FLOW CYTOMETRY AND SORTING

Second Edition

FLOW CYTOMETRY AND SORTING

Second Edition

Editors
MYRON R. MELAMED, M.D.
Departments of Pathology
Memorial Sloan-Kettering Cancer Center and
Cornell University Medical College
New York, New York

TORE LINDMO, Ph.D.
Biophysics Group
Division of Physics
University of Trondheim-NTH
Trondheim, Norway

MORTIMER L. MENDELSOHN, M.D., Ph.D.
Biomedical and Environmental Research
Lawrence Livermore National Laboratory
Livermore, California

WILEY-LISS
A JOHN WILEY & SONS, INC., PUBLICATION
NEW YORK • CHICHESTER • BRISBANE • TORONTO • SINGAPORE

Address all Inquiries to the Publisher
Wiley-Liss, Inc., 41 East 11th Street, New York, NY 10003

Printed in United States of America

First edition of **Flow Cytometry and Sorting** was published by John Wiley & Sons, Inc., 1979. Edited by Myron R. Melamed, Paul F. Mullaney, and Mortimer L. Mendelsohn.

Library of Congress Cataloging-in-Publication Data

Flow cytometry and sorting / editors, Myron R. Melamed, Tore Lindmo, Mortimer L. Mendelsohn.—2nd ed.
p. cm.
Includes bibliographical references.
ISBN 0-471-56235-1
1. Flow cytometry. I. Melamed, Myron R., 1922– II. Lindmo, T. III. Mendelsohn, Mortimer L.
[DNLM: 1. Cell Count—methods. 2. Cell Separation—methods. 3. Cytological Technics. QH 585.5 F644]
QH585.5.F56F56 1990
574.87′028—dc20
DNLM/DLC
for Library of Congress 89-13091
CIP

Contents

PART 8. APPLICATIONS IN ONCOLOGY

Contributors

Michael Andreeff, Memorial Sloan-Kettering Cancer Center and Cornell University Medical College, New York, NY 10021 [**697,725**]

Kenneth A. Ault, Maine Cytometry Research Institute, Portland, ME 04102 [**685**]

Bart Barlogie, Department of Hematology, University of Texas M.D. Anderson Cancer Center, Houston, TX 77030 [**745**]

A. Berge, SINTEF, Trondheim, Norway [**367**]

Craig F. Bohren, Department of Meteorology, Pennsylvania State University, University Park, PA 16802 [**81**]

J.G. Collard, Department of Cell Biology, Netherlands Cancer Institute, Antoni van Leeuwenhoekhuis, 1066 CX Amsterdam, The Netherlands [**195**]

L.S. Cram, Life Sciences Division, Los Alamos National Laboratory, Los Alamos, NM 87545 [**503**]

Harry A. Crissman, Cell Biology Group, Los Alamos National Laboratory, Los Alamos, NM 87545 [**227**]

Zbigniew Darzynkiewicz, Sloan-Kettering Institute for Cancer Research, New York, NY 10021 [**291,315,469**]

Phillip N. Dean, Biomedical Sciences Division, Lawrence Livermore National Laboratory, Livermore, CA 94550 [**171,415**]

F. Dolbeare, Biomedical Sciences Division, Lawrence Livermore National Laboratory, University of California, Livermore, CA 94550 [**445**]

Donald P. Evenson, Chemistry Department, Animal Science Complex, South Dakota State University, Brookings, SD 57007 [**531**]

Hugo Fellner-Feldegg, Max-Planck-Institut für Biochemie, D-8033 Martinsried, Federal Republic of Germany [**27**]

Michael H. Fox, Department of Radiology and Radiation Biology, Colorado State University, Fort Collins, CO 80523 [**633**]

S. Funderud, The Norwegian Radium Hospital, Oslo, Norway [**367**]

David W. Galbraith, Department of Plant Sciences, University of Arizona, Tucson, AZ 85721 [**633**]

Barton L. Gledhill, Biomedical Sciences Division, Lawrence Livermore National Laboratory, Livermore, CA 94550 [**531**]

J.W. Gray, Biomedical Sciences Division, Lawrence Livermore National Laboratory, University of California, Livermore, CA 94550 [**445,503**]

R.D. Hiebert, Cell Biology Group, Los Alamos National Laboratory, Los Alamos, NM 87545 [**127**]

Paul Karl Horan, Zynaxis Cell Science Inc., Malvern, PA 19355 [**397**]

James W. Jacobberger, Department of Genetics, School of Medicine, Case Western Reserve University, Cleveland, OH 44106 [**623**]

Ronald H. Jensen, Cytochemistry Section, Division of Biomedical Science, Lawrence Livermore National Laboratory, Livermore, CA 94550 [**553**]

James H. Jett, Life Sciences Division, Los Alamos National Laboratory, Los Alamos, NM 87545 [**381**]

Roger G. Johnston, Biochemistry/Biophysics Group, Life Sciences Division, Los Alamos National Laboratory, Los Alamos, NM 87545 [**81**]

Carl H. June, Naval Medical Research Institute, Bethesda, MD 20814 [**651**]

Volker Kachel, Max-Planck-Institut für Biochemie, D-8033 Martinsried, Federal Republic of Germany [**27,45**]

M.E. Kamarck, Molecular Therapeutics Inc., Miles Research Center, West Haven, CT 06516 [**563**]

Jan Kapuscinski, Sloan-Kettering Institute for Cancer Research, New York, NY 10021 [**291**]

Richard A. Keller, Chemistry and Laser Sciences Division, Los Alamos National Laboratory, Los Alamos, NM 87545 [**381**]

M. Lalande, Biotechnological Research Institute, National Research Council Canada, H4P 2R2 Montreal, Quebec, Canada; present address: Genetics Division, Children's Hospital, Boston, MA 02115 [**563**]

The numbers in brackets are the opening page numbers of the contributors' articles.

Richard G. Langlois, Biomedical Sciences Division, Lawrence Livermore National Laboratory, Livermore, CA 94550 [**249**]

Samuel A. Latt,† Genetics Division and Mental Retardation Center, The Children's Hospital, Boston, MA 02115 and Departments of Pediatrics and Genetics, Harvard Medical School, Cambridge, MA 02139 [**249**]

T. Lea, Rikshospitalet, University Hospital, 0172 Oslo, Norway [**367**]

James F. Leary, Department of Pathology, University of Rochester, Rochester, NY 14642 [**553**]

John M. Lehman, Department of Microbiology and Immunology, Albany Medical College, Albany, NY 12208 [**623**]

Tore Lindmo, Department of Biophysics, The Norwegian Radium Hospital, Montebello, 0310 Oslo 3, Norway; present address: Biophysics Group, Division of Physics, University of Trondheim-NTH, N-7034 Trondheim, Norway [**xi,145**]

Michael R. Loken, Becton Dickinson Monoclonal Center, San Jose, CA 95131 [**341**]

John C. Martin, Life Sciences Division, Los Alamos National Laboratory, Los Alamos, NM 87545 [**381**]

Myron R. Melamed, Departments of Pathology, Memorial Sloan-Kettering Cancer Center and Cornell University Medical College, New York, NY 10021 [**xi,1,755**]

Mortimer L. Mendelsohn, Biomedical and Environmental Research, Lawrence Livermore National Laboratory, Livermore, CA 94550 [**xi**]

Everhard Menke, Max-Planck-Institut für Biochemie, D-8033 Martinsried, Federal Republic of Germany [**27**]

Katharine A. Muirhead, Zynaxis Cell Science Inc., Malvern, PA 19355 [**397**]

Paul F. Mullaney,† Biophysics and Instrumentation Group, Los Alamos Scientific Laboratory, Los Alamos, NM 87544 [**1**]

Robert F. Murphy, Department of Biological Sciences and Center for Fluorescence Research in Biomedical Sciences, Carnegie Mellon University, Pittsburgh, PA 15213 [**355**]

Dinh C. Nguyen, Chemistry and Laser Sciences Division, Los Alamos National Laboratory, Los Alamos, NM 87545 [**381**]

K. Nustad, The Norwegian Radium Hospital, Oslo, Norway [**367**]

J.P. O'Connell, Becton Dickinson Research Center, Research Triangle Park, NC 27709 [**367**]

M.G. Pallavicini, Biomedical Sciences Division, Lawrence Livermore National Laboratory, University of California, Livermore, CA 94550 [**187,445**]

Donald C. Peters, Biomedical Sciences Division, Lawrence Livermore National Laboratory, Livermore, CA 94550 [**145**]

Daniel Pinkel, Biomedical Sciences Division, Lawrence Livermore National Laboratory, Livermore, CA 94550 [**531**]

Martin N. Raber, Department of Medical Oncology, University of Texas M.D. Anderson Cancer Center, Houston, TX 77030 [**745**]

Peter S. Rabinovitch, Department of Pathology, University of Washington, Seattle, WA 98195 [**651**]

A. Rembaum,† Jet Propulsion Laboratory, CALTECH, Pasadena, CA 91125 [**367**]

Gary C. Salzman, Biochemistry/Biophysics Group, Life Sciences Division, Los Alamos National Laboratory, Los Alamos, NM 87545 [**81**]

George C. Saunders, Life Sciences Division, Los Alamos National Laboratory, Los Alamos, NM 87545 [**381**]

Howard M. Shapiro, West Newton, MA 02165 [**1**]

Shermila Brito Singham, Biochemistry/Biophysics Group, Life Sciences Division, Los Alamos National Laboratory, Los Alamos, NM 87545 [**81**]

Sue E. Slezak, Zynaxis Cell Science Inc., Malvern, PA 19355 [**397**]

Lisa Staiano-Coico, Department of Surgery, Cornell University Medical College, New York, NY 10021 [**755**]

Harald B. Steen, Department of Biophysics, Institute for Cancer Research, The Norwegian Radium Hospital, Montebello, Oslo 3, Norway [**11,605**]

John A. Steinkamp, Cell Biology Group, Los Alamos National Laboratory, Los Alamos, NM 87545 [**227**]

Richard G. Sweet, Genetics Department, Stanford Medical Center, Stanford, CA 94305 [**145**]

I.W. Taylor, Queensland Institute of Medical Research, Brisbane, Queensland, Australia 4006 [**187**]

Frank Traganos, Sloan-Kettering Institute for Cancer Research and Investigative Cytology Laboratory, Memorial Sloan-Kettering Cancer Center, New York, NY 10021 [**469,773**]

A. Tulp, Department of Biochemistry, Netherlands Cancer Institute, Antoni van Leeuwenhoekhuis, 1066 CX Amsterdam, The Netherlands [**195**]

M.A. Van Dilla, Biomedical Sciences Division, Lawrence Livermore National Laboratory, Livermore, CA 94550 [**563**]

L.L. Vindelov, Medical Department, The Finsen Institute, DK-2100 Copenhagen, Denmark [**187**]

Jan W.M. Visser, Radiobiological Institute TNO, Rijswijk, The Netherlands [**669**]

Alan S. Waggoner, Department of Biological Sciences, Center for Fluorescence Research in Biomedical Sciences, Carnegie-Mellon University, Pittsburgh, PA 15213 [**209**]

Leon L. Wheeless, Jr., Analytical Cytology Unit, Department of Pathology and Laboratory Medicine, University of Rochester Medical Center, Rochester, NY 14642 [**109**]

† Deceased.

Preface

The first edition of *Flow Cytometry and Sorting* has been sold out for some years. Its usefulness is still very real, as judged by the dog-eared, shop-worn copies one sees wherever flow cytometry is practiced, and by the steady pressure on the editors to provide more copies or a new version of the book. With over a decade gone by, our choice is to provide a new edition.

Given the growth of the field and the explosive development of biologic techniques and applications, it should come as no surprise that nearly every chapter of this second edition has been greatly modified and that most are entirely rewritten.

There are new chapters on the study of higher plant cells, on flow cytometry of microorganisms, and on measurements of intracellular ionized calcium and membrane potential — each illustrating techniques of specimen preparation, measurement, and analysis not available at the time of the first edition. Also new are applications of flow cytometry to molecular genetics, to genetic toxicology, and to ultrasensitive analysis of molecules in solution. Fluorescent probes specifically designed for flow cytometry are treated in a separate new chapter, as are immunofluorescence techniques and ligand binding.

The continuing importance of flow cytometry for measurements of nucleic acids, chromatin, and DNA is evident in four new or rewritten chapters. Two other chapters dealing specifically with the cell cycle go beyond the traditional analysis of DNA histograms by using BrdU incorporation and DNA denaturability to more precisely identify and better study phases of the cell cycle, thus greatly enriching this important aspect of cell biology.

Chapters on the cytometry of sperm and other male germ cells, on the cytopathic effects of viruses, and on the technique and applications of flow cytogenetics have been entirely rewritten. Also rewritten are the chapters on clinical applications in hematology, immunology, clinical cytology, and the study of human solid tumors. Methods for the preparation of cellular or nuclear suspensions from solid tumors and methods for enrichment of cellular specimens are described in two new chapters. The chapter on effects of chemotherapeutic agents has been entirely rewritten to include new techniques and concepts.

While physical and engineering principles of flow cytometry were well established ten years ago, the great variety of custom and commercial instruments of that time has gradually given way to a smaller number of high-quality flow cytometers that each embody the best features of those early instruments. The chapter on hydrodynamic principles that is fundamental to the design of flow cytometers is now accompanied by a new chapter on electronics and signal processing, a new chapter on data processing, and an overview of the essential characteristics of flow cytometers. A single summary chapter on commercial instruments replaces six chapters that described the instruments of six different manufacturers in the first edition.

Physical methods of cell analysis are described in the chapter on electrical resistance pulse sizing and in updated chapters on light scattering, slit scanning, and the use of microspheres as immunoreagents. The chapter on flow sorters has been completely rewritten. Finally, a very important new chapter on standards and controls in flow cytometry is added.

As with the first edition, the book is divided into sections that bring together related chapters. We expect that it will be of value not only for those who work in the field, or intend to, but also for the many biologists, physicians, and others who are unfamiliar with flow cytometry and need to know whether this technology can help them in specific applications. The research instruments required for high-precision, multiparameter measurements and data analysis are still very expensive, and highly skilled operators are needed to assure optimum results. Thus, even frequent users may not themselves have "hands-on" experience with their instrument system, nor may they be aware of its many applications. This book should prove of value to them as well.

In a sense flow cytometry is a new kind of microscopy. As with the conventional light microscope, the flow cytometer may be applied to a wide range of cell studies, depending on the type of specimen, manner of specimen preparation, staining technique and staining specificity, and, in the case of flow cytometry, the method of analyzing data. With the passage of time we have witnessed a gratifying expansion of the techniques and applications of flow cytometry. The initial entirely experimental uses have broadened into clinical application that includes immunohematology and tentative steps in the analysis of solid tumors. The small fraternity of investigators who pioneered the development of flow cytometry have been joined and outrun by an ever-increasing number of recruits not only from engineering, physics, and chemistry but from almost every area of biology.

With the passage of time we also have been saddened by the death of friends and colleagues who contributed mightily in the early years, and to them we dedicate this book: Paul Mullaney who coedited the first edition of *Flow Cytometry and Sorting* and pioneered the studies of light scatter, Samuel Latt who devised methodology for analyzing DNA and chromosomal microstructure, Alan Rembaum who introduced microspheres as an immunochemical reagent for cell classification, and Tomas Hirschfeld who played a major role in developing multiple-beam flow cytometers and greatly extending the sensitivity of cytometry.

Myron R. Melamed
Tore Lindmo
Mortimer L. Mendelsohn

1

An Historical Review of the Development of Flow Cytometers and Sorters

Myron R. Melamed, Paul F. Mullaney, and Howard M. Shapiro

Department of Pathology, Memorial Sloan-Kettering Cancer Center, New York, New York 10021 (M.R.M.); Biophysics and Instrumentation Group, Los Alamos Scientific Laboratory, Los Alamos, New Mexico 87544 (P.F.M.) (deceased); Howard M. Shapiro, West Newton, Massachusetts 02165 (H.M.S.)

HISTORICAL OVERVIEW

Flow cytometry developed over the last 25 years from an effort initially intended to count and later to size particles, into a sophisticated analytic tool for rapidly quantitating multiple chemical and physical properties of the individual cells or cellular constituents of inhomogeneous populations. During most of this time, efforts were directed primarily toward the development of new and improved instrumentation, and this brief historical review is intended to recount those developments. We have drawn on personal discussions, internal publications, and patent disclosures, as well as formal publications in the scientific literature.

The first proposal to count cells automatically while in flow was reported by Moldavan in 1934 [54]. He described an apparatus in which a suspension of red blood cells or neutral red stained yeast is forced through a capillary glass tube on a microscope stage, and each passing cell registered (i.e., counted) by a photoelectric apparatus attached to the ocular. He noted problems in standardizing the capillary tube, ensuring proper focus, maintaining flow, and obtaining an appropriately sensitive photoelectric apparatus, and no further work was reported by him. In 1941, Kielland filed a patent for "Method and Apparatus for Counting Blood Corpuscles" [44], which appears to be the same as that described by Moldavan.

Problems with flow in the narrow channels required for flow cytometry, and frequent obstruction by large cells or cell clusters remained an important consideration in the design of later instruments. A solution to many of them was to come from the laminar sheath flow principle that Reynolds used in 1883 to study laminar flow and turbulence [58]. Gucker et al. incorporated this in a photoelectric counter for aerosol particles in 1947 [32]. Then, in 1953, Crosland-Taylor [14] applied the same principle to the design of a chamber for optical counting of red blood cells. An aqueous suspension of the cells was injected slowly into a faster flowing stream of fluid; the latter provided a laminar sheath surrounding and aligning the particles. This permitted wide-diameter channels to be used with a narrow central stream of the particles to be measured, and successfully accomplished two purposes: (1) it virtually eliminated any possibility of a narrow channel being blocked by large particles in the flow stream; (2) it made precise centering of the particle stream possible. Thus, the basis for hydrodynamic focusing was established. Almost all flow cytometry instruments today make use of the sheath-flow principle described by Crosland-Taylor.

In 1949, Wallace Coulter filed for a patent titled "Means for Counting Particles Suspended in a Fluid" [10]. In that patent he disclosed a means for accurately counting particles (blood cells) in suspension by causing the cells to pass through a constricted path (aperture), where the presence or absence of a particle gives rise to a detectable change in the electric characteristics of the path. It was necessary that there be a difference in electrical conductivity between the particles and their suspending medium, but for practical purposes this was easily achieved. Based on this principle a commercially successful blood counting instrument was developed (the "Model A" Coulter Counter®) [6,11]. Particle (cell) sizing also could be performed, since the amplitude of the signal produced [47] or area of the pulse [51] could be related to the volume of the particles, and sizing appeared independent of particle shape [48]. Kubitschek used a sliding, single-channel pulse-height analyzer for this purpose [47]. Multichannel pulse-height analyzers and nuclear pulse amplifiers were introduced later by Lusbaugh et al. [49] and Van Dilla et al. [66], greatly speeding data acquisition rates. In 1959, Coulter and co-workers filed a patent for improved electronic circuitry with a current-sensing amplifier that made the electrical pulses independent of electrolyte conductivity and improved particle-size measurements [12]. It led to the "Model B" Coulter Counter®. Short apertures, which were used to maximize the signal-to-noise ratio and reduce coin-

Flow Cytometry and Sorting, Second Edition, pages 1–9

cidence of particles within the orifice, resulted in nonuniformities of the electrical fields at the orifice entrance and exit, and caused an artefactual skewing of particle-size distributions. Van Dilla et al. [66] were able to correct this with long apertures, and Spielman and Goren by hydrodynamic focusing of the particle stream [63], but it remained for Grover et al. to describe the theoretical basis for secondary effects of this kind in the electrical sizing of particles in suspension [31].

Another patent filed by Coulter and Hogg, in 1966, proposed that two different radio frequencies (or radiofrequency and direct current) be used simultaneously to ascertain more than one physical characteristic [13]. Thus, particles of identical size but different substance might be differentiated by two different signals. During these years, Coulter Counters gradually replaced manual methods of blood cell counting in most large laboratories, and were widely used for research involving cell counting or cell sizing.

A photoelectric particle counter for counting opaque particles in the range of 50 to 700 μm was developed in 1954 by Bierne and Hutcheon, and published in 1957 [3]. In principle, the apparatus was similar to that of Moldavan. Particles were drawn in suspension through a capillary under the objective of a microscope, where they cut off light from a photomultiplier mounted above the eyepiece, producing electrical impulses that were amplified and counted electronically. Mechanical stirring devices were used to prevent settling of the particles. An effort was made to measure particle size by time duration of the pulse, but it was not possible to maintain a sufficiently steady flow rate. However, particle-size distributions could be obtained from the discriminator voltage measurements, and mean particle size from counts of the number of particles per gram.

Cornwall and Davison described a photoelectric counter for cells (apple cells 70–500μ length) in 1950 [9]. They stained the cells in dilute suspension with trypan blue. The cell suspension was then passed through a glass tube on a microscope stage and the stained cells counted by a phototransistor mounted above the eyepiece.

The first differential blood cell counting device was described by Parker and Horst in a patent application in 1953 [57]. A dilute suspension of blood in isotonic solution was stained such that the white blood cells were colored violet, indigo, or blue, while the red color of the red blood cells was retained or enhanced. The cell suspension was then made to flow (by gravity) through a narrow conduit with a constriction where the cells passed through a beam of light containing red and blue wavelengths. The transmitted light was separated into red and blue components, each going to a separate photocell. An electrical circuit (counter) was actuated by one of the photocells, according to whether a blue-absorbing or red-absorbing cell passed through the channel. The two independent photometric signals were not combined or quantitated. To minimize coincidence, the light beam was shaped to a slit, which was oriented perpendicular to the flow of the cells and had a width considerably less than the diameter of a single cell. This was accomplished by optically magnifying the constriction and inserting a mask in the magnified real image.

Two new concepts that greatly extended the potential applications of flow cytometry were introduced by Kamentsky et al. in 1965: one was the use of spectrophotometry to quantitate specific cellular constituents, and the other was that of cell classification by a combination of multiple simultaneous measurements of different cellular features [35–37]. In their initial report, ultraviolet absorption of nucleic acids and visible light scatter were measured simultaneously on unstained cells at rates of 500 cells/sec. Kamentsky et al. were the first to display and analyze multiparameter flow cytometry data by means of a two-dimensional histogram [35–37]. Subsequently, they reported a flow cytometer capable of carrying out up to four simultaneous measurements per cell, and were the first to record and analyze multiparameter data by an interfaced computer [37,39]. The Ortho Cytograf® and Cytofluorograf® were developed later from this instrument.

With these systems in which the suspension of cells passes through a constricted channel traversed by a beam of light orthogonal to the channel, the change in light intensity generated by each cell depends on the position of the cell in the channel with regard to the focal plane of the incident light. In 1965, Meyer-Doering and Knauer applied for a patent on a particle (blood cell) counter based on light scatter in which the cell stream flow was directed parallel to the beam of light through which it flows, thus requiring the cell to pass through the focal plane [53]. In 1968, Dittrich and Göhde applied for a patent on a system measuring fluorescence or phosphorence of particles moving in a flow stream parallel to the optical axis of a transmitted light microscope, and through the focal plane of a high numerical aperture objective [17,18]. Using Köhler illumination, their system provided for a homogeneously illuminated area filling the focused exit of the flow channel. Thus, no focusing problems were encountered. Shortly thereafter (1969), they applied for a patent that described means for deflecting the particle flow with a rinse fluid after passing the focal plane. Also, the same objective could be used for fluorescence excitation with Köhler illumination and for collecting emitted fluorescence [19]. By using high numerical aperture lenses, they achieved a large solid angle of excitation and fluorescence collection [27]. Thus, a conventional xenon or high-pressure mercury lamp could provide sufficient illumination, and excitation wavelength bands were available in the ultraviolet as well as the visible spectrum. In addition, the large solid angle of excitation and collection reduced or eliminated variation in signal due to different orientations of asymmetric cells. Göhde was able to achieve measurements of cellular DNA with extraordinarily low coefficients of variation. The Phywe Impulscytophotometer (ICP 22) became the commercial version of this instrument.

A stimulus for this early work, and much of what has been done since by others, came from efforts to automate the detection of cancer cells in clinical cytologic specimens. It was known that certain types of cancer cells are characterized by nuclear enlargement and hyperchromicity (due to increased or abnormal DNA), and this lent itself to possible identification by photometric techniques. Caspersson had shown that nucleic acids could be quantitated in unstained cells by ultraviolet absorption (260 μm) [7,8], and the first field trial of flow cytometry for detection of uterine cervical carcinoma made use of ultraviolet absorption to identify cells with high content of nucleic acids [46]. However, ultraviolet absorption did not distinguish DNA from RNA, and Kamentsky et al. subsequently described a technique for simultaneous measurements of DNA and protein per cell based on the absorption of specifically stained cells at two different wavelengths. They used the Feulgen reaction for DNA and counterstained with Naphthol Yellow S for protein, then measured absorption at 560 to 570 μm (Feulgen) and 430 μm (Naphthol Yellow S). The measurements were displayed as a two-dimensional histogram [38,39] or as a one-dimensional histogram of the ratio of the two measurements [42]. Göhde and Dittrich performed the first simultaneous staining of DNA and protein with fluorescent dyes. They stained DNA with the fluorescent dye ethidium bromide and counterstained for protein with the fluorescent dye DANS (1-dimethylamino-naphthalin-S-sulfochloride) [28] or FITC [20]. They then measured the ethidium bromide fluorescence at wavelengths higher than 590 nm and DANS or FITC stained protein fluorescence simultaneously at 520 or 510 nm, respectively, to obtain a two-parameter analysis of DNA and protein content per cell.

Fluorescent dyes provide important advantages over absorbing dyes. Properly used, they eliminate distributional error and greatly increase the signal-to-noise ratio. Interestingly, the first system proposed for cancer detection by automated cytologic examinations was a microfluorometer scanner described by Mellors and Silver in 1951 [52]. It was designed to measure nuclear fluorescence of cells spread on glass slides and stained by basic fluorescent dyes. In the case of flow cytometry systems, fluorescent stains were first used almost simultaneously by several different groups. In a 1967 conference of the New York Academy of Science, Kamentsky et al. reported three-parameter measurements of acridine orange-stained cells—fluorescence at two different wavelength bands and absorption at a third wavelength [38]; at the same time, Van Dilla et al. described fluorescent Feulgen measurements of human leukocytes and Chinese hamster ovary cells [67,68], and Göhde and Dittrich reported measurements of cellular DNA by quantitating fluorescence of ethidium bromide-stained cells [27]. A patent was filed by Wheeless et al. in September 1967 for method and apparatus to identify cells in static or flowthrough systems after staining with a fluorochrome, by measuring fluorescence at a plurality of separate wavelengths and determining size by light scatter or electrical resistance [69].

Illumination was orthogonal to flow in most of these flow cytometers. Van Dilla and the Los Alamos group developed the first instrument having orthogonal axes of flow, illumination, and detection [67,68]. Their system, which was not microscope-based, used a laminar flow chamber of the Crosland-Taylor design and introduced the argon-ion laser as a light source for flow cytometry. The orthogonal system promoted further development of fluorescence and scatter sensors, including combinations with the Coulter sensor. They were the first to demonstrate a linear relationship between ploidy and fluorescence intensity of cells stained by a fluorescent Feulgen reaction for DNA. They also were the first to produce DNA histograms that clearly defined G_1, S, and G_2 + M phases of the cell cycle, and to demonstrate quantitative cell kinetics with synchronized cultured cells [68]. Göhde and Dittrich were the first to undertake studies of drug effects on cell cycle kinetics by using flow cytometry measurements of DNA to determine changes in cell cycle kinetics [28,29].

A device for sorting cells in flow cytometry instruments was first reported by Fulwyler in 1965 [24] and an improved version in 1969 [26]. He adapted the electrostatic ink jet droplet deflection technique of Sweet [65] for use with a Coulter cell sizing instrument, and was able to sort cells according to their Coulter volume. Working independently and at the same time, Kamentsky filed a patent for a device to separate cells in flow after photometric or electrical sensing, using either pneumatic, hydraulic, or electrostatic techniques [40]. He later described results with a hydraulic cell separator after two-parameter photometric sensing [41]. Another fluidic switch cell sorter, patented by Friedman [23], diverted the cell stream by means of a sonic transducer that converted laminar to turbulent flow. However, the fastest and most efficient of the sorting devices still is the electrostatic sorter, and this is now the only one in general use. It was adapted for the separation of fluorescence-stained cells by Hulett et al. in 1969 [33]. Operation of the sorter was greatly simplified by Bonner et al. in 1972 [4,5], and improved by carrying out measurements of the cells in the fluid stream in air, after it left the nozzle of the flow chamber but before droplet formation occurred. Thus, there was minimal delay between cell measurement and droplet charging. The Becton-Dickinson Fluorescence Activated Cell Sorter (FACS)® was based on this instrument, and has been widely used in immunology to obtain pure populations of cells as identified by surface antigens [34].

Other modifications in the basic design of the flow cytometer are worthy of note. One of the most ambitious was an effort to obtain high-resolution morphologic information from the cells by scanning them in flow. In a patent filed in 1969, Ehrlich et al. described a system in which unstained biologic cells were made to flow in single file through a transparent tube while they were scanned with a mixture of ultraviolet and visible light using a flying spot scanner [21]. The apparatus measured nuclear diameter, size, and symmetry of the cytoplasmic shoulders on either side of the nucleus, ratio of nuclear to cellular size, and the product of the size and density of the nucleus. It also provided for a visual display of the scanned cells on a television type monitor. Similar features can be measured by the slit scan technique of Wheeless et al. [70,71]. They obtain low-resolution morphologic information by analyzing the fluorescent pulse contours of acridine orange stained cells flowing through a thin band of excitation illumination; but in this system measurements are affected by cell orientation.

Multiple stations in sequence for different measurements or combinations of measurements were described by Bonner et al. in 1972 [5] and Hulett et al. in 1973 [34], and developed independently at Block Engineering for their flow cytometry system, the Cytomat, in 1973 [61]. A fluorescent staining technique for use with this system was described in a patent by Kleinerman [45], and electronic circuitry for correlating and processing the signals from multiple stations was patented by Curbelo [15]. The Cytomat system is described in further publications by Curbelo et al. [16] and Shapiro et al. [62]. Other multistation instruments have been described by Stöhr [64] and by Fulwyler [25].

The Technicon Hemalog® was designed to perform different measurements or different combinations of measurements in parallel systems each with its own method of cell processing and its own measuring station, and each with an aliquot of the cell sample. The development of this system can be traced in a series of patents and publications [1,2,22,30,50,56,59,60].

Light scatter is measured in a number of instruments in addition to fluorescence and/or absorption. In theory, much of the morphologic information of each cell is contained in the scattered light, and there is a large body of knowledge that has accumulated in the study of aerosols and other applications of light scatter that could be applied to the problem of cell analysis by flow cytometry [43]. So far, light scatter has been used principally to measure cell size or dye uptake by the cell. In 1969, Mullaney et al. devised a flow cytometer for cell sizing with a helium neon laser as light source and a modified Crosland–Taylor flow channel [55]. They reported good agreement between cell size obtained by forward light scatter between 0.5 and 2.0°, Coulter volume measurements, and direct measurements of cell size by light microscopy.

During the early years of instrument development, there were few applications to problems of biologic interest. Most of the studies that were carried out were connected with the enumeration and characterization of blood cells, with the identification and classification of immunoreactive lymphocytes, or with the identification of cancer cells in clinical cytology specimens. Many experiments simply were designed to demonstrate the capabilities of newly built instrumentation. In the years that have followed, instrument development has continued, with corresponding progress in biologic preparatory techniques that are specifically designed for flow cytometry. These are reviewed below and detailed at some length in the chapters that follow.

DEVELOPMENTS IN FLOW CYTOMETRY SINCE 1980

In the decade since this historical review was published in the first edition of *Flow Cytometry and Sorting,* instrument development has matured, and the most exciting advances now focus on new methods of cell preparation, new fluorescent dyes and new markers of cell properties that lead toward an ever-widening spectrum of biologic and biomedical applications. This is not to diminish the critical role of the instrument manufacturers who have selected the best of the prototype and custom designs to provide state-of-the-art, rugged, and reliable high-precision instruments for research and clinical use. They have also taken advantage of the impressive technological advances in production of microprocessors and software to provide powerful, user-friendly data acquisition and analysis systems. Worthy of special mention is the development of high-speed cell sorting by Gray and others at the Lawrence Livermore and Los Alamos National Laboratories [84,85], providing a tool to sort chromosomes of a single type for gene mapping and construction of recombinant DNA libraries.

Almost from its beginning, flow cytometry has had obvious applications in cellular immunology. With the introduction of monoclonal antibody technology by Kohler and Milstein [92], an almost endless series of powerful, highly specific immunologic reagents became available for cell classification and study. These were applied first in flow cytometry by Reinherz et al. [93,101,102] to identify and subclassify T lymphocytes according to differences in cell surface antigens, and are now available not only to discriminate a bewildering array of mature and immature lymphocytes and other leukocytes, but also for the study of benign and transformed epithelial cells, and to measure an increasing number of cellular proteins. The latter include cytostructural proteins, cell surface receptors and other antigens, and protein products of gene (or oncogene) expression.

Simultaneous measurement and correlative studies of two antigens per cell using a single laser for excitation was first described by Loken et al. [94], who corrected electronically for the partially overlapping emission spectra of the two dyes used as fluorescent antibody labels: fluorescein and tetramethyl rhodamine. Dual antigen labelling with fluorescent dyes of nonoverlapping spectra was achieved by using fluorescein in conjunction with X RITC or Texas Red [111] but, because excitation as well as emission spectra are nonoverlapping, a dual-laser system is required. In 1982, Oi and co-workers made a major contribution by providing algae phycobiliproteins as fluorescent labels for antibodies and other molecules [96,107,108]. Phycobiliproteins, which function to transfer light energy in photosynthesis, have high absorption coefficients, high fluorescence quantum efficiency and relatively large Stokes shifts. The phycobiliprotein phycoerythrin contains 20 or more chromophores per molecule, can be excited with an argon laser and emits orange fluorescence which is readily separated optically from the green fluorescence of fluorescein. Thus, phycoerythrin labeled monoclonal antibodies, which are now available commercially, make dual-label immunofluorescence flow cytometry practical with single-beam argon laser instruments. Other phycobiliproteins—phycocyanin and allophycocyanin—absorb in the red and emit in the far red. Hardy et al. used a two-laser system to achieve three-color immunofluorescence with fluorescein, phycoerythrin, and allophycocyanin-conjugated antibodies [87].

Combined DNA and RNA measurement with the metachromatic fluorochrome acridine orange, described by Darzynkiewicz and Traganos and co-workers [76,112], is now an established method for measuring lymphocyte stimulation and monitoring cell cycle kinetics. It has been applied clinically by Andreeff to subclassify the acute leukemias [72] and monitor treatment effect. More recently, Shapiro described another technique for simultaneous DNA and RNA measurements using a dual beam flow cytometer to measure DNA staining by Hoechst 33342 and RNA staining by pyronin Y [103].

The most important clinical applications of flow cytometry are in hematology. Within a relatively few years flow cytometry has become the standard method of blood cell counting, including the differential counting of leukocytes and, with monoclonal antibodies, the subclassification of lymphocytes. Tanke et al. expanded this repertoire of measurements to include red blood cell reticulocyte counts based on RNA measurements [109].

Studies of cellular physiology by measurements of membrane potential changes using fluorescent cationic cyanine dyes were first proposed by Hoffman and Laris [89], and subsequently applied by Shapiro et al. [104] to assay lymphocyte activation by flow cytometry. More recently, Johnson et al. [91] reported specific staining of mitochondria in living cells by the membrane potential sensitive dye Rhodamine 123, which was adapted to flow cytometry by Darzynkiewicz et al. [77].

In 1985 a new fluorescent dye, Indo 1, introduced by Grynkiewicz et al. [86], made practical the precise measurement of intracellular calcium ion concentration by flow cytometry. The dye is excited by ultraviolet (UV) light and exhibits a large fluorescence emission wavelength shift in the presence of free calcium ion (485 nm) versus calcium complex (410 nm). The ratio of fluorescence intensities at the two different emission wavelength bands permits calculation of intracellular calcium ion concentration independent of dye concentration. Since changes in calcium ion concentration are a universal early marker of cell activation this dye can be used to identify and study that process by flow cytometry. Other physiologic measurements of cell function by flow cytometry include changes in intracellular pH, first described by Visser et al. [118], cell surface charge described first by Valet et al. [115], and redox state described by Thorell [110]. With time as an additional parameter [95], the kinetics of ligand binding, cell activation, enzyme reaction, drug uptake, and efflux from cells and other biologic processes can be measured over periods of seconds to minutes.

Measurements of cell proliferation have many applications in the study of drugs, growth factors, and other biologicals, as well as potential application in assessing the rate of growth of human tumors. Estimates of proliferation based on cell cycle distribution derived from DNA measurements are indirect, and do not distinguish quiescent (G_0) from cycling (G_1) cells. Further, there is mounting evidence that cells may enter quiescence from any phase of the cycle; thus it is possible to have quiescent S- or G_2-phase cells that cannot be distinguished on the basis of DNA content from proliferating S- or G_2-phase cells. Exact identification of DNA replicating cells was made possible by a technique first described by Gratzner et al. [82], in which Bromodeoxyuridine (BrdU) is incorporated into DNA-synthesizing cells grown in the presence of BrdU, and subsequently identified by an immunofluorescent technique using anti-BrdU antibody. BrdU may be given in vivo or added to cell cultures in vitro, and can be identified after even very short exposure by the high-affinity monoclonal antibodies that are now available [83].

One of the first, and still one of the most widely used, applications of flow cytometry is in measuring the DNA content of cells from solid tumors. Many but not all tumors contain populations of measurably aneuploid cells, and much controversy still centers on the possible significance of DNA aneuploidy. With the precision of modern flow cytometers even small differences in DNA content can be measured. Interestingly, Darzynkiewicz et al. showed that differences in staining of nuclear DNA by different DNA binding dyes under some conditions may be attributable not to differences in DNA content but to differences in chromatin structure affecting the accessibility of DNA to the dye, for example, in association with cell differentiation [78]. In 1983, Hedley and associates described a technique for DNA flow cytometry of tumor cell nuclei extracted from formalin fixed tissues embedded in paraffin blocks [88]. Since most pathology laboratories store these paraffin-embedded tissue blocks indefinitely, it is now possible to conduct retrospective studies of collected unusual tumors and tumors from patients with known clinical outcome. However, the effects of fixation, processing and storage, which are known to modify DNA stainability in at least some cases, still are not well understood.

Chromosomal karyotyping by flow cytometry (flow karyotyping) was first described by Gray et al., who used the DNA stain ethidium bromide [84]. Much better discrimination and a dramatic improvement in resolution of the human karyotype was subsequently achieved by Carrano et al. [75], who combined Hoechst 33258 and chromomycin A3 to stain AT and GC base-pair (bp) sequences differentially for flow karyotyping. Chromosome identification was based on the ratio AT/GC as well as total DNA.

Another approach to the study of chromosomal abnormalities was introduced by Trask and her colleagues who developed fluorescent genome probes that hybridize to repetitive DNA sequences characteristic of specific chromosomes. They were able to identify those chromosomes in interphase nuclei as well as metaphase spreads, and have successfully adapted the technique to flow cytometry [114]. There are now more than a dozen such probes for specific chromosomes, providing information on organization of the interphase nucleus as well as chromosomal abnormalities.

Among the more gratifying advances over the last decade have been the development and application of flow cytometry to an increasing variety of nonmammalian eukaryotic and prokaryotic cells. Bailey et al. [73] and Paau et al. [97,98] were the first to demonstrate the feasibility of flow cytometry of bacteria, which they used to study bacterial growth by protein, nucleic acid, and light-scatter measurements. Steen and Skarstad and their colleagues carried out DNA measurements of bacteria with coefficients of variation of 5% or less, described the bacterial cell cycle and also performed immunofluorescence flow cytometry studies of bacteria [105,106]. Ingram et al. proposed immunofluorescence flow cytometry for detection of the bacterium *Legionella pneumophilia* in clinical specimens [90]. Van Dilla et al. made use of combined AT-specific (Hoechst 33258) and GC-specific (chromomycin A3) stains to characterize bacteria by AT/GC ratio and total DNA content [117].

Price et al. were the first to study microalgae by flow cytometry, describing light scatter measurements [99]. Trask and Van den Engh and colleagues measured the chlorophyl fluorescence of algae by flow cytometry [113,116]; Bonaly and Mestre reported flow cytometry DNA measurements of the protozoan *Euglena gracilia* [74].

Studies of higher plant cells are complicated by the presence of a cellulosic cell wall, which must be removed under conditions of osmotic stability. Galbraith et al. described a general technique for this purpose involving chopping the tissue in an osmotically stabilized buffer with the detergent Triton X-100 [80]. The first flow cytometry studies of higher plant cells were described by Galbraith in a patent for identification and sorting of plant cell heterokaryons based on labeling of the original protoplasts with different fluorochromes [79]. Galbraith and Shields later used mithramycin to measure DNA changes of developing tobacco protoplasts in the presence of cell wall synthesis inhibitors [81]. At the same time Redenbaugh characterized and sorted plant protoplasts by fluorescence and light scatter after staining with various fluorochromes [100].

These wide ranging feasibility studies and early applications of flow cytometry ushered in a still continuing period of staining and techniques development that is distinguished from conventional cytochemistry and designed specifically for measurements by a flow cytometer. The chapters that follow provide detailed descriptions of the state of the art, that build on these innovative contributions. The methodology continues to improve, taking advantage of a growing number of ever more specific reagents. In this edition of *Flow Cytometry and Sorting*, biologic preparatory techniques begin to meet the expectations of earlier years as they match the highly sophisticated engineering of modern flow cytometers and computer data analysis.

REFERENCES

1. **Ansley H, Ornstein L (1970)** U.S. Patent No. 3,741,875, "Process and Apparatus for Obtaining a Differential White Blood Cell Count." Filed October 30, 1970. Issued June 23, 1973.
2. **Ansley H, Ornstein L (1971)** Enzyme histochemistry and differentiated white counts on the Technicon Hemalog™ D, in Technicon International Congress, 1970, "Advances in Automated Analysis," Vol. 1. Miami: Thurman Associates, pp 437–446.
3. **Bierne T, Hutcheon JM (1957)** A photoelectric particle counter for use in the sieve range. J. Sci. Instr. **34**:196–200.
4. **Bonner WA, Hulett HR, Sweet RG, Herzenberg LA**

(**1972**) Fluorescence activated cell sorting. Rev Sci Instr **43**:404–409.

5. **Bonner WA, Sweet RG, Hulett HR** (**1972**) U.S. Patent No. 3,826,364, "Particle Sorting Method and Apparatus." Filed May 22, 1972. Issued July 30, 1974.
6. **Brecher G, Schneiderman MA, William GE** (**1956**) Evaluation of an electronic red cell counter. Am. J. Clin. Pathol. **26**:1439–1449.
7. **Caspersson T** (**1936**) Uber den Chemischen Aufbau der Struktusen des Zellkernes. Skand Arch Physiol **73** (suppl. 8):1–151.
8. **Caspersson T** (**1950**) "Cell Growth and Cell Function." New York: Norton.
9. **Cornwall JB, Davison RM** (**1950**) Rapid counter for small particles in suspension J Sci Instr **37**:414–417.
10. **Coulter WH** (**1949**) U.S. Patent No. 2,656,508, "Means for Counting Particles Suspended in a Fluid." Filed August 27, 1949. Issued October 20, 1953.
11. **Coulter WH** (**1956**) High speed automatic blood cell counter and cell size analyzer. Proc Natl Electron Conf **12**:1034–1042.
12. **Coulter WH, Hogg WR, Moran JP, Claps WA** (**1959**) U.S. Patent No. 3,259,842 "Particle Analyzing Device." Filed August 19, 1959. Issued July 5, 1966.
13. **Coulter WH, Hogg WR** (**1966**) U.S. Patent No. 3,502,974, "Signal Modulated Apparatus for Generating and Detecting Resistive and Reactive Changes in a Modulated Current Path for Particle Classification and Analysis." Filed May 23, 1966. Issued March 24, 1970.
14. **Crosland-Taylor PJ** (**1953**) A device for counting small particles suspended in a fluid through a tube. Nature (Lond) **171**:37–38.
15. **Curbelo R** (**1975**) U.S. Patent No. 3,976,862, "Flow Stream Processor." Filed March 18, 1975. Issued August 24, 1976.
16. **Curbelo R, Schildkraut ER, Hirschfeld T, Webb RH, Block MJ, Shapiro HM** (**1976**) A generalized machine for automated flow cytology system design. J Histochem Cytochem **24**:388–395.
17. **Dittrich W, Göhde W** (**1968**) British Patent No. 1,300,585, "Automatic Measuring and Counting Device for Particles in a Dispersion." Filed December 18, 1968 in Germany. Issued December 20, 1972.
18. **Dittrich W, Göhde W** (**1969**) Impulsfluorometrie bei Einzelzellen in Suspensionen. Z. Naturforsch **24b**: 360–361.
19. **Dittrich W, Göhde W** (**1969**) British Patent No. 1,305,923, "Automatic Measuring and Counting Device for Particles in a Dispersion." Filed April 18, 1969 in Germany. Published February 7, 1973.
20. **Dittrich W, Göhde W, Severin E** (**1971**) Die Kern-Plasma-Relation in Der Impulscytophotometrie Des Cervical-Und Vaginalsmears. Fourth International Congress of Cytology, London, 1971 (abst).
21. **Ehrlich MP, Stoller M, Grand S, DeCote R** (**1969**) U.S. Patent No. 3,699,336, "Biological Cell Analysing System." Filed August 15, 1969. Issued October 17, 1972.
22. **Elkind A, Groner W, Saunders A** (**1970**) U.S. Patent No. 3,661,460, "Method and Apparatus for Optical Analysis of the Content of a Sheath Stream." Filed August 28, 1970. Issued May 9, 1972.
23. **Friedman M** (**1973**) U.S. Patent No 5,791,517, "Digital Fluidic Amplifier Particle Sorter." Filed March 5, 1973. Issued February 12, 1974.
24. **Fulwyler MJ** (**1965**) Electronic separation of biological cells by volume. Science **150**:910–911.
25. **Fulwyler MJ** (**1974**) U.S. Patent No 3,916,197, "Method and Apparatus for Classifying Biological Cells." Filed October 18, 1974. Issued October 28, 1975.
26. **Fulwyler MJ, Glascock RB, Hiebert, RD, Johnson NM** (**1969**) Device which separates minute particles according to electronically sensed volume. Rev. Sci. Instr. **40**:42–48.
27. **Göhde W, Dittrich W** (**1971**) Impulsefluorometrie-ein neuartiges Durchflussverfahren zur ultraschnellen Mengenbestimmung von Zillinhaltsstoffen. Presented at the "Symposion der Deutschen, Schweizerischen und Oesterreichhischen Gesellschaft fuer Histochemie," Graz (Austria) April 1969. Published in: Acta Histochem., Suppl. X, p. 42–51, 1971.
28. **Göhde W, Dittrich W** (**1970**) Simultane Impulsfluorimetrie des DANS—und Proteingehaltes von Tumorzellen. Z. Anal. Chem. **252**:328–330.
29. **Göhde W, Dittrich W** (**1971**) Die Cytostatische Wirkung von Daunomycin im Impulscytophotometrictest. Arzneim-Forsch (Drug Res.) **21**:1656–1658.
30. **Groner W, Kusnitz J, Saunders A** (**1970**) U.S. Patent No. 3,740,143, "Automation for Determining the Percent Population of Particulates in a Medium." Filed October 31, 1970. Issued June 19, 1973.
31. **Grover NB, Naaman J, Ben-Sasson S, Doljanski J** (**1969**) Electrical sizing of particles in suspension. I. Theory. Biophys. J. **9**:1398–1414, 1969.
32. **Gucker FT Jr, O'Konski CT, Pickard HB, Pitts JN Jr** (**1947**) A photoelectric counter for colloidal particles. J. Am. Chem. Soc. **69**:2442.
33. **Hulett HR, Bonner WA, Barret J, Herzenberg LA** (**1969**) Cell sorting: Automated separation of mammalian cells as a function of intracellular fluorescence. Science **166**:747–749.
34. **Hulett HR, Bonner WA, Sweet SG, Herzenberg LA** (**1973**) Development and application of a rapid cell sorter. Clin. Chem. **19**:813–816.
35. **Kamentsky LA, Melamed MR, Derman H** (**1965**) Spectrophotometer: New instrument for ultrarapid cell analysis. Science **150**:630–631.
36. **Kamentsky LA, Melamed MR, Derman H** (**1965**) A new ultra-rapid cell spectrophotometer: Preliminary observation of ultraviolet absorption in various human cells. IBM Research Report (RW 72) June 3, 1965.
37. **Kamentsky LA** (**1965**) Rapid biological cell identification by spectroscopic analysis. Proc. 18th Ann. Conf. on Engineering in Biol. and Med. **7**:178.
38. **Kamentsky LA, Melamed MR** (**1969**) Rapid multiple mass constitutent analysis of biological cells. Ann. NY Acad. Sci. **157**:310–323, 1969. (Presented June 6, 1967 at NY. Acad. Sci. Conf. on Data Extraction and Processing of Optical Images in the Medical and Biological Sciences, N.Y.C.)
39. **Kamentsky LA, Melamed MR** (**1969**) Instrumentation for automated examination of cellular specimens. IEEE **57**:2007–2016.
40. **Kamentsky LA** (**1965**) U.S. Patent No. 3,560,754,

"Photoelectric Particle Separator Using Time Delay." Filed November 17, 1965. Issued February 2, 1971.

41. **Kamentsky LA, Melamed MR (1967)** Spectrophotometer cell sorter. Science **156:**1364–1365.

42. **Kamentsky LA. Thorell B (1970)** Cell population identification studies. Acta Cytol. (Praha) **14:**307–312.

43. **Kerker M (1969)** "The Scattering of Light." New York: Academic Press.

44. **Kielland J (1941)** U.S. Patent No. 2,369,577, "Method and Apparatus for Counting Blood Corpuscles." Filed May 12, 1941. Issued February 13, 1945.

45. **Kleinerman M (1973)** U.S. Patent No. 3,916,205, "Differential Counting of Leukocytes and Other Cells." Filed May 31, 1973. Issued October 28, 1975.

46. **Koenig SH, Brown RD, Kamentsky L, Sedlis A, Melamed MR (1968)** A report of the efficacy of a rapid cell spectrophotometer in screening for cervical cancer. Cancer **21:**1091–1026.

47. **Kubitschek HE** (1958) Electronic counting and sizing of bacteria. Nature (Lond.) **182:**234–235.

48. **Kubitschek HE (1960)** Electronic measurement of particle size. Research (Lond.) **13:**128–135.

49. **Lusbaugh CC, Maddy JA, Basman NJ (1962)** Electronic measurement of cellular volumes. I. Calibration of the apparatus. Blood **20:**233–240.

50. **Mansberg HP, Saunders AM, Groner W (1974)** The Hemalog D white cell differential system. J. Histochem. Cytochem. **22:**711–724.

51. **Mattern CFT, Brackett FS, Olson BJ (1957)** The determination of number and size of particles by electrical gating: Blood cells. J. Appl. Physiol. **10:**56–70.

52. **Mellors RC, Silver R (1951)** A microfluorometric scanner for the differential detection of cells: Application to exfoliative cytology. Science **114:**356–360.

53. **Meyer-Doering H, Knauer F (1965)** U.S. Patent No. 3,412,254, "Apparatus for Counting Particles Suspended in Transparent Fluids." Filed June 4, 1965. Issued November 19, 1968.

54. **Moldavan A (1934)** Photo-electric technique for the counting of microscopical cells. Science **80:**188–189.

55. **Mullaney PF, Van Dilla MA, Coulter JR, Dean PN (1969)** Cell sizing. A light scattering photometer for rapid volume determination. Rev. Sci. Instr. **40:** 1029–1032.

56. **Ornstein L, Ansley HR (1974)** Spectral matching of classical cytochemistry to automated cytology. J. Histochem. Cytochem. **22:**453–469.

57. **Parker JC, Horst WR (1953)** U.S. Patent No. 2,875,666, "Method of Simultaneously Counting Red and White Blood Cells." Filed July 13, 1953. Issued March 3, 1959.

58. **Reynolds O (1883)** An experimental investigation of the circumstances which determine whether the motion of water shall be direct or sinuous, and of the law of resistance in channels. Phil. Trans. 935.

59. **Saunders AM (1973)** Hemalog D system—Recent developments. In "Advances in Automated Analysis." Vol. 3. Mediad, Inc., Tarrytown, New York.

60. **Saunders AM, Groner W, Kusnetz J (1971)** A rapid automated system for differentiating and counting white blood cells, in Technicon International Congress, 1970, "Advances in Automated Analysis." Vol. 1, Thurman Associates, Miami, 1971, pp 453–459.

61. **Schildkraut ER (1978)** Personal communication (March 22, 1978).

62. **Shapiro HM, Schildkraut ER, Curbelo R, Turner RB, Webb RH, Brown DC, Block MJ (1977)** Cytomat-R: A computer-controlled multiple laser source multiparameter flow cytophotometer system: J. Histochem. Cytochem. **25:**836–844.

63. **Spielman L, Goren SL (1968)** Improving resolution in Coulter counting by hydrodynamic focusing. J. Colloid Interface Sci. **26:**175–182.

64. **Stohr M (1976)** Double beam application in flow techniques and recent results. In Göhde W, Schumann J, Büchner TH (eds), "Pulse Cytophotometry." Second International Symposium. Ghent: European Press, pp 39–45.

65. **Sweet RG (1965)** High frequency recording with electrostatically deflected ink jets. Rev. Sci. Instr. **36:** 131–136.

66. **Van Dilla MA, Basman NJ, Fulwyler MJ (1964)** Electronic cell sizing. Annual Report, Biological and Medical Research Group (H-4) of the Health Division, LASL, July 1963–June 1964, pp 182–204 (LA-3132-MS).

67. **Van Dilla MA, Mullaney PF, Coulter JR (1967)** The fluorescent cell photometer: A new method for the rapid measurement of biological cells stained with fluorescent dyes. Annual Report, Biological and Medical Research Group (H-4) of the Health Division, LASL, July 1966–June 1967, pp 100–105 (LA-3848-MS).

68. **Van Dilla MA, Trujillo TT, Mullaney PF, Coulter JR (1969)** Cell microfluorometry: A method for rapid fluorescence measurement. Science **163:**1213–1214.

69. **Wheeless Jr LL, Wied GL, Patten Jr SF, Bahr GF** (1967) U.S. Patent No. 3,497,690, "Method and Apparatus for Classifying Biological Cells by Measuring the Size and Fluorescent Response Thereof." Filed September 21, 1967. Issued February 24, 1970.

70. **Wheeless LL, Patten SF (1971)** Definition of available cellular parameters with a slit-scan optical system. Acta Cytol. (Praha) **15:**111–112.

71. **Wheeless LL, Patten SF (1973)** Slit-scan cytofluorometry. Acta Cytol. (Praha) **17:**333–339.

72. **Andreeff M, Darzynkiewicz Z, Sharpless TK, Clarkson BD, Melamed MR (1980)** Discrimination of human leukemia subtypes by flow cytometric analysis of cellular DNA and RNA. Blood **55:**282–293.

73. **Bailey JE, Fazel-Madjlessi J, McQuitty DN, Lee Ly, Allred JC, Oro JA (1977)** Characterization of bacterial growth by means of flow microfluorometry. Science **198:**1175–1176.

74. **Bonaly J, Mestre JC (1981)** Flow fluorometric study of DNA content in nonproliferative *Euglena gracilis* cells and during proliferation. Cytometry **2:**35–38.

75. **Carrano AV, Van Dilla MA, Gray J (1979)** Flow cytogenetics: A new approach to chromosome analysis. In Melamed MR, Mullaney PF, Mendelsohn ML (eds), "Flow Cytometry and Sorting," 1st ed. New York: John Wiley & Sons, pp 447–449.

76. **Darzynkiewicz Z, Traganos F, Sharpless T, Melamed MR (1976)** Lymphocyte stimulation: A rapid multiparameter analysis. Proc. Natl. Acad. Sci. USA **73:** 2881–2884.

77. **Darzynkiewicz Z, Traganos F, Staiano-Coico L, Kapuscinski J, Melamed MR (1982)** Interactions of

Rhodamine 123 with living cells studied by flow cytometry. Cancer Res. **42**:799–806.

78. **Darzynkiewicz Z, Traganos F, Kapuscinski J, Staiano-Coico L, Melamed MR** (1984) Accessibility of DNA in situ to various fluorochromes: Relationship to chromatin changes during erythroid differentiation of Friend Leukemia cells. Cytometry **5**:355–363.
79. **Galbraith DW** (1981) Identification and sorting of plant heterokaryons. US Patent 4,300,310.
80. **Galbraith DW, Harkins KR, Maddox JM, Ayres NM, Sharma DP, Firoosabady E** (1983) Rapid flow cytometric analysis of the cell cycle in intact plant tissues. Science **220**:1049–1051.
81. **Galbraith DW, Shields BA** (1982) The effects of inhibition of cell wall synthesis on tobacco protoplast development. Physiol. Plant **55**:25–30.
82. **Gratzner HG, Leif RC, Ingram DJ, Castro A** (1975) The use of antibody specific for bromodeoxyuridine for the immunofluorescent determination of DNA replication in single cells and chromosomes. Exp. Cell. Res. **95**:88–84.
83. **Gray JW** (ed) (1985) Monoclonal antibodies against bromodeoxyuridine. Cytometry **6**:499–662.
84. **Gray JW, Carrano AV, Steinmetz LL, Van Dilla MA, Moore DH, Mayall BH, Mendelsohn ML** (1975) Chromosome measurements and sorting by flow systems. Proc. Natl. Acad. Sci. USA **72**:1231–1234.
85. **Gray JW, Dean PN, Fuscoe JC, Peters DC, Trask BJ, Van den Engh GJ, Van Dilla MA** (1987) High-speed chromosome sorting. Science **238**:323–329.
86. **Grynkiewicz G, Poenie M, Tsien RY** (1985) A new generation of Ca^{2+} indicators with greatly improved fluorescence properties. J. Biol. Chem. **260**:3440–3450.
87. **Hardy RR, Hayakawar K, Parks DR, Herzenberg LA** (1983) Demonstration of B-cell maturation in x-linked immunodeficient mice by simultaneous three-color immunofluorescence. Nature (Lond.) **306**: 270–272.
88. **Hedley DW, Friedlander ML, Taylor IW, Rugg CA, Musgrove EA** (1983) Method for analysis of cellular DNA content of paraffin-embedded pathological material using flow cytometry. J. Histochem. Cytochem. **31**:1333–1335.
89. **Hoffman JF, Laris PC** (1974) Determination of membrane potentials in human and amphiuma red blood cells by means of a fluorescent probe. J. Physiol (Lond.) **239**:519–552.
90. **Ingram M, Cleary TJ, Price BJ, Price RL, Castro A** (1982) rapid detection of *Legionella pneumophilia* by flow cytometry. Cytometry **3**:134–137.
91. **Johnson LV, Walsh ML, Chen LB** (1980) Localization of mitochondria in living cells with Rhodamine 123. Proc. Natl. Acad. Sci. USA **77**:990–994.
92. **Kohler G, Milstein C** (1975) Continuous cultures of fused cells secreting antibody of predefined specificity. Nature (Lond) **256**:495–497.
93. **Kung PC, Goldstein G, Reinherz EL, Schlossman SF** (1979) Monoclonal antibodies defining distinctive human T cell surface antigens Science **206**:347–349.
94. **Loken MR, Parks DR, Herzenberg LA** (1977) Two-color immunofluorescence using a fluorescence activated cell sorter. J. Histochem. Cytochem. **25**:899–907.
95. **Martin JC, Swartzendruber DE** (1980) Time: A new parameter for kinetic measurements in flow cytometry. Science **207**:199–201.
96. **Oi VT, Glazer AN, Stryer L** (1982) Fluorescent phycobiliprotein conjugates for analysis of cells and molecules. J. Cell. Biol. **93**:981–986.
97. **Pau AS, Cowles JR, Oro JA** (1977) Flow microfluorometric analysis of *Escherichia coli, Rhizobium meliloti* and *Rhizobium japonicum* at different stages in the growth cycle. Can. J. Microbiol. **23**:1165–1169.
98. **Pau AS, Lee D, Cowles JR** (1977) Comparison of nucleic acid content in free-living and symbiotic Rhizobium meliloti by flow microfluorometry. J. Bacteriol. **129**:1156–1158.
99. **Price BJ, Kollman VH, Salzman GC** (1978) Light-scatter analysis of microalgae: Correlation of scatter patterns from pure and mixed asynchronous cultures. Biophys. J. **22**:29–36.
100. **Redenbaugh K, Ruzin S, Bartholomew J, Bassham JA** (1982) Characterization and separation of plant protoplasts via flow cytometry and cell sorting. Z. Pflanzenphysiol. **107**:65–80.
101. **Reinherz EL, Kung PC, Goldstein G, Schlossman SF** (1979) A monoclonal antibody with selective reactivity with functionally mature thymocytes and all peripheral human T cells. J. Immunol. **123**:1312–1317.
102. **Reinherz EL, Kung PC, Goldstein G, Schlossman SF** (1979) Separation of functional subsets of human T cells by a monoclonal antibody. Proc. Natl. Acad. Sci. USA **76**:4061–4065.
103. **Shapiro HM** (1981) Flow cytometric estimation of DNA and RNA content in intact cells stained with Hoechst 33342 and Pyronin Y. Cytometry **2**:143–150.
104. **Shapiro HM, Natale PJ, Kamentsky LA** (1979) Estimation of membrane potentials of individual lymphocytes by flow cytometry. Proc. Natl. Acad. Sci. USA **76**:5728–5730.
105. **Skarstad K, Steen HB, Boye E** (1983) Cell cycle parameters of slowly growing *Escherichia coli* B/r studied by flow cytometry. J. Bacteriol. **154**:656–662.
106. **Steen HB, Boye E, Skarstad K, Bloom B, Godal T, Mustafa S** (1982) Applications of flow cytometry on bacteria; cell cycle kinetics, drug effects, and quantitation of antibody binding. Cytometry **2**:249–257.
107. **Stryer L, Glazer AN** (1985) Phycobiliprotein fluorescent conjugates. US Patent 4,542,104.
108. **Stryer L, Glazer AN, Oi VT** (1985) Fluorescent conjugates for analysis of molecules and cells. US Patent 4,520,110.
109. **Tanke HJ, Nieuwenhuis IAB, Koper GJM, Slats JCM, Ploem JS** (1981) Flow cytometry of human reticulocytes based on RNA fluorescence. Cytometry **1**:313–320.
110. **Thorell B** (1980): Intracellular red-ox steady states as basis for cell characterization by flow cytofluorometry. Blood Cells **6**:745–751.
111. **Titus JA, Haugland R, Sharrow SO** (1982) Texas red, a hydrophilic red-emitting fluorophore for use with fluorescein in dual parameter flow microfluorimetric and fluorescence microscopic studies. J. Immunol. Methods **50**:193–204.
112. **Traganos F, Darzynkiewicz Z, Sharpless T, Melamed MR** (1977) Simultaneous staining of ribonucleic and deoxyribonucleic acids in unfixed cells using acridine

orange in a flow cytofluorometric system. J. Histochem. Cytochem. 25:46–56.

113. **Trask BJ, Van den Engh GJ, Eigershuizen JHBW** (1982) Analysis of phytoplankton by flow cytometry. Cytometry 2:258–264.

114. **Trask B, Van den Engh G, Landegent J, Janssen in de Wal N, Van der Ploeg M** (1985) Detection of DNA sequences in nuclei in suspension by in situ hybridization and dual beam flow cytometry. Science **230:** 1401–1403.

115. **Valet G, Bamberger S, Hoffmann H, Schindler R, Ruhenstroth-Bauer G** (1979) Flow cytometry as a new method for the measurement of electrophoretic mobility of erythrocytes using membrane charge staining by fluoresceinated polycations. J. Histochem. Cytochem. **27**:342–349.

116. **Van den Engh GH, Trask BJ, Visser JWM** (1985) Flow cytometer for identifying algae by chorophyll fluorescence. US Patent 4,500,641.

117. **Van Dilla MA, Langlois RG, Pinkel D, Yajko D, Hadley WK** (1983) Bacterial characterization by flow cytometry. Science **222**:620–622.

118. **Visser JWM, Jongeling AAM, Tanke HJ** (1979) Intracellular pH determination by fluorescence measurements. J. Histochem. Cytochem. **27**:32–35.

2

Characteristics of Flow Cytometers

Harald B. Steen
Institute for Cancer Research, The Norwegian Radium Hospital, Montebello, Oslo 3, Norway

INTRODUCTION

The flow cytometer has been around for about two decades. During that time it has had significant impact on various fields of biology and medicine, including cell-cycle studies in relation to effects of drugs and radiation, immunology, ploidy determination in cancers, and studies of cellular parameters, such as intracellular pH and Ca^{2+} concentrations. In flow cytometry, cells are labeled with fluorescent molecules that bind specifically to the constituent(s) to be measured. For example, the DNA may be stained with propidium iodide or mithramycin, while other constituents may be labeled with (monoclonal) antibodies conjugated with some fluorescent dye such as FITC or phycoerythrin. Carried by a microscopic jet of water, the cells pass one by one through an intense beam of excitation light in the measuring region of the flow cytometer (Fig. 1). Each cell thereby produces a short flash of fluorescence, the intensity of which is proportional to the cellular content of the fluorescently labeled constituent. These flashes of fluorescence are collected by appropriate optics, which focus the light on a sensitive detector. The detector transforms the flashes of light into electrical pulses, which are measured and recorded by some electronics and a computer. Each cell also causes scattering of the excitation light. The intensity of this scattering is a function of the size, shape and structure of the cell. The resulting flash of scattered light is recorded by a separate detector. Thus, the cellular content of several constituents, labeled with dyes fluorescing at different wavelengths as well as size and shape or structure, are recorded for each individual cell. In contrast to other biomedical techniques that generally give averages over large numbers of cells, flow cytometry measures individual cells in large numbers. Hence this technique makes it possible to distinguish subpopulations of cells as, for example, in analyses of asynchronously growing cell cultures, where cells in the different phases of the cell cycle are readily distinguished, or in immunology, where the flow cytometer discriminates between different subsets of lymphocytes.

The flow cytometer is a remarkable and fascinating instrument from a technical point of view as well as with regard to performance. It employs a unique blend of modern technologies, including fluidics, lasers, optics, analogue and digital electronics, and computers and software. It can measure several parameters of each individual cell that is passed through its flow chamber to determine size, structure, and the precise contents of various cellular constituents. It can measure cells and other particles all the way down to submicroscopic sizes, that is, to ~ 0.1 μm. The sensitivity is sufficient to detect 10^{-18} g of a specific substance per cell. And such measurements can be carried out with a precision of a few percent and at a rate of several thousand cells per second. Some instruments even have the facility to sort cells "on line," that is, separate them physically, according to the values of the parameters measured, hence the name "cell sorter," which has wrongly been adopted also for the majority of flow cytometers that do not have the sorting facility. In fact, a more appropriate name is "flow microphotometer," since these instruments may be employed for analysis of all kinds of microscopic particles, not only biological cells.

This chapter describes the main components of flow cytometers, including light sources, optics, flow chambers, sample delivery systems, light detectors, and electronics, and how different types of instruments are configured to use these components optimally. The performance of flow cytometers in terms of sensitivity, measuring precision and rate of measurement, is analyzed and the benefits and shortcomings of different types of instruments are discussed. The emphasis is on general features and principles, whereas more in-depth and detailed treatment of the various subjects will be found in later chapters. This chapter is intended as an introduction and reference for the following chapters, which focus more closely on some of the various parts and functions of flow cytometers and on some of the many applications of this technique.

OPTICAL SYSTEMS

Optical Design of Laser-Based Instruments

Most laser-based flow cytometers have essentially the same optical layout, that is, an orthogonal configuration with the three main axes of the instrument, the sample flow, the laser beam, and the optical axis of fluorescence detection at right angles to each other as shown in Figure 2.

The laser light is focused on the sample flow by means of

Flow Cytometry and Sorting, Second Edition, pages 11–25

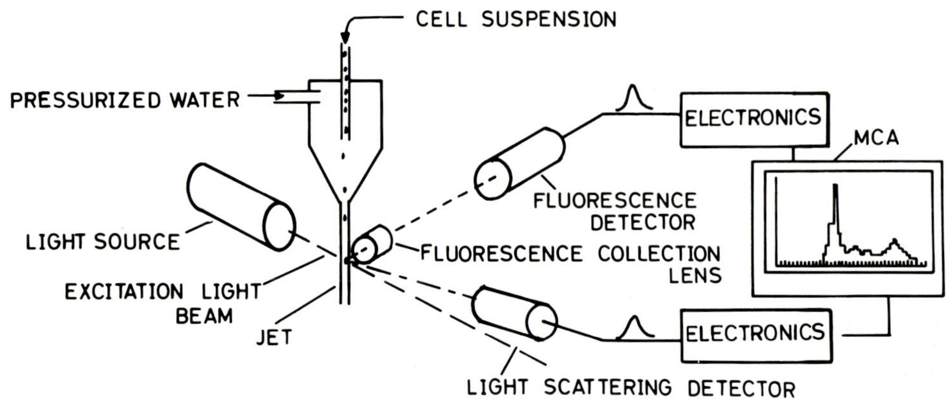

Fig. 1. Schematic representation of a flow cytometer with detectors for fluorescence and light scattering.

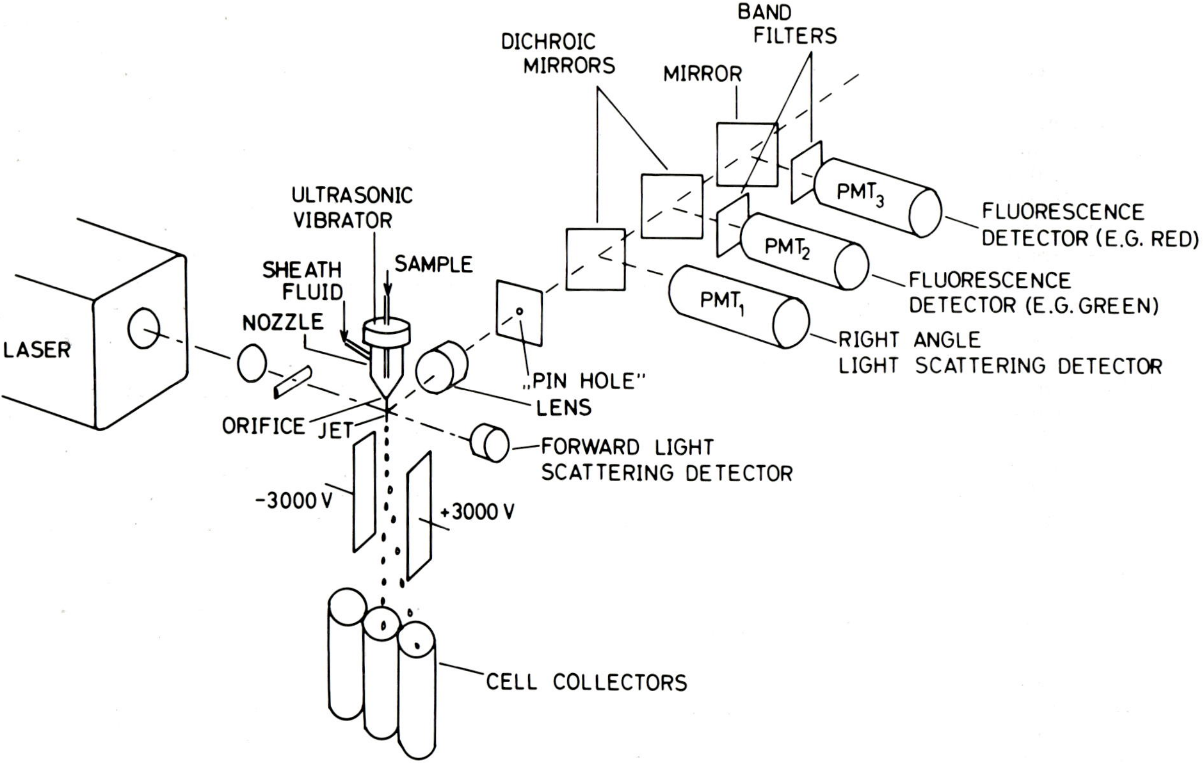

Fig. 2. Laser-based flow cytometer with orthogonal configuration. Right-angle light scattering and two fluorescence components may be detected through a microscope objective or equivalent lens perpendicular to the laser beam and the water jet carrying the cells. Forward-angle scattering is measured by means of a solid-state detector close to the laser beam. A piezoelectric crystal causes the nozzle to vibrate at a frequency of the order of 40 khz so that the water jet breaks into droplets for cell sorting by electrostatic deflection.

one or two lenses. At least one of these is a cylindrical lens that produces a flat focus with an elliptical cross section, the short axis of which is in the direction of sample flow. The intensity profile of the laser focus is approximately gaussian. The width (at l/e^2 of maximum intensity) of this focus perpendicular to the sample flow is typically in the range upward from 100 μm. The width of the focus is a compromise between the need for high sensitivity, which calls for a narrow focus to concentrate as much light as possible on the sample, and high resolution with less critical alignment of the cell stream, which requires a broader focus to allow a larger beam width and less precise positioning of the sample flow. For example, a width of 150 μm implies that the intensity is constant to within 1% over a region of 10 μm, assuming a gaussian intensity distribution. Hence, in order to be measured with this precision, the width of the track of cells through the excitation focus must be well within 10 μm. It should also be noted that for a "jet-in-air" flow chamber (see the section on Flow Chambers) a cylindrical jet of water has a significant focusing effect. For example, in an 80 μm-diameter jet the above 10 μm will be reduced to 8 μm at the center of the jet.

Some instruments employ two lasers. The laser beams are brought to closely adjacent but separate focuses so that the cells are excited sequentially with two different wavelengths. Hence it becomes possible to measure two cellular components with dyes that require different excitation wavelengths.

Fluorescence is detected at right angles to the laser beam through optics, which may be either a standard microscope objective or a specially designed lens. In order to collect as much fluorescence as possible the numerical aperture of these optics should be as large as possible. In instruments with a "jet in air" flow chamber the maximum practical numerical aperture is NA = 0.6. A lens with NA = 0.6 collects approximately 10% of the light emitted from an isotropic light source in its focus, if the light source is in air. When the light source is in water, however, as is the case in flow cytometers, collection efficiency drops to about 5%. Hence only a small fraction of the fluorescence is actually collected for detection in instruments having this type of optics.

An opaque screen with a small opening—the "pinhole"—is situated in the image plane of the detection optics. As explained below, this aperture has an important function, namely, to eliminate light from sources other than the cells and thereby increase the signal to noise ratio.

The fluorescence detection optics also collects scattered light from the cells, that is, light scattered at about right angles relative to the laser beam. Whereas the scattered light is of the same wavelength as the laser light, the fluorescence is always shifted toward higher wavelengths. Hence scattered light and fluorescence may be separated by a dichroic mirror, which is situated at 45° angle to the light beam behind the detection optics (Fig. 2). This mirror reflects the scattered light onto one detector, while the longer wavelength fluorescence is transmitted to another detector. Since the color separation of the dichroic mirror is not perfect, additional filters are used in front of each detector.

The same system is used for independent detection of two spectral components of the fluorescence, as is required for cells stained with two different dyes to be measured simultaneously. In that case the scattered light may be eliminated by a suitable longpass cutoff filter. Alternatively, all three components, that is, scattered light and two fluorescence colors, may be measured by adding a second dichroic mirror and a third detector (Fig. 2). In principle, such a stack of dichroic mirrors and detectors with specific filters may be extended to record any number of color components. In practice the number is usually limited to three.

In addition to fluorescence and light scattering from the cells, a background of light from other sources enters the detection optics. The major component of this background is laser light scattered off the cylindrical surface of the water jet, if a "jet-in-air" flow system is used, or from the surfaces of the flow channel if a closed flow chamber is used (see the section on Flow Chambers). In particular, the cylindrical jet surface reflects an enormous intensity of light in the plane perpendicular to the jet, as compared to the intensity of the light scattering and fluorescence of cells. To prevent this light from entering the detection optics, a field stop in the shape of a narrow strip of metal is placed horizontally across its front lens. Although this field stop reduces the effective aperture of the optics, it is a prerequisite to keep background intensity at tolerable levels in such systems.

Laser-based flow cytometers also have a detector for "low-angle" or "forward" light scattering measurement, that is, for light scattered within a relatively narrow cone around the laser beam (Fig. 2). Light scattering at low scattering angles, below 10°, is dominated by diffraction from the contour of the cell and thus depends primarily on its size. In contrast, the scattering at larger angles, for example, 90°, is a function also of the submicroscopic structure of the cell. That is, its intensity increases with what may be called the granularity of the cell.

Optical Design of Arc Lamp-Based Instruments

All arc lamp-based flow cytometers employ essentially the same optical system as that used in fluorescence microscopes with incident (epi-) illumination. As shown in Figure 3, epi illumination means that the fluorescence is collected through the same lens as that used to focus excitation light on the sample. The lens is usually a microscope objective. The numerical aperture (NA) of this lens must be as high as possible in order to concentrate as much light as possible on the target and collect a maximum of fluorescence. Hence oil or glycerin immersion optics with NA ≈ 1.3 is usually employed.

The amount of excitation light collimated by the microscope lens and the distribution of excitation light intensity in the object plane depends on the illumination system: Kohler or critical illumination. In Kohler illumination the arc is imaged in the back focal plane of the microscope lens, which produces a nearly constant excitation intensity across the object field. This makes the detection relatively insensitive to fluctuations in the position of the sample stream as well as to lamp flickering, that is, abrupt movements of the arc caused by deterioration of lamp electrodes. With critical illumination the arc is imaged in the object plane. This produces a smaller focal spot than does Kohler illumination and consequently a higher excitation intensity in the center of the object field through which the sample flow passes. Hence critical illumination facilitates a somewhat higher sensitivity than Kohler illumination. On the other hand, an instrument with critical illumination is more sensitive to sample stream fluctuations and lamp flickering.

In the epiillumination system excitation light and fluorescence are separated by a dichroic mirror behind the microscope lens (Fig. 3). Ideally, the dichroic mirror reflects all light below a certain wavelength and transmits all light above that wavelength. In reality, however, the transmission versus wavelength curve is not a step function, but rather a sigmoid curve having a 10–90% width of about 20 nm. Dichroic mirrors may also have significant transmission, 1–2%, even well below the threshold wavelength. Additional excitation and transmission filters are employed to avoid excitation light reaching the fluorescence detector. A high signal to noise ratio, that is, sensitivity, in this type of instrument is critically dependent on the smallest possible overlap between the transmission bands of the excitation and emission filters, as well as efficient blocking of wavelengths outside the transmission bands. As in laser-based instruments, a "pinhole" in the image plane of the detection optics is essential to eliminate background and thus increase signal to noise ratio. As discussed below, optimal size and positioning of the "pinhole" is of particular importance in arc lamp-based instruments.

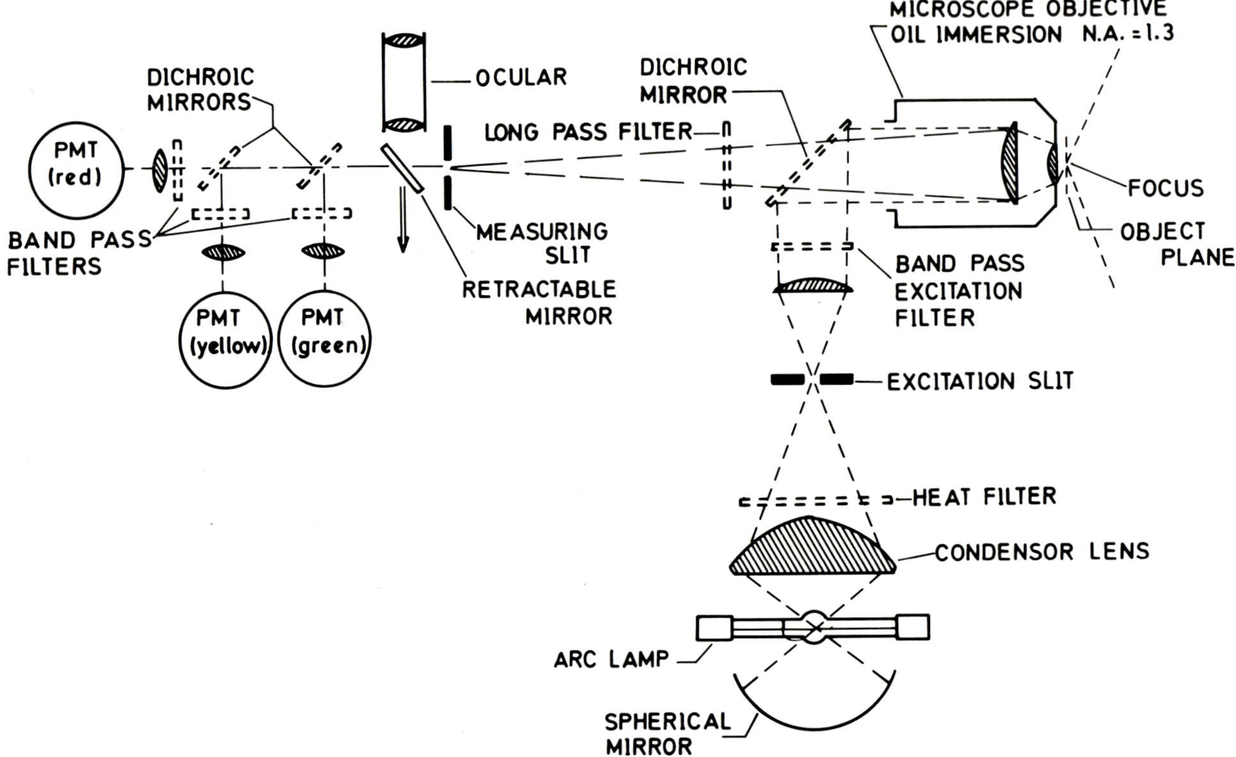

Fig. 3. Schematic representation of the optical design used in most arc lamp-based flow cytometers. In this epiillumination design the focus, through which the sample is passing, is illuminated through the same microscope objective that collects the fluorescence. Excitation light and fluorescence are separated by a dichroic mirror, which reflects the wavelengths of the excitation light while it transmits the longer wavelengths of the fluorescence. The long pass filter behind the dichroic mirror prevents traces of excitation light from entering the detection system. It is of crucial importance that there is no overlap between the transmission spectra of this filter and the excitation filter. The excitation slit serves to limit the excitation light to that portion of the object plane which contains the sample flow and thus reduce background light. Likewise, the measuring slit prevents background light from other parts of the object plane from reaching the detectors and thereby improves the signal to noise ratio of the detection.

Scattered light detection in arc lamp-based instruments. The use of immersion optics with high numerical aperture implies a large illumination field angle, about 140°, which makes light scatter detection in arc lamp-based instruments less straightforward than in laser-based instruments that employ an almost parallel excitation light beam. As shown in Figure 4, however, light scattering measurements in arc lamp-based instruments are feasible, provided critical illumination is employed. The dark field produced in this system has high contrast and allows detection optics for light scattering with NA = 0.4 or larger. The resulting sensitivity facilitates detection of particle sizes to about 0.1 μm. The system also facilitates separate detection of low-angle and large-angle scattering. Results obtained with simultaneous measurement of the two scattering components are closely similar to those obtained in laser-based instruments with forward- and right-angle scatter detection [8].

Background Rejection

As discussed in more detail below, the detection limit of flow cytometers is determined not only by the intensity of fluorescence and light scattering that can be excited in a cell and the efficiency by which this light can be collected and converted into electrical signals. It is equally dependent on the background of light from other sources from which the light pulses from the cells have to be distinguished. This background consists of excitation light scattered mainly from the jet or flow chamber, fluorescence and scattered light from optical components and filters, and Raman scattering from the water and the flow chamber material.

The detection limit depends on the number of molecules per cell producing light pulses that are sufficiently large to be discerned from the "noise" of the background. Thus, it is the "signal to noise ratio" that determines the detection limit. Suppression of all light other than that emitted by the cells is therefore essential.

Raman scattering is "inelastic scattering" of light, which means that Raman scattered light has a longer wavelength than the excitation light and therefore overlaps the fluorescence in many cases. For example, the main Raman scattering of water illuminated with 488 nm is concentrated in two narrow bands at about 529 and 584 nm. Hence the Raman scattering is not eliminated by the "longpass" cutoff filter which is used to block scattered light, and will therefore contribute to the background in the fluorescence detection if not eliminated by an appropriate filter. Fluorescence from the flow chamber material and other components exposed to excitation light will also add to this background.

In order to suppress background luminescence an opaque

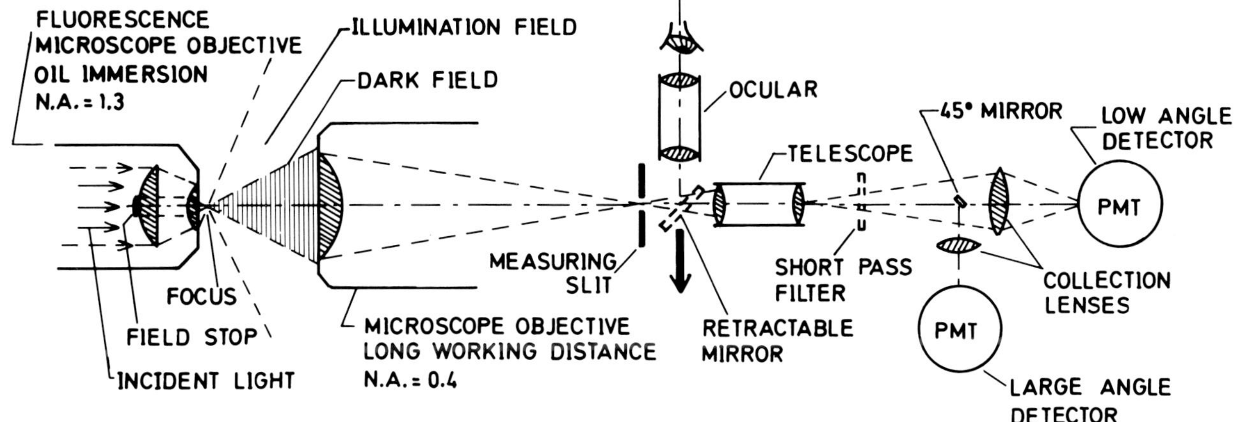

Fig. 4. Device for measuring light scattering in arc lamp-based flow cytometers with epiillumination through high numerical aperture (oil immersion) optics. A central field stop in the illumination lens produces a conical shadow in the illumination field which is pointed in the measuring focus through which the sample flow is passing. The shadow extending to the right of the focus contains only scattered light and fluorescence. The microscope lens looking into this shadow forms a scattered light image on the measuring slit which acts as a spatial filter preventing scattered light from outside the sample flow to reach the detector(s). This image can be viewed through an ocular. Whereas the periphery of the dark field is dominated by light scattered from low angles, the central core contains only light scattered from large angles, for example, above 18°. A telescope recreates the dark field behind the spatial filter. A 45° mirror on the axis of the field reflects the central core, that is the large-angle scattering, onto one detector, whereas the peripheral part, the low-angle scattering, is passed on to another detector, thereby facilitating separate detection of the two scattering components.

screen with a small opening, a so-called "pinhole," may be situated in the image plane of the fluorescence collecting optics. With proper alignment of the instrument the image of the illuminated part of the sample flow falls within this pinhole, whereas light from other sources is largely eliminated. Thus, the pinhole serves a very important function by reducing background and thereby enhancing the signal to noise ratio of the measurement.

Ideally, the pinhole should cover only the image of the illuminated part of the cell stream. However, that would make proper performance of the instrument too critically dependent on perfect alignment of the sample flow and the laser focus relative to the optical axis of the detection optics. Furthermore, the optimal size of the pinhole is different for different applications. Most instruments have a fixed pinhole, the size of which is considerably larger than that required for optimal background rejection. If needed, a significant increase of signal to noise ratio may be achieved by means of an adjustable pinhole, that is, a measuring slit that may be set for the particular application.

The efficiency of the pinhole also depends on the quality of the image formed by the detection optics. A blurred image implies that more background light is mixed into the fluorescence image of the cells. In this regard a microscope objective, or a similar high-quality lens, is preferable to simpler lenses for fluorescence and right-angle light scattering detection.

Filters

The quality of the filters used to eliminate stray light and isolate spectral components is essential to the performance of both laser-based and arc lamp-based flow cytometers. Desirable features are 1) high transmission inside the transmission band and high rejection outside the transmission band, 2) a sharply limited transmission band, 3) low fluorescence, and 4) good stability and light resistance. Two types of filters are used: color filters and interference filters. Color filters, which are plates of glass or polymer containing light-absorbing dyes, are commonly used as "longpass" cutoff filters, that is, filters absorbing all light below a certain wavelength. If not exposed to excessive light intensities or heat, that is, direct excitation light, color filters are generally quite stable. However, such filters are inherently fluorescent, although weakly, and should therefore be situated as far away from the detectors as possible. In particular the filter used to eliminate stray excitation light should be far in front of the pinhole. In order to avoid filter fluorescence, excessive heating and filter deterioration, the excitation filter in arc lamp-based instruments, which is exposed to very high intensities of excitation light, is usually an interference filter. Filter deterioriation is further prevented by the use of a heat filter, that is, a filter blocking the red and infrared part of the spectrum, situated between the lamp and the excitation filter (Fig. 3).

In contrast to color filters, interference filters and dichroic mirrors do not absorb the light they are not passing, but reflect it. Hence they are completely without fluorescence and can withstand much higher light intensities than color filters. They are therefore preferred in the excitation light path of arc lamp-based flow cytometers. The transmission band of interference filters can be made with steeper edges than is possible with color filters. The transmission band wavelength of such filters depends on the angle of incidence. To function optimally, such filters should always be situated in a near-parallel part of the light beam. In most interference filters the reflective layer(s) is confined between two glass plates.

However, in filters used as dichroic beam splitters the reflective surface is usually open and must be carefully protected from scratching. The reflective surface of the dichroic

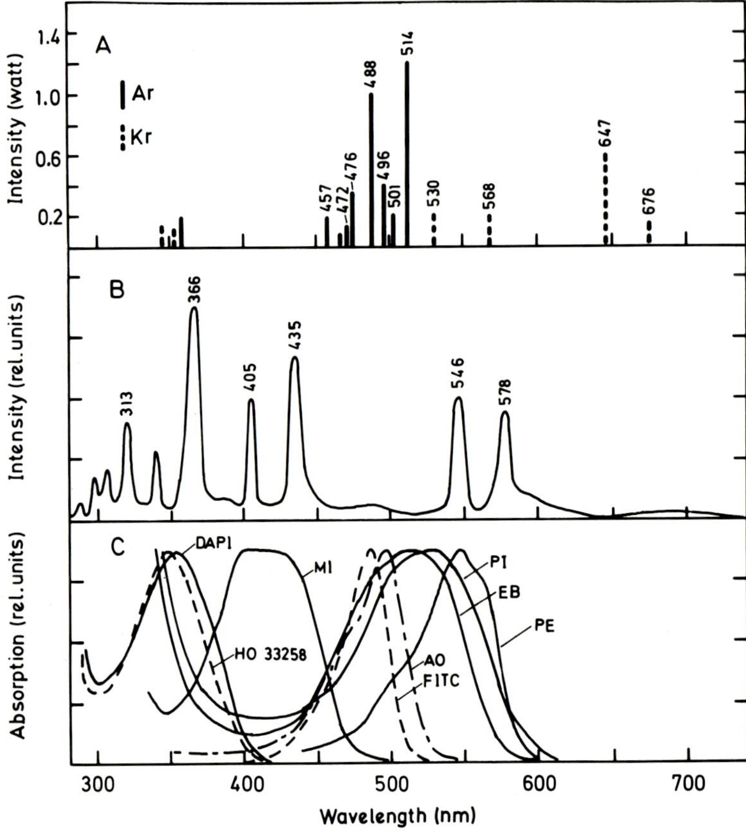

Fig. 5. **A:** The major lines of tunable argon and krypton lasers. Minor lines have been omitted. The lines around 350 nm are actually sums of several weaker lines. The intensity values refer to a 3-W argon laser and a 1.2-W krypton laser. (Data from Spectra Physics.) **B:** The emission spectrum of a high-pressure mercury arc lamp. Note the continuum between the lines. Most microscope optics have inadequate transmission below 350 nm. Hence these wavelengths are not utilized in flow cytometry. **C:** Absorption/fluorescence excitation spectra of some dyes commonly used in flow cytometry. All spectra have been normalized to a common peak value. The spectra of the DNA specific dyes: DAPI (4,6-diamidino-2-phenylindole), Hoechst 33258, Mi (mithramycin), AO (acridine orange), EB (ethidium bromide), and PI (propidium iodide), have been measured for the dyes bound to DNA. FITC (fluorescein-iso-thiocyanate) and PE (phycoerythrin) are the most commonly used fluorescent antibody conjugates.

beam splitter should always face the light source. Interference filters are susceptible to mechanical shock. Transmission may change with time and should be checked at regular intervals, for example, every 6 months. Some filters are a combination of interference and color filters. They should always be used with the reflective side facing the light source.

It is common for even high-quality interference band filters to have some transmission in the far-red and infrared. Arc lamps in particular, but also lasers to some extent, emit a lot of light at these wavelengths. Scattering of this light from the optics and flow chamber may cause additional background and reduce signal to noise ratio. This is remedied by a suitable shortpass filter in the fluorescence pathway.

LIGHT SOURCES

Lasers

The argon (Ar) laser is the standard in laser-based instruments. Most such instruments employ a water-cooled laser with a nominal output, that is, the combined maximum power of all emission lines, between 2 and 5 W. The main emission lines of argon lasers are depicted in Figure 5. Some flow cytometers have provisions for a second laser, which may either be another Ar laser or a Krypton (Kr) laser. As shown in Figure 5, the Kr laser facilitates excitation also in the yellow and red part of the spectrum. In both the Ar and Kr laser the emission lines in the UV require special optics.

These large, water-cooled lasers are expensive to buy and maintain, and are costly in operation. They require large amounts of cooling water (2–4 gal/min) and an electrical supply of 10–15 kW. A new generation of flow cytometers, designed primarily for detection of fluorescently conjugated antibodies on cells, but suitable for DNA measurements as well, are using small air-cooled Ar lasers with a typical output of 10 mW. These lasers are usually pretuned to deliver only the 488 line. Although not on par with the large water-cooled lasers, the stability of the air-cooled versions is adequate for most applications. The low intensity is at least partly compensated by the use of closed flow chambers and immersion optics.

Arc Lamps

The arc lamp used in flow cytometers is usually a 100-W high-pressure mercury (Hg) lamp. In such lamps the arc extends between two electrodes which are about 0.5 mm apart. Thus, the electrical effect is confined to a volume of about 0.02 mm^3, which yields the extremely high light densities of

TABLE 1. Excitation Intensities as Measured in an Arc Lamp-Based Flow Cytometer with Critical Illumination from a 100-W High-Pressure Hg lamp

Excitation bandpass (nm)	Typical application	Intensity (mW)
355–375	DAPI, Hoechst 33258	15
395–440	Mithramycin, chromomycin A_3	24
390–490	Acridine orange	28
450–490	Acridine orange	6
470–490	Fluorescein, FITC	3
520–560	Ethidium bromide, propidium iodide, phycoerythrin R	20
546/12	Phycoerythrin B	13
578/12	Texas red, 7-AMD, various cyanines for membrane potential and antibody conjugation	10

these light sources. These lamps must be operated on stabilized DC current in order to keep the light output constant. It should be noted that the excitation intensity cannot be increased by using a similar lamp with higher power, for example, a 200-W or a 500-W lamp. The reason is that the light density (flux) of the arc in the 100-W lamp is significantly greater than in the larger lamps of the same design; and, according to the laws of optics, it is not possible to focus a light source (except lasers) to a higher light density than that of the source itself. For some purposes, in particular excitation of fluorescein and fluorescein isothiocyanate (FITC), a somewhat higher excitation efficiency may be achieved by the use of a 75-W high pressure xenon (Xe) arc lamp. Again, larger lamps exhibit lower light densities and thus cannot produce excitation intensities comparable to the 75-W lamp.

The spectrum of the Hg arc lamp, shown in Figure 5B, consists of several narrow emission lines superimposed on a continuum. The total excitation intensity in an instrument with Kohler illumination has been measured to be about 10 mW when using an excitation filter, which transmits predominantly the 366 nm line and 30 mW with a filter transmitting in the region 340–500 nm [3]. Similar measurements on an instrument with critical illumination gave somewhat higher intensities as shown in Table 1. The spectrum of the Xe lamp is a continuum extending from UV through all of the visible spectrum. However, with regard to intensity it does not match the Hg lamp except in the region between 450 and 540 nm where the Hg lamp has no significant emission lines.

The light output of arc lamps decays gradually with time. The rate of decay for the Hg lamp is of the order of 2% per day (8 hours), while that of the Xe lamp is considerably less. The rated lifetime of the 100-W Hg lamp varies from 200 to 1,000 hours depending on the manufacturer. The corresponding value for Xe lamps ranges upward from 400 hours. For both types of lamps ignition of the lamp is expensive in terms of lifetime. These lamps should be turned off only when they are not to be used for at least 1 hour.

Arc lamps have a certain tendency to flicker, that is, occasional arc movements caused by deterioration of the lamp electrodes. Such flickering may cause abrupt changes in the excitation intensity of a few percent. As noted below, this problem can be remedied by a compensation device.

DETECTORS

Photomultiplier Tubes

The detectors used for fluorescence and right angle light scattering are photomultiplier tubes (PMT). The PMT is a vacuum tube with a light-sensitive photocathode that, with a certain probability (quantum efficiency), releases an electron when hit by a photon of light. Typically, the light pulses reaching the PMTs of a flow cytometer contain from a few tens to many thousands of photons. The electrons released by the light are multiplied when they travel down a chain of electrodes, driven by a voltage between the electrodes. The electrodes, so-called dynodes, are covered with an electron-emissive material that releases several electrons when struck by one. PMTs typically have on the order of 10 dynodes. The multiplication factor, which increases exponentially with the voltage across the dynode chain, usually goes to about 10^6. The PMT is a current amplifier which transforms pulses of light into equivalent electrical pulses. There are two main types of PMTs: "end-on" tubes that have a semitransparent, photosensitive cathode on the flat end of a cylindrical glass tube and "side-window" tubes that have an opaque photocathode behind a window at the side of a cylindrical glass tube. The most important difference between the two types with respect to flow cytometry is that the sensitivity of the end-on tubes is essentially constant across the photocathode, whereas the sensitivity of side-window tubes may be rather sharply peaked somewhere near the center of the cathode. Hence end-on tubes may be diffusely illuminated and are practically insensitive to the exact position of the tube relative to the entering light beam, whereas in order to fully utilize a side-window tube the light must be focused onto the most sensitive part of the photocathode, that is, on an area that is typically much less than 1 cm^2. This means that optimal functioning of a side-window tube is critically dependent on correct positioning relative to the entering light beam. Optimal positioning of such tubes should be checked, especially upon replacement.

PMTs for flow cytometry are selected for high cathode sensitivity and appropriate spectral range. End-on tubes for fluorescence detection in flow cytometers usually have an S-20 cathode. For the right-angle light scattering and blue fluorescence an S-11 cathode may also be used. Side window tubes usually have a GaAs cathode, which has a higher sensitivity than the S-20 type in the yellow and red part of the spectrum (see Fig. 6). Note that in the wavelength region where these PMTs are employed in flow cytometers, above 400 nm, the maximum cathode sensitivity, or "quantum efficiency," that is, the probability that a photon is actually detected, is below 20% in both types of tubes.

Cathode sensitivity should not be mixed up with "overall sensitivity," usually given in units of amp/lumen. This parameter is not critical for the performance of a PMT in a flow cytometer. "Dark current" is also immaterial, since it is normally too low to affect the measurement. What may be important, however, is "maximum current," which determines the maximum intensity of light (including background) that can be detected with linear response, that is, with an output signal proportional to light intensity. High background levels may cause the PMT to reach maximum current. In that case all light pulses above a certain intensity

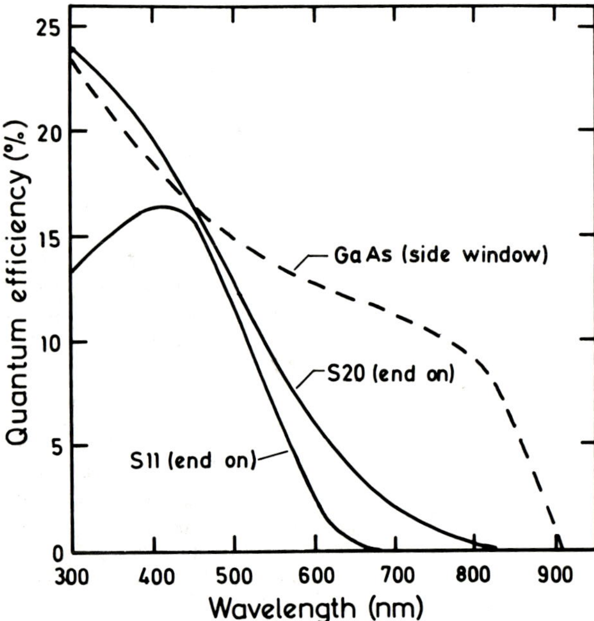

Fig. 6. The quantum efficiency, i.e., the probability that a photon will release an electron and hence be detected, of PMT photocathodes commonly used in flow cytometers. (Data from Hamamatsu Corp.)

will produce the same output pulse and hence be falsely detected as being equal.

PMTs may be permanently damaged if exposed to light intensities much in excess of that causing maximum current even if only for a few seconds. Otherwise they may be operated for years without deterioration.

Solid-State Detectors

The intensity of the low-angle scattering in laser-based instruments is generally much higher than that of the fluorescence and right-angle scattering. Low-angle scattering may be detected by a "solid state detector," usually configured as a flat, circular plate with its center in the laser beam. The laser beam itself is prevented from reaching the detector by means of a suitable field stop or "beam sink." Solid state detectors are less fragile, tolerate higher light intensities, and are less expensive than photomultiplier tubes. In fact, the sensitivity, that is, quantum efficiency, of solid-state detectors may be well above that of PMTs. However, the noise of these devices is so much larger than that of PMTs that the resulting signal to noise ratio, which is really the critical parameter, cannot match that of the PMT. The rapid development of solid-state detectors may change this situation within a not-too-distant future.

FLOW CHAMBERS

Flow Chambers for Laser-Based Flow Cytometers

All flow cytometers employ some type of nozzle with hydrodynamic focusing in order to make the cells follow the narrowest possible path through the measuring region (see Chapter 3, this volume). The bulk of the water that flows in

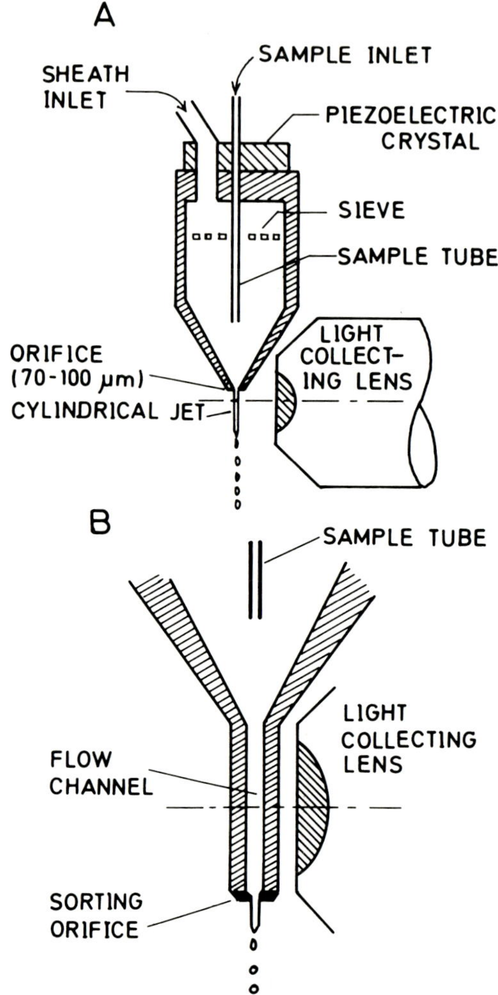

Fig. 7. Semischematic drawings of nozzles used in laser-based instruments. **A:** Jet in air type. **B:** The flow chamber and orifice of a closed chamber. In order to produce hydrodynamic focusing of the sample it is essential that the sheath flow be perfectly laminar. A sieve may thus be introduced to prevent any turbulence at the sheath inlet from reaching the conical part of the nozzle. The piezoelectric crystal vibrates the nozzle and thereby breaks the water jet into droplets for electrostatic cell sorting.

a laminar fashion through the nozzle, the "sheath flow," is usually supplied from a pressurized container, while the sample is introduced through a thin tube along the center of the stream in the nozzle. The sample is thereby confined to the central core of the laminar flow that is leaving the nozzle. This is called hydrodynamic focusing. The confinement of the sample flow allows the excitation focus to be narrowed and excitation intensity to be increased. The sheath flow, W, is typically around 5 ml/min, whereas the sample flow, w, is usually below 100 μl/min. Assuming perfect laminarity, the diameter of the sample flow, d, is given by Equation 1:

$$d = (w/W)^{1/2} \cdot D = 2(w/\pi v)^{1/2}, \qquad (1)$$

where D is the diameter of the flow leaving the nozzle and v the flow velocity. Thus, the width of the sample flow in-

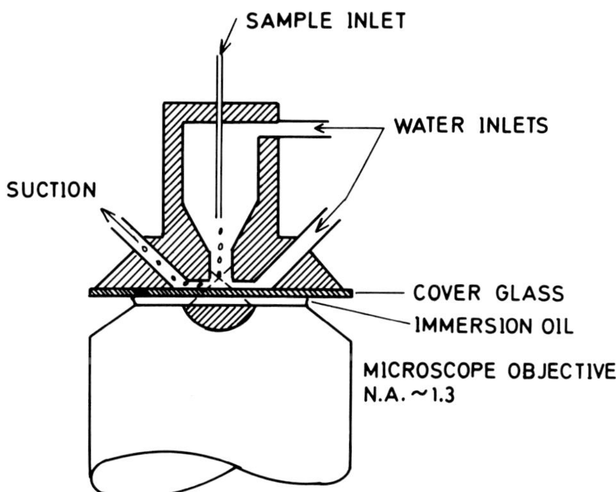

Fig. 8. Closed flow chamber used in arc lamp-based flow cytometer with epiillumination. The sample is introduced along the optical axis of the microscope objective through a nozzle with hydrodynamic focusing. After passing through the focus it is deflected from this path by a secondary flow of water running through a channel on the surface of the cover glass. The optics are coupled to the other side of the coverglass by means of immersion oil. The cells pass through the focus where they are measured while they are still moving on the optical axis. This feature, together with the high numerical aperture of the optics, makes this type of instrument insensitive to the shape of the cells and how they are oriented in the flow.

creases with the sample flow, and can be reduced by increasing the sheath flow.

In the jet-in-air type of flow chamber used in laser-based flow cytometers, the conical nozzle ends in an orifice from which emerges a laminar jet intersected by the laser beam just below the orifice (Fig. 7A). The orifice diameter is usually in the region 70–100 μm, but such nozzles work as well with larger diameters. In the closed flow chamber of laser-based instruments the nozzle leads into a tube, preferably of rectangular cross section both inside and outside (Fig. 7B). The tube is around 200 μm inside. If the nozzle is used in a sorting instrument, the tube ends in an orifice with a diameter of 70–100 μm. This orifice is essential for the formation of droplets in cell sorting.

Some closed flow chambers permit the use of immersion optics for fluorescence collection. Such optics typically have NA = 1.3 which implies that it collects about 10 times more fluorescence than do the optics used with jet in air systems where the aperture is limited to about NA = 0.6.

In order to maintain a laminar jet in air, its velocity cannot be reduced below a critical limit, which for a 100-μm orifice is about 3 m/sec (Chapter 3, this volume). A closed flow chamber has no such limit and may thus be operated at much lower flow velocities. It therefore facilitates higher sensitivity than the jet in air (see Eq. 6). A closed chamber, having a square cross section can also be used with immersion optics which further increase sensitivity. It also has the advantage that the intense reflection of excitation light from the cylindrical jet in air is avoided. On the other hand, contamination of the channel of the closed chamber may cause unstable flow as well as significant background of scattered light and fluorescence and is time-consuming to remedy.

Flow Chambers for Arc Lamp-Based Flow Cytometers

Both closed and open types of flow chambers are used with arc lamp-based flow cytometers. In one type of closed chamber the cells enter the object plane along the optical axis of the microscope objective (Fig. 8). After passing through the focus the cell path is deflected 90° by means of an additional flow perpendicular to the optical axis. In its present form this type of flow chamber does not facilitate detection of scattered light. In another type of closed chamber the cells are illuminated and measured while they are passing through a short tube of circular cross section. This tube is also used as a coulter orifice, which allows the electrical volume of cells to be measured simultaneously with fluorescence. It also permits detection of scattered light through a light guide situated outside the illumination field. The sensitivity of this light-scattering detection is relatively low, however, partly because of the small aperture of the light guide and partly because of the relatively high level of background light arising primarily from scattering off the tube wall.

An open type of flow chamber being used with an arc lamp-based instrument is shown in Figure 9 [5]. This type of flow chamber may be optically coupled to oil immersion microscope optics with NA = 1.3, thereby facilitating a high fluorescence collection efficiency. The open configuration of this flow chamber allows light-scattering detection with optics of fairly high aperture, for example, NA = 0.4 (Fig. 4). The flow on the glass surface, produced by a jet impinging at an oblique angle, is laminar, flat, and stable. Having a minimum number of reflecting surfaces, all of them perpendicular to the optical axis, this type of flow chamber yields very low background scattered light, which results in a high signal to noise ratio for both fluorescence and scattered light detection. It is also much simpler to clean than closed flow chambers.

This type of open flow chamber allows flow velocities below the critical limit if the nozzle orifice is situated close enough to the open glass surface to be directly connected to the flow on this surface by a meniscus of water. Thus, the flow velocity may be reduced from the usual 10–20 m/sec to a few cm/sec.

The flow chamber is a crucial part of any flow cytometer. Most instrument failures are due to malfunction of the nozzle caused by clogging, or a distortion of the flow produced by some particulate matter sticking in or close to the orifice, or to contamination of the flow channel of closed chambers. This causes breakdown of laminar flow and/or deviation of the sample flow, so that it is directed outside the central portion of the excitation focus. Increasing the diameter of the orifice makes the nozzle less susceptible to this problem though the consumption of water increases with the square of the orifice diameter, assuming constant velocity. On the other hand, it may be noted from Equation 1 that, for the same flow velocity, the diameter of the sample flow, that is the precision of the cell path, is independent of the orifice diameter.

SAMPLE INJECTION SYSTEMS

A prerequisite for optimal performance of a flow cytometer is that the sample be delivered to the nozzle at an even and constant rate. The optimal sample flow rate depends on the application. Hence it must be possible to vary the sample flow over a fairly wide range, say from 1 to 100 μl/min, preferably in calibrated steps. Furthermore, the dead volume

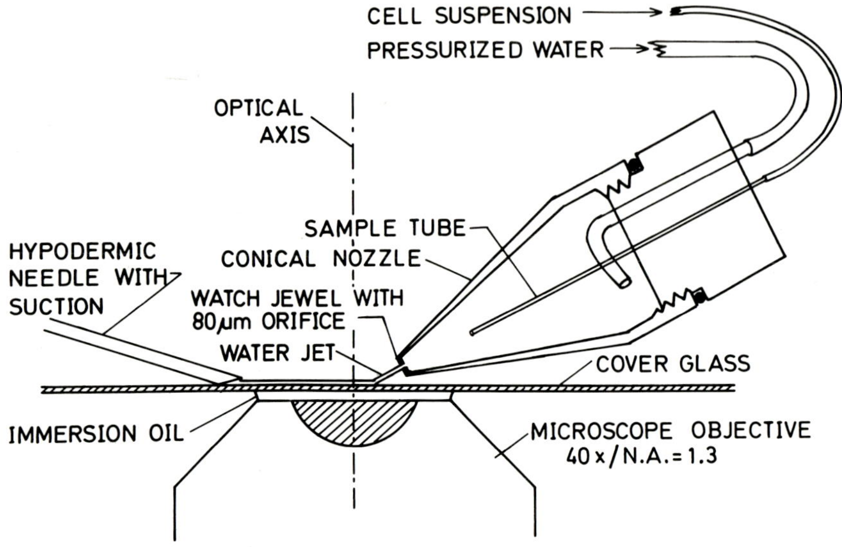

Fig. 9. Open flow chamber for arc lamp-based flow cytometers. A jet from a nozzle with hydrodynamic focusing impinges on the open surface of a microscope cover glass and thereby produces a flat, laminar flow across the surface of the glass with the cells confined to a narrow sector along the middle. This type of flow chamber allows light-scattering detection with high sensitivity when implemented with the optical system shown in Figures 3 and 4.

of the sample system, mainly the tubing carrying the sample to the nozzle, should be as small as possible to reduce loss of sample. Finally, it is important that the system may be efficiently flushed to minimize contamination of subsequent samples by residual material. Low sample flow velocities increase the tendency for cells to settle in the system and stick to the inside of the tubing. Hence tubing carrying sample should be as short as possible and have small inner diameter to increase the flow velocity. Furthermore, connectors and other devices where cells may hide should be avoided.

Differential Pressure Systems

Two main types of sample injection are in common use, the differential pressure system and volumetric injection. The working principle of the former type is shown in Figure 10. By means of a pressure regulator the sample container is kept at a slightly higher pressure than the sheath fluid container. The pressure difference determines the sample flow rate. In some instruments the sample flow is regulated also by a controlled constriction of the sample tube. The system is flushed by sheath fluid when the sample container is removed while the sheath fluid container is still pressurized. Thus, water flows back through the sample tubing from the nozzle. This system is simple to use and easy to flush between samples, but it has two significant disadvantages: It is difficult to maintain low sample flow rates, say below 5 µl/min, at a constant value because the pressure differential becomes too small to be regulated with sufficient precision. Also, the system cannot be set to calibrated values of sample flow. Hence the number of cells or particles per unit volume of sample cannot be determined accurately except by mixing into the sample a known density of a reference sample.

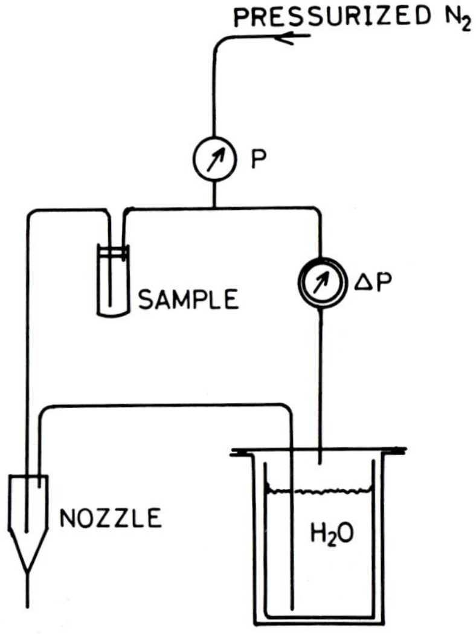

Fig. 10. Schematic representation of the differential pressure sample injection system. The sample flow is determined by a small pressure difference between the sheath fluid container and the sample tube, maintained by a gas pressure regulator.

Volumetric Sample Injection

The principle of volumetric sample injection is shown in Figure 11. The sample is drawn into a syringe, which is subsequently connected to the nozzle by means of a valve. The piston of the syringe is driven at a constant rate, by an electrical motor with variable speed. With this type of device the sample flow may be accurately regulated at least down to 0.2 µl/min. Furthermore, the sample flow rate can be determined directly from the rate of piston movement, so that the number of cells per unit sample volume can be readily calculated from the number of cells recorded per unit time. Volumetric sample injection systems require attention to flushing between samples to avoid leftovers in the syringe and valve.

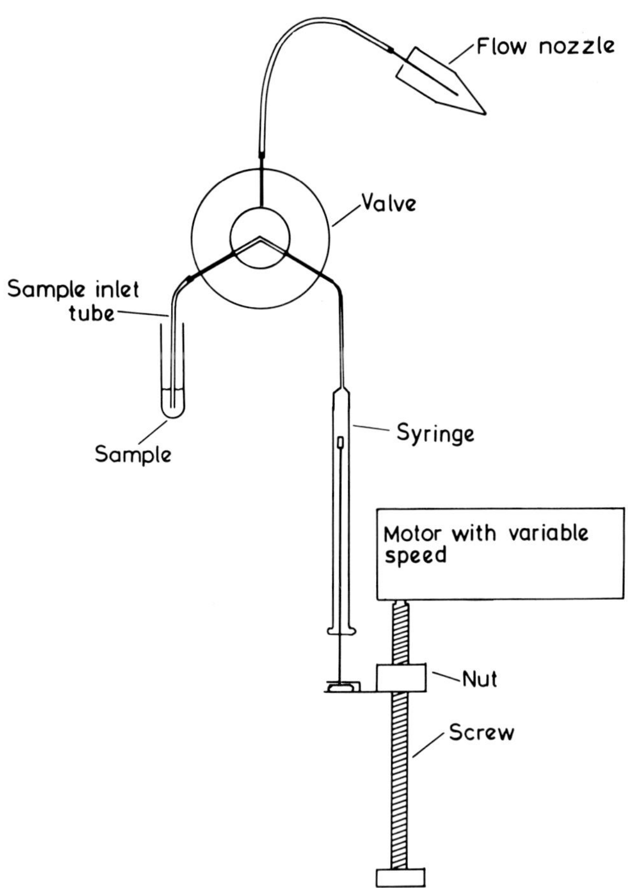

Fig. 11. System for volumetric sample injection. The sample may be drawn directly into the syringe or (in some systems) into a small chamber between the valve and the syringe. Turning the valve to connect the syringe and the nozzle, the sample is driven into the nozzle at a rate which is directly determined by the rate of the movement of the syringe piston.

ELECTRONICS

The electronics and data handling of flow cytometers are discussed in detail in Chapters 7 and 22 of this volume. In the present chapter we shall mention only the main elements and functions of these electronics, as shown schematically in Figure 12.

The pulses from the PMT detector are passed through a preamplifier located as close as possible to the base of the PMT. The preamplifier is essentially a current to voltage converter which provides the power required for the pulses to travel through a coaxial cable to the pulse amplifier. In contrast to the preamplifier, the pulse amplifier usually has a calibrated gain control. Thus, the user may control pulse amplification either through the high voltage of the PMT or by the pulse amplifier gain control. The pulse amplifier usually comprises both a linear and a logarithmic amplifier. The latter amplifier typically has a dynamic range of 3 to 4 decades and thus allows pulses of widely differing size to be accumulated within the same histogram.

As a result of the stochastic nature of the emission light as well as the electrons from the PMT photocathode, the pulses from a PMT always exhibit some high-frequency noise (1–10 MHz, typically). The pulse amplifier includes a filter that averages out this noise and thereby smooths the pulses. This filter may be either a simple RC element with a time constant approximately the same as the pulse width, or an active integrator. In instruments where the flow velocity, and hence the pulse width, can be varied, the time constant or integration time of the filter should be variable in order to optimize the signal to noise ratio.

The pulses from the pulse amplifier are passed on to a peak hold circuit which detects the peak voltage of each pulse and maintains that voltage at its exit for the period of time required for the analogue to digital converter (ADC) to assess this voltage and output the result in digital form, that is, as a number. The peak hold circuit is gated by a pulse from the pulse amplifier of one of the detectors, for example, that of the light-scattering detector. This pulse activates the peak hold circuit of all the detectors. The threshold for the signal required to trigger the gate pulse is adjustable. The threshold is typically set so that background noise and signals from irrelevant particles, such as cell debris, are not measured. The conversion time of the ADC, which may be in the range 10 to 50 μsec, obviously sets an upper limit to the measuring rate. The resolution of the ADC is typically 10 bits, which is to say that the pulses are measured with a precision of about 0.1%.

The data from the ADC are processed by the data system, which displays them in histograms of one or more parameters. The data may also be stored as such in list mode for later retrieval and analysis. The computer of the data system displays the histograms and performs various statistical analyses, curve fitting, etc.

PERFORMANCE CHARACTERISTICS OF FLOW CYTOMETERS

Flow cytometers exhibit remarkable performance with regard to sensitivity, reproducibility, and measuring rate. The detection limit is of the order $1 \cdot 10^3$ dye molecules, or roughly $1 \cdot 10^{-18}$ g of dye, per cell. (To appreciate this amount assume that 1 g of the dye is dissolved and evenly distributed in 3,000 km^2 (≈1000 sq. miles) of the North Sea where it is 300 m (≈1000 ft) deep. Thus, $1 \cdot 10^{-18}$ g is approximately the amount of dye that would be present in 1 g of this water.) This amount is measured in 1 μsec at an average rate of $1 \cdot 10^4$ cells/sec. Still the sensitivity may not be sufficient for all purposes, such as detection of virus, which has so far been carried out only in experimental systems [2] (see Chapter 20, this volume).

The performance of a flow cytometer is characterized primarily by the following parameters:

1. Sensitivity, which, with regard to fluorescence, may be expressed as the minimum number of given dye molecules per cell that can be measured with a certain precision (resolution), or, for light scattering, the smallest particle of a certain refractive index which can be measured with that precision.
2. Resolution, that is, the reproducibility of the signal produced by identical cells or particles.
3. Measuring rate, that is, the maximum average rate at which cells can be measured without exceeding a given frequency of coincidences (two cells detected as one).

Sensitivity

Sensitivity is a product of: 1) the excitation intensity, 2) the time spent by the cell in the excitation focus (i.e., the

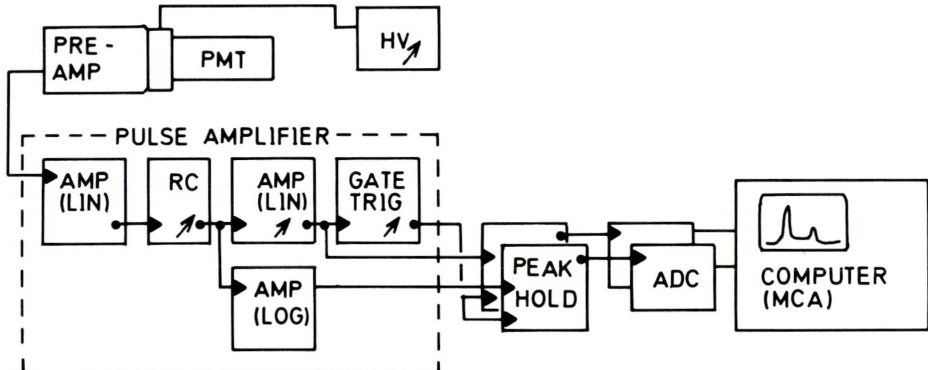

Fig. 12. Schematic representation of the main components of the electronics which amplify and transmit the signals from the photomultiplier detector to the data system. Each detector is supported by such a chain of electronics which feed the data into the data system. The data system may be either a specialized device called a multichannel pulse height analyzer (MCA) or a general computer which has been programmed to classify and analyze this type of data.

inverse flow velocity), 3) the effective numerical aperture (squared) of the fluorescence/light scattering collection optics, 4) The transmission of the fluorescence/light scattering pathway (in particular the transmission of filters), 5) the sensitivity of the detector(s), 6) the noise, resulting primarily from the background of scattered light and fluorescence reaching the detector(s), and 7) the overlap between the excitation wavelength and the absorption spectrum of the dye as well as its fluorescence quantum yield. As shown below, instrument sensitivity is closely related to resolution.

High excitation intensity is achieved by concentrating the excitation light into a very narrow focus. Whereas the laws of optics limit the light density in the focus from a conventional light source, such as a Hg arc lamp, to that of the source itself, the laser may be focused to a much smaller spot with correspondingly higher light density. However, the intensity profile of such a focus, whether the light comes from an arc lamp or a laser, is roughly bell-shaped. Since all cells must be illuminated by the same intensity, say to within 1%, in order to be measured reproducibly, the width of the focus must be much larger than the width of the cell path through the focus. In practice, therefore, the width ($1/e^2$) of the excitation focus in both laser- and arc lamp-based instruments is usually to the order of 100 to 200 μm.

Whereas laser light may be concentrated into such a focus by a simple lens, the arc lamp requires a complex lens, such as a microscope objective, to be utilized optimally. Furthermore, the intensity of excitation light in the arc lamp-based instrument increases with the square of the numerical aperture value of the optics. Hence, the importance of oil immersion optics with NA $\approx$ 1.3.

The amount of excitation light absorbed by a cell when it passes through the focus is obviously inversely proportional to the velocity of the cell, that is, the flow velocity. Hence, sensitivity may be increased by slowing down the sample flow (Eq. 6). As noted above, instruments with the excitation in a jet in air do not permit flow velocities below 2–3 m/sec, whereas there is almost no lower limit to the flow velocity in closed chambers. Hence instruments with closed flow chambers have a larger potential for high sensitivity in this regard.

As shown by Equation 1 a reduction of sheath flow velocity causes the width of the sample flow to increase, which may reduce the precision of the measurement if the sample flow is not reduced correspondingly. Furthermore, the maximum measuring rate is reduced in the same proportion as the flow velocity (Eq. 9).

As noted above, the detection limit of a flow cytometer is limited not only by the fraction of the emitted light that can be detected. The background of fluorescence and scattered light from the system itself is of equal importance. As discussed below, the very nature of light implies that such a background represents an optical noise which in many cases may completely swamp signals which would otherwise be easily detectable. In other words, the detection limit in flow cytometers may be limited by the signal to noise ratio rather than by sensitivity only.

Finally, it should be noted that the detection limit, in terms of fluorescent molecules per cell, depends on the overlap between the excitation wavelength and the absorption spectrum of the dye, that is, the extinction coefficient of the dye at the excitation wavelength. Furthermore, the fluorescence intensity is proportional to the fluorescence quantum yield, i.e., the probability that absorption of a photon will produce a photon of fluorescence. The detection limit is not only a function of instrumental factors, but depends also on the particular dye being used.

Resolution

Resolution is usually expressed as the coefficient of variation (cv), that is, the relative standard deviation of the signal produced by identical cells or particles. It is limited mainly by three factors: 1) the stability of the excitation light source, 2) the width and stability of the sample flow, and 3) the intensity of the signal and of the background luminescence, which determine the statistical fluctuation (noise) in the light detection. Whereas it is possible to reduce the two former factors to negligible proportions, statistical fluctuation is inherent to the nature of emission and detection of light, which are both stochastic processes, that is, random in time. Thus, the variation, in terms of relative standard deviation, in the number of photons, n_p, reaching the detector from identical particles, will be:

$$cv_p = 100\% \cdot n_p^{1/2}/n_p \cdot \qquad (2)$$

The light pulses produced by the cells are superimposed on the background luminescence which is subject to corresponding statistical fluctuation. If the average number of background photons reaching the detector during each pulse is n_b, the variation in the size of the pulses becomes

$$cv_s = 100\% \cdot (n_p + n_b)^{1/2}/n_p. \quad (3)$$

The detection of light depends on release of photoelectrons from the photocathode of the PMT. This is also a stochastic process with a variation given by

$$cv_d = 100\% \cdot [(n_p + n_b)\phi]^{1/2}/n_p \cdot \phi, \quad (4)$$

where ϕ is the quantum efficiency of the photocathode, that is, the probability that a photon will release a photoelectron. Taking into account variation which is due to statistical fluctuations associated with emission and detection of light, the total variation becomes

$$cv_t = 100\% \cdot (cv^2_s + cv^2_d)^{1/2} = 100\% \cdot (1 + 1/\phi)^{1/2} \cdot (n_p + n_b)^{1/2}/n_p. \quad (5)$$

This equation explains why background rejection is so important in flow cytometry. Typically, background intensity (n_b) exceeds the signal intensity (n_p) when one approaches the detection limit.

Equation 5 may be used to estimate the detection limit: the minimum number of dye molecules per cell that can be detected with a given resolution. Consider cells each carrying the same number N of fluorescent molecules. (In reality this number is also subject to variation from cell to cell, but for the sake of simplicity we shall neglect that variation here.) If we assume that each cell passes with velocity v (m/sec) through a gaussian laser focus having a width Δ (μm) and a power E (watts), the average number of fluorescence photons emitted per cell is given approximately by

$$n_f \approx 3.24\ 10^{-3}\ E \cdot \lambda \cdot \epsilon \cdot \phi_f \cdot N/\Delta \cdot v, \quad (6)$$

where λ (nm) is the excitation wavelength, $\epsilon(l \cdot M^{-1}\ cm^{-1})$ is the extinction coefficient of the dye at this wavelength and ϕ_f the fluorescence quantum yield of the dye. Taking as an example conjugated FITC, which has $\epsilon = 6 \cdot 10^4 l \cdot M^{-1} \cdot cm^{-1}$ and $\phi_f = 0.1$ (1), we calculate with $\lambda = 488$ nm, E = 1 watt, $\Delta = 130\ \mu$m, v = 10 m/sec:

$$n_f \approx 7.3N.$$

Since the fluorescence quantum yield, ϕ_f, the probability that absorption of a photon of excitation light causes emission of a photon of fluorescence, is 10%, this result implies that on the average each FITC molecule is excited 73 times while the cell is passing through the excitation focus. Assuming detection optics with NA = 0.6 (which means that about 5% of the fluorescence is collected) and a transmission factor T = 0.5 of the detection optics and filters, we obtain

$$n_p \approx 0.18N.$$

Assuming a PMT with a quantum efficiency of 15% and (unrealistically) that the background is negligible, Equation 5 gives

$$cv_t \approx 100\% \cdot 6 \cdot 5/N^{1/2} \quad (7)$$

or

$$N \approx 4.2 \cdot 10^5/cv_t^2. \quad (8)$$

If we require cv_t = 10%, the minimum number of FITC molecules per cell is $N \approx 5.5 \cdot 10^3$, whereas if we require cv_t = 50%, this number is reduced to $N \approx 1.7 \cdot 10^2$. (For gaussian histogram peaks the peak width at half-maximum is 2.35 cv_t.) In the real case background is not negligible, transmission of optics, filters, etc. is usually below 50%, and the laser focus is wider than 130 μm. Hence in most commercial instruments the detection limit for FITC is much higher than the values calculated above. However, if flow velocity can be substantially reduced, which is possible only in closed flow chambers, and/or if the laser is focused to a significantly smaller focus, the detection limit may be decreased below these values, provided background luminescence is sufficiently weak.

It is interesting to note that for a flow cytometer having oil immersion optics (NA = 1.3), a flow chamber which facilitates a flow velocity of 0.2 m/sec, and an arc lamp producing 3 mW of excitation light at the appropriate wavelength, that is the characteristics of the arc lamp-based instrument described above, Equations 6–8 give a somewhat lower detection limit, that is, higher sensitivity, than that calculated for the large laser-based instrument in the example above. For most other dyes the arc lamp yields a much higher excitation intensity whereas the Ar laser has significantly less power than at 488 nm. Hence, for these dyes the arc lamp-based flow cytometer should have a much higher sensitivity than the instrument using a large laser and a "jet-in-air" flow chamber.

From Equation 8, and the equations above, it may also be calculated that in order to achieve a CV_t of 1%, using a laser-based instrument as specified in the above example, the minimum number of dye molecules per cell must be $N \approx 4 \cdot 10^5$. (Assuming a dye with $\epsilon = 5 \cdot 10^4\ M^{-1}\ cm^{-1}$ and $\phi_f = 0.5$.) Again this is a minimum estimate. Although background is not as important at these higher fluorescence intensities, other factors, like variation in excitation light intensity, become relatively more important.

Measuring Rate

In the case of no aggregation, cells should come through the excitation focus randomly distributed in time. In this case the average measuring rate is given by

$$k_m \approx f_c/100 \cdot t_p, \quad (9)$$

where f_c is the percentage of coincidences, that is, when two cells follow each other so closely that they are detected as one, and t_p the pulse length. In a laser-based instrument, which may have a focus length (in the direction of the flow) of 10 μm and a flow velocity of 10 m/sec, t_p = 2 μsec assuming a 10 μm cell diameter. With f_c = 1%, Equation 9 yields $k_m = 5 \cdot 10^3$ cells/sec as the maximum measuring rate. In arc lamp-based instruments the typical pulse length may be around 10 μsec, which reduces the maximum measuring rate, with f_c = 1%, to $k_m = 1 \cdot 10^3$ cells/sec.

Laser-Based Versus Arc Lamp-Based Flow Cytometers

Each of the two main types of flow cytometers has some advantages and some disadvantages as compared to the other. Instruments having a laser with a total power above 1 W are superior with regard to excitation intensity at the wavelengths of the lasers main emission lines. However, sensitivity also depends on the overlap between the excitation wavelength and the absorption spectrum of the dye. Thus, the 488-nm line of the Ar laser is perfect for excitation of fluorescein and FITC (Fig. 5). With the UV option they are also better than the Hg arc for dyes like Hoechst 33342, Hoechst 33258, and DAPI. On the other hand, neither the Ar nor the Kr laser have suitable emission lines for chromomycin A_3 and mithramycin, which appear to be the most DNA specific dyes currently in use.

The Hg arc exhibits prominent lines to match the absorption spectrum of essentially all important dyes except fluorescein/FITC (Fig. 5). Even around 480 nm, however, the intensity of the arc lamp is sufficient for measuring FITC-conjugated antibodies on mammalian cells, although the sensitivity is lower than for some instruments that employ argon laser excitation. (In fact, the detection limit for FITC-conjugated antibodies is usually not determined by instrument sensitivity, but by the autofluorescence of the cells.) Detection of FITC-conjugated antibodies with arc lamp-based instruments is critically dependent on efficient rejection of background luminescence and on an absolute minimum of overlap between the transmission spectra of the excitation and emission filters. In order to minimize autofluorescence, which is excited predominantly at lower wavelengths, the bandwidth of the excitation filter must not be too broad. A 470–490 nm transmission band is about optimal. For the same reason, and to avoid the Raman scattering of water, the transmission of the emission filter should be limited to the region 510 to 560 nm. The problem with the low sensitivity of arc lamp-based instruments with regard to fluorescein/FITC is mitigated by the recent introduction of phycoerythrin conjugates. These conjugates are much brighter than FITC. Moreover, the absorption spectrum of phycoerythrin B and phycoerythrin R coincides perfectly with the intense Hg line at 546 nm (Fig. 5). Excitation at this wavelength has the further advantage that autofluorescence is much weaker than when the cells are excited below 500 nm. The detection limit of phycoerythrin-conjugated antibodies is well below that obtained with FITC conjugation.

Besides its lower intensity at some wavelengths the arc-lamp emission also has less stability than the laser. As noted above, arc lamps generally exhibit some degree of flickering. Furthermore, the intensity of arc lamps decreases gradually at a rate of the order of 2% per day. Both of these effects may be compensated for by means of an intensity monitor, the output of which is used to regulate the pulse amplifier(s) in such a way that the product of amplifier gains and excitation intensity remains constant [6]. Thus, measurement becomes insensitive to variations in excitation intensity.

The lower excitation intensity in the arc lamp-based flow cytometers is partly compensated by the larger aperture of the immersion optics used in these instruments. Thus, an oil immersion objective with NA = 1.3 collects about 10 times more light than the NA ≈ 0.6 objectives of most laser-based instruments. Arc lamp-based flow cytometers also have a relatively lower level of background luminescence than the laser-based instruments. The main reason for this is that with the epiillumination employed in arc lamp-based instruments fluorescence is detected in the backward direction where the intensity of scattered light is minimum. Furthermore, the high aperture of the objectives used in arc lamp-based instruments implies that the focus is quite shallow. Hence, the illuminated volume of water, which is the main source of Raman scattering, is relatively small.

An important difference between laser-based and arc lamp-based instruments is the illumination field angle. In laser-based instruments the target is illuminated by a nearly parallel beam of light. This gives rise to significant artifacts in the measurement of nonspherical objects, such as sperm cells, because the fluorescence and light-scattering intensities vary with the orientation of the object relative to the laser beam [9]. In contrast, the high aperture of the immersion optics in arc lamp-based instruments gives a very wide-angle illumination field, that is, about 150° for an objective with NA = 1.3. And fluorescence is collected over the same field. As a result nonspherical objects do not produce noticeable artifacts in such instruments. Pinkel et al. [4] demonstrated that by using a nozzle with a properly beveled sample inlet tube it was possible to orient nonspherical objects, so that this artifact could be greatly reduced in laser-based instruments.

The combination of a high degree of polarization of the laser light and the narrow illumination field makes the laser-based flow cytometers better suited for measurements of fluorescence polarization than arc lamp-based instruments where the wide field angle of the illumination and fluorescence detection diminish polarization quite considerably.

Perhaps the most important advantage of laser-based instruments, besides their greater sensitivity for fluorescein/FITC, is the possibility of cell sorting by electrostatic droplet deflection. A sorting device based on flow switching between channels in a closed flow chamber used in an arc lamp-based flow cytometer has been demonstrated [10]. At this writing it is not clear whether such a device is commercially available. Cell sorting is dealt with in Chapter 8, this volume.

Dual-focus excitation (by the use of two lasers) is also available only in laser-based flow cytometers.

The orthogonal configuration of laser-based flow cytometers implies that three axes—the laser beam, the water jet, and the optical axis of the fluorescence collection optics—have to be aligned so as to intersect a common point to within a few micrometers. In particular in some instruments having a rather complicated laser beam path, this alignment has been a significant problem. In arc lamp-based instruments with epiillumination, alignment is greatly simplified by the fact that illumination and fluorescence collection have a common optical axis and focus. Hence alignment is reduced to positioning the sample flow, that is, the flow chamber, relative to this focus. In one type of arc lamp-based instrument alignment can be monitored visually through an ocular which views the fluorescence (or light-scattering) image [5,6]. This ocular also facilitates correct setting of the adjustable measuring slit, the "pinhole", and makes it possible to see the sample flow at high (500×) magnification and thereby check its laminarity.

Finally, it should be mentioned that laser-based flow cytometers are more expensive than arc lamp-based ones. The cost of the laser contributes significantly to this difference. Furthermore, the plasma tube of the laser has a limited lifetime, typically 2 years. In addition to the large consumption of water and electrical power, this makes the operating expenses of laser-based instruments much higher than for instruments using arc lamp-based excitation.

FUTURE DEVELOPMENTS

Today's large laser-based flow cytometers are not much different from those produced more than a decade ago. However, a new generation of smaller instruments has recently been introduced. The instruments are simpler to operate and maintain, and significantly less expensive both to purchase and in operation. The lower cost is due partly to the use of less expensive, but also less powerful, light sources, both air-cooled lasers and arc lamps. Nevertheless, the performance of these smaller flow cytometers in many applications is on par with that of the large laser-based instruments. This is primarily due to new optical designs and new flow chambers, which facilitate improved light collection efficiency and background rejection.

Another important advance has been the introduction of new powerful PC computers and all the accompanying software tools. This development has provided inexpensive data-handling capacity which was previously available only with much larger, very expensive computers and custom programming. Further developments within the next few years can be expected to yield real time data analysis, much larger and safer storage of data, and real-time computer control of all instrument functions at about the same price as current PC-based systems.

Whereas PMTs are superior with regard to signal to noise as compared to photodiodes, the quantum efficiency of the latter is significantly higher than for PMTs. New advances in semiconductor technology may produce detectors which can compete with PMTs. Such detectors will be less expensive, require no high voltage supply and would facilitate more compact and less expensive instruments. We may also see small, powerful solid-state lasers with emission in the blue and even in UV wavelengths of the spectrum, which will replace the most expensive and voluminous component of today's large, laser-based instruments.

However, the most important prerequisite for more widespread use of flow cytometry, especially in routine applications, is increased reliability and ease of use. Self-testing routines for optical alignment and flow conditons coupled with automatic adjustments and calibration, facilitated by new computers and software, are likely to bring us further in this direction.

REFERENCES

1. **Fothergill JE, Ward HA, Nairn RC** (1972) "Fluorescent protein tracing." Edinburgh: Churchill, Livingstone, pp 39–67.
2. **Hercher M, Mueller W, Shapiro HM** (1979) Detection and discrimination of individual viruses by flow cytometry. J. Histochem. Cytochem. 27:350–352.
3. **Peters DC** (1979) A comparison of mercury arc lamp and laser illumination for flow cytometers. J. Histochem. Cytochem. 27:241–245.
4. **Pinkel D, Dean P, Lake S, Peters D, Mendelsohn M, Gray J, Van Dilla M, Gledhill B** (1979) Flow cytometry of mammalian sperm. Progress in DNA and morphology measurement. J. Histochem. Cytochem. 27: 353–358.
5. **Steen HB, Lindmo T** (1979) Flow cytometry: A high resolution instrument for everyone. Science 204:403–404.
6. **Steen HB** (1980) Further developments of a microscope-based flow cytometer: Light scatter detection and excitation intensity compensation. Cytometry 1: 26–31.
7. **Steen HB, Lindmo T** (1985) Differential light scattering detection in an arc lamp-based epi-illumination flow cytometer. Cytometry 6:281–285.
8. **Steen HB** (1985) Simultaneous separate detection of low angle and large angle light scattering in an arc lamp-based flow cytometer. Cytometry 7:445–449.
9. **Van Dilla MA, Gledhill BL, Lake S, Dean PN, Gray JW, Kachel V, Barlogie B, Göhde W** (1977) Measurement of mammalian sperm deoxyribonucleic acid by flow cytometry. Problems and approaches. J. Histochem. Cytochem. 25:763–773.
10. **Zold T** (1979) Method and device for sorting particles suspended in an electrolyte. U.S. Patent 4,175,662.

3

Hydrodynamic Properties of Flow Cytometry Instruments*

Volker Kachel, Hugo Fellner-Feldegg, and Everhard Menke
Max-Planck-Institut für Biochemie, D-8033 Martinsried, Federal Republic of Germany

All flow cytometry instruments used to analyze and classify biologic or other particles share a common feature: the particles are made to flow through a sensing region in which electrical resistance or optical properties are measured. The particles are suspended in a carrier fluid, usually physiologic saline, which is used to transport them through the measuring region. A knowledge of the principles governing this fluid flow is important for an understanding of the events that occur during measurement and of the effect of the flow on the data obtained.

In nearly all these instruments, the particles are transported by the carrier fluid through either short or long tubes; flow problems can be analyzed by studying fluid flow through equivalent tubes or pipes. These tubes serve different functions in the several different types of flow cytometry systems. In optical flow systems, relatively long tubes introduce the particles into the system and focus them so that they follow narrow, well-defined trajectories in the measuring region; in the instruments that measure particle volume by electrical resistance, the dimensions of the orifice, which may be considered a short tube, will have an effect on the resistance pulse that is measured.

This chapter considers the laminar flow properties of fluid both inside and outside the flow tubes, the behavior of particles transported in the flow stream, and the fluid jets and droplet formation in cell sorters.

FLOW CONDITIONS IN TUBES

Flow in tubes is either laminar or turbulent. Laminar flow is characterized by nonmixing of flow lines and flow layers in the system; that is, in a long straight tube in which laminar flow has been established, a dye marker threaded into the flow stream will remain undisturbed as a thin dye line within the stream, without mixing. By contrast, in turbulent flow, layers are mixed according to complicated laws, and the same dye marker would soon be distributed over the entire interior of the tube. In flow cytometry instruments, the suspending medium must transport the particles along well-defined paths through the sensing and measuring region. Thus, only laminar flow is suitable for use in these instruments, and our discussion is limited to laminar flow. Unless otherwise indicated, the flow laws referred to in this chapter are taken from a number of standard works on fluid mechanics [1,3,5,17,63,64,88].

THE REYNOLDS NUMBER

In 1883, Reynolds found that a dimensionless quantity, the Reynolds number could be used to characterize fluid flow:

$$\mathrm{Re} = \frac{d\,\rho\,\bar{v}}{\eta} = \frac{d\,\bar{v}}{\nu} \qquad (1)$$

This quantity describes the relationship among the diameter of the tube, the mean flow velocity, and the viscosity of the fluid. All real fluids in motion exhibit an internal friction or viscosity η. The viscosity unit is 1 poise (1 g/cm sec). It is usual, however, to consider the kinematic viscosity ν, which is the viscosity divided by the fluid density. For cytometric purposes, most flow instruments use water, saline, or other fluids with values of ν essentially the same as that of water. Most fluids have viscosities that show a strong temperature dependence, as shown in Table 1.

The critical Reynolds number $\mathrm{Re}_{kr} = 2300$ describes the transition between laminar and turbulent flow. Above this value, laminar flow is sometimes possible but small pressure disturbances, sharp edges, and so on, can cause a rapid deterioration from the laminar to turbulent condition. Below $\mathrm{Re} = 2300$ the flow is always laminar and, therefore, this is the region for which flow instruments should be designed.

The Reynolds number also defines the hydrodynamic law of similarities. All flow situations for which the Reynolds number values are equal are equivalent. Thus, flow through large-diameter tubes is similar to that through small-diameter tubes when the velocities of flow and the viscosities are appropriately adjusted. This principle is extremely important in modeling the flow of fluids.

*See abbreviations list at the end of the chapter.

Flow Cytometry and Sorting, Second Edition, pages 27–44

TABLE 1. Absolute and Kinematic Viscosity of Water as a Function of Temperature

T	273	283	293	303	313	323	Kelvin
η	1.79	1.31	1.00	0.80	0.68	0.55	Centipoise
ν	1.79	1.31	1.00	0.79	0.66	0.53	Centistokes

1 Poise = 1 g cm^{-1} sec^{-1}
1 Stokes = 1 cm^2/sec
273° Kelvin = 0° Celsius

THE LAW OF LAMINAR FLOW IN TUBES ACCORDING TO HAGEN POISEUILLE

Owing to viscous forces, real fluids form boundary layers at the walls. With laminar flow in a long straight tube a parabolic velocity profile is produced

$$v_r = 2\bar{v}\left(1 - \frac{r^2}{R^2}\right) \tag{2}$$

Here v_r is the flow velocity at a radial distance r from the tube axis. Along the tube axis, the velocity is $2\bar{v}$. The flow where such a parabolic velocity distribution exists is termed "completely developed laminar flow." The mean velocity in the tube is given

$$\bar{v} = Q/F \tag{3}$$

To pass a fluid at the rate Q through the tube, a pressure difference Δp must be established along its length. Assuming completely developed laminar flow,

$$Q = \frac{M}{t} = \frac{\Delta P \cdot \pi \cdot R^4}{8 \cdot \eta \cdot \ell} \tag{4}$$

or according to equation (2):

$$\bar{v} = \frac{\Delta p \cdot R^2}{8 \cdot \eta \cdot \ell} \tag{5a}$$

or solving for the pressure differences

$$\Delta p = \frac{8\,\ell\,\bar{v}}{R^2} \tag{5b}$$

Other relationships for the pressure differential are

$$\Delta p = \frac{64}{Re}\frac{\ell}{d}\frac{\rho}{2}\bar{v}^2 \tag{5c}$$

$$\Delta p = \lambda_0 \frac{\ell}{d}\frac{\rho}{2}\bar{v}^2 \tag{5d}$$

$$\xi = \lambda_0 \frac{\ell}{d} \tag{5e}$$

The dimensionless quantity in the last equation, $\lambda_0 = 64/Re$ is known as the tube friction number for laminar flow in cyclindrical tubes with circular cross section. Equation (5d) may also be used for turbulent flow, however, λ_0 assumes a more complicated form. Pressure differentials in tubes with other than circular cylindrical geometry may also be obtained from Equation (5d). For not fully developed flow ξ [eq. (5e)] is called "nozzle factor" (see jet formation below). In this case the diameter of the tube must be replaced by the "hydraulic diameter" $d_h = 4F/U$. In such cases, d_h is also used to calculate the Reynolds number. Laminar flow in tubes of rectangular cross section may also be considered by replacing λ_0 with $\lambda_\square = \phi\ 64/Re_\square$ where ϕ is taken from Table 2.

The major features of laminar flow in a long smooth tube have been considered; it is also important in flow instruments to examine the flow at points of change in tube diameter, both at the inlets and outlets of the system.

FLOW PROBLEMS AT THE INLET OF TUBES

Velocity Profile at the Inlet of Tubes

In long tubes with established laminar flow, the velocity profile is parabolic [cf. equation (2)]. However, this is not the case at the inlet to the tube. In flow instruments, short tubes may be used and there may also be transitions between tubes of differing diameters. In such cases, laminar flow may not fully develop and the velocity profile must be considered in the transition regions.

Consider the inlet from a large container into a tube with rounded edges. Several investigators [7,19,51,60,75,88] have considered the flow profile development at the tube inlet, and this has been reviewed by Smith [83]. If one neglects friction, a uniform velocity profile exists at the inlet of the tube with $v \simeq \bar{v}$.

As a result of the friction (viscosity) a boundary layer develops between the tube wall ($v = 0$) and the core of the flow which has a uniform velocity $v > \bar{v}$. With increasing distance into the tube from the inlet, the boundary layer thickness increases and that of the core decreases. The velocity profile changes into a parabolic one as the laminar flow completely develops, as was shown by Tatsumi [88]. He has formulated equations that, although criticized more recently [24], are in sufficient agreement with measurements for the purposes of the present discussion. These equations permit the calculation of the inlet profile at different distances $\times$ into the tube. This is done for three regions within the tube, two of which have uniform velocity profiles, and the third a parabolic profile. The velocity at a given element of area (x,r) is given

$$v = \bar{v}\,0.57\frac{y}{(1-2\delta)\delta} \qquad 0 \le y \le 0.63 \tag{6a}$$

$$v = \bar{v}\frac{1-(1.025-0.375\,y/\delta)}{1-2\delta} \qquad 0.63 \le y \le 2.87 \tag{6b}$$

$$v = \bar{v}\frac{1}{1-2\delta} \qquad 2.87\delta \le y \tag{6c}$$

TABLE 2. Values of ϕ for Rectangular Tubes*

h/b	0.1	0.2	0.3	0.4	0.5	0.6	0.7	0.8	0.9
ϕ	1.34	1.20	1.10	1.02	0.97	0.94	0.92	0.90	0.89

h/b is the ratio of height to width of the tube.

*From ref. [5].

The variable y, which is related to the distance from the axis of the flow, is calculated by

$$y = 1/2\,(1 - (r/R)^2) \tag{7}$$

The distance x is related to δ. For this case $0 < \delta < 0.15$ and this relationship is given by

$$\frac{x}{d\,Re} = \frac{1}{4}\left[-0.169 + \frac{0.718}{1 - 2\delta} + 1.268\ln(1 - 2\delta) - 0.550(1 - 2\delta)\right] \tag{8}$$

Typical values of these quantities are given in Table 3. Figure 1 shows the inlet profiles for the cases considered. Equation (6a) describes the flow adjacent to the walls; equation 6b the parabolic portion, and equation (6c) the core region of the flow. The type of flow profile established is a strong function of the value of the Reynolds number. An increase in Re inhibits the establishment of a parabolic profile. This is reasonable since a decrease in the viscosity causes the behavior of real fluids to approach the ideal, where a uniform velocity profile exists since friction is nonexistent in the ideal case. Due to the restriction $0 \leq \delta \leq 0.15$, Tatsumi's results are not applicable over the entire x distance from the inlet to the position where a true parabolic profile exists. However, according to Boussinesq (7) and Langhaar (51) the profile at the point

$$x = 0.06 \cdot d \cdot Re \tag{9}$$

is within 1% of the true parabolic profile. This equation better represents the real inlet length than that calculated by Schiller [75], who determined an inlet length of $x_e = 0.03 \cdot d \cdot Re$. In the present discussion, the inlet length x_e is the distance from the inlet of the tube to the point where laminar flow (a parabolic velocity profile) is established.

The Pressure Drop in the Inlet Length

Figure 2 shows the transition from a tube with a large diameter to one with a small diameter. The transition in tube diameter is considered to be gradual in order to maintain laminar flow. Now considering the transition shown by the solid line (curve) in Figure 2, two basic laws of flow mechanics may be demonstrated: the law of mass conservation and the equation of Bernoulli. The principle of mass conservation for incompressible fluids states that in closed tubes equal amounts of fluid must flow across all areas per unit time:

$$\bar{v}_1 F_1 = \bar{v}_2 F_2 \tag{10}$$

Bernoulli's equation is essentially a statement of energy conservation and relates the pressure to velocity at various zones within the flow:

TABLE 3. Relation Among δ, the Dimensionless Distance x/(d·Re) from the Tube Inlet, the Normalized Core Radius R_c/R, and the Normalized Core Velocity $v_c/\bar{v}$*

x/(d·Re)* 10^{-5}	δ	R_c/R	$v_c/\bar{v}$
4.33	0.05	0.84	1.11
20.5	0.06	0.81	1.14
41.0	0.07	0.77	1.16
67.0	0.08	0.74	1.19
105	0.09	0.70	1.22
139	0.10	0.65	1.25
186	0.11	0.61	1.28
244	0.12	0.56	1.32
313	0.13	0.50	1.35
443	0.14	0.44	1.39
485	0.15	0.37	1.43

*The values are calculated according to the Equations (6c), (7), and (8) (Tatsumi). For $\delta < 0.05$, x/(d·Re) is negative and therefore Equation (8) is no longer correct.

$$p_1 + \frac{\rho}{2}\bar{v}_1^2 = p_2 + \frac{\rho}{2}\bar{v}_2^2 \tag{11a}$$

$$\Delta p = p_1 - p_2 = \frac{\rho}{2}(\bar{v}_2^2 - \bar{v}_1^2) \tag{11b}$$

If there is any change in the level of the flow, the resulting potential energy difference must also be considered in equations 11a and 11b. If the flow is from a large reservoir into a smaller diameter tube, $v_1 \approx 0$ and

$$\Delta p = \frac{\rho}{2}\bar{v}_2^2 \tag{12}$$

For a real fluid, flowing from a large container into a smaller tube, friction is present and the situation is more complex. The requirements for Δp are then

1. Fluids entering the tube must be accelerated to the velocity v.
2. Losses due to friction at the walls and through the transition region must be considered.
3. Energy losses due to turbulence that can develop at sharp edges must be eliminated.

At inlets with rounded edges, a total pressure drop of $\Delta p = \Delta p_R + \Delta p_B$ occurs where Δp_R [cf. equation (5b)] is the pressure necessary to overcome the losses due to wall friction, Δp_B is the pressure differential required to produce the acceleration of the fluid and overcome the increased friction at the tube inlet prior to the development of a parabolic velocity profile. A tube that has a length longer than x_e [equation (9)] requires

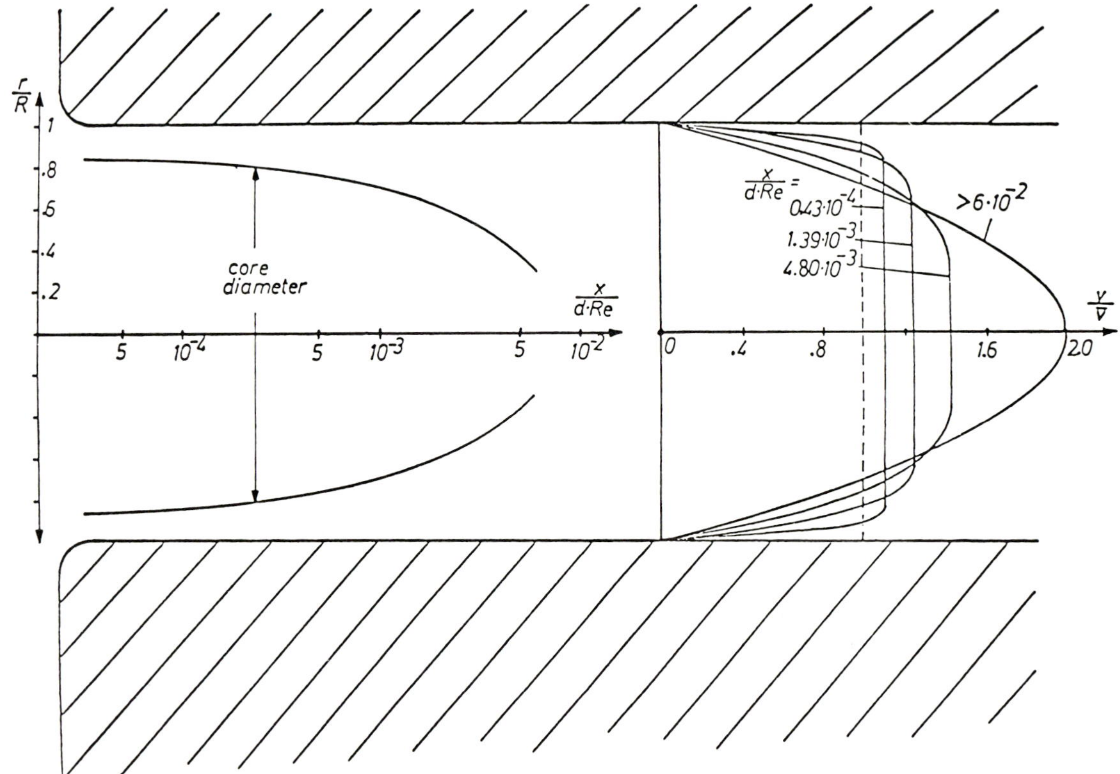

Fig. 1. Laminar velocity at the inlet of a circular tube according to the calculations of Tatsumi [88], which are summarized in Table 3. Flow direction is from left to right.

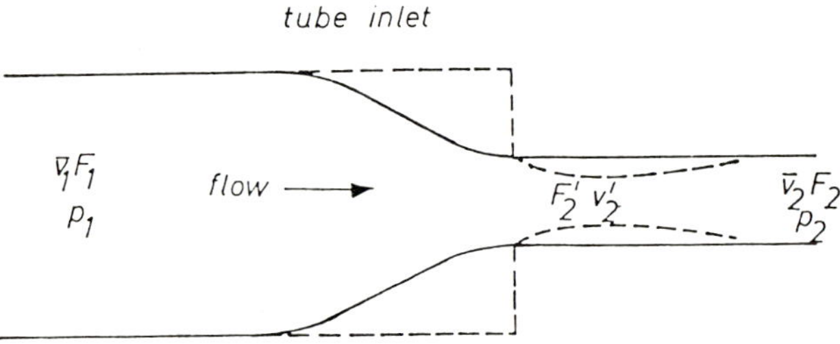

Fig. 2. Transition from a large-diameter tube to a small-diameter tube. If the transition between the cross-section F_1 and the cross-section F_2 is smoothed, the laminar flow remains undistorted. A sharp edge at the transition causes a constriction of the flow to the cross section F'_2 (broken lines).

$$\Delta p = \Delta p_R + \Delta p_B = \frac{8\,\eta\,\ell\,\bar{v}}{R^2} + m\frac{1}{2}\bar{v}^2 \tag{13}$$

Langhaar [51] calculates m = 2.28; however, other investigators have reported slightly different values ($2.11 \leq m \leq 2.41$).

If the length of the tube is less than x_e, the pressure drop in the tube may be taken from the Langhaar tables (51). Figure 3 depicts the pressure drop as a function of the length along the tube. The pressure drop near the tube inlet approaches asymptotically the Bernoulli pressure loss (equation 12):

$$\Delta p = 1/2\,\rho\,\bar{v}^2$$

If the parabolic profile is developed after a distance x_e, friction forces dominate [cf. equation (5b)]; the velocity profile developing in the transition region will become a straight line [described by equation (12)]. This situation in the transition region was confirmed experimentally by Gavis and Modan [26]. For a pressure drop Δp in a tube with the length x, the mean velocity $\bar{v}$ is determined from Figure 3. However, if the tube inlet is sharp rather than rounded, the flow will change behind the edge (see Fig. 2). Then the flow itself is responsible for the rounding of the edges. This immediately results in narrowing of the flow by the factor F'_2/F_2 with a dead water and turbulence zone between the tube wall and the narrowed flow. The effect will disappear with increasing length (x) down the tube. An additional loss is observed:

$$\Delta p_v = \frac{\rho}{2}(\bar{v}'_2 - \bar{v}_2)^2 \tag{14}$$

In this case, the total pressure necessary to establish the mean velocity $\bar{v}$ in the tube becomes

$$\Delta p = \Delta p_R + \Delta p_B + \Delta p_v$$

The velocity profile [cf. equations (6a,b,c)] that develops just beyond the tube inlet does not extend across the total tube diameter but is estimated across the diameter of the area F'_2. Between F'_2 and the tube wall there is no laminar flow and thus no defined velocity profile. Sharp edges, therefore, initiate turbulence and also causes losses in pressure. The mes-

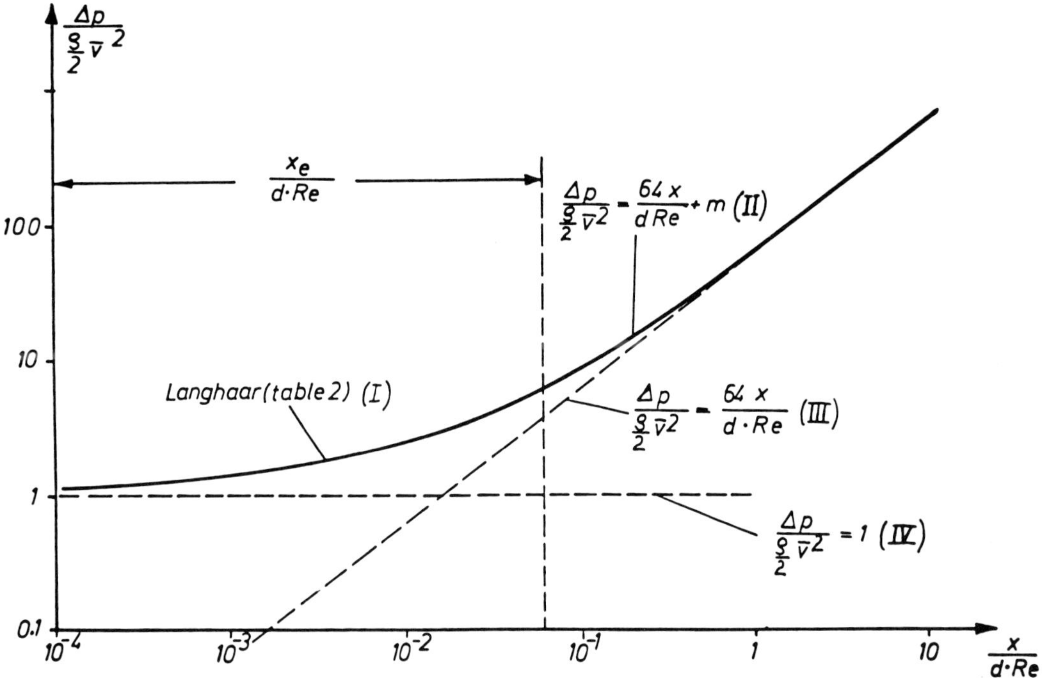

Fig. 3. The pressure in the inlet of a circular tube as a function of the dimensionless distance from the mouth. The pressure is normalized to the prssure drop ($1/2 \cdot \rho \cdot \bar{v}^2$) that occurs according to the Bernoulli equation (equation 12) in the mouth of the tube. The pressure curve is composed of the curves I and II, derived by Langhaar [51]. Near the tube mouth, this curve is asymptotically approximated to the Bernoulli pressure drop (IV) and with increasing x the curve is asymptotically approximated to curve (III), which describes the pressure loss in fully developed tube flow [equation (5c)].

sage of this section is that geometries with sharp edges must be eliminated or avoided in flow cytometric instrument design.

Flow Line Coordination

According to the potential flow theory, every flow is described by flow velocity lines and equipotential lines which are orthogonal to one another. In Figure 4 consider the two-dimensional situation where the flow is from a very large container of unlimited volume into a small diameter tube. The present analysis will be done based on potential flow theory.

Consider an area A in the reservoir that is located a distance s from the inlet to the tube of radius R. If $s >> R$, the flow is like that out of a sink, in the sense that the tube inlet is the center for a series of spherical areas (the equipotential areas) for the flow in the large container. The velocity lines then extend out radially from the tube inlet. At each area A, fluid passes through with the same velocity. As the fluid approaches and enters the tube inlet, the flow changes from one characterized by spherical equipotentials in the large container to equipotential areas that are parallel and that lie along the tube diameter. Conservation of mass [equation (10)] requires that the amount of fluid crossing a circular equipotential in the reservoir must equal the amount of fluid crossing a line in the tube. It follows that

$$\bar{v}_A \, 2 \cdot \pi \cdot s^2 = \bar{v} R^2 \cdot \pi \tag{15a}$$

$$\frac{\bar{v}_A}{\bar{v}} = \frac{R^2}{2\,s^2} \tag{15b}$$

Since the flow is considered to be laminar both in the reservoir as well as in the tube, the flow lines cannot cross or mix. Thus, large areas A in the reservoir must transform into small areas a in the tube. Fluid crossing through A will also cross the equivalent a within the tube. With a uniform velocity profile within the tube:

$$\frac{A}{a} = \frac{2\,s^2}{R^2} \qquad a = \frac{A\,R^2}{2\,s^2} \tag{16}$$

This is the principle of "hydrodynamic focusing" (84) which is used in most flow cytometric instruments. Since the requirement is that the flow remain laminar, the fluid flow lines from equivalent areas in the larger container transform to those in the tube or orifice with a corresponding increase in velocity. Large streams are focused into narrow threads. If, for example, the area A represents the cross-sectional area of an injection tube that contains cells, these cells would be focused into a small stream crossing the area a within the tube. The velocity v_A decreases with s^2; for large s, the fluid velocity and the particle sedimentation velocity may approach the same value. Moving the injection tube area A either toward or away from the flow axis of the system will change the position of the particle stream a within the tube. This can be verified experimentally as shown in Figure 5. Here a glass tube with a diameter of 120 μm extends into a fluid-filled region. In the tube $v \approx 3$ m/sec. The focusing effect is clearly obvious; the dye stream remains intact along the entire length of the tube, even when it is close to the tube wall. This experimental arrangement does not completely

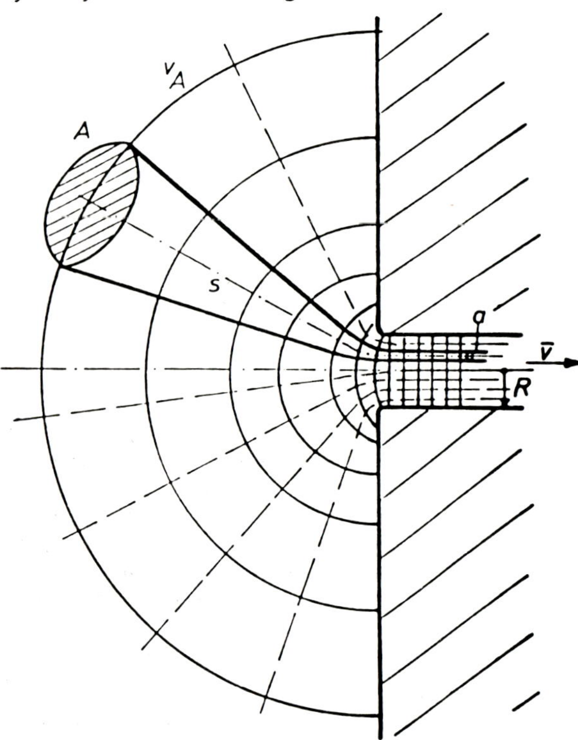

Fig. 4. The coordination of laminar flow lines outside and inside the inlet of a tube. The fluid passing through the cross section A in the distance s outside the tube inlet is focused to the cross section a in the tube.

agree with the theoretical considerations made for a tube inlet in a wall since the wall at the tube inlet is absent and fluid out of a larger angle enters the tube than that considered theoretically with the half sink (solid angle 180°); however, the basic effect is not altered. In contrast to predictions in the literature, no obvious constrictions of the flow are observed. In addition to the case considered here, fluid passing from a large-diameter to a small-diameter tube will undergo the same type of hydrodynamic focusing.

Problems at the Outlet of Tubes

Except for possible back-flow problems, where particles are recirculated into the sensing zone, the events that occur at the outlets of tubes are of little importance in analytical flow instruments. In cell sorters, however, the tube outlets deserve major consideration since they are responsible for the formation of the liquid jets in the air that transport the cells through the sorting steps. Sorters will be discussed below.

Tube outlets that do not produce jets are considered here, and the situation is depicted in Figure 6, where the flow is out of a tube of cross-sectional area F_2 into a larger tube of area F_3. The divergence angle is not allowed to exceed 6° to

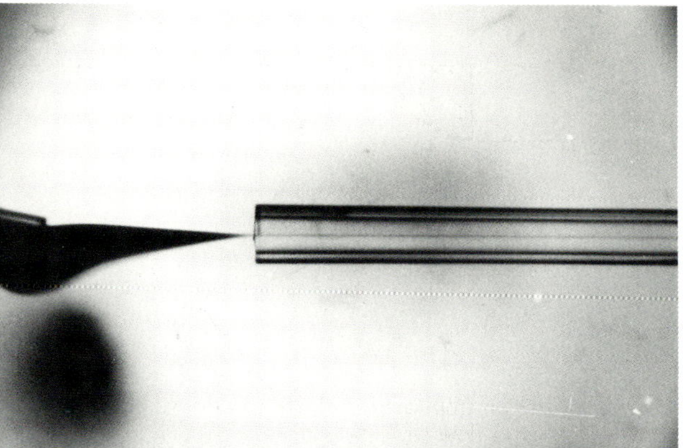

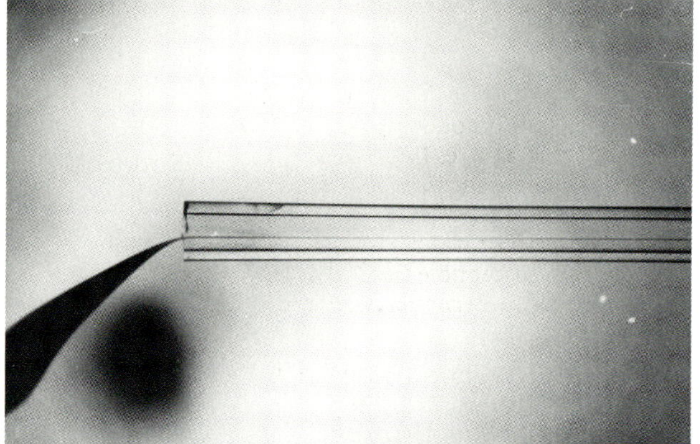

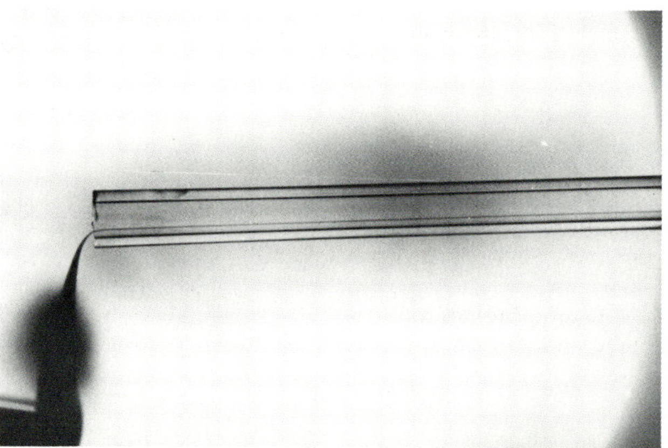

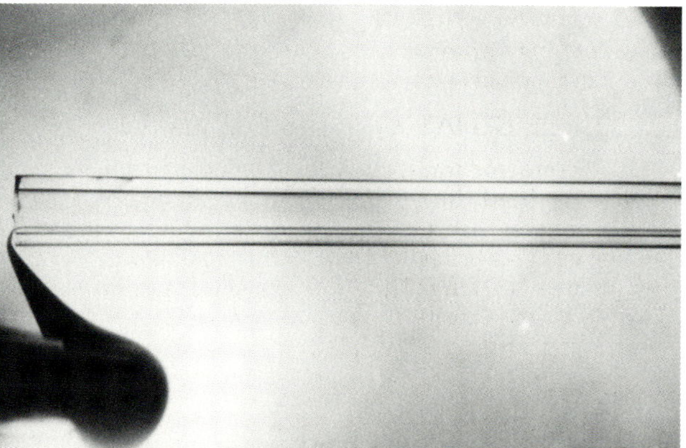

Fig. 5. Hydrodynamic focusing at the inlet of a circular glass tube with an inner diameter of 120 μm. The black ink is ejected by a glass tip of a diameter of about 150 μm. The path of the ink thread in the tube depends on the position of the ejecting glass tip outside the tube inlet. Mean flow velocity inside the tube 3 m/sec.

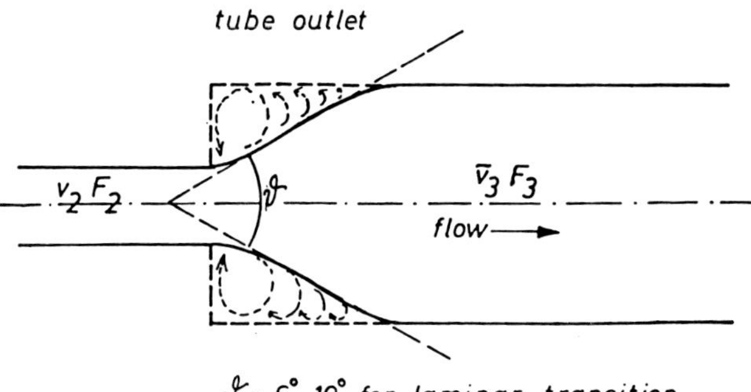

Fig. 6. Transition from a small-diameter tube to a large-diameter tube. To prevent flow separation at the widening tube wall, divergence should not exceed 6° to 10°. With larger angles ϑ, the flow is separated from the wall and an eddy zone exists behind the outlet of the small tube.

10° for laminar transformation. In the region of divergence, according to Bernoulli's principle [equation (11a)], kinetic energy is lost as a result of the velocity decrease and there is a corresponding increase in pressure. If ϑ is too large, flow separation at the walls occurs with a resulting loss in energy and formation of turbulence. The outlet transition region is much more susceptible to disturbances than the inlet region.

According to Bernoulli [equation (11b)], the pressure increase is

$$\Delta p = \frac{\rho}{2}(\bar{v}_3^2 - \bar{v}_2^2)$$

If $\vartheta = 180°$, the collision loss according to Carnot is

$$\Delta p_v = \frac{\rho}{2}(\bar{v}_2 - \bar{v}_3)^2 \tag{17}$$

The total increase in the pressure is $\Delta p' = \Delta p - \Delta pv$ and

$$\Delta p' = \rho\,\bar{v}_3\,(\bar{v}_2 - \bar{v}_3) \tag{18}$$

If the tube empties into a very large container, the pressure increase

$$\Delta p' = 0 \text{ since } v_3 = 0$$

Eddy currents form (see Fig. 6) and a backflow of part of the fluid ejected by the tube can occur. Particles suspended in the backflow may produce distorting pulses in resistance-pulse sizing instruments (see Chapter 4, this volume).

PARTICLE BEHAVIOR IN TUBE FLOW

So far only the transport media in flow instruments have been considered. In this section, the behavior of particles (biologic cells) that are transported by the carrier fluid will be treated. This behavior plays an important role in Coulter instruments (see ref. [43]) as well as optical instruments [27]. In general, the density of the cells is so close to that of the suspending medium that sedimentation of the particles is negligible.

Jeffery [40] demonstrated that rigid spheroids rotate differently in laminar flow, according to their shape. In addition, he found that particles flowing in a tube have a tendency to concentrate along the flow axis where there are paths of minimal energy dissipation due to the lack of shear forces. Saffman [72,73] reports, in contrast to Jeffery's work, that the forces on particles are not sufficient to explain the observed effect of particle migration toward the axis. He further postulates that the non-Newtonian characteristics of the carrier fluid are responsible for the effect. Segre and Silberberg [82] observed that rigid spherical particles do not concentrate along the axis in laminar flow of a carrier fluid with high viscosity, but rather they collect in a ringlike zone around the axis $r/R = 0.6$. Oliver [59] found that spheres rotating in flow behave according to the principles of Segre and Silberberg; nonrotating spheres, however, migrate closer to the tube axis. Goldsmith and Mason [29,30] investigated the motion of rigid spheres and fluid drops in relatively low Reynolds number flow, where rotation of the particles but no migration in the radial direction is observed. They also observed the migration of deformable drops toward the tube axis. Droplet deformation in shear flow was reported by Rumscheidt and Mason [69]. Goldsmith [31] showed that deformation of human erythrocytes in tube flow is similar to that of deformed drops [76]. These findings have only limited applicability to the behavior of particles in flow cytometry instruments since (1) the flow velocities are generally higher in these instruments than in the above cited experiments; and (2) the radial forces are so small that the requirement in flow cytometry for particles to be concentrated along a narrow well-defined path in a short time and distance along the flow is not fulfilled. Only the principle of hydrodynamic focusing will satisfy this requirement.

Particle Orientation and Deformation

From the preceding, it is apparent that there exists a flow in the region prior to the inlet of the tube where the flow velocity increases with decreasing distance from the inlet; i.e., $v\,(1/s)^2$. Our group has shown (41.42) that nonspherical cells always align in the direction of their longest dimension before they enter the tube (Fig. 7). This phenomenon is explained by the Bernoulli pressure–velocity relationship given in equation 11b and shown in Figure 8. The pressure difference between two equipotential regions is

$$\Delta_p = 1/2\,(\bar{v}_j^2 - \bar{v}_k^2)$$

By the law of mass conservation, the velocity of an equipotential region at a distance sj from the tube inlet is

$$vj = \frac{R^2\,\bar{v}}{2\,dj^2} \qquad \text{for } sj > R \tag{19}$$

The pressure difference between two equipotential areas at distances s_j and s_k from the inlet to the tube is

$$\Delta p_{j\mu} = \frac{\rho}{8} R^4\,\bar{v}^2 \left(\frac{1}{sj^4} - \frac{1}{sk^4}\right) \tag{20}$$

Now we define α as the angle between the particle axis and the tangent to the equipotential surface vj. If $\alpha = 0$, this is an unstable equilibrium, thus a net torque is applied to the particle and consequently the point B moves with a velocity that is greater than that experienced by the point A. The particle then rotates so that the axis A–B is aligned with the accelerating flow towards the tube inlet. From Equation (20), for

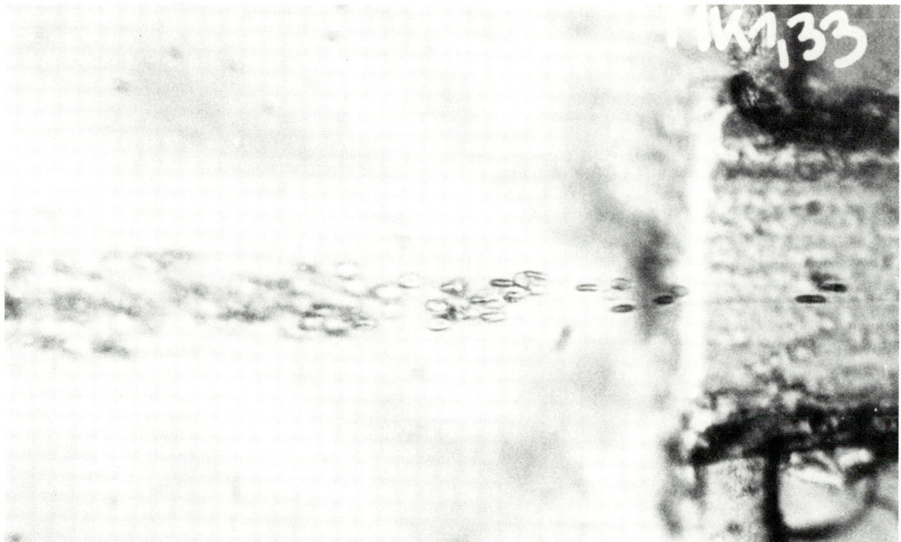

Fig. 7. Focused stream of human erythrocytes in front of a rectangular short tube. The cells are oriented and already deformed in front of the tube inlet. Flow velocity inside the orifice >3 m/sec.

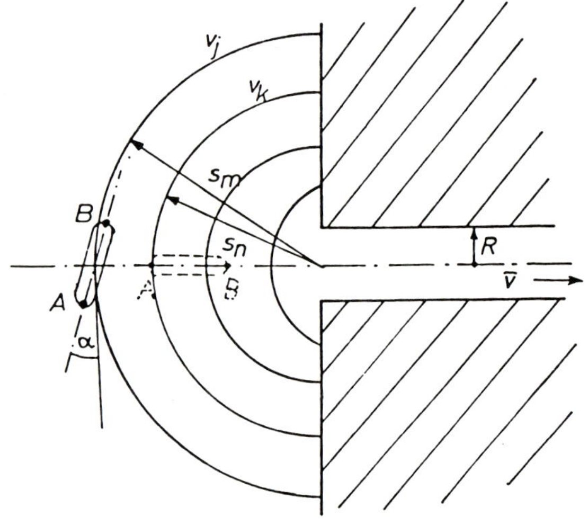

Fig. 8. Explanation of the orientation and deformation of particles in the accelerated flow in front of a tube inlet (see text).

a cell of axial dimension 10 μm located at a distance s = d where d = 60 μm, the pressure differential along the axis A-B is approximately 25 mbar. If the cell is flexible, like an erythrocyte, this is sufficient to cause flexing or other distortions [43] (see Fig. 7). As described above, there is a uniform velocity profile at the inlet to the tube. If the tube is short, this is still true except that a narrow boundary layer exists at the tube walls. The core flow is initially free of shear forces; these are only observed near the tube wall and are the result of friction at the wall. Consequently, the orientation of the particle is not altered if it flows in the core, where there are no shear forces. If the particle flows near the wall outside the core, the particle will experience some torque. If the tube has a sharp edge rather than a gradual transition zone at the inlet, additional disturbances are to be expected in the dead water region just behind the inlet. Furthermore, the flow near the container wall just before the tube inlet is not free of friction; hence the orientation of particles flowing in this region also will be altered.

Now, consider behavior in a short tube. Figure 9a shows native human erythrocytes on an off-axis trajectory in a short tube of rectangular cross-section. The relationships describing the flow in a rectangular tube are similar to those derived for cylindrical flow. The cells are still uniformly oriented; they will then follow an undisturbed path within the core of the flow as was the case in cylindrical flow. As the particles travel closer to the wall (Fig. 9b), the orientation will be disturbed. The cells are now in the turbulent zone just behind the sharp inlet edge. Figure 9b shows also the situation for human erythrocytes where turbulence exists close the wall and irregular deformation of these flexible cells can occur.

The situation is somewhat different in a longer tube. As the distance down the tube increases, the parabolic velocity profile develops with the result that there are shear forces present. Particle rotation results, as described by Goldsmith [31], for low Reynolds number flow. This is shown in Figure 10 for fixed chicken erythrocytes flowing in a tube 120 μm in diameter. In this case, if the particles flow along the axis there is no torque and hence no rotation. However, if the particles are located off the axis at the position r/R = 0.45, particle orientation is no longer uniform beyond x≈3d. At this point (Re = 170, v = 2.8 m/sec), the diameter of the shear free core has shrunk to about 35% of the tube diameter. Hence, there is cell rotaion in the shear zone outside the core. As mentioned above, the hydrodynamic focusing principle used in most flow cytometric instruments provides additional focusing features for aligning nonspherical particles. For some optical measurements of flat cells the lateral orientation of these cells influences the measured distribution curves [27]. A uniform lateral orientation of such cells is reached by constructing the focusing flow path such that symmetry of revolution no longer exists. The different constricting horizontal and vertical forces orient flat cells uni-

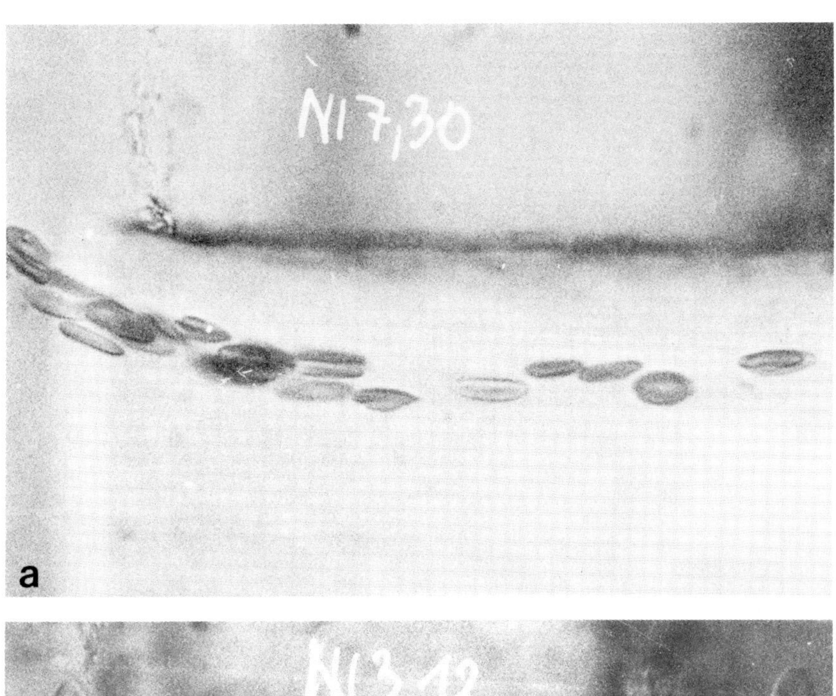

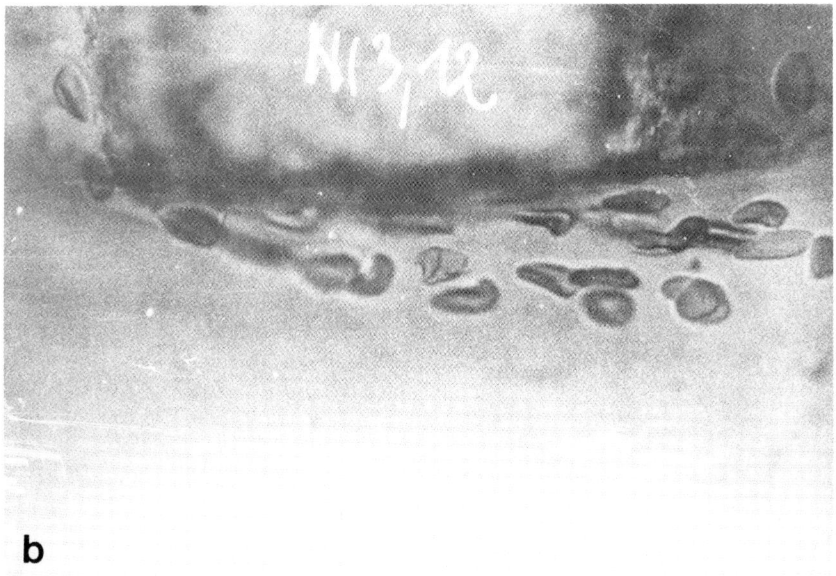

Fig. 9. **a:** Native human erythrocytes near the margin of the core stream of a short tube (orifice). The cells are uniformly oriented and elongated by the hydrodynamic forces of the inlet flow. **b:** In the turbulent flow near the tube wall, the cells are deformed and disoriented in a very individual way. $v > 3$m/sec.

formly in horizontal or vertical direction [44]. Figure 11 shows the axial view into a flow path which orients flat cells with their flat side in a horizontal direction. Horizontal and vertical orientation of chicken erythrocytes is demonstrated in Figure 12. Another hydrodynamically orienting device is described by Fulwyler [23].

FORMATION OF JETS AND DROPLETS IN CELL SORTERS

The hydrodynamic principles that determine the formation of jets and droplets deserve further consideration in regard to cell sorters. Initial studies on instability of fluid jets began with Savart [74] in 1833 and were pursued by Plateau, Rayleigh [67], Helmholtz, Boys [8], Bohr [6], Weber [90], and many others. Savart is responsible for the origin of the concept that uniform modulation of a stream can cause the breakup of the stream into uniform droplets at the modulation frequency. In the present century Dimmock [15] was probably the first to continue this work. The concept of jet formation and droplet generation are important in many technologies as well as cell sorting; that is, extrusion of synthetic fibers [26], determination of surface tension [6,47], ink jet printing [45,86], diagnosis of urinary tract diseases by drop spectrometry [48,92], production of aerosols [35,85], formation of uniform solid particles [36] and other applications [9,39,78].

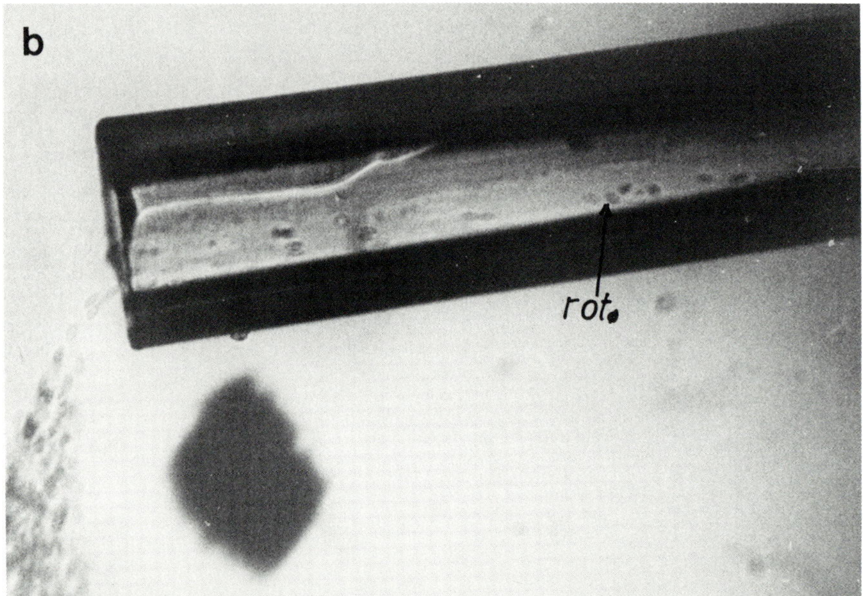

Fig. 10. **a:** Fixed chicken erythrocytes focused in a path at the axis of a longer circular tube. Over the whole visible length of the tube the cells remain uniformly oriented. **b:** Fixed chicken erythrocytes focused in a path at r/R 0.5. Beginning at x≈3d the orientation of the cells is no longer uniform. There the core region is reduced to about 35% of the tube diameter and the cells begin to rotate in the laminar shear field. Re = 170; v = 2.8 m/sec.

Minimum Pressure Conditions for Jet Formation

To form a stable jet, the rate of energy flow into the jet from the capillary tube feeding the jet should be greater than the rate of increase of surface energy required for the continuous formation of new surface on the jet [54,79]. Thus,

$$\bar{v} > 2 \left(\frac{2\sigma}{dj \cdot \rho} \right)^{1/2} \tag{21}$$

Assume that the jet diameter d_j is equal to the tube diameter d_o. The minimum pressure necessary to achieve stable jet formation is then calculated from equation (21) and the Hagen–Poiseuille equation (5d):

$$\Delta p > 4 \lambda_0 \frac{\ell \sigma}{{d_0}^2} \tag{22}$$

This holds only for long tubes (ℓ/d > 0.06 RE), where complete laminar flow has developed. For shorter nozzles, Δp

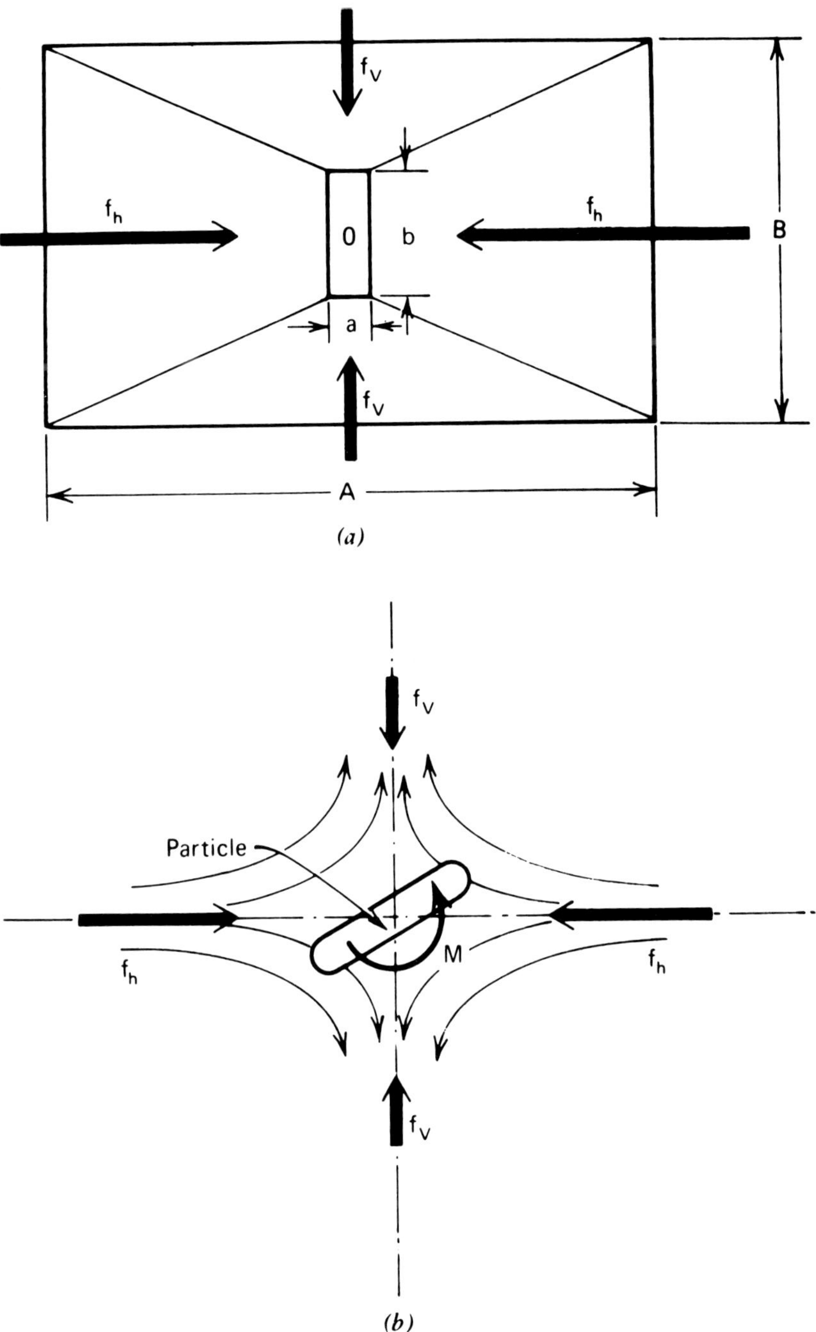

Fig. 11. **a:** Axial view into a flow path which orients flat cells laterally in the vertical direction. f_h: constricting flow forces that have a horizontally orienting effect; f_v: flow forces acting vertically; A: horizontal diameter of the flow path at its inlet; a: horizontal diameter of the flow path at the end of the constriction; B: vertical diameter of the flow path at its inlet; b: vertical diameter of the flow path at the end of the constriction. The dissimilar transformation of the cross-section causes preferential horizontal side forces that generate additional up- and down-flow components, which are responsible for the vertical orientation of flat cells. **b:** Detailed conditions of the flow around a particle. The main flow direction is normal to the plane of the paper. The decreased vertical force f_v enables a superpositioned flow in up-and-down direction at both sides of the vertical plane of symmetry. The torque M produced by these flow components (thin arrows) rotates flat cells to the stable vertical orientation.

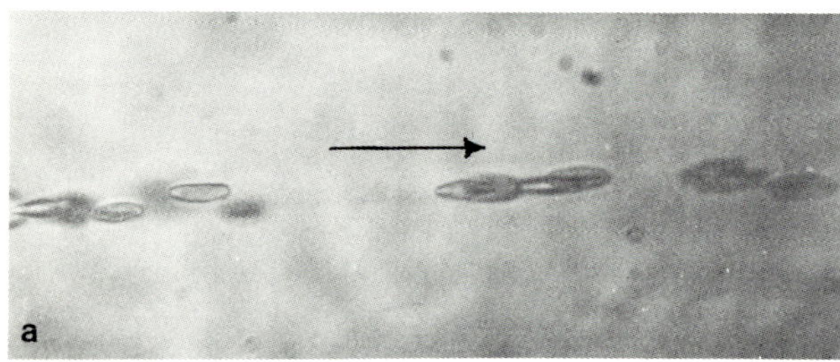

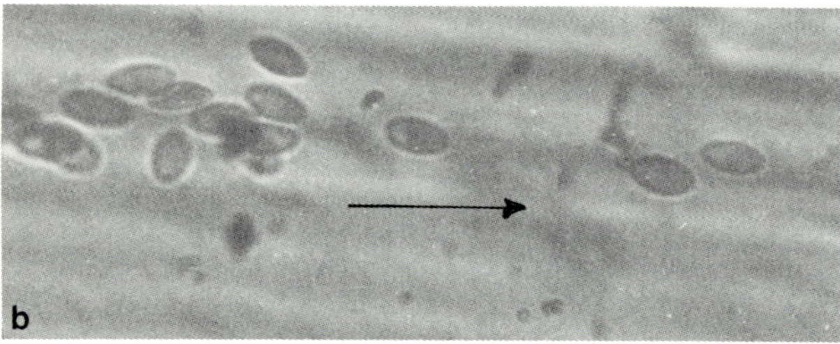

Fig. 12. Photographs of chicken erythrocytes moving with a velocity of 3 m/sec through lateral orienting flow paths. **a:** Vertically oriented cells. **b:** Horizontally oriented cells.

must be calculated from Langhaar [51] or taken from Figure 3. In the latter case, the tube length ℓ is substituted for the quantity x. In addition, there exists an internal jet pressure due to the surface tension, $p_j = 2\sigma/d_j$, which requires a correspondingly higher minimum pressure. If a large fluid droplet has formed at the exit nozzle, the jet can be destroyed. However, it cannot be reformed at pressures near the minimum pressure of equation (22), since in this situation the energy is dissipated in vortex formation within the drop.

CONTRACTION OF JETS

Jet contraction is an important consideration in the design of cell sorters. The jet diameter d_j defines the modulation frequencies that result in optimal droplet formation. The usual measurable parameter is the diameter of the outlet orifice d_o which is related to the jet diameter by $\chi = d_j/d_0$.

By applying the laws of momentum and energy conservation, Rayleigh [66] showed that for a cylindrical vessel with a short axial orifice the relationship between the cross-sections of the orifice, the vessel, and the final contracted jet is

$$\frac{2}{F_0} = \frac{1}{F_c} + \frac{1}{F_j} \tag{23}$$

The ratio F_jF_0 is 1/2 for jet formation in the case of a very large reservoir ($Fc \rightarrow \infty$, $1/Fc \rightarrow 0$). Hence, χ becomes

$$\chi = \frac{d_j}{d_0} = \frac{1}{\sqrt{2}} = 0.707$$

In fact, Gavis and Modan [26] reported that for short orifices and high Re flow, such sudden and sharp contractions occur.

Many investigators have studied jet contraction; Harmon [35] employed mass and momentum conservation at the point of jet exit from a tube with well-developed laminar flow, and at a point downstream on the jet at a distance greater than the length of the nozzle. The jet diameter d_j should always be determined at a point downstream, where the velocity profile is completely relaxed. Harmon [35] finds a coefficient of contraction, neglecting viscosity, $\chi = \sqrt{3/2} = 0.866$. By mass conservation (equation 10)

$$\bar{v}_j = \frac{F_0}{F_j}\bar{v} = \left(\frac{d_0}{d_j}\right)\bar{v}_0 = \frac{1}{\chi^2}\bar{v}_0 \tag{24}$$

Note that $vj > v_0$ by a factor $1/\chi^2 = 4/3$ when $\chi = 0.866$ [35]. However, this is only a minimum value for χ since the viscous forces (neglected by Harmon) dissipate some of the kinetic energy, thereby decreasing the mean velocity and increasing the final diameter of the jet. Under certain conditions, the effects of surface tension and viscosity can overcome the effects of inertia [26,33,57]. This effect is a strong function of the Reynolds number, and only for $Re > 16$ causes a contraction of jets ($\chi < 1$).

Gavis and Modan [26] considered the effect of tube length on jet contraction as well as the case of shorter tubes with non fully developed velocity profiles. These results are shown on the descending curve in Figure 13. The ascending branches in Figure 13 represent the effect of viscous forces. Surface tension showed no measurable effect [26]. In sorters, the values of $\ell/(d \cdot Re)$ that are used illustrate that internal forces predominate over the effect of the viscous forces represented by the ascending branches. The surrounding air also plays a role for jets with high velocities vj, where the effect of the air is to deaccelerate the jet, thus increasing d_j and χ [33].

Instability of Jets

A jet leaving a circular orifice can be regarded as a cylinder of liquid, once relaxation has occurred. Such cylinders have been shown to be unstable [74], and in order to minimize surface area the jet desintegrates into droplets under the influence of surface tension. Rayleigh [68] analyzed the problem by means of first order perturbation calculation. He showed that axissymmetrical perturbations grow only if their wavelength is longer than the circumference of the jet. Hence, the normalized wavenumber $kd_j/2 = \xi = \pi d_j/\lambda$ is < 1 for growing perturbations. The unstable waves grow exponentially with a growth rate α and drop break off occurs when perturbation amplitude gj becomes equal to the jet radius. The normalized perturbation amplitude δ is

$$\delta(t) = 2\Delta d_j/d_j = \delta(0) \exp(\gamma t) \tag{25}$$

where δ (0) is the excitation amplitude at the mouth of the nozzle.

The breakoff time T is given by

$$\delta(t) = 1 = \delta(0)\exp(\gamma T) \tag{26}$$

The break off distance L is then

$$L = vT$$

where v is the jet velocity.

Rayleigh's first order solution for γ is

$$\gamma^2 = \frac{8 \cdot \sigma}{\rho \cdot d_j^3}\,\xi\,(1 - \xi)\,I_1(\xi)/I_0(\xi) \tag{27}$$

σ is the surface tension, ρ the density, I_0 and I_1 are modified Bessel functions of the first kind [50]. γ has a maximum at ξ

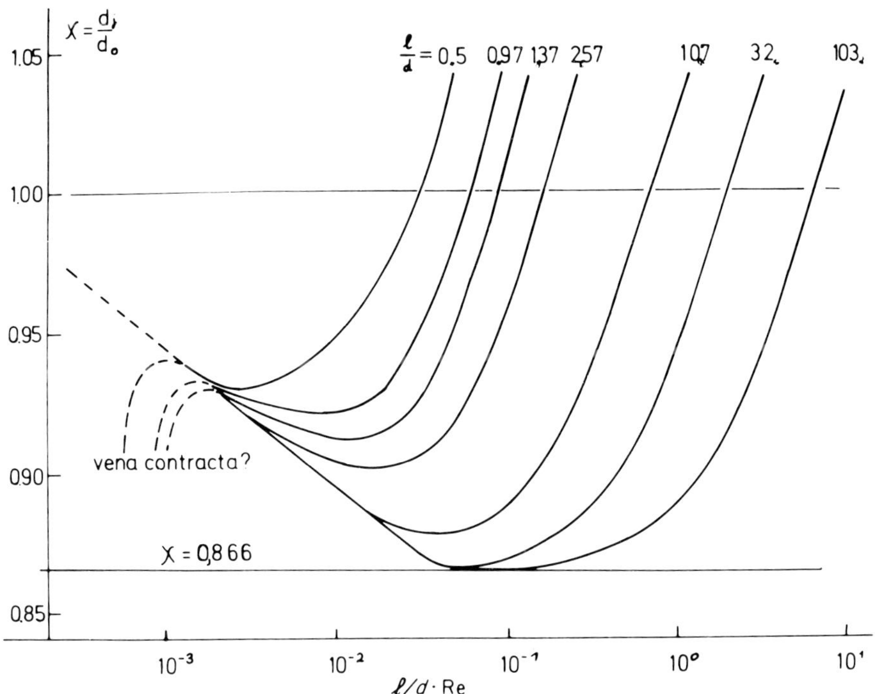

Fig. 13. Dependence between coefficient of contraction χ and the parameter (ℓ/d·Re) for various nozzle (tube) ℓ/d ratios, calculated from the experimental data of Figure 1 [26]. $\chi = 0.866$ is Harmon's [35] value. The descending line is due to the inertial forces; the ascending branches are due to the effect of viscosity and that of surface tension. The dotted parts have not been considered in detail [26].

= 0.697, indicating the waves with this wavenumber grow preferentially and the breakoff distance has a minimum at this point, corresponding to a wavelength

$$\lambda_{max} = \frac{\pi}{0.697} d_j = 4.508 d_j \qquad (28)$$

There is no theoretical upper limit for λ ($\xi \rightarrow 0$), although with increasing λ, γ decreases. However, under experimental conditions, there is a practical upper wavelength limitation for long-wavelength modulation based on noise considerations that might affect the jet. The longest wavelength reported is $\lambda = 18 d_j$ [79].

If the jet disturbances are applied periodically, as is done in cell sorters, the values of λ suitable for stable and effective modulation are found from Figure 14 and converted into modulation frequencies with equation (29):

$$f = \frac{\bar{v}_j}{\lambda} \qquad (29)$$

Other parameters that influence the stability of jets were not considered by Rayleigh [66] in the earlier theory:

Viscosity of the fluid. Viscocity of the fluid dampens the growth of disturbances [2,10,67]; thus, γ is less for viscous fluids than for those that are nonviscous. Smaller values of γ result in dispersion curves [34,89] that are below the Rayleigh curve but have identical end points ($\xi = 1$) with the maxima shifted to longer wavelengths (see Fig. 14). Experimental confirmation of this is evident in cases of spontaneous stream breakup; droplets in viscous fluids are larger than those in nonviscous fluids. The viscosity of some fluids is strongly temperature dependent, and temperature must be controlled if stable droplet formation is to be achieved.

Surrounding air. Weber [90] showed that the aerodynamic pressure is positive in the neck region and negative in the swell region of the disturbances, thereby increasing instability. Weber's dispersion curves are also shown in Figure 14 and should be compared with the experiments of Haenlein [34]. The effects of air on jets of higher velocity, e.g., cell sorters are as follows:

1. Increase in the growth rate γ
2. Decrease in the value of λ_{max} [cf. equation (28)] to $\lambda_{max} < 4.508 d_j$
3. Possibility of wave numbers $\xi > 1$. Under this condition, surface tension has a stabilizing effect [20] on the jet and disturbances tend to "heal."

Nonparallel flow components attributable to velocity profile relaxation have an effect on the jet breakup—especially in jets emerging from long tubes. This effect is most important at low Reynold's flow numbers [55] and is not significant at the high Re numbers encountered in sorters.

Formation of Droplets

For periodically applied stream perturbations, linear theory predicts the formation of uniform droplets, where the droplet diameter is [54]:

$$d_d = \left(\frac{3d_j}{2}\right)^{1/3} \qquad (30)$$

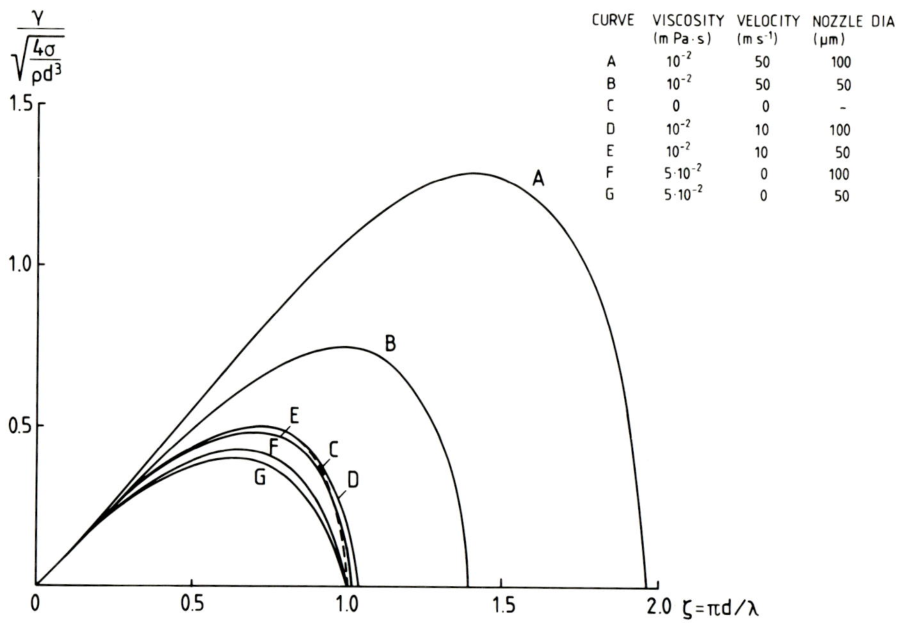

Fig. 14. Dispersion curve of the jet, according to Weber [90]. Dependence between the normalized growth rate γ and the wavenumber with viscosity η and jet velocity v as parameters.

with equations 10 and 29

$$d_d = \left(\frac{3d_j^2 v_j}{2f}\right)^{1/3} = \left(\frac{3d_0^2\, v_0}{2f}\right)^{1/3} \qquad (30a)$$

For $\lambda_{max} = 4.508\ d_j$, from the dispersion curve (see Fig. 14), with equation (30):

$$d_d = 1.89\ d_j \qquad (30b)$$

The droplet diameter is thus about twice the jet diameter.

Another useful form of equation (30) can be obtained from

$$\xi = \frac{\pi \cdot d_j}{\lambda}$$

with equation (30)

$$d_d = \left(\frac{3 \cdot \pi}{2\xi}\right)^{1/3} \cdot d_j \qquad (30c)$$

Rayleigh's uniform drop model does not explain smaller satellite droplets that occur under various excitation conditions [16,28]. Goedde [28] described the formation of satellite droplets from the neck region in a jet undergoing a periodic disturbance. As the neck thins into a ligament, the satellite droplet breaks free of the stream.

Different authorities [49,50,71] have extended Rayleigh's treatment to third order. With this extension, the formation of satellites, which is not predicted by Rayleigh's theory, can be explained. The satellites may be faster, slower than or synchronous with the main droplets [62]. For flow cytometry, quasi-synchronous satellites that have not yet merged with the main droplets when entering the deflection field are important, giving rise to a separation of the deflected stream.

Figure 15 shows the result of calculations of Lafrance [49] together with experimental results of Lafrance [50] and Rutland and Jameson [71]. According to Figure 15, no satellites should be produced with $\xi > 0.8$. However, the initial excitation amplitude $\delta(0)$ has also considerable influence on satellite formation. Curry and Portig [14] have experimentally shown with scaled up nozzles that satellite free operation can be achieved within certain excitation amplitude windows at $\xi \sim 0.45$— 0.70. Sharp entrance and exit edges of the nozzle reduce the satellite free window. Reynolds numbers of about 400 and Weber numbers ($W = \rho \cdot v^2 d/\sigma$) of the order of 100 gave satellite free operation from $\xi = 0.55$—0.70 over a wide range of excitation amplitudes. The nozzle aspect ratio ℓ/d was 0.995.

After their formation, spherules and droplets exhibit a damped oscillation [68]. The velocity decrease due to the aerodynamic drag is smaller for a droplet following another, than it would be for an isolated droplet. This effect, called channeling [85] is observed by deflecting a single droplet from the jet. This droplet moves at a slower speed as compared with the hole that was caused by the deflection.

The amplitude of the nozzle vibration can change by orders of magnitude when operating at a mechanical resonance frequency of the head drive system, keeping the transducer input voltage constant. Figure 16 shows such resonances at the nozzle tip of a FACS III-system (Becton-Dickinson).

The nozzle vibrations are transmitted into liquid jet giving rise to perturbations at the surface of the jet near the nozzle tip [δ (0) of equation (25)] that can be detected by changes in light reflection from a laser beam intersecting the liquid jet close to the nozzle. The amplitude of the AC component of this reflection is shown in Figures 17 and 18. Although the main features of the nozzle vibration resonances are preserved, there are significant changes in the resonance pattern

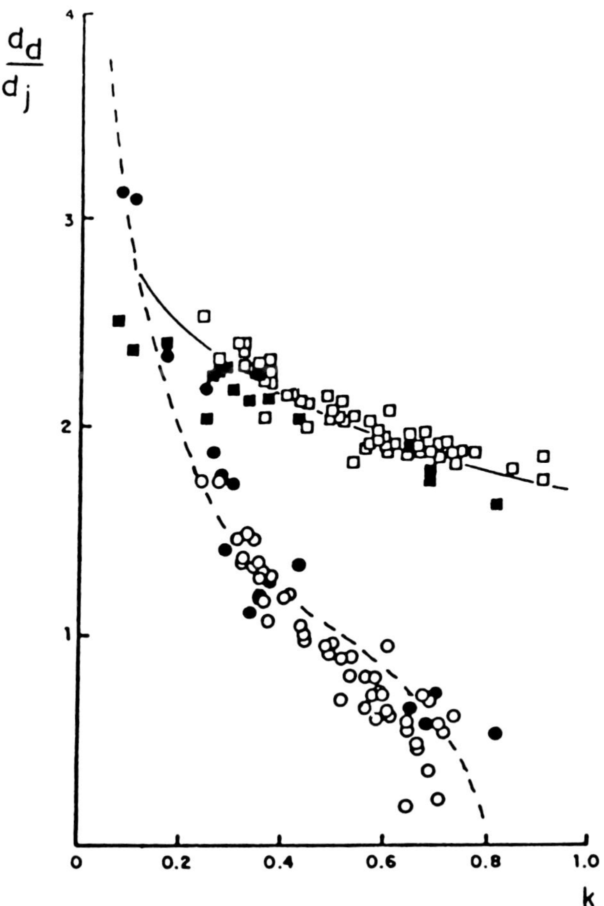

Fig. 15. Drop radius variation with wave number. Theory: ______main,------ satellite; Experiment: Lafrances's data □ main, ○ satellite; Rutland and Jameson ■ main, ● satellite;

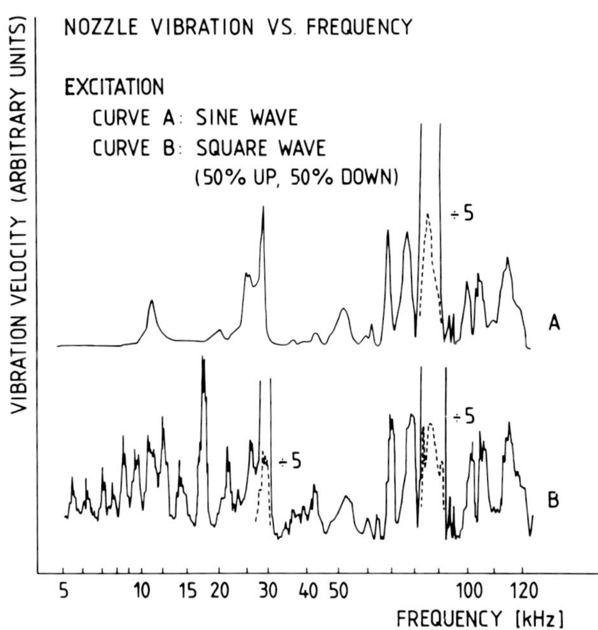

Fig. 16. Nozzle vibration versus frequency for a FACS III system (Beckton-Dickinson), measured with a moving coil transducer (Orthofon MC 2000 and transformer T 20) and a peak voltmeter (Brüel and Kjoer, type 2409). Transducer input voltage was 40 V peak to peak.

with different nozzle pressures. In general, it is advantageous to operate at a resonance frequency and to vary the sheath pressure in order to obtain optimum break off conditions at $\xi \sim 0.7$.

Scaling theory [14] requires for nozzles of different diameters that λ/d is constant. Using Weber's dimensionless number for free flow

$$We = \rho \cdot v^2/\sigma$$

gives with equations (5d) and (5e) $2\Delta P = \xi \cdot \rho \cdot v^2$

$$We = 2\Delta pd/\xi \cdot \sigma \qquad (31)$$

With the scaling factor $F = d_2/d_1$ and assuming ξ and σ = constant, the nozzle pressure ratio becomes

$$\Delta p_2/\Delta p_1 = F^{-1} \qquad (32)$$

Furthermore, since $\lambda = v/f$, where f is the head drive frequency, we get

$$\lambda^2/d^2 = v^2/f^2d^2 = \text{const.} \qquad (33)$$

and

$$We \cdot f^2d^3/v^2 = \rho \cdot f^2\, d^3/\sigma = \text{const.}$$

Keeping the sheath fluid unchanged (ρ/σ = const.) gives the frequency dependence

$$f_2/f_1 = F^{-3/2} \qquad (34)$$

Thus, large diameter nozzles require lower head drive frequency and lower nozzle pressure.

For example, using a resonance at 28 KHz with a 50-μm-diameter nozzle gives an optimum breakoff distance at 78-kPa nozzle pressure, while a 100-μm-diameter nozzle with the same nozzle aspect ratio ℓ/d operates best at a 10.5-kHz resonance with only 34-kPa head pressure. Large diameter nozzles operated at low pressure are advantageous for sorting of delicate biological material, such as plant protoplasts.

SUMMARY

Theoretical and experimental studies of hydrodynamic phenomena relevant to flow cytometric instrument design have been presented in this chapter, together with a survey of the appropriate literature describing

1. Laminar flow in tubes such as those used in flow instruments
2. The transition region in laminar flow—either from large-diameter flow to small-diameter flow, or the reverse
3. Behavior of particles in flow channels
4. Jet formation and its relationship to the frequency of the driving excitation
5. The formation of droplets from a jet stream as they apply to contemporary cell sorters

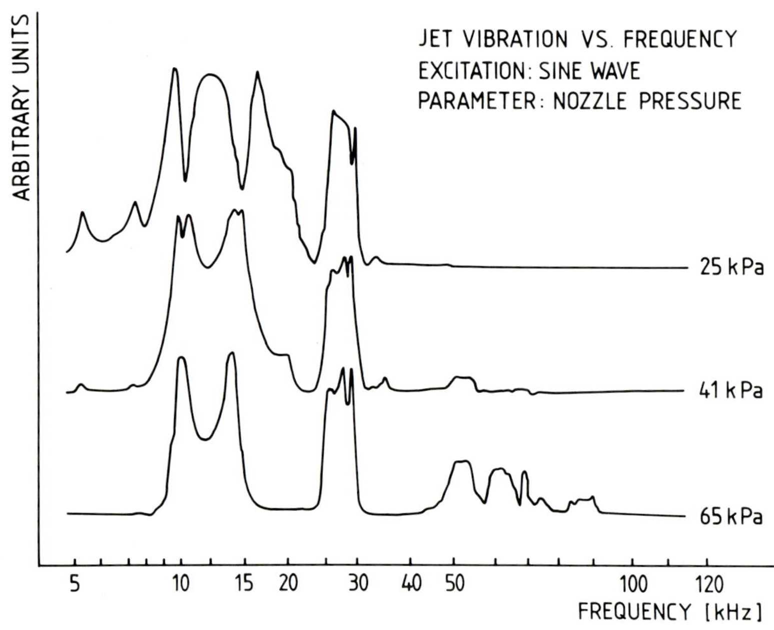

Fig. 17. Frequency dependence of the 90° reflection of a laser beam (488 nm) from the surface of a liquid jet of 92–98 μm diameter (depending on nozzle pressure), measured with a RMS voltmeter (Brüel and Kjoer, type 2409) at the output of a photomultiplier. Transducer input voltage was 40 V peak to peak.

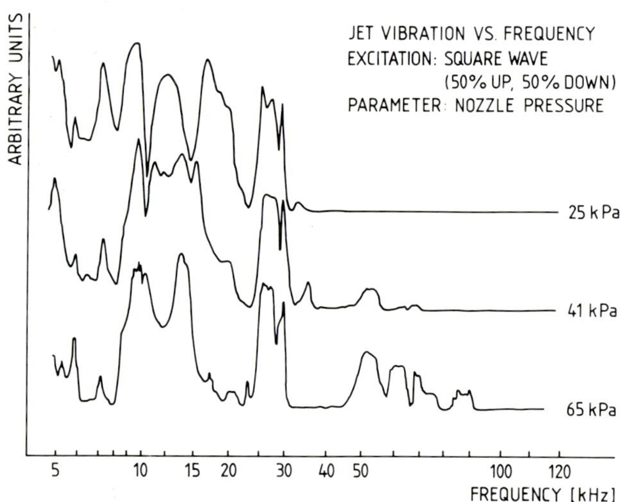

Fig. 18. Same as Figure 17, except for square wave transducer voltage, 40 V p-p, 50% up, 50% down.

Although no theory particularly designed for modern flow cytometric instruments or cell sorters is offered, it is clear that much of the past and present research in laminar flow or low Re flow is directly related to the present problems.

NOMENCLATURE

A,a	areas or equivelocity areas outside and inside a tube
d	tube diameter
d_h	hydraulic diameter
F	cross section of a tube, orifice
f	modulation frequency
ℓ	tube, orifice length
M	fluid volume
m	factor
p	pressure
Δp	pressure difference
Q	flow rate = fluid volume/time
R	tube radius
R_c	core radius
Re	Reynolds number
r	radius variable
s	distance form tube inlet or outlet
T	temperature
t	time
U	circumference
v	velocity
v_c	core velocity
$\bar{v}$	mean velocity
x_e	transition length between tube inlet and a fully developed parabolic profile
α	angle
δ	differences of diameters
η	viscosity
γ	growth rate
λ	wavelength
λ_0	tube friction number
ν	kinematic viscosity
ρ	density
σ	surface tension
ϕ	angle of tube dilation
χ	coefficient of contraction
ζ	normalized wave number for flow calculations in rectangular tubes
ξ	nozzle coefficient

INDICES USED

c	vessel
d	droplet
h	main droplet
s	satellite droplet
v	loss
o	tube, nozzle orifice
j	jet
m,n	arbitrary numbers
R	friction
B	acceleration

REFERENCES

1. **Albring W (1970)** "Angewandte Strömungslehre." Dresden: Steinkopff.
2. **Anno JN (1974)** Influences of viscosity on the stability of a cylindrical jet. AIAA J. 12:1137–1138.
3. **Bergmann-Schäfer (1974)** "Lehrbuch der Experimentalphysik." Vol. I. Berlin: W. de Gruyter.
4. **Bogy DB (1979)** Drop formation in a circular liquid jet. Annu. Rev. Fluid Mech. 11:207–228.
5. **Bohl W (1971)** "Technische Strömungslehre." Würzburg: Vogel Verlag.
6. **Bohr N (1909)** Determination of surface tension of water by the method jet vibration. Phil. Trans. A209: 281–317.
7. **Boussinesq J (1891)** C.R. Acad. Sci. (Paris) 113:9–49.
8. **Boys CV (1959)** "Soap Bubbles and the Forces which Mould Them." Garden City, N.Y.: Doubleday.
9. **Burkholder HC, Berg JC (1974)** Effect of mass transfer on laminar jet breakup. AIChE J. 20:863–880.
10. **Chandrasekar S (1961)** "Hydrodynamic and Hydromagnetic Stability." Oxford: Chalderon Press.
11. **Chaudhary KC, Maxworthy T (1980)** Nonlinear capillary instability of a liquid jet. 2. Experiments on jet behaviour before droplet formation. J. Fluid Mech. 96:275–286.
12. **Chaudhary KC, Maxworthy T (1980)** Nonlinear capillary instability of a liquid jet. 3. Experiments on satellite drop formation and control. J. Fluid Mech. 96: 287–297.
13. **Crane L, Birch S, McCormack PD (1964)** The effect of mechanical vibration on the break-up of a cylindrical water jet in air. Br. J. Appl. Phys. 15:743–750.
14. **Curry SA, Portig H (1980)** Scale model of an ink jet. IBM J. Res. Dev. 21:10–20.
15. **Dimmock NA (1950)** Production of uniform droplets. Nature (Lond.) 166:686.
16. **Donnelly RJ, Glaberson W. (1966)** Experiments on the capillary instability of a liquid jet. Proc. R. Soc. A290: 547–556.
17. **Eck B (1961)** "Technische Strömungslehre." Berlin: Springer.
18. **Entov V (1980)** Dynamical equations for a liquid jet. Fluid Dynamics 15:644–649.
19. **Erk S (1929)** Über die Zähigkeitsmessung nach der Kapillarmethode. Z. Tech. Phys. 452.
20. **Esmail MN, Hummel RL, Smith JW (1975)** Instability of two-phase flow in vertical cylinders. Phys. Fluids 18:508–516.
21. **Fillmore GL, Buehner WL, West DL (1977)** Drop charging and deflection in an electrostatic ink jet printer. IBM J. Res. Dev. 21:37–47.
22. **Fulwyler MJ (1965)** Electronic separation of biological cells by volume. Science 150:910–911.
23. **Fulwyler MJ (1977)** Hydrodynamic orientation of cells. J. Histochem. Cytochem. 25:781–783.
24. **Garg VK (1981)** Stability of developing flow in a pipe: Non-axisymmetric distrubances. J. Fluid Mech. 110: 209–216.
25. **Gavis J (1964)** Contribution of surface tension to expansion and contraction of capillary jets. Phys. Fluids 7:1097–1098.
26. **Gavis J, Modan M (1967)** Expansion and contraction of jets of Newtonian liquids in air: Effect of tube length. Phys. Fluids 10:487–497.
27. **Gledhill BL, Lake S, Steinmetz LL, Gray JW, Crawford JR, Dean PN, vanDilla MA (1976)** Flow microfluorometric analysis of sperm DNA content: Effect of shape on the fluorescence distribution. J. Cell. Physiol. 87:367–375.
28. **Goedde EF, Yuen MC (1969)** Experiments on liquid jet instability. J. Fluid Mech. 40:495–511.
29. **Goldsmith HL, Mason SG (1961)** Axial migration of particles in Poiseuille flow. Nature (Lond.) 190:1095–1096.
30. **Goldsmith HL, Mason SG (1962)** The flow of suspensions through tubes I. Single spheres, rods and discs. J. Coll. Sci. 17:448–476.
31. **Goldsmith HL (1971)** Deformation of human red cells in tube flow. Biorheology 7:235–242.
32. **Goren SL, Gottlieb M (1982)** Surface-tension-driven breakup of viscoelastic liquid threads. J. Fluid Mech. 120:245–266.
33. **Goren SL, Wronski S (1966)** The shape of low-speed capillary jets of Newtonian liquids. J. Fluid Mech. 25: 185–198.
34. **Hänlein A (1932)** Über den Zerfall eines Flüssigkeitsstrahles. Forsch. Ingenieurwes. 2:139–149.
35. **Harmon DB Jr (1955)** Drop sizes from low speed jets. J. Franklin Inst. 259:519–522.
36. **Hendricks CD, Babil S (1972)** Generation of uniform 0.5–10 μm solid particles. J. Phys. E5:905–910.
37. **Herzenberg LA, Sweet RG, Herzenberg LA (1976)** Fluorescence activated cell sorting. Sci. Am. 234:108–117.
38. **Hinch EJ:** The evolution of slender inviscid drops in an axissymmetric staining flow. J. Fluid. Mech. 191:545–553,1980.
39. **Iribarne JV, Klemes M (1975)** Electrification associated with droplet production from liquid jets. J. Chem. Soc. 70:1219–1227.
40. **Jeffery GB (1922)** The motion of ellipsoidal particles immersed in a viscous fluid. Proc. R. Soc. Lond. A102: 161–179.
41. **Kachel V, Metzger H, Ruhenstroth-Bauer G (1970)** Der Einfluβ der Partikeldurchtrittsbahn auf die Volumenverteilungskurven nach dem Coulter Verfahren. Z. Ges. Exp. Med. 153:331–347.
42. **Kachel V (1974)** Methodology and results of optical investigations of form-factors during the determination of cell volumes according to Coulter. Microsc. Acta 75:419–428.
43. **Kachel V (1990)** Electrical resistance pulse sizing (Coulter Sizing). Chapter 4, this volume.
44. **Kachel V, Kordwig E, Glossner E (1977)** Uniform lateral orientation of flat particles in flow through sys-

tems caused by flow forces. J. Histochem. Cytochem. 25:774–780.

45. **Kampfhoefner PJ (1972)** Ink jet printing. IEEE Trans. Elect. Dev. Ed. 19:584–593.
46. **Keller JB, Rubinov SI, Tu YO (1973)** Spatial instability of a jet. Phys. Fluids 16:2052–2055.
47. **Kochurova NN, Noskov BA, Rusanov AI (1974)** Taking account of the velocity profile in determining surface tension by the method of an oscillating jet. Coll. J. USSR 36:559–561.
48. **Lafrance P, Aiello G, Ritter RC, Trefil SJ (1974)** Drop spectrometry of laminar and turbulent jets. Phys. Fluids 17:1469–1470.
49. **Lafrance P (1974)** Nonlinear breakup of a liquid jet. Phys. Fluids 17:1913–1914.
50. **Lafrance P (1975)** Nonlinear breakup of a laminar liquid jet. Phys. Fluids 18:428–432.
51. **Langhaar HL (1942)** Steady flow in the transition length of a straight tube. J. Appl. Mech. 9:A55–A58.
52. **Lee HC (1974)** Drop formation in a liquid jet. IBM J. Res. Dev. 18:364–369.
53. **Levanoni M (1977)** Study of fluid flow through scaled-up ink jet nozzles. IBM J. Res. Dev. 21:56–68.
54. **Lindblad NR, Schneider JM (1965)** Production of uniform-sized liquid droplets. J. Sci. Instr. 42:635–638.
55. **Ling CM, Reynolds WC (1973)** Non-parallel flow corrections for the stability of shear flows. J. Fluid Mech. 49:571–591.
56. **Marston PL (1980)** Shape oscillations and static deformation of drops and bubbles driven by modulated radiation stresses—Theory. J. Acoust. Soc. Am. 67: 15–26.
57. **Middleman S, Gavis J (1961)** Expansion and contraction of capillary jets of Newtonian liquids. Phys. Fluids 4:355–359.
58. **Neukermans A (1973)** Stability criteria of an electrified liquid jet. J. Appl. Phys. 44:4769–4770.
59. **Oliver DR (1962)** Influence of particle rotation on radial migration in the Poiseuille flow of suspensions. Nature (Lond.) 194:1269–1271.
60. **Pfenninger W:** Further laminar flow experiments in a 40 foot long 2 inch diameter tube. Northrop Aircraft Co. Report AM 133.
61. **Pimbley WT (1976)** Drop formation from a liquid jet: A linear one-dimensional analysis considered as a boundary value problem. IBM J. Res. Dev. 20:180–196.
62. **Pimbley WT, Lee HC (1977)** Satellite droplet formation in a liquid jet. IBM J. Res. Dev. 21:21–30.
63. **Prandtl L, Tietjens OG (1934)** "Applied Hydro- and Aeromechanics." New York: McGraw-Hill.
64. **Prandtl L (1965)** Führer durch die Strömungslehre. Vieweg: Braunschweig.
65. **Punnis B (1947)** Zur Berechnung der laminaren Einlaufströmung im Rohr. Bericht 47 5/03, Max-Planck-Institute Strömungsforschung.
66. **Rayleigh JWS (1876)** Notes on Hydrodynamics: The contracted vein. Phil. Mag. 2:441–447.
67. **Rayleigh JWS (1892)** On the instability of a cylinder of viscous liquid under capillary force. Phil. Mag. 34: 145–154.
68. **Rayleigh JWS (1926)** "The Theory of Sound." Vol.II. London: Macmillan.
69. **Rumscheidt FD, Mason SG (1961)** Particle motions in sheared suspensions. J. Coll. Sci. 16:238–261.
70. **Rutland DF, Jameson GJ (1970)** Theoretical predictions of the sizes of drops formed in the breakup of capillary jets. Chem. Eng. Sci. 25:1689–1698.
71. **Rutland DF, Jameson GJ (1971)** A nonlinear effect in the capillary instability of liquid jets. J. Fluid Mech. 46:267–271.
72. **Saffman PG (1956)** The lift on a small sphere in a slow shear flow. J. Fluid Mech. 22:384–400.
73. **Saffman PG (1956)** On the motion of small spheroidal particles in a viscous liquid. J. Fluid Mech. 1:540–553.
74. **Savart F (1833)** Memoire sur la constitution des veines liquides lancées par des orifices circulaires en mince paroi. Ann. Chim. Phys. 53:337–386.
75. **Schiller L (1922)** Die Entwicklung der laminaren Geschwindigkeitsverteilung und ihre Bedeutung für die Zähigkeitsmessungen. Z. Angew. Math. Mech. 2:96–106.
76. **Schmid-Schönbein H, Wells RE, Goldstone J (1971)** Fluid drop-like behaviour of erythrocytes. Disturbance in pathology and its quantification. Biorheology 7: 227–234.
77. **Schneider JM (1964)** The stability of electrified liquid surfaces. Ph.D. thesis, University of Illinois.
78. **Schneider JM, Hendricks CD (1964)** Source of uniform-sized liquid droplets. Rev. Sci. Instr. 35:1349–1350.
79. **Schneider JM, Lindblad NR, Hendricks CD, Crowley JM (1967)** Stability of an electrified liquid jet. J. Appl. Phys. 38:2599–2605.
80. **Schröder A. (1957)** Durchflußmeßtechnik. In Hengstenberg J, Sturm B, Winkler O (eds), Messen und Regeln in der Chemischen Technik. Berlin: Springer-Verlag pp 160–266.
81. **Schuemmer P, Tebel KH (1982)** Production of monodispersed drops by forced disturbance of a free jet. Ger. Chem. Eng. 5:209–220.
82. **Segre G, Silberberg A (1961)** Radial particle displacements in Poiseuille flow of suspensions. Nature (Lond.) 189:209–210.
83. **Smith AMO (1960)** Remarks on transition in a round tube. J. Fluid Mech. 7:565–576.
84. **Spielman L, Goren SL (1968)** Improving resolution in Coulter counting by hydrodynamic focusing. J. Coll. Interface Sci. 26:175–182.
85. **Ström L (1969)** The generation of monodisperse aerosols by means of a disintegrated jet of liquid. Rev. Sci. Instr. 40:778–782.
86. **Sweet RG (1965)** High frequency recording with electrostatically deflected ink jets. Rev. Sci. Instr. 36:131–136.
87. **Suzuki M, Asano K (1979)** Mathematical model of droplet charging in ink jet printers. J. Phys. D. Appl. Phys. 12:529–537.
88. **Tatsumi T (1952)** Stability of the laminar inlet-flow prior to the formation of Poiseuille regime. I. J. Phys. Soc. Jpn. 7:489–495.
89. **Truckenbrodt E (1968)** "Strömungsmechanik." Heidelberg: Springer-Verlag.
90. **Weber C (1931)** Zum Zerfall eines Flüssigkeitsstrahles. Z. Angew. Meth. Mech. 11:136–154.
91. **Wetsel GC, Grover C Jr (1980)** Capillary oscillations on liquid jets. J. Appl. Phys. 51:3586–3592.
92. **Zimmer NR, Ritter RC, Sterling AM, Harding DC (1969)** Drop spectrometer. A non-obstructive non-interfering instrument for analyzing hydrodynamic properties of human urination. J. Urol. 101:914–198.

4

Electrical Resistance Pulse Sizing: Coulter Sizing*

Volker Kachel

Max-Planck-Institut für Biochemie, D-8033 Martinsried, Federal Republic of Germany

THE BASIC COULTER EFFECT

Efforts to count and size biologic cells and particles date from the invention of the light microscope in 1674 [89]. Until very recently, they required tedious visual examinations, and the resulting measurements had poor statistical reliability. In 1953, when Coulter described his invention, a "means for counting particles in a fluid" [24,25], he introduced an entirely new method of cell counting and cell sizing that was automatic, accurate, and reliable. Coulter's ingenious idea was to pass particles, suspended in a conducting fluid, through a small-diameter orifice of short length. The electrical resistance of the particles must be different from that of the electrolyte. A constant current was maintained across the orifice by two electrodes, one on either side of it. Then, as each particle traversed the orifice it displaced electrolyte, producing a change in electrical resistance. According to Ohm's law, this resistance pulse is observed as a voltage pulse. These signals are measured across the two electrodes, amplified, and counted with electronic scalers. Coulter stated that the pulses are directly proportional to the volumes of the particles, and thus the measured pulse height distributions should directly correspond to the volume distribution curve of the measured particles (Fig. 1).

SURVEY OF EXISTING LITERATURE

Many papers have been concerned with counting and sizing of red blood cells [1,5,10,11,14,15,19,20,37,50,59,64, 94,97,104–106,114,142,143,149–154,162,163,166,167], white blood cells [4,13,36,52], platelets [35,39,64,92,112, 127], bacteria [60,61,86,133,168], spermatozoa [18,139], nerve cells [23,66a], vaginal cells [21], virus [33], mitochondria [45], particles in milk [107], and other biologic or industrial particles [8,32,43,65,120,136,140,153,170]. More recently, resistance pulse sizing also has been incorporated within multiparameter flow cytometry analyzing and sorting systems [11a,44,63,77,98,126,137].

Historically, there are two main phases in the use of the Coulter method. Initially, most investigators accepted direct proportionality between pulse hight and volume. Differences (especially the skewness of blood cell distribution curves) were explained as biologic effects [37,50,94,163,167]. One of the first to question the direct pulse height to volume relationship was Mattern et al. [95], who presumed that the area under the pulse rather than its height should correlate with the volume of the particle. Coincidence problems were the only possible sources of error [27,117,118,138,160,161] considered in this period. In the second phase, artefacts of the Coulter method itself received increased consideration.

The first theoretical analysis of the resistance-change pulse height problem was by Kubitschek in 1960 [87]. He concluded that sizing of small particles is independent of particle shape. But, he points out, a more precise derivation could show that the pulse amplitude is not strictly independent of particle shape. In 1964, Van Dilla et al. [36] reduced the positive skewness of particle size distribution curves by the use of longer orifices and amplifiers whose frequency response matched that of the pulse. Gregg and Steidley [51], in 1965, introduced the shape factor 3/2 for spheres, and made the first model measurements in an enlarged electrolytic tank; but the field patterns they determined were not correct, and their treatment of nonspherical particles is questionable. Gutmann [57] explained the skewness of red cell distribution curves by the different orientation of the cells; only two discrete orientations of red cells were considered in his theory. Harvey and Marr [60,61] used an electronic integrating and differentiating device to influence the pulse shapes. They produced distribution curves of latex particles that were in conformity with curves determined by electron microscopy. The first experimental hydrodynamic focusing device was described by Spielmann and Goren [136], but without sufficient theoretical explanation. Shank at al. [130] reconsidered the bimodal erythrocyte distribution curves. On the basis of experiments with a "capillary directed flow system" the skewness and bimodality of erythrocyte curves were explained as resulting from different deformation of the cells at different radii in the orifice.

A comprehensive survey of the existing literature applicable to resistance pulse sizing, calculation of the electric field, shape factors, and hydrodynamic effects is given by Grover et al. [53]. Independently, Thom [143–147] and our group [69–76] investigated the Coulter method. Thom's model

*See abbreviations at the end of this chapter.

Flow Cytometry and Sorting, Second Edition, pages 45–80

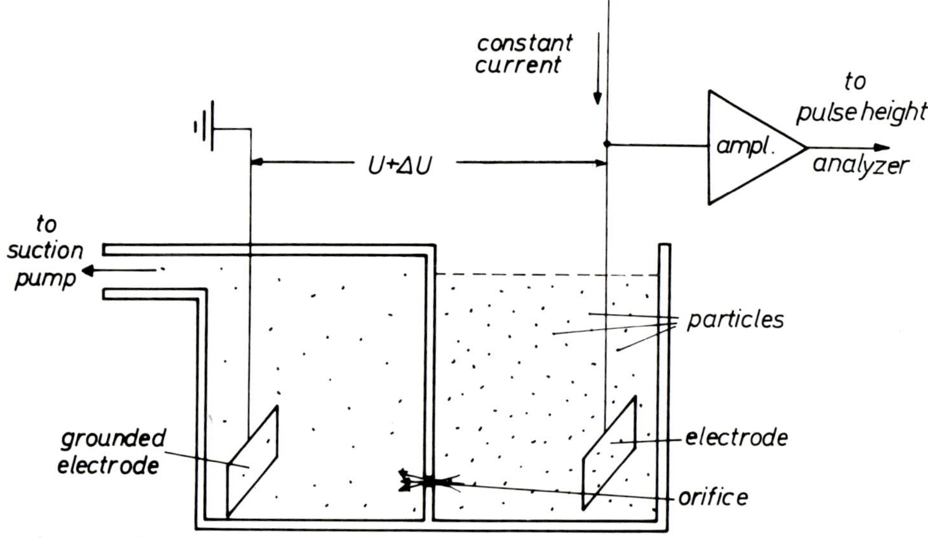

Fig. 1. Basic resistance pulse device according to Coulter.

experiments [144] made it evident that edge effects in the electric field strongly influenced volume distribution curves. Our group demonstrated these effects in a unique orifice and explained the relationship between pulse height and particle volume in short orifices [69,76]. Particle orientation, rotation, and deformation was demonstrated by a direct photographic technique [69,71,73,76,80,103]. Electronic models were developed and tested in order to reduce the edge artifacts [71,72], and a new instrument with electrical calibration [75,76,80] and a tubeless transducer [81] was designed. A focusing device including fluid resistors was described by Haynes [64]. Another edge artifact reducing transducer with a flow-collar was developed by Karuhn et al. [83]. Artifact peaks of particle size distributions have been investigated by Atkinson and Wilson [9], and Van der Plaats and Herps [156,157] investigated particle material and porosity.

Potential sensing transducers were described by Salzman et al. [123], Leif and Thomas [90], and Thomas et al. [148]. In this technique, well proven in other fields but new in resistance pulse sizing, the current electrodes and the potential sensing electrodes are separated from each other.

PARAMETERS THAT INFLUENCE THE SIZING RESULTS

Biologic particles will be considered nonconducting for these discussions. If conduction does occur, it is ionic conduction through the particle. By contrast, metal particles suspended in an electrolyte show a different behavior [56]; see Figure 2 for copper spheres suspended in NaCl solution. At first, with relatively small field strengths in the orifice, copper particles appear to be nonconducting, indicated by positive pulses. But above threshold field, a phenomenon like a breakdown occurs and the negative pulse originally expected of highly conductive metal appears. This phenomenon is explained by the energy transformation problems in the boundary layer between the ionic-conducting electrolyte and the electronic-conducting metal. Similar results are reported by others [156].

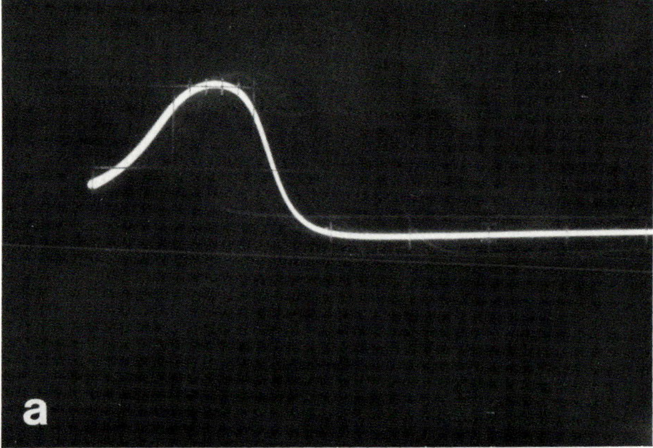

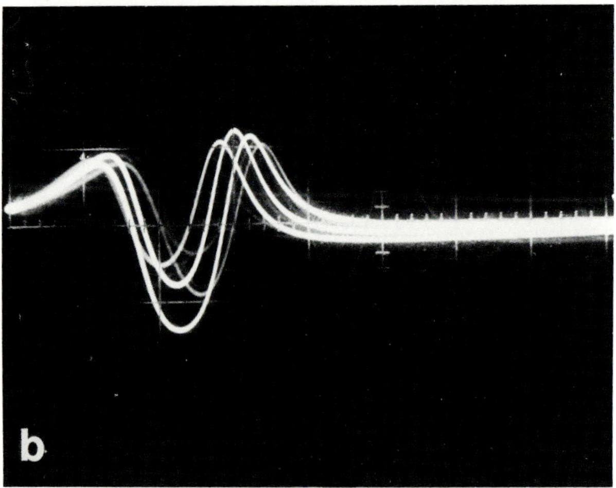

Fig. 2. Resistance pulses of electronically conducting copper spheres: (a) at low field-strength positive pulses like that of nonconducting particles are generated; (b) at high field-strength the internal electronic conduction causes inversion of the pulses.

MATHEMATICAL RELATIONSHIP BETWEEN PARTICLE VOLUME AND RESISTANCE PULSE HEIGHT

Maxwell [96], Fricke [41,42], and Velick and Gorin [158] derived equations for calculating the resistivity of an electrolyte in which differently shaped and oriented, reasonably conducting particles are suspended. With some transformations, these equations, which are also the ones used by Grover et al. [53], can be applied to calculate the resistance pulse produced by a particle passing through a Coulter orifice. With identical presumptions, the results of these investigators are in agreement with those found by Hurley [68], who calculated shape factors for nonconducting ellipsoids.

Consider now spheres and prolate oriented ellipsoids of revolution in cylindrical orifices. Ellipsoids with three different axes and other orientations are treated elsewhere [158]. For the mathematical treatment of reasonably conducting particles shaped like a prolate ellipsoid of revolution, we use the equation of Velick and Gorin:

$$\frac{1}{\rho} = \frac{1}{\rho_2} + \frac{P}{1 - P}\,\frac{2(1/\rho_1 - 1/\rho)}{2 + ab^2La\,(\rho_2/\rho_1 - 1)} \qquad (1)$$

The nomenclature used is explained at the end of the chapter (see abbreviations list). We assume here that the electric field in the orifice is uniform, and that the diameter of the particle is small as compared with the diameter of the orifice ($2b < 0.2d$). This ensures that the field distortion produced by the presence of a particle in the orifice (this distortion is the basis of the resistance change) has vanished at the orifice wall, and that the pulse produced by the particle is not influenced by the wall. The problem of larger particle diameter to orifice diameter ratios has been treated elsewhere [7,32,134,135]. We suppose that $V = q{\cdot}\ell$; ρ_1 and ρ_2 are known, and the resistance change during the passage of a particle is measured by $\Delta R_2 = \Delta u/i$. To introduce R_2 in equation 1, we set

$$R = R_2 + \Delta R_2 \qquad (2)$$

$$\frac{\rho{\cdot}\ell}{q} = \frac{\rho_2{\cdot}\ell}{q} + \Delta R_2$$

$$\rho = \rho_2 + \Delta R_2 \cdot \frac{q}{\ell} \qquad (3)$$

Replacing ρ by $\rho = \rho_2 + \Delta R_2 \cdot \alpha$

Equation 1 becomes

$$P = \frac{\Delta v}{V} = \frac{2[1 - (\rho/\rho_2)]}{1 - \rho/\rho_2 + \dfrac{2[(\rho/\rho_1) - 1]}{2 + ab^2La[(\rho_2/\rho_1) - 1\,]}} \qquad (4)$$

Assuming that $\Delta R_2 << R_2$, and introducing equation 3 into equation 4, we obtain with $R_2 = \rho_2/\alpha$ and $V = q{\cdot}\ell$:

$$\Delta V = \frac{\Delta R_2 q^2}{\rho_2}\;\frac{1}{\dfrac{2[1 - (\rho_2/\rho_1)]}{2 + ab^2La[(\rho_2/\rho_1) - 1]}} \qquad (5)$$

where

$$\frac{2[1 - (\rho_2/\rho_1)]}{2 + ab^2La[(\rho_2/\rho_1) - 1)} = f_e \qquad (6)$$

is the shape and conductivity factor of prolate ellipsoids of revolution. If the ellipsoids are nonconducting, f_e becomes

$$2/(ab^2La - 2) \qquad (7)$$

Replacing ΔR_2 by $\Delta U/i$, with equation 5 and equation 6:

$$\Delta V = \frac{\Delta U{\cdot}q^2}{\rho_2{\cdot}i{\cdot}f_e} \qquad (8)$$

with f_e from equation 6 or 7, where

$$ab^2La = 2 - \frac{2}{1 - m^2} + \frac{m \ln z}{(m^2 - 1)^{3/2}} \qquad (9)$$

$$z = \frac{m - (m^2 - 1)^{1/2}}{m + (m^2 - 1)^{1/2}} \qquad (10)$$

$$m = a/b \qquad (11)$$

as displayed in (158).

Equation 8 can also be used for approximated calculations of the shape and conductivity factor of spheres if $m = a/b \approx 1{\cdot}m = 1$—that is, $a = b$, leads to an indefinite solution of equation 8. In this case the equation of Maxwell can be used:

$$\rho = \frac{2\rho_1 + \rho_2 + p(\rho_1 - \rho_2)}{2\rho_1 + \rho_2 - 2p(\rho_1 - \rho_2)}\,\rho_2 \qquad (12)$$

With the above-introduced assumption and simplications we obtain for spheres from equation 12 :

$$\Delta V = \frac{\Delta U\,q^2}{\rho_2{\cdot}i}\;\frac{1}{\dfrac{3[1 - (\rho_2/\rho_1)]}{(\rho_2/\rho_1) + 2}} \qquad (13)$$

with

$$f_s = \frac{3(1 - (\rho_2/\rho_1))}{(\rho_2/\rho_1) + 2} \qquad (14)$$

$$\Delta V = \frac{+\,\Delta U\,{\cdot}q^2}{\rho_2{\cdot}i{\cdot}f_2} \qquad (15)$$

For nonconducting spheres $f_s = 3/2$. Equations 8 and 15 are identical except for factors f_s and f_e.

Figure 3 shows f_e as a function of the axial ratios of ellipsoids, the parameter being ρ_2/ρ_1. At high a/b ratios, f_e is near 1 for nonconducting ellipsoids. With $a/b = 0.001$, we find $f_e \approx f_s$. Figure 4 shows f_e as a function of the ρ_2/ρ_1 ratio, the axial ratio of the ellipsoids being used as a parameter.

At resistivity ratios below $\rho_2/\rho_1 = 0.001$, f_e is nearly constant—that is, the particle resistivity is only of increasing interest if it is less than 100 times the electrolyte resistivity.

If the original assumption that the particle is small with respect to the orifice diameter is not valid, the derived formulas are no longer accurate. From calculations by Smythe

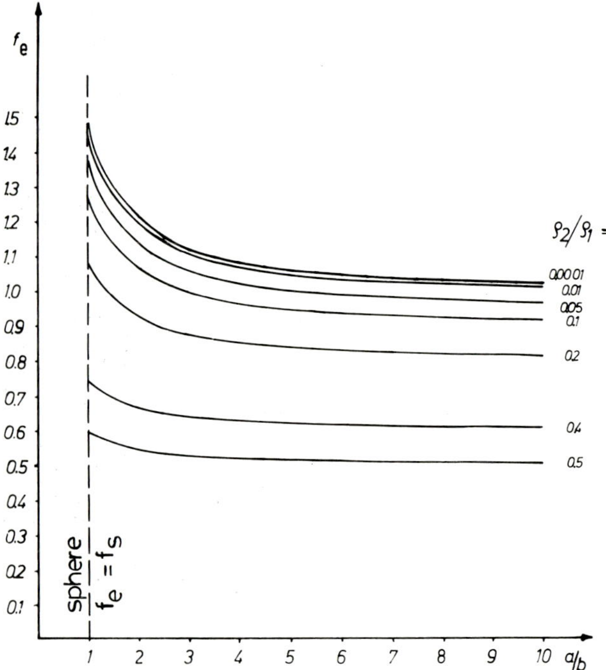

Fig. 3. Shape and conductivity factor f_e as a function of the axial ratios of prolate ellipsoids of revolution. The ratio of the electrolyte resistivity to the particle resistivity ρ_2/ρ_1 is used as a parameter.

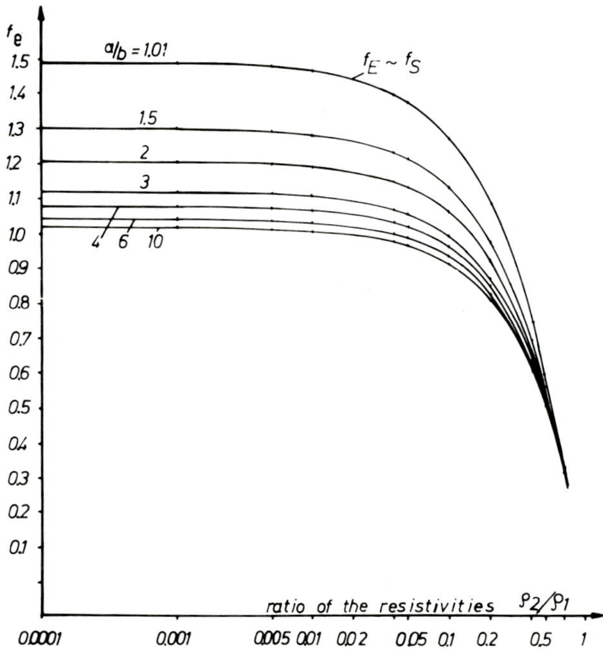

Fig. 4. Shape and conductivity factor f_e of prolate ellipsoids of revolution as a function of the ratio of the electrolyte resistivity to the particle resistivity ρ_2/ρ_1. The axial ratio of the ellipsoids is used as a parameter.

[134,135] the overestimated volume of large nonconducting particles can be corrected in the form of an increase of the shape factor of the particles.

Figure 5 shows the formal increase of the shape factor with increasing particle to orifice diameter ratio. With spheres and prolate ellipsoids (2 : 1), Smythe has shown that the increase depends on the shape of the particles. The curves of Figure 5 indicate that the error does not exceed 1% if the particle to orifice diameter ratio is below 0.2, and if the particles move near the axis of the orifice.

These equations indicate that the relationship between particle volume and pulse height is determined by (1) the electric current through the orifice, (2) the geometric dimensions of the orifice, (3) the geometric and electrical properties of the particles, and (4) the resistivity of the suspending medium. The pulse height is independent of the ℓ/d ratio of the orifice only if the electric field in the central area of the orifice is homogeneous. In short orifices, the pulse height is influenced by the inhomogeneity of the electric field, as will be shown below.

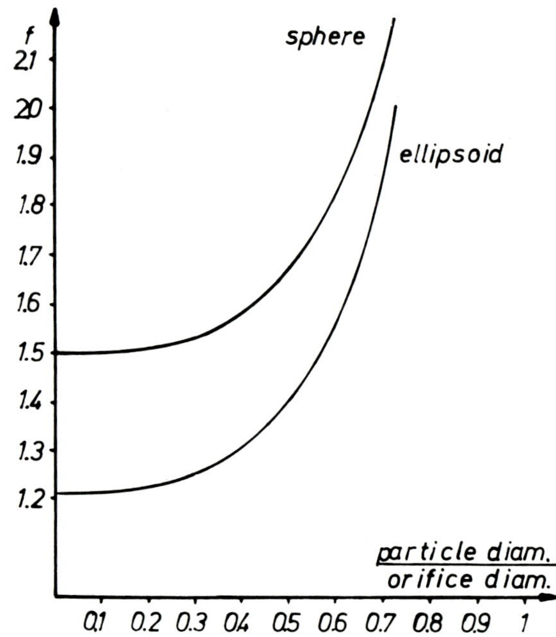

Fig. 5. Formal increase of the shape factor f_s and f_e of spheres and ellipsoids of revolution (axial ratio 2:1), respectively, as a function of the particle diameter to the orifice diameter ratio, derived from [135].

THE FLUID FLOW IN THE SENSING ZONE

Near the orifice, the flow is transformed from a sink trap flow into a tube flow inside the orifice. A relation can be derived that describes the hydrodynamic focusing features [69] (Fig. 6a,b):

$$\frac{B}{A} = \frac{r^2}{2s^2} \tag{16}$$

The significance of this focusing effect and general hydrodynamic problems is covered in detail in Chapter 3 (this volume).

The flow velocity is a function of the pressure difference across the orifice (see Fig. 20). Calculated Reynold numbers

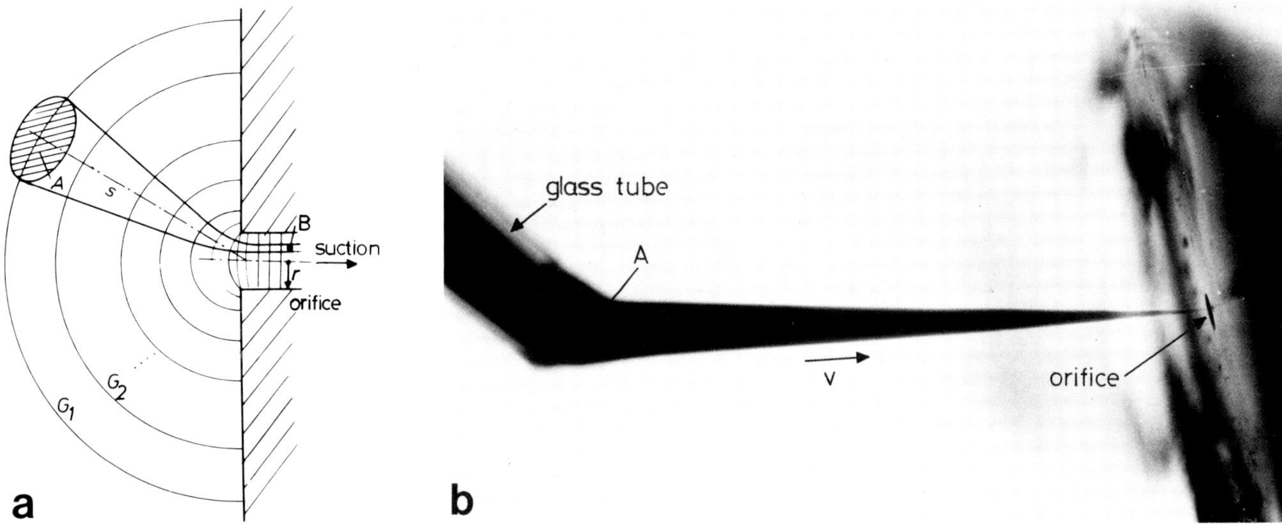

Fig. 6. (a) Sink-trap-flow tube flow system. Explanation of the focusing effect: G_i are equivelocity areas of the laminar flow. A is the cross section of the injecting tube, B the cross-section of the focused path, r the radius of the orifice, and s the distance of A from the inlet of the orifice. (b) Focusing of a dye stream ejected from a glass tip in front of a 100-μm diameter orifice.

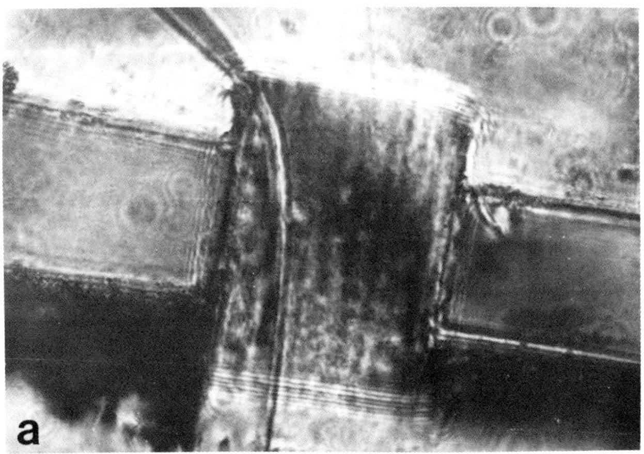

Fig. 7. Focused dye streams near the wall of a rectangular orifice (100-μm diameter, see Fig. 20a,c). (a) The stream is completely laminar. (b) The left side of the dye stream touches the dead water zone, the laminar flow is distorted, and the stream is broadened.

Re indicate that for all commonly used orifices and pressures, laminar flow exists. From the literature, it is known [88,122,132,141,162] that the velocity profile in a short tube such as a Coulter orifice is not parabolic but flat with a core region of linear velocity distribution. The radius of the core region is a function of the Reynold number and the distance from the inlet of the orifice. With increasing distance from the inlet, the velocity profile is transformed more and more to the parabolic Poiseuille profile. These profile calculations were made with the assumption that the entrance velocity across the inlet of the orifice is constant. This will be true only with smoothly rounded edges of an orifice. With sharp edges, the flow inside the orifice near the inlet is constricted to about 60% of the cross section of the orifice [38,115]. A dead zone with vortices exists in the outer 40% of the orifice. The constricted laminar flow has the tendency to dilate to the orifice walls again. The degree of constriction depends on the smoothing of the edges.

Figure 7a,b illustrates these effects with a dye stream in a rectangular orifice. Figure 7a illustrates the laminar flow situation; Figure 7b, the existence of the dead zone.

ELECTRIC CURRENT AND CURRENT DENSITY

The basic concept in resistance measurements of volume is to transport particles from a region of negligible current density through an orifice where the current density is high to a second region of low current density. Thus, for normal resistance pulse transducers where the fluid flow and current directions are the same, field inhomogeneities are inherent. Homogeneous electric fields will only exist in orifices of certain specified lengths. Other configurations can be imagined with current flow perpendicular to the fluid flow [47]. With the usual transducer, the electric field just outside the orifice inlet can be described as that associated with a spherical condenser, $E = \text{const}/s^2$, where s is the distance from the orifice inlet. The radially directed straight current flow lines of the sink-trap flow are transformed into a cylindrical parallel flow near and in the orifice (more details below). Field-strength E and current density j are related by $E = j\,\rho_2$; (ρ_2 =

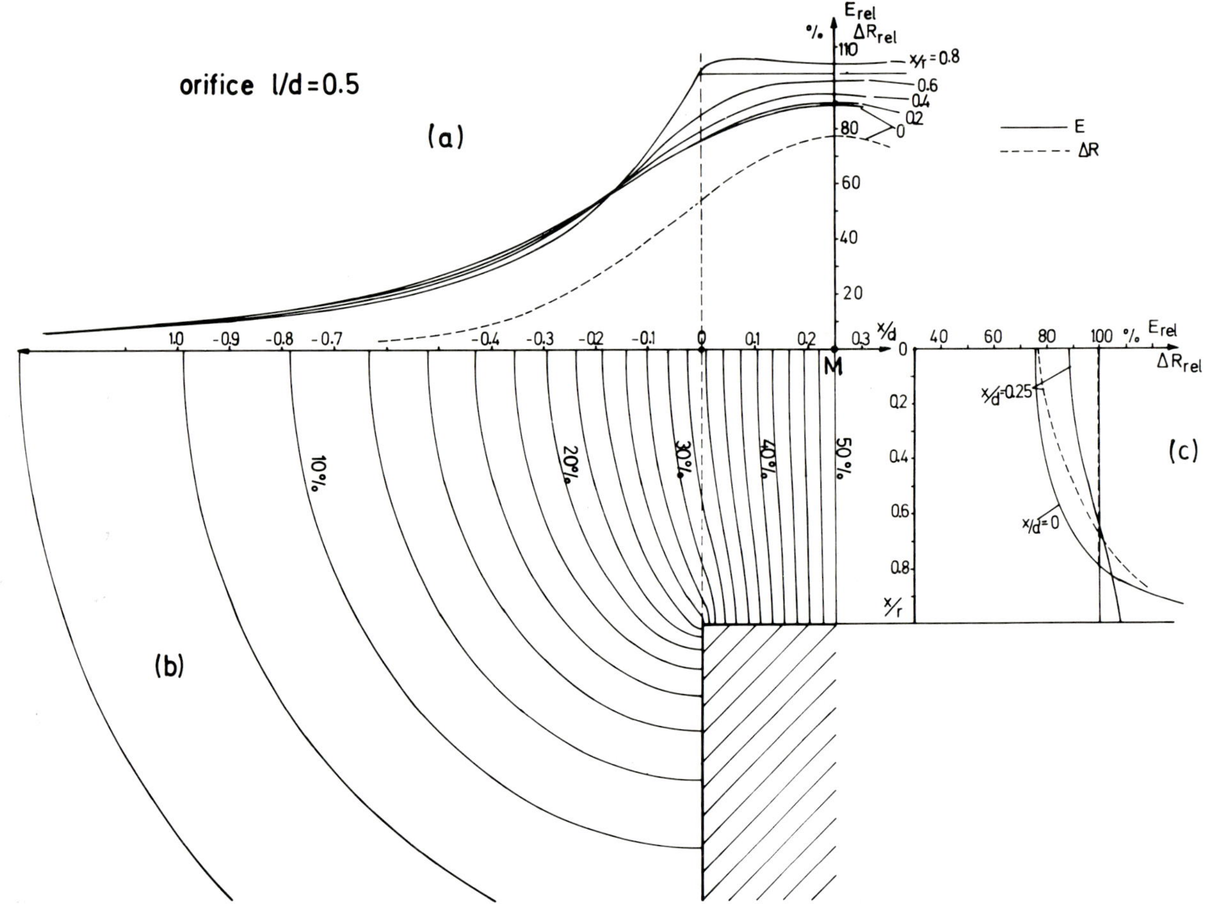

Fig. 8. Results of computer calculations of the electric field and of resistance pulse measurements in the enlarged model. Orifice of a length to diameter ratio of 0.5. (a) The electric field strength on paths parallel to the orifice axis at different distances x/r. The middle area of the orifice with 50% potential is located at x/d = 0.25 (point M). The origin of the x-axis has been arbitrarily placed on the inlet of the orifice. The units of the x-axis are expressed in terms of distances from the orifice inlet related to the orifice diameter. The field-strength values E_{rel} are related to the equivalent homogeneous field-strength in the orifice. The broken line is the resistance pulse height ΔR related to ΔR in the equivalent homogeneous field, measured with the model sphere. (b) Field of electric equipotential lines, which, by reasons of symmetry of revolution, represent equipotential areas. Because of the symmetry only one quarter of the orifice is plotted. The distance between the lines is 2% of the potential between the electrodes. The field-strengths at some x/r lines parallel to the orifice axis are plotted in (a). At the axis ≈ 32% of the potential drop is within the orifice proper. (c) The radial function of the field strength related to the equivalent homogeneous field-strength in the middle area (x/d = 0.25) and in the inlet area (x/d = 0) of the orifice. The broken line shows the radial function of ΔR, related to ΔR in the equivalent homogeneous field, measured with the model sphere at x/d = 0.25. The straight 100% lines represent the values in the equivalent homogeneous field.

resistivity of the electrolyte). Equation 8 and 15 show that pulse height and current i are proportional if the particle volume ΔV and the other parameters are kept constant. Therefore, high currents produce high-volume signals and high signal-to-noise ratios. But high currents cause increased heating of the fluid in the orifice, more electrolysis at the electrodes, and breakthrough effects of cell membranes [168,169]—all undesirable. Therefore, for cell sizing, the current should be chosen only as high as required for acceptable signal to noise ratio.

CURRENT DISTRIBUTION AND ITS EFFECTS

Distribution of the Electric Field in Cylindrical Orifices

The distributions of the electric field in cylindrical orifices have been determined in two ways: (1) measurement of the potential and the field strength in enlarged model orifices in an electrolytic tank [71], and (2) calculation of the potential and field strength distributions in cylindrical orifices of l/d ratios from 0.2 to 1.5, in steps of 0.1 [80,85]. Here [85], Laplace's equation is solved by the numerical relaxation method [164]. From these calculations, we obtain a complete potential and field strength distribution of high accuracy inside and outside the orifices. The previous model orifice measurements are in conformity with the calculated values within 3%. For all orifices, an equivalent homogeneous field strength with respect to current density was calculated by integrating the current densities and dividing the integral by the cross section of the orifice. The boundary conditions for these calculations were a half-spherical equipotential area with the potential 2% at a distance s = 3d from the orifice, and by reason of symmetry, the 50% equipotential area in

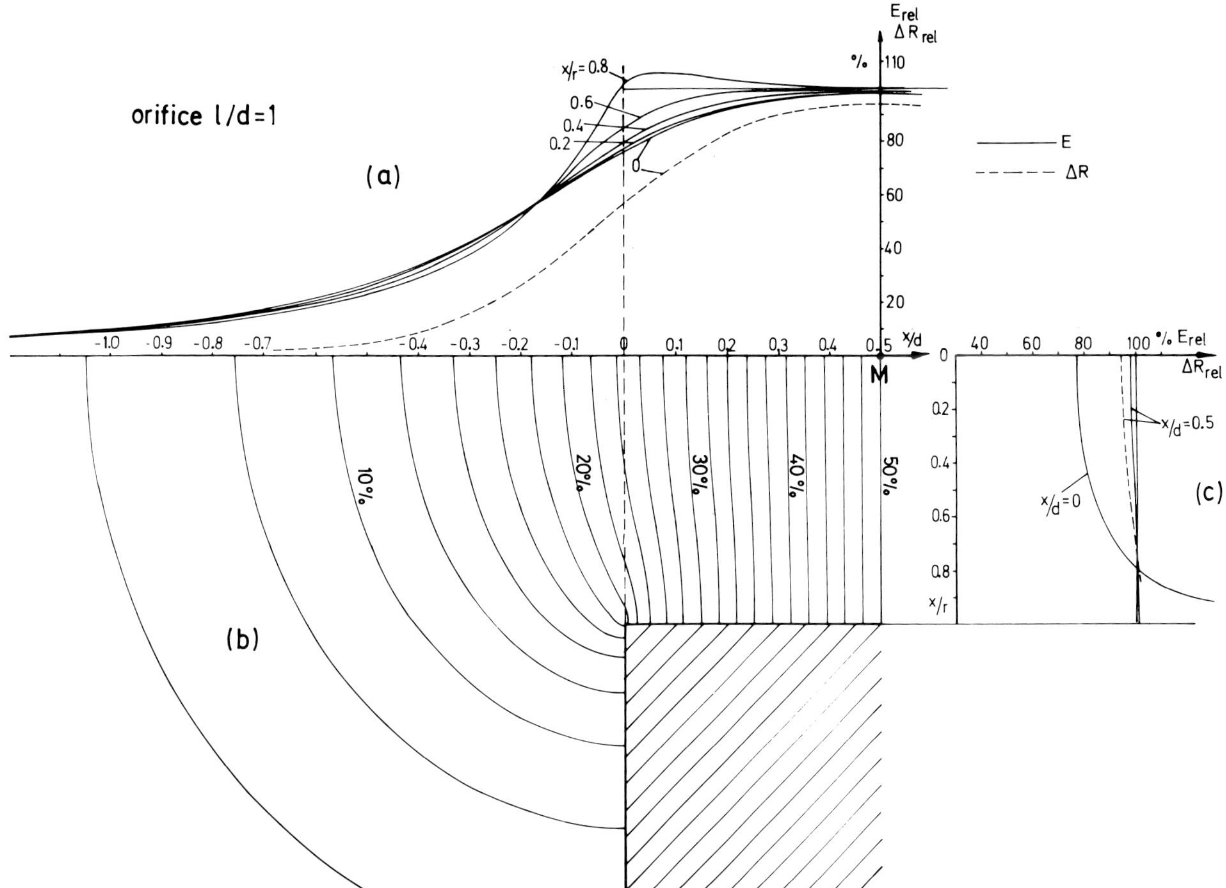

Fig. 9. Same as Figure 8, but the length to diameter ratio of the orifice is 1 and the middle area of the orifice (50% potential) is now at x/d = 0.5. (a) The field inhomogeneity in the middle area of the orifice is decreased, compared with the orifice l/d = 0.5. At all x/r values the field strength is near the equivalent homogeneous field strength. The ΔR curve measured at the orifice axis (x/r = 0) reaches 95% of ΔR in the equivalent homogeneous field. (b) At the axis now about 50% of the potential drop is within the orifice proper. (c) The radial change of E is within 2% and of ΔR within 5% between x/r = 0 and x/r = 0.8 in the middle area (x/d = 0.5).

the middle of the orifice perpendicular to the axis. The 2% area was determined by calculation of the equipotential areas with an electrode (potential 0%) far away from the orifice. The calculations were performed under the condition of identical resistivity of the electrolyte inside and outside the orifice. Figures 8 to 10 summarize the results of the field calculations. All field-strength values are related to the equivalent homogeneous field strength mentioned above.

The current streamlines (not shown) are trajectories orthogonal to the equipotential areas shown in Figures 8b to 10b.

On the basis of these calculations, we conclude that in long orifices the electric field strength is constant in the middle region in axial and radial directions. This is in general conformity with the results of Grover et al. [53]. However, their curves of x/R = 0 and x/R = 0.2 are inaccurate at the orifice inlet.

In short orifices, the field strength at any point of the orifice depends on both the axial and radial coordinates. At the edges of the inlet and outlet of cylindrical orifices there are regions of increased current densities with indefinite values exactly at the edges. The potential drop between inlet and outlet planes of the orifice is at most 60% of the total potential difference across the electrodes if $\ell/d < 1.5$, and depends on the length of the orifice and the radial distance from the axis.

Effect of the Orifice Edges on Volume Distribution Curves

Pulse height distributions and pulse shapes as a function of the radial distance from the orifice axis have been shown under original sizing conditions [69–71]. These experiments, performed with optical control of the particle path, will be described briefly. (The experiments were performed with the vertical chamber shown in Fig. 20b.)

A moveable sample tube injected a black-ink colored suspension of fixed erythrocytes into the focusing fluid stream above the orifice. By moving the sample tube, as described above, a narrow particle path, visible as a black dot, was generated at different points of the cross section of the orifice (Fig. 11). The upper part of Figure 11 shows four distribution curves of fixed erythrocytes, which were taken by keeping all parameters constant with the exception of the position of the ink-marked particle path through the orifice. In the four lower pictures of the 100-μm diameter orifice, the path

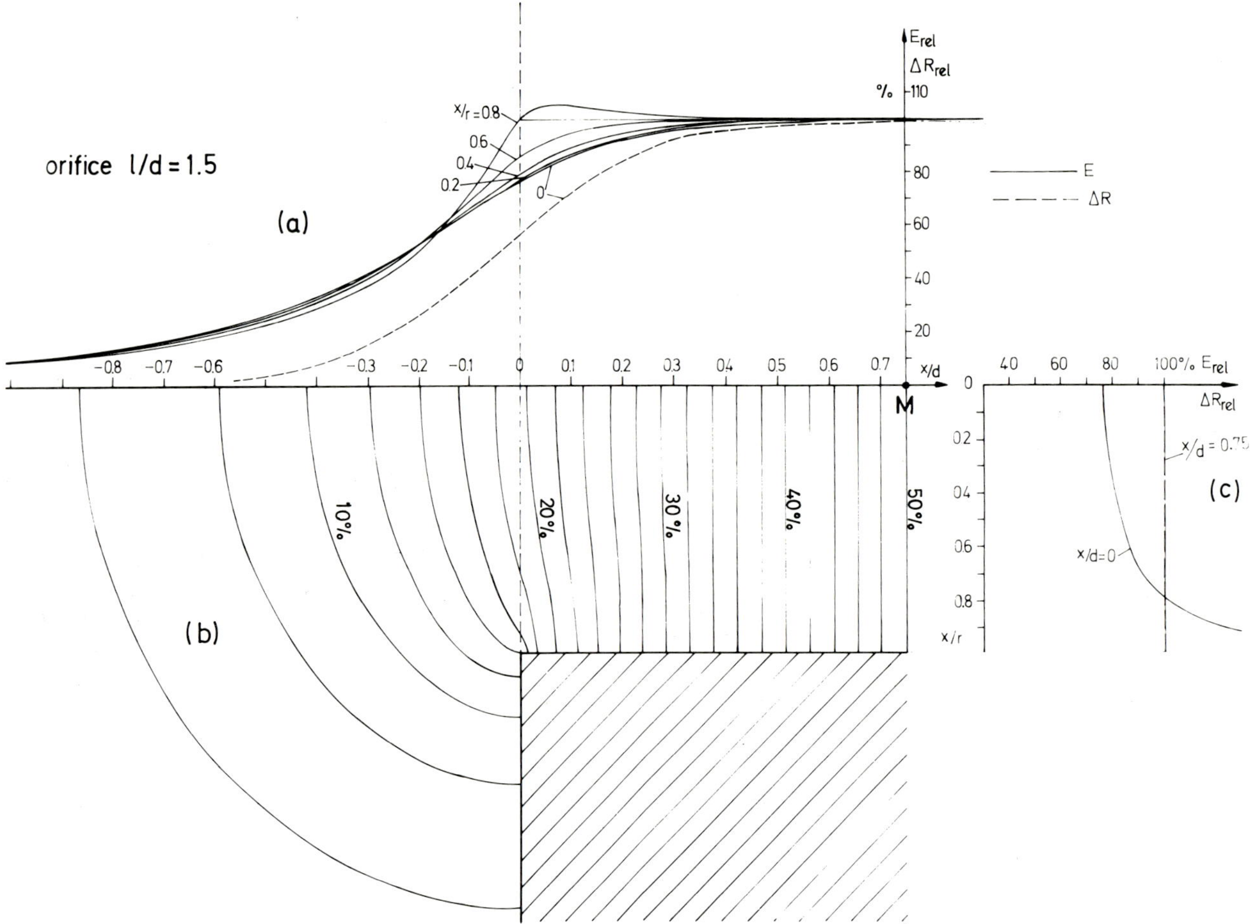

Fig. 10. Same as Figure 8, but the length to diameter ratio of the orifice is 1.5 and the middle area of the orifice (50% potential) is at x/d = 0.75. In the middle area at all x/d values E = 100%; i.e., there is a homogeneous field; therefore, ΔR at the axis is also 100%. E and ΔR are no longer depending on the radius x/r in the middle area (distortions by the wall are not considered here). At the axis 64% of the potential drop is within the orifice proper.

is represented by the black dot (indicated by an arrow). The middle pictures show the corresponding different pulse shapes generated by the particles in the indicated particle path. From these distribution curves, corresponding to a distinct path, an originally unfocused distribution curve can be reconstructed, which is skewed to the right [69].

Elimination of the Edge Distortions

Particle Path Method. The result of the particle path experiment suggests using only one defined particle path in the orifice. This focusing technique was first proposed—but without a clear explanation—by Spielman and Goren [136] and, in similar manner, was used by Thom et al. [143–147] and by Shank et al. [130]. To measure precise volume distribution curves, it is unnecessary to focus the particles exactly into the axial path of the orifice. In general, it is sufficient to size all particles at the same path in the core stream of the orifice [69]. The easiest way to control the particle path is by observations of the pulse shape with an oscilloscope. If the radial distance of the particle path from the axis does not exceed 60% of the radius of the orifice, bell-shaped or trapezoidal pulses are observed, provided the amplification is linear. Particle paths that produce M-shaped pulses should be avoided. In Figure 12, volume distributions of human erythrocytes are compared for both the focused and unfocused situations [145].

Electronic Elimination of Edge Field Distortions. The simplest electronic method to reduce the edge distortions is by use of long orifices and slow rise-time amplifiers [20,36,62]. The fast-rising first peaks of M-shaped pulses are suppressed by the integrating effect of the slow-rising amplifier. In addition the second peaks are reduced by the migration of the particles toward the orifice axis in long orifices as reported in [53]. Measurements with increased amplifier rise-time in long orifices are affected by a high coincidence probability and uncertainties resulting in pulse height variations induced by the flow velocity or particle path. The particle distributions measured by the integrating–differentiating method of Harvey [61] and Harvey and Marr [60] are also uncertain due to the arbitrary adjustment of the time constants.

The correlation between pulse height and pulse shape shown in Figure 11 suggests the use of a discriminator which accepts pulses based on electronic uniformity analysis. The following criteria can be employed:

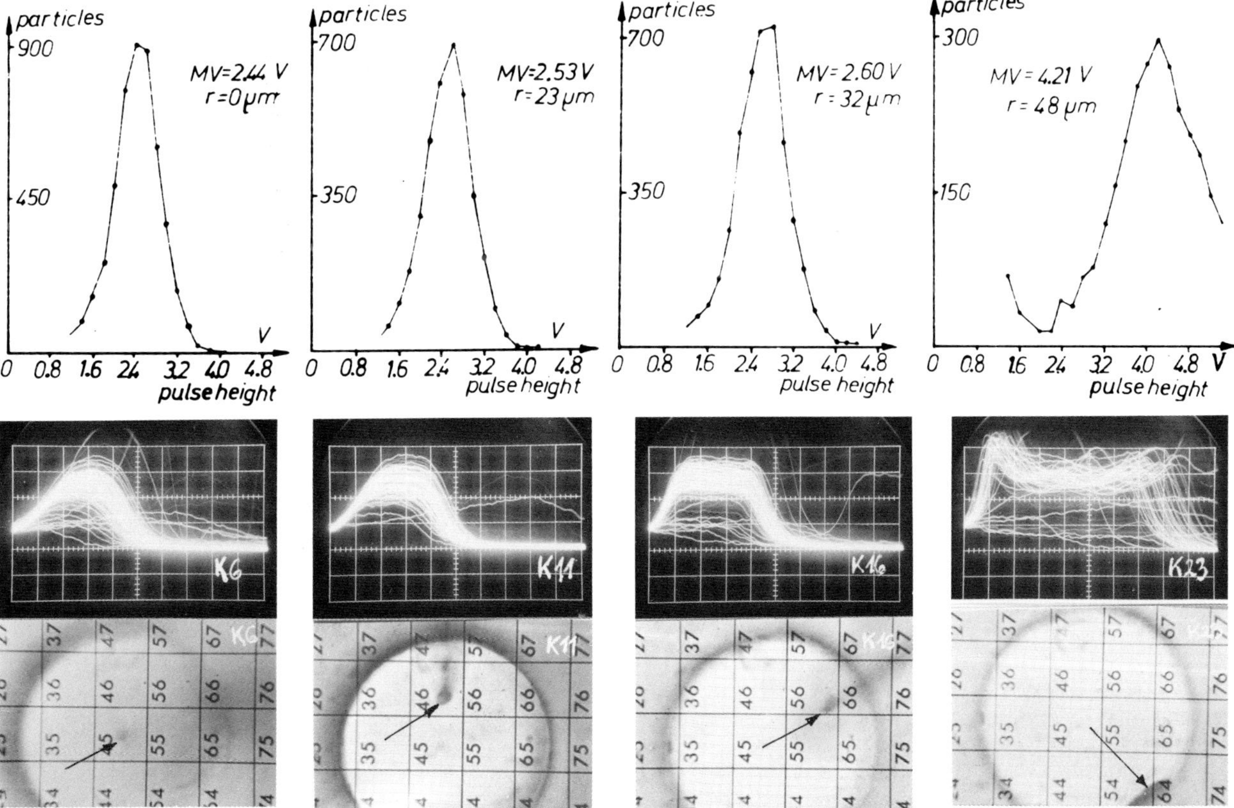

Fig. 11. Distribution curves and pulse shapes taken from four distinct particle paths. The lower pictures present axial views through the 100 µm diameter orifice taken just in the moment when the cells were sized. The suspension of the fixed erythrocytes is colored by black ink (indicated by the arrows). MV = mean value of the pulse heights in volts, r = distance of the particle stream from the orifice axis. Current through the orifice 0.46 mA, suction 85 mbar, time axis of the oscilloscope patterns 10 µsec/div.

1. Pulse length (duration): Pulses that are too long are rejected [28,29,71,162].
2. Time difference between pulse initiation and the first peak with M-shaped pulses: Pulses where this time is short are rejected [71,72].
3. Pulses with more than one peak: These are rejected [71].
4. Delayed peak evaluation: Pulses are evaluated only at the amplitude that corresponds to the middle region of the orifice [54,71].

The electronic distortion-eliminating instruments can be used without a focusing device. But in most cases this is not a striking advantage, because modern focusing devices are easy to use and require less sample than normal Coulter devices [12,64,75,76,147]. The pulse length, time difference and delayed evaluation methods are sensitive to flow velocity (pressure differential). The pulse height itself and the trigger threshold adjustment make the result uncertain. Attempts to compensate for such effects have been made by means of complicated electronic circuits [28,29]. Electronically improved volume distributions show decrease in skewness as compared with distributions obtained with the usual Coulter devices [69], but they are less accurate than the focusing technique [12,69].

Resistance Pulse Height and Its Relationship to the Field Strength in the Orifice

Pulse Height Determination by Model Experiments. The electric field-strength in the middle of long orifices ($\ell/d > 1$) is homogeneous, and the equations 1 to 15 are applicable for the calculation of absolute volumes. But in short orifices ($\ell/d < 1$) no homogeneous field exists; these equations are no longer applicable. The calculation of the particle volume-pulse height relationship for nonhomogeneous electric fields is a formidable problem in mathematical analysis; no such method exists at present. But it is desirable to use short orifices in order to reduce coincidence and noise (most of the original Coulter orifices have $\ell/d < 1$). Therefore, we carried out experiments in model orifices, to determine the relationship between field strength and pulse height for different ℓ/d ratios. These experiments are described in detail in ref. [71] and reported in ref. [80]; the salient results are given below.

The experiments were performed in an electrolyte tank (Fig. 13) and six cylindrical orifices with diameters of 11 cm and lengths between 2.75 and 22cm. Because of the symmetry, only the lower half of the orifice was filled with electrolyte. The model particles with identical volumes of 1 cm^3 (sphere, ellipsoid 4 : 1, erythrocyte disk) were immersed into

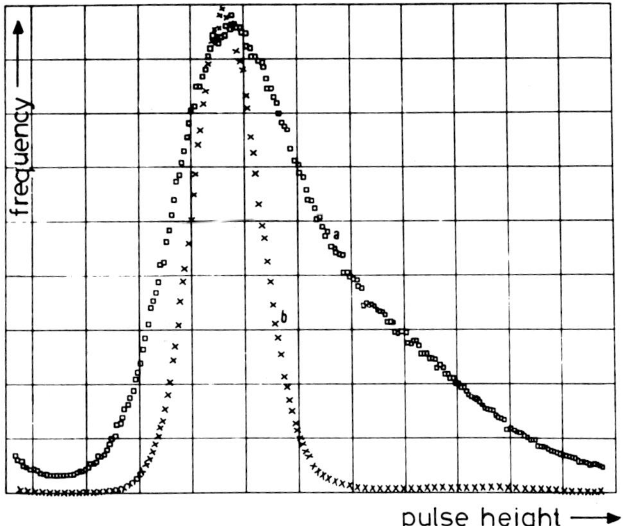

Fig. 12. Comparison of volume distribution curves of native erythrocytes measured with a normal Coulter Counter (a) and with a focusing device (b). Reproduced from [145], with permission of the publisher.

the electrolyte outside and inside the orifice. The field strength in the orifices was made homogeneous by placing parallel stainless steel plates at both ends of the orifices. This homogenized field strength corresponds to the so-called equivalent homogeneous field strength mentioned above. A constant alternating current source (800 Hz) generated a constant current through the orifice. Therefore, only the distribution and not the total amount of the current in the orifices was changed by the plate. The field strength measurements were performed with a double needle probe and a differential amplifier. These field measurements were confirmed by more recent computations. The resistance changes within the orifices due to the presence of the particles was calibrated by generating resistance changes with known resistors.

The apparatus was checked as follows: The resistance change produced by the model particles was measured in the homogeneous fields, and the particle volume calculated according to equations 3 and 15. The calculated values were within 3% of the a priori known volumes. In view of the focusing technique, the field strength E and the resistance pulse height ΔR were preferentially measured on the axial path and in the middle area of the orifices. In all orifices the particles were first measured in the natural field and then (under the same conditions) in the equivalent homogeneous field.

To generalize the results, all ΔR values are normalized to the ΔR values in the homogeneous field, and all field strength values E to the equivalent homogeneous field strength E_n; all axially directed distances (s) are related to the diameter (d) of the orifice.

The following results were obtained:

1. In long orifices, no difference between the measurements with and without the homogenizing plates was found. In both cases, the field is homogeneous in the central area of the orifice (x/d = 0.75, Fig. 10c).
2. With decreasing ℓ/d ratios, there is an increasing difference between the measured normalized ΔR values in the actual and homogenized field. The relative decrease of ΔR is not proportional to the relative decrease of the field strength. Therefore, the shape of the resistance pulses cannot be described by the function of the field strength along the particle path. The relation between the natural field-strength E and the resistance pulse values ΔR in the natural field varies with E^2 as is shown in Figure 14. This E^2 relation is a fundamental principle in resistance pulse sizing.

In Figures 8 to 10, the ΔR curves are drawn in broken lines. The rising and falling slopes of the ΔR curves are much steeper than the slopes of the E curves. At a distance of about

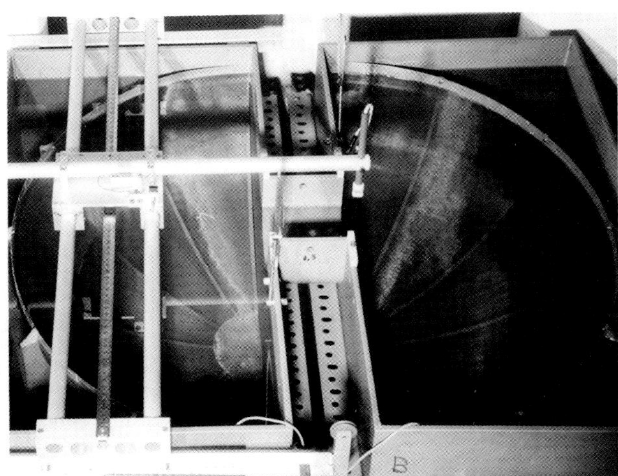

Fig. 13. The electrolytic tank model orifice with the half spherical electrodes forming an equipotential area of the natural field, and the measuring arm to hold the field measuring electrodes and model particles. Because of the symmetry, only a half-system is modeled, facilitating access from the upper side.

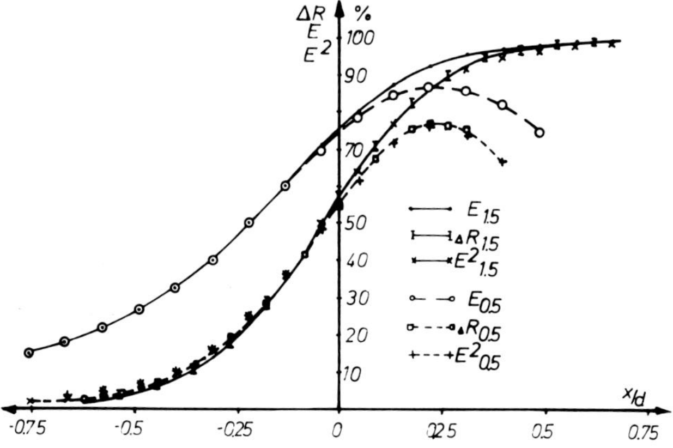

Fig. 14. Comparison of the axial curves of the field strength, the square of the field strength, and the resistance pulse height in two orifices of ℓ/d = 0.5 and ℓ/d = 1.5. All % values are related to the E and respective ΔR values in the equivalent homogeneous electrical field. The arbitrary origin of the x-axis lies at the inlet of the orifice and the distances x at the orifice axis are related to the diameter of the orifice. The ΔR values are measured with the sphere. In both orifices the ΔR and E^2 curves are in agreement.

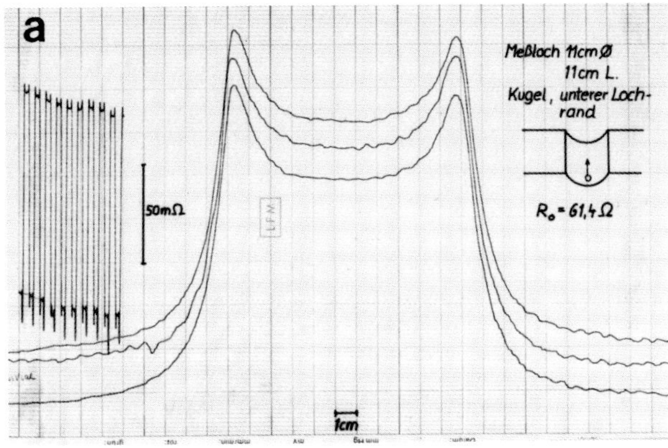

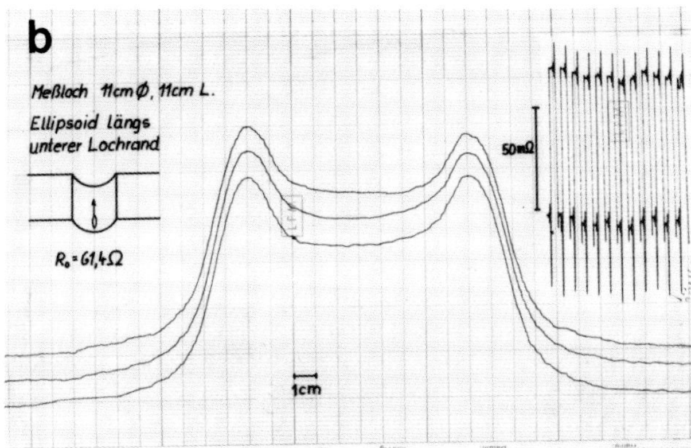

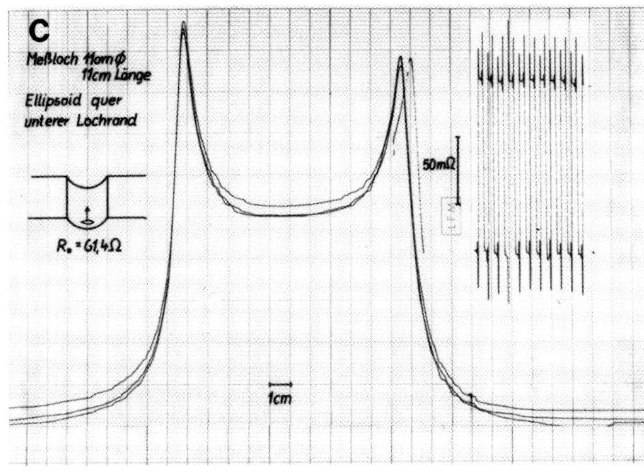

Fig. 15. Pulse shapes of model particles with identical volumes moving near the wall of the orifice with $\ell/d = 1$. (a) sphere, (b) ellipsoid (4:1) long axis in flow direction, (c) ellipsoid (4:1) long axis perpendicular to the flow axis. The shape of the M pulses is strongly influenced by the shape and orientation of the particles.

$x/d = -1$ outside the inlet of the orifice $E \approx 0.1\ E_{max}$; ΔR, however, has decreased to $\approx .01\ \Delta R_{max}$. In Figure 14, the E^2 and the ΔR curves all compared with good agreement.

According to Eq. 15:

$$\Delta U = \frac{\Delta V(\rho_2 \cdot i \cdot f)}{q^2}$$

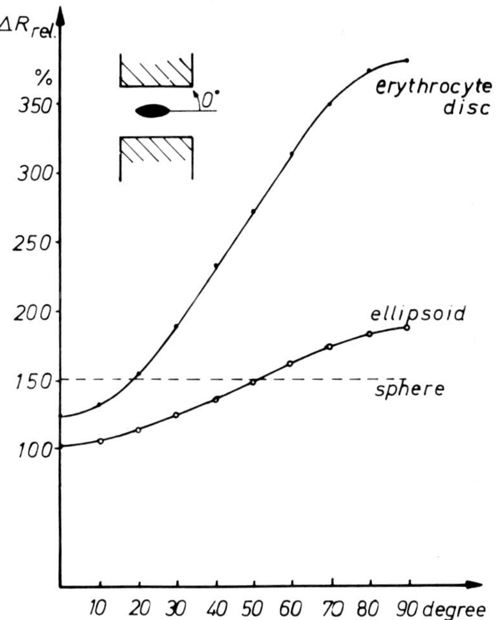

Fig. 16. Pulse heights measured in the middle area of a long model orifice ($\ell/d = 2$) as a function of cell rotation. O^0 when the long axes of the models are in the flow direction. The percentage values of the y-axis are related to the true volume of the particles.

Introducing the field-strength

$$E = i \cdot \rho_2/q \tag{17}$$

the pulse height

$$\Delta U = \frac{E^2 \cdot \Delta V \cdot f}{\rho_2 \cdot i} \tag{18}$$

proportional to E^2 is obtained at the electrodes of the transducer. In interpreting that pulse height/particle volume/field-strength relation one has to have in mind that field strength E and measuring current i are not independent; i.e., basically the field strength in the orifice is proportional to the current i. Increasing current i proportionally increases the amplitude of the pulse if a particle is in a distinct position somewhere in the sensing zone of the transducer.

By contrast, if current is constant and the particle is moving the pulse height generated by that particle is a function of the position of the particle and proportional to E^2. The E^2 relation can be used to calculate and construct pulse height topographies from measured or calculated fields of orifices of different ℓ/d ratios.

Figure 17a–c is calculated according to the E^2 relation from the orifice field strength values and demonstrate in 3 dimensional projections the local distributions of resistance pulse height in and around orifices of different ℓ/d ratios. By tracing particle trajectories through the orifice area, sizing pulses may be constructed (see Fig. 19). Furthermore, model experiments for particle paths along the orifice wall have shown that the height and shape of the M-pulses induced by the edge field are influenced by the volume, shape, and orientation of the particles. These edge-induced M-pulses generated by a sphere, by an ellipsoid with the long axis axially

l/d = 0,5

U_x/U_m (%)
60
40
20
0
x/d
1,5
1,0
0,5
0,25
0
−0,5
r/d
1,0
0,8
0,6
0,4
0,2
−1,0

a

l/d = 1,0

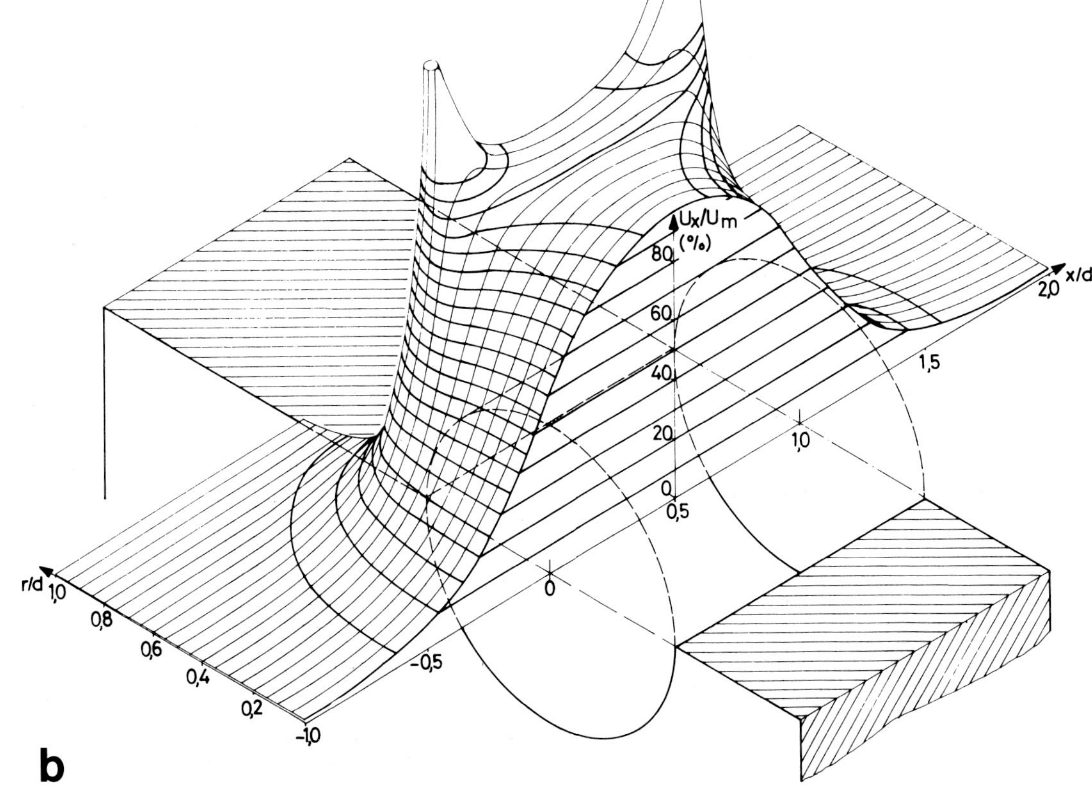

b

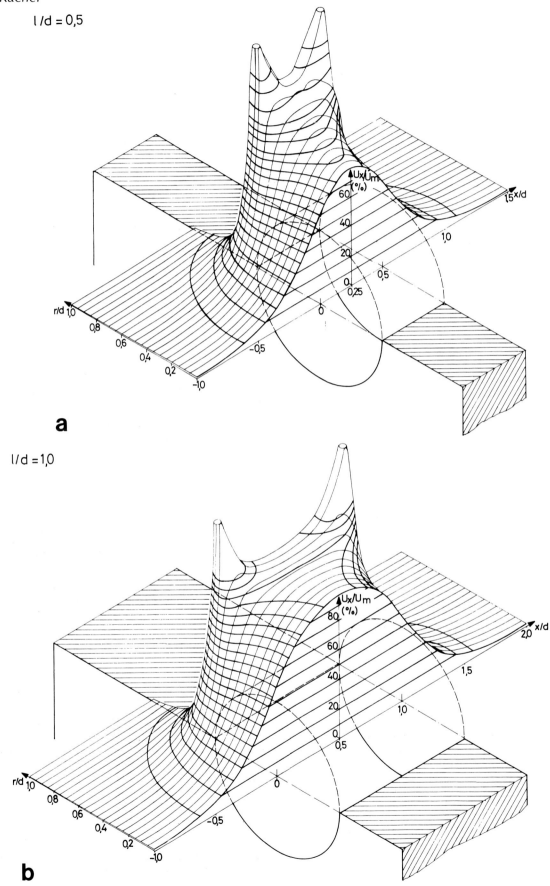

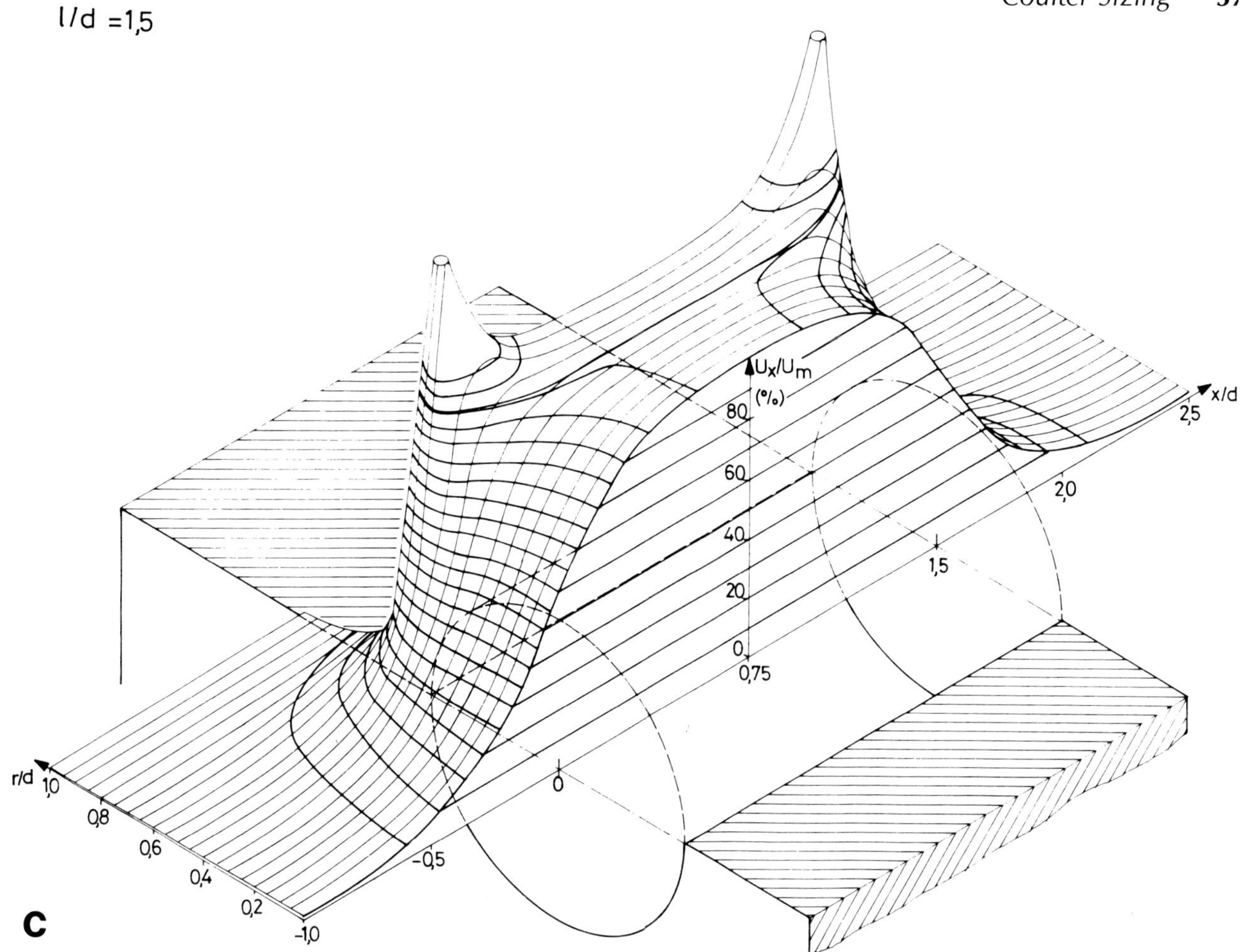

Fig. 17. Isometric view of the spatial distributions of the pulse heights in orifices of l/d ratios 0.5, 1 and 1.5. The patterns which by reason of radial symmetry are valid for all radial sections through the orifices are calculated from the field-strength distributions by applying the E square relation. Since there is symmetry the right half of the field is not shown. The pulse height values U_x are normalized to the pulse height U_m generated by an identical very small particle in the homogeneous electrical field of each orifice. Axial (x) and radial (r) dimensions are normalized to the orifice diameter. For each of the l/d relations extreme overestimation at the orifice edges is found. (a) short orifice of $\ell/d = 0.5$. The pulse height distribution inside the orifice is very inhomogeneous and U_x/U_m reaches in maximum 70% on the axial path. Deviations from the axial path result in a considerable increase of the pulse amplitude. (b) orifice of $\ell/d = 1$. This pulse height distribution pattern corresponds to the top view pattern of Figure 19a. U_x/U_m in the central plane (at $x/d = 0.5$) of the orifice reaches 95% and is much less dependent on deviations from the axial path. (c) long orifice of $\ell/d = 1.5$. U_x/U_m in the central plane at $x/d = 0.75$ is 99%. For such long orifices the electrical field in the central plane is nearly homogeneous. Deviations from the axial path have negligible influence on the pulse amplitudes with the exceptions of the edge distortion peaks.

oriented, and by an ellipsoid with the long axis transaxially oriented are shown in Figure 15. Figure 16 shows ΔR as a function of the angle of rotation of these objects in a long orifice of (ℓ/d = 2).

To test the predictions of the model experiments, fixed erythrocytes and latex particles suspended in a physiologic buffer solution were measured on the axial path of an orifice with a diameter of 100 μm and a length of 138 μm (ℓ/d = 1.38), and of an orifice with a diameter of 100 μm and a length of 27 μm (ℓ/d = 0.27), with a current of 0.52 mA and a suction pressure of 0.14 bar. The mean ΔU values of the distribution curves for the orifice of 27 μm reached only 55% to 60% of the mean values for the orifice of 138 μm in length. These results are in good agreement with the undersizing predicted by the model experiments.

The discrepancy of the model pulse height measurements with identical particles was in general 1% (standard deviation related to the values in the equivalent homogeneous field). The variations between different and differently oriented particles are in general within ± 3% (i.e., variations were found equivalent to ±3% of the ΔR values measured in the equivalent homogeneous field). In the case of (1), a prolate ellipsoid oriented with the major axis aligned with the flow; or (2) an erythrocyte disk oriented with the rotational axis perpendicularly to the flow, there is a tendency toward undersizing of the particles with short orifices. This effect requires further investigation.

Orifice Factor in Short Orifices. To apply equations 8 and 15 to calculations of absolute volumes of particles measured in short orifices, we introduce a so-called orifice factor f_k, which is derived from the inverse underestimation determined with the models. This factor, which corrects for this

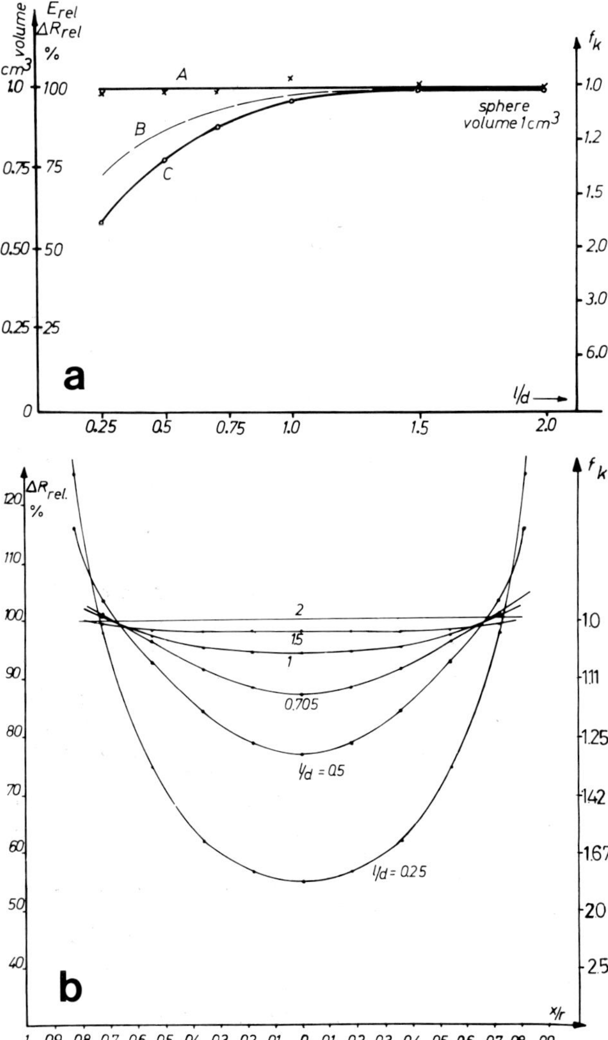

Fig. 18. (a) Resistance pulse height ΔR, volume, field strength E, and orifice factor f_k as functions of the length-to-diameter ratio of the cylindrical model orifices. The percentage values are related to the values in the equivalent homogeneous field and to the orifice axis. Curve A shows the field strength and volume of the 1 cm^3 sphere as a function of the l/d ratios in the equivalent homogeneous fields. The correct volume is always measured (± 3%); i.e., $f_k = 1$. Curve B shows the field-strength in the natural field. Curve C shows the measured volume of the sphere in the natural fields and the orifice factors f_k derived from these measurements. (b) Pulse heights and orifice factors f_k in the middle area of orifices of the different l/d ratios as functions of the radial distance from the axis (axis x/r = 0). The percentage scale is related to the ΔR values in the equivalent homogeneous field.

effect in short orifices, is valid only for a distinct particle path in an orifice of known l/d ratio. It is senseless to apply orifice factors without the focusing technique. In practice, equations 8 and 15 are complemented by the orifice factor f_k:

$$\Delta V = \frac{\Delta U \cdot q^2 \cdot f_k}{\rho_2 \cdot i \cdot f_e} \qquad (19)$$

For spheres f_e is replaced by f_s.

Figure 18a shows the percentage underestimation and the orifice factor f_k derived with spheres, as a function of the l/d ratios. In Figure 18b, the orifice factor f_k is shown as a function of the distance from the axis of the particle path. The descriptive parameter is the relative orifice length.

THE RECIRCULATION PROBLEM

The high-velocity flow of fluid out of the orifice generates a suction on the surrounding fluid according to the Bernouilli law (see Chapter 3, this volume). This suction results in a recirculating flow, returning the electrolyte and suspended particles, which have just been measured, back past the outlet region of the orifice. Owing to the current density at this region, spurious pulses are thereby generated by these recirculating particles (Fig. 19), the number of these, of course, depending on the number of particles in recirculation. These back-cursor pulses, as we will term them, are much longer than the original volume pulses due to their slow recirculating velocity as compared to the velocity of flow through the orifice. In many applications these back-cursor pulses can be identified by their very low amplitude and ignored, as is the case for erythrocyte distribution measurements. Figure 19a explains the generation of back-cursor signals.

The back-cursor pulses become important when both large and small cells are present in one suspension, especially if there are relatively few small cells. In this case, the back-cursor pulses created by the large cells will interfere with pulses of the small cells. Back-cursor pulses may be removed in two ways: (1) by electronically eliminating longer pulses, and (2) by hydromechanically preventing the recirculation of cells into the sensitive region at the low pressure side of the orifice.

The electronic elimination method is unreliable because low, back-cursor pulses may seem to be as short as the (original) volume pulses. This will occur if they are low amplitude and only exceed the triggering threshold for a short time (Fig. 19c). The hydromechanical method prevents generation of back-cursor pulses a priori by hindering cells from penetrating the sensitive region. One possible way to do this is by using two orifices, one after the other, whereby recirculation of cells after the first orifice is prevented by a second sheath flow that directs the cells through the second orifice [147] or a so-called catcher tube [64]. Another possible approach is to wash cells from the sensitive outlet region of the orifice with particle free electrolyte. Figure 19c shows pulses of 3.3 μm latex particles in a 30-μm orifice distorted by back-cursor pulses of fixed erythrocytes. In Figure 19c back-cursor pulses are prevented.

PROPERTIES OF THE PARTICLES

Factors f_e and f_s (see equations 6 and 14) describe the basic mathematical relations between the shape and the resistivity of the particles and the resulting pulse height. However, this mathematical description is not sufficient to determine the shape and resistivity factors of biologic particles. Such particles are not fixed in shape, they may be deformed or oriented by the hydrodynamic forces in the orifice. Their deformability and/or resistivity may be altered by swelling or shrinking processes, by the influence of chemical substances, or by electrical breakthrough effects of their membranes. Therefore, the actual problem is determination of the shape and conductivity of particles at the moment they are within the orifice.

DETERMINATION OF SHAPE AND ORIENTATION OF CELLS MOVING THROUGH THE ORIFICES

Direct Photographic Method

In order to observe and photograph the cells in the orifice in axial (horizontal chamber) and transaxial (vertical chamber) directions, two types of special transducer chambers have been developed. The results, described elsewhere [69–71,76,101] will be briefly summarized here. Transaxially directed photographs of cells passing on the axial path were also taken by Thom with giant pulse laser illumination [146,147].

Figure 20a,b shows schematic diagrams of the two chambers, Figure 20c,d examples of dye-colored particle streams, and Figure 20e the mean velocity to pressure relation in the orifices of both chambers. The optical apparatus consists of a Zeiss standard microscope with camera and a Nanolite 8N18D flashlamp with argon chamber and Nanolite Driver (illumination time 40 nsec). To photograph single cells, the amplified volume pulse of a cell activates a trigger generator which fires the flash with an adjustable delay. A symmetric double-stream device (Fig. 20c) was developed to simplify the comparison of various cell samples [71].

Figures 21 to 28 give some examples of cells photographed in the orifices. Untreated erythrocytes in the core stream of the orifice are deformed and always oriented as prolate ellipsoids of revolution (Fig. 21a) (see also refs. [6,48, 121,124,125]). Numerous pictures like that of (Fig. 21c), which were taken with the vertical chamber, show the circular cross section of native human erythrocytes. In contrast, fixed erythrocytes (Fig. 22) and nucleated avian erythrocytes (Fig. 23) show the biconcave or elliptical cross section. Note that the orientation of nonspherical particles occurs in front of the orifice (Fig. 24) in the highly accelerated fluid stream. The aligning forces in this region are explained in Figure 29, where the areas $v6 > v5 > v4 \ldots$ are equivelocity lines of increasing velocity with decreasing distance from the inlet of the orifice. The flow direction is perpendicular to these lines. If $v_B = v_A$, the position of the particles is not stable. A slight rotation from this position of unstable equilibrium shifts one of the points of the nonspherical particle to a position where $v_A <> v_B$. The forces produced by $v_{vA} - v_{vB}$ and by $v_{hA} - v_{hB}$ rotate the particle to the position of stable equilibrium where the difference between v_{vA} and v_{vB} is a maximum and the components v_{hA} and v_{hB} have vanished. Also, these torques tend to stretch deformable particles. At paths near the orifice wall, a constant flow velocity no longer exists. The flow velocity at the wall is reduced to zero by friction; the velocity gradient disturbs the alignment of the cells, some are in rotation (as shown with erythrocytes in Fig. 21b).

The rotation of cells can be recognized by the shape of the resistance pulse. Figure 30 shows pulse patterns of particles passing a 100 μm orifice near its wall; Figure 30a shows pulses of cup-shaped fixed erythrocytes, two pulses with a third peak caused by the shape factor change of rotating cells. The length of these three-peak pulses corresponds to that of the other two-peak pulses; therefore, coincidence may be excluded. For contrast, completely uniform pulses of latex spheres are shown in Figure 30b. Rotating cells and three-peak pulses were first photographed and explained elsewhere [69,101]. These observations were confirmed by Leif and Thomas [90]. (For shape determination by rotating cells, see next section.) The deformability of mammalian erythrocytes is strongly affected by the suspending electrolyte. Erythrocytes suspended in plasma show more deformability than those suspended in physiologic buffer solution. Thom found that the deformability of human erythrocytes is also influenced by the age of the cells [147]. The striking difference in deformability of differently treated human erythrocytes is shown in Figure 25, where cells from the same person suspended in plasma (lower stream) and in hypertonic NaCl solution are compared in the double-stream device. Figure 26 shows aggregated human erythrocytes passing through the orifice.

Nucleated cells show no visible deformation [4] (Figs. 27–28). From pictures of the cells their axial ratios are available, and by assuming that they have approximately a spherical or ellipsoidal shape, the shape factors can be calculated using equations 6 and 14. The deformed or natural cells shown here should not be considered as examples of general validity. This section shows that deformation must be investigated under sizing conditions of the cells to be sized.

Indirect Methods to Determine the Shape Factor

Attempts were made to obtain information about cell orientation and deformation without direct observation. Grover et al. [53,55] proposed that the minimum shape factor of cells and thus their axial ratios be determined by measuring the difference between the maximal and minimal pulse height produced by the rotation of cells in the orifice. Apart from the difficulty in producing distribution curves of cells with their long axis perpedicular to the orifice axis (it is possible only near the orifice wall, and there the pulse heights are distorted), this method would give true results only if the axial and transaxial shape of the cells is the same. Figure 21 demonstrates that an identical shape cannot be assumed for deformable cells moving in the core stream and others rotating near the wall; therefore, this method is not applicable to shape determinations of deformable cells. The minimal shape factor of nondeformable cells, which is necessary for volume calculations if the focusing technique is used, can easily be determined with stationary cells in a normal microscope.

The two-orientation method of shape determinations was also used by Thomas et al. [148] and Golibersuch [49], who interpreted the highest and lowest pulse amplitudes of presumably rotating cells. However, these investigators start from the errorneous assumption that native red blood cells maintain their biconcave shape; hence the disagreement with the present results. Thom [147] proposed a method for determining shape factors by the use of two orifices in sequence with different geometrical properties, comparing the pulse heights produced in these orifices by the same cell. For such quantitative measurements an exact knowledge of the shape factor to pulse height relationship in short orifices is required, and was not determined by Thom. The interpretation of his results is questionable because no currents are indicated, and it is possible that breakthrough effects were measured [146].

The method of Waterman et al. [162] for comparing the microhematocrit mean value with the resistance-pulse-measured mean value is correct if both measurements are performed in the same electrolyte, all other Coulter artifacts are eliminated, and there is no doubt about the flexibility of the investigated cells.

Deformability studies were carried out by Mel and Yee [97] with a mixture of fixed and native cells at slow and high flow rates. They draw conclusions about the deformability

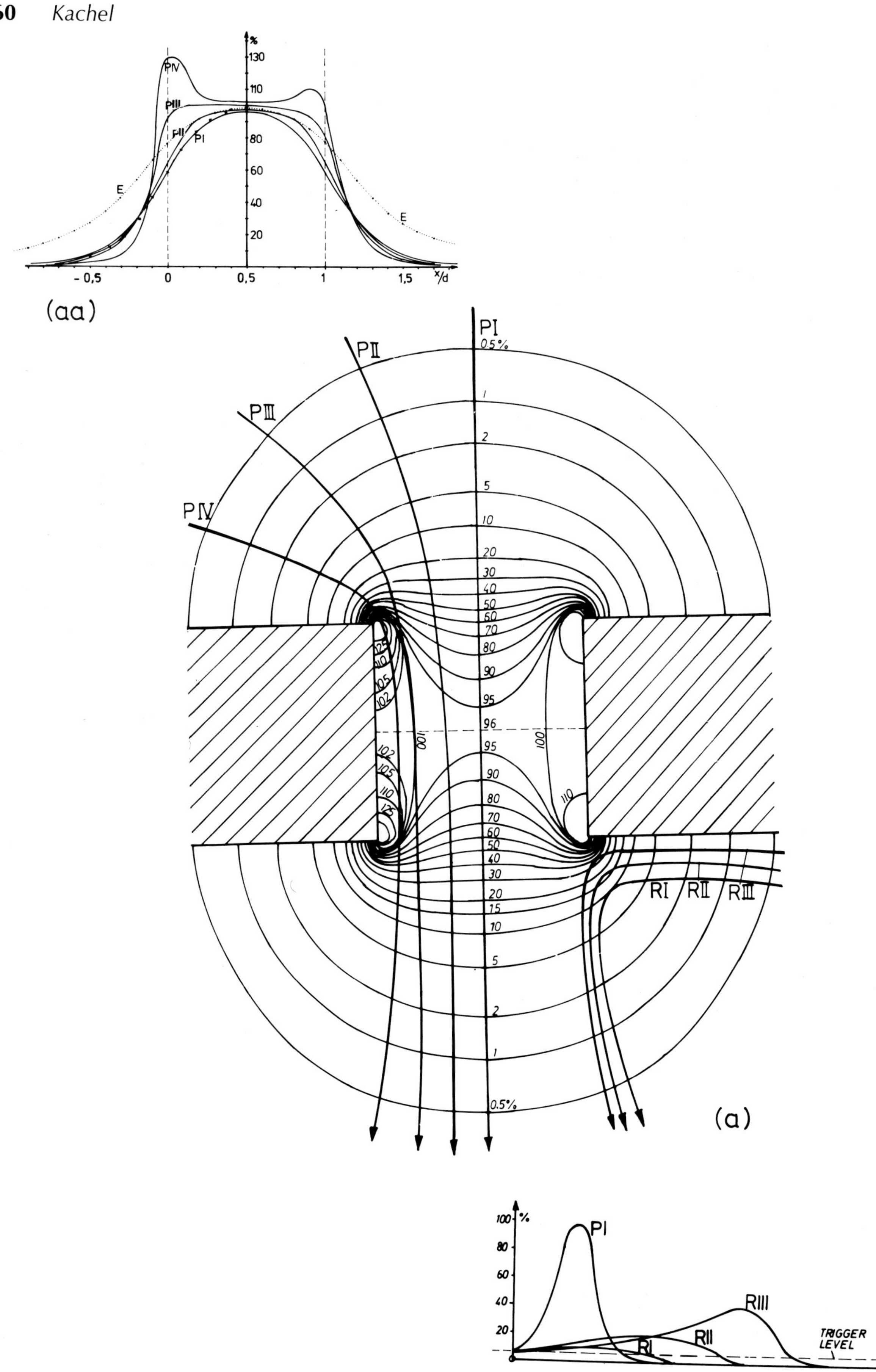
%
130
110
80
60
40
20
PIV
PIII
PII
PI
E
E
- 0,5
0
0,5
1
1,5
x/d
(aa)
PI
0.5%
PII
PIII
PIV
1
2
5
10
20
30
40
50
60
70
80
90
95
96
95
90
80
70
60
50
40
30
20
15
10
5
2
1
0.5%
100
100
110
125
110
105
102
102
105
110
125
RI
RII
RIII
(a)
100
%
80
60
40
20
0
PI
RI
RII
RIII
TRIGGER
LEVEL
t
(ab)

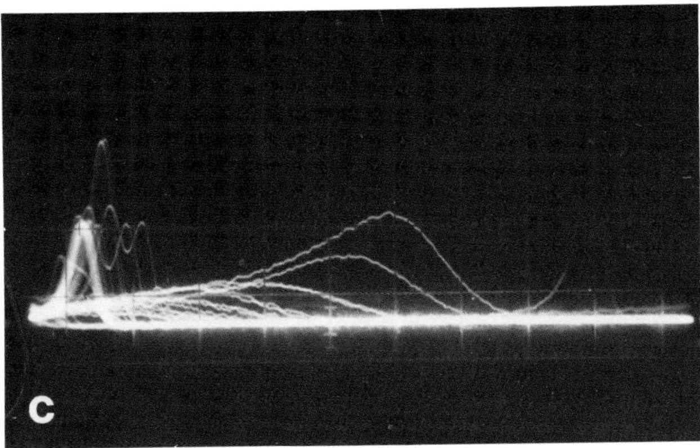

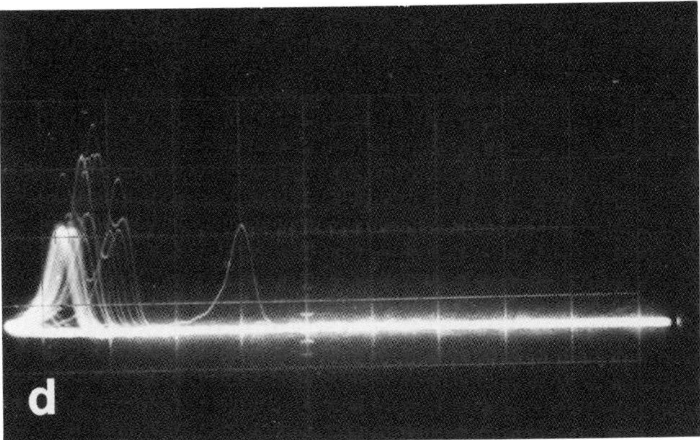

Fig. 19. Pulse generation and the origin of the back-cursor pulses. (a) Lines of identical pulse height inside and outside of an orifice of $\ell/d = 1$. The curves are derived from the calculated field-strength values (Fig. 9) by application of the E^2 relation, and the numbers indicate the pulse heights related to the pulse height in the equivalent homogeneous field. This pattern is another view of Figure 17b. PI to PIV are forward particle paths through the orifice, RI to RIII are back-cursor paths. (aa) Constructed volume pulses of identical particles on the paths PI to PIV, flow velocity neglected. (ab) Pulses coordinated to the forward-path PI and the recirculation-paths RI to RIII, time axis different from that of (aa). (c) Oscilloscope pattern of normal (type PI) and back-cursor pulses (type RI to RIII) supplied by a 30-μm orifice. The normal pulses are generated by latex particles (3.3-μm diameter), most of the back-cursor pulses by fixed erythrocytes that were added to the recirculating flow at the backside of the orifice. (d) Pulses from the same orifice as in (c), except that the antirecirculation flow is on. The back-cursor pulses are completely removed.

of the native cells from the peak difference between such measurements, but did not take into consideration possible orifice edge and cell rotation effects of their unfocused device.

Determination of the Electrical Resistivity of the Particles

The factors derived above also describe the influence of the resistivity of the particles, which, unlike the resistivity of the electrolyte, is difficult to determine. For practical purposes, it is usually sufficient to ensure that ρ_2/ρ_1 is below a certain value representing negligible error. In our laboratory two methods to determine the resistivity of particles have been developed. The first method, introduced by Metzger [102], involves suspending particles of unknown electric resistivity in electrolytes with increasing resistivities. With nonrigid biologic cells the change in osmolarity has to be compensated by the addition of nonionic substances. The suspensions are sized and the polarity of the volume pulses observed with an oscilloscope. When the amplitude of the pulses is reduced to zero and the first pulses with inverse polarity appear, the resistivities of the electrolyte and of the particles are nearly equal.

The second method is based on equations 8 and 15, respectively, which are provided to calculate particle volumes if the particle resistivity ρ_1 is known. But with unknown ρ_1 we have two unknown variables in the equation; the particle volume ΔV, and the resistivity ρ_1 of particles. We presume that the factors of the equation have been determined by methods described above or other well-known standard methods. The solution requires two independent equations. In order to obtain these equations, the cells are sized in two electrolytes of different specific resistivities ρ_2 and ρ_3, which are osmotically balanced by nonionic substances. If the volumes of the particles are unchanged, the volumes of both measurements are equalized. In the remaining equation (20), only the unknown resistivity ρ_1 of the cells remains. For these calculations, a characteristic value in the distribution curve is to be used that can be identified in the curves of both measurements, e.g., a mean value or the 50% value of the rising or falling slope of a peak in the distribution curves. The method fails if no characteristic point exists. Now consider prolate ellipsoids of revolution. According to equation 15 (long orifice with $f_k = 1$) we have

1st measurement electrolyte:

$$\Delta V_1 = \frac{\Delta U_1\, q^2\, [2 + ab^2\, La((\rho_2/\rho_1) - 1)]}{\rho_2 \cdot i \cdot 2(1-(\rho_2/\rho_1))} \tag{15a}$$

2nd measurement electrolyte:

$$\Delta V_2 = \frac{\Delta U_2\, q^2\, [2 + ab^2\, La\, ((\rho_3/\rho_1) - 1]}{\rho_3 \cdot i \cdot 2\, (1-(\rho_3/\rho_1))} \tag{15b}$$

ΔV_1 equals ΔV_2, hence:

$$\frac{\Delta U_1\{2 + ab^2\, La[(\rho_2/\rho_1) - 1]\}}{2[1 - (\rho_2/\rho_1)]} = \frac{\Delta U_2\, (2 + ab^2\, La[(\rho_3/\rho_1) - 1]}{2[1 - (\rho_3/\rho_1)]} \tag{20}$$

with

$$s = \frac{\Delta U_1}{\rho_2}$$

and

$$t = \frac{\Delta U_2}{\rho_3}$$

and $A = ab^2\, La$ [equations (9)–(11) we obtain:

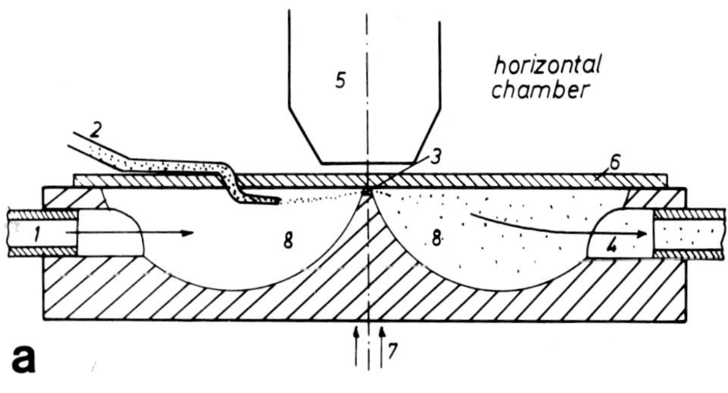

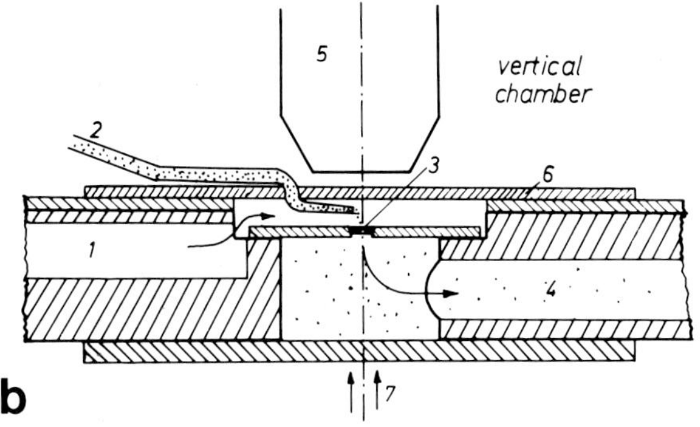

Fig. 20. (a,b) Schematic diagrams of the two optical chambers to observe cells moving through orifices. (1) supply of particle-free solution, (2) tube conducting the particle suspension, 3a) orifice with rectangular cross section (100-μm diameter), 3b) orifice with circular cross section (100-μm diameter), (4) tube connected to the pump, (5) objective of the microscope, (6) removable cover glass, (7) illumination, (8) semispherical excavations connected by the orifice. The electrodes are located outside of the tubes 1 and 4.

$$\rho_1 = \frac{-p \pm (p^2 - 4gc)^{1/2}}{2g} \qquad (21)$$

where

$$g = ((s/t) - 1)\,(2 - A)$$
$$p = \rho_3\{A[(s/t - 1) - 2(s/t)] + \rho_2(A[(s/t) - 1] + 2\}$$
$$c = A \cdot \rho_2\rho_3[1 - (s/t)]$$

Positive real solutions of equation 21 give the resistivity of the particles.

The volume of the reference point of the distribution curve is obtained by introducing ρ_1 into equation 15a or 15b:

$$\Delta V = s{\cdot}q^2 \frac{2 + A(1 - (\rho_2/\rho_1))}{i{\cdot}2{\cdot}((\rho_2/\rho_1) - 1)} \qquad (22)$$

The first method can be applied without knowledge of the distribution curve of the particles. For particles with relatively high resistivities, the field within the orifice ($E = i\,\rho_2/q$) increases with increasing electrolyte resistivity. Due to noise limitations, the current i cannot be decreased in proportion to increasing ρ_2, hence there is a probability that E will increase to the point where electrical breakthrough of the cell membrane will occur.

With the second method, the difference between ρ_2 and ρ_3 can be chosen such that the increase of the field in the orifice is compensated by a decrease in current, provided that breakthrough fields are avoided. With this method, it is at least

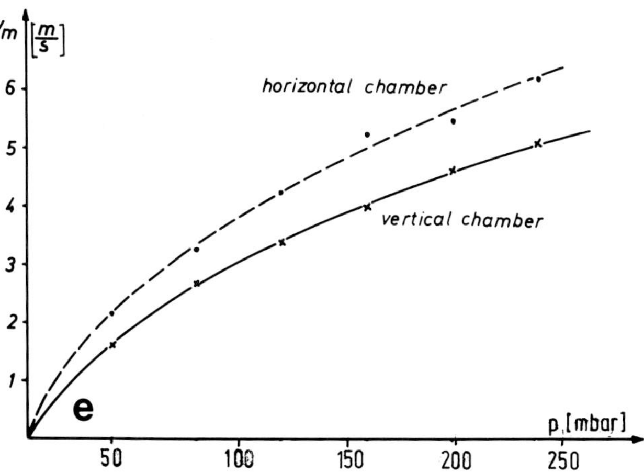

Fig. 20. (c) Particle streams in the horizontal chamber. The double-stream device is developed for simultaneous observation of two different samples. The black ink colored suspension is ejected from two glass tips at left (see also Figs. 7 and 25). (d) Particle stream near the wall of the vertical chamber. (e) The mean fluid flow velocities measured with water in the orifices of the two chambers as a function of the sucking pressure.

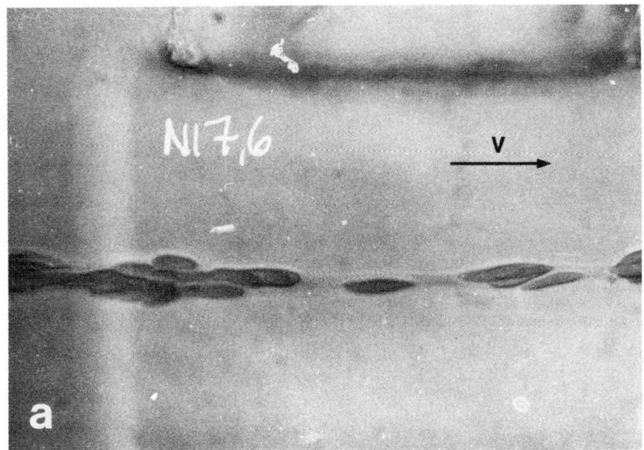

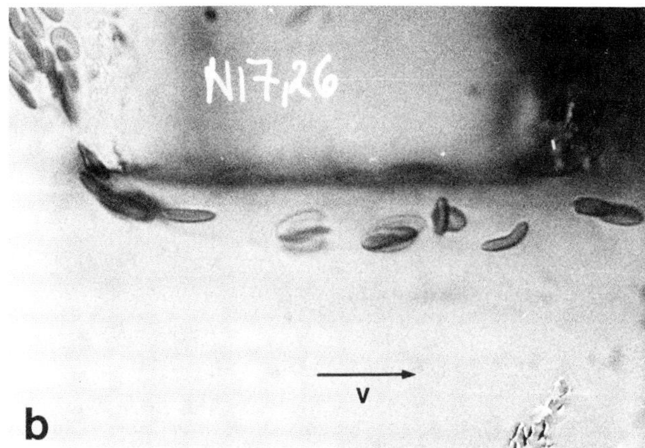

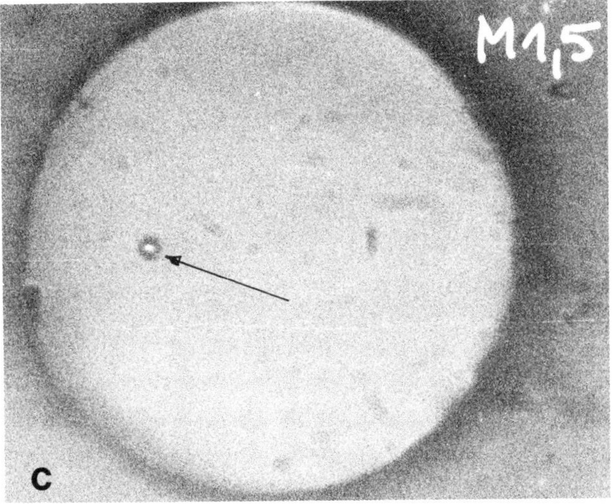

Fig. 21. (a) View perpendicular to flow axis of native human erythrocytes passing through the center of the horizontal orifice, $v_m = 3$ m/sec. The cells are oriented in the flow direction and deformed to ellipsoid-like bodies. (b) Native human erythrocytes passing the orifice near its wall, $v_m = 3$ m/sec. The orientation and deformation of the cells is no longer uniform. (c) Axial view of a native human erythrocyte passing the orifice of the vertical chamber. The circular cross-section due to the deformation of the cell is demonstrated, $v_m = 4$ m/sec. Arrows indicate flow direction (a,b).

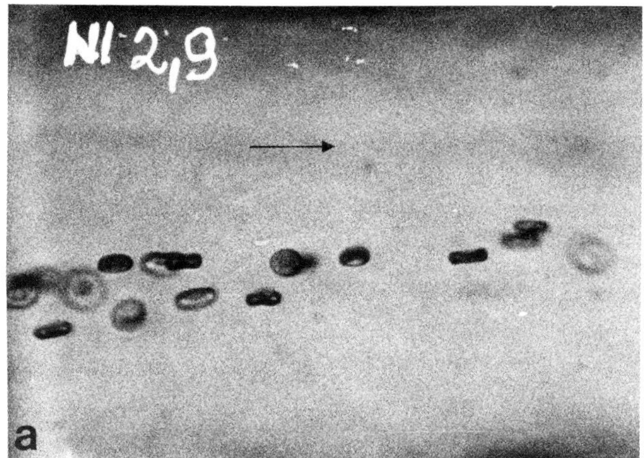

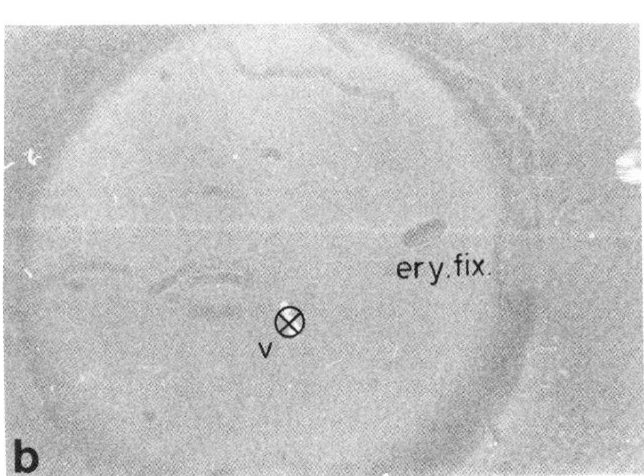

Fig. 22. (a) Biconcave fixed human erythrocytes passing the center of the horizontal chamber, $v_m = 2.5$ m/sec. The cells are oriented but not deformed. Arrow indicates flow direction. (b) Biconcave fixed human erythrocyte passing the orifice of the vertical chamber, $v_m = 4$ m/sec. In contrast to the native erythrocyte of Figure 21c the biconcave elliptical cross-section is demonstrated.

theoretically possible to detect resistivity variations within the same particle population. With rigid particles this method works with high reliability, but with living cells that are sensitive to changes of the electrolyte, uncertainties regarding volume changes of the particles will remain despite balancing the osmolarity of the electrolyte.

PROPERTIES OF THE ELECTROLYTE

Electrolytes used in sizing biologic cells have to fulfill the following conditions:

1. Neutrality in regard to the suspended biologic material; i.e., physiologic osmolarity and pH
2. No dirt or debris, which might block the orifice and/or produce noise
3. Negligible dissolved gases, to avoid distorting bubbles in the transducer
4. Equal resistivity of the electrolyte in the suspension and the sheath flow with focusing transducers.

A review of the most commonly used diluents for Coulter sizing is given by Helleman [66], who studied the effect of various solutions on blood cells. Other buffer solutions that

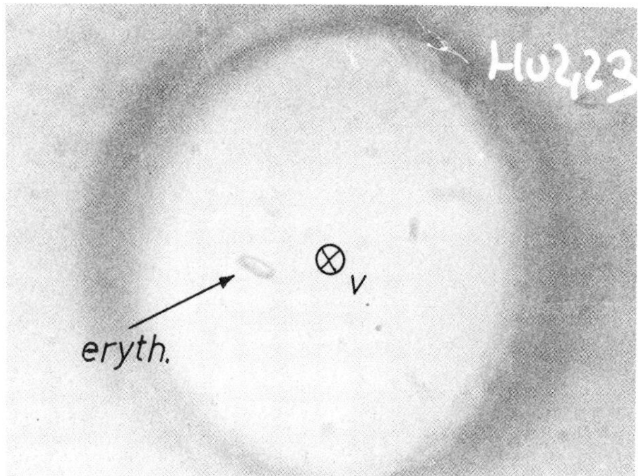

Fig. 23. Axial view of a chicken erythrocyte in the core stream of the orifice of the vertical chamber demonstrating the ellipsoidal cross-section of such nucleated cells during their passage in the orifice, v = 4 m/sec.

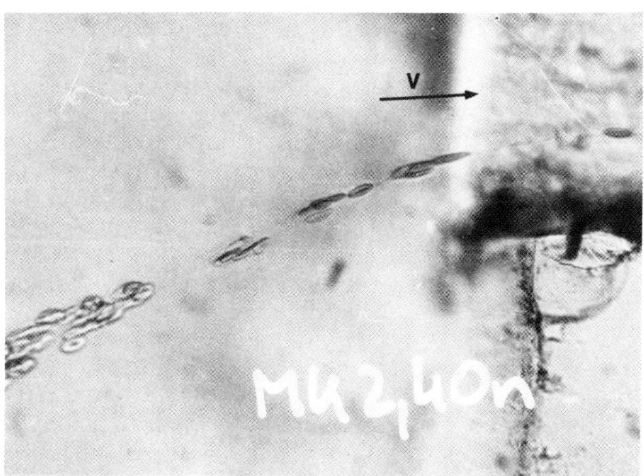

Fig. 24. A stream of native erythrocytes in front of the horizontal orifice, demonstrating that the cell orientation and deformation takes place outside the orifice, v_m = 4 m/sec. Arrow indicates flow direction.

can be used for sizing cells are described in refs. [30] and [46].

The resistivity of electrolytes is strongly dependent on temperature, thus it is advantageous to maintain constant temperatures while sizing. The pulse height generated in the orifice is proportional to the conductivity of the electrolyte (eq. 8). Conductivity-induced pulse height variations can be compensated by a current-sensing amplifier.

COINCIDENCE

Coincidence is due to the presence of two or more nonadherent particles in the orifice in such a manner that they are not resolved as single events in time. Coincidence may be caused by an analyzing device with a long dead time, or by too small a distance between the particles themselves (concentration too high). In the former case, fast electronic analyzers can reduce the electronic coincidence to neglibile values. In the latter case, Wales and Wilson [160] and Princen and Kwolek [117] described two types of coincidence: (1) the so-called horizontal interaction, where two particles enter the orifice in such a way that two pulses are generated, but only the larger one is evaluated, and (2) vertical interaction, where both particles form one pulse with a doubled pulse height.

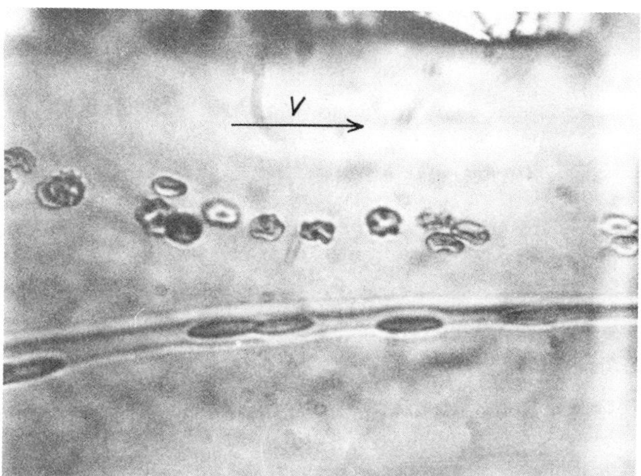

Fig. 25. Comparative study of differently treated human erythrocytes using the double-stream device. Lower stream: the cells suspended in their own plasma are distorted to ellipsoid-like bodies. The stria indicates the delimination of the plasma. Upper stream: The same cells suspended in 1.9% NaCl solution. The cells are osmotically distorted to crenated discs. A hydrodynamic deformation of these cells is not observed. Arrow indicates flow direction.

If particle counting is the sole measurement, there are methods of pulse processing with electronic differentiating circuits that can resolve coincidence due to horizontally interacting particles. Coincidence theories concerning the count loss of Coulter counters have been published [66,99,117,118,160,161]. In sizing particles, coincidence in all cases initially distorts the true pulse height of both particles by superposition (Fig. 31a). Thus the horizontal interaction as defined above is not completely correct.

With horizontal interaction, pulse heights lie between the true pulse heights of the single particles and the sum of both.

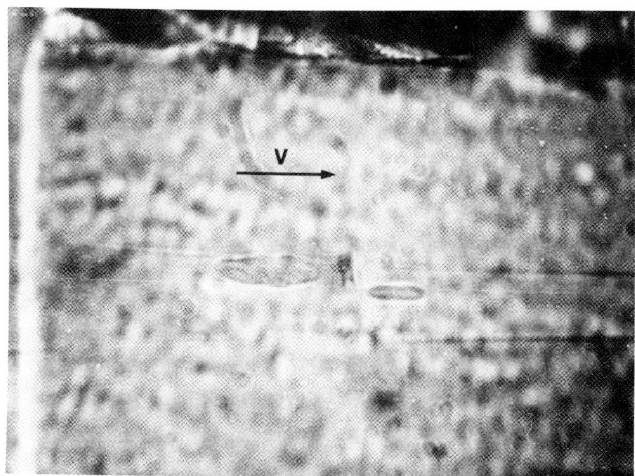

Fig. 26. Aggregate of four native human erythrocytes passing through the horizontal orifice, v_m = 3 m/sec. Arrow indicates flow direction.

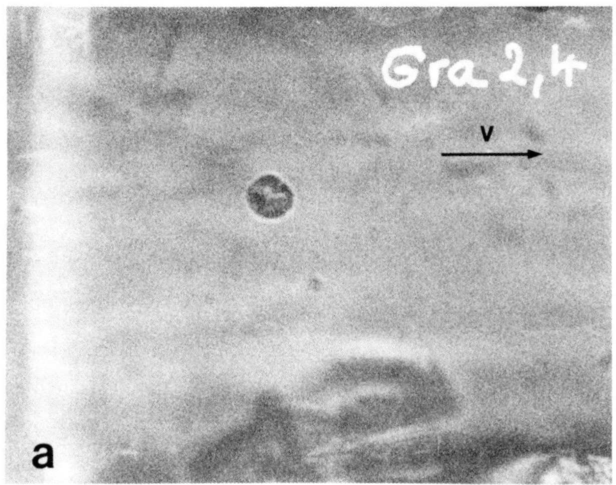

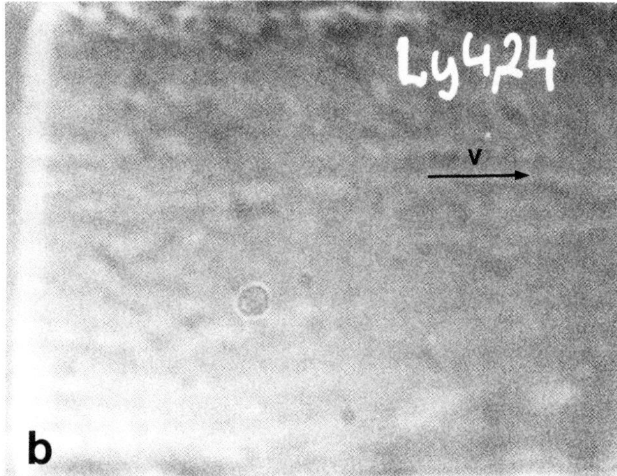

Fig. 27. (a) Native human granulocyte passing the horizontal orifice showing no distortion, $v_m = 3.5$ m/sec. Cells prepared according to [110]. (b) Native human lymphocyte passing the horizontal orifice showing no distortion, $v_m = 3.5$ m/sec. Cells prepared according to [109]. Arrows indicate flow direction.

Depending on the triggering circuits, either the first of the two or the greater pulse is evaluated. With vertical interaction, the sum of both pulses is always measured. In long orifices, there is increased probability of horizontal interaction. If we assume (1) a pulse rate of N particles per second that pass the orifice in a random distribution on a focused path with a mean velocity v_m, and (2) the sensitive length of the orifice (the physical length and the distance outside the orifice where the slope of the pulse is produced) is x; then $\gamma = x\ N/v_m$ particles are present in the orifice on the average.

The probability of the presence of n particles in the sensing zone is expressed by Poisson's law [160]:

$$p(n) = \gamma^n \cdot e^{-\gamma}/n! \qquad (23)$$

This coincidence probability depends on the flow velocity, the average number of particles passing through the orifice (which can be determined approximately by the counted pulse rate), and the length of the sensing zone. The flow

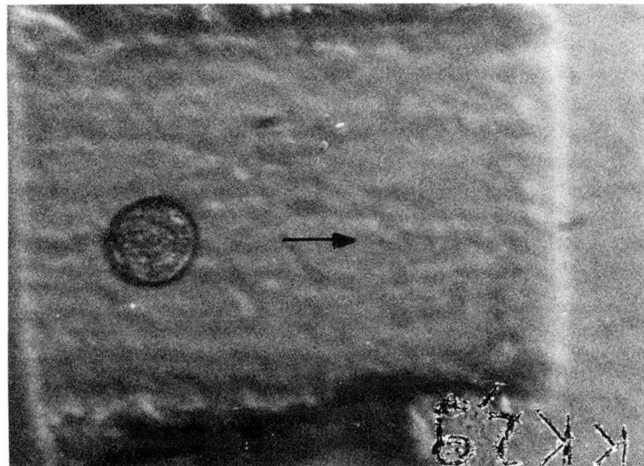

Fig. 28. Isolated liver cell passing the horizontal orifice showing no distortion, v = 3 m/sec. Cells prepared according to [67]. Arrow indicates flow direction.

velocity and the pulse rate depend on each other, and therefore the particle density related to the focused particle path is defined by $\beta = N/v_m$ (particles/length unit).

In Figure 31b, the ratio of the probability of two particles to the probability of one particle passing in the orifice p(2)/p(1) is plotted as a function of the particle density β. This percent ratio indicates how many doublets are expected if the probability of the singlets is assumed to be 100%. Thus, the coincidence probability may be estimated; e.g., with vm = 5 m/sec, N = 1,000 particles/sec and a sensing length of the orifice of 100 micrometer, the coincidence probability is near 1%.

Exact coincidence theories indicating the quantitative influence of the coincidence on the volume distribution curves are not known. Most of the coincidence pulses are characterized by an increased pulse length. An electronic time window may decrease the coincidence effect by rejecting long pulses. This method fails with particles of widely varied volume, or if cell aggregates are to be measured [14,15], since in such cases pulse length detected by the trigger device is also a function of pulse height (see Fig. 19ab).

INSTRUMENTATION

Many investigators have modified the commercial Coulter device or constructed instruments of their own. Doljanski et al. [37], Lusbaugh et al. [93], van Dilla et al. [36], and Paulus [112] equipped Coulter transducers with multichannel analyzers. Bull [20] equipped his Coulter counter with a delay circuit; and a delayed gating device was also used by Grover et al. (54) in an instrument of their own design. Other specially designed particle analyzers without focusing were used by Gutmann [57], Adams et al. [1], and Harvey [61]. Focusing devices are described by Spielman and Goren [136], Thom et al. [144–147], Shank et al. [130], our group [69–81], Merrill et al. [100], Steinkamp et al. [137], and v. Behrens and Edmonson [12]. A one-threshold analyzer was provided with a focusing device by Shuler et al. [131], but the focused suspension flow was not constant with time. A system with a so-called "flow directing collar" was developed by Karuhn et al. [83] and Davies et al. [31]. Electronic signal analysis techniques are described by our group [71,72], by Coulter patents [28,29], and by Waterman et al. [162].

A technique similar to the four-electrode impedance bridge

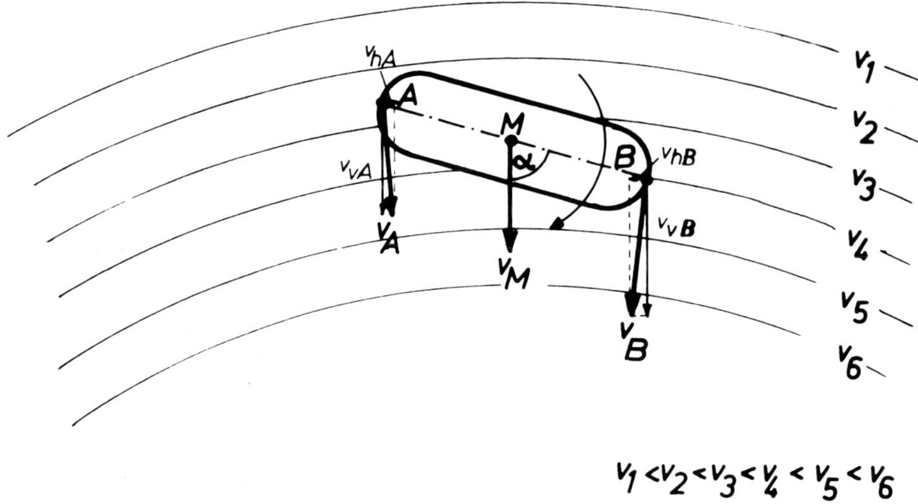

Fig. 29. Orienting forces exercised upon a nonspherical cell in the accelerating flow in front of an orifice. For explanation, see text.

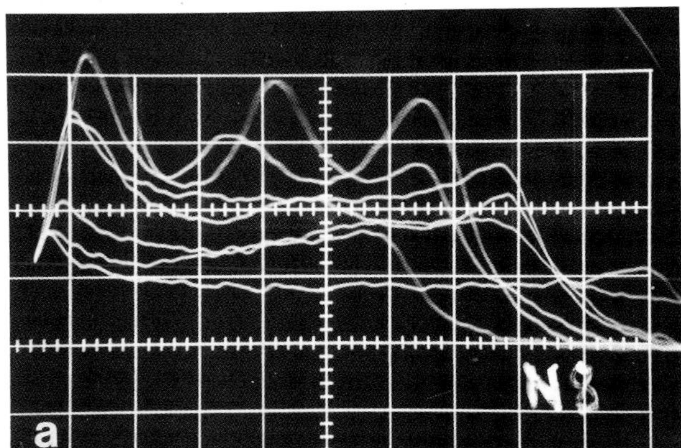

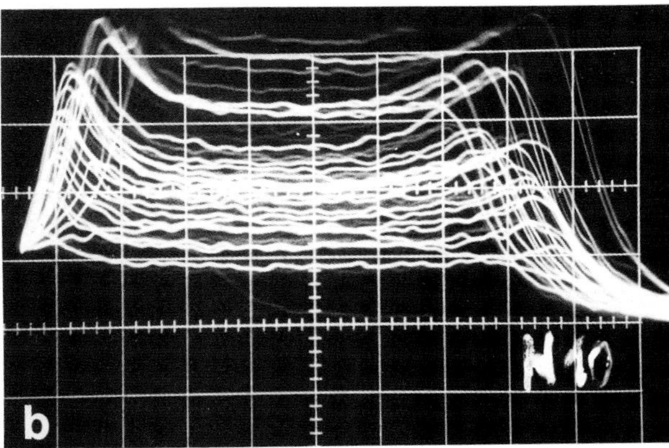

Fig. 30. Pulse patterns produced by particles passing through a 100-μm-diameter orifice near its wall. (a) Cup-shaped, fixed erythrocytes; the third peaks are produced by cell rotation, which changes the shape factor of the cells. (b) Spherical latex particles; particle rotation cannot change the shape factor, therefore uniform two-peak M pulses are observed.

devices [40] used to measure impedances of electrolytes and tissues, has been introduced into Coulter sizing by Salzman [123], Leif and Thomas [90], and Thomas et al. [148]. These so-called "potential sensing transducers" are reported to have improved frequency and noise response. Compared with a perfected two-electrode device with focusing, however, superior accuracy and resolution has yet to be demonstrated.

Multiparameter flow cytometric systems including electrical sizing are described by Steinkamp [137], our group [77], Thomas et al. [148a], Haskill et al. [63], and Becton Dickinson [11a]. It is not possible to describe and discuss all the devices mentioned. However, we can review some of the general features of the instrumentation from our point of view and on the examples of instruments developed in our institute. (Commercially available instruments are described in chapter 9, this volume.)

SENSOR SYSTEM

The transducer is the key element of any sizing instrument, and it determines the accuracy of the measurements and the convenience of the system. It should have the following features:

1. Hydrodynamic focusing, to avoid field distortion and cell-rotation artifacts
2. Simple mounting to permit quick and easy exchange of orifice by normal laboratory personnel
3. A simple means of cleaning:
   a. To remove dirt and bubbles from the system without disassembling.
   b. To avoid mutual contamination of different cell suspensions, and thus to permit quick exchange of probes.
4. Economic use of the cell suspension in controlled feeding to the transducer
5. Means to prevent signals from recirculating cells
6. A device for simultaneous counting of cells

ONE-PARAMETER CELL SIZING TRANSDUCER

A new transducer was developed for use in the instrument Metricell [75,76]. The electrolyte system consists of a storage tank containing electrolyte free of particles, a regulating

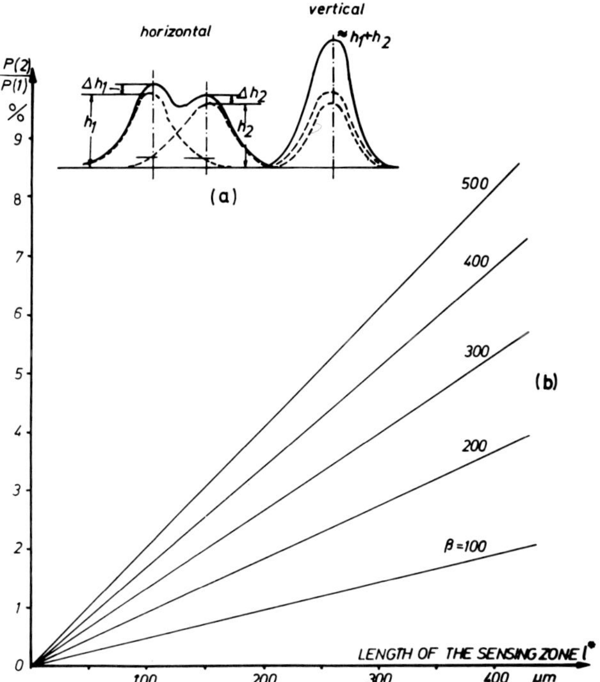

Fig. 31. (a) Explanation of the horizontal and vertical coincidence. With horizontal coincidence, each of the two pulses is affected by the rising or falling slope of the other pulse, resulting in a pulse height increase Δh. With vertical coincidence, the two peaks occur simultaneously and peak heights are summed. (b) The coincidence probability as a function of the length of the sensing zone of the orifice ℓ^*, which consists of the length of the orifice itself and the length of the region outside the orifice where the slopes of the pulses are generated. The particle density β at the focused path, measured in particles/m, is used as a parameter.

chamber that is adjustable in height for controlling the introduction of the cell sample, and an appropriate measuring chamber (Figs. 32,33). The suction, which causes flow through the orifice, is applied to the lower part of the measuring chamber containing the grounded electrode. A dropping chamber electrically separates the storage tank and the antirecirculation device from the lower part of the regulating chamber where a ball valve maintains a constant level of the electrolyte. The electrolyte then flows through a flexible tube to the upper part of the measuring chamber containing the electrode, which is connected to the current source and to the amplifier. Movable containers holding the particle suspension (pipette tips may be used as well) are connected to the particle tube, which terminates in a small tip above the orifice at a distance of about 5 to 10 orifice diameters from the orifice inlet. This tip injects the suspension into the focusing fluid stream. The lower part of the regulating chamber and the container for the cell suspension form a communicating fluid system.

Elevating or lowering the regulating chamber decreases or increases the flow of particles into the focusing particle-free fluid above the orifice. When there is no level difference Δh, the particle flow stops. To clean the upper side of the orifice, the suction is switched over from the lower part of the chamber to the cleaning tube.

For simultaneous counting of particles during sizing, an optional counting head has been developed that consists of a precision tube ⟨16⟩ that is connected to the upper side of the measuring chamber (Fig. 33). The tube is filled with the particle suspension to be counted. As the suspension of the tube empties into the measuring chamber, the meniscus passes a starting light barrier ⟨15⟩ and the refractive change activates the cell sizing measurements; likewise, when the meniscus passes a stop light barrier ⟨14⟩, sizing measurements are stopped. Thus, the volume distribution curve recorded represents the number of cells in the suspension contained between the two light barriers. The volume of the samples depends on the length and diameter of the precision tube. Measurements with a volume of 25 μl have shown a reproducibility of ± 3%.

Detail A of Figure 32 shows the device for removing recirculating cells [78]. Particle-free electrolyte is transported through channel ⟨9⟩ to the ring-shaped cup ⟨13⟩ by suction in the low pressure part of the chamber, and from there to the backside of the orifice. This particle-free flow completely prevents back-cursor pulses (see Fig. 19d). The antirecirculating flow may be switched off by closing tube ⟨11⟩ if the back-cursor pulse height lies outside the actual cell volume distribution. Figure 33 shows the standard Metricell transducer with counting head.

Figures 34 and 35 show a more recent design of a so-called tubeless transducer, where the suspension is directly fed from the removable container ⟨10⟩ via the hole ⟨3⟩ into the measuring orifice ⟨12⟩ [81]. This kind of transducer is of particular interest for kinetic measurements and quick exchange of samples.

The transducers are mounted in a shielded box. The grounded positive electrode is made of platinum, the negative electrode of stainless steel. Any particles from the current-free particle tube come into contact with the sheath flow, if at all, only a few microseconds before the sizing process. In this short time, cells should not be affected by electrode products as was reported by Alabaster et al. [5a].

TWO-PARAMETER CELL SIZING AND FLUORESCENCE TRANSDUCER

To perform combined cell volume and cell fluorescence investigations, the two-parameter transducer shown in Figure 36 [77,81] was developed. For the sizing portion, mainly the elements of the tubeless Metricell transducer were used. The additional cleaning flow on the grounded side of the orifice is also taken from the storage tank through an adjustable restrictor valve and a dropping chamber, to avoid an electrical short-circuit of the orifice. The volume and fluorescence signals are both taken from one orifice. Therefore, no correlation problems exist, as is true of systems where volume and fluorescence signals are taken from different places [137].

Figure 37 shows a two-dimensional cell volume and cell fluorescence histogram of blood cells from whole blood which were stained according to [155a] and measured with the two-parameter Fluvo-Metricell transducer [77] and a Cytomic 12 analyzer [79]. The x parameter shows cell volume, the y parameter fluorescence, both in log scale. Before plotting, the curve was electronically smoothed and the field is rotated to view from the back.

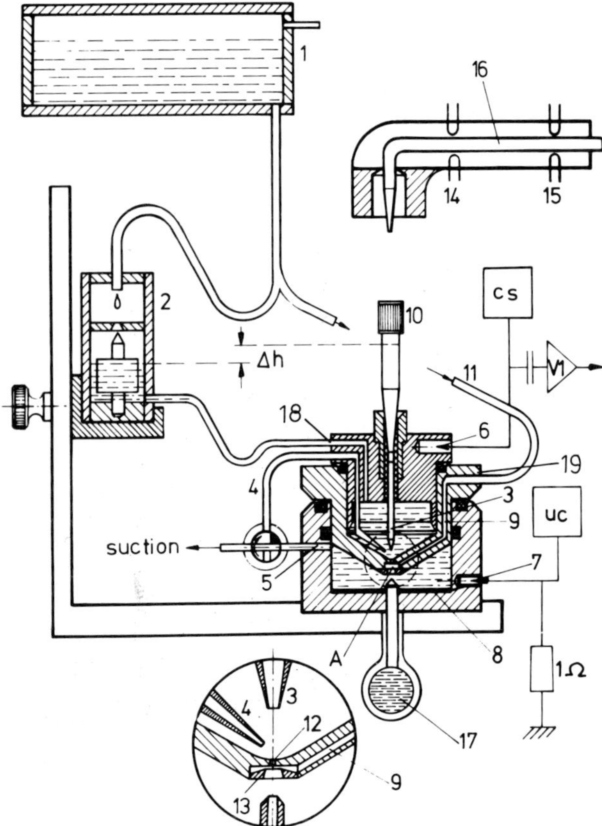

Fig. 32. Schematic diagram of the Metricell fluid system. Δh, adjusted by the regulating chamber, controls the particle flow. (1) storage tank of particle-free electrolyte, (2) regulating chamber, (3) particle tube, (4) cleaning tube, 5) suction connector, (6) electrode block, (7) ground connection with calibrator (uc) input, (8) ground electrode, (9) antirecirculation flow, (10) particle container, (11) connecting tube for antirecirculation flow, (12) orifice, (13) antirecirculation device, (14) stop light barrier, (15) start light barrier, (16) particle conting tube, (17) cleaning ball, (18) supply of sheath flow, (19) orifice holder. cs, current source; v1, pulse amplifier; uc, calibrator. The 1-ohm resistor is used to introduce the calibrator voltage.

ELECTRICAL INSTRUMENTATION

Current Source

The electric resistance R_o of the transducers (the resistance between the electrodes) is normally 5 to 50 kΩ, depending on the resistivity of the electrolyte and on the dimensions of the orifice. The natural signal limitation is due to the thermal noise generated by the resistors R_o. The lower limit of the noise signal in the orifice resistor is calculated near 10μV ($U = (4kTR_o f)$; with $R_o = 20$ kΩ, $f = 300$ kHz). It can be assumed that this noise is increased by the noise of the current transition from the electrodes to the electrolyte. The particle volume corresponding to a pulse height of 10μV is 0.08 μm, calculated from equation 15 by the assumption of a spherical particle in an orifice of 60 μm diameter and with i = 1 mA. If we intend to size particles under the indicated conditions with a lower noise limit at 1 μm of the volume scale, the current source may produce a minimum noise voltage of 20 μV on R_o of 20 kΩ, or a noise current of 10^{-6}mA. These hard conditions pertain only to the fast-oscillating or pulse-like short-time noise, and may be minimized by careful filtering of the supply voltages of the current source. The long-time stability conditions are less severe. Long-time current variations affect the pulse heights proportionally; i.e., a current source of 1% accuracy limits the pulse height accuracy to 1%. In our instrument, we use a simple constant current source equipped with field effect transistors (circuit Fig. 38). The long-time and short-time stability of the current depends on carefully filtered and regulated voltage U_B. The voltage range of U_o between the electrodes extends from 0 to 40 V with currents adjustable by R_c from 0.02 to 2 mA.

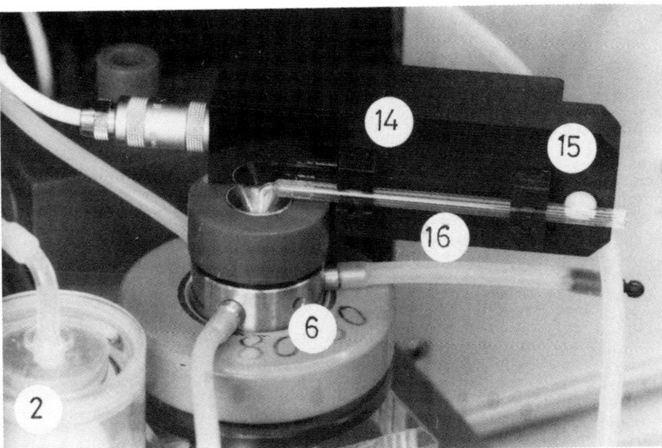

Fig. 33. Metricell transducer system. Version with the optional counting head for simultaneous sizing and counting cells. The suspension volume to be counted is determined by the two light barriers (14&15).

AMPLIFIERS, GENERAL RISETIME CONDITIONS

The rise time of amplifiers, especially of preamplifiers, should be chosen such that the signals to be amplified are transmitted without distortion. The rise times of resistance pulses can be estimated from the model pulse height patterns described below, which indicate that the pulse height at the distance of $x/d = -1$ (one orifice diameter outside the inlet of the orifice) is about 1%. From there to the point of maximal pulse height in the middle of the orifice requires a distance of about 1.5 orifice diameters with an orifice of $\ell/d = 1$. We calculate the half-spherical fluid volume enclosed by the equivelocity area that touches the 1% point at $x/d = -1$ with $U+ = 4/3\ \pi\ 8r^3/2$.

The time required to suck this fluid volume through the orifice to the middle area is approximately identical with the rise time of the pulse and is given by

$$t = \frac{U+}{v_m\, r^2 \pi} = \frac{1}{2\, v_m} \qquad (24)$$

where $v_m\ r^2\ \pi$ is the volume of fluid sucked through the orifice per time unit. $1/2v_m$ is the time required for the particle to move from the orifice inlet to the middle area of the orifice where the maximum pulse height is generated.

Assuming that $\ell/d = 1$; i.e., $\ell = 2r$, we obtain

$$t = \frac{16\, r}{3\, v_m} + \frac{r}{v_m} = 6.33\, \frac{r}{v_m} \qquad (25)$$

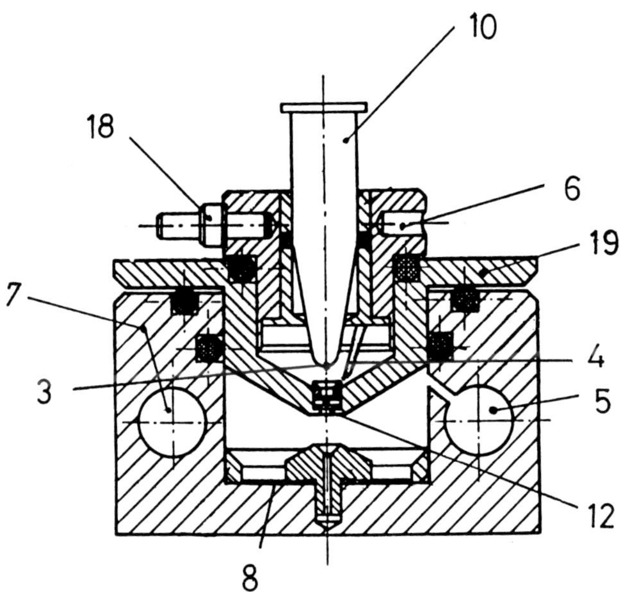

Fig. 34. A new type of tubeless transducer specially suited for kinetic experiments and quick exchange of cell samples [81]. The cell suspension is directly focused from the container 10 through the injection hole 3 into the orifice 12; 6 and 18 electrode block with electrical connection, 19 element bearing the orifice, 5 suction connector, 7 basic block, 4 sheath fluid supply, 8 ground electrode.

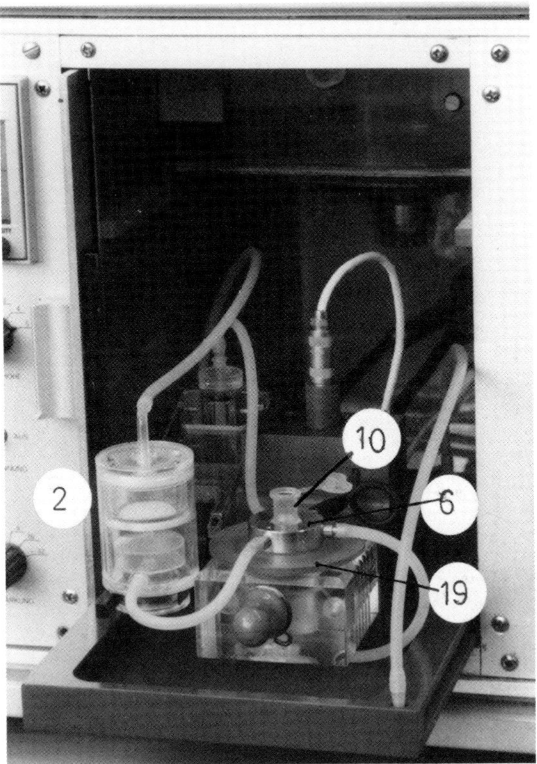

Fig. 35. New tubeless transducer. The numbers correspond to those in Figure 34.

With a 30-μm orifice ($\ell/d = 1$) and $v_m = 5$ m/sec, we obtain $t \approx 19 \mu s$. Rise times are usually defined as the time difference between 10% and 90% values of the pulse; therefore, pulse rise times under these conditions should not be below 10 μsec. With increased orifice diameter, the rise times are longer. If the amplifier rise time is faster by a factor 10 than the pulse rise time, the pulses will be amplified with negligible distortion. Hence the resistance pulse amplifiers should have rise times near 1 μsec.

PREAMPLIFIER

The preamplifier is directly coupled to the transducer and amplifies the weak resistance change signals ΔU from a range of about 0.1 to 5 mV, to U_A of about 50 to 1,000 mV. It should be composed of low-noise elements such as field effect transistors.

Two types of circuits may be used in the preamplifier stages. The amplification of the voltage amplifier $v = U_A/\Delta U$ is constant and defined by the internal components of the circuit. The frequency response plus rise time of the transducer-preamplifier system is influenced by the time constant of the resistance between the electrodes R_o, and by distorting capacitances of the first stage of the preamplifier and transducer, represented as being concentrated in C. Therefore, the rise time depends on the resistivity of the electrolyte and on the dimensions of the orifice. If the rise time and the pulse duration are the same order of magnitude, the pulse height to particle volume relation is distorted. The state-of-the-art preamplifier in resistance pulse sizing is the "current sensing" or "zero input" amplifier, schematically shown in Figure 39b. Here the output voltage U_A is fed back negatively to the input by the feedback resistor R_f. The open-loop amplification should be much higher than the intended amplification with feedback. With this condition, the amplification factor $v_{amp} = R_f/R_o$, v_{amp} increases with decreasing R_o and vice versa. This relation makes the output voltage of the transducer/current sensing amplifier system independent of variation of R_o. We obtain from equation 8

$$\Delta U = \frac{\Delta V \, \rho_2 \cdot i \cdot f \cdot \ell +}{q^2 \, \ell +} \qquad (26)$$

The resistance between the electrodes is $R_o = \rho_2 \cdot \ell + /q$, where $\ell +$ is an equivalent length which describes the real length of the orifice plus an additional fictive length of the same orifice which would be necessary to produce the resistance component of R_o from outside the actual orifice:

$$U_A = \Delta U \cdot v_{am}p = \frac{\Delta V \cdot i \cdot f \cdot R_f}{q \, \ell +} \qquad (27)$$

R_o is canceled; i.e., U_A is independent of R_o and of the resistivity of the electrolyte. By contrast, v_{amp} is a measure of the resistance between the electrodes independent of polarization like AC devices. This feature is used in the calibrating control device of our Metricell instrument described below. The second advantage of the zero-input amplifier is that, due to the negative feedback, the input voltage of the amplifier is kept near zero, and thus the distorting capacitances C_D cannot affect the rise-time response.

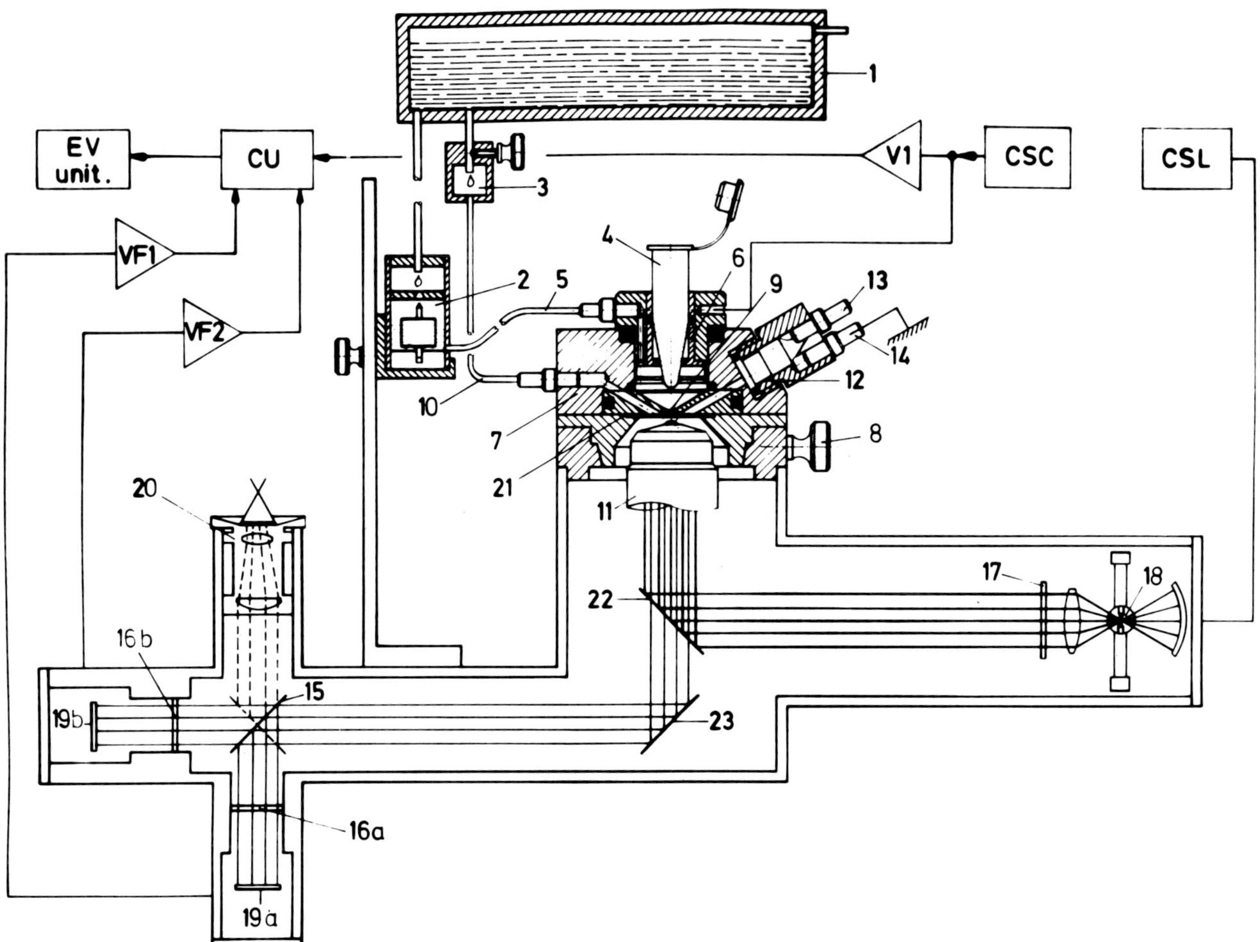

Fig. 36. Schematic diagram of a three-parameter Fluvo–Metricell transducer equipped with a tubeless particle injection. The transducer is able to measure simultaneously electrical volume and two fluorescences [81]. (1) storage tank of sheath fluid, (2) dropping chamber acting as control element, which adjusts the flow of suspension from vessel (4) into the sensing orifice, (3) dropping chamber isolating and adjusting the cleaning flow, (4) vessel containing and supplying the particle suspension, (5) supply of sheath fluid, (6) fitting for the particle container, (7) transducer block, (8) horizontal adjustment screw, (9) sensing orifice, (10) supply of cleaning fluid, (11) epi-illumination objective, (12) ground electrode, (13) suction connection, (14) ground connector, (15) dichroic mirror selecting the wavelength of the two fluorescence channels together with the two filters 16a,16b; (17) excitation filter of the high pressure lamp light source (18); (19a,b) photomultipliers for the two fluorescence channels; (20) eyepiece for controlling the orifice; (21) coverglass; (22) dichroic mirror reflecting the excitation light but being transparent for the fluorescence light emitted by the cells; (23) mirror; (CSL) current supply for the excitation lamp; (CSC) current source for electrical sizing; (CU) control unit preparing the signals from the amplifiers V1, VF1, and VF2 for the evaluation unit EV.

MAIN AMPLIFIER WITH PULSE HANDLING DEVICES

The main amplifier amplifies the pulse from the level of the preamplifier output to the input level of the evaluating unit. The amplification can be achieved by high-speed operational amplifiers, which are commercially available [159]. The time constants should be chosen such that the linear relation between pulse height and particle volume is not distorted.

A very useful device is a base line restorer if widespread pulse heights with high pulse rates are measured. The preamplifier is normally coupled to the transducer by a capacitor. The differentiating effect of such coupling capacitors produces a considerable undershoot, especially with high pulses. After such an undershoot, the AC-coupled base line needs a definite time to revert to the true zero level. All pulses that occur during this restoring time are evaluated with a long base line (Fig. 40). To shorten the duration of the undershoots, so-called base line restoration circuits are used. Such restorers specially developed for nuclear pulse height measurements are described in several publications [25, 111,119].

ANALOG-TO-DIGITAL CONVERSION AND EVALUATING UNITS

Modern data storing and processing units work digitally i.e. the analog pulses provided by the transducer have to be converted into digital signals. Pulse processing can be done by evaluation of pulse amplitude, pulse length or pulse area. Since resistance pulses represent cell volume best in their maximum amplitudes as is shown in the initial sections,

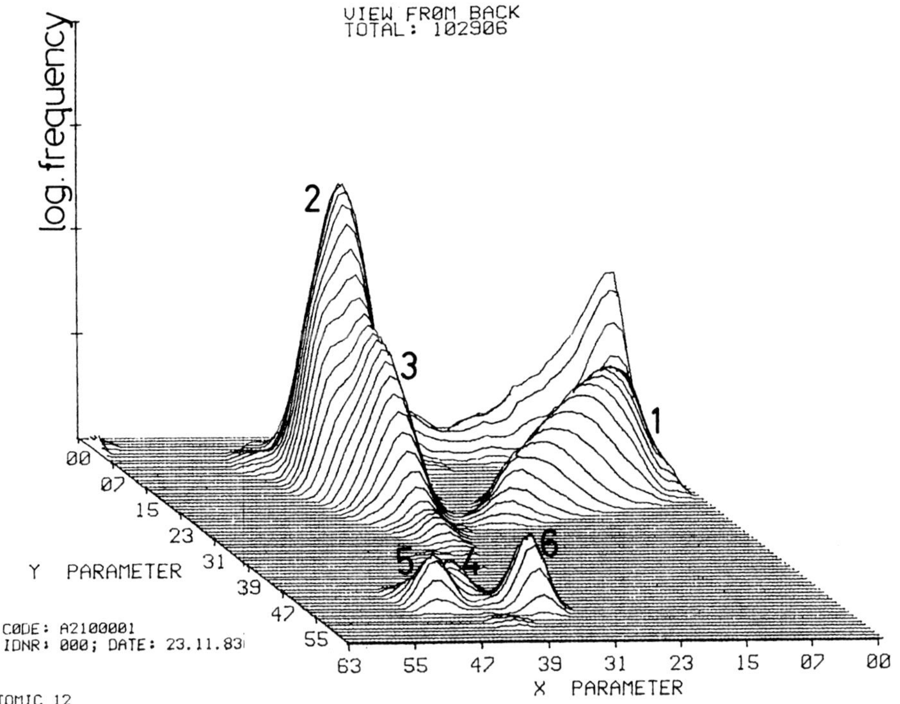

Fig. 37. Example of a two parameter distribution of blood cells. On the x-axis log cell volume, on y-axis log fluorescence, z-axis log frequency. The cells were stained according to (155a). Each axis has a resolution of 64 channels. The complete two parameter field which was electronically smoothed before plotting consists of 4096 channels. (1) thrombocytes, (2) erythrocytes, (3) reticulocytes, (4) lymphocytes, (5) granulocytes, (6) calibrator particles.

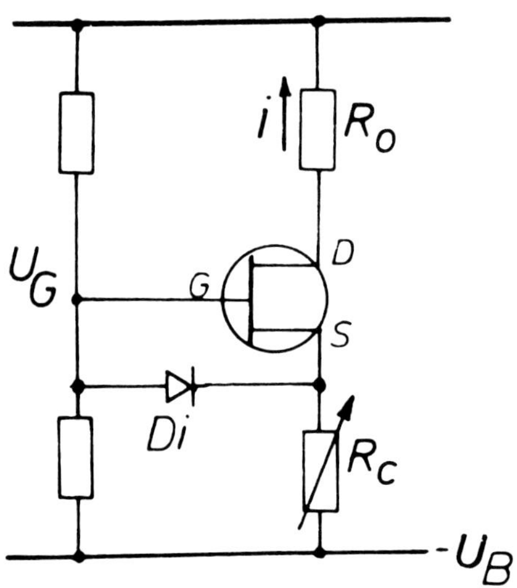

Fig. 38. Basic field effect transistor circuit of the constant current source; R_o resistance between the electrodes, $-U_B = 100$ V; gate voltage $U_G < -40$ V. The current is adjusted by R_C.

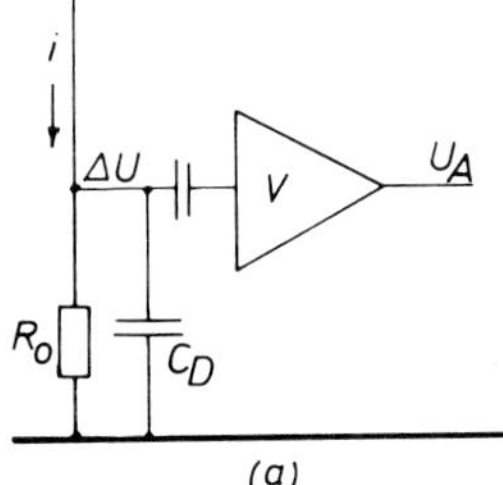

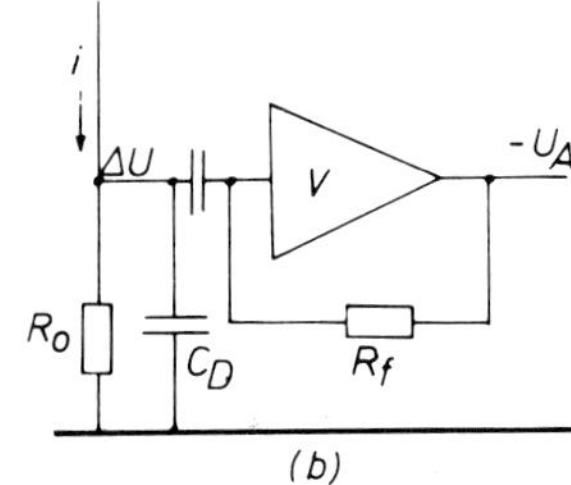

Fig. 39. Schematic circuits of preamplifier-orifice systems. R_o resistance between the electrodes, C_D capacity describing all distorting capacities. (a) Voltage amplifier, amplification $v = U_A/\Delta U$ is constant. (b) Current sensing amplifier, amplification $v = -U_A/\Delta U = R_f/R_o$ depends on R_o. U_A has the inverted polarity of ΔU.

pulse amplitude evaluation is the method of choice in resistance pulse sizing. Pulse length and pulse area evaluation may be distorted by change of flow velocities. The problems of data conversion, data storing and processing are treated in more detail in Chapter 22 (this volume).

The block circuit of Figure 41 gives an overview of the elements of flow cytometric data processing equipment. The amplified pulses are controlled by a user adjustable trigger circuit, which decides whether a pulse is accepted for evaluation. The accepted pulses are shown on an oscilloscope screen and counted by a rate meter in order to define the number of pulses accepted/time. The pulse rate may be used to estimate the coincidence probability. The peak detector following the trigger unit prepares the pulse amplitude for the analog to digital conversion by holding the pulse maximum until the analog-to-digital converter (ADC) has finished the conversion.

In modern instruments, the pulse height analyzer portion and computer portion are mostly one system that accepts the digital values, builds histograms that may be mathematically treated, numerically evaluated, stored on disk or tape, or printed or plotted.

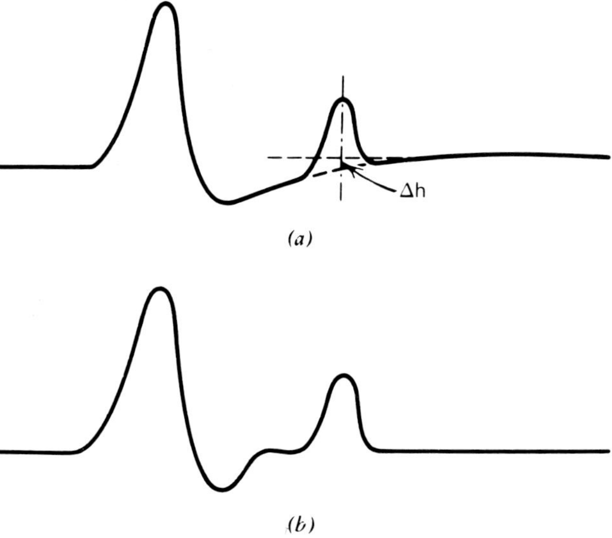

Fig. 40. Effect of baseline restoration. (a) A big pulse followed by a small pulse without baseline restoration, the small pulse is underevaluated by Δh. (b) A big pulse followed by a small pulse with baseline restoration, both pulses are correctly evaluated.

The accuracy of resistance pulse height measurements at present is not better than 1%. Therefore, a memory of 128 channels, or at maximum 256 channels for each histogram, is sufficient. More channels would provide more data to be handled and stored without a real increase of resolution and accuracy.

CALIBRATION OF ELECTRICAL SIZING INSTRUMENTS

Basically, there are two ways of calibrating: (1) the common method of calibration with particles of known volume, and (2) calibration with electrical pulses simulating resistance pulses of known height without the use of calibrating particles.

CALIBRATION WITH PARTICLES

It has already been shown how the particle volume and the pulse height are related and which parameters may distort an accurate result. If calibrating particles of the same shape, the same electrical resistivity, and an exactly known volume in the same range as the particles to be sized were available at low price, particle calibration would be ideal. But usually the correspondence of these parameters is not given, and conversion factors between the measured particles and the calibrating particles must first be determined by other methods. In addition, the commercially available uniform particles are very expensive, and it is difficult to verify their indicated volume specifications. If the suspension to be calibrated may not be contaminated, an additional calibrating measurement has to be done for each calibration.

ELECTRICAL CALIBRATION

Use for Actual Calibration

The considerations discussed above influenced us to develop a calibration method that was independent of calibrating particles. The concept of this electrical calibration was to simulate the volume pulses of particles quantitatively by electronic pulses, which are coupled to the transducers in series connection to the orifice (Fig. 42).

This new way of applying the calibrating pulses permits calibration during the actual sizing process. The uniform calibrator pulses generate calibrator peaks in the histogram. In equation (15) we introduce the pulse height of the calibrator pulses $\Delta U = i\Delta R$ instead of the voltage pulses generated by the particles. The shape and conductivity factor f_e or f_s, and the orifice factor f_k, must first be determined by the methods described above. This calculation attributes an absolute volume to the calibrator peak.

In practice, two calibrator peaks of known pulse height are

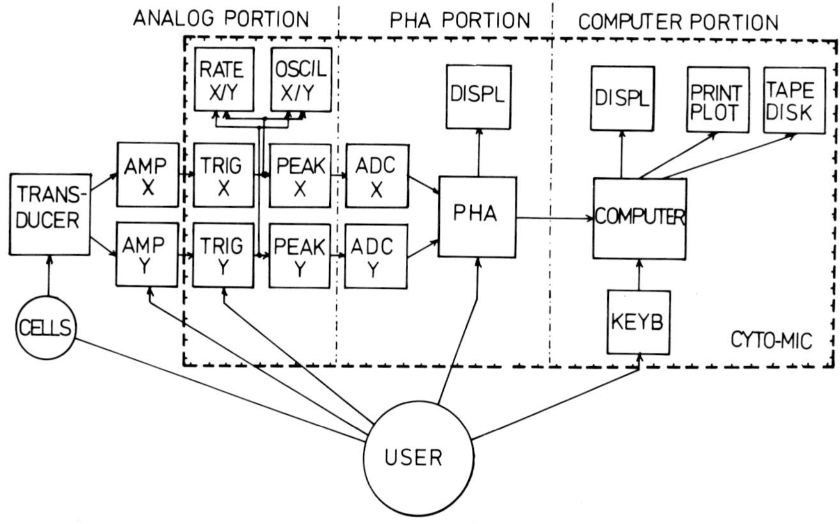

Fig. 41. Block diagram of the elements of a two parameter evaluation system consisting of an analog portion, a pulse height analyzer (PHA) portion and a computer portion for numerical evaluations and long time storing of data sets. The elements inside the broken line can be enclosed in a single compact data system [79]. TRIG = trigger device; PEAK = pulse peak detector; ADC = analog to digital converter.

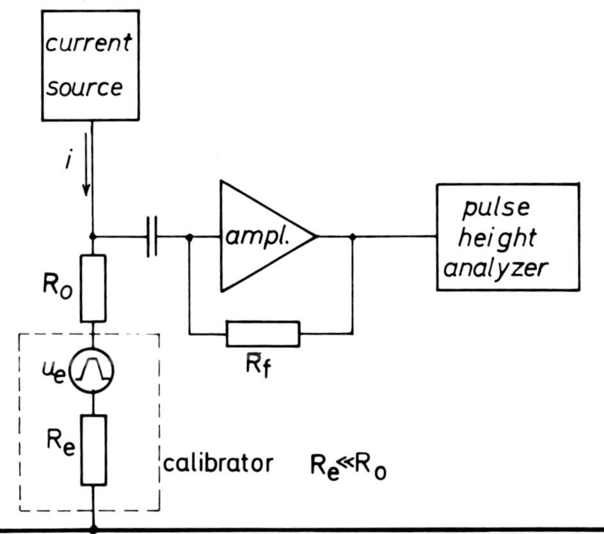

Fig. 42. Principle of feeding in the trapezoidal calibrator pulses into the orifice-transducer system. $R_e = 1$ Ohm. The calibrator pulses in the range of 0.5 – 5 mV are serially introduced via the calibrator resistor R_e. (See also Fig. 32.)

introduced into each volume distribution curve to be calibrated. By application of the equation of a straight line with these two points, a volume scale is generated, which is independent of the zero calibration of the pulse height analyzer:

$$\frac{\Delta V - V1}{X - X1} = \frac{V2 - V1}{X2 - X1} \tag{28}$$

X any point of interest in the volume distribution in channels
ΔV corresponding cell volume in the point of interest X
X1 mean of calibrator peak 1 in channels
X2 mean of calibrator peak 2 in channels
V1 cell volume corresponding to calibrator peak 1
V2 cell volume corresponding to calibrator peak 2

V1 and V2 are determined by introducing the calibrator voltages U1 and U2 into equation (19).

We obtain for the cell volume:

$$\Delta V = KX - KX1 + V1 \tag{29}$$

with

$$K = \frac{V2 - V1}{X2 - X1} \tag{30}$$

If in an offset free system instead of calibrator peak 1 the coordinate origin X1 = V1 = 0 is used, equation (29) is simplified to:

$$\Delta V = (V2/X2)\, X \tag{31}$$

With this calibration, the diameter of the orifice should be determined with high accuracy, because the radius of the orifice is involved in the calculation by r^4.

Figure 43 shows a volume distribution curve of human erythrocytes with two calibrator peaks inserted. Calibration by this method shows good reproducibility. The experimentally determined absolute volumes of native erythrocytes are about 10% smaller than reference values.

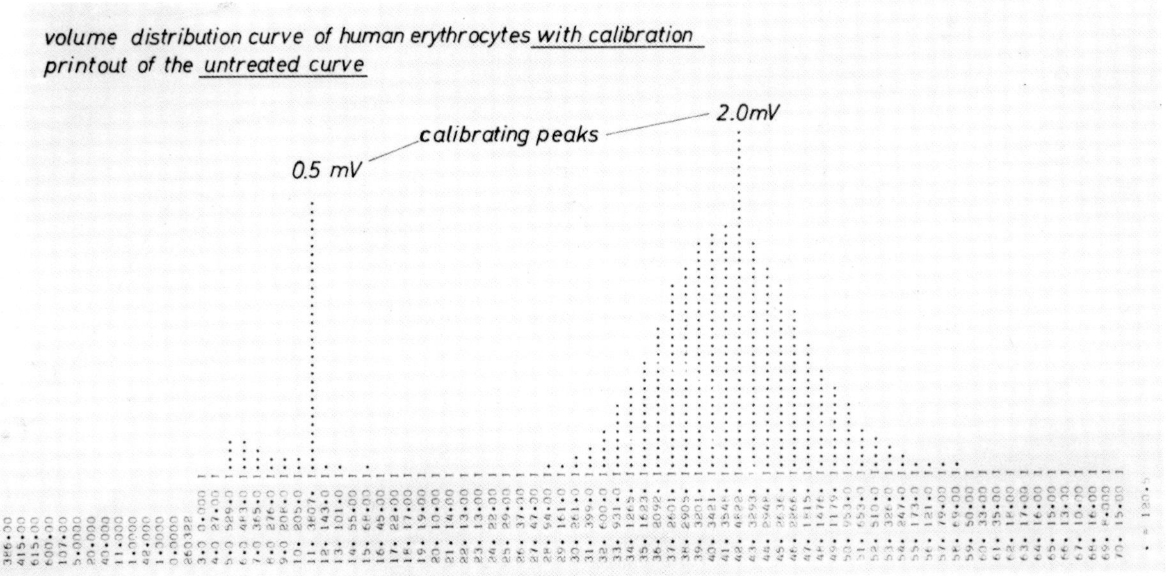

Fig. 43. Volume distribution curve of native human erythrocytes demonstrating the electrical calibration. Calibrator peaks in channel 11 (0.5 mV) and 42 (2 mV), which can be used for calculation of an absolute particle volume scale as described in the text. The coefficient of variation of the erythrocyte distribution is 12.2%. With orifice diameter 60 μm, electrolyte resistivity 61.5 Ω·cm, current i = 0.386, fk/fe = 1.37, the calibrator peak 0.5 mV corresponds with a cell volume of 23.06 μm and 2 mV with 92.2 μm. With the two calibrator points a linear volume scale can be calculated.

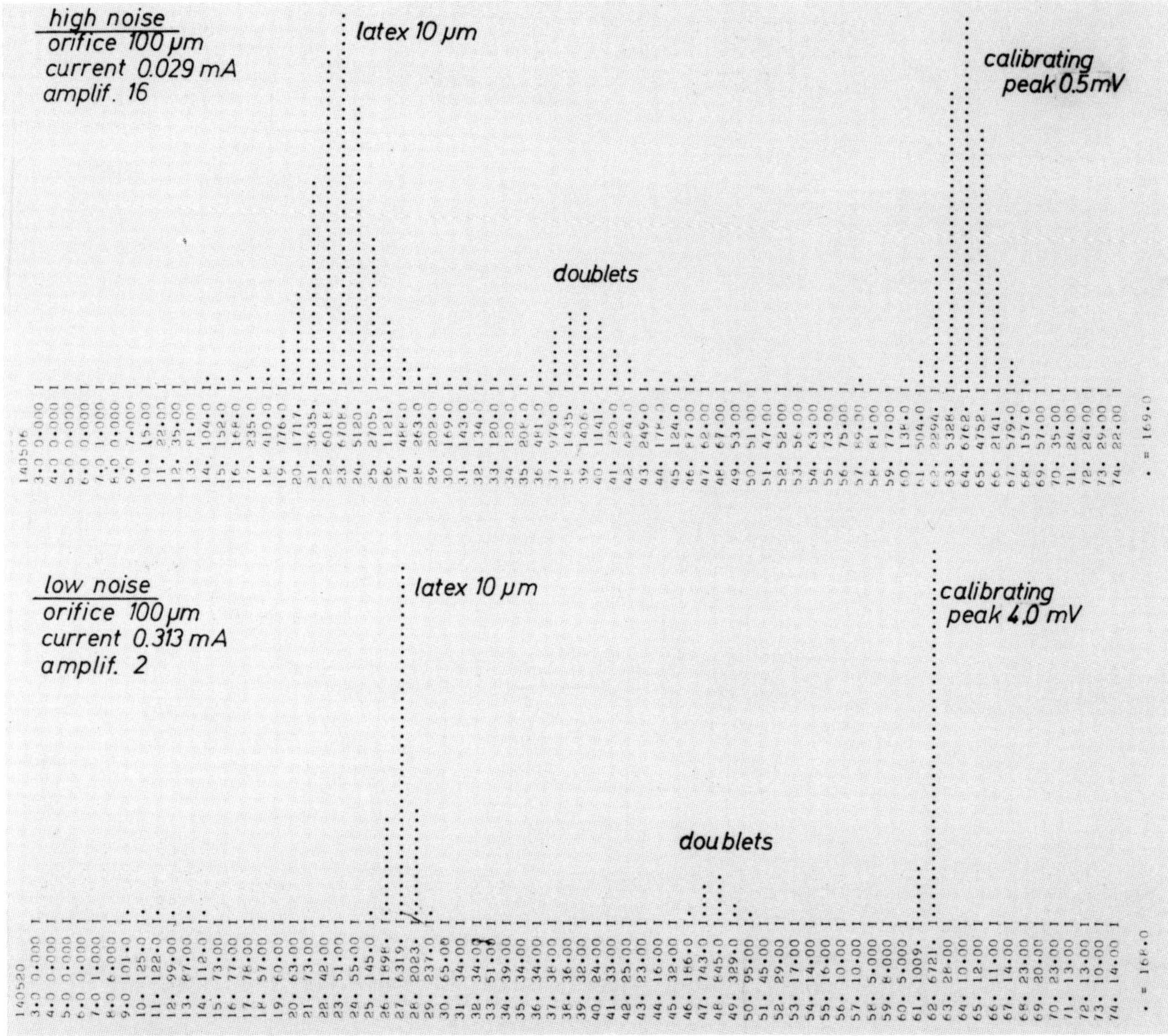

Fig. 44. Noise detection by means of electrical calibrator pulses. Upper: high noise level broadens all peaks, as revealed by wide calibrator pulse peak (C.V. latex = 8.36%,C.V. calibrator pulses 2.09%). Lower: low noise level as revealed by narrow calibrator pulse peak (C.V. latex = 2.85%, C.V. calibrator pulses = 0.54%).

USE OF THE CALIBRATOR AS A CONTROL UNIT

Analysis of Superimposed Noise

The described calibration method opens a way to recognize noise functions superimposed on the actual particle distribution. Because a priori uniform calibrating pulses are applied in a serial connection with the orifice, the noise function is also superimposed on these pulses. Therefore, the distribution curve of the calibrator pulses represents the noise function superimposed on the particle distribution. In Figure 44a, the size distribution of latex particles (diameter 10 μm) measured with relatively high noise, and the same particles measured with low noise are compared. The calibrating peak includes the noise function, which is also superimposed on the latex curve. This method detects errors that affect the base line of the sizing device: electric noise generated inside or outside the instrument, gas bubble motion, and so on. Distortions produced by particles moving on different paths in the orifice cannot be detected.

Control of the Effective Diameter of the Orifice

The fact that the amplification of the current sensing preamplifier of the Metricell depends on the ratio of the feedback resistor to the resistance between the electrodes is used for controlling the effective dimensions of the orifice or the resistivity of the electrolyte. If the resistivity of the electrolyte is changed, or if the orifice is partially blocked, the channel number of the calibrating peak is thereby changed. The ratio of the channel numbers corresponds to the ratio of the resistivities (Fig. 45).

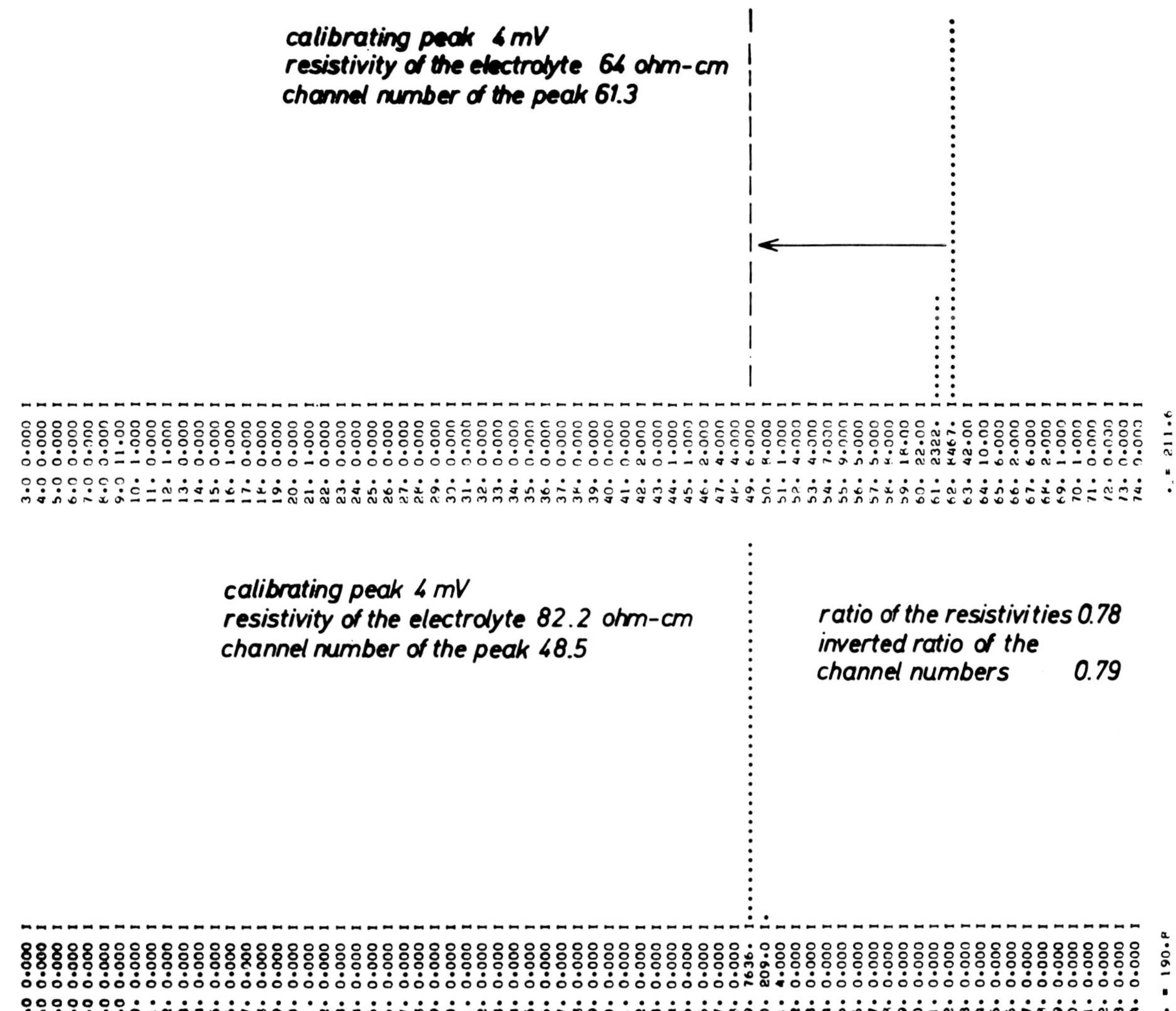

Fig. 45. Electrical calibrator senses electrolyte resistivity. The ratio of the resistivities of different electrolytes is proportional to the inverted ratio of the channel numbers of a distinct calibrator peak. Here ratio of the resistivities 0.78, inverted ratio of the channel numbers 0.79.

CONCLUSION

The resistance pulse technique has been known for many years. Though attempts were made to clarify the basic events in the Coulter orifice, sizing measurements by this technique were accepted with reservation for a long time because the phenomenon was not fully understood. Now the mathematical relationship between particle volume and pulse height and the effect of shape and conductivity of the particles on the measurements is known, and means for direct optical control of the behavior of the particles in the orifice are available; we know the precise distribution of the electrical potential and field strength inside and outside the orifice and the relation of pulse height to field strength. This theoretical background, together with a precise focusing technique, enables every user of the resistance pulse technique to obtain accurate and highly reliable size distribution curves of cells and other particles, and to interpret them in the right manner.

ACKNOWLEDGMENTS

I thank Professor Ruhenstroth-Bauer for promoting the basic investigations described in this article, Dr. N. Boss, Dr. E. Menke, and Dr. G. Valet for many helpful discussions, and Dr. Plies for his aid in the application of the field calculation computer program. The ingenious technical assistance of E. Glossner, H. Schneider, and E. Kordwig is gratefully acknowledged.

NOMENCLATURE

a	long axis of the ellipsoids
A	cross sectional area of the unfocused particle path
α	q/ℓ
B	area of the focused particle path in the orifice
b	short axis of ellipsoids of revolution
β	particle density N/vm
d	diameter of the orifice

E	field strength
E_h	equivalent homogeneous field strength
E_n	natural field strength
E_{rel}	field strength related to the field strength in the equivalent homogeneous field
f_e	shape and conductivity factor of ellipsoid-like cells
f_s	shape and conductivity factor of spherical cells
f_k	factor describing the influence of the orifice geometry
i	constant electric direct current through the orifice
j	current density
j,k	indices
k	Boltzmann constant
ℓ	length of the orifice
$\ell+$	fictive length of the orifice
N,n	number of particles or events
P	ΔV/V ratio of particle volume to orifice volume
p()	probability
q	cross section of the orifice
r	radius of the orifice
R	electric resistance of the orifice with particle
R_o	electric resistance between the electrodes
R_2	electric resistance of the orifice without a particle in it
ΔR_2	resistance change of the orifice caused by a particle
R_f	feedback resistor
R_e	calibrator resistor
ρ	specific resistivity of the suspension
ρ_1	specific resistivity of the particles
$\rho_{2,3}$	specific resistivities of electrolytes without particles
s	distance from the orifice inlet
t	time rise time
T	temperature
U	voltage
ΔU	pulse height in volts produced at the electrodes by a particle passing through the orifice
U+	fluid volume
U_A	output voltage of the amplifier
v_m	mean flow velocity
v_{amp}	amplification
V	volume of the orifice
ΔV	volume of the particle
ab^2 La	expression defined in equation 9

REFERENCES

1. **Adams RB, Voelker WH, Gregg EC (1967)** Electrical counting and sizing of mammalian cells in suspension. An experimental evaluation. Phys. Med. Biol. 12:79–92.
2. **Adams RB (1968)** Particle stream position effects in electrical sizing. Biophys. Soc. Abstr. 8:A122.
3. **Adams RB, Gregg EC (1972)** Pulse shapes from particles traversing Coulter orifice fields. Phys. Med. Biol. 17:830–842.
4. **Adell R, Skaklak R, Branemark PJ (1970)** A preliminary study of rheology of granulocytes. Blut 21:91–105.
5. **Akeson SP, Mel MC (1983)** Erythrocyte and ghost cytoplasmic resistivity and voltage-dependent apparent size. Biophys. J. 44:397–403.

5a. **Alabaster O, Glaubiger DL, Hamilton VT, Bentley SA, Shackney SE, Skramstad RS, Chen RF (1980)** Electrolytic degradation of DNA fluorochromes during flow cytometric measurement of electronic cell volume. J. Histochem. Cytochem. 28:330–334.

6. **Allan RS, Mason SC (1962)** Particle behavior in shear and electric fields. I. Deformation and burst of fluid drops. Proc. R. Soc. A267:45–76.
7. **Anderson JL, Quinn JA (1971)** The relationship between particle size and signal in Coulter-type counters. Rev. Sci. Instr. 42:1257–1258.
8. **Arvola L (1984)** A comparison of electronic particle counting with microscopic determinations of phytoplankton and chlorophyll A concentrations in 3 Finnish lakes. Ann. Bot. Fenn. 21:171–178.
9. **Atkinson CM, Wilson R (1983)** Artifact peaks in particle-size distributions measured by the electrical sensing zone (Coulter Counter) method. Powder Technol. 34:275–284.
10. **Bator JM, Groves MR, Price BJ, Eckstein EC (1984)** Erythrocyte deformability and size measured in a multiparameter system that includes impedance sizing. Cytometry 5:34–41.11a.
11. **Bauer J, Valet G (1983)** Cell-volume and osmotic properties of erythrocytes after complement lysis measured by flow-cytometry. J. Immunol. 130:839–844.

11a. **Becton Dickinson FACS Analyzer.** Becton Dickinson, Mountain View, CA.

12. **v. Behrens W, Edmondson S (1976)** Comparison of techniques improving the resolution of standard Coulter cell sizing systems. J. Histochem. Cytochem. 24:247–256.
13. **Ben Sasson S, Patinkin D, Grover NB, Doljanski F (1974)** Electrical sizing of particles in suspension. IV. Lymphocytes. J. Cell Physiol. 84:205–213.
14. **Boss N, Chmiel H, Kachel V, Ruhenstroth-Bauer G (1973)** Erythrocytenaggregation bei Nichtrauchern, Rauchern und Herzinfarkt-patienten. Blut 27:191–195.
15. **Boss N, Koenig S, Ruhenstroth-Bauer G (1975)** Die Erythrocyten-aggregation bei Menschen mit Risikofaktoren eines Herzinfarkts. Klin. Wochenschr. 53:385–389.
16. **Brecher, G, Jakobek EF, Schneidermann MA, Williams GZ, Schmidt PJ (1962)** Size distribution of erythrocytes. Ann. N.Y. Acad. Sci. 99:242–261.
17. **Breitmeyer MO, Lightfoot EN, Dennis WH (1971)** Model of red blood cell rotation in the flow toward a cell sizing orifice. application to volume distribution. Biophys. J. 11:146–157.
18. **Brotherton J (1975)** The counting and sizing of spermatozoa from 10 animal species using a Coulter counter. Andrologia 7:169–185.
19. **Buckhold BM, Murphy YR, Adams RB, Steidley KD (1969)** Erythrocyte deformability and the bimodal volume distribution. Biophys. Soc. Abstr. 8:A113–A113.
20. **Bull BS (1968)** On the distribution of red cell volumes. Blood 31:503–515.
21. **Cassidy M, Fowlkes BJ, Herman CJ (1975)** Electronic cell volume (Coulter volume) distribution of vaginal–cervical cytology samples. Acta Cytol. (Praha) 19:117–125.

22. **Chase RL, Poulo LR (1967)** A high precision dc restorer. IEEE Trans. Nucl. Sci. 167:83–88.
23. **Chaussy L, Baethmann A, Lubitz W (1981)** Electrical sizing of nerve and glia cells in the study of cell volume regulation. In Cervos-Navarro J, Fritschka E (eds), "Cerebral Microcirculation and Metabolism." New York: Raven Press, pp 29–40.
24. **Coulter WH:** Means for counting particles suspended in a fluid. U.S. Patent No. 2,656,508. Issued 1953.
25. **Coulter WH (1956)** High speed automatic blood cell counter and cell size analyzer. Proc. Natl. El. Conf. 12:1034–1040.
26. **Coulter WH, Hogg WR, Moran JP, Claps WA (1959)** Particle analyzing device. U.S. Patent No. 3,259,842.
27. **Coulter WH (1966)** Manual to Coulter Counter.
28. **Coulter Electronics (1971)** Deutsche Patentschrift 2,153,123 (Corresponds to the U.S. Patents 84, 440,101,352,113,165,113,920).
29. **Coulter Electronics (1972)** Deutsche Patentschrift 2,216,826 (Corresponds to the U.S. Patents 132, 771,142,531).
30. **Cutts JH (1970)** Balanced salt solutions. In Cutts JH (ed), "Cell Separation Methods in Hematology." New York: Academic Press, pp 169–174.
31. **Davies R, Karuhn R, Graf J (1975)** Studies on the Coulter counter. Part II. Investigations into the effect of flow direction and angle of entry of a particle volume and pulse shape. Powder Technol. 12:157–166.
32. **DeBloise RW, Bean CP (1970)** Counting and sizing of submicron particles by the resistive pulse technique. Rev. Sci. Instr. 41:909–916.
33. **DeBloise RW, Mayyasi SA, Schildlovsly G, Wesley R, Wolff JS (1974)** Virus counting and analysis by the resistive pulse (Coulter counter) technique. Proc. Am. Assoc. Cancer Res. 15:104–104.
34. **Dieckmann O, Hejmans HJ, Thieme HR (1984)** On the stability of the cell- size distribution. J. Math. Biol. 19:227–248.
35. **Diederich G, Krueger U, Haller H (1984)** Increased measuring sensitivity of Laboscale—A step forward in the exact conductometric counting and volume measurement of thrombocytes. Dtsch. Gesund. 39: 908–913.
36. **Van Dilla MA, Fulwyler MJ, Boone JU (1967)** Volume distribution and separation of normal human leukocytes. Proc. Soc. Exp. Biol. Med. 125:367–370.
37. **Doljanski F, Zajicek G, Naaman J (1966)** The size distribution of normal human red blood cells. Life Sci. 5:2095–2104.
38. **Eck B (1961)** "Technische Stroemungslehre." Berlin: Springer.
39. **Evans VJ, Glasser L (1981)** Accuracy of low electronic platelet counts using platelet distribution curves. Am. J. Med. Technol. 47:15–18.
40. **Ferris CD (1963)** Four-electrode electronic bridge for electrolyte impedance determinations. Rev. Sci. Instr. 34:109–111.
41. **Fricke H (1924)** A mathematical treatment of the electric conductivity and capacity of disperse systems. Phys. Rev. 24:575–587.
42. **Fricke H (1953)** The Maxwell-Wagner dispersion in a suspension of ellipsoids. J. Phys. Chem. 57:934–937.
43. **Freyer JP, Wilder ME, Raju MR (1984)** Coulter volume cell sorting to improve the precision of radiation survival assays. Radiat. Res. 97:608–614.
44. **Fulwyler MJ (1965)** Electronic separation of biological cells by volume. Science 150:910–911.
45. **Gebicki JM, Hunter FE (1964)** Determination of swelling and disintegration of mitochondria with an electronic particle counter. Biol. Chem. 293: 631–639.
46. **Wissenschaftliche Tabelle (1975)** (Geigy AG, Basel ed.) Stuttgart: Thieme Verlag.
47. **Goehde K, Goehde W (1972)** Deutsche Patentschrift 2.120.342.
48. **Goldsmith HL (1971)** Deformation of human red blood cells in tube flow. Biorheology 7:233–242.
49. **Golibersuch DC (1973)** Observation of asperical particle rotation in Poiseuille flow via the resistance pulse technique. Biophys. J. 13:265–280.
50. **Grant JL, Britton MC, Kurtz TE (1960)** Measurement of red blood cell volume with the electronic cell counter. Am. J. Clin. Pathol. 33:138–143.
51. **Gregg EC, Steidley KD (1965)** Electrical counting and sizing of mamalian cells in suspension. Biophys. J. 5:393–405.
52. **Grinstein S, Goetz JD, Furuya W, Rothstein A, Gelfand EW (1984)** Amiloride-sensitive Na^+-H^+ exchange in platelets and leukocytes—Detection by electronic cell sizing. Am. J. Physiol. 247:239–298.
53. **Grover NB, Naaman J, Ben Sasson S, Doljanski F (1969)** Electrical sizing of particles in suspension. I. Theory. Biophys. J. 9:1398–1414.
54. **Grover NB, Naaman J, Ben Sasson SF, Doljanski F, Nadev E (1969)** Electrical sizing of particles in suspension. II. Experiments with rigid spheres. Biophys. J. 9:1415–1425.
55. **Grover NB, Naaman J, Ben Sasson S, Doljanski F (1972)** Electrical sizing of particles in suspension. III. Rigid spheroids and blood cells. Biophys. J. 12: 1099–1117.
56. **Gunter RC Jr, Bamberger S, Valet G, Grossin M, Ruhenstroth-Bauer G (1978)** The trajectories of particles suspended in electrolytes under the influence of crossed electric and magnetic fields. Biophys. Struct. Mech. 4:87–95.
57. **Gutmann J (1966)** Elektronische Verfahren zur Ermittlung statistischer Masszahlen einiger medizinisch wichtiger Daten. Elektromedizin 11:62–79.
58. **Haigh GT (1973)** Current normalizer for particle size analysis apparatus. U.S. Patent No. 3,745,455.
59. **Hanser H, Valet G, Boss N, Ruhenstroth-Bauer G (1974)** Origin and regulation of the different erythrocyte volume populations in the newborn rat. XV Congr. Int. Soc. Hematology, Jerusalem, p. 318.
60. **Harvey RJ, Marr AG (1966)** Measurement of size distributions of bacterial cells. J. Bacteriol. 92:805–811.
61. **Harvey RJ (1968)** Measurement of cell volumes by electric sensing zone instruments. Methods Cell Physiol. 3:1–23.
62. **Harvey RJ (1969)** Effect of transducer length on volume measurements by electric sensing zone instruments. Rev. Sci. Instr. 40:1111–1112.
63. **Haskill S, Becker S, Johnson T, Marro D, Nelson K, Propst RM (1983)** Simultaneous 3-color and electronic cell-volume analysis with a single UV excitation source. Cytometry 3:359–366.
64. **Haynes JL (1980)** High-resolution particle analysis—Its application to platelet counting and suggestions

for further application in blood cell analysis. Blood Cells 6:201–213.

65. **Hazelton BJ, Torrance PM, George SL, Houghton PJ (1984)** Cloning Efficiency of cultured human-tumor cell-lines measured with the use of a Coulter particle counter. J. Natl. Cancer Inst. 73:555–563.

66. **Helleman PW (1972)** "The Coulter Particle Counter." De Bilt, Holland.

66a. **Heumann R, Kachel V, Thoenen H (1983)** Relationship between NGF-mediated volume increase and "priming effect" in fast and slow reacting clones of PC12 Pheochromocytoma cells. Exp. Cell Res. 145: 179–190.

67. **Howard RB, Pesch LA (1968)** Respiratory activity of intact isolated parenchymal cells from rat liver. J. Biol. Chem. 243:3105–3109.

68. **Hurley J (1970)** Sizing particles with a Coulter counter. Biophys. J. 10:74–79.

69. **Kachel V, Metzger H, Ruhenstroth-Bauer G (1970)** The influence of the particle path on the volume distribution curves according to the Coulter method. Z. Ges. Exp. Med. 153:331–347.

70. **Kachel V (1970)** Measuring chamber for cell volume according to Coulter. XIII Int. Congr. Hemat. Muenchen Abstr. 392–392.

71. **Kachel V (1972)** Methods of analysis and correction of operative errors in the electronic method of Coulter for particle sizing. Thesis Technische Universitaet Berlin D 83.

72. **Kachel V (1973)** The improvement of resolution in Coulter particle sizing by an electronic method. Blut 27:270–274.

73. **Kachel V (1974)** Methodology and results of optical investigations of form-factors during the determination of cell volumes according to Coulter. Microsc. Acta 75:419–428.

74. **Kachel V, Glossner E, Kordwig E (1974)** Vorrichtung zur Messung bestimmter Eigenschaften in einer Fluessigkeit suspendierter Partikel. German Patent 24, 620,63.1.

75. **Kachel V (1975)** An improved device according to Coulter to measure volumes of cells and particles equipped with a particle independent calibrating system. Biomed. Tech. 20(suppl. Vol. Mai):191–192.

76. **Kachel V (1976)** Basic principles of electrical sizing of cells and particles and their realization in the new instrument Metricell. J. Histochem. Cytochem. 24: 211–230.

77. **Kachel V, Glossner E, Kordwig E, Ruhenstroth-Bauer G (1977)** Fluvo-Metricell, a combined cell volume and cell fluorescence analyzer. J. Histochem. Cytochem. 25:804–812.

78. **Kachel V, Glossner E (1977)** Vorrichtung zur Messung bestimmter Eigenschaften in einer Partikelsuspension suspendierter Partikel. German Patent 27, 504, 47.8.

79. **Kachel V, Schneider H, Schedler K (1980)** A new flow cytometric pulse height analyzer offering microprocessor controlled data acquisition and statistical analysis. Cytometry 1:175–192.

80. **Kachel V (1982)** Sizing of cells by the electrical resistance pulse sizing technique. Methodology and application in cytometric systems. In Catsimpoolas N (ed), "Cell Analysis." Vol. 1, New York: Plenum, pp. 195–331.

81. **Kachel V, Glossner E, Schneider H (1982)** A new flow cytometric transducer for fast sample throughput and time resolved kinetic studies of biological cells and other particles. Cytometry 3:202–212.

82. **Kachel V, Schedler K, Schneider H, Haack L (1984)** "Cytomic" data system modules—Modern electronic devices for flow cytometric data handling and presentation. Cytometry 5:299–303.

83. **Karuhn R, Davies R, Kaye BH, Clinch MJ (1975)** Studies on Coulter counter. Part 1: Investigations into the effect of orifice geometry and flow direction on the measurement of particle volume. Powder Technol. 11:157–171.

84. **Kim YR, Ornstein L (1983)** Isovolumetric sphering of erythrocytes for more accurate and precise cell-volume measurement by flow-cytometry. Cytometry 3:419–427.

85. **Koller A:** POT 123 Numerische Berechnung von Potentialfeldern. Data Praxis, Siemens Publication, Bereich Datenverarbeitung, Muenchen.

86. **Kubitschek HE (1958)** Electronic counting and sizing of bacteria. Nature (Lond.) 182:234–235.

87. **Kubitschek HE (1960)** Electronic measurement of particle size. Research (Lond.) 13:128–135.

88. **Langhaar HL (1942)** Steady flow in the transition length of a straight tube. J. Appl. Mech. 9:A55–A58.

89. **Leeuwenhoek A (1674)** Microscopical observation concerning blood, milk, bones, the brain, spitle and cuticula. Phil. Trans. 121–130.

90. **Leif RC, Thomas RA (1973)** Electronic cell-volume analysis by use of AMAC I transducer. Clin. Chem. 19:853–870.

91. **Lewis HD, Goldman A (1965)** Proper analysis of Coulter counter data. Rev. Sci. Instr. 36:868–869.

92. **Lombarts AJ, Leijnse B (1983)** A stable human platelet white blood-cell control for the Coulter model S-Plus-II Clin. Chim. A 130:95–102.

93. **Lusbaugh CC, Maddy JA, Basman NJ (1962)** Electronic measurement of cellular volumes. I. Calibration of the apparatus. Blood 20:233–240.

94. **Lusbaugh CC, Basman NJ, Glascock B (1962)** Electronic measurement of cellular volumes. II. Frequency distribution of erythrocyte volumes. Blood 20:241–248.

95. **Mattern CFT, Brackett FS, Olson BJ (1957)** Determination of number and size of particles by electrical gating: Blood cells. J. Appl. Physiol. 10:56–70.

96. **Maxwell JC (1883)** "Lehrbuch der Elektrizitaet und des Magnetismus." Berlin: Springer-Verlag.

97. **Mel HC, Yee JP (1975)** Erythrocyte size and deformability studies by resistive pulse spectroscopy. Blood Cells 1:391–399.

98. **Menke E, Kordwig E, Stuhlmueller P, Kachel V, Ruhenstroth-Bauer G (1977)** A volume activated cell sorter. J. Histochem. Cytochem. 25:796–803.

99. **Mercer WB (1966)** Calibration of Coulter counters for particles ~ 1 diameter. Rev. Sci. Instr. 37:1515–1520.

100. **Merrill JT, Veizades N, Hulett HR, Wolf PL, Herzenberg LA (1971)** An improved cell volume analyzer. Rev. Sci. Instr. 42:1157–1163.

101. **Metzger H, Kachel V, Ruhenstroth-Bauer G (1971)** The influence of particle size, form and consistency on the right skewness of Coulter volume distribution curves. Blut 23:143–154.

102. **Metzger H:** Unpublished results.
103. **Metzger H, Valet G, Kachel V, Ruhenstroth-Bauer G** (1972) The calibration by electronic means of Coulter counters for the determination of absolute particle volumes. Blut 25:179–184.
104. **Miller RG, Wuest LJ, Cowan DH** (1972) Volume analysis of human red cells. I. The general procedures. Ser. Haematol. 52:105–127.
105. **Miller RG, Wuest LJ** (1972) Volume analysis of human red cells. II. The nature of the residue. Ser. Haematol. 52:128–141.
106. **Nevius DB** (1963) Osmotic error in electronic determinations of red cell volume. Am. J. Clin. Pathol. 39:38–41.
107. **Newbould FH** (1974) Electronic counting of somatic cells in farm bulk tank milk. J. Milk Food 37:504–510.
108. **Nosanchu JS** (1981) Comparison of hematocrit determinations by microhematocrit and electronic particle counter. Am. J. Clin. Pathol. 75:264–270.
109. **Otto F, Schmid DO** (1970) Lymphocytenisolierung aus dem Blut des Menschen und der Tiere. Blut 21: 118–122.
110. **Otto F** (1970) Granulocytenisolierung aus dem Blut der Menschen und der Tiere. Blut 21:290–294.
111. **Patzelt R** (1968) Improved base-line stabilization for pulse amplifiers. Nucl. Instr. Methods 59:283–288.
112. **Paulus JM** (1975) Platelet size in man. Blood 46:321–336.
113. **Poole RK** (1981) Mitochondria of tetrahymena-pyriformis—enumeration and sizing of isolated organelles using a Coulter-Counter and pulse height analyzer. J. Cell. Sci. 61:276–280.
114. **Porath-Furedi A** (1983) The mutual effect of hydrogen ion concentration and osmotic pressure on the shape of the human erythrocyte as determined by light scattering and by electronic cell volume measurement. Cytometry 4:263–267.
115. **Prantl L** (1965) Fuehrer durch die Stroemungslehre. Verlag F. Vieweg und Sohn Braunschweig.
116. **Price-Jones C** (1910) The variations in the sizes of red blood cells. Br. Med. J. 2:1418–1425.
117. **Princen LH, Kwolek WF** (1965) Coincidence corrections for particle size determinations with the Coulter counter. Rev. Sci. Instr. 36:646–653.
118. **Princen LH** (1966) Improved determination of calibration and coincidence correction constants for Coulter counters. Rev. Sci. Instr. 37:1416–1418.
119. **Robinson LB** (1961) Reduction of baseline shift in pulse amplitude measurements. Rev. Sci. Instr. 32: 1057–1057.
120. **Ruhenstroth-Bauer G, Valet G, Kachel V, Boss N** (1974) The electrical volume determination of blood cells in erythropoiesis, smokers, patients with myocardial infarction and leukemia and of liver cell nuclei. Naturwissenschaften 61:260–266.
121. **Rumscheidt FD, Mason SG** (1961) Particle motion in sheared suspensions. XII. Deformation and burst of fluid drops in shear and hyperbolic flow. J. Colloid Sci. 16:238–261.
122. **Sadikov IN** (1967) Motion of a viscous fluid in the initial section of a flat channel. Inzh.-Fiz. Zh. 12: 219–226.
123. **Salzman GC, Mullaney PF, Coulter JR** (1973) A Coulter volume spectrometer employing a potential sensing technique. Biophys. Soc. Abstr. 17:302a.
124. **Schmid-Schoenbein H, Wells RE, Goldstone J** (1971) Fluid-drop like behavior of erythrocytes. Disturbance in pathology and its quantification. Biorheology 7: 227–234.
125. **Schmid-Schoenbein H, Wells R** (1969) Fluid-drop like transition of erythrocytes under shear. Science 165:288–291.
126. **Schuette WH, Shackney SE, Plowman FA, Tipton HW, Smith CA, MacCollum MA** (1984) Design of flow chamber with electronic cell volume capability and light detection optics for multilaser flow-cytometry. Cytometry 5:652–656.
127. **Schulz J, Thom R** (1973) Electrical sizing and counting of platelets in whole blood. Med. Biol. Eng. 1973: 447–454.
128. **Schwartz A, Sugg H, Ritter TH, Fernandez-Repollet E** (1983) Direct Determination of cell diameter, surface area, and volume with an electronic volume sensing flow cytometer. Cytometry 3:456–458.
129. **Serwer P, Allen JL** (1983) Agarose-gel electrophoresis of bacteriophages and related particles. 4. An improved procedure for determining the size of spherical particles. Electrophoresis 4:273–276.
130. **Shank BB, Adams RB, Steidley KD, Murphy JR** (1969) A physical explanation of the bimodal distribution obtained by electronic sizing of erythrocytes. J. Lab. Clin. Med. 74:630–641.
131. **Shuler ML, Aris R, Tsuchiya HM** (1972) Hydrodynamic focusing and electronic cell sizing techniques. Appl. Microbiol. 24:384–388.
132. **Smith AMO** (1960) Remarks on transition in a round tube. J. Fluid Mech. 7:565–576.
133. **Smither R** (1975) Use of a coulter counter to detect discrete changes in cell numbers and volume during growth of *Escherichia coli*. J. Appl. Bacteriol. 39: 157–165.
134. **Smythe WR** (1961) Flow around a sphere in a circular tube. Phys. Fluids 7:756–759.
135. **Smythe WR** (1964) Flow around spheroids in a circular tube. Phys. Fluids 7:633–638.
136. **Spielman L, Goren SL** (1968) Improving resolution in Coulter counting by hydrodynamic focusing. J. Colloid Interface Sci. 26:175–182.
137. **Steinkamp JA, Fulwyler MJ, Coulter JR, Hiebert RD, Horney JL, Mullaney PF** (1973) A new multiparameter separator for microscopic particles and biological cells. Rev. Sci. Instr. 44:1301–1310.
138. **Strackee J** (1966) Coincidence loss in blood counters. Med. Biol. Eng. 4:97–99.
139. **Sundqvist T, Fjallbra. B, Magnusson KE** (1981) Computer-aided counting with the Coulter-Counter of low numbers of spermatozoa in human-semen. Int. J. Andr. 4:18–24.
140. **Talstad I** (1984) Electronic counting of spinal fluid cells Am. J. Clin. Pathol. 81:506–511.
141. **Tatsumi T** (1952) Stability of the laminar inlet-flow prior to the formation of Poiseuille regime. I. J. Phys. Soc. Jpn. 7:489–495.
142. **Tatsumi N** (1981) The size of erythrocyte ghosts. Bioc. Biop. A 641:276–280.
143. **Thom R** (1968) Rechtsschiefe der Erythrocyten Volumenverteilungskurven Coulter Symposium Bad Nenndorf 33–36.

144. **Thom R, Hampe A, Sauerbrey G (1969)** Die elektronische Volumenbestimmung von Blutkoerperchen und ihre Fehlerquellen. Z. Exp. Med. 151:331–349.

145. **Thom R, Kachel V (1970)** Fortschritte fuer die elektronische Groessenbestimmung von Blutkoerperchen. Blut 21:48–50.

146. **Thom R (1972)** Method and result by improved electronic cell sizing. In Izak G, Lewis SM (eds), "Modern Concepts in Hematology." New York: Academic Press, pp 191–200.

147. **Thom R (1972)** Vergleichende Untersuchungen zur elektronischen Zellvolumenanalyse. Telefunken Publikation N1/EP/V 1968.

148. **Thomas RA, Cameron BF, Leif RG (1974)** Computer based electronic cell volume analysis with the AMAC II transducer. J. Histochem. Cytochem. 22:626–641.

148a. **Thomas RA, Yopp TA, Watson BD, Hindman HK, Cameron BF, Leif SB, Leif RC, Roque L, Britt W (1977)** Combined optical and electronic analysis of cells with the AMAC transducers. J. Histochem. Cytochem. 25:827–835.

149. **Valet G, Metzger H, Kachel V, Ruhenstroth-Bauer G (1972)** The volume distribution curves of rat erythrocytes after whole body X-irradiation. Blut 24:274–282.

150. **Valet G, Metzger H, Kachel V, Ruhenstroth-Bauer G (1972)** The demonstration of several erythrocyte populations in the rat. Blut 24:42–53.

151. **Valet G, Hanser H, Metzger H, Ruhenstroth-Bauer G (1974)** Several erythrocyte populations in the blood of the newborn rat, mouse, guinea pig and in the human fetus. XVth Congr. Int. Soc. Hematol. Jerusalem, p 317.

152. **Valet G, Schindler R, Hanser H, Ruhenstroth-Bauer G (1975)** Several erythrocyte populations of different mean volume in the young sheep and rat with different electrophoretic mobilities. 3rd Meeting Europ. Afric. Div. Int. Soc. Hematol. London (abstr.) 18:2.

153. **Valet G, Silz S, Metzger H, Ruhenstroth-Bauer G (1975)** Electrical sizing of liver cell nuclei by the particle beam method. Mean volume, volume distribution and electrical resistance. Acta Hepato-Gastroenterol. 22:274–281.

154. **Valet G, Opferkuch W (1975)** Mechanism of complement-induced cell lysis demonstrating a three-step mechanism of EACI-8 cell lysis by C9 and of a non-osmotic swelling of erythrocytes. J. Immunol. 115: 1028–1033.

155. **Valet G, Hofmann H, Ruhenstroth-Bauer G (1976)** The computer analysis of volume distribution curves: Demonstration of two erythrocyte populations of different size in the young guinea pig and analysis of the mechanism of immune lysis of cells by antibody and complement. J. Histochem. Cytochem. 24:231–246.

155a. **Valet G (1984)** A new method for fast blood cell counting and partial differentiation by flow cytometry. Blut 49:83–90.

156. **Van der Plaats G, Herps H (1983)** A study on the sizing process of an instrument based on the electrical sensing zone principle. I. The influence of particle material. Powder Technol 36:131–136.

157. **Van der Plaats G, Herps H (1984)** A study on the sizing process of an instrument based on the electrical sensing zone principle. II. The influence of particle porosity. Powder Technol. 38:73–76.

158. **Velick S, Gorin M (1940)** The electrical conductance of suspensions of ellipsoids and its relation to the study of avian erythrocytes. J. Gen. Physiol. 23:753–771.

159. **Vordren A (1971)** Der Einsatz linearer integrierter Schaltungen in der Nuklear-Elektronik. Elektronikpraxis 1:25–30.

160. **Wales M, Wilson JN (1961)** Theory of coincidence in Coulter counters. Rev. Sci. Instr. 32:1132–1136.

161. **Wales M, Wilson JN (1962)** Coincidence in Coulter counters. Rev. Sci. Instr. 33:575–576.

162. **Waterman CS, Atkinson EE, Wilkins B, Fischer CL, Kimzey SI (1975)** Improved measurement of erythrocyte volume distribution by aperture-counter signal analysis. Clin. Chem. 21:1201–1211.

163. **Weed RJ, Bowdler AJ (1967)** The influence of hemoglobin concentration on the distribution pattern of the volumes of human erythrocytes. Blood 29:297–312.

164. **Wendt G (1958)** Elektrische Felder und Wellen. In Flügge S (ed), "Handbuch der Physik." Vol. XVI. Berlin: Springer-Verlag, pp 148–164.

165. **Wenger J, Nowak JS, Kai O, Franklin RM (1982)** Display and analysis of cell-size distributions with a Coulter-Counter interfaced to an Apple II microcomputer. J. Immunol. 54:385–392.

166. **Wilkins B, Fraudolig JE, Fischer CL (1970)** An interpretation of red cell volume distributions by pulse height analysis. J. Assoc. Adv. Med. Instr. 4:99–105.

167. **Winter H, Sheard RP (1965)** The skewness of volume distribution curves of erythrocytes. Aust. J. Exp. Biol. Med. Sci. 43:687–698.

168. **Zimmermann U, Schulz J, Pilwat G (1975)** Transcellular ion flow in escherichia coli B and electrical sizing of bacterias. Biophys. J. 13:1005–1013.

169. **Zimmermann U, Pilwat G, Riemann F (1974)** Dielectric breakdown of cell membranes. Biophys. J. 14: 881–899.

170. **Zucker RM, Whittington K, Price BJ (1983)** Differentiation of HL-60 cells—Cell-volume and cell-cycle changes. Cytometry 3:414–418.

5

Light Scattering and Cytometry

Gary C. Salzman, Shermila Brito Singham, Roger G. Johnston, and Craig F. Bohren

Biochemistry/Biophysics Group, Life Sciences Division, Los Alamos National Laboratory, Los Alamos, New Mexico 87545 (G.C.S., S.B.S., R.G.J.); Department of Meteorology, Pennsylvania State University, University Park, Pennsylvania 16802 (C.F.B.)

INTRODUCTION

Light scattering is a complex phenomenon that has found many uses in flow cytometry, particularly in cell discrimination. Each of the cellular components that has a refractive index different from that of the surrounding medium scatters light out of the incident beam. This collective response to an incident light beam has made it difficult to explain scattering in terms of specific cellular contents. Many apparently disparate theories have been proposed to solve parts of the scattering puzzle. This chapter begins with a general overview of the light scattering process to set the stage for a discussion of the theories that are relevant to various scattering phenomena. The specific theories are preceded by a brief discussion of scattering geometry and the mathematics of polarization. The theories are followed by a summary of work on polarization measurements. The various uses of scattering in flow cytometry are then presented followed by a summary and conclusion.

OVERVIEW OF SCATTERING

To understand light scattering it is important that the seemingly disparate phenomena that are fundamentally light scattering processes be brought together and cast into familiar terms. Simple theories are derived to explain observations at a macroscopic level. As more refined measurements are made, more detailed theories are required to explain the new observations. A successful theory can become so closely identified with a phenomenon that all reference to the phenomenon is in terms of the theory. This has happened with light scattering. Large particles are said to be Mie scatterers or are said to diffract light. Small particles are called Rayleigh scatterers. But all particles scatter light, and the various theories of scattering merely describe the process in greater or lesser detail.

Reflection and refraction, for example, are not separate processes, but rather separate names for scattering processes. The law of specular reflection is a convenient way of summarizing what is observed when a dense array of scatterers (i.e., atoms or molecules) composing what appears to be a smooth interface between apparently homogeneous media is illuminated. This law can be viewed as having been established by experiment, or derived from a theory in which the discrete nature of matter is ignored, or derived from a theory which explicitly recognizes that specular reflection is the combined effect of scattering by very many densely packed atoms or molecules. To go beyond the simple laws of reflection and refraction to describe the fraction of incident light reflected from a planar interface separating two different homogeneous media, the Fresnel equations [18] must be used. These are usually derived from continuum electromagnetic theory, a theory in which the discreteness of matter is hidden from view. Yet even the Fresnel equations can be made to betray their microscopic origins when properly interrogated. This was done by Doyle [38], who showed that the Fresnel equations can be factored into two terms: the scattering pattern of a single dipole and an array factor. The latter enters because the dipoles (atoms or molecules) form a phased array, by which is meant that there are definite phase relations among the waves radiated by all elements of the array. An illuminated pond or window pane may be looked upon as a phased array of a vast number of tiny antennas, each driven by the incident light and, as a consequence, radiating (i.e., scattering) waves in all directions. But the superposition of all these waves nearly cancels, except in two special directions: those given by the laws of reflection and refraction.

Implicit in the previous paragraphs is that what one observes when matter is illuminated depends as much on its disposition as on its composition. For example, an ordinary microscope slide scatters light that illuminates it mostly in the two directions given by the laws of reflection and refraction. When this same slide is etched, what is observed is different. Yet its chemical composition is unchanged. If the slide is smashed into small bits, the resulting pile of glass particles scatters light about equally in all directions and appears quite different from both the parent and etched slides. Again, nothing is different except the disposition of the glass. If illuminated matter is viewed as a collection of

Flow Cytometry and Sorting, Second Edition, pages 81–107

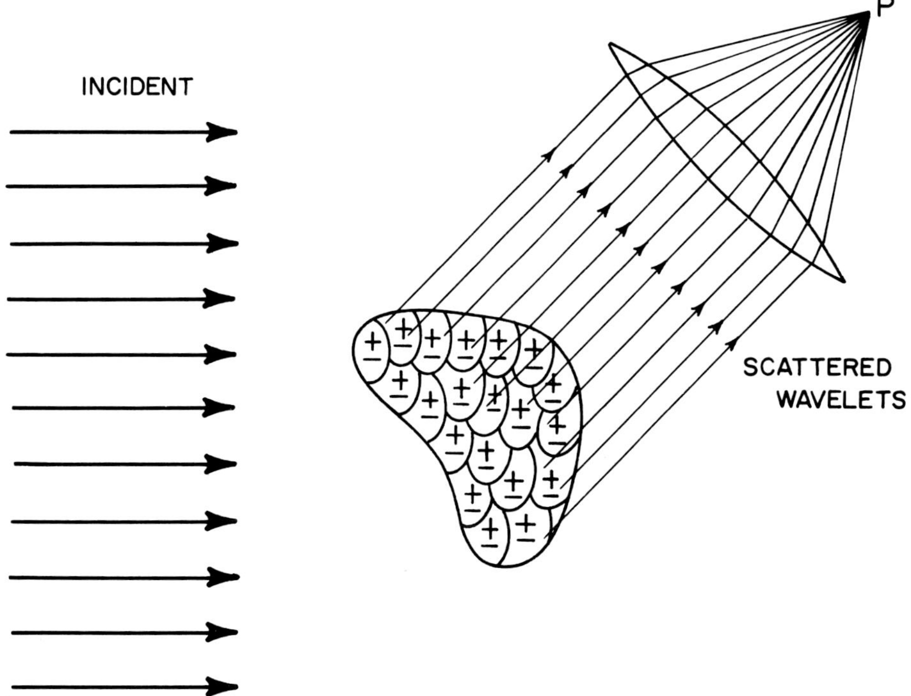

Fig. 1. The light scattered by a single particle is the sum of all the waves scattered by the dipoles of which the particle is composed. (Reproduced from ref. [17], with permission of the publisher.)

tiny dipolar antennas driven by the light incident on them, it is apparent that what is observed depends on how the antennas are arrayed. A smooth microscope slide is a more or less regular array of antennas. An etched slide is one in which those near the boundary are irregularly arrayed. And each particle in a pile of powdered glass is a clump of antennas, where the position of each clump is irregular. Imagine that a single one of these particles is plucked from the pile and illuminated.

Scattering by a Single Particle

Without invoking a specific theory or doing any calculations, insights can be obtained into scattering by a single particle by recognizing that it is an array of tiny dipolar antennas (Fig. 1) driven to radiate by the oscillating electric field of the light illuminating the particle. The light scattered by the particle is the sum of all the waves radiated (scattered) by each antenna. The value of this sum depends on all the phase relations among these radiated waves. In turn, these phase relations depend on the distances between antennas (relative to the wavelenth of the incident light) and the scattering angle (the angle between incident and scattered waves). Because these phase relations change with scattering angle, in general, the intensity of the light scattered by the particle changes with direction. Consider a simple example, two identical dipoles and two scattering directions, forward and backward (Fig. 2). To determine the total wave scattered by the two dipoles the waves that they scatter must be added, taking into account their phase difference.

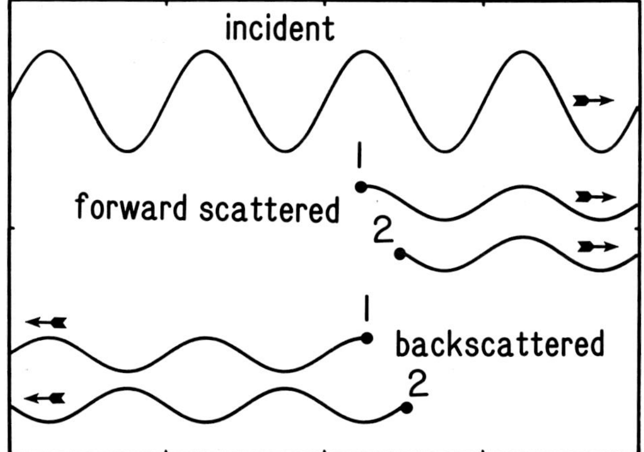

Fig. 2. The phase difference between the waves scattered by two dipoles excited by the same incident wave depends on their separation and the scattering direction. When the separation is one-quarter wavelength, the two waves scattered in the backward direction are exactly out of phase. In the forward direction, the two scattered waves are exactly in phase, which is true regardless of the separation. (Reproduced from ref. [16], p 139, with permission of the publisher.)

Because of the finite separation between the two dipoles they are excited out of phase. They also radiate out of phase, but in the forward direction this phase difference has the same magnitude and is opposite in sense. The net effect is that the two waves scattered in the forward direction are exactly in phase. Moreover, this is true regardless of the separation of the dipoles. And what is true for two dipoles is true for N: scattering by them is exactly in phase in one special direction, the forward direction. Interactions among the dipoles have been ignored. Each of them is excited not only by the incident wave but by the waves from all the other

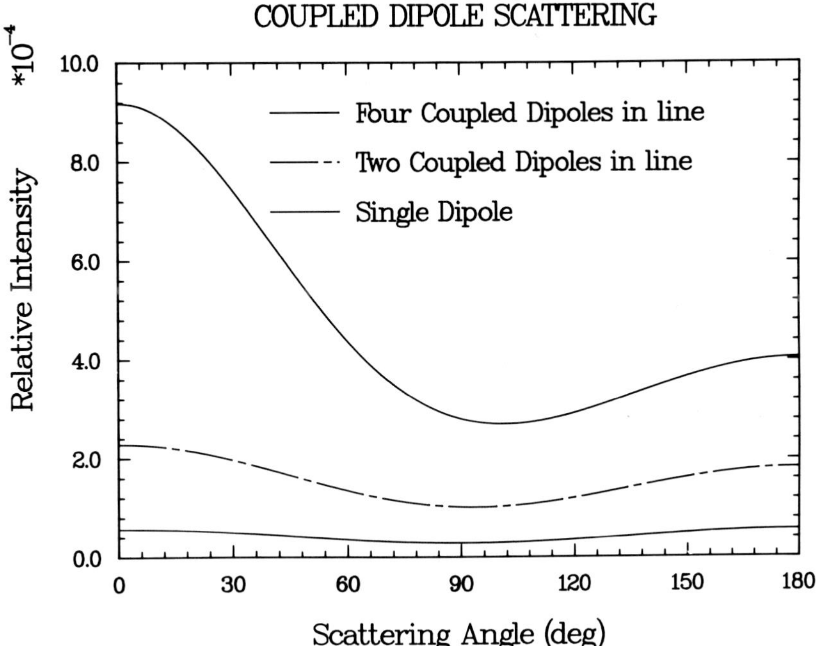

Fig. 3. Scattering by a line of interacting dipoles becomes more sharply peaked in the forward direction the greater the number of dipoles. For the example shown here the dipoles were separated by one wavelength and the intensity was averaged over all equally possible orientations of the line. (Reproduced from ref. [16], p 141, with permission of the publisher.)

dipoles. Accounting for interactions (which will be done in a further section) greatly complicates the mathematical formulation of scattering but it does not change the conclusion that the forward direction is special. Nor does it change the conclusion that the phase relations among the waves from an array of dipoles depend on the scattering direction. This is evident, for example, from Figure 2. When the two dipoles are separated by one-quarter wavelength, the waves radiated by them in the backward direction are (ignoring interactions) exactly out of phase, hence the two waves cancel in this direction.

The intensity of the light scattered by a particle is the square of the sum of the waves scattered by each of its constituent dipoles. Because these waves are exactly in phase in the forward direction, the scattering in this direction is expected to increase rapidly with the number of dipoles, that is, with the size of the particle. And because the forward direction is the only one for which all the dipoles radiate exactly in phase, the scattered intensity in this direction is expected to be a maximum. Since the number of dipoles is proportional to particle volume, the intensity of the light scattered in the forward direction should be proportional to the square of the particle volume. And indeed it is, at least for particles sufficiently small that the cumulative effect of all the dipolar interactions is negligible.

Regardless of particle size, scattering is greatest in the foward direction and becomes more sharply peaked in this direction as particle size is increased. An example of this is given in Figure 3, which shows scattering as a function of direction for one, two, and four interacting dipoles. Thus the increasing sharpness of scattering near the forward direction with increasing particle size is a general result that is a consequence of the phase relations among all the dipolar antennas composing an illuminated particle.

The quest for insight can be extended without entanglement in mathematical details. Suppose that a particle is small compared with the wavelength of the light illuminating it. In this instance all its constituent dipolar antennas radiate almost exactly in phase regardless of the scattering direction. Thus on physical grounds the scattering by such a particle should not be a complicated function of scattering direction. More detailed calculatons support this expectation.

Now suppose that the particle is such that the maximum separation between its constituent dipoles is approximately the wavelength of the incident light. It is possible that at some scattering angle many of the waves radiated by the dipoles will be almost exactly out of phase; thus, the scattered intensity will be a minimum. If the particle size is increased by an integral multiple, the number of minima in the scattering pattern should increase correspondingly. This conclusion is based on simple physical arguments and is supported by more detailed calculations and by experiments: the number of minima (and maxima) in the pattern of scattering by a particle increases in a more or less regular way with increasing size. The exact positions and magnitudes of these minima can be calculated (at least for spherical particles) using a complicated mathematical theory in which the underlying physics is obscured. Alternatively, the number and positions of these minima can be estimated using simple physical arguments about the phase relations among waves radiated by an array of dipoles.

Now consider a particle of fixed volume, that is composed of a fixed number of antennas. How does scattering by it vary as it is molded into different shapes? This question cannot be answered quantitatively. However, it can be said that as the particle's shape is changed, the relative positions of all its constituent dipolar antennas change, hence the phase relations among all the waves radiated by them change. The pattern of light scattered by a particle is expected to depend on its shape. This general, but rather vague statement can be made more precise.

From the previous paragraph it is clear that in or near the forward direction all the dipoles in a noninteracting array radiate exactly in phase regardless of their separation. In this direction, therefore, the scattered intensity does not change as the dipoles are moved about. Thus we expect scattering near the forward direction to be nearly independent of particle shape, which is consistent with both calculations and measurements (see, e.g., ref. [17], p. 400). A corollary of this is that if one wants to probe the details of particle shape using light scattering one should look far from the forward direction. For other than spherically symmetric particles, a change of orientation will cause scattering in a given direction to change because all the phase relations among the waves radiated by its constituent antennas change. And for the same reason that particle shape is expected to have the least influence on scattering near the forward direction, particle orientation is expected to have little effect on scattering near this particular direction.

Nothing has been said about the composition of particles, only their geometrical characteristics such as size, shape, and orientation. How does composition enter into these considerations? For simplicity, consider only particles of uniform composition. From a macroscopic optical point of view the material composition of such a particle is specified by its refractive index (relative to that of the surrounding medium). From a microscopic point of view, the relevant quantity is the polarizability (magnitude of the induced dipole moment for a unit exciting field) of the constituent dipolar antennas. These two quantities, refractive index and polarizability, must be related, although the precise relationship between them may not be known. Whatever this relationship, it is expected to be such that the greater the microscopic polarizability the greater the macroscopic refractive index. The greater the polarizability, the greater the amplitude of the waves radiated by the dipoles in the array making up a particle, hence the stronger the interactions among them. As a consequence, the general conclusions about scattering drawn on the basis of noninteracting dipoles are expected to be less valid as the relative refractive index of the particles increases. At visible wavelengths, the refractive indices of particles of biological origin are often not much different from those of the suspending medium, expecially the most common one, water. For many purposes, therefore, biological particles can be considered to be made up of weakly interacting dipolar antennas. But it would be unwise to conclude that the consequences of interactions are always negligible. What is or is not negligible depends on how deeply one probes. The noise in one experiment can be made the signal in another by a proper choice of apparatus.

The only characteristic of the light illuminating a particle that has been mentioned explicitly is its wavelength. But it has one more characteristic that determines how it is scattered by a given particle, namely, its state of polarization.

Consider a spherical particle that is much smaller than the wavelength of the light illuminating it. Such a particle may be looked upon as a single dipolar antenna. How this antenna radiates when excited by the oscillatory electric field of a light wave depends on the direction of the field (to be distinguished from the direction of the wave: fields of light waves are perpendicular to directions of propagation) and the direction of observation. When the incident field is linearly polarized perpendicular to a particular plane, so is the induced dipole moment, and the intensity of the light radiated by the dipole is the same for all directions lying in this plane (called the scattering plane, about which more will be said later). When the incident field is polarized parallel to the scattering plane, however, the intensity of the light radiated by the dipole is greatest at 0° and 180° and zero at a scattering angle of 90°.

Light of any polarization state can be decomposed into two components linearly polarized in orthogonal directions, in particular, parallel and perpendicular to the scattering plane. Since scattering by a dipole depends on the state of linear polarization of the field that excites it, scattering generally depends on the state of polarization of this field. In preceding paragraphs particles have been considered as arrays of dipoles. What is true for a single dipole in the array is true for many, namely, scattering by the array depends on the polarization state of the incident light.

A Hierarchy of Theories

A few simple physical arguments have provided some general understanding of scattering by particles. To go beyond this general understanding, that is, to obtain quantitative knowledge, specific theories must be invoked. It is important to understand the nature of theories. When something is said to be the result of reflection, shorthand is being used. What is meant is that the observed phenomenon can be explained at some level using only the theory of reflection. More precise observations require a better theory. This is, there is a hierarchy of theories for explaining any scatttering phenomenon. The rainbow is such a phenomenon, many of the features of which—its angular position, color separation, the existence of more than one bow—can be explained by invoking only geometrical optics and the laws of reflection and refraction. Yet the simple theory does not explain all features of rainbows. Geometrical optics predicts an infinite intensity at the rainbow angle; this intensity is not infinite. Geometrical optics predicts that all rainbows are identical regardless of the size of the drops that produce them; all rainbows are not identical. If an explanation of the rainbow that is more faithful to reality is desired, a better theory must be developed. And there is no end to this.

When a biological cell passes through the laser beam in a flow cytometer, the cell scatters light in all directions. The light is typically detected in the forward direction and in the angular region at right angles to the incident beam. A number of theories have been applied in attempts to describe the characteristics of the cells in terms of scattered irradiance and polarization measurements at one or more scattering angles. These theories vary in the precision with which they are able to describe the experimental data and in their complexity. The next section describes the scattering geometry and mathematical formalism needed to discuss the various theories in succeeding sections.

MATHEMATICAL FRAMEWORK FOR THE SCATTERING PROCESS

Figure 4 shows the geometry needed to describe the scattering of light by a small particle located at the origin of coordinates [17]. The incident beam comes from below and travels along the positive z axis. Some of the light is scattered by the particle along the direction given by the unit vector, $\hat{e}_r$, toward a detector located a distance r from the particle. The scattering direction is defined by the scattering angle, θ, and the azimuthal angle, φ. The scattering plane is determined by $\hat{e}_r$, the unit vector along the z axis, and $\hat{e}_r$. The electric field

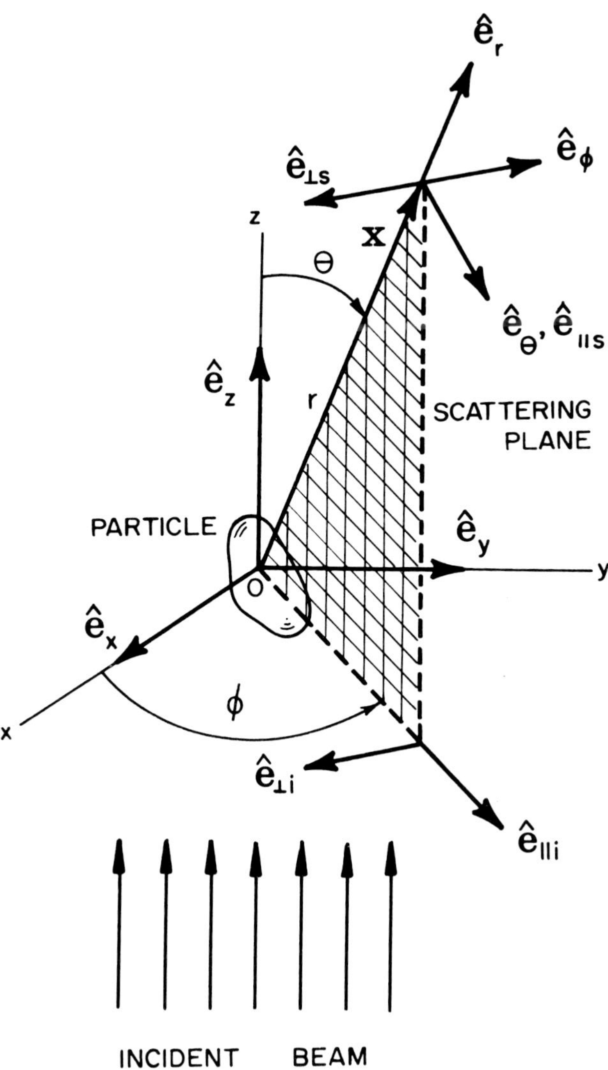

Fig. 4. Geometry of the scattering of light by a particle located at the origin. The incident light beam is parallel to the z-axis. A detector is located a distance r from the origin along the unit vector $\hat{e}_r$. (Reproduced from ref. [17], with permission of the publisher.)

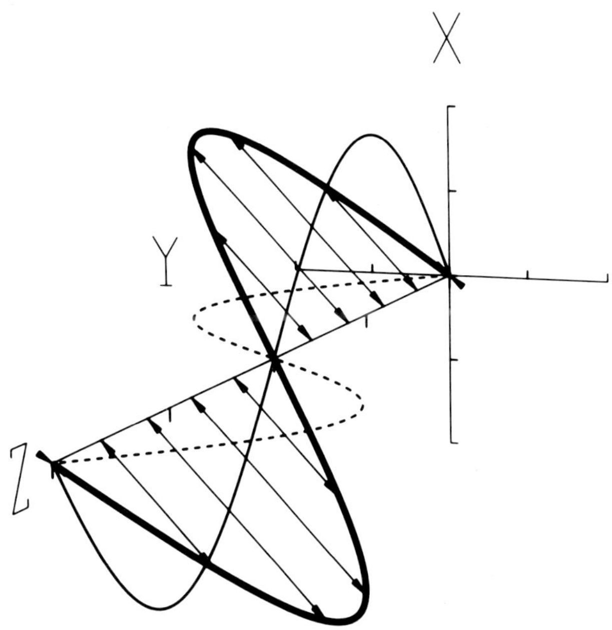

Fig. 5. Path of the electric field vector (arrows and heavy solid curve) for linearly polarized light traveling along the z axis. The plane of polarization is 45° from the x-z plane. The electric field can be resolved into a component parallel to the x-z plane (thin solid curve) and a component parallel to the y-z plane (dashed curve).

of the incident beam is in the x-y plane and can be resolved into components parallel, $E_{\parallel i}$, and perpendicular, $E_{\perp i}$, to the scattering plane. The incident electric field vector is given by [17]

$$\mathbf{E}_i = E_{\parallel i}\hat{e}_{\parallel i} + E_{\perp i}\hat{e}_{\perp i} \tag{1}$$

The irradiance or intensity of the incident light beam is given by

$$I_i = \langle E_{\parallel i}E^*_{\parallel i} + E_{\perp i}E^*_{\perp i}\rangle \tag{2}$$

where the asterisk denotes complex conjugation and the brackets denote a time average.

The electric field of the scattered light wave is perpendicular to $\hat{e}_r$ and can be resolved into components along the unit vectors $\hat{e}_{\parallel s}$ and $\hat{e}_{\perp s}$, which are parallel and perpendicular, respectively, to the scattering plane. The scattered electric field vector is given by

$$\mathbf{E}_s = E_{\parallel s}\hat{e}_{\parallel s} + E_{\perp s}\hat{e}_{\perp s} \tag{3}$$

The scattered and incident fields are defined with respect to different unit vectors. There is a linear relationship between the incident and scattered fields given by

$$\begin{bmatrix} E_{\parallel s} \\ E_{\perp s} \end{bmatrix} = \frac{e^{ik(r-z)}}{-ikr} \begin{bmatrix} S_2 & S_3 \\ S_4 & S_1 \end{bmatrix} \begin{bmatrix} E_{\parallel i} \\ E_{\perp i} \end{bmatrix} \tag{4}$$

where $k = 2\pi/\lambda$, $\lambda = \lambda_o/n_m$ is the wavelength in the medium, n_m is the refractive index of the medium, and λ_o is the wavelength of the light in vacuum. The complex numbers S_1, S_2, S_3, and S_4 are the elements of the amplitude scattering matrix. They each depend on θ and φ and contain information about the particle that scattered the light. Both amplitude and phase must be measured to quantitate the amplitude scattering matrix. Few attempts have been made to measure them. The work by Johnston et al. [62] to measure these elements directly is discussed later in this chapter.

Polarization refers to the pattern described by the electric field vector as a function of time at a fixed point in space. When the electric field vector oscillates in a single, fixed plane all along the beam, the light is said to be linearly polarized as shown in Figure 5. The heavy solid line is the plane of polarization of the linearly polarized wave traveling along the positive z axis. This linearly polarized wave can be resolved into a component parallel to the x-z plane (thin solid curve) and a component parallel to the y-z plane (dashed curve). If the x-axis is defined to be vertical, the thin solid curve represents vertical linearly polarized light and the

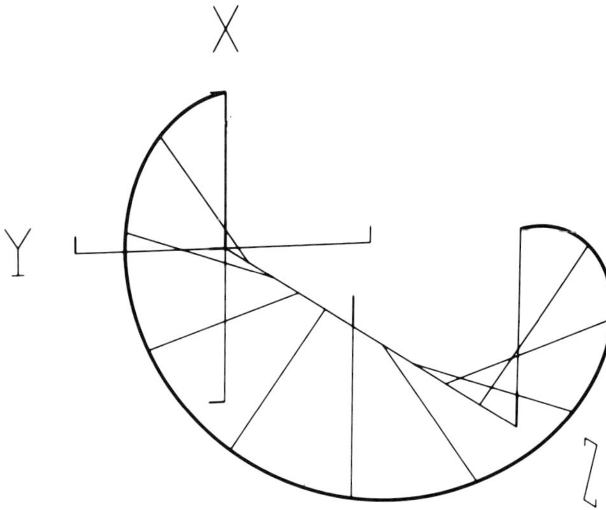

Fig. 6. Path of the tip of the electric field vector for a circularly polarized light wave traveling along the positive z-axis.

dashed curve represents horizontal linearly polarized light. If the plane of the electric field rotates, the light is said to be elliptically polarized, because the electric field vector traces out an ellipse at a fixed point in space as a function of time. If the ellipse happens to be a circle, the light is said to be circularly polarized. Figure 6 shows a circularly polarized wave traveling along the positive z-axis. The heavy solid curve shows the path traced out by the electric field vector as the wave moves along the positive z-axis. Viewed down the z-axis the path would be a circle. The connection between phase and polarization can be understood as follows: Circularly polarized light consists of equal amounts of vertically and horizontally polarized light where the vertical and horizontal electric fields oscillate exactly 90° out of phase. Light of arbitrary elliptical polarization consists of unequal amplitudes of vertically and horizontally polarized light. The electric fields for the two polarizations oscillate at the same sinusoidal frequency but have some constant phase difference.

Light of arbitrary polarization can be represented by four numbers known as the Stokes parameters, I, Q, U, and V. The first parameter, I, refers to the irradiance or intensity of the light. The parameters Q, U, and V represent the extent of horizontal linear, 45° linear, and circular polarization, respectively. The heavy solid curve in Figure 5 represents 45° linear polarization. Discussions in references 17 and 132 give quantitative explanations of the Stokes parameters. In terms of the electric field components the Stokes parameters are given by [17]

$$\begin{aligned} I &= \langle E_{\parallel}E_{\parallel}^{*} + E_{\perp}E_{\perp}^{*}\rangle \\ Q &= \langle E_{\parallel}E_{\parallel}^{*} - E_{\perp}E_{\perp}^{*}\rangle \\ U &= \langle E_{\parallel}E_{\perp}^{*} + E_{\perp}E_{\parallel}^{*}\rangle \\ V &= \langle E_{\parallel}E_{\perp}^{*} - E_{\perp}E_{\parallel}^{*}\rangle \end{aligned} \tag{5}$$

In what follows subscripts i and s refer to the incident and scattered light, respectively.

The relation between the Stokes parameters for the incident light and the Stokes parameters for the scattered light can be written [17]:

$$\begin{bmatrix} I_s \\ Q_s \\ U_s \\ V_s \end{bmatrix} = \frac{1}{k^2r^2} \begin{bmatrix} S_{11}S_{12}S_{13}S_{14} \\ S_{21}S_{22}S_{23}S_{24} \\ S_{31}S_{32}S_{33}S_{34} \\ S_{41}S_{42}S_{43}S_{44} \end{bmatrix} \begin{bmatrix} I_i \\ Q_i \\ U_i \\ V_i \end{bmatrix} \tag{6}$$

The matrix in equation (6) is known as the Mueller scattering matrix [17,101] or just scattering matrix. Hereafter, scattering matrix will be used. The 16 real matrix elements are generally functions of both wavelength (λ) and scattering angle (θ). They depend on the composition and geometry of the scatterer. These elements can be directly measured by experiments described later in the chapter. The Mueller matrix elements can be defined in terms of the elements of the amplitude scattering matrix. Several of these are given in Equation (7).

$$\begin{aligned} S_{11} &= 0.5(|S_1|^2 + |S_2|^2 + |S_3|^2 + |S_4|^2) \\ S_{12} &= 0.5(|S_2|^2 - |S_1|^2 + |S_4|^2 - |S_3|^2) \\ S_{34} &= \mathrm{Im}\{S_2S_1^* + S_4S_3^*\} \end{aligned} \tag{7}$$

Complete knowledge of the scattering by a sample requires that all 16 matrix elements be known, though not all the elements are independent [17,158]. There are only seven independent parameters in the scattering matrix of a single particle with fixed orientation. For scattering by a collection of randomly oriented ("orientationally averaged") particles, there are 10 independent parameters.

S_{11} is what is measured when the incident light is unpolarized. It provides only a fraction of the information theoretically available from scattering experiments [17,61,132]. S_{11} is much less sensitive to chirality [165] and long-range structure than some of the other matrix elements [13,14,17,28,57,69,92,116,123,133,134,138,139,153,170]. Note that the amplitude and phase of the incident and scattered electric fields do not appear explicitly in equation (6). The irradiance is much easier to measure than the electric field amplitude or phase [17,62].

THEORIES

Rayleigh Approximation for Small Particles

If a particle is small with respect to the wavelength of the incident light, its scattering can be described as if it were a single dipole [17,84–90,158]. Rayleigh arrived at this theory in an attempt to analyze why the sky is blue. Rayleigh theory is applicable under the condition that $|m|x \ll 1$ where m is the refractive index of the particle relative to that of the surrounding medium and x is the size parameter given by $x = 2\pi a/\lambda$, where a is the radius of the particle. If $x = 0.1$, the refractive index is 1.59 (polystyrene sphere) and the wavelength in vacuum is 488 nm, then the maximum particle radius is 4.9 nm for the Rayleigh theory to remain valid. The amplitude scattering matrix is given by [17]

$$\begin{bmatrix} E_{\parallel s} \\ E_{\perp s} \end{bmatrix} = \frac{e^{ik(r-z)}}{-ikr} \begin{bmatrix} 1.5a_1\cos\theta & 0 \\ 0 & 1.5a_1 \end{bmatrix} \begin{bmatrix} E_{\parallel i} \\ E_{\perp i} \end{bmatrix} \tag{8}$$

where

$$a_1 = -\frac{i2x^3}{3}\frac{m^2 - 1}{m^2 + 2} \tag{9}$$

The scattering matrix for the Rayleigh theory for spherical particles is given by [17]

$$\begin{bmatrix} I_s \\ Q_s \\ U_s \\ V_s \end{bmatrix} = \frac{9|a_1|^2}{4k^2r^2}\begin{bmatrix} 0.5(1+\cos^2\theta) & 0.5(\cos^2\theta-1) & 0 & 0 \\ 0.5(\cos^2\theta-1) & 0.5(1+\cos^2\theta) & 0 & 0 \\ 0 & 0 & \cos\theta & 0 \\ 0 & 0 & 0 & \cos\theta \end{bmatrix}\begin{bmatrix} I_i \\ Q_i \\ U_i \\ V_i \end{bmatrix} \tag{10}$$

For the Rayleigh theory the scattered irradiance for incident light is inversely proportional to λ^4 giving rise to the blueness of the sky. The Rayleigh theory also predicts that the scattered irradiance increases as a^6.

Rayleigh–Gans Approximation for Arbitrary Particles

The Rayleigh–Gans theory [70] addresses the problem of calculating the scattering by a special class of arbitrary shaped particles. If the refractive index of the particle is very close to that of the medium, the particle can be larger than allowed in the Rayleigh approximation. Specifically, the Rayleigh–Gans approximation requires that $| m - 1 | << 1$ and $kd | m - 1 | << 1$. The distance d is the largest dimension of the particle. These two conditions mean that the electric field inside the particle must be close to that of the incident field. If these conditions hold the particle can be viewed as a collection of independent dipoles that are all exposed to the same incident field. A biological cell might be modeled as a sphere of cytoplasm with a low refractive index relative to that of the surrounding water medium. If $m = 1.03$, $\lambda = 488$ nm, and $n_m = 1.33$, then $d = 0.585$ μm when the inequality above is interpreted to mean 0.03. The amplitude scattering matrix for the Rayleigh–Gans theory is given by [17]

$$\begin{bmatrix} E_{\|s} \\ E_{\perp s} \end{bmatrix} = \frac{e^{ik(r-z)}}{-ikr}\begin{bmatrix} S_2 & 0 \\ 0 & S_1 \end{bmatrix}\begin{bmatrix} E_{\|i} \\ E_{\perp i} \end{bmatrix} \tag{11}$$

where the amplitude scattering matrix elements are

$$\begin{aligned} S_1 &= -\frac{ik^3}{2\pi}(m - 1)\, vf(\theta,\varphi) \\ S_2 &= -\frac{ik^3}{2\pi}(m - 1)\, vf(\theta,\varphi)\cos\theta \end{aligned} \tag{12}$$

where v is the volume of the particle. The form factor $f(\theta,\varphi)$ involves the integration over the particle (see refs [17] or [70] for a detailed discussion). This theory is also called the Rayleigh–Debye theory [70]. Kerker [70] has extensively analyzed the range of validity of this theory. The Rayleigh–Gans approximation has been applied extensively to calculations of light scattering from suspensions of bacteria [30, 72,73,76,173]. Applications of light scattering in microbiology have been reviewed by Harding [51].

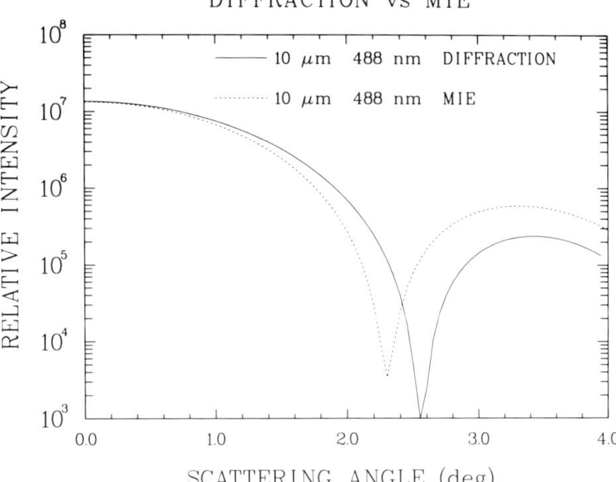

Fig. 7. Comparison of the diffraction approximation and Mie theory for the scattered irradiance at small angles for a 10 μm diameter particle illuminated by 488 nm light. The particle relative refractive index for the Mie theory solution is 1.03.

Diffraction Approximation for Near-Forward Scattering

The presence of a particle in a beam of light causes part of the incident beam to be scattered. In the vicinity of the particle edge the secondary waves add to the incident wave. This phenomenon can be described approximately by Fraunhofer diffraction theory. The particle is projected onto a plane perpendicular to the direction of the incident beam. By Babinet's principle [158], this planar object can be replaced by a hole of the same shape in a otherwise opaque screen. According to Fraunhofer diffraction theory the scattered light has the same polarization as that of the incident light and the scatter pattern is independent of the refractive index of the object. The Fraunhofer diffraction approximation is useful for large objects and small scattering angles. An application will be presented in the next section. Figure 7 shows a comparison of the relative irradiance of 488 nm light scattered by a 10 μm diameter homogeneous sphere for the Fraunhofer diffraction approximation and for Mie theory, an exact solution for homogeneous spheres. Near the forward direction Fraunhofer diffraction can represent accurately the change in irradiance as a function of particle size. Figure 8 shows a comparison of the Fraunhofer diffraction approximation for the relative scattered irradiance as a function of scattering angle for homogeneous spheres of two different sizes.

Hodkinson and Greenleaves [55] proposed a model in which the contributions of the theories of diffraction, refraction, and reflection are simply added together to obtain an approximation to the total scattered irradiance. Since this model ignores the phase relationships among the volume elements of the particle, it was qualified by the condition that the particle diameter must be larger than 3 or 4 wavelengths of light in suspensions in which there were a range of particle sizes having a diameter ratio of at least 2 to 1. Interference among the scattered waves is not taken into account. Mullaney [104] showed that this model cannot be used to describe small-angle scattering for objects with relative refractive index m less that 1.15 unless the phase shift $\rho = 2x | m-1 |$ is greater than 3.5. This is the phase difference between a ray that has traversed the particle along a diameter

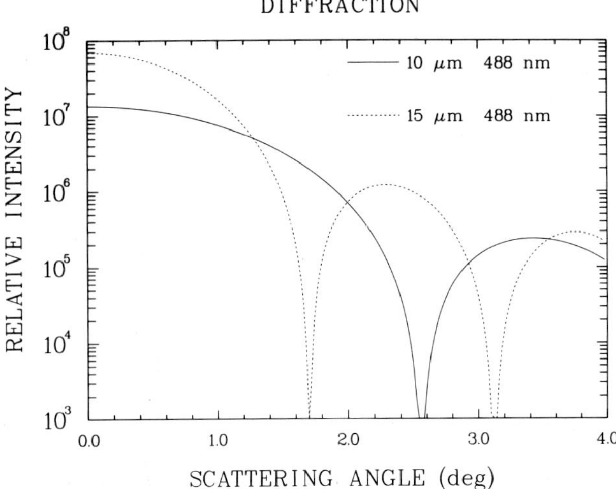

Fig. 8. Relative scattered irradiance as a function of scattering angle for two sizes of spheres according to the diffraction approximation.

and one that has covered the same distance in the surrounding medium. This phase shift requirement occurs because the refraction intensity contribution is a strong function of m as m approaches 1. If this condition is violated the refraction contribution will exceed that of the diffraction theory and the diffraction character of small-angle scattering will be lost. Hardy and Wheeless [50] applied diffraction theory to detector design for a flow cytometer. They placed masks shaped like an expected diffraction pattern in front of a forward scatter detector array. A signal on the detector meant that a particle with a scatter pattern represented by the expected diffraction pattern has passed through the light beam in the flow cytometer.

Fourier Transforms and the Fraunhofer Diffraction Approximation

In the previous section a particle was replaced by a planar object so that its scatter pattern in the near forward direction could be represented by a Fraunhofer diffraction pattern. Details in a particle (biological cell) such as nuclear texture and granules in the cytoplasm can be represented in the planar particle by variations in optical density. Optical density, D, is defined as $D = -\log_{10}T$, where T is the transmittance of light through the particle. For 100% transmittance, $T = 1$ and $D = 0$. For 10% transmittance, $T = 0.1$ and $D = 1$. The cell is essentially replaced by a positive photographic transparency. If a lens is placed between the transparency located in the object plane and a detector array such that the detector array is in the rear focal plane of the lens, the Fraunhofer diffraction pattern of the transparency appears on the detector array. This Fraunhofer diffraction pattern can be computed by a two-dimensional Fourier transform [20,48].

A Fraunhofer diffraction model for a lymphocyte might be a circle of high optical density for the nucleus surrounded by a larger circle of lower optical density to represent the cytoplasm. Spatial variations in optical density in the object plane are converted by a Fourier transform into spatial frequency variations in the Fourier transform plane in the rear focal plane of the lens. The Fourier transform is a mathematical model for the action of a lens placed appropriately between the object and the detector. To understand spatial frequency consider an optical density distribution in the object plane. If the optical density changes slowly across the object, the Fourier transform places most of the scattered light near 0° (low spatial frequency) in the Fourier transform plane. If the optical density changes rapidly across the object, the Fourier transform moves more of the energy to larger scattering angles (higher spatial frequency) in the Fourier transform plane. The Fourier transform for a cell model with clear cytoplasm (constant optical density) would be compact and located near the origin in the Fourier transform plane, while that for a cell model with highly granular cytoplasm (rapid changes in optical density across the cytoplasm) would be spread out in the Fourier transform plane (high spatial frequency). Spatial frequency increases from zero at the center of the two dimensional detector array to a maximum set by the maximum scattering angle subtended by the detector.

Kopp et al. [74,75], Pernick et al. [113,114], and Wohlers et al. [172] produced Fourier transforms of enlarged transparencies of human cervical cells. They showed that the best discrimination between normal and malignant cells occurred in the spatial frequency range from 20 to 40 cycles/mm (scattering angles of less than 3°). The malignant cells had condensed chromatin distributed in irregular patterns across the nucleus. The Fourier transforms of these cells showed complex patterns near 3°. Normal cells had smooth nuclei and produced smooth patterns with the diffraction pattern concentrated at small scattering angles. Rozycka et al. [119] used a similar approach to analyze normal and malformed myelin sheaths. There are many difficulties in carrying out Fourier transform measurements in static systems such as those described above. High background light levels can be introduced by dust on lens surfaces and by apertures along the optical path between the object and detector. If the incident light beam has a gaussian irradiance distribution and is cut off sharply by a small aperture in the object plane, the detector will record large irradiance variations due to the aperture alone. The irradiance distributions of most laser beams are approximately normally distributed. The gaussian mathematical function is an approximation to the normal distribution.

Pernick et al. [115] analyzed an enclosed flow chamber design and showed that Fourier transform techniques could be applied to the analysis of forward scatter in flow. They demonstrated that displacement of the cell from the center of the laser beam or displacement of the nucleus from the center of the cell had large effects on the small-angle scattering.

An optical Fourier transform is performed by a lens. The transform can be viewed in a microscope by imaging the rear focal plane of the objective. Seger et al. [127] and Turke et al. [157] developed a Fourier optical microscope for the examination of single cells in cervical smears. They found that the transforms of abnormal cells had significant high spatial frequency content compared to the transforms of normal cells. Genter and Salzman [44] developed a graphical technique for displaying these data and showed that two-population discriminant analysis methods were applicable to the data. Burger et al. [25] also developed a Fourier optical microscope and used it to extract shape and size information from single cells on slides.

Mie Theory

Mie theory [70,99] is an exact solution of Maxwell's electomagnetic field equations for a homogeneous sphere. The theory has been extended to arbitrary coated spheres [2,71] and to arbitrary cylinders. As a model for describing the scattering by biological cells Mie theory is an approximation. The amplitude scattering matrix is identical to that given in

Equation (11). S_1 and S_2 are given by complex series expansions. Programs are readily available for computing these amplitude scattering matrix elements [17,33]. The Mueller scattering matrix is given by [17]

$$\begin{bmatrix} I_s \\ Q_s \\ U_s \\ V_s \end{bmatrix} = \frac{1}{k^2r^2} \begin{bmatrix} S_{11} & S_{12} & 0 & 0 \\ S_{21} & S_{22} & 0 & 0 \\ 0 & 0 & S_{33} & S_{34} \\ 0 & 0 & -S_{34} & S_{44} \end{bmatrix} \begin{bmatrix} I_i \\ Q_i \\ U_i \\ V_i \end{bmatrix} \quad (13)$$

Brunsting and Mullaney [22,23] applied the coated sphere model to Chinese hamster tissue culture cells (CHO) because these cells are spherical and have concentric nuclei [24]. They measured both the nuclear and cytoplasmic refractive indexes for live CHO cells. The nucleus had a refractive index of 1.392. That for the cytoplasm was 1.3703. The medium was a 0.9% aqueous solution of sodium chloride with a refractive index of 1.3345. The measurements were made at 633 nm. They used the coated sphere theory of Aden and Kerker [2] to interpret the experimentally scattered light pattern for angles up to 15°. They showed that the coated sphere model gave significantly better results than the homogeneous sphere model.

The T-Matrix Method for Nonspherical Particles

The T-matrix method is based on an integral equation formulation of the scattering problem for an arbitrary particle. It was developed by Waterman [167,168]. Barber and Yeh [7] derived a related approach called the extended boundary condition method (EBCM). Wang and Barber [166] applied the method to spheroids of various eccentricities.

In the Mie theory solution for spheres the electromagnetic fields of the incident, internal and scattered waves are each expanded in a series. A linear transformation can be made between the fields in each of the regions. This approach can also be used for nonspherical objects such as spheroids. The linear transformation is called the transition matrix (T-matrix) [168]. The T-matrix for spherical particles is diagonal. Expressions for the T-matrix can be obtained by recasting problems from differential to integral form [150,162].

Computational problems make it difficult to obtain the T-matrix coefficients by integration if the aspherical particles have large aspect ratios. Iskander et al. [59,60] made improvements in the technique under the name iterated extended boundary condition method (IEBCM).

The T-matrix method has been applied to particles that are spheroids of revolution. Chebyshev polynomials have also been used as surfaces of revolution [102,103]. Work has also been done on several other approaches [3–5,41,42,169].

Coupled Dipole Approximation for Arbitrary Particles

The basis of the coupled dipole approximation is illustrated in Figure 9, which shows a collection of small spherical subunits representing a large sphere. Each subunit acts like a dipole oscillator with a polarizability which is related to the bulk dielectric constant of the particle. The dipoles are allowed to interact through their scattered fields. The field at a particular dipole is determined self-consistently as a sum of the incident field and the fields induced by all the other oscillators. The total scattered field at an external point is then the sum of the dipolar fields. This method was originally developed by Purcell and Pennypacker [118] to describe light scattering and absorption by nonspherical interstellar dust particles and was first applied to cells by Druger et al. [39]. There has been a great deal of theoretical work done on models of the scattering matrix for nonspherical chiral and nonchiral particles. Only a sample of the literature is cited here [11,26–29,52,53,133–139,154,174].

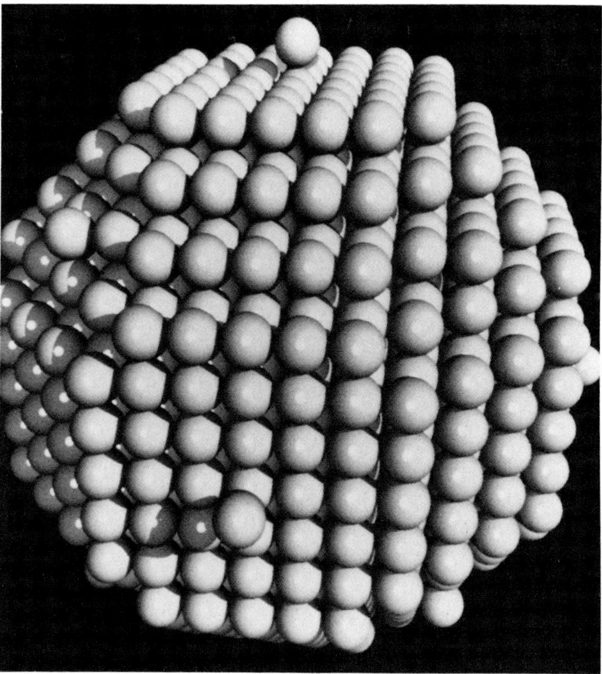

Fig. 9. Computer generated model showing a collection of small spherical subunits (dipoles) representing a solid sphere. Computer graphic courtesy Melvin Prueitt.

Equations (4) and (6) describe the amplitude and scattering matrices, respectively. The coupled dipole method is completely general and enables calculation of all the elements of the scattering matrix. Its validity is tested by comparison with Mie theory, which gives precise results for spheres, coated spheres and spherical shells.

Figure 10 shows a comparison between the coupled dipole approximation [139] and Mie theory for all of the independent scattering matrix elements for a sphere with size parameter $x = 1.5$. The 123 dipoles were placed on a cubic lattice with the maximum distance to a dipole from the dipole at the origin of 3 μm. For $x = 1.5$, $\lambda = 12.92$ μm, an appropriate isotropic polarizability is assigned to each dipole using the Clausius–Mosotti relation [61]. The solid curves represent the calculations using the coupled dipole approximation. The chain-dash curves are the Mie calculations. The dashed curves are the result of calculations using noninteracting dipoles. The agreement between the coupled dipole approximation and Mie theory is very good except for the S_{34} matrix element, which is very sensitive to the exact shape and optical properties of the particle. The lumpiness of the sphere contributes significantly to the error in S_{34}. The noninteracting dipole model gives 0 for S_{34}.

Figure 11 shows a comparison between the coupled dipole approximation [139] and Mie theory for a coated sphere. The coated sphere is represented in the coupled dipole approximation by a collection of dipoles placed in concentric

SPHERE x = 1.5

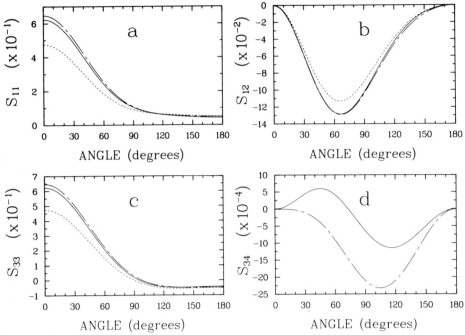

Fig. 10. Comparison between the coupled dipole approximation (sold curves) and Mie theory (chain—dash curves) for the independent Mueller matrix elements for a sphere with size parameter x = 1.5. The dashed curves are calculations for noninteracting dipoles. (Reproduced from ref. [139], with permission of the publisher.)

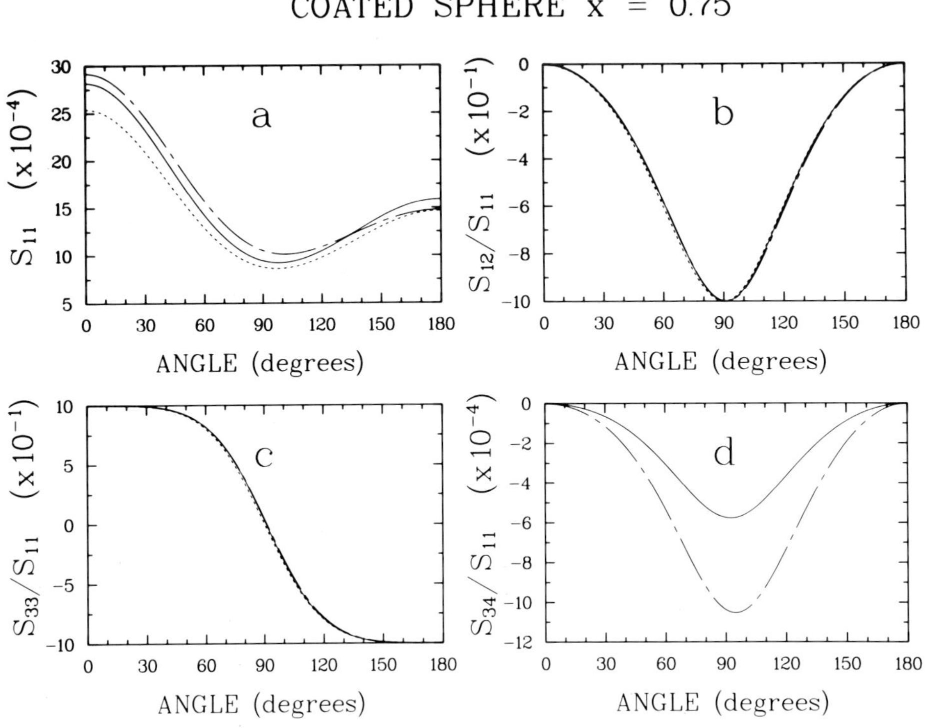

Fig. 11. Comparison between the coupled dipole approximation (solid curves) and Mie theory (chain—dash curves) for a coated sphere with x_{shell} = 0.75 and x_{core} 0.56. The dashed curves are for noninteracting dipoles. (Reproduced from ref. [139], with permission of the publisher.)

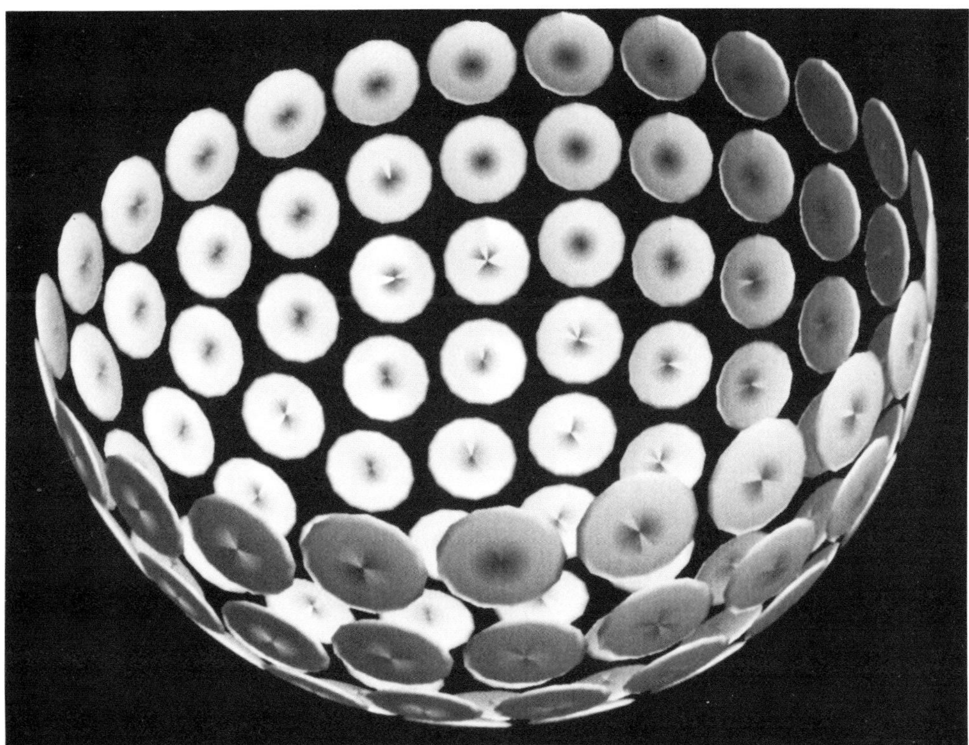

Fig. 12. Cutaway view of a spherical shell composed of oblate spheroidal dipoles. Computer graphic courtesy of Melvin Prueitt.

circles around an axis so that the distance of each dipole from the origin is 4 μm. A total of 149 dipoles is used and the distance between two dipoles on a circle is approximately 1 μm. Isotropic polarizabilities are used. Here the scattering matrix elements are normalized by the total scattered irradiance S_{11}. Only the S_{34} matrix element shows substantial disagreement. This, again, is due to the lumpiness of the coated sphere.

Isotropic polarizabilities for the dipolar subunits is appropriate for a sphere or coated sphere. This approach is impractical for thin shells or helical structures because the approach would require a very large number of spherical dipoles. A more useful method would be to use dipoles with anisotropic polarizabilities (i.e., ellipsoids) oriented appropriately along the particle. Figure 12 shows a cutaway view of a spherical shell composed of oblate spheroidal dipoles with anisotropic polarizabilities. Figure 13 shows a comparison between the coupled dipole approximation and exact theory for a thin shell with size parameter $x = 1.0$. A spherical shell is a coated sphere with vacuum for the core and the only material existing between the core radius and the shell radius. The coated sphere has a core radius of 3.443 μm and a shell radius of 3.444 μm. The wavelength is 21.6 μm. Dipolar subunits with anisotropic polarizabilities are particularly appropriate for irregularly shaped particles such as thin shells and helical structures.

Figure 14 shows a model for a thick helical structure whose subunits are prolate spheroids with anisotropic polarizabilities. The radius, R, and pitch, P, of this helix are 250 nm. The thickness, T, of the helix is 75 nm and there are 15 spheroidal dipoles per turn. Figure 15 shows the coupled dipole approximation for the total scattered irradiance, S_{11}, as a function of scattering angle for the helix in Figure 14 and several others. S_{11} is insensitive to helix pitch and thickness. It is sensitive to helix radius. Figure 16 shows the coupled dipole approximation for circular intensity differential scattering (CIDS) [6] from several helices. CIDS is the difference between scattering of left and right circularly polarized light. It is sensitive to helix radius, pitch and thickness. Figure 17 shows the coupled dipole approximation to the scattering matrix element S_{34} as a function of scattering angle for this same set of helices. S_{34} is insensitive to helix pitch but is quite sensitive to thickness and radius. The calculations in Figures 15–17 have been orientationally averaged, so they would be appropriate for scattering by a dilute suspension of such helices. Singham [135,136] addressed form and intrinsic optical activity in the scattering by chiral particles.

For N dipoles the coupled dipole approximation requires the solution of a set of 3N linear equations. If a matrix method is used for the solution, available computer memory quickly becomes a problem. Singham and Bohren [137] have taken steps toward solving this problem by reformulating the coupled dipole method in terms of internal scattering processes among the subunits. In this reformulation the field at the ith dipole is a sum of the field at the ith dipole due to the field of the incident beam, the field at the ith dipole due to single scattering by all the other dipoles, plus the field at the ith dipole due to double scattering by all the other dipoles, etc. Each of the terms is referred to as a scattering order. Within each scattering order, the dipoles that are to interact with the ith dipole are chosen. Only nearest neighbors can be considered or long range order can be taken into account by including dipoles far away from the ith dipole. For filamentary particles S_{11}, S_{12}, S_{33}, and S_{44} appear to be determined

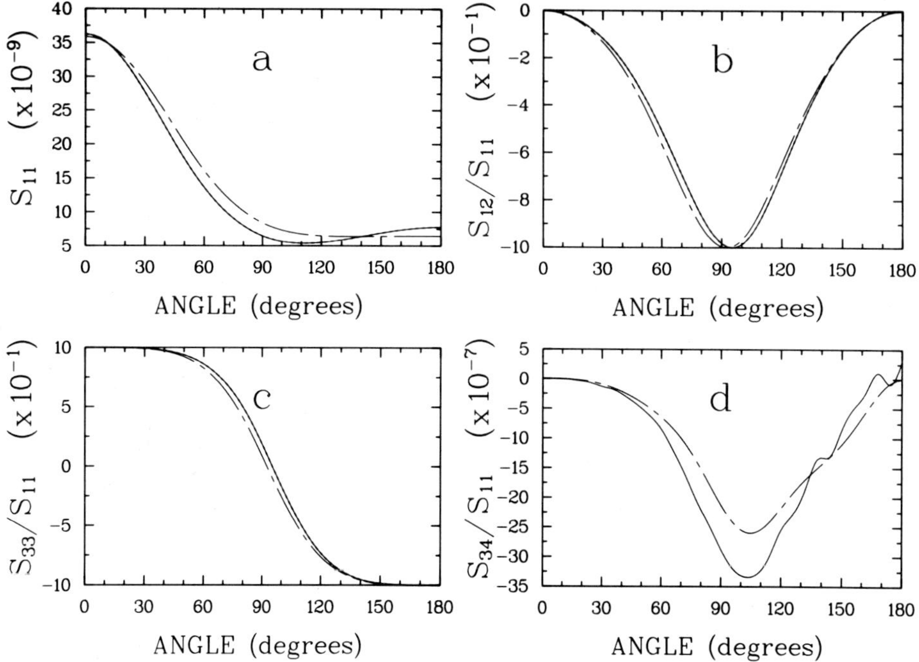

Fig. 13. Comparison between the coupled dipole approximation (solid curves) and Mie theory (chain—dash curves) for a very thin coated sphere with x_{shell} = 1.0 and x_{core} = 0.9997. The dashed curves are for noninteracting dipoles. (Reproduced from ref. [139], with permission of the publisher.)

by the incident field and single scattering from neighboring dipoles and S_{13}, S_{14}, S_{23}, S_{24}, and S_{34} appear to require higher scattering orders and long-range interactions.

POLARIZATION MEASUREMENTS

Scattering Matrix Measurements for Suspensions of Particles

Experimental measurements of the scattering matrix have largely concentrated on scattering by suspensions of particles [13,14,28,29,52,53,57,62,64,69,92,116,123,133,134,138, 139,153,170]. These measurements are difficult because many of the matrix elements are quite small when the scatterers are randomly oriented, as they are in an aqueous suspension. Several of the matrix elements that are sensitive to details of particle structure are typically 2 to 4 orders of magnitude smaller than S_{11}[63]. Small matrix elements include the set S_{13}, S_{14}, S_{23}, S_{24}, S_{34} and the elements symmetrically located across the diagonal. Large matrix elements include the set S_{11}, S_{12}, S_{21}, S_{33}, and S_{44}.

In theory, all the elements in the scattering matrix of Equation (6) can be measured using polarizers and retarders such as quarter-wave plates. Figure 18 shows the Mueller matrix representation for a beam of unpolarized light passing through a linear polarizer. As one looks into the beam the transmission axis of the polarizer, γ, is rotated 45° clockwise from the horizontal axis. The role of a retarder is to cause a phase shift between orthogonally polarized beams of linearly polarized light. A quarter-wave plate, for example, introduces a 90° phase difference between the electric field oscillations of vertically and horizontally polarized light. Figure 19 shows the Mueller matrix representation for the effect of a quarter wave retarder on a beam of linearly polarized light.

To measure a scattering matrix element using polarizers and fixed retarders, various combinations of the polarizers and retarders are inserted into the incident light beam to modify the incident polarization. Other polarizers and retarders are placed after the scattering suspension and used to analyze the polarization state of the scattered light. This approach is not generally practical [17,63]. The experimental artifacts caused by imperfections in the optical components and orientation and alignment errors restrict measurements to the large scattering matrix elements. Other major sources of error include light source fluctuations, DC gain instabilities, and the need to take differences between large signals to calculate some of the small matrix elements. Some of these experimental artifacts can be reduced by using rotating retarders. The rotation sinusoidally modulates the effective retardance [17,63,128].

More commonly, variable retarders such as photoelastic modulators or Pockels cells are used [17,130,131,171] for measurements of the scattering matrix elements. These devices have some of the same defects described above for randomly oriented scatterers. S_{34}, for example, can be measured quite accurately [13,14]. A photoelastic modulator consists of a block of amorphous quartz glued to a piezoelectric crystal. The piezoelectric crystal stresses the quartz block sinusoidally causing a small periodic change in the refractive index. The retardance is then modulated sinusoidally at the frequency of the driving crystal. Figure 20 shows the Mueller matrix of a photoelastic modulator acting on linearly polarized light. The result is elliptically polarized light. The retardance δ_1 oscillates sinusoidally with frequency f_1 and amplitude δ_0. Simultaneous measurement of eight of the Mueller matrix elements can be made using photoelastic modulators and the apparatus shown in Figure 21. The sample suspen-

Fig. 14. Computer graphic model of a thick helical structure whose subunit dipoles are prolate ellipsoids with anisotropic polarizabilities. Courtesy of Melvin Prueitt.

sion is illuminated with elliptically polarized light with the Stokes parameters U and V oscillating at 50 kHz due to a photoelastic modulator in the incident beam. The Stokes parameters of the scattered light are modulated at a frequency of 47 kHz by a second photoelastic modulator. Ideally, the eight matrix elements appear at different frequencies. However, errors in alignment and retardance of the various optical elements cause mixing of the matrix elements and large elements such as S_{12} tend to obscure the smaller elements such as S_{13} or S_{14}.

Measurement of S_{14} is difficult and often dominated by experimental artifacts [62,63,170]. S_{14} measurements on aqueous suspensions have been reported by Wells, et al. [170] and Katz et al. [69]. Misalignments, misadjustments, anharmonicities, and instabilities in the variable retarders can be the source of many experimental problems with these measurements [63,130,131]. Residual birefringence in the retarders and other optical components is especially troublesome.

Maestre et al. [92] and Katz et al. [69] developed an apparatus with which they have measured the CIDS ($-S_{14}/S_{11}$) of suspensions. They used a Pockels cell to produce an incident beam that alternated in square wave fashion between left and right circularly polarized light. They measured the CIDS of a suspension of the helical sperm head from the octopus Eledone cirrhosa (Fig. 22). The sperm head is a left handed helix with a pitch of 0.65 μm, an outside diameter of 0.60—0.65 μm and an inner diameter of 0.25 μm. Their measurements produced the results shown in Figure 23, which shows strong differential scattering of circularly polarized light. Wells et al. [170] carried out a detailed analysis of artifacts in CIDS measurements using these same cells. This technique may be a useful method for characterizing helical cells. Tinoco et al. [154] reviewed the imaging of biomolecular structures with polarized light.

Phase Differential Scattering Measurements

A new approach avoids the use of variable retarders and their associated artifacts. This interferometric technique is known as phase differential scattering (PDS) [62–64]. PDS uses a two-frequency, Zeeman effect laser [54,111] to produce two laser lines with orthogonal linear polarizations.

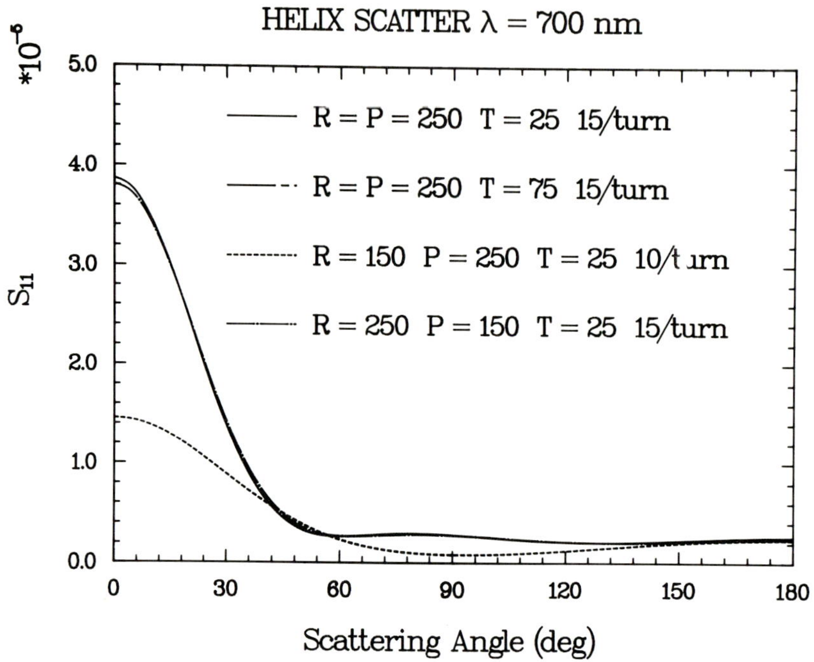

Fig. 15. Coupled dipole approximation for the total scattered irradiance as a function of scattering angle for the helix in Figure 14. S_{11} is sensitive to helix radius but insensitive to helix pitch and thickness.

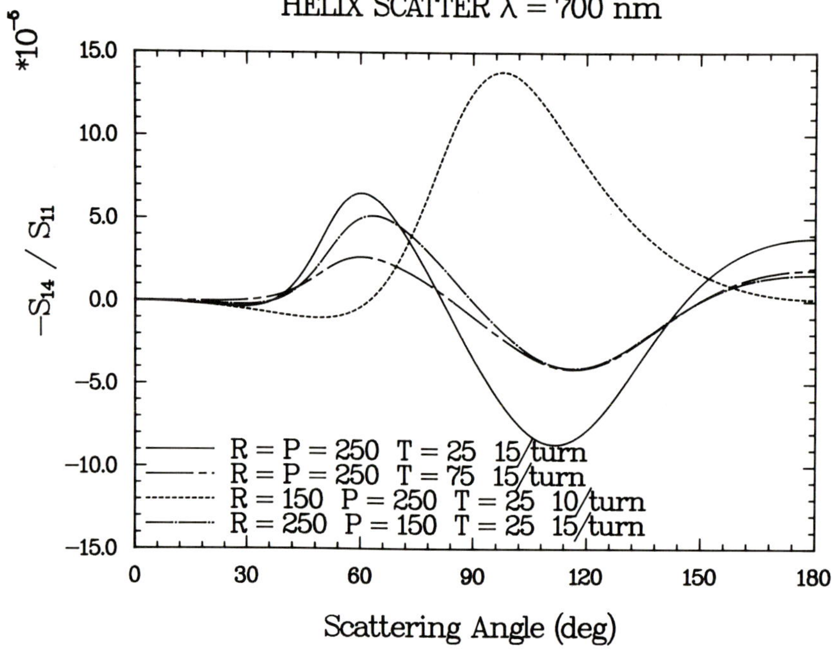

Fig. 16. Coupled dipole approximation for circular intensity differential scattering ($-S_{14}/S_{11}$) from several helical structures. R is radius, P is pitch and T is thickness of the helix.

The lines are near 632.8 nm and they differ in frequency by 250 kHz.

When the two collinear beams from the Zeeman laser pass through a properly oriented polarizer, they interfere on a photodetector and a 250 kHz beat frequency is detected. The phase of this 250 kHz signal is the optical phase difference between the two laser lines. The relative phase of two orthogonally polarized electromagnetic waves oscillating at ½ million GHz can thus be measured at an experimentally accessible 250 kHz.

When light from the Zeeman laser is scattered by a sample, the phase and amplitude of the 250 kHz beat frequency on

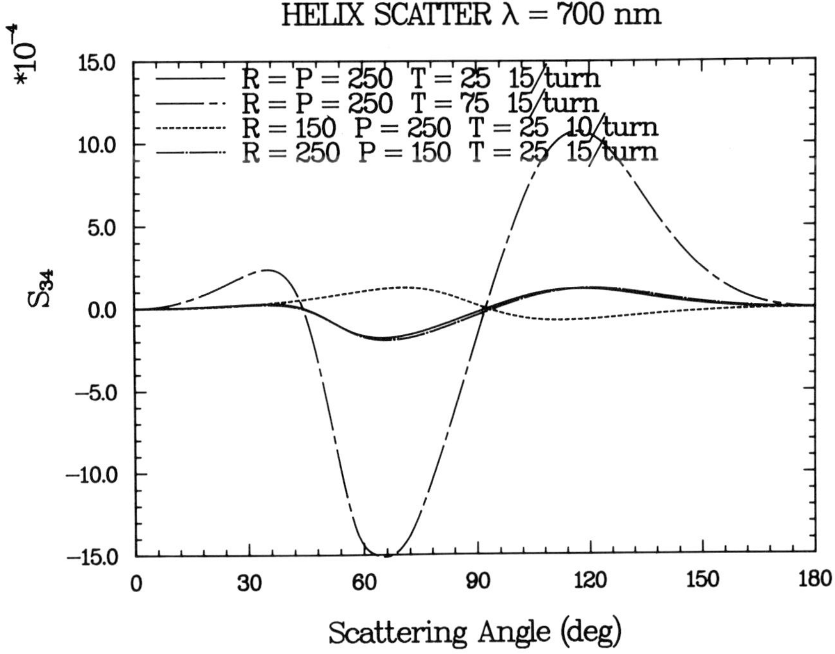

Fig. 17. Coupled dipole approximation to the S_{34} Mueller matrix element for several helices. S_{34} is insensitive to helix thickness but is quite sensitive to differences in helix pitch (P) and radius (R).

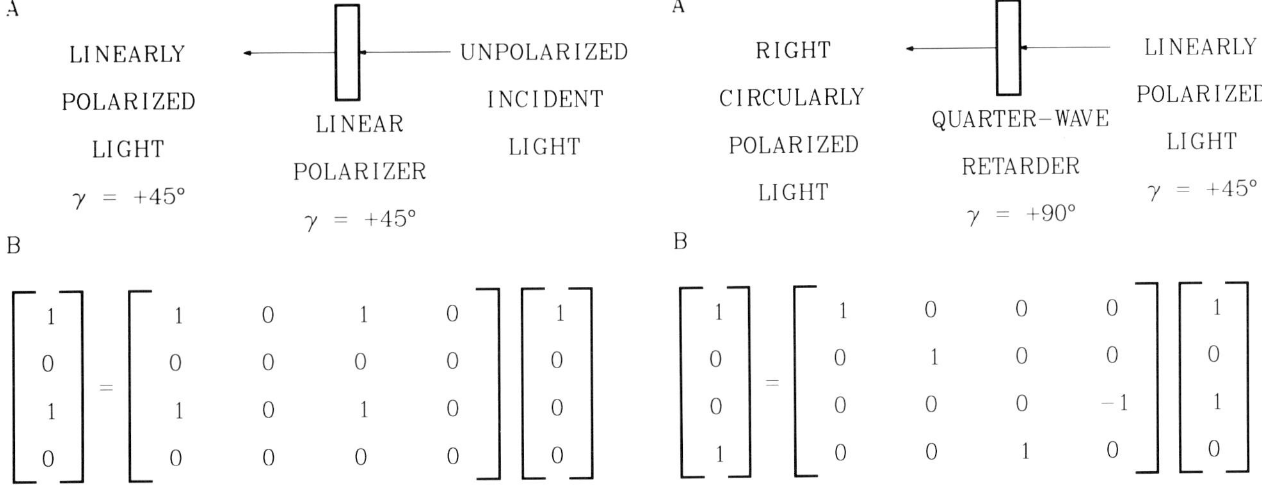

$$\begin{bmatrix}1\\0\\1\\0\end{bmatrix} = \begin{bmatrix}1&0&1&0\\0&0&0&0\\1&0&1&0\\0&0&0&0\end{bmatrix}\begin{bmatrix}1\\0\\0\\0\end{bmatrix}$$

Fig. 18. Mueller matrix formulation for the effect of a linear polarizer on a beam of unpolarized light. As you look into the beam the passing axis of the polarizer is rotated 45° clockwise from the horizontal.

$$\begin{bmatrix}1\\0\\0\\1\end{bmatrix} = \begin{bmatrix}1&0&0&0\\0&1&0&0\\0&0&0&-1\\0&0&1&0\end{bmatrix}\begin{bmatrix}1\\0\\1\\0\end{bmatrix}$$

Fig. 19. Mueller matrix formulation for the effect of a quarter wave plate retarder on a beam of linearly polarized light.

the photodetector contains structural information about the scatterer. The only optical component required for such PDS measurements on aqueous suspensions is an analyzing polarizer placed in front of the photodetector.

A theory for PDS has been derived [17,62,64]. At some scattering angle, the particle may retard the incident horizontal and vertical polarizations by different amounts, which will be apparent from the phase, α, of the 250 kHz beat frequency when the polarizer is oriented at $\gamma = 45°$. PDS for a suspension of particles can be analyzed by expressing α in terms of the Mueller matrix elements (S_{ij}) of Equation (6). For $\gamma = 45°$,

$$\alpha(45°) = \tan^{-1}\frac{S_{14} + S_{34}}{S_{13} + S_{33}} \qquad (14)$$

Figure 24 shows phase differential scattering by pure aqueous suspensions of several different viable bacteria. $\alpha(45°)$ is plotted as a function of scattering angle. Figure 25 shows relative phase, $\alpha(45°)$ as a function of scattering angle for

$$\begin{bmatrix} 1 \\ 0 \\ +\cos\delta_1 \\ +\sin\delta_1 \end{bmatrix} = \begin{bmatrix} 1 & 0 & 0 & 0 \\ 0 & 1 & 0 & 0 \\ 0 & 0 & +\cos\delta_1 & -\sin\delta_1 \\ 0 & 0 & +\sin\delta_1 & +\cos\delta_1 \end{bmatrix} \begin{bmatrix} 1 \\ 0 \\ 1 \\ 0 \end{bmatrix}$$

$$\gamma = +90°$$

$$\delta_1(t) = \delta_0\cos(2\pi f_1 t)$$

Fig. 20. Effect of a photoelastic modulator on a beam of linearly polarized light. The result is a beam of elliptically polarized light. The time dependence of the retardance δ_1 has amplitude δ_0 and frequency f_1.

encapsulated and unencapsulated cultures of Bacillus subtilis.

Aquaspirillum magnetotacticum is a spiral bacterium that deposits magnetite crytals and uses the earth's magnetic field to aid in maintaining its location near the bottoms of ponds [15]. Tomei et al. [155] measured PDS angular distributions for suspensions of these bacteria. They were able to distinguish readily between suspensions in which the bacteria were oriented in a magnetic field and suspensions in which no magnetic field was present. They were also able to distinguish between these two cases and a suspension of a culture of *A. Magnetotacticum* in which the bacteria deposited nonmagnetic amorphous iron.

FLOW CYTOMETRY

Sizing and Discrimination at Forward Angles

The scattered irradiance near the forward direction is expected to increase rapidly with particle size and to be approximately independent of particle refractive index and shape. Mullaney et al. [108] developed a flow cytometer with an annular forward scatter detector that collected light between 0.5° and 2°. The incident laser beam was stopped by a beam dump located in the center of the annular detector. Mullaney et al. [108] demonstrated that the irradiance of helium-neon laser light (633 nm) scattered by spheres with diameters between 6 and 14 μm is approximately linearly proportional to particle volume (diameter cubed). Steinkamp et al. [148] simultaneously measured Coulter volume and scattered 488 nm laser light in a forward angle range between 0.7° and 2.0°. They confirmed the diameter cubed dependence of the scattered irradiance for polystyrene spheres with diameters between 5.8 μm and 10.5 μm in agreement with the theoretical considerations of Mullaney and Dean [105,106]. These workers showed that this dependence changed to diameter squared beyond 10.5 μm. When they changed the beam shape and size from that produced by a 15 cm spherical lens to crossed cylindrical lenses, they observed diameter cubed dependence below 7.8 μm, followed by diameter squared, then diameter and finally a decrease for particles with diameters above 12 μm. The decrease in scattered irradiance for particles with diameters greater than 12 μm occurred because most of the light was scattered into a beam dump located in the core of the annular detector. Consistent cell sizing with forward angle light scattering can only be obtained if the detection geometry remains constant from one experiment to the next. With stream-in-air flow cytometers it is particularly important that the forward scatter beam dump remain in a constant position because it determines the minimum scattering angle viewed by the detector. A small change in the beam dump cross-sectional area will have a considerable effect on the detected scatter signal [125].

Julius et al. [66] showed that the collection of scattered light between 1° and 10° could be used to distinguish between live and dead murine thymocytes. The live cells were stained with fluorescein diacetate and fluorescence gated scatter was used to show that the live subpopulation scattered more light than the dead subpopulation. This discovery has been exploited by immunologists to remove the dead cells from the populations to be analyzed. Loken and Herzenberg [80] showed that the largest separation between live and dead cell scattered light pulse height distributions occurred with a collection angle of 0.5° to 12.5° and that the two peaks merged as the collection angle was decreased. Live and dead cells of the same type are likely to have the same cross sectional areas and would be expected to scatter the same amount of light at small scattering angles. The dead cells are likely to have torn membranes that would let the suspension medium permeate the cells reducing their relative refractive indices. The scattered irradiance at larger angles is more sensitive to refractive index differences. In a study of human lymphoma cell lines, Braylan et al. [21] showed that 50% ethanol fixation decreased the mean fluorescence activated cell sorter (FACS) forward light scatter (FLS) signal by 25% and increased the coefficient of variation (CV) from 15% to 22%. This could be caused by cell shrinkage and by reduction in the refractive index of the cytoplasm relative to that of the suspending medium. Jovin et al. [65] demonstrated that ratios at two angles of FLS in a flow cytometer could be used for particle sizing.

Forward light scatter at 488 nm has been used to gate fluorescence frequency histograms to obtain cell discrimination. Goldschneider et al. [45] used this approach with a FACS to select subpopulations of rat hematopoietic cells expressing Thy-1 antigen, to select pluripotent hematopoietic stem cells from rat bone marrow [46], and to isolate rat hematopoietic cells containing the intracellular enzyme terminal deoxynucleotidyl transferase [47].

Extinction in Flow

The signal on a small detector located at a 0° scattering angle is reduced by scattering and absorption of light along the path between the laser and the detector. This phenomenon is called extinction, which is the sum of absorption and scattering. Extinction has been exploited for cell sizing in flow cytometry [67,147]. Kamentsky and Melamed [68] and Adams and Kamentsky [1] were the first to use an extinction detector in a flow cytometer (Cytograf). Steinkamp [147] placed an optical fiber directly in the beam leaving the flow chamber. A cell passing through the beam absorbs some of the light and scatters some of the light. This produces a momentary decrease in irradiance at the fiber optic detector. Steinkamp showed that the height of this negative pulse was proportional to diameter squared for particles with diameters between 5 μm and 20 μm. Extinction measurements

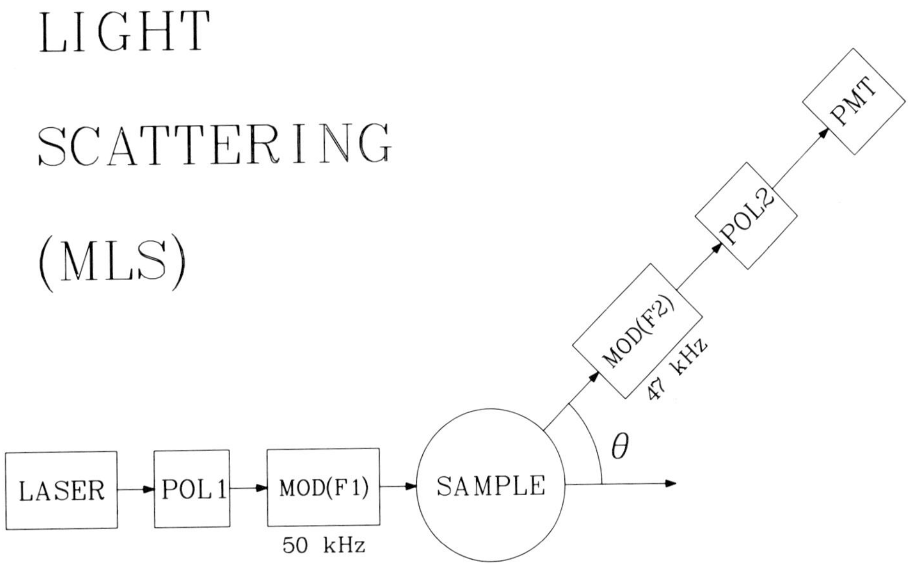

Fig. 21. Apparatus using two photoelastic modulators to measure 8 of the Mueller matrix elements for a particle suspension.

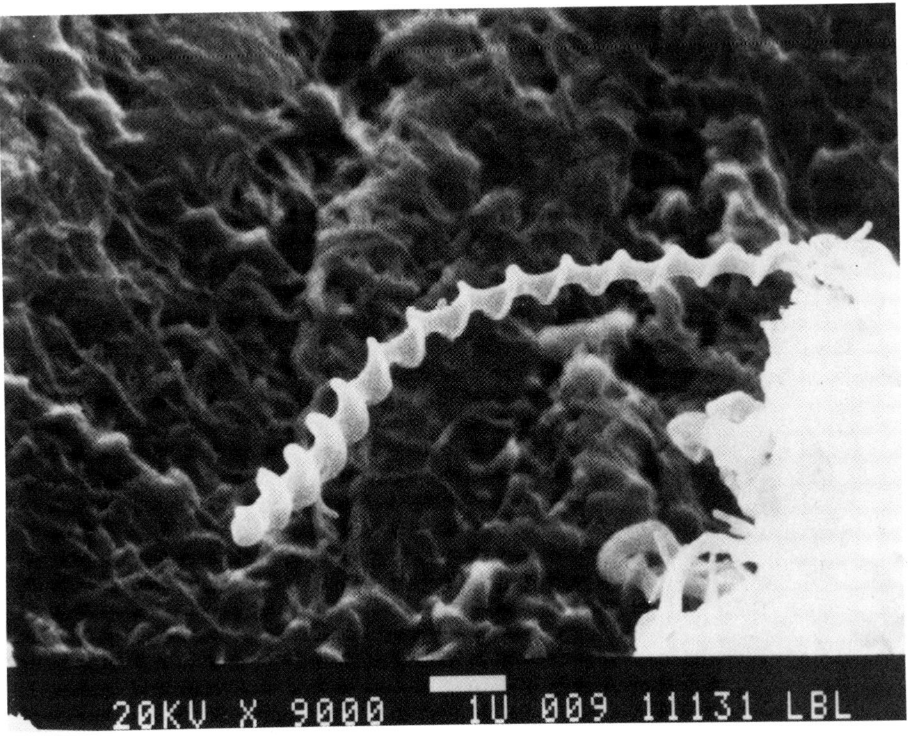

Fig. 22. Sperm head from the octopus Eledone cirrhosa. It is a left handed helix with a pitch of 0.6 μm. (Reproduced from ref. [92], with permission of the publisher.)

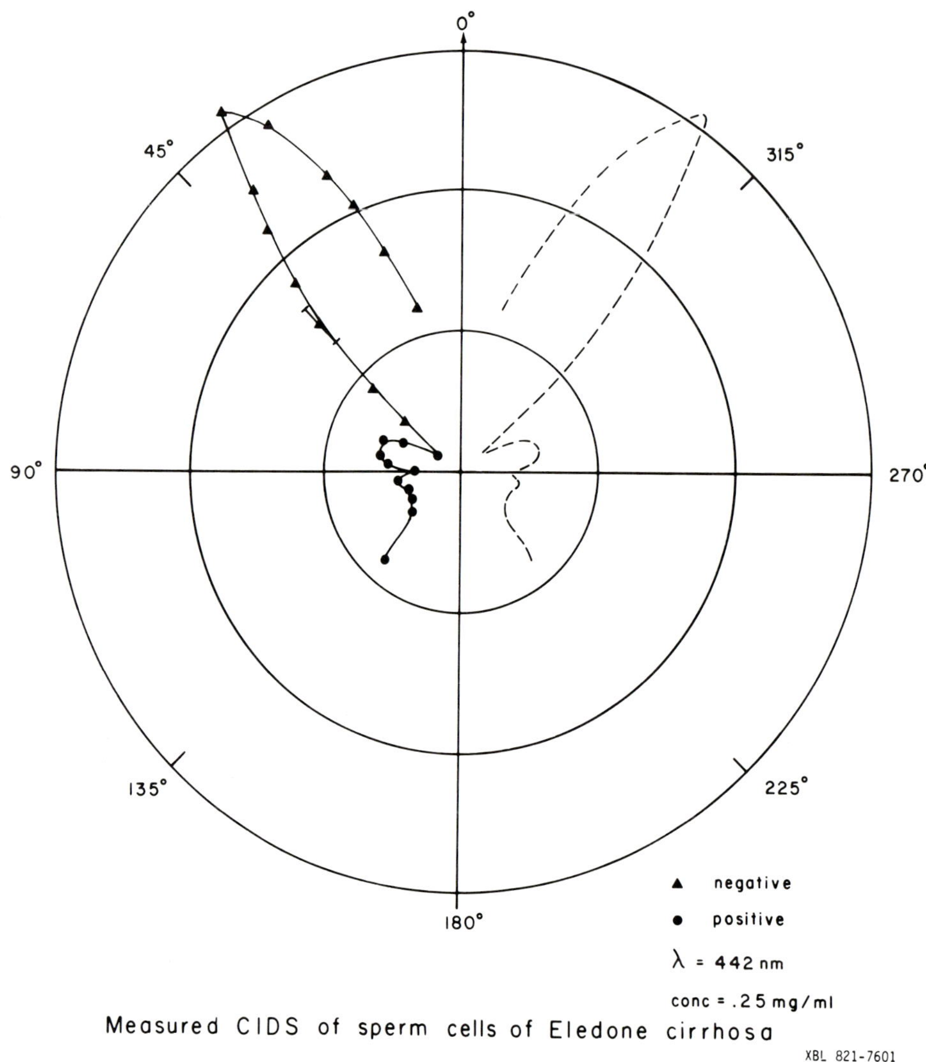

Fig. 23. Polar plot of the CIDS versus scattering angle for a suspension of sperm heads from Eledone cirrhosa. The triangles are for positive values and the solid circles are for negative values. (Reproduced from ref. [92], with permission of the publisher.)

have been combined with other flow cytometric parameters to assist in cell discrimination [78,79].

Forward and 90° Scatter

Salzman et al. [121] were the first to show that light scattered from unstained human leukocytes in a flow cytometer at 1° ± 0.1° (forward light scatter or FLS) in coincidence with that scattered at 90° ± 12.5° (perpendicular light scatter or PLS) could be used to identify three subpopulations consisting of segmented neutrophils, lymphocytes and monocytes. Logarithmic amplifiers were used for both parameters because of the large dynamic range of the signals. The significant discrimination observed has not been satisfactorily explained. The scattered irradiance at larger angles is much more sensitive to internal structural differences and to refractive index differences than is forward scattering [120]. Visser et al. [163] used this flow cytometer to show that NH_4Cl-treated mouse bone marrow cells could be sorted into four subpopulations based only on FLS and PLS. The four subpopulations were I(lobocytes), II(normoblasts and lymphocytes), III(metamyelocytes and lymphocytes), and IV-(granulocytes). Group I had the least FLS. Groups II, III, and IV had similar FLS. In order of increasing PLS were groups II, I, III, and IV. They showed that group III contained stem cells that gave rise to colonies of granulocytes and macrophages. van den Engh and Visser [159] and van den Engh et al. [160,161] used a FACS to extend this work to more classes of hematopoietic stem cells. This approach has been used for sorting mouse cytotoxic macrophages [94], for identifying four subpopulations in mouse islet cells [110], and for showing that there are two populations of human T8-positive cells [152]. Sklar et al. [140] showed that PLS decreases significantly when human neutrophils degranulate. McNeil et al. [95] used FLS and PLS to show that membrane ruffling by stimulated neutrophils caused increased FLS and decreased PLS. Doukas et al. [37] used scatter to study lymphocyte proliferation.

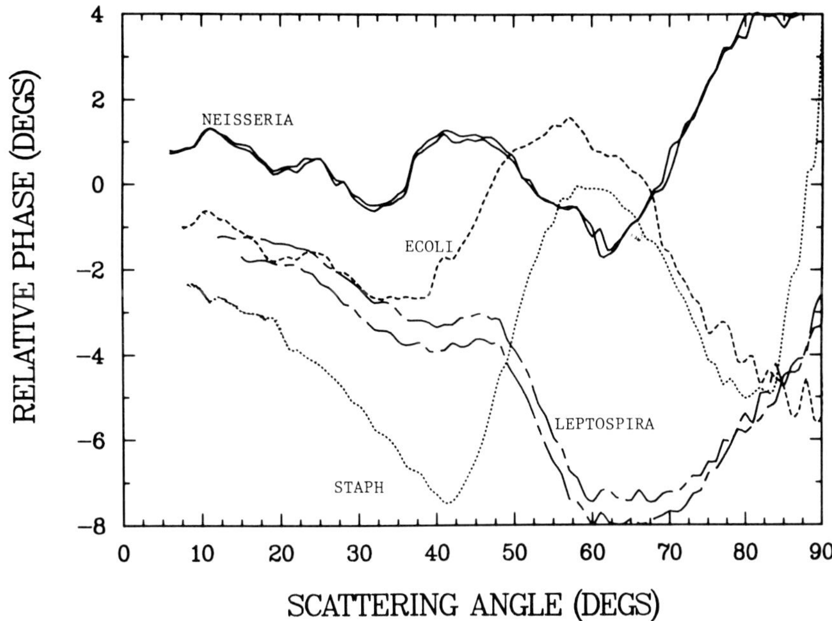

Fig. 24. Phase differential scattering (PDS) from pure aqueous suspensions of different viable bacteria. The relative phase α_2 (45°) is shown plotted as a function of scattering angle, θ. The bacteria are *Neisseria lactamica* (solid curves), *Leptospira biflexa* (chain—dash curves), *Escherichia coli* B (dashed curve), and *Staphylococcus aureus* (dotted curve), all at concentrations of approximately 5×10^7 bacterial cells per ml. The two Neisseria samples were aliquots from the same culture. They were run 5 hours apart. The two samples were prepared simultaneously, but the second sample was stored at room temperature for 5 hours before use. The two *Leptospira* samples were obtained from separate cultures grown three weeks apart. Details of the culture and preparation protocols are given in reference 63. Pure aqueous suspensions of bacteria have been distinguished by their PDS signatures at concentrations as low as 5×10^5 / ml with the 1 mW Zeeman effect helium neon laser. Bacteria can also be distinguished using backscattering measurements (not shown). (Reproduced from ref. [63], with permission of the publisher.)

Nicola et al. [109] combined FACS FLS, PLS, and fluoresceinated pokeweed mitogen fluorescence to obtain a 10- to 15-fold enhancement for in vitro colony forming cells from rat bone marrow. The three parameter measurements produced better enrichment than when only FLS and PLS were used.

Visser et al. [164] used red blood cells and suspending media of various osmolarities to demonstrate the sensitivity of FLS and PLS to cell shape. At low osmolarity, the red blood cells are swollen spherocytes and exhibit large FLS and small PLS. At high osmolarity, the red blood cells are shrunken and crenated and show small FLS and large and variable PLS. The scattered irradiance at angles away from the forward direction is much more sensitive to cell shape than is the forward scattered irradiance.

PLS has been shown experimentally to increase linearly with the area of the nucleus in four transplantable rat tumor lines [12]. The investigators were careful to choose cells for analysis that had similar levels of cytoplasmic granularity. They did not measure the area of the cytoplasm in the cells. The whole cell area may be important here. Mie theory [120] shows that the PLS detector response increases roughly linearly with whole cell cross sectional area for cell models with several different nucleus sizes.

Dubelaar et al. [40] used cyanobacteria with differing amounts of gas vacuoles to demonstrate effects on FLS and PLS in a flow cytometer. They showed that the presence of the gas vacuoles produced a 10-fold increase in PLS and a fivefold decrease in FLS. PLS is sensitive to changes in refractive index. Relative to the surrounding water medium the cell material has a refractive index of roughly 1.03, while that of the vacuoles is 0.75. The vacuoles represent a scattering object with a refractive index much different from that of the medium. Another way to view this is that the cell material adjacent to a vacuole has a large refractive index (1.37) relative to that of the vacuole. The presence of the vacuoles creates an environment in which the cell material has a high relative refractive index and thus high PLS. Without the vacuoles the cell material is surrounded only by water and has low relative refractive index and low PLS. The decrease in FLS with increasing volume of gas vacuoles inside the cell might be explained as follows: FLS is insensitive to refractive index but is senstive to cross sectional area of scattering material. The vacuoles may actually decrease the cross sectional area of the cell that is effective in scattering and lead to a reduction in FLS.

Steen and Lindmo [145] and Steen [142] developed a microscope-based flow cytometer with arc-lamp illumination (see Chapter 2, this volume). They demonstrated [143,146] that they could measure simultaneously FLS and PLS. They obtained reasonable discrimination between human lymphocytes, monocytes and granulocytes with their system. FLS in combination with DNA content measurements has been used to study cell cycle distributions in bacteria [19,144].

Hoffman et al. [56] were the first to combine immunofluorescence with FLS and PLS, making a significant practical advance in flow cytometric immunology. They used the scatter measurements to select only the lymphocyte fraction of the leukocytes. Single parameter fluorescence histograms were gated on the lymphocytes and used to identify T-lym-

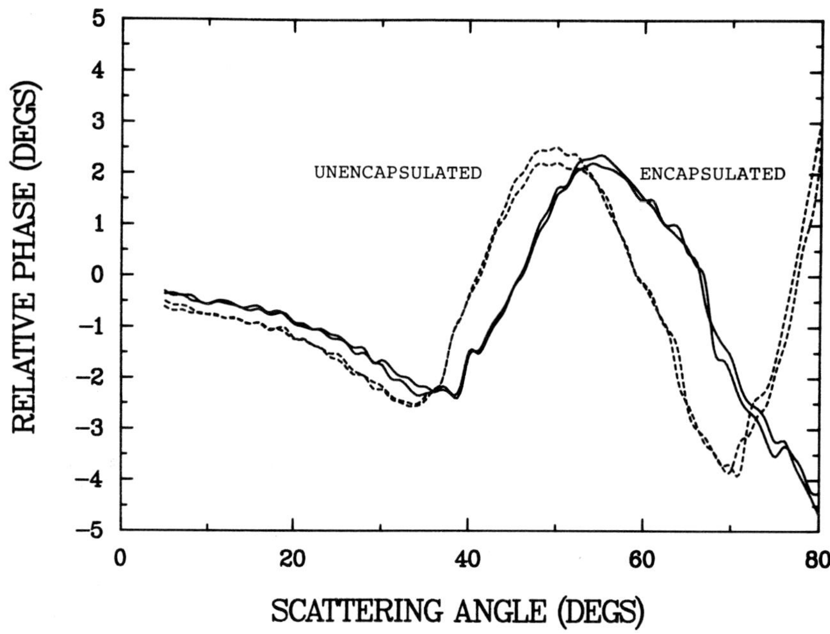

Fig. 25. Discrimination between encapsulated and unencapsulated bacteria based on α_2 (45°) versus scattering angle, θ. The viable bacteria are *B. subtilis* at a concentration of 5×10^7/ ml. The two solid curves represent replicate runs taken 20 min apart. The two dashed curves are for replicate runs taken 20 min apart for an identical sample of bacteria grown under conditions that preclude encapsulation.

phocyte subclasses by affinity for the various fluoresceinated monoclonal antibodies. Their technique is now used routinely in flow cytometric studies in immunology.

Spinrad and Brown [141] used FLS and PLS in a flow cytometer to estimate the relative real refractive indices of six species of phytoplankton cells. They showed that the ratio of PLS to FLS could be used to obtain the particle refractive index. The refractive indices were calibrated with glass beads and dispersions of oils with known refractive indices.

There have been numerous other uses for FLS and PLS. Some of these are mentioned briefly here. Sharpless et al. [129] showed that FLS could be used for size and refractive index measurements. Diamond and Braylan [36] compared FACS FLS with Coulter volume measurements on ethanol fixed cells and reported that Coulter volume was better for distinguishing non-Hodgkin's lymphomas from nonneoplastic cases. Barrett et al. [8,9] compared perpendicular light scatter measurements of cervical cells with microscopic planimeter measurements on sorted cells. They showed that PLS increased monotonically with cell cross sectional area. Frost et al.[43] used FACS FLS and PLS to enrich sputum samples for neoplastic cells. They obtained 10-fold enrichments and showed that malignant cells had scatter intensities similar to those of macrophages and squamous cells. Latimer et al. [77] calculated effects of asphericity on single particle scattering. Loken et al. [82] used a FACS to demonstrate the effects of cell asymmetry on light scattering. Meyer [96] investigated the theoretical dependence of backscattering on membrane thickness and refractive index. Meyer and Brunsting [97] examined several models of nucleated biological cells. Mullaney and Fiel [107] compared theory and experiment in studies of light scattering from suspensions of red cell ghosts.

Scattered Light Polarization Measurements

In a flow cytometer the direction of vibration of the electric field of the incident laser light is almost always vertical. If a very small homogeneous sphere passes through the laser beam the light scattered at 90° remains vertically polarized. For a larger particle, the direction of vibration of the electric field may be rotated away from the vertical. de Grooth et al. [34] devised a way to measure this depolarization in the PLS signal of a flow cytometer and have shown that scattered light depolarization can be used to distinguish among a number of leukocyte types. They used a 45° beam splitter in the PLS channel to direct the scattered light to two photomultiplier tubes. The light reflected from the beam splitter is proportional to the total amount of light scattered in the polar angular range between 65° and 115°. This signal is referred to as total orthogonal light scatter. The light transmitted by the beam splitter then passes through a polarizer with its transmission axis horizontal. This signal is referred to as depolarized orthogonal light scatter. Figure 26 contains three bivariate dot plots of human peripheral blood cells as a function of total orthogonal light scatter and depolarized orthogonal light scatter. The subpopulations were identified by sorting.

Multiwavelength Light Scattering

By illuminating cells in a flow cytometer with several wavelengths it is possible to improve the discrimination among cell types. Loken and Houck [81] illuminated the sample stream in a FACS using all the argon laser lines between 351 nm and 488 nm. They detected forward light scatter at UV wavelengths and at 488 nm. They showed that the single parameter frequency histogram resolution of mouse bone marrow cell subpopulations was better at 488

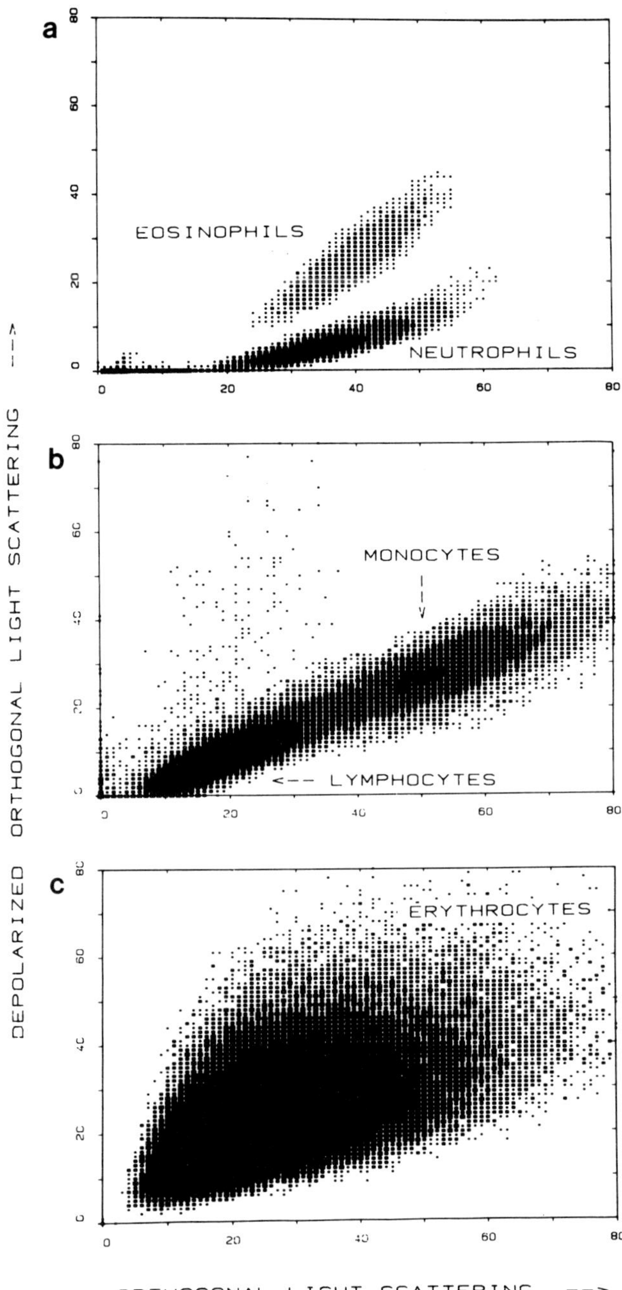

Fig. 26. Bivariate dot plots showing human blood cell number as a function of total PLS and depolarized PLS. a: Lysed human peripheral blood showing neutrophilic and eosinophilic granulocytes. b: Gain changed to show lymphocyte and monocyte subpopulations. c: Erythrocytes. (Reproduced from ref. [34], with permission of the publisher.)

nm than in the UV. Otten and Loken [112] showed that this same instrument could be used to distinguish among several mouse lymphocyte subpopulations.

In an approach similar to that of Curbelo et al. [32], de Grooth et al. [35] developed a flow cytometer in which a prism is used to disperse the light from a mercury arc lamp along the sample stream. In a study of lysed human blood cells, these investigators presented a bivariate plot of cell number as a function of PLS at 577 nm (x-axis) and PLS at 405 nm (y-axis). In order of increasing intensity, lymphocytes, monocytes and granulocytes fell along a straight line extending diagonally from the origin. The erythrocytes scattered 577 nm light the same as monocytes and 405 nm light the same as lymphocytes.

Multiangle Light Scattering

The successful use of forward light scattering for cell sizing and a desire to test theoretical models have led numerous investigators to develop both static and flow instruments for measuring light scattering at several angles. Brunsting and Mullaney [22] developed an instrument that enabled detection of scattered helium neon laser light between 2° and 25° from cuvette suspensions; 35 mm film was used to record the scattered light. A microdensitometer was used to read the irradiance distribution from the film. Brunsting and Mullaney [24] used this instrument to measure the scattered light angular distribution for Chinese hamster tissue culture cells (CHO) blocked in various phases of the cell cycle. They developed a computer program [23] to show that the coated sphere model of Aden and Kerker [2] was adequate to represent the data to a scattering angle of about 15°. In this same paper, they presented data on refractive index measurements of single CHO cells near 633 nm.

A 32-element concentric ring photodiode array was adapted to a flow cytometer for higher resolution forward scattering measurments between 0° and 21° [98,122]. The area of the rings increased rapidly with increasing scattering angle to partially compensate for the rapid decrease in scatter intensity with increasing angle. This system was used to discriminate among a variety of different cell types [31,117,126,149,150]. The largest differences among the subpopulations were in the angular range between 0.4° and 2°. Many of the signals were highly correlated.

In a system using a single detector with a FACS flow cytometer, Loken et al. [83] illuminated the stream with a slit of 488 nm laser light in which the slit was parallel to the stream. As a cell passed through the beam it scattered light onto a single detector at angles ranging from 1° to 49°. A waveform recorder captured the scatter pattern. Data were presented for several mouse thymocytes.

Ludlow and Kaye [91] developed a flow cytometer in which the scattered light from a single slowly moving particle can be measured at 1° intervals between 3° and 177° using fiber optic light guides, a rotating aperture and a single photomultiplier tube. They used the device and Mie theory to determine the refractive index and size of several bacterial spores.

Morris et al. [100] developed an instrument to study the dynamics of polymerization of microtubular protein and the osmotic lysis of chromaffin granules. The scattering from phospholipid vesicles and micelles in chromatographic column effluents were measured between 20° and 160° with a resolution of 0.25° using a cylindrical flow chamber and a photomultiplier tube that rotated around the cuvette at a frequency of 1 Hz.

Following a concept developed by an aerosol light scattering group at Indiana University [49,93,156], Bartholdi et al. [10] developed a flow cytometer using an ellipsoidal reflector and an annular ring photodiode array. The scattering angle ranged from 2.5° to 177.5° and 32 signals could be measured simultaneously for particle rates up to 1000 per second. The system was used to study polystyrene spheres with diameters between 1.1 μm and 19.5 μm.

Multiangle scattering flow cytometers have been most useful for studies involving comparisons with theory. For cell discrimination several carefully placed detectors are as good as any of the multiangle systems described above.

SUMMARY AND CONCLUSIONS

Good physical insight into the process of light scattering from biological cells can be obtained by considering a cell to be composed of many tiny dipolar antennas. The antennas radiate (scatter) in response to the electric field of the incident light beam. The irradiance at a detector is the sum of the wavelets incident on the detector taking into account the phase relations among the wavelets.

There are many theories that attempt to model scattering by cells. The theories are all approximations to reality. A particle should not be considered a "Mie scatterer" or a "Rayleigh particle." This chapter discusses the coupled dipole approximation for scattering by biological cells and subcellular organelles. This theory attempts to model an arbitrarily shaped scattering object as a composite of many tiny dipoles. Calculations are presented only for relatively small particles, but the theory looks promising as a method for approximating the scattering by complex biological particles.

Polarization measurements and calculations are becoming increasingly important as a tool for obtaining more information about cells. Nearly all biological molecules and structures are chiral, that is, they cannot be overlaid on their mirror images. This property guarantees that biological objects will differentially scatter light of different polarizations and thus provide information about their structure. The Mueller matrix formalism is presented to serve as a framework for a discussion of polarization. Phase differential scattering is presented as a new tool with which some types of polarization measurements may be made with very few artifacts. Scattered light polarization measurements in flow are discussed. They were used for distinguishing among some types of leukocytes.

The traditional uses of light scattering for cell sizing in flow cytometry are discussed. Extinction measurements are shown to be reliable for cell sizing over a broad range of cell diameters. Applications of combined forward (FLS) and 90° (PLS) scattering have become the predominant use of light scattering in flow cytometry. Multiangle and multiwavelength scattering are discussed, but have not been shown to provide much more discriminating power than that provided by combined FLS and PLS.

Further development of the coupled dipole theory may provide new methods for analyzing structural properties of cells, particularly by means of scattered light polarization. Experimental efforts to make flow cytometric polarization measurements are likely to increase and will provide new data to test the theoretical models. Light scattering has a bright future in flow cytometry.

ACKNOWLEDGMENTS

This work was performed under the auspices of the U.S. Department of Energy and was supported in part by the National Flow Cytometry and Sorting Resource (grant RR01315), the National Institute of General Medical Sciences (grant GM26857), and by the U.S. Army Chemical Research Development and Engineering Center. We would like to acknowledge Drs. Francisco Tomei and Marty Bartholdi for careful review of this manuscript. We also acknowledge Dr. Tomei and Ms. Cheryl Lemanski for microbiological assistance. This chapter is dedicated to Dr. Paul Mullaney, who pioneered flow cytometric light scattering studies and who provided visionary leadership for analytical cytology.

REFERENCES

1. **Adams LR, Kamentsky LA (1971)** Machine characterization of human leukocytes by acridine orange fluorescence. Acta Cytol. (Praha) 15:289–291.
2. **Aden AL, Kerker M (1951)** Scattering of electromagnetic waves from two concentric spheres. J. Appl. Phys. 22:1242–1246.
3. **Asano S (1979)** Light scattering properties of spheroidal particles. Appl. Opt. 18:712–723.
4. **Asano S, Sato M (1980)** Light scattering by randomly oriented spheroidal particles. Appl. Opt. 19:962–974.
5. **Asano S, Yamamoto G (1975)** Light scattering by a spheroidal particle. Appl. Opt. 14:29–48.
6. **Atkins PW, Barron LD (1969)** Rayleigh scattering of polarized photons by molecules. Mol. Phys. 16:453–466.
7. **Barber P, Yeh C (1975)** Scattering of electomagnetic waves by arbitrarily shaped dielectric bodies. Appl. Opt. 14:2864–2872.
8. **Barrett DL, King EB, Jensen RH, Merrill JT (1978)** Cytomorphology of gynecologic specimens analyzed and sorted by two-parameter flow cytometry. Acta. Cytol. (Praha) 22:7–14.
9. **Barrett DL, Jensen RH, King EB, Dean PN, and Mayall BH (1979)** Flow cytometry of human gynecologic specimens using log chromomycin A_3 fluorescence and 90° light scatter. J. Histochem. Cytochem. 27: 573–578.
10. **Bartholdi M, Salzman GC, Hiebert RD, Kerker M (1980)** Differential light scattering photometer for rapid analysis of single particles in flow. Appl. Opt. 19:1573–1581.
11. **Belmont A, Zietz S, Nicolini C (1985)** Differential scattering of circularly polarized light by chromatin modeled as a helical array of dielectric ellipsoids within the Born approximation. Biopolymers 24: 1301–1321.
12. **Benson BC, McDougal DC, Coffey DS (1984)** The application of perpendicular and forward light scattering to assess nuclear and cellular morphology. Cytometry 5:515–522.
13. **Bickel WS, Davidson JF, Huffman DR, Kilkson R (1976)** Application of polarizaton effects in light scattering: A new biophysical tool. Proc. Nat. Acad. Sci. USA 73:486–490.
14. **Bickel WS, Stafford ME (1981)** Polarized light scattering from biological systems: A technique for cell differentiation. J. Biol. Phys. 9:53–66.
15. **Blakemore RP (1982)** Magnetotactic bacteria. Annu. Rev. Microbiol. 36:217–238.
16. **Bohren CF (1987)** "Clouds in a Glass of Beer." New York: John Wiley & Sons.
17. **Bohren CF, Huffman DR (1983)** "Absorption and Scattering of Light by Small Particles." New York: John Wiley & Sons.
18. **Born M, Wolf E (1965)** "Principles of Optics." London: Pergamon Press.

19. **Boye E, Steen HB, Skarstad K (1983)** Flow cytometry of bacteria: A promising tool in experimental and clinical microbiology. J. Gen. Microbiol. 129:973–980.
20. **Bracewell BM (1965)** "The Fourier Transform and Its Applications." San Francisco: McGraw-Hill.
21. **Braylon RC, Benson NA, Nourse V, Kruth HS (1982)** Correlated analysis of cellular DNA membrane antigens and light scatter of human lymphoid cells. Cytometry 2:337–343.
22. **Brunsting A, Mullaney PF (1972)** A light scattering photometer using photographic film. Rev. Sci. Instr. 43:1514–1519.
23. **Brunsting A, Mullaney PF (1972)** Light scattering from coated spheres: models for biological cells. Appl. Optics 11:675–680.
24. **Brunsting A, Mullaney PF (1974)** Differential light scattering from spherical mammalian cells. Biophys. J. 14:439–453.
25. **Burger DE, Jett JH, Mullaney PF (1982)** Extraction of morphological features from biological models and cells by Fourier analysis of static light scatter measurements. Cytometry 2:327–336.
26. **Bustamante C, Maestre MF, Tinoco I Jr (1980)** Circular intensity differential scattering of light by helical structures. I. Theory. J. Chem. Phys. 73:4273–4281.
27. **Bustamante C, Maestre MF, Tinoco I Jr (1980)** Circular intensity differential scattering of light by helical structures. II. Applications. J. Chem. Phys. 73: 6046–6055.
28. **Bustamante C, Maestre MF, Keller D, Tinoco I Jr (1984)** Differential scattering (CIDS) of circularly polarized light by dense particles. J. Chem. Phys. 80: 4817–4823.
29. **Bustamante C, Tinoco I Jr, Maestre MF (1982)** Circular intensity differential scattering of light. IV. Randomly oriented species. J. Chem. Phys. 76:3440–3446.
30. **Cross DA, Latimer P (1972)** Angular dependence of scattering from Escherichia coli cells. Appl Optics 11: 1225–1228.Price BJ, Cram LS, Mullaney, PF
31. **Crowell JM, Hiebert RD, Salzma GC, Price BJ, Cram LS, Mullaney PF (1978)** A light scattering system for high speed cell analysis. IEEE Trans. Biomed. Eng. BME-25:519–526.
32. **Curbelo R, Schildkraut ER, Hirschfeld T, Webb RH, Block MJ, Shapiro H (1976)** A generalized machine for automated flow cytometry system design. J. Histochem. Cytochem. 24:388–395.
33. **Dave JV (1968)** Subroutine for computing the parameters of the electromagnetic radiation scattered by a sphere. IBM Scientific Center, Pal Alto, Calif., report 320-337. The program is available from G.C. Salzman at Los Alamos.
34. **deGrooth BG, Terstappen LWMM, Puppels GJ, Greve J (1987)** Light-scattering polarization measurements as a new parameter in flow cytometry. Cytometry 8:539–544.
35. **deGrooth BG, van Dam M, Swart NC, Willemsen A, Greve J (1987)** Multiple wavelength illumination in flow cytometry using a single arc lamp and a dispersing element. Cytometry 8:445–452.
36. **Diamond LW, Braylan RC (1980)** Flow analysis of DNA content and cell size in non-Hodgkin's lymphoma. Cancer Res. 40:703–712.
37. **Doukas JD, Ruckdeschel JC, Mardiney MR Jr (1977)** Quantitative and qualitative analysis of human lymphocytes proliferation to specific antigen in vitro by use of the helium neon laser. J. Immunol. Methods 15:229–238.
38. **Doyle WT (1985)** Scattering approach to Fresnel's equations and Brewster's law. Am. J. Phys. 53:463–468.
39. **Druger SD, Kerker M, Wang DS, Cooke DD (1979)** Light scattering by inhomogeneous particles. Appl. Optics 18:3888–3892.
40. **Dubelaar GBJ, Visser JWM, Donze M (1987)** Anomalous behavior of forward and perpendicular light scattering of a cyanobacterium owing to intracellular gas vacuoles. Cytometry 8:405–412.
41. **Fikioris JG, Uzunoglu NK (1979)** Scattering from an eccentrically stratified dielectric sphere. J. Opt. Soc. Am. 69:1359–1366.
42. **Fowler BW, Sung CC (1979)** Scattering of an electromagnetic wave from dielectric bodies of irregular shape. J. Opt. Soc. Am. 69:756–761.
43. **Frost JK, Tyrer HW, Pressman NJ, Albright CD, Vansickel MH, Gill GW (1979)** Automatic cell identification and enrichment in lung cancer. I. Light scatter and fluorescence parameters. J. Histochem. Cytochem. 27:545–551.
44. **Genter FC, Salzman GC (1979)** A statistical approach to the classification of biological cells from their diffraction patterns. J. Histochem. Cytochem. 27:268–272.
45. **Goldschneider I, Gordon LK, Morris RJ (1978)** Demonstration of Thy-1 antigen on pluripotent hemopoietic stem cells in the rat. J. Exp. Med. 148:1351–1366.
46. **Goldschneider I, Metcalf D, Battye F, Mandel T (1980)** Analysis of rat hemopoietic cells on the fluorescence activated cell sorter. J. Exp. Med. 152:419–437.
47. **Goldschneider I, Metcalf D, Mandel T, Bollum FJ (1980)** Analysis of rat hemopoietic cells on the fluorescence activated cell sorter. II. Isolation of terminal deoxynucleotidyl transferase-positive cells. J. Exp. Med. 152:438–446.
48. **Goodman JW (1968)** "Introduction to Fourier Optics." San Francisco: McGraw-Hill.
49. **Gucker FT, Tuma J, Lin H-M (1976)** Rapid measurements of light-scattering diagrams from single particles in an aerosol stream and determination of latex particle size. Aerosol. Sci. 55:624–636.
50. **Hardy JA, Wheeless LL (1977)** Application of Fraunhofer diffraction theory to feature specific detector design. J. Histochem Cytochem. 25:857–863.
51. **Harding SE (1986)** Applications of light scattering in microbiology. Biotechnol. Appl. Biochem. 8:489–509.
52. **Harris RA, McClain WM (1985)** On the manifestation of retardation effects in diagonally polarized light scattering. J. Chem. Phys. 82:658–663.
53. **Harris RA, McClain WM, Sloane CF (1974)** On the theory of polarized light scattering from dilute polymer solutions. Mol. Phys. 28:381–398.
54. **Hecht E, Zajac A (1974)** "Optics." Reading, Massachusetts: Addison-Wesley.

55. **Hodkinson JR, Greenleaves I (1963)** Computations of light-scattering and extinction by spheres according to diffraction and geometrical optics, and some comparisons with Mie theory. J. Opt. Soc. Am. 53: 577–588.
56. **Hoffman RA, Kung PC, Hansen WP, Goldstein G (1980)** Simple and rapid measurement of human T lymphocytes and their subclasses in peripheral blood. Proc. Nat. Acad. Sci. U.S.A. 77:4914–4917.
57. **Holland AC, Gagne G (1970)** The scattering of polarized light by polydisperse systems of irregular particles. Appl. Optics 9:1113–1121.
58. **Hunt AJ, Huffman DR (1973)** A new polarization-modulated light-scattering instrument. Rev. Sci. Instrum. 44:1753–1762.
59. **Iskander MF, Lakhtakia A, Durney CH (1983)** A new procedure for improving the solution stability and extending the frequency range of EBCM. IEEE Trans. Antennas Prop. AP-31:317–324.
60. **Iskander MF, Olson SC, Benner RE, Yoshida D (1986)** Optical scattering by metallic and carbon aerosols of high aspect ratio. Appl. Optics 25:2514–2520.
61. **Jackson JD (1975)** "Classical Electrodynamics." New York: John Wiley & Sons.
62. **Johnston RG, Singham SB, Salzman GC (1986)** Phase differential scattering from microspheres. Appl. Optics 25:3566–3572.
63. **Johnston RG, Singham SB, Salzman GC (1988)** Polarized light scattering. Comments on Molecular and Cellular Biophysics. 5:171–192.
64. **Johnston RG, Singham SB, Salzman GC (1987)** Zeeman laser scattering: A new light scattering technique. In Nicolini C (ed.), "Structure and Dynamics of Biopolymers." Boston: Martinus Nijhoff, pp 56–65.
65. **Jovin TM, Morris SJ, Striker G (1976)** Automatic sizing and separation of particles by ratios of light scattering intensities. J. Histochem. Cytochem. 24: 269–283.
66. **Julius MH, Sweet RG, Fatham CG, Herzenberg LA (1975)** Fluorescent-activated cell sorting and its applications. In Richmond CR, Petersen DF, Mullaney PF, et al. (eds.), "Mammalian Cells: Probes and Problems." ERDA Symposium Series CONF-731007, Technical Information Center, Oak Ridge, Tennessee, p 107.
67. **Kamentsky LA (1973)** Cytology automation. Adv. Biol. Med. Phys. 14:93–161.
68. **Kamentsky LA, Melamed MR (1965)** Spectrophotometer: New instrument for ultra-rapid cell analysis. Science 150:630–631.
69. **Katz JE, Wells S, Ussery D, Bustamante C, Maestre MF (1984)** Design and construction of a circular intensity differential scattering instrument. Rev. Sci. Instr. 55:1574–1579.
70. **Kerker M (1969)** "The Scattering of Light and Other Electromagnetic Radiation." New York: Academic Press.
71. **Kerker M, Cooke DD, Chew H, McNulty PJ (1978)** Light scattering by structured spheres. J. Opt. Soc. Am. 68:592–601.
72. **Koch AL (1968)** Theory of the angular dependence of light scattered by bacteria and similar-sized biological objects. J. Theoret. Biol. 18:133–156.
73. **Koch AL (1986)** Estimation of size of bacteria by low-angle light scattering measurements: Theory. J. Microbiol. Methods 5:221–235.
74. **Kopp RE, Lisa J, Mendelsohn J, Pernick B, Stone H, Wohlers R (1974)** The use of coherent optical processing techniques for the automatic screening of cervical cytologic specimens. J. Histochem. Cytochem. 22:598–604.
75. **Kopp RE, Lisa J, Mendelsohn J, Pernick B, Stone H, Wohlers R (1976)** Coherent optical processing of cervical cytologic samples. J. Histochem. Cytochem. 24: 122–137.
76. **Latimer P (1982)** Light scattering and absorption as methods of studying cell population parameters. Annu. Rev. Biophys. Bioeng. 11:129–150.
77. **Latimer P, Brunsting A, Pyle BE, Moore C (1978)** Effects of asphericity on single particle scattering. Appl. Optics 17:3152–3158.
78. **Lehnert BE, Steinkamp JA (1985)** Identification and isolation of subpopulations of pleural cells by multiparameter flow cytometry. Cell Biophys. 8:201–212.
79. **Lehnert BE, Valdez YE, Fillak DA, Steinkamp JA, Stewart CC (1986)** Flow cytometric characterization of alveolar macrophages. J Leukocyte Biol. 39:285–298.
80. **Loken MR, Herzenberg LA (1975)** Analysis of cell populations with a fluorescence activated cell sorter. Ann. N.Y. Acad. Sci. 254:163–171.
81. **Loken MR, Houck DW (1981)** Light scattered at two wavelengths can discriminate viable lymphoid cell populations on a fluorescence activated cell sorter. J. Histochem. Cytochem. 29:609–615.
82. **Loken MR, Parks DR, Herzenberg LA (1977)** Identification of cell asymmetry and orientation by light scattering. J. Histochem. Cytochem. 25:790–795.
83. **Loken MR, Sweet RG, Herzenberg LA (1976)** Cell discrimination by multiangle light scattering. J. Histochem. Cytochem. 24:284–291.
84. **Lord Rayleigh (1871)** On the light from the sky, its polarization and colour. Phil. Mag. 41:107–120.
85. **Lord Rayleigh (1871)** On the light from the sky, its polarization and colour. Phil. Mag. 41:274–279.
86. **Lord Rayleigh (1871)** On the scattering of light by small particles. Phil. Mag. 41:447–454.
87. **Lord Rayleigh (1881)** On the electomagnetic theory of light. Phil. Mag. 12:81–101.
88. **Lord Rayleigh (1910)** The incidence of light upon a transparent sphere of dimensions comparable with the wavelength. Proc. R. Soc. Lond. A84:25–46.
89. **Lord Rayleigh (1914)** On the diffraction of light by spheres of small relative index. Proc. R. Soc. Lond. A90:219–225.
90. **Lord Rayleigh (1918)** On the scattering of light by spherical shells, and by complete spheres of periodic structure, when the refractivity is small. Proc. R. Soc. Lond. A94:296–300.
91. **Ludlow IK, Kaye PH (1979)** A scanning diffractometer for rapid analysis of microparticles and biological cells. J. Colloid Interface Sci. 69:571–589.
92. **Maestre MF, Bustamante C, Hayes TL, Subirana JA, Tinoco I Jr. (1982)** Differential scattering of circularly polarized light by the helical sperm head from the octopus Eledone cirrhosa. Nature (Lond.) 298: 773–774.
93. **Marshall TR, Parmenter CS, Seaver M (1976)** Char-

acterization of polymer latex aerosols by rapid measurement of 360° light scattering patterns from individual particles. J. Colloid Interface Sci. 55:624–636.

94. **Martin A, Merdrignac G, Collet B, Genetet B (1983)** Characterization and sorting of mouse cytotoxic macrophages by their light scattering properties. Cytometry 4:250–253.

95. **McNeil PL, Kennedy AL, Waggoner AS, Taylor DL, Murphy RF (1985)** Light-scattering changes during chemotactic stimulation of human neutrophils: Kinetics followed by flow cytometry. Cytometry 6:7–12.

96. **Meyer RA (1979)** Light scattering from biological cells: Dependence of backscattering radiation on membrane thickness and refractive index. Appl. Optics 18:585–588.

97. **Meyer RA, Brunsting A (1975)** Light scattering from nucleated biolgical cells. Biophys. J. 15:191–203.

98. **Meyer RA, Haase SF, Podulso SE, McKhan GM (1974)** Light scattering patterns of isolated oligodendroglia. J. Histochem. Cytochem. 22:594–597.

99. **Mie G (1908)** Beitrage zur Optik truber Medien, speziell kolloidaler Metallosungen. Ann. Phys. 25: 377–445.

100. **Morris SJ, Schultens HA, Hellweg MA, Striker G, Jovin TM (1979)** Dynamics of structural changes in biological particles from rapid light scattering measurements. Appl. Optics 18:303–311.

101. **Mueller H (1948)** The foundations of Optics. J. Opt. Soc. Am. 38:661.

102. **Mugnai A, Wiscombe W (1980)** Scattering of radiation by moderately non-spherical particles. J. Atmos. Sci. 37:1291–1307.

103. **Mugnai A, Wiscombe WJ (1986)** Scattering from nonspherical Chebyshev particles. I: cross sections, single-scattering albedo, asymmetry factor, and backscattered fractions. Appl. Optics 25:1235–1244.

104. **Mullaney PF (1970)** Application of the Hodkinson scattering model to particles of low relative refractive index. J. Opt. Soc. Am. 60:573–574.

105. **Mullaney PF, Dean PN (1969)** Cell sizing: A small-angle light-scattering method for sizing particles of low relative refractive index. Appl. Optics 8:2361–2362.

106. **Mullaney PF, Dean PN (1970)** The small angle light scattering of biological cells. Biophys. J. 10:764–772.

107. **Mullaney PF, Fiel RJ (1976)** Cellular structure as revealed by visible light scattering: Studies on suspensions of red blood cell ghosts. Appl. Optics 15:310–311.

108. **Mullaney PF, Van Dilla M, Coulter JR, Dean PN (1969)** Cell sizing: A light scattering photometer for rapid volume determination. Rev. Sci. Instr. 40: 1029–1032.

109. **Nicola NA, Burgess AW, Staber FG, Johnson GR, Metcalf D, Battye FL (1980)** Differential expression of lectin receptors during hemopoietic differentiation: Enrichment for granulocyte-macrophage progenitor cells. J. Cell Physiol. 103:217–237.

110. **Nielson O, Larsen JK, Christensen IJ, Lernmark A (1982)** Flow sorting of mouse pancreatic B cells by forward and orthogonal light scattering. Cytometry 3:177–181.

111. **Optra, Inc., Peabody, Massachusetts.**

112. **Otten GR, Loken MR (1982)** Two color light scattering identifies physical differences between lymphocyte subpopulations. Cytometry 3:182–187.

113. **Pernick B, Jost S, Herold R, Mendelsohn J, Wohlers R (1978)** Screening of cervical cytological samples using coherent optical processing. Part 3. Appl. Optics 17:43–51.

114. **Pernick B, Kopp RE, Lisa J, Mendelsohn J, Stone H, Wohlers R (1978)** Screening of cervical cytologic samples using coherent optical processing. Part 1. Appl. Optics 17:21–34.

115. **Pernick B, Wohlers MR, Mendelsohn J (1978)** Paraxial analysis of light scattering by biological cells in a flow system. Appl. Optics 17:3205–3215.

116. **Perry RJ, Hunt AJ, Huffman DR (1978)** Experimental determination of Mueller scattering matrices for nonspherical particles. Appl. Optics 17:2700–2710.

117. **Price BJ, Kollman VH, Salzman GC (1978)** Light scatter analysis of microalgae: Correlation of scatter patterns from pure and mixed asynchronous cultures. Biophys. J. 22:29–36.

118. **Purcell EM, Pennypacker CR (1973)** Scattering and absorption of light by nonspherical dielectric grains. Astrophys. J. 186:705–714.

119. **Rozycka M, Lenczowski S, Sawicki W, Baranska W, Ostrowski K (1982)** Optical diffraction as a tool for semiautomatic, quantitative analysis of tissue specimens. Cytometry 2:244–248.

120. **Salzman GC (1982)** Light Scattering Analysis of Single Cells. In Catsimpoolas N (ed.), "Cell Analysis." New York: Plenum, pp 111–143.

121. **Salzman GC, Crowell JM, Martin JC, Trujillo TT, Romero A, Mullaney PF, LaBauve PM (1975)** Cell classification by laser light scattering: Identification and separation of unstained leukocytes. Acta. Cytol. (Praha) 19:374–377.

122. **Salzman GC, Crowell JM, Goad CA, Hansen KM, Hiebert RD, LaBauve PM, Martin JC, Ingram M, Mullaney PF (1975)** A flow-system multiangle light scattering instrument for cell characterization. Clin. Chem. 21:1297–1304.

123. **Salzman GC, Griffith JK, Gregg CT (1982)** Rapid identification of microorganisms by circular intensity differential scattering. Appl. Environ. Microbiol. 44: 1081–1085.

124. **Salzman GC, Mullaney PF, Price BJ (1979)** Light-scattering approaches to cell characterization. In Melamed MR, Mullaney PF, Mendelsohn MR (eds.), "Flow Cytometry and Sorting." New York: John Wiley & Sons. pp 108–124.

125. **Salzman GC, Wilder ME, Jett JH (1979)** Light scattering with stream-in-air flow systems. J. Histochem. Cytochem. 27:264–267.

126. **Schafer IA, Jamieson AM, Petrelli M, Price BJ, Salzman GC (1979)** Multiangle light scattering flow photometry of cultured human fibroblasts. Comparison of normal cells with a mutant cell line containing cytoplasmic inclusions. J. Histochem. Cytochem. 27: 359–365.

127. **Seger G, Achatz M, Heinze W, Sinsel F (1977)** Quantitative extraction of morphological cell parameters from the diffraction pattern. J. Histochem. Cytochem. 25:707–718.

128. **Sekera Z (1957)** Polarization of Skylight. In Flugge S (ed.), "Handbuch der Physik." Vol. 1. Berlin: Springer-Verlag. pp 288–328.

129. **Sharpless TK, Bartholdi M, Melamed MR** (1977) Size and refractive index dependence of simple forward-angle scattering measurements in a flow system using sharply focused illumination. J. Histochem. Cytochem. 25:845–856.
130. **Shindo Y, Nakagawa M** (1985) On the artifacts in circularly polarized emission spectroscopy. Appl. Spectrosc. 39:32–38.
131. **Shindo Y, Nakagawa M** (1985) Circular dichroism measurements. 1. Calibration of a circular dichroism spectrometer. Rev. Sci. Instr. 56:32–39.
132. **Shurcliff WA** (1962) "Polarized Light." Cambridge, Massachusetts: Harvard University Press.
133. **Singham MK, Singham SB, Salzman GC** (1986) The scattering matrix for randomly oriented spheroids. J. Chem. Phys. 85:3807–3815.
134. **Singham SB** (1986) Intrinsic optical activity in light scattering from an arbitrary particle. Chem. Phys. Lett. 130:139–144.
135. **Singham SB** (1987) Form and intrinsic optical activity in light scattering by chiral particles. J. Chem. Phys. 87:1873–1891.
136. **Singham SB** (1988) Coupled dipoles in light scattering by randomly oriented chiral particles. J. Chem. Phys. 88:1522–1527.
137. **Singham SB, Bohren CF** (1987) Light scattering by an arbitrary particle: A physical reformulation of the coupled dipole method. Opt. Lett. 12:10–12.
138. **Singham SB, Patterson CW, Salzman GC** (1986) Polarizabilities for light scattering from chiral particles. J Chem Phys 85:763–770.
139. **Singham SB, Salzman GC** (1986) Evaluation of the scattering matrix of an arbitrary particle using the coupled dipole approximation. J. Chem. Phys. 84: 2658–2667.
140. **Sklar LA, Oades ZG, Finney DA** (1984) Neutrophil degranulation detected by right angle light scattering: Spectroscopic methods suitable for simultaneous analyses of degranulation or shape change, elastase release, and cell aggregation. J. Immunol. 133:1483–1487.
141. **Spinrad RW, Brown JF** (1986) Relative real refractive index of marine microorganisms: A technique for flow cytometric estimation. Appl. Optics 25:1930–1934.
142. **Steen HB** (1979) Further developments of a microscope based flow cytometer: Light scatter detection and excitation intensity compensation. Cytometry 1: 26–31.
143. **Steen HB** (1986) Simultaneous separate detection of low angle light scattering in an arc lamp-based flow cytometer. Cytometry 7:445–449.
144. **Steen HB, Boye E** (1980) Escherichia coli growth studied by dual-parameter flow cytometry. J. Bacteriol. 145:1091–1094.
145. **Steen HB, Lindmo T** (1979) Flow cytometry: A high resolution instrument for everyone. Science 204: 403–404.
146. **Steen HB, Lindmo T** (1985) Differential light-scattering detection in an arc-lamp-based epi-illumination flow cytometer. Cytometry 6:281–285.
147. **Steinkamp JA** (1983) A differential amplifier circuit for reducing noise in axial light loss measurements. Cytometry 4:83–87.
148. **Steinkamp JA, Fulwyler MJ, Coulter JR, Hiebert RD, Horney JL, Mullaney PF** (1973) A new multiparameter separator for microscopic particles and biological cells. Rev. Sci. Instr. 44:1301–1310.
149. **Steinkamp JA, Hansen KM, Wilson JS, Salzman GC** (1977) Automated analysis and separation of cells from the respiratory tract: Preliminary characterization studies in hamsters. J. Histochem. Cytochem. 25:892–898.
150. **Strom S** (1975) On the integral equations for electromagnetic scattering. Am J. Phys. 43:1060–1069.
151. **Swartzendruber DE, Price BJ, Rall LB** (1979) Multiangle light-scattering analysis of murine teratocarcinoma cells. J. Histochem. Cytochem. 27:560–563.
152. **Terstappen LWMM, deGrooth BG, Nolten GMJ, ten Napel CHH, van Berkel W, Greve J** (1986) Physical discrimination between human T-lymphocyte subpopulations by means of light scattering, revealing two populations of T8-positive cells. Cytometry 7: 178–183.
153. **Thompson RC, Bottiger JR, Fry ES** (1980) Measurement of polarized light interactions via the Mueller matrix. Appl. Optics 19:1323–1332.
154. **Tinoco IJr, Mickols W, Maestre MF, Bustamante C** (1987) Absorption, scattering, and imaging of biomolecular structures with polarized light. Annu. Rev. Biophys. Biochem. 16:319–349.
155. **Tomei FA, Coulter KL, Horsten JL, Lemanski CL, Sebring RJ, Johnston RG** (1988) The study of bacterial morphology using phase differential light scattering. Los Alamos National Laboratory Report No. LAUR 88–130.
156. **Tuma J, Gucker FT** (1971) Design of an instrument for rapid measurement of light scattering diagrams from single aerosol particles. In Proceedings of the Second International Clean Air Congress. New York: Academic Press, pp 463–467.
157. **Turke B, Seger G, Achatz M, et al.** (1978) Fourier optical approach to the extraction of morphological parameters from the diffraction pattern of biological cells. Appl. Opt. 17:2754–2761.
158. **van de Hulst HC** (1957) "Light Scattering by Small Particles." New York: Wiley. (Reprinted by Dover, New York, 1981.)
159. **van den Engh G, Visser J** (1979) Light scattering properties of pluripotent and committed haemopoietic stem cells. Acta Haematol. 62:289–298.
160. **van den Engh G, Visser J, Trask B** (1979) Identification of CFU-s by scatter measurements on a light activated cell sorter. In Baum SJ, Ledney GD (eds.), "Experimental Hematology Today." New York: Springer-Verlag, pp 19–26.
161. **van den Engh G, Visser J, Bol S, Trask B** (1980) Concentration of hemopoietic stem cells using a light activated cell sorter. Blood Cells 6:609–623.
162. **Varadan VK, Varadan VV (eds.)** (1980) Acoustic, electromagnetic and elastic wave scattering—focus on the T-matrix approach. Oxford: Pergamon Press.
163. **Visser JWM, Cram LS, Martin JC, Salzman GC, Price BJ** (1978) Sorting of a murine granulocytic progenitor cell by use of laser light scattering measurements. In Lutz D (ed.), "Pulse Cytophotometry." Part III. Ghent, Belgium: European Press, pp 187–192.
164. **Visser JWM, van den Engh GJ, van Bekkum DW** (1980) Light scattering properties of murine hemopoietic cells. Blood Cells 6:391–407.

165. **Walker DC** (ed.) **(1979)** "Origins of Optical Activity in Nature." New York: Elsevier.
166. **Wang D-S, Barber PW (1980)** Scattering by inhomogeneous nonspherical objects. Appl. Optics 18: 1190–1197.
167. **Waterman PC (1965)** Matrix formulation of electromagnetic scattering. Proc. IEEE 53:805–812.
168. **Waterman PC (1971)** Symmetry, unitarity, and geometry in electromagnetic scattering. Phys. Rev. D3: 825–834.
169. **Weil H, Chu CM (1976)** Scattering and absorption of electromagnetic radiation by thin dielectric disks. Appl. Optics 15:1832–1836.
170. **Wells KS, Beach DA, Keller D, Bustamante C (1986)** An analysis of circular intensity differential scattering measurements: Studies on the sperm cell of Eledone cirrhosa. Biopolymers 25:2043–2064.
171. **Wilson J, Hawkes JFB (1983)** "Optoelectronics: An Introduction." Englewood Cliffs, New Jersey; Prentice-Hall. pp 93–109.
172. **Wohlers R, Mendelsohn J, Kopp RE, Pernick B (1978)** Screening of cervical cytological samples using coherent optical processing. Part 2. Appl. Optics 17: 35–42.
173. **Wyatt PF (1968)** Differential light scattering: A physical method for identifying living bacterial cells. Appl. Optics 7:1879–1896.
174. **Zeitz S, Belmont A, Nicolini C (1983)** Differential scattering of circularly polarized light as a unique probe of polynucleosome superstructure. Cell Biophys. 5:163–187.

6

Slit-Scanning

Leon L. Wheeless, Jr.

Analytical Cytology Unit, Department of Pathology and Laboratory Medicine,
University of Rochester Medical Center, Rochester, New York 14642

INTRODUCTION

Slit-scan cytofluorometry provides morphological information in addition to quantitative fluorescence measurements on cells and cellular organelles in flow. Originally developed to reduce the false-alarm rate in automated cytopathology screening instruments [41,42,44], slit-scanning is realizing broader application in flow cytometry. Current systems provide nuclear fluorescence; nuclear size and shape; nuclear–cell-diameter ratio; information on cell overlap and multinucleation for identification of abnormal cells with high sensitivity [34,40]; the recognition and exclusion from analysis of artifact, cell clumps, multinucleated cells, and debris [13,35]; chromosomes shape features, including length and centromeric index [7,9–11,14,23]; and head shape measurements of mammalian sperm [1,12].

Zero-resolution optical systems, in which the excitation and measuring apertures are larger than the cell of interest, permit measurements to be made on cells at rates up to several thousand cells per second. Multiparameter measurements may be realized by utilizing several detectors and possibly several excitation sources. These systems provide valuable information on specific cellular substances and parameter profiles on large populations of cells.

High-resolution scanning systems using a submicron scanning spot provide a data matrix ideally suited for detailed studies of individual cells. To date, scanning, computation, and focus requirements have constrained high-resolution analysis to fixed substrate systems.

The slit-scan technique provides a more complete set of cellular parameters than is available with a zero-resolution optical system without producing the large data matrix associated with a high-resolution system. Low-resolution flow systems, of which slit-scan is the prime example, employ field or measuring apertures larger than those used in high-resolution scanning systems, but smaller than the morphological features of interest of the cell. Current systems use effective aperture widths of 1–5 μm.

Like their zero-resolution counterparts, slit-scan systems may be multiparameter, employing light scatter, Coulter volume, multiwavelength fluorescence detection, or other measurements to expand the feature set.

Slit-scan systems may be fluorescence or absorbance based. This discussion focuses on fluorescence systems, as absorbance measurement has not yet been shown valuable in low-resolution flow analysis.

This chapter is intended as an introduction and overview of slit-scanning in flow. It covers the theory, practical considerations, and instrumentation of one-dimensional slit-scanning techniques and data processing. The effects of cell orientation are discussed. Finally, multidimensional slit-scanning and the extensive feature set available from these systems are considered. Segmented slit, swept-slit imaging, and three-multidimensional slit-scan instruments are described.

Slit-scan systems are useful for the analysis of cells, cellular organelles, and other objects in flow. For simplicity, the term "cell" will be used to refer to all the above objects unless a specific one is being discussed.

ONE-DIMENSIONAL SLIT-SCANNING

In a one-dimensional slit-scan system, the fluorescence from a narrow strip of a cell is collected at any given instant of time. Fluorescence measurements from sequential contiguous strips of a cell result in a slit-scan contour. Slit-scanning may be performed in a flow or static cell system. Cells may be sequentially illuminated as they flow through a thin "ribbon" or slit of fluorescence excitation illumination [36,37] (Fig. 1). Measurement of secondary fluorescence from the cell at discrete time intervals as the cell passes through the excitation slit yields a one-dimensional fluorescence intensity distribution or slit-scan contour (Fig. 2). Alternatively, the total cell may be illuminated, and scanning is accomplished by measuring the fluorescence intensity that passes through a slit placed in the image plane of the fluorescence collection lens. The slit-scan contour is processed to extract specific cell features of interest.

Theory

Slit-scanning represents a convolution of the instrument aperture (excitation beam profile or slit aperture) with the cell fluorescence emission along an axis perpendicular to the slit (usually the direction of flow) and integration along the

Flow Cytometry and Sorting, Second Edition, pages 109–125

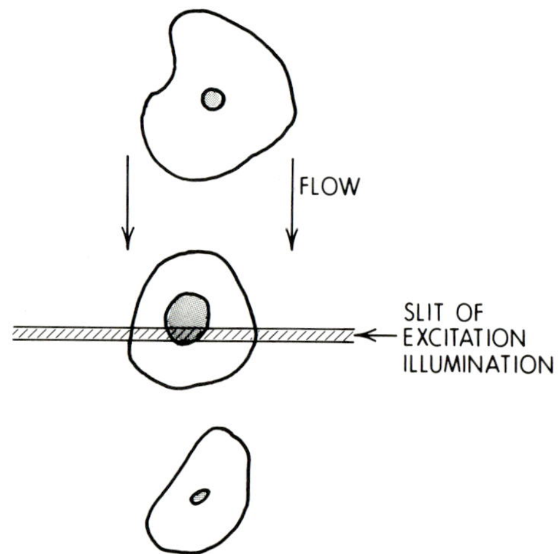

Fig. 1. One-dimensional slit-scanning. Cells are sequentially illuminated as they flow through a slit of excitation illumination.

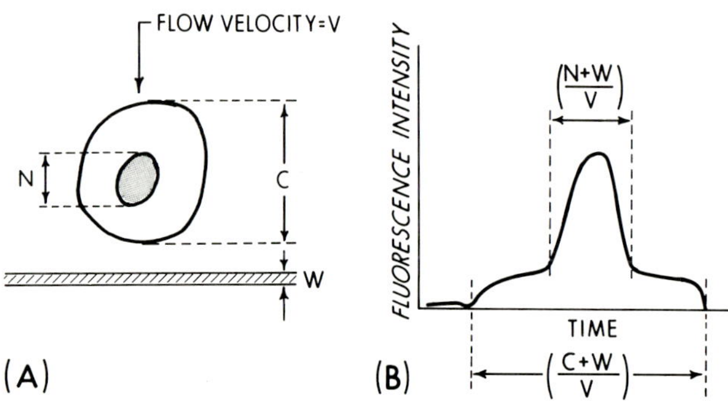

Fig. 2. **A:** Cell. **B:** Resulting slit-scan fluorescence contour.

other two axes (in the place of the slit). Mathematically, this is expressed by

$$S(vt) = \iiint C(x, y, vt - \xi)\, H(\xi) d\xi \, dx \, dy$$

where V is the velocity of the cell flow along the z axis, H(z) is the laser or slit irradiance profile as a function of z, and C(x,y,z) is the distribution of fluorescence emission sites in the cell. The slit or laser beam profile is the relevant instrument function for the measurement and determines the spatial resolution of the slit-scan contour.

Slit-scanning provides morphologic information about the cell only along the cell axis perpendicular to the slit. Consider a cell flowing with velocity V through a slit of thickness W (Fig. 2). Cell diameter C and nuclear diameter N are measured along the direction of flow. For a rectangular excitation beam profile, the length of the fluorescence contour would be (C + W)/V sec, while the length of the nuclear portion of the contour would be (N + W)/V sec.

Contour length, for other than spherical objects, is related to the projection of the object into the plane perpendicular to the slit (Fig. 3). An object of length L and angle α with the direction of flow (perpendicular to the slit) will produce a contour of length (L cos α + W)/V when passing through a slit of width W. These equations only hold for reasonably low values of α, for as α approaches 90°, it is the other dimension of the object (X) that determines contour length. Contour length is (X + W)/V, for α = 90 degrees. Maximum resolution is achieved when the axis containing the desired structural information is aligned perpendicular to the excitation slit. When the excitation slit is not placed 90 degrees to flow, contour length is given by (C + W)/Vsin θ (Fig. 4).

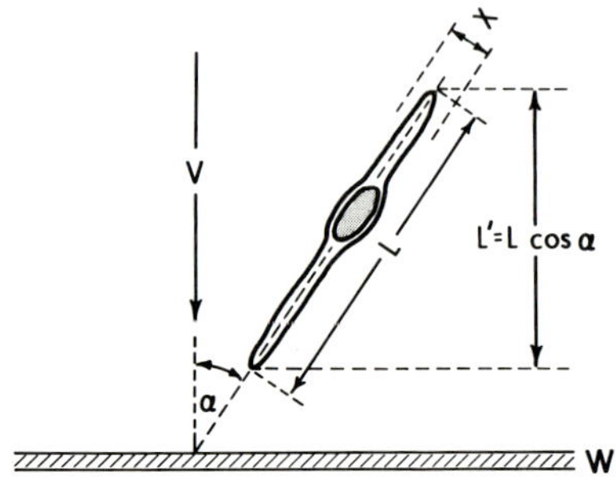

Fig. 3. Nonspherical cell at angle α with direction of flow.

The effects of low-resolution slit-scanning on the frequency content of the image may be determined by multiplying the cell image spectrum by the instrument aperture spectrum.

Practical Considerations

While rectangular aperture profiles may be used in static cell instruments, most flow systems use the gaussian profile of the laser excitation source (Fig. 5). A sheath flow geometry is used to confine cells to the center of the flow stream. Beam shaping optics spread the excitation beam across the specimen stream and compress the beam in the direction of fluid flow to achieve the desired beam thickness (slit width). Beam thickness, for a gaussian profile, is traditionally measured across the $1/e^2$ points. Secondary fluorescence is collected at 90° to the excitation beam to minimize blocking requirements of the barrier filter.

Figure 6A depicts frequency response data for a 5-μm rectangular measuring aperture or beam profile and for a 4.44 and 6.36-μm gaussian beam profile. Frequency data for a 32X, 0.4 NA objective is also depicted (Fig. 6B). The falloff in response of the objective at higher frequencies reduces high-frequency lobes of the rectangular beam profile. It may be concluded from these data that the frequency response of a 5-μm rectangular beam profile and a 6.36-μm gaussian are approximately the same out to 200 lines/mm (Hardy J, unpublished data). This permits comparison of data from static cell instruments incorporating a 5-μm rectangular beam profile with data from flow instruments using a 6.36-μm gaussian beam profile.

The extent to which gaussian beam thickness may be reduced is limited by specimen stream diameter and cell size. The larger the numerical aperture of the optics used to compress the excitation beam, the smaller the beam focus (thickness) in the center of the flow system, and the greater the divergence (spread) of the beam, (Fig. 7). This divergence places a limit on beam focus or resolution achievable in a system with finite specimen stream width.

Specimen stream width is usually set to equal or exceed the diameter of cells of interest. Divergence of the excitation beam results in an increase in beam thickness and corre-

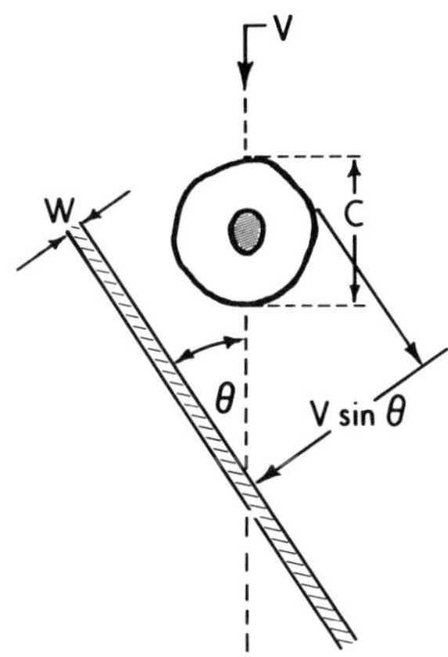

Fig. 4. Slit-scan relationships for a cell passing through an excitation slit that is not perpendicular to the flow stream.

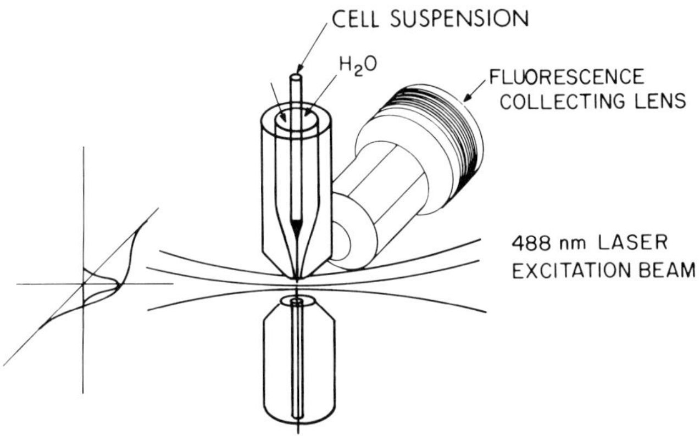

Fig. 5. Slit-scan system employing gaussian profile laser excitation beam. Cells flow in a sheath flow geometry through the excitation beam. Secondary fluorescence is collected at 90° to the excitation beam.

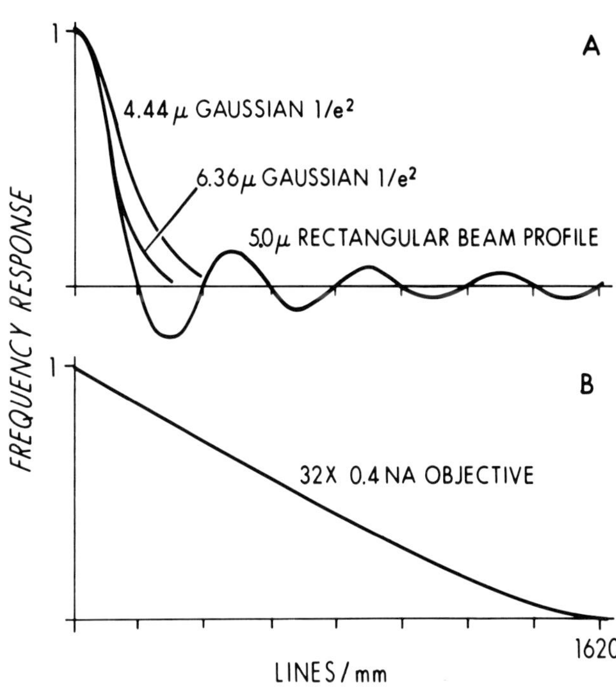

Fig. 6. **A:** Frequency response data for a 5-μm rectangular beam profile and for a 4.44 and 6.36-μm gaussian beam profile. **B:** Frequency response data for a 32×, 0.4 NA objective.

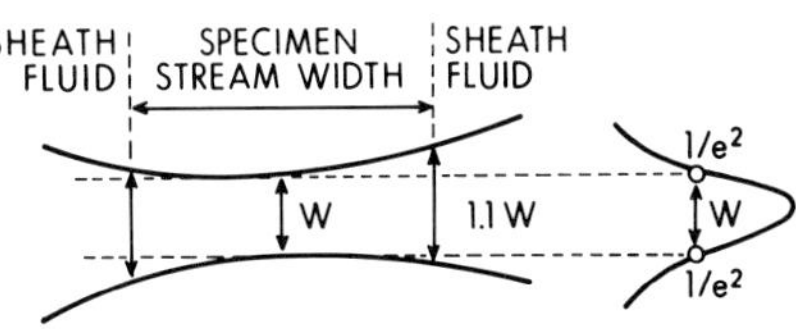

Fig. 7. Spread in excitation beam width across specimen stream. The smaller the beam thickness in the center of the flow stream (W), the greater the divergence of the beam, a 10% beam divergence across the specimen stream width is acceptable for most systems.

sponding decrease in resolution as one moves from center to edge of the specimen stream. To maintain an acceptable stream width while controlling resolution, we compromise with a 10% allowable beam divergence across the specimen stream width at our laboratory, when calculating beam and flow parameters. Figure 8 depicts beam focus (thickness) versus specimen stream diameter for a 10% beam divergence. Note from these data that for a prescreening system for cervical cytology where cell diameters may equal 50 μm, a beam thickness of 6 μm should theoretically be used. In practice, it has been found that a somewhat thinner beam may be used with acceptable results. To obtain increased resolution associated with a 4-μm gaussian beam, the specimen stream must be maintained at 25 μm. Likewise, 2-μm resolution demands a 5-μm stream. Finally, to achieve a 1-μm beam for slit-scanning chromosomes the specimen stream must be reduced to approximately 1.5 μm (assuming a single slit is used).

Signal-to-noise (S/N) ratio must also be considered when reducing beam focus (thickness). Fluorescence intensity is a function of many parameters. These include source radiance, excitation wavelength, excitation and light collection optical systems, flow velocity, temperature, stain quantum efficiency, half-life of the fluorochrome, and the number of sites for dye binding. Provided there are no fluorochrome stain saturation effects, fluorescence intensity should be independent of beam waist thickness over a reasonable range. This should be determined for each application.

Another source of signal noise must also be considered in a practical system as beam thickness is reduced. This is noise due to fluorescent particulate matter in the flow stream. System bandwidth is inversely proportional to beam thickness (slit width). Therefore, a decrease in beam thickness may pass additional spectral components of flow stream produced noise resulting in a decreased S/N ratio.

Other practical considerations include the coherent nature of the laser excitation source and achievement of uniform irradiance over the entire excitation volume. Because of the coherent nature of the excitation illumination, great care must be taken in preventing and controlling reflections and diffraction, and in the design of the beam-shaping optics to achieve the desired uniformity of the gaussian beam profile. Failure to do this results in "hot spots" in the beam profile [36]. The portion of the cell passing through the hot spot receives increased excitation, resulting in an increase in secondary fluorescence for that portion of the cell.

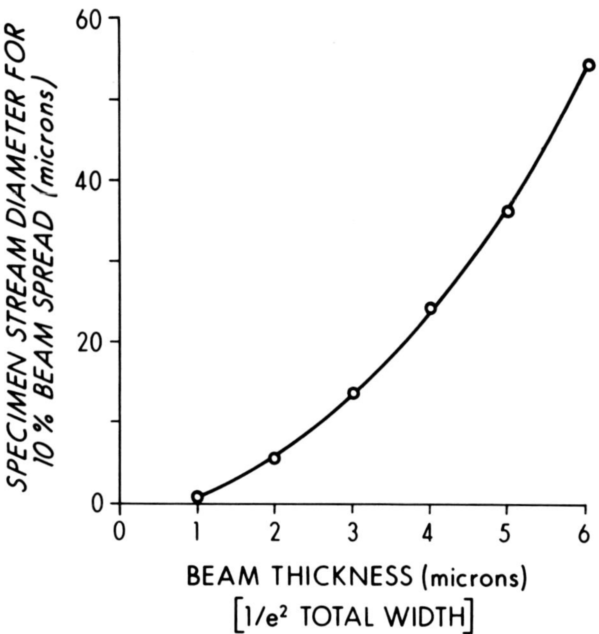

Fig. 8. Beam focus (thickness) versus specimen stream diameter for a 10% beam divergence.

Slit-Scan Optical Systems

Slit-scan optical systems are of two types distinguished by the effective placement of the measuring aperture. One design employs a slit-field stop, while the other design uses a laser beam focused into a ribbon of light to produce a slit of excitation illumination (Fig. 9).

Slit-scan using a slit-field stop. Figure 9A depicts a slit-scan system that employs a slit aperture in the image plane of the fluorescence collection lens. Light collected by the slit is bandpass filtered and detected by a photomultiplier tube. To provide the required resolution, a microscope objective is typically used to image the cell to the slit aperture. The numerical aperture (NA) of this objective must be sufficiently low to provide the depth of field required by the size of the cell and variation in position in the flow stream.

Slit-scan using a laser beam focused into a line. A one-dimensional slit-scan system employing a laser beam focused to a line to produce a "slit" of laser illumination is depicted in Figure 9B. A cell is sequentially illuminated as it flows through the slit of excitation illumination. Detection of the fluorescence emission provides the slit-scan contour. Use of high NA collection optics results in a S/N advantage over the slit-field stop approach.

One-dimensional slit-scan flow instrumentation. A one-dimensional slit-scan flow system was fabricated for the recognition of abnormal cells in cytologic specimens from the human female genital tract and the urinary bladder [36]. This instrument was based on initial studies with a static cell slit-scan cytofluorometer, documenting slit-scan cytofluorometry as a viable technique for separating acridine orange stained cells into pattern classes of normal and abnormal [6,43,44].

The slit-scan flow system configuration is illustrated in Figure 10. Light from an argon-ion laser (488 nm) is spatially filtered and shaped by optics to form a 4-μm ($1/e^2$ points) slit of excitation illumination across a 40-μm specimen stream in the gap region of the fluid-filled flow chamber. Cells flow in suspension, in a sheath flow geometry, through the slit of fluorescence excitation illumination, and are sequentially illuminated. A water-immersion objective positioned at 90° to the excitation beam is used to collect fluorescence and direct it to the detector assembly. The detector assembly contains a dichroic beam splitter, filters, and two photomultipliers for simultaneous recording of green (center frequency 540 nm, half-width 42 nm) and red (center frequency 655 nm, half-width 55 nm) fluorescence. Detectors are interfaced to a computer through a direct-memory access unit; 256 samplings are recorded as each cell passes through the excitation beam. Representative one-dimensional slit-scan contours recorded on normal epithelial cells are illustrated in Figures 11 and 12. Figure 11 presents contours from isolated cells, whereas Figure 12 depicts contours from multinucleate and overlapping cells. This ability to recognize multinucleate events in flow is a major advantage of the slit-scan technique. Contours are analyzed in real time at rates to 200 cells/sec for cellular morphological and quantitative fluorescence features.

Slit-scan cytofluorometry has also been applied to the analysis of chromosomes in flow. Instruments at Los Alamos and Livermore provide slit-scan contours as chromosomes flow lengthwise through a 2-μm-thick slit of argon-ion laser illumination [7,11]. The Los Alamos instrument (Fig. 13) contains a variable width slit in the image plane to define the measuring aperture. Resolution on the order of 1.8 μm is achieved. Recent studies have achieved a resolution of 0.8 μm using similar instrumentation. These low-resolution systems require exceptional stream stability to maintain flow of the chromosomes through the regions of highest resolution of the optical system. The slit-scan contours are recorded and analyzed "off-line" for length, centromere location and number, and total fluorescence. The instrument at Livermore has also been used to quantitate sperm shape [1,12].

Note that all slit-scan measurements are made in fluid-filled flow chambers. Diffraction-limited optical systems together with sample streams of well-defined position are required for optimal resolution and stable operation. Stream in air measurements must be avoided because of the complex and unstable optical characteristics of the air–water interface.

Contour Processing

Contours resulting from one-dimensional slit-scanning of cells or cellular organelles in flow may be processed for feature extraction by a variety of techniques. Both analog and digital signal processing are employed. Processing techniques are dependent on the features sought and whether the analysis is to be performed "on-line" in real time, or the contours are to be recorded for off-line analysis. Real-time analysis is constrained by the time available to analyze each contour. Processing time is minimized to obtain practical throughput rates, or permit the sorting of selected cells. In addition, real-time analysis provides information for instrument control.

Cells. Slit-scan contours of acridine orange stained cells from the female genital tract and the urinary bladder have been analyzed for the detection of abnormal cells [34,40]. Contour analysis was digital and performed in real time. Cell velocity was 0.5 m/sec, and the slit-scan contours were digitized at a 1-MHz rate to provide a spatial sampling interval of 0.5 μm. As the system resolution (beam waist) was 4 μm,

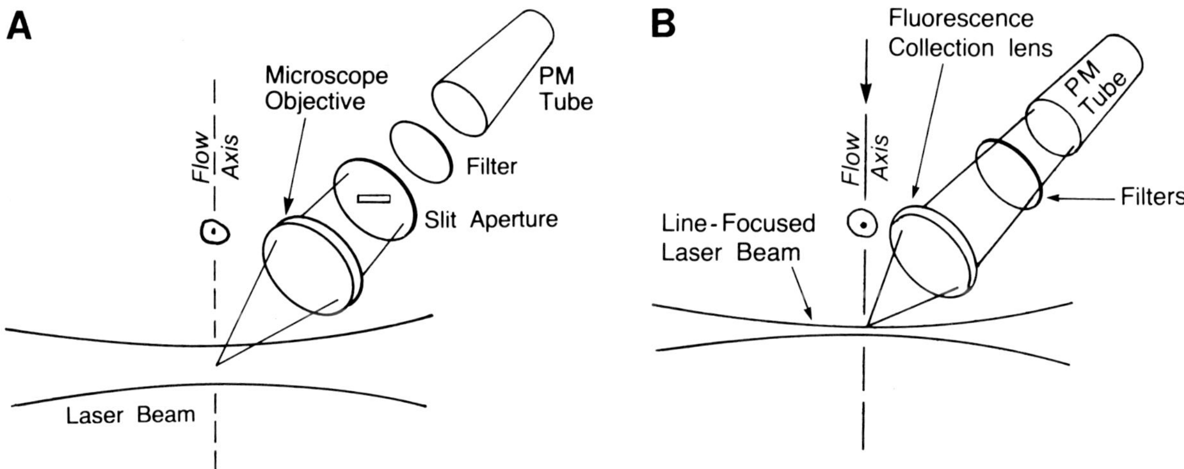

Fig. 9. Slit-scan optical system for flow geometry using **A:** A slit-field stop. **B:** A line-focused laser beam.

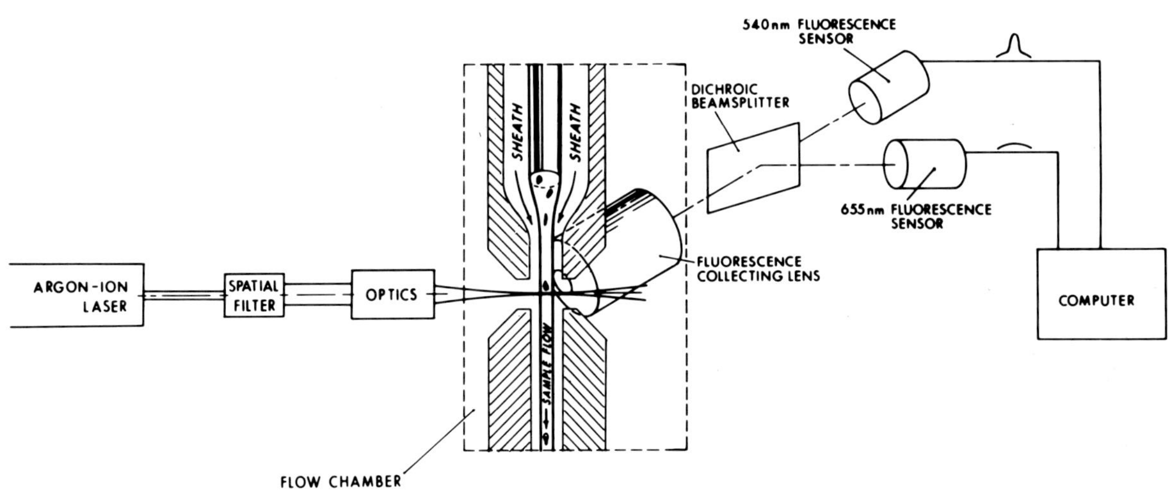

Fig. 10. Slit-scan flow system configuration (collecting lens is 90° to excitation axis).

this represented oversampling in terms of the Nyquist theorem but simplified feature extraction. Contours were analyzed to identify cell and nuclear boundaries and calculate desired cellular features (see Fig. 11). Algorithms detected cell boundaries principally through level detection. Nuclear boundaries were identified by shape changes within the contour. Algorithms reduced interference in cell and nuclear boundary detection from noise or small local peaks and disregarded cells not containing a nucleus or demonstrating insufficient fluorescence.

The following features were derived from each slit-scan contour at a rate exceeding 100 cells/sec: cell diameter, nuclear diameter, nuclear/cell-diameter ratio, nuclear fluorescence, cytoplasmic fluorescence, and certain other low-resolution morphological information such as information on binucleation and cell overlap. Partially overlapping cells and multinucleate cells with sufficient nuclear spacing along the direction of flow presented unique contours that permitted their recognition [2,4,36] (Fig. 12). Detection of fluorescence at multiple wavelengths was used if spectral information was desired. We found slit-scan contour analysis to be particularly important for use with the fluorochrome acridine orange in that epithelial cells stained with acridine orange by our method may exhibit considerable nonspecific cytoplasmic fluorescence. (Specificity of acridine orange staining is considered in Chapter 15, this volume.) A feature used for recognition of abnormal cells is the increased nuclear fluorescence of cells stained with acridine orange [6,43,44]. Measurement of total fluorescence would mask differences in nuclear fluorescence between normal and abnormal cells due to variability in cytoplasmic fluorescence. Likewise, identification of nuclear boundaries by level detection techniques is precluded by nonspecific staining, bacteria in the cytoplasm, or other factors contributing to variability in intensity of cytoplasmic fluorescence from cell to cell (see Fig. 11).

Chromosomes. One-dimensional slit-scan contour analysis of chromosomes provides the following features: chromosome length, total fluorescence, pulse-height fluorescence intensity along a chromosome arm, position of the centromere (centromeric index), and number of centromeres. The number of centromeres is used to distinguish aberrant from normal chromosomes.

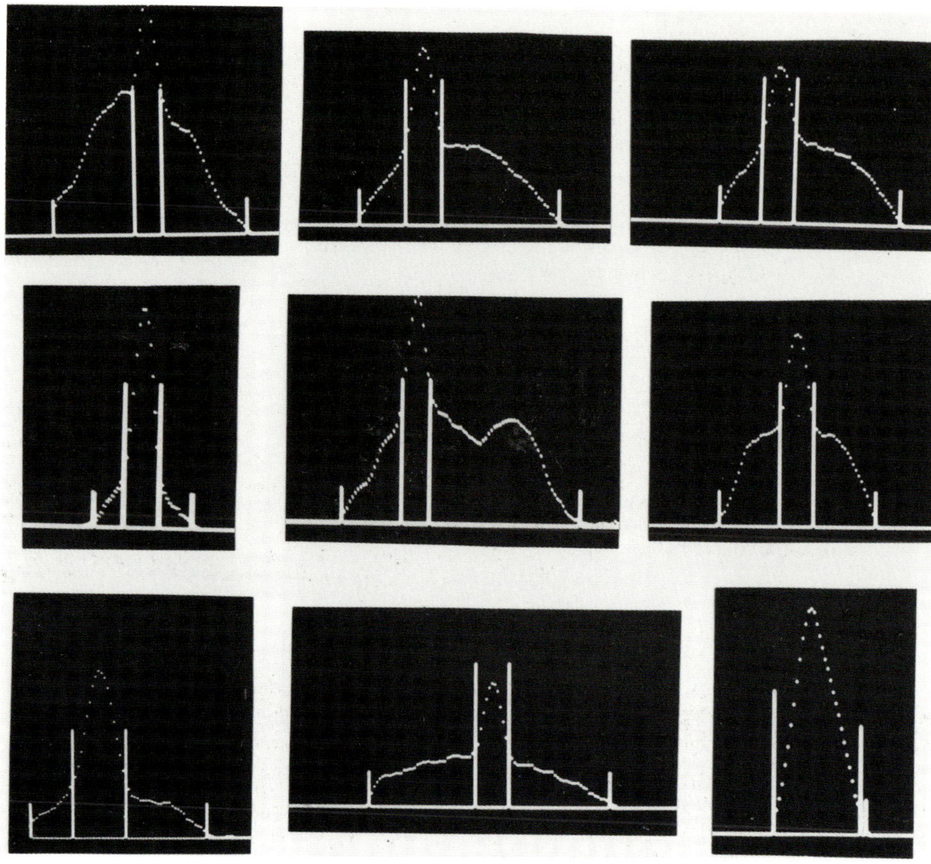

Fig. 11. Slit-scan flow contours recorded on normal squamous epithelial cells. Vertical lines indicate position of computer-determined cell and nuclear boundaries. Cells stained with acridine orange.

The centromere is detected by the presence of a dip, or a depression, in the chromosome slit-scan contour (Fig. 14). Typically, this is detected by comparing the slit-scan contour with a standard contour that is similar to the measured contour, but flat in the region of the centromere [7,10,23]. The centromere is then defined as the point of maximum difference between the two contours. An alternate algorithm determines the midpoint, maximum amplitude point, and edges of the contour. The half of the contour containing the point of maximum amplitude is then reflected about the contour midpoint producing a symmetrical contour. The points of maximum amplitude are joined to produce a centromere-free contour to which the chromosome contour is compared [7]. The key to accuracy in feature extraction is maintaining proper chromosome orientation in flow, achieving high spatial resolution, and maintaining an adequate S/N ratio.

Fourier analysis has been used to restore high spatial frequency information to the measured chromosome profile in an attempt to increase the information content available from flow slit-scan chromosome analysis [26]. This approach is based on the fact that the slit-scan contour is a convolution of the measuring aperture with the distribution of fluorescence emission along the chromosome. A high spatial resolution estimate of the chromosome slit-scan contour is theoretically obtainable from the low-resolution slit-scan contour and the shape of the measuring aperture by deconvolution. This technique has been used to recover microsphere doublet and chromosome slit-scan contours with a spatial resolution of about 1.6 μm from contours recorded in an instrument with a measuring aperture size of 2.5 μm. The spatial resolution that can be obtained by this technique is limited by the S/N ratio in the slit-scan contours and by the size of the measuring aperture.

An alternate approach produces a fringe pattern in the flow stream from the interference of two laser beams [25]. Contours are recorded as chromosomes pass through the fringe field. Precise knowledge of the fringe pattern permits deconvolution of the contours. This results in contours approximating those that would be obtained if the chromosomes passed through a single fringe. Fringe spacing can be less than one micron. This technique requires both stability of the fringe field and precise knowledge of the fringe pattern for deconvolution. Deconvolution is time intensive and must be performed off-line. Nevertheless, this technique does provide high-resolution slit-scanning with a depth of field unachievable by other flow techniques.

Pulse-width analysis. Pulse-width analysis [29,30] is a variation of contour analysis used in slit-scanning. Specificity of staining is required for its success. In pulse-width analysis, the width of the fluorescence pulse produced as a cell traverses the slit of excitation illumination is detected and used as a measure of the diameter of the cell plus slit width.

A pulse-width or "time of flight" technique used with two-color fluorescent staining to differentiate nucleus and cyto-

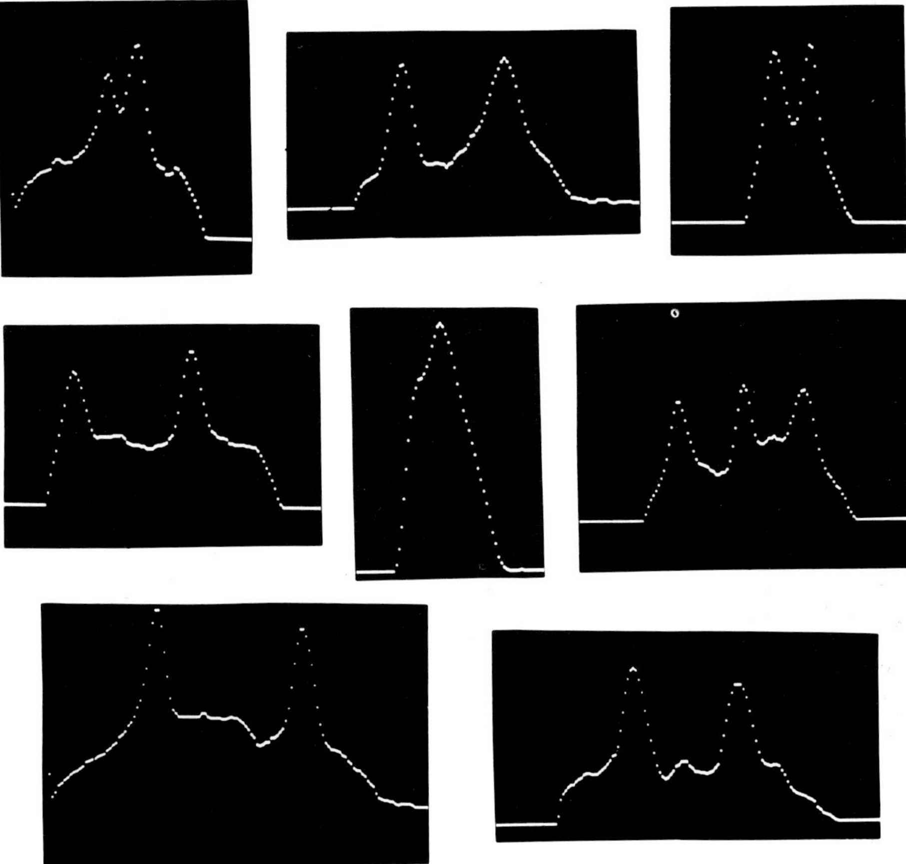

Fig. 12. Slit-scan flow contours recorded on normal squamous epithelial cells. These contours were rejected by contour analysis algorithms as representing multinucleate or overlapping cells.

plasm permits measurement of cell and nuclear diameters [31,32]. Alternatively, a single fluorochrome may be used to stain only the nucleus of the cell with determination of cytoplasmic boundaries or cell size derived from light-scatter measurements.

Slit-scan measurements are best carried out on systems using a thin slit of excitation illumination to achieve maximum resolution. Certain morphological information, however, is available by pulse-width or shape analysis from cells flowing through the larger excitation beams found in conventional flow instruments. Pulse shape approximates the beam intensity profile if cell diameter in the direction of flow is small compared with the beam width. If cell diameter is large compared with the beam, the pulse shape approximates the unidirectional image of the cell, and morphological information about the cell is available [29]. Techniques for amplitude-independent pulse-width measurement [28,30] and constant fraction rise time and pulse-width measurement [21] increase the accuracy of these techniques.

Cell orientation. Sheath-flow focusing maintains cells near the center of the flow stream, forcing them to pass through the desired geometric position within the detector assembly (see Chapter 3, this volume). The laminar flow in current flow cytometers tends to orient ovoid or elliptical cells, or rodlike objects such as chromosomes, with their long axis parallel to the direction of flow (Chapt. 3). Unfortunately, many cells and small chromosomes do not orient to the extent desired for slit-scan analysis [15,19,22,24]. Nor are elongate cells prevented from rotating about their long axis. Thus, the geometric relationship between the plane and the axis of the cell containing the desired morphologic information (typically the long axis), and the plane of the excitation beam will affect the resulting slit-scan contour for other than spherical cells. A prime example is epithelial cells which, because of their flattened shape, present a unique challenge. If cells pass through the laser excitation beam such that the plane of the cell is parallel to the plane of the beam, the resulting slit-scan contour will be compressed and contain inadequate morphologic information [35,36].

Figure 15 depicts two cell contours for comparison. The contour on the right suggests cell rotation orthogonal to the flow axis. The contour is compressed, and *cytoplasmic shoulders,* on either side of the nuclear peak, are nearly obscured by a loss in resolution and increased cytoplasmic fluorescence intensity per unit area. Several solutions exist to the problem of cell orientation in flow. One solution is control of cell orientation; another is multidimensional slit-scanning.

Cell orientation in flow has been studied by several investigators (see Chapt. 3). Kachel et al. [15] and Kay et al. [18,19] demonstrated that an asymmetrically converging nozzle could orient fixed chicken erythrocytes and epithelial cells. Fulwyler [8] showed that beveling the tip of a sample

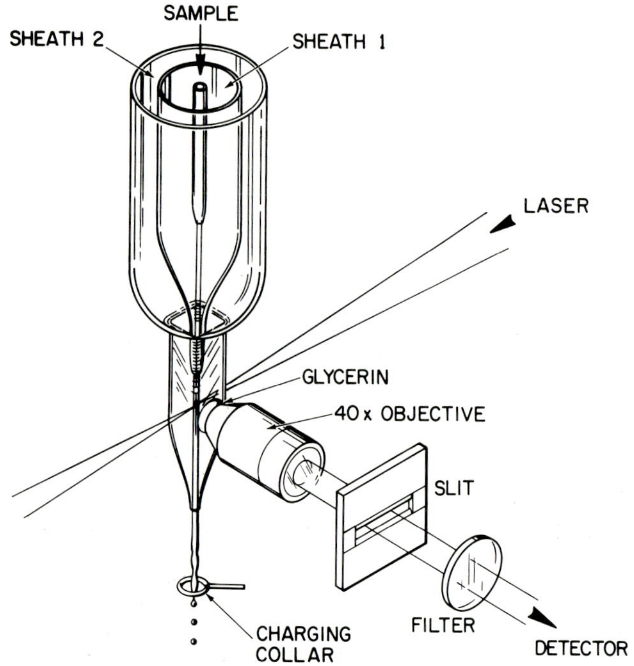

Fig. 13. Flow cytometer/sorter with image plane slit-scanning. The nozzle consists of three concentric glass channels that hydrodynamically constrain chromosomes to a sample core diameter of 2 μm within the square cross section. The 40× water-immersion objective (optically coupled to the glass channel with glycerine) collects fluorescence and forms an image of the object on the variable-width slit located at the image plane. (Reproduced from Cram et al. [7], with permission of the publisher.)

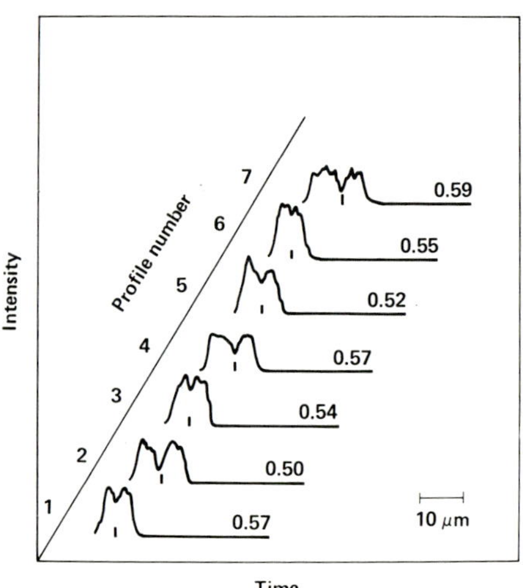

Fig. 14. Sequential pulse profiles recorded for human chromosomes 1 and 2 during object plane scanning. Centromeric locations are indicated by vertical bars. Centromeric indices calculated for each pulse profile are indicated to the right. (Reproduced from Cram et al. [7], with permission of the publisher.)

injection tube achieved a one dimensional convergence of the sheath flow, causing orientation of chicken erythrocytes. While these techniques work to a varying degree in selected applications, they do not provide the cell orientation required for detection of abnormal cells in cytologic specimens (rare event analysis). For these applications, multidimensional slit-scan analysis is required.

MULTIDIMENSIONAL SLIT-SCANNING

Variations of slit-scanning may be employed to extract additional features from cells in flow. Typically these are morphologic and quantitative fluorescence information along one or more of the other dimensions of the cell. Multidimensional slit-scan systems generating three orthogonal slit-scan contours provide information on the distribution of fluorescence throughout the volume of the cell. Features related to the three-dimensional shape of the cell and nucleus are also available, together with information on cell overlap and multinucleation independent of cell orientation.

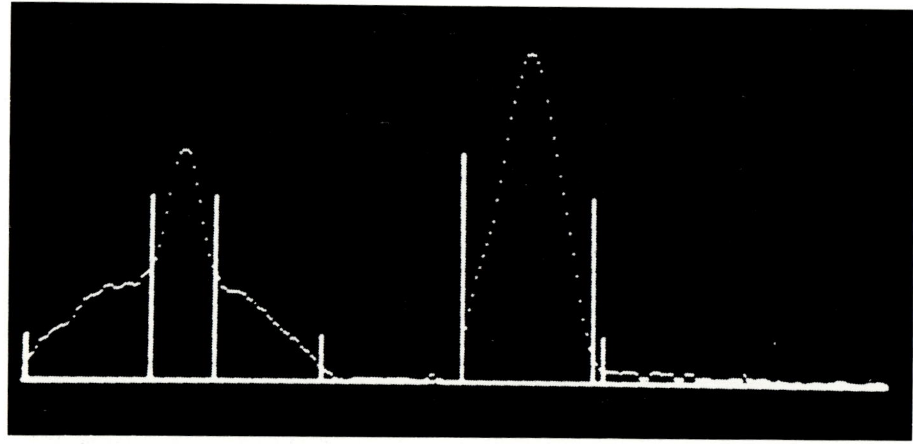

Fig. 15. Slit-scan flow contours recorded on squamous epithelial cells. Contour on the right suggests cell rotation orthogonal to the flow axis.

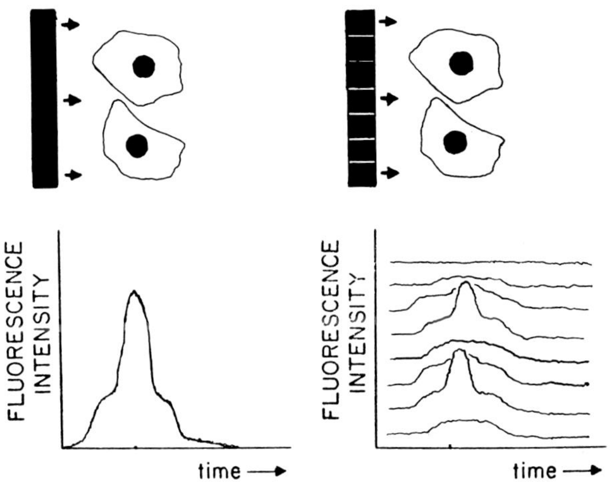

Fig. 16. Segmented slit analysis. Replacing the single detector (**left**) with a segmented detector or linear array (**right**) provides low-resolution morphologic information about cells and in this case discriminates two overlapping cells parallel to the excitation slit.

Segmented Slit

Through the use of imaging optics, the portion of the cell illuminated by the excitation slit may be imaged onto a detector. Replacing the single detector with a segmented detector or linear array provides low-resolution morphologic information about cells and cell overlap parallel to the excitation slit (Fig. 16). Sampling the output of each detector element (segment) as the cell traverses the excitation slit in flow provides a two-dimensional low-resolution matrix of cell fluorescence.

Swept-Slit Imaging in Flow

A dual-view swept-slit image correlation system has been developed to permit correlation of cell images and slit-scan contours in flow [16,17,38]. Using a 5-μm slit of excitation illumination, the system yields two-dimensional images (from two orthogonal directions) of cell fluorescence with a resolution of ~3 μm. Cells flow in suspension, in a sheath flow mode, through a fluid-filled flow chamber where they intersect the slit of argon-ion laser excitation illumination (Fig. 17). Secondary fluorescence is collected via a water immersion microscope ojective and directed through a dichroic beam splitter and filters to two photomultipliers. These photomultipliers detect green (540 nm) and red (655 nm) fluorescence. The detector outputs (slit-scan contours) are computer-analyzed to extract desired features.

A second optical system images the slit-scan region (where the excitation beam intersects the flowing objects) onto a SIT (silicon-intensified target) vidicon camera system. The SIT camera tube utilizes an S-20 multialkali photocathode that provides high sensitivity over the entire visible range. A scanning mirror in this path may be triggered to sweep the image of the slit-scan region across the camera tube at a rate related to the flow velocity of the object. Trigger information is derived from a probe laser beam and light-scatter detector, providing a signal to the computer 1 msec before an object's intersecting the slit-scan excitation beam. Video output from the SIT vidicon camera is digitized and circulated through a high-speed, 6-bit, shift-register memory capable of "freezing" one frame on receipt of a "hold" command. Image size is 200 × 256 pixels.

Figure 18 illustrates cell images recorded with the system together with the corresponding slit-scan contour. The cell was stained with acridine orange. Cell nucleus and cytoplasm are clearly visible in the photograph. Cell classification

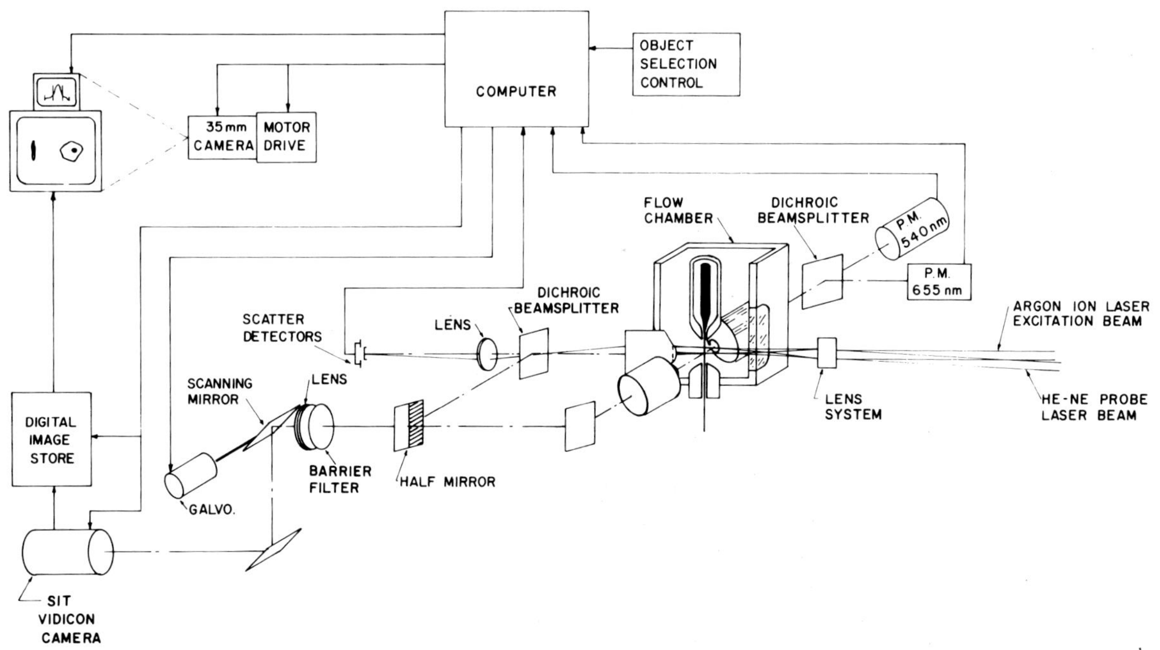

Fig. 17. Slit-scan correlation system showing dual-view slit-field imaging stations and Z slit-scan collection optics. Also shown are the image camera, storage, and display system which is under computer control. The computer-processed Z slit-scan contour and two images of a cell are displayed for photographic recording (see Fig. 18). (Reproduced from Kay et al. [17], with permission of the publisher.)

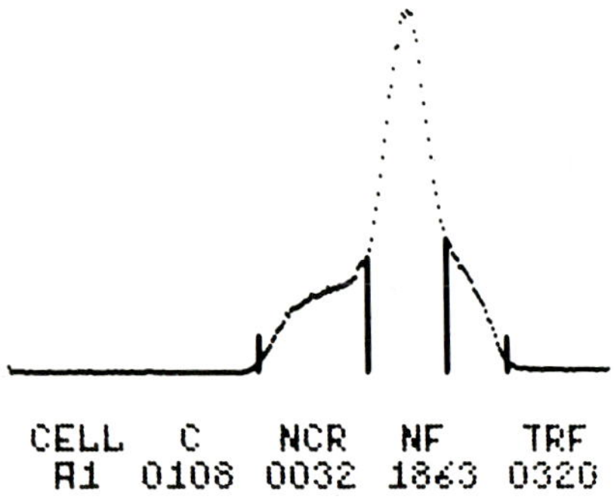

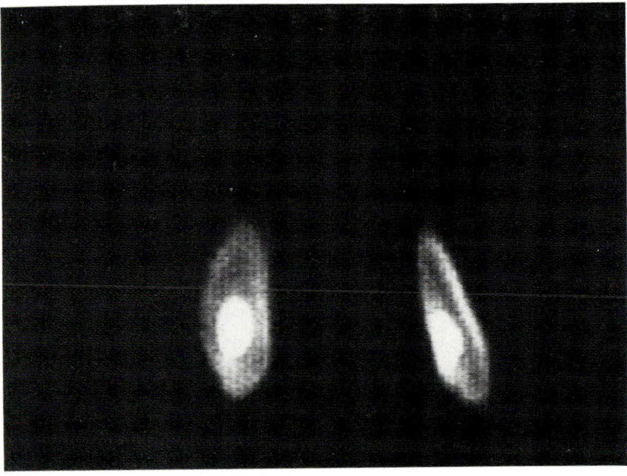

Fig. 18. Flow correlation cell images together with corresponding slit-scan contour. **Top:** Slit-scan contour with computer determined cell and nuclear boundaries. **Bottom:** Vidicon fluorescence image of two views of the cell recorded 90° apart.

was possible using the original vidicon images. The swept-slit image correlation system can be used in any application requiring low-resolution, dual-view image acquisition in flow. This system has been used for the study of false alarms in a one-dimensional slit-scan flow prescreening system [35,38]. It documents the necessity for a multidimensional slit-scan system.

Two- and Three-Dimensional Slit-Scan Systems

There are several techniques for deriving two- and three-dimensional slit-scan information in cells in flow. Using one or more excitation slits, these techniques provide slit-scan contours along two or three (typically orthogonal) dimensions of a cell. This section discusses these techniques and presents three instruments, which provide multidimensional slit-scan information in flow.

Multiple slit system. The use of three orthogonal slit excitation beams could provide three dimensional, low-resolution, slit-scan information on a cell [5,35,36]. While the three excitation slits could overlap (with imaging optics and masks used to define specific detector regions) a more practical solution would place the three slits sequential in flow. However, potential changes in cell orientation between slits and other complications make this technique less attractive than systems providing multidimensional slit-scan information from a single excitation station in flow.

X–Y–Z multidimensional slit-scan system

Instrumentation. A multidimensional, or X–Y–Z, slit-scan flow system has been fabricated using a single excitation slit with multiple detectors to provide three orthogonal slit-scan contours [3,39] (Fig. 19). This system employs slit-field imaging from two views at 90° angles to one another and the cell stream (X and Y contours), along with a conventional slit-scan fluorescence collecting system for the Z contour. The system concept is presented in Figure 20.

The system consists of a fluid-filled flow chamber employing a flow nozzle, a 350-μ-long gap region, an exit capillary, and three water-immersion objective lenses viewing the excitation beam in the gap region. The excitation beam is a 4-μm-thick slit of 488 nm light from an argon-ion laser. Fluorescence light for the Z contour is collected by one of the water-immersion objectives and measured by a photomultiplier tube. A conventional slit-scan contour (Z) is generated as a cell traverses the excitation beam.

The other two water-immersion objectives image the intersection region of the cell and excitation beam "edge-on" via anamorphic lenses onto the X and Y detectors. These detectors are charge-coupled photodiode (CCPD) linear arrays coupled to microchannel plate intensifier tubes. A small volume through the excitation region is imaged to each single element of the array. At the onset of the Z slit-scan contour, the image intensifiers are gated-on and the arrays begin to accumulate the imaged fluorescence. Summation for the duration of the Z contour provides volume integration of a vertical slice of cell fluorescence by each array pixel. At the end of the Z contour, the intensifiers are gated off, and the arrays are allowed to integrate the residual phosphorescence from the intensifier output phosphor. After this delay, the arrays are read out, providing the X and Y slit-scan contours.

Optical resolution is 4 μm in Z and 3 μm in X and Y. Representative multidimensional slit-scan contours from the analysis of a specimen from the human female genital tract are presented in Figure 21. Contour sets from each cell are analyzed in real time to provide a large feature set useful in the recognition of abnormal cells, as well as for other applications.

Contour processing. High-speed, real-time, analysis of the three slit-scan fluorescence contours from each cell is performed by a dedicated preprocessor, DMA interface and PDP-11/83 computer [27] (Fig. 22). The output of each detector is connected to a high-speed A/D converter and ring buffer module (Kinetic Systems 4010). These modules provide 10-bit sampling at rates to 2 MHz. A STOP signal stops the digitizing process after a preset number of samples. When enabled by the PDP-11/83, the Z INTERFACE preprocessor generates the STOP signal when a valid peak (cell with a nucleus) has been detected in the analog signal. The logic is also designed to ensure that sampling does not start in the middle of a contour. The occurrence of STOP in the Z channel generates a STOP signal for the X and Y channels after a preset delay; 256 samplings are recorded on each contour. The system can process three-dimensional slit-scan contour sets at rates exceeding 100 cells/sec while, in real time, checking each contour for multiple peaks, extracting 16 variables

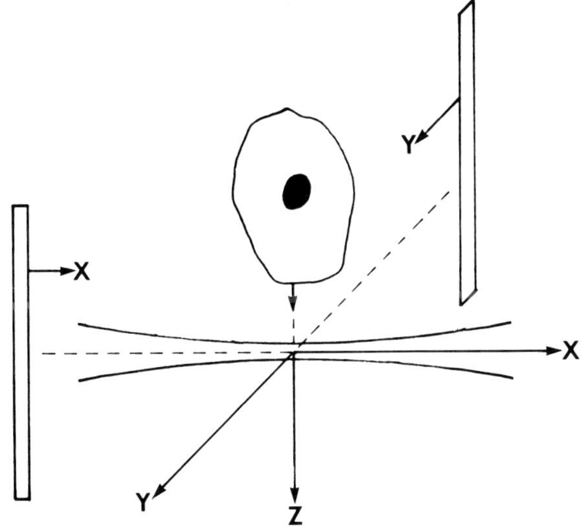

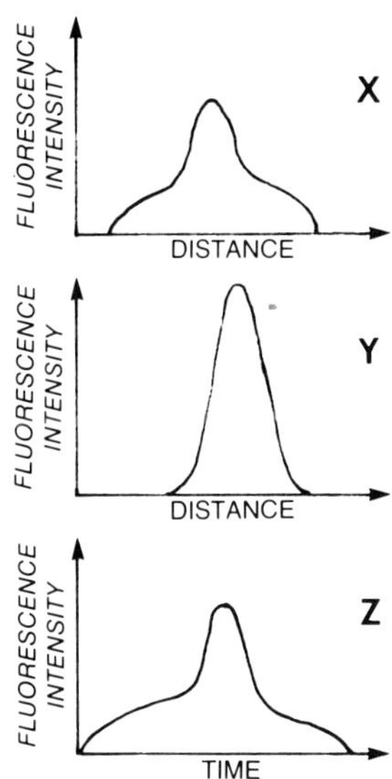

Fig. 19. Conceptual drawing of multidimensional slit-scanning. The multidimensional slit-scan flow system provides a one-dimensional fluorescence projection (slit-scan contour) along each of three orthogonal axes as a cell flows in suspension through a 4-μm slit of laser excitation illumination. The Z contour is obtained by sampling the output of a photomultiplier tube that collects 540-nm fluorescent light as the cell passes through the excitation slit. The X and Y contours are collected by semiconductor array detectors that integrate fluorescence from the cell-laser beam intersection region during the time the cell is passing through the excitation region.

from the primary contour, accumulating gated histograms of these data, and saving the derived data, selected contours, or both into list mode files on disk.

The system is being used to screen specimens from the female genital tract [40] and the urinary bladder [34] for cancer. Morphological information is derived from the three orthogonal slit-scan contours and can include cell shape and size, nucleus shape and size, N/C ratio, and the number of nuclei in the cell. These data are important in that they provide for the recognition and elimination from analysis, of most multinucleated, clumped, and overlapping cells. Such recognition is essential in increasing sensitivity for detection of small subpopulations of cells and in obtaining artifact-free DNA histograms. This system has provided abnormal cell detection with a sensitivity unmatched by other flow techniques since objects (such as debris particles, cell clumps, and binucleate cells) that often are classified incorrectly as abnormal cells in other systems are identified by multidimensional slit-scan analysis and eliminated.

Several new detectors have been added for current studies. They provide for one-dimensional slit-scan measurements at two additional wavelengths, together with low-angle forward and 90° slit-scan light-scatter contours, and axial extinction measurement. The system now permits, for example, the simultaneous measurement of DNA, antigen expression, and three-dimensional morphological features on cells in flow.

Two-station multidimensional slit-scan system. An alternate multidimensional slit-scan system has been designed to simplify the instrument by using only photomultipliers as detectors.

Instrumentation. The system employs two measurement stations [20] (Fig. 23). A conventional one-dimensional slit-scan measurement station having a high S/N ratio is followed immediately by a three-dimensional measurement station that provides a multidimensional slit-scan "second look" on selected or all cells. The first station employs a "slit" of laser excitation perpendicular to the direction of flow. Tilted slit-field stops imaged to the stream define mutually orthogonal detection slits at the second excitation station. A high numerical aperture collection lens is used at the first station for high S/N detection, whereas lower NA lenses are employed at the second station to provide the required depth of field for maintaining cells in focus across a specimen stream that can be as wide as 50 μm.

The first laser excitation station is identical to that of previous one-dimensional slit-scan systems. A 0.4 NA 20× water-immersion objective lens collects the fluorescence emissions from fluorochrome-stained cells as they traverse the slit defined by the 4-μm line-focused laser excitation beam. The fluorescence light is bandpass filtered and detected by a photomultiplier tube. This is the Z slit-scan contour.

Seven hundred microns downstream, the cell encounters the second laser excitation beam, which is conditioned to be collimated and nearly uniform in irradiance (±5%) over its central 100-μm diameter. Here, three water-immersion objective lenses (two 0.25 NA 10× lenses plus the aforementioned 0.4 NA 20× lens) image cell fluorescence onto slit-aperture field stops. The optical axes of the three lenses are separated by 120° about the flow axis (Fig. 23).

The slit-field stops in the image planes of the three water-immersion objective lenses restrict the X′, Y′, and Z′ optical channels to view only slit volumes of fluorescence of approximately 4 μm width. The field stops are oriented at 35.3° with respect to the flow axis. These angles provide an orthogonal measurement geometry as depicted in Figure 24. The lines of intersection of these slit volumes form equal angles with the flow axis (54.7°) and form a coordinate system that is rotated with respect to the axis of flow. The fluorescence emissions of these channels are bandpass-filtered and detected by photomultiplier tubes to provide the X′, Y′, and Z′ slit-scan signals.

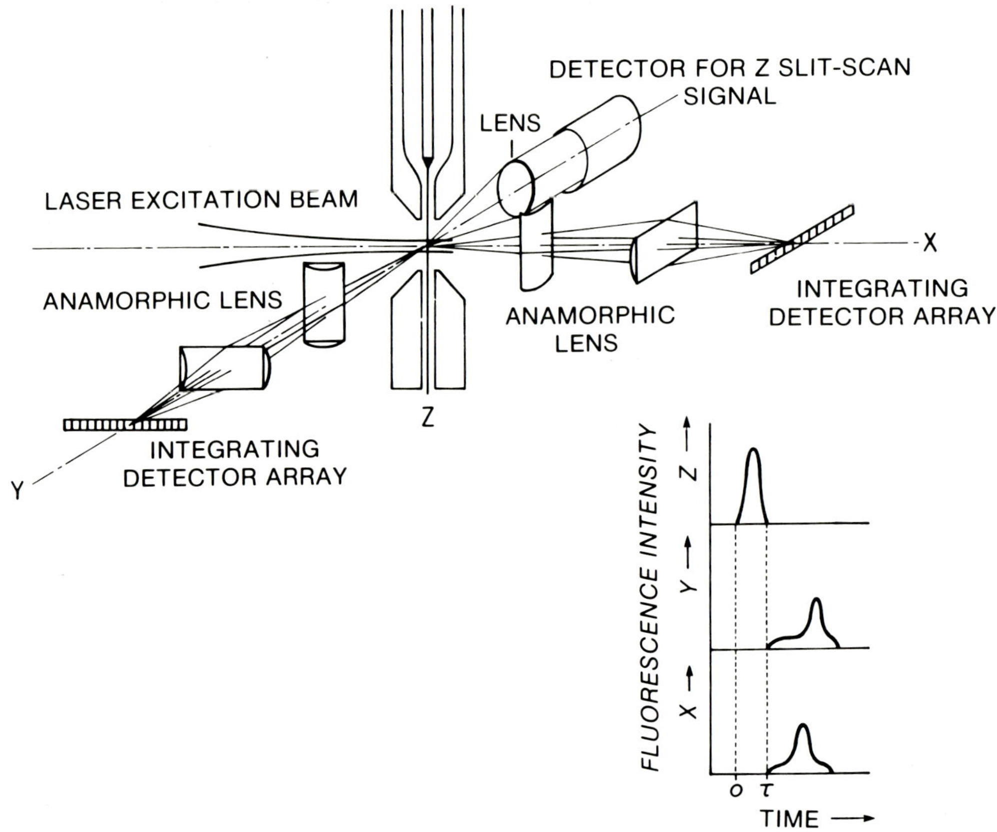

Fig. 20. X–Y–Z multidimensional slit-scan flow system. (See text.) (Reproduced from Cambier et al. [3], with permission of the publisher.)

An illumination system uses the X′ channel to introduce the laser excitation beams into the flow chamber. The two laser excitation beams are derived from the same argon-ion laser and conditioned by optics (not shown) before entering the dichroic beam splitter of the X′ channel. Ten percent of the 488 nm laser beam is diverted by a beam splitter and focused into a line before the 10× objective lens through which it is imaged into a focused line of excitation light at the first excitation station in the flow chamber. The remaining 90% of 488-nm laser excitation is converged to the front focal point of the 10× objective lens such that a collimated beam is directed to the second excitation station. Representative contours are depicted in Figure 25. The Z contour is generated at the first station, while the time-delayed Z′, X′, and Y′ contours are simultaneously generated at the second station.

Contour processing. High-speed, real-time analysis of the slit-scan contour sets from the two-station system is performed by a dedicated preprocessor–DMA interface and PDP-11/83 computer system (Fig. 26). The output of each detector is connected to a high-speed A/D converter and ring-buffer module (LeCroy 2262). This module provides 10-bit sampling at rates to 80 MHz, with 316 samplings recorded per contour. When enabled by the PDP-11/83, the Z INTERFACE preprocessor generates a STOP signal at the end of a cell containing a valid peak (nucleus). This STOP signal stops digitization in the Z converter. It also initiates a counter that, after a preset delay, enables the PRIME INTERFACE. This

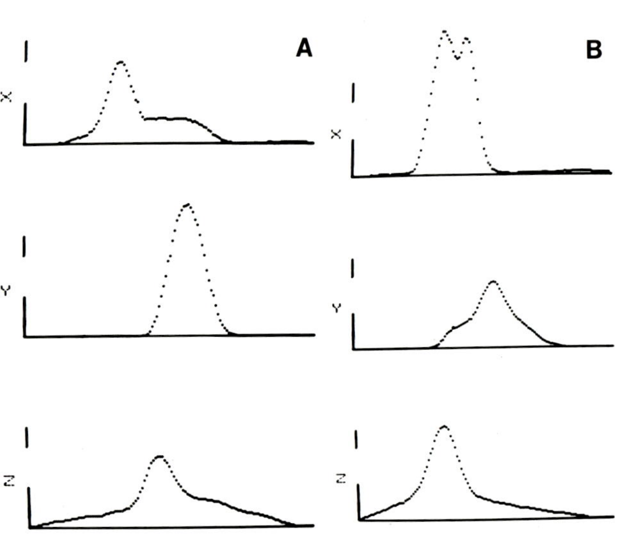

Fig. 21. Multidimensional slit-scan contour sets. **A:** Single normal epithelial cell. **B:** The Z contour appears to be from a single abnormal cell (elevated nuclear fluorescence). The double peak in the X contour, however, indicates that it is from two cells (or nuclei) passing through the slit simultaneously.

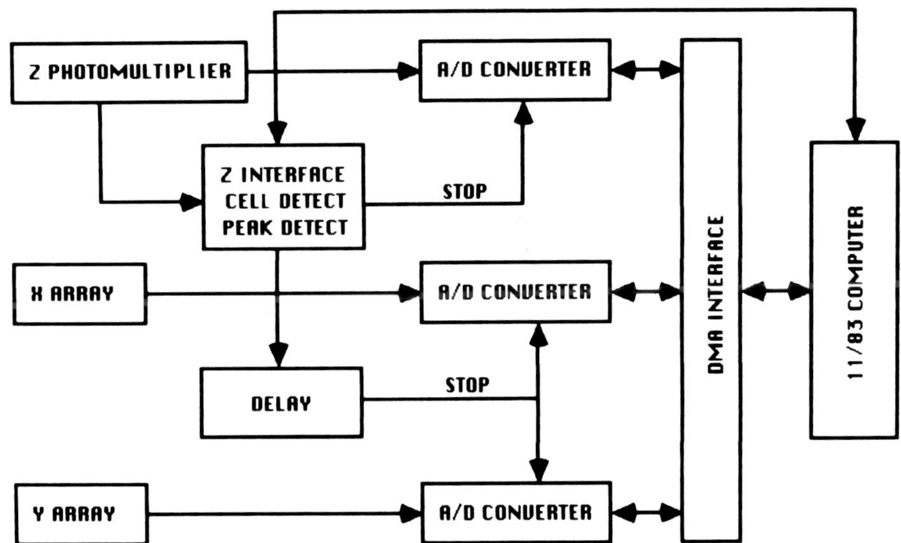

Fig. 22. Contour processing and data analysis system for the X–Y–Z multidimensional slit-scan flow system.

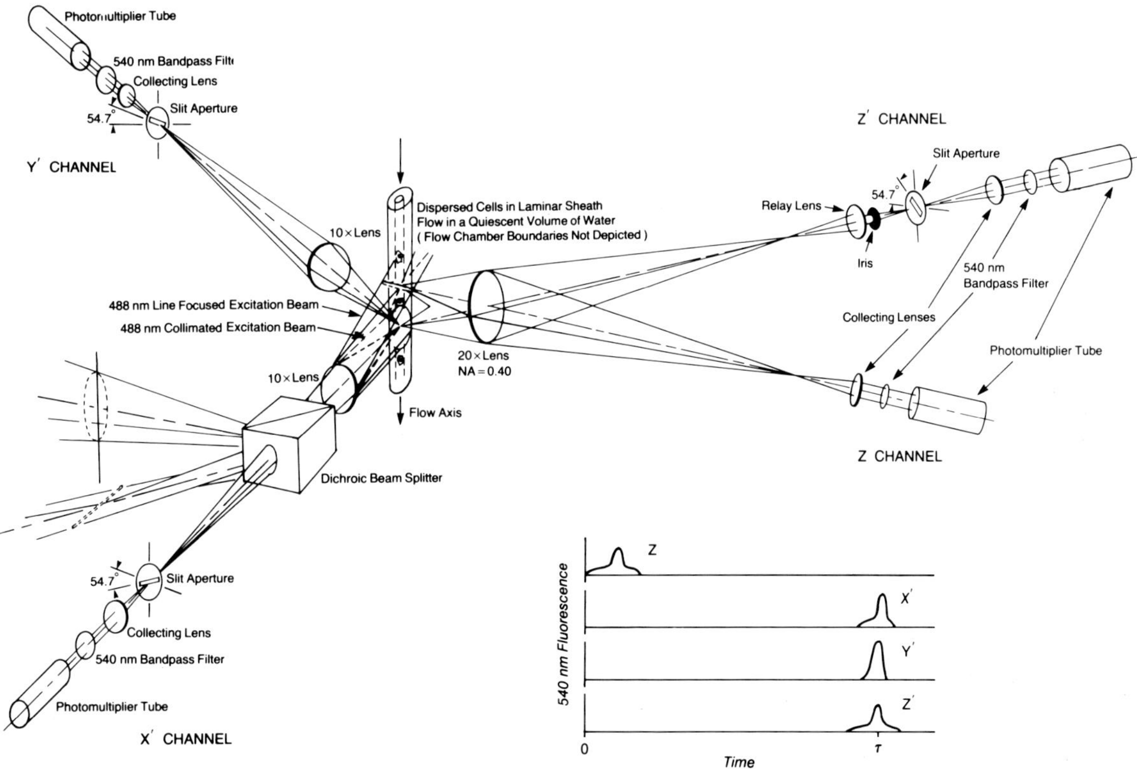

Fig. 23. Schematic diagram of the two-station multidimensional slit-scan optical system. The system employs epi-illumination to provide a 4-μm-thick line-focused beam at the first station and a collimated beam downstream at the second station. First-station fluorescence is collected by the 20× 0.4-NA objective lens and produces the Z-channel conventional slit-scan contour. Second-station fluorescence is collected by the 20× 0.40-NA lens and two other 10× 0.25-NA objective lenses and directed through slit field stops to produce the X′, Y′, and Z′ channel slit-scan contours. The slit-field stops are suitably tilted to provide an orthogonal slit-scan detection geometry at the second station. (Reproduced from Kay et al. [20], with permission of the publisher.)

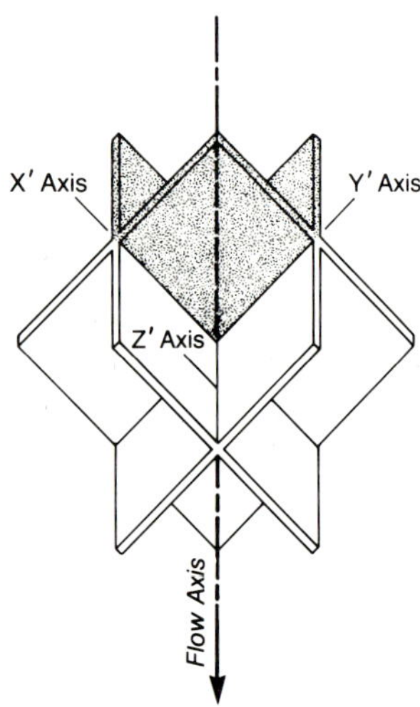

Fig. 24. Illustration of the orientation of the mutually orthogonal X'–Y'–Z' slit volumes at the flow stream, each generated by viewing the stream through the appropriate slit-field stop. (Reproduced from Kay et al. [20], with permission of the publisher.)

delay corresponds to the time required for a cell to travel from the first to second station. The PRIME INTERFACE generates a STOP signal for the prime channel A/D converters when a cell containing a valid peak is found. If no cell or peak is found within a preset time interval, a "no-cell" condition is sent to the PDP-11/83.

For some applications, multidimensional slit-scan data from all channels is used in the analysis of a cell. Other applications require slit-scan data from only some channels. For example, in the analysis of cytologic specimens stained with acridine orange from the female genital tract, the criterion for classification of individual cells as abnormal is an increase in 540-nm (green) nuclear fluorescence. Nuclear fluorescence for each cell is determined from the Z-channel contour. If the value is within normal limits (a normal cell), multidimensional slit-scan data from the second station are not required. If the cell is classified abnormal from data recorded at the first station, multidimensional slit-scan data from the second station are analyzed to detect and rule out false alarms. The additional analysis time is minimal, since only a few percent of events typically require multidimensional analysis. Multidimensional slit-scan morphological information is available on all cells, if desired.

Through exchange of filters and removal of slit apertures, the following measurement configurations are available at the second stage: (1) a set of three orthogonal slit-scan contours at one emission wavelength, (2) three one-dimensional slit-scan contours at three different emission wavelengths, (3) three zero-resolution fluorescence measurements at three different emission wavelengths, or (4) a combination of the above. Sixteen-parameter list-mode feature files may be recorded on each cell for each contour. Additional fluorescence detectors together with low-angle forward and high-angle slit-scan light-scatter detectors have been added to the system first stage to provide additional information. Both multidimensional slit-scan instruments have been designed to be versatile in hardware and software to permit their use in essentially any application.

Epi-illumination microscope objective slit-scan. An epi-illumination microscope objective slit-scan (EMOSS) flow system has been fabricated using two-dimensional slit-scanning with hydrodynamic sample stream focusing [33] (Fig. 27). Excitation radiation is generated by a 150 watt xenon arc lamp employing critical illumination. The hot spot of the arc is imaged to the flow plane, providing an illumination region through which the sample stream is hydrodynamically focused.

The excitation spot is imaged by a microscope objective to a region of approximately 150 μm in diameter within the specimen stream. Such a region is of suitable dimensions to allow total cell irradiation. Cellular fluorescence is collected by the microscope objective, transmitted by a dichroic mirror, and subsequently directed to an eyepiece and detection optics.

The fluorescent signal delivered to the detection portion of the system is bandpass-filtered and imaged to two detector channels by a lens and pellicle beam splitter. Each detector channel contains a slit-field stop in the image plane, a photomultiplier detector, and amplifier circuitry. Currently the slit-field stops being used have an effective width of 4 μm. The two slits are mutually orthogonal, each at 45° to the apparent direction of cell flow. Cells are slit-scanned in two orthogonal directions as they flow through the excitation

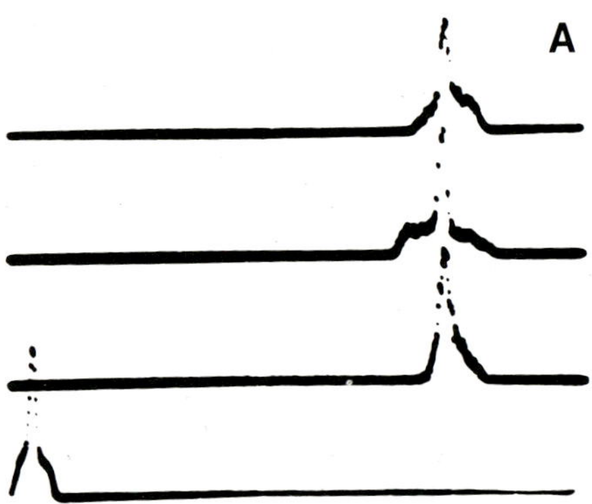

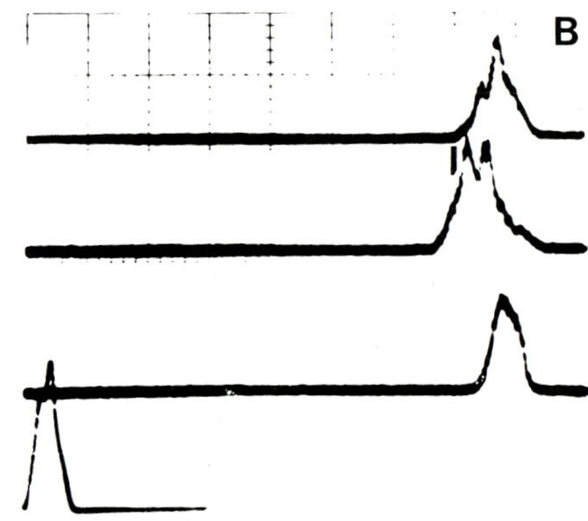

Fig. 25. **A:** Multidimensional slit-scan contours generated by a normal intermediate squamous cell. **B:** A false alarm. The Z contour appears to be produced by a high N/C ratio cell of elevated nuclear fluorescence, whereas the Z' and Y' contours depict the presence of two nuclei. (Reproduced from Kay et al. [20], with permission of the publisher.)

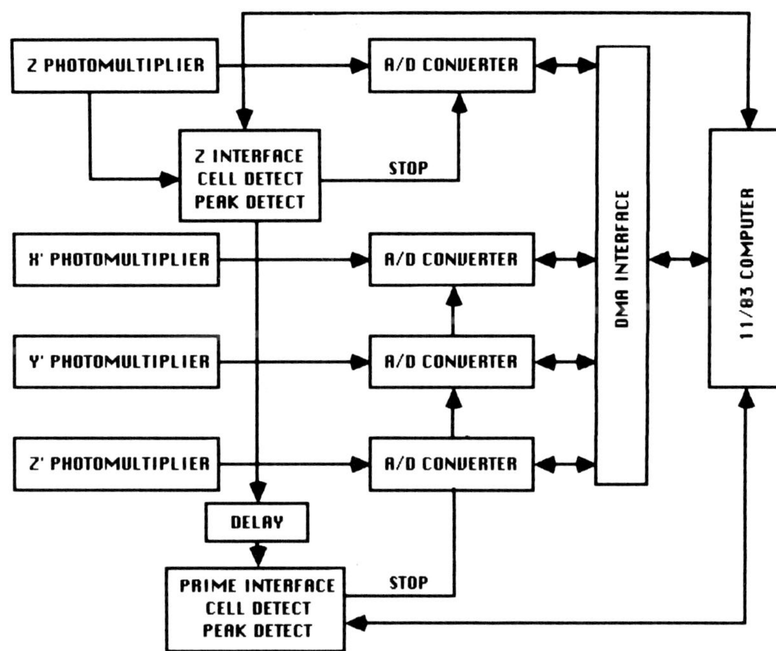

Fig. 26. Contour processing and data analysis system for the two station Z/X′–Y′–Z′ multidimensional slit-scan flow system.

region. This two channel system incorporates many features of multidimensional slit-scan instrumentation into a less complex design suitable for many applications.

CONCLUSION

Slit-scan flow cytometry has proved useful in many applications. The combination of morphological information, in addition to quantitative fluorescence measurements in flow, provides a feature set of great value. New multiparameter, multidimensional systems provide extensive information on cells in flow while permitting real-time data analysis and cell throughput rates exceeding 100 cells/sec.

Broader utilization of the slit-scan technology is dependent on several factors. The major factor is the development of less complex, less expensive, slit-scan instrumentation suitable for commercial production. Other factors include full use of the feature set available from multidimensional slit-scan systems, increased spatial resolution in flow, development of cell sorters operating with slit-scan systems, and continued development of dyes providing biological information based on their distribution within the cell.

Slit-scan instrumentation is currently quite complex. This is especially true for multidimensional slit-scan instrumentation and high speed chromosome slit-scan systems. Simpler, less complex instruments are required to extend this technology to the commercial market to make it available to more users. The EMOSS system is a step in this direction. Other designs are under development. Current studies in the author's laboratory permit off-line reanalysis of data with the opportunity to simulate results from X–Z and Y–Z instrumentation. Comparison with X–Y–Z results should document the extent to which three-dimensional slit-scan analysis is superior to two-dimensional analysis for a specific application. This will permit the specification of a minimum instrument configuration for a given level of performance.

Multidimensional slit-scan systems provide a much larger feature set than is currently used. Three orthogonal slit-scan contours provide (1) information on cell and nuclear size and shape, (2) number and location of nuclei within the cell, (3) N/C diameter, area and volume ratios, and (4) other low-resolution three-dimensional morphological information, together with quantitative information on the distribution of fluorescence throughout the cell. Full use of this feature set should open new applications.

The spatial resolution of slit-scan systems is set by the beam waist. Decreasing the beam waist increases resolution at the expense of a reduction in depth of field. Attempts are currently under way to design systems having increased resolution and to use increased numerical aperture optics to provide improved S/N ratio. The latter requires a full understanding of the effects of optical sectioning on fluorescence detection as only a portion of the cell is in focus. Fourier analysis and fringe scanning are being used in an attempt to increase the high spatial frequency information in chromosome profiles. Efforts will continue to increase spatial resolution and the information available from slit-scan systems.

Most slit-scan systems incorporating real-time contour analysis currently operate at flow velocities below those required for droplet sorting. Fluid cell sorters operating at these lower flow velocities, and producing no "upstream" effect during sorting, are required to permit the sorting of cells based on slit-scan features. Such a sorter is under development in the author's laboratory.

Most systems operating at higher flow velocities record contours for off-line analysis. An exception is the PRIME multidimensional slit-scan system in the author's laboratory, which has the capability of sampling fluorescence at rates to 80 MHz and operating at cell velocities to 10m/sec. Current data-analysis hardware and software, however, is designed for versatility, and not speed, and could not provide a sort

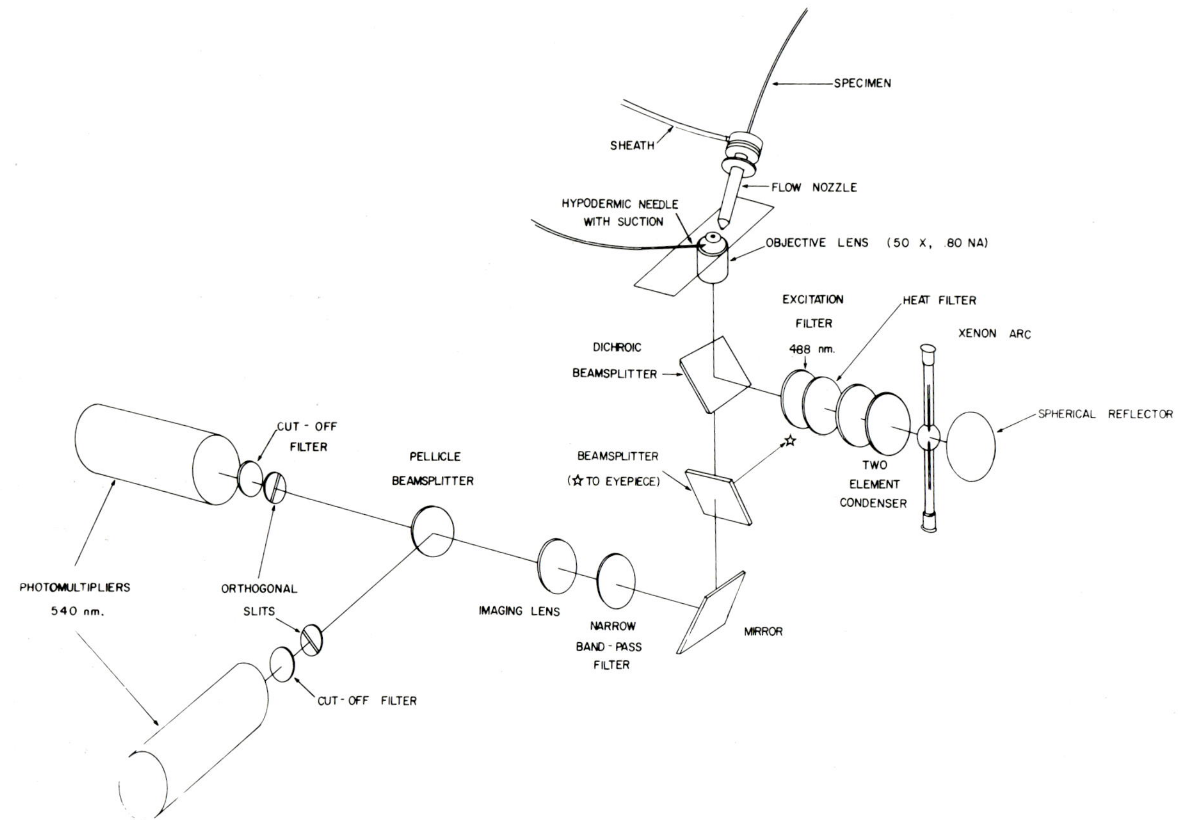

Fig. 27. Epi-illumination microscope objective slit-scan flow system (EMOSS). This system combines hydrodynamic sample stream focusing with two-dimensional slit-scanning. (Reproduced from Weller and Wheeless [33], with permission of the publisher.)

command within the required time. New hardwired microprocessors and parallel processing techniques are required for high-velocity systems capable of ballistic sorting.

Finally, new applications are dependent on the further development of fluorescent dyes providing biological information based on their distribution within and on a cell. Slit-scan flow cytometry provides a unique and powerful feature set with many proven and potential applications.

ACKNOWLEDGMENT

This work was supported by grants CA30582 and CA33148 from the National Cancer Institute under contract N01-CB-33862.

REFERENCES

1. **Benaron DA, Gray JW, Gledhill BL, Lake S, Wyrobek AJ, Young IT** (1982) Quantification of mammalian sperm morphology by slit-scan flow cytometry. Cytometry 2:344–349.
2. **Cambier JL** (1976) Binucleate cell recognition and prescreening system design in automated gynecologic cytopathology. Ph.D thesis, University of Rochester.
3. **Cambier JL, Kay DB, Wheeless LL** (1979) A multidimensional slit-scan flow system. J Histochem Cytochem 27:321–324.
4. **Cambier JL, Wheeless LL** (1975) The binucleate cell: Implications for automated analysis. Acta Cytol. (Praha) 19:281–285.
5. **Cambier JL, Wheeless LL** (1979) Predicted performance of single versus multiple-slit flow systems. J Histochem Cytochem 27:325–328.
6. **Cambier MA, Christy WJ, Wheeless LL, Frank IN** (1976) Slit-scan cytofluorometry: Basis for automated prescreening of urinary tract cytology. J Histochem Cytochem 24:305–307.
7. **Cram LS, Bartholdi MF, Wheeless LL, Gray JW** (1985) Morphological analysis by scanning flow cytometry. In Van Dilla MA, Dean PN, Laerum OD, Melamed MR (eds), "Flow Cytometry: Instrumentation and Data Analysis." New York: Academic Press, pp 163–194.
8. **Fulwyler MJ** (1977) Hydrodynamic orientation of cells. J Histochem Cytochem 25:781–783.
9. **Gray JW, Lucas J, Pinkel D, Peters D, Ashworth L, Van Dilla MA** (1979) Slit-scan flow cytometry: Analysis of Chinese hamster M3-1 chromosomes. In Laerum OD, Lindmo T, Thorud E (eds), "Flow Cytometry. Vol. IV. Oslo: Universitetsforlaget, pp 249–255.
10. **Gray JW, Lucas J, Yu LC, Langlois R** (1984) Flow cytometric detection of aberrant chromosomes. In Eisert WG, Mendelsohn MM (eds), "Biological Dosimetry." Berlin: Springer-Verlag, pp 25–35.

11. **Gray JW, Peters D, Merrill JT, Martin R, Van Dilla MA** (1979) Slit-scan flow cytometry of mammalian chromosomes. J Histochem Cytochem 27:441–444.
12. **Halamka J, Gray JW, Gledhill BL, Lake S, Wyrobek AJ** (1984) Estimation of the frequency of malformed sperm by slit-scan flow cytometry. Cytometry 5:333–338.
13. **Hardy J, Wheeless LL** (1977) Application of Fraunhofer diffraction theory to feature specific detector design. J Histochem Cytochem 25:857–863.
14. **Johnston RG, Bartholdi MF, Hiebert RD, Parson JD, Cram LS** (1985) A slit-scan flow cytometer for recording simultaneous waveforms. Rev Sci Instrum 56:691–695.
15. **Kachel V, Kordwig E, Glossner E** (1977) Uniform lateral orientation, caused by flow forces, of flat particles in flow through systems. J Histochem Cytochem 25: 774–780.
16. **Kay DB, Cambier JL, Wheeless LL** (1977) Imaging system for correlating fluorescence cell measurements in flow. Presented at the Proceedings of the Society of Photo-Optical Engineers, Clever Optics.
17. **Kay DB, Cambier JL, Wheeless LL** (1979): Imaging in flow. J Histochem Cytochem 27:329–334.
18. **Kay DB, Wheeless LL** (1976) Laser stroboscopic photography technique for cell orientation in flow. J Histochem Cytochem 24:265–268.
19. **Kay DB, Wheeless LL** (1977) Experimental findings on gynecologic cell orientation and dynamics for three flow nozzle geometries. J Histochem Cytochem 25: 870–874.
20. **Kay DB, Wheeless LL, Brooks CL** (1982) A two station multidimensional slit-scan flow system. IEEE Trans Biomed Eng 29:106–111.
21. **Leary JF, Todd P, Woods JCS, et al** (1979) Laser flow cytometric light scatter and fluorescence pulse width and pulse rise time sizing of mammalian cells. J Histochem Cytochem 27:315–320.
22. **Loken MR, Parks DR, Herzenberg LA** (1977) Identification of cell asymmetry and orientation by light scattering. J Histochem Cytochem 25:790–795.
23. **Lucas JN, Gray JW, Peters DC, Van Dilla MA** (1983) Centromeric index measurement by slit-scan flow cytometry. Cytometry 4:109–116.
24. **Lucas JN, Pinkel D** (1986) Orientation measurements of microsphere doublets and metaphase chromosomes in flow. Cytometry 7:575–581.
25. **Mullikin J, Norgren R, Lucas J, Gray J** (1988) Fringe-scan flow cytometry. Cytometry 9:111–120.
26. **Norgren RM, Gray JW, Young IT** (1982) Restoration of profiles from slit-scan flow cytometry. IEEE Trans Biomed Eng 29:101–105.
27. **Robinson RD, Wheeless DM, Hespelt S, Wheeless LL** (1988) A system for acquisition and real-time processing of multidimensional slit-scan flow cytometric data. (Submitted.)
28. **Sharpless TK, Bartholdi MF, Melamed MR** (1977) Size and refractive index dependence of simple forward angle scattering measurements in a flow system using sharply-focused illumination. J Histochem Cytochem 25:845–856.
29. **Sharpless TK, Melamed MR** (1976) Estimation of cell size from pulse shape in flow cytofluorometry. J Histochem Cytochem 24:257–265.
30. **Sharpless TK, Traganos F, Darzynkiewicz Z, et al** (1975) Flow cytofluorometry: Discrimination between single cells and cell aggregates by direct size measurements. Acta Cytol (Praha) 19:577–582.
31. **Steinkamp JA, Crissman HA** (1974) Automated analysis of DNA, protein and nuclear to cytoplasmic relationships in tumor cells and gynecologic specimens. J Histochem Cytochem 22:616–621.
32. **Steinkamp JA, Hansen KM, Crissman HA** (1976) Flow microfluorometric and light-scatter measurement of nuclear and cytoplasmic size in mammalian cells. J Histochem Cytochem 24:291–297.
33. **Weller LA, Wheeless LL** (1982) EMOSS: An epi-illumination microscope objective slit-scan flow system. Cytometry 3:15–18.
34. **Wheeless LL, Berkan TK, Patten SF, Eldidi MM, Hulbert WC, Frank IN** (1986) Multidimensional slit-scan detection of bladder cancer: Preliminary clinical results. Cytometry 7:212–216.
35. **Wheeless LL, Cambier JL, Cambier MA, et al** (1979) False alarms in a slit-scan flow system: Causes and occurrence rates (implications and potential solutions). J Histochem Cytochem 27:596–599.
36. **Wheeless LL, Hardy JA, Balasubramanian N** (1975) Slit-scan flow system for automated cytopathology. Acta Cytol (Praha) 19:45–52.
37. **Wheeless LL, Kay DB** (1985) Optics, light sources, filters, and optical systems. In Van Dilla MA, Dean PN, Laerum OD, Melamed MR (eds), "Flow Cytometry: Instrumentation and Data Analysis." New York Academic Press, pp 21–76.
38. **Wheeless LL, Kay DB, Cambier MA, et al** (1977) Imaging systems for correlation of false alarms in flow. J Histochem Cytochem 25:864–869.
39. **Wheeless LL, Kay DB, Cambier JL** (1979) Three dimensional slit-scan flow system. In Laerum OD, Lindmo T, Thorud E (eds), "Flow Cytometry." Vol. IV. Oslo: Universitetsforlaset, pp 45–48.
40. **Wheeless LL, Lopez PA, Berkan TK, Wood JCS, Patten SF** (1984) Multidimensional slit-scan flow prescreening system: Preliminary results of a single blind clinical study. Cytometry 5:1–8.
41. **Wheeless LL, Patten SF** (1971) Definition of available cellular parameters with a slit-scan optical system. Acta Cytol (Praha) 15:111–112.
42. **Wheeless LL, Patten SF** (1973) Slit-scan cytofluorometry. Acta Cytol (Praha) 17:333–339.
43. **Wheeless LL, Patten SF** (1973) Slit-scan cytofluorometry: Basis for an automated cytopathology prescreening system. Acta Cytol. (Praha) 17:391–394.
44. **Wheeless LL, Patten SF, Onderdonk MA** (1975) Slit-scan cytofluorometry: Data base for automated cytopathology. Acta Cytol (Praha) 19:460–464.

7

Electronics and Signal Processing

R. D. Hiebert
Cell Biology Group, Los Alamos National Laboratory, Los Alamos, New Mexico 87545

INTRODUCTION

Electronics for flow cytometers must be carefully selected or designed to cope with a wide variety of signal-processing needs. Early flow cytometric (FCM) instruments were assemblies of readily available general-purpose laboratory equipment, with the addition of a few special circuit designs to deal with unique signal features. Current systems have become sophisticated specialized designs capable of extracting the most information from sensor signals, while preserving high-quality data at high rates. Information gained from the measurement process is then strategically used to control cell sorting. This chapter describes the basic electronics features of many FCM systems with enough detail to give the reader an awareness of the strengths and limitations of the electronics components. Detailed designs of some building-block circuits have been given by Shapiro [80].

Figure 1 is an overall block diagram showing how the topics of this chapter are linked in a manner similar to the flow of signals in the processing circuits. The first sections describe how sensor signals are generated from the passage of cells in the flow stream. Electronic particle size transducers (Coulter principle) are discussed only to the extent that they influence front-end circuit design. More detailed treatment is given to photosensitive transducers. Descriptions of amplifier designs and analog pulse-processing and computational systems follow. Separate sections deal with digitization of the analog signals and with methods for pulse-shape measurements. The final topics discuss data acquisition interface systems for multistation multiparameter measurements and the unique electronics required of sorting systems; signal processing is taken to the point of interfacing to the computer. Other chapters describe data acquisition, processing, analysis, and display (see Chapter 22).

Noise control and suppression, an important design aspect of flow cytometers and many scientific instruments, is dealt with briefly in the sections on transducers, amplifiers, and analog processing.

TRANSDUCERS IN FLOW SYSTEMS

Cells in suspension are transported through the flow chamber of the flow cytometer that confines the stream to very small dimensions so that cells pass in single file through the measuring zones. Optical measurements are made by focusing one or more light sources onto the cell stream, thus generating observable phenomena such as fluorescence, absorption, light scatter, and extinction from individual cells. Optical transmitting systems project the light signals onto photosensitive transducers located around the flow chamber. An electronic cell-volume sensing orifice may be placed directly in the stream within the flow chamber. This transducer yields an electrical-resistance cell-size signal based on the Coulter principle [15] (see Chapter 4). The outputs from the cytometric transducers are electrical pulses that are processed and recorded by the electronics system.

Photodetectors

Photodetectors for flow cytometers can be placed into three categories [27]. First and foremost are photomultiplier tubes (PMT), which are most useful in applications in which light levels are very low, as in cell fluorescence measurements. Next are individual semiconductor photodiodes, which have many advantages for detection of brighter optical signals, as for light scatter or absorption. Finally, semiconductor photodetector arrays can give valuable spatial information in certain flow systems that yield moderately high levels of light.

Photomultiplier tubes. Fluorescence emitted from the intersection of the cell stream with the focused light source (laser or arc lamp) passes through an optical system of lenses and filters [89,91] and strikes the photocathode of one or more PMTs. A portion of the incident photons at a PMT are converted to photoelectrons, which are accelerated to a dynode structure where secondary-emission gain occurs. The electrons from the last dynode are collected by an anode that provides the output signal current. Multiplication noise and dark currents from PMTs seldom, if ever, limit sensitivity in flow cytometers.

Selection of a PMT type will depend on its physical characteristics, the wavelength and intensity of light to be measured, cost, and other factors. Table 1 lists characteristics of several tubes used in flow cytometers. The two basic physical configurations available depend on the direction of incident light. A head-on (end window) tube receives light along its axis and a side-on tube receives light perpendicular to its axis. A special version of the side-on tube is the dormer-

Flow Cytometry and Sorting, Second Edition, pages 127–144

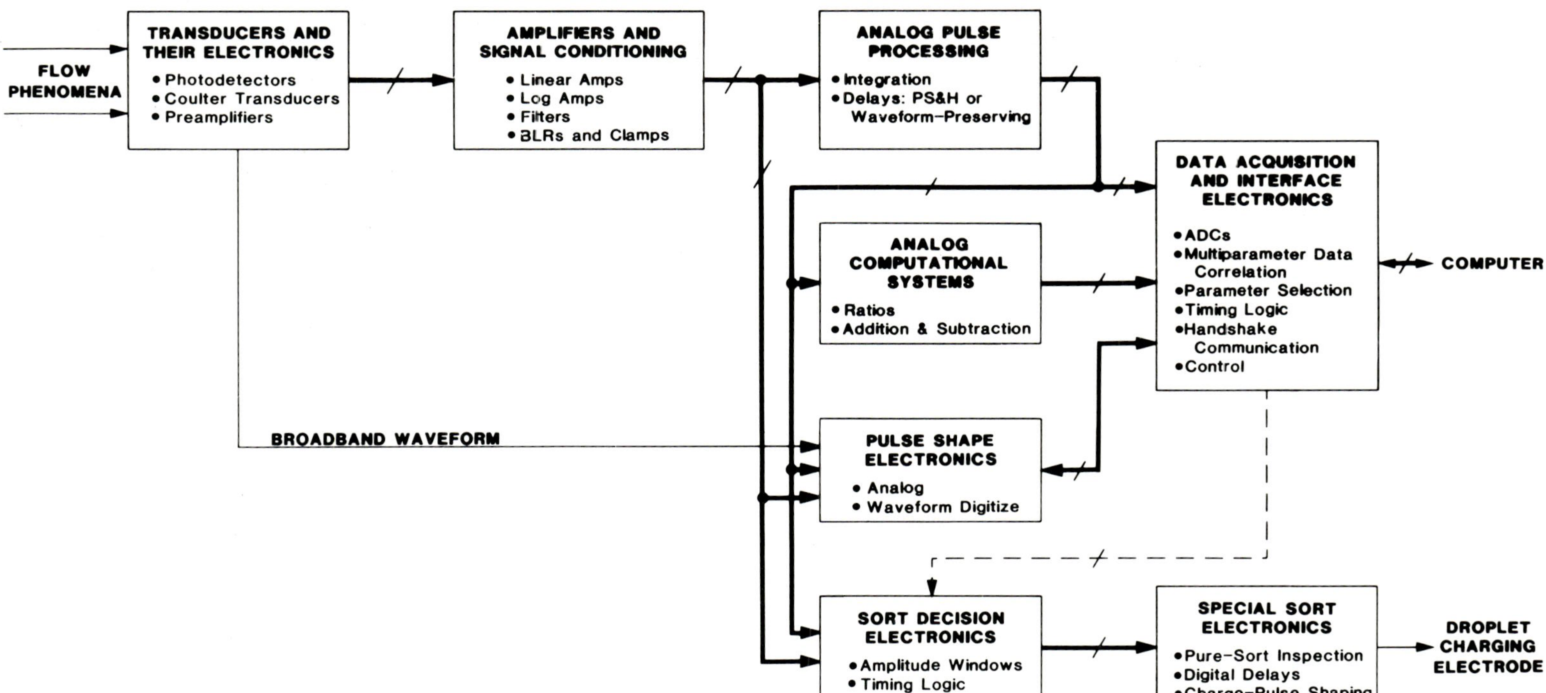

Fig. 1. Generalized diagram of electronics of a flow cytometer and cell sorter. The dashed line between data acquisition and sort decision electronics indicates linkage to correlate data displays and sort windows. Diagonal lines through interconnecting arrows indicate multiple connections.

TABLE 1 Photomultiplier Tubes Commonly Used in Flow Cytometers[a]

		Cathode characteristics			Dynode characteristics	
Tube type and mfgr	Physical characteristics	Useful wavelength λ (Peak λ)	QE at Peak λ (Elect/ Photon %)	Max current density (nA/cm^2)	Structure and number of stages	Max. pulse current (mA)[b]
4526-RCA	Dormer Window	300–800 (520)	20		Circ. focused 10 stage	20
8644-RCA	19 mm head on	300–800 (420)	19	250	Linear focused 10 stage	
7164R-RCA	50 mm Head On	300–800 (520)	9	250	Cric. focused 10 stage	20
1P28-RCA	Side On	240–600 (320)	19	5000	Circ. focused 9 stage	10
9798B-EMI	30 mm Head On	350–800 (400)	17	250	Box and grid 11 Stage	0.1
9524B-EMI	30 mm Head On	350–650 (400)	22	15	Box and grid 11 Stage	0.1
9924B-EMI	30 mm Head On	350–620 (380)	25	2.5	Box and grid 11 Stage	0.1
9558B-EMI	52 mm Head On	350–800 (400)	22	250	Venetian blind 11 Stage	2
9658B-EMI	52 mm Head On	350–870 (440)	28	250	Venetian blind 11 Stage	2
9659QB-EMI	52 mm Head On	200–900 (400)	19	250	Venetian blind 11 Stage	2
9818B-EMI	52 mm Head On	350–800 (400)	22	250	Linear focused 10 Stage	50
R928 Hamamatsu	Side On	200–900 (400)	21	5000	Circ. focused 9 Stage	
R1477 Hamamatsu	Side On	200–900 (450)	22	5000	Circ. focused 9 Stage	
R712 Hamamatsu	28 mm Head On	200–900 (640)	7	250	Box and grid 11 Stage	

[a]Detailed information on PMTs is available from RCA Electro Optics, Lancaster, PA 17604; EMI Gencom Inc., Plainview, NY 11803; and Hamamatsu Corp., Middlesex, NJ 08846.
[b]Peak anode currents for the EMI tubes are specified as those for which there is a 5% departure from linear amplification.

window arrangement with light entry near the end of the tube envelope. The photosensitive area of PMTs is ample for FCM applications.

The cathode response should match as closely as possible the wavelength of light to be detected. Table 1 shows the useful range of wavelengths with the peak-response wavelength indicated in parentheses. Tubes with particularly broad spectral responses are the head-on types 9659QB and R712, the side-on type R928, and its high-sensitivity variant R1477. Another variant of the R928 is the R955 (not listed in Table 1), with response extended down to 160 nm. Quantum efficiency (QE) is defined as the average number of electrons photoelectrically emitted from the photocathode per incident photon of a given wavelength. The QE should be as high as possible to minimize statistical effects in signal generation, which becomes a dominant factor influencing fluorescence intensity resolution for intermediate-brightness levels. Among several tube types with particularly good QE is the 9658B, which features a prismatic end plate on which the photocathode material is deposited; the resulting multiple reflections at longer wavelengths enhance the electron yield. The 9924B has a new bialkali cathode with higher blue–green sensitivity.

For high levels of fluorescence, the photocathode current density may become a factor because of the resistive nature of the thin layer of semitransparent cathode material in head-on tubes. As shown in Table 1, the commonly used multialkali (S20) cathodes (250 nA/cm^2 limit) are fairly conductive and are tolerant of moderate currents. Bialkali cathodes are the most vulnerable to defocusing effects from high currents.

In most cytometric applications, dynode characteristics are not a factor in selecting PMT. However, the user should be aware of maximum pulse current ratings for different dynode structures, as indicated in the table. The data-sheet specifications for maximum peak currents are usually vague, but fortunately, the preamplifier sensitivity is usually high enough that the pulse levels listed in the table need never be reached.

Gain stability of the PMT is very important, as it is in all other analog elements in a flow cytometer. The average anode current should be kept at 1 μA or less for utmost gain stability over long periods of time. The high-voltage power supply for the PMT must be of high quality. In some flow systems the light signals are bright enough that the needed gain of the PMT is less than the normal operating range of the tube. In these cases it is better to wire the PMT base for truncated operation (for example, 6 dynodes instead of 10) than to operate the full tube at voltages below manufacturer's limits. The unused dynodes may be connected together with the anode.

Photodiodes. Single silicon photodiodes are used as transducers for light signals of moderately high intensity; light scatter and absorption measurements are examples. Their advantages are small sensitive area, low-voltage operation, low cost, broad spectral response, high QE, stability, and resistance to effects from overloading. However, they have much higher dark currents than do PMT and, because there are usually no internal gain mechanisms in the devices, the low-output signal competes with signal-processing electronics noise.

Fiberoptic guides may be used to transmit the light from narrow collection angles to photodiode detectors [89,97]. Silicon avalanche photodiodes with gains as high as 1,000 are used in fiberoptic communications systems, and they can be used in cytometers if the small photosensitive areas can be tolerated.

Photodetector arrays. Multielement silicon photosensor arrays have been used in research instruments for multiangle light scattering analysis of single cells in flow [74]. The arrays can be divided into two general classes: those with individual photoelements that have separate connections for parallel data output, and those with individual elements that are integrated with circuitry to allow serial interrogation and readout.

Parallel-readout arrays of 50 photodiode elements arranged linearly on a common silicon substrate [21] have been used for absorption measurements, and full-circle arrays of 60 photodiode optochips mounted on a ceramic substrate [4] have been developed for differential scattering measurements over a wide range of angles. For weak scattering intensities, concentric fiberoptic rings have been used as light guides to individual PMT [22].

Monolithic serial-readout photodiode arrays consisting of photosensitive charge-coupled devices (CCD) have been used in simple light-scattering systems [3]; similar arrays are found in research systems for multidimensional slit-scan imaging of fluorescence in flow [43]. At these low light intensities, microchannel-plate image intensifiers must be used in front of the arrays.

Coulter-Effect Transducers

The basic Coulter transducer [15] consists of a small-diameter orifice placed in a conducting electrolyte solution that carries the cells to be measured. Electrodes are mounted on both sides of the orifice and connected to a power supply providing constant current flow from one electrode to the other through the orifice. The electrical resistance of the cells is different from that of the electrolyte, and as each cell passes through it, the resistance of the orifice is changed, generating a pulse. The pulse magnitude is a measure of cell size. The earliest cell sorters were based on measurements from Coulter-effect transducers [24]. Kachel [37] has given a comprehensive review of the Coulter principle (see Chapter 4), and Steinkamp [89] shows how the transducers are mounted in various flow chamber configurations. For kinetic studies of cells, special transducers have been made to minimize the delay from the sample source to the Coulter measuring zone [40].

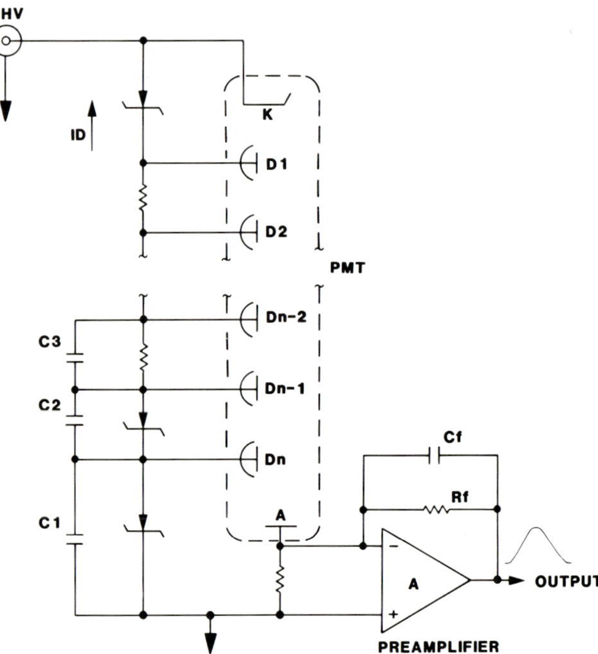

Fig. 2. Dynode-chain wiring of a PMT base for a flow cytometer. Note transimpedance preamplifier for converting PMT anode pulse current to output voltage. The zener diodes between dynodes may consist of several diodes in series to achieve the desired high voltage.

TRANSDUCER CIRCUITRY AND PREAMPLIFIERS

Circuits to be used with photodetectors and Coulter transducers must be designed carefully, with consideration of the operating conditions that occur in FCM systems. Figure 2 shows the dynode-chain wiring of a PMT base, with operational-amplifier symbolism for an associated preamplifier for the anode output pulses. The network providing interstage voltages to a PMT can take various forms, but the circuit shown in Figure 2 is applicable to almost any flow cytometer. Negative high voltage with respect to ground permits the anode (A) output to be at ground, allowing direct-current (dc)-coupled electronics in signal processing. Also, any noise on the high-voltage supply line is highly decoupled from the output signal. The disadvantage for head-on tubes is that, with the cathode (K) at high voltage with respect to ground, a voltage gradient will exist across the tube wall if there are close-fitting grounded structures in the mounting arrangement. This gradient can cause noise scintillations in the glass. Careful mounting or using shielding prescribed by PMT manufacturers prevents this problem.

The cathode-to-first-dynode voltage is very important to optimum operation of the PMT because it establishes the quality of the electron optics in the critical space between the photocathode and multiplier structure, and it defines the secondary emission gain at the first dynode. In applications where the overall voltage to the PMT is narrowly defined by the need for only small gain variability, the cathode-to-first-dynode voltage can be established with a fixed resistor. However, in many FCM applications in which the measured light level varies over wide extremes, requiring a broad range of applied high voltages, zener diode regulation should be used as shown in Figure 2. This voltage should be maintained at one-half to two-thirds the maximum cathode-to-dynode 1 ratings for the tube. Also, zeners at the last dynodes of the tube will preserve linearity for high output pulse current levels over a range of applied high voltages. Storage capacitors C1, C2, and C3 sustain high dynode pulse currents without using higher divider string current I_D.

Transducers in flow systems are current sources with relatively high output impedances, and current- or charge-sensitive preamplifiers are usually used. Therefore, the preamplifier must have a low-input impedance compared with the output impedance of the transducer. This is usually achieved as shown in Figure 2, wherein parallel feedback elements R_f and C_f are connected to a high-gain, low-noise voltage amplifier A. This circuit configuration is often called a transimpedance amplifier. The feedback to the inverting terminal of A forces that signal-input terminal to maintain a voltage that is nearly equal to that of the noninverting terminal, which is at ground. Thus the transducer output current is forced to flow through the feedback transimpedance elements. If these elements are chosen so that R_f is the dominant component and C_f is minimized, the output voltage signal will follow the time history of the input current pulse. By contrast, if C_f is made large and becomes the dominant component, the circuit becomes charge sensitive and integrates the input pulse current. Details for implementing these preamplifier circuits are described by Hiebert and Sweet [30]; the results of preamplifier pulse shaping are discussed in this chapter.

Photodiodes in cytometric applications are usually used in the photoconductive mode whereby external reverse bias voltage is applied to the junction. Connections to these devices are simple and the user should follow the manufacturer's recommendations. A current-sensitive preamplifier is almost always used.

Photodetector arrays are complicated assemblies requiring intricate peripheral circuitry. Electronics for parallel-readout arrays have been described by Bartholdi et al. [4] and Crowell et al. [17]. Support electronics for serial-readout photosensitive CCD arrays are available from the manufacturers (e.g., Fairchild CCD Imaging, 3440 Hillview Ave., Palo Alto, CA 94304; Reticon, 345 Potrero Ave., Sunnyside, CA 94086; Thomson-CSF Components, 301 Route Seventeen North, Rutherford, NJ 07070).

The constant-current source for Coulter orifice electrodes can be supplied by a simple circuit operating from a constant-voltage power supply [37,59]. The magnitude of current is frequently made adjustable. The preamplifier for a Coulter transducer should be a current-sensitive (transimpedance) type described earlier. This type of circuit maintains a virtual ground at its input, which makes the magnitude of the signal from the cell-resistance change in the orifice independent of the resistivity of the electrolyte. Also, this circuit permits a transient response (risetime) that is less dependent on the stray capacitances associated with the orifice.

Fluidic connections to the Coulter orifice are vulnerable to electromagnetic noise pickup that can degrade signal performance. Schuette et al. [75] describe a means for incorporating fluidic electrical noise-isolation resistance elements located in a small shielded enclosure mounted on top of the flow chamber.

AMPLIFIERS AND SIGNAL CONDITIONING

Main amplifier designs for flow cytometers can be classed as linear or logarithmic. The choice of the amplifier type and the signal conditioning (filtering, baseline control, noise rejection) will depend on the cytometric parameters to be measured.

Linear Amplifiers

Excellent commercial amplifiers are available in nuclear instrument module (NIM) form. They were used in early fluorescence-activated cell sorting (FACS) systems and continue to be used in multiparameter cell sorters [105] and measuring systems [102]. These amplifiers offer a wide range of gain control and filter time constants, as well as provision for baseline restoration (BLR). However, because their features are sometimes incompatible with cytometer requirements, and because of their relatively high costs, many laboratory and commercial systems use special designs.

Amplification can be achieved with high-speed operational amplifiers, and several options are available for organizing the gain and signal conditioning functions. Figure 3 shows how the arrangement might be for an amplifier for photodetector or Coulter-effect transducers. The dc-coupled input from a preamplifier branches to a gain-of-10 broadband stage and to a gain-of-100 stage to drive a dc panel meter that shows average anode current for PMT applications. The current sensitivity of this meter depends on the transresistance in the preamplifier. In Los Alamos instruments [60], the sensitivity can be switched between two ranges of 10 and 100 nA.

Output of the first pulse-gain stage feeds a single-pole switchable low-pass noise filter and a gain control circuit. This stage is followed by another gain-of-10 stage that may be operated differentially (dashed line) to minimize interference noise from the droplet generation action in sorting systems. A portion of the piezoelectric transducer (PZT) drive oscillator signal is passed through phase and amplitude conditioning circuits with controls to achieve a null for this systematic source of noise.

The next block in Figure 3 shows a BLR and clamp. Baseline restoration is important in amplifiers for Coulter transducers because they are almost always capacitively coupled to the preamplifier; this differentiating effect causes pulse undershoot that can take significant time to return to baseline [37]. Alternating-current (ac)-coupling may also be desired in photodetector systems that have low-frequency noise, and a BLR will give baseline control. The BLR may be a simple [69] or a sophisticated [14] circuit. At the output of the amplifier chain a pulse amplitude clamp may be furnished to provide abrupt transition from linear operation to pulse saturation at a level slightly lower than full-scale range on the pulse height analyzer (PHA) for the cytometer. This arrangement allows all the excess-amplitude pulses to pile up cleanly in a few channels near full scale so that they can be accounted for in cell-processing inventory. Examples of circuit components for a linear FCM amplifier are given by Hiebert and Sweet [30].

Logarithmic Amplifiers

Early FCM applications for logarithmic pulse amplifiers were in light-scattering instruments [71] that yielded a wide range of amplitudes from cell data. Logarithmic compression with relatively simple circuits permitted recording of these data without changing gain or resorting to higher-resolution digitizing devices. The value of logarithmic conversion of signals from Coulter transducers was recognized by Von Behrens and Edmondson [106]. A method of using data from logarithmically amplified Coulter signals to construct linear

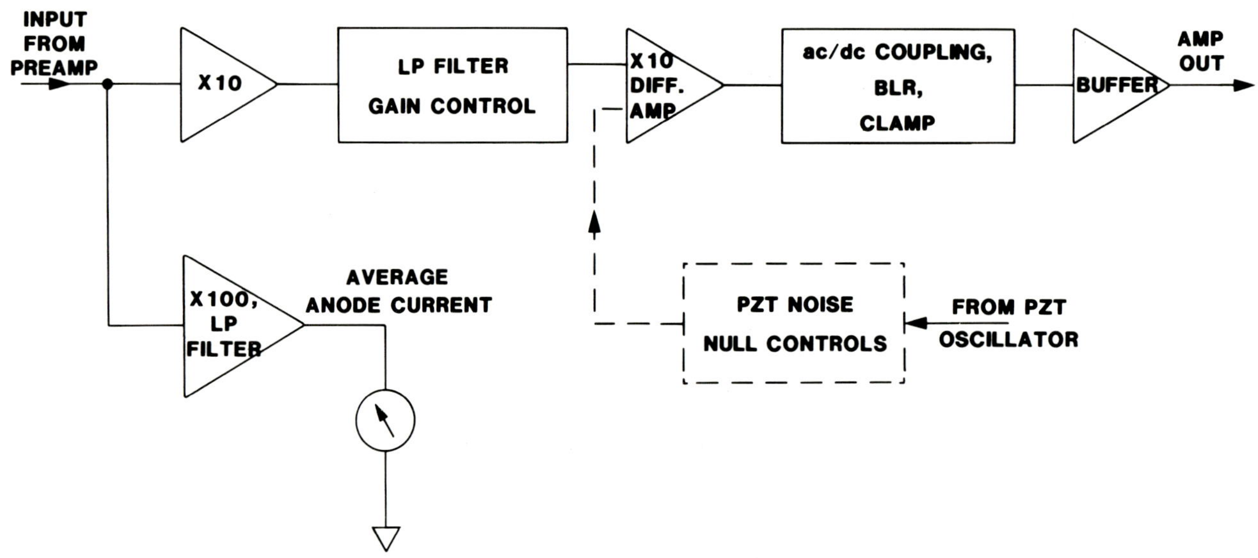

Fig. 3. General-purpose FCM linear amplifier, including a low-pass filter, a baseline restorer, and a differential stage to minimize noise that may arise from the piezo-electric transducer oscillator in sorting systems.

calibration plots for volume, surface area, and diameter of spherical cells is described by Schwartz et al. [76]. As FCM fluorescence techniques improved and the range of fluorescence intensities expanded, logarithmic amplifiers became useful for data display and analysis. The advantages of logarithmic compression of histograms collected from chromosome analysis by flow cytometry have been pointed out by Bartholdi [6]. Several ways to calibrate logarithmic fluorescence measurements to yield most meaningful data are given by Muirhead et al. [61] and by Parks and Peterson [63].

Figure 4 shows the basic elements of a logarithmic pulse amplifier. Input signals from the transducer enter one terminal of a differential preamplifier that feeds the logarithmic amplifier. A dc feedback system employing some means of baseline control connects the output of the logarithmic amplifier to the other input of the differential preamplifier. A postamplifier feeds the logarithmically amplified output pulses to processing elements that are described later.

The logarithmic amplifier may use monolithic integrated circuits to achieve ~21/2 decades of response [17]. These circuits have the advantage of a bandwidth or transient response yielding an output risetime (10–90% levels) of ~1 μs, and they are inexpensive. Other designs [100] use a commercial wideband logarithmic module with a specified minimum dynamic range of 70 dB (31/2 decades).

Stabilizing the baseline reference operating point of a logarithmic amplifier is extremely important. Any change in the quiescent input level shifts the operating point, and thus the small-signal gain of the system. If direct coupling is used, as for photodetector systems, this shift can be caused by changes in ambient light reaching the photodetector, by drift in the preamplifier, or by drift in the input circuitry of the logarithmic amplifier itself; ac-coupling of the signal to the logarithmic amplifier input will remove dc offsets produced in preceding components, but the no-signal input potential and the signal gain will then be a function of signal repetition rate and amplitude.

To attack the baseline stability problem, some form of feedback control (Fig. 4) is used. The simpler systems [17] use steady-state, direct-coupled feedback with an integrating operational amplifier. The system is equivalent to an ac-coupled amplifier with a low-frequency cutoff determined by the feedback integration time constant, and the resulting baseline is affected at high signal duty factors. More sophisticated amplifier designs with logarithmic modules incorporate gating action to switch off the feedback during the presence of signal so that the stabilized operating point is not affected by signal duty cycle [30].

The maximum signal range of a logarithmic system may be limited by the noise level in the preamplifier. This factor, along with amplitude accuracy and pulse recovery time, must be considered in choosing the optimum preamplifier and postamplifier bandwidths [30].

Alternatives to amplifiers for logarithmic conversion have been used with flow cytometers. Koper and Blanken [47] use a programmable read-only memory (ROM) circuit as the equivalent to a look-up table for the bit pattern emerging from an analog-to-digital converter (ADC). The rescaling algorithm converts an ADC with a conversion gain of 1024 channels to an 8-bit output with 1.3 decades of dynamic range. An ADC with 4,096 channels could be converted to a range of three decades. Whereas this circuit is placed between the ADC and the multichannel analyzer (MCA) storing the data, Brunk et al. [9] describe a microcomputer program that converts linear data acquired on a MCA to a logarithmic display. This scheme allows the combination of several sets of data into a single display of greater accuracy and dynamic range than could be achieved with any single linear analysis.

ANALOG PULSE PROCESSING

Most flow cytometers use special circuits to modify or manipulate the raw pulses that are generated at the transducers. In the previous section on amplifiers we discussed filters to limit the bandwidth of amplification. Low-pass filters are used to reject high-frequency electronic or systematic

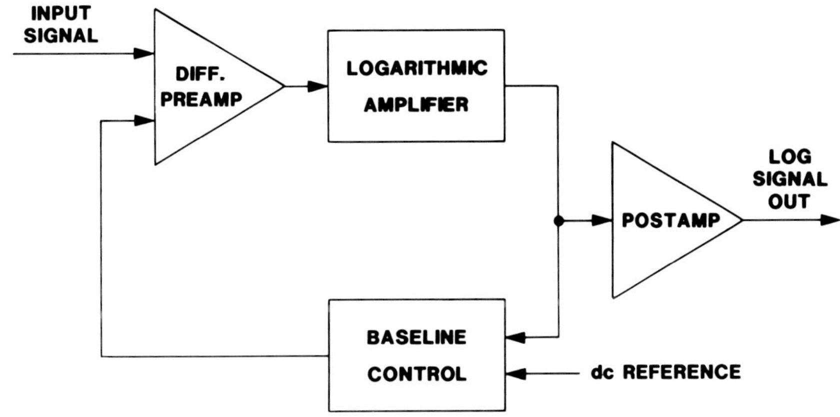

Fig. 4. Simplified diagram for FCM logarithmic pulse amplifier. A dc feedback loop to control the baseline operating point is very important to maintain predictable and stable gain for small pulse signals.

noise. High-pass filters (ac coupling) may be inherent to a system, as in Coulter-effect preamplifiers, or added as a selectable feature to allow rejection of system dc variations or low-frequency noise. As mentioned earlier, these high-pass filters are usually followed by BLR circuits to improve amplifier performance at high pulse rates. This section discusses the additional signal-processing features of analog pulse integration and analog signal delays.

Pulse Integration

Many fluorescence measurements of cells or chromosomes require a signal proportional to the total amount of light detected during the dwell time of the particle in the observation zone [5,49,68]. Similarly, absorption pulses can be integrated and combined with pulse shape measurements for particle sizing [21]. Time integration of scattered and absorption pulse signals has been correlated with width signals to determine size and refractive index of particles in flow [83].

Pulse integration can be made completely passive by collecting the photodetector current into a parallel resistor-capacitor (RC) network to ground and amplifying the pulse with a voltage-sensitive preamplifier. This method is generally not used because the amplitude and shape of the preamplifier signal depends on the transducer and stray capacitance to ground, which may not be constant for different experimental arrangements. Therefore, active pulse integration is preferred; Figure 5 shows two methods for achieving integration and pulse duration control.

Figure 5A shows a charge-sensitive-type preamplifier (see earlier section on transducer preamplifiers) with an R_fC_f product (which is time) that is much longer than the duration of the current signal from the photodetector. Thus, the duration of the input current is manifested in the output risetime of the preamplifier, and the crest value of the preamplified pulse is a measure of signal integral. These pulses must be shaped in the main amplifier to eliminate the long R_fC_f decay tail, or the system would be intolerably sensitive to pulse pileup. This shaping is gaussian or delay-line clipped in the NIM amplifiers discussed earlier, and the shaping time τ can be selected. In cytometers, the duration of the photocurrent pulse is variable because of such factors as particle shape, stream orientation, and velocity, and the resulting variation in risetime from the preamplifier must be considered in selecting τ. It must be long enough that cytometric duration variations do not affect the accuracy of the output pulse amplitude; typically τ is set for two to three times the nominal input current pulse duration. This method of integration has the advantage of wide dynamic range because it operates linearly on signals down to noise levels. Each pulse occupies a long baseline time due to the shaping action, which can be a disadvantage at high pulse rates. Also, time-resolved relationships are lost at the output of the amplifier.

A switched, or gated, integrator configuration, shown in Figure 5B, can be better for the variabilities in flow cytometers [30]. The preamplifier is a transimpedance type with R_f small enough to be the dominant component, and its voltage output follows the form of the current input. The main amplifier is a broadband type so that time-resolved pulses are available at its output for shape analysis measurements. The operational amplifier integrator is a three-mode type with two field-effect transistor switches S1 and S2 controlling the action as dictated by simple time-control circuits. In the quiescent reset condition, S1 is closed, S2 is open, C is discharged, and the output is at zero. When an incoming pulse crosses the noise-discriminator threshold in the time control circuits (not shown in Fig. 5B), S1 opens, S2 closes, and the circuit is in the integrate mode for the input pulse. When the input pulse recedes below threshold, S2 is opened and the circuit is kept in the hold mode for a short time before S1 closes to discharge C and complete the integration cycle. Minimum baseline time is used because each pulse is processed only for the time necessary for integration.

Analog Signal Delays

Flow cytometers having a single measuring station for multiparameter measurements or those having multiple measuring stations often use signal delays to satisfy system timing requirements. In systems involving many detectors at a single measuring station, the simultaneous detected pulses may have to be fed sequentially to a single ADC with peak sense-and-hold (PS&H) delays feeding an analog multiplexer [17,71]. A more common arrangement for measuring multiple cellular properties uses sequential measuring stations, and the signals produced at the stations must be coherently processed using either analog or digital delays for proper time alignment. In one method the crest values of pulses are temporarily stored in PS&H circuits and digital timing mark

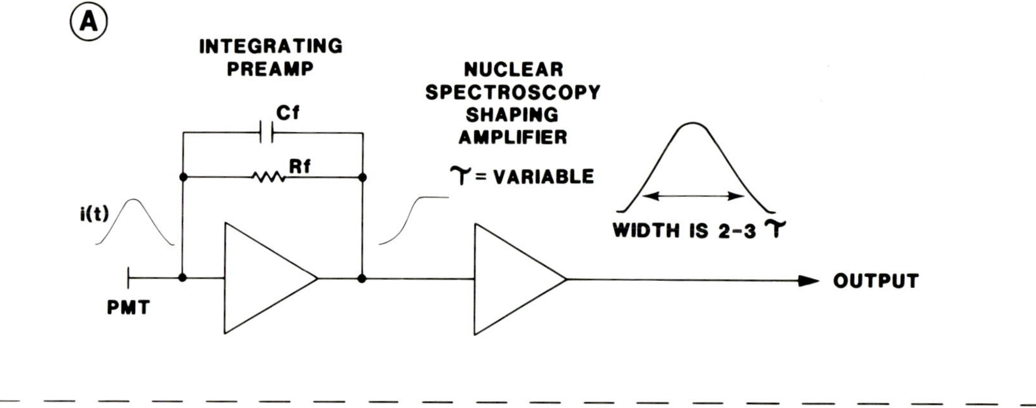

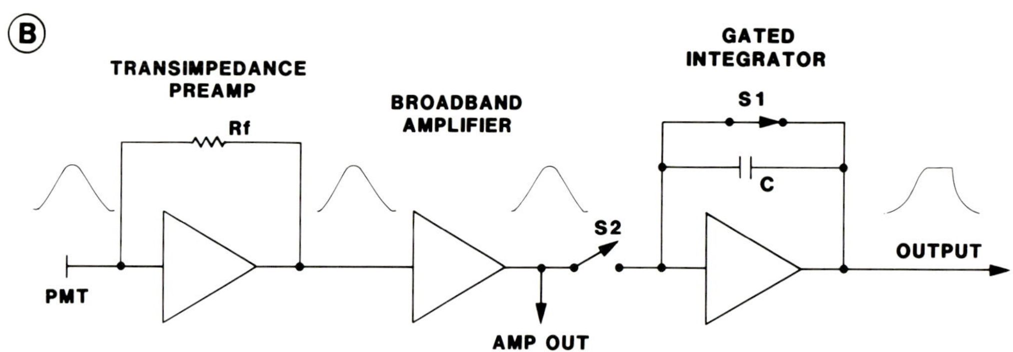

Fig. 5. Two methods for achieving integration of the pulse current from a PMT, using active circuits. In method A, a charge-sensitive preamplifier integrates the input current, and the main shaping amplifier generates a pulse whose crest value represents the integral, but the output pulse width has little relationship to the input current pulse shape. In method B, the current waveform is preserved in the voltage amplifiers, and integration is achieved with a gated (switched) operational-amplifier integrator.

signals are delayed appropriately for correlation with downstream measurements [2,72,77,87]. These systems have the disadvantages of not preserving the waveforms of the input pulses and of introducing dead time from the PS&H during its holding period. Waveform-preserving "pipeline" delays capable of acquiring and propagating many signals within the delay period can eliminate the hold dead time and thereby increase the cell measurement rate.

Short waveform-preserving delays can be achieved with passive electromagnetic delay lines. They are limited to maximum delays of about 10 μs for the bandwidths required for flow cytometric pulses, and they are difficult to terminate properly for distortion-free transmission. Longer waveform-preserving pipeline delays can be achieved using an integrated-circuit, charge-coupled device (CCD), which is an analog shift register with many elements that can be clocked at high frequencies. Figure 6 shows a typical application for these active delay elements in a flow cytometer [65]. Information from the first transducer in the flow stream, a Coulter orifice, is amplified and delayed in a single-channel Coulter-volume CCD pipeline delay. Pulses from two photodetectors positioned at the stream-laser 1 intersect point are amplified and delayed in a dual-channel photosignal pipeline delay. Photodetector pulses from the last measuring station at laser 2 are amplified and fed directly to the data acquisition system. The CCD delays are adjusted smoothly by varying the clock frequency that controls charge transport in the shift registers. The adjustments align the timing of the early signals to arrive for parallel data acquisition and processing coincidentally with the undelayed pulses from the last measuring station. The CCD integrated circuits are available as dual-channel devices driven by a single clock, which is convenient for delaying signals of two parameters equally from a single station.

Waveform-preserving delays can also be used in slit-scan flow cytometry in which simultaneous dual waveforms from a single measuring station can be recorded in a commercial broadband single-memory waveform recorder [34]. One channel of the CCD delay is used to delay one of the waveforms to a multiplexed switching network feeding the recorder. Homemade two-channel transient digitizers have been constructed for two-waveform slit-scan data acquisition and analysis [108,109,110].

ANALOG COMPUTATIONAL SYSTEMS

Analog computational preprocessing of signals before digitization is desired in many FCM systems. These real-time manipulations can result in higher quality data, and the higher speed of analog methods versus digital computation is important for making rapid decisions in sorting systems. These computational functions can be grouped into dividers for ratio-taking and adders and subtracters used as signal correctors. Multipliers used for special diagnostics, such as pulse-shape analysis, are not discussed here.

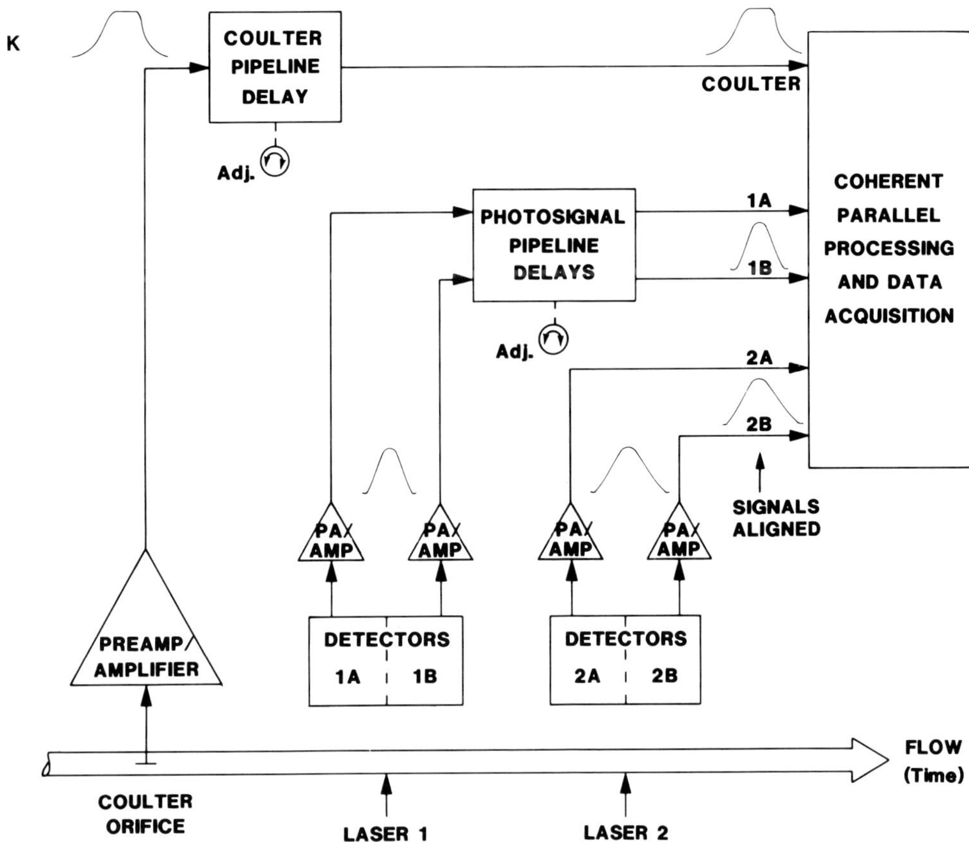

Fig. 6. Typical FCM system using adjustable charge-coupled-device analog pipeline delays to align multiparameter signals from two upstream measuring points (Coulter orifice and laser 1) with signals from the third measuring point (laser 2) for simultaneous processing. Direction of flow is shown along the X axis to more graphically portray passage of time (not to scale) as particles move with the stream. Coulter orifice signal delays can be many hundreds of μs, whereas photosignal delays are typically shorter, in the range of 30 to 200 μs.

Dividers

Ratios of measurements are useful in parameter normalization requirements, as in normalizing nuclear fluorescence to a cell size measurement [87]. A cytometer using commercial modular wideband multiplier/dividers to process microsecond-type pulses has been described [29], and somewhat slower integrated-circuit dividers have been added to commercial cell sorters [33]. Kachel et al. [38] described a hardwired digital calculation of ratios in which division is accomplished by multiplying data from one parameter by the inverted divisor data of the other. Values of the inverted divisor are tabulated in a ROM.

Analog dividers can be used to improve data affected by light source instabilities if the intensity of illumination the cell traverses can be accurately monitored for use as a denominator signal [30]. This technique has been used to improve cytometric data from arc lamp illumination [85].

Analog dividers are also useful for processing data to estimate fluorescence polarization [23,36]. Stewart et al. [93] developed a real-time analog polarization computer that uses a low-cost multiplier/divider integrated circuit. Keane and Hodgson [45] describe a fast analog divider system with careful attention to electronic and background offsets. Polarization calculations can also be done digitally after the data have been collected [51,67].

Signal Correctors

Sharpless [84] discussed the possibilities of analog computations to transform raw measurements for each cell into cell feature sets that have particular relevance to a cytometric problem. Numerous transformations are available, and the output data can be further processed by digital computer or used to establish unique analysis or sorting windows in the cytometer. We confine this discussion to linear combinations of sets of signals, with the objective of correcting or compensating for disturbing influences. Highly flexible modular circuits for making analog corrections have been presented by Steinkamp et al. [92].

In two-color immunofluorescence studies using a single wavelength of excitation, possible emission spectral overlap of the two dyes will affect the quantitative analysis of double-labeled cells. Loken et al. [53] described an electronic network in which a portion of each signal from the green or red channel is linearly cross-coupled to the other channel through a differential amplifier system. In this way, the component of unwanted fluorescence is electronically subtracted. This system has also been used for autofluorescence subtraction [1]. A more extensive system for analog signal compensation in four-color immunofluorescence measurements has been designed by Parks et al. [64].

Differential amplifiers have been used to reduce the effects

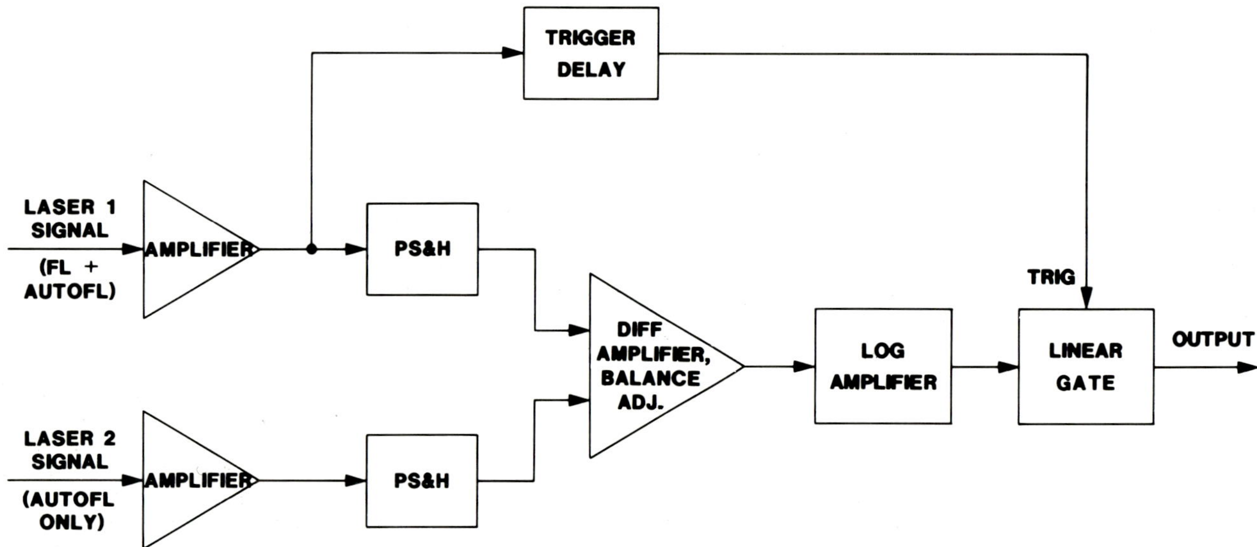

Fig. 7. Electronics arrangement for autofluorescence subtraction. Laser 1 excites fluorescence and autofluorescence, whereas laser 2 excites autofluorescence only. Crest values of the pulses are held in peak sense-and-hold circuits that feed a differential amplifier where subtractive correction occurs. After logarithmic amplification, the portion of the difference signal containing the fluorescence signal of interest is gated to the output.

of laser noise in axial light loss (ALL) measurements [88]. The laser is monitored with a photodiode to obtain a measure of laser excitation noise. This is carefully differentially balanced with the ALL photodetector output containing both ALL signal and laser noise to yield an output with a noise reduction of about 12 dB.

In cell sorters using enclosed flow chambers, an electro-optic monitoring system can be used to determine the transit time from the point of measurement of cell properties to the downstream region at which droplet charging occurs [56]. These delay monitor signals originating near the point of droplet breakoff contain the cell scattered-light pulses of interest along with contaminating waveforms caused by droplet formation. With careful differential mixing of signals from the droplet-generating oscillator, the pulses of interest can be extracted and the delay time measured.

Another application for differential amplifiers is in autofluorescence subtraction using two wavelengths of excitation (90). In some cell types, the autofluorescence signal is brighter than the fluorescence from the dye that is added as the cell marker, and the desired signal is obscured. Signal correction is based on the fact that autofluorescence emission and excitation spectral distributions are broad and relatively flat over a large wavelength range, whereas the dye is excited in a relatively narrow band. Sequential measurement stations are used. The first laser is tuned to excite in the spectral region for both desired fluorescence and autofluorescence, and the second laser is tuned to excite primarily the autofluorescence. As indicated in Figure 7, the signals from the two channels are amplified, and the crest values of the pulses are captured in PS&H circuits. The held levels are fed to a differential amplifier where subtraction of the two signals yields the desired fluorescence signal alone. The differential signal of interest does not appear until after the corrective action of laser 2, so a linear gate is added, with the gate trigger properly timed. A logarithmic amplifier is added to enhance the display of the difference signal.

ANALOG-TO-DIGITAL CONVERTERS

At the data acquisition interface (see Fig. 1), analog signals proportional to measured cellular features are digitized for storage, analysis, and display. The ADC is electronic circuitry that accomplishes the conversion process. If the ADC is incorporated in a complete integral instrument with digital memory, display, and control features, it is referred to as a pulse-height analyzer (PHA) or MCA (Multichannel Analyzer). For sorting decisions, the simplest form of ADC is used: the single-channel pulse-height analyzer (SCA), in which two electronic amplitude discriminators with preset thresholds become lower and upper bounds for a window of values representing a particular cell feature. The digital output pulse enters sort decision circuitry, as described in a later section.

The ADC process assigns the input pulse height (or analog level) to one of many channels. Ideally, all channels should have the same width so that the stored channel contents will yield a true pulse-height histogram. the degree to which all channel widths are not equal is a measure of differential nonlinearity; this digitization error is critical to the cumulative display requirements of a PHA. If the channel widths are not equal, spurious peaks and valleys will occur in the recorded frequency distributions.

Two types of ADC are used for FCM applications. The highest-quality histograms are obtained with the Wilkinson-type converter, which is widely used for recording nuclear pulse height spectra. With this technique, a capacitor is charged to the peak amplitude of the input pulse and linearly discharged by a constant current. During the discharge period, a high-frequency clock oscillator is gated to a counter that records the number of cycles of the clock. This number is proportional to the magnitude of the input pulse. With this conversion process, all channels are necessarily contiguous and, if the clock frequency is stable and the discharge linear, channel widths are nearly identical. Fast Wilkinson-type

ADC are relatively expensive, which can be a significant factor for multiparameter recording systems because they are not widely available in multiplexed configurations.

The second ADC is the successive-approximation type. The technique consists of comparing the input signal with a precisely generated internal voltage at the output of a digital-to-analog converter (DAC). The input to the DAC is a binary succession of test values, so that the error is halved at each approximation step. The process is continued until the last bit has been tried, at which time the output register keeping track of the succession of tests will show the conversion bit pattern. These ADC are very fast, capable of high precision, and relatively inexpensive. However, uniformity of channel width (i.e., equivalent analog size of the least significant bit) is poor, and rough histograms can result. Channel smoothing to improve effective differential nonlinearity can be achieved by several techniques. One method is to sum the input signal with the output from an offset DAC whose value increments with every conversion. This DAC output acts as a sliding-scale offset that is then subtracted from the ADC output after conversion [13]. Another technique is to use converters with excess precision and to discard some number of low-order bits from the recorded data. For example, one might truncate from 12 or 14 bits to 8 bits, which would result in a 256-channel histogram.

Completely self-contained MCA, most models based on microprocessors, are available from manufacturers of nuclear instruments. High-performance MCA systems based on personal computers are also available. For FCM applications, Kachel et al. [39] described a microprocessor-based PHA with additional capabilities for data processing and statistical calculations.

PULSE-SHAPE MEASUREMENTS

In many flow cytometers, the size of the illuminating beam in the direction of flow is larger than the cell. Furthermore, the limiting aperture in the detection channel permits collection of light from the entire illumination slit. Called a zero-resolution system [16], crude pulse-shape information can be extracted from the resulting low spatial resolution measurements. For example, in fluorescence measurements in which a specific component of the cell is stained, pulse height (maximum signal amplitude) measures the point at which the combination of fluorescent molecule density and laser intensity is a maximum. Pulse width measures the time during which the cell is in the beam and is a function of the length of the cell, the width of the measuring aperture, and size of the laser beam. Pulse area (pulse integral) is proportional to the total fluorescence emitted as the cell passes through the laser beam.

These zero-resolution measurements can readily be made if precautions for time-resolved amplification have been taken, as described earlier. A PS&H circuit will measure the crest value of the signal for pulse height. Pulse-width measurements can be made by recording the time that the pulse is above a fixed-amplitude threshold, as in the time-of-flight technique [89]. However, because this method depends on the pulse height of the signal, better data are acquired by using constant-fraction threshold measurements, which are independent of pulse amplitude. This has been demonstrated for widths from fluorescence [82] and from scatter and absorption [83]. Circuits for making constant-fraction width and risetime measurements are described by Hiebert and Sweet [30] and by Leary et al. [48]. An option in one of these circuits [30] permits sensing of the dip in a fluorescence pulse signal at the centromere of a chromosome. The timing of this dip is combined with the timing of the pulse crossing a lower fraction threshold to yield a measure of chromosome arm length and total length.

Slit-scan flow systems were developed [111] to reduce the size of the illuminating beam in order to obtain more detailed shape information. A summary of early slit-scanning systems and pulse analysis is given by Wheeless [112], and more recently in a comprehensive treatment of what is now termed a scanning flow cytometer (SFCM) [16] (see also Chapter 6). To record the complete pulse details from a SFCM, a waveform recorder is used in which the pulse profile is sampled at frequent intervals and the digitized values are stored in a computer. This is a time-consuming process, but the SFCM can be made more efficient by recording waveforms only if certain key features are detected by rapid analog means.

Cambier et al. [12] described such a combined analog and digital system for processing pulse profile signals in real time, using a custom-made waveform recorder. Figure 8 shows a similar Los Alamos system that uses some of the signal-processing elements described earlier in this chapter. The PMT anode feeds a broadband (50-MHz) hybrid-circuit transimpedance preamplifier [30] that produces a voltage pulse profile, faithfully showing the variation of fluorescence with time. The preamplifier pulse branches to an amplifier-integrator module [60] with amplifier bandwidth of 3 MHz for transmission of outstanding pulse features with good noise rejection, and to a commercial 25-MHz waveform recorder capable of rapidly digitizing the pulse in full detail at 10- or 20-ns intervals. The amplifier-integrator outputs enter two of the input ports of the data acquisition interface [28], where the crest values of the AMP OUT and pulse-area INTEG OUT signals are captured. The AMP OUT pulse also feeds an analog pulse-shape analyzer, whereby determinations of constant-fraction pulse width and pulse risetime are made and similarly sent to the data acquisition interface. If these analog data meet required time- and amplitude-window criteria as set up by the computer, a computer command triggers readout of the temporarily stored waveform from the waveform recorder. As a result of this multiparameter selection process, only waveforms likely to have features of interest are stored in full detail by the computer. Waveform recorders have large memories, and two (or more) simultaneous waveforms can be recorded, as for light scattering and fluorescence pulse profiles from individual metaphase chromosomes [34]. Slit-scan data acquisition and analysis systems have been used to measure multidimensional detail in real time [44,70]. A microprocessor-based system for acquiring complex slit-scan waveforms from chromosomes, and rapidly calculating the centromeric indexes has been described [7].

In some systems, particle shape is analyzed by later off-line means. Norgren et al. [62] have done Fourier analysis of slit-scan chromosome profiles to restore spatial frequency information. Lucas et al. [54] have described slit-scan centromeric index measurements using a profile "reflecting" algorithm. In this system waveform recording was triggered if the total fluorescence (pulse area) signal satisfied window discriminator bounds. Detailed size and shape information from cells in flow has been determined photographically in a system using Coulter aperture pulses to trigger a flashlamp illuminating the stream [8].

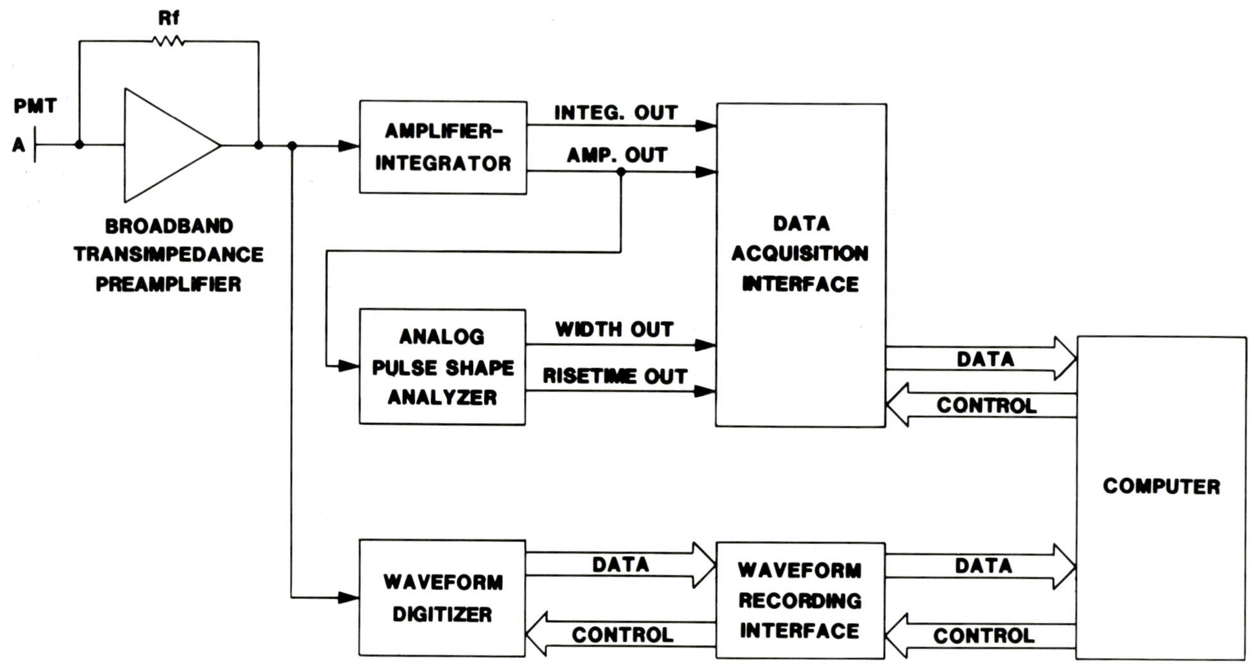

Fig. 8. Pulse-shape analysis system using analog electronics to record several FCM pulse features. If these characteristics meet preset criteria, full pulse-shape details are recorded from the waveform digitizer.

MULTIPARAMETER MEASUREMENTS

Flow cytometers with multiparameter capabilities generate a large amount of data as the single cells pass through one or more measurement zones along the flow stream. The computer-based system for data acquisition, storage, display, and control can take many configurations. Representative signal-processing arrangements are briefly described here; details of these complex systems are generously covered in the referenced literature.

The least complicated multiparameter systems are those using a single measurement station. The timing of pulse signals for such a system is usually simple because the signals are generated almost simultaneously in each transducer at the measurement zone. Separate ADC for each parameter are frequently used; this parallel processing decreases data acquisition time. The "Cytomic" system [41] is an example of a highly modularized design with each module made intelligent through the use of microprocessors. Another modular design with expansion capability to cope with two- or three-laser systems has been described by the Rijswijk group [95]. Stewart and Price [94] have developed a microprocessor-based system for realtime acquisition and display, analysis, and storage of correlated three-parameter data obtained from a commercial flow cytometer. Buican [10] has presented a flexible, high-speed data and control system based on CAMAC modules in a multiple bus arrangement. A system that consists mainly of commercial nuclear instrumentation elements is described by Turko and Leskovar [102]. The Los Alamos LACEL system [28] is an example of a modular approach based on the NIM system of hardware. LACEL, shown as an abbreviated block diagram in Figure 9, is arranged to receive inputs from four parameters. The two-parameter input processors detect and hold the crest values of input pulses and send timing pulses to the data acquisition coincidence logic module. If amplitude and time criteria are met, an ADC convert command signal is sent to the input processors by way of the computer; all inputs are simultaneously digitized and output words are made available to the computer. The four analog input lines are also connected to the sort module where amplitude window decisions are made and sort logic criteria are established. If conditions are satisfied, sort output pulses are generated for use in the sorting system.

Multistation multiparameter flow cytometers, which are more involved, have been thoroughly reviewed by Shapiro and co-workers [78,79]. Timing of signals is more complex because pulses are generated at different times due to the sequential measuring points along the flow stream. A number of papers have dealt with these signal-processing problems [2,20,25,50,72,77,86,87,95,105], and the earlier signal-delay section of this chapter discusses some of them. Careful use of waveform-preserving pipeline delays can make the correlated multistation parametric signals appear to be simultaneous at the time of cell passage at the last measuring station, which allows use of single-station processing techniques (Fig. 9).

Multiparameter instruments generate bursts of data that must be transferred rapidly to the data acquisition computer system. This is frequently done using direct memory access (DMA) controllers [55,105] or DMA interfaces [41] that transfer data directly to system memory and do not rely on time-consuming programmed interrupt routines [73].

SORTING SYSTEMS

Most flow sorter systems use the deflected-droplet technique. Each cell is isolated in a single droplet, the droplet is charged according to some measured cellular characteristic of interest and electrostatically deflected accordingly. The

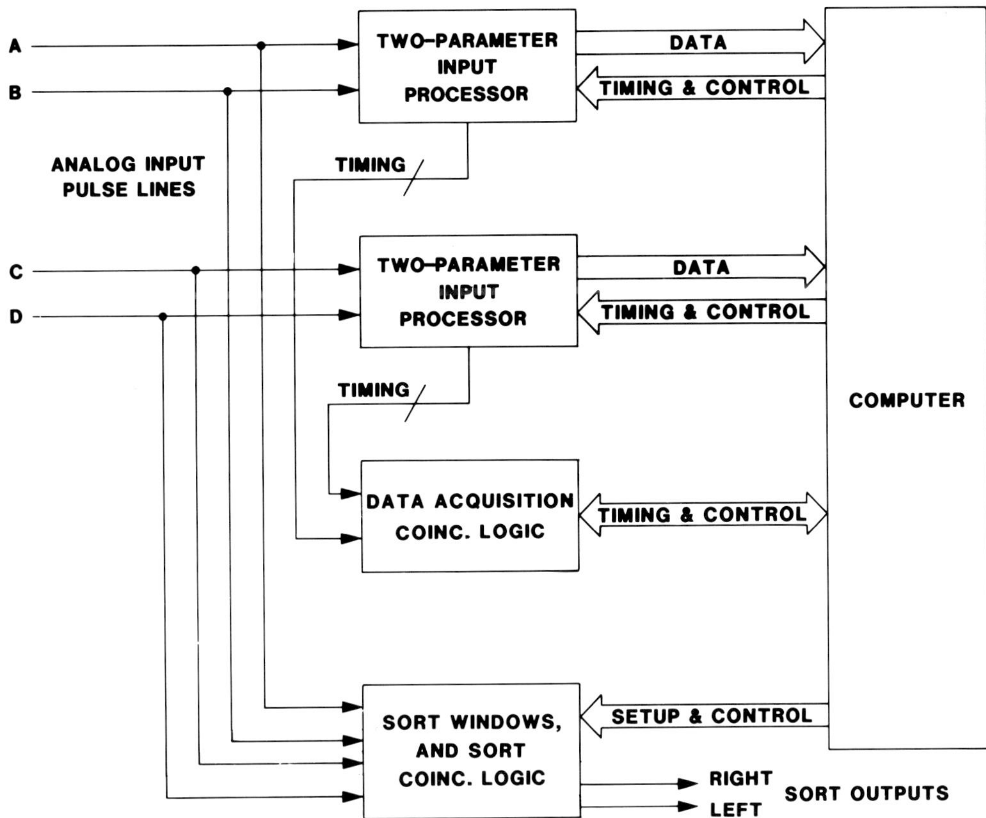

Fig. 9. Simplified diagram for a multiparameter data acquisition and sorting system for a single-measurement-station cytometer. Four analog input pulse lines (A, B, C, and D) are shown, and two digital sort output lines (right and left) to a separate sort chassis are indicated.

method of droplet formation, charging, and deflection is a modification of that developed for an ink-writing oscillograph [98], and there is a detailed description of deflected-droplet sorters for biological cells in Chapt. 8 of this book.

A general outline for a sorting system that is operated independently from the data acquisition system is shown in Figure 10. Two input channels from pulse amplifiers feed SCA discriminінators for amplitude bounds selection, followed by simple sort coincidence logic between channels. As indicated in Figure 9, these functions can be supplied for more than two inputs, and discriminator thresholds and timing logic can be controlled by the computer [28]. This computer control for the SCA thresholds can be through a look-up table classifier [46].

Sort coincidence logic yields valid pulses for individual cells, but additional circuits must be used to cancel sort action when unwanted and wanted particles are so close in time that both may be included in the deflected droplets. These "pure-sort" inspection schemes use digital shift registers to store sort command pulses temporarily for a selected number of droplet periods; if an unwanted event occurs in this period, the shift register is cleared and the sort pulse is not propagated. Figure 10 shows an "OR" input to the pure-sort block. This logic line can be used to inhibit any cell that is too close to a desired cell, or it can be set to cancel only when a particular unwanted cell classification is too close. Provision is usually made to disable pure-sort action in cases where contamination is not important.

The delay between the decision-making sorting circuitry and the generation of the droplet-charging pulse must be adjusted optimally. The delay is accomplished with shift registers clocked from the system master clock frequency, or a multiple from that frequency, that also drives the PZT for droplet generation. As indicated in Figure 10, a separate delay channel is provided for each sort direction. The delay is critical for few-droplet sorting, and an optimum system may require mechanical fine adjustment of the physical distance between the measurement laser intersection and droplet breakoff [52]. At the end of the delay period, high-level charging pulses must be generated. The duration of these pulses is usually separately controlled, adjusted to be shorter than the pure-sort checking period at the start of the delay [52], and crosslinked between the two channels to prevent application of two different polarities of charge at the same time. The sort-direction channels are combined into one charging-pulse generator line, with additional shaping controls to optimize droplet charging [99].

McCutcheon and Miller [57] described a very flexible arrangement for making sorting decisions using strategies stored in ROM. Their system can handle up to five independent input signals and can sort in up to four directions. Peters et al. [66] have given a detailed description of a high-speed sorter capable of droplet production rates and sort-processing rates 5 to 10 times greater than those of previous sorters, and the associated fast data-acquisition system has been described [104].

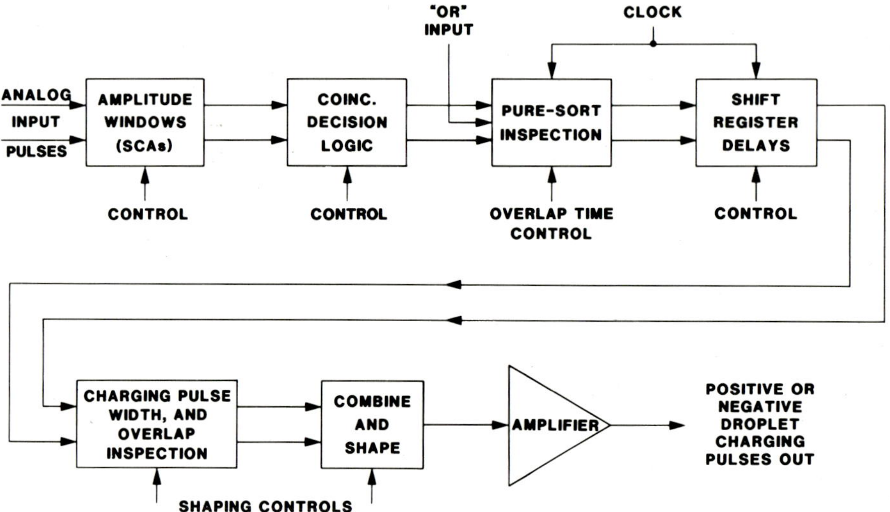

Fig. 10. Two-direction sort circuitry. Analog input pulses from two FCM amplifiers (not shown) must satisfy amplitude (SCA) and coincidence timing criteria. Emerging digital pulses are pure-sort inspected before they enter delay channels. At the end of the delays, there is further overlap inspection (see text) and charge-pulse shaping and amplification.

The sorting systems described above assign cells to collected subpopulations where they are pooled with other similar cells. To preserve the individual identity of processed cells, systems for individual cell sorting have been developed [19,58,96,103] in which cells are deposited in wells or in defined locations on a microscope slide. Data defining the cell characteristics are stored along with the cell's "address" (location). These systems require various electronics and mechanical refinements to increase droplet placement accuracy.

CONTEMPORARY AND FUTURE DIRECTIONS

New instrumentation hardware for flow cytometers will be developed as emerging technologies yield products of recognized value to cytometric problems. In general, the new devices used in FCM instruments are developed for other applications (e.g., communications, defense electronics, robotics, image processing, and general-purpose computing), and the designer must be aware of these modern elements and recognize their usefulness in cytometry. Forecasting is risky in the rapidly changing scientific world, but Kamentsky [42] predicted developments that are now commonplace in flow cytometry. Here, we point out some of the new technologies to watch.

Progress in semiconductor lasers is dramatic compared with that for other types of lasers [26,32]. Output of single GaAlAs devices (820-nm wavelength) has reached 200-mW continuous wave. Low-power lasers of the same material are available with fiberoptic output cables with 50-μm core diameter and integral optical feedback stabilization. Current technology limits diode-laser wavelengths to 700 nm or longer for reliable devices, so the burden of application to cytometry for these lasers may be transferred to developing the technology of making and using near-infrared dyes [107]. Shorter wavelengths can be obtained using a diode laser to pump a Nd : YAG laser, and doubling the output frequency through nonlinear optics. The devices typically produce 532-nm light with powers approaching the tens-of-milliwatts range [31]. The benefits of lower light-source costs, greater reliability, and reduced requirements for size, power, and cooling should stimulate the application of these lasers to flow cytometry.

New developments in semiconductor photodetectors may make these devices natural companions to diode-laser light sources. Hybrid modules containing an avalanche photodetector combined with a built-in input light pipe and integral broadband preamplifier are now available. Although they have high responsivity between 400 and 1,100 nm, the weakest optical signals will still require the superior detection features of PMT.

Rapid progress in reducing optical and electro-optical circuits to integrated-circuit size leads us to forecast miniaturized flow cells and associated electronics. This optical integrated circuit technology has brought speculation that a "flow cytometer on a chip" [81] will evolve.

Fabrication of digital and analog integrated circuits using automated software and fabrication tools allows the production of semicustom chips with fast turnaround services. Mask-programmable device arrays and standard cells can be assembled to produce intelligent analog circuits, mixing both analog and digital, so that a designer will have access to complex programmable linear functions. These products, aimed toward application-specific users, may be of interest to developers of cytometers for the clinical markets.

Advanced digital signal processing can be considered for the more sophisticated cytometric measurements. Large-scale digital integrated circuits, perhaps fabricated from GaAs for which speed demands are greatest, may use pipeline architecture to rapidly characterize analog waveform sources, as from SFCM signals from chromosomes. Preprocessing would include sampling, analog-to-digital conversion, shift register delay, digital filtering, and other functions of fixed format that are independent of the input data stream. The signal would then go to a programmable signal processor where signals would be transformed and categorized using high-speed arithmetic elements. Finally, the data processor/computer would interpret the signals and make system-level decisions.

The more conventional FCM signal-processing needs will continue to be served by microcomputers and powerful personal computers, and we will see more use of 32-bit microprocessors and array processors. Programmers will develop efficient processing algorithms and use modern numerical methods (e.g., matrix computation and clustering) for rapid analysis and interpretation of the complex data sets that can emerge from FCM systems.

Developments in new cytometric detection and analytical approaches also deserve mention. Methods for analyzing very small subpopulations of cells (rare-event analysis) place extreme requirements on instrument design and operation [18]. Measurements of the elastic properties of cells with acoustic excitation and sensing in flow [101] may add parameters to signal processing and analysis. Phase-sensitive detection methods will be increasingly used in flow cytometry. Buican [11] has described a spectral analyzer of single-cell emission based on an interferometer with phase-sensitive analysis. The technique of phase differential scattering using a Zeeman effect laser is applicable in flow cytometry and, compared with conventional intensity differential scattering measurements, the instrumentation is simple and relatively free of artifacts [35]. High-frequency phase-sensitive measurement of fluorescence emission is a promising technique for measuring and analyzing fluorescence lifetimes.

This chapter is aimed toward research laboratory workers who desire a high degree of flexibility and adaptability in their instruments and who may have assistance from dedicated instrument developers and operators who understand the complexities of their needs and can respond to requests for special measurement configurations. However, we recognize and give credit to manufacturers of commercial flow cytometers who have developed very sophisticated and effective instruments for the research and clinical markets. New machines are being designed to be very user-friendly, with a minimum of controls, with superior graphic displays, and with applications-oriented data analysis.

ACKNOWLEDGMENTS

I am grateful for support from the Los Alamos Flow Cytometry Resource (grant RR01315 from the National Institutes of Health) and the Los Alamos Institutional Supporting Research program for advanced analytical cytometry.

REFERENCES

1. **Alberti S, Parks DR, Herzenberg LA (1987)** A single-laser method for subtraction of cell autofluorescence in flow cytometry. Cytometry 8:114–119.
2. **Arndt-Jovin DJ, Grimwade BG, Jovin TM (1980)** A dual laser flow sorter utilizing a CW pumped dye laser. Cytometry 1:127–130.
3. **Bartholdi M, Salzman GC, Hiebert RD, Seger G (1977)** Single-particle light-scattering measurements with a photodiode array. Opt Lett 1:223–225.
4. **Bartholdi, M, Salzman GC, Hiebert RD, Kerker M (1980)** Differential light-scattering photometer for rapid analysis of single particles in flow. Appl Opt 19:1573–1581.
5. **Bartholdi MF, Sinclair DC, Cram LS (1983)** Chromosome analysis by high illumination flow cytometry. Cytometry 3:395–401.
6. **Bartholdi M (1985)** Logarithmic transformation of normal distributions. Int Conf Anal Cytol XI, Hilton Head, Abst 245.
7. **Bartholdi M, Meyne J, Parson J, Cram, LS (1987)** Chromosome pulse shape analysis and sorting. Cytometry Supplement 1, Abst 31:6.
8. **Bator JM, Groves MR, Price BJ, Eckstein EC (1984)** Erythrocyte deformability and size measured in a multiparameter system that includes impedance sizing. Cytometry 5:34–41.
9. **Brunk CF, Bohman RE, Brunk CA (1982)** Conversion of linear histogram flow cytometry data to a logarithmic display. Cytometry 3:138–141.
10. **Buican T (1987)** A multiple bus data acquisition, processing, and control system for flow cytometry. Cytometry 737 (Suppl 1):101 (abst).
11. **Buican TN (1987)** An interferometer for spectral analysis in flow. Cytometry 734(Suppl 1):101 (abst).
12. **Cambier JL, Kay DB, Wheeless LL (1979)** A multidimensional slit-scan flow system. J Histochem Cytochem 27:321–324.
13. **Casoli P, Maranesi P (1979)** Improved pulse-height store for A/D conversion. Nucl Instrum Methods 166:299–304.
14. **Chase RL, Poulo LR (1967)** A high precision dc restorer. IEEE Trans Nucl Sci NS-14:83–88.
15. **Coulter WH (1953)** Means for counting particles suspended in fluid. U.S. Patent 2,656,508.
16. **Cram LS, Bartholdi MF, Wheeless LL, Gray JW (1985)** Morphological analysis by scanning flow cytometry. In Van Dilla MA, Melamed M, Laerum OD, Dean PN (eds): "Flow Cytometry: Instrumentation and Data Analysis" San Diego: Academic Press, pp 163–194.
17. **Crowell JM, Hiebert RD, Salzman GC, Price BJ, Cram LS, Mullaney PF (1978)** A light-scattering system for high-speed cell analysis. IEEE Trans Biomed Eng BME-25:519–526.
18. **Cupp JE, Leary JF, Cernichiari E, Wood JCS, Doherty RA (1984)** Rare-event analysis methods for detection of fetal red blood cells in maternal blood. Cytometry 5:138–144.
19. **Dean PN, Peters DC, Mullikin J, Pallavicini, MG (1985)** Indexed sorting. Int Conf Anal Cytol XI, Hilton Head, abstr 315.
20. **De Grooth BG, van Dam M, Swart NC, Willemsen A, Greve J (1987)** Multiple wavelength illumination in flow cytometry using a single arc lamp and a dispersing element. Cytometry 8:445–452.
21. **Eisert WG, Nezel M (1978)** Internal calibration to absolute values in flowthrough particle size analyzers. Rev Sci Instrum 49:1617–1621.
22. **Eisert WG (1979)** Cell differentiation based on absorption and scattering. J Histochem Cytochem 27: 404–409.
23. **Fox MH, Delohery TM (1987)** Membrane fluidity measured by fluorescence polarization using an Epics V cell sorter. Cytometry 8:20–25.
24. **Fulwyler MJ, Glascock RB, Hiebert RD, Johnson NM (1969)** Device which separates minute particles according to electronically sensed volume. Rev Sci Instrum 40:42–48.
25. **Habbersett R, Steinkamp J, Hiebert R (1987)** Multiparameter flow cytometer (MPFC) provides unique capabilities for diverse biological applications. Cytometry 740(Suppl 1):102 (abst).
26. **Hecht J (1984)** Diode laser report. Lasers and Applications 3(11):89–92.
27. **Hiebert RD (1979)** Light sources, detectors, and flow

chambers. In Melamed MR, Mullaney PF, Mendelsohn ML (eds): "Flow Cytometry and Sorting." New York: John Wiley & Sons, pp 623–637.

28. **Hiebert RD, Jett JH, Salzman GC** (1981) Modular electronics for flow cytometry and sorting: The LACEL system. Cytometry 1:337–341.

29. **Hiebert RD, Steinkamp JA** (1982) Signal processor for a multiparameter flow cytometer using multilaser excitation. Los Alamos National Lab Rep LA-9066-MS.

30. **Hiebert RD, Sweet RG** (1985) Electronics for flow cytometers and sorters. In Van Dilla MA, Melamed M, Laerum OD, Dean PN (eds): "Flow Cytometry: Instrumentation and Data Analysis" San Diego: Academic Press, pp 129–162.

31. **Huth BG, Kuizenga D** (1987) Green light from doubled Nd : YAG lasers. Lasers Optron 6(10):59–61.

32. **Jacobs RR** (1984) Future lasers for flow cytometry and sorting. Int Conf Anal Cytol X, Asilomar, P5.4 (abst).

33. **Johnson GE, Dorman BP, Ruddle FH** (1979) Ratio of analog pulse height module for fluorescence activated cell sorter. Rev Sci Instrum 50:109–110.

34. **Johnston RG, Bartholdi MF, Hiebert RD, Parson JD, Cram LS** (1985) A slit-scan flow cytometer for recording simultaneous waveforms. Rev Sci Instrum 56:691–695.

35. **Johnston RG, Singham SB, Salzman GC** (1986) Phase differential scattering from microspheres. Appl Opt 25:3566–3572.

36. **Jovin TM** (1979) Fluorescence polarization and energy transfer: Theory and application. In Melamed MR, Mullaney PF, Mendelsohn ML (eds): "Flow Cytometry and Sorting." New York: John Wiley & Sons, pp 137–165.

37. **Kachel V** (1979) Electrical resistance pulse sizing (Coulter sizing). In Melamed MR, Mullaney PF, Mendelsohn ML (eds): "Flow Cytometry and Sorting." New York: John Wiley & Sons, pp 61–104.

38. **Kachel V, Meier H, Stuhlmuller P, Ahrens O** (1980) Ultra fast digital calculation of ratios of flow cytometric values. In: "Flow Cytometry." vol. IV. Bergen, Norway: Universitetsforlaget, pp 109–111.

39. **Kachel V, Schneider H, Schedler K** (1980) A new flow cytometric pulse height analyzer offering microprocessor controlled data acquisition and statistical analysis. Cytometry 1:175–192.

40. **Kachel V, Glossner E, Schneider H** (1982) A new flow cytometric transducer for fast sample throughput and time resolved kinetic studies of biological cells and other particles. Cytometry 3:202–212.

41. **Kachel V, Schedler K, Schneider H, Haack L** (1984) "Cytomic" data system modules—Modern electronic devices for flow cytometric data handling and presentation. Cytometry 5:299–303.

42. **Kamentsky** (1979) Future directions of flow cytometry. J Histochem Cytochem 27:1649–1651.

43. **Kay DB, Cambier JL, Wheeless LL** (1979) Imaging in flow. J Histochem Cytochem 27:329–334.

44. **Kay DB, Wheeless LL, Brooks CL** (1982) A two-station multidimensional slit-scan flow system: Concept and optical implementation. IEEE Trans Biomed Eng BME-29:106–110.

45. **Keane JP, Hodgson BW** (1980) A fluorescence polarization flow cytometer. Cytometry 1:118–126.

46. **Koper GJM, Christiaanse JGM** (1981) The look-up table: A classifier for cell sorters. Cytometry 1:394–396.

47. **Koper GJM, Blanken R** (1981) The look-up table: A logarithmic converter for cell sorters. Cytometry 2: 194–197.

48. **Leary JF, Todd P, Wood JCS, Jett JH** (1979) Laser flow cytometric light scatter and fluorescence pulse width and risetime sizing of mammalian cells. J Histochem Cytochem 27:315–320.

49. **Lebo RV, Bastian AM** (1982) Design and operation of a dual laser chromosome sorter. Cytometry 3: 213–219.

50. **Lebo RV, Bruce BD, Dazin PF, Payan DG** (1987) Design and application of a versatile triple-laser cell and chromosome sorter. Cytometry 8:71–82.

51. **Lindmo T, Steen HB** (1977) Flow cytometric measurement of the polarization of fluorescence from intracellular fluorescein in mammalian cells. Biophys J 18:173–187.

52. **Lindmo T** (1981) Protein synthesis as a function of protein content in exponentially growing NHIK 3025 cells studied by flow cytometry and cell sorting. Exp Cell Res 133:237–245.

53. **Loken MR, Parks DR, Herzenberg LA** (1977) Two-color immunofluorescence using a fluorescence-activated cell sorter. J Histochem Cytochem 25:899–907.

54. **Lucas, JN, Gray JW, Peters DC, Van Dilla MA** (1983) Centromeric index measurement by slit-scan flow cytometry. Cytometry 4:109–116.

55. **Malachowski GC, Hall BR, Ashcroft RG** (1984) "Cicero": Computerized instrument controller achieves 100% cell sorting efficiency at 25000 cells/s; doubles as 250 kHz data acquisition system and computing facility. Int Conf Anal Cytol X, Asilomar, p C39 (abst).

56. **Martin JC, McLaughlin SR, Hiebert RD** (1979) A real-time delay monitor for flow-system cell sorters. J Histochem Cytochem 27:277–279.

57. **McCutcheon MJ, Miller RG** (1982) Flexible sorting decision and droplet charging control electronic circuitry for flow cytometer-cell sorters. Cytometry 2: 219–225.

58. **Merrill JT, Dean PN, Gray JW** (1979) Investigations in high-precision sorting. J Histochem Cytochem 27: 280–283.

59. **Model 8107 Coulter Amplifier and Constant-Current Generator** (Los Alamos National Laboratory, Los Alamos NM), Drawing No. 4Y-89805.

60. **Model 8124A Amplifier-Integrator** (Los Alamos National Laboratory, Los Alamos, NM), Drawing No. 4Y-89839.

61. **Muirhead KA, Schmitt TC, Muirhead AR** (1983) Determination of linear fluorescence intensities from flow cytometric data accumulated with logarithmic amplifiers. Cytometry 3:251–256.

62. **Norgren RM, Gray JW, Young IT** (1982) Restoration of profiles from slit-scan flow cytometry. IEEE Trans Biomed Eng BME-29:101–106.

63. **Parks DR, Peterson LL** (1984) Calibration, comparison, and optimization of logarithmic amplifiers for flow cytometric fluorescence measurements. Int Conf Anal Cytol X, Asilomar, p B6 (abst).

64. **Parks DR, Hardy RR, Stovel RT** (1984) Design and

evaluation of a cell sorter system for four-color immunofluorescence measurements. Int Conf Anal Cytol X, Asilomar, p D42 (abst).

65. **Parson JD, Hiebert RD, Martin JC** (1985) Active analog pipeline delays for high signal rates in multistation flow cytometers. Cytometry 6:388–391.
66. **Peters D, Branscomb E, Dean P, Merrill T, Pinkel D, Van Dilla M, Gray J** (1985) The LLNL high-speed sorter: Design features, operational characteristics and biological utility. Cytometry 6:290–301.
67. **Pinkel D, Epstein M, Uekoff R, Norman A** (1978) Fluorescence polarimeter for flow cytometry. Rev Sci Instrum 49:905–912.
68. **Pinkel D, Steen HB** (1982) Simple methods to determine and compare the sensitivity of flow cytometers. Cytometry 3:220–223.
69. **Robinson LB** (1961) Reduction in baseline shift in pulse amplitude measurements. Rev Sci Instrum 32: 1057.
70. **Robinson RD, Wheeless DM, Wheeless LL** (1987) A system for real-time processing of multidimensional slit-scan flow cytometric data. Cytometry Supplement 1:34:7 (abst).
71. **Salzman GC, Crowell JM, Goad CA, Hiebert RD, Martin JC, Ingram ML, Mullaney PF** (1975) A flow-system multiangle light-scattering instrument for cell characterization. Clin Chem 21:1297–1304.
72. **Salzman GC, Hiebert RD, Crowell JM** (1978) Data acquisition and display for a high-speed cell sorter. Comput Biomed Res 11:77–88.
73. **Salzman GC, Wilkins SF, Whitfill JA** (1981) Modular computer programs for flow cytometry and sorting: The LACEL system. Cytometry 1:325–336.
74. **Salzman GC** (1982) Light scattering analysis of single cells. In Catsimpoolas N (ed): "Cell Analysis," Vol. 1. New York: Plenum, pp 111–143.
75. **Schuette WH, Shackney SE, Plowman FA, Tipton HW, Smith CA, MacCollum MA** (1984) Design of flow chamber with electronic cell volume capability and light detection optics for multilaser flow cytometry. Cytometry 5:652–656.
76. **Schwartz A, Sugg H, Ritter TW, Fernandez-Repollet E** (1983) Direct determination of cell diameter, surface area, and volume with an electronic volume sensing flow cytometer. Cytometry 3:456–458.
77. **Severin E, Ohnemus B, Kiegler S** (1983) A new flow chamber and processing electronics for combined laser and mercury arc illumination in an impulscytophotometer flow cytometer. Cytometry 3:308–310.
78. **Shapiro HM** (1983) Multistation multiparameter flow cytometry: A critical review and rationale. Cytometry 3:227–243.
79. **Shapiro HM, Feinstein DM, Kirsch AS, Christenson L** (1983) Multistation multiparameter flow cytometry: Some influences of instrumental factors on system performance. Cytometry 4:11–19.
80. **Shapiro HM** (1985) "Practical Flow Cytometry." New York: Alan R. Liss, Inc.
81. **Shapiro HM, Hercher M** (1986) Flow cytometers using optical waveguides in place of lenses for specimen illumination and light collection. Cytometry 7:221–223.
82. **Sharpless TK, Melamed MR** (1976) Estimation of cell size from pulse shape in flow cytofluorometers. J Histochem Cytochem 24:257–264.
83. **Sharpless TK, Bartholdi M, Melamed MR** (1977) Size and refraction index dependence of simple forward angle scattering measurements in a flow system using sharply-focused illumination. J Histochem Cytochem 25:845–856.
84. **Sharpless TK** (1979) Cytometric data processing. In Melamed MR, Mullaney PF, Mendelsohn ML (eds): "Flow Cytometry and Sorting." New York: John Wiley & Sons, pp 359–379.
85. **Steen HB** (1980) Further developments of a microscope-based flow cytometer: Light scatter detection and excitation intensity compensation. Cytometry 1: 26–31.
86. **Steinkamp JA, Fulwyler MJ, Coulter JR, Hiebert RD, Horney JL, Mullaney PF** (1973) A new multiparameter separator for microscopic particles and biological cells. Rev Sci Instrum 44:1301–1310.
87. **Steinkamp JA, Hiebert RD** (1982) Signal processing electronics for multiple electrical and optical measurements on cells. Cytometry 2:232–237.
88. **Steinkamp JA** (1983) A differential amplifier circuit for reducing noise in axial light loss measurements. Cytometry 4:83–87.
89. **Steinkamp JA** (1984) Flow cytometry. Rev Sci Instrum 55:1375–1400.
90. **Steinkamp JA, Stewart CC** (1986) Dual-laser, differential fluorescence correction method for reducing cellular background autofluorescence. Cytometry 7: 566–574.
91. **Steinkamp JA, Habbersett RC, Stewart CC** (1987) A modular detector for flow cytometric multicolor fluorescence measurements. Cytometry 8:353–365.
92. **Steinkamp JA, Hiebert RD, Habbersett RC** (1987) Dual ratio and sum/difference modules for flow cytometric applications. Cytometry 738(Suppl 1):101 (abst).
93. **Stewart SS, Miller RG, Price GB** (1980) A design for a real-time fluorescence polarization computer. Cytometry 1:204–211.
94. **Stewart SS, Price GB** (1986) Realtime acquisition, storage, and display of correlated three-parameter flow cytometric data. Cytometry 7:82–88.
95. **Stokdijk W, van den Engh GJ, van Dekken H** (1985) The electronics of the RELACS: the Rijswijk experimental light-activated cell sorter. Int Conf Anal Cytol XI, Hilton Head, abstr 318.
96. **Stovel RT, Sweet RG** (1979) Individual cell sorting. J Histochem Cytochem 27:284–288.
97. **Stovel RT, Parks DR, Nozaki T** (1984) A 130 degree light scatter detection system. Int Conf Anal Cytol X, Asilomar, p B23 (abst).
98. **Sweet RG** (1965) High frequency recording with electrostatically deflected ink jets. Rev Sci Instrum 36: 131–142.
99. **Sweet RG** (1979) Flow sorters for biologic cells. In Melamed MR, Mullaney PF, Mendelsohn ML (eds): "Flow Cytometry and Sorting." New York: John Wiley & Sons, pp 177–189.
100. **Sweet R, Parks D, Nozaki T, Herzenberg L** (1981) A 3 1/2 decade logarithmic amplifier for cell fluorescence data. Cytometry 2:130.
101. **Sweet RG, Fulwyler MJ, Herzenberg LA** (1984) Acoustic sensing in flow. Int Conf Anal Cytol X, Asilomar, p P1.2 (abst).

102. **Turko B, Leskovar B (1983)** Laser-based multiparameter system for microscopic particles. IEEE Trans Nucl Sci NS-30:605–616.
103. **Tyrer HW, Kunkel-Berkly C (1984)** Multiformat electronic cell sorting system. I. Theoretical considerations. Rev Sci Instrum 55:1044–1050.
104. **Van den Engh G, Stokdijk W, Peters D (1987)** A fast data-acquisition system for multiparameter flow analysis and sorting. Cytometry 33(Suppl 1):7 (abst).
105. **Voet L, Kurmann FJ, Hannig K (1982)** Data acquisition and control system for multiparameter cell sorting based on DEC LSI-11 microprocessor. Cytometry 2:383–389.
106. **Von Behrens W, Edmondson S (1976)** Comparison of techniques for improving the resolution of standard Coulter cell sizing systems. J Histochem Cytochem 24:247–256.
107. **Waggoner A (1987)** New red and infrared excited probes. Cytometry 158(Suppl 1):30 (abst).
108. **Weier H, Eisert WG (1986)** Two-parameter data-acquisition system for slit-scan chromosome analysis. Rev Sci Instrum 57:2902.
109. **Weier H, Eisert WG (1987)** Two-parameter data acquisition system for rapid slit-scan analysis of mammalian chromosomes. Cytometry 8:83–90.
110. **Weier H, Lucas JN, Mullikin JC, van den Engh G (1987)** Affordable two-parameter slit-scan data acquisition. Cytometry 744(Suppl 1):102 (abst).
111. **Wheeless LL, Hardy JA, Balasubramanian N (1975)** Slit-scan flow systems for automated cytopathology. Acta Cytol (Praha) 19:45–52.
112. **Wheeless LL (1979)** Slit-scanning and pulse width analysis. In Melamed MR, Mullaney PF, Mendelsohn ML (eds). "Flow Cytometry and Sorting." New York: John Wiley & Sons, pp 125–135.

8

Flow Sorters for Biological Cells

Tore Lindmo, Donald C. Peters, and Richard G. Sweet

Department of Biophysics, The Norwegian Radium Hospital, Montebello, 0310 Oslo 3, Norway (T.L.); Biomedical Sciences Division, Lawrence Livermore National Laboratory, Livermore, California 94550 (D.C.P.); Genetics Department, Stanford Medical Center, Stanford, California 94305 (R.G.S.)

INTRODUCTION

Flow sorters separate individual cells as a function of characteristics determined as they flow serially past one or more sensing devices. Classification and sorting are independent, sequential processes, in contrast to bulk sorters, such as centrifuges, where the separating forces act simultaneously on large numbers of cells and the sorting mechanism must be related to the characteristics used for classification.

A flow sorter combines one or more of the flow measurement methods described elsewhere in this book (i.e., light scattering, light absorption, electrical (Coulter) impedance, fluorescence intensity, or fluorescence polarization) with a method of physically isolating those cells having measured parameters in a specific range. The ability to make multiple measurements on each cell, together with the resolution and sensitivity attainable with such measurements in flow systems, makes possible the isolation of cell populations having a purity and specificity of function that can be obtained in no other way.

Because they process cells one by one, flow sorters have a throughput rate that is inherently much lower than that of bulk sorters. Bulk sorting techniques include methods based on selective destruction or stimulation, adherence, filtration, countercurrent distribution, electrophoresis, sedimentation and centrifugation. Recently, magnetic separation techniques have been improved through the availability of highly uniform, monodisperse, iron-containing particles (see Chapter 19, this volume). When coated with antibodies, these particles will bind to the appropriate cells and cell separation can be achieved in a sedimentation apparatus equipped with retaining magnets.

Bulk separation techniques can sometimes be used instead of flow cytometric sorting; in other cases, bulk separation techniques are useful as preparative steps to reduce the number of cells that must be serially processed in a flow sorter.

Two different techniques of flow cytometric cell sorting are described in this chapter. One technique is based on switching the flow between two outlets of the flow chamber, depending on the value measured as the cell passed through the sensing region. Such fluidic switching sorters are inherently slower than sorters based on the deflected droplet principle, in which the flow is broken up into a series of droplets that may be individually isolated. Fluidic switching sorters are only briefly described, since they have not been extensively used and until recently have not been commercially available. Most of the discussion is devoted to the design and operational characteristics of deflected droplet sorters. However, many of the operational characteristics discussed for deflected droplet sorters are also relevant for fluidic switching sorters. The emphasis in this chapter is on the design and operation of the components of sorting instruments that actually perform the sorting function. Cell analysis methods are discussed only to the extent that they influence sorter design or operation.

FLUIDIC SWITCHING SORTERS

An early flow sorter, developed by Kamentsky and Melamed [35], is illustrated in Figure 1. Cells are observed optically in a flow channel and arrive 2 msec later at a side port connected to a syringe operated by a stepper motor. An optical signal corresponding to a desired cell causes the stepper motor to increment and fluid to flow into the syringe for about 3 msec, diverting about 0.03 ml of fluid containing the selected cell. After a maximum of 300 selected cells have been isolated, the syringe is flushed through a Millipore filter, trapping the cells on the filter surface for subsequent staining and observation.

A fluidic switching sorter, developed by Friedman [17], uses acoustic energy to modify the flow stream and is illustrated in Figure 2. A laminarly flowing stream of cells, surrounded by sheath fluid, flows through an optical analyzer and then across a cavity to an outflow orifice. Cell flow to the outlet is interrupted by coupling acoustic energy to the cavity, causing turbulence that disperses the cell stream and diverts it to the side port. Switching from turbulent to laminar flow and back to turbulent flow takes 2 msec, a some-

Flow Cytometry and Sorting, Second Edition, pages 145–169

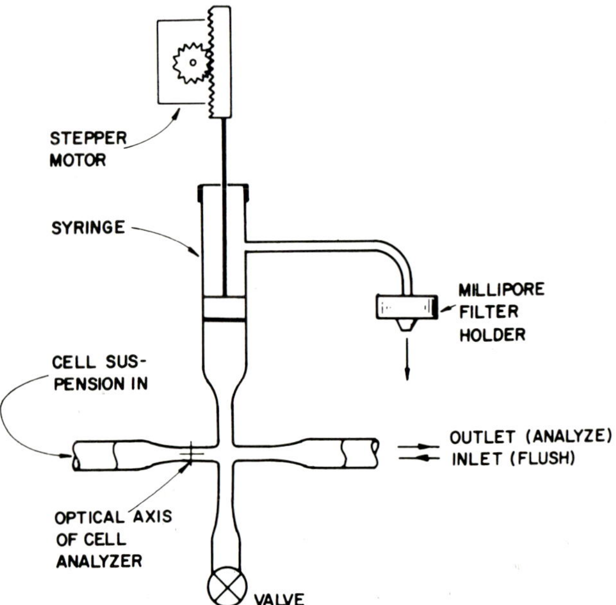

Fig. 1. Schematic drawing of early cell sorter developed by Kamentsky and Melamed [35]. Selected cells, classified by measuring optical properties in a narrow flow channel, are diverted from the main channel by a motor-driven syringe connected to a side port. After a number of cells have been selected, the syringe plunger is further withdrawn to uncover a side port in the syringe body. Wash fluid then reverse flushes the main channel and carries the selected cells to the surface of a Millipore filter.

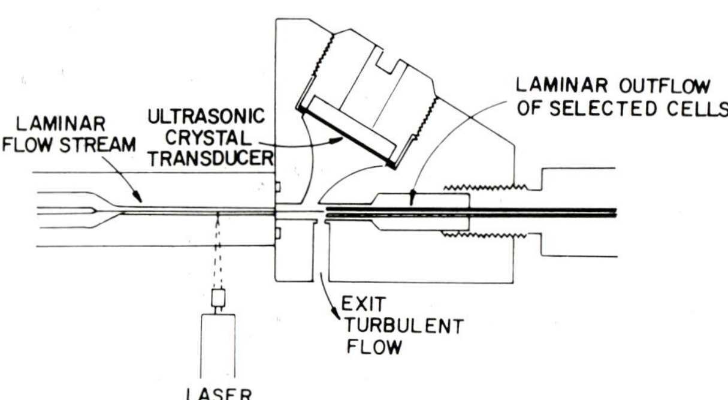

Fig. 2. Fluidic amplifier particle sorter developed by Friedman [17]. After classification, cells flow across a cavity to which acoustic energy can be coupled. When the sonic transducer is energized, turbulence in the cavity prevents cells from reaching the outflow orifice. The transducer is deactivated for a short time interval when a desired cell is detected, establishing laminar flow and allowing the selected cell to exit through the orifice on the flow axis, where it can be flushed to a collecting reservoir or slide.

what shorter response time than is achieved with the motor-driven syringe system. There is no fundamental limit to the number of cells that can be sorted in a single run.

Several principles for fluidic switching sorters have been developed at the Institut für Strahlenbiologie of the University of Münster. In these devices switching occurs between two outlet ports connected to a low vacuum. In one such sorter, one of the two outlet ports of the flow chamber is blocked by an electrolytically produced gas bubble each time an interesting cell is to be sorted [69]. In another design, one

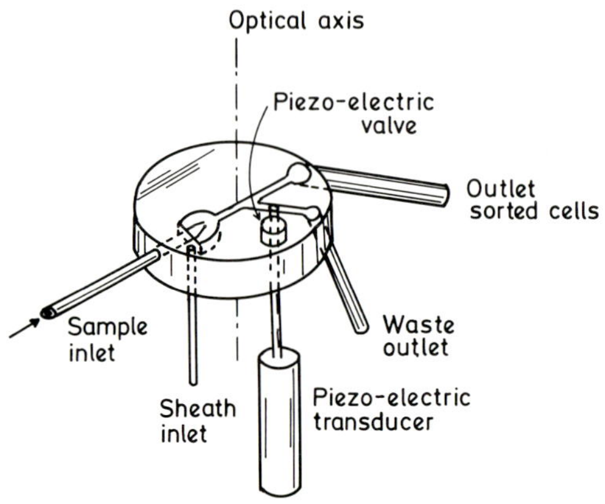

Fig. 3. Fluidic amplifier sorter developed by Goehde [20]. The flow chamber is incorporated into an epi-illumination configuration for fluorescence measurement with the optical axis perpendicular to the plane of the chamber. The flow channel branches after the sensing region, with most of the flow, including the sample stream, normally going to the waste outlet. A piezoelectric transducer in the waste outlet branch can be activated when a desired cell is to be sorted. This partly blocks the waste outlet and generates a pressure wave which, for a moment, deflects the sample flow over to the outlet for sorted cells.

output is blocked by activating a piezoelectric valve so that additional gas [14] or liquid [67] is drawn into the outlet channel, switching the fluid flow over to the other output port.

Another, similar, design that also has been developed into a commercially available sorter (PAS II, Partec AG, Arlesheim, Switzerland), is illustrated in Figure 3 [20]. The sample stream is hydrodynamically focused by the sheath flow in a tapered part of the flow chamber and passes through the measurement region toward a bifurcation where the waste channel branches off at an angle, leaving a straight flow path for the outlet of sorted cells. Normally, the laminar sheath flow splits at the bifurcation. About two-thirds of the liquid, including the sample flow, will pass through the waste channel due to the wall effect in this fluidistor-type chamber, and about one-third will flow through the outlet for sorted cells.

At the appropriate time after detecting an interesting cell to be sorted, the piezoelectric transducer is pulsed, generating a pressure wave that momentarily causes the flow path to switch over to the sort outlet. The piezoelectric transducer is capable of switching 5,000 times/sec, but realistic processing rates for this device are an input analysis rate of 1,000 cells/sec and a sorting rate of up to 500 cells/sec. This is partly dictated by the need for a certain time (about 40 μsec) between particles in order to avoid coicidences.

Estimating and checking the correct sorting delay are facilitated by a viewing ocular that is part of the epi-illumination optical geometry. The operator is thereby able to look down on the sorting chamber in high magnification and actually watch the sorted particles entering the sort channel when the delay has been correctly set. With appropriate optics, the possibility also exists for a second measurement station downstream in the sort channel, permitting direct verification of the sorted cells.

Fluid switching systems are closed systems that provide more protection against evaporation and contamination

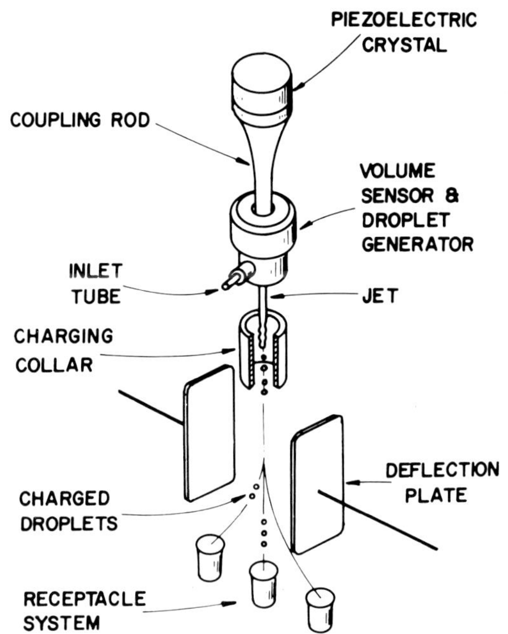

Fig. 4. Early deflected droplet sorter developed by Fulwyler [18]. A cell suspension enters the inlet tube under a pressure of 4 kg/cm^2, flows through a Coulter orifice, where cell volume is measured, and is projected into the atmosphere as a 36-μm-diameter jet travelling at 15 m/sec. A piezoelectric crystal is coupled to the fluid in the droplet generator, causing the jet to break up into precisely uniform droplets at a rate equal to the crystal excitation frequency of 72 kHz. Droplets containing cells in the desired volume range are charged as they form inside the charging collar and are then deflected by a steady electrostatic field into a collecting receptacle. Seven droplets are charged for each separated cell to allow for variation in cell transit time between detection and isolation in a droplet.

than do the deflected droplet systems described here. They are less complex than droplet deflection systems and are more easily adapted as an "add on" to existing flow analyzers. Their inherently low speed limits their use to the isolation of small numbers of cells for further analysis, to confirmation of flow analysis measurements, and to enriching for infrequently occurring cells in which the required separation rate, but not necessarily the input sample rate, is low.

DESIGN CHARACTERISTICS OF DEFLECTED DROPLET SORTERS

The flow sorting technique in most widespread use is the deflected droplet method originated by Fulwyler [18]. Instruments based on this technique have been commercially available for about 15 years and have established themselves as important tools in biomedical research.

Fulwyler's original system, shown in Figure 4, sorts cells as a function of volume. Cells, suspended in a conducting fluid, pass one by one through a Coulter orifice in which cell volume is measured electrically. The suspension then flows through a 36-μm-diameter nozzle to form a jet in air having a velocity of 15 m/sec. A piezoelectric crystal vibrating at 72 kHz is coupled to the fluid, causing the jet to break up a few millimeters from the nozzle into precisely uniform droplets at a rate of 72,000/sec. Each droplet, as it separates from the conducting jet, can be independently charged and subsequently deflected by a steady electric field to a specific collecting receptacle. Provided the cell flow rate is a small fraction of the repetition frequency, almost every cell is isolated in a separate droplet. When a cell in the desired volume range reaches the droplet-formation point, a group of several droplets is charged and deflected to the appropriate collector. The number of droplets deflected per cell is determined by uncertainty in cell position and is chosen to ensure deflection of the selected cell. The original system can analyze up to 1,000 cells/sec and separate up to 50% of these. The purity of the sorted sample, typically 95%, is limited by the finite probability of an unwanted cell occurring in the same group of droplets as a cell to be sorted.

A number of improved systems, based on Fulwyler's invention, have since been developed by Fulwyler et al. [19], Bonner et al. [9], Steinkamp et al. [60], Hulett et al. [31], and Arndt-Jovin et al. [2]. The design by Bonner et al. [9] was the first to implement fluorescence measurements as the basis for cell sorting. This design, shown schematically in Figure 5, is still representative of the principal components of today's deflected droplet sorters. All these systems include a droplet generator, a droplet-charging and -deflecting system, sample supply and collection components, sensors for measuring cell properties by fluorescence and light scatter, and electronic circuitry for processing the analyzer signals, for controlling the droplet formation, and for timing and generating the droplet charging pulses.

Droplet Generation Theory

The method of generating, charging, and deflecting fluid droplets is based on a technique originally developed by Sweet [64,65] as an ink jet recorder that relies, in turn, on phenomena described during the nineteenth century by Savart [57] and Lord Rayleigh [42]. A small fluid cylinder, formed by jetting fluid into the atmosphere from a circular orifice or nozzle, is unstable and decomposes into droplets having a smaller total surface area and lower surface energy. Vibrating the orifice or varying the fluid pressure at the proper frequency synchronizes this droplet formation to the disturbance frequency and produces a regular procession of droplets having remarkable uniformity in size, spacing, and formation point.

The droplet-forming transducer is generally a piezoelectric crystal acoustically coupled to the nozzle or to the fluid just before it issues from the nozzle. The droplet-forming force is provided by surface tension. The excitation energy, which only initiates and synchronizes the droplet-forming process, is very small, well below the level that would produce cell damage. Transducers with power outputs of less than 1 watt are usually adequate; only a small fraction of this energy is coupled to the fluid in the nozzle.

The transducer modulates the jet velocity, producing minute variations in the jet diameter. The undulation wavelength, λ, i.e., the distance between nodes of minimal diameter along the jet, is related to the transducer frequency f and the velocity of the jet v by

$$\lambda = v/f \tag{1}$$

If the frequency is chosen such that the wavelength exceeds the circumference of the jet (see Chapter 3, this volume), surface tension forces will cause the disturbance to grow exponentially with time until the jet is severed. For shorter wavelengths (higher frequencies), the surface tension of the jet exerts a dampening effect, causing high-frequency distur-

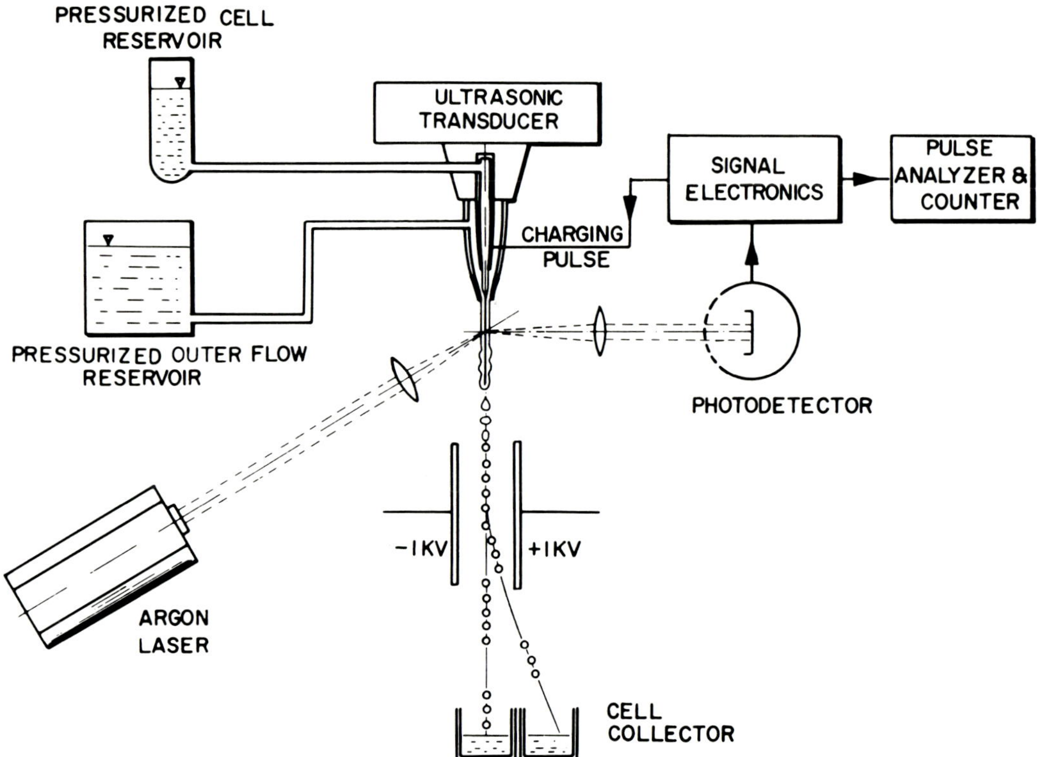

Fig. 5. Simplified diagram of droplet deflection cell sorter based on photometric cell measurements [9], representing the principal parts of present, commercially available cell sorters. Current instruments usually have more than one photodetector, thus measuring axial light loss, differential light scattering at various angles, and fluorescence in one, two or three colors. (Reproduced from ref. [9], with permission of the publisher.)

bances to "heal." Thus, droplet generation only occurs for frequencies corresponding to wavelengths longer than

$$\lambda_{min} = \pi \cdot d_j \tag{2}$$

where d_j is the diameter of the jet.

There is an optimal value of λ that causes the most rapid breakup into droplets, resulting in the shortest distance from the orifice to the droplet breakoff point. This optimal value has been determined as (see Chapter 3, this volume):

$$\lambda_{opt} = 4.508 \cdot d_j \tag{3}$$

Satisfactory droplet generation is generally limited to wavelengths of 4–8 jet diameters, although wavelengths up to 18 jet diameters have been reported [59].

A forming droplet, just before separation occurs, is connected to the parent jet by a small fluid ligament which usually breaks nearly simultaneously at both ends to form a small satellite droplet (see also Chapter 3, this volume). If the first break occurs on the orifice side, a "fast" satellite is produced that is accelerated by surface tension in a forward direction to merge with the major droplet preceding it. If the ligament first breaks on the downstream side, a "slow" satellite is produced that merges with the major droplet following it. Figure 6 shows a typical instantaneous profile of a regularly disturbed jet breaking up into droplets under conditions that result in fast satellites. The fast satellite mode is preferred, as discussed in the section on droplet charging.

Fig. 6. Scale drawing showing the breakup of a regularly distributed fluid jet. The small droplets are satellites formed from the thin liquid ligament that connects forming droplets just before separation. In this example, the ligament first breaks on the upstream side, producing "fast" satellites that merge with the major preceding droplets. Surface tension causes the ellipsoidal separated droplets to become spherical a short distance further downstream.

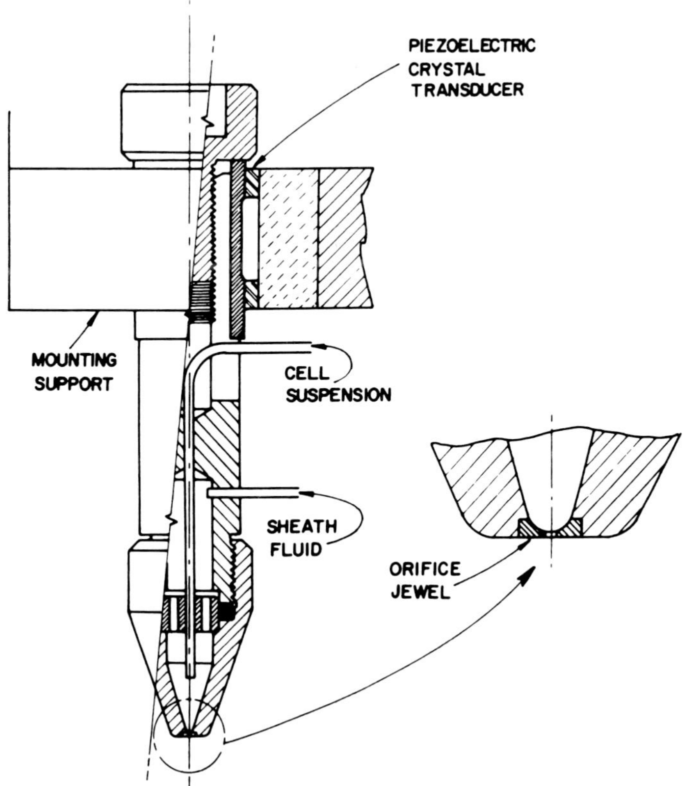

Fig. 7. Nozzle and transducer assembly of a commercially available deflected droplet cell sorter. The nozzle has a replaceable tip with a sapphire jewel orifice to produce jet diameters from 40 to 100 μm.

Operation in a mode where satellites never merge or merge erratically should be avoided as this results in inaccurate droplet charging and in collection of fluid where it is not wanted. Conditions affecting the generation of satellites such as excitation frequency and amplitude, are important in the field of ink jet printing and have been discussed in detail by Keur and Stone [33]. It has been reported that the addition of a third harmonic component to the sinusoidal waveform normally used for droplet generation might result in elimination of satellite droplets [10,52].

The Droplet Generator

Most systems use a coaxial flow arrangement, originated by Crosland-Taylor [13], in which the cell suspension constitutes only a small fraction of the total fluid in the flow system. The cell suspension is injected along the axis of a converging and accelerating sheath of cell-free fluid at a point where the fluid velocity is low and the cross-sectional area relatively high. Provided laminar flow and radial symmetry are maintained, cells are confined to a region centered on the flow axis that narrows to an area much smaller than the cross-sectional area of any tube or orifice through which the cells must pass. Cells are kept close to a specific path through the analysis section, minimizing analysis errors caused by position uncertainty and minimizing transit time variations between analysis and droplet separation, caused by velocity variations. Cells do not contact the nozzle wall, eliminating buildup of deposits of biological material that might affect jet or droplet generation.

Figure 7 shows a typical nozzle assembly for a system in which sorting is determined by optical measurements made through the cylindrical jet just after it emerges from the nozzle. The jet-forming nozzle has a replaceable tip with a watch jewel orifice. The two tubular piezoelectric transducers are longitudinally polarized; i.e., they have a permanent polarizing field in the direction of the cylinder axes. The ac excitation voltage is applied to cylindrical coatings bonded to the inside and outside surfaces of the tubes to provide an alternating field that is perpendicular to the polarizing field. The resulting shear forces produce axial motion of the inner surface with respect to the stationary outer surface of the tubes.

A relatively complex flow analysis chamber is shown in Figure 8 [60]. This system combines a Coulter orifice, coaxial flow system to center cells in the Coulter aperture, and viewing windows to permit optical measurements of cells before they issue from the jet-forming orifice. A second sheath fluid is introduced before jetting the cell stream into the atmosphere, where droplet formation and separation take place.

As illustrated by these examples, cells may be analyzed either before or after the fluid issues from the jet nozzle. Illuminating and viewing cells through optically flat windows before jet formation simplifies the optical design and, if a Coulter orifice is included, permits optical and Coulter volume measurements to be made nearly simultaneously. Errors in analysis caused by variations in cell trajectories through the optical sensing region and by optical misalignment can be minimized by uniformly illuminating and sensing an extended region that includes the volume through

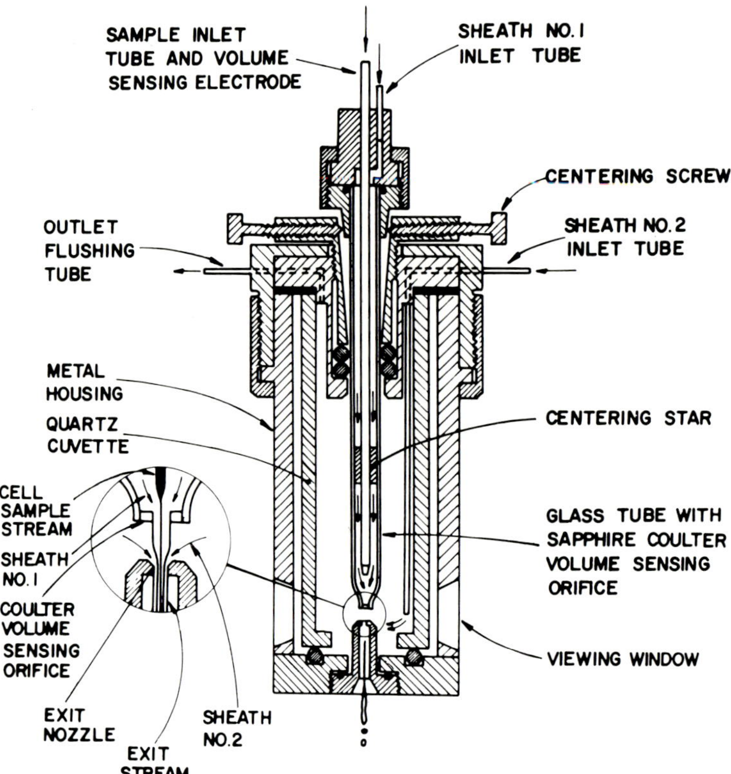

Fig. 8. Cell sorter flow chamber [60] incorporating a Coulter volume-sensing orifice and viewing windows through which optical measurements are made. The cell suspension first flows through the 75-μm-diameter exit orifice that forms the jet. A piezoelectric transducer (not shown) is coupled to the flow chamber and causes the jet to divide into droplets at the transducer excitation frequency of 45 kHz.

which all cells must pass. These advantages are offset by a longer cell transit time and transit time uncertainty between cell analysis and isolation in a droplet, and by the need to keep the windows of the viewing chamber scrupulously clean.

In instruments in which optical analysis follow jet formation, the fluid jet is an important optical component both of the illuminating and of the viewing systems. Light is refracted at the surface of the cylindrical jet, which acts as a very short focal length cylindrical lens (100 μm for a jet diameter of 50 μm). Optical signals, particularly scattered light, which is highly anisotropic, are thus critically dependent on cell position in the jet and on accurate alignment of the jet with the axis of the optical system. Coaxial flow systems, with high sheath-to-sample flow ratios, must be used to keep cells on the jet axis. The drop-forming transducer produces undulations in the jet surface which modulate light reflection and refraction by the jet. This limits the usable jet disturbance amplitude, hence the minimum distance from nozzle to drop separation point. Despite these drawbacks, most currently operating instruments use this approach because of its simplicity and relative ease of maintenance and cleaning.

In choosing parameter values for a system, one generally wishes to maximize the droplet frequency in order to maximize the cell separation rate. However, we see from Eqs. (1)–(3) that the maximum and optimal frequencies are given by

$$f_{max} = v/\lambda_{min} = v/3.14d_j \quad (4)$$

$$f_{opt} = v/\lambda_{opt} = v/4.5d_j \quad (5)$$

Thus, droplet frequency varies directly with jet velocity and inversely with jet diameter. The minimum jet diameter is determined by the size and other properties of the cells that must pass through the jet-forming orifice without significantly affecting droplet formation or causing stoppages by clumps. The maximum jet velocity is limited by the onset of turbulence at high velocities, which may cause wandering of the cell trajectory through the area in which analysis takes place, or instability in the drop-forming process. Turbulence is likely to occur if the Reynolds number (for water at 20°C, numerically equal to the product of nozzle diameter in micrometers and average velocity in meters per second see Chapter 3, this volume), exceeds the value 2,300, but this only applies to fully developed parabolic velocity profiles. Cell-sorter nozzles are made so short that fully parabolic profiles do not develop; laminar flow can then still exist for Reynolds numbers greater than 2,300 [51,52]. Increasing the velocity of the jet therefore represents an important potential for increasing the sorting rate, and this has been used in high-speed sorters (see section on high-speed sorting).

The initial jet disturbance amplitude, proportional to the droplet-forming transducer input amplitude, determines the distance from the nozzle to the droplet-separation point. The initial disturbance should not interfere with optical cell mea-

TABLE 1. Characteristic Parameters for Deflected Droplet Sorters used for Leukocyte Phenotyping in Immunology, General Purpose, Large-Particle Sorting, and High-Speed Sorting

	Immunology	General	Large cells	High speed
Max particle diam	12 μm	20 μm	50 μm	20 μm
Orifice diameter	50 μm	80 μm	155 μm	80 μm
Jet diameter	44 μm	68 μm	144 μm	65 μm
$lambda_{min}$	140 μm	215 μm	450 μm	205 μm
$lambda_{opt}$	200 μm	310 μm	650 μm	295 μm
Nozzle pressure	0.7 kg/cm^2	1 kg/cm^2	0.5 kg/cm^2	14 kg/cm^2
Jet velocity	10 m/sec	13 m/sec	8.5 m/sec	50 m/sec
Droplet frequency	40 kHz	34 kHz	6 kHz	220 kHz
Droplet period	25 μsec	29 μsec	167 μsec	4.5 μsec
lambda used	250 μm	400 μm	1400 μm	230 μm
Breakoff length	2.5 mm	6 mm	10 mm	20 mm
Typical delay	250 μsec	435 μsec	1170 μsec	495 μsec
in periods	10	15	7	110
Optimal input analysis rate*	8000 c/sec	6800 c/sec	1200 c/sec	44000 c/sec
Max acceptable dead time†	6.3 μsec	7.3 μsec	42 μsec	1.1 μsec

*Theoretical value for maximum sorting rate [m = 1/nT, see eq. (10)], using coincidence checking on 5 droplets, not taking into consideration possible limitations on analysis rate dictated by the requirements on purity of the sorted cells.
†Maximum acceptable dead time is indicated for 5% sort contamination at optimal input rate [eq. (14)].

surements but should, if possible, be large enough to dominate random disturbances and perturbations produced by cells. Typical parameter values for cell sorters often used for leukocyte phenotyping in immunology and for more general purpose cell sorters, are given in Table 1. These parameters can be modified substantially to handle cells or particles of different sizes. At the appropriate frequency, the droplet-formation technique will work for jet diameters in a range that extends at least from 10 to 1,000 μm.

In coaxial systems, cells are introduced through an inner nozzle or tube centered on the jet axis. Exit diameter and axial location of the inner nozzle are not critical, but radial symmetry is important. An inner nozzle with a 100-μm diameter exit orifice located 5 mm behind the outer nozzle exit is typical. The cross-sectional area of the outer nozzle is normally chosen such that at the cell injection point the velocities of sheath and sample are of the same order of magnitude. For a jet diameter of 50 μm and a typical sheath-to-sample flow ratio of 50 : 1, cell centers (not the entire cell) are confined to a 7-μm diameter core in the effluent jet and the cell suspension flow rate is approximately 0.4 μl/sec.

Fluid Flow and Cell Interaction

In the converging and accelerating flow region behind the jet nozzle exit, cells are subjected to forces that act to extend them in the direction of flow. Nonspherical cells tend to become aligned with the long axis parallel to the flow axis, and deformable cells can become considerably elongated in the direction of flow [34] (see Chapter 3, this volume).

Each cell, as it passes the nozzle exit, generates a perturbation on the jet that may increase or decrease the initial droplet-forming disturbance, depending on the relative phase relationship between the cell-induced and transducer-induced jet disturbances [62]. The result is a variation in the distance to the droplet-separation point that cannot be predicted, causing increased uncertainty in transit time between analysis and cell isolation in a droplet. This variation in transit time can be reduced by increasing the transducer-induced disturbance, thereby reducing the relative importance of the cell-induced disturbance, or by increasing the nozzle and jet diameter. For the system parameters given in Table I for sorters used in immunology, cell-induced disturbances become significant for cell diameters greater than about 10 μm. The cell-induced disturbance increases considerably if the cell path does not accurately follow the nozzle centerline. Cells can be kept close to the flow centerline by using a coaxial flow system having precise radial symmetry and a high sheath-to-sample flow ratio.

Larger cells can also modify the droplet size. This is apparently a different phenomenon, not caused by disturbance of the jet at the nozzle; it occurs perhaps when a large cell becomes positioned in or near the thin fluid ligament that connects a droplet in the final stage of formation to the parent jet. For the successful sorting of large particles, it is important to understand and try to minimize such particle-induced disturbances, as will be seen later in this chapter.

Cells are subjected to the pressure required to produce the desired jet velocity and to overcome friction in the nozzle until they approach the nozzle exit, where the static pressure quickly drops to zero as the jet enters the atmosphere. Cell damage will usually not be caused by this abrupt pressure change, but this will be of concern for fragile cells or in systems operating at high jet velocities and pressures (see section on high-speed sorting).

Droplet Charging

The droplet charging signal is applied between an electrode that contacts the fluid in the nozzle, or the metal fluid supply tube, and an electrode that surrounds, but does not touch, the jet at the droplet-separation point. The charging voltage establishes an electrostatic field and corresponding charge at the surface of the unbroken column of conducting saline fluid extending from the nozzle. To a first approximation, the charge on a separating droplet is proportional to the charging voltage at the instant of droplet separation. To facilitate observation of the jet and optical interrogation of cells in it, the electrode surrounding the jet may be omitted. The charging voltage is then the potential difference between the jet and nearby conducting surfaces. The voltage required for a given droplet charge is then increased and the droplet charge becomes sensitive to static charges on nearby surfaces and to the deflection plate voltages. The charging potential is typically 75 to 150 V peak for a 5- to 10-mm droplet deflection. The corresponding current represented by the flow of charge carried by the droplets is very small, e.g., 0.003 μA, if 10% of the droplets are charged. Significant current can flow, however, through the conducting fluid in the supply tubes if the charging potential establishes a voltage between the nozzle and fluid supply reservoirs. Direct currents must not be permitted in the fluid system, as they result in electrolysis, with erosion of metal parts and generation of corrosion products.

To charge a droplet precisely, charges induced by nearby previously charged droplets must be considered. A charged object induces a charge of opposite polarity on nearby conductors; therefore a charged droplet that has just separated has the effect of reducing the charge on the droplet behind it. If, following an uncharged series of droplets, several adjacent droplets are to be identically charged, the charging potential must be higher for the second droplet than for the first. Additional, but smaller, increases are needed for following droplets until a final equilibrium potential is reached that is typically 20 to 40% higher than that required for the first droplet. The required charging waveform can be approximated by a voltage that rises exponentially from initial to final value with a time constant of two to four droplet periods and can be generated by a simple resistance-capacitance network driven by a voltage step. Since the process is linear, the network provides the appropriate compensation for any number of drops and for any increase or decrease in charging voltage.

Generation of "fast" satellites does not alter droplet charging because the charging potential that determines satellite charge also determines the charge on the preceding droplet with which it merges. "Slow" satellites should be avoided as they are charged at a different time than are the following droplets with which they merge, complicating the charging waveform equalization and making droplet charging dependent on satellite geometry.

Droplet Deflection

After charging, the droplets pass through a constant transverse electrostatic deflection field established by two charged deflection plates. The deflection plate potentials are typically plus and minus 2,500 V, providing a field of 5 kV/cm for a 1-cm plate spacing. A high deflection potential is desirable to minimize the droplet charge for a given deflection, but it must be much less than the limit established by spark breakdown in air, which occurs at about 30 kV/cm. The deflection power supply system should have a low short-circuit current limit to minimize shock hazard. The best arrangement is a supply that regulates voltage accurately up to a current of 1 mA or so, to accommodate small leakage currents, but that cannot deliver more than a few milliamps maximum. Energy-storing filter capacitors across the output should be small.

Charged droplets will only follow one well-defined trajectory if they all have the same charge. This requires that the charging pulse have the correct shape, as discussed above, but also that it be in correct phase relative to the oscillator-generating droplet breakoff. Improper charge pulse shaping or incorrect charge pulse phasing will cause the individual droplets in a charged group to follow different trajectories, resulting in the spread in positioning usually called *fanning*. Just as the rising edge of the charge pulse is important for generating the correct charge on consecutive droplets to be deflected, the trailing edge is important for preventing induction of opposite charge on the first droplet following the charged group. Such induced charge can cause fanning of the main stream of undeflected droplets, necessitating larger deflection of sorted droplets to achieve reliable separation.

The trajectories of charged droplets are influenced by aerodynamic drag forces, acting along the line of flight, and by mutual electrostatic forces that tend to cause adjacent charged droplets to repel each other. Aerodynamic effects vary from droplet to droplet because the drag force is much greater on a droplet traveling in still air than on a droplet following in the wake of another droplet; the principal effect is to increase the transit time for deflected droplets and to cause adjacent droplets in a deflected group to run into each other and to merge. Variations in flight time are of no importance but, because slower droplets are in the deflection field longer, aerodynamic drag also causes deflection errors.

The overall result of these nonlinear processes is an uncertainty in droplet deflection that depends on the pattern of the droplets in flight. Because the cells to be sorted are randomly timed, the errors are difficult to compensate for, but are minimized by using low droplet charge, small maximum deflections, and a high deflection field established over as much of the flight path as possible. Increasing the deflection by increasing the distance between nozzle and collecting tubes increases the relative importance of errors caused by aerodynamic forces, while increasing deflection by increasing drop charge increases the relative importance of errors caused by mutual repulsion (proportional to charge squared) between adjacent droplets.

Droplet Charge Timing

Cells, classified in the analyzer section of the sorter, are isolated in a separating drop after a transit time determined principly by system geometry, droplet-separation point location, and flow velocity. A decision to sort must be delayed by this cell transit time before generating the droplet charging pulse. Figure 9 shows a timing diagram for initiation of droplet charging. The top trace shows the series of time points marking droplet breakoff. The sorting logic is clocked by synchronization pulses at the droplet frequency, but the phase relationship of these sync pulses to the droplet breakoff is unknown to the electronics. The passage of a particle is a random event relative to the cycle of one droplet generation. In Figure 9, detector pulse a is depicted as occurring early relative to the droplet break-off period, detector pulse b is close to the end of one break-off period. A certain classification time t_c is needed for analysis and sorting decision of each pulse. If the particle is to be sorted (a and b, but not c), an arming gate is set to initiate the charging pulse timing for that event. The charge gate switches on at the first sync pulse after the arming gate is activated. The arming gate has a set duration equal in periods to the number of droplets to be charged for each event. Termination of the arming gate is followed by termination of the charge gate at the next trigger. Relative to the charge gate the charge pulse has an adjustable phase delay t_p, which is set to bring droplet charging in phase with droplet generation. The charge pulse

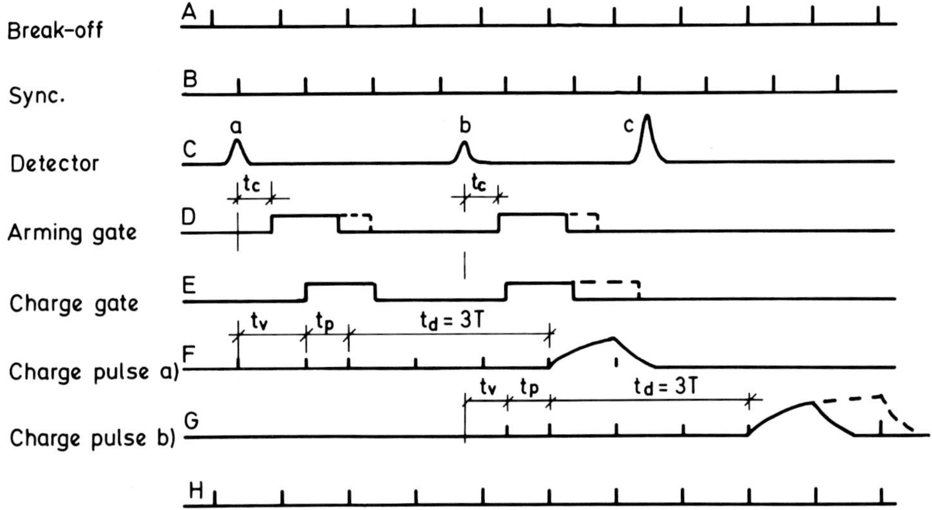

Fig. 9. Timing diagram for sort initiation and droplet charging. Trace A marks time point for droplet breakoff. Trace B shows synchronization pulses at the droplet frequency, having an unknown phase delay relative to droplet breakoff. Trace C shows primary fluorescence pulses detected from cells a, b, and c. Trace D shows the arming gate which is set as soon as pulse height analysis and sorting decision has been completed. The fully drawn arming gate has a duration of 1 droplet period. The situation for a duration of 1.5 droplet periods is shown with broken lines. Trace E shows the charge pulse gate which switches on at the first sync pulse after the arming gate is set, and switches off at the first sync pulse after the arming gate is reset. For an arming gate duration of 1,5 periods, the charge pulse gate for cell b lasts for 2 periods (compared with 1 for cell a) because cell b was detected very close to the end of the time interval corresponding to one droplet (compare trace A and C). The charge pulse gate signal is delayed by the appropriate number of periods, determined by the sort delay, before driving the charge pulse output amplifiers generating the charging pulse for cells a and b, shown separately on trace F and G, respectively. Note that in both cases droplet charging will be applied during the 5th droplet period after cell passage. Coincidence checking before and after the cells are detected is not included in the above diagram. The time intervals t_v, t_p, and t_d are explained in the text. Trace H is a repetition of trace A to facilitate timing comparisons.

is also delayed by the operator-determined sort delay time t_d before being applied. The total time from when the detector pulse occurred to when the charge pulse is applied is therefore $t_{total} = t_v + t_p + t_d$, of which t_v is randomly variable between t_c and $t_c + T$, where T is the droplet period. Note that the total delay time t_{total} is slightly longer than t_d, generally referred to as the sort-delay time.

Figure 9 also illustrates the case of an arming gate set equal to 1.5 droplet periods (broken line). Thus, the charge pulse gate varies according to when the particle was sensed relative to the time for droplet breakoff. If the particle was sensed late relative to the breakoff cycle, the arming gate will initiate the charging of two droplets (b) instead of one (a).

The sort delay is normally longer than that shown in Figure 9; therefore, sort decisions may be stored as digital signals in a shift-register delay line, with delay time adjustable in number of droplet periods to match the cell transit time, allowing new cells to be analyzed before previously analyzed cells have been isolated in charged droplets. The charging pulse duration is always an integral number of droplet periods, and that number of consecutive charged droplets always have the same charge and deflection pattern; thus, droplets cannot be deposited between collecting tubes by being indeterminately charged during charging pulse transitions.

Cell timing and droplet timing are uncorrelated; therefore, isolation of a cell in a separating droplet follows cell arrival at the droplet separation point by a random time interval between zero and one droplet repetition period. The arming gate thus has a minimum duration of one droplet period. Any uncertainty in the transit time, caused by variations in analyzer timing, flow velocity, or cell trajectory in the flow channel, requires a corresponding increase in arming gate width.

Timing errors in droplet separation, which may be caused by cell-induced disturbances at the nozzle, also require an increase in arming width to insure proper charging of the cell-containing droplet, but these errors also cause synchronization errors in droplet charge timing. Droplets not containing the cell may be partially charged if they separate during the charge pulse transitions.

Coincidence Detection

Cells that are too closely spaced to classify separately, or of different classes too closely spaced to isolate into separate deflected groups of droplets, should be detected by coincidence circuitry that then inhibits deflection of droplets containing those cells. The coincidence circuitry should operate on all detectable cells or particles that would be considered contaminants in the sorted fractions. If correct counting of the number of sorted cells is essential, the coincidence circuitry should cancel any sort decision in which a nearby cell might be deflected along with the desired cell, but if maximum sorting rate of desired cells is of interest, rejection due to coincidence should only take place if the nearby cell is an undesired cell. A switch for easy selection between these two operating modes should be provided on the instrument.

Some sorting errors caused by cell pairs too closely spaced to be detected as separate signals, or by cell doublets, are inevitable but may be reduced by careful sample preparation to minimize adhering doublets and by reducing cell flow rate to increase the average time spacing between cells. It is often possible to utilize characteristics of the analyzer signals to identify single-signal pulses that correspond to doublets or closely spaced pairs and to inhibit deflection of the corresponding droplets. Usually, the separated fraction containing "small" or "dull" cells has higher purity than does the "large" or "bright" fraction because unwanted cells are more likely to be deflected as undetected passengers along with cells producing large analyzer signals.

Sample Supply and Collection

Sheath and sample fluids are usually supplied from reservoirs pressurized with well-regulated air or nitrogen. The required fluid pressure equals the velocity head of the jet (0.5 kg/cm^2 or 7 psi for 10 m/sec) plus the pressure required to overcome friction losses in the nozzle and supply system (typically another 0.5 kg/cm^2). The differential pressure between sheath and sample determines the sheath-to-sample flow ratio and requires regulation to within a fraction of a centimeter of water. Pressure regulators with very low hysteresis are required. Inclusion of a capillary flow restrictor in the sample line to provide a friction pressure loss of a few cm of water at the required sample flow rate will make pressure regulation less critical, but this provides another undesirable component to trap cells. Alternatively, sheath and sample reservoirs may be pressurized from a common source, and differential pressure regulated by providing height adjustment for one of the reservoirs. Figure 10 shows a diagram for a typical fluid supply system that includes provisions for switching between samples and for flushing.

Positive displacement sheath supply systems based on gear or piston pumps are successfully being used in some flow cytometric analyzers, but their implementation in sorters may produce minute short-term pulsations or irregularities in the flow that affect droplet formation. Syringe driven sample supply systems are especially suitable for delivering small cell samples at precise, low flow rates, and are often optional on commercially available cell sorters. Valves and connecting tubing in the sample supply system should have minimum volume. The area of surfaces to which cells can adhere should be minimized, as should any crevices that might trap cells.

Charge carried by deflected droplets can cause static charge buildup on the collecting tubes that influences droplet trajectories. Fortuitous leakage paths are usually adequate to drain this charge but may require augmenting by addition of a conducting path between the collected fluid and ground or by employment of a radioactive static charge neutralizer.

For some live-cell sorting applications, sample and collecting tubes should be refrigerated by providing ice baths or cooling jackets. It is generally unnecessary to cool other components. The sample temperature can then approach the ambient room temperature during the travel time (typically about 1 min) from supply tube to nozzle before being recooled in the collecting tube. In general, this brief warming does not affect cell viability. There is little point in cooling the sheath fluid reservoir, since the sheath fluid will assume the uncooled nozzle assembly temperature before it merges with the sample.

For applications requiring sterility, all fluid-handling components that come into contact with the cell sample should be capable of being sterilized or should be sterile and disposable. If antibiotics alone are used to avoid sterilization in sorting live cells, resistant organisms will eventually colonize the system and contaminate the sorted sample. Components to be sterilized by autoclaving require attention to expansion coefficients and temperature stability of materials. Even

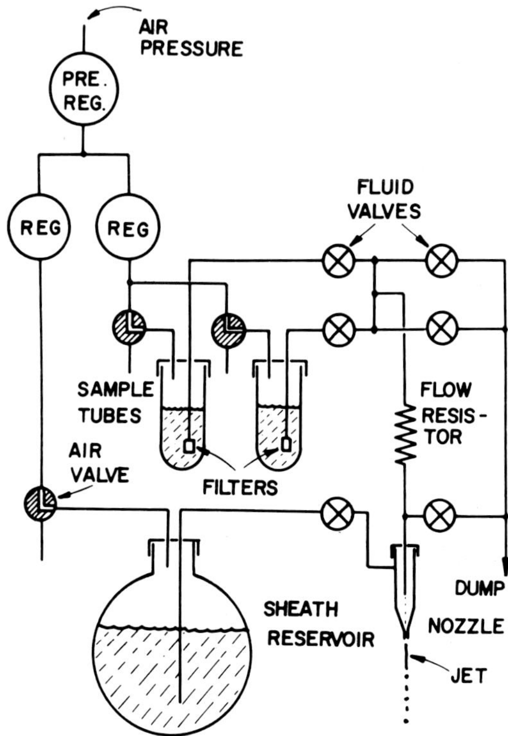

Fig. 10. Fluid supply system for a cell sorter. Sample and sheath fluid flow rates are determined by pressures separately set by precision air regulators. Valves in air and fluid lines are provided for sample selection, flushing, and dumping. Removing pressure from a sample tube allows sheath fluid to flow back from the nozzle and flush the sample supply line and filter. The optional flow resistor stabilizes the sheath to sample flow ratio against small differential changes between sheath and sample pressures or fluid heads. Sample cooling jackets are not shown.

though they withstand autoclaving without apparent damage, differential expansion and creeping often result in fluid leaks at interfaces between dissimilar materials after repeated sterilization.

Complete sterility requires sterility of the air through which the jet passes, which could be achieved by operating in a laminar flow hood or by completely enclosing the flow system in a sterile environment. However, culture experiments with sorted cells from droplets passed through nonsterile air indicate that the risk of contamination from this source is low.

However, unless precautions are taken, contamination of the air with the material being sorted can occur and this poses a possible hazard to operating personnel [46]. The two principal sources of contamination are (1) satellite droplets, which may drift away from the collection area if they do not merge with major droplets; and (2) mist, produced by droplet impact in the collecting tubes. Mist may be minimized by arranging the collecting system so that droplets impact on a surface that is nearly parallel to the flight direction. Contamination of the room atmosphere may be reduced by partially enclosing the volume around the droplet trajectories and flowing air through the volume and into a filter.

Collection vessels suitable for the various applications of flow cytometric cell sorting generally must be selected on the basis of the requirements for each particular run. Obviously, the need for unimpaired viability of live cells after sorting or demands for intact cellular morphology makes it harder to devise an optimal collection procedure than, for example, collecting a number of sorted cells pulse labeled for counting of radioactivity in the sorted fraction. In addition to descriptions in the original literature on various biological applications of flow cytometric cell sorting, there are some reports of special collection devices that may be found useful [1,15,30,45,48,49,54].

Monitoring Equipment

A sorter must incorporate procedures for monitoring the droplet-formation process, locating the droplet-separation point, determining the cell transit time to the separation point, and observing the trajectories of the deflected droplets. Droplet formation is conveniently monitored by observing the stroboscopically illuminated jet with a low-power (20 to 100×) measuring microscope. Typically, illumination is furnished by a light emitting diode driven by 1-μsec pulses synchronized with the droplet-forming transducer input and timed such that strobe pulses occur midway between droplet-charging transition points. With proper adjustment of the transducer input amplitude and relative phase, a droplet will appear to be just separating from the parent jet, and the location of this critical point can be determined relative to the nozzle tip or relative to the point at which cells are sensed by the analyzer.

Jet velocity is determined by multiplying the measured droplet-to-droplet spacing by the droplet frequency and then used to compute the time delay between cell analysis and cell arrival at the droplet-separation point. Satellite droplets, if they occur, can be checked for merging with the main droplets.

Droplet trajectories can be checked for uniformity of deflection and deflection amplitude by continuous illumination of the path so that light scattered by the droplets is visible. If the light is properly focused and the source masked from the viewing point, single droplets in flight can easily be seen. Using a test signal to simulate analyzer sorting decisions, the charging pulse amplitude and waveform shape are adjusted so that all deflected droplets follow the proper trajectory.

The effect on droplet trajectories by cell-induced disturbances modulating the droplet-separation point and the droplet size can be evaluated by comparing trajectories of cell-containing deflected droplets with trajectories of empty droplets that are deflected in response to test signals, or to the same cell-produced signals with an improperly set charging delay so that cells do not coincide with the deflected droplets.

Illumination for monitoring droplet formation and for monitoring droplet flight trajectories should be arranged so that it does not interfere with cell-generated optical signals. Subtle changes in nozzle flow conditions that affect droplet deflection, caused by drift in fluid or sample parameters or by accumulation of cell debris that does not stop jet flow, can then be checked during separation without interrupting operation.

Cell Classification for Sorting

Analysis for sorting requires timing information and classification speed not generally needed in instruments that perform analysis only. Unless more than one cell can be processed simultaneously, analysis and classification must be completed within a time that is short compared with the average interval between cells if sorting efficiency and purity is not to be reduced. Timing of the sorting signal must be related to the time the cell passes a specified point and must not depend on classification time. In a sorter, analysis should include detection of cell pairs too closely spaced to analyze or properly sort, and separation of these pairs into classified populations should be inhibited.

Cell classification is usually determined by electronically comparing the cell parameter signals with preset thresholds ("window" circuits or "single-channel analyzers") that specify upper and lower signal limits for a particular parameter. Window outputs are then combined with logic circuitry to generate sorting signals that define one or more cell classes to be sorted. Cell classes may also be defined by more complex relationships between cell parameters, for instance, specifying a range for the ratio of two parameters. Classification may be performed by analog voltage comparators or the cell parameter signals can be digitized and classification performed with digital logic. In multiparameter instruments, bit map techniques are often used to permit the sorting criteria to be defined as irregular regions of interest in multiparameter histograms.

Visualization of cell parameters to show bunching or clustering of subpopulations is an important aid to setting thresholds for sorting. Single- or two-parameter pulse-height distribution displays of cell parameters are often used. Another effective method displays two parameters as X,Y coordinates on a storage oscilloscope, with each cell recorded as an intensified dot. If two cell populations overlap (and they almost always do, to some extent), purer (but smaller) sorted populations can be obtained by setting thresholds to select only a portion of each population and discarding cells that lie in the overlapping area.

Frequently, thresholds must be set to sort a specified fraction of the input population. This can be accomplished with the aid of two counters that register total cell rate and sorted cell rate or with a ratio counter that indicates the fraction directly.

OPERATIONAL CHARACTERISTICS OF DEFLECTED DROPLET SORTERS

Ideally, cell sorting only requires that the user determine the sort delay, set the sort criteria that define the characteristic measure of desired cells to be sorted, and preset the number of sorted cells wanted. In practice, however, users have to think about how long this sorting procedure will take and must convince themselves that the number of actually sorted cells is correct and that the sorted cells satisfy the sorting criterion. Thus several operational characteristics such as sorting efficiency, and recovery and purity of the sorted cells, need to be considered.

Recovery

The number of cells or particles actually found in the sorted fraction at the end of a sort run, relative to the number of activated sort events, is usually called the recovery, expressed as a percentage. Some authors call this parameter sorting efficiency, but this term will be reserved for another aspect of cell sorting performance. Whether all the sorted cells are of the desired type as defined in the sort criteria, relates to the question of purity which will be discussed later in this section. Ideally, recovery is of course 100%, but it is often lower. If we take for granted that the instrument gives the correct count for the number of activated sort events, sorting recovery may be low for two principally different reasons: (1) The charged droplets never reached the collection vessel set up for the sorted fraction, and (2) The droplets reached the collection vessel, but contained no cell.

The first point may be checked, e.g., by calculating the expected weight of the liquid deflected into the sorted fraction for a certain number of sort events and comparing that to the amount of liquid recovered. Freezing the droplets as they are received in the collection vessel, may be used in order to obtain precise measurements of this aspect of sorting recovery, which might be called volume recovery.

Several factors may cause low volume recovery, such as improper charge pulse phasing in relationship to the oscillator of the droplet generator, or improper pulse shaping, both of which result in erroneous trajectories for the deflected droplets so that they will not reach the receptacle. Electrostatic buildup on the sorting vessel has been reported to cause arriving droplets to "bounce off" the top of the collection vessel. Air turbulence in the collection area could cause deflected droplets to miss the collection vessel.

If volume recovery is all right, cell recovery then only depends upon whether the cell destined for sorting actually was at the droplet breakoff point when the charging pulse was applied. Cell recovery therefore is critically dependent upon setting and maintaining the correct timing of the droplet charging pulse relative to the time point of cell sensing at the optical focus. Any variability in flow velocity between individual cells will introduce an uncertainty in transit time, and this uncertainty will increase with distance between the optical focus and the breakoff point. A short distance between sensing and breakoff therefore gives the best sorting recovery. Due to increasingly larger fluctuations of the surface of the liquid jet closer to the breakoff point, however, an optimal distance between sensing region and droplet breakoff will usually correspond to a delay of 10 to 20 droplet periods for sorters based on jet-in-air sensing.

Owing to uncertainty about particle position at the breakoff point, it is a common procedure to charge and deflect three droplets for each sort event, with the time delay set optimally for the center droplet. Setting of the correct sort delay will usually be a two-step procedure, first estimating the delay and then verifying the setting by test sorting. Estimates of the delay can be made by measuring out the distance from the optical focus along the jet to the droplet breakoff point, and determining how many droplet periods this corresponds to by counting the number of droplets over an equally long distance further down-stream. Such estimation is easily performed if the instrument is equipped with stroboscopic illumination and a viewing microscope containing an appropriate micrometer scale. In some instruments the optical sensing region is within the flow chamber where the bore diameter may be larger than that of the jet in air. This gives lower velocity within the chamber than in the jet and must be taken into account when the delay time is estimated [52].

The estimated delay is verified by performing test sorts with easily identifiable particles for the proposed delay setting and some settings above and below. The instrument is set up to sort, for example, 100 particles onto a microscope slide and, by counting the recovery, i.e., the number of particles found on the slide relative to the number of sort events, recovery profiles such as shown in Figure 11 are established. The open columns show the results for conventional three-droplet sorting. The optimal delay setting was in this case 13 periods, and the average recovery at that setting was found to be 98 ± 2%. Setting the delay one period shorter or

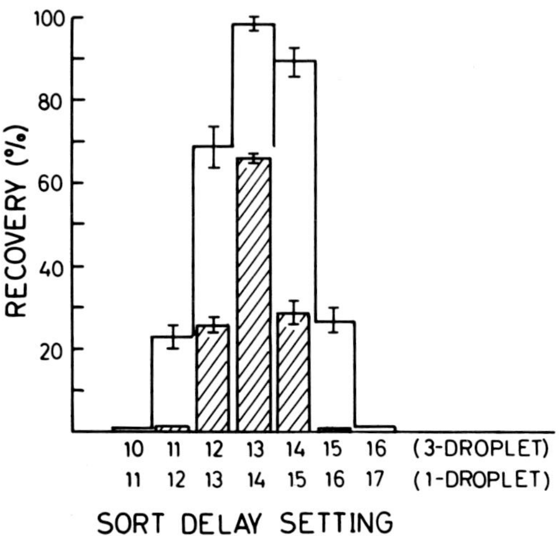

Fig. 11. Verification of the sort delay setting by scoring of the recovery obtained in test sorts performed for serial increments of the sort delay. The unshaded profile represents results obtained when 3 droplets were sorted for each event. The hatched columns represent data for single-droplet sorting. As will be clear from Figure 12, results as shown here (particularly for single-droplet sorting) can only be obtained after fine adjustment of the delay to within fractions of one period.

longer in this case also gave relatively high recoveries (70 and 90%, respectively) as might be expected for three-droplet sorting. It is seen from the figure that for three-droplet sorting, at least 5 different sort delays must be checked by microscopic counting in order to verify the optimal delay setting.

The hatched columns of Figure 11 show corresponding results for single-droplet sorting. The narrow time window for successful sorting in this case results in a sharper indication of the correct delay. Verification of a delay setting by single-droplet test sorting is therefore quickly done since test sorts at three different delays will be sufficient. The maximum recovery on single-droplet sorting will normally be below 100%, but once the correct delay is set and verified, the instrument can be switched to three-droplet sorting for processing of the desired samples at maximum recovery. In Figure 11, the delays for three-droplet sorting are indicated one period shorter than those for single-droplet sorting, since the delay circuitry sets the time from sensing to start of the charging pulse, rather than to the midpoint of the charging pulse (see Fig. 9).

Note that in distributions of recoveries such as Figure 11, there will be a few particles found also for delays outside the optimal region. This is due to the random probability of finding particles in any liquid segment equivalent to the number of droplets sorted. This probability is equal to the product of the number of droplets per sort event, the duration of one droplet period, and the average particle analysis rate [44].

It should also be noted that for a sorting recovery curve recorded at a series of integer settings of the sort delay, such as Figure 11 and 12, the integral under the curve ideally equals the number of droplets sorted for each event [16], i.e., 3.1 approximating 3 for the open columns of Figure 11, and 1.2 approximating 1 for the hatched columns. This is seen intuitively for one-droplet sorting since the recovery distri-

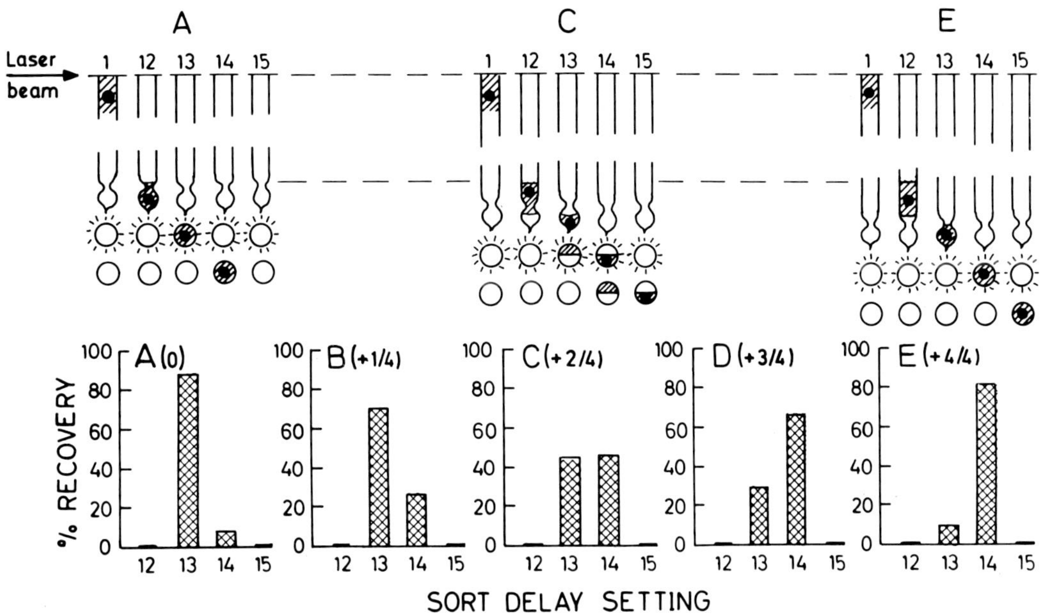

Fig. 12. Optimal sort delay setting requires fine adjustment to within fractions of one delay period, as revealed by single-droplet test sorts. The top panel shows schematically the liquid jet from the laser beam intersection to the droplet breakoff point. The top panels A, C, and E depict the situation at the end of sensing in a one-droplet segment of the liquid jet containing a particle of interest (hatched area), and the same after 12 to 15 periods. In A it is shown that after a delay of 13 periods, the liquid segment containing the desired particle is broken off as a charged (starlike) droplet. A delay of 12 or 14 would in this case result in charging the droplet before or after the desired one. In C the breakoff distance has been mechanically increased by half a period length, and in E the increase is by the full length corresponding to one period. The lower part of the figure shows the corresponding recoveries obtained by single-droplet test sorting for a series of integral settings of the sort delay. The sequence A–E represents increments of the physical distance between laser focus and droplet breakoff by the equivalent of 1/4 period for each new panel. The optimal recovery at delay 13 in panel A is gradually shifted to become optimal again at delay 14 in panel E as the vertical position of the flow chamber has been changed to increase the droplet breakoff distance by one period.

bution then shows the probability of finding the particle at each of the consecutive delays, the sum over all delay settings being equal to 1.

The symmetric distributions depicted in Figure 11 will not always be observed. In fact, Figure 11 shows the results that will be obtained only when the sort delay has been correctly fine-adjusted.

Droplet generation at the breakoff point may be thought of as a process in time where each droplet breakoff marks a time point. If these time points were also marked at the sensing region, they would cut the liquid jet into a series of adjacent liquid segments, each of which would constitute an entire droplet upon breakoff. For each droplet breakoff, optical sensing and classification of possible cells in that volume segment is completed a certain time before breakoff, and that is the exact time to be determined and set as the droplet charging delay after correcting for the duration of the charging pulse itself (see Fig. 9).

Generally, one should be able to fine-adjust the delay with a resolution of 1/5 to 1/10 of a period. However, many instruments can only set droplet delays in integer values of the droplet period. The correct timing is then obtained only if the time interval between sensing and breakoff is made to correspond to an integer number of periods by adjusting the physical distance from the sensing region to the droplet breakoff point (as was also assumed in the timing diagram of Figure 9). Such adjustment can be done either by changing the power to the droplet generator, or mechanically by microadjustment of the vertical position of the flow chamber. The latter approach was used on a laboratory-built sorting instrument by one of the present authors (T.L.) to obtain the data shown in Figure 12. The position of the flow chamber along the jet axis was adjusted by a micrometer driven positioning device that influenced no other geometrical setting of the flow cytometer. Thus the distance between the sensing region and the breakoff point could be varied over the length corresponding to one droplet period (0.4 mm). Only if the beginning and end of particle sensing for one "droplet equivalent" segment of the jet is in correct phase with the droplet breakoff, will the delay time from sensing to breakoff be an integer multiple of the droplet period, and optimal recovery will be seen as for delay 13 in Figure 12A and delay 14 in Figure 12E. If sensing and droplet breakoff is out of phase, sorted particles will be found at two neighbouring integer settings of the delay, and recovery will be lower than optimal, as shown in Figure 12B–D. Figure 12C illustrates the case where the delay setting of 13 periods is too short, only half the liquid segment sensed 13 periods earlier will be incorporated into the charged droplet, and the other half will be part of the next droplet. If the delay is set at 14 periods, half the liquid segment will be in the droplet before the charged one, thus only half of maximum recovery is found for both delay 13 and 14 in this situation.

A situation such as shown in Figure 12C can be optimized by changing the physical length of the liquid jet between sensing and droplet breakoff to correspond to an integer

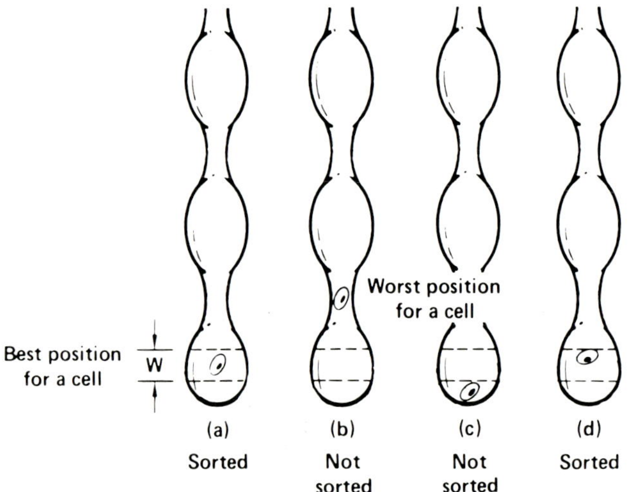

Fig. 13. Four examples of cell location in the jet and the sorting decision based on that location when position sensing is being used (47). The time window duration w is schematically represented as a region spanning the center of the forming droplet. Droplets with cells occurring inside the time window (a and d) are permitted to be sorted. Droplets with cells occurring outside the time window (b and c) are not sorted, i.e., aborted. (Reproduced from ref. [47], with permission of the publisher.)

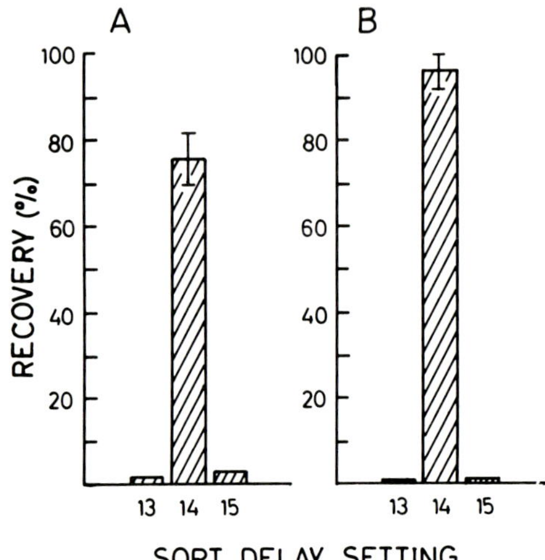

Fig. 14. Increased precision of single-droplet sorting by position sensing within the one-period liquid segment that will constitute one droplet. **A:** Recovery obtained for single-droplet sorting after optimal adjustment of the delay as illustrated in Figure 12. **B:** Results obtained with position sensing within the sensing volume. Sorting was allowed only if the cell was detected within the central 30 to 70% of the examined one-droplet liquid segment. The use of position sensing for single-droplet sorting has the same effect of bringing recovery up towards 100% as the usual procedure of increasing the number of droplets per sort event from one to three. Error bars indicate ±1 SD.

multiple of the generator period (as in Fig. 12A, 12E). Alternatively, the delay time can be set in fractions of one period to correspond to the correct mean transit time, which would be 13.5 periods in this case. Some instruments have a phase adjustment that alters the phase relationship between sensing (i.e., start of delay timing) and droplet generation but, depending on the electronics, the use of this facility may require readjustment of the charge pulse phasing relative to the generator in order to avoid fanning.

Owing to uncertainty in particle position caused by velocity variations, 100% recovery on single-droplet sorting will be difficult to achieve, since particles detected at the limits of the one-droplet time interval in some cases will end up in the droplet before or after the one being charged and deflected. This situation can be improved if sorting is allowed only when the timing of particle sensing is such that the particle is located in the central part of the liquid segment that will constitute one droplet, as illustrated in Figure 13a,d [47,62]. Figure 14 shows the result of an investigation by means of the position sensing device of an EPICS V flow cytometer. Panel A shows the conventional single-droplet sorting optimally adjusted so that a recovery of 76 ± 6% was obtained. Panel B shows the results obtained after switching on the position sensing circuitry that will permit sorting only if the particle was detected inside the 30 to 70% central interval of the one-droplet segment, resulting in recovery of 96 ± 4%.

Position sensing may be useful in cases where single droplet sorting with highest possible recovery is desired. Furthermore, the technique may be used to increase the positioning precision of sorted droplets, as demonstrated in Figure 15 [47].

Purity

Purity of the sorted cells is defined as the fraction of sorted cells that belong to the desired category. It is most practically defined relative to the number of recovered cells; i.e., if out

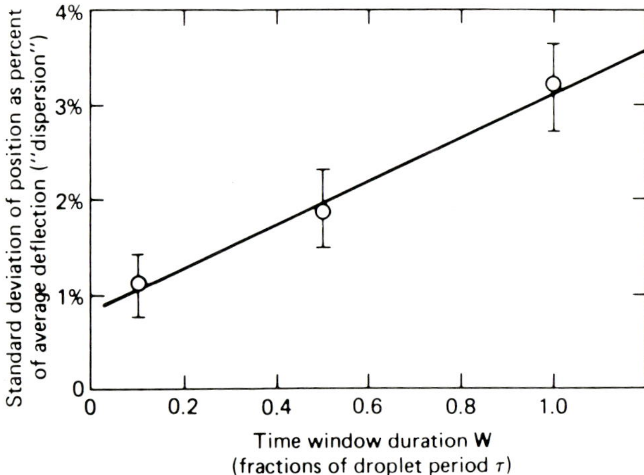

Fig. 15. Dispersion of sorted droplet positions on a microscope slide for various durations of the time window. The results were obtained sorting Lewis lung tumor cells using a 50-μm diameter nozzle. Each data point represents several hundred sorted droplets [47]. (Reproduced from ref. [47], with permission of the publisher.)

of 950 recovered cells 930 belong to the desired category, the purity is 100% × 930/950 = 98%. In cases where the analysis frequency is much lower than the droplet generator frequency, a recovery rate close to 100% guarantees that the purity will be close to 100%. The reason for this is that in the few cases where the desired cell is missed, the chance that the deflected droplets are empty is much larger than the possibility that the droplets contain an unwanted cell.

Generally, however, there is a need to use coincidence circuitry to check for the presence of unwanted cells in the liquid segment containing the desired cell to be sorted (see under Sorting Efficiency). A special problem arises if two cells are so close together that the instrument detector and electronics is unable to respond to the second cell; i.e., the second cell passes during the deadtime of the electronics. This time is typically around 20 μsec in flow cytometric analyzers, but special circuitry can bring this time down towards the actual duration of the primary pulse corresponding to the passage time of the particle through the optical focus. For an average analysis rate of 5,000 cells/sec, the probability of another cell passing within 20 μsec after an interesting cell is equal to $1\text{-}\exp(-5{,}000\times 20\times 10^{-6}) = 0.1$ [41]. This means that in 10% of cases, another cell is too close to the interesting cell to be detected and hence will be deflected along with the wanted cell. The reverse situation, that an interesting cell passes within 20 μsec after another cell, causes the interesting cell to go undetected and hence unsorted. This therefore does not influence the purity of sorted cells. The overall result is therefore that the number of cells in the sorted population is raised by 10%, and if these undetected cells all are unwanted cells, the purity will be reduced to 100/110 = 90% in this case. If it is generally assumed that any unknown "passenger" cell is considered a contamination of the sorted population, the requirement will be that for the fractional contamination to be c or less, the average analysis input rate m must be

$$m \leq c/t \tag{6}$$

where t is the dead time of the electronics. This approximation holds for values of mt so small that $1\text{-}\exp(-mt) \simeq mt$.

Correspondingly, the expression for the maximum achievable purity for given values of m and t is:

$$p_{max} = 1/(1+mt) \tag{7}$$

If the fraction of desired cells in the total population is significant and set equal to a, the probability that any passenger cell is an acceptable cell is also a, and the maximum purity will in this case be:

$$p_{max} = (1+amt)/(1+mt) \tag{8}$$

Any unwanted cells passing so far behind the desired one that it can be detected by the electronics, will activate the coincidence circuitry. Usually this circuitry will inhibit the charging pulse and thereby the sorting event if any other cell is detected close enough to the desired cell to be included in the droplets set for charging. If also the other cell detected belongs to the desired category, some instruments allow sorting of both cells, others may abort sorting also in this case and thus unnecessarily lose two wanted cells. Note that only if the latter approach is used, will counting of the number of sort events give a reliable count for the number of cells in the sorted fraction.

Because of uncertainty about cell positions at the breakoff point, it may be useful to set the coincidence checking gate wider than the sort gate (e.g., checking 5 droplets when sorting 3). This results in higher purity of the sorted cells, but reduces sorting efficiency, as seen below.

Sometimes purity of the sorted particles can be checked very easily, e.g. if the interesting particles are distinguishable from the rest of the population by microscopic examination. Thus, it is convenient to test the sorting performance by sorting one type of particles out of a mixture of easily distinguishable types.

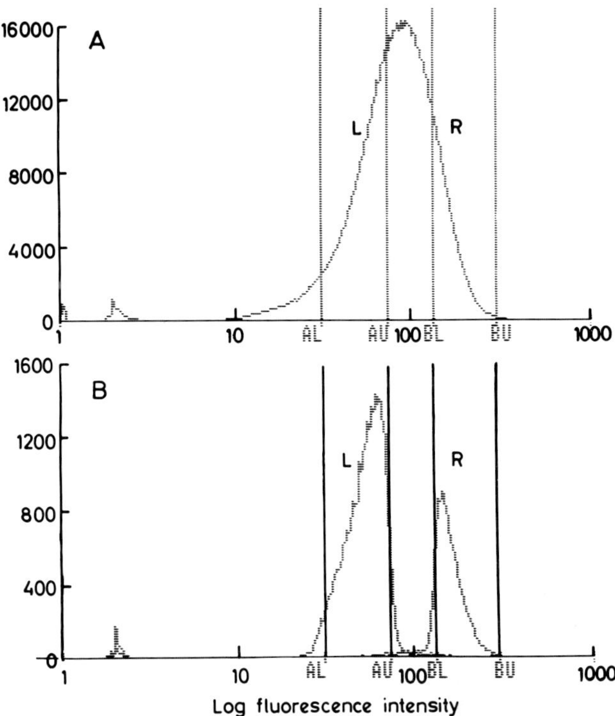

Fig. 16. Sorting of live cells (FME melanoma) stained for surface immunofluorescence [40]. **A:** Histogram recorded during the actual sort run, with upper and lower limits for the sort windows indicated for the left (L) and right (R) sorted fraction. **B:** Superposition of the two histograms obtained by reanalyzing the two sorted subpopulations L and R. The window limits were kept in the same positions to decide how many cells satisfied the original sort criterion upon reanalysis.

If the sorted cells cannot be identified by microscopic examination, one useful approach is to reanalyze the sorted fractions to determine what percentage of the sorted cells fall within the acceptance limits of the sort criterion upon reanalysis. An example of this is shown in Figure 16 [40]. In this case 90 and 80% of the left and right sorted fractions, respectively, fell within the sort windows originally used. In such a case various factors contribute to cause some cells to fall outside the sort windows upon reanalysis: (1) possible drift in the instrument or the cell staining during the period between sorting and reanalysis; (2) measurement uncertainty, characterized by a certain coefficient of variance, causing individual cells to be classified in a slightly higher or lower channel number upon reanalysis; and (3) high frequency of cells at (two of) the limiting channels used to define the sorting windows, thus the uncertainty caused by points 1 and 2 affects a relatively large number of the cells. In cases in which some of the above points can be eliminated, higher percentages of purity would be expected upon reanalysis of the sorted cells.

Efficiency

Since the particles arrive at the sensing region at random intervals, their occurrence can ideally be treated as a Poisson process. Normally, the requirement will be that the sorted droplets must contain no cell but the desired one. The prob-

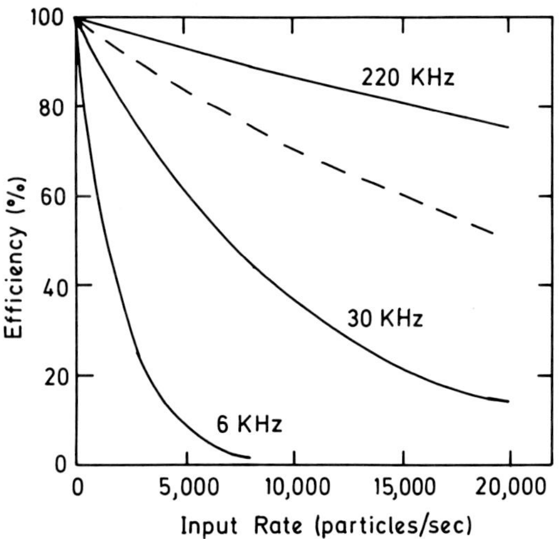

Fig. 17. Sorting efficiency comparison. Maintaining high-purity sorted populations requires that the probability of having more than one particle per sort event must be low. Since the particles occur at random intervals, their occurrence can be treated as a Poisson process [eq. (9)]. The efficiency when 3 droplets are used to sort a particle and check for coincidences is plotted versus the input particle rate for droplet frequencies of 6, 30, and 220 kHz. For the frequency of 30 kHz, the efficiency for one-droplet sorting and coincidence checking is also shown (broken line).

ability that the interval between two particles is larger than a certain time t is exp(−mt), where m is the average analysis input rate and mt is the average number of particles during the period [41]. For sorting to be allowed, there must be no other cell in the time interval t_1 before the interesting cell was detected, and no other cell in the time interval t_2 after the interesting cell. Thus, the probability for an allowed sort is

$$p(\text{sort}) = \exp(-mt_1) \cdot \exp(-mt_2) = \exp(-mnT)$$

where $nT = t_1 + t_2$ is the time duration corresponding to the number of droplets (n) deflected for each sorting event. With coincidence checking on the deflected droplets, this is also the probability that a desired particle will be sorted, therefore, it represents the sorting efficiency E:

$$E = \exp(-mnT) \qquad (9)$$

Figure 17 shows the sorting efficiency as a function of the input analysis rate for one and three droplet sorting using a 30 kHz droplet oscillator and coincidence check for the sorted droplets.

Obviously, if it is of paramount importance to obtain as many desired cells as possible out of a certain sample, sorting should take place at a low input rate, so that the rejection rate due to coincidences will be low. However, the situation will be different if the objective is to obtain a highest possible rate of sorted cells, as might be the case if there is a sufficient supply of sample and the aim is to reach a certain number of sorted cells in the shortest possible time.

If m is the input analysis rate and a is the fraction of wanted cells in the unsorted sample, ma will be the detection rate for wanted cells. The sorting efficiency will determine how many of these will actually be sorted, therefore the rate of wanted cells sorted will be:

$$R_s = ma \cdot \exp(-mnT). \qquad (10)$$

This expression has a maximum at m = 1/nT, i.e. the optimal input rate is to be such that on the average, one particle is detected within the time interval used for coincidence checking. For a three-droplet sorting at 30 kHz using coincidence check on the three-droplet window, the optimal input rate for sorting would then be 10,000 cells/sec. When sorting is performed at maximum rate, the sorting efficiency is only 37%, so nearly two-thirds of the sorting events will be aborted due to coincidences. If the interesting cells constituted 10% of the population, the theoretical maximum rate of sorted cells would then be

$$0.1 \times 10{,}000 \times 0.37 \text{ cells/sec} = 370 \text{ cells/sec}.$$

If the coincidence checking is set to cancel sort events only when the passenger cell is found to be an unwanted cell, the effective arrival rate of cells that cause cancellation is reduced to m(1−a), therefore the sorting efficiency in this case will be [52]:

$$E = \exp(-m(1-a)nT) \qquad (11)$$

and wanted cells will be separated out at a rate

$$R_s = ma \cdot \exp(-m(1-a)nT) \qquad (12)$$

If position sensing is being used, the sorting efficiency will be determined by the combined probability of the desired cell being in an acceptable position within the liquid segment and the interval between cells being large enough to avoid coincidence. The probability of acceptable position within the liquid segment is equal to the fraction of the acceptable part of this segment. Thus, if sorting only is allowed for cells in the central 30–70% of the segment constituting the sorted droplet, the corresponding probability w is 0,7−0,3 = 0,4, and the effective sorting rate [eq. (10)] would be modified by this factor:

$$R_s = maw \cdot \exp(-mnT) \qquad (13)$$

The problem of decrease in purity for high analysis rates due to cells passing undetected during the dead time of the electronics was neglected in the above analysis. The limitations on analysis rate set by the requirement for acceptable purity [eqs. (6–8)] will for many instruments set a lower maximum analysis rate than that indicated for maximum sorting rate [eq. (10)]. If a fraction c of contaminating cells is allowed [eq. (6)], the dead time requirement to enable analysis and sorting at the optimal input rate of m = 1/nT, would be:

$$t \leq cnT \qquad (14)$$

Thus, for a one-droplet sort with coincidence check on the charged droplet, the contaminating fraction would be c = t/T; i.e., as expected, the fraction of the droplet period in which the detector is unable to "see" another cell.

The above analyses were based on the assumption that the transit of cells through the sensing region could be modeled as a Poisson process. Measurement of time intervals between

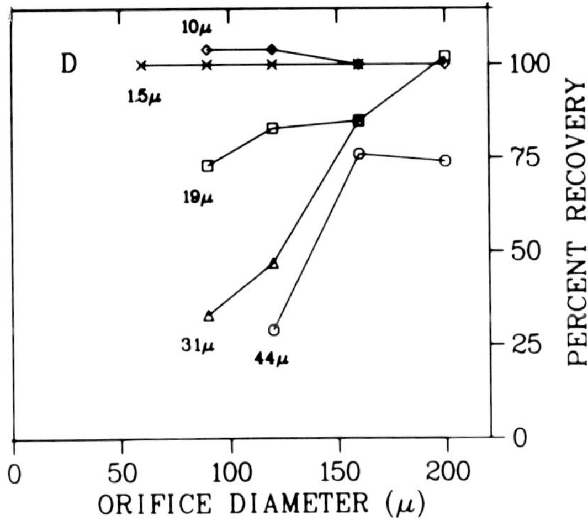

Fig. 18. Recovery percentage plotted as a function of orifice diameter for five sizes of microspheres. (Reproduced from ref. [32], with permission of the publisher.)

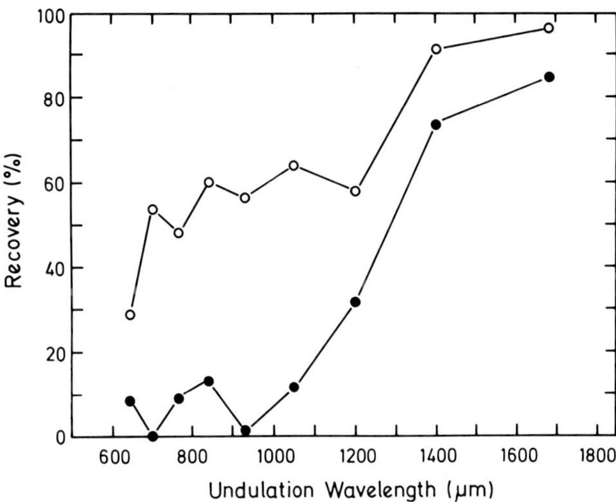

Fig. 19. Recovery percentage as a function of undulation wavelength for sorting of Carya pollen (open circles) and Zea pollen (closed circles) using a 155-μm nozzle. (Reproduced from ref [29], with permission of the publisher.)

particle arrivals in flow cytometry has confirmed that this often is a good representation, but has also documented situations where there was an overrepresentation of very short time intervals, thus leading to much higher rejection rates than expected from the analysis rate used [41]. In the reported case the cells were thought to have a tendency to "park" temporarily on the wall of the sample tubing, possibly due to electrostatic forces, and then be pushed along by another passing cell, the two cells thus passing through the sensing region closely together. The sample preparation also could have resulted in loosely adhering cell doublets and aggregates that were caused to separate by the acceleration forces in the hydrodynamic focusing region of the nozzle.

DEFLECTED DROPLET SORTING OF LARGE PARTICLES

The presence of a cell in the liquid jet disturbs the conditions for droplet breakoff, especially if it ends up in the neck region between two droplets [47,62]. Intuitively, one expects this effect to become more pronounced as the size of the particle increases, probably both relative to the jet diameter and relative to the droplet size, thus implicating dependence also on droplet frequency.

Established success in the sorting of 7 to 10-μm large lymphocytes in cell sorters using 50 to 80-μm orifices therefore does not guarantee equal success in sorting of large cells. This problem has become particularly apparent in the sorting of plant protoplasts, cells which typically have diameters in the range 30 to 100 μm. Protoplasts furthermore put the cell sorting technique to a difficult test since they are very fragile cells, obtained by enzymatic removal of the cell wall of plant cells (see Chapter 31, this volume).

Jett and Alexander [32] studied the effect of particle size on the performance of deflected droplet sorting with the view to find optimal conditions for protoplast sorting. As shown in Figure 18, their investigation using various orifice diameters demonstrated that although high recovery always was found when small (10 μm) particles were used, significantly reduced recovery was found for larger particles. They concluded that in order to obtain a reasonable recovery (>75%), the diameter of the sorted particle should be less than one quarter of the diameter of the orifice.

Freyer et al. [16] later found that with slight modifications of the same instrument and with carefully fine-adjusted droplet charging and delay setting, they succeeded in viable sorting of intact multicellular spheroids. Recoveries of 70 to 100% were obtained when sorting spheroids with mean diameters between 41 and 96 μm using a 200-μm orifice and droplet generation at 4.5 kHz.

Since the maximum drive frequency of a nozzle is inversely proportional to the diameter [eq. (4)], sorting of large cells using large nozzles inevitably will have to be at a relatively low rate. One would therefore like to be able to run close to the maximum frequency. However, in a study of factors influencing the sorting of large biological particles, Harkins and Galbraith [29] showed that recovery close to 100% could only be achieved at low oscillator frequencies. Figure 19 shows the result for Carya (50-μm diameter) and Zea (95-μm diameter) pollen cells using a 155-μm nozzle. Operated at a sheath pressure of 0.5 kg/cm^2, the jet diameter from this nozzle was 144 μm, indicating a minimum wavelength of 452 μm and an optimal wavelength of 650 μm, corresponding to frequencies about 19 and 13 kHz, respectively. To obtain good results (>75% recovery) for the larger particles, it was found necessary to reduce the frequency to about 6 kHz, corresponding to a wavelength of 1,400 μm (see Table 1). Using these conditions, Harkins and Galbraith showed that protoplasts could be successfully sorted with no significant impairment of cell integrity and cell viability.

Harkins and Galbraith interpreted the problems of large particle sorting to be due to the disturbance induced by the particle on the undulation giving rise to the droplet formation. The disturbance created by a given size particle will be larger for the generation of a small droplet (i.e., high frequency) than for the generation of a large droplet. This probably has to do with the proximity of the particle surface to the neck region of the undulation. Therefore, one might expect that position sensing, as explained in the previous section, could be used to increase recovery during sorting of large cells. Although the use of position sensing will proportionally reduce the efficiency of cell sorting as explained above, the advantage will be increased precision in depositing sorted cells [47] and more reliable and reproducible recovery.

HIGH-SPEED DEFLECTED DROPLET SORTING

Most commercially available cell sorters can scan several thousand particles per second, selecting from these the ones to be purified. Unfortunately, this processing rate still makes it extremely time consuming to purify bulk quantities of particles, such as human chromosomes of a single type, or to sort infrequently occurring particles, such as mutant erythrocytes. For example, with a conventional cell sorter processing 2,000 objects/second, over 120 hours of sorting time is required to purify 1 μg of DNA from the human Y chromosome [12] and almost 140 hours would be required to sort 100 mutant erythrocytes from a population in which the mutant cell frequency was 10^{-7}. This severely limits the utility of conventional sorting for the detection and isolation of rare event particles, and the utility of chromosome sorting for gene mapping or for production of recombinant DNA libraries. The time required for such tasks can be reduced five to eight fold at the same sorting efficiency by the use of a high speed (rather than a conventional) cell sorter.

The high-speed sorter differs from a conventional cell sorter mainly in an increased droplet production rate. Maintaining high purity sorted populations requires that the probability of having more than one particle per droplet must be low. Figure 17 shows the effect of input particle rate and droplet rate on sorting efficiency when 3 droplets are sorted and coincidence checked for each particle. Assuming an 80% sorting efficiency, the following comparison can be made. Conventional sorters, with droplet production rates from 30,000 to 40,000 per second, would be limited to particle processing rates of around 2,000 to 3,000 particles per second, respectively. The high speed sorter, with a droplet production rate of 220,000 per second, would be limited to somewhat over 16,000 particles per second. Higher processing rates for either type of sorter reduces the efficiency of sorting the desired particles to less than 80%.

The purpose of this portion of the chapter is to describe some of the theoretical, design, and developmental aspects of the high speed sorter that differentiate it from conventional cell sorters and to discuss its use in the above two examples of purifying human chromosomes for recombinant DNA libraries and rare event sorting. The high speed sorter, which was designed and developed at the University of California Lawrence Livermore National Laboratory, will be referred to as HiSS [24,51]. Some of the high-speed sorter description has been adapted from material published in ref. [50].

System Design Considerations

Theory. The nozzle diameter is a compromise between a small size for increased droplet frequency [eqs. (4) and (5)], and one large enough to produce little damage to cells or chromosomes and that, with some care, will rarely become plugged. An 80-μm-diameter nozzle is used on HiSS. Experiments indicated that stable droplet production could be achieved at droplet production frequencies up to 500 kHz by increasing the jet velocity to 100 m/sec. However, the droplet production stability was marginal at 500 kHz and precise droplet charging was impossible. Thus, somewhat arbitrarily, an intermediate velocity of 50 m/sec was selected for routine operation. The diameter of the jet emanating from the 80-μm diameter orifice is approximately 65 μm. With these data, eq. (5) would predict a frequency of 170 kHz for minimum jet length, and eq. (4) a maximum frequency of 240 kHz. HiSS operates at 220 kHz, close to the maximum, in order to maximize the particle processing rate.

According to Bernoulli's equation [eq. (11)] (Chapter 3, this volume), the velocity of the sorter jet is related to the pressure according to the equation

$$v = (2P/\rho)^{1/2} \tag{15}$$

where P is the water pressure and ρ is the density of the sheath liquid. Guided by this equation, the operating pressure on the sheath is adjusted to 14 kg/cm^2 (200 psi) to produce a 50 m/sec jet.

Flow chamber design, droplet charging, and deflection. The design of the flow chamber and nozzle assembly is critical for stable droplet production which again is a requirement for accurate sorting. Figure 20A shows the flow chamber design. A key feature of this chamber is the length to diameter ratio of the orifice (shown in the expanded view) forming the jet-in-air. The stability of droplet formation increases significantly as this ratio is reduced, thereby minimizing the turbulence induced by the interaction between the flowing liquid and the orifice walls [23]. The actual orifice is a synthetic ruby jewel through the center of which is a hemispherical depression centered on a 80-μm diameter hole. The side of the jewel opposite the hemispherical depression is polished to reduce the length of the 80-μm diameter region to about 40 μm. The jewel is expoxied, hemispherical depression down, into a cylindrical recess in the end of a square quartz flow chamber with a 250-μm square flow channel [52]. Cytometric measurements are performed while the particles are in the square cross section part of the chamber. This is advantageous because the high optical quality of this part of the chamber allows effective use of high quality optical components for laser beam shaping and fluorescence collection, and because the jet disturbances induced by the piezoelectric transducer to force accurate droplet production do not affect the dual beam flow cytometric measurements. In addition, the flow velocity in this region is only about one tenth of that in the jet-in-air. The relatively low flow velocity allows the cells to remain in the measurement region longer so more accurate measurements of total particulate fluorescence are possible.

The entire flow chamber is caused to vibrate at approximately 220 kHz by a piezoelectric crystal mounted axially on the top of the chamber, causing the jet to break into droplets of equal size and spacing. When particles to be sorted reach the end of the liquid jet, a 150-volt pulse is applied to the liquid jet for a time equal to that required for one or a few droplets to break off. Each droplet that breaks off during this time has a charge proportional to its capacitance to ground and to the voltage at the end of the jet. The droplets then continue downward through an electric field. The charged droplets containing particles of interest are deflected into a collection device. In HiSS, two classes of particles can be sorted by applying a negative charge to droplets containing one class and a positive charge to droplets containing the other class.

Droplet charging and deflection in HiSS is complicated by the shape of the jet at the breakoff point and by the velocity of the drops during passage through the deflecting electric field. Figure 21 shows a picture of the liquid jet in the vicinity

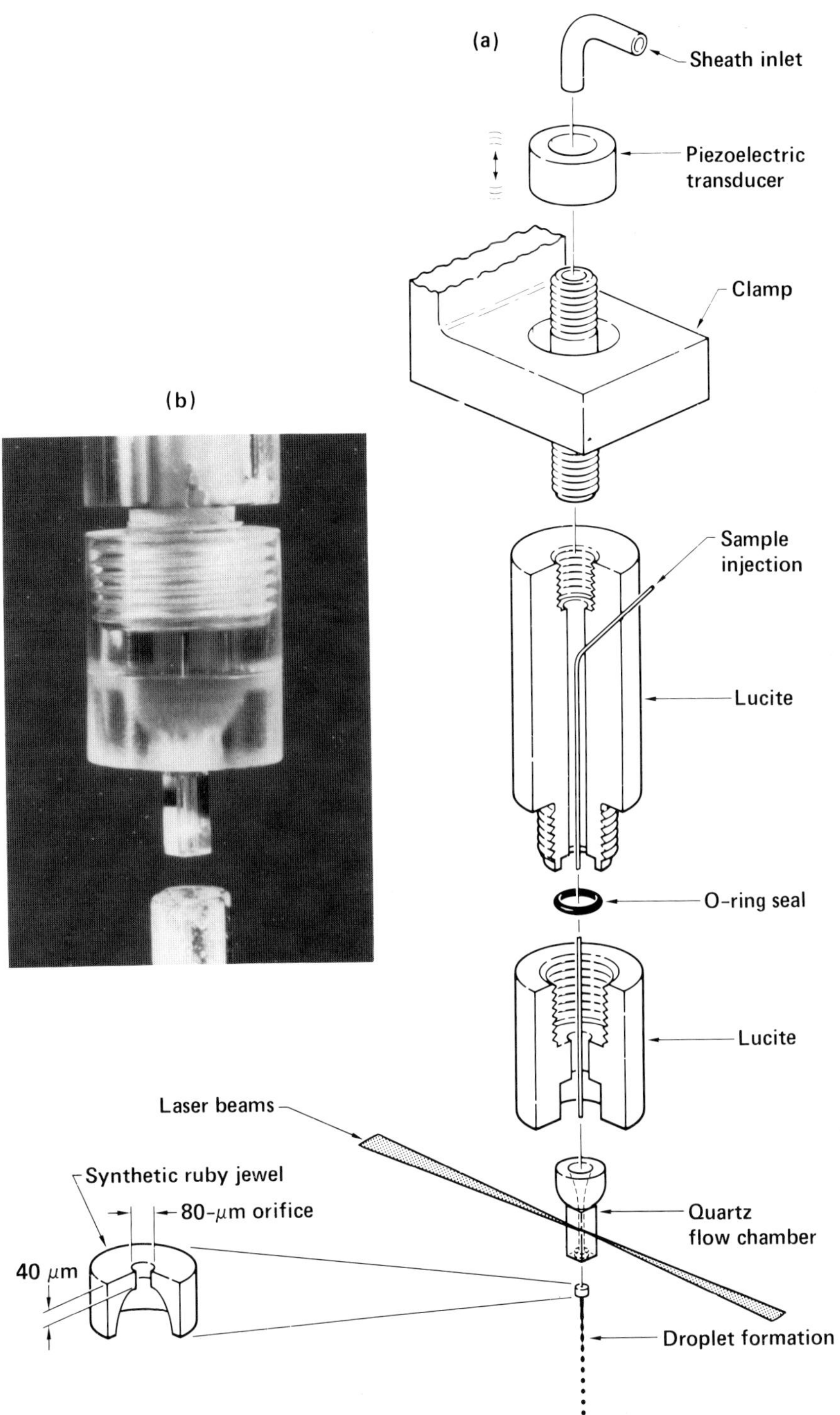

Fig. 20. Diagram of the flow chamber of the high speed cell sorter (HiSS). **a:** Sheath liquid is injected axially through a tube and through a piezoelectric crystal mounted on top of the flow chamber. The liquid flows downward into a tapered quartz section where the sample is injected. Particles are entrained in the liquid and flow in single file through a 250-μm square quartz channel where they intersect the two laser beams, vertically separated by 50 μm. The fluid then exits through a 80-μm diameter orifice. The enlargement shows the orientation and shape of the orifice that is formed by a synthetic ruby jewel, polished to produce an orifice length to diameter ratio of 0.5 and inserted cone side down. **b:** Shown here is a photograph of the lower portion of the flow chamber. Also shown is the upper portion of the grounding collar through which the liquid jet passes. (Reproduced from ref. [50], with permission of the publisher.)

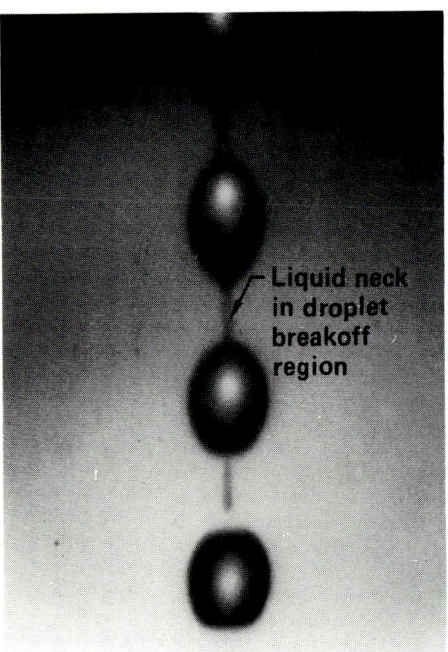

Fig. 21. The liquid jet at the droplet breakoff region. This photograph shows the long, thin neck that connects droplets prior to breakoff (cf. Fig. 6). The jet was illuminated by a pulsed light emitting diode (LED) driven at the droplet production frequency of 220 kHz. (Reproduced from ref. [50], with permission of the publisher.)

of the droplet breakoff region. This region is approximately 2 cm downstream from the point where the particles are measured and the transit time from the measurement region to the droplet breakoff point is about 500 μsec. The most important differences between HiSS and conventional sorters (see Fig. 6) in this region are: 1) the extended thin liquid neck between droplets in HiSS (see arrow in Figure 21) and 2) the higher jet velocity. The diameter of the jet in the neck region is only about 2 μm, which is sufficiently small to begin to interfere with the induction of charge on the separating droplets, and thereby interfering with proper droplet deflection during sorting. At higher jet velocity and droplet production frequencies, the neck diameter decreases still further and eventually prevents proper droplet charging. The high jet velocity results in rapid movement of the charged droplets through the deflecting DC electric field used to separate charged and uncharged droplets. To obtain sufficient droplet deflection, the deflection plate length was increased from the conventional length of about 4 cm to approximately 18 cm. In addition, the potential difference between the plates was set at 10 kV to maximize the deflecting force on each charged droplet. The plate separation is greater than 7 mm to minimize arcing between the plates.

Overall system. The overall high speed sorter system is diagrammed in Figure 22. The sorter is equipped with two argon ion lasers. To excite the two fluorescent dyes used for chromosome sorting, one beam is adjusted to operate in the visible portion of the spectrum, the other in the UV range. The minor dimension of the elliptical focus of the visible beam is about 7 μm, minimizing the possibility of multiple objects passing through the beam at the same time. The minor dimension of the UV beam is about 11 μm at focus, and the major dimensions of the two beam loci are 270 and 200 μm, respectively. The fluorescence and scattered light from the two measurement foci are imaged by a 40×, 0.60 N.A. microscope objective through two pinholes, one for each beam intersection point, and passed to two separate detector assemblies, each consisting of a beam splitter and a pair of photomultipliers with barrier filters. A lens in each of the detector assemblies forms an image of the back focal plane of the fluorescence pickup objective lens on the photomultiplier tube. This minimizes the movement of light on the photomultiplier tube photocathode (which may have a position sensitive response) due to the movement of particles through the laser beam.

The particles to be sorted are introduced onto the axis of the system flow chamber from a motor-driven syringe. They are surrounded with a conductive sheath liquid flowing from a stainless steel sheath container pressurized to 14 kg/cm^2 with dry nitrogen. The sheath stream carries the particles in the sample core through the square cross section flow channel (see Fig. 20A,B), where they are illuminated with the two laser beams.

The droplet breakoff region of the liquid jet is illuminated by a pulsed (200-nsec duration) light emitting diode driven at the droplet frequency of 220 kHz. The image of the stroboscopically illuminated jet is collected by a lens system and projected into a video camera to permit visual monitoring of the jet length stability on a video monitor. A ground reference collar, seen at the bottom of Figure 20B, surrounds most of the liquid jet between the end of the flow chamber and the droplet breakoff point. It shields the breakoff point from the electrostatic field of the deflection plates and increases the capacitive coupling of the jet to ground, thereby increasing the induced charge on drops separating from the jet during the application of the 150-volt charging pulse.

Electronics. The photomultiplier tubes convert the fluorescence intensity of each particle passing through the two laser beams into pulses that are amplified and actively integrated. The integrator is turned on and off by a threshold detector. The signal from one amplifier's threshold detector is designated as the system master signal and is assumed to be produced by every particle passing through the system. This signal is used to sense closely spaced objects and to inhibit sorting when an undesired object might contaminate a sort event. Signals from all active amplifiers are routed to window discriminators which are used to select particles to be sorted. Window discriminators, rather than a pulse height analyzer, are used to determine which events will be sorted in order to increase the rate at which objects can be processed. At the same time, the signals are routed to the analog-to-digital converter (ADC) inputs of a pulse height analyzer for digitization and accumulation into one-parameter or two-parameter distributions. At present, the system can analyze and sort on a maximum of two signals. The settings of the discriminators used to select particles for sorting are read by a computer and are transferred to the analyzer where they are displayed as intensified regions superimposed on the accumulating two-parameter distributions. Thus, the sorting windows selected by the discriminators are clearly visible on the two-parameter distributions thereby facilitating accurate window setting. The computer also is used for setting the delay between particle detection and droplet charging and for selecting the number of droplets to be charged per sorting event.

A new computer-based pulse-height analysis system is being designed which will make use of bit-mapped, or lookup table, sorting regions, rather than the present use of window

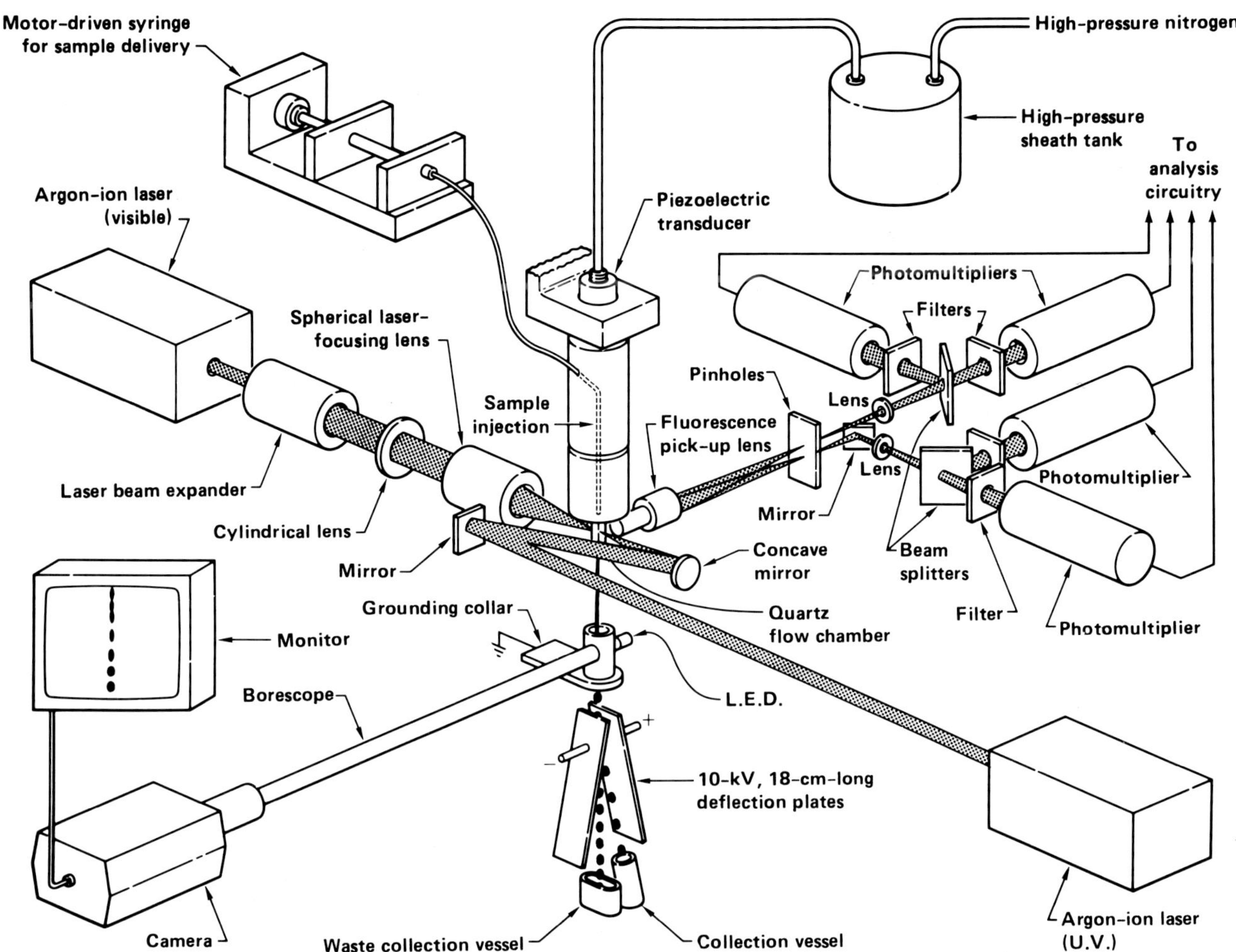

Fig. 22. Schematic diagram of HiSS. Shown are the principle system components, described in the text. (Reproduced from ref. [50], with permission of the publisher.)

discriminators to define sorting regions. The present system is confined to rectangular regions, while a lookup table based system will permit complex shapes, such as sort regions defined by contour levels on a two-parameter histogram peak. This will also permit for automation of the sort region generation and movement of the regions to track changes in the histogram features.

Performance Evaluations

A number of studies have been conducted to determine the operating characteristics of the high speed sorter. Specifically, its total fluorescence measurement resolution at several analysis and sorting rates, its sorting efficiency, and its physical effects on chromosomes, fixed cells and living cells have been determined.

System resolution. The total fluorescence measurement resolution of the system was estimated using 1.8-μm diameter fluorescent microspheres (Polysciences, Warrington, PA). One-parameter fluorescence distributions were measured for these microspheres at several analysis rates ranging from 2,500 cells/sec to 46,000 cells/sec. The diameter of the sample core was kept constant during these experiments and the measurement rate was varied by changing the concentration of the microspheres. Thus, these data illustrate the effect of increased measurement rates on the system resolution. The coefficients of variation increased from 1.2% to 1.4% as the rate increased from 2,500 to 46,000 microspheres/sec. The microspheres were excited at 458 nm (0.1 W) and fluorescence was collected through a colored glass filter (3-71; Corning Glass Works, Corning, NY), transmitting fluorescence at wavelengths longer than 520 nm, for these measurements.

The diameter of the sample core also affects the total fluorescence measurement resolution. The magnitude of this effect was estimated by analyzing fluorescence distributions for microsphere samples measured at several sample core diameters. The sample concentrations were adjusted during these studies to keep the measurement rate constant at about 2,500 c/sec. The CV increased from 1.1 to 1.2% as the sample stream diameter increased from 3.7 to 8.3 μm.

Recovery of sorted objects. The ultimate test of any sorter is whether objects of known characteristics can be selected accurately for sorting and effectively recovered. These aspects of HiSS were tested by counting the number of 1.8-μm microspheres collected on millipore filters after sorting known numbers (typically 20 to 50) directly onto the filters (0.8-μm pore size). The measured sorting efficiency ranged from 95 to 100% for both 1-droplet and 3-droplet sorting. The efficiency of the system was also evaluated by sorting microspheres onto microscope slides coated with a thin layer of grease. The slide was moved between each sort so that each deflected drop (the droplets of a 3-droplet sort appear to merge together) struck the slide in a new location. Each deflected drop left an impact crater in the grease thus allowing correlation of the sorting event (i.e., the appearance of a crater) with a proper sort (i.e., with the appearance of a fluorescent microsphere in the crater). On average, the ratio of craters to sort pulses was greater than 0.95 and, the ratio of microspheres to craters was also greater than 0.95. It was necessary, however, to insure good separation (7 mm) between the long deflection plates and the deflected stream to minimize the possibility of a charged droplet striking and adhering to a deflection plate.

There was also concern about the possible loss of sorted material due to splashing while sorting the high velocity drops into a test tube. The splashing loss was determined by comparing the mass of liquid collected while sorting a fixed number of drops with that expected. Using a 3-drop sort, the average number of sorted events needed to collect 1 g of liquid was 391,000. The variation was from −7,000 to +8,000 over 16 experiments. The expected mass of a droplet in HiSS was determined to be $8.5 \cdot 10^{-7}$ g by collecting a volume of sheath from the liquid jet for a preset length of time and dividing its weight by the number of droplet periods during that time. Thus, 394,000 events should produce a gram of liquid with a 3 drop sort. This value is in good agreement with the actual number per gram determined during sorting, indicating that splashing in the test tube is not a problem.

The ability to sort rare events in a concentrated mixture of unstained objects was also tested. Human erythrocytes fluorescently stained using a two step antibody procedure with the primary antibody specific for hemoglobin [5] were mixed with unstained human erythrocytes at a ratio of 1 to 10^7 and processed through HiSS at a rate of 10^6 per sec. In this mode of operation, the system did not "see" the nonfluorescent objects. In each of seven different trials, 100 fluorescent cells were sorted onto hot microscope slides and counted microscopically. Heating the slides speeded the evaporation of the sorted water and reduced the loss of cells by splashing. The fluorescent cell recovery was 88 ± 8% for the seven trials. Thus, the recovery was not significantly affected by the presence of unstained erythrocytes in most or all of the droplets.

Effects of sorting. The size of DNA fragments from isolated chromosomes before and after sorting was analyzed by agarose gel electrophoresis. Sorted and unsorted chromosomes stained with Propidium Iodide were pelleted by centrifuging at 40,000*g* for 30 min at 4°C. The supernatant was removed and the chromatin was hydrolyzed with proteinase K overnight at 37°C. The proteins were then extracted with phenol and chloroform and the purified DNA was loaded onto a 0.4% agarose gel for electrophoresis. The sorted chromosomes did not have shorter fragments than the unsorted ones. The DNA sizes ranged from 220 kb to greater than 500 kb for both the sorted and unsorted chromosomes. Thus, the chromosomal DNA was not damaged during passage through the sorter nozzle or upon impact in the collection vessel.

Ethanol-fixed Chinese hamster ovary (CHO) line cells were stained with the DNA specific dye CA3 and processed through the high speed sorter where all cells were sorted. The sorted cells were collected and processed through a conventional flow cytometer to determine the effect of the sorting on the total fluorescence distribution. The fluorescence distributions measured for cells before and after passage through HiSS were essentially identical, suggesting that HiSS does not disrupt fixed CHO line cells. Microscopic examination also confirmed that the sorted cells had the same morphology as unsorted ones.

Living CHO cells were also processed through HiSS. Exponentially growing CHO cells were resuspended in PBS and sorted according to their 90° light scatter intensity. Replicate sorts of 100 cells were made into petri dishes containing 5 ml MEM-a plus 10% FCS. These cultures were grown for 7 days at 37°C and then scored for colony formation. The average number of colonies per plate was 70.4 ± 9.6. This compares with about 85 colonies/plate in unsorted cells. A Coulter volume distribution of the sorted cells appeared normal. Thus, high speed sorting appears to cause little damage to living CHO cells. However, one attempt to culture colony forming units from murine bone marrow following high speed sorting was unsuccessful. This has not been pursued further. It is important to note that the sample must be injected with a syringe rather than from a gas-pressurized container. Living cells cannot be obtained after high speed sorting if the sample is pressurized with gas. The gas goes into solution at high pressure and is picked up by the living cells. These cells then rupture as they exit the sorting nozzle, when the dissolved gas comes back out of solution.

Experience

HiSS has been used by our laboratory for the sorting of individual human chromosomes for the National Gene Library Project [66]. Since the machine operates five- to eightfold faster than conventional sorters, it greatly reduced the time required to sort enough chromosomes for cloning. The number of sorted chromosomes required to provide enough DNA for one cloning experiment was estimated at about 10^6. This number was based on previous experience of other laboratories [66], and has turned out so far to be a reasonable estimate, although, as cloning efficiency improves, fewer chromosomes will be required. Since not all cloning attempts are successful, a safety factor of 4 was built into the sorting requirements, so that $4 \cdot 10^6$ chromosomes of each type was established as the target. Sorting time requirements for autosomes from human diploid fibroblast cell lines could then be estimated at about 9 working days for the conventional speed sorters and one day for HiSS. Experience over the past few years has generally followed these initial estimates. The course of an actual sorting run depends greatly on the quality and quantity of the chromosome preparation and on ease of machine alignment, stability, and freedom from nozzle plugs.

APPLICATION AREAS FOR FLOW CYTOMETRIC CELL SORTING

The unique features of flow cytometric cell sorting are the general applicability of the method and the precision by which it can perform. Cells can be sorted on the basis of any measurable quantity such as fluorescence, absorption, or light scattering, and be delivered in known numbers into exact locations. The processing speed of several thousand cells per second may seem impressive, but nevertheless is the weakest aspect of the technique, making it much slower than any bulk separation technique. The most important technical improvement in flow cytometric cell sorting over the 15 years since the technique was introduced, has therefore been the development of high speed cell sorting which has made possible an increase in processing rate of almost a factor of 10 for suitable samples.

The most general application of flow cytometric cell sorting is for identification of a particular subpopulation of cells defined by the flow cytometric analysis. Definition of the interesting subpopulation can be quite complex, based on multiparameter measurements and ratios or other derived measures of the parameters. For identification of cells or studies of morphology, the required number of sorted cells will usually be low, but other requirements such as need for intact morphology or viability of the sorted cells may have to be considered. An interesting variant of this approach is the so-called indexed sorting [63], i.e., sorting of individual cells into defined locations on a slide or in a microwell plate, with recording of the whole set of measurement values for each sorted cell along with its location or "address" in the receiving plate. In this case, the receiving plate is moved to a new position between each individual sort so that the sorted cells can be reprocessed afterward, e.g., by image cytometry.

Flow cytometric cell sorting offers the possibility of correlating flow cytometric quantities with various measurements based upon incorporation of radioactively labeled precursors [6–8,26,39,53]. This constitutes a powerful technique for correlating measurements of biochemical quantities by fluorescence staining with biochemical synthesis rate measurements based upon pulse labelling with radioactive precursors. In the field of DNA synthesis measurements, the need for cell sorting has, however, been elegantly eliminated by the use of fluorescent antibody staining for detecting BrdUrd incorporated after pulse labeling [21,22] (see Chapter 23, this volume).

Immunofluorescence is one of the major application areas of flow cytometry in general, but also for flow cytometric cell sorting. Leukocyte phenotyping with monoclonal antibodies has become perhaps the most widely recognized application field of flow cytometry, and flow cytometric cell sorting is playing an important role in the ongoing endeavour to correlate leukocyte phenotype subsets with functional properties of isolated subpopulations [36] (see Chapter 33, this volume). These studies have very important parallels in pathology, e.g. in the study of leukemias and AIDS-related disorders [43] (see Chapter 34, this volume).

In immunology the sorting criterion is sometimes qualitative rather than quantitative, i.e., requiring the separation of one well-defined class of cells from the other, where there is no significant overlap between the two classes. In such cases flow cytometry has been challenged by the technique of magnetic cell separation based on iron-containing microspheres labeled with the appropriate cell-binding antibodies (see Chapter 19, this volume). Magnetic cell separation offers much higher throughput rates than flow cytometric cell sorting, and is presently being investigated as a means of purifying autologous bone marrow for transplantation to cancer patients after high-dose radiotherapy. Although the cell sorting in this case is accomplished by magnetic separation, flow cytometry remains an important tool for analyzing the result of the separations.

Flow cytometry has proved a useful technique for microbiological strain improvement work in biotechnology [3]. By flow cytometric analysis and indexed sorting, the cellular property of interest can be measured and cells sorted out directly as a function of the interesting parameter. In some cases, flow cytometric cell sorting has been successfully used even if the desirable feature is not directly measurable by flow cytometry. The cells are then fractionated on the basis of some available correlated property, e.g., axial light loss or differential light scattering, and the sorted fractions later screened for the interesting product [3].

One inevitable requirement in applications of flow cytometric cell sorting for variant selection, strain improvement work or other functional studies, is that the analysis and sorting procedures must not compromise cell viability. Differential light scattering may be used to identify subpopulations for cell sorting, and immunofluorescence against cell surface markers causes no viability problems, but there are also vital staining techniques developed for intracellular components. Thus, although most DNA staining techniques are incompatible with cell viability, viable cell sorting on the basis of DNA staining with Hoechst 33342 has been achieved [28,55,58,61]. Other cell sorting applications based on vital fluorescence staining have used fluorescamine, acridine orange, and fluorogenic substrates such as fluorescein diacetate [28,38,61,68]. In cases in which the feature of interest is a cellular enzyme, a fluorogenic substrate makes an ideal marker. However, one may encounter a problem if the product, in this case the fluorescent label, leaks out of the cell as it is being produced. An interesting approach to the solution of this problem is the encapsulation of individual cells in microscopic agarose spheres. The fluorescent cellular product is retained in these spheres, which may be analyzed in the flow cytometer and subjected to sorting to isolate the better producers [56].

Flow cytometry has established itself as a very useful technique in molecular genetics, both as a preparative tool for making chromosome-specific DNA libraries [66] and as an elegant technique for mapping of chromosome loci by so-called spot-blot hybridization [4,11,27,37]. Genetic applications of flow cytometry and sorting are described in detail in Chapter 25 (this volume) and have also recently been reviewed [25].

Applications of flow cytometric cell sorting will continue to evolve. They will range from large projects like the chromosome sorting programs at Livermore and Los Alamos, through industrial applications like cell line and microbiological strain improvement programs, to applications in medicine and pure biological research. Flow cytometric cell sorting, although usually not the main methodology of an investigation, often allows the investigator to attack his problem in a more direct way.

REFERENCES

1. **Alberti S, Stovel R, Herzenberg LA (1984)** Preservation of cells sorted individually onto microscope slides with a fluorescence-activated cell sorter. Cytometry 5:644–647.
2. **Arndt-Jovin DJ, Jovin TM (1974)** Computer-controlled multiparameter analysis and sorting of cells and particles. J. Histochem. Cytochem. 22:622–625.
3. **Betz JW, Aretz W, Haertel W (1984)** Use of flow cytometry in industrial microbiology for strain improvement programs. Cytometry 5:145–150.
4. **Bianchi DW, Harris P, Flint A, Latt SA (1987)** Direct hybridization to DNA from small numbers of flow-sorted nucleated newborn cells. Cytometry 8:197–202.
5. **Bigbee WL, Branscomb EW, Weintraub HB, Papayannopoulou T, Stamatoyannopoulos G (1981)** Cell sorter immunofluorescence detection of human

erythrocytes labeled in suspension with antibodies specific for hemoglobin S and C. J. Immunol. Methods 45:117–127.

6. **Blair OC, Burger DE, Sartorelli AC (1982)** Analysis of glycosaminoglycans of flow sorted cells: Incorporation of ^{35}S-sulfate and ^{3}H-glucosamine into glycosaminoglycans of B16-F10 cells during the cell cycle. Cytometry 3:166–171.
7. **Blair OC, Roti Roti JL (1980)** Incorporation of ^{3}H-leucine, ^{3}H-lysine and ^{3}H-tryptophan during the cell cycle of Chinese hamster ovary cells. J. Histochem. Cytochem. 28:487–492.
8. **Bloch DP, Fu C-T, Dean PN (1981)** DNA and histone synthesis rate change during the S-period in Ehrlich ascites tumor cells. Chromosoma (Berl.) 82:611–626.
9. **Bonner WA, Hulett HR, Sweet RG, Herzenberg LA (1972)** Fluorescence activated cell sorting. Rev. Sci. Instr. 43:404–409.
10. **Chaudhary KC, Maxworthy T (1980)** The non-linear capillary instability of a liquid jet. Part 3. Experiments on satellite drop formation and control. J. Fluid Mech. 96:287–298.
11. **Collard JG, de Boer PAJ, Janssen JWG, Schijven JF, de Jong B (1985)** Gene mapping by chromosome spot hybridization. Cytometry 6:179–185.
12. **Cremer C, Gray JW, Ropers HH (1982)** Flow cytometric characterization of a Chinese hamster × man hybrid cell line retaining the human Y chromosome. Hum. Genet. 60:262–266.
13. **Crosland-Taylor PJ (1953)** A device for counting small particles suspended in a fluid through a tube. Nature (Lond.) 171:37–38.
14. **Dühnen J, Stegemann J, Wiezorek C, Mertens H (1983)** A new fluid switching sorter. Histochemistry 77:117–121.
15. **Folstad L, Look M, Pallavicini M (1982)** A polycarbonate filter technique for collection of sorted cells. Cytometry 3:64–65.
16. **Freyer JP, Wilder ME, Jett JH (1987)** Viable sorting of intact multicellular spheroids by flow cytometry. Cytometry 8:427–436.
17. **Friedman M (1973)** Digital fluidic amplifier particle sorter. US Patent No. 3.791.517.
18. **Fulwyler MJ (1965)** Electronic separation of biological cells by volume. Science 150:910–911.
19. **Fulwyler MJ, Glascock RB, Hiebert RD, Johnson NM (1969)** Device which separates minute particles according to electronically sensed volume. Rev. Sci. Instr. 40:42–48.
20. **Goehde H, Schumann J (1987)** Fluidic cell sorter. European Patent EP 0177.718.
21. **Gratzner H (1982)** Monoclonal antibody to 5-bromo and 5-deoxyuridine: A new reagent for detection of DNA replication. Science 218:474–475.
22. **Gray JW (guest ed) (1985)** Monoclonal antibodies against bromodeoxyuridine. Cytometry 6:499–674.
23. **Gray JW, Alger TW, Lord DE (1982)** Fluidic assembly for an ultra-high-speed chromosome flow sorter. US Patent No. 4.361.400.
24. **Gray JW, Dean PN, Fuscoe JC, Peters DC, Trask BJ, van den Engh GJ, Van Dilla MA (1987)** High speed chromosome sorting. Science 238:323–329.
25. **Gray JW, Langlois RG (1986)** Chromosome classification and purification using flow cytometry and sorting. Annu. Rev. Biophys. Biophys. Chem. 15: 195–235.
26. **Gray JW, Pallavicini MG, George YS, Groppi V, Look M, Dean PN (1981)** Rates of incorporation of radioactive molecules during the cell cycle. J. Cell. Physiol. 108:135–144.
27. **Grunwald D, Geffrotin C, Chardon P, Frelat G, Vaiman M (1986)** Swine chromosomes: Flow sorting and spot blot hybridization. Cytometry 7:582–588.
28. **Hamori E, Arndt-Jovin DJ, Grimwade BG, Jovin TM (1980)** Selection of viable cells with known DNA content. Cytometry 1:132–135.
29. **Harkins KR, Galbraith DW (1987)** Factors governing the flow cytometric analysis and sorting of large biological particles. Cytometry 8:60–70.
30. **Hinson WG, Pipkin JL, Hudson JL, Anson JF, Tyrer H (1982)** A microsample collection device for electrostatically sorted cells or particles and its preparative use for biomedical analysis. Cytometry 2:390–394.
31. **Hulett HR, Bonner WA, Sweet RG, Herzenberg LA (1973)** Development and application of a rapid cell sorter. Clin. Chem. 19:813–816.
32. **Jett JH, Alexander RG (1985)** Droplet sorting of large particles. Cytometry 6:484–486.
33. **Keur RI, Stone JJ (1976)** Some effects of fluid jet dynamics on ink jet printing. IEEE Trans. Ind. Appl. IA-12:86–90.
34. **Kachel V (1976)** Basic principles of electrical sizing of cells and particles and realization in the new instrument "Metricell." J. Histochem. Cytochem. 24:211–230.
35. **Kamentsky LA, Melamed MR (1967)** Spectrophotometric cell sorter. Science 156:1364–1365.
36. **Lanier LL, Engleman EG, Gatenby P, Babcock GF, Warner NL, Herzenberg LA (1983)** Correlation of functional properties of human lymphoid cell subsets and surface marker phenotypes using multiparameter analysis and flow cytometry. Immunol. Rev. 74: 143–160.
37. **Lebo RV, Tolan DR, Bruce BD, Cheung M-C, Kan YW (1985)** Spot-blot analysis of sorted chromosomes assigns a fructose intolerance disease locus to chromosome 9. Cytometry 6:478–483.
38. **Lessin SR, Abraham SR, Nicolini C (1982)** Biophysical identification and sorting of high metastatic variants from B16 melanoma tumor. Cytometry 2:407–413.
39. **Lindmo T (1986)** Protein synthesis as a function of protein content in exponentially growing NHIK 3025 cells studied by flow cytometry and cell sorting. Exp. Cell Res. 133:237–245.
40. **Lindmo T, Davies C, Fodstad Ø, Morgan AC (1984)** Stable quantitative differences of antigen expression in human melanoma cells isolated by flow cytometric cell sorting. Int. J. Cancer 34:507–512.
41. **Lindmo T, Fundingsrud K (1981)** Measurement of the distribution of time intervals between cell passages in flow cytometry as a method for the evaluation of sample preparation procedures. Cytometry 2: 151–154.
42. **Lord Rayleigh (1879)** On the capillary phenomena of jets. Proc. R. Soc. Lond. 29:71–97.
43. **Lovett EJ, Schnitzer B, Keren DF, Flint A, Hudson JL, McClatchey KD (1984)** Application of flow cytome-

try to diagnostic pathology. Lab. Invest. 50:115–140.

44. **McCutcheon MJ, Miller RG (1982)** Flexible sorting decision and droplet charging control electronic circuitry for flow cytometer cell sorters. Cytometry 2: 219–225.
45. **Meck RA, Benson NA, Ng AB, Brandon JP, Ingram M (1980)** Colloidon membrane secures cells sorted by flow cytofluorometry onto microscopic slides. Cytometry 1:84–86.
46. **Merrill JT (1981)** Evaluation of selected aerosol-control measures on flow sorters. Cytometry 1:342–345.
47. **Merrill JT, Dean PN, Gray JW (1979)** Investigations in high-precision sorting. J. Histochem. Cytochem. 27:280–283.
48. **Metezeau P, Bernheim A, Berger R, Goldberg ME (1984)** A simple device to obtain high local concentrations of material sorted by flow cytometry for biochemical or morphological analysis. Cytometry 5: 550–552.
49. **Patrick CW, Keller RH (1984)** A simple device for the collection of cells sorted by flow cytometry. Cytometry 5:308–311.
50. **Peters DC (1989)** Chromosome purification by high speed sorting. Gray JW (ed), In "Flow Cytogenetics." London: Academic Press, pp 211–224.
51. **Peters D, Branscomb E, Dean PN, Merrill T, Pinkel D, Van Dilla M, Gray JW (1985)** The LLNL high-speed sorter: Design features, operational characteristics, and biological utility. Cytometry 6:290–301.
52. **Pinkel D, Stovel R (1985)** Flow chambers and sample handling. In MA Van Dilla, PN Dean, OD Laerum, MR Melamed (eds), "Flow Cytometry: Instrumentation and Data Analysis." London: Academic Press, pp 77–128.
53. **Pipkin JL, Anson JF, Hinson WG, Schol H, Burns ER, Casciano DA (1986)** Analysis of protein incorporation of radioactive isotopes in the Chinese hamster ovary cell cycle by electronic sorting and gel microelectrophoresis. Cytometry 7:147–156.
54. **Pipkin JL, Hinson WG, Hunziker J (1980)** A collection device for small electronically sorted samples from flow cytometers. Anal. Biochem. 101:230–237.
55. **Rice GC, Dean PN, Gray JW, Dewey WC (1984)** An ultra-pure in vitro phase synchrony method employing centrifugal elutriation and viable flow cytometric cell sorting. Cytometry 5:289–298.
56. **Sahar E, Nir R, Yisraeli Y, Shabtai Y, Lamed R (1987)** A system for microbal strain improvement by product measurement and flow sorting of microcolonies. Cytometry 1(suppl.):93 (abst 599).
57. **Savart F (1833)** Memoire sur la constitution des veines liquides lancées par des orifices circulaires en mince paroi. Ann. Chim. Phys. 53:337–386.
58. **Schaap GH, Verkerk A, Vijg J, Jongkind JF (1983)** Flow sorting in the study of cell–cell interaction. Cytometry 3:408–413.
59. **Schneider JM, Lindblad NR, Hendricks CD, Crowley JM (1967)** Stability of an electrified liquid jet. J. Appl. Phys. 38:2599–2605.
60. **Steinkamp JA, Fulwyler MJ, Coulter JR, Heibert RD, Horney JL, Mullaney PF (1973)** A new multiparameter separator for microscopic particles and biological cells. Rev. Sci. Instr. 44:1301–1310.
61. **Storkus WJ, Balber AE, Dawson JR (1986)** Quantitation and sorting of vitally stained natural killer cell-target cells conjugates by dual beam flow cytometry. Cytometry 7:163–170.
62. **Stovel RT (1977)** The influence of particles on jet breakoff. J. Histochem. Cytochem. 25:813–820.
63. **Stovel RT, Sweet RG (1979)** Individual cell sorting. J. Histochem. Cytochem. 27:284–288.
64. **Sweet RG (1964)** High frequency oscillography with electrostatically deflected ink jets. Stanford Electronics Laboratory Technical Report No. 1722-1.
65. **Sweet RG (1965)** High frequency recording with electrostatically deflected ink jets. Rev. Sci. Instr. 36: 131–136.
66. **Van Dilla MA, Deaven LL, Albright KL, Allen NA, Aubuchon MR, Bartholdi MF, Browne NC, Campbell EW, Carrano AV, Clark LM, Cram LS, Fuscoe JC, Gray JW, Hildebrand CE, Jackson PJ, Jett JH, Longmire JL, Lozes CR, Luedemann ML, Martin JC, McNinch JS, Meincke LJ, Mendelsohn ML, Meyne J, Moyzis RK, Munk AC, Perlman J, Peters DC, Silva AJ, Trask BJ (1986)** Human chromosome-specific DNA libraries: Construction and availability. Biotechnology 4:537–552.
67. **Wiezorek C (1984)** Cell cycle dependence of Hoechst 33342 dye cytotoxicity on sorted living cells. Histochemistry 81:493–495.
68. **Wilson JS, Steinkamp JA, Lehnert BE (1986)** Isolation of viable type II alveolar epithelial cells by flow cytometry. Cytometry 7:157–162.
69. **Zöld T (1979)** Method and device for sorting particles suspended in an electrolyte. US Patent No: 4.175.662.

9

Commercial Instruments

Phillip N. Dean
Biomedical Sciences Division, Lawrence Livermore National Laboratory, Livermore, California 94550

INTRODUCTION

This chapter is intended to provide an overview of the variety of flow instruments available commercially. Characteristics of the various instruments are presented in tabular form, grouped by instrument function. The term flow cytometer includes a large array of devices, some of them having nothing to do with analytical cytology. Many kinds of flow cytometers have been manufactured by many companies for over 30 years, many of them available for only a short time. This chapter will discuss only those instruments that are commercially available at the time of this writing, as reported to the author by the manufacturers. Some instruments that have been advertised, particularly at international conferences, are not reported here due to lack of information from their manufacturers. Other instruments will probably come into existence after this book is published.

There are two major classes of instruments: flow cytometers and cell sorters. A flow cytometer is defined as a device that measures components and properties of cells and cell organelles that are suspended in a flowing liquid suspension. Cell sorters have the same capabilities as flow cytometers with the additional capability of selectively removing objects (e.g., cells, chromosomes) from the liquid suspension. Both classes of instruments operate at high measurement rates, typically several thousand per second. The tables that follow present these two classes separately. These tables are standardized to show the same information for each instrument. More information may be available directly from the manufacturers. Also, data are reported only for complete systems. There are several manufacturers that produce parts of a system, notably the data analysis systems. These companies are listed in Appendix A.

It is hoped that users and potential users of flow cytometers and cell sorters will find the information presented here useful both in selecting new instruments and in upgrading current ones.

INSTRUMENT CHARACTERISTICS

Lasers

Lasers provide four major advantages as sources of excitation light: (1) a wide variety of excitation wave lengths; (2) a light beam that can be easily transported and shaped; (3) the possibility of narrow angle light scatter measurements; and (4) the light is monochromatic. Currently offered are argon, krypton, xenon, and helium-neon ion lasers and dye lasers. The latter are usually used with Rhodamine 6G as the dye. Table 1 shows the wavelengths and powers available from several typical lasers. In some instances, such as for ultraviolet (UV) light, special optics are required and there is also usually a requirement for increased total power input to the laser power supply. Lasers are inefficient at this wavelength and a laser with a total power of several watts is usually required for stable operation. Lasers can be obtained with very large amounts of power although current instruments are designed for a limit of about 12 W. Disadvantages of lasers as excitation light sources are their size and cost, the latter including initial price, utilities, and repair.

The manufacturer usually specifies the type and model laser that is incorporated into an instrument. This is necessary because lasers from different manufacturers have different physical sizes and shapes. However, since it is possible to modify any instrument, it can be useful to know what parameters are important in making a selection. The following are suggested, not in any particular order.

1. Total light output (all lines)
2. Light output at desired single wavelength lines
3. Ease of cleaning and changing mirrors
4. Variety of mirrors provided
5. AC Line Voltage
6. Maximum current required
7. Type of cooling
8. Water flow rate (gal/min) for water cooled lasers
9. Repair record (talk to other users)

Arc Lamps

In many flow cytometers, the excitation source consists of high pressure mercury and xenon arc lamps. These lamps are specified by the amount of power they consume rather than the power they emit. Most lamps in use are rated at 100

Flow Cytometry and Sorting, Second Edition, pages 171–186
Published 1990 by Wiley-Liss, Inc.

TABLE 1. Power Levels (in watts) for Available Monochromatic Wavelengths for Several Lasers*

Wavelength (nm)	Argon (25 mW)	Argon (5 W)	Argon (12 W)	Argon (UV)	Krypton (750 mW)	He–Cd (40 mW)	He–Ne
325.0	—	—	—	—	—	7.5	0.01
351.1–363.8	—	0.060	0.400	1.5	0.075	—	—
413.1	—	—	—	—	0.060	—	—
441.6	—	—	—	—	—	30	0.05
454.5	—	0.12	0.60	0.60	—	—	—
457.9	0.001	0.35	0.95	0.95	—	—	—
465.8	—	0.20	0.35	0.35	—	—	—
472.7	—	0.30	0.55	0.55	—	—	—
476.5	0.002	0.75	1.95	1.95	—	—	—
488.0	0.015	1.5	4.7	4.7	—	—	—
496.5	—	0.70	1.75	1.75	—	—	—
514.5	0.009	2.0	5.2	5.2	—	—	—
528.7	—	0.34	0.80	0.80	—	—	—
632.8	—	—	—	—	—	—	0.002
647.1	—	—	—	—	0.500	—	—
676.4	—	—	—	—	0.120	—	—
752.5	—	—	—	—	0.100	—	—

*The powers shown are for typical lasers. Some wavelengths require special optics. The different lasers are distinguished by total power for all lines. For more precise and detailed information, please contact the manufacturers.

watts. See Figure 5, Chapter 2 (this volume) for the emission spectrum of a mercury arc lamp. Optical filters are used to select the range of excitation wavelength desired. For example, the UG1 excitation filter provides for illumination with 300 to 400 nm (UV) light. Total power provided by this combination is about 10 mW. Two advantages of arc lamps over lasers as excitation light sources are (1) they are much less expensive to purchase and to maintain; and (2) they are much smaller and easier to incorporate into instruments. Disadvantages include lower light output at all wavelengths, less selectivity of excitation wavelength (filters cannot produce monochromatic beams), more difficulty in shaping and transporting the beam, relatively short lifetimes for the lamp, instability of arc position and power, and no possibility for measuring light scatter at small angles. The only option for a user in the selection of arc lamps is the type of tube; there is quite a variation in price and dependability. Other users are a good source of information.

Flow Chambers

Flow Cytometers. Most flow chambers are constructed with square cross sections inside and outside, although a few are still available with a cylindrical cross section. The flat walls of the square cells provide a better surface for optimum focusing of the laser beam and for collecting the fluorescent light. If imaging of the detection region is desired, the flat walls are required. The fluid path in these instruments is usually totally enclosed. An important point in this regard is to have a method of collecting all fluid that passes through the chamber, for safety purposes. Most fluorescent dyes that are used are either outright or potential carcinogens or mutagens and must be carefully handled and disposed of.

The different commercial instruments tend to have different fluidic and optical arrangements. The following paragraphs give brief descriptions of the general design of several of the instruments, to illustrate the variety available. Design details are available from the manufacturers.

Figure 1 is a schematic drawing of the FACS Analyzer (Becton Dickinson Immunocytometry Systems, Mountain View, California), showing the combination of fluidics and optics used in the instrument. Figure 1A shows the arrangement of the mercury arc lamp used in an epi illumination configuration. Figure 1B shows the internal flow cell arrangement that permits the measurement of fluorescence and cell volume simultaneously.

Figure 2 shows a very different arrangement used by Kratel (Geneva, Switzerland) in their flow cytometer. In this instrument, the flow is horizontal and includes a laser as the illumination source, in an epi illumination arrangement. It also uses a split laser beam and two cell intersection points to control the flow velocity.

Figure 3 is a schematic drawing of the Skatron (Lier, Norway) flow cytometer. This instrument utilizes epi illumination for its fluorescence measurements and also measures forward angle light scatter. Its flow arrangement is a very novel one in that a nozzle is used to produce a flat, laminar flow across the open surface of a microscope cover glass, with the cells confined to a narrow sector along the middle. The cells pass through the illumination beam as they flow across the cover slip.

Flow Sorters. In flow sorters, the fluid stream carrying the cells is forced to pass through a small orifice (e.g., a sapphire watch jewel with 80-μm-diameter hole), forming a fluid jet (see also Chapter 8, this volume). Vibration of the chamber, also called the nozzle, results in the jet eventually breaking up into a stream of uniform droplets. In most instruments the drops are formed at a distance of about 6.5 mm from the orifice. In some sorters the laser beam used to excite the fluorescent dyes passes through the fluid stream shortly after it leaves the orifice. Measurements are made at this point. This is often referred to as the "sense in air" method. There is still sufficient time (about 150 μsec) to make a decision on

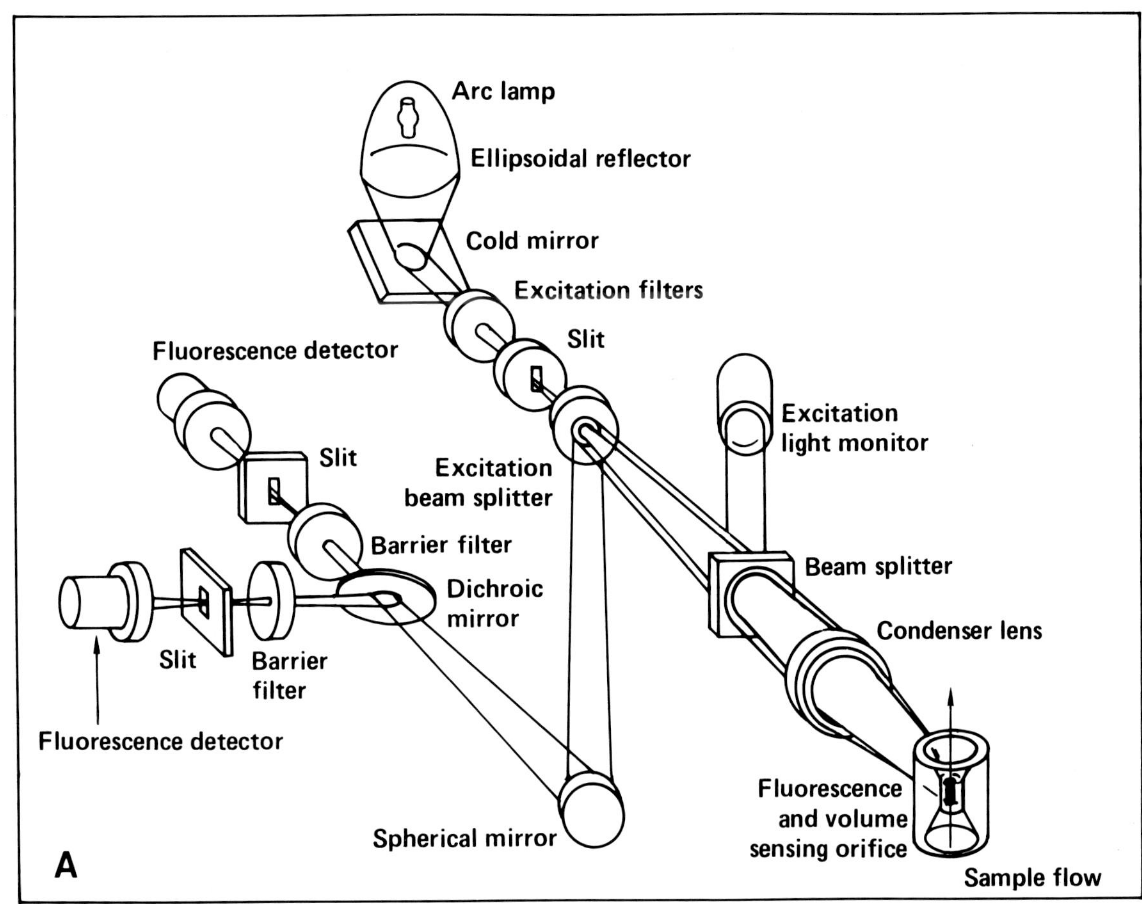

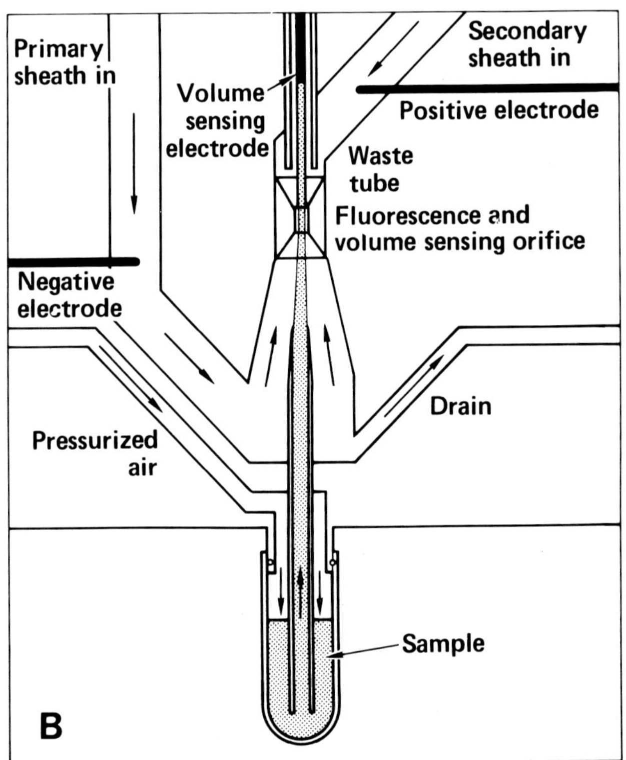

Fig. 1. **A:** Optical path of the Becton Dickinson FACS Analyzer. **B:** Fluidics path for the FACS Analyzer.

whether to sort the cell before the cell ends up in a drop. The advantage of this configuration is that this delay time is relatively short. For efficient sorting at high purity the time delay between cell detection and droplet charging must be very stable. Since no sorter is perfectly stable, the delay time should be kept as short as possible. Compensation for variation in delay time is made by sorting several drops for each cell. The major disadvantage for this type of detection/sorting arrangement is that the liquid stream acts as a very short focal length cylindrical lens. One result is that a large amount of the incident laser light is refracted and reflected into a plane of light that impinges onto the fluorescent light collection lens. To block this light, a thin strip of metal (obscuration bar) is placed across the middle of the fluorescent light collection lens. This results in lowered light collection efficiency. Since the obscuration bar cannot block all of the scattered laser light, greater demands are placed on the optical filters in front of the light detectors (photomultipliers). The lens effect of the stream also makes it impossible to accurately image the cell/laser intersection point at the back focal point of the light collection lens. As a result, the pin hole used to block unwanted light from impinging on the photomultipliers has to be larger than desired and can result in greater "noise" on the signal.

An alternate design uses the same type of square flow chamber as in a flow cytometer, with a watch jewel at the bottom. The laser beam passes through the flat walls of the chamber. The major advantages of this arrangement are (1) imaging of the measurement point is improved, (2) there is no plane of scattered laser light to increase background, (3)

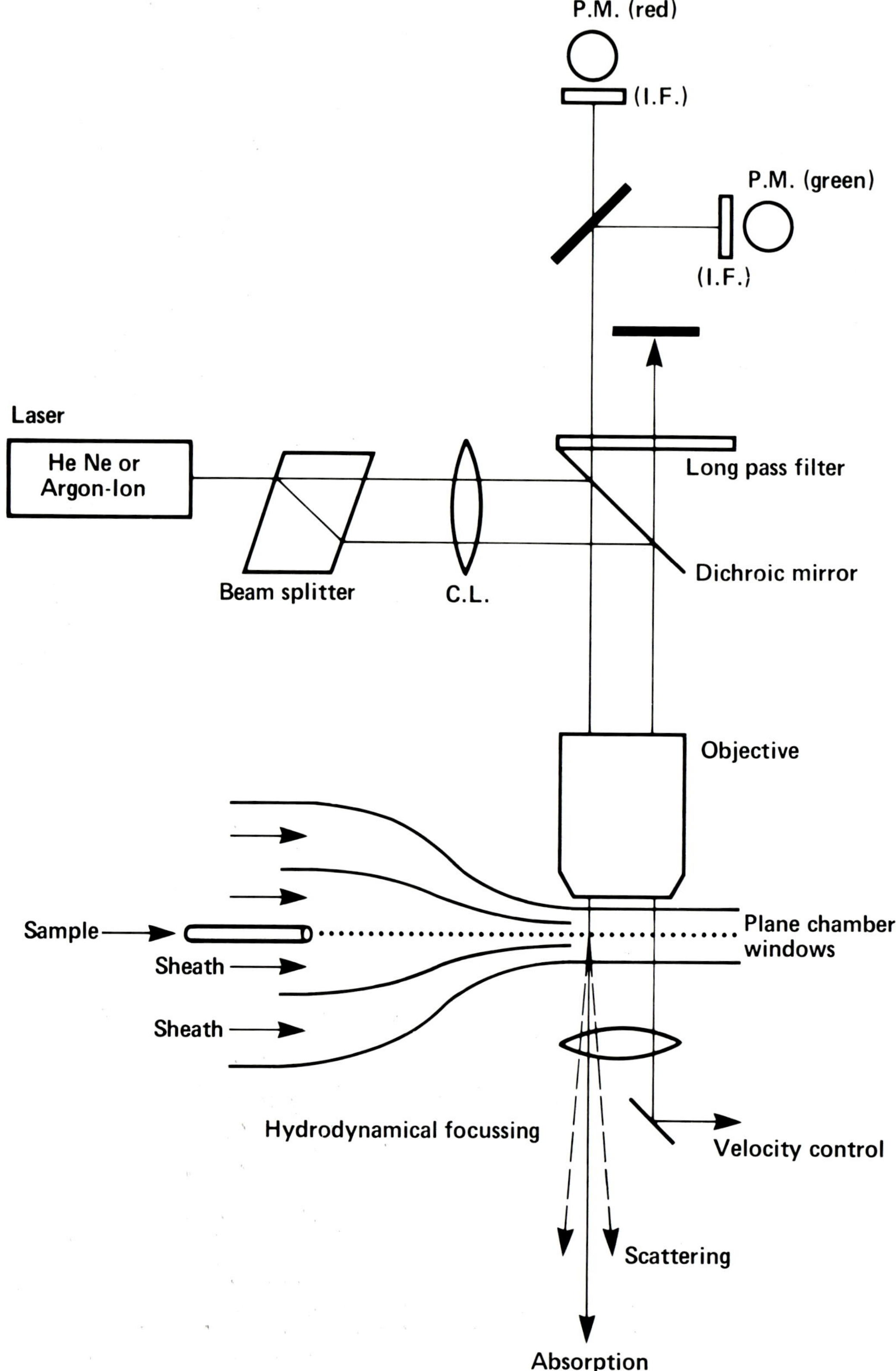

Fig. 2. Optical/fluidic arrangement of the flow cytometer produced by Kratel SA.

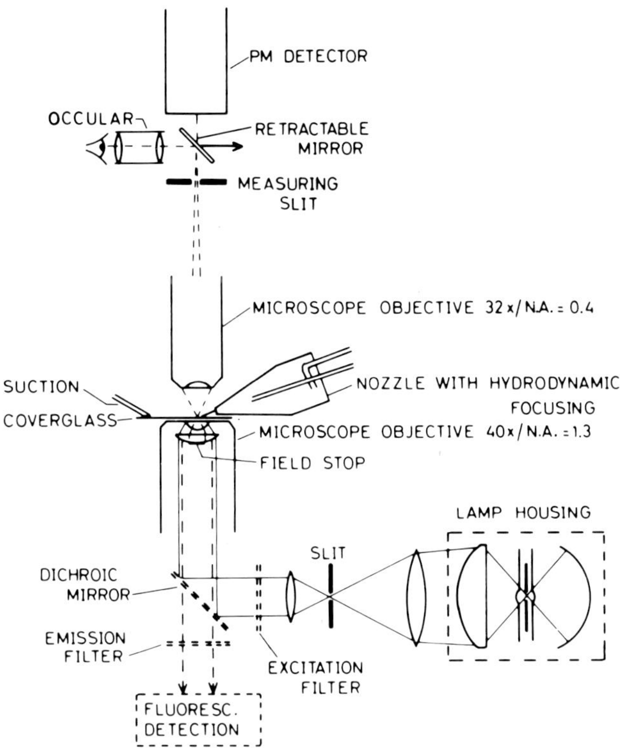

Fig. 3. Schematic drawing of the microscope based flow cytometer manufactured by Skatron A/S.

the light collection efficiency is higher since there is no obscuration bar, and (4) the flow velocity is lower, providing for more fluorescent light to improve measurement precision. Disadvantages are that the delay between cell detection and droplet charging is increased considerably, putting greater demands on flow stability, and a larger overall nozzle size, making the physical assembly larger.

Figure 4 shows the design of the Becton Dickinson FACS 440 cell sorter. This sorter is of the "sense in air" design described earlier, where the laser beam goes through the fluid stream after it leaves the nozzle. Only one laser is shown in this diagram. The instrument is available with two lasers. For the second laser, an additional pair of photomultipliers would be added, with the necessary electronics.

Figure 5 shows the design of the Coulter (Hialeah, Florida) EPICS 700 series flow sorter. This is a dual laser arrangement which includes four detectors at 90 degrees and a forward scatter photodetector. In this sorter the laser beams intersect the flow stream in air.

Figure 6 is a schematic diagram of the sorter manufactured by ODAM (Wissenbourg, France). It is a dual beam arrangement with three photomultipliers as fluorescence detectors. The light collection optics are unusual, using a toroidal mirror assembly to obtain high fluorescent light collection efficiency.

Figure 7 shows the arrangement used in the Partograph FMP Sorter manufactured by Kratel (Geneva, Switzerland). In this instrument, the measurement is made within a quartz cuvette. Laser excitation is made in an epi configuration, with forward and right angle detectors as well.

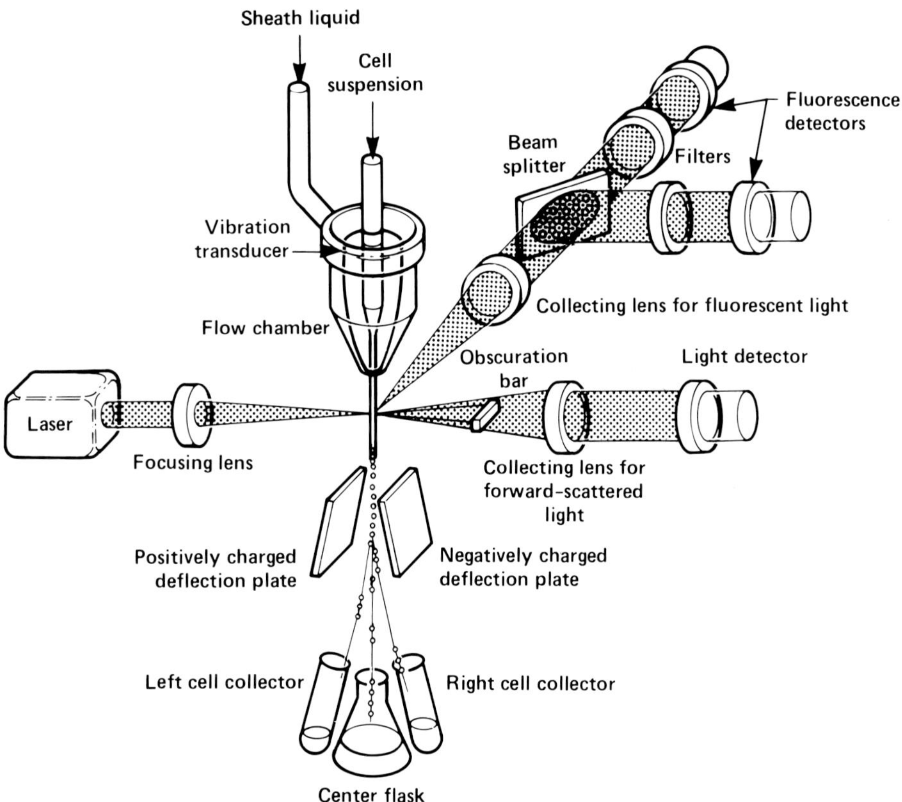

Fig. 4. Schematic drawing of the Becton Dickinson FACS 440 flow sorter.

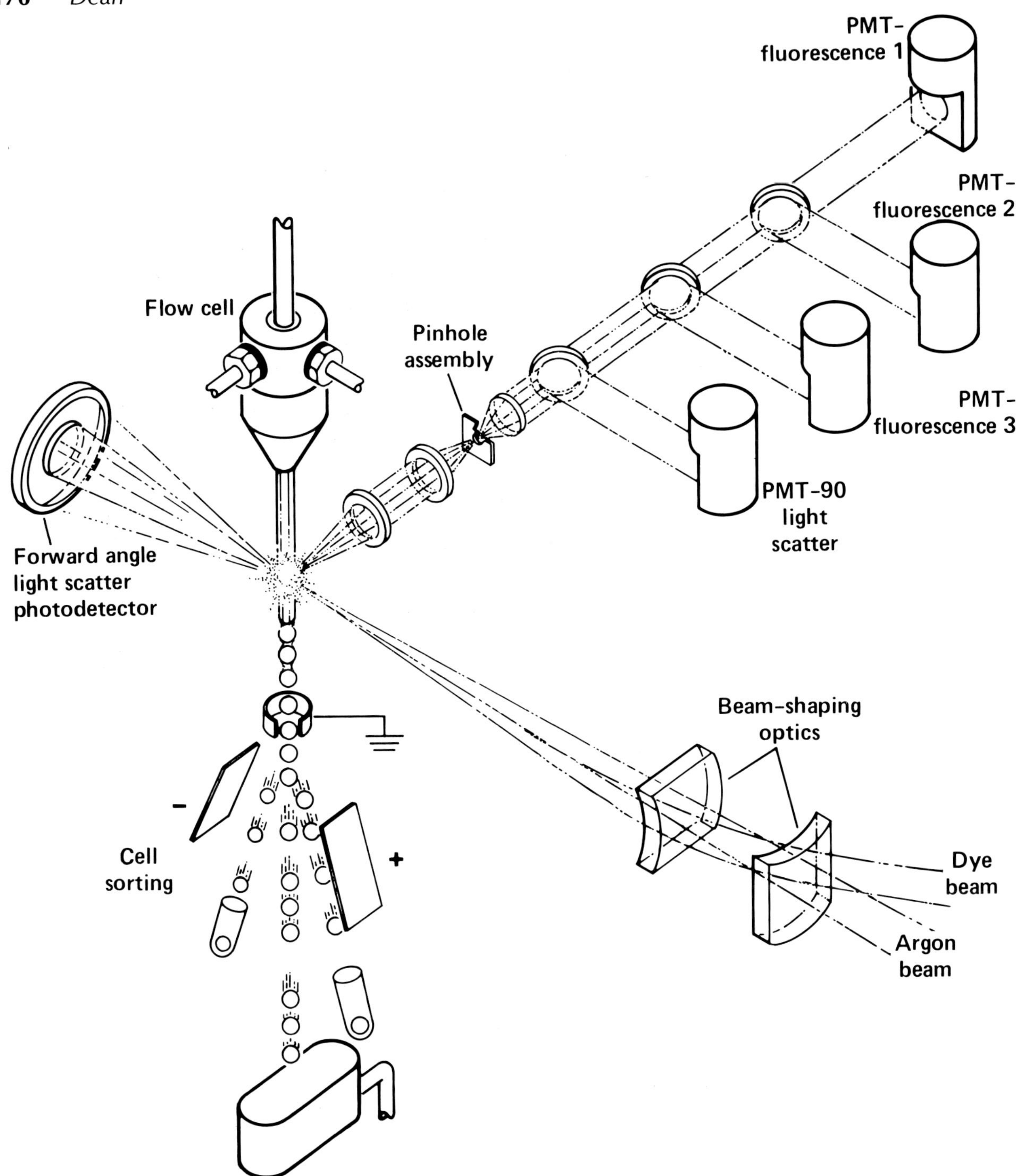

Fig. 5. Schematic drawing of the Coulter EPICS 750 series flow sorter.

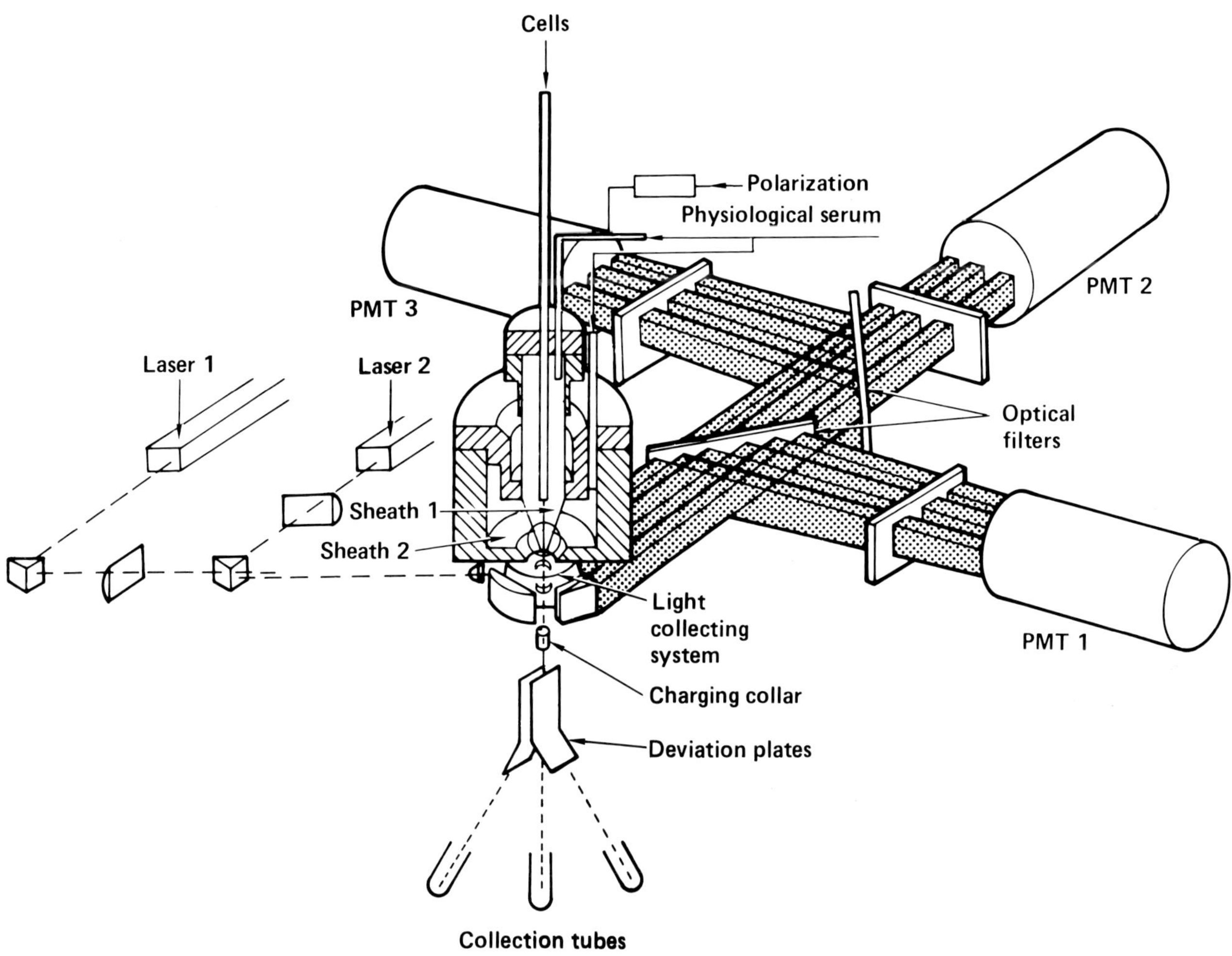

Fig. 6. Schematic drawing of the ODAM flow sorter.

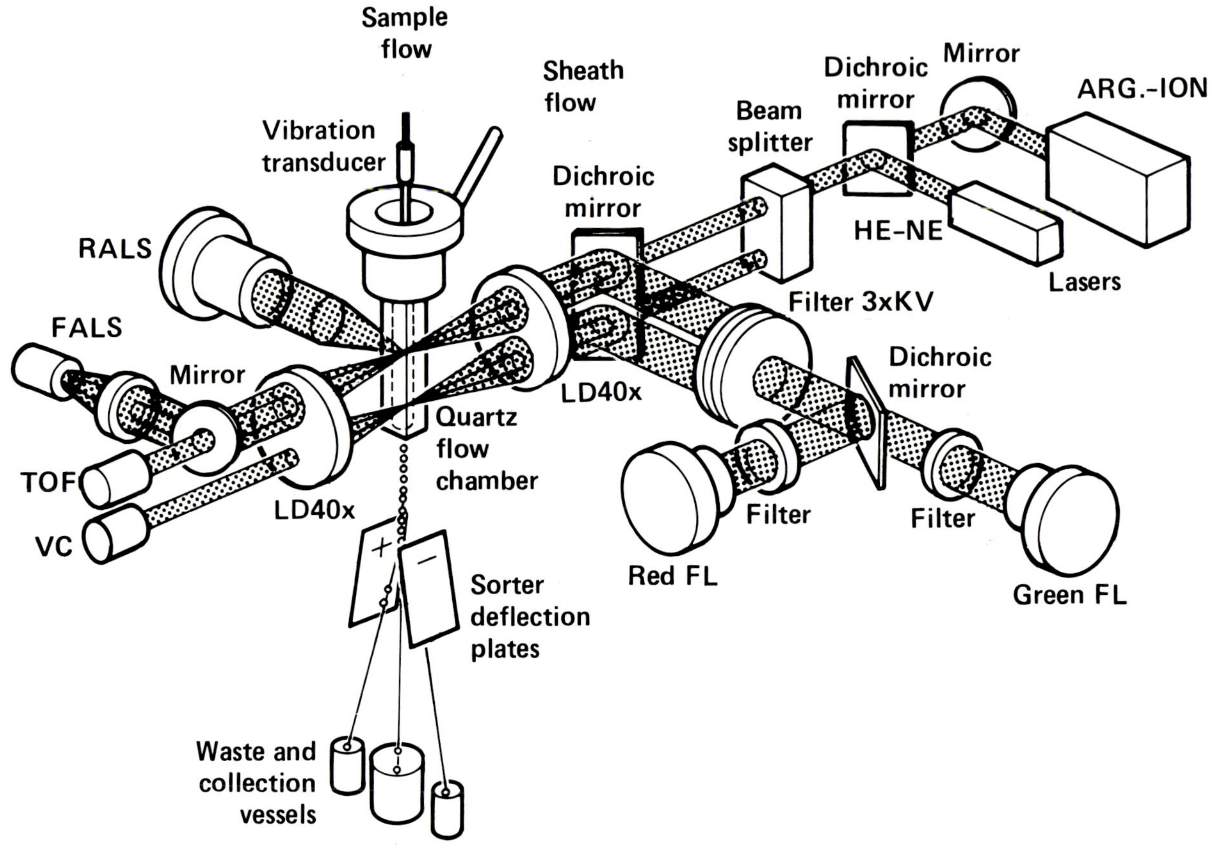

Fig. 7. Schematic drawing of the Kratel Partograph FMP Sorter.

Appendix A: Manufacturers

The following manufacturers, listed alphabetically, have contributed information to this chapter. For further or more detailed information on any of their products you are invited to contact them directly.

Becton Dickinson
Immunocytometry Systems
2375 Garcia Avenue * P.O. Box 7375
Mountain View, California 94039
(800) 821 9796 in CA (800) 223 8226

Epics Division
Coulter Corporation
P.O. Box 4486
Hialeah, Florida 33014-0486
(305) 885-0131

Kratel, SA
64 Ch. de St. Maurice/BP82
CH-1222 Geneve-Vesenaz
Switzerland (022)
52 33 74

ODAM
34, rue de L'Industrie
67160 Wissenbourg
France
(88) 94.99.32

Skatron A/S
P.O. Box 8
N-3401 Lier
Norway
(03)850770

The following manufacturers provide partial systems that can be attached in one way or another to complete systems. In some instances, for example with data analysis systems, a flow cytometer or cell sorter can be purchased without a data analysis capability and one of the competitor computer systems purchased and added to the instrument. Some of the companies provide alternatives in the hardware, for example the Multichannel Analyzer.

Data Acquisition/Analysis Systems

Brucker Spectrospin
34, Rue de L'Industrie
67160 Wissembourg
FRANCE

Catalyss Corporation
7400 South Tucson Way
Englewood, Colorado 80112
USA

Nuclear Data Incorporated
Golf and Meacham Roads
Schaumburg, Illinois 60196
USA

Data Analysis Systems

Cellsoft Biotechnology Systems
P.O. Box 13666
Research Triangle Park,
North Carolina 27709
USA

Oatka Software
P.O. Box 5
Scottsville, New York 14546

Verity Software House, Incorporated
P.O. Box 247
Topsham, Maine 04086
USA

Appendix B. Instrument Characteristics

Becton Dickinson Immunocytometry Systems

1. FACS Analyzer

Excitation Mercury arc lamp, 100 watts
Detectors Two fluorescence; one 90 deg. light scatter; volume sensor
Flow chamber 50–100 μm dia.; 75–225 μm long
Sample conc. 10^5 to 2×10^7 cells/ml
Data rate 1,000/sec
Resolution 2% CV on fluorescence, 2% CV on volume
Sensitivity 8,000 molecules of fluorescein
Signal proc. log/lin amplifiers; ratios; electronic compensation; list mode data acquisition
Data proc. uses the Consort 30 data management system
Utilities 115 VAC, 15 amps

2. FACS 440 Flow Sorter

Excitation two argon ion lasers, 2 watts standard
Detectors four photomultipliers; one photo diode
Flow chamber sense in air
Orifice 50–200 μm dia.
Sample conc. 10^5 to 2×10^7 cells/ml
Data rate 10,000/sec
Sort rate 5,000/sec
Resolution 2% CV on fluorescence
Sensitivity 2,000 molecules of fluorescein
Signal proc. log/linear amplifiers; ratios; region of interest sorting; list mode data acquisition; polarization anisotropy; dual fluorescence compensation
Data proc. accumulates up to sixteen 256 channel histograms or one 64 × 64 channel histogram; display 1 or 2 univariate histograms or bivariate data in a 3d perspective view; area of interest integration
Utilities 220 VAC, 50 amps, 5 gal/min water
Options 5 or 6 watt argon and krypton ion lasers, including uv enhanced; single cell deposition into 96 well trays; microsample delivery of 5–25 μl at 0.01–0.05 ml/min; enclosed sensing region; dye laser; helium–neon laser; universal data lister

3. FACStar Flow Sorter

Excitation 2 watt argon ion lasers
Detectors forward scatter photodiodes; three photomultipliers for fluorescence and side scatter
Flow chamber sense in air
Orifice 50–200 μm dia.
Sample conc. 10^5 to 2×10^7 cells/ml
Data rate 5,000/sec
Sort rate 7,000/sec
Sensitivity 2,000 molecules fluorescein
Signal proc. linear/log amplifiers, bivariate region of interest sorting with gating
Data proc. Hewlett Packard Model 310 microcomputer
Utilities 208 VAC at 50 amps, 5 gal/min water
Options automated cloning; signal area and width calculation; quartz flow cell; 5 watt UV enhanced argon ion laser

4. Consort 30 Data Management System

CPU HP 9920S, 64OKB RAM memory
Storage Dual floppy
Interaction menu driven
Acquisition univariate; bivariate; list mode
Data rate 5,000/sec at four variables
Programming HP BASIC and PASCAL 2.1

Display histograms; dot plots; contour plots
Analysis region of interest calculation for gated data; DNA phase fraction analysis
Options 14.5 MB fixed disk

5. Consort 40 Data Management System

CPU LSI 11/23, 256KB RAM memory
Storage dual floppy
Interaction menu driven
Acquisition up to four variables in list mode, simultaneous sort control and analysis of four variables
Data rate 5,000/sec at four variables
Programming FORTRAN
Display multiple univariate plots; bivariate contour and perspective plots
Analysis uses LACEL (Los Alamos National Laboratory) software that permits non rectangular region of interest analysis of multiple gated bivariate distributions; programmable analysis sequence
Options up to 31.4MB fixed disks; graphics hard copy

6. FACS Automate

Samples 96 well trays
Cycle time 2 minutes between samples
Volume 5–50 μl/sample
Rate 0.08–0.95 μl/sec
Interfaces FACS Analyzer and sorters

7. FACStar Plus

Excitation Argon, Dye, UV-Argon, He-Ne, Krypton lasers
Detectors six photodetectors, 8 variables in list mode
Flow chamber sense-in-air, quartz cuvette optional
Orifice 50–200 μm
Sample conc. 10^5 to 10^7 cells/ml recommended
Data rate event measurement time less than 30 μsec
Sort rate 7,000/sec
Sensitivity 2,000 molecules of fluorescein per cell
Resolution <2% on fluorescence
Signal proc. lin/log amplifiers; pulse height, width and area
Data proc. single and multiple histograms and contour plots; statistics; DNA curve fitting; histogram overlays; time mark on data
Utilities 100/120VAC, 25 amps, or 220/240VAC, 12 amps; laser 208VAC 3 phase, 50 amps; 5 gal/min water
Options automatic sampling; automatic cell deposition; refrigeration; aerosol removal; closed flow cell
Computer Hewlett Packard Model 310, 320; MicroVAX II
Storage choice of floppy disks, hard disks, magtape
Interaction menu-driven
Acquisition eight variables, list mode
Programming HP Pascal; DEC higher level languages
Display dual oscilloscopes; computer CRT; real time dot plots; histograms, pulse display, contour plots
Options mass storage devices; color graphics; printers; networking

8. FACScan

Excitation 15mW Ar-ion laser
Detectors 4 photomultipliers, 1 solid-state silicon detector
Flow chamber square quartz cuvette
Orifice 430μm × 180μm
Sample conc. 10^5 to 2×10^7 per ml
Sensitivity 1000 FITC molecules per particle
Resolution CV <2% for fluorescence

Signal proc. lin/log amplifiers; spectral overlap compensation
Data proc. single or multiple histograms, dot plots, contour plots; histogram overlays; nonparametric statistics; DNA analysis; immune monitoring software
Utilities 120 VAC, 20 amps
Options FACS AutoMATE automatic sample injection
CPU HP9000 Series, Model 310
Storage 20 Mbyte hard disk; microfloppy disk

Coulter Corporation

1. EPICS PROFILE Flow Cytometer

Excitation 150 mW air cooled argon ion laser/ one watt argon ion laser/ or three watt argon ion laser
Detectors three photomultipliers, one photodiode
Flow chamber 250 μm^2 quartz cuvette (125μm optional)
Sample conc. 5×10^6 cells per ml
Data rate 10,000 cells/sec
Sensitivity <1,000 molecules of fluorescein
Signal proc. One multiplexed successive approximation ADC with 1024 channel resolution; linear/log amplifiers; integral signals; signal overlap correction; amorphous gating windows; ratio of any two parameters; automatic computer control via test protocols
Data proc. high speed hardware data acquisition system; 32KB histogram memory; Intell 8088 16-bit microprocessor with 512 KB RAM memory; Intel 8087 floating point co-processor; 64-256-1024 channel histograms quad display screen; contour and dot plots; numerical and statistical analysis; automatic computer driven test protocols
Storage one 5 1/4-in. floppy disk drive with 360KB storage; one 20MB hard disk
Utilities 150mw laser 110 VAC, 15 amps, air cooled
1000mw laser 208 VAC, single phase/60 amps, 1.9 gal/min water
3000mw laser 208 VAC, 3 phase/40 amps,
2.2 gal/min water
Options streaming magnetic tape

2. EPICS CS Flow Sorter

Excitation 2 watt argon ion laser (5-watt optional)
Detectors three photomultipliers, one photodiode
Flow chamber sense in air standard (quartz flow cell optional)
Orifice 76um dia. (50 to 250 μm optional)
Sample conc. 5×10^6/ml
Data rate 10,000/sec
Sort rate 5,000/sec
Sensitivity <1,000 molecules of fluorescein
Signal proc. four ADCs; linear/log amplifiers; peak or integral signal; signal overlap correction; ratios of any 2 signals; 16 sided gating window; list mode; computer control via test protocols
Data proc. high speed hardware data acquistition system; 32KB histogram memory; Intel 8086 16-bit microprocessor with 512 KB RAM memory; Intel 8087 floating point coprocessor; 4 histograms per measurement; 64-256-1024 channel histograms; quad display screen, 3d perspective views of bivariate data; gated bivariate displays; contour and dot plots; nonparametric analysis of univariate data; numerical statistical analysis
Storage one 8-in. floppy disk drive, 1MB storage per disk; one 20MB hard disk drive
Utilities 208 VAC, 50 amps, 2.2 gal/min water
Options Coulter volume; Autoclone; microsampler delivery system; biohazard control system; upgrade to Model CD sorter; streaming tape drive; EASY 88 data analysis system; cytologic software

3. EPICS CD Flow Sorter

Same as EPICS CS, with dual laser bench for addition of second laser. Three photomultipliers, 2 photodiodes

4. EPICS 541 Flow Sorter

Excitation 2 watt argon ion laser (five watt optional)
Detectors three photomultipliers, one photodiode
Flow chamber sense-in-air standard (quartz flow cell optional)
Orifice 76um dia. standard (50 to 250 μm optional)
Sample conc. 5×10^6 cells/ml
Data rate 10,000/sec
Sort rate 5,000/sec
Sensitivity <1,000 molecules of fluorescein
Signal proc. 4 ADCs; linear/log amplifiers; signal amplitude or area; time; list mode; 16 sided gating window; signal overlap correction
Data proc. Intel 8085 microprocessor data acquisition system; 32 KB histogram memory; Intel 8086 microprocessor with 8087 floating point co-processor for online analysis with 512 KB RAM memory; up to 8 histograms per run; quad display screen; bivariate dot plots with windowing; 3D display of bivariate data; numerical statistical analysis; acquisition protocols; nonparametric analysis of univariate data
Storage two 8" floppy disk drives with 1 MB per disk
Utilities 208 VAC, 50 amps, 2.2 gal/min water
Options 2 additional ADC's; 1024 channel ADC's; emission anisotropy; polarization; time of flight; ratios of any two signals; 20MB or 40MB hard disk; microsampler delivery system; autoclone; autosort lock; biohazard control system; Coulter volume adapter; streaming tape drive; EASY 88 data analysis system; cytologic software

5. EPICS 751 sorter

Excitation 5 watt argon ion laser
Detectors 4 photomultipliers, 1 photodiode
Flow chamber sense in air standard (quartz flow cell optional)
Orifice 76 um dia. standard (50um to 250 um optional)
Sample conc. 5×10^6/ml
Data rate 10,000/sec
Sort rate 5,000/sec
Sensitivity <1,000 molecules fluorescein
Signal proc. 6 ADC's; linear/log amplifiers; pulse height and area; time of flight; polarization; emission anisotropy; 16 sided gating window; list mode; ratios of any two signals; signal overlap correction; time; microprocessor controlled
Data proc. Intel 8085 microprocessor data acquisition system; 32KB histogram memory; Intel 8086 microprocessor with 8087 floating point co-processor for online analysis with 512KB RAM memory; up to 8 histograms per run; quad display screen; bivariate dot plots with windowing; 3d display of bivariate data; numerical statistical analysis; acquisition protocols; nonparametric analysis of univariate data
Storage one 8-in. floppy disk drive with 1MB storage; 20MB hard disk
Utilities 208 VAC, 50 amps, 2.2 gal/min water
Options Microsample Delivery System; Auto-Clone System; EASY 88 data analysis system; 40MB hard disk; Biohazard Control System; Coulter Volume Adapter; Auto Sort Lock; streaming tape drive; cytologic software

6. EPICS 752 Flow Sorter

This sorter has the same features as the Model 751, with the addition of a dye laser to add more capability in the use of more than one fluorescent dye. The argon laser, operating in an all-lines mode, provides a 488-nm beam for direct cell excitation and a 514-nm beam for pumping the dye laser. The most frequently used dye is Rhodamine 6G. PRISM parameter standard for multicolor real time analysis and sorting

7. EPICS 753 Flow Sorter

This sorter has the same features as the Model 752, with the addition of a second argon ion laser to pump the dye laser. The second laser also provides for simultaneous operation of visible and uv beams.

8. Auto-Clone

A programmable automatic single-cell deposition system that sorts from 1 to 10 cells or multiples of 100 cells per well into multiwell microculture plates
Samples 24, 60, 96 well trays
Cells/well 1 to 10 per well, or multiples of 100
Timing 2 sec per well, 3.2 min for 96 well tray

9. Coulter Volume Adapter (No sorting)

The EPICS Coulter Volume Adapter (CVA) is an accessory that allows the volumetric measurement of cells and other particles using the electrical impedance method known as the Coulter Principle.
Aperture $50 \times 50 \times 50$ um
Current 50-500 μA
Dia. range 1-17 μm dia.
Flow rate 0-40 μl/min

10. Auto Sort Lock

This option, available for the 750 series of sorters, stabilizes the droplet breakoff point to a precision of ± 0.1 drops. It also includes ultrasonic excitation to help remove plugs in the orifice, controls the number of drops sorted and calculates the sort matrix for 1-, 2-, or 3-drop sorting

11. Microsample Delivery System

The MicroSampler Delivery System (MSDS) provides automated sample introduction and data acquisition for both single samples via standard 12 × 75 mm tubes and multiple samples via 96-well microtiter plates.
Samples 96 well microtiter trays and 12 × 75mm tubes, with temperature control (5±1°C), includes mix and wash cycles, selective sampling
Cycle time variable, depending on sample and protocol
Volume 10-100 μl/sample
Rate 10-250 μl/min delivery
Interfaces all EPICS sorters

12. Biohazard Containment System

The EPICS Biohazard Containment System (BCD) provides a totally enclosed environment for sample processing, which eliminates aerosols from hazardous materials. The system consists of a biohazard containment drawer (BCD and a 250 um square quartz flow cell for analysis only. The drawer houses a 3 liter waste container that can be incinerated for disposal.

13. EASY 88 Data Analysis System

CPU Intel 8088 with Intel 8087 floating point co-processor, MS-DOS operating system, Model 7220 graphics microprocessor
Storage 20MB hard disk, 10MB tape casette
Interaction Menu system
Acquisition Serial and parallel communications channels, EPINET network system
Programming Can run most commercial software written for use with the MS-DOS operating system
Display A 13" color monitor with 640 × 480 pixel resolution, with log or linear scaling of the y axis, zoom-in on specific areas of the data
Analysis EASY 88 provides a wide variety of numerical, statistical and graphical methods of analyzing data. Included are methods of comparing two or more histograms (graphical and statistical), list mode processing to produce multiple gated histograms, DNA phase fraction analysis
Options 20 MB hard disk, EPINET data links to sorters, floppy disk drives

14. Cytologic

Cytologic software is specifically designed for flow cytometry data analysis and graphics. It is compatible with IBM PC, XT, AT and equivalent systems. Written in Microsoft BASIC, it allows individuals to compile and link their own BASIC applications programs to the Cytologic package.
Two packages are available:
Cytologic Software for Immunofluorescence Applications
Cytologic software for DNA Applications

Kratel SA

1. Partograph FMP Analyzer

Excitation Argon ion and He–Ne, dual beam
Detectors 2 photomultipliers, 2 photodiodes
Flow chamber quartz flow cell, 250 × 250 μm
Sample conc. 10^7/ml
Data rate 10,000/sec
Sensitivity 3,000 molecules of fluorescein
Signal proc. log/linear amplifiers, pulse height, length and area, flow velocity
Data proc. microprocessor controlled (Z80) acquisition of all 4 signals, dual floppy disk system with BASIC and FORTRAN programming, real time graphical display with contour and 3d perspective plots of bivariate data
Utilities 220 VAC, 10 amps
Options mercury arc lamp, higher powered lasers, autosampler

2. CYTOMIC 12

CPU Z 80 microprocessor, 20KB ROM, 36KB RAM memory
Storage 74 univariate distributions of 128 channels, or 3 bivariate distributions of 4096 channels each
Interaction 20 key keyboard
Acquisition one or two variables, with gating
Display univariate distributions, with overlay of a second; bivariate data as contour or isometric plots, the latter with rotation
Analysis
Options floppy disk, holding 4800 univariate distributions or 132 bivariate distributions; printer

3. Partograph FMP Sorter

Excitation Argon ion and He–Ne, dual beam
Detectors 2 photomultipliers, 2 photodiodes
Flow chamber quartz flow cell, 250 × 250 μm, with 70 μm orifice
Sample conc. 10^7/ml
Data rate 10,000/sec
Sort rate 3,000/sec
Sensitivity 3,000 molecules of fluorescein
Signal proc. log/linear amplifiers, pulse height, length and area, flow velocity
Data proc. microprocessor controlled (IBM PC/AT 2) acquisition of all 4 signals, dual floppy disk system with BASIC and FORTRAN programming, hard disk, real time graphical display with contour and 3d perspective plots of bivariate data
Utilities 220 VAC, 10 amps
Options mercury arc lamp, higher powered lasers, autosampler

ODAM

1. ATC 3000

Excitation Argon and krypton ion lasers, dye lasers, dual system
Detectors 3 photomultipliers, 1 photodiode, 1 electrical cell volume detector, multiangle light collection optics
Flow chamber sense in a quartz cell, with 2 sheaths

Orifice 70 to 120 μm dia.
Sample 5×10^6/ml, with sample agitation and temperature control
Data rate 10,000/sec
Sort rate 5,000/sec
Sensitivity 2,000 molecules of fluorescein
Resolution 0.8% on fluorescence, 1.1% on scatter, 3% on volume
Signal proc. Intel 8085 analytical adjustments control processor, log/linear amplifiers, ratios, up to 8 measured and derived parameters are transferred into an 8 input ADC
Data proc. on-line processing detailed below
Utilities 380 VAC, 50 Hz, 70 amps for dual laser configuration; 18 liters per min cooling water
Options single cell deposition
CPU 24 bit bit-sliced central processor unit (AMD 2900)
Storage hard disk with 32, 80 or 160 MB capacity, and 512KB floppy
Interaction programmable
Acquisition 8 univariate and 8 bivariate distributions in realtime and list mode, 8 histograms with 256 channels per histogram, 8 cytograms with 256×256 points per cytogram
Display two graphics coprocessors, two color monitors, and one ASCII text display, dot plots, contour plots, 3d perspective views
Analysis 10 region of interest areas of any shape, (10 cursors in histograms, 10 windows in cytograms), data (curve) smoothing, comparison, subtraction, mode and CV calculation, calculation of cell cycle phase fractions, overlay of 3 histograms, specialized programs dealing with chromosome analysis and immunology, user programmability in PASCAL
Options 18MB magnetic tape storage unit, black and white ink jet printer, color ink jet printer, ethernet interface

Skatron A/S

1. Argus Flow Microphotometer

Excitation 100 watt Hg arc lamp, air cooled Ar-laser optional
Detectors 5 photomultipliers, for fluorescence and light scatter
Flow chamber Jet On Open Surface (JOOS) type
Orifice 70 or 100 um
Sample conc. Up to 10^9 per ml
Data rate 10^4 per sec
Resolution Fluorescence: CV = 1%, Light scatter: CV = 0.6%
Signal proc. Lin/Log amplifiers, electronic compensation
Data proc. Accumulates four 256 channel histograms and three 64×64 channel histograms simultaneously, plus list mode
Data proc. Region of interest, integrate, mode, CV, bivariate to univariate histogram collapse
Utilities 240 VAC or 110 VAC

10

Preparation of Cell/Nuclei Suspensions From Solid Tumors for Flow Cytometry

M. G. Pallavicini, I. W. Taylor, and L. L. Vindelov

Biomedical Sciences Division, Lawrence Livermore National Laboratory, Livermore, California 94550 (M.G.P.); Queensland Institute of Medical Research, Brisbane, Queensland, Australia 4006 (I.W.T.); Medical Department, The Finsen Institute, DK-2100 Copenhagen, Denmark (L.L.V.)

Flow cytometric analyses of cells from solid tissues in vivo have become increasingly prevalent in studies to relate characteristics measured in vitro to phenomena in vivo. For example, cytokinetic studies of tissues in vivo have been performed on both experimental and clinical samples in efforts to improve treatment protocols and to aid in diagnosis and prognosis. Univariate DNA distributions and more recently bivariate distributions of bromodeoxyuridine incorporated into DNA and DNA content have been used to measure the response of tumors to cancer chemotherapeutic agents or radiation [7,20]. DNA content distributions have also been used to detect aneuploid cell clones and the fraction of cells in S phase, both of which, in some cases, are useful for diagnosis and/or prognosis (see Chapters 23 and 24). The desirability of performing retrospective analyses on a large number of tumors of similar histological type, disease stage, and treatment course has led to the development of techniques to flow cytometrically analyze paraffin-embedded materials [10,11]. A common feature of all flow cytometric studies of tissues in vivo is the requirement for dispersal into single cells (or nuclei) before staining and analyses. The objective of this chapter is to review critically some of the commonly-used methods to disperse tumors (both fresh and paraffin-embedded samples) for flow cytometric cytokinetic analyses and to discuss inherent limitations in these techniques. Parts of this chapter have been presented in previous publications [21].

The aim of most flow cytokinetic studies is to quantitate the cell cycle properties of a particular tissue or subpopulation and to relate these measurements to phenomena in vivo. To achieve this goal, it is desirable that the cells in suspension have particular characteristics. These are summarized in Table 1. The dispersed cells should be representative of the tissue in situ (i.e., no preferential selection of differentiated versus undifferentiated, or proliferating versus quiescent cells, or cells in particular cell-cycle phases). The disaggregated cells (or nuclei) should exhibit minimal damage attributable to the dissociation, with minimal clumping and debris, and flow cytometric distributions of the measured variable (e.g., DNA profiles, bivariate DNA/BrdUrd distributions should be of high quality). In addition, retention of other molecular, biochemical and morphologic characteristics of the dispersed cells is useful for confirmation or evaluation of the subpopulation discrimination. Since cell clonogenicity is often used as a functional end point, retention of clonogenic capability is also advantageous. In studies requiring multiple measurements of several variables, a high degree of cell recovery per gram of tissue is important. Evaluation of cell yield and viability is relatively straightforward, whereas exclusion of preferential subpopulation selection requires specific protocols. Techniques to determine whether there is preferential selection of subpopulations have been described in detail [18,21] previously and are discussed only generally in the following sections.

The most commonly used techniques to disaggregate solid tumors for flow cytometric analyses can be categorized broadly into two groups: enzymatic procedures that produce single-cell suspensions, and detergent-based protocols that yield nuclear suspensions. Enzymatic disaggregation is often the method of choice for cytokinetic analyses which require preservation of cell morphology and function (i.e., clonogenicity) and high cell recovery. Other methods that yield suspensions of whole cells include chemical agents (e.g., ethylenediaminetetracetate, tetraphenylboron) and mechanical disruption. Generally, cells obtained with these latter methods are nonviable, the cell yield per gram of tissue is relatively low, and the suspensions are characterized by varying amounts of cellular debris and cell clumps. Disaggregation procedures that produce nuclear suspensions are suitable for flow cytometric analyses of DNA content, and are applicable to analyses of small clinical biopsied samples. Enzymatic and detergent-based protocols to disperse fresh and paraffin-embedded tissues into suspensions of whole cells or nuclei are described below.

Flow Cytometry and Sorting, Second Edition, pages 187–194

TABLE 1. Desirable Features of Tumor Cell Suspensions for Flow Cytometric Analyses

Representative of tissue in situ
High-quality suspensions characterized by minimal clumping, debris, low CV
Retention of cellular biochemical/morphologic characteristics
High cell yield

TABLE 2. Evaluation of Preferential Phase-Specific Cell Loss During Tumor Dispersal[a]

Hours after ^{3}HTdR injection	Labeling Index		
		Single Cells	
	Sections	Trypsin	Neutral Protease
0.5	27 ± 1 (4)	29 ± 1 (4)	32 ± 4 (8)
3	31 ± 2 (5)	32 ± 2 (5)	34 ± 2 (6)
13	58 ± 3 (5)	56 ± 1 (5)	55 ± 6 (7)

[a]Tumors were excised at either 0.5, 3, or 13 h after ^{3}H-TdR injection. Values represent the average ± 1 standard deviation of a number of samples (n) per group. (Reproduced from Pallavicini et al. [19], with permission of the publisher.)

DISAGGREGATION OF FRESH MATERIAL

Enzymatic Dispersal Into Single Cells

Disaggregation of tissues into single cells has been achieved with a variety of enzymes, including trypsin, neutral protease, collagenase, DNase, hyaluronidase and various combinations thereof. Examples of specific enzymatic dispersal procedures used for a variety of tumors in different laboratories can be found in Pallavicini [21]. The availability of enzymes with different specificities for intercellular components facilitates the design of dispersal protocols for specific purposes. For example, collagenase is useful to disperse tumors containing abundant amounts of collagen (e.g., carcinomas), whereas disaggregation of other types of tumors (i.e. lymphomas, some sarcomas) can often be achieved without exposure to collagenase. However, since the intercellular composition of individual tumors is rarely known, in practice several enzymes are usually used in combination. These enzyme cocktails provide relatively high cell yields but increase the potential for cell damage and alteration of cell-surface membrane components, hence should be used with caution in studies where retention of these characteristics is important.

We have used neutral protease and DNase–trypsin to disaggregate experimental mouse tumors into single cells for flow cytometric analysis. Both enzymatic procedures have been evaluated according to the criteria described (see Table 1). In the neutral protease technique [19], small pieces of excised tumor are finely minced with scissors and incubated with neutral protease for 1 hour; the resulting suspension is filtered through 37-μm nylon mesh before cell fixation for flow cytometric analyses. In the trypsin/DNase procedure [27], small pieces of tumor are first passed through a wire screen (50 mesh) with the aid of a syringe barrel. The resulting suspension contains single cells, clumps, and cellular debris. Treatment with a trypsin–DNase cocktail disaggregates most cell clumps and digests DNA released from broken cells. After the enzymatic treatment, the cells are passed through 37-μm nylon mesh and then fixed for subsequent staining.

To assess possible phase-specific selection of KHT tumor cells dispersed by either the neutral protease or the trypsin–DNase technique, tumor-bearing mice were injected with tritiated thymidine in protocols designed to label cells in S phase, G2 + M, and G2 + M + S + G1 phases [19]. One-half the tumor was dispersed with one of the enzymatic procedures and the remainder was fixed and histologic sections prepared. Autoradiography was used to compare the fraction of labeled cells in tissue sections and in cell suspensions. Preferential phase-specific selection as a result of tumor dispersal would be evidenced by differences in the fraction of labeled cells in the sections and suspensions. However, the labeled cell fraction (Table 2) is similar in suspensions prepared by either neutral protease or trypsin–DNase and does not differ significantly from that obtained in tumor sections, indicating that phase-specific selection has not occurred. Unfortunately, this methodology is not sensitive enough to reveal phase-specific selection of low-frequency subpopulations.

Additional comparison of the two enzymatic protocols indicates that neutral protease provides two-fold higher cell yield than trypsin/DNase, at least in the KHT tumor model. Both dispersal techniques are compatible with growth of tumor cells in either in vitro or in vivo clonogenic assays [19]. The DNA distributions of cells dispersed by either technique show minimal cell clumping and debris and coefficients of variation of 3–5% (Fig. 1). Thus, either dispersal protocol would be suitable for cytokinetic measurements of experimental KHT sarcomas by flow cytometry.

All proteolytic enzymatic dispersal procedures have potential limitations. Alteration of surface membrane components may affect discrimination of subpopulations, particularly when immunolabeling is used. For example, trypsin has been shown to release sialic acid-containing residues as well as specific cell-surface proteins [1,15,22,33]. Some of these alterations are reversible following extended culture periods. Also, enzymatic dispersal protocols, which are effective with some tumors, may be ineffective with others. For example, some sarcomas are easily digested by any number of single enzyme treatments, whereas carcinomas are much more difficult to disaggregate into suitable single cell suspensions. In these cases, other methods involving chemical (i.e., chelating agents) and/or mechanical methods in addition to enzyme digestion may be useful. Clinical tumor samples often require multiple enzyme treatments (e.g., trypsin, collagenase, hyaluronidase) [8] for adequate dispersal, and often cell yield is lower than that obtained with experimental tumors. Finally, enzymatic digestion can be expensive, particularly in studies involving large numbers of tumors. Generally, however, these limitations are outweighed by the advantages of enzymatic dispersal, which continues to be the most commonly used technique for solid tumors.

Dispersal Into Nuclear Suspensions

Preparation of nuclear suspensions is an alternative method to disaggregate tumors for analyses of nuclear DNA or cell cycle distribution. Several methods have been described to obtain suspensions of nuclei from solid tumors. These procedures, developed initially by Vindelov [28] and Fried et al. [2], use mechanical forces to disrupt the tissue matrix, and nonionic detergents, typically Triton-X-100 or

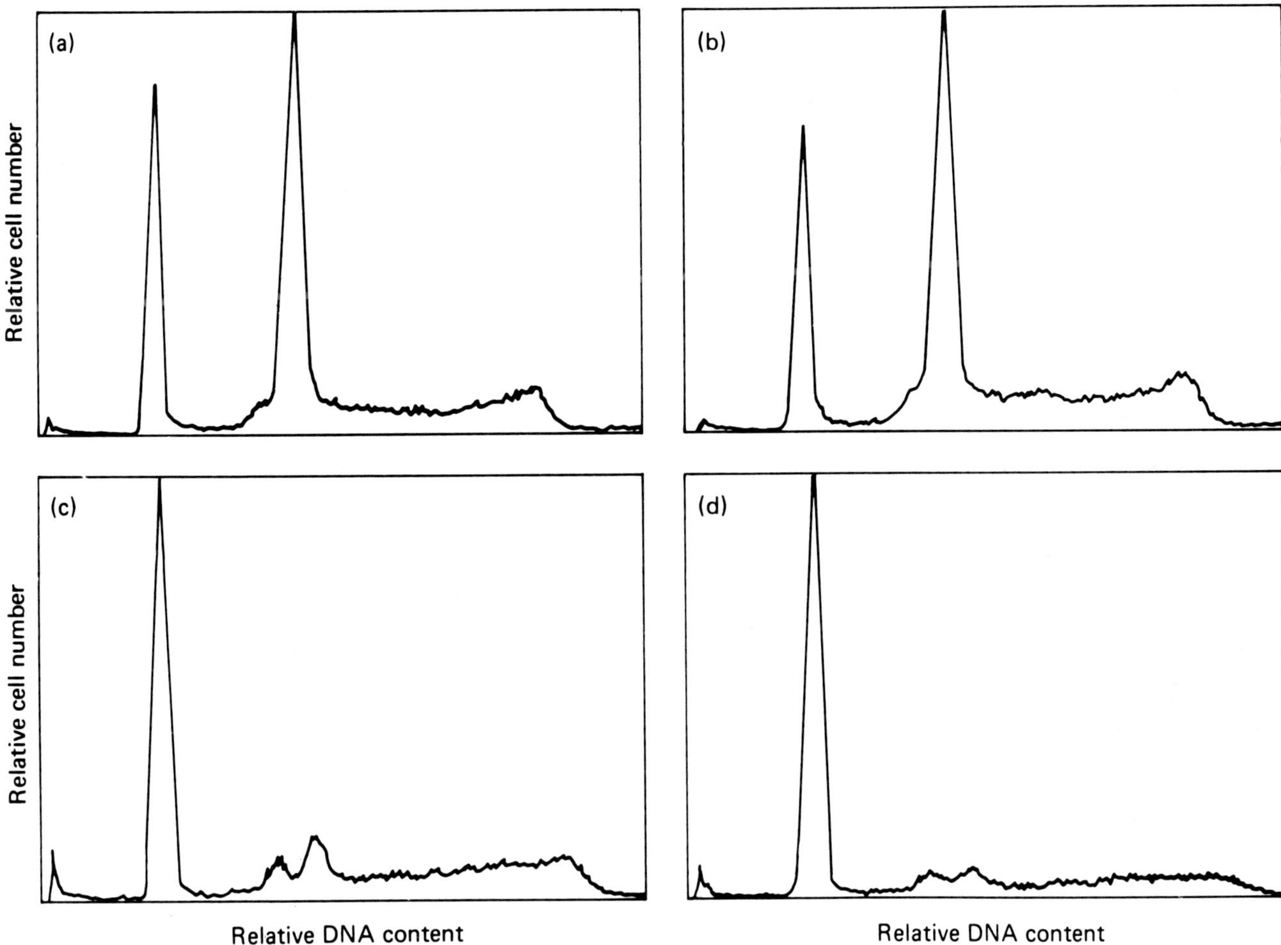

Fig. 1. DNA distributions of untreated KHT tumor cells dispersed by neutral protease **(a)** and trypsin–DNase **(b)**. **c,d:** show DNA distributions obtained from animals treated 12 hr prior to sacrifice with cytosine arabinoside (100 mg/kg). (Reproduced from Pallavicini et al. [19], with permission of the publisher.)

Nonidet P-40, to lyse cells and obtain nuclear suspensions. In a rapid, one-step procedure used by Taylor and colleagues [24,25], minced tumor pieces are simultaneously exposed to a nonionic detergent and a DNA fluorochrome (e.g., DAPI, ethidium bromide/mithramycin). The nuclei rapidly stain with the fluorochrome and are ready for analyses within 5 min. Although preferential subpopulation selection and nuclear yield were not directly evaluated with this procedure, the quality of the DNA distributions is good (CV of 2–3%), and ready detection of aneuploid clones is possible if appropriate standards are used. Figure 2 shows DNA distributions of primary breast tumors with different ploidy levels. The single-step technique has been used successfully with both experimental tumors and clinical material [3,5,9,26].

A multistep technique [29–32] to disperse fine-needle aspirates of solid tumors into single nuclei is shown in Figure 3. In this procedure, the aspirate is suspended in a small volume of buffer with citrate (anticoagulant) and dimethylsulfoxide, which permits long-term storage by freezing at −80°C. The subsequent addition of solutions A, B, and C lyse the cells and stain the nuclei. NP-40 lyses the aspirated cells, and spermine stabilizes the unfixed nuclei. The concentration of spermine is critical, since lower concentrations do not efficiently stabilize the nuclear membrane and higher concentrations result in contamination of the nuclei with cytoplasm and unlysed cells. Trypsin digests cytoplasmic fragments and minimizes differences in the fluorescence intensity of different types of cells (e.g., lymphocytes, granulocytes) that would otherwise be observed following DNA-specific staining. A trypsin inhibitor is used to inactivate trypsin and to obtain reproducible staining with propidium iodide (PI). Nonspecific staining of RNA with PI is eliminated by inclusion of RNase.

The multistep procedure was evaluated according to the criteria listed in Table 1. Since this procedure is most commonly used on clinical samples, direct measurement of possible selectivity in release or preservation of subpopulations by the techniques used for experimental tumors is difficult or impossible. Circumstantial evidence against any selectivity is derived from the observations that (1) no cell loss occurs during storage of ascites tumor cells at −80°C or by the detergent-trypsin procedure, (2) multiple analyses of nuclear suspensions of JB-1 ascites tumors and malignant melanomas grown in nude mice show highly reproducible DNA

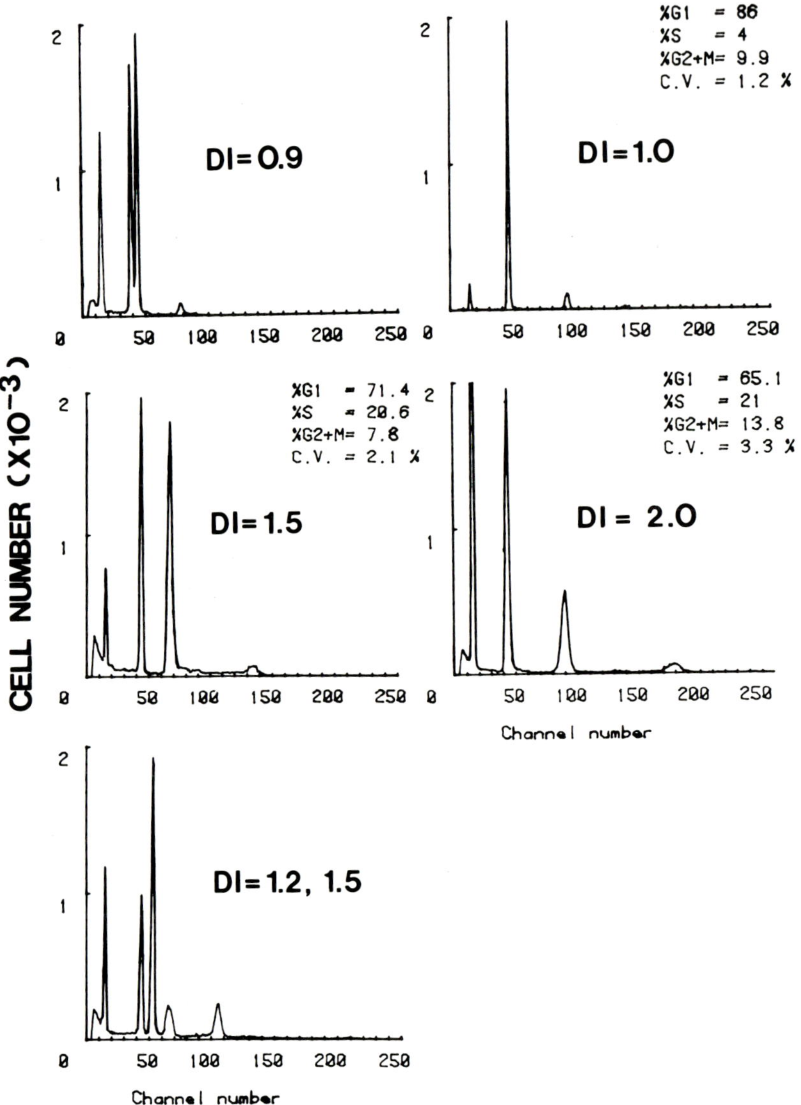

Fig. 2. DNA distributions of primary breast tumors obtained by the detergent-lysis technique of Taylor [26]. The peak between channels 10 and 20 in each histogram represents chicken red blood cells, which are used as an internal biological marker. (Reproduced from Taylor [26], with permission of the publisher.)

distributions (i.e., cell cycle phase fractions), and (3) DNA distributions of fine needle aspirates are similar to those obtained by disaggregation of tumors into single cells.

It is important to note that detergent lysis does not necessarily result in the loss of mitotic cells. The retention of mitotic figures in nuclear suspensions has been observed in several laboratories [2,9,14,16,28] and has been attributed to the existence of interchromosomal connections [23]. In general, DNA distributions of free nuclei show minimal cell clumping. Debris is a problem only if the tissue is very necrotic or if chemotherapy has induced extensive cell death. C.V.s of approximately 2% are obtained routinely. The DNA staining is highly quantitative and allows discrimination of small DNA differences, such as those caused by the human sex chromosomes. Although nuclear suspensions are not useful for subpopulation discrimination on the basis of surface antigens or other intracellular cytoplasmic markers, the nucleus does retain some structure as seen by phase-interference contrast microscopy and flow cyotmetric light-scatter measurements. The yield of nuclei from solid tissue using single or multistep detergent lysis techniques as described above has been difficult to evaluate. However, the capability to perform cytokinetic analyses on nuclei obtained from a single rat hair follicle (I. Taylor, unpublished data) indicates high recovery.

Flow cytometric DNA analyses have been performed on nuclei obtained from normal tissues (including liver, bone marrow, embryonic chick retina) and experimental and human tumors [3,6,17,26]. Some examples are shown in Fig. 4. Selected tissues (i.e., mouse sperm and highly necrotic tumors) have not been successfully analyzed. The fluorescence intensity signal of mouse sperm is much less than expected, and highly necrotic tumors contain fragmented and partially autolyzed nuclei. Nuclei derived from necrotic cells show

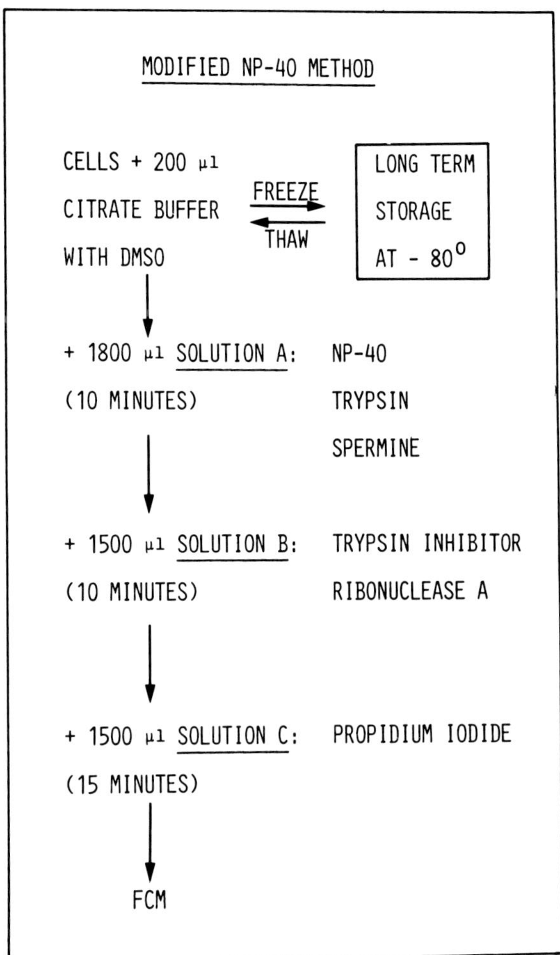

Fig. 3. Flow diagram and salient features of the multistep dispersal procedure.

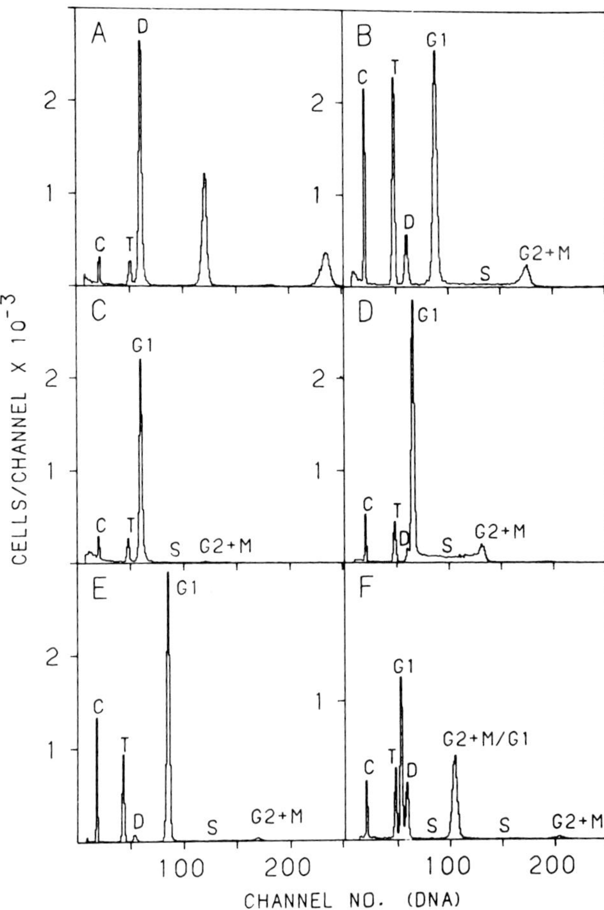

Fig. 4. Examples of DNA distributions obtained by the detergent–trypsin method for preparation of nuclei. The peaks produced by chicken and trout red blood cells are indicated by C and T, respectively. Diploid peaks are indicated by D. The parts of the DNA distribution corresponding to the different cell cycle phases are indicated by G1, S and G2 + M. **A:** Distribution of mouse liver nuclei. Diploid, tetraploid, and octaploid peaks are visible. CV of the G1 peak is 2.3%. **B,C:** DNA distributions of small cell carcinoma of the lung (CV = 2.2%) and oral squamous cell carcinoma (CV = 2.1%), respectively. **D,E:** Diffuse histiocytic lymphoma (CV = 2.1%) and carcinoma of the bladder (CV = 1.5%). **F:** DNA distribution of a mammary carcinoma containing two subpopulations. CV of the G1 peaks in each subpopulation are 1.7% and 1.6%.

higher light-scatter values and fluorescence intensities than do nuclei derived from viable cells. Debris, increasing exponentially to the left of the G1 peak, is invariably present in such distributions. In the single-step procedure [24,25,32] to obtain nuclei, the quality of the DNA distribution is highly dependent on the DNA to dye ratio. Too many cells per given volume of staining solution yields poor distributions. 1–3 $\times$ 10^6 cells/2 ml staining solution was found to be optimal. Other limitations of nuclear suspensions are the obvious loss of cell function, viability, and cell structure, which may be useful for subpopulation discrimination and clonogenic endpoints. The major advantages of nuclear suspensions are the ease with which they can be obtained, high quality DNA distributions, and application to small amounts of clinical material (i.e., fine needle aspirates).

DISPERSAL OF PARAFFIN-EMBEDDED MATERIAL

The availability of paraffin-embedded tumor samples in the archives of most hospitals and the desirability of retrospective analyses led to the development [10,11] of techniques to use these samples for flow cytometric analyses. In the method developed by Hedley and colleagues [10], thick sections (30–100 µm, depending on nuclear size) cut from paraffin blocks are dewaxed, rehydrated and suspended in an enzyme solution containing pepsin. After a short (30-min) incubation, the cell suspension is centrifuged and stained with a DNA fluorochrome (DAPI provides the best distributions although PI and mithramycin may also be used). RNase is added immediately before analysis. Stained samples can be kept for a few hours, but not overnight. This technique is suitable for dispersal of most tumor types in blocks up to ten years old. Failure to obtain sufficient cells for analysis or unacceptably high CV (10%) occurs in about 10–15% of all tissue blocks, independent of tumor type.

Flow cytometry of nuclear suspensions prepared from paraffin-embedded tumor has been carried out on several hundred samples. A comparison of the DNA distributions obtained from tumors before and after paraffin embedding is shown in Figure 5. Similar DNA distributions were obtained

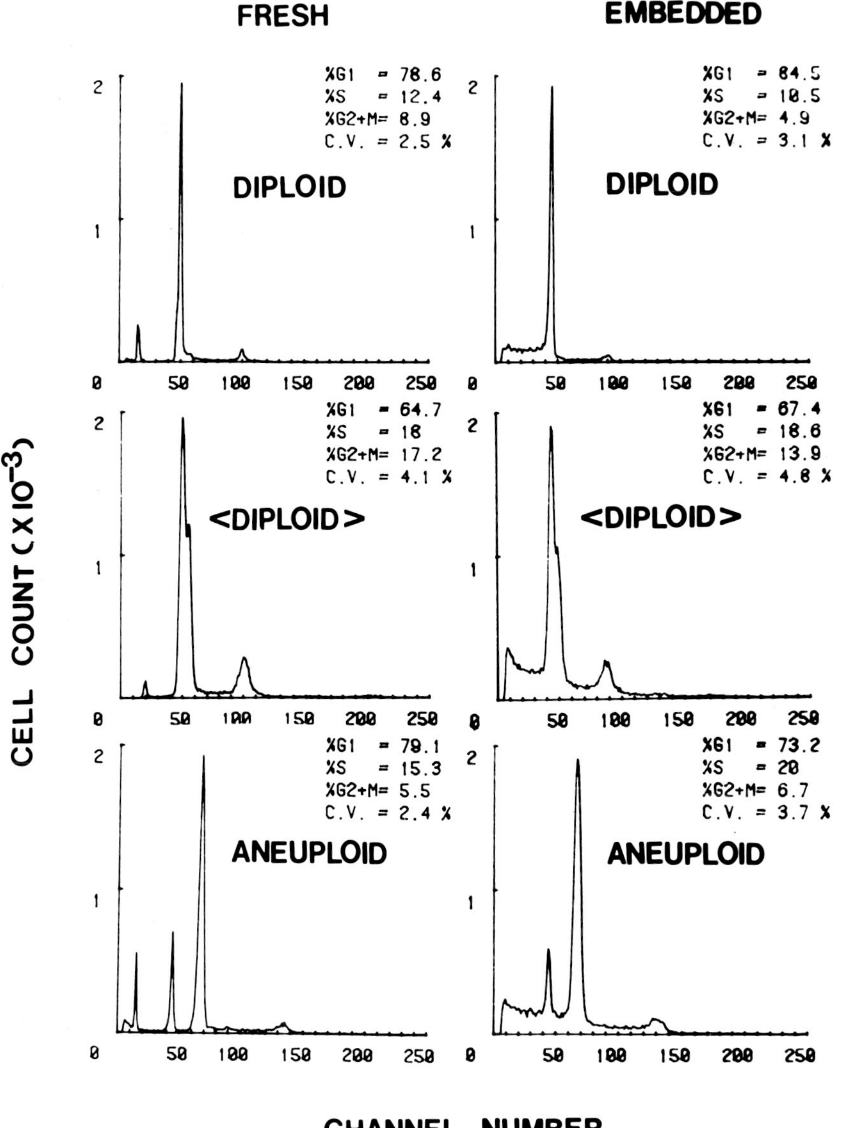

Fig. 5. Comparison of staining methods for nuclei derived from solid tumors. Samples of fresh tissue from three human tumors were divided into two equal parts. One aliquot, Fresh, was prepared for analysis by the detergent lysis technique described earlier [24], while the other was fixed and embedded in paraffin before being prepared for analysis using the method of Hedley et al. [10]. Chicken red blood cells were used as an internal standard in the fresh tissue distribution. (Reproduced from Hedley et al. [10], with permission of the publisher.)

from fresh and corresponding paraffin embedded material, suggesting that there was no selective cell loss during disaggregation of embedded material. However, cell-cycle phase-specific selection due to pepsin treatment of either paraffin-embedded or fresh tissue was not evaluated. Aneuploidy in 196 unfixed, deep frozen primary breast cancers stained with the detergent-lysis method [24] was compared with aneuploidy in 280 paraffin-embedded tumors disaggregated with pepsin, and the distribution of DNA index was almost identical (Fig. 6). Tumors with multiple aneuploid populations represented 15.9% and 15.2%, in unfixed and paraffin-embedded samples, respectively [13]. Collectively, these data suggest that at least the major populations are recovered following disaggregation of paraffin-embedded tissue. However, a slightly higher proportion of tumors with a DNA index of 1.0 (Fig. 6) was found in analyses of the embedded material, likely attributable to the higher CV (3.59 ± 1.05) for embedded versus fresh (2.73 ± 0.88) tumors. Although age of the paraffin block does not appear to influence the CV, it is likely that the quality of fixation and embedding of the original biopsy may be critical. An evaluation of different fixatives has shown that neutral formalin or formalin–acetic acid–acetone provides the best results, whereas Bouin's fixative and mercury-based fixatives are suboptimal. Poor tissue fixation resulting from poor penetration of thick

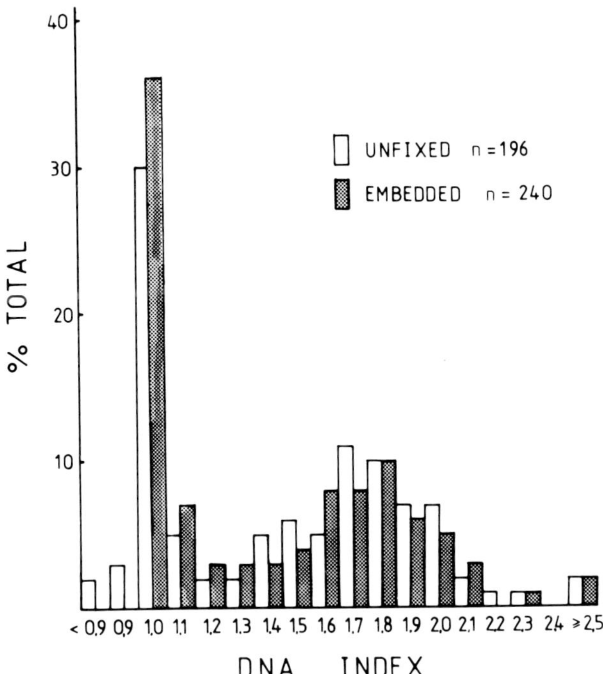

Fig. 6. Comparison of the distribution of DNA indices of stage II primary breast cancer from 196 samples of unfixed tissue and 240 samples of paraffin-embedded material. Multiple aneuploid tumors, which represented 15.9% and 15.2% of each group, respectively, are not included. We have used the convention of calling the G1 peak with the lowest DNA content "diploid" to facilitate comparison of distributions on paraffin-embedded samples. Hypodiploid tumors might therefore be represented as DNA index 1.1 or 1.2.

tissue blocks may contribute to the failure to obtain adequate DNA distributions on 10–15% of the paraffin-embedded samples.

Internal standards such as chicken red blood cells (CRBC) may be added to suspensions of cells (nuclei) from the ratio of diploid tumor G_1 peak to CRBC peak show a large variance for paraffin embedded (3.05 ± 0.41) and for fresh tissue (3.01 ± 0.13) stained with DAPI. Attempts to reduce the variance by using formalin-fixed and pepsin-treated CRBC have been unsuccessful. Fortunately, since most tumors have a diploid cell component, this subpopulation can be used as a reference for estimates of ploidy level. Thus, all but a few hypodiploid tumors can be identified with confidence.

The nuclear suspensions obtained from paraffin-embedded tumors meet few of the suggested criteria listed in Table 1. They are obviously not useful for functional studies or for subpopulation discrimination using cell-surface markers or other morphologic criteria. Nuclear yield is dependent on the size of the tissue section disaggregated and has not been critically quantitated. The DNA distributions suggest that there is loss of nuclear material, since a debris continuum is evident in most profiles. However, nuclear yield has been sufficient to permit successful analysis of embedded materials in several large clinical studies of breast cancer [12], ovarian cancer [6], and metastatic adenocarcinoma of unknown primary site [13]. In addition, the capability to obtain DNA distributions on samples taken 10 years previously is a powerful tool for retrospective studies of tumor diagnosis, treatment response, and prognosis.

SUMMARY

Dispersal of solid tumors for flow cytometric analyses can be achieved by a variety of methods. The choice of dispersal method depends on the type of material to be examined and the end point to be measured. For flow cytometric analyses of cell cycle distribution and aneuploidy, obtaining representative subpopulations in dispersed cell/nuclear suspension is as important as adequate yield and high-quality DNA measurements. Proper characterization of the cells/nuclei obtained from solid tumors is essential for accurate extrapolation of flow cytometric measurements in vitro to properties of the tumor in situ.

REFERENCES

1. **Cook GMW, Heard DH, Seamna GVF (1960)** A sialomucopetide liberated by trypsin from the human erythrocyte. Nature (Lond) 138:1011–1012.
2. **Fried J, Perez AG, Clarkson BD (1978)** A rapid hypotonic method for flow cytofluorometry of monolayer cell cultures: some pitfalls in staining and DNA analysis. J Histochem Cytochem 26:921–933.
3. **Friedlander ML, Taylor IW, Russell P, Musgrove EA, Hedley DW, Tattersall MHN (1983)** Ploidy as a prognostic factor in ovarian cancer. Int J Gynecol Pathol 2:55–63.
4. **Friedlander ML, Hedley DW, Taylor IW (1984)** The clinical and biological significance of aneuploidy in human tumors: A review. J Clin Pathol 37:961–974.
5. **Friedlander ML, Hedley DW, Taylor IW, Russell P, Coates AS, Tattersall MHN (1984)** The prognostic significance of cellular DNA content in advanced ovarian cancer. Cancer Res 44:397–400.
6. **Friedlander ML, Taylor IW, Russell P, Tattersall MHN (1984)** Cellular DNA content—A stable marker in epithelian ovarian cancer. Br J Cancer 49:173–181.
7. **Gray JW (1986)** Flow cytokinetics. In Gray JW, Darzynkiewicz, Z (eds), "Techniques for Analysis of Cell Proliferation." Clifton, NJ: Humana Press, pp 93–138.
8. **Hamburger AW, White CP, Tencer K (1982)** Effect of enzyme disaggregation on proliferation of human tumor cells in soft agar. J Natl Cancer Inst 68:945–949.
9. **Hedley DW, Joshua DE, Tattersall MHN, Taylor IW (1982)** Fine needle aspiration from the sternum. Lancet 2:415–416.
10. **Hedley DW, Friedlander ML, Taylor IW (1983)** Method for analysis of cellular DNA content of paraffin-embedded pathological material using flow cytometry. J. Histochem Cytochem 31:1333–1335.
11. **Hedley DW, Friedlander ML, Taylor IW, Rugg CA, Musgrove EA (1984)** DNA flow cytometry of paraffin-embedded tissue. (Letter to the editor.) Cytometry 5:660.
12. **Hedley DW, Rugg C, Taylor IW, Ng A (1984)** Influence of cellular DNA content on disease-free survival of Stage II breast cancer patients. Cancer Res 44: 5395–5398.
13. **Hedley DW, Friedlander ML, Taylor IW (1985)** Application of DNA flow cytometry to paraffin-embedded archival material for the study of aneuploidy and its clinical significance. Cytometry 6:327–333.
14. **Hymer EC, Kuff EL (1964)** Isolation of nuclei from

mammalian tissues through the use of Triton-X-100. J Histochem Cytochem 12:359–363.
15. **Milas L, Mujagic H (1973)** The effect of splenectomy on fibrosarcoma "metastasis" in lungs of mice. Int J Cancer 11:186–190.
16. **Morris VB, Taylor IW (1985)** Estimation of nonproliferating cells in the neural retina of embryonic chicks by flow cytometry. Cytometry 6:375–380.
17. **Murray JD, Berger ML, Taylor IW (1981)** Flow cytometric analysis of DNA content of mouse liver cells following *in vivo* infection by human adenovirus type 5. J Gen Virol 57:221–226.
18. **Pallavicini MG, Cohen AM, Dethlefsen LA, Gray JW (1978)** Dispersal of solid tumors for flow cytometer (FCM) analysis. In Lutz D (eds), "Pulse-Cytophotometry." Part III. Ghent, Belgium: European Press, pp 473–482.
19. **Pallavicini MG, Folstad LJ, Dunbar C (1981)** Solid KHT tumor dispersal for flow cytometric cell kinetic analysis. Cytometry 2:54–58.
20. **Pallavicini MG, Summers LJ, Dolbeare FD, Gray JW (1985)** Cytokinetic properties of asynchronous and cytosine arabinoside perturbed murine tumors measured by simultaneous bromodeoxyuridine/DNA analyses. Cytometry 6:602–610.
21. **Pallavicini MG (1986)** Solid tissue dispersal for cytokinetic analyses. In Gray JW, Darzynkiewicz Z (eds), "Techniques for Analysis of Cell Proliferation." Clifton, NJ: Humana Press, pp 139–162.
22. **Snow C, Allen A (1970)** The release of radioactive nucleic acid and mucoproteins by trypsin and ethylenediaminetetraacetate of baby hamster cells in tissue culture. Biochem J 119:707–714.
23. **Takayama S (1975)** Interchromosomal connections in squash preparations of L cells. Exp Cell Res 91:4018–412.
24. **Taylor IW (1980)** A rapid single step staining technique for DNA analysis by flow microfluorimetry. J Histochem Cytochem 28:1021–1024.
25. **Taylor IW, Milthrope BK (1980)** An evaluation of DNA fluorochromes, staining techniques, and analysis for flow cytometry. 1. Unperturbed cell populations. J Histochem Cytochem 28:1224–1232.
26. **Taylor IW, Musgrove EA, Friedlander ML, Foo MS, Hedley DW (1983)** The influence of age on the DNA ploidy levels of breast tumors. Eur J Cancer Clin Oncol 19:623–628.
27. **Thomson JE, Rauth AM (1974)** An *in vitro* assay to measure the viability of KHT tumor cells not previously exposed to culture conditions. Radiat Res 58: 262–276.
28. **Vindelov LL (1977)** Flow microfluorimetric analysis of nuclear DNA in cells from solid tumors and cell suspensions. Virchows Arch [Cell Pathol] 24: 227–242.
29. **Vindelov LL, Christensen IJ, Keiding N, Spang-Thomsen M, Nissen NI (1983)** Long term storage of samples for flow cytometric DNA analysis. Cytometry 3:317–322.
30. **Vindelov LL, Christensen IJ, Nissen NI (1983)** A detergent trypsin method for the preparation of nuclei for flow cytometric DNA analysis. Cytometry 3:323–327.
31. **Vindelov LL, Christensen IJ, Nissen NI (1983)** Standardization of high-resolution flow cytometric analysis by the simultaneous use of chicken and trout red blood cells as internal reference standards. Cytometry 3:328–331.
32. **Vindelov LL, Christensen IJ, Hensen G, Nissen NI (1983)** Limits of detection of nuclear DNA abnormalities by flow cytometric DNA analysis. Results obtained by a set of methods for sample storage, staining and internal standardization. Cytometry 3:332–339.
33. **Weiser RS, Heise E, McIvor K, Han S, Granger G (1969)** *In vitro* activities of immune macrophages. In Smith RT, Good RA (eds), "Cellular Recognition." E. Norwalk, CT: Appleton & Lange, pp 215–220.

11

Methods for Enrichment of Cellular Specimens for Flow Cytometry and Sorting

A. Tulp and J. G. Collard

Department of Biochemistry (A.T.) and Cell Biology (J.G.C.), Netherlands Cancer Institute, Antoni van Leeuwenhoekhuis, 1066 CX Amsterdam, The Netherlands

RATIONALE FOR ENRICHMENT PROCEDURES

The yield of any sorting method that processes cell by cell is limited by the cell measurement and sort time (*per definition*) and by dead time of the sorting decision. Quantitative yield benefits greatly from enrichment measures. These enrichment steps are generally carried out in batches and lend themselves to scaling-up whenever massive amounts are needed.

If in a mixture there are N particles, of which n are desirable for flow sorting, then contamination of the sorted pool by unwanted particles is given by

$$f = \frac{N - n}{N} \frac{3vt}{1 + v(3t - \tau)} \qquad (1)$$

in which f denotes fraction of contaminants, t drop generation time, τ dead time of analog electronics and v flow rate (particles/sec).

Equation (1) takes into account: 1° Poisson statistics; 2° dead time of the analog electronics; 3° three droplets are deflected per desired particle; 4° absence of overlap rejection [27]. It follows from equation (1) that a number of particles is never detected nor analyzed and that there is always a controversy of purity versus recovery. Under typical working conditions of a commercial flow sorter (τ = 20 μsec; t = 25 μsec and v = 1,000 particles/sec) equation (1) yields

$$f = 0.0714 \frac{N - n}{N} \qquad (2)$$

A numerical example for which n/N = 0.1, gives that the contamination is 6.4%. When an enrichment step augments n/N to 0.8, calculation using equation (1) shows that similar purity is obtained at particle flow rates as great as 5456 particles/sec; 0.8 × 5,456 = 4,365 desired particles per second are collected compared with 0.1 × 1,000 = 100 particles/sec from a nonenriched suspension. In this theoretical case, the sorting process is speeded up 44-fold. The lower the initial percentage of desired particles in the starting suspension, the more impressive the speeding-up factor. In general, when desired particles (at initial fraction a) are sorted at V_1 particles/sec, then enrichment of these particles to fraction b permits sorting at V_2 particles/sec to obtain a similar purity, and thus the speeding-up factor K equals

$$K = \frac{4(b - a) + \tau v_1(1 - a)}{v_1} \cdot \frac{2}{b} \qquad (3)$$

Theoretically it follows that the sorting of a particular, e.g., human chromosome (at a = 0.0435 and enriched to b = 0.5) can be speeded up 482-fold.

Apart from considerable speeding up, enrichment steps generally remove debris, large aggregates and dead cells. Removal of dead cells by equilibrium density centrifugation, since they are swollen and lighter than living cells, facilitates flow cytometry of the relevant DNA content of living cells after treatment by radiation or cytotoxic drugs. In chromosome sorting studies, debris may comprise 20% of the sorted material [14] and enrichment by velocity sedimentation removes debris almost completely. Thus, enrichment should also facilitate interpretation of flow histograms by avoiding the complication of a debris continuum under the peaks of the fluorescence frequency distribution. In analyzing DNA distributions, background due to cell debris and various aggregates of debris often extends over the whole histogram leading to wrong estimation of cell cycle compartments; enrichment avoids this. As for analytical purposes, mixed populations (e.g., from solid tumors) may have subdiploid cells

Flow Cytometry and Sorting, Second Edition, pages 195–208

that are difficult to demonstrate in low percent concentration, but may become obvious after enrichment procedures [56].

It is important that the desired particles (cells), that generally become diluted by the enrichment procedure(s), be easily washed and concentrated by sedimentation (centrifugation or at 1*g*) without major cell loss or damage due to pelleting or clump formation, and that they be brought as single cells to higher concentrations prior to flow cytometry or cell sorting. In the case of sticky particles like chromosomes this is obviously a difficult task.

PHYSICAL METHODS FOR THE ENRICHMENT OF CELLS AND LARGE CELL ORGANELLES

Enrichment procedures that exploit small differences in the physical properties of cells have the advantage that cells are not modified by staining or adherence of ligands that might initiate cellular processes. Intrinsic scalar parameters that can be used for cell separation purposes are size, specific gravity and surface charge. In this chapter, cell surface charge is only briefly touched upon; density and size are treated more extensively since these two parameters are more often the basis for enrichment procedures.

Electrophoresis

In electrophoresis, mobility is related linearly to particle surface charge (ζ-potential). Differences in surface charge are exploited for cell enrichment in the commercial (but costly) free-flow electrophoresis apparatus [16]. By injecting a continuous flow of particles into a thin vertically flowing curtain in a high voltage electrical field, cells carrying different surface charge are differentially deflected. Cells can also be separated in zones by electrophoresis stabilized in a stationary density gradient at low electrical field strength (1 to 6 V/cm) either in hand-made columns [42] or in the commercial Buchler PolyPrep (Büchler Instr., Fort Lee, NJ) column, as described in extenso by Platsoucas [33] and Tulp [43].

Two-Phase Partition

Two-phase partition offers a very gentle means of enriching cells that possess subtle cell surface differences. The cells are partioned over two aqueous polymer phases in equilibrium. There is then a propensity for them to enter one of the phases, depending on their cell surface properties. The method was most efficiently exploited in the counter current device of Albertsson [3]. This automatic apparatus processes large amounts of cells but due to the repetitive schedule of shaking and settling of the phases, the method is quite lengthy. The partition coefficient is related in an exponential form (which gives high selectivity) with the difference in electrostatic free energy, the difference in interfacial free energy between the phases and the surface area of the particle (cell) [57]. Covalent attachment of hydrophobic ligands to the polymers in the phases makes possible a separation of cells according to their surface hydrophobic properties.

Separation According to Specific Gravity

During density gradient centrifugation, particles equilibrate at their buoyant density irrespective of size and shape. Equilibrium density centrifugation, if properly done (see below) has a considerable cell capacity: 4×10^9 nucleated cells of different classes together with $2–4 \times 10^{10}$ erythrocytes can be separated with good resolution in density gradients of about 100 ml. There is a distinction between discontinuous density gradient separation, that by its very nature gives artifactually discrete populations at the interphase of the density discontinuities, and continuous density gradient separations. Separations may be performed in swing-out tubes, in angle head rotors (that both suffer from wall sedimentation) and in the reorienting zonal rotor (36). In this chapter use is made of a low g swing-out separation chamber (vide infra) according to Tulp et al. [48] for buoyant density separation of mammalian cells.

Size Separation by Velocity Sedimentation

According to Stokes' law, particles are sorted according to size when a thin zonal layer of particles enters into a viscous (turbulence-stabilized) medium under the influence of a gravity force. Separations are performed either at unit gravity or at higher g-force in (reorienting) zonal rotors, swing-out tubes, and in a centrifugal elutriation cell. Of these devices, swing-out tubes suffer from severe wall sedimentation in a radial centrifugal field; cell loss up to 75% of the desired cell types has been recorded (38). For that reason cylindrical swing-out tubes are not recommended for the separation of cells, nuclei, and chromosomes.

Reorienting zonal rotors (SZ-14, Du Pont, Newton, Conn.) function without rotating seals. These rotors are loaded while spinning by spraying a density gradient and the cell sample into the separation compartment(s). The rotor is unloaded at unit gravity after the liquid contents of the chamber are reoriented back to the horizontal. If the density difference between sample layer and top of the gradient is small, broad starting zones are produced [60] that are detrimental to good separations. It is noteworthy that Wells [59] who has a long-standing experience with the so called Reograd rotors, remarks that unit gravity separation of human lymphocytes and monocytes—although quite moderate in our opinion—is superior to that obtained in the zonal rotor [62]. Moreover, it has been demonstrated that the resolving power for red cells and lymphocytes was lower in the reorienting zonal rotor than it was in ordinary swing-out tubes [15]. We believe that the relatively low resolution is due to the peculiarities of the Reograd rotor.

Zonal rotors with rotating seals have been used for the separation of cells [58], chromosomes [5,8,35] and rat liver nuclei [20] with moderate to good success. The method is scarcely used in recent years probably because it is considered an expert's technique.

Of excellent resolving power and high capacity is the centrifugal elutriation rotor of Beckman Instruments. In this device, sedimentation of particles in a centrifugal field is counteracted by a flow of viscous medium that has access to the centrifugal chamber via a rotating seal, so that small, slowly sedimenting particles are entrained to be collected and larger particles remain stationary inside the chamber [25,28]. For high resolution, both buffer flow rate, temperature and rotor speed must be strictly controlled to within 0.5%. In the standard elutriation chamber (of 4.2 ml), about 10^9 cells (of about 10-μm diameter) can be processed.

The advantage of centrifugal elutriation is that simple buffers are used without addition of macromolecular solutes that may have adverse effects on cellular integrity. However, an inherent disadvantage is that elutriated fractions have rather large volumes (100 to 150 ml) irrespective of whether a large or small number of cells are processed. For exhaustive elutriation, fractions of 600 ml have to be collected [21]. There exist several methods to sweep cells from the elutriation boundary [44]: stepwise increase of elutriation flow at

Fig. 1. The low g sedimentation chamber. **Left:** the polycarbonate chamber is closed; a handle permits lifting the chamber from the centrifuge; the handle can be pressed into a groove in the top lid. The antivortex cross stands in front. **Right:** Chamber opened with the antivortex cross in position. In the top lid, a flow-deflector can be seen.

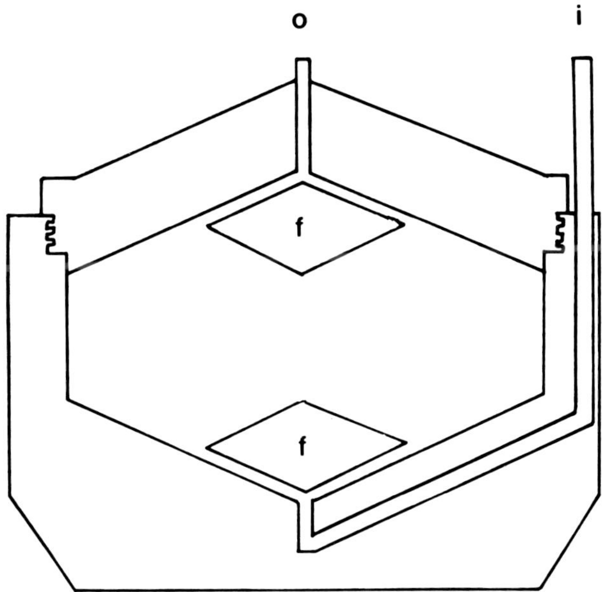

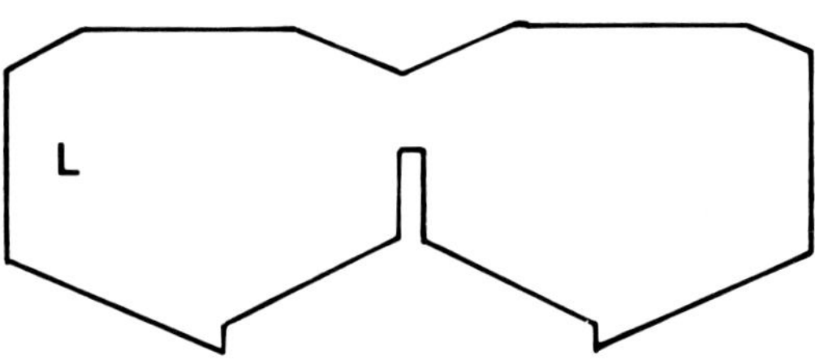

Fig. 2. Diagram of the low g sedimentation chamber, lateral view. f = flow-deflector; i = inlet; o = outlet; L = lamella of antivortex cross.

constant rotational speed; stepwise decrease of rotational speed at constant buffer flow; stepwise increase of the density of the elutriation buffer at constant flow and constant rotational speed. For its many applications the reader is referred to Beckman's own bibliography.

In this chapter, emphasis is laid upon a small (2 cm high, 50 cm^2) separation chamber that originated from a much larger unit gravity design [45,46]. It is operated at 10 to 130*g*. In its ease of manipulation it is applicable from bacteria and chromosomes up to large mammalian cells for velocity sedimentation of particles, and functions equally well for equilibrium density centrifugation of cells.

THE LOW GRAVITY SEDIMENTATION CHAMBER

The sedimentation chamber, originally made of Perspex and commercially available through Phywe (Göttingen, West Germany) is described in more detail in references [12,48]. Figure 1 shows a separation chamber made of the nylon "Macrolon" that is sterilizable. A diagram of the chamber is depicted in Figure 2. The effective internal cross-sectional area is 50 cm^2, and the effective internal height is 2 cm. The flow-deflectors in top and bottom cone guarantee undisturbed thin sample layering and undisturbed rapid fractionation. Samples of 0.3-mm thickness can be distributed uniformly over 50 cm^2. The stainless steel antivortex cross aborts eddy formation upon acceleration (linear in 1½ min to 310 rpm) and deceleration (to 0 rpm in 2 min), see discussion in reference (48). Briefly, in case cells are to be separated (for chromosomes, sucrose is used as a density solute [12]), the chamber is first filled completely with 70% Percoll (Pharmacia Fine Chem. Co., Uppsala) via the bottom inlet using a peristaltic pump. Next the direction of the peristaltic pump is reversed and a density gradient of 90 ml, dense end first, ranging linearly from 4 to 18% (w/v) Percoll is introduced via the top of the separation chamber within 5 minutes. The chamber is transferred in a 600 ml standard swinging bucket to the centrifuge (TGA-6 Damon/IEC, Needham Heights, Ma.). Then, 5 to 10 ml of a cell suspension is introduced into the chamber, via the top inlet, within 20 sec followed by an overlay of 22 ml buffer within 1 min, again by pumping cushion liquid from the chamber so that the sample just reaches the cylindrical (50 cm^2) part of the chamber. Centrifugation at 20*g* (310 rpm) is performed for time periods as described in the legends to the figures—70% Percoll functions as a cushion to collect cells that might otherwise escape from the 2-cm effective length of the chamber and collide with the bottom cone. After centrifugation, fractions of 4 ml are collected via the top cone by upward displacement as 70% Percoll is introduced via the bottom cone within about 5 minutes. Although not absolutely necessary, weak pulsations in the effluent due to the peristaltic pump may be suppressed by inserting a flow integrator (Cole-Palmer Instrument Co., Chicago, JL. cat. no. 7596-20) between peristaltic pump and separation chamber. All solutions contain PBS plus 0.5% BSA. All manipulations are carried out at 4°C.

Several factors render high resolving power to the device:

1. The flow-deflectors mounted in the top and bottom cone

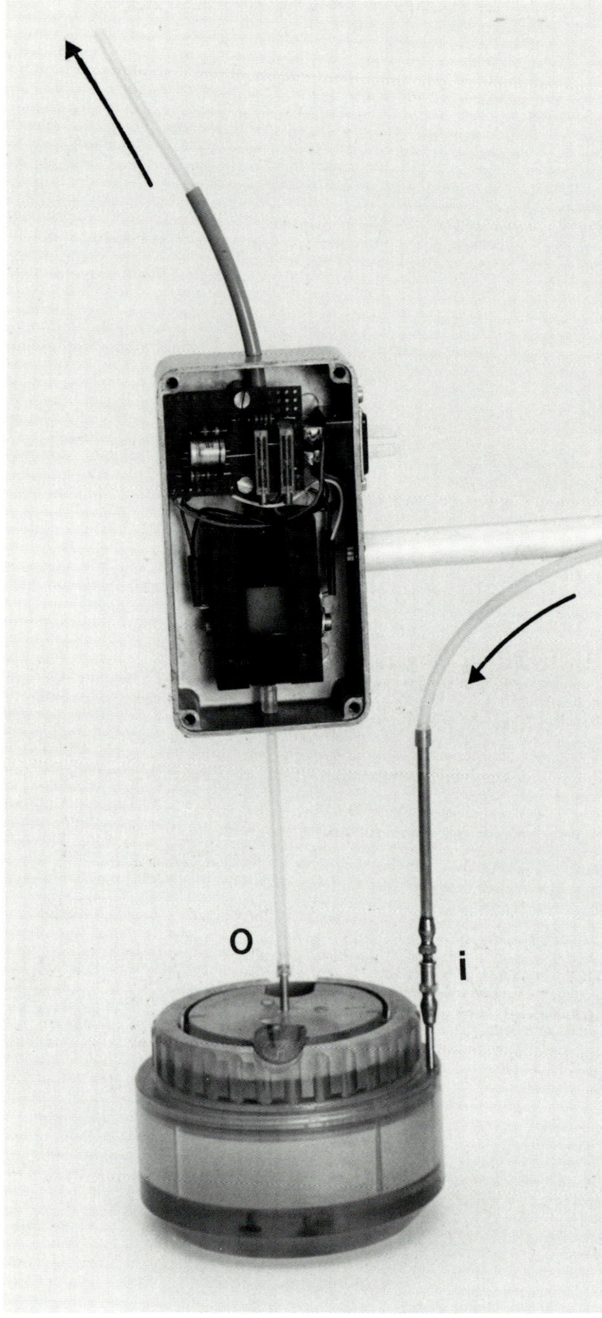

Fig. 3. Method for fractionation of the contents of the low g chamber following sedimentation. Dense cushion liquid (see arrows) is introduced via i into the chamber. The effluent, leaving the chamber at o, is monitored by a flow through cuvette (type 134-0s, Hellma, West Germany) equipped with a gallium arsenide phosphide red light emitting diode and a photodetector to measure changes in light scatter.

permit samples of 0.3-mm thickness to be distributed uniformly over a surface of 50 cm^2, of importance, since resolution in zonal velocity sedimentation is in part determined by the initial width of the sample layer.

2. At speed and upon complete swing-out of the sedimentation chamber, the sample layer is 21 cm from the axis of rotation. Since the chamber is only 2 cm high, it follows from geometry that there is almost no wall sedimentation (loss of particles) [47].
3. The liquid content of the chamber is prevented from swirling during acceleration and deceleration. The four perpendicular lamellae that form the antivortex cross (see Fig. 1) are sufficient to suppress eddy generation completely, the more so since a moderate, but absolutely necessary, acceleration-deceleration schedule ($\pm\Delta\omega$ = 200 rpm) goes with it.
4. It is a general rule that at higher gravity forces more cells can be applied to a density gradient than under unit gravity conditions (see [26] for theory). Thus, while 2×10^6 mononuclear cells per cm^2 can be layered and separated at 20g to 91% purity of monocytes, Wells [59] has shown that 0.2×10^6 mononuclear cells per cm^2 can be separated to 77% monocytic purity at unit gravity. In the present setup, cells are loaded at unit gravity, and then brought as rapidly as possible (without generating vortices) to the desired centrifugal speed, thus avoiding fluid instabilities at unit gravity due to overloading. These "streaming" phenomena develop usually after a few minutes at unit gravity.
5. The chamber being loaded at ambient gravity, liquid isodense layers which are initially flat turn into paraboloids of revolution upon centrifugation. This reorientation of isodense layers occurring twice (during acceleration and deceleration) is only marginal and easily tolerated by the density gradient used [49].

APPLICATIONS OF SIZE SEPARATION OF MAMMALIAN CELLS AND CELL ORGANELLES BY VELOCITY SEDIMENTATION AT MODERATE GRAVITY FORCES

Two exemplary size separation studies of well-defined cell preparations are given:

1. Mononuclear cells that are already separated from blood in the starting preparation
2. Cells enzymatically released from an intact organ

These two examples indicate that the method is generally applicable for enrichment of any desired cell type. In addition, an example of size separation is given that describes velocity sedimentation of metaphase chromosomes.

Separation of Monocytes and Lymphocytes

Human mononuclear cells were obtained from a Ficoll–Hypaque interphase according to Boyum [7]. Separation of lymphocytes and monocytes was effectuated at 20*g* for 10½ min only. For swift analysis, the effluent from the separation chamber can be monitored by optical density measurement. Fortuitously, monocytes have a prominent optical density profile that is not related to actual cell number but based upon the fact that the larger monocytes scatter more light than the small lymphocytes. Figure 4 shows the sedimentation profiles (expressed in optical density) of two separations from different donors indicating that preparations with varying amounts of monocytes yield clear cut peaks. Mere inspection of the optical density profiles allows one to allocate the populations of lymphocytes (L) and monocytes (M); separation takes place within a sedimentation path of only 1 cm underlining the fine resolving power of the device. The small

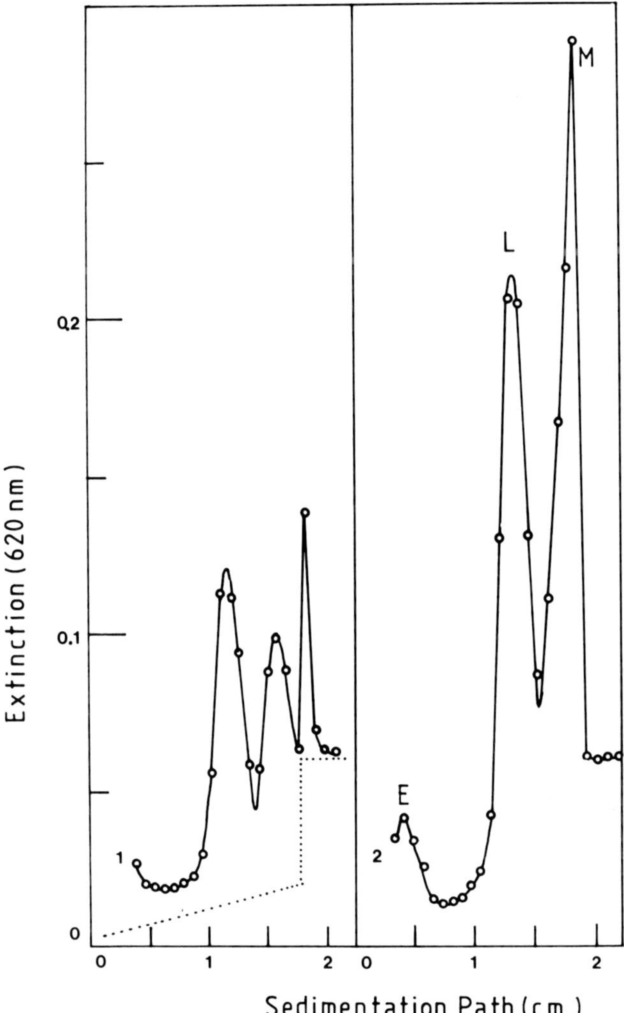

Fig. 4. Separation of human monocytes and lymphocytes by velocity sedimentation at 20*g*: 5 ml of a mononuclear cell suspension were separated for 10½ min at 4°C and 20 xg on a 90 ml linear density gradient (3.8 to 18% Percoll). For details, see text. Fractions of 4 ml each were collected and optical density of fractions was monitored at 620 nm. Curve 1 = 2×10^7 nucleated cells were loaded. Curve 2 = 5.8×10^7 nucleated cells were loaded. E = erythrocytes, L = lymphocytes, M = monocytes. The dotted line represents optical density of the Percoll gradient.

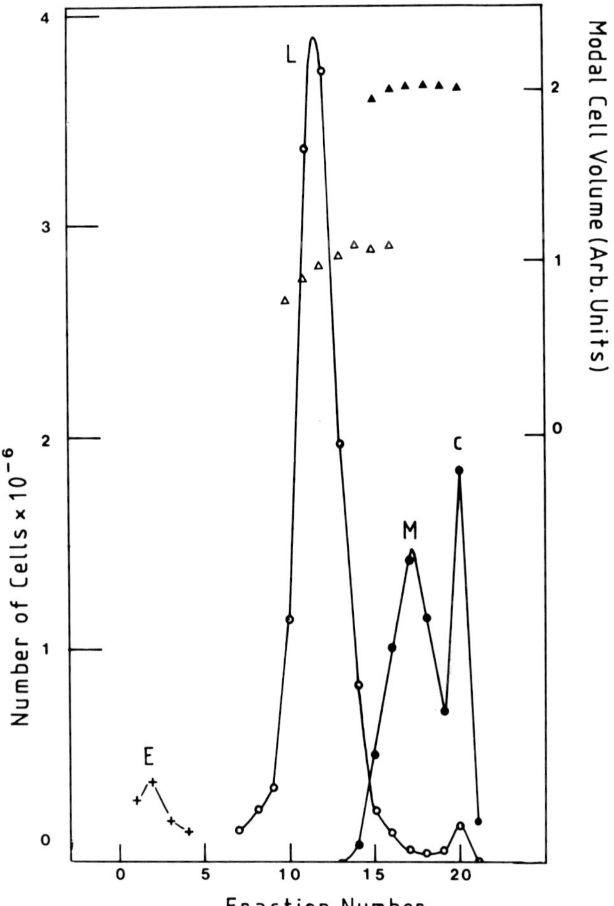

Fig. 5. Sedimentation profile of human mononuclear blood cells after velocity sedimentation at 20*g*. Experimental conditions were identical to those of curve 1 (Fig. 4). Modal cell volume was measured by electronic cell sizing [51]. (○——○) lymphocytes; (●——●) monocytes; (+——+) erythrocytes; (△——△) modal lymphocyte volume; (▲——▲) modal monocyte volume; c = cushion fraction.

population of contaminating erythrocytes (E) is at a comfortable distance from the lymphocyte population. Even at mononuclear cell loads of about 2×10^7 cells per 5 ml sample (curve 1) optical density measurement (at 1 cm optical path length) is feasible and facilitates rapid collection of the monocyte population upon fractionation: one starts to collect the monocytes right after the descending limb due to the lymphocytes is followed by an increase in optical density due to the appearance of the monocytes. Of necessity the optical density profiles depicted are more or less crude since aliquots of 4 ml (to limit the number of fractions) are collected. The sedimentation profile might be much more refined and sharper if there were continuous monitoring of the effluent. To that aim a simple glass flowthrough cuvette equipped with a LED and photodetector was constructed (see Fig. 3). Although not having the precision of sophisticated spectrophotometers the design met our needs rather well. With it, the monocyte population is easily traced during the fractionation procedure, adding to the rapidity of the separation method.

A cushion of 70% Percoll is present in the chamber to prevent cells from passing the effective part (2 cm) of the chamber, beyond which they may collide with the bottom cone. Figure 4 shows that in curve 1 the monocyte peak is at a small but significant distance from the cushion layer whereas in curve 2 the monocyte peak comes close to the cushion. These (small) differences might be due to the use of different donors or to lack of control of the exact centrifugal speed in the present set-up. Setting the centrifugal speed at 310 rpm (20*g*) might vary from 305 (19.4*g*) to 315 (20.6*g*) rpm.

Figure 5 shows the sedimentation profile (expressed in real cell number) of the separation study 1 of Figure 4. Erythrocyte (E), lymphocyte (L), and monocyte (M) populations stand well removed from each other. Some of the monocytes

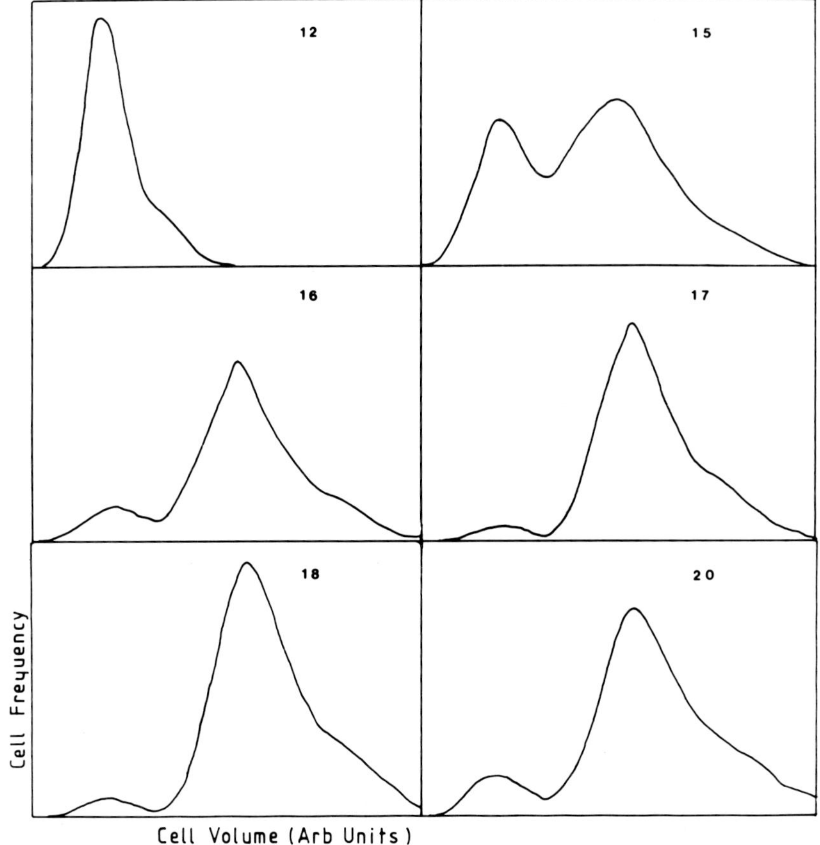

Fig. 6. Electronic cell sizing of mononuclear fractions as separated by velocity sedimentation at 20g. Experimental conditions were identical to those of Figure 5. Cell volume was analyzed with a Coulter counter equipped with a multichannelyzer [51]. Numbers denote fraction number.

ended up in the cushion liquid (C). Electronic cell sizing pertaining to separation study 1 is depicted in Figure 6. Only monocyte enriched fractions are given except for fraction 12; lymphocyte fractions that were >99% pure, are omitted. From these volume measurements it follows that modal lymphocyte volume (not depicted in Fig. 6) increased with increasing sedimentation path as is visualized by the symbols (△) relating modal cell volume to sedimentation path (see Fig. 5); modal cell volume of monocytes remains virtually constant, unlike findings we reported recently for equilibrium density centrifugation of monocytes [51]. Apparently constancy of volume represents the native state of monocytes remaining preserved at 4°C due to the short separation time. Prolonged stay of monocytes at room temperature promotes swelling of these cells probably due to pinocytic processes [55].

Table 1 presents the results of eight separations from different donors. Some features deserve attention: peak fractions of the monocyte population were 91.4 to 96.4% pure (Exp. 1–5). When the fraction(s) adjacent to the peak fraction was included 47.4 to 66.5% of all monocytes collected were in fractions that were 91 to 95.1% pure (exp. 1–5). A differentiation must be made between cell loads $<6\times10^7$ cells per 5 ml sample (expts. 1 to 5) and cell loads at about 10^8 cells per 5 ml (expt. 6). In the former case one stays well below the loading capacity of the chamber and high purities are produced, in the latter case one apparently exceeds the loading capacity so that streaming might ensue to the detriment of resolving power: purity of 81.8% only (expt. 6) was produced. At these high cell loads, it is preferable to increase sample layer volume so that one does not exceed the streaming limit [29,32]. Even though increasing the sample layer thickness twofold (to 10 ml) influences resolving power adversely, purities in the peak monocyte fractions of about 91% are still produced (expts. 7, 8) in contrast to the 81% purity one obtains at small sample layer volume (5 ml) (expt. 6). Total recoveries of nucleated cells vary from 88% to 95% (av. 90.3%). Viability of fractionated cells as measured by the trypan blue exclusion test was always in excess of 99%.

Separation of Kupffer Cells From Rat Liver

Sinusoidal lining cells were obtained after perfusion of a rat liver in situ with 0.1% pronase in Gey's balanced salt solution, subsequently mincing the tissue and then agitating for 1 hour at 37°C in the same pronase solution, exactly as described in ref. [22]. The enzymatic digest was centrifuged and freed from the majority of erythrocytes, debris and damaged cells by flotation in 17.5% (w/v) Metrizamide (Nyegaard, Oslo). Cells floating on top were composed of peripheral leukocytes, endothelial and Kupffer cells. Usually Kupffer cells are most conveniently separated by centrifugal elutriation [22,23]. The feasibility of separating rat liver en-

TABLE 1. Isolation of Human Blood Monocytes by Velocity Sedimentation at 20*g*

	Donor							
	1	2	3	4	5	6	7*	8*
Fraction[§]	Purity of monocytes (%)							
4.8	9.2		0	0	0		33.6	0
4.0	68.1	0	21.4	10.4	33.3	0	69.6	49.1
3.2	87.3	29.4	45.0	52.8	73.5	64.8	84.0	80.2
2.4	95.2	80.0	77.6	81.2	91.8	78.2	90.5	89.6
1.6	94.9	88.0	88.8	91.4	95.0	80.8	90.1	91.4
0.8	90.3	93.6	93.0	90.5	92.9	81.2	85.1	89.4
0(c)	89.9	96.4	93.7	84.9	85.5	81.8	89.6	83.4
No. of cells[†]	2.0	4.75	2.49	5.79	2.68	10	9.27	8.59
Recovery	92	90	90.8	94.6	90	87	89	88.9
Purity[‡]	82	85.1	77.6	83.3	82	77.6	78.1	81.2
m : n	47.4 : 94	66 : 95.1	64 : 93.4	53 : 91	66.5 : 93.3	70.2 : 81.5	37.4 : 90.3	38.9 : 90.5

*Sample volume = 10 ml; all others = 5 ml.
[†]Number of mononuclear cells added to the gradient $\times 10^{-7}$
[‡]Purity of monocytes collected in all these fractions.
[§] Distance (mm) from cushion fraction c; m = % of all monocytes in fractions that are n % pure.

dothelial and Kupffer cells using the small sedimentation chamber is demonstrated in Figure 7. Separation is effectuated by centrifugation at 20*g* for 7 min only. In this example, a cell load of 6×10^7 in 5 ml was processed. As clearly shown in Figure 7, the total chamber height was exploited for the concurrent separation of erythrocytes (E), leukocytes (L), endothelial (En), and Kupffer cells (K). Apart from the fact that the (still) contaminant population of erythrocytes is well separated from the nucleated cells, a pronounced shoulder—if not a solitary peak—in the endothelial cell population was due to peripheral leukocytes (not analyzed in detail). Fine resolution of the peaks is limited by the finite (4 ml) volume of the fractions. The Kupffer cell population is well removed from the endothelial cells. A number of Kupffer cells ended up in the interphase between density gradient and cushion liquid. At loads of 6×10^7 cells per 5 ml, purity of Kupffer cells approaching 99% was obtained. Because one is here very near the loading capacity, the descending limb of the endothelial cells progressed somewhat deeper into the Kupffer cell population to the effect that only 67.2% of all collected Kupffer cells were 99% pure in contrast to the results obtained at lower cell loads (Table 2). In summarizing results for relevant Kupffer cell enrichment, Table 2 shows that almost pure Kupffer cells are easily obtained at cell loads from 2 to 6×10^7 per 5 ml samples. Increasing sample layer thickness results in some loss of resolving power and, indeed, Table 2 shows that upon separating 10^8 cells in a 10-ml sample, purities of Kupffer cells up to 92% were obtained in most circumstances. This is sufficient to study those properties that make Kupffer cells different from endothelial cells. The present method compares favourably to centrifugal elutriation, by which method Kupffer cell purities of 84 to 90% are obtained [23]. A major advantage of centrifugal elutriation is that many more cells can be processed (10-fold in about 1 hour). On the other hand six small separation chambers can be run simultaneously in the centrifugal yoke.

Size-Separation of Chromosomes

Metaphase chromosomes are considerably smaller in size than the cells they originate from. Separation of intact cells usually requires approx. 100 to 200*g* min. For chromosome separations at unit gravity 52 hours and more are required. Gratifying resolution has been obtained by unit gravity sedimentation of Indian Muntjac [9] and Chinese hamster [10] chromosomes. This is undoubtedly because these chromosomes fall in rather discrete size classes and are large, so that Brownian movement (diffusion) is unimportant. However, chromosome separations at higher gravity forces are consistently better for two reasons. The short duration is beneficial for structural integrity of the chromosomes. Secondly, so-called "anomalous" band broadening [30,31,52] is strongly reduced at higher acceleration forces. At unit gravity, initially thin zones of particles (even at very low concentrations of 10^4/ml) are rather unstable in density gradients and are subject to considerable band broadening by a still unexplained mechanism. Miller [30,31] studied these 1 xg driven band instabilities and concluded that they are excessively large in weak density gradients, but relatively small in strong macromolecular gradients [52]. Our experience is that band broadening also occurs in strong sucrose gradients at 1*g* but is reduced at higher g forces. Therefore, reduction of the extent of "anomalous" band broadening at higher g forces is highly beneficial to the resolution of some of the more complex metaphase karyotypes separated in sucrose density gradients.

Previously, metaphase chromosomes have been separated in ordinary cylindric centrifugal tubes that suffer from severe wall sedimentation [34], in zonal rotors [5,8,35] and in the reorienting density gradient zonal rotor [17,39–41]. A comparison of these devices with the present small separation chamber points to an equal or even higher degree of resolving power for the sedimentation chamber [12]. Operating the chamber at 30 or 50*g*, chromosomes derived from rat [11], human [12], and mouse cells [18] have been separated. The separation of rat chromosomes was such that single copy genes could be allocated within chromosomal size classes by hybridizing gene specific probes to DNA isolated from the chromosomal fractions [11]. Figure 8 illustrates the separation of human chromosomes by velocity sedimentation at 52*g*. The upper panel (HFL) represents the flow karyotype of the unfractionated human chromosomes. The panels A-O

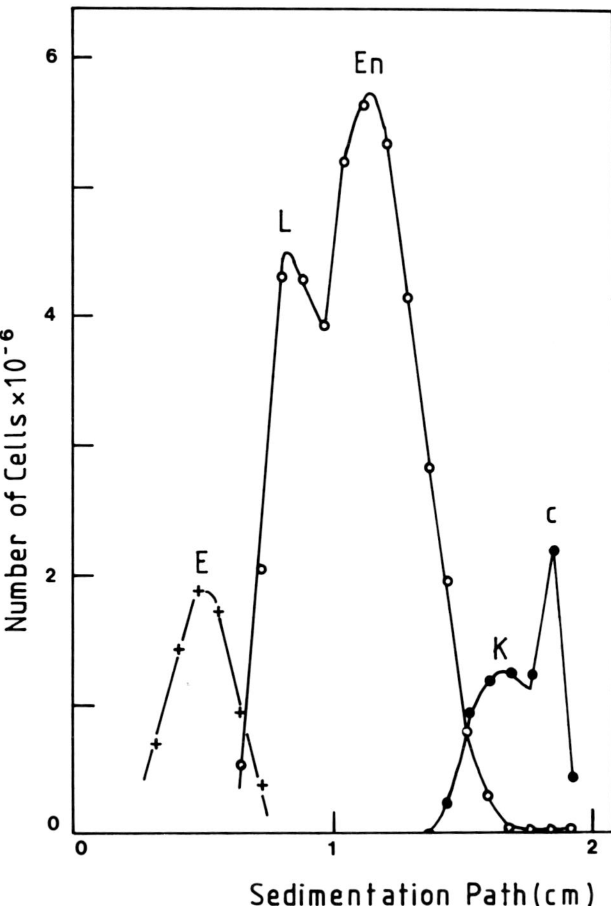

Fig. 7. Sedimentation profile of rat liver sinusoidal cells after velocity sedimentation at 20*g*; 5 ml rat liver cells containing 6×10^7 cells were separated for 7 min at 20*g* and 4°C on a 90 ml linear density gradient (3.8 to 18% Percoll). Fractions of 4 ml each were collected. Kupffer cells were determined by peroxidase staining [22]. (○—○) endothelial cells plus leucocytes L, En, (●—●) Kupffer cells K, (+—+) erythrocytes E, c = cushion fraction.

show the chromosomal composition of the fractions after sedimentation in a 5 to 15% sucrose gradient. The first fractions contain primarily the smallest chromosomes. Deeper in the gradient the middle-size class chromosomes are present while the last fractions contain virtually only the largest chromosomes. Thus, all fractions are highly enriched for certain chromosomes and are ideal starting material for fluorescence activated chromosome sorting. Following this procedure, the sorting rate for specific chromosomes can be speeded up by a factor of 5 to 10 [12]. Recently, we have demonstrated that chromosomes derived from the chromosomal fractions and spotted onto nitrocellulose filters can be used to assign unlocalized genes to chromosomal size classes [19]. The chromosomal DNA was hybridized with gene specific radioactively labeled DNA probes. This procedure may be applied to narrow the choice of chromosomes that need to be sorted for a precise gene localization on a specific chromosome. The chromosome spot hybridization technique is a rapid and simple method to detect gene specific hybridization to flow sorted or pre-enriched chromosomes [19,24]. As shown in Figure 9, sorting of 10,000 to 30,000 chromosomes (1 to 15 ng DNA) is sufficient to detect single-copy gene probe hybridization. The technique allows even subregional mapping of DNA probes by sorting translocated chromosomes. In this way, the human c-myb oncogene has been assigned to 6q22–q23 [19].

Another application of the sedimentation chamber is the separation of double minute chromosomes from a chromosomal suspension. The population of double minutes present in a chromosome preparation of Colo 320 cells [4] sedimented apart from the smallest human chromosomes. By means of spot hybridization, the amplification of the c-myc oncogene on the double minutes could be confirmed (unpublished results). Furthermore, chromosome separation by velocity sedimentation has been applied to construct a chromosome specific DNA library of the human chromosome 21 [18].

After separation of the chromosomes derived from a mouse–human hybrid cell line (SCC-16-5), containing hu-

Table 2. Isolation of Rat Liver Kupffer Cells by Velocity Sedimentation at 20*g*

	Experiment			
	1	2	3	4*
Fraction§	% of Kupffer cells			
8	0	0	0	0
7.2	2	0	0	0
6.4	28	0	0	0
5.6	29	3	1	0
4.8	98	50	2	0
4.0	99	81	8	10
3.2	100	94	53	16
2.4	98	95	80	38
1.6	99	98	97	90
0.8	98	96	99	92
0(c)	96	99	99	89
Load†	2.0	4.0	6.0	10.0
Recovery	91	90.5	90	90
Purity‡	54.5	69.4	40	42.8
m : n	85.4:98.5	65.5:96.2	67.2:99	59.9:90.4

*sample volume = 10 ml; all others 5 ml.
†Number of cells added to the gradient $\times 10^{-7}$.
‡Purity of Kupffer cells collected in all of these fractions.
§Distance (mm) from cushion fraction c; m = % of all Kupffer cells in fractions that are n % pure.

Fig. 8. Flow cytometric histograms of human chromosomes as separated by velocity sedimentation at 52*g*. Chromosomes from human lung fibroblasts were separated for 60 min at 52*g* and 22°C. Fractions of 4 ml each were collected. For details, see ref. [12]. Chromosomes were stained with mithramycin and ethidium bromide and analysed with a FACS IV flow sorter. Abcissa: channel number (relative amount of DNA per chromosome); ordinate: relative number of chromosomes per channel. Panel HFL = unfractionated chromosomes; Panels A–O: chromosomal composition of velocity sedimentation fractions. A is at 0.5 cm in the sedimentation chamber; each following fraction differs 0.8 mm in sedimentation path. Numbers 1–22 and X and Y represent chromosomal classification.

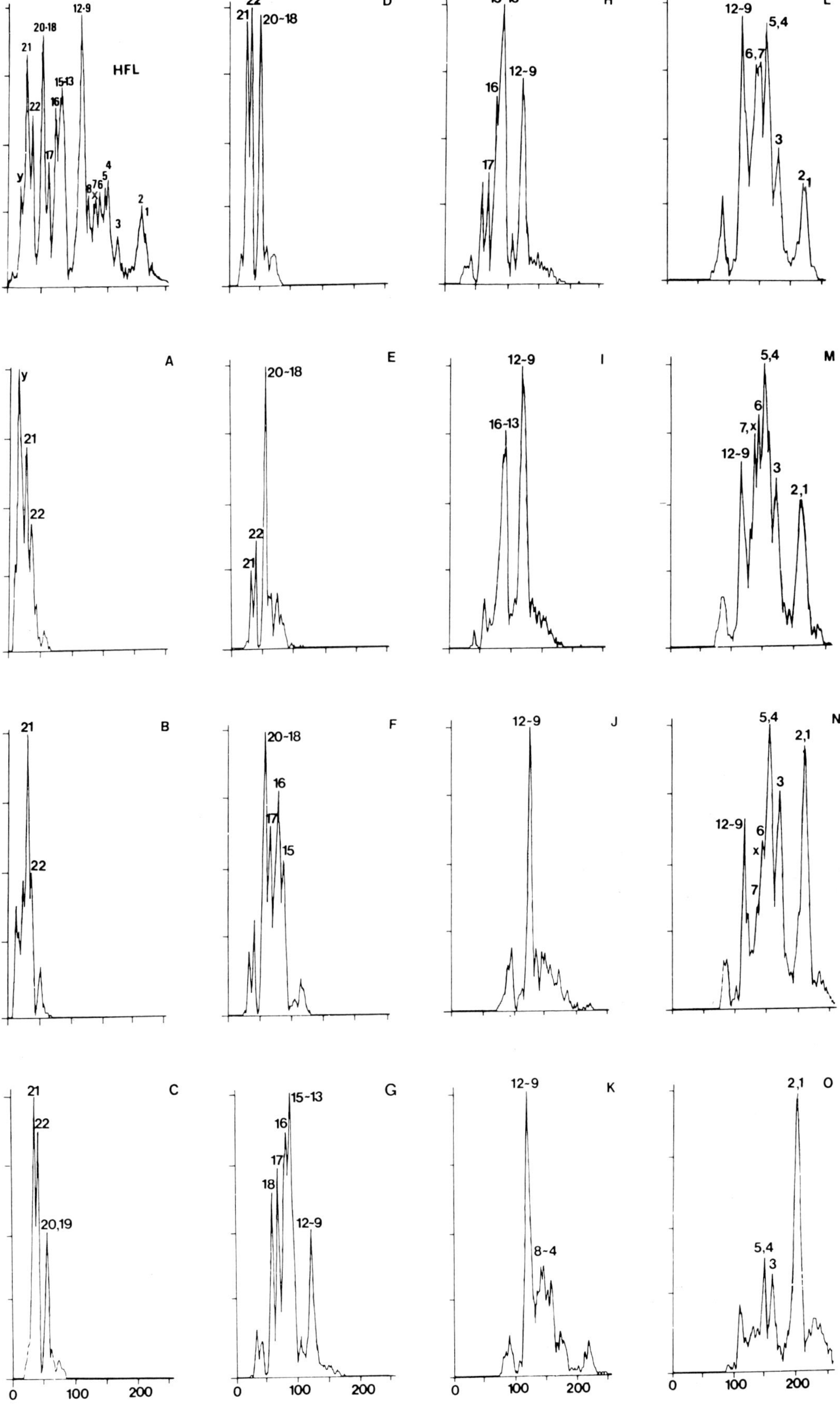
HFL
y
21
22
20-18
17
16
15-13
12-9
8
7
6
x
5
4
3
2
1
A
B
C
D
E
F
G
H
I
J
K
L
M
N
O
20,19
16-13
8-4
6,7
5,4
7,x
2,1
0
100
200

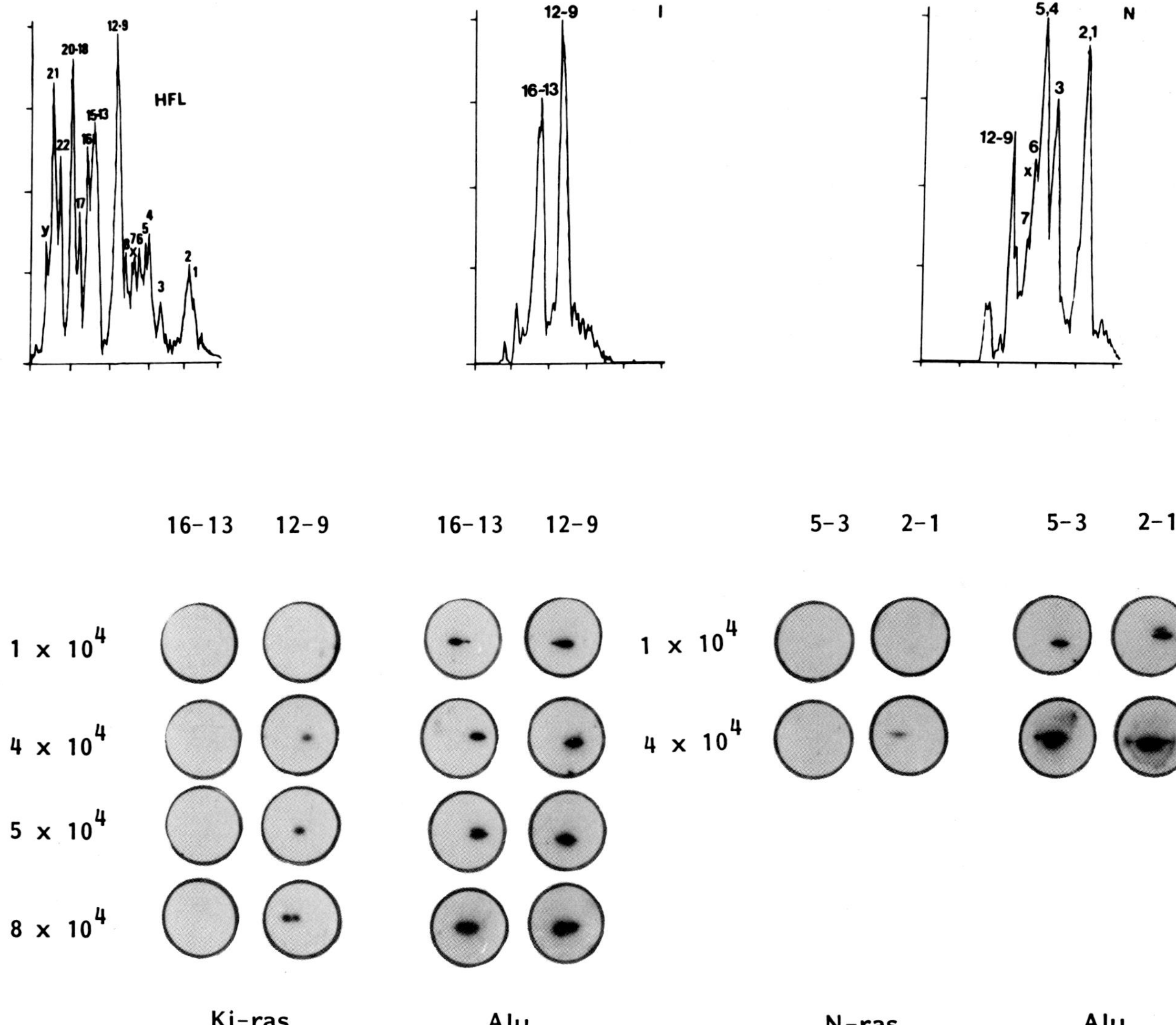

Fig. 9. Detection level of chromosome spot hybridization with single-copy gene-specific DNA probes. Increasing amounts of chromosomes were sorted onto nitrocellulose filters and chromosomal DNA was subsequently hybridized with radioactive DNA probes. For details see [13]. Top: flow histograms of human chromosomes. Conditions were identical to that of Figure 8. Panel HFL = unfractionated chromosomes. Panels I and N: chromosomal composition of two velocity sedimentation fractions. Bottom: chromosome spot hybridization. The 8 filters on the left contain increasing amounts (as indicated) of sorted chromosomes 9–12 and 13–16 derived from fraction I. Filters were hybridized with c-Ki-*ras* probe and subsequently, as a control, with the human Alu-repeat probe. Ki-*ras*-2 has been mapped to chromosome 12. Specific hybridization is detected only on filters containing chromosomes 9–12. 10,000 Ki-*ras* gene copies (on chromosome 12), corresponding with 40,000 chromosomes of the group 9–12, are sufficient for detection. The four filters on the right contain increasing amounts (as indicated) of sorted chromosomes 5-3 and 1,2 respectively derived from fraction N. The filters were hybridized with the N-*ras* probe and subsequently, as a control, with the Alu-probe. The N-*ras* gene, localized on chromosome 1, is detected after sorting 40.000 chromosomes 1 and 2, indicating that 20.000 chromosomes 1 are sufficient for detection.

man chromosome 21 only, we were able to obtain chromosomal fractions consisting of 50% mouse DNA and 50% human chromosome 21 DNA. This DNA was partially digested with MboI, size fractionated on an NaCl gradient and cloned in the EMBL-3 phage vector. The employed cloning strategy resulted in a human chromosome 21 specific phage library containing 50% recombinants with human DNA inserts of 15 to 20 kb in size and 50% recombinants with mouse DNA inserts [18].

EQUILIBRIUM DENSITY CENTRIFUGATION OF MAMMALIAN CELLS

Sorting of cells according to their buoyant density can be quite rapid using the present separation chamber. Thanks to the small height of the chamber, cells need to travel only a short distance to their equilibrium position so that separation time is limited to about 20 min at 130*g* and room temperature, in particular when the low viscosity solute Percoll is used. Essentially an equilibrium method, the final buoyant density distribution is independent of the way the cells are initially distributed along the density gradient. One may load cells as a zonal layer either on top of the gradient or at the bottom in a medium of the appropriate density. In both cases, however, one is faced with overloading phenomena such as droplet sedimentation [29] during all manipulations that take place at unit gravity. Also, crowding of cells may occur at the density interphases upon centrifugation leading to distribution artifacts. It is therefore advisable to distribute the cells over the total of the gradient when it is loaded into the separation chamber. One may opt for uniform distribution of the cells over the gradient by adding equal amounts of cells to the two (communicating) vessels of the gradient mixing device, that contain the light and dense gradient solutions respectively. Alternatively, and preferentially, one may add all the cells to the gradient mixing vessel that contains the dense gradient solution so that the cells are introduced as a linear gradient into the separation chamber. Thus, without even the slightest trace of the streaming phenomenon and without mutual hindrance massive amounts of cells can be loaded and separated: 4×10^9 nucleated cells plus 1 to 4×10^{10} erythrocytes are still well resolved into the constituent cell types. The high loading capacity of equilibrium density centrifugation may be exploited to enrich a desired cell type before it is purified by size separation from smaller (larger) cells by velocity sedimentation with its essentially lower loading capacity. Such a combination of purification steps has enabled us to procure 86.4% pure myeloblasts plus promyelocytes together with 4.2% myelocytes from human bone marrow [53,54].

To be described herein is the concomittant separation of the constituent cell types of peripheral human blood for which the total height of the separation chamber was used. Usually, however, when one is aiming for one particular cell type it is recommended that a density gradient of limited range be loaded that encompasses the density variation of that particular cell type, over the total height of the chamber, letting other cells collect in bulk in the cushion liquid.

Separation of Peripheral Blood Cells

Specific gravity is a rather characteristic property of the various cell types that constitute the compartment of peripheral blood. Prior to subjecting the cells to the equilibrium density centrifugation procedure, the bulk of erythrocytes is removed by adding dextran T-500 (Pharmacia) [37,51] to a final concentration of 1% to 100 ml of diluted blood. Interestingly, the small separation chamber functions well for the bulk removal of erythrocytes within a short time at unit gravity. The artificially produced (by dextran) agglomerations of red cells are separated within 5 min at 1*g* from the nucleated cells by introducing the blood mixture into the chamber. A clear-cut sedimenting layer of erythrocyte rouleaux is already formed within 1 min at unit gravity over the 50 cm^2 cross section of the chamber. After 5 min, the contents of the chamber are rapidly (within 30 sec) collected from the top of the chamber by introducing dense cushion liquid into the bottom of the chamber. The supernatant containing the white cells is collected within 0.1 ml accuracy before the dense layer of erythrocytes leaves the chamber: 97% of the erythrocytes are removed while 96% of the nucleated cells are recovered, in separate fractions. A more detailed description of the removal of erythrocytes at unit gravity is to be found in ref. [51].

Figure 10 shows the buoyant density profile of peripheral

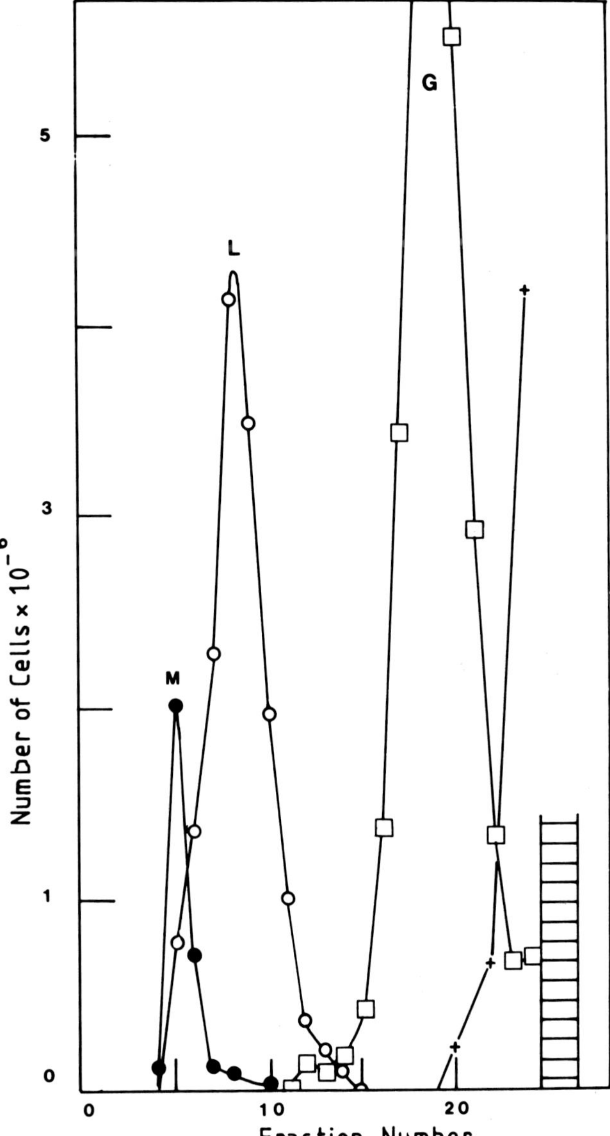

Fig. 10. Equilibrium density centrifugation of human peripheral blood cells. Blood cells, freed from the majority of erythrocytes, were separated for 50 min at 75g and 22°C on a 90 ml (42 to 70%) linear Percoll gradient. Fractions of 4 ml each were collected. (+—+) erythrocytes, (○—○) lymphocytes L, (●—●) monocytes M, (□—□) granulocytes G. In the hatched area the bulk of erythrocytes was collected. Fraction 18 contained 6.5×10^6, fraction 19:7.4×10^6 granulocytes.

blood cells from a normal donor on a linear Percoll (42 to 70%) density gradient of 90 ml. The very sharp peaks are indicative of good resolution. Major peaks of monocytes (M), lymphocytes (L), and granulocytes (G) respectively can be observed. In the hatched area the bulk of erythrocytes (often occurring in clumps) is collected. Flow cytometry analysis by staining the cells with acridine orange in hypotonic reagent and measuring green and red fluorescence [2] corroborated the progressive enrichment of sub-classes of leukocytes by the buoyant density centrifugation procedure

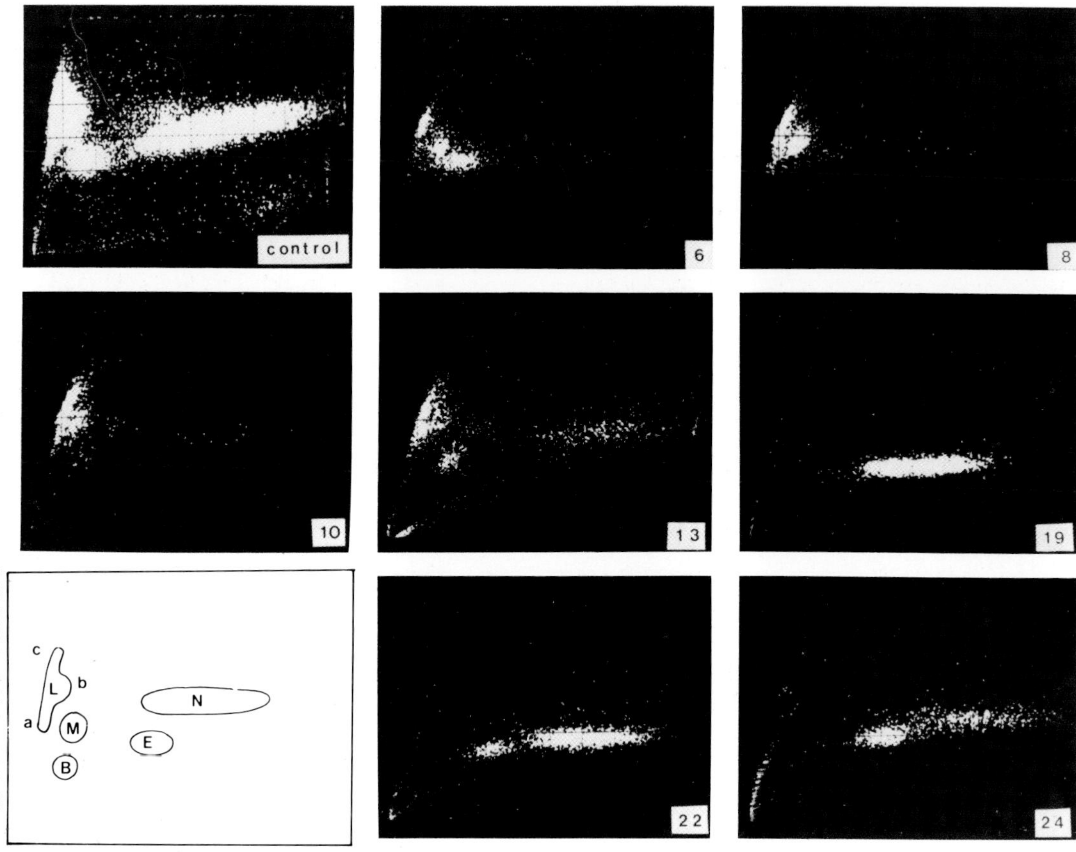

Fig. 11. Identification by flow cytometry of blood cells as separated by buoyant density centrifugation. Experimental conditions were identical to that of Figure 10. Red versus green fluorescence of acridine orange stained cells was recorded with a Cytofluorograf (FC-200). Numbers denote fraction number. L = lymphocytes (a, "null" cells, b, B cells; c, T-cells); M, monocytes, B, basophils; E, eosinophils; N, neutrophils.

(Fig. 11). In agreement with the findings of other investigators [1,6], the lymphocyte cluster was found to be subdivided into B, T, and nonrosetting cells. According to this analysis, fraction 6 in Figure 11 was found to contain B and T cells as well as monocytes; "null" cells were almost absent. In fraction 8, B cells and only a few monocytes were present, whereas in fraction 10 the B cell contribution was diminished and the monocyte percentage was very low. Furthermore, Figure 11 shows that in fraction 13, B-cell content was low and basophils were considerable enriched (12% according to microscopic analysis). Fractions 16 to 18 (not shown) and 19 consisted almost completely of neutrophil granulocytes. In fraction 22, the eosinophil cluster becomes visible, to be highly enriched in fraction 24. Whereas the density gradient in Figure 10 runs from 42 to 70% Percoll to encompass all cell types, a more refined selection of cells can be obtained by altering slope and compass of the density gradient. If monocytes are required one may expand the light part of the density gradient (35 to 45% Percoll) so that it takes up the largest part of the chamber and granulocytes may be collected in a cushion of Percoll. Indeed, under such conditions 99% pure monocytes can be obtained and 98% of all monocytes collected are purified to 92.5% [51]. As for the separation of B and T cells, the resolution was improved by centrifuging a mononuclear cell suspension (which was depleted from monocytes after phagocytosis of iron particles) in a linear 41 to 49% Percoll (90 ml) gradient. Thus, fractions were obtained that contained 70% pure B cells that gradually, at higher densities, diminished to 0%, while the T cells in that order increased from 12 to 73% [48].

ACKNOWLEDGMENTS

We gratefully acknowledge the contributions of Mrs. G. Meijerink for preparing the manuscript; Mrs. M. G. Barnhoorn, J. F. Schijven, and E. Philippus for technical assistance; Dr. M. Stukart and Dr. J. W. G. Jansen for their contribution in some of the experiments; and W. van Es and C. Slee for construction of the flow-through cuvette.

REFERENCES

1. **Adams LR (1977)** Staining for the cytograf and cytofluorograf. J. Histochem. Cytochem 25:965–975.
2. **Adams LR, Kamentsky LA (1974)** Fluorometric characterization of six classes of human leucocytes. Acta Cytol. (Praha) 18:389–391.
3. **Albertsson PA, Anderson B, Larsson C, Akerlund HE (1982)** Phase partition—A method for purification and analysis of cell organelles and membrane vesicles. In Glick D (ed), "Methods of Biochemical Analysis." Vol. 28, New York: John Wiley & Sons, pp 115–150.

4. **Alitalo K, Schwab M, Lin CC, Varmus HE, Bishop JM (1983)** Homogeneously staining chromosomal regions contain amplified copies of an abundantly expressed cellular oncogene (c-myc) in colon carcinoma. Proc. Natl. Acad. Sci. USA 80:1707–1711.
5. **Benz RD, Burki J (1978)** Preparation and identification of partially fractionated Chinese hamster chromosomes. Exp. Cell. Res. 112:143–153.
6. **Beveridge RA, Super BS, Chretien PB (1979)** Identification and quantitation of B- and T-cells by cytofluorographic analysis. J. Immunol. Methods 26:47–60.
7. **Boyum A (1977)** Separation of lymphocytes, lymphocyte subgroups and monocytes. Lymphology 10:71–76.
8. **Burki JH, Regimbal TJ, Mel HC (1973)** Zonal fractionation of mammalian metaphase chromosomes and determination of their DNA content. Prep. Biochem. 3:157–182.
9. **Collard JG, Tulp A, Hollander JH, Bauer FW, Boezeman J (1980)** Separation of large quantities of chromosomes by velocity sedimentation at unit gravity. Exp. Cell. Res. 126:191–197.
10. **Collard JG, Tulp A, Stegeman J, Boezeman J, Bauer FW, Jongkind JF, Verkerk A (1980)** Separation of large quantities of Chinese hamster chromosomes by velocity sedimentation at unit gravity followed by flow sorting. Exp. Cell. Res. 130:217–227.
11. **Collard JG, Schijven J, Tulp A, Meulenbroek M (1982)** Localization of genes on fractionated rat chromosomes by molecular hybridization. Exp. Cell. Res. 137:463–469.
12. **Collard JG, Philippus E, Tulp A, Lebo RV, Gray JW (1984)** Separation and analysis of human chromosomes by combined velocity sedimentation and flow sorting applying single- and dual-laser flow cytometry. Cytometry 5:9–19.
13. **Collard JG, de Boer PAJ, Jansen JWG, Schijven JF, de Jong B (1985)** Gene mapping by chromosome spot hybridization. Cytometry 6:179–185.
14. **Gray JW, Lucas J, Pinkel D, Peters D, Ashworth L, Van Dilla MA (1980)** Slit-scan flow cytometry: Analysis of Chinese hamster M3-1 chromosomes. In Laerum OD, Lindmo T, Thorud E (eds), "Flow Cytometry IV." Oslo: Universitetsforlaget, pp 249–255.
15. **Green CL, Pretlow TP, Tucker KA, Bradley EL, Cook WJ, Pitts AM, Pretlow TG (1980)** Large capacity separation of malignant cells and lymphocytes from the Furth mast cell tumor in a reorienting zonal rotor. Cancer Res. 40:1791–1796.
16. **Hannig K (1972)** Separation of cells and particles by continuous free-flow electrophoresis. In Glick D, Rosenbaum RM (eds), "Techniques of Biochemical and Biophysical Morphology, Vol I". New York: Wiley–Interscience, pp 191–232.
17. **Hughes SH, Stubblefield E, Payvar F, Engel JD, Dogson JB, Spector D, Cordeel B, Schimke RT, Varmus HE (1979)** Gene localization by chromosome fractionation: globin genes are on at least two chromosomes and three estrogen-inducible genes are on three chromosomes. Proc. Natl. Acad. Sci. USA 76:1348–1352.
18. **Janssen JWG, Collard JG, Tulp A, Cox D, Millington-Ward A, Pearson P (1986)** Construction and analysis of an EMBL-3 phage library representing partially digested human chromosome 21 specific DNA inserts. Cytometry 7:411–417.
19. **Janssen JWG, Vernole P, de Boer PAJ, Oosterhuis JW, Collard JG (1986)** Sublocalization of c-myb to 6q21-q23 and c-myb expression in a human teratocarcinoma with 6q rearrangements. Cytogenet. Cell Genet. 41:129–135.
20. **Johnston IR, Mathias AP, Pennington F, Ridge D (1968)** The fractionation of nuclei from mammalian cells by zonal centrifugation. Biochem. J. 109:127–135.
21. **Keng PC, Li CKN, Wheeler KT (1981)** Characterization of the separation properties of the Beckman elutriator system. Cell. Biophys. 3:41–56.
22. **Knook DL, Sleyster EC (1976)** Separation of Kupffer and endothelial cells by centrifugal elutriation. Exp. Cell. Res. 99:444–449.
23. **Knook DL, Sleyster EC (1977)** Preparation and characterization of Kupffer cells from mouse and rat liver. In Wisse E, Knook DL (eds), "Kupffer Cells and other Liver Sinusoidal Cells." Amsterdam: Elsevier/North-Holland Biomedical Press, pp 273–288.
24. **Lebo RV, Gorin F, Fletterick RJ, Kao F-T, Cheung M-C, Bruce BD, Kan YW (1984)** High resolution chromosome sorting and DNA spot-blot analysis assign McArdle's syndrome to chromosome 11. Science 225:57–59.
25. **Lindahl PE (1956)** On counter streaming centrifugation in the separation of cells and cell fragments. Biochim. Biophys. Acta 21:411–415.
26. **Mason DW (1976)** A diffusion driven instability in systems that separate particles by velocity sedimentation. Biophys. J. 16:407–416.
27. **McCutcheon MJ, Miller RG (1982)** Flexible sorting decision and droplet charging control electronic circuitry for flow cytometer-cell sorters. Cytometry 2: 219–225.
28. **McEwen CR, Stallard RW, Juhos ET (1968)** Separation of biological particles by centrifugal elutriation. Anal. Biochem. 23:369–377.
29. **Mel HC (1964)** Stable-flow free boundary migration and fractionation of cell mixtures. J. Theoret. Biol. 6:181–200.
30. **Miller RG (1973)** Separation of cells by velocity sedimentation. In Pain RH, Smith BJ (eds), "Biophysics Cell Biology." New York: John Wiley & Sons, pp 87–112.
31. **Miller RG (1984)** Separation of cells by velocity sedimentation. Methods Enzymol 108:64–87.
32. **Miller RG, Phillips RA (1969)** Separation of cells by velocity sedimentation. J. Cell. Physiol. 73:191–202.
33. **Platsoucas CD (1983)** Separation of cells by preparative density gradient electrophoresis. In Pretlow TG, Pretlow TP (eds), "Cell Separation." Vol. 2. Orlando, Florida: Academic Press, pp 145–182.
34. **Salzman NP, Moore DE, Mendelsohn J (1966)** Isolation and characterization of human metaphase chromosomes. Proc. Natl. Acad. Sci. USA 56:1449–1456.
35. **Schneider EL, Salzman NP (1970)** Isolation and zonal fractionation of metaphase chromosomes from human diploid cells. Science 167:1141–1143.
36. **Sheeler P, Gross DM, Wells JR (1970)** Zonal centrifugation in reorienting density gradients. Biochim. Biophys. Acta 237:28–42.
37. **Skoog WA, Beck WS (1956)** Studies on the fibrinogen, dextran and phytohemagglutinin methods of isolating leukocytes. Blood 11:436–454.

38. **Stewart MJ, Pretlow TG, Hiramoto R (1972)** Separation of ascites myeloma cells, lymphocytes and macrophages by zonal centrifugation on an isokinetic gradient. Am. J. Pathol. 68:163–182.
39. **Stubblefield E, Cram S, Deaven L (1975)** Flow microfluorometric analysis of isolated Chinese hamster chromosomes. Exp. Cell. Res. 94:464–468.
40. **Stubblefield E, Wray W (1978)** Isolation of specific human metaphase chromosomes. Biochem. Biophys. Res. Commun. 83:1404–1414.
41. **Stubblefield E, Oro J (1982)** The isolation of specific chicken macrochromosomes by zonal centrifugation and flow sorting. Cytometry 2:273–281.
42. **Todd P, Milito RP, Boltz RC, Gaines RA (1979)** Cell electrophoresis. In Melamed MR, Mullaney PT, Mendelsohn ML (eds), "Flow Cytometry and Sorting." New York: John Wiley & Sons, pp 217–229.
43. **Tulp A (1984)** Density gradient electrophoresis of mammalian cells. In Glick D (ed), "Methods of Biochemical Analysis, Vol. 30. New York: Interscience, pp 141–198.
44. **Tulp A (1985)** Centrifugal elutriation of mammalian cells. In Kurstak J (ed), "Techniques in Cell Biology." Vol. 1. Amsterdam: Elsevier, pp 1–10.
45. **Tulp A, Welagen JJMN, Westra JG (1978)** Binding of the chemical carcinogen N-hydroxy-acetylaminofluorene to ploidy classes of rat liver nuclei as separated by velocity sedimentation at unit gravity. Chem. Biol. Interactions 23:293–303.
46. **Tulp A, Collard JG, Hart AAM, Aten JA (1980)** A new unit gravity sedimentation chamber. Anal. Biochem. 105:246–256.
47. **Tulp A, De Leng P, Barnhoorn MG (1980)** A sedimentation chamber to sort cells, nuclei and chromosomes at ten times gravity. Anal. Biochem. 107:32–43.
48. **Tulp A, Kooi MW, Kipp JBA, Barnhoorn MG, Polak F (1981)** A separation chamber to sort cells, nuclei and chromosomes at moderate g-forces. II. Studies on velocity sedimentation and equilibrium density centrifugation of mammalian cells. Anal. Biochem. 117:354–365.
49. **Tulp A, Aten JA, Barnhoorn MG, Van Beek WP, Collard JG, Lutter R, Westra JG (1982)** Separation of cells and cell organelles by weak physical forces. IV. Applications. In Akoyunoglou G (ed), "Cell Function and Differentiation." New York: Alan R. Liss, Inc., pp 105–114.
50. **Tulp A, Barnhoorn MG, Collard JG, Kraan W, Sluyser M, Polak F, Aten JA (1982)** Separation of mammalian cells and cell organelles at low g-forces. In Reid E, Cook G, Morré DJ (eds), "Methodological Surveys." Vol. II. New York: John Wiley & Sons, pp 158–170.
51. **Tulp A, Barnhoorn MG (1984)** A separation chamber to sort cells and cell organelles by weak physical forces. V. A sector-shaped chamber and its application to the separation of peripheral blood cells. J. Immunol. Methods 69:281–295.
52. **Tulp A, Hart AAM, Barnhoorn MG and Stukart MJ (1985)** Separation of cells and cell organelles by low gravity forces. In Kurstak J (ed), "Techniques in Cell Biology." Vol. I. Amsterdam: Elsevier, pp 1–36.
53. **Van Beek WP, Tulp A, Egbers-Boogaards M, Roozendaal KJ, Smets LA (1982)** Continuous expression of cancer-related fucosyl glycopeptides on the surface of human promyelocytic leukemia cells (HL-60) following terminal differentiation in vitro. Cancer Res. 42: 5222–5230.
54. **Van Beek W, Tulp A, Bolscher J, Blanken G, Roozendaal KJ, Egbers-Boogaards M (1984)** Transient versus permanent expression of cancer-related glycopeptides on normal versus leukemic myeloid cells coinciding with marrow egress. Blood 63:170–176.
55. **Wakefield JJ, Gale JS, Berridge MV, Jordan TW, Ford HC (1982)** Is Percoll innocuous to cells? Biochem. J. 202:795–797.
56. **Walle A, Kodama T, Melamed MR (1983)** A simple density gradient for enriching subfractions of solid tumor cells. Cytometry 3:402–407.
57. **Walter H (1982)** Separation and subfractionation of blood cell populations based on their surface properties by partitioning in two-polymer aqueous phase systems. In Pretlow TG, Pretlow TP (eds), "Cell Separation." Vol. I. New York: Academic Press, pp 261–299.
58. **Warmsley AMH, Pasternak CA (1970)** The use of conventional and zonal centrifugation to study the life cycle of mammalian cells. Biochem. J. 119:439–499.
59. **Wells JR (1982)** A new approach to the separation of cells at unit gravity. In Pretlow TG, Pretlow TP (eds), "Cell Separation." Vol. 1. Orlando, Florida: Academic Press, pp 169–189.
60. **Wells JR, James TW (1972)** Cell cycle analysis by culture fractionation. Exp. Cell. Res. 75:465–474.
61. **Wells JR, Sheeler RS, Gross D (1972)** A reorienting density gradient rotor for zonal centrifugation. Anal. Biochem. 46:7–18.
62. **Wells JR, Opelz G, Cline MJ (1977)** Characterization of functionally distinct lymphoid and myeloid cells from human blood and bone marrow. II. Separation by velocity sedimentation. J. Immunol. Methods 18: 79–93.

12

Fluorescent Probes for Cytometry

Alan S. Waggoner

Department of Biological Sciences, Center for Fluorescence Research in Biomedical Sciences, Carnegie-Mellon University, Pittsburgh, Pennsylvania 15213

INTRODUCTION

Fluorescent probes provide a sensitive method with which to obtain information about the structure, function, and health of cells. In fact, the method can be sensitive in two ways. First, it is possible to detect very few fluorescent molecules in a cell. Fluorescent chromophores for labeling antibodies, DNA, and lipids are designed for high detection sensitivity [26]. The ideal labels have large extinction coefficients and high quantum yields and are insensitive to pH and the polarity of their local molecular environments.

The second way fluorescence can be sensitive is in the ability of probe molecules to respond spectroscopically to subtle changes in their molecular environment. The dependence of absorption wavelength, extinction, emission wavelength, quantum yield, excited state lifetime, and emission polarization on fluorescent probe microenvironment provides a powerful method for understanding the behavior of cells. One of the key requirements for exploiting the sensitivity of probes microenvironment is the careful design of the fluorophore structure so that it can be targeted to the site of interest in the cell, e.g., plasma membrane versus endoplasmic reticulum, or mitochondria versus cytoplasm, or one class of lysosomes versus another. Therefore, the skills of the organic chemist must be blended with those of the spectroscopist, biochemist, and cell biologist to exploit this potential of fluorescent probes optimally.

OPTICAL PROPERTIES OF IDEAL FLUORESCENT PROBES

Large Extinction Coefficient (ϵ) at the Wavelength of Excitation

The value of ϵ, which characterizes the light-absorbing power of the fluorophore, should be as large as possible. Extinction coefficients and other spectroscopic parameters for a variety of fluorescent probes are listed in Table 1. Fluorescein has an ϵ ~67,000 liters/mole · cm at its wavelength of maximum absorption (pH 7.0). The multichromophore phycobiliproteins have exceedingly large extinction coefficients (~2×10^6 liters/mole · cm), which contributes to their value as very fluorescent antibody-labeling reagents. There are useful fluorescent probes with lower excitation coefficients but, to be detected by flow cytometry, they must have high quantum yields and be at high concentrations within cells.

High Quantum Yield (ϕ)

ϕ should be large when the probe is bound to the target and is in the solvent environment where the fluorescence measurement is to be made. Fluorescein-labeled proteins can have quantum yields near 0.5 to 0.7 at pH 8, but ϕ drops rapidly with decreasing pH and is lower when more than one fluorescein is bound to the protein. Rhodamine is much less sensitive to pH and is still fluorescent in acidic compartments of cells or in the presence of acidic fixatives. If a large amount of a probe is associated with each cell, e.g., with propidium staining of DNA, the limitations due to small ϵ and ϕ are not as severe. The fluorescence intensity from a fluorophore is proportional to the product of ϵ and ϕ.

Optimal Excitation Wavelength

Cells excited at wavelengths below 500 nm produce considerable autofluorescence that arises mainly from flavins, flavoporteins, and NADH [3,8]. In situations in which autofluorescence can swamp the probe fluorescence, it is useful to have the probe absorb at wavelengths above 500 nm.

Photostability

Fluorescein can survive between 10^4 and 10^5 excitations before decomposing [57]. As Mathies and Stryer discuss, photostability is important for detecting a small number of probes in solutions. Photostability is also a concern for visualization of tagged materials with conventional fluorescence microscopes. For example, the fading of fluorescein fluorescence under intense microscopic illumination is well known, and numerous chemical approaches have been taken with various degrees of success to reduce fading [10,60]. Removal of oxygen is the most effective step that can be taken but this is impossible in studies of most living cells. Photofading is generally not a problem in flow cytometry because the stained sample remains in the laser beam only a short time.

Flow Cytometry and Sorting, Second Edition, pages 209–225

Table 1. Spectral Properties of Selected Fluorescent Probes[a]

PARAMETER	PROBE (a)	ABSORPTION MAXIMUM (b)	EXTINCTION MAX. (c)	EMISSION MAXIMUM (b)	QUANTUM YIELD	MEASURMENT CONDITIONS	REFERENCES
Covalent labeling reagents	FITC-NH-CH3	490	67	520	0.71	pH7, PBS	[32], W, MP
	FITC-NH-Ab	490	67	520	0.1-0.4 (d)	pH7, PBS	W
	TRITC-amines	554	85	573	0.28	pH7, PBS	[32], MP
	XRITC-NHCH3	582	79	601	0.26	pH7, PBS	W
	XRITC-NH-Ab	580		604	0.08 (h)	pH7, PBS	W
	Texas Red-amines	596	85	620	0.51 (w)	pH7, PBS	[94], W, MP
	Texas Red-NH-Ab	596	85	620	0.01 (g)	pH7, PBS	W
	Phycoerythrin-R	480-565	1960	578	0.68	pH 7, PBS	[67]
	Allophycocyanine	650	700	660	0.68	pH7, PBS	[67]
DNA-RNA content	Hoechst 33342	340	120	450	0.83	+DNA	W
	DAPI	350		470		+DNA	W
	Ethidium Bromide	510	3.2	595		+DNA	[71]
	Propidium Iodide	536	6.4	623	0.09	+DNA	W
	Acridine Orange	480		520		+DNA	[43, 78]
		440-470		650		+RNA	[43, 78]
	Pyronine Y	549-561	67-84	567-574	0.04-0.26	+ds DNA (e)	[17, 42]
		560-562	70-90	565-574	0.05-0.21	+ds RNA (e)	[17, 42]
		497	42	563	LOW	+ ss RNA	[17, 42]
	Thiazole Orange	453	26	480	0.08	RNA	[52]
Membrane potential	diO-Cn-(3)	485	149	505	0.05	MeOH	[82], W
	diI-Cn-(3)	548	126	567	0.07	MeOH	[82], W

Excited State Lifetime (τ)

In flow cytometry the greatest sensitivity can be achieved with a combination of high laser power and probes with short fluorescence lifetimes. Under saturation conditions, the strong illumination of a focused laser spot can excite dye molecules immediately after they have fluoresced or relaxed from a previous excitation. Thus, as a cell passes through the beam the greatest number of excitations can occur when a dye has a very short τ. The maximum number of excitations (and photons emitted) that are possible with a saturating laser beam is approximately given by the time the fluorescent molecule is in the beam divided by τ. Most highly fluorescent molecules do not have lifetimes much shorter than a nanosecond. Fluroescein and pyrene have emission lifetimes of ~4 and 100 nsec, respectively.

In addition to the ideal optical properties listed, there are other considerations, e.g., solubility, chemical stability, and photostability. Also, the probe should not perturb the function of the cell, organelle, or target molecule by reacting with key groups in active sites or by causing steric perturbations because of its size. The probe should not be phototoxic. Few fluorescent probes have all the ideal properties, but a number have proved to be very valuable and are widely used. They are discussed more fully later.

FLUORESCENT LABELS

Reactive fluorescent chromophores can be used to tag proteins, polynucleotides, lipids, or other biological molecules, which can in turn be used as biological probes. The structures of common covalent labeling reagents are shown in Fig. 1 and their optical properties are given in Table 1. Ideal labeling reagents generally have appropriate selectivity and modest reactivity. If too reactive, the tagging reagent hydrolyzes before binding to the protein. Isothiocyanates, chlorotriazinyl derivatives, and succinimide active esters are the most common functional groups that permit chromophores to be attached covalently to primary amino groups of proteins. Generally, the reactions are carried out in aqueous solutions at pH 8.9 to 9.5. Iodoacetamido and malemido functional groups on chromophores can be used to form linkages to protein sulfhydryl groups. This chemistry has been reviewed by Haugland [32].

TABLE 1. *Continued*

PARAMETER	PROBE (a)	ABSORPTION MAXIMUM (b)	EXTINCTION MAX. (c)	EMISSION MAXIMUM (b)	QUANTUM YIELD	MEASURMENT CONDITIONS	REFERENCES
	dl-Cn-(5)	646	200	668	0.4	MeOH	[82], W
	diBA-Isopr-(3)	493	130	517	0.03	MeOH	[82], W
	Rhodamine 123	511	85			EtOH	EK
Lipid content and fluidity	Nile Red	485		525		Heptane	[30]
		530		605		Acetone	[30]
	Diphenylhexatriene (DPH)	330, 351, 370	77 (351nm)	430		Hexane	MP
	dil-C18-(3)	546	126	565	0.07	MeOH	W
	NBD phosphatidylethanolamine	450	24 (f)	530		Lipid	[85]
	Anthroyl stearate	361, 381	8.4, 7.5	446		MeOH	[99]
pH	6-Carboxyfluorescein	495		520		High pH	MP
		450				Low pH	MP
	BCECF (i)	505		530		High pH	MP
		460				Low pH	MP
	DCDHB (j)	340-360		500-580		High pH	[97], MP
		340-360		420-440		Low pH	[97], MP
Calcium	Fura 2	335	33	512-518	0.23	Low Calcium	[31]
		360	27	505-510	0.49	High Calcium	[31]
	Indo 1	330	34	390-410	0.56	High Calcium	[31]
		350	34	482-485	0.38	Low Calcium	[31]
Enzyme substrates	Rhodamine-di-arg-CBZ substr.	-	Low at 495nm	532	0.09	Hepes pH7.5+15% EtOH	[53]
	Product of rxn. (rhodamine)	495	67	523	0.91	Hepes pH7.5+15%EtOH	[53]
	Coumarin-glucoside substr.	316	13	395		Acetate pH 5.5+1%Lubrol	W
	Product of rxn (hydroxy coumarin)	370	17	450		Glycine pH10+1%Lubrol	W

[a]Key: EK = Eastman Kodak Chemical Catalog; MP = Molecular Probes, Inc catalog (or personal communication); W = Waggoner laboratory determination; (a) Ab = antibody; (b) nanometers; (c) in thousands (liters/mol · cm); (d) dye–antibody ratio 2–5; (e) base pair dependent; (f) value for NBD–ethanolamine in MeOH, which has an abs. max at 470 nm and an emission max at 550 nm [4]; (g) dye–antibody ratio 1.2; (h) dye–antibody ratio 2.5; (i) BCECF = 2′,7′-bis(2-carboxyethyl)-5(and 6)carboxyfluorescein; (j) DCDHB = 2,3-dicyano-1,4-dihydroxybenzene.

Some tagging reagents, such as fluorescein, are soluble in aqueous solutions and are easily conjugated to proteins. Others, such as the hydrophobic rhodamines, are less soluble and not only can they precipitate from the labeling reaction mixture but they can also cause precipitation of more heavily labeled antibodies (even with only 2 to 3 fluorochrome molecules per protein molecule).

Certainly the most commonly used fluorescent tag is fluorescein isothiocyanate. It has been attached to antibodies, lectins, avidins, hormones, lipids, protein analogues, and other biological molecules [32]. Properties that make fluorescein popular are its reasonably high extinction coefficient and quantum yield, its water solubility, ideal reactivity, and emission at a perfect wavelength (515 nm) for detection by the human eye and by photodetectors. Antibodies, avidin and lectins can be labeled with 2 to 8 fluoresceins before fluorescence quenching, which occurs when fluorophores are in close proximity, reduces the brightness of the labeled species. The disadvantages of fluorescein are its relative photoinstability [57], its loss of fluorescence when the pH is below 8, and its wavelength of excitation, which is in a region that produces autofluorescence.

Other fluorescent labels have been developed either to improve on the properties of fluorescein or to provide additional reagents that absorb and emit at different wavelengths and can therefore be detected simultaneously. Rhodamines, for example, are generally more photostable than fluorescein, are pH insensitive under physiological conditions, and excite in the 500- to 600-nm range, where less autofluorescence is generated. However, the quantum yield of rhodamine on labeled antibodies is generally less than that of fluorescein at similar dye/protein ratios and, as mentioned earlier, problems with the limited water solubility of rhodamines make these reagents less attractive as labeling reagents. Lissamine rhodamine and Texas Red are more soluble rhodamines and are excited at longer wavelengths.

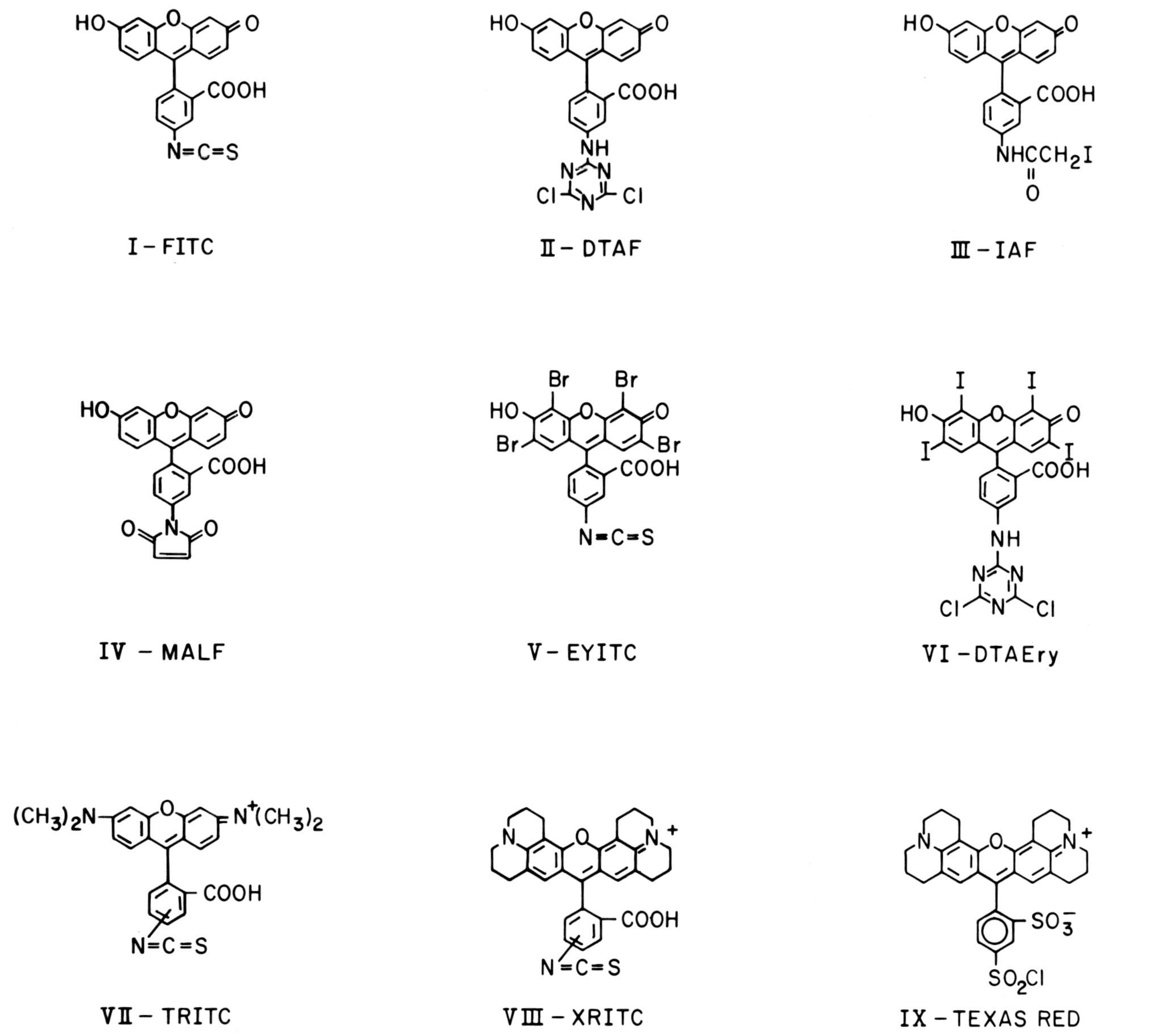

I - FITC

II - DTAF

III - IAF

IV - MALF

V - EYITC

VI - DTAEry

VII - TRITC

VIII - XRITC

IX - TEXAS RED

Fig. 1. Fluorescent labeling probes. I, Fluorescein isothiocyanate; II, Dichlorotriazinylamino-fluorescein; III, Iodoacetamidofluorescein; IV, Maleimidofluorescein; V, Eosin Y isothiocyanate; VI, Dichloro-triazinylaminoerythrosin; VII, Tetramethylrhodamine isothiocyanate; VIII, Rhodamine isothiocyanate analog; IX, Texas Red is a trademark of Molecular Probes, Inc.

The phycobiliproteins are unique in their strong absorbing and fluorescing capabilities [29,57,67]. Methods were developed for coupling these large protein–chromophore complexes to antibodies, and the phycobiliprotein-tagged antibodies have become very popular for flow cytometric studies of cell surface antigens. Of particular value are phycoerythrin-labeled antibodies, which can be used simultaneously with fluorescein-labeled antibodies for two-color analysis of cells [45]. Both tags can be excited at 488 nm, but the phycoerythrin has a large Stokes shift and its orange emission can be detected independently of the yellow green fluorescein emission.

More labeling reagents need to be developed. For example, it would be valuable to have a substitute for fluorescein that is as easy to couple to antibodies as fluorescein isothiocyanate but which is more photostable, less sensitive to pH, and has a higher ϵ and ϕ when bound to proteins. Probes that absorb and emit further to the red (even into the near-infrared) than those available now would be valuable for a number of reasons. Excitation of such dyes would produce far less autofluorescence, and inexpensive HeNe lasers could be used for excitation sources. It may be possible to develop fluorescent labels that can be excited in the 670- to 820-nm range and can be used in flow cytometers "built on chips" that incorporate laser diodes for excitation and solid-state detectors. Cyanine, merocyanine, and oxonol dyes are being exploited as labeling reagents in the laboratory of the author. These chromophores have very large extinction coefficients and can be synthesized with many different structures, charges, and wavelengths. The cyanines appear particularly promising as antibody tags.

Fluorescent labels can also be used as probes to localize and monitor interesting biological processes in living cells. Endocytic processing of receptors and ligands, e.g., insulin and EGF, which have been tagged with fluorescein and other markers, have been studied [64,65].

Detection of specific genes in situ with biotin derivatized DNA probes and fluorescein-tagged avidin first occurred in 1981 [51]. Subsequently, hapten molecules were attached to DNA probes so that they could be detected with fluorescent monoclonal antibodies. The result is a powerful technology for detection of infectious agents, oncogenes, and genetic defects in tissues, cells and fluids [12]. Efforts are being made in a number of laboratories to develop procedure for in situ hybridization with cells and chromosomes to be analyzed by flow cytometry. For analysis of cells by this method it is likely that multiple gene copies will have to be present or else new fluorescence amplification methods will have to be devised to provide sufficient signal to overcome auto fluorescence and other sources of noise. For flow cytometric analysis of chromosomes, hybridization methods will have to be developed that preserve chromosome morphology so that flow karyotyping and gene detection can be done simultaneously.

Another important advance was the development of fluorescein-tagged antibodies to quantify incorporation of bromodeoxyuridine into DNA [23] (see also Chapter 23). This reagent is useful in identifying S-phase cells in cell kinetic analysis. A complete volume of Cytometry [Vol. 6, No. 6, 1985] is devoted to this subject.

There is no doubt that there is a bright future for the application of fluorescent probes for analysis of antigens, genes, and biochemical processes in living and fixed cells and tissues. Still needed is further development of fluorescent chromophores with extended spectroscopic properties that can be covalently attached to biological molecules without interfering with their function. Interested organic and inorganic [34] chemists: take note!

NONCOVALENTLY ASSOCIATING FLUORESCENT PROBES

Because of their particular molecular composition, fluorescent probes in this class associate noncovalently with special structures in cells. Structures that can be visualized with these probes include DNA, RNA, lipid, electrically negative or positive compartments such as mitochondria, and compartments with a low pH.

Probes of DNA and RNA Content

A wide variety of fluorescent probes have been developed for quantification of the DNA and RNA contents of cells. The structures of the more commonly used probes are shown in Figure 2. The major applications involve cell cycle analysis (see Chapter 24), chromosome analysis (see Chapter 25), and detection of aneuploid cells in tumor samples (see Chapter 37). Other applications can be found in other reviews [46,63,78,84].

There are a number of important factors to consider when selecting a DNA or RNA content probe. The first is specificity. Some probes, such as Hoechst 33342 and the Feulgen stain, interact preferentially with DNA, whereas intercalating fluorophores such as propidium and acridine orange bind to double-stranded RNA as well as DNA [46,63,78,84]. In order to use one of the nonselective probes to measure only DNA content, prior treatment of the sample with ribonuclease is generally required to remove RNA.

It is interesting that the DNA-intercalating probe, acridine orange, will also bind to single stranded RNA, but shows a red-shifted emission spectrum when this occurs. Darzynkiewicz, Traganos, and Melamed [18] and their colleagues have taken advantage of the green DNA fluorescence and the red RNA fluorescence from acridine orange to develop an important method for simultaneously quantifying DNA and RNA in cells by flow cytometry (see Chapter 15). Kapuscinski et al. [43] have elucidated the mechanism for the spectral shift. This group has also recently analyzed the interaction of pyronin Y with DNA and RNA, and has been able to account for absorption and fluorescence properties of these complexes [17,42].

Shapiro showed that it is possible to combine probes, Hoechst 33342 and pyronin Y, which have differential selectivity for DNA and RNA to measure DNA and RNA simultaneously on single cells [79]. Thioflavin T [73] and $diOC_1$ (3) [38] and thiazole orange [52] have been used to estimate RNA content of reticulocytes.

Some probes have selectivity for AT-rich or GC-rich regions of DNA [46,63,78]. Hoechst 33342, Hoechst 33258, DAPI, and DIPI (4-6-bis[2-imidazxolinyl-4H,5H]-2-phenylindol) prefer AT-rich regions, but mithramycin, chromomycin A3, and olivomycin select GC-rich regions of DNA. This selectivity is useful for analysis of bacterial samples which have a wide range of AT/GC ratios. Measurement of this parameter using appropriate AT/GC-selective probes can provide a method for bacterial classification by flow cytometry [98] (see also Chapter 29). Two-color analysis using probes selective for AT-rich or GC-rich regions is also valuable for karyotype analysis of chromosomes (see Chapter 25). Chromosomes that have near-identical total DNA content and that are not separable in a single-parameter DNA histogram often are sufficiently different in AT/GC base content to be discriminated by appropriate base-selective DNA stains used in dual parameter histograms.

The second factor to be considered in DNA and RNA probe selection is plasma membrane and nuclear permeability. Hoechst 33342 is the only commonly used DNA probe that will stain living cells. Thus, living cells in different stages of the cell cycle can be sorted and analyzed on the basis of DNA content determined with the use of Hoechst 33342 [2]. Other stains, such as propidium, are too highly charged or too polar to cross the membrane of a living cell. (Propidium readily crosses the membranes of "dead" cells to make them highly fluorescent. This property is useful in immunology for cell-killing assays [63].) Darzynkiewicz and colleagues [19] found significant differences in the accessibility of cellular DNA to different fluorochromes. Studies of these differences may lead to a better understanding of chromatin structure.

The third factor is the spectral properties of the probe. Hoechst 33342, DAPI, and a number of other DNA/RNA probes are optimally excited in the ultraviolet (UV) region of the spectrum. This is an advantage when dual laser instruments are used to quantify DNA with UV laser excitation and simultaneously to measure immunological properties of the cells using a laser emitting visible light. In other cases, it is preferable to excite DNA probe fluorescence in the visible spectrum because less autofluorescence is produced and because laser tubes operating in the visible range generally last longer. Propidium is popular for blue light excitation at 488 nm. Shapiro and Stephens [81] discovered several helium neon-laser-excitable (633-nm) dyes that may be useful for

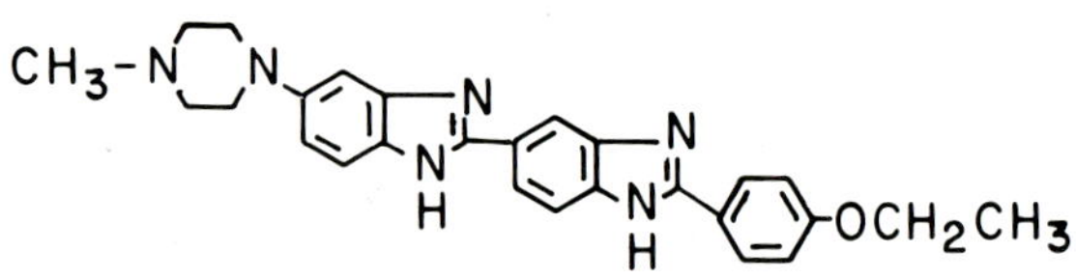

I-HOECHST 33342

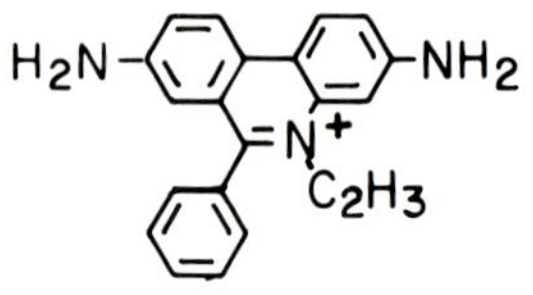

II- ETHIDIUM

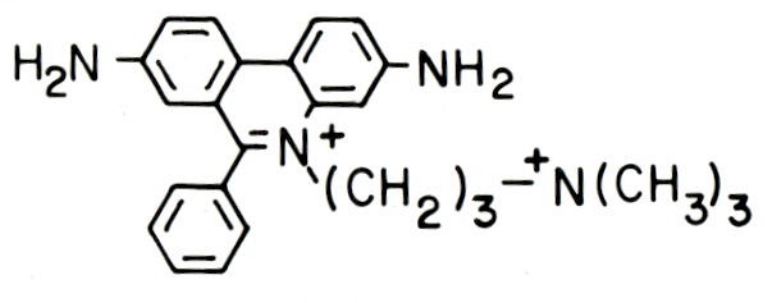

III PROPIDIUM

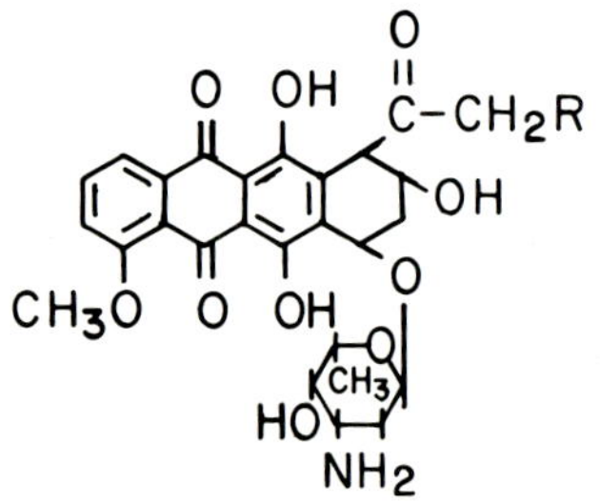

IV ANTHRACYCLINE DYES

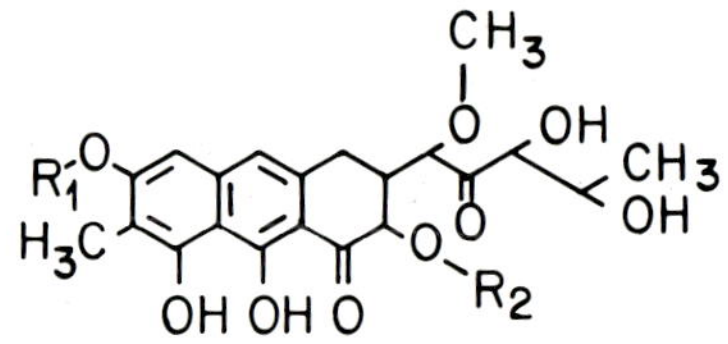

V MITHRAMYCIN

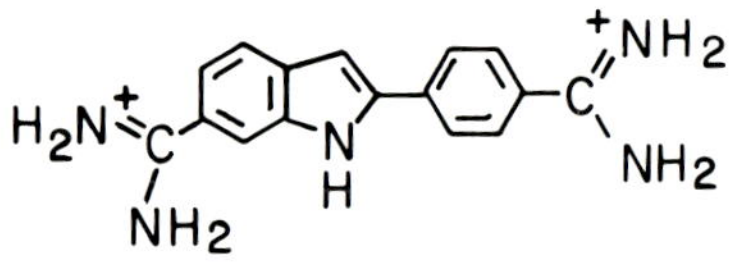

VI DAPI

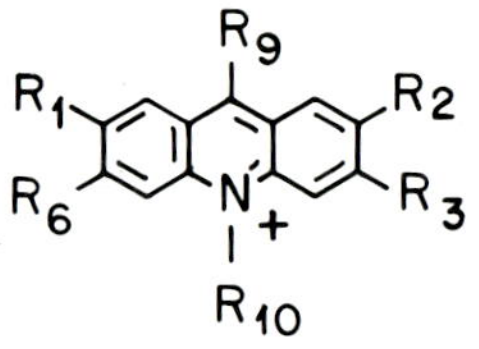

VII ACRIDINE DYES

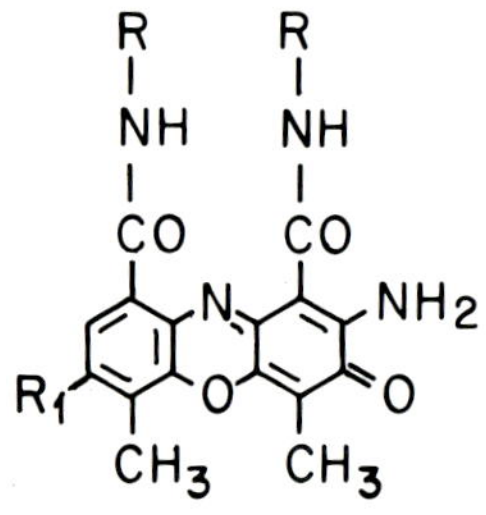

VIII ACTINOMYCINS

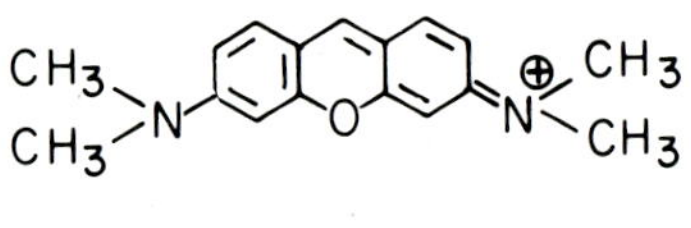

PYRONIN Y

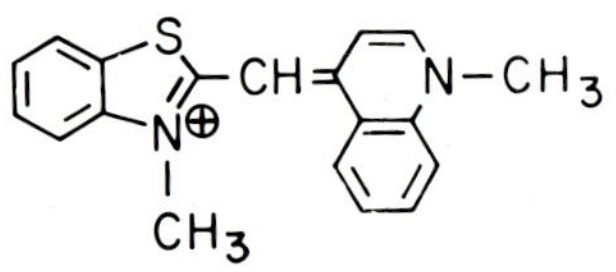

THIAZOLE ORANGE

Fig. 2. Fluorescent probes that bind noncovalently to DNA or RNA. For the anthracycline dyes (IV) the R group is -H on Daunomycin and -OH on Adriamycin. The R groups of mithramycin (V) are two- and three-unit oligosaccharides. The acridines (VII) have a wide variety of R-groups including alkyl, amino, alkylamino, methoxy, and halogen substituents. The R groups on Actinomycin D (VIII) are cyclic polypeptide chains and R_1 is a hydrogen.

quantifying DNA. This advance would leave available the shorter regions of the spectrum for measurement of other cell parameters with flow cytometers.

Membrane Potential Probes

Membrane potential is an important property of cells and some organelles. The electrical potential difference across the plasma membrane is directly involved in the transport of ions and nutrients into and out of the cell. Modulation of the cell membrane potential is the mechanism of nerve conduction and a part of sensory transduction. The electrical potential difference works together with the pH gradient across the mitochondrial membrane to couple energy released during electron transport to the phosphorylation of ADP to form ATP. Quantification of membrane potentials can be accomplished with microelectrodes in large cells, but in small cells and organelles, membrane potentials are estimated from the transmembrane distribution of membrane permeant cations and anions that are either radioactive or fluorescent. Cyanine dyes (Fig. 3) were the first fluorescent membrane potential probes to have had their mechanism of action explained [82], and they have been used extensively since [100,101]. The cationic cyanine dyes are accumulated in compartments with a negative potential, such as cell interiors and mitochondria. Depolarization of the cell membrane or uncoupling of mitochondria leads to a loss of fluorescence as cyanine dye leaks out. Rhodamine 123 is also a membrane potential stain, exploited by Chen and colleagues for visualization of mitochondria in living cells under the fluorescence microscope [39]. Rhodamine 123 has been termed a mitochondrial specific dye because the mitochondria are so brightly stained. Cyanine dyes also yield bright mitochondria in healthy cells. Undoubtedly, the high fluorescence of mitochondria arises because they have a negative membrane potential. Uncoupling agents and inhibitors reduce mitochondrial fluorescence. The large membrane surface area in the mitochondrial matrix may contribute to the staining by binding large amounts of accumulated probe. Rhodamine 123 retention by tumor cell lines has been studied and related to anticancer drug sensitivity [50]. Other uses of this mitochondrial stain have been reviewed recently [102].

Cyanine dyes and other membrane potential probes have been used in flow cytometry experiments to study changes in neutrophils [76,77] and lymphocytes [60,61,80,103,104] following stimulation of the cells with formylated peptides and mitogens, respectively. The fluorescent changes are substantial, but the mechanisms are not well understood [14,91]. A major complication is the possible contribution of mitochondria to the total cell fluorescence change [11,60, 61,101–104]. Chused et al. [15] show that membrane-permeant anionic oxonol dyes may provide answers to this question. Mitochondria tend to exclude anionic dye, decreasing the contribution by this organelle. Anionic dyes also distribute across the plasma membrane according to the membrane potential, but the total cell fluorescence will be small because these dyes are excluded from the negatively charged (−50 to −60 mV) cells.

The large fluorescence changes that occur when these blood cells are stimulated in the presence of these probes undoubtedly reflect important physiological changes taking place in cells. It will be interesting to determine the nature of these changes eventually as well as their relationship to probe fluorescence alterations.

Probes for Visualization of Membranes and Lipid Compartments

Fluorescent probes for lipid bilayers (Fig. 4) have been available for some time [99]. Many are uncharged hydrocarbons like diphenyl hexatriene (DPH). Others are charged but have long hydrocarbon chains attached (e.g., $diIC_{18}[3]$). Since not all lipid membranes in a cell are alike, it is not surprising that fluorescent lipid probes with different structures partition between the different membranes in interesting ways. For example, Pagano and Sleight [68] made extensive use of fluorescent lipid analogues to trace pathways of membrane movement during endocytosis and recycling. Unfortunately, flow cytometry gives little information about the spatial distribution of these probes within cells. However, lipid probes are taken up differently by cells that differ in physiology, and flow cytometry can be used to quantify the uptake. For example, the hydrophobic dye Nile Red has been shown to be highly specific for lipid oil droplets in cells [30]. With 488 nm excitation, fluorescence measurements in the 515- to 560-nm range give good discrimination of acylated low-density lipoprotein uptake by macrophages.

pH Probes That Partition Between Compartments with Different pH

There are two classes of pH probes. One class of fluorophores have spectral properties that change with pH (e.g., fluorescein). These fluorophores are discussed in the next section, which centers on environment-sensitive fluorescent probes.

A second class of pH probes partition differently into cell compartments with different internal pH. The pH of each compartment can be estimated by the amount of fluorescent weak acid or weak base that distributes into that compartment. Weak bases cross membranes as the conjugate base (usually the whole molecule is neutral) but become trapped in acidic compartments when protonated. The converse is true for weak acids. Acridine orange is accumulated in neutrophil granules [1] and acidic lysomal compartments presumably for this reason. These authors showed that the kinetics of degranulation following stimulation of the neutrophils with the calcium ionophore A23187 could be followed by flow cytometry. Degranulation leads to loss of red fluorescence as the acridine orange (in an aggregated state?) in the lysosomes is released into the extracellular space. Apparently, the lysomal fluorescence is not masked or complicated by acridine orange fluorescence from cellular RNA in the live cell. The probe 9-amino acridine also is accumulated in acidic compartments [20] but this probe has evidently not been useful in flow cytometry experiments. Unless this probe is used at extremely low concentrations, the fluorescence of accumulated 9-amino acridine molecules is actually reduced as a result of concentration quenching inside the compartment. It would be useful if there were specific fluorescent probes for acidic and basic compartments. This should provide another area for collaboration of organic chemists and spectroscopists.

FLUORESCENT PROBES THAT ARE SENSITIVE TO THEIR MICROENVIRONMENT

Certain fluorescent chromophores can be used to estimate the following properties of their local environment: pH, calcium concentration redox potential, polarity, fluidity, and the presence of other ions or molecules that can deactivate the excited state of the chromophore by energy transfer or

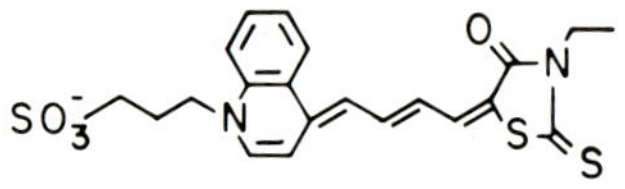

WW375

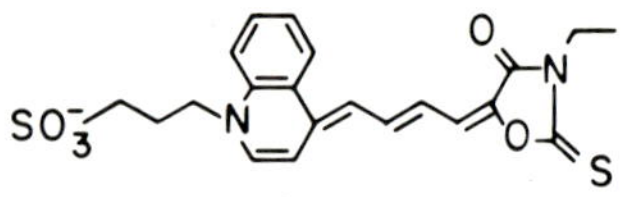

NK2367

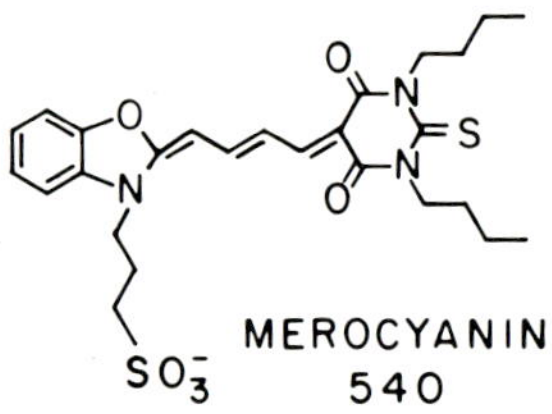

MEROCYANINE 540

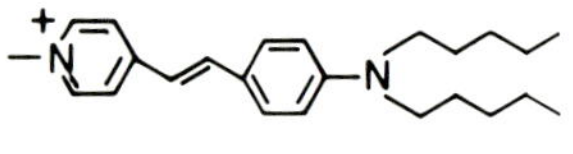

di-C_5-ASP

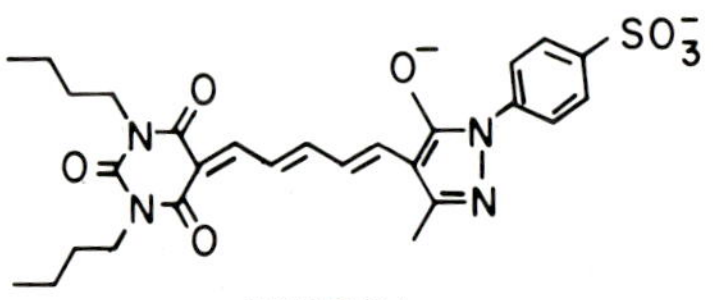

WW781

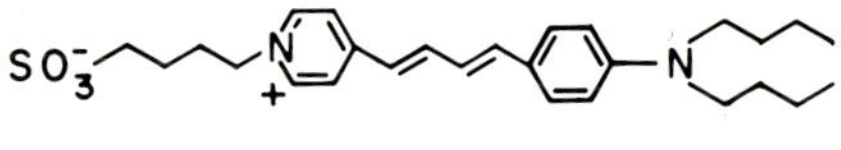

RH160

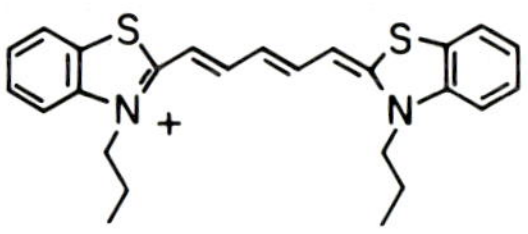

diS-C_3-(5)

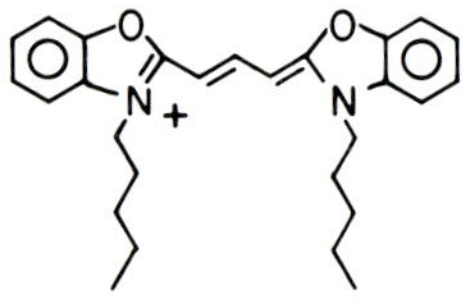

diO-C_5-(3)

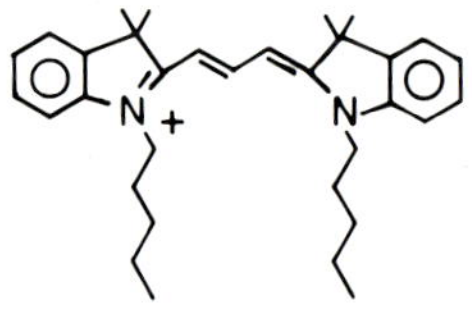

diI-C_5-(3)

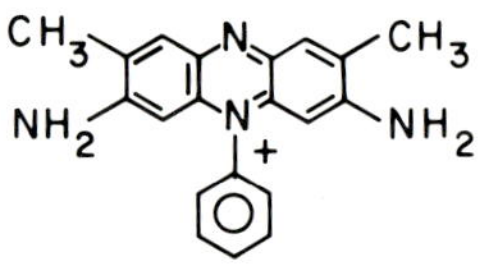

SAFRANIN O

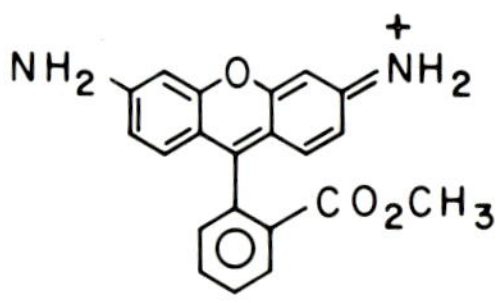

RHODAMINE 123

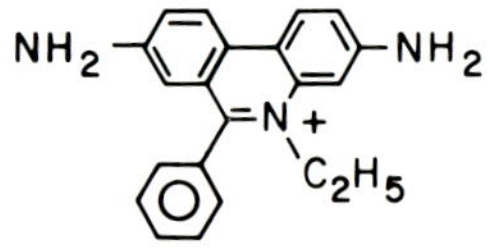

ETHIDIUM

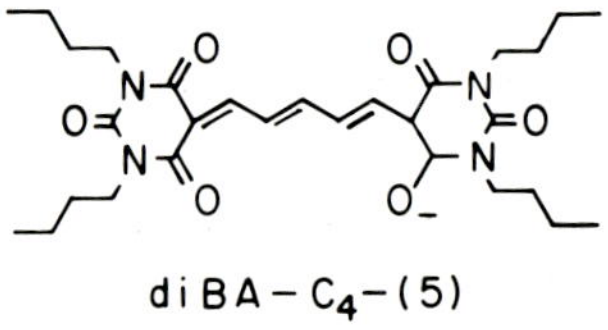

diBA-C_4-(5)

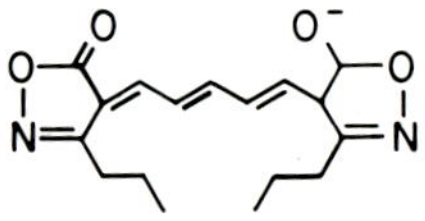

OX-VI

Fig. 3.

charge transfer. The major challenge in making use of these dyes is in finding a method, by synthetic modification of the probe or otherwise, to target the probe to the region of interest within the cell.

pH Probes That Respond with Spectral Changes

Hydrogen ion or metal ion binding changes the electronic structure of a number of dyes. The absorption properties, the fluorescence properties, or both, of the molecule can be affected, depending on whether the hydrogen or metal ion concentration is near the dissociation constant of the ground state or the excited state, respectively [75]. The rates of complexation and decomplexation relative to the excited state lifetime of the probe also play a role in the sensitivity of fluorescence to pH or to pION [−log(cation or anion)]. An example of a probe with an absorption spectrum sensitive to pH changes in the physiological range is 6-carboxyfluorescein, which has a ground-state pKa near 6.5. The absorption spectrum shifts from the 440 to 450 region at more acid pH to near 490 as the pH is raised [92]. Emission is not especially affected for reasons discussed by Martin and Lindquist [56]. Therefore, pH measurements are made by calculating the ratio of the emission produced with 496 excitation and the emission produced with 452 excitation. Ohkuma and Poole [66] used a standard fluorometer to measure pH in endosomes. Heiple and Taylor [33] and Tanasugarn et al [90] demonstrated how this method can be used to determine cellular pH with a fluorescence microscope. Murphy et al. [64] used an interesting variation of this technique to measure endosomal pH of single cells with a dual-laser excitation flow cytometer. Instead of exciting fluorescein at two wavelengths, a mixture of endocytosed fluorescein and rhodamine labeled insulin molecules were excited. PH-insensitive rhodamine fluorescence provided a measure of the amount of both ligands in each cell. The ratio of fluorescein emission to rhodamine emission gave the pH of the cell. (See Chapter 18 for further discussion of pH measurements by flow cytometry.)

There is also a useful pH probe that changes its *emission* properties in the physiological pH range. Valet et al. [97] showed that 2,3-dicyano-1,4-dihydroxybenzene can be excited at a single wavelength, and the emission spectrum shifts from 450 to 384 nm as pH increases [47]. The excited state pKa of this molecule is in the physiological range, and proton exchange can occur within its excited-state lifetime. The probe is loaded into cells in the diacetyl form. Nonspecific esterases expose the pH-sensitive hydroxyl groups. Thus, this probe is expected to report the pH of esterase-rich regions, including the cytoplasmic compartment. The ester-loading trick can be used with 6-carboxyfluorescein diacetate [92]. Unfortunately, these probes may not be limited to any one compartment of most cells.

Fig. 3. Membrane potential probes. The top six structures are fast dyes that respond to membrane potential changes that take place in microseconds. These probes are membrane impermeant and are useful for detecting action potentials in excitable membranes. The work of London et al. [55] illustrates the use of these probes. The bottom eight structures are slow dyes that redistribute across membranes over a period of milliseconds to seconds and accumulate in cells or cellular compartments according to the membrane potential difference. Chused et al. [15] discuss the application of several of these probes, as does Waggoner [100,101].

Hydroxypyrene trisulfonate is pH sensitive in the physiological range [16]. The molecule has the advantage that it does not leak out of cells but has the disadvantage that loading methods [59] are required to get the probe into cells. The relevance of cellular pH to function has been reviewed [13].

Calcium Probes

Because of its important role in regulating many cellular processes, calcium has received considerable attention in recent years [14,96] (see Chapter 32). The indicator Quin II, developed by Tsien [95], provided the first fluorescence method for detecting changes in cytoplasmic calcium [96]. The structure of Quin II is shown in Figure 6. Its acetoxy ester precursor is membrane premeant. Cytoplasmic esterase activity liberates the carboxylate groups, which limit the escape of the probe into the medium and provide chelation groups for binding the calcium. Bound calcium shifts the absorption peak toward the red and measurements are usually made with excitation near 340 nm. Calcium determinations with Quin II and its improved analogue fura-2 [31] are best made by a ratio technique whereby emission is measured at the two optimal excitation wavelengths for probe with and without bound calcium. Quin II is sufficiently nonfluorescent, so that cells must be loaded with 0.1- to 1.0-mM probe. The large Quin II concentration has worried some investigators because the probe is such a strong calcium buffer. The dissociation constant of the probe calcium complex is below 10^{-6} M. Thus, makeup calcium must enter the cytoplasm, which normally contains less than 10^{-6} M calcium, before the cytoplasmic Quin II is at equilibrium and calcium measurements can begin. Makeup calcium is provided in the medium during experiments. Several recent reports describing internal calcium concentration changes in platelets [40] and 3T3 fibroblasts [58,62] suggest that Quin II signals may not accurately indicate the magnitude or kinetics of the changes. Thus Quin II may be best for qualitatively detecting calcium level changes. The improved probe from Tsien's laboratory, fura-2, is more sensitive and can be used at lower concentrations [31]. Another advance is the synthesis of indo-1 which shows an emission shift upon calcium binding [31]. In flow cytometry, this is a definitive advantage since indo-1-stained cells can be excited with a single laser line and calcium concentration determined from the ratio at two emission wavelengths [15]. Care must be exercised in applying and interpreting the fluorescence signals of the new generation of fluorescent calcium indicators.

Fluorescent Redox Potential Probes

Neutrophils release superoxide and hydrogen peroxide when they are stimulated with chemotactic factors. Several fluorescent probes have been used to monitor the kinetics of this process in normal and diseased cells. Dichlorofluorescin (Fig. 7) can be loaded into cells in its acetylated form and trapped as a result of esterase activity. Bass and associates [7] showed that H_2O_2 released from stimulated neutrophils converts the nonfluorescent reduced form into dichlorofluorescein thereby making stimulated cells fluorescent. Cellular NADH and NADPH are intrinsic fluorescent probes of the redox state of cells and have also been used to monitor the energy state of cells and tissues [93].

Fluorescent Probes for Measuring Enzyme Activities

Fluorogenic enzyme substrates are chromophores converted by specific enzymes into products that have either

C_6 NBD PA

NBD PE

diI C_{18}–(3)

TRANS-PARANARIC ACID

$CH_3-(CH_2)_m-\underset{\underset{\underset{\underset{\text{anthryl}}{C=O}}{|}}{O}}{C}-(CH_2)_n-COO^-$

(m,n)-ANTHROYLSTEARATE

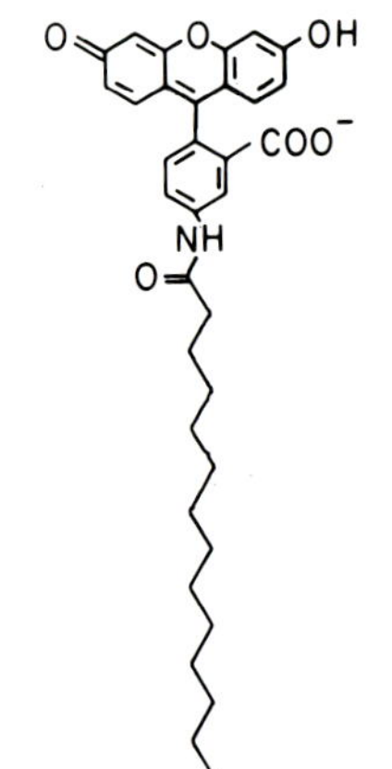

ACYL AMINOFLUORESCEIN

4-ALKYL UMBELLIFERONE

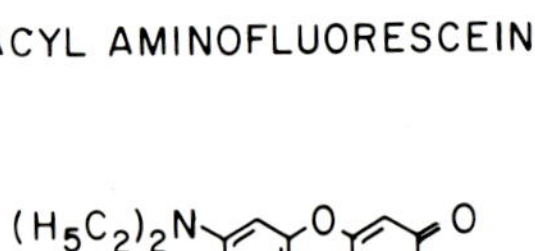

NILE RED

DPH

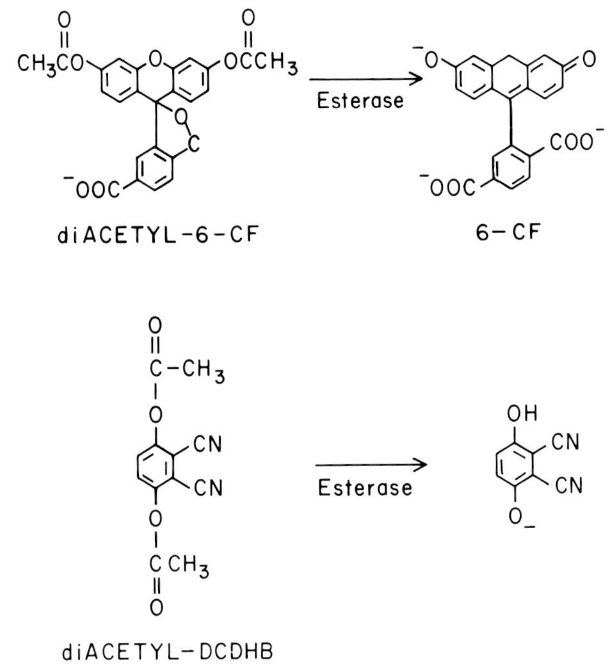

diACETYL-6-CF → Esterase → 6-CF

diACETYL-DCDHB → Esterase →

Fig. 5. pH probes. Two reactions illustrate the conversion, by cellular esterase enzymes, of membrane permeant ester molecules into fluorescent pH probes that are membrane impermeant because of their increased charge. The trapped products are 6-carboxyfluorescein (top) and 2,3-dicyano-1,4-dihydroxybenzene (bottom).

increased fluorescence or shifted spectra. These substrates can therefore be said to be sensitive to the presence of certain enzymes in their microenvironment. Dolbeare [21,22] provided excellent reviews of the use of fluorogenic substrates in flow cytometry.

Hydrolase enzymes play crucial roles in cell function. Lysosomal proteases for example are responsible for breakdown of intracellular proteins and peptides as well as polypeptide material that has been endocytosed [6,36]. Other proteases are bound to the external surfaces of cells or are secreted in soluble form. The latter enzymes breakdown extracellular protein molecules [70,72,86,107]. The breakdown of extracellular macromolecules is required for movement of phagocytic cells like macrophages and neutrophils through tissues [35,105]. It is also thought that the extracellular proteases play a significant role in tumor invasion, rheumatoid arthritis, and other disease states [86]. Thus, because of their significant roles in cell function, it is important to be able to measure activity and localization of enzymes such as proteases, glycosidases, esterases, phosphodiesterases, phosphatases, and sulfatases, in tissues, cells, organelles, and fluids.

Antibodies against proteases have come into wide use during the past 5 years for localizing and quantifying the enzymes in fixed tissues and cells and in serum. However, antibodies do not indicate whether the enzymes they bind to are functional (since the epitope on the enzyme is usually not associated with the active site of the enzyme) nor are antibodies sensitive to features involved in allosteric regulation of enzyme activity.

Most fluorogenic substrates have been constructed from chromophores that have either an amino or a hydroxyl group that participates in the electronic conjugation (Fig. 8). When an appropriate amino acid sequence is attached by an amide bond to an aminochromophore, the molecule usually becomes much less fluorescent. Cleavage of the amide bond of the substrate by a peptidase that will act upon the particular chromophore–peptide complex releases the more fluorescent primary amine chromophore. Naphthylamine, amino coumarins, amino quinolines, and rhodamine have served as chromophore bases for fluorogenic peptidase substrates [21,53].

Esterase, glycosidase, sulfatase, phosphatase, and phosphodiesterase substrates can be made by attaching the appropriate functional group to a hydroxyl-containing chromophore, as shown in Figure 8. Hydroxy coumarins and fluorescein have most often been used for construction of these substrates.

Fluorogenic substrates are now routinely used to measure activities of enzymes in such fluids as blood plasma [83]. However, these substrates have found little use in studies of enzyme activities in intact cells because they do not usually diffuse readily into cells and the fluorescent products rapidly leak out of the cells [21,25]. A variety of coupling reagents have been developed that convert products into insoluble colored substances that can be measured by absorbance techniques, but only one coupling reaction is available that results in an insoluble fluorescent product that stays within the cell. The latter involves coupling of nitrosalicylaldehyde to the methoxynaphthylamine product that is liberated by peptidase activity. This reaction, developed by Dolbeare and Smith [24] is illustrated in Figure 8. It is surprising that little use has been made of the coupling reaction, since it provides a way to produce fluorescence in cells that is proportional to enzyme activity.

A second disadvantage of available substrates is that most need to be excited with UV light, so their signals may be obscured by cell autofluorescence. The most commonly used peptidase substrates are based on the coumarin and naphthylamine chromophores that are excited near 365 nm. A recent and significant improvement was made when peptidase substrates based on the rhodamine chromophore were developed by Leytus et al. [53]. These substrates are excited at wavelengths near 490 nm. A limitation of presently available fluorogenic substrates is that none have been attached to peptide hormones or to phagocytosable particles like viruses, bacteria, and parasites so that the kinetics of lysosomal processing of these materials by living cells could be followed by sensitive fluorescent methods.

Certainly if the technical problems with fluorogenic substrates can be overcome there is potential for their wide use. Many hydrolases exit in cells and substrates need to be developed for the measurement of their activities in fluids, tissues, and cells.

Fig. 4. Lipid probes. The top two probes have been exploited by Pagano and colleagues [68]. Barak and Webb [5] used $diIC_{18}$-(3) to study the low-density lipoprotein receptor in cultured cells. Paranaric acid is discussed by Hudson et al. [37]. The anthroyl stearates have been used to study lipid fluidity and organization as a function of depth in bilayer membranes [9]. The pH-sensitive fluorescein and umbelliferone chromophores with long alkyl chains can be used to study the effects of membrane surface potential on the pH of aqueous regions close to the membrane surface [28]. Nile red is a lipid droplet probe [30]. Diphenylhexatriene, DPH, is a membrane lipid fluidity probe discussed by Lakowicz [48,49].

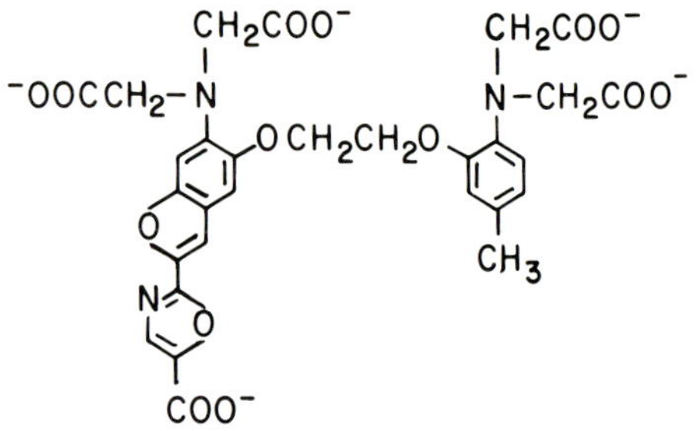

FURA-2

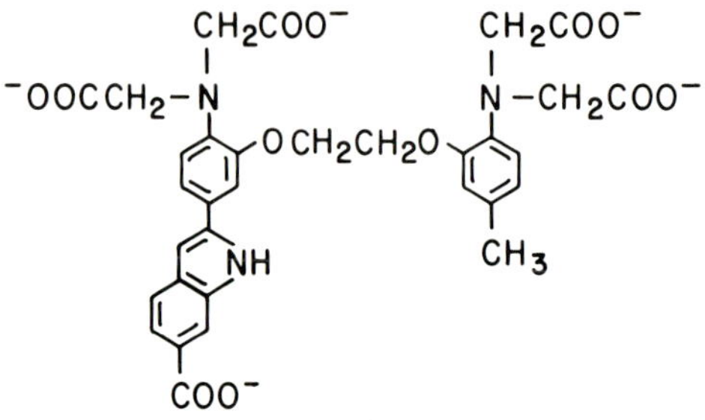

INDO-1

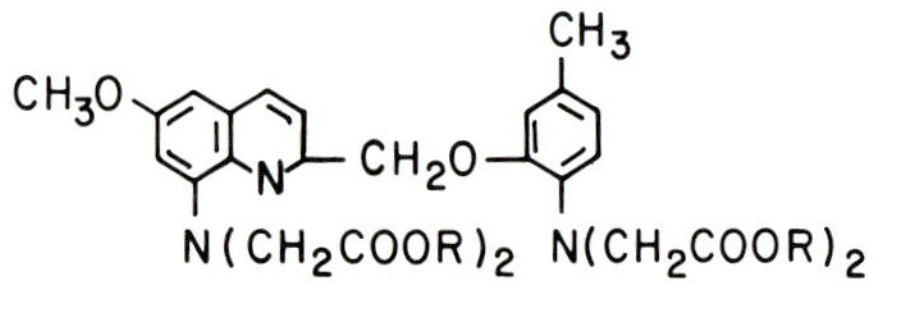

QUIN II

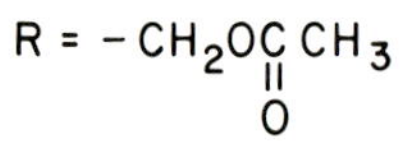

$R = -CH_2OCCH_3$ (with =O on the C)

ACETOXY-DERIVATIVE

Fig. 6. Calcium probes. Fura-2 and indo-1 can be obtained as acetoxyester derivatives, similar to the derivatized form of Quin II shown on the bottom left. The ester derivatives are membrane permeant. Cellular esterases convert the derivatives into the free-acid forms of the probes, which are trapped inside cells and are able to chelate calcium.

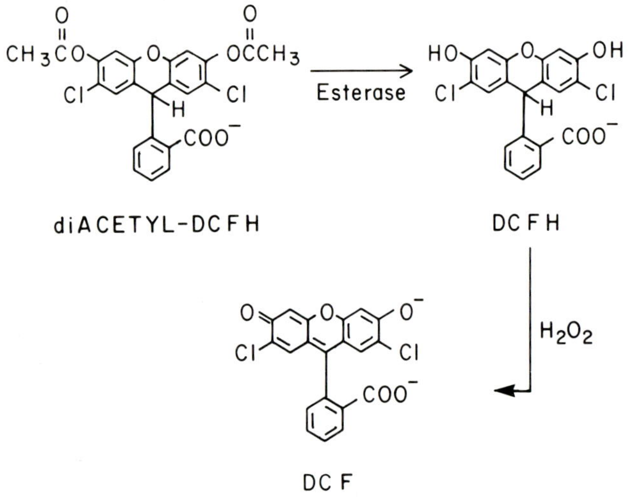

diACETYL-DCFH

Esterase

DCFH

H_2O_2

DCF

Fig. 7. Peroxide probe. Membrane permeant diacetyl-dichlorofluorescin is converted by cellular esterases to dichlorofluorescin. Peroxides released upon neutrophil stimulation oxidize the nonfluorescent dichlorofluorescin to the fluorescent dichlorofluorescein. Details can be found in Bass et al. [7].

Energy Transfer

Fluorescent chromophores in the excited state are sensitive to the presence of nearby chromophores that can act as acceptors for resonance energy transfer. Energy transfer can occur between two chromophores separated by distances of tens of Angstroms provided that the donor fluorescence spectrum overlaps significantly with the acceptor absorption spectrum. The relative orientation of the optical transition moments of the two chromophores also affects the efficiency of transfer. Energy transfer is a nonradiative process. That is to say, there is not absorption by the acceptor of a photon emitted by the donor. If the acceptor is fluorescent, however, it can emit a photon as a result of excitation of the donor. The theory and early uses of energy transfer to study protein and membrane structure have been reviewed elsewhere [68,106]. More recently, the technique has been used to detect aggregation of cell surface components which were separately labeled with fluorescein (donor) and rhodamine (acceptor) [27,89]. Energy transfer might also be used to look at self-assembly or binding events even inside living cells [91], since large numbers of cells can be loaded with fluorescent analogs by bulk loading methods [59].

Energy transfer can be a nuisance as well. For example, in experiments designed to quantify cell-surface markers with antibodies tagged with different-colored fluorophores, extensive capping or patching may, depending on the extent of tagging and the size of the antibody–antigen complex, result in energy transfer. The quantity of marker tagged with the energy donor would be underestimated in this circumstance. This possibility should not be overlooked in multicolor flow cytometry experiments.

In addition to energy transfer, it is possible for certain molecules and ions to quench fluorescence directly by deactivating the excited state. Direct quenching of probe fluorescence can occur in the presence of oxygen, heavy atoms, and molecules that form charge transfer complexes [49]. The latter processes can in principle be used to learn something about biophysical properties of cells, such as accessibility of dissolved oxygen to structures containing fluorescent probes. This potential has not been exploited with flow cytometry.

Solvent Polarity

Almost all fluorescent molecules are sensitive to some degree to the dynamic polarity of their solvent environment. It has been known for decades that the quantum yield of anilino-naphthaline sulfonate (ANS) is high in nonpolar solvents but more than 30-fold lower in water [87]. A large red shift in the emission wavelength of ANS also occurs in nonviscous polar solvents. This happens because energy of the

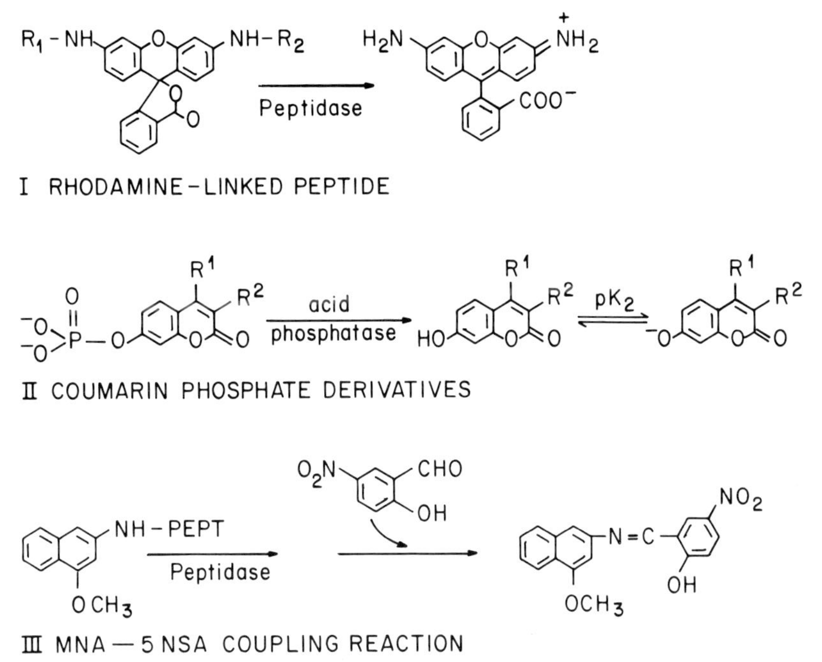

I RHODAMINE-LINKED PEPTIDE

II COUMARIN PHOSPHATE DERIVATIVES

III MNA — 5 NSA COUPLING REACTION

Fig. 8. Fluorogenic enzyme substrates. I, Rhodamine-linked peptides developed by Leytus et al. [53]. The R groups are peptides specific for the enzyme of interest. II, Acid phosphatase derivatives based on the coumarin structure. The R groups are chosen to shift the fluorescence of these derivatives toward the visible region of the spectrum [44]. For I and II, the products are more fluorescent than the substrates. III, The coupling reaction devised by Dolbeare and Smith [24]. The free amino group, which is made available when the peptide bond of peptidylmethoxynaphthylamine is hydrolyzed by a peptidase, couples with 5-nitrosalicylaldehyde, which is present during the reaction, to form the insoluble red-orange-fluorescing product.

excited state is lost to solvent when polar solvent molecules reorient around the more dipolar structure of excited ANS shortly after excitation. Reduction of the distance between the excited and ground state means that emitted photons will have lower energy, i.e., be red-shifted.

Sensitivity to solvent polarity is an important component of the mechanism of certain probes. For example, propidium is more fluorescent when protected from water by DNA intercalation. Also, many potential sensitive dyes are more fluorescent in hydrocarbon environments such as membranes and therefore show fluorescence changes when driven by membrane potential changes to compartments of different polarity [101].

Sometimes it is better if a probe is not sensitive to solvent polarity. Quantification of the binding of fluorescent labeled antibodies may be difficult if the probe is sensitive to the polarity of the cellular environment near the antigen or to different aqueous and nonaqueous preserving agents or to air drying. It complicates physiological interpretations when pH probes, calcium probes, energy-transfer probes, and distribution-type membrane potential probes also change their spectral properties because of their solvent environment changes rather than because of a change in the property they are intended to monitor.

Probes of Lipid Microviscosity

There is a direction defined on each chromophore—the transition moment—for which polarized light will be optimally absorbed or emitted. Absorption and emission transition moments are not always in the same direction but often they are. This property can be used to find the orientation of probes bound to biological structures illuminated with polarized light. The property can also be used to determine the rotational mobility of probes and therefore determine the microviscosity of their environment or the binding of the probe. The light emitted from a population of fixed fluorophores that have been excited with a beam of polarized light is highly polarized. However, if the fluorophores are rapidly rotating relative to their excited state lifetime, the emitted light will be depolarized. The rate of rotation can be determined by measuring the rate of depolarization of fluorescence after pulse excitation [106], by phase-modulation techniques [49], and by measuring the steady-state fluorescence polarization of probe molecules excited with continuous polarized light [106].

The theory and instrumentation for flow cytometric measurements of fluorescence polarization have been available for a number of years [41,54,108]. However, relatively few biological experiments have been carried out with single cell polarization measurements. Schaap et al. [74] determined the fluorescence polarization of 6 membrane probes associated with embryonic carcinoma cells after differentiation. Diphenylhexatriene (DPH) (see Fig. 4) and a charged analogue of DPH showed polarization increases upon differentiation, but anthroyl stearate analogues (see Fig. 4) gave no polarization change. Fox and Delohery [108] recently detected a difference in the DPH fluorescence from CHO cells grown in different media.

The value of fluorescence polarization measurements ultimately depends on having probes that selectively bind to cell structures or components in which rotational motion or microviscosity is involved in an interesting cell function. As often is the case in flow cytometry, the instrumentation capabilities outstrip the availability of ideal probes.

FUTURE FLUORESCENT PROBES FOR FLOW CYTOMETRY

The availability of new specific and sensitive fluorescent probes is the key to advances in the power of flow cytometry. New labeling reagents that fluoresce in all regions of the spectrum from the UV to the near-infrared (IR) have been developed that match new excitation wavelengths from lasers and laser diodes. These advances increase the flexibility a researcher has in choosing combinations of probes and provide additional capabilities for multicolor analysis of single cells. Furthermore, new probes with extended Stokes shifts will permit simultaneous excitation of two or more probes simultaneously with one laser, just as is done presently with the phycoerythrin-fluorescein combination. All these new labeling reagents to come will not only be used for tagging monoclonal antibodies but will also be used in conjunction with DNA and RNA hybridization probes both to detect sequences in individual chromosomes and to quantify the RNA content of single cells.

Fluorescent ion indicators are another class of probes certain to expand because of the driving need of biologists to detect ion concentration changes in single living cells. Besides long-wavelength excitable calcium indicators, sodium, potassium, and magnesium indicators are needed. Anion and trace metal indicators would also be valuable.

Additional and improved fluorogenic enzyme substrates for detection of enzyme activities are needed to study growth control and other functions of single living cells. Organic chemists who understand the needs of the biologists and the capabilities of the flow cytometer will need to overcome the disadvantages of available substrates that were mentioned earlier.

Altogether, the co-development of fluorescent probes, instrumentation, monoclonal antibodies, and nucleic acid hybridization technologies promises an exciting and fruitful future for flow cytometry.

REFERENCES

1. **Abrams WR, Diamond LW, Kane AB (1983)** A flow cytometric assay of neutrophil degranulation. J Histochem Cytochem 31:737–744.
2. **Arndt-Jovin DJ, Jovin TM (1977)** Analysis and sorting of living cells according to DNA content. J Histochem Cytochem 25:585–589.
3. **Aubin J (1979)** Autofluorescence of viable cultured mammalian cells. J Histochem Cytochem 27:36–43.
4. **Barak LS, Yocum (1981)** 7-Nitrobenz-2-oxa-1, 3-diazole (NBD)-Phallacidin: Synthesis of a fluorescent actin probe. Anal Biochem 110:31–38.
5. **Barak LS, Webb WW (1981)** Fluorescent low density lipoprotein for observation of dynamics of individual receptor complexes on cultured human fibroblasts. J Cell Biol 90:595–604.
6. **Barrett AJ (1979)** "Proteinases in Mammalian Cells and Tissues." New York: Elsevier North-Holland.
7. **Bass DA, Parce JW, Dechatelet LR, Szejda P, Seeds McThomas M (1983)** Flow cytometric studies of oxidative product formation by neutrophils: A graded response to membrane stimulation. J Immunol 130: 1910–1917.
8. **Benson R, Meyer RA, Zaruba M, McKhann G (1979)** Cellular Autofluorescence. Is it due to Flavins? J Histochem Cytochem 27:44–48.
9. **Blatt E, Sawyer WH (1985)** Depth-dependent fluorescence quenching in micelles and membranes. Biochim Biophys Acta 822:43–62.
10. **Bock G, Hilchenbach M, Schauenstein K, Wick G (1985)** Photometric analysis of antifading reagents for immunofluorescence with laser and conventional illumination sources. J Histochem Cytochem 33: 699–705.
11. **Brand MD, Felber SM (1984)** Membrane potential of mitochondria in intact lymphocytes during early mitogenic stimulation. Biochem J 217:453–459.
12. **Brigatti DJ, Myersow D, Leary JJ, Spalhotz B, Travis SZ, Fong CYK, Hsiung GD, Ward DC (1983)** Detection of viral genomes in cultured cells and paraffin embedded tissue sections using biotin-labeled hybridization probes. Virology 126:32–50.
13. **Busa WB, Nuccitelli R (1984)** Metabolic regulation via intracellular pH. Am J Physiol 246:R409–R483.
14. **Campbell AK (1983)** "Intracellular Calcium." New York: John Wiley & Sons.
15. **Chused TM, Wilson HA, Seligmann BE, Tsien RY (1986)** Probes for use in the study of leukocyte physiology by flow cytometry. In Taylor DL, Waggoner AS, Murphy RF, Lanni F, Birge R (eds), "Applications of Fluorescence in the Biomedical Sciences," New York: Alan R. Liss, pp 531–544.
16. **Clement NR, Gould MJ (1981)** Pyranine (8-hydroxy-1,3,6- pyrenetrisulfonate) as a probe of internal aqueous hydrogen ion concentration of phospholipid vesicles. Biochemistry 20:1534–1538.
17. **Darzynkiewicz Z, Kapuscinski J, Traganos F, Crissman HA (1987)** Application of Pyronin Y(G) in cytochemistry of nucleic acids. Cytometry 8:138–145.
18. **Darzynkiewicz Z, Traganos F, Melamed M (1980)** New cell cycle compartments identified by multiparameter flow cytometry. Cytometry 1:98–108.
19. **Darzynkiewicz Z, Traganos F, Kapuscinsky J, Staiano-Coico L, Melamed M (1984)** Accessibility of DNA in situ to various fluorochromes: Relationship to chromatin changes during erythroid differentiation of Friend leukemic cells. Cytometry 5:355–363.
20. **Deamer DW, Prince RC, Crofts AR (1972)** The response of fluorescent amines to pH gradients across liposome membranes. Biochim Biophys Acta 274: 323–335.
21. **Dolbeare F (1981)** Fluorometric quantification of specific chemical species in single cells. In Wehry EL (ed), "Modern Fluorescence Spectroscopy." New York: Plenum Press, pp 251–293.
22. **Dolbeare F (1983)** Flow cytoenzymology—An update. In Fishman WH (ed), "Oncodevelopmental Markers. Biologic, Diagnostic, and Monitoring Aspects." Orlando, FL: Academic Press, pp 207–217.
23. **Dolbeare F, Gratzner H, Pallovicini MG, Gray JW (1983)** Flow cytometric measurement of total DNA content and incorporation of bromodeoxyuridine. Proc Natl Acad Sci USA 80:5573–5577.
24. **Dolbeare FA, Smith RE (1977)** Flow cytometric measurement of peptidases with use of 5-nitrosalicylaldehyde and 4-methoxy-β-naphthylamine derivatives. Clin Chem 23:1485–1491.

25. **Dolbeare F, Vanderlaan M** (1979) A fluorescent assay of proteinases in cultured mammalian cells. J Histochem Cytochem 27:1493–1495.

26. **Dovichi NJ, Martin JC, Jett JH, Trkula M, Keller RA** (1984) Laser-induced fluorescence of flowing samples as an approach to single molecule detection in liquids. Anal Chem 56:348–354.

27. **Fernandez SM, Berlin RD** (1976) Cell surface distribution of lectin receptors determined by resonance energy transfer. Nature (Lond) 264:411–415.

28. **Fromhertz P, Masters B** (1974) Interfacial pH at electrically charged lipid monolayers investigated by the lipoid pH-indicator method. Biochim Biophys Acta 356:270–275.

29. **Glazer AN, Stryer L** (1984) Phycofluor probes. TIBS: 423–427.

30. **Greenspan P, Mayer EP, Fowler SD** (1985) Nile Red: A selective fluorescent stain for intracellular lipid droplets. J Cell Biol 100:965–973.

31. **Grynkiewicz G, Poenie M, Tsien RY** (1985) A new generation of Calcium indicators with greatly improved fluorescence properties. J Biol Chem 260: 3440–3450.

32. **Haugland RP** (1983) Covalent Fluorescent Probes. In Steiner RF (ed), "Excited States of Biopolymers." New York: Plenum Press, pp 29–58.

33. **Heiple JM, Taylor DL** (1980) Intracellular pH in single motile cells. J Cell Biol 86:885–890.

34. **Hemmila I, Dakubu S, Mukkala V-M, Siitari H, Lovgren T** (1984) Europium as a label in time-resolved immunofluorometric assays. Anal Biochem 137: 335–343.

35. **Herscowicz H, Holden H, Bellante J, Ghaffor A** (1981) "Manual of Macrophage Methodology." New York: Marcel Dekker.

36. **Holtzman E** (1976) "Lysosomes: A Survey." New York: Springer-Verlag.

37. **Hudson B, Harris DL, Ludescher RD, Ruggiero A, Cooney-Freed A, Cavalier SA** (1986) Fluorescent probe studies of proteins and membranes. In Taylor DL, Waggoner AS, Murphy RF, Lanni F, Birge R (eds), "Applications of Fluorescence in the Biomedical Sciences." New York: Alan R. Liss, pp 159–202.

38. **Jacobberger JW, Horan PK, Hare JD** (1984) Flow cytometric analysis of blood cells stained with DiOC1(3): Reticulocyte quantification. Cytometry 5: 589–600.

39. **Johnson LV, Walsh ML, Bockus BJ, Chen LB** (1982) Monitoring of relative mitochondrial membrane potential in living cells by fluorescence microscopy. J Cell Biol 88:526–535.

40. **Johnson PC, Ware JA, Cliveden PB, Smith M, Dvorak AM, Salzman EW** (1985) Measurement of ionized calcium in blood platelets with the photoprotein aquorin. Comparison with quin II. J Biol Chem 260:2069–2076.

41. **Jovin TM** (1979) Fluorescence polarization and energy transfer: Theory and application. In Melamed MR, Mullaney PF, Mendelsohn ML (eds), "Flow Cytometry and Sorting." New York: John Wiley & Sons, pp 137–165.

42. **Kapuscinski J, Darzynkiewicz Z** (1987) Interactions of Pyronin Y(G) with nucleic acids. Cytometry 8: 129–137.

43. **Kapuscinski J, Darzynkiewicz Z, Melamed M** (1982) Luminescence of the solid complexes of acridine orange with RNA. Cytometry 2:201–211.

44. **Koller E, Wolfbeis OS** (1985) Synthesis and spectral properties of long wavelength absorbing and fluorescing substrates for direct and continuous kinetic assay of carboxyesterases, phosphatases, and sulfatases. Monatsschr Chem 116:65–75.

45. **Kronick MN, Grossman PD** (1983) Immunoassay techniques with fluorescent phycobiliprotein conjugates. Clin Chem 29:1582–1586.

46. **Kruth HS** (1982) FLow cytometry: Rapid biochemical analysis of single cells. Anal Biochem 125:225–242.

47. **Kurtz I, Balaban RS** (1985) Fluorescence emission spectroscopy of 1,4-dihydroxyphthalonitrile. Biophys J 48:499–508.

48. **Lakowicz JR** (1986) Biochemical applications of frequency-domain fluorometry. In Taylor DL, Waggoner AS, Murphy RF, Lanni F, Birge R (eds), "Applications of Fluorescence in the Biomedical Sciences." New York: Alan R. Liss, pp 225–244.

49. **Lakowicz JR** (1983) "Principles of Fluorescence Spectroscopy." New York: Plenum Press.

50. **Lampidis TJ, Bernal SD, Summerhays IC, Chen LB** (1982) Rhodamine 123 is selectively toxic and preferentially retained in carcinoma cells in vitro. Ann NY Acad Sci 397:299–302.

51. **Langer PR, Waldrop AA, Ward D** (1981) Enzymatic synthesis of biotin labeled polynucleotides: Novel nucleic acid affinity probes. Proc Natl Acad Sci USA 78:6633–6637.

52. **Lee LG, Chen C-H, Chiu LA** (1986) Thiazole Orange: A new dye for reticulocyte analysis. Cytometry 7:508–517.

53. **Leytus SP, Patterson WL, Mangel WF** (1983) New class of sensitive and selective fluorogenic substrates for serine proteinase. Biochem J 215:253–260.

54. **Lindmo T, Steen HB** (1977) Flow cytometric measurement of the polarization of fluorescence from intracellular fluorescein in mammalian cells. Biophys J 18:173–187.

55. **London J, Zecevic D, Loew LM, Orbach HS, Cohen LB** (1986) Optical measurement of membrane potential in simple and complex nervous systems. In Taylor DL, Waggoner AS, Murphy RF, Lanni F, Birge R (eds), "Applications of Fluorescence in the Biomedical Sciences." New York: Alan R. Liss, pp 423–447.

56. **Martin MM, Lindquist L** (1975) The pH dependence of fluorescein fluorescence. J Lumin 10:381–390.

57. **Mathies RA, Stryer L** (1986) Single-molecule fluorescence detection: A feasibility study using phycoerythrin. In Taylor DL, Waggoner AS, Murphy RF, Lanni F, Birge R (eds), "Applications of Fluorescence in the Biomedical Sciences." New York: Alan R. Liss, pp 129–140.

58. **McNeil PL, McKenna MP, Taylor DL** (1985) A transient rise in cytosolic calcium follows stimulation of quiescent cells with growth factors and is inhibitable with phorbol myristate acetate. J Cell Biol 101:372–379.

59. **McNeil PL, Murphy RF, Lanni F, Taylor DL** (1984) A method for incorporating macromolecules into adherent cells. J Cell Biol 98:1556–1564.

60. **Monroe JG, Cambier JD** (1983) B cell activation. I. Antiimmunoglobulin-induced receptor crosslinking

results in a decrease in plasma membrane potential of murine B lymphocytes. J Exp Med 157:2073–2089.

61. **Monroe JG, Cambier JC (1983)** B cell activation. II. Receptor crosslinking by thymus-independent and thymus-dependent antigens induces a rapid decrease in the plasma membrane potential of antigen-binding B lymphocytes. J Immunol 131:2641–2644.
62. **Moolenaar WH, Tertoolen LGJ, deLaat SW (1984)** Growth Factors immediately raise cytoplasmic free Ca^{++} in human fibroblasts. J Biol Chem 259:8066–8069.
63. **Muirhead K, Horan PK, Poste G (1985)** Flow cytometry: Present and future. Bio/Tech 3:337–356.
64. **Murphy RF, Powers S, Cantor CR (1984)** Endosome pH measurement in single cells by dual fluorescence flow cytometry: Rapid acidification of insulin to pH 6. J Cell Biol 98:1757–1762.
65. **Murphy RF, Roederer M (1986)** Flow cytometric analysis of endocytic pathways. In Taylor DL, Waggoner AS, Murphy RF, Lanni F, Birge R (eds), "Applications of Fluorescence in the Biomedical Sciences." New York: Alan R. Liss, pp 545–566.
66. **Ohkuma S, Poole B (1978)** Fluorescence probe measurement of the intralysosomal pH in living cells and the perturbation of pH by various agents. Proc Natl Acad Sci USA 75:3327–3331.
67. **Oi V, Glazer AN, Stryer L (1982)** Fluorescent phycobiliprotein conjugates for analysis of cells and molecules. J Cell Biol 93:981–986.
68. **Pagano RE, Sleight RG (1985)** Defining lipid transport pathways in animal cells. Science 229:1051–1057.
69. **Picciolo GL, Kaplan DS (1984)** Reduction of fading of fluorescent reaction product for microphotometric quantification. Adv Appl Microbiol 30:197–234.
70. **Pietras RJ, Szego CM, Roberts JA, Seelser BJ (1981)** Lysosomal cathepsin B-like activity: Mobilization in prereplicative and neoplastic epithelial cells. J Histochem Cytochem 29:440–450.
71. **Pohl FM, Jovin TM, Baehr W, Holbrollk JJ (1972)** Ethidium bromide as a cooperative effector of DNA structure. Proc Natl Acad Sci USA 69:3805–3809.
72. **Poole A, Mort J (1981)** Biochemical and immunological studies of lysosomal and related proteinases in health and disease. J Histochem Cytochem 29:494–500.
73. **Sage BH Jr, O'Connell JP, Mercolino TJ (1983)** A rapid vital staining procedure for flow cytometric analysis of human reticulocytes. Cytometry 4:222–227.
74. **Schaap GH, de Josselin de Jong JE, Jongkind JF (1984)** Fluorescence polarization of six membrane probes in embryonic carcinoma cells after differentiation as measured on a FACS II cell sorter. Cytometry 5:188–193.
75. **Schulman SG (1976)** Acid-base chemistry of excited singlet states. In Wehry EL (ed), "Modern Fluorescence Spectroscopy," Vol. 2. New York: Plenum Press, pp 239–275.
76. **Seligman B, Chused T, Gallin JI (1984)** Differential binding of chemoattractant peptide to subpopulations of human neutrophils. J Immunol 133:2641–2645.
77. **Seligman GE, Gallin JI (1983)** Comparison of indirect probes of membrane potential utilized in studies of human neutrophils. J Cell Physiol 115:105–115.
78. **Shapiro H (1985)** "Practical Flow Cytometry." New York: Alan R. Liss.
79. **Shapiro HM (1981)** Flow cytometric estimation of DNA and RNA content in intact cells stained with Hoechst 33342 and Pyronin Y. Cytometry 2:143–150.
80. **Shapiro HM, Natale PJ, Kamentsky L (1980)** Estimation of membrane potential of individual lymphocytes by flow cytometry. Proc Natl Acad Sci USA 76:5728–5730.
81. **Shapiro HM, Stephens S (1986)** Flow cytometry of DNA content using oxazine 750 or related laser dyes with 633 nm laser excitation. Cytometry 7:107–110.
82. **Sims PJ, Waggoner AS, Wang CH, Hoffman JF (1974)** Studies on the mechanism by which cyanine dyes measure membrane potential in red blood cells and phosphatidylcholine vesicles. Biochemistry 13: 3315–3330.
83. **Smith RE (1983)** Contributions of histochemistry to the development of the proteolytic enzyme detection system in diagnostic medicine. J Histochem Cytochem 31:199–209.
84. **Steinkamp JA (1984)** Flow cytometry. Rev Sci Instrum 55:1375–1400.
85. **Struck DK, Hoekstra D, Pagano RE (1981)** Use of resonance energy transfer to monitor membrane fusion. Biochemistry 20:4093–4099.
86. **Struli P, Barrett AJ, Bauci A (1980)** "Proteinases and Tumor Invasion." New York: Raven Press.
87. **Stryer L (1986)** Fluorescence spectroscopy of proteins. Science 162:526–533.
88. **Stryer L (1978)** Fluorescence energy transfer as a spectroscopic ruler. Annu Rev Biochem 47:819–846.
89. **Szollosi J, Trons L, Damjanoviek S, Hellewell SH, Arndt-Jovin D, Jovin TM (1984)** Fluorescence energy transfer measurements on cell surfaces: A critical comparison of steady state fluorimetric and flow cytometric methods. Cytometry 5:210–216.
90. **Tanasugarn L, McNeil P, Reynolds GT, Taylor DL (1984)** Microspectrofluorometry by digital image processing: Measurement of cytoplasmic pH. J Cell Biol 98:717–724.
91. **Taylor DL, Reidler J, Spudich, Stryer L (1981)** Detection of actin assembly by fluorescence energy transfer. J Cell Biol 89:362–367.
92. **Thomas JA, Buschbaum RN, Zimmick A, Racker E (1979)** Intracellular pH measurements in Ehrlich ascites tumor cells utilizing spectroscopic probes generated in situ. Biochemistry 18:2210–2218.
93. **Thorell B (1983)** Flow cytometric monitoring of intracellular flavins simultaneously with NAD(P)H levels. Cytometry 4:61–65.
94. **Titus JA, Haugland RP, Sharrow SO, Segal DM (1982)** Texas Red, a hydrophilic red-emitting fluorophore for use with fluorescein in dual parameter microfluorometry and fluorescence microscopic studies. J Immunol Methods 50:193–204.
95. **Tsien RY (1980)** New calcium indicators and buffers with high selectivity against magnesium and protons: Design, synthesis, and properties of prototype structures. Biochemistry 19:2396–2404.
96. **Tsien RY, Pozzan T, Rink TJ (1984)** Measuring and

manipulating cytosolic Ca^{+2} with trapped indicators. Trends Biochem Sci 9:263–266.

97. **Valet G, Raffael A, Moroder L, Wunsch E, Ruhenstroth-Bauer G** (1981) Fast intracellular pH determination in single cells by flow cytometry. Naturwissenschaften 68:265–266.
98. **Van Dilla MA, Langlois RG, Pinkel D, Yajko D, Hadley WK** (1983) Bacterial characterization by flow cytometry. Science 220:620–621.
99. **Waggoner AS, Stryer L** (1970) Fluorescent probes of biological membranes. Proc Natl Acad Sci USA 67: 579–589.
100. **Waggoner AS** (1979) Dye indicators of membrane potential. Annu Rev Biophys Bioeng 8:47–68.
101. **Waggoner AS** (1985) Dye probes of cell organelle, and vesicle membrane potentials. In Martonosi A (ed), "The Enzymes of Biological Membranes." New York: Plenum Press, pp 313–331.
102. **Weiss MJ, Chen LB** (1984) Rhodamine 123: A lipophilic mitochondrial-specific vital dye. Kodak Lab Chem Bull 55:1–4.
103. **Wilson HA, Chused TM** (1985) Lymphocyte potential and Ca^{++} sensitive potassium channels described by oxonol dye fluorescence measurements. J Cell Physiol 125:72–81.
104. **Wilson HA, Seligmann BE, Chused TM** (1985) Voltage sensitive cyanine dye fluorescence signals in lymphocytes: Plasma membrane and mitochondrial components. J Cell Physiol 125:57–71.
105. **Wright DG** (1982) The neutrophil as a secretory organ in host defense. In Gallin JI, Fauci AS (eds), "Host Defense Mechanisms," Vol 1. New York: Raven Press, pp 75–110.
106. **Yguerabide J, Yguerabide EC** (1984) Nanosecond fluorescence spectroscopy. In Rousseau DL (ed), "Optical Techniques in Biological Research." Orlando, FL: Academic Press, pp 181–290.
107. **Zucker-Franklin D, Lavi D, Franklin E** (1981) Demonstration of membrane-bound proteolytic activity on the surface of mononuclear leukocytes. J Histochem Cytochem 29:451–456.
108. **Fox MH, Delohery TM** (1987) Membrane fluidity measured by fluorescence polarization using EPICS V cell sorter. Cytometry 8:20–25.

13

Cytochemical Techniques for Multivariate Analysis of DNA and Other Cellular Constituents

Harry A. Crissman and John A. Steinkamp
Cell Biology Group, Los Alamos National Laboratory, Los Alamos, New Mexico 87545

DEVELOPMENT OF QUANTITATIVE CYTOCHEMISTRY

The early studies by Casperson and co-workers [13,14] employing the then infant science of quantitative cytophotometry first demonstrated the potential for quantitative measurement of biochemical constituents in single cells. These studies showed the capability for performing population biochemical analysis on a cell-by-cell basis so that distinct subpopulations could be discriminated and identified. Such methods as first reported by Casperson and Schultz [14] and others (see review by Swift [126]) showed the potential of ultraviolet (UV) (260 nm) absorption cytophotometry for determining cellular nucleic acid content. Kamentsky et al. [70], in the earliest report on flow cytometry, utilized this analytical technique along with light-scattering measurement to quantify nucleic acid content and cell size. Quantitative cytochemical analysis using colorometric procedures such as the Feulgen reaction [46] allowed for microspectrophotometric assay of DNA content in single cells. Pyronin Y was used first by Brachet [11] and later by Kurnick [78] for measuring RNA content. The quantitative cytophotometric methods employed in these early studies [13] accurately demonstrated that both nucleic acid and protein contents were elevated in rapidly growing cells. These findings are in agreement with biochemical events well established for the cell cycle.

Flow cytometry (FCM) incorporates many of the principles established in the previous cytophotometric studies. In addition, FCM provides ease, speed, and statistical accuracy of the measurements as well as other desirable features. For example, when cellular constituents are labeled stoichiometrically with fluorochromes, subsequent fluorescence measurements by flow cytometry are less sensitive than absorption cytophotometry to "distributional errors" [97] caused by the inhomogeneous distribution of the materials within the cells. Furthermore, the loss of quantitative accuracy due to fluorescence fading does not appear to limit the precision of flow measurements, since stained cells are exposed uniformly, and only briefly (i.e., 3 to 5 μsec) to high intensity excitation sources. Through the use of multicolor staining techniques, it is possible to analyze several biochemical constituents on a cell-to-cell basis. When such data are acquired and stored in a computer in correlated list mode fashion [110], subsequent reprocessing of the data allows for studying the interrelationships of the various biochemical moieties, including DNA. In this way, other cellular parameters such as RNA and protein content, for example, can, by "gated analyses," be investigated for cells in specific phases of the cell cycle (see Chapter 22, this volume).

The similarity of design, and the widespread availability of FCM instruments, along with ease and reproducibility of staining methods for DNA and other cellular parameters, have allowed for more direct comparison of data from laboratories throughout the world. For the most part, the versatility and acceptance of FCM technology have increased as a function of the advances in development of cell preparation and staining methodology. Except for recent specific designs for detailed chromosome analysis [17], flow cytometric instruments have, in principle, remained much the same for the past 15 to 20 years.

This chapter summarizes the cytochemical techniques and approaches for analysis of DNA content and various other cellular constituents which are potentially useful for elucidating biochemical events involved in control of cell cycle progression and proliferation. Data presented demonstrate the feasibility of the methods and illustrate the value of the additional information that can be obtained from multivariate analysis for cell cycle studies.

CHARACTERISTICS OF FLUORESCENT DYES AND ENVIRONMENTAL EFFECTS

Excitation and emission spectra of dyes in solution are analyzed in a fluorescence spectrophotometer. However, the spectral properties of the free dye are influenced by the nature of the solvent. Changes in the spectral characteristics of the free dye can occur when it binds to a particular biochemical. Fluorochromes are therefore most often analyzed in the solution designed for cell staining and bound either to the biochemical of interest in solution or in cells. Upon binding to DNA, many fluorochromes exhibit a significant increase

Flow Cytometry and Sorting, Second Edition, pages 227–247

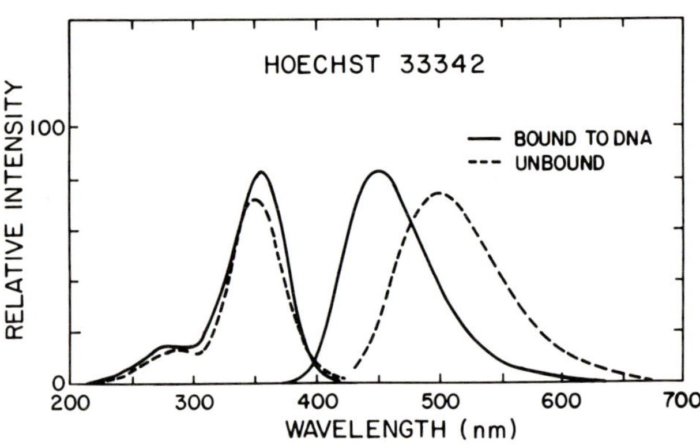

Fig. 1. Excitation and emission spectra for the fluorescent dye, Hoechst 33342. The dye exhibits a dramatic shift in emission to the lower wavelength when complexed with DNA (——), compared with the unbound state (-----). The excitation peak value of Hoechst 33342–DNA solution in the UV range is elevated approximately 10 nm with respect to the free dye.

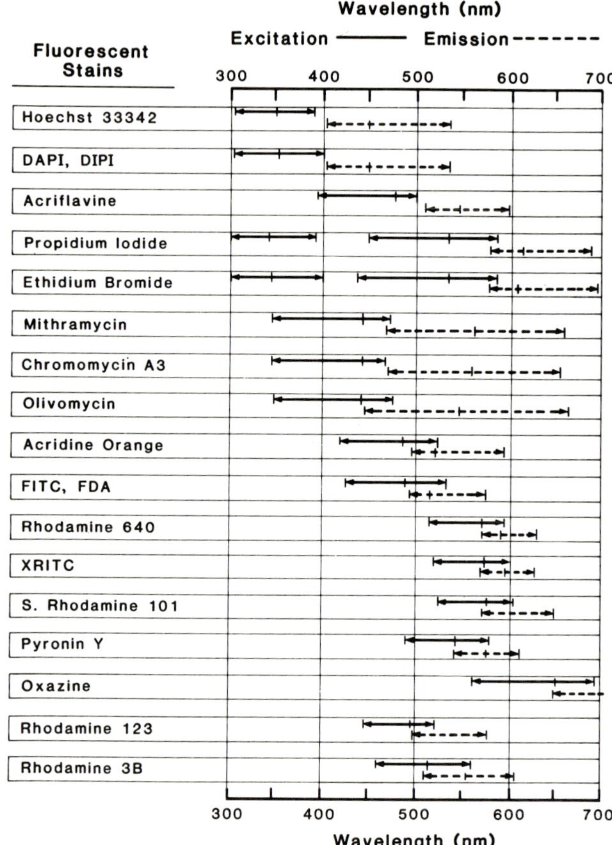

Fig. 2. Fluorescence excitation (<——>) and emission (<--->) wavelength ranges for selected fluorescent dyes. Short vertical lines designate the position of the relative intensity peak in the spectral curves.

in their emission intensity (i.e., increased quantum yield) which may be accompanied by a shift in either or both the excitation and emission wavelength spectra. For example, complexes of DNA and Hoechst 33342 in phosphate-buffered saline (PBS) exhibit a significant decrease in emission peak wavelength value compared with the free dye (460 nm versus 500 nm (Fig. 1), while complexes of DNA with ethidium bromide (EB) or propidium iodide (PI) both show a shift of approximately 50 nm to the longer wavelength in the excitation spectrum as compared with the free dyes [25]. When bound to DNA, the emission peak wavelength position of these two dyes shifts only slightly. Preliminary fluorescence analysis is critical for determining the appropriate excitation and emission wavelengths required for staining cells with a combination of dyes. The aim is to achieve appropriate excitation and optimal color separation of the emitted fluorescence for quantitative accuracy. Spectral characteristics of dyes frequently used in FCM are shown in Figure 2. Reference to specific fluorochromes is made throughout this chapter.

PRINCIPLES OF FLUORESCENCE ANALYSIS

Flow cytometric analysis of cellular biochemical content is achieved by labeling or staining the moiety of interest with a fluorochrome. During analysis the bound fluorochrome in stained cells is excited at an appropriate wavelength and light of equal intensity is emitted in all directions. The excitation wavelength and color of emitted fluorescence is characteristic of each fluorochrome. If the dye is bound specifically to the particular cellular biochemical of interest, and the staining reaction is stoichiometric, ideally the fluorescence intensity reflects, quantitatively, the cellular biochemical content. This scheme is an oversimplification; however, in principle, it provides the systematic, logical basis for development of quantitative cytochemistry for flow cytometry.

SPECIFICITY AND STOICHIOMETRY OF FLUOROCHROME REACTIONS

Fluorometric measurements represent indirect assays. The flow instrument measures fluorescence, not the biochemical moiety, directly; therefore, the assay method relies on the accuracy of the fluorochrome labeling technique. Establishing optimal specificity and stoichiometry of the staining method are critical to the measurements.

The degree of specificity of staining is often established on the basis of results obtained either following enzymatic degradation of the cellular substrate or by chemically blocking the molecular sites to be labeled by the dye. Subsequent cell staining should be negative or significantly diminished. Such tests are performed on ethanol-fixed cells with permeabilized membranes, since enzymes do not penetrate viable cells. Formalin fixation is avoided, since some aldehyde complexes of DNA, RNA, and protein are often not readily accessible to the particular enzymes.

Stoichiometry of DNA stains can initially be established from the ratio of the G2 + M and G1 peak values in the DNA histograms. Ideally, this ratio should yield a value of 2.0. Also the computer fit of the DNA histogram should provide nearly the same percentage of G1, S, and G2 + M phase cells as calculated by an independent method with [^{3}H]thymidine [39,56]. Coulson et al. [16] tested the stoichiometry of various DNA staining reactions using cells from different animal species which varied in DNA content. Of the methods tested, propidium iodide (PI) staining of ethanol-fixed cells following RNase treatment [23] gave the more linear results. The method used by Coulson et al. [16] is one of the best approaches for establishing linearity of staining with cellular content. Similar studies can be designed for establishing stoichiometry of staining reactions for other cellular constituents (e.g., RNA, protein) by the appropriate choice of several cell populations whose cellular content for a given moiety differ over a given range as determined by an alternative independent method.

ENERGY TRANSFER

Energy of one excited dye molecule may be transmitted and absorbed by another molecule under appropriate conditions. For example, when two fluorochromes are bound in close proximity (i.e., generally within 100 Å) and the emission spectrum of one dye (donor) overlaps significantly the excitation spectrum of the second dye (acceptor), energy transfer can occur [108]. An enhancement in fluorescence intensity of the acceptor molecule is observed coincidental to a diminution or quenching of fluorescence of the donor. Therefore, the extent of this phenomenon depends upon the steric orientation of the dyes and their fluorescence properties (see review by Jovin [69]). Energy transfer can also occur between a fluorescent molecule such as a Hoechst DNA reactive dye and a nonfluorescent molecule such as BrdU, in which case only quenching of the Hoechst fluorescence is observed. Such a phenomenon has been exploited in some cell cycle studies mentioned later in this chapter. The method can also provide information on the proximity of cellular moieties within the cell, such as cell surface receptor clustering after ligand interaction.

The potential for energy transfer must be taken into consideration when selecting dyes for multiple fluorochrome staining. In some cases energy transfer may significantly interfere with the stoichiometry of analysis, particularly if the fluorescence of one dye is increased only at the expense of the donor dye. Interpretation of data should take into account the potential for energy transfer that might exist for combined fluorochrome staining.

INTRINSIC FLUORESCENCE OF UNSTAINED CELLS (AUTOFLUORESCENCE)

Autofluorescence, in contrast to luminescence, is a term commonly used in FCM to denote fluorescence excited in unstained cells. It is known that naturally occurring fluorescent materials, generally pyrimidines and flavin nucleotides [8] are found in many cell types. However, phagocytes such as macrophages also often contain ingested fluorescent particles which can produce "autofluorescence" when excited at appropriate wavelengths. When the levels of autofluorescence are far below that produced from cell staining, analytical problems are not encountered. However, difficulties involving cellular autofluorescence can arise when (1) the number of bound dye molecules is small, such as for antibodies against cell surface antigens, (2) the quantum yield of the dye is small, or (3) the autofluorescence spectra overlap that of the labeling fluorochrome. Under any of these conditions, the resolution of fluorochrome labeling and quantitative analysis may be impaired. Fluorescence analysis of unstained cells at the excitation wavelength, laser power, and electronic gain setting predetermined for the fluorochrome-labeled studies, will potentially detect and determine the degree of analytical distortion due to autofluorescence. The use of a fluorochrome with spectral properties different from cellular autofluorescence can often minimize the problem.

Alternatively, instrumental electronic analog [2,118] and computerized [105] methods have been devised to compensate for autofluorescence on a cell-by-cell basis. These methods each provide a means for subtracting the autofluorescence component from the total fluorescent component, resulting in measurement of fluorescence of only the fluorochrome label.

Autofluorescence can also be a useful parameter in examining certain cell populations. For example, autofluorescence intensity analysis has been used to differentiate specific subpopulations of viable cell types in rat lung lavage samples [90]. Metabolic activation in cytocholasin B-stimulated neutrophils has been assessed by monitoring the autofluorescence produced during the reduction of NAD(P)H [135]. In that study, time [96] was recorded as a second parameter, so that the kinetics of the reaction could be assessed by bivariate analysis of autofluorescence vs time.

FLUORESCENCE POLARIZATION

When polarized illumination is used to excite a fluorochrome-labeled cell some of the fluorescence is emitted as polarized light, but some depolarization of the emitted light occurs as well. The extent of depolarization reflects the molecular fluidity or mobility of the dye-bound molecule. Such measurements have been used in cell membrane studies where changes in fluidity were used to assess the functional state of the cells. Diphenylhexatriene (DPH) is a dye commonly used in such studies [5]; however, polarization measurements of intracellular exogenous fluorescein have also been reported [92]. These studies represent more specialized usage of particular fluorochromes. A review of fluorescence polarization was presented by Jovin [69] in the first edition of this book.

CELL PREPARATION AND FIXATION

Flow systems that are designed to perform rapid and precise measurements on individual cells cannot be totally relied on to distinguish fluorescent cellular debris and cell clumps from properly stained single cells. The production of good-quality single-cell suspensions is often the key to successful FCM analysis. Methods for preparing cell suspensions will vary in difficulty. Certain biological considerations must be taken into account in establishing criteria for preparation as well as fixation so that the cellular constituent of interest is optimally preserved for specific fluorochrome labeling. For analysis of cell membrane components, mild dispersal methods are required to preserve intact cells. However, for DNA content analysis, only the intact nuclei must be maintained. Likewise, paraformaldehyde fixatives are preferable to ethanol for preserving membrane antigens; however, paraformaldehyde often interferes with DNA dye binding, which is best achieved in ethanol-fixed preparations. For these reasons each investigator must make the final decisions on the choice of preparation and fixation as based on the initial biological material, the cell constituent to be preserved, and the fixative that is most compatible for both preserving the cell constituent and subsequently providing optimal fluorochrome-labeling. We have previously provided details on cell preparation and fixation [21]. Here we emphasize some general considerations of the methodology.

Plasma membranes of viable cells exclude many of the fluorochromes currently used in FCM. For rapid cell staining, membranes are permeabilized by brief treatment with nonionic detergents [36,51,129,138], or with hypotonic solutions [74], proteolytic enzymes [144] and/or low pH [36]. Such techniques are useful for rapid nuclear staining and

FCM cell cycle analysis; however, cell membrane components and cytoplasmic constituents can be lost from analysis by these treatments.

Alternatively, cells can be treated with ethanol which perforates cell membranes but seems to preserve most cytoplasmic materials. Ethanol (70%) does not appear to impair seriously fluorochrome binding to cellular DNA, RNA, and proteins. No detailed studies have been done, however, to determine the extent of preservation of these and other cellular constituents after fixation in ethanol. Such studies generally involve radioisotope labeling with precursors specific for the constituent of interest and then subsequent quantitative determination of the cellular loss of radioactivity during fixation and staining.

For ethanol fixation we routinely harvest cells from culture or from tissue dispersal solutions by centrifugation. Cells are then resuspended in one part cold "saline GM" (g/L: glucose 1.1; NaCl 8.0; KCl 0.4; Na_2 HPO4.12 H_2O, 0.39; KH_2PO_4, 0.15) containing 0.5 mM EDTA for chelating free calcium and magnesium ions. The addition of three parts cold 95% nondenatured ethanol to the cell suspension, with mixing, produces a final ethanol concentration about 70%. It is most important that the cells be thoroughly resuspended before the addition of alcohol, to prevent cell aggregation. Ethanol fixation can be used in conjunction with most of the cell staining procedures employed in flow cytometry, except for the Feulgen procedure and for labeling of the cell membrane. For most cell cycle analysis studies, ethanol is recommended because, unlike the aldehyde fixatives, ethanol does not crosslink molecules, hence does not interfere with subsequent DNA staining by most dyes currently used in FCM.

Before Feulgen-DNA staining, cells are fixed for at least 12 hours in saline GM containing 10% formalin (i.e., 3.7% formaldehyde). Fresh formalin (pH 7.0), free of precipitate, is recommended for cell fixation. For preserving cell membranes, the use of cacadylate buffers (0.05 to 1.0 M) containing 1.0% paraformaldehyde is suggested. Electronic cell volume distributions of fixed cells are nearly identical to those obtained for viable cells at the same electronic gain setting. Cells fixed in glutaraldehyde or formalin sometimes show intense autofluorescence when excited at 488 nm.

In general, paraformaldehyde is a fixative reserved for use in special studies, particularly for investigation of cell membrane properties (i.e., antigenic sites). C.C. Stewart (personal communication) has found it useful in some studies to first react unfixed cells with appropriate fluorochromes or fluorochrome-labeled antibodies to cell surface antigens. Cells are then rinsed and fixed 2 hours in paraformaldehyde, rinsed again, and stored for up to 1 month in Ca and Mg/free phosphate-buffered saline.

Recently Hedley et al. [59] introduced a new method for preparing nuclei from formalin-fixed paraffin-embedded tissue. Following removal of paraffin with xylene or the less toxic commercial product "Histoclear" [58], tissue samples are rehydrated, treated 30 min with aqueous 0.5% pepsin-HC1 and subsequently stained for DNA with DAPI. DNA histograms compared well with histograms obtained from fresh material stained with ethidium bromide–mithramycin (EB-MI). The authors suggest that pepsin digestion provides good nuclei suspensions and also apparently breaks cross linkages in chromatin produced during initial formalin fixation. The method has proven successful for recovery, staining, and FCM analyses of nuclei from many types of tumor tissues [58].

CURRENT FLOW-SYSTEM MODIFICATIONS FOR MULTIVARIATE ANALYSIS

Multiple analyses of several cell constituents in cells is achieved by staining of the moeities with fluorochromes of different colors (i.e., different spectral properties). As can be seen in Figure 2, the fluorescence spectra of many stains overlap to some extent. When fluorescence spectra overlap, precise quantitative determinations must rely on analytical capabilities that separate the emission of each dye. These requirements have led to modifications in flow systems, which specifically accommodate excitation and emission analysis of multicolor stained cells. For example, dual- or triple-beam flow systems have been developed by several laboratories [3,40,112,117,119–121]. In many of these systems, the arrangement of the excitation source and the flow chamber are similar to that illustrated graphically in Figure 3 [114,119]. An essential feature of this arrangement is the spatial separation of three laser beams, each of which is tuned to a near-optimal wavelength for excitation of at least one dye [119]. Appropriate color-separation filters ensure fluorescence measurement over a wavelength range specifically selected to measure the emission of only one of the dyes at each point of excitation on the cell stream. In this flow system, five fluorescence signals from each cell can be collected and stored by computer in list mode fashion [110], and measurements can be correlated subsequently.

Cell volume, light scattered by cells at three different wavelengths, and axial light loss (ALL) can be measured in the system and correlated with the fluorescence measurements. Analog electronic capabilities allow for on-line ratio determinations of any two parameters [23], as well as addition or subtraction of signals. Ratio analysis of fluorescence to cell volume is useful for determining the density of a dye-bound molecule as a function of cell size. Also by electronically converting the volume signal to the 2/3 power [116] a signal proportional to cell surface area can be generated and the density per unit surface area of a labelled surface antigen can be determined by ratio analysis and correlated, for example, with cell cycle phase (DNA content). Subtraction has been used to remove the autofluorescence component from the combined signal generated by bound fluorochrome plus autofluorescence [2,105,118].

Cell volume is a good indicator of the relative protein content in viable cells, and neither electronic volume analysis nor electronic sorting by FCM appear to reduce long-term cell viability as determined by the plating (colony-forming) assay [49]. Light scatter can also be used as a basis for viable cell sorting. When scattering from white blood cells is analyzed at 0.5 to 2.0° (forward) and at 90° to the laser beam three subpopulations of cells can be differentiated [109]. Visser has used similar analyses to detect the stem cells from mouse bone marrow [140]. With correlated light scatter measurements at 351 and 488 nm, Otten and Loken [101] were able to discriminate subsets of viable lymphoid cells. Axial light loss measurements determine the amount of light the cell removes from the laser beam. Lehnert et al. [90] have used this parameter to discriminate subpopulations of viable cells in rat lung lavage preparations. Kaplow et al. [71] used the absorptive and light-scattering properties to analyze enzyme content and classify stained human leukocytes. In most instances these techniques are used in conjunction with fluorescent probes, and they provide additional valuable parameters for multivariate cell analysis (see review by Steinkamp [114]).

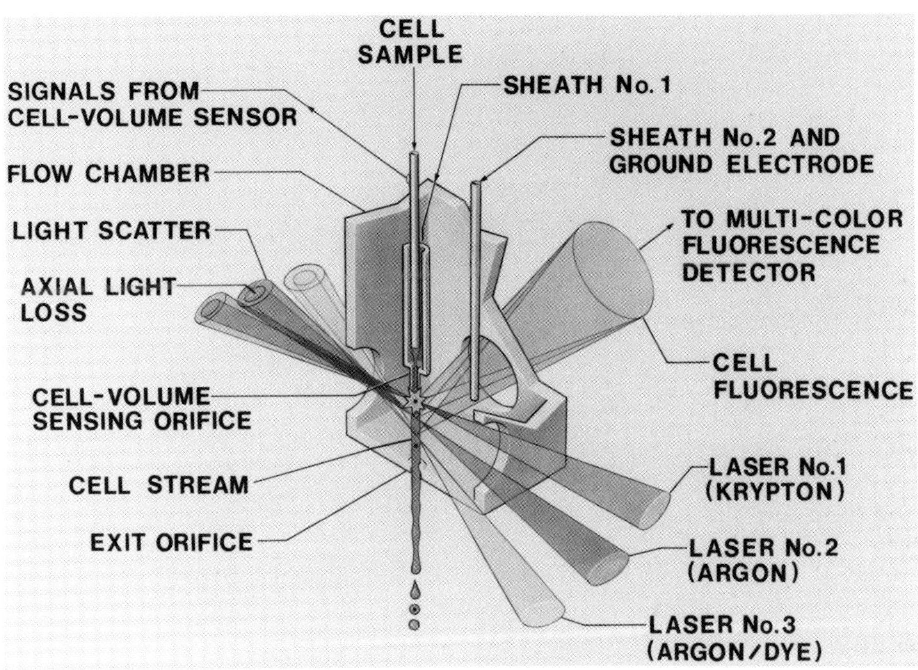

Fig. 3. Diagrammatic cutaway display of the flow chamber in the three-laser excitation FCM system. The system has capabilities for analysis of cell volume, light scattering at three wavelengths, and five colors of fluorescence. The three laser beams are spatially separated 250 μM apart to provide for sequential excitation and selective emission wavelength analysis of stained cells.

A unique flow cytometer that determines the fluorescence emission spectrum from each cell has recently been developed by Buican [12]. Because spectral resolution is significantly improved, the instrument has the potential for expanding the number of fluorochromes that can be combined for cell staining as compared to current FCM systems that rely mainly on colored filters to obtain fluorescence discrimination.

DNA CONTENT ANALYSIS BY FLOW CYTOMETRY

Most of the early studies in flow cytometry (FCM) involved quantitative fluorescence measurement of DNA-bound dyes. Analysis of the generated histograms provided the relative frequency of cells in various phases of the cell cycle. Comparisons of computer-fit analyses of the FCM derived fluorescence histograms and data obtained from conventional assays of DNA replication involving autoradiographic detection of [^{3}H]thymidine confirmed the accuracy and precision of the FCM approach to cell cycle analysis [39]. These early studies provided the credibility necessary to rapidly and firmly establish flow cytometry as an important analytical tool for biological analyses. After nearly 20 years, cell staining with DNA-specific fluorochromes for FCM cell cycle analysis still remains one of the most frequent applications of the technology.

The attractive features of the methodology include the rapidity and ease of both cellular DNA staining and FCM procedures. Other advantages, discussed previously [130], are the abilities to (1) monitor cell cycle distributions in ongoing experiments, with the added option of altering an experiment in progress in response to a population change; (2) localize cells within S phase, and distinguish between early-, mid-, and late S phase, etc.; (3) analyze populations with radiolabeled RNA and/or protein; (4) monitor populations composed of slowly progressing or arrested cells; (5) analyze populations devoid of cells in the S or mitotic phases; (6) analyze populations containing cells unable to transport, incorporate, or metabolize [^{3}H]thymidine; (7) detect abnormalities in progression through mitosis such as nondisjunction or polyploidization [73]; and (8) distinguish between intact and fragile (dying) cells in drug-treated or virus-infected cells. In addition there is also the option of sorting cells based on DNA content for identification and further biochemical analysis. It is important, however, to recognize that FCM cell-cycle analysis has certain limitations.

Flow cytometric DNA content analysis alone fails to provide information relating to the metabolic state of cells such as the cycling capacity and the rate at which cells will traverse the cell cycle. By analogy, the cell cycle frequency histogram is equivalent to a "snapshot" photograph of individuals in a race. The position of all runners in the course of the race at any instance is evident; however, the speed and the capacity of each runner to successfully complete the race is indeterminant. Such is also the case for cells represented in a FCM cell-cycle distribution. For instance, two exponentially growing cell populations may have very different doubling times (i.e., 15 hours for Chinese hamster ovary (CHO) cells versus 24 to 30 hours for HL-60 cells) but still yield similar DNA content distributions if the duration of the individual phases of the cell cycle in proportion to the cycling time is similar. Also under experimental conditions, it is difficult to ascertain proportions of cycling and noncycling cells within the various phases.

DNA content analysis has also been useful for detecting aneuploid subclones within a given population. From a clinical standpoint such determinations can provide information of prognostic value [82]. The terms aneuploid or heteroploid, which refer to chromosome number, have occasionally been used to specify subclones within the DNA content histogram. In a recent publication [60], guidelines are proposed to standardize nomenclature and procedural descriptions for DNA content analyses.

In recent years, flow cytometry has been used to elucidate physiological aspects which, in addition to DNA metabolism, regulate and control cell proliferation. Many current studies now include staining and measurement of other cellular constituents such as protein and RNA simultaneously with DNA. Cellular levels of such descriptors and others are known to be important indicators of cell cycle progression capacity.

We present some current methods and criteria for multicolor fluorescent staining and analysis of those cellular biochemicals that, in addition to DNA content, can yield information on cell cycle traverse potential. In most instances, the techniques involve modification of preexisting staining methods that will allow for the appropriate combination of the dyes and near-optimal quantitative analysis.

DNA SPECIFIC CELL STAINING

In previous publications, we [21,22,27,28] and others [65,66,80,84,87,128] have reviewed various dyes, their mode of binding, and methods for their usage as DNA stains in FCM studies. This chapter addresses only those aspects that are important for combining these dyes in protocols for multicolor staining and correlated analysis of several biochemical compounds and organelles in cells.

Feulgen–DNA Reaction

The Feulgen procedure [46] represents one of the earliest and most well established cytochemical procedures for labeling and quantitating cellular DNA content. Modifications of the Culling and Vassar adaptation [30] of the procedure were used in earlier FCM studies employing the fluorochromes auromine 0 [132,134] or acriflavine [21,131]. The use of the Feulgen method is somewhat limited since it is time consuming, produces cell loss and nonspecific staining in many cell types. In addition, HCl hydrolysis, an essential step in the protocol, depurinates DNA, removes RNA and histones, and possibly damages other cellular components, including membranes. Therefore, we discuss more recently available DNA dyes that can be used rapidly and without the loss of other cellular constituents. Several of these dyes have different fluorescence and DNA-binding properties so there are more options for selecting a DNA stain with spectral characteristics different from the other dyes used in multicolor staining.

DNA-Reactive Hoechst Dyes; DAPI and DIPI, and LL 585

The Hoechst dyes are benzimidazole derivatives which emit blue fluorescence when excited by UV light at about 350 nm (see Fig. 2). They have a high specificity for DNA and bind preferentially to A-T base regions but do not intercalate [100]. The Hoechst 33258 derivative was originally used by Hilwig and Gropp [61] in mouse chromosome banding studies. Latt [83] showed that fluorescence of this dye was quenched when bound to BrdU-substituted DNA and used the phenomenon in a technique for detecting regions of sister chromatid exchange in metaphase chromosomes. Using flow cytometric analysis, Arndt-Jovin and Jovin [4] first demonstrated the use of both Hoechst 33258 and 33342 for quantitative DNA staining and sorting of viable cells. The structural and spectral properties of DAPI and DIPI are similar and each can be used interchangeably with the Hoechst 33258 or 33342 derivatives for staining ethanol-fixed cells.

Stock solutions of the Hoechst dyes, DAPI or DIPI, prepared in distilled water (1.0 mg/ml) and refrigerated in dark-colored containers, may be used for at least 1 month without noticeable degradation. These dyes tend to precipitate in PBS at concentrations above 50 μg/ml. Fluorescence of ethanol-fixed CHO cells stained 15 to 20 min at room temperature with Hoechst 33342 (0.5 to 1.0 μg/ml) in PBS remains stable for at least 5 to 6 hours. Stained cells usually are analyzed in equilibrium with the dye, using a uv laser beam or mercury-arc lamp excitation source. Dye concentrations for fixed cells may vary slightly, depending on the cell type and flow instrumentation used for analysis.

Latt et al. [86] recently examined the binding properties and cell-staining characteristics of several fluorescent compounds produced by the Eastman Kodak Company. Of three compounds studied, one, designated LL 585, appears most useful for DNA-specific staining. The spectral characteristics of LL 585 in the visible range (data not shown) are similar to those of propidium iodide (see Fig. 2); however, as with the Hoechst dyes, it binds preferentially to A-T sites and does not seem to intercalate into double stranded DNA. In contrast to HO 33342, LL 585 does not readily penetrate viable cells even after a 3-hour staining period. Human lymphoblasts, permeabilized with 0.1% Triton ×100 and stained with LL 585 yielded typical cell cycle distributions (coefficient of variation 2.9%) when analyzed by FCM using the 514-nm line of an argon-ion laser beam. Cells treated with RNase gave essentially the same DNA histograms. The usefulness of the other two Eastman Kodak dyes examined by Latt et al. [86] remains to be further demonstrated.

Viable Cell Staining for DNA with Hoechst 33342

Synchronization protocols potentially perturb the cell cycle. DNA-specific staining of viable cells, followed by FCM analysis and cell sorting, provides an alternative approach for selecting and recovering living cells from various phases of the cell cycle. In experimental studies, the sorted cells may be cultured and examined with regard to long-term viability, functional activity, immunological properties, and other physiological features. Such studies can significantly advance the understanding and interpretation of cellular changes that regulate and control cell cycle progression and cell proliferation.

Unfortunately, most DNA-specific fluorochromes are not easily transported into viable cells, and thus intracellular concentration levels sufficient for optimal staining and analysis are rarely attained. Some dyes, such as DAPI, can penetrate membranes of some cell types, but they are very cytotoxic. Of the fluorochromes used in FCM, HO 33342 remains the preferred dye for viable cell staining. However, experimental usage of this compound also has limitations as will be mentioned below.

Viable cell staining for DNA is usually performed by direct addition of HO 33342 (final concentration 2.0 to 5.0 μg/ml) to cells in culture medium (37°C) for incubation periods of 30 to 90 min, depending on the cell type. Samples stained at 4°C yield poor DNA distributions, possibly indicating that conditions favoring active dye transport and retention are required. However, even at optimal conditions, dye uptake, cytotoxicity, DNA binding and analytical resolution, as judged by coefficients of variation (CV) in the FCM–DNA histograms are cell type dependent. The permeability and/or retention factors are extremely variable and CVs for stained cell populations can range from 4.6% (i.e., cultured colon 26 cells) to 14.0% (i.e., CHO cells). The accuracy of sorting cells from specific cycle phases is seriously impaired when the quality (i.e., CV value) of the DNA content distribution is poor. Also the stoichiometry of the staining reaction is questionable in such cases.

If the cell membrane is perforated with nonionic detergents (i.e., NP40, Triton × 100) or ethanol, staining in most cell types is rapid (i.e., 5 to 10 min) even at dye concentrations of 0.5 μg/ml. Untreated cells with damaged membranes also stain rapidly. Fixed CHO cells stained with HO 33342 generally yield DNA content histograms with coefficient of variation (CV) values of about 4 to 5%. These results indicate that membrane permeability of different cell types is responsible for observed variations in staining quality and probably not DNA binding per se.

Heterogeneous cell populations, as contained in bone marrow samples, present a special problem because of the variability in HO 33342 uptake by the different cell types. In bivariate DNA content and cellular light-scatter studies, Visser [139] demonstrated that many of the large myeloid progenitor cells did not stain as intensely as the smaller lymphocytes, even after several hours staining at pH 6.7. Later Visser et al. [140] improved dye uptake in the larger cells by increasing the pH to 7.2; however, cell survival at the elevated pH, even in the absence of HO 33342, was significantly decreased. (J.W.M. Visser, personal communication). Interestingly, Loken [93] took advantage of differences in HO 33342 stainability (i.e., membrane transport) to differentiate viable T and B cells in spleen samples (i.e., the T cells were less fluorescent). These differences were not apparent in the fixed spleen sample. Lalande et al. [79] demonstrated that the degree of Colcemid-resistance in different clones of CHO cells, a diminished drug uptake phenomenon, somewhat paralleled the uptake of HO 33342; i.e., resistant clones took up less dye.

Dyes such as HO 33342, which bind to DNA, can potentially interfere with cell replication and thereby impair viability. Fried et al. [50] studied the effects of the dye on survival of various cell types and demonstrated that HeLa S-3 cells were highly resistant but SK-DHL2 cells were highly sensitive to cytotoxic affects of the fluorochrome. Pallavicini et al. [102] showed that X-irradiated cells are much more sensitive to HO 33342 than unirradiated cells, indicating a possible synergistic cytotoxic effect. However, two very recent studies show that treatment of viable cells with membrane interacting agents may improve Hoechst uptake and possibly improve survival.

We have recently reported [20] that the membrane potential mitochondrial stain, DiO-C5-3, when applied to viable CHO cells in conjunction with Hoechst 33342, increased cellular uptake of the Hoechst dye twofold and provided coefficient of variation values of about 3.0% compared with 8.3% for cells treated with Hoechst alone when stained cells were excited with 500 mW of UV illumination (Fig. 4). Following cell sorting, population survival values (colony-forming assay) of Hoechst-stained cells either treated with DiO-C5-3 or untreated were near 100%. However DiO-C5-3 treatment did not increase Hoechst uptake in cells, such as L1210 cells or human skin fibroblasts, that routinely stain well with Hoechst. By comparison, the more commonly used mitochondrial stain, rhodamine 123 (R123) did not improve the uptake of Hoechst 33342 by CHO cells. Krishan [75] has also recently shown that calcium channel blocking agents, such as verapamil, can also increase Hoechst stainability in viable cells normally refractory to Hoechst uptake. Krishan [75] speculated that rapid metabolic dye efflux could account for poor Hoechst stainability. Calcium channel blocking agents have been shown to improve Adriamycin (ADR) retention and increase cytotoxicity in cells that are usually resistant to ADR [76]. Collectively these data provide sufficient evidence to stimulate new lines of research aimed at improving both dye uptake and survival of Hoechst-stained populations.

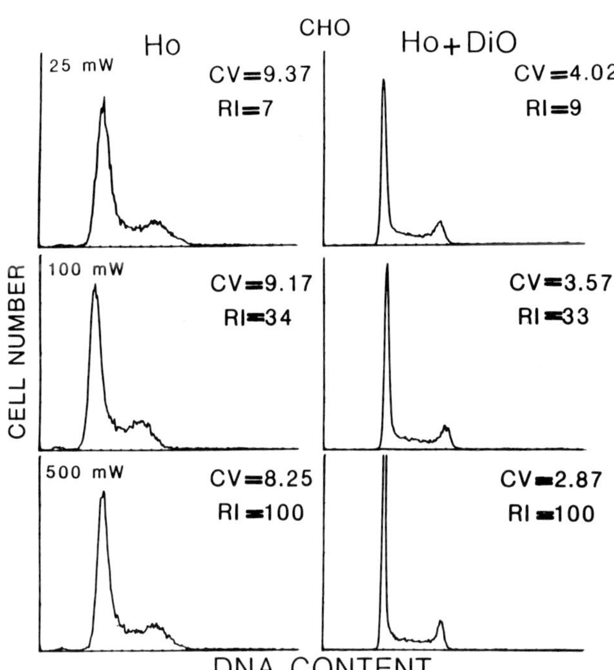

Fig. 4. DNA content frequency histograms obtained for viable CHO cells stained with Hoechst 33342 (HO, 5.0 μg/ml) with and without DiO-C5-3 (DiO, 0.3 μg/ml) and excited with a UV laser adjusted to 25-, 100-, or 500-mW power. RI, relative intensity of G1 cells based on a value of 100 for cells excited at 500 mW. The relative intensity of cells stained with both HO and DiO was twofold that obtained for cells stained only with HO. CV, coefficient of variation of the G1 subpopulations.

A useful technique, developed by Stohr and Vogt-Schaden [122], combines two dyes, HO 33342 and propidium iodide (PI), to differentiate viable and dead cells. Viable cells exclude PI and fluoresce blue while dead cells (i.e., cells with damaged membranes) fluoresce both red and blue. Hoechst fluorescence in dead cells is significantly quenched due to energy transfer to PI and therefore well separated from the Hoechst-DNA histogram of viable cells.

In summary, the use of HO 33342 for staining and recovery of viable cells within specific cell cycle phases is useful in some studies providing that control studies are designed to test for any potential adverse drug effects.

Mithramycin, Chromomycin, and Olivomycin Cell Staining

Mithramycin (MI) (Pfizer Co., Groton, CT) is a green-yellow fluorescent DNA-reactive antibiotic similar in structure and dye-binding characteristics to chromomycin A3 [65,143] and olivomycin (see Fig. 2). Ward et al. [142] have

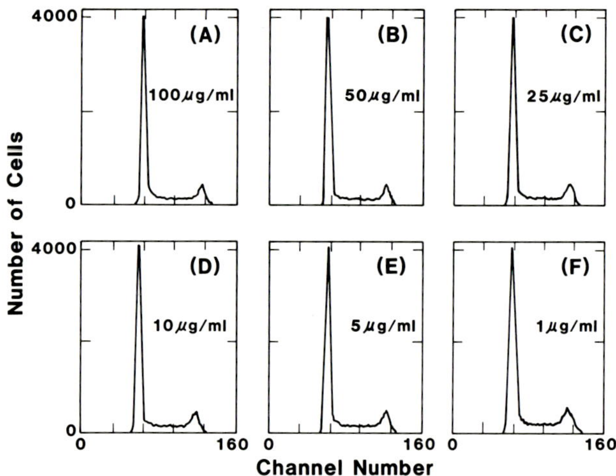

Fig. 5. DNA content frequency distribution histograms for CHO cells stained with different mithramycin concentrations in Tris–HCl (10 mM) buffer, (pH 7.2) containing 20 mM $MgCl_2$. Data obtained by computer-fit analysis are shown in Table 1.

shown that these compounds, when complexed with magnesium ions, preferentially bind to G–C base regions by nonintercalating mechanisms. Mithramycin has been studied extensively with regard to its spectral characteristics, specificity for cellular DNA, and as a quantitative DNA stain for use in flow cytometry [27]. Spectrofluorometric analysis of mithramycin or chromomycin A3–Mg complexes bound to DNA in PBS show two excitation peaks, a minor peak in the UV region (about 320 nm) and a major peak at about 445 nm. A broad green-yellow emission spectrum is observed with a peak at 575 nm. By comparison, olivomycin has slightly lower wavelength excitation and emission characteristics (see Fig. 2).

Our original procedure [28] was designed for rapid MI-labeling and analysis of unfixed cells. Ethanol (25%) was included in the staining solution specifically to permeabilize cell membranes. More recently, DNA histograms with improved CV values of 3.0% versus 6.0% for the above method have been obtained when CHO cells are first fixed with 70% ethanol, and then the ethanol removed by centrifugation prior to staining and analysis [24,27]. Fixation periods as short as 30 min are adequate. Also, CHO samples stored in fixative for more than 1 year at 4°C will yield MI-DNA fluorescence histograms of good quality.

Ethanol-fixed CHO cells are routinely stained for 20 min at room temperature and analyzed at 457 nm in solutions of mithramycin (100 μg/ml) in 10 mM Tris–HCl (pH 7.4) containing 20 mM $MgCl_2$. In a recent study, we found that concentrations of mithramycin as low as 5 μg/ml could be used to obtain acceptable FCM cell cycle frequency resolution (Fig. 5, Table 1). Staining with higher mithramycin concentrations improved the coefficient of variation slightly. We have no explanation for these recent results compared with our previous dye concentration studies of more than 10 years ago, from which we concluded that 100 μg/ml mithramycin was optimal for good DNA content resolution. Possibly the antibiotic is now prepared in a more purified form. From an economic standpoint, significant financial benefits can be gained by using the lower dye concentrations.

Mithramycin–DNA cell staining has been used following fixation with agents other than ethanol. Previously we showed [27] by FCM analysis that glutaraldehyde-fixed cells bound only about one-half the MI of ethanol-fixed, MI-stained cells. Although the CV values were 6.4% and 4.5%, respectively, the cell cycle frequency distributions obtained by computer-fit analysis of the respective DNA histograms were quite similar. Larsen et al. [81] showed that cells in mitosis could be discriminated from cells in other phases of the cell cycle when FCM analysis was applied to nuclear suspensions prepared with nonionic detergent, fixed in formalin and stained with MI, as well as PI or EB. Formaldehyde quenched the fluorescence of these dyes in interphase cell nuclei to a greater extent than mitotic nuclei having MI fluorescence intensities 20 to 40% greater but light scattering 30 to 60% lower than G_2 cell nuclei. Pierrez et al. [103] showed that peripheral blood and bone marrow samples could be stained with MI following picric acid–alcohol fixation. Fixation neither prevented subsequent Saponin lysis of red blood cells (RBC) nor significantly impaired MI staining.

In another recent study, van Kroonenburgh et al. [136] were able to distinguish eight different cell populations in MI-stained rat testis cell suspensions. The subpopulations were identified using bivariate analysis of the peak amplitude of the fluorescence signal versus the total fluorescence intensity integrated over time. Cell sorting permitted morphologic identification of the subpopulations.

Experimental drugs that interact directly with DNA or that interfere with DNA metabolism can have effects on subsequent staining. In in vivo studies, Alabaster et al. [1] found differences in MI–DNA stainability of untreated (control) L1210 ascites cells and cells treated with a combination of cytosine arabinoside (ara-c) and Adriamycin. An unexpected finding reported by Cunningham et al. [31] was more related to the conditions of ethanol fixation prior to MI staining. They observed an aberrant peak, slightly elevated above the G_1 peak position, in the DNA histograms of human lymphocytes fixed continuously at 4°C. By contrast, this spurious peak was absent or diminished in samples first fixed for 7 hours at room temperature or for 2 hours at 37°C before refrigeration. Similar reports for other cell types have not been found in the literature indicating that human lymphocytes may pose a unique problem.

Propidium Iodide, Ethidium Bromide, and Hydroethidine Staining

Propidium iodide (PI) and ethidium bromide (EB) are red fluorescing compounds that have similar chemical structures and DNA dye-binding properties. Both dyes intercalate between base pairs of both double-stranded DNA and RNA, without base specificity [91]. The dyes can be used as DNA-specific stains following pretreatment of fixed cells with RNase.

Ethidium bromide (EB) was used by Caspersson et al. [15] to determine DNA content in single chromosomes by scanning microspectrofluorometry. Dittrich and Gohde [41] demonstrated the use of the dye for cell cycle analysis by flow cytometry. PI, a dye used by Hudson et al. [62] in a buoyant density procedure for isolating closed circular DNA, was first introduced to FCM in a double staining technique for analysis of DNA and protein, in which fluorescein isothiocyanate was used for protein labeling [23]. Krishan [74], Fried et al. [51], and Vindelov et al. [138] have since developed methods for rapid staining of unfixed cells with PI. Techniques vary with procedures used for permeabilizing cell membranes and/or for tissue disaggregation. Vindelov et

TABLE 1. Comparison of Cell Cycle Frequency Distributions†, Relative Fluorescence Intensity,* Coefficient of Variation (CV), and $G_2 + M$ to G_1 Peak Value Ratio for CHO Cells Stained with Different Concentrations of Mithramycin

Dye concn (μg/ml)	Percentage of Cells G_1	S	$G_2 + M$	Relative intensity (AU)	CV (%)	Ratio $G_2 + M$ to G_1
100	58.2	32.4	9.4	100	3.2	1.99
75	62.6	30.1	7.4	91	3.5	1.99
50	58.3	33.0	8.8	82	3.1	2.00
25	57.1	33.3	9.6	68	3.6	2.01
10	57.6	33.5	8.9	52	3.6	2.02
5	56.2	34.6	9.2	38	3.8	2.03
3	55.6	34.4	10.0	29	3.9	2.04
1	52.0	38.3	9.7	16	4.5	2.08

*Based on the relative G_1 mean channel value obtained at the same electronic gain setting and laser power.
†Obtained by computer fit of DNA histograms [39].

al. [137] also proposed the use of PI-stained nucleated trout and chick red blood cells as internal standards for assessing relative DNA content. Alternatively, Frankfurt [47] and Wallen et al. [141] measured double-stranded RNA content in fixed cells stained with PI following DNase pretreatment.

Hydroethidine (HE) is a fluorescent compound produced by chemical reduction of EB. Gallop et al. [52] demonstrated, microscopically, that HE is taken up by viable cells. In the cell, HE is enzymatically dehydrogenated, in part, to form ethidium, which then intercalates into double-stranded DNA and RNA and fluoresces red. Nonreacted HE in the cytoplasm fluoresces blue when excited at 370 nm. Using 535-nm excitation, only red (ethidium) is seen. Luce et al. [94] used HE in conjunction with sulfofluorescein diacetate to enumerate cytotoxic cell and target cell interactions by FCM.

Flow cytometric analysis of PI stained cells yields DNA (fluorescence) histograms comparable to those obtained for cells stained with several other DNA stains [27]. Spectral studies on PI bound to calf thymus DNA in PBS show two excitation peaks, a minor but substantial peak in the uv region (340 nm) and a major peak at about 540 nm. One emission peak was observed at 615 nm. Solutions of PI (5 to 20 μg/ml) in PBS containing 50 μg/ml RNase (Worthington code R) are used to stain ethanol-fixed CHO cells for 30 min at 37°C before FCM analysis at 488 nm excitation wavelength [23,24]. PI does not penetrate viable cells sufficiently for cell-staining purposes. Cells with damaged membranes stain readily with PI so the dye is often used for differentiating and quantitating viable and dead cells in a given population.

Mazzini and Giordano [98] have shown that the quantum efficiency of PI and EB are significantly enhanced in solutions of deuterium oxide. Dye concentrations of 1.0 μg/ml were adequate for cell staining and FCM analysis. This technique can be used for flow measurements providing there is not significant mixing of the deuterium oxide stain solution in the cell stream with the saline or water in the sheath fluid (G. Mazzini, personal communication).

Additional DNA-Binding Fluorochromes

A number of other DNA-specific dyes have been introduced into flow cytometry. 7-Amino-actinomycin D (7-Act D), synthesized by Modest and Sengupta [99], is a red fluorescent analog of actinomycin D (excitation 540 nm) that intercalates in the G–C regions on DNA. Gill et al. [53] examined the spectral properties of this dye; later, Darzynkiewicz et al. [35] examined the accessibility of DNA to the fluorochrome during DMSO-induced, erythroid differentiation of Friend leukemia cells. Evenson et al. [45] used 7-Act D for studying changes in DNA accessibility to various fluorochromes during spermatogenesis. In that study another intercalating dye, ellipticine, was also used. Zelenin et al. [145] have recently provided a detailed study on the use of 7-Act D for cell staining and FCM analysis. Shapiro and Stephens [113] have used several laser dyes, oxazine 750, LD700, and rhodamine 800, as DNA-specific stains in FCM studies. The advantage in using these dyes is that they can be excited with a low-power, relatively inexpensive, helium–neon laser (633 nm). Fluorescence measured above 665 nm provided good DNA content histograms for stained CCRF–CEM cells.

Multiple Fluorochrome-Labeling of DNA

Although DNA specific stains are most often used as single dye agents, several FCM studies have demonstrated advantages in combining some of these dyes. Berkhan [9] used HO in combination with EB and both Zante et al. [144], and Barlogie et al. [7] used MI and EB in protocols designed to take advantage of energy transfer from HO or MI to EB. Under these conditions, there was an increase in the red fluorescence signal, thereby increasing resolution (i.e., reduced CV) in the single-color fluorescence (DNA) histograms. Gray et al. [57] double-stained chromosomes with HO 33258 and chromomycin A-3 (CA3) and used bivariate FCM karyotyping analyses to resolve some human chromosomes having similar DNA content but different base composition, demonstrated by the proportion of A-T-bound Hoechst and G-C bound CA3 fluorescence. Van Dilla et al. [133] also used differences in DNA base composition to distinguish three bacterial strains stained with HO and CA3, and Dean et al. [38] used this same dye combination to compare cell cycle analysis results directly from cells stained with the two DNA stains simultaneously.

We recently used a combination of HO, MI, and PI, in a three-color staining and FCM analysis technique designed to

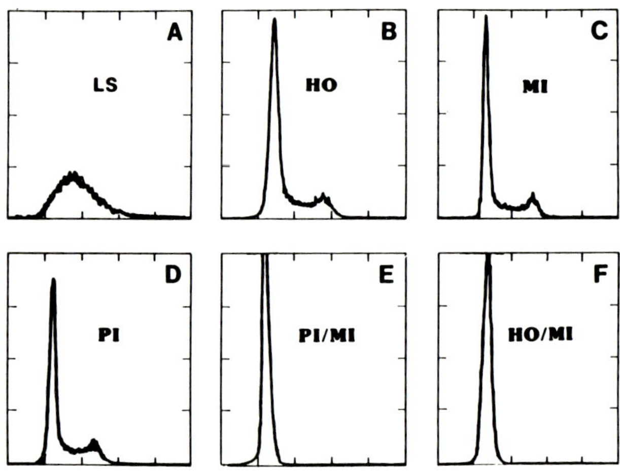

Fig. 6. Single-parameter frequency histograms of light scatter, LS **(A)**, Hoechst 33342, HO **(B)**, mithramycin, MI **(C)**, and propidium iodide, PI **(D)** fluorescence and fluorescence ratios of PI/MI **(E)**, and HO/MI **(F)** obtained by analysis of three color-stained CHO cells. Exponentially growing CHO cells were fixed in 70% ethanol and subsequently stained with a solution containing all three DNA stains. Sequential excitation and analysis were performed as described in the text. Coefficients of variation and cell cycle frequency distributions obtained by computer fit of the DNA content histograms are shown in Table 2.

detect experimentally or physiologically induced modifications in chromatin structure. Such structural modifications are reflected in the redistribution of specific regions on DNA available for binding by each of the three dyes. The analytical design shown in Figure 3 allows for sequential excitation and emission analysis, and alleviates potential difficulties in obtaining spectral resolution of HO 33342 (blue), MI (green), and PI (red) fluorescence emissions. This approach takes advantage of the differences in excitation and/or emission characteristics of each dye (see spectra, Fig. 2). For example, at the position of the UV beam in Figure 3, excitation of HO in stained cells is optimal, PI near-optimal, and MI suboptimal. However, appropriate filters that transmit only blue fluorescence (400 to 485-nm range), predominantly pass fluorescence from HO 33342. Exposure of stained cells to the second 457-nm beam excites MI optimally and PI suboptimally; however, only green-yellow fluorescence over the 510 to 575-nm range (predominantly from MI) is analyzed. At 530 nm (beam 3), only PI is excited and red fluorescence (580 to 800 nm) is analyzed. Ratio analysis [23] of HO/MI (blue/green) fluorescence allows for direct comparison of AT/GC regions available for dye binding in DNA on a cell-to-cell basis. Detailed descriptions of experimental design for cell staining, sequential excitation, and fluorescence analysis will be described in a separate publication (manuscript in preparation).

Near-optimal simultaneous three-color staining and analytical conditions for exponentially growing cells should ideally provide three fluorescence (DNA) histograms, one for each dye, with relatively low CV. Also, all histograms generated by FCM analysis of three-color stained cells should be comparable with respect to the frequency distributions throughout the cell cycle. Such results have been obtained from populations of CHO cells (Fig. 6, Table 2) as well as HL-60 and Friend erythroleukemia cells. In particular, the

TABLE 2. Comparison of Computer-Fit Analysis of DNA Frequency Distribution Histograms Shown in Figure 6

		Percentage of cells			Ratio
	CV (%)	G_1	S	G_2 + M	G_2 + M to G_1
HO	8.5	60.2	32.1	7.7	2.02
MI	6.6	59.2	34.0	6.8	1.98
PI	5.9	60.4	33.8	5.8	1.98

percentages of cells in G1, S, and G2 + M (Table 2) are in good agreement and the CV values are acceptable for accurate computer analysis. Compared with cells stained separately with each dye, three-color-stained cells exhibit a 75 to 80% decrease in HO fluorescence, a 35 to 40% decrease in MI, and a decrease of 20 to 25% in PI fluorescence. This could be due to either energy transfer and/or nonspecific quenching by unbound dye in stain solution surrounding cells during analysis.

The correlation of dye binding by the various dyes (same data as for Fig. 6) is illustrated in the computer-generated bivariate displays shown in Figure 7A–E. In addition Figure 8B shows a PI/MI fluorescence ratio versus HO for a population of CHO cells treated for 4 hours with Colcemid to accumulate a subpopulation of cells in mitosis. Compared with the PI/MI ratio distribution for the exponentially growing untreated cells (Fig. 7E), cells in mitosis (Fig. 8B) with highly condensed chromatin exhibit a decrease in stainability of PI compared with MI and thus have a decreased PI/MI ratio with respect to the G_2 cells. This subpopulation of mitotic cells is also observed in two parameter MI versus PI distributions (Fig. 8A).

Detection of BrdU-Labeled Cells by Differential Fluorescence Analysis of DNA Fluorochromes

Cells cultured in medium containing BrdU will incorporate the analogue in place of thymidine during the DNA synthetic (S) phase. Latt [83] showed that the fluorescence of HO 33258 was quenched when bound to BrdU-substituted DNA in chromosomes. Later, Latt et al. [85] used FCM analysis and demonstrated the detection of cycling BrdU containing cells as based on BrdU quenching of HO 33342. Swartzendruber [124,125] observed that CHO cells cultured in 30 μM BrdU for one- and two-cell cycle division periods showed an increase in MI fluorescence intensity of 25 and 40%, respectively. It was concluded that the BrdU-induced alterations in native chromatin, previously demonstrated by Latt and Wohlleb [88], had increased the availability of MI binding sites, since total DNA content did not increase nor did very brief addition of BrdU to cells affect MI stainability. Based on the spectral properties of BrdU and MI, and the distant proximity of G–C-bound MI to BrdU sites on DNA, it is difficult to attribute increased MI fluorescence to energy transfer from BrdU.

Recently Dolbeare et al. [42] developed a sensitive method for labeling both BrdU and DNA using FITC conjugated to a monclonal antibody to BrdU (anti-BrdU) and PI. The antibody assay method is similar to that described previously by Gratzner [55]. PI–DNA staining and two-color bivariate analysis (488-nm excitation) showed that the antibody was bound to cells in S phase. Dolbeare et al. [42] used only short (10 to 30-min) periods for exposure of cells in culture to

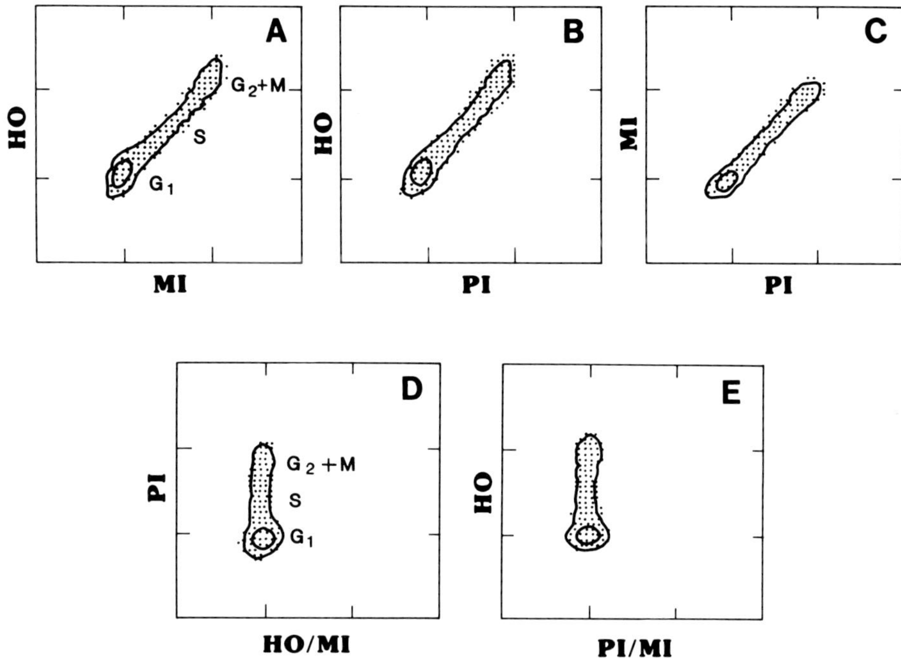

Fig. 7. Bivariate contour displays showing the correlation of the various DNA dyes **(A–C)** in three color-stained CHO cells and the ratios for HO/MI **(D)** and PI/MI **(E)** throughout the cell cycle as derived by the PI **(D)** and HO **(E)** fluorescence. Data are the same as used for the single-parameter distributions shown in Figure 6.

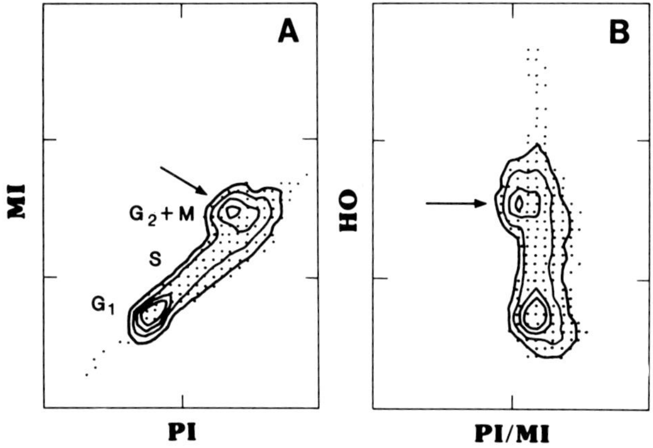

Fig. 8. Detection of mitotic cells in populations fixed in ethanol and stained with three DNA dyes. CHO cells were cultured for 4 hours in medium containing 0.1 μg/ml Colcemid to accumulate cells in mitosis. Mitotic cells (indicated by arrow) show a decrease in PI stainability as compared with MI **(A)** and are thus easily detected by bivariate ratio analysis **(B)**.

BrdU (10-μM concentrations). This illustrates that the method is very sensitive and comparable to the tritiated-thymidine assay (see Chapter 23, this volume, for further details and examples of this technique).

We recently described a sensitive two-color flow cytometric method for detecting BrdU-labeled DNA in cells treated for 30 min or less [26]. The technique uses two nonintercalating DNA-specific fluorochromes: Hoechst 33342 (HO) and mithramycin (MI), a dye whose fluorescence in the presence of BrdU remains stoichiometric to DNA content under the conditions employed. Using dual-wavelength excitation, the blue (HO) and green-yellow (MI) fluorescence emissions are measured. A differential amplifier [118] subtracts the blue fluorescence from the green-yellow fluorescence signal amplitude on a cell-by-cell basis, and the resulting difference signal is amplified. Cells in S phase exhibit a significant BrdU/Hoechst quenching and produce a greater differential fluorescence signal compared with that of cells in the G_1 and the G_2 + M phase that, except for minor differences in HO and MI stainability, show relatively small (near zero) fluorescence differences. The technique is simple, rapid, and requires only one-step staining. It is mild and therefore minimizes cell loss and loss of other important cellular markers such as DNA and/or chromatin and RNA. Recently, we included pyronin Y in the staining procedure in order to examine RNA content of cycling and noncycling cells (manuscript in preparation).

Untreated Chinese hamster ovary (line CHO) cells examined by this technique show an equal affinity for HO and MI as seen by the linear relationship in staining throughout the cell cycle in Figure 9A. However, slight differences in stainability between the two dyes are detected with increased sensitivity and amplified in the MI–HO fluorescence difference versus DNA content (mithramycin) profile (Fig. 9B). By contrast, a population of CHO cells treated for 30 min with 30 μM BrdU shows a significant quenching of HO fluorescence in S phase, as seen in Figure 9C. The G_1 and G_2 + M subpopulations are essentially unaffected. Comparison of the MI–DNA distributions in Figure 9A and 9C indicates that mithramycin fluorescence was not affected by the BrdU-substituted DNA nor by the quenching of the Hoechst fluores-

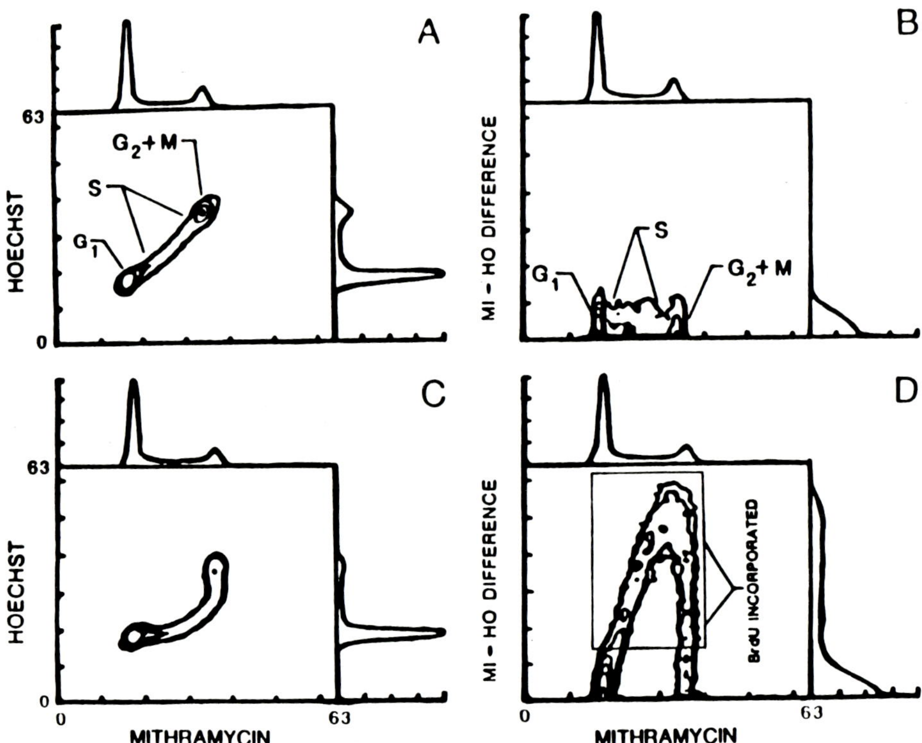

Fig. 9. DNA content (HO and MI) and MI–HO signal difference frequency distribution histograms and the corresponding bivariate contour diagrams for untreated CHO cells **(A,B)** and for CHO cells treated in culture with 30 μM BrdU for 30 min **(C,D)**. The X and Y axes are linear relative units. The MI–HO difference signal amplitudes from G_1, S, and G_2 + M phase control (untreated) cells are shown in the bivariate diagram **(B)**. The difference signals from cells with zero or slightly negative difference values, due to stainability, were accumulated along the X axis of this display, since the negative outputs from the signal difference amplifier cannot be visualized in the bivariate diagram unless a positive offset voltage is added to the amplifier output as shown in Figure 10. **D:** The cells that have incorporated BrdU are easily visualized, whereas the data counts from most G_1 and G_2 + M phase cells remain along the X axis. The difference amplifier gain was initially adjusted and fixed to give a maximum expansion of the fluorescence difference signal range from cells that had incorporated BrdU. The untreated cell population was analyzed at this same instrument gain setting.

cence. Profiles obtained for untreated and BrdU-treated CHO populations stained with MI, in the absence of HO (data not shown), were the same as shown in Figure 9A and 9C. The G1 peak positions in the single-parameter HO–DNA profile are also the same in Figure 9A and 9C. However, the magnitude of the MI–HO fluorescence differences, reflecting BrdU/Hoechst quenching (Fig. 9D) is most significant for cells in S phase that have incorporated BrdU during the 30-min pulse-labeling period. The percentage of BrdU-containing S-phase cells detected by this technique (i.e., relative fraction of cells in the boxed region in Fig. 9D) was 35%. This value is in good agreement with the flow cytometric (FCM) computer-fit S-phase calculation of 37% and also with a 33% thymidine labeling index derived by standard autoradiography. The bivariate contour distribution for the BrdU-treated population (Fig. 9D) is also similar to distributions obtained for BrdU-treated CHO cells using the BrdU-antibody technique as shown previously by Dolbeare et al. [42].

The sensitivity and stoichiometry of the technique were examined using L1210 cells in culture. The results showed that the amount of BrdU detected by the differential fluorescence measurement was proportional to the period of BrdU labeling, as demonstrated by analysis of L1210 cells treated with 10 μM BrdU for 30-, 15-, or 5-min durations, respectively (Fig. 10A–C). The bivariate profiles in Fig. 10 are displayed with a slight zero axis offset so that the actual range in fluorescence differences of most G_1 and G_2 + M phase cells could be visualized (see legend to Fig. 10). Cells in S phase that incorporated BrdU are clearly separated from the G_1 and G_2 + M subpopulations. Details of studies on stoichiometry and sensitivity are presented elsewhere (manuscript in preparation).

Using this technique, we have also obtained similar results in detecting BrdU incorporation into normal human skin fibroblasts, Friend erythroleukemia cells, CHO–K1 cells, whole Chinese hamster embryo cells from primary culture, and phytohemagglutinin (PHA)-stimulated human lymphocytes. Currently, we are also examining the cycling S-phase compartment in populations of cells from solid tumors of mice injected with BrdU.

Several investigators [10,44,77] have successfully used a staining combination of ethidium bromide and Hoechst 33258 and FCM analyses to detect BrdU-containing cells. However, differential fluorescence analyses were not used in those studies, and in vitro BrdU labeling periods of 6 hours or longer were required before BrdU/Hoechst quenching in cells could be detected in the bivariate displays.

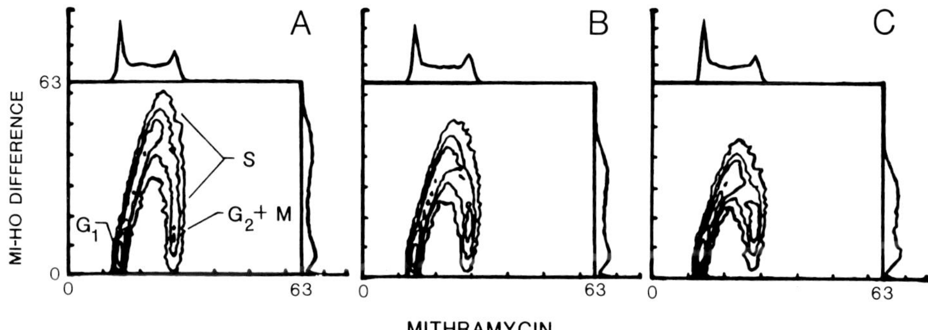

Fig. 10. DNA content (MI) and MI–HO signal difference frequency distribution histograms and the corresponding bivariate contour for L1210 treated in culture with 10 μM BrdU for **(A)** 30 min, **(B)** 15 min, or **(C)** 5 min. Stained cells were analyzed as described in the text and in Figure 9, except that a small positive differential offset bias voltage was added to the difference amplifier output so that the zero and slightly negative MI–HO difference signals associated with G1 and G2 + M cells could be visualized.

TWO-COLOR DNA AND PROTEIN STAINING

Simultaneous FCM analysis of DNA and protein allows for direct assessment of the cell growth at various phases of the cell cycle. Within heterogeneous populations, cell types with distinct protein content variations can also be identified. Such determinations have been used in a number of studies involving a wide variety of cell types. For most studies, two-color staining of DNA and protein for cells in suspension has been accomplished using a method combining ethidium bromide (EB) and fluorescein isothiocyanate (FITC) [54] or propidium iodide (PI) and FITC [23,24]. Stohr et al. [123], evaluated several other dye combinations. Of these, DAPI and sulfarhodamine 101 (SR101) was most rapid and gave the best results. Recently, Roti-Roti et al. [106] have used a combination PI-FITC to examine and directly compare nuclear protein and DNA contents of cells in various phases of the cell cycle. By modifying our earlier PI-FITC technique we recently demonstrated a rapid, one-step staining procedure for cellular DNA-protein determinations by single or dual laser beam FCM [24]. Methods were demonstrated for optimizing dye concentrations of a staining solution that contained PI, FITC and RNase. FCM analysis was performed using single laser excitation at 488 nm. Alternatively the DNA stains, MI or HO 33342, were applied in combination with either of the red protein stains, X-rhodamine isothiocyanate (X-RITC) or rhodamine 640 (R-640) without RNase and sequential dual laser FCM analysis. Rhodamine-640 and HO 33342 were also used to stain unfixed cells. In all cases, the dye concentrations were adjusted to permit equilibrium staining without the usual washing and centrifugation steps. Cells were then analyzed in the stain solution. For comparative purposes the cell concentration in the staining solution was kept constant from sample to sample. The fluorescence properties of PI-bound to calf thymus DNA and FITC are shown in Figure 2.

We have previously shown the advantage of simultaneous DNA-protein analysis for detecting myeloma cells in human bone marrow [29]. Recently we used sequential dual laser excitation (i.e., UV and 488 nm) for DNA and protein analysis of cells stained with HO 33342 and FITC. Using this method, we were able to resolve two subpopulations of cycling cells in samples from non-Hodgkin's lymphoma patients that showed bimodal protein contents [25]. Bivariate analysis was required to detect the subpopulations, since the two DNA distribution patterns were overlapping. Ffrench et al. [48] have recently used PI-FITC staining and FCM in analysis of leukoblasts from acute myeloid leukemia patients. In that study, they observed G_1 cells with low protein content (i.e., G_0/G_{1A}) that did not appear to cycle. The relative frequency of these low protein content cells was significantly greater in the complete remission group.

Cell populations stained for DNA and protein can be analyzed by various FCM and computer techniques to provide information on the ratio of protein to DNA content of cells in different phases of the cell cycle, and the ratio of protein to cell size [23], as well as the nuclear to cytoplasmic ratio, a parameter that is often of importance in clinical diagnosis. Ratio analysis can be performed on-line, electronically [23] or by subsequent computer analysis, if data are collected in correlated list mode [110]. (See also Chapt. 22) Using computer gating techniques across the DNA content histogram the protein to DNA content ratios can be generated for cells in various phases of the cell cycle.

Nuclear to cytoplasmic ratio analysis requires the use of a FCM system in which the laser beam is focused to a thin elliptical slit that is smaller than the diameter of the nucleus [114]. When a PI–FITC-stained cell crosses the illuminating slit both the time duration of the green (cytoplasm) and the red (nucleus) fluorescence are measured and provide a relationship proportional to the cytoplasmic and nuclear diameter of the cell. A ratio of the red to green time duration signals thus provides information on the relative nuclear to cytoplasmic relationship [115]. Data obtained from analysis of PI–FITC-stained CHO cells are shown in Figure 11. The narrow, unimodal histogram in Figure 11F, demonstrates the good correlation in nuclear to cytoplasmic relationship that is usually observed in exponentially growing, homogeneous cell populations.

RNA-BINDING FLUOROCHROMES AVAILABLE FOR USE IN FLOW CYTOMETRY

In Chapter 15 (this volume), Darzynkiewicz and Kapuscinski describe the use of acridine orange (AO) for staining and FCM analysis of DNA versus RNA. Data presented show the interaction of dye and nucleic acids in cell cycle related events. In addition to acridine orange, several other fluorochromes have been used for flow cytometric analysis of RNA content. DiO-C1(3) [63], thioflavin T [107], and thiazole orange [89] have been used for viable reticulocyte analysis, and oxazine 1 [111] for total RNA content in viable

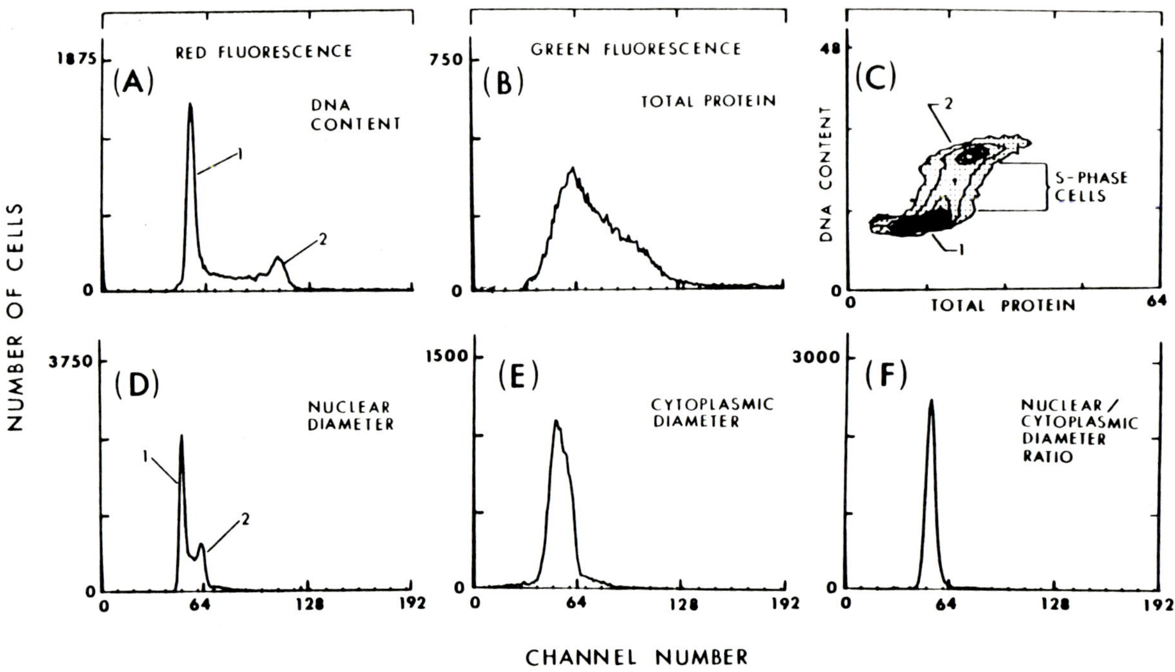

Fig. 11. Frequency distribution histograms of CHO cells fixed in 70% ethanol and stained with both propidium iodide (PI) (DNA content) and fluorescein isothiocyanate (FITC) (total protein). **A:** DNA content (red fluorescence). **B:** Total protein (green fluorescence). **C:** DNA versus protein two-parameter contour diagram. **D:** Nuclear diameter. **E:** Cytoplasmic diameter. **F:** Nuclear-to-cytoplasmic diameter ratio distribution. The horizontal axes (channel number) are proportional to red fluorescence signal amplitude **(A)**; green fluorescence signal amplitude **(B)**; red fluorescence signal time duration (nuclear diameter) **(D)**; green fluorescence signal time duration (cytoplasmic diameter) **(E)**; and the ratio of red-to-green fluorescence signal time duration **(F)**. The vertical and horizontal axes of **(C)** are proportional to red fluorescence (DNA content) and green fluorescence (protein) amplitudes, respectively.

CCRF–CEM cells. In those studies, thiazole orange and DiO C1(3) were excited at 488 nm, thioflavin T at 457 nm, and oxazine 1 at 633 nm (using a helium–neon laser). The fluorochrome, pyronin Y, has been used in FCM studies by several investigators [104,111,127]. Recently, Darzynkiewicz et al. [33] presented data from detailed studies on the application of pyronin Y (PY) in cytochemistry of nucleic acids, and Kapuscinski and Darzynkiewicz [72] have characterized the fluorescent properties of complexes of PY with natural and synthetic nucleic acids. These studies showed that under appropriate conditions pyronin Y is quite specific for RNA, and that it is quite reliable for FCM-RNA content analysis.

CORRELATED ANALYSIS OF DNA, RNA, AND PROTEIN

There appears to be a close coupling of the metabolic patterns of protein with RNA and DNA in maintaining the state of balanced growth. However, under experimental conditions, such as drug treatment, the DNA synthetic and cell division patterns are often grossly perturbed, resulting in an uncoupling of transcriptional and translational activity. Correlated studies on cellular DNA, RNA, and protein can be extremely useful in detecting such abnormalities in the cell cycle as well as elucidating normal cell growth and cell cycle progression patterns.

We recently developed a technique for simultaneous analysis of DNA, RNA, and protein [19] based on modifications of FCM methods described by Arndt-Jovin and Jovin [4] for HO 33342–DNA staining; by Tanke et al. [127], Shapiro [111], and Pollack et al. [104] for pyronin Y–RNA staining; and by Gohde et al. [54] and us [23,24] for FITC protein staining. Sequential excitation of stained cells (UV, 458 and 530 nm) is followed by fluorescence analysis of HO–DNA (blue), FITC–protein (green) and pyronin Y–RNA (red), respectively. Fluorescence ratio analysis permits direct correlation of RNA/protein (red/green) and RNA/DNA contents (red/blue) throughout the cell cycle.

Comparisons of results obtained from analyses of CHO cells stained with each dye alone versus results for cells stained with all three dyes simultaneously (Fig. 12) demonstrate the good resolution obtainable with the staining protocol combining these three dyes. Hoechst-DNA fluorescence is essentially unchanged (compare Fig. 12A,D), whereas the slight decrease in FITC fluorescence (Fig. 12B,D) is probably due to energy transfer to PY. Comparison of Figure 12C and Fig. 12D shows a slight increase in PY fluorescence in the three-color stained cell population. In general, however, the shape of the distributions are similar demonstrating adequate fluorescence resolution to quantify DNA versus RNA versus protein.

This staining and analysis technique was used to correlate cellular DNA, RNA, and protein as well as the ratios of RNA/DNA and RNA/protein throughout the cell cycle of CHO cell populations in the exponential growth phase (Fig. 13A) and to examine these parameters in populations perturbed by the chemotherapeutic agent adriamycin (Fig. 13B) (see Crissman et al. [18] for details). It is evident that G1 cells in the untreated, exponentially growing population (Fig. 13A) with a low RNA or protein content do not enter the S phase. Detailed characterization of the critical RNA and pro-

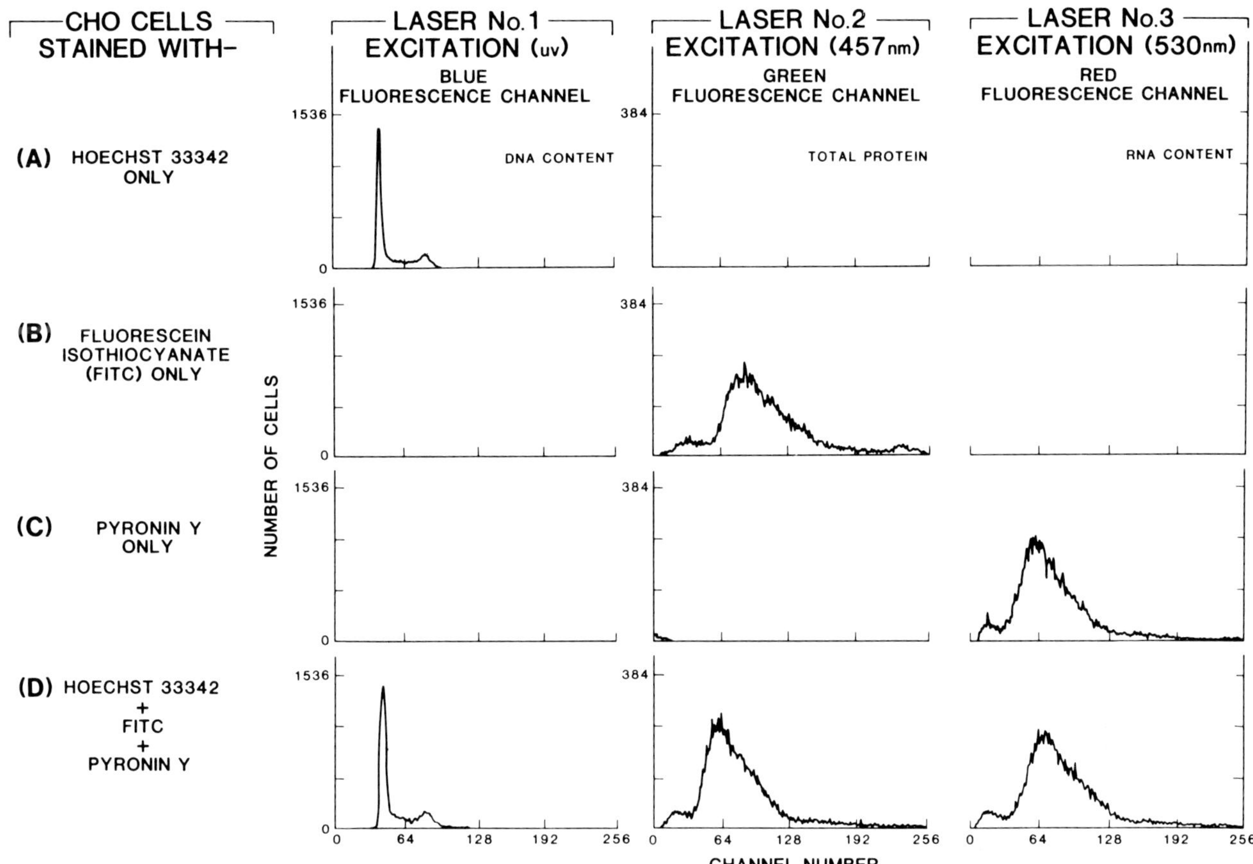

Fig. 12. Single-parameter frequency distributions for populations of ethanol-fixed CHO cells stained with only Hoechst 33342 (0.5 μg/ml) for DNA **(A)**; with fluorescein isothiocyanate (0.08 μg/ml) for protein **(B)**; with pyronin Y (1.0 μg/ml) for RNA **(C)**, or with a combination of all three dyes **(D)**. All populations **(A–D)** were analyzed in the three-laser system at all three wavelengths indicated, and fluorescence was monitored in each channel.

tein thresholds for entry of G1 cells into S phase are presented in Chapter 24 (this volume). The DNA to protein and DNA to RNA patterns are similar, and the RNA/protein ratio throughout the cell cycle remains quite constant as would be expected for cell populations in a balanced growth state.

Analysis of the cellular RNA/DNA ratio in relation to DNA content reveals a characteristic pattern reflecting changing rates of DNA replication and transcription during the cell cycle. Thus, during G1 when DNA content is stable, cells accumulate RNA but at different rates so that the G1 phase is quite heterogeneous with respect to RNA. However, during progression through S phase, the rate of DNA replication exceeds RNA accumulation giving rise to a nonvertical, negative slope of the S phase cell cluster. Cells in G2 + M have an RNA/DNA ratio in the same range as the majority of the G1 cells.

Adriamycin (AdR) treatment induced a differential response in metabolism of DNA, RNA, and protein in exponentially growing cells (Fig. 13B). Analysis of the single parameter DNA profiles at 15 hours after drug treatment (Fig. 13B, left panel) showed a large accumulation of cells from the exponential culture arrested in the G2 + M phase (i.e., 85%). Based on numerical data this subpopulation had mean ratio values for RNA to DNA and RNA to protein that were 44% and 31% elevated, respectively, above control ratio values. Few cells are observed in S phase, but the cells remaining in G1 phase (i.e., 15%) had RNA to DNA and RNA to protein ratio values almost identical to control values. Survival studies showed that compared with the exponential control CHO cell population the AdR-treated exponential population had only a 12% surviving fraction. Analysis with Trypan Blue revealed that, at the time the survival studies were performed (15 hours after AdR treatment), more than 97% of the drug-treated cells excluded the dye. The method represents a very useful approach for studying mechanisms and control of the cell cycle. In addition to the study described briefly above, we have used this technique to (1) characterize granulosa cell subpopulations from preovulatory avian follicles [95]; (2) correlate DNA, RNA, and protein contents in normal and rickettsia-infected L929 cells [6]; (3) study the effects of conditioned medium on proliferation and antimetabolite sensitivity of promyelocytic leukemia (HL-60) cells in vitro [43]; and (4) determine the effects of three different drugs on the entry of synchronized CHO G_1 cells into S phase [32].

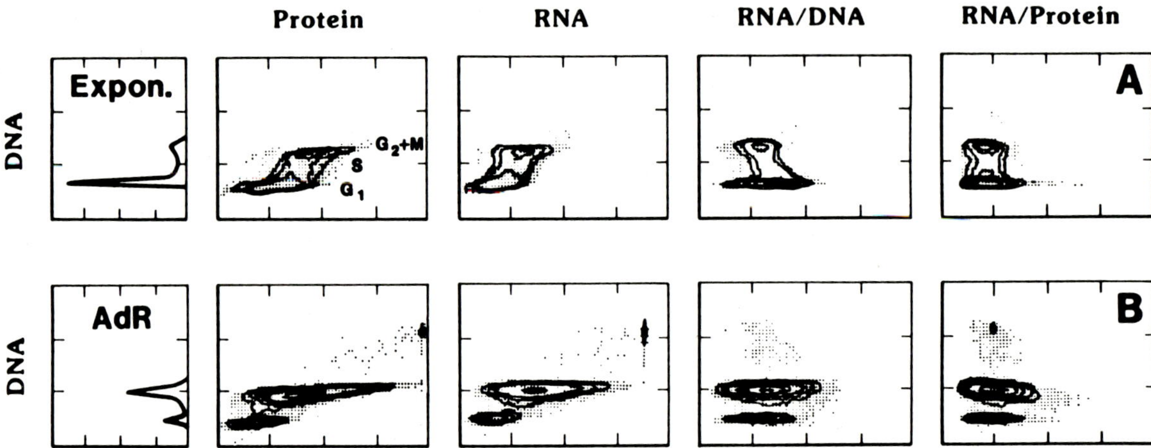

Fig. 13. **A:** Single-parameter DNA distribution (left panels) and bivariate contour density profiles for DNA (Y axis) versus protein and RNA as well as the RNA to DNA and RNA to protein ratios (left to right) for control (untreated) exponentially growing CHO cells. **B:** Corresponding population of AdR-treated cells at 15 hours after drug treatment (i.e., 6 μg/ml AdR for 2 hours.)

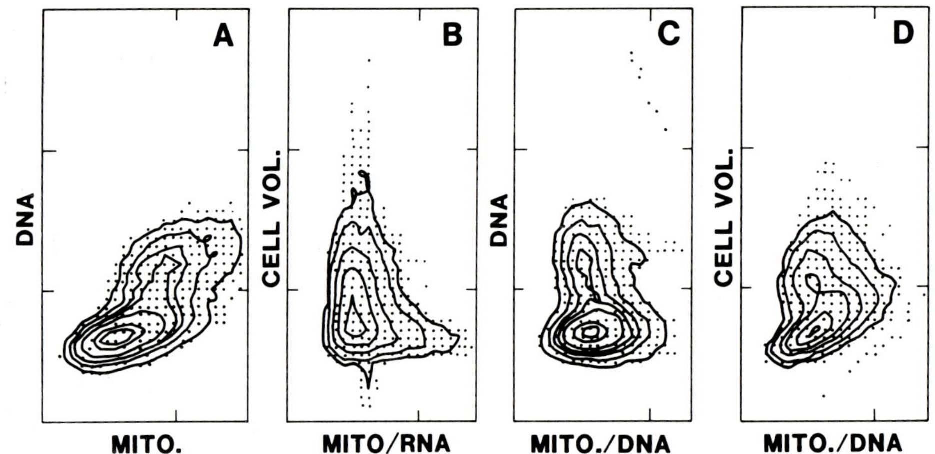

Fig. 14. Bivariate distributions illustrating the correlation of DNA, RNA, and cell volume to mitochondrial fluorescence. Growing CHO cells in suspension were stained in culture for 1 hour with HO 33342 (5.0 μg/ml) and PY (2.0 μg/ml); 3,3 dipentyloxacarbocyanine (DiO-C5-3, 0.3 μg/ml) was then added for an additional 30 min before analysis in the dye-containing medium. Cells were analyzed in the three-laser system set at the excitation wavelengths stated in the text.

CORRELATED ANALYSIS OF DNA, RNA OR PROTEIN, AND MITOCHONDRIA

Johnson et al. [68] first reported the use of rhodamine 123 (R123) as a fluorescent probe for selective staining of mitochrondria in viable cells. In subsequent studies, Johnson et al. [67] characterized mitochrondrial specificity of R123 and several other rhodamine and cyanine dyes, including Di0-C5-3. These cationic dyes accumulate due to the electronegativity of the inner mitochrondrial membranes. Loss of the negative charge or interference with electron transport results in a loss or decrease in fluorescence intensity. These findings show that these cationic dyes do not attach to specific molecules as in the case of DNA, RNA or protein stains. Therefore an increase in fluorescence of stained cells is a function of the increase in mitochondrial number and/or electronegativity.

In FCM studies, Darzynkiewicz et al. [34] showed a dramatic increase in R123 fluorescence of PHA-stimulated lymphocytes concomitant to lymphoblast activation. Based on these studies, they suggested the use of R123 as a probe for distinguishing between cycling and quiescent cells. James and Bohman [64] also used FCM techniques to study proliferation of mitochondria during the cell cycle of HL-60 cells. Darzynkiewicz et al. [37] later showed decreased mitochondrial (R123) fluorescence in differentiating Friend Leukemia cells as well as in stationary L1210 and CHO cells compared with the respective exponentially growing populations.

In an effort to correlate DNA, RNA or protein, and mitochondrial activity, we incorporated the green fluorescent DiO-C5-3 dye [67] into the protocol described by Shapiro [111] for viable cell staining of DNA and RNA content respectively with HO 33342 and pyronin (PY). Alternatively, for protein content analysis the red fluorescent rhodamine 640 (R640) was applied to the viable cells in place of pyronin Y. The three-laser system was employed as described using the UV laser line for HO 33342 excitation, the 457-nm laser line for DiO-C5-3 excitation, and the 528-nm line for PY or the 568-nm line for R640 excitation (see fluorescence spectra, Fig. 2). DiO-C5-3 has excitation and emission peak values at about 470 and 500 nm, respectively. Frequency histograms for volume, DNA and RNA contents, mitochrondrial DiO-C5-3 fluorescence and ratios of mitochondrial/DNA (green/blue) and mitochondrial/RNA (green/red) fluorescence for CHO cells were similar to those obtained previously with R123 [68], except for the decreased coefficient of variation in the Hoechst–DNA histograms. For cells in balanced growth, the protein content (R640 analysis) did not significantly differ from the PY–RNA content distributions shown here. Mitochondria fluorescence correlated well, but not directly with cell volume and RNA content. However, it was least correlated with DNA content. Bivariate analysis (Fig. 14) permits direct comparison of the mitochondrial activity as a function of the cell cycle (Fig. 14A and 14C) or as related to cell volume (Fig. 14B,D). The mitochondrial/RNA ratio appears to remain constant with increasing cell size (Fig. 14B). However, the mitochondrial/DNA ratio shows two slopes with increasing cell size (Fig. 14D). The negative slope in larger cells is presumably due, in part, to the more rapid increase in DNA content during S phase. Extension of these preliminary studies will prove useful in examining populations of drug treated and irradiated cells. Care should be taken in interpretation of mitochondrial fluorescence, since agents that are potentially cytotoxic may induce alterations in membrane potential that alter mitochondrial fluorescence without any real fluctuation in mitochondrial number and/or size.

CONCLUSION

The development and availability of new fluorescent biomolecular probes will provide the potential for unprecedented multiparameter FCM investigations on the interrelationships among cell cycle, cell metabolism, and chromatin structure. Such studies will yield new insight on the control and regulation of immunological properties and functional capacities of cells. Advances in multiple-labeling techniques will permit even more definitive and objective approaches for rapid detection and sorting of discrete subpopulations of cells. Successful application of these combined labeling methods will rely upon preliminary control studies, directed in a discriminate manner to insure reliability, accuracy and reproducibility. In this chapter, we have provided guidelines and suggested approaches to address these criteria.

It is important that fluorochromes be developed with high quantum yields comparable to the phycobiliproteins, but with greater stability than with these compounds. Such labels can be used in FCM systems with relatively inexpensive low-power laser systems or, in the case of the infrared dyes, with diode–laser systems. Finally, advances in Fourier transform not only will expand the number of dyes that can be coupled for cell labeling but will also provide a means for obtaining spectral information relevant to molecular structure and orientation. It is certain that current and potential flow system capabilities far exceed the limited applications of fluorochrome cytochemistry and more emphasis will need to be placed in this area to exploit fully the potential of FCM systems.

REFERENCES

1. **Alabaster O, Tannenbaum E, Habbersett MC, Magrath I, Herman C** (1978) Drug-induced changes in DNA fluorescence intensity detected by flow microfluorometry and their implications for analysis of DNA content distributions. Cancer Res 38:1031–1035.
2. **Alberti S, Parks DR, Herzenberg LA** (1987) A single laser method for subtraction of cell autofluorescence in flow cytometry. Cytometry 8:114–119.
3. **Arndt-Jovin DJ, Grimwade BG, Jovin TM** (1980) A dual laser flow sorter utilizing a CW pumped dye laser. Cytometry 1:127–131.
4. **Arndt-Jovin DJ, Jovin TM** (1977) Analysis and sorting of living cells according to deoxyribonucleic acid content. J Histochem Cytochem 25:585–589.
5. **Arndt-Jovin DJ, Ostertag W, Eisen H, Klimek F, Jovin TM** (1976) Studies of cellular differentiation by automated cell separation. J Histochem Cytochem 24:332–347.
6. **Baca OG, Crissman HA** (1987) Correlation of DNA, RNA, and protein content by flow cytometry in normal and *Coxiella burnetii*-infected L929 cells. Infect Immun 55:1731–1733.
7. **Barlogie B, Spitzer G, Hart JS, Johnston D, Buchner T, Schumann J** (1976) DNA histogram analysis of human hemopoietic cells. Blood 48:245–258.
8. **Bensen RC, Meyer RA, Zaruba ME, McKhann GM** (1979) Cellular autofluorescence—Is it due to flavins? J Histochem Cytochem 27:44–48.
9. **Berkhan E** (1975) Pulse-cytophotometry as a method for rapid photometric analysis of cells. in Haanen CAM, Hillen HFP, Wessels JMC (eds), "Pulse Cytophotometry." Ghent: European Press, pp 15–21.
10. **Bohmer RM, Ellwart J** (1981) Cell cycle analysis by combining 5-bromodeoxyuridine/33258 Hoechst technique with DNA-specific ethidium bromide staining. Cytometry 2:31–34.
11. **Brachet J** (1940) La détection histochimique des acides pintose nucléiques. CR Soc Biol 133:89–90.
12. **Buican TN** (1987) Spectral analysis in flow: Fourier transform flow cytometry. In Burger G, Ploem JS, Goerttler K (eds), "Clinical Cytometry and Histometry." London: Academic Press, pp 67–71.
13. **Caspersson T, Santesson L** (1942) Studies on protein metabolism in the cells of epithelial tumors. Acta Radiol (Stockh) 46(Suppl):1–105.
14. **Caspersson T, Schultz J** (1938) Nucleic acid metabolism of the chromosomes in relation to gene reproduction. Nature (Lond) 142:294–295.
15. **Caspersson T, Zech L, Modest J, Foley GE, Wagh U, Simonsson E** (1969) DNA-binding fluorochromes for the study of the organization of the metaphase nucleus. Exp Cell Res 58:141–152.
16. **Coulson PA, Bishop AO, Lenarduzzi R** (1977) Quantitation of cellular deoxyribonucleic acid by flow cytometry. J Histochem Cytochem 25:1147–1153.
17. **Cram LS, Bartholdi MF, Wheeless LL, Gray JW.** (1985) Morphological analysis by scanning flow cytometry. In Van Dilla MA (ed), "Flow Cytometry Instrumentation and Data Analysis." New York: Academic Press, pp 163–194.
18. **Crissman HA, Darzynkiewicz Z, Tobey RA, Steinkamp JA** (1985) Normal and perturbed CHO cells: Correlation of DNA, RNA and protein by flow cytometry. J Cell Biol 101:141–147.
19. **Crissman HA, Darzynkiewicz Z, Tobey RA, Steinkamp JA** (1985) Correlated measurements of

DNA, RNA and protein in individual cells by flow cytometry. Science 228:1321–1324.

20. **Crissman HA, Hofland MH, Stevenson AP, Wilder ME, Tobey R A (1988)** Use of DiO-C_5-3 to improve Hoechst 33342 uptake resolution of DNA content and survival of CHO cells. Exp Cell Res 174:388–396.
21. **Crissman HA, Mullaney PF, Steinkamp JA (1975)** Methods and applications of flow systems for analysis and sorting of mammalian cells. In Prescott DM (ed), "Methods in Cell Biology," Vol. 9. New York: Academic Press, pp 179–246.
22. **Crissman HA, Orlicky DA, Kissane RJ (1979)** Fluorescent DNA probes for flow cytometry. J Histochem Cytochem 27:1652–1654.
23. **Crissman HA, Steinkamp JA (1973)** Rapid simultaneous measurement of DNA, protein, and cell volume in single cells from large mammalian cell populations. J Cell Biol 59:766–771.
24. **Crissman HA, Steinkamp JA (1982)** Rapid, one step staining procedures for analysis of cellular DNA and protein by single and dual laser flow cytometry. Cytometry 3:84–90.
25. **Crissman HA, Steinkamp JA (1986)** Multivariate cell analysis: Techniques for correlated measurements of DNA and other cellular constituents. In Gray JE, Darzynkiewicz Z (eds), "Techniques in Cell Cycle Analysis." Clifton, NJ: Humana Press, pp 163–206.
26. **Crissman HA, Steinkamp JA (1987)** A new method for rapid and sensitive detection of bromodeoxyuridine in DNA replicating cells. Exp Cell Res 173:256–261.
27. **Crissman HA, Stevenson AP, Kissane RJ, Tobey RA (1979)** Techniques for quantitative staining of cellular DNA for flow cytometric analysis. In Melamed MR, Mullaney PF, Mendelsohn ML (eds), "Flow Cytometry and Sorting." New York: John Wiley & Sons, pp 243–261.
28. **Crissman HA, Tobey RA (1974)** Cell cycle analysis in 20 minutes. Science 184:1297–1298.
29. **Crissman HA, Von Egmond JV, Holdrinet RG, Pennings A, Hannen C (1981)** Simplified method for DNA and protein staining of human hematopoietic cell samples. Cytometry 2:59–62.
30. **Culling C, Vassar P (1961)** Desoxyribose nucleic acid. A fluorescent histochemical technique. Arch Pathol Lab Med 71:88–92.
31. **Cunningham RE, Skramstad KS, Newburger AE, Schackney SE (1982)** Artifacts associated with mithramycin fluorescence in the clinical detection and quantitation of aneuploidy by flow cytometry. J Histochem Cytochem 30:317–322.
32. **D'Anna J, Crissman HA, Jackson PJ, Tobey RA (1985)** Time dependent changes in H1 content turnover, DNA elongation, and the survival of cells blocked in early S phase by hydroxyurea, aphidicolin or 5-fluorodeoxyuridine. Biochemistry 24:5020–5026.
33. **Darzynkiewicz Z, Kapuscinski J, Traganos F, Crissman HA (1987)** Application of pyronin Y (G) in cytochemistry of nuclei acids. Cytometry 8:138–145.
34. **Darzynkiewicz Z, Staiano-Coico L, Melamed MR (1981)** Increased mitochondrial uptake of rhodamine 123 during lymphocyte stimulation. Proc Natl Acad Sci USA 78:2383–2387.
35. **Darzynkiewicz Z, Traganos F, Kapuscinski J, Staiano-Coico L, Melamed MR (1984)** Accessibility of DNA in situ to various fluorochromes: Relationship to chromatin changes during erythroid differentiation of friend leukemia cells. Cytometry 5:355–363.
36. **Darzynkiewicz Z, Traganos F, Sharpless T, Melamed MR (1976)** Lymphocyte stimulation: A rapid, multiparameter analysis. Proc Natl Acad Sci USA 76:358–362.
37. **Darzynkiewicz Z, Traganos F, Staiano-Coico L, Kapuscinski J, Melamed MR (1982)** Interaction of rhodamine 123 with living cells studied by flow cytometry. Cancer Res 42:799–806.
38. **Dean PN, Gray JW, Dolbeare FA (1982)** The analysis and interpretation of DNA distributions measured by flow cytometry. Cytometry 3:188–195.
39. **Dean PN, Jett JH (1974)** Mathematical analysis of DNA distributions derived from flow microfluorometry. J Cell Biol 60:523–527.
40. **Dean PN, Pinkel D (1978)** High resolution dual laser flow cytometry. J Histochem Cytochem 26: 622–627.
41. **Dittrich W, Gohde W (1969)** Impulse fluorometry with single cells in suspension. Z Naturforsch 24B: 360–361.
42. **Dolbeare F, Gratzner HG, Pallavicini MG, Gray JW (1983)** Flow cytometric measurement of total DNA content and incorporated bromodeoxyuridine. Proc Natl Acad Sci USA 80:5573–5577.
43. **Elias L, Wood A, Crissman HA, Ratliff R (1985)** Effects of conditioned medium upon proliferation, deoxynucleotide metabolism and antimetabolite sensitivity of promyelocytic leukemia cells *in vitro*. Cancer Res 45:6301–6307.
44. **Ellwart J, Dohmer P (1985)** Effect of 5-fluoro-2′-deoxyuridine (FdUrd) on 5-bromo-2 deoxyuridine (BrdUrd) incorporation into DNA measured with a monoclonal BrdUrd antibody and by BrdUrd/Hoechst quenching effect. Cytometry 6:513–520.
45. **Evenson DE, Darzynkiewicz Z, Jost L, Janca F, Ballachey B (1986)** Changes in accessibility of DNA to various fluorochromes during spermiogenesis. Cytometry 7:45–53.
46. **Feulgen R, Rossenbeck H (1924)** Mikroskopisch-chemischer Nachweis einer Nucleinsaure von Typus der Thymonucleinsaure und die darauf beruhende elektive Farbung von Zellkerzen in mikroskopischen Praparaten. Z Physiol Chem 135:203–244.
47. **Frankfurt OS (1980)** Flow cytometric analysis of double-stranded RNA content distributions. J Histochem Cytochem 28:663–669.
48. **Ffrench M, Bryon PA, Fiere D, Vu Van H, Gentilhomme O, Adeleine P, Viola JJ (1985)** Cell cycle, protein content, and nuclear size in acute myeloid leukemia. Cytometry 6:47–53.
49. **Freyer JP, Wilder ME, Raju MR (1984)** Cell volume sorting to improve the precision of radiation survival assay. Radiat Res 97:608–614.
50. **Fried J, Doblin J, Takamoto S, Perez A, Hansen H, Clarkson B (1982)** Effect of Hoechst 33342 on survival and growth of two tumor cell lines and on hematopoietically normal bone marrow cells. Cytometry 3:42–47.
51. **Fried J, Perez AG, Clarkson BD (1976)** Flow cytofluorometric analysis of cell cycle distributions using

propidium iodide. Properties of the method and mathematical analysis of the data. J Cell Biol 71: 174–181.

52. **Gallop PM, Paz MA, Henson SA, Latt SA (1984)** Dynamic approaches to the delivery of reporter agents into living cells. Biotechniques 1:32–36.
53. **Gill JE, Jotz MM, Young SG, Modest EJ, Sengupta SK (1975)** 7-amino-actinomycin D as a cytochemical probe. I. Spectral properties. J Histochem Cytochem 23:793–799.
54. **Gohde W, Spies I, Schumann J, Buchner T, Klein-Dopke G (1976)** Two parameter analysis of DNA and protein content of tumor cells. In Buchner T, Gohde W, Schumann J (eds), "Pulse-Cytophotometry." Ghent:European Press, pp 27–32.
55. **Gratzner HG (1982)** Monoclonal antibody to 5-bromo- and 5-iododeoxyuridine: A new reagent for detection of DNA replication. Science 218:474–475.
56. **Gray JW (1976)** Cell cycle analysis of perturbed cell populations: Computer simulation of sequential DNA distributions. Cell Tissue Kinet 9:499–516.
57. **Gray JW, Langlois RG, Carrano AV, Burkhart-Schultz K, Van Dilla MA (1979)** High resolution chromosome analysis: One and two parameter flow cytometry. Chromosoma 73:9–27.
58. **Hedley DW, Friedlander ML, Taylor IW (1985)** Application of DNA flow cytometry to paraffin embedded archival material for the study of aneuploidy and its clinical significance. Cytometry 6:327–333.
59. **Hedley DW, Friedlander ML, Taylor IW, Rugg CA, Musgrove EA (1985)** Method for analysis of cellular DNA content of paraffin embedded pathological material using flow cytometry. J Histochem Cytochem 31:1333–1335.
60. **Hiddemann W, Schumann J, Andreeff M, Barlogie B, Herman CJ, Jeff RC, Mayall B, Murphy RF, Sandberg A (1981)** Convention on nomenclature for DNA cytometry. Cytometry 5:445–446.
61. **Hilwig I, Gropp A (1973)** Decondensation of constitutive heterochromatin in L cell chromosomes by a benzimidazole compound ("33258 Hoechst"). Exp Cell Res 81:474–477.
62. **Hudson B, Upholt WB, Deninny J, Vinograd J (1969)** The use of an ethidium bromide analogue in the dye-bouyant density procedure for the isolation of closed circular DNA. The variation of the superhelix density of mitochondrial DNA. Proc Natl Acad Sci USA 62: 813–820.
63. **Jacobberger JW, Horan PK, Hare JD (1984)** Flow cytometric analyses of blood cells stained with the cyanine dye DiO C_1 (3): Reticulocyte quantification. Cytometry 5:589–600.
64. **James TW, Bohman R (1981)** Proliferation of mitochondria during the cell cycle of the human cell line (HL-60). J Cell Biol 89:256–260.
65. **Jensen RH (1977)** Chromomycin A_3 as a fluorescent probe for flow cytometry of human gynecological samples. J Histochem Cytochem 25:573–579.
66. **Jensen RH, Langlois RG, Mayall BH (1977)** Strategies for choosing a DNA stain for flow cytometry of metaphase chromosomes. J Histochem Cytochem 25: 954–964.
67. **Johnson LV, Walsh ML, Bockus BJ, Chen LB (1981)** Monitoring of relative mitochondrial membrane potential in living cells by fluorescence microscopy. J Cell Biol 88:526–535.
68. **Johnson LV, Walsh ML, Chen LB (1980)** Localization of mitochondria in living cells with rhodamine 123. Proc Natl Acad Sci USA 77:990–994.
69. **Jovin TM (1979)** Fluorescence polarization and energy transfer: Theory and applications. In Melamed MR, Mullaney PF, Mendelsohn ML (eds), "Flow Cytometry and Sorting." New York: John Wiley & Sons, pp 137–165.
70. **Kamentsky LA, Melamed MR, Derman H (1965)** Spectrophotometer: New instrument for ultrarapid cell analysis. Science 150:630–631.
71. **Kaplow LS, Dauber H, Lerner E (1976)** Assessment of monocyte esterase activity by flow cytometry. J Histochem Cytochem 24:363–372.
72. **Kapuscinski J, Darzynkiewicz Z (1987)** Interactions of pyronin Y(G) with nucleic acids. Cytometry 8: 129–137.
73. **Kraemer P, Deaven L, Crissman H, Van Dilla M (1972)** DNA constancy despite variability in chromosome number. In DuPraw EJ (ed), "Advances in Cell and Molecular Biology," Vol. 2. New York: Academic Press, pp 47–108.
74. **Krishan A (1975)** Rapid flow cytophotometric analysis of mammalian cell cycle by propidium iodide staining. J Cell Biol 66:188–193.
75. **Krishan A, Swagata N, Gordon K (1987)** Blocking of flurochrome efflux can improve intracellular quantitation of calcium and DNA by laser flow cytometry Cytometry 60A(Suppl I):60 Abst. 337.
76. **Krishan A, Sridkar KS, Davila E, Vogel C, Sternheim W (1987)** Patterns of anthracycline retention modulation in human tumor cells. Cytometry 8:306–314.
77. **Kubbies M, Rabinovitch PS (1983)** Flow cytometric analysis of factors which influence the BrdUrd Hoechst quenching effect in cultivated human fibroblasts and lymphocytes. Cytometry 3:276–281.
78. **Kurnick NB (1952)** Histological staining with methylgreen-pyronin. Stain Technol 27:233–242.
79. **Lalande ME, Ling V, Miller RG (1981)** Hoechst 33342 dye uptake as a probe of membrane permeability changes in mammalian cells. Proc Natl Acad Sci USA 78:363–367.
80. **Langlois RG, Jensen RH (1979)** Interaction of DNA specific fluorescent stains bound to mammalian cells. J Histochem Cytochem 27:72–79.
81. **Larsen JK, Munch-Petersen B, Christiansen J, Jorgensen K (1986)** Flow cytometry discrimination of mitotic cells: Resolution of M, as well as G_1, S and G_2 phase nuclei with mithramycin, propidium iodide, and ethidium bromide after fixation with formaldehyde. Cytometry 7:54–63.
82. **Latreille J, Barlogie B, Johnston D, Drewinko B, Alexanian R (1982)** Ploidy and proliferative characteristics in monoclonal gammopathies. Blood 59:43–51.
83. **Latt SA (1973)** Microfluorometric detection of deoxyribonucleic acid replication in human metaphase chromosomes. Proc Natl Acad Sci USA 70: 3395–3399.
84. **Latt SA (1979)** Fluorescent probes of DNA microstructure and synthesis. In Melamed MR, Mullaney PF, Mendelsohn ML (eds), "Flow Cytometry and Sorting." New York, John Wiley & Sons, pp 263–284.

85. **Latt SA, George YS, Gray JW** (1977) Flow cytometric analysis of bromodeoxyuridine-substituted cells stained with 33258 Hoechst. J Histochem Cytochem 25:927–934.
86. **Latt SA, Marino M, Lalande M** (1984) New fluorochromes compatible with high wavelength excitation for flow cytometric analysis of cellular nucleic acids. Cytometry 5:339–347.
87. **Latt SA, Stetten G** (1976) Spectral studies on 33258 Hoechst and related bisbenzimidazole dyes for fluorescent detection of deoxyribonucleic acid synthesis. J Histochem Cytochem 24:24–33.
88. **Latt SA, Wohlleb JC** (1975) Optical studies of the interaction of 33258 Hoechst with DNA, chromatin, and metaphase chromosomes. Chromosoma 52:297–316.
89. **Lee LG, Chen CH, Chiu LA** (1986) Thiazole orange: A new dye for reticulocyte analysis. Cytometry 7: 508–517.
90. **Lehnert BE, Valdez YE, Fillak DA, Steinkamp JA, Stewart CC** (1986) Flow cytometric characterization of alveolar macrophages. J Leukocyte Biol 39:285–298.
91. **LePecq JB, Paoletti C** (1967) A fluorescent complex between ethidium bromide and nucleic acids. J Mol Biol 27:87–106.
92. **Lindmo T, Steen H** (1977) Flow cytometric measurement of the polarization from intracellular fluorescein in mammalian cells. Biophys J 18:173–187.
93. **Loken MR** (1980) Separation of viable T and B lymphocytes using a cytochemical stain, Hoechst 33342. J Histochem Cytochem 28:36–39.
94. **Luce GG, Sharrow SO, Shaw S** (1985) Enumeration of cytotoxic cell-target conjugates by flow cytometry using internal fluorescent stains. Biotechniques 3: 270–272.
95. **Marrone BL, Crissman HA** (1988) Characterization of granulosa cell subpopulations from avian preovulatory follicles by multiparameter flow cytometry. Endocrinology 122:651–658.
96. **Martin JC, Swartzendruber DE** (1980) Time: A new parameter for kinetic measurements in flow cytometry. Science 207:199–201.
97. **Mayall BH, Mendelsohn ML** (1970) Errors in absorption cytophotometry: Some theoretical and practical considerations. In Wied GL, Bahr GG (eds), "Introduction to Quantitative Cytochemistry," Vol. II. New York: Academic Press, pp 171–208.
98. **Mazzini G, Giordano P** (1979) Effects of some solvents on fluorescence intensity of phenoantridinic derivatives-DNA complexes: Flow cytometric applications. In Laerum OD, Lindmo T, Thorud E (eds), "Flow Cytometry," Vol. IV, Bergen, Norway: Universitetsforlaget, pp 74–77.
99. **Modest EJ, Sengupta SK** (1974) 7-Substituted actinomycin D (NSC-3053) analogs as fluorescent DNA-binding and experimental antitumor agent. Cancer Chemother Rep 58:35–48.
100. **Mueller W, Gautier F** (1975) Interaction of heteroaromatic compounds with nucleic acids. A T-specific nonintercalating DNA ligands. Eur J Biochem 54: 385–394.
101. **Otten GR, Loken MR** (1982) Two color light scattering identifies physical differences between lymphocyte subpopulations. Cytometry 3:182–187.
102. **Pallavicini MG, Lalande ME, Miller RG, Hill RP** (1979) Cell cycle distribution of chronically hypoxic cells and determination of the clonogenic potential of cells accumulated in G_2 + M phases after irradiation of a solid tumor *in vivo*. Cancer Res 39:1891–1897.
103. **Pierrez J, Guerci A, Guerci O** (1987) Technique and staining optimization leucoconcentration. Cytometry 8:529–533.
104. **Pollack A, Prudhomme DL, Greenstein DB, Irvin GL III, Claflin AJ, Block NL** (1982) Flow cytometric analysis of RNA content in different cell populations using pyronin Y and methyl green. Cytometry 3:28–35.
105. **Roederer M, Murphy RF** (1986) Cell by cell autofluorescence correction for low signal-to-noise systems: Application to epidermal growth factor endocytosis by 3T3 fibroblasts. Cytometry 7:558–565.
106. **Roti-Roti JL, Higashikubo R, Blair OC, Vygur H** (1982) Cell cycle position and nuclear protein content. Cytometry 3:91–96.
107. **Sage BH, O'Connell JP, Mercolino TJ** (1983) A rapid vital staining procedure for flow cytometric analyses of human reticulocytes. Cytometry 4:222–227.
108. **Sahar E, Latt SA** (1978) Enhancement of banding patterns in human metaphase chromosomes by energy transfer. Proc Natl Acad Sci USA 75: 5650–5654.
109. **Salzman GC, Crowell JM, Martin JC, Trujillo TT, Romero A, Mullaney PF, La Bauve PM** (1975) Cell classification by laser light scattering: Identification and separation of unstained leucocytes. Acta Cytol (Praha) 19:374–377.
110. **Salzman GC, Wilkins SF, Whitfill JA** (1981) Modular computer programs for flow cytometry and sorting: The LACEL system. Cytometry 1:325–336.
111. **Shapiro HM** (1981) Flow cytometric estimation of DNA and RNA content in intact cells stained with Hoechst 33342 and pyronin Y. Cytometry 2:143–150.
112. **Shapiro H, Schildkraut R, Curbelo R, Brough-Turner R, Webb R, Brown D, Block M** (1977) Cytomat R: A computer-controlled multiple laser source multiparameter flow cytophotometer system. J Histochem Cytochem 25:836–844.
113. **Shapiro H, Stephens S** (1986) Flow cytometry of DNA content using oxazine 750 or related laser dyes with 633 nm excitation. Cytometry 7:107–110.
114. **Steinkamp JA** (1984) Flow cytometry. Rev Sci Instrum 55:1375–1400.
115. **Steinkamp JA, Crissman HA** (1974) Automated analysis of deoxyribonucleic acid, protein, and nuclear to cytoplasmic relationships in tumor cells and gynecologic specimens. J Histochem Cytochem 22: 616–621.
116. **Steinkamp JA, Kraemer PM** (1974) Flow microfluorometric studies of plant lectin binding to mammalian cells. II. Estimation of the surface density of receptor sites by multiparameter analysis. J Cell Physiol 84:197–204.
117. **Steinkamp JA, Orlicky DA, Crissman HA** (1979) Dual-laser flow cytometry of single mammalian cells. J Histochem Cytochem 27:273–276.
118. **Steinkamp JA, Stewart CC** (1986) Dual laser, differential fluorescence correction method for reducing

cellular background autofluorescence. Cytometry 7: 566–574.

119. **Steinkamp JA, Stewart CC, Crissman HA** (1982) Three-color fluorescence measurements on single cells excited at three laser wavelengths. Cytometry 2:226–231.

120. **Stohr M** (1976) Double beam application in flow techniques and recent results. In Gohde W, Schumann J, Buchner T (eds), "Pulse Cytophotometry." Ghent: European Press, pp 39–45.

121. **Stohr M, Eipel H, Goerttler K, Vogt-Schaden M** (1977) Extended application of flow microfluorometry by means of dual laser excitation. Histochemistry 51:305–313.

122. **Stohr M, Vogt-Schaden M** (1979) A new dual staining technique for simultaneous flow cytometric DNA analysis of living and dead cells. In Laerum OD, Lindmo T, Thorud E (eds), "Flow Cytometry," Vol IV. Bergen; Norway: Universitetsforlaget, pp. 96–99.

123. **Stohr M, Vogt-Schaden M, Knobloch M, Vogel R, Futterman G** (1978) Evaluation of eight fluorochrome combinations for simultaneous DNA-protein flow analyses. Stain Technol 53:205–215.

124. **Swartzendruber DE** (1977) A bromodeoxyuridine–mithramycin technique for detecting cycling and noncycling cells by flow microfluormetry. Exp Cell Res 109:439–443.

125. **Swartzendruber DE** (1977) Microfluorometric analysis of cellular DNA following incorporation of bromodeoxyuridine. J Cell Physiol 90:445–454.

126. **Swift H** (1966) Analytical microscopy of biological materials. In Weid GL (ed), "Introduction to Quantitative Cytometry." New York: Academic Press, pp. 1–39.

127. **Tanke HJ, Nieuwenhuis AB, Koper GJM, Slats JCM, Ploem JS** (1981) Flow cytometry of human reticulocytes based on RNA fluorescence. Cytometry 1:313–320.

128. **Taylor IW, Milthorpe BK** (1980) An evaluation of DNA fluorochromes, staining techniques and analysis for flow cytometry. J Histochem Cytochem 28: 1224–1232.

129. **Thornthwaite JT, Sugarbaker EV, Temple WJ** (1980) Preparation of tissues for DNA flow cytometric analysis. Cytometry 1:229–237.

130. **Tobey RA, Crissman HA** (1975) Unique techniques for cell cycle analysis utilizing mithramycin and flow microfluorometry. Exp Cell Res 93:235–239.

131. **Tobey RA, Crissman HA, Kraemer PM** (1972) A method for comparing effects of different synchronizing protocols on mammalian cell cycle traverse. J Cell Biol 54:638–645.

132. **Trujillo TT, Van Dilla MA** (1972) Adaptation of the fluorescent Feulgen reaction to cells in suspension for flow microfluorometry. Acta Ctyol (Praha) 16:26–30.

133. **Van Dilla MA, Langlois RG, Pinkel D** (1983) Bacterial characterization by flow cytometry. Science 220: 620–622.

134. **Van Dilla MA, Trujillo TT, Mullaney PF, Coulter JR** (1969) Cell microfluorometry: A method for rapid fluorescence measurement. Science 163:1213–1214.

135. **Van Epps DE, Bender JG, Steinkamp JA, Chenoweth DE** (1985) Modulation of neutrophil-reduced pyridine nucleotide content following stimulation with phorbol myristate acetate and chemotactic factors. J Leukocyte Biol 38:586–601.

136. **van Kroonenburgh MJ, Beck JL, Scholtz JW, Hacker-Klom V, Herman CJ** (1985) DNA analysis and sorting of rat testis cells using two-parameter flow cytometry. Cytometry 6:321–326.

137. **Vindelov LL, Christensen IJ, Jensen G, Nissen NI** (1983) Standardization of high resolution flow cytometric DNA analysis by the simultaneous use of chicken and trout red blood cells as internal reference standards. Cytometry 3:328–331.

138. **Vindelov LL, Christensen IJ, Nissen NI** (1983) A detergent–trypsin method for the preparation of nuclei for flow cytometric DNA analysis. Cytometry 3: 323–327.

139. **Visser JWM** (1979) Vital staining of hemopoietic cells with the fluorescent bis-benzimidazole derivatives Hoechst 33342 and 33258. In Laerum OD, Lindmo T, Thorud E (eds), "Flow Cytometry," Vol IV. Bergen, Norway: Universitetsforlaget, pp 86–90.

140. **Visser JWM, Bol SJL, Van den Engh GJ** (1981) Characterization and enrichment of murine hemopoietic stem cells by fluorescence activated cell sorting. Exp Hematol 9:644–655.

141. **Wallen CA, Higashikubo R, Dethlefson LA** (1982) Comparison of two flow cytometric assays for cellular RNA–acridine orange and propidium iodide. Cytometry 3:155–160.

142. **Ward DC, Reich E, Goldberg IH** (1965) Base specificity in the interaction of polynucleotides with antibiotic drugs. Science 149:1259–1263.

143. **Waring M** (1970) Variation of the supercoils in closed circular DNA by binding of antibiotics and drugs: Evidence for molecular models involving intercalation. J Mol Biol 54:247–279.

144. **Zante J, Schumann J, Barlogie B, Gohde W, Buchner T** (1976) Preparation and staining procedures for specific and rapid analysis of DNA distributions. In Gohde W, Schumann J, Buchner T (eds), "Pulse Cytophotometry." Ghent: European Press, pp 97–106.

145. **Zelenin AV, Poletaev AI, Stephanova NG, Barsky VE, Kolesnikov VA, Nikitin SM, Zhuze AL, Gnutchev NV** (1984) 7-Amino actinomycin D as a specific fluorophore for DNA content analysis by laser flow cytometry. Cytometry 5:348–354.

14

Fluorescent Probes of DNA Microstructure and DNA Synthesis

Samuel A. Latt and Richard G. Langlois

Genetics Division and Mental Retardation Center, The Children's Hospital, Boston, Massachusetts 02115; Departments of Pediatrics and Genetics, Harvard Medical School, Cambridge, Massachusetts 02139 (S.A.L.) (Deceased); Biomedical Sciences Division, Lawrence Livermore National Laboratory, Livermore, California 94550 (R.G.L.)

FEATURES MEASURABLE

A wide range of structural and functional aspects of DNA are measurable, given the appropriate fluorescent probe and, for the most part, using instrumental capabilities present or easily added to existing flow cytometers. The availability of fluorescent probes continues to be the major "rate" or progress-limiting component. However, adding dye laser capability to existing line emission lasers puts most dyes, once identified, within reach of some instrument-compatible light source. Also, a number of independent excitation sources can be used in sequence, each eliciting two or more emission readouts from each cell, covering different wavelength ranges, which are selected with appropriate filters [289]. Thus it is possible to make multiple measurements, including some dependent on the electronic interactions between multiple dyes [396]. Ultimately, a moderate resolution excitation or emission spectrum should be measurable on each cell (or particle) examined. Appropriate use of polarizers can provide additional information, both about dye-dye interactions and about rotational mobility of dyes and the cellular moieties to which they are attached. Means of synthesizing such information with data on light scatter and possibly light absorption continue to be developed, and there is the interesting possibility of examining such features in the wavelength range of probe optical transitions to detect dispersive effects associated with electronic transitions.

The time during which individual objects traverse a light beam in the flow cytometer (typically in the few-microsecond range) imposes certain constraints, primarily on the emission kinetics of acceptable probes. For example, probes with very long emission lifetimes are relatively less useful, and gathering information about spectral distributions requires very fast detectors and processing methods. There are constraints also on various spatially sensitive excitation/emission configurations that can provide information about fluorophore distribution within objects. Improvements in excitation intensity, emission detection, and probe absorption and fluorescence efficiency in principle can reduce the number of detectable molecules from the present range of 10^3–10^4/object, bringing measurements of individual chromosomes and perhaps specific gene sequences within theoretical range. Examples of the current status and likely prospects for future measurement are described below.

Amount of DNA

Measurement of DNA is rarely a problem with whole human cells (containing approximately 6×10^9 base pairs (bp) DNA per diploid, G1 genome), or even with isolated human metaphase chromosomes (the smallest of which contains almost 10^8 bp DNA). Most DNA dyes bind, at saturation, at ratios of at least one dye molecule per 50 base pairs, with the resulting fluorescence of chromosomes well within detection limits. Pinkel and Steen [341] have described methods to determine and compare instrument sensitivity. DNA measurements of small objects such as yeast [1,32], parasites [171,178,463], the slime mold *Physarum* [317], phytoplankton [420], or even bacteria [281,394] (approximately 3×10^3 kb) are possible, including response of bacteria to antibiotics [395]. Dye combinations can provide information about the base composition of bacterial DNA [432]. Internal standards (e.g., chick and trout RBC; resting human lymphocytes) [180,399,438] may be used to obtain absolute values for DNA or to standardize and calibrate instrumentation or assess sensitivities, for example, in detecting small differences between the DNA contents of different cells [47,439]. DNA measurements also have been used to achieve and refine cell synchrony [87,353].

Detection sensitivity can become a problem if, for example, the minute amount of DNA within different regions of a chromosome is to be measured, for example, via a slit scan arrangement. It can become even more severe when microbiological samples are to be examined. Absorption measurements can help standardize flow cytometry measurements of

Flow Cytometry and Sorting, Second Edition, pages 249–290

chromosome DNA [294], since most fluorescent DNA dyes have some base composition bias [226]. The minimum amount of DNA per object that can be detected depends, of course, on excitation and emission geometry, detection efficiency, and, in certain wavelength ranges, autofluorescence [11,20,299,416]. If the detection limit is in the 10^3 per object range [271], then even many bacteriophage should be sensed by existing equipment. For specific DNA sequences, as will be discussed later, some means of artificially amplifying the number of chromophores per sequence may prove necessary. At the other extreme, in the presence of many chromophores, there may be internal absorption of fluorescence perhaps accompanied by energy transfer with or without fluorescence reemission at lower energy levels. However, this does not appreciably alter the linearity between fluorescence and DNA content [199] except at very high dye concentration and absorption [408].

Measurement of DNA in the presence of RNA can be effected either with a DNA-specific fluorophore or by first treating cells with RNase before staining with less specific dyes. Conversely, measurement of RNA in the presence of DNA requires either DNase treatment before staining [124] or staining with a dye such as pyronine, which preferentially binds to single-stranded nucleic acids after DNA binding sites are occupied by another fluorescing or spectrally noninterfering DNA-specific dye [346,385] (see also Chapter 15, this volume). Pyronine (or thioflavine [364]) or the cyanine dye $DiOC_1$ [179] alone will suffice for RNA measurement in reticulocytes, which have no appreciable DNA [407].

Composition of DNA

The sensitivity of various DNA-binding fluorochromes to binding and/or quantum yield variation, DNA base composition, and perhaps methylation, is well known. These dyes are amenable to use by flow cytometry. Furthermore, as will be detailed later, appropriate pairs of dyes serving as energy donor-acceptor pairs can accentuate differences in regions of DNA enriched for one base pair or another. The fluorescence depolarization that accompanies energy transfer can be used to enhance this effect (by a theoretical factor as high as 2–3), although the additional instrumental problems associated with polarization measurements have thus far inhibited exploitation of this possibility.

The ultimate biologic components of DNA are the specific genetic sequences, which, in principle, can be detected once cytochemical techniques are developed to exploit the many available cloned nucleic acid probes. Polymeric dyes constructed of subunits with strong base specificity might also provide a means of "reading" DNA sequences, with the advantage that DNA denaturation would not be a necessary pretreatment. An intermediate step in developing DNA probe technology will, no doubt, be the detection of highly repeated sequences numbering 10^3–10^4 or more per object measured. Another intermediate solution could be based on the spatial distribution of chromatin taking up a high concentration of specific base composition-sensitive dyes. Approaches (and problems) for accomplishing this will be discussed later.

Conformation of DNA

There is increasing evidence for multiple conformations of isolated DNA, including that of reverse chirality. Perhaps the most widely discussed of this latter class is Z-DNA [189,354,360,461], which has been defined in isolated nucleic acid but whose status in vivo remains an object of intense investigation. Differential dye binding was involved in initial studies of alternative DNA conformations [345], and it stands to reason that certain dyes will ultimately prove to be useful detectors of DNA conformation. This is subject to the caveat that the very structural specificity that influences dye binding confers on the dye the potential to alter the DNA conformation to be measured. Less esoteric aspects of DNA conformation detectable by dyes include DNA strandedness (i.e., native or denatured) and, especially for those dyes that bind to DNA by intercalation, torsion constraints, or supercoiling of DNA induced by its site in a cell or an isolated chromosome.

Accessibility and Condensation of DNA

Intuitively, one would expect chromosomal proteins to reduce dye binding stoichiometry and affinity, and that is often the case. The details depend greatly on the specific dye involved. In spite of increasing knowledge of chromosomal subunit (nucleosome) structure, rationalization of these details is still difficult, perhaps because of higher order chromatin packing. Energy transfer between bound dyes, under appropriate conditions, can serve as a measure of DNA accessibility. However, the relatively low values of dye pair 50% energy transfer efficiency distances ($R_0 \leq 50$–100 Å) [402] compared even with nucleosomal dimensions (>100 Å in diameter) [355] place limits on the use of this approach except as an empirical readout of DNA condensation. However, with unfixed cells or nuclei and on a sufficiently rapid time scale, such measurements might still prove of great practical use. Alternatively, one can compare the fluorescence signals on cells with identical DNA content but different chromosome condensation [69,84] or during differentiation [94]. Finally, one can use proteins, with or without DNA-binding dyes, to protect regions of DNA from specific nucleases [362] or even to promote decondensation of specific chromosomal regions.

Interaction of DNA With Other Cellular Components

The types of interactions detectable, in addition to those with proteins, include those with RNA or with exogenously added agents, for example, carcinogens. One can rely either on extrinsic DNA probes, optical properties of the interacting species (e.g., certain carcinogens fluoresce), or spectroscopic tags placed on the interacting species (e.g., RNA or protein) [79–81,136,137,254,361,421]. For the latter, were it possible to extend flow cytometric excitation to the UV level (e.g., 270–280 nm), the tryptophan fluorescence in nonhistone proteins and the tyrosine fluorescence of histones might ultimately be exploited. Importantly, most DNA specific dyes do not alter the electrophoretic behavior of proteins subsequently isolated from chromatin [468].

Rates of DNA Synthesis

One can use a variety of DNA base analogues and analogue detection systems to follow DNA synthesis. The time scale of interest can range from instantaneous rates (a few minutes of labeling) to a retrospective look at the number of total DNA replication cycles undergone by subpopulations of cells on which other features are also measured. The dynamic potential of flow cytometry places detection of chemicals influencing cellular proliferation, for example, antimetabolites, within the range of readily measurable entities.

TYPES OF SAMPLES AND APPLICATIONS

In the previous section, the most common samples, for example, cells, isolated metaphase chromosomes, bacteria or viruses were enumerated. Other entities, related to or entirely different from the above, for example, gametes, can be detected subject to constraints of DNA amount and, in some instances, object geometry [340]. Detailed criteria will be given for cells and chromosomes.

Cells

The major discriminating feature among DNA-binding dyes is the extent to which they will permeate cells and bind stoichiometrically and specifically to cellular DNA. Freedom from toxic effects, protection from photosensitization, and reversibility of dye binding are additional relevant features.

In the case of fixed cells, particularly those treated to denature the DNA, any change in optical properties of the dye dependent on the extent and distribution of native and denatured DNA becomes important. Also, it is important to retain binding sites of cell surface antigens [272].

Throughout, dye stoichiometry and compatibility with other types of measurements must be considered. Dye uptake and optical signals can be correlated with the entities that they presumably detect if objects (cells) of specified fluorescence are sorted and subjected to appropriate chemical analysis. Alternatively, signals attributed to DNA or RNA can be compared before and after treatment with specific nucleases [343]. Obviously, such correlations depend on consistent methods of cell preparation.

Chromosomes

Here the problem is due to the constraints of signal sensitivity [252], object preparation, and preharvest treatment (influencing object condensation) [44,45]. Still, with dual laser instruments, one can resolve all but a few human chromosomes (#9–#12 and at times #14–#15). By adding a third dye, using slit scan or other instrumental modifications [72,275,300] or by taking advantage of chromosomal structural variants or rearrangements, virtually complete flow karyotyping is within the range of possibility [251]. Heteromorphic variations between homologues, especially for certain chromosomes, can also be measured to a high degree of precision [227]. Variations in karyotype due to intrinsic factors (e.g., mosaicism, rearrangement) or to external effects (e.g., DNA damage by mutagenic carcinogens) can be detected relative to an appropriate control [51]. Quantitation of such changes, including the production of micronuclei [174,324] or chromosome rearrangements [73,115,329,330] requires some standardization and/or calibration of preparative procedures [199,438]. Preparative sorting of chromosomes can be facilitated by preceding flow sorting with velocity sedimentation as a rough size fractionation step [62].

Additional studies, now performed in many laboratories, include preparative sorting of chromosomes for specific DNA sequence analysis (via Southern or dot blots) or for the construction of chromosome-enriched libraries. With the use of additional probes, including anticentromeric antibodies [40,46,71,109] and repetitive DNA chromosome-specific probes [177,451,465,469], the speed and resolution with which variations in chromosome structure and sequence content can be studied by flow cytometry should improve.

GENERAL CRITERIA FOR PROBE SELECTION

There are a large number of fluorescent dyes that bind to DNA, differing in spectral properties, modes of binding, and interaction with other biological components of cells or chromosomes. Choosing among them depends on the biological material that is being analyzed, the properties that are to be measured, and the characteristics of the measurement system. A consideration of the cytochemical characteristics of DNA probes should facilitate probe selection and provide insight into the biochemical factors that could affect the fluorescence signals produced by each probe.

One important criterion for selection of probes is their specificity for DNA. Solution studies have shown that many DNA dyes also form fluorescent complexes with RNA. For intercalating dyes such as the phenanthridines this interaction in cells is primarily with double-stranded hairpin regions in RNA, yielding fluorescence shifts similar to that of DNA [258]. The acridine dyes can complex directly with single-stranded RNA, but the fluorescence properties of these complexes often differ from those of acridine-DNA complexes. For most applications in which DNA in fixed cells or isolated chromosomes is studied, RNase treatment is usually effective in removing RNA. For applications in which both DNA and RNA are to be measured, the bisbenzimidazole and chromomycinone dyes, as well as the red fluorescing dye LL585, have all been shown to be useful for DNA detection, since they have minimal affinity for RNA [161,216,247,443].

Interactions of these dyes with other cellular components have been observed. At low ionic strength, ethidium bromide has been reported to bind with high affinity to hydrophobic regions on cell membranes [143], but such binding is generally not dominant in fixed cells stained at physiological salt concentrations. The cationic dye acridine orange also will bind by electrostatic forces to polyanions, such as polysaccharides or glycosaminoglycans, which can affect the analysis of cell types such as mast cells [35,298,462]. Substantial cytoplasmic fluorescence has also been observed when fixed cells are stained in suspension with quinacrine, but the cell components responsible for this staining have not been clearly identified. Methods for verifying that cellular fluorescence primarily results from dye-DNA complexes include microfluorometric comparison of nuclear versus whole-cell fluorescence [181,182], elimination of cellular fluorescence by DNase treatment [142], or flow cytometric confirmation that G2-phase cells have twice the fluorescence intensity of G1-phase cells [262].

The stoichiometry of dye-DNA complexes in cytological samples is determined by the dye and cell concentrations, the affinity constant between the dye and DNA, solvent conditions, and the presence of secondary modes of binding between the dye and DNA. In general, optimum cell to cell uniformity is obtained when the cells are suspended at equilibrium with a dye concentration that is sufficiently high to saturate the primary binding sites on the DNA, so that subtle differences among cells in affinity for dye do not affect the observed fluorescence intensities. Since dye dissociation constants typically range from 10^{-4} to 10^{-7} M^{-1}, near stoichiometric binding can be expected for cells at a concentration of 10^7/ml (nucleotide concentrations of 2×10^{-4} M). For bacteria, however, 10^7 organisms/ml corresponds to approximately 10^{-7} M nucleotides, so that a large molar excess of dye is often required to saturate the primary binding sites.

Determination of the equilibrium concentrations of bound and free dye are complicated by secondary modes of binding, since the fluorescence properties of dyes often differ between primary and secondary binding sites. With both Hoechst

33258 and ethidium bromide, the presence of dye in secondary binding sites can reduce the fluorescence quantum yield and lifetime of dyes bound to primary sites [43,236,257]. Since most secondary dye binding modes result from relatively low-affinity electrostatic interactions with DNA phosphate groups, they can be reduced by the use of high ionic strength and low excess dye concentrations. Dye and cell concentrations should be carefully controlled for studies that depend on measurements of dye bound in two different modes, that is, DNA and RNA determinations with acridine orange [418], particularly since secondary binding of this dye is highly cooperative, further complicating the equilibria [192] (see Chapter 15, this volume).

Major additional criteria for dye selection are that the optical properties of the dye be compatible with the wavelength and sensitivity limitations of the measuring system, and that the emission spectrum be separable from that of other fluorescent labels if used for multiparameter measurements. Existing DNA probes provide a wide range of excitation and emission wavelengths, which allow considerable flexibility in combining these probes with other fluorescent labels [386]. Examples of useful probes (to be detailed subsequently) with different spectral properties are the following: bisbenzimidazole dyes or related compounds (UV excitation, blue emission); the acridine and chromomycinone dyes (blue excitation, green-yellow emission); phenanthridine and anthracycline dyes, and LL585 (green excitation, red emission). For a given measurement system, the fluorescence intensity is proportional to the product of the dye extinction coefficient at the excitation wavelength, the fluorescence quantum yield, and the number of dye binding sites per nucleotide on DNA. Calculations of the expected signal intensity, using a typical laser-based flow cytometer, show that the signal varies widely among dyes, with Hoechst 33258, for example, yielding more than 20 times the signal of chromomycin A_3 per unit DNA [182]. Since many reports indicate comparable resolution is obtained for mammalian cells stained with these two dyes, signal intensity does not appear to make a significant contribution to measurement precision with whole cells. However, for particles with a small DNA content, such as chromosomes or bacteria, statistical fluctuations in the fluorescence photon flux can make a significant contribution to measurement resolution [341].

Dye extinction coefficient also plays a role in determining both the linearity of fluorescence with DNA content and the sensitivity of dye fluorescence to environmental factors. For Feulgen stains such as acriflavin and pararosanaline, high extinction and large numbers of dyes bound per nucleotide result in substantial absorbance at the excitation wavelength so that fluorescence is reduced by inner filter effects, which vary depending on the size of the cell [139,408]. Inner filter effects are generally negligible for the commonly used noncovalently bound dyes [199,228]. For a given fluorescence quantum yield, the fluorescence lifetime of a dye is often inversely correlated with dye extinction coefficient [334]. This approximation of course is sensitive to prefluorescent relaxation of excited dyes to an energetically lower state from which emission occurs. Still, low extinction dyes are thus generally well suited for studies of rotational depolarization or solvent quenching, while high extinction dyes, with shorter emission lifetimes, are less sensitive to these environmental factors. One additional optical characteristic that can affect the utility of dyes is the relative quantum yield of bound versus free dye. Background fluorescence due to free dye in the solution or dye nonspecifically absorbed to non-DNA components will clearly be more of a problem for dyes that have reduced quantum yield when bound to total cellular DNA (i.e., quinacrine [233] or adriamycin [422]) than for dyes with quantum yield enhancements when bound to virtually any natural DNA (i.e., Hoechst 33258 and ethidium bromide [4,236]). Other characteristics that are important for probe selection are the sensitivity of the dyes to DNA composition, structure, and the presence of chromosomal proteins, as well as the uptake and toxicity of dyes for studies on living cells.

Solution studies have shown that the binding or fluorescence of many stains is affected by the average base composition of the DNA. Initial studies that utilized bacterial DNAs differing in base composition or synthetic polynucleotides of known base composition determined that there are two general mechanisms responsible for dye-base specificity. Dyes such as Hoechst 33258 and chromomycin A_3 preferentially bind to DNA regions that are rich in A-T or G-C base pairs [18,236,314], while quinacrine and the anthracycline dyes have little base preference in binding but their fluorescence quantum yield varies with base composition [184,233]. More recently, reports of 1) electrophoretic mobility shifts of DNA restriction fragments in the presence of dyes [316], 2) dye-induced shifts in DNA buoyant density [159], and 3) footprinting studies of the specific sequences protected from cleavage by bound dye [291,434,435] have shown that for the first class of dyes mentioned, that is, bisbenzimidazoles or chromomycinones, specific sequences typically 3–6 base pairs long are required for dye binding. Probe fluorescence can also be affected by the presence of modified nucleotides in DNA. Incorporated bromodeoxyuridine has been shown to both enhance or quench the fluorescence of different dyes, as will be discussed later (see Chapter 23, this volume). Less is known about the sensitivity of probes to naturally occurring modifications, such as methylated bases. Further studies of dye interactions with modified bases would be very useful because of the importance of methylation in gene regulation in eukaryotes [27] and in the restriction-modification systems in microorganisms [5].

As mentioned in the previous section, DNA conformation and the presence of chromosomal proteins can also affect the fluorescence from DNA probes. While double-stranded nucleic acid regions provide the primary binding site for most probes, dyes such as acridine orange have been used to label single- as well as double-stranded DNA [89,356] (see also Chapter 15, this volume). Conformational factors, such as the presence of Z-DNA or superhelical DNA, may also affect dye binding, particularly for intercalating dyes that are sensitive to topological constraints. The presence of chromosomal proteins generally leads to a reduction in dye binding sites [41], and these proteins can also affect dye binding affinities [226].

Dye transport through cell membranes and dye toxicity affect the utility of DNA probes for studies involving living cells. Since propidium iodide is not appreciably taken up by living cells, this dye has been useful in discriminating live from dead cells [88,168]. Certain bisbenzimidazole dyes, such as Hoechst 33342, in contrast are taken up by living cells [238] and have become important probes for vital staining of cells [6,125]. Toxicity to cells can result from both dye uptake and photochemical damage occurring during flow analysis. Both problems can be reduced by low stain concentrations and reduced illumination intensity.

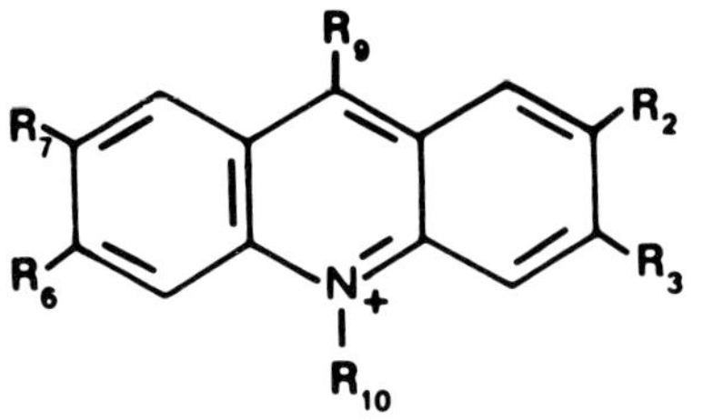

	R_2	R_3	R_6	R_7	R_9	R_{10}
ACRIDINE ORANGE	H	$(CH_3)_2N-$	$(CH_3)_2N-$	H	H	H
PROFLAVINE	H	H_2N-	H_2N-	H	H	H
ACRIFLAVINE	H	H_2N-	H_2N-	H	H	CH_3-
QUINACRINE	H	Cl–	H	CH_3O-	HN-X	H
2,7-DI-T-BUTYL-PROFLAVINE	Y–	H_2N-	H_2N-	Y–	H	H

$X = -CH(CH_3)(CH_2)_3N(C_2H_5)_2$ $\quad Y = (CH_3)_3C-$

Fig. 1. The chemical structures of the acridine dyes: acridine orange, proflavine, acriflavine, quinacrine, and 2,7-di-t-butylproflavine. (Reproduced from ref. [243], with permission of the publisher.)

EXAMPLES AND SPECIFICITIES OF NUCLEIC ACID PROBES POTENTIALLY USEFUL IN FLOW CYTOMETRY

In the previous edition of this book, an entire chapter was devoted to DNA probes that might be utilized in flow cytometry [243]. Many of the basic considerations of that chapter remain unchanged. However, a subset of the dyes described has proved or should soon prove to be of greater utility in flow cytometry than the others, in large part because they satisfy the criteria delineated earlier in this chapter. Also, methods for measuring RNA in the presence of DNA have evolved, some utilizing combinations of probes. It is on the more useful probes that the greatest emphasis of this section will be given.

Intrinsic Probes

Intrinsic probes of DNA content and synthesis, such as formycin (7-amino-3-(beta-D-ribofuranosyl) (pyrazolo-(4,3-d)pyrimidine) [444,445] and 7-deazanebularin (pyrollo(4,3-d)-pyrimidine-3-beta-D-ribofuranoside) [39,156,446] are limited by low wavelength emission, low quantum yield, and toxicity. Similarly, the intrinsic fluorescent properties of 5-methyldeoxycytidine [138] have been of little use in studying cells, although they might eventually find use in HPLC detection of DNA methylation in studies of gene control [29,68,350,466,470]. Also, while additional variations on ethenyl derivatives of adenine and cytidine, produced, for example, by reaction with chloracetaldehyde, continue to be developed [13,205,256,379], none thus far have overcome the disruptive effects of these derivatives on DNA base pairing. One possible use of the agents forming these derivatives, for example, chloracetaldehyde or bromoacetaldehyde [453], is in exploiting this latter feature to detect regions of chromatin that are single-stranded or deformed so as to lead to a single-stranded conformation. Similar use of these chemical agents in principle could be in flow cytometry.

Extrinsic Fluorescent Probe of DNA or RNA

As in the previous edition of this volume, this chapter will not address fluorescent Feulgen reagents, which are considered in detail elsewhere [141,262], but will focus on those families of extrinsic nucleic acid-binding dyes that continue to be of high utility.

Acridine Dyes. Acridine dyes continue to be widely used as fluorescent stains of nucleic acids. A rich variety of electronic properties have been ascribed to different types of acridine-DNA complexes. However, precise interpretation of the structural basis underlying measured optical properties remains limited by an incomplete understanding of the number and molecular features of dye-binding interactions.

Acridine orange (3,6-dimethylaminoacridine) (Fig. 1) is perhaps the most popular acridine dye in cytology, although its interactions are among the most complex [28,89,356] (see Chapter 15, this volume). It possesses both basic and aromatic features, and both electrostatic and hydrophobic forces influence its binding to DNA. Dilute neutral aqueous solutions of the free dye exhibit an absorption maximum at 492 nm [37] and green fluorescence. These are replaced at higher dye concentrations by a new band (near 450 nm) and red fluorescence, presumed to reflect dye-dye interactions.

The optical properties of acridine orange-polynucleotide complexes exhibit marked saturation dependence. The complex predominating at low dye/phosphate ratios possesses an absorption maximum near that of dilute free dye and exhibits green fluorescence [28,452]. This complex has been hypothesized to involve isolated [35], intercalated [397], dye molecules. At higher dye/phosphate ratios, the absorption maximum shifts to lower wavelengths and red fluorescence appears. Previous interpretations of the general feature of this form of the dye spectrum, however attained, considered that it reflected multiple acridine dye molecules in close proximity. Recently, Kapuscinski and associates [194,195] have suggested a different model for this red fluorescence, which better explains variations between dye concentrations and dye-polymer spectra, increased DNA accessibility, and the occurrence of dye-polymer precipitates [92]. It is argued that this spectrum can be derived, at least in many cases, from a dye-nucleic acid precipitate that also tends to denature nucleic acids [192–195]. Perhaps the greatest difference in the Kapuscinski-Darzynkiewicz model compared with earlier models is that the high dye/DNA structure involves dye intercalation (i.e., dye-base) rather than direct dye-dye interactions.

The structure of the low dye/phosphate complex was already convincingly associated with intercalative binding. Lerman [260,261] showed that binding of quinacrine, proflavine, or acridine yellow to linear DNA molecules (at low dye/phosphate ratios) increased the apparent solution viscosity. Moreover, fluorescence polarization measurements on flow-oriented dye-DNA complexes indicated that the in-plane acridine transitions were orthogonal to the flow direction. Both of the above observations are consistent with intercalation. Other phenomena supportive of intercalative binding include alterations in the supercoiling of closed circular DNA molecules, which are detectable by sedimentation velocity measurements [448] and a parallel alignment of dye and DNA base planes, the latter discernible from electric birefringence measurements [170]. Neville and Davies [318] obtained x-ray fiber diffraction patterns from DNA-acridine orange complexes, which indicated that at least some of the dye was bound by intercalation, while Rich and associates [380] as well as Sobell et al. [392] demonstrated that dyes such as 9-aminoacridine intercalated between the base pairs formed in crystals, for example, of adenylyl-3′-5′-uridine. As mentioned above, additional mode(s) of dye binding, suggested by optical measurements to prevail at dye/phosphate ratios greater than about 0.2 [28,35,36,356,397] are less well characterized and contribute to some of the difficulties in interpreting experiments using acridine orange.

Indirect evidence that secondary binding of acridine orange as well as proflavine involved the major groove of the DNA helix derives from observations of Galley and Purkey [131] that the excited-state triplet lifetime of such dye when bound to poly(dA-BrdU) is less than that when bound to poly(dA-dT). Consistent with these results, which are thought to reflect excited-state interactions between the dye and Br atoms, the fluorescence of cytological chromosome preparations stained with acridine orange is quenched by substitution of BrdU for dT in chromosomal DNA [90, 105,123,196,237]. The observation of red dye-nucleic acid precipitates at high saturations, with emission sensitive to DNA base composition, further attests to the role of red fluorescence in aggregates between denatured stretches of DNA fully intercalated with dye. This should be testable by optical activity and crystallographic experiments [192,194].

The binding affinity [312], fluorescence wavelength [280], and fluorescence efficiency [356,454] of acridine and its derivatives are relatively insensitive to the DNA A + T/G + C ratio, although some preference for 5′-pyrimidine-purine sites in dinucleotides or short oligonucleotides doublets has been observed [392]. However, the maximum number of DNA sites capable of binding acridine orange is reduced by chromosomal proteins [126]. Changes in nuclear morphology related to genetic activity (e.g., the response to phytohemagglutinin stimulation [357,358]) or chick erythrocyte maturation [48] are associated with an increase or decrease (respectively) in acridine orange fluorescence intensity, which could reflect variations in dye binding and/or quantum efficiency.

Variations in acridine substituent groups that cause relatively small spectral shifts may markedly alter both the binding characteristics and base composition dependence of dye fluorescence. Unless specifically blocked (e.g., by bulky ring substituents [311]), the primary dye binding mode is intercalation, but the relative importance of secondary binding may vary widely. Intercalative acridine binding exhibits relatively little base composition specificity [233,312], although both the affinity and stoichiometry of binding are reduced by chromosomal proteins [33,86,126,234]. The fluorescence intensity of some acridine dyes, for example, quinacrine (Fig. 1) increases with the A + T/G + C ratio of the DNA to which they are bound [58,64,232,233,239,302,321,381, 382,423,454–456], most likely due to increases in bound dye fluorescence quantum yield. Quinacrine has also been found to exhibit specificity for 5′-pyrimidine-purine doublets [12]. Unfortunately, some of these dyes, such as quinacrine, which are very useful in identifying metaphase chromosomes [52–55,111], have proved less useful in flow studies because of a decrease in overall fluorescence when binding to most DNA and high fluorescence when bound to RNA or even membranes [292]. Other new substituted acridines [339] may, with further study, prove to have specific uses in flow cytometry.

The binding of proflavine (3,6-diaminoacridine) (Fig. 1) to nucleic acids has been extensively studied by optical and viscometric methods, which provide evidence for intercalative binding [374,448]. The high-wavelength absorption band of proflavine shifts from 445 nm to 460 nm upon binding to poly(dA-dT) [26] but changes little upon binding to poly(dG)-poly(dC). Binding to either polymer reduces the dye absorption amplitude in this region. Fluorescence of the poly(dA-dT) complex is much greater than that of the poly(dG)-poly(dC) complex [26,331,415], and thermal destabilization of A-T base pairs in calf thymus DNA results in a redistribution of dye from fluorescent to nonfluorescent, presumably G-C-rich binding sites. Binding of proflavine to DNA is also associated with an increase in fluorescence polarization and a reduction in the susceptibility to quenching by added iodide ion [26]. This latter property, which is shared by other dyes capable of binding to DNA, such as quinacrine [415] and 33258 Hoechst [236], constitutes in principle a convenient way of selectively suppressing free dye fluorescence in cell or chromosome suspensions. Deubel and Leng [99] characterized the interaction of proflavine with double-helical polyribonucleotides. As with the deoxypolymers, G-C base pairs quenched proflavine fluorescence, and dye bound to poly(rI)-poly(BrrC) fluoresced less efficiently than that bound to the unbrominated polymer.

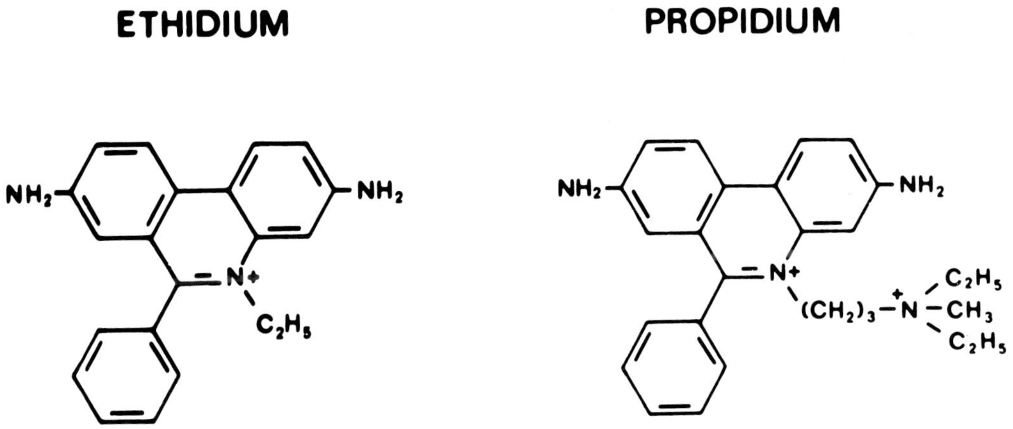

ETHIDIUM **PROPIDIUM**

Fig. 2. The chemical structures of ethidium and propidium. (Reproduced from ref. [243], with permission of the publisher.)

The properties of proflavine-DNA complexes have been characterized by fluorescence and circular dichroism measurements. The ellipticity of proflavine depends markedly on the dye/phosphate ratio [86]. Since this dependence is less for chromatin than for DNA, it was suggested that chromosomal proteins might restrict the structural alterations induced in DNA by intercalated dye. Such a curtailment of dye-induced conformation changes would provide an indirect mechanism by which chromosomal proteins could reduce the apparent binding affinity of an intercalating dye.

The discovery of Caspersson and associates that quinacrine (Fig. 1) and related alkylating derivatives produced a banded fluorescence pattern when used to stain metaphase chromosomes [52–55] stimulated interest in the optical properties of quinacrine and its complexes with DNA or chromatin. At low dye/DNA ratios, quinacrine intercalates with DNA [261]. As with other intercalating acridine derivatives, the isotherms of quinacrine-DNA binding are compatible with a one dye/4 phosphate limiting stoichiometry [28,64,233]. Curvature in standard reciprocal binding plots is expected because of the excluded site-binding interactions [38,82,230,296,475] and does not of itself establish the existence of secondary binding modes.

The intensity of quinacrine fluorescence increases with DNA A + T content [64,66,111,233,331,454–456], but quinacrine binding affinity shows little or no dependence on DNA base composition [233,312], except for the previously mentioned pyr-pur preferences [12]. Thus, the marked increase in quinacrine fluorescence in the presence of A-T-rich DNA almost certainly reflects a change in dye fluorescence quantum yield. In complexes of quinacrine with natural DNA, an increasing dye/phosphate ratio is associated with reduced fluorescence intensity, polarization, and lifetime [233] that has been associated with energy transfer between bound quinacrine molecules, a conclusion that has received subsequent support [12,244,245,365]. Chromosomal proteins decrease both the affinity and stoichiometry of quinacrine binding to DNA [33,233]. Moser et al. [309] have shown that the cytologic appearance of quinacrine-stained nuclei reflects chromatin condensation, and Therman and associates [413] have detected dull fluorescence, which they suggest to be a center involved in human X-chromosome inactivation. While quinacrine could in principle also prove useful for sorting similarly sized chromosomes according to A + T composition, it has a high background fluorescence, low extinction coefficient, and low quantum yield that, taken together, have reduced its use for this purpose relative to that of other dyes (e.g., bisbenzimidazoles) with appreciable but differently based A-T specificity.

Phenanthridinium Dyes. Ethidium bromide (2,7-diamino-9-phenyl-10-ethylphenanthridinium bromide) (Fig. 2) and related phenanthridinium dyes illustrate another pattern of dye-nucleic acid interactions. The primary, intercalative mode of dye binding depends on double-helical polynucleotide structure. Such binding, which is associated with a shift in dye absorption from 480 nm to 520 nm and a marked increase in fluorescence efficiency [134,257,447], can occur with either RNA or DNA. Ethidium-induced unwinding of DNA has been estimated by fluorescence energy transfer measurements [135], hydrodynamic analyses [373,448], and electron microscopy [269]. In addition to intercalative binding, ethidium can undergo an external electrostatic interaction with double helices. This secondary dye binding, which tends to suppress overall fluorescence, can be reduced by working in solutions of moderate ionic strength [257]. Unlike acridine orange, this secondary binding mode of ethidium does not cause DNA denaturation or precipitation [194,195], and it may indeed reflect dye-dye interactions.

Intercalative ethidium bromide depends little on nucleic acid base composition. Mueller and Crothers [312] obtained binding competition data with natural DNA samples suggesting a weak G + C preference for this dye. Similarly, there are few data to indicate a marked variation in dye fluorescence quantum yield due to DNA base composition. Chromosomal proteins reduce the apparent accessibility of ethidium to DNA [3,41,226,295] by a mechanism that may reflect not only physical coverage of DNA by protein but also protein-induced conformational changes in DNA [41]. Propidium diiodide (3,8-diamino-5-diethylmethylamino-propyl-6-phenylphenanthridium diiodide) (Fig. 2), a dye structurally related to ethidium, was originally used because of its ability to shift differentially the buoyant density of linear and closed circular DNA [173]. This dye, which absorbs and fluoresces at slightly higher wavelengths than ethidium when complexed with DNA (a feature that is not desirable when also utilizing phycoerythrin probes), has become a popular DNA stain in flow fluorometry (to be detailed in the subsequent section).

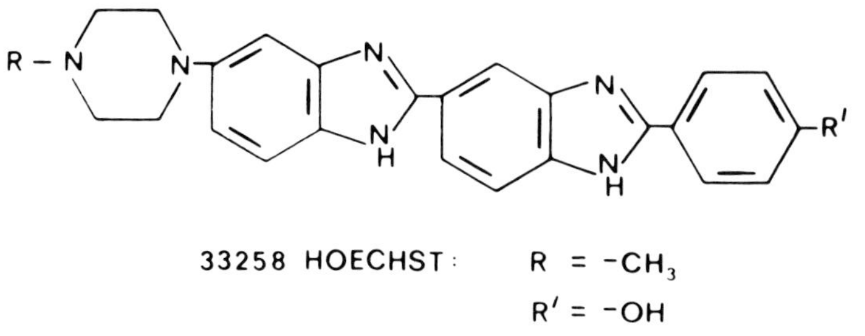

33258 HOECHST: $R = -CH_3$

$R' = -OH$

Fig. 3. Bisbenzimidazole dye exhibiting BrdU-sensitive fluorescence. (Reproduced from ref. [243] with permission of the publisher.)

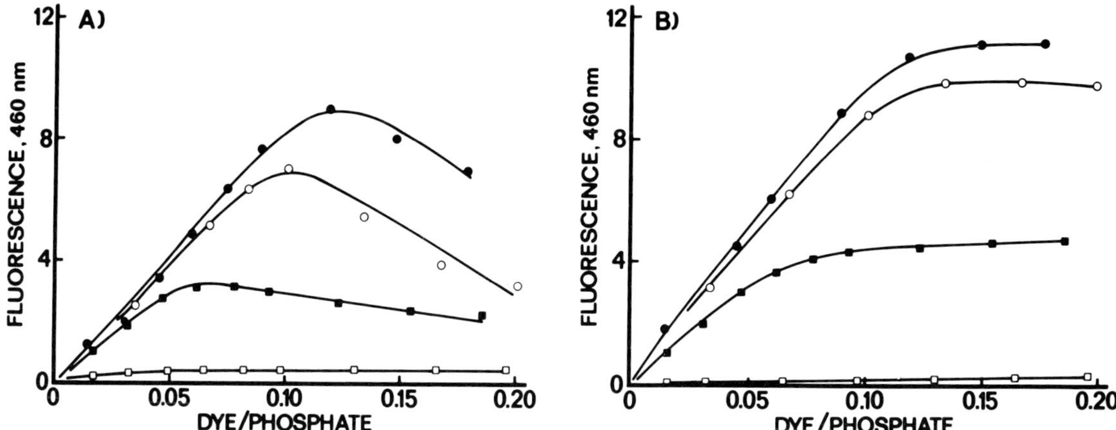

Fig. 4. Fluorescence titration of DNA and polydeoxy nucleotides with 33258 Hoechst. Increments of 33258 Hoechst were added to 3×10^{-5} M solutions of poly(dA)-poly(dT) (●), poly(dA-dT) (○), calf thymus DNA (■), or poly(dG-dC) (□) in 0.005 M Hepes, pH 7.0, with 0.01 M NaCl **(A)** or 0.40 M NaCl **(B)**. The fluorescence amplitude, in arbitrary units, at 460 nm (at or near the peak of the uncorrected fluorescence emission spectrum) is plotted versus the amount of added dye/phosphate. The baseline for these measurements refers to the buffer alone. The free dye fluorescence peaks at about 505 nm and contributes negligibly to all amplitudes in this figure except those determined in the presence of poly(dG-dC). (Reproduced from ref. [238], with permission of the publisher.)

Bisbenzimidazole Dyes. The bisbenzimidazole dye 33258 Hoechst (Fig. 3) is one of several A-T-specific molecules that bind tightly to DNA without intercalation [34,236, 246,303,314]. Addition of this dye to closed circular DNA molecules does not induce major changes in hydrodynamic properties [314], and electrooptical studies [34] indicate that the major in-plane transitions of the dye are not polarized perpendicular to the DNA helix axis. The latter measurements are consistent with binding of dye along the grooves of the helix, a possibility previously suggested by measurements of BrdU-dependent quenching of bound dye fluorescence [231].

The free dye possesses a long-wavelength absorption maximum near 340 nm at pH 7, with fluorescence emission peaking slightly above 500 nm [65,236,270,457]. Upon binding to DNA, the dye absorption undergoes a bathochromic shift of approximately 15 nm, and the wavelength of maximum fluorescence decreases to near 470 nm. Fluorescence of the free dye is very weak, while emission intensity of dye-DNA complexes increases with DNA A + T content (Fig. 4). The dependence of 33258 Hoechst DNA optical properties on dye-DNA ratios (Fig. 5) [238] is another variable whose use in flow cytometry is being exploited [400].

The absorption, fluorescence, and circular dichroic properties of 33258 Hoechst DNA complexes depend both on the DNA base composition and on the dye/DNA ratio. Spectroscopic [236] and electrooptic data [34] are consistent with the existence of two modes of nonintercalative dye binding. A strong binding mode with marked A-T base preference is associated with intense fluorescence and a circular dichroism extremum near 360 nm. Binding saturation occurs at one dye per three to four A-T base pairs. A second binding mode, which can be selectively suppressed by increasing the solution ionic strength, exhibits very little fluorescence (confirmed by Cowell and Franks [70] and by Stokke and Steen [400]) and relatively weak ellipticity. This secondary binding, which is more important at high dye/phosphate ratios, is associated with a small hypsochromic shift in the wavelength of the absorption band located near 350 nm. The A-T specificity of 33258 Hoechst has proved of use in isolating human Y chromosomes [114], and it may bias analyses of DNA synthesis rates from flow histograms [24,95,367,371] because later replicating DNA is slightly more A-T-rich than earlier replicating DNA [208].

33258 Hoechst and related compounds were synthesized by Dr. H. Loewe [270] during a systematic search for antifilariasis drugs. Prolonged exposure of cells produced a significant increase in chromosome aberrations [163], and 33258 Hoechst has not been used for medicinal purposes. However, in the course of its initial evaluation, the dye was found to fluoresce, and it was observed that live cells could be stained simply by adding the dye directly to cell cultures

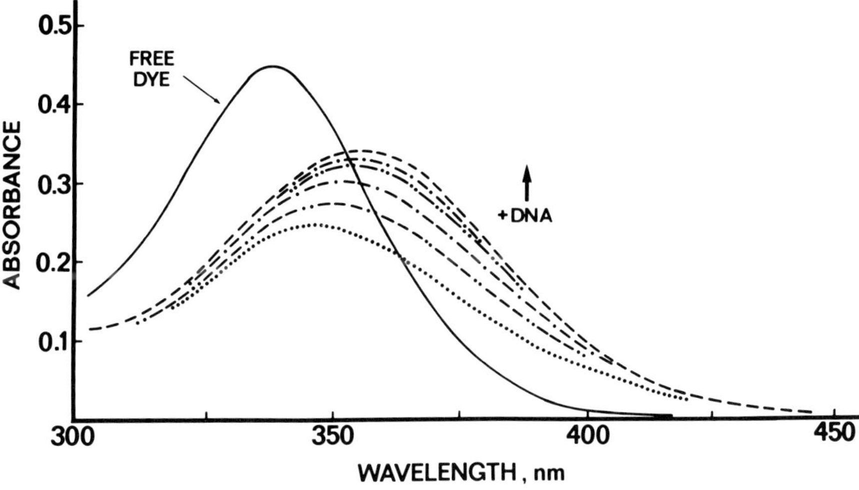

Fig. 5. Effect of calf thymus DNA on the absorption of 33258 Hoechst. The measured absorption spectrum of 1.06×10^{-5} M 33258 Hoechst in 0.01 M NaCl, 0.05 M Hepes, pH 7.0, in a 1-cm pathlength cuvette is plotted versus wavelength for the free dye (—) and after the addition of calf thymus DNA to a DNA/dye ratio of 3.4 (···), 6.7 (··—), 10 (·–·–·–), 13 (··—··—), 17 (·—·—), and 24 (—) (Data from [238]).

[163,221]. Hilwig and Gropp [164] subsequently demonstrated that 33258 Hoechst stained cytologic preparations of mouse chromosomes, with especially bright fluorescence coinciding with the centromeric locations of mouse satellite DNA. These same authors also demonstrated that addition of high concentrations of dye to mouse cells before harvest for chromosome analysis selectively decondensed the centromeric regions of chromosomes [165]. This type of experiment, with or without utilization of a BrdU-dependent increase in dye binding affinity has now been repeated in many different variations [236,285–287,293]. Both the fluorescence and chromatin decondensation regional specificity of 33258 Hoechst, as well as that of the related dye DAPI [347] are compatible with the known A-T binding preference of the dye.

Additional interest in 33258 Hoechst has centered around its use as a fluorescent probe of DNA synthesis. Dye bound to DNA or chromatin exhibits reduced fluorescence if the base analogue 5-bromodeoxyuridine (BrdU) is substituted for thymidine [231,236]. Such an effect presumably reflects dye binding in the DNA major groove near the bromine atom, which in principle can quench fluorescence because of its high atomic number and/or polarizability [131]. The fluorescence at pH 7 of the complex of 33258 Hoechst with poly(dA-BrdU) is one fourth that with poly(dA-dT) at low (0.01) ionic strength, and one tenth that of the poly(dA-dT) complex at higher (0.4) ionic strength [236]. Fluorescence quenching is not observed if the pH is reduced to 4 [234,236]. Dye fluorescence is also suppressed when bound to polynucleotides in which IdU is substituted for dT [237]. Moreover, at 0.4 ionic strength, at which secondary dye binding is largely suppressed, the fluorescence quenching is associated with a marked reduction in fluorescence lifetime [236,239]. The last observation is consistent with the hypothesis that much of the quenching takes place at the first excited singlet state and is thus compatible with (although it does not constitute absolute proof for) an intersystem crossing quenching mechanism. 33258 Hoechst binds weakly to RNA, especially at moderate ionic strength, and exhibits less fluorescence than when bound to DNA [238]. The dye apparently becomes a better stain for RNA in fixed cells if the ambient pH is lowered [166].

A number of other bisbenzimidazole dyes, in addition to 33258 Hoechst, have been prepared by Dr. Loewe [270]. Many of these have been examined and found to exhibit absorption spectra and BrdU-dependent fluorescence quenching similar to that of 33258 Hoechst [238,240]. Some have larger alkyl groups as ring amino substituents, while others have an alkoxy rather than a hydroxy aromatic ring substituent. As anticipated [238], these changes appear to increase membrane permeability of the dyes [6], and the latter substitution results in a loss of alkaline pK_a, which can be associated with marked changes in dye fluorescence.

Additional A-T Base-Specific Dyes. It has been noted [101] that the chromosome staining properties of 33258 Hoechst are mimicked by a nonintercalating proflavine derivative, 2,7-di-t-butylproflavine [311]. Another dye, DAPI (4′,6-diamidino-2-phenylindole) [363] (Fig. 6) possesses absorption and fluorescence spectra resembling those of 33258 Hoechst [267,268]. Like 33258 Hoechst, DAPI has been used as a stain for mycoplasma [363]. Neither tert-butylproflavine nor DAPI exhibit less fluorescence when bound (at pH 7) to poly(dA-BrdU) than when bound to poly(dA-dT), although the fluorescence of DAPI is reduced by BrdU at pH 11 [267,268]. A procedure utilizing DAPI to detect DNA by fluorescence in agarose gels, which apparently overcomes the high background fluorescence seen with 33258 Hoechst [248], has been published by Kapuscinski and Yanagi [191].

Two nonfluorescent dyes, netropsin and distamycin A (Fig. 7), share the nonintercalative A-T-specific binding characteristics of 33258 Hoechst [210,211,276,314,450,477]. However, the binding of netropsin, unlike that of 33258 Hoechst [236], is unaffected by BrdU [450], and it has been suggested that the dye binds in the minor groove of DNA. All three dyes can act as A + T-specific ligands for reducing DNA buoyant density in CsCl [314]. The binding of netropsin and distamycin A as well as other derivatives has been studied by NMR [129], optical rotation [277,478], re-

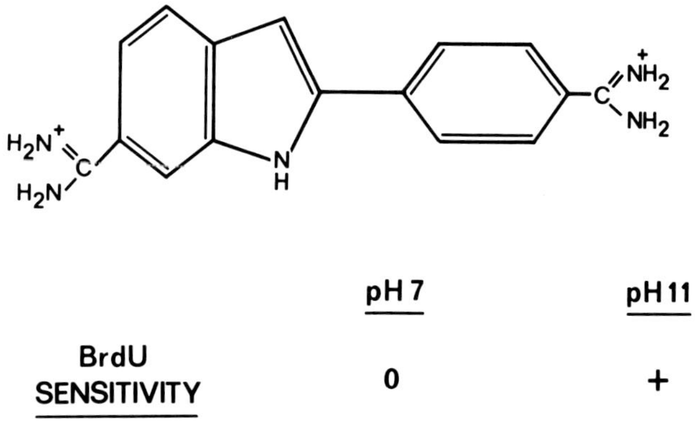

	pH 7	pH 11
BrdU SENSITIVITY	0	+

Fig. 6. The chemical structure of 4′,6-diamidino-2-phenylindole (DAPI) [373a].

striction enzyme protection (footprinting) [85,121,149,433], and crystallographic procedures [206]. Binding of dye in the minor groove is confirmed, with slight differences in protection seen for each dye. Preference for A-T base pairs (approximately 4) may be driven by the greater water of hydration of A-T versus G-C base pairs [425,474], a mechanism previously invoked to explain other A-T-specific binding (e.g., of tetramethylammonium ion) [230].

In the case of distamycin A, intermediate H bonds are formed between adjacent chains in the DNA minor groove. Interestingly, steric hindrance of the CH_3 of the pyrolle residues seems to block G-C binding, which Kopka et al. [206] predict would be reversed by substituting an imidazole for the methylpyrolle. The ultimate application of this phenomenon could be the synthesis of distamycin analogues with strings of predetermined methylpyrolles and imidazoles, terminated by a fluorescent dye, that would recognize a desired DNA sequence *without* DNA denaturation and *without* competing with histones, which dominate the major groove [175].

Actinomycins. Actinomycin D (Fig. 8) is a DNA ligand that has been employed more for its binding than for its spectroscopic properties. The dye exhibits an absorption maximum at 440 nm that shifts to approximately 460 nm when complexed with DNA [56,310,443]. However, antinomycin D is essentially nonfluorescent. The hallmark of actinomycin D binding to DNA is its G-C specificity [310,443,460]. The antibiotic does not bind to poly(dA-dT) or to RNA. Loss of the 2-NH_2 group from guanine (e.g., poly(dI-dC) abolishes actinomycin D binding, while the dye binds tightly to a synthetic polydeoxyribonucleotide containing the 2-NH_2 derivative of adenine, 2,6-diaminopurine [56]).

Actinomycin D consists of a phenoxazine ring to which are attached two cyclic pentapeptides [443]. The peptide rings serve to limit dye binding to less than one molecule per six DNA base pairs. Both spectroscopic data on actinomycin-arylsulfonate interactions [310] and crystallographic analyses of actinomycin-dinucleotide complexes [391] indicate that the aromatic ring of the antibiotic intercalates between G-C base pairs. The data on protection of DNA from cleavage by methidium propyl-EDTA Fe(II) [433] show that actinomycin D covers 4–16 bp regions centered around one or more G-C base pairs. The data are compatible with a binding azimuth that places the peptide rings in the DNA minor groove. Actinomycin D binding stoichiometry is restricted by chromosomal proteins [21,203,359], and undergoes a cyclic variation in synchronized HeLa cells, peaking near the G_1-S boundary [336,337].

While actinomycin D is nonfluorescent, a few fluorescent analogues of the drug have been described. Mueller and Crothers reported the preparation and binding properties of a number of actinomycin derivatives, including 7-aminoactinomycin C_2 and C_3 [310]. Subsequently, Modest and Sengupta [306] synthesized 7-aminoactinomycin D (C_1). This derivative exhibits an absorption maximum at 505 nm, which shifts to approximately 550 nm when the dye binds to different types of DNA [140]. The free dye exhibits a fluorescence emission peaking at 657 nm in DNA complexes. Like actinomycin D itself, 7-aminoactinomycin D can bind to DNA in intact cells and is then useful in a number of flow fluorometric experiments that exploit both the base specificity and cell-cycle sensitivity of its interaction with DNA [476]. This dye has also proved especially useful when combined with bisbenzimidazole dyes in energy transfer studies [244,245,365]. The latter permit detection of clusters of G-C base pairs in regions (e.g., Q-dull bands) in which the average G-C content [208] and actinomycin binding [42] show very little variation.

Anthracyclines and Chromomycinones. Ward [443], Kersten [200], and their coworkers, as well as Chaires et al. [57], have summarized the base-specific properties of a number of other DNA-binding antibiotics. Daunomycin (Fig. 9) and chromomycin A_3 (Fig. 10) are polycyclic dyes to which single or multiple sugar groups are attached, respectively. The antineoplastic anthracycline adriamycin is a monohydroxy derivative of daunomycin [266]. Daunomycin alters the sedimentation coefficient of closed circular DNA in a manner consistent with intercalation [448]. This same analysis indicates that the chromomycinone dyes chromomycin A_3 and mithramycin do not intercalate. RNA synthesis inhibition studies indicate that daunomycin can interact with polyribonucleotides, while mithramycin, chromomycin, and olivomycin are more specific for DNA and exhibit a G + C preference [443] (Fig. 11).

Mithramycin and related dyes possess absorption maxima at 335 and 395 nm and fluoresce near 550 nm (or slightly higher, depending on spectral correction [225] when bound to intact cells. Olivomycin, which is structurally similar to mithramycin and chromomycin A_3, has been shown to bind preferentially to G-C-rich DNA and to produce a banded metaphase chromosome pattern that is virtually the converse of that found with quinacrine [430]. It is important to remember with these dyes that they bind to DNA with 1:1 stoichiometry with a bivalent cation, for example, Mg^{2+} [443]. One can either add dye to target plus Mg^{2+} or first add dye to Mg^{2+} [183,429] and then add this to DNA. A 20- to 30-min waiting time is needed for the first approach, though perhaps not for the second. The chromomycin A_3-Mg^{2+} complex bound to DNA appears to be kinetically stable, although with excess Mg^{2+} overall fluorescence of chromomycinone-stained DNA-containing specimens increases after several hours [183,220].

Miscellaneous Heteroaromatic Dyes. A nonfluorescent dye with a complicated, bisintercalative binding to DNA, echinomycin (Fig. 12) [449,255], which seems to migrate to specific sites over intervals of several minutes [121], has some G + C-dependent specificity evident in solution (Fig. 13). From footprinting [435] or crystallographic [426,442] work, this appears to be a CpG specificity, with surrounding A and T base pairs enhancing binding. Echinomycin can be useful

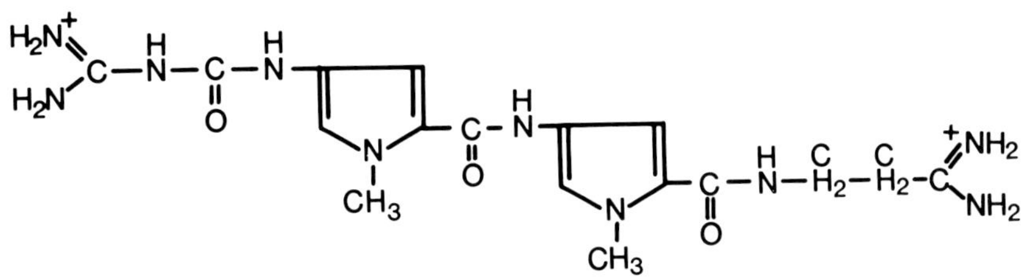

Netropsin

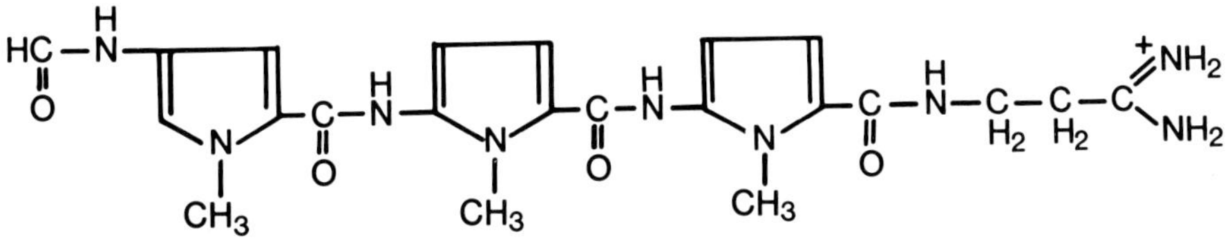

Distamycin A

Fig. 7. The chemical structures of netropsin and distamycin A [477].

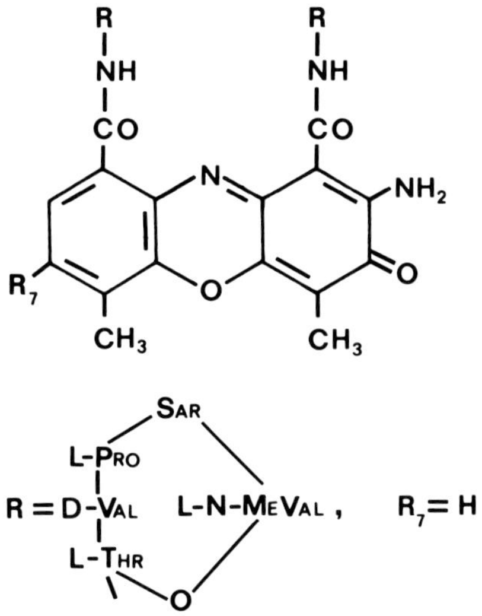

Fig. 8. The chemical structure of actinomycin D (C_1). Actinomycin C_2 and C_3 have different peptide ring substituents (R), while R_7 = NH_2 for 7-aminoactinomycins [243,306,310]. (Reproduced from ref. [243], with permission of the publisher.)

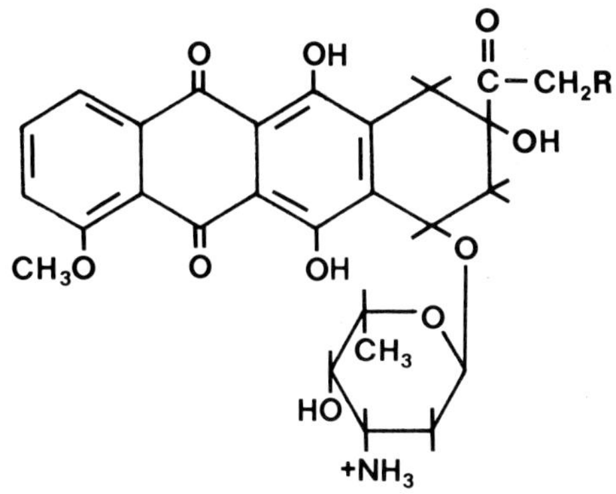

	R
DAUNOMYCIN	H
ADRIAMYCIN	OH

Fig. 9. The chemical structures of the anthracycline dyes daunomycin and adriamycin. (Reproduced from ref. [243], with permission of the publisher.)

with bisbenzimidazole dyes to generate new fluorescent patterns [244,245].

Mueller and associates surveyed the properties of a number of fluorescent and nonfluorescent DNA binding dyes [312,314,315] which were divided into two classes. One group of dyes, which bound by intercalation, exhibited G + C specificity [312,313]. The extent of this specificity increased as the absorption maximum of the dye shifted to longer wavelengths, an observation interpreted as correlating with excited-state polarizabiity [312]. Neutral red exhibited a very high G + C specificity, which was extended to specificity for adjacent G + C base pairs by addition of a phenyl residue [313]. Another G + C-specific DNA ligand, PNR (2-methyl-3-amino-7-dimethylamino-5-phenylphenanzinium cation) was used by Pakroppa and Mueller [315,332] to promote early elution of G + C-rich DNA from hydroxyapatite columns.

The second class of dyes was composed of nonintercalat-

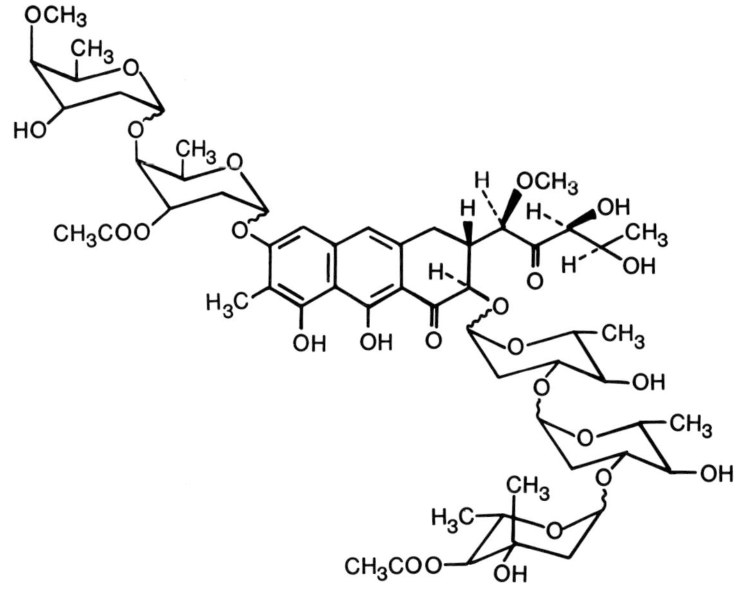

Fig. 10. The chemical structure of chromomycin A_3. (From The Merck Index, 1968, p. 258.)

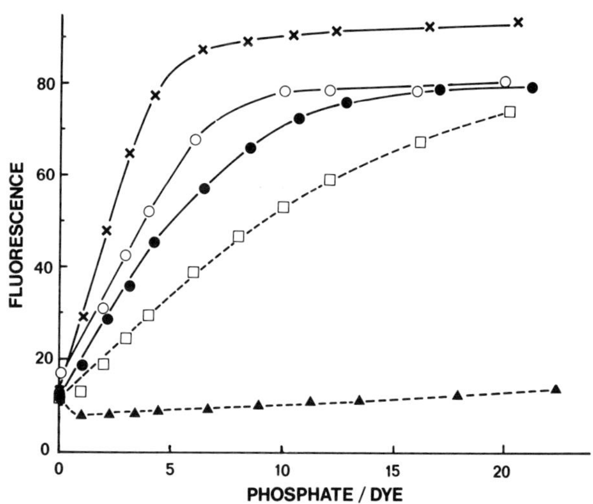

Fig. 11. Effect of different types of DNA on the fluorescence of chromomycin A_3. Aliquots of poly(dG-dC) (X). *M. lysodeikticus* DNA (○), calf thymus DNA (●). *C perfringens* DNA (□) or poly(dA-dT) (▲) were added to 8×10^{-6} M chromomycin A_3 plus 10^{-2} M $MgCl_2$ in 0.15 M NaCl, 0.005 M Hepes, pH 7. The fluorescence amplitude of chromomycin A_3 (545 nm, uncorrected) following excitation at 425 nm is plotted versus DNA phosphate/dye ratio. (Reproduced from ref. [239].)

ing A + T-specific ligands [312]. Included in this group, in addition to 33258 Hoechst, were methyl green, bis(benzamidine) compounds, 2,7-di-t-butylproflavine, crystal violet, and auramine O. The basis of the A + T specificity exhibited by these dyes is unknown. One possible mechanism would exploit the greater hydration of A-T base pairs, which such dyes might recognize [230,425]. Mueller has further considered constructing compound dyes with increased specificities, and has made polyethylene glycol derivatives of these dyes to facilitate electrophoretic separation of DNA segments differing in base content [316].

Nonaromatic Cationic Ligands. Nonaromatic cationic compounds have been used as relatively nonspecific DNA ligands. A prototype of this class is polylysine, which can be tagged with dyes while retaining its binding affinity for DNA [113]. By analogy with unsubstituted polylysine [61] such conjugates probably reflect some feature in chromatin related to DNA accessibiity [232,288]. The utility of such conjugates with intact cells may be restricted both by limited membrane permeabiity of these polycations and by nonspecific electrostatic binding to other cellular components. Quinacrine mustard derivatives of oligolysine [232] or synthetic dimeric quinacrines with a known spacer [431] or other bisacridines [8,9] may prove much better for cell staining. Gabbay and associates developed a set of chromophore-substituted diamines that bind to nucleic acids, apparently in the minor groove of the double helix [116,130]. Similar fluorescent diamines might serve an analogous purpose in cytofluorometry.

Fluorescent lanthanide ions constitute a different class of DNA ligands. The principal transitions of these ions originate from 4f electrons [169], which are relatively shielded from external interactions and are insensitive to bleaching. Many lanthanide ions exhibit fluorescence that can be excited by weak, narrow absorption transitions [335,351], one of Tb^{3+}, for example, coinciding with an argon ion laser line (488 nm) [169]. One can effect a photosensitization of lanthanide fluorescence via organic ligands [108], aromatic amino acids [278], or DNA [117]. Terbium appears particularly useful for detecting G-C base pairs. Appropriately tuned, highly intense laser excitation may overcome problems associated with the low extinction coefficients of the intrinsic transitions, while at the same time exploiting the narrow bandwidth for selective excitation. A potentially useful feature of lanthanide ion–DNA interactions is the ability of chromosomal proteins to restrict ion binding [471]. A potential drawback is the long emission lifetime (because the

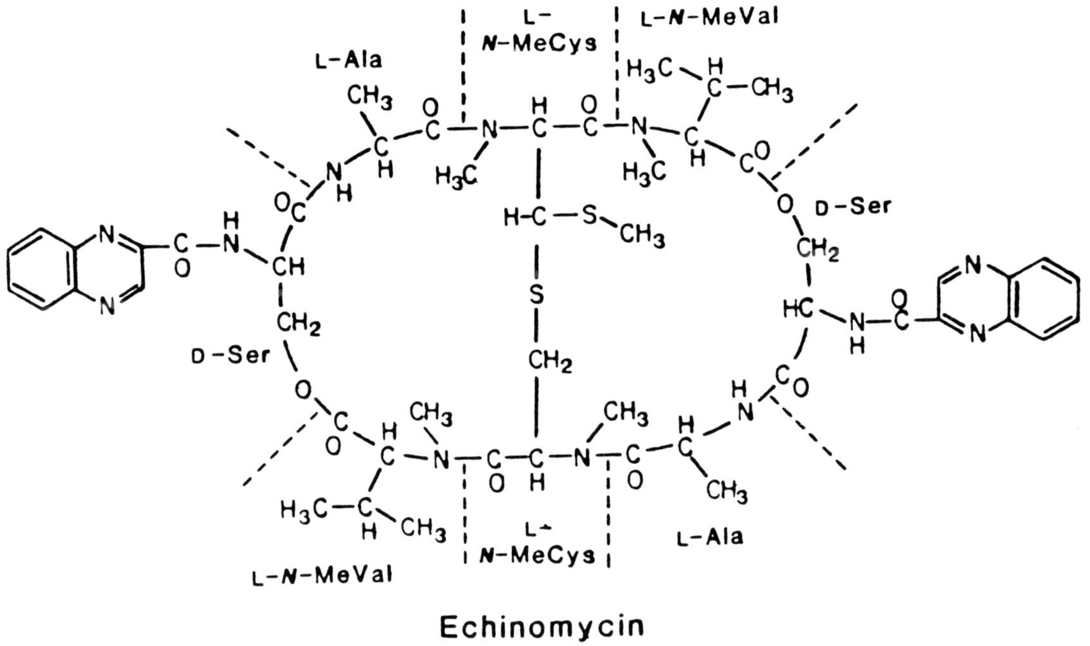

Echinomycin

Fig. 12. The chemical structure of echinomycin [449].

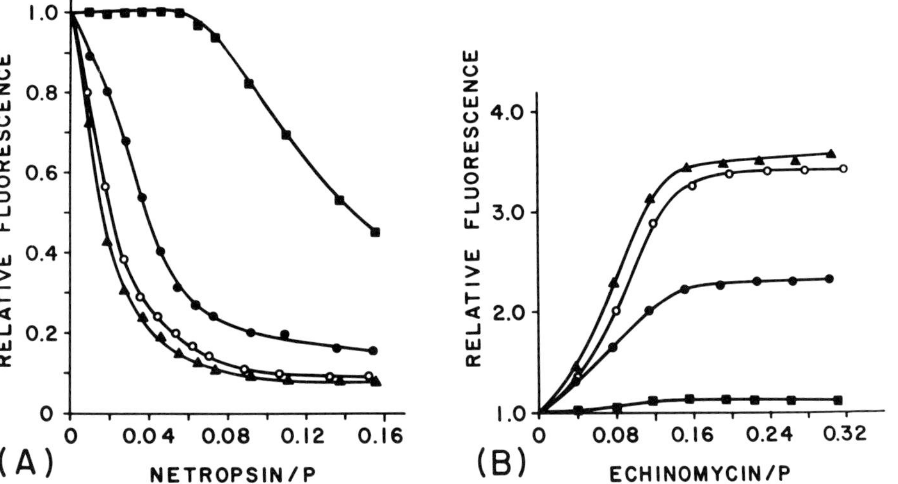

Fig. 13. The dependence on quinacrine-DNA complex fluorescence of the addition of netropsin **(A)** or echinomycin **(B)**. The relative fluorescence of a quinacrine/phosphate = 0.02 complex is plotted versus added netropsin or echinomycin. 0.01 M NaCl, 0.005 M Hepes, pH 7, 2 × 10^{-4} M DNA. DNA:Poly(dA-dT) (■), *C. perfringens* DNA (●), calf thymus DNA (○), *E. coli* DNA (▲). Displacement of quinacrine from A-T- or G-C-rich sites, with associated decrease or increase in fluorescence, respectively, is used to monitor the binding preferences of netropsin or echinomycin. (S.A. Latt, unpublished data.)

transitions are "spin-forbidden"), with the result that little of the emission is detected during dye-laser interception time.

Immunofluorescent Techniques. Immunological techniques constitute perhaps the most flexible approach for creating base-specific fluorescent tags. Antibodies are prepared against antigens consisting of ribonucleosides conjugated as haptens to serum albumin [112,204]. These antibodies can then interact with DNA, which must be in a denatured state. Antibody detection is usually via fluorescein [127] or horseradish peroxidase-conjugated [274] heterospecific antiimmunoglobulin antibody, although direct labeling is also possible. Fluorescent latex spheres [307] attached to antigammaglobulin antibodies might offer increased detection sensitivity (see Chapter 19, this volume).

In addition to serving as probes of DNA denaturation in interphase nuclei [14,264,335], antinucleoside antibodies have been used to stain metaphase chromosomes [279]. Immunological techniques appear to be especially useful as a qualitative method for localizing rare or modified bases generally present in small amounts. A highly successful example is the localization of 5-methylcytosine to certain heterochromatic regions of metaphase chromosomes [274,305]. Antibodies against thymine dimers [375] have also been employed, and detection of specific di- and trinucleotides by similar methods appears feasible. Sawicki et al. [368] prepared antibodies against 5-bromo- and 5-iodouracil, although cross-reactivity with thymine limited their use.

More recently, Gratzner and coworkers have utilized affinity-purified polyclonal antibodies [150–152] and mouse monoclonal antibodies [102,153] for highly specific labeling of 5-bromouracil. These antibodies, localized either by fluorescence or immunoperoxidase activity have been used to detect BrdU incorporation into cytologic material in a manner similar to previous BrdU-fluorochrome and modified Giemsa techniques [342]. Antibodies against 5-bromouracil might also be employed to study RNA synthesis. The immunological approach for detecting nucleic acid synthesis possesses at least two specific advantages: 1) it produces a positive rather than a negative readout of BrdU, enhancing sensitivity, and 2) it can be enzymatically amplified leading to the deposition of electron-dense products when coupled with peroxidase. However, permeability factors as well as the need for DNA denaturation restricts its use with intact cells, and quantitative measurements are subject to the steric interference discussed above. Since the previous edition of this volume, these antibodies have become generally available and used, first with dansyl hydrazine [152], and more precisely with propidium in fixed cells [96,102,308], to permit effective measurement of DNA synthesis at different points in S phase and to simplify quantitation of S-phase cells (see Chapter 23, this volume). A remaining goal is the use of anti-BrdU to detect duplex DNA. Similar studies with anti-AAF, which thus far have been restricted to in situ hybridization, should also be possible [223,412].

APPLICATIONS

Single Dyes, Cells, Chromosomes

Given the biochemical complexities of most biological samples, interpretation of the results from DNA probes often depends on both characterization of dye interactions with pure DNA in solution as well as characterization of dye interactions in situ in cells or chromosomes. Solution studies with purified DNA, and chromatin, as described in the previous section, have been useful for defining the spectroscopic properties of free and bound dye as well as for defining the specificities of dyes for DNA samples differing in primary or secondary structures. Solution studies have also been useful for determining the effects of environmental factors such as ionic strength, pH, and the presence of specific ions on the formation of dye-DNA complexes. Most of the dyes mentioned in the previous section have applications in flow cytometry. A few illustrative applications will be mentioned in this and the following section.

Acridine Dyes. Among the acridine dyes, acridine orange has been the most useful flow cytometric probe.

A seminal application of acridine orange fluorescence was in the attempted differentiation between normal and cancerous human uterine cells. Early work indicated that red cytoplasmic fluorescence was much more pronounced in tumor cells than in normal cells [23]. Differentiation between cell types was improved subsequently by slit scan measurements of the ratio of nuclear and cellular diameters and total nuclear fluorescence at 540 nm [464]. Wheeless and associates have now developed two-dimensional slit scan (at 4 μm effective width) as a further refinement of this approach [459]. Standardized microfluorometric analyses indicated that the nuclei of unfixed malignant cells exposed to high acridine orange concentrations exhibited greater fluorescence at 530 nm than the nuclei of corresponding normal cells [148]. While these observations may reflect total nucleic acid content, they might also be influenced by nucleic acid conformation and associated protein molecules (see Chapters 15 and 16, this volume).

Differential staining of G2 from M-phase cells after ribonuclease and low pH treatment (the mitotic cells more readily increase red fluorescence), and moderate sensitivity to BrdU [89–91,93,237] are additional properties of acridine dyes, which Daryzynkiewicz, Traganos, Melamed, and associates [91] have exploited with great effectiveness. The subdivision of G1 into a stochastic, RNA-poor phase (A) and a deterministic RNA-rich phase (B), which fits with the Smith and Martin [389] model of G1, is especially interesting, as is the correlation of *increased* acid denaturability with *increased* chromatin condensation and quiescence. This somewhat counterintuitive observation may be explained by Kapuscinski's data [192]: if the increased condensation brought denatured DNA strands into greater proximity, increasing potential precipitation and accounting for red fluorescence at high dye concentration.

The cell sorting capability of flow systems provides a unique opportunity to establish the basis for differences in acridine orange fluorescence of various cell types. It should be possible to compare fluorescence intensities with the DNA, RNA, and dye content of sorted cells. Alternative experiments have shown that most if not all DNA or RNA signals can be abolished by appropriate nucleases [83,84]. Exceptions include red-fluorescing glycosaminoglycans [343]. Moreover, cells differentiated by fluorescence might subsequently be examined in culture for proliferative abnormalities. Possible complications for these latter studies include mutagenesis [328] and photosensitization [22], both properties of acridine dyes.

Phenanthridine Dyes. Phenanthridine dyes are useful both for quantitating DNA and for differentiating dead cells from cells with an intact membrane. The DNA staining is limited primarily by the need to permeabilize cells and the need to deal with the ability of these dyes to stain RNA as well as DNA. On the other hand, these dyes can be excited by easily attainable visible light sources (e.g., 488-nm or 514-nm argon ion laser lines), and procedures have been described for permeabilizing cells and minimizing RNA staining. Krishan [212] developed a rapid, useful procedure based on propidium fluorescence for obtaining relative cellular DNA content. Dye is added directly to cells, which are made permeable by being suspended in a hypotonic sodium citrate solution. Salt can be added to suppress nonspecific binding [410] and cells treated with RNase to make the signal specific for DNA. Alternatively, in DNase-treated cells, or in yeast in which RNA >> DNA, the propidium signal can reflect RNA [1]. Propidium iodide has also been amalgamated into multistain techniques, which afford simultaneous flow fluorescence analysis of DNA and protein [78–81].

Since the binding of these dyes can be affected by the presence of chromosomal proteins on DNA, flow measurements have been used to quantify these affects on biological samples. In Figure 14, for example, flow measurements of the fluorescence intensity of isolated chromosomes as a function of ethidium bromide concentration were used to assess the binding affinity and number of binding sites on the chro-

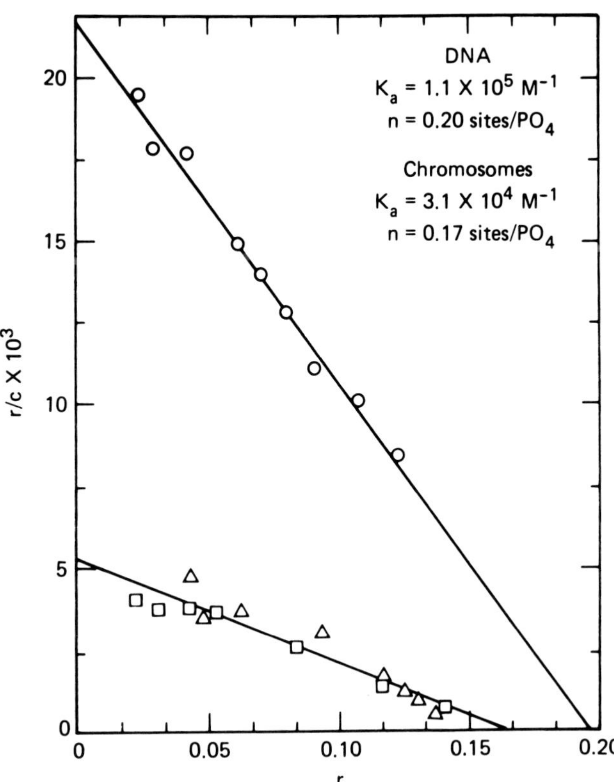

Fig. 14. Scatchard [369] plots of the binding of ethidium bromide to DNA and Chinese hamster chromosomes. DNA binding was measured using 20 μM ethidium bromide and DNA concentrations from 42 to 840 μM (○). Chromosome binding was measured under two sets of conditions. In one experiment, the chromosome concentration was constant at 5 × 10^6 sheared cells/ml (200 μM DNA), and the ethidium concentration was varied from 10 to 250 μM (□). In the other experiment, both stain and chromosome concentration were varied keeping the dye to phosphate ratio (D/P) constant. The ethidium bromide concentrations ranged from 10 to 400 μM with a constant D/P = 0.5 (△). Binding parameters were calculated from least-square fits to the data. The two sets of chromosome data were included in the same fit because both gave nearly identical binding parameters. (Reproduced from ref. [226], with permission of the publisher.)

mosomes for comparison with measurements of DNA in solution. The results indicate that presence of chromosomal proteins leads to a reduction in both the binding affinity and the number of dye binding sites. Addition of ethidium and other intercalators to chromosomes isolated for sorting can also maintain them in an extended form, which facilitates their subsequent identification [44,45].

As stated above, ethidium is not readily taken up by live cells [139,168], making it a good stain of dead cells. It can be used together with fluorescein diacetate hydrolysis/retention which is a live cell stain [185], as can Hoechst 33342 plus fluorescein diacetate [160]. The pink dye erythrosin [209] can substitute for phenanthridines (or trypan blue, which does not fluoresce) for tagging of dead cells, although there are implications of at least some penetration of living cells by the dye (e.g., trypanocidal properties); addition of the dye to cell cultures produces chromosomal banding [427] and aberrations due to chromosomal decondensation [172, 176,297].

As will be discussed later, reversible chemical modification of ethidium does permit its uptake by live cells. Staining of fixed cells by ethidium or propidium is strong [78], as is staining of suspensions of isolated metaphase chromosomes [50,51,388,403]. At dye/DNA ratios and ionic strengths reducing the likelihood of secondary binding, ethidium fluorescence should serve as a fairly accurate estimate of total DNA content. Illumination of cells or chromosomes sorted according to ethidium bromide fluorescence should be minimized, since ethidium bromide can apparently photosensitize single polynucleotide chain cleavage in DNA [98].

Bisbenzimidazole Dyes. Bisbenzimidazole dyes are used primarily as DNA stains (with or without membrane permeabilization) and as stains for DNA synthesis, the latter via reduced fluorescence following BrdU incorporation. The DNA specificity of 33258 or 33342 Hoechst makes them excellent stains for cell or chromosome DNA content (e.g., Fig. 15). Addition of 0.1% Triton X-100 guarantees uniform cellular permeabilization, a treatment less necessary with alkyl derivatives of the dye phenol moiety, for example, 33342 Hoechst [238].

The use of bisbenzimidazole dyes, for example, 33342 Hoechst [6,238] for staining live cells, while extremely useful, is not without toxicity [104,128,333], depending to some extent on the type of cell. Stoichiometric detection of DNA requires high dye concentrations (5–6 μg/ml), long (90 min) incubations (at 37°) [218,436], and is subject to passive membrane permeability (which can be reduced by mutation in a manner similar to Colcemide) [218] plus an energy-dependent dye extrusion. At low 33342 Hoechst concentrations (e.g., 1 μg/ml), mouse T and B cells can be differentiated (B cells fluoresce slightly more) [272,273]. Activated T cells show a ≥2-fold fluorescence increase [217], and 33342 Hoechst uptake rate appears to increase during S phase [125]. The coefficients of variation (CVs) of fluorescence of cells of a given (e.g., G1) DNA content obtained with live cells and 33342 Hoechst tend to be broad [6,133], although often sufficient for desired use. Also, use of this dye in the absence of triton does not abolish surface antigen staining with specific antibodies [272,273]. Otherwise, complicated fixation procedures are needed to keep surface antibody bound before staining cells for DNA [322]. If cells are permeabilized (e.g., 0.1% Triton-X100), 33342 Hoechst gives flow histograms with relatively tight CVs [190].

The specificity and mode of binding of 33258 Hoechst with cytological samples is affected not only by 33258 Hoechst concentration, but also by environmental factors such as solution pH. The spectra obtained from stained cells and chromosomes illustrate this sensitivity (Fig. 16). At low pH the spectra of cells and chromosome differ, indicating dye interaction with nonchromosomal components. In addition, the emission spectra of both cells and chromosomes varies with excitation wavelength, suggesting multiple modes of dye binding to both cells and chromosomes. At pH 8, in contrast, the emission spectra is the same for cells and chromosomes, and the emission maximum is the same as observed with purified DNA. This suggests that at pH 8, the fluorescence detected is primarily of 33258 Hoechst bound to cellular DNA in a similar environment as when bound to DNA in solution.

For flow studies of metaphase chromosomes with 33258 Hoechst, the relative peak positions are determined by base composition as well as DNA content because of the A + T specificity of this dye. The effective base specificity of Hoechst can vary with staining conditions. Figure 17 shows the

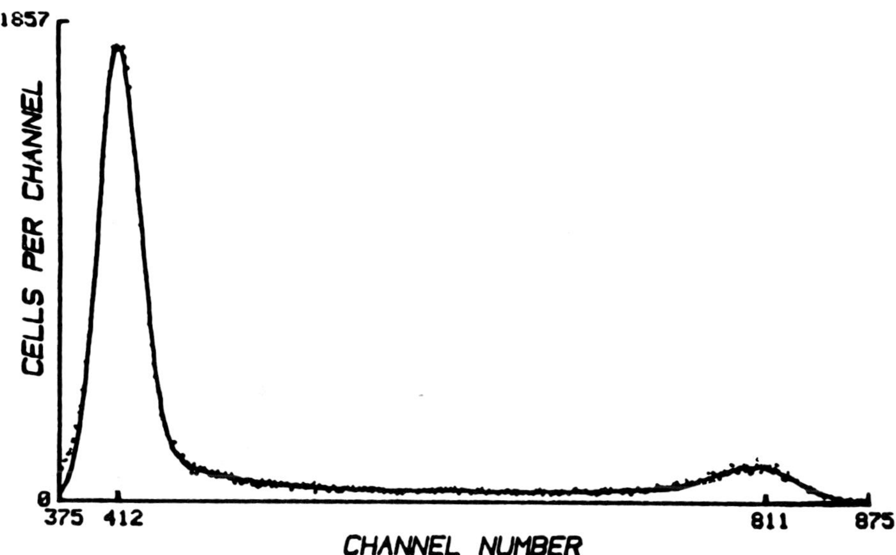

Fig. 15. DNA flow histogram of Chinese hamster cells stained with 10^{-5} M 33342 Hoechst after permeabilization with 0.1% triton X-100 (E. Sahar, M.L. Wage, and S.A. Latt, unpublished data).

results of flow analysis of human chromosomes equilibrated with different concentrations of 33258 Hoechst. While the absolute intensity and resolution varies with dye concentration, the relative position of most peaks is remarkably constant. Shifts in relative peak position are observed for some of the smaller chromosomes, particularly the Y chromosome, which increases in relative fluorescence as the Hoechst concentration is decreased. This indicates that the binding affinity of Hoechst is higher for the Y chromosome than for other chromosomes, which may be due to the presence of large amounts of repetitive DNA [67], some of which is rich in A + T base pairs [186].

The quenching of 33258 Hoechst fluorescence by biosynthetically incorporated BrdU continues to be of practical use in cytology. This effect can be detected in fixed interphase nuclei [234] and metaphase chromosomes [231], and fluorescence quenching or related modified Giemsa techniques have been used as an alternative to autoradiography to study DNA replication kinetics and sister chromatid exchange [105,167,201,207,338,467]. More useful from the standpoint of fluorescence flow systems, BrdU-dependent fluorescence quenching can be observed in unfixed cells after direct addition of the dye to the culture medium [238,241]. Exposure of cells to BrdU plus dye is not of itself cytotoxic, although the combination of dye plus BrdU is a powerful photosensitizer [398]. The precise toxicity at the light dosage expected in a fluorescence flow system remains to be determined. The single-stranded nicks caused by this photoreaction permit simple analysis of the replication kinetics of DNA sequences replicating at specific times [87] or differentiation between single- and double-strand BrdU substitution [247]. In addition to analysis and sorting of intact cells according to the extent of BrdU incorporation, 33258 Hoechst fluorescence methodology has also facilitated the purification of individual chromosomes, such as the late-replicating hamster Y chromosome [74,75], and it might also serve for the human X, based on DNA synthesis kinetics.

Chromomycinone Dyes. Chromomycinone dyes, especially mithramycin, though highly G-C-specific, are also highly DNA-specific, and hence have proved useful as stains for flow fluorometric analysis of DNA content [77,78,417]. Additional uses of these dyes are realized when they are used in combination with other (e.g., A-T-specific) dyes (see below).

Dye Combinations: Energy Transfer

The selectivity of individual dyes can be combined to form compound molecules with useful properties. Dye combinations can also provide information about proximity relationships and DNA accessibility in chromatin. Fluorescence specificity can be enhanced with multiple chromophore–polycation combinations. The backbone, such as diamine [49,259] or polylysine [232] presumably binds to DNA without marked specificity, while conjugated dyes (e.g., quinacrine mustard derivatives) exhibit appreciable fluorescence dependent on the coincidence of multiple dyes interacting with appropriate sites. For example, LePecq et al. [259] described diacridine derivatives in which the components fluoresced brightly when bound near A + T base pairs, but binding of one dye moiety near a G-C pair was sufficient to quench the other even if bound to a fluorescent site, presumably by electronic excitation energy transfer [119, 229,396,401,402]. Van de Sande et al. [431] have described a similar and useful "dimeric quinacrine," a spermidine bisacridine called "$(CMA)_2S$."

Energy transfer between hydrocarbon chromophores covalently linked to DNA was reported by Pochon et al. [344]. Methylbenz(alpha) anthracene (absorption 250–350 nm, fluorescence 400–450 nm) served as an acceptor for energy absorbed by acetylaminofluorene (absorption 250–300 nm), which itself did not fluoresce. Energy-transfer detection was based on enhancement of fluorescence excitation in the vicinity of donor absorption and by changes in fluorescence polarization. Energy transfer in this system was abolished by DNA hydrolysis or denaturation.

Consideration of the effects of energy transfer is particularly important when pairs of dyes are used at near-saturating concentrations to stain cells or chromosomes for flow cytometry. At saturation, dye-dye distances are frequently

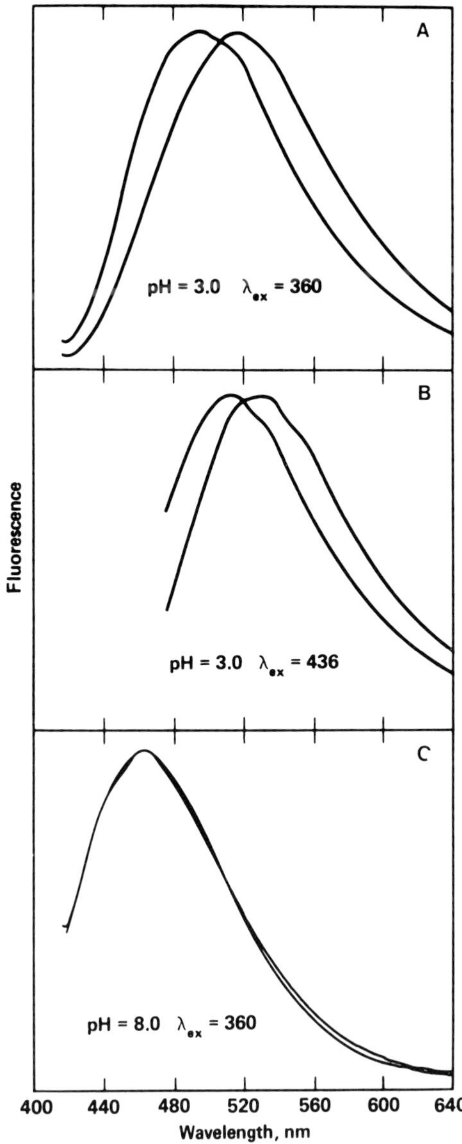

Fig. 16. Corrected emission spectra of Hoechst 33258 bound to mitotic cells or chromosomes. Conditions of staining are 3.7×10^6 M Hoechst 33258 with either 3.3×10^5 Chinese hamster mitotic cells (right curves) per milliliter, or 5×10^5 sheared mitotic cells per ml (left curves) in: **A,** 5×10^{-2} M citrate pH 3, Hoechst filter combination (excitation ~360 nm); **B,** 5×10^{-2} M citrate pH 3, chromomycin A_3 filter combination (excitation ~436 nm); **C,** 5×10^{-2} M Tris pH 8, Hoechst filter combination (excitation ~360 nm). All spectra were measured with the microspectrofluorometer. At low pH, cell spectra (right curves) differ from chromosome spectra (left curves) and the emission maxima of both particles varies with excitation wavelength. At high pH, the same spectrum is seen for both cells and chromosomes, and the observed emission maxima are the same for Hoechst bound to purified DNA in solution. (Reproduced from ref. [182], with permission of the publisher.)

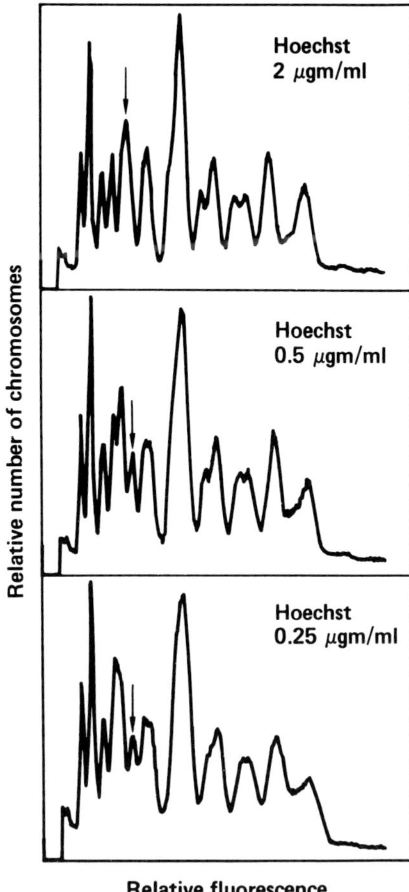

Fig. 17. Flow distributions of isolated metaphase chromosomes from human fibroblasts stained in suspension with different concentrations of 33258 Hoechst. All distributions were normalized to the same intensity for the large central peak corresponding to chromosomes 9–12. Shifts in relative peak position are clearly seen for the smaller chromosomes. The Y chromosome (indicated by the arrows) has the same intensity as chromosomes 16 and 18 in the top panel, but its relative intensity shifts to higher values as the dye concentration is decreased. (R.G. Langlois, unpublished data.)

less than the critical distance for energy transfer. The efficiency of energy transfer can vary dramatically because of the inverse sixth-power dependence of transfer with distance. The primary consequence of energy transfer is that excitation energy absorbed by the donor is emitted at the same wavelength as energy absorbed by the acceptor so that the fluorescence of the two stains cannot be independently measured using two different emission filters. Both microspectrofluorometry [182,225] and flow cytometry [226] have been used to demonstrate the existence of energy transfer in cytological preparations stained in suspension. These studies have also shown that the transfer efficiency can be very high (i.e., greater than 90% of the donor energy transferred to the acceptor), and that fixation can increase the transfer efficiency presumably due to extraction of some chromosomal proteins increasing the binding of acceptor molecules.

Energy transfer can also complicate the interpretation of experiments where it is important to independently measure the fluorescence of two dyes. Examples of this are measurements of DNA and RNA using one DNA-specific dye and one dye for total nucleic acids, or measurement of bromodeoxyuridine incorporation using one dye whose fluorescence is affected by BrdU and a second that is unaffected by this substitution. While the fluorescence of the two stains cannot be separated using two different emission filters, dual wave-

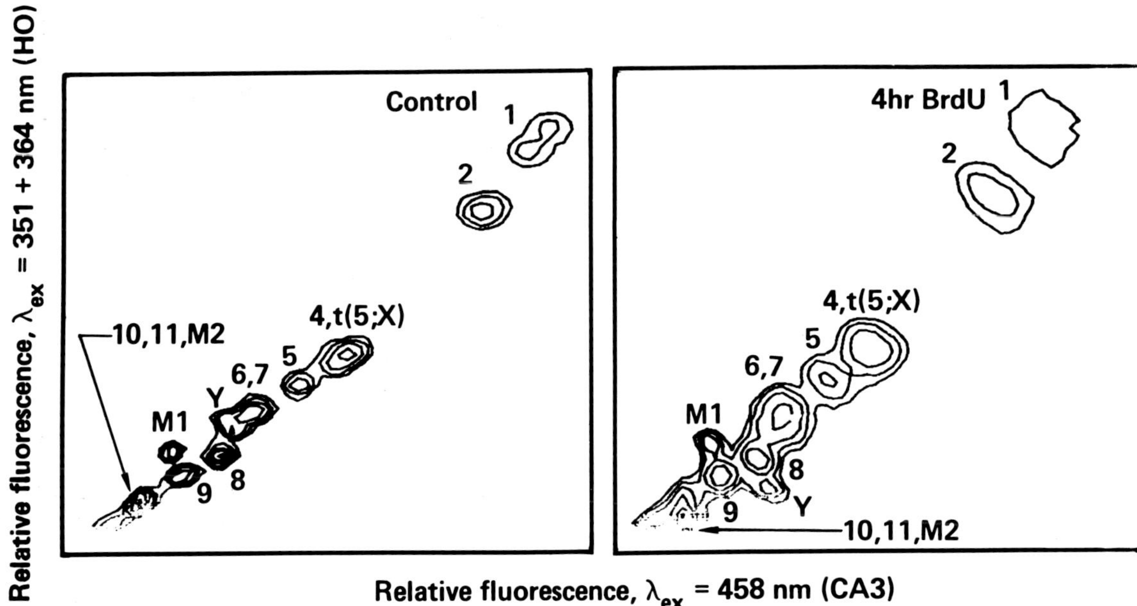

Fig. 18. Bivariate flow distributions of Chinese hamster M3-1 chromosomes stained with 33258 Hoechst and chromomycin A_3 with and without a 3½-hour terminal pulse of BrdU in late S phase. The late-replicating chromosomes 10, 11, M1, and Y show decreased Hoechst fluorescence due to the presence of incorporated BrdU (right panel). (Data from [74].) (See also Chapter 25, this volume.)

length excitation can be used to selectively excite each stain. In this case, the total fluorescence resulting from donor excitation is approximately proportional to bound donor concentration independent of energy transfer, while acceptor excitation yields the concentration of bound acceptor. Figure 18 illustrates the application of dual wavelength excitation for measurement of BrdU incorporation and DNA content on isolated chromosomes. Figure 19 illustrates the use of these dyes on cytological preparations of chromosomes, the latter providing additional evidence that the major factor responsible for reduced 33258 Hoechst fluorescence is BrdU incorporation into DNA and not indirect effects mediated by chromosomal proteins.

Energy transfer can also be exploited to alter the specificities or spectral characteristics of dyes, enhancing their utility for cytological analyses. One example of this is the use of the dye pair mithramycin and ethidium bromide for DNA content measurements [147,411]. This dye combination is particularly well suited for measurements on flow systems utilizing a mercury arc illumination source [394]. Microspectrofluorometric measurements [180] on cells stained with the related dye pair chromomycin A_3 and ethidium bromide showed efficient energy transfer for this pair with almost 80% of the chromomycin excitation energy being transferred to ethidium and emitted as ethidium fluorescence (Fig. 20). While chromomycin A_3 has high specificity for DNA and is efficiently excited by the 436-nm emission line from a mercury lamp, its low quantum yield of approximately 0.05 [182] limits the fluorescence intensity from stained cells. Ethidium bromide, in contrast, has a higher quantum yield (approximately 0.2), but it binds to RNA and excites in a wavelength region where there are no strong mercury emission lines. Thus, excitation at 436 nm preferentially excites the DNA specific chromomycin with high efficiency, and the resulting excitation energy is emitted with

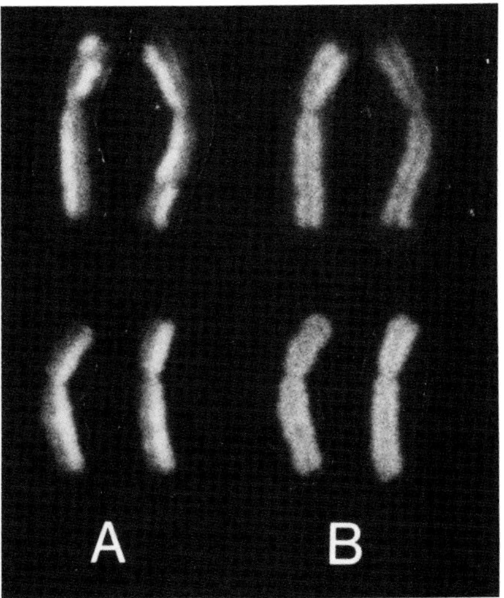

Fig. 19. Fluorescence of metaphase chromosomes from human lymphocytes cultivated two cycles in medium containing BrdU and stained with both 33258 Hoechst and ethidium bromide. Chromosomes from human peripheral lymphocytes that had undergone two cycles of BrdU incorporation were stained first with 10^{-6} M 33258 Hoechst and then 10^{-6} M ethidium (in pH 7.5 McIlvaine's buffer, the mounting solution). Fluorescence predominantly due to 33258 Hoechst **(A)** was detected by exciting between 360–400 nm and observing emission above 460 nm; sister chromatid differentiation is apparent. Ethidium fluorescence was selectively excited between 500 and 550 nm in **(B).** Ethidium fluorescence is not sensitive to BrdU, and sister chromatids fluoresce with essentially equal intensity. (Reproduced from ref. [242], with permission of the publisher.)

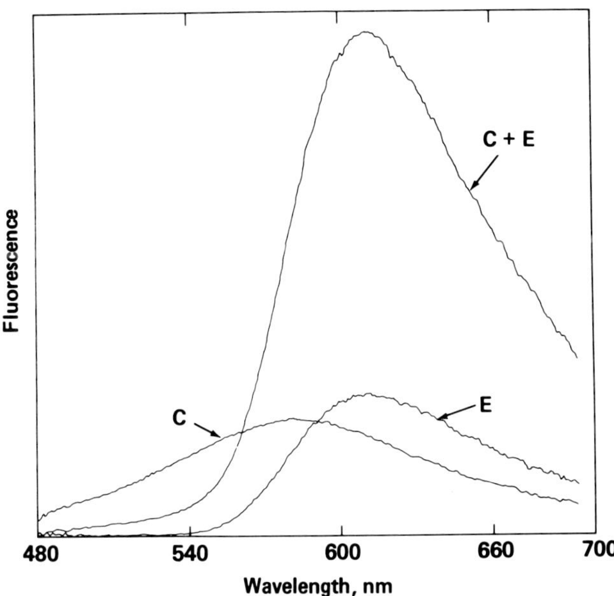

Fig. 20. Corrected emission spectra of individual Chinese hamster mitotic cells stained in suspension with chromomycin A_3 alone, ethidium bromide alone, and with both dyes. The vertical scale is the same for all three spectra. The reduction of chromomycin fluorescence (C) and the enhancement of ethidium fluorescence (E) in the double-stained sample results from energy transfer. (Data from [182].)

increased efficiency by the ethidium bromide. An additional benefit of this dye pair is that the effective Stokes shift between excitation and emission is increased, simplifying the separation of fluorescence from scattered excitation light.

Noncovalent energy donor-acceptor combinations have now been used extensively. Gursky et al. [157] employed acridine orange as an energy donor with actinomycin D, methylene blue, or ethidium bromide as energy acceptors. Brodie et al. [41] utilized ethidium as an acceptor for energy transferred from quinacrine, acridine orange, or 33258 Hoechst. In both sets of experiments, binding of increments of acceptor to a donor-DNA complex progressively quenched donor fluorescence as the average donor-acceptor distance decreased. Under conditions such that virtually all added dye molecules bound, the amount of acceptor required for a given reduction in donor fluorescence increased in proportion to the amount of DNA. Since accessible DNA was being measured, its occlusion by chromosomal proteins could be detected by enhanced energy transfer. However, the definition of "accessible" is operational, reflecting chromatin conformation, correlations in dye-binding selectivity, and base-specific variations in fluorescence efficiency [233]. Thus, while this approach is probably not an accurate gauge of DNA occlusion or exposure in chromatin, it might serve, under appropriate conditions, as a rapidly measurable index of chromatin condensation.

The use of energy transfer for analyzing chromatin structure and its application to cytology continues to expand. Shapiro [383] incorporated energy transfer from a cytoplasmically bound stilbene disulfonic acid derivative to nuclear bound ethidium into a fluorochrome system for blood cell counting and classification, and he has more recently sung the praises of other dyes for this purpose [384]. In other multistain systems for DNA and protein analysis [78], energy transfer between the stains might convey valuable additional information. Combinations of fluorescence intensity, to which polarization measurements might eventually be added, have proved particularly useful in studying fixed metaphase chromosomes [242,365] and individual cells and chromosomes [182,220,225,227,300].

The general idea behind the application of energy transfer between dyes (or dye dimers [430]), to study chromosomes, as reviewed by Latt et al. [244,245], is that an energy donor with a binding or fluorescence efficiency specific for one type of DNA base pair will be quenched by the presence of an energy acceptor nearby, yet will be relatively resistant to quenching in regions containing short clusters compatible with donor fluorescence but not with acceptor binding (Fig. 21). As mentioned above, the primary feature permitting energy transfer between arrays of dye molecules bound to DNA is spectral overlap between donor fluorescence and acceptor absorption [120,402]. Energy transfer varies as the inverse sixth power of donor-acceptor separation [401,402], and, for pairs of dyes with good spectral overlap, the "critical distance" at which energy transfer is 50% efficient is in the 20–50 Å range [41,182,225,242,244,245,366], that is the distance spanning 5–15 base pairs in helical DNA. Thus, energy transfer can, in principle, be used to highlight chromosomal regions containing clusters of at least 10–30 base pairs of a particular type. Based on theoretical considerations, and measurements on soluble dye-DNA complexes, a number of dye pairs satisfy the above criteria. All have been observed to enhance the general type of banding expected. Additional dye pairs between which energy transfer is possible have also been described (Fig. 22) [245,247,248]. Also, energy transfer between 33342 Hoechst and hydrolyzed fluorescein diacetate [160] can help detect nucleated, viable cells.

It should be noted that energy transfer-dependent effects on chromosome staining are distinct from effects due primarily to binding competition between different dye pairs. With the former, donor quenching can be shown to be accompanied by sensitization of acceptor fluorescence, and a reduction in donor fluorescence lifetime [365,366], while for the latter, induction of specific patterns of donor fluorescence is not associated with complementary changes in acceptor fluorescence (following donor excitation) (Figs. 23 and 24). Unlike energy transfer, binding competition, which has also been utilized in chromosome staining [187,376–378], does not require spectral overlap between the two dye types employed, and the fluorescence patterns induced rarely have the same generalized contrast as those produced by energy transfer. Examples of dye pairs relying primarily on binding competition are 33258 Hoechst or DAPI (4′6-diamidinoindolephenol) plus netropsin or distamycin A [377] (similar but nonidentical binding) and chromomycin A_3 plus netropsin [187] (reverse binding specificity). Netropsin and distamycin A compete with 33258 Hoechst or DAPI for binding to DNA [244,245,366] with less competition for heterochromatic regions, which also stain with antibodies to 5-MedC. This combination has proved especially useful in studies of chromosome #15 [244,245,366,377,378]. Echinomycin is another dye that can be used as a non-energy transfer-dependent counterstain [244,245]. Notably, all three dyes, that is, distamycin A, netropsin, and echinomycin, absorb at wavelengths far below the emission of the fluorescent dyes with which they are used as counterstains.

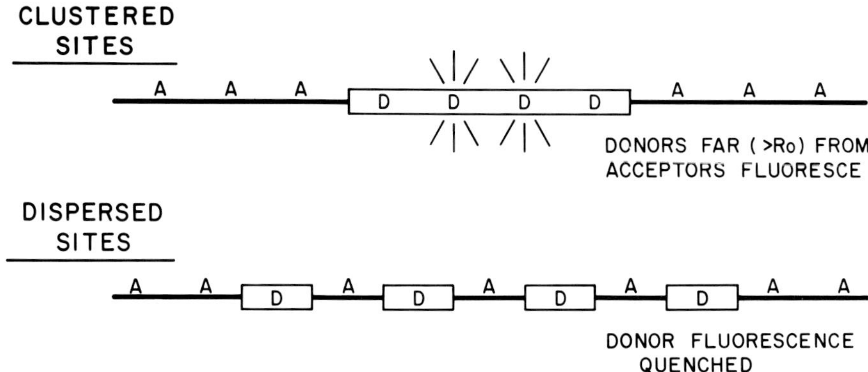

Fig. 21. Diagrammatic representation of the effect of DNA base pair clustering on energy transfer-enhanced chromosome banding. Sites appropriate for energy donors (D) are those specific for donor binding and/or efficient fluorescence. For the energy acceptor (A) only binding specificity is important, and this specificity should, with respect to base composition, be the reverse of that of the donor. Donors separated from acceptors by more than R_0 (a distance typically equivalent to 5–15 base pairs) will be resistant to energy transfer-dependent quenching. A chromosome region containing many (perhaps $\geq 10^5$–10^6) such clusters, within which the donor can bind and fluoresce but the acceptor cannot bind, will (in the presence of both donor and acceptor) exhibit brighter donor fluorescence than do the surroundings, even if the donor or acceptor alone exhibits little differential fluorescence. (Reproduced from ref. [244], with permission of the publisher.)

Use of these nonfluorescent counterstains has been detailed in the previous section. However, some of the best banding described thus far for pairs of dyes, chosen because of binding properties, involved pairs in which a nonfluorescent member incidentally satisfied spectral overlap criteria for efficient energy transfer, for example, 33258 Hoechst plus actinomycin D [187].

A third use of dye combinations involves 33258 Hoechst to detect DNA synthesis, via BrdU incorporation, and ethidium, to quantitate DNA. In this case, the energy transfer between 33258 Hoechst and ethidium (or propidium) is either noncontributory or a phenomenon that tends to reduce the 33258 Hoechst fluorescence, requiring greater UV excitation. When combined with chromomycinones or phenanthridines as stains for total DNA, the bisbenzimidazole dyes can be used to detect DNA replication and content (Figs. 18, 19) [31,74,242,320,323] and cell cycling [214,348].

It is important to note that the assumption made in some of these studies [17,31,323] using 33258 Hoechst, that DNA substituted with BrdU in one strand has no fluorescence, is probably incorrect [231].

Thus far, two types of dyes, 33258 Hoechst or DAPI and quinacrine, have been used as the A-T-specific (via binding or quantum yield, respectively) donor component of energy transfer pairs. Counterstaining with the G-C-specific [306] fluorochrome 7-aminoactinomycin D, in theory a good energy acceptor for 33258 Hoechst fluorescence [140, 239,240], enhances the otherwise indistinct Q-type banding of the Hoechst dye (Fig. 23). These brightly fluorescing regions presumably contain clusters of A-T base pairs. In contrast, 33258 Hoechst has little effect on the pattern of 7-aminoactinomycin D fluorescence. As mentioned above, enhancement of a similar banding pattern with 33258 Hoechst as a primary stain has also been observed using the G-C-specific [243,325] fluorochrome chromomycin A_3 as an energy acceptor. Chromomycin A_3, which exhibits marked G-C preference (Fig. 11), has proved especially useful when combined with 33258 Hoechst, as a dual fluorochrome stain of fixed (Fig. 24) or isolated chromosomes [76,154,225, 227,365,428] (Fig. 25), cells [181] (Fig. 26), or bacteria [432] (Fig. 27).

The functional relationship between base composition and the fluorescence signals from 33258 Hoechst and chromomycin A_3 is determined by a number of factors. Recent studies have shown that both dyes bind with highest affinity to clusters of A-T or G-C base pairs, but that weaker binding to less base-specific regions is observed at high dye concentrations [236,239,291,434]. Thus, the apparent base specificity of both dyes varies with stain concentration. Energy transfer can further enhance the base specificity for this stain pair since the acceptor chromomycin A_3 (CA_3) has a lower quantum yield than the donor 33258 Hoechst (Ho). The relatively low Hoechst fluorescence from G-C-rich regions will be further reduced by efficient energy transfer to the relatively high concentration of acceptor molecules in these regions. This effect will be most dramatic when the average transfer efficiency is near 50%, and when the Hoechst detector is more sensitive to donor fluorescence than acceptor fluorescence.

By choosing staining conditions that maximize base specificity, subtle differences in base composition can be detected. Individual human chromosome types have been shown to have subtle differences in base composition (A-T contents of approximately 56–63% [208]). Figure 25 demonstrates that these small differences can be clearly detected by flow analysis with Hoechst and chromomycin A_3. Flow analysis of interphase nuclei shows that the Hoechst to chromomycin ratio also varies during S phase, suggesting that early replicating DNA is relatively G-C-rich (Fig. 26) [95]. This correlates nicely with observations that quinacrine-dull bands, which are presumably G-C-rich, are replicated early in S phase [107,235]. Theoretical calculations based on these data suggest that a 1% increase in A-T content yields a 10% increase in the ratio of 33258 Hoechst to chromomycin A_3 fluorescence [226].

For flow analysis of bacterial species, which differ widely in base composition, staining conditions can be adjusted to minimize base specificity. Staining at very high dye to base pair ratios decreases the intrinsic specificity of both stains and increases the average energy transfer efficiency. The en-

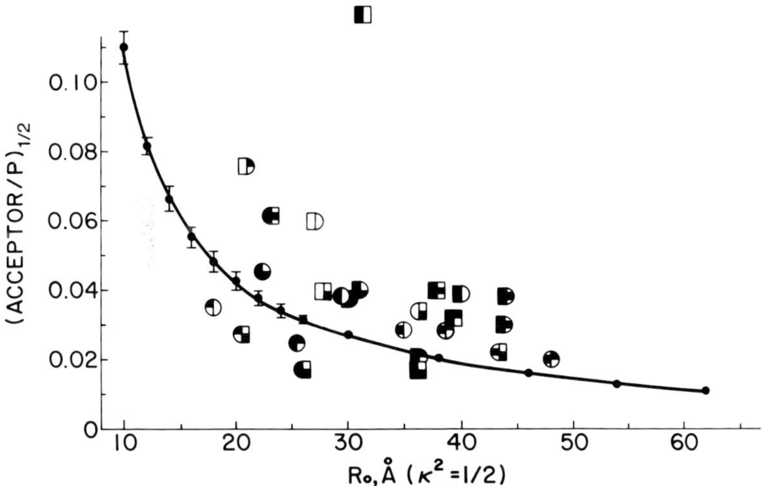

Fig. 22. Dependence of energy transfer quenching efficiency on donor-acceptor R_0 in calf thymus DNA complexes. Quenching efficiency is expressed on the ordinate as the acceptor/P ratio needed for a 50% reduction in donor fluorescence. Theoretical values (see Latt et al. [245]) are plotted as a continuous curve. The vertical limits bracketing the theoretical curve indicate the range of values expected for variations in the angle between the donor and acceptor transition dipoles (when donor and acceptor are bound at the same site) of 0, 30, or 60°. Experimental values are expressed as compound symbols, the left half of which identifies the energy donor and the right half identifies the energy acceptor. Symbols for individual dyes are: 33258 Hoechst (■), chromomycin A_3 (□), quinacrine (●), 2,7-di-t-butylproflavine (⬒), actinomycin D (⊟), acridine orange (⊖), ethidium (◒), daunorubicin (◓), 7-aminoactinomycin D (○), and methyl green (⬓). (Reproduced from ref. [245], with permission of the publisher.)

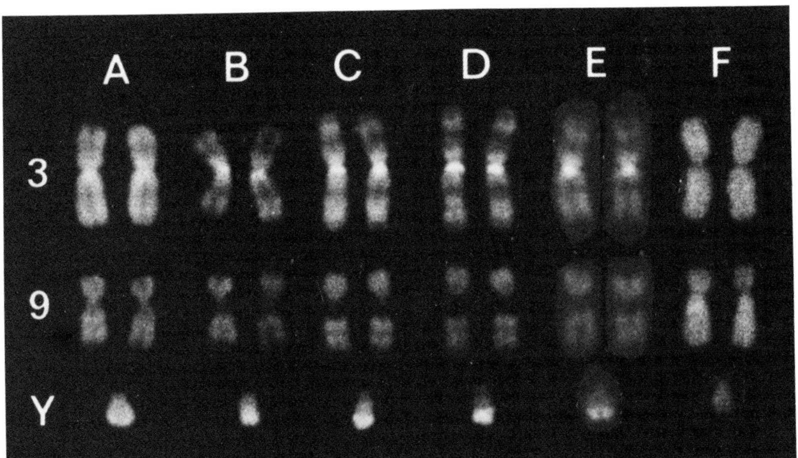

Fig. 23. Effect of 7-aminoactinomycin D on the fluorescence patterns of human chromosomes stained with 33258 Hoechst. Chromosomes were stained with 0.2 μM 33258 Hoechst in 0.14 M NaCl/0.004 M KCl/0.01 M sodium phosphate buffer at pH 7 and observed either directly (column A) or after counterstaining with 7-aminoactinomycin D at 0.5 μM (column B), 1 μM (column C), 3 μM (column D), or 10 μM (column E) in pH 7.5 McIlvaine's buffer. Fluorescence predominantly due to 33258 Hoechst was observed (columns A–E) by exciting with 360- to 400-nm light and viewing the chromosomes through a 430-nm high-pass filter. 7-Aminoactinomycin D fluorescence (column F), with 3 μM of this dye as a counterstain, was observed by exciting with 500- to 550-nm light and viewing through a 580-nm high-pass filter [365].

hancement of base specificity by energy transfer is further reduced by using an emission filter on the Hoechst detector that primarily passes acceptor fluorescence. Figure 27 shows the results of flow analysis of a mixture of six bacterial species with base compositions varying from 32 to 72% A-T content, and illustrates the wide range of base compositions that can be measured with these staining conditions. For these conditions, a 1% increase in A-T content leads to only a 3% increase in fluorescence ratio [432].

Careful standarization is required before base composition values can be derived from these fluorescence parameters because both Hoechst and chromomycin require clusters of A-T or G-C base pairs to form a binding site. In unique sequence DNA the frequency of such clusters is nonlinearly related to average base composition. In repetitive DNA, the frequency of dye binding sites can be primarily determined by the sequence of the repeat unit rather than the average base composition. Hoechst bright heterochromatic regions

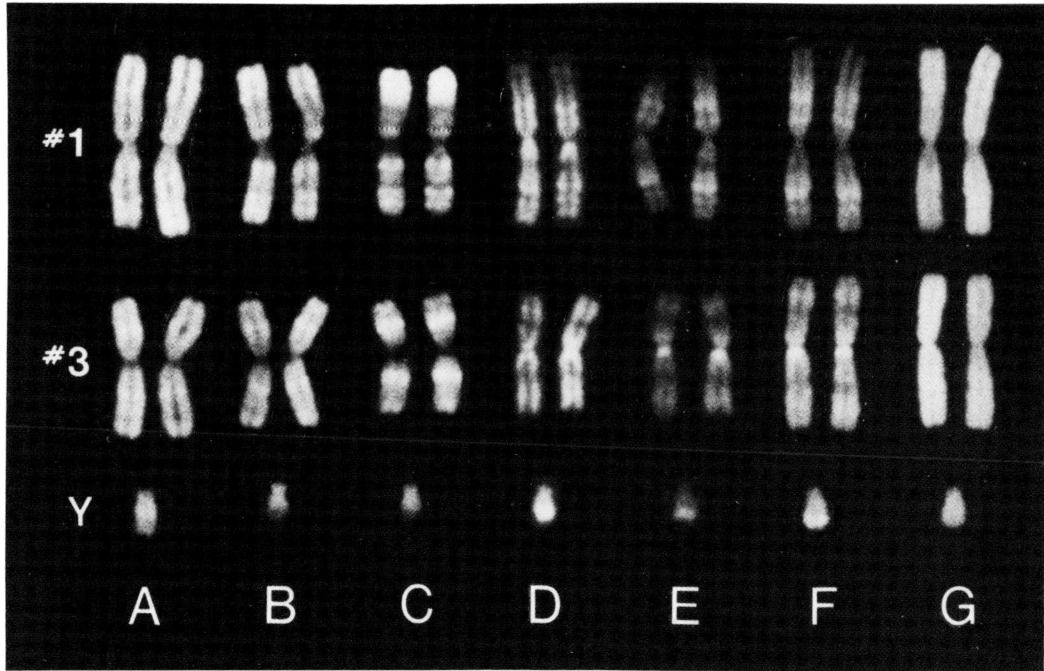

Fig. 24. Dependence of fluorescence contrast enhancement by DNA-binding counterstains on relative dye binding specificities and spectral overlap. Shown are human 1,3 and Y chromosomes from different cells of the same individual. Chromosomes in columns A and B were stained only with chromomycin A_3 at 500 μM (A) or 4 μM (B), in 0.14 M pH 6.8 phosphate buffer plus $MgCl_2$. Chromosomes in column C were stained with 500 μM chromomycin A_3 followed by 100 μM methyl green, the latter in 0.15 M NaCl/0.005 M Hepes, pH 7.0. Chromosomes in column D were stained with 0.4 μM 33258 Hoechst, in 0.14 M NaCl/0.004 M KCl/0.01 M phosphate, pH 7.0, while those in columns E, F, and G were stained with 0.4 μM 33258 Hoechst, followed by 500 μM chromomycin A_3. Fluorescence from doubly stained chromosomes was due to chromomycin A_3 in C and G and to 33258 Hoechst in E and F. The mounting medium in E was pH 7.5 McIlvaine's buffer. In all other cases, glycerol, which stabilizes chromomycin A_3 fluorescence, was used. Chromomycin A_3 staining solutions contained equimolar amounts of the dye and magnesium [365].

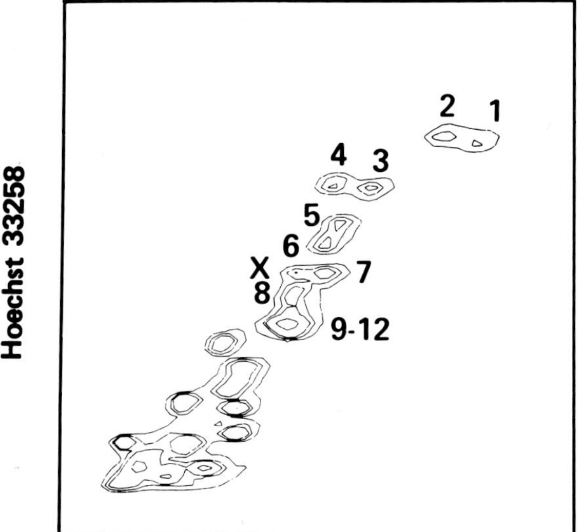

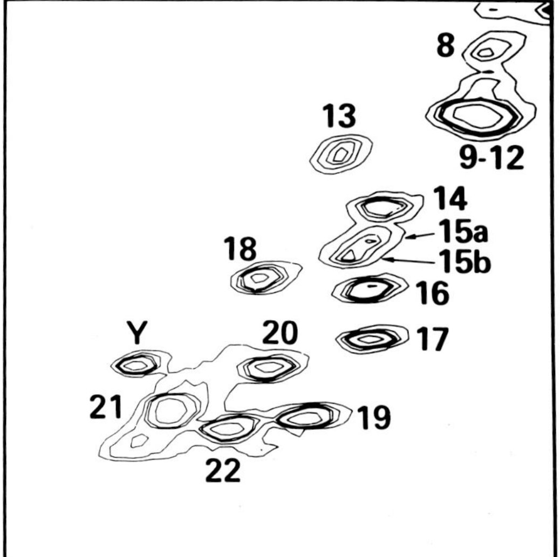

Fig. 25. Bivariate flow distributions of human chromosomes derived from PHA-stimulated peripheral lymphocytes stained with 33258 Hoechst and chromomycin A_3. The left panel shows the whole karyotype while the right panel shows an expanded view of the smaller chromosome types. The individual homologues of chromosome 15 are separately resolved with this donor. (Reproduced from ref. [154a], with permission of the publisher.)

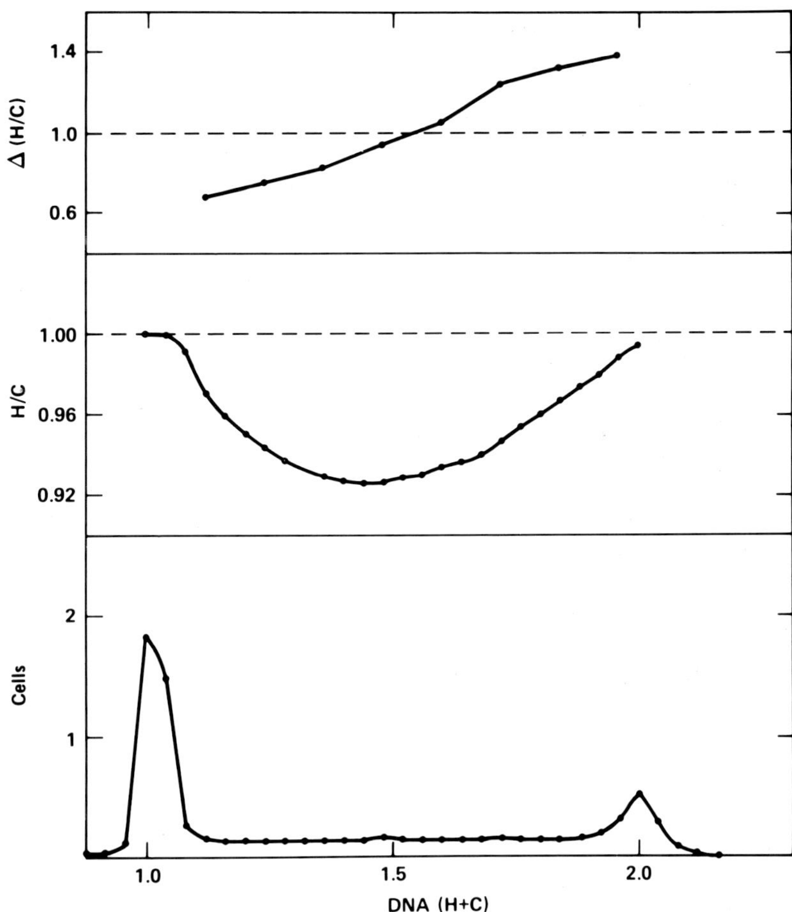

Fig. 26. Staining characteristics of interphase nuclei from PHA-stimulated human lymphocytes stained with 33258 Hoechst and chromomycin A_3. High-resolution list mode data were used to calculate the DNA content of each cell (approximated by the sum of the two fluorescence signals) and the ratio of the Hoechst to chromomycin fluorescence (H/C) for each cell. The number of cells (bottom panel) and the fluorescence ratio per cell (middle panel) are shown for different intervals in the cell cycle. The H/C ratio for the DNA synthesized in each interval of S phase (Δ (H/C) in the top panel) was determined by the difference in the two fluorescence intensities from the beginning to the end of the interval. These results are consistent with early replicating regions being relatively G-C-rich (low H/C ratio). (R.G. Langlois, unpublished data.)

on mouse chromosomes, for example, contain satellite DNA which has clusters of A-T base pairs in its repeat unit [165,387] A-T-rich polymorphic chromosomal regions can also be highlighted by the combination of 33258 Hoechst and actinomycin D [244,378] (Fig. 28).

Further utilization of chromomycinone dyes would benefit from a more complete analysis of the geometry, base composition, and chromosomal protein dependence of both binding and fluorescence properties. An additional feature of chromomycinones, namely, a 15–25% enhancement in fluorescence following BrdU incorporation, has been used by Swartzendruber and associates [404,405] to follow DNA synthesis. BrdU incorporation was also shown by this group [406] to influence teratocarcinoma cell development, a result consistent with other effects of BrdU. The anthracyclines have weak but measurable fluorescence (excitation at 505 nm [with strong absorption 20–30 nm lower]) (emission near 600 nm), especially daunomycin, when added to cells [409], permitting direct use [393] or use as a quencher of more brightly fluorescing dye [110,213].

Additional resolution of chromosomes can be achieved by substituting the DAPI-derivative, DIPI [253] or even adding netropsin or distamycin A to 33258 Hoechst plus chromomycin A_3 [220,300]. Use of fluorescent or nonfluorescent actinomycin counterstains with quinacrine as an energy donor enhances Q-banding contrast and can be made to highlight quinacrine-bright polymorphic regions (Fig. 28). As suggested by other work (cited earlier) these regions probably contain clusters of A-T base pairs.

Reverse banding patterns can be generated if G-C-specific dyes are used as energy donors with an A-T-specific energy acceptor. For example, methyl green (the commonly used name for a dye, CI 42590, that is actually ethyl green) [265], serving in the latter capacity, can enhance contrast in chromosomes stained with chromomycin A_3 [365]. With human chromosomes, certain telomeric regions tend to be accentuated, and the combination of standard staining techniques plus chromomycin A_3-methyl green banding permits very sensitive detection of certain reciprocal translocations (e.g., Figs. 29 and 30). The chromomycin A_3-methyl green pair also provides for convenient and sensitive detection of certain acrocentric chromosome polymorphisms (Fig. 31), described

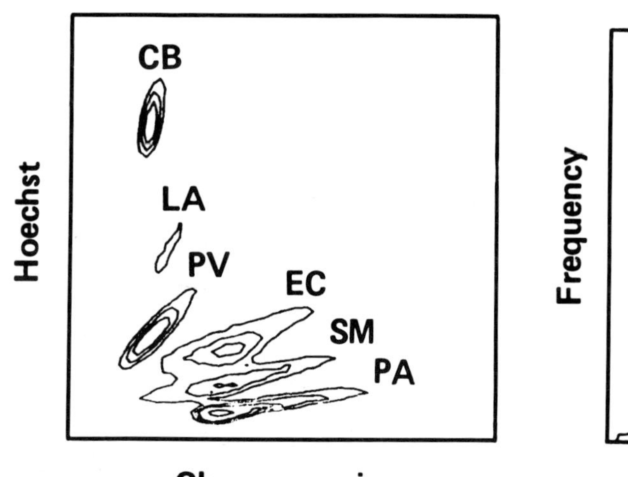

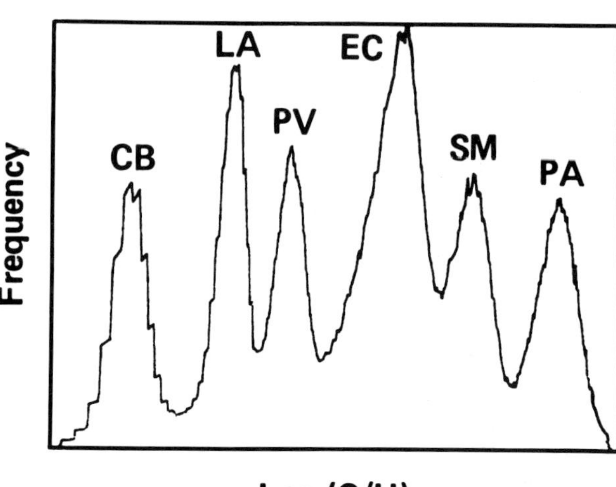

Fig. 27. Bivariate flow distribution of a mixture of six species of fixed bacteria stained with 33258 Hoechst and chromomycin A_3 (left panel). The right panel shows the same data as a frequency distribution of number of cells versus the log of the chromomycin to Hoechst ratio. The following are the species analyzed (with percent G + C content): (PA) *Pseudomonas aeruginosa* (68), (SM) *Serratia marcescens* (57.5), *Escherichia coli* (EC) (50), (PV) *Proteus vulgaris* (37), (LA) *Lactobacillus acidophilus* (33), and (CB) *Clostridium butyricum* (28). For experimental details see Van Dilla et al. [432]. (R.G. Langlois, W.F. Hadley, D. Yajko, unpublished data.)

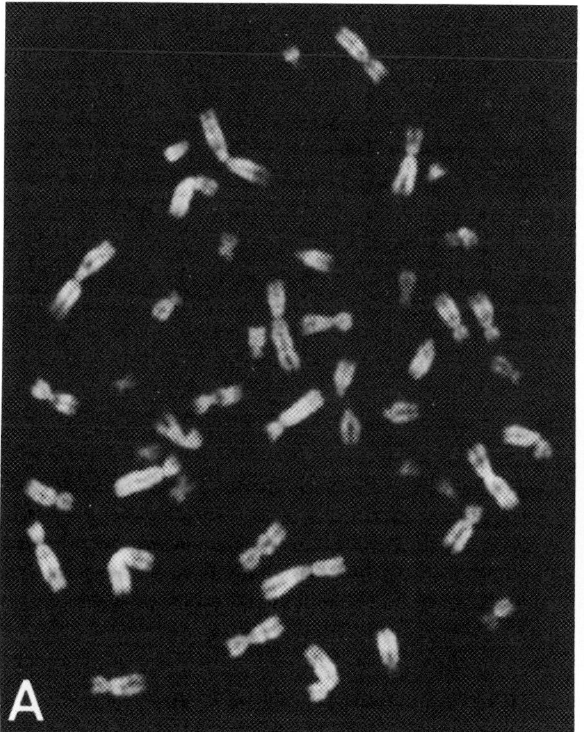

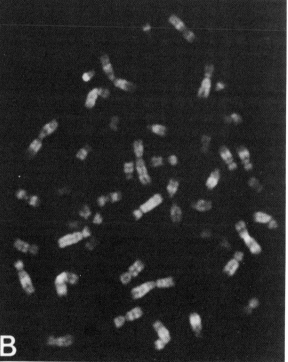

Fig. 28. Effect of actinomycin D on the fluorescence of human metaphase chromosomes stained with 33258 Hoechst. Metaphase chromosomes of a 46,XY individual were stained with 33258 Hoechst, mounted in McIlvaine's pH 7.5 citrate-phosphate buffer, photographed **(A)**, counterstained with 3×10^{-5} M actinomycin D and rephotographed **(B)**. (S.A. Latt, unpublished data.)

with other R-banding methods [437], but distinct from those observed with Q banding. Methyl green absorption also overlaps 7-aminoactinomycin D fluorescence, and this pair of dyes can generate weak R banding [244,245]. However, there exists the need for a more highly fluorescent G-C-specific fluorochrome for full utilization of this approach.

NEW DIRECTIONS

Red Dyes

A number of advantages accrue to the development and use of new, red nucleic acid-specific fluorochromes. These include a lower background due to intrinsic cell fluorescence,

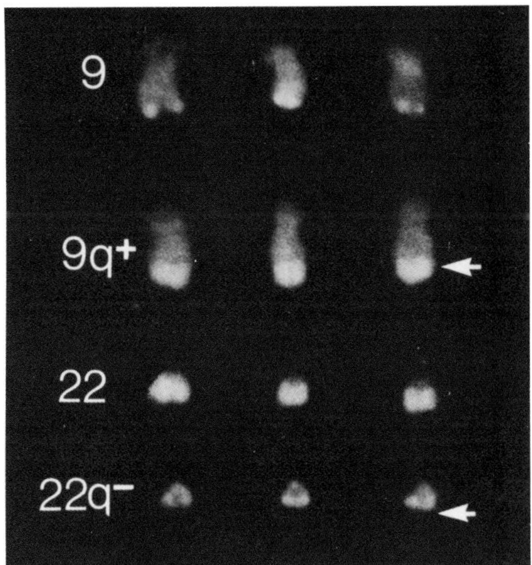

Fig. 29. Appearance of t(9;22) Ph′ translocation following chromomycin A_3-methyl green staining. Slides of lymphocytes from a patient with chronic myelogenous leukemia were stained and photographed by the chromomycin A_3-methyl green procedure. Chromosome identification was confirmed by subsequent quinacrine staining. Shown are structurally normal and abnormal #9 and #22 chromosomes from three different cells. Note in particular the brightly staining chromosome #22 material, which retains its pronounced (chromomycin A_3) fluorescence when translocated to the distal part of the long arm of #9. The present pictures do not permit a definitive assessment of the extent of the (reciprocal) translocation from 9q to 22q, although the size of the bright material present on the long arm of the 22q is about the same size as the bright material at the long arm terminus of the normal #9. (Reproduced from ref. [244], with permission of the publisher.)

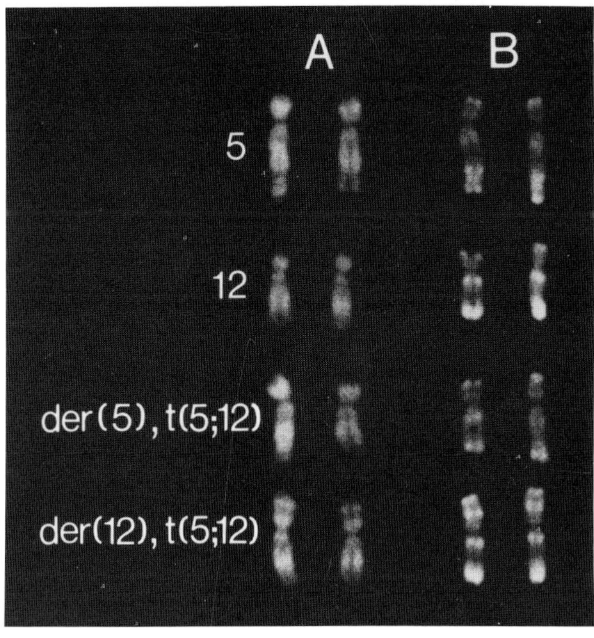

Fig. 30. Analysis of a reciprocal translocation by quinacrine (A) and chromomycin A_3-methyl green (B) fluorescence. The chromosomes shown are from cells of a fetus who had a parent with a reciprocal translocation between chromosomes #5 and #12, i.e., rcp (t(5q; 12p). The chromosomes (from two separate cells) shown were stained with quinacrine or with chromomycin A_3 plus methyl green. The normal chromosomes are in the top two rows, while the translocation chromosomes are in the last two rows. The fetus had the translocation in a balanced form and was born phenotypically normal. (Reproduced from ref. [367a], with permission of the publisher.)

access to high-wavelength light sources and detectors, independence of low-wavelength dyes, or, in appropriate combinations, extended multidye energy donor-acceptor chains. At least three different types of dyes have been introduced for this use.

1. High-wavelength, cellular-permeable fluorochromes. One of a number of dyes supplied by Eastman Kodak, a thioimidazole dye (LL585) (Fig. 32), has a number of desirable properties. It has high absorption near 550 nm, emits in the red (Fig. 33), has little fluorescence when free, permeates live cells without high toxicity, is highly specific for DNA versus RNA, and can be used to detect DNA in agarose gels [248]. While, in principle, LL585 should have a use in flow cytometry of live cells, and it has been useful in permeabilized cells [247], thus far it has yielded only poor-resolution DNA flow histograms with live cells. Perhaps it binds to non-nucleic acid components or unevenly permeates cells. This latter point can be examined by making and testing more hydrophobic derivatives of LL585. At present, LL585 can be used instead of propidium, without RNase, for DNA staining in permeabilized cells, for detecting nucleated cells or isolated DNA (under conditions not requiring high resolution), and for staining metaphase chromosomes. Zelenin et al. [476] describe alternative desirable properties (G-C specificity, emission >650 nm but low quantum yield and high G1 CV) for 7-aminoactinomycin D, which should serve a similar purpose. They have stated that 7-aminoactinomycin D is insensitive to different states of chromatin condensation. In contrast, Darzynkiewicz et al. [94] find 7-aminoactinomycin D binding to change dramatically with differentiation, and especially if histones are removed (perhaps in part transiently) by 0.1 N HCl treatment followed by dilution with neutral buffer.*
2. A different approach to effect the incorporation of cationic dyes into live cells is the reversible reduction of these dyes; this approach has been developed by Gallop et al. [133] with ethidium, which can be reduced to hydroethidine with sodium borohydride (Fig. 34). Ethidium itself is so much more permeable to dead cells than live cells that it can serve in an assay for the former [168]. The blue fluorescent, colorless hydroethidine, which in previous studies failed to bind tRNA (and presumably DNA) [414], permeates cells. At least part is reoxidized to a racemic mixture, including the parent compound, ethidium, which then stains cellular nucleic acids yielding red fluorescence [133]. Previous studies [319] had also indi-

*Shapiro and Stephens [386a] have recently described three laser dyes, all with substantial absorption above 600 nm (i.e., oxazine-750, LD-700, and rhodamine-800), which appear to be extremely promising as cellular DNA stains. A major virtue of these dyes, besides their DNA specificity, is their compatibility with low-power, high-wavelength excitation, such as that provided by the 633-nm emission of a helium neon laser.

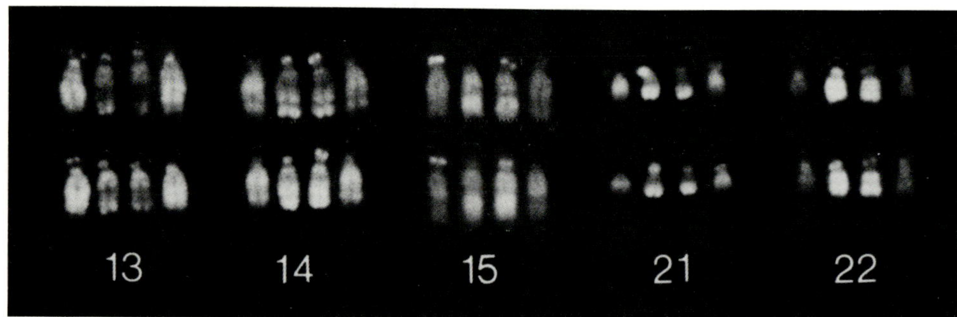

Fig. 31. Identification of polymorphic centromeric regions in human acrocentric chromosomes by staining with quinacrine (outer member of each pair) or chromomycin A_3 plus methyl green (inner member of each pair). (Reproduced from ref. [245], with permission of the publisher.)

S
N
S
N
CH=CH
$N(CH_3)_2$
CH_3-⊙-SO_3^-
C_2H_5

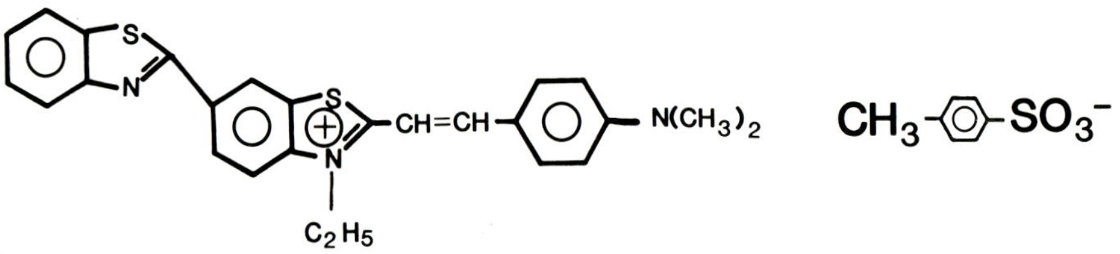

Fig. 32. Chemical formula of the dye LL585. (Reproduced from ref. [248], with permission of the publisher.)

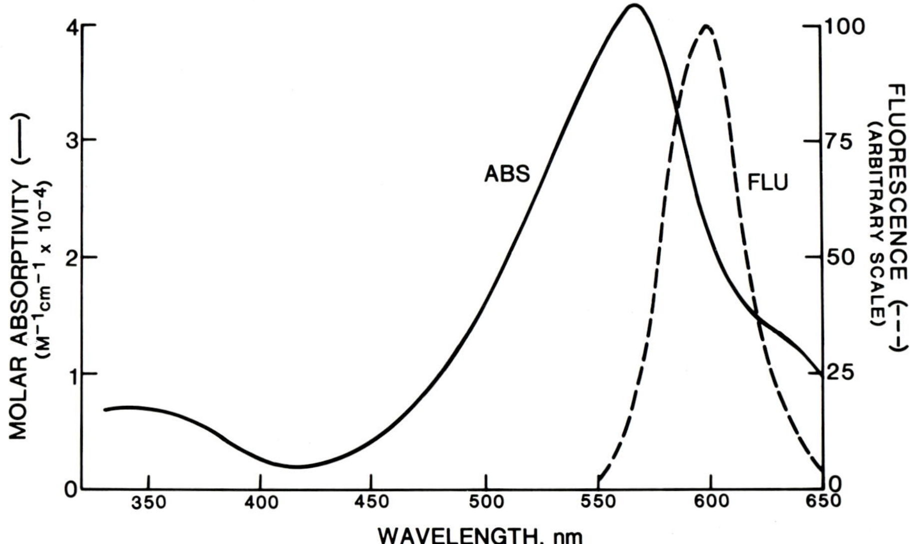

Fig. 33. Absorption and fluorescence of the dye LL585 complexed with calf thymus DNA. 0.15 M NaCl, 0.005 M Hepes, pH 7.0, dye/DNA = 10^{-2} (DNA = 10^{-3} M, absorption; 5 × 10^{-4} M, fluorescence). (Data from [247].)

cated that the antitumor effects of reduced ethidium were comparable to those of ethidium, although without considering the oxidation of reduced ethidium upon traversing the cell membrane. The generality of this approach for delivering different dyes into cells [30] or across the blood-brain barrier is apparent.

3. Phycoerythrins, plant pigments with a wide range of red fluorescence, have been introduced as cellular stains by Oi, Glazer, and Stryer [145,326]. While initially used for cell surface staining (reviewed in Chapter 17, this volume), such dyes, their derivatives [327], including tandem pairs [144], could in principle be used as part of nucleic acid-specific stains, with the advantage of multiple chromophores and increased signal per tag.

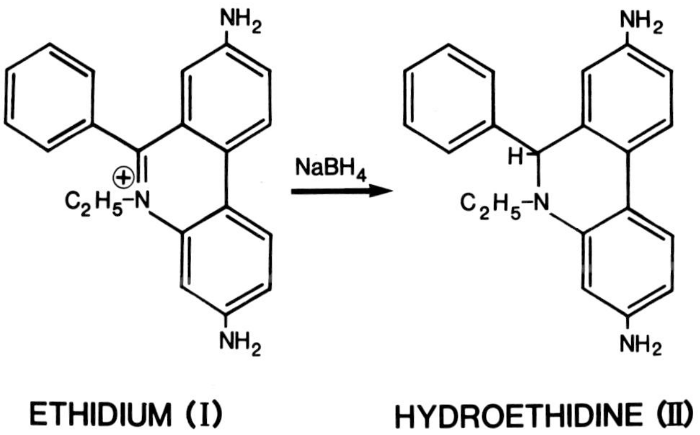

ETHIDIUM (I) NaBH$_4$ HYDROETHIDINE (II)

Fig. 34. Reduction of ethidium bromide to hydroethidine by $NaBH_4$. (Reproduced from ref. [133], with permission of the publisher.)

New Permeability Modes

1. *Boradeption:* Gallop and associates have utilized reversible complexation of boronic acid conjugated dyes with a dihydroxy buffer or complexation of dihydroxy dyes with boronic acids to transfer these dyes into live cells [132,133]. Thus far, the dyes used have themselves had a wide range of specificity and vivid, complex staining patterns have been produced. Specificity of binding for cis-hydroxyls, for example, ribose, is expected of the first type of boronic dyes but has not yet been established.
2. Gallop and associates have also exploited noncovalent complexes of aromatic acids (e.g., 3-amino, 6-7′, 7′, 8′, 8′-tricyanoquinodimethane-phenylboronic acid) with other dyes, for example, acriflavines [133], which then permeate and stain live cells. Variation of this approach to highlight specific nucleic acid components should be possible in the future.

Different DNA structures: B versus Z, etc.

Early circular dichroic data suggesting the existence of a form of poly(dG-dC) in high salt with a chirality opposite to that of B-DNA [345] was corroborated and extended by the x-ray crystallographic data of Rich and associates [354,441], establishing the existence of Z DNA in a crystalline state. The existence, induction, localization, and significance of this DNA in vivo, and hence prospects for its immunofluorescent detection, remains a matter of debate [162, 189,321,461]. The initial Phol and Jovin studies demonstrated that intercalating dyes retarded formation of this new structure. Hence one would expect such dyes to bind less tightly to Z versus B DNA. Nonintercalators, such as 33342 or 33258 Hoechst, do not bind tightly to Z-DNA [189]. Much more work is needed to define the conditions promoting dye interactions with this almost certainly important DNA conformation.

Enzyme-Linked Readouts

For stains of DNA, occurring at 6×10^9 bp per diploid human cell, dye fluorescence sensitivity is rarely a limiting factor. However, when attempting immunological detection of a specific sequence or family of sequences [16,118] detection sensitivity becomes all-important. Anticipating success in other components of this system, one can anticipate that some means of amplifying immunochemical signals will be needed. Enzyme-linked assays, such as the polyalkaline phosphatase methodology of Ward and associates with biotin conjugates [224,250] should be considered a forerunner of this approach.

Other Optical Modalities

In addition to fluorescence intensity, other modalities of fluorescence can in principle lend themselves to flow cytometry. Two limiting factors are the time during which dye fluorescence is usually observed (5×10^{-6} sec) and dye emission lifetimes (τ) (10^{-9}–10^{-8}-sec range). In principle, one might be able to obtain lifetime data directly (using appropriate excitation and processing of 10^3 photons per cell) [290]. An easier approach might be to compute the ratio of lifetime (τ) to rotational relaxation time via measurement of steady-state polarization. Depending on the readout geometries (180° or 90° apart) [197,370], horizontal or vertically polarized light is needed to balance channels for $I_{\parallel}$ and $I_{\perp}$. Dye binding, alignment, and energy transfer (i.e., proximity) are all features that can influence such measurements, in addition to lifetime (τ). Spectra of cells in a flow cytometer have been obtained using a vidicon detector [440], although such spectra were typically averaged over multiple cells, since nearly 0.5 msec was needed to obtain a single spectrum. Very recently, a method was described for fluorescence detection of the circular dichroism of ethidium bromide, which in turn is highly sensitive to changes in dye geometry [222]. Application of this methodology to flow cytometry was suggested by the authors. As described, this would require averaging of signals derived from multiple cells, but the instrumentation could in principle be modified to obtain data on individual cells, at least at a small number of exciting wavelengths.

In the case of spin-forbidden transitions (e.g., phosphorescence or lanthanide emission) [169], the time scale of emission is increased by orders of magnitude. If emission is delayed too much, that is, $\tau > 10^{-6}$ sec, then intensity per cell becomes limiting and emission is spread out over a large-stream distance. If this is increased it may be necessary to reduce particle flow rate. But there should be an intermediate range in which spin-forbidden transitions and flow cytometry are compatible.

Fixation of Probes In Situ

Optical substituents, for example, diazo groups, can permit fixation of optical probes in cells for subsequent chemical localization following cell sorting. Similarly, immunofluorescent methodology is amenable to this approach. While largely for the future, it is likely to assume greater importance as the crucial problem of deducing three-dimensional spatial relationships of components within the nucleus is approached [2,155,282–284]. Slit scanning of somatic or germ cells or gametes [19,459] can further resolve spatial detail of fluorescence.

Sequence Specific Probes

This is perhaps the ultimate goal of optical probing of nuclei or chromosomes, at least from the perspective of specificity, which is inherent in nucleic acid sequences and exceeds that attainable by dyes. Some specificity is achieved by restriction enzyme digestion of chromosomes, which can produce banding [25,301,304] or specific labeling [198, 215]. If done with enzymes such as MboI, which generate clonable DNA ends, this might both help resolve chromosomes and provide DNA for subsequent cloning. Alternatively, pretreatment of cells with dyes, such as distamycin A [347,372] or compounds such as aphidicoin [146] that bring out fragile sites [473], might help isolate individual chromosomes. Recently published oligonucleotide-ferrous-chelate compounds may provide a new approach to cleave denatured DNA at specific sites; subsequent recognition of these sites by appropriate stains, or even attachment of dyes to the oligonucleotides might provide a yet newer approach to sequence-specific staining.

One might include repeated, centromeric-specific [177, 451,465,469] or other [67,158,249,349] chromosome-specific repeat sequences, and single-copy DNA sequences as well as RNA sequences as potential targets. Studies with anti-BrdU and propidium diiodide have demonstrated that partial, controlled denaturation of cellular DNA can be achieved and the cells still studied by cytometry. The extension of this to specific DNA sequences is more difficult, although, if highly repeated sequences are studied, only partial denaturation of cellular DNA should be necessary. Alternatively, one might rely on proteins, for example, repressors [202,263,458] or synthetic distamycin analogues [206] with high specificity for nondenatured DNA, followed by secondary immunochemical detection of the probe. The chemistry to add probes to oligonucleotides certainly exists [10,59, 380].

Detection of RNA poses less of a preparative problem, in that the RNA may already be single-stranded and complementary labeled probes can easily be synthesized. Stability and localization of the RNA may prove to be the limiting feature. Whatever the interaction, one will still be facing a photon-limited situation. The original sensitivity estimate in flow of 3×10^3 fluorescein molecules [271] can be reduced only slightly by more powerful light sources. Better chromophores, for example, phycobiliproteins, or amplified, indirect labeling might reduce the number of primary determinants to the 10^2 range. If these determinants reflect nucleic acid substituents at high density, for example, 1/20 bases, one might then detect 2 kb of hybridized sequence. This size would have to be multiplied by the inverse of hybridization efficiency. However, for repeated DNA sequences or highly expressed RNA, detection in cells should be possible, with methodological improvement required for single-copy sequences. For isolated chromosomes, maintenance of chromosome morphology under denaturing conditions becomes a problem. However, one could envision detection by DNA hybridization to chromosome repeat (condition-dependent annealing) sequences. Given enough chromosome-specific sequences, acquired from chromosome-enriched recombinant libraries produced by flow sorting alone [15,76, 97,100,219,246,472] or flow sorting in combination with velocity sedimentation [63], multiple probes per chromosome should be possible via nucleic acid-hybridization [253].

ACKNOWLEDGMENTS

Those parts of this work described as from the authors' laboratories were supported by grants from the National Institutes of Health: GM21121/33579 (S.A.L.) and HD17665 (R.G.L.).

REFERENCES

1. **Agar DW, Bailey JE (1982)** Cell cycle operation during batch growth of fission yeast populations. Cytometry 3:123–128.
2. **Agard DA, Sedat JN (1983)** Three-dimensional architecture of a polytene nucleus. Nature (Lond) 302: 676–681.
3. **Angerer LM, Moudrianakis EN (1972)** Interaction of ethidium bromide with whole and selectively deproteinized deoxynucleoproteins from calf thymus. J. Mol. Biol. 63:505–521.
4. **Angerer LM, Georghiou S, Moudrianakis EN (1974)** Studies on the structure of deoxyribonucleoproteins. Spectroscopic characterization of the ethidium bromide binding sites. Biochemistry 13:1075–1082.
5. **Arber W (1974)** DNA modification and restriction. In Choen WE (ed), "Progress in Nucleic Acid Research and Molecular Biology," vol. 14. Orlando, FL: Academic Press, pp 1–37.
6. **Arndt-Jovin DJ, Jovin TM (1977)** Analysis and sorting of living cells according to deoxyribonucleic acid content. J. Histochem. Cytochem. 25:585–589.
7. **Arndt-Jovin DJ, Latt SA, Striker G, Jovin TM (1979)** Fluorescence decay analysis in solution and in a microscope of DNA and chromosomes stained with quinacrine. J. Histochem. Cytochem. 27:87–95.
8. **Assa-Munt N, Denny WA, Leupin W, Kearns DR (1985)** ^{1}H NMR study of the binding of bis (acridines) to $d(AT)_5$ $d(AT)_5$ I. Mode of binding. Biochemistry 24:1441–1449.
9. **Assa-Munt N, Leupin W, Denny WA, Kearns DR (1985)** ^{1}H NMR study of the binding of bis (acridines) to $d(AT)_5$ $d(AT)_5$ II. Dynamic aspects. Biochemistry 24:1449–1460.
10. **Asseline U, Delarue M, Lancelot G, Toulme F, Thuong NT, Montenay-Garestier T, Helene C (1984)** Nucleic acid-binding molecules with high affinity and base sequence specificity: Intercalating agents covalently linked to oligodeoxynucleotides. Proc. Natl. Acad. Sci. USA 81:3297–3301.
11. **Aubin JE (1978)** Mammalian cell autofluorescence. "Abstracts, Engineering Foundation Conference on Automated Cytology VI." Schloss Elmau, Germany, April 1978.
12. **Baldini G, Doglia S, Dolci S, Sassi G (1981)** Fluorescence-determined preferential binding of quinacrine to DNA. Biophys. J. 36:465–477.
13. **Barrio JR, Secrist JA, Leonard NJ (1972)** Fluorescent adenosine and cytidine derivatives. Biochem. Biophys. Res. Commun. 46:597–604.
14. **Bases R, Mendez F, Hsu KC (1975)** Immunoreactivity antinucleoside antibodies persists during G-2 arrest in X-irradiated HeLa cells. Exp Cell Res 92:505–509.
15. **Baum R (1985)** Chromosome-specific human gene libraries made available. Chem. Eng. News March 4:24–25.
16. **Bauman JGJ, Wiegant J, van Duijn P (1983)** The development, using poly(Hg-U) in a model system, of a new method to visualize cytochemical hybridization in fluorescence microscopy. J. Histochem. Cytochem. 31:571–578.
17. **Beck H-P (1981)** Proliferation kinetics of perturbed cell populations determined by the bromodeoxyuridine-33258 technique: Radiotoxic effects of incorporated [^{3}H] thymidine. Cytometry 2:170–174.
18. **Behr W, Honikel K, Hartmann G (1982)** Interaction of the RNA polymerase inhibitor chromomycin with DNA. Eur. J. Biochem. 9:82–92.
19. **Benaron DA, Gray JW, Gledhill BL, Lake S, Wyrobek AJ, Young IT (1982)** Quantification of mammalian sperm morphology by slit-scan flow cytometry. Cytometry 2:344–349.

20. **Benson RC, Meyer RA, Zaruba ME, McKhann GMT** (1979) Autofluorescence of viable cultured mammalian cells. J. Histochem. Cytochem. 27:36–43.
21. **Berlowitz L, Pallotta D, Sibley CH** (1969) Chromatin and histones; binding of tritiated actinomycin D to heterochromatin in mealy bugs. Science 164:1527–1529.
22. **Berns MW, Floyd AD** (1971) Chromosomal microdissection by laser. Exp. Cell Res. 67:305–310.
23. **Bertalanaffy FD** (1962) Evaluation of the acridine-orange fluorescence microscope method for cytodiagnosis of cancer. Ann. NY Acad. Sci. 93:717–750.
24. **Bertuzzi A, Gandolfi A, Germani A, Spano M, Starace G, Vitelli R** (1984) Analysis of DNA synthesis rate of cultured cells from flow cytometric data. Cytometry 5:619–628.
25. **Bianchi MS, Bianchi NO, Pantelias GE, Wolff S** (1985) The mechanism and pattern of banding induced by restriction endonucleases in human chromosomes. Chromosoma 91:131–136.
26. **Bidet R, Chambron J, Weill G** (1971) Hétérogénéite des sites de fixation de la proflavine sur le DNA. Biopolymers 10:225–242.
27. **Bird AP** (1984) DNA methylation—How important in gene control? Nature (Lond) 307:503–504.
28. **Blake A, Peacocke AR** (1968) The interaction of aminoacridines with nucleic acids. Biopolymers 6:1225–1253.
29. **Blasi F, Toniolo D** (1983) DNA methylation and X-chromosome inactivation. Mol. Biol. Med. 1:271–274.
30. **Bodor N, Farag HH** (1983) Improved drug delivery through biological membranes. II. A redox chemical drug-delivery system and its use for brain-specific delivery of phenylethylamine. J. Med. Chem. 26:313–318.
31. **Bohmer R-M, Ellwart J** (1981) Cell cycle analysis by combining the 5-bromodeoxyuridine/33258 Hoechst technique with DNA-specific ethidium bromide staining. Cytometry 2:31–34.
32. **Bonaly J, Mestre JC** (1981) Flow fluorometric study of DNA content in nonproliferative Euglena gracilis cells and during proliferation. Cytometry 2:35–38.
33. **Bontemps J, Fredericq E** (1974) Comparative binding study of the interaction of quinacrine and ethidium bromide with DNA and nucleohistone. Biophys. Chem. 2:1–22.
34. **Bontemps J, Houssier C, Fredericq E** (1975) Physicochemical study of the complexes of "33258 Hoechst" with DNA and nucleohistone. Nucleic Acids Res. 2: 971–984.
35. **Bradley DF, Wolf MK** (1959) Aggregation of dyes bound polyanions. Proc. Natl. Acad. Sci. USA 45: 944–946.
36. **Bradley DF, Felsenfeld G** (1959) Aggregation of an acridine dye on native and denatured deoxyribonucleates. Nature (Lond) 184:1920–1922.
37. **Bradley DF** (1962) Molecular biophysics of dye-polymer complexes. Trans. NY Acad. Sci. Ser. II 24: 64–74.
38. **Bradley DF, Lifson S** (1967) Statistical mechanical analysis of binding of acridines to DNA. In Pullman B (ed), "Molecular Associations in Biology." Orlando, FL: Academic Press, pp 261–270.
39. **Brdar B, Reich E** (1972) 7-Deazanebularin metabolism in cultures of mouse fibroblasts and incorporation into cellular and viral nucleic acids. J. Biol. Chem. 247:725–730.
40. **Brenner S, Pepper D, Berns MW, Tan E, Brinkley BR** (1981) Kinetochore structure, duplication and distribution in mammalian cells: Analysis by human autoantibodies from scleroderma patients. J. Cell Biol. 91:95–102.
41. **Brodie S, Giron J, Latt SA** (1975) Estimation of accessibility of DNA in chromatin from fluorescence measurements of electronic excitation energy transfer. Nature (Lond) 253:470–471.
42. **Brothman AR, Lindell TJ** (1984) Actinomycin D in low concentrations binds uniformly to human chromosomes. Exp. Cell Res. 151:252–257.
43. **Burns VWF** (1969) Fluorescent decay time characteristics of the complex between ethidium bromide and nucleic acids. Arch. Biochem. Biophys. 133: 420–424.
44. **Buys CHCM, Koertz T, Aten JA** (1982) Well identifiable human chromosomes isolated from mitotic fibroblasts by a new method. Hum. Genet. 61:157–159.
45. **Buys CHCM, Koertz T, van der Veen AY** (1984) Banding of unfixed mitotic chromosomes in suspension after release from human lymphocytes and fibroblasts. Hum. Genet. 66:361–364.
46. **Calarco-Gillam PD, Siebert MC, Hubble R, Mitchison T, Kirschner M** (1983) Centrosome development in early mouse embryos as defined by an autoantibody against pericentriolar material. Cell 35: 621–629.
47. **Callis J, Hoehn H** (1976) Flow fluorometric diagnosis of euploid and aneuploid human lymphocytes. Am. J. Hum. Genet. 28:577–584.
48. **Campbell G LeM, Gledhill BL** (1973) Chromatin of primitive erythroid cells from the chick embryo. I. Changes in acridine orange binding and the sensitivity to thermal denaturation during maturation. Chromosoma 41:385–394.
49. **Canellakis ES, Shaw YH, Hanners WE, Schwartz RA** (1976) Diacridines: Bifunctional intercalators. I. Chemistry, physical chemistry and growth inhibitory properties. Biochim. Biophys. Acta 418:277–289.
50. **Carrano AV, Gray JW, Moore DH II, Minkler JL, Mayall BH, Van Dilla MA, Mendelsohn ML** (1976) Purification of the chromosomes of the Indian Muntjac by flow sorting. J. Histochem. Cytochem. 24: 348–354.
51. **Carrano AV, Gray JW, Van Dilla MA** (1978) Flow cytogenetics: Progress towards chromosome aberration detection. In Evans HJ, Lloyd DC (eds), "Mutagen-Induced Chromosome Damage in Man." Edinburgh: Edinburgh University Press, pp 326–338.
52. **Caspersson T, Farber S, Foley GE, Kudynowski J, Modest EJ, Simonsson E, Wagh U, Zech L** (1968) Chemical differentiation along metaphase chromosomes. Exp. Cell Res. 49:219–222.
53. **Caspersson T, Zech L, Modest EJ, Foley GE, Wagh U, Simonsson E** (1969) DNA-binding fluorochromes for the study of the organization of the metaphase nucleus. Exp. Cell Res. 58:141–152.
54. **Caspersson T, Lomakka G, Zech L** (1971) The 24 fluorescence patterns of the human metaphase chro-

mosomes—distinguishing characters and variability. Hereditas. 67:89–102.
55. **Caspersson T, Lindsten J, Lomakka G, Moller A, Zech L** (1972) The use of fluorescent techniques for the recognition of mammalian chromosomes and chromosome regions. Int. Rev. Exp. Pathol. 11:1–72.
56. **Cerami A, Reich E, Ward DC, Goldbert IH** (1967) The interaction of actinomycin with DNA: Requirement for the 2-amino group of purines. Proc. Natl. Acad. Sci. USA 57:1036–1042.
57. **Chaires JB, Dattagupta N, Crothers DM** (1982) Studies on interaction of anthracycline antibiotics and deoxyribonucleic acid: Equilibrium binding studies on interaction of daunomycin with deoxyribonucleic acid. Biochemistry 21:3933–3940.
58. **Chan LM, Van Winkle Q** (1969) Interaction of acriflavine with DNA and RNA. J. Mol. Biol. 40:491–495.
59. **Chollet A, Kawashima EH** (1985) Biotin-labeled synthetic oligodeoxyribonucleotides: Chemical synthesis and uses as hybridization probes. Nucleic Acids Res. 13:1529–1541.
60. **Chu BCF, Orgel LE** (1985) Nonenzymatic sequence-specific cleavage of single-stranded DNA. Proc. Natl. Acad. Sci. USA 82:963–967.
61. **Clark RJ, Felsenfeld G** (1971) Structure of chromatin. Nature New Biol. 229:101–106.
62. **Collard JG, Philippus E, Tulp A, Lebo RV, Gray JW** (1984) Separation and analysis of human chromosomes by combined velocity sedimentation and flow sorting applying single- and dual-laser flow cytometry. Cytometry 5:9–19.
63. **Collard JG, de Boer PAJ, Janssen JWG, Schijven JF, de Jong B** (1985) Gene mapping by chromosome spot hybridization. Cytometry 6:179–185.
64. **Comings DE, Kovacs BS, Avelino E, Harris DC** (1975) Mechanisms of chromosome banding V. Quinacrine banding. Chromosoma 50:111–145.
65. **Comings DE** (1975) Mechanism of chromosome banding. VIII. Hoechst 33258-DNA interaction. Chromosoma 52:229–243.
66. **Comings DE, Drets ME** (1976) Mechanisms of chromosome banding IX. Are variations in DNA base composition adequate to account for quinacrine, Hoechst 33258 and daunomycin banding? Chromosoma 56:199–211.
67. **Cooke H, Schmidke J, Gosden JR** (1982) Characterization of a human Y chromosome repeated sequence and related sequences in higher primates. Chromosoma 87:491–502.
68. **Cooper DN** (1983) Eukaryotic DNA methylation. Hum. Genet. 64:315–333.
69. **Cowden RR, Curtis SK** (1981) Microfluorometric investigations of chromatin structure. Histochemistry 72:11–23.
70. **Cowell JK, Franks LM** (1980) A rapid method for accurate DNA measurements in single cells in situ using a simple microfluorimeter and Hoechst 33258 as a quantitative fluorochrome. J. Histochem. Cytochem. 28:206–210.
71. **Cox JV, Schenk EA, Olmsted JB** (1983) Human anticentromere antibodies: Distribution, characterization of antigens, and effect on microtubule organization. Cell 35:331–339.
72. **Cram LS, Arndt-Jovin DJ, Grimwade B, Jovin TM** (1979) Analysis of chromosomes by fluorescence polarization, fluorescence inhibition, time of flight and image slit scanning in a flow system. J. Histochem. Cytochem. 27:445–453.
73. **Cremer C, Cremer T, Gray JW** (1982) Induction of chromosome damage by ultraviolet light and caffeine: Correlation of cytogenetic evaluation and flow karyotype. Cytometry 2:287–290.
74. **Cremer C, Gray JW** (1982) Application of the BrdU/thymidine method to flow cytogenetics: Differential quenching/enhancement of Hoechst 33258 fluorescence of late-replicating chromosomes. Somat. Cell Genet. 8:319–327.
75. **Cremer C, Gray JW** (1983) Replication kinetics of Chinese hamster chromosomes as revealed by bivariate flow karyotyping. Cytometry 3:282–286.
76. **Cremer C, Rappold G, Gray JW, Muller CR, Ropers H-H** (1984). Preparative dual-beam sorting of the human Y chromosome and in situ hybridization of cloned DNA probes. Cytometry 5:572–579.
77. **Crissman HA, Tobey RA** (1974) Cell-cycle analysis in 20 minutes. Science 184:1297–1298.
78. **Crissman HA, Oka MS, Steinkamp JA** (1976) Rapid staining methods for analysis of deoxyribonucleic acid and protein in mammalian cell. J. Histochem. Cytochem. 24:64–71.
79. **Crissman HA, van Egmond J, Holdrinet RS, Pennings A, Haanen C** (1981) Simplified method for DNA and protein staining of human hematopoietic cell samples. Cytometry 2:59–62.
80. **Crissman HA, Steinkamp JA** (1982) Rapid, one step staining procedures for analysis of cellular DNA and protein by single and dual laser flow cytometry. Cytometry 3:84–90.
81. **Crissman HA, Darzynkiewicz Z, Tobey RA, Steinkamp JA** (1985) Correlated measurements of DNA, RNA and protein in individual cells by flow cytometry. Science 228:1321–1324.
82. **Crothers DM** (1968) Calculation of binding isotherms for heterogeneous polymers. Biopolymers 6: 575–584.
83. **Curtis SK, Cowden RR** (1981) Four fluorochromes for the demonstration and microfluorometric estimation of RNA. Histochemistry 72:39–48.
84. **Curtis SK, Cowden RR** (1983) Evaluation of five basic fluorochromes of potential use in microfluorometric studies of nucleic acids. Histochemistry 78:503–511.
85. **Dabrowiak JC** (1983) Minireview: Sequence specificity of drug–DNA interactions. Life Sci. 32:2915–2931.
86. **Dalgleish DG, Dingoyr E, Peacocke AR** (1973) A comparison of the interactions of proflavine with DNA and with deoxyribonucleohistone using circular dichroism spectroscopy. Biopolymers 12:445–457.
87. **D'Andrea AD, Tantravahi U, Lalande M, Perle MA, Latt SA** (1983) High resolution analysis of the timing of replication of specific DNA sequences during S phase of mammalian cells. Nucleic Acids Res. 11: 4753–4774.
88. **Dangl JL, Parks DR, Oi VT, Herzenberg LA** (1982) Rapid isolation of cloned isotype switch variants using fluorescence activated cell sorting. Cytometry 2: 395–401.

89. **Darzynkiewicz Z, Traganos F, Arlin ZA, Sharpless T, Melamed MR (1976)** Cytofluorometric studies on conformation of nucleic acids in situ. II. Denaturation of deoxyribonucleic acid. J. Histochem. Cytochem. 24:49–58.
90. **Darzynkiewicz Z, Andreff M, Traganos F (1978)** Discrimination of cycling and noncycling lymphocytes by BUdR-suppressed acridine orange fluorescence in a flow cytometric system. Exp. Cell Res. 115:31–35.
91. **Darzynkiewicz Z, Traganos F, Xue S, Staiano-Coico L, Melamed MR (1981)** Rapid analysis of drug effects on the cell cycle. Cytometry 1:279–286.
92. **Darzynkiewicz Z, Evenson D, Kapuscinski J, Melamed MR (1983)** Denaturation of RNA and DNA in situ induced by acridine orange. Exp. Cell Res. 148:31–46.
93. **Darzynkiewicz Z, Traganos F, Melamed MR (1983)** Distinction between 5-bromodeoxyuridine labelled and unlabelled mitotic cells in flow cytometry. Cytometry 3:345–348.
94. **Darzynkiewicz Z, Traganos F, Kapuscinski J, Staiano-Coico L, Melamed MR (1984)** Accessibility of DNA in situ to various fluorochromes: Relationship to chromatin changes during erythroid differentiation of friend leukemia cells. Cytometry 5:355–363.
95. **Dean PN, Gray JW, Dolbeare FA (1982)** The analysis and interpretation of DNA distributions measured by flow cytometry. Cytometry 3:188–195.
96. **Dean PN, Dolbeare F, Gratzner H, Rice GC, Gray JW (1984)** Cell cycle analysis using a monoclonal antibody to BrdUrd. Cell Tissue Kinet. 17:427–436.
97. **Deaven L (1984)** Construction of human chromosome-specific DNA libraries from flow-sorted chromosomes. Am. J. Hum. Genet. 36:Abstract #396, 134S.
98. **DeNiss IS, Morgan AR (1970)** Studies on the mechanism of DNA cleavage by ethidium. Nucleic Acids Res. 3:315–323.
99. **Deubel V, Leng M (1974)** Interaction between proflavine and double stranded polynucleotides. Biochemie. 56:641–648.
100. **Disteche CM, Kunkel LM, Lojewski A, Orkin SH, Eisenhard M, Sahar E, Travis B, Latt SA (1982)** Isolation of mouse X-chromosome specific DNA from an X-enriched lambda phage library derived from flow sorted chromosomes. Cytometry 2:282–286.
101. **Disteche D, Bontemps J (1974)** Chromosome regions containing DNAs of known base composition, specifically evidenced by 2,7-di-t-butyl proflavine. Chromosoma 47:263–281.
102. **Dolbeare F, Gratzner H, Pallavicini MG, Gray JW (1983)** Flow cytometric measurement of total DNA content and incorporated bromodeoxyuridine. Proc. Natl. Acad. Sci. USA 80:5573–5577.
103. **Dreyer GB, Dervan PB (1985)** Sequence-specific cleavage of single-stranded DNA: Oligodeoxynucleotide-EDTA FE(II). Proc. Natl. Acad. Sci. USA 82: 968–972.
104. **Durand RE, Olive PL (1982)** Cytotoxicity, mutagenicity and DNA damage by Hoechst 33342. J. Histochem. Cytochem. 30:111–116.
105. **Dutrillaux MB, Laurent C, Couturier J, LeJeune J (1973)** Coloration des chromosomes humains par l'acridine orange après traitement par le 5-bromodeoxyuridine. CR Acad. Sci. Paris 276:3179–3181.
106. **Dutrillaux MB, Fosse AM, Prieur M, LeJeune J (1974)** Analyse des echanges de chromatides dans les cellules somatiques humaines. Chromosoma 48:327–340.
107. **Dutrillaux B, Couturier J, Lise-Richer C, Viegas-Pequignot E (1976)** Sequence of DNA replication in 277 R- and Q-bands of human chromosomes using a BrdU treatment. Chromosoma 58:51–61.
108. **Dyer DL, Mori K (1969)** Fluorescent nuclear staining with europium thenolytrifluoroacetonate. J. Histochem. Cytochem. 17:755–757.
109. **Earnshaw WC, Rothfield N (1985)** Identification of a family of human centromere proteins using autoimmune sera from patients with scleroderma. Chromosoma 91:313–321.
110. **Egorin MJ, Clawson RE, Cohen JL, Ross LA, Bachur NR (1980)** Cytofluorescence localization of anthracycline antibiotics. Cancer Res. 40:4669–4676.
111. **Ellison JR, Barr HJ (1972)** Quinacrine fluorescence of specific chromosome regions, late replication and high AT content in Samoaia leonensis. Chromosoma 36:375–390.
112. **Erlanger BF, Beiser SM (1964)** Antibodies specific for ribonucleosides and ribonucleotides and their reaction with DNA. Proc. Natl. Acad. Sci. USA 52:68–70.
113. **Evett J, Isenberg I (1969)** DNA polylysine interaction as studied by polarization of fluorescence. Ann. NY Acad. Sci. 158:210–222.
114. **Fantes JA, Green DK, Cooke HJ (1983)** Purifying human Y chromosomes by flow cytometry and sorting. Cytometry 4:88–91.
115. **Fantes JA, Green DK, Elder JK, Malloy P, Evans HJ (1983)** Detecting radiation damage to human chromosomes by flow cytometry. Mutation Res. 119: 161–168.
116. **Farber J, Baserga R, Gabbay EJ (1971)** The effect of a reporter molecule on chromatin template activity. Biochem. Biophys. Res. Commun. 43:675–681.
117. **Formoso C (1973)** Fluorescence of nucleic acid-terbium (III) complexes. Biochem. Biophys. Res. Commun. 53:1084–1087.
118. **Forster AC, McInnes JL, Skingle DC, Symons RH (1985)** Non-radioactive hybridization probes prepared by the chemical labelling of DNA and RNA with a novel reagent, photobiotin. Nucleic Acids Res. 13:745–761.
119. **Forster T (1948)** Zwischenmolekulare Energiewanderung und Fluoreszenz. Ann. Physik. 2:55–75.
120. **Forster T (1965)** Delocalized excitation and excitation transfer. In Sinanoglu O (ed), "Modern Quantum Chemistry," vol. 3. Orlando, FL: Academic Press, pp 43–75.
121. **Fox KR, Waring MJ (1984)** DNA structural variations produced by actinomycin and distamycin as revealed by DNAse I footprinting. Nucleic Acids Res. 12:9271–9285.
122. **Fox KR, Waring MJ (1985)** Kinetic evidence that echinomycin migrates between potential DNA binding sites. Nucleic Acids Res. 13:595–603.
123. **Franceschini P (1974)** Semiconservative DNA duplication in human chromosomes treated with BUdR

and stained with acridine orange. Exp. Cell Res. 89: 420–421.

124. **Frankfurt OS (1980)** Flow cytometric analysis of double-stranded RNA content distributions. J. Histochem. Cytochem. 28:663–669.

125. **Frankfurt OS (1983)** Increased uptake of vital dye Hoechst 33342 during S phase in synchronized HeLa S3 cells. Cytometry 4:216–221.

126. **Fredericq E (1971)** The chemical and physical properties of nucleohistones. In Phillips DMP (ed), "Histones." New York: Plenum Press, pp 135–186.

127. **Freeman MVR, Beiser SM, Erlanger BF, Miller OJ (1971)** Reaction of antinucleoside antibodies with human cells in vitro. Exp. Cell Res. 69:345–355.

128. **Fried J, Doblin J, Takakmoto S, Perez A, Hansen H, Clarkson B (1982)** Effects of Hoechst 33342 on survival and growth of two tumor cell lines and on hematopoietically normal bone marrow cells. Cytometry 3:42–47.

129. **Fritzsche H, Crothers DM (1983)** ^{1}H NMR study of the interaction of the peptide antibiotic netropsin with the miniduplexes d(pApT)$_3$ d(pApT)$_3$ and d(pA)$_6$ d(pT)$_6$. Studia Biophys. 97:43–48.

130. **Gabbay EF (1969)** Topography of nucleic acid helices in solutions. XII. The origin of the oppositely-induced circular dichroism of reporter molecules bound to ribo- and deoxyribonucleic acid. J. Am. Chem. Soc. 91:5136–5150.

131. **Galley WC, Purkey RM (1972)** Spin-orbital probes of bimolecular structure. A model DNA-acridine system. Proc. Natl. Acad. Sci. USA 69:2198–2202.

132. **Gallop PM, Paz MA, Henson E (1982)** Boradeption: A new procedure for transferring water-insoluble agents across cell membranes. Science 217:166–169.

133. **Gallop PM, Paz MA, Henson E, Latt SA (1984)** Dynamic approaches to the delivery of reported reagents into living cells. Biotechniques: J. Lab. Technol. Bio-Res. Jan/Feb: 32–36.

134. **Gatti G, Houssier C, Fredericq E (1975)** Binding of ethidium bromide to ribosomal RNA. Biochim. Biophys. Acta. 407:308–319.

135. **Genest D, Wahl P, Auchet JC (1974)** The fluorescence anistropy decay due to energy transfer occurring in the ethidium bromide-DNA complex. Determination of the deformation angle of the DNA helix. Biophys. Chem. 1:266-278

136. **Gerdes J, Schwab U, Lemke H, Stein H (1983)** Production of a mouse monoclonal antibody reactive with a nuclear antigen associated with cell proliferation. Int. J. Cancer 31:13–20.

137. **Gerdes J, Lemke H, Baisch H, Wacker HH, Schwab U, Stein H (1984)** Cell cycle analysis of a cell-proliferation-associated human nuclear antigen defined by the monoclonal antibody Ki-67. J. Immunol. 133: 1710–1715.

138. **Gill JE, Mazrimas JA, Bishop CC (1974)** Physical studies on synthetic DNAs containing 5-methyl-cytosine. Biochim. Biophys. Acta. 335:330–348.

139. **Gill JE, Jotz MM (1974)** Deoxyribonucleic acid cytochemistry for automated cytology. J. Histochem. Cytochem. 22:470–472.

140. **Gill JE, Jotz MM, Young SG, Modest EJ, Sengupta SK (1975)** 7-Amino-actinomycin D as a cytochemical probe I. Spectral properties. J. Histochem. Cytochem. 23:793–799.

141. **Gill JE (1979)** Principles of cytochemistry. In Melamed M, Mendelson M, Mullaney P (eds), "Flow Cytometry and Sorting." New York: Wiley, pp 233–242.

142. **Gill JE, Wheeless LL, Hanna-Maden C, Marisa RJ (1979)** Cytofluorometric and cytochemical comparisons of normal and abnormal human cells from the female genital tract. J. Histochem. Cytochem. 27: 591–595.

143. **Gitler C, Rubalcava B, Caswell A (1969)** Fluorescence changes of ethidium bromide on binding to erythrocyte and mitochondrial membranes. Biochim. Biophys. Acta 193:479–481.

144. **Glazer A, Stryer L (1983)** Fluorescent tandem phycobiliprotein conjugates. Biophys. J. 43:383–386.

145. **Glazer AN, Stryer L (1984)** Phycofluor probes. Trends Biochem. Sci. 9:423–427.

146. **Glover TW, Berger C, Coyle T, Echo B (1984)** DNA polymerase alpha inhibition by aphidicolin induces gaps and breaks at common fragile sites in human chromosomes. Hum. Genet. 67:136–142.

147. **Gohde W, Schumann J, Buchner T, Otto F, Barlogie B (1979)** Pulse cytophotometry: Application in tumor cell biology and clinical oncology. In Melamed MR, Mullaney PF, Mendelsohn ML (eds), "Flow Cytometry and Sorting," New York: Wiley pp 599–620.

148. **Golden JR, West SS, Echols CK, Shingleton HM (1976)** Quantitative fluorescence spectrophotometry of acridine orange-stained unfixed cells. Potential for automated detection of human uterine cancer. J. Histochem. Cytochem. 24:315–321.

149. **Goppelt M, Langowski J, Pingoud A, Haupt W, Urbanke C, Mayer C, Maass G (1981)** The effect of several nucleic acid binding drugs on the cleavage of d(GGAATTCC) and pBR322 by the EcoRI restriction endonuclease. Nucleic Acids Res. 9:6115–6127.

150. **Gratzner HG, Leif RC, Ingram MJ, Castro A (1975)** The use of antibody specific for bromodeoxyuridine for the immunofluorescent determination of DNA replication in single cells and chromosomes. Exp. Cell Res. 95:88–94.

151. **Gratzner HG, Pollack A, Ingram DJ, Castro A (1976)** Deoxyribonucleic acid synthesis in single cells and chromosomes detected by immunological techniques. J. Histochem. Cytochem. 24:34–39.

152. **Gratzner HG, Leif RC (1981)** An immunofluorescence method for monitoring DNA synthesis by flow cytometry. Cytometry 1:385–389.

153. **Gratzner HG (1982)** Monoclonal antibody to 5-bromo- and 5-iododeoxyuridine: A new reagent for detection of DNA replication. Science 218:474–475.

154. **Gray JW, Langlois RG, Carrano AV, Burkhart-Schultz K, Van Dilla MA (1979)** High resolution chromosome analysis: One and two parameter flow cytometry. Chromosoma 73:9–27.

154a. **Gray JW, Langlois RG (1986)** Chromosome classification and purification using flow cytometry and sorting. Ann. Rev. Biophys. Biophys. Chem. 15:195–235.

155. **Gregoire M, Hernandez-Verdun D, Bouteille M (1984)** Visualization of chromatin distribution in living PTO cells by Hoechst 33342 fluorescent staining. Exp. Cell Res. 152:38–46.

156. **Grunberger D, Ward DC, Reich E (1972)** 7-Deaza-

nebularin: Coding properties of triplets and polynucleotides. J. Biol. Chem. 247:720–724.

157. **Gursky GV, Zasedatelev AS, Strelzov SA (1973)** Energy transfer between dye molecules bound to DNA: Application to determine the structure of complexes between small molecules and DNA. Stud. Biophys. 40:105–108.

158. **Gusella JF, Jones C, Kao F-T, Housman D, Puck TT** (1982) Genetic fine-structure mapping in human chromosome 11 by use of repetitive DNA sequences. Proc. Natl. Acad. Sci. USA 79:7804–7808.

159. **Guttann T, Votavova H, Pivec L (1976)** Base composition heterogeneity of mammalian DNAs in CsCl-netropsin density gradient. Nucleic Acids Res. 3: 835–845.

160. **Hamori E, Arndt-Jovin DJ, Grimwade BG, Jovin TM** (1980) Selection of viable cells with known DNA content. Cytometry 1:132–135.

161. **Hill BT, Whatley S (1975)** A simple, rapid microassay for DNA. FEBS Lett. 56:20–23.

162. **Hill RJ, Stollar BD (1983)** Dependence of Z-DNA antibody binding to polytene chromosome on acid fixation and DNA torsional strain. Nature (Lond) 305:338–340.

163. **Hilwig I (1970)** On the influence of a benzimidazol derivative (fluorochrome) on cell lines in tissue culture. Z. Zellforsch. 104:127–137.

164. **Hilwig I, Gropp A (1972)** A staining of constitutive heterochromatin in mammalian chromosomes with a new fluorochrome. Exp. Cell Res. 75:122–126.

165. **Hilwig I, Gropp A (1973)** Decondensation of constitutive heterochromatin in L cell chromosomes by a bisbenzimidazole compounds ("33258 Hoechst"). Exp. Cell Res. 81:474–477.

166. **Hilwig I, Gropp A (1975)** pH dependent fluorescence of DNA and RNA in cytologic staining with "33258 Hoechst." Exp. Cell Res. 91:457–460.

167. **Holmquist G, Comings D (1975)** Sister chromatid exchange and chromosome organization based on a bromodeoxyuridine Giemsa-C-banding technique (TC banding). Chromosoma 52:245–259.

168. **Horan PK, Kappler JN (1977)** Automated fluorescent analysis for cytotoxicity assays. J. Immunol. Methods 18:309–316.

169. **Horrocks W DeW, Sudnick DR (1981)** Lanthanide ion luminescence probes of the structure of biological macromolecules. Acc. Chem. Res. 14:384–392.

170. **Houssier C, Fredericq E (1964)** Electrooptical effects on nucleic acids and nucleoproteins. Biochim. Biophys. Acta. 88:450–452.

171. **Howard RJ, Battye FL, Mitchell GF (1979)** Plasmodium-infected blood cells analysed and sorted by flow fluorimetry with the deoxyribonucleic acid binding dye 33258 Hoechst. J. Histochem. Cytochem. 27:803–813.

172. **Hsu TC, Pathak S, Kusyk CJ (1975)** Continuous induction of chromatid lesions by DNA-intercalating compounds. Mutat. Res. 33:417–420.

173. **Hudson B, Upholt WB, Devinny J, Vinograd J (1969)** The use of an ethidium analogue in the dye-buoyant density procedure for the isolation of closed circular DNA: The variation of the superhelix density of mitochondrial DNA. Proc. Natl. Acad. Sci. USA 62: 813–820.

174. **Hutter KJ, Stohr M (1982)** Rapid detection of mutagen-induced micronucleated erythrocytes by flow cytometry. Histochemistry 75:353–362.

175. **Igo-Kemenes T, Horz W, Zachav HG (1982)** Chromatin. Ann. Rev. Biochem. 51:89–121.

176. **Ikeuchi T (1984)** Inhibitory effect of ethidium bromide on mitotic chromosome condensation and its application to high-resolution chromosome banding. Cytogenet. Cell Genet. 38:56–61.

177. **Jabs EW, Wolf SF, Migeon BR (1984)** Characterization of a cloned DNA sequence that is present at centromeres of all human autosomes and the X chromosome RNA shows polymorphic variation. Proc. Natl. Acad. Sci. USA 81:4884–4886.

178. **Jacobberger JW, Horan PK, Hare JD (1983)** Analysis of malaria parasite-infected blood by flow cytometry. Cytometry 4:228–237.

179. **Jacobberger JW, Horan PK, Hare JD (1984)** Flow cytometric analysis of blood cells stained with the cyanine dye $DiOC_1[3]$: Reticulocyte quantification. Cytometry 5:589–600.

180. **Jakobsen A (1983)** The use of trout erythrocytes and human lymphocytes for standardization in flow cytometry. Cytometry 4:161–165.

181. **Jensen RH (1977)** Chromomycin A_3 as a fluorescent probe for flow cytometry of human gynecologic samples. J. Histochem. Cytochem. 25:573–579.

182. **Jensen RH, Langlois RG, Mayall BH (1977)** Strategies for choosing a deoxyribonucleic acid stain for flow cytometry of metaphase chromosomes. J. Histochem. Cytochem. 25:954–964.

183. **Johannisson E, Thorell B (1977)** Mithramycin fluorescence for quantitative determination of deoxyribonucleic acid in single cells. J. Histochem. Cytochem. 25:122–128.

184. **Johnston FP, Jorgenson KF, Lin CC, van de Sande JH** (1978) Interaction of anthracyclines with DNA and chromosomes. Chromosoma 68:115–129.

185. **Jones KH, Senft JA (1985)** An improved method to determine cell viability by simultaneous staining with fluorescein diacetate-propidium iodide. J. Histochem. Cytochem. 33:77–79.

186. **Jones KW, Purden IF, Prosser J, Cornero G (1974)** The chromosomal localization of human satellite DNA. Chromosoma 49:161–171.

187. **Jorgenson KF, van de Sande JH, Lin CC (1978)** The use of base pair specific DNA binding agents as affinity labels for the study of mammalian chromosomes. Chromosoma 68:287–302.

188. **Jotz MM, Gill JE, Davis DT (1976)** A new optical multichannel microspectrofluorometer. J. Histochem. Cytochem. 24:91–99.

189. **Jovin TM, van de Sande JH, Zarling DA, Arndt-Jovin DJ, Ekstein B, Fuldner HH, Greider C, Greiger I, Hamort E, Kalisch B, McIntosh LP, Robert-Nicoud M (1982)** Generation of left-handed Z DNA in solution and visualization in polytene chromosomes by immunofluorescence. Cold Spring Harbor Symp Quant Biol. 47:143–154.

190. **Kaiser TN, Lojewski A, Dougherty C, Juergens L, Sahar E, Latt SA (1982)** Flow cytometric characterization of the response of Fanconi's anemia cells to mitomycin C treatment. Cytometry 2:291–297.

191. **Kapuscinski J, Yanagi K (1979)** Selective staining by 4′,6- diamidine-2-phenylindole of nanogram quanti-

ties of DNA in the presence of RNA on gels. Nucleic Acids Res. 6:3535–3542.

192. **Kapuscinski J, Darzynkiewicz Z, Melamed MR** (1982) Luminescence of the solid complexes of acridine orange with RNA. Cytometry 2:201–211.
193. **Kapuscinski J, Darzynkiewicz Z** (1983) Increased accessibility of bases in DNA upon binding of acridine orange. Nucleic Acids Res. 11:7555–7568.
194. **Kapuscinski J, Darzynkiewicz Z** (1984) Condensation of nucleic acids by intercalating aromatic cations. Proc. Natl. Acad. Sci. USA 81:7368–7372.
195. **Kapuscinski J, Darzynkiewicz Z** (1984) Denaturation of nucleic acids induced by intercalating agents. Biochemical and biophysical properties of acridine orange-DNA complexes. J. Biomol. Struct. Dynamics 1:1485–1499.
196. **Kato H** (1974) Spontaneous sister chromatid exchanges detected by a BUdR labelling method. Nature 251:70–72.
197. **Keene JP, Hodgson BW** (1980) A fluorescence polarization flow cytometer. Cytometry 1:118–126.
198. **Kerem B, Goitein R, Richler C, Marcus M, Ledar H** (1983) In situ nick-translation distinguishes between active and inactive X chromosomes. Nature 304:88–90.
199. **Kerker M, Van Dilla MA, Brunsting A, Kratohvil JP, Hsu P, Wang DS, Gray JW, Langlois RG** (1982) Is the central dogma of flow cytometry true: that fluorescence intensity is proportional to cellular dye content? Cytometry 3:71–78.
200. **Kersten W, Kersten H, Szybalski N** (1966) Physicochemical properties of complexes between deoxyribonucleic acid and antibiotics which affect ribonucleic acid synthesis (actinomycin, daunomycin, cinerubin, nogalamycin, chromomycin, mithramycin, and olivomycin). Biochemistry 5:236–244.
201. **Kim MA** (1974) Chromatid exchange and heterochromatin alteration of human chromosomes with BUdR-labelling demonstrated by benzimidazol fluorochrome and Giemsa stain. Humangenetik 25:179–188.
202. **Kim R, Kim SH** (1982) Direct measurement of DNA unwinding angle in specific interaction between LAC operator and repressor. Cold Spring Harbor Symp. Quant. Biol. 47:451–454.
203. **Kleiman L, Huang RCC** (1971) Binding of actinomycin D to calf thymus chromatin. J. Mol. Biol. 55: 503–521.
204. **Klein WJ Jr, Beiser SM, Erlanger BF** (1967) Nuclear fluoresence employing antinucleoside immunoglobulins. J. Exp. Med. 125:61–70.
205. **Kochetkov NK, Shibaev VN, Kost AA** (1971) New reaction of adenosine and cytosine derivatives, potentially useful for nucleic acids modification. Tetrahedron Lett. 22:1993–1996.
206. **Kopka ML, Yoon C, Goodsell D, Pjura P, Dickerson RE** (1985) The molecular origin of DNA-drug specificity in netropsin and distamycin. Proc. Natl. Acad. Sci. USA 82:1376–1380.
207. **Korenberg JR, Freedlender E** (1974) Giemsa technique for detection of sister chromatid exchanges. Chromosoma 48:355–360.
208. **Korenberg JR, Engels WR** (1978) Base ratio, DNA content and quinacrine brightness of human chromosomes. Proc. Natl. Acad. Sci. USA 75:3382–3386.
209. **Krause AW, Carley WW, Webb WW** (1984) Fluorescent erythrosin B is preferable to trypan blue as a vital exclusion dye for mammalian cells in monolayer culture. J. Histochem. Cytochem. 32:1084–1090.
210. **Krey AK, Hahn FE** (1970) Studies on the complex of distamycin A with calf thymus DNA. FEBS Lett. 10: 175–178.
211. **Krey AK, Allison RG, Hahn FG** (1973) Interactions of the antibiotic, distamycin A, with native DNA and with synthetic duplex polydeoxyribonucleotides. FEBS Lett. 29:58–62.
212. **Krishan A** (1975) Rapid flow cytofluorometric analysis of mammalian cell cycle by propodium iodide staining. J. Cell Biol. 66:188–193.
213. **Krishan A, Ganapathi R** (1979) Laser flow cytometry and cancer chemotherapy: Detection of intracellular anthracyclines by flow cytometry. J. Histochem. Cytochem. 27:1655–1657.
214. **Kubbies M, Rabinovitch P** (1983) Flow cytometric analysis of factors which influence the BrdU-Hoechst quenching effect in cultured human fibroblasts and lymphocytes. Cytometry 3:276–281.
215. **Kuo MT, Plunkett W** (1985) Nick translation of metaphase chromosomes. In vitro labelling of nuclease hypersensitive regions in chromosomes. Proc. Natl. Acad. Sci. USA 82:854–858.
216. **Labarca C, Paigen K** (1980) A simple, rapid, and sensitive DNA assay procedure. Anal. Biochem. 102: 344–352.
217. **Lalande M, Miller RG** (1979) Fluorescence flow analysis of lymphocyte activation using Hoechst 33342 dye. J. Histochem. Cytochem. 27:394–397.
218. **Lalande M, Ling V, Miller RG** (1981) Hoechst 33342 dye uptake as a probe of membrane permeability changes in mammalian cells. Proc. Natl. Acad. Sci. USA 78:363–367.
219. **Lalande M, Kunkel LM, Flint A, Latt SA** (1984) Development and use of metaphase chromosome flow-sorting methodology to obtain recombinant phage libraries enriched for parts of the human X chromosome. Cytometry 5:101–107.
220. **Lalande M, Schreck RR, Hoffman R, Latt SA** (1985) Identification of inverted duplicated #15 chromosomes using bivariate flow cytometric analysis. Cytometry 6:1–6.
221. **Lammler G, Schutze HR** (1969) Vital-fluorochromierung tierischer Zellkerne mit einem neuen Fluorochrom. Naturwissenschaften 56:286.
222. **Lamos ML, Turner DH** (1985) Fluorescence-detected circular dichroism of ethidium in vivo and bound to deoxyribonucleic acid in vitro. Biochemistry 24: 2819–2822.
223. **Landegant JE, Jansen in de Wal N, Baan RA, Hoeijmakers JHJ, van der Ploeg M** (1984) 2-Acetylaminofluorene-modified probes for the indirect hybridocytochemical detection of specific nucleic acid sequences. Exp. Cell Res. 153:61–72.
224. **Langer-Safer PR, Levine M, Ward DC** (1983) Immunological method for mapping genes on Drosophila polytene chromosomes. Proc. Natl. Acad. Sci. USA 79:4381–4385.
225. **Langlois RG, Jensen RH** (1979) Interactions between pairs of DNA-specific fluorescent stains bound to mammalian cells. J. Histochem. Cytochem. 27:72–79. Erratum p 1559.

226. **Langlois RG, Carrano AV, Gray JW, Van Dilla MA** (1980) Cytochemical studies of metaphase chromosomes by flow cytometry. Chromosoma 77: 229–251.
227. **Langlois RG, Yu L-C, Gray JW, Carrano AV** (1982) Quantitative karyotyping of human chromosomes by dual beam flow cytometry. Proc. Natl. Acad. Sci. USA 79:7876–7880.
228. **Langlois RG, Gray JW, Kerker M** (1983) The central dogma of flow cytometry (Letter to the Editor). Cytometry 3:459–460.
229. **Latt SA, Cheung AHT, Blout ER** (1965) High resolution analysis of the timing of replication of specific DNA sequences during S phase of mammalian cells. Nucleic Acids Res. 11:4753–4774.
230. **Latt SA, Sober HA** (1967) Protein-nucleic acid interactions. II. Oligopeptide-polynucleotide binding studies. Biochemistry 6:3293–3306.
231. **Latt SA** (1973) Microfluorometric detection of deoxyribonucleic acid in human metaphase chromosomes. Proc. Natl. Acad. Sci. USA 70:3395–3399.
232. **Latt SA, Gerald PS** (1973) Staining of human metaphase chromosomes with fluorescent conjugates of polylysine. Exp. Cell Res. 81:401–406.
233. **Latt SA, Brodie S, Munroe SH** (1974) Optical studies of complexes of quinacrine with DNA and chromatin: Implications for the fluorescence of cytological chromosome preparations. Chromosoma 49:17.
234. **Latt SA** (1974) Detection of DNA synthesis in interphase nuclei by fluorescence microscopy. J. Cell Biol. 62:546–550.
235. **Latt SA** (1975) Fluorescence analysis of late DNA replication in human metaphase chromosomes. Somat. Cell Genet. 1:293–321.
236. **Latt SA, Wohlleb JC** (1975) Optical studies of the interaction of 33258 Hoechst with DNA, chromatin and metaphase chromosomes. Chromosoma 52:297–316.
237. **Latt SA** (1976) Longitudinal and lateral differentiation of metaphase chromosomes based on the detection of DNA synthesis by fluorescence microscopy. In Pearson P, Lewis K (eds), "Chromosomes Today." New York: Wiley, pp 367–394.
238. **Latt SA, Stetten G** (1976) Spectral studies on 33258 Hoechst and related bisbenzimidazole dyes useful for fluorescent detection of deoxyribonucleic acid synthesis. J. Histochem. Cytochem. 24:24–33.
239. **Latt SA** (1977) Fluorescent probes of chromosome structure and replication. Can. J. Genet. Cytol. 19: 603–623.
240. **Latt SA** (1977) Fluorometric detection of deoxyribonucleic acid synthesis: Possibilities for interfacing bromodeoxyuridine dye techniques with flow fluorometry. J. Histochem. Cytochem. 25:913–926.
241. **Latt SA, George YS, Gray JW** (1977) Flow cytometric analysis of BrdU-substituted cells stained with 33258 Hoechst. J. Histochem. Cytochem. 25:927–934.
242. **Latt SA, Sahar E, Eisenhard ME** (1979) Pairs of fluorescent dyes as probes of DNA and chromosomes. J. Histochem. Cytochem. 27:65–71.
243. **Latt SA** (1979) Fluorescent probes of DNA microstructure and synthesis. In Melamed M, Mendelson M, Mullaney P (eds), "Flow Cytometry and Sorting." New York: Wiley, pp 263–284.
244. **Latt SA, Juergens LA, Matthews DJ, Gustashaw KM, Sahar E** (1980) Energy transfer-enhanced chromosome banding. An overview. Cancer Genet. Cytogenet. 1:187–196.
245. **Latt SA, Sahar E, Eisenhard ME, Juergens LA** (1980) Interactions between pairs of DNA-binding dyes: Results and implications for chromosome analysis. Cytometry 1:2–11.
246. **Latt SA, Kunkel LM, Tantravahi U, Aldridge J, Lalande M** (1983) Construction, analysis, and utilization of recombinant phage libraries obtained using fluorescence activated flowsorting. In Messer A, Porter IH (eds), "Recombinant DNA and Medical Genetics," Orlando, FL: Academic Press, pp 35–47.
247. **Latt SA, Marino M, Lalande M** (1984) New fluorochromes, compatible with high wavelength excitation, for flow cytometric analysis of cellular nucleic acids. Cytometry 5:339–347.
248. **Latt SA, Lalande M, Kunkel LM, Schreck R, Tantravahi U** (1985) Application of fluorescence spectroscopy to molecular cytogenetics. Biopolymers 24:77–95.
249. **Lau Y-F, Huang JC, Dozy AM, Kan YW** (1984) A rapid screening test for antenatal sex determination. Lancet i:14–16.
250. **Leary JJ, Brigati DJ, Ward DC** (1983) Rapid and sensitive colorimetric method for visualizing biotin-labeled DNA probes hybridized to DNA or RNA immobilized in nitrocellulose: Bio-blots. Proc. Natl. Acad. Sci. USA 80:4045–4049.
251. **Lebo RV** (1982) Chromosome sorting and DNA sequence localization. Cytometry 3:145–154.
252. **Lebo RV, Bastian AM** (1982) Design and operation of a dual laser chromosome sorter. Cytometry 3: 213–219.
253. **Lebo RV, Gorin F, Fletterick RJ, Kao FT, Cheung MC, Bruce BD, Kan YW** (1984) High resolution chromosome sorting and DNA spot-blot analysis assign McArdle's syndrome to chromosome 11. Science 225:57–59.
254. **Lee JCK, Bahr GF** (1983) Microfluorometric studies on chromosomes. Quantitative determination of protein content of Chinese hamster chromosome 1 in situ with and without trypsin digestion. Chromosoma 88: 374–376.
255. **Lee JS, Waring MJ** (1978) Bifunctional intercalation and sequence specificity in the binding of quinomycin and triostin antibiotics to deoxyribonucleic acid. Biochem. J. 173:115–128.
256. **Leonard NJ** (1985) Adenylates: Bound and unbound. Biopolymers 24:9–28.
257. **Le-Pecq JB, Paoletti C** (1967) A fluorescent complex between ethidium bromide and nucleic acids. J. Mol. Biol. 27:87–106.
258. **Le-Pecq JB** (1971) Use of ethidium bromide for separation and determination of nucleic acids of various conformational forms and measurement of their associated enzymes. In Glick D (ed), "Methods of Biochemical Analysis," vol. 20. New York: Wiley, p 41–86.
259. **Le-Pecq J, LeBret ML, Barbet J** (1975) DNA polyintercalating drugs: DNA binding of diacridine derivatives. Proc. Natl. Acad. Sci. USA 72:2915–2919.
260. **Lerman LS** (1961) Structural considerations in the interaction of DNA and acridines. J. Mol. Biol. 3: 18–30.

261. **Lerman LS (1963)** The structure of the DNA-acridine complex. Proc. Natl. Acad. Sci. USA 49:94–102.
262. **Levinson JW, Langlois RS, Maher VM, McCormick JJ (1978)** An improved acriflavine-feulgen reagent for quantitative DNA cytofluorometry. J. Histochem. Cytochem. 26:680–684.
263. **Lewis M, Jeffrey A, Wang J, Ladner R, Ptashne M, Pabo CO (1982)** Structure of the operator-binding domain of bacteriophage lambda repressor: Implications for DNA recognition and gene regulation. Cold Spring Harbor Symp. Quant. Biol. 47:435–440.
264. **Liebeskind D, Hsu KC, Erlanger B, Bases R (1974)** Immunoreactivity to antinucleoside antibodies in X-irradiated HeLa cells. Exp. Cell Res. 83:399–405.
265. **Lillie RD, Conn's HJ (1969)** "Biological Stains," 8th ed. Baltimore, MD: Williams and Wilkins, p 195.
266. **Lin CC, van de Sande JH (1975)** Differential fluorescent staining of human chromosomes with daunomycin and adriamycin—the D bands. Science 190:61–63.
267. **Lin MS, Alfi OS (1976)** Detection of sister chromatid exchanges by 4′-6′diamidino-2-phenyl indole fluorescence. Chromosoma 57:219–225.
268. **Lin MS, Comings DE, Alif OS (1977)** Optical studies of the interaction of 4′-6-diamidinio-2-phenylindole with DNA and metaphase chromosomes. Chromosoma 60:15–25.
269. **Liu LF, Wang JC (1975)** On the degree of unwinding of the DNA helix by ethidium II. Studies by electron microscopy. Biochim. Biophys. Acta. 395:401–412.
270. **Loewe H, Urbanietz J (1974)** Basisch substituierte 2,6-Bis-benzimidazol-derivate, eine neue chemotherapeutisch active Körperklasse. Arzneim. Forsch. 24: 1927–1933.
271. **Loken M, Herzenberg LA (1975)** Analysis of cell populations with a fluorescence-activated cell sorter. Ann. NY Acad. Sci. 254:163–171.
272. **Loken Mr (1980)** Simultaneous quantitation of Hoechst 33342 and immunofluorescence on viable cells using a fluorescence activated cell sorter. Cytometry 1:136–142.
273. **Loken MR (1980)** Separation of viable T and B lymphocytes using a cytochemical stain, Hoechst 33342. J. Histochem. Cytochem. 28:36–39.
274. **Lubit BW, Schreck RR, Miller OJ, Erlanger BF (1974)** Human chromosome structure as revealed by an immunoperoxidase staining procedure. Exp. Cell Res. 89:426–429.
275. **Lucas JN, Gray JW, Peters DC, Van Dilla MA (1983)** Centromeric index measurement by slit-scan flow cytometry. Cytometry 4:109–116.
276. **Luck G, Triebel H, Waring M, Zimmer CH (1974)** Conformation dependent binding of netropsin and distamycin to DNA and DNA model polymers. Nucleic Acids Res. 1:503–530.
277. **Luck G, Zimmer C, Reinert KE, Arcamone F (1977)** Specific interactions of distamycin A and its analogues with (A T) rich and (G C) rich duplex regions of DNA and deoxypolynucleotides. Nucleic Acids Res. 4:2655–2670.
278. **Luk CK (1971)** Study of the nature of the metal-binding sites and estimate of the distance between the metal-binding sites in transferrin using trivalent lanthanide ions as fluorescent probes. Biochemistry 10: 2838–2843.
279. **Mace ML Jr, Tevethia SS, Brinkley BR (1972)** Differential immunofluorescent labelling of chromosomes with antisera specific for single strand DNA. Exp. Cell Res. 75:521–523.
280. **MacInnes JW, McClintock M (1970)** Differences in fluorescence spectra of acridine orange-DNA complexes related to DNA base composition. Biopolymers 9:1407–1411.
281. **Mansour JD, Robson JA, Arndt CW, Schulte TH (1985)** Detection of Escherichia coli in blood using flow cytometry. Cytometry 6:186–190.
282. **Manuelidis L (1982)** Repeated DNA sequences and nuclear structure. In Dover GA, Flavell RB (eds), "Genome Evolution." London: Academic Press, pp 263–285.
283. **Manuelidis L (1984)** Active nucleolus organizers are precisely positioned in adult central nervous system cells but not in neuroectodermal tumor cells. J. Neuropathol. Exp. Neurol. 43:225–241.
284. **Manuelidis L, Ward DC (1984)** Chromosomal and nuclear distribution of the HindIII 1.9 kb human DNA repeat segment. Chromosoma 91:28–38.
285. **Marcus M, Nattenberg A, Goitein R, Nielsen K, Gropp A (1979)** Inhibition of condensation of human Y chromosome by the fluorochrome Hoechst 33258 in a mouse-human cell hybrid. Hum. Genet. 46:193–198.
286. **Marcus M, Goitein R, Gropp A (1979)** Condensation of all human chromosomes in phase G2 and early mitosis can be drastically inhibited by 33258 Hoechst treatment. Hum. Genet. 51:99–105.
287. **Marcus M, Spering K (1979)** Condensation-inhibition by 33258-Hoechst of centromeric heterochromatin in prematurely condensed mouse chromosomes. Exp. Cell Res. 123:406–411.
288. **Marfey SP, Li HG (1974)** Relationship between tritiated poly-L-lysine binding and template activity of human chromosomes. Nature (Lond) 249:559–560.
289. **Martin CL, Rolland JM, Nairn RC, Muller HK (1982)** Selection of optical filters for laser flow cytometry using fluorescent conjugates. Cytometry 2: 374–382.
290. **Martin JC, Wilder ME (1978)** Nanosecond spectroscopy. In "Abstracts, Engineering Foundation Conference on Automated Cytology VI," Schloss Elmau, Germany, April 1978.
291. **Martin RF, Holmes N (1983)** Use of an ^{125}I-labelled DNA ligand to probe DNA structure. Nature 302: 452–454.
292. **Massari S, Dell'Antone P, Colonna R (1974)** Mechanism of atebrin fluorescence changes in energized submitochondrial particles. Biochemistry 13:1038–1046.
293. **Matsukuma S, Utakoji T (1978)** Asymmetric decondensation of the L cell heterochromatin by Hoechst 33258. Exp. Cell Res. 113:453–455.
294. **Mayall BH, Carrano AV, Moore DH II, Ashworth LK, Bennett DE, Mendelsohn ML (1984)** The DNA-based human karyotype. Cytometry 5:376–385.
295. **Mazzini G, Giodano P, Riccardi A, Montecucco CM (1983)** Flow cytometry study of the propidium diodide staining kinetics of human leukocytes and its relationship with chromatin structure. Cytometry 3: 443–448.
296. **McGhee JD, von Hippel OH (1974)** Theoretical as-

pects of DNA-protein interactions: Co-operative and non-co-operative binding of large ligands to a one-dimensional homogeneous lattice. J. Mol. Biol. 86: 469–489.

297. **McGill M, Pathak S, Hsu TC (1974)** Effects of ethidium bromide on mitosis and chromosomes: A possible material basis for chromosome stickiness. Chromosoma 47:157–166.

298. **Menter JM, Golden JF, West SS (1978)** Kinetics of fluorescence fading of acridine orange-heparin complexes in solution. Photochem. Photobiol. 27:629–633.

299. **Meyer RA, Benson RC, Zaruba ME, McKhanu GMT (1979)** Cellular autofluorescence: Is it due to flavins? J. Histochem. Cytochem. 27:44–48.

300. **Meyne J, Bartholdi MF, Travis G, Cram LS (1984)** Counterstaining human chromosomes for flow karyology. Cytometry 5:580–583.

301. **Mezzanotte R, Ferrucci L, Vanni R, Bianchi U (1983)** Selective digestion of human metaphase chromosomes by Alu I restriction endonuclease. J. Histochem. Cytochem. 31:553–556.

302. **Michelson AM, Monny C, Kovoor A (1972)** Action of quinacrine mustard on polynucleotides. Biochimie. 54:1129–1136.

303. **Mikhailov MV, Zasedatelev AS, Krylov AS, Gurskii GV (1981)** Mechanism of the recognition of AT pairs in DNA by molecules of the dye Hoechst 33258. Molec. Biol. USSR 15:541–553.

304. **Miller DA, Choi YC, Miller OJ (1983)** Chromosome localization of highly repetitive human DNA's and amplified ribosomal DNA with restriction enzymes. Science 219:395–397.

305. **Miller OJ, Schnedl W, Allen J, Erlanger BF (1974)** 5-methylcytosine localized in constitutive heterochromatin. Nature (Lond) 251:636–637.

306. **Modest EJ, Sengupta SK (1974)** 7-substituted actinomycin D (NSC-3053) analogs as fluorescent DNA-binding and experimental antitumor agents. Cancer Chemother. Rep. 58:35–48.

307. **Molday RS, Dreyer WJ, Rembaum, Yen SPS (1975)** New immunolatex spheres; visual markers of antigens on lymphocytes for scanning electron microscopy. J. Cell Biol. 64:75–88.

308. **Morstyn G, Hsu SM, Kinsella T, Gratzner H, Russo A, Mitchell JB (1983)** Bromodeoxyuridine in tumors and chromosomes detected with a monoclonal antibody. J. Clin. Invest. 72:1844–1850.

309. **Moser GC, Muller H, Robbins E (1975)** Differential nuclear fluorescence during the cell cycle. Exp. Cell Res. 91:73–78.

310. **Mueller W, Crothers DM (1968)** Studies of the binding of actinomycin and related compounds to DNA. J. Mol. Biol. 35:251–290.

311. **Mueller W, Crothers DM, Waring MJ (1973)** A non-intercalating proflavine derivative. Eur. J. Biochem. 39:223–234.

312. **Mueller W, Crothers DM (1975)** Interactions of heteroaromatic compounds with nucleic acids. 1. The influence of heteroatoms and polarizability on the base specificity of intercalating ligands. Eur. J. Biochem. 54:267–277.

313. **Mueller W, Bunemann H, Dattagupta N (1975)** Interactions of heteroaromatic compounds with nucleic acids. 2. Influence of substituents on the base and sequence specificity of intercalating ligands. Eur. J. Biochem. 54:279–291.

314. **Mueller W, Gautier F (1975)** Interactions of heteroaromatic compounds with nucleic acids; A-T specific non-intercalating DNA ligands. Eur. J. Biochem. 54:385–394.

315. **Mueller W, Pakroppa W, Bunemann H, Eigel A, Dattagupta N, Flossdor J (1975)** Fractionierung von DNA mit Hilfe von Basen und sequenzspezifischen Komplexbildern. Hoppe Seylers Z. Physiol. Chem. 356:256.

316. **Muller W, Hatteshol I, Schuetz H-J, Meyer G (1981)** Polyethylene glycol derivatives of base and sequence specific DNA ligands: DNA interaction and application for base specific separation of DNA fragments by gel electrophoresis. Nucleic Acids Res. 9:95–119.

317. **Murphy RF, Daban J-R, Cantor CR (1981)** Flow cytofluorometric analysis of the nuclear division cycle of Physarum polycephalum plasmodia. Cytometry 2: 26–30.

318. **Neville DM Jr, Davies DR (1966)** The interaction of acridine dyes with DNA: An X-ray diffraction and optical investigation. J. Mol. Biol. 17:57–74.

319. **Nishiwaki H, Miura M, Imai K, Ohno R, Kawashima K, Ezaki K, Veda R, Yoshikawa H, Nagata K, Takeyama H, Yamada K (1974)** Experimental studies on the antitumor effect of ethidium bromide and related substances. Cancer Res. 34:2699–2703.

320. **Noguchi PD, Johnson JB, Browne W (1981)** Measurement of DNA synthesis by flow cytometry. Cytometry 1:390–393.

321. **Nordheim A, Pardue ML, Lafer EM, Moller A, Stollar BD, Rich A (1981)** Antibodies to left-handed DNA Z bind to interband regions of Drosophila polytene chromosomes. Nature (Lond) 294: 417–422.

322. **Noronha A, Richman DP (1984)** Simultaneous cell surface phenotype and cell cycle analysis of lymphocytes by flow cytometry. J. Histochem. Cytochem. 32:821–826.

323. **Nusse M (1981)** Cell cycle kinetics of irradiated synchronous and asynchronous tumour cells with DNA distribution analysis and BrdUrd-Hoechst 33258-technique. Cytometry 2:70–79.

324. **Nusse M, Kramer J (1984)** Flow cytometric analysis of micronuclei found in cells after irradiation. Cytometry 5:20–25.

325. **Nyak R, Sirsi M, Posser SK (1975)** Spectrophotometric studies on the interaction of chromomycin A_3 with DNA and chromatin of normal and neoplastic tissue. Biochim. Biophys. Acta. 378:195–204.

326. **Oi VT, Glazer AN, Stryer L (1982)** Fluorescent phycobiliprotein conjugates for analyses of cells and molecules. J. Cell Biol. 93:981–986.

327. **Ong LJ, Glazer A, Waterbury JB (1983)** An unusual phycoerythrin from a marine cyanobacterium. Science 224:80–83.

328. **Orgel A, Brenner S (1961)** Mutagenesis of bacteriophage T4 by acridines. J. Mol. Biol. 3:762–768.

329. **Otto FJ, Oldiges H (1980)** Flow cytogenetic studies in chromosomes and whole cells for the detection of clastogenic effects. Cytometry 1:13–17.

330. **Otto FJ, Oldiges H, Gohde W, Jain VK (1981)** Flow cytometric measurement of nuclear DNA content

variations as a potential in vivo mutagenicity test. Cytometry 2:189–191.

331. **Pachmann U, Rigler R (1972)** Quantum yield of acridines interacting with DNA of defined base sequence. Exp. Cell Res. 72:602–608.

332. **Pakroppa W, Muller W (1974)** Fractionation of DNA on hydroxylapatite with a base-specific complexing agent. Proc. Natl. Acad. Sci. USA 71:699–703.

333. **Pallavicini MG, Lalande ME, Miller RG, Hill RP (1979)** Cell cycle distribution of chronically hypoxic cells and determination of the clonogenic potential of cells accumulated in G_2 + M phases after irradiation of a solid tumor in vitro. Cancer Res. 39:1891–1897.

334. **Parker CA (1968)** "Photoluminescence of Solutions With Applications to Photochemistry and Analytical Chemistry." Amsterdam: Elsevier, p 22.

335. **Pauli F (1971)** Microspectrofluorometry in studies of absorption complexes between biopolymers and lanthanides. Microscopy 27:257–266.

336. **Pederson T (1972)** Chromatin structure and the cell cycle. Proc. Natl. Acad. Sci. USA 69:2224–2228.

337. **Pederson T, Robbins E (1972)** Chromatin structure and the cell division cycle-actinomycin binding in synchronized HeLa cells. J. Cell Biol. 55:322–327.

338. **Perry P, Wolff S (1974)** New Giemsa method for the differential staining of sister chromatids. Nature (Lond) 251:156–158.

339. **Petschel K, Naujok A, Kempter P, Seiffert W, Zimmermann HW (1984)** Über eine Bertalanffy-analoge Fluorochromierung mit 3-Dimethylamino-6-methoxyacridin. Histochemistry 80:311–321.

340. **Pinkel D, Lake S, Gledhill BL, Van Dilla MA, Stephenson D, Watchmaker G (1982)** High resolution DNA content measurements of mammalian sperm. Cytometry 3:1–9.

341. **Pinkel D, Steen H (1982)** Simple methods to determine and compare the sensitivity of flow cytometers. Cytometry 3:220–233.

342. **Pinkel D, Thompson LH, Gray JW, Vanderlaan M (1984)** Induction of sister chromatid exchanges (SCEs) by BrdUrd: Determination of exchange frequency at low incorporation levels using a monoclonal antibody to BrdUrd. Abstract #D41. Analytical Cytology X. Asilomar, CA, June.

343. **Piwnicka M, Darzynkiewicz Z, Melamed MR (1983)** RNA and DNA content of isolated cell nuclei measured by multiparameter flow cytometry. Cytometry 3:269–275.

344. **Pochon F, Kapuler AM, Michelson AM (1971)** Energy transfer in modified DNA. Proc. Natl. Acad Sci. USA 68:205–209.

345. **Pohl F, Jovin T, Baehr W, Holbrook JJ (1972)** Ethidium bromide as a cooperative effector of a DNA structure. Proc. Natl. Acad. Sci. USA 69:3805–3809.

346. **Pollack A, Prudhomme DL, Greenstein DB, Irvin GL III, Claflin AJ, Block NL (1982)** Flow cytometric analysis of RNA content in different cell populations using pyronin Y and methyl green. Cytometry 3:28–35.

347. **Prantera G, Pimpinelli S, Rocchi A (1979)** Effects of distamycin A on human leukocytes in vitro. Cytogenet. Cell Genet. 23:103–107.

348. **Rabinovitch PS (1983)** Regulation of human fibroblast growth rate by both non-cycling cell fraction and transition probability is shown by growth in 5-bromodeoxyuridine followed by Hoechst 33258 flow cytometry. Proc. Natl. Acad. Sci. USA 80:2951–2955.

349. **Rappold GA, Cremer T, Hager HD, Davies KE, Muller CR, Yang T (1984)** Sex chromosome positions in human interphase nuclei as studied by in situ hybridization with chromosome specific DNA probes. Hum. Genet. 67:317–325.

350. **Razin A, Riggs AD (1980)** DNA methylation and gene function. Science 210:604–610.

351. **Reisfeld R, Greenberg E, Velapoldi R (1972)** Luminescence quantum efficiency of GD and TB in borate glasses and the mechanism of energy transfer between them. J. Chem. Phys. 56:1698–1705.

352. **Rembaum A (1979)** Microspheres as immunoreagents for cell identification. In Melamed M, Mendelson M, Mullaney P (eds), "Flow Cytometry and Sorting." New York: Wiley, pp 335–350.

353. **Rice GC, Dean PN, Gray JW, Dewey WC (1984)** An ultra-pure in vitro phase synchrony method employing centrifugal elutriation and viable flow cytometric cell sorting. Cytometry 5:289–298.

354. **Rich A (1982)** Right-handed and left-handed DNA: Conformational information in genetic material. Cold Spring Harbor Symp. Quant. Biol. 47:1–13.

355. **Richmond TJ, Finch JT, Rushton B, Rhodes D, Klug A (1984)** Structure of the nucleosome core particles at 7A resolution. Nature (Lond) 311:532–537.

356. **Rigler R Jr (1966)** Microfluorometric characterization of intracellular acids and nucleoproteins by acridine orange. Acta. Physiol. Scand. 67(Suppl 267):5–122.

357. **Rigler R, Killander D (1969)** Activation of deoxyribonucleoprotein in human leukocytes stimulated by phytohemagglutinin. II. Structural changes of deoxyribonucleoprotein and synthesis of RNA. Exp. Cell Res. 54:171–180.

358. **Rigler R, Killander D, Bolund L, Ringertz NR (1969)** Cytochemical characterization of deoxyribonucleoprotein in individual cell nuclei. Exp. Cell Res. 55: 215–224.

359. **Ringertz NR, Bolund L (1969)** Actinomycin binding capacity of deoxyribonucleoprotein. Biochim. Biophys. Acta. 174:147–154.

360. **Robert-Nicoud M, Arndt-Jovin DJ, Zarling DA, Jovin TM (1984)** Immunological detection of left handed Z DNA in isolated polytene chromosomes. Effects of ionic strength, pH, temperature and topological stress. EMBO J. 3:721–731.

361. **Roti Roti JL, Higashikubo R, Blair OC, Uygur N (1982)** Cell-cycle position and nuclear protein content. Cytometry 3:91–96.

362. **Roti Roti JL, Wright WD, Higashikubo R, Dethlefsen LA (1985)** DNase I sensitivity of nuclear DNA measured by flow cytometry. Cytometry 6:101–108.

363. **Russel WC, Newman C, Williamson DH (1975)** A simple cytochemical technique for demonstration of DNA in cells infected with mycoplasma and viruses. Nature 253:461–462.

364. **Sage BH, O'Connell JP, Mercolino TJ (1983)** A rapid, vital staining procedure for flow cytometric analysis of human reticulocytes. Cytometry 4:222–227.

365. **Sahar E, Latt SA (1978)** Enhancement of banding

patterns in human metaphase chromosomes by energy transfer. Proc. Natl. Acad. Sci. USA 75:5650–5654.

366. **Sahar E, Latt SA (1980)** Energy transfer and binding competition between dyes used to enhance staining differentiation in metaphase chromosomes. Chromosoma 79:1–28.

367. **Sahar E, Wage ML, Latt SA (1983)** Mutation rates and transition probabilities of cycling cells. Cytometry 4:202–210.

367a. **Sandstrom MMcH, Beauchesne MT, Gustashaw KM, Latt SA (1982)** Prenatal cytogenetic diagnosis. Methods in Cell Biol. 26:35–66.

368. **Sawicki DL, Erlanger BF, Beiser SM (1971)** Immunochemical detection of minor bases in nucleic acids. Science 174:70–72.

369. **Scatchard G (1949)** The attractions of proteins for small molecules and ions. Ann. NY Acad. Sci. 51: 660–663.

370. **Schaap GH, de Josselin de Jong JE, Jongkind JF (1984)** Fluorescence polarization of six membrane probes in embryonal carcinoma cells after differentiation as measured on a FACS II cell sorter. Cytometry 5:188–193.

371. **Scheck LE, Muirhead KA, Horan PK (1980)** Evaluation of the S phase distribution of flow cytometric DNA histograms by autoradiography and computer algorithms. Cytometry 1:109–117.

372. **Schmid M, Hungerford DA, Poppen A, Enger W (1984)** The use of distamycin A in human lymphocyte cultures. Hum. Genet. 65:377–384.

373. **Schmir M, Revet BM, Vinograd J (1974)** Dependence of the sedimentation coefficient of denatured close circular DNA in alkali on the degree of strand interwinding. The absolute sense of supercoils. J. Mol. Biol. 82:35–45.

373a. **Schnedl W, Dann O, Schweitzer D (1980)** Effects of counterstaining with DNA binding drugs on fluorescent banding patterns of human and mammalian chromosomes. Eur. J. Cell Biol. 20:290–296.

374. **Schoentjes M, Fredricq E (1972)** Proflavine binding to yeast and RNA and ribosomes as related to structure. Biopolymers 11:361–374.

375. **Schreck RR, Erlanger BF, Miller OJ (1974)** The use of antinucleoside antibodies to probe the organization of chromosomes denatured by ultraviolet irradiation. Exp. Cell Res. 88:31–39.

376. **Schweizer D (1976)** Reverse fluorescent chromosomes banding with chromomycin and DAPI. Chromosoma 58:307–324.

377. **Schweizer D, Ambros P, Andrle M (1978)** Modification of DAPI banding on human chromosomes by prestaining with a DNA binding oligopeptide antibiotic, distamycin A. Exp. Cell Res. 111:327–332.

378. **Schweizer D (1981)** Counterstain-enhanced chromosome banding. Hum. Genet. 57:1–14.

379. **Secrist JA, Barrio JR, Leonard NJ, Weber G (1972)** Fluorescent modification of adenosine-containing coenzymes. Biological activities and spectroscopic properties. Biochemistry 11:3499–3506.

380. **Seeman MC, Day RO, Rich A (1975)** Nucleic acid-mutagen interactions: crystal structure. Nature (Lond) 253:324–327.

381. **Selander RK, de la Chapelle A (1973)** The fluorescence of quinacrine mustard with nucleic acids. Nature New Biol. 245:240–244.

382. **Selander RK (1973)** Interaction of quinacrine mustard with mononucleotides and polynucleotides. Biochem. J. 131:749–755.

383. **Shapiro HM, Schildkraut EF, Curbelo R, Laird C, Turner B, Hirschfeld T (1976)** Combined blood cell counting and classification with fluorochrome stains and flow instrumentation. J. Histochem. Cytochem. 24:396–411.

384. **Shapiro HM (1977)** Fluorescent dyes for differential counts by flow cytometry: Does histochemistry tell us much more than cell geometry? J. Histochem. Cytochem. 25:976–989.

385. **Shapiro HM (1981)** Flow cytometric estimation of DNA and RNA content in intact cells stained with Hoechst 33342 and pyronin Y. Cytometry 2:143–150.

386. **Shapiro HM (1983)** Multistation multiparameter flow cytometry: A critical review and rationale. Cytometry 3:227–243.

386a. **Shapiro HM, Stephens S (1986)** Flow cytometry of DNA content using oxazine 750 or related laser dyes with 633 nm excitation. Cytometry 7:107–110.

387. **Shmookler Reis RJ, Biro PA (1978)** Sequence and evolution of mouse satellite DNA. J. Mol. Biol. 121: 357–374.

388. **Sillar R, Young BD (1981)** A new method for the preparation of metaphase chromosomes for flow analysis. J. Histochem. Cytochem. 29:74–78.

389. **Smith JA, Martin L (1973)** Do cells cycle? Proc. Natl. Acad. Sci. USA 70:1263–1267.

390. **Smith LAM, Fung S, Hunkapiller TJ, Hood LE (1985)** The synthesis of oligonucleotides containing an alliphatic amino group at the 5′ terminus: Synthesis of fluorescent DNA primers for use in DNA sequence analysis. Nucleic Acids Res. 13:2399–2412.

391. **Sobell HM, Jain SC (1972)** Stereochemistry of actinomycin binding to DNA. II. Detailed molecular model of actinomycin-DNA complex and its implications. J. Mol. Biol. 68:21–34.

392. **Sobell HJM, Sakore TD, Jain SC, Banerjee KK, Bhandary KK, Reddy BS, Lozansky ED (1982)** Beta-kinked DNA—a structure that gives rise to drug intercalation and DNA breathing—and its wider significance in determining the premelting and melting behavior of DNA. Cold Spring Harbor Symp. Quant. Biol. 47:293–314.

393. **Speth PAJ, Linssen PCM, Boezeman JBM, Wessels HMC, Haanen C (1985)** Quantitation of anthracyclines in human hematopoietic cell subpopulations by flow cytometry correlated with high pressure liquid chromatography. Cytometry 6:143–150.

394. **Steen HB, Boye E (1980)** Bacterial growth studied by flow cytometry. Cytometry 1:32–36.

395. **Steen HB, Boye E, Skarstad K, Bloom B, Godal T, Mustafa S (1982)** Applications of flow cytometry on bacteria: Cell cycle kinetics, drug effects, and quantitation of antibody binding. Cytometry 2:249–257.

396. **Steinberg IZ (1971)** Long-range nonradiative transfer of electronic excitation energy in proteins and polypeptides. Ann. Rev. Biochem. 40:83–114.

397. **Steiner RF, Beers RF (1959)** Polynucleotides. V. Titration and spectro-photometric studies upon the in-

teraction of synthetic polynucleotides with various dyes. Arch. Biochem. Biophys. 81:75–92.

398. **Stetten G, Latt SA, Davidson RL (1976)** 33258 Hoechst enhancement of the photosensitivity of bromodeoxyuridine-substituted cells. Somat. Cell Genet. 2:285–290.
399. **Stewart CC, Steinkamp JA (1982)** Quantitation of cell concentration using the flow cytometer. Cytometry 2:238–243.
400. **Stokke T, Steen HB (1985)** Multiple binding modes for Hoechst 33258 to DNA. J. Histochem. Cytochem. 33:333–338.
401. **Stryer L, Haugland R (1967)** Energy transfer: A spectroscopic ruler. Proc. Natl. Acad. Sci. USA 58:719–726.
402. **Stryer L (1978)** Fluorescence energy transfer as a spectroscopic ruler. Annu. Rev. Biochem. 47:819–846.
403. **Stubblefield E, Cram S, Deaven L (1975)** Flow microfluorometric analysis of isolated Chinese hamster chromosomes. Exp. Cell Res. 94:464–468.
404. **Swartzendruber DG (1977)** Microfluorometric analysis of cellular DNA following incorporation of bromodeoxyuridine. J. Cell Physiol. 90:445–454.
405. **Swartzendruber DG (1977)** A bromodeoxyuridine (BUdR)-mithramycin technique for detecting cycling and noncycling cells by flow microfluorometry. Exp. Cell Res. 109:439–443.
406. **Swartzendruber DE, Travis GL, Martin JC (1980)** Flow cytometric analysis of the effect of 5-bromodeoxyuridine on mouse teratocarcinoma cells. Cytometry 1:238–244.
407. **Tanke HJ, Nieuwenhuis IAB, Koper GJM, Slats JCM, Ploem JS (1980)** Flow cytometry of human reticulocytes based on RNA fluorescence. Cytometry 1:313–320.
408. **Tanke HJ, van Oostveldt P, van Duijn P (1982)** A parameter for the distribution of fluorophores in cells derived from measurements of inner filter effect and reabsorption phenomenon. Cytometry 2:359–369.
409. **Tapiero H, Fourcade A, Vaigot P, Farhi JJ (1982)** Comparative uptake of adriamycin and daunorubicin in sensitive and resistant friend leukemia cells measured by flow cytometry. Cytometry 2:298–302.
410. **Tate EH, Wilder ME, Cram LS, Wharton W (1983)** A method for staining 3T3 cell nuclei with propidium iodide in hypotonic solution. Cytometry 4:211–215.
411. **Taylor IW (1980)** A rapid single step staining technique for DNA analysis by flow microfluorometry. J. Histochem. Cytochem. 28:1021–1024.
412. **Tchen P, Fuchs RPP, Sage E, Leng M (1984)** Chemically modified nucleic acids as immunodetectable probes in hybridization experiments. Proc. Natl. Acad. Sci. USA 81:3466–3470.
413. **Therman E, Sarto GE, Palmer CG, Kallio H, Denniston C (1979)** Position of the human X inactivation center on Xq. Hum. Genet. 50:59–64.
414. **Thomas G, Roques B (1972)** Proton magnetic resonance studies of ethidium bromide and its sodium borohydride reduced derivative. FEBS Lett. 26:169–175.
415. **Thomas JC, Weill G, Daune M (1969)** Fluorescence of proflavine-DNA complexes: heterogeneity. Biopolymers 8:647–659.
416. **Thorell B (1981)** Flow cytometric analysis of cellular endogenous fluorescence simultaneously with emission from exogenous fluorochromes, light scatter and absorption. Cytometry 2:39–43.
417. **Tobey RA, Crissman HA (1975)** Unique techniques for cell cycle analysis utilizing mithramycin and flow microfluorometry. Exp. Cell Res. 93:235–239.
418. **Traganos F, Darzynkiewicz Z, Sharpless T, Melamed MR (1977)** Simultaneous staining of ribonucleic and deoxyribonucleic acids in unfixed cells using acridine orange in a flow cytofluorometric system. J. Histochem. Cytochem. 25:46–56.
419. **Traganos F, Darzynkiewicz Z, Melamed MR (1982)** The ratio of RNA to total nucleic acid content as a quantitative measure of unbalanced cell growth. Cytometry 2:212–218.
420. **Trask BJ, van den Engh GJ, Elgershuizen JHBW (1982)** Analysis of phytoplankton by flow cytometry. Cytometry 2:258–264.
421. **Trask B, van den Engh G, Gray J, Vanderlaan M, Turner B (1984)** Immunofluorescent detection of histone 2B on metaphase chromosomes using flow cytometry. Chromosoma 90:295–302.
422. **Tsou KC, Yip KF, Go KJ (1976)** Fluorescence in human metaphase chromosomes produced by adriamycin with and without deoxyribonuclease treatment. J. Histochem. Cytochem. 24:752–756.
423. **Tubbs FR, Ditmars WE Jr, Van Winkle Q (1964)** Heterogenicity of the interaction of DNA with acriflavine. J. Mol. Biol. 9:545–557.
424. **Tuffanelli DL, McKeon F, Kleinsmith DM, Burnham TK, Kirschner M (1983)** Anticentromere and anticentriole antibodies in the scleroderma spectrum. Arch. Dermatol. 119:560–566.
425. **Tunis M-JB, Hearst JE (1968)** On the hydration of DNA. II. Base composition dependence of the net hydration of DNA. Biopolymers 6:1345–1353.
426. **Ughetto G, Wang AHJ, Quigley GJ, van der marel G, van Boom J, Rich A (1985)** A comparison of the structure of echinomycin and triostin A complexed to a DNA fragment. Nucleic Acids Res. 13:2305–2323.
427. **Unakul W, Hsu TC (1973)** Induction of chromosome banding in early stages of spermatogenesis. Chromosoma 44:285–290.
428. **Van den Engh G, Trask B, Cram S, Bartholdi M (1984)** Preparation of chromosome suspensions for flow cytometry. Cytometry 5:108–117.
429. **Van den Engh GJ, Trask BJ, Gray JW, Langlois RG, Yu L-C (1985)** Preparation and bivariate analysis of suspensions of human chromosomes. Cytometry 6: 92–100.
430. **Van de Sande JH, Lin CC, Jorgenson KF (1977)** Reverse banding on chromosomes produced by a guanosine-cytosine specific DNA binding antibiotic: Olivomycin. Science 195:400–402.
431. **Van de Sande JH, Lin CC, Deugau KV (1979)** Clearly differentiated and stable chromosome bands produced by a spermine bis-acridine, a bifunctional intercalating analogue of quinacrine. Exp. Cell Res. 120:439–444.
432. **Van Dilla MA, Langlois RG, Pinkel D, Yajko D, Hadley WK (1983)** Bacterial characterization by flow cytometry. Science 220:620–622.
433. **Van Dyke MW, Hertzberg RP, Dervan PB (1982)** Map of distamycin, netropsin, and actinomycin binding sites on heterogeneous DNA: DNA cleavage-in-

hibition patterns with methidiumpropyl-EDTA-Fe(II). Proc. Natl. Acad. Sci. USA 79:5470–5474.

434. **Van Dyke MW, Dervan PB (1983)** Chromomycin, mithramycin, and olivomycin binding sites on heterogeneous deoxyribonucleic acid. Footprinting with (methidiumpropyl-EDTA) iron(II). Biochemistry 22: 2373–2377.

435. **Van Dyke MW, Dervan PB (1984)** Echinomycin binding sites on DNA. Science 225:1122–1127.

436. **Van Zant G, Fry CG (1983)** Hoechst 33342 staining of mouse bone marrow: Effects on colony-forming cells. Cytometry 4:40–46.

437. **Verma RS, Dosik H, Lubs H (1977)** Frequency of RFA colour polymorphisms of human acrocentric chromosomes in Caucasians: Interrelationships with QFQ polymorphisms. Ann. Hum. Genet. (Lond) 43: 257–267.

438. **Vindelov LL, Christensen IJ, Nissen NI (1983)** Standardization of high resolution flow cytometric DNA analysis by the simultaneous use of chicken and trout red blood cells as internal reference standards. Cytometry 3:328–331.

439. **Vindelov LL, Christensen JJ, Jemsen G, Nissen NI (1983)** Limits of detection of nuclear DNA abnormalities by flow cytometric data analysis. Cytometry 3:332–339.

440. **Wade CG, Rhyne RH, Woodruf WH, Bloch DP, Bartholemew JC (1979)** Spectra of cells in flow cytometry using a vidicon detector. J. Histochem. Cytochem. 27:1049–1052.

441. **Wang AHJ, Quigley GJ, Kolpak FJ, Crawford JL, van Boom HJ, van der Marel G, Rich A (1979)** Molecular structure of a left-handed double helical DNA fragmet at atomic resolution. Nature 282:680–686.

442. **Wang AHJ, Ughetto G, Quigley GJ, Hakoshima T, van der Marel GA, van Boom JH, Rich A (1984)** The molecular structure of a DNA-triostin complex. Science 225:1115–1121.

443. **Ward DC, Reich E, Goldberg IH (1965)** Base specificity in the interaction of polynucleotides with antibiotic drugs. Science 149:1259–1261.

444. **Ward DC, Reich E, Stryer L (1969)** Fluorescence studies of nucleotides and polynucleotides. Formycin, 2-aminopurine riboside, 2,6-diaminopurine riboside, and their derivatives. J. Biol. Chem. 244:1228–1237.

445. **Ward DC, Cerami A, Reich E, Acs G, Altweger L (1969)** Biochemical studies of the nucleoside analogue, formycin. J. Biol. Chem. 244:3243–3250.

446. **Ward DC, Reich E (1972)** Fluorescence studies of nucleotides and polynucleotides II. 7-Deazanebularin: Coding ambiguity in transcription with base pairs containing fewer than two hydrogen bonds. J. Biol. Chem. 247:705–719.

447. **Waring MJ (1965)** Complex formation between ethidium bromide and nucleic acids. J. Mol. Biol. 13: 269–282.

448. **Waring M (1970)** Variation of the supercoils in closed circular DNA by binding of antibiotics and drugs: Evidence for molecular models involving intercalation. J. Mol. Biol. 54:247–279.

449. **Waring MJ, Wakelin LPG (1974)** Echinomycin: A bifunctional intercalating antibiotic. Nature 252: 653–657.

450. **Wartell RM, Larson JE, Wells RD (1974)** Neotropsin—a specific probe for A-T regions of duplex deoxyribonucleic acid. J. Biol. Chem. 249: 6719–6731.

451. **Waye JS, Willard H (1985)** Chromosome-specific alpha satellite DNA: Nucleotide sequence analysis of the 2.0 kilobase pair repeat from the human X chromosome. Nucleic Acids Res. 13:2731–2743.

452. **Weill G, Calvin M (1963)** Optical properties of chromophore macromolecule complexes: Absorption and fluorescence of acridine dyes bound to polyphosphates and DNA. Biopolymers 1:401–417.

453. **Weintraub H (1983)** A dominant role for DNA secondary structure in forming hypersensitive structures in chromatin. Cell 32:1191–1203.

454. **Weisblum B, deHaseth PL (1972)** Quinacrine, a chromosome stain specific for deoxyadenylate-deoxythymidylate-rich regions in DNA. Proc. Natl. Acad. Sci. USA 69:629–632.

455. **Weisblum B (1973)** Why centric regions of quinacrine-treated mouse chromosomes show diminished fluorescence. Nature (Lond) 246:150–151.

456. **Weisblum B (1974)** Fluorescent probes of chromosomal DNA structure: three classes of acridines. Cold Spring Harbor Symp. Quant. Biol. 38:441–449.

457. **Weisblum B, Haenssler E (1974)** Fluorometric properties of the bisbenzimidazole derivative Hoechst 33258, a fluorescent probe specific for AT concentration in chromosomal DNA. Chromosoma 46: 255–260.

458. **Weiss MA, Eliason JL, States DJ (1984)** Dynamic filtering by two-dimensional ^{1}H NMR with application to phage lambda repressor. Proc. Natl. Acad. Sci. USA 87:6019–6023.

459. **Weller LA, Wheeless LL Jr (1982)** EMOSS: An epi-illumination microscope objective slit-scan flow system. Cytometry 3:15–18.

460. **Wells RD, Larson JE (1970)** Studies on the binding of actinomycin D to DNA and DNA model polymers. J. Mol. Biol. 49:319–342.

461. **Wells RD, Brennan R, Chapman KA, Goodman TC, Hart PA, Hillen W, Kellogg DR, Kilpatrick MW, Klein RD, Klysik J, Lambert PF, Larson JE, Miglietta JJ, Neuendorf SK, O'Connor TR, Singleton CK, Stirdivant SM, Veneziale CM, Wartell RM, Zacharias W (1982)** Left-handed DNA helices, supercoiling and the B-Z junction. Cold Spring Harbor Symp. Quant. Biol. 47:77–84.

462. **West SS, Lorincz AE (1973)** Fluorescent molecular probes in fluorescence microspectrophotometry and microspectrofluorometry. In Thaer AA, Sernetz M (eds), "Fluorescence Techniques in Cell Biology." New York: Springer-Verlag, p 395.

463. **Whaun JM, Rittershaus C, Ip SHC (1983)** Rapid identification and detection of parasitized human red cells by automated flow cytometry. Cytometry 4: 117–122.

464. **Wheeless LL, Patten SF (1974)** Slit-scan cytofluorometry: Basis for automated cytopathology prescrenning system. Acta. Cytol. 17:391–394.

465. **Willard HF, Smith KD, Sutherland J (1983)** Isolation and characterization of a major tandem repeat family from the human X chromosome. Nucleic Acids Res. 11:2017–2033.

466. **Wolf SF, Jolly DJ, Lunnen KD, Friedmann T, Migeon BR (1984)** Methylation of the hypoxanthine phosphoribosyl transferase locus on the human X chro-

mosome: Implications for X-chromosome inactivation. Proc. Natl. Acad. Sci. USA 81:2806–2810.

467. **Wolff S, Perry P** (1974) Differential staining of sister chromatids and the study of sister chromatid exchanges without autoradiography. Chromosoma 48: 341–353.

468. **Wray W, Wray VP** (1980) Proteins from metaphase chromosomes treated with fluorochromes. Cytometry 1:18–20.

469. **Yang TP, Hansen SK, Oishi KK, Ryder OA, Hamkalo BA** (1982) Characterization of a cloned repetitive DNA sequence concentrated on the human X chromosome. Proc. Natl. Acad. Sci. USA 79:6593–6597.

470. **Yen PH, Patel P, Chinault AC, Mohandas T, Shapiro LJ** (1984) Differential methylation of hypoxanthine phosphoribosyl transferase genes on active and inactive human X chromosomes. Proc. Natl. Acad. Sci. USA 81:1759–1763.

471. **Yonuschot G, Mushrush GW** (1975) Terbium as a fluorescent probe for DNA and chromatin. Biochemistry 14:1677–1681.

472. **Young BD** (1984) Chromosome analysis by flow cytometry: A review. Basic Applied Histochem. 28:9–19.

473. **Yunis JJ** (1984) Constitutive fragile sites and cancer. Science 226:1199–1204.

474. **Zakrewska K, Lavery R, Pullman B** (1983) The solvation contribution to the binding energy of DNA with non-intercalating antibiotics. Nucleic Acids Res. 11:8825–8839.

475. **Zasadatelev AS, Gursky GV, Volkenshtein MV** (1971) Theory of one dimensionial absorption. I. Absorption of small molecules on a polymer. Mol. Biol. (USSR) 5:245–251.

476. **Zelenin AV, Poletaev AI, Stepanova NG, Barsky VE, Kolesnikov VA, Nikitin SM, Zhuze AL, Gnutchev NV** (1984) 7-Amino-actinomycin D as a specific fluorophore for DNA content analysis by laser flow cytometry. Cytometry 5:348–354.

477. **Zimmer CH, Reinert KE, Luck G, Wahnert U, Lober G, Thrum H** (1974) Interaction of the oligopeptide antibiotics netropsin and distamycin A with nucleic acids. J. Mol. Biol. 58:329–348.

478. **Zimmer CH, Luck G, Nuske R** (1980) Protection of (dA-dT) cluster regions in the DNAseI cleavage of DNA by specific interaction with netropsin. Nucleic Acids Res. 8:2999–3010.

15

Acridine Orange: A Versatile Probe of Nucleic Acids and Other Cell Constituents

Zbigniew Darzynkiewicz and Jan Kapuscinski
Sloan-Kettering Institute for Cancer Research, New York, New York 10021

INTRODUCTION

Spectrum of Applications of Acridine Orange

Acridine dyes, having different affinity to various cell constituents, are widely used as fluorescent stains (fluorochromes) in histochemistry and cytochemistry. Among them acridine orange (AO) is the most popular and the most versatile. No other dye approaches AO in the number of applications and diversity of staining reactions. This fluorochrome has found application in several protocols developed to stain nucleic acids for their quantitative analysis, to distinguish RNA from DNA, to measure the extent of DNA denaturation (single versus double strandedness), to selectively stain certain glycosaminoglycans, and as a lysosomal probe [reviews 69,191,214,235,266]. These protocols were often used in different fields and in different biological assays. For instance, in immunology AO was introduced to discriminate different phases of lymphocyte stimulation and to quantify the stimulation process [91]. In cytology or cell biology, AO was applied to recognize quiescent versus cycling cells, enumerate cells in different phases of the cell cycle and study cell kinetics [83,97], or discriminate between live and dead cells [83]. The AO-based methodology enables one to distinguish potentially subfertile spermatozoa from normal spermatozoa, which has found an application in the fertility clinic [116]. Staining with AO also makes it possible to assess DNA damage and repair, which is of interest in radiation biology [246]. Because chromosome banding can be generated by techniques employing AO, this methodology is of practical use in cytogenetics [44,110,111]. In analytical hematology, AO is applied to obtain differential white blood cell counts [189,190]. AO applications in cell physiology involve the use of this dye to study ion transport across membranes [197], membrane structure [166], or lysosomal activity [319]. In clinical oncology, AO has found numerous applications, either as a diagnostic [14,62,204,208,283,306] or prognostic [12,26,159,185] tool. The above examples, although incomplete, illustrate the extent of range of applications of this universal dye.

Despite extensive literature, many uncertainties regarding interactions between AO and particular cell constituents remain. Also, because of the extreme sensitivity of this dye to any change in equilibrium conditions and complexity of the interactions, AO has the reputation of a dye that is difficult to use. We hope that this review will clarify some misconceptions and will encourage still wider use of AO. With better understanding of the mechanisms of cell staining with AO, the data obtained employing this dye can yield a plethora of information about content, molecular structure, conformation, and environment of many cell constituents in situ that no other dye can provide.

Interest in studying interactions between acridine orange or other acridine dyes [250] with cellular constituents stems not only from the use of these dyes in cytochemistry, but also from the fact that they exhibit strong biological activity. These dyes are known as antibacterial [105], antimalarial [228], mutagenic, carcinogenic [68,135,272], or antitumor [68,107,135,228,294] agents. Also, their structure and mode of interactions with nucleic acids resemble other biologically active chemicals, such as carcinogens (e.g., polycyclic hydrocarbons, benzacridines), nucleic acid derivatives (e.g., purines), certain antibiotics (e.g., actinomycins), or other dyes (e.g., pyronin Y(G), toluidine blue, methyl green, ethidium). Furthermore, these or similar dyes may be used to modify the conformation of nucleic acids in a desired way to obtain information relating to the structure of the latter [31,164,210].

Historical Perspective

Acridine orange was introduced as a fluorescent stain for biological structures, independently by Strugger [268], and by Bukatsch and Haitinger [53]. Strugger [268,269] observed the metachromatic change of fluorescence in AO-stained dead cells and discussed this phenomenon in terms of increasing concentration of the dye ("concentration effect"), proposing use of AO for the differentiation of living and dead cells. The metachromatic behavior of AO was later explained by Zanker [316–318], who has shown that dye–

Flow Cytometry and Sorting, Second Edition, pages 291–314

dye interactions occurring at increasing AO concentration (dye aggregation, polymeric forms) are responsible for the metachromatic shift toward red luminescence (phosphorescence). By contrast, the monomeric form of the dye prevailing at low AO concentration is characterized by green fluorescence. By providing a basis for the differential staining of various classes of substances in the cell, this metachromatic property of AO was the main cause contributing to its wide use in cytochemistry.

In early cytochemical studies, even prior to the understanding of the mechanisms of interaction between AO and nucleic acids, the green orthochromatic fluorescence of cell nuclei and red metachromatic luminescence of the cytoplasm were attributed to differential stainability of DNA and RNA, respectively [19,20,39,192]. Subsequently, more and more studies have been appearing in which AO was used to stain DNA and RNA [36–38,109,175,186,311]. Most of these studies were done on clinical material, predominantly in exfoliative cytology where AO was applied as an alternative to the Papanicolaou technique (36–39], or in microbiology, to detect viral inclusions in the cells [19,20,186]. Metachromatic staining of nucleic acids was also proposed as a method to distinguish single stranded denatured sections of DNA from the native DNA within the same cell [200,235]. AO was also used to stain DNA in gels following its denaturation by glyoxal [55].

Before the advent of flow cytometry, almost all the cytochemical studies with AO were done on preparations of cells attached to glass and fixed. Cells were stained at relatively high dye concentrations and washed before fluorescence measurement, to remove the unbound and weakly bound dye fraction. In contrast, measurements by flow cytometry are done in equilibrium with the dye. There are significant differences between these two methods. Thus, while staining of cells in equilibrium enables one to observe weak reversible interactions between the dye and the cell constituents, the diffusion of AO from stained cells in a nonequilibrium state makes these interactions undetectable and often results in uncontrolled changes in cell stainability [172]. Also, the phenomenon of altered cell stainability due to variation in local cell densities [45,237] (discussed later) is nonexistent in the case of cells studied by flow cytometry. Furthermore, under equilibrium conditions, staining is extremely sensitive to concentrations of free AO in solution, which in turn, in some situations, may be sensitive to the molar ratio of the dye to the potential binding sites [86,94].

The objective of the present review is to summarize and discuss the studies in which AO has been used as a probe of nucleic acids in situ. Special attention is given to flow cytometric studies and, when possible, comparisons are made between equilibrium versus nonequilibrium staining. Mechanisms of interactions between AO and nucleic acids are also described; understanding of these interactions is essential for practical use of this dye in cytochemistry. This chapter updates and extends the earlier review [69].

INTERACTIONS OF AO WITH NUCLEIC ACIDS

Properties of AO in Solution

3,6-(Dimethylamino)acridine (acridine orange, AO) in aqueous solution at pH around 7 and at room temperature is singly protonated (pK_a = 10.5) [317]. The dye molecules self-associate to form dimers and higher aggregates [241,242,317,318]. Because the process of self-association is anticooperative [242], in diluted solution of AO (below 10^{-5} M in 0.15 M NaCl, pH 7.0), the concentration of AO oligomers higher than dimers is negligible [145,241]. The concentration of the dye can be measured colorimetrically at isosbestic points (in which the absorption of monomer and dimer are equal) (Fig. 1), e.g., at 470 nm using extinction coefficient E = $4.33 \times 10^4\ M^{-1}\ cm^{-1}$ [287]. The respective concentrations of AO monomer and dimer in solutions can also be assayed colorimetrically [145,160,287]. AO dimerization constant is salt concentration dependent; an empirical formula illustrates this relation:

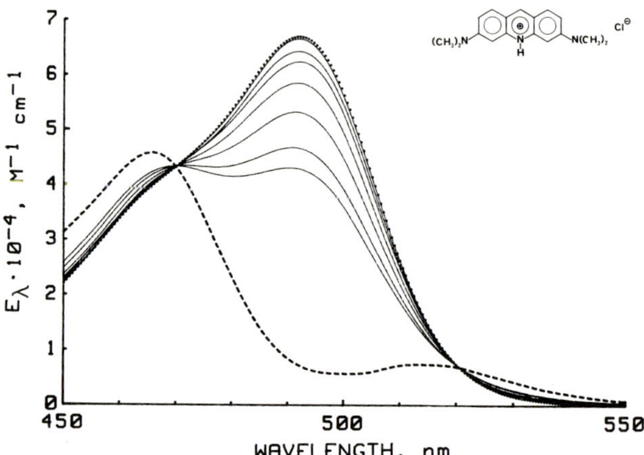

Fig. 1. Absorption spectra of AO in 0.15 M NaCl, pH 7.0 at 25°C. (——) The spectra of the samples containing AO at concentration: 1.0, 1.9, 2.8, 5.2; 10, 19, and 28 μM; their peaks heights are in descending order (from 1.0 to 28 μM, respectively). (...) The calculated spectra of AO monomer and (----) dimer are also shown. Reproduced from [145], with permission of the publisher. (Insert) chemical formula of acridine orange (AO).

$$K_D = (1.08 + 7.37\,[Na^+]) \times 10^4 \pm 350\ M^{-1}\ [145]$$

The presence of organic solvents such as methanol, ethanol, or dioxane in solution markedly decreases self-aggregation of AO and changes its absorption spectrum slightly [148,241].

Luminescence of AO depends strongly on its concentration (Fig. 1) and temperature. Diluted aqueous solutions containing predominantly monomer fluoresce green with maximum emission at 525 nm [47,145,317]. This is true fluorescence emission with short (2 nsec) lifetime [47] and is a result of excited single-state (S_1) to ground single-state (S_0) transition ($S_1 \rightarrow S_0$). At high concentration and at low temperature (−150°C) a long life (>0.1 sec) red emission with maximum at 605 nm can be observed [316]. This is a phosphorescence emission resulting from excited triplet (T_1) to ground-state transition ($T_1 \rightarrow S_0$), characteristic of AO aggregates whose absorption properties favor intersystem crossing transition $S_2 \rightarrow T_1$ if excited at wavelength <500 nm [145, and references cited therein]. In diluted aqueous AO solutions the long lifetime T_1 state is very effectively relaxed in a nonradiative way by collision with oxygen or solvent molecules (collision quenching [147]).

Binding of AO to Double-Stranded Nucleic Acids

Double-stranded nucleic acids form two types of nonbinding complexes with small molecules such as AO. The first

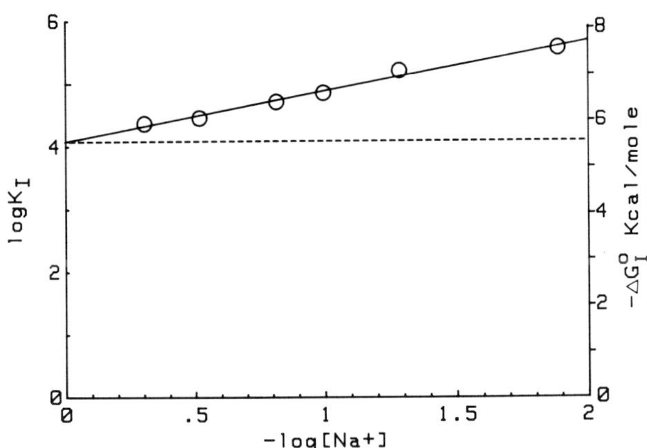

Fig. 2. Correlation between NaCl concentration and association constant (K_I, left scale) or free energy change ($\Delta G^o{}_I$, right scale) of intercalative AO binding to calf thymus DNA (——). (---) Intrinsic component of intercalation (K_i or $\Delta G^o{}_i$), independent of salt concentration. Data from [145].

type involves only phosphate groups of the polymer, which electrostatically interact with organic or inorganic cations present in solution. The counterion condensation theory developed by Manning [180] and Record et al. [225] provides quantitative analysis and conceptual understanding of this process. Simplifying, only a few parameters, i.e., charge density of the polymer, valence of the ligand, and concentration of the counterion in the solution (e.g., Na^+ or Mg^{2+}), are necessary to estimate the free energy change (or association constant) of such interactions. These assumptions bear important practical implications for assessment of pure ionic (electrostatic) ligand–nucleic acid interactions. Thus, the interactions are not specific with regard to nucleic acid base composition and type of ligand (except its charge) but are inversely depend on ionic strength. At Na^+ concentration = 1 M (standard state), the pure electrostatic interactions became unfavorable; for monocationic ligand, such as AO, at 0.15 M Na^+ the association constant of its electrostatic association with ds DNA K_{el} = 5.4 M^{-1} (calculated based on Manning and Record model). This is a rather low value corresponding to free energy change $\Delta G^o{}_{el}$ = 1.0 kcal/mole. At 10 mM Na^+ the corresponding values increase to K = 57.5 M^{-1} and $\Delta G^o{}_{el}$ = 2.4 kcal/mole, respectively (Fig. 2).

For a long time the electrostatic AO–ds DNA interactions were held responsible for metachromatic (red) luminescence observed at high ligand concentrations. This dye–dye stacking model assumed the binding of AO aggregates to DNA or RNA (weak interactions or type II binding). According to this model, the polymer supports the formation of long stacks of the dye outside the double helix [50,213,309]. However, the recently developed base–dye stacking model explains the red luminescence of ds nucleic acids by AO-induced denaturation of the double helix and subsequent condensation of the complexes of AO with ss polymers in which not only ionic but also specific base–AO interaction takes place [143,145]. This model is described in more detail further in this chapter.

The intercalation is a second type of nonbonding AO–ds nucleic acid interaction (strong or type I interaction) in which planar dye molecules (monomers) are inserted between the neighboring base pairs. The plane of the intercalated molecule is parallel to that of the base pair. Intercalation extends and locally unwinds the helix of DNA. In the case of AO, the unwinding angle (unwinding of the double helix after insertion of a single AO molecule) is about 20° [257]. Both electrostatic and hydrophobic (π–π) interactions take part in the complex formation. This type of interaction, involving AO or other aromatic cations, originally described by Lerman [167,168], has been extensively studied since then [reviewed in 42,43,50, 219,294,309]. The base specificity, affinity and mode of AO binding to ds DNA were reevaluated recently using computer-based techniques and taking into account the possibility of DNA denaturation and condensation by AO [145]. The main features of intercalative AO binding based on data from that paper and references cited thereof are summarized in Table 1 and discussed below.

The intercalative AO–DNA complexes show bathochromic (red) shift and hypochromicity in relation to the spectrum of AO monomer (Fig. 3). The degree of the bathochromic shift and hypochromicity vary somewhat, depending on the primary DNA structure. This indicates interaction of the dye's chromophore with DNA bases. Interestingly, not only base composition but also base sequence influences spectral properties of the chromophore.

Excitation spectra generally overlap absorption spectra. All emission maxima of the AO–DNA complexes are shifted to blue, compared with emission of the AO monomer. The lifetime of green fluorescence is 5 nsec as compared to 2 nsec of the free dye [47]. The short lifetime is indicative of true fluorescence ($S_1 \rightarrow S_0$ transition) of both dye monomer and its intercalative form. As expected, the increase in lifetime of AO in the complex corresponds to 2.5-fold increase in fluorescence quantum yield [47]. The data of relative quantum yield Q_R of AO intercalated into natural DNA and various synthetic polymers are listed in Table 1.

Stoichiometry of AO–ds DNA interaction depends on ionic strength. While theoretical maximum binding density (bound dye/phosphate) of intercalation is 0.25 (i.e., one AO molecule per two base pairs), only one-half this value can be achieved experimentally. Namely, during the titration of polymer with AO, with progression of the polymer saturation, concentration of free dye in equilibrium with the complex rises rapidly and leads to the destabilization of the double helix, as will be described in detail later in the chapter. This destabilization is salt-dependent and, e.g., is reflected in the increase in AO binding site size (n) from n = 2 to n = 2.4 bp in 0.1 M and 0.01 M NaCl, respectively. The increase in salt concentration, however, has an adverse effect on the association constant of the AO–DNA system.

Both the affinity and accessibility of intercalators to nucleoproteins are generally reduced compared to free nucleic acids in solutions [81,114,211]. Nevertheless, it should be noted that high-affinity sites for intercalators have been detected in chromatin and ribosomes; they consist, however, of only a small portion of the total accessible sites of nucleic acids in cells [150,211].

AO base specificity is low when measured as affinity of the dye for synthetic DNAs of different base composition and structure (see Table 1). An exception is poly(dA) · poly(dT); the association constant of AO to this polymer is severalfold lower and the n-value larger, compared with other polymers and natural DNA. A similar pattern is evident when fluorescence quantum yield (Table 1, Fig. 3) is taken as a criterion for base specificity. Consequently, the assay of DNA content based on fluorimetry of its complex with AO may not be

TABLE 1. Properties of AO and Its Complexes With Nucleic Acids

	Absorption		Emission			Affinity	
	λ_{max} (nm)	$E_{max} \cdot 10^{-4}$ (M^{-1} cm^{-1})	λ_{max} (nm)	Φ_{525}	Q_R	n Base pairs	$K_I \times 10^{-4}$ (M^{-1})
ds DNAs[a]							
Poly(dA) · poly(dT)	502	5.13	524	1.62	1.42	19.3	1.2
Poly(dA–dT) · poly(dA-dT)	504	6.23	522	2.63	2.13	2.0	9.2
Poly(dG) · poly(dC)	506	4.33	523	3.04	2.38	2.4	7.6
Poly(dG–dC) · poly(dG–dC)	503	6.64	522	2.59	2.06	2.0	7.9
Poly(dA–dC) · poly(dG–dT)	500	5.17	520	2.31	1.83	2.0	7.4
Calf thymus DNA	502	5.85	522	2.64	2.23	2.0	5.0
RNAs[b]							
Poly(rA)	457[c]	4.76[c]	630[d]	0	[e]	1[f]	20.6[c,g]
Poly(rI)	458[c]	3.09[c]	630[d]	0	[e]	1[f]	13.9[c,g]
Poly(rC)	426[c]	1.62[c]	644[d]	0	[e]	1[f]	11.5[c,g]
Poly(rU)	438[c]	2.28[c]	643[d]	0	[e]	1[f]	3.6[c,g]
Natural RNA	—	—	638[d]	0	[e]	1[f]	—
Free AO							
AO monomer	492	6.85	525	1.00	1.00		
AO dimer	466	4.57	—	<0.01	<0.01		
Crystals[b]	—	—	618[d]	0	[e]		

λ_{max}, maximum wavelength; E_{max}, molar extinction coefficient at maximum; Φ_{525}, increase in emission intensity at 525 nm as compared with AO monomer; Q_R, relative quantum yield as compared with AO monomer; n, binding site size; K_I, association constant.
[a]Measured in 0.15 M NaCl, pH 7.0 at 25°C, data from [145].
[b]Suspensions in 0.1 M NaCl, pH 7.0, data from [147,148].
[c]In 10 mM phosphate buffer, pH 7.0, containing 25% (v/v) of ethanol, data from [148].
[d]Noncorrected for emission monochromator and photomultiplier response.
[e]Phosphorescence.
[f]Nucleotides.
[g]Cooperative association constant (M^{-1} nucleotides).

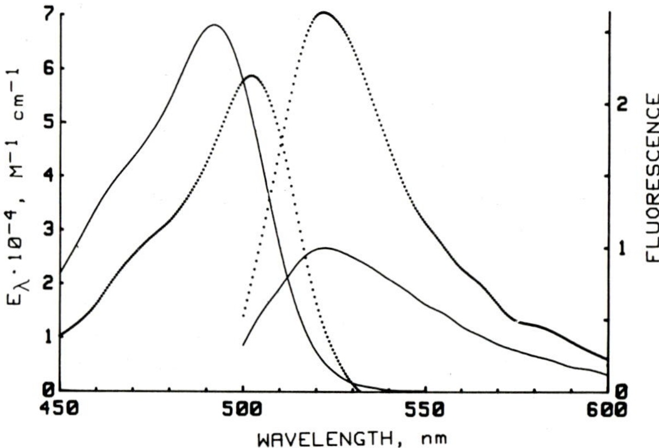

Fig. 3. Excitation and emission spectra of AO monomer (——) in comparison with the spectra of AO–calf thymus DNA intercalative complex (...). The intensities of the excitation spectra were normalized to the values of the corresponding maxima of absorption spectra. Reproduced from [145], with permission of the publisher.

accurate when comparing DNA differing in AT/GC ratio and containing long sections of AT sequences.

AO–base pair interactions are manifested in the nonelectrostatic (intrinsic) component of the free energy change of intercalation (ΔG^o_i). The intrinsic component is not salt dependent, and $\Delta G^o_i = -5.6 \pm 0.3$ kcal/mole (corresponding to $K_i = 1.2 \times 10^4$ M^{-1} at 25°C). This energy represents π–π interactions of aromatic systems of the dye and base pairs (dispersion forces). The electrostatic (ionic) component of AO intercalation (ΔG^o_{el} or K_{el}) is salt dependent and can be calculated based on the counterion condensation theory [180,225]. For monocationic salt (e.g., NaCl) $\Delta G^o_{el} = 1.2$ 1og[Na^+] kcal/mole (corresponding to $K_{el}=10^{-0.88\log[Na^+]}$ at 25°C). Total free energy change or association constant can be estimated from the equations: $\Delta G^o_I = \Delta G^o_i + \Delta G^o_{el}$ and $K_I = K_iK_{el}$ (see Fig. 2).

The effect of AO on the stability of the double helix depends on its concentration. At low concentration of free dye and D/P<0.5, the intercalative binding is favorable. This type of binding stabilizes the double-stranded structure of DNA. At higher AO concentration, a destabilization of the double helix occurs [142,143,148]. This process is described further in this chapter.

Binding of AO to Single-Stranded Nucleic Acids

Final products of AO interaction with ss nucleic acids are very different from those with ds DNA at low D/P ratio. Thus, their stoichiometry at saturation is one dye molecule per nucleotide. The absorption spectrum of these complexes is shifted to blue and emission far to the red as compared to the AO monomer or its complexes with dsDNA (see Table 1). The shift in emission to red received the name of metachromasia in cytochemistry. The lifetime of the red emission at room temperature is >20 nsec [47]. Thus, spectral properties of AO complexes with ss nucleic acids or other polyanions at high binding density resemble, to some extent, those of dye aggregates.

The classic dye–dye stacking model of AO interaction with ss nucleic acids developed by Bradley and Wolf [50] assumed binding of ligand molecules to each other on the external surface (backbone) of the polymer resulting in the formation of long dye stacks bound electrostatically to phosphates of the polymer. The nucleic acid was believed to only provide support (lattice) for AO aggregates, and the model did not postulate any base–ligand interaction [50,192,213]. Because interaction of AO with polyanions (DNA, RNA, polynucleotides, heparin, polyphosphate) was observed to be cooperative, Bradley and Wolf [50] proposed the stacking coefficient to characterize the cooperativity of the reaction. Significant differences in cooperativity between different polyanions were observed, which were explained as due to a flexibility of the polymers. Attempts were later made to explain why the noncooperative aggregation of free AO became highly cooperative in the presence of polyanions [48,49,241,242]. However, despite extensive studies, the mechanism of interaction between AO and ss nucleic acids remained unclear and controversial [reviews in refs. 42,174,212,219].

Significant differences in emission spectra of AO complexes with ss nucleic acids of different base composition [235] (see Table 1) and observations of the interaction between AO and purine and pyrimidine derivatives [33,136,149,265,287,300] led to modifications of the original model of AO–ss nucleic acid interactions. Two such modifications [138,287] assumed transient insertion of dye molecule between the adjacent bases of the ss polymer; the final product, however, was proposed to have the same structure as described by Bradley and Wolf [50]. Perhaps the most interesting structure of the intercalator DNA complex was considered by Pritchard et al. [220; see also 212]. Their model, developed as an alternative to Lerman's model of intercalation [167] of acridines to ds DNA, assumed stacking interactions of the ligand with bases instead of interactions with base pairs. This explained apparently similar reactivities of denatured and native DNA with acridine dyes [220] but left unexplained metachromasia of AO complex with ss nucleic acids.

Unfortunately, two important features of AO–ss polyanion interactions, i.e., (1) insolubility of the complex in aqueous media, and (2) the fact that metachromatic red emission is a phosphorescence and not a fluorescence, were ignored in early attempts to describe the nature of the complexes and their spectral properties (metachromasia). Although poor solubility of AO-polyanion complexes has occasionally been reported, e.g. [33,201,235], it was treated as a nuisance in experiments and never investigated in detail. The red luminescence of AO complexes with polyanions is still labeled fluorescence in the literature despite early reports that AO aggregates and its complexes with ATP or adenine forms red phosophorescent complexes [149,317]. Only recently has a model of AO interactions with ss nucleic acids, taking into account the above features, been proposed [143,147,148]. This new model can be summarized as follows:

1. Both electrostatic and dye–base stacking interactions are involved in the AO–ss nucleic acid complex formation. The π–π interactions of the dye chromophore with different bases (Fig. 4) manifest in spectral differences of the complexes, depending on their base composition (see Table 1).
2. Spectral properties of dye–base stacks, similar to those of dye–dye stacks, indicate that the intersystem crossing transition, i.e., phosphorescence ($T_1 \rightarrow S_0$) emission, is favored. The long lifetime T_1 excited state is very effectively relaxed in a nonradiation way by interaction with oxygen or solvent (collision quenching) and, therefore, cannot be observed in solutions, when excited molecules are exposed. The condensation and precipitation of the complex, as well as freezing [316], facilitate the red emission by partially eliminating the quenching.
3. There is a 1 : 1 stoichiometry of AO binding per nucleotide, and the interaction is highly cooperative. Affinity parameters differ for different homopolymers. The affinity of AO is higher for purine than pyrimidine homopolymers (see Table 1) and for ribo- as compared with deoxyribohomopolymers. Similar to intercalation, both electrostatic (ionic) and π–π interaction of the dye and the bases participate in complex formation. The free energy change of AO binding to ss nucleic acids, calculated per dye molecule, is about one-half of that for intercalation to ds nucleic acids.
4. Because of repulsive ionic forces, polyanion (nucleic acid) molecules in solution are extended (coil, wormlike structure). The charge neutralization by cationic ligands like AO leads to condensation (collapse) of the polymer and formation of compact forms (tori?, spheres?). The condensation can be observed as an increase in light scattering or appearance of red luminescence during titration of nucleic acids with AO. The condensed molecules have a tendency to aggregate and form precipitates. It is conceivable that, in addition to electrostatic and dye–base stacking interactions, favorable nonelectrostatic forces resulting from the propensity of the polymer itself to collapse [181] are involved in AO–ss nucleic acid complex formation. The latter component is most likely responsible for differences in reactivity of ssRNA versus ssDNA and for cooperativity of AO–ss nucleic acid interaction. This conclusion is based on observations that condensation of nucleic acids by ligands deprived of aromatic system (e.g., by spermine^{4+} or Co^{3+}) are also cooperative [144].

Denaturation of Double-Stranded Nucleic Acids by AO

Earlier cytochemical studies suggested that AO may exert a denaturing effect on ds nucleic acids or prevent renaturation of ss polymers [95,96,137]. These observations were unexplained and neglected at that time, especially in light of evidence that intercalation stabilizes the double helix. It was recently shown, however, that while at low ligand concentration and low D/P ratio, AO indeed intercalates and stabilizes the double helix; at increased D/P, binding of AO to ss nucleic acids becomes thermodynamically preferable [148]. Namely, at a certain point of titration, the weaker but more numerous 1 : 1 (dye–phosphate) interactions of AO with ss polymer dominate thermodynamically over the stronger but less numerous (1 : 4) intercalative binding to the ds form [148], as outlined in Figure 4. These data thus provide an explanation for the observed denaturation of nucleic acids by AO. A mathematical model that may be pertinent to describe denaturating properties of AO was proposed by McGhee [187]. Although developed for large ligands, this model is also accurate in describing the complicated equilibrium interactions of intercalators with ds nucleic acids.

Denaturation and condensation of ds DNA became evident during its stepwise titration in solution with AO. At the

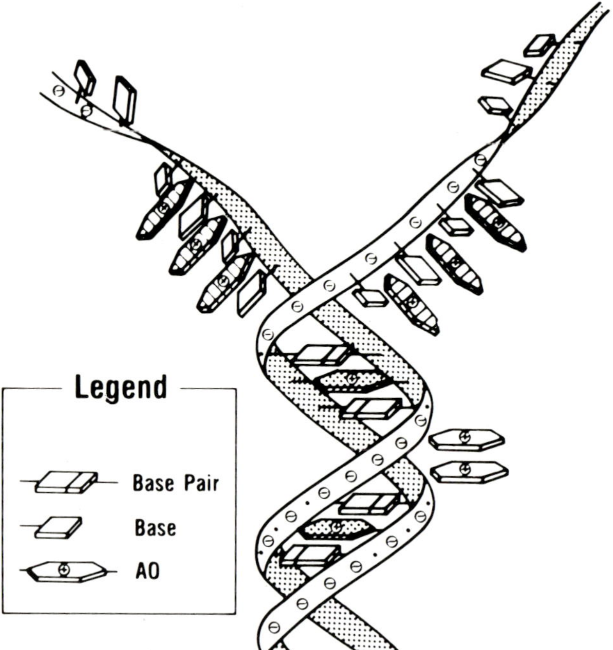

Fig. 4. Schematic representation of different modes of AO binding to nucleic acids, based on data from [142–145,147,148]. At low dye concentration and D/P ratio, AO binds to ds nucleic acids by intercalation (lower part of the structure) and the dye fluoresces green (AO molecules filled with dots). At increased D/P ratio, binding of AO to ss nucleic acids becomes thermodynamically preferable and, thus, the double helix is destabilized (upper part of the structure, AO molecules dashed). Both electrostatic and dye–base stacking interactions are involved in the AO–ss nucleic acids complex formation and the process is highly cooperative. The high binding density of AO binding to single strand (one AO molecule per nucleotide) results in polymer charge neutralization which in turn leads to condensation of the product (not shown). In condensed form the complex luminesces red; see the text for further explanation. Nucleic acids can also bind AO molecules or AO aggregates by purely ionic interactions. Such a binding of AO dimer outside of the double helix is presented in the figure (lower part of the structure). In solution, red luminescence of such aggregates is quenched by collisions with solvent molecules. Reproduced from [146a], with permission of the publisher.

onset of titration, the green fluorescence of the AO–DNA complex is higher than that of free dye, a finding consistent with the higher quantum yield of fluorescence of an intercalated versus free dye monomer. During titration the green fluorescence of the AO–DNA complex decreases, indicating DNA denaturation, and the red luminescence increases sharply. The sigmoidal transition, characteristic of the cooperative process, is followed by light-scattering increase [142,143]. This increase in light scattering is an indication of polymer condensation and the condensed product tends to aggregate and precipitate. During formation or dissociation of the condensed products, accessibility of DNA bases for external probes increases, indicating that ds structure of DNA undergoes destabilization (denaturation, strand separation). The condensed product is, in fact, the complex of AO with ss DNA [142,143]. It should be noted that destabilization of the double helix is not a unique feature of AO, but was observed for other intercalators as well [142–144,146]. An exception is ethidium bromide which, perhaps because of its bulky structure, may not form 1 : 1 complexes with ss nucleic acids.

The cooperative interaction of the ligand with ss nucleic acids is a driving force of intercalator-induced double-helix destabilization. It is not surprising, therefore, that there is strong selectivity in this process with regard to nucleic acid sugar (RNA versus DNA) and base composition, reflecting the affinity of the dye to ss polymers [144–146]. In addition, the stability of the double helix itself influences its sensitivity to undergo denaturation in the presence of the intercalator.

SUPRAVITAL CELL STAINING WITH AO

Viability Tests, AO Cytotoxicity

Interactions of AO with unfixed cells, living or dead, are very complex and often the subject of different interpretation. The pattern of cell stainability under these conditions depends on type and physiological state of the cell, AO concentration, incubation time, pH, and ionic strength of the environment. Because it is impossible to assess specificity of staining of cells with nondisrupted membranes by such means as pretreatment with nucleases or specific extractions, little is known about intracellular targets for AO and chemistry of the staining reaction. However, because there are numerous practical uses of AO involving supravital cell staining, it is necessary to discuss it even if there is uncertainty about AO binding mechanisms.

Historically, AO was first applied in cytochemistry to discriminate viable and dead cells [268] and, because of low cytotoxicity, AO was often used as a vital fluorochrome [101,126,248]. The dead cells were reported to stain metachromatically red, whereas live cells exhibited green luminescence [268]. While this pattern of staining is indeed reproducible [258], it is also possible to stain cells that, either live or dead, may luminesce exclusively in green or red or in any intermediate hue, depending on concentration and pH of the dye solution, cell type, and the kind of cellular injury leading to cell death [310]. Generally, at low concentration of the dye ($<10^{-5}$ M), only live cells can accumulate AO in lysosomes (which luminesce red); thus, only live cells exhibit red luminescence. Dead cells stain uniformly green. By contrast, at high AO concentration ($>10^{-4}$ M), dead cells stain rapidly and uniformly in red (nucleus and cytoplasm), whereas living cells still have only red lysosomes but green nuclei and cytoplasm. With time, however, when the intracellular concentration of AO increases, live cells also stain uniformly red. Caution should therefore be exercised in interpretation of the data on supravital cell staining with AO, and often independent tests of cell viability should be done in parallel.

Low concentrations of AO ($<10^{-6}$ M) appear to be nontoxic to normal fibroblasts in cultures [310]. Their growth remains unaltered for about 20 generations despite the presence of AO in the cytoplasm detected by ultraviolet (UV) microscopy. The phototoxic effect, however, becomes apparent following exposure of AO-stained cells to light [310]. Among a variety of photodynamic effects, damage to nucleic acids photosensitized by AO has been observed [10,11]. AO appears to be selectively toxic to tumor cells. The antitumor effects of this dye were reported first by Lewis and Goland [169] and acridine analogs are common antitumor drugs [review 228]. Like most intercalating antitumor drugs, AO is also a mutagen [51]. Exposure of bacteria to AO sensitizes them to lysogenic activity of phages [18,301] and prevents

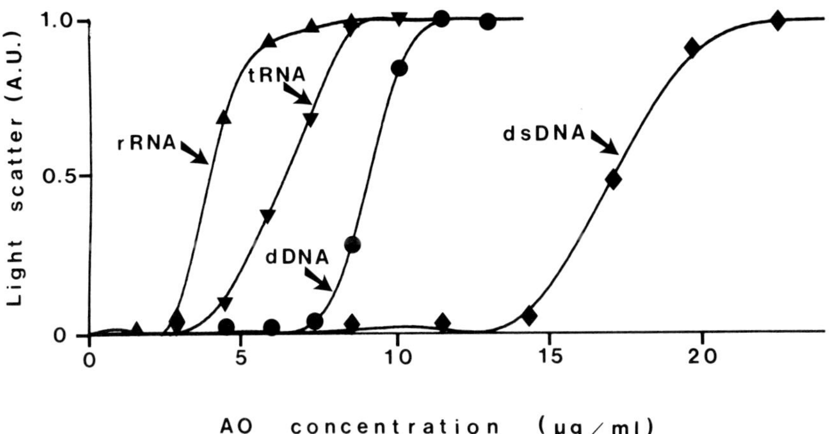

Fig. 5. Different sensitivity of nucleic acids to condensation induced by AO as a principle of differential stainability of RNA versus DNA. During titration of rRNA, tRNA, denatured (d) and native (ds) DNA with AO (in 0.15 N NaCl), an increase in light scatter occurs indicating condensation of the products. As is evident, the AO–RNA complexes undergo condensation at lower concentration of the ligand compared to AO–DNA products. Because AO luminesces red in the condensed state, the differential staining, e.g., rRNA (red luminescence) and ds DNA (green fluorescence), can be obtained at AO concentration range from 5 to 12 μg/ml; i.e., under conditions in which rRNA–AO is already in the condensed form, while ds DNA–AO (the dye bound by intercalation) is still in solution. Reproduced from [80], with permission of the publisher.

the appearance of resistance to drugs [134]. In vitro effects of AO intercalation into supercoiled DNA on its transcription were investigated by Greene et al. [129], who observed suppression of transcription in proportion to the supercoil unwinding.

It should be stressed, however, that most studies on the cytotoxicity of AO and other in vivo effects of this dye have been done using standard, nonpurified preparations of AO (C.I. grade). AO in this form often contains up to 50% of other, unspecified contaminants. It is possible, therefore, that the cytotoxic effects may be due as well to these contaminants rather than to AO itself and caution should be exercised in interpretation of these data. Needless to say, application of low purity AO in cytochemistry may also result in irreproducible results if the contaminants interfere with the binding sites for AO.

Lysosomal Uptake of AO

In living cells exposed to relatively low concentration of AO ($<10^{-5}$ M), the dye is preferentially accumulating in lysosomes where it luminesces red [4,239,240,243,245,319]. The uptake of AO, which is a weak base, is a consequence of low pH within the lysosomes. The cell membrane is permeable to the uncharged form of AO; the dye once inside becomes protonated and therefore trapped in lysosomes, due to high proton concentration in these organelles. Thus AO behaves as a lysosomotrophic substance [102]. Uptake of AO is also due to pinocytosis. The concentration of AO in lysosomes is predominantly determined by the pH gradient across the membrane of these organelles [102,239], which is a reflection of the efficiency of the ATP-dependent proton pump [315]. In living cells incubated with AO, therefore, intensity of red luminescence is a reflection of total lysosomal volume and the capacity of the lysosomal membrane to maintain proton gradient.

Granulocytes supravitally concentrate AO in azurophilic granules, which, when observed under the UV microscope, luminesce intensively red [1]. This luminescence, similar to luminescence of lysosomes in other cell types, requires intact membranes. Cell fixation or permeabilization by detergents abolishes metachromatic stainability of these organelles.

Because of its affinity to lysosomes, AO is often used to study the proton pump in isolated lysosomes in suspension [197,198]. When suspensions of lysosomes are incubated with AO, the overall luminescence rapidly drops due to the uptake of AO by these organelles. Namely, the dye's luminescence is quenched at low pH inside the lysosomes, compared to that in solution at higher pH. Uptake of AO by isolated lysosomes estimated this way, i.e., expressed by the degree of luminescence suppression, serves as a sensitive assay of the lysosomal proton pump. This assay is widely used in cell physiology to study mechanisms maintaining a proton gradient across the lysosomal membrane. In viable cells, activation of lysosomes measured by increased uptake of AO was reported to occur as early as 3 hours following stimulation of lymphocytes by mitogens, and the assay based on uptake of this dye was proposed as a method of early detection of lymphocyte stimulation [279].

It is unclear why AO luminesces red in lysosomes. It is possible that during incubation with AO, high concentration of the dye is generated in these organelles, and the dye molecules are then in the stacked, aggregate form, which is known to luminesce red [147,316–318]. The second explanation of the metachromatic stainability of lysosomes is that AO binds therein to some polyanionic substrate and the complex, when condensed, luminesces red. No studies were done, however, to detect such a complex and nothing is known about the possible AO binding sites in lysosomes.

Staining of Other Cell Constituents

In addition to lysosomes, other cell constituents stain supravitally with AO. Thus, nuclei generally exhibit green fluorescence, and a uniform green or yellow background of the cytoplasm is apparent as well. In all probability, this is a reflection of AO binding to DNA (nuclei) and RNA (nucleoli, cytoplasm). Attempts have been made to extract information about the nature of AO binding products from spectral measurements of the stained cells [127]. In specialized cells containing polyanions, such as hyaluronic acid (fibroblasts), chondroitin sulfate (chondrocytes), or heparin (mast cells), complexes of AO with these polyanions show red luminescence regardless of whether the cells are stained supravitally or after fixation. Keratohyalin granules of differentiating epidermal cells containing dermatan sulfate also exhibit luminescence having both green and red components [113,155]. Neurosecretory material of nerve cells has also been reported to stain metachromatically at relatively high (0.4 mM) AO concentration [32].

Acridine orange has relatively low affinity to mitochondria of live cells. Hydrophobic acridine dyes, however, such as 10 *n*-nonylacridinium orange-chloride, show high specificity towards these organelles [252] and can be applied in flow cytometry as mitochondrial probes [223].

Flow Cytometry of Supravitally Stained Cells

Flow cytometric measurements of supravitally stained cells are relatively scarce. Melamed et al. [189,190] have developed differential staining of lymphocytes versus monocytes versus granulocytes based on different red luminescence (content of lysosomes) of these cells following a short incubation with 3×10^{-6} M AO. This principle of staining has found an application in automated white blood cell differential counting instruments. These investigators also reported that uptake of AO by granulocytes varies with the presence of infection [189]. Wilson et al. [308] were able to discriminate type II alveolar epithelial cells from lavaged rat lungs based on their unique uptake of AO. Flow cytometry has also been used by Abrams et al. [1] to measure degranulation of neutrophils induced by calcium ionophore A23187.

Wheeless et al. [54,304–306] employed AO to stain cytological material of the cervix supravitally in cancer screening. In these studies, however, cells were postfixed in glutaraldehyde before measurements were made by flow cytometry. This resulted, most likely, in differential staining of RNA and DNA, although DNA was not stained stoichiometrically and appeared to be more accessible to AO in tumor cells [124]. These measurements, in conjunction with measurements of the nucleus-to-cytoplasm ratio, were adequate to obtain good differentiation of normal versus tumor cells [54,304–306]. The cells could later be sorted and counterstained according to Papanicolaou [35].

DIFFERENTIAL STAINING OF DNA AND RNA

Staining with AO Under Nonequilibrium Conditions

The possibility of differential staining of DNA versus RNA with AO on cytological preparations has been demonstrated by several investigators [19,20,39,118,249,311]. In these earlier studies based on nonequilibrium staining, the choice of a method of cell fixation, the concentration of AO, ionic strength, ion composition and pH during staining are all listed as critical factors essential for specificity of the reaction. Under nonequilibrium conditions, however, i.e., when the prestained cells are then mounted in the dye-free media, a continuous leakage of the dye from the stained structures takes place, which makes the reaction unstable and unsuitable for quantitative cytometry. The present review thus concentrates on the technique of cell staining in equilibrium with the dye and describes the procedures of cell preparation and staining found to be the most satisfactory for quantitation of DNA and RNA.

Equilibrium Staining: Cell Permeabilization

To suppress accumulation of AO in lysosomes and ensure specific and stable staining of DNA and RNA under equilibrium conditions, the cells have to be permeable to the dye. To this end, the cells are either fixed or permeabilized with a detergent. Fixation in ethanol or in ethanol–acetone mixture, used in the early studies, is adequate [96], but somewhat better resolution of the DNA measurements (lower c.v. values of the G_1 peak) is achieved by staining cells permeabilized by nonionic detergents [91]. The detergent procedures, however, require special caution to ensure that cells remain unbroken. Low pH, low temperature (0–4°C), the presence of proteins in the cell suspension and avoidance of rapid mechanical agitation have been recommended as preventing disintegration of the cell membrane and leakage of the cytoplasmic constituents, as well as suppressing activity of endogenous nucleases [84,91,96,279]. Another advantage of the detergent treatment combined with cell staining is that the procedure is rapid, avoids cell centrifugation, and disperses cell aggregates [84].

The flow cytometric method using AO to stain DNA and RNA differentially in the detergent-permeabilized cells has been developed in our laboratory [84,91,96,279]. The method was later tested with respect to stoichiometry of the reaction [28,64] and compared with other methods measuring DNA and/or RNA [76,255,292]. The method was also extended to express the ratio of RNA to DNA per cell in relationship to cell position in the cell cycle, which was useful to measure the degree of unbalanced growth during cell treatment with various drugs [276]. This procedure was also adapted to isolated cell nuclei [217], modified to have on-line addition of the reagents [215], and combined with FITC-labeled antibodies to measure in parallel the cell surface antigens [203]. Characteristics of this method with respect to stainability of RNA and DNA, as well as its applications, are described below.

Stainability of RNA

More than 80% of total cellular RNA consists of rRNA. Native RNA in the cell has mixed conformation: a large portion in situ (>40%) is ds and thus, like native DNA, can bind AO by intercalation and stain orthochromatically green [96]. To differentially stain DNA versus RNA, it is therefore necessary to selectively denature ds RNA and transform the AO–RNA complexes to the condensed, luminescing red form. Owing to higher affinity of AO to ssRNA than to ssDNA [143,145,148], it is possible at certain AO concentration (~20 μM at 0.15 N NaCl) to denature dsRNA selectively, leaving DNA at ds conformation. AO thus acts not only as a stain but, because of its higher affinity to ssRNA compared with ssDNA, also as a selective RNA-denaturing agent. Significantly higher AO concentration (>50 μM) is required to denature DNA [145,148]. Denaturation of RNA in situ by AO was found to be enhanced by EDTA [96]. Apparently, chelation of divalent cations weakens the RNA–protein interaction in ribosomes and makes rRNA more amenable to denaturation by AO [96]. Thus, at approximately 20 μM AO (in 0.15–0.25 N NaCl) in the presence of 1 mM EDTA, all cellular RNA undergoes denaturation, and the condensed RNA–AO complexes luminesce red, while DNA still remaining in ds conformation upon intercalation of AO fluoresces green [91,96,279]. The reaction is extremely sensitive to AO concentration. Any diffusion of the dye during measurement of fluorescence, which occurs in some cell sorters, necessitates use of increased AO concentration in the staining mixture to compensate for it [28]. Likewise, the procedure is sensitive to a decrease in D/P ratio below a certain threshold and thus the staining protocols are formulated to have a limited number of cells per volume of staining solution [91,96,279]. Under these conditions the reaction of AO with RNA, resulting in red luminescence, is highly stoichiometric to RNA content [28]. To minimize spectrum overlap from AO–DNA interactions, it is advisable to measure the red luminescence of AO–RNA product in the cell using the long pass filters transmitting at relatively long wavelength (630 or 640 nm).

The AO–RNA condensed complexes have high electron opacity and can be visualized by electron microscopy; in unfixed cells treated with 1 mM AO, the complexes forming in place of ribosomes undergo transposition toward the nucleus [73]. Insolubility of the condensed and precipitated products in aqueous solution is perhaps the main reason why the AO-stained RNA does not leak out of the permeabilized cell. This ensures stoichiometry of the reaction in the cell.

Stainability of DNA

DNA in situ is only partially accessible to AO. Removal of histones, e.g., by treatment with acid or at high salt concentration, increases binding of AO to nuclear DNA threefold [81]. The mechanism of restriction of AO binding is thus associated with maintenance of the nucleosomal structure of chromatin. The topological rigidity of the DNA helix wound around the core histones restricts intercalation of AO, perhaps because unwinding of the helix and its elongation are needed to accommodate intercalating AO molecules, and nucleosomal DNA at low ionic strength resists such unwinding. Increased salt concentration (above 0.8 N NaCl) of the staining solution or removal of histones by acid makes DNA more accessible to intercalate AO and thus increases DNA stainability [81]. It is therefore likely that in nuclear chromatin spacer DNA is more accessible to AO compared with DNA associated with the core histones of the nucleosomes.

Additional restriction of AO binding was observed during differentiation of Friend erythroleukemia cells [81] and in the course of spermatogenesis [115,125,320]. The character of changes in nuclear chromatin that may be responsible for variation in restriction of AO intercalation to nuclear DNA, or binding of other dyes, is discussed elsewhere [70,81], and is also the subject of Chapter 16 of this volume. It should be mentioned here, however, that numerous reports about variation in accessibility of DNA in situ to bind AO in relation to "genome activation" observed under nonequilibrium conditions [e.g., 154,233,234] were later shown to be due to uncontrolled variation in local cell densities unrelated to genome activity [45,171,172,236]. Cell prefixation in formaldehyde or studies of DNA accessibility to AO at certain specific ionic environments or pretreatments may result, however, in different AO binding related to differences in chromatin structure [66]. In most situations, however, binding of AO to nuclear chromatin at low dye concentration and in equilibrium, regardless of the cell type, is proportional to cell ploidy, and fluorescence intensity measured around 530 nm correlates stoichiometrically with DNA content of the cells [64,66,278]. In the flow cytometric method designed to stain RNA and DNA differentially, however, most histones are removed before staining by cell exposure to HCl, which also ensures stoichiometry of the reaction [81].

Applications

The flow cytometric method of differential staining of DNA and RNA [91,96,279] has found wide application in different fields of biology and medicine. In an early study [91], it was observed that the noncycling (G_0) lymphocytes can be distinguished from their cycling counterparts based on differences in RNA content. It was later found that RNA content can also discriminate cells in several distinct subcompartments of the cell cycle, including G_{1A} versus G_{1B}, S versus S_Q, or G_2 versus G_{2Q}, having different metabolic and kinetic properties [75,77,83]. Details about distinction of these phases of the cell cycle are presented in Chapter 24 of this volume. This approach to discriminating quiescent cells was used by numerous workers investigating a variety of cell types. Thus, the plateau phase, or noncycling and nonclonogenic EMT6 cells have been reported by Watson and Chambers [296] to have markedly lower RNA content compared with exponentially growing EMT6 cells. Later, these investigators studied RNA content changes during the progression of EMT6 cells through interphase [297] or mitosis [298]. Quiescent EMT6 cells from plateau cultures having low RNA content have also been studied with respect to their sensitivity to radiation [176]. Based on differences in RNA content, Ashihara et al. [21] recognized quiescent AF8 cells and compared them with the postmitotic cells. By the same parameter, Bauer and Dethlefsen [29] identified quiescent HeLa 53 cells in plateau phase or unfed cultures; Bauer et al. [30] and Luk and Sutherland [177] distinguished G_0 cells in multicellular tumor spheroids; Wallen et al. [293] recognized quiescent cells in murine mammary tumors in vitro; Dethlefsen et al. [108] and Pande et al. [209] characterized the heterogeneity of cell populations in solid tumors. By analysis of the RNA content of isolated nuclei of plant protoplasts, Bergounioux et al. [34] were able to identify quiescent and differentiated protoplasts.

Correlation between RNA content of individual cells and their rate of progression through the cell cycle have been investigated for CHO cells [72,75,282], PHA-stimulated lymphocytes [74], 3T3 and SV40-3T3 [2,253,254], V79 [120] and HeLa S3 cells [121]. Simultaneous measurements of DNA and RNA have also been helpful to estimate cytotoxicity of glucocorticoids on lymphoid cell lines in relation to the cell ploidy and cell-cycle position [112].

The flow cytometric method of RNA and DNA detection has also found an application in studies of cell differentiation. Thus it was observed that the erythroid differentiation of Friend leukemia cells is associated with a dramatic decline in RNA content [280]. Marked changes in RNA content were also observed during differentiation of HL-60 leukemic cells [57], as well as normal human [113,155] or mouse keratinocytes [140].

Simultaneous staining of DNA and RNA has also proved useful in studies of drug effects on various cell types [e.g., 97–99,122,256,259] (reviewed in Chapter 39, this volume). The advantage of this method is in the ability to discern drug effects on DNA replication or inducing a block in the cell cycle (including the G_0 to G_1 transition arrest) versus effects on DNA transcription or RNA turnover, the latter two reflected by changes in accumulation of RNA. By measurement of the RNA/DNA ratios [276], it was also possible to estimate the severity of the growth imbalance induced by antitumor drugs of various classes.

The method of differential staining of DNA and RNA with AO has been most extensively used in studies of lymphoid cells. Nonstimulated normal T lymphocytes have been observed to have elevated RNA content compared with B lymphocytes [13]. The ability to discern G_0 (G_{1Q}) cells from the cycling population and among the latter to recognize G_{1T}, G_{1A}, G_{1B}, S and G_2 + M compartments offered by this method [83,91] have proven to be of great value in detailed evaluation of the process of lymphocyte stimulation induced by polyvalent mitogens or antigens [17,40,41,56,58,100, 106,123,128,133,139,151,152,156–158,163,182,193–196, 206,207,221,226,231,232,244,263,264,273,274,288,295, 299,302,307,313,314]. A wealth of information was obtained in these studies regarding the relationship between immunological properties of stimulated lymphocytes and their rate of progression through different compartments of the cell cycle. The scope of this review does not permit discussion of these findings. The RNA/DNA staining method has now become a classic method of approach among cell immunologists to study the cell-cycle progression of lymphocytes following their stimulation.

There are several advantages to the flow cytometric assay of lymphocyte stimulation based on differential staining of DNA and RNA over the traditional method employing [^{3}H]thymidine uptake. One of the advantages is the ability to detect the stage of lymphocyte stimulation that does not involve cell entrance to the S phase, namely, G_0 (G_{1Q}) to G_1 transition. This transition, characterized by elevation in RNA content and nondetectable by the [^{3}H]thymidine uptake assay, appears to be of importance in several clinical situations. For instance, the G_0 to G_1 transition was observed among lymphocytes from cerebrospinal fluid (CSF) of all patients with multiple sclerosis, regardless of the stage; cell entrance to S phase, however, was associated with exacerbation of the disease [204]. Likewise, the increased RNA content in the absence of DNA replication was found to characterize lymphocytes from synovial fluid of some of rheumatoid arthritis patients, while other patients had increased RNA and DNA content; it is unclear whether this distinction correlates with severity of the disease [46]. A similar pattern was observed in peripheral blood lymphocytes of patients with systemic lupus erythematosus [3] and in bronchoalveolar lymphocytes of patients with Sarcoidosis alveolitis [199]. The increased RNA content indicating G_0–G_1 transition of peripheral blood lymphocytes was also noted among patients with kidney transplants before graft rejection, and this parameter was proposed as a rejection prognostic marker [290]. The ability and kinetics of lymphocytes to enter G_1 compartment in response to polyvalent mitogens, as compared with their progression through S phase, was also investigated as a function of aging [264]. The above examples indicate that discrimination of two stages of lymphocyte stimulation, one characterized by the G_0–G_1 transition (rise in RNA content), and another involving interleukin-2 (IL-2) and cell proliferation (entrance to S phase), offered by this method may have diagnostic and prognostic value in various clinical situations.

In addition to lymphocytes, stimulation of macrophages can also be analyzed by simultaneous measurements of DNA and RNA [261,262]. Differences in increase in RNA content following stimulation make it possible to distinguish different subpopulations of macrophages responding to different stimuli [262]. It should be noted, however, that stimulated macrophages exhibit a significant component of green and red luminescence unrelated to RNA and DNA; some of it may be attributable to their autofluorescence.

RNA and DNA content measurements have also been useful in clinical hematology, especially in analysis of leukemias, lymphomas, and myelomas (see Chapters 35 and 36). The studies published by Andreeff and colleagues [12–16,227], Barlogie and his associates [22–26,260], Walle [289], Mauro et al. [185], and Zittoun et al. [322] clearly indicate that DNA/RNA flow cytometry can be used as a diagnostic tool in some leukemias and as a prognostic marker in nearly all types of neoplastic hematological disease. Actually, RNA content appears to be one of the best parameters correlated with prognosis and prediction of remission during therapy. The mouse models of leukemia also indicate a correlation between RNA content of the preleukemic or leukemic cells and their proliferative potential [141,206].

Simultaneous staining of DNA and RNA was applied to identify malaria *(Plasmodium falciparum)*-infected human red blood cells [303] and to study growth and replication of this parasite [131,132,229]. Among other uses, the method also served to identify reticulocytes in peripheral blood [247,251] (see Chapter 33).

A note of caution should be added regarding the quantitative aspect of this methodology. In most cell systems following staining with AO, intensity of green fluorescence is proportional to DNA content and red luminescence to RNA content. However, the stoichiometry may be perturbed when cells are pretreated with agents having high affinity to DNA and RNA or modifying these polymers. Thus, for instance, stainability of RNA is severely suppressed in cells treated with the intercalating antitumor drug, ditercalinium, which in living cells has the propensity to bind to RNA and to condense this polymer [275]. Decreased DNA stainability was also observed in cells treated with some *N*-alkyl derivatives of adriamycin [281], as well as following incorporation of BrdUrd [71,85,111]. In the latter case, most likely, the energy transfer of the excited AO molecules to BrdUrd takes place, resulting in fluorescence quenching similar to fluorescence quenching of Hoechst dyes by this precursor [161,162]. The phenomenon of quenching of AO luminescence by BrdUrd has been applied in flow cytometry to detect cycling cells independently of their RNA content [71,185]. Displacement of AO from DNA by metal mutagens and carcinogens was detected by fluorescence polarization measurements [230], which would indicate that DNA stainability of cells exposed to such mutagens may also be altered. Also, there may be other constituents in the cell, especially in differentiated cells, that can stain with AO.

APPLICATION OF AO IN STUDIES OF DNA DENATURATION

Detection of DNA Denaturation by Static Microfluorometry

Native nuclear DNA is double stranded. During treatment with heat, alkali or acids, the two strands can separate. This phenomenon is known as denaturation, melting, or helix–coil transition and is the result of destruction of hydrogen bonding between the paired bases. Denaturation of free DNA depends exclusively on its primary structure. The melting temperature (T_m) of any DNA relates directly to its guanine–cytosine content, since this base pair confers added stability due to an additional hydrogen bond in comparison with adenine–thymine.

Free DNA in solution is stabilized by counterions that neutralize DNA phosphates. At low concentrations of counterion, the melting temperature is low and the width of the transition (range of temperature over which DNA denatures) is broad. The transition becomes narrow and occurs at higher temperatures when the concentration of counterions is increased.

DNA in chromatin in situ is stabilized by interactions with histones; its stability is also modulated by the supranucleosomal levels of chromatin structure [review in Ref. 70]. The ability to measure DNA denaturation in situ therefore offers an insight into chromatin structure allowing one to discern the interactions that stabilize the double helix. Although there are biochemical methods based on measurements of UV light absorption to study DNA denaturation in chromatin, they involve chromatin isolation, shearing and solubilization, which destroys higher orders of chromatin structure and limits their application in investigations of DNA in situ [70]. Attempts, therefore, have been made to use the metachromatic properties of AO to stain double-stranded versus denatured DNA differentially following partial denaturation in situ. Such measurements were initiated by Nash and Plant [200], who observed that *Drosophila* chromosomes mounted on glass slides heated to over 75°C and then stained with AO exhibited red luminescence while the nonheated chromosomes fluoresced green. Since then, AO was widely used in cytogenetics to discriminate denatured and nondenatured sections of DNA in chromosomes following their treatment with heat, acids, alkali, or other dyes [44,63,65, 103,104,110,173,267,284–286,312]. Such treatments often induced chromosome banding. Apart from the practical application of banding for chromosome identification, the AO-induced bands, when compared with banding generated by Giemsa, quinacrine, or other methods of staining, offered clues to the mechanisms responsible for the banding in relation to the molecular structure of chromosomes. Pretreatment of live cells with AO was observed to enhance the resolution of banding by Giemsa [184].

Rigler and his colleagues [233–235] introduced a quantitative method for assaying DNA denaturation based on AO staining and microfluorometry. The extent of DNA denaturation was demonstrated as the spectrum of AO luminescence changed from green to red when the dye was bound to ss-rather than ds DNA. The proportion of denatured DNA was represented by the ratio of intensity of red-to-green luminescence, and the transition curve showing the ratio changes as a function of increasing temperature during cell heating indicating the cooperative helix-to-coil transition. To omit interference of RNA, the cells were pretreated with RNase prior to heating. Rigler's technique has been widely applied in a variety of cell systems [6–9,153,170,216,236,238,321]. In most of these studies, cells were heated while immersed in solutions of relatively high ionic strength (0.15 N NaCl) containing formaldehyde, which was included to prevent DNA renaturation; in its absence renaturation was believed to occur immediately after cell cooling [233–235]. Since formaldehyde, in addition to preventing DNA renaturation, induces extensive changes in nuclear chromatin affecting stability of DNA, the results obtained with this agent cannot be compared with the data obtained in its absence [reviewed in 69,88,277]. Furthermore, because of the complexity of interactions involving formaldehyde, such results are difficult to interpret, and they offer limited information regarding molecular interactions responsible for DNA stability in situ.

In general, all studies in which denaturation was assayed in the presence of formaldehyde indicate that there are significant differences between various cell types in the stability of DNA to heat, and that the stability of DNA is directly related to the degree of chromatin condensation or the transcriptional activity of the cell ("genome activation"). DNA of active cells and/or cells with a large portion of euchromatin denatures at lower temperature than DNA of dormant cells and/or cells with condensed chromatin (heterochromatin) [6–9,125,153,170,216,236,238,321]. These results remain in contrast to some of the data provided by biochemical methods [5,270] or the in situ studies in the absence of formaldehyde (see further), which indicate that DNA in condensed chromatin, rather than in euchromatin, is more sensitive to denaturation. However, an increase in AO binding and a change in the color of nuclei (from green to yellow), in the absence of formaldehyde, was observed by Maroudas and Wray [183] during stimulation of muscle nuclei, suggesting that in the stretch-dependent tissue, activation of nuclear chromatin may indeed decrease stability of DNA. Recently, Raap et al. [222] applied AO to investigate the degree of DNA denaturation in isolated nuclei induced by heat, acids or alkali, in the absence of formaldehyde, and correlated it with the ability of such DNA to hybridize in situ with the complementary strands.

Studies of DNA denaturation in situ by measurement of the thermal depolarization of fluorescence of AO bound to DNA were attempted by MacInnes and Uretz [178,179]. The assumption was made that the intercalated AO molecules, being markers of ds sections of DNA, will be characterized by highly polarized fluorescence, whereas AO associated with ss regions was expected to have higher rotational freedom and thus lower degree of fluorescence polarization. Indeed, during exposure of chromosomes to heat, depolarization of fluorescence was apparent, but the results were complicated by changes in polarization also occurring as a result of dissociation of proteins during heating and by dynamic instabilities ("breathing") of the double helix [179].

DNA Denaturation Measurements by Flow Cytometry

A method has been developed in our laboratory to study DNA denaturation in situ by flow cytometry [87,95]. Originally the method was applied in studies of DNA denaturation induced in fixed cells, or isolated nuclei, by heat. The RNase-treated cells or nuclei were suspended in media of relatively low ionic strength, heated at various temperatures, stained with AO under conditions (AO concentration, pH, ionic strength) favoring differential staining of ds versus ss DNA, and measured in equilibrium with AO by flow cytometry [87,95,97]. Fluorescence measurements were performed with standardized settings of the photomultiplier sensitivities such that the green fluorescence of the nonheated cells (nearly all DNA is ds) had the same numerical value as the red luminescence of the same cells heated to 100°C under conditions of maximal DNA denaturation (nearly all DNA is ss) [87]. This standardization was done in order to obtain a quantitative representation of fluorescence intensities in green and red wavelengths that would be proportional to the quantities of the stainable ds and ss DNA, respectively.

The extent of DNA denaturation at a given temperature (αt) was proposed as a ratio of red luminescence intensity to total (red plus green) cell luminescence intensity [$\alpha t = F_{>600}/(F_{>600} + F_{530})$], which thus represents the portion of DNA that is denatured and luminesces red in relationship to the total DNA that stains with AO [87].

Interestingly, despite the lack of formaldehyde during cell heating and cooling, no DNA renaturation was observed [87,95], which remains in contrast to earlier observations

[69,234,235]. However, because the flow cytometric measurements were done under conditions of cell equilibrium with AO, the lack of renaturation can be explained by the presence of ss DNA in the condensed, luminescing red complexes with AO; such DNA cannot, of course, reform base-pairing without a prior dissociation from AO. As in the case of differential stainability of RNA versus DNA with AO, the most critical factor in obtaining the best discrimination between ds versus denatured DNA is choice of the optimal concentration of the dye.

The flow cytometric method of DNA denaturation measurements was widely applied in studies of chromatin structure. It was observed that DNA denaturation in situ was multiphasic; the heat-sensitive fraction most likely represented spacer DNA, while the resistant fractions, the nucleosomal particle associated DNA [89]. Because extraction of basic proteins at low pH lowered the melting temperature of DNA to the point at which free DNA melts, DNA in situ appears to be stabilized mostly by histones [87,89,97]. It was also observed that proteins extractable from nuclei with 0.25–0.35 N NaCl, i.e., predominantly HMG proteins, provided additional stabilization of the heat-sensitive fraction and lowered the melting point of the resistant fraction [90]. These data suggested that HMG proteins bind to spacer DNA and may also decrease association of DNA with the core particle histones.

Significant differences in DNA sensitivity to heat denaturation were observed between various cell types [78,88] and in cells progressing through various phases of the cell cycle [92,93]. Generally, a correlation was observed between the degree of chromatin condensation and sensitivity of DNA to denaturation; mitotic and G_0 cells having the most condensed chromatin exhibited the most sensitivity, whereas cycling cells, especially in late G_1 and early S phase, had DNA the most resistant. An exception was sperm cells; DNA in mature spermatozoa was remarkably resistant to denaturation [116]. However, in cases of abnormal sperm maturation associated with infertility, DNA of spermatozoa was sensitive to denaturation [116]. Sensitivity of DNA to denaturation thus offers a new marker to detect certain forms of male infertility related to defective chromosome condensation [116,117] (see Chapt. 26).

Denaturation of DNA in situ by alkali revealed by AO staining was proposed by Rydberg [246] as a method to quantify ssDNA breaks caused by ionizing radiation. The method offers the advantage of correlation of the ssDNA breaks with the cell position in the cell cycle.

The pattern of sensitivity of DNA to denaturation by heat, showing dependence on the degree of chromatin condensation (i.e., DNA is less stable in condensed chromatin), is similar to that induced by acid [79,82,92,94]. DNA denaturation by acid, however, as compared with heat denaturation, offers many advantages, chiefly simplicity of the reaction and lack of DNA loss that occurs at high temperatures. The method is based on subjecting cells, prefixed in ethanol or ethanol–acetone mixture, to treatment with RNase and then to 0.1 N HCl, followed by staining with AO at pH 2.6 [79,94]. It is not entirely clear why staining with AO at such low pH levels following the acid treatment provides the best discrimination between ss versus dsDNA. It is possible that staining at low pH, i.e., when DNA bases are more protonated, precludes rapid renaturation of the denatured portion of DNA (before the reaction with AO, i.e., condensation of ss DNA, is completed) that otherwise may occur when pH of the staining solution is higher. To explain higher sensitivity of DNA in condensed chromatin to denaturation, it was suggested that following acid treatment (which removes histones from chromatin) the repulsive sources of the charged DNA polyanion, supercoiled, compressed and maintained at high spatial density via interactions with the scaffold (matrix) proteins induce topological strain which destabilizes the double helix [79]. This instability is also reflected by increased sensitivity of DNA to the single strand specific S_1 nuclease [79]. In contrast to the condensed chromatin, DNA compression and topological strain are less in euchromatic regions of nuclei.

The technique of cell staining, based on acid-induced DNA denaturation, allows one to discriminate cells in G_0 (G_{1Q}), G_{1A}, G_{1B}, S, G_2, and M compartments of the cell cycle, often to recognize noncycling cells with an S- or G_2-phase-DNA content (S_Q, G_{2Q} cells) and to discern differentiating or dead cells [67,79,82,92,94]. This is because of the differences in both both DNA content (which is measured as total cell luminescence, red plus green) and sensitivity of DNA to denaturation (expressed as αt) between cells in the respective compartments of the cell cycle.

The ability to discriminate various compartments of the cell cycle contributed to application of this technique in cell biology [27,29,52,79,94,98,188,205,291]. Especially useful in these studies was the possibility of detection of noncycling (G_{1Q}) cells, characterized by high sensitivity of DNA to denaturation. A correlation was observed between position of the cells in the compartments discriminated by this method and expression of different proliferation-specific antigens [27,59–61,123]. It was also observed that sensitivity of DNA to acid denaturation measured in isolated nuclei from normal human colon epithelium, papillomas and colon carcinoma can be a useful diagnostic and prognostic marker [159]. The technique, in conjunction with stathmokinetic experiments arresting cells in mitosis, has also found an application in analysis of the cell cycle kinetics and its perturbation by various drugs [98, review 82]; see also Chapt. 39).

Denaturation of DNA In Situ Induced by AO

Acridine orange can denature nucleic acids. Because the affinity of AO to ss RNA is higher than that to ss DNA, RNA is denatured at lower AO concentration (~20 μM at 0.15 N NaCl) than DNA (>50 μM at 0.15 N NaCl). The products undergo condensation and luminesce red. This denaturation and condensation of nucleic acids can be detected in situ; stepwise titration of cells with increasing concentration of AO results in a loss of green fluorescence and an increase of red luminescence intensity; the changes occur in a cooperative manner [73,80]. Because the products scatter light, their appearance in cells can also be detected by right angle scatter measurements by flow cytometry [80]. The condensed products are electron opaque and can be visualized by electron microscopy [73,80]. Sensitivity of DNA in situ to denaturation induced by AO, as in the case of denaturation by heat or acid, is also higher in condensed chromatin compared to euchromatin [80]. Thus, denaturation and condensation of DNA induced by AO, either in isolated nuclei or in viable, permeabilized or fixed cells, provide still another approach to discriminate cell subpopulations with different chromatin structure by flow cytometry [80]. This approach was used in the human fertility clinic, where a correlation was observed between the sensitivity of spermatozoal DNA to denaturation induced by AO and the donor's fertility [271] (see Chapter 26).

A note should be added about the condensation of DNA

induced by AO in living cells, which are then fixed in glutaraldehyde, digested with DNase and observed by electron microscopy [119,130,165,224]. As originally reported by Frenster [119], such treatment resulted in the formation of electron-opaque granular products localized preferentially in euchromatin and considered to represent sections of transcriptionally active DNA. Interpretation of these data, however, is difficult because of extensive translocation of nucleic acids due to the condensation that occurs when live cells are treated with comparable (1 mM) concentration of AO [73,80]. Thus, the original localization of the DNA found in these products is uncertain. Furthermore, the role of DNase in generating the pattern is unexplained. The relationship of the condensed products to active DNA transcription therefore remains to be proven.

PRACTICAL NOTES ON APPLICATION OF AO IN FLOW CYTOMETRY

Certain critical points in the method of differential staining of RNA and DNA [73,91,279] or acid-induced DNA denaturation [92] are of practical importance. The most critical in the application of AO is the selection of proper concentration of the dye. Unlike staining techniques employing other dyes, the reaction with AO is very stringent, requiring a precise concentration of free dye in solution. This is due to the fact that the dye serves two functions: it denatures and condenses RNA and also stains differentially ss- and ds-nucleic acids. Thus, unlike other dyes, the AO methodology is unforgiving to even minor variation in the dye concentration from the optimum. Unfortunately, in some cell sorters or flow cytometers having long channels, rapid diffusion of AO from the sample core flow to the sheath flow takes place, lowering the AO concentration in the cell and breaking the staining equilibrium. In such instruments, higher AO concentration in the original staining solution is required to compensate for the AO loss due to diffusion. To this end, pilot experiments should be done, in which several different concentrations of AO are tested to select the optimal one for that instrument. The optimal concentrations of AO found by several investigators for particular types of flow cytometers or sorters are provided in the original publications [e.g., 22–30,52,59–61,205,291–293].

Another problem that relates to AO concentration is encountered during use of this dye for microscopy. Namely, equilibrium staining of cells or chromosomes mounted on microscopic slides requires significantly higher concentrations of dye compared to staining for flow cytometry. This is because AO is absorbed on glass or plastic surfaces; the surface-to-volume ratio of a flat chamber made of the coverslip mounted on the microscope slide is much higher than when cells are stained in the test tube. To compensate for AO absorption, either higher AO concentration should be used or the chamber should be prewashed with the AO solution to saturate the dye-absorbing sites.

Regarding the effects of AO concentration, there is a misconception that the reaction with AO follows exactly the mass action law according to the simple formula R = μM AO/μM DNA [202]. The mass action law is not applicable in this situation because one of the products undergoes phase transition (condensation and precipitation) [144,146]. Furthermore, at the ionic strength which is generally used, it is the concentration of free AO in solution and not the D/P ratio that is critical for the differential stainability of ss- versus ds-nucleic acids [94]. It is a simple task to test this, i.e., by lowering the number of cells per given volume of the staining solution of constant AO concentration; the decrease in cell number (increase in D/P) by several orders of magnitude) neither changes green nor red luminescence intensity. The D/P ratio becomes of importance only when there is such an excess of cells in a given volume that due to uptake of AO by the cells, concentration of free AO drops significantly. Therefore, there should not be more than approximately 10^5 cells per 1 ml of the final staining solution containing 20 μM AO [73,91,279]. Unfortunately, this restricts the use of high cell densities for high flow rates, unless an adjustment is made in total AO concentration. However, at cell densities below this value, there is such an excess of free AO that the drop of AO concentration attributable to uptake by cells is of no significance for the staining reaction.

Another point of practical importance is standardization of the data, which is required, e.g., to estimate the DNA index or compare RNA content of cell populations between day-to-day measurements, different cell types or different laboratories. Standardization of DNA measurements is in principle similar to that of other DNA dyes and requires standards such as normal diploid lymphocytes or other cells of known DNA content. To standardize RNA content, particular cells should be measured before and after treatment with RNase to establish the value of the RNA-specific red luminescence. The remaining red luminescence is due mostly to spectrum overlap from AO bound to DNA and to cell autoluminescence [73]; its extent can vary depending on the cell type, spectral characteristics of the phototube and the emission filters. To minimize this, long-pass filters transmitting above 630 or 640 nm are recommended. The mean RNase-specific value of red luminescence of a given cell population should be compared with and expressed as a multiple of the RNase-specific fraction of red luminescence of normal, nonstimulated lymphocytes. To be even more precise, pure B or T lymphocytes ought to be used as standard since they are more uniform with respect to RNA content [13]. Expressing the RNA index simply as a ratio of red luminescence of given cells to standard cells (i.e., lymphocytes) without subtracting the RNase-resistant fraction of the emission is inaccurate and does not allow for comparison between different cell types or instruments.

Standardization of DNA denaturation is relatively simple because it deals with standard ratios of luminescence intensities (αt) rather than their absolute values, and it was described in detail previously [87,89]. In the technique of acid-induced DNA denaturation, a useful standard again is nonstimulated lymphocytes. Adjustment of photomultiplier sensitivities to obtain the same numerical values of red and green luminescence of these cells (i.e., to have the cluster of lymphocytes located on the diagonal axis of the green/red bivariate distribution display) assumes that 50% of stainable DNA is denatured. This can be verified by subjecting these cells, following removal of histones, to maximal DNA denaturation by heat (at 100°C), and measuring αt changes as described [87]. In some situations the standard may be normal cells present in tumors to which the value of αt of tumor cells is related [159].

A trivial point that should be mentioned relates to the widespread belief that use of AO irreversibly contaminates the tubing in flow systems, hindering further measurements of any weak fluorescence. A strongly fluorescing cationic dye, AO indeed adheres to various surfaces and as in the case of other dyes rigorous washings are needed to remove it and prevent cross-contamination. Generally, sequential washes with ethanol (50%), detergent and bleach (Clorox) remove the dye quite efficiently, enabling measurement of weak immunofluorescence rather shortly (~30 min) following AO. In some instruments it may be a simpler task to replace the sample-flow tubing and keep one set of tubing devoted to AO only. The adherence problems, however, are not unique to AO and some other dyes (i.e., rhodamine 123) are more troublesome in this respect.

Similar to most dyes interacting with DNA, AO is a mutagen [68,135,272] and, needless to say, requires careful handling. Potentially, a danger exists when it is used in open sorters and the aerosol generated during sorting can spread through the laboratory.

CONCLUDING REMARKS

As reviewed in this chapter, AO is already widely applied in flow cytometry in a variety of diverse cytochemical reactions. Considering the potential of these methods in cell biology and clinically, the methodology is still greatly underutilized. This is true especially in the clinic where it is already evident that the RNA content per cell is one of the best, if not the best, prognostic parameter in leukemias, lymphomas and myelomas [12–16,22–26,227]. Considering the close relationship that exists between nucleolar/cellular RNA content and cell proliferation or transformation, this is not surprising, and the RNA parameter is expected to have similar prognostic value in the case of solid tumors. Although there are other methods of RNA detection [218,255], only the AO-based technique provides stoichiometric estimates of total cellular RNA and, simultaneously, the cell cycle distribution [76]. There is also a large, unexplored field in the application of other AO-employing techniques, especially those related to studies of chromatin structure or lysosomal function in clinical material.

The main reason hindering even wider application of AO in flow cytometry is the general perception of difficulty in its use and complexity of molecular interaction with cell constituents. The latter limits data interpretation. Ironically, complexity of interactions is also responsible for the versatility of AO, which is used in so many diverse applications. The apparent difficulty in applying AO stems from the extreme sensitivity to even minor changes in dye concentration. Finally, interpretation of the data suffered from the old classical model describing dye-dye stacking interactions [48–50] as responsible for the metachromatic property of this dye in reactions with nucleic acids. Such a model could not explain many phenomena, especially the specificity of staining of RNA versus DNA. It is hoped that this chapter, by reviewing earlier data, by providing some practical details related to its use and, above all, in describing the new model of AO–nucleic acid interactions that explains denaturation and condensation of the latter, can be of help in further applications of this unique, highly versatile cytochemical probe.

ACKNOWLEDGMENTS

We thank Ms. Rose Vecchiolla for typing the manuscript. This work was supported by grants R37 CA23296 and R01 CA28704 from the National Cancer Institute.

REFERENCES

1. **Abrams WR, Diamond LW, Kane AB (1983)** A flow cytometric assay of neutrophil degranulation. J Histochem Cytochem 31:737–744.
2. **Adam G, Steiner U, Seuwen K (1983)** Proliferative activity and ribosomal RNA content of 3T3 and SV40-3T3 cells. Cell Biol Int Rep 7:955–962.
3. **Alarcon Seqovia D, Llorente L, Fishbein E, Diaz Jouanen E (1982)** Abnormalities in the content of nucleic acids of peripheral blood mononuclear cells from patients with systemic lupus erythrematosus. Relationship to DNA antibodies. Arthritis Rheum 23:304–317.
4. **Allison AC, Young MR (1964)** Uptake of dyes and drugs by living cells in culture. Life Sci 3:1407–1414.
5. **Almagor M, Cole RD (1987)** A high melting (105°C) form of chromatin characterizes the potential of cells for mitosis. J Cell Biol 262:15071–15075.
6. **Alvarez MR (1973)** Microfluorometric comparisons of chromatin thermal stability in situ between normal and neoplastic cells. Cancer Res 33:786–790.
7. **Alvarez MR (1974)** Early nuclear cytochemical changes in regenerating mammalian liver. Exp Cell Res 83:225–230.
8. **Alvarez MR (1975)** Microfluorometric comparisons of heat-induced nuclear acridine orange metachromasia between normal cells and neoplastic cells from primary tumors of diverse origin. Cancer Res 35:93–98.
9. **Alvarez MR, Truitt AJ (1977)** Rapid nuclear cytochemical changes induced by dexamethasone in thymus lymphocytes of adrenalectomized rats. Exp Cell Res 106:105–110.
10. **Amagasa J (1986)** Mechanisms of photodynamic inactivation of acridine orange-sensitized transfer RNA: Participation of singlet oxygen and base damage leading to inactivation. J Radiat Res (Tokyo) 27:339–351.
11. **Amagasa J (1986)** Binding of acridine orange to transfer RNA and photodynamic inactivation. J Radiat Res (Tokyo) 27:325–338.
12. **Andreeff M, Assing G, Cirrincione C (1986)** Prognostic value of DNA/RNA flow cytometry in myeloblastic and lyphoblastic leukemia in adults: RNA content and S-phase predict remission duration and survival in multivariate analysis. Ann NY Acad Sci 486:386–407.
13. **Andreeff M, Beck JD, Darzynkiewicz Z, Traganos F, Gupta S, Melamed MR, Good RA (1978)** RNA content in human lymphocyte subpopulations. Proc Natl Acad Sci USA 75:1938–1942.
14. **Andreeff M, Darzynkiewicz Z, Sharpless T, Clarkson BD, Melamed MR (1980)** Discrimination of human leukemia subtypes by flow cytometric analysis of cellular RNA and DNA. Blood 55:282–293.
15. **Andreeff M, Gaynor J, Chapman D, Little C, Gee T, Clarkson BD (1987)** Prognostic factors in acute lymphoblastic leukemia in adults: The Memorial Hospital experience. Haematol Blood Transf 30:111–124.
16. **Andreeff M, Hansen H, Cirrincione C, Filippa D, Thaler H (1986)** Prognostic value of DNA/RNA flow cytometry on B-cell non-Hodgkin's lymphoma: Development of laboratory model and correlation with four taxonomic systems. Ann NY Acad Sci 486:368–386.
17. **Antel JP, Oger JJF, Dropcho E, Richman DP, Kuo HH, Arnason BGW (1980)** Reduced T-lymphocyte cell reactivity as a function of human aging. Cell Immunol 54:184–192.
18. **Arditti RR, Coppo A (1965)** Effect of acridines and

temperature on a strain of Bacillus megaterium lysogenic for phage. Virology 25:643–649.

19. **Armstrong JA** (1956) Histochemical differentiation of nucleic acids by means of induced fluorescence. Exp Cell Res 11:640–643.
20. **Armstrong JA, Niven JSF** (1957) Histochemical observations on cellular and virus nucleic acids. Nature (Lond) 180:1335–1338.
21. **Ashihara T, Traganos F, Baserga R, Darzynkiewicz Z** (1978) A comparison of cell cycle related changes in post-mitotic and quiescent AF8 cells as measured by flow cytometry after acridine orange staining. Cancer Res 38:2514–2518.
22. **Barlogie B, Alexanian R, Dixon D, Smith L, Smallwood L, Delasalle K** (1985) Prognostic implications of tumor cell DNA and RNA content in multiple myeloma. Blood 66:338–341.
23. **Barlogie B, Alexanian R, Gehan EA, Smallwood L, Smith T, Drewinko B** (1983) Marrow cytometry and prognosis in myeloma. J Clin Invest 72:853–861.
24. **Barlogie B, Maddox AM, Johnston DA, Raber MN, Drewinko B, Keating MJ, Freireich EJ** (1983) Quantitative cytology in leukemia research. Blood Cells 9:35–55.
25. **Barlogie B, McLaughlin P, Alexanian R** (1987) Characterization of hematologic malignancies by flow cytometry. Anal Quant Cytol 9:147–155.
26. **Barlogie B, Alexanian R, Gehan EA, Smallwood L, Smith T, Drewinko B** (1983) Marrow cytometry and prognosis in myeloma. J Clin Invest 72:853–861.
27. **Bauer KD, Clevenger CV, Williams TJ, Epstein AL** (1986) Assessment of cell cycle-associated antigen expression using multiparameter flow cytometry and antibody-acridine orange sequential staining. J Histochem Cytochem 34:245–250.
28. **Bauer KD, Dethlefsen LA** (1980) Total cellular RNA content: Correlation between flow cytometry and ultraviolet spectroscopy. J Histochem Cytochem 28:493–498.
29. **Bauer KD, Dethlefsen LA** (1981) Control of cellular proliferation in HeLa-53 suspension cultures. Characterization of cultures utilizing acridine orange staining procedures. J Cell Physiol 108:99–112.
30. **Bauer KD, Keng PC, Sutherland RM** (1982) Isolation of quiescent cells from multicellular tumor spheroids using centrifugal elutriation. Cancer Res 42:72–78.
31. **Bauer W, Vinograd J** (1968) The interaction of closed circular DNA with intercalating dyes. I. The superhelix density of SV40 DNA in the presence and absence of the dye. J Mol Biol 33:141–171.
32. **Beattie TM** (1970) Vital staining of neurosecretory material with acridine orange in the insect Periplaneta americana. Experientia 27:110–110.
33. **Beers RF, Armilei G** (1965) Heterogeneous binding of acridine orange by polyribonucleotides. Nature (Lond) 208:466–468.
34. **Bergounioux C, Perennes C, Brown SC, Gadal P** (1988) Nuclear RNA quantification in protoplast cell-cycle phases. Cytometry 9:84–87.
35. **Berkan TK, Reeder JE, Lopez PA Jr, Gorman KM, Wheeless LLJr** (1986) A protocol for Papanicolaou staining of cytologic specimens following flow analysis. Cytometry 7:101–103.
36. **Bertalanffy FD** (1962) Evaluation of the acridine-orange fluorescence microscope method to cytodiagnosis of cancer. Ann NY Acad Sci 93:715–750.
37. **Bertalanffy LV** (1963) Acridine orange fluorescence in cell physiology, cytochemistry and medicine. Protoplasma 57:51–83.
38. **Bertalanffy LV, Bertalanffy FD** (1960) A new method for cytological diagnosis of pulmonary cancer. Ann NY Acad Sci 84:225–228.
39. **Bertalanffy LV, Bickis I** (1956) Identification of cytoplasmic basophilia (ribonucleic acid) by fluorescence microscopy. J Histochem Cytochem 4:481–493.
40. **Betel I, Martijnse J, van der Western G** (1979) Mitogenic activation and proliferation of mouse thymocytes. Exp Cell Res 124:329–337.
41. **Bettens F, Kirstensen F, Walker C, Bonnard GD, DeWeck AL** (1984) Lymphokine regulation of human lymphocyte proliferation: Formation of resting G0 cells by removal of interleukin 2 in cultures of proliferating T lymphocytes. Cell Immunol 86:337–346.
42. **Blake A, Peacocke AR** (1968) The interaction of aminoacridines with nucleic acids. Biopolymers 6:1225–1253.
43. **Bloomfield VA, Crothers DM, Tinoco I Jr** (1974) Binding of small molecules. In "Physical Chemistry of Nucleic Acids." New York: Harper & Row, pp 374–477.
44. **Bobrow M, Madan K** (1973) The effects of various banding procedures on human chromosomes studied with acridine orange. Cytogenet Cell Genet 12:145–153.
45. **Bolund L, Darzynkiewicz Z, Ringertz NR** (1970) Cell concentration and the staining properties of nuclear deoxyribonucleoprotein. Exp Cell Res 62:76–81.
46. **Bonvoisin B, Cordier G, Revillard JP, Bouvier M** (1984) Increased DNA and/or RNA content of synovial fluid cells in rheumatoid arthritis; a flow cytometric study. Ann Rheum Dis 43:222–228.
47. **Borisova OF, Tumerman LA** (1964) Luminescence of complexes of acridine orange with nucleic acids. Biofizika 9:537–544.
48. **Bradley DF** (1961) Molecular biophysics of dye–polymer complex. Trans NY Acad Sci 24:64–73.
49. **Bradley DF, Lifson S** (1968) Statistical mechanical analysis of binding of acridines to DNA. In Pullman B (ed): "Molecular Associations in Biology." New York: Academic Press, pp 261–270.
50. **Bradley DF, Wolf MK** (1959) Aggregation of dyes bound to polyanions. Proc Natl Acad Sci USA 45:944–952.
51. **Brenner S, Barnett L, Crick FHC, Orgel A** (1961) The theory of mutagenesis. J Mol Biol 3:121–121.
52. **Brock WA, Swartzendruber DE, Grdina DJ** (1982) Kinetic heterogeneity in density-separated murine fibrosarcoma subpopulations. Cancer Res 42:4499–5003.
53. **Bukatsch F, Haitinger M** (1940) Beitraqe zur fluoreszenzmikroskopischen Darstellung des Zellinhaltes insbesondere des Cytoplasmas und des Zellkernes. Protoplasma 34:515–523.
54. **Cambier MA, Wheeless LL, Patten SF** (1979) A post-staining fixation technique for acridine orange. Quantitative aspects. Anal Quant Cytol 1:57–57.
55. **Carmichael GG, McMaster GK** (1980) The analysis

of nucleic acids in gels using glyoxal and acridine orange. Methods Enzymol 65:380–391.

56. **Carotenuto P, Pontesili O, Cambier JC, Hayward AR** (1986) Desteroxamine blocks IL2 receptor expression on human T lymphocytes. J Immunol 136: 2342–2347.

57. **Cayre Y, Raynal MC, Darzynkiewicz Z, Dorner MH** (1987) Model for intermediate steps in monocytic differentiation: c-myc, c-fms and ferritin as markers. Proc Natl Acad Sci USA 84:7619–7623.

58. **Clark EA, Ledbetter JA** (1986) Activation of human B cell mediated through two distinct cell surface differentiation antigens Bp35 and Bp50. Proc Natl Acad Sci USA 83:4494–4498.

59. **Clevenger CV, Bauer KD, Epstein AL** (1987) Modulation of the nuclear antigen p105 as a function of cell cycle progression. J Cell Physiol 130:336–343.

60. **Clevenger CV, Epstein AL, Bauer KD** (1985) A method for simultaneous nuclear immunofluorescence and DNA content quantitation using monoclonal antibodies and flow cytometry. Cytometry 6: 208–214.

61. **Clevenger CV, Epstein AL, Bauer KD** (1987) Quantitative analysis of a nuclear antigen in interphase and mitotic cells. Cytometry 8:280–286.

62. **Collste LG, Darzynkiewicz Z, Traganos F, Sharpless T, Sogani P, Grabstald H, Whitmore WF Jr, Melamed MR** (1980) Flow cytometry in cancer detection and evaluation using acridine orange metachromatic nucleic acid staining of irrigation cytology specimens. J Urol 123:478–485.

63. **Comings DE** (1973) Biochemical mechanisms of chromosome banding and color banding with acridine orange. In Caspersson T (ed): "Chromosome Identification. Techniques and Applications in Biology and Medicine." Academic Press, New York, pp 292–306.

64. **Coulson PB, Bishop AO, Lenarduzzi R** (1977) Quantitation of cellular deoxyribonucleic acid by flow microfluorometry. J Histochem Cytochem 25: 1147–1153.

65. **Couturier J, Dutrillaux B, Jejeune J** (1973) Etude de fluorescences specifiques des bandes R et des bandes Q des chromosomes humains. CR Acad Sci Paris Ser D 276:339–342.

66. **Cowden RR, Curtis SK** (1976) Some quantitative aspects of acridine orange fluorescence in unfixed sucrose-isolated mammalian nuclei. Histochem J 8:45–49.

67. **Creasey AU, Bartholomew JC, Merigan TC** (1981) The importance of G_0 in the site of action of interferon in the cell cycle. Exp Cell Res 134:155–160.

68. **Crick FHC, Barnett L, Brenner S, Watts Tubin RJ** (1961) General nature of genetic code for proteins. Nature (Lond) 192:1227–1230.

69. **Darzynkiewicz Z** (1979) Acridine orange as a molecular probe in studies of nucleic acids in situ. In Melamed MR, Mullaney PF, Mendelsohn ML (eds): "Flow Cytometry and Sorting." New York: John Wiley & Sons, pp 285–316.

70. **Darzynkiewicz Z** (1986) Cell growth and division cycle. In Dethlefsen LA (ed): "Cell Cycle Effects of Drugs. International Encyclopedia of Pharmacology and Therapeutics," Section 121. Oxford: Pergamon Press, pp 1–43.

71. **Darzynkiewicz Z, Andreeff M, Traganos F, Sharpless T, Melamed MR** (1978) Discrimination of cycling and noncycling lymphocytes by BUdR-suppressed acridine orange fluorescence in a flow cytometric system. Exp Cell Res 115:31–35.

72. **Darzynkiewicz Z, Crissman H, Traganos F, Steinkamp J** (1982) Cell heterogeneity during the cell cycle. J Cell Physiol 112:465–474.

73. **Darzynkiewicz Z, Evenson D, Kapuscinski J, Melamed MR** (1983) Denaturation of RNA and DNA in situ induced by acridine orange. Exp Cell Res 148:31–46.

74. **Darzynkiewicz Z, Evenson D, Staiano-Coico L, Sharpless T, Melamed MR** (1979) Relationship between RNA content and progression of lymphocytes through the S phase of the cell cycle. Proc Natl Acad Sci USA 76:358–362.

75. **Darzynkiewicz Z, Evenson DP, Staiano-Coico L, Sharpless T, Melamed MR** (1979) Correlation between cell cycle duration and RNA content. J Cell Physiol 100:425–438.

76. **Darzynkiewicz Z, Kapuscinski J, Traganos F, Crissman HA** (1987) Application of pyronin Y(G) in cytochemistry of nucleic acids. Cytometry 8:138–145.

77. **Darzynkiewicz Z, Sharpless T, Staiano-Coico L, Melamed MR** (1980) Subcompartments of the G1 phase of the cell cycle detected by flow cytometry. Proc Natl Acad Sci USA 77:6696–6700.

78. **Darzynkiewicz Z, Traganos F, Arlin Z, Sharpless T, Melamed MR** (1976) Cytofluorometric studies on conformation of nucleic acid in situ. II. Denaturation of deoxyribonucleic acid. J Histochem Cytochem 24: 49–58.

79. **Darzynkiewicz Z, Traganos F, Carter SP, Higgins PJ** (1987) In situ factors affecting stability of DNA helix in interphase nuclei and metaphase chromosomes. Exp Cell Res 172:168–179.

80. **Darzynkiewicz Z, Traganos F, Kapuscinski J, Melamed MR** (1985) Denaturation and condensation of DNA in situ induced by acridine orange in relation to chromatin changes during growth and differentiation of Friend erythroleukemia cells. Cytometry 6:195–207.

81. **Darzynkiewicz Z, Traganos F, Kapuscinski J, Staiano-Coico L, Melamed MR** (1984) Accessibility of DNA in situ to various fluorochromes: Relationship to chromatin changes during erythroid differentiation of Friend leukemia cells. Cytometry 5:355–363.

82. **Darzynkiewicz Z, Traganos F, Kimmel M** (1988) Assay of cell cycle kinetics by multivariate flow cytometry using the principle of stathmokinesis. In Gray JE, Darzynkiewicz Z (eds): "Techniques in Cell Cycle Analysis." Clifton, NJ: Humana Press, pp 291–336.

83. **Darzynkiewicz Z, Traganos F, Melamed MR** (1980) New cell cycle compartments identified by flow cytometry. Cytometry 1:98–108.

84. **Darzynkiewicz Z, Traganos F, Melamed MR** (1981) Detergent treatment as an alternative to cell fixation for flow cytometry. J Histochem Cytochem 29:329–330.

85. **Darzynkiewicz Z, Traganos F, Melamed MR** (1983) Distinction between 5-bromodeoxyuridine labelled and unlabelled mitotic cells by flow cytometry. Cytometry 3:345–348.

86. **Darzynkiewicz Z, Traganos F, Sharpless T, Friend C, Melamed MR** (1976) Nuclear chromatin changes during erythroid differentiation of Friend virus induced leukemic cells. Exp Cell Res 99:301–309.
87. **Darzynkiewicz Z, Traganos F, Sharpless T, Melamed MR** (1975) Thermal denaturation of DNA in situ as studied by acridine orange staining and automated cytofluorometry. Exp Cell Res 90:411–428.
88. **Darzynkiewicz Z, Traganos F, Sharpless T, Melamed MR** (1976) Cytofluorometric studies on conformation of nucleic acids in situ. II. Denaturation of DNA. J Histochem Cytochem 24:49–58.
89. **Darzynkiewicz Z, Traganos F, Sharpless T, Melamed MR** (1976) DNA denaturation in situ: Effect of divalent cations and alcohols. J Cell Biol 68:1–10.
90. **Darzynkiewicz Z, Traganos F, Sharpless T, Melamed MR** (1976) Effect of 0.25 M sodium chloride treatment on DNA denaturation in situ in thymus lymphocytes. Exp Cell Res 100:393–396.
91. **Darzynkiewicz Z, Traganos F, Sharpless T, Melamed MR** (1976) Lymphocyte stimulation: A rapid, multiparameter analysis. Proc Natl Acad Sci USA 73: 2881–2884.
92. **Darzynkiewicz Z, Traganos F, Sharpless T, Melamed MR** (1977) Cell cycle related changes in nuclear chromatin of stimulated lymphocytes as measured by flow cytometry. Cancer Res 37:4635–4640.
93. **Darzynkiewicz Z, Traganos F, Sharpless T, Melamed MR** (1977) Different sensitivity of DNA in situ in interphase and metaphase chromatin to heat denaturation. J Cell Biol 73:128–138.
94. **Darzynkiewicz Z, Traganos F, Sharpless T, Melamed MR** (1977) Interphase and metaphase chromatin. Different stainability of DNA with acridine orange after treatment at low pH. Exp Cell Res 110:201–214.
95. **Darzynkiewicz Z, Traganos F, Sharpless TK, Melamed MR** (1974) Thermally-induced changes in chromatin of isolated nuclei and of intact cells as revealed by acridine orange staining. Biochem Biophys Res Commun 59:392–399.
96. **Darzynkiewicz Z, Traganos F, Sharpless TK, Melamed MR** (1975) Conformation RNA in situ as studied by acridine orange staining and automated cytofluorometry. Exp Cell Res 95:143–153.
97. **Darzynkiewicz Z, Traganos F, Xue S, Melamed MR** (1984) Effect of n-butyrate on cell cycle progression and in situ chromatin structure of L1210 cells. Exp Cell Res 136:279–239.
98. **Darzynkiewicz Z, Traganos F, Xue SB, Staiano Coico L, Melamed MR** (1981) Rapid analysis of drug effects on the cell cycle. Cytometry 1:279–286.
99. **Darzynkiewicz Z, Williamson B, Carswell EA, Old LJ** (1984) The cell cycle specific effects of tumor necrosis factor. Cancer Res 44:83–90.
100. **Davis L, Lipsky PE** (1986) Signals involved in T cell activation. II. Distinct roles of intact accessory cells, phorbol esters and interleukin 1 in activation and cell cycle progression of resting T lymphocytes. J Immunol 136:3588–3596.
101. **De Bruyn PPH, Robertson RC, Farr RS** (1950) In vivo affinity of di-aminoacridines for nuclei. Anat Rec 108:279–295.
102. **de Duve C, de Barsy T, Poole B, Trouet A, Tulkens P, Van Hoof F** (1974) Lysosomotrophic agents. Biochem Pharmacol 23:2495–2531.
103. **de la Chapelle A, Schroder J, Selander RK** (1973) In situ localization and characterization of different classes of chromosomal DNA: Acridine orange and quinacrine mustard fluorescence. Chromosoma 40: 347–360.
104. **de la Chapelle A, Schroder J, Selander RK, Stenstrand K** (1973) Differences in DNA composition along mammalian metaphase chromosomes. Chromosoma 42:365–382.
105. **Dean ACR, Hinselwood CD** (1964) "Growth, Function and Regulation in Bacterial Cells." Oxford: Clarendon Press.
106. **DeFranco AL, Raveche E, Paul WE** (1985) Separate control of B lymphocyte early activation and proliferation in response to anti-IgM antibodies. J Immunol 135:87–94.
107. **Denny WA, Baguley BC, Cain BF, Waring MJ** (1983) Antitumor acridines. In Neidle S, Waring MJ (eds): "Molecular Aspects of Anticancer Drugs." Weinheim: Verlag-Chemie, pp 1–34.
108. **Dethlefsen LA, Bauer KD, Riley RM** (1980) Analytical cytometric approaches to heterogeneous cell populations in solid tumors. A review. Cytometry 1:89–97.
109. **Dukes CD, Parson JL, Stephens CAL** (1969) Use of acridine orange in lymphocyte transformation test. Proc Soc Exp Biol Med 131:1168–1170.
110. **Dutrillaux B, Covie M** (1974) Etude de facteurs influençant la denaturation thermique menagée. Exp Cell Res 85:143–153.
111. **Dutrillaux B, Laurent C, Couturier J, Lejeune J** (1973) Coloration par l'acridine orange de chromosomes préalablement traités par le 5-bromodeoxyuridine (BUdR). CR Hebd Seances Acad Sci 276:3179–3181.
112. **Dyson JED, Quirke P, Bird CC, McLaughlin JB, Surrey CR** (1984) Relationship between cell ploidy and glucocorticoid induced death in human lymphoid cell lines. Br J Cancer 49:731–738.
113. **Eisinger M, Lee JS, Hefton JM, Darzynkiewicz Z, Chiao JW, de Harven E** (1979) Human epidermal cell cultures: Growth and differentiation in the absence of dermal components or medium supplements. Proc Natl Acad Sci USA 76:5340–5345.
114. **Elson D, Spitnik-Elson P, Avital S, Abramovitz R** (1979) Binding of magnesium ions and ethidium bromide: Comparison of ribosomes and free ribosomal RNA. Nucleic Acids Res 7:465–480.
115. **Evenson DP, Darzynkiewicz Z, Melamed MR** (1980) Comparison of human and mouse sperm chromatin structure by flow cytometry. Chromosoma 78:225–238.
116. **Evenson DP, Darzynkiewicz Z, Melamed MR** (1980) Relation of mammalian sperm chromatin heterogeneity to fertility. Science 210:1131–1131.
117. **Evenson DP, Klein FA, Whitmore WF, Melamed MR** (1984) Flow cytometric evaluation of sperm from patients with testicular carcinoma. J Urol 132:1220–1225.
118. **Franklin AL, Filion WG** (1981) Acridine orange–methyl green fluorescent staining of nucleoli. Stain Technol 56:343–348.
119. **Frenster JH** (1971) Electron microscopic localization

of acridine orange binding to DNA within human leukemic bone marrow cells. Cancer Res 31:1128–1133.

120. **Fujikawa Yamamoto K (1982)** RNA dependence in the cell cycle of V79 cells. J Cell Physiol 112:60–66.

121. **Fujikawa Yamamoto K (1983)** The relation between length of the cell cycle duration and RNA content in HeLa S3 cells. Cell Struct Funct 8:303–308.

122. **Genovesi E, Collins JJ (1983)** In vitro growth inhibition of murine leukemia cells by antibody specific for the major envelope glycoprotein (gp71) of Friend leukemia virus. J Cell Physiol 117:215–229.

123. **Gerdes J, Lemke H, Baisch H, Wacker HH, Schwab U, Stein H (1984)** Cell cycle analysis of a cell proliferation-associated human nuclear antigen defined by the monoclonal antibody Ki-67. J Immunol 133: 1710–1715.

124. **Gill JE, Wheeless LL Jr, Hanna Madden C, Marisa RJ, Horan PK (1978)** A comparison of acridine orange and Feulgen cytochemistry of human tumor cell nuclei. Cancer Res 38:1893–1898.

125. **Gledhill BL, Gledhill MP, Rigler P Jr, Ringertz NR (1966)** Changes in deoxyribonucleoprotein during spermiogenesis in the bull. Exp Cell Res 41:652–665.

126. **Golden JF, West SS (1979)** Fluorescence spectroscopic and fading behavior of Ehrlich's hyperdiploid mouse ascites tumor cells supravitally stained with acridine orange. J Histochem Cytochem 22: 495–505.

127. **Golden JF, West SS, Echols CE, Shingleton HM (1976)** Quantitative fluorescence spectrophotometry of acridine orange-stained unfixed cells. Potential for automated detection of human uterine cancer. J Histochem Cytochem 24:315–321.

128. **Gordon J, Walker L, Guy G, Brown G, Rowe M, Rickinson A (1986)** Control of human B-lymphocyte replication. II. Transforming Epstein–Barr virus exploits three distinct viral signals to undermine three separate points in B-cell growth. Immunology 58: 591–595.

129. **Greene RS, Alderfer J, Munson BR (1983)** In vitro effects of acridine intercalation on RNA polymerase interactions with supercoiled DNA. Int J Biochem 15: 1231–1239.

130. **Hara H, Moriki T, Hiroi M, Yamane T (1984)** Electron microscopic localization of acridine orange binding to DNA within rat astrocytoma C6 cells. Acta Pathol Jpn 34:1049–1057.

131. **Hare JD (1986)** Two-color flow-cytometric analysis of the growth cycle of Plasmodium falciparum in vitro: Identification of cell cycle compartments. J Histochem Cytochem 34:1651–1658.

132. **Hare JD, Bahler DW (1986)** Analysis of Plasmodium falciparum growth in culture using acridine orange and flow cytometry. J Histochem Cytochem 34:215–220.

133. **Hawrylowicz CM, Klaus GGB (1984)** Activation and proliferation signals in mouse B cells. IV. Concanavalin A stimulates B cells to leave G_0 but not to proliferate. Immunology 53:703–711.

134. **Heller CS, Sevag MG (1966)** Prevention of the emergence of drug resistance in bacteria by acridines, phenothiazines and dibenzocycloheptenes. Appl Microbiol 14:879–885.

135. **Hirohita Y (1960)** The effect of acridine dyes on mating type factors in Escherichia coli. Proc Natl Acad Sci USA 46:57–64.

136. **Ichimura S, Zama M, Fujita H (1971)** Quantitative determination of single-stranded sections in DNA using the fluorescent probe acridine orange. Biochim Biophys Acta 240:485–495.

137. **Ichimura S, Zama M, Fujita H, Ito T (1969)** The nature of strong binding between acridine orange and deoxyribonucleic acid as revealed by equilibrium dialysis and thermal renaturation. Biochim Biophys Acta 190:116–125.

138. **Imae T, Hayashi S, Ikeda S, Sakaki T (1981)** Interaction between acridine orange and polyriboadenylic acids. Int J Biol Macromol 3:259–266.

139. **Jelinek DF, Lipsky PE (1985)** The roles of T cell factors in activation, cell cycle progression, and differentiation of human B cells. J Immunol 134:1690–1701.

140. **Jensen PKA, Pedersen S, Bolund L (1985)** Basal-cell subpopulations and cell cycle kinetics in human epidermal explant cultures. Cell Tissue Kinet 18:201–215.

141. **Joshi DS, Goyal U, Phondke GP (1983)** Flow cytometric studies on lymphoid cells in murine leukemia. Cell Mol Biol 29:39–47.

142. **Kapuscinski J, Darzynkiewicz Z (1983)** Increased accessibility of bases in DNA upon binding of acridine orange. Nucleic Acids Res 11:7555–7568.

143. **Kapuscinski J, Darzynkiewicz Z (1984)** Denaturation of nucleic acids induced by intercalating agents. Biochemical and biophysical properties of acridine orange–DNA complexes. J Biomol Struct Dyn 1:1485–1499.

144. **Kapuscinski J, Darzynkiewicz Z (1984)** Condensation of nucleic acids by intercalating aromatic cations. Proc Natl Acad Sci USA 81:7368–7372.

145. **Kapuscinski J, Darzynkiewicz Z (1987)** Interactions of acridine orange with double stranded nucleic acids. Spectral and affinity studies. J Biomol Struct Dyn 5:127–143.

146. **Kapuscinski J, Darzynkiewicz Z (1987)** Interactions of pyronin Y (G) with nucleic acids. Cytometry 8: 129–137.

146a. **Kapuscinski J, Darzynkiewicz Z (1989)** Structure destabilization and condensation of nucleic acids by intercalators. In Sarma RH, Sarma MH (eds): "Biological Structure, Dynamics, Interactions, and Stereodynamics." Schenectady, NY: Adenine Press. (In press.)

147. **Kapuscinski J, Darzynkiewicz Z, Melamed MR (1982)** Luminescence of the solid complexes of acridine orange with RNA. Cytometry 2:201–211.

148. **Kapuscinski J, Darzynkiewicz Z, Melamed MR (1983)** Interactions of acridine orange with nucleic acids. Properties of complexes of acridine orange with single stranded ribonucleic acids. Biochem Pharmacol 32:3679–3694.

149. **Kareeman G, Mueller H, Szent-Györgyi A (1957)** Competitive binding of ATP and acridine orange by muscle. Proc Natl Acad Sci USA 45:373–379.

150. **Kean JM, White SA, Draper DE (1985)** Detection of high-affinity intercalator sites in a ribosomal RNA fragment by the affinity cleavage intercalator methidiumpropyl–EDTA–Iron(II). Biochemistry 24:5062–5070.

151. **Kehrl JH, Muraguchi A, Fauci AS** (1984) Differential expression of cell activation markers after stimulation of resting human B lymphocytes. J Immunol 132:2857–2861.
152. **Kenter AL, Watson JV, Azim T, Rabbitts TH** (1986) Colcemid inhibits growth during early G1 in normal but not in tumorigenic lymphocytes. Exp Cell Res 167:241–251.
153. **Kernell AM, Bolund L, Ringertz NR** (1971) Chromatin changes during erythropoiesis. Exp Cell Res 65:1–6.
154. **Killander D, Rigler R** (1969) Activation of deoxyribonucleoprotein in human leukocytes stimulated by phytohemagglutinin. Exp Cell Res 54:163–170.
155. **Kimmel M, Darzynkiewicz Z, Staiano Coico L** (1986) Stathmokinetic analysis of human epidermal cells in vitro. Cell Tissue Kinet 19:289–289.
156. **Klaus GGB, Hawrylowicz CM, Carter CJ** (1985) Activation and proliferation signals in mouse B cells. VI. Anti-Ig antibodies induce dose-dependent cell cycle progression in B cells. Immunology 55:411–418.
157. **Klaus GGB, Hawrylowicz CM, Holman M, Keller KD** (1984) Activation and proliferation signals in mouse B cells. III. Intact (IgG) anti-immunoglobulin antibodies activate B cells but inhibit induction of DNA synthesis. Immunology 53:693–701.
158. **Kristensen F, Walker C, Bettens F, Joncourt F, DeWeck AL** (1982) Assessment of interleukin 1 and interleukin 2 effects on cycling and noncycling murine thymocytes. Cell Immunol 74:140–149.
159. **Kunicka JE, Darzynkiewicz Z, Melamed MR** (1987) DNA in situ sensitivity to denaturation: A new parameter for flow cytometry of normal human colonic epithelium and colon carcinoma. Cancer Res 47: 3942–3947.
160. **Lamm ME, Neville DM Jr** (1965) The dimer spectrum of acridine orange hydrochloride. J Phys Chem 69:3872–3877.
161. **Latt SA** (1973) Microfluorometric detection of deoxyribonucleic acid replication in human metaphase chromosomes. Proc Natl Acad Sci USA 70: 3395–3399.
162. **Latt SA** (1979) Fluorometric detection of deoxyribonucleic acid synthesis: Possibilities of interfacing bromodeoxyuridine dye techniques with flow fluorometry. J Histochem Cytochem 25:913–926.
163. **Le Clercq L, Cambier JC, Mishal Z, Julius MH, Theze J** (1986) Supernatant from a cloned helper T cell stimulates most small resting B cells to undergo increased I-A expression, blastogenesis, and progression through the cell cycle. J Immunol 136:539–545.
164. **Le Pecq JB** (1976) Cationic fluorescent probes of polynucleotides. In Chen RF, Edelhoch H (eds): "Biochemical Fluorescence Concepts." New York: Marcel Dekker, pp 711–736.
165. **Lehman R, Slavkin HC** (1976) Localization of "transcriptionally active" cells during odontogenesis using acridine orange ultrastructural cytochemistry. Dev Biol 49:438–356.
166. **Lelkes G, Fodor I, Hollan S** (1986) The mobility of intramembrane particles in nonhemolyzed human erythrocytes. Factors affecting acridine orange-induced particle aggregation. J Cell Sci 86:57–63.
167. **Lerman LS** (1961) Structural considerations in the interaction of DNA and acridines. J Mol Biol 3:18–30.
168. **Lerman LS** (1963) The structure of the DNA–acridine complex. Proc Natl Acad Sci USA 49:94–94.
169. **Lewis MR, Goland PG** (1948) In vivo staining and retardation of tumours in mice by acridine compounds. Am J Med Sci 215:282–282.
170. **Liedeman R, Bolund L** (1976) Acridine orange binding to chromatin of individual cells and nuclei under different staining conditions. II. Thermodenaturation of chromatin. Exp Cell Res 101:175–183.
171. **Liedeman R, Bolund L** (1976) Acridine orange binding to chromatin of individual cells and nuclei under different staining conditions. I. Binding capacity of chromatin. Exp Cell Res 101:164–174.
172. **Liedeman RR, Matveyeva NP, Vostricova SA, Prilipko LL** (1976) Extrinsic factors affecting the binding of acridine orange to DNP complex of cell nuclei in different physiological states. Exp Cell Res 90:105–110.
173. **Lin CC, Jorgenson KF, van de Sande JH** (1980) Specific fluorescent bands on chromosomes produced by acridine orange after prestaining with base specific nonfluorescent DNA ligands. Chromosoma 79:271–286.
174. **Lochmann E-R, Michler A** (1973) Binding of organic dyes to nucleic acids and the photodynamic effect. In Duchesne J (ed): "Physico-chemical Properties of Nucleic Acids," Vol. 1. London, Academic Press, pp 223–267.
175. **Loeser CN, West SS** (1962) Cytochemical studies and quantitative television fluorescence and absorption spectroscopy. Ann NY Acad Sci 97:346–357.
176. **Luk CK, Keng PC, Sutherland RM** (1985) Regrowth and radiation sensitivity of quiescent cells isolated from EMT6/RO-fed plateau monolayers. Cancer Res 45:1020–1025.
177. **Luk CK, Sutherland RM** (1986) Influence of growth phase, nutrition and hypoxia on heterogeneity of cellular buoyant densities in in vitro tumor model systems. Int J Cancer 37:883–890.
178. **MacInnes JW, Uretz RB** (1966) Organization of DNA in dipteran polytene chromosomes as indicated by polarized fluorescence microscopy. Science 151: 689–691.
179. **MacInnes JW, Uretz RB** (1967) Thermal depolarization of fluorescence from polytene chromosomes stained with acridine orange. J Cell Biol 33:597–604.
180. **Manning GS** (1978) The molecular theory of polyelectrolyte solutions with application to the electrostatic properties of polynucleotides. Q Rev Biopys 11:179–246.
181. **Manning GS** (1980) Thermodynamic stability theory for DNA doughnut shapes induced by charge neutralization. Biopolymers 19:37–59.
182. **Marder P, Schmidtke JR** (1983) Effects of methylprednisolone on Concanavalin A-induced human lymphocyte blastogenesis: A comparative analysis by flow cytometry, volume determination and ^{3}H-thymidine incorporation. Immunopharmacology 6:155–166.
183. **Maroudas NG, Wray S** (1985) Activation of nuclear chromatin in stretch-dependent growth of tissues. Connect Tissue Res 13:217–225.
184. **Matsubara T, Nakagone Y** (1983) High-resolution

banding by treating cells with acridine orange before fixation. Cytogenet Cell Genet 35:148–151.

185. **Mauro F, Teodori L, Schumann J, Gohde W** (1986) Flow cytometry as a tool for the prognostic assessment of human neoplasia. Int J Radiat Oncol Biol Phys 12:625–636.

186. **Mayor HD** (1963) The nucleic acids of viruses as revealed by their reaction with fluorochrome acridine orange. In Richter GM, Epstein HA (eds): "International Review of Experimental Pathology," Vol. 2. New York: Academic Press, pp 1–45.

187. **McGhee JD** (1976) Theoretical calculations of the helix–coil transition of DNA in the presence of large, cooperatively binding ligands. Biopolymers 15: 1345–1375.

188. **Melamed J, Darzynkiewicz Z** (1985) RNA content and chromatin structure of CHO cells arrested in metaphase by Colcemid. Cytometry 6:381–385.

189. **Melamed MR, Adams LR, Traganos F, Kamentsky LA** (1974) Blood granulocyte staining with acridine orange changes with infection. J Histochem Cytochem 22:526–530.

190. **Melamed MR, Adams LR, Traganos F, Zimring A, Kamentsky LA** (1972) Preliminary evaluation of acridine orange as a vital stain for automated differential leukocyte counts. Am J Clin Pathol 57:95–95.

191. **Melamed MR, Darzynkiewicz Z** (1981) Acridine orange as a quantitative cytochemical probe for flow cytometry. In Stoward PJ, Polak JM (eds): "Histochemistry: The Widening Horizons." New York, John Wiley & Sons, pp 237–261.

192. **Michaelis L** (1947) The nature of interaction of nucleic acids and nuclei with basic dye stuffs. Cold Spring Harbor Symp Quant Biol 12:131–142.

193. **Monroe JG, Cambier JC** (1983) Level of mla expression on mitogen-stimulated murine B lymphocytes is dependent on position in cell cycle. J Immunol 130: 626–631.

194. **Monroe JG, Cambier JC** (1983) Sorting of B lyphoblasts based upon cell diameter provides cell populations enriched in different stages of cell cycle. J Immunol Methods 63:45–56.

195. **Monroe JG, Havran WL, Cambier JC** (1982) Enrichment of viable lymphocytes in defined cycle phases by sorting on basis of pulse width of axial light extinction. Cytometry 3:24–27.

196. **Monroe JG, Kass MJ** (1985) Molecular events in B cell activation. I. Signals required to stimulate G_0 to G_1 transition of resting B lymphocytes. J Immunol 135:1674–1681.

197. **Moriyama Y, Takano T, Ohkuma S** (1982) Acridine orange as a fluorescent probe for lysosomal proton pump. J Biochem 92:1333–1336.

198. **Moriyama Y, Takano T, Ohkuma S** (1984) Solubilization and reconstitution of lysosomal H^+ pump. J Biochem (Tokyo) 93:927–930.

199. **Mornex JF, Cordier G, Pages J, Revillard JP** (1983) Sarcoidosis alveolitis: A classification based on flow cytometric studies of T lymphocyte activation. Monogr Allergy 18:178–180.

200. **Nash D, Plant W** (1964) On the denaturation of chromosome DNA in situ. Proc Natl Acad Sci USA 51:731–735.

201. **Neville DM Jr, Davies DR** (1966) The interaction of acridine dyes with DNA: An X-ray diffraction and optical investigation. J Mol Biol 17:57–74.

202. **Nicolini C, Belmont A, Parodi S, Lessin S, Abraham S** (1979) Mass action and acridine orange staining: Static and flow cytofluorometry. J Histochem Cytochem 27:102–113.

203. **Noronha A, Richman DP** (1984) Simultaneous cell surface phenotype and cell cycle analysis of lymphocytes by flow cytometry. J Histochem Cytochem 32: 821–826.

204. **Noronha ABC, Richman DP, Arnason BGW** (1980) Detection of in vivo stimulated cerebrospinal-fluid lymphocytes by flow cytometry in patients with multiple sclerosis. N Engl J Med 303:713–717.

205. **Nusse M, Egner HJ** (1984) Can nocodazole, an inhibitor of microtubule formation, be used to synchronize mitotic cells. Cell Tissue Kinet 17:13–23.

206. **O'Donnell PV, Traganos F** (1986) Changes in thymocyte proliferation at different stages of viral leukemogenesis in AKR mice. J Immunol 136:720–727.

207. **Oliver K, Noelle RJ, Uhr JW, Krammer PH, Vitetta ES** (1985) B-cell growth factor (B-cell growth factor I or B-cell-stimulating factor, provisional 1) is a differentiation factor for resting B cells and may not induce cell growth. Proc Natl Acad Sci USA 82:2465–2467.

208. **Olszewski W, Darzynkiewicz Z, Rosen P, Schwartz MK, Melamed MR** (1981) Flow cytometry of breast carcinoma. I. Relation of DNA ploidy level to histology and estrogen receptor. Cancer 48:985–988.

209. **Pande G, Joshi DS, Sundaram K, Das MR** (1986) Isolation and characterization of the two subpopulations of cells with different lethalities from Zajdela ascitic hepatoma. Cancer Res 46:1673–1678.

210. **Paoletti C, Le Pecq JB, Lehman IR** (1971) The use of ethidium bromide-circular DNA duplexes for the fluorometric analysis of breakage and joining of DNA. J Mol Biol 55:75–100.

211. **Paoletti C, Magee BB, Magee PT** (1976) The structure of DNA in native chromatin as determined by ethidium bromide binding. In Cohn WE, Volkin E (eds): "Progress in Nucleic Acid Research and Molecular Biology," Vol. 19. New York: Academic Press, pp 373–377.

212. **Peacocke AR** (1973) The interaction of acridines with nucleic acids. In Acheson RM (ed): "Acridines." 2nd ed. New York: Interscience, pp 723–757.

213. **Peacocke AR, Skerrett JNH** (1956) The interaction of aminoacridines with nucleic acids. Trans Faraday Soc 52:261–279.

214. **Pearse AGE** (1972) "Histochemistry Theoretical and Applied," Vol. 2. Edinburgh: Churchill Livingstone, pp. 1185–1192.

215. **Pennings A, Speth P, Wessels H, Haanen C** (1987) Improved flow cytometry of cellular DNA and RNA by on-line reagent addition. Cytometry 8:335–338.

216. **Piesco N, Alvarez MR** (1972) Nuclear cytochemical changes in onion roots stimulated by kinetin. Exp Cell Res 73:129–139.

217. **Piwnicka M, Darzynkiewicz Z, Melamed MR** (1983) RNA and DNA content of isolated nuclei measured by multiparameter flow cytometry. Cytometry 3: 269–275.

218. **Pollack A, Prudhomme DL, Greenstein DB, Irwin GL III, Claflin AJ, Block NL** (1982) Flow cytometric analysis of RNA content in different cell populations

using pyronin Y and methyl green. Cytometry 3:28–35.

219. **Porumb H (1978)** The solution spectroscopy of drugs and drug–nucleic acid interactions. Prog Biophys molec Biol 34:175–195.
220. **Pritchard NJ, Blake A, Peacocke AR (1966)** Modified intercalation model for the interaction of amino acridines and DNA. Nature (Lond) 212:1360–1361.
221. **Proust JJ, Chrest FJ, Buchholz MA, Nordin AA** (1985) G_0 B cells activated by anti-u acquire the ability to proliferate in response to B-cell-activating factors independently from entry into G_1 phase. J Immunol 135:3056–3061.
222. **Raap AK, Marijnen JGJ, Vrolijk J, van der Ploeg M** (1986) Denaturation, renaturation and loss of DNA during in situ hybridization procedures. Cytometry 7:235–242.
223. **Ratinaud MH, Leprat P, Julien R (1988)** In situ flow cytometric analysis of nonyl acridine orange stained mitochondria from splenocytes. Cyometry 9: 206–212.
224. **Rechter L, Chan H, Sykes JA (1973)** Comparative study of microspherules and acridine orange reaction products. J Ultrastruct Res 44:347–354.
225. **Record MTJr, Lohman TM, de Haseth P (1976)** Ion effects on ligand–nucleic acid interactions. J Mol Biol 107:145–158.
226. **Redelman D, Wormsley S (1986)** The induction of the human T-cell growth factor receptor precedes the production of RNA and occurs in the presence of inhibitors of RNA synthesis. Cytometry 7:453–462.
227. **Redner A, Melamed MR, Andreeff M (1986)** Detection of central nervous system relapse in acute leukemia by multiparameter flow cytometry of DNA, RNA and CALLA. Ann NY Acad Sci 486:241–255.
228. **Remers WA (1984)** Chemistry of antitumor drugs. In Remers WA (ed): "Antineoplatic Agents." New York, John Wiley & Sons, pp 83–269.
229. **Richards DF, Hunter DT, Janis B (1969)** Detection of plasmodia by acridine orange stain. Am J Clin Pathol 51:280–283.
230. **Richardson CL, Verna J, Schulman GE, Shipp K, Grant AD (1981)** Metal mutagens and carcinogens effectively displace acridine orange from DNA as measured by fluorescence polarization. Environ Mutagen 3:545–553.
231. **Richman DP (1980)** Lymphocyte cell-cycle analysis by flow cytometry. Evidence for a specific postmitotic phase before return to G_0. J Cell Biol 85:459–465.
232. **Rigby WFC, Moelle RJ, Krause K, Fanger WF (1985)** The effects of 1,25-dihydroxyvitamin D_3 on human T lymphocyte activation and proliferation: A cell cycle analysis. J Immunol 135:2279–2286.
233. **Rigler R, Killander D (1969)** Activation of deoxyribonucleoprotein in human leukocytes stimulated by phytohemagglutinin. II. Structural changes of deoxyribonucleoprotein and synthesis of RNA. Exp Cell Res 54:171–179.
234. **Rigler R, Killander D, Bolund L, Ringertz NR (1969)** Cytochemical characterization of deoxyribonucleoprotein in individual cell nuclei. Exp Cell Res 55: 215–224.
235. **Rigler R Jr (1966)** Microfluorometric characterization of intracellular nucleic acids and nucleoproteins by acridine orange. Acta Physiol Scand 67(Suppl 267):1–122.
236. **Ringertz NR, Bolund L (1969)** Activation of hen erythrocyte deoxyribonucleoprotein. Exp Cell Res 55:205–214.
237. **Ringertz NR, Bolund L, Darzynkiewicz Z (1970)** AO binding to intracellular nucleic acids in fixed cells in relation to cell growth. Exp Cell Res 63:233–238.
238. **Ritzen M, Carlsson SA, Darzynkiewicz Z (1972)** Cytochemical evidence of nucleoprotein changes in rat adrenal cortex following hypophysectomy or dexamethasone suppression. Exp Cell Res 70:417–421.
239. **Robbins E, Marcus PI (1963)** Dynamics of acridine orange–cell interaction. I. Interrelationships of acridine orange particles and cytoplasmic reddening. J Cell Biol 18:237–250.
240. **Robbins E, Marcus PI, Gonatas NK (1964)** Dynamics of acridine orange–cell interaction. II. Dye-induced ultrastructural changes in multivesicular bodies (acridine orange particles). J Cell Biol 21:49–62.
241. **Robinson BH, Loffler A, Schwarz G (1973)** Thermodynamic behaviour of acridine orange in solution. Model system for studing stacking and charge-effects on self-aggregation. J Chem Soc Faraday Trans 1 69: 56–69.
242. **Robinson BH, Seelig-Loffler A, Schwarz G (1975)** Kinetic and amplitude measurements for the process of association of acridine orange studied by temperature-jump relaxation spectroscopy. J Chem Soc Faraday Trans 1 71:815–830.
243. **Rolland JM, Ferrier GR, Nairn RC, Cauchi MN** (1976) Acridine orange fluorescence cytochemistry for detecting lymphocyte immunoreactivity. J Immunol Methods 12:347–254.
244. **Roska AK, Lipsky PE (1985)** Dissection of the functions of antigen-presenting cells in the induction of T cell activation. J Immunol 135:2953–2961.
245. **Rundquist I, Olsson M, Brunk U (1984)** Cytofluorometric quantitation of acridine orange uptake by cultured cells. Acta Pathol Microbiol Immunol Scand 92:303–309.
246. **Rydberg B (1984)** Detection of DNA strand breaks in single cells using flow cytometry. Int J Radiat Biol 46:521–527.
247. **Schmitz FJ, Werner E (1986)** Optimization of flow-cytometric discrimination between reticulocytes and erythrocytes. Cytometry 7:439–444.
248. **Schummelfeder N (1950)** Zur Morphologie und Histochemie nervoser Elemente. I. Mitt Die Fluorochromie-rung merkhaltiger Nervenfasein mit Acridinorange. Virchows Arch [Cell Pathol] 319:294–298.
249. **Schummelfelder N, Ebschner KJ, Krogh RE (1957)** Die Grundlage der different Fluoromierung von Ribo- und Desoxyribonucleinsauren mit Acridinorange. Naturwissenschaften 44:467–468.
250. **Schwarz G, Wittekind D (1982)** Selected aminoacridines as fluorescent probes in cytochemistry in general and in the detection of cancer cells in particular. Anal Quant Cytol 4:44–54.
251. **Seligman PA, Allen RH, Kirchanski SJ, Natale PJ** (1983) Automated analysis of reticulocytes using fluorescent staining with both acridine orange and immunofluorescence technique. Am J Hematol 14:57–66.
252. **Septinus M, Berthold T, Naujok A, Zimmerman HW**

(1985) Uber hydrophobe Acridinfarbstoffe zur Fluorochromierung von Mitochondrien in lebenden Zellen. Histochemistry 82:51–51.

253. **Seuwen K, Adam G (1983)** Only one of the two signals required for initiation of the cell cycle is associated with cellular accumulation of ribosomal RNA. Biochem Biophys Res Commun 117:223–230.

254. **Seuwen K, Steiner U, Adam G (1984)** Cellular content of ribosomal RNA in relation to the progression and competence signals governing proliferation of 3T3 and SV40-3T3 cells. Exp Cell Res 154:10–24.

255. **Shapiro HM (1981)** Flow cytometric estimation of DNA and RNA content in intact cells stained with Hoechst 33342 and pyronin Y. Cytometry 2:143–159.

256. **Smith PJ, Anderson CO, Watson JV (1985)** Effects of X-irradiation and sodium butyrate on cell-cycle traverse of normal and radiosensitive lymphoblastoid cells. Exp Cell Res 160:331–342.

257. **Sobel HM, Sakore TD, Jain SC, Banerjee A, Bhandary KK, Reddy BS, Lozansky ED (1982)** Beta-kinked DNA-A structure that gives rise to drug intercalation and DNA breathing and its wider significance in determining the premelting and melting behavior of DNA. Cold Spring Harbor Symp Quant Biol 47:293–314.

258. **Soderstrom KO, Parvinen LM, Parvinen M (1977)** Early detection of cell damage by supravital acridine orange staining. Experientia 33:265–266.

259. **Song P, Li S, Xue S, Li Z, Lou Z, Sun R, Han R, Shen W, Chen R, Rong Z (1987)** Effect of acclacinomycin (ACM) on cell cycle of L1210 leukemia cells by dual parameter flow cytometer. Acta Biophys Sin 3:111–119.

260. **Srigley J, Barlogie B, Butler JJ, Osborne B, Blick M, Johnston D, Kantarjian H, Reuben J, Batsakis J, Freireich E (1985)** Heterogeneity of non-Hodgkin's lymphoma probed by nucleic acid cytometry. Blood 65:1090–1096.

261. **Stadler BM, DeWeck AL (1978)** Cytofluorometric analysis of macrophages activated in vivo and in vitro. Eur J Immunol 8:243–246.

262. **Stadler BM, DeWeck AL (1980)** Flow-cytometric analysis of mouse peritoneal macrophages. Cell Immunol 54:36–48.

263. **Stadler BM, Kirstensen F, DeWeck AL (1980)** Thymocyte activation by cytokines: Direct assessment of G0–G1 transition by flow cytometry. Cell Immunol 55:436–443.

264. **Staiano-Coico L, Darzynkiewicz Z, Melamed MR, Weksler M (1984)** Immunological studies of aging. Impaired proliferation of T lymphocytes in elder humans by flow cytometry. J Immunol 132: 1788–1792.

265. **Steiner RF, Beers RF (1959)** Polynucleotides. V. Titration and spectrophotometric studies upon the interaction of synthetic polynucleotides with various dyes. Arch Biochem Biophys 81:75–92.

266. **Stockert JC (1985)** Cytochemistry of nucleic acids. Microsc Elect Biol Cell 9:89–131.

267. **Stockert JC, Lisanti JA (1972)** Acridine-orange differential fluorescence of fast- and slow-reassociating chromosomal DNA after in situ DNA denaturation and reassociation. Chromosoma 37:117–130.

268. **Strugger S (1940)** Fluoreszenz-mikroskopische Untersuchungen uber die Aufnahme und speicherung des Akridinorange durch lebende und fote Pflanzenzellen. Jen Z Med Naturwiss 73:97–112.

269. **Strugger S (1949)** "Fluoreszenzmikroskopie und Mikrobiologie." Hannover: Schaper.

270. **Subirana JA (1973)** Studies on thermal denaturation of nucleohistone. J Mol Biol 74:363–381.

271. **Tejada TI, Mitchell JC, Norman A, Marik JJ, Friedman S (1984)** A test for the practical evaluation of male fertility by acridine orange (AO) fluorescence. Fertil Steril 42:87–91.

272. **Terzaghi E, Okada Y, Streisinger G, Emrich J, Inouye M, Tsuqita A (1966)** Change of a sequence of amino acids in phage T4 lysosyme by acridine orange induced mutations. Proc Natl Acad Sci USA 56:500–507.

273. **Thompson CB, Schaefer ME, Finkelman FD, Scher I, Farrar J, Mond JJ (1985)** T cell-derived B cell growth factor(s) can induce stimulation of both resting and activated B cells. J Immunol 134:369–374.

274. **Thompson CB, Scher I, Schaefer ME, Lindsten T, Finkelman FD, Mond JJ (1984)** Size-dependent B lymphocyte subpopulations: Relationship of cell volume to surface phenotype, cell cycle, proliferative response, and requirements for antibody production to TNP–Ficoll and TNP–BA. J Immunol 133: 2333–2342.

275. **Traganos F, Bueti C, Kapuscinski J, Darzynkiewicz Z (1989)** The antitumor intercalating drug ditercalinium binds preferentially to RNA in Friend erythroleukemia cells. Leukemia 3:522–530.

276. **Traganos F, Darzynkiewicz Z, Melamed MR (1982)** The ratio of RNA to total nucleic acid content as a quantitative measure of unbalanced cell growth. Cytometry 2:212–218.

277. **Traganos F, Darzynkiewicz Z, Sharpless T, Melamed MR (1975)** Denaturation of deoxyribonucleic acid in situ: Effect of formaldehyde. J Histochem Cytochem 23:431–438.

278. **Traganos F, Darzynkiewicz Z, Sharpless T, Melamed MR (1977)** Nucleic acid content and cell cycle distribution of five human bladder cell lines analysed by flow cytofluorometry. Int J Cancer 20:30–36.

279. **Traganos F, Darzynkiewicz Z, Sharpless T, Melamed MR (1977)** Simultaneous staining of ribonucleic and deoxyribonucleic acids in unfixed cells using acridine orange in a flow cytofluorometric system. J Histochem Cytochem 25:46–56.

280. **Traganos F, Darzynkiewicz Z, Sharpless T, Melamed MR (1979)** Erythroid differentiation of Friend leukemia cells as studied by acridine orange staining and flow cytometry. J Histochem Cytochem 27:382–289.

281. **Traganos F, Israel M, Silber R, Seshadri R, Kirchenbaum, Potmesil M (1985)** Effects of new N-alkyl analogues of adriamycin on in vitro survival and cell cycle progression of L1210 cells. Cancer Res 45: 6273–6279.

282. **Traganos F, Kimmel M, Bueti C, Darzynkiewicz Z (1987)** Effects of inhibition of RNA or protein synthesis on CHO cell cycle progression. J Cell Physiol 133:277–287.

283. **Tyrer HW, Golden JF, Vansickel MH, Echols CK, Frost JK, West SS, Pressman NJ, Allbright CD, Adams LA, Gill GW (1979)** Automatic cell identification and enrichment in lung cancer. II. Acridine or-

ange for cell sorting of sputum. J Histochem Cytochem 27:552–556.

284. **Verma RS, Lubs HA (1975)** A simple R banding technique. Am J Hum Genet 27:110–117.

285. **Verma RS, Lubs HA (1976)** Description of the banding patterns of human chromosomes by acridine orange reverse banding (RFA) and comparison with the Paris banding diagram. Clin Genet 9:553–557.

286. **Verma RS, Lubs HA (1976)** Additional observations on the preparation of R banded human chromosomes with acridine orange. Can J Genet Cytol 18:45–50.

287. **von Tscharner V, Schwarz G (1979)** Complex formation of acridine orange with single-standed polyriboadenilic acids and 5′-AMP: cooperative binding and intercalation between bases. Biophys Struct Mech 5:75–90.

288. **Walker C, Kristensen F, Bettens F, DeWeck AL (1983)** Lymphokine regulation of activated (G_1) lymphocytes. J Immunol 130:1770–1773.

289. **Walle AJ (1985)** Distribution and content of nuclear and cellular RNA among cell populations of acute lymphoblastic and nonlymphoblastic leukemia. Cancer Res 45:5193–5195.

290. **Walle AJ, Wong GY, Suthanthiran M, Rubin AL, Stenzel KH, Darzynkiewicz Z (1988)** Altered nucleic acids contents of mononuclear cells in blood of renal allograft recipients. Cytometry 9:177–182.

291. **Wallen AC, Higashikubo R, Dethlefsen LA (1984)** Murine mammary tumour cells in vitro. I. The development of a quiescent state. Cell Tissue Kinet 17: 65–77.

292. **Wallen CA, Higashikubo R, Dethlefsen LA (1982)** Comparison of two flow cytometric assays for cellular RNA–acridine orange and propidium iodide. Cytometry 3:155–160.

293. **Wallen CA, Higashikubo R, Dethlefsen LA (1984)** Murine mammary tumour cells in vitro. II. Recruitment of quiescent cells. Cell Tissue Kinet 17:79–89.

294. **Waring MJ (1981)** DNA modification and cancer. Annu Rev Biochem 50:159–192.

295. **Warner GL, Lawrence DA (1986)** Cell surface and cell cycle analysis of metal-induced murine T cell proliferation. Eur J Immunol 16:1337–1342.

296. **Watson JV, Chambers SH (1977)** Fluorescence discrimination between diploid cells on their RNA content: A possible distinction between clonogenic and nonclonogenic cells. Br J Cancer 36:592–600.

297. **Watson JV, Chambers SH (1978)** Nucleic acid profile of the EMT6 cell cycle in vitro. Cell Tissue Kinet 11:415–422.

298. **Watson JV, Chambers SH (1978)** Inference on RNA synthesis rates during mitosis from flow cytofluorimetric RNA histograms. J Histochem Cytochem 26: 691–695.

299. **Waxdal MJ (1983)** An early biochemical pathway of transmembrane signaling in the stimulation of lymphocytes. In Parker JW, O'Brien RL (eds): "Intercellular Communication in Leukocyte Function." New York: John Wiley & Sons, pp 413–418.

300. **Weill G, Calvin M (1963)** Optical properties of chromophore macromolecule complexes: Absorption and fluorescence of acridine dyes bound to polyphosphates and DNA. Biopolymers 1:401–417.

301. **Weiss DL, Nitzkin P (1969)** Phage lysis of Staphylococcus aureus exposed to acridine orange. Am J Clin Pathol 51:379–383.

302. **Wetzel GD, Swain SL, Dutton RD, Kettman JR (1984)** Evidence for two distinct activation states available to B lymphocytes. J Immunol 133:2327–2332.

303. **Whaun JM, Rittershaus C, Ip SHC (1983)** Rapid identification and detection of parasitized human red cells by automated flow cytometry. Cytometry 4: 117–122.

304. **Wheeless LL, Berkan TK, Patten SF, Reeder JE, Robinson RD, Eldidi MM, Hulbert WC, Frank IN (1986)** Multidimensional slit-scan detection of bladder cancer. Preliminary clinical results. Cytometry 7:212–216.

305. **Wheeless LL, Cambier MA, Kay DB, Wightman LL, Patten SF (1979)** False alarms in a slit-scan flow system: Causes and occurrence rates. Implications and potential solutions. J Histochem Cytochem 27:596–596.

306. **Wheeless LL, Patten SF, Berkan TK, Brooks CL, Gorman KM, Lesh SR, Lopez PA, Wood JCS (1984)** Multidimensional slit-scan prescreening system: preliminary results of a single blind clinical study. Cytometry 5:1–8.

307. **Williams JM, Ransil BJ, Shapiro HM, Strom TB (1984)** Accessory cell requirement for activation antigen expression and cell cycle progression by human T lymphocytes. J Immunol 133:2986–2995.

308. **Wilson JS, Steinkamp JA, Lehnert BE (1986)** Isolation of viable type II alveolar epithelial cells by flow cytometry. Cytometry 7:157–162.

309. **Wilson WD, Jones RL (1982)** Intercalation in biological systems. In Whittingham MS, Jacobson AJ (eds): "Intercalation Chemistry." New York: Academic Press, pp 445–501.

310. **Wolf MK, Aronson SB (1960)** Growth, fluorescence and metachromasy of cells cultured in the presence of acridine orange. J Histochem Cytochem 9:22–29.

311. **Wolstenholme DR (1965)** The distribution of DNA and RNA in salivary gland chromosomes of chrionomus tentans as revealed by fluorescence microscopy. Chromosoma 17:219–229.

312. **Wyandt HE, Vlietinck RF, Magenis RE, Hecht F (1974)** Colored reverse-banding of human chromosomes with acridine orange following alkaline/formalin treatment: Densitometric validation and applications. Humangenetic 23:119–130.

313. **Yagita H, Hashimoto Y (1986)** Monoclonal antibodies that inhibit activation and proliferation of lymphocytes. II. Requisite role of the monoclonal antibody-defined antigen systems in activation and proliferation of human and rat lymphocytes. J Immunol 136:2062–2068.

314. **Yagita H, Masuko T, Takahashi N, Hashimoto Y (1986)** Monoclonal antibodies that inhibit activation and proliferation of lymphocytes. I. Expression of the antigen on monocytes and activated lymphocytes. J Immunol 136:2055–2061.

315. **Yamashiro DJ, Fluss SR, Maxfield FR (1983)** Acidification of endocytic vesicles by an ATP-dependent proton pump. J Cell Biol 97:929–934.

316. **Zanker V (1952)** Quantitative Absorptions- und Emissionsmessungen am Acridinorangekation bei Normal- und Tieftemperatur im organishen Lo-

sungsmittel und ihr Beitrag zur Deutung des metachromatischen Fluoreszensproblems. Z Phys Chem 200:250–292.

317. **Zanker V (1952)** Uber den Nachweis definierter reversibler Assoziate (reversibler Polymerisate) des Acridinorange durch Absorptions und Fluoreszenzmessungen in wassriger Losung. Z Phys Chem 199: 225–258.

318. **Zanker V, Held M, Rammensee H** (1959) Neuere Ergebnisse der Absorptions,—Fluoreszenz- und Fluoreszenz-Polarization-grad-Messungen am Akridinorange-Kation, ein Weiterer Beitrang zum Metachromasie-Problem dieses Vitalfarbstoffs. Z Naturforschg 146:789–801.

319. **Zelenin AV** (1966) Fluorescence microscopy of lysosomes and related structures in living cells. Nature (Lond) 212:425–426.

320. **Zelenin AV, Shapiro IM, Kolesnikov VA, Senin VM** (1974) Physicochemical properties of chromatin of mouse sperm nuclei in heterokaryons with Chinese hamster cells. Cell Diff 3:95–101.

321. **Zelenin AV, Vinogradova NG** (1973) Influence of partial deproteinization on the cytochemical properties of DNP of lymphocyte nuclei. Exp Cell Res 82: 411–414.

322. **Zittoun R, Pochat L, Mathieu M, Pochat R (1981)** Coloration simultanee de l'ADN et de l'ARN par l'acridine orange pour cytophotometrie en flux liquide. Etude de la fiabilité de la measure de l'ADN. Pathol Biol 29:317–320.

16

Probing Nuclear Chromatin by Flow Cytometry

Zbigniew Darzynkiewicz

Sloan-Kettering Institute for Cancer Research, New York, New York 10021

INTRODUCTION

Replication and transcription of nuclear DNA, the two most fundamental processes of living cells, are essential for cell proliferation and differentiation. Although influenced by the cytoplasmic environment, they take place exclusively in the cell nucleus. The cell constituents directly involved in these events, such as enzymes, coenzymes, structural, or regulatory macromolecules, reside within the nucleus. The structure of nuclear chromatin undergoes extensive modulation in preparation for and parallel to DNA replication, DNA packing into chromosomes, or cell differentiation. Chromatin constituents are specifically modified by addition or removal of various chemical moieties, and proportions of the constituents are changed and interactions between them altered [60]. Taking into account the above, it is not surprising that analysis of nuclear constituents with respect to their content, modifications and conformation provides clues as to the functional state of the cell, especially regarding the regulatory mechanisms associated with its proliferative or differentiation status.

Whereas basic chromatin structure, at nucleosomal or subnucleosomal level, is already quite well characterized [60,122,144,163], the nature of higher orders of DNA packing in the interphase nucleus or in mitotic chromosomes (supranucleosomal structure, "superstructure") remains still largely unknown [80,178]. One of the major roadblocks to studies of chromatin superstructure is the paucity of adequate methods. Biochemical procedures rely on solubilization of the component, which, in the case of the nucleus, is equivalent to destruction of the superstructure. Furthermore, DNA and some other constituents in situ in native configuration are in a condensed, solid-state form. Their analysis in solutions introduces methodological bias. This perhaps is the reason why biochemical techniques have failed so far to resolve the chromatin superstructure. By contrast, electron microscopy cannot provide information on functional relationships between the constituents.

Flow cytometry appears to be ideally suited to the study of nuclear chromatin. Isolated nuclei, permeabilized or prefixed cells can be measured under conditions in which nuclear constituents are in the native state with their three-dimensional structure preserved. Furthermore, because populations of individual nuclei are investigated, it is possible to obtain information on internuclear variability or detect the presence of distinct subpopulations. This is clearly an advantage over bulk measurements in solution. Because DNA content is one of the nuclear features that can be measured, and provides information on cell position in the cell cycle, there is no need for prior cell synchronization or separation by chemical or physical means in order to relate the findings to cell progression through the cycle. The latter techniques are cumbersome and often induce undesirable side effects.

In addition to being used as a tool for basic research on nuclear chromatin flow cytometry can be applied to more practical tasks, i.e., as a means to identify cells with different chromatin structure. In pathology, identification and classification of tumor cells are often based on differences in nuclear morphology. Introduction of objective criteria for the evaluation of nuclei as offered by flow cytometry can be an aid in the identification and classification of tumor cells and perhaps in tumor prognosis. With further development of the technique and application of new probes this approach may yield information about proliferative potential, differentiation or drug sensitivity of tumor cells. Because isolation of nuclei from solid tumors is often a much simpler task than isolation of intact cells, rapid progress is expected in the development of methods exploring nuclear chromatin and their application in pathology.

This chapter describes techniques used in the study of nuclear chromatin applicable to flow cytometry. While basic studies especially in relation to the cell cycle are advanced, clinical applications as yet are rare. This imbalance is reflected in this review which deals predominantly with changes in chromatin during the cell cycle. Studies in which flow sorting is used only to separate cells (e.g., based on DNA content) for further biochemical analysis of nuclear chromatin are not covered by this review.

ACCESSIBILITY OF DNA IN SITU TO VARIOUS FLUOROCHROMES

Among studies of nuclear chromatin, investigations of DNA accessibility to different fluorescent ligands are the most numerous. The results of these investigations are also the most controversial and often appear to be contradictory.

Flow Cytometry and Sorting, Second Edition, pages 315–340

This review aims toward an explanation for the controversies and stresses the data obtained by flow cytometry under conditions of cell equilibrium with the fluorochromes.

Before the advent of flow cytometry, investigations of DNA accessibility were by cell measurements in static cytofluorometers. The initial experiments employed predominantly the intercalating dye acridine orange (AO), which was reported to bind DNA in situ in proportion to the degree of "genome activation" [139,140,215]. Large differences in AO binding were observed between noncycling cells characterized by a low level of transcription and condensed chromatin (e.g., nonstimulated lymphocytes) and their cycling counterparts (mitogen-stimulated cells). The change in binding occurred immediately after addition of the mitogen, even prior to an increase in the rate of RNA synthesis [140]. Since then, there have been numerous reports indicating that indeed DNA in condensed chromatin of noncycling cells binds less of the intercalating dyes. Most of these early studies were done on cells in smears, stained, rinsed and then measured in the absence of the dye, i.e., under nonequilibrium conditions. It was later observed that this difference in AO binding between active and inactive cells was an artifact of the nonequilibrium staining due to the bias in cell density of smears rather than the state of metabolic activity or chromatin structure [20,158,220]. Further studies using static cytofluorometry, but careful adjustment of cell densities or equilibrium conditions, showed no differences in staining between active and inactive cells differing in chromatin structure [157].

There are two cell systems, however, in which the density artifacts could not explain the observed difference in AO binding; i.e., under carefully controlled conditions, a decrease in DNA accessibility to AO was observed during spermiogenesis [105] and erythropoiesis [137]. Biochemical data on isolated chromatin in solution [230] and on avian erythrocyte nuclei reactivated in heterokaryons [219] confirm that during differentiation of erythrocytes a progressive restriction in accessibility of DNA to some intercalating dyes indeed does occur.

Using static microfluorometry, Cowden and Curtiss [48] compared the binding of several dyes to DNA in condensed chromatin of thymocytes versus DNA in diffused chromatin of hepatocytes. Their data indicated that after removal of RNA, accessibility of DNA to mithramycin, proflavin, 7-amino actinomycin D and berberine sulfate was reduced in thymus nuclei as compared with hepatocytes, whereas binding of pyronin Y and quinacrine mustard was increased. It is difficult to compare these results with other studies, however; or even to make comparisons of the binding of individual dyes because conditions of staining (equilibrium versus nonequilibrium, pH, ionic strength, dye concentration) were different for each of the dyes. Furthermore, the HMG proteins and perhaps other constituents were extracted from these nuclei by pretreatment with 5% perchloric acid [90,120].

The introduction of flow cytometry made fluorescence measurements simpler and more accurate. In most situations the cells are measured in equilibrium with the dyes. Generally, there is good stoichiometry in DNA stainability with various dyes, regardless of the structure of chromatin. It is a common observation, for instance, that DNA stainability of highly condensed chromatin in mitotic cells is similar to that in G_2 cells. Exceptions, however, were reported. For instance, Bonaly et al. [21] noticed that the transition of *Euglena* cells from stationary phase to exponential growth is paralleled by a decrease in DNA accessibility to several intercalating dyes but not to Hoechst 33258. Extraction of basic proteins from nuclei abolished the difference and raised the overall DNA stainability [21]. Smets [240] observed that DNA accessibility to AO in 3T3 cells is decreased when cells enter quiescence. Mazzini et al. [162] described nonstoichiometric stainability of DNA with propidium in relationship to differences in chromatin structure. In the latter two reports, however, the effects could be seen only at very low dye concentration. Studies on dye binding under these conditions are very difficult because of the excess of potential binding sites per dye molecule. It requires extreme care in estimating and adjusting cell densities to have identical dye/DNA-P molar ratios in the compared samples.

Larsen et al. [153] recently reported that following fixation with formaldehyde, DNA is more accessible to mithramycin, propidium or ethidium (but not to DAPI or Hoechst 33342) in condensed chromatin of metaphase cells compared with cells in interphase. The difference was large enough to enable these authors to quantify mitotic cells by flow cytometry. Apparently crosslinking of chromatin constituents by formaldehyde introduced more restriction in DNA accessibility in interphase nuclei than in metaphase chromosomes, enhancing the difference in chromatin structure. Similarly, Stokke and Steen [246,247] observed that after formaldehyde fixation and permeabilization, monocytes bound 30 to 130% more 7-amino actinomycin D than did lymphocytes. By contrast, when the cells were permeabilized before fixation, monocytes and lymphocytes bound similar amounts of the fluorochrome.

Accessibility of DNA in situ to several dyes is altered as a result of chromatin changes during cell differentiation in two cell systems, namely, during erythroid differentiation and spermiogenesis. This was documented in repeated studies in equilibrium with the dyes and also when fluorescence was measured by flow cytometry [41,71,73,93,94]. Our early observations indicated that erythroid differentiation of Friend leukemia (FL) cells induced by dimethyl sulfoxide (Me_2SO) is accompanied by a progressive reduction in accessibility of DNA to AO [73]. This phenomenon was studied in detail recently, e.g., with respect to a variety of dyes, both intercalators and those binding externally to DNA [71]. In these experiments, exponentially growing and fully differentiated FL cells were stained with each of 10 different dyes before and after extraction of nuclear protein with 0.1 N HCl. The acid extraction removes, or at least dissociates, histones, HMG proteins, and other basic proteins from DNA [119]. The results are summarized in Figure 1 and described in detail in [71].

It is evident that accessibility of DNA to DAPI, ellipticine, acridine orange, and 7-amino actinomycin D is significantly diminished in differentiated cells (Fig. 1). With other dyes the difference is minor and cannot be proven statistically. Another interesting feature of this data is the difference in the extent of DNA in chromatin that is accessible to each of the dyes. Thus, for instance, whereas in exponentially growing cells nearly 70% DNA is accessible to DAPI, less than 10% is accessible to 7-amino actinomycin D. Approximately half of the nuclear DNA is accessible to the externally binding dyes mithramycin, chromomycin A_3 and Hoechst 33342.

Changes in accessibility of DNA during spermatogenesis were studied recently by us [93] and the results are summarized in Figure 2. In comparison with diploid cells in testis (spermatogonia, interstitial cells) which stain uniformly, the tetraploid cell population shows somewhat decreased stainability with most of the dyes; the statistical evidence for this

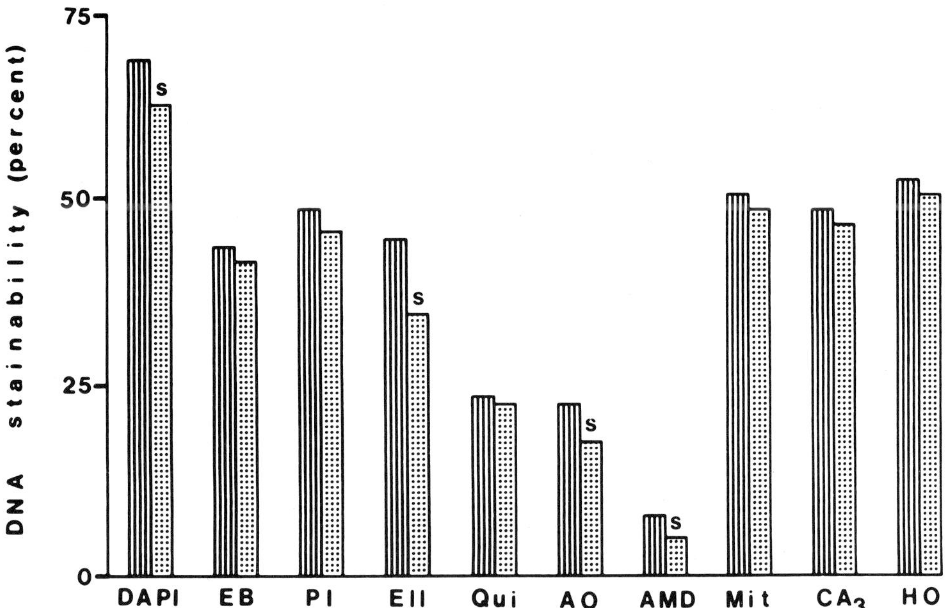

Fig. 1. Accessibility of DNA in situ in Friend erythroleukemia (FL) cells to various fluorochromes. The vertical bars represent the portion of nuclear DNA in exponentially growing (striped bars) and differentiated (Me_2SO-induced erythroid differentiation; dotted bars) FL cells accessible to each fluorochrome. The data are expressed as percent of stainable DNA relative to DNA in nuclei of exponentially growing FL cells from which the nuclear proteins have been extracted with 0.1N HCl prior to staining. The "s" above the dotted bars indicates that DNA accessibility in differentiated cells is significantly lower (statistically demonstrated) compared with exponentially growing cells for a particular dye. The fluorochromes from left to right are DAPI (4'6-diamidino-2-phenyl indole), ethidium bromide, propidium idodide, ellipticine, quinacrine, acridine orange, 7-amino actinomycin D, mithramycin, chromomycin A_3 and Hoechst 33342. (For details see [71].)

decrease was documented only for DAPI, AO, and Hoechst 33342. The round spermatids, however, have DNA more accessible to all the dyes, with the highest increase observed for propidium, ellipticine, mithramycin, chromomycin A_3 and DAPI.

Very interesting is the behavior of DNA in elongated spermatids. Namely, DNA accessibility in these cells to intercalating dyes (propidium, ellipticine, AO, 7-amino actinomycin D) is increased; by contrast, accessibility to dyes binding externally (mithramycin, chromomycin A_3 and Hoechst 33342) is decreased, both relative to DNA of diploid cells. DNA accessibility to DAPI is also decreased. There is a controversy regarding the mechanism of DAPI binding to DNA, whereas some reports suggest external binding, others indicate intercalation [135]. If DAPI is an intercalator, however, in contrast to other intercalators its ability to unwind DNA (unwinding angle) is minimal.

Sperm cells bind significantly less of each of the fluorochromes per unit of DNA compared to cells representing all earlier stages of spermatogenesis. The lowest accessibility of sperm cell DNA is toward externally binding dyes (only 10% DNA is stainable compared with diploid cells), the least restricted is binding of some intercalators (e.g., ellipticine, more than 60%). These results [93] (Fig. 2) conform with the observation of Clausen et al. [41], who reported an increase in accessibility of DNA to ethidium at the stage of elongating spermatids, and a decrease in sperm cells.

It is interesting to consider mechanisms that may restrict accessibility of DNA to the fluorochromes and to correlate the observed changes in DNA accessibility with chromatin modifications known to occur during erythroid differentiation of FL cells or spermatogenesis. In the case of small intercalators the restriction is most likely related to the maintenance of DNA wound around the octamer of the nucleosome core histones in the superhelical conformation [15]. DNA–histone contacts confined to the inner surface of the superhelix are made on every turn of the DNA double helix and involve approximately 20% of phosphates of the core particle DNA, which makes the DNA–histone complex stable at physiological ionic strength [213]. The strength of histone–DNA interactions in the core particle, however, varies depending on the degree of histone modifications, e.g., as caused by treatment with n-butyrate, which induces their hyperacetylation [214,263]. The topological rigidity of the superhelix constrains unwinding and elongation of the DNA helix, which is necessary for acceptance of every intercalating molecule, especially one characterized by a large unwinding angle. The presence of closed DNA loops may provide additional limitation in the number of molecules that can intercalate. The observation that removal of histones (HCl extraction) enhances binding of intercalating dyes more than externally binding dyes (see Fig. 1) favors this interpretation. The latter is also supported by the observation that intercalators, such as ethidium, dissociate core particle histones from DNA at relatively low (0.8 N NaCl) ionic strength [15]; ethidium thus is synergistic with NaCl in dissociation of nucleosomes. DAPI, whose binding does not involve any significant unwinding, is less restricted by the HCl-soluble proteins than other intercalators.

The size of the fluorochrome and involvement of the minor groove of the double helix plays a major role in dye binding to chromatin, as illustrated by the example of 7-

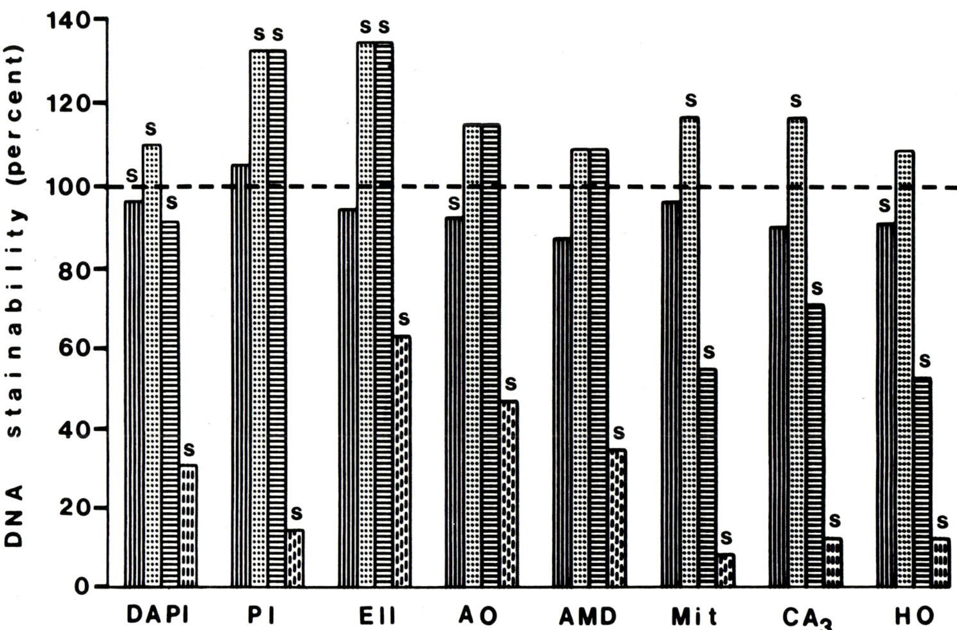

Fig. 2. Changes in DNA accessibility to various dyes during spermatogenesis in mice. Testicular cells were measured by flow cytometry in equilibrium with the dyes (see legend to Fig. 1 for the abbreviations) and DNA stainability of individual cell types (recalculated per unit of DNA) is expressed as a percent of that of diploid cells in testis (i.e., G_1 phase spermatogonia, interstitial cells, lymphocytes, histiocytes). Thus, the stainability of DNA in diploid cells was used as a yardstick of stainability for each of the dyes (broken line = 100%). Vertically striped bars—cells with tetraploid DNA content (predominantly spermatocytes), dotted bars—round spermatids; horizontally striped bars—elongated spermatids; broken vertical stripes—vas deferens sperm cells. The "s" above each bar indicates that DNA stainability of these cells is significantly different from that of the diploid cells. Methods of cell preparation, staining, measurements and identification are described in ref. [93].) See also Chapter 15 (this volume). (Reproduced from ref. [93], with permission of the publisher.)

amino actinomycin D. The latter dye binds by intercalation of the chromophore between the G-C bases, while the bulky peptide rings are located in the minor groove. This dye exhibits the greatest difference in binding to DNA in chromatin of differentiated versus nondifferentiated cells [61], and the highest degree of restriction in binding by the acid extractable proteins.

In the case of FL cells, the most prominent chromatin modifications during differentiation involve the appearance of the histone H1 variant, H1° (IP 25) [143,190], a change in histone H2A variants' proportions (179), and the extent of H2A phosphorylation [179]. Because accessibility of the core particle DNA to intercalators is lower than that of the spacer DNA (see under Nuclease Accessibility), any change in the ratio of spacer/nucleosomal core DNA will be reflected as a change in accessibility of nuclear DNA to these dyes. Likewise, modifications of histones H1 (and perhaps HMG1 and 2 proteins) which either weaken or enhance the strength of locking of DNA in the nucleosome at the point of its entry and exit (stability of "chromatosomes") [22] will be expected to alter the capacity of DNA to accept the intercalators which unwind the polymer. Apparently such changes occur during differentiation of FL cells.

An entirely different pattern of changes is observed during spermatogenesis. Thus, there is an increase in accessibility of DNA to all dyes that takes place at the stage of round spermatids. The increase is followed by a decrease in accessibility to externally binding dyes only (elongated spermatids) and then by a dramatic decrease in accessibility to all dyes (sperm cells). The testis-specific variants of histone H1 are replacing somatic H1 early during spermatogenesis; such modification is believed to promote extended or flexible chromatin conformation [156]. Extensive acetylation of histones, primarily H4 [109], and their substitution by the basic transition proteins occurs at the stage of spermatids [8,109], with elongated spermatids having no more detectable histones [109]. It appears that the presence of the testis-specific variants of H1 and the displacement of histones by transition proteins coincide in time with increased accessibility of DNA in situ to all the dyes. By contrast, the DNA in chromatin already containing transition proteins instead of histones is more accessible to intercalators and less accessible to dyes binding externally, and this property is reversed by acid extraction of nuclear proteins [93]. Finally, substitution of transition proteins by protamines, which results in maximal chromatin condensation (sperm cells), coincides with severalfold restriction in DNA accessibility to nonintercalating dyes and somewhat lesser restriction to intercalators.

Rabinovitch et al. [207] designed an interesting and potentially very useful approach to study lymphocyte stimulation that combines DNA staining with 7-amino actinomycin D and two-color surface immunofluorescence. An increase in cell stainability with 7-amino actinomycin D coincides with stimulation of the lymphocyte which permits detection of stimulated lymphocytes and characterization of their cell cycle distribution in addition to measurements of the surface antigens of the stimulated and nonresponding cells [207].

While actinomycin D indeed appears to be the best probe

for sensing changes in nuclear chromatin [61], when used supravitally its binding is similar to the binding of other dyes and may be affected by cell membrane permeability. In live cells, membrane permeability to the dye, as well as the efflux pump [146,147], are often the rate-limiting factors of binding. Some dyes are taken up by pinocytosis and then the binding reflects membrane pinocytic activity [164], whereas in the case of certain cationic dyes uptake reflects electronegativity of cellular or mitochondrial membranes [47,80,233]. Thus, while DNA accessibility to fluorochromes in live cells can be modulated by nuclear proteins that may indeed play a role in the binding, there is an uncertainty with respect to other factors, which restricts interpretation of such data in terms of modulation of chromatin structure. Furthermore, it is impossible to assess specificity of staining by treatment of viable cells with exogenous nucleases because of the enzymes' exclusion by the intact cell membrane. Studies on DNA stainability in live cells, therefore, are not reviewed in this chapter.

Conclusions

Results of the studies on DNA in situ accessibility to different dyes can be summarized as follows:

1. DNA in situ is only partially accessible to the dyes. In somatic cells, the portion of accessible DNA varies from less than 12 to 70%, depending on the dye. These are minimal estimates based on removal of nuclear proteins with 0.1 N HCl (destruction of nucleosomal structure).
2. Approximately 50% DNA is accessible to each of the externally binding dyes (mithramycin, chromomycin A_3, Hoechst 33342). Accessibility to intercalators varies markedly. Some correlation with the unwinding angle and involvement of the minor groove of the double helix is apparent.
3. The accessibility varies with chromatin changes during cell differentiation (erythroid differentiation, spermatogenesis), but less with chromatin condensation during the cell cycle or G_1–G_0 transition; an exception to the latter appears to be *Euglena* cells. However, following fixation with formaldehyde, accessibility of DNA to several dyes is increased in mitotic cells compared with cells in interphase.
4. Generally, binding of 7-amino actinomycin D appears to be the most sensitive probe of chromatin conformation.
5. Modifications of nuclear chromatin during erythroid differentiation of Friend leukemia cells (appearance of H1°, hyperacetylation of H4, change in proportion of the H2A variants) are paralleled by decreased accessibility of DNA to some intercalators (7-amino actinomycin D, acridine orange, ellipticine, DAPI), but not to externally binding dyes.
6. During spermatogenesis, initiation of histone replacement by the transition protein is paralleled by increased accessibility of DNA to all dyes. Later, when histones are predominantly replaced by transition proteins, DNA accessibility to the intercalators is increased, whereas to nonintercalators it is decreased. Finally, association of protamines with DNA is accompanied by a decrease in DNA accessibility to all dyes, more to those that are externally binding than to intercalators.
7. Dye-accessibility studies appear to provide information primarily on changes in relationship to the extent of histone modifications, replacement of histones by protamines, changes in histone H1 variants, and on DNA spacer length.
8. Because, regardless of the dye, only a portion of DNA in situ is stainable, caution should be exercised in interpretation of the cytometric data on DNA stainability as reflecting the absolute DNA content per cell or per chromosome, e.g., when defining cell aneuploidy. This is especially critical in the analysis of DNA content of formaldehyde prefixed, paraffin-embedded cells [116], inasmuch as formaldehyde alters the stainability of DNA with different dyes in relationship to chromatin structure [153].

ACCESSIBILITY OF DNA IN SITU TO EXOGENOUS NUCLEASES

Sensitivity of DNA in situ to exogenous nucleases varies in relationship to the primary (nucleosomal) and higher orders of nuclear structure [181,187,277]. Generally, DNA in the spacer ("linker") region is more sensitive to nucleases than the DNA of the core particle. The latter, although located on the periphery of the particle, is protected via strong interactions with the core histones. Thus, for instance, due to preferential hydrolysis of spacer DNA, extensive digestion with endonucleases results in a release of individual core particles. This approach has become a routine method for obtaining isolated core particles, e.g., for biochemical studies.

Kinetic studies of the stepwise digestion of chromatin with nucleases reveal additional features of chromatin structure when transient-size particles are produced before the release of the core particles. Initially, a nucleosome is released containing both spacer and core particle DNA associated with cell histones including H1. With further digestion, a "chromatosome" can be obtained that contains 168 base pairs of DNA. In comparison with the core particle, a chromatosome has 10 more base pairs each of spacer DNA exiting and entering the particle (e.g., additional 20 base pairs per particle), which are protected against the nuclease by association with histone H1 [1,22,218]. The core particle itself can be further probed with nucleases; its digestion reveals periodic sites of increased DNA sensitivity 10 base pairs apart. These sites are believed to reflect "kinking" or similar topological changes of DNA wound around the core histones [51,217].

There is a difference in sensitivity to DNase I between DNA in transcriptionally active versus inactive chromatin, the former being "hypersensitive" to the nuclease [276,277]. Association of HMG14 and 17 proteins with nucleosomes, characteristic of the active chromatin, appears to render DNA in those nucleosomes increasingly sensitive to DNase I [93]. Although molecular mechanisms explaining hypersensitivity of DNA in transcriptionally active chromatin are still not entirely clear, the phenomenon is widely used to explore gene activity in the interphase nuclei. Progression of cells through the cell cycle is also associated with variation in DNA sensitivity to DNase I, with DNA in G_1 cells having the most and in mitosis the least sensitivity to the enzyme [205].

Probing sensitivity of DNA in situ to exogenous nucleases by flow cytometry is a simple procedure. Isolated nuclei or permeabilized cells are exposed to low concentration of nucleases for different time intervals, then stained with DNA-specific dyes and measured. As a result of sequential hydrolysis of the most sensitive sections of DNA in nuclei, stainability of total nuclear DNA drops and the kinetics of changes in nuclear fluorescence can be measured (Fig. 3). We have studied and compared DNA sensitivity in situ to DNase I and micrococcal nuclease in thymocytes [34] and L1210 cell nuclei in cells growing exponentially versus those treated with n-butyrate [81]. Butyrate treatment suppresses the activity of endogenous deacetylase and results in the accumulation of multiacetylated forms of histones H3 and H4 [60,214,263].

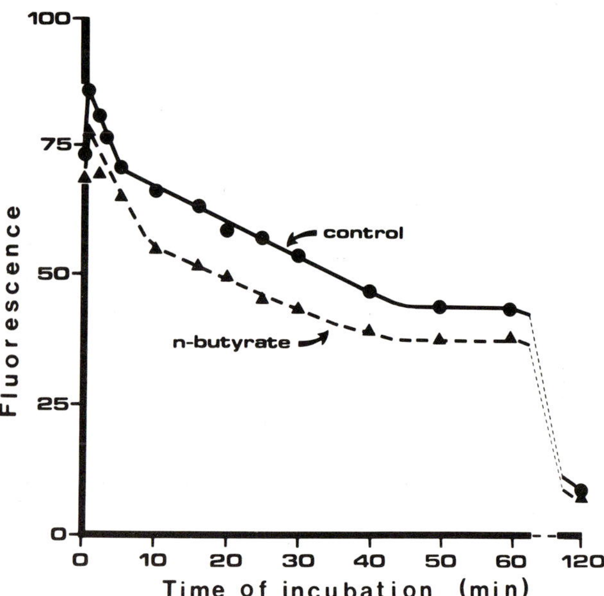

Fig. 3. Sensitivity of DNA in situ to DNase I. Nuclei were isolated from L1210 cells growing exponentially (control) or in the presence of 1 mM n-butyrate [55] treated with RNase and suspended in 10 mM Tris, 2 mM $MgCl_2$ and 0.25 M sucrose at pH 7.4 at 20°C. DNase I was added at 0 time (10 μg/ml) and at various time intervals samples were withdrawn and stained with 20 μM (final concentration) of ethidium bromide. Fluorescence measurements were completed 20 sec after the sample withdrawal. A decrease in nuclear fluorescence reflects the rate of DNA digestion. An increase in the DNase concentration to 100 μg/ml and incubation at 37°C (broken line) results in near total loss of nuclear fluorescence. (Reproduced from ref. [81], with permission of the publisher.)

Four phases can be distinguished from the kinetic curve of DNase digestion (Fig. 3). Immediately after addition of DNase a transient increase in fluorescence (up to 15%) is evident. This increase apparently is the result of the initial nicking by the enzyme of any closed DNA loops or superhelical turns, thus releasing the constraint against intercalation of the dye which requires unwinding and elongation of the double helix for its binding. A similar increase in fluorescence is observed after chromatin treatment with X-rays to induce DNA breaks [225]. Between 1 and 10 min there is a rapid decrease in fluorescence, followed by a phase in which the decrease rate is moderate. About half of the DNA stained with ethidium resists DNase at this low concentration. All the phases are seen in both nuclei from control cells and from cells treated with n-butyrate. However, DNA sensitivity towards the enzyme, especially during the rapid phase, is increased in nuclei from n-butyrate-treated cells. The DNase I-resistant portion of DNA is 15% higher in control cells.

The phases as shown in Figure 3 were also demonstrated in elegant experiments by Roti Roti et al. [225] in nuclei of HeLa and murine mammary carcinoma cells. In the latter studies a difference in DNA accessibility was also observed between exponentially growing versus quiescent cells, and a change was noted in cells exposed to heat shock [225].

A somewhat different approach to assay the sensitivity of DNA in situ to DNase I was described by Szabo et al. [251]. These workers subjected methanol: acetic acid prefixed cells to DNase I. The enzyme produced nicks in DNA which were then labeled with Brd-5′triphosphate and subsequently detected with FITC-labeled BrdUrd antibodies by flow cytometry. Using this procedure, Szabo and co-workers [251] observed a decrease in sensitivity of DNA to DNase I during myeloid differentiation of HL-60 cells induced by dimethylsulfoxide. In contrast to Prentice et al. [205], however, they did not see a significant variation in DNA sensitivity during cell progression through the cell cycle. This may be due to removal of histones by acetic acid in their fixation procedure.

It is also possible to detect by flow cytometry the sections of DNA that are selectively sensitive to the single-strand specific nucleases [69]. This approach is discussed in further detail under DNA In Situ Denaturation.

Conclusions

Although investigations on DNA accessibility to nucleases by flow cytometry are as yet scarce, the few existing reports together with extensive biochemical evidence indicate that such studies can provide new insights into chromatin structure. The main findings are as follows:

1. Four phases of the reaction with DNase I can be distinguished:
   a. An initial increase in DNA stainability may reflect DNA nicking and subsequent elongation of the double helix in superhelical loops, which can then unwind to accommodate more intercalator molecules.
   b. The rapid phase of the decrease in stainability may be a reflection of the digestion of transcriptionally active, "hypersensitive" DNA in chromatin.
   c. The slow phase of the decrease may represent hydrolysis of most of the spacer DNA.
   d. The plateau of the reaction reflects stability of DNA of the core particles in the inactive chromatin.
2. Approximately 50% of DNA stainable with AO or EB is relatively inaccessible to micrococcal nuclease or DNase I. This is, in all probability, the DNA of the core particles. Because in most cell types the core particle:spacer DNA ratio is 7:3 [122,144,163], the data indicates that the core particle DNA is approximately 2.3 times less stainable with these intercalators than the spacer DNA.
3. Histone modifications (e.g., H4 hyperacetylation) and changes in nonhistone proteins are reflected in the kinetics of the reaction with DNase I.
4. Changes in DNA in situ sensitivity to DNase I accompany cell differentiation, transition to quiescence, or response to heat shock.

NUCLEAR PROTEIN CONTENT AND NUCLEAR SIZE

Measurements of nuclear protein or nuclear weight ("dry mass") were initiated well before the advent of flow cytometry [4,6,30,63,221] with the use of microspectrophotometric or microinterferometric techniques. These early studies indicated that nuclear protein content is doubled during the cell cycle and that protein content of quiescent cell nuclei is very low compared with cycling cells. One of the observed early events of activation of quiescent cells was an increase in nuclear protein content [4,5,221]. Influx of protein from the cytoplasm during activation correlated with decondensation of nuclear chromatin and preceded a rise in DNA transcription [6,221,222]. This was especially evident in nuclei of quiescent cells activated in heterokaryons following cell fusion [221,222]. These data also suggested that the ability of cells to enter S phase depends on attainment of a critical size (protein content) of the cell nucleus.

Nuclear size can be estimated by flow cytometry either by light scattering [283], Coulter volume [125], pulse width [78,125], or protein content [201,224] measurements. The initial reports indicated that cell entrance to S phase is not correlated with nuclear size estimated by light scatter [283] or pulse width [125]. However, extremely high heterogeneity of cell populations was evident in these studies, and it is likely that the heterogeneity was the result of rather low resolution of the light scatter or pulse width measurements. This could obscure detection of the exact relationship between nuclear size and entrance to S phase. Furthermore, the relationship between forward light scatter and nuclear size is not directly proportional because the intensity of the light scatter signal, in addition to the nuclear size, may be modulated by its structure (granularity, refractiveness).

Simultaneous measurements of forward and right angle light scatter can be used to probe the internal structure of isolated nuclei [195–197]. Using this approach, Papa and associates were able to distinguish parenchymal cell nuclei from liver [196] and study the structure and protein associations in the nuclear matrix [195,197].

The method of simultaneous measurement of DNA versus protein developed by Crissman and Steinkamp [53,54] was adapted for measurement of isolated nuclei by Pollack et al. [201] and Roti Roti et al. [224]. Their studies brought several interesting observations. The data of Roti Roti et al. [224] indicate that exponentially growing HeLa cells attain a critical threshold of nuclear protein prior to entrance to S phase. The prethreshold G_1 subpopulation (G_{1A}) is similar to the G_{1A} subpopulation discriminated by either low RNA content or DNA sensitivity to denaturation (see Chapter 24, this volume).

After lysis of the mitotic cell, metaphase chromosomes remain attached to each other and, despite a lack of nuclear envelope, the amount of DNA and protein associated with these chromosome clusters can be estimated. Although these structures contain full 4C DNA content, they can be distinguished from G_2 cells due to their lower protein content or low light scatter properties [153,224]. Thus, the amount of protein associated with chromosomes in metaphase is significantly lower compared to that present in nuclei of G_2 cells. This observation was confirmed by Pollack et al. [201] on lymphocytes and by us on CHO cells (Fig. 4). The threshold protein content in G_1 and diminished protein content of the chromosome aggregates as compared with G_2 cells are evident in Figure 4.

The data of Pollack et al. [201] also confirm the microspectrophotometric observations of Auer and colleagues [4–6] that nuclei of quiescent cells have minimal protein content. Based on differences in nuclear protein content, Pollack et al. [201] were able to discriminate cells in deep quiescence (G_{1Q}), recognize G_{1A} versus G_{1B} populations, and observe an increase in nuclear protein during G_2. The latter made it possible to differentiate between early G_2 (G_{2A}) and late G_2 (G_{2B}) populations. In a subsequent study, these investigators [202] observed that irradiation of cells with tritium from incorporated thymidine arrests them in G_2 phase, with nuclear protein content characteristic of G_{2B} cells. Subdivision of the G_2 phase into G_{2A} and G_{2B} compartments, as proposed by Pollack et al. [201,202], is therefore helpful in more accurate characterization of cells with respect to their cell cycle position during normal growth and following perturbation of the cell cycle.

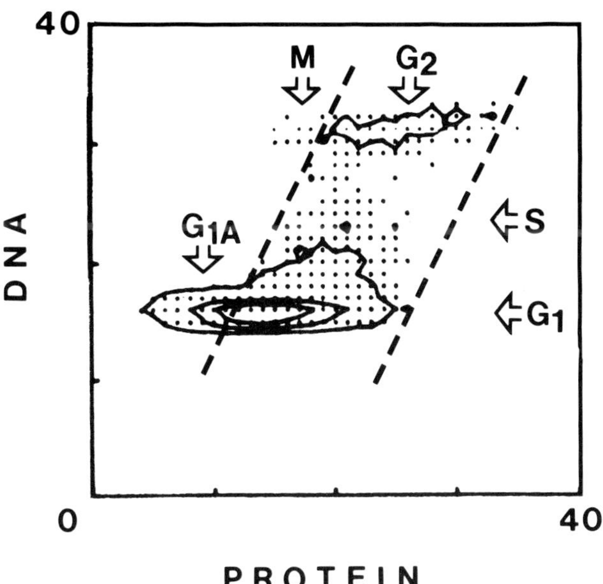

Fig. 4. Correlated measurements of DNA and protein content of nuclei isolated from exponentially growing CHO cells. Fluorescence of nuclei stained with Hoechst 33342 and FITC was measured using two laser excitation, as described [52,53]. Under the assumption that histone to DNA ratio remains essentially constant during the cell cycle, variability in the protein to DNA ratio reflects changes in the nonhistone protein content in the nuclei. Such a change (a decrease, seen in the cells falling outside the area contained within the broken lines) is observed for aggregates of mitotic chromosomes (M) and early postmitotic (G_{1A}) cells.

Nuclear chromatin consists of DNA, RNA, and two classes of protein, the histone and nonhistone proteins. The ratio of histones to DNA remains rather constant throughout the cell cycle, inasmuch as the prevalent portion of histones is synthesized during S phase [60,122,144,163]. By contrast, nonhistone proteins are produced independently of DNA replication, and their content relative to DNA changes during the cell cycle, contributing to the observed changes in total nuclear protein:DNA ratio. Nonhistones represent a heterogeneous group of proteins; the function of most of them has not yet been identified [120]. As is evident from the flow cytometric data (Fig. 4), the nuclear protein portion which varies during G_1 or G_2 [5,6,201], i.e., belonging to the category of nonhistone nuclear proteins, is therefore responsible for discrimination of G_{1A} versus G_{1B} or G_{2A} versus G_{2B} cells.

Perhaps the most interesting observation from the flow cytometric studies is the loss of nonhistone proteins from chromosome aggregates during mitosis and their reaccumulation in nuclei prior to cell entrance to S. These proteins behave identical to several proliferation-associated antigens, e.g., the *c-myc* proto-oncogene transcripts [278], which also dissociate from chromosomes and reappear in the G_1 nucleus shortly after its formation following mitosis. It is possible that this phenomenon is part of the regulatory mechanisms of cell replication; i.e., just before and during mitosis some classes of nonhistone proteins are lost from the chromatin and their loss may be essential for chromatin to condense. Their influx during G_1 (G_{1A}) correlates with, and perhaps causes, decondensation of chromatin; DNA replication can occur only when the decondensation has passed the threshold value (see further). The influx of nonhistone proteins during G_1 may therefore play a critical role in the preparation of chromatin for replication and the rate of influx may be regulated by specific signals.

Oud et al. [191,192] measured the protein : DNA ratio of normal and malignant human endometrium. They observed that the ratio for malignant or normal proliferating endometrium is significantly higher than for normal nonproliferating (postmenopausal) tissue. These results support the interpretation above of changes in the protein to DNA ratio during the cell cycle, which may be a useful parameter in pathology to discriminate the cycling cell populations. The value of nuclear size in discrimination of cycling cells in human acute myeloid leukemias was supported by studies of Ffrench et al. [100], who described nuclear size changes during maturation and differentiation of myeloid blasts.

There are technical problems related to protein estimation in isolated nuclei. One of the problems stems from the fact that stainability of nuclear proteins, like that of DNA, is based not only on their quantity but on accessibility of the fluorochrome-reactive groups as well. Thus, for instance, we observed an increase rather than a decrease in stainability of nuclei with FITC following the extraction of HMG proteins with 0.35 N NaCl [62]. Most likely "loosening" of chromatin with 0.35 N NaCl, due to dissociation of some protein–protein or protein–DNA ionic bonds, contributed to increased protein stainability which more than offset their partial removal. Dyson et al. [89] studying the effects of hyperthermia, irradiation, or antitumor drugs, also observed an increase in stainability of nuclear proteins with FITC due to an increase in accessibility of the protein-reactive groups to the fluorochrome. These investigators proposed to utilize this change as a probe to discern modulations in chromatin structure that occur during experimental manipulation of cells in vitro or treatment of tumors in vivo [89]. In this respect, the SH-reactive fluorochromes may be even more attractive as probes, since they may specifically stain the transcriptionally active portions of chromatin known to have uncovered SH groups of the histone H3 of the nucleosome core particles [243].

Another technical problem relates to the nuclear isolation procedure, which may either be incomplete, leaving tags of cytoplasm attached to the envelope, or may result in inadvertent extraction of nuclear proteins. Protein extraction is especially extensive in solutions of low pH. The choice of proper isolation technique and observation of isolated nuclei by electron or light microscopy to assess the completeness of isolation are imperative in these studies.

Conclusions

1. On the basis of nuclear protein content it is possible to discriminate G_{1Q} versus G_{1A} versus G_{1B} cells, to distinguish mitotic from G_2 cells and even to recognize early versus late G_2 (G_{2A} versus G_{2B}) cells.
2. Because the DNA to histone ratio remains relatively constant during the cell cycle, the differences in DNA to protein ratio reflect the changing ratio of the nonhistone : histone proteins; this ratio is the lowest in G_{1Q} and M cells.
3. Nonhistone chromosomal protein content is decreased during mitosis and these proteins accumulate in the nucleus early in G_1 (G_{1A}); DNA replication is initiated in cells that attain the critical threshold of these proteins.
4. Accessibility of the nuclear protein-reactive groups to different fluorochromes may provide still another probe of chromatin structure.

NUCLEAR RNA CONTENT AND NUCLEOLUS

The flow cytometric methods that allow one to obtain correlated measurements of DNA and RNA were developed and originally applied in studies of whole cells [52,75, 77,232,258]. The data on whole-cell RNA provided interesting information regarding the relationship between RNA content and cell cycle progression, differentiation or quiescence (see Chapter 24, this volume). The techniques were also applied in clinical cytology [3,141].

Several lines of evidence suggest that RNA content of isolated nuclei may be an equally or perhaps even more interesting feature than the whole cell RNA. First, different RNA species with different functions and turnover rates are located in the nucleus and cytoplasm, respectively [114,171, 285]. Thus, it is of interest to relate cell position in the mitotic cycle with nuclear RNA content and compare it with whole cell RNA inasmuch as the latter reflects predominantly (~80%) rRNA. Furthermore, because nuclear RNA is very closely related to the rate of DNA transcription, RNA processing and RNA transport, it may be an even more sensitive marker of the changes in genome transcription, and may react more rapidly to any metabolic change affecting transcription (e.g., drug or hormone induced) than does whole-cell RNA content. From a practical point of view, isolation of nuclei from solid tumors is often more successful than isolation of whole, well-dispersed cells [145]. Preferably, the nuclei should be isolated using solutions of low pH to decrease the possibility of RNA leakage during isolation or storage.

Nuclear RNA is mainly localized in nucleoli; relatively little RNA is in nucleoplasm. It is well established that nucleolar morphology and staining properties correlate with cell proliferation, cell cycle phases and neoplasia [28,239]. Cells with dense nucleoli are typical of a rapidly proliferating tumor population, whereas cells with trabecular or ring-shaped nucleoli do not cycle or cycle at very slow rates [203]. Also, changes in nucleoli (e.g., "segregation") are often among the earliest effects in cells exposed to antitumor drugs, especially intercalators [236]. These morphological changes may correlate with alterations of nucleolar RNA content, which may serve as a marker of cell sensitivity to these drugs. In summary, nuclear RNA content may indeed be one of the most sensitive parameters of cell metabolism and drug sensitivity that can be explored by flow cytometry.

Studies of nuclear RNA content are sparse. The technique of differential staining of DNA versus RNA, based on the use of acridine orange [75,77,258], was recently adapted to isolated nuclei [199]. It was observed that depending on the cell type, between 10 and 20% of total cellular RNA is localized in nuclei. Generally, nuclear RNA content correlated well with whole cell RNA and it was possible, based on the former, to discriminate quiescent (G_{1A}) cells from the cycling population as well as to recognize G_{1A} versus G_{1B} cells [149]. On the basis of the RNA content of isolated nuclei, Higgins et al. [118] were able to distinguish cycling hepatocytes in regenerating liver and study the kinetics of regeneration, and Walle [264] found nuclear RNA content to be a useful parameter to identify and characterize L3 leukemia and Burkitt's lymphoma cells. Analysis of RNA content during the cell cycle indicates that following mitosis, when the nucleus reforms early in G_1, it contains very low RNA content. The RNA content increases markedly during G_1 before the cells

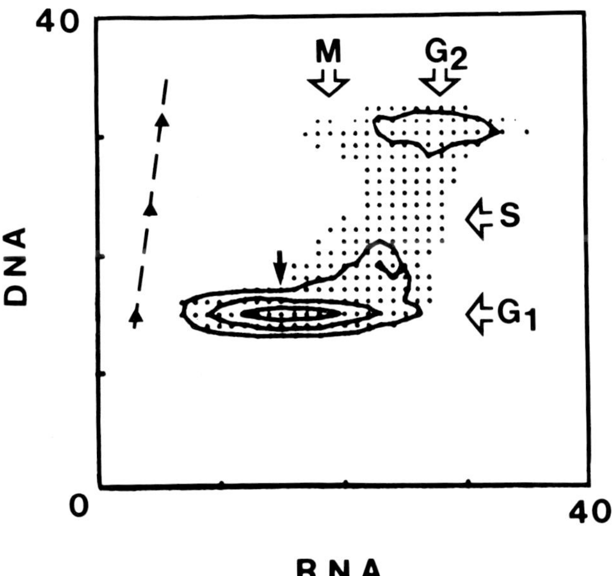

Fig. 5. Correlated measurements of DNA and RNA content of nuclei isolated from exponentially growing CHO cells. Fluorescence of nuclei stained with Hoechst 33342 and pyronin Y was measured using two laser-excitation [52]. The line to the left of the contour plot indicates the level of the pyronin Y-fluorescence after incubation of the nuclei with RNase (the nonspecific component of staining). The data shows that RNA content of metaphase chromosome aggregates (M) is lower than that of nuclei of G_2 cells, and that there is a threshold RNA nuclear content in G_1 prior to cell entrance to S phase.

are able to enter S phase, and the increase contributes to the large heterogeneity of the G_1 population with respect to nuclear RNA [62] (Fig. 5).

RNA content of isolated nuclei, measured after staining with AO, was found by Bergounioux et al. [16] to correlate with initiation of proliferation of plant (*Petunia hybrida*) protoplasts in tissue culture. These authors observed that an increase of RNA above a critical, well-defined threshold was required for cells to be able to initiate DNA replication, and that G_2 nuclei had an RNA to DNA ratio similar to that of G_1 nuclei, whereas the quiescent cells had minimal RNA [16].

The major drawback to nuclear RNA staining with AO is relatively low specificity in discrimination of DNA versus RNA; i.e., because of low RNA content in nuclei compared with whole-cell RNA, the nonspecific component of red luminescence, due to spectrum overlap from the emission of AO bound to DNA, becomes significant and lowers resolution of the method [199]. To decrease the contribution of the green "tail" of the AO emission spectrum representing its binding to DNA and thus to raise the sensitivity and specificity of RNA detection in nuclei, the use of long pass filters transmitting emission of AO above 640 nm is recommended.

Measurements based on multiple laser, different wavelength excitation are superior for the detection of small amounts of RNA. We attempted to adapt the technique for simultaneous measurements of DNA versus RNA versus protein with Hoechst 33342, pyronin Y and FITC, respectively, from whole cells [52] to isolated nuclei [62]. The data on nuclei of exponentially growing CHO cells are shown in Figure 5. Differential staining of DNA and RNA with Hoechst 33342 and pyronin Y, respectively, offers higher sensitivity for RNA detection than the AO-staining technique and thus may be applied to nuclei with very low RNA content. However, because staining of RNA with pyronin Y is not stoichiometric [65], interpretation of the results employing this dye in quantitative terms is limited. The virtues and limitations of the techniques of RNA measurement employing AO versus pyronin Y are presented elsewhere [65].

Conclusions

1. Nuclear RNA content represents predominantly nucleolar RNA and may be a valuable marker of nucleolar function.
2. Changes in nuclear RNA content during the cell cycle are similar to changes in whole-cell RNA content. Based on nuclear RNA content, it is possible, e.g., to distinguish G_{1Q}, G_{1A}, and G_{1B} compartments in the G_1 phase of the cell cycle.
3. Measurement of nuclear RNA content may be a sensitive and useful probe to discriminate normal versus tumor or quiescent versus cycling cells as well as to estimate the effects of certain antitumor drugs.

DNA IN SITU DENATURATION

The stability of free nucleic acids in solution varies with the nature and sequence of the bases involved in base-pairing and the composition of the backbone chain. The melting temperature of DNA is related to its G-C content, since this pair confers extra stability on the molecule.

Stability of the helix is also influenced by its environment. Fewer counterions are associated with denatured than with native DNA. Therefore, because during denaturation the counterions associated with the polymer are released into solution, the equilibrium between native and denatured DNA can be influenced by changing the activity of the counterion, as would be the case with any chemical equilibrium [211]. An increase in the concentration of the counterion in solution shifts equilibrium toward helix stability, and higher temperatures are then needed to induce denaturation. An additional element contributing to stability of the helix is screening of its phosphate charges by the counterions, originally thought to be the primary mechanism of stabilization [211].

Stability of DNA in chromatin is also influenced by local counterions provided by histones and nonhistone proteins [248]. The strength and extent of interactions between DNA and these local counterions, as well as any covalent association or crosslinking which further stabilizes the double helix, may be evaluated by analysis of the profiles of DNA denaturation ("melting curves"). Modifications of histones, which weaken their interactions with DNA (phosphorylation, acetylation), are expected to decrease the stability of DNA in chromatin [260].

Biochemical studies on DNA stability in chromatin are done by measurement of light absorption at 260 nm ("hyperchromicity changes") in soluble preparations of nucleohistones heated at increasing temperatures. Nuclear breakage and solubilization of chromatin, however, destroys higher orders of chromatin structure, often displaces histone H1 and HMG proteins, and removes other constituents of chromatin. Such studies therefore yield information pertinent only to the structure of the core particles (e.g., modification of the core histones), but not to the higher orders of chromatin organization [260,282]. The in situ studies on DNA stability in isolated nuclei or permeabilized cells are,

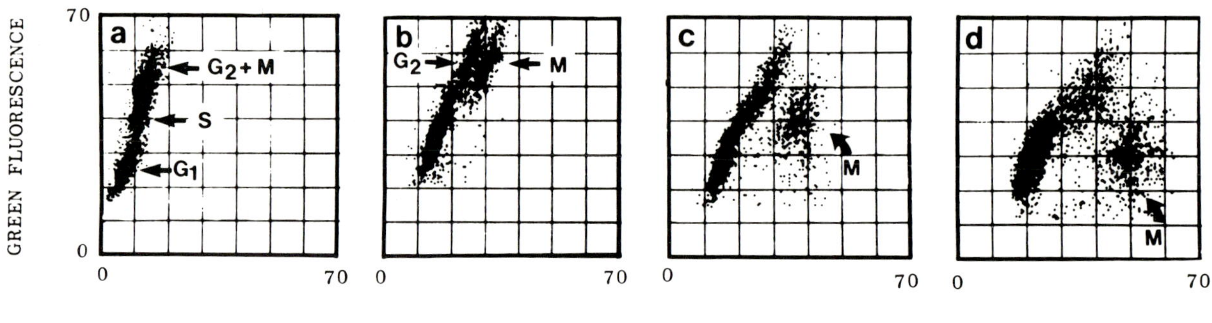

Fig. 6. Heat-induced denaturation of DNA in situ. Prefixed and RNase-treated Friend leukemic cells, nonheated **(a)** and heated at 50°C **(b)**, 85°C **(c)** or 95°C **(d)** were stained with the metachromatic dye AO under conditions when double-stranded DNA stains in green, whereas denatured DNA reacts with AO giving red luminescence (see Chapter 15, this volume). Details of the experiment, cell classification, etc. are described in ref. [76]. As is evident, the extent of DNA denaturation is higher in mitotic (M) cells than in cells in interphase.

therefore, of great interest inasmuch as they offer insight into the native conformation of chromatin.

In situ studies on DNA denaturation were initiated by Nash and Plaut [177], who used the metachromatic properties of AO to differentially stain double-stranded versus denatured DNA in preparations of *Drosophila* chromosomes heated on coverslips (see Chapter 15, this volume). These investigators assayed the transition by visual discrimination of the color change from green to red luminescence. Using static microfluorometry, Rigler et al. [216] introduced quantitative estimates of the transition by measurement of the intensity of green versus red luminescence of AO-stained cells. Unfortunately, in these earlier studies [2,138–140,216], DNA denaturation was induced at relatively high ionic strength and in the presence of formaldehyde. Because detection of protein–DNA ionic interactions is obscured at high ionic strength and formaldehyde intensively crosslinks chromatin constituents, also per se enhancing DNA denaturation, it is difficult, if not impossible, to interpret these early data in the context of the molecular structure of chromatin.

Flow cytometric assays of DNA denaturation also make use of AO to differentially stain native versus denatured polymer but under equilibrium conditions [68,74,76,79]. Either isolated nuclei [74] or permeabilized cells are incubated with RNase and then heated or treated with acid buffers to partially denature DNA. Subsequent staining with AO reveals the extent of denatured DNA, which stains metachromatically red. Denaturation by acid causes a lesser loss of DNA from nuclei compared to thermal DNA denaturation [206] and, therefore, as a method it offers better analytical resolution. Cell staining is done at ambient temperature (following heating) or at higher pH (after acid denaturation); hence, the technique reveals the extent of strand separation rather than the initial denaturation (base unstacking due to base-pairing breakage) as is generally assayed by the hyperchromicity changes. The latter can remain undetected by flow cytometry because of the possibility of fast reassociation of the nonseparated DNA strands during staining. Under appropriate staining conditions, the ratio of red to total luminescence (α_t) represents the fraction of denatured DNA while the total luminescence is proportional to total DNA content [76,79]. The technique was widely used to analyze differences between cells at various phases of the cell cycle, to differentiate between cycling and noncycling cells [9,16,60,125,167,168] (see Chapter 24, this volume), as well as to detect drug-induced changes in nuclear chromatin [19] (see Chapter 39, this volume).

Heat-induced denaturation of DNA in situ is revealed in the form of cooperative transition curves representing an increase in α_t as a function of rise in temperature [76,79]. When cells are heated while immersed in solutions of ionic strength above 0.1 N NaCl, a more than twofold increase in total fluorescence (AO binding to DNA) precedes DNA denaturation, reflecting most likely thermal disruption of the nucleosomal structure [74]. This phenomenon is being currently studied in our laboratory in order to reveal whether or not it might be used as a probe of chromatin structure. The thermal DNA denaturation in situ is generally biphasic. The heat-sensitive fraction is believed to represent spacer DNA, while the resistant fraction DNA is associated with the core particle histones [60]. Further discussion on the type of changes in nuclear chromatin that can be analyzed by studying thermal denaturation of DNA in situ is presented in ref. [60] and, in part, in Chapter 15 (this volume).

There is a striking correlation between sensitivity of DNA to heat or acid-induced denaturation and the degree of chromatin condensation. Thus, the most sensitive to denaturation is DNA in metaphase chromatin, less in condensed chromatin of quiescent cells, and the least sensitive is DNA in interphase nuclei of cycling cells (Fig. 6). Even interphase cells differ with respect to DNA sensitivity; the late G_1 (G_{1B}) and early S cells have DNA the most resistant, whereas DNA in early G_1 (G_{1A}) and G_2 is the most sensitive to denaturation [68,79]. DNA in condensed chromatin of pyknotic nuclei (dying cells) has extreme sensitivity to denaturation, even higher than DNA of mitotic cells [83].

The above pattern of varying DNA sensitivity to denaturation across the cell cycle resembles very much the pattern of varying content of nonhistone proteins in the nucleus (see Fig. 4). Namely, the cells that have the highest protein to DNA ratio (early S, G_{1B}) have the least condensed chromatin and are the most resistant to denaturation of DNA. Conversely, G_0 and M cells have minimal protein to DNA ratio, maximally condensed chromatin, and their DNA is the most sensitive to denaturation.

Using this principle of cell staining, it is possible to distinguish G_0 (G_{1Q}) versus G_{1A} versus S versus G_2 versus M cells, to recognize S- and G_2-phase cells with condensed chromatin

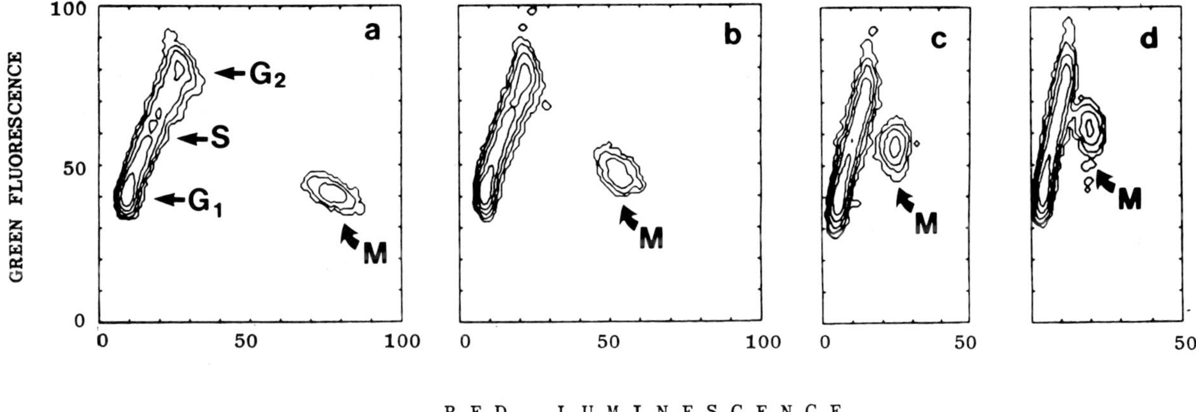

Fig. 7. DNA denaturation in situ induced by acid treatment. L1210 cells were fixed, incubated with RNase, then treated with 0.1N HCl–KCl buffer and stained with AO at pH 2.6. **a:** Under these conditions, DNA in mitotic (M) cells is more extensively denatured and, in reaction with AO, exhibits decreased green and increased red luminescence relative to cells in other phases of the cell cycle. **b–d:** Same cells pretreated with the single-stranded-specific S_1 nuclease (at pH 2.6) for 20, 60, and 120 min, respectively, prior to staining with AO. The decrease in red luminescence of M cells provides additional evidence that DNA in mitotic cells is more extensively denatured.

that do not progress through the cell cycle (S_Q, S_{2Q} cells) as is occasionally seen in tumors [72], to study the stability of cells arrested in metaphase [165], to discriminate dying cells with pyknotic nuclei [83], and to observe specific effects of antitumor drugs (e.g., intercalators, polyamine-depleting drugs) on nuclear chromatin [18,82]. There are tissue-specific differences in DNA denaturation as well, which in some situations enable us to discriminate normal versus tumor cells of the same ploidy level and the same phase of the cell cycle.

The technique of acid-induced denaturation has recently been applied in pathology to characterize human benign and malignant colonic epithelium [149]. A correlation was observed in this study between DNA sensitivity to denaturation and the pathologically determined stage of the disease, the most resistant being DNA in the invasive Dukes' B and C/D tumors. Thus, the sensitivity of DNA to denaturation appears to have a prognostic value in clinical pathology.

A difference in DNA sensitivity to denaturation was also observed between fertile and nonfertile sperm cells and the assay of DNA in situ denaturation was proposed to estimate sperm fertility [9,97] or to study the effects of mutagens on spermatogenesis [96] (see Chapter 26, this volume).

It should be stressed that the distinction of various cell classes with different kinetic properties, or tumor versus normal cells, as discussed above, is based on the rather nonspecific phenomenon of DNA denaturability correlating with an equally nonspecific marker, chromatin condensation. The situation may exist, therefore, when functionally distinct cells can still exhibit similar DNA denaturability. Such an example was provided by Brock et al. [27], who observed that anoxic cells from tumors did not progress through the cell cycle but had DNA sensitivity to denaturation similar to that of the cycling cells.

What are the molecular mechanisms responsible for the observed differences between the cell types or phases of the cell cycle in DNA in situ sensitivity to denaturation? Several lines of evidence indicate that the differences are unrelated to histone H1 phosphorylation, modifications (acetylation, phosphorylation, poly(ADP)-ribosylation) of the core particle histones or the presence of S-S bonds in chromatin, but require the preservation of higher orders of chromatin structure such as exist in whole, undisrupted nuclei [69]. Thus, nuclear disruption and studies of chromatin in solution reveal only minimal differences in DNA denaturability between interphase and metaphase chromatin [235]. Furthermore, acid-induced DNA denaturation (Fig. 7) involves pretreatment of nuclei with low pH buffer (0.1 N HCl) which extracts most, if not all, histones. Yet, such pretreatment renders DNA in condensed chromatin of metaphase cells much more denatured when assayed at pH 2.6 in comparison with DNA in the diffuse chromatin of interphase cells. Dissociation of histones also precedes the heat-induced denaturation of DNA in situ.

Our recent data indicate that the residual proteins extractable with ≧1.2 N NaCl, most likely involved in anchoring DNA loops to the matrix (scaffold) proteins, are responsible for the observed differences in DNA in situ sensitivity to denaturation [69]. Their role is illustrated in Figure 8, which schematically presents the sequence of events occurring during denaturation of DNA in situ. The release of histones that occurs during treatment with acid (or during heating preceding DNA denaturation [74]) leaves loops of free DNA attached to the nuclear matrix or chromosome scaffold proteins. In condensed chromatin, the repulsive forces of the charged DNA polyanion, supercoiled, compressed, and maintained at high spatial density via interactions with the matrix proteins, induce topological stress, destabilizing the double helix. The destabilized sections of DNA can be detected by their metachromatic stainability with AO at pH 2.6, sensitivity to single-stranded specific nucleases or full melting (strand separation) after heating and staining with AO at pH 6.0 [69]. Interestingly, such stressed DNA also appears to react with anti-Z-DNA antibodies [223]. In contrast to condensed chromatin, DNA in euchromatic regions of the nucleus is packaged in a less dense form and thus, following dissociation of histones, the spatial density of the polyanion is lower and the repulsive forces causing the torsional stress are weaker. Loosening of the DNA–matrix (scaffold) interactions by extraction of the residual proteins at ≧1.2 N NaCl swells nuclei and chromosomes, makes DNA more resistant to denaturation, and abolishes the dif-

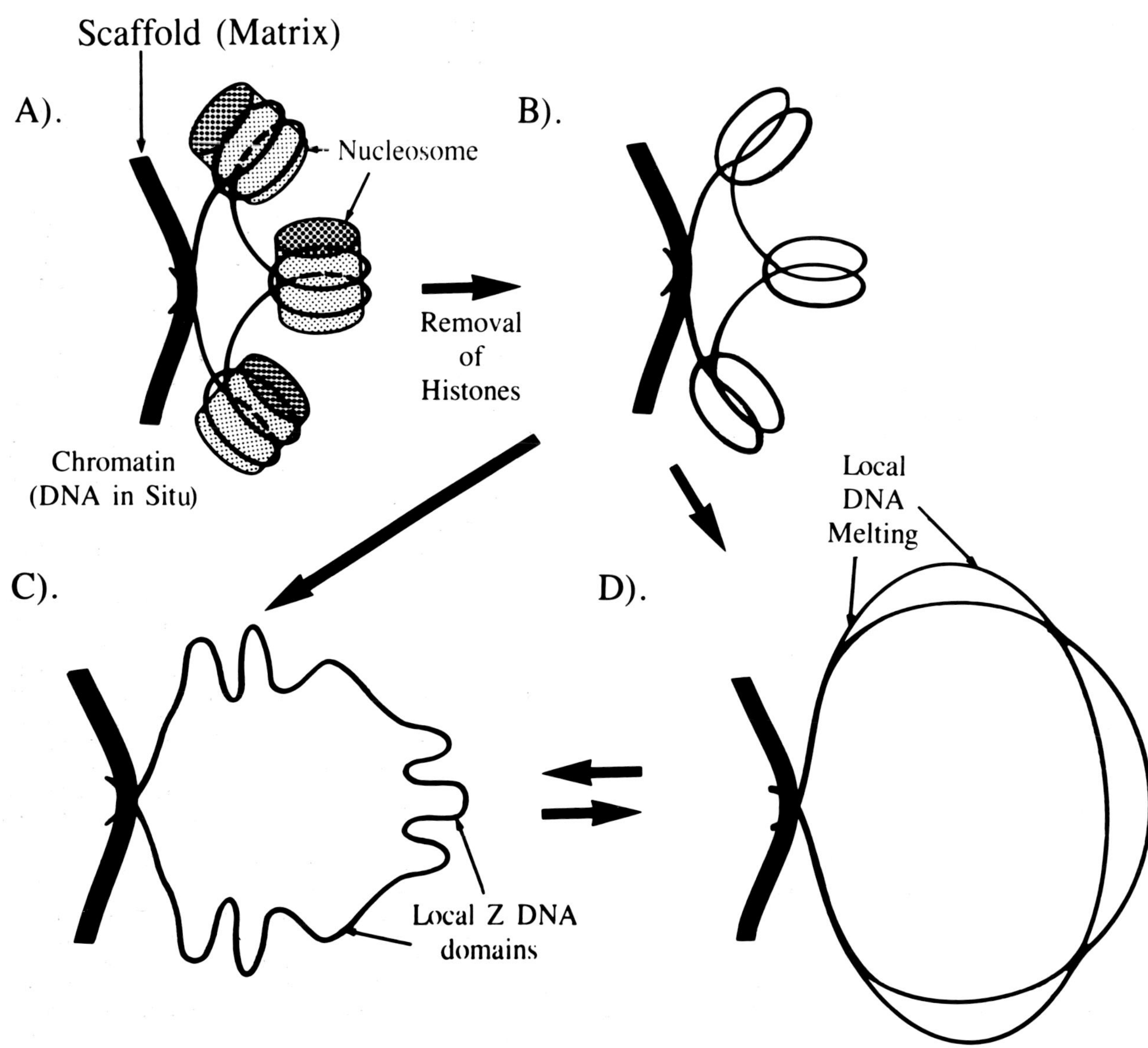

Fig. 8. Schematic representation of chromatin changes during acid- or heat-induced DNA denaturation. **A:** DNA in situ is organized in loops attached to the matrix in the interphase nucleus or to the chromosome scaffold proteins. DNA in double-stranded conformation is wound around the core particle histones, interacting electrostatically with their amino groups. Histone H1, which is located between the nucleosome core particles, interacting with the spacer (linker) sections of DNA, is not shown. **B:** During treatment with acid or heat prior to DNA denaturation, histones dissociate. Loops of free supercoiled DNA are still attached to the matrix (scaffold) proteins. **C,D:** Subsequent heating at higher temperatures, or staining with AO at low pH, results in DNA denaturation (melting). Regions of single-stranded DNA can be detected by single-strand-specific nucleases, or by metachromatic staining with AO [69]. Under the same conditions (removal of histones with acids, assay at low pH), such DNA reacts also with antibodies against Z genus DNA [223]. The differential sensitivity of DNA in condensed versus diffuse chromatin to denaturation is believed to be due to differences in the topological (torsional) stress of the double helix following dissociation of histones. In condensed chromatin the repulsive forces of the charged DNA polyanion supercoiled, compressed, and at high spatial density contribute to the increased topological stress of the superhelix compared to DNA in euchromatin [69].

ferences in DNA denaturability otherwise related to the degree of chromatin condensation. The above discussed mechanisms (torsional stress of the DNA superhelix causing destabilization of the double-stranded structure) are believed to play a role in regulation of transcriptional activity [121].

As mentioned, apart from the use of AO to detect DNA denaturation in situ, this transition can also be assayed by single-strand specific enzymes such as S_1 or mung bean nucleases [69]. Actually, it was the use of S_1 nuclease that made it possible to reveal the stress on the DNA helix, as illustrated in Figure 7; i.e., the release of the stress manifested as an increase in double-strandedness (higher green fluorescence) following the initial nicking of DNA by the S_1 nuclease [69]. The application of these specific nucleases makes it possible to use DNA dyes other than AO to detect DNA denaturation in situ [69].

Conclusions

1. Sensitivity of DNA in situ to heat- or acid-induced denaturation closely correlates with the degree of chromatin condensation and is inversely correlated with the transcriptional activity of chromatin.

2. Based on DNA in situ sensitivity to denaturation, it is possible to distinguish cells in G_{1Q}, G_{1A}, G_{1B}, early-, mid-, and late-S, G_2, and M phases, to discriminate quiescent cells with condensed chromatin in S and G_2 phase (S_Q and G_{2Q}) and recognize cells with pyknotic nuclei.

3. Sensitivity of DNA in situ to denaturation may be a probe of chromatin changes induced by DNA-intercalating or other chromatin modifying drugs and thus be used to reveal cell sensitivity to these drugs.

4. Differences in DNA sensitivity to denaturation cannot be detected in chromatin (nucleohistone) extracted from nuclei in solution, suggesting that the phenomenon is associated with chromatin packing (higher order structure) into the nucleus rather than with changes at the nucleosomal level.

5. The phenomenon appears to be unrelated to phosphorylation, acetylation, or poly(ADP) ribosylation of nuclear proteins as well as the presence of HMG proteins or S-S bonds in chromatin.

6. A close correlation is observed between the content of nonhistone proteins (protein to DNA ratio) and DNA stability to denaturation, and similar subphases can be distinguished based on nonhistone protein content.

7. The structural proteins extractable with $\geqq 1.2$ N NaCl are involved in anchoring DNA to the nuclear matrix or chromosome scaffold and may be responsible for maintaining a high degree of chromatin compaction in situ, such as observed in metaphase or G_0 cells chromatin. These proteins confer the differential sensitivity of DNA in situ to denaturation.

DENATURATION AND CONDENSATION OF NUCLEIC ACIDS IN SITU INDUCED BY INTERCALATING LIGANDS

We recently reported that AO or other intercalating cations can induce denaturation and condensation of nucleic acids [64,70,130–134]. This phenomenon was observed in preparation of free nucleic acids in solution [130–134], as well as in situ, in intact or permeabilized cells [64,70]. Denaturation was reflected as the loss of the green fluorescence characteristic of AO intercalation into double-stranded nucleic acids [38,49] and by increased accessibility of DNA bases to external probes, such as formaldehyde or diethyl pyrocarbonate [130]. The thermodynamics of AO binding to single-stranded versus double-stranded nucleic acids [90], and suppression of renaturation of nucleic acids by AO [74,76], provided additional evidence for the denaturing properties of the ligand.

The condensation (collapse) of nucleic acids manifested as an increase in light scatter during titration of nucleic acids with intercalators in solutions [134] and the appearance of visible precipitates, which, in the case of AO, exhibited red luminescence [132]. In viable cells exposed to AO the condensed complexes of AO-nucleic acids were electron-opaque and could be observed by electron microscopy [64,70]. They formed preferentially at sites of ribosomes, in nucleoli and in areas of heterochromatin in nuclei.

Studies on permeabilized RNase-treated cells or isolated nuclei made it possible to correlate sensitivity of DNA in situ to AO-induced denaturation occurring during the cell cycle and differentiation [70]; i.e., the AO-induced denaturation and condensation was studied by flow cytometry in nuclei of Friend erythroleukemia cells from exponentially growing, differentiated or quiescent cells. The DNA in nuclei of quiescent cells was the most sensitive to denaturation. It denatured (measured as changes in luminescence, at the midpoint of the transition) at 70 μM AO. DNA in nuclei of differentiated cells was more resistant, denaturing at 77 to 83 μM, whereas DNA in exponentially growing cells was the most resistant, denaturing at 86 μM (Fig. 9). Extraction of basic proteins with 0.1N HCl abolished the differences between the cells and shifted the transition to a lower AO concentration, comparable to that causing the transition of free DNA in solution.

Condensation of nucleic acids in viable cells, as condensation in solutions, can be monitored by changes in light scatter [70]. The decrease in forward light scatter observed during this reaction is, in all probability, due to the retraction of ribosomes, polysomes and cytoplasmic organelles towards the nucleus, caused by the collapse of the cytoplasmic RNA upon binding AO. The rise of right-angle light scatter can be explained as due to the increased refractive index (density) of the AO–nucleic acid complexes, compared to the native ribo- or deoxyribonucleoprotein.

The mechanisms responsible for the observed differences in DNA sensitivity to denaturation/condensation in quiescent versus cycling cells are unexplained. However, it is already evident that the phenomenon can find an application in flow cytometry as a probe discriminating cells with different chromatin structure. Thus, for instance, in addition to our observation, Mitchell et al. [170] exploited it for analysis of sperm cell chromatin; the DNA in sperm cells of infertile donors exhibited markedly increased sensitivity to AO-induced condensation (see Chapter 26, this volume). The method appears to be useful for male fertility tests.

Not only intercalators, but the dyes presumed to bind externally to nucleic acids appear to induce condensation of the latter. Namely, Stokke and Steen [245,246] recently reported that Hoechst 33258 condenses DNA and, like AO, upon condensation the spectral properties of the dye are changed (red-shifted). Condensation could be detected by right angle light scatter [246]. This phenomenon is also very likely responsible for the differential staining of lymphocyte subpopulations with Hoechst dyes in relationship to differences in chromatin structure, as recently reported by Watson et al. [269]. Similarly, Smith et al. [241] observed spectral changes in permeabilized human tumor cells titrated with Hoechst 33342. These changes were not sensitive enough, however, to discriminate the differences in chromatin structure that could be detected by DNA sensitivity to DNase I. The changes in spectrum of Hoechst 33342 that may be associated with DNA condensation by this dye also were observed by Sandhu et al. [227] during titration of supercoiled or relaxed plasmid pBR322 DNA. Although the mechanism of nucleic acid-induced condensation by externally binding dyes is even less understood than condensation caused by intercalators, the reaction appears to be sensitive to differences in chromatin structure and thus of potential value in flow cytometry.

Conclusions

1. Denaturation and condensation of nucleic acids induced by intercalating ligands can be detected and measured as spectral changes in luminescence or as light scatter increase,

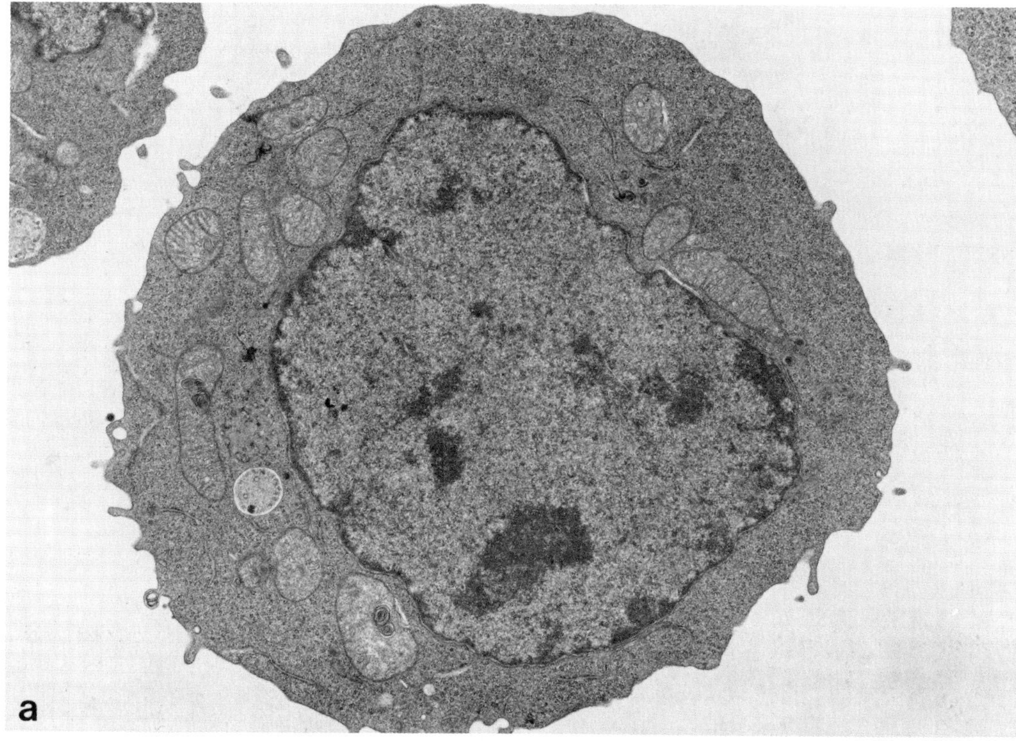

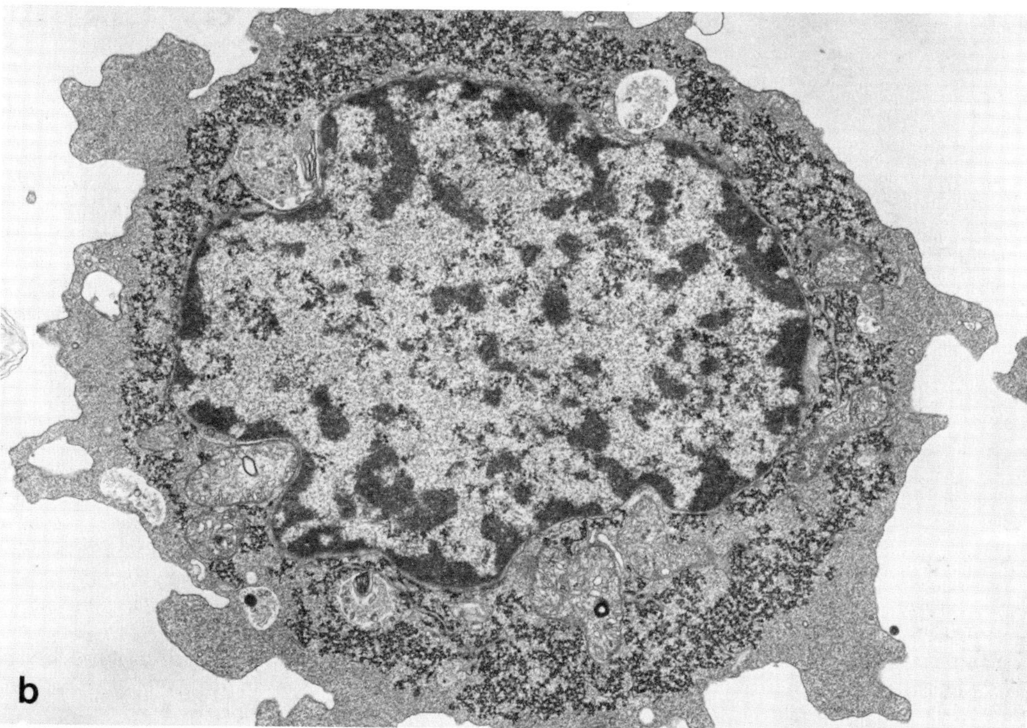

Fig. 9. AO-induced denaturation and condensation of nucleic acids in situ. Electron micrographs of Friend erythroleukemia cells untreated (**a**) and treated with 0.4 mM AO for 20 min (**b**). The dense complexes of AO with RNA and DNA are localized at sites of ribosomes in the cytoplasm and in nuclear heterochromatin. These electron-dense complexes represent nucleic acids denatured and condensed by AO [64,70]. Note that the ribosomes undergo retraction towards the nucleus upon reaction with AO. This phenomenon can be measured by flow cytometry as a decrease in green fluorescence of nucleic acids during titration with AO (denaturation) and increase of red luminescence (condensation). Condensation is also reflected by an increase in the right angle light scatter and a decrease of the forward light scatter [70].

both in solution and in individual cells, the latter by flow cytometry.

2. Because of the high base and sugar specificity of the reaction, the method makes it possible to distinguish nucleic acids differing in primary (e.g., RNA versus DNA, repetitive DNA sections) and secondary (single- versus double-stranded) structure.

3. In cells, the reaction is sensitive to differences in chromatin structure and may be used as a probe of chromatin conformation.

4. The condensed complexes of ligand-nucleic acid are electron-opaque and can be observed by electron microscopy.

DETECTION OF SPECIFIC NUCLEAR CONSTITUENTS

The possibility of detection of specific marker constituents, unique for a particular phase of the mitotic cycle or associated with cell proliferation, differentiation or quiescence, is of great importance. Such constituents may play a regulatory or other essential role in these processes and their presence, quantity or altered conformation may have prognostic value, e.g., with respect to cell kinetics, tumor progression, and sensitivity to drugs. Several products that manifest such specificity have recently been detected in nuclei. Most of them are proteins or peptides; the antibodies raised against them or other means of their cytochemical detection can serve as specific probes of nuclear chromatin and be markers of the cell proliferative potential. The most notable constituents of chromatin which show these characteristics are the following:

Products of *c-myc, c-myb,* and *c-fos* Oncogenes

Several oncogene products, which appear to be implicated in the regulation of cell proliferation, are localized in cell nuclei. Their function in chromatin is still unclear [12,62,80,269]. The most extensively studied is the *c-myc* oncogene product, which appears to be a 65 to 68-kD nuclear phosphoprotein [198] of short half-life [112], closely associated with the nuclear matrix [91], having high nonspecific affinity for double-stranded DNA [273], and believed to promote DNA replication [123]. It dissociates from chromatin during mitosis [91,278], and its level remains unchanged during the cell cycle [113,256]. Expression of this oncogene is induced shortly after stimulation of G_0 cells by mitogens [212]. Several studies suggest that within the framework of the competence-progression model of the cell cycle [200, 231], *c-myc* expression is associated with attainment of cell competence [127,136]. Induction of *c-myc* expression alone, however, is not sufficient for acquisition of competence to progress through the cell cycle and some additional inducible functions must be present in order for the cell to proliferate [238,279]. The transcription of the *c-myc* oncogene and level of the transcripts are markedly suppressed during cell quiescence [136,212] and at certain stages during induction of differentiation [31,101]. There is no correlation, however, between cell proliferation and *c-myc* expression [31,154, 155,279]. Thus, for example, Larsson et al. [154] observed that the expression of *c-myc* in chronic lymphocytic leukemia cells induced to differentiate by phorbol esters is greatly enhanced, yet the cells arrest in the G_{1A} or G_{1D} compartment and do not enter the S phase.

Recent observations by Watson and his associates [207, 270–272] indicate that the immunochemical detection of *c-myc* proteins by flow cytometry provides valuable clinical information on the growth of solid tumors in humans. These authors observed that in some tumor types (testicular tumors), the *c-myc* product accumulation correlated with cell differentiation rather than with proliferation [270]. Given the above, it appears that the cellular content of *c-myc,* per se, cannot be a parameter predicting kinetic properties of the cell because elevated levels are observed when proliferation is stimulated as well as when suppressed (e.g., during differentiation).

The product of *c-myb* oncogene is a 75-kD protein, it also binds to DNA in vitro [142]; its expression is often correlated with the entrance of cells into the cell cycle. During stimulation of quiescent cells to proliferate, it peaks after the peak of expression of *c-myc* but prior to expression of the histone H2b gene, i.e., prior to cell entrance of S phase [157]. In comparison with *c-myc,* its expression appears to correlate somewhat better with cell proliferation, at least in the system of monocytic cell differentiation [155].

The *c-fos* oncogene product of 40 kD has the ability to bind to DNA [226] and is phosphorylated [55]. It is located in the cell nucleus, interacting with a 39-kD protein [56]. During stimulation of quiescent cells a transient peak of expression of *c-fos* occurs very early [107], preceding the *c-myc* peak [172,193]. However, similar to *c-myc, c-fos* expression is also often enhanced during cell differentiation when proliferation is downregulated [155,174], and the cellular level of this oncogene product alone cannot be a marker of cell proliferation.

It should be pointed out, however, that because regulation of the oncogenes' expression is often altered during neoplasia [99,237], estimates of their cellular content may have diagnostic and perhaps prognostic value in some tumors.

The Proliferating Cell Nuclear Antigen (PCNA) Cyclin

The PCNA, described by Tan and colleagues [252], and cyclin, described by Bravo and Celis and their associates [32,34], are identical proteins. This protein of 36-kD is present in small amounts in normal noncycling cells. In some types of transformed cells, as well as in normal proliferating cells, the level of this protein is markedly elevated, especially during S phase [32,34,173]. The S-phase cells show variable patterns of distribution of cyclin, both with respect to intensity of the staining with antibodies and localization, depending on the degree of progression through S [33].

Two distinct populations of cyclin can be detected in the nucleus [24]. One, localized diffusely in nucleoplasm, is seen in quiescent cells and is easily extracted with detergent. Another is bound to nuclear structures that resemble replicon clusters and is seen in the S-phase nucleus [24]. Cyclin/PCNA was recently recognized as an auxiliary factor of DNA polymerase δ [25,204]; its level in cycling and noncycling cells was measured by flow cytometry [150]. It is clear that the recently developed monoclonal antibodies to this protein [185] represent a powerful reagent to immunochemically discriminate the DNA-replicating cells.

The Family of Proteins of Approximately 53 kD ("p53 proteins")

This protein(s) is associated with virus-induced cell transformation and cell proliferation [50]. Monoclonal antibodies against p53 are available and are used to study the regulation of cell proliferation [85,87,189]. As recently shown by Mercer et al. [166,168], microinjection of p53 antibodies inhibits

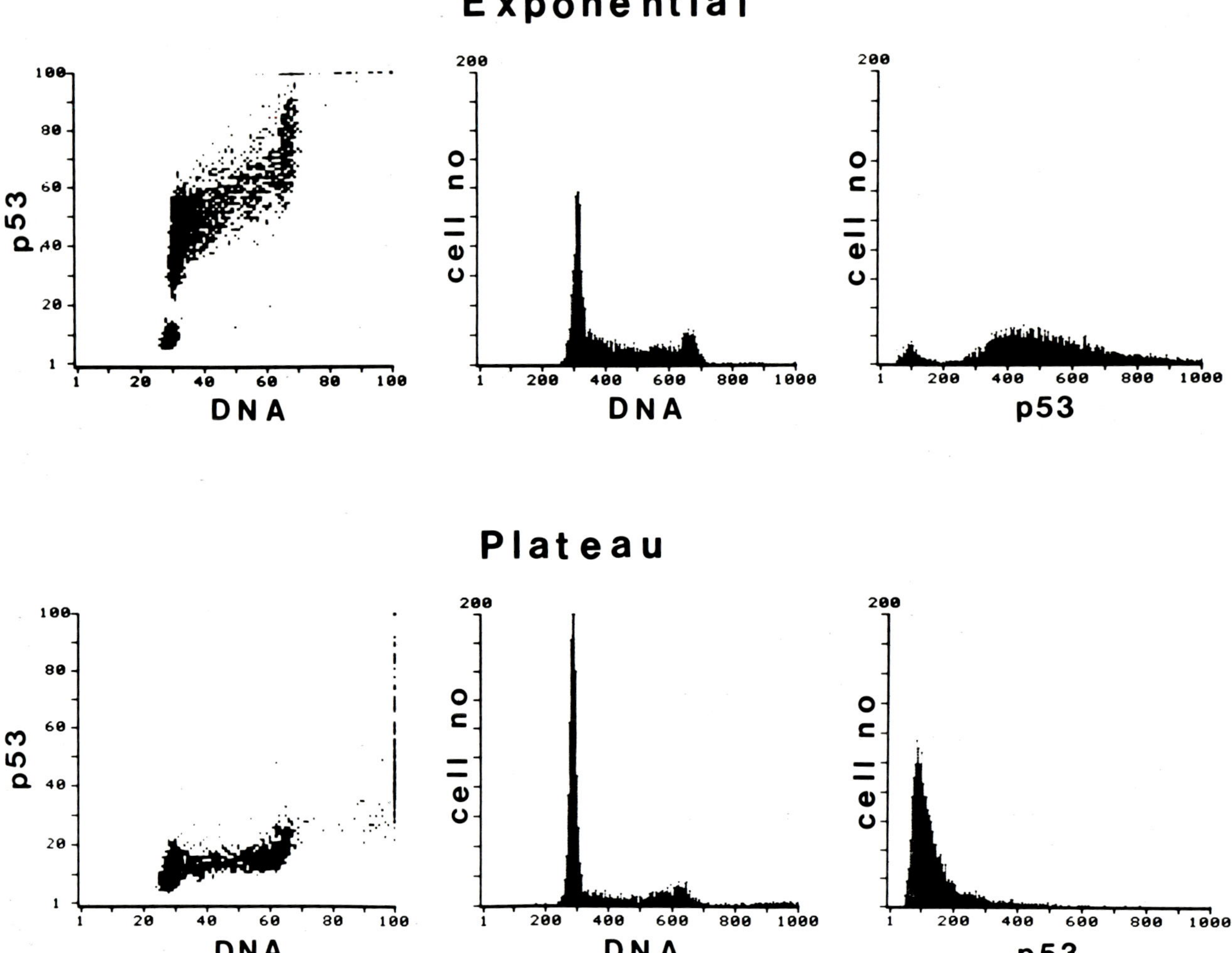

Fig. 10. Correlated measurement of DNA and p53 in exponentially growing and plateau-phase L1210 cells. The data demonstrate immunochemical detection of the specific nuclear protein (p53), expression of which appears to correlate with cell proliferation [50], in combination with DNA measurements. Suppression of the proliferation in plateau cultures is correlated with a decrease of p53 content in all cells regardless of the phase of the cell cycle (see ref. [67] for details).

the transition of 3T3 cells from G_0 to S but has no effect on the cell progression through G_1 and S. Induction of cell differentiation correlates with a decline of p53 [234]. Flow cytometric measurements of p53, in conjunction with DNA and RNA content [67], in cycling and noncycling cells indicate that expression of p53 is closely correlated with accumulation of cellular RNA (Fig. 10). These data, as well as observations by Wynford-Thomas et al. [281] suggest that p53 may play a role during cell growth in G_1 and attainment of competence rather than be directly associated with cell progression through particular phases of the cell cycle. Mercer and Baserga [167] and Kaczmarek et al. [126], however, while demonstrating a role for p53 as a competence factor, suggest that expression of this protein correlates more with DNA replication than with growth in size. Interestingly, p53, like *c-myc* product, is absent in mitotic, or very early postmitotic chromatin, and appears in the nucleus early in G_1, migrating from the cytoplasm [188]. Thus, although the role of p53 in cell growth or proliferation is not yet clear [188], it appears that its detection in the nucleus may be indicative of its proliferative potential, or correlate with neoplastic transformation. There are exceptions, however, as in the case of other oncogenes, when its content is not correlated with cell proliferation [84].

Nuclear Antigen Detected by Ki-67 Antibodies

The monoclonal antibodies against this protein, induced by immunizing animals with nuclei of L-428 cells, were reactive toward Hodgkin lymphoma cells as well as normal proliferating cells in interphase and mitosis, but not G_0 cells [103]. The function of the protein detected by these antibodies is unknown. Interestingly, during lymphocyte stimulation cells passing the early events of mitogen-induced transition from G_0 (G_{1Q}) to G_1; i.e., G_{1T} and G_{1A} lacked the Ki-67 nuclear antigen, whereas G_1 cells after mitosis were Ki-67 positive [102]. A growing number of publications indicate that the use of Ki-67 antibodies (often in conjunction with flow cytometry) offers a practical means for detecting proliferating cell populations [7,209,228,229].

Nonhistone Proteins Released from Chromatin by Limited Digestion with DNase I

Proteins of 37 and 100 kD were detected in proliferating cells, whereas the levels of proteins with molecular weight 52 and 75 kD were higher in quiescent cells [26]. These proteins were released from nuclei after incubation with low concentration of DNase I. Using the same strategy (DNase digestion), Goldberger et al. [106] detected proteins associated with terminal cell division and myeloid differentiation of human leukemic cell lines. Little more is known about these proteins.

Nuclear Proteins Associated with Interchromatin Granules; p105 Antigen

Monoclonal antibodies against several nuclear proteins were induced by immunization of mice with preparations of broken nuclei, and immunofluorescence data indicate that certain antibodies recognize cell-specific antigens [41–43]. The most explored was an interchromatin-associated antigen designated p105 [12,44–46]. Its expression correlates well with cell proliferation. Thus, stimulation of lymphocytes with mitogens causes a severalfold increase in expression of p105, which occurs even before cell entrance to S phase [45]. Further increase in p105 parallels cell progression through S, and a 10-fold increase is observed during transition from G_2 to M [46]. An increase in expression of this protein is closely associated with modulations of chromatin condensation that can be detected by changes in sensitivity of DNA in situ to acid-induced denaturation [45]. The antigen is stable and can be detected in fixed and paraffin-embedded nuclei, which makes it possible to study its content in archival pathological material [12]. The p105 antigen, similar to that of Ki-67 or cyclin, now appears to be one of the most specific markers of proliferating cells.

Nuclear Antigen of 55 kD:PSL

This protein accumulates specifically in the nucleus during S phase and appears to be distinct from p53 or cyclin [10]. Recently, Barque et al. [11] noted that PSL is composed of two acidic proteins, p55A and p55B. However, these proteins are also elevated during granulocytic differentiation of HL-60, much above the level seen in proliferating cell populations [11]. This again imposes limitation on the use of this antigen as a marker of cycling cells.

Highly Phosphorylated Nonhistone Proteins B23 and C23

These proteins are present in nucleoli or in nucleolar organizer regions of metaphase chromatin and are associated with rDNA-containing structures. Location, quantity and degree of phosphorylation of their serine and threonine residues show clear cell cycle phase dependency [183,242]. The role of these proteins in nucleologenesis was recently studied by Ochs et al. [184].

Adenosine (5′)tetraphospho(3′) adenosine (Ap_4A)

Although its function in cellular metabolism is not yet fully understood, this constituent is interesting for several reasons. The quantity of Ap_4A varies over a 1,000-fold range, depending on cell proliferation, growth, transformation, or differentiation [275]. Ap_4A binds to DNA polymerase α [210], primes DNA replication [284], triggers DNA replication, and may serve as the "second messenger" in regulation of the cell cycle [111,275]. Its level peaks at the initiation of S phase of the cell cycle [275]. However, because of its low molecular weight, Ap_4A cannot be retained in fixed cells. This restricts the possibility of cytochemical localization.

Enzymes Associated with DNA Replication

Cycling cells have increased activity of *DNA polymerase* α [58,254,280] and β [38]. Because the biotinylated dUTP is now availabe [152], it is possible to measure DNA polymerase α activity using fluorescent probes. DNA polymerase α can also be directly visualized by reaction with monoclonal antibodies; the enzyme is detectable in nuclei of cycling cells but not in condensed chromatin of quiescent or mitotic cells [14,161,176].

Dihydrofolate reductase and the multifunctional protein associated with the "housekeeping" functions of the cell are both cell cycle regulated [151]. The dihydrofolate reductase can be cytochemically detected by labeled methotrexate which binds with extremely high affinity to the enzyme [66]. Because dansylated fluorescent analogues of methotrexate were synthesized [148], it is now possible to measure the content of this enzyme also by flow cytometry.

Measurements of *DNA topoisomerases* are now possible, and the enzyme has been found to be localized predominantly in the nucleolus [175]. Relative to total DNA, ribosomal DNA sequences are 20-fold enriched in topoisomerase. This enzyme expression, therefore, should correlate with transcription of rDNA and be a marker of cell growth. The recent study of Heck and Earnshaw [115] indicates that topoisomerase II appears not to be required for transcription, but may function during DNA replication, and may be a specific marker of proliferating cells. Observations of Sullivan et al. [249], however, indicate that topoisomerase II content is proliferation-dependent only in some cell types, e.g., exponentially growing and plateau phase L1210 cells exhibit similar levels of the enzyme. Their studies, in addition [249], suggest that topoisomerase II content may be predictive of the cell's sensitivity to certain antitumor drugs. Tricoli et al. [259] also did not observe a difference between proliferating and confluent mouse fibroblasts in topoisomerase II activity, but they noted a fourfold rise in topoisomerase I activity in proliferating cells. In addition to its role in DNA replication, topoisomerase II may also be involved in the chromatin condensation cycle [262].

Ribonucleotide reductase is another enzyme associated with DNA replication. Its activity is cell cycle-dependent, rising dramatically in S phase [92]. *Thymidylate kinase* genes are also cell cycle regulated, and the enzyme is maximally expressed in S phase [29,255]. In some cell types, however [124], the level of mRNA for this enzyme appears to be invariant throughout the cycle. Still another DNA-replication-associated enzyme, *ornithine decarboxylase*, shows cell-cycle-related expression as well [129].

A modest variation during the cell cycle is in the enzyme likely involved in DNA repair, *ADP-ribosyltransferase*, which is most active in S phase [110] and more expressed in proliferating than in quiescent cells [250]. The enzymes associated with *biosynthesis of polyamines* are induced specifically during G_0–G_1 transition and remain constant during G_1 [243].

Nonproliferating Cell-Specific Protein: Statin

Statin, a 57 kD protein, present in nuclei of quiescent cells, was identified by monoclonal antibodies established after immunization of mice with extracts of in vitro aged human fibroblasts [267,268]. These antibodies appear to be a good marker of quiescent and perhaps senescent cells and can be used in flow cytometry [268].

Variants of Histone H1

Cell differentiation or quiescence is associated with a change in the proportion of histone H1 variants. Thus, for instance, noncycling cells have increased content of histone H1° [190,194]. During erythropoiesis histone H1 is replaced by the tissue-specific histone H5 [261]. Antibodies against these variants can also be useful in detecting quiescent or differentiating cells.

Other Cell-Cycle-Specific Proteins

The list of proteins that are either cell cycle phase specific or described as unique for proliferating cell populations is growing almost daily. Some of the most interesting proteins are (1) *numatrin*, a 40-kD nuclear matrix protein, which is associated with induction of proliferation of B lymphocytes [99]; (2) *dividin*, a 54-kD nuclear basic phosphoprotein, maximally expressed during S [36]; (3) nuclear *proteins of 110 and 85 kD*, characteristic of proliferating cells, with 85 kD having nucleolar localization [17]; (4) *calmodulin*, the intracellular Ca^{2+} receptor, which increases markedly during stimulation of quiescent cells to proliferation [38]; (5) *progressin*, a 33-kD protein whose expression peaks in mid- to late-S phase (37); (6) *primate-specific nuclear proteins* (76 kD) of varying charge, described by Celis et al. [35], which are expressed at a higher level in transformed and proliferating cells; (7) a *heat shock protein* (HSP 70), which increases 10 to 15-fold upon entry to S phase and declines in G_2 [169]; (8) the *microtubule-associated protein-1* (MAP-1), present in interphase cells but not in G_0 [186]; (9) *tubulin*, which is synthesized before cell division [39]; and (10) several *proteins coded by genes activated during stimulation of lymphocytes,* described by Kaczmarek et al. [104,126].

Conclusions

The list of constituents of nuclear chromatin, which specifically vary in quantity or quality during particular phases of the cell cycle, quiescence or differentiation, is growing rapidly. Several (e.g., certain oncogene products) appear to be implicated in regulation of the cell cycle. Antibodies to these constituents are now becoming available. Their use in flow cytometry, which has just begun, will yield a wealth of information of interest in basic research and could be of help in tumor diagnosis and prognosis in the clinic.

Extensive studies, however, are needed to (1) develop optimal methods of cell fixation and permeabilization for quantitative detection of these antigens; (2) reveal how universal the expression of particular nuclear antigens may be, depending on the cell type; and (3) correlate expression of nuclear constituents with other markers of the cell phenotype and with cell kinetics.

ACKNOWLEDGMENTS

I thank Mrs. Sally Carter and Mrs. Rose Vecchiolla for their assistance in the preparation of the manuscript. This work was supported by grants R37 CA 23296 and RO1 CA 28704 from the National Cancer Institute.

REFERENCES

1. **Allan J, Hartman PG, Crane-Robinson C, Aviles FX (1980)** The structure of histone H1 and its location in chromatin. Nature (Lond.) 288:675–679.
2. **Alvarez MR (1975)** Microfluorometric comparisons of heat-induced nuclear acridine orange metachromasia between normal cells and neoplastic cells from primary tumors of diverse origin. Cancer Res. 35: 93–98.
3. **Andreeff M, Darzynkiewicz Z, Sharpless T, Clarkson BD, Melamed MR (1980)** Discrimination of human leukemia subtypes by flow cytometric analysis of cellular DNA and RNA. Blood 55:282–293.
4. **Auer G (1972)** Nuclear protein content and DNA–histone interaction. Exp. Cell. Res. 75:231–236.
5. **Auer G, Ono J, Caspersson T (1983)** Determination of the fraction of G_0 cells in cytological samples by means of simultaneous DNA and nuclear protein analyses. Anal. Quant. Cytol. 5:1–4.
6. **Auer G, Zetterberg A (1972)** The role of nuclear proteins in RNA synthesis. Exp. Cell. Res. 75:245–253.
7. **Baisch H, Gerdes J (1987)** Simultaneous staining of exponentially growing versus plateau phase cells with the proliferation-associated antibody Ki-67 and propidium iodide: analysis by flow cytometry. Cell Tissue Kinet. 20:387–391.
8. **Balhorn R, Gledhill BL, Thomas C, Wyrobek A (1984)** DNA packaging in mouse spermatids. Exp. Cell. Res. 150:298–308.
9. **Ballachey BE, Miller HL, Jost LK, Evenson DP (1986)** Flow cytometry evaluation of testicular and sperm cells obtained from bulls implanted with zeranol. J. Anim. Sci. 63:995–1004.
10. **Barque JP, Danon F, Peraudeau L, Yeni P, Larsen CJ (1983)** Characterization by human autoantibody of a nuclear antigen related to the cell cycle. EMBO J. 2:743–749.
11. **Barque JP, Lagaye S, Ladoux A, Della Valle V, Abita JP, Larsen CJ (1987)** PSL, a nuclear cell-cycle associated antigen is increased during retinoic acid-induced differentiation of HL-60 cells. Biochem. Biophys. Res. Commun. 147:993–999.
12. **Bauer KD, Clevenger CV, Endow RK, Murad T, Epstein AL, Scarpelli DG (1986)** Simultaneous nuclear antigen and DNA content quantitation using paraffin-embedded colonic tissue and multiparameter flow cytometry. Cancer Res. 46:2428–2434.
13. **Bauer KD, Dethlefsen LA (1981)** Control of cellular proliferation in HeLa-53 suspension cultures. Characterization of cultures utilizing acridine orange staining procedures. J. Cell. Physiol. 108:99–112.
14. **Bensch KG, Tanaka S, Hu SZ, Wang TSF, Korn D (1982)** Intranuclear localization of human DNA polymerase with monoclonal antibodies. J. Biol. Chem. 257:8391–8396.
15. **Benyajati C, Worcel A (1976)** Isolation, characterization and structure of folded interphase genome of *Drosophila melanogaster.* Cell 9:393–407.
16. **Bergounioux C, Perennes C, Brown SC, Gadal P (1988)** Nuclear RNA quantification in protoplast cell-cycle phases. Cytometry 9:84–87.
17. **Bishop JM (1983)** Cellular oncogenes and retroviruses. Annu. Rev. Biochem. 52:301–354.
18. **Black A, Freeman JW, Zhou G, Busch H (1987)**

Novel cell cycle-related nuclear proteins found in rat and human cells with monoclonal antibodies. Cancer Res. 47:3266–3272.

19. **Block AL, Bauer KD, Williams TJ, Seidenfeld J** (1987) Experimental parameters and a biological standard for acridine orange detection of drug-induced alterations in chromatin condensation. Cytometry 8:163–169.
20. **Bolund L, Darzynkiewicz Z, Ringertz NR** (1970) Cell concentration and the staining properties of nuclear deoxyribonucleoprotein. Exp. Cell. Res. 62:76–89.
21. **Bonaly J, Bre MH, Lafort-Tran M, Mestre JC** (1987) A flow cytometric study of DNA staining *in situ* in exponentially growing and stationary *Euglena gracilis*. Cytometry 8:42–45.
22. **Boulikas T, Wiseman JM, Garrard WT** (1980) Points of contact between histone H1 and the histone octamer. Proc. Natl. Acad. Sci. USA 77:127–131.
23. **Bravo R** (1986) Synthesis of the nuclear protein cyclin (PCNA) and its relationship with DNA replication. Exp. Cell. Res. 163:287–293.
24. **Bravo R, Macdonald-Bravo H** (1986) Existence of two populations of cyclin/proliferating cell nuclear antigen during the cell cycle: Association with DNA replication sites. J. Cell. Biol. 105:1549–1554.
25. **Bravo R, Frank R, Blundell PA, Macdonald-Bravo H** (1987) Cyclin-PCNA is the auxillary protein of DNA polymerase-δ. Nature (Lond.) 326:515–517.
26. **Briggs RC, Brewer G, Goldberger A, Wolff SN, Hnilica LS** (1983) Antigens in chromatin associated with proliferating and nonproliferating cells. J. Cell. Biochem. 21:249–252.
27. **Brock WA, Swartzendruber DE, Grdina DJ** (1982) Kinetic heterogeneity in density-separated murine fibrosarcoma subpopulations. Cancer Res. 42:4499–5003.
28. **Busch H, Smetana K** (1970) "The Nucleolus." New York: Academic Press.
29. **Cadman E, Heimer R** (1986) Levels of thymidylate synthetase during normal culture growth of L1210 cells. Cancer Res. 46:1195–1198.
30. **Caspersson T** (1961) Nucleo-cytoplasmic relations in normal and malignant cells. In Allen V (Ed), "The Molecular Content of Cellular Activity." New York: McGraw-Hill, pp 127–141.
31. **Cayre Y, Raynal MC, Darzynkiewicz Z, Dorner MH** (1987) Model for intermediate steps in monocytic differentiation: *c-myc, c-fms* and ferritin as markers. Proc. Natl. Acad. Sci. USA 84:7619–7623.
32. **Celis JE, Bravo R, Larsen PM, Fey SJ** (1984) Cyclin: A nuclear protein whose level correlates directly with the proliferative state of normal as well as transformed cells. Leukemia Res. 8:143–157.
33. **Celis JE, Celis A** (1985) Cell cycle dependent variations in the distribution of the nuclear protein cyclin proliferating cell nuclear antigen in cultures cells: Subdivision of S phase. Proc. Natl. Acad. Sci. USA 82:3262–3266.
34. **Celis JE, Fey SJ, Larsen PM, Celis A** (1984) Expression of the transformation-sensitive protein "cyclin" in normal human epidermoid basal cells and Simian virus 40-transformed keratinocytes. Proc. Natl. Acad. Sci. USA 81:3128–3132.
35. **Celis JE, Madsen P, Nielsen S, Ratz GP, Lauridsen JB, Celis A** (1987) Levels of synthesis of primate-specific nuclear proteins differ between growth-arrested and proliferating cells. Exp. Cell. Res. 168: 389–401.
36. **Celis JE, Nielsen S** (1986) Proliferation-sensitive nuclear phosphoprotein "dividin" is synthesized almost exclusively during S phase of the cell cycle in human AMA cells. Proc. Natl. Acad. Sci. USA 83: 8187–8190.
37. **Celis JE, Ratz GP, Celis A** (1987) Progressin: A novel proliferation-sensitive and cell cycle-regulated human protein whose rate of synthesis increases at or near the G_1/S transition border of the cell cycle. FEBS Lett. 223:237–242.
38. **Chafouleas JG, Lagace L, Bolton WE, Boyd III AE, Means AR** (1984) Changes in calmodulin and its mRNA accompany reentry of quiescent (G_0) cells into the cell cycle. Cell 36:73–81.
39. **Chang MT, Dove WF, Laffler TG** (1983) The periodic synthesis of tubulin in the *Physarum* cell cycle. J. Biol. Chem. 258:1352–1356.
40. **Chin TW, Baril EF** (1975) Nuclear DNA polymerases and the HeLa cell cycle. J. Biol. Chem. 250: 7951–7957.
41. **Clausen OAF, Parvinen M, Kirkus B** (1982) Stage-related variations in DNA fluorescence distribution during rat spermatogenic cycle measured by flow cytometry. Cytometry 2:421–425.
42. **Clevenger CV, Epstein A** (1984) Identification of nuclear protein component of interchromatin granules using a monoclonal antibody and immunogold electron microscopy. Exp. Cell. Res. 151:194–207.
43. **Clevenger CV, Epstein AL** (1984) Use of immunogold electron microscopy and monoclonal antibodies in the identification of nuclear substances. J. Histochem. Cytochem. 32:757–763.
44. **Clevenger CV, Bauer KD, Epstein AL** (1985) A method for simultaneous nuclear immunofluorescence and DNA content quantitation using monoclonal antibodies and flow cytometry. Cytometry 6: 208–214.
45. **Clevenger CV, Epstein AL, Bauer KD** (1987) Modulation of the nuclear antigen p105 as a function of cell-cycle progression. J. Cell. Physiol. 130:336–343.
46. **Clevenger CV, Epstein AL, Bauer KD** (1987) Quantitative analysis of a nuclear antigen in interphase and mitotic cells. Cytometry 8:280–286.
47. **Cohen RL, Muirhead KA, Gill JE, Waggoner AS, Horan PK** (1981) A cyanine dye distinguishes between cycling and noncycling fibroblasts. Nature (Lond.) 290:593–595.
48. **Cowden RR, Curtis SK** (1976) Some quantitative aspects of acridine orange fluorescence in unfixed sucrose-isolated mammalian nuclei. Histochem. J. 8: 45–49.
49. **Cowden RR, Curtis SK** (1981) Microfluorometric investigations of chromatin structure. I. Evaluation of nine DNA-specific fluorochromes as probes of chromatin organization. Histochemistry 72:11–23.
50. **Crawford L** (1983) The 53,000 dalton cellular protein and its role in transformation. Int. Rev. Exp. Pathol. 25:1–50.
51. **Crick FHC, Klug A** (1975) Kinky helix. Nature (Lond.) 255:530–533.
52. **Crissman HA, Darzynkiewicz Z, Tobey RA, Steinkamp JA** (1985) Correlated measurements of

DNA, RNA and protein content in individual cells by flow cytometry. Science 228:1321–1324.

53. **Crissman HA, Steinkamp JA (1973)** Rapid simultaneous measurement of DNA protein and cell volume in single cells from large mammalian cell populations. J. Cell. Biol. 59:766–772.

54. **Crissman HA, Steinkamp JA (1982)** Rapid one step staining procedures for analysis of cellular DNA and protein by single and dual laser flow cytometry. Cytometry 3:84–90.

55. **Curran T, Miller DA, Zokas L, Verma IM (1984)** Viral and cellular *fos* proteins: A comparative analysis. Cell 36:259–268.

56. **Curran T, Van Beveren C, Ling N, Verma IM (1985)** Viral and cellular *fos* proteins are complexed with a 39,000 dalton protein. Mol. Cell. Biol. 5:167–172.

57. **D'Anna JG, Gurley LR, Becker RR (1981)** Histones H1° and H1°b are the same as CHO histones H1 (III) and H1 (IV): New features of H1° phosphorylation during the cell cycle. Biochemistry 20:4501–4505.

58. **Darzynkiewicz Z (1973)** Detection of DNA polymerase activity in fixed cells. Exp. Cell. Res. 80:483–486.

59. **Darzynkiewicz Z (1979)** Acridine orange as a molecular probe in studies of nucleic acids *in situ*. In Melamed MR, Mullaney PF, Mendelsohn ML (eds); "Flow Cytometry and Sorting." New York: John Wiley & Sons, pp 285–316.

60. **Darzynkiewicz Z (1986)** Cell growth and division cycle. In Dethlefsen LA (ed); "Cell Cycle Effects of Drugs." International Encyclopedia of Pharmacology and Therapeutics, Section 121. Oxford: Pergamon Press, pp 1–43.

61. **Darzynkiewicz Z, Bolund L, Ringertz NP (1969)** Actinomycin D binding of normal and phytohemagglutinin stimulated lymphocytes. Exp. Cell. Res. 55: 120–126.

62. **Darzynkiewicz Z, Crissman H, Steinkamp JA, Traganos F (1990)** (In preparation.)

63. **Darzynkiewicz Z, Dokov V, Pienkowski M (1967)** Dry mass of lymphocytes during transformation after stimulation by phytohemagglutinin. Nature (Lond.) 214:1265–1266.

64. **Darzynkiewicz Z, Evenson D, Kapuscinski J, Melamed MR (1983)** Denaturation of DNA and RNA *in situ* induced by acridine orange. Exp. Cell. Res. 148:31–46.

65. **Darzynkiewicz Z, Kapuscinski J, Traganos F, Crissman HA (1987)** Application of pyronin Y(G) in cytochemistry of nucleic acids. Cytometry 8:138–145.

66. **Darzynkiewicz Z, Rogers AW, Barnard EA, Wong DH, Werkheiser WC (1966)** Autoradiography with tritiated methotrexate and the cellular distribution of folate reductase. Science 151:1528–1530.

67. **Darzynkiewicz Z, Staiano-Coico L, Kunicka JE, DeLeo AB, Old LJ (1986)** p53 content in relation to cell growth and proliferation in murine L1210 leukemia and normal lymphocytes. Leukemia Res. 10:1383–1389.

68. **Darzynkiewicz Z, Traganos F, Andreeff M, Sharpless T, Melamed MR (1979)** Different sensitivity of chromatin to acid denaturation in quiescent and cycling cells as revealed by flow cytometry. J. Histochem. Cytochem. 27:478–485.

69. **Darzynkiewicz Z, Traganos F, Carter SP, Higgins PJ (1987)** *In situ* factors affecting stability of the DNA helix in interphase nuclei and metaphase chromosomes. Exp. Cell. Res. 172:168–179.

70. **Darzynkiewicz Z, Traganos F, Kapuscinski J, Melamed MR (1985)** Denaturation and condensation of DNA *in situ* by acridine orange in relation to chromatin changes during growth and differentiation of Friend erythroleukemia cells. Cytometry 6: 195–207.

71. **Darzynkiewicz Z, Traganos F, Kapuscinski J, Staiano-Coico L, Melamed MR (1984)** Accessibility of DNA *in situ* to various fluorochromes. Relationship to chromatin changes during erythroid differentiation of Friend leukemia cells. Cytometry 5:355–363.

72. **Darzynkiewicz Z, Traganos F, Melamed MR (1980)** New cell cycle compartments identified by flow cytometry. Cytometry 1:98–108.

73. **Darzynkiewicz Z, Traganos F, Sharpless T, Friend C, Melamed MR (1976)** Nuclear chromatin changes during erythroid differentiation of Friend virus induced leukemic cells. Exp. Cell. Res. 99:301–309.

74. **Darzynkiewicz Z, Traganos F, Sharpless T, Melamed MR (1974)** Thermally-induced changes in chromatin of isolated nuclei and of intact cells as revealed by acridine orange staining. Biochem. Biophys. Res. Commun. 59:392–399.

75. **Darzynkiewicz Z, Traganos F, Sharpless T, Melamed MR (1975)** Conformation of RNA *in situ* as studied by acridine orange staining and automated cytofluorometry. Exp. Cell. Res. 95:143–153.

76. **Darzynkiewicz Z, Traganos F, Sharpless T, Melamed MR (1975)** Thermal denaturation of DNA *in situ* as studied by acridine orange staining and automated cytofluorometry. Exp. Cell. Res. 90:411–428.

77. **Darzynkiewicz Z, Traganos F, Sharpless T, Melamed MR (1976)** Lymphocyte stimulation: A rapid, multiparameter analysis. Proc. Natl. Acad. Sci. USA 73: 2881–2884.

78. **Darzynkiewicz Z, Traganos F, Sharpless T, Melamed MR (1977)** Cell cycle-related changes in nuclear chromatin of stimulated lymphocytes as measured by flow cytometry. Cancer Res. 37:4635–4640.

79. **Darzynkiewicz Z, Traganos F, Sharpless T, Melamed MR (1977)** Different sensitivity of DNA *in situ* in interphase and metaphase chromatin to heat denaturation. J. Cell. Biol. 73:128–138.

80. **Darzynkiewicz Z, Traganos F, Staiano-Coico L, Kapuscinski J, Melamed MR (1982)** Interactions of rhodamine 123 with living cells studied by flow cytometry. Cancer Res. 42:799–806.

81. **Darzynkiewicz Z, Traganos F, Xue SB, Melamed MR (1981)** Effect of n-butyrate on cell-cycle progression and *in situ* chromatin structure of L1210 cells. Exp. Cell. Res. 136:279–293.

82. **Darzynkiewicz Z, Traganos F, Xue SB, Staiano-Coico L, Melamed MR (1981)** Rapid analysis of drug effects on the cell cycle. Cytometry 1:279–286.

83. **Darzynkiewicz Z, Williamson B, Carswell EA, Old LJ (1984)** Cell cycle specific effects of tumor necrosis factor. Cancer Res. 44:83–90.

84. **Davis D, Wynford-Thomas D (1986)** Heterogeneity in distribution and content of p53 in SV40-transformed mouse fibroblasts. Exp. Cell. Res. 166: 94–102.

85. **DeLeo A, Jay G, Apella E, Dubois G, Law LW, Old**

JL (1979) Detection of transformation-related antigen in chemically induced sarcomas and other transformed cells of mouse. Proc. Natl. Acad. Sci. USA 76:2420–2424.

86. **Dethlefsen LA, Bauer KD, Riley RM (1980)** Analytical cytometric approaches to heterogenous cell populations in solid tumors. A review. Cytometry 1:89–97.
87. **Dippold WG, Jay G, DeLeo AB, Khoury G, Old LJ** (1981) p53 transformation-related protein. Detection by monoclonal antibody in mouse and human cells. Proc. Natl. Acad. Sci. USA 7:1695–1699.
88. **Duesberg PH (1983)** Retroviral transforming genes in normal cells. Nature (Lond.) 304:219–226.
89. **Dyson JEF, McLaughlin JB, Surrey CR, Simmons DM, Daniel J** (1987) Effects of hyperthermia, irradiation, and cytotoxic drugs on fluorescein isothiocyanate staining intensity for flow cytofluorometry. Cytometry 8:26–34.
90. **Einck L, Bustin M** (1985) The intracellular distribution and function of the high mobility group of chromosomal proteins. Exp. Cell. Res. 156:295–310.
91. **Eisenman RN, Tachibana CY, Abrane HD, Hann SR** (1985) *v-myc* and *c-myc*-encoded proteins are associated with the nuclear matrix. Mol. Cell. Biol. 5: 114–126.
92. **Eriksson S, Graslund A, Skog S, Thelander L, Tribukait B** (1984) Cell cycle-dependent regulation of mammalian ribonucleotide reductase. J. Biol. Chem. 259:1695–1700.
93. **Evenson DP, Darzynkiewicz Z, Jost L, Janca F, Bellachey B** (1985) Changes in accessibility of DNA *in situ* to various fluorochromes during spermatogenesis. Cytometry 7:45–53.
94. **Evenson DP, Darzynkiewicz Z, Melamed MR** (1980) Comparison of human and mouse sperm chromatin structure by flow cytometry. Chromosoma 78:225–238.
95. **Evenson DP, Darzynkiewicz Z, Melamed MR** (1980) Relation of mammalian sperm chromatin heterogeneity to fertility. Science 210:1131–1133.
96. **Evenson DP, Higgins PJ, Grueneberg D, Ballachey BE** (1985) Flow cytometric analysis of mouse spermatogenic function following exposure to ethylnitrosourea. Cytometry 6:238–253.
97. **Evenson DP, Klein FA, Whitmore WF, Melamed MR** (1984) Flow cytometric evaluation of sperm from patients with testicular carcinoma. J. Urol. 133:1220–1225.
98. **Fahrlander PD, Marcu KB** (1986) Regulation of *c-myc* expression in normal and transformed mammalian cells. In Kahn P, Graf T (eds), "Oncogenes and Growth Control." Berlin: Springer-Verlag, pp 264–270.
99. **Feuerstein N, Moud JJ** (1987) "Numatrin," a nuclear protein associated with induction of proliferation in B lymphocytes. J. Biol. Chem. 262:11389–11397.
100. **Ffrench M, Fiere PA, Van HV, Gentilhomme O, Adeleine P, Viala JJ** (1985) Cell cycle, protein content and nuclear size in acute myeloid leukemia. Cytometry 6:47–53.
101. **Filmis J, Buick RN** (1985) Relationship of *c-myc* expression to differentiation and proliferation of HL-60 cells. Cancer Res. 45:822–825.
102. **Gerdes J, Lemke H, Baisch H, Wacker HH, Schwab U, Stein H** (1984) Cell cycle analysis of a cell proliferation-associated human nuclear antigen defined by the monoclonal antibody Ki-67. J. Immunol. 133: 1710–1713.
103. **Gerdes J., Schwab U, Lemke H, Stein H** (1983) Production of a mouse monoclonal antibody reactive with a human nuclear antigen associated with cell proliferation. Int. J. Cancer 31:13–20.
104. **Gibson CW, Rittling Sr, Hirschorn RR, Kaczmarek L, Calabretta B, Stiles CD, Baserga R** (1986) Cell cycle dependent genes inducible by different mitogens in cells from different species. Mol. Cell. Biochem. 71:61–69.
105. **Gledhill BL, Gledhill MP, Rigler R Jr, Ringertz NR** (1966) Changes in deoxyribonucleoprotein during spermiogenesis in the bull. Exp. Cell. Res. 41:652–665.
106. **Goldberger A, Brewer G, Hnilica LS, Briggs PC** (1984) Nonhistone protein antigen profiles of five leukemic cell lines reflect the extent of myeloid differentiation. Blood 63:701–710.
107. **Greenberg ME, Ziff EB** (1984) Stimulation of 3T3 cells induces transcription of the *c-fos* proto-oncogene. Nature (Lond.) 311:433–437.
108. **Grimes SR, Henderson N** (1983) Acetylation of histones during spermatogenesis in the rat. Arch. Biochem. Biophys. 221:108–116.
109. **Grimes SR, Meistrich ML, Platz RD, Hnilica LS** (1977) Nuclear protein transitions in rat testis spermatids. Exp. Cell. Res. 110:31–39.
110. **Grobner P, Loidl P** (1985) ADP-ribosyltransferase in isolated nuclei during the cell cycle of *Physarum polycephalum*. Biochem. J. 232:21–24.
111. **Grummt F** (1985) Diadenosine tetraphosphate (Ap_4A): A putative chemical messenger of cell proliferation control and inducer of DNA replication. Plant Mol. Biol. 2:41–44.
112. **Hann SR, Eisenman RN** (1984) Proteins encoded by the human *c-myc* oncogene. Differential expression in neoplastic cells. Mol. Cell. Biol. 4:2486–2497.
113. **Hann SR, Thompson CB, Eisenman RN** (1985) *c-myc* oncogene protein synthesis is independent of the cell cycle in human and avian cells. Nature (Lond.) 314:366–369.
114. **Harris H** (1974) "Nucleus and Cytoplasm." Oxford: Clarendon Press.
115. **Heck MMS, Earnshaw WC** (1986) Topoisomerase II: A specific marker for cell proliferation. J. Cell. Biol. 103:2569–2581.
116. **Hedley DW, Friedlander ML, Taylor IW, Rugg CA, Musgrove EA** (1983) Method for analysis of cellular DNA content of paraffin embedded pathological material using flow cytometry. J. Histochem. Cytochem. 31:1333–1335.
117. **Heldin CH, Westermark B** (1984) Growth factors: Mechanism of action and relation to oncogenes. Cell 37:9–20.
118. **Higgins PJ, Piwnicka M, Darzynkiewicz Z, Melamed MR** (1984) Multiparameter flow cytometric analysis of hepatic nuclear RNA and DNA of normal and hepatotoxin-treated mice. Am. J. Pathol. 115:31–35.
119. **Hnilica LS** (1972) "The Structure and Biological Function of Histones." Cleveland: CRC Press.
120. **Hnilica LS** (1983) "Chromosomal Nonhistone

Proteins." Vols. I–IV. Boca Raton, Florida: CRC Press.
121. **Hutchison N, Weintraub H (1985)** Localization of DNase I-sensitive sequences to specific regions of interphase nuclei. Cell 43:471–482.
122. **Igo-Kemenes T, Horz W, Zachan HG (1982)** Chromatin. Annu. Rev. Biochem. 51:89–121.
123. **Iguchi-Ariga SMM, Itani T, Kiji Y, Ariga H (1987)** Possible function of the *c-myc* product: Promotion of cellular DNA replication. EMBO J. 6:2365–2371.
124. **Iman AMA, Crossley PH, Jackman AL, Little PFR (1987)** Analysis of thymidylate synthase gene amplification and of mRNA levels in the cell cycle. J. Biol. Chem. 262:7368–7373.
125. **Johnson TS, Swartzendruber DE, Martin JC (1981)** Nuclear size of G_1/S transition cells measured by flow cytometry. Exp. Cell. Res. 134:201–205.
126. **Kaczmarek L, Calabretta B, Baserga R (1985)** Expression of cell cycle dependent genes in phytohemagglutinin-stimulated human lymphocytes. Proc. Natl. Acad. Sci. USA 82:5375–5379.
127. **Kaczmarek L, Hyland JK, Watt R, Rosenberg M, Baserga R (1985)** Microinjected *c-myc* as a competence factor. Science 228:1313–1315.
128. **Kaczmarek L, Oren M, Baserga R (1986)** Cooperation between the p53 protein tumor antigen and platelet-poor plasma in the induction of cellular DNA synthesis. Exp. Cell. Res. 162:268–272.
129. **Kahana C, Nathans D (1983)** Isolation of cloned cDNA encoding ornithine decarboxylase. Proc. Natl. Acad. Sci. USA 80:3645–3649.
130. **Kapuscinski J, Darzynkiewicz Z (1983)** Increased accessibility of bases in DNA upon binding of acridine orange. Nucleic Acids Res. 11:7555–7567.
131. **Kapuscinski J, Darzynkiewicz Z (1984)** Denaturation of nucleic acids induced by intercalating agents. Biochemical and biophysical properties of acridine orange–DNA complexes. J. Mol. Struct. Dyn. 1:1485–1499.
132. **Kapuscinski J, Darzynkiewicz Z, Melamed MR (1982)** Luminescence of the solid complexes of acridine orange with RNA. Cytometry 2:201–211.
133. **Kapuscinski J, Darzynkiewicz Z, Melamed MR (1983)** Interactions of acridine orange with nucleic acids: Properties of complexes of acridine orange with single stranded ribonucleic acid. Biochem. Pharmacol. 32:3679–3694.
134. **Kapuscinski J, Darzynkiewicz Z (1984)** Condensation of nucleic acids by intercalating aromatic cations. Proc. Natl. Acad. Sci. USA 81:7368–7372.
135. **Kapuscinski J, Skoczylas B (1978)** Fluorescent complexes of DNA with DAPI, 4′6-diamidine-2- phenylindole, 2 HCl or DCl, 4′6-dicarboxyamide-2-phenylindole. Nucleic Acids Res. 5:3775–3799.
136. **Kelly K, Cochran BH, Stiles CD, Leder P (1983)** Cell-specific regulation of the *c-myc* gene by lymphocyte mitogens and platelet-derived growth factor. Cell 35: 603–610.
137. **Kernell AM, Bolund L, Ringertz NR (1981)** Chromatin changes during erythropoiesis. Exp. Cell. Res. 65:1–6.
138. **Kernell AM, Ringertz NR (1972)** Cytochemical characterization of deoxyribonucleoprotein by UV-microspectrophotometry in heat-denatured cell nuclei. Exp. Cell. Res. 72:240–251.
139. **Killander D, Rigler R Jr (1965)** Initial changes of deoxyribonucleic and synthesis of nucleic acids in phytohemagglutinin stimulated human leukocytes *in vitro*. Exp. Cell. Res. 39:701–712.
140. **Killander D, Rigler R (1969)** Activation of deoxyribonucleoprotein in human leukocytes stimulated by phytohemagglutinin. Exp. Cell. Res. 54:163–170.
141. **Klein FA, Melamed MR, Whitmore WF, Herr HW, Sogani PC, Darzynkiewicz Z (1982)** Characterization of papilloma by two-parameter DNA–RNA flow cytometry. Cancer Res. 42:1094–1097.
142. **Klempnauer KH, Sippel A (1986)** Subnuclear localization of proteins encoded by the oncogene *v-myb* and its cellular homolog *c-myb*. Mol. Cell. Biol. 6: 62–69.
143. **Koppel F, Allet B, Eisen H (1977)** Appearance of a chromatin protein during erythroid differentiation of Friend virus-transformed cells. Proc. Natl. Acad. Sci. USA 74:653–656.
144. **Kornberg RD (1977)** Structure of chromatin. Annu. Rev. Biochem. 46:931–954.
145. **Koss LG, Wolley RC, Schreiber K, Mendecki J (1977)** Flow-microfluorometric analysis of nuclei isolated from various normal and malignant human epithelial tissues. A preliminary report. J. Histochem. Cytochem. 25:565–572.
146. **Krishan A (1987)** Flow cytometric studies on intracellular drug resistance. In Gray JW, Darzynkiewicz Z (eds), "Techniques in Cell Cycle Analysis." Clifton, NJ: Humana Press, pp 337–366.
147. **Krishan A (1987)** Effect of drug blockers on vital staining of cellular DNA with Hoechst 33342. Cytometry 8:642–645.
148. **Kumar AA, Kempton RJ, Anstead GM, Freishen JH (1983)** Fluorescent analogues of methotrexate: Characterization and interaction with dihydrofolate reductase. Biochemistry 22:390–395.
149. **Kunicka JE, Darzynkiewicz Z, Melamed MR (1987)** DNA *in situ* sensitivity to denaturation: A new parameter for flow cytometry of normal colonic epithelium and colon carcinoma. Cancer Res. 47:3942–3947.
150. **Kurki P, Vanderlaan M, Dolbeare F, Gray J, Tan EM (1986)** Expression of proliferating cell nuclear antigen (PCNA)/cyclin during the cell cycle. Exp. Cell. Res. 166:209–219.
151. **LaBella F, Brown EH, Basilico C (1983)** Changes in the levels of viral and cellular gene-transcripts in the cell cycle of SV40 transformed mouse cells. J. Cell. Physiol. 117:62–68.
152. **Langer PR, Waldrop AA, Ward AC (1981)** Enzymatic synthesis of biotin-labeled polynucleotides. Novel nucleic acid affinity probes. Proc. Natl. Acad. Sci. USA 78:6633–6637.
153. **Larsen JK, Munch-Petersen B, Christiansen J, Jorgensen K (1986)** Flow cytometric discrimination of mitotic cells: Resolution of M, as well as G_1, S and G_2 phase nuclei with mithramycin, propidium iodide, and ethidium bromide after fixation with formaldehyde. Cytometry 7:54–63.
154. **Larsson LG, Gray HE, Tofferman T, Pettersson U, Nilsson K (1987)** Drastically increased expression of MYC and FOS protooncogenes during *in vitro* differentiation of chronic lymphocytic leukemia cells. Proc. Natl. Acad. Sci. USA 84:223–227.

155. **Lee J, Mehta K, Blick MB, Gutterman JU, Lopez-Berenstein G (1987)** Expression of c-fos, c-myb and c-myc in human monocytes: Correlation with monocytic differentiation. Blood 69:1542–1545.

156. **Lennox RW, Cohen LH (1984)** The alterations in H1 histone complement during mouse spermatogenesis and their significance for H1 subtype function. Dev. Biol. 103:840–845.

157. **Liedeman R, Bolund L (1976)** Acridine orange binding to chromatin of individual cells and nuclei under different staining conditions. I. Binding capacity of chromatin. Exp. Cell. Res. 91:164–174.

158. **Liedeman RR, Matveyeva NP, Vostricova SA, Prilipko LL (1975)** Extrinsic factors affecting the binding of acridine orange to the DNP complex of cell nuclei in different physiological states. Exp. Cell. Res. 90: 105–110.

159. **Manzini G, Barcellona ML, Avitabile M, Quadrifoglio F (1983)** Interaction of diamidino-2-phenylindsole (DAPI) with natural and synthetic nucleic acids. Nucleic Acids Res. 11:8861–8876.

160. **Mathews MB, Bernstein RM, Franza RB Jr, Garrels HI (1984)** Identity of the proliferating cell marker antigen and cyclin. Nature (Lond.) 309:374–376.

161. **Matsukage A, Yamamoto S, Yamagushi M, Kauakabe M, Takahashi T (1983)** Immunocytochemical localization of chick DNA polymerases and α.β. J. Cell. Physiol. 117:266–271.

162. **Mazzini G, Giordano P, Riccardi A, Montecucco MC (1983)** A flow cytometric study of the propidium iodide staining kinetics of human leukocytes and its relationship with chromatin structure. Cytometry 3: 443–448.

163. **McGhee JD, Felsenfeld G (1980)** Nucleosome structure. Annu. Rev. Biochem. 49:1115–1156.

164. **Melamed MR, Adams LR, Traganos F, Zimring A, Kamentsky LA (1972)** Acridine orange metachromasia for characterization of leukocytes in leukemia, lymphoma and other neoplasms. Cancer 29:1361–1368.

165. **Melamed J, Darzynkiewicz Z (1985)** RNA content and chromatin structure of CHO cells arrested in metaphase by colcemid. Cytometry 6:381–385.

166. **Mercer WE, Avignolo C, Baserga R (1984)** Role of the p53 protein in cell proliferation as studied by microinjection of monoclonal antibodies. Mol. Cell. Biol. 4:278–281.

167. **Mercer WE, Baserga R (1985)** Expression of the p53 protein during the cell cycle of human peripheral blood lymphocytes. Exp. Cell. Res. 160:31–46.

168. **Mercer WE, Nelson D, DeLeo AB, Old LJ, Baserga R (1982)** Microinjection of monoclonal antibody to protein p53 inhibits serum-induced DNA synthesis in 3T3 cells. Proc. Natl. Acad. Sci. USA 79:6309–6312.

169. **Milarski KL, Morimoto RI (1986)** Expression of human HSP70 during the synthetic phase of the cell cycle. Proc. Natl. Acad. Sci. USA 83:9517–9521.

170. **Mitchell C, Norman A, Tejada R (1984)** Acridine orange denaturation of human sperm DNA. In Proceedings of the International Conference of Analytical Cytology, June 3–8, Asiolomar, Pacific Grove, California (abst. 28).

171. **Monahan JJ (1978)** The structure, origin and function(s) of RNA in the nuclei of eukaryotic cells. Int. Rev. Cytol. 8(suppl.):229–245.

172. **Moore KS, Sullivan K, Tan EM, Prystowsky MB (1987)** Proliferating cell nuclear antigen/cyclin is an interleukin 2-responsive gene. J. Biol. Chem. 262: 2447–2450.

173. **Moore JP, Todd JA, Hasketh TR, Metcalfe JC (1986)** *c-fos* and *c-myc* gene activation, ionic signals and DNA synthesis in thymocytes. J. Biol. Chem. 261: 8158–8162.

174. **Muller R (1986)** Cellular and viral *fos* genes: Structure, regulation of expression and biological properties of their encoded products. Biochim. Biophys. Acta 822:207–225.

175. **Muller MT, Pfund WP, Mehta VB, Trask DK (1985)** Eukaryotic type I topoisomerase is enriched in the nucleolus and catalytically active on ribosomal DNA. EMBO J. 4:1237–1243.

176. **Nakamura H, Morita T, Masaki S, Yoshida S (1984)** Intracellular localization and metabolism of DNA polymerase α in human cells visualized by monoclonal antibody. Exp. Cell. Res. 151:123–133.

177. **Nash D, Plaut W (1964)** On the denaturation of chromosomal DNA *in situ*. Proc. Natl. Acad. Sci. USA 51:731–735.

178. **Nelson WG, Pienta KJ, Barrack ER, Coffey DS (1986)** The role of the nuclear matrix in the organization and function of DNA. Annu. Rev. Biophys. Chem. 15:457–475.

179. **Neuman JR, Housman D, Ingram UD (1978)** Nuclear protein synthesis and phosphorylation in Friend erythroleukemia cells stimulated with DMSO. Exp. Cell. Res. 111:277–284.

180. **Newport JW, Forbes DJ (1987)** The nucleus: Structure, function, and dynamics. Annu. Rev. Biochem. 56:535–565.

181. **Null M (1974)** Subunit structure of nuclear chromatin. Nature (Lond.) 251:249–251.

182. **Nusse M, Egner HJ (1984)** Can nocodazole, an inhibitor of microtubule formation, be used to synchronize mammalian cells? Cell Tissue Kinet. 17:13–23.

183. **Ochs R, Lischwe M, O'Leary P, Busch H (1983)** Localization of nucleolar phosphorylations B23 and C23 during mitosis. Exp. Cell. Res. 146:139–149.

184. **Ochs RL, Lischwe MA, Shen E, Carrol RE, Busch H (1985)** Nucleologenesis: Composition and fate of prenucleolar bodies. Chromosoma 92:330–336.

185. **Ogata K, Kurki P, Celis JE, Nakamura RM, Tam EM (1987)** Monoclonal antibodies to a nuclear protein (PCNA/cyclin) associated with DNA replication. Exp. Cell. Res. 168:475–486.

186. **Ohno T, Kako R, Sato C, Ohkawa A (1986)** Distinction of G_0 cells from senescent cells in cultures of non-cycling human fetal lung fibroblasts by anti-MAP-1 monoclonal antibody staining. Exp. Cell. Res. 163:309–316.

187. **Oosterhof D, Hozier J, Ri UR (1975)** Nuclease action on chromatin. Evidence for discrete repeated nucleoprotein units along chromatin fibrils. Proc. Natl. Acad. Sci. USA 72:633–637.

188. **Oren M (1986)** p53: Molecular properties and biological activities. In Kahn P, Graf T, (eds), "Oncogenes and Growth Control." Berlin: Springer-Verlag, pp 284–289.

189. **Oren M, Reich NC, Levine AJ (1982)** Regulation of the cellular p53 tumor antigen in teratocarcinoma

cells and the differentiated progeny. Mol. Cell. Biol. 2:443–449.

190. **Osborne HB, Chabanas A** (1984) Kinetics of H1° accumulation and commitment to differentiation of murine erythroleukemia cells. Exp. Cell. Res. 158: 449–458.

191. **Oud PS, Katzko MW, Pahlplatz MMM, Vooijs GP** (1987) Evaluation of DNA and nuclear protein features for their use in describing normal and malignant endometrium. Cytometry 8:453–460.

192. **Oud PS, Reubsaet-Veldhuizen JAM, Hendrik JBJ, Pahlplatz MMM, Hermkens HG, James JTJ, Vooijs GP** (1986) DNA and nuclear protein measurement in isolated nuclei of human endometrium. Cytometry 7:318–324.

193. **Palumbo AP, Rossino P, Comoglio PM** (1986) Bombesin stimulation of *c-fos* and *c-myc* gene expression in cultures of Swiss 3T3 cells. Exp. Cell. Res. 167:276–280.

194. **Panyim S, Chalkley R** (1969) A new histone found only in mammalian tissues with little cell division. Biochem. Biophys. Res. Commun. 37:1042–1043.

195. **Papa S, Billi AM, Martelli AM, Capitani S, Miscia S, Maraldi NM, Manzoli FA** (1987) The application of flow cytometry to the study of nuclear matrix. A multiparametric analysis. J. Submicrosc. Cytol. 19:257–263.

196. **Papa S, Capitani S, Matteuci A, Vitale M, Santi P, Martelli AM, Maraldi NM, Manzoli FA** (1987) Flow cytometric analysis of isolated rat liver nuclei during growth. Cytometry 8:595–601.

197. **Papa S, Caramelli E, Billi AM, Santi P, Capitani S, Manzoli FA** (1986) Three-parameter flow cytometric characterization of rat liver nuclear matrix. Cell Biol. Int. Rep. 10:271–276.

198. **Perrson H, Leder P** (1984) Nuclear localization and DNA binding properties of a protein expressed by human c-myc oncogene. Science 225:718–720.

199. **Piwnicka M, Darzynkiewicz Z, Melamed MR** (1983) RNA and RNA content of isolated cell nuclei measured by multiparameter flow cytometry. Cytometry 3:269–275.

200. **Pledger WJ, Stiles CD, Scher CD** (1978) An ordered sequence of events is required before BALB/c-3T3 cells become committed to DNA synthesis. Proc. Natl. Acad. Sci. USA 75:2839–2843.

201. **Pollack A, Moulis H, Block NL, Irvin III, GL** (1984) Quantitation of cell kinetic responses using simultaneous flow cytometric measurements of DNA and nuclear protein. Cytometry 5:473–481.

202. **Pollack A, Moulis H, Greenstein DB, Block NL, Irvin III GL** (1985) Cell kinetic effects of incorporated ^{3}H-thymidine on proliferating human lymphocytes: Flow cytometric analysis using the DNA/nuclear protein method. Cytometry 6:428–436.

203. **Potmesil M, Goldfeder A** (1981) Nucleolar morphology, nucleic acid synthesis and growth rates of experimental tumors. Cancer Res. 31:789–797.

204. **Prelich G, Tan CK, Kostura M, Mathews MB, So G, Downey KM, Stillman B** (1987) Functional identity of proliferating cell nuclear antigen and a DNA polymerase-δ auxillary protein. Nature (Lond.) 326:517–520.

205. **Prentice DA, Tobey RA, Gurley LR** (1985) Cell cycle variations in chromatin structure detected by DNase I. Exp. Cell. Res. 157:242–252.

206. **Raap AK, Marijnen JGJ, Vrolijk J, van der Ploeg M** (1986) Denaturation, renaturation and loss of DNA during *in situ* hybridization procedures. Cytometry 7:235–242.

207. **Rabbitts PH, Watson JV, Lamond A, Forster A, Stinson MA, Evan G, Fisher W, Atherton E, Sheppard R, Rabbits TH** (1986) Metabolism of *c-myc* gene products: *c-myc* mRNA and protein expression in the cell cycle. EMBO J. 4:2009–2015.

208. **Rabinovitch P, Torres RM, Engel D** (1986) Simultaneous cell cycle analysis and two-color surface immunofluorescence using 7-amino-actino-mycin D and a single laser excitation: Applications to study of cell activation and the cell cycle of murine LY-1 cells. J. Immunol. 136:2769–2775.

209. **Ralfkiaer E, Steintl, Bosq J, Gatter KC, Ralfkiaer N, Wantzin GL, Mason DY** (1986) Expression of a cell cycle-associated nuclear antigen (Ki-67) in cutaneous lymphoid infiltrates. Am. J. Dermatol. 8:37–43.

210. **Rapaport E, Zamecnik PC, Baril EF** (1981) Association of diadenosine 5′,5″-p p^4-tetraphosphate binding protein with HeLa cell DNA polymerase. J. Biol. Chem. 256:12143–12151.

211. **Record TM Jr, Anderson CF, Lohman TM** (1978) Thermodynamic analysis of ion effects on the binding and conformational equilibria of proteins and nucleic acids: The roles of ion association or release, screening and ion effects on water activity. Q. Rev. Biophys. 11:103–178.

212. **Reed JC, Alpers JD, Nowell PC** (1987) Expression of c-myc proto-oncogene in normal human lymphocytes. Regulation by transcriptional and post-transcriptional mechanisms. J. Clin. Invest. 80:101–106.

213. **Richmond TJ, Finch JT, Rushton B, Rhodes D, Klug A** (1984) Structure of nucleosome core particle at 7A resolution. Nature (Lond.) 211:532–537.

214. **Riggs MG, Whittaker RG, Newman JR, Ingram VM** (1977) N-butyrate causes histone modification in HeLa and Friend erythroleukemia cells. Nature (Lond.) 268:462–464.

215. **Rigler R Jr** (1966) Microfluorometric characterization of intranuclear nucleic acids and nucleoproteins by acridine orange. Acta Physiol. Scand. 67(suppl 267):1–122.

216. **Rigler R, Killander D, Bolund L, Ringertz NR** (1969) Cytochemical characterization of deoxyribonucleoprotein in individual cell nuclei. Exp. Cell. Res. 55: 215–224.

217. **Riley D, Weintraub H** (1978) Nucleosomal DNA is digested to repeats of 10 bases by exonuclease III. Cell 13:281–293.

218. **Ring D, Cole RD** (1979) Chemical cross-linking of histone H1 to the nucleosomal histones. J. Biol. Chem. 254:11688–11695.

219. **Ringertz NR** (1969) Cytochemical properties of nuclear proteins and deoxyprotein complexes in relation to nuclear fusion. In Lima-de-Faria A (ed), "Handbook of Molecular Cytology." Amsterdam, North-Holland, pp 657–684.

220. **Ringertz NR, Bolund L, Darzynkiewicz Z** (1970) AO binding of intracellular nucleic acids in fixed cells in relation to cell growth. Exp. Cell. Res. 63:233–238.

221. **Ringertz NR, Carlsson SA, Ege T, Bolund L** (1971)

Detection of human and chick nuclear antigens in nuclei of chick erythrocytes during reactivation in heterokaryons with HeLa cells. Proc. Natl. Acad. Sci. USA 68:3228–3232.

222. **Ringertz NR, Savage RE (1976)** "Cell Hybrids." New York: Academic Press.

223. **Robert-Nicoud M, Arnd-Jovin DJ, Zarling DA, Jovin TM (1984)** Immunological detection of left-handed Z DNA in isolated polytene chromosomes. Effects of ionic strength pH, temperature and topological stress. EMBO J. 3:721–731.

224. **Roti Roti JL, Higashikubo R, Blair CC, Uygur N (1982)** Cell-cycle position and nuclear protein content. Cytometry 3:91–96.

225. **Roti Roti J, Wright WD, Higashikubo R, Dethlefsen LA (1985)** DNase I sensitivity of nuclear DNA measured by flow cytometry. Cytometry 6:101–108.

226. **Sambucetti L, Curran T (1986)** The Fos protein complex is associated with DNA in isolated nuclei and binds to DNA cellulose. Science 234:1417–1419.

227. **Sandhu CL, Warters RL, Dethlefsen LA (1985)** Fluorescence studies of Hoechst 33342 with supercoiled and relaxed plasmid pBR322 DNA. Cytometry 6: 191–194.

228. **Sasaki K, Murakami T, Kawasaki M, Takahashi M (1987)** The cell cycle associated change of the Ki-67 reactive nuclear antigen expression. J. Cell. Physiol. 133:579–584.

229. **Schwarting R, Gerdes J, Niehus J, Jaeschke L, Stein H (1986)** Determination of the growth fraction in cell suspensions by flow cytometry using the monoclonal antibody Ki-67. J. Immunol. Methods 90:65–70.

230. **Seligy VL, Lurquin PF (1973)** Relationship between dye binding and template activity of isolated ovian chromatin. Nature (New Biol.) 243:20–21.

231. **Seuwen K, Steiner U, Adam G (1984)** Cellular content of ribosomal RNA in relation to the progression and competence signals governing proliferation of 3T3 and SV40-3T3 cells. Exp. Cell. Res. 154:10–24.

232. **Shapiro HM (1981)** Flow cytometric estimation of RNA content in intact cells stained with Hoechst 33342 and pyronin Y. Cytometry 2:143–150.

233. **Shapiro HN, Natale PJ, Kamentsky LA (1979)** Estimation of membrane potentials of individual lymphocytes by flow cytometry. Proc. Natl. Acad. Sci. USA 76:5728–5730.

234. **Shen DW, Real FC, DeLeo AB, Old LJ, Marks PA, Rifkind RA (1983)** Protein p53 and inducer-mediated erythroleukemia cell commitment to terminal cell division. Proc. Natl. Acad. Sci. USA 80:5919–5922.

235. **Shih TY, Lake RS (1972)** Studies on structure of metaphase and interphase chromatin of Chinese hamster cells by circular dichroism and thermal denaturation. Biochemistry 11:4811–4826.

236. **Simard R, Langelier Y, Mandevill R, Maestracci N, Royal A (1974)** Inhibitors as tools in elucidating the structure and function of the nucleus. In Busch H (ed), "The Cell Nucleus." New York: Academic Press, pp 447–487.

237. **Slamon DJ, Boone TC, Murdock DC, Keith DE, Press MF, Larson RA, Souza LM (1986)** Studies on the human c-myb gene and its product in human acute leukemias. Science 233:347–351.

238. **Smeland E, Godal T, Rand E, Beiske K, Funderud S, Clark EA, Pfeifer-Ohlsson S, Ohlsson R (1985)** The specific induction of myc protooncogene expression in normal human B cells is not a sufficient event for acquisition of competence to poliferate. Proc. Natl. Acad. Sci. USA 82:6255–6259.

239. **Smetana K, Busch H (1974)** The nucleolus and nucleolar DNA. In Busch H, (ed), "The Cell Nucleus." Vol. 1. New York: Academic Press, pp 73–143.

240. **Smets LA (1973)** Activation of nuclear chromatin and the release from contact-inhibition of 3T3 cells. Exp. Cell. Res. 79:239–243.

241. **Smith PJ, Nakeff A, Watson JV (1985)** Flow-cytometric detection of changes in fluorescence emission spectrum of a vital DNA-specific dye in human tumor cells. Exp. Cell. Res. 159:37–46.

242. **Spector DL, Ochs RL, Busch H (1984)** Silver staining immunofluorescence and immunoelectron microscopic localization of nucleolar phosphoprotein B23 and C23. Chromosoma 90:139–148.

243. **Sterner R, Boffa LC, Chen TA, Allfrey VG (1987)** Cell cycle-dependent changes in conformation and composition of nucleosomes containing human histone gene sequences. Nucleic Acids Res. 25:4375–4391.

244. **Stimac E, Morris DR (1987)** Messenger RNAs coding for enzymes of polyamine biosynthesis are induced during G_0–G_1 transition but not during traverse of normal G_1. J. Cell. Physiol. 133:590–594.

245. **Stokke T, Steen HB (1985)** Multiple binding modes for Hoechst 33258 to DNA. J. Histochem. Cytochem. 33:333–338.

246. **Stokke T, Steen HB (1986)** Binding of Hoechst 33258 to chromatin *in situ*. Cytometry 7:227–234.

247. **Stokke T, Steen HB (1987)** Distinction of leukocyte classes based on chromatin-structure-dependent DNA binding of 7-aminoactinomycin D. Cytometry 8:576–583.

248. **Subirana JA (1973)** Studies on the thermal denaturation of nucleohistone. J. Mol. Biol. 74:363–385.

249. **Sullivan DM, Latham MD, Ross WE (1987)** Proliferation-dependent topoisomerase II content as a determinant of antineoplastic drug action in human, mouse and Chinese hamster ovary cells. Cancer Res. 47:3973–3979.

250. **Sweigert SE, Petzold SJ, Surowy CS, Berger SJ, Dethlefsen LA, Berger NA (1986)** Poly(ADP) polymerase activity in proliferating and quiescent murine mammary carcinoma cells. Radiat. Res. 205:219–226.

251. **Szabo G, Damjanovich S, Sumegi J, Klein G (1987)** Overall changes in chromatin sensitivity to DNase I during differentiation. Exp. Cell. Res. 169:158–168.

252. **Takasaki Y, Fishwild D, Tan EM (1984)** Characterization of proliferating cell nuclear antigen recognized by autoantibodies in lupus sera. J. Exp. Med. 159:981–992.

253. **Tan EM, Ogata K, Takasaki Y (1987)** PCNA/cyclin: A lupus antigen connected with DNA replication. J. Rheumatol. 14(suppl. 13):89–96.

254. **Taylor GR, Lagosky PA, Storms RK, Haynes RH (1987)** Molecular characterization of the cell cycle-regulated thymidylate synthase gene of *Saccharomyces cerevisiae*. J. Biol. Chem. 262:5298–5307.

255. **Thommes P, Reiter T, Knippers R (1986)** Synthesis of DNA polymerase α analyzed by immunoprecipitation from synchronously proliferating cells. Biochemistry 25:1308–1314.

256. **Thompson CB, Challoner PB, Neiman PE, Groudine M** (1985) Levels of *c-myc* oncogene mRNA are invariant throughout the cell cycle. Nature (Lond.) 314:363–366.

257. **Tompson CB, Challoner PB, Neiman PE, Groudine M** (1986) Expression of the *c-myb* proto-oncogene during cellular proliferation. Nature (Lond.) 319: 374–380.

258. **Traganos F, Darzynkiewicz Z, Sharpless T, Melamed MR** (1977) Simultaneous staining of ribonucleic and deoxyribonucleic acid in unfixed cells using acridine orange in a flow cytofluorometric system. J. Histochem. Cytochem. 25:46–56.

259. **Tricoli JV, Sahai BM, McCormick PJ, Jarlinski S, Bertram JS, Kowalski D** (1965) DNA topoisomerase I and II activities during cell proliferation and the cell cycle in cultured mouse embryo fibroblast (C^3H 10T1/2) cells. Exp. Cell. Res. 158:1–14.

260. **Tsai YH, Ansevin AT, Hnilica LS** (1975) Association of tissue-specific histones with deoxyribonucleic acids. Thermal denaturation of native partially dehistonized and reconstituted chromatin. Biochemistry 14:1257–1265.

261. **Tsai YH, Hnilica LS** (1975) Tissue-specific histones in the erythrocytes of chicken and turtle. Exp. Cell. Res. 91:107–112.

262. **Uemura T, Ohkura H, Adachi Y, Morino K, Schiozaki K, Yanagida M** (1987) DMA topoisomerase II is required for condensation and separation of mitotic chromosomes in *S. pombe.* Cell 50:917–925.

263. **Vidali G, Boffa LC, Bradbury EM, Allfrey V** (1978) Butyrate suppression of histone deacetylation leads to accumulation of multiacetylated forms of histone H3 and H4 and increase DNase I sensitivity of the associated DNA sequences. Proc. Natl. Acad. Sci. USA 75:2239–2243.

264. **Walle AJ** (1986) Identification of L3 leukemia and Burkitt's lymphoma cells by flow cytometric quantitation of nuclear and cellular RNA and DNA content. Leukemia Res. 10:303–312.

265. **Wallen CA, Higashikubo R, Dethlefsen LA** (1984) Murine mammary tumour cells *in vitro*. I. The development of a quiescent state. Cell Tissue Kinet. 17: 65–77.

266. **Wallen CA, Higashikubo R, Dethlefsen LA** (1984) Murine mammary tumour cells *in vitro*. II. Recruitment of quiescent cells. Cell Tissue Kinet. 17:78–89.

267. **Wang E** (1985) A 57,000-mol-wt protein uniquely present in nonproliferating cells and senescent human fibroblasts. J. Cell. Biol. 100:545–551.

268. **Wang E** (1987) Contact-inhibition-induced quiescent state is marked by intense nuclear expression of statin. J. Cell. Physiol. 133:151–157.

269. **Watson JV** (1986) Oncogenes, cancer and analytical cytology. Cytometry 7:400–410.

270. **Watson JV, Nakeff A, Chambers SH, Smith PJ** (1985) Flow cytometric fluorescence emission spectrum analysis of Hoechst-33342-stained DNA in chicken thymocytes. Cytometry 6:310–315.

271. **Watson JV, Sikora KE, Evan GI** (1985) A simultaneous flow cytometric assay for *c-myc* oncoprotein and DNA in nuclei from paraffin embedded material. J. Immunol. Methods 83:179–185.

272. **Watson JV, Stewart J, Evan GI, Ritson A, Sikora K** (1986) The clinical significance of flow cytometric *c-myc* oncoprotein quantitation in testicular cancer. Br. J. Cancer 53:331–337.

273. **Watt RA, Shatzman AR, Rosenberg M** (1985) Expression and characterization of the human *c-myc* DNA-binding protein. Mol. Cell. Biol. 5:448–456.

274. **Weinmann-Dorsch C, Hedl A, Grummt I, Albert W, Ferdinand FJ, Friis RR, Pierron G, Moll W, Grummt F** (1984) Drastic rise of intracellular adenosine (5′) tetraphospho (5′) adenosine correlation with onset of DNA synthesis in eukaryotic cells. Eur. J. Biochem. 138:179–185.

275. **Weinmann-Dorsch C, Pierron G, Wick R, Sauer H, Grummt F** (1984) High diadenosine tetraphosphate (Ap_4A) level at initiation of S phase in the naturally synchronous mitotic cycle of *Physarum polycephalum.* Exp. Cell. Res. 155:171–177.

276. **Weintraub H, Groudine M** (1976) Chromosomal subunits in active genes have an altered conformation. Science 193:848–856.

277. **Weisbrod S** (1982) Active chromatin. Nature (Lond.) 287:289–295.

278. **Winqvist R, Saksela K, Alitalo K** (1984) The myc proteins are not associated with chromatin in mitotic cells. EMBO J. 3:2947–2950.

279. **Womer RB, Krick K, Mitchell CD, Ross AH, Bishayee S, Scher CD** (1987) PDGF induces c-myc mRNA expression in MG-63 human osteosarcoma cells but does not stimulate cell replication. J. Cell. Physiol. 132:62–72.

280. **Wong SW, Wahl AF, Yuan PM, Arai N, Pearson BE, Arai K, Korn D, Hunkapiller MW, Wang TSW** (1988) Human DNA polymerase α gene expression is cell proliferation dependent and its primary structure is similar to both prokaryocytic and eukaryocytic replicative DNA polymerases. EMBO J. 7:37–47.

281. **Wynford-Thomas D, LaMontagne A, Marin G, Prescott DM** (1985) The role of p53 in growth of mouse fibroblast lines. Exp. Cell. Res. 159:191–200.

282. **Yau P, Thorne AW, Imai BS, Matthews HR, Bradbury EM** (1982) Thermal denaturation studies of acetylated nucleosomes and digonucleosomes. Eur. J. Biochem. 129:281–289.

283. **Yen A, Pardee AB** (1979) Role of nuclear size in cell growth initiation. Science 204:1315–1317.

284. **Zamecnik AC, Rapaport E, Baril EF** (1982) Priming of DNA synthesis by diadenosine 5′5‴p p^1,p^4 tetraphosphate with double-stranded octodecamer as a template and DNA polymerase. Proc. Natl. Acad. Sci. USA 79:1791–1794.

285. **Zieve G, Penman S** (1981) Subnuclear particles containing a small nuclear RNA and heterogenous RNA. J. Mol. Biol. 1345:501–519.

17

Immunofluorescence Techniques

Michael R. Loken

Becton Dickinson Monoclonal Center, San Jose, California 95131

The value of immunofluorescence in analytical cytometry is in studying specific gene products rather than entire chemical species such as DNA or RNA. During the process of differentiation and maturation of normal (and neoplastic) cells, genes are expressed that are required for the proper function of the cell. Some of the specific gene products are displayed on the cell surface, while others are intracellular. The cell-surface molecules may play a role in cell–cell recognition for homing and attachment to specific tissues [13,71]; they may be receptors for specific ligands such as hormones or other growth regulatory signals [76,78]; they may be molecules that mediate that cell's particular function [81]; or they may be a component of cellular structure [2]. Since different cell types have different cell-surface proteins, these provide the basis for distinguishing one cell as different from another using immunofluorescence techniques.

A short definition of terms may be useful in understanding the basic concepts of immunofluorescence. An immune response is mounted by an animal when a foreign substance called an antigen enters the body. As part of this immune response, a large number of B cells (lymphocytes of the B lineage) are stimulated to produce proteins called antibodies (immunoglobulins), which aid in neutralizing the foreign substance. Although an individual B lymphocyte produces only one type of antibody, other B lymphocytes can secrete different antibodies with different binding specificities. The immune response is a summation of the response of all the B cells stimulated by the antigen. An antiserum is the mixture of proteins (including the immunoglobulins) remaining in a blood sample after the red cells (erythrocytes) and white cells (leukocytes) are removed by letting the blood clot.

The value of immunoassays was greatly enhanced by the development of monoclonal antibodies by Kohler and Milstein [33]. A monoclonal antibody is secreted by the progeny of a single B lymphocyte that has been immortalized by fusion with a malignant plasma cell [33]. The new antibody-secreting cell, called a hybridoma, grows indefinitely and can be frozen and recovered like other tissue culture cell lines.

The two ends of an antibody have quite different functions. There are two identical specificity-conferring combining sites on an immunoglobulin molecule at the amino-terminal end. The amino acid sequences that make up these combining sites fold to form a region that is highly variable and determines the nature of the antigen recognized (V regions). These are connected to the constant region (C region), which is essential for the elimination of recognized foreign antigen. The same combining sites can be attached to several different types of constant regions, which produce different effects in the immune response. The eight (in the mouse) different constant regions (immunoglobulin heavy-chain classes: IgM, IgD, IgG1, IgG2a, IgG2b, IgG3, IgE, and IgA) arise from closely linked genes [1,22].

Antibodies have such exquisite specificity that they can distinguish minor differences between molecules of similar structure [18,19,77]. This ability to bind only to a single three-dimensional structure is exploited in the immunofluorescence techniques. The specificity is conferred by the antibody. The chromophore, which is conjugated to the antibody, acts simply as a tag or reporter molecule. The specificity of binding to the target molecule is not intrinsic to the dye as it is with DNA or RNA specific dyes. By separating the binding specificity from the chromophore, it is possible to use the same dye molecule with an infinite number of specificity-conferring antibodies. This allows standardization of the optical filters in a flow cytometer, since different cellular characteristics can all be assessed using the same filters. Conversely, it is possible to couple any given antibody to different chromophores. This provides flexibility in selecting spectral characteristics of the dye molecule independent of the binding characteristics of the antibody. By selecting fluorochromes that have optically separable spectral characteristics, it is also possible to study multiple antibodies, each coupled with a different chromophore, in the same experiment. Two, three, or four immunofluorescence tags have been used simultaneously in a single experiment, and even more may be possible [48,49,64,77] (also R.R. Hardy and D.R. Parks, personal communication, 1986). The general methodologies and strategies used in immunofluorescence may also be applicable to other techniques in which the reporter molecule is coupled to a specificity-conferring molecule [26,75,79].

Immunofluorescence has been most widely used to identify and quantify a specific population of cells in a heterogeneous mixture of other cells. These studies have focused pri-

Flow Cytometry and Sorting, Second Edition, pages 341–353

marily on cell-surface characteristics; however, recent applications of immunofluorescence techniques have begun to probe intracellular antigens [9,24,69]. Since the binding of immunofluorescent molecules to cell surfaces does not in itself kill the cell, a desired population not only can be identified but can be sorted for further functional or biochemical study [27] (also R.R. Hardy and D.R. Parks, personal communication, 1986).

The ability to discriminate one cell type from another depends on several factors. The relative frequency of the desired cell subpopulation to the rest of the cells places limitations on which techniques can be used. If the cells represented $1/10^5$ of the entire population, different strategies would be needed than if the cells were 1/20.

The ability to distinguish one cell type from another also depends on how different they are. The total amount of each antigen expressed on the cell is genetically controlled and varies considerably depending on the antigen. Some antigens may have as few as 100 molecules per cell, while other antigens are more frequent with as many as 10^5 molecules expressed on the cell surface [40]. A further complication arises from the fact that all cells of a given type do not have the same amount of antigen on their cell surfaces. Unlike the constant amount of DNA per cell, there may be a 10 to 100-fold difference in the expression of individual cell-surface components even on cloned cell populations [41]. The degree of variability is dependent on which cell-surface molecule is assessed, since the expression of each molecule is regulated independently [73]. Glycophorin A, present on human erythrocytes, has only a 10% variation from cell to cell within an individual [36]. By contrast, a common leukocyte antigen identified by CD45 varies in expression over a range of 10^3 [47].

The strategy used for immunofluorescence is also dependent on how accurate the result must be. Assessing the presence or absence of tumor cells to determine minimal residual disease in leukemia, for example, requires more stringent procedures than determining whether the proportion of T lymphocytes is within the normal range.

This chapter addresses the strategies and techniques of maximizing the immunofluorescence signal on the cells of interest while minimizing the noise (or background) contributed by the remainder of the cell population. The brightness of the signal is related to the specificity of the reagents, the techniques used to label the cells and the chromophores used to detect antibody binding. The reduction of background is dependent on decreasing nonspecific binding, reducing intrinsic fluorescence of the cells (autofluorescence) and decreasing the noise inherent in the instrument itself. Pre-enrichment of the desired cell population can also increase the effective discrimination of this cell.

MAXIMIZING THE IMMUNOFLUORESCENCE SIGNAL

The usefulness of immunofluorescence to discriminate cell populations is directly related to how specific the antibodies are in distinguishing one cell from another. The antibodies used in immunoassays originate from two sources; antisera and monoclonal antibodies. The general characteristics of these two sources of reagents can affect the sensitivity of the immunofluorescence assay.

Antisera

An antiserum is a very heterogeneous mixture of antibodies, only some of which have the desired binding characteristics. Comparison of the steps required to produce antisera and monoclonal antibodies illustrates their basic differences (Fig. 1). The reactivity observed with a particular serum is, in essence, the summation of all the individual antibodies present at the time the serum was collected. In different animals, different B cells respond to the same antigen preparation even if the animals are genetically identical. This means that the serum in each animal contains a different spectrum or repertoire of antibodies. In addition, this spectrum of antibodies can change within an animal during the course of immunization. The limited availability of antibodies of defined activity is a constant problem with conventional antisera.

Once a desired antiserum is generated, the antibodies must be purified. The antibodies found in that serum are heterogeneous with respect to their binding ability, and most antibodies are reactive with antigens other than the one used for immunization. It is first necessary to make the serum specific for a single antigen. This is accomplished by extensive absorption of the serum to remove all possible other reactivities that might interfere with the immunoassay to be performed. Once the serum has been shown to be monospecific, it is preferable to isolate the antibodies that have the desired binding. Affinity purification is used to isolate the specific antibodies from all the rest of the antibodies by coupling the purified antigen to a solid support medium such as Sepharose [51]. The serum is then passed over the antigen in a column so that the specific antibodies are retained while all the contaminating proteins are washed away. The specific antibodies can then be eluted with acid or high concentrations of salt and collected for further processing. The purified antibodies can then be conjugated with a dye for immunofluorescence studies.

An advantage of antisera over monoclonal antibodies results from the heterogeneity of the antibodies. Since one antigen may have several sites that can be recognized by antibodies, an antiserum usually contains a mixture of antibodies that bind to different parts of the antigen molecule. By contrast, a monoclonal antibody is a single antibody that will bind to only one determinant on the antigen. When using an antiserum, multiple sites of the antigen may be bound by different antibodies simultaneously. As a result, the signal produced by an antiserum is often brighter than that produced by a monoclonal antibody. Although monoclonal antibodies have supplanted antisera for many immunofluorescence applications, affinity-purified antisera are used extensively as second-step reagents in order to amplify the immunofluorescent signal.

Monoclonal Antibodies

Hybridoma technology focuses primarily on the cells secreting antibodies, rather than on the antibodies themselves (see Fig. 1). With monoclonal antibodies, the immune response is separated into its component parts. The cells secreting the antibodies rather than the antibodies themselves are the focus of attention (illustrated as circles in the schematic drawing). The cells responding to the antigenic challenge are fused with continuously growing myeloma cells. These hybridoma cells are separated by cloning into individual wells so that each antibody secreting cell is isolated from the rest. Only the cells that have reactivity with antigen 1 are selected for further growth, while the rest are destroyed. The selected cells are usually cloned a second or third time and are frozen in liquid nitrogen for storage and retrieval at a later time. The antibodies secreted from these hybridoma cell

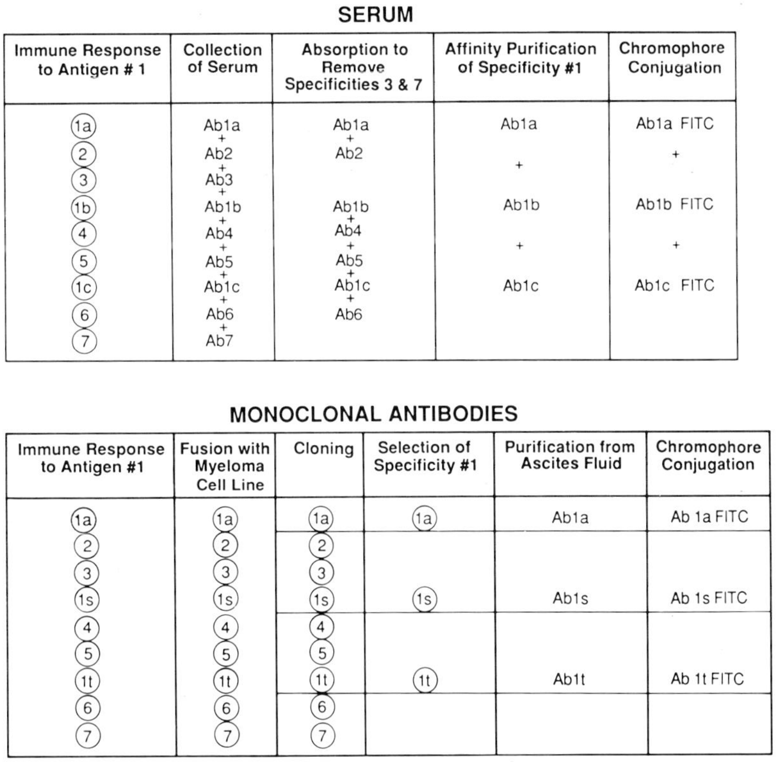

SERUM

Immune Response to Antigen # 1	Collection of Serum	Absorption to Remove Specificities 3 & 7	Affinity Purification of Specificity #1	Chromophore Conjugation
(1a)	Ab1a +	Ab1a +	Ab1a	Ab1a FITC
(2)	Ab2 +	Ab2	+	+
(3)	Ab3 +			
(1b)	Ab1b +	Ab1b +	Ab1b	Ab1b FITC
(4)	Ab4 +	Ab4 +	+	+
(5)	Ab5 +	Ab5 +		
(1c)	Ab1c +	Ab1c +	Ab1c	Ab1c FITC
(6)	Ab6 +	Ab6		
(7)	Ab7			

MONOCLONAL ANTIBODIES

Immune Response to Antigen #1	Fusion with Myeloma Cell Line	Cloning	Selection of Specificity #1	Purification from Ascites Fluid	Chromophore Conjugation
(1a)	(1a)	(1a)	(1a)	Ab1a	Ab 1a FITC
(2)	(2)	(2)			
(3)	(3)	(3)			
(1s)	(1s)	(1s)	(1s)	Ab1s	Ab 1s FITC
(4)	(4)	(4)			
(5)	(5)	(5)			
(1t)	(1t)	(1t)	(1t)	Ab1t	Ab 1t FITC
(6)	(6)	(6)			
(7)	(7)	(7)			

Fig. 1. Schematic comparison of antisera and monoclonal antibodies. Many B cells respond to the antigenic challenge with the specific responses illustrated as 1a, 1b, and 1c. The collected serum can be made specific by absorption. The immunoglobulins can then be purified and conjugated with fluorochromes. During monoclonal antibody production, the individual secreting plasma cells are immortalized by fusion with a myeloma cell line. Unique antibodies can be isolated by cloning these cells and selecting the appropriate specificity.

lines can be collected from the tissue-culture supernatant or from ascites fluid grown in mice. These antibodies can be purified and conjugated with a chromophore for immunofluorescent studies.

Each monoclonal antibody selected may be different in its binding characteristics. Some of the antibodies identified in the antiserum, as noted by Ab1a, may be found in the hybridomas generated by the fusion. Other monoclonal antibodies, Ab1s and Ab1t, may not be identified in the serum either because they represent such a small portion of the immune response or because they were not stimulated by the antigen at that particular time. The lymphocytes of the immunized animal responding to the antigen are fused with a continuously growing myeloma cell that "immortalizes" the antibody-secreting cell. The cells are isolated into individual tissue culture wells, thereby segregating the immune response into its component parts. Through a rigorous process of selection, only the cells secreting antibodies with the desired characteristics are saved. These separate cell lines, each secreting a single antibody, can be stored indefinitely in liquid nitrogen. The antibodies can be purified from the tissue culture fluid in which the cells are grown or can be isolated from ascites fluid produced in mice. The details of producing, screening and maintaining monoclonal antibodies have been described in several reviews [11,17,30,34,66].

Once a useful monoclonal antibody has been generated, an infinite supply of the identical reagent can be made for use by any laboratory. The monoclonal antibodies can be purified from ascites fluid using conventional biochemical techniques rather than by affinity chromatography. This is important since many antigens may not be abundant enough or in a form to be able to make an affinity column. Since it is possible to work with homogeneous antibodies, the relative amount of nonspecific reactivity can be reduced.

Using hybridoma technology, it is possible to generate reagents that could not be made in a conventional manner. Pure antibodies can be generated from impure antigens because the immune response is separated into its individual component parts. Even if multiple antigens are presented to the animal, each monoclonal antibody will recognize only one of the many antigens used in the immunization procedure. With proper screening procedures, it is possible to identify monoclonal antibodies produced by relatively weak immunogens, even in the presence of very immunogenic material. For antisera to be specific, the immunizing antigen must be pure in order to eliminate contaminating antibodies that might interfere with the specificity required of the reagent. For monoclonal antibodies, this requirement is eliminated, since all unwanted reactivities are ignored and only the antibodies with the desired characteristics are preserved. It should be noted that the repertoire of individual antibodies isolated from the hybridomas may not reflect the repertoire of the antibodies which comprise an antiserum. The specificity of a particular monoclonal antibody may not be identifiable in the antiserum, since that single antibody may comprise a small part of the total antibody response. Once an antibody is obtained, it can be used to purify the antigen from impure starting material by affinity chromatography. This purified antigen might then be used for subsequent immunizations and production of second generation monoclonal antibodies [77].

Usually more than one antibody from more than one hybridoma with a defined specificity is obtained from a single fusion. This permits the selection of the antibody with the best characteristics for further use. Some properties that may affect usefulness include stability in solution, stability after freezing and thawing, ease of conjugation with chromophores, ability to be used for complement-mediated cell killing, or ability to bind to specific phagocytic cells through the C region of the antibody (Fc binding).

The affinity of the antibody for the antigen is another property that affects the usefulness of the antibody. When antibodies bind to the antigen with both combining sites, the effective affinity (termed avidity) is much higher than when only one binding site is used. The affinity becomes critical when only a single combining site is used. An illustration of the effect of affinity on immunofluorescence comes from the reactivity of different anti-BrdU antibodies with BrdU-substituted DNA [4,80]. When DNA is synthesized in the presence of high concentrations of BrdU, the substituted bases are clustered. Both high- and low-affinity anti-BrdU monoclonal antibodies can be used to detect the thymidine analogue. However, when BrdU is mixed with thymidine in a ratio of 1 : 100, the substituted bases are physically separated so that an antibody can bind only with a single combining site. Under these conditions, only anti-BrdU antibodies with high affinities can be retained on the sparsely substituted DNA.

A hybridoma can be driven to produce an antibody with the desired characteristics. Mutations occur within a population of hybridoma cells at a rate of 10^{-6}–10^{-7} [53]. These mutations can affect either the combining site or the heavy-chain class of the antibody [83]. These mutant cells, or variants, can be sequentially selected by cell-sorting techniques [10,32,42,55]. A switch of heavy-chain class can be performed by isolating the very few cells in the hybridoma population that bind an antibody reactive with the desired heavy-chain class. Variants that have a changed affinity can also be selected by successive sorting under the appropriate experimental conditions [83]. Strategies for isolation of these rare cells are discussed in Chapter 34.

Labeling Techniques

The procedure used for cell staining, that is, whether to use direct or indirect staining procedures, can affect the intensity of the resulting signal. With direct staining, the primary antibody is conjugated directly to a chromophore. This procedure yields less intense staining but also gives less nonspecific binding. With indirect staining, the primary antibody is identified by secondary reagents that generate brighter cell stains. But the use of a second step to develop the primary antibody is limited by an increase in nonspecific binding. A balance between increased signal and increased background must be maintained. Detailed staining protocols have been described elsewhere [25].

Several types of secondary reagents can be used to enhance the immunofluorescence signal. The most frequently used second-step reagents are antisera that recognize species differences between immunoglobulins (heteroantisera). These sera require absorption to be specific and should be affinity purified to avoid increasing the background staining. The removal of the constant regions by enzymatic digestion to form the divalent $F(ab')_2$ molecule reduces the nonspecific binding of antisera [61]. Monoclonal antibodies which recognize species differences in either the heavy or light chains of the immunoglobulins can be used as secondary reagents [31,38]. Sera or monoclonal antibodies that identify genetic differences between immunoglobulins produced from different mice (antiallotypes) can also be used as second-step antibodies [60]. The monoclonal antibodies are not as bright as antisera but usually have low background staining. It is possible to mix several different monoclonal antibodies together, simulating an antiserum having defined composition. These mixtures produce a bright signal with low nonspecific binding.

Small molecules, called haptens, can be coupled to the primary antibody and then become the targets of secondary antibodies. Arsinylate and trinitrophenol have been used as haptens to produce sensitive, low background immunofluorescence staining [7]. The disadvantage is the complexity of the procedure required for conjugation of the haptens to the primary antibody. Biotin is another hapten that is used for amplification of the immunofluorescence signal [3]. Biotin is easy to conjugate to antibodies and forms a stable complex with avidin [6]. Avidin is a tetramer that is readily coupled with fluorochromes without losing its capacity to bind to biotin. Avidin is purified from two sources: hen egg albumen and *Streptomyces avidini.* With some preparations of avidin obtained from hen egg albumen, there is high nonspecific binding to monocytes, lymphocytes, and dead cells. This nonspecific binding is greatly reduced when Strept–avidin is used as a second step (an example is shown later in Fig. 4).

An enormous amplification of the immunofluorescence can be obtained when antibodies are coupled to a fluorescent bead [62]. This technique has been used to identify rare positive cells in a background of negative cells. The disadvantage of this procedure is loss of the quantitative relationship between fluorescence and antigen. The use of fluorescent microspheres in analytical cytometry is discussed in Chapter 19.

Selection of Fluorochromes

The spectral properties of the chromophores affect their usefulness and sensitivity as immunofluorescent tags. First, the absorption spectrum must match the spectral output of the light source. Fluorescein, with an absorption maximum at 495 nm, is well matched to the 488-nm line obtained from the argon ion laser. Rhodamine is efficiently excited by the 546-nm light from a mercury arc lamp but is poorly stimulated by an argon ion laser. One way of assessing the relative sensitivities of chromophores as immunofluorescence labels is to compare their extinction coefficients at the wavelength of excitation. The extinction coefficient is a measure of the probability of absorbing a photon of light. While fluorescein has an extinction coefficient of 86,000 at 488 nm, R-phycoerythrin has an extinction coefficient of 2,000,000 when measured at 488 nm [21] (Fig. 2).

This difference in extinction coefficient is reduced by conjugating multiple fluorescein molecules to a single antibody molecule. Dye to protein ratios of 5 are not uncommon for fluorescein conjugated reagents. Since the phycobiliproteins are as large or larger than immunoglobulins, there can be several antibodies coupled to a single phycoerythrin molecule, resulting in dye–antibody ratios of <1.

Once the light energy has been absorbed by the dye, the energy must be efficiently converted to emitted light (fluo-

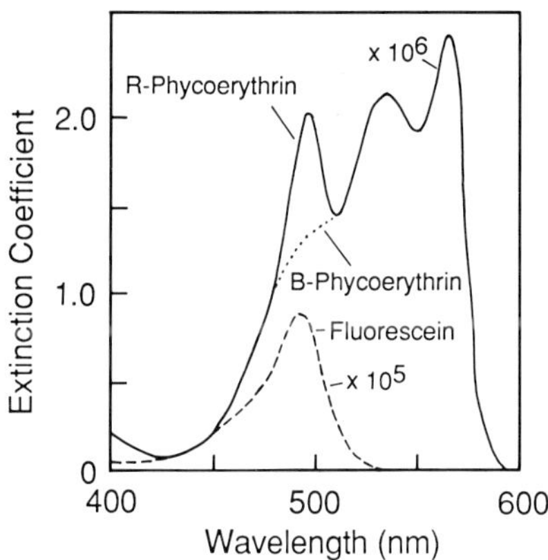

Fig. 2. Comparison of absorption spectra of fluorescein, B phycoerythrin, and R phycoerythrin. The scale for the two phycoerythrin spectra is 10 times greater than the scale for fluorescein, illustrating that these two molecules are very efficient at absorbing blue-green light. The primary difference between B and R phycoerythrin is in the additional peak at 490 nm observed with R phycoerythrin.

rescence) for optimal use as an immunofluorescent probe. A measure of the efficiency of converting absorbed photons of light to emitted photons of light is the quantum yield. The quantum yield is defined as the probability of emitting a photon once a photon of light is absorbed. B phycoerythrin (quantum yield of 0.98) is extremely efficient in energy conversion and will emit a photon essentially every time it absorbs a photon [21]. Fluorescein (quantum yield 0.5), will emit an average of one photon for every two photons absorbed [21]. Therefore, phycoerythrin is a better immunofluorescent tag as compared with fluorescein, with respect to both extinction coefficient and quantum yield.

The emission spectrum of the dye also contributes to its sensitivity. There must be a sufficient difference between the exciting and emitting wavelengths of light (Stokes shift) to permit easy discrimination of relatively dim fluorescent light from the intense excitation light. The difference in sensitivity of the various chromophores is a combination of all these spectral factors. Five dyes have been used extensively as immunofluorescence probes, each with its own advantages and disadvantages (for a comparison of the spectra of these chromophores, see Parks et al. [63].

1. *Fluorescein:* The extensive experience with this dye makes it the standard of immunofluorescence probes. It is easy to conjugate to proteins via the isothiocyanate derivative [16]. The absorption maximum at 495 nm is well matched to the 488-nm output of the argon ion laser. The quantum yield is high, at 0.5. When used in fluorescence microscopy, fluorescein fades quickly.
2. *Tetramethyl rhodamine:* This dye has been almost completely replaced in flow cytometric applications by phycoerythrin. It is inefficiently excited by ion lasers since the absorption spectrum does not match the light output of laser light sources. It is much better suited for fluorescence microscopy or for flow cytometers using halogen lamps, since the light output from a mercury arc (546-nm) light is at the peak of absorption of this dye. In contrast with fluorescein, rhodamine does not fade quickly in the fluorescent microscope.
3. *Phycoerythrin:* This large fluorochrome (240,000 daltons) is a member of a family of natural dyes, the phycobiliproteins, derived from photosynthetic cyanobacteria and red algae [15]. The two forms, B and R, are obtained from different species and have almost identical emission properties. R-phycoerythrin has an extra absorption peak at 490 nm as compared with B phycoerythrin, making it more efficient for use with 488-nm light from an argon laser (see Fig. 2). Phycoerythrin is almost ideal for two-color fluorescence when matched with fluorescein, since both can be excited with a single laser at 488 nm and the emission of the dyes are well separated (575 and 530 nm, respectively [44,59]. The absorption peak at 546 nm makes phycoerythrin a very bright dye to be used with mercury arc excitation. It does, however, fade quickly in the fluorescent microscope. For use as a single color of immunofluorescence excited at 488 nm, the sensitivity of the dyes can be ranked in the following order: R phycoerythrin > B phycoerythrin > fluorescein. Maximal sensitivity using phycoerythrin can be achieved by tuning the argon ion laser to 514 or 528 nm for excitation. (Two-color immunofluorescence in combination with fluorescein, however, is not possible at these wavelengths.) The conjugation of phycoerythrin to proteins is not as simple as FITC. Crosslinking groups such as SPDP or SMPB have been used for stable conjugate formation [21]. These conjugates are stable at 4°C for months or even a year.
4. *Texas Red:* This rhodamine derivative is more soluble than a similar compound, X-RITC, and has an absorption maximum at 596 nm [74]. Although it originally was used for two-color immunofluorescence excited with a krypton laser (576 nm), it is more efficiently excited with a dye laser tuned to 595 nm. This dye can be used in combination with fluorescein and phycoerythrin to permit three-color immunofluorescence, or it can be paired with allophycocyanin to perform four-color immunofluorescence [48] (also R.R. Hardy and D.R. Parks, personal communication). Texas Red conjugated to avidin can be purchased commercially (Molecular Probes, Junction City, OR). The direct conjugation of Texas Red to monoclonal antibodies has been described [74]; however, the general use of direct conjugates has been questioned [63].
5. *Allophycocyanin:* This is another member of the phycobiliprotein family, which absorbs maximally at 650 nm and emits at 660 nm. This molecule is smaller than phycoerythrin (104,000 daltons) with a lower maximum extinction coefficient, 700,000 at 650 nm [21]. The quantum yield is also slightly lower than phycoerythrin (0.68). This dye can be excited by a dye laser (600–640 nm), a helium–neon laser (633 nm), or a krypton laser (647 nm) [43,70]. It has been used in both three- and four-color immunofluorescence experiments matched with fluorescein, phycoerythrin, and Texas Red [64] (also R.R. Hardy and D.R. Parks, personal communication). The procedure used to conjugate this dye to proteins is the same as for phycoerythrin.

The size of the conjugate is also an important consideration, especially working with the phycobiliproteins. The IgG antibody (150,000 daltons) is smaller than phycoerythrin (240,000 daltons) so that careful consideration must be given to the stoichiometry of the reaction since the complex may be so large that it is no longer soluble. Phycoerythrin conjugates of IgM monoclonal antibodies (1,000,000 daltons) have not been successful because the conjugate is so large it is no longer soluble.

The stoichiometry between antibody and dye does affect the final sensitivity of the immunofluorescent reagent. Increasing the number of fluorescein molecules on an antibody does increase the fluorescence of the conjugate. However, dye–dye interactions that can cause quenching of fluorescence or that change the emission spectrum of the conjugate, can be observed at high dye-to-antibody ratios [3]. Aggregates of antibodies can also form at high conjugation ratios that increase the nonspecific binding of the reagent. This means that for each antibody, there is an optimum number of fluorochromes that can be attached that will yield the greatest fluorescence with the smallest amount of nonspecific binding.

Once the positive cells have been made as bright as possible, the next step in improving their discrimination from the rest of the cells lies in the reduction of background or noise coming from the unstained cells.

MINIMIZING SAMPLE BACKGROUND AND NOISE

In most immunofluorescence systems, the limiting factor of sensitivity is not the instrument but the sample (D. Recktenwald, personal communication, 1986). This means that the fluorescence associated with the unstained cells in the sample is greater than the optical and electronic noise. This fluorescence results from two sources: nonspecific binding of the reagents to the negative cells and endogenous chromophores within the unstained cells (termed autofluorescence). The techniques used to enhance the signal to make the positive cells brigher may increase the nonspecific staining of the negative cells. A real enhancement of signal to background requires the reduction of the background staining and inherent fluorescence of the negative cells in concert with the amplification of the positive signal.

Minimizing Background

The exquisite specificity of antibodies targeting fluorochromes to specific cells is dependent on the binding taking place through the combining site. If the antibody binds to cells by any other part of the molecule, the specificity is lost. The most common type of nonspecific binding is through the constant portion of the molecule, the Fc region. Certain cells, such as macrophages, monocytes, and neutrophils, as well as B cells, natural killer (NK) cells, and some T cells, have receptor molecules for the Fc region of specific subclasses of antibodies. These receptor molecules are required for the elmination of the antigen–antibody complexes from the body. The Fc receptor molecules bind antibodies tightly through the constant region without regard to the specificity of the combining site. Dead cells also bind antibodies nonspecifically. Since these cells are often included in the heterogeneous populations studied in blood and tissue, the reagent binding may appear to define a particular subset of cells. However, the staining may just reflect the inclusion of the Fc receptor-bearing leukocytes or dead cells.

Several steps can be taken to reduce this nonspecific binding. With antisera it is essential to purify the desired antibodies as much as possible. Thus, affinity purification before conjugation to the dye is highly desirable. Since most antibodies in sera are not of the desired specificity, removing them from the reagent will reduce their contribution to nonspecific binding. It is also essential to remove other serum proteins since they also may bind to cells and confuse the specific reactivity identified by the antibodies.

An increase in nonspecific binding of the reagents may also arise during the purification and conjugation procedures where antibodies may form aggregates. Although these complexes may greatly increase the signal on positive cells, they also have increased binding to Fc receptors. It is important to remove this aggregated material from reagents before use by centrifugation for 10–15 min at 20,000–50,000*g* in a microfuge or airfuge [25].

The constant region can be cleaved from the rest of the molecule by digestion with thiol-free papain or pepsin, in order to eliminate Fc binding [61]. This leaves a divalent $F(ab')_2$ molecule with binding characteristics similar to those of the whole molecule. This smaller molecule can then be coupled with dyes for immunofluorescence studies. The procedure can be used both with antisera and with purified monoclonal antibodies.

A different strategy can be used with monoclonal antibodies for reducing Fc binding. It is possible to change or to "switch" the Ig constant region to one having more desirable characteristics. By isolating the rare variants that have changed their C region, new clones can be derived that may be more useful [32]. It is also possible to generate mutants that have lost segments of the constant region [83]. These mutants lose biological activities, such as Fc receptor binding and complement fixation associated with the deleted part of the molecule, while retaining their antigen-binding ability. Monoclonal antibodies also lend themselves to manipulation using genetic engineering techniques. It is possible to attach specific combining sites of one species with the constant regions of another species [52,57]. By combining the specificity conferring V regions derived from murine immunoglobulin with human immunoglobulin C regions, a new "chimeric" antibody is formed. This hybrid antibody is expected to have reduced immunogenicity when used in immunotherapy.

Genetic engineering can be taken one step further. Once the sequence of amino acids of an antibody has been determined, it is possible to synthesize just the combining site and place it in a framework set of amino acids that retain the required three-dimensional structure [58]. It can be envisioned that specific combining sites could be attached to any molecule [58]. This synthetic molecule could then be targeted to a cell type without the requirement for an antibody, with its potential nonspecific reactivities.

An assessment of positive versus background staining can only be made by comparing the labeled cells to proper control samples. Such controls must be run with any quantitative immunofluorescence experiment. When using a direct immunofluorescence staining procedure, the staining observed with the experimental antibody must be compared with the fluorescence from unstained cells and with the fluorescence from cells reacted with a fluorochrome-conjugated irrelevant antibody of the same isotype as the experimental antibody. The indirect procedure requires a different set of controls. The experimental stained cells must be compared with unstained cells and to cells exposed to nonreactive antibody of the same isotype as the experimental antibody followed by the second-step reagent. It is also advisable to add a sample of cells stained with the second step reagent alone as a control. Most nonspecific staining is caused by the second step rather than by the first step.

When performing multicolor immunofluorescence studies, it is also necessary to include controls for assessing reagent and dye interactions. The correction required for spectral overlap in two-color immunofluorescence studies (compensation) makes the assumption that the binding and associated fluorescence from each reagent are independent [49].

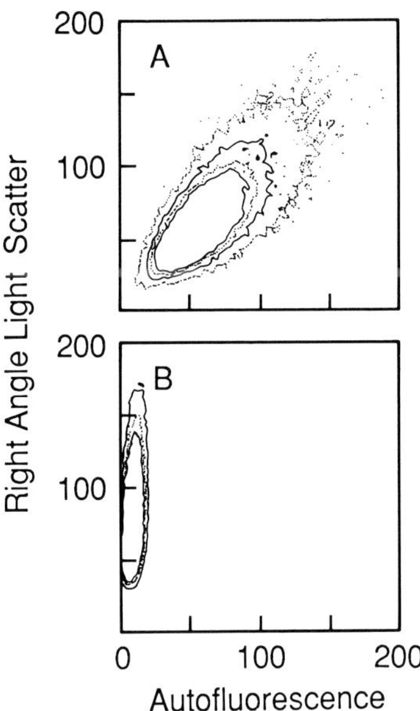

Fig. 3. Autofluorescence from the murine cell line, 1591. **A:** A positive correlation exists between the amount of autofluorescence excited at 488 nm and right-angle light scattering. **B:** This fluorescence was quenched by fixing the cells with OsO_4. The fixation did not change the right-angle light-scattering signal, indicating that the autofluorescence was not the result of light scattered from the cells.

This means that there is no steric hinderance or enhancement of binding of one reagent with another. There also must be no dye–dye interactions such as energy transfer or quenching between the two reagents. The independence of reagent binding can be determined by staining the cells with each reagent alone as well as in combination with the others. The histogram of the single-color labeled cells should be identical to the histogram of that same color when a second (or more) reagent is added. By comparing the histograms from each (the single stains with those of the multiple stains), it is possible to identify changes in intensity that result from reagent interactions.

Reducing Autofluorescence

Once nonspecific reagent binding has been effectively reduced or eliminated from the experiments, cellular autofluorescence becomes the sensitivity limiting factor in immunofluorescence studies. Autofluorescence is observed on unstained cells and has been shown to be true fluorescence, since it can be quenched and is not a result of light scattering (Fig. 3). The fluorescence of live cells is significantly above the noise observed with nonfluorescent, OsO_4 fixed cells (1% for 15 min on ice). The amount of autofluorescence is dependent on the cell type and the activation state of the cell. Certain tissue culture cell lines and macrophages have intensities of autofluorescence that are so high that it is almost impossible to discriminate positive staining above the background.

The fluorescence observed with unstained cells has been attributed to flavins and other coenzymes [5]. The emission spectra of autofluorescence is extremely broad in comparison to the emission from the chromophores used in immunofluorescence. Therefore, the contribution of specific fluorescence can be maximized by using appropriate bandpass filters centered on the peak emissions of the fluorochromes rather than using longpass filters. The bandwidth of the optical filter should be selected to maximize the fluorescence of the chromophore while minimizing the autofluorescence.

The amount of autofluorescence from the cells is also dependent upon the wavelength of excitation. The greatest autofluorescence is observed when ultraviolet (UV) or blue light is used to excite the chromophores. Less fluorescence from unstained cells is detectable with green or red excitation. The study of alveolar macrophages has been limited because of the intense autofluorescence observed on unstained cells. This fluorescence can be dramatically reduced by using red absorbing and emitting allophycocyanin rather than fluorescein or phycoerythrin (Fig. 4).

A technique for subtracting the contribution of autofluorescence from a cell population has been described [72]. This requires one laser tuned to a wavelength that excites autofluorescence separately from immunofluorescence. The autofluorescence contribution is then subtracted from the total fluorescence (autofluorescence + immunofluorescence) stimulated by the second laser. An assumption in applying this technique is that the proportionality of the autofluorescences stimulated at the two wavelengths does not change from cell to cell.

Once nonspecific binding and autofluorescence have been eliminated, instrument noise becomes the sensitivity limiting factor in flow cytometry. Although a complete discussion of the effects of instrument noise on fluorescence is beyond the scope of this chapter, there are two sources of noise that can significantly alter immunofluorescence data: inefficient optical filters and incorrect compensation of spectral overlap. The correct selection of optical filters is essential not only for maximal sensitivity but also for complete elimination of the exciting laser light. Without the proper optical filters, light scattered by the cells could enter the fluorescence channel. Since this scattered light differs for different cell types, subpopulations of cells might be identified by their light scattering properties rather than by specific immunofluorescence. A reduction of the exciting light by a factor of 10^{-5} is required to separate scattered light from immunofluorescence [46].

In multicolor immunofluorescence, it is necessary to compensate properly for overlapping emission spectra that cannot be eliminated by optical filters [49]. This is accomplished by electronically subtracting the fluorescence contribution of each fluorochrome identified in the "wrong" fluorescence channels. The compensation network must be tested with single-color samples each time the instrument is used, in order to establish the accuracy of the settings. Insufficient compensation results in the identification of single-labeled cells as double labeled (Fig. 5). With too much compensation, the reverse is observed: double-labeled cells are counted as having only one color.

OTHER TECHNIQUES FOR INCREASING SIGNAL TO NOISE

Increasing the relative frequency of the desired cells in a heterogeneous population enhances their discrimination from the remainder of the cells. This can be accomplished by using a second or third parameter on the flow cytometer to aid in identifying the cells under study, or by purifying the cells prior to immunofluorescence studies (see Chapter 11). In either procedure, the background contributed by the unwanted cells is decreased.

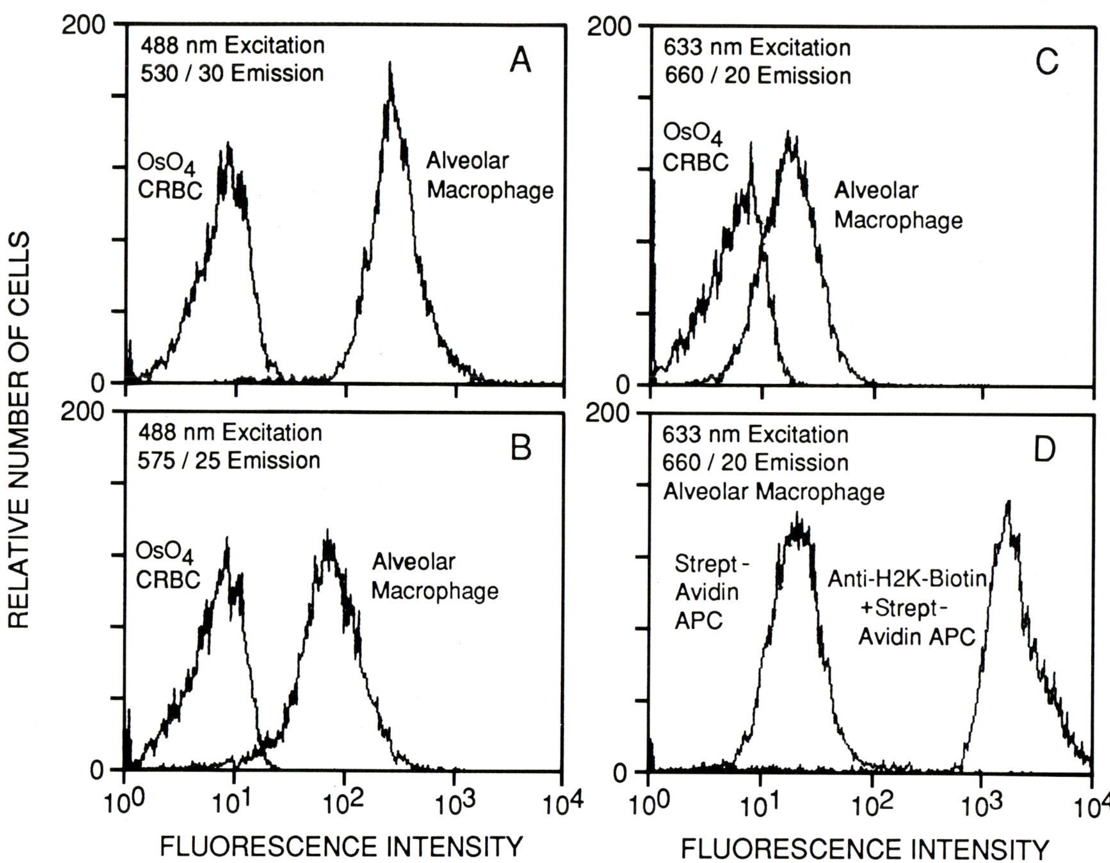

Fig. 4. Reduction of autofluorescence by high wavelength excitation. Highly fluorescent murine alveolar macrophages that had not been stained were compared with nonfluorescent OsO_4-fixed chicken red blood cells. (OsO_4 CRBC). The excitation was 200 mW of 488-nm light, while the emission was detected at 530 nm for fluorescein **(A)** and 575 nm for phycoerythrin **(B)**. The amount of autofluorescence was reduced in the phycoerythrin channel as compared with the fluorescein channel but was still brighter than some positive immunofluorescence stains. **C:** A similar comparison between unstained alveolar macrophage and OsO_4-fixed CRBC using a dye laser tuned to 633 nm with 200-mW power. The emission was detected at 660 nm,where allophycocyanin has peak fluorescence. Under these conditions, the fluorescence of the unstained alveolar macrophage is just above the nonfluorescent OsO_4-fixed CRBC. **D:** A positive immunofluorescent stain. The alveolar macrophages were reacted with anti-H2K–biotin followed by Strept–Avidin conjugated with allophycocyanin or with just the Strept–Avidin–APC second step alone. The fluorescence of the alveolar macrophages exposed to the second step alone had the same fluorescence as the unstained cells indicating that nonspecific reactivity of the second step was undetectable. The positively stained cells were clearly separable from the unstained cells using allophycocyanin as the immunofluorescent dye.

Multiparameter Analysis

Physical characteristics such as size and granularity are easily assessed on the flow cytometer by collecting light scattered by the cell as it passes through the laser beam. Although the cell size, as identified by forward light scattering, is a complex function of refractive index and cell diameter, it has been useful as an independent parameter to discriminate the cells of interest based on relative size [54]. Another measure of size that has been used on flow cytometers is electrical resistance [29]. Like light scattering, electrical resistance is used most frequently as a relative, rather than absolute, size measurement. (See Chapters 4 and 5, this volume.)

Right-angle light scattering is another physical parameter that can be assessed without staining the cell population [68]. Light scattered perpendicular to the laser beam measures primarily reflection from the cell rather than size or refractive index. In combination with either forward light scattering or electronic volume, it is possible to discriminate lymphocytes, monocytes, and granulocytes solely on the basis of these physical differences. Immunofluorescence can then be correlated with one of these cell lineages without physical separation.

The discrimination between cell types is particularly important in studying infrequent populations using immunofluorescence. The nonspecific binding of antibodies to monocytes is much higher than with lymphocytes. Contamination of these cells in the preparation will increase the background staining and may obscure the specific immunofluorescence. In a similar manner, the autofluorescence of monocytes and

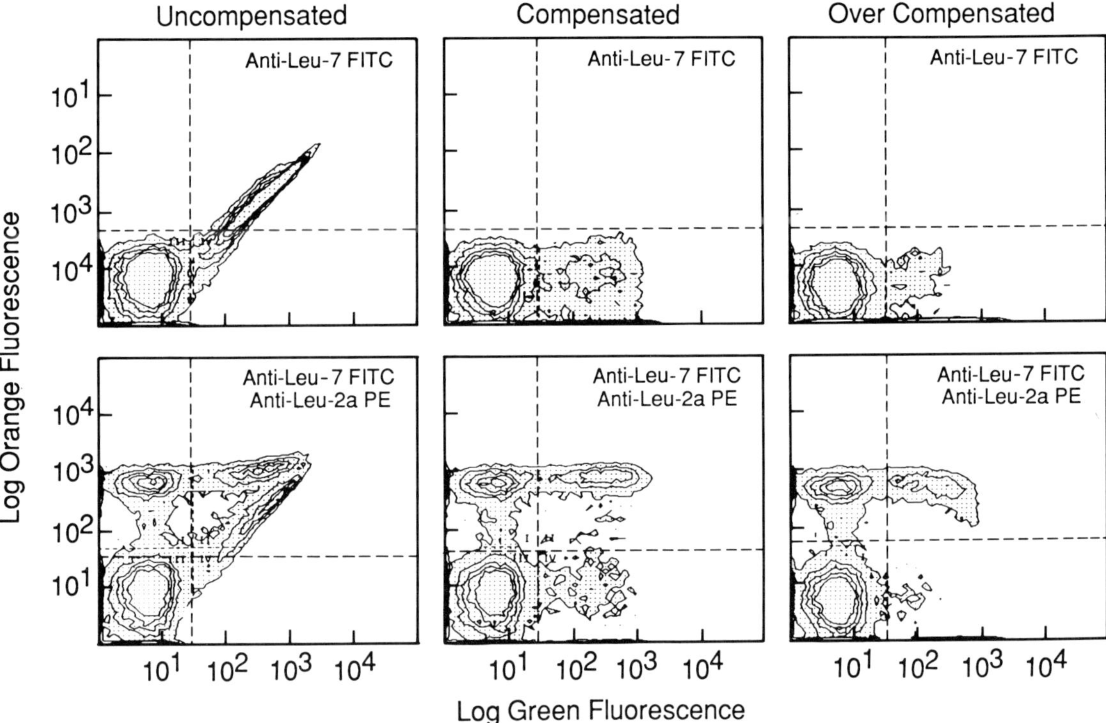

Fig. 5. Effect of incorrect setting of the spectral compensation network for two-color immunofluorescence. Peripheral blood mononuclear cells were reacted with anti-Leu-7 FITC alone (top panels) or with the combination of anti-Leu-7 FITC and anti-Leu-2a PE (bottom panels). These two cell populations were analyzed using three different settings of the fluorescence compensation network: uncompensated, correctly compensated, or over compensated. Using the 530- and 575-nm optical filters in the fluorescence channels, the compensation must be made only in one direction (fluorescein in the phycoerythrin channel) as demonstrated, in the panels designated Uncompensated. When the circuit is set for less compensation than is needed, single fluorescein-labeled cells are detected as being double labeled. With correct compensation the single FITC-labeled cells are in a horizontal line with the unstained cells in that sample. With overcompensation, some of the double-labeled cells are detected as single-labeled, while the single-labeled cells are off scale (i.e., seemingly have less fluorescence than unstained cells). (Data kindly provided by Dr. L. Lanier.)

granulocytes is much higher than that of lymphocytes. The autofluorescence coming from these cells may hide specific staining of a minor population of cells [8]. It is also possible that a contaminating cell type, such as monocytes, may express the antigen to be studied on lymphocytes [82]. Distinction between the two cell types must be made in order to achieve accurate assessment of the number of lymphocytes stained.

The inclusion of dead cells in a sample can also obscure specific labelling of a minor cell population. When working with unfixed cell samples forward light scattering can be used to discriminate live and dead lymphocytes; however, the collection angle of the scatter detector must be set very accurately (D. Recktenwald, personal communication), and this technique cannot be used with cell populations that are heterogeneous in size distribution. Dead cells can also be identified by staining with propidium iodide [35]. This dye is excluded from live cells (with intact cell membranes) but is free to diffuse into dead cells and bind to the nucleic acids. Using this technique, only the immunofluorescence of propidium iodide-negative cells is collected. Once the cells have been fixed, however, the propidium iodide technique does not discriminate between cells that had been living or dead at the time of immunofluorescence staining.

The combination of multiple cell surface markers has greatly enhanced the study of lymphocyte populations in the mouse and human. A particular cell set can be identified using one antibody, while a second, third, or fourth antibody can be used to define subsets within that set of cells. This can be performed in combination with the physical parameters of forward- and right-angle light scattering. Using these techniques it is possible to identify very infrequent cells. These multiparameter experiments require list-mode data collection to permit the correlation of multiple characteristics at one time. In list-mode data collection, all parameters for each cell are collected in a long list. The experiment can be replayed at a later time and the correlations made between different parameters of the cells. The analysis and interpretation of multidimensional data are beyond the scope of this discussion (see Chapter 22).

Enrichment of Desired Cell Type

The discrimination of leukocytes from erythrocytes in blood has been a problem for the analysis of lymphocyte subsets by immunofluorescence techniques. In contrast to the procedure just described, the most common techniques provide for physical separation of the leukocytes or lysis of the erythrocytes before analysis [8]. Erythrocytes can be lysed before or after immunofluorescence staining by using ammonium chloride [51]. Buoyant density differences can also be used to separate the cell types found in blood or bone marrow using Ficoll–Hypaque (Pharmacia, Piscataway, NJ),

Mono/Poly resolving media (Flow Laboratories, McLean, VA), or LEUCO-prep (Becton Dickinson Immunocytometry Systems, Mountain View). However, it is well to remember that abnormalities of buoyant density or fragility of the blood cells in diseased patients may lead to an incorrect assessment of cell types using these procedures.

Buoyant density has also been used to fractionate mononuclear cell preparations rather than simply separating erythrocytes and granulocytes from lymphocytes. Percoll gradients have been used to enrich the NK cells from human blood [65]. The low-density lymphocytes, which comprise about 20–25% of the mononuclear cells, contain most of the NK activity. The cell-surface phenotype of these cells can be more readily identified by enriching them from the more frequent dense cells. By performing two- and three-color immunofluorescence on the NK-enriched fraction, the identity and functional activity of minor subpopulations of both T cells and NK cells have been determined [37]. Since these cells occur at such low frequency in the blood, they could not be identified without an enrichment step.

Velocity sedimentation, which separates cells by size rather than by density, has also been used extensively as a preliminary enrichment step before immunofluorescence [12]. Size separation can be accomplished by unit gravity separation [50], by brief centrifugation in a gradient-stabilized centrifuge tube [45], or by countercurrent flow in a large centrifuge (elutriation) [67]. The fractions isolated by these procedures can then be processed for immunofluorescence studies.

SAMPLE STORAGE

Once the sample has been labeled for immunofluorescence, it must either be analyzed immediately or be fixed for later analysis. Because viable cells can undergo modulation of the cell-surface-bound antibodies, the cells must be stored on ice or in the presence of 0.1% NaN_3, or both, to inhibit the loss of the immunofluorescent tag [63]. Cell death can be reduced by suspending the cells in nutrient media in the presence of serum or serum proteins. Under these conditions, the cells remain viable for several hours. Propidium iodide can be used to discriminate live and dead cells in these samples.

Alternatively, the cells can be fixed and stored for analysis at a later time. Fixation has the additional advantage of reducing the biohazard of human samples. The fixation procedure that preserves the cells best for flow cytometry uses 1% paraformaldehyde in buffered saline [39]. The fluorescences of fluorescein, rhodamine, phycoerythrin, Texas Red, and allophycocyanin are almost unchanged from fresh samples after several days fixation when cells are stored at 4°C in the dark. (A significant increase in autofluorescence is observed after several weeks of storage [63].) This technique also preserves the forward- and right-angle light-scattering properties of the cells as well as electronic volume.

Paraformaldehyde does not destroy the lipid structure of the cells, so nucleic acid stains that require permeablization (e.g., propidium iodide) do not stain the cells quantitatively. Low concentrations of nonionic detergents have been used to disrupt the cell membrane [9,24]. The permeablization after paraformaldehyde fixation permits entrance of large molecules such as antibodies and nucleases, which can then react with the intracellular components.

SUMMARY

The study of heterogeneous populations of cells by flow cytometry is becoming easier with improved techniques for distinguishing differences between cells. These require maximizing the signal to background particularly when using immunofluorescence stains. The positive signal is completely dependent on the specificity of the reagents used. Thus, it is necessary to characterize the antibodies especially for reactivities that will interfere with the identification of the desired cell populations. The specificity of reagents has dramatically improved with the advent of monoclonal antibodies to study cell-surface antigens. Once the proper reagent has been identified, fluorescence can be maximized with indirect staining, or by mixing several monoclonal antibodies together, or by the proper selection of chromophores.

The background staining can be reduced by removing aggregates, using affinity purified antisera, selecting monoclonal antibodies with low Fc binding and/or removal of the C region by variant selection or enzymatic digestion. When working with highly autofluorescent cells it may be necessary to reduce the emission of unstained cells by excitation at long wavelengths or by subtracting the autofluorescence using multiple laser systems. The background contributed by extraneous cells can be reduced by multiparameter analysis or by physical separation of the cells before analysis.

Since each immunofluorescent stain reacts differently with a given population of cells, a systematic study of each of these procedures will help refine the method best suited for detection of the cells reactive with each stain and will increase the usefulness of that reagent.

REFERENCES

1. **Adams JM, Kemp DJ, Bernard O, Gough N, Webb E, Tyler B, Gerondakis S, Cory S (1981)** Organization and expression of murine immunoglobulin genes. Immunol Rev 59:5–32.
2. **Anstee DJ, Mawby WJ, Tanner MJ (1982)** Structural variations in human erythrocyte sialoglycoproteins. In Martonosi AN (ed), "Membranes and Transport." Vol. 2 New York: Plenum Press, pp 427–451.
3. **Bayer E, Wilchek M (1978)** The avidin–biotin complex as a tool in molecular biology. Trends Biochem Sci 3:257–260.
4. **Beisker W, Dolbeare F, Gray JW (1987)** An improved immunocytochemical procedure for high-sensitivity detection of incorporated bromodeoxyuridine. 8:235–242.
5. **Benson RC, Meyer RA, Zaruba ME, McKhann GMJ (1979)** Cellular autofluorescence—Is it due to flavins? Histochem Cytochem 27:44–55.
6. **Bonnard C, Papermaster DS, Kraehenbuhl J-P (1984)** The streptavidin–biotin bridge technique: Application in light and electron microscope immunocytochemistry. In Polak R, Varndell M (eds), "Immunolabelling for Electron Microscope." New York: Elsevier Science, pp 95–110.
7. **Cammisuli S, Wofsy L (1976)** Hapten-sandwich labelling. III. Bifunctional reagents for immunospecific labeling of cell surface antigens. J Immunol 117:1695–1707.
8. **Civin CI, Loken MR, Strauss LC, Hess A (1987)** Flow cytometric characterization of MY-10 positive progenitor cells. Antigenic analysis of hematopoiesis. VI. Exp Hematol 15:10–17.
9. **Clevenger CV, Bauer KD, Epstein AL (1985)** A method for simultaneous nuclear immunofluorescence and DNA content quantitation using monoclonal antibodies and flow cytometry. Cytometry 6:208–214.

10. **Dangl JL, Parks DR, Oi VT, Herzenberg LA (1982)** Rapid isolation of cloned isotype switch variants using fluorescence activated cell sorting. Cytometry 2:395–401.
11. **Eshhar Z (1985)** Monoclonal antibody strategy and techniques. In Springer TA (ed), "Hybridoma Technology in the Biosciences and Medicine." New York: Plenum Press, pp 3–41.
12. **Fathman CG, Small M, Herzenberg LA, Weissman IL (1975)** Thymus cell maturation. II. Differentiation of three "mature" subssclasses in vivo. Cell Immunol 15: 109–114.
13. **Galatin WM, Weissman IL, Butcher EC (1983)** A cell-surface molecule involved in organ-specific homing of lymphocytes. Nature (Lond) 304:30–35.
14. **Galfre G, Milstein C (1981)** Preparation of monoclonal antibodies: Strategies and procedures. Methods Enzymol 73:3–46.
15. **Glaser AN (1984)** Phycobilisome. A macromolecular complex optimized for light energy transfer. Biochem Biophys Acta 768:29–51.
16. **Goding JW (1976)** Conjugation of antibodies with fluorochromes: Modification of the standard methods. J Immunol Methods 13:215–236.
17. **Goding JW (1983)** "Monoclonal Antibodies: Principles and Practice." New York: Academic Press.
18. **Haber E, Margolies MN (1984)** Combining site specificity and idiotypy: A study of antidigoxin and antiarsonate antibodies. In Greene MI, Nisonoff A (eds), "The Biology of Idiotypes." New York: Plenum Press, pp 141–170.
19. **Haber E, Novotny J (1985)** The antibody combining site. In Springer TA (ed), "Hybridoma Technology in the Biosciences and Medicine." New York: Plenum Press, pp 57–76.
20. **Hardy RR (1986)** Purification and coupling of fluorescent proteins for use in flow cytometry. In Weir DM, Blackwell CC, Herzenberg LA, Herzenberg LA (ed), "Handbook of Experimental Immunology." Edinburgh: Blackwell Publications, p 31.1–31.12.
21. **Haugland RP (1985)** "Handbook of Fluorescent Probes and Research Chemicals." Junction City, OR: Molecular Probes.
22. **Honjo T, Kataoka T (1978)** Organization of immunoglobulin heavy chain genes and allelic deletion model. Proc Natl Acad Sci USA 75:2140–2145.
23. **Horan PK, Loken MR (1985)** A practical guide for the use of flow systems. In Van Dilla MA, Dean PN, Laerum OD, Melamed MR (eds), "Flow Cytometry: Instrumentation and Data Analysis." New York: Academic Press, pp 259–280.
24. **Houck DW, Loken MR (1985)** Simultaneous analysis of cell surface antigens, BrdU incorporation and DNA content. Cytometry 6:531–538.
25. **Jackson AL, Warner NL (1986)** Preparation, staining and analysis by flow cytometry of peripheral blood leukocytes. In Rose NR, Friedman H, Fahey JL (eds), "Manual of Clinical Laboratory Immunology." 3rd ed. Washington DC: American Society of Microbiology, p 226–235.
26. **Johnston RN, Beverley SM, Schimke RT (1983)** Spontaneous DHFR gene amplification shown by cell sorting with FACS. Proc Natl Acad Sci USA 80:3711–3714.
27. **Julius MH, Herzenberg LA (1974)** Isolation of antigen binding cells from unprimed mice. J Exp Med 140: 904–1004.
28. **Julius MH, Masuda T, Herzenberg LA (1972)** Demonstration that antigen binding cells are precursors of antibody producing cells after purification using a fluorescence activated cell sorter. Proc Natl Acad Sci USA 69:1934–1938.
29. **Kachel V (1979)** Electrical resistance pulse sizing (Coulter Sizing). In Melamed MR, Mullaney PF, Mendelsohn ML (eds), "Flow Cytometry and Sorting." New York: John Wiley & Sons, pp 61–104.
30. **Kennett RH, Mckearn TJ, Bechtol KB (eds) (1980)** "Monoclonal Antibodies and Hybridomas: A New Dimension in Biological Analyses." New York: Plenum Press.
31. **Kincade PW, Lee G, Lee L, Watanabe T (1981)** Monoclonal rat antibodies to murine IgM determinants. J Immunol Methods 42:17–25.
32. **Kipps TJ, Herzenberg LA (1986)** Hybridoma immunoglobulin isotype switch variant selection using the fluorescence activated cell sorter. In Weir DM, Blackwell CC, Herzenberg LA, Herzenberg LA (eds), "Handbook of Experimental Immunology." Edinburgh: Blackwell Publications, pp 109.1–109.8.
33. **Kohler G, Milstein C (1975)** Continuous cultures of fused cells secreting antibody of predefined specificity. Nature (Lond) 256:495–497.
34. **Kohler G (1981)** The technique of hybridoma production. In Lefkowits I, Pernis B (eds), "Immunological Methods." Vol 2. New York: Academic Press, pp 285–298.
35. **Krishan A (1975)** Rapid flow cytometric analysis of mammalian cell cycle by propidium iodide staining. J Cell Biol 66:188–192.
36. **Langlois RG, Bigbee WL, Jensen RH (1985)** Flow cytometric characterization of normal and variant cells with monoclonal antibodies specific for glycophorin A. J Immunol 134:4009–4014.
37. **Lanier LL, Phillips JH (1986)** Human cytotoxic lymphocytes. Immunol Today 7:132–134.
38. **Lanier LL, Guttman GA, Lewis DE, Griswald ST, Warner NL (1982)** Monoclonal antibodies against rat immunoglobulin kappa chains. Hybridoma 1:125–134.
39. **Lanier LL, Warner NL (1981)** Paraformaldehyde fixation of hematopoietic cells for quantitative flow cytometry (FACS) analysis. J Immunol Methods 47:25–30.
40. **Ledbetter JA, Frankel AE, Herzenberg LA, Herzenberg LA (1981)** Human Leu T cell differentiation antigens: Quantitative expression on normal lymphoid cells and cell lines. In Hammerling G, Hammerling U, Kearney J (eds), "Monoclonal Antibodies and T Cell Hybridomas, Perspectives and Notes." New York: Elsevier, pp 16–22.
41. **Leibson PJ, Loken MR, Panem S, Schreiber H (1979)** Clonal evolution of myeloma cells leads to quantitative changes in immunoglobulin secretion and surface antigen expression. Proc Natl Acad Sci USA 76:2937–2941.
42. **Liesegang B, Radbuch A, Rajewsky K (1978)** Isolation of myeloma variants with predefined variant surface immunoglobulin by cell sorting. Proc Natl Acad Sci USA 75:3901–3905.
43. **Loken MR, Keij J, Kelley KA (1987)** A comparison of

Helium–Neon and Dye lasers for excitation of allophycocyanin. Cytometry 8:96–100.
44. **Loken MR** (1986) Cell surface antigen and morphological characterization of leukocyte populations by flow cytometry. In Beverly P (ed), "Methods in Hematology." London: Churchill Livingstone, pp 132–144.
45. **Loken MR, Kubitschek HE** (1984) Constancy of cell buoyant density for cultured murine cells. J Cell Physiol 118:22–26.
46. **Loken MR** (1980) Evaluating optical filter efficiency in a flow cytometer. J Histochem Cytochem 28:1136–1137.
47. **Loken MR, Shah VO, Dattilio KA, Civin CI** (1987) Flow cytometric characterization of human bone marrow. I. Normal erythroid development. Blood 69: 255–263.
48. **Loken MR, Lanier LL** (1984) Three color immunofluorescence of Leu antigens on human peripheral blood using two lasers on a fluorescence activated cell sorter. Cytometry 5:151–158.
49. **Loken MR, Parks DR, Herzenberg LA** (1977) Two color immunofluorescence using a fluorescence activated cell sorter. J Histochem Cytochem 25:899–906.
50. **Miller RG, Phillips RA** (1969) Separation of cells by velocity sedimentation. J Cell Physiol 73:191–201.
51. **Mishell BB, Shiigi SM** (1980) "Selected Methods in Cellular Immunology." San Francisco: WH Freeman.
52. **Morrison SL, Johnson MJ, Herzenberg LA, Oi VT** (1984) Chimeric human antibody molecules: Mouse antigen-binding domains with human constant region domains. Proc Natl Acad Sci USA 81:6851–6855.
53. **Morrison SL, Scharff MD** (1981) Mutational events in mouse myeloma. CRC Crit Rev Immunol 3:1–22.
54. **Mullaney PF, Van Dilla MA, Dean PN** (1969) Cell sizing: A light scattering photometer for rapid volume determination. Rev Sci Instrum 40:1929–1934.
55. **Muller CE, Rajewsky K** (1983) Isolation of immunoglobulin heavy chain class switch variants from hybridoma lines secreting anti-idiotype antibodies by sequential sublining. J Immunol 131:877–881.
56. **Neuberger MS, Williams GT, Mitchel EB, Jouhal SS, Flanagan JG, Rabbitts TH** (1985) A hapten-specific chimaeric IgE antibody with human physiological effector function. Nature (Lond) 314:268–270.
57. **Neuberger MS, Williams GT, Fox RO** (1984) Recombinant antibodies possessing novel effector functions. Nature (Lond) 312:604–608.
58. **Neuberger MS, Jones PT, Dear J, Foate MS, Winter G** (1986) Replacing the complementarity determining regions in a human antibody with those from a mouse. Nature (Lond) 321:522–525.
59. **Oi VT, Glazer AN, Stryer L** (1982) Fluorescent phycobiliprotein conjugates for analysis of cells and molecules. J Cell Biol 93:981–986.
60. **Oi VT, Herzenberg LA** (1979) Localization of murine Ig-1b and Ig-1a (IgG2a) allotypic determinants defined with monoclonal antibodies. Mol Immunol 16:1005–1018.
61. **Parham P** (1986) Preparation and purification of active fragments from mouse monoclonal antibodies. In Weir DM, Blackwell CC, Herzenberg LA, Herzenberg LA (eds), "Handbook of Experimental Immunology." St. Louis: CV Mosby, pp 14.1–14.23.
62. **Parks DR, Bryan VM, Oi VT, Herzenberg LA** (1979) Antigen specific identification and cloning of hybridomas with a fluorescence activated cell sorter. Proc Natl Acad Sci USA 76:1962–1967.
63. **Parks DR, Lanier LL, Herzenberg LA** (1986) Flow cytometry and fluorescence activated cell sorting (FACS). In Weir DM, Blackwell CC, Herzenberg LA, Herzenberg LA (eds), "Handbook of Experimental Immunology." Edinburgh: Blackwell Publications, pp 29.1–29.21.
64. **Parks DR, Hardy RR, Herzenberg LA** (1984) Three-color immunofluorescence analysis of mouse B lymphocyte subpopulations. Cytometry 5:159–164.
65. **Phillips JH, Warner NL, Lanier LL** (1984) Correlation of biophysical properties and cell surface antigenic profile of Percoll gradient-separated human natural killer cells. Natl Immunol Cell Growth Reg 3:73–81.
66. **Pollock RA, Teillaud J-L, Scharff MD** (1984) Monoclonal antibodies: A powerful tool for selecting and analyzing mutations in antigens and antibodies. Annu Rev Microbiol 38:389–417.
67. **Pretlow TG, Pretlow TP** (1979) Centrifugal elutriation (counter streaming centrifugation) of cells. Cell. Biophys 1:195–205.
68. **Salzman GC, Crowell JM, Martin JC, Trujillo A, Romero A, Mullaney PF, LaBauve PM** (1975) Cell classification by laser light scattering: Identification and separation of unstained leukocytes. Acta Cytol (Praha) 19:374–377.
69. **Schroff RW, Corazon CD, Klein RA, Farrel MM, Morgan AC** (1984) Detection of intracytoplasmic antigens by flow cytometry. J Immunol Methods 70: 167–177.
70. **Shapiro HM, Glazer AN, Christenson L, Williams JM, Strom TB** (1983) Immunofluorescence measurement in a flow cytometer using low-power Helium–Neon laser excitation. Cytometry 4:276–279.
71. **Springer TA, Anderson DC** (1985) Functional and structural interrelationships among the Mac-1, LFA-1 family of leukocyte adhesion glycoproteins, and the deficiency in a novel heritable disease. In Springer TA (ed), "Hybridoma Technology in the Biosciences and Medicine." New York: Plenum Press, pp 191–200.
72. **Steinkamp JA, Stewart CC** (1986) Dual laser, differential fluorescence correction method for reducing cellular background autofluorescence. Cytometry 7: 566–574.
73. **Taupier MA, Kearney JF, Leibson PJ, Loken MR, Schreiber H** (1983) Nonrandom escape of tumor cells from immune lysis due to intra-clonal, non-heritable fluctuations in antigen expression. Cancer Res 43: 4050–4056.
74. **Titus JA, Haugland R, Sharrow SV, Segal DM** (1982) Texas Red, a hydrophilic red-emitting fluorophore for use with fluorescein in dual parameter flow microfluorometric and fluorescence microscopic studies. J Immunol Methods 50:193–201.
75. **Trask B, van den Engh G, Landegent J, Jansen in de Wal N, van der Ploeg M** (1985) Detection of DNA sequences in nuclei in suspension by in situ hybridization and dual beam flow cytometry. Science 240: 1401–1403.
76. **Trowbridg IS, Omary MB** (1981) Human cell surface glycoprotein related to cell proliferation in the receptor for transferrin. Proc Natl Acad Sci USA 78:3039–3043.
77. **Trowbridge IS, Lopez F** (1982) Monoclonal antibody

to transferrin receptor blocks transferrin binding and inhibits human tumor cell growth in vitro. Proc Natl Acad Sci USA 79:1175–1179.

78. **Uchiyama T, Broder S, Waldman TA (1981)** A monoclonal antibody (Anti-TAC) reative with activated and functionally mature human T cells. I. Production of anti-TAC monoclonal antibody and distribution of TAC(+) cells. J Immunol. 126:1393–1401.
79. **Van NT, Raber M, Barrows GH, Barlogie B (1984)** Estrogen receptor analysis by flow cytometry. Science 224:876–879.
80. **Van Der Laan M, Watkins B, Thomas C, Dolbeare F, Stanker L (1986)** Improved high affinity monoclonal antibodies to iododeoxyuridine: Cytometry 7:449–507.
81. **Warner NL (1974)** Membrane immunoglobulins and antigen receptors on B and T lymphocytes. Adv Immunol 19:67–89.
82. **Wood GS, Warner NL, Warnke RA (1983)** Anti-Leu-3/T4 antibodies react with cells of monocyte/macrophage and Langerhans lineage. J Immunol 131:212–217.
83. **Yelton DE, Scharff MD (1982)** Mutant monoclonal antibodies with alterations in biological functions. J Exp Med 156:1131–1148.

18

Ligand Binding, Endocytosis, and Processing

Robert F. Murphy
Department of Biological Sciences and Center for Fluorescence Research in Biomedical Sciences, Carnegie Mellon University, Pittsburgh, Pennsylvania 15213

INTRODUCTION

During the last 20 years, major advances have been made in our understanding of the role of membrane receptors in stimulus–response coupling through the use of ligand analogues, ultrastructural localization, membrane solubilization, and cell fractionation. These by now classic techniques employ both radioactive and electron-dense substances to produce ligand conjugates that retain biological activity and can be used to determine the biochemical characteristics of receptors. More recently, fluorescent derivatives have been used to provide additional information about the location and environment in which ligand–receptor complexes are contained. Such fluorescent derivatives can be used effectively in combination with flow cytometry to provide answers to questions that are difficult, if not impossible, to address using classic techniques. This chapter discusses the ways in which flow cytometry has been used to analyze the pathways followed by ligands, receptors, and other surface molecules during endocytosis.

LIGAND BINDING

In a sense, analysis of cells stained by immunofluorescence techniques is the earliest application of flow cytometry to the measurement of ligand binding. However, using this approach, fluorescent ligands (antibodies) are mainly used to identify cell populations, rather than to determine the characteristics of the binding site. Since the many ways in which flow cytometry can be applied in immunology have been thoroughly reviewed [31,38] (see also chapters 17 and 34, this volume), they are not discussed further here.

The initial application of flow cytometry to the analysis of ligand binding per se took place in the Biomedical Research Group at Los Alamos Scientific Laboratory. Kraemer et al. [37] used a fluorescein conjugate to measure the number of concanavalin A (Con A) binding sites on Chinese hamster ovary (CHO) cells. Cells were incubated with labeled Con A for various periods of time, washed, fixed, and analyzed either with a microfluorometer or a cell volume spectrometer. From these measurements, Kraemer and co-workers were able to measure the kinetics of binding of the ligand at 0°C. Con A binding was found to be approximately 75% specific, based on competition experiments with the inhibitor α-D-mannopyranoside. In addition, they measured the number of binding sites and the cell volume for synchronized cells at various times after mitotic selection and found that the average number of binding sites per unit surface area remained constant throughout the cell cycle. Steinkamp and Kraemer [74] reported similar findings. No attempt was made to convert measured fluorescence values into number of molecules of Con A bound per cell.

Ligand Binding at Equilibrium

A major advance occurred with the description by Bohn [5] of equilibrium ligand-binding experiments on viable cells. Bohn described measurements of Con A binding to Morris hepatoma cells and also described procedures for calculating the number of bound ligands from the measured fluorescence values. Using these methods, Bohn determined the K_A to be $4.1 \times 10^7\ M^{-1}$ (K_d of 24 nM) and the number of binding sites at saturation to be 6.6×10^6 molecules/cell. While he observed that, at high ligand concentrations (above 100 nM), the binding became linearly dependent on ligand concentration, there was no explanation for this observation. Bohn also demonstrated that ligand binding by dead cells could be eliminated by using a cell viability marker, such as fluorescein diacetate in combination with a tetramethyl rhodamine-conjugated ligand.

Bohn and Manske [7] later recognized that, at high ligand concentrations, the fluorescence from free ligand in the stream surrounding the cell becomes significant. This explains the increase in apparent binding at ligand concentrations above 100 nM observed previously [5]. These investigators measured Con A and *Ricinus communis* agglutinin binding to mouse thymoma cells; the number of binding sites ranged from 4.2×10^6 to 10.6×10^6 per cell, and the K_d ranged from 0.48 to 12 nM.

Steinkamp and Kraemer [75] determined the number of Con A-binding sites on CHO cells to be approximately 1.5×10^7 per cell; they first described the method for measurement of ligand-binding kinetics in which histograms are recorded at selected time points for a sample continuously incubated with a labeled ligand. Bohn [6] gave an excellent

Flow Cytometry and Sorting, Second Edition, pages 355–366

description of the methods for acquiring and analyzing continuous ligand binding measurements.

Murphy et al. [58] used a different approach to analyze the binding of FITC-insulin to adherent mouse 3T3 fibroblasts. Monolayer cultures were incubated with various concentrations of ligand at 0°C and removed from the tissue culture plate by incubation in PBS containing 0.5 mM EDTA; these live cells were then analyzed in flow. A low degree of specificity of binding was observed, because of both the high ligand concentrations used and the possibility of dissociation of ligand from receptor during the removal and analysis steps.

Sklar and Finney [71] used the continuous analysis methods described by Bohn [5] and Steinkamp and Kraemer [75] to analyze the binding of a fluoresceinated derivative of the peptide CHO-Nle-Leu-Phe-Nle-Tyr-Lys. These workers determined the number of receptors on human neutrophils to be approximately 70,000, with a K_d of 0.4 nM. Their results are an excellent example of how high-affinity binding of a ligand to a relatively small number of sites (<100,000 per cell) can be measured using flow cytometry. Flow cytometric analysis of the binding of fluorescein derivatives of the anaphylatoxins C5a and C3a to formaldehyde-fixed human neutrophils and monocytes has also been described [86].

Flow cytometric measurements have not been limited to the binding of peptides and proteins to cell surfaces. Notter et al. [60] used flow cytometry to measure the binding of Rous sarcoma virus to chicken embryo fibroblasts and correlated the amount of virus binding with susceptibility to viral infection. Osband et al. [63] described analysis of histamine receptors; Van et al. [84] measured the specific binding of a fluorescent analogue of estradiol to human tumor cells and found acceptable agreement with results obtained from radioreceptor assays. Measurements of binding of steroids to intracellular receptors using fluorescein conjugates are difficult to interpret, since the presence of the probe hinders passage of the ligand through cell membranes. Thus, these apparent ligand-binding rates measured using fluorescein conjugates may actually reflect an additional step of endocytosis.

Physical Basis for Discrimination of Bound and Free Ligand

The unique advantages of flow cytometry for the analysis of equilibrium binding are illustrated in Figure 1. With the appropriate combination of moderate to high ligand affinity, low free ligand concentration, and low sample volume flow rate, the fraction of total fluorescence resulting from free ligand in the stream outside the cell becomes negligible. The

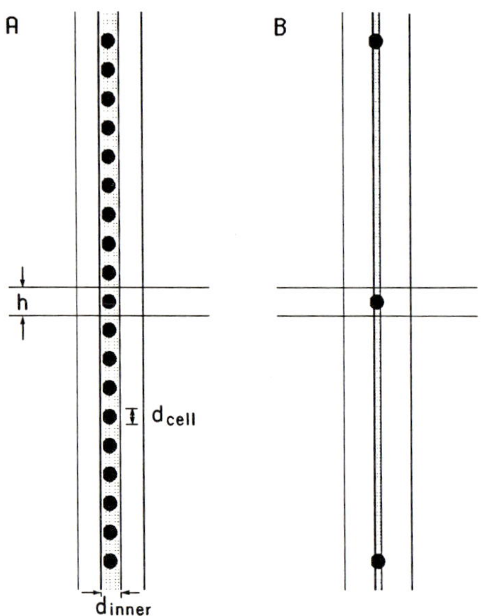

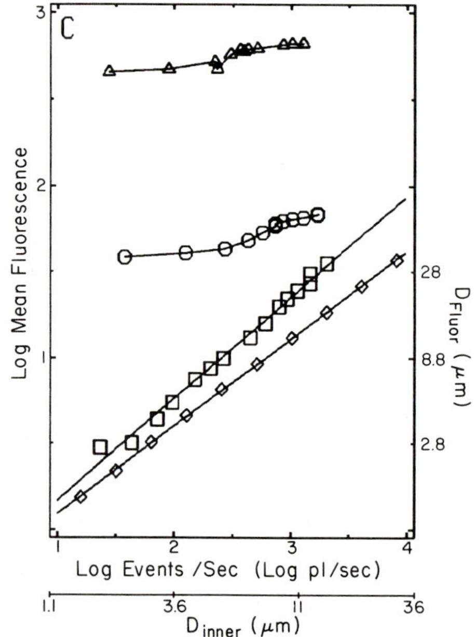

Fig. 1. Effect of flow rate on signal to noise (S/N) ratio for equilibrium ligand-binding studies using stream-in-air flow cytometry. **A:** Fast-flow configuration. The inner stream diameter is larger than the cell diameter. **B:** Slow-flow configuration. The inner stream has narrowed so that the cells appear as beads-on-a-string. Drawn to scale for the dimensions as given in the text and a 50μm nozzle, with the exception that the distance between the cells (if they are at a concentration of 10^6/ml) is actually 283 times that shown. **C:** Fluorescence from free fluorochrome as a function of sample volume flow rate; 2.02-μm microspheres (□), 7.5-μm microspheres (○) and trypsinized 3T3 cells (△) were suspended at a concentration of 10^6 per ml in PBS containing 5 μg/ml FITC–dextran (0.55 μM fluorescein). List mode data for 10,000 events were recorded on a FACS 440 using a 50-μm nozzle at various settings of the flow rate. The 3T3 cell samples were run at 0°C to prevent endocytosis of FITC–dextran. The mean fluorescence of the singlet peak was calculated for each condition. The sample volume flow rate was calculated from the observed particle flow rate; at 10^6 particles per ml, or 1 particle per pl, the sample volume flow rate (pl/sec) is equal to the event rate (particles/sec). The inner stream diameter shown on the abscissa, d_{inner}, was calculated from the sample volume flow rate by assuming a stream velocity of 10 m/sec. The inner stream diameter shown on the ordinate, d_{fluor}, was calculated from the volume necessary to produce a given fluorescence signal (using a conversion factor of 63.7 fluorescein equivalents per channel, determined using Becton Dickinson Research Center beads). The slope of the least-squares fit line for the 2.02-μm beads is close to 0.59, showing the effect of the AC-coupling circuit in the FACS (see text). Values predicted for the 2.02 μm beads (see text) are also shown (◇). Note the close agreement between these two curves.

signal-to-noise (S/N) ratio can be determined for the two cases shown:

$$\frac{S}{N} = \frac{N_b}{\pi\left(hd_{inner}^2/4 - d_{cell}^3/6\right)C_{free}10^{-15}N_0} \text{ for } d_{inner} > d_{cell} \quad (1)$$

and

$$\frac{S}{N} = \frac{N_b}{\pi/4d_{inner}^2(h - d_{cell})C_{free}10^{-15}N_0} \text{ for } d_{inner} < d_{cell} \quad (2)$$

where N_b is the number of molecules of ligand bound per cell, N_0 is Avogadro's number, h is the height of the laser beam (μm), d_{inner} is the diameter of the inner (sample) stream (μm), d_{cell} is the diameter of a cell (μm), and C_{free} is the concentration of free ligand (moles per liter). This assumes that (1) the laser beam is wider than the inner stream in all cases, (2) the sheath fluid and unlabeled cell have no fluorescence, and (3) the analog-to-digital electronics convert the entire fluorescence within the illuminated area (i.e., the pulse height (or area) determination is not AC coupled). The smallest number of receptors of a given K_d that can be detected with a given S/N ratio can be calculated if the dimensions of the cell and stream are known. Assuming a beam height of 20 μm, a cell diameter of 10 μm, and an inner stream diameter of 15 μm, Eq. (1) reduces to

$$S/N = 5.5 \times 10^{-13}N_b/C_{free} \quad (3)$$

When half-maximal binding occurs, $N_b = 0.5\ N_t$ (N_t is the total number of receptors per cell) and $C_{free} = K_d$ (K_d is the dissociation equilibrium constant). Thus, to achieve a signal to noise ratio of at least 10, N_t must be at least 3.6×10^{13} K_d. For example, a ligand with a K_d of 10 nM should have at least 360,000 receptors per cell to be detectable under these conditions. [Using Eq. (3) and the data of Bohn [5], the estimated S/N for measuring Con A binding to Morris hepatoma cells is 70.]

The effect of reducing the inner stream diameter (by decreasing the sample flow rate), can be seen by assuming the same beam height and cell diameter, but assuming an inner stream diameter of 5 μm. Equation (2) then reduces to

$$S/N = 8.5 \times 10^{-12}N_b/C_{free} \quad (4)$$

and N_t must be at least 2.4×10^{12} K_d for a S/N of 10. Under these conditions, 700,000 receptors for a ligand with a K_d of 300 nM or as few as 24,000 receptors for a ligand with a K_d of 10 nM can be detected. In practice, the autofluorescence from unlabeled cells (typically 10,000–100,000 fluorescein equivalents, depending on cell type) may become the limiting factor at these low numbers of molecules of ligand per cell (see below for discussion of a method for partially compensating for this).

Whether Eq. (1) or (2) applies in a given experiment is a function of flow cytometer type. This can be demonstrated by measurements of fluorescence from an external volume marker as a function of sample flow rate. A useful method is to mix a known number of nonfluorescent beads with a known concentration of a fluorescent probe. The average fluorescence per bead and the number of seconds required to acquire a given number of beads are determined for various settings of the sample flow rate. The fluorescence from the illuminated volume can then be plotted versus the calculated sample volume flow rate (beads per second divided by beads per unit volume) (Fig. 1C). The diameter of the inner stream can also be calculated if the stream velocity is known. The observed fluorescence should be proportional to the sample volume flow rate.

Many flow cytometers are equipped with an AC-coupling (or baseline-restoring) circuit to aid in discriminating cell-associated fluorescence from free fluorescence. This has the effect that the observed fluorescence will be (on average) the sum of the cell-associated fluorescence and the square root of the constant inner stream fluorescence. Equations (1) and (2) may be modified to take this into account by simply taking the square root of the denominator in both equations. Figure 1C shows that for small beads (2.02-μm diameter), the slope of the log–log plot is close to 0.5, as would be expected for AC coupling. For comparison, a predicted curve generated using the denominator of Eq. (1) (with the square root taken to reflect AC coupling) is also shown. This curve was generated independently from the data, using only the geometry shown in Figure 1B, the stream velocity and the fluorescence detection efficiency (fluorescein equivalents per channel, estimated using beads with known amounts of covalently-bound fluorescein).

When the square root effect is taken into account, Eq. (4) (for the slow flow case) is replaced by

$$S/N = 3 \times 10^{-6}N_b/C_{free}^{0.5} \quad (5)$$

and 680 receptors for a ligand with a K_d of 10 nM should be detectable with a S/N of 10, assuming that noise from free ligand were the only limiting factor.

It is often desirable to generate a similar curve with the cells of interest and a nonbinding probe to more closely approximate the experimental conditions. The results in Figure 1C for the larger beads and cells show a much smaller dependence of fluorescence on flow rate. One possible explanation for this observation is that the larger particles perturb the inner stream as they leave the 50-μm nozzle and that an additional volume of sample (not dependent on flow rate) is therefore present in the illuminating beam. Some of this constant fluorescence may result from nonspecific binding of FITC–dextran to the cell or bead surface; this would appear to be likely for the 3T3 cells. Since the fluorescence signal resulting from either stream perturbations or nonspecific binding is present only when the cell is in the illuminating beam, it would be converted at full efficiency (i.e., the AC-coupling circuit would not reduce it to a square root). In addition, if the effective laser beam width is actually narrower than the cell width, the dependence of total fluorescence on flow rate would be reduced. A combination of these three factors is responsible for the slopes of the curves for 7.5-μm beads and 3T3 cells.

To summarize, a number of variables affect flow cytometric discrimination of bound and free ligand. These include (1) *cellular autofluorescence,* the use of probes that are excited in the red or autofluorescence correction may reduce this problem; (2) *laser beam height and width,* decreasing either to the cell size while maintaining the same light density increases the S/N; (3) *sample volume flow rate,* decreasing the flow rate to decrease the inner stream diameter, thereby increasing the S/N; (4) *pulse detection electronics,* use of AC-coupling increases the S/N; and (5) *nozzle size,* increasing the nozzle size decreases the effectiveness of the hydrodynamic focusing but reduces perturbations induced by passage of the cell through the nozzle.

Autofluorescence Correction

Measurements of fluorescence from bound (or internalized) ligand are frequently corrected by subtracting the mean autofluorescence of unstained cells from the mean fluorescence of stained cells. The variation in the resulting difference is mainly a function of the width of the distributions whose means are being subtracted. This is often a major problem, since wide variations in cell size, which can affect both number of receptors and amount of autofluorescence, are frequently observed with cultured cells such as fibroblasts; thus, a small cell with a number of ligands bound often shows less total fluorescence than does a large unstained cell. What is needed is a method for subtracting the autofluorescence of each individual cell from the total measured fluorescence for that cell. When the autofluorescence is much greater than the signal, any attempt to make this correction becomes impossible. However, when the fluorescence above background for a labeled sample is of the same order of magnitude as the autofluorescence, it is possible to use either light-scattering measurements or measurements of autofluorescence at a wavelength at which fluorescence from the ligand is small to calculate a corrected fluorescence on a cell by cell basis [67].

An illustration of this approach for a FITC conjugate of epidermal growth factor (EGF) internalized by mouse 3T3 fibroblasts is shown in Figure 2; this approach is equally applicable to binding experiments. With 488-nm excitation and a 530-nm bandpass emission filter (30-nm bandwidth), the autofluorescence of these cells is equivalent to 20,000–35,000 molecules of fluorescein, depending on whether Becton Dickinson Research Center beads or Fluorotrol are used for the calibration. The number of EGF receptors is 13,000–22,000. Because autofluorescence shows wide excitation and emission spectra, it is possible to analyze a sample of cells and collect a parameter containing only autofluorescence and not signal from the labeled ligand. In this example, the autofluorescence is detected by a 625-nm bandpass filter (bandwidth 35 nm). 3T3 monolayers were incubated with or without 10 nM FITC–EGF for 30 min at 37°C in the presence or absence of 1-μM unlabeled EGF and then trypsinized. Forward scatter, right-angle (side) scatter, fluorescence at 530 nm [F(488,530)], and 625 nm [F(488,625)] were measured with excitation by an argon laser at 488 nm (400 mW); 10,000 live cells (selected with a forward-scatter threshold) were analyzed for each sample.

Measurements of strongly fluorescent samples (e.g., FITC-conjugated beads or cells labeled with FITC–dextran), demonstrate that only a small amount of FITC fluorescence is observed in the F(488,625). In order to correct for the amount of autofluorescence in the F(488,530) signal, therefore, only one assumption need be made: that the autofluorescence at 625 nm is proportional to the autofluorescence at 530 nm. To obtain this constant of proportionality, a control cell sample (with no FITC-EGF) is analyzed (Fig. 2A). A straight line with a correlation coefficient of 0.824 has been observed to relate the emission at 530 nm and that at 625 nm. This constant is then applied to subsequent experiments with fluorescent signal to correct for the autofluorescence. In particular, the amount of signal in the autofluorescence parameter is used to guage the amount of autofluorescence in the fluorescence parameter:

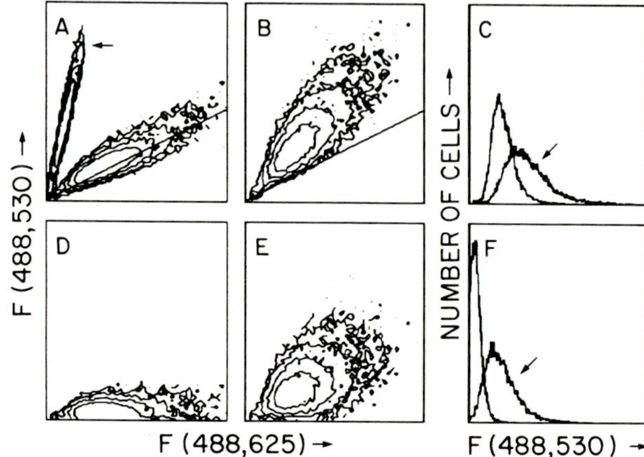

Fig. 2. Autofluorescence correction method. Dual-parameter histograms are shown for two samples, control cells **(A,D)** and cells incubated with 10 nM FITC–EGF **(B,E)**. **A:** Correlation between autofluorescence at 530 nm and at 625 nm for the unstained cells. The line shows the best-fit line relating these parameters (it is reproduced in **B** for reference). The slope of this line predicts the proportion of 625-nm fluorescence that will be present in the 530-nm signal, in addition to any "real" 530-nm fluorescence that is present. **B:** Most of the labeled cells are above this line. By correcting for autofluorescence (see text), the control distribution is shifted down to the horizontal axis, and the vertical axis becomes "real" 530-nm fluorescence **(D,E)**. Single-parameter histograms are shown for the uncorrected **(C)** and corrected **(F)** F(488,530) for the same samples (arrows indicate the histogram for the labeled sample). Note the decrease in overlap between the control and the labeled samples in the corrected histograms compared with the uncorrected histograms. The second set of contours shown in **A** (arrow) shows data for cells labeled with FITC–dextran, indicating the amount of spillover of fluorescein fluorescence into the autofluorescence channel that is present. (Reproduced from Roederer and Murphy [67], with permission of the publisher.)

$$F_{real} = F(488{,}530) - F(488{,}625) \cdot S_{625,530} \qquad (6)$$

where F_{real} is the amount of FITC fluorescence that would have been observed in the absence of any autofluorescence, F(488,530) and F(488,625) are the measured fluorescence values as described above, and $S_{625,530}$ is the spillover between them calculated from the control. This calculation is done on a cell-by-cell basis for list-mode data for the samples containing FITC–EGF; the results of all cells are combined to obtain corrected histograms (Fig. 2C,F). The success of the correction is shown by the reduced overlap between the histograms.

The degree to which noise is reduced should be related to the correlation between the actual autofluorescence at 530 nm and the parameter being used to estimate it in the control sample. For instance, when forward-angle scatter, right-angle scatter, and F(488,625) were compared as measures of the autofluorescence at 530 nm, the correlation was found to be highest with F(488,625), followed by side scatter and forward scatter. This order was observed in the quality of the correction.

The one-way correction method may be improved by also correcting for spillover from F(488,530) to F(488,625) (two-way correction), and has been extended to simultaneous correction for spillover and autofluorescence for more than one fluorochrome [67].

Calculation of Receptor Density

If binding measurements are limited to total number of ligands bound per cell, potential differences in surface receptor density may be difficult to detect. Steinkamp and Kraemer [74] described the use of a $X^{2/3}$ generator circuit to convert electronic cell volume signals into values proportional to cell surface area; this conversion is only strictly correct for spherical cells with no surface protrusions. In combination with an analog ratio circuit, this permitted them to measure the surface density of Con A sites (fluorescence/surface area) for synchronized cells at various times after mitosis. Steinkamp and Kraemer concluded that the surface receptor density remains constant during the cell cycle.

Rudolph et al. [68] used light-scatter pulse width to perform a similar normalization for transferrin receptors on K562 cells. By making simultaneous measurements (in list mode) of light-scatter pulse width, receptor number and DNA content, they found that histograms of surface transferrin receptor density for cells in G1, S, and G2 + M were essentially identical. However, these investigators showed that the receptor density was not constant within any one of these populations and concluded that receptor *number* rather than density was regulated during the cell cycle.

Receptor Clustering

As initially described by Chan et al. [10], fluorescence energy-transfer measurements by flow cytometry can be used to follow the clustering of ligand–receptor complexes, which frequently follows ligand binding. This approach has been described in detail recently [78,80] and has been used to measure changes in lectin-binding site proximity during differentiation of HL-60 leukemia cells [32]. The applications of this technique to date have all used lectins, since (1) close packing of lectins is facilitated by their multivalent nature, and (2) the large number of binding sites (typically $>10^6$ per cell) produce sufficiently large fluorescence signals to permit accurate calculation of the energy transfer efficiency. The successful application of this technique to other ligand–cell systems should permit the determination of whether specific ligands cluster before endocytosis.

LIGAND INTERNALIZATION

The earliest use of flow cytometry to study ligand internalization was in the analysis of the nonspecific adsorptive endocytosis of fluorescein-conjugated histone by CHO cells [53,54]. In order to study the fate of histone internalized by this adherent cell line, monolayer cultures were incubated with labeled histone for various periods of time, trypsinized and analyzed by flow cytometry. Trypsinization was used both to remove the cells from the culture plate and to remove surface-bound histone. (However, complete removal of all surface histone under the conditions used was not rigorously demonstrated.) Using this approach, the half-time to reach steady-state levels of internalized histone was found to be approximately 50 min. Pulse-chase experiments were used to show that the exocytosis rate was related to the length of the initial incubation with ligand; histone was released with a half-time of 30 min after a 1-hour incubation, and with a half-time of 11 hours after a longer incubation (24–48 hours). This 22-fold difference in rate is evidence that exocytosis can occur from two different compartments.

A similar approach was used to study the endocytosis of FITC–insulin and FITC–dextran by mouse 3T3 fibroblasts [58]. Internalization was assayed using trypsin, and the combination of binding and internalization was assayed using EDTA to remove the cells from the plate without removing surface label. The trypsinization conditions were shown to remove essentially all surface label. Under the conditions used (1 μM FITC–insulin), approximately 25% of the insulin binding and endocytosis occurred through receptor-mediated endocytosis. Measurements at low insulin concentrations, at which higher specificity would be expected, were made unreliable by the high level of autofluorescence observed with unlabeled cells, part of which can be eliminated using higher-quality optical filters (see above for a discussion of digital autofluorescence correction).

These methods were used to study the endocytosis of dextran, histone, insulin, and α_2-macroglobulin by fibroblast cell lines differing in growth requirements [57]. The amount of insulin endocytosis was found to be correlated with the insulin requirement for growth.

Accessibility of a ligand to an exogenous antibody can also be used as an assay for internalization [48]. The "ligands" in this case were rabbit antimouse immunoglobulin G (IgG) antibodies, and the "receptors" were mouse immunoglobulins on the surface of B lymphocytes. Approximately 40% of the rabbit antibodies bound at 0°C were endocytosed during 5–30 min at 37°C, while the remaining 60% was on the surface even after 90 min. Some of this remainder might include ligand that has been internalized and recycled to the surface.

Finney and Sklar [18] introduced an excellent technique for discriminating surface bound from internalized ligand. Their method, which can be used for ligands conjugated with a pH-sensitive fluorochrome such as fluorescein, relies on a rapid external pH shift to quench surface ligand. These workers applied this technique to the internalization of a fluoresceinated chemotactic peptide by human neutrophils and found that internalization of ligand began approximately 1 min after ligand addition and that more than 50% of the occupied receptors are internalized within 3 min. Both spectroscopic and flow cytometric measurements of chemotactic factor binding and internalization by human neutrophils have been reviewed by Sklar et al. [72].

A potential problem with the pH-drop method is that the fraction of surface ligand must be extrapolated back to the time of the pH shift, since internal ligand is also quenched (at a slower rate). Part of this problem can be overcome by acquiring kinetics data in list mode rather than using sequential histograms.

van Deurs et al. [85] used flow cytometry to measure the kinetics of endocytosis and exocytosis of FITC–dextran by mouse L-cells. They reported biphasic exocytosis kinetics, in agreement with previous results [1,3,54]. Unfortunately, they did not control for quenching of FITC–dextran by acidification. This makes it difficult to interpret their observed rates of internalization and exocytosis.

LIGAND ACIDIFICATION

Endocytosis, Acidification, and Differential Processing

One of the most interesting questions currently being addressed by cell biologists regards the existence, mechanism, and role of differential endocytic processing of surface molecules. A major stimulus to the development of this field has been the publication of a number of elegant studies on toxin [15,39,69] and virus [29,40,91] endocytosis. The results of these studies indicated a major role for acidification of endocytosed ligand in the biological activity of that ligand, in this case, achieving entry into the cytoplasm. It was also suggested that diphtheria toxin may enter the cell from nonlysosomal acidic vesicles [15,39] and that endosome acidification may play a role in recycling of some receptors [25,79,87]. Recent results have been reviewed by Steinman et al. [77], Helenius et al. [30], and Pastan and Willingham [64].

Role of Internalization and Acidification in Growth Stimulation

Insulin. While many effects of insulin on mammalian cells have been described, a classification of these effects can be made based on the cellular process affected and the kinetics with which that effect is produced [35]. At physiological concentrations (0.1–1 nM) and short times (seconds to minutes), insulin stimulates glucose transport and glucose oxidation. At higher concentrations and longer times, insulin stimulates DNA and RNA synthesis and cell multiplication. The specific binding of insulin to cells or cell membranes can also be divided into two classes: high affinity and low affinity. The latter class almost certainly represents receptors for the insulin-like growth factors (IGF). Since anti-insulin receptor antibodies, which can block insulin binding to the high-affinity receptor, can also stimulate glucose oxidation [33], it is widely and reasonably held that the short-term metabolic effects of insulin are mediated by the high-affinity receptor. However, these antibodies cannot stimulate cell growth [36], and it is not currently known whether antibodies to the IGF receptor would be able to produce this stimulation. It should therefore be stressed that insulin growth promotion may require interaction of insulin or anti-insulin receptor antibodies with intracellular sites or with the plasma membrane receptor inside an endocytic vesicle. A number of intracellular binding sites and effects of insulin have been demonstrated [24,66].

EGF. A similar resolution into early and late effects can be made for EGF [9]. The late effects may require proteolytic processing of the EGF receptor, as suggested by Das and Fox [11]. The two effects may be unrelated, since a nonmitogenic analogue of EGF can induce early, but not late, responses [90]. Monoclonal antibodies against EGF receptors induce both early and late effects [70]; however, the late effects may result from the interaction of the anti-EGF receptor antibody with intracellular sites.

Effects of lysosomotropic amines. The effect of lysosomotropic amines on endocytosis and growth factor action has been the subject of a considerable controversy in recent years. Maxfield et al. [42] and Davies et al. [12] reported that amines inhibited receptor clustering and endocytosis of α_2M and EGF and Maxfield et al. [41] reported that these compounds potentiated EGF stimulation of DNA synthesis. More recent experiments by Haigler et al. [26] and Yarden et al. [89] led to the conclusion that amines do not affect surface receptor clustering or internalization but do inhibit ligand degradation. As the latter investigators have pointed out, clustering observable by videomicroscopy is caused by fusion of endocytic vesicles within the cell rather than by movements on the cell surface, therefore, the effects of the amines on clustering are probably attributable to the inhibition of this fusion event. While the effects of primary amines on "clustering" and internalization are now understood, their effects on growth factor action are not. In contrast to Maxfield et al. [41], King et al. [34] reported that methylamine, lidocaine, chloroquine, and dansyl cadaverine inhibited EGF stimulation of DNA synthesis by confluent monolayers. Cain and Murphy [8] demonstrated that growth stimulation of subconfluent, serum-starved cells by 10% serum is inhibited by chloroquine, methylamine, ammonium chloride, tributylamine, and benzylamine and that this inhibition can be directly correlated with effect on intravesicular pH, as measured by flow cytometry. The inhibition by tributylamine and benzylamine are of particular interest since these weak bases can neutralize acidic vesicles without causing vacuolation [65].

Measurement of Acidification by Flow Cytometry

Amine-ratio method. As originally demonstrated by Ohkuma and Poole [62], fluorescein conjugates provide an excellent means of measuring the pH value of compartments containing endocytosed material. They used the ratio of fluorescein emission when excited by 495-nm and 450-nm light to measure the pH of macrophage lysosomes containing endocytosed FITC–dextran. A number of investigators have used this method in combination with spectrofluorimetry or microspectrofluorimetry (reviewed by Nuccitelli [61]).

Given the limited excitation wavelengths available in laser-based flow cytometers, the dual-excitation method is not practical. To circumvent this limitation, Murphy et al. [54] used a lysosomotropic amine, chloroquine, to enable measurement of the total amount of internalized probe. Figure 3 illustrates the basic principle of this method. Cells are allowed to internalize a fluorescein-conjugated probe for various periods of time and are then trypsinized to remove surface-bound probe. A histogram of FITC fluorescence is then acquired, using appropriate light-scattering criteria to limit measurements to viable cells. A weak base, such as chloroquine or methylamine, is then added, at a concentration sufficient to neutralize acidic compartments, to the remaining cells and a second histogram is acquired. The ratio of the mean fluorescence from these two histograms is proportional to the average pH before amine treatment. This method was used to demonstrate that ligand acidification occurs rapidly: both histone (in CHO cells [54]) and insulin (in 3T3 cells [58]) were acidified within 10 min after ligand addition. This method was also used to demonstrate that insulin is internalized and acidified by activated, but not resting, T lymphocytes [52].

Some care must be taken in using the amine-ratio method. While the method was originally developed using chloroquine, possible interactions between chloroquine and the fluorophore can make interpretation of results difficult. While chloroquine did not appear to affect the fluorescence of IAAF-histone [54] or FITC–insulin [58], significant quenching of FITC–dextran fluorescence was observed at millimolar chloroquine concentrations by Ohkuma and Poole [62]. These interactions are not observed for simpler amines, such as methylamine or ammonia. Care must also be taken in calibrating the pH using the amine-ratio method. The amine addition must reproducibly raise the pH to a known value. This value is best determined by comparison with an in vivo pH calibration curve. Such a curve can be generated by incubating cells with probe for a fixed period of time, such as 30 min, and then measuring FITC fluorescence for (1) an aliquot incubated with the neutralizing agent (e.g., 200 mM methylamine), and (2) aliquots incubated with metabolic inhibitors (200 mM 2-deoxyglucose and 40 mM sodium azide), strong buffers of known pH and an equilibrating agent such as nigericin or monensin. Differences between optimal conditions for neutralization of acidic vesicles may

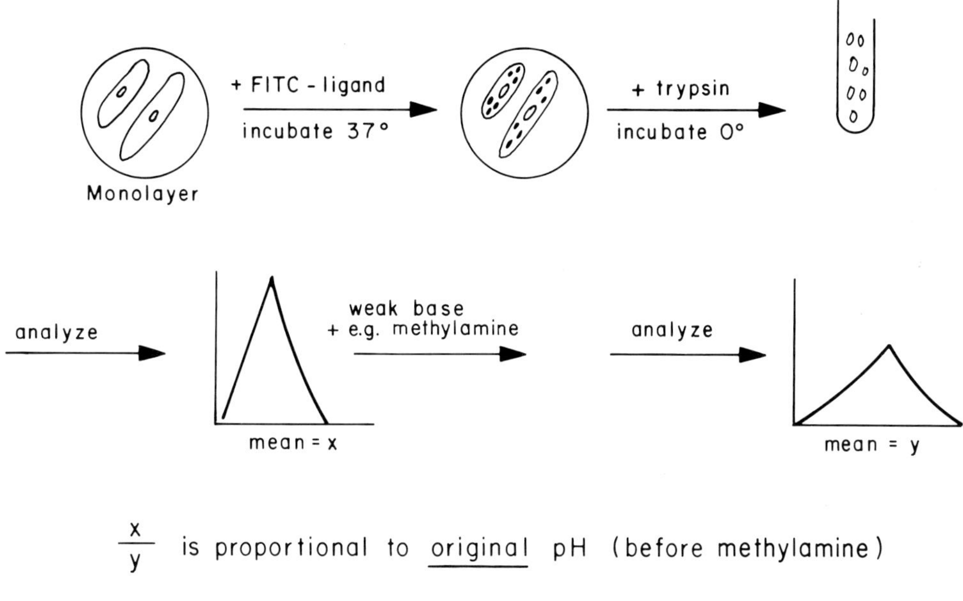

Fig. 3. Schematic for endocytic vesicle pH measurements using the amine ratio method. Note that each sample must be analyzed twice in order to obtain an estimate of the average intravesicular pH.

exist between cell types, but such conditions can normally be found for both amines and ionophores.

Dual-fluorescence method. Since sequential measurements of cell populations are required, the amine-ratio method is not capable of determining the pH distribution within a cell population. This limitation can be overcome by using the dual fluorescence (rhodamine-fluorescein) method developed by Murphy et al. [55,56]. Figure 4 illustrates this method, which uses a combination of ligands conjugated with pH-sensitive (e.g., fluorescein) and pH-insensitive (e.g., rhodamine) fluorochromes. Once again, calibration with buffers of known pH is extremely important. With careful calibration and correction for autofluorescence, in vivo and in vitro curves of fluorescein/rhodamine fluorescence versus pH should be identical. Using the dual fluorescence method, insulin was observed to be rapidly acidified to pH 6 within 10 min, and then to reach pH 5 at 30–120 min after the addition of ligand.

Tse and Pernis [81] demonstrated that the mouse histocompatibility antigen, H-2K, is internalized spontaneously by activated T lymphocytes. The dual fluorescence method was used to analyze the acidification of a monoclonal antibody directed against H-2K [59]. Activated T lymphocytes were shown to internalize and acidify this ligand with biphasic kinetics; there is an initial acidification to around pH 6, followed by a further acidification to pH 5. Similar results were obtained using the amine-ratio technique. By contrast, activated B lymphocytes did not acidify this ligand.

Metezeau et al. [49] recently described flow cytometric measurements of acidification, based on the method of Ohkuma and Poole. Successive measurements of fluorescein fluorescence excited at 458 nm and 488 nm were made, and changes in the ratio were used as a reflection of pH changes; no attempt to calibrate the ratio was made. Unfortunately, 458-nm excitation is significantly pH dependent (as compared with either 450-nm excitation, or the use of a non-pH-sensitive fluorochrome such as rhodamine); therefore, it is difficult to convert the measured ratios to accurate pH values. Because of the loss of both signals observed at low pH, it is also suitable only for highly fluorescent samples.

ANALYSIS OF ENDOCYTIC VESICLES

Fluorescein-labeled ligands and flow cytometry have been used to demonstrate that the acidification process occurs much more rapidly than had previously been believed [54,58]. Tycko and Maxfield [82] presented evidence that α_2-macroglobulin may be acidified before entry into lysosomes. Since α_2-macroglobulin is rapidly degraded in lysosomes, interpretation of these results is somewhat difficult. By contrast, transferrin is not appreciably degraded by some cell types. Van Renswoude et al. [88] demonstrated that transferrin-containing vesicles are nontheless mildly acidified, and Harding et al. [27] presented electron microscopic evidence of recycling of the transferrin receptor. To determine whether any nonlysosomal endocytic vesicles could be identified, Merion and Sly [47] used Percoll gradients to resolve homogenates from cells that had endocytosed FITC–dextran on the basis of their density. Galloway et al. [21] then found that two distinct fractions from Percoll gradients were both capable of acidifying their contents in vitro. Merion et al. [46] described a mutant CHO cell line that fails to acidify its endosomes, but not its lysosomes. To complicate matters, both fractions of clathrin-coated vesicles [20] and Golgi vesicles [23] have been demonstrated to contain proton pumps. While, like α_2-macroglobulin, EGF has been demonstrated to be degraded in lysosomes [9], at least some EGF is internalized into nonlysosomal vesicles [50]. Using dual-fluorescence flow cytometry, Murphy et al. [56] demonstrated two phases of acidification of insulin by 3T3 cells. Mellman et al. [45] and Mellman and Plutner [44] demon-

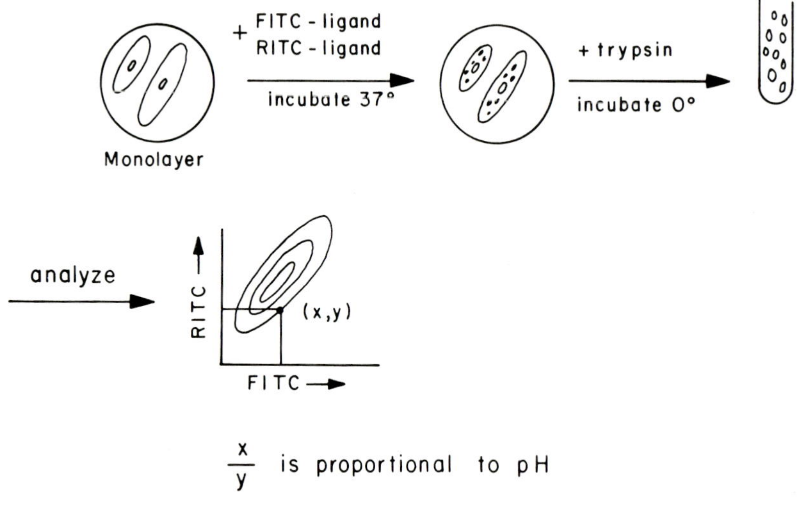

Fig. 4. Schematic for endocytic vesicle pH measurement using dual fluorescence method. Note that the RITC–ligand can be replaced by any pH insensitive conjugate which is distinguishable from FITC (e.g., XRITC, Texas Red). Note also an extension of this method permits measurements of cytoplasmic pH by introducing inert conjugates (e.g., FITC–dextran and RITC–dextran) into the cytoplasm [17].

strated that complexes formed by monovalent, but not polyvalent, antibodies against the macrophage Fc receptor are recycled.

Taken together, the data for a variety of cell types and a number of ligands indicate that at least two types of endocytic compartments (endosomes and lysosomes) are acidic and that some receptor–ligand complexes (e.g., those involving transferrin and monovalent antibodies to Fc receptor) may be endocytosed and recycled without encountering lysosomes. At least two types of endosomes may also exist.

Many of the details of endocytic pathways are not yet known because of the limitations of the methods used. For example, morphological electron microscopic studies have the advantage that spatial and structural information may be obtained. However, the method is inherently static, sample sizes are ordinarily limited, and it is difficult to quantitate enzymatic and biochemical properties. Density-gradient centrifugation has been very useful because of the possibility of performing biochemical analyses on the resulting fractions. However, the length of time required for fractionation introduces the possibility of changes in vesicle properties, as might be expected from proteolysis or loss of ion gradients. In addition, bulk fractionation techniques are unable to address questions which require particle by particle analysis, such as whether different endocytic probes or different enzymes are contained in the same compartment.

Clearly, a technique that permits measurements on individual vesicles would be desirable. Flow cytometric analysis of unfractionated cell lysates can be used to demonstrate that fusion of endocytic vesicles must occur [51]. Examples of the results of this type of analysis are shown in Figures 5 and 6. Three populations of endocytic vesicles are observed in lysates from cells incubated for various periods with FITC–dextran; a progression from dim to intermediate to bright can be observed with increasing time of incubation (Fig. 5B–D). The enzymatic contents of endocytic compartments can also be determined by this technique. Figure 6 shows histograms for vesicles from cells labeled with a fluorogenic substrate (*N*-carbobenzyloxy-ala-arg-arg-4-methoxy-β-naphthylamine) for cathepsin B. The flow cytometric vesicle analysis technique offers the advantages of rapid measurement of large numbers of individual vesicles, the ability to measure simultaneously more than one vesicle property, and the possibility of sorting individual vesicles for further analysis.

PHAGOCYTOSIS

For many of the reasons outlined above for ligand binding and internalization, flow cytometry has advantages over other techniques for the analysis of phagocytosis. This approach has been used to study the phagocytosis in vitro of FITC-labeled heat-killed *Staphylococcus aureus* by human neutrophils [2]. Bassøe et al. [2] demonstrated agreement with manual counts of Giemsa-stained smears and found that the modal fluorescence was linearly proportional to incubation time at 15–60 min. These workers also demonstrated that phagocytosis measurements on clinical samples were in agreement with expectations based on the clinical diagnosis.

Valet et al. [83] measured phagocytosis in vitro of rhodamine-stained 1.8 μm Latex microspheres by guinea pig peritoneal macrophages and demonstrated an increased phagocytic activity following stimulation by lymphokine-containing supernatants. This increase was correlated with an apparent repolarization of the plasma membrane 1–2 hours after stimulation.

Dunn and Tyrer [16] used 2-μm green-fluorescing Fluoresbrite microspheres from Polysciences to measure phagocytic uptake by human neutrophils and correlated the mea-

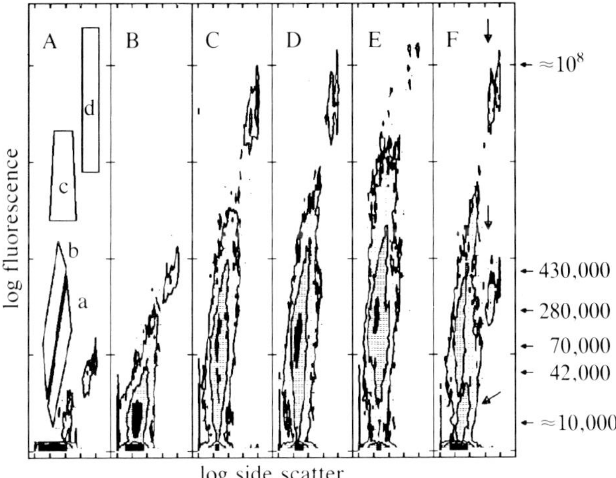

Fig. 5. Detection of endocytic vesicles in cell lysates using flow cytometry. Cell monolayers were incubated in the absence of FITC–dextran (**A**), or in the presence of 50 mg/ml FITC–dextran for 5 sec (**B**), 20 min (**C**), 60 min (**D**), and 180 min (**E**). Crude cell lysates were prepared by homogenization in ice cold 0.25 M sucrose, 2 mM EDTA, 10 mM Hepes, pH 7.4 and analyzed by flow cytometry. Each panel represents 50,000 events. The polygons in **A** show the regions in **C–E**, where vesicles showing weak (a), intermediate (b), and bright (c) fluorescence are found. The position of intact fluorescent cells (d) is also shown. The arrows show the average fluorescence of four types of calibration microspheres, and the extrapolated positions corresponding to 10,000 and 100,000,000 molecules of fluorescein per event. A mixture of equal volumes of the samples in **A** and **D** was also analyzed (**F**). Note the presence of a population of unlabeled vesicles and two populations of intact cells in this mixture (large arrows). Contours are drawn at 3, 20, and 80 events per bin. Tics are placed at approximately one log intervals. (Reproduced from Murphy [51].)

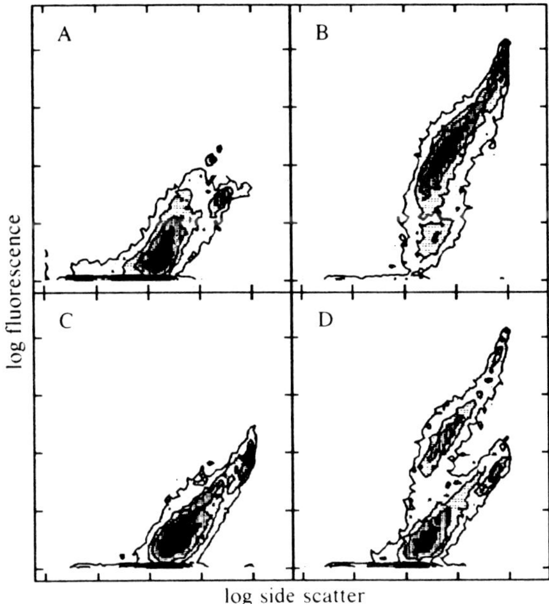

Fig. 6. Detection of vesicles containing cathepsin B activity. Flow cytometric analysis for lysates from cells incubated with (**B,C**) and without (**A**) fluorogenic substrate and with (**C**) or without (**A,B**) 100-μg/ml leupeptin. A mixture of equal volumes of the samples shown in **B** and **C** was also analyzed (**D**). Contours are drawn at 5, 20, 40, and 80 events per bin. Tics are placed at approximately one log intervals. (Reproduced from Murphy [51].)

surements with manual counts by fluorescence microscopy. Steinkamp et al. [76] also used Latex microspheres from Polysciences (green-fluorescing 1.83-μm microspheres) to measure phagocytosis both in vivo and in vitro by rat alveolar macrophages. These workers showed that the size distribution of the phagocytosing cells matched that of the whole macrophage population and that the number of cells with associated microspheres could be reduced 95% by exposure to sodium azide.

Haskill et al. [28] combined enzyme assays based on the techniques of Dolbeare and Smith [13] and Dolbeare and Vanderlaan [14] with measurements of zymosan particle phagocytosis and DNA; this combination is useful for the analysis of complex tumor cell populations. Bjerknes and Bassøe [4] used quenching of fluorescein fluorescence by Trypan blue (which does not penetrate viable cell membranes) to aid in distinguishing surface-bound from internalized FITC-conjugated zymosan particles. This method has excellent potential for following the internalization of small ligands as well.

Two points may be made about phagocytosis measurements. First, the apparent uptake must be demonstrated to be inhibitable by sodium azide, or other inhibitors such as iodoacetate. Second, if a pH-sensitive fluorochrome is used to label the particles to be internalized (e.g., bacteria or yeast), an equilibrating agent such as methylamine or nigericin should be added shortly before the measurement to eliminate quenching.

An interesting application of flow cytometry to the analysis of phagocytosis was described by Geisow et al. [22]. They labeled mouse peritoneal macrophages with FITC–dextran for 16–24 hours followed by a 2-hour chase. A parallel culture was allowed to ingest 5-μm Latex spheres. Partially purified homogenates from these two cultures were then mixed for 60 min at 37°C and analyzed by flow cytometry. Phagosomes were detected by light scattering, and fusion of FITC-dextran containing vesicles with phagosomes was detected by measuring the number of phagosomes showing fluorescence above control values.

LIGHT-SCATTERING CHANGES

Since flow cytometry can measure multiple parameters on individual cells with excellent temporal resolution, it is possible to correlate ligand binding and internalization kinetics with changes in other cell properties. One of the best examples of this approach is the analysis of degranulation and shape changes by human neutrophils upon addition of chemotactic factor [19,43,73]. Two changes in light-scattering properties can be measured by flow cytometry. In the absence of degranulating stimuli, a decrease in side (90°) scatter and an increase in forward scatter are observed within 20 sec of the addition of chemotactic factor. Starting 2–6 min after stimulation, side scatter increases to 5–15% above the prestimulation value, and forward scatter returns to its initial level. McNeil et al. [43] suggested that these two phases correspond to surface ruffling and cell elongation and demonstrated that they are correlated temporally with binding and internalization of the chemotactic peptide, respectively.

In the presence of cytochalasin B to facilitate degranulation, the elongation phase is replaced by a marked decrease in side scatter. The kinetics of this decrease is side scatter are well correlated with independent measurements of enzyme release [19,73]. Changes in membrane potential have also been observed during stimulation [19].

SUMMARY

The results reviewed illustrate how flow cytometry can be used to probe one of the fundamental components of cell physiology and metabolism, the endocytic/exocytic pathway. Measurements can be made in real time on living cells, and can also be made on individual organelles. Thus, in addition to its often-stated potential for clinical applications, flow cytometry should play a major role in basic research into the ways in which ligand binding and processing are coupled with cellular responses.

ACKNOWLEDGMENTS

I thank Robert Bowser, Amy Kennedy, Gregory LaRocca, Robert Mays, and Daniel Shmorhun for excellent technical assistance, and Mario Roederer, David Sipe, D. Lansing Taylor, and Alan Waggoner for stimulating discussions and critical reading of this manuscript. This work was supported in part by research grant GM32508 from the National Institutes of Health, and by Presidential Young Investigator Award DCB-8351364 from the National Science Foundation, with matching funds from Becton Dickinson Monoclonal Center, Inc.

REFERENCES

1. **Adams CJ, Maurey KM, Storrie B** (1982) Exocytosis of pinocytic contents by Chinese hamster ovary cells. J Cell Biol 93:632–637.
2. **Bassøe C-F, Solsvik J, Laerum OD** (1980) Quantitation of single cell phyagocytic capacity by flow cytometry. In Leaerum OD, Lindmo T, Thorud E (eds), "Flow Cytometry." Vol. IV. Oslo: Universitetsforlaget, pp 170–174.
3. **Besterman JM, Airhart JA, Woodworth RC, Low RB** (1981) Excytosis of pinocytosed fluid in cultured cells: Kinetic evidence for rapid turnover and compartmentation. J Cell Biol 91:716–727.
4. **Bjerknes B, Bassøe C-F** (1984) Phagocyte C3-mediated attachment and internalization: Flow cytometric studies using a fluorescence quenching technique. Blut 49:315–323.
5. **Bohn B** (1976) High-sensitivity cytofluorometric quantitation of lectin and hormone binding to surfaces of living cells. Exp Cell Res 103:39–46.
6. **Bohn B** (1980) Flow cytometry: A novel approach for the quantitative analysis of receptor–ligand interactions on surfaces of living cells. Mol Cell Endocrinol 20:1–15.
7. **Bohn B, Manske W** (1980) Application of flow cytofluorometry to ligand binding studies on living cells: Practical aspects and recommendations for calibration and data processing. In Laerum OD, Lindmo T, Thorud E (eds), "Flow Cytometry." Vol. IV. Oslo: Universitetsforlaget, pp 227–232.
8. **Cain CC, Murphy RF** (1986) Growth inhibition of 3T3 fibroblasts by lysosomotropic amines: Correlation with effects on intravesicular pH but not vacuolation. J Cell Phys 129:65–70.
9. **Carpenter G, Cohen S** (1979) Epidermal growth factor. Annu Rev Biochem 48:193–216.
10. **Chan SS, Arndt-Jovin DJ, Jovin TM** (1979) Proximity of lectin receptors on the cell surface measured by fluorescence energy transfer in a flow system. J Histochem Cytochem 27:56–64.
11. **Das M, Fox CF** (1978) Molecular mechanism of mitogen action: Processing of receptor induced by epidermal growth factor. Proc Natl Acad Sci USA 75: 2644–2648.
12. **Davies PJA, Davies DR, Levitzki A, Maxfield FR, Milhaud P, Willingham MC, Pastan IH** (1980) Transglutaminase is essential in receptor-mediated endocytosis of alpha$_2$-macroglobulin and polypeptide hormones. Nature (Lond.) 283:162–167.
13. **Dolbeare FA, Smith RE** (1977) Flow cytometric measurement of peptidases with use of 5-nitrosalicylaldehyde and 4-methoxy-beta-naphthylamine derivatives. Clin Chem 23:1485–1491.
14. **Dolbeare F, Vanderlaan M** (1979) A fluorescent assay of proteinases in cultured mammalian cells. J Histochem Cytochem 27:1493–1495.
15. **Draper RK, Simon MI** (1980) The entry of diphtheria toxin into the mammalian cell cytoplasm: Evidence for lysosomal involvement. J Cell Biol 87:849–854.
16. **Dunn PA, Tyrer HW** (1981) Quantitation of neutrophil phagocytosis, using fluorescent latex beads. Correlation of microscopy and flow cytometry. J Lab Clin Med 98:374–381.
17. **Fechheimer M, Denny C, Murphy RF, Taylor DL** (1986) Measurement of cytoplasmic pH in *Dictyostelium discoideum* by using a new method for introducing macromolecules into living cells. Eur J Cell Biol 40:242–247.
18. **Finney DA, Sklar LA** (1983) Ligand/receptor internalization: A kinetic, flow cytometric analysis of the internalization of N-formyl peptides by human neutrophils. Cytometry 4:54–60.
19. **Fletcher MP, Seligmann BE** (1985) Monitoring human neutrophil granule secretion by flow cytometry: Secretion and membrane potential changes assessed by light scatter and a fluorescent probe of membrane potential. J Leukocyte Biol 37:431–447.
20. **Forgac M, Cantley L, Wiedenmann B, Altstiel L, Branton D** (1983) Clathrin-coated vesicles contain an ATP-dependent proton pump. Proc Natl Acad Sci USA 80: 1300–1303.
21. **Galloway CJ, Dean GE, Marsh M, Rudnick G, Mellman I** (1983) Acidification of macrophage and fibroblast endocytic vesicles *in vitro*. Proc Natl Acad Sci USA 80:3334–3338.
22. **Geisow M, D'Arcy Hart P, Young MR** (1982) Extracellular fusion of macrophage phagosomes with lysosomes. Cell Biol Int Rep 6:361–367.
23. **Glickman J, Croen K, Kelly S, Al-Awqati Q** (1983) Golgi membranes contain an electrogenic H^+ pump in parallel to a chloride conductance. J Cell Biol 97: 1303–1308.
24. **Godlfine ID** (1981) Interaction of insulin, polypeptide hormones, and growth factors with intracellular membranes. Biochim Biophys Acta 650:53–67.
25. **Gonzalez-Noriega A, Grubb JH, Talkad V, Sly WS** (1980) Chloroquine inhibits lysosomal enzyme pinocytosis and enhances lysosomal enzyme secretion by impairing receptor recycling. J Cell Biol 85: 839–852.
26. **Haigler HT, Willingham MC, Pastan I** (1980) Inhibi-

tors of ^{125}I-epidermal growth factor internalization. Biochem Biophys Res Commun 94:630–637.

27. **Harding C, Heuser J, Stahl P (1983)** Receptor-mediated endocytosis of transferrin and recycling of the transferrin receptor in rat reticulocytes. J Cell Biol 97: 329–339.
28. **Haskill S, Kivinen S, Nelson K, Fowler WC Jr (1983)** Detection of intratumor heterogeneity by simultaneous multiparameter flow cytometric analysis with enzyme and DNA markers. Cancer Res 43: 1003–1009.
29. **Helenius A, Kartenbeck J, Simons K, Fries E (1980)** On the entry of Semliki Forest virus into BHK-21 cells. J Cell Biol 84:404–420.
30. **Helenius A, Mellman I, Wall D, Hubbard A (1983)** Endosomes. Trends Biochem Sci 8:245–249.
31. **Herzenberg LA, Herzenberg LA (1978)** In Weir DM (ed), "Handbook of Experimental Immunology." Oxford: Blackwell, pp 22.1–22.21.
32. **Jenis DM, Stephanowski AL, Blair OC, Burger DE, Sartorelli AC (1984)** Lectin receptor proximity on HL-60 leukemia cells determined by fluorescence energy transfer using flow cytometry. J Cell Physiol 121:501–507.
33. **Kahn CR, Baird KL, Jarrett DB, Flier JS (1978)** Direct demonstration that receptor crosslinking or aggregation is important in insulin action. Proc Natl Acad Sci USA 75:4209–4213.
34. **King AC, Hernaez-Davis L, Cuatrecasas P (1981)** Lysosomotropic amines inhibit mitogenesis induced by growth factors. Proc Natl Acad Sci USA 78: 717–721.
35. **King GL, Kahn CR (1981)** Non-parallel evolution of metabolic and growth-promoting functions of insulin. Nature (Lond.) 292:644–646.
36. **King GL, Kahn CR, Rechler MM, Nissley SP (1980)** Direct demonstration of separate receptors for growth and metabolic activities of insulin and multiplication-stimulating activity (an insulin-like growth factor) using antibodies to the insulin receptor. J Clin Invest 66:130–140.
37. **Kraemer PM, Tobey RA, Van Dilla MA (1972)** Flow microfluorometric studies of lectin binding to mammalian cells. I. General features. J Cell Physiol 81: 305–314.
38. **Loken MR, Stall AM (1982)** Flow cytometry as an analytical and preparative tool in immunology. J Immunol Methods 50:R85–112.
39. **Marnell MH, Stookey M, Draper RK (1982)** Monensin blocks the transport of diphtheria toxin to the cell cytoplasm. J Cell Biol 93:57–62.
40. **Marsh M, Bolzau E, Helenius A (1983)** Penetration of Semliki Forest virus from acidic prelysosomal vacuoles. Cell 32:931–940.
41. **Maxfield FR, Davies PJA, Klempner L, Willingham MC, Pastan I (1979)** Epidermal growth factor stimulation of DNA synthesis is potentiated by compounds that inhibit its clustering in coated pits. Proc Natl Acad Sci USA 76:5731–5735.
42. **Maxfield FR, Willingham MC, Davies PJA, Pastan I (1979)** Amines inhibit the clustering of alpha$_2$-macroglobulin and EGF on the fibroblast cell surface. Nature (Lond.) 277:661–663.
43. **McNeil PL, Kennedy AL, Waggoner AS, Taylor DL, Murphy RF (1985)** Light-scattering during chemotactic stimulation of human neutrophils: Kinetics followed by flow cytometry. Cytometry 6:7–12.
44. **Mellman I, Plutner H (1984)** Internalization and degradation of macrophage Fc receptors bound to polyvalent immune complexes. J Cell Biol 98:1170–1177.
45. **Mellman I, Plutner H, Ukkonen P (1984)** Internalization and rapid recycling of macrophage fc receptors tagged with monovalent antireceptor antibody: Possible role of a prelysosomal compartment. J Cell Biol 98:1163–1169.
46. **Merion M, Schlesinger P, Brooks RM, Moehring JM, Moehring TJ, Sly WS (1983)** Defective acidification of endosomes in chinese hamster ovary cell mutants "cross-resistant" to toxins and viruses. Proc Natl Acad Sci USA 80:5315–5319.
47. **Merion M, Sly WS (1983)** The role of intermediate vesicles in the adsorptive endocytosis and transport of ligand to lysosomes by human fibroblasts. J Cell Biol 96:644–650.
48. **Metezeau P, Djavadi-Ohaniance L, Goldberg ME (1982)** The kinetics and homogeneity of endocytosis of a receptor-bound ligand in a heterogenous cell population studied by flow cytofluorometry. J Histochem Cytochem 30:359–363.
49. **Metezeau P, Elguindi I, Goldberg ME (1984)** Endocytosis of the membrane immunoglobulins of mouse spleen B-cells: A quantitative study of its rate, amount and sensitivity to physiological, physical and cross-linking agents. EMBO J 3:2235–2242.
50. **Miskimins WK, Shimizu N (1984)** Uptake of epidermal growth factor into a lysosomal enzyme-deficient organelle: Correlation with cell's mitogenic response and evidence for ubiquitous existence in fibroblasts. J Cell Physiol 118:305–316.
51. **Murphy RF (1985)** Analysis and isolation of endocytic vesicles by flow cytometry and sorting: Demonstration of three kinetically distinct compartments involved in fluid-phase endocytosis. Proc Natl Acad Sci USA 82: 8523–8526.
52. **Murphy RF, Bisaccia E, Cantor CR, Berger C, Edelson RL (1984)** Internalization and Acidification of Insulin by Activated Human Lymphocytes. J Cell Physiol 121: 351–356.
53. **Murphy RF, Jorgensen ED, Cantor CR (1981)** Internalization of exogenous histone measured by flow cytofluorometry. Cytometry 2:116 (abst).
54. **Murphy RF, Jorgensen ED, Cantor CR (1982)** Kinetics of histone endocytosis in Chinese hamster ovary cells. A flow cytofluorometric analysis. J Biol Chem 257:1695–1701.
55. **Murphy RF, Powers S, Bisaccia E, Edelson RL, Cantor CR (1983)** Dual fluorescence flow cytometric analysis of endosome acidification: Application to insulin internalization by fibroblasts and activated lymphocytes. J Cell Biol 97:110 (abst).
56. **Murphy RF, Powers S, Cantor CR (1984)** Endosome pH measured in single cells by dual fluorescence flow cytometry: Rapid acidification of insulin to pH 6. J Cell Biol 98:1757–1762.
57. **Murphy RF, Powers S, Cantor CR, Pollack R (1984)** Reduced insulin endocytosis in serum-transformed fibroblasts demonstrated by flow cytometry. Cytometry 5:275–280.
58. **Murphy RF, Powers S, Verderame M, Cantor CR, Pollack R (1982)** Flow cytofluorometric analysis of insu-

lin binding and internalization by Swiss 3T3 cells. Cytometry 2:402–406.

59. **Murphy RF, Tse DB, Cantor CR, Pernis B** (1984) Acidification of internalized class I major histocompatibility complex antigen by T lymphoblasts. Cell Immunol 88:336–342.
60. **Notter MFD, Leary JF, Balduzzi PC** (1982) Adsorption of Rous sarcoma virus to genetically susceptible and resistant chicken cells studied by laser flow cytometry. J Virol 41:958–964.
61. **Nuccitelli R, Deamer DW** (1982) "Intracellular pH: Its Measurement, Regulation and Utilization in Cellular Functions." New York: Alan R. Liss.
62. **Ohkuma S, Poole B** (1978) Fluorescence probe measurement of the intralysosomal pH in living cells and the perturbation of pH by various agents. Proc Natl Acad Sci USA 75:3327–3331.
63. **Osband ME, Cohen EB, McCaffrey RP, Shapiro HM** (1980) A technique for the flow cytometric analysis of lymphocytes bearing histamine receptors. Blood 56: 923–925.
64. **Pastan I, Willingham MC** (1983) Receptor-mediated endocytosis: Coated pits, receptosomes, and the Golgi. Trends Biochem Sci 8:250–252.
65. **Poole B, Ohkuma S** (1981) Effect of weak bases on the intralysosomal pH in mouse peritoneal macrophages. J Cell Biol 90:665–669.
66. **Purrello F, Vigneri R, Clawson GA, Goldfine ID** (1982) Insulin stimulation of nucleoside triphosphatase activity in isolated nuclear envelopes. Science 216:1005–1007.
67. **Roederer M, Murphy RF** (1986) Cell-by-cell autofluorescence correction for low signal-to-noise systems: Application to EGF endocytosis by 3T3 fibroblasts. Cytometry 7:558–565.
68. **Rudolph NS, Ohlsson-Wilhelm BM, Leary JF, Rowley PT** (1985) Single-cell analysis of the relationship among transferrin receptors, proliferation, and cell cycle phase in K562 cells. Cytometry 6:151–158.
69. **Sandvig K, Olsnes S** (1980) Diphtheria toxin entry into cells is facilitated by low pH. J Cell Biol 87:828–832.
70. **Schreiber AB, Lax I, Yarden Y, Eshhar Z, Schlessinger J** (1981) Monoclonal antibodies against receptor for epidermal growth factor induce early and delayed effects of epidermal growth factor. Proc Natl Acad Sci USA 78:7535–7539.
71. **Sklar LA, Finney DA** (1982) Analysis of ligand–receptor interactions with the fluorescence activated cell sorter. Cytometry 3:161–165.
72. **Sklar LA, Finney DA, Oades ZG, Jesaitis AJ, Painter RG, Cochrane CG** (1984) The dynamics of ligand–receptor interactions. Real-time analyses of association, dissociation, and internalization of an N-formyl peptide and its receptors on the human neutrophil. J Biol Chem 259:5661–5669.
73. **Sklar LA, Oades ZG, Finney DA** (1984) Neutrophil degranulation detected by right angle light scattering: Spectroscopic methods suitable for simultaneous analyses of degranulation or shape change, elastase release, and cell aggregation. J Immunol 133:1483–1487.
74. **Steinkamp JA, Kraemer PM** (1974) Flow microfluorometric studies of lectin binding to mammalian cells. J Cell Physiol 84:197–204.
75. **Steinkamp JA, Kraemer PM** (1979) Quantitation of lectin binding by cells. In Melamed MR, Mullaney PF, Mendelsohn ML (eds), "Flow Cytometry and Sorting." New York: John Wiley & Sons, pp 497–504.
76. **Steinkamp JA, Wilson JS, Saunders GC, Stewart CC** (1982) Phagocytosis: Flow cytometric quantitation with fluorescent microspheres. Science 215:64–66.
77. **Steinman RM, Mellman IS, Muller WA, Cohn ZA** (1983) Endocytosis and the recycling of plasma membrane. J Cell Biol 96:1–27.
78. **Szollosi J, Tron L, Damjanovich S, Helliwell SH, Arndt-Jovin D, Jovin TM** (1984) Fluorescence energy transfer measurements on cell surfaces: A critical comparison of steady-state fluorimetric and flow cytometric methods. Cytometry 5:210–216.
79. **Tietze C, Schlesinger P, Stahl P** (1980) Chloroquine and ammonium ion inhibit receptor-mediated endocytosis of mannose-glycoconjugates by macrophages: Apparent inhibition of receptor recycling. Biochem Biophys Res Commun 93:1–8.
80. **Tron L, Szollosi J, Damjanovich S, Helliwell SH, Arndt-Jovin DJ, Jovin TM** (1984) Flow cytometric measurement of fluorescence resonance energy transfer on cell surfaces. Quantitative evaluation of the transfer efficiency on a cell-by-cell basis. Biophys J 45: 939–946.
81. **Tse DB, Pernis B** (1984) Spontaneous internalization of class I major histocompatibility complex molecules in T lymphoid cells. J Exp Med 159:193–207.
82. **Tycko B, Maxfield FR** (1982) Rapid acidification of endocytic vesicles containing alpha-2-macroglobulin. Cell 28:643–651.
83. **Valet G, Jenssen HL, Krefft M, Ruhenstroth-Bauer G** (1981) Flow cytometric measurements of the transmembrane potential, the surface charge density and the phagocytic activity of guinea pig macrophages after incubation with lymphokines. Blut 42:379–382.
84. **Van NT, Raber M, Barrows GH, Barlogie B** (1984) Estrogen receptor analysis by flow cytometry. Science 224:876–879.
85. **van Deurs B, Ropke C, Thorball N** (1984) Kinetics of pinocytosis studied by flow cytometry. Eur J Cell Biol 34:96–102.
86. **Van Epps DE, Chenoweth DE** (1984) Analysis of the binding of fluorescent C5a and C3a to human peripheral blood leukocytes. J Immunol 132:2862–2867.
87. **Van Leuven F, Cassiman J-J, Van Den Berghe H** (1980) Primary amines inhibit recycling of alpha-2-macroglobulin receptors in fibroblasts. Cell 20:37–43.
88. **Van Renswoude J, Bridges KR, Harford JB, Klausner RD** (1982) Receptor-mediated endocytosis of transferrin and the uptake of Fe in K562 cells: Identification of a nonlysosomal acidic compartment. Proc Natl Acad Sci USA 79:6186–6190.
89. **Yarden Y, Gabbay M, Schlessinger J** (1981) Primary amines do not prevent the endocytosis of epidermal growth factor into 3T3 fibroblasts. Biochim Biophys Acta 674:188–203.
90. **Yarden Y, Schreiber AB, Schlessinger J** (1982) A nonmitogenic analogue of epidermal growth factor induces early responses mediated by epidermal growth factor. J Cell Biol 92:687–693.
91. **Yoshimura A, Kuroda K, Kawasaki K, Yamashina S, Maeda T, Ohnishi S** (1982) Infectious cell entry mechanism of influenza virus. J Virol 43:284–293.

19

Microspheres as Immunoreagents for Cell Identification and Cell Fractionation

T. Lea, J. P. O'Connell, K. Nustad, S. Funderud, A. Berge, and A. Rembaum

Rikshospitalet, University Hospital, 0172 Oslo, Norway (T.L.); Becton Dickinson Research Center, Research Triangle Park, North Carolina 27709 (J.P.O.); The Norwegian Radium Hospital, Oslo, Norway (K.N. and S.F.); SINTEF, Trondheim, Norway (A.B.); Jet Propulsion Laboratory, CALTECH, Pasadena, Califonia 91125 (A.R., deceased)

INTRODUCTION

A detailed characterization of the various components making up the cell surface is fundamental to an understanding of how cells respond to stimuli, how they function, and how they communicate with each other. Many techniques have been devised to visualize cell-surface determinants. The best established methods involve the use of fluorochrome or radiolabeled antibodies, lectins, hormones, or other tracer molecules binding to receptors or ligands present in the cell membrane. The same tracers can be conjugated with gold, ferritin, hemocyanin, peroxidase, or other enzymes, making possible the detection of target structures with equipment of varying sophistication. In microbiology and immunology, antibodies have been coupled with cells or microorganisms such as erythrocytes, bacteria, yeast, or viruses that served as markers. Most of these techniques have serious drawbacks, however, making them unsuitable as general methods in the study of cell membrane morphology. The major limitations are attributable to lack of reagent stability, the need for special equipment or test conditions, and nonspecific interactions between the tracer molecule and the cell surface.

For these reasons, investigations were undertaken to develop better systems for sensitive and specific detection of membrane structures. Because of the unique reactivity and specificity of the monoclonal antibodies made available during the late 1970s, there was strong motivation for polymer chemists to construct carriers that would permit easy coupling of these monoclonal antibodies as tracer molecules and at the same time would be convenient, sensitive, and have high specificity for the study of biological specimens. The most promising were polymeric microspheres that could be produced in many sizes for various applications. Such microbeads can be synthesized according to different principles, and of several starting materials. Accordingly, they have different properties. They vary in ease of preparation, in density, in biocompatibility, and in reactive groups available for covalent coupling of tracer molecules or for modification with compounds enhancing their detection. During the last few years, for example, magnetic polymer particles have been created.

These microspheres are of potential value in the study of cell surface structures, their distribution and movement in the cell membrane and in the process of phagocytosis. Microspheres have also been used for cell separation and enrichment. This chapter summarizes some of the properties and discusses some of the potential applications of polymeric microspheres in cell biology and immunology. Most of the applications are directly related to flow cytometry. However, the development of magnetic microspheres greatly simplifies the isolation of highly purified cell populations to be studied by surface markers or in functional assays. Thus, immunomagnetic cell separation in combination with flow cytometry has become a powerful tool in many research areas. The first part of this chapter concentrates on microparticles and their possibilities in flow cytometry, the second on immunomagnetic beads in cell separation. Besides giving detailed examples based on our own experiences, other interesting methodological developments are presented.

PROPERTIES OF POLYMERIC MICROSPHERES

Depending on their purpose, the properties of polymeric microparticles should satisfy several specific requirements. One important criterion is biocompatibility. Polymer beads that are to be used in the study of living cells should not contain compounds that can affect cell function in any way. It is well documented that some polymerization initiators, terminators, and detergents have toxic effects on mammalian cells. Although beads suffering from these drawbacks may be used for special purposes, they will not find general applica-

Flow Cytometry and Sorting, Second Edition, pages 367–380

tion. Also, the polymer used to build the particles has to be inert, in the sense that it should not interact with the cell suface, leading to irrelevant responses in addition to those under study.

The density of the particles is another important parameter. When microspheres coupled with specific tracer molecules are to react with cells in suspension, particle density should approximate that of the cells' own density, so that their similar sedimentation rates assure the necessary contact between beads and cells. Alternatively, the suspension of particles and cells must be rotated continuously, a process that may have deleterious effects on cell viability. Another possibility that has been used to advantage in hemagglutination techniques is to increase cell concentrations or even pellet the particles and cells by a brief centrifugation to establish the proper contact. It is also conceivable that particle density will influence the kinetics of the reaction between the tracer and the target molecule, depending on the avidity of their interaction. Finally, high particle density may increase the shear forces on the cell membrane leading to particle detachment.

Uniformity of particle size is not an absolute prerequisite. However, a narrow size distribution will provide homogenous sedimentation and more predictable behaviour in solution. In general, the size range should not exceed mean ±20%. Besides, monodispersity can be an advantage in many experimental situations. Thus, it is possible to visualize two markers simultaneously by employing monodisperse particles of different sizes.

A problem often encountered when working with polymeric microspheres is their tendency to agglomerate. Such particle aggregates can usually be dispersed by vortexing or by brief sonication. However, depending on the physicochemical properties of the particles, they may soon reaggregate, a process which in most instances limits their applicability. The analysis of cell populations by flow cytometry using these microspheres can only be carried out successfully with aggregate-free particle suspensions, since clumps of microspheres may completely mask the cells to be studied.

Other important properties relate to particle charge and whether they are hydrophobic. Under most experimental conditions, the interaction between beads and cells should be determined only by the binding of tracer to its specific ligand. Because of the lipids making up biological membranes, hydrophobic microspheres tend to bind nonspecifically to cell surfaces. There are several ways to increase the hydrophilic properties of particle surfaces. The easiest is to couple hydrophilic compounds such as carbohydrates or some irrelevant protein to the beads. Because of the presence of terminal sialic residues in the carbohydrates of the membrane glycocalyx, there will be a net negative charge on the cell suface under physiological conditions. Through the introduction of charged groups on the surface of the microspheres, the tendency to nonspecific binding via hydrophobic interactions can be balanced and controlled.

The particle surface should also possess chemical groups permitting efficient coupling of the various tracer molecules by standard techniques. Since the tracers in most instances will be proteins, hydroxyl-, amino-, or sulfhydryl groups are especially useful. Protein may then be coupled to the particle surface after activation of the beads with cyanogen bromide or sulfonyl chlorides or by coupling via a series of crosslinkers such as carbodiimides, glutardialdehyde, and N-succinimidyl-3-(2-pyridyldithio) propionate (SPDP).

For several reasons, a smooth surface will be an advantage. Porous microspheres may permit coupling of tracer molecules inside the pores, making them unavailable for binding to target structures. An irregular surface may likewise reduce the amount of interactions possible between microspheres and cells. Accordingly, it has been reported that spacer molecules increasing the distance between the particle surface and the tracers will improve binding efficiency. Thus, in immunomagnetic cell separation, particle binding directed by murine monoclonal antibodies of the immunoglobulin G (IgG) isotype benefits considerably from prior coupling of a polyclonal anti-mouse immunoglobulin antibody to the beads [16,34]. Alternatively, direct coupling of monoclonal antibodies of the IgM isotype have proved to be equally useful [10]. In this instance, the large size of the pentameric IgM molecule provides the necessary distance to the particle surface for efficient binding to take place. When indirect technique is used, the quality of the secondary antibodies is of utmost importance. Depending on the system under study, secondary antibodies should be properly adsorbed and tested in a sensitive assay to confirm species-specificity before use; affinity-purification ensures necessary immune reactivity.

The principles of affinity chromatography or immunosorbent chromatography also hold for the coupling of proteins to microspheres. However, performance does not necessarily depend on the number of tracer molecules coupled to the particle surface. Thus, the optimal amount of tracer should be determined by careful titration. In the case of antibodies it is also important that activation of the beads prior to coupling be regulated so that multipoint attachment is reduced to a minimum. This will ensure optimal antibody function.

In general, coupling of a secondary antibody directly to the beads results in satisfactory performance. If antibody orientation is of special importance, however, a primary layer of staphylococccal protein A may improve the reactivity of the particles. Protein A, which binds to the Fc portion of IgG molecules from many species, will direct the antigen-binding sites outward from the particle surface. Owing to the high-avidity interaction between biotin and avidin (10^{15} M^{-1}), it has also been suggested that avidin, or streptavidin, may act as an efficient linker between the microspheres and the antibodies. Biotinylation is a gentle process and most of the antibody activity will be retained. Efficient biotinylation may be obtained by reacting protein either with the hydroxysuccinimide ester of biotin [27], or with biotin hydrazide [37]. In the latter case, biotin is conjugated to the antibodies via carbohydrates located in the Fc region of the IgG molecules, contributing to favorable orientation of the antibodies. The biotin-avidin interaction may be used for binding any protein to the particle surface.

FLUORESCENT POLYMERIC MICROSPHERES AS CELL LABELS

Small polymeric microspheres were first used during the early 1970s as cell-labeling reagents in scanning electron microscopy (SEM). The submicron styrene Latex particles were targeted with antibody (usually attached by physical adsorption), exposed to the appropriate cell types, and examined with the SEM. The specificity of the reagents could then be determined visually [17,26]. The development of flow cytometry during the early 1970s and monoclonal antibodies in 1975 made possible the use of similar, highly specific fluorescent microspheres for quantitative analysis of cell-surface markers [20,25,26]. The obvious advantage of this method is that a high concentration of fluors can be contained within the microspheres and targeted with specific antibodies or other ligand-binding substances such as lectins. They make it possible to identify even small subpopulations of cells for further examination and quantification.

The "ideal" bead should be of proper size, have the dye homogeneously distributed throughout the bead volume, have reactive functional groups on the particle surface, and, perhaps most important, show negligible nonspecific interaction with cells. As to the preparation of fluorescent polymer particles, in principle, three different methods are used to get the dye into the particles:

1. *Solvent dyeing.* With oil-soluble dyes, the particles are first made to swell in a dye–solvent solution. The solvent is then distilled off, leaving the dye stranded within the particles. Crosslinked particles, for example from styrene–divinylbenzene, may be introduced as dry powder into the dye solution, provided their size is not too small, >3–5 μm. Smaller particles and those without crosslinking stay in an aqueous phase in contact with the organic solvent-dye phase. A relatively detailed description of the procedure for solvent dyeing is given by Bangs [1].
2. *Covalent coupling.* Polymer particles may be prepared with various types of functional groups, such as epoxy-, hydroxy-, amino-, and carboxylic acid. Fluorescent dyes may be coupled covalently through such functional groups. For example, fluorescein isothiocyanate (FITC) and tetramethylrhodamine isothiocyanate (TRITC) react readily with amino groups.
3. *Electrostatic binding.* Valet et al. [32] showed that the pH-sensitive fluorescent dye 2,3-dicyano hydrochinon (DCH) may be bound electrostatically to cross-linked acrylate particles with functional amino groups. DCH lost its pH indicator properties by binding but remained fluorescent.

For use in immunofluorescence, monosized particles of various sizes are clearly advantageous. A convenient method for the preparation of suitable monosized particles has been developed by Ugelstad et al. [30]. By this method, one may prepare highly monodisperse polymer particles of predetermined size, both compact as well as porous, with different functional groups. Fluorescence may be introduced by any of the methods mentioned above.

Fluorescent beads for coupling of antibodies are commercially available from several different sources: Dyno Particles A.S. (Lillestrøm, Norway), Duke Scientific Laboratories (Palo Alto, CA), Pandex Laboratories (Mundelein, IL), Polysciences, Inc. (Warrington, PA), and Seragen Diagnostics Laboratory (Indianapolis, IN). Most fulfill only some of the requirements mentioned above. In addition, many producers do not give precise information about particle composition, the dye used and properties of the particle surface.

In the work described here, we demonstrate the use of monoclonal antibody-targeted fluorescent microspheres as cell labels in flow cytometry. The microspheres contain the equivalent of 1 million fluorescein molecules per microsphere, leading to an amplification factor of about 1,000 versus what is achievable with conventional staining reagents.

We investigated the use of fluorescent microspheres in the determination of allotypes of IgD heavy chains on the surface of mouse B cells. This model system was attractive for various reasons. First, highly specific monoclonal antibodies were available to each of the heavy-chain allotypes (Becton Dickinson, Mountain View, CA). Mice of each allotype were readily available and were well characterized. We used C57Bl/6 (Igh-5^{b+}, 5a$-$) and BALB/c (Igh-5^{b-}, 5^{a+}) (Jackson Laboratories, Bar Harbor, ME) as positive and negative controls. Furthermore, conventional staining methods and reagents yielded low fluorescent signals, even when amplified with the biotinylated antibody-fluorescent avidin method. We then extended our studies to human T-lymphocyte subsets stained with fluorescent microspheres and quantitated with a flow cytometer.

Microsphere Preparation

Four sizes of microbeads, 0.1, 0.4, 0.7, and 1.0 μm in diameter, were synthesized for these studies. They contained 78%, 2-hydroxyethyl methacrylate, 10% methacrylic acid, 10% acrylamide, and 2% *N*,*N*-bis-acrylamide as the major co—monomers (BAH-microspheres). Allyl fluorescein and allyl tetramethylrhodamine were synthesized by reacting allyl amine with isocyanates of the respective dyes. The allyl dyes were combined with the co-monomers before polymerization, allowing the dye to become covalently linked within the resulting microspheres. The co-monomers were dissolved in a 0.4% aqueous solution of polyethylene oxide (~100,000 M_r), which acts as a stabilizer during polymerization. The size of the microspheres was controlled by the concentration of co—monomers in solution. Polymerization was initiated by 4.0 mrad of γ-radiation from a ^{60}Co source at a dose rate of 0.5 mrad/hour. Additional details of microsphere preparation may be found in reports published by Rembaum and co-workers [23,25,26]. Following the cleanup procedure, which involved multiple centrifugations and resuspensions, samples were taken for size and distribution characterization. The mean diameter was determined by SEM on samples critically dried from ethanol. The size and fluorescence distributions were determined on the FACS IV and found to have a coefficient of variation of 15–25%.

Antibody Conjugation Methods

The procedure used to attach antibody to the microspheres was a two-step carbodiimide method. Two general methods were used. In some experiments, the monoclonal antibody was directly attached to the microsphere (direct method). Alternatively, the microspheres were first reacted with an affinity-purified goat antimouse IgG (Fc specific), and subsequently labelled with the monoclonal antibody (indirect method). This method was particularly useful when the monoclonal antibody was stabilized with high concentrations of other proteins such as gelatin or bovine serum albumin. In either case, the antibody was attached to the microsphere using the following method: the stock 0.7-μm diameter BAH-allylfluorescein microspheres were washed several times in low pH buffer (0.01 *N* 2-hydroxyethyl-piperazine-N^1-2-ethane sulfonic acid, Hepes, pH 4.5) by repeated centrifugation at 13,000*g*, and resuspended to a concentration of 1 mg/ml (dry weight) in the 4.5 pH buffer. The carbodiimide (EDCI, 1-ethyl-3-3-dimethyl aminopropyl carbodiimide·HCl) was added to the bead suspension at 0.5 g per 15 mg microbeads, and the mixture was incubated at room temperature for 4 hours. The activated microspheres were washed three times by centrifugation; the final pellet was resuspended in 1 ml 0.01 M Hepes, pH 7.0, containing 100 μg of the protein to be attached. The protein and activated beads were rotated together overnight at 4°C, washed several times to remove any unattached protein, and resuspended to 1 mg/ml in the pH 7.0 Hepes buffer. Microspheres labeled with the goat antimouse antibody were reacted for

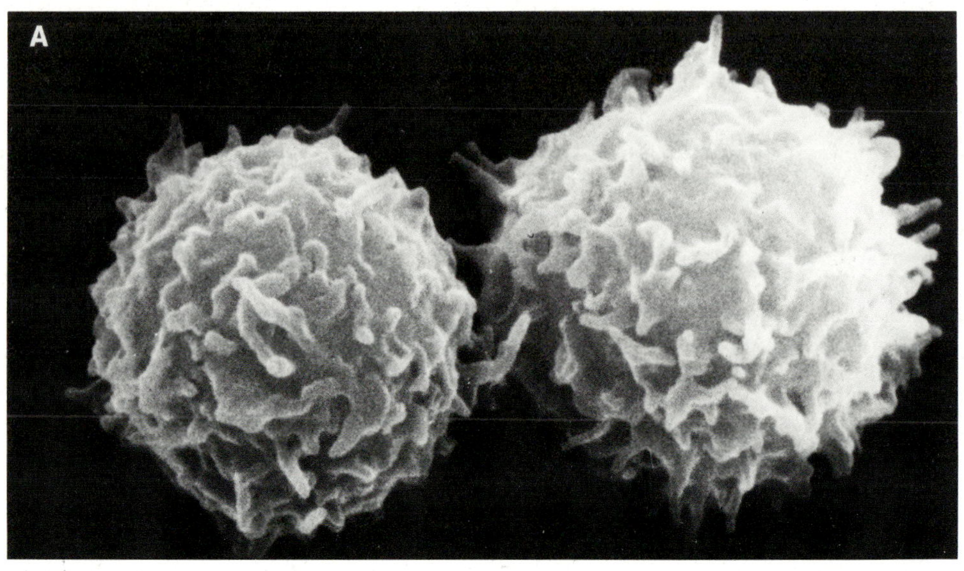
A

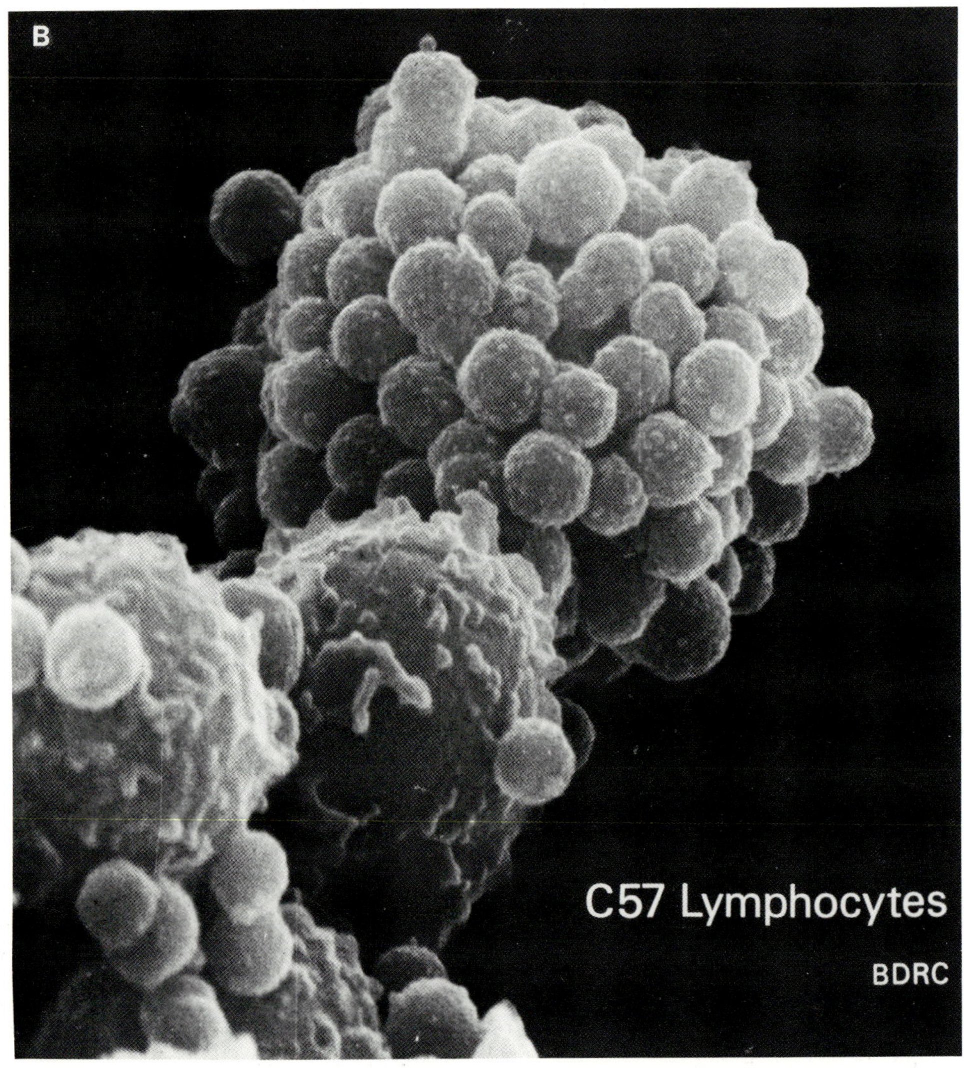
B
C57 Lymphocytes
BDRC

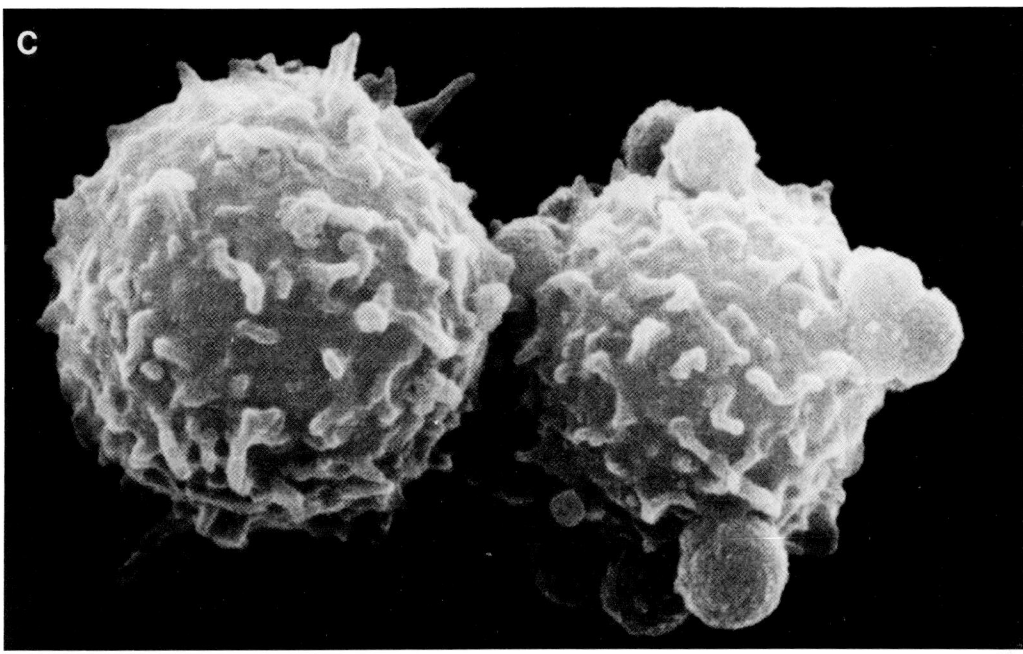

Fig. 1. Scanning electron micrographs of BAH microsphere-stained mouse lymphocytes. **A:** Negative control cells (BALB/c) stained with BAH microspheres conjugated with antimouse IgD^{5b} antibody. **B:** Lymphocytes from a C57Bl/6 mouse. **C:** Stained and unstained lymphocyte from a C57Bl/6 mouse. On the basis of flow cytometric analysis, the average positive cell was stained with about 40 microspheres. $\times$18,000.

16 hours at 4°C with the monoclonal mouse antibody at the rate of 1 mg goat antimouse microspheres for each 100 μg monoclonal antibody. The microspheres were washed several times by centrifugation and stored at 1 mg/ml at 4°C.

Mouse spleen cell preparation. Female mice about 6 weeks old were used in these experiments. They were killed by cervical dislocation, and the spleens were removed. The spleen was then minced with fine scissors and the cells expressed between the frosted ends of two microscope slides. Large pieces of debris were removed by filtration through a fine mesh in a minimum amount of Hanks' balanced salt solution (HBSS). The lymphocytes were collected on a Ficoll–Hypaque gradient and washed several times with HBSS and resuspended in HBSS at a concentration of 2×10^6 cells in 50 μl.

Human peripheral blood lymphocytes. Whole blood was collected from healthy donors in heparin- or EDTA Vacutainer™ brand evacuated blood collection tubes. Lymphocytes were obtained by standard separation on Ficoll–Hypaque gradients. The cells were then handled as described above for mouse cells. The antibodies used in these studies were anti-Leu 1, anti-Leu 3a and anti-Leu 4 (Becton Dickinson, Mountain View CA).

Cell staining. Fifty microliters of HBSS containing 2×10^6 cells were placed in one well of a standard 96-well microtiter plate; 10 μg of prepared beads was added to the cells in the well, and incubated for 1 hour at 37°C. The cells were washed three times by centrifugation at 800*g* with HBSS; the final cell pellet was resuspended in 50 μl HBSS and carefully layered on 2 ml fetal calf serum (FCS) in a 6 × 50-mm glass tube. The cells were centrifuged at 500*g* for 5 min; the pellet was then resuspended in 100 μl HBSS and placed on ice until they were analyzed on the flow cytometer.

Analysis. Cells were analyzed on a FACS IV (Becton Dickinson) equipped with logarithmic amplifiers. We used the 488-nm line of the argon ion laser at 550 mW; the photomultiplier tube voltage was 550 V. Control cells were stained with commercially available reagents (either fluorescein-conjugated antibody or biotinylated antibody coupled with fluorescein-labeled avidin). Because of the greatly amplified signal obtained with the fluorescent microspheres, a 2.0 neutral density (ND) filter was required to keep the labeled cells on scale.

The lymphocyte population was detected using forward-angle light scatter and the fluorescence (either one or two channels) was determined on a cell-by-cell basis. Scatter thresholds were generally set at channels 110–255 and fluorescence at channels 85–255 (with the 2.0 ND filter in the light path).

Conjugation of antibody to the 0.7-μm BAH-allyl fluorescein microbeads. The two-step carbodiimide procedure yielded the most immunoreactive microspheres. A high concentration of carbodiimide (e.g., 2.0 M) at pH 4.5 for 4 hours at room temperature produced microspheres that bound significant amounts of antibody. The amount of antibody attached per unit weight of microbeads was essentially linear with the concentration of carbodiimide used, up to the limits of the carbodiimide. Similarly, we determined the effect of antibody concentration on the uptake of protein by the activated microspheres. Protein concentrations of 25–100 μg/mg microspheres were tested and showed a linear increase in protein uptake as protein concentration increased. Cell-labeling experiments with microspheres carrying varying amounts of antibody showed increased reactivity as a function of increased antibody content. These results are summarized in Table 1.

TABLE 1. Effect of Carbodiimide Concentration and Protein Concentration on Percentage of Cells Labeled with Fluorescent Microspheres[a]

Concentration of EDCI used to activate microspheres	Concentation of anti-IgD^{5b} reacted per mg of microspheres	% Cells stained C57Bl/6 (+)	BALB/c (−)
0	25	0.4	0.3
0	50	0.5	0.3
0	100	0.5	0.3
1.0	25	14.0	1.3
1.0	50	17.0	1.1
1.0	100	18.0	1.6
2.5	25	24.3	1.1
2.5	50	28.8	1.4
2.5	100	32.0	1.3
Soluble staining control (biotinylated anti-IgD^{5b}-fluoresceinated avidin)		32.4	4.6

[a]Cells were labeled with 10 μg microspheres per 2 × 10^6 cells in 50 μl of medium.

Size of microspheres used as cell labels. In earlier experiments, we investigated microsphere size and its effect on the efficiency of cell labeling. Microspheres with diameters of 0.1, 0.4, 0.7, and 1.0 μm were synthesized and evaluated as cell-labeling reagents in the IgD5a/5b model system. In our hands, the 0.7-μm microspheres showed the greatest sensitivity as cell labels and were easy to work with.

Microspheres of this type below 0.5-μm diameter require high speed ultracentrifugation to separate the unbound antibody fraction from the microspheres, and the amount of available fluorescent signal was greatly decreased. Microspheres larger than 1.0 μm tended to be removed from the labeled cells during the washing procedures, presumably because of the increased shear forces encountered by the larger-sized microsphere. Scanning electron micrographs of mouse lymphocytes labelled with 0.7-μm BAH-microspheres are illustrated in Figure 1a–c.

Cell staining with fluorescent microspheres. In the studies that follow, we compared cell staining with both fluorescein- and rhodamine-containing fluorescent microspheres. They were essentially of the same size (0.7 μm) and were each conjugated to an affinity-purified Fc-specific goat antimouse IgG under the same conditions. In other experiments, we had seen essentially the same cell-staining characteristics with microspheres containing either the fluorescein or rhodamine conjugates. Control cells were stained with a biotinylated antibody followed by reaction with fluorescein-conjugated avidin. All cells were analyzed on the FACS IV using the 488-line from the argon-ion laser. Although this is far from the optimum excitation wavelength of rhodamine, sufficient signal was present to yield a minimum 20-log channel difference between the positively stained cells and the autofluorescent cells (Fig. 2c). The amplification obtained with the fluorescein-containing microspheres is most apparent by comparing Figure 2a and 2b. Cells labeled with the conventional staining reagents (Fig. 2a) are very poorly separated from the autofluorescent cells, and nonspecific staining of the negative control cells was 5%. Cells from the same mouse are clearly differentiated from the autofluorescent cells with the fluorescein-containing microspheres (Fig. 2b). The two peaks are separated by more than 50-log channels, with a 2.0-ND filter in the light path. On the basis of the ND filter alone, this represents an amplification of at least 100× versus the conventional reagents. The BALB/c cells showed very low levels of nonspecific staining (2–3%) with either microsphere preparation. The average number of microspheres per cell in both histograms is about 40, with a maximum of 80 in the highest channel counted. We consider a cell labeled with one microsphere to be specifically labeled; however, fewer than 5% of the stained cells are labeled with one microsphere. We found almost no coincidence counting when the cells were washed properly.

Staining of human peripheral blood lymphocytes with fluorescent microspheres. We also evaluated the use of fluorescent microspheres as cell-staining reagents for human peripheral T cells. Again, using the 0.7-μm BAH-microsphere conjugated with affinity-purified Fc-specific goat antimouse IgG reacted with specific monoclonal antibody, we found highly specific staining of these human cells as well.

Anti-Leu 3a, a monoclonal antibody specific for the helper/inducer subset of human T cells, was attached to the microspheres via the goat antimouse IgG conjugated to the microspheres and used to stain human peripheral blood lymphocytes. Figure 3c shows the histogram of cells stained with a fluorescein conjugate of the antibody; 52% of the cells are stained. Figure 3d shows the histogram of cells stained with fluorescent microspheres; 50% of these cells are stained. Control microspheres conjugated with antimouse IgD^{5b} were exposed to the human cells under the same conditions and stained 2.3% of the cells (Fig. 3b). As would be expected, there is a wide separation between the specifically labeled and the autofluorescent cells as was seen with the mouse cells.

The same staining characteristics were obtained with other antihuman T-cell antibodies tested. Cells stained with goat antimouse microspheres conjugated with anti-Leu 1 (a pan T-cell reagent that reacts with greater than 95% of human peripheral T cells) and with the fluorescein conjugate of the same antibody are shown in Figure 4a and 4b. The same population of cells was stained by both reagents, with the microsphere-labeled cells showing clear differentiation from the unstained, unreactive cells.

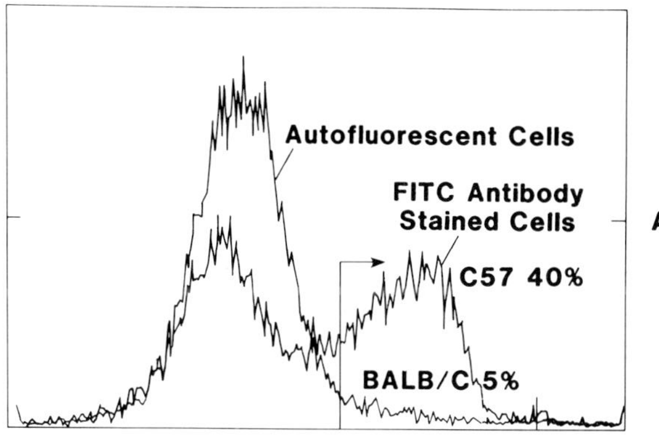

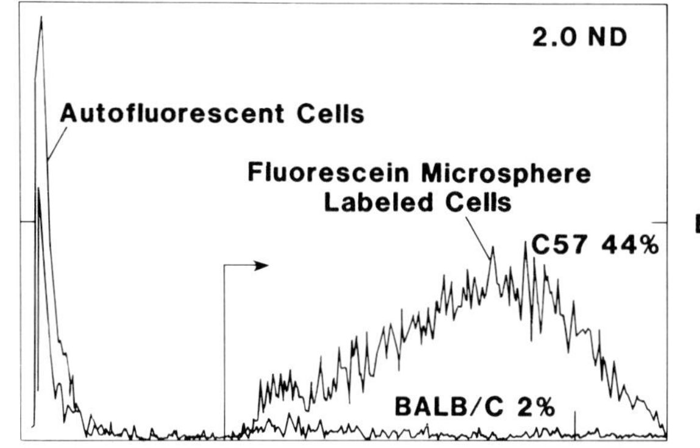

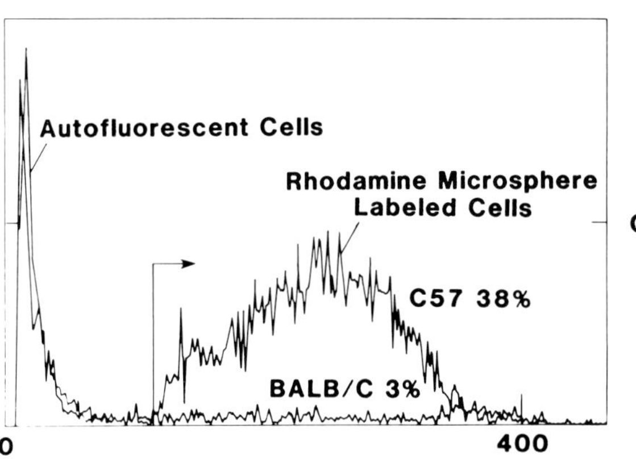

Fig. 2. Histograms of mouse lymphocytes stained with **(A)** conventional staining reagents, **(B)** BAH-allyl fluorescein microspheres, and **(C)** BAH-allyl rhodamine microspheres. A 2.0 neutral-density filter was in the light path for histogram 2b. All experiments employed the 488 nm line of the argon-ion laser. The maximum excitation of the rhodamine derivative used was 560 nm.

These types of microspheres were also used in 2-color cell staining with human peripheral T cells. In one experiment, anti-Leu 1 was attached to the rhodamine-containing microspheres. Anti-Leu 2a (a T-cell-specific monoclonal antibody specific for the suppressor/cytotoxic subpopulation of human peripheral T cells) was attached to the identical fluorescein-containing microspheres. The prepared human peripheral blood lymphocytes were reacted with each type of microsphere either separately, or with a 1 : 1 mixture of both types of labeled microspheres. In the first trial, the results given in Table 2 were obtained. When the microspheres were used separately, 64% of the cells showed positive staining with the anti-Leu 1 (red) microspheres, and 25% positive staining with the anti-Leu 2a (green) microspheres. The mixture of the two microspheres showed slightly higher (e.g., about 5%) results with the dual-parameter analysis, indicating a probable overlap of fluorescence spectra. All Leu 2a-positive cells are also Leu 1-positive (although the converse is not true), and this experiment showed that both cell types could be detected accurately with one sample and one reagent (that is, a mixture of the two different-colored microspheres.

We then repeated this experiment, but in this case we evaluated each monoclonal antibody on each type of microsphere, both separately and mixed. The results indicated that in spite of some overlap in the fluorescence spectra of the two fluors, 2-color staining with microspheres loaded with fluorescein or rhodamine is possible and yields results that compare favorably with results obtained with soluble antibody staining. (Table 3)

Fluorescent microspheres are easy to use as cell-labeling reagents and offer great signal amplification. In some cases, for example, background staining was lower with the microsphere method than with conventional staining reagents, even though the amount of signal was increased by a factor of about 1,000. The method appears to be a general one, applicable to both T and B lymphocytes, and is presumably adaptable to many other cell types. We have conjugated a number of different monoclonal and affinity-purified polyclonal antibodies to the microspheres with the methods described here. The microspheres can be loaded with a variety of fluorescent dyes and used in two-color staining protocols. In general, fluorescent microspheres can be effectively used as cell-labeling reagents with a high degree of sensitivity and specificity.

OTHER APPLICATIONS OF MICROPARTICLES

The high signal obtained with fluorescent microspheres makes them particularly useful when trying to analyze rare cells among large populations of other cells. Cupp and collaborators [6] took advantage of this in their studies of transplacental passage of fetal blood into maternal circulation. They reported detecting one Rh-positive erythrocyte among 100,000 that were Rh-negative because it bound fluorescent microspheres that gave more than 100 times higher signal than ordinary indirect immunofluorescence technique, with an increase in the signal-to-noise (S/N) ratio. Likewise, signal amplification was important for the detection of Fcε receptors and IgE-binding factors on human lympoid cells and lymphoid cell lines as reported by Bonnefoy et al. [4].

Nonfluorescent microparticles have been employed to establish an assay for circulating immune complexes [18]. In this system, the beads are coated with C1q and incubated with serum samples from patients and controls. Immune complexes binding to the beads are then visulized by addition of fluorochrome-conjugated antibodies and are finally analyzed by flow cytometry. This assay proved to be several orders of magnitude more sensitive than a standard C1q-binding assay.

Flow cytometry also offers possible light-scatter signals for marker identification. Böhmer and King [5] employed colloidal gold particles to increase the 90° sidescatter signal. The magnitude of the amplification was satisfactory for quanti-

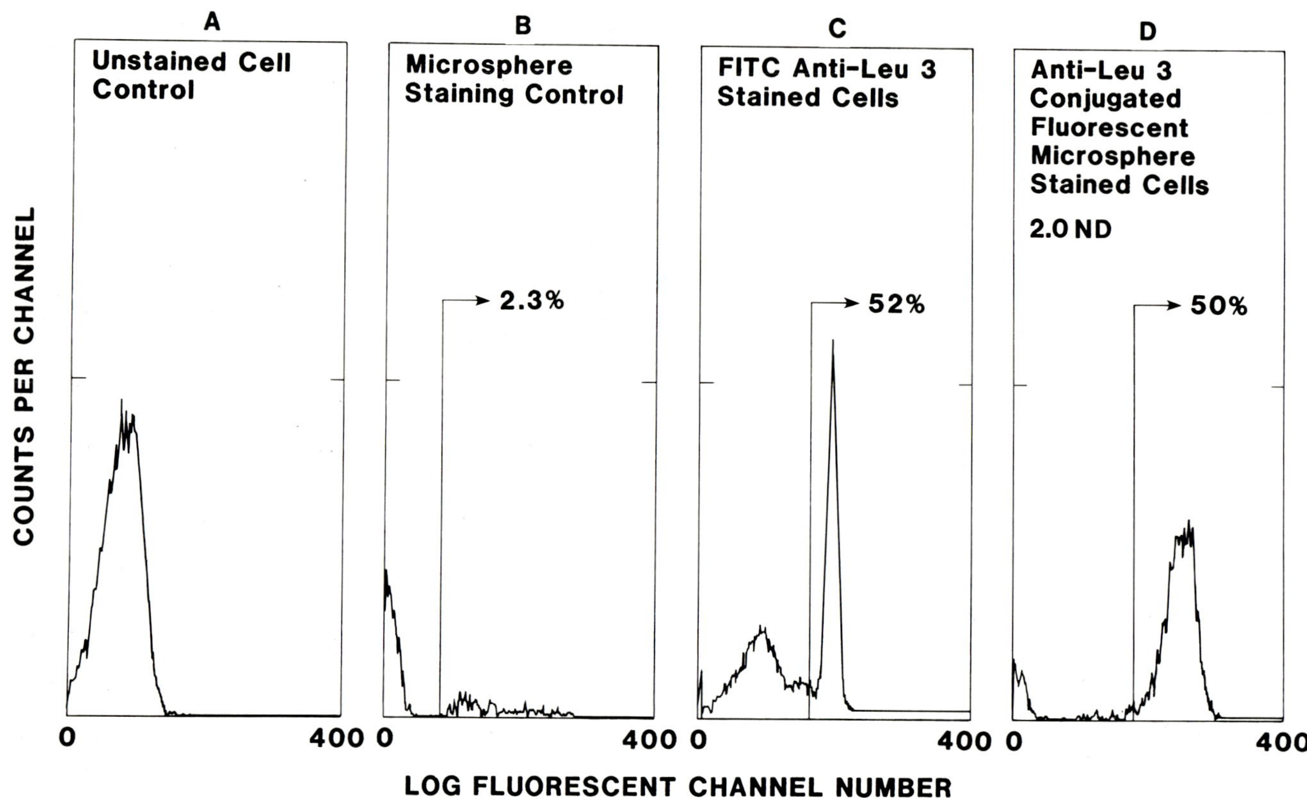

Fig. 3. Histograms of human lymphocytes stained with FITC-anti-Leu 3 reagents. **A:** Unstained control cells showed essentially no fluorescent cells beyond channel 95. **B:** Microsphere-stained control cells were stained with equivalent amounts of microspheres labeled with antimouse IgD^{5b}; 2.3% of these cells were stained. **C:** Cells stained with the FITC conjugate of the monoclonal antibody directly showed 52% positive cells, with a shoulder of questionable cells between the two peaks. **D:** Cells from the same sample stained with the fluorescent microspheres. A 2.0 ND filter was in the light path; 50% of the cells were stained, and the stained cell peak is clearly separated from the autofluorescent cell peak.

tative discrimination between positive and negative cells. They also combined fluorescein- and colloidal gold-labeled antibodies and found that these labels did not interfere with each other. Thus, colloidal gold may be used as an additional label for dual-parameter analysis and cell sorting.

The above studies were recently extended by Festin and collaborators [9]. Using a single laser flow cytometer, they demonstrate that immunogold staining of cells can be combined with ordinary two-color staining, employing standard red and green fluorochrome-conjugated antibodies, for the simultaneous detection of cells expressing three different surface markers. This methodology offers interesting possibilities in the identification of discrete steps in cell-differentiation pathways.

IMMUNOMAGNETIC CELL SEPARATION

Cell sorting by flow cytometry has long been the method of choice for the isolation of specific cell populations by surface markers. This technique has made possible functional studies of highly purified cell populations in vitro. However, cell sorting by flow cytometry suffers from several drawbacks, especially low speed and sterility problems. Also, the equipment is very costly, making the technique available only to a small number of laboratories.

During the past few years, isolation of cells by antibody-coupled magnetic beads has been developed into a reliable tool for the purification and characterization of lymphoid cell populations and subpopulations [10,16,34]. This technique can handle large numbers of cells, is very fast, and introduces no problems regarding sterility. The purity of the isolated cells compares well with electronic cell sorting. The success of this technique is due to the development of the monodisperse superparamagnetic microspheres by Ugelstad et al. [31].

Magnetic microspheres were first introduced by Guesdon and Avrameas in 1977 [11], who were able to prepare magnetite-containing particles by polymerization of acrylamide and agarose. Because of the acrylamide-derived amino

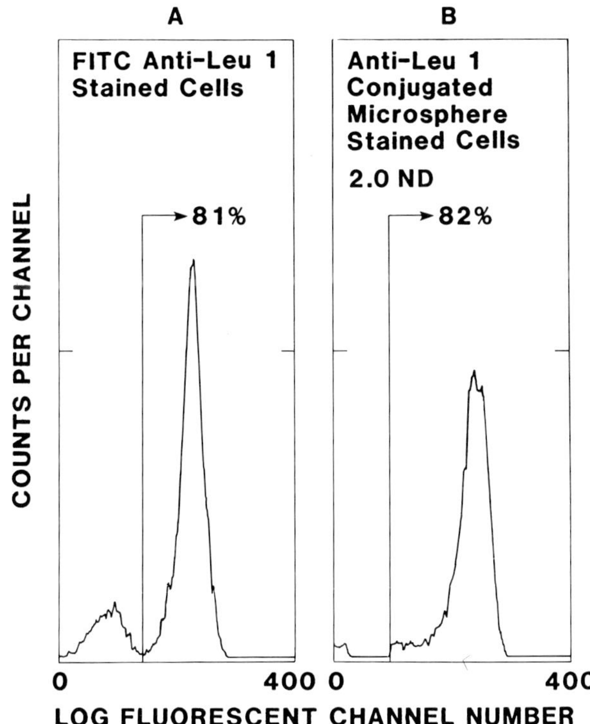

Fig. 4. Human lymphocytes stained with **(A)** FITC-anti-Leu 1 showed 81% positive cells. In **(B)** BAH-allyl fluorescein microspheres labelled with anti-Leu 1 used as a staining reagent yielded 82% positive cells, with a nearly 800 fold increase in fluorescence intensity; see text for details.

groups, protein could easily be coupled to the beads by crosslinking with glutardialdehyde.

Magnetic microspheres have also been prepared from starch, dextran, agarose, and albumin for specific purposes, but they have never come into general use. In 1980, Kronick reported magnetic particles prepared by polymerization of acrylates, methacrylates, and styrene [13]; Rembaum et al. [24] later described several techniques to prepare magnetic microspheres.

Monosized, Magnetic Microspheres

Monosized, magnetic microspheres are prepared from highly monodisperse macroreticular polymer particles according to the method devised by Ugelstad and collaborators [28–30]. The particles are then made magnetic by a process in which magnetic iron oxides are deposited as small grains evenly throughout the macroporous beads. The particles prepared in this way are highly porous, with a surface area of approximately 100 m^2/g. To reduce the surface, the pores are finally filled with polymeric material. In this process the particles may be equipped with a variety of chemical groups, such as hydroxyl, epoxy, isocyanate, and anhydride groups, which can later be used for covalent coupling of various ligands.

The particles that so far have been most successfully used for cell-separation purposes are commercially available as Dynabeads M-450 (Dynal A/S, Oslo, Norway). They have a diameter of 4.5 μm with an iron content of around 20%. The iron is present in the particles as maghemite, γ-Fe_2O_3. A property of major importance is their lack of magnetic memory. When cells are rosetted with magnetic particles and collected by a suitable magnet, they should be easy to redisperse for efficient washing. This is usually not the case with magnetic microspheres. However, magnetization curves of the M-450 beads showed no hysteresis and consequently, both coercivity and remanence are zero.

Because they are hydrophobic, the M-450 particles will bind protein by physical adsorption only. Thus the easiest way to bind antibodies to the beads is to mix protein and beads in the proper ratio and incubate overnight in a cold room [16]. Excess protein-binding sites should then be blocked by addition of an irrelevant protein such as human serum albumin or FCS. Antibodies adsorbed to the particle surface in this way result in a highly functional bead that will retain its activity for months when stored in the cold.

The M-450 particles also have some residual hydroxyl groups which can be used for covalent coupling of proteins after activation with either sulfonyl chlorides [22] or 2-fluoro-1-methylpyridinium toluene-4-sulfonate [21].

Sensitization of the Particles with Antibodies

So far, mostly antibodies have been used to bind the M-450 beads to their targets. Sensitization with antibodies can be done in principally three different ways: (1) by coupling monoclonal antibodies of the IgM isotype directly to the beads [10]; (2) by first sensitizing the particles with a polyclonal antibody specific for mouse IgG and then add the IgG monoclonal antibodies as a second layer [34]; and (3) by sensitizing the cells with monoclonal antibodies, washing them, and then adding secondary antibody-coupled magnetic beads [15,16]. In general, comparable results are obtained with all three techniques, although sensitization according to the third method would be expected to result in optimal orientation and the highest amount of monoclonal antibodies available for binding.

After appropriate blocking, such antibody-coupled M-450 beads show negligible nonspecific binding to proteins and cells. Thus, their interaction with the targets is wholly determined by the specificity of the monoclonal antibodies.

In many research laboratories, immunomagnetic cell separation has become a valuable tool for isolating homogeneous populations of cells for functional studies. Owing to the availability of a series of monoclonal antibodies against membrane markers of the cell types participating in the immune response, this system is the best characterized, so far. Figure 5 shows the binding of Dynabeads M-450 to the surface of a human T lymphocyte via monoclonal antibodies against the CD2 pan T marker.

Depending on the purpose of the experiments, positive or negative selection methods can be employed. Positive selection means that a specific cell population is directly isolated from a heterogeneous mixture of cells by monoclonal antibody-coupled magnetic beads. Negative selection means that a cell population is obtained after immunomagnetic depletion of contaminating cells. Although positive selection, resulting in single step purification of a particular cell population, may seem attractive, there are limitations to this approach. This is largely due to problems in removing the beads afterward. However, depending on the surface marker selected for, overnight culture will lead to detachment of a substantial fraction of the magnetic beads. Thus, it has been possible to isolate human T lymphocytes in high yield using

TABLE 2. Two-Color Staining of Human Peripheral Blood Mononuclear Cells with Anti-Leu 1- and Anti-Leu 2a-Coupled Fluorescent Microspheres

Microsphere preparation	Antibody attached	% cells stained: Microspheres separate[a]	% cells stained: Microspheres mixed[b]
BAH–rhodamine	Anti-Leu 1	64	69
BAH–fluorescein	Anti-Leu 2a	25	29

[a]Single-parameter analysis.
[b]Dual-parameter analysis.

TABLE 3. Two-Color Staining of Human Peripheral Blood T Lymphocytes with Soluble Monoclonal Antibodies Compared with Monoclonal Antibody-Coupled Fluorescent Microspheres

		% cells stained			
		Microsphere separate		Microsphere mixed	
Antibody	Soluble Antibody	Fluorescein	Rhodamine	Fluorescein	Rhodamine
Anti-Leu 1	44	51	48	46	44
Anti-Leu 2a	28	24	26	25	30

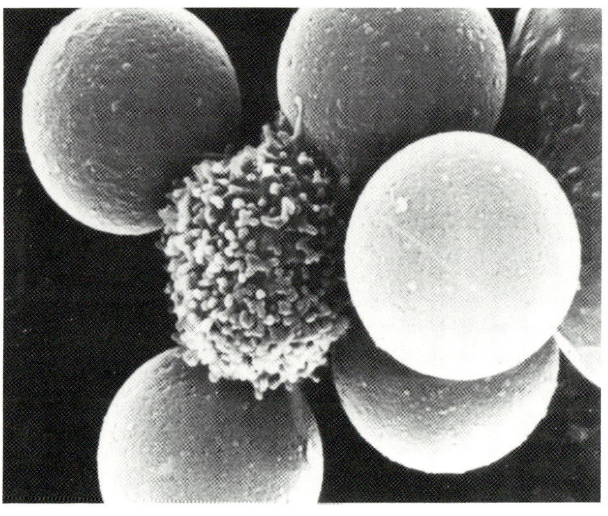

Fig. 5. Scanning electron micrograph of human T lymphocyte sensitized with monoclonal anti-CD2 antibodies and rosetted with Dynabeads M-450 coupled with rabbit antimouse IgG antibodies.

positive selection techniques and monoclonal antibodies against the CD2, CD3, CD4, CD6, and CD8 markers [10,15]. Likewise, activated T lymphocytes expressing the CD25 antigen, the receptor for interleukin 2 (IL-2), can be obtained in close to quantitative yields [15].

When using the positive immunomagnetic cell separation technique, one should be aware of the possibility that the marker selected for might be involved in signal transduction processes. Thus, the cells might not be in their resting state after isolation. This problem is circumvented by negative selection procedures. In the latter case, however, it might be difficult to obtain a completely homogeneous cell population due to lack of monoclonal antibodies against all contaminating cells.

The following discussion demonstrates how immunomagnetic beads can be used to obtain highly purified cell suspensions for further studies in vitro. Judged by relevant markers, the purity obtained is regularly better than 99%. In many cases this situation is of vital importance, especially if the purpose of the study is to monitor events that could be influenced by direct contact with other, contaminating cell types or soluble factors from the same cells.

To illustrate the versatility of immunomagnetic cell separation, we have chosen to present data from both positive and negative selection experiments. First, we tried to establish sublines of human T-cell lines (TCL) according to their expression of several membrane phenotypic markers to study events in the activation of human T lymphocytes. Second, we used monoclonal antibodies against the CD37 pan B lymphocyte marker to isolate a homogenous population of resting human peripheral blood B lymphocytes. The purity and the functional characteristics of these cells will be described.

Establishment of T-Cell Lines of Defined Phenotypes

In many experimental situations, knowledge of essential phenomena in cell biology and immunology has been obtained through studies of malignant cell lines and how they respond to various stimuli in vitro. To delineate early events in the activation of human T lymphocytes, important information has been obtained from experiments with the Jurkat cell line [14,35,36]. Most available cell lines adapted to grow in vitro have been cultured under variable conditions in laboratories all over the world. Through the years, they have attained different surface phenotypes. In the T-cell system, several of the known surface phenotypic markers have been suggested to take part in the activation process. Thus, both the CD2, CD3, CD4, CD5, and CD8 molecules have been assigned roles in the transduction of signals affecting the stimulation of human T lymphocytes [2,3,7,19,33].

To be able to better define the functional role of the various T-lymphocyte-specific membrane molecules, we have isolated several sublines of established human T-cell lines. These cell lines were either homogeneous regarding the expression of one or several of these membrane markers or completely devoid of the same markers. This situation allowed us to carry out experiments under defined conditions

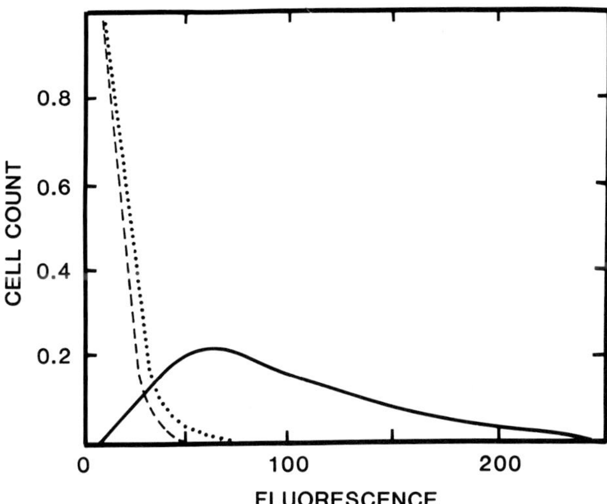

Fig. 6. Flow cytometric analysis of positively selected CD2⁺ T cells from the CEM cell line (—). (. . . .) designates the same T-cell line after immunomagnetic depletion. (— — —) indicates control ascitic fluid matched for isotype. Indirect immunofluorescence technique was used throughout; 50,000 cells were counted in each instance. Histograms were normalized and superimposed on each other to facilitate comparison.

and with the necessary homogeneous populations of control cells.

The surface phenotype of the various cell lines was established by indirect immunofluorescence technique and flow cytometry. The following monoclonal antibodies were used: OKT11 (anti-CD2, Ortho Diagnostic Systems, Raritan, NJ), anti-Leu 4 (anti-CD3, Becton Dickinson, Mountain View, CA), anti-Leu 3a (anti-CD4, Becton Dickinson), anti-Leu 1 (anti-CD5, Becton Dickinson), T12 (anti-CD6, Coulter Immunology, Hialeah, CA), BMA 0112 (anti-CD7, Behringwerke, Marburg, West Germany), and 2D4 (anti-CD8, produced in our own laboratory). Staining protocol was as given above. The flow cytometric analysis was carried out on an Argus flow cytometer equipped with a mercury arc lamp (Skatron A/S, Lierbyen, Norway).

Immunomagnetic Cell Separation

To obtain sublines devoid of specific markers, thorough depletion procedures were carried out using the M-450 beads sensitized with rabbit antimouse Ig antibodies in the first layer and by the relevant monoclonal antibodies given above, in the second. A high ratio of particles to cells (30 : 1) was used to achieve complete depletion. A typical result from such a negative selection experiment is shown in Figure 6. Before separation, 45% of the cells expressed the CD2 marker. After a single depletion step carried out in less than 20 min, the cell suspension was close to devoid of cells expressing the CD2 marker. The homogeneity of this subline is further corroborated by the fact that it has demonstrated the same phenotype after more than 4 months of continuous culture.

To isolate cells carrying the CD2 marker, the positive selection technique was employed. To pick out only those cells with the highest expression of this particular membrane molecule, a low ratio (3 : 1) of beads to cells was used. After collecting the rosetting cells on a suitable magnet, they were thoroughly washed and resuspended in medium for overnight culture. The day after, and the following days, cells free of magnetic beads were collected and pooled. The positively selected cells were then expanded in bulk culture for analysis by flow cytometry. The established cell line was strongly positive for the selected marker (Fig. 6).

Isolation of Human B Lymphocytes by Positive Selection

Human peripheral blood leukocytes (PBL) were isolated by density-gradient centrifugation on Lymphoprep (Nycomed A/S, Oslo, Norway) according to standard procedure. The cells were adjusted to a cell concentration of 5×10^6/ml in cold HBSS, and a saturating amount of monoclonal antibodies specific for the CD37 pan B marker (clone HH1, S. Funderud, The Norwegian Radium Hospital, Oslo, Norway) was added. The cell suspension was incubated for 45 min at 4°C and then thoroughly washed in cold HBSS. Cell concentration was then adjusted to 3×10^7 cells/ml in HBSS containing 1% FCS; M-450 beads, coupled with rabbit antimouse Ig antibodies, were added in amounts corresponding to two to three beads per anticipated B cell. The cell–particle suspension was placed on ice to avoid phagocytosis of the beads by monocytes. During a 20-min incubation period the culture tube was gently swirled every 5 min. The suspension was then diluted 10 times with HBSS containing 1% FCS; the tube was placed on a suitable magnet to attract the rosetted cells and the surplus of unreacted beads. Unrosetted cells were poured off, and the rosetted cells were washed five times by redispersion in a large volume of HBSS–1% FCS followed by collection on the magnet. After the final wash the cells were counted and suspended in culture medium (RPMI 1640 with 10% FCS) in a concentration of 10^6/ml and put in a CO_2-incubator overnight.

The following day, a substantial fraction of the cells were free of magnetic beads and could be directly poured off after the culture flask was placed on a magnet. The rest of the cell–particle suspension was diluted in a small volume of culture medium and pipetted through a narrow-bore Pasteur pipet a few times. Then the suspension was diluted 10 times once again, placed on the magnet, and particle-free cells poured off. In the particular experiment described here, total recovery of 85% of the B lymphocytes was achieved.

Characterization of the Isolated B Lymphocytes by Flow Cytometry

Indirect immunofluorescence technique was used throughout. The cells were stained according to standard techniques with the following monoclonal antibodies: AB1 (anti-CD19, pan B marker) (S. Funderud, The Norwegian Radium Hospital, Oslo); OKT11 (anti-CD2, pan T marker) (Ortho Diagnostic Systems, Raritan, NJ); 1D5 (monoclonal antibody shown to react with the majority of peripheral blood monocytes) (G. Gaudernack, Rikshospitalet, Oslo). FITC-conjugated rabbit antimouse Ig antibodies (F/P ratio = 3.5) were employed as second layer. Fluorescence was monitored with an Epics V flow cytometer (Coulter Electronics, Hialeah, FL) equipped with an argon laser.

The isolated B lymphocytes were further characterized for expression of the 4F2 activation marker [8], after being stimulated with $F(ab')_2$ fragments of polyclonal rabbit antihuman IgM antibodies coupled to agarose beads (anti-μ) (Immunobeads, BioRad Laboratories, CA). Staining and analysis of activated cells were carried out as described above.

Flow cytometric analysis of the isolated B lymphocytes

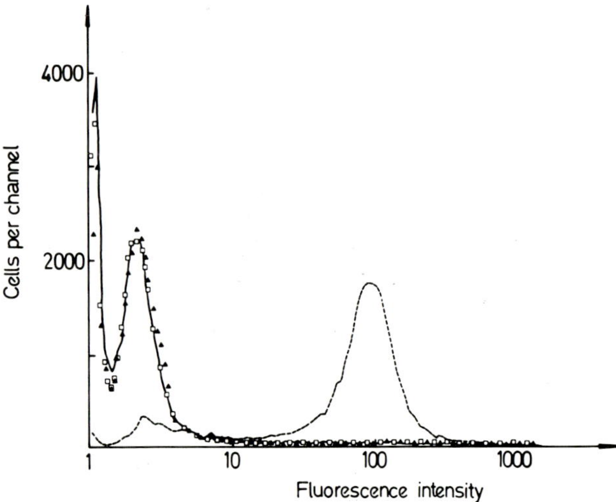

Fig. 7. Flow cytometric analysis of human B lymphocytes positively selected with immunomagnetic beads and HH1 monoclonal antibodies (anti-CD37). The cell population was stained by indirect technique using the monoclonal antibodies given below and FITC-conjugated rabbit antimouse IgG antibodies in the second layer: — — —, AB1 (anti-CD19 pan B-cell antibody); ▲, OKT3 (anti-CD3 pan T-cell antibody; □, 1D5 (monoclonal antibody defining peripheral blood monocytes). The continuous line gives the background staining with control ascitic fluid. The histograms represent the logarithmic fluorescence intensity from 40,000 cells.

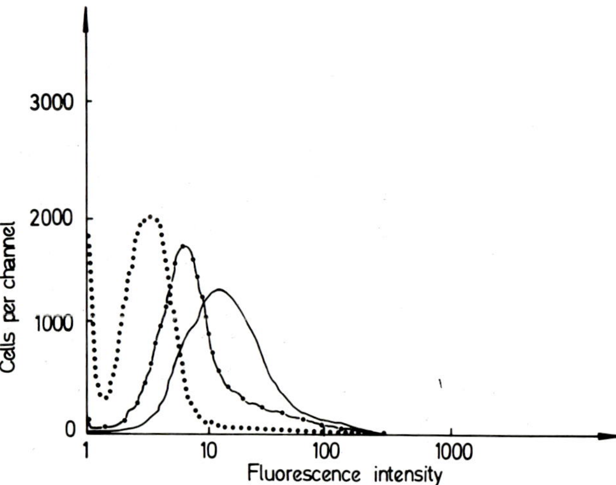

Fig. 8. Cellular fluorescence distribution of positively isolated B cells stained with 4F2 monoclonal antibodies and indirect technique as described above. . . . , background control; _ . _ . _, cells after 20-hour incubation in medium alone; —, cells after stimulation with anti-μ reagent for 20 hours.

showed no contaminating T lymphocytes or monocytes, as shown in Figure 7. Staining with anti-CD19 antibodies resulted in a brightly fluorescent, homogeneous peak of B lymphocytes.

After activation of the cells with anti-μ-coupled agarose beads, expression of the 4F2 marker was monitored during consecutive intervals. As demonstrated in Figure 8, the cells showed increased expression of this particular activation antigen, as expected. These data also indicated that the isolation procedure by itself did not induce 4F2 expression.

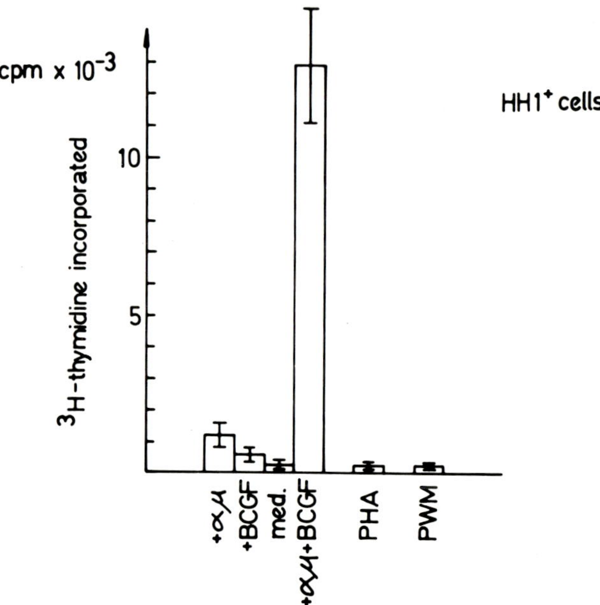

Fig. 9. DNA synthesis data showing the response of positively isolated HH1-positive B cells to the listed stimuli. The histograms represent means ±SD of five experiments.

Functional Properties of the Isolated B Lymphocytes

The cells were activated with anti-μ-coupled agarose beads in the presence and absence of a BCGF-1 preparation (Cytokine Technologies, NY). The cells were seeded in the wells of 96-well tissue culture plates giving 75×10^3 cells per well. Culture medium was used throughout. Titrated amounts of anti-μ-coupled beads were added. The plates were incubated for 48 hours. Then 0.5 μCi [^{3}H]methyl thymidine was added to each well, and cultivation continued for 24 hours before harvesting. Cell proliferation was measured by liquid scintillation counting.

The ability of the isolated B lymphocytes to respond to a polyclonal activator such as anti-μm antibodies in the presence and absence of BCGF-1 is demonstrated in Figure 9. The cells did not respond with proliferation either to stimulation with anti-μ or BCGF on their own, but added together strong synergy appeared. The lack of response of the cells to potent mitogens like phytohemagglutinin (PHA) and pokeweed mitogen (PWM) is also shown. The latter data add further proof of the purity of the cells under study, since even slight contamination of T lymphocytes and accessory cells would have resulted in a detectable proliferative response with PHA.

These few examples demonstrate the versatility of immunomagnetic cell separation in biomedical research. The procedures are fast and specific and can handle large amounts of cells. In many situations, the cells can be isolated in a single step directly from blood or other biological fluids. The purity obtained is consistently better than 99% and makes possible studies of homogeneous cell populations under well-defined conditions. As a complement to other techniques, such as flow cytometry, immunomagnetic particle technology offers important advantages, particularly in the study of phenomena related to cell communication and stimulation.

REFERENCES

1. **Bangs LB** (1984) Uniform latex particles. Indianapolis: Seragen Inc., pp 40–42.
2. **Bank I, Chess L** (1985) Perturbation of the T4 molecule transmits a negative signal to T-cells. J Exp Med 162:1294–1303.
3. **Biddison WE, Rao PE, Talle MA, Goldstein G, Shaw S** (1982) Possible involvement of the OKT4 molecule in T-cell recognition of class II HLA antigens. Evidence from studies of cytotoxic T lymphocytes specific for SB antigens. J Exp Med 156:1065–1076.
4. **Bonnefoy JY, Banchereau J, Aubry JP, Wijdenes J** (1986) A flow cytometric method for the detection of Fcε receptors and IgE binding factors using fluorescent microspheres. J Immunol Methods 88:25–32.
5. **Böhmer R-M, King, NJC** (1984) Immuno-gold labelling for flow cytometric analysis. J Immunol Methods 74:49–58.
6. **Cupp JE, Leary JF, Cernichiari E, Wood JCS, Doherty RA** (1984) Rare-event analysis methods for detection of fetal red blood cells in maternal blood. Cytometry 5:138–144.
7. **Engleman E, Benike CJ, Glickman E, Evans RL** (1981) Antibodies to membrane structures that distinguish suppressor/cytotoxic and helper T lymphocyte subpopulations block the mixed leukocyte reaction in man. J Exp Med 153:193–198.
8. **Fauci AS, Muraguchi A, Kehrl JH, Butler JL** (1984) Activation and immunoregulation of human B lymphocyte function. J Cell Biochem 8A (suppl):96.
9. **Festin R, Björklund B, Tötterman TH** (1987) Detection of triple antibody-binding lymphocytes in standard single laser flow cytometry using colloidal gold, fluorescein and phycoerythrin as labels. J Immunol Methods 101:23–28.
10. **Gaudernack G, Leivestad T, Ugelstad J, Thorsby E** (1986) Isolation of pure functionally active CD8+ T-cells. Positive selection with monoclonal antibodies directly conjugated to monosized magnetic microspheres. J Immunol Methods 90:179–188.
11. **Guesdon JL, Avrameas S** (1977) Magnetic solid-phase enzyme immunoassay. Immunochemistry 14:443–447.
12. **Khaw BA, Scott J, Fallon JT, Cahill SL, Haber E, Homcy C** (1982) Myocardial injury. Quantitation by cell sorting initiated with anti-myosin fluorescent spheres. Science 217:1050–1053.
13. **Kronick PL** (1980) Magnetic microspheres in cell separation. In Catsimpoolas N (ed), "Methods of Cell Separation," Vol. 3. New York: Plenum Press, pp 115–139.
14. **Landegren U, Andersson J, Wigzell H** (1985) Analysis of human T lymphocyte activation in a T-cell tumor model system. Eur J Immunol 15:308–311.
15. **Lea T, Smeland E, Funderud S, Vartdal F, Davies C, Beiske K, Ugelstad J** (1986) Characterization of human mononuclear cells after positive selection with immunomagnetic particles. Scand J Immunol 23:509–519.
16. **Lea T, Vartdal F, Davies C, Ugelstad J** (1985) Magnetic, monosized polymer particles for fast and specific fractionation of human mononuclear cells. Scand J Immunol 22:207–216.
17. **Lobuglio A, Rhinehart J, Balcerzak S** (1972) A new immunologic marker for scanning electron microscopy. In "Scanning Electron Microscopy." Part II. Chicago: IIT Research Institute, pp 313–320.
18. **McHugh TM, Stites DP, Casavant CH, Fulwyler MJ** (1986) Flow cytometric detection and quantitation of immune complexes using human C1q-coated microspheres. J Immunol Methods 95:57–61.
19. **Meuer SC, Hussey RE, Fabbi M, Fox DA, Acuto O, Fitzgerald KA, Hodgdon JC, Protentis JP, Schlossman SF, Reinherz EL** (1984) An alternative pathway of T-cell activation: A functional role for the 50 kD T11 sheep erythrocyte receptor protein. Cell 36:897–909.
20. **Molday RS, Dreyer WJ, Rembaum A, Yen SPS** (1975) Latex spheres as markers for studies of cell surface receptors by scanning electron microscopy. J Cell Biol 64:75–88.
21. **Ngo TT** (1986) Facile activation of Sepharose hydroxyl groups by 2-fluoro-1-methylpyridinium toluene-4-sulfonate: Preparation of affinity and covalent chromatography matrices. Biotechnology 4:134–137.
22. **Nustad K, Danielsen H, Reith A, Funderud S, Lea T, Vartdal F, Ugelstad J** (1988) Monodisperse polymer particles in immunoassays and cell separation. In Rembaum A, Tokes ZA (eds), "Microspheres: Medical and Biological Applications." Boca Raton, Florida: CRC Press, pp 53–75.
23. **Rembaum A, Yen SPS, Cheong E, Wallace S, Molday RS, Gordon IL, Dreyer WJ** (1976) Functional polymeric microspheres based on 2-hydroxy-ethylmethacrylate for immunochemical studies. Macromolecules 9:328–336.
24. **Rembaum A, Yen RCK, Kempner D, Ugelstad J** (1982) Cell labelling and magnetic separation by means of immunoreagents based on polyacrolein microspheres. J Immunol Methods 52:341–347.
25. **Rembaum A** (1979) Microspheres as immunoreagents for cell identification. In Melamed MR, Mullaney PF, Mendelsohn ML (eds), "Flow Cytometry and Sorting." New York: John Wiley & Sons, pp 335–347.
26. **Rhinehart J, Balcerzak S, Lobuglio A** (1971) Study of the malaria-red cell relationship with the use of a new immunologic marker. J Lab Clin Med 78:167–171.
27. **Subba Rao PV, McCartney-Francis NL, Metcalfe DD** (1983) An avidin–biotin microELISA for rapid measurement of total and allergen-specific human IgE. J Immunol Methods 57:71–85.
28. **Ugelstad J, Mørk PC, Mfutakamba HR, Soleimany E, Nordhuus I, Schmid R, Berge A, Ellingsen T, Aune O, Nustad K** (1983) Thermodynamics of swelling of polymer, oligomer and polymer/oligomer particles. Preparation and application of monodisperse polymer particles. In Poehlein GW, Ottewill RH, Goodwin JW (eds), "Science and Technology of Polymer Colloids." NATO ASI Series I. Boston: M. Nijhoff, pp 51–99.
29. **Ugelstad** (1982) Processes for preparing an aqueous emulsion or dispersion of a partly water-soluble material, and optionally further conversion of the prepared dispersion or emulsion to a polymer dispersion when the partly water-soluble material is a polymerizable monomer. U.S. Patent 4,336,173.
30. **Ugelstad J, Mørk PC, Kaggerud KH, Ellingsen T, Berge A** (1980) Swelling of oligomer-polymer particles. New methods of preparation of emulsions and polymer dispersions. Adv Colloid Interfac Sci 13:101–140.

31. **Ugelstad J, Söderberg L, Berge A, Bergström J** (1983) Monodisperse polymer particles—A step forward for chromatography. Nature (Lond) 303:95–96.
32. **Valet G, Ugelstad J, Berge A** (1982) In "Combined International Conference on Analytical Cytology and Cytometry, and the Sixth International Symposium on Flow Cytometry," Elmau, West Germany.
33. **van Wauve JP, De Mey JR, Goossens JG** (1980) OKT3: A monoclonal anti-human T lymphocyte antibody with potent mitogenic properties. J Immunol 124:2708–2713.
34. **Vartdal F, Kvalheim G, Lea T, Bosnes V, Gaudernack G, Ugelstad J, Albrechtsen D** (1987) Depletion of T lyphocytes from human bone marrow. Use of magnetic monosized polymer microspheres coated with T lymphocyte specific monoclonal antibodies. Transplantation 43:366–371.
35. **Weiss A, Stobo JD** (1984) Requirement for the coexpression of T3 and the T-cell antigen receptor on a malignant human T-cell line. J Exp Med 160:1284–1299.
36. **Weiss A, Wiskocil RL, Stobo JD** (1984) The role of T3 surface molecules in the activation of human T-cells: A two-stimulus requirement for IL 2 production reflects events occurring at a pre-translational level. J Immunol 133:123–128.
37. **Wilchek M, Bayer EA** (1987) Labeling glycoconjugates with hydrazide reagents. Methods Enzymol 138: 429–442.

20

Ultrasensitive Molecular-Level Flow Cytometry

James H. Jett, Richard A. Keller, John C. Martin, Dinh C. Nguyen, and George C. Saunders

Chemistry and Laser Sciences Division (R.A.K., D.C.N.) and Life Sciences Division (J.H.J., J.C.M., G.C.S.), Los Alamos National Laboratory, Los Alamos, New Mexico 87545

The traditional domain of flow cytometry has been the measurement of properties of discrete biological particles—primarily cells or chromosomes. As detailed in other chapters in this book, the number and variety of cellular properties that can be measured by flow cytometry is increasing at a rapid rate. In actuality, most flow cytometric fluorescence measurements of cellular properties are determinations of the amount of a specific type of molecule in a cell, such as DNA, or a cell-surface antigen. Since the inception of flow cytometry, there has been a continuing quest for more sensitive instruments and for techniques to detect and quantitate low levels of fluorescence. A generally accepted value for the sensitivity of detection of fluorescein on the surface of a cell is 3,000–5,000 molecules [21]. This value can be attained only for cells with low levels of autofluorescence.

Often, fundamental discoveries have followed upon major advances in detection sensitivity. Techniques to determine the concentration of molecules in solution, along lines to be described later, have led to the detection of a single molecule as it passes through the illuminated volume of a flow cytometer. This chapter describes some of those recent advances in the application of flow cytometry technology to sensitive molecular detection in solutions.

The techniques developed to date can be divided into two categories: direct solution measurements and microsphere-based assays. By adapting flow cytometers to measure the level of fluorescence from a continuous dye stream, measurements of the number of fluorescent molecules in solution have been made with sensitivities greater than those achieved by any reported technique. The ultimate goal of the solution measurements is to increase the sensitivity to the point that a single fluorescent molecule with one chromophore can be detected with high efficiency as it passes through the probe volume.

In the microsphere-based assays, the sphere can be thought of as a reagent carrier, the function of which is to localize the reaction to a small volume that is subsequently probed in a flow cytometer. By employing the specificity of antibody–antigen interactions specifically to bind small fluorescent microspheres to large nonfluorescent microspheres, new homogeneous immunoassays have been developed with unparalleled sensitivities.

The direct solution measurements require that new techniques for signal processing and detection be developed. Until one approaches the level of single molecule detection, the fluorescence from the sample stream can be considered continuous. Thus, unless one modulates the exciting laser beam, the signal to be quantitated is continuous as contrasted to signals obtained from particulate measurements that are typically 2–5 μsec wide. This necessitates the implementation of new (to flow cytometry) signal-processing electronics using phase-sensitive detection techniques. As the single molecule level of detection is approached, another change in detection scheme becomes necessary. To achieve this level of sensitivity, photon-counting techniques are being employed. A continuously scaling electronic module employing event detection based on the total number of photoelectrons detected during a molecular transit time has been constructed.

There are several properties of flow systems in general which make them ideally suited for sensitive molecular detection. The primary attribute of flow cytometers is the small probe volume that is created by hydrodynamic focusing of the sample stream. The hydrodynamic focusing combined with the focused laser beam produces a probe volume typically on the order of a few picoliters and can be as low as 50 fl [36]. Since, in cuvette flow systems with appropriate spatial filtering, the confining walls of the cuvette are relatively far removed from the detection volume, the volume probed is effectively windowless. Thus, background due to scattered excitation light from index of refraction mismatches at liquid–glass interfaces is greatly reduced.

Techniques commonly employed in flow cytometers have also been employed in chromatographic systems. Hydrodynamic focusing has been used to produce a small sample stream detector for liquid chromatography systems [16] and for flow injection analysis [19]. Both enclosed flow chambers [16,19a] and stream-in-air interrogation [11] systems have been developed. To avoid the intense plane of scattered light in the stream-in-air systems, the detection optics viewed the laser beam sample stream intersection from 30° above the

Flow Cytometry and Sorting, Second Edition, pages 381–396

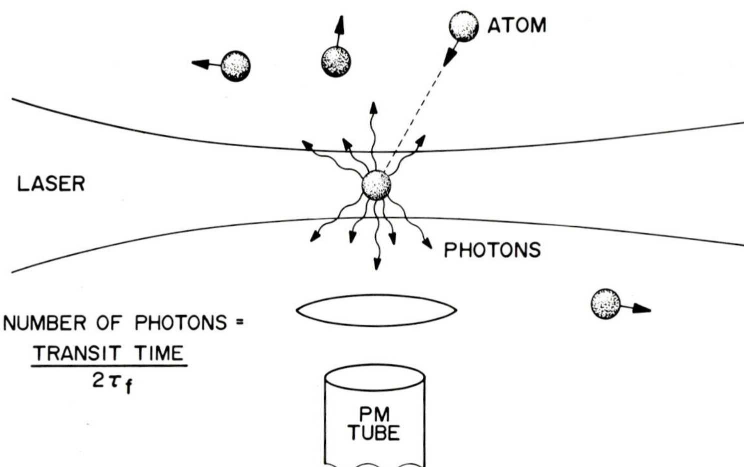

Fig. 1. Schematic representation of photon-burst detection of single atoms as they cross a focused laser beam.

plane of scattered light. This geometry results in a reduction in the background due to scattered excitation light as large as 10^6. The best sensitivity reported [11] was a detection limit of 1.8×10^6 molecules of fluoranthene for an approximately half-second measurement. In general, as the probe volume decreased, the detection limit decreased [5,6]. Recent work using flow techniques reported a detection limit of 3×10^4 molecules of phycoerythrin [22].

Our earlier work reported for direct solution measurements using a more or less standard flow cytometer achieved a detection limit of 22,000 molecules of rhodamine 6G for a 1-sec measurement [5,6]. The hybrid system employed an Ortho flow cell mounted on a FACS II optical bench. The laser beam, modulated at 10 kHz, was focused by a 5-cm focal length spherical lens. Quantitation of the preamplifier signal was accomplished by integrating the output of a lock-in amplifier for 1 sec. Significant improvements in these results are discussed in detail below.

The microsphere-based assay systems use flow cytometers to make measurements of the fluorescence of cellular-sized objects. These assays are homogeneous. That is, to perform the assay, one does not have to separate the bound from the free label. It is the small probe volume inherent in a flow cytometer combined with the ability to trigger data collection by a light scatter signal that makes this type of assay possible. The assays can be performed with a conventional flow cytometer. The sensitivity achieved is equal to or greater than that achieved by conventional radioimmunoassay without the problems associated with separation chemistry or with radiolabeled compounds.

In this chapter, we discuss three areas in ultrasensitive molecular level flow cytometry. First, recent results using flow systems for sensitive molecular detection are described. The next section describes the microsphere-based assays—both competitive binding assays and sandwich assays—that depend on antibody specificities. The third section describes other types of molecular assays that could potentially be performed by flow cytometers.

CONTINUOUS-FLOW MOLECULAR DETECTION

There is continued interest in improving the sensitivity of fluorescence detection. We have demonstrated that the use of focused flow techniques, as practiced in flow cytometry, can significantly improve detection limits in fluorescence analysis [6,5,25,27]. Calculations based on experimental observations and improved apparatus designs allow us to project much better sensitivity, ultimately to the single chromophore limit.

Photon-burst techniques have been used successfully to detect single atoms [1,8,14,29]. An atom traversing a focused laser beam, tuned to a resonance transition, continually recycles between the ground and excited state emitting many photons (Fig. 1). Under optical saturation conditions, the number of photons emitted, N_p, is given by

$$N_p = t/2\tau, \tag{1}$$

where t is the transit time of the atom across the laser beam and τ is the fluorescence lifetime. Only a fraction of these photons is detected. During the mid-1970s, single atom detection was claimed for sodium atoms when the concentration of sodium at the detection limit was such that the probability of an atom being in the detection volume at any instant of time was less than unity [8]. True single-atom detection was demonstrated in the early eighties by directly associating observed photon bursts with individual atoms traversing the focused laser beam [14].

Molecules are harder to detect by photon-burst techniques because:

1. The population of gas-phase molecules is distributed over many rotation and vibration levels. This means that a narrow band laser, tuned to one (or a few) rotational/vibrational transition(s), can excite only a small fraction of the molecules. For this reason, the sensitivity is reduced by ~1/partition function ~(1/1,000).
2. The recycling process, so important for photon burst detection, is interrupted because, in general, molecular emission places molecules in rotation/vibration levels that are not in resonance with the excitation laser. Since the molecule is trapped in these pseudo-metastable levels, only a few photons are emitted during the molecule's transit across the laser beam.
3. The large number of vibration and rotation levels dilutes

the strength of the electronic transition; consequently, molecular excitation cross sections are much smaller than atomic excitation cross sections, again by ~1/partition function.

For these reasons, the best molecular fluorescence detection limits in the gas phase are in the range of 10^4 molecules/cc *in a particular rotation/vibration level.*

Most of the problems associated with the detection of small, gas-phase molecules can be circumvented by studying large molecules in solutions. The energy levels of a large molecule in liquid solution are homogeneously broadened into a pseudo-continuum, resulting in optical transition widths of tens of nanometers. At the same time, energy relaxation in the ground electronic state following molecular emission occurs in picoseconds. These two factors contribute to the elimination of problems (1) and (2) listed above. A narrow-band laser can now address all of the molecules in its focused probe volume, and molecular fluorescence selection rules do not put molecules into states inaccessible to the laser. We expect that $\sim 10^3$ photons will be emitted as a single molecule transits the focused laser beam [27]. The third problem can be reduced if we choose a molecule with a large optical excitation cross section; rhodamine 6G, with a molar extinction coefficient ~100,000 liter · $mole^{-1}$ · cm^{-1}, appears ideal. Although the optical absorption cross section is not as large as that of a fully allowed atomic transition, available laser powers (~1 W focused to a beam waist ~10 μm) are sufficient to saturate the optical transition of rhodamine 6G [5,27] or phycoerythrin [22]. Equally important in rhodamine 6G are the high fluorescence quantum yield (~0.8) and the short fluorescence lifetime (~3 nsec).

Extensive calculations, incorporating the principle of focused flow, convinced us that a properly designed apparatus would have the potential for detecting single molecules. Particularly important attributes of focused flow are (1) the elimination of scattering at index gradients in the laser beam path from the field of view of the detection optics, (2) creation of a small probe volume (10^{-12} – 10^{-15} liter) to enhance fluorescence signals relative to the background and to permit tight focusing of the laser to attain high irradiance, (3) focused flow of all sample molecules through the probe volume, and (4) reduced adsorption of sample molecules onto the walls of the flow cuvette. Although we realized that standard flow cytometry apparatus did not have the required sensitivity for single molecule detection, measurements made on such a system served as a check of our calculations and formed a basis for projection of future improvements. These measurements are described below.

Experimental

The apparatus for solution fluorescence detection is shown in Figure 2. Approximately 1 W of cw laser radiation at 514.5 nm was focused to a diameter of 12 μm in the center of an Ortho Instruments System 50 flow cell. Hydrodynamic focusing confined the rhodamine 6G sample, traveling at a velocity ~0.14 m/sec, to a 20-μm-diameter stream in the center of the sheath flow. Molecular diffusion increased the sample stream diameter to ~40 μm at the point of observation. Fluorescence photons were collected with a microscope objective (40×, 0.55 NA), spectrally filtered, and impinged onto the photocathode of a cooled photomultiplier tube. The amplified photoelectron pulses were counted for 1 sec. Means and standard deviations were calculated by averaging 5–10 1-sec counts for each dye concentration.

High-purity water, treated and filtered through a 0.2-μm pore membrane, was used for both sheath fluid and sample preparation. Rhodamine 6G solutions were prepared within 48 hours of the measurements by serial dilution of a stock 1.2×10^{-6} mol/liter solution. Samples were contained in glass flasks and test tubes during the measurements. Adsorption of the rhodamine 6G solution onto the glass walls was negligible. Solutions were allowed to flow for several minutes to establish equilibrium before measurements were taken. Blank measurements were made with the sample stream off and with only pure water flowing in the sheath.

Rhodamine 6G Results

Figure 3 (originally in color) is a multiple-exposure photograph of the emission from a thousand molecules in the probe volume of the focused laser beam. The spot in the center (bright yellow in the original) is rhodamine 6G fluorescence. The surrounding red emission (shown as lighter, surrounding arcs) is Raman scatter from the stretching modes (centered at 625 nm) of the water solvent. This Raman scatter is removed by optical filters. There is a much weaker Raman scatter, from the water-bending mode (centered at 562 nm), masked by the strong Raman scatter and the molecular fluorescence. It is the *fluctuations* in the Raman scatter from this weak bending mode and background fluorescence from impurities that ultimately determine the detection limit.

Viewing the emission with the detection system described above leads to the results summarized in Table 1. The fluorescence signal as a function of rhodamine 6G concentration is shown in Figure 4. Below 10^{-11} mol/liter, we have difficulty making and handling standard reference solutions. Following standard analytical practice, the detection limit was set at the concentration that would lead to a signal twice the standard deviation of the fluctuations in the background signal. Our results are compared with those of other workers in Table 2. On the basis of the number of fluorophors, the data demonstrate greater than two orders of magnitude higher sensitivity than previous works employing a flowing fluid system. More importantly, we expect to do considerably better with the modifications described below.

At our calculated detection limit, the probability of a rhodamine 6G molecule being in the probe volume is ~0.06. By some definitions, this is single-molecule detection. In truth ~800 molecules pass through the probe volume during our 1-sec apparatus integration time. Since the noise in the background varies as the square root of the measurement time, this detection limit can be used to calculate the number of fluorescent tags necessary on a single species in order for us to detect it with the present apparatus during the 85-μsec transit time through the laser beam. The effective fluorophore detection limit, Ω, is given by

$$\Omega = 800\ (85\ \mu\text{sec}/1\ \text{sec})^{1/2} \cong 8 \qquad (2)$$

Thus, we could presently detect a species containing the equivalent of eight rhodamine 6G fluorophores. We are attempting to reduce this number to one.

In order to minimize the contribution from the background emission, it is important to minimize the probe volume and maximize the molecular transit time. Because fluctuations in the background signal are proportional to the square root of the signal, increasing both the fluorescence signal and the background signal is just as beneficial as reducing the background with respect to the fluorescence.

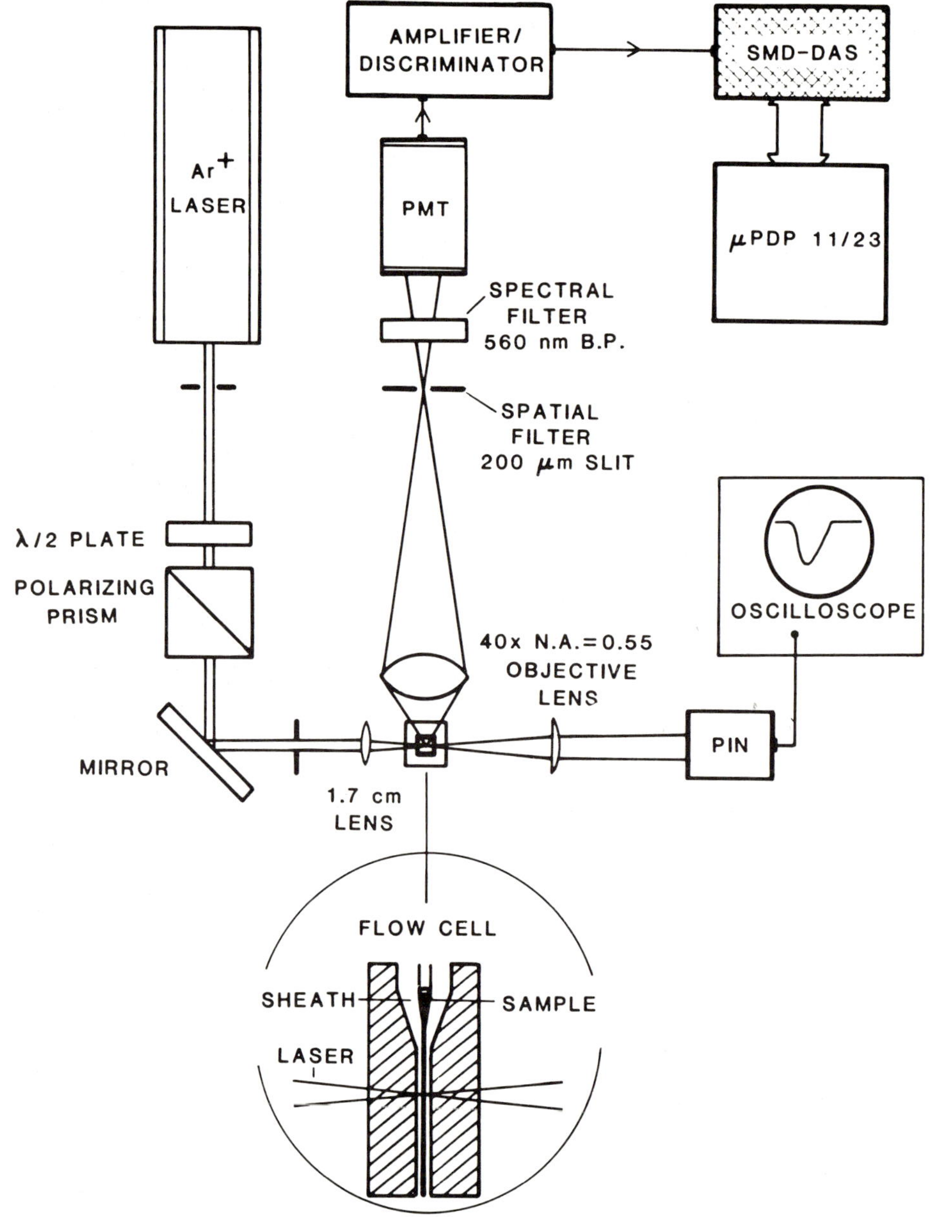

Fig. 2. Schematic drawing of experimental apparatus for experiments achieving a detection limit of 800 molecules of rhodamine 6G.

Phycoerythrin Results

The detection limit described above (a species containing the equivalent of 8 rhodamine 6G chromophores) was derived by extrapolation of the solution measurement results. In order to verify this detection limit, we investigated the detection of individual phycoerythrin molecules. B-phycoerythrin is a large phycobiliprotein containing 34 bilin chromophores [12,13]. When the molar absorptivity at 515.5 nm (1.4×10^{-6} liters · mol^{-1} · cm^{-1}) and the fluorescence quantum yield (0.98) of B-phycoerythrin are taken into account, B-phycoerythrin fluoresces as strongly as 25 rhodamine 6G molecules under saturated irradiation conditions. These properties, coupled with a low photodegradation quantum yield (1.1×10^{-5}) [22], make B-phycoerythrin an excellent candidate for single large-molecule detection.

Experimentally, the apparatus used was as described above and diagrammed in Figure 2 with the following changes. The laser power was 0.1 W. The sheath flow rate was reduced such that the molecular transit time through the laser beam was 180 μsec. With these flow conditions, the sample stream diameter was 21.6 μm, resulting in 33% of the sample passing through the 1.1-pl probe volume defined by the intersection of the laser beam and the detector field of view. Data collection was accomplished with a computer-controlled CAMAC-based multichannel scaler system [26].

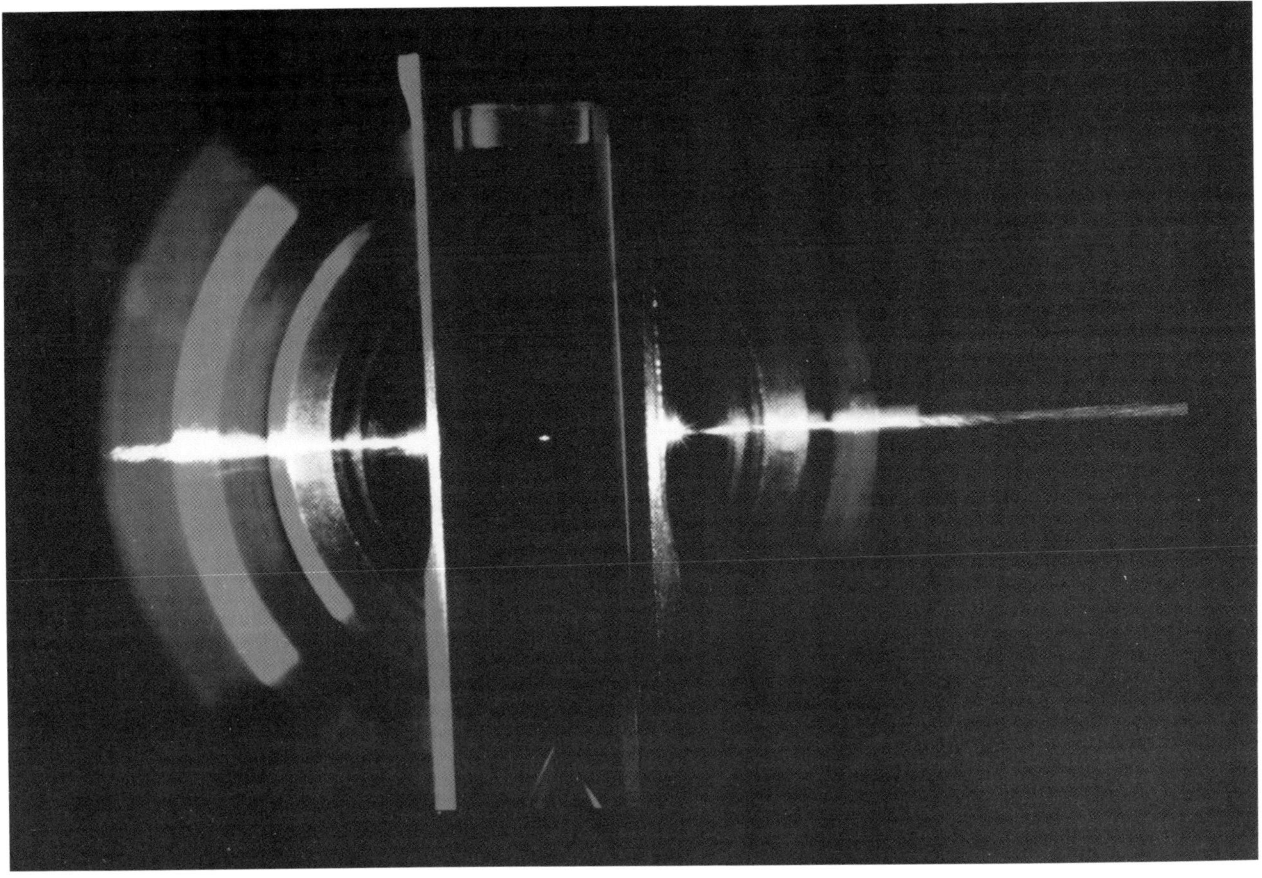

Fig. 3. Photograph of fluorescence from 1,000 molecules of rhodamine 6G (color not shown). The rhodamine fluorescence is the dot in the center of the photograph (yellow in the original), surrounded by wings of light (red in the original) due to Raman scattering of the laser beam by sheath water.

Each data set consisted of the number of photons detected during each of 4,000, 20-μsec time segments.

Data sets with a 10^{-12} M concentration B-phycoerythrin and a blank were recorded and analyzed in the following manner. The sum of the number of photons detected in nine 20-μsec segments corresponds to the total number of photons detected during a molecular transit time. A histogram of the distribution of nine segment sums is shown in Figure 5a. The average number of background counts during a transit time was 4.7. Using the results described above for the 1 second measurements of rhodamine 6G [27], the average number of photons predicted to be detected during the transit of a molecule of B-phycoerythrin through the laser beam is 14 [25]. Figure 5b shows the histograms of the nine segment sums derived from two data sets for a 10^{-12} solution of B-phycoerythrin. As can be seen, the mean number of photons detected during a transit time has shifted upward to yield a mean of 12–14. Events following a Poisson distribution with an estimated mean of 14 will have a 65% probability of having 12 or more counts. Hence, nine segment sums above 12 are attributed to single molecules of B-phycoerythrin. Complete details of the analysis are given by Nguyen et al. [25].

Three criteria were developed to confirm the detection of individual molecules of B-phycoerythrin [25]. First, the number of counts observed per molecule is in agreement with the expected number. Second, based on the sample concentration and the sample flow rate, the number of events with sums of 12 or greater was estimated to be 320. The number observed, 126, is in good agreement with the expected number considering the approximate nature of the calculated number of molecules. Finally, the distribution of the time intervals between events agrees with the calculated interval distribution. In order to detect molecules such as rhodamine 6G with close to 100% efficiency, several improvements in the apparatus must be made. These modifications are discussed in the next section.

Improved Apparatus

The most important improvements that we have made in the design of a more sensitive apparatus are related to the increase in photon collection efficiency and data-handling procedures. The standard Ortho flow cell was modified by grinding away most of the glass on the detection side, leaving only 100 μm (coverslip thickness) between the sheath and the microscope objective. The microscope objective (40×, 0.55 NA) will be replaced by an oil immersion type microscope objective (100×, 1.3 NA), which permits a closer working distance and thereby increased light collection efficiency. A specially designed spherical mirror will be used to reflect photons initially radiating away from the microscope objective back through the source. Index matching fluid will

TABLE 1. Experimental Conditions for Sensitive Fluorescence Measurements[a]

Characteristics	Value
Flow velocity	0.14 m/sec
Laser beam waist, diameter	12 μm
Sample stream diameter (diffusion corrected)	42 μm
Width of detection window image at probe volume	5.0 μm
Probe volume	0.6 pL
Transit time	85 μsec
Laser irradiance	0.7 MW/cm^2
Fluorescence transmitted through spectral filters	0.49
Detection efficiency (pe$^-$/photon emitted)	0.0015
Number of molecules crossing probe volume during 1-sec integration time at detection limit	800
Probability of molecule in probe volume at any instant of time at detection limit	0.06
Number of fluorescence photocounts per molecule during transit time	1.0
Number of background photocounts during transit time	14.5
Number of tags necessary to see a single species at current detection limit	8[b]

[a]See ref [27] for details.
[b]See text for details of calculation.

Table 2. Comparison Laser-Induced Fluorescence Detection Limits

Technique and Reference	Molecule	Year	Number of Molecules
Stain on microscope slide [16A]	FITC	1976	80
Effluent chromatography [4A]	aflatoxin	1977	5×10^7
Focused flow [19]	rhodamine 6G	1981	7×10^6
Effluent chromatography [11]	fluoranthene	1982	1.8×10^6
Focused flow [5]	rhodamine 6G	1984	2.2×10^4
Adsorbed on surface of small glass beads [19A]	rhodamine 6G	1985	8×10^3
Effluent chromatography[a] [24]	coumarin 440	1985	3.4×10^4
Flow[b] [22]	phycoerythin	1986	2.6×10^4
Focused flow [27]	rhodamine 6G	1987	800
Focused flow [25]	phycoerythrin	1987	1

[a]Adjusted to S/N = 2.
[b]Adjusted to correspond to the one second integration time.

be inserted between the Ortho cell, the mirror, and the microscope objective to reduce reflection losses associated with index mismatches. We are now using a photon-counting photomultiplier tube (RCA C31034A34) selected for a quantum efficiency of 24% at 560 nm.

It is important to optimize the optical filters for maximum transmission and maximum ratio of fluorescence to Raman and Rayleigh scattering. The filter set used to obtain the results described above performed well but was not optimized for this task. The increase in the efficiency of the new filter set is given in Table 3.

The flow, laser beam waist, and spatial filtering will be adjusted to obtain a probe volume ~40 fl with a flow velocity of 7 cm/sec corresponding to a molecular transit time of 170 μsec. It is unlikely that much slower flow velocities

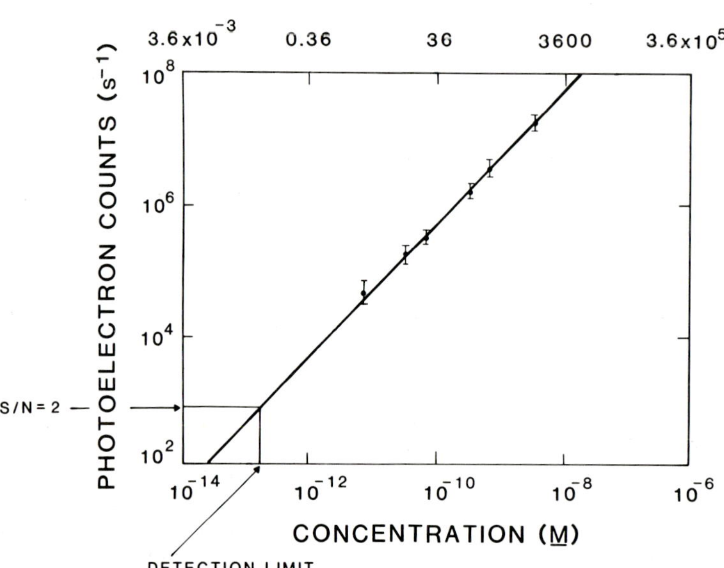

Fig. 4. Photoelectron counts as a function of the concentration of rhodamine 6G. The detection limit was determined as described in the text.

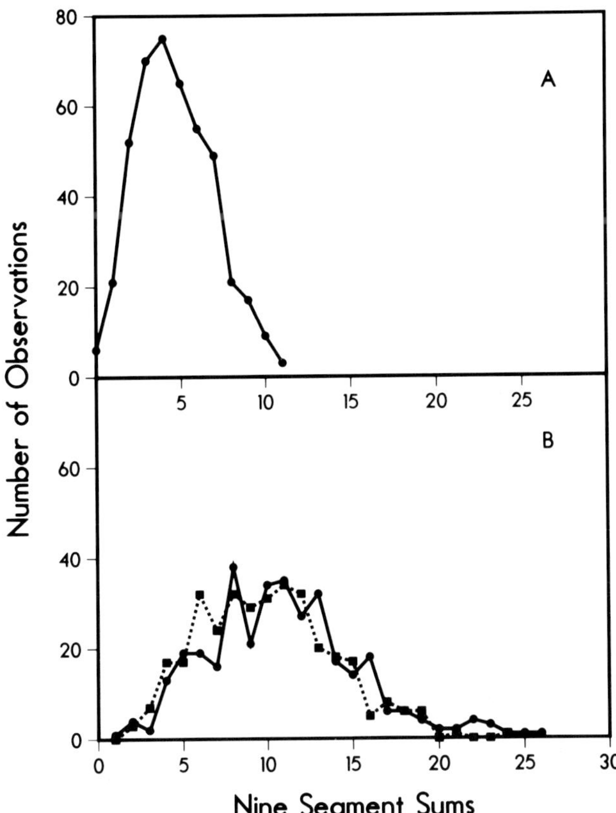

Fig. 5. Histograms of nine segment sums for background **(A)** and for two data sets with a 10^{-12} M solution of B phycoerythrin flowing **(B)**.

will be practical because we are beginning to observe diffusional mixing between the sheath and sample streams.

The capability of detecting single species is significantly enhanced if photon counting and sophisticated signal processing are used to observe photon burst time correlations in the molecular fluorescence [14]. Although both molecular fluorescence and background scatter are detected by the photomultiplier tube, only fluorescence has a time response correlated with the passage of a single molecule across the laser beam. The known transit time is divided into several (10–20 μsec) intervals and a photon count is continuously recorded for every interval using a fast counter. The counts are then summed over the transit time. Since background photons are statistically uncorrelated, their counts average to a DC level with a superimposed fluctuation that has been reduced by the square root of the average. The strongly correlated photons from a single molecule's fluorescence add to give a sum larger than the background level. A threshold can be set so that the probability of detecting a single molecule in the probe volume is high while the frequency of recording a false event can be made quite low [5,27]. A computer-controlled data acquisition system has been designed, built and tested to perform these operations [26].

The laser irradiance of 0.7 MW/cm^2 used in our current measurements is already above the saturation irradiance. Further increase in the laser irradiance will not lead to improved signal-to-noise ratio (S/N).

The modifications discussed above are used to calculate the projected S/N for single molecule detection. The results of this calculation are listed in Table 3. Our projected S/N = $76/\sqrt{415} = 3.7$ for single-molecule detection encourages us to continue in this direction.

TABLE 3. Experimental Modifications to Enhance Detection Sensitivity

Improvements	Background[a]	Fluorescence[a]
Using a 100X, NA = 1.3 immersion-type objective to increase geometric collection and reduce reflection loss	7.0	7.0
Using a concave mirror to retroreflect photons traveling away from the detector	2.0	2.0
Replacing the existing filter with a 560-nm FWHM = 60 nm bandpass filter	1.5	1.6
Reducing the sample flow rate by a factor of 2, to double the transit time	2.0	2.0
Reducing the window width and sample stream diameter to 2 μm	0.4	1.0
Increasing PMT quantum efficiency from 0.16 to 0.24 and reducing reflection loss due to PMT housing	1.7	1.7
Total enhancement	28.6	76.2
Projected no. of photocounts during transit time	415	76

[a]Multiplicative factors.

A powerful technique, not considered in our projected detection limit, that would significantly decrease the Raman background, would be to use a mode-locked Ar^+ pump laser and gated, delayed detection [30]. Another technique is to fold the laser beam so that molecules have to pass through it twice. By correlating the two photon bursts, the discrimination against false events will be greatly improved.

Applications

An increase in the sensitivity of fluorescence detection will have significant impact on many fields.

1. *Sensitive cellular measurements:* In some cases, cell labeling and analysis is limited by the number of receptor sites or by steric hindrance from the antibodies used in present tagging procedures [9]. The ability to detect single (or a few) chromophores should be a significant improvement in this situation.
2. *General molecular detection:* The specificity of antibody–antigen interactions is being combined with the sensitivity of fluorescence detection to provide analytical techniques with the potential for detecting species ranging from chemical agents and hazardous molecules to pathogens. The sensitivity of the antibody/antigen techniques will be increased if individual chromophores can replace the relatively large tags used. For example, phycoerythrin is 1,000 times smaller than the 0.1-μm fluorescent microspheres currently used. In addition, multiple color tagging

and interrogation could lead to schemes with sensitivity perhaps below the 10^{-15} mol/liter level.

3. *High-performance liquid chromatography (HPLC):* HPLC [28] and flow injection analysis [31] have become important tools in analytical and separation chemistry. Current sensitivity and selectivity are limited by fluorescence probe volumes at the exit of the column [28]. Typical probe volumes for state-of-the-art HPLC are 8–10 μl, while probe volumes as small as 50×10^{-15} liter l[36] have been reported for focused flow techniques. Some progress has been made in the mating of these two techniques [19,24], but there is room for further work in this area.
4. *Cellular autofluorescence discrimination:* Often the limit in the number of molecules that can be detected on a cell is not sensitivity but interference from autofluorescence. We are developing techniques for distinguishing fluorescence associated with the tags from autofluorescence [4,30].

AMPLIFIED FLOW CYTOMETRIC FLUOROIMMUNOASSAYS

The introduction of radioimmunoassay (RIA) in 1959 by Yallow and Berson [35] led to a revolution in the quantitative measurement of biochemical compounds. Substances previously unmeasurable by conventional techniques could now be quantitated with high sensitivity and specificity. However, problems associated with RIA, especially the licensing and bookkeeping requirements along with the instability of reagents due to isotope decay, soon led to a search for other types of sensitive non-isotopic methods, including fluorescence-based approaches.

Fluorescence immunoassays (FIA) have the potential for wide use if they offer the sensitivity, precision, versatility, and cost competitiveness of RIA [10]. However, previous use of fluorescent rather than radioactive probes has not proved very satisfactory where assay sensitivity greater than 10^{-10} mol/liter is required. To a large degree, the lack of sensitivity is due to the fluorescent "noise" inherent in the sample being analyzed, e.g., serum [15]. The intensity of this "noise," which is really a background against which the measurement is made, is affected by the number and amount of fluorescing compounds (e.g., proteins, bilirubin) present in the sample. The presence of these compounds also increases the Rayleigh and Raman scattering by the sample. The increased scattering reduces the S/N ratio and results in a loss of sensitivity.

In 1982 Lisi et al. [20] described a FIA that employed flow cytometric (FCM) detection. For this assay, antibody-coated microspheres, sample, and soluble fluorescent antibody are reacted together as in a conventional double-antibody sandwich assay. Laser flow cytometric measurement allowed the omission of separation and washing steps. Gating the collection of fluorescence intensity data by detection of light scattered by the microspheres results in discrimination against fluorescence not associated with a microsphere. Lisi et al. developed an assay for human IgG that could be performed in a 1 : 10 diluted serum matrix and achieved a sensitivity of 6×10^{-11} mol/liter. This assay represented a step forward, but still lacked the sensitivity (10^{-12}–10^{-13} mol/liter) of similar radioactive isotope-based sandwich assays.

At Los Alamos, we have developed separation free, dual microsphere, flow cytometric competitive binding and sandwich FIA with one to three orders of magnitude greater sensitivity than those reported by Lisi et al. [20].

The remainder of this section discusses this technology. As in the Lisi technique, a relatively large (3.5–10-μm diameter) nonfluorescent microsphere is used as an antibody carrier. The essential difference between the two techniques is that instead of using a soluble fluorescent label, we employ very-small volume (0.05–0.15-μm diameter) labeled fluorescent microspheres to which either antigen (competitive binding) or antibody (sandwich) is attached. Since each microsphere contains many tens of molecules of fluorochrome, the fluorescence associated with each antibody–antigen bond that couples a small sphere to a larger sphere is amplified.

The large antibody-coated microsphere serves as the reaction vehicle, while the antigen- or antibody-coated small fluorescent particle serves as the amplified label. In these assays, the number of large particles is limiting (~100,000/ml), while for each large particle present at least 10,000 small particles are made available for interaction. Each large particle provides the vehicle for a separate microassay. Thus the individual FCM measurement of several thousand of these solid-phase carriers in each sample tube yields results of very high precision. Properties inherent to flow systems also allow these microassays to be analyzed without separating free from bound label. The absence of a separation step both facilitates the mechanics of the assay and improves precision, because equilibrium conditions are maintained throughout the measurement process.

The sensitivity and stability of the fluorescence detection system along with the hydrodynamic focusing principle employed in the design of flow cytometers makes these assays both possible and practical. Hydrodynamic focusing, which creates an extremely narrow sample stream (10 to 15-μm diameter) with the resulting very small probe volumes (6 pl), is partially responsible for the separation free attribute of these assays. While upward of 3×10^9 small fluorescent particles may be present in 1 ml of sample, there are less than 20 nonbound spheres present in the probe volume at any time. The small probe volume also reduced background problems due to the presence of fluorescent sample constituents (e.g., bilirubin), which, because of their spectral properties, have previously been reported to interfere with FIA [3]. Also, AC coupling of the fluorescence signal amplifying electronics together with the ability to trigger on the light scatter signal from the large spheres allows the detection electronics to ignore the low level of DC background fluorescence from the steady-state concentration of free beads and other fluorescent molecules, and record only the fluorescence associated with the relatively large (3.5- to 15-μm diameter) particles as they pass individually through the probe volume.

Flow Cytometry

The fluorescence of the microsphere aggregates was measured with a modified Becton-Dickinson FACS II (Becton-Dickinson Immunocytometry Systems, Mountain View, CA). The fluidics including a nozzle tip with a 60-μm-diameter exit orifice and light-collection components were standard. The microspheres were excited by the 457 line from an argon laser, and the fluorescence was selected by two longpass filters (510-nm dielectric and 520-nm colored glass). Fluorescence signals were amplified, integrated, and recorded by the computer-based data acquisition system. Data collection was triggered by forward-angle light scatter detected in the standard FACS II geometry.

Immunomicrosphere Preparation

Polystyrene microspheres have been used in these assays because of the very tight noncovalent bond that occurs between their surface and protein molecules under suitable adsorptive reaction conditions. Microsphere reagents prepared in the manner described below have remained stable for at least 1 year in buffer at 4°C. It is important that the large microsphere carrier be both of uniform diameter and have low autofluorescence. For our experiments, we have used nonfluorescent microspheres obtained from the Dow Chemical Co. (Indianapolis, IN). Small microspheres, volume-labeled with various fluorochromes are available as a 2.5% suspension from Polysciences, Inc. (Warrington, PA). Most of our work has been done using 10-μm diameter large spheres and 0.10-μm-diameter small yellow-green fluorescent spheres.

Before labeling, all particles should be washed at least four times in coating buffer (0.01 mol/liter carbonate buffer, pH 9.5). To wash the small particles, a centrifuge capable of at least 25,000*g* is required. After washing, 5×10^7 large particles are added to 1 ml of coating buffer in which is dissolved 25–100 μg of the specific antibody of interest. After 24 hours at 4°C, the particles are washed three times in coating buffer and are then resuspended to a concentration of 2×10^6/ml in 0.1 mol/liter phosphate buffer (pH 8.3) containing 1% bovine serum albumin (BSA) and 0.1% sodium azide. Washed small fluorescent particles are suspended (1% v/v) in coating buffer into which 0.1–1.0 mg of the desired labeling protein has been dissolved. The particle–protein solution is immediately sonicated for 1 min in an ultrasonic cleaner to disperse aggregated particles and to facilitate protein binding. After 24 hours at 4°C, the particles are washed four times in coating buffer and resuspended in BSA–phosphate buffer as above. All labeled immunobead reagents are stored at 4°C in the BSA–phosphate buffer and appear to remain stable indefinitely.

In practice, the amount of protein bound to the small particles must be optimized for each assay. A limitation in the competitive binding assays could occur where purified antigen is difficult or expensive to obtain. A 1% suspension of 0.10-μm-diameter particles contains more than 10^{13} particles per ml. Each particle has sufficient surface area to bind more than 1,000 globular 40,000–dalton molecules. Therefore, to saturate a 1-ml suspension completely, 0.6 mg of the protein would be required. In practice, only 10–40 molecules are bound to each sphere. Even at this level, significant amounts of protein are required. For the sandwich assay, which employs antibody on both large and small particles, the labeling requirement can be met more easily, since large amounts of relatively pure antibody can usually be obtained. We have not attempted to bind haptens to small microspheres but believe this should be possible via linker molecules such as hapten–protein or hapten–poly-L-lysine [7].

Reaction Buffer

At low protein concentrations nonspecific interactions of small and large particles could be high, especially in the presence of NaCl. These interactions are minimal or absent in the presence of at least 2.5% protein. Therefore, the reaction buffer contains 5% BSA in a 0.1 mol/liter phosphate buffer, pH 8.2–8.4, plus 0.1% sodium azide. Preliminary experiments also indicate that whole serum, the matrix for many clinical analyses, may be used.

Performing the Competitive Assay

The principle and binding sequence of the dual sphere flow cytometric competitive binding assay are illustrated in Figure 6. Briefly, soluble unlabeled analyte (antigen) and labeled antigen compete for the limiting number of binding sites available on the large particles. In the absence of soluble antigen, maximum binding of antigen-coated fluorescent particles occurs. As the concentration of soluble antigen increases, fewer sites are available for the binding of the antigen-coated fluorescent spheres. Thus, the fluorescence of the large particles decreases as the soluble antigen concentraton increases.

The assay can be performed in either a single or double step. In the single step method both soluble and labeled antigen are added together, while in the double-step method the soluble antigen is first added for time t_1, after which the labeled antigen is added for time t_2. Although the two-step process has been reported to increase the sensitivity of competitive binding RIA protein assays by factors of 2–4 [17], we observe virtually no difference in sensitivity between these two types of assay procedures [32].

Incubation times and small-to-large bead ratios must be optimized for each assay. Both the desired assay dynamic range and the antibody affinity affect these parameters. In a prototype assay for the enzyme horseradish peroxidase (HRP), we desired a concentration of HRP-coated-particles which yielded 40–60% saturation at equilibrium with HRP–antibody-coated 10-μm particles in the absence of soluble HRP. To determine the desired ratio, varying amounts of small HRP-coated particles were incubated with a constant number of large particles for 48 hours at room temperature. Without separating free from bound immunobead reagent, the mean fluorescence of about 5,000 large particles was determined by a flow cytometric [32] measurement and plotted as a function of small bead concentration (Fig. 7). The concentration that yielded 60% saturation was chosen for use in the assay.

The time required for the chosen concentration of HRP-coated fluorescent spheres to reach equilibrium with antibody-coated spheres was then determined (Fig. 8). The reaction was 93% complete by 14 hours; therefore, a minimum time of 14 hours was subsequently allowed for this interaction. Once the equilibrium has been established it is quite stable such that the assay can be flow cytometrically analyzed any time (even weeks) after equilibrium has occurred.

Standard Competitive Binding Displacement Curves

The details of generating a standard competitive binding displacement curve are reported elsewhere [32]. In summary, serial dilutions of soluble antigen are made from a standard stock solution. One tube in each series receives no antigen. A constant number ($\sim 10^5$) antibody-coated large spheres are added to each assay tube. In a two step assay, the tubes are then incubated at ambient temperature for time t_1 before adding the antigen-coated particles for time t_2. In a single-step assay, both soluble and microsphere-labeled antigen are added simultaneously for time t_1. After equilibrium has occurred, each tube is directly analyzed flow cytometrically for fluorescence with data collection triggered by light-scatter detection of the large spheres. The mean of each fluorescence histogram is plotted as a function of antigen concentration. Examples of standard displacement curves for the antigen

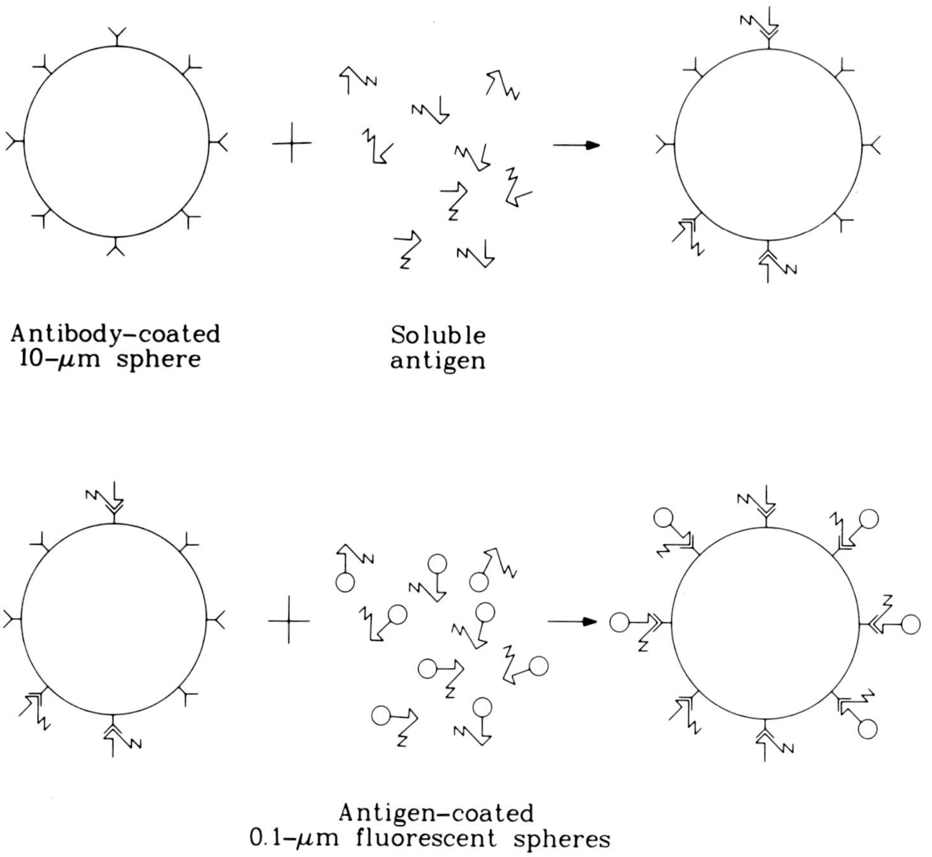

Antibody–coated
10–μm sphere

Soluble
antigen

Antigen–coated
0.1–μm fluorescent spheres

Fig. 6. In the binding sequence of the dual-sphere competitive binding fluoroimmunoassay, the Ab-coated large sphere is incubated with unknown or standard concentrations of soluble Ag for time t_1. Excess Ag-coated 0.1 μm fluorescent spheres are added for time t_2, after which the fluorescence of the large spheres (resulting from big bead–little bead interactions) is measured with a flow cytometer. The fluorescence intensity is inversely proportional to the soluble Ag concentration.

HRP are shown in Figure 9. These curves illustrate both the sensitivity (10^{-12} mol/liter) and the precision of the methodology.

Sandwich Assay

The principles and binding sequence of the dual sphere flow cytometric sandwich assay are shown in Figure 10. The sandwich assay is applicable where the analyte of interest has at least two well-separated antibody-binding sites. The potential for extremely sensitive and specific assays exists where two high-affinity monoclonal antibodies with different epitope specificity on the analyte molecule are available. One monoclonal antibody would be put on the large sphere and the other would be placed on the small sphere. Our prototype assays developed to date have employed only polyclonal antibodies. Even here, remarkable sensitivity has been attained. Where two distinct antibodies were used, a single step assay could be done as a matter of course. With polyclonal antibodies, a two-step assay as illustrated in Figure 10 was used.

As in the competitive binding assay, one must optimize the incubation times and bead ratios to suit assay requirements. Figure 11 illustrates an example of how the ratio of small to large spheres can influence the sensitivity of alphafetoprotein detection.

A standard curve for HRP (Fig. 12) illustrates the best sensitivity we have obtained to date, 10^{-14} mol/liter. Each assay is performed with 10^5 large particles per ml of volume assayed. At the sensitivity level of 10^{-14} mol/liter, assuming that every HRP molecule originally in solution is bound to a large sphere (a very unlikely condition), an average of 60 molecules of HRP are bound to each 10-μm-diameter particle! (One ml of a 10^{-14} mol/liter solution contains 6.023×10^6 molecules, or about 60 molecules per large sphere.) Thus, at the detection limit, each large sphere is read as positive when fewer than 60 molecules of analyte are bound to it.

To define the real potential value of these assays to clinical medicine more work needs to be done using a variety of clinical samples to quantitate various analytes present at ultralow concentrations. Nevertheless, based on our early results with α-fetoprotain, we are very optimistic that these techniques will prove both useful and practical. Meantime, the technology will be immediately useful to those in the research community who have access to a flow cytometer and who are involved in work in which the sensitive quantitation of hormones, growth factors, lymphokines, and so forth in experimental systems is important.

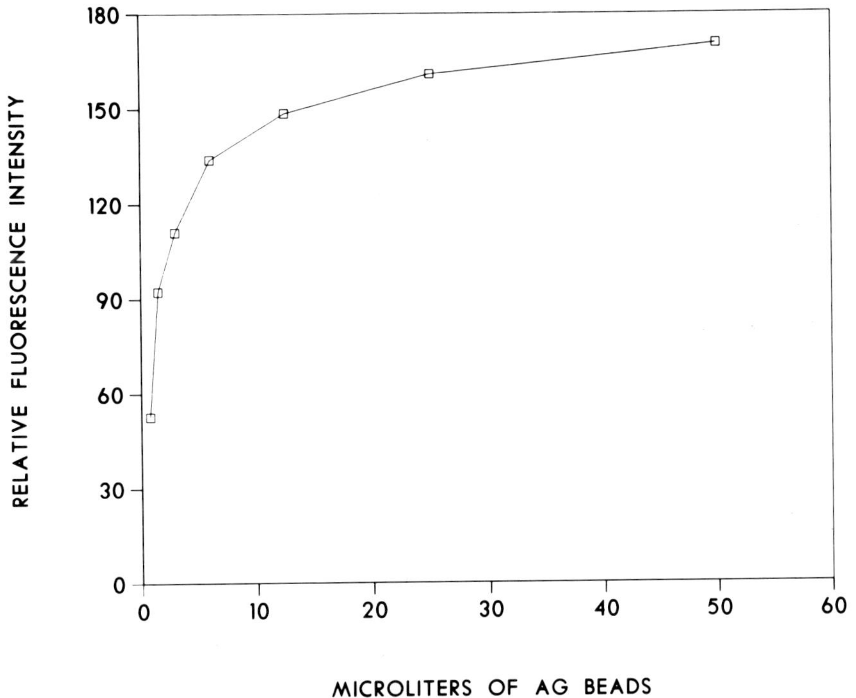

Fig. 7. Saturation of Ab-binding sites on large spheres by Ag coated fluorescent small spheres. See text for details. The concentration of the Ag bead stock solution was 0.05%, or about 1×10^{12}/ml, with a constant number of 10^5 Ab-coated spheres in a 1 ml reaction volume; 60% saturation occurred with 3 μl stock solution.

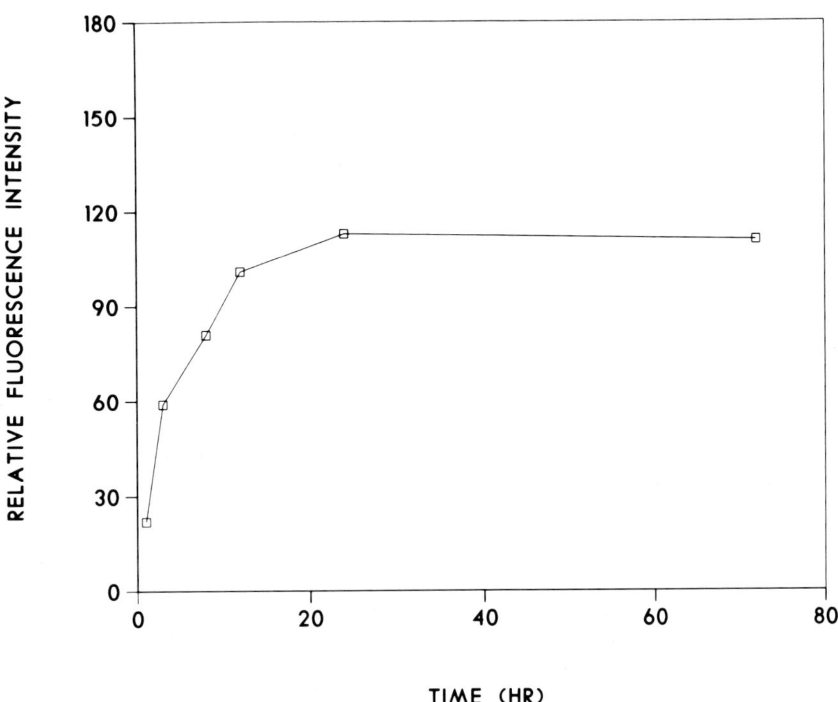

Fig. 8. Time to reach equilibrium when 10^5 Ab-coated spheres are incubated with 3×10^9 Ag coated spheres in a total volume of 1 ml.

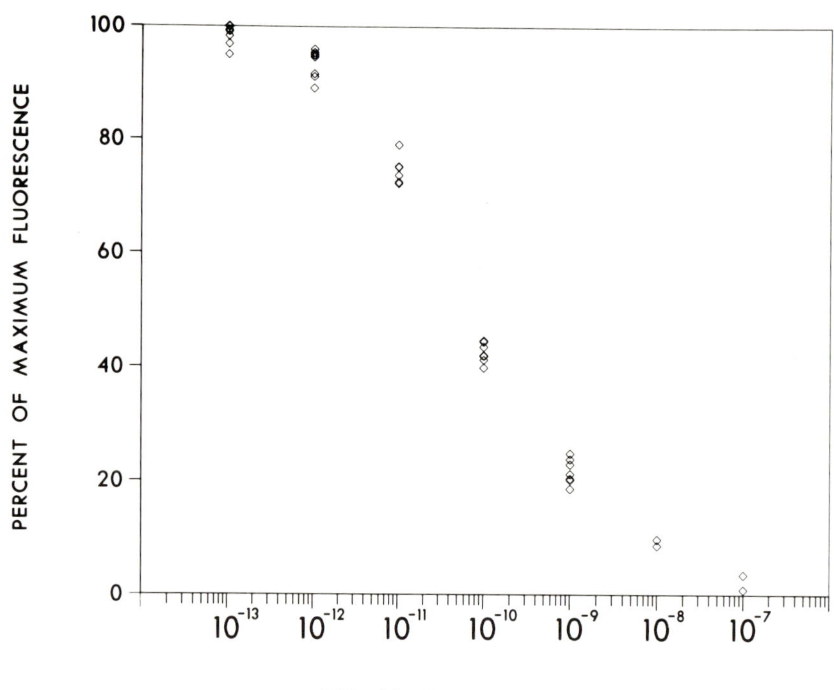

Fig. 9. HRP displacement curves obtained from a series of 7 experiments performed over a 1-month period. Five of the experiments cover a HRP molar range of 10^{-9}–10^{-13}; two experiments span a HRP molar range of 10^{-7}–10^{-13}. The reproducibility is excellent. The volume of sample analyzed in the flow cytometer is less than 0.1 ml; at 10^{-12} mol/liter concentration this translates to less than 10^{-16} moles of HRP.

OTHER FLOW CYTOMETRIC MOLECULAR ASSAYS

A major objective of developing fluorescence based immunoassays is to take advantage of the strong signals from fluorescent labels and to devise clever techniques for reducing background contributions and nonspecific labeling. In recent years many fluoroimmunoassay techniques have been developed, several of which are being routinely used in clinical applications. A recent review describing a variety of fluoroimmunoassays is given by Hemmila [15]. Some of these other techniques could be applied to the sensitive molecular assays described above to increase the selectivity of the measurement of fluorescent probes that would reduce background fluorescence, further increasing sensitivity. This section describes several other techniques for molecular detection. Some of these methods have been realized in flow cytometry; others are yet to be accomplished in flow cytometry.

A single microbead fluorescence sandwich assay reported by Lisi et al. [20] was described earlier. It is also possible to perform single microbead assays in a competitive binding mode. Both actinomycin D and mithramycin can be quantitated in a system in which DNA is bound to a 10-μm-diameter nonfluorescent microsphere. Because actinomycin D and mithramycin bind to similar parts of the DNA molecule, their competition for available DNA binding sites can be measured. Mithramycin bound to DNA is highly fluorescent, while actinomycin D is not. By using a constant amount of either chemical, the amount of the other can be determined by measuring the fluorescence of the DNA coated particles. In preliminary experiments, we have achieved a detection limit of ~10^{-8} mol/liter [33]. We believe that immunoassays in which high sensitivities are not required could also be developed using similar principles. Other applications include the competitive binding of any two molecules (one of which is fluorescent) to a third molecule that has been immobilized onto a microsphere. For example, the competition between two enzymes for a common substrate could be used as the basis for an assay.

Quenching or enhancement of fluorescence from molecules, either in solution or attached to microsphere carriers could provide the basis of a new type of assay. An example of the use of this type of signature is energy transfer [18,34]. In a system using two fluorochromes in which the emission spectrum of one fluorochrome (donor) overlaps the excitation spectrum of the other fluorochrome (acceptor) it is possible, under certain conditions, to obtain a transfer of energy from an excited state of a donor molecule to that of an acceptor molecule. This energy transfer can occur when the separation between the two molecules is in the range of 1–6 nm and can be observed as a loss of characteristic fluorescence from the donor molecules and an increase in the characteristic fluorescence of the acceptor molecules. The most commonly used donor–acceptor pair is fluorescein–rhodamine. It is also possible to use a nonfluorescent acceptor, in which case the measurement consists of recording changes in the donor fluorescence. Energy transfer can be used in immunofluorescence assays when two different antibodies to the same antigen are labeled with a donor–acceptor pair. A sandwich assay for a bivalent antigen would then use a donor–acceptor pair as a binding indicator for both types of antibodies.

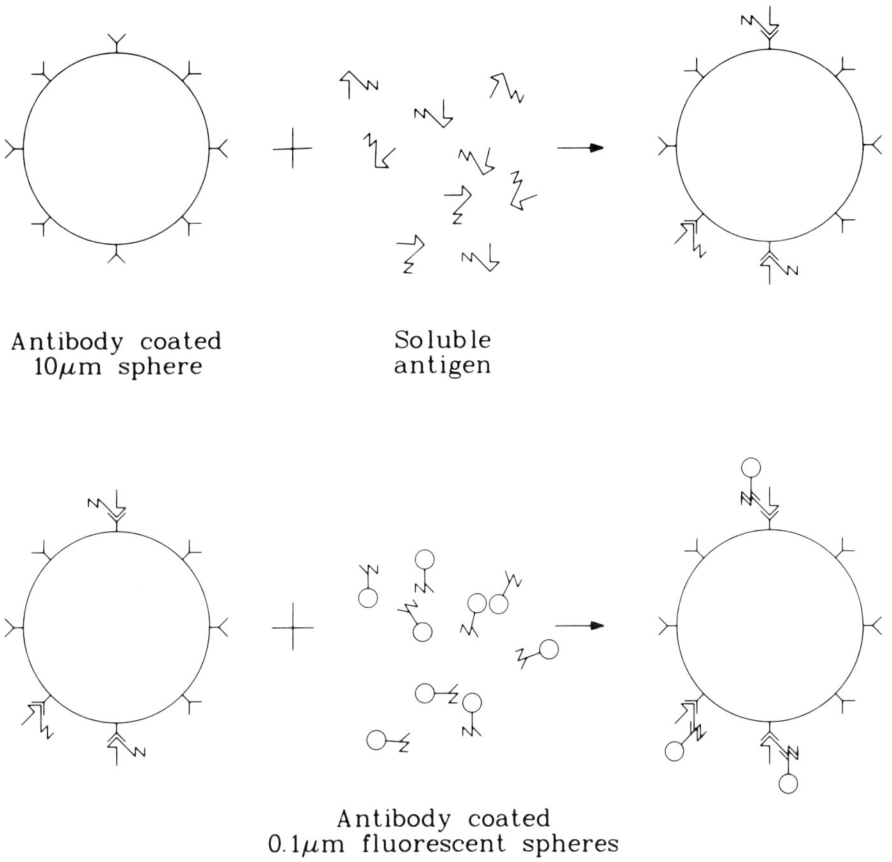

Antibody coated 10μm sphere

Soluble antigen

Antibody coated 0.1μm fluorescent spheres

Fig. 10. In the binding sequence of the dual-sphere sandwich-type fluorescent immunoassay, both large and small spheres are coated with Ab. This type of assay is useful for quantifying Ag with multiple binding sites for Ab. Both small and large Ab-coated spheres bind to the Ag to make a sandwich, with the Ag as a filling. In this assay, the fluorescence of the large sphere is directly proportional to soluble Ag concentration.

When fluorescent probe molecules are excited by polarized light, the emission from probe molecules bound to a specific matrix exhibits enhanced polarization, whereas the emission from unbound probe molecules is less polarized. This effect is due to differences in rotational relaxation times between the bound and unbound states.

By taking advantage of the differences in the time dependent nature of fluorescence and scatter, scatter backgrounds can be reduced or eliminated in measurements using either pulsed or modulated excitation sources. Time-resolved detection requires discrimination between the prompt scatter signals and the delayed fluorescence emission. Most fluorescent dye molecules have fluorescence lifetimes in the range of 1–20 nsec. Time-resolved measurements of such dye molecules require a pulsed excitation source having nanosecond, or shorter, optical pulse duration and photodetectors and electronics capable of detecting and processing nanosecond pulses. Pulsed or mode-locked lasers are the most commonly used light sources for this application. Fast correlation techniques have been used to discriminate between fluorescein and bilirubin fluorescence [2]. These fast techniques are available to the research laboratory but at the present time may not be cost effective in the clinical laboratory. An alternative is to use the relatively long excited state lifetimes (10–1,000 μsec) found in many rare earth chelates. In these time-resolved measurements, lanthanide labels are excited by short duration optical pulses (10–20 nsec) from a pulsed arc lamp. Gated photon counting of the fluorescence is begun after a delay of several microseconds and is continued for several tens or hundreds of microseconds. In this way, it is relatively easy to make time resolved fluorescence immunoassay measurements using rare earth labels [15]. A time-resolved fluoroimmunoassay instrument has recently been developed and is being marketed for use in several clinical assays.

In addition to the pulsed time-resolved fluorescence method, it is relatively easy to modulate the light from a continuous source at high frequency in order to make time-resolved fluorescence measurements using phase sensitive detection techniques similar to those described in the introduction. Phase-sensitive detection of fluorescence emission can discriminate against both Rayleigh and Raman scatter backgrounds [4]. Phase-sensitive fluorescence detection of admixtures of fluorescein and fluorescein isothiocyanate have been made using 18-MHz modulation of the excitation light [23].

CONCLUSIONS

This chapter describes a newly emerging field of application for flow cytometric techniques. Direct molecular detection has led to new sensitivities that are unparalleled. The

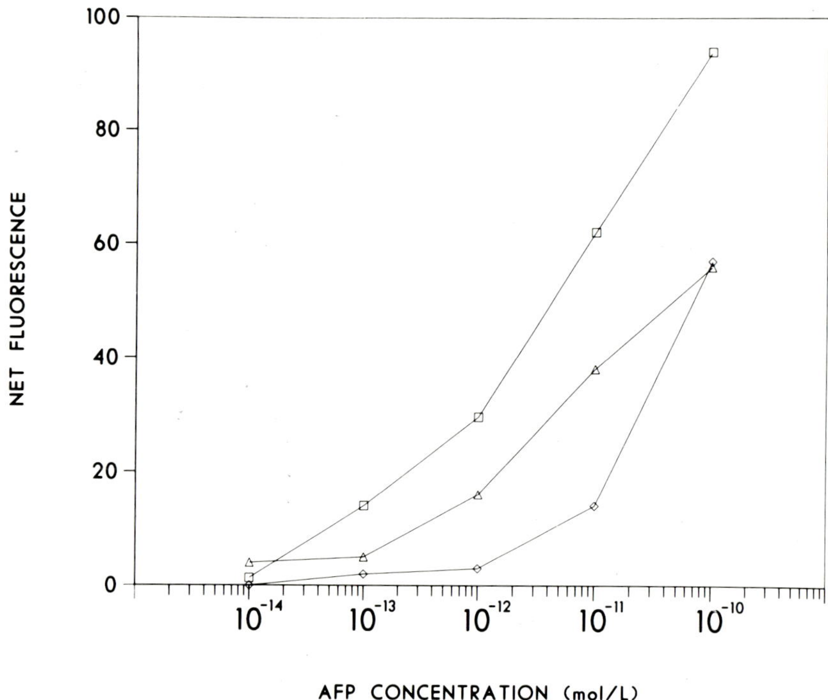

Fig. 11. α-fetoprotein detection in a sandwich-type assay. Various quantities of a 0.025% suspension of fluorescent spheres were used to generate standard AFP curves using a constant number of 10^5 large spheres in a volume of 1 ml. The volumes of small spheres used were 30 μl (□), 20 μl (△), and 10 μl (◇). In this experiment, the most usable curve was generated with 30 μl of small spheres.

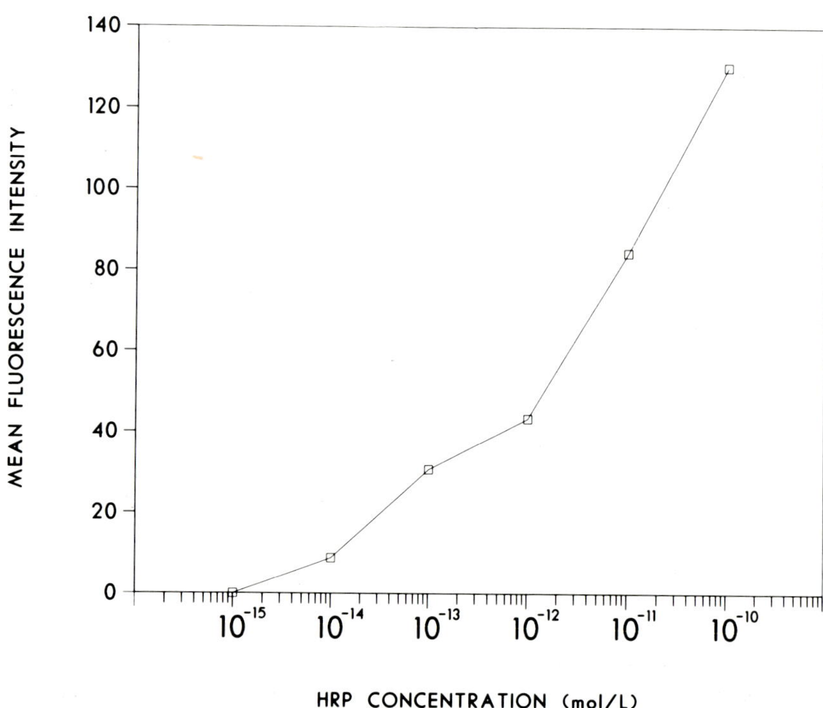

Fig. 12. HRP concentration versus mean fluorescence intensity for a two-step HRP sandwich assay where both small and large spheres are coated with the same HRP Ab used in the previously illustrated competitive binding curves (Fig. 6). There is an apparent increase in sensitivity of two orders of magnitude in the sandwich-type assay.

single molecule detection capability will find use in new assays employing antibody–antigen interactions in solution and in providing a new detection methodology for high-performance liquid chromatography assays. The microsphere-based assays of both types are providing new sensitivities that are directly applicable to clinical medicine.

NOTE ADDED IN PROOF

Since writing this review, the development of a new application of the single molecule detection capability to DNA sequencing has been initiated. Briefly, the approach is to synthesize, using a single strand template, the complementary strand of DNA with fluorescently tagged precursors, attach the labeled duplex molecule to a microsphere, suspend the complex in a flow system capable of single molecule detection, sequentially cleave the labeled bases into the flow from the 3′ end of the molecule with an exonuclease, and identify the fluorescent label attached to the base as it passes through the laser beam. Based on molecular detection times and enzyme cleavage rates, sequencing rates between 100 and 1000 bases per second for long strands of DNA are projected. Details are contained in: Jett JH, Keller RA, Martin JC, Marrone BL, Moyzis RK, Ratliff RL, Seitzinger NK, Shera EB, and Stewart CC (1989) High-speed DNA sequencing: An approach based upon fluorescence detection of single molecules. Journal of Biomolecular Structure and Dynamics (in press).

ACKNOWLEDGMENTS

This work was supported by the Los Alamos National Laboratory, by the U.S. Department of Energy, and by NIH grant RR01315 for the National Flow Cytometry Resource.

REFERENCES

1. **Balykin VI, Letokhov VS, Mishin VI (1980).** Multiphoton fluorescence detection of single atoms by laser radiation. Appl Phys 22:245–250.
2. **Bright FV, Vickers GH, Hieftje GM (1986).** Use of time resolution to eliminate bilirubin interference in the determination of fluorescein. Anal Chem 58: 1225–1227.
3. **Dandliker WB, Hsu ML, Levin J, Rao BR (1980).** Equilibrium and kinetic assays based upon fluorescence polarization. Methods Enzymol 74:3–28.
4. **Demas JN, Keller RA (1985).** Enhancement of luminescence and Raman spectroscopy by phase-resolved background suppression. Anal Chem 57:538–545.

4a. **Diebold GJ, Zare RN (1977).** Laser Fluorimetry: Subpicogram detection of aflatoxins using high-pressure liquid chromatography. Science 196:1439–1441.

5. **Dovichi NJ, Martin JC, Jett JH, Trkula M, Keller RA (1984).** Laser induced fluorescence of flowing samples as an approach to single molecule detection in liquids. Anal Chem 56:348–354.
6. **Dovichi NJ, Martin JC, Jett JH, Keller RA (1983).** Attogram detection limit for aqueous dye samples by laser-induced fluorescence. Science 219:845–847.
7. **Ekele GI, Exley D (1981)** The assay of steroids by fluoroimmunoassay. In Pal SB (ed), "Enzyme labelled immunoassay of hormones and drugs." de Gruyter, Berlin, pp 195–205.
8. **Fairbank WM Jr, Hansch TW, Schawlow AL (1975)** Absolute measurement of very low sodium-vapor densities using laser resonance fluorescence. J Opt Soc Am 65:199–204.
9. **Feinstein A, Beale D (1977)** Models of immunoglobins and antigen–antibody complexes. In Glynn LE, Steward MW (eds), "Immunochemistry: An Advanced Textbook." John Wiley & Sons, pp 263–306.
10. **Focus: Clinical Immunoassay Industrial Markets (1984)** Anal Chem 56:1420A–1422A.
11. **Folestad S, Johnson L, Josefsson B, Galle B (1982)** Laser induced fluorescence detection for conventional and microcolumn liquid chromatography. Anal Chem 54:925–929.
12. **Glazer AN (1977)** Structure and molecular organization of the photosynthetic accessory pigments of cyanobacteria and red algae. Mol Cell Biochem 18:125–140.
13. **Glazer AN, Streyer L (1984)** Phycofluor probes. Trends Biochem Sci 9:423–427.
14. **Greenless GW, Clark DL, Kaufman SL, Lewis DA, Tonn JF, Broadhurst JH (1977)** High resolution laser spectroscopy with minute samples. Opt Comm 23: 236–239.
15. **Hemmila I (1985)** Fluoroimmunoassays and immunofluorometric assays. Clin Chem 31:359–370.
16. **Hershberger LW, Callis JB, Christian GD (1979)** Submicroliter flow-through cuvette for fluorescence monitoring of high performance liquid chromatographic effluents. Anal Chem 51:1444–1446.

16a. **Hirschfeld T (1976)** Optical microscopic observation of single small molecules. Appl Opt 15:2965–2966.

17. **Hunter WM (1978)** Radioimmunoassay. In Weir DW (eds), "Handbook of Experimental Immunology." 3rd ed. Oxford: Blackwell Scientific, pp 14.24–14.25.
18. **Jovin TM (1979)** Fluorescence polarization and energy transfer: Theory and application. In Melamed MR, Mullaney PF, Mendelsohn ML (eds), "Flow Cytometry and Sorting." John Wiley & Sons, pp 137–165.
19. **Kelly TA, Christian GD (1981)** Fluorometer for flow injection analysis with application to oxidase enzyme dependent reactions. Anal Chem 53:2110–2114.

19a. **Kirsch B, Voigtman E, Winefordner JD (1985)** High-sensitivity laser fluorometry. Anal Chem 57:2007–2009.

20. **Lisi PJ, Huang CW, Hoffman RA, Teipel JW (1982)** A fluorescent immunoassay for soluble antigens employing flow cytometric detection. Clin Chim Acta 120: 171–179.
21. **Loken MR, Herzenberg LA (1975)** Analysis of cell populations with a fluorescence activated cell sorter. Ann NY Acad Sci 254:163–171.
22. **Mathies RA, Stryer L (1986)** Single-molecule fluorescence detection: A feasibility study using phycoerythin. In Taylor DL, Waggoner AS, Murphy RF, Lanni F, Birge RR (eds), "Applications of Fluorescence in the Biomedical Sciences." New York: Alan R Liss, pp 129–140.
23. **McGowan LB (1984)** Phase-resolved fluorimetric determination of two albumin-bound fluorescein species. Anal Chem Acta 157:327–332.
24. **McGuffin VL, Zare RN (1985)** Laser fluorescence detection in microcolumn liquid chromatography: Application to derivatized carboxylic acids. Appl Spect 39: 847–853.
25. **Nguyen DC, Keller RA, Jett JH, Martin JC (1987)** Detection of single molecules of phycoerythrin in hy-

drodynamically focused flows by laser-induced fluorescence. Anal Chem 59:2158–2161.

26. **Nguyen DC, Keller RA, Parson JD, Jett JH, Martin JC** (1990) Single molecule detection of fluorophors in flowing samples using photon-burst detection techniques. (in preparation).
27. **Nguyen DC, Keller RA, Trkula M** (1987) Ultrasensitive laser -induced fluorescence detection in hydrodynamically focused flows. J Opt Soc Am B 4:138–143.
28. **Novotny M** (1981) Microcolumns in liquid chromatography. Anal Chem 53:1294A–1308A.
29. **Pan CL, Prodan JV, Fairbank WM Jr, She CY** (1980) Detection of individual atoms in helium buffer gas and observation of their real-time motion. Opt Lett 5: 459–461.
30. **Russo RE, Hieftje GM** (1982) A new instrument for time-resolved reduction of scattered radiation in fluorescence measurements. Anal Chem Acta 134:13–19.
31. **Ruzicka J** (1983) Flow injection analysis from test tubes to integrated microconduits. Anal Chem 55: 1040A–1053A.
32. **Saunders GC, Jett JH, Martin JC** (1985) Amplified flow-cytometric separation-free fluorescence immunoassays. Clin Chem 31:2020–2023.
33. **Saunders GC, Martin JC, Jett JH, Perkins A** (1990) Flow cytometric competitive binding assays: Determination of actinomycin-D concentrations. Cytometry (in press).
34. **Streyer L** (1978) Fluorescence energy transfer as a spectroscopic ruler. Annu Rev Biochem 47:819–846.
35. **Yallow RS, Berson SA** (1960) Immunoassay of endogenous plasma insulin in man. J Clin Invest 39:1157–1175.
36. **Zarrin F, Dovichi NJ** (1985) Sub-picoliter detection with the sheath flow cuvette. Anal Chem 57:2690–2692.

1478

21

Standards and Controls in Flow Cytometry

Paul Karl Horan, Katharine A. Muirhead, and Sue E. Slezak
Zynaxis Cell Science Inc., Malvern, Pennsylvania 19355

INTRODUCTION

As used in the literature, the terms STANDARD and CONTROL have a variety of meanings. Therefore, an appropriate place to begin is with our definitions for these terms. For this chapter, we shall define STANDARD as a stable material that has or may be assigned values for fluorescence or light scatter intensity in one or more spectral windows. Standards are typically used to optimize and/or calibrate instrumentation. Examples of standards are microspheres having a known number of fluorochromes per microsphere or microspheres having an unknown but stable number of dye molecules, which can be assigned a fluorescence value using a specific filter arrangement, laser intensity, and detector high voltages and gain settings. We shall define CONTROL as a material that gives an expected result, although it may lack the long term stability or assigned value required of a standard. Controls may be either positive or negative and are used to assure that reagents and/or cell preparation methods are working as expected. A control often used for immunofluorescence analysis is blood from a healthy normal donor. Stained with a reagent expected to bind, it serves as a positive control; stained with a reagent not expected to bind, it serves as a negative control. Similarly, normal lymphocytes are often used as a control for DNA analysis; they are expected to give a single peak with minimal heterogeneity and an intensity corresponding to diploid DNA content under the staining conditions employed.

Why do we need standards and controls when running flow cytometric analyses? Because they give us the ability to make accurate comparisons of our data: from cell to cell within a sample, from sample to sample on a given day, from day to day within a laboratory, and from laboratory to laboratory. The last is in some ways most difficult. Different laboratories use instruments from different manufacturers which differ in terms of light sources, illumination optics and intensities, light collection optics and efficiencies, and so on. Even in two laboratories using instruments from the same manufacturer, it is impossible to use identical filters since every filter has slightly different spectral properties. Proper choice of standards and controls allows normalization of instrument-related differences and more accurate comparisons between laboratories. However, it is unreasonable to make comparisons with another laboratory unless we have first insured that our own data is reliable. Standards can be used to check instrument alignment and/or calibration on a daily basis, to monitor sorting efficiency, and to monitor instrument performance during a single run, as discussed below.

Problems of selecting standards to assure instrument performance and reproducibility are common to all types of applications, while selection of appropriate controls is more dependent on the particular application of interest. In this chapter, we will focus primarily on instrument standards, but will also review a variety of commonly used controls for DNA cell cycle analysis and immunofluorescence analysis.

INSTRUMENT STANDARDS

Instrument standards are used for a variety of purposes: optimizing alignment, reproducing specific operating conditions, calibrating intensity scales, comparing sensitivities, monitoring instrument performance, etc. Different types of standard particles have varying advantages and disadvantages relative to different purposes, as discussed below. Some are commercially available; some are made by the user; some come with assigned values for intensity or fluorochrome number; some are assigned a value by the user under defined operating conditions. Commercial sources of standard particles often used in flow cytometry include Becton Dickinson Immunocytometry Systems (Mountain View, CA), Coulter Electronics Inc. (Hialeah, FL), Duke Scientific Laboratories (Palo Alto, CA), Dynal Inc. (Great Neck, NY), Flow Cytometry Standards Corporation (Research Triangle Park, N.C.), Pandex Laboratories (Mundelein, IL), Polysciences Inc. (Warrington, PA), and Seragen Diagnostics Laboratory (Indianapolis, IN).

Alignment and Performance Standards

Let us begin with how standard particles are used for instrument alignment, which typically involves assignment of values by the user. For a given set of operating conditions, optimum alignment should produce the maximum possible signal, assessed as mean intensity or channel number, and minimum variability, assessed as coefficient of variation (CV = standard deviation divided by mean). Commercially purchased fluorescent spheres may come with some indica-

Flow Cytometry and Sorting, Second Edition, pages 397–414

tion of heterogeneity (SD or CV of diameter or volume) but, with the exception of calibration kits for specific fluorochromes, rarely have assigned intensity values. When running alignment particles for the first time, you typically adjust system optics to maximize the mean intensity and minimize the CV, and then use the optimized values as target values in further runs of the same microspheres under the same conditions on your instrument. For example, Figure 1A shows a histogram obtained from some Coulter grade II "full bright" microspheres, under instrument conditions which gave a mean fluorescence intensity of 177 channels and a CV of 2.27%. If these are the highest mean and lowest CV values obtained by your most expert operator running the same particles over several days using the same laser power, filters, and detector gains, these values are then assigned to the alignment particles under those conditions. To use this standard from day to day, each operator begins by determining whether microsphere values under current running conditions agree well with the assigned values. If not, it is assumed that the discrepancy is due to some alteration in instrument conditions and/or alignment and these are adjusted until acceptable agreement is obtained. Clearly, some consensus on what constitutes "acceptable agreement" is also required.

Use of alignment standards as described above provides a sensitive means of detecting both flow-related and optics-related changes in signal intensity from day to day. Cross-calibration with a new lot of microspheres before the old one is depleted allows a long term record of instrument performance to be maintained (see also section F on Quality Control Records). Note that it is important to determine the effect of changes in alignment on *all* of the signals to be measured in the biological system of interest. If the optics used in a flow cytometer are not completely color-corrected, different wavelengths of light may be focused at slightly different places. Therefore it is possible to optimize the fluorescence signal at the expense of the light scatter signal, or to optimize one color of fluorescence at the expense of another. Obviously, the compromise you choose will depend on your particular application.

From the above description of how alignment standards are used, it is apparent that several properties are desirable in any material to be used for this purpose. First, it should have a stable signal over a time frame of months to years. Second, it should be as sensitive as possible to altered alignment resulting from changes in flow or optical properties of the system. Third, it should be similar in size to the cells of interest, so that it will be handled similarly by both fluidic and optical systems. Finally, it should be as inexpensive as possible consistent with all of the foregoing properties.

Glutaraldehyde-fixed chicken red blood cells (gCRBC) have been popular alignment particles because they meet many of the above criteria. They are inexpensive, easily prepared in your own laboratory, and stable over long periods of time so long as contamination with microorganisms is avoided. Because they are ellipsoidal in shape, their forward (low angle) light scatter signal is multimodal and very sensitive to variations in system fluidics and in geometry of the light scatter detector [8]. This may be an advantage in your own system, where you have learned what is expected under optimal conditions, but presents some problems in comparing system to system. Figures 1B and 1C show forward light scatter histograms obtained for glutaraldehyde-fixed chick red cells on EPICS 753 and FACS II systems, respectively. Differences in scatter detector geometry (e.g., width of obscuration bar, angle of detection) lead to significant differences in the histograms obtained even when both instruments are optimized. An advantage of using gCRBC is that their fluorescence intensity is much closer to that of immunofluorescently stained lymphocytes than that of the high intensity commercial microspheres, allowing the optimization to be carried out at instrument settings similar to those used for running the cells. A disadvantage is that they are very heterogeneous and give broad peaks, making it harder to estimate visually whether a change in instrument conditions has improved or worsened the alignment with respect to fluorescence.

Another cell-based standard that can be used to assess instrument performance is FluoroTrol-GF [3], shown in Figure 1D. FluoroTrol-GF is a three part mixture of calf thymocyte nuclei, unstained or stained to two different levels of fluorescence with fluorescein isothiocyanate. As with gCRBC, fluorescence intensities are similar to those obtained with direct or indirect immunofluorescence staining and the peaks are quite broad. However, only a single light scatter peak is obtained from the spherical nuclei. To use the FluoroTrol-GF as an alignment standard, mean intensity is maximized and CV is minimized as previously described for other alignment standards. However, because it consists of mixed populations with differing intensities, this standard can also be used to assess relative instrument sensitivity, i.e., how well unstained cells can be resolved from dimly stained or brightly stained cells. Ability to resolve all three peaks will be affected by laser power, residence time in the beam, efficiency of collecting optics, choice of filters, etc., as well as instrument alignment. The levels of autofluorescence and dim or bright staining in FluoroTrol-GF were chosen to represent "typical" values for immunofluorescence work, with the goal of designing a standard specifically capable of asking "Is Instrument performance adequate for immunofluorescence analysis?" as opposed to "Is instrument performance optimal?"

It should be clear from the above examples of standards used to monitor instrument alignment and performance that it is essential to choose some standard which is run in your laboratory on a daily basis. Daily monitoring provides both reassuring evidence of the reliability of properly functioning instrumentation and a sensitive indicator of malfunction. It should also be obvious that there is no single standard which we can demand be used in every laboratory. The "right" standard is the one which best fits the systems, instruments, and applications in your laboratory and which allows you to assess whether instrument performance criteria are acceptable before you begin taking data.

Instrument Calibration Standards

Flow cytometric data have generally been expressed in terms of relative rather than absolute intensities, i.e., as channel numbers rather than number of photons emitted per cell. Note that many of the factors which influence relative intensity (excitation intensity, flow rate, signal collection efficiency, filter choice, detector gains and spectral response curves, etc.) are biologically uninteresting and difficult to reproduce from system to system, day to day, or run to run. For many applications it is adequate to adjust our sliding intensity scale so that all cells are on scale and displayed with acceptable resolution. However, for others we may wish to know the correlation between a particular channel number and given number of photons, a given number of fluorochromes, or a given number of cellular binding sites. Standards may be used to carry out instrument calibration at a variety of levels, the complexity of the calibration procedure

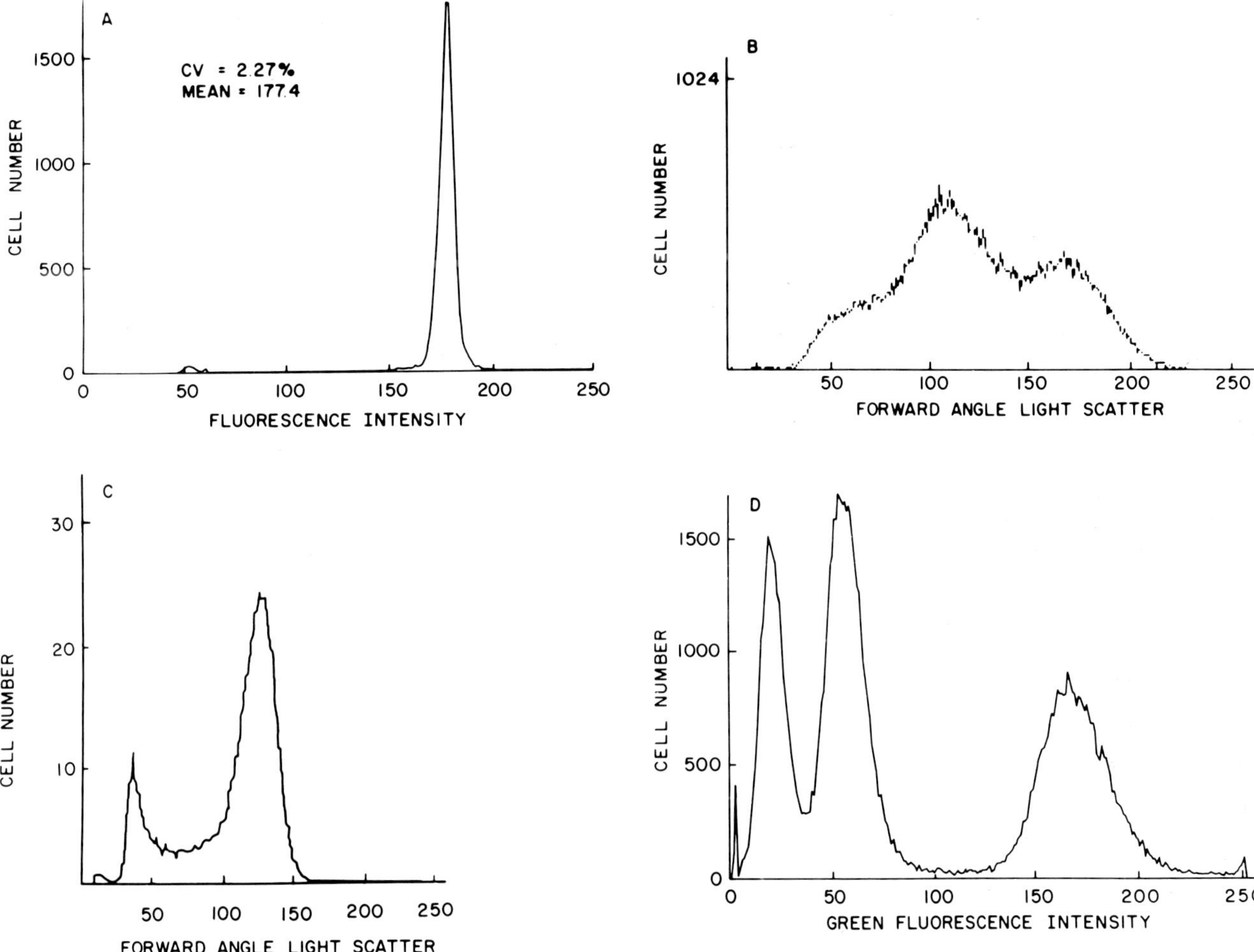

Fig. 1. Some commonly used instrument alignment/performance standards **A:** Polystyrene spheres (Coulter Electronics, grade II full bright): high intensity fluorescence relative to cellular autofluorescence, very homogeneous. **B:** Glutaraldehyde fixed chick red blood cells: autofluorescence intensity comparable to immunofluorescently stained cells, shape of heterogeneous forward angle light scatter distribution sensitive to instrument alignment and exact scatter detector geometry. Light scatter distribution shown obtained on optimally aligned Coulter EPICS 753. **C:** Light scatter distribution for glutaraldehyde-fixed chick red blood cells obtained on optimally aligned FACS II (courtesy of M. Loken): different scatter detector geometry results in significantly different distribution shape. On a given instrument, the characteristic shape of the light scatter distribution can be used as a sensitive indicator of misalignment. **D:** FluoroTrol-GF: three part mixture of unstained and fluorescein isothiocyanate stained calf thymocyte nuclei with forward light scatter intensity similar to lymphocytes. Fluorescence intensities are similar to lymphocyte autofluorescence, dim immunofluorescence or bright immunofluorescence (left to right). Position and resolution of peaks can be used to assess instrument alignment and performance.

generally being directly proportional to the absoluteness of the calibration desired.

The simplest level of calibration is standardization of the intensity scale by adjusting detector high voltage and/or gain to reproduce the mean intensity of a stable standard from day to day. This is equivalent to re-setting the instrument to a constant relative intensity scale, something which cannot necessarily be accomplished simply by re-establishing a constant set of instrument parameters; laser power in particular is difficult to reproduce with sufficient exactness. Standardization of intensity scale(s) can be done using FluoroTrol-GF, fluorescent microspheres, gCRBC, other stable autofluorescent particles, etc. Criteria for the standard particles include long term stability (months to years), availability of reproducibly similar lots, and signal intensities similar to samples of interest in the spectral windows of interest. By using similar signal intensities we avoid readjustment of instrument parameters between running the standard and running the sample, especially helpful for parameters which have a nonlinear effect on channel number (e.g., photomultiplier high voltage).

The converse of using standards to fix the intensity scale is using them to calibrate the effect of changing instrument conditions on measured channel number. In some instruments overall system gain is a function only of photomultiplier high voltage while in others it is determined by the combination of high voltage and amplifier gain setting. However, the effects of these two types of gain adjustment on channel number are quite different. Changes in photomultiplier high voltage have nonlinear (exponential) effects on channel number; changes in amplifier gain (whether stepwise or continuously variable) typically produce linearly pro-

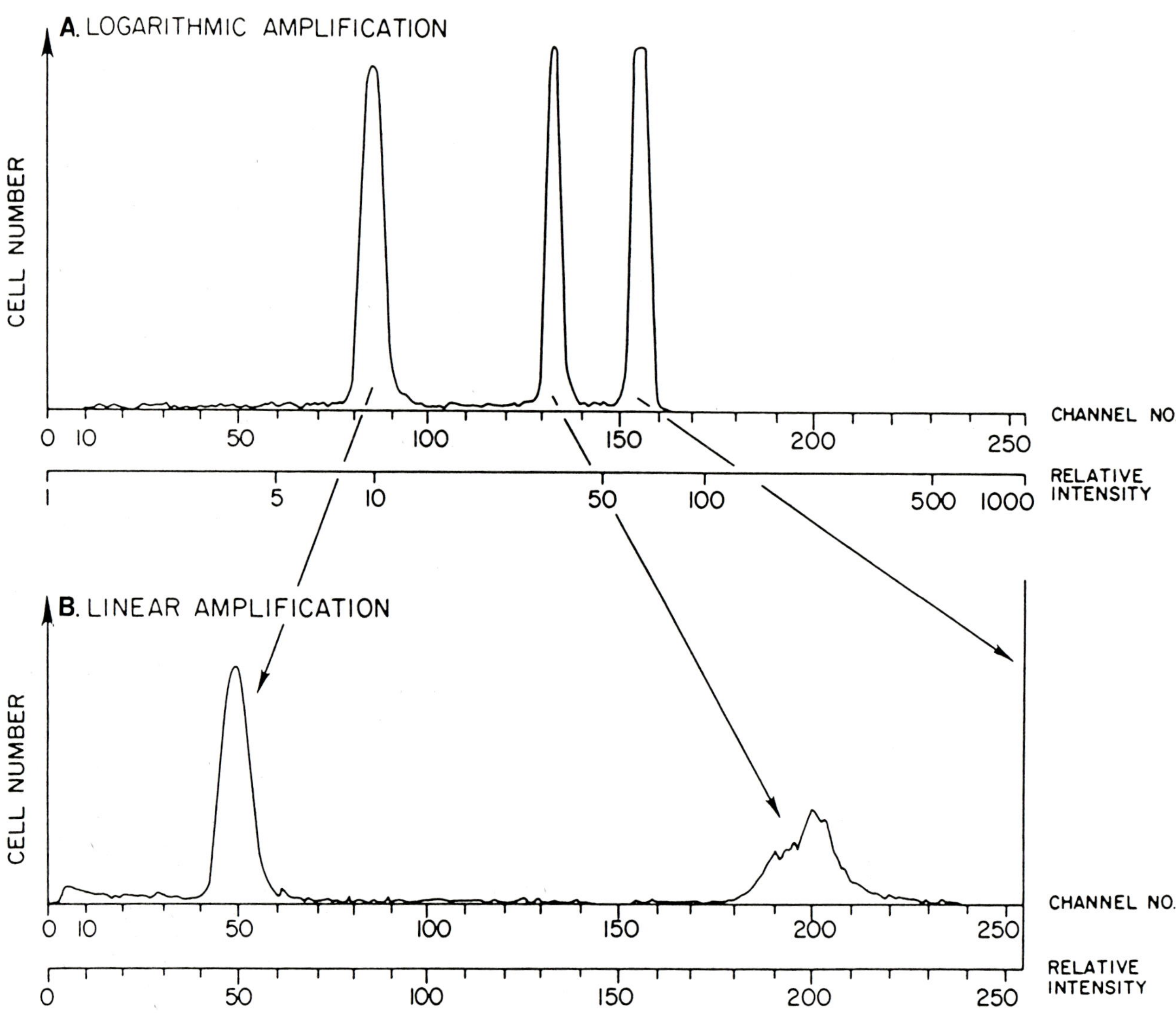

Fig. 2. Comparison of logarithmic and linear signal amplification. **A:** Logarithmic amplification condenses a wide intensity range (here 1,000-fold) into a fixed number of channels (here 256) and therefore is often used in collecting immunofluorescence data. The channel number value assigned to a cell or particle is proportional to the logarithm of its original signal voltage, resulting in different signal increments per channel number at different positions on the scale (compare channel number and relative intensity scales). Relative linear intensities of the three fluorescent microsphere populations shown are 1:3.7:7.4 (left to right). **B:** Linear amplification results in a channel number directly proportional to original signal voltage and equal signal increments per channel number over all positions on the scale. This type of amplification is typically used in collecting cell cycle data, where the intensity range is more limited. Note that the highest intensity population, which fell onscale with logarithmic amplification, is now offscale (*i.e.* outside the dynamic range of the linear amplifier).

portional changes in channel number. Comparison of samples run at fixed high voltage but different gains is done using the gain ratio as the intensity correction factor. Comparison of samples run at different high voltages requires constructing a calibration curve: mean channel number is plotted versus high voltage setting for standard particles run over the range of settings of interest [6]. A word of caution: the shape of the calibration curve may be a function of the laser power used, since laser power determines the number of photons incident on the detector from a fixed number of fluorochromes, and one cannot assume the same curve shape at all laser powers [1].

Another type of calibration is cross-calibration of logarithmic and linear intensity scales, both of which are available on many instruments. When a logarithmic amplifier is used, the resulting channel number is proportional to the logarithm of the original signal intensity rather than to the signal intensity itself. This type of signal transformation is particularly useful in analysis of heterogeneous popuations with a wide range of signal intensities, e.g., immunofluorescently stained samples. However, as can be seen in Figure 2, drastically different increments in relative signal intensity are represented by the same increment in channel number at different positions on the logarithmic scale. Effectively, we have expanded the

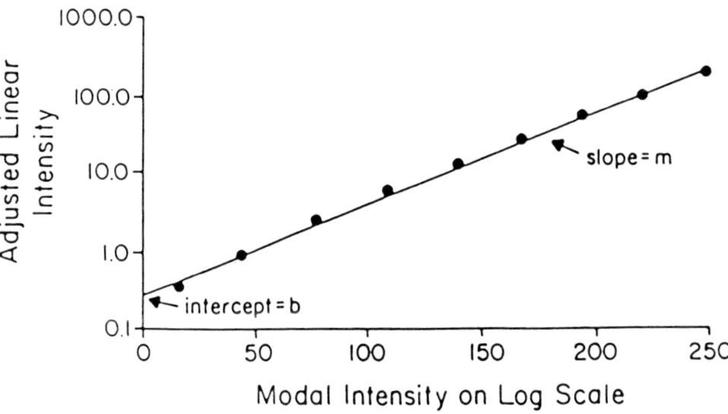

Fig. 3. Cross-calibration of logarithmic and linear intensity scales. Microsphere or other standards are run using a range of instrument settings which span the dynamic range of the logarithmic amplifier. At each setting the modal channel on the logarithmically amplified histogram (x axis) is plotted against the modal channel on the linearly amplified histogram after the latter is adjusted to account for any gain change necessary to keep the population onscale (y axis).

TABLE 1. Data for Cross-Calibration of Linear and Logarithmic Amplifiers

HV	Gain (G)	Log mode (LG)	Lin mode (LI)	Adj lin (ALI)	Calc. value
300	100	16	33	0.33	0.77
350	100	43	86	0.86	1.47
400	50	76.5	115	2.3	3.3
450	20	109	114	5.7	7.1
500	10	140	131.5	13.2	14.9
550	5	168	136	27.2	29.1
600	2	194.5	110	55	54.7
650	2	221	212	106	103
700	1	249	200	200	201

lower portion of the intensity scale at the expense of compressing the upper portion. Therefore it is no longer valid to compare the relative intensities of two populations by simply comparing their mean channel numbers, as we could on the linear intensity scale. Instead, we must convert channel numbers from logarithmically and linearly amplified scales to a common relative intensity scale. This type of calibration is illustrated in Figure 3 and Table 1. A sample of microspheres or some other appropriate standard is run several times, using a range of instrument high voltage/gain settings chosen so that the resulting peak positions span as much as possible of the full intensity scale on the logarithmic histogram. At each instrument setting, both logarithmically amplified and linearly amplified histograms are accumulated and the modal (peak) channels on each scale are recorded. Where amplifier gain settings are altered to keep the peak onscale in the linear histogram, the adjusted linear intensity is calculated as the observed modal channel divided by the gain setting. This corresponds to normalizing all the values to those that would have been observed at a gain setting of 1. A plot of adjusted linear intensity versus log scale modal channel (Fig. 3) allows calculation of system constants relating log and linear intensity scales and determination of the linear intensity corresponding to any given channel on the logarithmic scale. Plotted on the semilogarithmic scale shown, this corresponds to fitting a straight line of the following form to the data and determining its slope m and intercept b:

$$\text{Adjusted linear intensity} = 10^{\text{m(modal log scale channel)} + \text{b}}$$

Because the relationship between logarithmic and linear scales is a system parameter determined by the electronic characteristics of the particular amplifiers, the slope and intercept determined by the above calibration procedure are also valid for particles, excitation wavelengths, emission filters, etc., other than those used in the original calibration. However, lest you believe it is always best to take your data on the log scale and convert back when you wish to make intensity comparisons, we must mention that there are some drawbacks to this procedure. In the log-to-linear conversion we are effectively interpolating between channel numbers at the high end of the logarithmic scale, which represents much more widely spaced channel numbers on the linear scale. One price we pay for the greater dynamic range obtained on a logarithmic scale is that information about the wide range of linear intensities lumped into one log scale channel is lost and cannot be reconstructed by the log-to-linear conversion.

For some applications we may wish to calibrate either linear or logarithmic histograms in terms of numbers of fluorochromes, numbers of antibodies, or numbers of receptors, rather than simply in units of relative intensity. If the fluorochrome is fluorescein, we can use commercially available standards such as FluoroTrol-GF or Fluorescein Quantitation Kit (Flow Cytometry Standards Corp.) to calibrate our intensity scale(s) in terms of free fluorescein equivalents. In such standards, cells or particles are covalently labeled to different levels with fluorescein. The number of particles required to give the same level of fluorescence as a known concentration of free fluorescein molecules in solution is determined using a spectrofluorometer. This allows calculation of the average number of free fluorescein equivalents per particle ($\approx 3\times10^4 - 1.3\times10^5$ equivalents for FluoroTrol-GF, $\approx 6\times10^3 - 1\times10^6$ for Fluorescein Quantitation kits). Plotting free fluorescein equivalents versus peak channel number allows readout of either linear or logarithmic scales in terms of free fluorescein equivalents and expression of sample intensities in these terms (also cross-calibration of log and linear scales if desired). Note, however, that free fluorescein equivalents are *not* the same as number of fluoresceins. This is because fluorescein fluorescence is generally quenched by conjugation to proteins or particles, but to varying degrees depending on the particular protein and on the level of labelling; "typical" extent of quenching is 60 to 70%. Translation from free fluorescein equivalents per cell to number of antibodies bound per cell requires some way of estimating the degree of quenching for the particular antibody (or other ligand) of interest. This could be done either by assuming some "typical" quenching value or more accurately by determining the concentration of antibody which gives fluorescence equal to that of a known concentration of free fluorescein, assessed spectrofluorimetrically. F/P ratios based on absorbance measurements are not accurate for this purpose.

What about absolute calibrations when the fluorochrome is not fluorescein? Could we just refer everything back to free fluorescein equivalents, or perhaps refer all determinations to some common calibration particles with a broad spectrum dye useful over a larger color range than fluorescein itself? This is possible, but cross-calibration between dyes will be a strong function of excitation wavelength and emission filters used, again requiring additional determinations or assump-

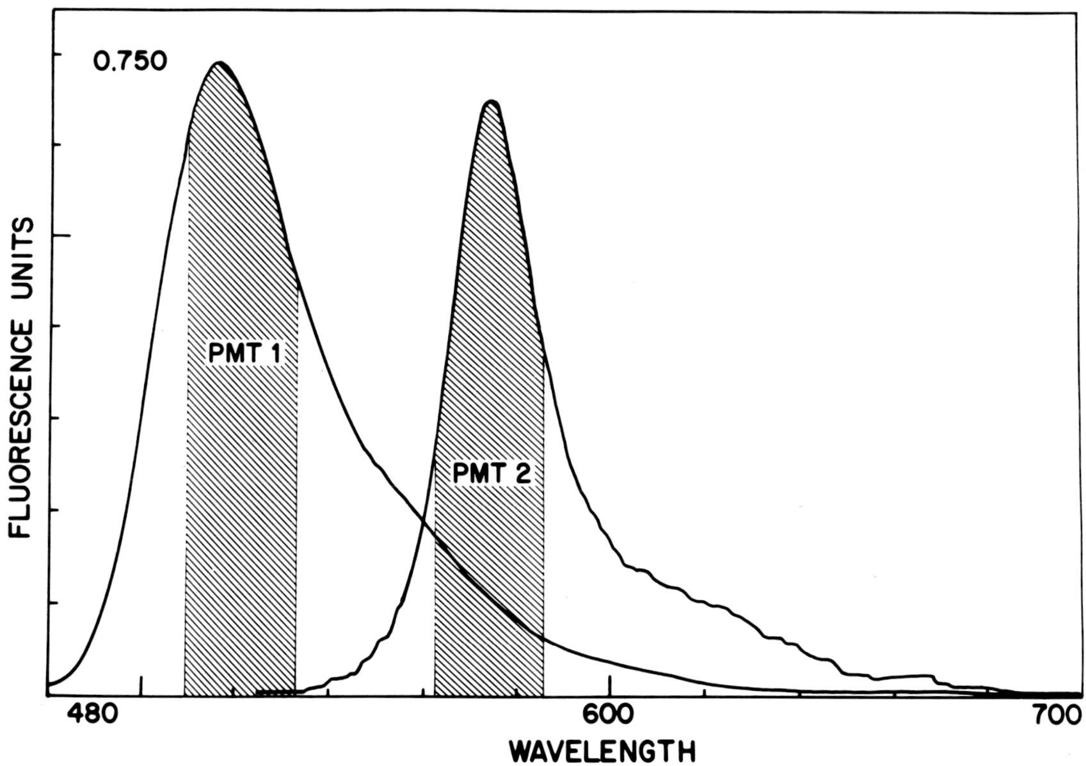

Fig. 4. Spectral overlap between fluorescein and phycoerythrin fluorescence. Emission spectra for fluorescein (left hand curve) and phycoerythrin (right hand curve) show significant overlap in the 550–600 nm range. If filter windows corresponding to the shaded regions are chosen, some fluorecein emission will be detected in both the "fluorescein channel" (PMT1) and the "phycoerythrin channel" (PMT2), resulting in the histogram of Figure 5A for cells stained with both dyes.

tions for accrate results. An experimental calibration particle (Simply Cellular) newly available from Flow Cytometry Standards Inc. offers a different way to solve this problem when the fluorescent reagents of interest are mouse monoclonal antibodies. Polyclonal goat anti-mouse serum is covalently conjugated to uniform size particles and the number of binding sites for mouse monoclonal antibody is determined by incubation with saturating amounts of a radiolabeled mouse monoclonal of known specific activity. To use as calibration standards the beads are incubated in saturating concentration of the fluorescent monoclonal of interest. Because the number of binding sites for murine antibody per bead is known, the resulting fluorescence intensity can then be related back to a known number of antibody molecules using your own fluorochrome, excitation wavelengths, and emission filters.

As stated initially, the complexity of calibration procedures increases in proportion to the absoluteness of the calibration desired, i.e. how much we wish to express our results in terms of biologically meaningful units versus relative intensity units. However, use of at least the basic levels of intensity standardization/calibration discussed above allows us to ask questions about the biological significance of quantitative as well as qualitative differences between cell types.

Color Compensation Standards

In most two color flow cytometric applications, our goal is to label cells with two distinct fluorochromes reflecting discrete cellular properties and to detect a signal proportional to one fluorochrome in one detector, unbiased by any contribution from the second fluorochrome, and vice versa. The discussion which follows uses fluorescein and phycoerythrin as sample fluorochromes, but similar standardization procedures are needed for any other two color combination. As illustrated in Figure 4, although we think of fluorescein fluorescence as "green", its emission spectrum is extremely broad, and in fact, a significant fraction of the emitted light lies in the "red" window where we wish to detect phycoerythrin emission. Similarly, our "red" dye, phycoerythrin, emits some light below 560nm in the "green" range. Thus, when choosing band pass filters to select the color ranges for analysis, it is impossible to choose wavelength bands which will accept all the light emitted by one dye and exclude all the light emitted by the other. If we use the filter windows indicated in Figure 4 to analyze a sample stained only with fluorescein, some of the emitted light will be detected at PMT2 and therefore called "phycoerythrin", even though none is present. The result is the histogram shown in Figure 5A; its diagonal character indicates that as we increase the fluorescein signal a proportional increase in the "phycoerythrin" fluorescence occurs. In a double-labeled cell, the total "phycoerythrin" signal is therefore the sum of the true phycoerythrin signal and the fluorescein-related "phycoerythrin" signal (see also Chapt. 17). Most systems provide some way to electronically or computationally subtract the fluorescein crossover signal from the total "phycoerythrin" signal to estimate the true phycoerythrin signal. To do this we take advantage of the fact that the crossover signal is pro-

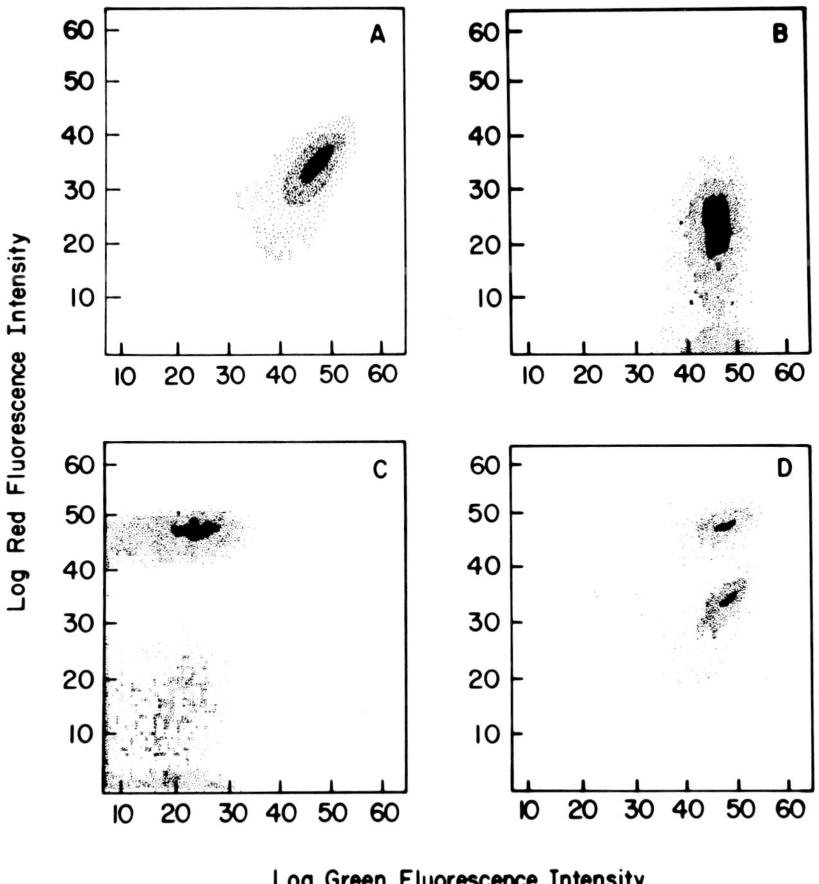

Fig. 5. Correction for color crossover in two color immunofluorescence. **A:** Cells stained only with fluorescein analyzed using the type of filter windows shown in Figure 4, without compensation for fluorescein crossover signal measured in the red (PMT2 of Fig. 4). Increasing fluorescein signal causes a proportional increase in "phycoerythrin" signal, giving the histogram a diagonal character. **B:** Second aliquot of cells stained only with fluorescein, analyzed using electronic subtraction to correct for fluorescein crossover signal. A fixed fraction of the fluorescein signal measured in the green (PMT1 of Fig. 4) is subtracted from the total signal measured in the red (PMT2 of Fig. 4), on a cell by cell basis. The appropriate fractional value is chosen so that the distribution becomes nondiagonal in character; i.e., the level of red fluorescence becomes independent of the level of fluorescein staining. At this point the level of red fluorescence should be similar to that of an unstained control sample (not shown). **C:** Third aliquot of cells stained only with phycoerythrin and analyzed using electronic subtraction to correct for phycoerythrin crossover measured in the green. **D:** Fourth aliquot of cells stained with both fluorescein and phycoerythrin, analyzed using the same electronic compensation settings chosen in Panels B and C. If filters or detector high voltage/gain are changed, the levels of compensation must be reset (i.e., the samples of B and C must be rerun).

portional to the size of the true fluorescein signal (cf. Fig. 5A) and that for each cell the amount to be subtracted from the phycoerythrin channel is therefore a fixed percentage of the fluorescein signal. The exact percentage is determined by the filter windows used, i.e., the relative areas under PMT1 and PMT2 regions for the fluorescein emission curve in Figure 4. Similar reasoning is followed for correction of phycoerythrin crossover into the fluorescein channel. Note that with the filter windows and relative dye concentrations shown, the correction for phycoerythrin fluorescence in the fluorescein channel is only a small percentage of the total phycoerythrin signal. However, this is *not* the same as saying that the correction is small compared with total signal in the fluorescein channel, especially if the true fluorescein signal is much smaller than the true phycoerythrin signal.

Obviously, real filters do not cut off as precisely as shown in Figure 4; we may not know the exact shape of our filter windows; and we usually do not know the relative strengths of fluorescein and phycoerythrin signals expected from cells. Therefore, it is difficult to predict the degree of color correction needed, and some type of standards must be used to empirically establish the appropriate degree of subtraction under our instrumental and biological conditions. Traditionally the standards used have been cells stained with either fluorescein alone or phycoerythrin alone. Cells labeled with fluorescein alone are run and color compensation is adjusted until the resulting histogram resembles that of Figure 5B. Note that the diagonal character seen in the uncorrected histogram (Figure 5A) is gone and the green fluorescence distribution is vertical in nature, i.e., the level of "red" staining is independent of the intensity of fluorescein staining. Note also that the level of red staining has not been reduced to zero, indicating that the cells have significant autofluorescence in the red channel. With correctly set compensation, the "phycoerythrin" distribution from cells stained only with fluorescein should be nearly identical to that obtained from unstained cells run under the same conditions. In parallel fashion, cells stained with phycoerythrin alone are run and compensation adjusted to eliminate any crossover signal in the green channel, giving the histogram shown in Figure 5C. The distribution of red fluorescence for positive cells is horizontal, indicating that the green channel intensity is not a function of phycoerythrin intensity. Note that again there is significant autofluorescence, this time in the green.

After correction for crossover of fluorescein fluorescence into the phycoerythrin channel and vice-versa, we are finally ready to run samples stained with both fluorochromes. The resulting histogram is shown in Figure 5D. However, if we wish to adjust the detector high voltages or try a different filter combination the corrections for emission overlap must be performed again. By now we may be running out of the samples stained with only one fluorochrome! One solution to this problem is to use beads which resemble phycoerythin-stained or fluorescein-stained cells, available from Becton Dickinson or Flow Cytometry Standards. Because the beads are used to establish the correct percentage (not absolute amount) of fluorescein signal to be subtracted from the phycoerythrin channel and vice-versa, they need not be exactly the same intensity as cells so long as they fall onscale under the same instrument conditions. However, it *is* critical that the beads have emission spectra exactly like fluorescein and phycoerythrin, so that altering filters or detector gains has exactly the same effect on their crossover fluorescence as it would for cells. One caveat in using beads to establish compensation settings is that their autofluorescence may be less than that of cells under some excitation conditions. If this is the case, correct compensation for beads may represent overcompensation for cells.

Finally, the problems discussed here are not peculiar to the combination of fluorescein and phycoerythrin. When fluorescein and propidium iodide are used for combined DNA/immunofluorescence analysis (e.g., anti-BrdU determination of S phase cells), crossover of "red" propidium iodide signal into the "green" fluorescein channel is the major problem. It cannot be too strongly emphasized that it is imperative to run the appropriate single fluorochrome standards, be they beads or cells, to establish valid crossover compensation settings for whatever dye combination is being used.

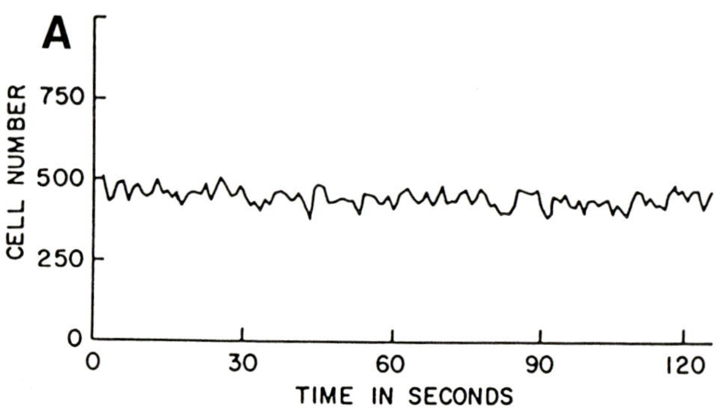

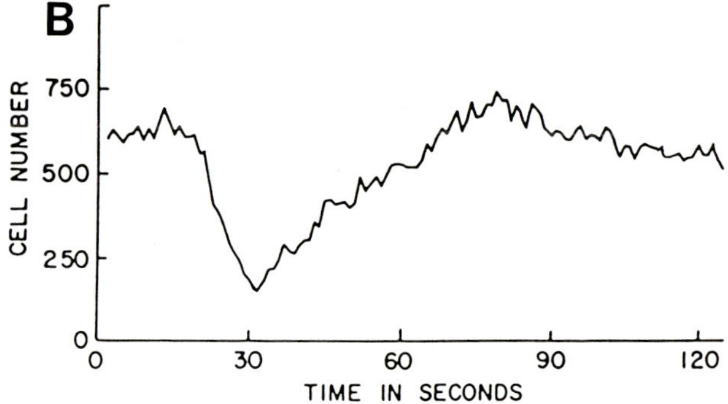

Fig. 6. Real-time monitoring without internal microsphere standards. Rate of data acquisition (number of cells analyzed per second) was monitored as a function of time, where zero time represents 1 minute after sample startup. **A:** Stable analysis rate indicates no significant change in rate of sample introduction or rate of exit from flow chamber. **B:** Sharp reduction in analysis rate followed by relatively quick return to original rate typical of that induced by formation and clearance of partial blockage of sample introduction tube or exit from flow chamber. Slow steady decrease in analysis rate may indicate settling of cells in sample tube and decreasing concentration of cells in suspension.

Standards for Real Time Monitoring

There are situations in the operation of a flow cytometry facility when we need to assure that instrument performance is constant throughout a period of analysis or sorting. Under such conditions it is important to be able to monitor instrument performance while the sorting functions or measurement or analysis functions are actually being carried out. One way to achieve this is to make periodic measurements on some internal standard and to use those measurements as a guide to continuing or discontinuing sorting, measurement or analysis. Such "on the fly" determinations are referred to as real time monitoring. One way to achieve this on-line monitoring of the flow system is to add a microsphere standard to the sample. (Note that unless the standard has been washed free of any suspending detergent, cell fatalities may result.) The microsphere fluorescence signal is acquired at one photomultiplier tube and used to monitor shifts in alignment, which are in practice usually flow related. If a standard with a known mean fluorescence intensity and CV is used, it is possible to calculate these values periodically and compare them to determine whether instrument performance is equivalent, better, or worse in succeeding time intervals during the run. Normally, acceptable ranges for mean intensity and/or CV are entered into the data acquisition computer by the user. Suppose the microsphere mean intensity is initially channel 170 and you wish to terminate sorting or analysis if this value fluctuates by more than 10%. Each time a specified number of microspheres (typically 100–1,000) has been accumulated, their mean is calculated and if it falls outside the range 153–187, data acquisition or sorting is halted. If your software permits, skewness of the peak may also be useful as an on-line performance criterion.

Figure 6 illustrates a different means of real-time monitoring, one which does not require addition of microsphere standards to the sample and can therefore be used under circumstances where microspheres do not give a discrete signal or might affect the cellular properties of interest. Many instrument problems which invalidate analysis or sorting results are associated with changes in the rate at which cells flow by the laser beam. Monitoring number of cells analyzed as a function of time helps to detect changes in flow rate and serves as an alert for related problems. Data collection in Figure 6 was begun one minute after starting the sample to avoid looking at fluctuations in flow rate associated with sample startup. For the run shown in Figure 6A, the analysis rate was approximately 500 cells per second and did not deviate over the 120 seconds of total analysis time. However, the run shown in Figure 6B began with an initial data acquisition rate of approximately 600 cells per second but showed a significant reduction in acquisition rate approximately 30 seconds into the run, followed by a return to the initial rate. This type of change in sample acquisition rate suggests that a partial plug or other flow problem altered the rate at which cells were introduced into or exited from the flow chamber. As with mean intensity or CV, it is possible to monitor cell acquisition rate and to terminate analysis or sorting functions if this rate changes beyond acceptable limits.

Standards for Sorting

One of the most critical aspects of flow cytometry is verification that the sorting function is performing correctly. The validity of our conclusions about the cells sorted depends on the purity of our sort, while the time it takes us to sort the desired cells depends on the sorting efficiency. Recall that at commonly used analysis and sorting rates there is approximately one cell-containing droplet for every 60 droplets formed, i.e., the large majority of the drops do not contain any cells at all. Thus unless the time delay between signal acquisition and charging of the drop(s) to be sorted is extremely accurate, we are very likely to sort empty droplets and reduce our effective sorting rate. One way to check the time delay might be to sort a stated number of cells onto a microscope slide and then verify sorting recovery by counting them in the microscope. However, cells may be lysed upon striking the microscope slide and/or destroyed by salt deposits as the droplet dries on the slide. Because of such problems, most laboratories use fluorescent microspheres as standards to optimize or monitor sorting functions. Fluorescence and/or light scatter histograms are accumulated for the microspheres and appropriate sorting gates or windows are set up. The operator then sorts a stated number of microspheres, typically 100, onto a glass microscope slide and allows the drops to dry. As shown in Figure 7, the microspheres tend to line up at the periphery of the crystalized droplet, making them very easy to count. If the sort logic indicates that 100 droplets were originally sorted but only 95 microspheres were found on the slide, this indicates that 95% of the droplets sorted contained a microsphere (assuming that all the droplets actually hit the slide).

Many sorting applications use 3-droplet sorting, i.e. at the appropriate time following identification of a cell of interest, 3 adjacent drops are charged and deflected. This is because some uncertainty arises about the exact location of the cell during the time it travels from the analysis point to the droplet charging point. Even if we are not quite sure that it will be in the central drop of the trio, we can be much more sure that it will be in either the drop immediately before or the one immediately after, and therefore that our sorting recovery will be relatively high. (The potential risk associated with this method for increasing recovery is that there may be an unwanted cell in one of the adjacent droplets, which will reduce the sorting purity.)

The length of time we wait before initiating droplet charging is referred to as the delay time or droplet delay. Delay times may be measured either in microseconds or droplet numbers, depending on the particular instrument you are using: at a droplet formation rate of 32,000 per second, 16 drops is equivalent to 5×10^{-4} seconds or 500 μsec. Table 2 illustrates the relationship between delay time and sorting recovery for 3-droplet sorting on an instrument which measures delay in drop numbers. At each of several different delay settings, 100 sort decisions were made and the number of microspheres actually deposited on a slide were counted.

TABLE 2. Recovery Optimization for 3-Droplet Sorting

Droplets switch setting	Particles counted	Droplets sorted								
		14	15	16	17	18	19	20	21	22
13	0	0	0	0						
14	5		0	0	5					
15	95			0	5	90				
16	100				5	90	5			
17	95					90	5	0		
18	5						5	0	0	
19	0							0	0	0

A switch setting of 13 (column 1) means "wait 13 drops, then charge and sort drops 14–16". At this setting, no microspheres were found on the slide, meaning that none were present in drops 14, 55, or 16. Using a switch setting of 14, 5 microspheres were found. These must have been in drop 17 since drops 15–17 are the ones sorted at this setting and we know from the previous run that no microspheres will be found in drops 14 and 15. Similarly, when we find 95 microspheres sorted at a switch setting of 15 (drops 16–18 sorted), we expect that 5 of them were in drop 17 and the other 90 in drop 18, etc., etc.. By now it should be obvious why another name for this type of table is "sorting confusion matrix"! It should also be obvious that anything which alters the flow rate and therefore the time it takes for a cell to go from analysis point to droplet breakoff point will have a drastic effect on sorting accuracy.

How does all of this translate to sorting recovery? At a setting of 14, we retrieved 5 microspheres out of 100 actually observed and requested, for a recovery of 5%, while at a setting of 16 we retrieved 100/100 for a recovery of 100%. Note that the recovery would be different if we were doing 1-droplet sorting. If we said "Wait 17 drops, then charge and deflect drop 18" the best we could do would be 90 microspheres for 100 sort decisions, or a recovery of 90%. Sorting purity in the 3-droplet case would be a function of how often we would expect the events in adjacent drops to be wanted or unwanted, in turn a function of the relative frequency of wanted events and sample concentration. Where you are trying to balance considerations of recovery and purity, it may pay to use a mixture of microspheres differing in color or size to select optimum sorting conditions.

Common practice is to use microsphere standards to establish that sorting recovery is at least 95% at the beginning of a sort run and then to continue sorting the sample without any further checks unless we notice some change in the stream dynamics or droplet formation point. Recovery should also be checked at any time during the sorting operation after a problem is encountered, and at the end of the run. In a typical nonplugging run this operation may be carried out as many as three or four times. Real time monitoring using internal standards or acquisition rate (see "Standards for Real Time Monitoring," above) can provide more continuous reassurance of proper instrument performance. In a more demanding sorting experiment where it is critical to know the exact number of cells sorted into a tissue culture dish (e.g., determination of subpopulation cloning recovery), one may sort cells to the left and microspheres to the right. Thus, to get an accurate determination of the sorting recovery live cells are sorted to the left into a culture

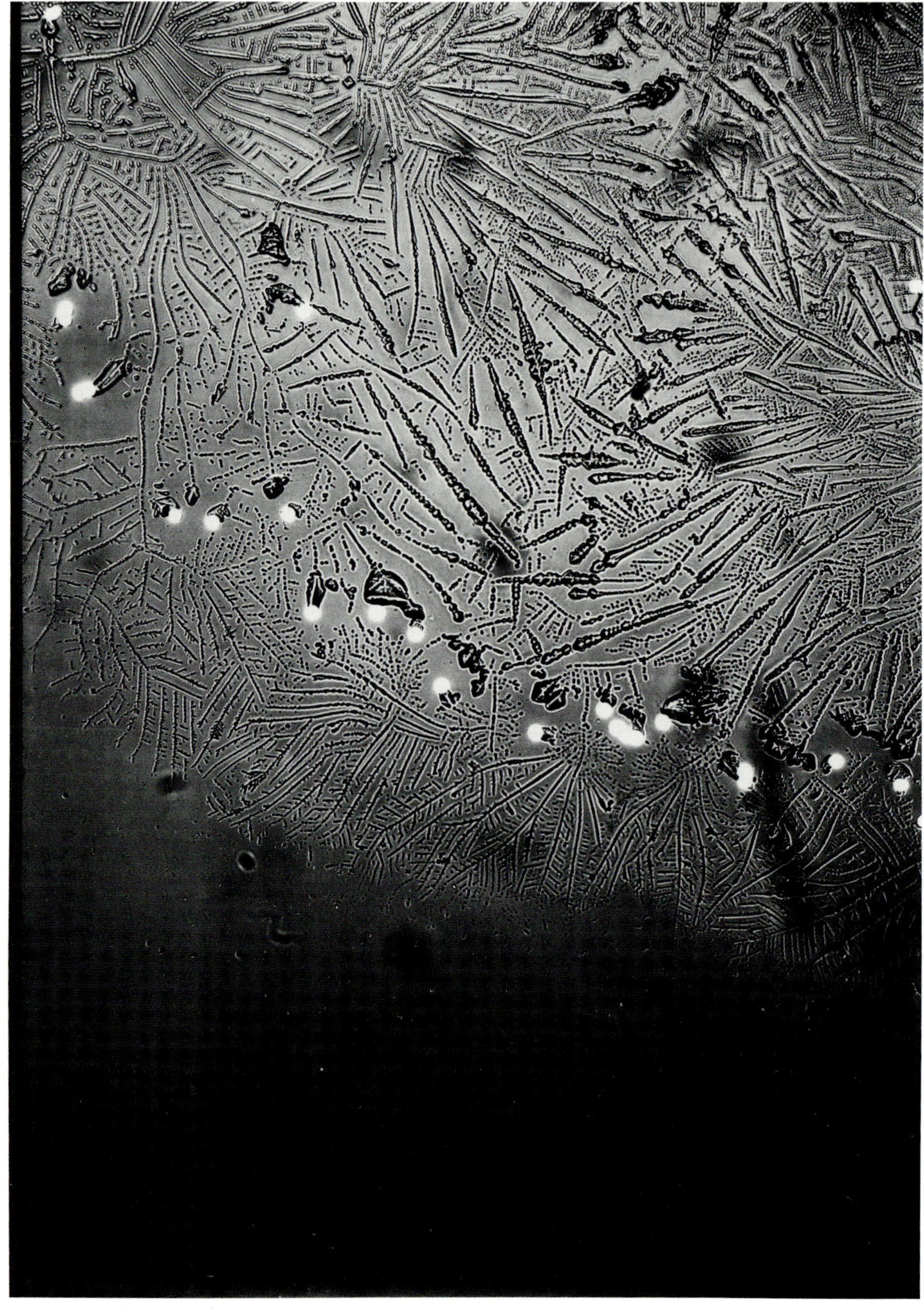

Fig. 7. Microsphere standards for monitoring sorting efficiency. Because of their stability to impact and high salt concentrations, fluorescent microspheres are excellent standards to compare number of particles theoretically sorted with those actually recovered.

vessel and microspheres are sorted to the right onto microscope slides. The number of microspheres sorted onto the slide is counted and the sorting recovery determined. If microsphere sorting recovery is 90% and 1500 sort decisions were made on cells going into the culture dish, then we calculate that the actual number of cells sorted into the tissue culture dish is $0.9 \times 1500 = 1350$. Internal sorting standards can also help resolve the situation where cell recovery is unexpectedly low but the reasons are unknown. If sorting recovery for cells is only 50% but that for microspheres is 95%, it suggests that you should be looking for stain stability problems or improvements in cell recovery methods rather than altering the delay time (see also Chapt. 8).

Standards for Quality Control Records

The use of flow cytometry in clinical laboratories is becoming more and more commonplace. One of the major requirements in running clinical samples is the assurance of proper instrument performance on a day-to-day basis. This is particularly important where laboratories are relatively new to the use of this technology; since experience cannot be used as a basis for assessing instrument reliability, some objective criteria must be available for this purpose. One way of following the instrument performance and monitoring changes in that performance is to run some standard on a daily basis, recording mean intensity and CV as performance parameters. Figure 8 shows such a quality control record for two laboratories. Both laboratories had the same model instrument and each optimized daily, keeping instrument parameters constant and recording mean and CV for both fluorescence and light scatter channels. Within each laboratory, the same lot of microspheres was used for optimization over the time period shown. In laboratory 1 (panel A) fluctuations were relatively minimal, but much poorer reproducibility was found in laboratory 2 (panel B). The general downward trend seen for the "green" fluorescence signal in laboratory 2 suggests some systematic change, perhaps filter degradation or fluorescence collection optics becoming dirty, and the need for further investigation. There is no reason why a properly functioning instrument purchased from any manufacturer cannot provide the kind of reproducibility shown Figure 8A. Note also that systematic trends are usually easier to spot in a running plot of optimization data than in a column of logbook entries.

Cell-based standards may also be used in quality control procedures and sometimes provide additional information not available from microsphere standards. We have already mentioned the use of FluoroTrol-GF or gCRBC for intensity standardization (see above). Variation in detector high voltage required to reset the standard peak to a particular channel, excitation power and emission filters remaining fixed, provides an index of instrument performance under the same conditions used for running samples. Figure 9 shows the detector settings required to reproduce FluoroTrol-GF peak positions on both "green" (panel A) and "red" (panel B) channels, after optimization of alignment using microspheres. Instrument performance was very reproducible (dotted lines indicate ±2 SD), with the notable exception of the "red" channel on day 34. However, the "green" channel setting for FluoroTrol-GF and all means and CVs for microspheres were well within formal limits. Further investigation revealed that low levels of propidium iodide (or propidium iodide stained cells) remained in the sample pickup tubing from a previous day's DNA analysis run and caused staining of the thymocyte nuclei used in FluoroTrol-GF. The fact that the shutdown method supposed to remove adsorbed dye had in fact failed to do so did not influence the microspheres (which have very little DNA!), but would have resulted in artifactual data for immunofluorescence analysis of fixed cells.

We have described two different ways of monitoring instrument reproducibility. The first, illustrated in Figure 8, keeps instrument conditions constant and monitors variations in peak position of the standard. The second, illustrated in Figure 9, keeps peak position constant and monitors variations in instrument conditions (detector high voltage). Is one better than the other? No, although it is perhaps easier to understand how much of a shift in instrument performance is indicated by a 10% shift in peak intensity than by a 10% shift in detector high voltage, because the latter has a nonlinear effect on measured intensity. Far more important than the exact method used is that *some* method be used consistently and with a clear definition of the acceptable range of performance parameters. With this kind of monitoring, instrument performance can be observed on a day-to-day basis and anomalous results obtained on patient or research samples can often be avoided and/or traced to problems in instrument performance.

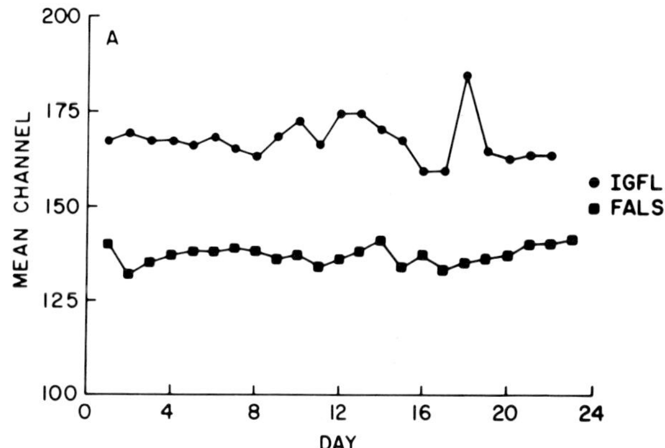

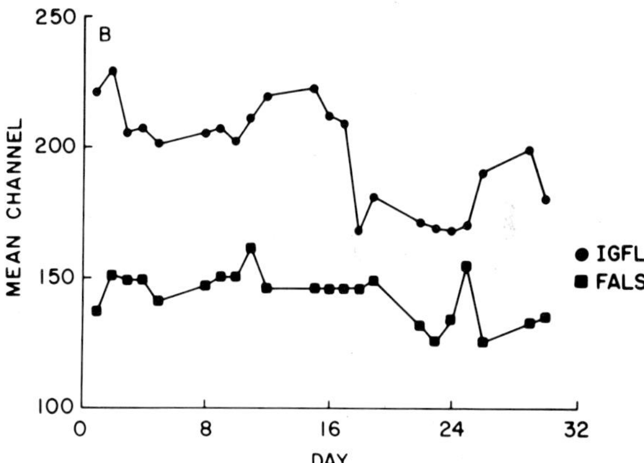

Fig. 8. Quality control using microsphere standards. Instrument performance was monitored under fixed conditions (see text) by recording mean channel number (A, B) and C.V. (not shown) for relevant parameters, here forward angle scatter and linear green fluorescence. **A:** Laboratory 1 shows relatively minor fluctuations and no steady trends; this type of record can be used to establish acceptable ranges for parameter values. **B:** Laboratory 2 shows significantly larger fluctuations and a general downward trend in green fluorescence intensity; this type of record indicates the need for corrective maintenance.

CONTROLS

While standards are used primarily to monitor instrument performance, controls are used to monitor reagent performance, quality of cell preparation methods, etc. Because the definition of appropriate controls is so dependent on the particular application in question and what experience tells us are critical steps to monitor, we cannot cover all possible controls. Instead we shall focus on those used in two of the most common applications, namely, DNA analysis and im-

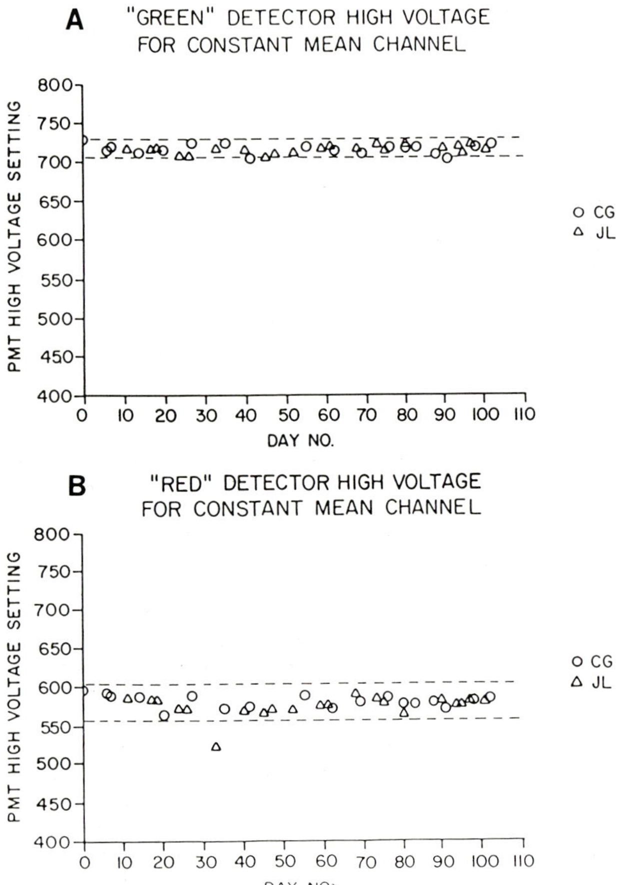

Fig. 9. Quality control using cell-based standards. Instrument performance was monitored under fixed conditions (see text) by recording detector high voltage/gain setting required to reset highest intensity peak of FluoroTrol-GF (Fig. 1D) to a fixed channel number. Instrument alignment had previously been checked by method of Figure 8. **A:** Only minor fluctuations detected in "green" channel (530±15nm) over a 3-month period by two different operators; dashed lines represent ±2 SD from the mean setting. **B:** Significant fluctuation observed on day 34 in "red" channel (>570nm), although "green" channel and microspheres on same day showed no deviation from acceptable range. Propidium iodide from a previous run had not been completely removed and was concentrated by thymocyte nuclei, increasing their intensity and decreasing the high voltage setting required to obtain the standard channel number.

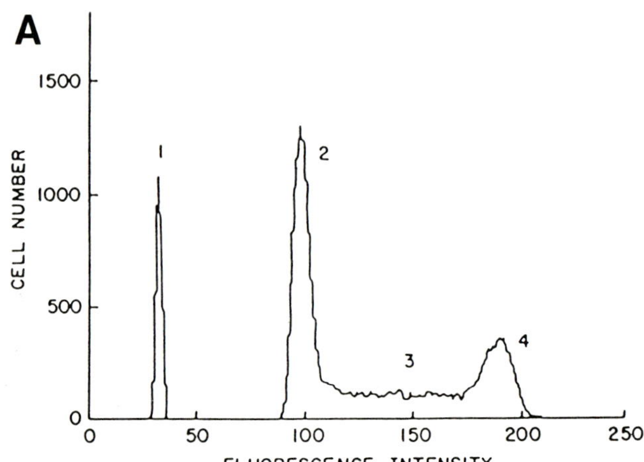

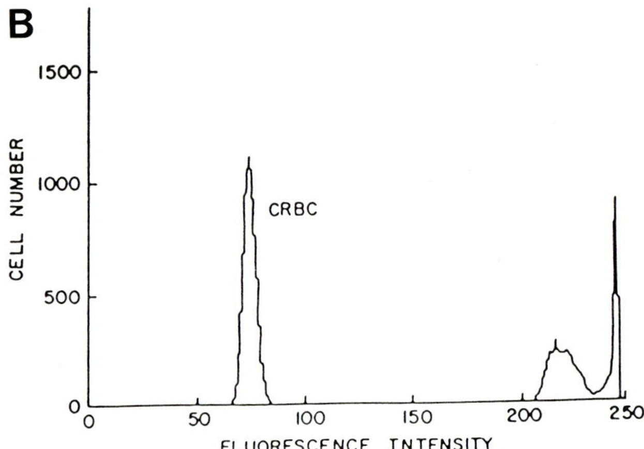

Fig. 10. Internal stain standards for DNA fluorescence analysis. **A:** Chick red cell nuclei (nCRBC, region 1) added to tumor cells (regions 2–4) prior to staining to allow assessment of sample to sample variability due to differences in stain concentration which are not possible using separate tubes of tumor cells and nCRBC standards (see text). **B:** Improved detection of small intensity shifts can be obtained by processing the signal used to generate Panel A simultaneously through a second amplifier set at higher gain to more sensitively monitor changes in the position of peak 1. The histogram of A is still available for calculation of cell cycle phase distributions of the tumor population.

munofluorescence analysis. Perhaps the most general statements that can be made about controls are that you usually need both positive and negative ones, that they should be as sensitive as possible to the parameter(s) being monitored, and that they should be as similar as possible to the biological sample being analyzed.

Controls for DNA and Cell Cycle Analysis

Most DNA analysis is carried out to determine whether a population of cells has normal or abnormal DNA content, to determine the distribution of cells in different proliferative compartments (G_0, G_1, S, G_2, M), or to determine both of the foregoing. The first application typically involves comparing the DNA-related fluorescence of the test population to that of a control. The second involves resolving cell cycle compartments from each other, either on the basis of differences in DNA content alone or in combination with some additional parameter (RNA content, anti-BrdU immunofluorescence, etc.). Controls to assess homogeneity, reproducibility and linearity of staining are important in either case.

Figure 10A shows a DNA histogram obtained from a mixture of tumor cells and chick red cell nuclei (nCRBC). Peak 1 is due to nCRBC, which give DNA fluorescence approximately one-third that of normal diploid mammalian cells. Peak 2 is G_1 phase tumor cells, region 3 is S phase tumor cells, and peak 4 is G_2 and M phase tumor cells. In evaluating the tumor cell population, we consider relative percentages of G_1, S, and G_2+M phase cells and the relative position of the G_1 peak. Suppose we run a replicate tumor sample and find that its G_1 mean fluorescence is 110 rather than 100. Several explanations are possible: concentrations of stain in the two samples were slightly different, there was a partial obstruction to flow or other problem while sample number two was running which caused increased fluorescence to be measured for all cells; or the two samples came from different parts of the tumor and represent populations with different DNA content. Use of instrument standards and stain controls can help us decide among these possibilities. If the problem is an instrumental one, an internal microsphere standard would also give proportionally increased fluorescence. If the difference results from staining variability, microspheres will not reflect it. However, an internal stain control such as the nCRBC included in the same tube as the tumor cells will also give altered intensity relative to the microspheres. If we find that both the microspheres and nCRBC intensities remain constant, we can conclude that the differences in staining intensity are due to some legitimate difference between tumor cell populations, although we have not yet eliminated differences due to variations in cell preparation methods. Note that simply running tumor cells plus microspheres in one tube and nCRBC plus microspheres in another tube will *not* give us equivalent information: the stained cells and the staining control must be exposed to the same staining conditions for the staining control to be an accurate reference point.

Figure 10B illustrates a method for more sensitively detecting small changes in intensity of the stain control in cases where the control differs considerably in staining intensity from the cells being monitored. The same signal used to generate the histogram in panel A was split and simultaneously processed through a different amplifier set at a higher gain, causing the nCRBC peak to appear at approximately channel 75 instead of 35. On this scale a 10% difference in staining would result in a 7-8 channel shift in peak position, while on the scale shown in panel A it would cause only a 3-4 channel shift. Obviously, in panel B we have pushed the tumor cells offscale by increasing the gain, so we must use the histogram shown in panel A for estimation of cell cycle phase distributions.

A variety of different cell types have been used by different investigators as staining controls. Human lymphocyte nuclei and mouse spleen or thymocyte nuclei have mammalian diploid DNA content, are available in large quantities, and give tight homogeneous staining distributions characteristic of noncycling G_1 cells. Avian erythrocyte nuclei have a lower DNA content (fluorescence $\simeq 1/3$ that of diploid mammalian nuclei), are readily obtainable in very large quantities, give tight G_1 distributions, and are quite stable once prepared and fixed [10]. Note, however, that whenever fixed cells are used as DNA staining controls, care must be taken to insure that doublets and larger multiplets are dispersed prior to addition to the sample, since such peaks may overlap with and distort the test population if present in significant numbers.

Choice of cell type for use as a staining control depends on the purpose of the DNA analysis. If the control is simply being used to correct for differences in staining among samples, any of the above will do as long as they are distinct from the test population in DNA staining intensity. However, in cases where a DNA Index is being calculated (G_0-G_1 mean of test population divided by G_0-G_1 mean of control population) and used to determine whether the test population has abnormal DNA content, it is generally agreed that the reference cells must be normal diploid cells from the same species as the test population [7]. In cases where abnormal cells with very near normal DNA content are admixed with cells of normal DNA content, the result may be a broadening of the G_1 peak without the appearance of discrete normal and abnormal peaks. Some investigators have used the ratio of test cells G_1 CV to nCRBC G_1 CV to detect slight broadening which may indicate the presence of a subpopulation of abnormal cells. However, this is not generally regarded as proof of DNA "aneuploidy" unless a second parameter can be used to separate the two populations and demonstrate that their two DNA fluorescence values are indeed different [2]. Note that strictly speaking, the term aneuploidy refers to an abnormal karyotype, which may or may not be associated with an abnormal DNA content (e.g., a balanced translocation gives an abnormal karyotype but normal DNA content). The older literature tends to use these two terms interchangeably and can be somewhat confusing; hence the convention on nomenclature [7] suggests the term "abnormal DNA stemline."

Staining controls can also be used to optimize staining conditions. In this case, optimization means minimizing CV while maintaining staining linearity. If binding sites for stain are not saturated, cell to cell variation and increased CVs may result from differences in chromatin structure and/or affinity, decreasing the accuracy of cell cycle compartment analysis and our ability to distinguish normal from abnormal DNA content. If stain concentration is too high, the linear relationship between increased stain and increased fluorescence is lost, distorting the relationship between G_2+M and G_1 intensities and subsequent cell cycle compartment analysis. The simplest way to determine optimum staining conditions is to choose a cultured cell with G_1 DNA content similar to the test cells of interest and characterize the effect of variations in stain concentration on relevant histogram parameters. Optimal staining is that which gives minimum G_1 CV consistent with maximum G_1 mean intensity and maximum G_2+M/G_1 intensity ratio; usually this will represent a concentration range as opposed to a sharp optimum. Over the optimum range there should be only minor variations in the cell cycle compartment analysis.

Where DNA Index and/or %S phase are being determined as part of a clinical assessment, some method of quality control is helpful in monitoring reagents, cell preparation procedures, and instrument optimization on a daily basis. The following system, designed for use in conjunction with DNA analysis of frozen breast tumor specimens [11], illustrates the use of both normal and abnormal controls and some of the considerations involved. Freshly isolated peripheral mononuclear cells from a healthy donor serve as the normal control. Their measured CV (1.86 ± 0.20 for 21 preparations, see Table 3 for method) and $\%S+G_2+M$ (should be 0 since they are noncycling cells) can be used to monitor the success of staining, instrument optimization, and doublet correction techniques. Hyperploid cultured human colon carcinoma cells (LOVO) propagated, aliquoted, and frozen at $-70°C$. serve as the abnormal control. An aliquot is thawed, processed, and stained in parallel with each batch of tumor specimens, with or without prior admixture of normal mononuclear cells.

Figure 11 illustrates the type of DNA histogam obtained from admixed normal and abnormal controls. Variations in cell preparation/staining or instrument optimization affect CV, DNA Index, and calculated %S phase: therefore these parameters can be used to monitor overall system performance and to assist in troubleshooting when problems arise. As illustrated in Table 3, a single batch of LOVO cells gave results which were quite stable over a 5 month period: %S phase varies somewhat ($\pm10\%$) from batch to batch but DNA Index is highly reproducible. Data of this type can be

TABLE 3. Stability of LOVO Quality-Control Specimens for DNA Analysis*

Date	DNA index†	CV‡	%S (PARA1)§	%S (SFIT)#
10-30	1.11	2.71	21.35	23.10
11-04	1.11	2.45	23.33	20.10
11-17	1.12	2.33	24.38	22.30
11-25	1.13	2.61	22.92	22.90
12-09	1.11	2.36	23.49	22.70
12-11	1.12	2.48	23.48	22.10
11-19	1.13	2.27	22.58	22.10
12-16	1.13	2.31	22.65	23.30
1-06	1.14	2.05	23.98	20.70
2-05	1.14	2.57	25.76	24.37
1-28	1.11	2.86	20.75	22.52
1-13	1.16	2.24	22.99	9.58
1-29	1.16	2.45	26.12	22.62
1-20	1.16	2.24	22.99	24.41
2-10	1.13	2.30	22.89	25.91
2-17	1.15	2.31	26.57	23.20
2-24	1.15	2.24	20.58	22.56
3-10	1.12	1.80	25.20	24.35
3-24	1.13	1.76	28.07	25.99
3-25	1.13	1.80	25.52	23.10
3-31	1.14	2.09	23.98	27.01
Mean	1.13	2.30	23.79	23.06
SD	0.02	0.28	1.90	1.44
N	21	21	21	21

*Staining method after K.D. Bauer with some modifications: incubate 45 min. @ 20°C in hypotonic stain (50μg/ml propidium iodide in 3.6 mM citrate, pH 7.8, containing 3% w/v PEG 6000, 0.1% Triton X-100, and 800 units/ml RNase) at 2×10^6/ml; add equal volume of hypertonic stain (50 μg/ml propidium iodide in 0.375M NaCl containing 3% w/v PEG 6000 and 0.1% Triton X-100). Store at 4°C for >1 hour prior to analysis; stable for 3 days.
†LOVO G_0-G_1 mean ÷ lymphocyte G_0-G_1 mean.
‡LOVO G_0-G_1 peak.
§G_0-G_1 and G_2-M areas determined by peak reflection, S phase by subtraction [4].
#Second-degree polynomial fit to S phase; G_0-G_1 and G_2-M by subtraction [4].

used to set confidence limits which must be met by the quality control sample before any patient specimens are analyzed.

LOVO are certainly not perfect models for analysis of solid tissue preparations. They have a higher %S and lower proportion of debris than typical human neoplasms. They are also unlikely to be appropriate controls for combined DNA/immunofluorescence staining using breast-tumor- specific markers. However, they do allow us to monitor our ability to reliably detect a DNA Index ≥1.1 and the consistency of %S phase estimates obtained using two different non-iterative analysis algorithms. They also allow estimation of effects of other variables such as specimen storage temperature: CVs increased to >5% after 3 weeks at −20°C [11]. As was the case with standards, there is no one system we can recommend that fits all applications. However, it is clearly important that some control which reflects the significant variables and/or parameters for your particular application be chosen (or designed) and run regularly.

Controls for DNA analysis can also be used to assess the sensitivity and reproducibility of compartmental analysis, i.e., to determine the level of variation among replicate samples and the degree of perturbation in cell cycle distribution which can reliably be detected. Cultured cells treated with agents which block cells in different compartments (e.g., vinblastine, thymidine) can be used to establish that the DNA staining method used is capable of detecting the type and level of perturbation of interest. The importance of running replicate samples is emphasized by a study which found that variations of less than 5% in cell cycle compartment analysis are unlikely to be significant and that the same batch of cells stained with different DNA stains gave surprisingly different compartmental analysis results [5]. It is always important to remember that we are really measuring DNA-related fluorescence intensity, not DNA content, and to be somewhat cautious in the conclusion we draw from "DNA" histograms.

Controls for Immunofluorescence Analysis

The goal of most immunofluorescence analysis is to use specific staining with antibody reagents recognizing distinct cellular antigens to identify and/or enumerate subpopulations of cells. "Specific" in this context means selectively identifying only the cells and/or the antigen of interest. Therefore, we need two different types of controls for immunofluorescence. Negative controls, or specificity controls, ensure that the antibody reagent is not interacting with cells other than those of interest. Positive controls insure that antibody is capable of recognizing the cells of interest. The level of negative control fluorescence is used to help decide what level of fluorescence we wish to call "positive" (i.e., specifically stained), while the positive control establishes that positive cells should be detected if present.

Let us start with negative controls and some of the types of nonspecific binding which we wish to detect and/or minimize. As cartooned in Figure 12, an antibody is composed of two distinct regions with very different functions. The $F(ab')_2$ portion is responsible for antigen recognition and binding; this activity is retained even if the $F(ab')_2$ portion is enzymatically cleaved and separated from the rest of the immunoglobulin molecule. The Fc portion of the molecule is responsible for bringing the bound antigen to the attention of immune cells by binding to Fc receptors on their surfaces. However, when we are interested in $F(ab')_2$-mediated binding to cellular antigen, Fc binding becomes by definition "nonspecific." $F(ab')_2$ fragments cannot bind to Fc receptors, but many reagents of interest are only available as intact immunoglobulin because of the significant losses entailed in their preparation.

Figure 13 illustrates that Fc-mediated binding can result in significant staining, especially in activated cells which have enhanced levels of Fc receptor expression. HL60 cells, either unstimulated or γ-interferon stimulated (500 units/ml for 48 hours), were incubated with an aggregated mouse IgG_{2a} antibody which did not recognize any HL60 antigens. Therefore the only binding observed was due to "nonspecific" Fc mediated binding. Fluorescein conjugated goat antimouse IgG was used to detect Fc mediated binding of the aggregated IgG_{2a}. Significant aggregate binding was observed even in unstimulated cells and markedly enhanced binding is evident in the stimulated cells. When using immunologic reagents to detect antigens of interest on the surface of the cell, it is highly likely that denatured antibodies or antibody aggregates in suspension may in fact bind to cells via the Fc portion of the molecule and not by the $F(ab')_2$. The flow cytometer, however, detects only total fluorescence intensity, and

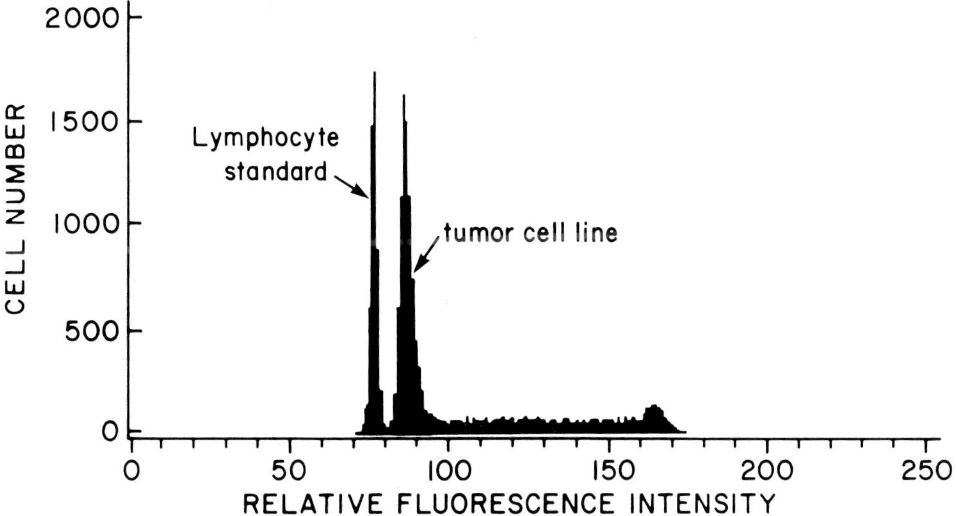

Fig. 11. Normal and abnormal controls in DNA analysis. Fresh normal human lymphocytes admixed with human colon carcinoma cells (DNA index 1.13 ± 0.02) can be used for quality control of reagents, cell preparation procedures, and instrument performance (see text). Aliquots of frozen cells provide reproducible specimens with which to monitor system performance over time (see Table 3).

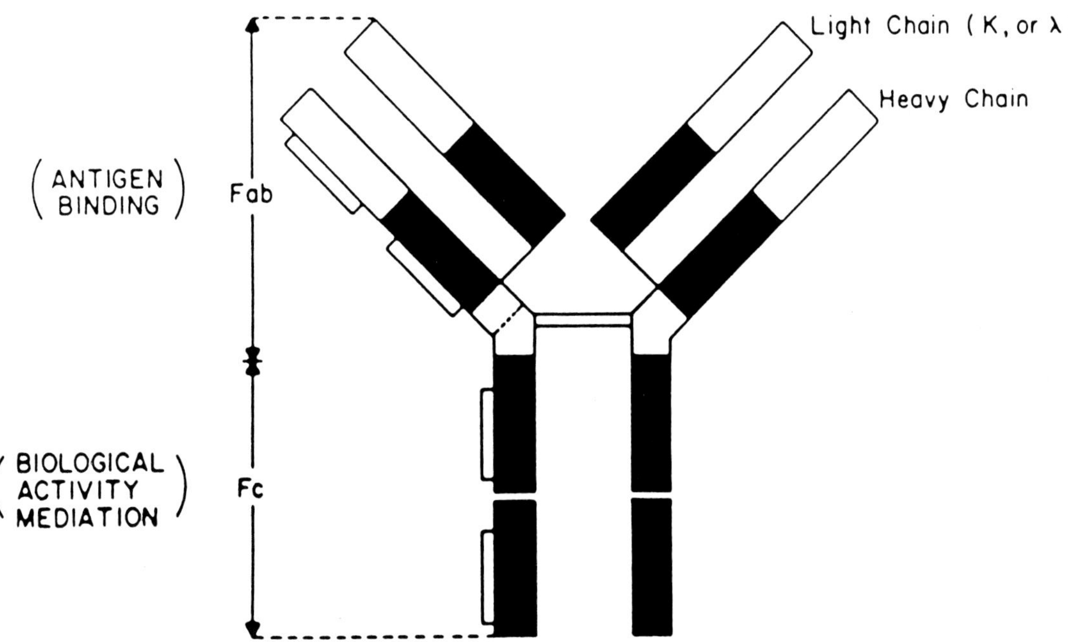

Fig. 12. Schematic structure of an IgG molecule. Antibody binding via the Fc portion is important for mediation of biological responses but nonspecific and may interfere with detection of immunologically specific $F(ab')_2$-mediated binding to cellular antigens (see text).

the operator must therefore determine whether the binding is via the antigen binding site or by the Fc portion of the antibody.

The first step in solving problems of nonspecific binding is identifying how severe they are. One negative control which serves as a good reference point is a completely unstained control; any fluorescence from this sample must be autofluorescence. Other appropriate negative controls vary somewhat depending on whether cells are being labeled using direct or indirect immunofluorescence staining (Fig. 14A or B, respectively). For indirect immunofluorescence, a second reagent control in which cells are exposed to labeled second antibody in the absence of primary antibody is a bare minimum. Increased fluorescence intensity of the second reagent control as compared with the unstained control suggests Fc-mediated or some other type of nonspecific binding of the secondary reagent; in this context, unwanted cross-reactivity due to $F(ab')_2$ binding of the second antibody would also be defined as nonspecific binding. If the primary antibody is a polyclonal antiserum, the ideal negative control would be to

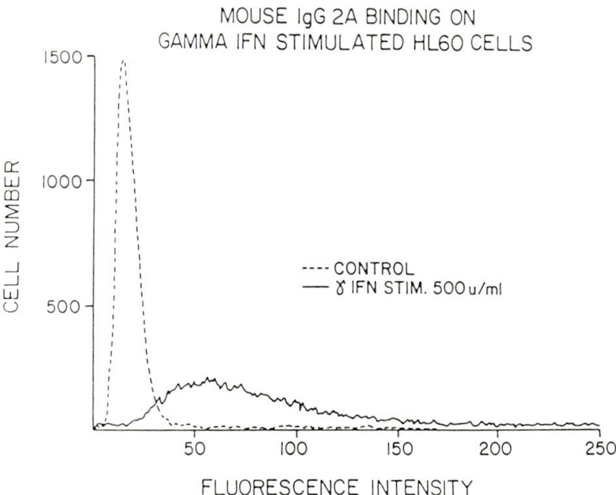

Fig. 13. Enhanced Fc receptor-mediated staining in activated cells. γ-Interferon activated HL60 cells display marked increases in level of Fc-mediated binding of mouse IgG_{2a}. Similar increases are observed in human peripheral monocytes treated with γ-interferon (not shown). Where Fc receptor binding represents "nonspecific" background, appropriate reagents and/or controls must be used to distinguish it from specific binding.

incubate with a pre-immune serum (taken from the same animal prior to immunization) followed by fluorescent secondary antibody. This attempts to mimic any nonspecific binding found with the primary immunoglobulin as well as that of the secondary antibody. Pre-immune serum from the same animal may not be available (as with commercially purchased primary antibody) or may be in short supply (when you have immunized the mouse yourself). In this case, a logical substitute would be normal (unimmunized) serum from the appropriate species, even if not the same animal. If the primary antibody is a mouse monoclonal antibody, it may be more appropriate to use a different antibody of the same isotype which is not thought to recognize any antigens on the cell of interest, the so-called "isotype control." Negative controls for direct immunofluorescence staining would certainly include the isotype control, but in this case it is also important to know that the fluorochrome:protein ratio of the specific and the nonspecific reagent are similar. Otherwise, similar levels of nonspecific binding can give rise to quite different levels of cellular fluorescence. Nonreactive cell types can serve as negative controls, in addition to nonreactive reagents. For example, when staining with a reagent specific for the $H\text{-}2K^k$ histocompatibility antigen in one strain of mice, cells from a mouse which is genetically identical except for that histocompatibility antigen (e.g., $H\text{-}2D^d$) can be used to assess nonspecific binding.

Dead cells represent another source of nonspecific binding. In the microscope they can often be distinguished by virtue of a diffuse cytoplasmic staining pattern, but the flow cytometer simply measures total fluorescence and we must be able to eliminate dead cells from analysis on the basis of nonmorphological criteria. Microscopically, dead cells frequently display decreased refractility, the flow cytometric correlate of which is decreased light scatter. For a homogeneous cell population, exclusion of cells with low light scatter can be a reasonably effective method of eliminating dead cells from consideration. However in heterogeneous populations, where large dead cells and small live cells may give equivalent scatter signals, this criterion may be quite ineffective [8]. A more robust technique is to use propidium iodide, which enters dead cells (i.e., those with permeable membranes) and stains their nuclei intensely red. Gates or windows may then be set to exclude those cells with high red fluorescence from the analysis. One caveat in using this technique is that when doing two-color immunofluorescence with phycoerythrin, it must be established that there are no degenerating cells which take up very low levels of propidium iodide and therefore appear live but phycoerythrin positive. This can be determined by running a sample stained with propidium iodide only.

Alternatively, live cells may be recognized by their ability to trap probes such as fluorescein diacetate (FDA) or BCECF [9]. Esterified probes readily cross cell membranes, the esters are cleaved by nonspecific esterases, and the resulting charged fluorochromes are not expected to leak back out across intact cell membranes. The validity of the latter assumption must be tested over the time scale of your experiment, but is more likely to be valid for probes with multiple charges. Gates or windows may then be set to include only live (positively stained) cells. One disadvantage of positive staining for live cells is that such stains are relatively bright and typically not compatible with quantitation of immunofluorescence in the same spectral window. However, this may be the method of choice where there are significant contributions to nonspecific staining from non-nucleated sources (e.g., red cells, platelet aggregates, enucleated cells, or large organic debris).

Negative controls of the type just described will help determine how much of a problem nonspecific staining poses. If it appears to be significant, there are two philosophies as to how to solve the problem. One is to "control" it, i.e. to run another sample which we assume imitates the level and type of nonspecific staining found in the test sample of interest, and to use that control to choose the level of fluorescence required to identify cells exhibiting specific binding. The problem with this approach is that it is often difficult, if not impossible, to know that the level and type of nonspecific staining are in fact well matched in control and test samples. A more conservative approach is to try to eliminate the nonspecific staining. If we think it is due to Fc binding, we may try high-speed centrifugation to remove any aggregates of denatured antibody. If we think it is due to some other immunoglobulin–cell interaction, we may try adding normal serum from an appropriate species during the incubation with specific antibody. If we think it is due to cross-reactivity with a related antigen/epitope, we may try to absorb out or titer out the undesired reactivity. If we think it is due to dead cell staining, we may try improving sample viability or eliminating dead cells by propidium iodide gating. The most difficult case is clearly a small unimodal shift in intensity, where there is considerable overlap between the unstained control and the stained sample. If all manipulations fail to eliminate the staining, our belief that the intensity shift represents specific but weak staining is increased but not necessarily proven.

Positive controls are also essential in immunofluorescence analysis, to assure that positive cells will in fact be identified. For example, a very commonly used positive control in lymphocyte subset analysis is the so-called "normal control." Peripheral blood from a healthy individual is stored and prepared in parallel with the test samples and stained with the same reagents. If such a control gives the results expected on

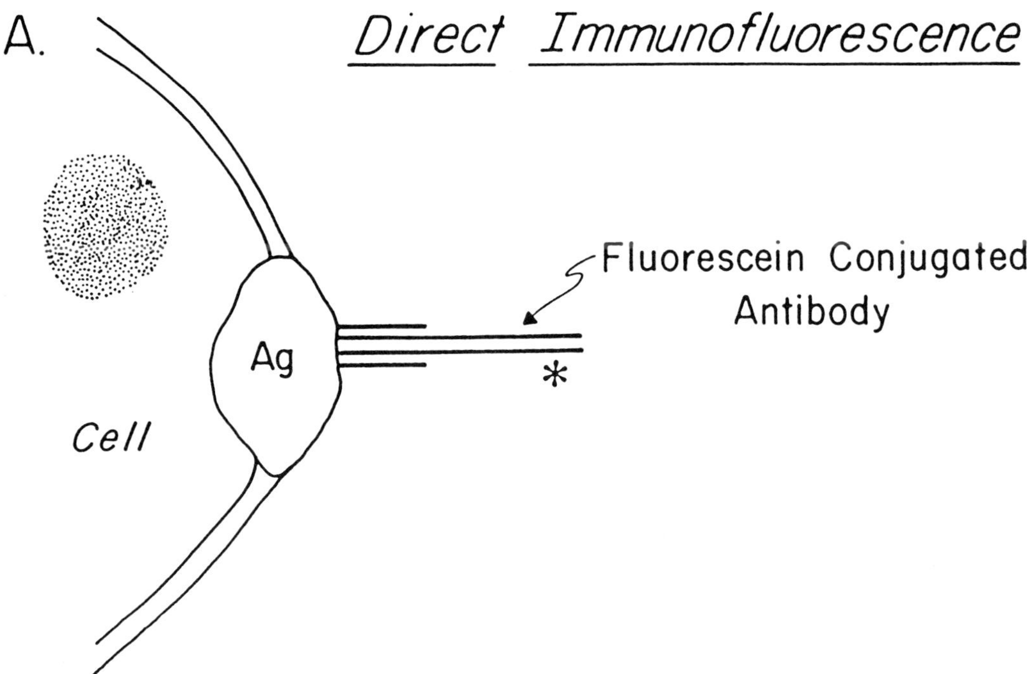

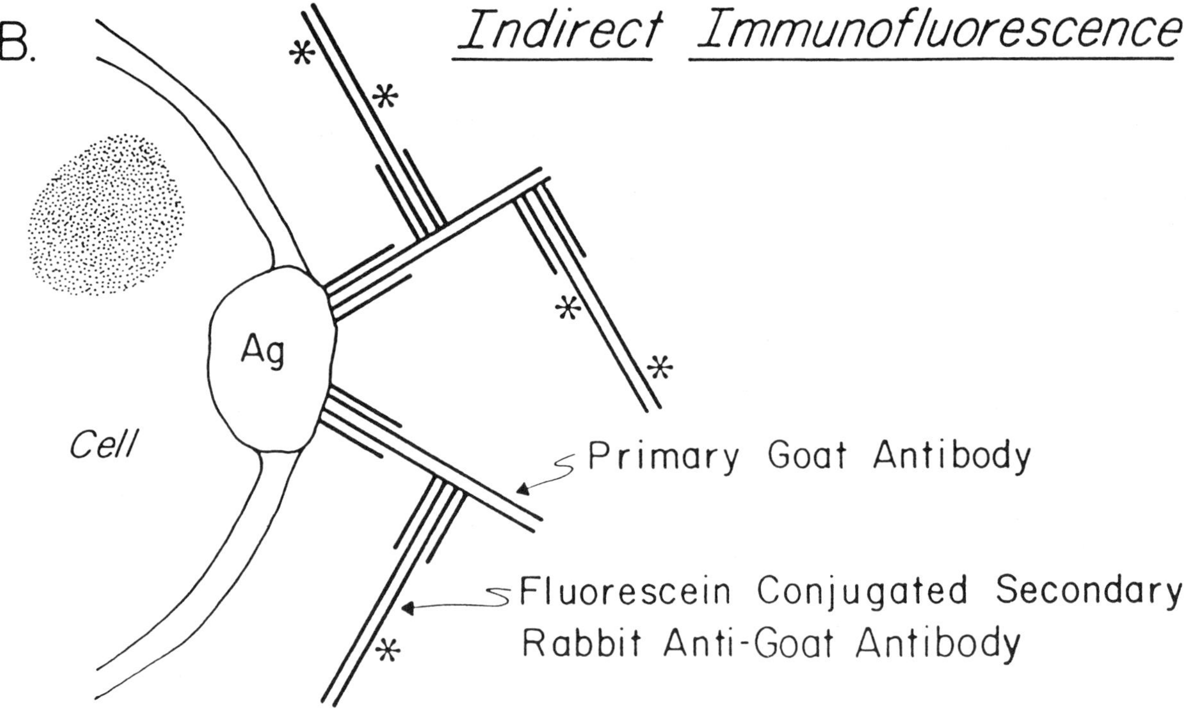

Fig. 14. Direct and indirect methods for immunofluorescence staining. **A:** Direct immunofluorescence uses fluorochrome (*) directly conjugated to the antibody recognizing the cellular antigen of interest and a single step staining procedure. Negative controls for this type of staining attempt to mimic the fluorochrome to protein ratio and level of nonspecific binding occurring with positive reagent. **B:** Indirect immunofluorescence requires two-step staining but generally results in higher fluorescence intensity per cell. An incubation with unlabeled primary antibody recognizing the cellular antigen of interest is followed by incubation with a labeled secondary antibody recognizing the primary antibody. Negative controls include incubation with unlabeled irrelevant antibody from the same species as the primary antibody and/or with the labeled secondary antibody in the absence of the primary antibody (see text).

the basis of our own or others previous experience with similar cells and reagents, we feel confident that positively stained cells in the test samples will be detected. What do we use for a positive control when the cell type of interest is not a cell type found in readily accessible samples from a healthy individual or animal? Cultured cells of the appropriate phenotype can indicate that the antibody reagent is working properly but may not detect problems with loss of antigen due to poor sample storage conditions or preparation procedures. For abnormal cell types, such as leukemias, it is sometimes possible to obtain and freeze relatively large quantities from a patient to use as later staining controls. Or if samples positive for the phenotype of interest come into the laboratory almost every day, a previously analyzed sample can also be used as a positive control. Note, however, that to serve in this fashion it must be re-prepared and restained, not just rerun.

Obviously if one is looking for a large number of unusual phenotypes, running all the positive controls may require a significant effort (and expense!). Are positive controls really needed? We would say emphatically "Yes". If a test sample does not stain with a given reagent and no positive control was run, it is impossible to say whether that sample was truly negative or whether there was a staining problem of some sort.

CONCLUSION

Potential causes of variability and lack of reproducibility in flow cytometric data are numerous, and may be either instrumental or biological in nature. Instrument standards allow us to minimize and/or compensate for instrument-related sources of variation, while negative and positive controls allow us to minimize those sources of biological variability which are not of interest and to focus on those that are. The more creative we are in designing and using appropriate standards and controls, the more successful we will be in applying the full quantitative capabilities of flow cytometry to measurement of biologically relevant parameters.

REFERENCES

1. **Bohmer R-M, Papaioannou J, Ashcroft RG (1985)** Flow-Cytometric determination of fluorescence ratios between differently stained particles is dependent on excitation intensity. J. Histochem. Cytochem. 33: 974–976.
2. **Braylan RC, Benson NA, Nourse VA (1984)** Cellular DNA of Human Neoplastic B-cells Measured by Flow Cytometry. Cancer Res. 44:5010–5016.
3. **Brown MC, Hoffman RA, Kirchanski S (1986)** Controls for flow cytometers in hematology and cellular immunology. Ann. N.Y. Acad. Sci. 468:93–103.
4. **Dean PN, Gray JW, Dolbeare FA (1982)** The Analysis and Interpretation of DNA Distributions Measured by Flow Cytometry. Cytometry 3:188–195.
5. **Dean PN (1985)** Methods of data analysis in flow cytometry. In Van Dilla MA, Dean PN, Laerum OD, Melamed MR (eds) "Flow Cytometry: Instrumentation and Data Analysis." New York: Academic Press, pp 195–221.
6. **Durand RE (1981)** Calibration of flow cytometry detector systems. Cytometry 2:192–193.
7. **Hiddeman WH, Schumann J, Andreeff M, Barlogie B, Herman CJ, Leif RC, Mayall BH, Murphy RF, Sandberg AA (1984)** Convention on Nomenclature for DNA Cytometry. Cytometry 5:445–446.
8. **Horan PK, Loken MR (1985)** A practical guide to the use of flow systems. In Van Dilla MA, Dean PN, Laerum OD, Melamed MR (eds) "Flow Cytometry: Instrumentation and Data Analysis." New York: Academic Press, pp 260–280.
9. **Dive C, Watson JC, Workman P (1988)** Polar fluorescein derivatives as improved substrate probes for flow cytoenzymological assay of cellular esterases. Mol. Cell. Probes 2:131–145.
10. **Vindelov LL, Christensen IJ, Jensen G, Nissen NI (1983)** Limits of detection of nuclear DNA abnormalities by flow cytometric DNA analysis. Results obtained by a set of methods for sample-storage, staining and internal standardization. Cytometry 3:332–339.
11. **Wallace PK, Muirhead KA, Horan PK (1987)** Specimen Preparation and Quality Control for Determination of DNA Ploidy and % S Phase in Breast Cancer (manuscript in preparation). Cytometry (Suppl) 1:53.

22

Data Processing

Phillip N. Dean

Biomedical Sciences Division, Lawrence Livermore National Laboratory, Livermore, California 94550

INTRODUCTION

Flow cytometers have the capability of rapidly overwhelming a user with data. The amount of data available, and the precision and accuracy available in its acquisition, provides researchers with enhanced capabilities in their experimental programs, but also puts great demands on their data processing methods. Most flow cytometers generate histograms of at least 256 channels, with a lot of attention currently being given to acquiring data of two or more variables. Bivariate histograms usually utilize 4,096 channels (64 × 64). Consequently, the bivariate measurement of 100 samples in a day, not unusual at many laboratories, generates 409,600 channels of information. The acquisition of list mode data, where all variables are stored for each cell measured, provides even more data and can easily overwhelm storage devices. This large amount of data presents a challenge to users of flow instruments, in both managing and analyzing such data. Data base management, providing for optimized storage and access to the data, is often neglected in the operation of a laboratory. This is an especially important consideration when more than the raw data are stored, for example data about the type of sample and its method of preparation, instrument parameters, and user and operator names.

Without rapid and efficient methods of analysis, investigators would soon be relying on only subjective judgments of their data. Laboratories that operate several instruments with active research programs have learned that a thorough, well maintained data processing operation is mandatory. Since the types of data obtained vary greatly, a large armamentarium of methods of analyzing data must be available. This chapter presents the various types of data that are obtained from flow cytometers, and the methods of data analysis that are currently available, paying special attention to those that are available online utilizing microcomputers. This chapter also emphasizes the practical aspects of data analysis, such as: (1) which method is "best" for a particular type of data; (2) what kinds of information and insights can be gained from different approaches; (3) are there data that cannot reasonably be analyzed with any method; and (4) what kinds (size) of computers are required with the various methods.

UNIVARIATE DATA

Nonparametric Analysis

All flow cytometric data can be processed through graphical and mathematical/statistical procedures. The graphical techniques discussed below are not restricted to univariate data but can be applied to bivariate data as well. Many of the analytical methods are also applicable to both types of data.

The simplest, fastest, and most useful technique in data analysis is to look at the data. For univariate data this requires only that the data be plotted as a histogram (frequency distribution); number of counts per channel versus the channel number. As described elsewhere in this book (Chapter 9, this volume), pulse height analyzers, also called multichannel analyzers, are used to acquire the data. The amplitude (pulse height) of the measured variable is digitized (converted to a digital number) and becomes represented as the "channel number"; the greater the magnitude of the variable, the higher the channel number. Thus the abscissa (X) of the plot represents the magnitude of the variable. For example, in DNA distributions the channel number is proportional to and represents DNA content. The analyzers have a finite number of channels, often 256, and the digitized signal amplitudes must fit into them. Thus, each channel covers a small range of amplitudes. This "binning" of the data determines the ultimate resolution of a measurement: it can never be smaller than one channel. For example, if a peak is positioned with its mean in channel 50, the best possible measurement resolution is 2% (1 channel out of 50). This is an important consideration when designing an experiment. If the experimental system, including instrument and staining protocols, performs at high resolution, one should be certain to use a large number of channels in the pulse height analyzer. For example, one should not use 100 channels to measure the DNA content of chromosomes stained with Hoechst 33258. With 23 chromosome types, the peaks due to some of the smaller chromosomes would be spread over as few as three channels, making mathematical analysis of the data impossible. On the other hand, for many immunological measurements, where the resolution is 20% or greater, 100 channels would be adequate. The ordinate (Y) of the plot is

Flow Cytometry and Sorting, Second Edition, pages 415–444
Published 1990 by Wiley-Liss, Inc.

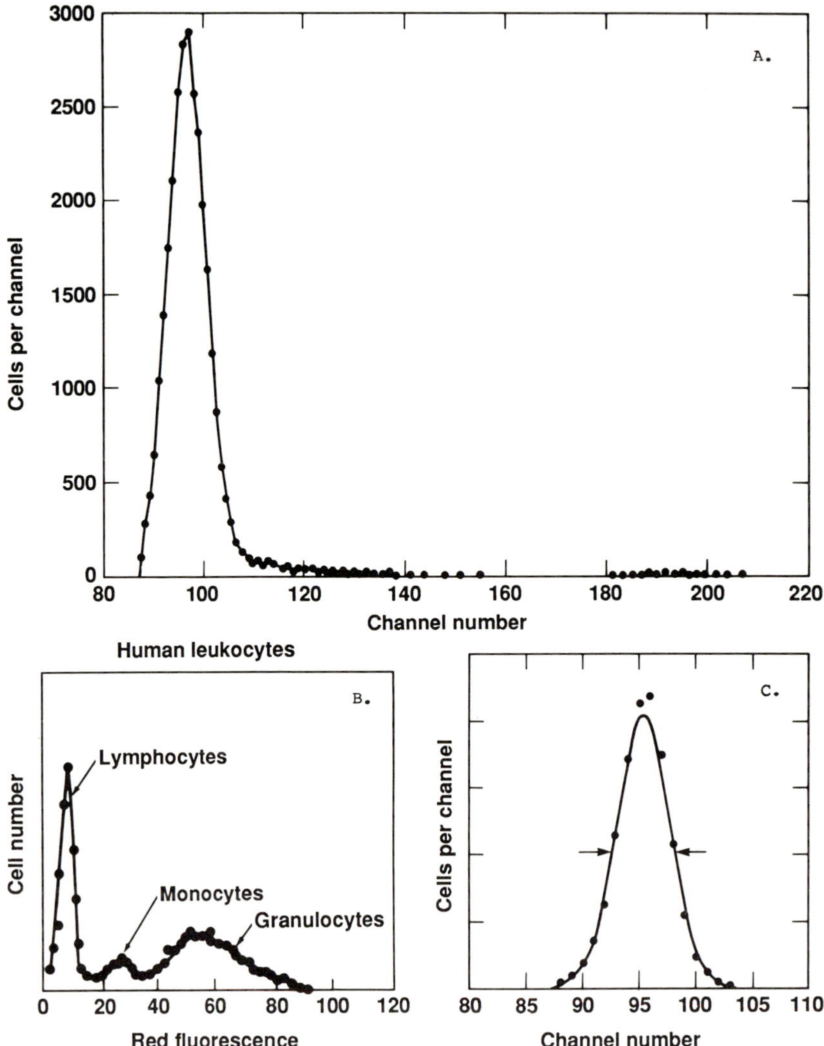

Fig. 1. **A:** DNA content distribution for Chinese hamster ovary (CHO) cells in culture, having reached "confluency." The cells have stopped proliferating (moving through the cell cycle), being suspended primarily in G0/G1-phase. **B:** Normal human leukocytes supravitally stained with acridine orange, revealing three subpopulations. **C:** Fluorescence distribution for hamster sperm cells stained by the acriflavine-Feulgen method. The arrows indicate the point at which the full-width-half-maximum is measured. The solid line is the result of fitting a gaussian function to the data.

the number of objects (e.g., cells or chromosomes) that have an amplitude that falls within the channel width, producing a frequency distribution.

There are many kinds of univariate data, and many ways of analyzing them. The first major application of flow cytometers was in the measurement of a single variable, usually DNA. These measurements are still very popular although the measurement of immunological probes has become very popular as well. Often a plot of the data is all that is required, subjective analysis being sufficient to yield the desired information. Figure 1 illustrates this. Figure 1A shows the DNA distribution for CHO cells growing in culture. The question being asked in this measurement is: "Are the cells in exponential growth or has the culture grown to confluence?" This amounts to asking whether or not there are cells in S-phase. Since about 40% of the population would be in S-phase for actively growing cells, it is clear that the cells of Figure 1A have reached confluency; almost all are in G0/G1-phase. Although this result is not quantitative, it is adequate to answer the question. In some cases this is not true. For example, in the treatment of leukemia it is important to know whether there are any cells in S-phase, at the few percent level. Treatment of the patient depends on such quantitative results and subjective analysis is not acceptable. In these cases more rigorous methods of analysis are required, as discussed later. Figure 1B shows data for which the objective is to determine whether there is more than one subpopulation of cells in a sample. These data are for human leukocytes, stained by an acridine orange method [43]. Several different staining protocols have been used in an effort to find one that can distinguish the individual cell types in the mixture. The protocol used for the cells in Figure 1B clearly resolves at least three subpopulations. Although the procedure does not identify the subpopulations, sorting or calibra-

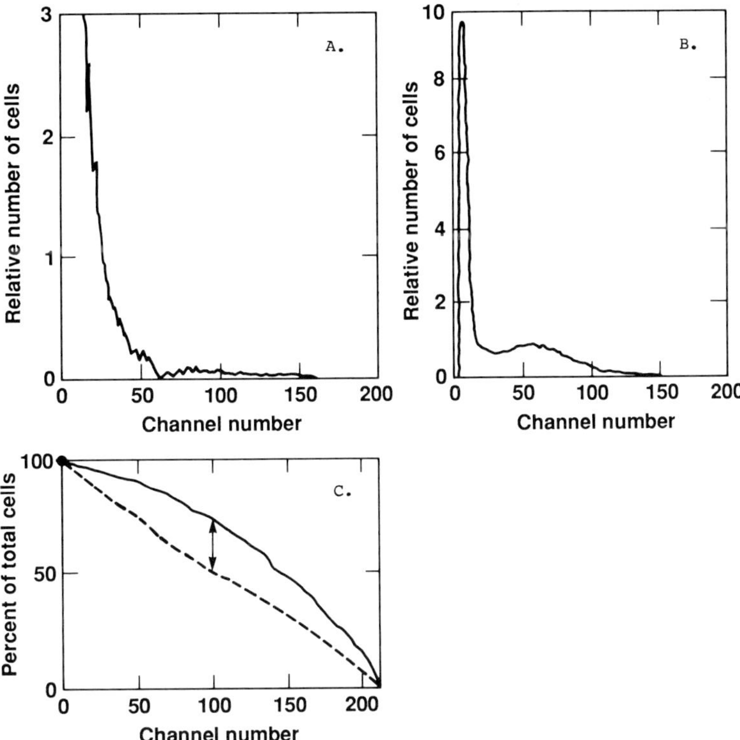

Fig. 2. **A:** human peripheral blood lymphocytes, stained with an arsanilated anti-HLA-2 antiserum followed by fluoresceinated rabbit antiarsanilic acid. **B:** Mouse spleen cells stained with a fluoresceinated rabbit antimouse Fab serum, showing two subpopulations. **C:** Percent of total distribution that has a fluorescence intensity (channel number x) greater than that channel number (x). The solid line is for the unknown sample and the dashed line is the negative control.

tion techniques would be required to do this, it is adequate for the purpose.

Many of the data obtained from immunologically marked cells are difficult to analyze quantitatively. The amount of fluorescently labelled antibody that can be attached to a cell is usually quite low, containing a few thousand molecules of dye. The fluorescence intensity of this amount of dye (e.g. fluorescein) is often not much greater than the cell's natural fluorescence. Figure 2A illustrates the problem [34]. Human peripheral lymphocytes were analyzed for HLA type by staining with an arsanilated anti-HLA-2,28 antiserum followed by fluoresceinated rabbit antiarsanilic acid. The question is, are there any positive cells? They would be represented as a peak somewhere to the right (beyond channel 50) on the histogram. The large peak at low fluorescence intensity is due to the natural fluorescence of the cells. A subpopulation can usually be easily identified if it contains a large number of cells or if there is a large number of binding sites per cell (a large number of fluorescein molecules per cell), producing a signal that is large compared with the natural fluorescence of the cell. Where neither condition exists one must use a "calibration" histogram. In this technique, a second measurement is made of the same cells without the added fluorescein, constituting a background histogram. The two histograms are then normalized to the same total number of cells and the background subtracted from the unknown. If the difference yields a statistically significant peak, then the sample is positive. The Kolmogorov-Smirnov [51] and the Gosset t-tests [6] are two methods of determining whether two frequency distributions differ significantly. These procedures usually require a computer or sophisticated calculator to accomplish.

For histogram subtraction to yield accurate results, especially when there are peaks in the data, the histograms must first be aligned (corresponding peaks put into the same channels in the two histograms). The method usually used for the translation will be illustrated with two examples. In the first example, the simpler one, a peak in channel 100 is to be translated (moved) to channel 50. This is a factor of two reduction, so data in channels 100 and 101 are added together and placed in channel 50 in the new histogram. Data in channels 102 and 103 are summed and placed in channel 51. Data in channels 98 and 99 are summed and placed in channel 49. The procedure is repeated for all channels in the original histogram. In the second example, a peak in channel 103 is to be moved to channel 90. This is a factor of 1.1444. Channel 90, actually from 90 to 91, a width of 1 channel, will receive data from channel 103 to 104.14. This means summing channel 103 (1.1444×90) and 14% of channel 104 ($91 \times 1.144 = 104.14$). Channel 95 will receive 86% of channel 104 (100-14) and 28% of channel 105 ($92 * 1.1444 = 105.28$). This procedure is repeated for all channels of the original histogram.

The critical assumption in the translation procedure just described is that the data are distributed evenly across the width of a channel. If narrow peaks are present in the distribution and/or the channels are relatively wide (few channels cover the peak), the assumption is clearly wrong. However, since the true distribution is usually unknown, and large numbers of channels are usually used, the procedure is adequate for the type of analysis being performed.

Often an experimenter will simply use minimum fluorescence intensity and minimum number of cells criteria to make a judgement. In the data of Figure 2A this consists of calculating the fraction of the population that lies above a brightness of channel 50. This minimum brightness for a positive cell is established using a standard histogram for unlabelled cells. In this example the cells are HLA negative. Figure 2B illustrates a positive sample. In this example, mouse spleen cells are shown to contain two major subpopulations by staining with a directly fluoresceinated rabbit antibody directed against the Fab fragment of mouse immunoglobulin. Positive cells, having a fluorescence intensity above channel 25, comprise 49% of the population.

An additional method [39] has been described that is reported to be more accurate than the simple subtraction method. In this method, two histograms are generated, as shown in Figure 2C. The ordinate (y-axis) consists of the number of cells with a brightness greater than the corresponding channel number. Thus, in channel one the y value is 100% and in channel 230 the y value is 0%. The upper curve, solid line, is for the "unknown" population. The lower curve, dashed line, is for the control population. The largest difference between these curves, indicated by the arrow, is the percent positives (20%). It can be shown that this procedure is rigorous mathematically, though it suffers from the problem of validity of the control data. A real advantage over the subtraction method is that no normalization is required, except for the x-axis, fluorescence intensity.

Comparing histograms through subtraction and ratios is sometimes improved if the data are first "smoothed." This is especially true when the histograms contain small numbers of cells per channel (poor counting statistics). Smoothing procedures must be used with great care. Done improperly, valid information (structure in the data) can be lost rather than gained. Smoothing procedures have been described in great detail in many publications (e.g., refs [11,13,15]) and is described here only briefly. The smoothing process attempts to find the middle of the measured distribution. It does this through an averaging process, using data on both sides of the point (channel) being smoothed. The simplest method of smoothing simply sums the contents of the channel being smoothed and one channel on either side. Dividing the sum by 3 yields the smoothed point. The average can be formed over any number of channels. However, as the number of channels is increased, fine structure in the data (e.g., peaks) is lost. In a considerably more sophisticated procedure, using robust locally weighted regression [15], also known as the lowess procedure, each data point is replaced by a value computed from a weighted least-squares fit to the data region around the point. The weighting factor is symmetrical about the point, being greatest at the point in question and zero outside the region. For example, nine points could be included in a calculation, the principal data point and four on either side, the weighting factors could be derived from a cosine function, and the nine points fitted by weighted least squares with a second degree polynomial. This calculation is performed for every data point, producing the smoothed distribution. Both distributions being compared are smoothed using this procedure; then, the subtraction or ratio is taken with the smoothed distributions.

A somewhat simpler smoothing process, lying between the two already described in complexity, smoothes data by averaging over adjacent channels using a binomial distribution to weight the averages. For three-channel averaging, the function for channel x is

$$y(x) = \tfrac{1}{4}\,y(x-1) + \tfrac{1}{2}\,y(x) + \tfrac{1}{4}\,y(x+1) \qquad (1)$$

and for five-channel averaging the function is

$$y(x) = \tfrac{1}{16}\,y(x-2) + \tfrac{1}{4}\,y(x-1) + \tfrac{3}{8}\,y(x) + \tfrac{1}{4}\,y(x+1) + \tfrac{1}{16}\,y(x+2) \qquad (2)$$

This method is faster than the lowess procedure and preserves the data structure better than the simple averaging process.

Methods of comparing two histograms to determine whether two populations of cells are different have other applications. For example, the direct subtraction method has been used to distinguish type T and B human peripheral lymphocytes [6] by comparing DNA distributions of cells stimulated with phytohemagglutinin, with and without the addition of 200 μg of thymocyte extract.

Logarithmic transformation of data can be very useful, in extending the dynamic range of a measurement and in some cases making analysis simpler. Figure 3 illustrates the first point. Figure 3A is a histogram of mouse bone marrow cells stained with wheat germ agglutinin labeled with phycoerythrin. Note that there is one well defined peak at low fluorescence values and a very broad undefined region at higher fluorescence values. Figure 3B shows the same cells with the fluorescence intensity measured with a logarithmic amplifier. The broad ridge of Figure 3A, spread over more than 150 channels, is now a relatively narrow peak covering only about 30 channels. The logarithmic plot covers a total of about 2½ decades in fluorescence intensity. In these two forms of representing data, changes in signal intensity have very different effects. In the linear plot of Figure 3A, a doubling of the amount of dye will double the peak mean (channel number), as well as the peak width (standard deviation). In the logarithmic display, the entire histogram shifts evenly to the right ($\log 2x = \log 2 + \log x$); peak widths and shape of the distribution do not change. The only way to change the widths of the peaks, and their separation, is to change the conversion gain of the amplifier (number of channels in the logarithmic plot that correspond to a doubling in fluorescence intensity). It is important to keep these properties of logarithmic transformations in mind if one is developing a mathematical model with which to analyze the data.

For immunological data, where the range in amount of bound fluorescent antibody is very large, this type of data processing is very effective and useful. The danger in using this transformation lies in the interpretation of results. Since the transformation is logarithmic, a symmetrical peak in the log plot will not be symmetrical in the linear plot. This also means that the standard deviation (or CV) of a peak in the log plot will not be the same in the linear plot. Peak means and ratios of peak means will also be different in the two plots. It is possible to convert parameters from one domain to the other. The following equations convert the mean and standard deviation of a peak in the log plot to what they would be in the linear plot. If μ_y and σ_y are the mean and standard deviation in the log domain, then μ_x and σ_x would be the mean and standard deviation in the linear domain.

$$\mu_x = \exp(\mu_y + \tfrac{1}{2}\sigma_y^2) \qquad (3)$$

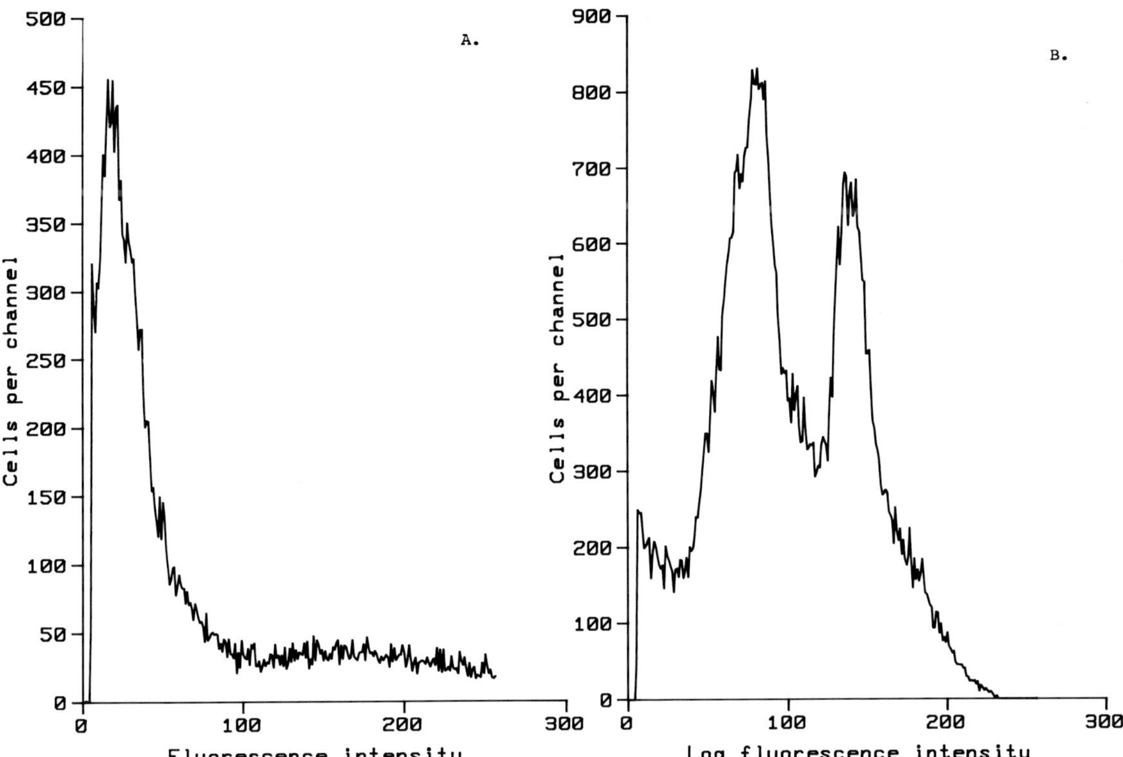

Fig. 3. Mouse bone marrow cells stained with wheat germ agglutinin that was labeled with phycoerythrin. **A:** Data collected using a linear amplifier; channel number is proportional to wheat germ content. **B:** Data collected using a logarithmic amplifier with a conversion gain where seven channels represent a doubling of fluorescence intensity; channel number is proportional to logarithm of wheat germ content.

$$\sigma_x = \exp(2\cdot\mu_y + 2\cdot\sigma_y^2) - \mu_x^2 \qquad (4)$$

Parametric Analysis

Simple models. In many cases, mathematical processing of the data is required. This processing can vary greatly in complexity. Figure 1C shows the fluorescence distribution for mouse sperm cells stained by the acriflavin–Feulgen method. Changes in sperm cells are being used as indicators of genetic damage in humans [42]. The indicator of damage is variability in the DNA content of the sperm, due possibly to nondisjunction in the replication process. This variability in DNA content would be reflected in variability in stain uptake. The width of the peak is determined by the instrument resolution, stain reproducibility, and DNA variability. If the first two items are constant, possible in good systems, then the last item can be measured. The peak width can be determined in several ways. The simplest is to measure the width of the peak at half its height, usually called the full width at half-maximum (FWHM). A more quantitative method that is more reproducible is to calculate the standard deviation of the mean of the distribution. The equation used for the mean, XMEAN, is

$$\text{XMEAN} = \sum(x\cdot y)/\sum(y) \qquad (5)$$

where x is the channel number and y is the number of counts in channel x. The standard deviation, STD, is then

$$\text{STD}^2 = \sum[y\cdot(\text{XMEAN}-x)^2]/\sum(y) \qquad (6)$$

These equations are easily solved using pocket calculators which have the functions built in. The calculations are also available online in most commercially available instruments. An even better method of analysis, which can provide good estimates even when the counting statistics are poor (low number of cells) is to "fit" the data with a normal distribution function (a Gaussian). The function used is

$$y = A/(\sqrt{2\pi})/\sigma\cdot\text{EXP}\{-0.5\cdot[(x-\mu)/\sigma]^2\} \qquad (7)$$

where A is the area of the peak, μ is it's mean channel number and σ is its standard deviation. This function is "fit" to the data by the method of least squares. Basically, the objective is to minimize the sum of the squares of the differences between the data, y, and the value, y_c, computed from Eq. 7.

$$S = \sum[y_c-y)^2] \qquad (8)$$

First, an estimate is made for each parameter (A, μ, and σ) and S computed. The three parameters are then changed slightly and S recomputed. If S decreases, then the parameters are changed and S computed once again. The parameter magnitudes may be increased or decreased. In general, the procedure can be strongly affected by the accuracy (nearness to the "true" answer) of the initial estimates; the better the

TABLE 1. Analysis of the Fluorescence Distribution of Mouse Sperm Cells Stained by the Acriflavin–Feulgen Method

Method of Analysis	Mean channel	Standard deviation	Percent CV*
Gaussian fit	95.4	2.32	2.43
FWHM†	95 4	2.25	2.36
Eqns. 5 & 6	95.4	2.36	2.47

* CV = Mean/Standard deviation (SD).
†FWHM = full width at half-maximum.

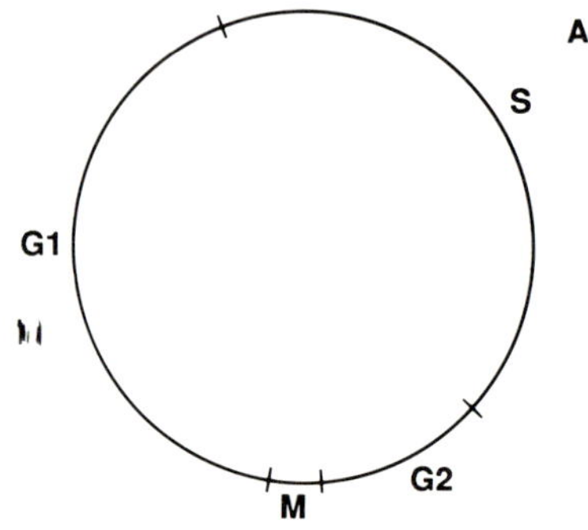

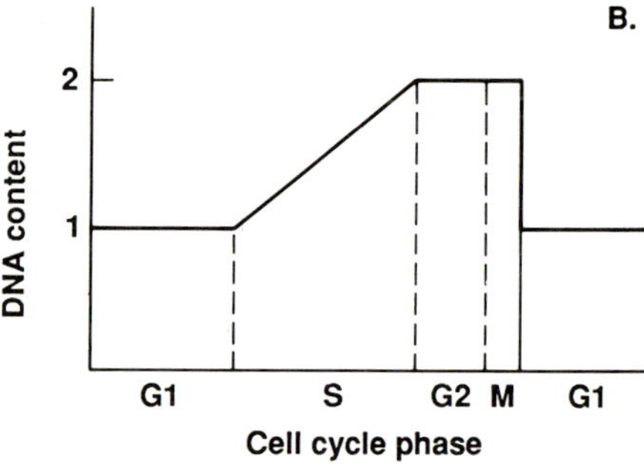

Fig. 4. **A:** Model of the life cycle of a cell. The four basic phases of the cycle are labelled G1, S, G2, and M. **B:** Same model, related to DNA content.

estimate the faster and more accurate the final result. There is a very extensive literature covering the procedure for knowing when to change a parameter and by how much. See Bevington [11] for a good discussion of this topic. This process is continued until the fractional change in S between two iterations is small (less than some arbitrary value, usually 0.0001).

Figure 1C also shows the data analyzed by the methods discussed above. The arrows indicate the points at which the FWHM is measured. The solid curve is the gaussian function fit to the data by the method of least squares (using the Marquardt algorithm described in Bevington). Table 1 presents the numerical results of these two methods of analysis, as well as the values obtained using equations (5) and (6). Included in the results is the CV, a popular measure of system performance. It is the coefficient of variation, defined as the standard deviation of the peak divided by the mean. To obtain this parameter from the FWHM we assume that the data are normally distributed. It can then be shown that

$$CV = 0.425 \cdot FWHM/mean. \tag{9}$$

It must be noted that references to the CV of a peak assume that the peak is symmetrical, which is not always the case. For example, the G1 peak of a DNA distribution is usually skewed to the right due to overlap with S-phase cells. The FWHM is larger than the true G1 peak standard deviation and consequently the computed standard deviation and the CV will be overestimated.

DNA Distribution Analysis. Measurement of the DNA content of cells was an early use of flow cytometers (49). The first application was in monitoring the ploidy of cells in suspension culture, where the assumption was that changes in the ploidy of growing cells would be reflected as a change in the shape of the distribution. This assumption turned out to be incorrect [27] but the development of the technology led to other applications. The shape of the distribution, as will be seen later, does reflect the distribution of cells around the cell cycle. According to the cell cycle model introduced by Howard and Pelc [24], the life cycle of a cell can be divided into four basic phases, G1, S, G2, and M. In the life of a cell two major events occur, the duplication of its DNA (in S-phase) and its division into two daughter cells (in mitosis). G1- and G2-phases were gaps in knowledge of the life cycle. A cell begins life at the beginning of G1-phase, with unit DNA content maintained throughout this phase. As a cell enters S-phase it begins the replication process, which is completed at the end of S-phase (beginning of G2-phase). The cell proceeds through G2-phase with two units of DNA and enters M-phase (mitosis), at the end of which it divides. This entire process is illustrated in Figure 4. This model, developed in 1953, is still the basic model used today, with some modifications that will be described later (see Chapters 23 and 24, this volume).

As is evident throughout this book, staining protocols and flow cytometers are not perfect. Ideal DNA distributions are never obtained and methods of data analysis must be able to process data such as is shown in Figure 5. The objective is to obtain the fraction of the population in the different phases of the cell cycle. G2- and M-phases are combined into G2M-phase, since they cannot be distinguished on the basis of DNA content alone. The combined errors in a measurement cause the ideal DNA distribution of Figure 5A to be "broadened" by a gaussian function into the distribution of Figure 5B. This distribution immediately suggests a mathematical model of the analysis of the data. Since the G1 and G2M peaks result from gaussian broadening, they can be represented by such functions. The S distribution is not so simple. It must be represented by a function that can follow its shape throughout S-phase. The first function suggested for such use was a second-degree polynomial [17], broadened in the same way as for the G1 and G2M peaks. The entire function used for the analysis is made up of three parts:

$$G1 = A_1/(\sqrt{2\pi})/\sigma_1 \cdot EXP(-0.5((x-\mu_1)/\sigma_1)^2) \tag{10}$$

$$G2M = A_2/(\sqrt{2\pi})/\sigma_2 \cdot exp(-0.5((x-\mu_2)/\sigma_2)^2) \tag{11}$$

$$S = \sum_{j=\mu_1}^{\mu_2} \{(a+bx+cx^2) \cdot G_j\} \tag{12}$$

$$Y = G1 + G2M + S \tag{13}$$

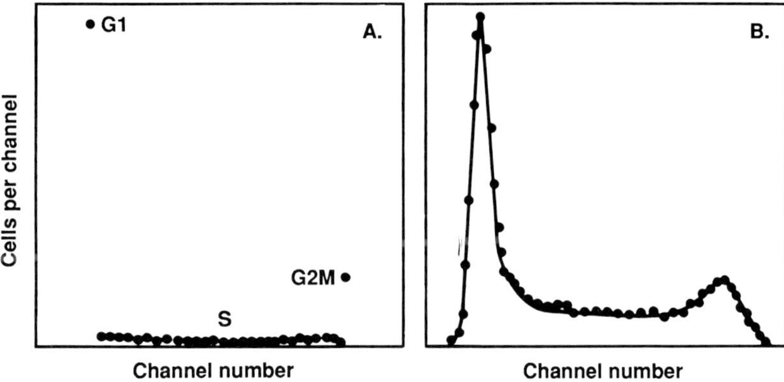

Fig. 5. **A:** DNA distribution for cells in asynchronous growth, measured in a perfect system; no staining or measurement errors. The shape of S-phase is determined by the rate of DNA synthesis through S-phase. **B:** DNA distribution for the cells of A, including measurement and staining errors.

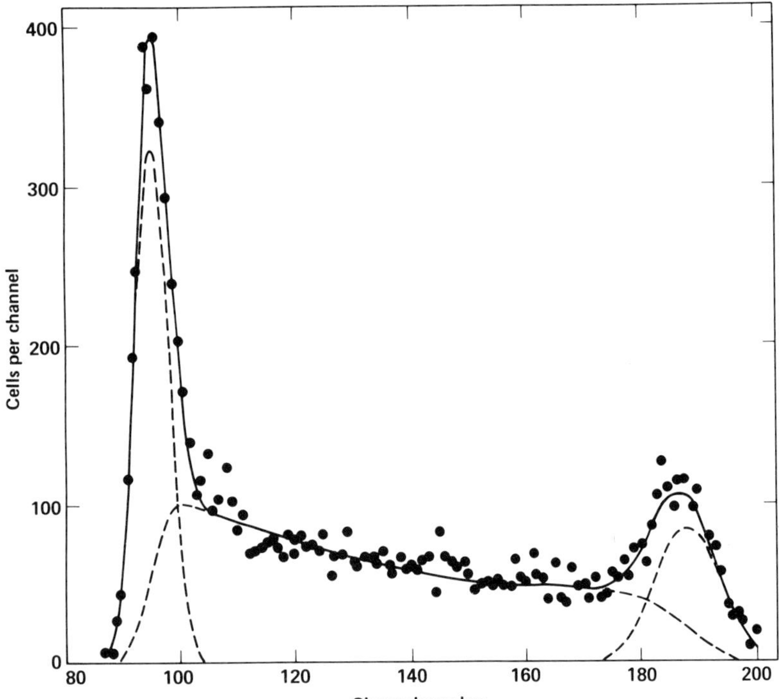

Fig. 6. DNA distribution of L1210 cells stained with chromomycin A3 and analyzed for phase fractions by the broadened polynomial method. The solid circles are the measured data, the solid line is the total equation fitted to the data, and the dotted lines show the broadened polynomial fit to S-phase, and the gaussian functions fit to the G1 and G2M peaks.

This function is fitted to the distribution by the method of least squares. The parameters are the areas (A_i), standard deviations (σ_i) and means (μ_i) of the peaks and a, b, and c are the coefficients of the polynomial. The gaussian function used to broaden the polynomial, G_i in Eq. (12), uses a constant CV. The area of (number of cells in) S-phase is obtained by integrating the polynomial between the G1 and G2M peak means.

The phase fractions are the areas (number of cells) of each phase divided by the total number of cells in the distribution. This method of analysis also yields the means and standard deviations of the G1 and G2M peaks. The standard deviations are useful in evaluating the adequacy of the measurement and the peak means can be used to monitor stability (repeatability) of the staining procedure and instrument performance. Figure 6 shows the method applied to L1210 cells in asynchronous growth. Table 2 shows the numerical results. Any good data analysis procedure requires some method of determining how accurately the data were analyzed, a "goodness-of-fit" criterion. The weighted variance, WTVAR in Table 2, is a good measure of how well the mathematical model fits the data:

TABLE 2. Results of the Analysis of the Data of Figure 5, Using the Broadened Polynomial Model for S-Phase*

G1	
Area	2,277 cells (24.2 ± 0.6%)
Mean	95.4 ± 0.1 channels
SD	2.79 ± 0.02 channels
CV	2.92%
G2M	
Area	1,174 cells (12.5 ± 0.6%)
Mean	188.0 ± 0.3 channels
SD	5.54 ± 0.04 channels
CV	2.94%
S	
Area	5,943 cells (63.3 ± 1.3%)

*Weighted variance = 1.81.

$$\text{WTVAR} = \text{SQRT}\{\sum[1/W\cdot(y-y_c)^2]\}/\text{ndf} \qquad (14)$$

where ndf is the number of degrees of freedom (number of data points minus the number of parameters), y_c is the computed number of cells and W is the weighting factor. If the so-called statistical weighting factor $[\surd(Y_i)]$ is used, and the only error in the data were due to sampling statistics, then a perfect correspondence between the data and the model would yield a WTVAR of exactly 1. In practice, nothing is perfect and WTVARs up to about 10 are quite acceptable. If it gets much higher than this, either the model or the adequacy of the data should be questioned.

The model just described works very well with cells in asynchronous growth, where the S-phase distribution (S-phase part of the curve in Fig. 6) has very little structure and the second degree polynomial can easily follow it. This is not always the case, as can be seen in Figure 7 for CHO cells measured 6 hours after treatment with the drug, cytosine arabinoside (ara-c). Ara-c kills cells in S-phase and causes a block at the G1/S boundary. At 6 hours post-drug treatment, all the cells originally in S-phase are gone and a large number of G1-phase cells have entered S-phase. In order to analyze this distribution, a different mathematical model must be used. In the preferred model, introduced by Bagwell [5], S-phase is described with a series of connecting rectangles, beginning with the mean of the G1 peak and ending with the mean of the G2M peak, as illustrated in Figure 8A. As in the polynomial model, to account for measurement errors the rectangles must be "broadened" with a gaussian function. Also, as in the polynomial model, the G1 and G2M peaks are represented by gaussian functions. The three components of the model are

$$\text{G1} = \text{gaussian 1} \qquad (15)$$

$$\text{G2M} = \text{gaussian 2} \qquad (16)$$

$$\text{S} = \sum\{G_j\cdot R_j\} \qquad (17)$$

and the total function is

$$\text{Y} = \text{G1} + \text{G2M} + \text{S} \qquad (18)$$

In Eq. (17), G_j is the gaussian used to broaden rectangle j, represented by R_j. The actual method of performing the broadening cannot easily be described since it involves the accumulation of several sums. For more information, readers are advised to read the ref. [5]. The solid lines in Figure 8 show the overall fit to the data and the broadened rectangles that fit S-phase. An improvement in this model, at some cost in complexity and in computational time, is obtained by us-

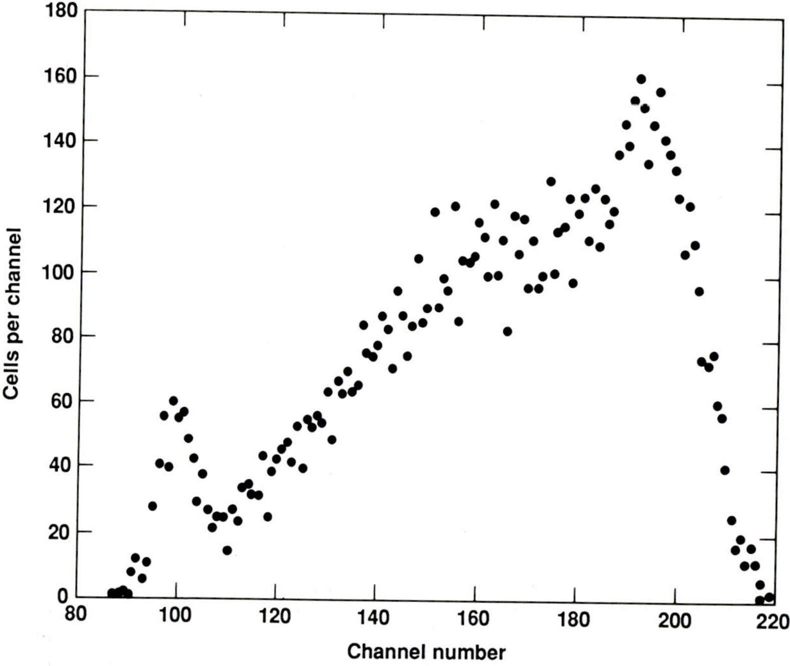

Fig. 7. DNA distribution for CHO cells 6 hours after the addition of ara-c.

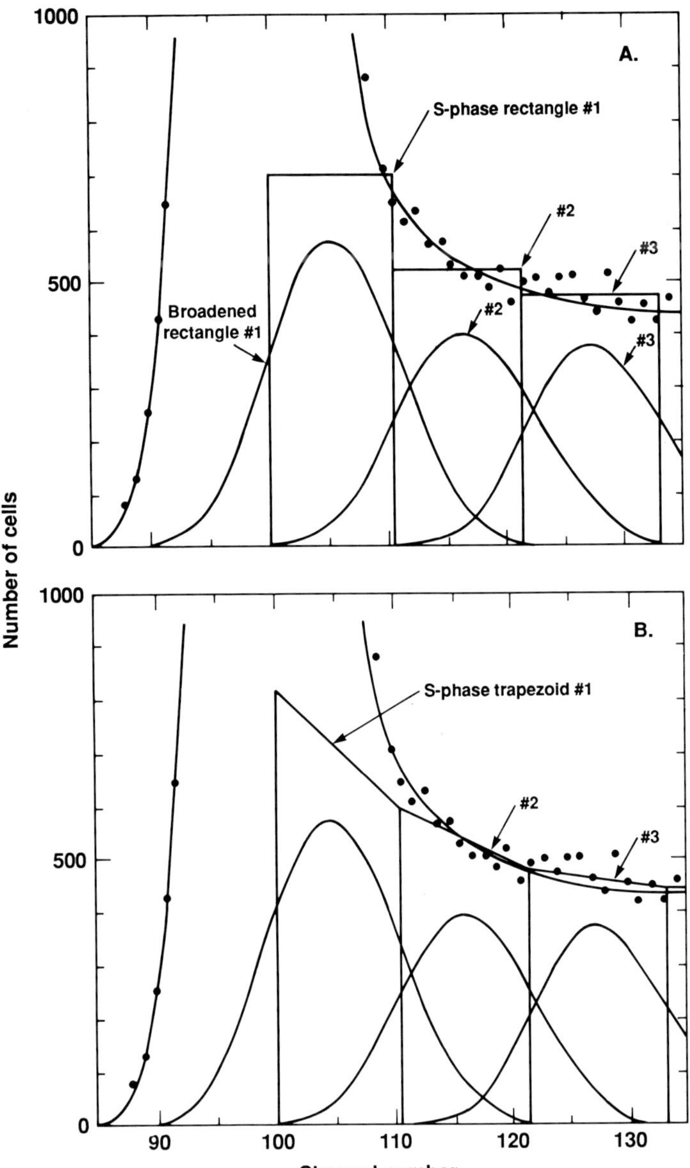

Fig. 8. **A:** Broadened rectangle model. S-phase is divided into series of rectangles, each of which is broadened with a gaussian function with constant coefficient of variation. The solid lines show the rectangles, and the peaks that result from the broadening effect. **B:** The broadened trapezoid model. S-phase is divided into a series of trapezoids, with connected corners. As with the rectangle model, each trapezoid is broadened with a gaussian function, as illustrated with the solid lines.

ing connecting trapezoids rather than rectangles, as illustrated in Figure 8B. This model, which provides for a continuous S-phase, is the recommended one where sufficient computational power is available. Figure 9 shows the results of analyzing the data of Figure 7 with the multiple rectangle model. The numerical results are given in Table 3. Note that more than phase fractions and G1 and G2M peak parameters are obtained from this type of analysis. Included are the areas of all peaks used to fit S-phase, providing the distribution of cells throughout S-phase. In addition, these programs estimate the error associated with each parameter calculation, based on the statistics of the process. These error estimates should not be ignored as they can often show that a calculation has produced results with little confidence. This is particularly true for data with low G2M-phase fractions and/or low S-phase fractions.

If a sequence of measurements is made at various times after drug treatment, it could be possible to study the progression of cells through S-phase. DNA distributions of perturbed cells can sometimes be very difficult to analyze, due to large populations in early or late S-phase. In these cases, there is no clear G0/G1 (or G2M) peak and the computer programs can have considerable difficulty in locating them. The programs can also have difficulty determining the standard deviation of the Gaussian functions that describe the peaks. In such cases, it is necessary to provide this information to the program, which can then hold these parameters constant, not let them vary during the calculations. The bet-

TABLE 3. Results of the Analysis of Data of Figure 9, Using the Broadened Rectangle Model for S-Phase*

G1	
Area	495 cells (5.1 ± 0.4%)
Mean	99.8 ± 0.2 channels
SD	3.7 ± 0.1 channels
CV	3.7%
G2M	
Area	1,315 cells (13.5 ± 0.9%)
Mean	199.6 ± 0.4 channels
SD	7.4 ± 0.2 channels
CV	3.7%

Peak #	Area (cells)
S-Phase	
1	53
2	298
3	478
4	802
5	1,141
6	1,517
7	1,595
8	2,083

*Weighted variance = 1.04.

ter programs allow one to "fix" any parameter in the function.

The two methods of analysis described above require significant computational power. On a PDP-11 class computer (e.g., PDP-11/44) the analysis requires about 20 to 40 seconds. On a VAX class machine (e.g., MicroVAX II), the analysis takes about 3 to 5 seconds. These models are usually not appropriate for present microcomputers (personal computers), since they would require 20 minutes to more than an hour per analysis, depending on the data. This situation is changing rapidly, however, and some of the more powerful personal computer models are now capable of making the calculations in a reasonable time.

Simplified mathematical models have been developed for use on microcomputers. Their main feature is that they require operator intervention in the analysis procedure, or very specific types of data, in particular for asynchronously growing cells with a low S-phase fraction. These models are characterized as "linear" in that the function is linear in all parameters. Figure 10 illustrates the use of a polynomial, which is such a function, in analyzing data from CHO cells [19]. In this method the middle of S-phase, the part that is not overlapped by either the G1 or G2M peaks, is assumed to be representative of S-phase. That is, it is appropriate to extend to the G1 and G2M peak means, a polynomial fit to this part of the data. The fitting is performed by the method of linear least squares. The polynomial is integrated to obtain the S-population and subtracted from the data to obtain the G1 and G2M populations. This method has been applied to many distributions and found to be adequate when the CV is low (4 to 5% or less) and the S-fraction less than about 40% [19]. Another method assumes that the left side of the G1 peak and the right side of the G2M peak are not significantly affected by S-phase. It sums these areas and doubles them to get the respective populations and obtains S-phase by subtraction [9]. Other methods draw straight lines through S-phase, where the histogram makes this feasible [7,8]. All these methods, and many similar ones, can be used only for very specific types of data. The more complete models described earlier can be used for a much wider range of data.

Not all data are as easy to analyze as those used to illustrate the methods described. Often data include a "background" contribution due to cell and other fluorescent debris, usually in the form of a continuum underlying the data of interest. Two methods of dealing with the background problem will be described. The first and simplest method is to draw a straight line from the left side of the G1 peak to the right side of the G2M peak, at the presumed background level at each position. The data below the straight line are subtracted from the total distribution (sometimes called a "stripping" process) and the "pure" DNA distribution then analyzed by one of the methods already described. An alternative to the straight line is an exponential curve. In this case, the subtraction is more easily carried out if the data are first converted to a semilog plot, log number of counts versus channel number. The disadvantage of this method of accounting for the background component in a DNA distribution is that it is entirely subjective, and carries the risks associated with such approaches. The accuracy of results is very dependent on the expertise of the person performing the analysis and on the quality of the data; the shape of the background is usually determined by very few data points. A much more satisfactory approach to the background problem is to include a background function in the mathematical model used to analyze the distribution. For example, an exponential function, $B = aEXP(bX)$, could be added to Eq. (18). The parameters of B would be obtained in the same least-squares procedure that provides all of the other parameters. The function could also be a second degree polynomial or some other function that is found to accurately model the background. The advantage of this procedure is that it is objective, is statistically valid if the uncertainty in sampling the background is the same as the uncertainty in sampling the cells, and makes use of all of the data.

Another problem that can occur with the analysis of DNA distributions is caused by cell clumping. Two G1-phase cells that are stuck together have the total DNA content of a G2M-phase cell and will be measured as one in a flow cytometer. This will result in an artificially high G2M-phase fraction. Nothing can be done mathematically to correct for this problem, since the number of cells that stick together is unknown. Some instruments utilize special electronic circuits, called doublet discriminators, to reject such events; two G1-phase cells stuck together are larger (longer) than a G2M-phase cell.

For rigorous least-squares fitting of mathematical functions to data, smoothing of the data first is not recommended. However, having made the requisite disclaimer, it must be said that smoothing can sometimes be beneficial. When the function being used does not exactly match the data, for example when data are distributed in a Poisson distribution but fit with a gaussian distribution, the resulting parameter estimates are incorrect. The estimates can be improved if the minimum chi square can be reduced. Smoothing the data by averaging over adjacent channels does not change the total number of cells in the distribution. It will, however, reduce the chi square in a least-squares fit (by an order of magnitude or so) and thereby improve the estimate of the area. Although this type of smoothing will improve the estimate of the area, estimates of other parameters, e.g., stan-

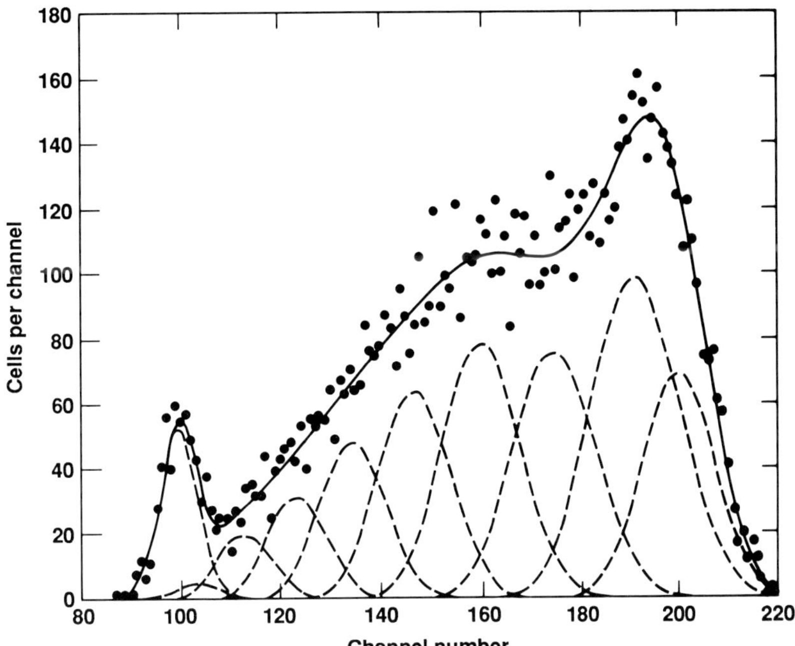

Fig. 9. Broadened rectangle model applied to the data of Fig. 6. The solid circles are the data points and the dashed lines are the G1 and G2M gaussian functions and the broadened rectangles fit to S-phase.

dard deviation, will be less accurate. The smoothing procedure was described earlier, in its application to nonparametric analysis.

It should be realized that a flow cytometric measurement does not guarantee good data. The quality of the data depends completely on the preparation and staining of the cells (in some instances on their source), and on the performance of the instrument. Indeed, there are a lot of data that cannot be analyzed by any method with any degree of confidence. Figure 11 illustrates this fact. These data were obtained from cells that had deteriorated, losing some of their DNA. The CV is about 20%, and the G1 and G2M peaks are overlapped. There is simply no way to accurately calculate any phase fractions, nor detect an S-phase fraction. In some cases, particularly for cells obtained from tissues, cell preparation methods produce significant amounts of debris, some of which contains DNA. This problem and artifacts due to improperly tuned instruments can produce histograms that abrogate the assumption that the G1 and G2M peaks can be represented by gaussian functions. In these cases, any analysis becomes suspect and results must be carefully evaluated by the experimenter.

Histograms resulting from the measurement of tumor cell populations are a special case. The difficulty arises because of the presence of two distinct populations, tumor cells and normal diploid cells. If the tumor cells are of identical ploidy to the normal cells, they cannot be distinguished on the basis of DNA content. If there is a difference in ploidy, analysis may be possible. Figure 12A shows data obtained from in vivo KHT cells. The tumor population is slightly hypertetraploid, with a small amount of overlap between the normal G2M peak and the tumor G1 peak. These data were analyzed using the broadened polynomial method, with the mathematical function

$$Y_c = G1_n + S_n + G2M_n + G1_t + S_t + G2M_t \quad (19)$$

Referring to Eq. (10) to (13), the subscript n designates the normal cells and t designates tumor cells. A total of 18 parameters are involved in the calculation, with the results shown in Table 4. Figure 12B is from human tumor cells, stained with chromomycin A3. Since the tumor cells are only slightly hyper diploid, S-phase overlaps for both populations and it is impossible to resolve them. The best that can be done with this kind of data is to compute the G1-phase populations. This is accomplished by fitting the two peaks with gaussian functions, using an analysis region restricted to the two peaks; channels 80 to 130 in Figure 12B. The mean channel numbers were found to be 114.5 and 92.0. The ratio is 1.24, also called the DNA index, and the CV was 2.5%. The G1 normal cells were 61% of the population and the tumor G1 cells were 29%. Another way to approach such data is the use of control populations. If one can obtain a sample of pure normal cells and can infer that this distribution of normal cells is the same within the tumor, analysis may be possible. The phase fractions obtained from analyzing the control population can be used to establish the shape of the normal cell distribution. This reduces the number of parameters to 10 and makes it possible to analyze almost any tumor cell distribution. The requirement of constancy of the normal cell distribution, if one can even be obtained, remains an obstacle.

Many mathematical models have been proposed for the analysis of DNA distributions. Only three, the ones the author prefers, have been discussed in any detail. Interested readers are directed toward several recent reviews on the subject [8,19,31] for more detailed information.

Kinetic Analysis Models. Using the methods described earlier for the analysis of single cell DNA content distributions, one can obtain some information about the life cycle of the cells being measured, namely, the relative number of cells in each phase and thereby in most cases the relative phase durations. For many applications this level of information is adequate.

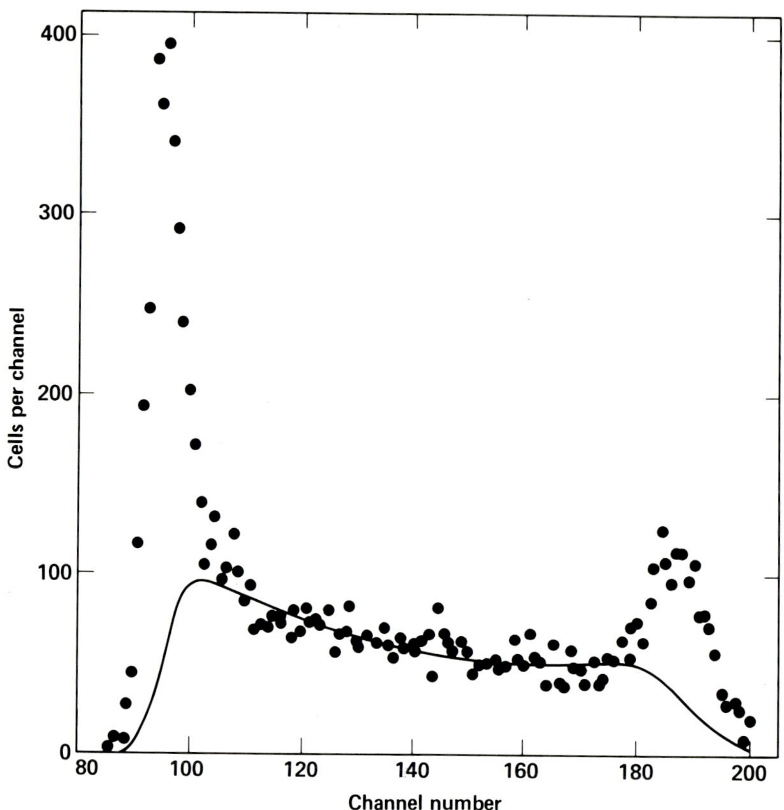

Fig. 10. DNA distribution for CHO cells, analyzed with a linear model, as described in the text. The polynomial used to fit S-phase is shown as a solid line.

However, in the treatment of cancer it fails to provide the information necessary for the use of many anticancer drugs, specifically those that exhibit cell cycle phase specific effects. Some drugs are known to be toxic only to cells in S-phase (e.g., ara-C), others are known to inhibit cell cycle traverse at specific points of the life cycle.

Tumor and normal cells are known to grow at different rates (32). If one knew the cycle and phase times of the two cell populations (normal and tumor) and they were sufficiently different, the application of blocking agents followed at the appropriate times by phase specific drugs could potentially be successful in eliminating the tumor cells. For example, if a drug were administered that would inhibit progression into S-phase (prevent DNA synthesis), all of the cells would eventually end up in late G1-phase. Suppose that the normal cells had an S-phase duration of 6 hours and the tumor cells one of 9 hours (in general tumor cells mature slower than normal cells). Then, 6 hours after the introduction of the blocking drug, all normal cells would have left S-phase and the application of a drug that kills cells in S-phase would kill only tumor cells. Tumor cells that were not in S-phase would not be affected. When the blocking drug is removed, or becomes metabolized in the in vivo situation, the cells again move through the cell cycle and in 6 hours the drugs are applied once more. In principle, this procedure would be repeated until all the tumor cells are killed.

There are two major difficulties with the procedure just described.

First, the absolute phase times (for all phases) must be known. Second, the phase times are known to vary from cell to cell. All the normal cells will not have left S-phase in 6 hours, and some of them will be killed; the drugs are not cell type specific. Consequently, the application of the drugs has to be made in an optimized manner to produce a maximum effect on the tumor cells and a minimum effect on the normal cells. To do this, one must know the phase durations and the dispersions in them. One must develop a model that describes the experimental data ("fits" them) and predicts the response to different drugs and their administration protocols. This is best done mathematically and a great deal of effort has been spent in developing such models. Chapter 39 (this volume) discusses this subject in more detail. The interested reader is advised to read this chapter and its references. To illustrate what is involved in developing a kinetic model, this chapter describes one very simple model that makes use of flow cytometric data.

Cell kinetic models can be simple or complex. The simplest model is one that describes how the number of cells in a population changes with time. Let us assume that each cell has a life time of T_c hours, at the end of which it divides into two daughter cells. If the number of cells at time 0 is N_0, then the number in T_c hours is

$$N_1 = 2\,N_0$$
$$N_2 = 2\,N_1$$

etc

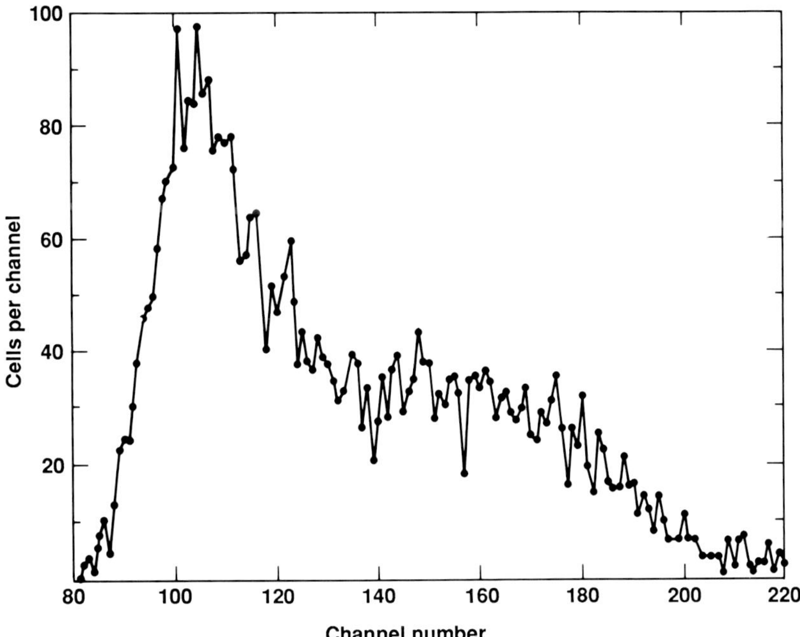

Fig. 11. Chromomycin fluorescence distribution for tumor cells. The low number of cells in the distribution, along with a CV of about 12%, makes analysis of these data virtually impossible.

TABLE 4. Results from Analysis of Data of Figure 12A*

Normal cells	Tumor cells
G1	G1
Area 14,191 cells (63.3 ± 0.6%)	Area 25,759 cells (45.1 ± 1.0%)
Mean 51.98 ± 0.03 channels	Mean 113.55 ± 0.06 channels
SD 2.82 ± 0.02 channels	SD 3.76 ± 0.04 channels
G2M	G2M
Area 3,166 cells (14.1 ± 2.2%)	Area 3,495 cells (6.1 ± 0.4%)
Mean 103.1 ± 0.9 channels	Mean 220.2 ± 0.3 channels
SD 5.5 ± 0.7 channels	SD 7.2 ± 0.1 channels
S	S
Area 5,071 cells (23.6 ± 0.2%)	Area 27,893 cells (48.8 ± 1.0%)

*SD = standard deviation of the gaussian peak; weighted variance = 2.1

$$N_i = 2\ N_{i-1} = N_0\ 2^i$$

or

$$= N_0\ 2^{(t/T_c)} \qquad t = i \cdot T_c$$

and

$$N(t) = N_0\ e^{\ell n2/T_c t} \tag{20}$$

This is the exponential growth curve. If one measures the density of cells growing in culture as a function of time, and plots the log of the density versus time, the slope of the line will reveal the doubling time. This is illustrated in Figure 13 for CHO cells. In Figure 13A the cell number is plotted versus time. The model assumes that all of the cells are in cycle and that each cell produces exactly two daughter cells. Figure 13 shows that these two conditions are not always true. When the cells are first placed into growth medium, they do not always immediately start to grow but have a "startup" period. They then go through a period of exponential growth. Finally, as the concentration of cells becomes high and the cells have difficulty in obtaining enough of the nutrients to sustain growth, the cells move into what is termed plateau phase. Figure 13 includes the G1-phase fraction for samples removed from the culture at various times, showing the accumulation of cells in G1-phase as the cells enter plateau phase. If cells from the culture are to be used in experiments where asynchronous growth is required, then one needs to be careful when the sample is taken.

The exponential growth model described above is useful for cells in culture (in vitro). However, for in vivo cases such as tumors, it is well known that the tumor does not grow uniformly. In particular, the center of a tumor mass tends to have a poor blood supply and cells in this region grow poorly or die. A simple correction that accounts for the death of cells in a population is as follows. Suppose the birth rate is α and the death rate is β per unit of population.

$$N(t) = \text{size of population at time t}$$
$$N(t+\Delta t) = N(t) + \alpha N(t)\Delta t - \beta N(t)\Delta t$$

$$\frac{N(t+\Delta t)-N(t)}{\Delta t} = (\alpha-\beta)\ N(t)$$

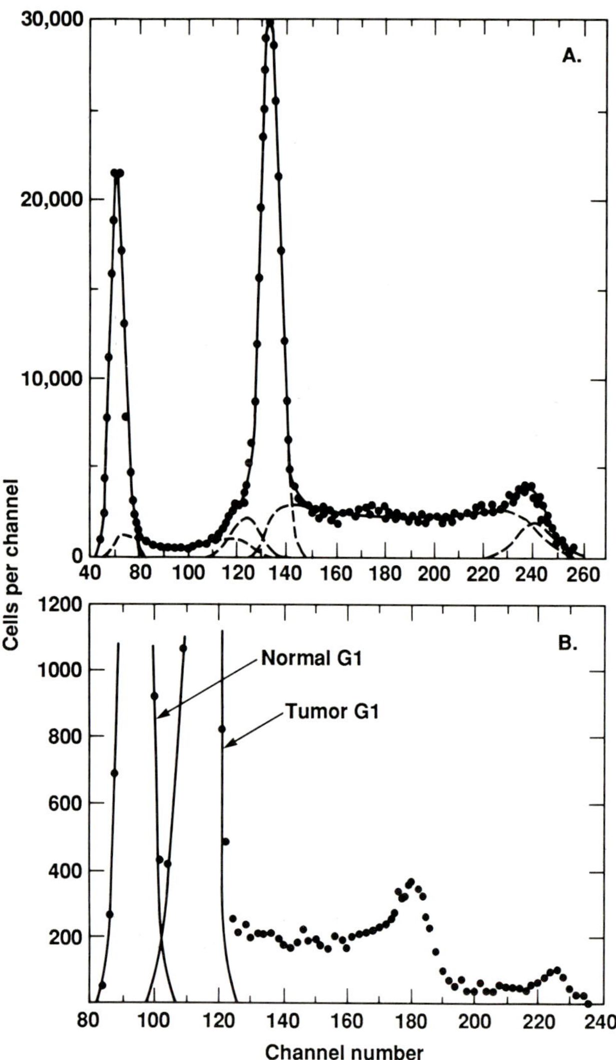

Fig. 12. **A:** DNA distribution of in vivo KHT cells, analyzed with a broadened polynomial model. The tumor population is slightly hypertetraploid. **B:** DNA distribution of a patient whose tumor population is only slightly hyperdiploid. Only the G1 peak means and phase fractions can be computed.

as Δt approaches zero

$$dN/dt = (\alpha - \beta)\, N(t)$$

$$N(t) = N(0) \exp(\mu t) \qquad \mu = \alpha - \beta \tag{21}$$

This equation is essentially the same as Eq. (20), with a different growth constant. However, with this model, the tumor can grow, remain constant in size, or decrease in size, according to the relative magnitude of α and β.

The models just described are not full kinetic models, since they describe only the behavior of the population as a whole, and provide no information about individual cells or life cycle phases. They cannot be used in the scheduling of cancer therapy. To be useful in studying the response of cell systems to drugs, a model must describe the distribution of cells around the life cycle, the phase time distributions, allow for cells to enter and leave the cycle within any phase, and provide for blocking cell maturation at any point in the cycle.

BIVARIATE DATA

Nonparametric Analysis

At many laboratories bivariate methods of data analysis have become more important than those for univariate data (see Chapters 23,24). Even in DNA analysis the current method of choice uses two dyes, propidium iodide bound to DNA and fluorescein bound to a monoclonal antibody against bromodeoxyuridine (BrdUrd). In this procedure, cells are grown in the presence of BrdUrd, which is an analogue of Thymidine. Cells which are in S-phase, replicating DNA, will incorporate some of the BrdUrd. The cells are then exposed to a fluorescently labeled monoclonal antibody against the BrdUrd. Thus, only those cells that were in S-phase will be labeled with the antibody and thus contain fluorescein. All cells are labeled with the propidum iodide, which binds to all DNA. The cells are measured in a flow cytometer utilizing blue-green light (488 nm from an argon ion laser) for excitation. Both the fluorescein and the propidium iodide are excited at this wavelength but emit at different wavelengths. Figure 14 shows data for human ALL cells, grown in vitro. The x-axis is DNA content (propidium iodide) and the y-axis is BrdUrd content (fluorescein). Figure 14A shows cells in asynchronous growth and Figure 14B shows cells after treatment with the drug cytosine arabinoside (ara-c). As shown in Figure 14B, ara-c kills cells in S-phase. In analyzing such data, most of the graphical/numerical methods used with univariate data are applicable, with slight modification. These histograms are analyzed with a "windowing" technique, which consists of drawing windows (also called areas of interest) around the features of interest and summing the number of cells within the windows. For rectangular windows, the analysis can be performed with a listing of the data and a hand calculator. If a computer, even a personal computer, is available, more complex windows can be used. In this procedure, the data are displayed as a contour plot (lines drawn at equal numbers of cells per channel) on a graphics terminal and then a cursor is used to define the vertices of the window, as illustrated in Figure 14A. This process can be carried out online or offline, in seconds.

The method of analysis just described, windowing, is not restricted to the DNA/BrdUrd data shown but can be used with any bivariate data. It is often used with data from cells labeled with two monoclonal antibodies. This is illustrated in Figure 15 for human thymocytes stained with fluorescein isothiocyanate (FITC)-labeled anti-Leu-6 and phycoerythrin (PE)-labeled anti-Leu-4 [30]. To cover a wide range of antigen contents, the fluorescence intensity is plotted on a logarithmic scale. In this example, there are four areas, which are delineated with two thresholds, channel 10 on the abscissa and 10 on the ordinate. These regions are usually spoken of as containing cells with high or low amounts of the antibodies, or positive/negative. The results are often quoted as A+B- for a subpopulation high in antibody A and low in antibody B. For the example shown in Figure 15, there are four subpopulations. The A+B+ population (positive for both Leu-4 and Leu-6) comprised most of the thymocytes (61%), with less than 10% of the cells expressing neither marker (A-B-).

The nonparametric methods of data comparison (e.g., differences, ratios, between two or more histograms) described earlier for univariate data can also be used with bivariate

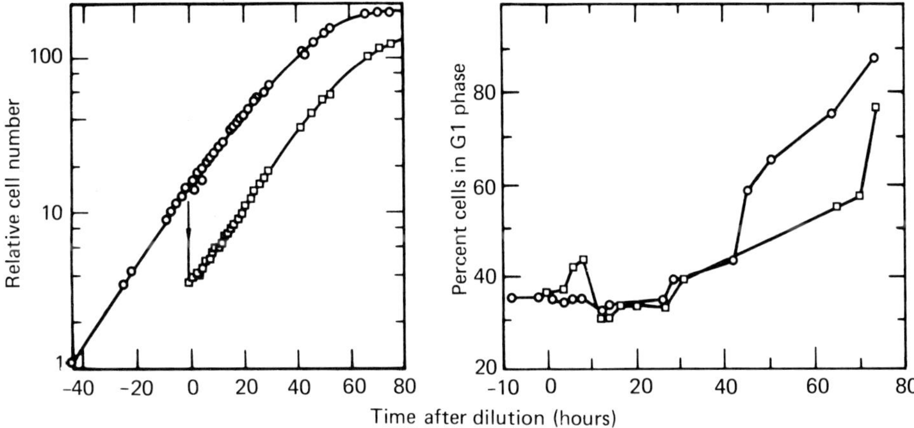

Fig. 13. CHO cells in suspension culture. The left panel shows the logarithm of cell number versus time, showing the culture going into plateau phase at about 50 hours after dilution. The right panel shows the percent of the cell population in G1-phase at various sampling times.

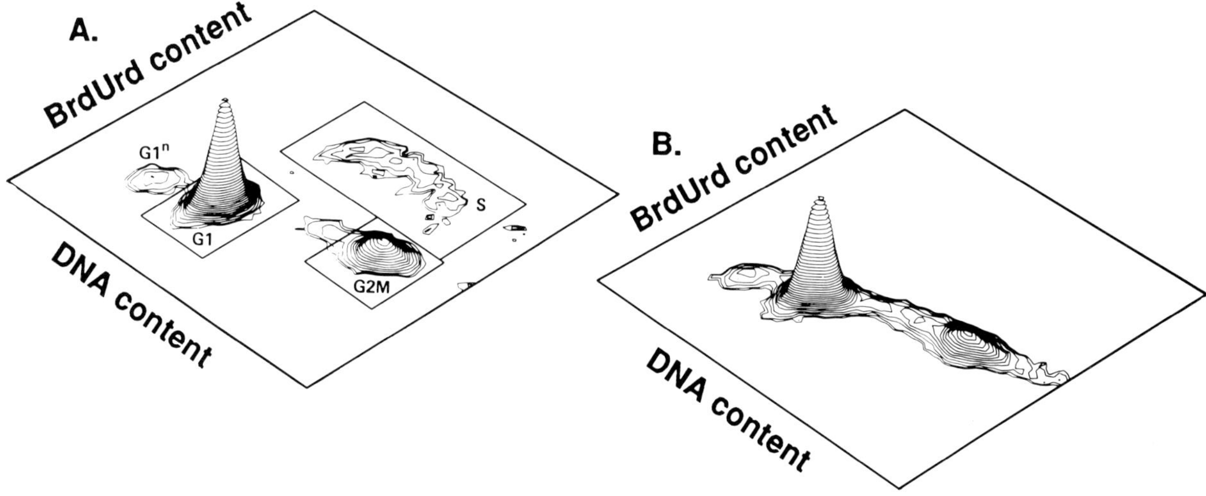

Fig. 14. **A:** Bivariate distribution (contour plot) of DNA content versus incorporated BrdUrd for human ALL cells grown in vitro. **B:** The same type of plot as for panel A, after the cells were treated with ara-c, showing the loss of cells from S-phase.

data. Bennett and Mayall [10] applied one such method to cervical cells measured for DNA content and light scatter at 90 degrees. These two variables have been shown to distinguish several cell types in cervical samples. The objective was to find a method of distinguishing normal from abnormal (cancer) patients. First, distributions were measured for seven normal women, with no history of dysplasia. These data were then aligned such that inflammatory cells fell in the same channel in each histogram, normalized such that each histogram had the same number of transition zone cells (diploid DNA content), and the average histogram computed. The latter involved computing the mean and standard deviation of the number of cells in each channel. Data from women who were being evaluated for possible dysplasia were also aligned and normalized and then subtracted channel by channel from the average distribution. The probability, p, that the unknown was significantly different from the average normal was computed and evaluated for diagnosis by applying the one-sided tolerance test statistic to each data point.

Parametric Analysis

Mathematical models are also used in the analysis of some bivariate data, as illustrated in Figure 16 for mouse gut cells measured for DNA content (chromomycin A3) and cell length (length of scatter signal) [19]. The objective was to distinguish crypt cells (reproducing) from villus cells (differentiated). If the S-phase fraction can be determined, one can study the growth kinetics of the crypt cells in the presence of chemotherapeutic drugs. In a two-dimensional analog to the DNA analysis methods described earlier, the procedure used here was to describe the crypt G1- and G2M-phase cells with three-dimensional gaussian functions of the form

$$G(x,y) = A\, e^{-\frac{1}{2}\left[((x-\mu_x)/\sigma_x)^2 + ((y-\mu_y)/\sigma_y)^2\right]} \qquad (22)$$

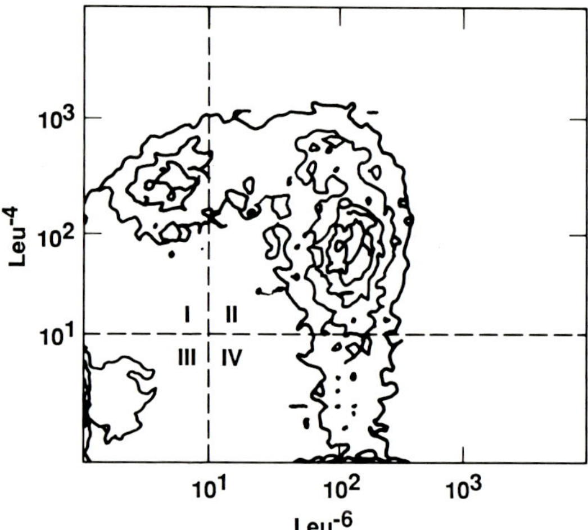

Fig. 15. Human thymocytes stained with phycoerythrin and fluorescein, bound to monoclonal antibodies against Leu-4 and Leu-6 antigens. The logarithm of fluorescence intensity is plotted on both axes.

where μ_x and μ_y are the coordinates of the mean of the peak and sigx and sigy are the standard deviations in each direction. S-phase is represented with a series of these functions. Since typically 4096 channels are used in one of these histograms, and a large number of peaks, 10 in this example, the computation is best done on a large computer (e.g., VAX, Digital Equipment Corporation). In Figure 16, the left diagram shows the experimental data and the right one shows the result of the analysis. This method is not restricted to gaussian functions whose coordinates are parallel to the x and y axes. The functions can be written such that the peaks can have any orientation.

Figure 17 shows an application that combines the windowing technique with least-squares analysis. The data are for human chromosomes dual stained with Hoechst-33258 and chromomycin A3 (see Chapter 25, this volume). These dyes bind preferentially to AT and GC-rich regions of DNA, respectively [29]. As is evident in Figure 17A, most of the chromosomes are clearly delineated. In one application at this Laboratory we wish to sort large numbers of specific chromosomes, in an effort to develop libraries of specific gene segments of them [50]. This is done by drawing the "sort windows" around the peaks due to the chromosomes of interest; two chromosome types can be sorted simultaneously. In Figure 17B, window A is drawn around chromosome 17 and window B is drawn around chromosome 16. The instrument will sort all chromosomes that produce signals falling within these windows. Since the sorted chromosomes will be used to produce, ultimately, a gene library, it is very important to know the purity of the sorted chromosomes. In Figure 17B, the peaks in window C seem to be well separated. This illustrates one of the difficulties with contour plots. The degree of separation of features in such plots (the peaks in this case) depends very strongly on the contour levels selected. If high levels are used, e.g., 50% of the peak height, the peaks will appear to be well separated. Conversely, 10% contours may show extensive overlap between neighboring peaks. Rather than depend on the contour plot in trying to estimate the amount of contamination in the sorted fraction, we use two different procedures. In the first, simpler procedure, we draw a window around the peaks being sorted, and around their neighbors, parallel to one of the axes. This is window C in Figure 17B. In this example the data within this window are then projected onto the y-axis, as shown in Figure 18. In the projection process, for a given channel in y (Hoeschst fluorescence), the contents are summed for all channels in x (CA3 fluorescence) that are within the window. The result is the univariate distribution of Figure 18. Each peak in the plot, four in the example, is then fit by least squares with a gaussian function, represented by the solid lines. The presence of background in the histogram is accounted for by adding a polynomial, exponential, or gaussian function to the series of gaussian functions representing the peaks. The user selects the function interactively, the decision being based on the specific data set and chromosome peaks being analyzed. The fraction of each peak that falls within the sort window, channels 32 to 37 and 39 to 43, is then computed. Table 5 shows the results. The chromosome 16 sort window includes no contamination from number 17 chromosomes and the chromosome 17 sort window includes 1.0% contamination from number 16 chromosomes. The windows used in this type of procedure can have any shape (not restricted to rectangular) and the enclosed data can be "collapsed" onto a line at any angle. Statistically, this procedure works best if more channels are used than in the 64×64 histogram shown. This can be accomplished easily by using list mode data, a procedure that will be discussed in more detail below.

The second method of analysis of chromosome data involves much more computation than in the process just described but also provides much more capability. In this procedure, bivariate gaussian functions are fit to each chromosome peak. The function used is G(x,y):

$$G(x,y) = \frac{V}{2\pi\,\sigma_x\,\sigma_y\sqrt{(1-\rho^2)}}\exp$$

$$-\left[\frac{(x-\mu_x)^2}{\sigma_x^{\,2}} - 2\,\rho\,\frac{(x-\mu_x)(y-\mu_y)}{\sigma_x\quad\sigma_y} + \frac{(y-\mu_y)^2}{\sigma_y^{\,2}}\right]/2(1-\rho^2) \tag{23}$$

where μ_x is the mean value of x, μ_y is the mean value of y, σ_x is the standard deviation in x, σ_y is the standard deviation in y, and V is the volume of the peak. The parameter ρ is the correlation coefficient between the variables x and y and can have values in the range -1 to 1. Chromosome distributions usually contain background. In this application, a bivariate polynomial function is used:

$$B(x,y) = \;= a + bx + cy + dxy + ex^2 + fy^2 \tag{24}$$

In some cases, the polynomial function does not fit the data well and other functions are used. A set of gaussian functions, one for each chromosome peak, and the background function are fit to a data set by the method of least squares. Owing to the large number of calculations involved in this procedure, not all of the peaks can be handled simultaneously. To keep the computation time reasonably short (a few minutes on a VAX computer), a maximum of six peaks are included in a single fitting procedure. An example of using this analysis technique is presented in Figure 19. Frame A shows the original data, for human chromosomes 9 to 22, stained with both chromomycin A3 and Hoeschst 33258.

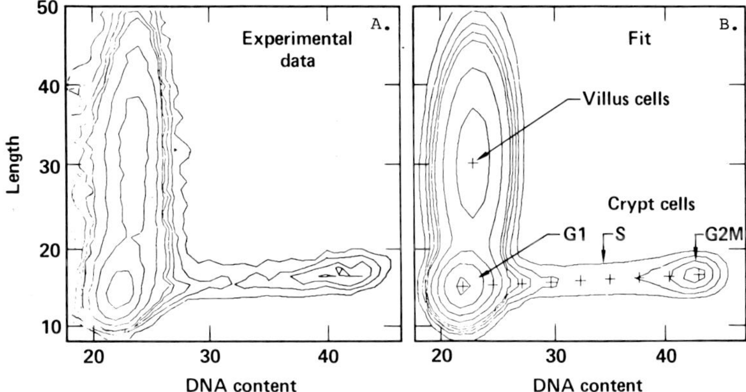

Fig. 16. **A:** Contour plot of mouse gut cells measured for DNA content and cell length. **B:** Calculated distribution for the data of A, where a series of bivariate gaussian functions is used to model the data.

These are the same data as in Figure 18. Frame B shows the gaussian functions fit to chromosomes 14 to 17 and Frame C shows the background function. This type of analysis procedure can be used to obtain the sorting purity results described earlier. It can also be used to analyze data for the relative volume of each peak (number of chromosomes of each type), for the possible detection of chromosome abnormalities such as extra or missing chromosomes, and for translocations. It can also be used for data other than from chromosomes; the only requirement is that there be peaks that can be represented by gaussian functions.

MULTIVARIATE DATA

The human brain still has a superior capability to group and classify data. This capability, together with the ability to judge the quality of the data analysis results, will continue to make human inspection and interaction a key part of multivariate data processing. Data display and analysis techniques that provide for maximal human interaction are therefore extremely valuable. Bivariate analysis of flow cytometric data has the major advantage that a human can see and understand the data and the results of the analysis in complete detail. Also, a large amount of experience has accumulated with the processing of bivariate data, as discussed earlier. This means that the various analysis techniques can be applied in particularly relevant ways after the data have been examined and the strengths and weaknesses of the analysis techniques have been taken into account.

Despite the very desirable aspects of bivariate data display and analysis techniques, there are also significant advantages to techniques that deal with more than two variables. The simultaneous evaluation of more than two variables may provide important additional biological insight. Mathematically, a major advantage of using more than two variables is that clusters of data elements can be separated more easily. This is illustrated as follows. Suppose that a set of data elements are described by the two variables DNA content and cell size. Also assume that there are two clusters of data elements in this data set. If the centroids of the clusters are separated by one unit of measurement in DNA content and also by one unit of measurement in cell size, then by using the conventional euclidean distance measure, the centroids will be separated by a distance of approximately 1.41. However, if the same data set is described by the original two variables plus one additional variable, membrane potential, and if the centroids of the two clusters are separated by one unit of measurement in membrane potential, then the centroids will be separated by approximately 1.73. Thus the separation between the centroids has been increased by more than 20%.

The disadvantage of data display and analysis techniques which can deal with more than two variables is obvious. Humans can not directly perceive objects in spaces of dimension greater than three. This means that the human capacity for grouping and classifying data, as well as evaluating the results of data analysis, can not be effectively used for large sets of data.

Graphical Methods

A large number of methods display multivariate data graphically [1,2,13,14,15,21,35,36,45]. This chapter presents only a few of the more popular ones.

Frequency Distributions. Graphically, only one and two variables can be plotted as frequency distributions, as shown in the previous sections. For data of more than two variables, multiple histograms can be used. For example, for data of three variables, one can generate 6 plots, as illustrated in Figure 20. The variables are 90-degree light scatter, cyanine dye fluorescence (a measure of membrane potential) and Hoeschst 33342 (DNA content), measured on murine bone marrow cells [50]. Each possible bivariate distribution is plotted in a three-dimensional perspective view, along with the three univariate distributions. If one can also rotate the bivariate distributions, preferably in real time as is possible in some commercial pulse height analyzers (e.g., Nuclear Data model 620, model 6660), small subpopulations can be distinguished. The bivariate views could also be presented as contour plots.

This approach can be used with any number of variables. One method is to plot n-variable data as a nxn square matrix of scatter (dot) plots [45]. This provides all possible views, including both axis orientations, of bivariate combinations. For large numbers of variables, this method can require a lot of effort to interpret the histograms and find correlations in the data. The method can be enhanced considerably through a technique called "brushing," especially when a graphics work station type of computer is used. These computers, including many personal computers, have "bit-mapped" memory, which means that a portion of the computer memory can be continuously displayed, providing for very fast graphics response. In "brushing" the data, a window of vir-

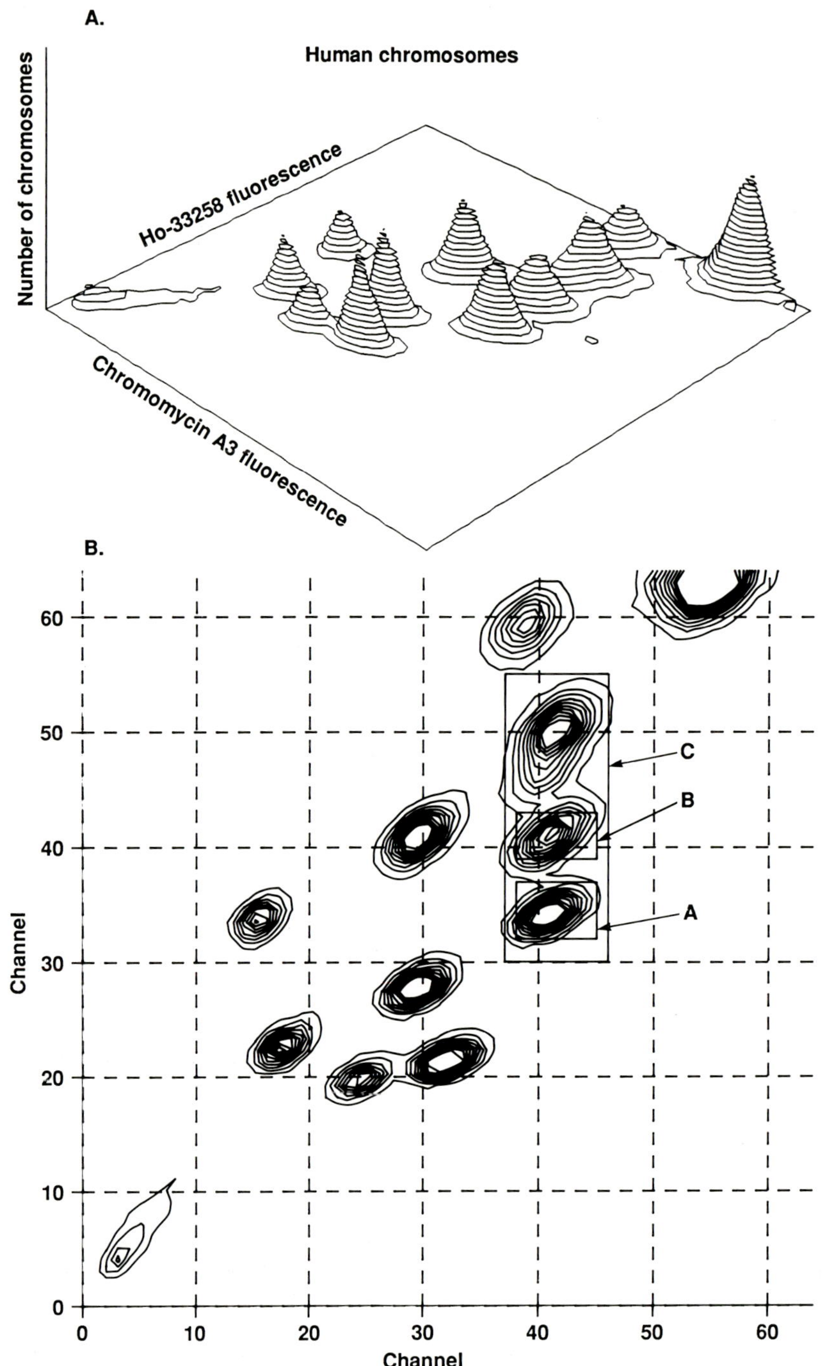

Fig. 17. **A:** A three-dimensional perspective view of human chromosomes 9 to 22, dual stained with Hoechst 33258 and chromomycin A3. Each peak corresponds to one chromosome type, with the exception of the 9 to 12 peak. **B:** Contour plot of the data of panel A. Boxes labeled A and B establish sort windows. All chromosomes that have the measured Ho and CA3 content to place them in box A will be sorted left and those in box B will be sorted right. The boxes specify mostly the number 16 and 17 chromosomes. Box C is drawn to include all chromosomes that may fall within either box.

tually any shape is drawn in one of the bivariate histograms, e.g., variable 1 versus variable 2. The data points enclosed in the window are then highlighted in all other histograms. The highlighting can be in the form of increased intensity, a changed symbol, or best of all, color. As with other methods that use scatter plots, a limited number of points can be used or the histograms become saturated, limiting the usefulness of the method in the detection of rare populations. This depends on the resolution of the display (e.g., the number of pixels on each axis).

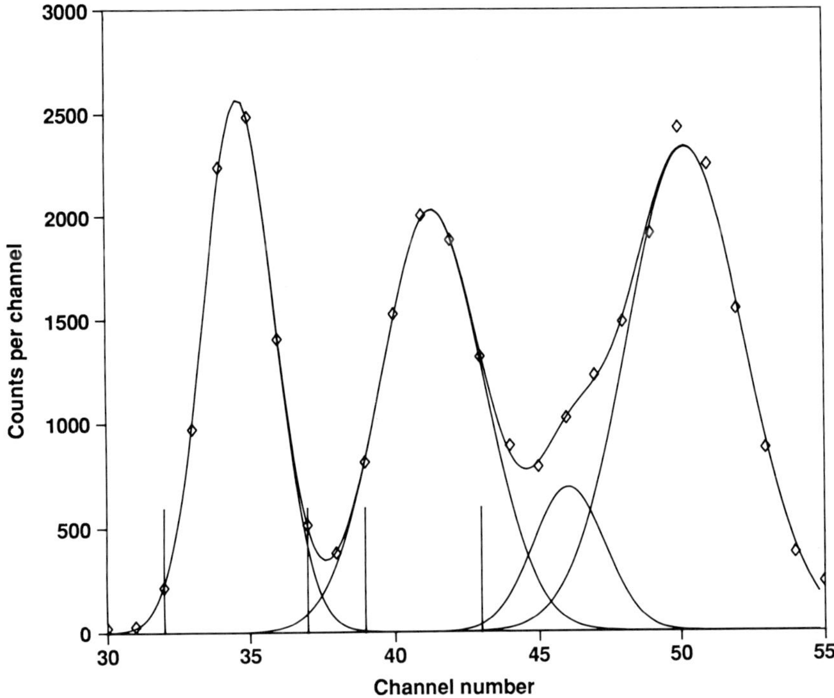

Fig. 18. Hoechst 33258 distribution for data from Figure 16B that fall within box C. The diamonds are the data points and the solid lines are the total fit of four Gaussian functions to the data.

TABLE 5. Results of Analysis of Data of Figure 18; Hoeschst Fluorescence Distribution of Four Human Chromosomes, Fit with Four Gaussian Functions

					Sort left‡		Sort right‡	
Peak No.*	Area	Mean	CV (%)	Fraction†	N	%	N	%
1	7,860	34.7	3.5	0.25	7,535	99.0	3	0.0
2	8,948	41.4	4.2	0.29	49	1.0	6,670	99.7
3	2,118	46.0	2.7	0.07	0	0.0	13	0.2
4	12,147	50.2	4.1	0.39	0	0.0	2	0.0

†Fraction = fraction of chromosomes analyzed.
*Peak 1, chromosome 17; peak 2, chromosome 16; peaks 3 and 4, chromosomes 14 and 15.
‡Left sort window is 95.9% of peak 1; right sort window is 74.6% of peak 2.

Three variables can also be displayed as a "dot plot" or "scattergram", where each cell is plotted as a single point in a three-dimensional perspective view. However, there is no depth perception and the data are difficult to interpret. This can be overcome by graphical techniques such as reducing the size of a point as the y-coordinate value of the point increases; it is farther from the viewer. As the number of points plotted increases, the value of this method decreases; the plot can become virtually solid. Also, for effective use the number of channels in the display should be large. Otherwise, with a relatively poor resolution of 64 channels, many points can fall into the same channel and be represented as a single cell. Structure within the data can be made more visible through rotation of the data within the plot. To make this method of display more useful, especially where there is a limited number of channels, one can generate a three-dimensional contour plot, as illustrated in Figure 21 [46,48] for ovarian ascites cells measured for esterase activity, volume, and DNA content [48]. Each line in the graph represents channels of equal density (number of cells), in this case those channels whose values are 20% of the most frequent channels. Five subpopulations were detected. This method of analysis has also been applied to colorectal tumor cells measured for DNA content, volume and carcinoembryonic antigen (CEA) [47].

There is an additional method of displaying three variable data as a contour plot. The two axes show two of the variables. The third variable can be represented by the size of the plotting symbol, e.g., a circle. The greater the magnitude of the variable, the larger the circle. A series of symbols, e.g., numerals, could also be used.

Another method of processing three- and higher-variable data is called LIST processing. The name comes from the fact that the data are stored sequentially in a list, on magnetic tape or disk. All variables are recorded for each cell, usually at the full measured resolution. For example, univariate data

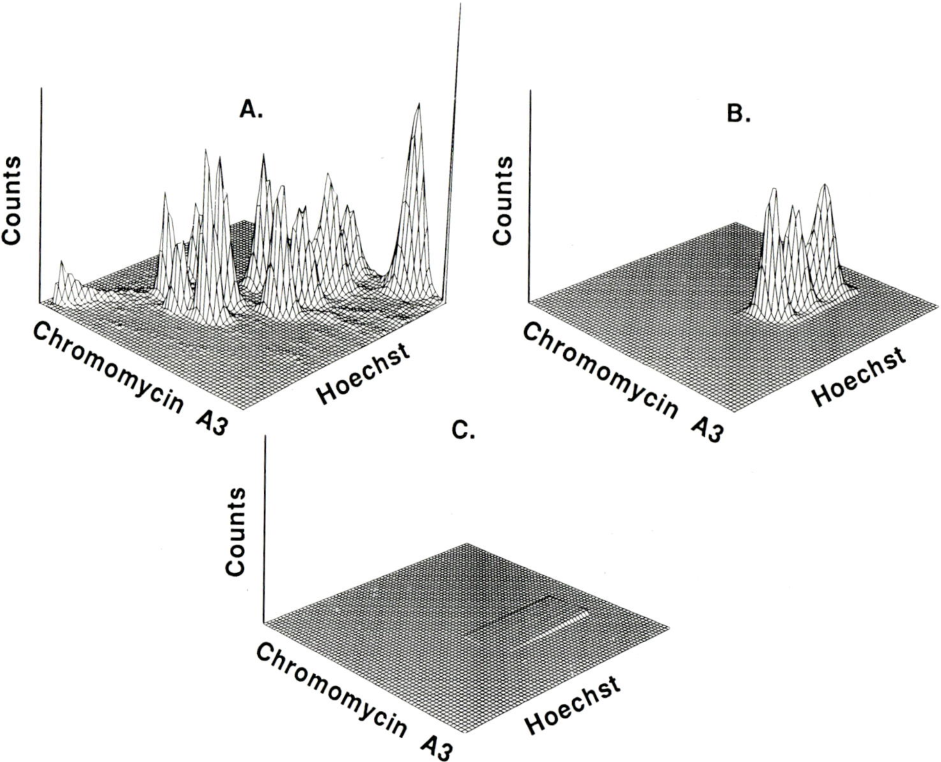

Fig. 19. **A:** 3-d perspective view of human chromosomes 9 to 22 stained with Hoechst 33258 and chromomycin A3. **B:** Theoretical distibution of chromosomes 14 to 17 obtained by fitting bivariate gaussian functions to the data of A. **C:** Bivariate polynomial distribution included as background in the analysis of the data of A.

are usually measured and stored at a resolution of 256 channels, bivariate data at a resolution of 64 channels (64 × 64). At this laboratory, we store list data at the measured resolution of 4,096 channels. This allows us to manipulate (e.g., scale, normalize, translate, transform) the data before displaying or otherwise analyzing them. Analysis can be very simple, e.g., multiple plots using methods already described, or complex, e.g., cluster analysis.

LIST processing permits the use of complex windowing schemes, such as that illustrated in Figure 22. These data are for mouse bone marrow cells, with the variables small angle light scatter, orthogonal light scatter, and Hoechst fluorescence [40]. The objective of the measurements was to identify colony-forming cells in general and individual stem cell types in particular. Sorting was used to establish the window (area of interest) in Figure 22B such that it included the majority of the blast cells. Figure 22A shows the DNA distribution for all cells in the bone marrow. Cells corresponding to the area of interest yielded the Hoechst fluorescence distribution shown in Figure 22C. Further sorting of these cells based on their fluorescence intensity showed that the low fluorescence region (below the apparent G1 peak) yielded a 110-fold enrichment of CFU-GM cells and 100 fold enrichment of CFU-S cells. The cells with an apparent S-phase DNA content yielded few colony-forming cells. This study is continuing with the addition of a fourth probe, a fluoresceinated monoclonal antibody, one that hopefully will label a specific type of colony-forming cell. If the process is successful, the cell will meet several tests: (1) have a low angle/wide angle light scatter signal that falls within the area of interest; (2) have a Hoechst fluorescence signal between channels 10 and 60; and (3) a fluorescein intensity greater than about channel 120. If the antibody helps but is not conclusive, then another antibody probe tagged with another dye, e.g., Texas red, could be added and an additional test performed.

Another application of LIST processing that also involves the use of scatter plots and the use of color is shown in black and white in Figure 23. The data, collected at a resolution of 4,096 channels as described above, are for mouse bone marrow cells. The data are plotted as "scatter grams" or "dot plots". This means that every measured cell is plotted individually; these are not frequency distributions (number of cells vs measured variable). Approximately 20,000 cells are plotted in each quadrant of Figure 23 (20,000 dots in each plot). The variables are low angle light scatter (X), othogonal light scatter (Y), and a monoclonal antibody against a cell surface marker (color; not shown). The four plots show respectively a mixture of antibodies, the 9A4 and BMW7 antibodies, and a depleted population. The color bar (here in black and white) is intended to show the conversion from magnitude of the variable to color of the point. When viewed, the colors clearly delineate several subpopulations, based on the magnitude of the monoclonal antibody signal. The use of color in data analysis is expected to increase dramatically in the near future.

When the number of measured variables exceeds three, iterative visual analysis of the data becomes progressively more difficult. Although helpful, multiple views, even with

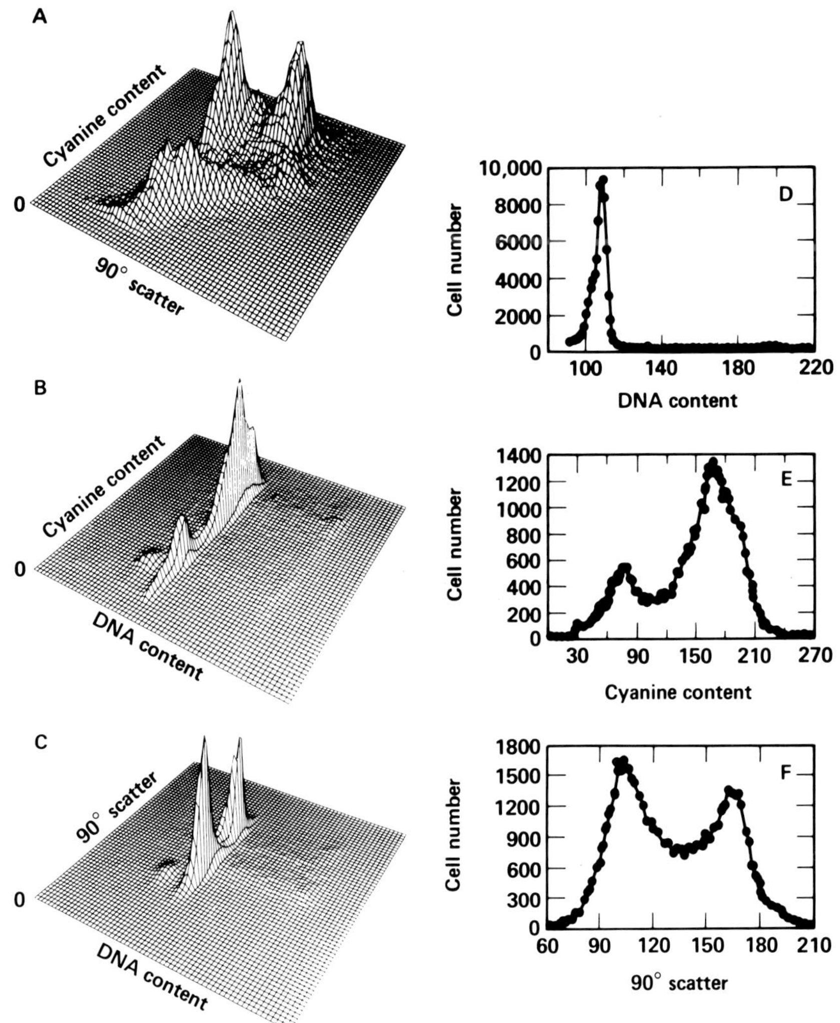

Fig. 20. Orthogonal light scatter (size), Hoechst fluorescence (DNA content), and cyanine fluorescence (membrane potential) for mouse bone marrow cells. Panels A, B, and C are the three possible bivariate distributions, D, E, and F are the univariate distributions.

the use of gating methods, are not adequate, and more sophisticated statistical methods of analysis are required. A common approach is to use clustering algorithms.

Symbols. Methods of displaying up to three variables in histogram form have been described. When four or more variables are measured, different methods are required. Many methods have been developed [13,15,45] in which each data point is represented by a symbol whose shape is determined by the magnitudes of the variables. Some of the methods are illustrated in Figure 24. In Figure 24A, each data point of n variables is represented by a polygon of n sides. The length of the vector from the center of the circle to each vertex is proportional to the magnitude of a variable. This procedure is also called a star plot. Figure 24B illustrates a popular method using faces to represent each data point [14]. Magnitudes of the variables determine the facial characteristics, such as diameter of the head, size of the eyes, shape of the mouth. Since we are all taught (or quickly learn) to recognize individual people quickly, this method is very useful in resolving data sets where there are substantial subpopulations. There are very little data shown in Figure 24A,B, only enough to illustrate the principle. However, in Figure 24B there are enough to clearly identify four subpopulations.

In Figure 24C, each measurement is plotted as a series of connected points, one point for each variable, sometimes called a profile symbol plot [13,16]. In such a plot, clusters of cells can be distinguished. Figure 24D shows the graph created by mapping the multivariate data into trigonometric functions (2), f(x):

$$f(x,t) = x_1/\sqrt{2} + x_2 \sin(t) + x_3 \cos(t) + x_4 \sin(2t) + x_5 \cos(2t) \qquad (25)$$

where the x's are the variables (four in this example) and t has the range $-\pi<t<\pi$. One function is plotted for each measurement. Those curves that lie close together at some value of t represent clusters.

Plots of symbols, such as shown in Figure 24A,B, are limited to about 100 measurements (cells) per graph. Conse-

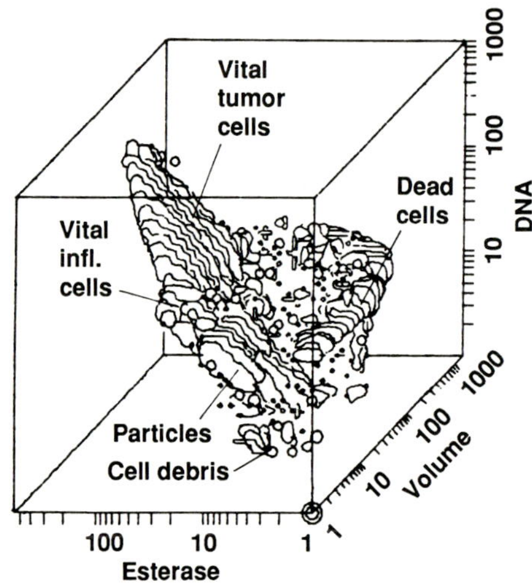

Fig. 21. Three-dimensional perspective view of ovarian ascites cells measured for esterase activity, volume, and DNA content. The data are plotted in a contour format where each line represents channels of equal density (number of cells). Five subpopulations of cells have been identified from these data.

quently, these methods are limited to situations where the subpopulations to be distinguished are relatively abundant; they are not useful for detecting clusters of rare cells. In Figure 24C,D, if enough cells are plotted to clearly distinguish clusters, small numbers of rare cells are masked by the noise (variability) in the data.

Principal Components

In a multivariable measurement, one hopes to measure enough variables to resolve all of the subpopulations in a set of objects (e.g., cells). This means that there should be a great deal of variability in the data. Otherwise, only some of the subpopulations (in the limit only one) will be resolved. The intent of a principal component analysis is to reduce the dimensionality of a data set by determining if a few linear combinations of the variables can account for nearly as much of the variability of the data as the entire original collection of variables: Can the data set be resolved into its components using less than the measured number of variables? Are any of the variables superfluous? The mathematical heart of the analysis is an eigenvector and eigenvalue analysis of the correlation matrix of the data. Along with a reduction in the dimensionality, a principal component analysis also produces completely uncorrelated new variables. This can make the later task of clustering considerably more intuitive and can also be useful in interpreting the biological significance of the original variables.

Principal component analysis can be understood in more detail by looking at how it could be applied to a relevant data set. Consider a data set to be used to selectively identify stem cells in bone marrow. In this case, suppose that three measurements will be made on each cell as it passes through the flow cytometer. Let D be the DNA content of the cell, let S be the amount of small angle scattered light, and let L be the amount of large angle scattered light. Thus for each cell the collection of measurements (D, S, L) are made. This collection can be represented as the vector $X = (D, S, L)'$, where $'$ means the transpose operation on vectors. X can be thought of as a random vector each of whose components is a random variable which models a measurement made on the cells.

The following notation is necessary to make the notion of principal components precise. Let E(v) and Var(v) represent the mean and variance respectively of a random variable v. The same notation will be used to denote the mean and variance of random vectors as well. The covariance of the random variables v and w is denoted by Cov(v,w). This same notation will also be used for random vectors. Finally, let Z be the covariance matrix for the random vector X.

The idea in principal component analysis is to find a new random vector whose components are uncorrelated and have variances which are as large as possible. To do this, we define three new random variables A, B, and C which are linear combinations of the previously defined random variables D, S and L. That is, for any numbers a1, a2, a3, b1, b2, b3, c1, c2, and c3 we have:

$$A = a1\ D + a2\ S + a3\ L \tag{26}$$

$$B = b1\ D + b2\ S + b3\ L \tag{27}$$

$$C = c1\ D + c2\ S + c3\ L \tag{28}$$

The random variable A is called the first principal component if the numbers a1, a2, and a3 are chosen so that Var(A) is a maximum subject to the constraint that the vector (a1, a2, a3) has unit length. Similarly B is called the second principal component if b1, b2, and b3 are chosen to maximize Var(B) subject to the constraints that Cov(A,B) = 0 and that the vector (b1, b2, b3) has unit length. Finally, C is called the third principal component if c1, c2, and c3 are chosen to maximize Var(C) subject to the constraint that Cov(A,C) = 0, Cov(B,C) = 0, and the vector (c1,c2,c3) has unit length.

Without knowing the details of the entire data set it is not possible to give an intuitive explanation of what the principal components mean. However, it is useful to describe some simple cases that could happen and which will serve to build up intuition. For example, if the computations produced a1 = 1, a2 = 0, a3 = 0, b1 = 0, b2 = 1, b3 = 0, c1 = 0, c2 = 0, c3 = 1, Var(A) = 100, Var(B) = 50 and Var(C) = 5, then the first principal component would just be D, the second principal component would be S and the third principal component would be L. In this case it would seem that that no information has been gained from the principal analysis. Actually, a very useful piece of information has been learned, namely that L is not contributing to the data set in a significant way (low variance) and therefore can be ignored.

As another intuitive example consider the result a1 = 1, a2 = 0, a3 = 0, b1 = 0, b2 = 0.707, b3 = 0.707, c1 = 0.577, c2 = 0.577, c3 = 0.577, Var(A) = 200, Var(B) = 160 and Var(C) = 2. In this case the first principal component is again D. However, the second principal component is a combination of of S and L, and the third principal component is a combination of D, S, and L. The variances show that only the first two principal components are important. Therefore, this principal component analysis shows us that there are only two quantities of interest in this data set, D and a combination of S and L. All future data analysis should be done in terms of these quantities. An additonal ramification of this analysis is that the researcher should spend some time looking for an interpretation of the combination of S and L, perhaps in terms of cell size and internal structure.

The definition of principal components given above does

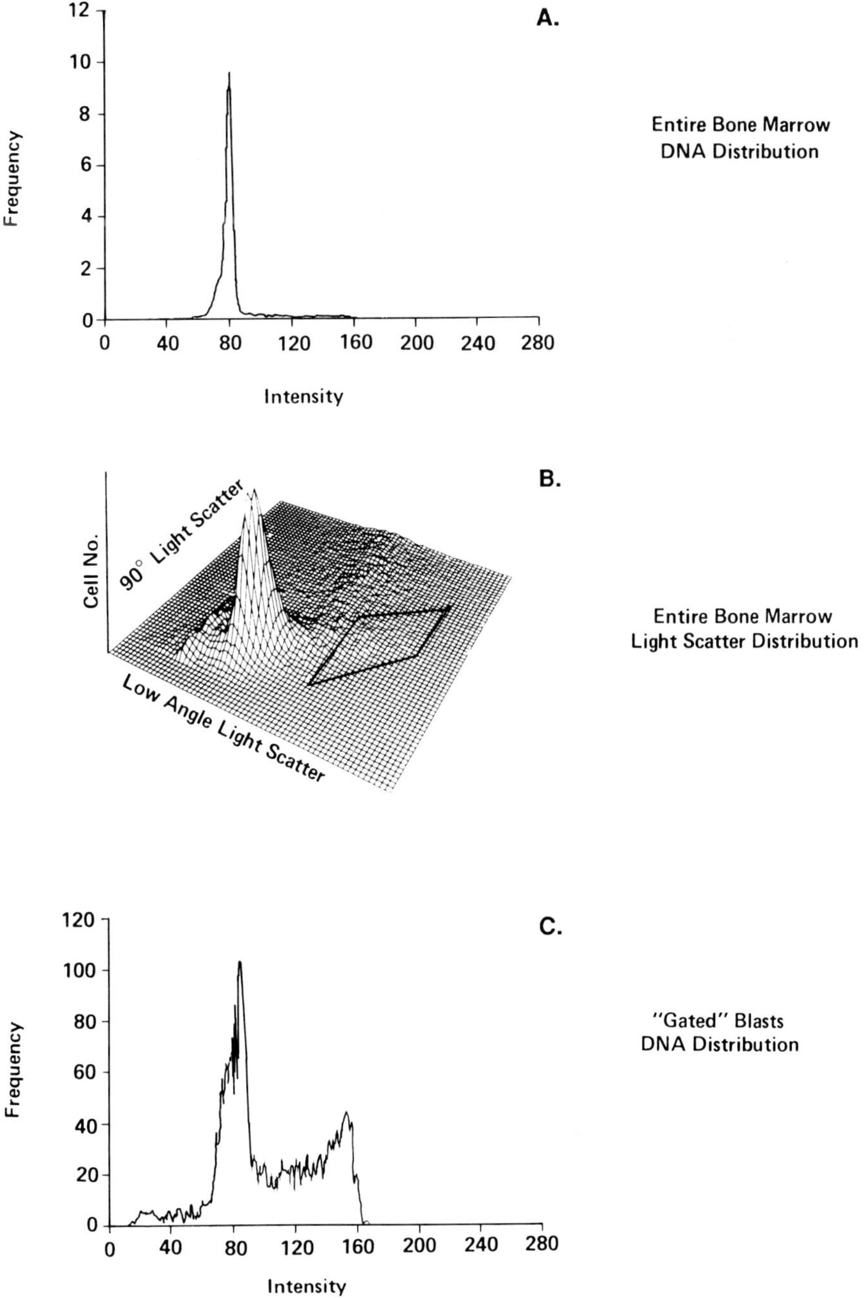

Fig. 22. Orthogonal light scatter, narrow angle light scatter, and Hoechst fluorescence for mouse bone marrow cells. **A:** Hoechst 33258 distribution for entire cell population. **B:** Bivariate distribution of low angle versus 90-degree light scatter, for total population. **C:** Hoechst fluorescence distribution for cells within the "blast" window shown in Panel B.

not provide an easy indication of how the numbers a1, a2, a3, b1, b2, b3, c1, c2, and c3 are to be found. The general mathematical tools needed to show how to find these numbers are results of the maximation of quadratic forms on the unit sphere. Without going into the details of these mathematical tools, the end result is that a1, a2, and a3 are the components of the normalized eigenvector corresponding to the largest eigenvalue of the covariance matrix Z. This eigenvalue is the variance of the first principal component. Similarly, b1, b2, and b3 are the components of the normalized eigenvector corresponding to the next largest eigenvalue of the covariance matrix Z. This eigenvalue is again the variance. The same result holds for the third principal component and would also hold for higher dimensional situations.

A principal component analysis is a very common procedure in multivariate statistical analysis for which a number of commercial packages are available. The BMDP function BMDP4R (BMDP Statistical Software, Los Angeles, Califor-

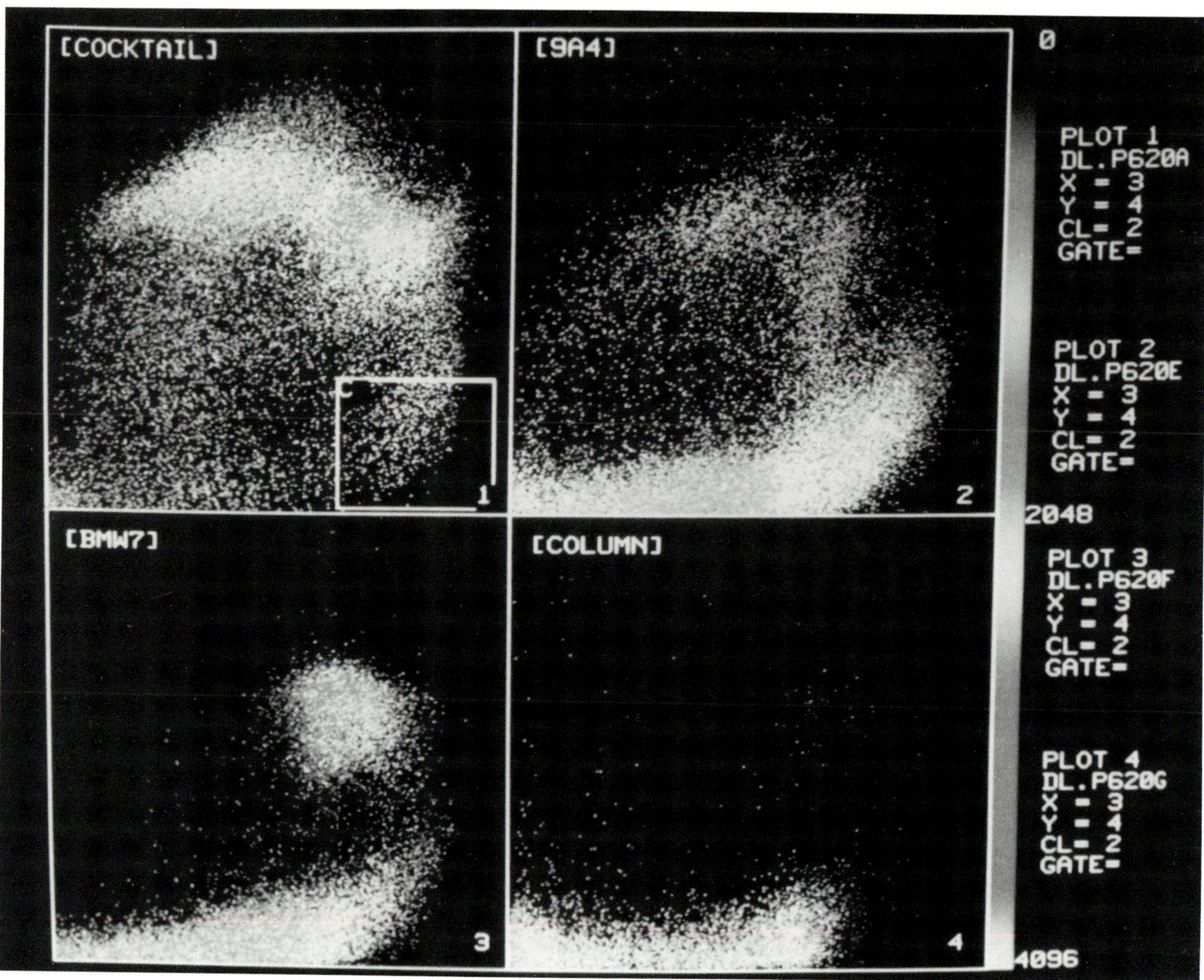

Fig. 23. Scatter plot for mouse bone marrow cells measured for two light scatter signals (X and Y coordinates) and for surface markers using monoclonal antibodies (in color in the original, here shown in black and white). Plot 1 is for a mixture of antibodies, plots 2 and 3 are for 9A4 and BMW7 antibodies, and plot 4 is for a depleted population. The magnitude of the antibody signal varies from 0 (blue) to 4095 (pink), as defined by the color bar on the right (shown here in black and white).

nia), the SAS function FACTOR (SAS Institute, Cary, North Carolina), and the SPSS-X function FACTOR (SPSS Inc., Chicago, Illinois) all carry out principal component analyses.

Clustering

The objective of clustering is to place all data elements into an arbitrary number of groups. Arbitrary in that the number and characteristics of the groups are determined from the data: they are not predetermined. After clustering, the data elements in each group are defined to be similar. It is then up to the researcher to determine the biological or biochemical significance of each group. The hope would be that each group would define a different group of cells.

The process of clustering is very heuristic. There is very little mathematical theory to justify the various clustering algorithms. There is however a reasonably complete classification scheme for clustering methods, both hierarchical and nonhierarchical. Hierarchical methods can be further broken down into agglomerative and divisive methods. The idea behind both methods is to systematically process the data set to find reasonable clusters without having to examine all possible configurations of data elements. Agglomerative hierarchical methods start with as many groups as there are data elements. Clusters that are most similar are then merged to form new clusters. Clusters are then successively divided to form new clusters that are more reasonable. The process is stopped when there are only a few reasonable clusters.

Quantitative measures of the similarity of clusters are almost always determined in the following way. First, a measure of how similar any two data points are must be decided. This is usually done via some type of distance measure in euclidian space. Next, a measure of how similar any two clusters are must be determined. This involves picking some linkage scheme.

The determination of the proper distance measure to use

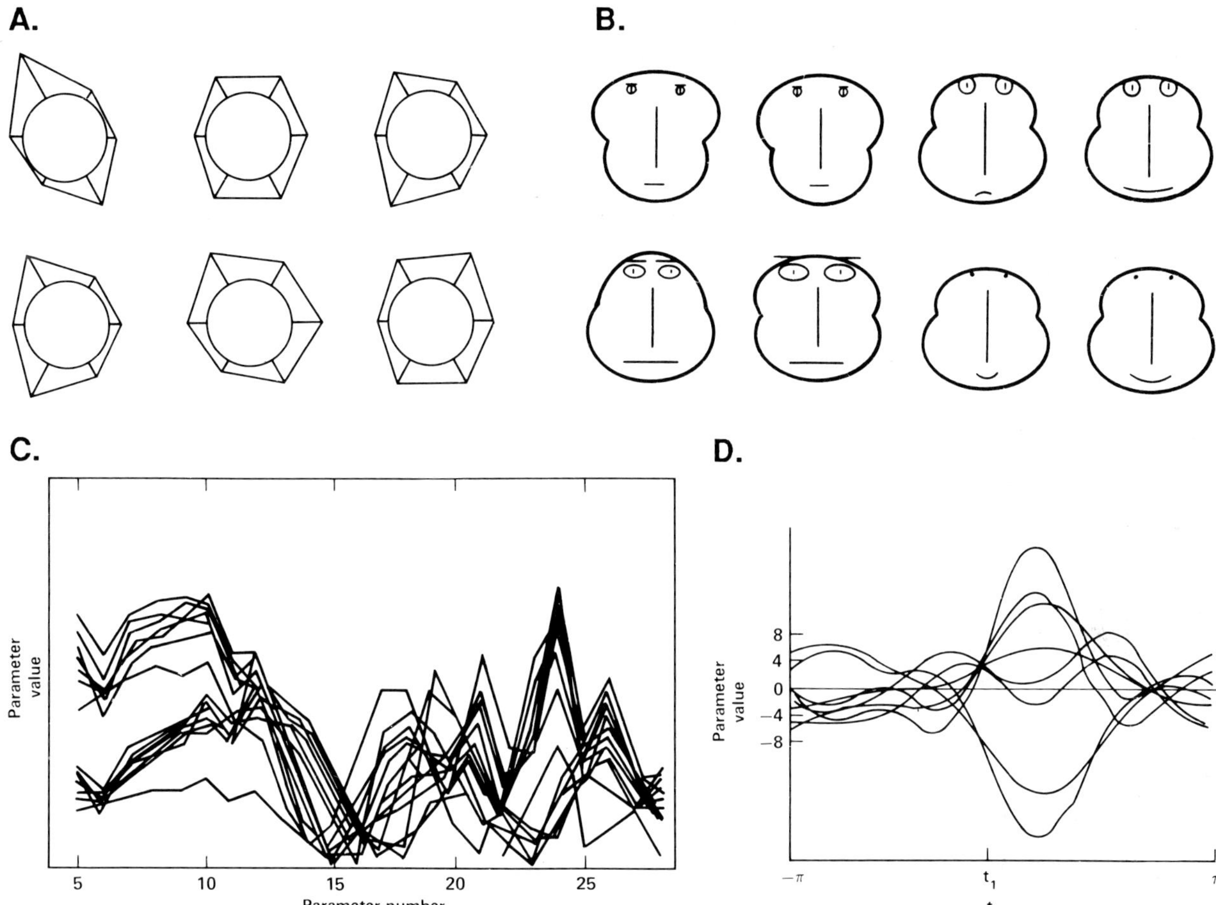

Fig. 24. Multivariate data plotted in four different ways. **A:** "Star" or polygon plot where the length of a ray is proportional to the magnitude of the corresponding variable. Data of six variables are represented here. **B:** Stylized face is used to represent each data point. Each feature of the face (e.g., size of head, size of eyes, curvature of mouth) represents a variable and its magnitude. **C:** Profile plot where the magnitudes of each variable of the measurement are connected by straight lines. **D:** Each data point (measurement) is represented as a Fourier series.

to quantify the similarity of individual pairs of data points is the subject of a significant fraction of the literature on clustering. The standard euclidean distance measure is often not the proper one to use. If the measurements are correlated, a statistical distance measure is often appropriate. However, if a principal component analysis has been done before clustering, the measurements will not be correlated and the simple intuitive distance measures will probably be adequate.

Determining a cluster linkage scheme means choosing a way of combining the distance measures between the data elements in different clusters to come up with a single number quantifying the distance between the clusters. Three common linkage schemes are called single, complete, and average. All of these schemes use the set of all distances between data points in different clusters. In the single linkage scheme, the minimum distance between data points in different clusters is defined to be the distance between clusters. In the complete linkage scheme, the maximum distance is used. In the average linkage scheme, the average distance is used. The clustering behavior is significantly affected by which linkage scheme is chosen. The most dramatic effect is that highly non-elliptical groups can be created by the single linkage scheme.

The main non-hierarchical clustering method in use is the K-means method. In order for this method to be used, the final number of clusters must be specified in advance. In this method the data elements are initially distributed arbitrarily into K clusters. The elements in the data set are then repeatedly examined. A data element is reassigned to the cluster whose centroid is closest. This assignment process continues until no further reassignments are made. If the correct number of clusters is not known in advance, the K-means algorithm must be run a number of times with different values of K. The user must then decide what the best number of clusters is.

Clustering is also a very common procedure in multivariate statistical analysis. A number of commercial software packages are available for doing agglomerative hierarchical clustering and K-means clustering. Divisive hierarchical clustering is seldom done in practice. The BMDP function BMDP2M, and the SAS (SAS Institute, Cary, North Carolina) process CLUSTER both do agglomerative hierarchical cluster analysis. The BMDP function BMDPKM and the SAS process FASTCLUS both carry out K-means clustering.

The application of clustering methods of analysis to flow cytometric data has been demonstrated [16,38]. In a recent

TABLE 6. Class Definitions (Upper and Lower Values for Each Variable) for Helper/Suppressor Subsetting Using Cluster Analysis

Class	Variables							
	FS		Leu-2A (FITC)		Leu-3 (PE)		SS	
No.	Lower	Upper	Lower	Upper	Lower	Upper	Lower	Upper
1	10.4	17.6	11.4	155.7	1.8	2.2	11.1	112.0
2	32.1	186.7	162.0	228.5	2.3	177.7	15.2	72.0
3	199.0	255.0	229.7	232.0	178.1	255.0	74.0	255.0
4	—	—	235.0	255.0	—	—	—	—

FS = forward light scatter; FITC = fluorescein isothiocyanate; PE = phycoerythrin; SS = slit scan.

study, Murphy [38] studied the reproducibility of a cluster from sample to sample, as a criterion for whether the cluster represents a significant population. In this study, data for eight samples of RBC-lysed whole peripheral blood from four donors were split into two portions, four of which were stained with FITC-conjugated-anti-Leu2A (anti-T-suppressor) and phycoerythrin-conjugated-anti-Leu3 (anti-T-helper) antibodies, and four were unstained. In addition to the two immunological variables, forward and slit scan light scatter were also measured. A modified K-means clustering program was used to identify clusters within each sample, and then the resulting clusters were compared using a simple categorization program. The range (defined as a user-specified number of standard deviations on either side of the mean) for each cluster was compared for each variable. If the ranges of two clusters overlapped, a composite range was created consisting of the smaller of the two lower bounds and the larger of the two upper bounds. This process was continued until no overlap between ranges remained. For each variable, the resulting ranges were ordered, and given class numbers starting with one. A class number then corresonds to a range in a given variable. For each cluster, a 4-digit classification could then be assigned by combining the class numbers for each variable for that cluster. For example, a classification of 1,123 means that variables 1 and 2 (FS and Leu-2A) fell within class (window) 1 for both variables, Leu-3 fell within its class 2 group, and SS within its class 3 group.

This procedure was applied to the first 4,000 detected events to define the clusters. All 20,480 events for each sample were then assigned to the nearest cluster. The results are presented in Tables 6 and 7. For each sample, all clusters with the same classification were combined. Table 6 shows the range of each variable used to define each class. Class four was established only for variable 2, LEU2A. Table 7 shows the class assignments for cells from the 4 donors. The minus (−) and (+) columns refer to whether or not the cells received the antibodies. Twenty clusters were detected, with class numbers ranging from 1123 to 3433 (column 1). The frequencies are given as the fraction of the sample that was assigned to each cluster. Populations corresponding to unlabeled small cells (e.g. B cells; class 2122) and large cells (e.g., granulocytes, monocytes; class 2123) were found in similar frequencies in all samples. Two populations, corresponding to T helper cells (class 2132) and T suppressor (class 2222) cells were found in all samples that received the antibodies. The average percent of helper T cells was 8.52 ± 1.33, and the average percent suppressor T cells was 3.79 ± 0.38, giving an average helper/suppressor ratio of 2.24 ± 0.21.

Classification

Classification is also a very common mathematical procedure in multivariate statistical analysis. The basic idea is to develop a classification criterion using a well understood set of data and then apply this criterion to additional unknown data elements to classify them. In terms of bone marrow data, which was used as an example in discussing principal component analysis, the idea would be to put known types of cells into the flow cytometer so that a training data set could be produced. From this data set a classification criterion would be developed so that any future cells processed by the system can be classified into one of the known groups of cells used to produce the training data set.

The principal classification technique is called discriminant analysis in which a discriminant function is developed using the training set. Many different discriminant functions have been developed. An early example is Fischer's linear discriminant function, which was developed around the idea of transforming multivariate data elements into univariate data elements such that the univariate data elements are as separated as possible. More recent discriminant functions such as Anderson's classification function and various quadratic discrimination functions were obtained by applying criteria such as minimizing the expected cost of misclassification to normally distributed data.

A recent procedure that holds some promise is called CART (Classification And Regression Trees) [12] and is particularly well suited to computer-based multivariate data analysis. In this method, the training data set is used to partition the measurement space into smaller pieces. Each step in the partitioning involves the splitting of a piece of the measurement space into two smaller pieces. In this way, a binary tree is built up. The partitioning criteria involve simple questions about the ranges in which the components of the data elements lie. The partitioning process is stopped when all the data elements in each terminal node belong to only one class. The classification procedure for a new data element then consists of determining which terminal node of the classification tree the data element will end up in.

Several commerical packages exist for doing discriminant analysis. The BMDP function BMDP7M, the SAS functions DESCRIM and STEPDISC and the SPSS-X function DISCRIMINANT are all reasonably complete and employ many modern extensions to the original basic procedures.

A disadvantage of discriminant analysis is that its theoretical justification relies upon normality of the data. There is no reason to assume that the flow cytometer data are nor-

TABLE 7. Class (Cluster) Frequencies for 20,480 Events from Measurement of Cells from Four Donors, with (+) and without (−) Addition of Fluorescently Labeled Antibodies*

	Donors							
	1		2		3		4	
Class	−	+	−	+	−	+	−	+
1 1 2 3:		0.015						
2 1 1 1:								1.646
2 1 1 2:								3.042
2 1 2 2:	30.222	13.535	23.882	9.956	27.798	15.356	33.970	32.739
2 1 2 3:	66.802	68.413	72.661	69.766	68.750	63.989	62.102	44.604
2 1 3 2:		8.496		9.082		9.810		6.699
2 1 3 3:		0.274		0.894		0.088		
2 2 2 2:		3.579		4.194		4.009		3.374
2 2 2 3:		1.367		0.684		0.107	0.024	0.103
2 2 3 2:						0.112		
2 2 3 3:				0.210				
2 3 3 2:			0.176					
2 3 3 3:			0.083					
2 4 2 3:		0.083						
2 4 3 3:						0.024		
3 1 2 3:	0.371		0.405			0.908	0.190	
3 1 3 3:		0.806						
3 1 2 3:	0.010	0.015			0.337		0.181	
3 1 3 3:						0.127		0.142
3 4 3 3:				0.459				
"Noise"	2.598	3.418	2.793	4.756	3.115	5.469	3.511	7.651

*Variables measured and the class codes are defined in Table 6.

mally distributed and in fact there is some graphical evidence to the contrary.

STANDARDS

With the great proliferation of flow cytometers in recent years, in number and application, more and more investigators are generating data which they would like to share with their colleagues, both within and outside their Laboratory. This is highly desirable in order to share interesting data with collaborators, as well as to enlist aid in the analysis of such data. This data interchange has been made difficult to accomplish, since individual investigators (and instrument manufacturers) tend to vary greatly in their selection of computers and operating systems. There is no uniformity in the format of flow cytometry data, that would make it convenient, or in some cases possible, for the recepient of the data to be able to enter them directly into local data analysis programs.

This problem also existed in the personal computer world, where users wished to transmit data between spreadsheet, data base, statistics, and graphics programs, not necessarily all from the same company. This requirement led to the adoption of the Data Interchange Format (DIF) standard [26].

In the DIF (trademark of Software Arts Inc.) standard, data are arranged in tables, using the terms "vector" for column and "tuple" for row. All tuples in a table have equal length and all vectors have equal length. Every DIF file has two parts: a header section containing descriptive information, and a data section. Each field in the header section contains four items: (1) the name of the header item; (2) the vector number; (3) the vector value; and (4) a "string" of characters. An example of a header entry is

TABLE
0,1
"Chromosomes"

The vector number of 0 indicates that the title "Chromosomes" refers to the entire table, and the value of 1 is the version number, not important in this entry. Required entries in the header section are TABLE, VECTORS, TUPLES, and DATA. Items VECTORS and TUPLES define the numbers of each in the table. The keyword LABEL is used to describe each vector. DATA is the last entry in the header section and indicates that all the following items are data values. Each data entry consists of three fields on two lines. The first line includes a type indicator (numeric, string or special) and the value (integer, real or floating point). If the indicator is 1, the data consists of a string of characters, which is contained on the second line. At the end of the data section, the string "EOD" is entered. This data format is now used by most cmpanies that write data processing software for personal computers.

In 1984 Murphy and Chused (37) proposed a standard specific to flow cytometric data. Their proposal was similar to the DIF standard in that keywords are used to designate specific fields within the data file. The file is divided into three segments, TEXT, DATA and ANALYSIS. The TEXT segment is in ASCII format and consists entirely of keywords followed by their values, set off by a "separator" character. The proposal incudes 35 standard keywords; additional keywords can be defined by the user. For example, $EXP/P Dean/ specifies the name of the experimenter, $CELLS/Lymphocytes/ specifies the type of cells being measured, and the sequence $MODE/U/$P1R/256/$P1B/16/ specifies a univariate histogram of 256 16-bit words for variable number 1. In the examples, the "/" character delimits the value of the keyword. The DATA segment consists of a list of the data values, whose length in bits was specified by a keyword in the TEXT segment. The ANALYSIS segment contains the results of processing the data, for example phase fractions from analyzing a DNA content distribution. This data file standard has been implemented on several different kinds of computers and is available to some extent on commerical flow cytometers as the FCS file format.

In an effort to establish a formal standard that would be supported by industry as well as the user community, in the fall of 1987 the Society for Analytical Cytology formed a Committee on Flow Cytometry Data Standards. This Committee is charged with the responsibility of defining a data file standard that is both comprehensible and flexible, and suitable for use with a wide variety of computers and operating systems. The standard is to include standard nomenclature to the extent that it has been established [23].

ACKNOWLEDGMENTS

I would like to acknowledge the very valuable assistance and advice from many colleagues at laboratories throughout the world. As can be seen from the great variety of topics in this chapter, data analysis is a complicated and diverse subject. I have been fortunate in the friends I have made who have been willing to share their knowledge, data, and experience with me.

SOFTWARE SOURCES

Software such as that described in this chapter is available from both commercial and private sources. Some of these packages are available for the price of the transport medium (e.g., magnetic tape), and some carry a substantially higher fee. Anyone interested in obtaining one of the packages should contact the source directly and make the appropriate arrangements. The list presented below contains all the sources known to the author as of the writing of this chapter.

Commercial Sources

Company	Computers
Becton Dickinson Immunocytometry systems 2375 Garcia Ave. P.O. Box 7375 Mountain View, California 94039	DEC LSI-11 VAX Hewlett Packard
Catalyss Corporation 7400 South Tucson Way Englewood, Colorado 80112	PDP-11/73
Cellsoft Biotechnology Systems P.O. Box 13666 Research Triangle Park, North Carolina 27709	IBM/PC
EPICS Division Coulter Corporation P.O. Box 4486 Hialeah, Florida 33014-0486	IBM/PC
Oatka Software P.O. Box 5 Scottsville, New York 14546	
Verity Software House, Inc. P.O. Box 247 Topsham, Maine 04086	IBM/PC

Private Sources

Dr. James Jett Los Alamos National Laboratory MS-M888 Los Alamos, New Mexico 87545	PDP-11 VAX (FORTRAN)
David Ow Biomedical Sciences Division Lawrence Livermore National Laboratory P.O. Box 5507 L-452 Livermore, California 94550	LSI-11 PDP-11/44 VAX (FORTRAN)
Dr. Robert Murphy Carnegie Mellon University Dept. of Biological Sciences 4400 Fifth Street Pittsburgh, Pennsylvania 15213	VAX (FORTRAN)
Dr. Roy Robinson Dept. of Pathology University of Rochester Medical Center Rochester, New York 14642	PDP-11 (FORTRAN)
James Herman Chemistry Department Colorado State University Fort Collins, Colorado 80523	IBM/PC

REFERENCES

1. **Afifi AA, Clark V** (1984) "Computer Aided Multivariate Analysis." Belmont, California: Lifetime Learning Publications.
2. **Andrews DF** (1972) Plots of high dimensional data. Biometrics 28:125–136.
3. **Arvesty J** (1973) Tumor growth and chemotherapy: Mathematical models, computer simulations, and experimental foundations. Math. Biosci. 17:243–300.
4. **Atchley WR, Bryant EH (eds.)** (1975) "Multivariate Statistical Methods: Among-Groups Covariation." Stroudsburg, Pennsylvania: Dowden, Hutchinson & Ross, Inc.
5. **Bagwell CB** (1979) "Theory and Application of DNA Histogram Analysis." Ph.D. thesis, University of Miami, Florida.
6. **Bagwell CB, Hudson JL, Irvin GL II** (1979) Non parametric flow cytometry analysis. J. Histochem. Cytochem. 27:293–296.
7. **Baisch H, Göhde W, Linden WA** (1975) Analysis of PCP-data to determine the fraction of cells in the various phases of the cell cycle. Radiat. Environ. Biophys. 12:31–39.
8. **Baisch H. Beck H-P, Christensen IJ, Hartmann NR, Fried J, Dean PN, Gray JW, Jett JH, Johnston DA, White RA, Nicolini C, Zeitz S, Watson JV** (1982) A comparison of mathematical models for the analysis of DNA histograms obtained by flow cytometry. Cell Tissue Kinet. 15:235–249.
9. **Barlogie B, Drewinko B, Johnston DA, Buchner T, Hauss WH, Freidreich EJ** (1976) Pulse cytophotomet-

ric analysis of synchronized cells in vitro. Cancer Res. 36:1176–1181.

10. **Bennett DE, Mayall BH** (1979) Interactive display and analysis of data from bivariate flow cytometers. J. Histochem. Cytochem. 27:579–583.
11. **Bevington P** (1969) "Data Reduction and Error Analysis for the Physical Sciences." New York: McGraw-Hill.
12. **Breiman L, Friedman JH, Olshen RA, Stone CJ** (1984) "Classification and Regression Trees." Belmont, California: Wadsworth International Group.
13. **Chambers JM, Cleveland WS, Kleiner B, Tukey PA** (1983) "Graphical Methods for Data Analysis." Boston: Duxbury Press.
14. **Chernoff H** (1973) The use of faces to represent points in k-dimensional space graphically. J. Am. Stat. Assoc. 68:361–368.
15. **Cleveland WS** (1979) Robust locally weighted regression and smoothing scatterplots. J. Am. Stat. Assoc. 74:829–836.
16. **Crowell JM, Hiebert RD, Salzman GB, Price BJ, Cram LS, Mullaney PF** (1978) A light-scattering system for high-speed cell analysis. IEEE Trans. Biomed. Eng. BME-25:519–526.
17. **Dean PN, Jett JH** (1974) Mathematical analysis of DNA distributions derived from flow microfluorometry. J. Cell Biol. 60:523–527.
18. **Dean PN, Anderson EC** (1974) The rate of DNA synthesis during S-phase by mammalian cells in vitro. In Hannen CAM, Hillen HFP, Wessels JMC (eds.) "Pulse Cytophotometry." Ghent, Belgium: European Press Medikon, pp 77–86.
19. **Dean PN** (1985) Methods of data analysis in flow cytometry. In Van Dilla MA, Dean PN, Laerum OD, Melamed MR (eds.) "Flow Cytometry: Instrumentation and Data Analysis." London: Academic Press, pp 195–220.
20. **Dean PN** (1987) Data analysis in cell kinetics research. In Gray JW, Darzynkiewicz Z (eds.) "Techniques in Cell Cycle Analysis." Clifton, New Jersey: Humana Press, pp 207–253.
21. **Gabriel KR** (1971) The biplot graphic display of matrices with application to principal component analysis. Biometrika 58:453–467.
22. **Gray JW, Dean PN** (1980) Display and analysis of flow cytometric data. Annu. Rev. Biophys. Bioeng. 509–539.
23. **Hiddeman W, Schumann J, Andreeff M, Barlogie B, Herman CJ, Leif RC, Mayall BH, Sandberg AA** (1984) Convention on nomenclature for DNA cytometry. Cytometry 5:445–446.
24. **Howard A, Pelc SR** (1953) Synthesis of deoxyribonucleic acid in normal and irradiated cells and its relation to chromosome breakage. Heredity 6(suppl.):261–273.
25. **Johnson RA, Wichern DW** (1982) "Applied Multivariate Statistical Analysis." Englewood Clifts, New Jersey: Prentice-Hall.
26. **Kalish CE, Mayer MF** (1981) DIF: A format for data exchange between applications programs. BYTE Nov: 174–206.
27. **Kraemer PM, Deaven LL, Crissman HA, Van Dilla MA** (1972) DNA constancy despite variability in chromosome number. In Du Praw EJ (ed), "Advances in Cell and Molecular Biology." Vol. 2. New York: Academic Press, pp 47–108.
28. **Langlois RG, Carrano AV, Gray JW, Van Dilla MA** (1980) Cytochemical studies of metaphase chromosomes by flow cytometry. Chromosoma 37:229–251.
29. **Langlois RG, Yu L-C, Gray JW, Carrano AV** (1982) Quantitative karyotyping of human chromosomes by dual beam flow cytometry. Proc. Natl. Acad. Sci. USA 79:7876–7880.
30. **Lanier LL, Allison, JP, Phillips JH** (1986) Correlation of cell surface antigen expression on human thymocytes by multi-color flow cytometric analysis; Implications for differentiation. J. Immunol. 137:2501–2507.
31. **Lindmo T, Aarnaes E** (1979) Selection of optimal model for the DNA histogram by analysis of error of estimated parameters. J. Histochem. Cytochem. 27: 297–304.
32. **Lipkin M, Bell B, Sherlock P** (1963) Proliferation in the gastrointestinal tract of man. I. Cell renewal in colon and rectum. J. Clin. Invest. 6:767–776.
33. **Loken MR, Herzenberg LA** (1975) Analysis of cell populations with a fluorescence-activated cell sorter. Ann NY Acad. Sci. 254:163–171.
34. **Loken MR, Stout RD, Herzenberg LA** (1979) Lymphoid cell analysis and sorting. In Melamed MR, Mullaney P, Mendelsohn ML (eds.) "Flow Cytometry and Sorting." New York: John Wiley & Sons, pp 505–528.
35. **Mann RC, Papp DM, Hand RE** (1984) The use of projections for dimensionality reduction of flow cytometric data. Cytometry 5:304–307.
36. **Mann RL** (1987) On multiparameter data analysis in flow cytometry. Cytometry 8:184–189.
37. **Murphy RF, Chused TM** (1984) A proposal for a flow cytometric data file standard. Cytometry 5:553–555.
38. **Murphy RF** (1985) Automated identification of subpopulations in flow cytometric list mode data using cluster analysis. Cytometry 6:302–309.
39. **Overton WR** (1985) Analysis of immunofluorescence histograms by cumulative subtraction. Abstract, Analytical Cytology X, Asilomar, California, 3–8 June 1987.
40. **Pallavicini MG, Summers LJ, Dean PN, Gray JW** (1985) Enrichment of murine hemopoietic clonogenic cells by multivariate analysis and sorting. Exp. Hematol. 13:1173–1181.
41. **Pallavicini MG, Summers LJ, Giroud FJ, Dean PN, Gray JW** (1985) Multivariate analysis and list mode processing of murine hemopoietic subpopulations for cytokinetic studies. Cytometry 6:539–549.
42. **Pinkel D, Gledhill BL, Van Dilla MA, Lake S, Wyrobek AJ** (1983) Radiation-induced DNA content variability in mouse sperm. Radiat. Res. 95:550–565.
43. **Steinkamp JA, Romero A, Van Dilla MA** (1973) Multiparameter cell sorting: Identification of human leukocytes by acridine orange fluorescence. Acta Cytol. (Praha) 17:113–117.
44. **Steinkamp JA, Romero A** (1974) Identification of discrete classes of normal human peripheral lymphocytes by multiparameter flow analysis. Proc. Soc. Exp. Biol. Med. 146:1061–1066.
45. **Tukey JW, Tukey PA** (1983) Some graphics for studying four-dimensional data. In "Computer Science and Statistics: Proceedings of the Fourteenth Symposium

on the Interface." New York: Springer-Verlag, pp 60–66.

46. **Valet G** (1980) Graphical representation of three-parameter flow cytometer histograms by a newly developed FORTRAN IV computer program. In Laerum O (ed.) "Flow Cytometry IV." Oslo: Universitetsforlaget, pp 125–129.
47. **Valet G, Rüssmann L** (1984) Automated flow-cytometric identification of colo-rectal tumor cells by simultaneous DNA, CEA-antibody and cell volume measurements. J. Clin. Chem. Clin. Biochem. 22:935–942.
48. **Valet G, Warnecke HH, Kahle H** (1984) New possibilities of cytostatic drug testing on patient tumor cells by flow cytometry. Blut 49:37–43.
49. **Van Dilla MA, Trujillo TT, Mullaney PF** (1969) Cell microfluorometry: A method for rapid fluorescence measurement. Science 163:1213–1214.
50. **Van Dilla MA** (1986) Human chromosome-specific DNA libraries: Construction and availability. Bio/Technology 4:537–552.
51. **Young IT** (1977) Proof without prejudice. J. Histochem. Cytochem. 25:935–941.

23

Quantitative Cell-Cycle Analysis

J. W. Gray, F. Dolbeare, and M. G. Pallavicini

Biomedical Sciences Division, Lawrence Livermore National Laboratory, University of California, Livermore, California 94550

INTRODUCTION

Quantitative information about the proliferative characteristics (hereafter called cytokinetic characteristics) of cells and cell populations is essential for a variety of biological and biomedical studies. Knowledge of the fraction of cells at risk to cell cycle specific cytotoxic agents is important to oncologists seeking to improve cancer therapy by maximizing tumor cell kill while minimizing toxicity to normal cells [1,12,13,33,51,76,80,88,95,100]. The fractions of cells in the G1, S, and G2M phases of the cell cycle and the fraction of actively proliferating cells (i.e., the growth fraction) are especially important in this regard. The rate at which cells traverse the various cell-cycle phases before and during therapy may also be prognostically important. Information about the rate at which cells traverse the cell cycle also is fundamental to studies of the developmental biology of complex tissues such as the hematopoietic system and the intestinal epithelium [3,75,77,87]. Biochemists interested in the mechanisms of action of cell cycle phase specific agents require information about the effects of these agents on cell cycle traverse (e.g., the phase(s) in which cell cycle traverse is inhibited or in which cell kill occurs) [14,18,59,64,67,92]. In addition, the cytokinetic status of a population of cells may influence its response to mutagenic and/or carcinogenic agents [2,68,92,104].

This chapter describes a battery of techniques based on flow cytometry and sorting that facilitate accumulation of cytokinetic information such as that described above. An increasing number of distinct cytokinetic conditions have been proposed in recent years [2,20–23]. For the purposes of this chapter, we consider the movement of cells through physiologically distinct conditions called states and through four phases of the cell cycle [42,44]. We postulate that cells may reside in one of four states: Proliferating and capable of forming multicell colonies (clonogenic) (PC), proliferating and nonclonogenic (PD), not actively proliferating but potentially clonogenic (QC) and not actively proliferating and not clonogenic (QD). In addition, actively proliferating cells pass through four phases designated G1, S, G2, and M as they replicate their DNA and divide. Figure 1 suggests the variety of states and phases in which cells may reside. Cytokinetic analysis involves quantitative determination of the distribution of cells among these phases and states and the rates at which cells move among them. Cytokinetic information commonly required includes the fractions of cells in the G1, S, G2, and M phases of the cell cycle, the fraction of actively proliferating cells in the entire population (i.e., the fraction of cells in the PC + PD states), the fraction of clonogenic cells (i.e., the fraction of cells in the PC + QC states), the duration of the G1, S, G2, and M phases (designated T_{g1}, T_s, T_{g2}, and T_m, respectively), the rate of DNA synthesis and/or the rates at which cells enter or leave the PC and PD states.

During the past 15 years, numerous flow cytometric techniques have been developed that facilitate accumulation of this information. These methods include (1) DNA distribution analysis for assessment of the fractions of cells in the various phases of the cell cycle and for assessment of the effects of perturbing agents on cell cycle traverse, (2) the use of flow cytometry and/or sorting to monitor the cell cycle traverse of cohorts of cells labeled with DNA precursors (e.g., bromodeoxyuridine, BrdUrd; or tritiated thymidine, [^{3}H]-TdR), and (3) multivariate analysis to allow cytokinetic analysis of cytometrically distinct subpopulations. This chapter describes these methods conceptually, experimentally and analytically as they apply to asynchronously growing populations, to perturbed populations and to complex tissues. The biological and mathematical models presented here were chosen to be illustrative and should not be considered to be the models of choice for kinetic studies. Other models and approaches are referenced throughout the chapter. One important application of flow cytometry in cytokinetic analysis is not described in this chapter; namely, multivariate analysis of cytokinetically important subpopulations defined according to their RNA content, chromatin structure, etc. This application is described in detail in Chapter 24 (this volume).

ASYNCHRONOUS POPULATION ANALYSIS

DNA Distributions

DNA Distributions and Cytokinetic Analysis. The importance of DNA distribution analysis to cell cycle studies, illustrated in Figure 2, comes from the fact that cellular DNA

Flow Cytometry and Sorting, Second Edition, pages 445–467
Published 1990 by Wiley-Liss, Inc.

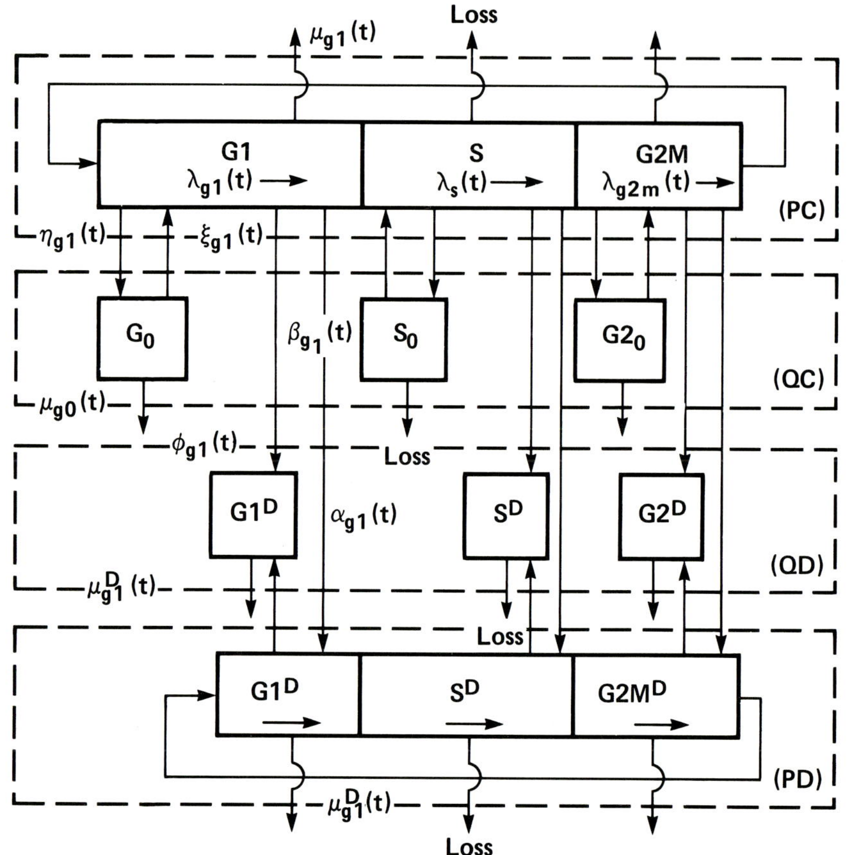

Fig. 1. Schematic representation of various cytokinetic states and phases through which cells may pass during proliferation and/or during response to treatment with chemotherapeutic agents. Four states are defined: PC (proliferating and clonogenic), PD (transiently proliferating but not clonogenic), QC (quiescent but clonogenic if stimulated into proliferation), and QD (not proliferating and not clonogenic). The cells in any state may reside in any of several compartments within the G1, S, and G2M phases. Coefficients λ, η, β, α, and ξ describe the movement of cells through the PC compartment, from PC to QC, from PC to QD, from PC to PD and from QC to PC; respectively. In addition, μ describes the loss of cells. These coefficients may vary with time in perturbed populations.

content is an unambiguous marker for cellular maturity. That is, cells in G1 phase have unit DNA content, cells in S-phase increase their DNA content continuously as they mature until they enter the G2 and M phases where they have twice the G1-phase DNA content (Figs. 2a,b). Theoretically, the fractions of cells in each of these phases can be obtained from the DNA distribution (Fig. 2c). In practice, DNA content cannot be measured perfectly by flow cytometry because of imperfect staining and imperfect measurement. Typically, the magnitude of such errors is only a few percent, so that DNA distributions like that shown in Figure 2d are obtained routinely. As a result, mathematical analysis is often required for accurate estimation of the fractions of cells with G1-, S-, and G2M-phase DNA contents.

Experimental Techniques. The accuracy of DNA distribution analysis depends on several aspects of the analysis procedure including cell dispersal, staining, and flow cytometric analysis.

Cell dispersal. The cells must be reduced to a suspension of single, intact cells that is representative of the original population or tissue under study. That is, selective cell loss (i.e., preferential loss of a specific cell type or loss of cells within a particular state or phase) must be minimized. Hemopoietic cells are especially fragile and often are lost in harsh preparative procedures. Likewise, mitotic cells may be lost in procedures that yield suspensions of cell nuclei. Cell dispersal procedures and procedures to monitor cell loss are described in detail by Pallavicini [78] and by Pallavicini et al. in Chapter 10 (this volume). It is also important to minimize disruption of cell nuclei during dispersal to avoid generation of a "debris continuum" in the DNA distributions. Such continua are the result of nuclear fragments in the dispersed cell population. These continua are usually smoothly varying; highest near the origin and decreasing with increasing fluorescence intensity.

Staining. Cells must be stained with fluorescent dyes that bind (or can be made to bind) specifically and stoichiometrically to cellular DNA. Commonly used dyes for this purpose include ethidium bromide (EB), propidium iodide (PI), Hoechst 33258 (HO58), Hoechst 33342 (HO42), chromomycin A3 (CA3) and mithramycin (MI). These dyes and some of their characteristics are listed in Table 1. The reader is referred to Crissman and Steinkamp [18] and to Chapters 13 and 14 (this volume) for more information on the cytochemistry for DNA distribution analysis.

Flow cytometry. Typically, cells to be analyzed are stained

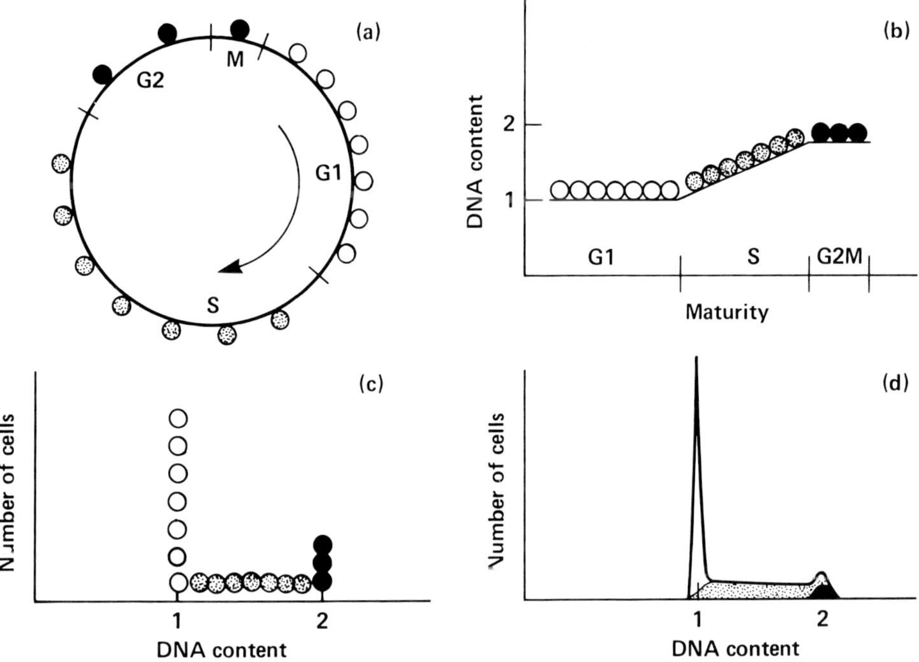

Fig. 2. Schematic illustration of the generation of a DNA distribution. **a:** Distribution of cells around the cell cycle for an asynchronous population. **b:** Illustration of the relation between DNA content and cell cycle position for an asynchronous population. **c:** DNA distribution of the hypothetical population illustrated in b. **d:** Broadened DNA distribution for the hypothetical distribution in c. Broadening may result from variability in cell staining or in fluorescence measurement.

TABLE 1. DNA-Specific Dyes and Selected Characteristics Important for DNA Distribution Analysis

Dye	Excitation maximum (nm)	Emission maximum (nm)	Cytochemical characteristics
EB	525	620	No DNA base composition preference; binds to all double-stranded nuclei acids [63]
PI	535	630	Similar to EB [30]
CA3	430	585	Binds preferentially to GC-rich double-stranded DNA (dsDNA); Requires Mg^{2+} [54]
MI	430	585	Similar to CA3 [19]
HO58	350	450	Binds preferentially to AT-rich dsDNA [52]
HO42	350	450	Similar to HO58, except it can be used to stain living cells [56]

with one of these dyes and processed flow cytometrically where fluorescence intensity is recorded as a measure of relative DNA content. The results from several hundred thousand measurements are accumulated to form a histogram showing the distribution of DNA among the cells of the population (called the DNA distribution). One figure of merit in DNA distribution analysis is the accuracy with which DNA content can be measured. The coefficient of variation of the G1-phase portion of the DNA distribution (CV_{g1}) is usually given as an indicator of measurement accuracy. DNA distributions measured for fixed, cultured cells typically have CV_{g1} values of less than 0.03 when measured using modern instruments. CV_{g1} values for viable cells stained with HO42 range from 0.04 to 0.15, depending on the cell type, stain concentration and staining duration and temperature. The effect of the CV on DNA distribution shape is illustrated in Figure 3, which shows simulated DNA distributions in which the fractions of cells in the G1, S, and G2M phases are 0.62, 0.29, and 0.09, respectively. Figure 3a–d shows DNA distributions in which the measurement CV increases from 0.02 to 0.12.

Phase Fraction Estimates. Several analytical techniques have been developed in recent years to estimate the fractions of cells in the G1, S, and G2M phases from DNA distributions measured for asynchronously growing cells (i.e., cells distributed randomly around the cell cycle). The goal of these meth-

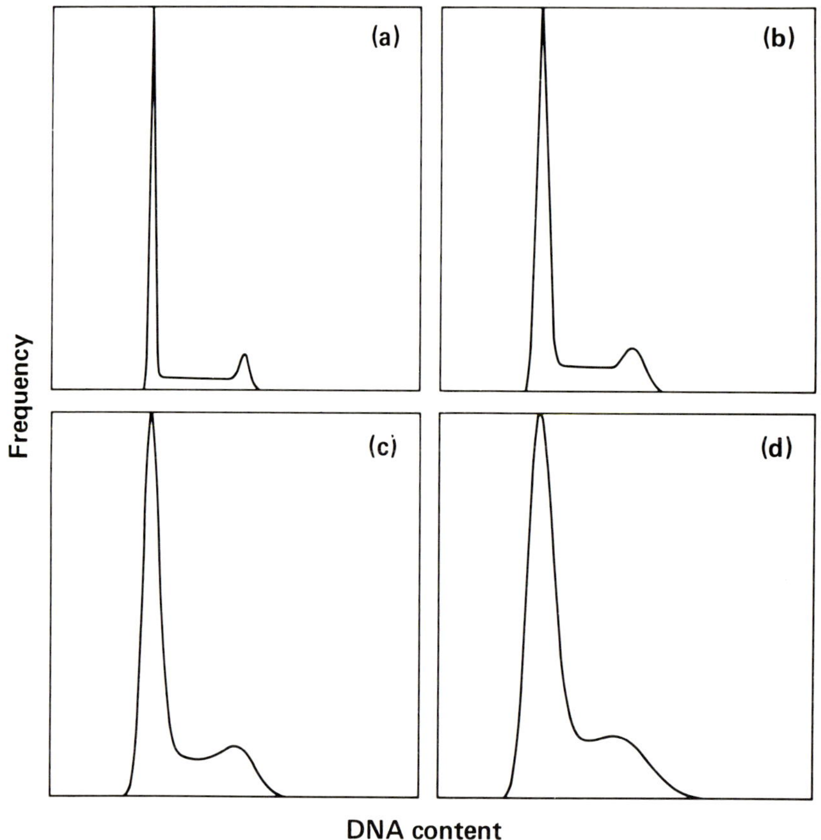

Fig. 3. Illustration of the effect of measurement CV on the shape of DNA distributions. All DNA distributions were generated for a hypothetical asynchronous population in which the fractions of cells in the G1, S, and G2M phases were 0.62, 0.29, and 0.09. Measurement CVs was as follows: **a:** 0.02. **b:** 0.05. **c:** 0.09. **d:** 0.12.

ods is to take proper account of the overlap between the G1- and early S-phase cells and between late-S-phase and G2M-phase cells so that accurate estimates for the fractions of cells in each phase can be made. Figure 4 illustrates two approaches to phase fraction estimation. Both of these approaches rely on development of functions that approximate the G1-, S-, and G2M-phase distributions. Figure 4a shows a graphic approximation of the S-phase distribution as a rectangular distribution. The G1-phase area is estimated as proportional to the area to the left of the S-phase distribution and the G2M-phase area is estimated as proportional to the area to the right of the S-phase distribution. This method works best for distributions in which the measurement CV is low. Figure 4b shows a more sophisticated approach to phase-fraction estimation. In this approach, the G1 and G2M phases are approximated as normal distributions. That is,

$$f_{g1}(x) = \frac{A_{g1}}{\sqrt{2\pi}\sigma_{g1}} e^{-.5\left(\frac{x - \mu_{g1}}{\sigma_{g1}}\right)^2} \qquad (1a)$$

and

$$f_{g2m}(x) = \frac{A_{g2m}}{\sqrt{2\pi}\sigma_{g2m}} e^{-.5\left(\frac{x - \mu_{g2m}}{\sigma_{g2m}}\right)^2} \qquad (1b)$$

where A_{g1} and A_{g2m} are the areas, α_{g1} and σ_{g2m} are the standard deviations and μ_{g1} and μ_{g2m} are the means of the G1- and G2M-phase distributions, respectively. The S-phase distribution is described as the sum of a series of N regularly spaced normal distributions whose areas are constrained by a second order polynomial. That is, the S-phase distribution, $f_s(x)$ is defined as

$$f_s(x) = \sum_{j=1}^{N} \frac{1}{\sqrt{2\pi}\sigma_j} e^{-.5\left(\frac{x - \mu_j}{\sigma_j}\right)^2} (a + b + cx^2) \qquad (1c)$$

where σ_j and μ_j are the standard deviation and mean of the jth normal distribution. The sum of these three distributions, $S(x) = f_{g1}(x) + f_s(x) + f_{g2m}(x)$, is adjusted using a least squares-best-fit procedure so that S(x) matches the experimental distribution as closely as possible. Parameters adjusted during the fit include: A_{g1}, A_{g2m}, σ_{g1}, σ_{g2m}, μ_{g1}, μ_{g2m}, a, b, and c. The μ_j and σ_j may be constrained by the other parameters. For example,

$$\mu_j = \left(\frac{\mu_{g2m} - \mu_{g1}}{N}\right) j \qquad (2a)$$

and

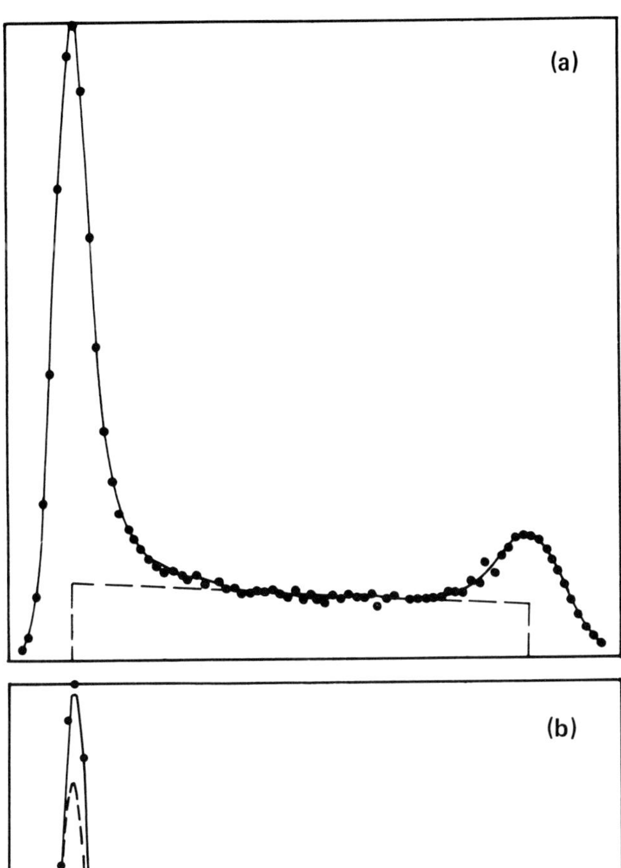

Fig. 4. Techniques for quantitative DNA distribution analysis. **a:** Illustration of the use of a trapezoidal approximation for the S-phase distribution. **b:** Illustration of the use of normal distributions for approximation of the G1- and G2M-phase distributions and a broadened second-order distribution as an approximation for the S-phase distribution.

$$\sigma_j = \frac{\mu_j}{2}\left(\frac{\sigma_{g1}}{\mu_{g1}} + \frac{\sigma_{g2m}}{\mu_{g2m}}\right) \qquad (2b)$$

The parameters that result in the best fit are taken as appropriate descriptors of the experimental distribution. Thus, the fractions of cells in the G1-, S-, and G2M-phases are

$$F_{g1} = \int_0^\infty f_{g1}(x)\, dx/\text{Total} = A_{g1}/\text{Total} \qquad (3a)$$

$$F_s = \int_{\mu_{g1}}^{\mu_{g2m}} (a + b + cx^2)\, dx/\text{Total}$$

$$= \frac{\left[\left(a\mu_{g2m} + \frac{b}{2}\mu_{g2m}^2 + \frac{c}{3}\mu_{g2m}^3\right) - \left(a\mu_{g1} + \frac{b}{2}\mu_{g1}^2 + \frac{c}{3}\mu_{g1}^3\right)\right]}{\text{Total}} \qquad (3b)$$

$$F_{g2m} = \int_0^\infty f_{g2m}(x)\, dx/\text{Total} = A_{g2m}/\text{Total} \qquad (3c)$$

with

$$\text{Total} = A_{g1} + A_{g2m} + \left[\left(a\mu_{g2m} + \frac{b}{2}\mu_{g2m}^2 + \frac{c}{3}\mu_{g2m}^3\right]\right.$$

$$\left. - \left[\left(a\mu_{g1} - \frac{b}{2}\mu_{g2m}^2 + \frac{c}{3}\mu_{g2m}^3\right)\right]\right.$$

and the CV describing the measurement resolution for the G1-phase cells is

$$CV_{g1} = \sigma_{g1}/\mu_{g1}$$

The key to accurate phase fraction estimation from DNA distributions is in use of the proper functions to describe the overlap between the S-phase distribution and the G1- and G2M-phase distributions. Both of the approaches described here work well when the measurement resolution is high (e.g., when CV_{g1} is less than 2%). However, the rectangular approximation for S phase becomes inaccurate when CV_{g1} increases to the point when the G1- and G2M-phase distributions overlap most of S phase. If the S-phase distribution is not smoothly varying across S phase so that it can be approximated by a second-order polynomial, both methods become inaccurate and a different functional approximation must be used. During the past several years, numerous mathematical procedures have been developed for analysis of DNA distributions. Several of these are listed in Table 2 along with references to which the reader is referred for more information. In addition, the reader is referred to Baisch et al. [5] and Dean [28,29] for detailed reviews of the various analysis procedures (see Chapter 22, this volume). The accuracy with which phase fractions can be estimated from DNA distributions depends on the CV of the DNA distribution, on the magnitude of any debris continuum, on the accuracy of the model used to parameterize the distributions and on whether the DNA distribution is representative of the original population. For example, phase fraction estimation can be compromised by (1) a high debris continuum that obscures the S-phase portion of the distribution, (2) clumping of G1-phase cells to mimic G2M-phase cells, (3) preferential loss of cells within one phase, and (4) skewing of the DNA content measurement (e.g., skewing caused by nonspecific binding of the fluorescent dye to cytoplasmic components) so that the peaks cannot be accurately approximated using normal distributions. Studies of the accuracy of phase

TABLE 2. Analytical Procedures for Phase Fraction Estimation from DNA Distributions

Method	Description	References
Rectangular S phase	S phase approximated by a trapezoid truncated at the G1 and G2M means.	6
Peak reflect	G1- and G2M-phase distributions are generated by reflecting the left and right portions of the G1 and G2M phase peaks about their modes. S phase is what remains.	8
Polynomial S phase	The S-phase distribution is approximated by a broadened second-order polynomial and G1- and G2M-phase distributions are approximated by normal distributions.	24
Multiple distributions	The G1- and G2M-phase distributions are approximated by normal distributions. S phase is approximated by a series of normal distributions or broadened rectangles or trapezoids.	4,36

fractions estimated from DNA distributions have shown that the fraction of cells in the G1 phase can be estimated most accurately and the fraction in G2M phase least accurately [5,28,29]. It is not unusual for the G2M phase fraction estimates to be in error by more than 25%. However, the fraction of cells with G1-phase DNA content can usually be estimated to within 10% using these methods. It is important to note, however, that the S-phase fraction estimated from a DNA distribution is the fraction of cells with S-phase DNA content. It is impossible to determine from a single DNA distribution whether these cells were actually synthesizing DNA at the time of sampling.

Rate of DNA Synthesis. The shape of a DNA distribution depends on the distribution of cells around the cell cycle and on the variation in the rate of DNA synthesis across S-phase. Figure 5, for example, shows hypothetical distributions to be expected for asynchronous, exponentially growing populations in which the rate of DNA synthesis across S phase was highest in late-S-phase (Fig. 5a), constant rate across S phase (Fig. 5b) or oscillatory across S phase (Fig. 5c). Quantitative analysis procedures have been developed to estimate the variation in the rate of DNA synthesis across S phase from the shape of DNA distributions [16,25]. These methods assume that the population is growing asynchronously so that the function describing the distribution of cells around the cell cycle is known. In addition, they assume that all cells with S-phase DNA content are actively proliferating and that the dyes used to stain for DNA content are not dependent on DNA base composition. Information about the variation in rate of DNA synthesis across S phase may be inaccurate for drug treated populations or when HO58, CA3 or similar base-composition-dependent dyes are used [26].

Radioactive DNA Precursor Incorporation

RC Analysis of Cell Cycle Traverse. Classically, cell cycle traverse studies have relied on introduction of a radioactive DNA precursor (e.g., [^{3}H]-TdR) into S-phase cells and on determination of G1-, S-, and G2M-phase durations from the rate at which the labeled cohort of cells traversed these phases. For example, in the fraction of labeled mitosis (FLM) method [50,53,81,85], cells in S phase are pulse labeled with [^{3}H]-TdR and the population is sampled periodically thereafter. Cell-cycle traverse information is estimated from the rate at which the labeled cells pass through mitosis by scoring the fraction of labeled mitoses detected autoradiographically. One alternative to the FLM procedure is to replace the autoradiographic ^{3}H-detection scheme with cell sorting (to define the cell cycle location) and liquid scintillation counting (to determine the presence of the labeled cells). In this procedure [45,74,90], the cells are labeled and sampled in the same manner as for FLM analysis and then processed through a sorter where cells in mid-S, G1, and possibly G2M phase are sorted for liquid scintillation counting (Fig. 6b). Cell cycle traverse information is estimated from the variation in the Radioactivity per Cell (RC) in the various sort windows as illustrated in Figure 6c, which shows the variations in the RC in mid-S (RCS_i) and in G1 (RCG1) expected for a population in which all cells cycle at the same rate. Figure 6d shows RCS_i and RCG1 curves measured for exponentially growing Chinese hamster ovary (CHO) cells [45].

Quantitative estimates of the G1-, S-, and G2M-phase durations and coefficients of variation thereof can be obtained from mathematical analyses of the RCS_i and RCG1 curves [45]. In this process, a model describing the movement of a labeled cohort of cells around the cell cycle is developed and used to simulate RCS_i and RCG1 curves. The cytokinetic parameters of the model such as phase durations and dispersions therein are varied until the simulations match the experimental data as closely as possible. One commonly used model divides the cell cycle into sequential compartments as illustrated in Figure 7 [40,90,91]. The multicompartment model for data analysis is used to illustrate the concept of mathematical analysis. Numerous other mathematical models have been proposed to describe cell cycle traverse [35,49,83,84,86,89,107,108]. Description of these various models is beyond the scope of this review. The mathematical formalism describes the movement of cells from compartment to compartment. For example, the rate of change in the number of cells in compartment i at time t, $N_i(t)$, might be described as

$$\frac{dN_i(t)}{dt} = \lambda_{i-1}N_{i-1}(t) - \lambda_i N_i(t) \tag{4a}$$

where λ_i defines the rate at which cells leave compartment i and move to compartment i+1. For compartment 1, the equation becomes

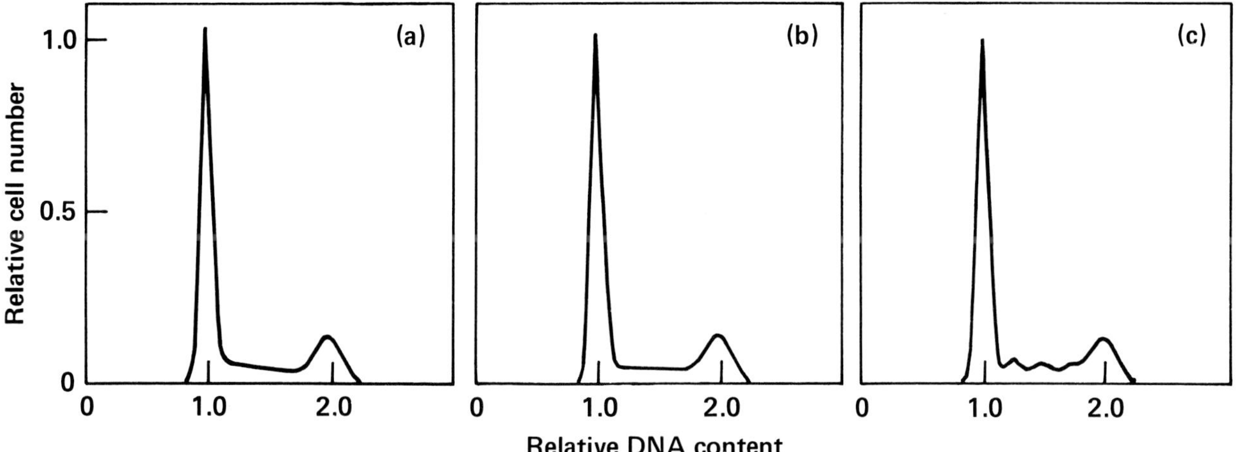

Fig. 5. DNA distributions showing the effect of variation in the rate of DNA synthesis across S phase. **a:** DNA distribution for a population in which the rate of DNA synthesis was highest in late S phase. **b:** DNA distribution for a population in which the rate ofDNA synthesis was constant across S phase. **c:** DNA distribution in which the rate of DNA synthesis varied in an oscillatory fashion across S phase. The distribution of cells around the cell cycle was the same for all distributions.

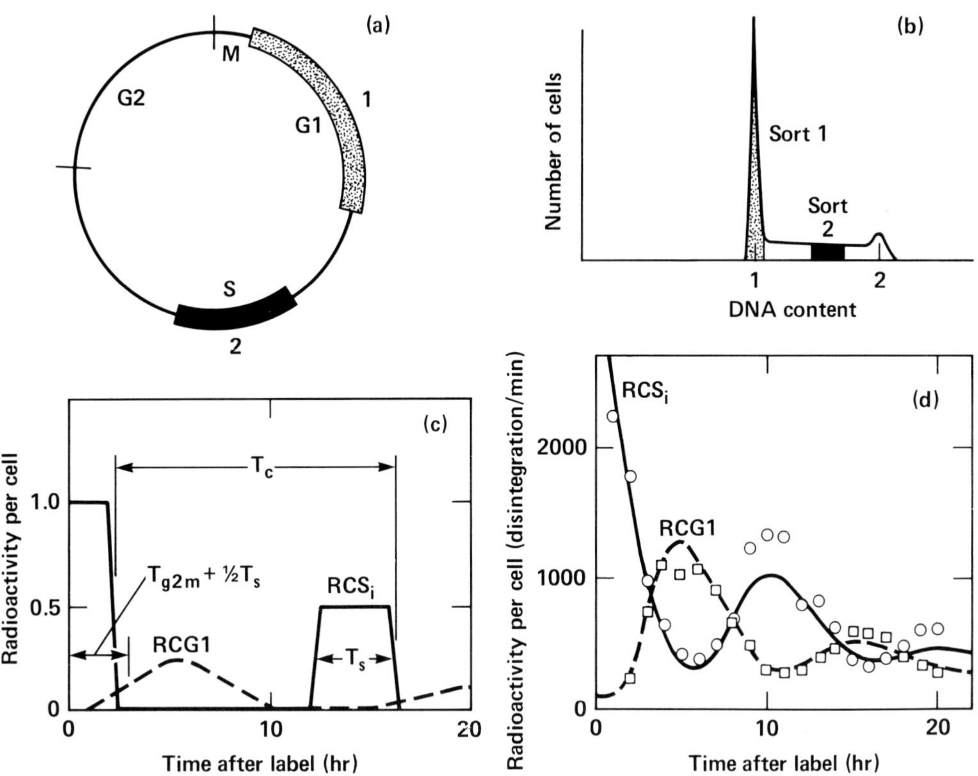

Fig. 6. Schematic representation of RC analysis. **a:** Regions of the cell cycle from which cells were sorted for RC analysis. **b:** Comparable windows in the DNA distribution from which cells were sorted for analysis of the Radioactivity per Cell in G1 (RCG1) and mid S phase (RCS_i). **c:** Hypothetical distributions showing RCG1 and RCS_i curves for a population in which all cells traverse the cell cycle at the same rate. Also shown is the relationship among G1-, S-, and G2M-phase duration and the RC curves. **d:** Actual RCG1 and RCS_i curves measured for a Chinese hamster ovary (CHO) cell population labeled for 30 min with [^{3}H]-TdR. ○, Experimental RCG1 data. □, RCS_i data—Computer-Generated RCG1 and RCS_i curves generated during quantitative analysis of these data. The simulated curves were generated assuming G1-, S-, and G2M-phase durations of 2.4 hours, 6.5 hours, and 1.6 hours, respectively.

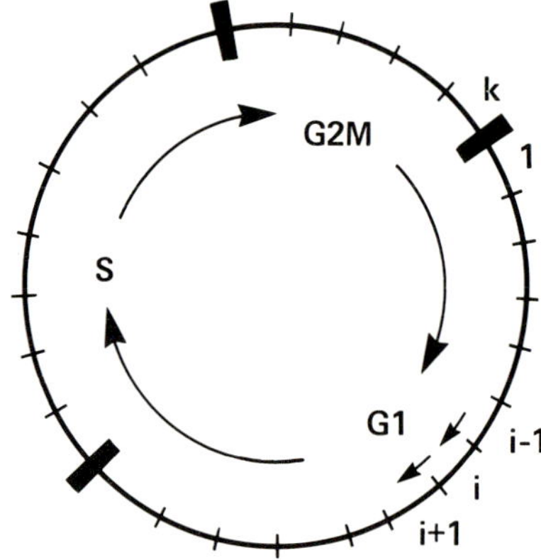

$$\frac{dN_i(t)}{dt} = \lambda_{i-1} N_i(t) - \lambda_i N_i(t)$$

where $N_i(t)$ is the number of cells in compartment i at time t, and λ_i determines the rate at which cells move from compartment i to i+1.

Fig. 7. Schematic illustration of the division of the cell cycle into a series of sequential compartments. Also illustrated are mathematical relations that define the manner in which cells move from compartment to compartment.

$$\frac{dN_1(t)}{dt} = 2\lambda_k N_k(t) - \lambda_1 N_1(t) \qquad (4b)$$

where k is the number of the last compartment in the cycle and the factor of 2 takes account of the fact that the cell number doubles as cells move from G2M to G1 phase.

Equations such as these can be defined for each compartment in the cell cycle and, given the number of cells in each compartment at $t = 0$, can be solved simultaneously to yield the number of cells in any compartment at time t. Usually, the number of cells in each compartment at $t = 0$ is chosen to simulate the cell cycle distribution for asynchronously growing cells.

Similar equations also can be defined to describe the radioactivity per cell in each compartment. That is

$$\frac{dRC_i(t)}{dt} = \lambda_{i-1} RC_{i-1}(t) - \lambda_i RC_i(t) \quad \text{for all } i \qquad (5)$$

where $RC_i(t)$ is the radioactivity associated with compartment i at time t. The factor of 2 is not needed for the first equation, since the radioactivity per compartment does not increase at cell division. Initially, the radioactivity per compartment is defined to be proportional to the number of cells in that compartment for all S-phase compartments and zero for the rest (thereby simulating a pulse label at $t = 0$). That is,

$RC_i(0) = 0$ for all i not in S phase

$RC_i(0) = CN_i(0)$ for all i in S phase with C being an arbitrary constant

RCS_i and RCG1 curves can then be simulated according to the equations

$$RCS_i(t) = \sum_{i \text{ in } S_i} RC_i(t) \Big/ \sum_{i \text{ in } S_i} N_i(t) \qquad (6a)$$

$$RCG1(t) = \sum_{i \text{ in } G_i} RC_i(t) \Big/ \sum_{i \text{ in } G_i} N_i(t) \qquad (6b)$$

These simulated curves are matched to the experimental curves by varying the λ_i values and the number of compartments in each phase. This permits estimation of the phase durations, since the λ_i values are related to the phase durations and coefficients of variation thereof by the equations

$$\overline{T}_{phase} = \frac{\#\text{ components in phase}}{\lambda_{phase}};$$

assuming all λ within phase are the same

$$CV_{phase} = [\#\text{ components in phase}]^{-1/2}$$

RC Analysis of the Rate of DNA Synthesis. RC analysis has also proved useful for analysis of the rate of DNA synthesis (actually the rate of incorporation of radioactive DNA precursor) across S phase [60]. In these studies, the population to be analyzed is pulse labeled (e.g., treated with [^{3}H]-TdR for 30 min) dispersed and stained with a DNA-specific dye. Cells are sorted from several windows across S phase, as indicated in Figure 8a. The RC value for each window is measured as an estimate of the rate of DNA synthesis by cells with that DNA content. This method is only accurate to the extent that the rate of precursor incorporation is a good measure of the rate of DNA synthesis. Figure 8b shows the RC values measured for Lewis Lung tumors pulse labeled in vivo. One advantage of the method is that it can be applied to populations in vivo or in vitro; either asynchronous or perturbed. Information gained in this way confirms earlier observations that the rate tends to be low in early and late S phase and maximal in mid-S phase. The accuracy of the method is diminished near the G1- and G2M-phase peaks where the overlap between the S- and G1- and G2M-phase cells is substantial. In these regions, G1- or G2M-phase cells may be sorted and scored as S-phase cells but may contain no radioactivity. Thus, the rate of precursor incorporation will appear abnormally low in these regions.

Halogenated Pyrimidine Incorporation

Bromodeoxyuridine (BrdUrd) Quenching of HO58 Fluorescence. BrdUrd is an analog of thymidine in which a methyl group in thymidine has been replaced by bromine to form BrdUrd. Several studies have shown that BrdUrd is incorporated into cellular DNA with about the same efficiency as thymidine [32] and that it is relatively nontoxic [37,57,71]. Indeed, BrdUrd has been approved for use in humans in some circumstances [106]. Others have demonstrated that BrdUrd incorporated into cellular DNA quenches the fluorescence from HO58 or HO34 bound to DNA [10,11, 58,61,62]. Furthermore, the extent of HO58 fluorescence quenching increases with increasing BrdUrd substitution. BrdUrd induced HO58 fluorescence quenching is the basis for one technique that allows estimation of cell cycle traverse rates in cells that can be heavily labeled with BrdUrd [10,11,58]. This approach is illustrated in Figure 9. BrdUrd is administered continuously at a concentration selected so that the substitution level will be sufficient to keep the HO58

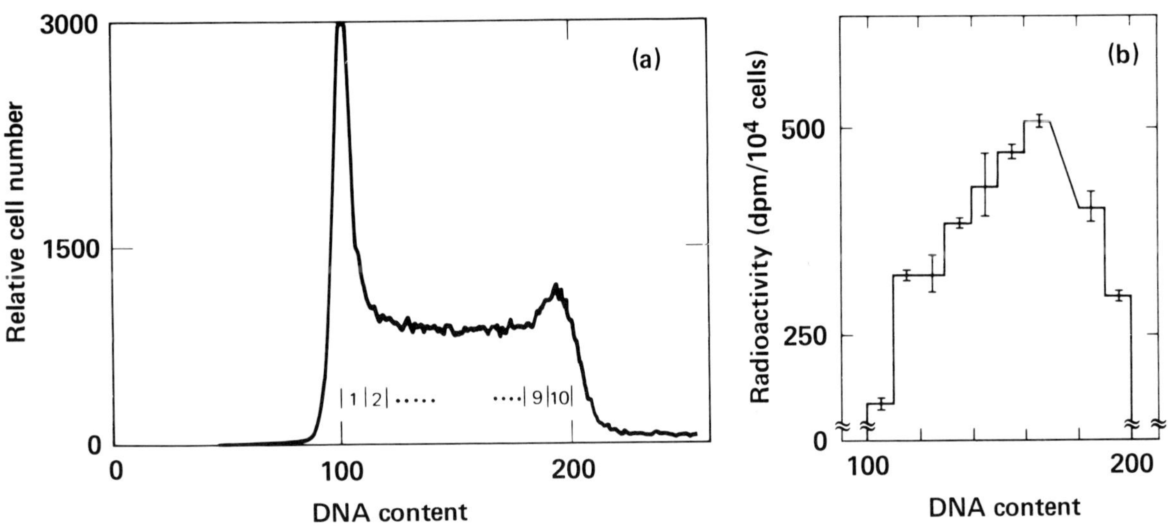

Fig. 8. Application of RC analysis to determination of the variation in the rate of DNA synthesis across S phase. **a:** A series of DNA content windows from which cells were sorted for RC analysis. **b:** RC values measured for the cells sorted from the various windows shown in a.

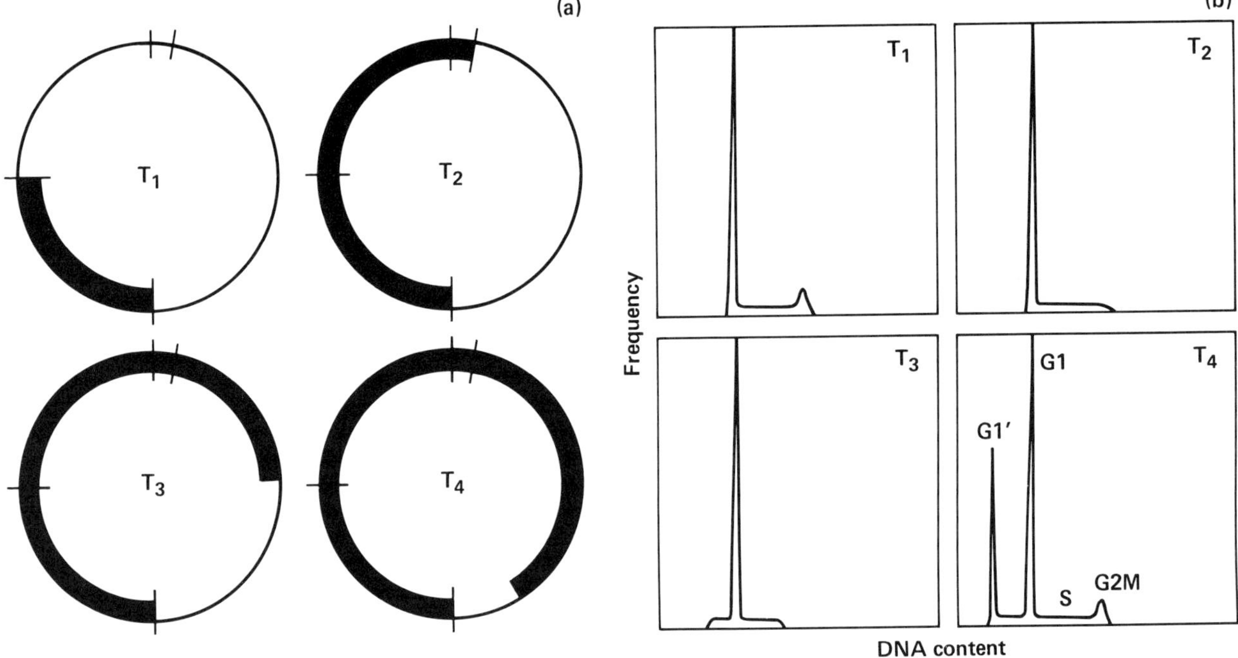

Fig. 9. Cell cycle labeling patterns **(a)** and HO58 fluorescence distributions **(b)** for hypothetical cell populations at four times ($T_1 > T_2 > T_3 > T_4$) after initiation of growth in medium containing BrdUrd. The BrdUrd labeling was assumed to be sufficient to quench fluorescence from all newly synthesized DNA. The filled regions in a indicate the cells that have synthesized DNA in the presence of BrdUrd. The G1′ portion of the distribution in b is produced by cells that have completed one round of DNA synthesis in the presence of BrdUrd and subsequently divided.

fluorescence intensity approximately constant as DNA is synthesized. Thus, for example, the fluorescence intensity of cells in mid-S-phase at the time of BrdUrd administration will remain unchanged until the cells divide. At this point, their fluorescence intensity will decrease twofold. Phase durations can be estimated from the rate at which the HO58 fluorescence distribution changes (e.g., the rate at which cells enter the G1′ portion of Fig. 9).

Phase durations can be estimated quantitatively using mathematical models similar to those described above for RC analysis. For example, two sets of compartments may be defined; those carrying labeled cells and those carrying unlabeled cells. The numbers of labeled and unlabeled cells in compartment i at times t are defined as $L_i(t)$ and $U_i(t)$, respectively. The transport cells among these compartments can then be described as

For the labeled cells:

$$\frac{dL_i(t)}{dt} = \lambda_{i-1}L_{i-1}(t) - \lambda_i L_i(t) \qquad (7a)$$

for all i except i=1 and for the first compartment in S−phase (designated j)

$$\frac{d\,L_i(t)}{dt} = 2\lambda_k L_k(t) - \lambda_1 L_i(t) \quad \text{for } i = 1 \qquad (7b)$$

and

$$\frac{dL_j(t)}{dt} = \lambda_{j-1}[L_{j-1}(t) + U_{j-1}(t)] - \lambda_j L_j(t) \qquad (7c)$$

for the first compartment in S-phase.

The last equation demands that all unlabeled cells become labeled as they enter the first compartment in S-phase.

$$\frac{dU_i(t)}{dt} = \lambda_{i-1}U_{i-1}(t) - \lambda_i U_i(t) \qquad (8a)$$

for all i in G1- and G2M-phase except for compartment 1 and for the S-phase compartments.

$$\frac{dU_i(t)}{dt} = 2\lambda_k U_k(t) - \lambda_1 U_1(t) \qquad \text{for } i = 1. \qquad (8b)$$

$$\frac{dU_i(t)}{dt} = -\lambda_i U_i(t) \qquad \text{for all i in S-phase.} \qquad (8a)$$

Initially,

$$\left.\begin{aligned} L_i(0) &= N_i(0) \\ U_i(0) &= 0 \end{aligned}\right\} \text{for all compartments in S-phase}$$

and

$$\left.\begin{aligned} L_i(0) &= 0 \\ U_i(0) &= N_i(0) \end{aligned}\right\} \text{for all other compartments}$$

The fraction of cells in G1′, designated F_{g1}', can then be calculated as

$$F_{g1}'(t) = \sum_{i \text{ in } G1} L_i(t) \Big/ \sum_{\text{all } i} [L_i(t) + U_i(t)] \qquad (9)$$

The λ_i values (related to the phase durations and dispersions as described above) are adjusted until $F_{g1}'(t)$ matches the experimental measurements as closely as possible for all t. Similar equations may be written to describe the fractions of cells in S′ and G2M′.

The principal disadvantage of this method is that rather high levels of BrdUrd must be administered over an extended period so that the cell cycle traverse characteristics of the labeled cell may be changed. In addition, sufficiently high levels of BrdUrd may not be possible to generate in vivo or may be toxic to hematopoietic cells.

Antibody Detection of BrdUrd Incorporation. Quantification of BrdUrd incorporation can be accomplished with greater sensitivity using antibodies against BrdUrd in cellular DNA [31,32,38,39,46,69,70,76,82,93,98,99].

Immunochemistry and Flow Cytometry. Figure 10 illustrates the procedures required to stain cells to permit flow cytometric analysis of cellular BrdUrd content and DNA content. In this procedure, cellular DNA is denatured to expose the BrdUrd for antibody binding. This is necessary since the antibodies now available recognize BrdUrd only in single-stranded DNA (ssDNA). The degree of anti-BrdUrd antibody binding is demonstrated by immunofluorescent labeling with, for example, fluorescein labeled goat anti-mouse antibody. Alternately, the anti-BrdUrd antibody can be labeled with a fluorescent molecule directly. The intensity of fluorescence resulting from these labeling schemes is taken as a measure of the amount of BrdUrd in the cellular DNA. In most studies, the cells are counterstained with a DNA specific dye that fluoresces at a wavelength that allows it to be distinguished from the immunofluorescence [31,46]. Typically, fluorescein is used in the immunochemical procedure and PI is used as the DNA stain. In this case, the cells are excited at 488 nm during flow cytometry; green fluorescence (510 to 530 nm) is recorded as a measure of relative BrdUrd incorporation and red fluorescence (>620 nm) is recorded as a measure of relative DNA content. The most critical aspect of this immunochemical procedure is the technique used for

Partially denature cellular DNA

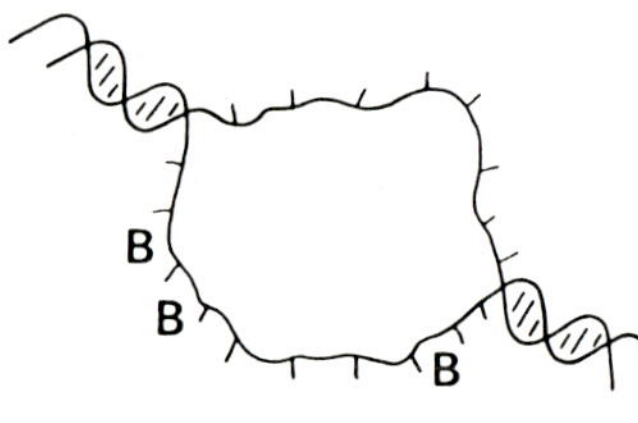

Immunochemically detect the incorporated BrdUrd and suppress nonspecific staining and/or autofluorescence

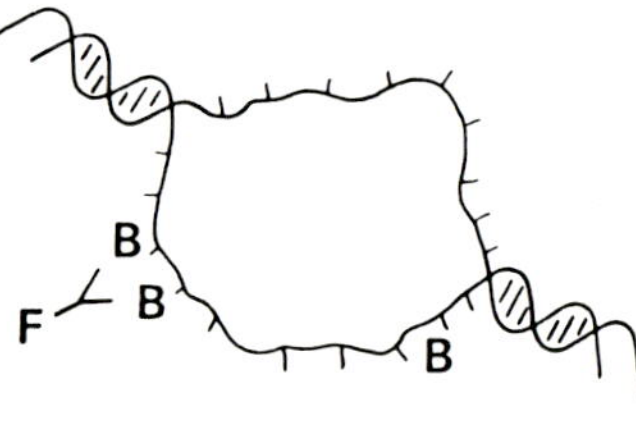

Fluorescently stain the remaining dsDNA

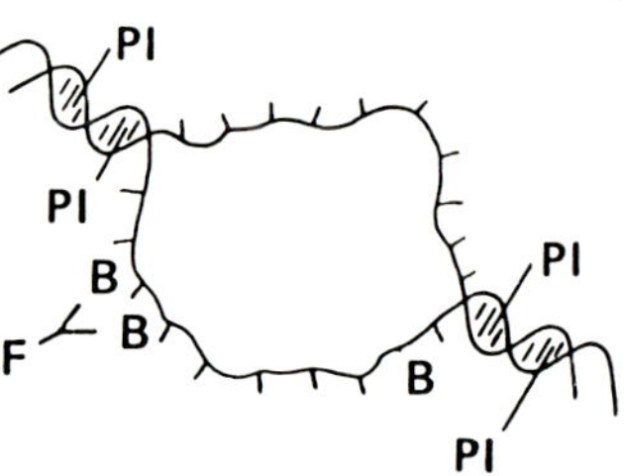

Fig. 10. Cytochemistry for BrdUrd/DNA analysis. In this procedure, the DNA is partially denatured to expose BrdUrd (labeled B) in single-stranded DNA (ssDNA). The exposed BrdUrd is immunochemically stained with fluorescein (F ─<; fluorescing in the green) using a monoclonal antibody that binds preferentially to BrdUrd in ssDNA. The remaining dsDNA is stained using propidium iodide (PI; fluorescing in the red).

TABLE 3. DNA Denaturation Procedures for BrdUrd Analysis

Technique	Comments	References
1. HCl denaturation	Extent of DNA denaturation is moderate. Little preferential cell loss occurs. Cell-surface antigens may be lost.	[31]
2. Histone extraction followed by denaturation at 80°C in 50% formamide	Substantial DNA denaturation. Loss of hematopoietic cells may be severe.	[32]
3. Histone extraction followed by denaturation at 100°C in water followed by $NaBH_4$ treatment	Excellent denaturation. Moderate hematopoietic cell loss. Highest discrimination between label and unlabeled cells.	[9,69]
4. Histone extraction followed by exposure of ssDNA using endonuclease and exonuclease	Good exposure of ssDNA. Minimal hematopoietic cell loss.	[32a]

DNA denaturation. If too harsh, substantial cell loss may occur or too little dsDNA may remain to permit PI binding. If too mild, insufficient BrdUrd will be exposed for immunochemical analysis or the BrdUrd may be exposed nonrandomly so that the amount of subsequent antibody binding may not be strictly proportional to the amount of BrdUrd. Table 3 compares several DNA denaturation schemes that have been evaluated in recent years.

Phase Fraction Analysis. An important application of BrdUrd incorporation analysis is estimation of the fraction of labeled cells (if BrdUrd alone is measured) or estimation of the fractions of cells in the G1, S, and G2M phases (if BrdUrd content and DNA content are measured simultaneously). In this approach, BrdUrd is administered during a short interval, cells are harvested and stained and processed as described above. Figures 11a,b shows typical results. Figure 11a shows a univariate BrdUrd distribution and Figure 11b shows a bivariate BrdUrd/DNA distribution for human leukemia cells labeled in vitro with 10 μM BrdUrd for 30 min. Also shown are regions defining the G1, S-, and G2M-phase cells. The fractions of cells in these regions can be determined simply by summing the number of cells in each region and dividing by the total. The discrimination between the labeled and unlabeled cells and normal and leukemic cells in Figure 11b is sufficiently large that the regions defining the leukemic G1-, S-, and G2M-phase populations can be drawn with little ambiguity.

Two important advantages of the BrdUrd/DNA analysis procedure are (1) that distinction can be made between cells with S-phase DNA content that are actively synthesizing DNA and those that are not [30,76,104], and (2) very small amounts of BrdUrd incorporation are required because of the high sensitivity of the immunochemical detection procedure. For example, the amount of BrdUrd incorporated by CHO cells during a 30-sec exposure to medium containing 10-μM BrdUrd can be readily detected [9].

Phase Duration Estimates. The BrdUrd/DNA procedure is well suited to study of cell cycle traverse. The analysis procedure is similar to that used for RC analysis [30,47,76]. That is, BrdUrd is given as a short pulse to label cells in S-phase. The cell population is then sampled periodically for BrdUrd/DNA analysis. Figure 12 shows serial BrdUrd/DNA distributions measured for CHO cells labeled for 10 minutes with BrdUrd. The progression of labeled cells from S-phase through G2M-phase and into G1-phase can be readily followed. Estimates for the G1-, S-, and G2M-phase durations and coefficients of variation thereof can be calculated by using the same mathematical techniques that have been developed for RC analysis as described above. In this approach, curves showing the BrdUrd content per cell in mid-S phase (BCS_i) and in G1 phase (BCG1) are simulated and matched to the experimental data by adjusting cytokinetic parameters such as phase durations and dispersions. The BCS_i and BCG1 values for KHT tumor cells labeled in vivo are shown in Figure 13 along with the computer analysis of these data. Note that BCS_i and BCG1 are exact analogs of the RCS_i and RCG1 values described above and have been analyzed using the same model.

Growth Fraction. Information about the fraction of actively proliferating cells in a population can be obtained by analysis of BrdUrd/DNA distributions measured periodically during continuous labeling with BrdUrd [7]. Combination of the rate at which cells become labeled during continuous labeling with information about the rate at which pulse-labeled cells traverse the G1, S, and G2M phases permits estimation of the growth fraction for the population and the G1-, S-, and G2M-phase durations for the actively proliferating cells.

PERTURBED POPULATION ANALYSIS

DNA Distribution Analysis

Kinetic information about perturbed populations can be derived from DNA distributions measured periodically after the perturbing event. Figure 14, for example, shows cell cycle and DNA distributions for a hypothetical cell population initially synchronized in early G1 phase. The rate of movement of cells from G1 through the S and G2M phases is clearly reflected in the DNA distributions which show the expected increase in DNA content as the cells move through these phases. The actual G1-, S-, and G2M-phase durations (and variability therein) are determined from the rates at which cells move through these phases.

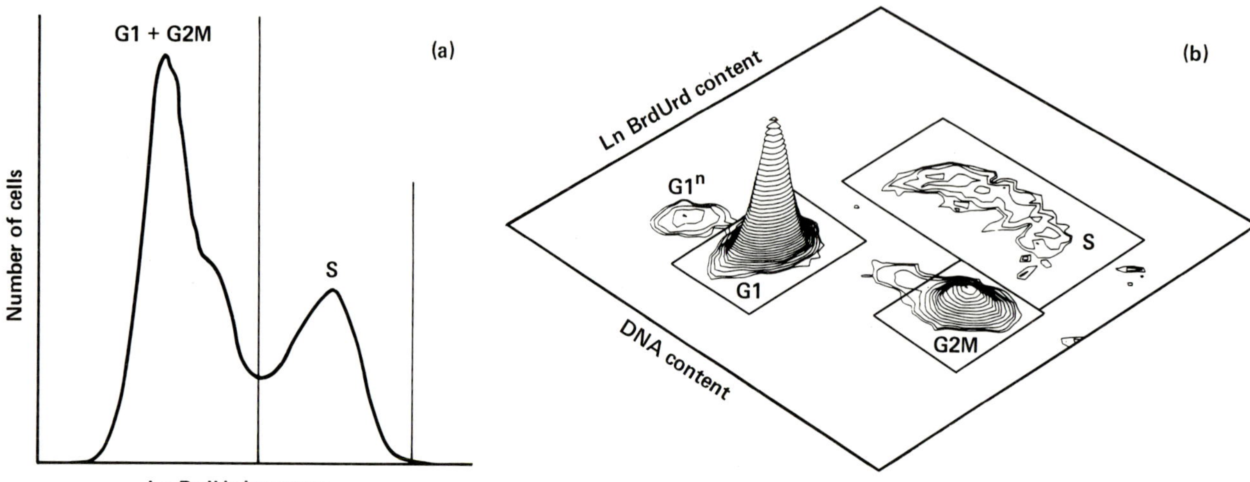

Fig. 11. Fluorescence distributions measured for cells from the bone marrow of a patient with leukemia following immunochemical staining for incorporated BrdUrd. The bone marrow cells were grown for 1 hour in vitro in medium containing BrdUrd. **a:** A univariate fluorescence distribution of BrdUrd-linked immunofluorescence intensity. The vertical line divided the labeled and unlabeled cells in the distribution. **b:** Bivariate BrdUrd/DNA distribution for the same cells. The G1-, S-, and G2M-phase leukemic cells are clearly and distinctly resolved in the distribution. Also visible are unlabeled normal bone marrow cells (labeled $G1^n$).

Two approaches to quantitative analysis of DNA distributions have been widely used: (1) mathematical approximation of the fractions of cells in one or a few DNA content ranges (e.g., the fractions of cells in G1 phase and/or in S phase [34,48,108], or (2) mathematical approximation of the DNA distributions themselves [40,76]. The underlying model of the cell cycle may be the same for both approaches. For complex populations in which cells are found in several states, the models may be mathematically formidable, since a large number of parameters must be included to describe the movement of cells among the various possible states and phases. Worse still, many of the parameters may be time dependent to account for the fact that the cytokinetic characteristics of a population may change as the population (and perhaps the host) responds to the perturbing influence. Equation (10), for example, describes the movement of proliferating, clonogenic (PC) cells into and out of compartment i as suggested in Figure 1.

$$\frac{dPC_i(t)}{dt} = \lambda_{i-1}(t)\,PC_{i-1}(t) + \xi_i(t)\,QC_i(t) - [\eta_i(t) + \beta_i) + \alpha_i(t)]\,PC_i(t) \qquad (10)$$

where $PC_i(t)$ and $QC_i(t)$ are the numbers of cells in the ith compartment in PC and QC, respectively, and where $\lambda_i(t)$ determines the rate of movement of PC cells through the ith compartment. $\xi_i(t)$ determines the rate at which cells in the ith compartment move from QC to PC at time t, $\eta_i(t)$ determines the rate at which cells in the ith compartment move from PC to QC at time t, $\alpha_i(t)$ determines the rate at which cells in the ith compartment move from PC to PD and $\beta_i(t)$ determines the rate at which cells in the ith compartment move from PC to PD. μ_i determines the rate at which cells in the ith compartment of PC are lost as a result of cell lysis or migration. These equations, together with an estimate of the distribution of cells among the states and phases of the cycle at the beginning of the experiment (all in early G1 phase in the hypothetical example), can be solved to predict the distribution at any subsequent time(s). The calculated cell cycle distributions are then converted to simulated DNA distributions or G1-, S-, and G2M-phase fractions for comparison with the experimental data. The conversion process required for DNA distribution simulation, illustrated schematically in Figure 15, requires information about the rate of DNA synthesis across S-phase. Such information can be estimated as described earlier in this chapter or from a BrdUrd/DNA distribution.

The analysis of DNA distribution sequences or G1-, S-, and G2M-phase fractions then proceeds in the usual fashion. The cell cycle traverse parameters, e.g., $\lambda_i(t)$, $\xi_i(t)$, $\eta_i(t)$, $\alpha_i(t)$, $\mu_i(t)$, and $\beta_i(t)$, are adjusted until the simulated DNA distributions or phase fractions match the experimental data as closely as possible. For example, Figure 16 shows the match of a model to experimental DNA distributions measured for Chinese hamster ovary cells synchronized to be in late G1 phase at the beginning of the experiment. In this relatively simple biological situation, the G1-, S-, and G2M-phase durations that resulted in the match were 4.5, 4.8, and 2.8 hours, respectively. Furthermore, it was not necessary to include cell loss and recruitment phenomena in the analysis. As a result, a complete cell cycle traverse analysis was possible using only the data contained in the DNA distribution sequence.

RC and BrdUrd/DNA Analysis

DNA Synthesis Following Perturbation. Most perturbed populations are sufficiently complex kinetically that the information contained in a DNA distribution sequence is insufficient to permit complete, unambiguous determination of all the important kinetic parameters. In these situations, additional information may increase the accuracy of the analysis. One important piece of information in this regard is the fraction of cells actually engaged in DNA synthesis and the rate of DNA synthesis. The fraction of DNA synthesizing cells may be estimated by measuring number of cells that

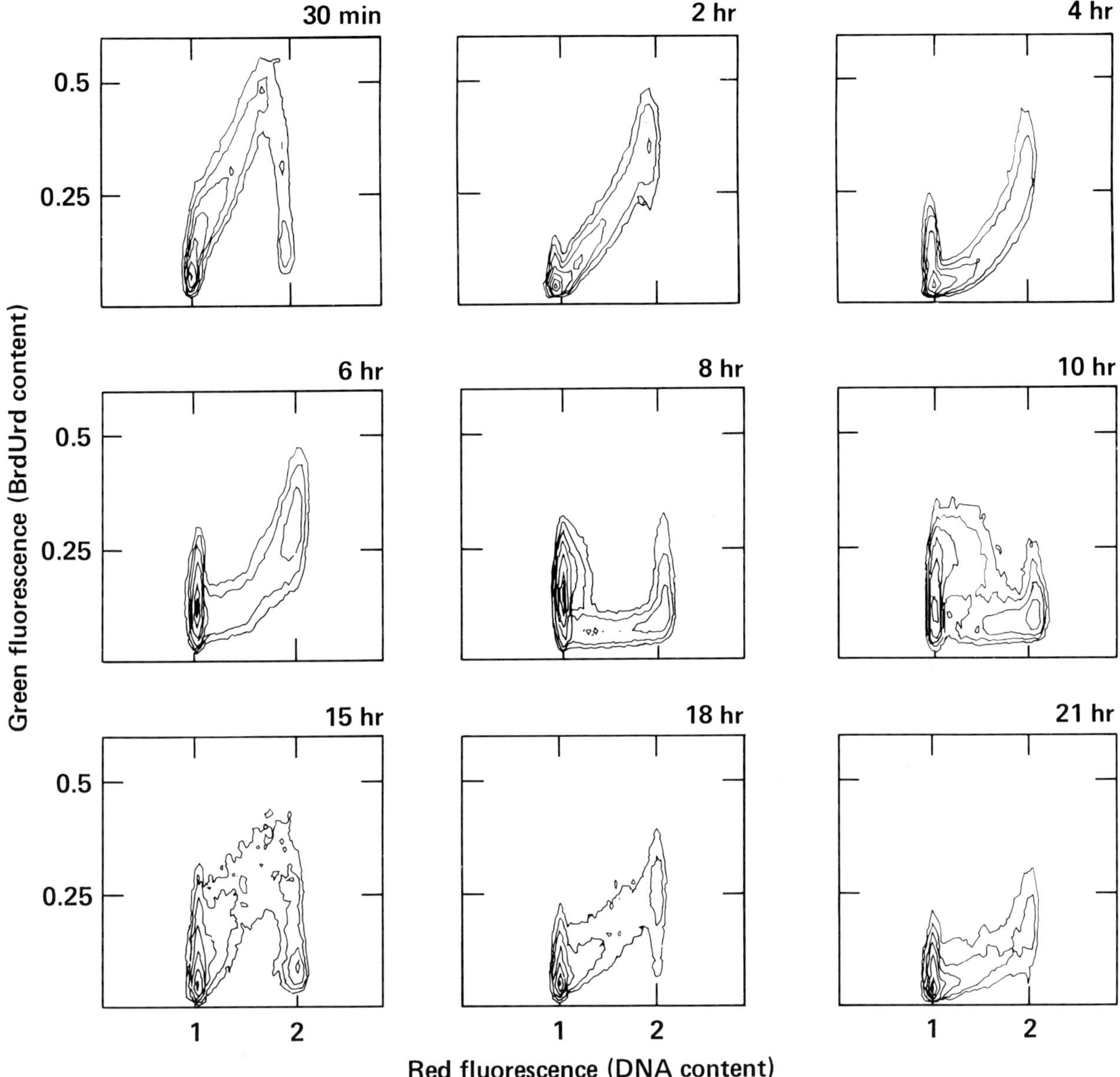

Fig. 12. Serial BrdUrd/DNA distributions measured for Chinese hamster ovary (CHO) cells labeled in vitro with BrdUrd for 10 min. Cells were taken for BrdUrd/DNA analysis 30 min, 2 hours, 4 hours, 6 hours, 8 hours, 10 hours, 15 hours, 18 hours, and 21 hours after BrdUrd labeling. Distributions for samples taken at each of these times.

incorporate DNA precursors like [^{3}H]-TdR or BrdUrd administered at selected times after perturbation. Figure 17 shows BrdUrd/DNA distributions measured for CHO cells following treatment with cytosine arabinoside (ara-c). In this study, BrdUrd was administered to replicate CHO cell cultures 30 min before harvest at several times after ara-c treatment. The suppression of DNA synthesis 2 hours after ara-c is clear as is the increase in the rate of BrdUrd incorporation beginning 6 hours after ara-c treatment.

The fraction of cells synthesizing DNA may be included in the quantitative analysis process by relating this experimental variable to the mathematical description of the distribution of cells among the various phases and states of the cell cycle. In the model used in this chapter, the fraction of cells synthesizing DNA may be calculated as follows:

$$F_s(t) = \frac{\sum_{i \text{ in } S}[PC_i(t) + PD_i(t)]}{\sum_{i \text{ in } S}[PC_i(t) + PD_i(t) + QC_i(t) + QD_i(t)]} \quad (11)$$

where $PC_i(t)$ and $QC_i(t)$ are as defined above and $PD_i(t)$ and $QD_i(t)$ are the numbers of cells in compartment i in states PD and QD at time t. This information makes it possible to adjust the mathematical model to simultaneously match DNA distributions and the fractions of cells synthesizing DNA measured at several times after perturbation.

Cytokinetic Properties of S-Phase Cells. The BrdUrd/DNA and RC analysis techniques can also be used to generate information about the cell cycle traverse or cell loss that

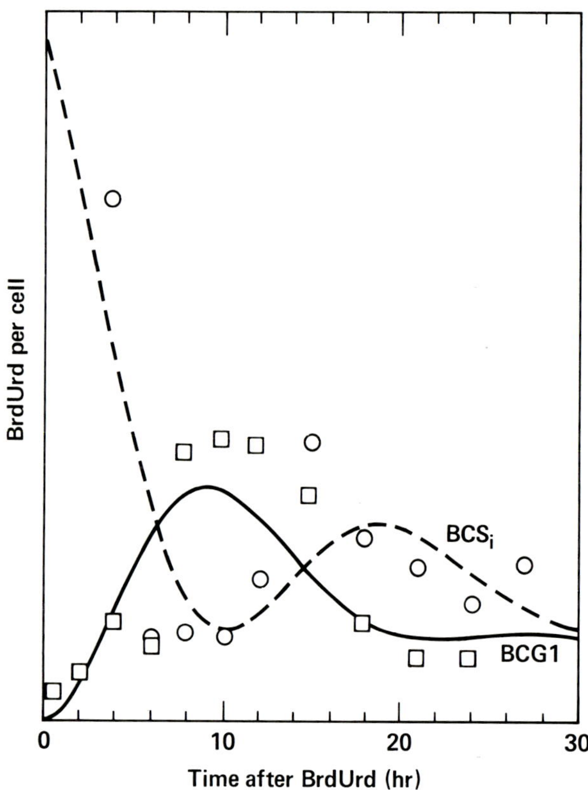

Fig. 13. BrdUrd content per cell curves for KHT tumor cells determined from BrdUrd/DNA distributions measured following pulse labeling with BrdUrd in vivo. The points on the BCG1 and BCS curves represent the mean BrdUrd contents for labeled cells with G1- and mid-S-phase DNA contents at several times after BrdUrd labeling. The solid lines show a computer analysis of the data generated assuming that the G1-, S-, and G2M-phase durations were 6.6 hours, 10.4 hours, and 2.9 hours, respectively.

occurs in S-phase populations exposed to toxic levels of S-phase cytotoxic agents such as ara-c and hydroxyurea. The S-phase cells may be marked with BrdUrd or [^{3}H]-TdR before treatment with the cytotoxic agent and followed after treatment. For example, Figure 18 shows BrdUrd/DNA distributions measured for CHO cells at several times after ara-c treatment. These cells were labeled with BrdUrd 30 min before treatment with ara-c. Two populations with S-phase DNA content are visible beginning about 6 hours after ara-c treatment; BrdUrd-labeled cells that were in S phase at the time of ara-c treatment and unlabeled cells that were in G1 or G2M phase at the time of ara-c treatment. The BrdUrd-labeled cells progressed slowly out of S phase and eventually were lost from the population. The unlabeled cells were unaffected by the toxic ara-c treatment and progressed through S phase at the normal rate [83].

MULTIVARIATE CYTOKINETIC ANALYSIS

Multivariate Analysis Concept

Tissues in vivo comprise multiple subpopulations, many proliferating with different cytokinetic characteristics. Substantial effort has been devoted to development of techniques to allow discrimination of selected subpopulations [55, 60,65,66,97,101–103]. Subpopulations within heterogeneous tissues can be studied by purifying the subpopulation of interest prior to the cytokinetic analyses (e.g., using velocity sedimentation, density gradients, sorting). Alternatively, interesting subpopulations may be distinctly labeled so that flow cytokinetic analyses can be performed directly on the labeled subpopulation(s). The latter technology is rapidly evolving as unique subpopulation markers/probes (i.e., monoclonal antibodies, lectins, enzymes, physiologic, and intracellular probes) are identified. Fluorophores with different spectral characteristics are used to label each probe so that several properties for each cell can be measured simultaneously during flow cytometry. This approach is particularly useful for analyses of low frequency subpopulations which are not easily purified. Simultaneous discrimination of flow cytometrically distinct subpopulations and analysis of their cytokinetic properties is readily achieved by multivariate flow cytometry.

List Mode Data Processing and Acquisition. Multivariate flow cytometric measurements and data analyses are facilitated by list mode data processing capabilities. In list mode data acquisition, cells stained with multiple probes are analyzed flow cytometrically, and the fluorescent characteristics of each cell are recorded in a list. For example, if four variables (e.g., v1,v2, v3, and v4) are measured for each cell, the list for three cells (e.g., i, j, and k) would read

$$v1_i,\ v2_i,\ v3_i,\ v4_i;\ v1_j,\ v2_j,\ v3_j,\ v4_j;\ v1_k,\ v2_k,\ v3_k,\ v4_k$$

In a cytokinetic experiment, v1 and v2 might define a particular subpopulation, v3 might be BrdUrd content and v4 might be DNA content. The recorded list can then be processed to display data as univariate or bivariate distributions (e.g., v1 versus v2 or v3 versus v4). In addition, specific information (e.g., mean fluorescence intensity or frequency) on selected flow cytometrically distinct subpopulations may be calculated. Multivariate gating techniques can be used to define clusters of cells and to determine their characteristics. For example, in the four-variable analysis described above, subpopulations may be inscribed in a bivariate distribution of v1 vs. v2 and the bivariate BrdUrd/DNA (v3 versus v4) distribution may be generated for cells falling in the inscribed region. Thus, BrdUrd/DNA distribution analyses may be conducted for each cytometrically distinct population in the v1 vs. v2 distribution. This approach to data acquisition permits repeated processing of multivariate data for better subpopulation definition.

Cytokinetic Analysis of Murine Hematopoietic Subpopulations. Multivariate cytokinetic analysis is illustrated here by an example showing the analysis of two cytometrically distinct hematopoietic subpopulations in murine bone marrow. In these studies, bone marrow cells were obtained from mice after a brief treatment with BrdUrd [77]. These cells were stained for DNA content (HO42), BrdUrd incorporated into DNA (anti-BrdUrd monoclonal antibody labeled indirectly with fluorescein) and cell surface N-acetylneuraminic acid residues [wheat germ agglutinin (WGA) conjugated with rhodamine]. The fluorescent characteristics of each cell were recorded in list mode fashion as it passed through two laser beams of the flow cytometer. One laser was adjusted to emit in the UV range to excite HO42 and the resulting blue fluorescence intensity was recorded as a measure of cellular DNA content. The other was adjusted to emit at 488 nm to excite fluorescein and rhodamine; green fluorescence was recorded as a measure of BrdUrd content and red fluores-

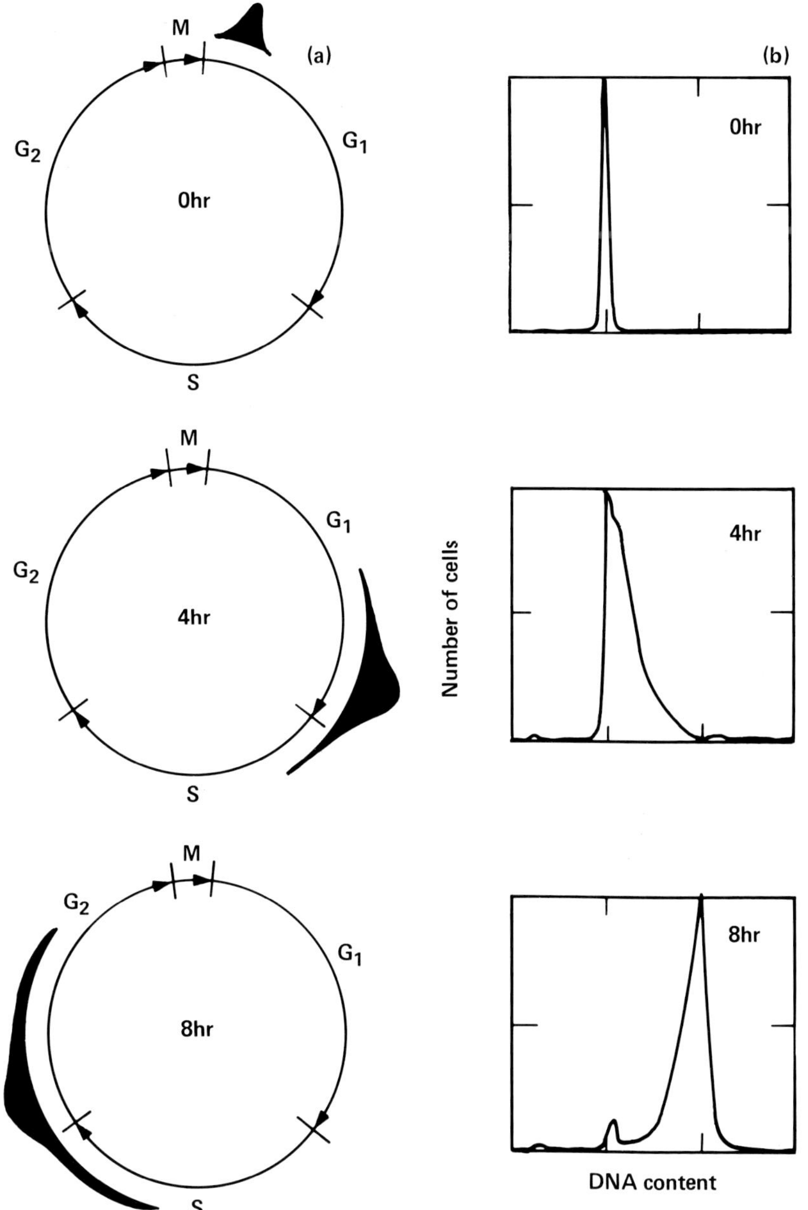

Fig. 14. Hypothetical cell cycle and DNA distributions for a population initially synchronized in early G1 phase. **a,b:** Cell cycle and DNA distributions predicted before and 4 and 8 hours after release from synchrony.

cence was recorded as a measure of WGA binding. Figure 19 shows several representations of the data. Univariate distributions of DNA content and WGA binding are shown in Figures 19a and 19b, respectively. Bivariate distributions of DNA content versus incorporated BrdUrd and DNA content versus WGA binding are shown in Figure 19c and 19d, respectively. The DNA profiles and bivariate BrdUrd/DNA content distribution are typical of whole bone marrow containing approximately 10 to 15% cells in S phase. Two subpopulations with different WGA binding characteristics are visible in Figure 19b,d. The bivariate distribution of DNA content versus WGA binding also shows two inscribed regions; one around the low WGA binding cells and the other around the high WGA binding cells. Figures 19e and 19f show bivariate BrdUrd/DNA distributions for the cells in the low and high WGA binding cells inscribed in Figure 19d. These distributions show that S-phase cells in both the low and high WGA binding populations can incorporate BrdUrd. Since cell kinetic information can be extracted from the BrdUrd/DNA content distributions, this technique permits estimation of the cell kinetic properties of the two flow cytometrically distinct subpopulations without purification of the subpopulations. This approach has been used to efficiently determine the cytokinetic properties of murine bone

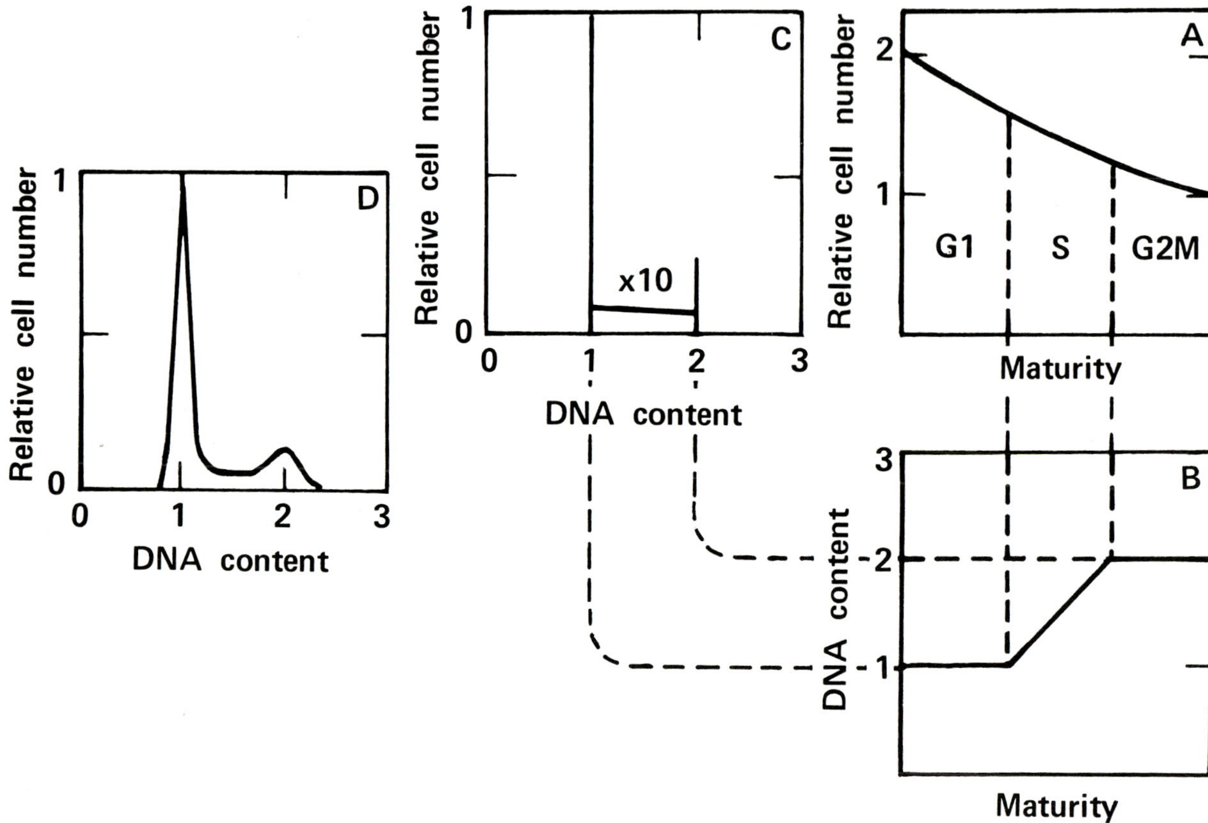

Fig. 15. Schematic diagram showing the process by which DNA distributions can be calculated from information about the distribution of cells around the cell cycle and about the rate of DNA synthesis.

marrow cells before and after treatment with chemotherapeutic agents such as ara-c.

Other Subpopulation Probes. The usefulness of multivariate cytokinetic analyses is dependent on several factors, perhaps the most important being the capability to define a homogeneous subpopulation. Although WGA clearly defines two subpopulations in mouse bone marrow, neither of these subpopulations is homogeneous. Erythroid, myeloid, and lymphoid cells are contained in the low WGA subpopulation, whereas the high WGA binding population contains predominantly, although not exclusively, myeloid cells. Thus kinetic analysis of the two subpopulations is still complicated by the presence of cells of different lineages, stages of differentiation and cytokinetic characteristics, and the kinetic estimates thus obtained reflect those of the predominant cell type. Probes to label unique subpopulations are continually and rapidly emerging. Monoclonal antibodies directed against cell surface antigens are the most widely used probes, since they are easily produced and labeled with a variety of fluorophores. In addition, these probes are generally compatible with cell viability; a wide selection of targets are potentially available [55,66,97,105]. Physiologic probes for cell function such as membrane potential [15,17], intracellular pH [72,73,96], membrane permeability [60,65,101] and RNA [20–33] have also been used with varying degrees of success to discriminate subpopulations (see also Chapter 32, this volume). Since cell differentiation is a gradual process in all heterogeneous tissues it is likely that combinations of probes will be required to define homogeneous subpopulations adequately. In addition, techniques are now developing for quantitative analysis of cellular levels of specific nucleic acids [79,94]. Flow cytometric techniques are already available for quantitative analysis of the amounts of specific DNA sequences in cell nuclei. These techniques may be extended to permit analysis of specific mRNA levels in the future. This capability would provide a powerful tool for subpopulation discrimination and for analysis of the events associated with or controlling cell cycle traverse.

ACKNOWLEDGMENTS

This chapter was prepared under the auspices of the U.S. Department of Energy, contract W-7409-ENG-48, with support from grants CA 14533 and CA 28752 from the National Institutes of Health. We thank Ms. Lil Mitchell and Mr. Rick Wooten for expert help in the preparation of this manuscript.

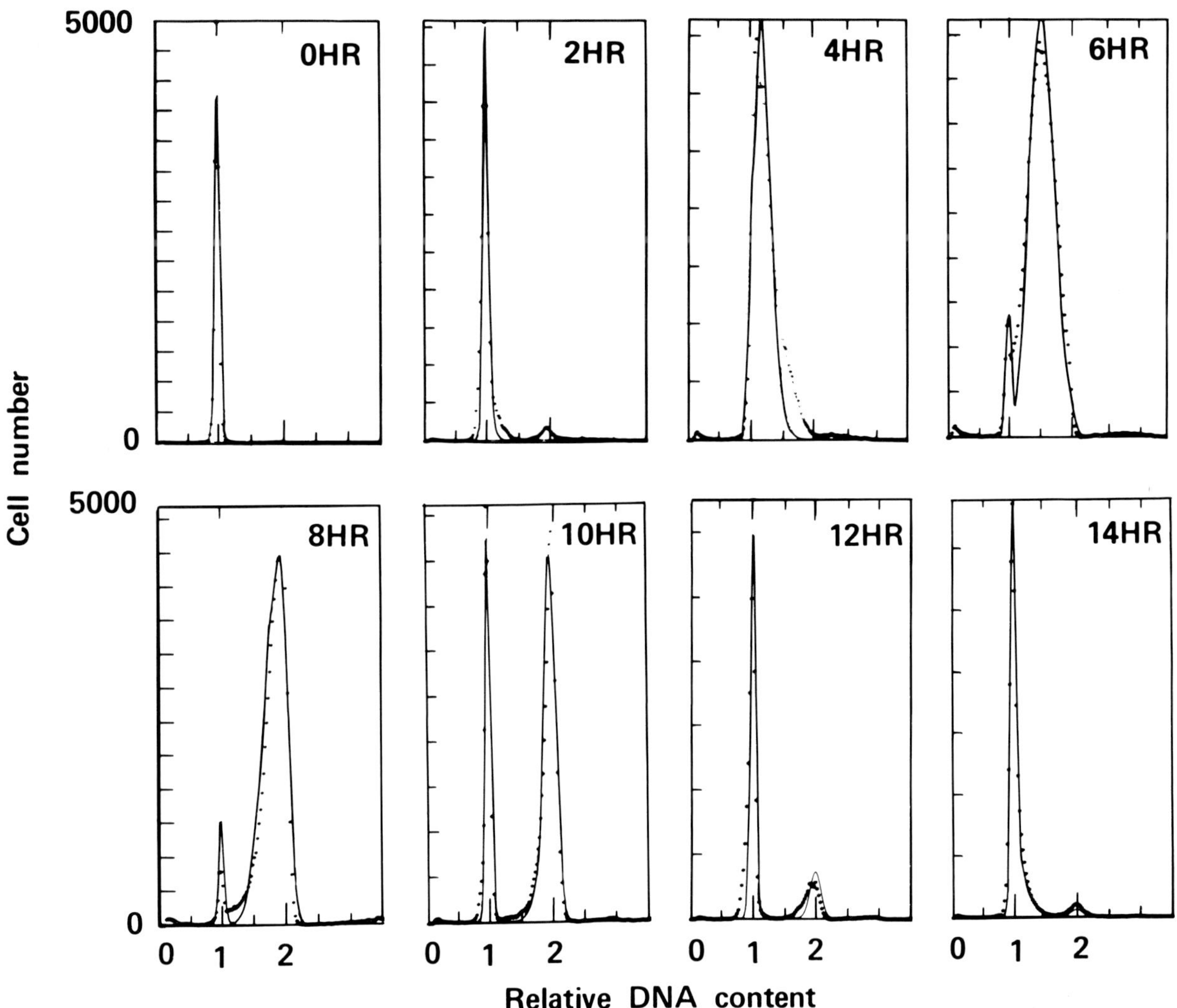

Fig. 16. Sequential DNA distributions measured for Chinese hamster ovary (CHO) cells at several times after release from synchrony in early G1 phase. The points show the experimental measurements and the solid lines show the DNA distributions simulated assuming initial synchrony in early G1 phase and G1-, S-, and G2M-phase durations of 4.5, 4.8, and 2.8 hours, respectively.

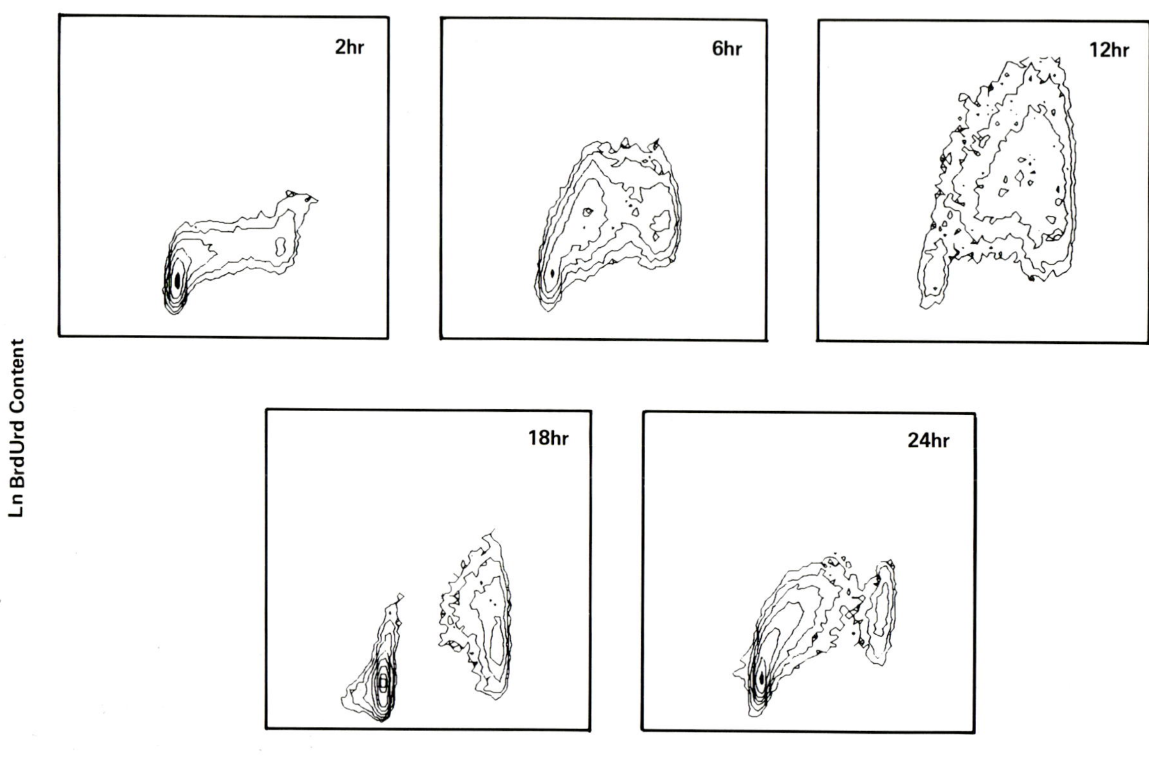

Fig. 17. Sequential BrdUrd/DNA distributions measured for Chinese hamster ovary (CHO) cells at 2 hours, 6 hours, 12 hours, 18 hours, and 24 hours after treatment with ara-c. BrdUrd was administered to each culture 30 min before harvest.

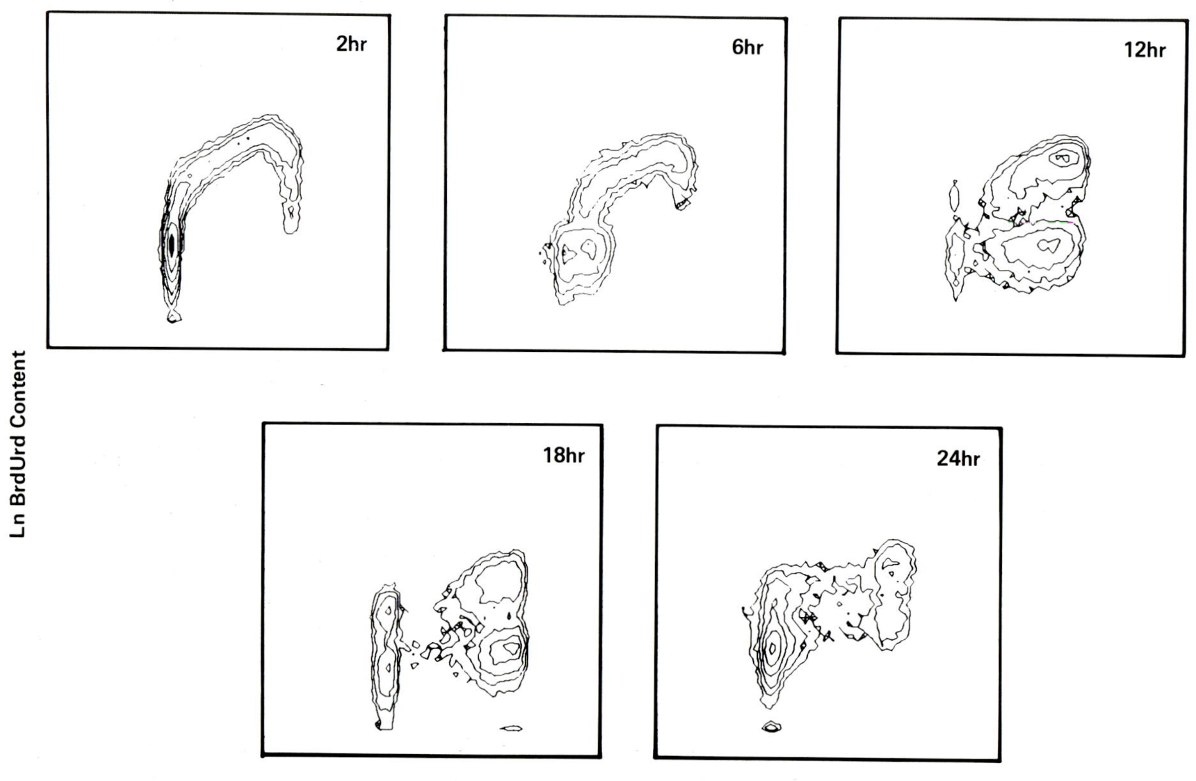

Fig. 18. Sequential BrdUrd/DNA distributions measured for Chinese hamster ovary (CHO) cells responding to treatment with ara-c. These distributions were measured for CHO cells treated with BrdUrd 30 min before treatment with ara-c and then sampled for BrdUrd/DNA analysis at 2 hours, 6 hours, 12 hours, 18 hours, and 24 hours after treatment with ara-c.

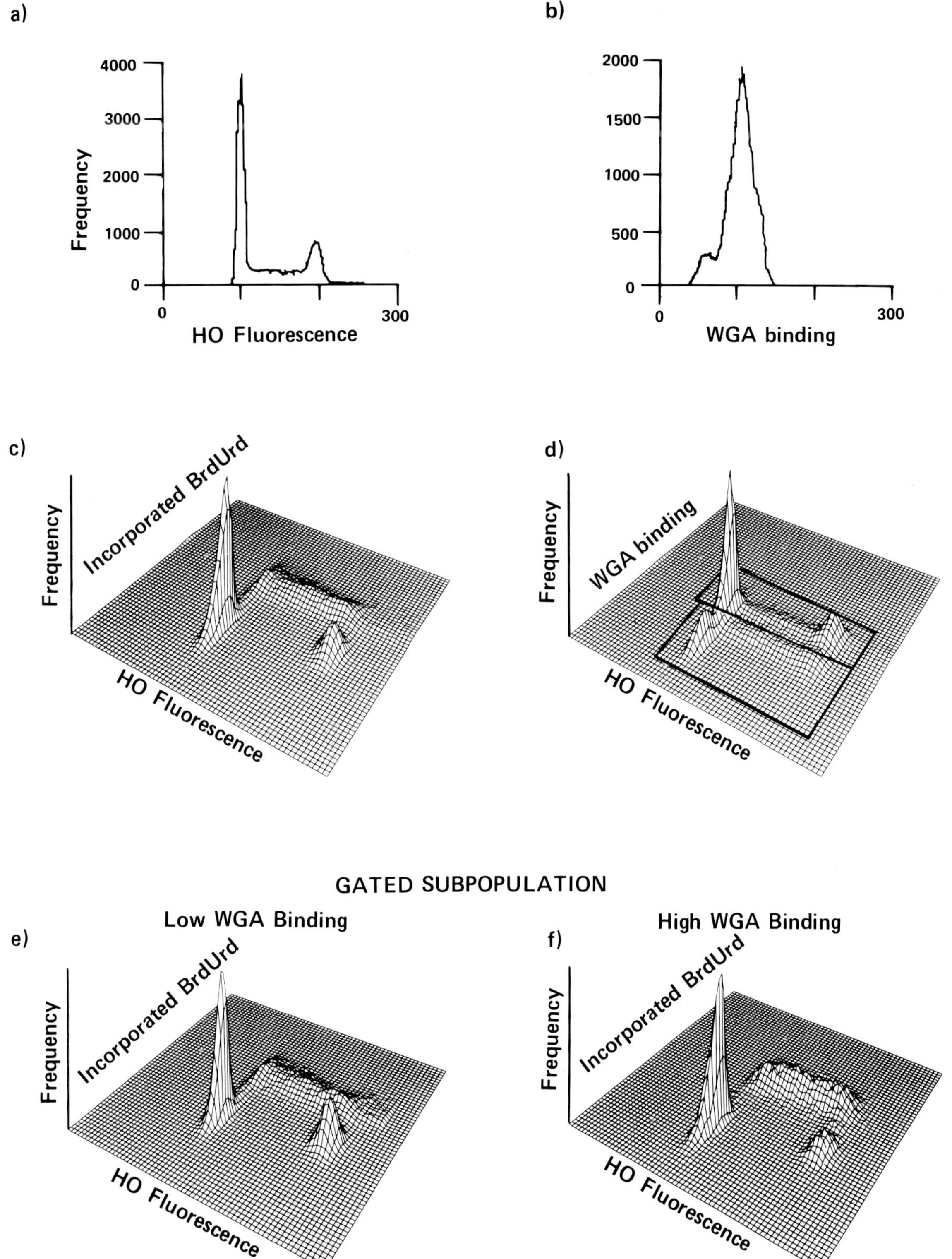

Fig. 19. Multivariate analyses of murine bone marrow. **a:** Univariate DNA distribution measured for total bone marrow. **b:** WGA binding distribution for total bone marrow. **c:** BrdUrd/DNA distribution for total bone marrow. **d:** BrdUrd/WGA distribution. The inscribed regions define populations of low and high WGA. **e:** BrdUrd/DNA distribution for low WGA binding bone marrow cells. **f:** BrdUrd/DNA distribution measured for high WGA binding bone marrow cells.

REFERENCES

1. **Aglietta M, Sonneveld P (1978)** The relevance of cell kinetics of optimal scheduling of 1-β-D-arabinofuranosyl cytosine and methotrexate in a slow growing acute myeloid leukemia (BNML). Cancer Chemother. Pharmacol. 1:219–223.
2. **Alabaster O, Bunnag B (1976)** Flow microfluorimetric analysis of sensitive and resistant leukemia L1210 following 1-β-D-arabinofuranosylcytosine *in vivo*. Cancer Res. 36:2744–2749.
3. **Appleton D, Sunter J, Watson A (1980)** "Cell Proliferation in the Gastrointestinal Tract." Kent, England: Pitman Medical Limited.
4. **Bagwell C (1979)** Theory and application of DNA histogram analysis. PhD thesis, University of Miami.
5. **Baisch H, Beck H, Christensen I, Hartmann N, Fried J, Dean P, Gray J, Jett J, Johnston D, White R, Nicolini C, Zietz S, and Watson J (1982)** A comparison of mathematical models for the analysis of DNA histograms obtained by flow cytometry. Cell Tissue Kinet. 15:235–249.
6. **Baisch H, Gohde W, Linden W (1975)** Analysis of PCP-data to determine the fraction of cells in the various phases of the cell cycle. Radiat. Environ. Biophys. 12:31–39.
7. **Barcellos-Hoff M, Gray J, Marton L, Deen D** Cell cycle parameters of 9L rat brain tumors in monolayer and multicellular spheroids using a bromodeoxyuridine labeling technique. (manuscript in preparation).
8. **Barlogie B, Spitzer G, Hart J, Johnston D, Buchner T, Schumann J, Drewinko B (1976)** DNA-histogram analysis of human hemopoietic cells. Blood 48:245–258.
9. **Beisker W, Dolbeare F, Gray JW (1987)** An improved procedure for BrdUrd/DNA analysis. Cytometry 8: 235–239.
10. **Böhmer R (1979b)** Flow cytometric cell-cycle analysis using the quenching of 33258 Hoechst fluorescence by bromodeoxyuridine incorporation. Cell Tissue Kinet. 12:101–112.
11. **Böhmer R, Ellwart J (1981)** Combination of BUdR-quenched Hoechst fluorescence with DNA-specific ethidium bromide fluorescence for cell cycle analysis with a two-parameter flow cytometer. Cell Tissue Kinet. 14:653–658.
12. **Braunschweiger P (1986)** Tumor growth fractions: Estimation, perturbation and prognostication. In Gray JW, Darzynkiewicz Z (eds), "Techniques in Cell Cycle Analysis." Clifton, New Jersey: Humana Press, pp 47–71.
13. **Clarkson B, Dowling M, Gee T, Cunningham I, Burchenal J (1975)** Treatment of acute leukemia in adults. Cancer 36:775–795.
14. **Coffino P, Gray J (1978)** Regulation of S49 lymphoma cell growth by cyclic adenosine 3′:5′-monophosphate. Cancer Res. 38:4285–4288.
15. **Cohen R, Muirhead K, Gill J, Waggoner A, Noron P (1981)** A cyanine dye distinguishes between cycling and monocycling fibroblasts. Nature (Lond.) 290: 593–595.
16. **Collins J (1978)** Rates of DNA synthesis during the S-phase of HeLa cells. J. Biol. Chem. 253: 8570–8577.
17. **Collins J, Foster K (1983)** Differentiation of promyelocytic (H6-60) cells into mature granulocytes: Mitochondrial-specific rhodamine 123 fluorescence. J. Cell Biol. 96:94–99.
18. **Crissman H, Steinkamp J (1986)** Multivariate cell analysis: Techniques for correlated measurements of DNA and other cellular constituents. In Gray J, Darzynkiewicz Z (eds), "Techniques in Cell Cycle Analysis." Clifton, New Jersey: Humana Press, pp 163–206.
19. **Crissman H, Tobey R (1974)** Cell cycle analysis in 20 minutes. Science 184:1297–1298.
20. **Darzynkiewicz Z, Traganos F, Melamed M (1980)** New cell cycle compartments identified by multiparameter flow cytometry. Cytometry 1:98–108.
21. **Darzynkiewicz Z (1984)** Metabolic and kinetic compartments of the cell cycle distinguished by multiparameter flow cytometry. "Growth, Cancer and the Cell Cycle." Clifton, New Jersey: Humana Press.
22. **Darzynkiewicz Z (1986)** Cytochemical probes of cycling and quiescent cells applicable for flow cytometry. In Gray JW, Darzynkiewicz Z (eds), "Techniques in Cell Cycle Analysis." New Jersey: Humana Press, pp 255–290.
23. **Darzynkiewicz Z, Traganos F, Kimmel M (1986)** Assay of cell cycle kinetics by multivariate flow cytometry using the principle of stathmokinesis. In Gray JW, Darzynkiewicz Z (eds), "Techniques in Cell Cycle Analysis." Clifton, New Jersey: Humana Press, pp 291–336.
24. **Dean P, Jett J (1974)** Mathematical analysis of DNA distributions derived from flow microfluorometry. J Cell Biol 60:523–527.
25. **Dean P, Anderson E (1975)** The rate of DNA synthesis during S phase by mammalian cells *in vitro*. In Haanen C, Hillen H, Wessels J (eds), "Pulse-Cytophotometry." European Press, Ghent, Belgium, pp 77–86.
26. **Dean P, Gray J, Dolbeare F (1982)** The analysis and interpretation of DNA distributions measured by flow cytometry. Cytometry 3:188–195.
27. **Dean P, Dolbeare F, Grazner H, Rice G, Gray J (1984)** Cell-cycle analysis using a monoclonal antibody to BrdUrd. Cell Tissue Kinet. 17:427–436.
28. **Dean P (1985)** Methods of data analysis in flow cytometry. In Van Dilla M, Dean P, Laerum D, Malamed M (eds), "Flow Cytometry: Instrumentation and Data Analysis." Orlando, Florida: Academic Press, 195–221.
29. **Dean P (1986)** Data analysis in cell kinetics research. In Gray J, Darzynkiewicz Z (eds): "Techniques in Cell Cycle Analysis." Clifton, New Jersey: Humana Press, pp 207–254.
30. **Dittrich W, Göhde W (1969)** Impulsfluorometrie bei Einzelzellen in Suspensionen. Z. Naturforsch. 24: 360–361.
31. **Dolbeare F, Gratzner H, Pallavicini M, Gray J (1983)** Flow cytometric measurement of total DNA content and incorporated bromodeoxyuridine. Proc. Natl. Acad. Sci. USA 80:5573–5577.
32. **Dolbeare F, Beisker W, Pallavicini M, Vanderlaan M, Gray J (1985)** Cytometry for BrdUrd/DNA analysis: Stoichiometry and sensitivity. Cytometry 6:521–530.

32a. **Dolbeare F, Gray J (1988)** Use of restriction endonucleases and exonuclease III to expose halogenated pyrimidines for immunochemical staining. Cytometry 9: 631–635.

33. **Dosik G, Barlogie B, Johnston D, Mellard D, Freireich E (1981)** Dose-dependent suppression of DNA synthesis in vitro as a predictor of clinical response in adult myeoblastic leukemia. Eur. J. Cancer 17:549–555.
34. **Dosik G, Barlogie B, White A, Göhde W, Drewinko V (1981)** A rapid automated stathmokinetic method for determination of in vitro cell cycle transit times. Cell Tissue Kinet. 14:121–134.
35. **Eisen M (1979)** "Mathematical Models in Cell Biology. Lecture Notes in Biomathematics." New York: Springer-Verlag.
36. **Fried J (1976)** Method for the quantitative evaluation of data from flow microfluorometry. Comp. Biomed. Res. 9:263–276.
37. **Goz B (1978)** The effects of incorporation of 5-halogenated deoxyuridines into the DNA of eukaryotic cells. Pharmacol Rev. 19:249–272.
38. **Gratzner H, Leif R (1981)** An immunofluorescence method for monitoring DNA synthesis by flow cytometry. Cytometry 1:385–389.
39. **Gratzner H (1982)** Monoclonal antibody against 5-bromo- and 5-iododeoxyuridine: A new reagent for detection of DNA replication. Science 218:474–475.
40. **Gray J (1976)** Cell-cycle analysis of perturbed cell population. Computer simulation of sequential DNA distribution. Cell Tissue Kinet. 9:499–510.
41. **Gray J, Carver J, George Y, Mendelsohn M (1977)** Rapid cell cycle analysis by measurement of the radioactivity per cell in a narrow window in S phase (RCS_i). Cell Tissue Kinet. 10:97–107.
42. **Gray J, Pallavicini M (1981)** Quantitative cytokinetic analysis reviewed. In Rotenberg M (ed), "Biomathematics and Cell Kinetics." New York: Elsevier/North-Holland Biomedical Press, pp 107–123.
43. **Gray J, Pallavicini M, George Y, Groppi V, Look M, Dean P (1981)** Rates of incorporation of radioactive molecules during the cell cycle. J. Cell Physiol. 108: 135–144.
44. **Gray J (1983)** Quantitative cytokinetics:Cellular response to cell cycle specific agents. Pharmacol. Ther. 22:163–197.
45. **Gray J, Bogart E, Gavel D, George Y, Moore DH II. (1983)** Rapid cell cycle analysis. II. Phase durations and dispersions from computer analysis of RC curves. Cell Tissue Kinet. 16:457–471.
46. **Gray J, Mayall B (1985)** "Monoclonal Antibodies Against Bromodeoxyuridine." New York: Alan R. Liss Inc.
47. **Gray J, Dolbeare F, Pallavicini M, Vanderlaan M (1985)** Flow cytokinetics. In Gray JW, and Darzynkiewicz Z (eds), "Techniques in Cell Cycle Analysis." Clifton, New Jersey: Humana Press, pp 93–138.
48. **Grdina D, Meistrich M, Meyn R, Johnson T, White R (1986)** Cell synchrony: A comparison of methods. In Gray JW, Darzynkiewicz Z (eds), "Techniques in Cell Cycle Analysis." Clifton, New Jersey: Humana Press, pp 367–402.
49. **Hahn G (1966)** State vector description of the proliferation of mammalian cells in tissue culture. I. Exponential growth. Biophys. J. 6:275–286.
50. **Hartmann N, Gilbert C, Jansson B, MacDonald P, Steel G, Valleron A (1977)** Mitoses curves. Cell Tissue Kinet. 8:119–124.
51. **Hill B (1978)** Cancer chemotherapy: The relevance of certain concepts of cell cycle analysis. Biochem. Biophys. Acta 5516:389–417.
52. **Hillwig I, Gropp A (1972)** Staining of constitutive heterochromatin in mammalian chromosomes with a new fluorochrome. Exp. Cell Res. 75:122–126.
53. **Howard A, Pelc S (1953)** Synthesis of deoxyribonucleic acid in normal and irradiated cells and its relation to chromosome breakage. Heredity 6(suppl.): 261–273.
54. **Jensen R (1977)** Chromomycin A3 as a fluorescent probe for flow cytometry of human gynecologic samples. J. Histochem. Cytochem. 25:573–579.
55. **Johnson G, Nicola N (1984)** Characterization of two populations of CFU-S fractionated from mouse fetal liver by fluorescence-activated cell sorting. J. Cell. Physiol. 118:45–52.
56. **Jovin D, Jovin T (1977)** Analysis and sorting of living cells according to DNA content. J. Histochem. Cytochem. 25:585–589.
57. **Kriss J, Revesz L (1961)** The distribution and fate of bromodeoxyuridine and bromodeoxycytidine in the mouse and rat. Cancer Res. 22:254–265.
58. **Kubbies M, Rabinovitch P (1983)** Flow cytometric analysis of factors which influence the BrdUrd–Hoechst quenching effect in cultured human fibroblasts and lymphocytes. Cytometry 3:276–281.
59. **Kufe D, Major P, Munroe M, Herrick D (1983)** Relationship Between Incorporation of 9-β-D-arabinofuranosylcytosine in L1210 DNA and Cytotoxicity. Cancer Res. 43:2000–2004.
60. **LaLande M, Ling V, Miller R (1981)** Hoechst 33342 dye uptake as a probe of membrane permeability changes in mammalian cells. Proc. Natl. Acad. Sci. USA 78:363–367.
61. **Latt S (1973)** Microfluorometric detection of DNA replication in human metaphase chromosomes. Science 70:3395–3399.
62. **Latt S, Wollheb J (1975)** Optical studies of the interaction of 33258 Hoechst with DNA, chromatin and metaphase chromosomes. Chromosoma 52:297–316.
63. **Le Pecq J, Paoletti C (1967)** A fluorescent complex between ethidium bromide and nucleic acids. J. Mol. Biol. 27:87–106.
64. **Linfoot P, Gray J, Dean P, Marton L, Deen D (1986)** Effect of cell cycle position on the survival of 9L cells treated with nutrosoureas that alkylate, crosslink and carbamoylate. Cancer Res. 46:2402–2406.
65. **Loken M (1980)** Separation of viable T and B lymphocytes using a cytochemical stain, Hoechst 33342. J. Histochem. Cytochem. 28:36–39.
66. **Loken M, Dessner-DeJose D, vanZant G, Goldwasser E (1983)** Characterization of murine hemopoietic cells using rat anti-mouse monoclonal antibodies. Hybridoma 2:55–68.
67. **Major P, Egan E, Beardsley M, Minden M, Kufe D (1981)** Lethality of human myeloblasts correlates with the incorporation of arabinofuranosylcytosine into DNA. Proc. Natl. Acad. Sci. USA 78:3235–3239.
68. **Morales-Ramires R, Vallarino-Kelly T, Rodriques-Reyes R (1983)** Effects of BrdUrd and low doses of

gamma radiation on sister chromatid exchange, chromosome breaks, and mitotic delay in mouse bone marrow cells in vivo. Environ. Mutagen. 5:589–602.

69. **Moran R, Darzynkiewicz Z, Staiano-Coico L, Melamed M (1985)** Detection of 5-bromodeoxyuridine (BrdUrd) incorporation by monoclonal antibodies: Role of the DNA denaturation step. J. Histochem. Cytochem. 33:821–827.
70. **Mortsyn G, Hsu S, Kinsella T, Gratzner H, Russo A (1983)** Bromodeoxyuridine in tumors and chromosomes detected with a monoclonal antibody. J. Clin. Invest. 71:1844–1850.
71. **Mortsyn G, Kinsella T, Hsu S, Russo A, Gratzner H, Mitchell J (1984)** Identification of bromodeoxyuridine in malignant and normal cells following therapy: Relationship to complications. Int. J. Radiat. Oncol. Biol. Phys. 10:1441–1445.
72. **Murphy RF, Powers S, Cantor CR (1984)** Endosome pH measured in single cells by dual fluorescence flow cytometry. J. Cell Biol. 98:1757–1762.
73. **Musgrove E, Rugg C, Hedley D (1986)** Flow cytometric measurement of cytoplasmic pH: A critical evaluation of available fluorochromes. Cytometry 7: 347–355.
74. **Pallavicini M, Gray J, Folstad L (1982)** Quantitative analysis of the cytokinetic response of KHT tumors in vivo to 1-β-D-arabinofuranosylcytosine. Cancer Res. 42:3125–3131.
75. **Pallavicini M, Ng C, Gray J (1984)** Bivariate flow cytometric analysis of murine intestinal epithelial cells for cytokinetic studies. Cytometry 5:55–62.
76. **Pallavicini M, Summers L, Dolbeare F, Gray J (1985)** Asynchronous and ara-c perturbed cytokinetics using simultaneous BrdUrd/DNA analysis in a murine tumor in vivo. Cytometry 6:602–610.
77. **Pallavicini M, Summers L, Giroud F, Dean P, Gray J (1985)** Multivariate analysis and list mode processing of murine hematopoietic subpopulations for cytokinetic studies. Cytometry 6:539–549.
78. **Pallavicini M (1986)** Solid tissue dispersal for cytokinetic analyses. In Gray JW, Darzynkiewicz Z (eds), "Techniques in Cell Cycle Analysis." Clifton, New Jersey: Humana Press, pp 139–162.
79. **Pinkel D, Straume T, Gray J (1986)** Cytogenetic analysis using quantitative, high sensitivity, fluorescence hybridization. Proc. Natl. Acad. Sci. USA 83:2934–2938.
80. **Preisler H, Bjornsson S, Henderson E (1977)** Adriamycin–cytosine arabinoside therapy for acute myelocytic leukemia. Cancer Treat. Rep. 61:89–92.
81. **Quastler H, Sherman F (1959)** Cell population kinetcs in the intestinal epithelium of the mouse. Exp. Cell Res. 17:420–438.
82. **Raza A, Preisler H, Myers G, Bankert R (1984)** Rapid enumeration of S-phase cells by means of monoclonal antibodies. N. Engl. J. Med. 310:991.
83. **Rotenberg M (1981)** "Biomathematics and Cell Kinetics." New York: Elsevier/North-Holland Biomedical Press.
84. **Roti Roti J, Okada S (1973)** A mathematical model of the cell cycle of L51784. Cell Tissue Kinet. 6:111–124.
85. **Shackney S, Ritch P (1986)** Percent labeled mitosis curve analysis. In Gray J, Darzynkiewicz Z (eds), "Techniques in Cell Cycle Analysis." New Jersey: Humana Press, pp 31–45.
86. **Sharpless T, Schlesinger F (1982)** Flow cytometric analysis of G1 exit kinetics in asynchronous L1210 cell cultures with the constant transition probability model. Cytometry 3:196–200.
87. **Siegers M, Feinendegen L, Lahiri S, Cronkite E (1979)** Relative number and proliferation kinetics of hemopoietic stem cells in the mouse. Blood Cells 5: 211–236.
88. **Skipper H, Schabel F, Wilcox W (1967)** Experimental evaluation of potential anticancer agents. XXI. Scheduling of arabinofuranosylcytosine to take advantage of its S-phase specificity against leukemia cells. Cancer Chemother. Rep. 51:125–165.
89. **Smith J, Martin L (1973)** Do cells cycle? Proc. Natl. Acad. Sci. USA 70:1263–1267.
90. **Takahashi M (1966)** Theoretical basis for cell cycle analysis. J. Theoret. Biol. 13:202–211.
91. **Takahashi M, Hogg J, Mendelsohn M (1971)** The automatic analysis of FLM curves. Cell Tissue Kinet. 4:505–518.
92. **Tobey R, Crissman H (1972)** Use of flow microfluorometry in detailed analysis of the effects of chemical agents on cell cycle traverse. Cancer Res. 32: 2726–2732.
93. **Traincard F, Ternyck T, Danchin A, Avrameas S (1983a)** Une technique immunoenzymatique pour la nise en evidence de l'hybridization moleculaire entre acides nucléiques. Ann. Immunol. (Inst. Pasteur) 134D:399–405.
94. **Trask B, van den Engh G, Landegent J, in de Wal N, van der Ploeg M (1985)** Detection of DNA sequences in suspension by *in situ* hybridization and dual beam flow cytometry. Science 230:1401–1403.
95. **Valeriote F (1979)** The use of cell kinetics in the development of drug combinations. Pharmacol. Ther. 4:1–33.
96. **Valet G, Raffael A, Moroder L, Wunsch E, Rukenstroth-Bauer G (1981)** Fast intracellular pH determination in single cells by flow cytometry. Naturwissenschaften 68:265–266.
97. **van den Engh G, Bauman J, Mulder D, Visser J (1983)** Measurement of antigen expression of hemopoietic stem cells and progenitor cells by fluorescence activated cell sorting. In Killmann SA, Cronkite EP, Muller-Berat CN (eds), "Haemopoietic Stem Cells," Copenhagen: Munsgaard, pp 59–74.
98. **Vanderlaan M, Thomas C (1985)** Characterization of monoclonal antibodies to bromodeoxyuridine. Cytometry 6:501–505.
99. **Vanderlaan M, Watkins B, Thomas C, Dolbeare F, Stanker L (1986)** Improved high-affinity monoclonal antibody to iododeoxyuridine. Cytometry 7:499–507.
100. **Vaughan W, Burke P (1983)** Development of a cell kinetic approach to curative therapy of acute myelocytic leukemia in remission using the cell cycle-specific drug 1-β-D-arabinofuranosylcytosine in a rat model. Cancer Res. 43:2005–2009.
101. **Visser J (1979)** Vital staining of hemopoietic cells with the fluorescent bis-benzimidazole derivatives Hoechst 33342 and Hoechst 33258. In Laerum OD, Lindmo T, Thorud E (eds), "Fourth International Symposium on Flow Cytometry: Pulse Cytophotom-

etry." Bergen-Oslo-Tromsø: Universitetsforlaget, pp 86–90.

102. **Visser J, van den Engh A, and van Bekkum D** (1980) Light scattering properties of murine hemopoietic cell. Blood cells 6:391–407.

103. **Visser J, Bauman J, Mulder A, Eliason J, deLeeuw A** (1984) Isolation of murine pluripotent hemopoietic stem cells. J. Exp. Med 159:1576–1590.

104. **Waldman F, Dolbeare F, Gray J** (1985) Detection of ara-c resistant cells at low frequency using the Brd-Urd/DNA assay. Cytometry 6:657–662.

105. **Watt S, Metcalf D, Gilmore D, Stenning G, Clark M, Waldman F** (1983) Selective isolation of murine erythropoietin-responsive progenitor cells (CFU-E) with monoclonal antibodies. Biol. Med. 1:95–115.

106. **Wilson G, McNally N, Dunphy E, Kärcher H, Pfranger R** (1985) The labelling index of human and mouse tumors assessed by bromodeoxyuridine staining in vitro and in vivo and flow cytometry. Cytometry 6:641–647.

107. **Yanagisawa M, Dolbeare F, Todoroki T, Gray J** (1985) Cell cycle analysis using numerical analysis of bivariate DNA/bromodeoxyuridine distributions. Cytometry 6:550–562.

108. **Zietz D** (1980) FPI analysis. I. Theoretical outline of a new method to analyze time sequences of DNA histograms. Cell Tissue Kinet. 13:461–471.

24

Multiparameter Flow Cytometry in Studies of the Cell Cycle

Zbigniew Darzynkiewicz and Frank Traganos
Sloan-Kettering Institute for Cancer Research, New York, New York 10021

INTRODUCTION

Multiparameter flow cytometry allows one to estimate, with high accuracy, relative quantities of a variety of cell constitutents simultaneously [13,145,226,238,257]. When the measurements are recorded in list mode fashion, it is possible to attribute each of the several measured features to a particular cell and thus to obtain correlated measurements of these features on a cell by cell basis. Multivariate analysis of such data yields a plethora of information on the relationship between the parameters measured. In contrast to biochemical analysis in bulk, cellular heterogeneity can be estimated and cell subpopulations with distinct characteristics can be discriminated by multiparameter flow cytometry.

DNA content is the most commonly measured cell feature. This measurement allows one to position the cell within the traditional phases of the cell cycle, that is, G_1 versus S versus G_2 + M. Thus, correlated bi- or multivariate analysis of the data, in which one parameter is cellular DNA content, provides information on the quantity of a second constituent relative to the cell's position in the cell cycle. Multiparameter flow cytometry is, therefore, the method of choice to study the relationship between cell growth (e.g., expressed as RNA or protein content) or differentiation (e.g., measured by content of a differentiation specific product) and progression through the cell cycle. The method can also estimate the time of appearance and content of a particular, specific constituent, such as an oncogene product or other molecules with a presumed regulatory function, at a given point of the cell cycle. The list of antibodies, predominantly of monoclonal origin, which can detect the latter constituents, is rapidly growing, and their application for cell characterization and in cell cycle studies is becoming widespread.

Three different strategies are used in applying multiparameter flow cytometry to studies of the cell cycle. The first is the "snapshot" or static analysis of the populations. This, so far, is the most common approach, which, in a single measurement, can reveal relationships between the cell's metabolic state and cell-cycle phase that apply to a variety of different cell types [60,74]. These features can be used as metabolic markers distinguishing different kinetic compartments within the traditional phases of the cell cycle [60,74].

The second strategy represents a combination of multiparameter flow cytometry with kinetic measurements of the cell population. Thus, the "snapshot" observations, which do not contain any kinetic information per se, can be correlated with the *rate* of cell progression through the cell cycle. One such experimental design involves the use of synchronized cells. The progression of synchronized populations (e.g., cells released from G_1 arrest) can be measured sequentially in time such that the rates of cell entrance into S, S traverse, or entrances into G_2 can be estimated. This approach, for instance, has been used to measure cell kinetics in relation to RNA content in cultures of stimulated lymphocytes [66] or CHO cells synchronized by selective detachment at mitosis [65,264]. Another experimental design is based on the principle of stathmokinesis. This design has been introduced to measure the rates of cell traverse through several points of the cell cycle, simultaneously, and has found application in studies of drug effects on the cell cycle [73, 83,90] (see Chapter 39, this volume).

A third strategy in the application of multiparameter flow cytometry for cell-cycle analysis is based on incorporation of the thymidine analogue, 5-bromodeoxyuridine (BrdUrd). When incorporated into DNA during replication, this parameter alters DNA stainability with several fluorochromes [157] (see Chapters 13 and 14, this volume). The presence of incorporated BrdUrd can be also detected immunochemically [107] (see Chapter 23, this volume). Thus, cells that undergo DNA replication in the presence of BrdUrd can be recognized, making it possible to study their rate of entrance into S, or to discriminate cycling from noncycling cells. Multiparameter analysis of cells that incorporate BrdUrd can be accomplished by combination of Hoechst 33358 and propidium iodide (PI) [26], by simultaneous staining of DNA versus RNA with acridine orange (AO) [61] or differential staining of double-stranded versus denatured DNA with AO to distinguish BrdUrd labeled- from unlabeled-mitotic cells [76]. Recently, Dolbeare et al. [88] introduced a method of simultaneous analysis of DNA content (PI-staining) and

Flow Cytometry and Sorting, Second Edition, pages 469–501

BrdUrd incorporation based on the use of anti-BrdUrd antibodies. The latter technique, due to its sensitivity, has already gained widespread use in studies of cell kinetics not only in vitro but also in vivo and is separately reviewed in Chapter 23 of this volume.

The present review describes and discusses findings resulting from application of these multiparameter techniques to cell cycle studies. Special attention is given to the association between cell growth and progression through the mitotic cycle. Also described are characteristic metabolic states of the cell associated with cell quiescence, normal or unbalanced growth, progression through particular phases of the cell cycle, and differentiation. As previously mentioned, identification of those states allows us to subclassify cells into a variety of compartments recognized by multiparameter flow cytometry. Information predictive of cell cycle kinetic behavior can be obtained from a single measurement based on this classification. This review updates earlier publications in which some of the above topics were presented and discussed [59,60,70,74]. The application of multiparameter flow cytometry to studies of cell cycle-specific drug effects is described in Chapter 39 (this volume). Studies in which isolated nuclei or chromatin structures were investigated by flow cytometry, often in relation to the cell cycle, are reviewed separately (Chapter 16 in this volume). For recent reviews of the literature directly related to the cell cycle, the reader is advised to turn to the book by Baserga [20] and the article by Prescott [205]. Chromatin changes during the cell cycle were reviewed in ref. [59].

Static, "Snapshot" Measurements

DNA Content and Cell Size. It is generally accepted that noncycling "G_0" cells (classified here as G_{1Q}, see further) are smaller than their cycling counterparts [64,160]. The diameter of a nonstimulated lymphocyte, for instance, is between 8 and 12 μm, whereas the diameter of mitogen-stimulated cycling lymphoblasts varies between 20 and 30 μm [160]. The increase in size of both the nucleus and cytoplasm contribute to cell enlargement.

Extensive literature exists describing the changes in cell size during progression through the cell cycle, especially in relation to attainment of a "critical" size, or mass prior to entrance into S phase [18,19,60,177,203]. There is no general consensus, however, as to whether or not attainment of a critical mass is a prerequisite for initiation of DNA replication [19,20]. In detailed studies on this subject, Killander and Zetterberg [131,132] observed, by combined interferometry, microspectrophotometry, and time-lapse cinematography, that, during nonperturbed exponential growth, individual cells initiate DNA synthesis only after attainment of a critical mass. This topic will be discussed further in this chapter when characterizing the cell-cycle compartments identified by flow cytometry.

Cell size can be estimated in flow by physical methods such as Coulter volume [121], forward light scatter [219], axial light loss [126,145], or pulse-duration [229] measurements. Whereas reports of cell size estimates in relation to cell cycle position based on single-parameter analysis are common [166,180,237,295], multiparameter measurements (i.e., cell size and a second parameter) are relatively scarce. Most studies have been done on lymphocytes; their stimulation by mitogens, which represents a transition from the resting (G_0) to the cycling state, is accompanied by cell enlargement [64]. This transition has been measured by Coulter volume [102,236], axial light loss [92,290], or light scatter

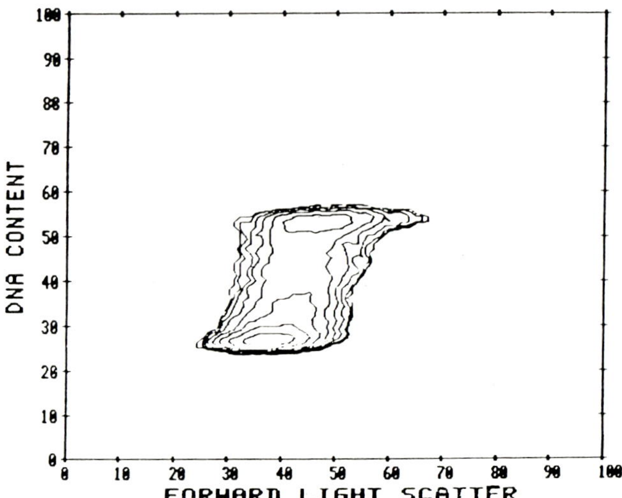

Fig. 1. Contour map of DNA versus forward-angle (2–19°) light scatter of exponentially growing CHO cells.

[262] estimates; these methods were proposed (as an alternative to [^{3}H]-Tdr incorporation) for a rapid assay of lymphocyte stimulation.

The DNA content and forward light scatter values of exponentially growing CHO cells are shown in Figure 1. An increase in light scatter is observed during progression through S, inasmuch as cells in G_2 have higher scatter values than cells in G_1. The scatter values of the G_1 cell population, however, are unimodally distributed and are similar to those of early-S cells. Thus, based on light scatter measurements alone, no distinct subpopulations in G_1 can be discriminated, and there is no evidence of a critical threshold prior to entrance to S phase.

Figure 2 illustrates the distribution of cells from mixed allogeneic lymphocyte cultures with respect to their forward light scatter and DNA values. It is apparent that cycling cells scatter more light than do G_0 cells. The progression of cells through S and G_2 phase in this culture, however, is not associated with an additional increase in light scatter (note the vertical position of the S + G_2 + M cluster). Nonviable cells have lowered DNA stainability and low scatter values.

The light scatter signal is, in addition to cell size, influenced by the refractive index of cell constituents, light reflection from the cell surface and internal structure, as well as cell shape [219]. It cannot, therefore, be considered a direct measure of cell size. In addition, fixed and/or permeabilized cells have altered scatter signals such that results on viable cells cannot always be compared with those on fixed or detergent-treated cells. Another source of variability are differences in the geometry of light scatter detection between different instruments.

Light scatter measured simultaneously at several different angles provides more information on cell morphology than does a single measurement in the forward direction [219]. Using multiangle measurements, Visser and his colleagues [268,269] were able to distinguish the pluripotent stem cells in the bone marrow and to subdivide the differentiating cells of bone marrow into several groups. These measurements, however, were not correlated with cell cycle progression.

DNA and RNA Content. Simultaneous staining and correlated measurements of DNA and RNA can be accom-

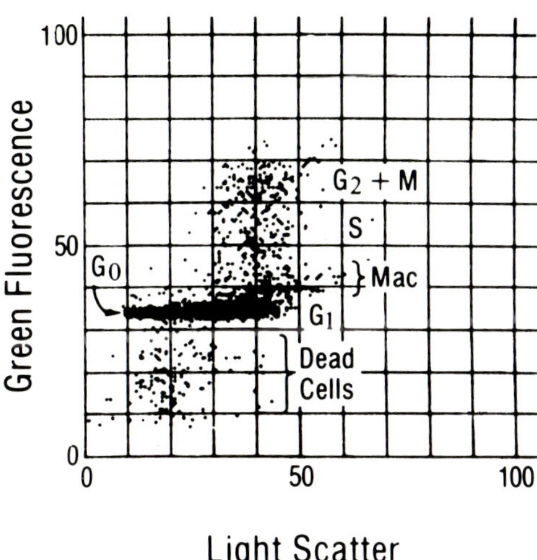

Fig. 2. Cytogram of DNA (acridine orange green fluorescence) versus forward-angle (2–19°) light scatter of human lymphocytes from a 6-day mixed lymphocyte culture reaction [262]. There is a population of G_1 cells of monocyte origin (Mac) showing somewhat higher DNA stainability than lymphocytes. Note, there is no increase in light scatter as cells proceed from G_1 through S to G_2 + M phase.

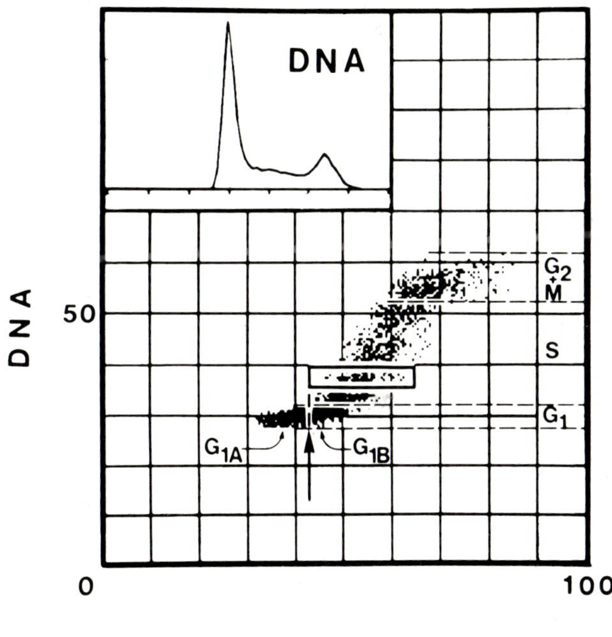

Fig. 3. Cytogram of exponentially growing CHO cells stained with acridine orange to measure relative DNA and RNA content. The G_{1A} compartment is established by selecting a slice of early S phase cells (see box) and calculating the mean and standard deviation (SD) of RNA content for that slice. The line separating G_{1A} from G_{1B} is placed at 3 SD below the mean RNA content of those early-S-phase cells.

plished by taking advantage of the metachromatic properties of acridine orange (AO; see Chapter 15, this volume) or by using a combination of the dyes Hoechst 33342 and pyronin Y [201,225,247]. Both methods require permeabilized or fixed cells, inasmuch as supravital staining with AO results in accumulation of the dye in lysosomes (Chapter 15, this volume), whereas pyronin Y has an affinity to mitochondria in viable cells [50,67]. Furthermore, pyronin Y, as a fluorochrome, stains only double-stranded RNA [67].

Results of correlated measurements of DNA and RNA content of exponentially growing CHO cells are presented in Figure 3. The G_1 population as seen in the figure has a characteristically broad RNA distribution. The arrow indicates a threshold dividing G_1 cells into a subpopulation in which cells have an RNA content similar to early-S cells (G_{1B}) and another subpopulation, containing cells with a lower RNA content than any S-phase cell (G_{1A}). Since this type of RNA distribution has been observed in over 40 different cell systems studied [60,74], whether derived from normal or tumor material, it is apparent that, during unperturbed growth, cells do accumulate a critical amount of RNA prior to entrance into S phase. The fact that the RNA distribution of G_1 cells in exponentially growing cultures is generally unimodal [66,74] suggests that the G_{1A} and G_{1B} transition is continuous.

To have an objective criterion to discriminate between G_{1A} and G_{1B} cells, a gating window can be located around the early one third of S phase (based on DNA content) and the cells measured within this window to obtain the mean RNA value and its standard deviation. The RNA threshold is then established at the level of the mean value of RNA minus three standard deviations of these early-S cells as shown by the arrow in Figure 3. By definition, therefore, the G_{1A} cells have an RNA content significantly lower than early-S cells, whereas the G_{1B} cells overlap in RNA content with early-S cells. We have emphasized in earlier publications [60,63, 66,74,107], and will discuss further in this chapter, that the kinetic properties of the G_{1A} and G_{1B} populations appear to be different.

A technical point should be stressed that is important in discriminating the G_{1A} and G_{1B} populations. Namely, the presence of any dead or broken cells in the sample precludes detection of the threshold because the broken S-phase cells depleted of all or a portion of cytoplasm will be located on cytograms below this RNA threshold. The dead cells, however, can be removed by preincubation with trypsin and DNase I prior to fixation and staining [85]. Detection of the threshold also requires good accuracy (resolution); when the cv value of the mean DNA content of G_1 is higher than 6–8% the distinction between G_{1A} and G_{1B} becomes obscured unless the G_{1A} population is very prominent.

Changes in the DNA and RNA distribution of cells undergoing transition to quiescence are shown in Figure 4. This figure provides an example of cell arrest in the G_{1A} compartment, as induced in L1210 leukemia cells whose proliferation has been suppressed by growth in the presence of 1 mM n-butyrate [68]. The number of cells in S, G_2, and M is diminished, and most G_1 cells have an RNA content typical of a G_{1A} population. Extended growth of L1210 cells in the presence of n-butyrate results in loss of viability rather than a further reduction in RNA content. Cell arrest in G_{1A} was also observed in plateau-phase tumor cell cultures [60,74] as well as in cultures of tumor cells exposed to the mitochondrial cationic probe, rhodamine-123 [82].

The "depth" of quiescence can vary [10], and Figure 5 illustrates the RNA content of 3T3 cells in two different states of quiescence. Exponentially growing 3T3 cells (Fig.

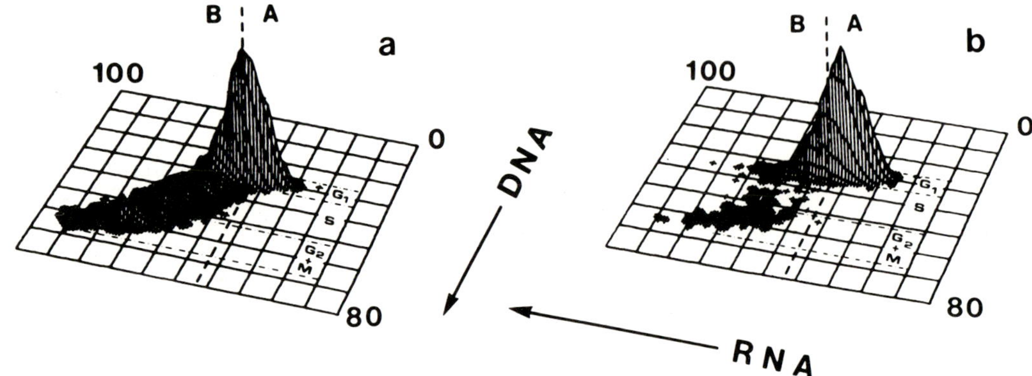

Fig. 4. Two-dimensional (DNA versus RNA) histograms of control (left) and 1 mM sodium butyrate (right) treated L1210 cells. Sodium butyrate inhibits cell proliferation and causes an increase in the proportion of G_{1A} cells [84].

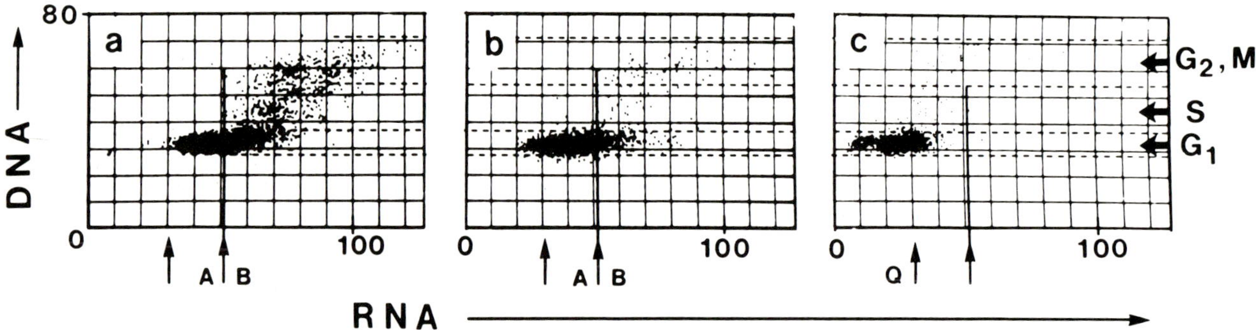

Fig. 5. DNA versus RNA distribution of 3T3 cells during **(a)** exponential growth, **(b)** growth for 3 days in 0.5% serum, and **(c)** growth for 7 days in 0.5% serum [68]. The G_{1A} (A) versus G_{1B} (B) discrimination was performed as in Fig. 3. Note that in **(c)**, 3T3 cells enter a state (G_{1Q}) in which their RNA content is significantly lower than that of G_{1A} cells.

5a) show a well-defined RNA threshold in G_1, which allows the discrimination of G_{1A} and G_{1B} subpopulations, as discussed above for CHO cells. When cells are maintained in 0.5% serum for 72 hours (Fig. 5b), a marked suppression in the number of S and G_2 + M cells is apparent. Most cells are arrested in G_1 and are characterized by low RNA values typical of G_{1A} cells. Thus, by the criterion of RNA content these cells are arrested in G_{1A}. Following addition of serum most of these G_{1A}-arrested cells enter S phase after a 12-hour delay. The cells enter a deeper state of quiescence when maintained at confluence for an extended period of time (Fig. 5c). Namely, following an additional 4 days at confluence, the RNA content of these G_1 cells falls below the level of the G_{1A} population. These cells are viable and, when trypsinized and replated at lower cell densities, resume progression and enter S phase after a delay of approximately 16 hours. Such deeply quiescent G_1 cells can thus be distinguished as a separate category (G_{1Q} cells) having distinct metabolic (very low RNA content) and kinetic (long delay before entrance into S phase) properties.

Nonstimulated peripheral blood lymphocytes are another example of G_{1Q} cells. These cells have minimal RNA content and following stimulation by mitogens require at least 24 hours to enter S phase. Changes in the RNA and DNA content of stimulated lymphocytes are shown in Figure 6. As described in detail [60,74,78] several compartments representing different phases of cell growth and progression through the mitotic cycle can be distinguished during lymphocyte stimulation. Prior to stimulation nearly all cells are in a deep, quiescent phase (G_0, G_{1Q}). During the first hours of stimulation, for example, with phytohemagglutinin (PHA), an increase in RNA content is observed in a subpopulation of cells, although the RNA level of these cells is still less than the minimal RNA content of "cycling" lymphocytes. These cells were classified as undergoing transition from the resting state to the mitotic cycle (G_{1T}). On the second day of stimulation there are a large number of cells with increased RNA content, but, as yet, few enter S phase. Maximal stimulation, both with respect to an increase in RNA content and in the number of cycling cells (S and G_2 + M phase cells), is observed on the third day after addition of PHA. In the cultures shown in Figure 6, both B and T lymphocytes undergo stimulation; stimulated B cells have a higher RNA content than do T lymphocytes. This tissue-specific difference in RNA contributes to a wide distribution of stimulated cells with respect to their RNA content, as is evident in Figure 6, panel c. This heterogeneity obscures discrimination of G_{1A} versus G_{1B} cells, which otherwise is possible in cultures of stimulated lymphocytes. Different flow cytometric techniques that may be used to assay lymphocyte stimulation are discussed later in this chapter.

Multicellular spheroids represent another cell system in which quiescent cells with minimal RNA content can be detected [22]. Minimal RNA content is also observed in several

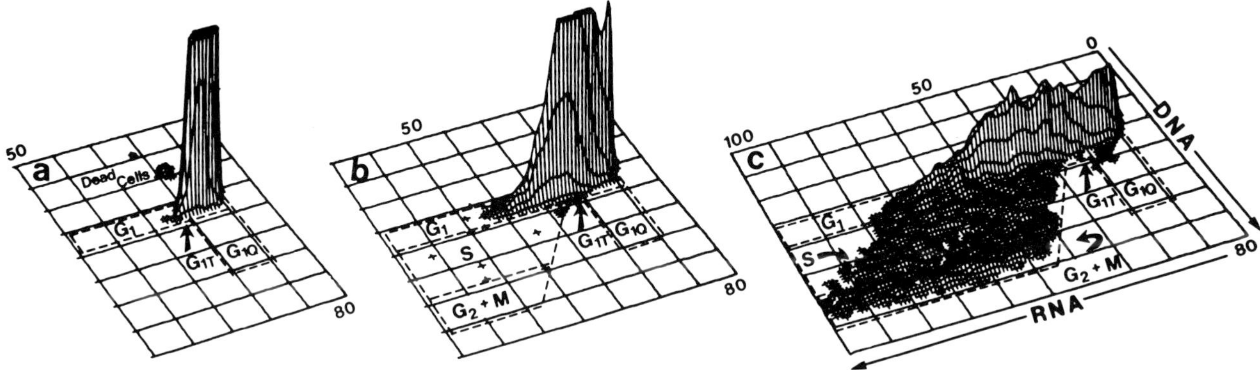

Fig. 6. DNA versus RNA distribution of human lymphocytes incubated for **(a)** 0, **(b)** 1, **(c)** 3 days with the polyvalent mitogen phytohemagglutinin (PHA). Stimulated cells are identified by their increase in RNA content first as cells in transition (G_{1T}) and then as G_{1A} and G_{1B} cells in 1-day-old cultures. In 3-day-old cultures, there are cells in G_1, S, and G_2 + M phase. Identification of S and G_2 + M phase cells is based on both increased RNA and DNA content and was confirmed by comparison with synchronized cultures [78].

cell lines when deprived of nutrients [21,163,273,274,279]. In contrast to normal cells, however, tumor cells or established tumor cell lines cannot remain viable for an extended period of time in the G_{1Q} state.

Although in the majority of the cases reported quiescent cells were characterized by a G_1-DNA content (G_{1A}-arrested, G_{1Q} cells), in some situations the noncycling cells or cells cycling at very slow rates were observed to have an S- or G_2-DNA content [60,74]. The terms S_Q and G_{2Q}, respectively, were proposed to describe such cells [60,74]. The autoradiographic evidence for quiescent G_2 cells (G_{2Q}) was provided by Gelfant more than two decades ago [99]. S_Q and G_{2Q} cells have been observed during late stages of erythroid differentiation of Friend leukemia cells in cultures treated with dimethylsulfoxide [260] or retinoic acid [263], in chronic myeloid leukemia during blastic crisis [60,74], in multicellular spheroids [5] and in solid tumors [48]. It appears that, under certain adverse conditions, S-phase cells may decrease or perhaps even stop synthesizing DNA and, if catabolic processes prevail, lose RNA.

One striking example in which loss of RNA content can be demonstrated to be linked to loss of proliferative potential even among cells in S or G_2 phase is illustrated in Figure 7. In the presence of dimethylsulfoxide (DMSO), a classical inducer of Friend erythroleukemia cell differentiation, a population of cells evolves with dramatically lower RNA content (Fig. 7B). This population of cells, in addition to manifesting differentiation-specific proteins (e.g., hemoglobin), has lost its proliferative potential, although a small proportion of cells (approximately 15%) have left the cycle with an S or G_2-DNA content [263]. Other agents, such as the synthetic vitamin A analog retinoic acid (RA), also induce a population of cells that have lost their ability to proliferate [263]. However, FL cells do not respond uniformly to RA (Fig. 7A), so that, while a population appears with an RNA content equivalent to DMSO-differentiated cells, there also exists a cell population with an intermediate RNA content. RA-treated cells with the lowest RNA content (G_{1Q}, S_Q, G_{2Q} cells) can be shown to have lost their ability to proliferate [263]. However, RA-treated cells with an intermediary RNA content (G_{1T}, S_T, G_{2T} cells) continue to proliferate, albeit more slowly than exponentially growing FL cells and, upon removal of RA, return to a normal growth rate with normal cellular levels of RNA [263].

A second example in which differentiation in vitro has been linked to alterations in RNA content is illustrated in Figure 8. Myeloid differentiation of HL-60 human leukemic cells induced by dimethylsulfoxide is accompanied by a dramatic reduction in RNA content and a parallel suppression of cell proliferation (Fig. 8b,c). Thus, HL-60, like FL, exhibits a differentiated phenotype characterized by changes in morphology, cell function, and synthesis of differentiation-specific proteins concomitant with the change (lowering) in RNA content and loss of proliferative potential. To help discriminate between instances in which the loss of RNA content is associated with an irreversible transition to a differentiated state, as opposed to a quiescent but generally reversible state, it is suggested that the subscript D (e.g., G_{1D}) be used rather than Q [60,74]. However, it should be noted that the differentiated phenotype is not always characterized by lowered RNA content, as we see in the example of keratinocytes described below.

Keratinocytes, which differentiate in vitro, exhibit a unique pattern of cell distribution with respect to their RNA and DNA values (Fig. 9). This cell system was studied by us in detail, and the kinetic properties of the subpopulations that can be distinguished by multiparameter flow cytometry [90,134,234] were established in stathmokinetic experiments [134,234]. As is evident in Figure 9, three distinct subpopulations can be discriminated based on differences in RNA content. The first subpopulation is composed of low-RNA content cells with a G_1, S, and G_2 + M DNA content, which, after sorting and visual inspection [234], resemble basal keratinocytes of the epidermis and show characteristics of stem cells [134]. These cells cycle slowly, having a mean generation time of about 100 hours [134]. The second population consists of G_1, S, and G_2 + M cells with a higher RNA content; these cells morphologically resemble stratum spinosum cells and have an average in vitro generation time of approximately 40 hours. The third subpopulation consists of the most differentiated (noncycling) cells with a G_1-DNA content and cytochemical features similar to those of cells in the granular layer of the epidermis (G_{1D} cells). In this instance, keratinocyte differentiation is associated with an increased rather than decreased cellular RNA content. Due to the "yellow" fluorescence of keratohyalin granules following staining with AO, these cells form a slanted cluster on cytograms (Fig. 9). The kinetic data regarding the generation

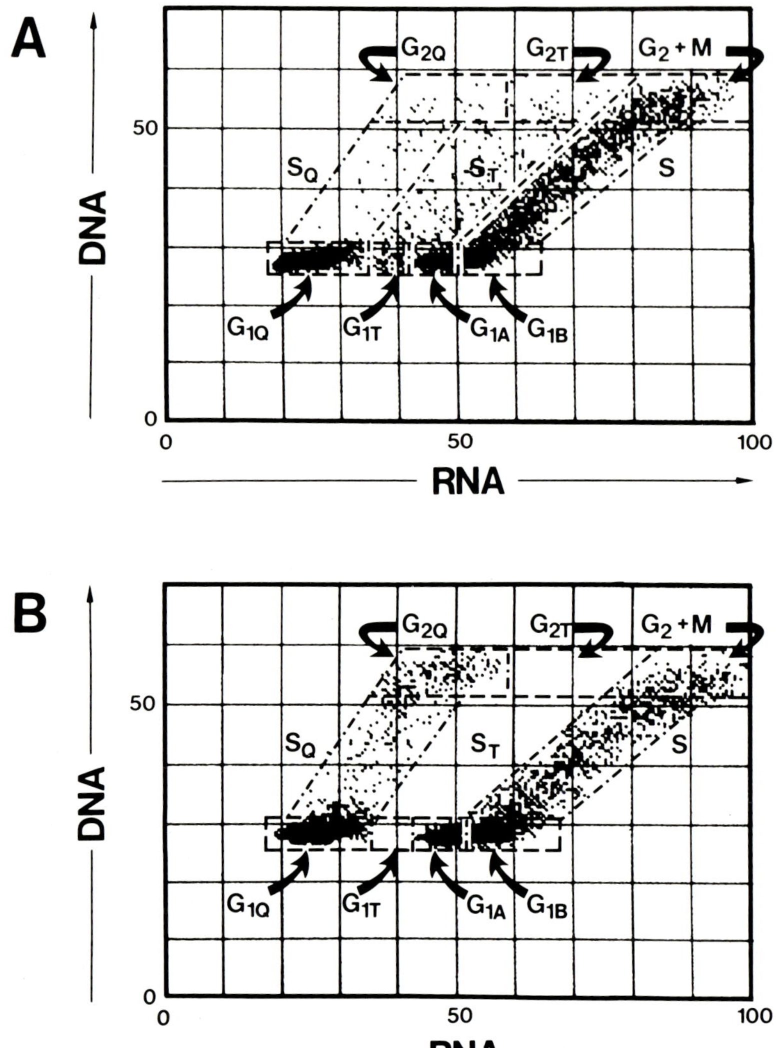

Fig. 7. DNA versus RNA distributions of exponentially growing Friend leukemia (FL) cells mixed with cells treated with **(A)** 5 × 10^{-6}M retinoic acid (RA) or **(B)** 2% dimethylsulfoxide (DMSO). Exponentially growing FL cells, having the highest RNA content, are represented by the clusters in the far right of (A) and (B). RA-treated cultures **(A)** are characterized by two populations differing in RNA content; a nonproliferating population with low RNA content and a slowly proliferating population with intermediate RNA values [263]. The latter population (G_{1T}, S_T, and G_{2T} cells), upon removal of RA, returns to exponential growth. In the presence of DMSO, cells fully differentiate **(B)** and form a completely separate population with lowered RNA content [260]. These cells have lost their ability to proliferate such that G_{1Q} (G_{1D}) cells cannot enter S phase while cells in S_Q (S_D) and G_{2Q} (G_{2D}) have either ceased progression or progress very slowly through the cycle.

times of the respective populations [134,234] conform with earlier observations made by single-parameter flow cytometry and autoradiography, which indicated the presence of two subpopulations cycling at distinctly different rates [40,202]. The data in Figure 9 indicate that the transition from the low RNA (A) to high RNA (B) compartment is abrupt, inasmuch as there are only a few cells in transition, that is, with intermediate RNA values.

A word of caution is necessary regarding the relationship between RNA content and cell quiescence. Whereas in most cell types studied to date a good correlation has been observed between these two parameters, exceptions have been noted. Thus, for instance, Brock et al. [29] observed noncycling (anoxic) fibrosarcoma populations with a RNA content similar to that of cycling cells. Likewise, in the case of keratinocytes the small, low-RNA cells do progress through the cell cycle though at a very slow rate. In the latter case, the low-RNA containing keratinocytes are most likely stem cells, inasmuch as small keratinocytes are truly clonogenic, whereas large keratinocytes have a limited potential to divide, forming only small colonies [17]. Although growth and proliferation are normally correlated, there are examples in which induction of cell proliferation of quiescent cells (e.g., cell entrance into S) can be dissociated from an increase in RNA content [19]. This subject is discussed later in this chapter.

The cell cycle-related changes in RNA content discussed above, should be distinguished from tissue-specific variations

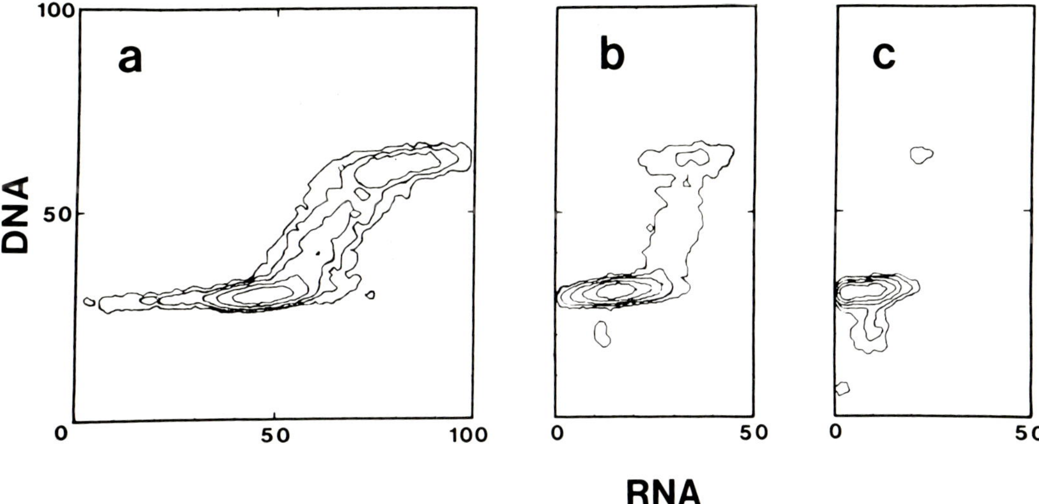

Fig. 8. DNA versus RNA content of HL-60 human leukemic cells before **(a)**, 3 **(b)**, and 5 **(c)** days after induction of differentiation by 1.5% DMSO. The decrease in RNA content and proliferating cells observed is clearly dependent upon the duration of the differentiation process. By 5 days **(c)**, the mean RNA content of all cells is markedly decreased with the vast majority of the cells confined to G_1 phase; these cells manifest a differentiated cell phenotype (G_{1D} cells).

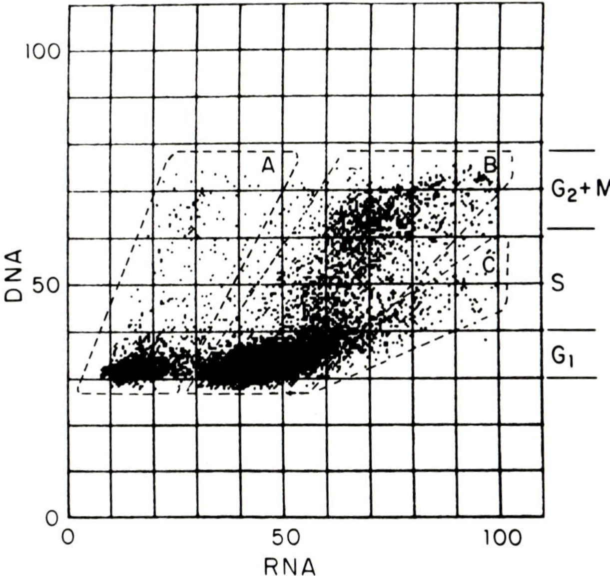

Fig. 9. DNA versus RNA cytogram of an exponentially growing keratinocyte culture stained with acridine orange. Compartments A and B contain cycling cells. Population A consists mainly of basal keratinocytes which show characteristics of stem cells while population B resembles stratum spinosum cells from epidermis and cycle approximately 2.5 times faster than cells in population A (100 versus 40 hr, respectively) [134]. Cells in population C do not cycle and appear to represent differentiated keratinocytes containing keratohyalin granules [134]. Population C cells detach during culturing and thus are few in number when only cells remaining attached are collected by trypsinization as in this example [134,234].

in RNA content. Thus, for instance, stimulated B lymphocytes have considerably higher RNA content than stimulated T cells, and this difference is unrelated to cell-cycle progression [62]. Likewise, there is a difference between nonstimulated (G_{1Q}) B versus T lymphocytes, with the latter having higher RNA values [6], and between blast cells of different leukemia types [7] (see Chapter 35, this volume).

DNA and Protein Content; Correlated Measurements of DNA, RNA, and Protein. Numerous methods have been developed to differentially stain cellular DNA and proteins [56–58,104,240]. Regardless of the cell type or the method of staining, the DNA versus protein distribution, as shown in Figure 10, is consistently observed for exponentially growing cells. This pattern resembles the DNA versus RNA distributions observed above (Fig. 3). This similarity is rather expected inasmuch as cellular protein and RNA (rRNA) values should be correlated. Indeed, such a correlation was observed, at least in the case of exponentially growing cells, when simultaneous measurements of DNA, RNA, and protein were obtained in the same cells.

Cells in deep quiescence are small, and their dry weight is lower than that of their cycling counterparts [64]. Therefore, the protein content of quiescent cells should be equally diminished, and this parameter may serve as a nonspecific flow cytometric marker of quiescent cells.

The method for simultaneous, direct determination and correlation of DNA, RNA, and protein was recently developed based on the use of a three-laser flow cytometer [52,53]. In addition to these three parameters, the RNA/DNA and RNA/protein ratios can also be measured and recorded in list mode so that each of these five measurements can be correlated with any other for a given subpopulation of cells. An example of such measurements performed on CHO cells is shown in Figure 11. Analysis of these multiparametric measurements yields many details of the relationship between cell growth and position in the cell cycle.

The DNA versus RNA and DNA versus protein cell distributions in Figure 11 are similar to each other and to the data shown in Figures 3 and 4. Judging from the width of the G_1, S, or G_2 + M clusters and from the standard deviation of the mean values, the populations are more heterogeneous when an analysis is based on protein as compared to RNA

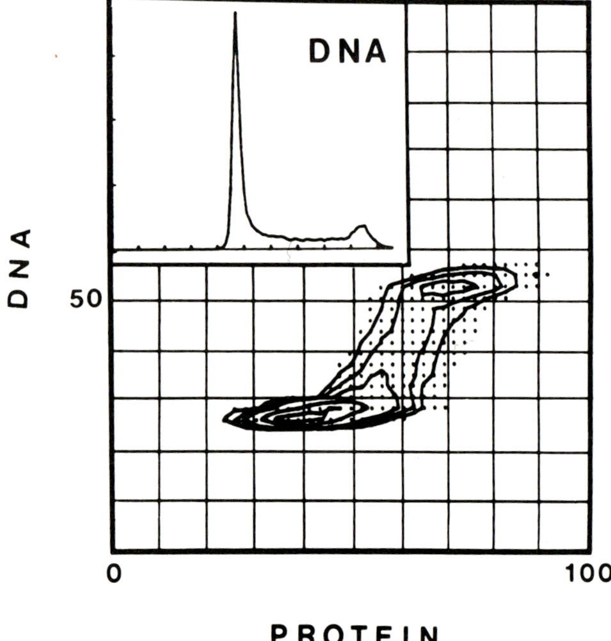

Fig. 10. Contour map of DNA versus protein content of exponentially growing CHO cells stained with Hoechst 33342 and rhodamine 640 [56]. Note that the distribution is similar to that obtained for DNA versus RNA content in Figure 3 and that a G_{1A} compartment can be clearly identified.

content [53,63]. The aforementioned thresholds for RNA or protein content in G_1 are also clearly manifested (arrows).

Multiparameter correlated data, as shown in Figure 11, offer more than just the distribution of cells with respect to DNA versus RNA and DNA versus protein. Thus, it is possible to estimate whether the low-RNA G_1 cells (G_{1A}) also have low protein content, that is, whether the G_{1A} population discriminated either by RNA or protein content are identical. Indeed, by gating the low RNA containing G_1 cells based on RNA content (G_{1A} cells) and replotting the gated population with respect to DNA versus protein, we observed that the G_{1A} populations discriminated by RNA and protein content are nearly identical [53].

Analysis of the RNA/DNA ratio in relation to DNA content (Fig. 11e) reveals a characteristic pattern of changeable rates of DNA replication versus transcription during the mitotic cycle. Thus, during G_1, while DNA content remains constant, cells accumulate increasing quantities of RNA, and, as described above, G_1 phase is observed to be heterogeneous with respect to the RNA/DNA ratio. The critical ratio (arrow) reflects the same RNA threshold discussed above. During S phase, the rate of DNA replication exceeds RNA accumulation giving rise to a slanted, negatively skewed slope for the S cluster. Cells in G_2 + M have an RNA/DNA ratio similar to that of the majority of G_1 cells as judged from the peak values on contour maps.

The ratio of RNA/protein is a novel parameter, and it reflects the relationship between cellular transcriptional versus translational activity. During exponential growth this ratio remains remarkably constant and uniform throughout the cell cycle with only a few G_1 cells having elevated ratios contributing to the skewed distribution of the G_1 population (Fig. 11f). The RNA and protein content of individual cells are highly correlated (Fig. 11c). The multiparameter analysis as shown in Figure 11 is of great value in estimating unbalanced growth, for example, as induced by antitumor drugs or cell synchronization procedures, especially in situations where the factors inducing unbalanced growth have different effects on RNA versus protein synthesis or their respective turnover rates.

DNA and Nuclear Size, Protein, or RNA Content. Studies on isolated nuclei are often more advantageous than analysis of whole cells. From a practical point, isolation of nuclei from solid tissues is a simpler task than obtaining a preparation of well-dispersed intact cells suitable for flow cytometry. Isolated nuclei, therefore, may be preferable for analysis of solid tumor samples in pathology. Indeed, isolated nuclei are commonly used to estimate DNA index and cell cycle distribution by single-parameter flow cytometry [103,142, 250,255].

Nuclear constituents are directly involved in DNA replication, transcription, and chromosome packaging. Therefore, multiparameter studies performed on isolated nuclei [198,199,218] are likely to be more sensitive to any cell heterogeneity occurring during progression through the mitotic cycle than are measurements of gross changes in cytoplasmic constituents.

Different tissues require somewhat different methods of nuclear isolation for optimal yield and purity [204]. The main problem is uncontrolled loss of nuclear constituents. Generally, the detergent-based techniques result in a loss of various proteins and low molecular weight RNA. It should be mentioned, however, that nonionic detergents can be used at certain pHs and in the presence of proteins to permeabilize the cells rather than to isolate nuclei [75,78]. The citric acid-based procedures do not result in loss of nucleic acids, but certain proteins (especially histone H1 and HMG proteins) are removed. The nonaqueous techniques [292] or isolation of nuclei in hypertonic sucrose may be performed to obtain nuclei with most of the constituents preserved [204]. Total removal of the adjacent cytoplasm ("cytoplasmic tags"), which can be monitored by phase or interference microscopy, is essential if the data are to be interpreted as relating to isolated nuclei.

The methods of measurement of nuclear size, RNA, or protein content are reviewed in Chapter 16 of this volume. A relationship between nuclear size, protein, or RNA content versus cell-cycle progression is also discussed. In the case of exponentially growing cells, it is possible to distinguish G_{1A} and G_{1B} compartments based on differences in nuclear protein or RNA, as is the case for total cellular protein or RNA content (Fig. 3 and 10). Furthermore, because the amount of protein or RNA associated with metaphase chromosomes is lower than that of the G_2 nuclei following cell lysis (during mitosis some nonhistone proteins and RNA dissociate from chromatin), the aggregate of metaphase chromosomes can be distinguished from nuclei of G_2 cells due to this deficit, and thus the proportion of cells in mitosis can be estimated [199,218]. Following cytokinesis the dissociated constituents do not immediately reappear in G_1 nuclei inasmuch as early G_1 (G_{1A}) nuclei show a deficit in protein or RNA content when compared to cells in late G_1 (G_{1Q}), S, or G_2. The G_{1A} cells are characterized by a lower ratio of nuclear protein or RNA per unit of DNA than cells in G_{1B}, S, or G_2. It may be concluded that an influx of these constituents from the cytoplasm to the nucleus takes place during G_{1A}.

Recognition of cells in deep quiescence (G_{1Q}) is also pos-

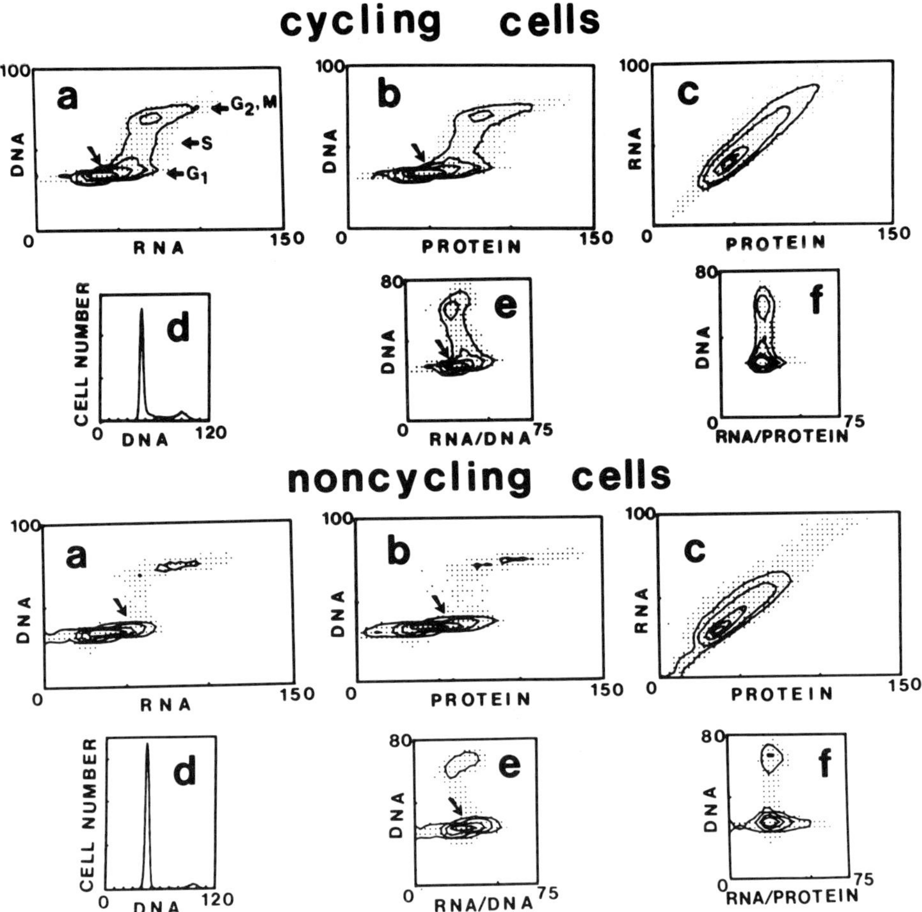

Fig. 11. Simultaneous staining of DNA, RNA, and protein for a population of cycling and noncycling CHO cells. CHO cells obtained from exponentially growing cultures (top; 57% G_1, 24% S, 19% G_2 + M cells) and cultures deprived of isoleucine for 30 hr (bottom; 95% G_1, 2% S, 6% G_2 + M cells) were fixed and stained with Hoechst 33342, pyronin Y and fluorescein isothiocyanate [53]. The correlated measurements of DNA versus RNA (a), DNA versus protein (b) and RNA versus protein (c) were obtained with a three-laser flow cytometer [52,53]. List mode recording of the data provides the ratios of DNA versus RNA/DNA (e) and DNA versus RNA/protein (f). The single-parameter DNA distributions are shown in (d). The arrows indicate the threshold RNA or protein content of G_1 cells below which cells do not immediately enter S phase.

sible based on RNA (Fig. 12) or protein [199] content of isolated nuclei. Figure 12 shows DNA and RNA values of nuclei from the liver of a young mouse. It is evident that cells entering S phase originate from the high-nuclear RNA population. In older mice, when proliferation of hepatocytes is minimal, nearly all diploid nuclei belong to the low-RNA (G_{1Q}) population, whereas stimulation of hepatocyte proliferation in regenerating liver coincides in time with disappearance of the G_{1Q} population [112]. In comparison with whole-cell RNA, nuclear RNA content, reflecting predominantly accumulation of rRNA precursor in nuclei, is expected to be a more sensitive marker of changes in genome transcription and also to react more rapidly to any metabolic change associated with the cell cycle. No studies have been done, however, to compare the kinetics of changes in nuclear versus whole cell RNA content (e.g., in cells stimulated to enter the cell cycle or undergoing transition to quiescence) to prove such a relationship.

DNA and Membrane Potential-Sensing Probes. There are several cationic fluorescent probes that show affinity toward mitochondria of living cells [137,270]. Among such probes, rhodamine-123 (R123) is the most commonly used [119]. Uptake of this fluorochrome depends both on the number (mass) of mitochondria per cell and the electronegativity of the mitochondrial membrane. Uptake of R123 has been studied in relation to cell position in the cell cycle, quiescence, and differentiation [69,82,117].

An example of changes in R123 uptake during the transition of quiescent cells to the cell cycle is illustrated in Figure 13. In this analysis, the second parameter is forward light scatter. Nonstimulated lymphocytes (G_{1Q}) are very uniform, having low light scatter and R123 fluorescence values. The population of PHA-stimulated lymphocytes consists of responding (cycling) and nonresponding (G_{1Q}) subpopulations. Discrimination of these subpopulations is possible based on both R123 fluorescence and light scatter; fluorescence, however, appears to offer better discrimination. A close correlation was observed between the increase in RNA content of stimulated lymphocytes and R123 uptake [69]. Simultaneous measurements of DNA and uptake of mito-

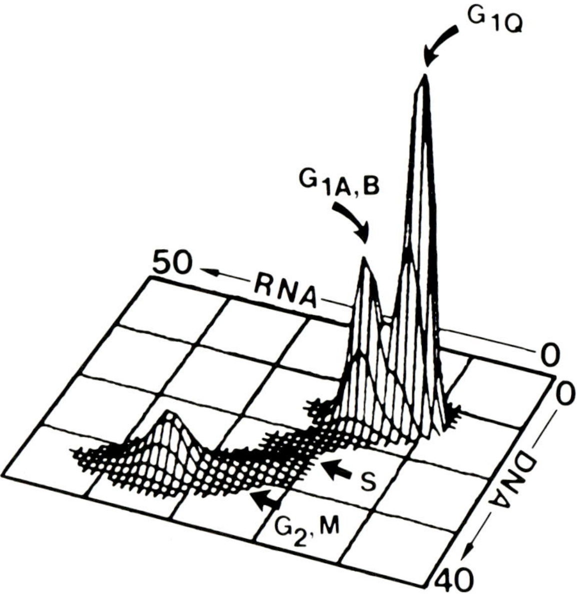

Fig. 12. DNA versus RNA content of mouse hepatocyte nuclei. The nuclei were isolated from the liver of a 14-week-old mouse and stained with acridine orange [112].

chondrial probes (R123, cyanine dyes) requires two-laser sequential excitation; such data on mitogen-stimulated lymphocytes was presented by Shapiro [226].

Friend erythroleukemia cells induced to differentiate in the presence of dimethylsulfoxide exhibit markedly diminished uptake of R123 in comparison with their exponentially growing counterparts [82]. Also, L1210, CHO, and Friend erythroleukemia cells in stationary cultures bind approximately half the R123 of their counterparts in exponentially growing cultures. Because of high intercellular variability in the binding of R123 in most cell types, the discriminatory power of this fluorochrome is lower than, for example, that of cellular RNA content.

Some mitochondrial (or membrane potential sensing) probes modify the uptake of DNA fluorochromes by living cells. Thus, for instance, Crissman et al. [54] have recently reported that preincubation of CHO cells with the membrane potential dye DiO-C5-3 (3,3-dipentyloxacarbocyanine), but not with R123, enhances the uptake and stainability of DNA with Hoechst 33342 and has a positive effect on cell survival. Although the mechanism of this phenomenon is unclear, the combination of Hoechst 33342 and DiO-C5-3 may be optimal for simultaneous measurements of DNA content and membrane potential in living cells. Because it is unknown to what extent pyronin Y stains mitochondria versus RNA in live cells [50,67] a combination of Hoechst 33342 and pyronin Y, although producing excellent DNA staining [225,226], may yield results that are difficult to interpret.

Sensitivity of DNA In Situ to Denaturation. Profound changes occur in both the gross and molecular structure of nuclear chromatin during the cell cycle as well as during cell transition to quiescence or differentiation [59] (see also Chapter 16 in this volume). We have developed a technique to study these chromatin changes based on differences in sensitivity of DNA in situ to denaturation. Briefly, RNase-treated cells are subjected to heat or acid treatment followed by staining with the metachromatic fluorochrome, acridine orange. After partial denaturation of DNA by heat or acid, acridine orange stains the nondenatured DNA sections green, whereas interactions of the dye with the denatured sections results in red luminescence. Thus, the relative proportions of red and green luminescence of acid or heat-treated cells represent the proportions of denatured and native DNA, respectively.

In general, the sensitivity of DNA to denaturation in the majority of cell types analyzed correlates with the degree of chromatin condensation. DNA in mitotic and in quiescent cells is characterized by condensed chromatin (G_{1Q}), and is most sensitive to denaturation. The most resistant DNA is in late G_1 (G_{1B}) and early-S-phase cells. This pattern of staining is shown by exponentially growing cells in Figure 14 and during mitogenic stimulation of lymphocytes in Figure 15. Analyzing exponentially growing populations, it is possible to distinguish G_1 versus S versus G_2 + M cells based on total (red plus green) luminescence, which, under appropriate conditions of cell staining and measurement, is proportional to total DNA content. The ratio of red to total luminescence (α_t), on the other hand, allows one to discriminate mitotic from G_2 cells and to recognize G_{1A} and G_{1B} subpopulations. Here, as in the case of cellular RNA (Fig. 3) or protein (Fig. 10), the threshold discriminating G_{1A} versus G_{1B} cells is equivalent to the maximal α_t value (mean value plus three standard deviations) of the early-S cells. By definition, therefore, the G_{1B} cells have α_t values similar to those of early-S cells, whereas G_{1A} cells are significantly different from the latter.

Mechanisms related to changes in chromatin structure during the cell cycle that are responsible for the observed differences in sensitivity of DNA to denaturation are discussed in Chapter 16 of this volume.

Quiescent, nonstimulated lymphocytes are very uniform, having approximately equal proportions of green and red luminescence (Fig. 15a). Lymphocytes stimulated for 18 hours with PHA (Fig. 15b) consist mostly of cells undergoing transition to the cell cycle (G_{1T}) prior to entrance to S. These cells have wider distribution with the major subpopulation characterized by increased green and lowered red luminescence, in comparison with G_{1Q} cells. In lymphocyte cultures stimulated by PHA for 3 days (Fig. 15c,d) subpopulations of cells in G_{1Q}, $G_{1A,B}$, G_2, and M can be distinguished. The number of cells in M is increased concomitant with a decrease in the proportion of G_1 cells in culture treated with Colcemid (Fig. 15d). The peaks and ridges ascribed to particular subphases as marked on the histogram (Fig. 15a–d) were identified by studies of synchronized populations [79].

Cell Cycle Specific Constituents Detected Immunochemically. Until quite recently, the predominant view was that, with the exception of histones, the rates of synthesis of other proteins were invariable during the cell cycle [94,175]. However, the past several years have seen the discovery of cell products, predominantly proteins, whose transcription or synthesis and accumulation vary sharply during the cell cycle. These constituents can be broadly categorized into three groups. The first are those that are suspected of having regulatory function during the cell cycle. Some oncogene products are believed to belong to this category. The second are proteins or peptides directly involved in cell proliferation, for instance, enzymes and cofactors of DNA replication, microtubule-associated or other structural proteins. The third

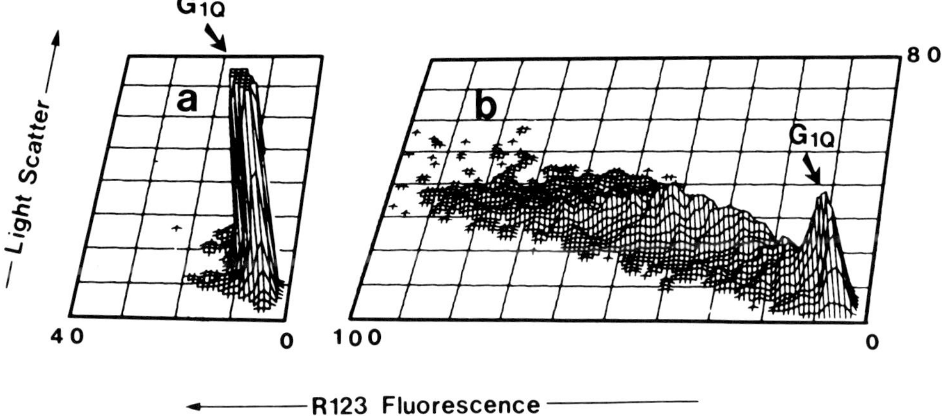

Fig. 13. Forward angle (2–19°) light scatter versus green fluorescence of unstimulated **(a)** and PHA-stimulated **(b)** human lymphocytes supravitally stained with the mitochondrial probe rhodamine 123 (R123). Unstimulated lymphocytes (G_{1Q}) are uniform with respect to size (light scatter) and fluorescence. After PHA stimulation (3 days), a responding population becomes apparent having both increased size (light scatter) and mitochondrial stainability (R123 fluorescence); the nonresponding cells in the stimulated culture (G_{1Q} cells) show no change in light scatter or fluorescence.

group are the cell surface proteins or glycoproteins that exhibit cell cycle (phase) dependency.

Detection of these constituents by cytochemical or immunological techniques may provide new markers of cell proliferation. Their use, especially in multiparameter analysis, in conjunction with other markers of cell metabolism (cellular RNA or protein content, nuclear protein content, mitochondrial activity) and with DNA content measurements could be of great value in providing a better understanding of the biology of the cell cycle and as possible prognostic tools in clinical oncology.

A list of the cell constituents that are best characterized with respect to their cell cycle specificity is given in Table 1. Several of the products listed have not yet been characterized with respect to their possible function. Some of the proteins were analyzed by correlating time of transcription with particular phases of the cell cycle, although little is known about their synthesis and turnover rates. It is possible that post-transcriptional regulation of the expression of these genes could shift appearance of their products to other phases of the cell cycle. However, a large number of the constituents listed in Table 1 have already been investigated by single or multiparameter flow cytometry in model systems and on clinical material. The products listed in Table 1, which are constituents of nuclear chromatin, are reviewed in more detail in Chapter 16 of this volume. More extensive review of protooncogenes and other proliferation-associated cellular constituents can be found in [121,125,189,215].

While flow cytometric analysis of cell cycle-specific protein products could be of great value in numerous situations, there are several potential problems and difficulties associated with the use of antibodies to quantify cell constituents. First, because the majority of the products are localized intracellularly, cells have to be permealized to allow access of the primary and then secondary antibodies. Unfortunately, fixation does not always ensure retention of the products to be assayed; some or all of the material can leak out of the cell during fixation or subsequent treatments [164]. Furthermore, fixation may cause chemical (e.g., reaction of active groups with aldehydes or other crosslinking fixatives) or biophysical (precipitation of the products by anhydrous fixatives) changes in the constituents, which can affect their affinity in reaction with antibodies. Therefore, fixation should be individually tailored for particular antigens to obtain both specific and quantitative estimates of their expression [116]. Isolation of unfixed nuclei (to detect nuclear antigens) may also result in loss of various constituents. In addition, nuclei tend to aggregate with repeated washings, which hinders their analysis by flow cytometry.

A second, major limitation of quantitative immunocytochemistry involves the possible variations in accessibility of the antigen. The binding of large molecules, such as antibodies, is obviously affected if the antigen is in a complex with other constituents, that is, present but sequestered from "view." Furthermore, allosteric changes in the conformation of the antigen could affect accessibility to and affinity for the antibody. Antibodies are often designed to recognize a specific epitope on an antigen. Allosteric changes in the conformation of the antigen are likely to take place at such regions (e.g., phosphorylation) changing the affinity for the antibody. Thus, the antigen may be present but no longer recognized by that particular antibody.

Still another point that requires critical interpretation is any increase in the quantity of a particular constituent observed during the G_0–G_1 transition. It is well established that G_0 cells are several times smaller than their cycling counterparts. For instance, during mitogenic stimulation of lymphocytes their dry mass [64], RNA [69,78], or mitochondrial content [69] all increase up to 6- to 7-fold. Considering the above, an increase in the amount of any constituent, unless it exceeds the above values, must be considered just a reflection of cell growth in size and cannot be interpreted as a specific indication of the G_0–G_1 event. This situation is similar to the twofold increase of any constituent in G_2 + M related to early-G_1-phase cells. To document that an increase of a particular constituent is more *specific* than cell growth in size, a rise in the *ratio* of that constituent to total protein or RNA content should be demonstrated.

In summary, caution should be exercised when the data obtained by immunocytochemical methods are interpreted as representing changes in the quantity of an intracellular antigen. Independent biochemical evidence, using immunopre-

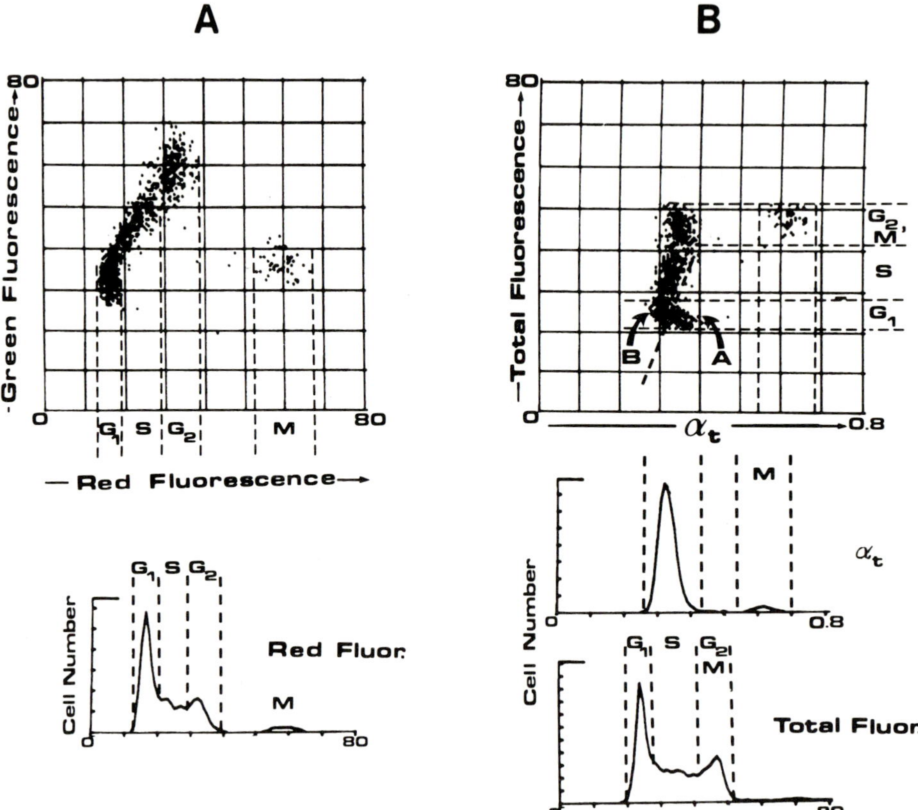

Fig. 14. Cytograms of L1210 leukemic cells stained with acridine orange following partial acid-induced denaturation of DNA in situ. Exponentially growing L1210 cells were fixed, treated with RNase, exposed to acid (pH 1.3) and stained with acridine orange [83]. The relative intensities of green and red fluorescence **(A)** correlated with the extent of DNA in the native and denatured state, respectively [71,83]. The single-parameter red fluorescence histogram (A, bottom) often provides complete separation of G_2 and M cells allowing discrimination of G_1-, S-, G_2- and M-phase cells. Transformation of the data to total fluorescence (red + green fluorescence) and the ratio α_t relating to the sensitivity of cells to denaturation (red divided by red plus green fluorescence) is shown in **(B)**. Total fluorescence and α_t single parameter histograms (B, bottom) provide the DNA distribution and percentage of mitotic cells directly. The G_1 cluster can be further subdivided into G_{1A} and G_{1B} cells based on α_t values as shown in **(B)**.

cipitation techniques, should confirm quantitative flow cytometric data before accepting as meaningful.

Kinetic Analysis of Cell Populations

Synchronized Populations. Many of the techniques found most useful for obtaining synchronized populations of pro- or eukaryotic cells have recently been reviewed by Lloyd et al. [162]. Generally, synchronization is meant to imply a relatively uniform population of cells in a particular cell cycle phase, that is, according to DNA content. However, cell synchrony is not necessarily synonomous with metabolic uniformity, since, for example, even immediately postmitotic G_1 daughter cells are heterogeneous with respect to RNA, protein, or mitochondria [63]. Nevertheless, as a practical matter, a population of cells of equivalent chronological age (i.e., time since division) are considered to be synchronized.

The various techniques used to obtain a synchronous cell population fall into three categories: 1) chemical, 2) nutritional, and 3) mechanical [162].

The use of chemical agents involves blocking cell cycle progression with an inhibitor of DNA synthesis or mitosis. Thus, agents such as thymidine (at high levels), hydroxyurea, aphidicolin, some antimetabolites, and folic acid inhibitors will, by a variety of mechanisms, block cells at the G_1-S boundary, whereas agents such as Colcemid (colchicine), the vinca alkaloids (vinblastine, vincristine), epipodophyllotoxins (VM 26, VP 16-213), or nocodazol accumulate cells in mitosis. Not all these agents provide a uniformly synchronous population when applied to a culture of synchronously growing cells. This is especially true of DNA synthesis inhibitors, which, in addition to accumulating cells at the G_1-S border, will often block cell progression throughout S phase. Varying degrees of toxicity are associated with many of these agents. For example, extended incubation with mitotic inhibitors often leads to cell death [183] and/or formation of polyploid cells as a result of endoreduplication [217,293]. The action of most of these agents is rarely completely reversible and often associated with a lag before "normal" cell cycle progression rates are reestablished.

Nutritional synchronization procedures generally involve removal of essential growth factors from the medium either in part or totally. This form of synchronization is only useful in obtaining cells in a metabolic subphase of G_1 that is usually equated with quiescence (e.g., G_{1Q}). Again, except for normal (e.g., nontransformed) fibroblasts, the technique rarely provides complete synchronization, since cells often

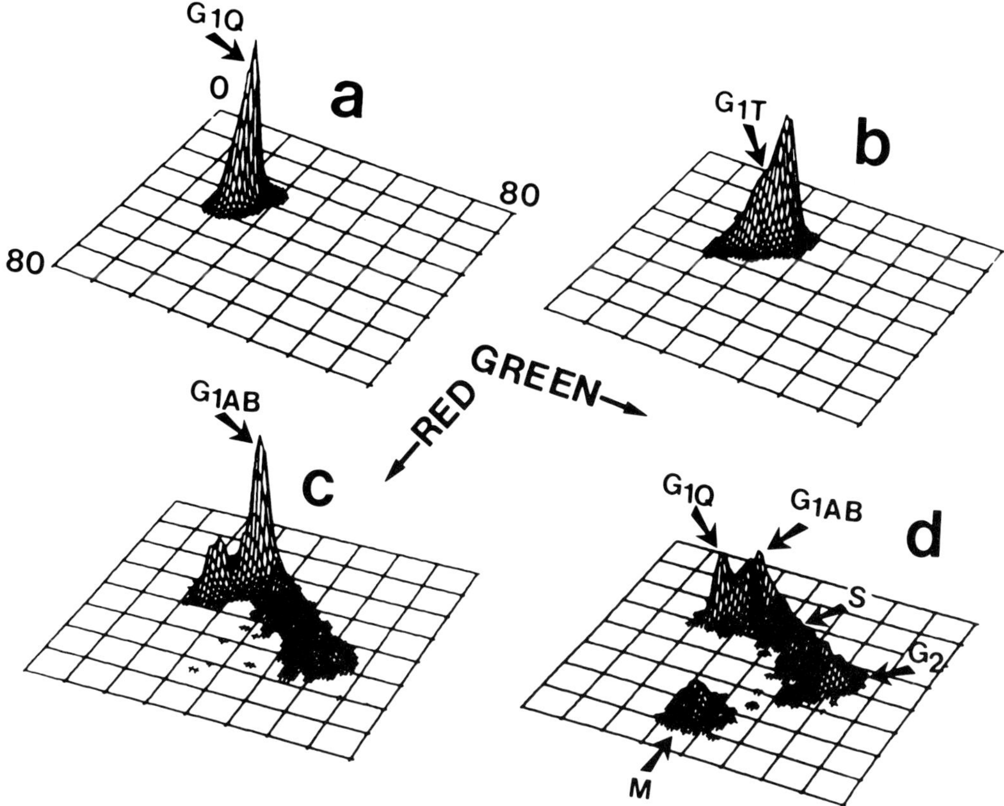

Fig. 15. DNA acid denaturation pattern of human unstimulated and PHA-stimulated lymphocytes. Unstimulated lymphocytes **(a)** form a uniform population. A short time after PHA stimulation (18 hours; **b**) cells undergoing transition to proliferation (G_{1T}) are characterized by increased green (native DNA) and decreased red (denatured DNA) luminescence. During maximal stimulation after 3 days with PHA **(c)** in addition to G_{1Q}, G_{1A}, and G_{1B} populations there are also cells in S, G_2, and M. Treatment with Colcemid for 8 hours **(d)** caused an accumulation of mitotic cells and depletion of cells from the post-mitotic G_1 peak.

either remain in or continue to slowly progress through S and G_2 for days after removal of serum [9]. Reversal from this block is often accompanied by a delay in the onset of entrance into S phase that is correlated with the length of nutrient deprivation [19].

Mechanical synchronization involves either some form of velocity-sedimentation (e.g., gradient centrifugation, differential sedimentation at unit gravity, continuous flow centrifugation, centrifugal elutriation) or mechanical dislodging of mitotic cells, which tend to be less tightly adherent to the culture vessel than interphase cells [197,251]. Several of these techniques require large numbers of cells, fail to provide tight synchrony, are time-consuming, and/or are not very efficient (large cell loss). However, centrifugal elutriation and mitotic detachment can provide good numbers of highly synchronized cells sufficient even for biochemical experimentation. Mechanical techniques in general, in contrast to chemical or nutrient-deprivation methods, have fewer problems with toxicity, reversibility or lag-time asynchrony.

Multiparameter flow cytometry provides a unique means to both monitor attempts at cell synchronization and to use synchronized populations to follow the kinetics of cell proliferation either under ideal or perturbed conditions. Two examples, representing different methods of synchronization, are presented below. They illustrate the type of kinetic information that this approach can provide.

Chemical Synchronization. Using a chemical inhibitor, in this case hydroxyurea (HU), mitogen-stimulated lymphocytes were allowed to undergo the transition from G_0 to G_1, but were blocked at the beginning of S phase (Fig. 16). Stimulated lymphocytes normally exhibit a large variation in RNA content, and, with extended treatment with HU (18–24 hours), the variation in RNA content was increased to the point where some of the cells attained RNA levels equivalent to that normally found in G_2 cells [65,66]. Release from the block allowed initiation of DNA synthesis among the lymphocytes synchronized at the entrance to S phase. It is clear from the bivariate distribution in Figure 16 that cells with the highest RNA content were the first to progress through S phase. Analysis of the rate of S-phase traverse as a function of RNA content demonstrated that these cells in the highest and lower quartile with respect to RNA content reached G_2 phase in 5 or 7 hours, respectively [66]. It is also interesting to note that those cells in G_1 with very high RNA levels equivalent to normal G_2 lymphocytes did not, as would normally be the case, increase their RNA content as they traversed S phase (Fig. 16). Thus, "overloading" cells with RNA during their arrest in G_1 not only alters their subsequent kinetics of progression through S phase, but also mobilizes some compensatory-regulatory mechanisms precluding additional increase in RNA content that otherwise occurs during S phase.

Mechanical Synchronization. A second kind of synchronization experiment is illustrated in Figure 17. In this instance

TABLE 1. Cell Constituents Showing Cell Cycle Specificity

Substance	Nature and MW	Localization, affinity	Cell cycle phase specificity, role
c-*myc* product	64 kD phosphoprotein [111,280]	Nuclear matrix [92] ds-DNA [280]	Peaks early during G_0–G_1 transition [129]; invariable during the cell cycle [252]; high in tumor cells [111,278], generally low in differentiated cells [96,152], occasionally increases during differentiation [156]; role in acquisition of cell competence [124]; promotion of DNA replication [114]
c-*fos* product(s)	40–72 kD protein(s) [144,186,200]	Nucleus	Peaks very early during G_0–G_1 transition, precedes c-*myc* appearance [109,144,186]
c-*myb* product	75 kD protein [140]	Nucleus, DNA [140]	Peaks during G_0–G_1 transition, following c-*myc* and c-*ras* [253]
c-*ras* product	21 kD protein [249]	Cell membrane [289]; GTP-binding threonine kinase [159]	Peaks during G_0–G_1 transition, following maximal expression of c-*myc* [32,105,106,184]; increased expression during differentiation [241]
c-*src* product	60 kD phosphoprotein [30]	Cell membrane [49]; tyrosine kinase [47]	Serine-phosphorylated very early during G_0–G_1 transition [246]; invariable during the cell cycle [46]; role in differentiation [3]
"cyclin" [35] "proliferating cell nuclear antigen" [245]	36 kD acidic protein [168]	Nucleus [35]	High level in cycling cells [35]; peaks during S [36]; low in senescent cells [34]; role in DNA replication [28]
"p53 antigen" [87]	48–53 kD proteins [87]	Nucleus	High in transformed cells [51,87]; low in quiescent [176] and differentiated cells [230]; complements c-*myc* [291]; role during G_0–G_1 transition [172]
"p105 antigen" [43]	105 kD protein [43]	Nucleus, euchromatin, interchromatin	Low in G_0 cells, high in cycling cells [42]; highest in M
"Ki67 antigen" [101]	Unknown	Nucleus	Low in G_0 cells, high and invariable in cycling cells [100,101]
statin [275]	57 kD protein	Nucleoplasm, nuclear envelope	High in senescent and quiescent fibroblasts [275,277]; lost during G_o–G_1 transition [276]
"p55 antigen" (PSL)	53 kD acidic proteins [14,15]	Nucleus	High during S, distinct from p53 and cyclin [14,15]
B23, C23	Acidic phosphoproteins	Nucleoli, nucleolar organizer [191,233]	High in transcriptionally active cells [233]; differences between G_0 versus G_1 (?)
Ap 4A	Adenosine (5′) tetra phospho (3′) adenosine	Binds to DNA polymerase[210]	Low in quiescent and differentiated cells, peaks at the onset of S [110,282], primes DNA synthesis [296]
DNA polymerase α	Enzyme, 140 kD sub units 76 and 66 kD	Nucleus, active during DNA replication	Minimal in quiescent cells, peaks during S [23,169] and G_2 [187]; Igs against the subunits available
Dihydrofolate reductase	Enzyme 23 kD	Nucleus	Low in quiescent cells, high during S [151,161]; detected by binding of dansylated methotrexate [152]
DNA topoisomerase I	Enzyme 100 kD	Nucleus, nucleolus [185]	Fourfold higher in cycling cells relative to G_0 [185]
Thymidilate synthetase	Enzyme 75 kD	Intracellular	High in cycling cells, peaks in mid-S
Ribonucleotide reductase	Enzyme, 60 kD	Intracellular	3- to 7-fold increase during transition from G_1-S [95]; Igs against MC subunit available [95]
Thymidine kinase	Enzyme, 55 kD	Intracellular	Increase during stimulation of quiescent cells [161]
Calmodulin	Protein 25–30 kD	Intracellular, nuclear receptor [171]	Increase during G_1–S transition [37], and G_1–G_0 transition [38]
Actin	Structural protein 42 kD	Intracellular	Increases during G_0–G_1 transition [127,211,212]; β- and γ-actin mRNAs increase 3- 6-fold early during G_0–G_1 transition [170]
Tubulin	Structural protein 50 kD	Intracellular	Increased synthesis in G_2 [39,284]; 10-fold rise in mRNAs during G_0–G_1 transition [170]

(Continued)

TABLE 1. Cell Constituents Showing Cell Cycle Specificity (Continued)

Substance	Nature and MW	Localization, affinity	Cell cycle phase specificity, role
Microtubule-associated antigens	Structural proteins, 250 kD [188], 340–370kD [222]	Nucleus, cytoskeleton [223]	Increase during G_0–G_1 transition; decrease in normal but not tumor cells during S [222,223]
SV40 T antigen	Protein, 94 kD	Nucleus, nuclear matrix [235]	Present in certain cell lines only during G_2 [115]; stimulates DNA replication and rRNA synthesis [113,256]
p21	Pair of 21 kD proteins pI 5.5	Intracellular	Synthesized and present only during G_2 and M [284]
Histone H1 variants	H1° [194,195], H5 [267]	Nucleus, binds to spacer DNA	H1°-increased in G_0 [194,195]; H5-increased in differentiated cells [267]; H1/H1° ratio changes during G_1
Nucleosome core histones	H2A, H2B, H3, and H4 histones and their variants	Nucleus, bind to DNA in the core particle	Synthesized primarily during S except the "basal" histones
4F2 antigen [128]	Glycoprotein	Cell membrane	Appears during G_0–G_1 transition [128]
5E9 antigen [128]	Glycoprotein	Cell membrane	Appears after 4F2 in G_0–G_1 transition [128]
p34 ("Nuclear LDH")	34 kD protein, lactic dehydrogenase activity, and homology [196]	Nucleus binds to ss DNA [196]	Low in G_0; nuclear function unknown [31]
2-5A synthetase	Enzyme 42 kD	Intracellular polymerizes ATP into oligomers (2-5A system)	Increases 10-fold in late S [283]; 2-5A activates specific RNase, which cleaves mRNA and rRNA, and phosphodiesterase, which degrades 2-5A

CHO cells were synchronized in mitosis by mechanical detachment [197]. Within about 30 min, nearly all cells enter the immediately postmitotic G_1 phase (Fig. 17, 0 hour). Cell exit from G_1 begins at about 4–5 hours and can be observed to occur first among those cells with the highest G_1 RNA content. Progression through S to G_2 phase is accompanied by a continuous increase in RNA content. However, there is some loss of synchrony observable as a tail of cells trailing the major population back to the G_1 cluster. This effect is not due to bias in the method but rather demonstrates that exit of cells from G_1 occurs stochastically with the result that synchronized cells eventually reestablish an asynchronous growth distribution. Cultures synchronized by this technique are particularly useful for studying the sensitivity of cells toward drugs in relation to their cell cycle phase or the effect of unbalanced growth on cell cycle progression (see Chapter 39 in this volume).

Stathmokinetic Approach

The stathmokinetic or metaphase arrest technique was originally developed by Puck and Steffen [206] to estimate the rate of entry of cells into mitosis ("mitotic rate") or the cell birth ("production") rate [2,206,248,293,294]. The technique is based upon addition of agents that arrest exponentially growing asynchronous cell populations in metaphase (see above). The slope of the plot representing the accumulation of cells arrested in mitosis versus time after addition of the stathmokinetic agent provides an estimate of the rate of cell entrance into mitosis. Assuming the cells are proliferating and the growth fraction is known, the slope of the mitotic collection curve describes the "cell cycle time" [2,293]. Extensive mathematical analysis and interpretation of such data have been published elsewhere [73,90,135, 136,165,228,248,294]. Reviews detailing the advantages, limitations, and pitfalls of the stathmokinetic technique have appeared by Wright and Appleton [293] and Darzynkiewicz et al. [73].

Clearly, the quantitation of mitotic figures by light microscopy is both cumbersome and time-consuming. By contrast, flow cytometry offers a rapid, unbiased and accurate estimation of the distribution of cells throughout the entire cell cycle. Early flow cytometric approaches applied to stathmokinetic experiments generally employed a single-parameter DNA measurement to quantitate the accumulation of cells in the G_2 + M compartment [11,12,138]. Dosik et al. [90] extended the analysis to include the cell number in $G_{0/1}$ and S. This enabled them to estimate the rate of exit from $G_{0/1}$ and to observe cell transit through S phase. By adding a chemotherapeutic agent to the culture in addition to the stathmokinetic agent, Langen et al. [154] utilized single-parameter flow cytometric analysis to quantitate drug effects on cell cycle kinetics in vitro.

The combination of multiparameter flow cytometry with stathmokinesis offers a more comprehensive analysis of cell cycle kinetics [59,60,70,73,74]. The dye acridine orange provides simultaneous measurements of cellular DNA and RNA content in individual cells, but in the absence of cellular RNA (i.e., following removal with RNase) can provide a measure of the extent of double- versus single-stranded DNA. These latter measurements are carried out on fixed cells in which the DNA in situ has been partially denatured by heat or acid [71,72,77,80,81]. As discussed previously, the ratio of denatured to total DNA in the cell, expressed as α_t, indicates the degree of chromatin condensation in vitro [54]. The combi-

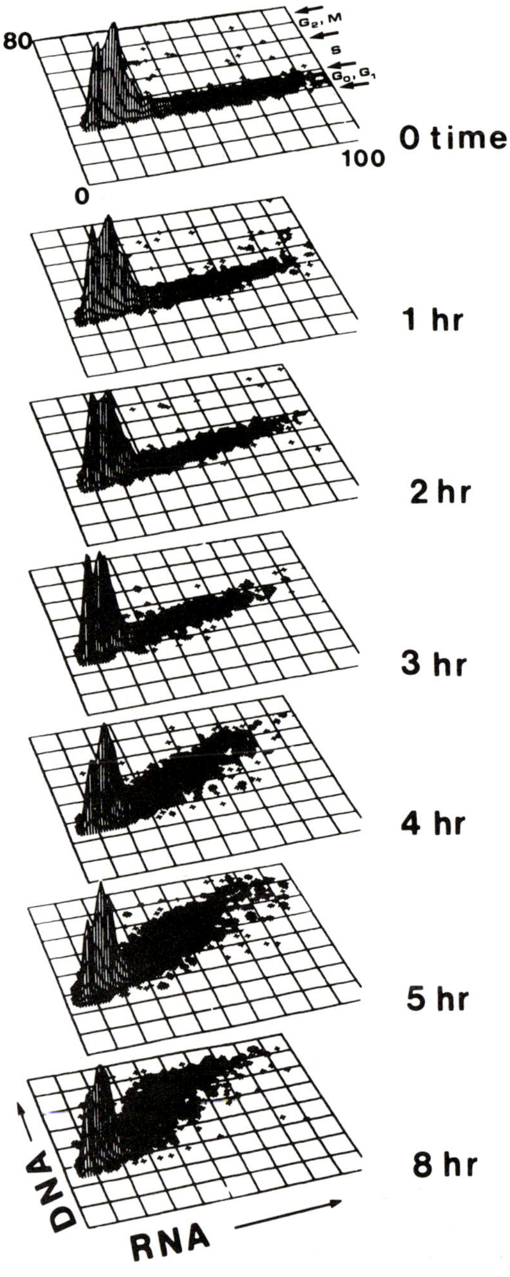

Fig. 16. S-phase progression of stimulated lymphocytes in relation to their RNA content. PHA-stimulated human lymphocytes were arrested at the G_1/S boundary by hydroxyurea and then released from the block [66]. Their progression during the subsequent 8 hours was studied on sequential histograms as shown in the figure. During the final 3 hours, vinblastine was added to prevent dividing cells from reentering G_1. The rate of cell progression through S is highly correlated with their RNA content.

nation of total fluorescence (denatured plus native DNA) versus α_t provides a means of discriminating mitotic cells as a separate population and permits identification of early-G_1 (G_{1A}) cells, late-G_1 (G_{1B}) cells and, in certain instances, cells with a G_1-DNA content that are quiescent (e.g., G_0, G_{1Q}, G_{1D}).

Figure 18 illustrates the staining pattern of CHO cells obtained following partial acid-induced denaturation and staining with acridine orange. DNA content (total fluorescence) provides discrimination of $G_{0/1}$, S, and G_2 + M cells, whereas the α_t parameter allows the discrimination of M, G_{1A}, G_{1B}, and G_2 cells. Upon addition of a stathmokinetic agent, the mitotic peak increases, while the percentage of cells in interphase decreases; the disappearance of cells from the G_{1A} compartment is especially obvious at early times (Fig. 18, top). For each time point (generally every 30 min or 1 hour after addition of the stathmokinetic agent) the percentage of cells in the various cell-cycle compartments can be determined by multivariate analysis and plotted as in the bottom of Figure 18.

Plot A (bottom of Fig. 18) illustrates the cell accumulation curves for the M and G_2 + M phases when a stathmokinetic agent is added at t = 0 to an asynchronously growing culture. The curves are exponential (straight line on a scale of log (1 + fx), where x is the fraction of cells in M or G_2 + M) and their slopes reflect the rate of entrance into M or G_2 + M [206]. Occasionally an early lag in the accumulation curve can be observed (dashed lines) that may reflect the time required for the mitotic inhibitor to penetrate the cell membrane and interact with the mitotic spindle. A deviation from the exponential accumulation of cells in mitosis can indicate that the motitic block is leaky, cell growth is not exponential, or there is selective cell death in some part of the cycle. An early change in slope of the mitotic accumulation curve might indicate perturbation in transit through the G_2 phase [73]. A negative slope may indicate disintegration of mitotic cells, for example, as a result of the drug treatment, or at longer culture times due to the toxicity of the stathmokinetic agent [183].

Since cell reentry into G_1 phase is prevented by the stathmokinetic agent, the rate of emptying of the G_1 compartment can be measured [84,261,265,266]. The curve representing cell exit from G_1 has two distinct slopes (Fig. 18B). On a semilogarithmic scale, the first portion of the curve has a concave shape, whereas the second is linear. The exponentially declining, linear slope provides an estimate of the half-time of cell residence in G_1 for a subpopulation of cells exhibiting stochasticlike G_1 exit kinetics. Within the framework of the probablistic model of the cell cycle [232], this parameter represents the probability of cell passage from the indeterminate ("A") to the deterministic ("B") compartment. The mean duration of G_1 transit times can be estimated based on the detection of the first time-point on the early portion of the G_1 curve where the data represented by the concave portion of the curve departs from the upward extension of the exponential portion of the curve (T_1). The time difference between the G_1 and G_{1A} exit curves (T_2) provides similar information (Fig. 18b).

Cell progression through S phase can also be directly analyzed in a stathmokinetic experiment by counting the number of cells progressing through various S-phase windows. The latter are established on the basis of DNA content, for example, at early-, mid-, and late-S phase (Fig. 18C). The number of cells measured in the window(s) increase at early times during stathmokinesis, reflecting the exponential age distribution of individual cells. Then the slope declines in a manner similar to the G_1 or G_{1A} exit curves (Fig. 18B,C). The length of the ascending portion of the curve provides a measure of the combined duration of the linear portion of G_1 phase plus that portion of S phase that precedes the lower threshold of the window. By comparing the changes in deflection points and the slopes of the descending portions of

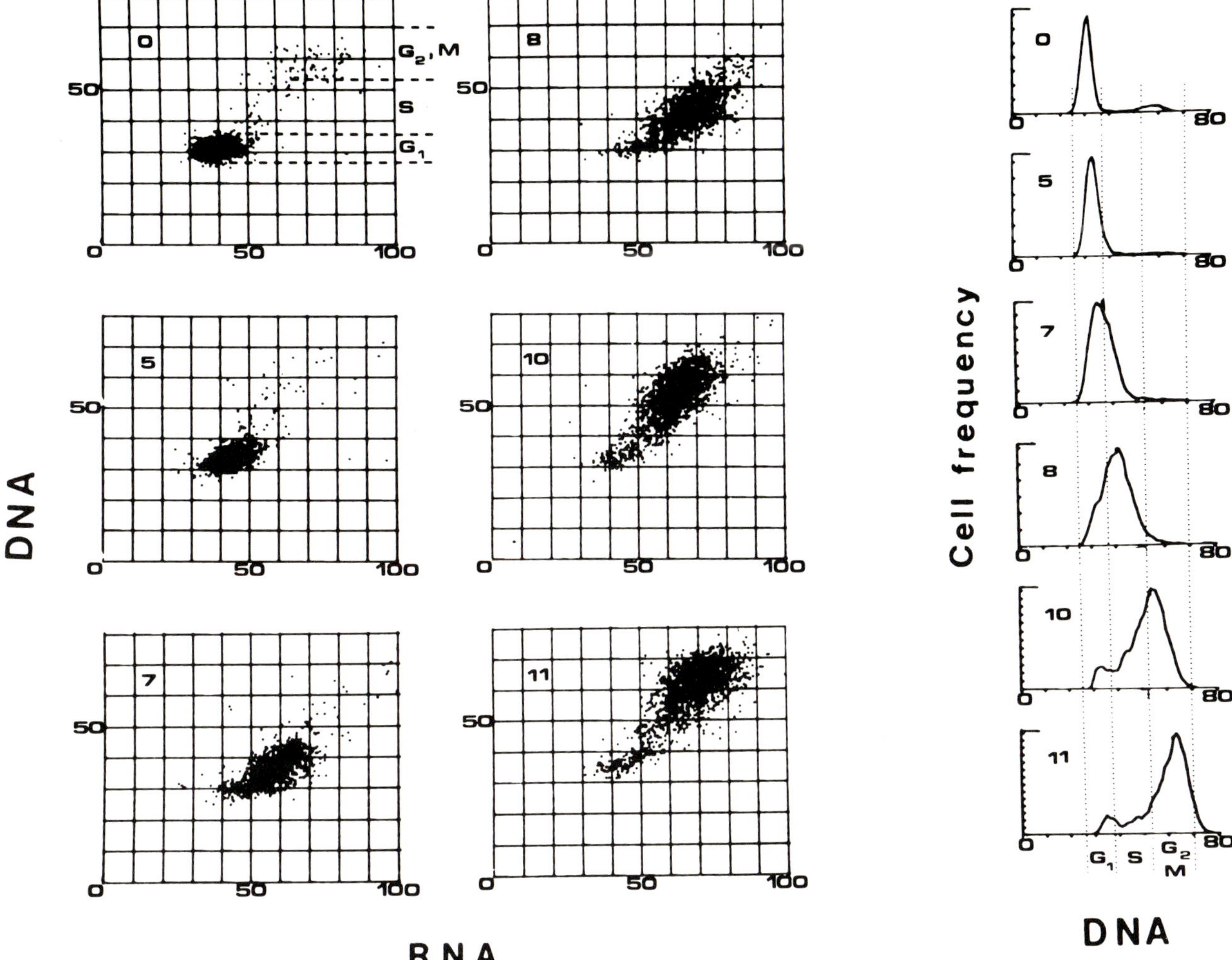

Fig. 17. Progression of CHO cells, initially synchronized by selective detachment in mitosis, through the cell cycle. An increase in RNA content parallels progression through the cycle. The stochastic nature of the G_1 exit is evident as a trail of cells remains in G_1 even 11 hours after division. The experiment was designed to correlate cell progression rate through particular phases of the cell cycle with their RNA content [65].

the S-phase transit curves from S-phase windows spaced throughout S phase, a perturbation in cell transit through S can be localized and measured [261,265,266]. For a specific example of the use of the stathmokinetic experiment to analyze cell cycle-phase perturbations induced by a chemotherapeutic agent, refer to Chapter 39 in this volume. Other details of the stathmokinetic method employing AO are presented elsewhere [73]. Mathematical analysis of stathmokinetic experiments can also be found in ref. [73] and the application of such analysis to practical situations in ref. [135].

Another approach to analyze cell kinetics that is similar, in principle, to stathmokinesis, employs drugs that prevent cytokinesis leading to cell growth at higher ploidy levels. One such agent is the metal chelating antitumor drug ICRF 159, which, at certain concentration, completely precludes reentrance of cells to G_1 [258]. Addition of this drug into exponentially growing cultures and sequential analysis of the cell-cycle distribution, allows one to estimate the rate of cell exit from G_1, as in the case of stathmokinesis induced by mitotic blockers [258]. This approach has been applied in vivo to distinguish cells that do not exit G_1 for an extended period of time and are therefore presumed to be noncycling [182].

Detection of BrdU Incorporation

The thymidine analogue 5-bromodeoxyuridine (BrdUrd), when incorporated in DNA, can be detected flow cytometrically by a variety of techniques. This approach, in many cases, can replace the laborious, expensive, and time-consuming techniques of [^{3}H]-Tdr autoradiography. First applied to the field of cytogenetics [157,158] (see Chapter 14, this volume), fluorescence quenching or enhancement caused by the analog was then measured flow cytometrically. The dyes Hoechst 33258 [25,27,147] and acridine orange [61] show reduced fluorescence upon binding to DNA in which thymidine has been substituted by BrdUrd, while mithramycin appears to exhibit enhanced fluorescence (20–25%) within BrdUrd-substituted cells [243,244] (see Chapter 13, this volume).

The quenching of 33258 Hoechst by BrdUrd has been

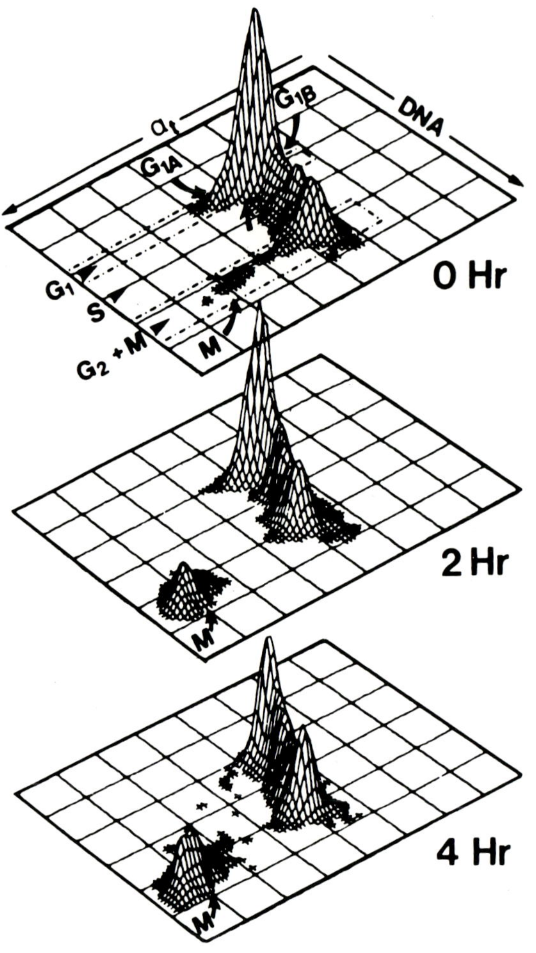
α
G1B
DNA
G1A
G1
S
G2+M
M
0 Hr
2 Hr
M
4 Hr
M

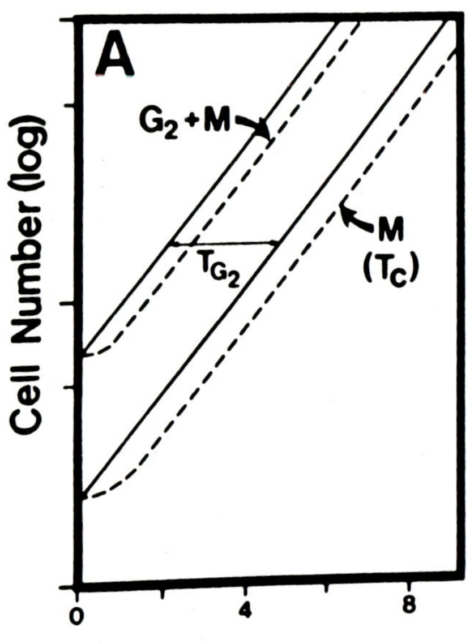
A
G2+M
M
(Tc)
Cell Number (log)
0
4
8

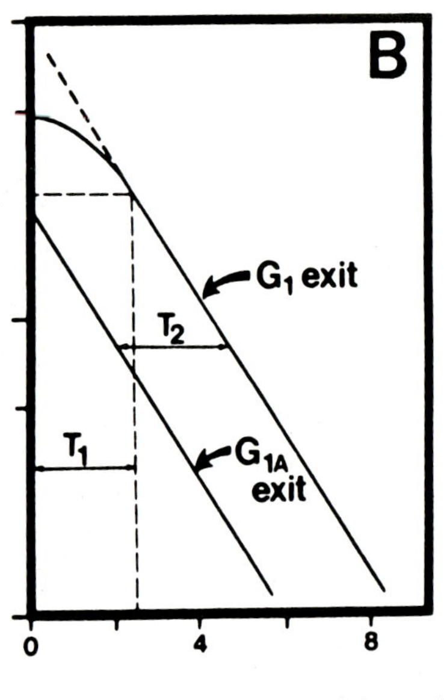
B
G1 exit
T2
T1
G1A
exit
0
4
8
Hours of Stathmokinesis

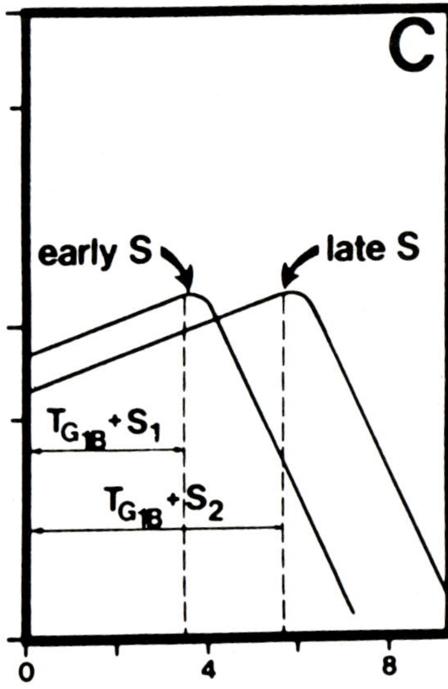
C
early S
late S
0
4
8

combined with staining of total DNA with ethidium bromide [27,147] to allow analysis of cells over several cycles. This method was perfected by Kubbies, Rabinovich, and their associates to estimate the cell transit rates through G_1, S, and G_2 phase and was successfully applied by these authors to various cell systems [148,149,207].

A very clever method of detection of BrdUrd-incorporating cells was recently developed by Crissman and Steinkamp [55]. Namely, the authors stained cells with Hoechst 33342 and mithramycin and subtracted the blue fluorescence signal of Hoechst 33342 (quenched by BrdUrd) from the green-yellow fluorescence signal of mithramycin (enhanced by BrdUrd incorporation. The S-phase cells produced fluorescence difference signals that were significantly greater compared to cells that did not incorporate the precursor [54]. This method is very sensitive, allowing recognition of DNA-synthesizing cells following as little as a 5-min pulse of BrdUrd (see Chapter 13 in this volume).

By combining BrdUrd quenching of acridine orange fluorescence with the acid denaturation of DNA in situ, it is possible to automatically identify mitotic cells that have and have not incorporated BrdUrd; this technique provides a flow cytometric equivalent to the autoradiographic procedure for obtaining the fraction of labeled mitoses [76]. The techniques described above lack sensitivity because they require that a significant portion of thymidine be replaced by BrdUrd molecules. For instance, an entire round of DNA synthesis in the presence of BrdUrd decreases green fluorescence of DNA stained with acridine orange by approximately 40% [61].

An example of the use of the fluorescence-quenching properties of BrdUrd-substituted DNA is presented in Figure 19. The upper three panels (Fig. 19a,b,c) illustrate the RNA versus DNA distribution of PHA-stimulated lymphocytes grown in the absence of BrdUrd for 4 (a), 6 (b), and 8 (c) days. The cultures manifest maximal stimulation on day 4 as characterized by the proportion of cycling (G_1, S, G_2 + M) cells and their high RNA content. At longer culture times the percentage of cycling cells decreases and their RNA content is reduced toward that of the G_0, unstimulated lymphocytes (Fig. 19b,c). The three lower panels (Fig. 19d–f) illustrate parallel cultures exposed to BrdUrd for approximately one cell generation. Thus, the culture in (d) is parallel to the one in (a) but BrdUrd was present for the final 20 hours prior to harvesting. Cultures (b) and (d) are parallel except that the cells in (d) were grown for 20 hours in the presence of BrdUrd. Cultures (c) and (f) are also parallel; however, (f) was treated for 20 hours with BrdUrd on the sixth day and harvested as (c) on day 8. Those cells having incorporated BrdUrd for approximately one full S phase show, following staining with acridine orange, a 40% decrease in DNA fluorescence though RNA fluorescence was relatively unaffected. The above example illustrates that the BrdUrd technique can be used to: 1) recognize the cycling population during proliferation; 2) prelabel the cycling cells for their later identification when they enter quiescence; and 3) correlate metabolic parameters such as RNA with proliferation [60].

Recently, an immunochemical probe has been prepared against BrdUrd, which, when tagged with a fluorochrome, provides a positive fluorescence signal for the identification of incorporated BrdUrd and is very sensitive [107,108]. A pulse as short as 30 min may be sufficient for the detection of DNA-synthesizing cells [107]. The technique has one drawback; it is necessary to fix the cells and partially denature the DNA, since the antibody will only bind to single-stranded regions of DNA. A comprehensive description of the flow cytometric techniques utilizing anti-BrdUrd monoclonal antibodies is presented in Chapter 23 in this volume.

Fig. 18. Kinetic experiment demonstrating measurements of cell exit rates from G_1, progression through S and entrance to M. Several compartments of the cell cycle can be distinguished following partial DNA denaturation by acid and subsequent staining with acridine orange (top; see also Fig. 14). During stathmokinesis induced by addition of Colcemid or other mitotic blockers, changes in cell number within these comparments can be measured to obtain kinetic information. **A:** In exponentially growing cultures, an increase in the number of cells in mitosis (M) is represented by a linear slope on the log scale; this slope provides an estimate of cell doubling time (Tc). Cell accumulation in G_2 + M is represented by a similar slope; the time distance between the slopes measures the average duration of G_2. Broken lines indicate data when a time lag between addition of the stathmokinetic agent and onset of mitotic arrest is apparent. **B:** Emptying of the G_1 and G_{1A} compartments. Cell exit from G_{1A} is represented by a linear slope on the log scale; the slope provides a measure of the half-time of cell residence in G_{1A}. The linear portion of the G_1 exit is similar to that of G_{1A}. The convex-shape of the G_1 curve indicates the exit of a cohort of cells having linearly distributed residence time in G_1. T_1 and T_2 are estimates of the duration of G_{1B} calculated in two different ways. **C:** Cell number measurements in "windows" in S phase provide curves. An initial increase in cell number in the windows represents the log age distribution of cells in exponentially growing cultures. The declining slope is due to the cell deficit resulting from their arrest in mitosis. The time of the slope reverse in the first window represents the duration of G_1 plus that portion of S which preceeds the window location. If the first and second window in S are very close to the entrance and exit of S, respectively, the time distance between the slope-reverse in these windows is a measure of the minimal duration of S phase. For details see refs. [70,73,83].

Metabolic and Kinetic Compartments of the Cell Cycle Distinguished by Multiparameter Flow Cytometry

Examples of cell subpopulations showing different metabolic properties that can be distinguished by flow cytometry were illustrated in the first portion of this chapter. Based on these differences, we have proposed to subdivide the cell cycle into several distinct subcompartments [60,74]. Such subdivision offers higher accuracy of cell classification than the traditional four main phases of the cell cycle. A summary of the characteristics and typical examples of these compartments are given below; justification of the subdivision and possible relationships to specific molecular events in the cell cycle will be discussed further.

1. *G_{1A} compartment* is detected during exponential cell growth. It is comprised of G_1 cells, which, in addition to DNA content, are significantly different from early-S-phase cells by other parameters. Thus, the G_{1A} cells are predominantly early G_1, postmitotic cells that have an RNA or protein content distinctly lower than cells in early S phase. The chromatin structure of G_{1A} cells is also distinct from that of cells in early-S; the difference is manifested as increased sensitivity of DNA in situ to acid-induced denaturation (increased chromatin condensation) and decreased nonhistone protein content. Their nuclear RNA content is also very low.

Kinetically, the cell residence times in G_{1A} have a characteristic, exponential-like component that may be responsible

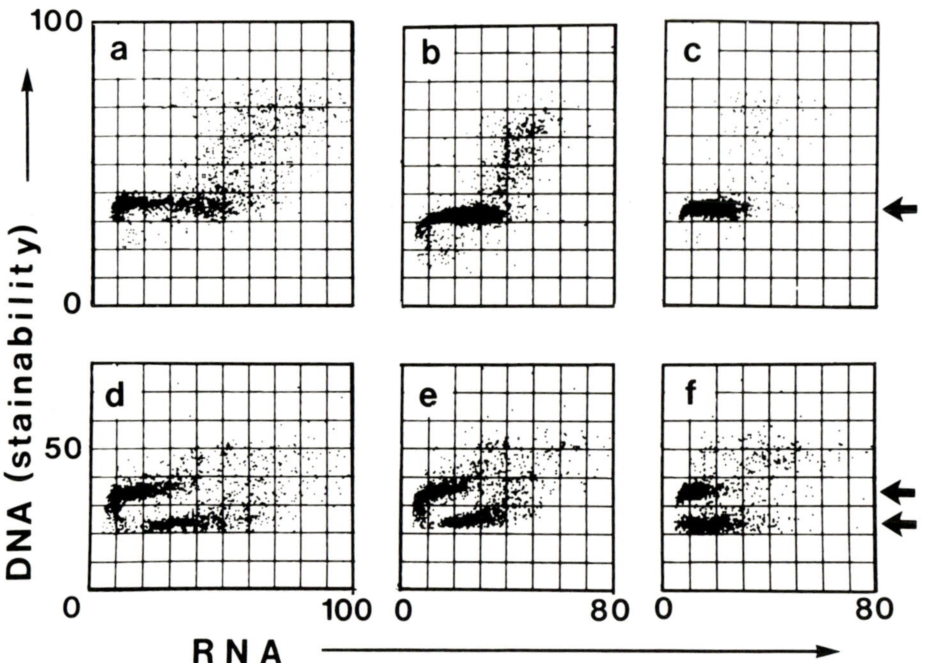

Fig. 19. RNA and DNA stainability with acridine orange of PHA-stimulated lymphocytes grown in the absence (**a–c**) and presence of BrdUrd (**d–f**); for an explanation see text and ref. [60].

for heterogeneity of cell generation times [63]. Within Mitchison's framework of the cell cycle [177] the G_{1A} phase represents a gap in the DNA replication/division cycle. During this phase, cells are in the growth cycle, accumulating certain metabolic constituents up to a critical threshold. Thus, G_{1A} can be also characterized as a "growth" or "equalization" phase [63]. Cell residence times in G_{1A} appear to be inversely proportional to the amount of the metabolic constituents that they inherit during asymmetric cytokinesis [63,133].

There is a paucity of nonhistone proteins and RNA in nuclei of G_{1A} compared to early-S-phase cells. Both constituents dissociate from chromatin during mitosis and their level is minimal in newly formed postmitotic nuclei (see Chapter 16 of this volume). Considering the nuclear events, therefore, the G_{1A} period represents the time during which nonhistone proteins and RNA migrate from the cytoplasm to the nucleus, the nucleolus is reformed and rRNA synthesis is initiated; DNA replication starts when a threshold quantity of these constituents is accumulated in the nucleus. It is likely that the nonhistone proteins are involved in or perhaps essential for decondensation of the postmitotic chromatin. As an analogy to the observed changes in G_{1A} nuclei, a critical level of chromatin decondensation, critical content of nonhistone proteins and development of nucleoli are also required for initiation of DNA replication during reactivation of quiescent nuclei in cell heterokaryons [214,215]. Sensitivity of DNA in situ to acid denaturation, which also shows a threshold in G_1, strongly correlates with chromatin condensation and thus may be a direct reflection of the presence of nonhistone proteins.

The role of the G_{1A} compartment in cell cycle kinetics is schematically presented in Figure 20. The upper panel (A) shows a classical cell age distribution diagram of an exponentially growing, asynchronous population. As is evident, this distribution cannot account for an exponential-like compartment of cell residence times and assumes a linear cell progression through each phase. The lower portion (Fig. 20B) shows a distinction between the cell growth (equalization) subphase (G_{1A}) and the remainder of the cell cycle. It was experimentally demonstrated by time-lapse cinematography [131,132] and flow cytometry [63] that cytokinesis is asymmetric, resulting in two daughters cells of unequal size. The circle in Figure 20B symbolizes the inequality of cell division; the width of the circle is proportional to cell heterogeneity measured as the coefficient of variation of mean RNA or protein content of a synchronous subpopulation located within the narrow windows of the cell cycle marked by the double arrows. Early postmitotic cells are the most heterogeneous, whereas cells progressing through S are the most uniform. The equalization point is symbolically marked E; this point by definition is at the transition between G_{1A} and G_{1B}, inasmuch as it represents attainment of a threshold in cell size (rRNA content). Cells with subthreshold size values do not enter S phase, at least during the exponential phase of cell growth. The exponential-like component of cell residence times in G_{1A}, that is, the equalization times, are symbolized by the asymptotic curve of the G_{1A} distribution. As a result of cytokinesis (due to heterogeneity of mother cells plus asymmetric division), the postmitotic cells may be located anywhere in G_{1A}; the largest cells may be directly located in G_{1B}, thus rapidly entering S and bypassing the "growth" phase. The very smallest G_{1A} cells have extremely long equalization times and for all practical purposes may be considered to be noncycling, hence the asymptotic character of the curve. Our experimental data on cell heterogeneity and transit times [63] and the kinetic model data [133] all support the above characteristics of G_{1A}.

It should be stressed that, although the G_{1A} compartment or the "prethreshold" populations discriminated either by RNA, protein, or chromatin differences, all consist predom-

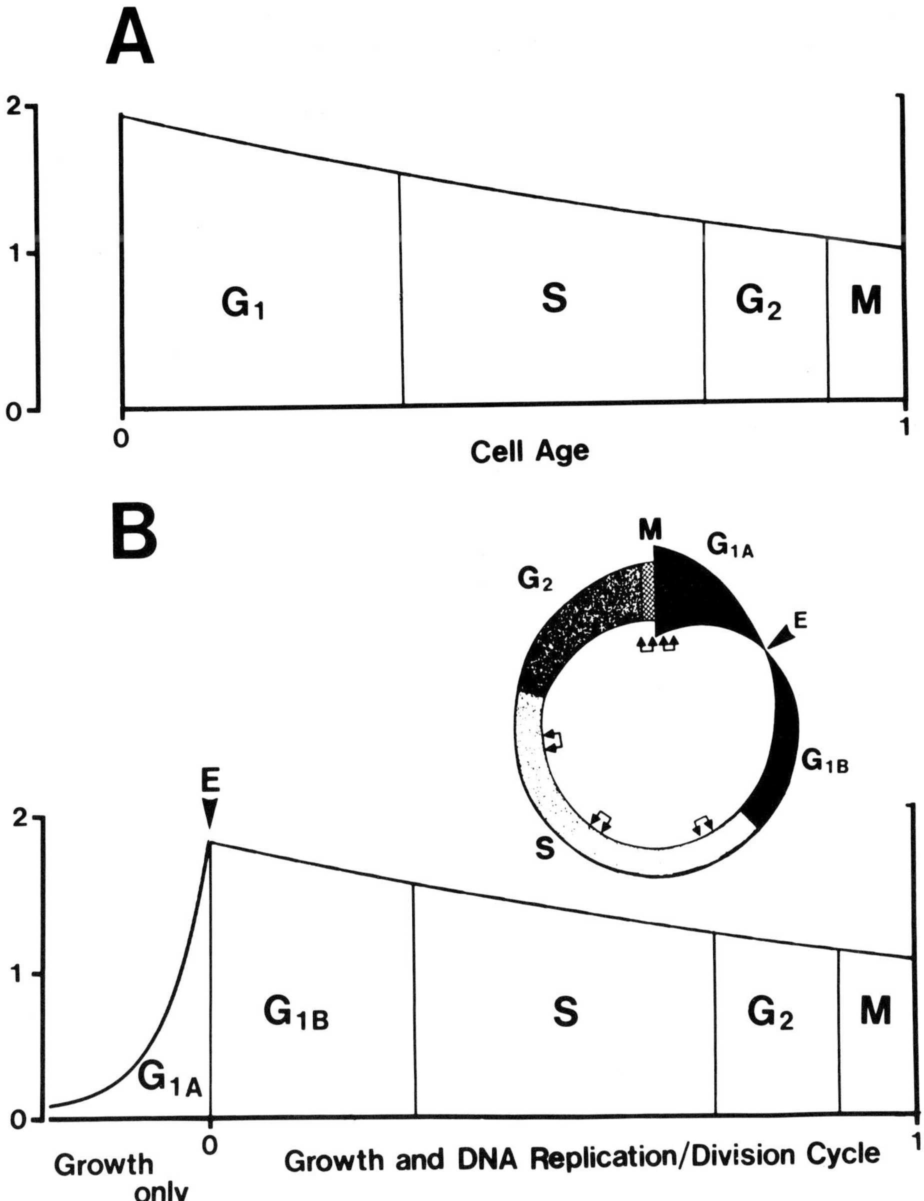

Fig. 20. Graphic illustration of the role of the G_{1A} compartment in the cell cycle. **A:** The cell age distribution in exponentially growing populations according to the traditional subdivision of the cell cycle. **B:** Subdivision of the cell cycle in which the G_{1A} compartment represents the growth (equalization) phase during which cells attain a threshold size (RNA, protein content) prior to entering the cell cycle. Residence times in G_{1A} have an exponential component. The circle symbolizing the cell cycle represents cell heterogeneity; its width is proportional to the coefficient of variation of the mean RNA content measured within narrow windows of the cell cycle [63].

inantly of early G_1 cells and have a common distinct kinetic feature (exponential component in transit times), they may not be identical [63]. Further studies are needed to determine the relationship between all these metabolic features that discriminate G_{1A} populations and the cells' kinetic properties.

2. *G_{1A}-arrested cells* are metabolically similar to G_{1A} cells. In contrast to the latter, however, they do not progress toward S. 3T3 cells maintained in cultures, deprived of serum [68], or L1210 cells growing in the presence of sodium butyrate [84], show characteristics of G_{1A} cells. Upon release from the arrest, a delay of 10–12 hours is observed prior to their entrance into S.

3. *G_{1B} cells* are present in exponentially growing populations. The RNA or protein content of G_{1B} cells is similar to that of early-S-phase cells. Likewise, the chromatin structure of G_{1B} cells, reflected by nonhistone protein content and DNA in situ sensitivity to denaturation, resembles that of early-S-phase cells. Nuclear RNA content, representing most likely the development and activity of nucleoli, is also similar to that of early-S-phase cells. It is quite possible that G_{1B} cells are committed to enter S phase and are in a preparatory state

for DNA replication. G_{1B} and early-S cells exhibit the highest resistance to DNA denaturation in situ and the highest nonhistone/histone protein content ratio in nuclei.

4. *Traditional S phase* comprises cell-replicating DNA. As discussed, in unperturbed exponentially growing populations, a threshold RNA or protein content is required prior to entrance into S. However, the cells may bypass this requirement and under certain conditions enter S phase with a "subthreshold" quantity of RNA or protein [19]. It is unknown at present whether or not cells can enter S phase with a "subthreshold" nonhistone protein content. A progressive increase in DNA sensitivity to denaturation occurs during S manifested as a skew in the axis of the S-cell cluster with respect to the α_t coordinate (Fig. 14B). Multiparameter flow cytometric data clearly indicate that there is a close relationship between RNA content (number of ribosomes) and the rate of cell progression through S phase [65,66,97]. Most of the latter data, however, were obtained using hydroxyurea to arrest cells and induce unbalanced growth; the results thus await confirmation by other methods, inasmuch as hydroxyurea can affect the duration of the subsequent phases by means unrelated to RNA content.

5. *Cells in G_2 phase* accumulate additional quantities of RNA and protein and initiate chromatin changes preparatory to mitosis. The sensitivity of DNA in situ in G_2 cells to denaturation is similar to that of G_{1A} cells and is clearly higher than that of S-phase cells. Additional increases in the level of nonhistone proteins above that seen in late-S phase makes it possible to discriminate between G_{2A} and G_{2B} cells [200].

6. *Cells in mitosis* (late prophase, anaphase, metaphase, telophase) can be recognized as having the most condensed chromatin that coincides with the highest degree of sensitivity of DNA in situ to denaturation. Following fixation with formaldehyde, the DNA of mitotic cells has increased accessibility to several dyes, as compared to the DNA of interphase cells. This feature also allows identification of mitotic cells by flow cytometry [155]. Cytokinesis is unequal, producing two daughter cells with different quantities of RNA or protein; this primarily generates heterogeneity in the cell cycle [63,133]. A mathematical model of the cell cycle, based on unequal distribution of cell constituents having regulatory functions during mitosis, has recently been proposed by us [133]. In this model the randomness generated by asymmetric cytokinesis is reflected in the stochastic-like progression through G_{1A}. Thus, the probablistic-like distribution of generation times [33,232] can be explained by the "randomness" of cell division.

7. *G_{1Q} cells* are characterized by minimal metabolic activity. They have low RNA and protein content, a minimal content of nuclear RNA, a low ratio of nonhistone/histone proteins, a high sensitivity of DNA in situ to acid- or heat-induced denaturation and a low number of mitochondria. By all these criteria G_{1Q} cells are significantly different from their exponentially growing counterparts and thus can be easily recognized by flow cytometry. Upon stimulation, these cells require a long (16–24 hours) prereplicative phase during which they synthesize RNA and proteins, develop new mitochondria, and decondense chromatin. Nonstimulated peripheral blood lymphocytes or 3T3 cells maintained at confluency for an extended time are examples of G_{1Q} cells. In most cell systems, G_{1Q} cells are equivalent to "G_0" or "dormant" cells, as originally defined by Lajtha [153].

8. *S_Q cells* are characterized by an S-phase DNA but low RNA content and have DNA more sensitive to denaturation than cycling S-phase cells. S_Q cells, however, either do not progress through the cell cycle, or do so very slowly. Cells with these characteristics were observed in human leukemia cell lines coincubated with activated macrophages, or in bone marrow of patients with chronic myeloid leukemia during blastic crisis [60]. The presence of such cells may cause a discrepancy between quantitative estimates of S-phase cells based on single-parameter DNA histograms versus thymidine- or BrdUrd-incorporation assays [4,220].

9. *G_{2Q} cells* share characteristics of S_Q cells except they have a G_2-DNA content. The presence of quiescent G_2 cells was recently documented in spheroids [5] and in solid tumors [48].

10. *S-arrested cells* have all the metabolic features of S-phase cells except they do not replicate DNA. In certain tumors, anoxic cells exhibit these characteristics [29]. Cells arrested in S by most antitumor drugs (e.g., araC, hydroxyurea) develop signs of unbalanced growth. The degree of unbalanced growth can be estimated from the deviation in the ratio of RNA/DNA or protein/DNA from normal values. When cells are in negative unbalanced growth (e.g., after treatment with actinomycin D or cycloheximide) they resemble S_Q cells. Examples of positive and negative unbalanced growth are described in ref. [259].

11. *G_2-arrested cells* are similar by metabolic criteria to G_2 cells. Many antitumor drugs, predominantly intercalators, arrest cells in G_2. Such G_2-arrested cells often develop signs of unbalanced growth as described above for S-arrested cells.

12. *G_{1D} cells* are fully differentiated and have a G_1-DNA content. Differentiated cell phenotypes may be characterized by low (granulocytes, nucleated erythrocytes; see Figs. 7 and 8), moderate (macrophages, hepatocytes), or high (plasma cells, keratinocytes; see Fig. 9) RNA or protein content. In addition, G_{1D} cells contain differentiation-specific products. These phenotypic markers can be detected in conjunction with DNA content by flow cytometry using a variety of staining methods [146,189,287]. Unlike G_{1Q} cells, G_{1D} cells cannot enter the cell cycle except under unusual circumstances (e.g., cell fusion).

13. *Cells in transition:* When cells undergo transition, for example, from the G_{1Q} state to proliferation, for a certain period of time they possess intermediate metabolic features. Peripheral blood lymphocytes, stimulated by mitogens during the initial 12–24 hours, exhibit characteristics of cells in transition from quiescence to proliferation (G_{1T}). Depending on the DNA content cells manifest during their transition, they can be subdivided in G_{1T}, S_T, or G_{2T} [74].

The above classification represents an attempt to subdivide cells into distinct kinetic and metabolic compartments that can be objectively determined using rather simple flow cytometric procedures. Because the metabolic (biochemical) properties of individual cells differ significantly between these compartments, cell sensitivities to various factors, including antitumor drugs, are expected to be different. This classification therefore provides a more accurate characterization of tumor cell populations with respect to their kinetics and drug sensitivity than the traditional subdivision into G_1, S, and G_2 + M phase cells and could have prognostic value in clinical oncology. Thus, for instance, the possibility of estimating the proportion of noncycling cells (i.e., cells in deep quiescence) could influence a decision as to whether cell cycle specific drugs should be used. Likewise, the proportion of G_{1A} versus G_{1Q} cells may be indicative of whether the populations will be more sensitive to drugs that affect cell growth (e.g., actinomycin D) versus drugs that affect cell progression through the mitotic cycle.

The proposed classification does not take into account any specific molecular events but rather is based upon gross metabolic cell features, predominantly related to growth in size. While in most situations cell growth and proliferation are closely coordinated, several exceptions have been noted, and the possibility of their dissociation should be considered [19,20,173]. Of the parameters described in this chapter, changes in chromatin (nonhistone protein content, stability of DNA in situ) appear to be less related to cell growth and thus perhaps more germane to cell-cycle progression. Further studies, however, are needed to observe how strictly these features correlate with cell progression through the mitotic cycle, especially in situations in which cell growth or proliferation is suppressed by antitumor drugs.

Discovery of cell cycle-specific molecular events such as transcription or translation of protooncogenes or other similar nuclear protein constituents, discussed earlier in this chapter, is quite recent. In most instances, the temporal sequence of oncogene expression, their temporal relationship to each other and their role in cell progression are not as yet fully understood. However, it is already apparent that several molecular events can be located with rather good precision within the framework of the classification proposed above. Thus, for instance, quiescent fibroblasts require platelet-derived growth factor (PDGF), which is reasonably homologous to the product of the c-*sis* oncogene [89,239], to initiate the sequence of molecular events leading to their entrance into the mitotic cycle [141]. The receptor for PDGF has a tyrosine-specific protein kinase activity and may be a product of c-*src* or c-*abl* [239]. Formation of the PDGF-receptor complex triggers a kinase activity, which in turn activates the "competence genes" c-*myc* and c-*fos* [45,109, 129,144]. Acquisition of "competence" is associated with growth in size, and in particular, with increases in rRNA content [1,224,242]. Expression of c-*scr* and c-*abl* takes places in G_{1Q} cells, whereas the "competence gene family" [239] is expressed in G_{1T}, that is, during transition from G_{1Q} to G_{1B}.

The sequential activation of oncogenes and other "proliferative antigens" has also been demonstrated during the mitogenic activation of peripheral blood lymphocytes [123, 174]. Thus, polyclonal mitogens, such as PHA, trigger very early expression of the c-*myc* oncogene [129], appearance of the receptors for interleukin-2, and subsequently an increase in the quantity of the proliferation-related protein p53 [174]. Here again, within the framework of the proposed classification scheme, the c-*myc* oncogene would be G_{1T} phase-specific, whereas p53 is expressed during the proliferative phase (G_{1A}, G_{1B}, S, G_2).

Stimulation of Lymphocytes

Stimulation of lymphocytes by polyvalent mitogens or antigens is an example of transition of cells from deep quiescence (G_0, G_{1Q}) into the cycling compartment and subsequent progression through the cell cycle. This is a complex multiphase process. During the initial stage of stimulation (the G_{1Q}–G_{1A} transition), the appearance of receptors to interleukin-2 takes place [122]. This stage can be induced in lymphocyte cultures by mitogens in the absence of interleukin-2. Cell-cycle progression, on the other hand, requires the presence of interleukin-2 as a growth factor, and occurs only after completion of the G_{1Q}–G_{1A} transition. A model of B-cell stimulation by polyclonal activators, as suggested, for example, by Wetzel et al. [285], identifies the G_{1A} stage as the one during which cells become receptive to signals delivered by interleukins which, in turn, triggers cell progression to G_{1B} and further through the cell cycle.

Up until now, the prevalent assay of this widely studied process (especially in the clinical setting) involved bulk measurements of radioactivity in the insoluble fraction of cultures incubated with [^{3}H]-Tdr, reflecting DNA replication. This assay is of limited value for several reasons. For instance, the quantity of incorporated [^{3}H]-TdR depends on the pool of endogenous precursors, which can vary in an uncontrolled way. Release of the endogenous precursors from even a few dead cells that may be present in cultures may markedly dilute [^{3}H[-TdR, decreasing its specific activity and lowering incorporation of tritium in such cultures. Furthermore, bulk assays do not provide information about individual cells. As such, they cannot discriminate between situations in which a few cells are heavily labeled as compared to those in which many cells are moderately labeled. Also, the time of availability of [^{3}H]-TdR in cultures is generally short and may vary. Since the [^{3}H]-TdR incorporation assay generally involves a rather long (6–18 hours) pulse, heavily labeled cells may become arrested in G_2 due to tritium radiation [41,200], whereas less labeled cells can divide and progress to the next S phase during the pulse. DNA damage and DNA repair synthesis occur during the assay, and it is unknown to what extent these radiobiological effects influence incorporation of the precursor.

The most critical limitation of the [^{3}H]-TdR incorporation assay, however, is the lack of information it provides concerning the early events of stimulation prior to the entrance of cells into S phase. It becomes evident that there are numerous important immunological situations in which lymphocytes are triggered to progress from G_0 to G_1 but do not enter S phase [208,272]. Such stimulation is, of course, undetectable by the traditional assay. In summary, thus, [^{3}H]-TdR incorporation provides a rather remote approximation of lymphocyte stimulation.

Multiparameter flow cytometry is ideally suited to measure the process of lymphocyte stimulation. Two procedures that can reveal several steps in lymphocyte stimulation were described earlier in this chapter. Simultaneous measurement of RNA and DNA (Fig. 6) makes it possible to discriminate all the compartments of the cell cycle described above and offers great precision in defining the degree of cell progression through the stimulation process. Furthermore, because stimulated B and T lymphocytes have different RNA content [62] they can be discriminated from each other at the peak of stimulation. Addition of BrdUrd during culturing also makes it possible to identify the cells incorporating this precursor in relation to their RNA content (Fig. 19). Because the procedure is very simple and rapid (one step, no centrifugation, cells can be measured 1 min after removal from the culture) this technique appears to be optimal for the routine assay of lymphocyte stimulation. To recapitulate, the method provides an estimate of the proportions of cells in G_{1Q}, G_{1T}, G_{1A}, G_{1B}, S, and G_2 + M (and dead cells). When sequential measurements are made during stimulation, it is possible to measure the rates of transition from G_{1Q} to G_1 and cell progression through S. If cultures are treated with a stathmokinetic agent, additional kinetic parameters, related to the rate of progression through all of the compartments listed above, can be obtained. It is also possible to obtain the absolute number of cells in cultures that respond to mitogen by increasing RNA content (G_0–G_1 transition) and estimate how many of them then enter and progress through S phase.

Flow cytometric analysis of lymphocyte stimulation based

on DNA/RNA measurements [61,74,78,262] has gained wide acceptance in immunology. Pioneering studies employing this method in various fields of immunology are described in references [8,24,86,100,118,123,139,143,167,178,179,181, 193,213,231,254,259,285,286,288]. Additional references are presented in Chapter 15 in this volume, describing applications of AO. The method, applied in a variety of immunological studies both in vivo and in vitro, contributed to a better understanding of the process of lymphocyte stimulation. In particular, it proved useful in discriminating between the states of stimulation that involve only G_{1Q}–G_{1A} transition versus those that are associated with further cell progression through the mitotic cycle [123,181,213,231,281,286].

Stimulation of lymphocytes can also be assayed by the method based on measurements of DNA denaturation (Fig. 15). The advantage of this method over the RNA/DNA technique is the ability to discriminate between G_2 and M cells. This is important in stathmokinetic experiments utilizing M-phase blockers and provides very accurate estimates of the rates of cell progression through each of the compartments of the cell cycle in control and in perturbed cultures [73]. In comparison with the DNA/RNA method, the DNA denaturation technique is more cumbersome, as it requires cell fixation and digestions with RNase [71].

Simultaneous measurements of DNA content and cell volume, light scatter or protein content can perhaps provide information similar to that obtained with the DNA/RNA technique. Cell volume, however, requires cells with intact membranes, which imposes a limitation on the staining of DNA. Light scatter, which should provide information on cell size that in turn is expected to correlate with RNA content, is influenced by the reflective and refractive properties of the cell and thus its relation to cell size may not always be straightforward. Hence the bivariate distributions of DNA/scatter may differ from those of DNA/RNA.

Correlated measurements of DNA and protein should, in principle, yield similar data to DNA and RNA. However, because ribosomes are synthesized prior to bulk protein synthesis, early in stimulation an increase in RNA content may precede the increase in protein content [160]. The kinetic events of stimulation (G_{1Q}–G_{1A} transition) may differ somewhat when assayed by protein, compared to RNA, content.

Instead of staining all cellular proteins, it is possible, using fluorescence-labeled immunoglobulin antibodies, to specifically stain immunoglobulins within permeabilized (e.g., ethanol-fixed) cells. Simultaneous staining of DNA (e.g., with PI) and immunoglobulins (e.g., with FITC-labeled antiimmunoglobulin antibodies) makes it possible to recognize differentiated B cells that synthesize immunoglobulins and locate them within the cell cycle [98]. As mentioned, these cells have higher RNA content than stimulated T cells [62].

A potentially useful technique that permits simultaneous measurements of two-color surface immunofluorescence and DNA content was recently proposed by Rabinovitch et al. [209]. These authors combined staining of permeabilized cells for DNA using 7-aminoactinomycin D (red) with FITC (green) and phycoerythrin (orange) fluorescence-labeled antibodies. Using a single-laser excitation, they were able to discriminate the cell-cycle distributions of various subsets of stimulated lymphocytes from mouse spleen and human peripheral blood. Interestingly, differences in the stainability of DNA with 7-aminoactinomycin D related to differences in chromatin structure makes it possible to distinguish G_0 from stimulated cells. This approach will certainly be of value in immunology.

The methods described above detect rather "late" stages of stimulation (e.g., significant accumulation of RNA occurs only 6–8 hours after the event triggering stimulation [78]). However, there are two events occurring earlier that can be measured by flow cytometry. One is a change in the membrane potential (depolarization) that takes place within the first 2 hours of stimulation [130,227]. Shapiro et al. [227] demonstrated that the membrane potential measurements of stimulated lymphocytes by flow cytometry can be used to recognize mitogen-responsive cells. The second event is a rapid rise in intracellular Ca^{2+}. Rabinovitch et al. [120,208] combined measurements of intracellular Ca^{2+}, using indo-1, with surface-immunofluorescence detection to identify subpopulations of cells responding rapidly to various mitogens. The latter two techniques will undoubtedly find wide use in immunological studies.

As briefly described above, numerous flow cytometric methods already exist that allow the measurement of sequential steps of lymphocyte stimulation. Especially fruitful in further immunological studies will be their combined application, for example, when subpopulations of cells responding to early events (Ca^{2+} increase, transmembrane potential change) are compared with those responding by RNA increase, chromatin changes, and progression through the cell cycle. This will allow detection and characterization of heterogeneity in the states of stimulation that have so far escaped recognition by traditional techniques employing [^{3}H]-TdR.

Application of multiparameter flow cytometry to studies of the cell cycle has a rather short history. Yet it has already provided a plethora of information contributing to a better understanding of biological processes associated with the cell cycle, quiescence, and differentiation. Availability of molecular probes and further development of instrumentation guarantee future progress in this field. Application of multiparameter flow cytometry in the clinic is expected to provide new information on the kinetic behavior of tumor cells, which may contribute to better designs of therapeutic strategies (see Chapter 39, this volume).

ACKNOWLEDGMENTS

This work was supported by U. S. Public Health Service Grants CA23296 and CA28704 from the National Cancer Institute. The authors greatly appreciate the assistance provided by Ms. Robin Nager in the preparation of this manuscript.

REFERENCES

1. **Adam G, Steiner U, Seuwen K (1983)** Proliferative activity and ribosomal RNA content of 3T3 and SV40-3T3 cells. Cell. Biol. Int. Rep. 7:955–962.
2. **Aherne WA, Camplejohn RS, Wright NA (1977)** "An Introduction to Cell Population Kinetics." London: Edward Arnold.
3. **Alema S, Tato F, Boettiger D (1985)** *myc* and *src* oncogenes have complementary effects on cell proliferation and expression of specific extracellular matrix components in definitive chondroblasts. Mol. Cell. Biol. 5:538–544.
4. **Allison DC, Ridolpho PF, Anderson S, Bose K (1985)** Variations in the (^{3}H)thymidine labelling of S-phase cells in solid mouse tumors. Cancer Res. 45:6010–6016.
5. **Allison DC, Yuhas JM, Ridolpho PF, Anderson SL, Johnson TS (1983)** Cytophotometric measurement of

the cellular DNA content of (^{3}H)thymidine-labelled spheroids. Demonstration that some non-labelled cells have S and G_2 DNA content. Cell Tissue Kinet. 16:237–246.

6. **Andreeff M, Beck JD, Darzynkiewicz Z, Traganos F, Gupta S, Melamed MR, Good RA (1978)** RNA content in human lymphocyte subpopulation. Proc. Natl. Acad. Sci. USA 75:1938–1942.
7. **Andreeff M, Darzynkiewicz Z, Sharpless TK, Clarkson BD, Melamed MR (1980)** Discrimination of human leukemia subtypes by flow cytometric analysis of cellular DNA and RNA. Blood 55:282–293.
8. **Antel JP, Oger JF, Dropcho E, Richman DP, Kuo HH, Arnason GBW (1980)** Reduced T-lymphocyte cell reactivity as a function of human aging. Cell Immunol. 54:184–192.
9. **Ashihara T, Traganos F, Baserga R, Darzynkiewicz Z (1978)** A comparison of cell cycle-related changes in postmitotic and quiescent AF8 cells as measured by cytofluorometry after acridine orange staining. Cancer Res. 38:2514–2518.
10. **Augenlicht LH, Baserga R (1974)** Changes in the G_0 state of WI fibroblasts at different times after confluence. Exp. Cell. Res. 89:255–262.
11. **Barfod IJ, Barfod NM (1980)** Cell-production rates estimated by the use of vincristine sulphate and flow cytometry. I. An in vitro study using murine tumour cell lines. Cell Tissue Kinet. 13:1–8.
12. **Barfod IJ, Barfod NM (1980)** Cell-production rates estimated by the use of vincristine sulphate and flow cytometry. II. Correlation between the cell-production rates of aging ascities tumours and the number of S phase tumour cells. Cell Tissue Kinet. 13:9–19.
13. **Barlogie B, Raber MN, Schumann J, Johnson TS, Drewinko B, Swartzendruber DE, Gohde W, Andreeff M, Freireich EJ (1983)** Flow cytometry in clinical cancer research. Cancer Res. 43:3982–3997.
14. **Barque JP, Danon F, Perandean L, Yeni P, Larsen CJ (1983)** Characterization by human antibody of a nuclear antigen related to the cell cycle. EMBO J 2: 743–749.
15. **Bargue JP, Lagaye S, Bendayan M, Purion-Dutilleul F, Danon F, Larsen CJ (1985)** PSL, an S phase-related p53 nuclear antigen associates transiently with chromatin. Exp. Cell Res. 157:8–14.
16. **Barque JP, Lagaye S, Ladoux A, Della Valle V, Abita JP, Larsen CJ (1987)** PSL, a nuclear cell-cycle associated antigen is increased during retinoic acid-induced differentiation of HL-60 cells. Biochem. Biophys. Res. Commun. 147:993–999.
17. **Barrandon Y, Green H (1985)** Cell size as a determinant of the clone-forming ability of human keratinocytes. Proc. Natl. Acad. Sci. USA 42:5390–5394.
18. **Baserga R (1976)** "Multiplication and Division in Mammalian Cells." New York: Marcel Dekker.
19. **Baserga R (1984)** Growth in size and cell DNA replication. Exp. Cell Res. 151:1–5.
20. **Baserga R (1985)** "The Biology of Cell Reproduction." Cambridge, Massachusetts: Harvard University Press.
21. **Bauer KD, Dethlefsen LA (1981)** Control of cellular proliferation of HeLa-S3 suspension cultures. Characterization of cultures utilizing acridine orange staining procedures. J. Cell. Physiol. 108:99–112.
22. **Bauer KD, Keng PC, Sutherland RM (1982)** Isolation of quiescent cells from multicellular spheroids using centrifugal elutriation. Cancer Res. 42:72–78.
23. **Bensch KG, Tanaka S, Hu S-Z, Wang SF, Korn D (1982)** Intracellular localization of human DNA polymerase with monoclonal antibodies. J. Biol. Chem. 257:8391–8396.
24. **Bettens F, Kristensen F, Walker C, Bonnard GD, DeWeck AL (1984)** Lymphokine regulation of human lymphocyte proliferation: Formation of resting G_0 cells by removal of interleukin 2 in cultures of proliferating T lymphocytes. Cell. Immunol. 86:337–346.
25. **Bohmer RM (1979)** Flow cytometrical cell cycle analysis using quenching of 33258 Hoechst fluorescence by BUdR incorporation. Cell Tissue Kinet. 12:101–105.
26. **Bohmer RM, Ellwart J (1981)** Combination of BUdR-quenched Hoechst fluorescence with DNA-specific ethidium bromide fluorescence for cell cycle analysis with two parametrical flow cytometer. Cell Tissue Kinet. 14:653–658.
27. **Bohmer RM, Ellwart J (1981)** Cell cycle analysis by combining the 5-bromodeoxyuridine/33258 Hoechst technique with DNA-specific ethidium bromide staining. Cytometry 2:31–34.
28. **Bravo R, Frank R, Blundell PA, Macdonald-Bravo H (1987)** Cyclin/PCNA is the auxillary protein of DNA polymerase-δ. Nature (Lond.) 326:515–517.
29. **Brock WA, Swartzendruber DE, Grdina DJ (1982)** Kinetic heterogeneity in density-separated murine fibrosarcoma subpopulations. Cancer Res. 42:4499–5003.
30. **Bruge JS, Steinbough PJ, Erikson RL (1978)** Characterization of the avian sarcoma virus protein $p60^{src}$. Virology 91:130–140.
31. **Callissano P, Volonte C, Biocca S, Cattaneo A (1985)** Synthesis and content of a DNA-binding protein with lactic dehydrogenase activity are reduced by nerve growth factor in the neoplastic cell line PC12. Exp. Cell. Res. 161:117–129.
32. **Campisi J, Gray HE, Pardee AB, Dean M, Sonenshein GE (1984)** Cell-cycle control of c-*myc* but not c-*ras* expression is lost following chemical transformation. Cell 36:241–247.
33. **Castor LN (1980)** A G_1-rate model accounts for cell cycle kinetics attributed to "transition probability." Nature (Lond.) 287:76–79.
34. **Celis JE, Bravo R (1984)** Synthesis of the nuclear protein cyclin in growing, senescent and morphologically transformed human skin fibroblasts. FEBS Lett. 165:21–25.
35. **Celis JE, Bravo R, Larsen PM, Fey SJ (1984)** Cyclin: A nuclear protein whose level correlates directly with the proliferation state of normal as well as transformed cells. Leukemia Res. 8:143–157.
36. **Celis JE, Celis A (1985)** Cell cycle-dependent variations in the distribution of the nuclear protein cyclin proliferating cell nuclear antigen in cultured cells: Subdivision of S phase. Proc. Natl. Acad. Sci. USA 82:3262–3266.
37. **Chafouleas JG, Bolton WE, Hidaka H, Boyd AE III, Means AR (1982)** Calmodulin and the cell cycle: Involvement in regulation of cell cycle progression. Cell 28:41–50.

38. **Chafouleas JG, Legace L, Bolton WE, Boyd AE III, Means AR (1984)** Changes in calmodulin and its mRNA accompany reentry of quiescent (G_0) cells into the cell cycle. Cell 36:73–81.
39. **Chang MT, Dove WF, Laffler TG (1983)** The periodic synthesis of tubulin in the *physarum* cell cycle. Characterization of *physarum* tubulins by affinity for monoclonal antibodies and by peptide mapping. J. Biol. Chem. 258:1353–1356.
40. **Clausen OPF, Kjell E, Kirkhus B, Petersen S, Bolund L (1983)** DNA synthesis in mouse epidermis: S phase cells that remain unlabelled after pulse labelling with DNA precursors progress slowly through S. J. Invest. Dermatol. 81:545–549.
41. **Cleaver JE (1967)** "Thymidine Metabolism and Cell Kinetics." Amsterdam: North-Holland.
42. **Clevenger CV, Bauer KD, Epstein AL (1986)** Modulation of the nuclear antigen p105 as a function of cell cycle progression. J. Cell. Physiol. 130:336–343.
43. **Clevenger CV, Epstein AL (1984)** Identification of a nuclear protein component of interchromatin granules using a monoclonal antibody and immunogold electron microscopy. Exp. Cell. Res. 151:194–207.
44. **Clevenger CV, Epstein AL, Bauer KD (1987)** Quantitative analysis of nuclear antigen in interphase and mitotic cells. Cytometry 8:280–286.
45. **Cochran BH, Zallo J, Verma DM, Stiles CD (1984)** Expression of the c-*fos* and of a *fos* related gene is stimulated by platelet-derived growth factor. Science 226:1080–1082.
46. **Collet MS, Bruge JS, Erikson RL (1979)** Characterization of a normal avian cell protein related to the avian sarcoma virus transforming gene product. Cell 15:1363–1369.
47. **Collett MS, Erikson E, Erikson RL (1979)** Structural analysis of the avian sarcoma virus transforming protein: Sites of phosphorylation. J. Virol. 29:770–781.
48. **Coninx P, Liautaud-Roger F, Boisseau A, Loirette M, Cattan A (1983)** Accumulation of non-cycling cells with a G_2 DNA content in aging solid tumors. Cell Tissue Kinet. 16:505–515.
49. **Courtneidge SA, Levinson AD, Bishop JM (1980)** The protein encoded by the transforming gene of avian sarcoma virus ($pp60^{src}$) and a homologous protein in normal cells ($pp60^{proto\text{-}src}$) are associated with the plasma membrane. Proc. Natl. Acad. Sci. USA 77:3783–3787.
50. **Cowden RR, Curtis SK (1983)** Supravital experiments with pyronin Y, a fluorochrome of mitochondria and nucleic acids. Histochemistry 77:535–542.
51. **Crawford L (1983)** The 53,000 dalton cellular protein and its role in transformation. Int. Rev. Exp. Pathol. 25:1–50.
52. **Crissman HA, Darzynkiewicz Z, Tobey RA, Steinkamp JA (1985)** Normal and perturbed Chinese hamster ovary cells: Correlation of DNA, RNA and protein by flow cytometry. J. Cell. Biol. 101:141–147.
53. **Crissman HA, Darzynkiewicz Z, Tobey RA, Steinkamp JA (1985)** Correlated measurements of DNA, RNA and protein in individual cells by flow cytometry. Science 228:1321–1324.
54. **Crissman HA, Hofland MH, Stevenson AP, Wilder ME, Tobey RA (1988)** Use of DiO-C5-3 to improve Hoechst 33342 uptake, resolution of DNA content and survival of CHO cells. Exp. Cell. Res. 174:388–396.
55. **Crissman HA, Steinkamp JA (1987)** A new method for rapid and sensitive detection of bromodeoxyuridine in DNA-replicating cells. Exp. Cell. Res. 173: 256–261.
56. **Crissman HA, Van Egmond JV, Hodrinet RS, Pennings A, Haanen C (1981)** Simplified method for DNA and protein staining of human hematopoietic cell samples. Cytometry 2:59–62.
57. **Crissman HA, Steinkamp JA (1973)** Rapid simultaneous measurement of DNA, protein and cell volume in single cells from large mammalian cell populations. J. Cell. Biol. 59:766–769.
58. **Crissman HA, Steinkamp JA (1982)** Rapid, one-step staining procedure for analysis of cellular DNA and protein by single and dual laser flow cytometry. Cytometry 3:84–90.
59. **Darzynkiewicz Z (1986)** Cell growth and division cycle. In Dethlefsen LA (ed), "Cell Cyclew Effects of Drugs. International Encyclopedia of Pharmacology and Therapeutics," Section 121. Oxford: Pergamon Press, pp 1–43.
60. **Darzynkiewicz Z (1984)** Metabolic and kinetic compartments of the cell cycle distinguished by multiparameter flow cytometry. In Skehan P, Friedman SJ (eds), "Growth, Cancer and the Cell Cycle." Chifton, NJ: Humana Press, pp 249–278.
61. **Darzynkiewicz Z, Andreeff M, Traganos F, Melamed MR (1978)** Discrimination of cycling and noncycling lymphocytes by BUdR-suppressed acridine orange fluorescence in a flow cytometric system. Exp. Cell. Res. 115:31–35.
62. **Darzynkiewicz Z, Andreeff M, Traganos F, Melamed MR (1980)** Proliferation and differentiation of normal and leukemic lymphocytes as analyzed by flow cytometry. In Laerum OD, Lindmo T, Thorud E (eds), "Flow Cytometry IV," Bergen, Norway: Universitetsforlaget, pp 392–397.
63. **Darzynkiewicz Z, Crissman HA, Traganos F, Steinkamp JA (1982)** Cell heterogeneity during the cell cycle. J. Cell. Physiol. 113:465–474.
64. **Darzynkiewicz Z, Dokov V, Pienkowski M (1967)** Dry mass of lymphocytes during transformation after stimulation by phytohemagglutinin. Nature (Lond.) 214:1265–1266.
65. **Darzynkiewicz Z, Evenson DP, Staiano-Coico L, Sharpless T, Melamed MR (1979)** Correlation between cell cycle duration and RNA content. J. Cell. Physiol. 100:425–438.
66. **Darzynkiewicz Z, Evenson DP, Staiano-Coico L, Sharpless T, Melamed MR (1979)** Relationship between RNA content and progression of lymphocytes through the S phase of the cell cycle. Proc. Natl. Acad. Sci. USA 76:358–362.
67. **Darzynkiewicz Z, Kapuscinski J, Traganos F, Crissman HA (1987)** Application of pyronin Y (G) in cytochemistry of nucleic acids. Cytometry 8:138–145.
68. **Darzynkiewicz Z, Sharpless T, Staiano-Coico L, Melamed MR (1980)** Subcompartments of the G_1 phase of cell cycle detected by flow cytometry. Proc. Natl. Acad. Sci. USA 77:6696–6700.
69. **Darzynkiewicz Z, Staiano-Coico L, Melamed MR (1981)** Increased mitochondrial uptake of rhodamine

123 during lymphocyte stimulation. Proc. Natl. Acad. Sci. USA 78:2383–2387.

70. **Darzynkiewicz Z, Traganos F (1981)** RNA content and chromatin structure in cycling and noncycling cell populations studied by flow cytometry. In McCarthy KS Sr, Padilla GM (eds), "Genetic Expression in the Cell Cycle." Orlando, FL: Academic Press, pp 103–128.
71. **Darzynkiewicz Z, Traganos F, Andreeff M, Sharpless T, Melamed MR (1979)** Different sensitivity of chromatin to acid denaturation in quiescent and cycling cells as revealed by flow cytometry. J. Histochem. Cytochem. 27:478–485.
72. **Darzynkewicz Z, Traganos F, Arlin Z, Sharpless T, Melamed MR (1976)** Cytofluorometric studies on conformation of nucleic acids in situ. II. Denaturation of deoxyribonucleic acid. J. Histochem. Cytochem. 24:49–58.
73. **Darzynkiewicz Z, Traganos F, Kimmel M (1986)** Assay of cell cycle kinetics by multivariate flow cytometry using the principle of stathmokinesis. In Gray JW, Darzynkiewicz Z (eds), "Techniques in Cell Cycle Analysis." Clifton, New Jersey: Humana Press, pp 292–336.
74. **Darzynkiewicz Z, Traganos F, Melamed MR (1980)** New cell cycle compartments identified by multiparameter flow cytometry. Cytometry 1:98–108.
75. **Darzynkiewicz Z, Traganos F, Melamed MR (1981)** Detergent treatment as an alternative to cell fixation for flow cytometry. J. Histochem. Cytochem. 29: 392.
76. **Darzynkiewicz Z, Traganos F, Melamed MR (1983)** Distinction between 5-bromodeoxyuridine labeled and unlabeled mitotic cells by flow cytometry. Cytometry 3:345–348.
77. **Darzynkiewicz Z, Traganos F, Sharpless T, Friend C, Melamed MR (1976)** Nuclear chromatin changes during erythroid differentiation of Friend virus induced leukemic cells. Exp. Cell. Res. 99:301–309.
78. **Darzynkiewicz Z, Traganos F, Sharpless T, Melamed MR (1976)** Lymphocyte stimulation: A rapid multiparameter analysis. Proc. Natl. Acad. Sci. USA 73: 2881–2884.
79. **Darzynkiewcz Z, Traganos F, Sharpless T, Melamed MR (1977)** Cell cycle related changes in nuclear chromatin of stimulated lymphocytes as measured by flow cytometry. Cancer Res. 37:4635–4640.
80. **Darzynkiewicz Z, Traganos F, Sharpless T, Melamed MR (1977)** Different sensitivity of DNA in situ in interphase and metaphase chromatin to heat denaturation. J. Cell. Biol. 73:128–138.
81. **Darzynkiewicz Z, Traganos F, Sharpless T, Melamed MR (1978)** Differential stainability of M vs G_2 and G_0 vs G_1 cells. In Lutz D (ed), "Third International Symposium on Pulse Cytophotometry." Ghent, Belgium: European Press, pp 267–273.
82. **Darzynkiewicz Z, Traganos F, Staiano-Coico L, Kapuscinski J, Melamed MR (1982)** Interactions of rhodamine 123 with living cells studied by flow cytometry. Cancer Res. 42:799–806.
83. **Darzynkiewicz Z, Traganos F, Xue S-B, Staiano-Coico L, Melamed MR (1981)** Rapid analysis of drug effects on the cell cycle. Cytometry 1:279–286.
84. **Darzynkiewicz Z, Traganos F, Xue S-B, Melamed MR (1981)** Effect of n-butyrate on cell cycle progression and in situ chromatin structure of L1210 cells. Exp. Cell. Res. 136:279–293.
85. **Darzynkiewicz Z, Williamson B, Carswell EA, Old LJ (1984)** Cell cycle-specific effects of tumor necrosis factor. Cancer Res. 44:83–90.
86. **DeFranco AL, Raveche ES, Paul WE (1985)** Separate control of B lymphocyte early activation and proliferation in response to anti-IgM antibodies. J. Immunol. 135:87–94.
87. **DeLeo AB, Jay G, Appella E, Dubois G, Law LW, Old LJ (1979)** Detection of a transformation-related antigen in chemically induced sarcomas and other transformed cells of the mouse. Proc. Natl. Acad. Sci. USA 76:2420–2424.
88. **Dolbeare F, Gratzner H, Pallavicini M, Gray JW (1983)** Flow cytometric measurements of total DNA content and incorporated bromodeoxyuridine. Proc. Natl. Acad. Sci. USA 80:5573–5577.
89. **Doolittle RF, Hunkapiller MW, Hood LE, Devare SG, Robbins KC, Aaronson SA, Antoniades HN (1983)** Simian virus *onc* gene, v-sis is derived from the gene (or genes) encoding a platelet-derived growth factor. Science 221:275–277.
90. **Dosik GM, Barlogie B, White AR, Gohde W, Drewinko B (1981)** A rapid automated stathmokinetic method for determination of in vitro cell cycle transit times. Cell Tissue Kinet. 14:121–134.
91. **Doukas JG, Ruckdeschel JC, Mardiney MR (1977)** Quantitative and qualitative analysis of human lymphocyte proliferation to specific antigen in vitro by use of the helium neon laser. J. Immunol. Methods 15:229–238.
92. **Eisenman RN, Tachibana CY, Abrane HD, Hann SR (1985)** v-*myc* and c-*myc*-encoded proteins are associated with the nuclear matrix. Mol. Cell. Biol. 5: 114–126.
93. **Eisinger M, Lee JS, Hofton JM, Darzynkiewicz Z, Chiao JW, DeHarven E (1979)** Human epidermal cell cultures: Growth and differentiation in the absence of dermal components or medium supplements. Proc. Natl. Acad. Sci. USA 76:5340–5344.
94. **Elliot SG, McLoughlin C (1978)** Rate of macromolecular synthesis through the cell cycle of the yeast Saccharomyces cerevisiae. Proc. Natl. Acad. Sci. USA 75:4384–4388.
95. **Eriksson S, Graslund A, Skog S, Thelander L, Tribukait B (1984)** Cell cycle-dependent regulation of mammalian ribonucleotide reductase. J. Biol. Chem. 259:11695–11700.
96. **Filmus J, Buck RN (1985)** Relationship of c-*myc* expression to differentiation and proliferation of HL-60 cells. Cancer Res. 45:822–825.
97. **Fujikawa-Yamamoto K (1982)** RNA dependence in the cell cycle of V79 cells. J. Cell. Physiol. 112:60–66.
98. **Garner JG, Colvin RB, Schooley RT (1984)** A flow cytometric technique for quantitation of B-cell activation. J. Immunol. Methods 67:37–51.
99. **Gelfant S (1981)** Cycling-noncycling cell transitions in tissue aging, immunological surveillance, transformation and tumor growth. Int. Rev. Cytol. 70:1–25.
100. **Gerdes J, Lemke H, Baisch H, Wacker H-H, Schwab U, Stein H (1984)** Cell cycle analysis of a cell proliferation-associated human nuclear antigen defined by the monoclonal antibody Ki-67. J. Immunol. 133: 1710–1715.

101. **Gerdes J, Schwab U, Lemke H, Stein H** (1983) Production of a mouse monoclonal antibody reactive with a human nuclear antigen associated with cell proliferation. Int. J. Cancer. 31:13–20.
102. **Gibbs JH, Brown RA, Robertson AJ, Potts RC, Swanson Beck J** (1979) A new method of testing for mitogen-induced lymphocyte stimulation: Measurement of the percentage of growing cells and of some aspects of their cell kinetics with an electronic particle counter. J. Immunol. Methods 25:147–158.
103. **Goerttler K, Ehemann V, Tschanargane C, Stoehr M** (1977) Monodispersal and deoxyribonucleic acid analysis of prostatic cell nuclei. J. Histochem. Cytochem. 25:560–564.
104. **Gohde W, Spies I, Schumann J, Buchner T, Kliene-Dopke G** (1978) Two parameter analysis of DNA and protein content of tumor cells. In Gohde W, Schumann J, Buchner T (eds), "Proceedings of the 2nd International Symposium on Flow Cytometry." Ghent, Belgium: European Press, pp 27–32.
105. **Goyette M, Pedropoulos CJ, Shank PR, Fausto N** (1983) Expression of a cellular oncogene during liver regeneration. Science 219:510–512.
106. **Goyette M, Pedropoulos CJ, Shank PR, Fausto N** (1984) Regulated transcription of c-Ki-*ras* and c-*myc* during compensatory growth of rat liver. Mol. Cell. Biol. 4:1493–1498.
107. **Gratzner HG, Leif RC** (1981) An immunofluorescence method for monitoring DNA synthesis by flow cytometry. Cytometry 1:385–389.
108. **Gratzner HG, Pollack A, Ingram DJ, Leif RC** (1976) Deoxyribonucleic acid replication in single cells and chromosomes by immunologic techniques. J. Histochem. Cytochem. 24:34–39.
109. **Greenberg ME, Ziff EB** (1984) Stimulation of 3T3 cells induces transcription of the c-*fos* proto-oncogene. Nature (Lond.) 311:433–437.
110. **Grummt F** (1985) Diadenosine tetraphosphate (Ap4A): A putative chemical messenger of cell proliferation control and inducer of DNA replication. Plant Mol. Biol. 2:41–44.
111. **Hann SR, Eisenman RN** (1984) Proteins encoded by the human c-*myc* oncogene. Differential expression in neoplastic cells. Mol. Cell. Biol. 4:2486–2497.
112. **Higgins PJ, Piwnicka M, Darzynkiewicz Z, Melamed MR** (1984) Multiparameter flow cytometric analysis of nuclear RNA and DNA of normal and hepatotoxin-treated mice. Am. J. Pathol. 115:31–35.
113. **Ide T, Whelly S, Baserga R** (1977) Stimulation of RNA synthesis in isolated nuclei by partially purified preparations of SV40 T-antigen. Proc. Natl. Acad. Sci. USA 74:3189–3192.
114. **Iguchi Ariga SMM, Itani T, Kiji Y, Ariga H** (1987) Possible function of the c-*myc* product: Promotion of cellular DNA replication. EMBO J 6:2365–2371.
115. **Imbert J, Lawrence JJ, Birg F** (1983) Simian virus 40 T antigen is detected only in cells in the G_2 phase of the cell cycle in one group of rat transformants. Virology 126:711–716.
116. **Jacobberger JW, Fogelman D, Lehman JM** (1986) Analysis of intracellular antigens by flow cytometry. Cytometry 7:356–364.
117. **James TW, Bohman R** (1981) Proliferation of mitochondria during the cell cycle of human cell line (HL-60). J. Cell. Biol. 89:256–260.
118. **Jelinek D, Lipsky PE** (1985) The roles of T-cell factors in activation, cell cycle progression and differentiation of human B cells. J. Immunol. 134:1690–1701.
119. **Johnson LV, Walsh ML, Chen LB** (1980) Localization of mitochondria in living cells with rhodamine 123. Proc. Natl. Acad. Sci. USA 77:990–994.
120. **June CH, Rabinovitch PS, Ledbetter JA** (1987) CDS antibodies increase intracellular ionized calcium concentration in T cells. J. Immunol. 138:2782–2792.
121. **Kachel V** (See Chapter 4, this volume.)
122. **Kaczmarek L** (1986) Protooncogene expression during the cell cycle. Lab. Invest. 54:365–376.
123. **Kaczmarek L, Calabretta B, Baserga R** (1985) Expression of cell-cycle dependent genes in phytohemagglutinin-stimulated human lymphocytes. Proc. Natl. Acad. Sci. USA 82:5376–5379.
124. **Kaczmarek L, Hyland JK, Watt R, Rosenberg R, Baserga R** (1985) Microinjected c-*myc* as a competence factor. Science 228:1313–1315.
125. **Kahn P, Graf T (Eds)** (1986) "Oncogenes and Growth Control." Berlin: Springer-Verlag, pp 1–369.
126. **Kaplow LS, Eisenberg M** (1975) Leukocyte differentiation and enumeration by cytochemical-cytographic analysis. In Haanen CAM, Hillen HFP, Wessels JMC (eds), "Pulse Cytophotometry." Ghent, Belgium: European Press, pp 262–274.
127. **Kecskemethy N, Schafer KP** (1982) Lectin-induced changes among polyadenylated and non-polyadenylated mRNA in lymphocytes: mRNAs for actin, tubulin and calmodulin respond differently. Eur. J. Biochem. 126:573–582.
128. **Kehrl JH, Muraguchi A, Fauci AS** (1984) Differential expression of cell activation markers after stimulation of resting human B lymphocytes. J. Immunol 132:2857–2861.
129. **Kelly K, Cochran BH, Stiles CD, Leder P** (1983) Cell-specific regulation of the c-*myc* gene by lymphocyte mitogens and platelet-derived growth factor. Cell 35: 603–610.
130. **Kiefer H, Blume AJ, Kaback HR** (1980) Membrane potential changes during mitogenic stimulation of mouse spleen lymphocytes. Proc. Natl. Acad. Sci. USA 77:2200–2204.
131. **Killander D, Zetterberg A** (1965) A quantitative cytochemical investigation of the relationship between cell mass and initiation of DNA synthesis in mouse fibroblasts in vitro. Exp. Cell. Res. 40:12–20.
132. **Killander D, Zetterberg A** (1965) Quantitative cytochemical studies on interphase growth. I. Determination of DNA, RNA and mass content of age determined mouse fibroblasts in vitro and of intercellular variation in generation time. Exp. Cell. Res. 38:272–284.
133. **Kimmel M, Darzynkiewicz Z, Arino O, Traganos F** (1984) Analysis of a cell cycle model based on unequal division of metabolic constituents to daughter cells during cytokinesis. J. Theoret: Biol. 110:637–664.
134. **Kimmel M, Darzynkiewicz Z, Staiano-Coico L** (1986) Stathmokinetic analysis of human epidermal cells in vitro. Cell Tissue Kinet. 19:289–304.
135. **Kimmel M, Traganos F** (1985) Kinetic analysis of drug-induced G_2 block in vitro. Cell Tissue Kinet. 18:91–110.

136. **Kimmel M, Traganos F, Darzynkiewicz Z (1983)** Do all daughter cells enter the "intermediate" ("A") state of the cell cycle? Analysis of stathmokinetic experiments on L1210 cells. Cytometry 4:191–201.

137. **Kinnally LW, Tedeschi H, Maloff BL (1978):** Use of dyes to estimate the electrical potential of the mitochondrial membrane. Biochemistry 17:3419–3428.

138. **Kipp JBA, Jongsma APM, Barendsen GW (1979)** Cell cycle phase durations derived by a flow cytometric method using the mitotic inhibitor vinblastine. In Laerum OD, Lindmo T, Thorud E (eds), "Flow Cytometry IV." Oslo: Universitetsforlaget, pp 341–344.

139. **Klaus GGB, Hawrylowicz CM, Carter J (1985)** Activation and proliferation signals in mouse B cells. II. Anti-Ig antibodies induce dose dependent cell cycle progression in B cells. Immunology 55:411–418.

140. **Klempnauer KH, Sippel A (1986)** Subnuclear localization of proteins encoded by oncogene *v-myb* and its cellular homolog *c-myb*. Mol. Cell. Biol. 6:62–69.

141. **Kohler N, Lipton A (1984)** Platelets as a source of fibroblast growth-promoting activity. Exp. Cell. Res. 87:279–301.

142. **Koss LG, Wolley RC, Schreiber K, Mendecki J (1977)** Flow microfluorometric analysis of nuclei isolated from various normal and malignant human epithelial tissues. A preliminary report. J. Histochem. Cytochem. 25:565–570.

143. **Kristensen F, Walker C, Bettens F, Joncourt F, DeWeck AL (1982)** Assessment of interleukin 1 and interleukin 2 effects on cycling and noncycling murine thymocytes. Cell. Immunol. 74:140–149.

144. **Kruijer W, Cooper JA, Hunter T, Verma IM (1984)** Platelet-derived growth factor induces rapid but transient expression of the c-*fos* gene and protein. Nature (Lond.) 312:711–716.

145. **Kruth HS (1980)** Flow cytometry: Rapid biochemical analysis of single cells. Ann. Biochem. 125:225–242.

146. **Kruth HS, Braylan RC, Benson NA, Nourse VA (1981)** Simultaneous analysis of DNA and cell surface immunoglobin in human B-cell lymphomas by flow cytometry. Cancer Res. 412:4885–4891.

147. **Kubbies M, Rabinovitch PS (1983)** Flow cytometric analysis of factors which influence the BrdUrd-Hoechst quenching effects in cultured human fibroblasts and lymphocytes. Cytometry 3:276–281.

148. **Kubbies M, Schindler D, Hoehn H, Rabinovitch PS (1985)** Cell cycle kinetics by BrdU-Hoechst flow cytometry: An alternative to the differential metaphase labelling technique. Cell Tissue Kinet. 18:551–562.

149. **Kubbies M, Schindler D, Hoehn H, Rabinovitch PS (1985)** BrdU-Hoechst flow cytometry reveals regulation of human lymphocyte growth by donor-age-related growth factor and transition rate. J. Cell. Physiol. 125:229–234.

150. **Kumar AA, Kempton RJ, Anstead GM, Freishen JH (1983)** Fluorescent analogues of methotrexate: Characterization and interaction with dihydrofolate reductase. Biochemistry 22:390–395.

151. **LaBella F, Brown EH, Basilico C (1983)** Changes in the levels of viral and cellular gene-transcripts in the cell cycle of SV 40 transformed mouse cells. J. Cell. Physiol. 117:62–68.

152. **Lachman HM, Skoultchi AI (1984)** Expression of c-*myc* changes during differentiation of mouse erythroleukemia cells. Nature (Lond.) 310:592–594.

153. **Lajtha LG (1963)** On the concept of the cell cycle. J. Cell. Comp. Physiol. 62(Suppl 1):142–145.

154. **Langen P, Graetz H, Lehmann E (1979)** The use of cytosine arabinoside and colchicin in FCM analysis of cell cycle effects of inhibitors in an asynchronous cell culture. In Laerum OD, Lindmo T, Thorud E (eds), "Flow Cytometry IV." Oslo: Universitetsforlaget, pp 362–366.

155. **Larsen JK, Munch-Petersen B, Christiansen J, Jorgensen K (1986)** Flow cytometric discrimination of mitotic cells. Resolution of M as well as G_1, S and G_2 nuclei with mithramycin, propidium iodide and ethidium bromide after fixation with formaldehyde. Cytometry 7:54–63.

156. **Larsson LG, Gray HE, Tofferman T, Pettersson U, Nilsson K (1987)** Drastically increased expression of MYC or FOS protooncogenes during in vitro differentiation of chronic lymphocytic leukemia cells. Proc. Natl. Acad. Sci. USA 84:223–227.

157. **Latt SA (1973)** Microfluorimetric detection of deoxyribonucleic acid replication in human metaphase chromosomes. Proc. Natl. Acad. Sci. USA 70:3395–3399.

158. **Latt SA (1977)** Fluorometric detection of deoxyribonucleic acid synthesis; possibilities for interfacing bromodeoxyuridine dye techniques with flow fluorometry. J. Histochem. Cytochem. 25:913–926.

159. **Lautenberger JA, Ulsh L, Shih TY, Papas TS (1983)** High level expression in Escherichia cells of enzymatically active Harvey murine sarcoma virus $p21^{ras}$ protein. Science 281:858–860.

160. **Ling MR, Kay JE (1965):** "Lymphocyte Stimulation." New York: Elsevier.

161. **Lin HT, Gibson CW, Hirschorn RR, Sittling S, Baserga R, Mercer WE (1985)** Expression of thymidine kinase and dihydrofolate reductase genes in mammalian ts mutants of the cell cycle. J. Biol. Chem. 260:3269–3274.

162. **Lloyd D, Poole RK, Edwards SW (1982)** "The Cell Division Cycle." Orlando, FL: Academic Press.

163. **Luk CK, Keng PC, Sutherland RM (1985)** Regrowth and radiation sensitivity of quiescent cells isolated from EMT6/Ro-fed plateau monolayers. Cancer Res. 45:1020–1025.

164. **Mann GJ, Dyne M, Musgrove EA (1987)** Immunofluorescent quantification of ribonucleotide reductase M1 submit and correlation with DNA content by flow cytometry. Cytometry 8:509–517.

165. **Macdonald PDM (1981)** Towards an exact analysis of stathmokinetic and continuous-labelling experiments. In Rottenberg M (ed), "Biomathematics and Cell Kinetics." Amsterdam: Elsevier, pp 125–142.

166. **MacDonald HR, Zaech P (1982)** Light scatter analysis and sorting of cells activated in mixed leukocyte culture. Cytometry 3:55–58.

167. **Marder P, Schmidtke JR (1983)** Effects of methylprednisolone on Concanavalin A-induced human lymphocyte blastogenesis: A comparative analysis by flow cytometry, volume determination and ^{3}H-thymidine incorporation. Immunopharmacology 6:155–166.

168. **Mathews MB, Bernstein PM, Franza RB Jr, Garrels**

JI (1984) Identity of the proliferating cell nuclear antigen and cyclin. Nature (Lond.) 309:374–376.

169. **Matsukage A, Yamamoto S, Yamaguchi M, Kusakabe M, Takahashi T** (1983) Immunocytochemical localization of chick DNA polymerase α* and β⁺. J. Cell. Physiol. 117:266–271.

170. **McCarins E, Fahey D, Muscat GE, Murray M, Rowe P** (1984) Changes in levels of actin and tubulin mRNAs upon the activation of lymphocytes. Mol. Cell. Biol. 4:1754–1760.

171. **Means AR, Dedman JR** (1980) Calmodulin—An intracellular calcium receptor. Nature (Lond.) 285:73–77.

172. **Mercer WE, Avignolo C, Baserga R** (1984) Role of the p53 protein in cell proliferation as studied by microinjection of monoclonal antibodies. Mol. Cell. Biol. 4:276–281.

173. **Mercer WE, Avignolo C, Galanti N, Rose KM, Hyland JK, Jacob ST, Baserga R** (1984) Cellular DNA replication is independent of the synthesis or accumulation of ribosomal RNA. Exp. Cell. Res. 150: 118–130.

174. **Mercer WE, Baserga R** (1985) Expression of the p53 protein during the cell cycle of human peripheral blood lymphocytes. Exp. Cell. Res. 160:31–46.

175. **Milcarek C, Zahn N** (1978) The synthesis of ninety proteins including actin throughout the HeLa cell cycle. J. Cell. Biol. 79:833–838.

176. **Milner J, Milner S** (1981) SV40-53K antigen: A possible role for 53K in normal cells. Virology 112:785–788.

177. **Mitchison JM** (1971) "The Biology of the Cell Cycle." Cambridge: The University Press.

178. **Monroe JG, Cambier JC** (1983) Sorting of B lymphoblasts based upon cell diameter provides cell proliferation enriched in different stages of cell cycle. J. Immunol. Methods 63:45–56.

179. **Monroe JG, Cambier JC** (1983) Level of mla expression on mitogen-stimulated murine B lymphocytes is dependent on position in cell cycle. J. Immunol. 130: 626–631.

180. **Monroe JG, Havran WL, Cambier JC** (1982) Enrichment of viable lymphocytes in defined cycle phases by sorting on the basis of pulse width of axial light extinction. Cytometry 3:24–27.

181. **Monroe JG, Kass MJ** (1985) Molecular events of B cell activation. I. Signals required to stimulate G_0 to G_1 transition of resting B lymphocytes. J. Immunol. 135:1647–1681.

182. **Morris VB, Taylor IW** (1985) Estimation of nonproliferating cells in the neural retina of embryonic chicks by flow cytometry. Cytometry 6:375–380.

183. **Morris WT** (1967) In vivo studies on optimum time for the action of colchicine on mouse lymphoid tissue. Exp. Cell. Res. 48:209–215.

184. **Mulcahy LS, Smith MR, Stacey DW** (1985) Requirement for *ras* protooncogene function during serum-stimulated growth of NIH 3T3 cells. Nature (Lond.) 313:241–243.

185. **Muller MT, Pfund WP, Mehta VB, Trask DK** (1985) Eukaryotic type I topoisomerase is enriched in the nucleolus and catalytically active on ribosomal DNA. EMBO J. 4:1237–1243.

186. **Muller R, Bravo R, Burckhardt J, Curran T** (1984) Induction of c-*fos* gene and protein by growth factors precedes activation of c-*myc*. Nature (Lond.) 312: 716–720.

187. **Nakamura H, Morita T, Masaki S, Yashida S** (1984) Intracellular localization and metabolism of DNA polymerase α in human cells visualized with monoclonal antibody. Exp. Cell. Res. 151:123–133.

188. **Newmeyer DD, Ohlsson-Willhelm BM** (1985) Monoclonal antibody to a protein of the nucleus and mitotic spindle of mammalian cells. Localization and synthesis throughout the cell cycle. Chromosoma 92: 297–303.

189. **Noronha A, Richman DP** (1984) Simultaneous cell surface phenotype and cell cycle analysis of lymphocytes by flow cytometry. J. Histochem. Cytochem. 32:821–829.

190. **Noronha ABC, Richman DP, Arnason BGW** (1980) Detection of in vivo stimulated cerebrospinal fluid lymphocytes by flow cytometry in patients with multiple sclerosis. N. Engl. J. Med. 303:713–717.

191. **Ochs R, Lischwe M, O'Leary P, Busch H** (1983) Localization of nucleolar phosphoproteins B23 and C23 during mitosis. Exp. Cell. Res. 146:139–149.

192. **Ohlsson RI, Pfeifer-Ohlsson SB** (1987) Cancer genes, proto-oncogenes and development. Exp. Cell. Res. 173:1–16.

193. **Oliver K, Noelle RJ, Uhr JW, Krammer PH, Vittetta ES** (1985) B-cell growth factor (B-cell growth factor I or B-cell-stimulating factor, provisional 1) is a differentiation factor for resting B cells and may not induce cell growth. Proc. Natl. Acad. Sci. USA 82:2465–2467.

194. **Osborne HB, Chabanas A** (1984) Kinetics of H1° accumulation and commitment to differentiation of murine erythroleukemia cells. Exp. Cell. Res. 158: 449–458.

195. **Panyim S, Chalkley R** (1969) A new histone found only in mammalian tissues with little cell division. Biochem. Biophys. Res. Commun. 37:1042–1048.

196. **Patel GL, Reddigari S, Williams KR, Baptist E, Thompson PE, Sisodia S** (1983) A rat liver helix-destabilizing protein: Properties and homology to LDH-5. In Skehan P, Friedman SJ (eds), "Growth, Cancer and the Cell Cycle." Clifton, New Jersey: Humana Press, pp 41–58.

197. **Petersen DF, Anderson EC, Tobey RA** (1968) Mitotic cells as a source of synchronized cultures. In Prescott DM (ed), "Methods in Cell Physiology, vol. III." Orlando, FL: Academic Press, pp 347–370.

198. **Piwnicka M, Darzynkiewicz Z, Melamed MR** (1983) RNA and DNA content of isolated cell nuclei measured by multiparameter flow cytometry. Cytometry 3:269–275.

199. **Pollack A, Maulis H, Block NL, Irwing GL III** (1984) Quantitation of cell kinetic responses using simultaneous flow cytometric measurements of DNA and nuclear protein. Cytometry 5:473–481.

200. **Pollack A, Moulis H, Greenstein DB, Block NL, Irvin III GL** (1985) Cell kinetic effects of incorporated ^{3}H-thymidine on proliferating human lymphocytes: Flow cytometric analysis using the DNA/nuclear protein method. Cytometry 6:428–436.

201. **Pollack A, Prudhomme D, Irwin GL III, Claffin AJ, Block NL** (1981) Flow cytometric analysis of RNA content per cell using pyronin Y and methyl green. Cytometry 2:122.

202. **Potten CS, Wichman HE, Loeffler M, Dobek K, Major D** (1982) Evidence for discrete cell kinetic subpopulations in mouse epidermis based on mathematical analysis. Cell Tissue Kinet. 15:305–329.
203. **Prescott D** (1976) The cell cycle and the control of cellular reproduction. Adv. Genet. 18:99–177.
204. **Prescott DM (ed)** (1977) "Methods in Cell Biology XVI. Chromatin and Chromosomal Protein Research I." Orlando, Florida, Academic Press, pp 1–123.
205. **Prescott DM** (1987): Cell reproduction. Int Rev Cytol 100:93–128.
206. **Puck TT, Steffen J** (1963) Life cycle analysis of mammalian cells. I. A method for localizing metabolic events within the life cycle and its application to the action of colcemid and sublethal doses of X-irradiation. Biophys. J. 3:379–397.
207. **Rabinovitch PS, Kubbies M, Chen YC, Schindler D, Hoehn H** (1988) BrdU-Hoechst flow cytometry: A unique tool for quantitative cell cycle analysis. Exp. Cell. Res. 174:309–318.
208. **Rabinovitch PS, June CH, Grossman A, Ledbetter JA** (1986) Heterogeneity among T cells in intracellular free calcium responses after mitogen stimulation with PHA or anti-CD3. Simultaneous use of indo-1 and immunofluorescence with flow cytometry. J. Immunol. 137:952–961.
209. **Rabinovitch PS, Torres RM, Engel D** (1986) Simultaneous cell cycle analysis and two-color surface immunofluorescence using 7-amino-actinomycin D and single laser excitation: Applications to study cell activation and the cell cycle of murine Ly-1 B cells. J. Immunol. 136:2769–2775.
210. **Rapaport E, Zamecnik PC, Baril EF** (1981) Association of diadenosine 5′, 5′′-p-p^4-tetraphosphate binding protein with HeLa cell DNA polymerase. J. Biol. Chem. 256:12143–12151.
211. **Riddle VGH, Dubrow R, Pardee AB** (1979) Changes in the synthesis of actin and other cell proteins after stimulation of serum-arrested cells. Proc. Natl. Acad. Sci. USA 76:1298–1302.
212. **Riddle VGH, Pardee AB** (1980) Quiescent cells but not cycling cells exhibit enhanced actin synthesis before they synthesize DNA. J. Cell. Physiol. 103:11–15.
213. **Rigby WFC, Noelle RJ, Krause K, Fanger MW** (1985) The effects of 1,25-dihydroxyvitamin D_3 on human on human T lymphocyte activation and proliferation: A cell cycle analysis. J. Immunol. 135: 2279–2286.
214. **Ringertz NR, Bolund L** (1974) Reactivation of chick erythrocyte nuclei by somatic cell hybridization. Int. Rev. Exp. Pathol. 8:83–116.
215. **Ringertz NR, Savage RE** (1976) "Cell Hybrids." Orlando, Florida: Academic Press.
216. **Ritling SR, Baserga R** (1987) Regulatory mechanisms in the expression of cell cycle dependent genes. Anticancer Res. 7:541–522.
217. **Rizzoni M, Palitti F** (1973) Regulatory mechanisms of cell division. I. Colchicine-induced endoreduplication. Exp. Cell. Res. 77:450–458.
218. **Roti Roti JL, Higoshikubo P, Blair CC, Uyger N** (1982) Cell-cycle position and nuclear protein content. Cytometry 3:91–96.
219. **Salzman GC, Singham SB, Johnston RG, Bohren CF** (See Chapter 5, this volume.)
220. **Sambucetti L, Curran T** (1986) The Fos protein complex is associated with DNA in isolated nuclei and binds to DNA cellulose. Science 234:1417–1419.
221. **Sasvari-Szekely M, Szabo G, Staub M, Spasokukotskaja T, Antoni F** (1983) Discrepancies between flow cytometric analysis and (^{3}H)-thymidine incorporation in stimulated lymphocytes. Biochim. Biophys. Acta 762:452–457.
222. **Sato C, Nishizawa K, Nakamura H, Ueda R** (1984) Nuclear immunofluorescence by a monoclonal antibody against microtubule-associated protein-1 as it is associated with cell proliferation and transformation. Exp. Cell. Res. 155:33–42.
223. **Sato C, Tanabe K, Nishizawa K, Nakayma T, Kobayashi T, Nakamura H** (1985) Localization of 350K molecular weight and related proteins in both cytoskeleton and nuclear flecks that increase during G_1 phase. Exp. Cell. Res. 160:206–220.
224. **Seuwen K, Steiner V, Adam G** (1984) Cellular content of ribosomal RNA in relation to the progression and competence signals governing proliferation of 3T3 and SV40-3T3 cells. Exp. Cell. Res. 154:10–24.
225. **Shapiro HM** (1981) Flow cytometric estimation of DNA and RNA content in intact cells stained with Hoechst 33342 and pyronin Y. Cytometry 2:143–150.
226. **Shapiro HM** (1985) "Practical Flow Cytometry." New York: Alan R. Liss, Inc.
227. **Shapiro HM, Natale PJ, Kamensky LA** (1979) Estimation of membrane potential of individual lymphocytes by flow cytometry. Proc. Natl. Acad. Sci. USA 76:5728–5732.
228. **Sharpless TK, Schlesinger FH** (1982) Flow cytometric analysis of G_1 exit kinetics in asynchronous L1210 cell cultures with the constant transition probability model. Cytometry 3:196–200.
229. **Sharpless T, Traganos F, Darzynkewicz Z, Melamed MR** (1975) Flow cytofluorimetry: Discrimination between single cells and cell aggregates by direct size measurements. Acta Cytol. (Praha) 19:577–581.
230. **Shen DW, Real FX, DeLeo AB, Old LJ, Marks PA, Rifkind RA** (1983) Protein p53 and inducer-mediated erythroleukemia cell commitment to terminal cell division. Proc. Natl. Acad. Sci. USA 80:5919–5922.
231. **Smeland E, Godal T, Rund E, Beiske K, Funderud S, Clark EA, Pfeifer-Ohlsson S, Ohlsson R** (1985) The specific induction of *myc* protooncogene expression in normal human B cells is not a sufficient event for acquisition of competence to proliferate. Proc. Natl. Acad. Sci. USA 82:6255–6259.
232. **Smith JA, Martin L** (1973) Do cells cycle? Proc. Natl. Acad. Sci. USA 70:1263–1267.
233. **Spector DC, Ochs RL, Busch H** (1984) Silver staining, immunofluorescence and immunoelectron microscopic localization of nucleolar phosphoproteins B23 and C23. Chromosoma 90:139–148.
234. **Staiano-Coico L, Higgins PJ, Darzynkiewicz Z, Kimmel M, Gottlieb AB, Pagan-Charry I, Madden MK, Finkelstein JL, Hefton JM** (1986) Human keratinocyte culture. Identification and staging of epidermal cell subpopulations. J. Clin. Invest. 77:396–404.
235. **Staufenbiel M, Deppert W** (1983) Different structural systems of the nucleus are targets for SV40 large T antigen. Cell 33:173–181.
236. **Steen HB, Lindmo T** (1978) The effect of colchicine

and colcemid on the mitogen-induced blastogenesis of lymphocytes. Eur. J. Immunol. 8:667–671.

237. **Steen HB, Lindmo T (1978)** Cellular and nuclear volume during the cell cycle of NHIK 3025 cells. Cell Tissue Kinet. 11:69–81.

238. **Steinkamp JA (1984)** Flow cytometry. Rev. Sci. Instrum. 55:1375–1400.

239. **Stiles CD (1985)** The biological role of oncogenes-insights from platelet-derived growth factor. Rhoads Memorial Award Lecture. Cancer Res. 45:5215–5218.

240. **Stohr M, Vogt-Schaden M, Knobloch M, Vogel R, Futterman G (1978)** Evaluation of eight fluorochrome combinations for simultaneous DNA-protein flow analyses. Stain Technol. 53:205–212.

241. **Studzinski GP, Brelvi ZS (1987)** Increased expression of oncogene c-HA-ras during granulcytic differentiation of HL60 cells. Lab. Invest. 56:499–504.

242. **Sturani E, Toschi L, Zippel R, Mortegani E, Aberghina L (1984)** G_1 phase heterogeneity in exponentially growing Swiss 3T3 mouse fibroblasts. Exp. Cell. Res. 153:135–144.

243. **Swartzendruber DE (1977)** A bromodexoyuridine-mithramycin technique for detecting cycling and non-cycling cells by flow microfluorometry. Exp. Cell. Res. 109:439–443.

244. **Swartzendruber DE (1977)** Microfluorometric analysis of cellular DNA following incorporation of bromodeoxyuridine. J. Cell. Physiol. 90:445–454.

245. **Takasaki Y, Fishwild D, Tan EM (1984)** Characterization of proliferating cell nuclear antigen recognized by autoantibodies in lupus sera. J. Exp. Med. 159:981–992.

246. **Tamura T, Friis RR, Bauer H (1984)** pp60$^{c\text{-}src}$ is a substrate for phosphorylation when cells are stimulated to enter the cycle. FEBS Lett. 177:151–155.

247. **Tanke HJ, Niewenhuis IAB, Koper GJM, Slats JCM, Ploem JS (1980)** Flow cytometry of human reticulocytes based on RNA fluorescence. Cytometry 1:313–320.

248. **Tannock IF (1967)** A comparison of the relative efficiencies of various metaphase arrest agents. Exp. Cell. Res. 47:345–356.

249. **Taparowsky E, Shimizu K, Goldfarb M, Wigler M (1983)** Structure and activation of the human N-ras gene. Cell 34:581–586.

250. **Taylor IW (1980)** A rapid single-step staining technique for DNA anlysis by flow microfluorimetry. J. Histochem. Cytochem. 28:1021–1028.

251. **Terasima T, Tolmach LJ (1963)** Growth and nucleic acid synthesis in synchronously dividing populations of HeLa cells. Exp. Cell. Res. 36:344–362.

252. **Thompson CB, Challoner PB, Neiman PE, Groudine M (1985)** Levels of c-*myc* oncogene mRNA are invariant throughout the cell cycle. Nature (Lond.) 314:363–369.

253. **Thompson CB, Challoner PB, Neiman PE, Groudine M (1986)** Expression of the *c-myb* proto-oncogene during cellular proliferation. Nature (Lond.) 319: 373–380.

254. **Thompson CB, Schaefer ME, Finkelman FD, Scher I, Farrar J, Mond JJ (1985)** T cell-derived B cell growth factor(s) can induce stimulation of both resting and activated B cells. J. Immunol. 134:369–374.

255. **Thornthwaite JT, Sugerbaker EV, Temple WJ (1980)** Preparation of tissues for DNA flow cytometric analysis. Cytometry 1:299–306.

256. **Tijan R, Fey G, Graessman A (1978)** Biological activity of purified simian virus 40 T antigen proteins. Proc. Natl. Acad. Sci. USA 75:1275–1283.

257. **Traganos F (1984)** Flow cytometry. Principles and application. II. Cancer Invest. 2:239–258.

258. **Traganos F, Darzynkiewicz Z, Melamed MR (1981)** Effects of L-isomer (+)-1,2-bis(3,5,-dixoypiperazine-1-yl) propane on cell survival and cell cycle progression of cultured mammalian cells. Cancer Res. 41: 4566–4575.

259. **Traganos F, Darzynkiewicz Z, Melamed MR (1982)** The ratio of RNA to total nucleic acid content as a quantitative measure of unbalanced cell growth. Cytometry 2:212–218.

260. **Traganos F, Darzynkiewicz Z, Sharpless T, Melamed MR (1979)** Erythroid differentiation of Friend leukemia cells as studied by acridine orange staining and flow cytometry. J. Histochem. Cytochem. 27:382–389.

261. **Traganos F, Evenson DP, Staiano-Coico L, Darzynkiewicz Z, Melamed MR (1980)** Action of dihydroxyanthraquinone on cell cycle progression and survival of cultured mammalian cells. Cancer Res. 40:671–681.

262. **Traganos F, Gorski AJ, Darzynkiewicz Z, Sharpless T, Melamed MR (1977)** Rapid multiparameter analysis of cell stimulation in mixed lymphocyte culture reactions. J. Histochem. Cytochem. 25:881–887.

263. **Traganos F, Higgins PJ, Bueti C, Darzynkiewicz Z, Melamed MR (1984)** Effects of retinoic acid versus dimethyl sulfoxide on Friend erythroleukemia cell growth. II. Induction of quiescent, nonproliferating cells. J. Natl. Cancer Inst. 73:205–218.

264. **Traganos F, Kimmel M, Bueti C, Darzynkiewicz Z (1987)** Effects of inhibition of RNA or protein synthesis on CHO cell cycle progression. J. Cell. Physiol. 133:277–287.

265. **Traganos F, Staiano-Coico L, Darzynkiewicz Z, Melamed MR (1981)** Effects of dihydro-5-azacytidine on cell survival and cell cycle progression of cultured mammalian cells. Cancer Res. 41:780–789.

266. **Traganos F, Staiano-Coico L, Darzynkiewicz Z, Melamed MR (1981)** Effects of aclacinomycin on cell survival and cell cycle progression of cultured mammalian cells. Cancer Res. 41:2728–2737.

267. **Tsai YH, Hnilica LS (1975)** Tissue-specific histones in the erythrocytes of chicken and turtle. Exp. Cell. Res. 91:107–112.

268. **Van den Engh G, Visser J, Bol S, Trask B (1980)** Concentration of hemopoietic stem cells using a light-activated cell sorter. Blood Cells 6:1–12.

269. **Visser JWM, Cram LS, Martin JC, Salzman GC, Price BJ (1978)** Sorting of a murine progenitor cell by use of laser light scattering measurements. In Lutz D (ed), "Pulse Cytophotometry, VI." Ghent, Belgium: European Press, pp 178–192.

270. **Waggnoner AS (1979)** Dye indicators of membrane potential. Annu. Rev. Biophys. Bioeng. 8:47–68.

271. **Walker C, Kristensen F, Bettens F, DeWeck AL (1983)** Lymphokine regulation of activated (G_1) lymphocytes. I. Prostaglandin E2-induced inhibiton of interleukin 2 production. J. Immunol. 130:1770–1773.

272. **Walle AJ, Wong GY, Suthanthiran M, Rubin AL,**

Stenzel KH, Darzynkiewicz Z (1988) Altered nucleic acid contents of mononuclear cells in blood of renal allograft recipients. Cytometry 9:177–182.

273. **Wallen CA, Higashikubo R, Dethlefsen LA (1984)** Murine mammary tumor cells in vitro. I. The development of quiescent state. Cell Tissue Kinet. 17:65–77.

274. **Wallen CA, Higashikubo R, Dethlefsen LA (1984)** Murine mammary tumor cells in vitro. II. Recruitment of quiescent cells. Cell Tissue Kinet. 17:79–89.

275. **Wang E (1985)** A 57,000-mol-wt protein uniquely present in nonproliferating cells and senescent human fibroblasts. J. Cell. Biol. 100:545–551.

276. **Wang E (1985)** Rapid disappearance of statin, a nonproliferating and senescent cell-specific protein upon reentering the process of cell cycling. J. Cell. Biol. 101:1695–1701.

277. **Wang E, Krueger JG (1985)** Application of a unique monoclonal antibody as a marker of nonproliferative subpopulations of cells in some tissues. J. Histochem. Cytochem. 33:587–594.

278. **Watson J (1985)** Simultaneous assay of DNA and c-*myc* oncoprotein in archive material. "Proceedings of the Flow Cytometry Conference, NIMR Mill Hill, London, April 18,19."

279. **Watson JV, Chambers SH (1977)** Fluorescence discrimination between diploid cells based on their RNA content: A possible distinction between clono-genic and non-clonogenic cells. Br. J. Cancer 36:592–600.

280. **Watt RA, Shatzman AR, Rosenberg M (1985)** Expression and characterization of the human c-*myc* DNA binding protein. Mol. Cell. Biol. 5:448–456.

281. **Waxdal MJ (1983)** An early biochemical pathway of transmembrane signaling in the stimulation of lymphocytes. In Parker JW, O'Brien RL (eds), "Intercellular Communication in Leukocyte Function." New York: Wiley & Sons, pp 413–418.

282. **Weinmann-Dorsch C, Pierron G, Wick R, Sauer H, Grummt F (1984)** High diadenosine tetraphosphate (Ap4A) level at initiation of S phase in the naturally synchronous mitotic cycle of Physarum polysephalum. Exp. Cell. Res. 155:171–177.

283. **Wells V, Malluci L (1985)** Expression of the '2-5A system during the cell cycle. Exp. Cell. Res. 159:27–36.

284. **Westwood JT, Wagenaar EB, Church RB (1985)** Synthesis of unique low molecular weight proteins during late G_2 and mitosis. J. Biol. Chem. 260:695–698.

285. **Wetzel GD, Swain SL, Dutton RW, Kettman JR (1984)** Evidence for two distinct activation states available to B lymphocytes. J. Immunol. 133:2327–2332.

286. **Williams JM, DeLoria D, Hansen JA, Dinarello CA, Loertscher R, Shapiro HM, Strom TB (1985)** The events of primary T-cell activation can be staged by use of sepharose-bound anti-T3 (64.1) monoclonal antibody and purified interleukin. J. Immunol. 135:2249–2255.

287. **Williams JM, Loertscher R, Cotner R, Reddish HM, Shapiro HR, Carpenter CB, Strominger JL, Strom TB (1984)** Dual parameter flow cytometric analysis of DNA content, activation, antigen expression and T cell subset proliferation in human mixed lymphocyte reaction. J. Immunol. 132:2330–2336.

288. **Williams JM, Ransil BJ, Shapiro HM, Strom TB (1984)** Accessory cell requirement for activation antigen expression and cell cycle progression by human T lymphocytes. J. Immunol. 133:2986–2995.

289. **Willingham MC, Pastan I, Shih TY, Scolnick EM (1980)** Localization of the *src* gene product of the Harvey strain of MSV on the plasma membrane of transformed cells by electron microscopic immunocytochemistry. Cell 19:1005–1014.

290. **Winchurch RA, Cimino E (1978)** Analysis of variability of in vitro lymphocyte responses as measured by laser cytometry and thymidine incorporation. J. Immunol. Methods 23:269–274.

291. **Wolf D, Rotter V (1985)** Major deletions in the gene encoding the p53 tumor antigen cause lack of p53 expression in HL-60 cells. Proc. Natl. Acad. Sci. USA 82:790–794.

292. **Wray W, Stubblefield E (1970)** A new method for the rapid isolation of chromosomes, mitotic apparatus or nuclei from mammalian fibroblasts at near neutral pH. Exp. Cell. Res. 59:469–481.

293. **Wright NA, Appleton DR (1980)** The metaphase arrest technique. A critical review. Cell Tissue Kinet. 13:643–663.

294. **Wright NA, Britton DC, Bone G, Appleton DR (1977)** A stathmokinetic study of cell proliferation in human gastric carcinoma and gastric mucosa. Cell Tissue Kinet. 10:429–437.

295. **Yen A, Fried J, Clarkson B (1977)** Alternative modes of population growth inhibition in human lymphoid cell line growing in suspension. Exp. Cell. Res. 107:325–341.

296. **Zamecnik AC, Rapaport E, Baril EF (1982)** Priming of DNA synthesis by diadenosine $5'$-$5''$-p$'$,p^4 tetraphosphate with double-stranded octodecomer as a template and DNA polymerase. Proc. Natl. Acad. Sci. USA 79:1791–1794.

25

Flow Karyotyping and Chromosome Sorting

J. W. Gray and L. S. Cram
Biomedical Sciences Division, Lawrence Livermore National Laboratory, Livermore, California 94550 (J.W.G.). Life Sciences Division, Los Alamos National Laboratory, Los Alamos, New Mexico 87545 (L.S.C.)

INTRODUCTION

Flow cytogenetics is defined as the use of flow cytometry and sorting for classification and purification of chromosomes isolated from mitotic cells. The utility of flow cytogenetics has increased substantially since the first reports of the possibility of flow cytometric chromosome analysis in 1975 [53,54,151]. For example, it is now possible to classify chromosomes according to their DNA content, base composition, protein content and/or shape, using multivariate flow cytometry. As a result, flow cytometric chromosome classification, hereafter called flow karyotyping, has proved useful for analysis of chromosomes from Chinese hamster [8], muntjac [22], mouse [42], swine [67], rat kangaroo [149], chicken [152] and rodent/human hybrid [35,163] cells. In addition, it has been applied extensively to classify chromosomes from human fibroblasts [21], lymphocytes [98,119,170,172], amniocytes [63,161] and chorionic villus cells (J.W. Gray, private communication, 1986). Flow karyotyping of these cells has proved especially useful for quantitative detection of aberrant chromosomes. Translocations, deletions, additions, and numerical aberrations have been detected in samples where they occur homogeneously (i.e., in most or all of the cells from which the chromosomes were isolated [30,56,63]. In addition, the frequency of heterogeneous aberrations (i.e. occurring randomly and at low frequency in the cells from which the chromosomes were isolated) such as deletions, dicentrics and other structural aberrations have been quantified as an indication of the degree of induced genetic damage [56,60,82].

Flow cytogenetics has also allowed purification of chromosomes to a high degree by sorting since any chromosomes that can be distinguished flow cytometrically can be purified by sorting. Sorter-purified chromosomes have proved useful for gene mapping [17,24,25,87,105,107–110], analysis of protein content [169] and as a source of DNA for production of recombinant DNA libraries [13,37,83,84,86,125,171]. The molecular weight of the DNA extracted from sorted chromosomes is sufficiently high that it may also be suitable for analysis on pulsed field gel electrophoresis thereby allowing construction of large fragment maps of selected chromosomes [39,48,55].

This chapter reviews the principles and techniques necessary for flow cytogenetics, including chromosome preparation and staining, flow cytometric chromosome analysis, quantitative analysis of flow karyotypes, and purification by sorting. In addition, we describe the applications of flow cytogenetics to analysis of homogeneous and heterogeneous aberrations and to purification of chromosomes for gene mapping and for library production.

INSTRUMENTATION

Instruments currently in use for chromosome classification and purification include single and dual laser sorters, slit-scan flow cytometers, and high-speed sorters. The aspects of these instruments that are important for chromosome analysis and purification are presented for each instrument type.

Single-Laser Chromosome Analysis

Single-laser flow cytometers are commonly used for classification and purification of chromosomes stained with a single DNA specific fluorescent dye such as Hoechst 33258 (Ho58), propidium iodide (PI), ethidium bromide (EB), or chromomycin A3 (CA3). In this approach, illustrated in Figure 1, chromosomes are classified according to the intensity of fluorescence that they emit as they flow through one laser beam whose wavelength is adjusted for efficient excitation of the dye. The results from several hundred thousand chromosome measurements are accumulated to form a univariate fluorescence distribution like that shown for human chromosomes in the insert to Figure 1. These distributions are characterized by several peaks, each produced by one or a few chromosome types with similar DNA content (dye content). The mean of each peak is proportional to the dye content of that chromosome type. The relative area of each peak is proportional to the frequency of that chromosome type in the population [54,63,122,123,146].

High measurement resolution (i.e. measurement coefficients of variation (CV) lower than ~0.02) is essential in flow cytogenetics because differences between chromosomes

Flow Cytometry and Sorting, Second Edition, pages 503–529
Published 1990 by Wiley-Liss, Inc.

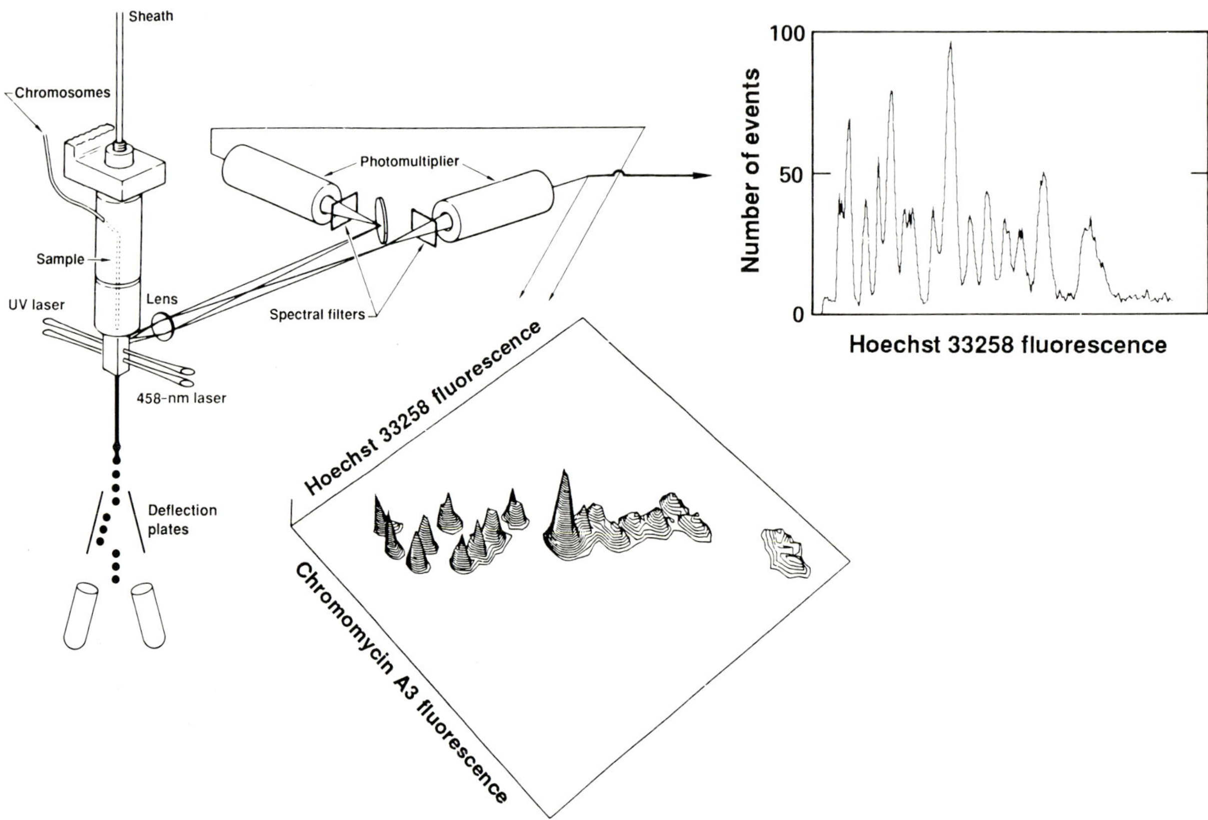

Fig. 1. Instrumentation for chromosome analysis and sorting. Chromosomes flow one at a time through laser beams adjusted to efficiently excite the dye(s) with which the chromosomes have been stained. One laser is used for univariate flow karyotyping and two lasers are employed for bivariate flow karyotyping. The fluorescence emitted as the chromosomes pass through the laser beam(s) is collected by a lens and projected through spectral filters to define the wavelength of light to be measured and onto photomultipliers that produce electrical pulses proportional in amplitude to the amount of light reaching them. The amplitudes of the pulses from the detectors are then amplified and digitized. The results are stored as a univariate (see right insert) or bivariate (see lower left insert) distribution showing the distribution of fluorescence intensities among the objects of the population. Homogeneously staining chromosomes produce peaks in the fluorescence distributions (flow karyotypes). Chromosomes are purified according to their location in the flow karyotypes. Chromosomes with fluorescence intensities that place them within a predefined region of the flow karyotype are electronically selected for sorting. That is, these chromosomes travel from the measuring region in a jet in air to the point where the liquid jet breaks into droplets. The droplets containing the chromosomes selected for sorting are charged as they separate from the liquid column. Thus the chromosomes to be sorted reside in charged droplets. The other chromosomes are in uncharged droplets. The charged droplets are separated from the others during passage through a high voltage electric field. Two classes of chromosomes may be separated simultaneously by charging the droplets carrying one class negatively and charging the droplets carrying the other class positively.

of different types are small. This is illustrated in Figure 2, which shows the DNA content distributions expected for human chromosomes assuming measurement CV of 0.04, 0.02, and 0.01. It is clear that CVs lower than 0.04 are required for resolution of even a few single chromosome types. Two aspects of flow cytometry are crucial for high resolution chromosome analysis: (1) collection of sufficient photons from each chromosome to permit accurate determination of the chromosomal stain content, and 2) uniform illumination of the chromosomes so that the emitted fluorescence is proportional to the total chromosomal dye content. The number of photons collected from each chromosome increases with increasing intensity of illumination and with increasing efficiency of the optical system used for fluorescence collection. In addition, the number of photons collected can be increased by adjusting the laser wavelength to match the absorption maximum of the dye, by selecting spectral filters to pass as much fluorescence as possible without passing scattered laser light and by using photomultipliers with maximum sensitivity in the spectral range over which the dye fluoresces. At least 10,000 photoelectrons should be produced in the detector for each chromosome for 1% resolution. Most commercially available flow systems equipped with lasers that emit several hundred milliwatts are capable of meeting this requirement if careful attention is given to dye and filter selection. Uniform illumination of the chromosomes requires that the chromosomes be forced to flow through a part of the laser beam that varies in intensity by less than -1%. In most flow systems, this requires that the diameter of the sample stream in which the chromosomes flow be kept to less than a few micrometers. For example, the chromosomes must be constrained to flow in a cylindrical region whose diameter is only $\sim 3.5\ \mu m$ in a system in which the width of the laser beam (full width of $1/e^2$ points) is $50\ \mu m$. The diameter of the sample stream can be made larger if the width of the laser beam is made correspondingly

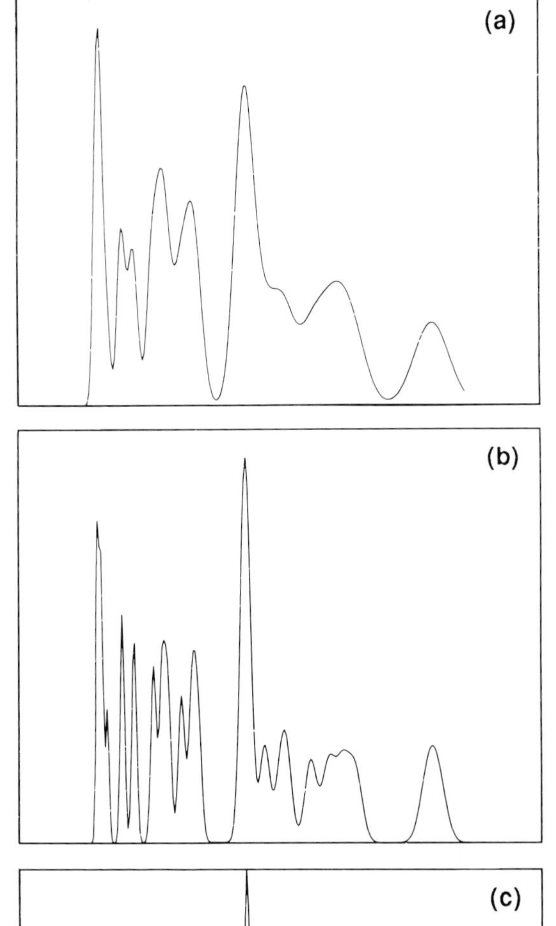

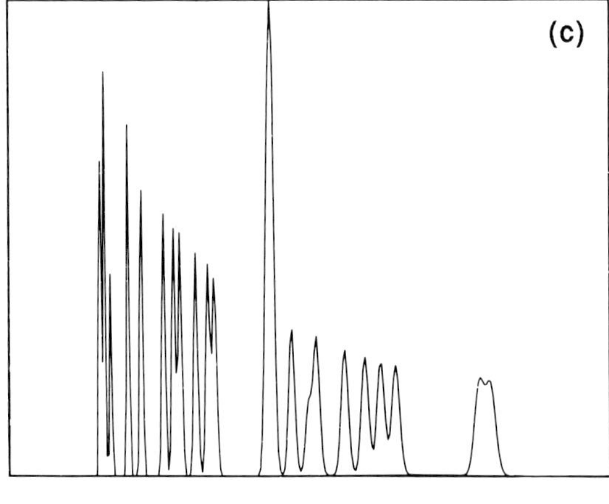

Fig. 2. DNA distributions for human chromosomes. **a–c:** DNA distributions expected for human chromosomes assuming measurement coefficients of variation of 0.04, 0.02 and 0.01, respectively.

larger. However, this can be achieved only at the expense of illumination intensity. The chromosome concentration should be higher than 10^7/ml for flow analysis in order to maintain high throughput at high resolution. To see this, consider that the chromosome analysis rate would be 1,000/sec at a sample concentration of 10^7/ml assuming a flow velocity of 10 m/sec and a sample stream diameter of 3.5 μm [56].

Dual Laser Flow Cytometry

Dual laser flow cytometry is usually employed for analysis of human chromosomes stained with two DNA specific dyes like Ho58 (or DAPI) and CA3 (or mithramycin). Ho58 binds preferentially to adenine-thymine (AT) rich DNA [73,100,102–104] while CA3 binds preferentially to guanine-cytosine (GC)-rich DNA [11,77,100,160,161,164] so that chromosomes can be classified according to their DNA content (roughly proportional to the Ho58 plus CA3 content) and DNA base composition (related to the Ho58/CA3 ratio). During dual laser flow cytometry (see Fig. 1), the chromosomes pass sequentially through two laser beams; one adjusted to the UV to efficiently excite Ho58 and the other adjusted to efficiently excite CA3. The fluorescence intensities for several hundred thousand chromosomes are accumulated to form a bivariate flow karyotype like that shown in the insert to Figure 1. Each peak is produced by one or a few chromosome types that have similar Ho58 and CA3 contents. All human chromosomes except chromosomes 9 through 12 (and sometimes 14 and 15) can be resolved in bivariate Ho58/CA3 flow karyotypes [24,57,97,98,162]. Thus, the degree of human chromosome discrimination is higher than in univariate flow karyotypes measured at comparable resolution because the peaks are spread out over two dimensions. Other dye combinations may be used to further improve the discrimination between chromosomes that are difficult to distinguish using Ho and CA3 [108,121].

Chromosome Sorting

Chromosomes that can be distinguished flow cytometrically can be purified by sorting (see Fig. 1). The rate at which chromosomes can be purified depends on the sorter droplet production rate (~20,000/sec in conventional sorters). Typically, chromosomes are processed at approximately 5% of the droplet rate (~1,000/sec in conventional sorters). These rates suggest that human chromosomes of a single type can be purified at ~40/sec (i.e., 1/25th of 1,000 chromosomes/sec). In practice, the sorting rates can drop to about half the theoretically expected rate, since many of the objects through the system are not chromosomes. Thus, approximately 7 hours is required for purification of 10^5 chromosomes, 50 hours to purify 1 μg of DNA from larger human chromosomes and as much as 250 hours to purify 1 μg of DNA from one of the smaller chromosomes [6,36,37,40,42,56,163]. These long sorting times stimulated the development of the high speed sorter [55,138]. This instrument, which operates at ~200 psi, produces droplets at approximately 200,000/sec so that chromosomes can be purified at rates up to 10 times faster than in conventional sorters. This sorter permits purification of several hundred human chromosomes of one type each second, thereby speeding purification of microgram amounts of chromosomal DNA.

The purity with which chromosomes can be sorted depends on the degree to which the chromosome of interest can be distinguished from chromosomes of similar size, from clumps of chromosomes and from DNA-containing debris that may be produced during chromosome isolation. The bivariate Ho58 versus CA3 flow karyotypes shown in Figure 1 and Figure 3a show that most of the human chromosomes smaller than chromosome 12 can be distinguished from each other and can be sorted at high purity. Chromosomes 9 through 12 cannot be distinguished from each other. Chromosomes 14 through 17 and 8 through 5 can be distinguished only partially. Chromosomes 1 through 4 are often superimposed on

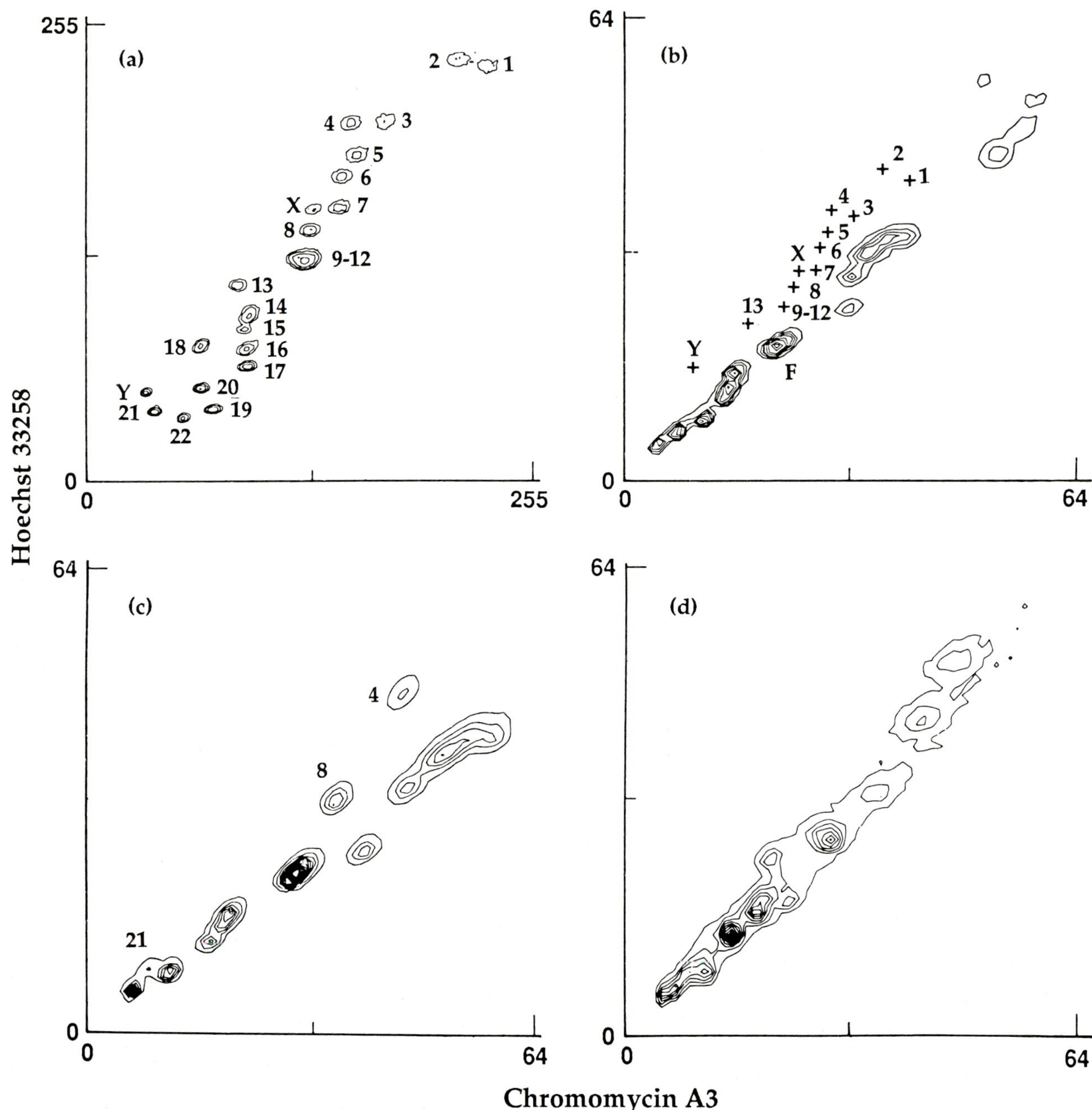

Fig. 3. Bivariate flow karyotypes for several cell types: **a:** Flow karyotype for human chromosomes isolated from karyotypically normal human lymphoblastoid cells. The chromosomes producing each peak are indicated in the figure. **b:** Flow karyotype for a hamster cell line. The crosses indicate the locations where human chromosomes 1 through 13 and Y would appear in a human hamster hybrid cell line made using this hamster line as one of the parents. **c:** Flow karyotype for a human × hamster cell line carrying human chromosomes 4, 8, and 21. The human chromosome peaks are numbered in the figure. The flow karyotype of the hamster cell line used to produce this hybrid is shown in b. **d:** Flow karyotype for mouse chromosomes.

a continuum produced by clumps of smaller chromosomes. Several approaches have been used to facilitate purification of chromosomes that are not fully resolved. In one approach, human cell lines are selected that carry polymorphisms, structural rearrangements or numerical aberrations that serendipidously increase the distinctiveness with which one or more chromosome types can be resolved [39,48,50,65,68,109, 170]. Another approach is to sort from human × hamster hybrid cells carrying one or a few human chromosomes [36, 40,58,163]. Figure 3c illustrates the ease with which the larger human chromosomes can be separated from Chinese hamster chromosomes in a hybrid line carrying human chromosomes 4, 8, and 21. Human chromosomes 4 and 8 are especially well resolved and can be easily purified by sorting. The National

Laboratory Gene Library Project has relied on several human cell lines and on a panel of hybrid cells for purification of human chromosomes of each type. Measured sorting purity ranges from 70% to 90% for most chromosome types [40,163]. Other dye combinations may also allow selected chromosomes to be resolved [105]. For example, use of DAPI in place of Ho58 often allows human chromosome 9 to be resolved from chromosomes 10 through 12.

Several applications of sorted chromosomes require that the molecular weight of the DNA therein be high. Such applications include production of large insert recombinant DNA libraries or pulsed field gel electrophoresis following digestion with restriction enzymes such as Not 1, which cut human DNA infrequently. The DNA fragment lengths from sorted chromosomal DNA can be larger than 500 kb. It is important in studies requiring high molecular weight DNA to minimize the time between chromosome isolation and sorting, to minimize DNA degradation by nucleases released from the cells during chromosome isolation [39,50,66,106] and to minimize shear forces during chromosome concentration and DNA extraction.

Slit-Scan Flow Cytometry

Chromosome shape has also proved a useful chromosome discriminant. The centrometric index (CI), the ratio of the DNA content of the long arm of a chromosome divided by its total DNA content, is the most commoniy measured shape descriptor. Chromosome shape analysis has also proved useful in the analysis of the extent of induced genetic damage. In this assay, the frequency of chromosomes with two centromeres (dicentric chromosomes) is measured as an indication of the extent of damage. Scanning flow cytometry has been developed to allow measurement of chromosome shape [32,59–61,76,115,116,130]. In this approach, illustrated in Figure 4, the chromosomes stained with a DNA-specific fluorescent dye are forced to flow lengthwise through a thin laser beam. Fluorescence intensities are measured periodically (e.g. every 10 nsec) as a chromosome passes through the beam so that a digital profile of the distribution of the dye (DNA) along that chromosome is recorded. Chromosomal DNA content is lower around the centromere so that the digital profiles show a dip or reduced fluorescence intensity in the segment produced by passage of a centromere through the scanning region. Figure 5 shows slit-scan profiles measured for normal and dicentric chromosonies. Centromeric dips are clearly visible in these profiles. Dicentric chromosomes are recognized as producing profiles with two centromeric dips.

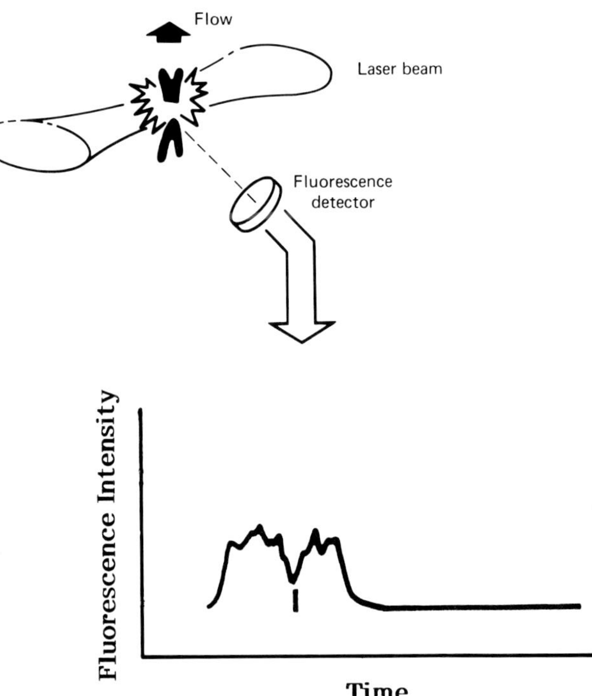

Fig. 4. Slit-scan flow cytometry. Chromosomes are scanned as they flow lengthwise through a thin (e.g., 1.5 μm thick) scanning laser beam. The intensity of fluorescence coming from the laser beam chromosome intersection is measured periodically (e.g., every 10 nsec) to produce a digital profile showing the distribution of fluorescence dye along the length of the chromosome (see also Chapter 6, this volume).

CHROMOSOME ISOLATION

The single most important step in flow karyotyping is the isolation of well-dispersed intact metaphase chromosomes. Chromosome isolation requires that cells be blocked in mitosis, swelled and mechanically or chemically ruptured to release the chromosomes into an aqueous buffer selected to maintain chromosomal integrity [2,15,54,147,150,159,161, 170,172]. This section focuses both on cell culture methods and on chromosome isolation techniques. Our goals are to describe the existing techniques and the rationale behind these procedures; their limitations, relative advantages, and disadvantages; and qualitative descriptors that provide a measure of the quality of a chromosome suspension. These procedures are designed to produce suspensions of isolated chromosomes that contain minimal debris and nuclei, morphologically recognizable chromosomes, and chromosomes that can be efficiently and uniformly stained with DNA specifc dyes, as well as having high-molecular-weight DNA. Minimization of the selective loss of small, or large chromosome types is also important. The staining (cytochemistry) employed is designed to allow one chromosome type to be resolved from another during flow cytometry. The staining procedure is an integral part of the isolation procedure. Indeed, cytochemical demands often dictate the isolation protocol. The optimal chromosome isolation procedure may vary depending on the experimental endpoint. For example, the isolation buffer that maintains the best morphology for banding may not be the buffer preferred for highest resolution flow karyotype analysis.

Cell Culture Methods

A key aspect of successful chromosome isolation is the development of a culture environment in which a large number of cells can be collected in mitosis during a few hour treatment with an agent such as colcemid. To accomplish this, a large fraction of the cell population must be stimulated into active proliferation and the cell cycle traverse rate should be maximal. The exact culture procedures are necessarily different for cells that grow in suspension or that grow in monolayer cultures.

Monolayer Culture. Chromosomes have been successfully isolated from mitotic cells from cultures of Chinese hamster, Indian muntjac, mouse, rat kangaroo, and chimpanzee cells, human fibroblasts and amniocytes and human carcinomas.

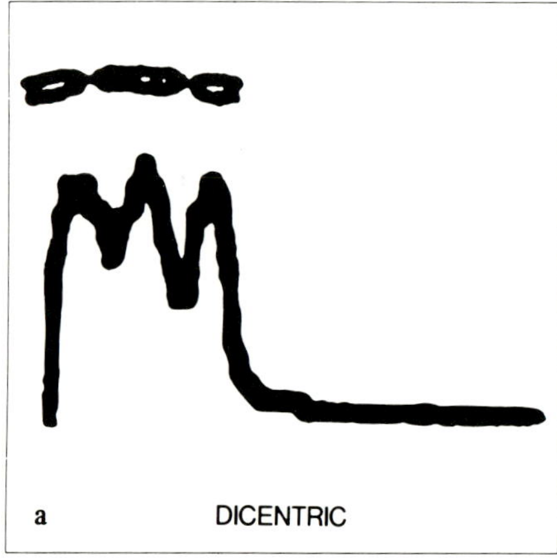

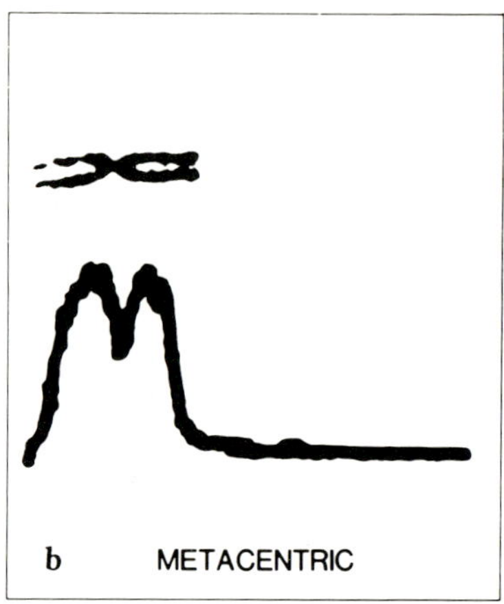

Fig. 5. Slit-scan profiles for metaphase chromosomes. **a:** Profile measured for a chromosome thought to be dicentric. **b:** Profile measured for a normal metacentric chromosome.

Monolayer cultures are usually seeded at relatively high density. This ensures that the efficiency of the culture procedure is high (i.e., the largest mitotic cell to cell culture surface area ratio is achieved). Colcemid or some other mitotic inhibitor is then added to block a substantial fraction of the cell population in mitosis. Treatment times usually range from 3 to 4 hours for rapidly growing cultures (e.g., for those with total cell cycle durations of 12 to 15 hours) to as much as 12 hours for more slowly growing cells (e.g., for those with total cell cycle durations of 24 to 36 hours). To date it has been almost impossible to produce useful chromosome suspensions from cultures with total cell cycle durations greater than 48 hours. In most monolayer cultures, mitotic cells can be removed by shake selection. It may be advantageous to change the medium and vigorously shake the cultures at the beginning of the colcemid treatment to remove dead or dying cells that may be disrupted during subsequent chromosome isolation. Mitotic indices ranging from 50% to >90% are achieved using this technique. Up to 10^6 mitotic cells can be obtained from a single T75 flask for a rapidly growing cell line. In general, monolayer culture leads to flow karyotypes of high quality (high measurement precision with minimal debris [2,8,159].

Suspension Cultures. Chromosomes have been successfully isolated from stimulated human peripheral blood lymphocytes [159,161,172], Chinese hamster cells [7,31,142], murine bone marrow cells and lymphoblastoid cells grown in suspension culture [15,37,91,105,147,170]. Culture procedures designed to produce rapidly growing, high density cultures are important for both suspension and monolayer cultures. It is especially important to maximize cell viability in suspension since dead and dying cells cannot be removed using a shake selection process as is possible with monolayer cultures. Mitotic cells are accumulated by treatment with inhibiting agents like colcemid. The mitotic index achieved may be as high as 40% for rapidly proliferating cultures in which the fraction of the proliferating cells is high. The mitotic index may be increased further by synchronizing the culture prior to colcemid treatment or by enriching for mitotic cells using a size selection procedure such as velocity sedimentation. In general, suspension culture is best suited for the efficient production of large numbers of chromosomes required for flow sorting since as many as 10^8 mitotic cells can be obtained from a single suspension culture.

Isolation Protocols

Isolation of chromosomes from mitotic cells is accomplished in three steps: (1) cell swelling, (2) chromosome stabilization, and (3) cell shearing to release the chromosomes.

Cell Swelling. This is a critical step in the isolation of chromosomes. When mitotic cells are placed in a hypotonic environment, the cell membrane swells allowing the mitotic chromosomes to separate. The success with chromosome disentanglement can be determined by microscopic observation of a drop of the cell suspension on a microscope slide to which Hoechst 33342 and PI have been added. Hoechst 33342 (10 μM final concentration) will diffuse into cells with intact membranes and PI (50 μg/ml final concentration) will be excluded. The swelling and separation of the blue fluorescing Hoechst stained chromosomes can be easily monitored. At the point that the membrane becomes permeable, the diffusion of PI into the cell can be observed. This double staining procedure provides a means of monitoring if satisfactory cell swelling is occurring prior to membrane lysis. Ideally, a swelling procedure can be achieved in which a majority of the mitotic cells are swollen and contain well separated blue fluorescing chromosomes. The frequency of small (unswollen) red fluorescing cells with little physical separation between chromosomes should be minimized. These cells have lost membrane integrity prior to swelling.

Chromosome Stabilization. The four most commonly used chromosome stabilization buffers are often referred to by the component of the buffer that acts as the stabilizing agent. Those agents are hexylene glycol, propidium iodide, magnesium sulfate and polyamines. Table 1 lists the advantages and disadvantages and some of the characteristics of each of these procedures.

Hexylene glycol: This procedure was developed as a way to isolate chromosomes in suspension for subsequent fractionation on a sucrose gradient [153,168]. Chromosomes are swelled in hypotonic KCl and resuspended in Hepes buffer containing 0.75 M hexylene glycol. The mechanism of action of hexylene glycol as a stabilizing agent [54,57,151,168] is not clear. Chromosomes isolated in this buffer have good morphology and can be Q-banded [57]. In addition, isolated chromosomes stored at −70°C retain their flow karyotype. However, the molecular weight of chromosomes isolated in

TABLE 1. Comparison of Chromosome Isolation Procedures

Common name	Buffer components	Chromosome morphology	Protocol of choice for	Ref.
Hexylene glycol	KCl Trisma base Hexylene glycol* $CaCl_2$	Normal length and morphology	Chromosome identificaton by G-banding, long-term sample storage	151,168
Propidium iodide	KCL Triton X-100 Propidium iodide* RNase	Excellent morphology—slightly extended	High-resolution, univariate analysis, fast and easy to reproduce	2,15,19,80
Magnesium sulfate	$MgSO*_4$ HEPES Dithiothreitol RNase Triton X-100	Slightly contracted, difficult to recognize	High-resolution, allows dual staining , small sample size	159,161
Polyamine	Tris-HCl EDTA EGTA KCl Spermine* Spermidine* Digitonin Mercaptoethanol	Slightly contracted, difficult to recognize	Cells in suspension, high resolution, allows dual staining, maintains high molecular weight	18,147

*Designates purported stabilizing agent.

this buffer may not be as high as that achieved using other procedures.

Propidium iodide: In this procedure, the mitotic cells are swelled in hypotonic KCl. Propidium iodide and detergent are then added. Propidium iodide intercalates in the DNA to both stabilize and stain the chromosomes [2,15,19,80]. Chromosome morphology is excellent and limited G-banding can be done with these preparations. pH control is critical and can be a problem because of the small sample volumes involved. RNase is effective at reducing small particle fluorescent debris that results, in part, from propidium iodide binding to double-stranded RNA. The procedure has the virtue of simplicity and high resolution [2]. However, the propidium iodide does not permit effective use of dyes such as Hoechst 33258 and chromomycin A3.

Magnesium sulfate: The $MgSO_4$ procedure is a variation on the PI procedure [155,159,161]. Mitotic cells are swelled in hypotonic KCl and resuspended in a buffer containing $MgSO_4$, detergent, and dithiothreitol. Magnesium ions are added to stabilize the chromosomes and to facilitate the binding of the chromomycin family of dyes to DNA. Hepes is added as a pH buffer (pH 8.0). Dithiothreitol is added as a reducing agent and acts to reduce chromosome clumping. Trition X-100 is the detergent most commonly used with the maganesium sulfate procedure. This procedure is well suited to bivariate flow karyotyping since it supports the simultaneous use of the base composition-dependent dyes such as Hoechst 33258 and chromomycin A3, and it can be successfully applied to small numbers of mitotic cells. Chromosomes isolated using this procedure can be banded, although with some difficulty.

Polyamine: The polyamine procedure was developed to permit rapid isolation of chromosomes containing high-molecular-weightDNA[18,147]. When applied to chromosome isolation for flow karyotyping, swollen mitotic cells are resuspended in a buffer containing chelating agents and the polyamines spermine and spermidine. The detergent digitonin is added later. The polyamines stabilize chromosome structure [18,92] and heavy metal chelators reduce nuclease activity. The molecular weight of the DNA in chromosomes isolated using this procedure is high (greater than 5×10^8 daltons). The disadvantage of this method is that chromosomes are highly condensed and are therefore difficult to G-band [19,37]. However, in situ hybridization with chromosome specific probes alleviates this problem by allowing identification of sorted chromosomes according to their DNA sequence [105,108,124,139].

CYTOCHEMISTRY

The success with which chromosomes can be distinguished using flow cytometry depends on the binding characteristics of the fluorescent dye(s) used to stain them [75]. The dyes that have proved most useful to date bind preferentially to nucleic acids [56]. However, some dyes that quench fluorescence from DNA-specific dyes have also proved useful [144] as have immunofluorescence staining techniques [155,158].

DNA-Specific Dyes

Three broad classes of DNA specific dyes have proved useful in chromosome staining: (1) propidium iodide (PI) or ethidium bromide (EB) binds to double-stranded nucleic acids by intercalating between the base pairs [47,111,131,136]. These dyes show little DNA base specificity; (2) Hoechst 33258 (Ho58) and Hoechst 33342 (Ho42) bind in the minor groove of double-stranded DNA (dsDNA) and bind preferentially to ademine-thymine-rich DNA [73,79,88,102,104,113,140]. (DAPI and DIPI also bind preferentially to AT-rich DNA and may have binding mechanisms similar to those of the Hoechst dyes [20,113,135]); and (3) chromomycin A3 (CA3) and mithramycin (Mith) bind preferentially to guanine-cytosine rich DNA [11,75,77,97,100,102,104,117,164]. The spectral characteristics of the three classes of dyes are different. The intercalating dyes excite preferentially at ~500 nm and fluoresce at ~650 nm. Hoechst and related dyes excite at ~350 nm and fluoresce at ~450 nm. Chromomycin-like dyes excite around 420 nm and fluoresce around 520 nm.

The dependence of some of the dyes on the base composition of the DNA in the chromosomes permits some chromosome types to be better resolved by one dye type than another. In general, the largest number of chromosome groups can be resolved using Hoechst, DAPI, DIPI, or PI. PI has the advantage that it can be excited at 488 nm, while the others excite in the ultraviolet (UV) range. CA3 and Mith do not allow resolution of as many chromosome groups [57,72]. However, they do permit resolution of chromosome types that are not well resolved using Ho58 and related dyes. This proves advantageous for bivariate flow karyotyping as described below. The chromosome types producing the peaks in the flow karyotypes have been determined by banding analysis of chromosomes sorted from each peak [57], by comparison of the peak DNA contents with DNA contents measured from banded chromosomes using quantitative image analysis [69], and by in situ hybridization with mapped probes to sorted chromosomes [105]. It is important to note that the locations of some peaks may vary as a result of chromosome polymorphisms. Figure 6 shows univariate flow karyotypes measured for human chromosomes stained with PI, Ho58 and CA3.

Dual Staining. Hoechst-like dyes and chromomycin-like dyes have proved useful chromosome stains when used together [5,57,103]. These dyes do not appear to compete for binding sites in DNA. Furthermore, they excite and emit at different wavelengths. Thus, essentially independent measurements of the Ho58 (or DAPI, DIPI, or Ho42) and CA3 (or mith) contents of the chromosomes can be measured by passing them through a dual beam flow system with one beam adjusted to the UV (351 nm + 363 nm) to excite Ho and the other adjusted to 458 nm to excite CA3. Figure 3a shows the bivariate flow karyotype for human chromosomes stained with Ho58 and CA3. The fluorescence emitted following UV excitation is a measure of chromosomal Ho binding and the fluorescence emitted following 458-nm excitation is a measure of chromosomal CA3 binding. However, substantial energy transfer occurs from Ho to CA3 [98,160]. Some dyes have slightly different DNA base binding characteristics that prove useful in flow karyotyping. For example, DAPI and Ho58 bind differently to AT-rich heterochromatic DNA. Thus, chromosomes that cannot be resolved in Ho vs CA3 flow karyotypes (e.g., chromosome 9) may be resolved in DAPI vs CA3 flow karyotypes [105] (see Fig. 7a).

Modifiers. The fluorescence from selected chromosomes can be modified by adding agents such as netropsin and distamycin that modify the Ho fluorescence from selected chromosomes [92,121,144]. Both these compounds bind preferentially to adenine-thymine-rich DNA. Binding appears to be in the minor groove of the DNA double helix. These dyes, when used in conjunction with Ho58 or Ho42 enhance the centromeric fluorescence intensity of human chromosomes such as 1,9,15,16, and Y with large amounts of centromeric heterochromatin [121]. Figure 7b shows a bivariate flow karyotype for human chromosomes stained with Ho58, distamycin, and CA3.

Ho Quenching. The fluorescence from Ho is quenched by the incorporation of halogenated pyrimidines such as bromodeoxyuridine (BrdUrd) into chromosomal DNA [20,99]. This, coupled with the fact that not all chromosomes replicate at the same rate during DNA synthesis has been used for preferential staining of chromosomes that replicate during a limited portion of S-phase [33,34,145]. For example, Figure 7 shows bivariate Ho versus CA3 distributions measured for Chinese hamster chromosomes grown normally (Fig. 7c) or grown so that BrdUrd was present in the growth medium during the last 4 hours of S-phase (Fig. 7d). Thus, the late replicating Y chromosome incorporated BrdUrd. The relative Ho fluorescence of the labeled Y chromosome is lower than that for the comparably sized unlabeled chromosomes.

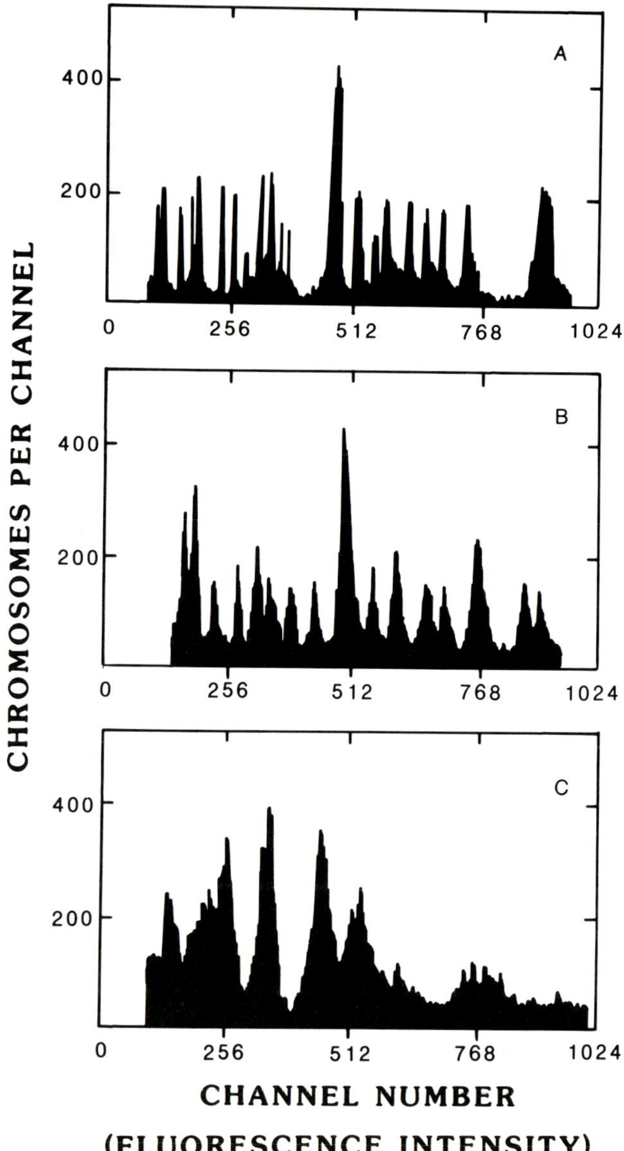

Fig. 6. Univariate flow karyotypes measured for human chromosomes. **a:** Following staining with PI. **b:** Following staining with Ho58. **c:** Following staining with CA3.

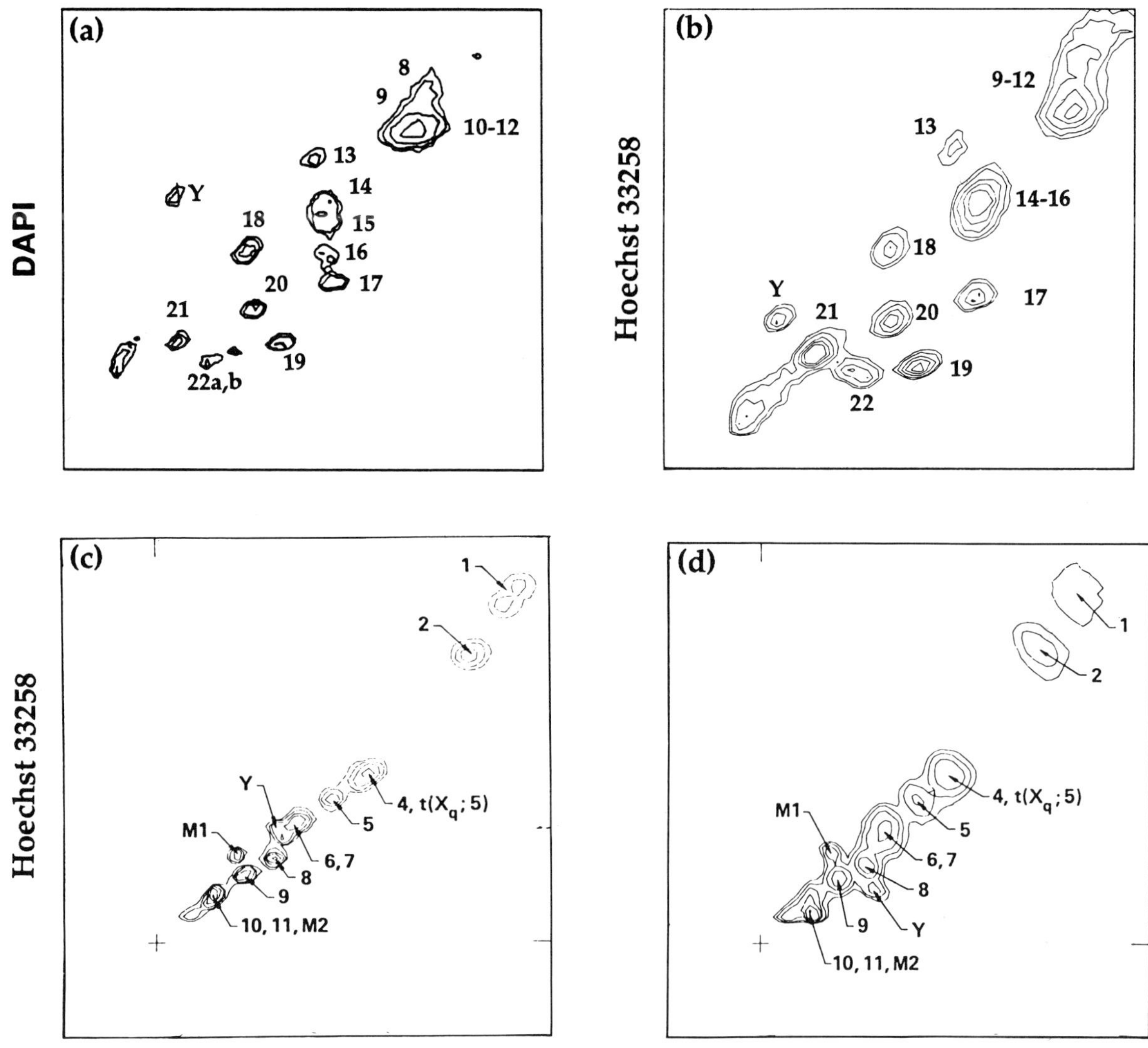

Fig. 7. Bivariate flow karyotypes. **a:** Human chromosomes stained with DAPI and CA3. **b:** Human chromosomes stained with Ho58, distamycin A and CA3. **c:** Chinese hamster chromosomes stained with Ho58 and CA3. **d:** Chinese hamster cells stained as in c following treatment with BrdUrd so that the late replicating Y chromosome contains substantial BrdUrd.

Immunofluorescence

Antihistone Staining. The possibility of staining chromosomes using immunochemical techniques has been demonstrated using antibodies against histone H2B [155,158]. In these studies, human and hamster chromosomes were incubated in buffer containing a mouse anti-H2B antibody, washed, incubated in fluorescein-labeled goat antimouse antibody and counterstained with PI. Figure 8 shows a bivariate PI versus immunofluorescence flow karyotype measured for a mixture of hamster and human chromosomes. Two points are evident: (1) chromosomes can be immunofluorescently stained without destroying chromosome integrity, and (2) the anti-H2B antibody binds with different affinity to the human and hamster chromosomes. However, this difference disappears if the human and hamster chromosomes come from a hybrid cell line. Thus, the difference in binding affinity probably comes from differences in chromatin structure rather than a difference in histone H2B. This seems likely, since histone H2B is highly conserved among mammals. Further evidence for the dependance of immunofluorescence on chromatin structure comes from the observation that the intensity of immunofluorescence following staining with anti-H2B is reduced approximately fivefold by staining with PI prior to immunofluorescence staining. This effect is thought to be due to the unwinding of DNA due to the intercalation of PI [155,158].

Antikinetochore Stain. Quantification of the frequency of dicentric chromosomes as a measure of induced genetic damage depends critically on being able to detect chromosome

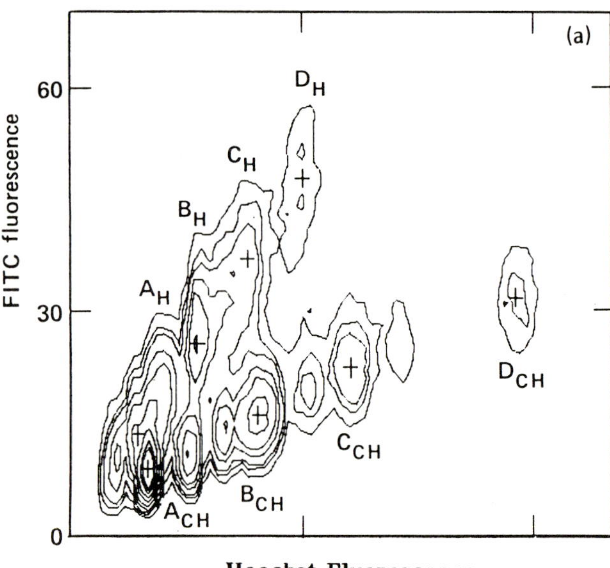

Fig. 8. Bivariate flow karyotype for a mixture of human and hamster chromosomes. The chromosomes were immunofluorescently stained with an antibody against human histone H2B. Fluorescein was used to detect the antibody binding and the chromosomes were counterstained with Ho58 [52,63]. Peaks produced by the human chromosomes are labeled A_H through D_H. Peaks produced by the Chinese hamster chromosomes are labeled A_{CH} through D_{CH}.

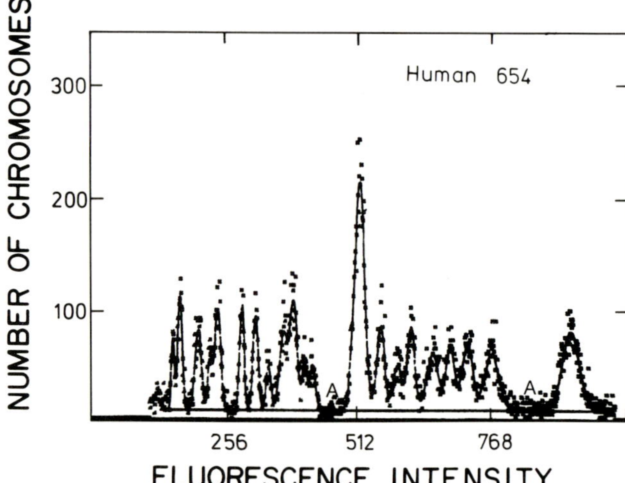

Fig. 9. Least squares-best-fit-analysis of a human univariate flow karyotype. The lines indicate the fit. The data are shown as individual points.

centromeres. Positive staining of the kinetochores of human chromosomes using CREST serum from patients with scleroderma has been proposed as one means of increasing the distinctness of centromere staining [64]. In this approach, chromosome suspensions are incubated in CREST serum, washed, and incubated in fluorescein-labeled antihuman antibody. The chromosomes are then counterstained for total DNA content with PI or Ho58 as described above. The number of centromeres per chromosome may be determined by slit scanning or by measuring total centromere-linked immunofluoresce [64].

DATA ANALYSIS

Flow karyotypes consist of several more or less distinct peaks produced by intact, homogeneously staining chromosomes, sometimes superimposed on a continuum produced by nuclear and cytoplasmic debris and chromosome clumps. Flow karyotypes contain quantitative information about the chromosome population under analysis: (1) the mean of each peak defines the chromosome type that produces that peak; (2) the area (or volume) of the peak is proportional to the frequency of occurrence of that chromosome group; (3) the coefficient of variation (CV) of each peak describes the precision of chromosome staining and flow cytometric analysis; and (4) the area under the debris continuum is a measure of the frequency of the debris and/or random aberrant chromosomes and may be used as an indication of the extent of nuclear or chromosomal disruption (e.g., damage caused by the shearing necessary to release chromosomes during chromosome isolation) or induced chromosomal damage (e.g., aberrant chromosomes induced by clastogenic agents like radiation). Accurate information about peak means and areas is important for chromosome classification and aberrant chromosome detection, since the occurrence of numerical and structural aberrations characteristically causes changes in peak volumes and/or means. Information about peak means, CV, areas, and the magnitude of the debris continuum is important for chromosome sorting since those variables affect sorting purity. In general, the highest sorting purity will be obtained when the CV and the magnitude of the debris continuum are low. The CV and the magnitude of the debris continuum are also useful indicators of the degree of damage induced by clastogenic agents since the magnitude of the debris continuum and the peak CV increase with increasing chromosome damage. Several analysis schemes developed for estimation of these parameters are described in the following section.

Univariate Analysis

Partitioning. The simplest scheme for determination of peak means, areas and CV is accomplished interactively. That is, a peak is selected interactively. The mean, area and CV for the peak are calculated as described by Equations (1):

$$\text{Mean} = \sum_i (Y_i x_i) / \text{Area} \tag{1a}$$

$$\text{Area} = \sum_i Y_i \tag{1b}$$

$$\text{CV} = \sqrt{\sum_i (i - \text{mean})^2 \, Y_i} \Big/ \text{mean} \tag{1c}$$

where Y_i is the number of events in channel i. The CV of a peak may also be estimated from an estimate of the full width of the peak at half of its maximum height (FWHM) as:

$$\text{CV} = \frac{\text{FWHM}}{2.35 \cdot \text{Mean}}$$

The magnitude of the debris continuum may be estimated by determining the height of the distribution at points where it might be expected to be near zero in the absence of debris

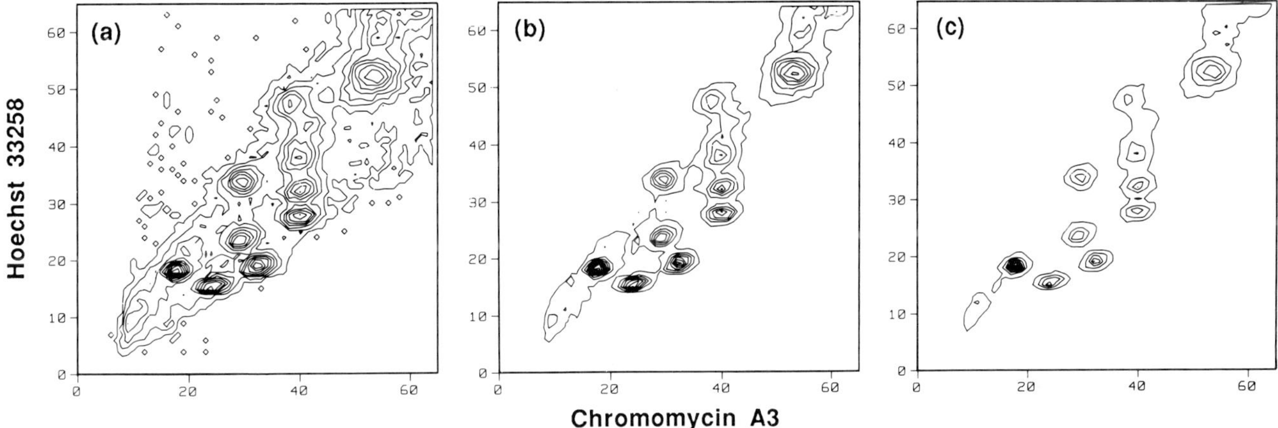

Fig. 10. Contour plots of human chromosomes. **a:** The contour spacing increases geometrically (e.g., contours at 2, 4, 8, 16, . . .). **b:** The contour spacing is constant (e.g., contours at 10, 20, 30, . . .). **c:** The contour spacing decreases linearly (e.g., contours at 100, 300, 475, 625, 750, . . .).

(e.g., at point A in Figure 9) [45]. These techniques work well only for flow karyotypes where the CV are low, so that the peaks are well separated and where the debris continuum is low so that it does not significantly affect the estimation of the peak parameters using Equations (1).

Least-Squares Fitting. More accurate estimation of the peak areas can be accomplished by approximating each peak in the distribution with a normal distribution and by approximating the debris continuum with a smooth function during a least-squares best fit. In this procedure, the sum of the normal distribution and the debris continuum is matched to the flow karyotype by adjusting the parameters of the functions until the sum of these functions matches the experimental flow karyotype as closely as possible [54,55,122]. A typical function f(x) used to fit univariate flow karyotypes is given in Equation 2.

$$f(x) = \sum_j \frac{A_j}{\sqrt{2\pi}\,\sigma_j} e^{-1/2[(x-\mu_j)^2/\sigma_j^2]} + b(x) \qquad (2)$$

where A_j, μ_j, and σ_j are the area, mean and standard deviation of the jth peak and b(x) is a function describing the debris continuum. The magnitude of the debris continuum from X_1 to X_2 is

$$\text{Mag} = \int_{x_1}^{x_2} b(x)\,dx \qquad (3)$$

A flow karyotype for human chromosomes and the function fitted to it is displayed in Figure 9.

The purity of a sort can be estimated by calculating the fraction of the desired chromosome type that lies in the sort window. For example, the fraction of chromosomes from peak k that lie in the region $x_1 \leq X \leq x_2$ can be calculated as:

$$\text{Frac}(k) = \int_{x_1}^{x_2} \frac{A_k}{\sqrt{2\pi}\,\sigma_k} e^{-1/2[(x-\mu_k)^2/\sigma_k^2]}\,dx$$

$$\div \text{ (total area between } x_1 \text{ and } x_2)$$

The fractions of the chromosomes on either side of peak k that lie in the sorting window is

$$\text{Frac}(k-1;k+1) = \left\{ \int_{x_1}^{x_2} \frac{A_{k-1}}{\sqrt{2\pi}\,\sigma_k} e^{-1/2[(x-\mu_{k-1})^2/\sigma_{k-1}^2]}\,dx + \int_{x_1}^{x_2} \frac{\hat{A}_{k+1}}{\sqrt{2\pi}\,\sigma_{k+1}} e^{-1/2[(x-\mu_{k+1})^2/\sigma_{k+1}^2]}\,dx \right\} \qquad (4)$$

$$\div \text{ (Total area between } x_1 \text{ and } x_2).$$

and the fraction of the sorted chromosomes that are debris particles is

$$\text{Frac (debris)} = \int_{x_1}^{x_2} b(x)dx \Big/ \text{(Total area between } x_1 \text{ and } x_2) \qquad (5)$$

The relative purity of the sort can be increased in this example by decreasing the width of the sorting window. This comes at the expense of sorting efficiency.

Bivariate Analysis

Display. Bivariate flow karyotypes may be displayed as contour plots, dot plots, or as isometric displays. The contour plot has the advantage that all of the data can be viewed in one plot and it is simple to interact with the data during analysis (ie. to use cursors to interactively circumscribe a peak). However, unless care is taken, contour plots can give an improper impression about the height of the peaks in a distribution, about the apparent separation between peaks and about the magnitude of the debris continuum. Figure 10, for example shows three contour displays for the same human flow karyotype. In 10b, the contours are spaced equally. This method displays the data accurately but may give little information about the existence of a debris continuum if the lowest contour is higher than the debris continuum. In 10a, the contours are spaced so that spacing increases with increasing peak height. This method clearly shows the debris continuum but may provide a misleading impression about the relative heights of the peaks. In 10c, the contour spacing decreases with increasing peak height. This method accentuates the peaks but provides almost no information about the debris continuum. In general, several displays may be necessary to show all aspects of a flow karyotype.

Boxing. The most simple method for interacting with bivariate flow karyotypes is by interactively defining a peak or region about which information is desired [63]. The bivariate mean, CV, and volume can then be calculated for the data in the circumscribed region using conventional statistical procedures. Alternately, a region of a flow karyotype may be circumscribed and the data within the circumscribed region may be collapsed onto the X or Y axes [55,161]. The mean and CV along the X or Y axes may be calculated as indicated in the section on univariate analysis. The area of each peak in this univariate distribution is equal to the volume of the corresponding peak in the bivariate flow karyotype.

Least-Squares Best-Fit Analysis. The peaks in a bivariate flow karyotype and the debris continuum may be approximated using a least-squares best-fit technique similar to that used for analysis of univariate flow karyotypes [98,123]. In this case, each peak in the bivariate flow karyotype may be approximated by a bivariate normal distribution of the form:

$$f(x,y) = \frac{U}{2\pi\sigma_x\sigma_y(1-\rho^2)} \exp\left\{ -\frac{1}{2(1-\rho^2)} \times \left[\frac{(x-\mu_x)^2}{\sigma_x^2} - \frac{2\rho(x-\mu_x)(y-\mu_y)}{\sigma_x\sigma_y} + \frac{(y-\mu_y)^2}{\sigma_y^2}\right]\right\} \quad (6)$$

where U is the peak volume, σ_x and σ_y are standard deviations in the x and y directions, and ρ is a cross-correlation coefficient. The bivariate mean of the bivariate normal distributions is μ_x; μ_y. The complete function to be matched to the data is the sum of a series of bivariate normal distributions (one for each peak) plus a distribution to match the debris continuum. Selection of the function for the debris continuum is often difficult since the continuum forms a ridge with a Ho/CA3 ratio approximately equal to that defined by the peaks in the flow karyotype. The ridge is high at low Ho and CA3 values and decreases in magnitude at higher Ho and CA3 values. This ridge has been approximated most successfully using a bivariate normal distribution whose covariance term is selected to make the distribution run along the desired diagonal. This function is clearly ad hoc, however, and the accuracy of the resulting fit depends heavily on the degree to which this function is an accurate approximation of the debris continuum.

In general, a complete least-squares best fit analysis of all of the peaks in a bivariate flow karyotype measured for human chromosomes is not feasible because of the computational complexity of the fitting procedure. Such flow karyotypes typically contain 4096 elements and are made up of ~20 peaks. Thus, a complete analysis may require simultaneous adjustment of as many as 120 variables (bivariate mean, CV, volume for each peak). This is a challenging task even for modern supercomputers. Thus, a more reasonable task is to circumscribe several of the peaks in a flow karyotype and to fit these. Alternately, univariate distributions that are derived from collapsing the contents of circumscribed regions onto the abscissa or ordinate may be analyzed using the univariate analysis programs described above.

Slit-Scan Analysis

Chromosome Shape Analysis.

Centromeric Index (CI) Calculation. Chromosomes analyzed using slit-scan flow cytometry are usually classified according to their total fluorescence and CI [32,59,61,114–116]. The CI is a shape parameter derived from the profile recorded during slit scanning. The CI is calculated from the recorded profile as the ratio of the fluorescence from the long arm of the chromosome to its total fluorescence. The fluorescence of the long arm of a chromosome is estimated from the slit-scanned profile by locating the dip in the profile produced when the centromere with its reduced DNA content traverses the laser beam. The areas under the profile to either side of the centromeric dip are then calculated and the CI is estimated as the ratio of the larger area to the total profile area. Accurate location of the centromeric dip is essential to accurate CI calculation. This is accomplished by locating the two edges of the profile and the profile maximum. The part of the profile containing the maximum is then reflected about its median value and a horizontal line is drawn between the two highest points. The experimental profile is subtracted from this "standard" profile. The point of maximum difference is taken as the location of the centromere. The areas on either side of the centromere are obtained by simple summation for Cl calculation. This algorithm works even for profiles in which the centromeric reduction in the profile is not a distinct dip [115].

Dicentric Chromosome Detection. Detection of profiles produced by dicentric chromosomes is important to efforts to estimate induced genetic damage. Such profiles typically show three peaks separated by two centromeric dips (see Figure 5a. The general approach is to classify as dicentric profiles only those for which the three peak heights are normal and approximately equal. In addition, the two centromeric dips should be distinct and approximately equal. Dicentric chromosome profiles produced by smaller or misoriented chromosomes may be missed by this technique. Thus, the goal in dicentric chromosome quantification by slit-scan flow cytometry is to determine a number proportional to the number of dicentric chromosomes in the population. The damage index is taken as the ratio of the number of profiles scored as dicentric to the number of normal profiles of one or a few chromosome types (e.g., the number of chromosomes with total fluorescence expected for the number one and two human chromosomes).

KARYOTYPING HOMOGENEOUS POPULATIONS

A chromosome aberration that is present in a high percentage of cells and is the same from cell to cell is referred to in this chapter as a homogeneous aberration. Such aberrations may be either numerical (e.g., trisomies and monosomies) or structural (e.g., translocations, deletions, additions, inversions). Homogeneous aberrations are clinically important for prenatal diagnosis of genetic disease and for screening populations for inherited disease-linked abnormalities. Analysis of karyotype stability is also important during the extended culture of mammalian cells, since a change from normal indicates the clonal emergence of a genetically altered cell population.

Univariate Karyotyping

Application to Several Species. Univariate flow karyotyping has proven remarkably effective for resolving a large number of chromosome types in cells from several animal species [3,22,30,54,65]. Human chromosomes, for example, can be resolved into 18 to 20 groups in high resolution flow karyotypes. The success of univariate flow karyotype analysis is based on the fact that (1) the DNA content of an individual chromosome type is extremely well controlled, and (2) DNA specific fluorescent dyes like PI bind specifically and stoichiometrically to double-stranded nucleic acids. Typically, chromosomes differing in DNA content by less than 2% (i.e., by 5 fg DNA) can be resolved as separate peaks in the flow karyotypes. In addition changes in chromosomal DNA content less than 0.5% (i.e., ~1 fg) can be readily detected. Figure 11 shows univariate flow karyotypes for several species. Figure 11a shows the 24 types of human chromosomes resolved into 18 groups. Figure 11b shows the 12 types of Chinese hamster groups resolved into 13 groups (the two number 9 homologs are separately resolved because of a subtle polymorphism in one chromosome [142]). Of course, chromosomes can be resolved only if they differ sufficiently in DNA content. Figure 11c shows the 40 types of mouse chromosomes resolved only into 10 groups [3,42].

Univariate flow karyotyping has been found useful when (1) only a single laser is available, (2) the adenine : thymine (AT) to guanine : cytosine (GC) base ratio of the chromosomes is constant or nearly so for all the chromosome types, and (3) it is important to have the lowest coefficient of variation possible or (4) the chromosome types differ by more than about 3 percent in DNA content. For Chinese hamster, mouse, and Muntjac chromosomes, univariate analysis is generally the method selected.

Karyotype Interpretation. The identification of the chromosome types that produce the peaks in a flow karyotype can be determined by comparing peak means (closely related to chromosome DNA content) with chromosome DNA contents or lengths measured by image analysis, from estimates of peak areas (proportional to the frequency of occurrence of that chromosome type) and by visual or molecular biological analysis of chromosomes sorted from a particular peak.

Sorted chromosomes may be analyzed visually following banding, or after in situ hybridization with a chromosome specific DNA probe [21,49,105,124,139]. Classical banding techniques are difficult to interpret when applied to sorted chromosomes because the complete normal complement of chromosomes from a single cell is not available for comparison and the chromosomes may be poorly banded due to a high degree of condensation. In situ hybridization of a chromosome-specific repeated sequence DNA probe to sorted chromosomes has been used to identify contaminating hamster chromosomes mistakenly sorted with human chromosomes from a Chinese hamster/human somatic cell hybrid [40,163]. In this case, a whole genomic Chinese hamster probe was used to label contaminating Chinese hamster chromosomes present in the sorted fraction. This technique is particularly suited for accurate determination of low levels of contamination. The technique can be extended to analysis of chromosomes sorted from human cells in those cases where a suitable chromosome-specific probe is available. Repetitive sequence probes that are chromosome specific have been used for this purpose [55,124] (M. Van Dilla and L. Deaven, private communication). Peak identities also may be determined by in situ hybridization with chromosome specific probes to chromosomes sorted onto nitrocellulose filters [105,108,109]. Probes suitable for this purpose are readily available since any previously mapped sequence may be used.

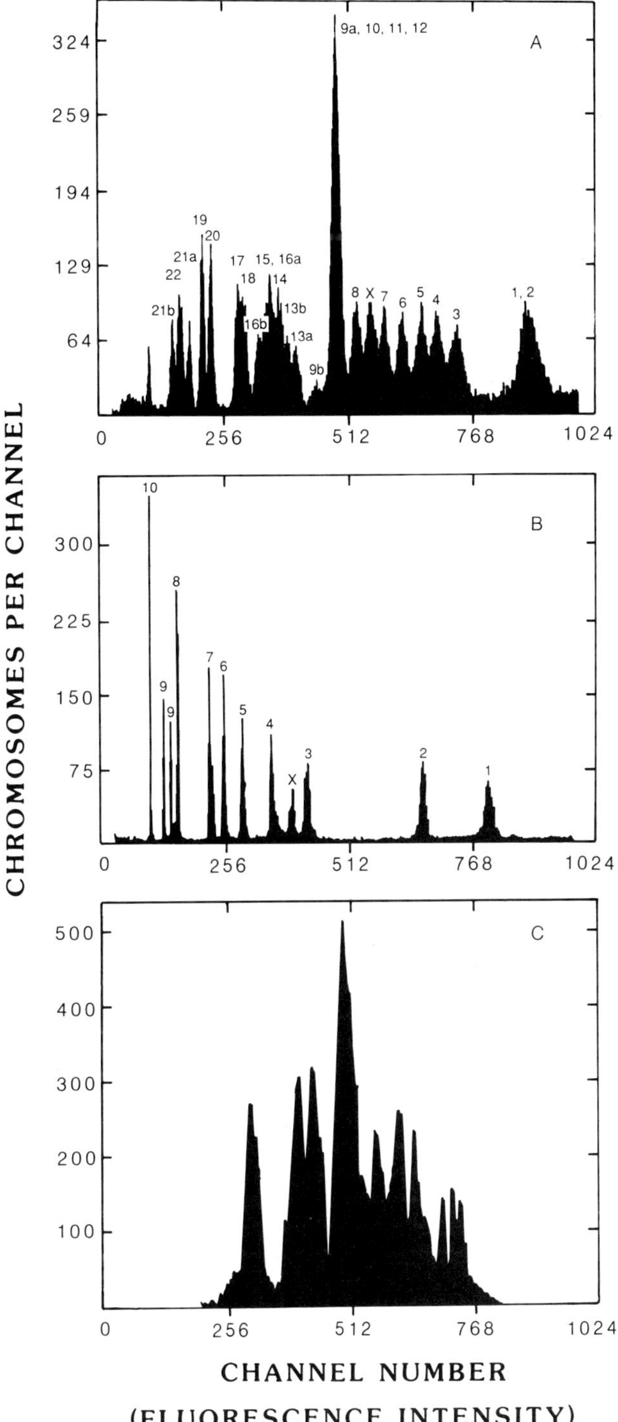

Fig. 11. Univariate flow karyotypes measured for several cell types following staining with PI. **a:** Human chromosomes. **b:** Hamster chromosomes. **c:** Mouse chromosomes.

Univariate flow karotype interpretation also may be accomplished by comparing the flow karyotype for a new cell line to a flow karyotype whose peak identities have been determined as described above. However, errors in peak identification can result from using this approach due to chromosomal polymorphisms. A large number of chromosomal polymorphisms are known to occur in the human karyotype [65,170]; polymorphisms in other species such as the Chinese hamster have also been identified [30,142]. Figure 11b shows a chromosomal polymorphism in the number 9 Chinese hamster chromosome that causes separation of the two homologues.

Cell Types That Have Been Flow Karyotyped

Mammalian Cells. Cells from both in vivo and in vitro sources have been successfully flow karyotyped. Cells successfully flow karyotyped include peripheral lymphocytes, amniocytes, bone marrow, tumor tissue, and numerous cultured mammalian cells [7,62,63,80,167,170,172]. Species that have been flow karyotyped include human, Chinese hamster, mouse, chimpanzee, orangutan, chicken, muntjac, gorilla, swine, and rat kangaroo [4,9,17,21,22,42,67,80, 133,149,152]. The quality of a flow karyotype depends on the relative homogeneity of the cell population and the degree to which the isolation procedure has been adapted to the cells from which the chromosomes are to be isolated [8]. Populations containing several different cell types may be difficult to flow karyotype, since some of the cell types may respond differently to the chromosome isolation protocol resulting in fewer single chromosomes, chromosomal damage, and/or chromosome aggregation. Heterogeneous populations of cells from fresh tumor tissue or whole embryonic tissue are often placed in culture for a short period of time to increase the mitotic index and to reduce the degree of cell heterogeneity. Figure 12 compares flow karyotypes obtained from whole Chinese hamster embryo cells at passage 7 (Fig. 12a) and after crisis has occurred at passage 19 (Fig. 12b). Passage 7 cells are a heterogeneous mixture of cell type while at passage 19 the culture was karyotypically homogeneous by visual inspection. This type of improvement between the two flow karyotypes is reproducible [8] and is characterized by a reduced debris continuum and lower coefficient of variation of the peaks.

Plant Cells. The analysis and sorting of plant chromosomes is of considerable economic interest and has been accomplished using cell suspensions of *Haplopappus gracilis* [41]. As is the case for mammalian chromosomes, flow karyotyping and chromosome sorting provides the opportunity for gene mapping and the construction of chromosome specific libraries. Because of the economic interest in manipulating plant genetic material, additional opportunities in chromosome transplantation via fusion techniques and microinjection are also likely applications for sorted plant chromosomes. Advances in plant chromosome sorting have been slowed by the difficulty of obtaining well-synchronized cultures. The existence of cell walls in plants also presents problems for chromosome isolation. These must be dissolved by enzymes. Alternatively, the problem can be avoided by working with protoplasts. *Haplopappus gracilis* is comparatively easy to flow karyotype, since it is karyotypically simple and it responds to hydroxyurea synchronization and colchicine mitotic arrest. The important cereal grains are much more difficult to work with and have not been effectively flow karyotyped to date.

Genome Stability

The precise role of karyotype instability in the neoplastic process is uncertain but critical. It has been shown to be one of the earliest events in what appears to be the stepwise progression of events that culminate in a tumorigenic cell.

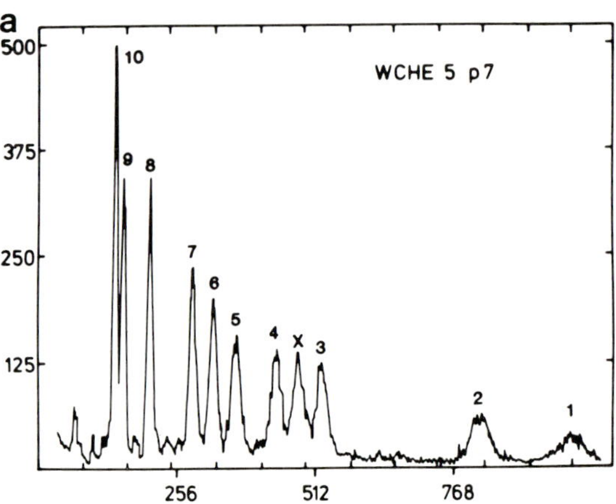

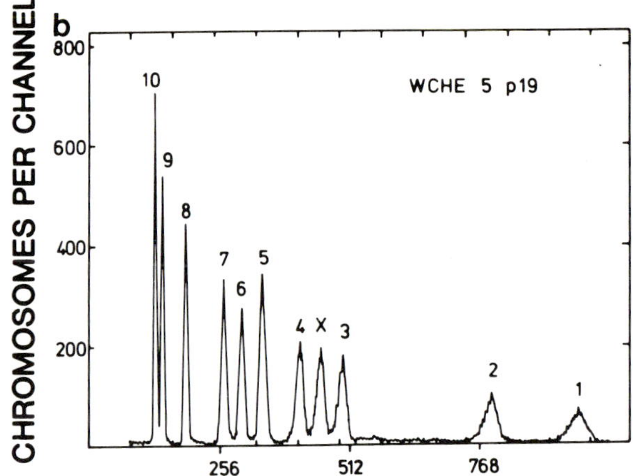

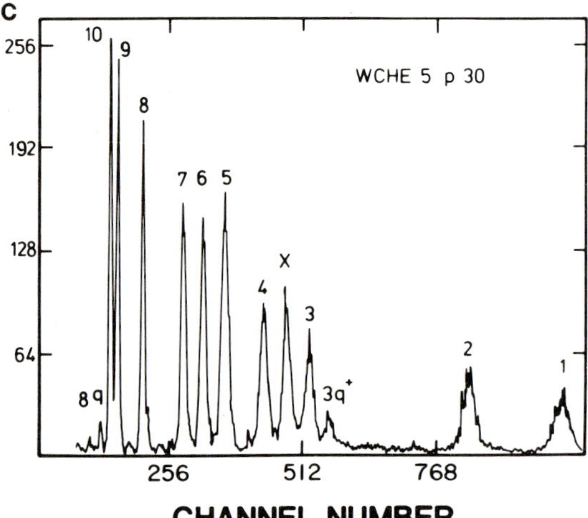

Fig. 12. Univariate flow karyotypes for PI-stained Chinese hamster embryo cells. **a:** After 7 passages in vitro. **b:** After 19 passages in vitro. **c:** After 30 passages in vitro. Cells were prepared as described by Aten et al. [2].

Univariate flow karyotype analysis has played a significant role in analysis of karyotypic changes associated with the spontaneous transformation of Chinese hamster (WCHE/5) fibroblasts and with the directed transformation of human (KD) fibroblasts [30,31,38,81,143]. Karyotype stability is also an important monitor of genetic stability in the production of biologic standards. For example, in the production of viral biologics the loss of karyotype stability is of major concern when using mammalian tissue culture as the host in which virus is replicated.

Flow karyotypes are particularly sensitive to the occurrence of nonrandom chromosome changes. Chromosome rearrangements that occur randomly and at low frequency will contribute to the background continuum. However, as the frequency of an altered chromosome increases, it appears as a peak above the continuum [59]. Such chromosomes are referred to as *marker chromosomes*. Such marker chromosomes are often easier to detect using flow karyotyping than conventional banding analysis. Banding analysis is especially hard to apply to the detection of aberrant chromosomes that occur nonrandomly in cells from a population in which random aberrations are present at high frequency. A comparison of flow karyotypes obtained at sequential stages of culture progression can be made quickly and easily thereby facilitating study of the development of karyotypic instability [9,143]. Figure 12 shows flow karyotypes measured for WCHE/5 fibroblasts at several stages of transformation. Flow karyotyping was advantageous in this study, since it provided (1) analysis of a large number of chromosomes within a short period of time, (2) identification of statistically significant nonrandom aberrations, and (3) quantitative information about the aberrant chromosome(s). The ability to monitor the progressive changes that occur in a culture over a period of time was particularly useful. Once the peaks of a normal euploid flow karyotype had been identified, univariate analysis of sequentially derived flow karyotypes measured at regular intervals allowed identification of the appearance of aberrant chromosomes and the detection of chromosome loss. Figure 12 illustrates some of the numerical and structural changes that occurred in Chinese hamster (WCHE/5) cells at several points in time after the culture was started from embryonic tissue [31]. The first event was a trisomy of the #5 chromosome. This was followed by the appearance of an insertion element in the long arm of the #3 chromosome resulting in a new peak in the flow karyotype located between the #2 and #3 Chinese hamster chromosomes. This aberrant chromosome was first detected when present in only 10 percent of the cells. The position of the peak in the flow karyotype focused the cytologic search for the aberrant chromosome. Analysis of the larger chromosomes in G-banded metaphase spreads showed the aberrant chromosome to be a 3q+. G-banding suggested variability in the size of the insertion element in the long arm of chromosome 3 (3q+). Flow karyotyping clearly indicated that this was not the case. The peak for this chromosome was well defined and not skewed, thus showing that this karyotype aberration was the same in most or all of the cells in the population. It may have been missing in some cells but the chromosome DNA content did not vary from cell to cell [31].

Flow karyotyping also has been used to monitor the loss of karyotype stability in human cells and to correlate the appearance of a cytogenetic abnormality with expression of a mutant form of β-actin. Flow karyotypes were measured at each step of incremental tumorigenicity, starting with a normal euploid cell and concluding with the emergence of a tumorigenic cell. The rate of synthesis of a mutant β-actin doubled in the tumorigenic cell at the same time that trisomy of chromosome 7 was detected by flow karyotyping [9]. The actin gene was subsequently mapped to human chromosome 7 [129].

BIVARIATE FLOW KARYOTYPING

Flow Karyotype for Several Species

Bivariate flow karyotyping has proved superior to univariate flow karyotyping for discrimination between chromosomes in those species whose chromosomes differ in DNA base composition. Figure 3, for example, shows bivariate (Ho versus CA3) flow karyotypes measured for chromosomes isolated from mouse, human, and hamster cells. These flow karyotypes are characterized by a band of peaks whose means fall more or less along a single diagonal defined by a fixed Ho/CA3 ratio. The peaks that lie off the diagonal differ in base composition and are likely to be better resolved in the bivariate flow karyotype than in univariate flow karyotypes. Figure 3a shows that human chromosomes (especially those smaller than chromosome 12) differ substantially in DNA base composition and thus are better resolved in bivariate than in univariate flow karyotypes. The X chromosome in the Chinese hamster flow karyotypes also falls off the constant Ho/CA3 diagonal and thus is better resolved in the bivariate flow karyotype. By contrast, mouse chromosomes usually fall more or less on the diagonal, indicating that they have similar DNA base composition ratios. In this situation, little is gained in the bivariate flow karyotype [30].

Bivariate flow karyotypes also are useful in resolving selected human chromosomes from hamster chromosomes in chromosome suspensions isolated from human hamster hybrid cells [35,40,56,163]. Figure 3c shows flow karyotypes measured for a human × hamster hybrid containing human chromosomes 4, 8, and 21. The larger human chromosomes can be distinguished easily from the hamster because they have higher Ho/CA3 ratios than do the hamster chromosomes. The Ho-CA3 ratios of the smaller human chromosomes do not differ from the hamster so markedly and thus cannot be distinguished from the hamster chromosomes as well. This is summarized in Figure 3b which shows the locations of peaks for human chromosomes 1 through 13 and Y relative to the hamster chromosomes from a hamster cell line commonly used to produce human × hamster hybrids. In general, human chromosomes larger than 13 are well resolved from hamster chromosomes. This fact is particularly useful to investigators interested in purifying the larger human chromosomes for production of recombinant DNA libraries [39,40,48,50,55,163] or for chromosome gene mapping [105–110].

Classification of Human Chromosomes and Detection of Aberrant Chromosomes

Chromosome Classification. Human chromosomes can be classified into approximately 20 groups in bivariate Ho vs CA3 flow karyotypes [57,63,97,98,105,161,173]. Chromosomes 9 through 12 are consistently difficult to resolve. In addition, some chromosome types (e.g., 14 and 15) may or may not be resolved because of small changes in their DNA content and/or DNA base composition as a result of normal polymorphism. The identities of the chromosomes producing each peak shown in Figure 3a have been determined by banding analysis and DNA probe analysis of chromosomes sorted from each peak [57,105]. Use of other dye pairs (e.g.,

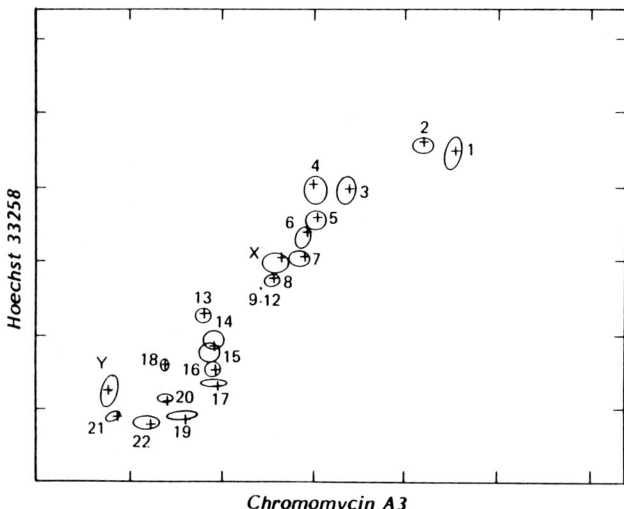

Fig. 13. Variability in the means of the peaks in Ho versus CA3 bivariate flow karyotypes measured for human amniocyte chromosomes. The ellipses show the 95% tolerance regions for the various peaks estimated from flow karyotypes measured for 50 different, karyotypically normal individuals. The crosses show the bivariate means for human lymphocyte chromosomes.

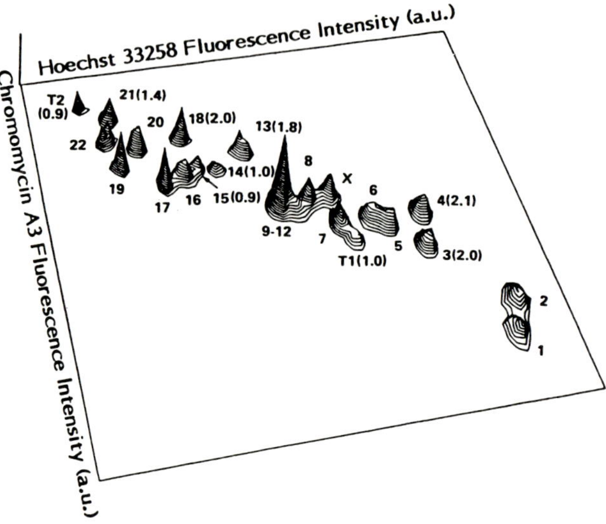

Fig. 14. Bivariate Ho versus CA3 flow karyotype measured for an amniocyte culture carrying the translocation, t(14;15). The numbers indicate the locations of selected normal and derivative chromosomes. The numbers in parentheses indicate the relative volumes of selected peaks.

DIPI versus CA3 or Ho plus neutropsin versus CA3) sometimes improve resolution of chromosomes such as 1, 9, 16, and Y that contain large polymorphic heterochromatic regions [96,108,121]. Chromosome 9 may be resolved from chromosomes 10 through 12 using these dyes.

Variability in Peak Means and Areas The peak locations of chromosomes in bivariate flow karyotypes have proved to be remarkably constant in flow karyotypes measured for the same chromosomes prepared on different days, for flow karyotypes measured for chromosomes from different tissues, and for chromosomes isolated from different individuals [63,98]. Figure 13 shows the variability found in the relative peak means for chromosomes isolated from 50 different primary amniotic cultures. The variability observed in this study was caused by normal polymorphic variability among normal individuals and by variability in the estimation of peak means [63].

Bivariate flow karyotyping has been applied to classification of chromosomes isolated from a variety of human tissues including fibroblasts [57,161], lymphoblasts [97,161], lymphoblastoid cell lines [163], human carcinoma cell lines [56], and amniocytes and chorionic villus biopsies [63,161]. The flow karyotypes are similar for all cell sources. For example, Figure 13 shows the means of the peaks in flow karyotypes measured for 10 normal human lymphoblast cell cultures [97] superimposed on the 95% tolerance regions measured for 50 normal human amniocyte cultures [63]. In all cases, the lymphocyte means fall inside the 95% tolerance regions even though the tissues, chromosome isolation procedures and staining conditions were different in the two studies. The magnitude of the variability observed among flow karyotypes is about 5% of the DNA content of any average-sized chromosomes (i.e., approximately 1/250th of the total genome DNA content). This is comparable in DNA content to that associated with a single chromosome band (assuming 300 bands per mitotic cell). Thus, a change in chromosomal DNA content or base composition caused by the gain or loss of a single band can often be detected in a flow karyotype and should be distinguishable from a normal polymorphism. Figure 14 shows a flow karyotype measured for a 46XY, t(14:15) amniocyte culture. Two new derivative chromosome peaks are evident in the flow karyotype; one due to the chromosome that lost DNA content as a result of the translocation and the other to the chromosome that gained DNA content. In this case, the translocation was balanced so that the DNA lost by one chromosome was gained by the other [61].

The variability in peak volumes is considerably larger than for the peak means. Figure 15, for example, shows the variability in peak volumes calculated for the smaller human chromosomes in 21 flow karyotypes measured for karyotypically normal cell samples [61]. In general, the peak volumes, normalized so that the volume of a peak produced by two identical homologous chromosomes is 2, differ from the expected volumes by less than 0.25. Thus, the frequency of a chromosome in the population can be estimated by measuring the peak volume and rounding to the nearest whole number. Trisomies can be detected as producing peaks that are approximately 50% larger than normal. Figure 16a shows flow karyotypes measured for the smaller chromosomes from a normal 46XY amniocyte culture and Figure 16b shows a similar flow karyotype measured for a 47XY, +21 amniocyte culture. The volume for the peak for chromosome 21 is more than 50% larger in the trisomic sample [63].

KARYOTYPING HETEROGENEOUS POPULATIONS

The detection of chromosome aberrations and determination of the frequency of these aberrations is important for analyzing extent of genetic damage and for study of the mechanism(s) of aberration formation. The application of flow cytometry to aberration detection is discussed in two

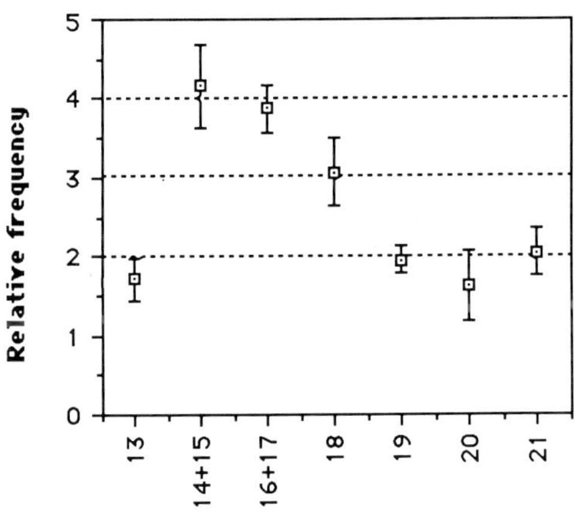

Fig. 15. Variabilities in peak volumes for the smaller human chromosomes. This peak shows the mean volumes (and standard deviations therein) for human chromosomes 13 through 21. These estimates were made from bivariate flow karyotypes measured for amniotic samples in which chromosome 18 was trisomic. The volumes for peaks 14 and 15 were combined as were the volumes for peaks 16 and 17. The horizontal dashed lines show the peak volumes expected for normal homologous chromosomes (chromosomes 13, 19, and 21), trisomic chromosomes (chromosome 18) and pairs of homologous chromosomes (chromosomes 14+15 and 16+17).

parts; (1) using conventional flow cytometers, and (2) using scanning flow cytometers.

Random Aberration Analysis via Conventional Flow Cytometry

Random chromosome aberrations that can be detected in conventional flow karyotypes result from structural rearrangements that lead to chromosomes with abnormal DNA contents and/or base compositions. These aberrant chromosomes and fragments may occur anywhere in the flow karyotype. However the frequency of smaller chromsomes or fragments tends to be higher than the frequency of larger objects. Thus, these objects produce a rapidly decreasing continuum that underlies the peaks in a flow karyotype. The magnitude of the continuum has been estimated by fitting to it, a decreasing function and taking the area under the fitted functions as an estimate of the magnitude of the genetic damage [122]. Alternately, the height of the continuum at several points between peaks has been used as a measure of the debris continuum [45]. There is also some evidence that the widths of the peaks in a flow karyotype increase with increasing genetic damage [132–134]. Figure 17 shows increases in the magnitude of the debris continuum and peak CVs from flow karyotypes measured for rodent cells following exposure to increasing doses of ionizing radiation. Both variables clearly increase with increasing dose.

Fourier analysis also has been applied to quantitation of clastogen induced changes in flow karyotypes. In this approach, a Fourier transformation is made of each flow karyotype. The low frequency components of the Fourier transform increase with an increase in the magnitude of the debris continuum and/or with an increase in the peak CVs. Thus, the magnitude of the low frequency portions of the transform has been shown to be a useful measure of the extent of clastogen induced chromosomal damage [2,165].

The main problem associated with flow karyotypic analysis of random chromosome aberration is its low sensitivity. Most studies to date suggest that damage produced by radiation doses below 25 to 50 rads is probably not detectable with this method. The sensitivity of flow karyotyping is limited by the chromosomal and nuclear fragments that result during the chromosome isolation process. These chromosome fragments cannot be distinguished from true aberrant chromosomes. Furthermore, the frequency of such fragments cannot be tightly controlled in biologically interesting cell populations such as stimulated peripheral lymphocytes. It is difficult to detect induced damage when the frequency of the resultant aberrant chromosomes (marker chromosomes) is lower than the frequency of isolation-induced DNA fragments. In spite of these limitations, flow karyotyping has been used to demonstrate the direct relationship between chromosome damage and cell survival for both low and high LET radiations [45,60,165].

Detections of Structurally Aberrant Chromosomes Using Scanning Flow Cytometry

Slit-scan flow cytometry has proved useful for detection of clastogen induced structurally distinct aberrant chromosomes such as dicentrics. Dicentric chromosomes stained with DNA specific dyes produce distinctive profiles with two dips (produced by the reduced DNA content of centromeres) as they pass through the scanning region (see Figure 5a). The frequency of such profiles has been measured as an indication of the extent of radiation-induced genetic damage in hamster and human cells [32,60].

Figure 18 shows the result of applying slit-scan flow cytometry to quantification of the frequency of dicentric chromosomes in Chinese hamster cells following irradiation with x-rays of doses ranging up to 2 Gy. The frequency of dicentrics measured by slit-scan analysis in this study was shown to be proportional to the frequency of dicentrics measured by conventional karyotyping [60]. Recent experiments with human lymphoblast cells suggest that this method is now capable of detecting radiation-induced increases in the dicentric frequency at doses below 0.5 Gy (J.N. Lucas et al., manuscript in preparation). A principal advantage of slit-scan flow cytometry for measurement of radiation-induced chromosome damage is that it speeds measurement of a cytogenetic endpoint whose cytogenetic significance is already known. The most significant limitation of dicentric analysis by slit-scan analysis is that current systems allow detection of only a small fraction of the total dicentric chromosomes (only dicentrics with centrally located centromeres spaced more than 1 μm apart can be detected with current systems). The dicentric chromosome detection efficiency may be as low as 0.003 for human chromosomes.

The efficiency with which dicentric chromosomes can be detected by slit-scanning may be increased significantly by developing techniques to selectively stain chromosome centromeres. This should allow, for example, detection of centromeres located near the ends of chromosomes that are missed when scanning chromosomes stained with DNA specific dyes. Recent reports have suggested that antibodies against kinetichore proteins may be used to stain chromosomes immunofluorescently for slit-scan analysis [64].

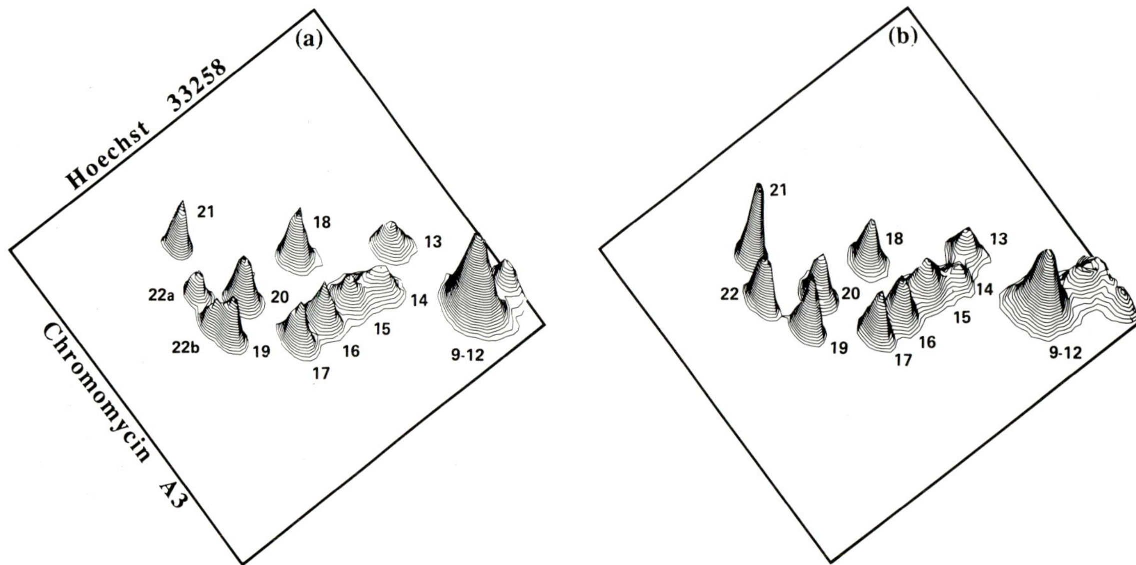

Fig. 16. Bivariate Ho versus CA3 flow karyotypes for normal and trisomic human amniocyte chromosomes. **a:** A 46XY culture. **b:** A 47XY, +21 culture.

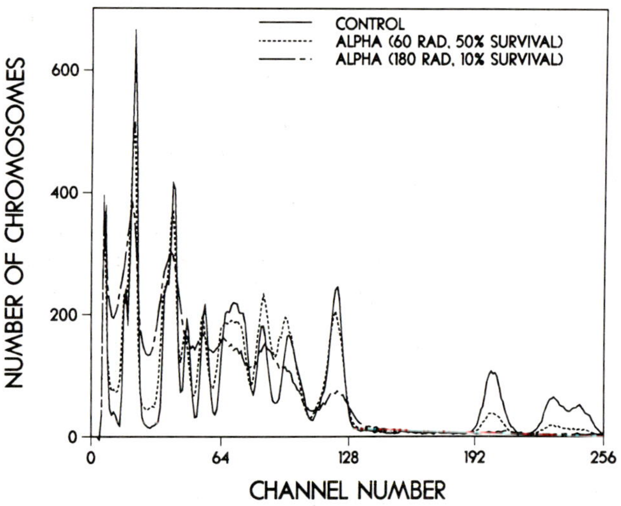

Fig. 17. Chromosome damage estimated from the magnitude of the debris continuum in univariate flow karyotypes measured for α-irradiated Chinese hamster chromosomes. Chromosomes were isolated and stained using the PI procedure described in Table 1.

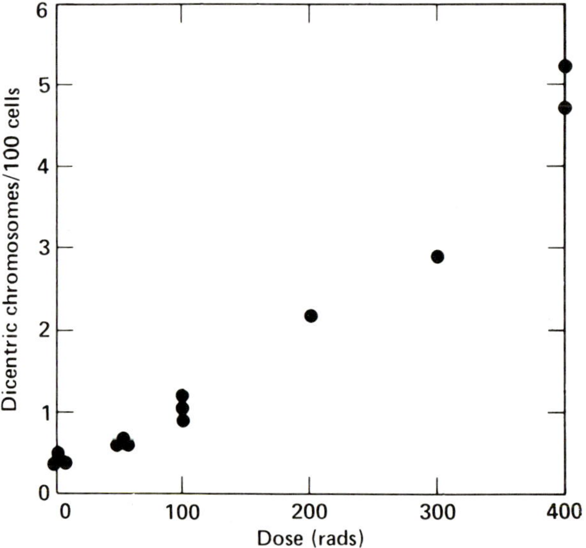

Fig. 18. Slit-scan estimates of dicentric chromosome frequencies in Chinese hamster cells after X-irradiation at doses from 0 to 4 Gy.

CHROMOSOME SORTING

Study of the genetic and physical properties of selected mammalian and plant genomes may be simplified considerably if the study can be confined to only a subset of the genome. Chromosomes provide a natural basis for subdividing the genome. In the human, for example, genetic studies of a single chromosome type are, on average, only 1/24th as complex as the entire genome. Chromosome purification by sorting has facilitated genetic studies by providing chromosomal DNA from only one or a few chromosome types. In recent years, chromosomes purified by sorting have proved useful for gene mapping and for production of recombinant DNA libraries. The libraries are simplifying efforts to order the human genome and to identify disease-linked restriction fragment length polymorphisms (see Chapter 28, this volume).

Technical Considerations

Sorting Purity. The degree to which chromosomes can be purified by sorting is determined by how well the chromosome(s) of interest can be resolved in a flow karyotype. Thus, the smaller human chromosomes (e.g., 18–22 and Y) can be sorted with high purity from a suspension of human chromosomes while chromosomes 10–12 can only be purified as a group. The purity of sorting also may be reduced by the occurrence of chromosome clumps that are indistinguishable in the flow karyotype from the chromosome of interest. These clumps will be sorted along with the target chromosome(s) thereby reducing the purity of the sorted fraction. Debris produced during chromosome isolation may be indistinguishable from the smaller chromosomes and may also reduce the purity of these chromosomes during sorting. It may be desirable to shear heavily when sorting larger chromosomes (to reduce clumping) and gently when sorting smaller chromosomes (to reduce debris).

In general, sorting purity is highest when the chromosome type of interest is cleanly resolved from surrounding chromosomes. Thus, human chromosomes sorted using bivariate flow cytometry (eg. of chromosomes stained with Ho and CA3) are likely to be more pure than those sorted following univariate flow cytometry since the discrimination based on DNA content and base composition is likely to be greater than discrimination according to DNA content alone. Use of other dyes (e.g., DIPI, neutropsin) [96,98,105,107,121] or measurement of additional chromosome parameters (e.g., shape, [114,115,127]) may further increase the purity with which chromosomes can be sorted. Figure 7, for example, shows that the staining with DIPI or DAPI in place of Ho often improves discrimination of human chromosome 9. Purification of this chromosome from chromosomes isolated from human cells is not possible using Ho and CA3.

The purity with which chromosomes can be sorted may also be increased by judicious selection of the cell line from which chromosomes are to be sorted [68]. The reason is that many chromosome types (especially human chromosomes) may vary in DNA content and/or DNA base composition among normal individuals. This polymorphic variability, usually the result of an increase or decrease in the amount of heterochromatin in one or a few chromosome types, may allow a particular chromosome type to be distinguished more readily in one cell type than in another. The purity of sorting can also be increased by sorting from a chromosome suspension prepared from a hybrid cell line containing only a few chromosomes from the species of interest. Figure 3c, for example, shows a flow karyotype measured for a human × hamster hybrid cell line carrying human chromosomes 4, 8, and 21, along with a full complement of hamster chromosomes. The human chromosomes are sufficiently well resolved from each other and from the hamster chromosomes that chromosomes 4 and 8 can be purified.

Sort Rate. The rate at which chromosomes can be purified by sorting depends on the relative frequency of that chromosome type in the population, the throughput rate of the sorter and the fraction of the chromosomes of the type to be sorted that actually falls in the sort window. In a conventional sorter, approximately 1,000 fluorescent objects can be processed each second. This processing rate would allow sorting of approximately 40 human chromosomes of a single type each second (assuming a starting population of isolated human chromosomes in which the chromosome of interest was present at a frequency of 1/23). Thus, approximately 10 hours would be required to purify 1×10^6 chromosomes of a single type. Of course, the sort rate will be reduced substantially by the presence of debris in the population and by the use of a sort window that contains only a fraction of the chromosome type to be sorted. In practice, purification of 20–30 human chromosomes per second has been achieved routinely [40,163].

High-speed sorting has recently been developed to facilitate chromosome sorting [55,58,138]. High speed sorting is conceptually similar to conventional sorting except that the system is operated at higher pressure and droplet production frequency so that the sorting rate is increased approximately 10-fold. Chromosome analysis rates as high as 20,000/sec are routine in high speed sorting as is purification of chromosomes of a single type at rates up to 200/sec. The purity achieved using high speed sorting is comparable to that achieved by conventional sorting [55].

The speed with which chromosomes of a single type can be purified can be increased by increasing the frequency of the desired chromosome type in the population from which chromosomes are sorted. This may be accomplished by prepurifications using velocity sedimentation (i.e. enrichment according to the rate at which the chromosomes sediment in a sucrose gradient) [27,28]. Another approach is to isolate chromosomes from cells in which the desired chromosome frequency is higher. For example, the frequency of the Y chromosome in the cell line (49XYYYY) is four times higher than normal and the rate at which the chromosome can be purified by sorting is increased correspondingly.

Gene Mapping

The assignment of cloned DNA sequences to a single chromosome type (or subregion) has been simplified by chromosome sorting [13,14,25,46,51,71,74,94,105,108,110,112, 126,128,137,166,167]. In this application ~20,000 chromosomes of a single type are sorted onto a small spot on nitrocellulose paper. Spots are prepared for each different chromosome type. The DNA on the nitrocellulose paper is then denatured and hybridized to radioactively labeled, denatured probe DNA. The probe and sorted chromosomal DNAs are allowed to reanneal under conditions where the probe will bind only to DNA sequences to which it has high homology. The hybridized probe is then detected autoradiographically. Unique sequence probes are mapped by hybridizing radioactively labeled probe DNA to filters containing DNA of each chromosome type. Unique sequence probes bind only to one sorted spot. Probes can be mapped to chromosome subregions by hybridizing against DNA obtained by sorting derivative chromosomes that carry only a small part of the whole chromosome to which the probe has been mapped. Such derivative chromosomes result from structural aberrations such as translocations.

Library Production

Chromosome sorting has allowed production of recombinant DNA libraries for each human chromosome type [10,36,37,39,40,48,50,83,85,86,89,90,91,93,120,125, 148, 163,170,171], for selected aberrant chromosomes [12,43,70, 78,101,118] and for single rodent chromosome types [3, 4,17,66,84]. This has been accomplished by purifying chromosomes of each type by sorting. DNA is then extracted from 10^5 to 10^6 sorted chromosomes, digested with a restriction enzyme and ligated into a recombinant DNA vector (typically a lambda phage derivative). The recombinant phage DNA is then packaged to form infective phage and amplified by growth on a bacterial lawn.

Most of the libraries produced to date are complete digest libraries. That is, the DNA was digested to completion prior to ligation to the vector so that the recombinant DNA fragment in each phage is only a few kilobases in length. The complete digest HindIII and EcoR1 (the restriction endonucleases used to cut the DNA) libraries for each human chromosome type produced by the National Laboratory Gene Library Project are now available from the American Type Culture Collection (Camden, NJ). Only a few large insert libraries have been produced [106]. These are more difficult since more sorted DNA is required for each library (see Chapter 28, this volume).

The quality of the libraries produced from sorted DNA depends on the purity with which chromosomes can be sorted. The purity of sorted chromosomal material has been estimated by analysis of the flow karyotype produced during sorting [55], cytologic analysis of the sorted chromosomes [21], analysis of sorted material by hybridization with chromosome specific probes to chromosomes on slides [163] or on nitrocellulose filters [106] and by mapping probes selected from recombinant DNA libraries [163]. In general, about 50–80% of the DNA in a library can be expected to be from the intended chromosome type.

FUTURE DIRECTIONS

The success of flow karyotyping is due to the accuracy and precision of chromosome classification and the sensitivity and speed with which chromosome aberrations can be detected. Chromosome sorting allows collection of chromosomes of a single type with purity that cannot be achieved by any other technique. These attributes have made flow cytogenetics a valuable tool for studies in the fields of molecular genetics, carcinogenesis, radiation biology and clinical cytogenetics. However, flow cytogenetics still suffers from limitations in several areas. First, it is difficult to isolate chromosomes without production of debris fragments and/or chromosome clumps. These fragments and clumps often cannot be distinguished from normal chromosomes. This leads to contamination of sorted chromosome fractions. These objects also may be mistaken for aberrant chromosomes in studies designed to assess the frequency of random aberrations as a measure of clastogen-induced genetic damage. Second, the cytochemical differences between chromosomes that are the basis for chromosome classification are small so that high analysis precision is required for complete chromosome discrimination or purification. Third, molecular biological studies often require microgram or larger quantities of sorted chromosomes. Purification of sufficient chromosomes for these studies is time consuming. Future work in the field of flow cytogenetics will be directed at minimization or elimination of these limitations and may come in several areas.

Chromosome Isolation

Major advances have come in recent years as procedures for chromosome isolation have been improved to allow isolation of chromosomes with reduced amounts of debris and chromosome clumps. In general, debris seems to be produced during cell lysis to release chromosomes into suspension, and cell clumps result from incomplete cell swelling and chromosome separation prior to shearing. Figure improvements may come from improved cell swelling and membrane treatment so that all cells can be broken with the same, minimal amount of shearing. Improved chromosome stabilization (e.g., by treating with DNA or protein cross linking agents) may also reduce debris by increasing the amount of manipulation that the chromosomes can tolerate. Chromosome preparations may also be improved by development of procedures such as differential centrifugation, velocity sedimentation [26,28] or enzymatic digestion of free DNA in solution prior to chromosome isolation to reduce debris clumping. Improvements may also come from development of improved culture techniques that lead to a faster cell cycle traverse rate and/or to a larger fraction of cycling cells so that the mitotic index can be increased. This should increase the efficiency of chromosome isolation and should minimize debris from interphase nuclei that disintegrate during cell shearing. Development in these areas is essential if flow cytogenetic techniques are to be applied successfully to tissues such as human solid tumors that currently are difficult to culture in vitro.

Cytochemistry

Chromosome classification and sorting can be improved by increasing the distinctness with which chromosomes can be stained. This may be achieved by increasing the precision of chromosome staining as has been achieved by adding millimolar amounts of sodium bisulfite and sodium citrate to Ho and CA3 stained chromosomes just prior to analysis [159,161,162]. The precision of chromosome analysis can also be increased by development of dyes that have increased quantum efficiency increases upon binding to chromosomal DNA [131]. Dyes of this sort will have a reduced effect on the chromosomal measurement when free in solution. A replacement for CA3 would be especially useful since its quantum efficiency increases only about 4-fold upon binding to DNA. Improvements in chromosome staining would also result from chemical modifications that enhance the sensitivity of the DNA specific dyes to DNA base composition. Such enhancement may be achieved by altering the isolation buffers to change the amount of energy transfer between dyes such as Ho and CA3. The potential is large here since the amount of energy transfer between Ho and CA3 may be as high as 0.75 [56,97,100]. The use of DNA specific dyes like DAPI and DIPI or modifying agents such as distamycin and netropsin also may be employed to enhance discrimination between selected chromosome types.

Fluorescence hybridization with chromosome-specific probes may be the most powerful procedure for increasing the distinctness with which selected chromosome types can be stained [93,95,124,139,141,154]. In this approach, the DNA in the target chromosomes is denatured and incubated with chemically modified (eg. biotinylated), chromosome specific DNA probes. The target chromosomes and probe are allowed to reanneal under conditions such that the probes bind only to the chromosomes to which they have high homology. The chemically modified probes are detected using a fluorescent reagent that binds only to the chemically modified probes (e.g., fluorescein-labeled avidin to detect biotinylated probes). In this way, the chromosomes may be stained according to their DNA sequence. This approach has already been used successfully to stain selected chromosomes in metaphase spreads on slides and in nuclei on slides and in supension [124,139,156,157]. In addition, some evidence has been presented that isolated chromosomes can be stained in suspension using fluorescence hybridization [44]. Chromosome-specific repeat sequence probes suitable for fluorescence hybridization chromosome staining have already been discovered for over half of the human chromosomes [49] (Hunt Willard, Dan Pinkel, private communication). Composite chromosome-specific probes consisting of multiple unique sequences have also been shown to be suitable for fluorescence hybridization [49]. Thus, it seems likely that fluorescence hybridization with chromosome specific probes may contribute substantially to flow cytogenetic studies in the next few years.

Antibodies against chromosome specific proteins may also allow improved chromosome staining [155,158]. Antibodies against kinetochore proteins have been already used successfully to stain chromosomal centromeres [64]. The existence of specific telomeric and centromeric proteins seems likely [16]. Thus, the development of antibodies against these proteins offers another option for specific chromosome staining.

Instrumentation

Enhanced collection of fluorescence emission, more uniform illumination across a wide sample stream, increased sorting speed and new electro-optical techniques may improve existing instruments used for chromosome analysis and sorting. The efficiency with which fluorescence is collected in most current flow systems is only about 15%. As a result, high power lasers must be employed to stimulate sufficient fluorescence that chromosomal dye contents can be measured with high accuracy (10^4 photoelectrons must be produced for each chromosome to allow 1% measurement precision). Increasing the numerical aperture of the collection optics through the use of reflection optics allows measurement of flow karyotypes for Chinese hamster chromosomes that are nearly the same quality as that illustrated in Figure 12 using only 15 mW of 488 nm, excitation. It seems likely in the future, that the efficiency of light collection can be increased as high as 75% so that air cooled lasers can be employed for high resolution chromosome excitation.

Improvements may also be made in the uniformity with which broad (e.g., 6 to 8 μm diameter) sample streams can be illuminated. The rate of chromosome processing is limited currently by the diameter of the sample stream and by the concentration of the chromosomes in suspension. Chromosome concentrations higher than 5×10^7 are difficult to achieve. This concentration permits analysis of only 1,500 chromosomes per second in a conventional system in which the sample stream diameter is 2 μm and the flow velocity is 10 m/sec. Increasing the sample stream diameter to 8 μm would allow analysis of 24,000 chromosomes/sec. Increased throughput is especially important in high speed chromosome sorting where the sort rate is determined at least partly by the rate at which chromosomes can be analyzed. Unfortunately, the diameter of the illuminating laser beam must be increased if the sample stream diameter is increased. This can be achieved using Gaussian cross section beams only by increasing the lateral dimension of the beam. This, in turn, reduces the intensity of illumination and thus reduces the measurement precision. Future work to produce laser beams with more rectangular profiles (e.g., by laser beam apodization) may allow use of larger sample streams without loss of illumination intensity.

The rate at which chromosomes can be purified continues to limit accumulation of sufficient chromosomal DNA and proteins for molecular biological studies of individual chromosomes. High speed sorting [55] has been developed to reduce this limitation. In the future, four and six way sorting may be developed to further increase the rate of chromosome purification [1]. Four way sorting, allowing simultaneous purification of four chromosome types, uses two different levels of negative and positive charging to establish four separate sorted droplet trajectories. Four way sorting (versus two way) doubles the rate at which different types of chromosomes can be sorted if several different chromosome types are to be purified. The rate at which DNA from one chromosome type can be purified also may be increased by prefractionation (e.g., by using velocity sedimentation). This approach has been shown to increase the representation of one chromosome type in a population of human chromosomes from 1/24th to about 1/6th. This will allow one of the enriched chromosomes to be sorted four times faster. This procedure will be practical, however, only if the chromosomes can be concentrated after prefractionation so that chromosomes can be processed at the normal rate.

Analysis based on chromosome shape may be improved by increasing the rate at which chromosomes can be processed using slit-scan flow cytometry so that chromosome sorting by shape becomes possible. In addition, fringe-scan flow cytometry may be employed to increase the scanning resolution. In this approach, a series of interference fringes is generated by the intersection of two laser beams [127]. The profile of a chromosome traversing the fringe field is recovered by mathematical deconvolution of the profile recorded during the passage of the chromosome through the fringe field. Slit-scan techniques can resolve features as small as 0.8 μm and fringe scanning has a theoretical resolution as low as 0.3 μm [32,127].

Instruments may also be developed to classify chromosomes according to differences in their fluorescence emission anisotropies [29]. This technique may be most effective when combined with slit-scanning to permit classification according to the differences in emission anisotropy along the chromosomes. This technique may allow detection of aberrant chromosomes that have discontinuities such as breaks or homogeneously staining regions. Chromosomes may also be classified according to differences in their energy transfer characteristics or their fluorescence emission lifetimes. Phase sensitive detection (J. Steinkamp, J. Martin, private communication, Los Alamos National Laboratory) is being developed at the Los Alamos National Laboratory to distinguish between fluorescent molecules that have similar emission spectra but different fluorescence lifetimes. This technique may be applied to chromosome analysis to allow independent analysis of multiple fluorescent molecules that have independent binding characteristics.

ACKNOWLEDGMENTS

Work performed under the auspices of the U.S. Department of Energy by the Lawrence Livermore National Laboratory under contract number W-7405-ENG048, with support from Grant number HD 17665, and the Los Alamos National Laboratory with support from the National Institutes of Health Division of Research Resources, Grant number RR 01315.

REFERENCES

1. **Arndt-Jovin D, Jovin TM (1978)** Automated sorting with flow systems. Annu. Rev. Biophys. Bioeng. 70: 527–558.
2. **Aten JA, Kipp JBA, Barendsen GW (1980)** Flow cytofluorometric determination of damage to chromosomes from X-irradiated Chinese hamster cells. In Laerum OD, Lindmo T, Thorud E (eds.), "Flow Cytometry IV. Proceedings of the Fourth International Symposium on Flow Cytometry." Oslo: Universitetsforlaget, pp 485–491.
3. **Baron B, Metezeau P, Hatat D, Roberts C, Goldberg ME, Bishop C (1986)** Cloning of DNA libraries from mouse Y chromosomes purified by flow cytometry. Som. Cell Mol. Genet. 12:289–295.
4. **Baron B, Metezeau P, Kelly F, Bernheim A, Berger R, Guenet JL, Goldberg ME (1984)** Flow cytometry isolation and improved visualization of sorted mouse chromosomes. Exp. Cell Res. 152:220–230.
5. **Bartholdi MF (1985)** DNA base composition of hu-

man chromosomes. J Coll. Interface Sci. 105:416–434.

6. **Bartholdi MF, Meyne J, Albright K, Luedemann M, Campbell E, Deaven LL, van Dilla M, Chritton D, Cram LS (1986)** Chromosome sorting by flow cytometry. In Goettesman M (ed.), Methods in Enzymology. vol. 76: "Molecular Genetics of Mammalian Cells." Orlando, Florida: Academic Press, pp 252–267.
7. **Bartholdi MF, Ray FA, Cram LS, Kraemer PM (1987)** Karyotype instability of Chinese hamster cells during in vivo tumor progression. Somat. Cell Mol. Genet. 13:1–10.
8. **Bartholdi MF, Ray FA, Jett JH, Cram LS, Kraemer PM (1984)** Flow karylogy of serially cultured Chinese hamster cell lineages. Cytometry 5:534–538.
9. **Bartholdi MF, Travis GL, Cram LS, Porreca P, Leavett J (1986)** Flow karyology of neoplastic human fibroblast cells. Ann. NY Acad. Sci. 468:339–349.
10. **Bartlett RJ, Pericak-Vance MA, Yamaoka L, Gilbert J, Herbstreith M, Hung WY, Lee JE, Mohandas T, Bruns G, Laberge C, Thibault MC, Ross D, Roses AD (1987)** A new probe for the diagnosis of myotonic muscular dystrophy. Science 235:1648–1650.
11. **Behr W, Honikel K, Hartmann G (1969)** Interaction of the RNA polymerase inhibitor chromomycin with DNA. Eur. J. Biochem. 9:82–92.
12. **Benedict WF, Murphee AL (1983)** Chromosomal and genetic basis for cloning the retinoblastoma gene within 13q14. In "Banbury Report 14: Recombinant DNA Applications to Human Disease." Cold Spring Harbor, New York: Cold Spring Harbor Laboratory, pp 135–140.
13. **Bernheim A, Metezeau P, Guellaen G, Fellows M, Goldberg ME, Berger R (1983)** Direct hybridization of sorted human chromosomes: Localization of the Y chromosome by flow karyotypes. Proc. Natl. Acad. Sci. USA 80:7571–7575.
14. **Bernheim A, Metezeau P, Guellaen G, Weissenbach J, Fellous M, Goldberg M, Berger R (1982)** Identification of molecular hybridization of human chromosomes after cell sorter purification. C.R. Seanc. Acad. Sci. (III) (Fr.) 295:439–43.
15. **Bijman JT (1983)** Optimization of mammalian chromosome suspension preparations employed in a flow cytometric analysis. Cytometry 3:354–358.
16. **Blackburn EH, Challoner PB (1984)** Identification of a telomeric DNA sequence in *Trypanosoma brucei.* Cell 36:447–457.
17. **Blin N, Stohr M, Hutter KJ, Alonso A, Goerttler K (1982)** Assignment of snRNA gene sequences to the large chromosomes of rat kangaroo and Chinese hamster isolated by flow cytometric sorting. Chromosoma 85:723–733.
18. **Blumenthal AD, Dieden JD, Kapp LN, Sedat JW (1979)** Rapid isolation of metaphase chromosomes containing high molecular weight DNA. J. Cell. Biol. 81:255–259.
19. **Buys CH, Koerts T, Aten JA (1982)** Well-identifiable human chromosomes isolated from mitotic fibroblasts by a new method. Hum. Genet. 61:157–159.
20. **Buys CHCM, Mesa J, van der Veen AY, Aten JA (1986)** A comparison of the effect of 5-bromodeoxyuridine substitution on Hoechst 33258 and DAPI fluorescence on isolated chromosomes by bivariate flow karyotyping. Histochemistry 84:462–470.
21. **Carrano AV, Gray JW, Langlois RG, Burkhart-Schultz KJ, Van Dilla MA (1979)** Measurement and purification of human chromosomes by flow cytometry and sorting. Proc. Natl. Acad. Sci. USA 76: 1382–1384.
22. **Carrano AV, Gray JW, Moore DH II, Minkler JL, Mayall BH, Van Dilla MA, Mendelsohn ML (1976)** Purification of the chromosomes of the Indian muntjac by flow sorting. J. Histochem. Cytochem. 24: 348–54.
23. **Carrano AV, Lebo RV, Yu LC, Kan YW (1981)** Regional gene mapping of human chromosomes purified by flow sorting. In Neth R, Gallo R, Graf T, Mannweiler K, Winkler K (eds), "Modern Trends in Human Leukemia." Vol. IV. Berlin: Springer-Verlag, pp 156–159.
24. **Carrano AV, Van Dilla MA, Gray JW (1979)** Flow cytogenetics: A new approach to chromosome analysis. In Melamed MR, Mullaney PF, Mendelsohn ML (eds), "Flow Cytometry and Sorting." New York: John Wiley & Sons, pp 421–451.
25. **Collard JG, de Boar PAJ, Janssen JWG, Schijven JF, deJong B (1985)** Gene mapping by chromosome spot hybridization. Cytometry 6:179–185.
26. **Collard J, Phillipus E, Tulp A, Lebo R, Gray JW (1984)** Separation and analysis of human chromosomes by combined velocity sedimentation and flow sorting: Applying single and dual laser cytometry. Cytometry 5:9–19.
27. **Collard JG, Tulp A, Hollander JH, Bauer FW, Boezeman J (1980)** Separation of large quantities of chromosomes by velocity sedimentation at unit gravity. Exp. Cell. Res. 126:191–197.
28. **Collard JG, Tulp A, Stegeman J, Boezeman J, Bauer FW, Jongkind JF, Verkerk A (1980)** Separation of large quantities of Chinese hamster chromosomes by velocity sedimentation at unit gravity followed by flow sorting. (1980) Exp. Cell. Res. 130:217–228.
29. **Cram LS, Arndt-Jovin DJ, Grimwalde BG, Jovin TM (1979)** Fluorescence polarization and pulse width analysis of chromosomes by a flow system. J. Histochem. Cytochem. 27:445–453.
30. **Cram LS, Bartholdi MF, Ray FA, Habbersett MC, Kraemer PM (1988)** Univariate flow karyotype analysis. In Gray JW (ed), "Flow Cytogenetics." Orlando, Florida: Academic Press, pp. 113–135.
31. **Cram LS, Bartholdi MF, Ray FA, Travis GL, Kraemer PM (1983)** Spontaneous neoplastic evolution of Chinese hamster cells in culture: Multistep progression of karyotype. Cancer Res. 43:4828–4837.
32. **Cram LS, Bartholdi MF, Wheeless LL, Gray JW (1985)** Morphological analysis by scanning flow cytometry. In Van Dilla MA, Dean PN, Laerum OD, Melamed MR (eds), "Flow Cytometry Instrumentation and Data Analysis." Orlando, Florida: Academic Press, pp. 163–194.
33. **Cremer C, Gray J (1982)** Application of the BrdU/thymidine method to flow cytometry: Differential quenching/enhancement of the Hoechst 33258 fluorescence of late replicating chromosomes. J. Somat. Cell Genet. 8:319–327.
34. **Cremer C, Gray JW (1983)** Replication kinetics of Chinese hamster chromosomes as revealed by bivariate flow karyotyping. Cytometry 3:282–286.
35. **Cremer C, Gray JW, Ropers HH (1982)** Flow cyto-

metric characterization of a Chinese hamster × man hybrid cell line retaining the human Y chromosome. Hum. Genet. 60:262–266.

36. **Cremer C, Rappold G, Gray JW, Muller CR, Ropers HH (1984)** Preparative dual beam sorting of the human Y chromosome and in situ hybridization of cloned DNA probes. Cytometry 5:572–579.
37. **Davies KE, Young BD, Elles RG, Hill ME, Williamson R (1981)** Cloning of a representative genomic library of the human × chromosome after sorting by flow cytometry. Nature (Lond.) 293:374–376.
38. **Deaven LL, Cram LS, Wells RS, Kraemer PM (1981)** Relationships between chromosome complement and cellular DNA content in tumorigenic cell populations. In Arrighi FE, Rao PN, Stubblefield E (eds), "Genes, Chromosomes and Neoplasia." New York: Raven Press, pp 419–449.
39. **Deaven LL, Hildebrand CE, Fuscoe JC, Van Dilla MA (1986)** Construction of human chromosome specific DNA libraries: The National Laboratory Gene Library Project. In Setlow JK, Hollaender A (eds), "Genetic Engineering." Vol. 8. New York: Plenum, pp 317–332.
40. **Deaven LL, Van Dilla MA, Bartholdi MF, Carrano AV, Cram LS, Fuscoe JC, Gray JW, Hildebrand CE, Moyzis RK, Perlman J (1986)** Construction of human chromosome-specific DNA libraries from flow sorted chromosomes. Cold Spring Harbor Symp. Quant. Biol. 51:159–168.
41. **de Laat AMM, Blaas J (1984)** Flow-cytometric characterization and sorting of plant chromosomes. J. Theor. Appl. Genet. 67:463–467.
42. **Disteche CM, Carrano AV, Ashworth LK, Burkhart-Schultz K, Latt SA (1981)** Flow sorting of the mouse Cattanach × chromosome, T(X;7) 1 ct. in an active or inactive state. Cytogenet. Cell. Genet. 29: 189–197.
43. **Donlon TA, Lalande M, Wyman A, Gruns G, Latt SA (1986)** Isolation of molecular probes associated with the chromosome 15 instability in the Prader–Willi syndrome. Proc. Natl. Acad. Sci. USA 83: 4408–4412.
44. **Dudin G, Cremer T, Schardin M, Hausmann M, Bies F, Cremer C (1987)** A method for nucleic acid hybridization to isolate chromosomes in suspension. Hum. Genet. 76:290–292.
45. **Fantes JA, Green DK, Elder JK, Malloy P, Evans HJ (1983)** Detecting radiation damage to human chromosomes by flow cytometry. Mutat. Res. 119:161–168.
46. **Fojo A, Lebo R, Shimizu N, Chin JE, Roninson IB, Merlino GT, Gottesman MM, Pastan I (1986)** Localization of multidrug resistance-associated DNA sequences to human chromosome 7. Somat. Cell Mol. Genet. 12:415–420.
47. **Fried J, Perez AG, Clarkson BD (1976)** Flow cytofluorometric analysis of cell cycle distributions using propidium iodide. J. Cell. Biol. 71:172–181.
48. **Fuscoe JC, Clark LM, Van Dilla MA (1986)** Construction of fifteen human chromosome-specific DNA libraries from flow-purified chromosomes. Cytogenet. Cell Genet. 43:79–86.
49. **Fuscoe J, Collins C, Pinkel D, Gray JW (1989)** An efficient method for selecting unique-sequence clones from DNA libraries and its application to selecting fluorescent staining of human chromosome 21 using in situ hybridization. Genomics 5:100–109.
50. **Fuscoe J, Van Dilla MA, Deaven LL (1986)** Construction and availability of human chromosome-specific gene libraries. In Ramel C, Lambert B, Agnusson J (eds), "Genetic Toxicology of Environmental Chemicals. Part A: Basic Principles and Mechanism of Action." New York: Alan R. Liss, pp 465–472.
51. **Gartler SM, Riley DE, Lebo RV, Cheung M-C, Eddy RL, Shows TB (1986)** Mapping of human autosomal phosphoglycerate kinase sequence to chromosome 19. Somat. Cell Mol. Genet. 12:395–401.
52. **Gray JW, Carrano AV, Langlois R, Lucas J, Yu LC, Van Dilla MA (1982)** Flow cytogenetics: Chromosome classification and purification by flow cytometry and sorting. In "Galjaard: The Future of Prenatal Diagnosis." Edinburgh: Churchill Livingstone, pp 33–40.
53. **Gray JW, Carrano AV, Moore DH II, Steinmetz LL, Minkler J, Mayall BH, Mendelsohn ML, Van Dilla MA (1975)**. High speed quantitative karyotyping by flow microfluorometry. Clin. Chem. 21:1258–1262.
54. **Gray JW, Carrano AV, Steinmetz LL, Van Dilla MA, Moore DH II, Mayall BH, Mendelsohn ML (1975)** Chromosome measurement and sorting by flow systems. Proc. Natl. Acad. Sci. USA 72:1231–1234.
55. **Gray JW, Dean PN, Fuscoe JC, Peters DC, Trask BJ, van den Engh GJ, Van Dilla MA (1987)** High-speed chromosome sorting. Science 238:323–329.
56. **Gray JW, Langlois RG (1986)** Chromosome classification and purification using flow cytometry and sorting. Annu. Rev. Biophys. Chem. 15:195–235.
57. **Gray JW, Langlois RG, Carrano AV, Burkhart-Schultz K, Van Dilla MA (1979)** High resolution chromosome analysis: One and two parameter flow cytometry. Chromosoma 73:9–27.
58. **Gray JW, Lucas J, Peters D, Pinkel D, Trask B, van den Engh G, Van Dilla M (1986)** Flow karyotyping and sorting of human chromosomes. Proc. Cold Spring Harbor Symp. Quant. Biol. 51:141–150.
59. **Gray JW, Lucas J, Pinkel D, Peters D, Ashworth L, Van Dilla MA (1980)** Slit-scan flow cytometry: Analysis of Chinese hamster M3-1 chromosomes. In Laerum OD, Lindmo T, Thorud E (eds), "Flow Cytometry. IV. Proceedings of the Fourth International Symposium on Flow Cytometry." Oslo: Universitetsforlaget, pp 485–491.
60. **Gray JW, Lucas J, Yu L-C, Langlois R (1984)** Flow cytometric detection of aberrant chromosomes. In Eisert WG, Mendelsohn ML (eds), "Biological Dosimetry: Cytometric Approaches to Mammalian Systems." Heidelberg: Springer-Verlag, pp 25–35.
61. **Gray JW, Peters D, Merrill JT, Martin R, Van Dilla MA (1979)** Slit-scan flow cytometry of mammalian chromosomes. Cytochemistry 27:441–444.
62. **Gray JW, Pinkel D, Trask B, van den Engh G, Pallavicini M, Fuscoe J (1988)** Analytical cytology applied to detection of prognostically important cytogenetic aberrations. In "Current Status of Future Directions in Proceedings of Prediction of Tumor Treatment Response, Banff, 1987."
63. **Gray JW, Trask B, van den Engh G, Silva A, Lozes C, Grell S, Schoenberg S, Yu L-C, Golbus M (1988)**

Application of flow karyotyping in prenatal detection of chromosome aberrations. Am. J. Hum. Genet. 42: 49–59.

64. **Fantes J, Green DK, Malloy P, Sumner AT (1989)** Flow cytometry measurements of human chromosome kinetochore labeling. Cytometry 10:134–142.
65. **Green DK, Fantes JA, Buckton KE, Elder JK, Malloy P, Carothers A, Evans HJ (1984)** Karyotyping and identification of human chromosomes polymorphisms by single fluorochrome flow cytometry. Hum. Genet. 66:143–146.
66. **Griffith JK, Cram LS, Crawford BD, Jackson PJ, Schilling J, Schimke RT, Walters RA, Wilder ME, Jett JH (1984)** Construction and analysis of DNA sequence libraries from flow-sorted chromosomes: Practical and theoretical considerations. Nucleic Acids Res. 12:4019–4034.
67. **Grunwald D, Geffrotin C, Chardon P, Frelat G, Vaiman M (1986)** Swine chromosomes: Flow sorting and spot blot hybridization. Cytometry 7:582–588.
68. **Harris P, Boyd E, Ferguson-Smith MA (1985)** Optimizing human chromsome separation for the production of chromosome-specific DNA libraries by flow sorting. Hum. Genet. 70:59–65.
69. **Harris P, Boyd E, Young BD, Ferguson-Smith MA (1986)** Determination of the DNA content of human chromosomes by flow cytometry. Cytogenet. Cell. Genet. 41:14–21.
70. **Harris P, Morton CC, Guglielmi P, Li F, Kelly K, Latt SA (1986)** Mapping by chromosome sorting of several gene probes, including c-myc, to the derivative chromosomes of a 3;8 translocation associated with familial renal cancer. Cytometry 7:589–594.
71. **Harris A, Young BD, Griffin BE (1985)** Random association of Epstein–Barr virus genomes with host cell metaphase chromosomes in Burkitt's lymphoma-derived cell lines. J. Virol. 56:328–332.
72. **Hauser-Urfer I, Leemann U, Ruch F (1982)** Cytofluoremetric determination of the DNA base content in human chromosomes with quinicrine mustard, Hoechst 33258, DAPI, and mithramycin. Exp. Cell. Res. 142:455–459.
73. **Hilwig I, Gropp A (1975)** pH-Dependent fluorescence of DNA and RNA in cytologic staining with "33258 Hoechst." Exp. Cell. Res. 91:457–460.
74. **Janssen JWG, Vernole P, Boer PAJ, Oosterhuis JW, Collard JG (1986)** Sublocalization of c-myb to 6q21-q23 by in situ hybridization and c-myb expression in a human teratocarcinoma with 6q rearrangements. Cell. Genet. 41:129–135.
75. **Jensen RH, Langlois RG, Mayall BH (1977)** Strategies for choosing a deoxyribonucleic acid stain for flow cytometry of metaphase chromosomes. J. Histochem. Cytochem. 25:954–964.
76. **Johnston R, Bartholdi M, Hiebert R, Parson J, Cram LS (1985)** A slit-scan flow cytometer for recording simultaneous waveforms. Rev. Sci. Ins. 56:691–695.
77. **Kamiyama M (1968)** Mechanisim of action of chromomycin A3. J. Biochem. 63:566–572.
78. **Kanda N, Schreck R, Alt F, Bruns G, Baltimore D, Latt S (1983)** Isolation of amplified DNA sequences from IMR-32 human neuroblastoma cells: Facilitation by fluorescence-activated flow sorting of metaphase chromosomes. Proc. Natl. Acad. Sci. USA 80:4069–4073.
79. **Kapuscinski J, Skoczylas B (1977)** Simple and rapid fluorimetric method for DNA micro assay. Anal. Biochem. 83:252–257.
80. **Kooi MW, Aten JA, Stap J, Barendsen GW (1984)** Preparation of chromosome suspensions from cells of solid tumors for measurement by flow cytometry. Cytometry 5:547–549.
81. **Kraemer PM, Deaven LL, Crissman HA, Van Dilla MA (1972)** DNA constancy despite variability in chromosome number. In DuPraw EJ (ed), "Advances in Cell and Molecular Biology." vol. 2. New York: Academic Press, pp 47–108.
82. **Kraemer PM, Ray FA, Bartholdi MF, Cram LS (1987)** Spontaneous in vitro neoplastic evolution: Selection of specific karyotype in Chinese hamster cells. Cancer Genet. Cytogenet. 27:273–287.
83. **Krumlauf R, Jeanpierre M, Young BD (1981)** Construction and characterization of genomic libraries from specific human chromosomes. Proc. Natl. Acad. Sci. USA 79:2971–2975.
84. **Krumlauf R, Jeanpierre M, Young BD (1982)** Isolation of mouse X-chromosome specific DNA from an X-enriched lambda phage library derived from flow sorted chromosomes. Cytometry 2:282–286.
85. **Kunkel LM, Lalande M, Monaco AP, Flint A, Middlesworth W, Latt SA (1985)** Construction of a human X-chromosome-enriched phage library which facilitates analysis of specific loci. Gene 33:251–258.
86. **Kunkel LM, Tantravahi U, Eisenhard M, Latt SA (1982)** Regional localization on the human X of DNA segments cloned from flow sorted chromosomes. Nucleic Acids Res. 10:1557–1578.
87. **Kunkel LM, Tantravahi U, Eisenhard M, Latt SA (1982)** Regional localization of the human X of DNA segments cloned from flow sorted chromosomes. Nucleic Acids Res. 10:1557–1578.
88. **Labarca C, Paigen K (1980)** A simple, rapid and sensitive DNA assay procedure. Anal. Biochem. 102: 344–352.
89. **Lalande M, Donlon T, Petersen RA, Liberfarb R, Manter S, Latt SA (1986)** Molecular detection and differentiation of deletions in band 13q14 in human retinoblastoma. Cancer Genet. Cytogenet. 23:151–157.
90. **Lalande M, Dryja TP, Schreck RR, Shipley J, Flint A, Latt SA (1984)** Isolation of human chromosome 13-specific DNA sequences cloned from flow sorted chromosomes and potentially linked to the retinoblastoma locus. Cancer Genet. Cytogenet. 13:283–295.
91. **Lalande M, Kunkel LM, Flint A, Latt SA (1984)** Development and use of metaphase chromosome flow-sorting methodology to obtain recombinant phase libraries enriched for parts of the human X chromosome. Cytometry 5:101–107.
92. **Lalande M, Schreck RR, Hoffman R, Latt SA (1985)** Identification of inverted duplicated #15 chromosomes using bivariate flow cytometric analysis. Cytometry 6:1–6.
93. **Landegent J, Jansen in de Wal J, Baan R, Hoeijmakers J, van der Ploeg M (1984)** 2-Acetylaminofluorene-modified probes for the indirect hybridocytochemical detection of specific nucleic acid sequences. Exp. Cell. Res. 153:61–72.
94. **Langer G, Blin N, Stoehr M (1984)** Chromosomes

for molecular hybridization: Assignment of repetitive and single copy genes using a rapid filter-fixation method. Histochemistry 80:469–473.

95. **Langer-Safer P, Levine M, Ward D** (1982) Immunological methods for mapping genes on *Drosophila* polytene chromosomes. Proc. Natl. Acad. Sci. USA 79:4381–4385.

96. **Langlois RG** (1988) DNA stains as cytochemical probes for chromosomes. In Gray JW (ed), "Flow Cytogenetics." Orlando, Florida: Academic Press, pp 61–81.

97. **Langlois RG, Carrano AV, Gray JW, Van Dilla MA** (1980) Cytochemical studies of metaphase chromosomes by flow cytometry. Chromosoma 77:229–252.

98. **Langlois RG, Yu LC, Gray JW, Carrano AV** (1982) Quantitative karyotyping of human chromosomes by dual beam flow cytometry. Proc. Natl. Acad. Sci. USA 79:7876–7880.

99. **Latt SA** (1973) Microfluorometric detection of DNA synthesis of human chromosomes. Proc. Natl. Acad. Sci. USA 70:3395–3399.

100. **Latt SA** (1977) Fluorescent probes of chromosome structure and replication. Can. J. Genet. Cytol. 19: 603–623.

101. **Latt SA, Alt FA, Schreck RP, Kanada N, Baltimore D** (1982) The use of chromosome flow sorting and cloning to study amplified DNA sequences. In Schimke RT (ed), "Gene Amplification." Cold Spring Harbor, New York: Cold Spring Harbor Laboratory, pp 283–290.

102. **Latt SA, Sahar E, Eisenhard ME** (1979) Pairs of fluorescent dyes as probes of DNA and chromosomes. J. Histochem. Cytochem. 27:65–71.

103. **Latt SA, Sahar E, Eisenhard ME, Juergens LA** (1980) Interactions between pairs of DNA-binding dyes: Results and implications of chromosome analysis. Cytometry 1:2–12.

104. **Latt S, Wohleb JC** (1975) Optical studies of the interaction of 33258 Hoechst with DNA, chromatin, and metaphase chromosomes. Chromosoma 52:297–316.

105. **Lebo RV** (1982) Chromosome sorting and DNA sequence localization. Cytometry 3:145–154.

106. **Lebo R, Anderson L, Lau Y-F, Flandermeyer R, Kan YW** (1986) Flow sorting analysis of normal and abnormal human genomes. Cold Spring Harbor Symp. Quant. Biol. 51:169–176.

107. **Lebo RV, Carrano AV, Burkhart-Schultz K, Dozy AM, Yu L, Kan YW** (1979) Assignment of human β-, γ-, and δ-globin genes to the short arm of chromosome 11 by chromosome sorting and DNA restriction enzyme analysis. Proc. Natl. Acad. Sci. USA 76: 5804–5808.

108. **Lebo RB, Gorin F, Fletterick RJ, Kao FT, Cheung MC, Bruce BD, Kan YW** (1984) High-resolution chromosome sorting and DNA spot-blot analysis assign McArdle's syndrome to chromosome 11. Science 225:57–59.

109. **Lebo RV, Kan YW, Cheung MC, Carrano AV, Yu LC** (1982) Assigning the polymorphic human insulin gene to the short arm of chromosome 11 by chromosome sorting. Hum. Genet. 60:10–15.

110. **Lebo R, Tolan DR, Bruce BD, Cheung MC, Kan YW** (1985) Spot-blot analysis of sorted chromosomes assigns on fructose intolerance disease locus to chromosome 9. Cytometry 6:478–483.

111. **Le Pecq JB, Paoletti C** (1967) A fluorescent complex between ethidium bromide and nucleic acids, physical–chemical characterization. J. Mol. Biol. 27:87–106.

112. **Liao Y-CJ, Lebo RV, Clawson GA, Smuckler EA** (1986) Human prion protein cDNA: Molecular cloning, chromosomal mapping, and biological implications. Science 233:364–367.

113. **Lin MS, Comings DE, Alfi OS** (1977) Optical studies of the interaction of 4′-6-diamidino-Z-phenylindole with DNA and metaphase chromosomes. Chromosoma 60:15–25.

114. **Lucas JN, Gray JW** (1987) Centromeric index versus DNA content flow karyotypes of human chromosomes measured using slit-scan flow cytometry. Cytometry 8:273–279.

115. **Lucas JN, Gray JW, Peters DC, Van Dilla MA** (1983) Centromeric index measurement by slit-scan flow cytometry. Cytometry 4:109–116.

116. **Lucas JN, Peters D, Van Dilla MA, Gray JW** (1981) The application of chromosome stretching to slit-scan flow cytometry. Cytometry 2:113.

117. **Martin RF, Holmes N** (1983) Use of an ^{125}I-labeled DNA ligand to probe DNA structure. Nature (Lond.) 302:452–454.

118. **Mathieu-Mahul D, Caubet JF, Bernheim A, Mauchauffe M, Palmer E, Berger R, Larsen C-J** (1985) Molecular cloning of a DNA fragment from human chromosome 14 (14q11) involved in T-cell malignancies. EMBO J. 4:2427–2433.

119. **Matsson P, Rydberg B** (1981) Analysis of chromosomes from human peripheral lymphocytes by flow cytometry. Cytometry 1:369–372.

120. **McDermid HE, Duncan AMV, Higgins MJ, Hamerton JL, Rector E, Brasch KR, White BN** (1986) Isolation and characterization of an α-satellite repeated sequence from human chromosome 22. Chromosoma 94:228–234.

121. **Meyne J, Bartholdi MF, Travis G, Cram LS** (1984) Counterstaining human chromosomes for flow karyology. Cytometry 5:580–583.

122. **Moore DH II** (1979) A template method for decomposing flow cytometry histograms of human chromosomes. J. Histochem. Cytochem. 27:305–310.

123. **Moore D II** (1988) Mathematical analysis of flow karyotypes. In Gray JW (ed), "Flow Cytogenetics." Orlando, Florida: Academic Press, pp 83–111.

124. **Moyzis RK, Albright KL, Bartholdi MF, Cram LS, Deaven LL, Hildebrand CE, Joste NE, Longmire JL, Meyne J, Schwarzacher-Robinson T** (1987) Human chromosome-specific repetitive DNA sequence: Novel markers for genetic analysis. Chromosoma 95: 375–386.

125. **Mueller CR, Davies K, Cremer C, Rappold G, Gray JW, Ropers H** (1983) Cloning of genomic sequences from the human Y chromosome by combined velocity sedimentation and flow sorting: Applying single and dual laser cytometry. Hum. Genet. 64:110–115.

126. **Muller U, Donlon TA, Harris P, Rose E, Hoffman E, Bruns GP, Latt SA** (1987) Highly polymorphic DNA sequences in the distal region of the long arm of human chromosome 18. Cytogenet. Cell Genet. 45:16–20.

127. **Mullikin J, Norgren R, Lucas J, Gray JW (1988)** Fringe-scan flow cytometry. Cytometry 9:111–120.
128. **Murray SS, Deaven LL, Burton EW, O'Connor DT, Mellon PL, Deftos LJ (1987)** The gene for human chromogranin A (CgA) is located on chromosome 14. Biochem. Biophys. Res. Commun. 142:141–146.
129. **Ng S-Y, Gunning P, Eddy R, Poute P, Leavitt J, Shows T, Keds L (1985)** Evolution of a functional human β-actin gene and its multipseudogene family: Construction of noncoding regions and chromosomal dispersion of pseudogenes. Cell. Biol. 5:2720–2732.
130. **Norgren RM, Gray JW, Young IT (1982)** Restoration of profiles from slit-scan flow cytometry. IEEE Trans. Biomed Eng. 29:101–106.
131. **Olmsted J III, Kearns DR (1977)** Mechanism of ethidium bromide fluorescence enhancement on binding to nucleic acids. Biochemistry 16:3647–3654.
132. **Otto FJ, Oldiges H (1980)** Flow cytogenetic studies in chromosomes and whole cells for the detection of clastogenic effects. Cytometry 1:13–17.
133. **Otto FJ, Oldiges H, Gohde W, Barlogie B, Schumann J (1980)** Flow cytogenetics of uncloned and cloned Chinese hamster cells. Cytogenet. Cell Genet. 27:52–56.
134. **Otto F, Oldiges H, Gohde W, Dertinger H (1980)** Flow cytometric analysis of mutagen induced chromosomal damage. In Laerum OD, Lindmo T, Thorud E (eds), "Flow cytometry. IV. Proceedings of the Fourth International Symposium on Flow Cytometry." Oslo: Universitetsforlaget, pp 485–491.
135. **Otto F, Tsou KC (1985)** A comparative study of DAPI, DIPI, and Hoechst 33258 and 33342 as chromosomal DNA stains. Stain Technol. 60:7–11.
136. **Paoletti J, Le Pecq JB (1971)** Resonance energy transfer between ethidium bromide molecules bound to nucleic acids. Does intercalation wind or unwind the DNA helix? J. Mol. Biol. 59:43–62.
137. **Parker RC, Mardon G, Lebo RV, Varmus HE, Bishop JM (1985)** Isolation of duplicated human c-src genes located on chromosomes 1 and 10. Mol. Cell. Biol. 5:831–838.
138. **Peters D, Branscomb E, Dean P, Merrill T, Pinkel D, Van Dilla M, Gray JW (1985)** The LLNL high-speed sorter: Design features, operational characteristics, and biological utility. Cytometry 6:290–301.
139. **Pinkel D, Straume T, Gray JW (1986)** Cytogenetic analysis using quantitative high-sensitivity fluorescence hybridization. Proc. Natl. Acad. Sci. USA 83: 2934–2938.
140. **Pjura P, Grzeskowiak K, Dickerson R (1987)** Binding of Hoechst 33258 to the Minor groove of β-DNA. J. Mol. Biol. 197:257–271.
141. **Rappold GA, Cremer T, Hager H, Davies KE, Muller CR, Yang T (1984)** Sex chromosome positions in human interphase nuclei as studied by in situ hybridization with chromosome specific probes. Hum. Genet. 67:317–325.
142. **Ray FA, Bartholdi MF, Kraemer PM, Cram LS (1984)** Chromosome polymorphism involving discrete heterochromatic blocks in Chinese hamster number nine chromosome. Cytogenet. Cell Genet. 38:257–264.
143. **Ray FA, Bartholdi MF, Kraemer PM, Cram LS (1986)** Spontaneous in vitro neoplastic evolution: Recurrent chromosome changes of newly immortalized Chinese hamster cells. Can. Gen. Cytogenet. 21:35–51.
144. **Sahar E, Latt SA (1980)** Energy transfer and binding competition between dyes used to enhance staining differentiation in metaphase chromosomes. Chromosoma (Berl.) 79:1–28.
145. **Severin E, Ohnemus G (1982)** Flow cytometric analysis of chromosomes and cells using a modified BrdU–Hoechst method. Histochemistry 76:113–122.
146. **Shay JW, Cram LS (1985)** Cell fusion and chromosome sorting. In Goettsman MM (ed), "Molecular Cell Genetics. The Chinese Hamster Cell." New York: John Wiley & Sons, pp 155–180.
147. **Sillar R, Young BD (1981)** A new method for the preparation of metaphase chromosomes for flow analysis. J. Histochem. Cytochem. 29:74–78.
148. **Stewart GD, Harris P, Galt J, Ferguson-Smith MA (1985)** Cloned DNA probes regionally mapped to human chromosome 21 and their use in determining the origin of nondisjunction. Nucleic Acids Res. 13: 4125–5132.
149. **Stohr M, Hutter K, Frank M, Futterman G (1980)** A flow cytometric study of chromosomes from rat kangaroo and Chinese hamster cells. Histochemistry 67: 179–190.
150. **Stohr M, Hutter KJ, Frank M, Goerttler K (1982)** A reliable preparation of mono-dispersed chromosome suspensions for flow cytometry. Histochemistry 74: 57–61.
151. **Stubblefield E, Cram S, Deaven L (1975)** Flow microfluorometric analysis of isolated Chinese hamster chromosomes. Exp. Cell Res. 94:464–468.
152. **Stubblefield E, Ors J (1982)** Isolation of specific chicken macrochromosomes by zonal centrifugation and flow sorting. Cytometry 2:273–281.
153. **Stubblefield E, Wray W (1978)** Isolation of specific human metaphase chromosomes. Biochem. Biophys. Res. Commun. 83:1404–1414.
154. **Tchen P, Fuchs R, Sage E, Leng M (1984)** Chemically modified nucleic acids as immunodetectable probes in hybridization experiments. Proc. Natl. Acad. Sci. USA 81:3466–3470.
155. **Trask B, van den Engh G, Gray JW, Vanderlaan M, Turner B (1984)** Immunofluorescent detection of histone 2B on metaphase chromosomes using flow cytometry. Chromosoma 90:295–302.
156. **Trask B, van den Engh G, Landegent J, Jansen in de Wal N, van der Ploeg M (1985)** Detection of DNA sequences in nuclei in suspension by in situ hybridization and dual beam flow cytometry. Science 230: 1401–1403.
157. **Trask B, van den Engh G, Pinkel D, Mullikin J, van Dekken H, Gray JW (1988)** Fluorescence in situ hybridization to interphase cell nuclei in suspension allows flow cytometric analysis of chromosome content and microscopic analysis of nuclear organization. Hum. Genet. 78:251–259.
158. **Turner B, Keohane A (1987)** Antibody labeling and flow cytometric analysis of metaphase chromosomes reveals two discrete structural forms. Chromosoma 95:263–270.
159. **van den Engh G, Trask B, Cram S, Bartholdi M**

(1984) Preparation of chromosome suspensions for flow cytometry. Cytometry 5:105–117.

160. **van den Engh GJ, Trask BJ, Gray JW** (1986) The binding kinetics and interaction of DNA Fluorochromes used in the analysis of nuclei and chromosomes by flow cytometry. Histochemistry 84:501–508.

161. **van den Engh GJ, Trask GJ, Gray JW, Langlois RG, Yu L-C** (1985) Preparation and bivariate analysis of suspensions of human chromosomes. Cytometry 6: 91–100.

162. **van den Engh G, Trask B, Landsdorp P, Gray JW** (1988) Improved resolution of flow cytometric measurements of Hoechst-chromomycin-stained human chromosomes after addition of citrate and sulfite. Cytometry 9:266–270.

163. **Van Dilla MA, Deaven LL, Albright KL, Allen NA, Aubuchon MR, Bartholdi MF, Browne NC, Campbell EW, Carrano AV, Clark LM, Cram LS, Crawford BD, Fuscoe JC, Gray JW, Hildebrand CE, Jackson PJ, Jett JH, Longmire JL, Lozes CR, Leudemann ML, Martin JC, McNinch JS, Meincke LJ, Mendelsohn ML, Meyne J, Moyzis RK, Munk AC, Perlman J, Peters DC, Silva AJ, Trask BJ** (1986) Human chromosome-specific DNA libraries: Construction and availability. Biotechnology 4:537–552.

164. **Van Dyke MW, Dervan PB** (1983) Chromomycin, mithramycin, and olivomycin binding sites on heterogeneous deoxyribonucleic acid footprinting with (methidiumpropyl-EDTA iron II). Biochemistry 22: 2373–2377.

165. **Welleweerd J, Wilder ME, Carpenter SG, Raju MR** (1984) Flow cytometric determination of radiation-induced chromosome damage and its correlation with cell survival. Radiat. Res. 99:44–51.

166. **Wen D, Dittman WA, Ye RD, Deaven LL, Majerus PW, Sadler JE** (1987) Human thrombomodulin: Complete cDNA sequence and chromosome localization of the gene. Biochemistry 26:4350–4357.

167. **Wirschubsky Z, Ingvarsson S, Carstenssen A, Wiener F, Klein G, Sumegi J** (1985) Gene localization on sorted chromosomes: Definitive evidence on the relative positioning of genes participating in the mouse–plasmacytoma-associated typical translocation. Proc. Natl. Acad. Sci. USA 82:6975–6979.

168. **Wray W, Stubblefield E** (1970) A new method for the rapid isolation of chromosomes, mitotic apparatus, or nuclei for mammalian fibroblasts at near neutral pH. Exp. Cell Res. 59:469–478.

169. **Wray W, Wray VP** (1980) Proteins from metaphase chromosomes treated with fluorochromes. Cytometry 1:18–20.

170. **Young BD, Ferguson-Smith, MA, Sillar R, Boyde E** (1981) High resolution analysis of human peripheral lymphocyte chromosomes by flow cytometry. Proc. Natl. Acad. Sci. USA 78:7727–7731.

171. **Young BD, Jeanpierre M, Hoyns MH, Krumlauf R** (1983) Construction and characterization of chromosomal DNA libraries. Hematol. Blood Trans. 28: 301–310.

172. **Yu L-C, Aten J, Gray J, Carrano AV** (1981) Human chromosome isolation from short-term lymphocyte culture for flow cytometry. Nature (Lond.) 293:154–155.

173. **Yu L-C, Gray JW, Langlois R, Van Dilla MA, Carrano AV** (1984) Human chromosome karyotyping and molecular biology by flow cytometry. In Sparks RS, de la Cruz FF (eds), "Research Perspectives in Cytogenetics." Baltimore: University Park Press, pp 63–74.

26

Flow Cytometry and Sorting of Sperm and Male Germ Cells

Barton L. Gledhill, Donald P. Evenson, and Daniel Pinkel

Biomedical Sciences Division, Lawrence Livermore National Laboratory, Livermore, California 94550 (B.L.G.; D.P.); Chemistry Department, Animal Science Complex, South Dakota State University, Brookings, South Dakota 57007 (D.P.E.)

The potential for reproductive ill effects from environmental, occupational, accidental, and therapeutic exposure to noxious agents is a major, growing public concern. Similarly, there is increased interest in clinical reproductive matters, such as fertility assessment and genetic counseling, and in veterinary reproductive techniques, such as fertility prediction and preselection of gender of offspring. These different concerns have stimulated a substantial portion of the flow cytometric studies of spermatogenesis.

Spermatogenesis is a sensitive process with an enormous capacity for amplifying physiological or pharmacological perturbations. Isolation of the many types of testicular cells by flow cytometry offers unlimited potential for understanding this process. Multiparameter analysis and powerful staining methods have been developed that allow study of the various phases of the complex process of sperm development in both normal and perturbed circumstances. The dramatic changes in germ-cell structure and biochemistry during spermatogenesis offer numerous additional possibilities to measure features related to cell differentiation (e.g., cytoskeletal proteins, internal pH, and receptors).

In this chapter, we describe the different flow cytometric techniques available to study spermatogenesis. We discuss the progress in quantifying sperm morphology and physiology, and the alterations of sperm chromatin following in vivo exposure to mutagens and carcinogens. In addition, we review the use of flow cytometry (FCM) to achieve significantly accurate DNA measurements to resolve the two peaks corresponding to the slight difference in DNA content of X- and Y-chromosome-bearing sperm populations, a resolution that was not possible as recently as 1979 [23]. Finally, we summarize recent developments in understanding semen quality and fertility, testicular neoplasia, and gender preselection.

DISCRIMINATION OF CELL POPULATIONS

Germ Cells of the Testes

Mammalian spermatogenesis is a continuum of dramatic biochemical events and morphological alterations leading from diploid spermatogonia to mature haploid sperm. Spermatogenesis (Fig. 1) is generally divided into three major phases: (1) spermatogonial proliferation and renewal, (2) mitosis and meiosis, and (3) spermiogenesis [38]. The sequential progression of these phases during puberty has been studied by FCM of testicular biopsy material in mice [31] and rats [6,7,36]. The many parameters of proliferation and differentiation in adult male germ cells analyzed by FCM, either directly or indirectly, include DNA content [23,37, 44,48], ploidy and chromatin condensation [41,42], size [13,54,59], shape [3,13,30,48,59], RNA content [19], exchange of histones for sperm specific transition proteins [13], sperm-associated immunoglobulins [26,27], and level of disulfide bonding between protamine-SH groups (D.P. Evenson, unpublished data).

Given the many dramatic biochemical changes that occur during spermatogenesis, interpreting the intensity of staining for any particular cellular constituent requires great care, especially when comparing between cell types. Certain stains provide information both on the structure of the cell as well as the amount of target for the stain. These properties are most pronounced in mature sperm [24] but are present in all cell types. The fact that one stain provides two different types of information may have caused problems in the initial measurements on testis cells [41,42]. On a basis of cellular DNA content, one expects to see spermatogonia and spermatocytes cycling between 2C and 4C equivalents of DNA content and all spermatids and sperm stages at 1C. In fact, seven distinct peaks are distinguished and sorted, corresponding to diploid and tetraploid cells, round and elongated spermatids, and sperm. (This finding agrees with data previously obtained with quantitative fluorescence microscopy measurements of DNA stain uptake by elongating and elongated spermatids and sperm [22].)

More complex measurement schemes permit a more precise discrimination of cell type. Dual parameter measurements of DNA versus RNA [10] or peak of the DNA fluorescence pulse versus area of the pulse (total emitted

Flow Cytometry and Sorting, Second Edition, pages 531–551
Published 1990 by Wiley-Liss, Inc.

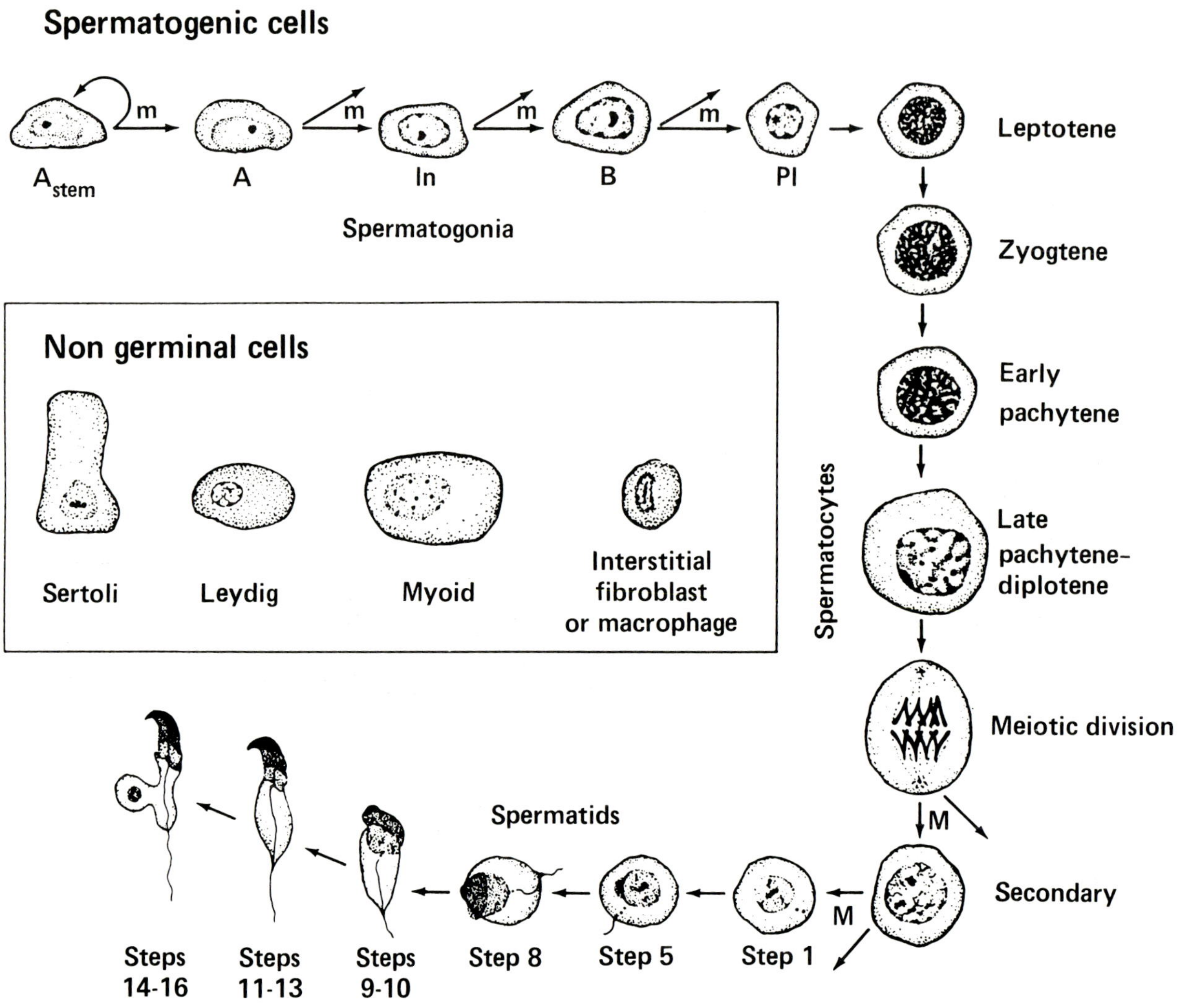

Spermatogenic cells

m m m m

A_{stem} A In B Pl

Spermatogonia

Leptotene

Zyogtene

Early pachytene

Late pachytene–diplotene

Meiotic division

M

Secondary

Spermatocytes

Non germinal cells

Sertoli Leydig Myoid Interstitial fibroblast or macrophage

Spermatids

M

Steps 14-16 Steps 11-13 Steps 9-10 Step 8 Step 5 Step 1

Fig. 1. Cells of mouse testis. Mitotic and meiotic divisions are indicated by m and M, respectively. In, Intermediate; Pl, Preleptotene. (Redrawn from Meistrich [38].)

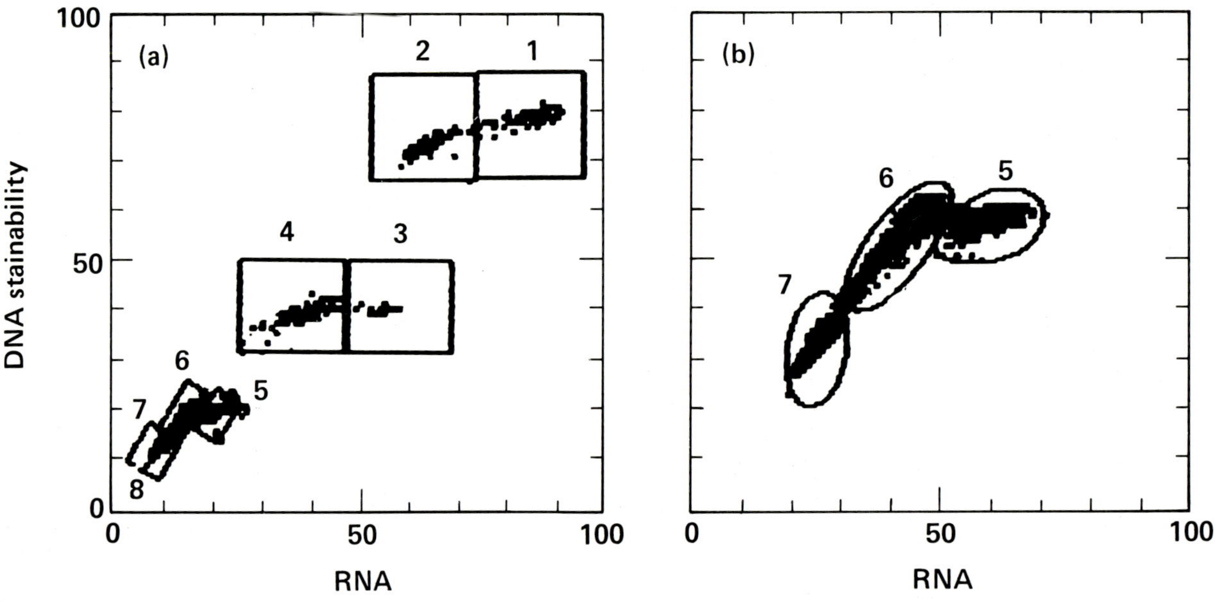

Fig. 2. Cytograms of red fluorescence (RNA) versus green fluorescence (DNA) signals of AO stained adult mouse testicular cells. **a:** Distribution of 4n (boxes 1 and 2), 2n (boxes 3 and 4), and 1n (boxes 5,6,7) cells. **b:** Computer enhancement of 1n populations showing increased resolution between cells in boxes 5–7. (Reproduced from Evenson et al. [12], with permission of the publisher.)

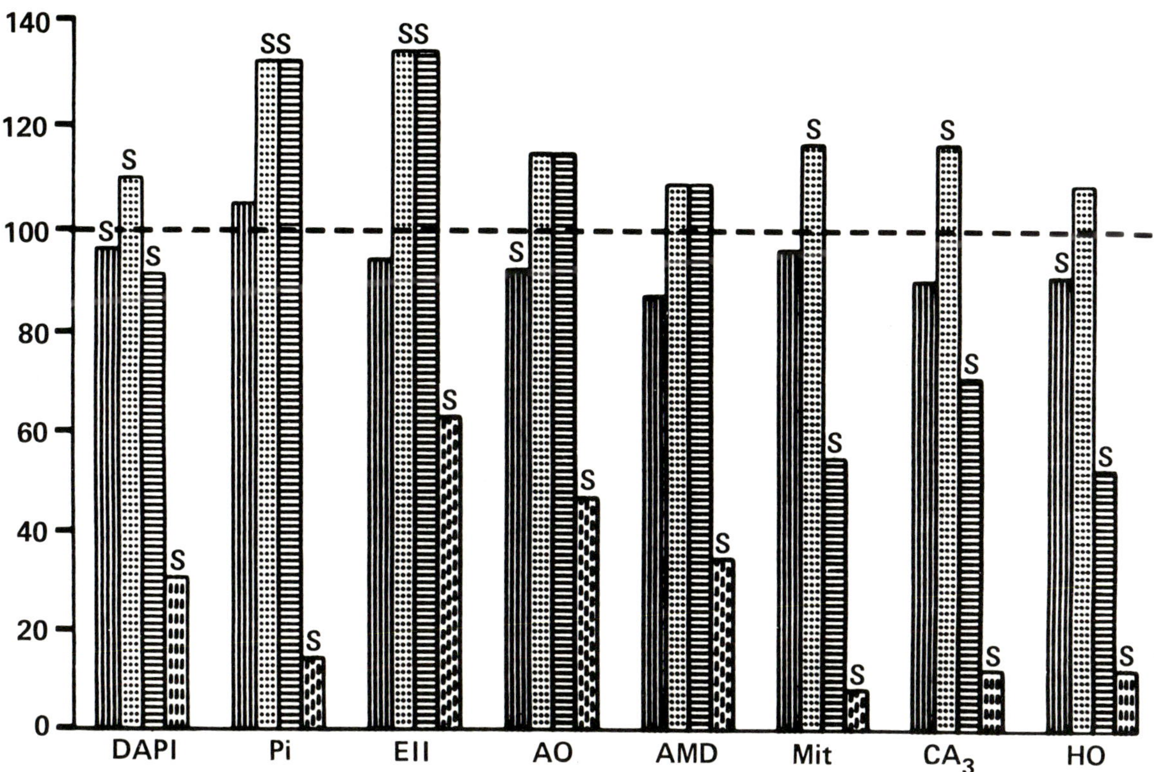

Fig. 3. Relative DNA stainability of fresh, Triton X-permealized, non-HCl-treated (one-step procedure) mouse testicular cells and vas deferens sperm stained with eight different DNA dyes (see text): DAPI, PI, EII, AO, AMD, Mit, CA3, and HO. Peak staining values of all cell types are normalized to unit content of DNA with diploid cell values set to 100%. Vertical lines = tetraploid cells; dots = round spermatids; horizontal lines = elongated spermatids; dashed vertical lines = vas sperm. Each bar represents the mean peak staining values; 'S' at the top of bars indicates statistical significance as defined by being 2 SD away from the diploid value. (Reproduced from Evenson et al. [13], with permission of the publisher.)

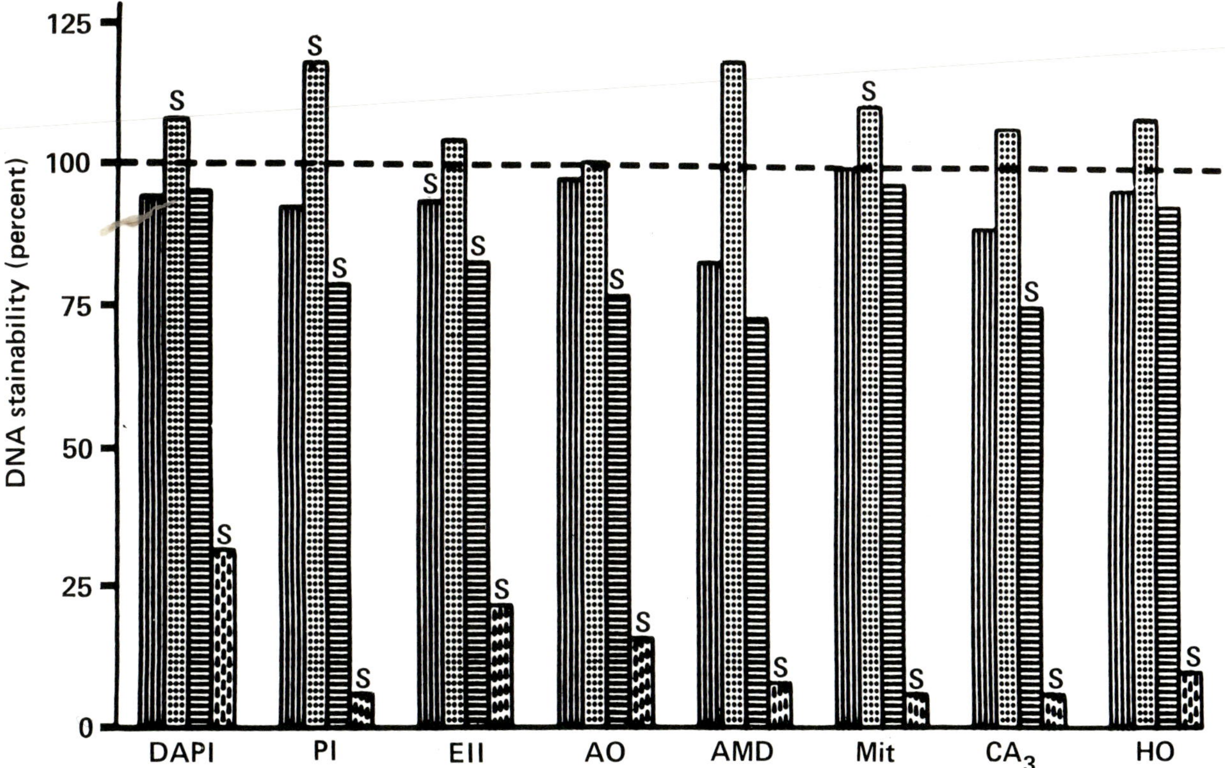

Fig. 4. Relative DNA stainability of fresh, Triton X-permealized and HCl-treated (two-step procedure) mouse testicular cells and vas deferens sperm stained with eight different dyes. The key and abbreviations are the same as in Figure 3. (Reproduced from Evenson et al. [17], with permission of the publisher).

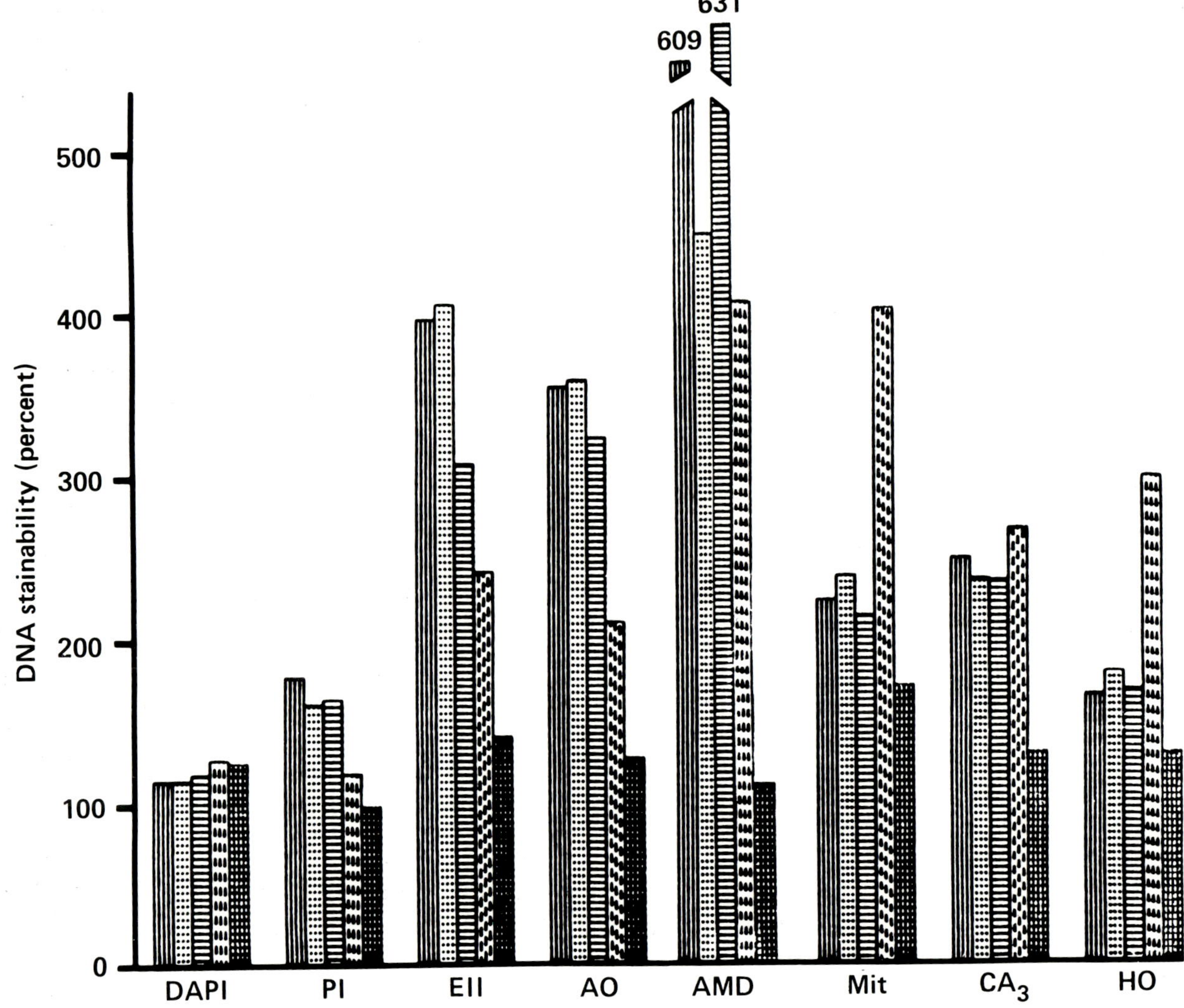

Fig. 5. Percent DNA stainability of HCl-treated and -stained mouse testicular cells and vas deferens sperm cells relative to the same cell types stained without acid pretreatment. Each value was derived from the mean of all peak fluorescence values obtained from each dye and cell type treated with HCl divided by those not treated with HCl. Vertical lines, diploid cells; dots, tetraploid cells; horizontal bars, round spermatids; dashed vertical lines, elongated spermatids; grid lines, vas sperm. (Reproduced from Evenson et al. [17], with permission of the publisher.)

fluorescence) [12,59] identify seven or eight testicular cell types. Figure 2a is a two-parameter (DNA, RNA) scatterplot distribution (cytogram) with each dot representing a single acridine orange-stained mouse testicular cell. Tetraploid (boxes 1,2) and diploid (boxes 3,4) populations each have two relatively distinct subpopulations differing in amount of red fluorescence (RNA content). The haploid population contains three subpopulations (boxes 5–7), an expanded view of which is shown in Figure 2b. Loss of RNA from round spermatids (box 5) as they mature into elongating spermatids (box 6) and elongated spermatids (box 7) is readily seen in Figure 2b. A sharp decrease in DNA stainability is seen as the cells progress from round spermatids with somatic-like histones to elongated spermatids with a mixture of protein types. Another sharp drop in DNA stainability occurs as cells progress from the elongated spermatid stage to mature sperm containing condensed, protamine-complexed DNA.

The inherent complexity of staining data is illustrated in Figures 3–5. Figures 3 and 4 are based on different sample preparation procedures and show the results obtained with several nominally DNA-specific stains; Figure 5 illustrates a comparison of the preparation procedures [13]. In Figure 3, mouse testicular cell suspensions were stained by a one-step procedure that utilizes a mixture of the dye and a permealizing detergent (0.1% Triton X-100) in the staining buffer, while a two-step procedure that includes a 30-s pretreatment with acid to dissociate histones and other basic proteins prior to staining was used to generate the data in Figure 4. The dyes used were of several general classes: (1) intercalators, i.e., acridine orange (AO), propidium iodide (PI), and ellipticine (Ell); (2) external DNA-binding dyes, i.e., mithramycin (Mit), Chromomycin A3 (CA3), and Hoechst 33342 (HO); and (3) dyes with other possible modes of binding, i.e., 4,'6-diamidino-2-phenylindole (DAPI) and 7-amino-actinomycin D (AMD). Intercalating dyes do not distinguish between round and elongated spermatids unless acid treatment is used. For the histone-containing tetraploid, diploid, and round spermatid cells, HCl extraction of nuclear proteins causes an approximately four- to sixfold increase of AMD stainability, but has essentially no effect on DAPI stainability (Fig. 5). By contrast, HCl treatment of vas sperm does not increase the staining level of AMD, DAPI, or PI but does increase the staining level for the other intercalating dyes and external dyes. The elongated spermatid is the cell type most variable in response to different dyes and the effect of acid treatment. Its nucleus contains a mixture of protein types, including histones, transition proteins, and protamines [1,4,55]. Some of the differential stainability of late stage spermatids and extratesticular sperm is due to the presence of intra- and intermolecular disulfide bonding of proteins [16]. Consequently, a combination of carefully selected dyes could be used to detect abnormalities in protein transition during spermatogenesis.

Certain other stains offer the possibility of directly probing the biochemistry of male germ-cell differentiation. As sperm undergo the final stages of development in the testis, leave the testis and traverse the epididymis, free sulfhydryl groups (−SH) on chromatin protamines are oxidized to form intermolecular disulfide bonds (S−S) which couple adjacent protamine molecules around the DNA helix [1,4]. Dyes that fluoresce only after coupling with free −SH groups provide a means for their quantification. Pellicciari et al. [45], using such dyes and microspectrofluorometry, were able to examine the loss of free −SH groups as sperm pass through the epididymis. Flow cytometric determinations of free sylfhydryl groups with the fluorescent probe CPM, 7-diethylamino-3-(4′-maleimidylphenyl)-4-methylcoumarin, made by D.P. Evenson (unpublished data) showed a sharp decrease in relative staining values between corpus and cauda epididymidal sperm in the rat and mouse. When rat sperm were detained in passage by ligatures proximal and distal to each major epididymal segment, FCM measurements of free −SH groups in nuclei showed that formation of S−S bonds is temporally controlled and is independent of regional environment.

High-Resolution DNA Content Measurements of Spermatids and Sperm

Mammalian spermatids and sperm usually contain either the X or Y chromosome. Because these chromosomes generally are different in size, haploid cells comprise two populations which differ in DNA content. In most species, the difference in DNA content ranges between 3% and 5%. This difference has served as a benchmark for establishing DNA measurement precision. Resolution of the two populations requires a measurement coefficient of variation on the order of 1.5%.

Resolution of X and Y populations was accomplished first in spermatids using enzymatic treatment to assure stoichiometric stain uptake [39,40]. A combination of special staining and measurement techniques is required to achieve this accuracy in sperm measurements. The sperm nucleus is dominated by the highly condensed, tightly bound, DNA–protamine complex and is roughly planar in shape. Condensation hampers stoichiometric penetration of nuclear stains and also produces an index of refraction for the nucleus that is much higher than for water. The mismatch between nuclear and sample-sheath indices of refraction, coupled with the nonspherical nuclear shape, produces asymmetric emission of fluorescence. More light is emitted in the plane of the sperm than normal to the flat faces [24]. Thus, the measured intensity depends not only on stain content of the sperm, but also on the details of nuclear shape and the orientation of the nucleus to both the excitation beam and the detection optics. Consequently, accurate DNA-content measurements of sperm depend on adequate access of stain to the nuclei and use of flow cytometers to control the nuclear orientation.

Accurate sperm–DNA staining was originally accomplished by slightly swelling the nucleus: the disulfide crosslinks in the protamines surrounding the DNA were broken and the nucleus subjected to mild proteolytic digestion [44]. More recently, it has been found that mild sonication followed by long incubations in the DNA dye, Hoechst 33342, gives quantitative staining [13].

The optical difficulties have been overcome with two types of flow cytometers [52], Figure 6. In one, the sperm nuclei are measured as they flow along the optical axis of the instrument, which is basically an epi-illumination fluorescence microscope with the microscope slide replaced by the flow chamber. The same lens is used to focus the excitation light and collect the fluorescence. Hydrodynamic forces orient the long axis of the nuclei parallel to the flow direction. There is no control over cell–cell variations in the orientation of the plane of the nuclei about the flow axis or over which end of the nucleus first enters the chamber. However, since the flow is parallel to the optical axis of the instrument, the signals are insensitive to the rotational distribution of the sperm. In the other type of flow cytometer, the orientation of the plane of the nuclei about the flow direction is controlled by shaping the tip of the sample injection tube to give a ribbonlike flow, and sometimes by using a rectangular nozzle orifice. The sperm nuclei are illuminated by a laser beam incident on one of their flat faces and fluorescence emitted normal to the opposite face is measured by a detector aimed directly into the excitation beam. Both instruments are capable of achieving coefficients of variation below 1.5% for DNA-content measurements, which is sufficient to discriminate the X and Y populations in most domestic animals. However, resolution of the X and Y population is not routinely possible in human sperm, presumably due to the relatively great heterogeneity of chromatin composition; nonetheless, it has been achieved [44].

MEASUREMENT OF TOXIC EFFECTS

Alterations in Testis–Cell Kinetics

Testis–cell proliferation invokes the complex expression of many genes to coordinate numerous physiological and morphological events. Exposure to agents that interfere with these processes most likely will have severe consequences, such as killing or blocking cells at specific stages of development [28,59], and will alter the normal ratio of cell types present. Interpretation of testis–cell kinetic data is complicated in some species by marked seasonal variation in testicular functions [53].

The effect of chemical exposures can be investigated by obtaining sequential samples from mice treated at several dose levels with perturbing agents. Figure 7 shows the relative percentage of germ cells present in mouse testes at three different times after administration of fractionated doses of thiotepa [12]. At 7 days after the last exposure to 0.5 mg/kg thiotepa, a significant decrease in the percentage of round spermatids and a relative increase in the percentage of elongating spermatids is observed. A decreased percentage of 4N cells also is observed with a complete loss of cell types falling

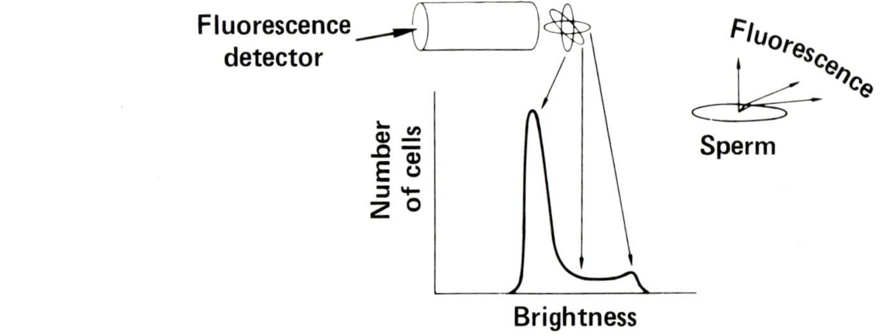

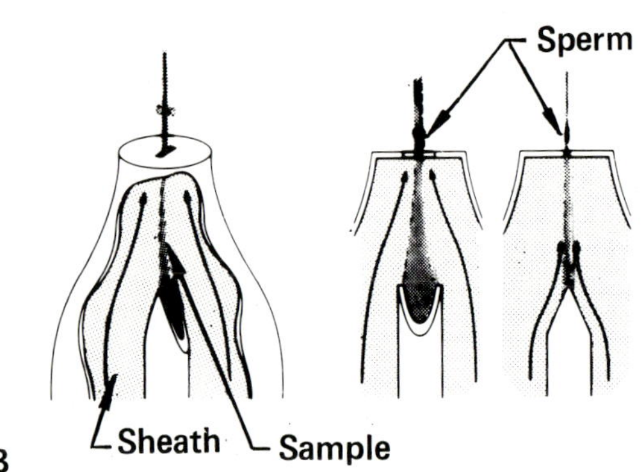

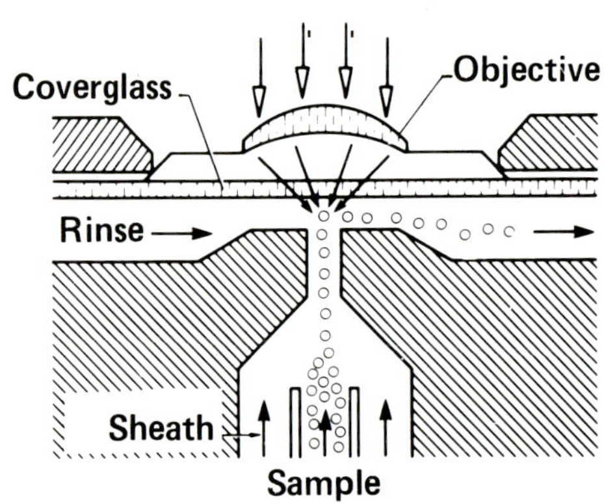

Fig. 6. Special flow-cytometric measurement techniques are required for sperm. **A:** Measuring DNA content of flat, condensed mammalian sperm in orthogonal flow cytometers results in distorted frequency distributions [24]. When the edge of the sperm is toward the detector a high fluorescence is recorded; when the flat side is toward the detector, a low fluorescence is recorded; intermediate orientations produce intermediate values. **B:** Flow chamber of the orienting flow cytometer. Shaping the end of the sample injection tube and using a rectangular flow orifice cause the sample stream to be drawn into a thin ribbon. The hydrodynamic forces encountered by the flat sperm cause them to be preferentially oriented in the plane of the ribbon. The output of the orifice enters a cylindrical quartz tube of quiescent liquid (not shown) where the cells are illuminated by a laser beam. **C:** Flow chamber of the epi-illumination flow cytometer (Ortho ICP22). The sperm flow upward along the optical axis towards the microscope objective (N.A.–1.25). Hydrodynamic forces cause them to orient with their longitudinal axis parallel to the flow. The emitted light reaching the photomultiplier is not affected by the random rotational orientation of the nuclei because the optics are radially symmetric. A maximum in the fluorescence signal occurs as the nuclei move through the focal plane and is the basis of the photometric measurement. The rinse fluid rapidly removes the nuclei from the chamber after measurement. (Adapted from Pinkel [47], with permission of the publisher.)

into box 2 of the cytogram. Samples obtained 28 days after last exposure to a dose rate of 2.5 mg/kg show a nearly total elimination of 1n cell types. At 67 days after the last exposure germ cell populations are recovering from the chemically induced damage.

The cytotoxicity of ionizing radiation has been studied with murine spermatogenesis as an in vivo biologic dosimeter [29]. Changes in the distribution frequency of cellular DNA content of whole testis preparations were analyzed by FCM. Observations from this analysis include a linear increase in the coefficient of variation of DNA content of spermatids irradiated as spermatocytes, a dose-dependent arrest of differentiated spermatogonia, and an induction of diploid sperm.

Alterations of Sperm

DNA content. Caudal epididymal mouse sperm collected 35 days after acute localized exposure of testes to x rays show dose-dependent increases in the CV of fluorescence distributions of DNA content (Fig. 8). Equivalent results were obtained with several staining protocols and with both of the high resolution measurement techniques discussed previously (see Fig. 6), which tends to confirm that the wider distributions are due to variability induced in DNA content [51]. The data suggest that the fluorescence distributions from the exposed animals have two components, one consisting of cells with normal DNA content, which is bimodal due to the presence of the X and Y subpopulations, and the other consisting of cells with abnormal DNA content, whose fluorescence distribution is broader. In the dosage range 0–600 rads, the dose dependence of the square of the CV of DNA content, CV_D^2, is described by

$$CV_D^2 = Bx + Cx^2$$
$$\text{with} \quad 0 < B < 0.23 \times 10^{-2}$$
$$\text{and} \quad C = (0.44 \pm 0.06) \times 10^{-4}$$

The dose x is measured in rads, and CV_D^2 expressed in per-

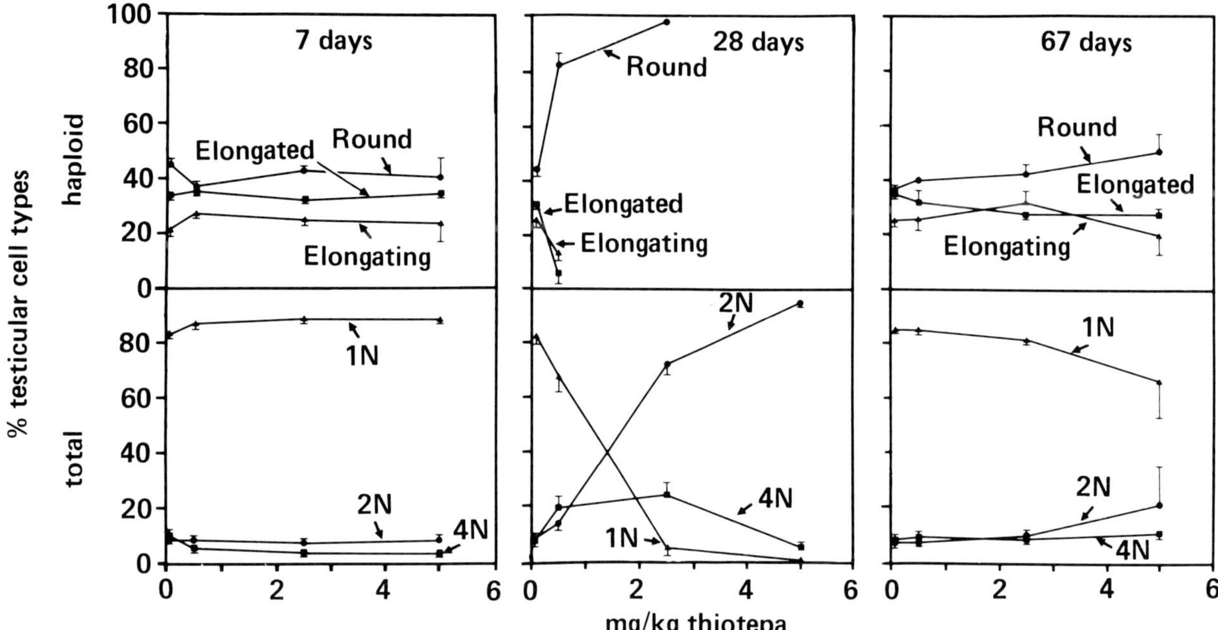

Fig. 7. Effect of various doses of thiotepa on the presence of 5 major testicular cell types at 7, 28, and 67 days after last exposure. Each point represents mean values from at least three individual mice from each time/dosage group derived from raw data such as seen in Figure 2. The vertical bars show standard deviations. (Reproduced from Evenson et al. [12], with permission of the publisher.)

cent. Computer modeling of the shapes of the fluorescence distributions at a dosage of 600 rads reveals that 30–40% of the sperm have abnormal DNA content. Some have deviations as large as two whole chromosomes, but it is not clear whether they are due to whole chromosome nondisjunction, fragmentation of the genome, or perhaps a general effect of the irradiation on the stoichiometry of stain. Two chemical agents that produce abnormally shaped sperm did not cause variability in DNA content. Because the x-ray dose response has a small slope at low doses, the sensitivity of this technique is low. Similarly its use in detecting mutagen exposure will be limited to agents that produce substantial aneuploidy in sperm. Use of another species with fewer chromosomes than the mouse might help optimize detection of exposure-induced errors in sperm DNA content [47].

Chromatin structure. The complex process of sperm cell development is very sensitive to perturbing agents. The sperm chromatin structure assay (SCSA) [10] is based on the concept (see Chapter 16, this volume) that the DNA in cells has a differential susceptibility to acid- or heat-induced denaturation in situ depending on the state of cell growth and differentiation or influence of perturbing agents.

In the SCSA, whole sperm or isolated nuclei are subjected to heat or acid to induce partial denaturation of DNA in situ and then stained with acridine orange (AO). AO complexed with native DNA fluoresces green and can be distinguished by FCM from the red-fluorescing AO complexed to denatured DNA (in the absence of RNA). The degree of denaturation is measured using the index

$$\alpha_t = \frac{\text{red flourescence}}{\text{(red + green) flourescence}}$$

which can range between 0 and 1 [8].

Early reports expressed the degree of sperm chromatin-structure abnormality as the ratio of α_t values in heat-denatured samples compared to α_t values in unheated samples [15]. However, because later studies showed that highly abnormal sperm could have altered α_t values in unheated samples, current studies use the ratio α_t experimental values compared to α_t control values as a measure of abnormality. Current studies also use acid rather than heat for induction of DNA denaturation because it appears equally efficient for detecting chromatin abnormalities, is less time consuming, and allows for use of samples with low concentrations of sperm.

Figure 9 shows the effect of various concentrations of intraperitoneally administered ENU (ethylnitrosourea, a strong direct-acting alkylating agent) on the susceptibility of mouse sperm chromatin to acid denaturation [17]. Twenty-eight days after the last exposure, caudal epididymidal sperm were measured by the SCSA. Although the total amount of DNA stained (red plus green fluorescence) does not change significantly, there is a marked dose–response shift to higher values of α_t.

Figure 10 presents the CV of α_t from acid treatment as a function of ENU dosage. Nearly identical plots resulted when nuclei were heated prior to AO staining, indicating that acid- or heat-induced partial DNA denaturation elicit similar changes in sperm chromatin structure [17].

In the mouse, a strong correlation has been noted [12,17] between toxin-induced abnormalities of sperm-head shape [58,60,62], and both the standard deviation (SD) and the CV of α_t of acid- or heat-treated epididymal sperm measured by the SCSA. Figure 11 shows the effect of thiotepa on SD α_t and epididymal–sperm-head morphology in mice. No significant effect was observed at 7 days. At 28 and 67 days after the last exposure to 2.5 mg/kg, the SD α_t and abnormal morphology data ponts are all greater than 1 SD from the

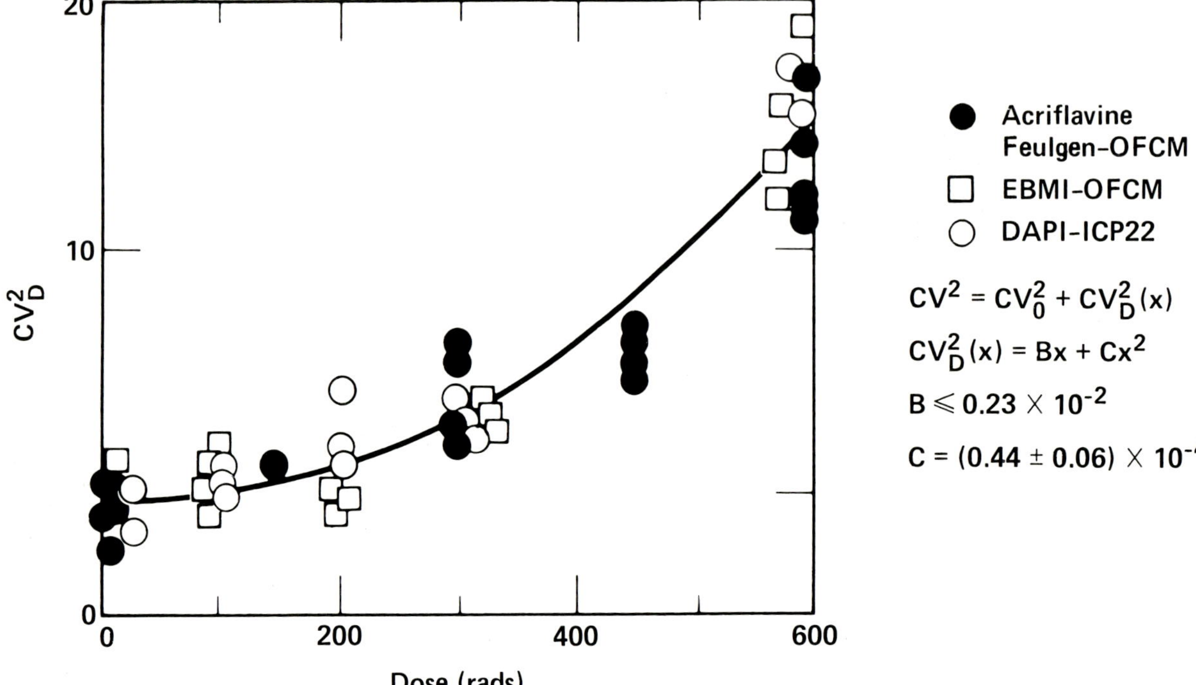

Fig. 8. Increased DNA-content variability as a function of radiation dose. The observed X-ray dose response is the same for three DNA measurement techniques. The square of the coefficient of variation of the fluorescence distribution increases with radiation dose (OFCM = orienting flow cytometer). Each symbol represents an independent determination. The solid line is the least-squares fit of a second-order polynomial (second equation) to the data. (Redrawn from Pinkel [47].)

control mean; at 28 and 67 days after the last exposure to 5.0 mg/kg, the SD α_t is greater than 2 SD from the control mean and the abnormal morphology value is >1 SD from the mean. These nearly parallel slopes of dose response for sperm-head morphology and SD α_t are similar to those observed after exposure to ENU and eight other chemicals including alkylating agents and metabolic inhibitors [17]. In each instance, the minimal effective dose needed to alter sperm-head morphology was approximately equivalent to that which altered sperm chromatin structure detected as an increase of SD α_t.

The desire to increase growth rates in animals used for meat production has led to use of exogenous promoters that alter hormonal regulation of growth. The promoter Zeranol, a resorcyclic acid lactone, when implanted subcutaneously in bulls at 1 month or at 1 and 4 months of age has a deleterious influence on spermatogenesis at 15 months of age, with a decrease in numbers of elongated spermatids frequently observed [2]. Perhaps of greater interest, sperm obtained from the vas deferens of bulls treated with zeranol demonstrated increased susceptibility to acid-induced DNA denaturation as determined by the SCSA protocol. Zeranol causes alterations in hormonal regulation that apparently result in the side-effect of abnormal differentiation of sperm chromatin.

Altered nuclear morphology. In all species of mammals so far examined, the frequency of sperm with malformed heads increases after exposure to most mutagens, carcinogens, and teratogens; this suggests that malformed sperm may indicate induced genetic damage [61,62]. The usual method for assessment of morphological defects, visual examination of morphological features of sperm mounted on microscope slides, is subjective. While individuals may score reproducibly, substantial variability exists among technicians and laboratories. Quantitative procedures are needed to provide objectivity and to improve reproducibility. Flow cytometry enhances speed, provides standardization, and should lead to improved archival libraries, data exchange, and retrieval systems.

Shape information is available from flow cytometry by slit scanning the sperm heads [3]. Sperm nuclei (approximately 7 μm long) are stained with a DNA-specific fluorescent dye and forced to flow through a 2.5-μm-thick laser beam. Fluorescence is recorded as each sperm flows across the beam. The time course of the signal from the fluorescence detector, the fluorescence profile, is a measure of the shape of the sperm nucleus. The slit-scan flow-cytometric (SSFCM) fluorescence procedure can distinguish profiles for sperm from rabbits, mice, hamsters, and bulls (Fig. 12).

Benaron et al. [3] developed computer algorithms to distinguish profiles of malformed mouse sperm and to determine their frequency of occurrence. By adjusting data-analysis parameters in an experiment involving x-ray irradiation of mice, they showed that the frequencies of malformed sperm in the samples as determined by computer analysis of the SSFCM-generated profiles correlated very well (r = 0.99) with the frequencies of malformed sperm estimated by visual analysis of samples from the same mice. In a subsequent study, Halamka et al. [30] sought to validate the SSFCM technique by analyzing sperm samples from untreated mice and from mice treated with methyl methane sulfonate (MMS) or procarbazine without prior reference to the results obtained by visual analysis of samples taken from the same mice. Visual and SSFCM analyses were conducted in a dou-

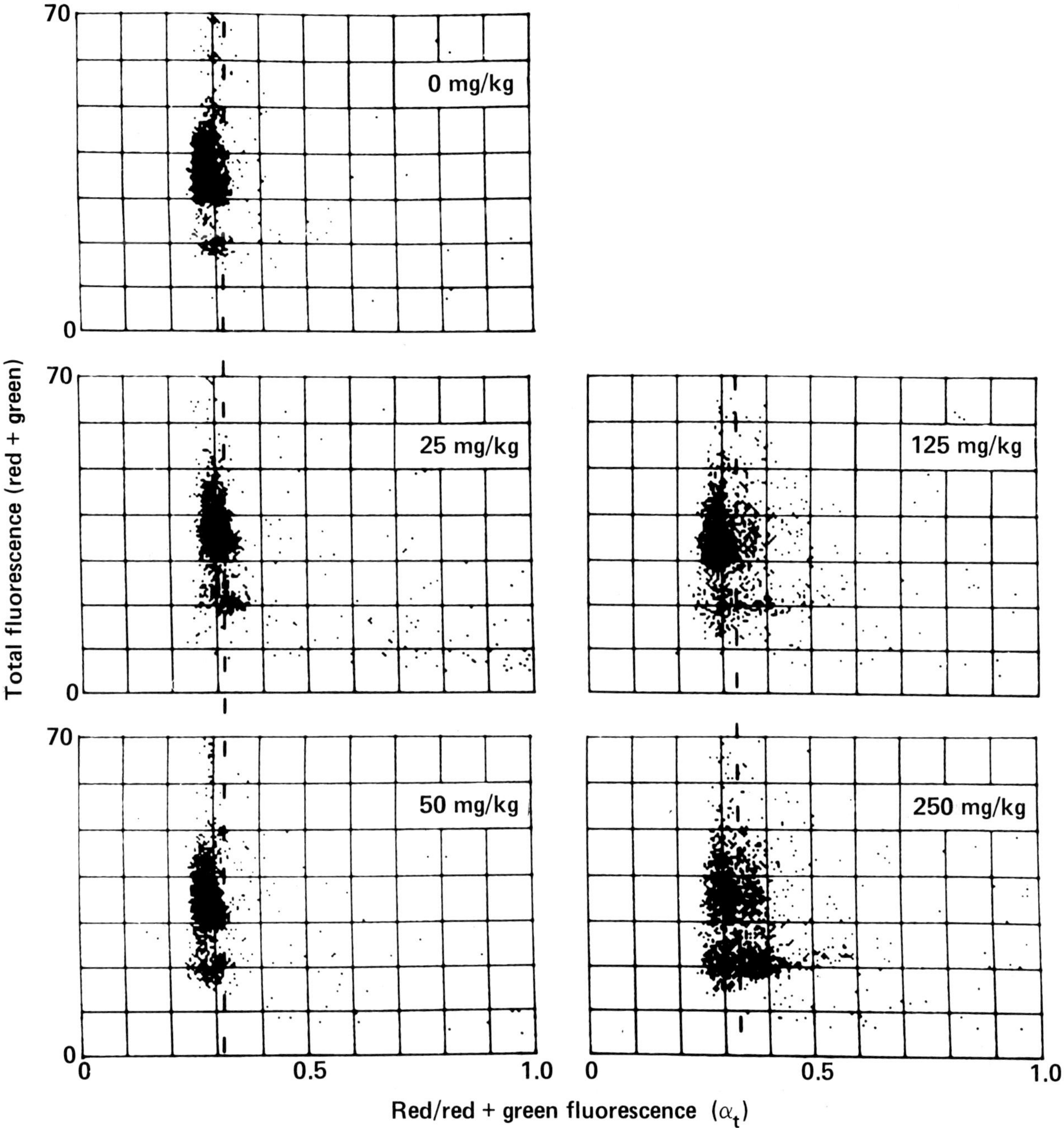

Fig. 9. Effect of acid denaturation on mouse epididymidal sperm nuclear DNA in situ following exposure to ENU. Five mice in each of five groups were injected i.p. with Hank's balanced salt solution (HBSS) alone or containing various concentrations of ENU. (Reproduced from Evenson et al. [17], with permission of the publisher.)

ble-blind fashion and the results were compared only at the end of the study. Three significant conclusions were reached from this work: (1) the SSFCM technique can be used to quantify morphological changes in sperm heads induced by physical and chemical agents, and dose-effect plots can be obtained; (2) the frequencies of aberrant sperm determined visually and by SSFCM correlate well (r = 0.83), although the concordance between the visual and SSFCM estimates of abnormal sperm is not one-to-one for all samples; and (3) the present SSFCM hardware and software are adequate for analyzing 500 profiles per sample in a reproducible manner.

Sperm sorting based on slit-scan analysis has so far not been implemented, but it is technically possible. If such an instrument were built, flow sorting of large numbers of normal and malformed sperm would furnish material for biochemical analyses that might provide insight into mecha-

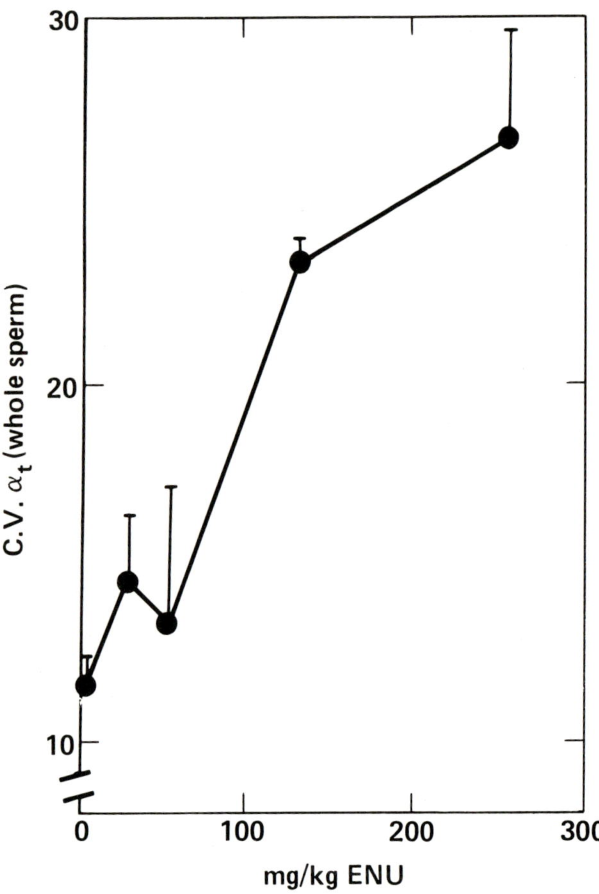

Fig. 10. Coefficient of variation of α_t from acid treatment as a function of exposure level to ENU. (Reproduced from Evenson et al. [17], with permission of the publisher.)

nisms of sperm shaping and into the consequences of morphological abnormalities.

SEMEN QUALITY AND FERTILITY

Mammalian ejaculates frequently contain more than a hundred million sperm; normally only one fertilizes an ovum. Establishing the effect of a particular sperm abnormality on fertility requires study of many sperm from multiple ejaculates. Study of sperm by FCM has the potential to provide adequate characterization of several semen and/or sperm traits that would be useful for diagnostic and prognostic applications. Studies on humans are often complicated by the extreme heterogeneity of sperm in human ejaculates. Furthermore, semen analysis is rarely done before a clinical problem exists. Consequently, longitudinal, quantitative FCM studies of sperm from normal men for an extended period are urgently needed.

Garner et al. [21] studied sperm from men, bulls, boars, dogs, horses, and mice using a fluorogenic stain consisting of the membrane-permeant substrate, carboxyfluorescein diacetate (CFDA), and the relatively membrane-impermeant DNA stain, propidium iodide (PI). PI significantly stains only the DNA of cells that are dead or that have damaged membranes. When excited it produces a red fluorescence. Although CFDA is a nonfluorescent compound, hydrolysis of its ester bonds produces a highly fluorescent, membrane-impermeant green fluorophore that is trapped intracellularly by intact membranes. Microscopic examination revealed three distinct populations of sperm from each species. Motile sperm retained the green fluorescent hydrolytic products of CFDA. The second population of sperm exhibited red nuclei and green acrosomes. The third population, consisting of red fluorescent nuclei, were presumed to be degenerated sperm. These three differing populations of sperm were quantified with dual-parameter flow cytometry in 14 samples of cryopreserved bovine sperm. Flow-cytometric analyses correlated well with standard measurements of seminal quality.

A similar type of study was conducted by Evenson et al.

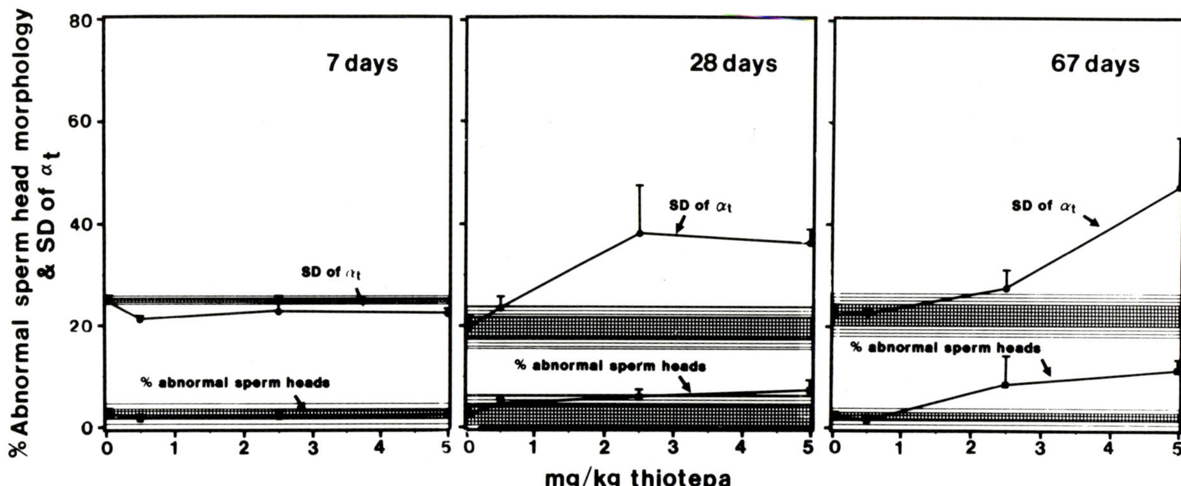

Fig. 11. Relation of mouse epididymal sperm head morphology to standard deviation (SD) of α_t at 7, 28, and 67 days after the last exposure to thiotepa. Each point represents mean values from at least three individual mice from each time/dosage group. The hatched and horizontal-lined boxes represent one and two standard deviations from the control values (Drawn from data in Evenson et al. [12].)

[14] using FCM to measure sperm mitochondrial function and cell viability simultaneously. They used double staining with Rhodamine 123 (R123) and ethidium bromide (EB). EB is virtually excluded from cells with intact membranes but penetrates damaged membranes and stains the nuclear DNA. R123 is a cationic dye specific for mitochondria because of the high electronegative charge on these energy-producing organelles [32]. The results indicated that the R123 fluorescence intensity of sperm mitochondria is generally proportional to sperm motility. Figure 13 shows measurements of a single semen sample obtained from a fertile donor stained with R123 and EB. At 4 hours after collection, the majority of sperm cells have a relatively homogeneous high green and low red fluorescence. A quantum loss of green fluorescence occurs as cells age in vitro. After 30 hours, all cells have reduced R123 staining with an increased EB staining.

In addition to the assays just described, the SCSA may have clinical utility in human [11,18] and veterinary medicine. The relationship between heterogeneity of sperm nuclear chromatin structure and bull fertility was evaluated in two groups of Holstein bulls (D.P. Evenson, unpublished data). Strong correlations between α_t and fertility ratings suggest that the SCSA can identify sires with low-fertility and poor-quality semen samples (Fig. 14).

Human semen aliquots can be fruitfully analyzed by the SCSA protocol. Dramatic differences are often seen between semen samples from fertile donors and infertility patients. Staining patterns of untreated and denatured nuclei do not significantly differ for individuals of proven fertility [9,15]. Thus, for fertile donors, a relatively homogeneous chromatin structure must exist in sperm that is resistant to DNA denaturation in situ.

Infertile patients show more variability in sperm chromatin structure. For example, a patient that had worked in the chemical industry for about 20 years has been studied. The first 10 years he worked with chemicals without benefit of protective hoods, gloves, and/or masks. The next 10 years he used protective clothing and hoods. The patient's wife experienced 16 miscarriages during the 20 years. Sperm from this patient and a fertile donor were examined using the SCSA protocol. The two AO-stained populations of sperm are seen in Figure 15; this figure shows differing sensitivity to acid-denaturing conditions, one that is close to normal and a second one that is quite abnormal.

Figure 16 shows SCSA derived data on sperm obtained from a fertile donor and five patients from an infertility clinic. In sharp contrast to the relatively normal homogeneous staining pattern seen for the fertile donor, a significant AO staining heterogeneity exists between unheated and heated samples for the patients. The sperm cells from the patients had a larger range of green fluorescence values in unheated samples than the control. Cells with 5–10 times the green fluorescence values of normal sperm may be round spermatids and diploid cells, respectively. Higher values may also indicate the presence of immature cells lacking normal condensation and/or the exchange of histones for prota-

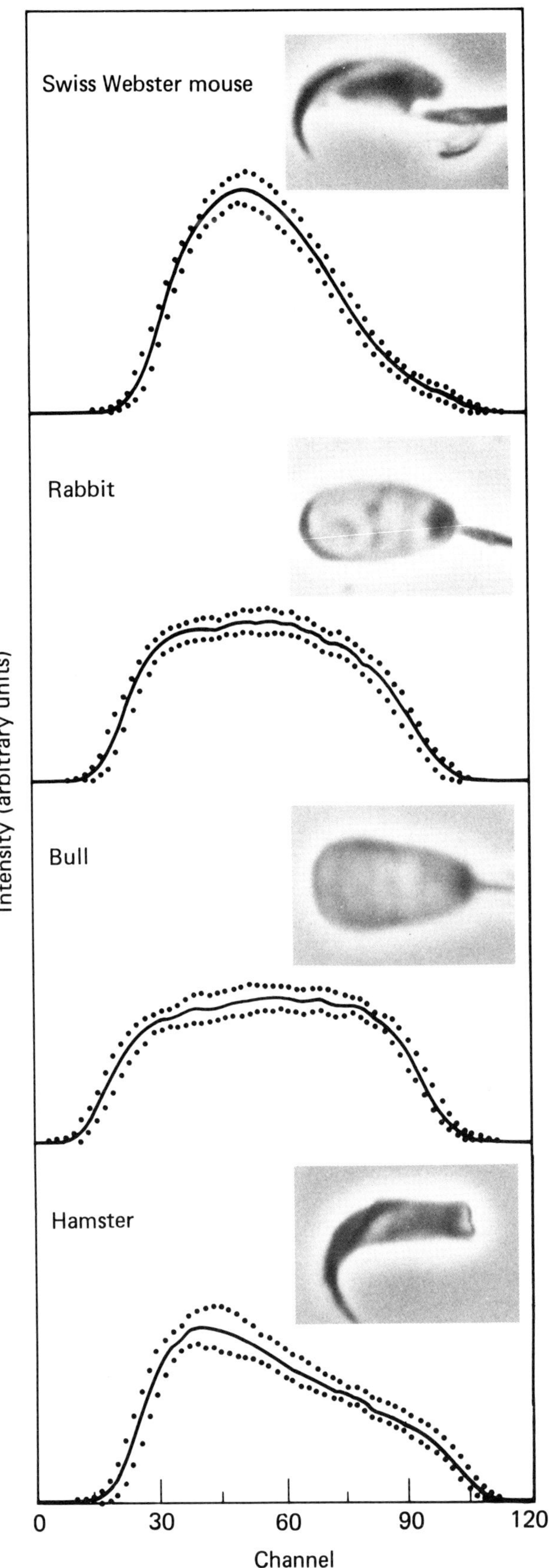

Fig. 12. Average slit-scan flow-cytometric profiles for sperm from four species of mammals.——, sample average, · · · , 1 SD from the average. Sperm from each animal yield an average profile with a distinct length and shape. (Reproduced from Benaron et al. [3], with permission of the publisher.)

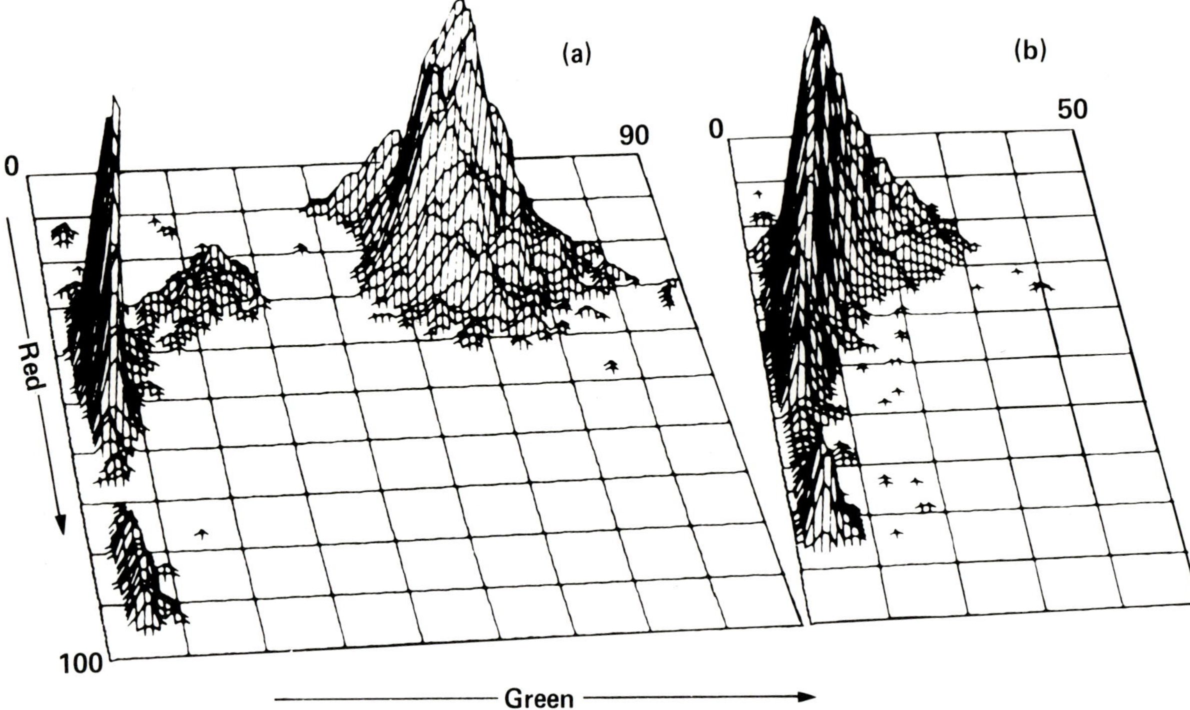

Fig. 13. Two-parameter frequency histograms demonstrating the stainability of human sperm cells with rhodamine 123 (R123, green fluorescence) and ethidium bromide (EB, red fluorescence). The height of the peaks corresponds to relative cell numbers. **a:** Measurements 4 hours after collection. **b:** Measurements 30 hours after collection. (Reproduced from Evenson et al. [14], with permission of Elsevier Science Publishing Co., Inc. Copyright 1982 by the Histochemical Society, Inc.)

mines. Cells with lower green fluorescence values may be dying or dead cells, or cells with overly condensed DNA or less DNA. An even greater level of heterogeneity is evident in the patients' sperm after heat-induced DNA denaturation; that is a decreased green fluorescence and an increased red fluorescence is seen. Particularly interesting for clinical applications is the observation that samples appearing normal without heating (patients 3 and 4 and, to a lesser degree, patient 5), demonstrated a significant level of DNA denaturation after heating. Thus, these FCM techniques may detect sperm abnormalities not identified by routine methods.

Tejada et al. [56] adapted the FCM based SCSA technique [17] for use in a fluorescent microscope to determine the relationship between sperm chromatin structure and clinical infertility in a study of 89 men. Sperm were applied to a glass microscope slide, air dried, fixed in an acid fixative, and stained with a high concentration of AO (which may also enhance DNA denaturation). The sperm nuclei were then visually scored for red and green sperm chromatin. An "effective sperm count" (total sperm count × % green fluorescent sperm nuclei) of less than 5×10^6 was seen in 60 of 61 infertility patients, while 27 of 28 fertile donors had an effective sperm count of more than 40×10^6.

TESTICULAR NEOPLASIA

The effect of testicular neoplasia, systemic disease, and hormonal imbalance on human spermatogenesis has been demonstrated by FCM measurements of cell suspensions of testis biopsies [5,9,19,35,46,57,63]. For ethical reasons, such investigations usually are limited to patients with andrological irregularities or the symptoms of testicular cancer. Other diseases are usually studied with postmortem specimens [35].

Figure 17 illustrates the results of FCM measurements of semen from two testicular-neoplasia patients. The diagnosis of embryonal cell testicular carcinoma had been established for patient A. A semen sample was obtained 6 months after unilateral orchiectomy and 3 months following chemotherapy. Most of the cells in the suspensions made from a testicular biopsy of the opposite testis from this patient appeared to be round spermatids. Those sperm morphologically identifiable in the biopsy had, relative to control samples, a higher level of both green and red fluorescence indicating a lack of normal chromatin structure. Patient B had mixed testicular cancer with teratoma, embryonal cell carcinoma, and seminoma. The semen sample was obtained 11 months following unilateral orchiectomy. The FCM distribution of AO-stained semen cells was very broad, ranging from nearly normal sperm to diploid cells. Light microscopy revealed the presence of abnormally shaped sperm, immature germ cells, and somatic cells [19].

Figure 18a–c shows measurements of AO-stained biopsy cells from a normal testis, a testis with carcinoma but exclusive of the tumor area, and the carcinoma, respectively. The biopsy obtained from the macroscopically normal region of the tumor-containing testis was virtually devoid of tetraploid cells and had an abnormally high ratio of diploid to round spermatid cells. This ratio demonstrates abnormal kinetics of spermatogenesis, perhaps due to the presence of the tumor and other disease symptoms. In this particular tumor, only diploid cells were detected; however, aneuploid cells commonly found in tumors can be easily recognized by flow cytometry [63].

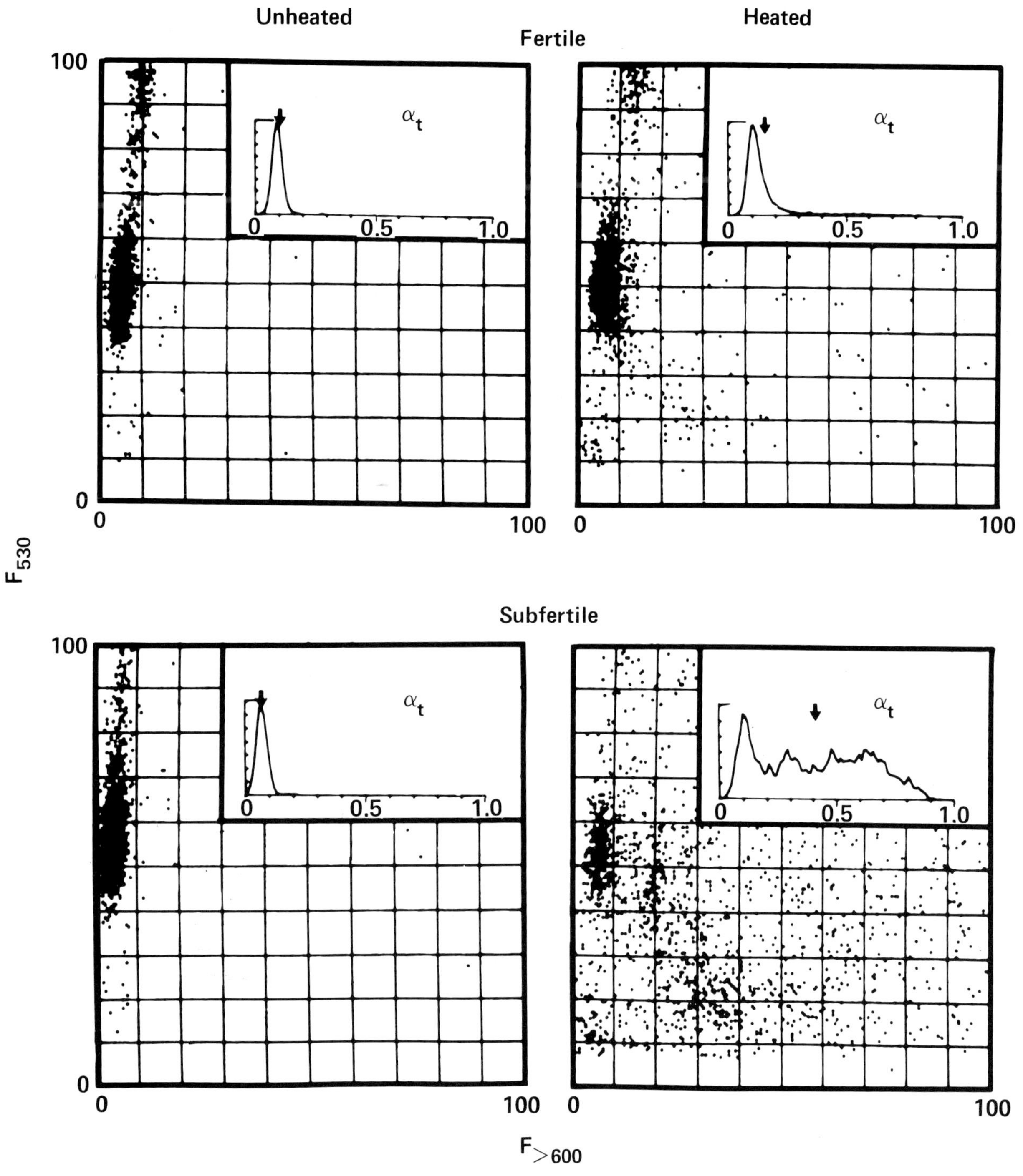

Fig. 14. Relationship between fertility of two Holstein bulls and susceptibility of sperm nuclear DNA to heat-induced DNA denaturation in situ as determined by the SCSA. The AO staining profile and the corresponding α_t values were similar between the unheated and heated samples from the fertile bull while the staining distribution shifted significantly for the heated sample from a subfertile bull.

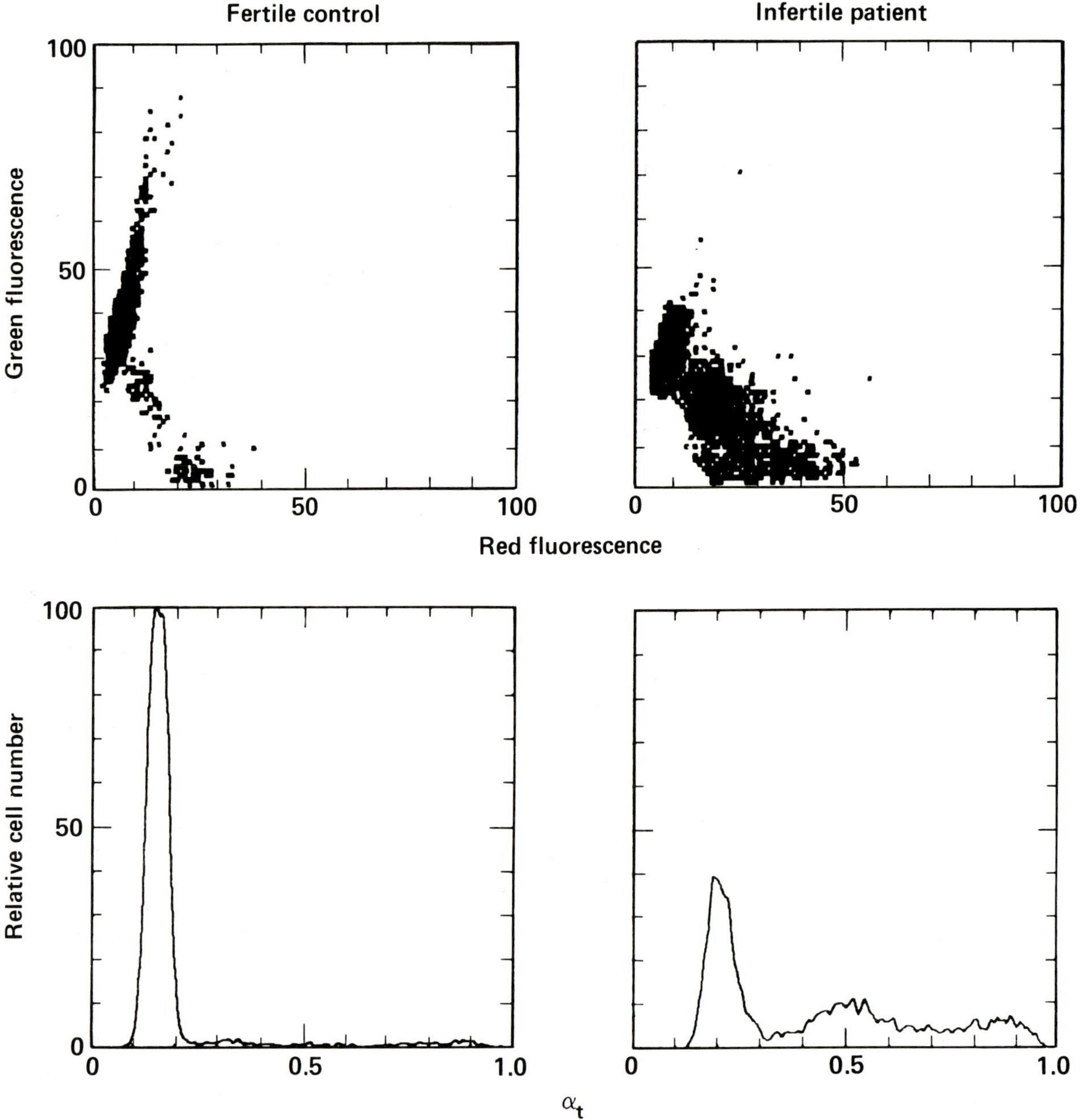

Fig. 15. Relationship between fertility of two human males and susceptibility of sperm nuclei DNA to acid-induced DNA denaturation in situ as determined by the SCSA. **Upper:** AO-staining distribution of semen cells obtained from a fertile donor and a patient exposed to unknown industrial chemicals. **Lower:** Frequency histogramplot of α_t. For the fertile control, 8% of the cells have a staining distribution outside of the main population. For the suspected infertile patient, not only does the main population have a lower green fluorescence and high α_t value, but 55% of the cells have an AO staining distribution that is considered to be of low quality.

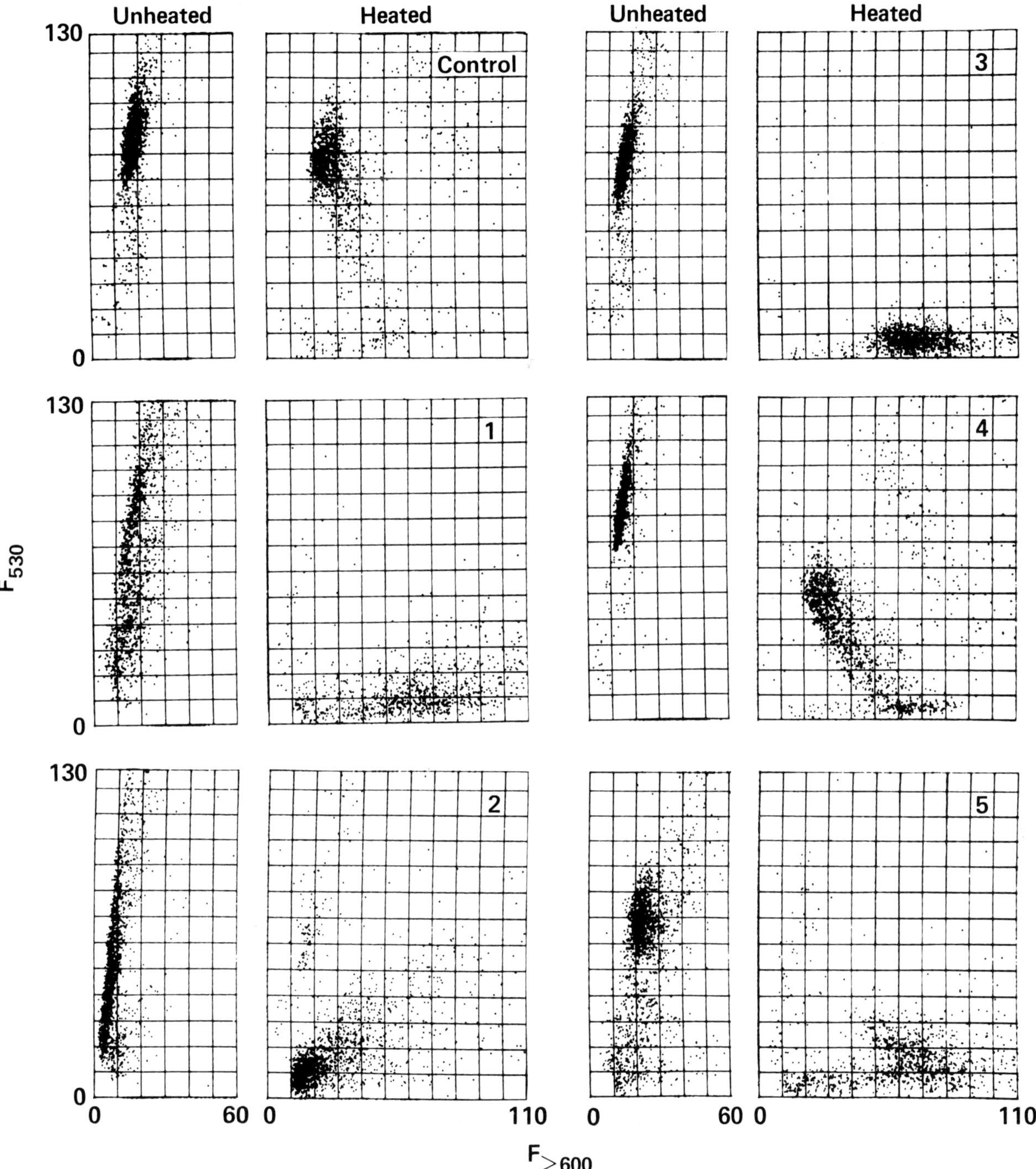

Fig. 16. Computer-drawn scattergrams derived from SCSA measurements on sperm nuclei from one fertile (control) donor and five patients attenting an infertility clinic. (Reproduced from Evenson et al. [13], with permission of the publisher.)

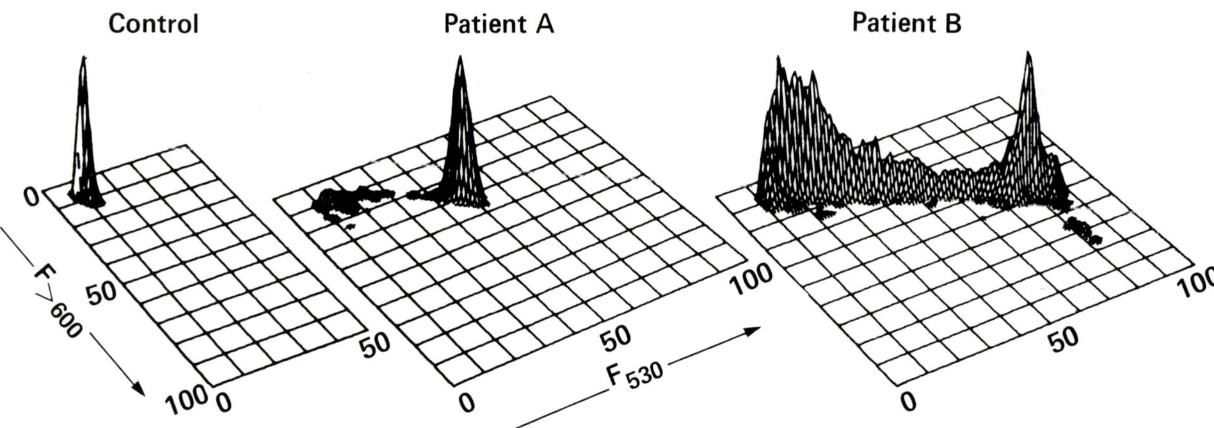

Fig. 17. Computer-drawn two-parameter (F530 versus F>600) histogram representing the distribution of AO-stained human semen cells obtained from a healthy fertile donor (control) and two cancer patients, one treated (A) and one not treated (B) with chemotherapy agents. (Reproduced from Evenson and Melamed [19], with permission of Elsevier Science Publishing Co., Inc. Copyright 1983 by the Histochemical Society, Inc.)

GENDER PRESELECTION OF OFFSPRING

The ability to resolve the peaks due to the X- and Y-chromosome-bearing sperm populations has important applications to control of the gender of offspring. Any method for influencing the sex ratio of agriculturally important animals would have a major impact on the genetic improvement of animals and should produce substantial economic benefits. Gender preselection for humans would have important sociological implications and would raise ethical questions for society to address.

In mammals, information for sexual differentiation is carried in the sperm, so gender-control schemes have concentrated on semen manipulations. One can imagine strategies based on physical separation of the X- and Y-sperm populations or on altering their relative fertilizing capacities. In spite of a great deal of effort, there is currently no prefertilization gender-selection technique available that is generally recognized as effective [25]. Flow-cytometric work associated with gender selection has concentrated on two areas, assaying the effectiveness of a variety of techniques for sperm separation and sorting the nonviable sperm from the two populations for subsequent study.

Sperm Analysis

Current FCM techniques have been used successfully to resolve the X- and Y-sperm populations. This success extends to essentially every species in which the techniques have been applied except for the human, where they work in only a small proportion of the samples [44]. The lack of a reliable method for use in humans has been attributed to increased variability in chromatin structure, but this has not been firmly established.

Techniques have been developed so that X- and Y-population measurements can be made on sperm obtained from the epididymis, from ejaculates, and after cryopreservation [20]. Measurement of a sample results in a bimodal histogram, such as Figure 19, consisting of two closely spaced peaks. Analysis of this data by fitting a pair of gaussian distributions yields the proportion of cells in each of the populations and the coefficient of variation of the measurement. The separations in the peak means have ranged from just over 9% in the vole *Microtus oregoni* (Fig. 20) to about 3% in the mouse [50,52]. The separations for most animals of agricultural interest fall in the range of 3.5–4.5%. These numbers are in agreement with expectations based on independent measures of the DNA content of the X and Y chromosomes in these species.

Flow cytometric measurements were applied to bull semen samples that were processed by a number of methods purported to enrich one of the sperm populations [49]. The techniques used for the separations ranged from gradient methods to proprietary chemical and electrical approaches. In all cases, the flow analysis found equal proportions of X and Y sperm. The only samples in which separation has been confirmed are those separated by flow sorting.

Sperm Separation by Sorting

Once the two sperm populations can be resolved they can be sorted. This was first done [50] with sperm of the vole, *Microtus oregoni*, which was chosen because the separation of the two populations, 9%, is unusually large. In fact, no X chromosomes appear in the sperm of this animal; one population contains the Y chromosome and the other has no sex chromosome [43]. The separation technique was also used for the chinchilla (separation 7%) and the bull [33]. Both FCM investigations [33,50] employed a flow chamber with the sample tube modified as shown in Figure 6b followed by a standard sorting orifice. With sample-flow rates on the order of 100 cells per second, sufficiently precise orientation is achieved. In addition, a fluorescence detector in the forward direction must be used (Fig. 6a). These modifications can be added to commercially available instruments [33]. The purity of the separated fractions, as determined by reanalysis, is routinely greater than 80% and can be well above 90% [50]. The results of a sort on vole sperm are shown in Figure 21. The two histograms show the analysis of the sorted fractions; the lower histogram shows a photographic superposition of the sorted histograms similar in shape to the separations before sorting.

Recently, repetitive nucleic acid probes with base sequences that allow them to bind specifically to individual human chromosomes have been discovered. These permit labeling of chromosomes and nuclei, including human sperm, by in situ hybridization. Thus, although flow DNA-

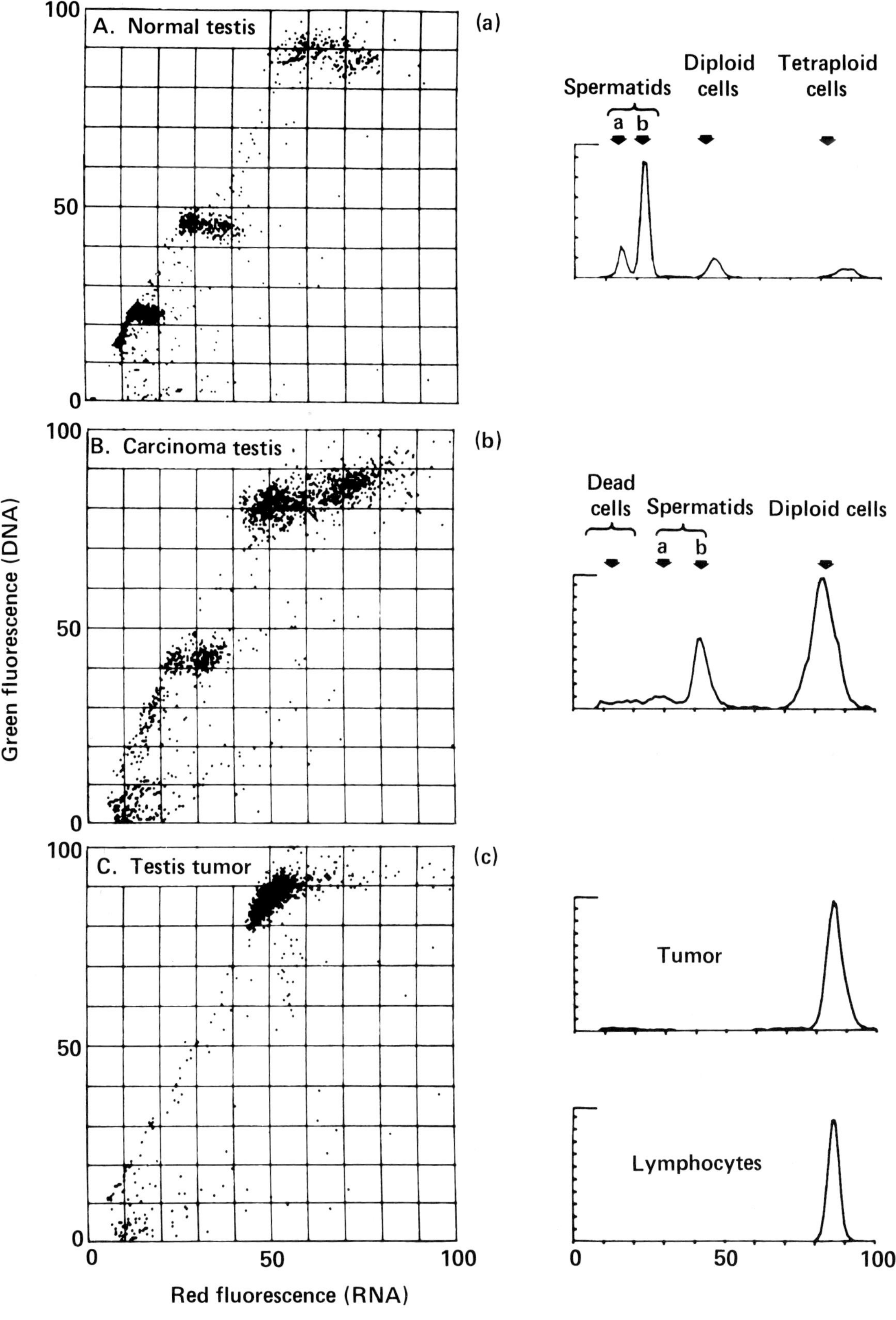

Fig. 18. Computer-drawn scattergrams of the distribution of testicular biopsy cells from a normal (A) and tumor-bearing testis (B and C) according to their fluorescence intensities. To the right of each scattergram are the green fluorescence frequency histograms. The bottom frequency histogram represents the green fluorescence of AO-stained peripheral blood lymphocytes used as a marker for normal diploid cell DNA stainability. (Reproduced from Evenson et al. [13], with permission of the publisher.)

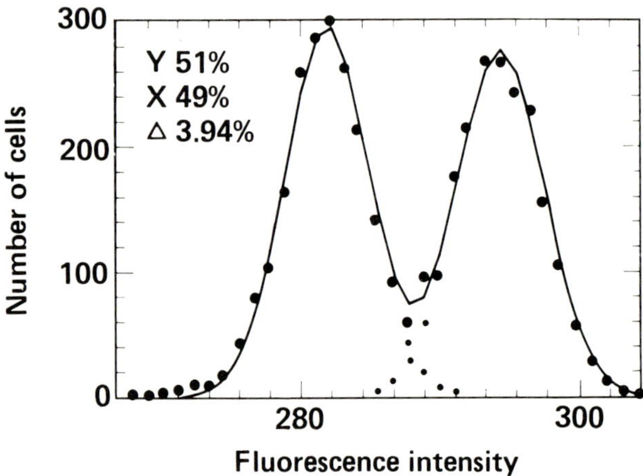

Fig. 19. Resolution of bull X- and Y-sperm populations. For computer analysis, the distribution obtained from flow-cytometric measurement is truncated to the channels of fluorescence intensity shown and fitted with a pair of gaussian distribution. Each of the two distributions is represented by small dots (···) and their sum by the solid line (——). The actual number of sperm per channel are shown as large dots (●). The only restriction placed on the computer fit is that the coefficients of variation of the two gaussian distributions be identical. There are 51% of the sperm in the Y peak (lower intensity) and 49% in the X peak. The difference in modal fluorescence intensity of the two peaks is 3.94%. (Redrawn from Garner et al. [20].)

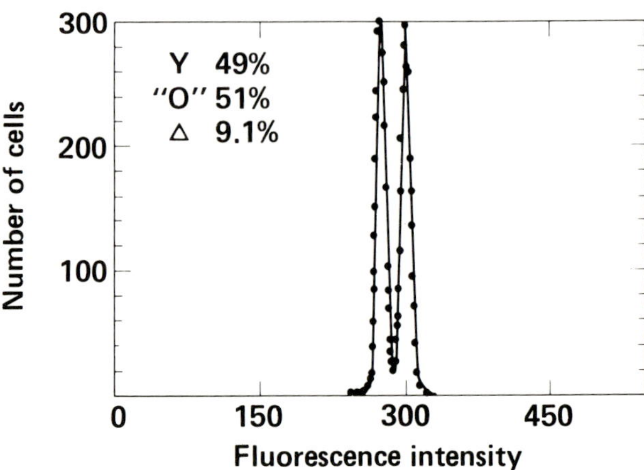

Fig. 20. The "O"- and Y-sperm populations are clearly resolved in *M. oregoni* sperm and the 9.1% difference in modal fluorescence intensity of the two peaks corresponds closely to the 8.8% that was expected based on length of chromosomes.

content measurements cannot currently be used to analyze human sperm for their sex-chromosome content, it should be possible to distinguish the two sperm types with probes specific for the X and Y chromosomes. Similar probes also may be eventually discovered for other species. Their use probably would be preferential to flow measurements, once they are available. It should be noted that techniques for fluorescently labeling probes have been developed and their applicaton to flow is rapidly developing. Thus, flow measurements of sperm may remain important, but the staining may shift from general DNA dyes to DNA-sequence-specific probes.

So far no viable sperm have been sorted.* In fact, it is difficult to maintain viability of unstained cells on passage through a sorter. Nonetheless, sorted cells may be useful for analysis of antigenic surface components that might allow bulk separation of X and Y sperm if a marker is found. Additionally, the DNA in the sorted sperm may still be active, and might be useful for in vitro fertilization.

SUMMARY AND CONCLUSIONS

In the past decade significant refinements of flow cytometric techniques for analyzing male germ cells have occurred. Detailed information has been provided on the complex differentiation pathways of spermatogenesis and the striking uniformity of mature sperm. Chemical staining and highly specific antibodies permit discrimination of various cell types and analysis of their constituents on a cell-by-cell basis. Dose–effect responses of the reproductive cells from individuals ill or exposed to toxic agents have been determined. Further development of hardware, software, and protocols for staining will establish rapid and routine methods for quantification of aberrant sperm and should lead to standardization of measurement criteria promoting the creation of data bases easily exchangeable between laboratories. Precise DNA content measurements now permit accurate analysis of the proportions of X- and Y-chromosome-bearing sperm in the semen of domestic animals. Sorting these cells may eventually lead to methods for controlling the sex of offspring. The recent development of DNA probes specific for single chromosomes combined with fluorescence methods to measure their hybridization to cells opens the possibility of rapid genetic analysis of germ cells and sperm using flow systems or conventional microscopy.

ACKNOWLEDGMENTS

We thank our colleagues, too numerous for mention, for their many contributions and their steadfast encouragement. This work was supported in part by grants CR-810991 and R-812363 from the Environmental Protection Agency, grant R01 ES03035 from the National Institutes of Health, grant 86-CRCR-1-2107 from the U.S. Department of Agriculture, and by the South Dakota State University Experimental Station. This work was also performed under the auspices of the U.S. Department of Energy, Office of Health and Environmental Research, by the Lawrence Livermore National Laboratory under contract number W-7405-ENG-48.

DISCLAIMER

This document was prepared as an account of work sponsored by an agency of the United States Government. Neither the U.S. Government nor the University of California nor any of their employees, makes any warranty, express or implied, or assumes any legal liability or responsibility for the accuracy, completeness, or usefulness of any information, apparatus, product, or process disclosed, or represents that its use would not infringe privately owned rights. Reference

*Note added in proof: Intact, viable X and Y chromosome-bearing sperm of the rabbit have been separated according to DNA content using a flow sorter (Johnson LA, Flook JP, Hawk HW (1989) Sex preselection in rabbits: live births from X and Y sperm separated by DNA and cell sorting. Biol Reprod 41). The phenotypic sex ratio at birth was accurately predicted from the FCM measured proportion of X- and Y-bearing sperm used for artificial insemination.

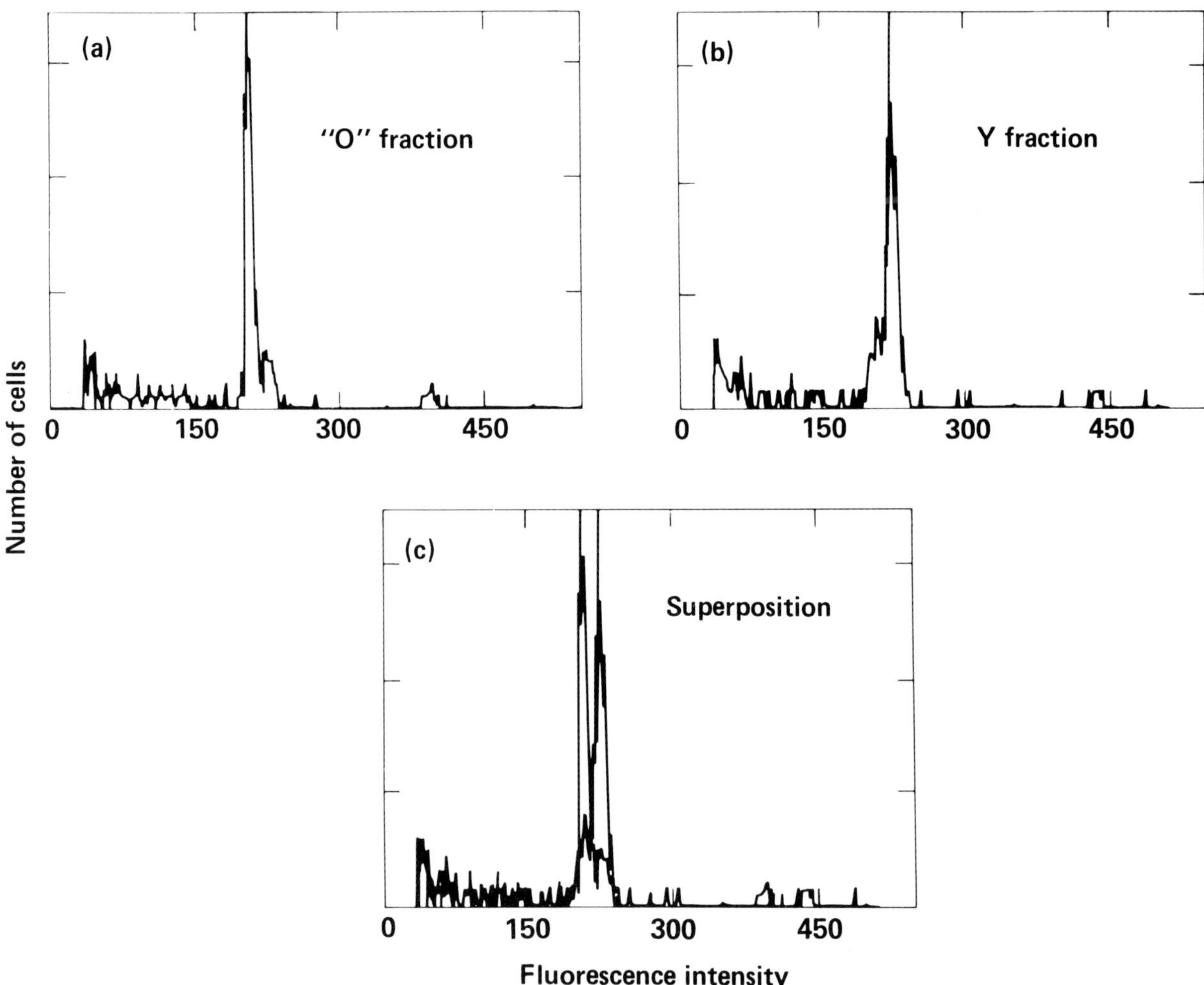

Fig. 21. Verification of sorting purity of *M. oregoni* sperm. Sperm were sorted into two fractions. After sorting, the putatively enriched fractions were restained and analyzed in ICP22. Analysis of the "O" and Y fractions are shown in (a) and (b), respectively. The photographic superposition (c) of the peaks is very similar in shape to that shown in Figure 20. (Redrawn from Pinkel et al. [50].)

herein to any specific commercial products, process, or service by tradename, trademark, manufacturer, or otherwise, does not necessarily constitute or imply its endorsement, recommendation, or favoring by the United States Government or the University of California. The views and opinions of authors expressed herein do not necessarily state or reflect those of the United States Government thereof and shall not be used for advertising or product endorsement purposes.

REFERENCES

1. **Balhorn R** (1982) A model for the structure of chromatin in mammalian sperm. J Cell Biol 93:298–305.
2. **Ballachey BE, Miller HL, Jost LK, Evenson DP** (1986) Flow cytometry evaluation of testicular and sperm cells obtained from bulls implanted with zeranol. J Animal Sci 63:995–1004.
3. **Benaron DA, Gray JW, Gledhill BL, Lake S, Wyrobek AJ, Young IT** (1982) Quantification of mammalian sperm morphology by slit-scan flow cytometry. Cytometry 2:344–349.
4. **Calvin HI, Yu CC, Bedford JM** (1973) Effects of epididymal maturation, zinc(II), and copper(II) on the reactive sulfhydryl content of structural elements in rat spermatozoa. Exp Cell Res 81:333–341.
5. **Clausen OPF, Abyholm T** (1980) Deoxyribonucleic acid flow cytometry of germ cells in the investigation of male infertility. Fertil Steril 34:369–374.
6. **Clausen OPF, Parvinen M, Kirkhus B** (1982) Stage-related variations in DNA fluorescence distribution during rat spermatogenic cycle measured by flow cytometry. Cytometry 2:421–425.
7. **Clausen OPF, Purvis K, Hansson V** (1977) Application of micro-flow fluorometry to studies of meiosis in the male rat. Biol Reprod 17:555–560.
8. **Darzynkiewicz Z, Traganos F, Sharpless T, Melamed MR** (1975) Thermal denaturation of DNA in situ as studied by acridine orange staining and automated cytofluorometry. Exp Cell Res 90:411–428.
9. **Evenson DP, Ballachey BE, Jost LK, Baer RK** (1986) Male germ cell analysis by flow cytometry: Effects of cancer, chemotherapy and other factors on testicular function and sperm chromatin structure. In Andreef

MA (ed), "Proceedings of the Conference on Clinical Cytometry." New York: New York Academy of Science, pp 350–367.

10. **Evenson DP (1988)** Flow cytometry assays of male fertility. In Darzynkiewicz Z, Crissman H (eds), Methods in Cell Biology, Vol. 33: "Flow Cytometry." Academic Press, New York.
11. **Evenson DP, Arlin Z, Welt S, Claps ML, Melamed MR (1984)** Male reproductive capacity may recover following drug treatment with the L-10 protocol for acute lymphocytic leukemia. Cancer 53:30–36.
12. **Evenson DP, Baer RK, Jost LK, Gesch RW (1986)** Toxicity of thiotepa on mouse spermatogenesis as determined by dual-parameter flow cytometry. Toxicol Appl Pharmacol 82:151–163.
13. **Evenson D, Darzynkiewicz Z, Jost L, Janca F, Ballachey B (1986)** Changes in accessibility of DNA to various fluorochromes during spermatogenesis. Cytometry 7:45–53.
14. **Evenson DP, Darzynkiewicz Z, Melamed MR (1982)** Simultaneous measurement by flow cytometry of sperm cell viability and mitochondrial membrane potential related to cell motility. J Histochem Cytochem 30:279–280.
15. **Evenson DP, Darzynkiewicz Z, Melamed MR (1980)** Relation of mammalian sperm chromatin heterogeneity to fertility. Science 210:1131–1133.
16. **Evenson DP, Darzynkiewicz Z, Melamed MR (1980)** Comparison of human and mouse sperm chromatin structure by flow cytometry. Chromosoma 78:225–238.
17. **Evenson DP, Higgins PJ, Grueneberg D, Ballachey BE (1985)** Flow cytometric analysis of mouse spermatogenic function following exposure to ethylnitrosourea. Cytometry 6:238–253.
18. **Evenson DP, Klein FA, Whitmore, WF, Melamed MR (1984)** Flow cytometric evaluation of sperm from patients with testicular carcinoma. J Urol 132: 1220–1225.
19. **Evenson DP, Mclamed MR (1983)** Rapid analysis of normal and abnormal cell types in human semen and testis biopsies by flow cytometry. J. Histochem Cytochem 31:248–253.
20. **Garner DL, Gledhill BL, Pinkel D, Lake S, Stephenson D, Van Dilla MA, Johnson LA (1983)** Quantification of the X- and Y-chromosome-bearing spermatozoa of domestic animals by flow cytometry. Biol Reprod 28: 312–321.
21. **Garner DL, Pinkel D, Johnson LA, Pace MM (1986)** Assessment of spermatozoal function using dual fluorescent staining and flow cytometric analysis. Biol Reprod 34:127–138.
22. **Gledhill BL, Gledhill MP, Rigler R Jr, Ringertz NR (1966)** Changes in deoxyribonucleoprotein during spermiogenesis in the bull. Exp Cell Res 41:652–665.
23. **Gledhill BL, Lake S, Dean PW (1979)** Flow cytometry and sorting of sperm and other male germ cells. In Melamed MR, Mullaney PF, Mendelsohn ML (eds), "Flow Cytometry and Sorting." New York: John Wiley & Sons, pp 471—484.
24. **Gledhill BL, Lake S, Steinmetz LL, Gray JW, Crawford JR, Dean PN, Van Dilla MA (1976)** Flow microfluorometric analysis of sperm DNA content: Effect of cell shape on the fluorescence distribution. J Cell Physiol 87:367–376.
25. **Gledhill BL (1983)** Control of mammalian sex ratio by sexing sperm. Fertil Steril 40:572–574.
26. **Haas GG Jr, Cunningham ME (1984)** Identification of antibody-laden sperm by cytofluorometry. Fertil Steril 42:606–613.
27. **Haas GG Jr, Cunningham ME, Culp L (1984)** The effect of freezing on sperm-associated immunoglobulin G (IgG). Fertil Steril 42:761–764.
28. **Hacker U, Schumann J, Gohde W (1980)** Effects of acute gamma-irradiation on spermatogenesis as revealed by flow cytometry. Acta Radiol [Oncol] 19: 361–368.
29. **Hacker U, Schumann J, Gohde W, Muller K (1981)** Mammalian spermatogenesis as a biologic dosimeter for radiation. Acta Radiol [Oncol] 20:279–282.
30. **Halamka J, Gray JW, Gledhill BL, Lake S, Wyrobek AJ (1984)** Estimation of the frequency of malformed sperm by slit scan flow cytometry. Cytometry 5:333–338.
31. **Janca FC, Jost LK, Evenson DP (1986)** Mouse testicular and sperm cell development characterized from birth to adulthood by dual parameter flow cytometry. Biol Reprod 34:613–623.
32. **Johnson LV, Walsh ML, Chen LB (1980)** Localization of mitochondria in living cells with rhodamine 123. Proc Natl Acad Sci USA 77:990–994.
33. **Johnson LA, Flook JP, Look MV, Pinkel D (1987)** Flow sorting of X and Y chromosome-bearing spermatozoa into two populations. Gamet Res 16:1–9.
34. **Johnson LA, Pinkel D (1986)** Modification of a laser-based flow cytometer for high-resolution DNA analysis of mammalian spermatozoa. Cytometry 7:268–273.
35. **Klein R, Pfitzer P (1984)** Flow cytometry of postmortem human testicular tissues in cases of atherosclerosis. Cytometry 5:636–643.
36. **Libbus BL, Schuetz AW (1978)** Analysis of the progression of meiosis in dispersed rat testicular cells by flow cytofluorometry. Cell Tissue Kinet 11:377–391.
37. **McLean-Grogon W, Farnham WF, Sabau JM (1981)** DNA analysis and sorting of viable testis cells. J Histochem Cytochem 29:738–746.
38. **Meistrich ML (1977)** Separation of spermatogenic cells and nuclei from rodent testes. In Prescott DM (ed), "Methods in Cell Biology." Vol. 15: Orlando, FL: Academic Press, pp 15–54.
39. **Meistrich ML, Gohde W, White RA, Longtin JL (1979)** "Cytogenetic" studies of spermatids of mice carrying Cattanach's translocation by flow cytometry. Chromosoma 74:141–151.
40. **Meistrich ML, Gohde W, White RA, Schumann J (1978)** Resolution of X and Y spermatids by pulse cytophotometry. Nature (Lond) 274:821–823.
41. **Meistrich ML, Lake S, Steinmetz LL, Gledhill BL (1978)** Increased variability in nuclear DNA content of testis cells and spermatozoa from mice with irregular meiotic segregation. Mutat Res 49:397–405.
42. **Meistrich ML, Lake S, Steinmetz LL, Gledhill BL (1978)** Flow cytometry of DNA in mouse sperm and testis nuclei. Mutat Res 49:383–396.
43. **Ohno S, Jainchill J, Stenius C (1963)** The creeping vole as a gonosomic mosaic. The OY/XY constitution in the mole. Cytogenetics 2:232–239.
44. **Otto FJ, Hacker U, Zante J, Schumann J, Gohde W,**

Meistrich ML (1979) Flow cytometry of human spermatozoa. Histochemistry 61:249–254.

45. **Pellicciari C, Hosokawa Y, Fukuda M, Manfredi Romanini MG** (1983) Cytofluorometric study of nuclear sulphydryl and disulphide groups during sperm maturation in the mouse. J Reprod Fertil 68:371–376.
46. **Pfitzer P, Gilbert P, Roly G, Vyska K** (1982) Flow cytometry of human testicular tissue. Cytometry 3: 116–122.
47. **Pinkel D** (1984) Cytometric analysis of mammalian sperm for induced morphologic and DNA content errors. In Eisert WG, Mendelsohn ML (eds), "Biological Dosimetry: Cytometric Approaches to Mammalian Systems." Berlin: Springer-Verlag, pp 111–126.
48. **Pinkel D, Dean P, Lake S, Peters D, Mendelsohn M, Gray J, Van Dilla M, Gledhill B** (1979) Flow cytometry of mammalian sperm: Progress in DNA and morphology measurement. J Histochem Cytochem 27: 353–358.
49. **Pinkel D, Garner DL, Gledhill BL, Lake S, Stephenson D, Johnson LA** (1985) Flow cytometric determination of the proportions of X- and Y-chromosome-bearing sperm in samples of purportedly separated bull sperm. J Anim Sci 60:1303–1307.
50. **Pinkel D, Gledhill BL, Lake S, Stephenson D, Van Dilla MA** (1982) Toward sex preselection in mammals? Separation of sperm bearing Y and "O" chromosomes in the vole *Microtus oregoni*. Science 218: 904–906.
51. **Pinkel D, Gledhill BL, Van Dilla MA, Lake S, Wyrobek AJ** (1983) Radiation-induced DNA content variability in mouse sperm. Radiat Res 95:550–565.
52. **Pinkel D, Lake S, Gledhill BL, Van Dilla MA, Stephenson D, Watchmaker G** (1982) High resolution DNA content measurements of mammalian sperm. Cytometry 3:1–9.
53. **Smith AJ, Clausen OPF, Kirkhus B, Jahnsen T, Moller OM, Hansson V** (1984) Seasonal changes in spermatogenesis in the blue fox (*Alopex lagopus*) quantified by DNA flow cytometry and measurements of soluble Mn^{2+}-dependent adenylate cyclase activity. J Reprod Fertil 72:453–461.
54. **Spano M, Calugi A, Capuano V, deVita R, Gohde W, Hacker-Klom U, Maistro A, Mauro F, Otto F, Pacchierotti F, Rocchini P, Schumann J, Teodori L, Vizzone A** (1984) Flow cytometry and sizing for routine andrological analysis. Andrologia 16:367–375.
55. **Stanker LH, Wyrobek AJ, Balhorn R** (1987) Monoclonal antibodies to human protamines. Hybridoma 6:293–303.
56. **Tejada RI, Mitchell JC, Norman A, Marik JJ, Friedman S** (1984) A test for the practical evaluation of male fertility by acridine orange (AO) fluorescence. Fertil Steril 42:87–91.
57. **Thorud E, Clausen OPF, Abyholm T** (1981) Fine needle aspiration biopsies from human testes evaluated by DNA flow cytometry. In Lareum O, Lindmo T, Thorud E (eds), "Flow Cytometry." Vol. IV. Oslo: Universitetsforlaget, pp 175–177.
58. **Topham JC** (1980) The detection of carcinogen-induced sperm head abnormalities in mice. Mutat Res 69:149–155.
59. **van Kroonenburgh MJ, Beck JL, Scholtz JW, Hacker-Klom U, Herman CJ** (1985) DNA analysis and sorting of rat testis cells using two-parameter flow cytometry. Cytometry 6:321–326.
60. **Wyrobek AJ, Bruce WR** (1975) Chemical induction of sperm abnormalities in mice. Proc Natl Acad Sci USA 72:4425–4429.
61. **Wyrobek AJ, Gordon LA, Burkhart JG, Francis MW, Kapp RW Jr, Letz G, Malling HV, Topham JC, Whorton MD** (1983) An evaluation of the mouse sperm morphology test and other sperm tests in nonhuman mammals: A report of the U.S. Environmental Protection Agency Gene-Tox Program. Mutat Res 115:1–72.
62. **Wyrobek AJ, Gordon LA, Burkhart JG, Francis MC, Kapp RW Jr, Letz G, Malling HV, Topham JC, Whorton MD** (1983): An evaluation of human sperm as indicators of chemically induced alterations of spermatogenic function: A report of the U.S. Environmental Protection Agency Gene-Tox Program. Mutat Res 115:73–148.
63. **Zimmermann A, Truss F** (1980) The prognostic power of flow-through cytophotometric DNA determinations for testicular diseases. Anal Quant Cytol 2:247–251.

Mutagenesis as Measured by Flow Cytometry and Cell Sorting

Ronald H. Jensen and James F. Leary

Cytochemistry Section, Division of Biomedical Science, Lawrence Livermore National Laboratory, Livermore, California 94550 (R.H.J.); Department of Pathology, University of Rochester, Rochester, New York 14642 (J.F.L.)

During the past two decades, there has been increasing concern over the genetic hazards from chemicals introduced into our environment. Much information has been obtained from a large number of mutagenicity tests in different organisms with a wide array of chemicals. These tests have been graded into a three-tier approach. Tier I tests are inexpensive and usually use prokaryotes; tier II tests are more difficult and usually use cultured mammalian cells; tier III tests require long-term animal exposure studies to attempt confirmation and quantitation of genotoxicity of particular compounds [3,36]. These tests have been extremely informative about the heterogeneity of genetic hazard of different chemicals, as well as the heterogeneity of response of different organisms. Because of the large differential effects between organisms, it has proved extremely difficult to determine the genotoxic response in one organism and use those results to predict the response in another organism. Therefore, we cannot always use the results of the three tiers of mutagenicity tests to make informed estimates about the genotoxic risk to humans from many of these chemicals.

Recently there has been growing interest in developing new tests for detecting mutations and analyzing mutagenesis in eukaryotes, and particularly in humans [9,27]. By using such tests, we may be able to determine the genotoxic effects of exposure on different mammalian species and also on individual humans. Many of these efforts are aimed toward detecting mutations in individual cells rather than examining the progeny of these cells or monitoring the progeny of whole eukaryotic organisms for heritable phenotypes to confirm a genotoxic effect.

VALUE OF FLOW CYTOMETRY AND CELL SORTING IN RARE EVENT DETECTION

Because many mutagenesis studies require detection of very rare occurrences, large sample sizes often are necessary to ensure statistical significance of data generated from the analyses. Flow cytometry and cell sorting are being applied to such problems to take advantage of the high speed with which flow cytometers analyze the properties of individual cells. Billions of cells can be screened to identify subpopulations at frequencies as low as 10^{-5}. The identity of the particles that generate signals at these low frequencies can be confirmed by cell sorting followed by some other analysis (e.g., microscopic, microbiochemical, biological, immunological, or enzymatic).

A valuable feature of flow cytometry is the ability to correlate multiple parameters per cell. This is especially important in situations in which two or more parameters are tightly correlated in a mutagenic event. One case is that of cell transformation assays in which more than one cellular property must change, minimizing errors due to random variations in measuring a single property. Other chapters of this book illustrate the many different cellular parameters that can be measured by flow cytometry. For example, fluorescent DNA stains can be used to measure changes in DNA content or composition. Light-scatter intensities at different angles relative to the light source can be used to measure cell size, shape, and viability. Fluorescent polynucleotide probes can be used to measure particular DNA sequences in cells or chromosomes. Fluorescent antibodies (particularly monoclonal antibodies) provide exquisitely specific probes for antigens contained in different cell subpopulations. All these parameters have been used in different flow cytometric methods for study of mutagenesis, and it is probable that other parameters and multiparameter analysis with several parameters will be used in the near future.

IN VITRO

A powerful advantage of flow sorting for the study of mutagenesis of cultured cells lies in the ability to select for variant phenotypes without applying selective growth conditions. Often growth conditions that are selective for desirable mutations cannot be obtained, particularly for diploid cell systems. Even recessive mutants that are at a disadvantage under selective growth conditions may grow well since the unmutated allelic gene can provide growth requirements

Flow Cytometry and Sorting, Second Edition, pages 553–562

for the cell. Therefore, selective systems are generally limited to genes that map in hemizygous regions of cellular genomes [35]. Genes on the sex chromosomes are naturally haploid, whereas haploid genes on autosomal chromosomes must be induced by tissue culture manipulation such as cross-species cell hybridization or subcloning of cells that have acquired chromosome segregation errors during division. Such techniques result in cells that are atypical of the natural diploid state of some genes, and even though mutagenesis by growth selection may be performed, there remains concern that such events are not representative of the natural process in normal diploid cells. Indeed, many genes of interest, even when they occur in a hemizygous state, do not control functions for which selective growth conditions can be generated. One solution to these selection difficulties is to label variant cells so that flow cytometers can discriminate between the many nonvariant cells and an exceedingly low frequency of variants. Then high-speed cell sorting can be used to separate variant cells from the normal population and to culture them independently.

Isotype Switch Variants

To date, the most common means of labeling for variant cells has been to use antibodies that bind particular variant cell-surface antigens. A good example of this technique is the use of specific antibodies against mouse immunoglobulin isotypes to label viable hybridoma cells for the presence of variant isotype-producing cells [29]. The eight classes of immunoglobulin isotypes—IgM, IgD, IgG3, IgG1, IgG2a, IgG2b, IgE, and IgA—are defined by the primary sequence of the constant regions of their heavy chains. Expression of heavy chain results from intragenomic DNA rearrangements in which a V (variable) gene segment is joined with other gene segments (D and J) as an early step in completing the immunoglobulin gene assembly [39]. In vivo, this VDJ assembly first is attached to the μ gene segment to code for synthesis of an IgM heavy chain. During cellular differentiation, VDJ will translocate to a position 5′ to one of the other heavy-chain gene segments (e.g., δ, $\gamma 3$, $\gamma 1$, $\gamma 2b$, $\gamma 2a$, ϵ or α) and code for a different immunoglobulin isotype. This chain class switch phenomenon is a form of DNA reorganization that is specifically built into the cellular differentiation scheme. Therefore, it is not really a mutagenic event, but rather a developmental event.

Cell sorting has been used to detect a very similar DNA reorganization in cells growing in culture, linked to mutational disturbance [8,24]. A typical experimental sequence is illustrated in Figure 1, which shows the results of the Herzenberg group's attempt to isolate a clone of IgG2b-producing hybridoma cells from an IgG1-producing hybridoma cell culture. Fluorescent antibodies specific for isotype IgG2 were added to hybridoma cells to label any rare cells that have experienced a chain class switch and started to produce an IgG2 isotype. Flow cytometric analysis showed that most of the cells remained unstained. However, sorting of cells that displayed fluorescence intensity high enough to be IgG2 producers, resulted in a small yield of viable cells. These cells were expanded in culture and selected for IgG2 producers by a repetition of the same flow sorting procedure. This resulted in a high enough concentration of the variant cell type to perform cloning of the new cell type, an IgG2b producer. Detailed analyses of the DNA sequence in cell types isolated by these techniques show significant differences between the in vivo differentiation sequence and the in vitro spontaneous class switch sequence [32]. The best hypothesis is that the in vitro occurrence is the expression of a transchromosomal recombinatorial event. Thus, the flow-sorting technique is detecting events that are a result of mutational disturbance and expressed as cell-surface variant protein.

Gene Amplification

Another phenomenon that can be studied effectively using flow cytometry and cell sorting is the occurrence of gene amplification in cultured cells. The system studied most intensively is the amplification of the dihydrofolate reductase gene (DHFR) in Chinese hamster ovary (CHO) cells. Expression of the genetic change was first detected by challenging the cells with methotrexate (MTX), an inhibitor of DHFR. After such selective pressure was applied, cells accumulated that were resistant to the drug. Analysis of these cells showed that many copies of the DHFR gene had accumulated in viable cells, enabling them to synthesize large amounts of the enzyme, making them resistant to MTX inhibition. A major question in this phenomenon was whether gene amplification was a consequence of the selective conditions. Since MTX causes damage and fragmentation to chromosomes [13], it seemed possible that it caused deranged DNA synthesis resulting in amplification of the DHFR gene. Alternatively, gene amplification may have occurred spontaneously, with no MTX treatment, and drug selection merely displayed the presence of resistant cells. The classic Luria–Delbruck fluctuation analysis [25] has been used to show a low frequency (ca. 4×10^{-5} per generation) of cells resistant to high levels of MTX are present as variants in an untreated population of cells [36]. The fluctuation analysis, however, is insensitive to some features of gene amplification in that it is most suitable for detecting rare, random, stable mutations that confer high resistance to a drug. Gene amplification resistance often is unstable and usually confers graded resistance to selective agents [20]. Therefore, Terzi and Hawkins [37] may have determined an inaccurate frequency of resistant cells by misaccounting for intermediate-resistant or unstable variant cells.

A method for isolating and independently culturing drug-resistant cells without using selective pressure was developed using a high-speed cell sorter [19]. The method requires using fluoresceinated MTX as a probe to stain cells under nonselective conditions. Cells containing high levels of DHFR bind high levels of the fluoresceinated MTX and could be collected using flow sorting, even though their presence was not obvious in fluorescence histograms. These sorted cells were then cultured under nonselective conditions, stained, and sorted again in the same manner. After several such sequential sorts and growths, the histogram of such cultures showed high levels of fluoresceinated MTX bound to a significant fraction of the cell population [19]. Cells derived from a 10-fold sorting series showed 50-fold more fluorescence when stained with fluoresceinated MTX, and these cells were nearly 100-fold more resistant to MTX than were the original cultured cells. In addition, such cells showed about a 40-fold increase in amount of DNA that hybridizes with DHFR complementary DNA (cDNA), indicating extensive gene amplification.

Kaufman and Schimke [20] also selected a highly resistant clone from this growth and performed the reverse flow-sorter selection technique for low fluorescent staining to show that cells with less copies of the DHFR gene (de-amplified cells) could be found in the amplified clone.

These cell-sorter results clearly demonstrated that gene amplification and de-amplification could occur without selective pressure. In addition, mathematical analysis of the flow histograms was used to calculate the rate of accumulation of variant cells in each population. From these analyses, they calculated that 8×10^{-4} gene-amplified resistant cells were generated in each cell division and as many as 3×10^{-2} de-amplified sensitive cells were generated per cell division in the drug-resistant population. Thus, while these are rare events, the frequency is surprisingly high and increases as gene copy number increases. If these frequencies of genome alteration are typical of much of the genome, the genome may be much more variable than previously thought.

Flow cytometry and cell sorting for detecting and studying gene amplification also can be applied to other gene loci. No toxic chemical treatment is required, only a fluorescent probe that binds differentially to cells depending on their extent of gene expression. An example of such selection techniques for gene amplification is the study by Kavathas and Herzenberg [21] using flow sorting to select a murine–human hybrid cell that expresses increased levels of human T-cell differentiation antigen, Leu-2. A population of growing hybrid cells was stained with fluoresceinated monoclonal anti-Leu-2, and the cells labeled most brightly were sorted, grown, relabeled and re-sorted in an analogous fashion to that described above. After repeating this procedure six times, the resulting population stained forty times brighter than the original; the cellular DNA transformed other cells 20 times more efficiently for Leu-2; and the cells contained 50-fold more human DNA sequences than the original. Thus, by using flow sorting, gene amplification of the Leu-2 gene in this hybrid cell line has been found.

An exciting new direction to carry this approach would be to use flow cytometry and cell sorting to select for cells that contain amplified oncogenes. In various types of mammalian tumors and tumor lines, oncogenes have been found in an amplified state and increased expression of the cellular homolog of known viral oncogenes has been detected [22,34]. If a sensitive fluorescent polynucleotide probe were used for labeling cell populations, flow cytometry should be capable of detecting subpopulations of cells with genomes containing amplified oncogenes. In some cases, the activation of an oncogene involves a point mutation in the cell homolog oncogene and a resultant amino acid substitution in the encoded polypeptide. Thus, if a very sensitive technique were developed to label cells with antibodies for the presence of such an altered polypeptide, amplified gene expression might be detectable on a cellular level, and cell sorting could be performed on cell populations to select out cells that exhibit high expression of an oncogene.

CLASTOGENICITY ASSAY

While flow cytometry is not used extensively in assays designed to screen chemical compounds for their mutagenic properties, the capability to analyze individual cells at high speed makes it potentially extremely useful for these assays. One application toward this end is the use of flow cytometry to screen for mutagens using the mouse micronucleus assay [33].

The mouse micronucleus assay is based on the observation that mitotic cells with chromatid breaks or exchanges tend to suffer from disturbances in the distribution of chromatin during anaphase. After telophase, a sizable proportion of this displaced chromatin is not included within the nuclei of daughter cells but instead forms single or multiple nuclei (micronuclei) in the cytoplasm of these cells. These micronuclei generally contain only 1–10% of diploid DNA. Testing for the presence of micronuclei represents a method of screening for clastogenic (chromosome breaking) agents. Because the chemical exposure is to whole animals, this assay is categorized in tier III of screening assays, and is deemed to be a reasonable estimate of a chemical's clastogenic risk to humans.

During normal erythropoiesis, erythrocytes in the bone marrow expel their nuclei. If a mouse is first exposed to a clastogenic agent, this normal differentiation pattern is disturbed and a micronucleus remains behind in the erythrocyte. A manual mouse blood micronucleus test has been used for many years as a short-term test for clastogenic agents. This conventional method involves microscopic detection and enumeration of the frequency of micronucleated cells in an early reticulocyte subpopulation of blood cells. The manual scoring process is influenced by technician experience and bias and provides marginal statistical significance because of the low frequency of micronucleated polychromatic erythrocytes and the relatively small sample sizes (usually 1,000 reticulocytes are scanned). In the bone marrow of mice treated with clastogenic agents, micronucleated polychromatic reticulocytes occur at a frequency of approximately 1–10%, whereas in the bone marrow of unexposed mice the frequency is typically 0.1–0.5%. A major improvement in the manual method was achieved by staining with acridine orange [11] or Hoechst 33258/pyronin Y [26]. These fluorescent DNA and RNA-staining procedures have made manual counting easier and less subjective, but processing of slides for testing a single chemical compound still can require 15–25 hours of manual cell counting with a fluorescent microscope—an obviously slow, labor-intensive, and costly process.

Recently, flow cytometry was used to automate the mouse micronucleus assay [14] by staining micronucleated erythrocytes with a DNA-specific fluorophor, DAPI (4′-6′-diamidino-2-phenylindole) and also a protein-specific fluorophor, SR101 (sulforhodamine 101). Two-parameter flow cytometry was used to identify red blood cells by the intensity of SR101 fluorescence and micronucleated cells by DAPI fluorescence of DNA. Animals treated with the clastogenic agents cyclophosphamide or mitomycin C were examined for increases in frequency of micronucleated erythrocytes caused by these agents. The frequency data showed a strong linear correlation with a high degree of statistical significance (correlation coefficient $r = 0.96$) between manual microscopic and flow cytometric counts of micronucleated erythrocytes. Several other research groups are now attempting to develop this flow cytometric assay into a routine screening test for potentially clastogenic chemicals in the environment.

MUTAGENESIS OF HUMAN SOMATIC CELLS IN VIVO

Because of the connection between mutagenesis and carcinogenesis, it would be desirable to measure the level of mutagenic injury in somatic cells of humans. If such a measurement can be obtained, then a correlation between mutagenic burden and risk of cancer may be possible. Unfortunately, most somatic cells replicate poorly in culture. Thus, clonogenic assays of human somatic cells are not generally available.

Hypoxanthine Guanine Phosphoribosyl Transferase

One assay that has been developed (1) selects for T lymphocytes that have lost the ability to synthesize the purine

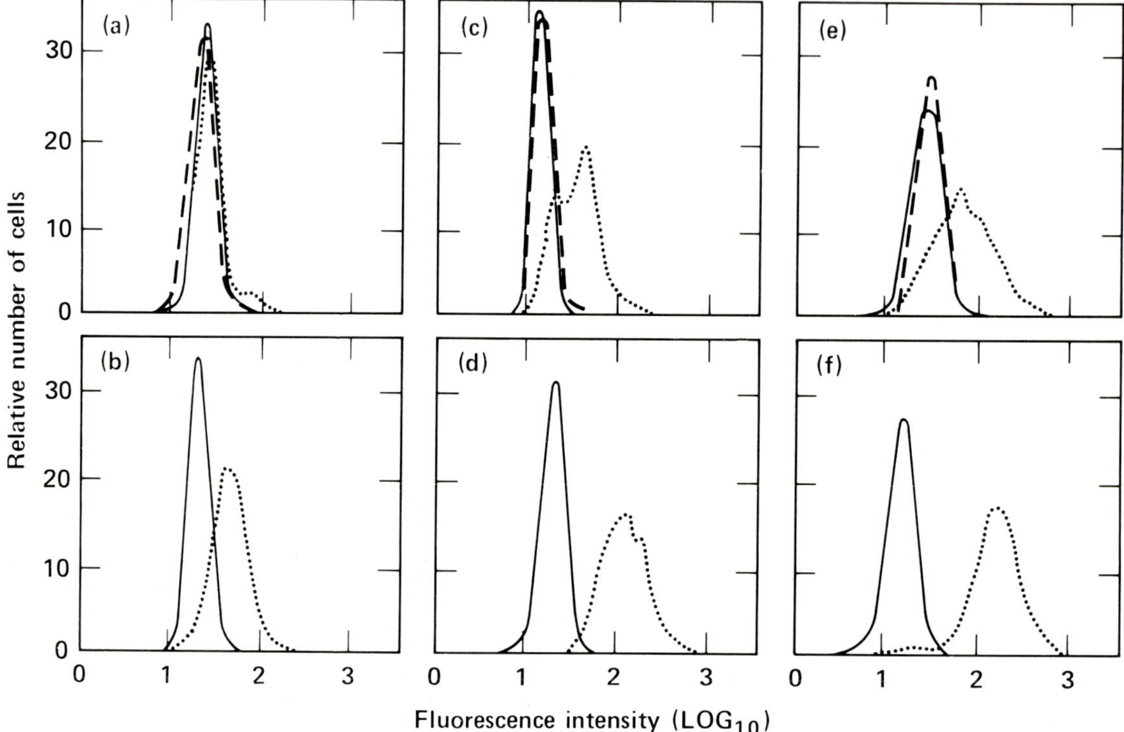

Fig. 1. Selection of isotype switch variant cells using flow sorting. **a, c, e:** Bulk-type sorting enrichment. **b, d, f:** Single-cell sorting and cloning. Three different mouse hybridoma cell lines were grown in culture, and an aliquot of each was suspended in a solution of fluorescein-conjugated antibody that is specific for a different isotype from the one produced by the cell line being analyzed. **a, b:** The cell line produces IgG1 and the antibody is IgG2 specific. **c, d:** The cell line produces IgG2b, and the antibody is IgG1 specific. **e, f:** The cell line produces IgG2a, and the antibody is IgE specific. The aliquots were analyzed by flow cytometry (solid lines in all panels). In the bulk sorting enrichment (**a, c, e**, the brightest 0.5% of the cells were sorted and regrown. This new population was again labeled with the same isotype-specific antibody, reanalyzed (dashed lines), and again the brightest 0.5% of the cells were sorted. These cells also were grown, labeled with the sample isotype-specific antibody, and reanalyzed (dotted lines). **b, d, f:** Cells grown after the first enrichment in **a, c, e** were sorted as individual cells into individual wells of a 96-well plate and grown as clones. **b, d, f:** One of the clones from each 96-well plate was labeled with the same isotype-specific antibodies and reanalyzed (dotted lines). (Reproduced from Dangl et al. [8], with permission of the publisher.)

metabolizing enzyme, hypoxanthine-guanine phosphoribosyl transferase (HGPRT). This enzyme converts a synthetic purine, 6-thioguanine (TG), into a toxic metabolite so that cells lacking HGPRT are resistant to the toxic effects of 6-thioguanine (TG^r). Since the HGPRT gene is located on the X chromosome, its expression is haploid. Thus, a single mutation in a somatic cell can result in complete loss of function of the enzyme. The original assay for TG^r lymphocytes used [^{3}H]thymidine incorporation and autoradiography in the presence and absence of TG to enumerate the cells that continue to synthesize DNA in the presence of TG and thus are viable, TG^r variants. Because this was a null assay, the overriding concern was that nonmutational events could also lead to lack of expression of the HGPRT and appear as variants in the assay. These variants, called phenocopies, would be falsely identified as mutated cells; thus, there were continuing efforts to develop a clonogenic assay [2]. Because of recent technical advances in the culture of somatic cells, clones of human T cells have been developed using complex culture media, interleukin-2 (IL-2) priming, and feeder layers [28]. This clonogenic assay is being tested as a means for determining the frequency of TG^r T cells in human blood samples. Early results suggest that it will function as a mutagenesis assay, but requires long periods of time to grow generations of cells and to detect the variant phenotype in progeny.

In parallel with the development of the clonogenic assay, the original autoradiographic TG^r assay has been improved by flow cytometric enumeration of labeled cells [4,40]. Cells deficient in HGPRT activity are resistant to the toxic effects of TG and can synthesize DNA, while the HGPRT proficient cells are sensitive and stop DNA synthesis early in S phase. Thus, after incubating in the presence of TG, the resistant cells should have greater DNA content than the sensitive cells. If at this stage of the assay, cells are labeled with a DNA-specific fluorophor, a DNA histogram can be obtained that relates the frequency of cells to their phase in the cell cycle. A lymphocyte population that has been stimulated to grow will show a modest number of S-phase and G2-phase cells, while a population that is stimulated in the presence of toxic amounts of TG will be inhibited by the toxin (Fig. 2). In theory a sorting window set to collect cells only in the S and G2 phase will isolate only cells that are TG^r and therefore $HGPRT^-$. In practice there are a small number of cells that nonspecifically attach to one another and these pairs contain twice the DNA of a single cell. They also appear in

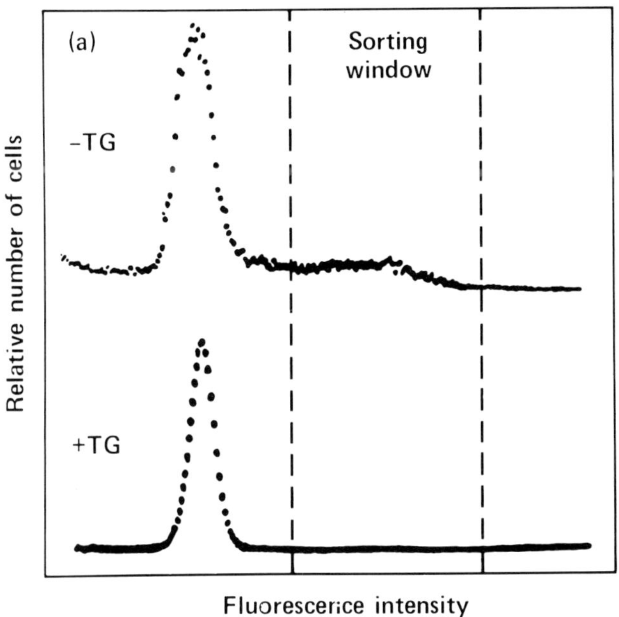

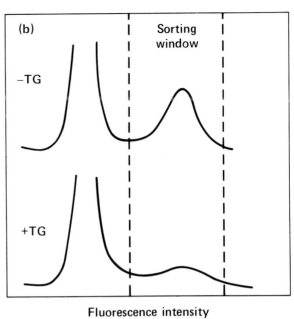

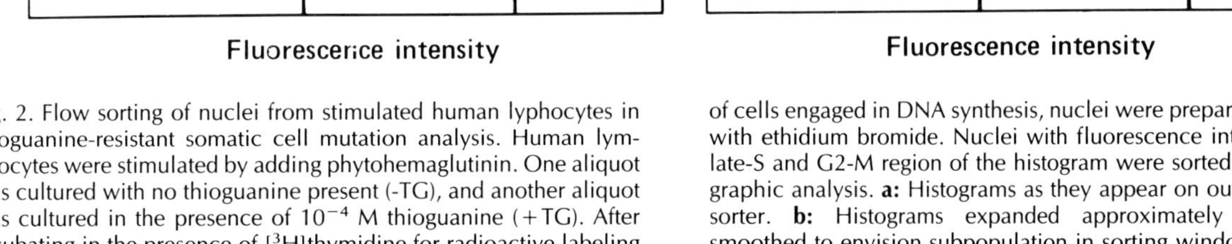
Fig. 2. Flow sorting of nuclei from stimulated human lyphocytes in thioguanine-resistant somatic cell mutation analysis. Human lymphocytes were stimulated by adding phytohemaglutinin. One aliquot was cultured with no thioguanine present (-TG), and another aliquot was cultured in the presence of 10^{-4} M thioguanine (+TG). After incubating in the presence of [^{3}H]thymidine for radioactive labeling of cells engaged in DNA synthesis, nuclei were prepared and stained with ethidium bromide. Nuclei with fluorescence intensities in the late-S and G2-M region of the histogram were sorted for autoradiographic analysis. **a:** Histograms as they appear on output from flow sorter. **b:** Histograms expanded approximately 25-fold and smoothed to envision subpopulation in sorting window.

the sorting window even though they are HGPRT$^+$. The enrichment of DNA-synthesizing cells can be determined by autoradiography of cells that were first [^{3}H] thymidine labeled exactly as in the original procedure, then sorted by the above procedure. With this two-step approach enrichment of tritiated thymidine-labeled cells is at least 2,000-fold. Therefore, the cell-sorting procedure does not replace the original autoradiographic procedure but makes it several orders of magnitude easier to perform.

Hemoglobin Variants

Flow cytometers and cell sorters can be used in several other ways to detect and count mutant somatic cells in populations of cells obtained from humans. The most direct approach is to label cells with antibodies that bind specifically to mutationally modified variant protein. Even proteins that differ in their structure by a single amino acid may be distinguished by specific antibodies. Jensen and colleagues [6] depended on immunologic labeling for flow cytometry detection of mutated human erythrocytes containing hemoglobin with a single amino acid substitution. The rare individuals carrying a heritable single amino acid substitution in a hemoglobin gene show that such hemoglobin is expressed in red cells circulating in the peripheral blood. If the mutational event occurs somatically in bone marrow cells, progeny of such cells would express heterozygous phenotypes for their hemoglobin and appear in the peripheral blood. The frequency of such cells in the peripheral blood should be an indicator of the frequency of somatic mutations in the bone marrow population and thus be a measure of the mutagenic burden that is carried by that individual.

To test the validity of this hypothesis requires highly specific antibody, high-speed cell analysis, and cell-sorting capability. Antibodies specific for variant human hemoglobins have been purified from serum samples [30] and isolated as monoclonal antibodies [15]. One variant hemoglobin that often is used as a phenotypic marker for the somatic mutants is sickle cell hemoglobin (hemoglobin S). To label cells for hemoglobin type, it is possible to use a membrane permeable crosslinking reagent, dimethylsuberimidate, to fix the intracellular hemoglobin into a stable, cell-sized pellet. The resulting "hard cells" are resistant to hypotonic shock, but their membranes can be made permeable by enzyme degradation, organic chemical extraction, or modest heating. In general, the heated cells are best for flow cytometry since they aggregate less than cells treated with enzymes or organic extracts. Heated hard cells that contain hemoglobin S bind antihemoglobin S antibodies specifically and exhibit bright fluorescence for detection in the flow cytometer. A histogram that displays this specificity of fluorescent labeling is shown in Figure 3.

Manual fluorescence microscopic analysis of normal human blood stained with hemoglobin S-specific antibody has been performed, and the frequency of stained cells was found to be 1–10 × 10^{-8} [35]. This frequency of variant cells is small enough that normal flow cytometric procedures cannot be used for enumeration of the labeled cells. Typical flow rates and cell concentrations result in processing 100–1,000 cells/sec. To enumerate 100 variant cells that occur at a frequency of 10^{-7} would require hundreds of hours of continuous flow cytometry. Flow cytometry can be performed at much higher rates by increasing the cell concentration (to about 5 × 10^8 cells/ml) and increasing the rate of flow in the cytometer or sorter by about 10-fold [31]. Performing both of these changes results in processing at 1–2 × 10^6 cells/sec. Thus, a sample of 10^9 cells can be analyzed in 20 min. Be-

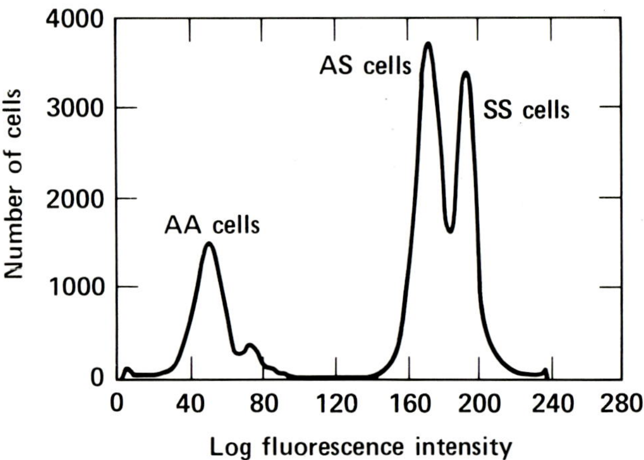

Fig. 3. Flow histogram of erythrocytes labeled with anti-hemoglobin S monoclonal antibody, HuS-1. A mixture of three different blood specimens was fixed, stained with HuS-1, and analyzed using flow cytometry as described in Jensen et al. [15]. The three blood types were normal human blood containing hemoglobin A (AA cells), blood from a homozygous sickle hemoglobin individual (SS cells), and blood from a heterozygous sickle trait individual (AS cells). Logarithmic amplification of the fluorescence intensity results in doubling of intensity every 28 channels on the abscissa.

cause the cell concentration is very high, there are several cells in the light beam at all times. If a large fraction of the cells were labeled, there would be continuously overlapping signals, and the measuring system would not function correctly. At the low frequency of variant somatic cells, no great difficulties with signal coincidence occur.

The primary difficulty in measuring events that occur at the low frequency of fluorescently labeled somatic variant red cells is the occurrence of false alarms. Whenever blood samples are labeled with fluorescent antibodies, some cells will bind antibody nonspecifically (e.g., white blood cells will bind antibodies at the Fc receptors, or poorly fixed cells will entrap antibody by hydrophobic binding). In addition, antibody may adsorb to small contaminant particles that occur in the suspension, or there may be precipitated antibody aggregates. By any of these means, a significant number of particles occur that give signals bright enough to be measured as variant cells. In fact, the frequency of such events is higher than the variant cell frequency by about 100-fold. Even extraordinary care in sample preparation and handling does not decrease this false alarm frequency to less than the variant cell frequency.

The best way to solve the false-alarm problem is to perform multiparameter flow sorting with subsequent analysis of the sorted cells. The first step is to double-stain cells, not only with monospecific antibody but also with a DNA-specific fluorophor. Then a two-parameter cytometric analysis is performed. One parameter is fluorescence from antibody corresponding with mutant protein and the other is from the nucleic acid-specific fluorophor, corresponding with DNA content. Using these two parameters with cell sorting, one can select for cells that bind antibody and also select against cells that contain DNA (such as white blood cells or platelets, which take up antibody nonspecifically). Objects that fluoresce brightly enough to be variant cells, but contain no DNA are potentially variant red cells and are sorted. From a sample with 10^9 red cells, about 10^4 such objects will be sorted. This sample can then be flow sorted once again, this time at a cell concentration low enough to analyze fluorescence and light scatter from every particle in the suspension. Most of the objects found in this sample are small, highly fluorescent particulate debris and are not sorted with the labeled cells because they display low intensity light scatter. To confirm the variant red cell origin of signals that give correct fluorescence and light scatter, the droplets are sorted onto microscope slides and analyzed by fluorescence microscopy. When this two-step sorting and microscopic analysis is performed, fluorescently labeled cells in the final sample occur at frequencies such that the original variant cell frequency is calculated to be $1–10 \times 10^{-8}$. This complex sorting and counting routine is being performed on a number of blood samples to determine reproducibility and correlation with exposure to mutagenic chemicals. The long range objective is to develop an assay of the mutagenic burden of individuals that will indicate their risk of neoplasia.

Glycophorin A Variants

One somatic cell mutation assay has been devised that depends on immunofluorescent flow cytometry to detect cells that have lost the expression of a gene [23]. This assay is based on detecting the red cell surface glycoprotein, glycophorin A (GPA), which carries the antigenic structures to determine the M and N blood groups. These blood group determinants are defined by a polymorphism in the amino acid sequence of the protein encoded by a pair of codominantly expressed alleles located on chromosome 4. The polymorphic amino acid sequence occurs near the amino terminus of glycophorin A and is shown in Figure 4. Except for the amino acid differences at positions one and five, the two glycoproteins are identical, both in amino acid sequence and in saccharide side chain sequences [38]. Individuals homozygous for the M or N allele synthesize only the A(M) or the A(N) sequence respectively, while heterozygous individuals have equal amounts of both glycoproteins on every erythrocyte.

The null mutation flow cytometric assay is performed on blood from GPA heterozygotes to detect rare erythrocytes that fail to express one or the other of the two allelic forms of GPA. To detect these functionally hemizygous cells using flow cytometry, monospecific antibodies that bind to only one of the two allelic forms of GPA were isolated [5]. Such antibodies were then labeled with different colored fluorophors and incubated with the blood from a heterozygote. Variant cells were detected by two-parameter flow cytometry as illustrated in Figure 5. Such cells would be missing one of the two allelic forms, but must express the other of the two forms. Phenocopies that lost the ability to express either of the GPA are not included in the variant cell enumeration.

Because the genetic target is larger than the single amino acid substitution assay described previously, the frequency of mutated cells is about 100-fold higher than in the hemoglobin studies. As a result, flow cytometry can be performed on cell suspensions at low enough concentration to analyze signals from every cell. This null mutation assay with the built in control for eliminating phenocopies, gives histograms in which the frequency of somatic null mutations can be measured reliably (Fig. 6).

Systematic evaluation of the feasibility of this assay is ongoing. Preliminary tests have been performed on cancer patients undergoing chemotherapy with mutagenic compounds

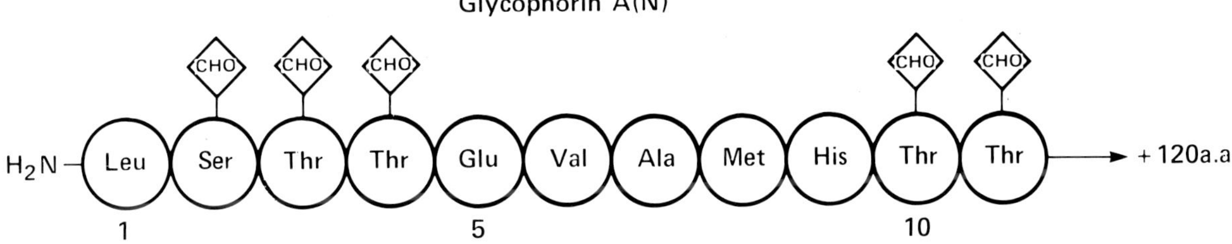

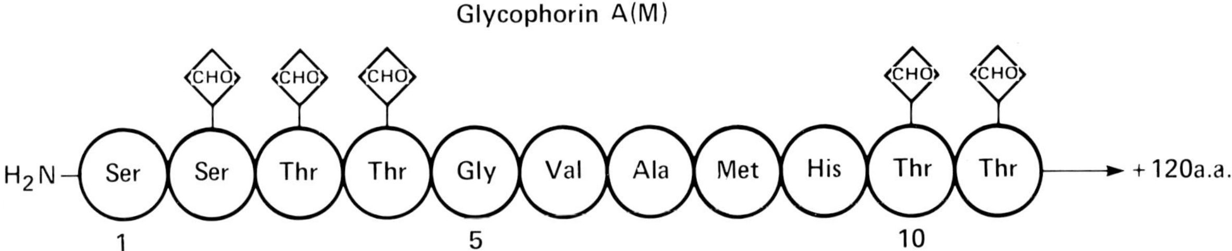

Fig. 4. Amino acid sequence of amino-terminus of the two allelic forms of human glycophorin A. Boxes labeled CHO signify carbohydrate side chains, all of which are identical tetrasaccharide units. These red cell surface antigens are imbedded through the cell membrane at amino acid position 73 and continue into the cell cytoplasm.

[16]. Longitudinal studies on patients as they are treated and after cessation of treatment have indicated that significant increases in variant erythrocyte frequency occur during the first few months of treatment, but persistence of the genotoxic burden from such treatment appears to be short. Within one red cell lifetime after treatment (120 days), variant cell frequencies return to normal [17]. This transient effect may be due to the high doses administered during chemotherapy, or it may be dependent on the particular mutagenic chemical to which individuals are exposed.

A more persistent effect has been observed from exposure of individuals to ionizing radiation. Samples obtained from survivors of the Atomic bomb over Hiroshima showed a correlation of variant cell frequency with the estimated radiation dose each individual received from the A bomb in 1945 [18]. This persistent effect more than 40 years after exposure indicates that stem cells in the bone marrow were mutated at the GPA locus and continue to produce variant progeny today. Although it is still early in the development of this assay, it appears to show promise as a lifelong cumulative biodosimeter for each person's radiation mutagenic burden.

INSTRUMENTATION ADVANCES FOR RARE EVENT DETECTION AND SORTING

While it is beyond the scope of this chapter to detail the complex and competing factors responsible for instrumental limitations on high-speed analysis and sorting, a few of these factors can be discussed and possible solutions outlined. One of the biggest problems in high speed flow cytometry is that cells do not flow synchronously (evenly spaced) in the sample stream, but instead are distributed randomly according to Poisson statistics. The magnitude of this problem can be addressed in two separate functions: (1) cell analysis problems resulting from incorrect single cell measurements or erroneous enumeration of cells analyzed, and (2) cell-sorting prob-

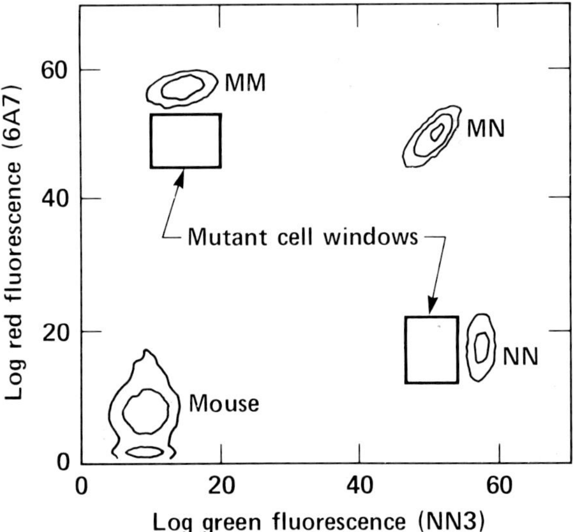

Fig. 5. Two-parameter flow histogram of mixed cell population labeled with two different monoclonal antibodies. Equal numbers of erythrocytes from four blood samples: human bloods; homozygous M, homozygous N, and heterozygous MN; mouse red cells (contain no glycophorin A), were mixed and stained with two different monoclonal antibodies, one of which (6A7) is specific for glycophorin A(M) and the other of which (NN3) is specific for glycophorin A(N). 6A7 was conjugated with biotin for fluorescent labeling with Texas Red-labeled avidin. NN3 is an IgM, and fluoresceinated goat antibody specific for mouse IgM was used for fluorescent labeling of NN3. Logarithmic amplification of the signals was used such that doubling of intensity corresponds to seven channels on either axis. Mutant cell windows are shown at fluorescence intensities such that they correspond to expected intensities from cells that express glycophorin A of the type NO or MO.

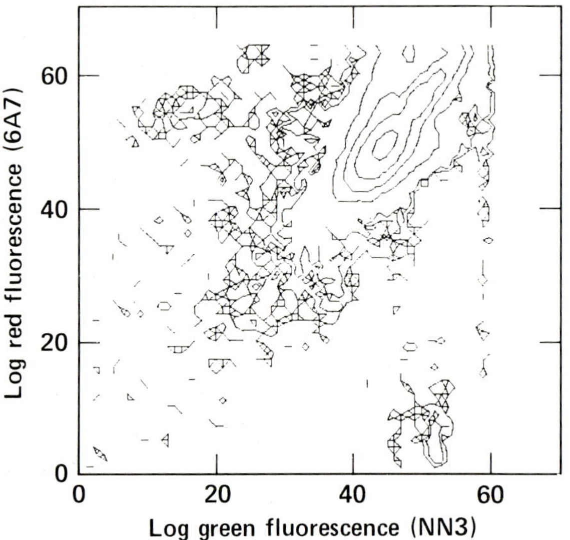

Fig. 6. Two-parameter flow histogram of cell sample from human glycophorin A(MN) heterozygote. Cell labeling was performed as described for Figure 5, and two million cells were analyzed by flow cytometry. The contours in this plot are drawn at logarithmically increasing frequencies of cells. The contour levels are 1, 3, 30, 300, 3,000, and 30,000. Cell frequency at the modal peak channel is 57,494 cells. Variant cell populations appear in the areas at x = 15, y = 48 and x = 48, y = 5, as indicated in Figure 5. Phenocopies appear along the diagonal from 0,0 toward the main cell population at x = 48, y = 48. Frequency of variant cells determined for this sample was 12 × 10^{-6}.

lems resulting from incorrect sorting of individual cells that were analyzed correctly.

In general, the analysis of rare cellular events has fewer problems than cell sorting of low frequency subpopulations. One reason for this difference is that errors in cell analysis will be passed directly to the sorting logic and result in subsequent sorting errors. Also, since analysis requires much less time than sorting, cell–cell coincidence in the excitation beam is much less probable than such coincidence in each sort function. For example, in a typical commercially available cell sorter with a 12-μm-wide light beam, the sample stream runs at 10 m/sec, droplets are generated at a rate of 32,000 per second, and three droplets per sorting event are standard. If a sample of 8-μm diameter cells is analyzed at 20,000 cells per second, the probability of coincidence of two or more cells in the light beam at one time is only 0.08%. However, the time required for most detector systems to complete the measurement and be ready for the next is typically 10 μsec. At 20,000 cells per second, 17% of the cells will not be analyzed during this instrumental "dead time." If higher rates of analysis are desired, say 100,000 cells per second, the coincidence frequency in the light beam is 1.8%, and loss due to dead time is up to 50%.

Even though the above "facts of life" for high-speed analysis must be traded off against one another in a complex fashion, some useful progress has been made in attempts to improve the instrumentation for rare event detection. One interesting change in concept is that most cells need not be analyzed. Only cells with high fluorescence are potential positive events. By setting a high fluorescence threshold for cell enumeration, individual cells can be counted (but not analyzed) rapidly without requiring the signal to be restored to zero between cells. Several inexpensive modifications [7] and more recent additions to commercial cell sorters at the University of Rochester have permitted instrument dead times to be reduced to 0.63 microseconds for counting rates as high as 189,000 cells per second with detection of subpopulations as small as 10^{-4} at rates above 20,000 cells per second. Ultimately, one is limited by the spot size of the laser beam, the physical dimensions of the cell, and the degree to which cells can be concentrated and still maintain a single-cell suspension.

The larger errors obtained by cell sorting can be illustrated with the 20,000-cell-per-second model introduced above. If this same sample is sorted with a typical commercially available cell sorter, the purity of the sorted cell fraction depends on the number of droplets deflected in each sort pulse. Using queuing theory [10] and Poisson statistics, one can calculate that a sort in which three droplets are deflected at each positive sort signal would result in only 49.8% sorted cell purity. If two droplets are deflected, this increases to 60% purity and single droplet deflection gives 75% purity. These calculated error rates have been confirmed by performing sorts at the rates indicated. While the purity of sorting increases significantly by sorting fewer droplets per event, control of this precise sorting mode is extremely difficult.

Intensive efforts at Lawrence Livermore National Laboratory to perform high-speed sorting of chromosomes and rare variant cells have yielded important technical advances [31]. The overall design of this ultrahigh-speed sorter is similar to commercial dual-beam cell sorters, except that the speed of droplet formation is increased tenfold. To perform this feat, the pressure for sample and sheath stream injection is increased 10- to 20-fold over commercial instruments to 200 psi. As a result, a number of apparently small changes in instrumental design had important effects on sorter performance. The nozzle assembly was changed significantly to minimize turbulence induced by the interaction between flowing liquid and orifice walls. Analysis with the two laser beams occurs in a 250-μm square flow channel, allowing effective use of high-quality optical components for beam shaping and fluorescence collection. The outlet orifice is a synthetic ruby with an 80-μm circular hole in the center, so that linear flow rate increases significantly as the sample leaves the cytometric region and enters the droplet formation region. This allows the cells to pass through the analysis region just as in a conventional sorter, and yet be sufficiently separated from each other in the sorting function because of the 10-fold increased droplet rate. Other important design changes are the elongated droplet deflection plates (18 cm) and a high electric field (10 kV) to maximize the deflecting force. These are necessary to obtain sufficient deflection of droplets traveling at speeds 10 times faster than in commercial sorters. As a result of these instrumental improvements, sorting of the rare hemoglobin-based somatic cell variants has been accomplished as described in the section on Mutagenesis of Human Somatic Cells In Vivo.

Perhaps as more flow cytometers and cell sorters are designed specifically for analyzing and sorting rare cell subpopulations, we will see very different approaches not only in instrumental design, but also in data analysis methods. For example, perhaps the wisdom of digitizing rare-event data into many channels should be questioned. This approach was created for an entirely different set of flow cytometry needs and may not be appropriate for analysis of rare cell subpopulations. Analysis time and computer space are largely wasted by digitizing and storing hundreds of millions of data points on the normal, nonmutant cells in a population. Leary's group at Rochester has taken the approach of counting all cells (to preserve original frequency information) but not digitizing signals from cells lacking any fluorescence. A small aliquot of sample from a pure negative population is digitized to allow comparison with samples containing rare fluorescent positive cells. Data from the pure negative sample and the mixture with an unknown number of positive cells are then both normalized to the same number of cells over a portion of the curve where only negative cells are assumed to be present and channel-by-channel subtractions of the data from the two samples is performed to extract the rare positive cells. This approach has been used successfully to provide a relatively inexpensive and easy to operate system for detection and sorting of live rare cells.

FUTURE PROSPECTS

Difficulties encountered in variant cell analysis and mutant cell sorting are enough different from those generally seen in flow cytometry that many of the solutions found will be applicable to other rare subpopulation studies. Present efforts to apply solutions to mutagenesis are motivated mainly by an interest in developing mutagenesis assays for eukaryotic systems. Some examples of other problems that are benefiting from rare event detection by flow cytometry are (1) detection of fetal blood cells in maternal peripheral blood [7,12], (2) monitoring the fate of transfused blood cells using post-transfusion antibody labeling of transfused cells, and (3) detecting small sub-populations of neoplastic cells in biopsies or cytology specimens for diagnosis or prognosis by flow cytometry (see Chapters 37 and 38).

ACKNOWLEDGMENTS

This work was performed under the auspices of the U.S. Department of Energy by the Lawrence Livermore National Laboratory under Contract W-7405-ENG-48 with financial support of grants R-808642-03 and R-811819-01 from the U.S. Environmental Protection Agency.

REFERENCES

1. **Albertini RJ (1982)** Studies with T-lymphocytes: An approach to human mutagenicity monitoring. In Bridges BA, Butterworth BE, Weinstein B (eds), "Indicators of Genotoxic Exposure." Cold Spring Harbor, NY: Cold Spring Harbor Laboratory, pp 393–412.
2. **Albertini RJ, Castle KL, Borcherding WR (1982)** T-cell cloning to detect the mutant 6-thioguanine-resistant lymphocytes present in human peripheral blood. Proc Natl Acad Sci USA 79:6617–6621.
3. **Ames BN, Durston WE, Yamasaki E, Lee FD (1973)** Carcinogens are mutagens: A simple test system combining liver homogenates for activation and bacteria for detection. Proc Natl Acad Sci USA 70:2281–2285.
4. **Amneus H, Matsson P, Zetterberg G (1982)** Human lymphocytes resistant to 6-thioguanine; restrictions in the use of a test for somatic mutations arising in vivo studied by flow-cytometric enrichment of resistant cell nuclei. Mutat Res 106:163–178.
5. **Bigbee WL, Langlois RG, Vanderlaan M, Jensen RH (1984)** Binding specificities of eight monoclonal antibodies to human glycophorin A: Studies with M^cM, and $M^kEn(UK)$ variant human erythrocytes and M- and MN^v-type chimpanzee erythrocytes. J Immunol 133:3149–3155.
6. **Bigbee WL, Branscomb EW, Jensen RH (1984)** Detection of mutated erythrocytes in man. In de Serres FJ, Pero RW (eds), "Individual Susceptibility to Genotoxic Agents in the Human Population." New York: Plenum Press, pp 249–266.
7. **Cupp JE, Leary JF, Cernichiari E, Wood JCS, Doherty RA (1984)** Rare-event analysis methods for detection of fetal red cells in maternal blood. Cytometry 5:138–144.
8. **Dangl JL, Parks DR, Oi VT, Herzenberg LA (1982)** Rapid isolation of cloned isotype switch variants using fluorescence activated cell sorting. Cytometry 2:395–401.
9. **de Serres FJ, Sheridan W (eds) (1983)** "Utilization of Mammalian Specific Locus Studies in Hazard Evaluation and Estimation of Genetic Risk." New York: Plenum Press.
10. **Graybeal WJ, Pooch UW (1979)** "Simulation: Principles and Methods," Cambridge, MA: Winthrop Publishers.
11. **Hayashi M, Sofune T, Ishidate M (1983)** An application of acridine orange fluorescent staining to the micronucleus test. Mutat Res 120:241–247.
12. **Herzenberg LA, Bianchi DW, Schroder J, Cann HM, Iverson MG (1979)** Fetal cells in the blood of pregnant women: detection and enrichment by fluorescence-activated cell sorting. Proc Natl Acad Sci USA 76:1453–1455.
13. **Hittelman WN (1973)** The type and time of occurrence of aminopterin-induced chromosome aberrations in cultured Potorous cells. Mutat Res 18:93–102.
14. **Hutter KJ, Stohr M (1982)** Rapid detection of mutagen induced micronucleated erythrocytes by flow cytometry. Histochemistry 75:353–362.
15. **Jensen RH, Vanderlaan M, Grabske RJ, Branscomb EW, Bigbee WL, Stanker LH (1985)** Monoclonal antibodies specific for sickle cell hemoglobin. Hemoglobin 9:349–362.
16. **Langlois RG, Bigbee WL, Jensen RH (1986)** Determination of somatic mutations in human erythrocytes. Hum Genet 74:353–362.
17. **Bigbee WL, Langlois RG, Jensen RH, Wyrobek AW, Everson RB (1987)** Chemotherapy with mutagenic agents elevates the in vivo frequency of glycophorin A "null" variant erythrocytes. Environ Mutagen 9(suppl 8):14–15.
18. **Langlois RG, Bigbee WL, Kyoizumi S, Nakamura N, Bean MA, Akiyama M, Jensen RH (1987)** Evidence for increased somatic cell mutations at the glycophorin A locus in atomic bomb survivors. Science 236:445–448.
19. **Johnston RN, Beverley SM, Schimke RT (1983)** Rapid spontaneous dihydrofolate reductase gene amplification shown by fluorescence-activated cell sorting. Proc Natl Acad Sci USA 80:3711–3715.
20. **Kaufman RJ, Schimke RT (1981)** Amplification and loss of dihydrofolate reductase genes in a Chinese hamster ovary cell line. Mol Cell Biol 1:1069–1076.
21. **Kavathas P, Herzenberg LA (1983)** Amplification of a gene coding for human T-cell differentiation antigen. Nature (Lond) 306:385–387.
22. **Kohl NE, Gee CE, Alt FW (1984)** Activated expression of the N-myc gene in human neuroblastomas and related tumors. Science 226:1335–1337.

23. **Langlois RG, Bigbee WL, Jensen RH** (1985) Flow cytometric characterization of normal and variant cells with monoclonal antibodies specific for glycophorin A. J Immunol 134:4009–4017.
24. **Liesegang B, Radbruch A, Rajewsky D** (1978) Isolation of myeloma variants with predefined variant surface immunoglobulin by cell sorting. Proc Natl Acad Sci USA 75:3901–3905.
25. **Luria SE, Delbruck M** (1943) Mutation of bacteria from virus-sensitive to virus-resistant. Genetics 28: 491–511.
26. **MacGregor JT, Wehr CM, Langlois RG** (1983) A simple fluorescent staining procedure for micronuclei and RNA in erythrocytes using Hoechst 33258 and pyronin Y. Mutat Res 120:269–275.
27. **Mendelsohn ML** (1985) Prospects for cellular mutational assays in human populations. In Woodhead AD, Shellabarger CJ, Pond V, Hollaender A (eds), "Assessment of Risk from Low-Level Exposure to Radiation and Chemicals." New York: Plenum Press, pp 319–328.
28. **Morley AA, Trainor KJ, Seshadri R, Ryall RG** (1983) Measurement of in vivo mutations in human lymphocytes. Nature (Lond) 302:155–156.
29. **Morrison SL, Scharff MD** (1981) Mutational events in mouse myeloma cells. CRC Crit Rev Immunol 3:1–22.
30. **Papayannopoulou T, McGuire TC, Lim G, Garzel E, Nute PE, Stamatoyannopoulos G** (1976) Identification of haemoglobin S in red cells and normoblasts, using fluorescent anti-Hb S antibodies. Br J Haematol 34: 25–31.
31. **Peters D, Branscomb E, Dean P, Merrill T, Pinkel D, Van Dilla M, Gray JW** (1985) The LLNL high speed sorter: Design features, operational characteristics and biological utility. Cytometry 6:290–301.
32. **Sablitzky F, Radbruch A, Rajewsky K** (1982) Spontaneous immunoglobulin class switching in myeloma and hybridoma cell lines differs from physiological class switching. Immunol Rev 67:59–71.
33. **Schmid W** (1975) The micronucleus test. Mutat Res 31:9–15.
34. **Seeburg PH, Colby WW, Hayflick JS, Capon DJ, Goeddel DV, Levinson AD** (1985) Biological properties of human c-Ha-rasl genes mutated at codon 12. Nature (Lond) 312:71–75.
35. **Siminovich L** (1976) On the nature of heritable variation in cultured somatic cells. Cell 7:1–11.
36. **Stamatoyannopoulos G** (1979) Possibilities for demonstrating point mutations in somatic cells, as illustrated by studies of mutant hemoglobins. In Berg K (ed), "Genetic Damage in Man Caused by Environmental Agents." New York: Academic Press, pp 49–62.
37. **Terzi M, Hawkins TS** (1974) Chromosomal variation, cellular aging and resistance to anti-folate analogs in a Chinese hamster cell line. Biochem Exp Biol 11:245–254.
38. **Tomita M, Furthmayer H, Marchesi VT** (1978) Primary structure of human erythrocyte glycophorin A. Isolation and characterization of peptides and complete amino acid sequence. Biochemistry 17:4756–4770.
39. **Tonegawa S** (1983) Somatic generation of antibody diversity. Nature (Lond) 302:575–581.
40. **Woodhead AD, Shellabarger CJ, Pond V, Hollaender A (eds)** (1985) "Assessment of Risk from Low-Level Exposure to Radiation and Chemicals." New York: Plenum Press.
41. **Zetterberg G, Amneus H** (1982) Use of flow cytometry to concentrate 6-thioguanine-resistant variants of human peripheral blood lymphocytes. In Bridges BA, Butterworth BE, Weinstein B (eds), "Indicators of Gentoxic Exposure." Cold Spring Harbor, NY: Cold Spring Harbor Laboratory, pp 413–421.

28

Applications of Flow Cytometry and Sorting to Molecular Genetics

M.A. Van Dilla, M.E. Kamarck, and M. Lalande

Biomedical Sciences Division, Lawrence Livermore National Laboratory, Livermore, California 94550 (M.A.V.D.); Molecular Therapeutics Inc., Miles Research Center, West Haven, Connecticut 06516 (M.E.K.); Biotechnological Research Institute, National Research Council Canada, H4P 2R2 Montreal, Quebec, Canada (M.L.)

INTRODUCTION

Flow cytometry has an impact on molecular genetics through both its analysis and sorting functions. The analysis function is useful for studying both isolated chromosomes and also for detecting and quantitating surface (or internal) cellular molecules coded by specific genes. Flow sorting of specific chromosomal types is useful for mapping DNA sequences to their chromosomal location, or to chromosomal subregions taking advantage of translocations involving a portion of the chromosome carrying the sequence to be mapped, and for construction of chromosome-specific gene libraries. These and other applications of flow cytometry to the study of the structure and function of genes are discussed in this chapter. The material to be covered also falls into three interrelated areas.

The first area deals with the construction of gene libraries from individual chromosomal types purified by flow sorting. Gene libraries are important tools in molecular genetics for several reasons. They are the raw material from which DNA fragments containing complete genes or segments thereof can be isolated (easier from chromosome-specific libraries than from total genomic libraries). This is crucial for the study of gene structure and expression, a fundamental problem in biology. They are also important in medical genetics, as a rich source of chromosome-specific probes for DNA polymorphisms linked to a genetic disease locus. The objective here is diagnosis of the disease by detecting the linked DNA polymorphism in cases in which there are no symptoms but the person is known to be at risk. A good example of this medical application is Huntington's disease, known to be caused by a defective (probably single) gene near the end of the short arm of chromosome #4. Nothing is known about the nature of the gene or the defect, or about the biochemical nature of the disease. Nor is a cytogenetic marker linked to the locus. And the individual is usually symptom free until midlife, after which the disease gradually develops and runs its fatal course. There is an intense effort to find such diagnostic molecular markers, and then to home in on the defective gene and clone it for study of its expression. After reviewing the development of the technology making gene library construction possible, we trace the development of libraries produced from total genomic DNA, from DNA obtained from flow sorted chromosomes, and finally describe in some detail the National Laboratory Gene Library Project—the objective of which is the construction of libraries for each of the 24 human chromosomal types.

The second area deals with the study of genes that code for cell surface molecules, and exploits the use of fluorescently labeled probes that are specific for the surface molecule being studied. A good example comes from immunological research. The genes coding for the histocompatibility antigens have been studied by flow cytometric measurement of these surface antigens as labelled by fluorescent monoclonal antibodies. Flow sorting and somatic cell genetic techniques provide general methods for studying not only the genes of the immunological system, but any gene whose protein product appears on the cell surface, such as receptors for growth factors and transferrin. These genes can now be cloned using the powerful combination of cell-sorting and DNA-mediated gene transfer, as discussed in this chapter. Genes whose protein products are cytoplasmic can also be studied by these methods. An example of this, which also involves the phenomenon of gene amplification, is the gene which codes for the purine biosynthesis pathway enzyme dihydrofolate reductase. Exposure of cells to methotrexate, a drug which binds to the enzyme and blocks it, causes amplification of the gene. This can be quantitated flow cytometrically using a fluorescent derivative of methotrexate. In addition, sorting normal cells has been used in the absence of drug selection to isolate a sub-population with high fluorescent methotrexate binding and hence amplification of the gene, showing that gene amplification occurs normally and not just as the result of the drug insult.

Flow Cytometry and Sorting, Second Edition, pages 563–603
Published 1990 by Wiley-Liss, Inc.

The third area deals with oncogenes and chromosomal rearrangements in neoplasia. The human genome contains DNA sequences, termed cellular proto-oncogenes or c-onc genes, which are homologous to sequences in retroviruses highly tumorigenic in other species, such as birds. These c-onc genes are prime candidates for a key role in human carcinogenesis and are under intense study—although the story is not clear yet. Human tumors are frequently accompanied by chromosomal rearrangements, and these may involve c-onc genes. An example is Burkitt's lymphoma, in which the specific c-onc, termed c-myc, on the #8 chromosome is translocated to the region of the immunoglobulin genes on either chromosome #2, 14, or 22. There are variants in the breakpoint such that the c-myc gene stays on #8, as demonstrated by flow sorting the rearranged chromosomes and analyzing the DNA by hybridization techniques using a radioactively labelled probe for c-myc. Flow cytometry has also been used effectively in the study of other human neoplasias, including neuroblastoma, retinoblastoma, and Wilms' tumor. For example, retinoblastoma may be accompanied by deletion of a band (q14) on the long arm of chromosome #13. This has been studied using probes from a gene library constructed of flow sorted #13 chromosomes, and two probes that map to band 13q14 have been identified and are being used to study this locus at the molecular level.

Thus, it is clear that flow cytometry is versatile enough to be useful in both basic studies of normal gene structure and function, and abnormalities leading to medical problems. What follows is a more detailed discussion of the three areas just outlined, all of which represent a marriage of recombinant DNA and flow cytometry and sorting technologies.

GENE LIBRARIES

Molecular Cloning and Flow Cytometry and Sorting

Recombinant DNA Technology. The origins of gene libraries can be traced to the beginnings of recombinant DNA technology, in the early 1970s. There were two key developments. The first was the discovery of restriction endonucleases that cut the DNA molecule at sequence-specific sites [264]; the first one was isolated from *Hemophilus influenzae* bacterial cells, a blunt end 6-cutter now called HindIII. The second key development was the discovery of methods of using these endonucleases in conjunction with DNA ligase to insert foreign DNA into circular replicating units such as SV40 [114] and the plasmid pSC101 [31], and to introduce these hybrid replicating units into cells for cloning. For the first time, it became possible to cut genomic DNA into fragments of reproducible size, insert the fragments into DNA molecules (vectors) that would replicate in host cells, and thus amplify the number of copies of each fragment considerably (by a factor of about 10^6). Some of the amplified fragments contain complete genes, sections of genes, or other important sequences. This collection of cloned fragments containing all (or most) of the original genomic DNA, having lost its order along the chromosome, is called a gene library.*

Genomic and Enriched Libraries. The first use of gene libraries to map DNA sequences in the chromosomes of *Drosophila melanogaster* quickly followed these pioneering developments [287]. The *Drosophila* DNA was sheared to a mean length of 8 kb, and extended 3′ dT ends were produced by use of lambda exonuclease followed by terminal transferase and dTTP. The tetracycline-resistant vector pSC101 was linearized with EcoR1 and extended 3′ dA ends were produced similarly. These two DNAs were then mixed, allowed to anneal, and used to transform a suitable tetracycline-sensitive *Escherichia coli* host (strain h303). Easily visible colonies grew in agar plates in about 2 days, each representing a clone of cells transformed to antibiotic resistance by a specific hybrid plasmid (now highly amplified). Homogeneous populations of hybrid DNAs were isolated from the clones of transformants and studied in detail. Three showed no repetitive sequences detectable by reassociation kinetics and were mapped to polytene chromosomes by in situ hybridization.

Technical improvements in molecular cloning techniques followed. Derivatives of bacteriophage lambda were developed as cloning vectors convenient to use while insuring biological containment [19,187]. Rapid screening methods for examining very large numbers of bacterial colonies or phage plaques for specific sequences were developed [11,93]. In vitro packaging of recombinant lambda vector molecules into infectious phage particles by supplying bacterial extracts containing packaging proteins and preheads [109] made cloning more efficient. As a result of this improved technology, libraries could be constructed from a large mammalian genome (the rabbit), from which structural genes for beta-globin could be isolated [193], and from human embryonic DNA from which the linked delta- and beta-globin genes were isolated and characterized [168].

Libraries or other DNA preparations enriched for specific chromosomal sequences have many advantages over their total genomic counterparts, and reports of both biochemical and recombinant DNA approaches appeared in the late 1970s and early 1980s. The biochemical methods are based on DNA hybridization kinetics and require two identical sources of DNA, except that one contains the desired chromosome and the other does not. Typical sources are DNA from a human–mouse hybrid cell line containing a single human chromosome (such as the X) and DNA from normal mouse cells. Repeated hybridization of a small amount of labelled DNA from the hybrid cells with large excesses of unlabeled mouse DNA permits isolation of the single-stranded human DNA on a hydroxyapatite column [249]. Variants of the approach have been used in other studies of the human X chromosome [209] and in the study of the human Y chromosome [154]. The other enrichment approach involves construction of a genomic DNA library from human–Chinese hamster hybrid cells carrying a portion of the human genome (e.g., a chromosome 11 fragment) as the only human material [95]. This library was screened with labeled human genomic DNA under conditions of stringency such that only human clones containing repetitive DNA will hybridize. Since highly repetitive DNA is widely dispersed throughout the human genome and the average insert size

*Nomenclature in this field is somewhat confusing, and not standardized. The term *library* is generally used, but sometimes the term *bank* is encountered. Both are somewhat misleading since they imply order, which is completely lacking in the unordered collection of cloned DNA fragments that constitute a library. The term *gene library* is widely used, but is not precise since a given gene may be present only in fragmented form rather than as a continuous DNA sequence. The term *DNA library* is also used and is more accurate. Also used are such terms as genomic library, recombinant DNA library, recombinant phage library. We use the terms *gene library* and *DNA library* interchangeably.

was 15 to 20 kb, most human inserts should contain one or more such sequences and hence hybridize. Fifty human clones were isolated; five were characterized in detail and mapped to chromosome 11 regions using a panel of DNAs from hybrid cell lines carrying various 11p deletions. These approaches are useful but limited; a more direct and general approach uses chromosomes purified by flow sorting as the input material for molecular cloning.

Libraries from Flow-Sorted Chromosomes. For the first gene library constructed from chromosomes enriched by flow sorting [45], a FACS-2 sorter purified X chromosomes from the human lymphoblastoid line GM 1416 (48.XXXX). A polyamine procedure for chromosome isolation, ethidium bromide staining, and flow cytometric techniques developed for high resolution analysis of the human karyotype [261,295] were used in this work. 2×10^6 X chromosomes (400 to 500 ng of DNA) were sorted at a rate of 17/second, so that 33 hours of sorting time were required. The DNA was extracted, purified, digested extensively with EcoR1, ligated to the arms of the λ vector λgtWES.λB (4 to 14 kb acceptance range), packaged in vitro, and grown up on the bacterial host; 50,000 recombinants were obtained with a background of parental phage (containing no inserts) of less than 5%. Reports of other gene libraries constructed from flow-purified chromosomes quickly followed, including libraries from an HSR-bearing #1 chromosome from the human neuroblastoma line IMR-32 [165], from mouse X-chromosome specific DNA [49,50], from human X-chromosomes sorted from the human lymphoblastoid line GM 1202 (49.XXXXY) [155], and from human #21 and #22 chromosomes sorted from the same lymphoblastoid cell line used by the Davies group [149]. Interest in chromosome-specific gene libraries steadily increased thereafter, with further work on amplified DNA sequences from an HSR-bearing #1 library [129], and the construction of a library using human Y chromosomes sorted from a Chinese hamster–human hybrid line bearing the human Y-chromosome only [201]. Soon thereafter came the construction of libraries from about 10^6 Chinese hamster #1 and #2 chromosomes, which are very favorable for sorting [90], a library enriched by flow sorting for parts of the human X-chromosome [157], a library from human #13 chromosomes for use in retinoblastoma studies [158], and the first library from flow sorted chromosomes (the human X-chromosome) cloned into a λ phage vector, Charon 30, that accepts large inserts (mean insert size was 13.7 kb; [152]. But the facilities for flow karyotyping and sorting around the world fell short of the demand for libraries enriched for each of the 24 human chromosomal types, leading to a systematic approach that we shall now describe.

The National Gene Library Project

Strategy. The two National Laboratories at Livermore and Los Alamos have played a prominent role in the development and application of flow cytometry and sorting to chromosome classification and purification. Both laboratories began to receive numerous requests for specific human chromosomal types purified by flow sorting for gene library construction, but these requests were difficult to satisfy owing to time and personnel constraints. The Department of Energy, through its Office of Health and Environmental Research, has a longstanding interest in the human genome in general and in the mutagenic and carcinogenic effects of energy-related environmental pollutants in particular. Hence, it was decided early in 1983 to use the flow cytometric and molecular biological skills at both laboratories to construct chromosome-specific gene libraries to be made available to the genetic research community. This National Laboratory Gene Library Project was envisioned as a practical way to deal with requests for sorted chromosomes, and also as a way to promote increased understanding of the human genome and the effects of mutagens and carcinogens on it. The strategy for the project was developed with the help of an advisory committee* and input from the genetics community.

The first goal (phase I) is the construction of small insert chromosome-specific gene libraries for each of the 24 human chromosomal types. These are intended mainly for the medical genetics community for use in the study and diagnosis of genetic diseases by the restriction fragment length polymorphism method. For such studies, short DNA probes on the order of 1 kb in size are ideal, and such small insert libraries are the easiest to produce from the limited amounts of chromosomal DNA available from flow sorting. The second goal (phase II) is the construction of large insert libraries (20 to 40 kb) in phage and cosmid vectors, better suited to the study of gene structure and function. Phase II is more demanding than phase I because larger amounts of chromosome-specific DNA are required for large-insert libraries, and cloning procedures are more difficult. Work began on phase I in mid-1983 and gained full momentum in 1984 and was completed in 1987.

The specific strategy of phase I is outlined in the schematic diagram of Figure 1, using the HindIII library constructed from sorted #18 chromosomes as an example. The human DNA source is cultured cells, either normal, diploid human fibroblasts (as for the #18 library in Fig. 1), human lymphoblastoid cells, or human–hamster hybrids having a reduced number of human chromosomes. Metaphase chromosomes are isolated, stained with DNA-specific fluorescent dyes, and processed through a flow sorter to yield enough material for cloning (nominally 4×10^6 chromosomes, enough for several clonings). The chromosomal DNA is then extracted, purified, and digested to completion with restriction enzyme. The vector DNA is cut with the same restriction enzyme and the arms are dephosphorylated with calf alkaline phosphatase (to inhibit re-ligation). The chromosomal DNA fragments and the vector arms are mixed so that complementary (sticky) ends can anneal and join covalently in the ligation reaction. The resulting recombinant DNA molecules are packaged in vitro into infective virions using complementary packaging extracts from two mutant lysogenic bacterial strains, allowed to infect a lawn of *E. coli* cells (strain LE392) and to multiply; the resulting phage lysate is the amplified #18 gene library.

Both laboratories use the same bacteriophage lambda vector, Charon 21A, which has been used successfully to construct chromosome-specific libraries by others [156,159]. Charon 21A is an insertion vector with a single restriction site for Hind III and another for EcoR1. It accepts inserts of size up to 9.1 kb. Each laboratory undertook cloning all 24 human chromosomal types with Livermore using HindIII

*Membership: Dr. P. Berg (Stanford), Dr. T. Maniatis and Dr. S.A. Latt (Harvard University), Dr. A.G. Motulsky (University of Washington), Dr. W.J. Rutter (University of California, San Francisco), Dr. C.W. Schmid (University of California at Davis), Dr. T.B. Shows (Roswell Park Memorial Hospital), Dr. C.T. Caskey (Baylor University), Dr. F.R. Blattner (University of Wisconsin), Dr. M.H. Edgell (University of North Carolina), and Dr. R.E. Gelinas (University of Washington).

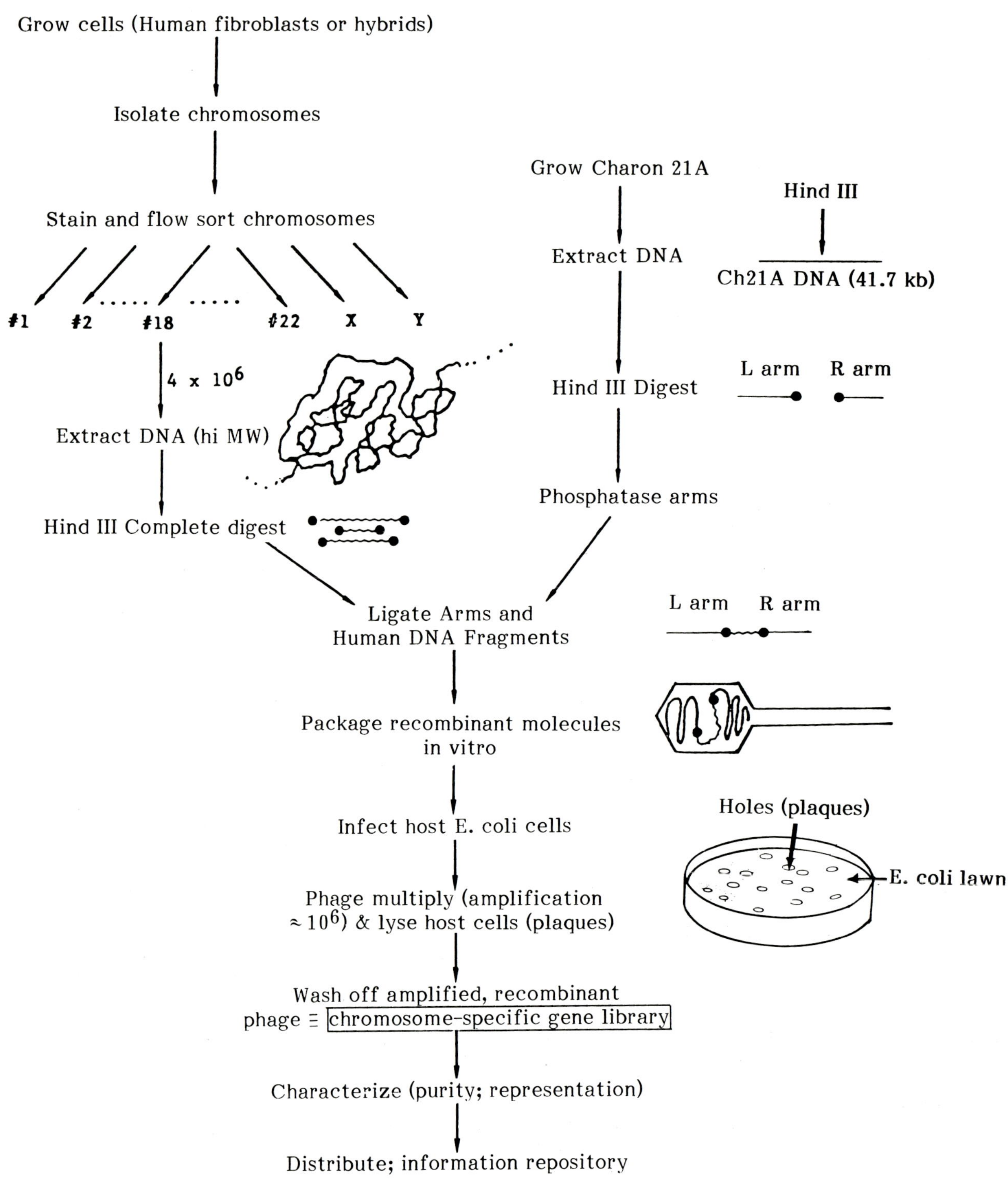

Fig. 1. The steps in the molecular cloning of DNA from flow-sorted human chromosomes are outlined. The vector is an engineered version of bacteriophage lambda called Charon 21A, which has both HindIII (shown here) and Eco R1 cloning sites. In this example, #18 chromosomes were sorted from a human fibroblast line. Both the human and lambda DNA are extracted, purified, and digested to completion (with HindIII in this example), and then brought together in the ligation reaction to form recombinant DNA molecules. In vitro packaging into infective phage particles which multiply in the *Escherichia coli* host results in the final product, the chromosome-specific gene library.

and Los Alamos using Eco R1. In this way it is likely that sequences larger than 9.1 kb that do not cut with one restriction enzyme will be cut with the other and hence be cloned.

Technologies Required

Flow Cytometry. A crucial requirement for chromosome-specific library construction is the ability to discriminate each of the 24 human chromosomal types by flow cytometry with sufficient precision so that they can be purified by flow sorting. This technology has been developed extensively during the past decade and is described elsewhere in this book (see Chapter 25, this volume); only the aspects particular to library construction are discussed here.

Flow Sorting. The flow instrumentation used for most of the work was a dual laser flow cytometer and a high speed sorter at Livermore, and a modified FACS 11 and an EPICS V sorter at Los Alamos (see Chapters 8 and 9, this volume). It takes about an 8-hour day of machine-on time to sort the target number of chromosomes (4×10^6) under optimal conditions, or a few days under average conditions [278].

Cell Culture and Chromosome Isolation. These factors turn out to be critical for successful sorting and cloning, and are the bottleneck in the project. Cell lines are chosen largely on the basis of their flow karyotype, that is, the degree to which the peak generated by the desired chromosomes is isolated from nearby peaks, and also on the basis of minimal underlying background of chromosomal debris or clumps. Also important are good growth in mass culture, ability to produce a high mitotic index, and stability of karyotype. These considerations led to the decision to use human cells (either normal foreskin fibroblast or lymphoblastoid lines) for the smaller chromosomes (13–22, Y) and human × Chinese hamster hybrid cell lines containing a reduced number of human chromosomes for the larger chromosomes (1–12). The smaller human chromosomes have more base composition variation than the larger ones, hence are well resolved in the bivariate flow karyotype of human lines, using Hoechst 33258 and chromomycin A3 staining. The larger human chromosomes are less well resolved, and, indeed, the 9–12 group shows almost complete overlap for most human cell lines. These larger human chromosomes have base compositions sufficiently different from hamster chromosomes so that they are well resolved in human × hamster hybrids having few, or only one, human chromosome. An example is the hybrid line UV20HL21-27, containing the human #4, 8 and 21 human chromosomes only; the #4 and 8 chromosomes produce peaks well resolved from each other and from the hamster background. This is not true for the #21 peak, but human cell lines work well for this chromosomal type. The cell lines used in this project are listed and described in [278].

A problem encountered with some of the hybrid lines is karyotype instability, that is, a tendency for the cell population to lose human chromosomes during proliferation. This seems to be a random process, and varies from cell to cell line; it can be very frustrating if the chromosome of interest is lost during the proliferation necessary for mass culture. If the desired chromosome has a selectable marker, then the line can be grown in selection medium and the resultant population will contain the chromosome. Examples are human-mouse hybrid lines [4] containing single human chromosomes in a mouse background (HPRT deficient) in which the human chromosomes contain the Ecogpt gene (for XPRT) introduced by recombinant DNA techniques. When grown in medium containing mycophenolic acid and xanthine, only cells retaining the human chromosome will survive and proliferate. Unfortunately, mouse and human chromosomes are not well resolved flow karyotypically. However, these marked human chromosomes can be transferred to a hamster background, in which case a very advantageous line would result that is both stable and sortable.

Mass cell culture procedures are necessary to produce enough metaphase chromosomes for sorting. For human diploid fibroblast cell lines growing in monolayer culture, up to 100 T150 flasks may be necessary to produce the desired number of mitotic cells (about 10^8) from the colcemid block by shakeoff. Cells that grow in suspension (lymphoblastoid and hybrid lines) may be cultured in glass bottles or in closed T150 flasks more easily and economically but will have a smaller mitotic index after the colcemid block. One liter of culture is generally sufficient to produce 10^8 mitotic cells.

Both the $MgSO_4$ and the polyamine methods [261,276, 277] have been used in this work to produce metaphase chromosome suspensions suitable for staining with the DNA dyes Hoechst 33258 and chromomycin A3 and for sorting. Both yield preparations of high quality, that is, good peak resolution and minimal debris. The size of the DNA isolated from chromosomes prepared by either method is large enough for small-insert library construction. The chromosomes are stable morphologically for several months at 4°C. The main disadvantage is that the isolated chromosomes cannot be readily banded, which would be very useful in verifying the purity of a sort. Staining is carried out as for chromosome analysis as described in Chapter 25 (this volume).

Molecular Cloning. The cloning strategy has been outlined in Figure 1, and the steps are shown in more detail in Table 1. The cloning protocol leans on prior experience in cloning

TABLE 1. Isolation and Cloning of Human DNA from Flow-Sorted Metaphase Chromosomes

1. Extract and purify the sorted chromosomal DNA
   - Pellet chromosomes at 40,000g, 30 min, 4°C*
   - Hydrolyze protein with proteinase K overnight 37°C
   - Extract proteins with phenol and chloroform†
   - Dialyze in 0.5 ml volume against Tris–EDTA†
2. Digest to completion with HindIII or EcoR1
   - Cleave with HindIII or EcoR1
   - Another protein extraction and dialysis against H_2O†
3. Ligate vector arms with chromosomal DNA fragments
   - Add Ch21A arms and concentrate under vacuum†
   - Precipitate DNA with ethanol†
   - Wash DNA pellet with 70% ethanol
   - Ligate Ch21A–human DNA mixture†
4. Package recombinants and amplify
   - Package recombinant DNA into viable phage particles
   - Infect *E. coli* strain LE392 to amplify clones

*For chromosomes isolated by the polyamine procedure, omit centrifugation step.

†Check DNA molecular weight; run small aliquot on minigel, Southern blot, and probe with labeled total human DNA. This provides valuable diagnostic information on each step of this procedure.

both genomic and sorted DNA (see above), and well-worked-out procedures were used whenever possible in the interests of minimizing library production time. A complete description has been published [81], and will not be given here. However, in the course of library construction, experience with manipulating submicrogram amounts of DNA stimulated the development of two useful procedures.

Cloning reliability and efficiency are very important, since sorted chromosomes can be produced in only limited quantities after a labor intensive cell culture and chromosome isolation process. Whereas total genomic library construction may start with 50 to 100 μg of DNA—easy to obtain—only 0.1 to 1.0 μg of sorted DNA is generally available. The two procedures introduced to increase the reliability and efficiency of cloning are indicated in Table 1. The first is a mini-blot hybridization procedure to analyze the efficiency of the cloning steps to be sure that the reactions were actually taking place as expected. This useful diagnostic procedure is applied at several points in the cloning process as indicated in Table 1. A small DNA aliquot (of the order of 1 ng) is electrophoresed through a 1% agarose minigel, transferred to a nitrocellulose filter, and hybridized with a probe consisting of total human DNA labeled by nick-translation with ^{32}P. Autoradiography reveals the size distribution of the DNA, yielding valuable information on whether the procedure is working. For example, digestion to completion with the restriction enzyme is crucial, and if cutting is inefficient for some reason, the resulting library may be too small to be useful. This can happen, and is revealed by a large size distribution for the DNA in step 1 of Table 1, and a similar distribution for step 2. The chromosomal DNA in step 1 is expected to be large (25 to 50 kb or more), but after digestion it should be degraded to a broad exponential distribution with an average size of about 4 kb extending down below this size with increasing frequency. The mini-blot procedure clearly shows the size distribution of the chromosomal DNA after the digestion step, and so gives positive evidence as to whether this step is working.

The digestion step, in fact, proved to be very variable in efficiency in cloning experiments. Several techniques were tried to optimize it; the best was a simple microdialysis step before cleavage. This was accomplished by transferring the DNA (after the degraded proteins were removed by extraction with phenol and chloroform) to a mini-collodion membrane for dialysis against a Tris–EDTA buffer. Incorporation of this step into the standard protocol greatly increased the reliability and efficiency of cloning.

Further cloning efficiency increases were obtained by fine-tuning the ligation reaction conditions, which affect not only library size (number of recombinants) but also the background of nonrecombinants (which result in spite of the fact that the vector arms are dephosphorylated with calf alkaline phosphatase). Good libraries should have a large number of independent recombinants, to maximize the probability of a given sequence being present, but only a small fraction of nonrecombinants, to simplify screening—and it would be ideal to achieve this with the minimum number of starting chromosomes. The first libraries were made with 10^6 chromosomes and 2 μg of vector arms, and very large libraries resulted once the dialysis and mini-blot procedures were introduced. This encouraged experiments with fewer starting chromosomes (down to 2×10^5), which still give large libraries, but with excessive backgrounds of nonrecombinants (10 to 60%). Reducing the amount of vector arms and reaction volume, proportional to the reduction in number of starting chromosomes, corrected this situation and resulted in large libraries with low backgrounds (<1 to 10%).

Library Construction

Library Properties. The characteristics of the complete set of Phase I small-insert libraries are listed in Tables 2 and 3. All 24 human chromosomal types have been cloned, and two versions are available—one cloned into the EcoR1 site of Charon 21A and the other cloned into the HindIII site of Charon 21A. Note that normal human fibroblasts or lymphoblastoid lines were used for the smaller human chromosomes (13–22, Y) and human–hamster hybrids for the larger chromosomes (1–12,X) almost exclusively. The amount of starting material was generally about 10^6 chromosomes, although as techniques improved as few as 0.1 to 0.2×10^6 sufficed. The number of recombinant DNA molecules resulting from the ligation and packaging reactions varied by two orders of magnitude, as did the frequency of nonrecombinants. This variability is probably inherent in cloning very small amounts of DNA. It should be noted that the frequency of nonrecombinants is based on packaging a mock ligation without insert DNA; hence, it does not directly measure the nonrecombinants in the library, but rather gives an estimate. The efficiency of converting sorted chromosomal DNA into phage clones is low. At best, it is $354/1 \times 10^6 = 3.5 \times 10^{-4}$ for the LL*20*NS01, and about two orders of magnitude less for some. This is common experience in library construction, the ligation and packaging steps probably having the lowest efficiency. Recombinant phage growth on the *Escherichia coli* lawn results in an amplification of 10^5–10^6, so that the final library ends up with about as much human DNA as was contained in the original starting chromosomes. In several cases, especially for large libraries with chromosome equivalents > 5, only a fraction of the packaging reaction was amplified, so that the remainder is available for future amplification if needed. The intention has been to achieve a number of chromosome equivalents of at least 5, generally considered a "complete" library with the statistical probability of any sequence being present of 0.99. This has been accomplished in most cases, but even a small library, such as the LL*04*NS01 in Table 2, has many probes (about 10^4) for this chromosome and is being extensively used in attempts to find RFLP markers for Huntington's disease and to eventually clone the gene involved [85].

Sorting Purity. Table 4 also lists estimates of sorting quality, which refers to the resolution of the peak due to the chromosomal type being sorted from nearby peaks of other chromosomes, and to the amount of background continuum present in the flow karyotype (due to fluorescent debris and chromosome clumping). These factors depend on the cell line being used, the quality of the chromosome isolation and staining, and on sorter performance. In some cases the best cell line available may be less than ideal with respect to the separation between the desired chromosome and neighboring ones, as was the case for the #11 sort from line UV20HL4. Here the #11 peak was well separated from other human peaks, but there was some overlap with a large nearby hamster peak. In addition to this inherent source of impurity, the quality of chromosome preparations is hard to control and shows considerable variability. Machine performance is never perfect over the hours or days of a sorting run, another source of purity loss. Visual inspection of the set of flow karyotypes recorded during the sorting run will give a qualitative idea of sorting purity, as indicated in Table 4. More quantitative estimates of sorting purity can come from computer analysis of the bivariate histogram by summing

TABLE 2. Twenty-Eight Chromosome-Specific DNA Libraries (Charon 21A; HindIII)

Library ID No.*	Independent recombinants	Frequency of nonrecombinants	Chromosome equivalents†	Chromosome source‡	Starting chromosomes (× 10^6)	Starting DNA (ng)
LL *01* NS01	8.3×10^4	0.04	2.1	UV24HL10-12	0.5	270
LL *01* NS02	1.3×10^6	0.01	32 (20)	UV24HL6	0.5	270
LL *02* NS01	6.6×10^5	0.08	16 (5)	UV24HL5	0.5	270
LL *03* NS01	1.6×10^5	0.10	4.8	314-1b	0.5	220
LL *04* NS01	2.3×10^4	0.02	0.8	UV20HL21-27	1.0	400
LL *04* NS02	5×10^5	0.27	10	UV20HL21-27	0.5	210
LL *05* NS01	3.4×10^6	0.26	113 (30)	640-12	0.5	200
LL *06* NS01	7.6×10^5	0.06	27 (20)	UV20HL15-33	0.4	135
LL *07* NS01	3×10^5	0.01	11.5	GM131	0.9	310
LL *08* NS02	2.2×10^6	0.03	93 (20)	UV20HL21-27	0.7	210
LL *09* NS01	3.0×10^5	0.02	13	UV41HL4	0.4	120
LL *10* NS01	2.4×10^5	0.01	10.6	762-8A	0.5	150
LL *11* NS01	1.1×10^5	0.05	4.9	UV20HL4	0.2	50
LL *12* NS01	7.5×10^5	0.01	34 (20)	81P5D	0.4	120
LL *13* NS01	2.2×10^4	0.04	1.3	761	1.0	240
LL *13* NS02	8.5×10^5	0.03	47 (20)	GM131	0.5	120
LL *14* NS01	2.3×10^6	0.06	135	GM131	0.5	110
LL *45* NS01	2.6×10^6	0.02	152 (30)	811	1.0	210
LL *15* NS01	7.0×10^4	0.06	4.4	GM131	1.0	110
LL *16* NS03	7.6×10^5	0.02	51 (20)	HSF7	0.5	100
LL *17* NS02	3.4×10^5	0.02	24 (20)	HSF7	0.5	95
LL *18* NS01	8.9×10^5	0.13	72	761	1.0	170
LL *19* NS01	1.5×10^6	0.02	145 (10)	811	1.0	130
LL *20* NS01	3.9×10^6	0.01	354 (20)	811	1.0	140
LL *21* NS02	4.7×10^5	0.34	60 (20)	811	0.5	45
LL *22* NS01	6.1×10^5	0.05	71 (17)	811	0.5	50
LL *OX* NS01	2.1×10^6	0.33	84 (30)	UV24HL5	0.5	170
LL *OY* NS01	2.5×10^5	0.02	27	811	1.0	115

*The ID Code consists of 8 alphanumeric items. The first two items indicate which laboratory made the library, i.e., LL = Lawrence Livermore National Laboratory and LA = Los Alamos National Laboratory. The next 2 items are italicized and indicate chromosome type (in the one case of a mixed 14/15 library, chromosome type is designated as 45). The fifth item is a letter indicating chromosome status, i.e., N for normal, T for translocation, etc. The sixth item is either S (for small insert, complete digest libraries) or L (for large insert, partial digest libraries). The final 2 items represent library construction number.

†Number of recombinants for 1 chromosome equivalent = $(3 \times 10^9)(0.65)(f)/4100$ where 3×10^9 bp is the size of the human haploid genome; 0.65 is the clonable fraction; f is the fraction of cellular DNA in particular chromosome; 4100 bp is the average fragment size. Numbers in parentheses refer to representation of amplified library; for very large libraries, only a fraction of the packaging reaction was amplified.

‡Cell lines from which metaphase chromosomes were isolated:

a. Normal diploid human fibroblast lines 761, 811, HSF7
b. Apparently normal human lymphoblastoid line GM131
c. CHOx human lymphocyte lines (human chromosome content)—
    UV24HL10-12 (#1, 2 del, 3, 11, 13, 19)
    UV24HL6 (#1, 2del, 3?, 11–13, 14?, 18?, 19)
    UV24HL5 (#2, X)
    314-1b (#3)
    UV20HL21–27 (#4, 8, 21)
    640-12 (#5, 9, 12)
    UV20HL15-33 (#6, 9, 13, 15, 17, 20, 21)
    UV41HL4 (#6, 9, 13, 16, 18, Y)
    762-8A (#10, Y)
    UV20HL4 (#1, 4–6, 11, 14–16, 19,21)
    81P5D (#12, 15, X)

events in the sorting window and its vicinity, and by collapsing the sorted peak and its neighbors on a convenient axis for univariate gaussian fitting [47]. These methods are currently being applied and show that sorting purity of 95% or more is expected in the favorable case of a well-isolated peak, a good chromosome preparation, and good machine operation.

The type of impurity present in sorted chromosomes depends on the cell line. When sorting from human-hamster hybrid lines with few human chromosomes, impurities will be almost entirely of hamster origin and can be detected by species-specific hybridization. Contamination by unwanted human material will be negligible. However, sorts from human lines will contain only human impurities, which can

TABLE 3. Twenty-eight Chromosome-Specific DNA Libraries (Charon 21A; Eco R1)

Library ID No.*	Independent recombinants	Frequency of nonrecombinants	Chromosome equivalents†	Chromosome source‡	Starting Chromosomes ($\times 10^6$)	Starting DNA (ng)
LA *01* NS01	1.3×10^6	<0.01	31	UV24HL10–12	0.5	250
LA *02* NS01	7.0×10^4	0.04	1.8	UV24HL5	0.5	260
LA *03* NS01	2.7×10^4	0.04	0.8	314-1b	0.4	160
LA *03* NS02	2.1×10^5	0.08	6.4	314-1b	0.4	180
LA *04* NS01	5.1×10^4	<0.01	1.6	UV20HL21–27	0.9	370
LA *04* NS02	7.4×10^4	<0.01	2.3	UV20HL21–27	0.9	370
LA *05* NS01	1.3×10^6	<0.01	43	640-12	0.4	130
LA *06* NS01	4.8×10^4	0.06	1.7	UV20HL15–33	1.9	700
LA *07* NS01	2.4×10^6	0.06	9.2	MR3.316TG6	0.1	40
LA *08* NS04	3.6×10^4	0.10	1.5	UV20HL21–27	0.5	150
LA *09* NS01	1.6×10^5	0.07	7	HSF-7	0.3	90
LA *10* NS01	4.0×10^5	<0.01	18	762-8A	0.3	90
LA *11* NS02	6.2×10^4	0.17	2.8	80H10	0.4	100
LA *12* NS01	6.0×10^5	<0.01	27	81P5d	1.5	210
LA *13* NS03	7.5×10^4	0.14	4.2	HSF-7	0.4	80
LA *14* NS01	6.1×10^5	<0.01	36	762-8A	0.5	115
LA *15* NS02	4.0×10^5	0.06	20	80H10	0.3	71
LA *15* NS03	3.3×10^5	0.09	4	81P5d	0.4	82
LA *16* NS01	1.1×10^5	0.08	2	1634	3.0	590
LA *16* NS03	1.6×10^4	0.03	21.7	HSF-7	0.5	100
LA *17* NS03	1.1×10^6	0.01	7.9	HSF-7	0.6	110
LA *18* NS04	2.5×10^5	0.02	19	GM130A	0.5	81
LA *19* NS03	1.1×10^5	0.16	11	HSF-7	0.5	65
LA *20* NS01	1.6×10^4	0.24	1.5	HSF-7	0.5	72
LA *21* NS01	1.1×10^6	<0.01	137	HSF-7	0.9	87
LA *22* NS03	9.3×10^4	0.19	11	HSF-7	0.5	55
LA *OX* NS01	2.1×10^5	0.02	8.5	81P5d	1.2	380
LA *OY* NS01	1.1×10^5	0.10	11.5	HSF- 7	0.5	64

*The ID Code consists of 8 alphanumeric items. The first two items indicate which laboratory made the library, i.e., LL = Lawrence Livermore National Laboratory and LA = Los Alamos National Laboratory. The next 2 items are italicized and indicate chromosome type (in the one case of a mixed 14/15 library, chromosome type is designated as 45). The fifth item is a letter indicating chromosome status, i.e., N for normal, T for translocation, etc. The sixth item is either S (for small insert, complete digest libraries) or L (for large insert, partial digest libraries). The final 2 items represent library construction number.

†Number of recombinants for 1 chromosome equivalent = $(3 \times 10^9)(0.65)(f)/4100$ where 3×10^9 bp is the size of the human haploid genome; 0.65 is the clonable fraction; f is the fraction of cellular DNA in particular chromosome; 4100 bp is the average fragment size. Numbers in parentheses refer to representation of amplified library; for very large libraries, only a fraction of the packaging reaction was amplified.

‡Cell lines from which metaphase chromosomes were isolated:

a. Normal diploid human fibroblast lines 761, 811, HSF7
b. Apparently normal human lymphoblastoid line GM131
c. CHOx human lymphocyte lines (human chromosome content)—
   - UV24HL10-12 (#1, 2 del, 3, 11, 13, 19)
   - UV24HL6 (#1, 2del, 3?, 11–13, 14?, 18?, 19)
   - UV24HL5 (#2, X)
   - 314-1b (#3)
   - UV20HL21–27 (#4, 8, 21)
   - 640-12 (#5, 9, 12)
   - UV20HL15-33 (#6, 9, 13, 15, 17, 20, 21)
   - UV41HL4 (#6, 9, 13, 16, 18, Y)
   - 762-8A (#10, Y)
   - UV20HL4 (#1, 4–6, 11, 14–16, 19,21)
   - 81P5D (#12, 15, X)

come from nearby peaks that partially overlap the sort window, from elsewhere in the karyotype as fragments or clumps that fall in the sort window, or from failure of the purity monitor circuits to abort sorting when the near-coincidence of a small fragment and the desired chromosome occurs. These sources of impurity are all minimal for good chromosome preparations with high resolution and little background continuum, emphasizing the importance of good input material to the sorting process.

Molecular Characterization. Molecular analysis of library DNA yields information on purity and other library properties. In the case of libraries constructed from Chinese ham-

TABLE 4. Molecular Characterization of 14 Libraries

Library*	Sorting quality[†]	Mapping unique sequence clones	Estimated hamster impurity (%)	Average insert size (kb)	Reference**
LA *03* NS01	+ + +		23		Hildebrand et al. (1986)
LL *04* NS01	+ + +	47/47→#4	~20	2.0	85
			17		215
LA *04* NS01	+ + +		8		Hildebrand et al. (1986)
LL *06* NS01	+ + +		23		215
LA *07* NS01		18/19→#7			Scambler et al. (1986)
		53/55→#7	20–30		Barker et al. (1986)
LL *08* NS02	+ + +		16		215
LL *09* NS01	+ +		38		215
LL *11* NS01	+ +		52		215
LA *11* NS01		6/6→#11	40–50		Shows (1986)
LA *11* NS02	+ +		20		Hildebrand et al. (1986)
LL *45* NS01	+ + +	3/4 → #14/15	NA[‡]	3.5	Van Tuinen (1986)
LL *17* NS01	+	10/20→#17	NA[‡]	2.0	Van Tuinen (1986)
LL *18* NS01	+ + +	3/3→#18	NA[‡]	1.9	215
LL *20* NS01	+ + +	5/6→#20	NA[‡]	3.1	215

*See first footnote to Table 2 for explanation of the library identification code (ID code). Italicized numbers indicate chromosomal type.
[†]Estimated from resolution of sorted peak from neighboring peaks and amount of debris present. Best is + + +.
[‡]Not applicable, since human cell lines were used.
**Personal communication, except for refs. 85 and 215.

ster–human lymphocyte hybrid cells, contaminating sequences will most likely be hamster DNA, since few human chromosomes are present—and they produce peaks in the flow karyotype distant from the one being sorted. Indeed, some lines contain only a single human chromosome, such as line 762-8A, containing #10 only. Since species-specific hybridization can distinguish hamster from human sequences [95], several libraries constructed from hybrid cells were analyzed for purity by hybridizing random plaques with probes of whole genomic DNA from human or hamster cells. If the chromosome-specific libraries were free of sequences from hamster chromosomes, they should contain no clones that hybridize to the hamster total DNA probe, but should contain many clones which hybridize to the human total DNA probe. If all clones in the libraries contained either a hamster or a human repetitive DNA element, then the fraction containing contaminating hamster DNA could be directly determined by this approach. Actually, it is necessary to correct for the presence of clones with no inserts, or with inserts too small to contain repetitive DNA sequences [215]. The results are given in Table 4.

In the case of the HindIII #18 and 20 libraries constructed from normal human fibroblast cells, inserts of individual clones were characterized with regard to size, whether they contain repetitive or strictly unique sequences, and whether the unique sequences map to the desired chromosome [215]. This more extensive characterization involves the following steps:

1. Isolation and characterization of individual clones
2. Preparation of DNA from a panel of human–hamster hybrid cell lines containing overlapping sets of chromosomes which allow chromosomal assignment of the cloned single-copy sequence.
3. Hybridization of single-copy clones with the DNA panel to verify expected chromosomal assignment.

To characterize the chromosome 18 library, 20 clones were picked at random. Upon screening for inserts containing repetitive sequences using the method of Benton and Davis [11] with total human DNA as a probe, six clones hybridized and therefore contained repetitive DNA. The remaining 14 clones were propagated in the bacterial host, and phage DNA was isolated, digested with HindIII to cut inserts from the vector, and subjected to agarose gel electrophoresis using standard techniques [192]. The insert sizes were determined to be less than 300 bp in six of the clones, too small to be visible on usual gels or useful as probes. The remaining eight clones contained inserts of 0.7 to 5.4 kb in length, average 1.9 kb; three of them contained either two or three inserted HindIII fragments. Each of these eight clones (phage plus insert) was used as a probe on a panel of DNAs from hamster cells, human cells, and two human–hamster hybrid cell lines, one containing only three human chromosomes and the other the same three plus the #18 chromosome [274]. The DNA of this panel had been prepared by digestion of about 2 μg DNA from each cell line with HindIII, electrophoresis on an agarose minigel, and transfer to nitrocellulose filters [167]. The eight clones were labeled with ^{32}P by nick-translation and used in the hybridization step; this and subsequent treatment of the filters, i.e., washing and exposure of X-ray film, were by standard techniques [192].

Five of the eight clones hybridized in smears along the lengths of all lanes containing human DNA, showing that they contain repetitive sequences. The repeats are presumably short or present in moderate copy number or both since they were only revealed by their use as a purified probe. The other three clones, including one with a triple insert, contained only unique sequences. The three HindIII fragments in the triple insert are contiguous on the chromosome, as shown by using that clone as a probe against total human DNA digested with three other restriction enzymes that cut relatively unfrequently [215]. The triple insert, then, is a product of incomplete HindIII digestion. All three unique clones mapped to the human #18 chromosomes, i.e., three out of three unique sequence insert fragments are #18 DNA.

This indicates high purity, although statistical precision is low.

Similar experiments were carried out to characterize the HindIII #20 library. Twenty clones were picked that did not hybridize with total human DNA in a primary screen. Further analysis yielded 8 clones free of repetitive DNA. DNA from these eight unique sequence clones was isolated and characterized as above; the insert sizes (0.3–7 kb; average 3.1 kb) were significantly larger than in the clones from the HindIII #18 library, and there were no multiple inserts. The reason for the differences in insert size and number between the two libraries is unknown, and may be simply due to chance. The six clones with sufficiently large inserts were used as probes on a panel of DNAs that can map sequences specific to the #20 chromosome. All are human sequences; five map to chromosome 20 and one does not.

Summarized in Table 4 are (1) the results on hamster contamination, (2) the results of the characterizations of the HindIII #18 and #20 libraries, and (3) data from user groups. The mapping data are extensive for the HindIII #4 and EcoR1 #7 libraries, which are being used in a search for DNA polymorphisms linked to the defective gene in Huntington's disease and cystic fibrosis, respectively. The human inserts map almost exclusively to the proper human chromosome (#4 or #7); this is expected since the only other human chromosomes in the hybrid lines used in sorting are not close in the flow karyotype.

Purity is high in general, although there are some surprises. The 20% hamster contamination in the #4 library was unexpected, since the HindIII #4 peak was very well isolated from the hamster background and should have negligible overlap by the nearest hamster peaks in the flow karyotype. This discrepancy is unexplained. In the case of the EcoR1 #7 library, hamster contamination is a greater possibility because the EcoR1 #7 peak is close to the hamster background. Several other libraries (HindIII #6, 8, 9, 11) show hamster impurities in the 15 to 50% range, in general correlated with flow karyotype quality, i.e., impurity level is low for high-quality flow karyotypes. In the case of the HindIII #17 library (LL*17*NS01), the mapping data show that one-half the unique sequence inserts map to chromosome #17 and one-half to other human chromosomes. This library was one of the first constructed. The low purity is due to the fact that the #17 peak was not well isolated from the neighboring #16 peak in the flow karyotype of human fibroblast line used, and the chromosome preparation contained much debris and yielded poor resolution. In addition, this library was small (number of chromosome equivalents <5). This library has been recloned from better starting material and is listed in Table 2 as LL*17*NS02; it has much better purity than the old one, which it replaces in the ATCC repository.

The average insert size for five libraries is somewhat less than the average size of 4.1 kb predicted for a six-cutter and a random distribution of restriction sites. This is perhaps not surprising, since the assumption of randomness is not correct (because of repetitive DNA, for example) and the screening procedures for unique clones tend to select for smaller insert sizes.

Storage. During the course of library construction at both National Laboratories, libraries were stored in 4 ways:

1. Unamplified in 7% DMSO at −80°C
2. Plate lysates at 4°C
3. Plate lysates in 7% DMSO at −80°C
4. In CsCl at 4°C

Early experience with the Hind III libraries showed that method [2] worked well. For example, the first library (#18) was constructed in March 1984 and stored as a plate lysate over chloroform at 4°C. Its titer was monitored over a period of a few months and was found to be relatively stable. In fact, the titer dropped only 5.5-fold in 20 months (3.3-fold/year). Similar results were obtained with most of the other libraries. The titers of 10 of the 15 Hind III libraries dropped less than fivefold/year when stored in this simple manner. The titer of another library (#4) dropped about ninefold/year.

Four of the HindIII libraries, however, showed titer losses of greater than 10-fold/year. These were the #6 (13-fold/year drop), #13 (31-fold/year drop), #17 (12.5-fold/year drop) and Y (15-fold/year drop). Similar titer losses occurred for 7 of the EcoR1 libraries. The reasons for the rapid loss in these libraries are not known, as they were amplified and stored just like the other libraries. Another method of storage was then tried for the #13, #17 and Y libraries. They were frozen in 7% DMSO at −80°C in an attempt to stabilize them. After 8 months of storage under these conditions, the titers fell only about twofold (threefold/year). Thus, the sharp loss of the library viability could be arrested by low-temperature storage.

Nine of the HindIII libraries were extremely large and portions of them were not amplified. These nonamplified libraries are the primary packaged recombinant DNA and titers were highly stable. Thus, it seemed that impurities in the plate lysates must be the cause of the titer loss in the amplified libraries. Therefore, portions (about one-half) of 10 of the HindIII libraries were purified by CsCl density-gradient centrifugation. Briefly, the phage were precipitated with polyethylene glycol, resuspended in a small volume and centrifuged to equilibrium on a CsCl gradient. Fractions were collected to ensure that phage with both small and large inserts were represented. These are currently being held at 4°C and closely monitored. Similarly, 17 EcoR1 libraries were purified by CsCl density gradient centrifugation. In February 1986, major fractions of all libraries were transferred to the American Type Culture Collection, Rockville, Maryland, which now serves as a repository with long-term storage at liquid nitrogen temperatures.

Availability. Libraries are available to the general scientific research community, both national and international. Close to 1,200 library aliquots have been sent to about 300 laboratories worldwide from Livermore and Los Alamos through February 1986. The repository and distribution functions were transferred on that date to American Type Culture Collection, Rockville, Maryland; the repository is sponsored by the Division of Research Resources, National Institutes of Health (NIH). It stores library stocks from Livermore and Los Alamos frozen in liquid nitrogen, handles distributions and receives and assembles characterization information from users into an information database open to the general scientific community.

Large Insert Libraries. Phase II of the National Laboratory Gene Library Project is the construction of a set of large insert (20 to 40 kb) human chromosome-specific libraries begun October 1987.

Polyamine chromosome isolation is being used, since DNA of very high molecular weight (≥200 kb) is needed for efficient cloning. Lymphoblastoid cells are used as a source for the smaller chromosomes because of the relative economy with which they can be grown. In addition, certain of these lines show good resolution of some large chromosomes. In this way, reliance on hybrid lines with their problem of chromosome instability is reduced. Although some attempt was made to avoid the use of lymphoblastoid lines transformed by Epstein-Barr virus (EBV) for phase I constructions, their use for phase II is necessary because of the larger numbers of sorted chromosomes required. The fact that some gene rearrangements accompany transformation should not be a detriment to most user groups, since only a very small fraction of the genome appears to be affected by the integrated EBV. Human–hamster hybrid lines will be used when favorable for sort purity, as in the case for the larger human chromosomes. These lines also have only one human homolog present, advantageous for clone ordering and physical mapping.

An important consideration for phase II is the choice of vector. Lambda replacement vectors are widely used; several convenient, recently engineered versions are available. Insert size for these vectors is about 20 kb. Cosmid vectors are a newer development, and accept larger insert size (about 40 kb). This is advantageous for the study of gene structure and function and for physical mapping of large genomes. Cosmid cloning is more difficult than lambda cloning, and not as widely used as lambda cloning. For maximum utility in genetic research, our goal is the cloning of each chromosomal type into both types of vector. Charon 40 [63] will be the lambda vector used; it has an acceptance range of 10 to 24 kb, and incorporates numerous useful features. Experiments with several cosmid vectors are now being carried out. These double cos site cosmid vectors are c2RB [8], a modification of a Lorist vector [92] by Pieter de Jong (LLNL), p cos 5 [226] and s cos 5 [280].

GENE EXPRESSION

Introduction

During the past two decades, a number of methods have been developed for the introduction of genes into somatic cells. These methods are whole cell hybridization, microcell-mediated gene transfer (MMGT), chromosome-mediated gene transfer (CMGT), and DNA-mediated gene transfer (DMGT). For human genetic studies host cells are usually mouse or hamster cells. The method utilized in transferring human genes determines the fraction of the human genome introduced (Table 5). Whole cell hybridization results in the production of human × rodent hybrids, which may contain from one up to a dozen or more intact human chromosomes, while MMGT produces somatic cell hybrids with one or a few human chromosomes. When purified metaphase chromosomes are transferred into host cells by CMGT, the transfectants often contain visible chromosome fragments. In contrast, DNA-mediated gene transfer has been used to introduce genomic DNA or purified, cloned genes into host somatic cells [290]. In all of the above methods, the expression of human genes in the host cell facilitates human genetic studies on gene location, function and structure (Table 5).

A growing number of investigators have combined these somatic cell genetics methods with flow cytometry to study genes which code for cell surface antigens. The expression of a donor surface antigen on a host somatic cell serves as a marker for the genetic material that codes for it. Since cell surface antigens can be labeled on single, viable cells by fluorescent antibodies, the expression of genetic material that codes for donor antigens can be monitored and selected using fluorescence-activated cell sorting (FACS). In this way, flow cytometry has made a significant contribution to the study of genes that code molecules of the cell surface.

TABLE 5. Introduction of Portions of the Human Genome into Recipient Cells*

Whole cell hybridization
Retention of 1–15 chromosomes
$6–30 \times 10^4$ kbp of DNA per chromosome
Mapping genes to chromosomes
Microcell-mediated gene transfer
Retention of 1–3 chromosomes
$6–30 \times 10^4$ kbp of DNA per chromosome
Mapping and isolation of chromosome-specific DNA
Chromosome-mediated gene transfer
Retention of chromosome fragments
$1 \times 10^2 - 5 \times 10^4$ kbp of foreign DNA introduced
Subchromosomal gene localization
DNA-mediated gene transfer
Transfer of whole-cell DNA or cloned genes
10^1 to 10^4 kbp of foreign DNa introduced
Gene expression and gene cloning

*Somatic cell genetic methods employed for fractionating the human genome into rodent recipient cells. The human haploid genome contains approximately 3×10^6 kbp of DNA. The fraction of this material transferred by each method is indicated in the table. Applications of each method in human gene mapping are summarized; see text for details.

Hybrid Cells and Human Genetics

Human chromosomes can be introduced into host mouse cells by the method of whole cell hybridization. Sendai virus or polyethylene glycol induce fusion by breaking down cellular membranes transiently and allowing the contents of two somatic cells to mix (Fig. 2). Initially these hybrids exist as heterokaryons which contain the separate intact nuclei of both cells [103]. But, once nuclear fusion takes place human chromosomes begin to segregate from the hybrids due to their reduced stability during cell division [237]. In human × mouse hybrids, human chromosome loss is more or less a random phenomenon. Some genomic stabilization occurs in tissue culture, but even cloned lines display marked genetic heterogeneity [205]. For this reason, the development of efficient selection systems for the retention of specific human genetic material remains a major goal in human × rodent hybrid cell analysis.

Intact human chromosomes can also be introduced into rodent recipient cells by the method of microcell-mediated gene transfer (MMGT). This method involves the formation of nuclear membrane bounded micronuclei following extensive colcemid treatment [64]. Each of these structures surrounds several condensed metaphase chromosomes. These micronuclei are then isolated by centrifugation and are fused with host cells using polyethylene glycol. This results in the formation of hybrid cells containing one or several human chromosomes [78]. Recent technological advances have allowed the application of these techniques of microcell preparation to primary human fibroblasts [77]. Although substantially reducing the subset of the human genome in a hybrid, fragmentation of the transferred chromosomes has

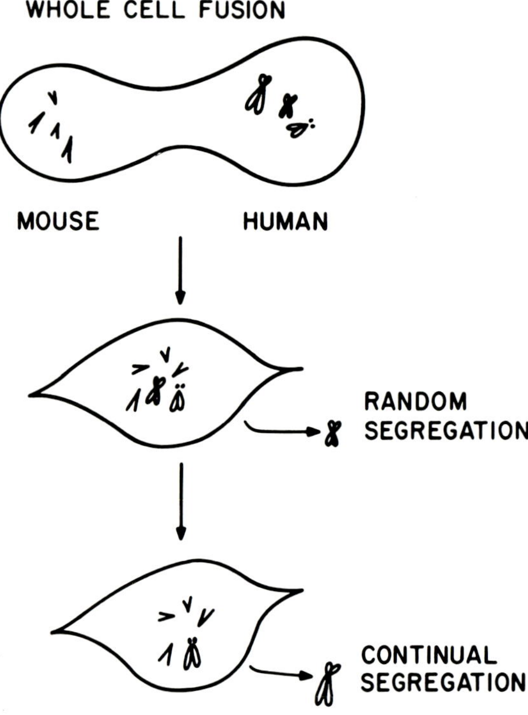

Fig. 2. Somatic cell hybrids. Human chromosomes are gradually lost from human × mouse hybrids formed from whole cell fusion as the hybrid cells are expanded in culture. Human chromosome segregation is rapid, continual, and random, so that even cloned hybrid cell lines are only partly stable.

been observed following this procedure. But, as in the case with whole cell hybridization, human chromosomes transferred to the host must be retained by selection pressure if the cell line is to maintain genetic stability.

Hybrid cell lines produced by whole cell hybridization or microcell-mediated gene transfer contain a subset of human chromosomes, and are useful for mapping human genes to these chromosomes. First, accurate determination of human chromosomes or chromosome-fragments retained by hybrid cells must be made. Donor chromosomes are typically characterized by isoenzyme and karyotype analysis. In isoenzyme analysis, the species-specific differences in enzyme mobility during electrophoresis can be used to distinguish human and rodent enzyme activities in cell lines. At least one isoenzyme activity has been localized to each human chromosome, and is indicative of its presence in a hybrid cell [102]. Isoenzyme analysis does not reveal the presence of fragmented or translocated chromosomes. For this reason it is important to perform Giemsa banding of metaphase chromosomes to confirm the isoenzyme data. This method produces a unique banding pattern on each metaphase chromosome which allows it to be distinguished from other human and rodent chromosomes. It is also possible to distinguish human genetic material in hybrid cells by Giemsa staining at pH 11 [79], or by in situ hybridization with fluoresceinated human repetitive sequence probes [223]. In both cases, the human genetic material is colored or stained differently than the rodent host chromosomes.

Chromosome assignment of an expressed donor gene is accomplished using a group of hybrid cell lines in which each

TABLE 6. A Minimal Hybrid Panel Needed to Map Genes to One of Eight Chromosomes*

Hybrid Panel Member	Chromosome No.							
	1	2	3	4	5	6	7	8
A	0	0	0	0	+	+	+	+
B	0	0	+	+	0	0	+	+
C	0	+	0	+	0	+	0	+

*The three hybrid members in the panel are designated A, B and C, and the chromosomes present in each hybrid are indicated by a + (chromosome present) or a 0 (chromosome absent). Unambiguous assignment of a donor phenotype or DNA sequences to one of the eight chromosomes can be made using this panel.

hybrid member contains unique and overlapping subsets of the human genome [35]. For instance, the panel shown in Table 6 could be used to map a gene to one of eight chromosomes unambiguously. One merely correlates a phenotype assay, such as the expression of a surface antigen, with the donor chromosome complement of each hybrid to obtain a chromosome assignment. A given phenotype pattern (e.g., +0+ for hybrids, A,B,C) assigns the gene coding for that phenotype to one and only one chromosome (chromosome 6 in this example). One would require a similar hybrid panel containing at least five members to map genes to 1 of the 24 human chromosomes.

Hundreds of single-copy human DNA sequences have already been isolated either in the study of specific human genes or to provide markers for genetic linkage studies [240,293]. A human × mouse hybrid cell panel can be used to map these fragments to chromosomes. In the case of probes generated by chromosome sorting, hybrid cells may be used to confirm the expected map position. DNA is first isolated from the characterized hybrids, cut with a specific restriction enzyme, and the different-sized fragments resolved by gel electrophoresis. These genomic DNAs are then hybridized with the cloned, labeled probe using standard Southern blot technology. This approach is schematically illustrated and described in Figure 3. Although human and rodent sequences often cross-hybridize, species differences in endonuclease restriction sites within a gene produce species-specific gene fragments of different sizes. These size differences are resolved by electrophoresis and make it possible to distinguish the human genes. Hundreds of cloned human DNA sequences have already been assigned chromosome map positions, and provide genetic markers for use in generating a human genetic linkage map [20].

Improved technologies for in situ hybridization have recently made it possible to localize single-copy DNA fragments directly to Giemsa-banded metaphase chromosomes [100]. This technology is directly applicable to metaphase spreads of human cells. But, if hybrid panels are first used to map the fragment to a specific chromosome, in situ hybridization can not only confirm this assignment, but also localize the DNA fragment to a small region of the chromosome.

Selection for Somatic Cell Hybrids Containing Specific Human Chromosomes

Biochemical Selection. Random continual human chromosome segregation poses the major obstacle to somatic cell hybrid preparation and analysis. The retention of specific chromosomes or their fragments in somatic cell hybrid populations may be accomplished by biochemical selection me-

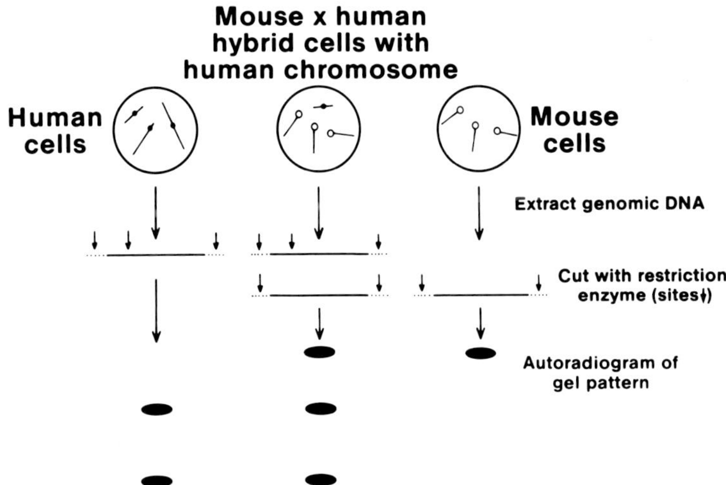

Fig. 3. Restriction fragment mapping. This simplified example shows how a cloned human DNA fragment can be mapped to a specific human chromosome using mouse × human hybrids. The fragment is complementary to a specific region of the human genome (heavy line) with noncomplementary flanking sequences (dotted lines) on each side. The mouse genome has a similar, but not identical sequence (heavy line). Restriction sites in and flanking this sequence for a particular restriction enzyme are shown by arrows, the only difference being the absence of an internal site in the mouse DNA. The total genomic DNA from the human cells, human × mouse hybrid cells containing a single human chromosome, and mouse cells is extracted, purified, cut with the restriction enzyme, and subjected to gel electrophoresis to separate the DNA fragments according to size. These DNA fragments are transferred to a nitrocellulose filter, hybridized with the ^{32}P-labeled human DNA clone, and autoradiographed. The resulting band pattern shows that the hybrid cell DNA has a composite structure corresponding to a mixture of mouse and human DNA. This establishes that the cloned human DNA sequence maps to the human chromosome contained in the hybrid.

dia based on genetic complementation of enzyme defects in the host cell. For example, hypoxanthine, aminopterin, thymidine (HAT) supplemented media select for somatic cell hybrids derived from host variants defective in either hypoxanthine phosphoribosyl transferase (HPRT) or thymidine kinase (TK) [185]. Only hybrids containing either human HPRT from the X-chromosome or TK from chromosome 17 will survive the selection. A limited number of other rodent auxotrophic mutants have been developed that can be used in conjunction with biochemical selection schemes to drive the retention of specific human chromosomes in these hybrids, and these systems have been extensively characterized [130].

An alternate system of hybrid selection has been developed that extends this biochemical selection to other chromosomes. A dominant selectable marker (e.g., the bacterial xanthine-guanine phosphoribosyl transferase or neomycin resistance genes) can be integrated at random into the chromosomes of human cells by DNA-mediated gene transfer. These marked chromosomes are then transferred to mouse cells by whole cell or microcell fusion. Hybrids containing the marked chromosomes are then isolated in selection media. Effectively, this introduces selectable biochemical markers into chromosomes which do not contain their own [127].

Surface Antigen Selection and Chromosome Mapping. The first human × rodent somatic cell hybrid line containing functioning genes was produced by Weiss and Green in 1967 [286]. Using a mixed cell agglutination assay these workers were able to demonstrate the expression of human-specific surface antigens on these hybrids. They concluded that surface markers of both species are codominantly expressed and that the human antigen serves as a marker for the chromosome that codes it. Antibodies can readily be generated to these human-specific cell surface determinants and can be used in indirect immunofluorescence assays to label hybrid cells which express the antigen. When the expression of this surface antigen is examined across a hybrid panel, it is possible to map surface antigen genes to specific chromosomes. In conjunction with immunoselection procedures these antibodies also can be used to select for [53] or against [120] hybrid cells expressing these determinants. The fluorescence-activated cell sorter provides a powerful tool for immunoselection because it is capable of simultaneously selecting cells that both express and lack a given antigen. Studies of the human 4F2 determinant demonstrate the use of the FACS for both gene mapping and hybrid chromosome selection [221].

Monoclonal antibodies to human specific-surface determinants provide an important tool for the identification and study of surface molecules critical to cell function, growth and differentiation. The monoclonal antibody 4F2 was isolated in a study of human T-lymphocyte antigens, and was characterized as a marker of human lymphocyte activation [105]. The 4F2 antibody defines a human cell surface determinant that is expressed on fibroblasts and on a variety of hematopoietic cell lines, monocytes and on activated T and B cells, but is absent from nonactivated peripheral B and T cells. It recognizes a determinant on a large peptide of 100,000 M_r which is noncovalently associated with a smaller subunit of 40,000 M_r on the cell surface [108]. The expression of the 4F2 antigen was examined on a panel of human × mouse hybrid cell lines and a number of lines demonstrated 4F2 antigen subpopulations of various sizes [221]. One line in particular (FRY 5) displayed marked heterogeneous staining (Fig. 4a), and 4F2 positive (Fig. 4c) and negative (Fig. 4b) cells were sorted using the FACS.

The expectation that sorting based on surface antigen ex-

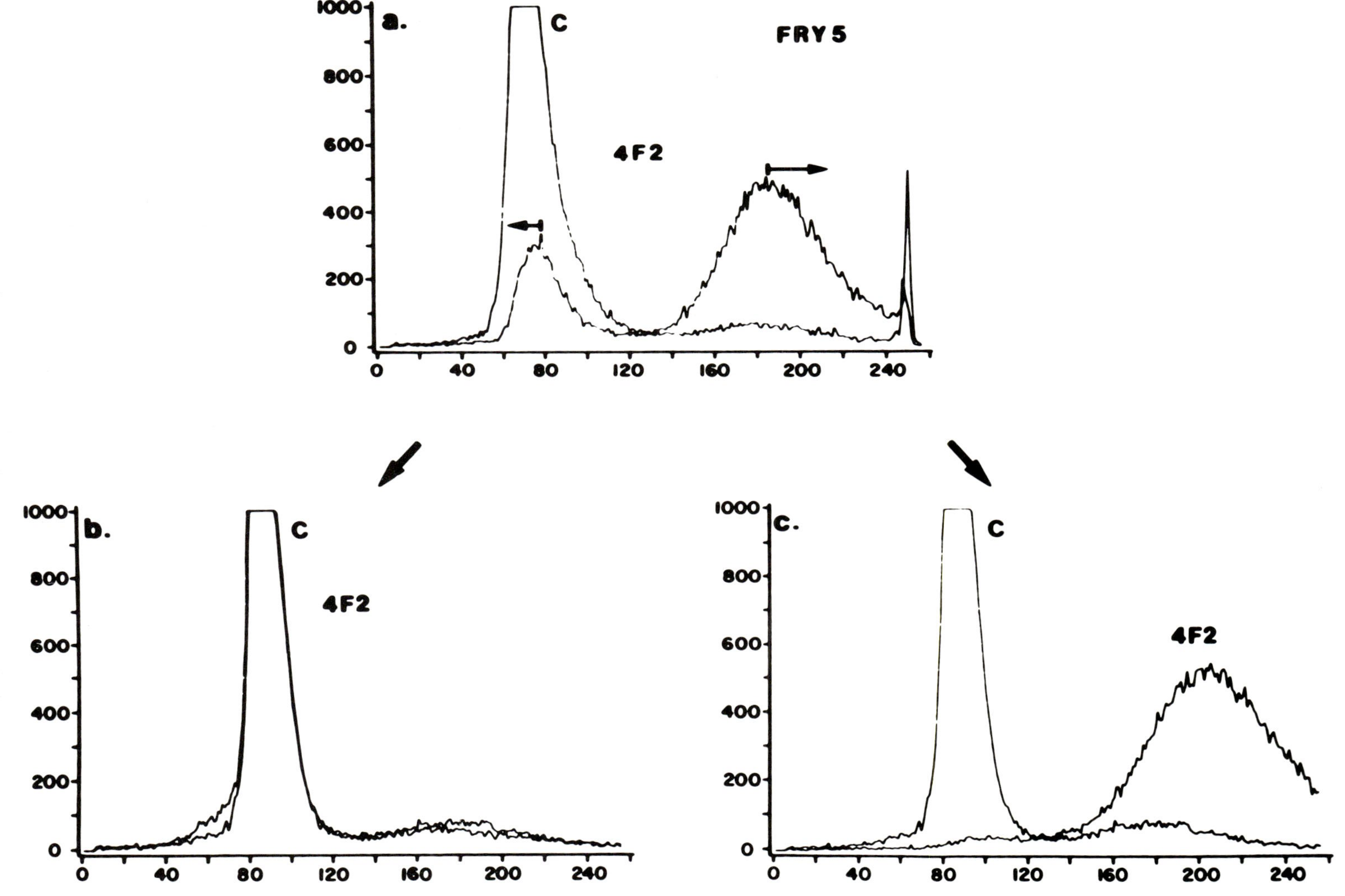

Fig. 4. Analysis and cell sorting of human × mouse hybrid line Fry 5 for expression of the 4F2 antigen. Cells were stained by indirect immunofluorescence using monoclonal antibody 4F2 or control supernatants (C) and analyzed using the FACS. Photomultiplier voltages were converted by logarithmic amplification for display on the multichannel analyzer. Cells that stained specifically with the 4F2 antibody were sorted above the threshold level indicated by the right arrow. Cells that did not stain specifically with the 4F2 antibody were sorted below the threshold level indicated by the left arrow. (a) FRY 5; (b) FRY 5.4F2 1-; (c) FRY 5.4F2 2+ [221].

TABLE 7. Isozyme Analysis of a Human–Mouse Hybrid and FACS-Derived Sublines*

	Human chromosome																						
Cell lines	1	2	3	4	5	6	7	8	9	10	11	12	13	14	15	16	17	18	19	20	21	22	X
FRY5 unsorted		X			X	X		X		X	X	X	X			X	X			X			
		X																					
FRY5 4F2-		X				X		X		X		X	X			X	X			X			
		X																					
FRY5 4F2+		X				X		X		X	X	X	X			X	X			X			
		X																					

*Cell preparations, starch gel electrophoresis, and cellogel techniques were performed as described by Harris and Hopkinton [102]. Cell sorter derivation of homogeneous sublines is described in text. Isozyme assay for all human chromosomes was performed. Presence of a chromosome is indicated by X. The chromosome containing the 4F2 locus must be present in antigen-positive populations and absent from antigen-negative populations. Chromosome 11 is the only chromosome fitting the required pattern [221].

pression would yield cell populations characterized by specific chromosome retention or segregation patterns was tested by analyzing the sorted hybrid sublines for the presence of human chromosomes. A summary of this study performed by isoenzyme analysis for each human chromosome is tabulated in Table 7. It is clear that chromosome 11 alone segregates concordantly with the expression of surface antigen 4F2, indicating that the gene for this antigen is on chromosome 11. This analysis was confirmed by antigen/genotype analysis of the other members of the hybrid cell panel including a hybrid cell line which contains only human chromosome 11 [221].

A large number of human genes coding chromosome-specific surface antigens already have been identified and include those on chromosomes 1, 2, 3, 5, 6, 7, 11, 12, 15, 17, 21, 22, and X [214]. This list will continue to expand with the application of monoclonal antibody technology to the genetic dissection of the human cell surface. As demonstrated by the 4F2 study, a surface antigen serves as a constitutive marker for the chromosome that codes it. Thus, antibodies to a species-specific determinant can be used in conjunction with the FACS to select hybrid cells that retain or have segregated a specific human chromosome.

It also has been possible to regionally map human genes using cell hybrids which contain fragments of human chromosomes [120]. Such fragments can be obtained by fusion of the rodent parent with human cells which contain intraspecies chromosome translocations. Alternatively, regional mapping can be performed with cell hybrids containing human × mouse interspecies translocations. Interspecies translocations are randomly produced in hybrid cell lines which have been in culture for an extended time. They also occur following chromosome-mediated gene transfer (CMGT) of calcium phosphate precipitated donor metaphase chromosomes into host cells [194]. CMGT donor fragments are initially unstable, but some undergo chromosome integration, inserting relatively large contiguous fragments of donor DNA into the host genome [140]. Kamarck et al. examined the surface expression of the human X-coded S11 antigens on a panel of hybrid lines which had fragmented the X-chromosome either during chromosome-mediated gene transfer or hybrid cell expansion [128]. One chromosome-mediated gene transfer line was identified that expressed S11 antigens and the two X-linked isozymes G6PD and HPRT, indicating that these three markers were closely linked. It also was demonstrated that a hybrid cell line had segregated the S11 antigen from a fraction of cells during expansion, while maintaining the selectable marker HPRT. These two hybrid populations were separated by cell sorting based on the expression of S11. Sorted cells were analyzed both by isozyme and by karyotype methods. Three independent interspecies translocations were characterized from the sorted populations, two of which were present in the antigen negative sorted population. By correlating data provided from these sorted hybrids it was possible to map the S11 antigen gene(s) to the X chromosome region Xq27-28 between the genes coding the isoenzymes HPRT and G6PD (Fig. 5) [128].

Intraspecies Cell Hybrids

Somatic cell hybrids can be classified as being either interspecies, such as the human × mouse hybrids already described, or intraspecies, when cells originate from the same species. In both cases whole cells are fused, but intraspecies hybrids are characterized by a high degree of chromosome stability. In human genetic studies, such hybrids are uniquely suited to the studies of genetic complementation between two mutants. An example of this approach can be found in work on the genetic complementation of lysosomal enzyme deficiencies [121]. This has involved the fusion of fibroblast cytoplasts (anucleate cells) with fibroblasts to give rise to cytoplasmic hybrids between two mutants.

Jongkind and co-workers approached the problem of creating such hybrids by using the cell sorter [122]. Two days before fusion, fluorescing microspheres were added to nonconfluent cell cultures. Cells to be enucleated were labeled with green-fluorescing beads, while the other parental cell received red-fluorescing microspheres. Nonspecific uptake of the beads by both lines resulted in cells that could be distinguished based on the wavelength of fluorescence emission. Following fusion hybrid cells containing red and green microspheres were identified by the FACS and sorted, producing a homogeneous population of intraspecies hybrids.

An experimental fusion was performed between β-galactosidase deficient cells derived from a patient with GM1-gangliosidosis type I and cells from an individual with a combined β-galactosidase and neuraminidase deficiency. Genetic complementation between these two mutant cells resulted in the restoration of β-galactosidase activity, and pro-

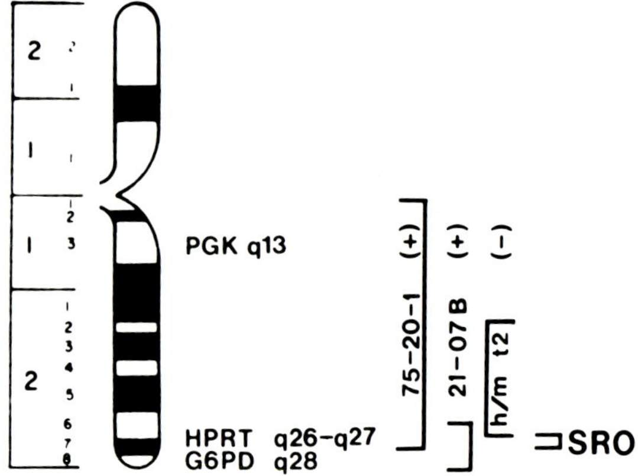

Fig. 5. Regional mapping of the human S11 gene on the X chromosome. Analyzed lines include two hybrid clones (75-20-1 and H/M T2) which spontaneously translocated part of the human X–chromosome to a mouse chromosome. A third translocation (not shown) did not contain a visible human chromosome fragment. Cell line 21-07B was produced by chromosome mediated gene transfer of HeLa chromosomes into a mouse L cell host. Brackets indicate the genetic region included by each cell line, and the (+) or (−) indicates the expression of S11 antigens on the line. The bracketed region designated SRO indicates the shortest region of overlap which includes the gene position [128].

vided valuable information about the genetic basis of each defect [121,122]. The formation of such intraspecies hybrids with the FACS is generally applicable to the hybridization of cell populations where drug resistance selection systems are not available.

Somatic Cell Genetic Variants

Although the use of hybrid cell lines facilitates genetic studies such as chromosome mapping and genetic complementation, specific gene alterations and mutation can be studied directly in unfused somatic cells. Sensitive biochemically based selection systems have been devised to isolate rare variants from a population with altered gene function [29]. The application of fluorescence-activated cell sorting to the isolation of rare somatic cell variants was first made by Liesegang et al. [183] who selected variants of myeloma MPC-11. This mouse myeloma cell expresses an immunoglobulin of the IgG2b subclass on the cell surface. To isolate isotype switch variants expressing different immunoglobulin heavy chain constant regions, these workers stained the cell population by indirect immunofluorescence with both anti-IgG1 and anti-IgG2a reagents. These reagents bind specifically to the 2 different heavy chain constant regions coded by the Y_1 and Y_2 genes (Fig. 6). The brightest 0.5 to 2.0% of the cells were isolated by sorting, grown up and resorted. After three rounds of enrichment sorting, the positive cells were cloned. Clones were isolated which expressed the IgG2a, but not IgG1 heavy-chain isotype, while retaining the idiotypic determinants present on the original parental cell. The frequency of these cells in the original population was estimated to be 10^{-4} [183].

The same workers extended this observation with studies of the isotype switch variants that could be isolated from myeloma X63, which expresses an IgG1 immunoglobulin [231]. From this work [183,206,231] and that of Herzenberg and coworkers [43] a pattern of isotype switch variants has emerged. The order in which isotype changes could be isolated corresponded to the order of genes from 5′ to 3′ in the immunoglobulin heavy chain locus (Fig. 6). In other words, while isotypes could readily be switched to a locus downstream (3′), those isotype genes that were 5′ to the original heavy chain isotype of the myeloma have already been deleted during immunoglobulin gene rearrangement. For example, by starting with a hybridoma producing IgG1 antibodies to the hapten dansyl, Dangl et al. successfully isolated the heavy chain "family" of antidansyl antibodies consisting of the same variable region genes on molecules with heavy chain isotypes IgG1, IgG2b, IgG2a, and IgE [43]. This application of somatic cell variant analysis provides biological material for studies of the molecular events involved in heavy chain switching, and the functional differences between immunoglobulins with different heavy chains.

The isolation of rare (frequency = 10^{-4} to 10^{-6}) heavy chain switch variants involved important technical modifications aimed at reducing the number of nonspecifically sorted cells, and visualization of rare positive cells [42,213]. These developments included the use of a logarithmic amplifier to handle the wide range of fluorescence signal amplitudes, the use of highly specific monoclonal antibodies, and gating out nonspecifically stained dead cells. These cells can be distinguished by their inability to exclude the vital fluorescent dye propidium iodide (PI). Dead cells treated with PI take up the dye, fluoresce at its characteristic red wavelength, and are excluded from analysis and sorting by "gating."

In addition to immunoglobulins, other cell surface antigens have served as a target for somatic cell variant selection using the FACS. Rajewsky and co-workers studied the expression of mouse histocompatibility antigen (H-2) genes with altered epitopes (i.e., the part of the molecule that the antibody recognizes). The extensive antigenic polymorphism of the histocompatibility antigens provides the immune system with its major barrier to tissue transplantation. These studies focused on the H-$2K^k$ molecule, which reacts with a number of distinct monoclonal antibodies [111]. Rajewsky et al. selected cells that reacted with one anti-H-$2K^k$ monoclonal but that no longer competed for binding of a second anti-H-$2K^k$ antibody (Fig. 7). Variant cells (10^{-5} to 10^{-6}) expressing an altered H-$2K^k$ gene product were isolated by this method. By imposing positive FACS selection for antigen expression with the first monoclonal, this approach was successful in excluding variants that had simply lost expression of the entire surface antigen. This provides a general method for the generation of structural mutants in cell surface expressed molecules.

DNA-Mediated Gene Transfer

Isolated fragments of DNA can be introduced into host cells by the method of DNA-mediated gene transfer (DMGT) [246]. Cloned genes whose size may be on the order of tens of kilobases can be efficiently transferred into host cells and permit assessment of gene expression. DNA-mediated gene transfer also can be used for the transfer of human whole cell DNA into mouse hosts as a method for fragmentation of the human genome [242].

Several systems have been developed for the introduction of foreign, donor DNA into recipient cells. The most widely

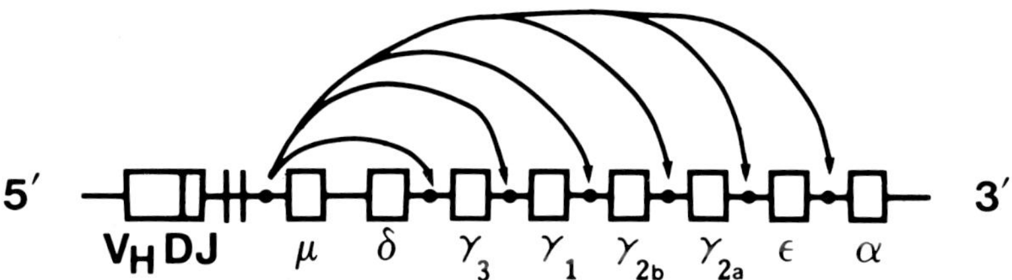

Fig. 6. Structure of the mouse heavy-chain immunoglobulin gene locus. Variable region exons (V_H) joining region exons (D_H and J_H) and heavy chain genes (μ, δ, Y_3, Y_1, Y_{2b}, Y_{2a}, ε, α) are indicated by boxes. The assembly of V,D, and J segments can be associated sequentially with different constant region chains as indicated by the arrows resulting in the deletion of the intervening DNA. Class switch variants can therefore be isolated 3′ to the starting heavy chain.

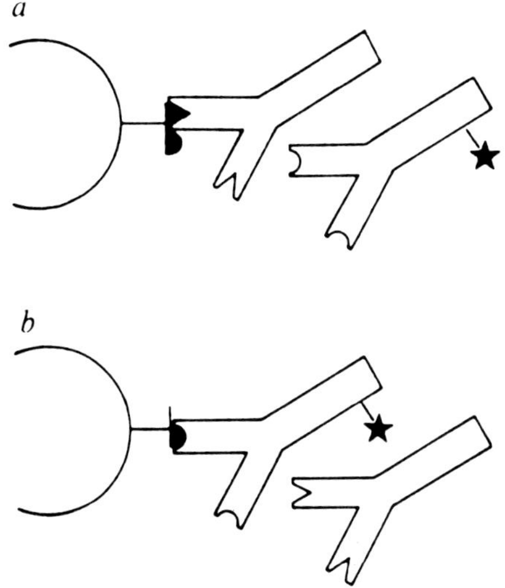

Fig. 7. Principle of variant selection. **a:** The labeling of cells carrying wild-type H-$2K^k$ by a fluorescent (*) monoclonal antibody is inhibited by an unlabeled second antibody that binds to a neighboring antigenic determinant. **b:** Cells expressing variant H-$2K^k$ molecules that have lost the neighboring determinant are successfully stained in the same experimental conditions. In reality, the unlabelled antibody is present in a 1,000-fold excess [111].

used approach has been the DNA/calcium phosphate precipitation technique [87]. Basically, DNA is coprecipitated with calcium phosphate to form microscopic aggregates in the 0.1 to 1.0 μm size range. These bind efficiently to the cell surface, and are ingested endocytotically. In readily transfectable cell populations such as NIH/3T3 and L cells, the majority of cells ingest large amounts of precipitate within a few hours [189]. Shortly after uptake of DNA into recipient cells, it is covalently joined together to form high molecular weight molecules [220]. These giant concatamers have been termed transgenomes (Fig. 8). Although the mechanism of their formation is not understood, it is believed that the process occurs in the nucleus and may represent some form of DNA repair activity. Transgenomes are unstable following transfection, and it has been proposed that the transgenome is not initially integrated into a recipient chromosome [245]. However, after several weeks of selection, transformed cells invariably give rise to stable subpopulations. The stable phenotype correlates with transgenome integration into a single random chromosomal site (Fig. 8) [235].

Cell populations are usually transfected with bacterial or

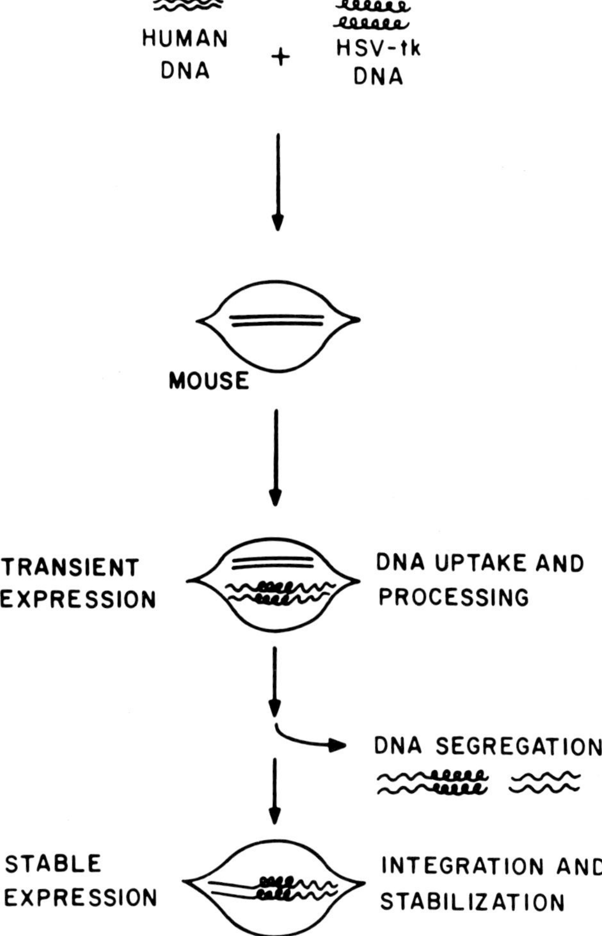

Fig. 8. DNA-mediated gene transfer. The uptake, segregation, and stabilization of human DNA and herpes simplex virus thymidine kinase DNA in a mouse-cell host following DNA-mediated gene transfer. Following DNA uptake transient expression of the gene probably corresponds to the presence of an extrachromosomal transgenome. Stable expression of transferred genes has been shown to be correlated with the integration and stabilization of the introduced gene into a host chromosome. Details of this process are described in the text.

viral genes containing selectable markers. These genes then provide the selective basis with which transfected cells can be isolated from the vast majority of cells which were not competent for stable transfection. The most commonly used gene for this purpose is the herpes simplex virus-thymidine kinase gene (HSV–tk) [289]. When this gene has been transferred into L cells which are thymidine kinase deficient (tk^-) the viral gene complements the genetic defect and cells transfected with the gene survive HAT selection medium. In a similar fashion mammalian cells that have been transfected with the bacterial gene that confers neomycin resistance can survive the addition of the neomycin analogue G418 [204]. The advantage of this second system is that it is more general; it does not require genetic complementation, and so does not necessitate the selection of a host cell mutant. Finally, Look and co-workers [188] have transfected NIH3T3 cells with the oncogene v-fms, and selected transfectants based on their increased growth rates and altered morphology. This characteristic provides the major advantage of this selection system, which is the more rapid growth and expansion of transfectants.

Wigler and co-workers [290] demonstrated that if nonselectable cloned genes are mixed with the selectable gene (i.e., tk) a high percentage of the thymidine kinase positive transfectants also contain the cotransferred sequences. This phenomenon of cotransfection can be explained by the extensive amount of DNA processing that takes place during the formation of transgenomes. Cotransfection frequencies are high because the non-selected passenger gene sequence is joined during the formation of concatamers to the selectable gene. The selectable gene becomes integrated as a consequence of direct selection pressure, whereas the passenger gene is maintained indirectly, because of its attachment to the selectable marker. Integration into a chromosomal site renders both markers stable even in the absence of selection pressure. This process is highly efficient, because of the large amount of DNA normally incorporated into a transgenome. It should be noted that genomic DNA as well as cloned genes can be equally well cotransferred in this manner [242].

The application of gene transfection techniques have been limited by the relatively few cell lines which can serve as hosts, and the efficiency with which genes can be transferred into these cells. Recent developments in the methodology of gene transfer include the use of retroviral vectors with expanded host range, and the use of electroporation to allow more efficient gene uptake in cells with low transfection frequency [28,225]. Both of these methods promise to extend the utility of DMGT in the study of gene expression.

DNA-Mediated Gene Transfer and Cloned Surface Antigen Genes

A number of workers have used DNA-mediated gene cotransfer to transfect cells with cloned surface antigen genes. The fluorescence-activated cell sorter has been invaluable in quantitation of gene expression following transfection.

Expression of Surface Antigen Genes. The histocompatibility antigens form the major barrier to tissue transplantation and play a crucial role in cell to cell interactions in the immune system [224]. A number of laboratories isolated cDNA clones representing both the HLA (human) and H-2 (mouse) gene products. Southern blot analysis of human and mouse genomic DNA, hybridized with these cloned probes, revealed 15 to 20 or more HLA or H-2 related gene sequences, all of which mapped to the histocompatibility locus [23,180]. To identify which members of this multigene family were the functioning genes, as opposed to pseudogenes, individual clones were transferred into mouse recipient cells by DNA-mediated gene transfer and gene expression was examined by flow cytometry or other quantitative methods.

TABLE 8. Expression of $H2L^d$ on a Transfected Cell Line*

		Mean fluorescence	
Specificity	Antibody	DAP (parent)	T1.1.1 (transformed)
K^k	36.7.5 S	<u>620.5</u>	<u>162.8</u>
I-A^b, Ia^d	25.9.17 S	118.8	104.2
L^d/R^d	28.14.8 S	118.3	<u>459.6</u>
L^d	20.5.7 S	125.0	<u>492.2</u>
D^d	34.5.8 S	129.8	107.7
—	No Ab	110.0	98.0

*Mouse L cells (DAP-3) and a cell line transformed with the H-$2L^d$ gene (T1.1.1) were analyzed for the expression of H-2 antigens. The cells (10^6) were incubated with hybridoma culture supernatant containing monoclonal antibodies specific for H-2 antigens. After reaction with fluorescein-conjugated goat $F(ab')_2$ anti-mouse F_c, cells were fractionated by flow cytofluorimetry. Mean relative fluorescence at constant gain (normalized to 8) was determined for 10^5 viable DAP-3 (L-cell parent) or T1.1.1 cells. The SD varied from 37 to 60 fluorescence units. Fluorescence values over background have been underlined, and show that H-$2L^d$ is expressed on the surface of T1.1.1 transfectants. Four additional experiments with this transformed cell line gave similar results. In addition to clone T1.1.1, three additional transformation experiments with this gene yielded transformed cells that produce H-$2L^d$ determinants [72].

The first use of flow cytometry for this purpose was the characterization of the Balb/c H-2L gene [72]. Mouse L cells were cotransfected with cloned genes and the HSV-tk gene and selected on HAT media. These cells were then analyzed by flow cytometry using indirect immunofluorescence with a panel of monoclonal antibodies. Cells transfected with the functioning gene reacted strongly with the specific monoclonal antibodies recognizing the H-$2L^d$ antigen [72] (Table 8). Using the identical strategy with other assay methods other groups discriminated the functional genes for the H-$2L^d$ and porcine SLA^d histocompatibility antigens from pseudogenes containing homologous sequences [86,262].

Ruddle and co-workers first characterized functioning genes for human histocompatibility antigens through the use of transfection, indirect immunofluorescence and the FACS [6]. From a genomic library, they selected a panel of 18 phage clones that showed varying degrees of hybridization with an HLA-B7 cDNA probe. These genes were transfected into mouse L cells with HSV-tk as described. Using specific monoclonal antibodies and flow cytometry these workers were able to identify transfectants expressing HLA-A2 and HLA-B7, and hence the specific phage clones carrying that genetic information. In a similar fashion, Jordan and co-workers identified the genes coding human HLA-A3 and HLA-CW3 [179,180,191]. As was the case with mouse histocompatibility genes, the vast majority of these human gene sequences did not code functioning genes, indicating that these mammalian genomes contain multiple copies of homologous pseudogenes. It has been suggested that these nonfunctional genes may play a role in generating and maintaining the high degree of polymorphism at this locus by serving as genetic repositories for variation.

Ruddle and co-workers examined the expression of HLA-A,B,C antigens on recipient cells at 60 hours after gene trans-

TABLE 9. Surface Antigens Selected from Genomic Transfections

Antigen	Tissue specificity
Mouse	
Ly-1(112)	T-cell and B-cell subset
Ly-2(112)	T-cell
BLA-1(112)	B-cell
30-Fl(112)	B-cell
BLA-2(112)	B-cell
Th-B(112)	B-cell and immature T-cell
Fc-R(112)	B-cell and killer cell
Thy-1(112)	T-cell, brain
L3T4a(112)	T-cell
$H\text{-}2^d$ (112)	Most cells
T-200(112)	Lymphocytes
Human	
Leu1(T4)(132)	T-cell and B-cell subset
Leu2(T8)(134)	T-cell and NK cell
Nerve GF Rcptr(28a)	Nerve tissue
Rhinovirus Rcptr (89a)	Many cell types
gp150(188)	Myeloid cells
Transferrin Rcptr(151,207)	Most cells
HLA-A,B,C(132,150)	Most cells
B_2M(132)	Most cells
4F2(150)	Fibroblasts, T-cells
S11	Fibroblasts
Rat	
Nerve GF Rcptr(112)	Nerve tissue

fer before selection had been imposed [6]. Using flow cytometry following transfer of functional genes, it was possible to quantitate the transient expression of HLA determinants on 1 to 25% of the recipient cells (Fig. 9). This transient expression probably results from transcriptional activity of the gene before its degradation or chromosomal integration. This expression disappears by 84 hours posttransfer, but allows a more rapid assessment of gene expression than by waiting for gene integration into chromosomes. It has also been possible to sort cells based on transient expression and, after numerous selection passes over a period of time, to recover stable surface antigen transfectants without the use of any biochemical selection pressure [12,126].

Cloned surface antigen genes have been engineered by exon shuffling in an effort to dissect the function of different coding domains. Unique restriction sites in the introns between coding domains of a surface antigen gene are used to cut the gene, and recombine with homologous domains from a second molecule. The expression of these genes and the presence of specific epitopes (antibody-binding sites) has been characterized using flow cytometry [73,195]. Genetic engineering was used to generate in vitro genetic recombinants between the genes for two mouse H-2 antigens, involving the first two coding exons (N and C_1 in Fig. 10), and the remainder of the molecule. These engineered genes were then transferred into L cells as has been described. Monoclonal antibodies and flow cytometry were used to characterize 13 epitopes on these hybrid molecules. This analysis allowed the mapping of monoclonal antibody defined epitopes to different domains on the H-2 molecules (Fig. 10). Hybrid molecules expressed on L cells have also been used to define functional domains of the H-2 molecule. In this way, the major epitopes recognized by cytotoxic T cells have been shown to be controlled by the extensively polymorphic first two coding exons (N and C_1) of the genes for the Class I antigens [211].

Finally, it has been possible to construct recombinant genes for surface molecules composed of both H-2 and the I-A surface antigens, transfect them into L cells, and to examine epitope expression with flow cytometry [94]. These studies suggest the broad application that recombinant DNA construction, transfection and flow analysis will have in the study of the serology and function of surface antigen gene domains.

Regulation of Surface Antigen Genes. Flow cytometry also has been useful in the quantitation of gene expression following DMGT. The ability of the flow cytometer to quantitate fluorescence accurately, and therefore antigen levels, has made it possible to assess the transcriptional activity of these transfected cells. Ruddle and coworkers used flow cytometry to quantitate HLA antigen expression on cloned transfected cell lines. They were able to correlate antigen expression with the number of gene copies which had been integrated into these lines, indicating that the HLA antigen genes are constitutively expressed in fibroblasts [6]. Several groups have assessed the transcriptional activity of introduced genes in different cellular environments. It had been shown that the treatment of human cells with human interferon, or mouse cells with mouse interferon, causes an increase in the number of histocompatibility antigens available for antibody binding at the cell surface [76,237]. Rosa et al. [237] transfected the human HLA-A3 gene into mouse cells, and demonstrated increased gene transcription upon treatment of transfectants with mouse interferon. This increase in transcriptional activity was monitored by Northern Blot hybridization for HLA mRNA, immunoprecipitation of antigen, radioimmunoassay, and quantitative flow cytometryc analysis.

Shulman et al. [126,260] performed similar studies on L-cells and transfectants which contained either the HLA-A2 or HLA-B7 genes. All three cell lines were treated with mouse β-interferon and monitored by the FACS for expression of the mouse and human antigens. In all three lines, the mouse histocompatibility antigens $H\text{-}2K^k$ and $H\text{-}2D^k$ were induced, although $H\text{-}2K^k$ was consistently induced at a higher level. The human HLA antigens were also increased on mouse cells in response to mouse β-interferon. As was the case with the endogenous histocompatibility antigens, these antigens were disproportionately induced: HLA-A2 increased 2.7-fold over 36 hours, while HLA-B7 increased only 2.2-fold. These results were confirmed on a transfected cell line that had been transfected with both HLA-A2 and HLA-B7 [260].

Evans et al. [71] studied the expression of the Thy-1.2 glycoprotein gene in different genomic environments. This antigen has been used as a tissue specific marker normally found in high levels on T-lymphocyte and neuronal cells. Using DNA-mediated gene transfer, these investigators introduced the surface antigen gene into cell lines derived from either fibroblast, lymphoid or neuronal tissues. Although each transfectant could be shown to contain similar copy numbers of the gene, different cell lines expressed vastly differing amounts of cell-surface Thy-1 as monitored by the FACS. Cells that normally express high levels of Thy-1 antigens also expressed high levels of the transfected genes [71]. This result suggested that the Thy-1 differentiation antigen gene is under tissue-specific regulation by the promoter present on the gene clone.

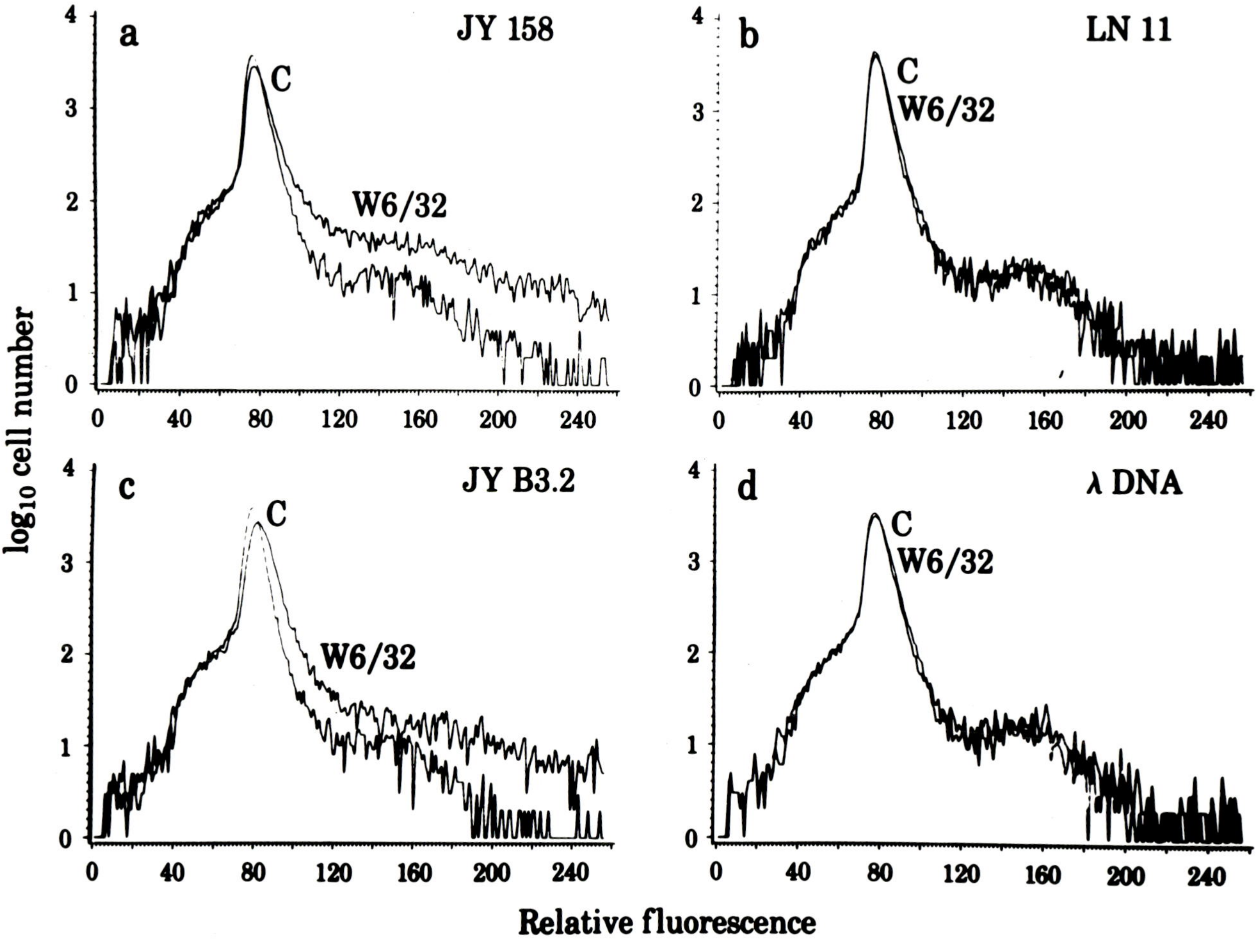

Fig. 9. Transient gene expression. Indirect immunofluorescence and FACS analysis of HLA expression at 60 hours after transfection with human genomic clones. Each panel presents a population of transfected Ltk^- cells stained with monoclonal antibody W6/32 compared with control staining with myeloma supernatants (C). Cell number was converted to logarithmic values to facilitate visualization of the results. Transfection with: a) JY158 (HLA-B7 gene); b) LN11 (HLA pseudogene); c) JY B3.2 (HLA-A2 gene); d) lambda DNA (control) [6].

DNA-Mediated Gene Transfer and Gene Cloning

DNA-mediated gene transfer serves as an efficient method for the fractionation of the human genome in host mouse cells. Following DMGT with both a selectable marker and genomic human DNA, each of the resulting transfectants contains a small fraction of the human genome. Evidence from a variety of sources indicates that the human DNA present in each cell may range from 10^3 to 10^4 kb pairs [150]. This transfected human DNA can be distinguished from mouse DNA by the presence of species-specific repetitive sequences such as those defined by the restriction endonuclease Alu [116]. These repetitive sequences are estimated to occur at 1- to 10-kb intervals throughout the human genome, and thus serve as a convenient endogenous marker sequence for the identification of human DNA [95].

Isolation of Genomic Transfectants for Gene Cloning. A number of investigators have used DMGT and selection for gene expression as a first step in cloning human gene sequences. Human Harvey-ras oncogene transfectants were identified by the transformation of NIH3T3 cells which resulted in an increased growth rate [257]. Both TK and HPRT genomic transfectants have been isolated by genetic complementation of host L cell mutants selected by HAT selection [21,119,184]. Following the isolation of these primary transfectants, an additional round of DMGT is performed using primary cell DNA, producing secondary transfectants. This serves to further dilute out extraneous donor sequences, as the original transgenome is further fragmented with host cell carrier DNA [242].

DNA from this secondary transformant usually contains between 30 and 100 kilobase pairs of the original donor DNA, and includes the selected gene. This DNA can be used directly for the construction of phage libraries, using human repetitive sequences to identify and isolate specific phage

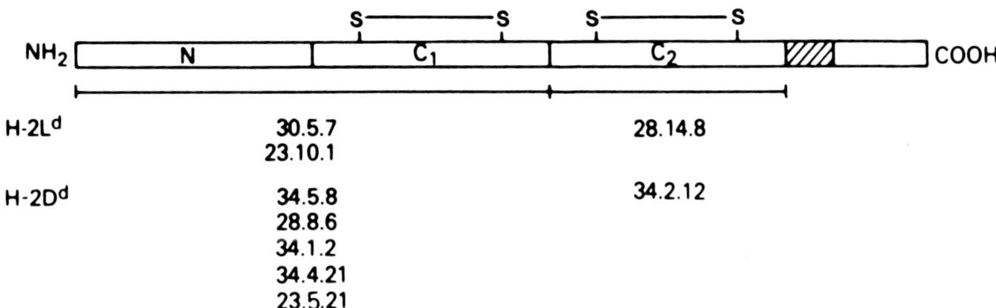

Fig. 10. Localization of polymorphic determinants on the external domains of H-2 proteins. This diagram summarizes the reactivity of $H\text{-}2L^d$ and $H\text{-}2D^d$-specific monoclonal antibodies with hybrid H-2 gene products. The protein domains of the H-2 molecule are summarized at the top of the diagram. The hatched region indicates the region of the hybrid gene resulting from exchange between the $H\text{-}2L^d$ and $H\text{-}2D^d$ genes. The reacting monoclonal antibody is listed in relation to the globular subunit of the protein to which it binds. The determinants recognized by monoclonal antibodies 30.5.77 and 28.14.8 have previously been shown to correlate with classical serological specificities [73].

clones with human inserts. The presence of the selected gene in these phage clones is assayed once again by transfection. Finally, single copy coding sequences in this genomic DNA are used to isolate a cDNA clone from a human cDNA library which codes the gene. Both the human TK and Ha-ras genes have been cloned in this manner [21,184,257].

Isolation of Surface Antigen Transfectants. A number of workers have used this strategy for the isolation of surface antigen transfectants, and ultimately gene cloning (Fig. 11). Kavathas and Herzenberg [132] first demonstrated the efficacy of this method in isolating genomic transfectants expressing human cell surface antigens (Fig. 11). Mouse Ltk- cells were cotransfected with human genomic DNA and HSV-tk. Transfectants containing both human DNA and HSV-tk were isolated with HAT selection. Using indirect immunofluorescence and flow sorting they were able to select those transfectants expressing human surface antigens HLA, β_2M, Leu 1 (T4) and Leu 2 (T8) [132]. These transfectants occurred in the population at frequencies of 10^{-3} to 10^{-4}, consistent with estimates of the fraction of the human genome transfected into each cell. As was the case in the isolation of immunoglobulin switch variants, the recovery of these rare cells was facilitated by gating for living cells. Shortly thereafter, other laboratories isolated HLA 4F2 and transferrin receptor transfectants using the same methods [150,151,207].

Characterization of Surface Antigen Transfectants. Primary and secondary surface antigen transfectants can be characterized by biochemical, immunological, functional, and genetic criteria. The original transfectants of Kavathas and Herzenberg [132] were analyzed biochemically by immunoprecipitation and gel electrophoresis. It was demonstrated that the molecular weights of HLA, Leu 1 (T4) and Leu 2 (T8) molecules on the transfectants were identical to those on cells that normally express the antigen. The transfected Leu 2 (T8) antigen was also identified by a second anti-Leu 2 monoclonal antibody which had not been used in the selection process. This implied that the antigen epitopes had not been altered by the transfection process.

Trowbridge and co-workers [207] characterized transferrin receptor transfectants by immunological and functional criteria. Their transfectants reacted with four unique antihuman transferrin receptor monoclonal antibodies, suggesting that all the receptor epitopes were intact. In addition, these workers provided evidence that human transferrin receptors on transfected cells were capable of normal function. Monoclonal antibodies were available that specifically blocked the uptake of transferrin-bound Fe mediated by both the human and mouse transferrin receptors. The data presented in Figure 12 establishes that the human transferrin receptors expressed on the transfected mouse cells (J4) can transport transferrin-bound Fe across the cell membrane. On transfected cells, the human-specific and mouse-specific antibodies each partially block Fe uptake. The two antibodies together however, are capable of reducing Fe uptake to background levels [207]. The isolation of these transfectants

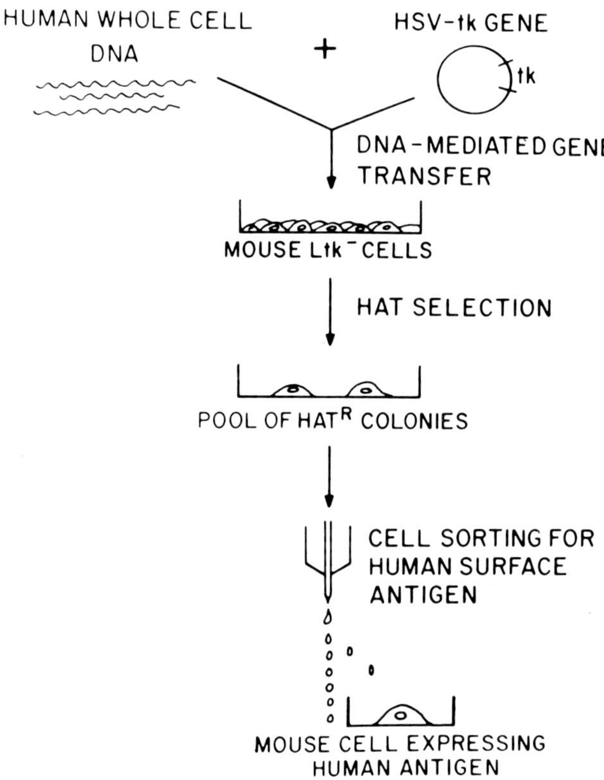

Fig. 11. Isolation of mouse transfectants expressing human cell surface antigens following DMGT with whole cell DNA. Following the selection of L-cell transfectants with HAT medium, these cells are surface labeled and isolated aseptically with the FACS.

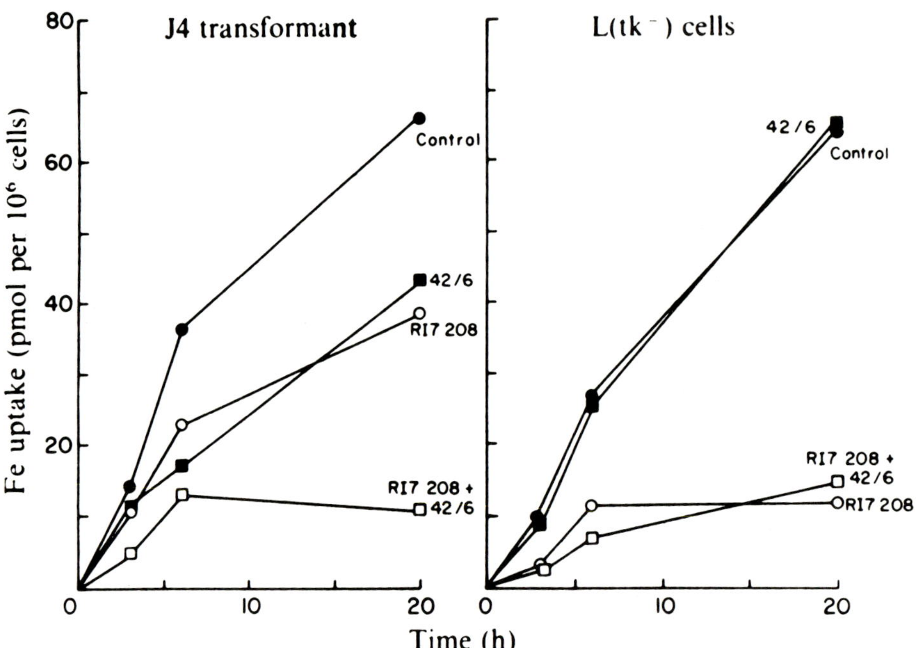

Fig. 12. Function of transferrin receptors on transfectants. Measurement of Fe uptake in the J4 transfectant in the presence of monoclonal antibodies against human and murine transferrin receptors. ^{59}Fe-labelled human transferrin (specific activity 2 × 10^4 cpm/μg) was prepared using established methods. Ltk- and J4 cells were plated at 6 × 10^5 cells per 35 mm tissue culture dish in DMEM supplemented with 10% horse serum; the following day the cell monolayers were washed extensively with serum-free DMEM at 37°C and incubated for 30 minutes in DMEM containing an excess of purified monoclonal anti-human transferrin receptor (42/6), monoclonal anti-murine transferrin receptor (R17 208), or both antibodies together. After preincubation with the antibodies, ^{59}Fe-human transferrin was added in 50 μl of DMEM to give a final concentration of 5 μg/ml. At 0 min and various times thereafter, duplicate dishes from each treatment group were washed three times with 2 ml of cold 0.15 M NaCl–0.015 M NaN^3 0.01 M Na phosphate buffer containing 0.1% BSA. The cells were then removed from the dish with versene and ^{59}Fe radioactivity was measured in a gamma counter [207].

thus provided a valuable system for other functional studies of the human transferrin receptor.

The genetic constitution of surface antigen transfected cells can be assessed by DNA hybridization. Kuhn et al. [150] demonstrated that genomic cell transfectants expressing HLA antigens contained the human genes coding the antigen. They also assessed the quantity of human DNA in both primary and secondary transfectants using Southern blot hybridization with human repetitive sequence probes. They showed that while primary transfectants contained large amounts of human DNA (many in excess of 3,000 kb), isolated secondary transfectants for 4F2 and the transferrin receptor contained less than 50 kilobase pairs of human genetic material [150]. More importantly, it was shown that independent secondary transfectants shared specific restriction bands which contained the gene coding sequences. This demonstrated the feasibility of cloning these surface antigen genes.

Expression of Transfected Surface Antigen Genes. Human surface antigens HLA β_2M, 4F2, and transferrin receptor are expressed on a wide variety of normal cells and also on human × mouse hybrid cell lines. These antigens are also readily expressed on transfectants following genomic gene transfer. These constitutively expressed antigens can be contrasted with another class of surface antigens whose expression is limited to cells of specific lineage, such as T and B lymphocytes. Lymphocyte-specific surface antigen expression on human × mouse hybrid cell lines has been reported, and these lines have been used to map differentiation antigen genes to chromosomes [222,250]. However, the hybrids used in these studies had mouse and human lymphoid cell parents and segregated human chromosomes extremely rapidly.

Herzenberg and coworkers have done extensive work on the expression of lymphoid differentiation antigens on transfected cells. Kavathas and Herzenberg [133] first demonstrated that T-lymphocyte markers Leu 1 (T4) and Leu 2 (T8) could be stably expressed on transfected cells. This suggested that gene transfection had altered the cell lineage specific restrictions on gene expression. These workers went on to show that a wide variety of differentiation antigens, both human and mouse, could be effectively transferred and expressed in mouse fibroblast L cells [112]. Table 8 lists the genes that have been successfully transferred and expressed in mouse cell hosts, and includes many differentiation antigens. Significantly, the source of genomic DNA used in transfer does not affect the ability to transfer these differentiation antigens. Fibroblast DNA is as effective in the transfection of T-cell differentiation antigen genes as is T-lymphocyte DNA [112]. This suggests that there are no permanent modifications made in the donor DNA to restrict its expression. It has not been possible, however, to transfer the HLA genes from trophoblast cell DNA. This is a tissue which must remain immunologically neutral to the maternal immune system during pregnancy. It may be inferred that, in this particular tissue, permanent alterations in the HLA sequence have rendered these genes refractory to gene transfer and expression.

Gene Amplification. It has been demonstrated that the expression of some genes may be increased as a result of spontaneous amplification of gene copy number [248,268]. Gene amplification is thought to be an abnormal process in mammalian cells, which is revealed by specific cell selection with growth inhibiting drugs. Among the mechanisms that mediate drug resistance is an elevation in the activity of the protein which is the direct target of the inhibitor. There are now about 20 different examples of cells resistant to drugs due to the overaccumulation of a protein. In more than one-half of these examples, gene amplification has been shown to be the basis of this resistance [268]. In general, these duplicated gene sequences may either occur as concatamerized additions to existing chromosomes or as small autonomously replicating "double minute" chromosomes. The selection of amplificants occurs in a stepwise fashion, where each succeeding selection involves increased levels of the inhibitor drug, and increased amplification of the gene.

The two best studied cases of drug resistance in cultured cells are the amplification of the CAD gene when selected by the drug PALA and the amplification of dihydrofolate reductase (DHFR) as a consequence of methotrexate selection. In the latter case, the fluorescence-activated cell sorter has contributed to studies of the dynamics of gene amplification [118,131]. Because of the stoichiometric binding of inhibiting drug methotrexate to DHFR and its active export from the cell, equilibrium measurements of methotrexate in the cell are proportional to the quantity of dihydrofolate reductase within the cell [131]. Schimke and coworkers have treated cells with a fluorescent derivative of methotrexate (MTX-FI) and used the FACS to quantitate cells based on cellular fluorescence. They were able to show a linear relationship between fluorescence ligand bound within a cell and the DHFR specific activity. These workers then used the FACS to examine the events involved in the selective multiplication and loss of dihydrofolate reductase genes during the emergence and loss of methotrexate resistance. Specifically, by sorting for cells with high MTX-FI binding in the absence of drug selection these workers were able to select a cell population with amplified DHFR genes [118]. This provided direct proof that gene amplification is a spontaneous event, and not a consequence of DNA damage or alterations caused by drug treatment. The FACS has also been used in this system to study the loss of amplified sequences, especially those carried in double minute chromosomes during cell division. Because the FACS permits analysis of single cells, quantitative data can be used to accurately calculate the rate of gene amplification or deletion.

Several investigators have observed amplification of surface antigen genes following gene transfection. Herzenberg and co-workers first observed the spontaneous amplification of the Leu 2 gene in transfectants following cell sorter selection for the surface antigen [133]. By continual selection for the highest expressing cells, it was possible to obtain amplified antigen expression up to 50 times the original level [134] (Fig. 13). This stepwise selection, involving higher and higher selection stringency, is similar to the process of increased gene amplification in the DHFR or CAD system revealed by increasing doses of inhibitor drug. Surface antigen amplification for a transfected myeloid antigen (gp150) [188], the human X-linked antigen S11 and the mouse Lyt-2 surface antigen [112] have also been reported. It is possible that gene transfer may permit amplification of transfected genes, depending on the location of integration into the host genome [268], or that each gene may carry flanking sequences that enable it to amplify spontaneously. It is clear that the FACS provides a direct method for examining the mechanism of surface antigen amplification.

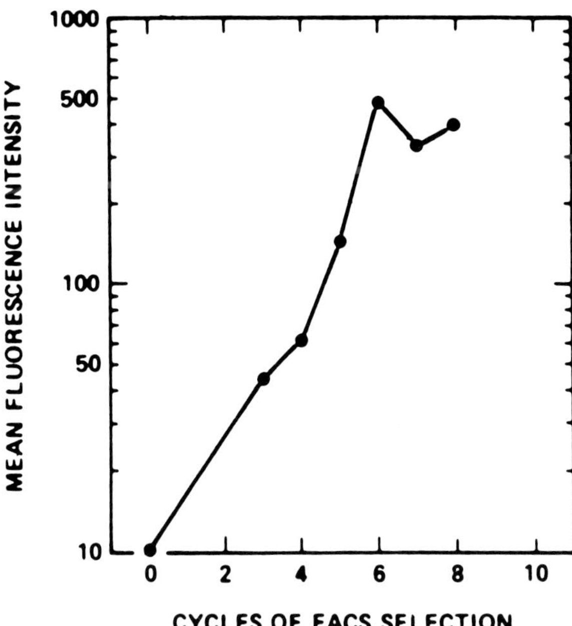

Fig. 13. Amplification of LEU 2 gene expression in transfectants. Mean fluorescence of Leu 2+ transferent, J10, after each round of sorting the brightest cells [133]. Starting with a cloned transferent J10, the cells were stained with fluorescein-conjugated anti-Leu 2a monoclonal antibody and the brightest 0.3% of the cells were sorted aseptically by the FACS. The cells were regrown, stained, analysed and resorted as described elsewhere except that propidium iodide was not used. The amount of anti-Leu 2a was sufficient for maximum staining. The mean fluorescence is calculated as the geometric mean of the distribution and expressed relative to 1.83 um green fluorescent beads (Polysciences Inc., Warrington, PA) which were used as a standard of 100 fluorescence units. As the fluorescence of the cells increased the voltage of the photomultiplier was reduced or a neutral density filter was inserted [133].

Cloning Surface Antigen Genes. Secondary transfectants contain 30 to 100 kb of human genomic donor DNA, a quantity of DNA that can be managed by prokaryotic cloning vectors. Genomic libraries of secondary transfectant DNA are generated by amplifying cosmid or lambda phage vectors carrying transfectant DNA inserts in host E. coli cells. Colonies are screened for human inserts using human repetitive DNA sequences. Positive phage are isolated, colony purified and mapped with respect to each other using restriction endonuclease analysis. Individual phage DNA is then transfected back into L cells and monitored for its ability to direct surface antigen expression. If the intact gene is not contained within a single phage, they can be transferred in pairwise combination to achieve functional transformation by recombination. This approach was successfully employed by Ruddle and co-workers in the cloning of both the thymidine kinase and transferrin receptor genes [151,184].

In some cases, human repetitive sequences may not be found sufficiently close to the gene of interest to be used in the screening of phage libraries. The transfectants may still be used in gene cloning by a method described as subtractive hybridization [134,186]. Basically, the method depends on the isolation of messenger RNA from both the transfectants and parental L cell. It is reasoned that these message populations will differ by only a few species, one of them being the messenger that codes for the surface antigen of interest. Complementary DNA copies of transfectant cell message are made, and hybridized with an excess of L cell mRNA. The cDNAs which are not hybridized can be separated, and are candidates for antigen gene coding sequences. The first application of this methodology in surface antigen gene cloning was for the amplified Leu 2 (T8) transfectants already described [134].

Other strategies for cloning surface antigen genes can be foreseen. New prokaryotic vectors should improve the efficiency with which new genes can be cloned by gene transfer methods. For example, cosmid vectors have been developed which can accommodate DNA inserts of up to 45 kb, and which contain selectable markers both for growth in bacteria and for the transfection of mammalian cells [166]. These "shuttle vectors" can be used directly to transfer individual cloned genes or entire gene libraries into a variety of different cell types, obviating the need for cotransfection. In addition, the specificity of the phage in vitro packaging system for rescuing cosmids provides an elegant way of recovering selected genes from the host cell genome. This will eliminate the need to construct and screen recombinant libraries. Another alternative to the cloning of amplified gene sequences is the direct isolation of the double minute chromosomes from transfected cells. As the technology and staining protocols improve for chromosome sorting [223,275], it may be possible to isolate double minute chromosomes directly by flow sorting rather than sucrose density gradients. The understanding of the structure of this autonomously replicating molecule may lead to its use directly as a cloning/expression vector.

The techniques of transfection and FACS selection can be extended to other genes which are not accessible to antibodies on the cell surface. As suggested by the work with methotrexate and DHFR, it may be possible to fluoresceinate a substrate or ligand and use it in a direct assay for its corresponding enzyme or receptor expressed by a cell. In theory, any ligand for a donor cell receptor that is absent on host cells could be fluoresceinated and used in the isolation of gene transfectants. For example, human interferon displays species-specificity in binding, and could be used to label mouse transfectants for the isolation of human interferon receptor transfectants. The binding of fluoresceinated lectins which are human specific, such as wheat germ agglutinin, could also be used for the isolation of mouse transfectants expressing specific carbohydrate moieties. In this way the FACS could be used in the cloning of glycosyl transferase enzymes.

A future approach may be to make the expression of an internal protein accessible to binding by fluorescence probes at the cell surface. Using recombinant DNA vectors which contain leader and transmembrane sequences, it may be possible to present the products of transferred genes at the surface of the host cell, and in conjunction with specific monoclonal antibodies and the FACS select specific transfectants [94].

GENE MAPPING

Introduction

The mapping of a gene or arbitrary DNA segment to a specific chromosomal region can be performed either by determining the extent of its genetic linkage to a marker* of known chromosomal location or by using a physical mapping technique.† DNA sequence mapping is crucial for determining the order of genes such as disease loci on chromosomes and inferring evolutionary relationships between genes or families of genes in different species.

The advent of recombinant DNA technology has greatly accelerated the construction of a genetic linkage map in man [20,288]. Such a map will be constructed using chromosome-specific DNA probes that detect DNA sequence polymorphisms and are spaced at roughly equal (40 centimorgan)‡ intervals throughout the human genome. A cloned single copy DNA fragment of unknown map position and which detects a restriction fragment length polymorphism can be assigned a chromosomal localization on the basis of its linkage to one or more of such DNA marker loci [20,288]. The construction of chromosome-specific recombinant DNA libraries using flow cytometric techniques provides a useful source of DNA probes for the massive undertaking which the construction of human genetic map represents. The chromosomal localization of many of the linkage map DNA markers will be accomplished using physical mapping techniques. The molecular hybridization of cloned, unmapped DNA probes to genomic DNA samples containing a single or limited number of human chromosomes permits the identification of the chromosome which contains the genomic DNA sequence, i.e., maps the probe. Such chromosome-enriched DNA samples can be obtained either by segregating a single or a few chromosomes of one species in cells of another species by somatic cell hybridization [233] or by chromosome flow sorting [88]. Hybridization of a cloned DNA probe to a metaphase chromosome spread in situ [84, 100,101,190] can be used for a high resolution mapping of single copy DNA sequences to a specific region of a chromosome.

Hybrid Cell Lines

Hybrid cell lines have been isolated by fusing human and mouse cell lines and propagating in tissue culture those hybrid cells which retain a single or a limited number of human chromosomes and a complete set of mouse chromosomes [241]. DNA is extracted from a panel of such hybrid cell lines; each cell line will yield a DNA sample that represents a different human chromosome set in a mouse chromosomal background. The DNA is then cleaved with the appropriate enzyme; the resulting fragments are separated on the basis of size by electrophoresis in agarose gels. The size fractionated DNA is then transferred to and immobilized on nitrocellulose filter paper (see Fig. 3). The DNA probe to be mapped is tagged with a tracer molecule such as a radioactive isotope

*The marker (i.e., a detectable chromosomal locus) can be a region on a metaphase chromosome of distinctive staining or morphology or a DNA sequence polymorphism.

†A gene or DNA fragment is said to be physically mapped if its location along the nucleotide sequence of a specific chromosome has been determined.

‡In classical genetics, the distance between two genes on the same chromosome (i.e., linked genes) is expressed in terms of the frequency of crossing-over during meiosis. If this frequency = 0.01, the distance is 1 centimorgan named after Thomas Hunt Morgan, who pioneered gene mapping in the fruit fly *Drosophila* during the 1930s. A centimorgan is roughly 10^6 bp. For the human genetic linkage map, roughly 100 DNA polymorphisms (for the haploid DNA content of 3.3×10^9 bp) are required, not an impossibly large number.

or modified with a hapten such as biotin [163,169] or 2-acetylaminofluorene [162], which can then be labeled with a fluorescent or other cytochemically detectable molecule. The filter paper is then immersed in a hybridization solution which contains the labeled probe. If sufficient homology exists between the DNA probe and, the genomic DNA immobilized to the filter paper, hybrids between the molecules are formed. The hybridization of the labeled probe is detected in single or multiple discrete size regions of the genomic DNA which are termed hybridization bands (Fig. 14). In the example shown in Fig. 14, Southern blot hybridization of a human chromosome #13-specific DNA probe will be detected only in the hybrid cell line DNA samples that bear human chromosome #13, i.e., that contain the DNA sequences homologous to the DNA probe. By examining the pattern of positive and negative hybridization in all hybrid cell line DNAs of the panel, the human chromosomal location of the DNA probe can be determined [38]. This method, which is not technically difficult if the hybrid cell lines are available, is widely used for gene and DNA sequence mapping. In particular, the characterization of the phage inserts from recombinant libraries constructed from several flow sorted chromosomes [45,129,155] has been greatly facilitated by the availability of human–rodent hybrid cell line DNA samples.

Chromosome Fractionation

Another technique available for gene mapping involves the extraction of DNA from chromosome populations that have been fractionated by physical separation procedures. Rate zonal centrifugation [113], velocity sedimentation at 30*g* [33] and flow sorting [172] have all been employed for the enrichment of specific chromosomes in gene mapping studies. The degree of enrichment that can be achieved using the physical fractionation technique will determine the precision of the chromosomal assignment. The high resolution of the flow cytometric technique [171] is advantageous with respect to the specificity of the chromosomal assignment as compared with the other two separation procedures. Human globin [172] and insulin [175] genes have been assigned to human chromosome 11 following chromosome flow sorting, chromosomal DNA isolation, restriction enzyme digestion, and Southern blotting of the fractionated chromosome populations. The amount of sorted material required for such analysis is relatively large; consequently, the flow sorting times are long. This problem has been circumvented, however, with the use of chromosomal DNA spot blot procedures which were first used for the identification of Y chromosome-specific repetitive sequences [14,15]. In this procedure, chromosomes are flow sorted directly onto nitrocellulose filter paper. Following denaturation of the chromosomal DNA that has accumulated as a spot on the filter paper, a cloned DNA probe is hybridized to the filter (Fig. 16). Using this technique, only 30,000 chromosomes [174] are required to obtain a positive hybridization signal of a single copy DNA probe in the appropriate chromosomal DNA fraction. DNA blot analysis has now been used in several laboratories for single-copy gene mapping [115, 123,125,134,174,176,212], and has been greatly enhanced by instrumentation capable of high resolution bivariate chromosome analysis in which roughly 20 distinct chromosome subpopulations are identified [160,171,174,197]. The high resolution of the dual laser flow-sorting techniques permits isolation of structurally abnormal chromosomes such as those bearing translocations and, consequently, gene mapping assignments can be refined to the level of chromosomal subregions [32,123,174].

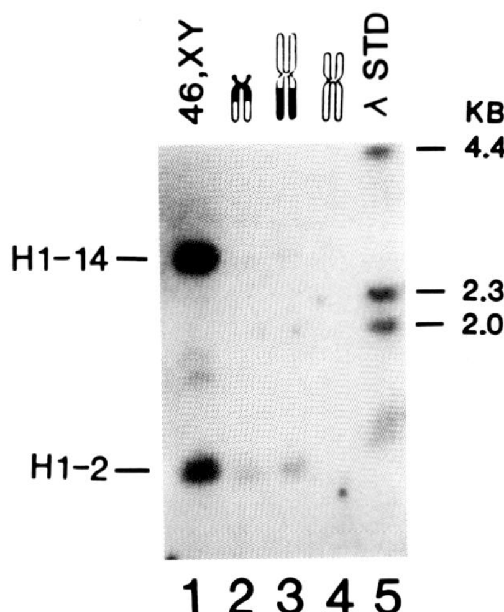

Fig. 14. Mapping of probes using human–rodent cell hybrid DNA samples. Hind III-digested genomic DNA samples from cell lines differing in human chromosome 13 content were separated by electrophoresis in agarose gels and transferred to nitrocellulose paper. The region of human chromosome 13 contained in each hybrid cell (lanes 2,3) corresponds to the darkened area of each chromosome idiogram whereas the blank areas (lanes 2,3,4) of the idiograms indicate the human X chromosome content of the hybrid cell lines. The DNA samples are derived from: (lane 1) 46,XY human lymphoblasts: (lane 2) a human–mouse hybrid cell line containing a 13/X translocated (13pter- 13q22::Xq22-Xqter) chromosome; (lane 3) a human–mouse hybrid cell line bearing an X/13 translocated (Xqter-Xp22::13q12-13qter) chromosome; (lane 4) a human–hamster hybrid cell line containing an intact human X chromosome; (lane 5) Hind III-restricted lambda DNA size standards. Autoradiographs of the hybridization of radioactively-labeled Hind III fragments isolated from a chromosome 13-enriched library. The 2.7 kb fragment (H1–14) does not appear to be chromosome 13-specific since it shows no hybridization in the hybrid DNA samples containing human chromosome 13 (lanes 2,3). The pattern observed for the 1.0 kb (H1–2) insert is consistent with localization to human chromosome 13 since hybridization bands are observed in human 13-containing DNA samples (lane 2,3). The H1–2 insert can also be tentatively sublocalized to the 13q12 to 13q22 region since this is the only part of 13 shared by the two chromosome 13-containing hybrid cell DNA samples (lanes 2,3). Based on the data of Lalande et al. [159].

Mapping the Homeo Box

The importance of gene mapping in studies of the evolutionary conservation of DNA sequences as well as a comparison of the different physical mapping techniques can be demonstrated by the localization of the homeo box locus in humans and mice [123,230]. The highly conserved homeo box is a 180 base pair (bp) DNA sequence that has been identified in homeotic genes in Drosophila. The homeotic genes play a crucial role in regulating *Drosophila* embryonic development. The identification of homeo box sequences in mammals confirms that this genetic locus is strongly conserved in evolution. The role, if any, of such sequences in mammalian development remains to be elucidated. A cloned human DNA fragment, homologous to the Drosophila homeo box, [181] was assigned to the long arm of human chromosome 17 (17q) [123] using dual laser chromosome flow sorting and DNA spot blot analysis [174] of chromosomes isolated from cell lines bearing normal and translocated chromosomes 17 [123]. The homologous mouse sequence was localized to mouse chromosome 11 by using the human DNA fragment to probe the DNA extracted from a panel of rat–mouse or hamster–mouse somatic hybrid cell lines. The results confirm not only that the human and mouse homeo box loci show extensive homology to each other as well as the *Drosophila* locus but also demonstrate that the human and mouse loci map to chromosomal regions which are tested evolutionarily, (i.e., human chromosome 17 and mouse chromosome 11). In a separate study confirming this result [230], a recombinant phage insert containing a mouse DNA sequence homologous to the *Drosophila* homeo box was mapped to mouse chromosome 11 using a panel of DNA samples extracted from different mouse–hamster hybrids. In addition, a regional mapping assignment for the human homeo box DNA sequence [181] was obtained by in situ hybridization [223]. Using this technique, in which a single copy DNA probe is hybridized in situ to banded metaphase chromosomes [84,100,101,190], the human homeo box locus was mapped to the 17q11–17q22 subregion [230]. The results of the two studies [123,230] on the gene mapping of the mammalian homeo box loci exemplifies the usefulness and complementarity of the different physical techniques of gene mapping. In both studies, somatic cell hybrid mapping panels were used in the preliminary mapping assignments indicating the widespread application and acceptance of this procedure. In one study [123], the human homeo box DNA probe was rapidly assigned to 17q using the cell sorter and DNA spot blot analysis whereas the refined chromosomal localization to the 17q11 to q22 subregion was obtained by in situ hybridization in the other investigation [230]. Thus mapping by the use of somatic cell hybrids can be confirmed and refined using either the rapid flow cytometric approach or the more precise in situ procedure. Preliminary results [275] indicate that it may be possible to perform in situ hybridization using fluorescence flow cytometry and, therefore, develop a method for gene mapping that combines the advantage of these two techniques.

ONCOGENES, CHROMOSOME REARRANGEMENTS, AND HUMAN NEOPLASIA

Introduction

The cytologic observations that certain defined chromosomal aberrations are consistently associated with specific human neoplasia [198,296] supports a widely held view that such abnormalities play a role in the initiation and/or maintenance of tumor growth. The DNA rearrangements that were grossly apparent in certain cases by chromosome banding studies are now the subject of investigations at the molecular level using recombinant DNA technology. Such techniques have permitted a much refined analysis of the structure of chromosome abnormalities such as translocations and the involvement of oncogenes in B-cell neoplasia; amplification of chromosomal DNA in neuroblastoma, and DNA deletions in Wilms' tumor and retinoblastoma.

Chromosome Translocations Involving Proto-oncogenes

Burkitt's Lymphoma. The tumorigenic retroviruses bear transforming oncogenes which share striking homology with DNA sequences in normal mammalian cells [16,17,62,285]. Retroviral oncogenes probably originated from the integration and mutation of the cellular sequences (proto-oncogenes) into the retroviral genome. The development of certain neoplasias might, therefore, involve the activation of cellular proto-oncogenes (c-onc), although the occurrence and mechanism of such a process remains controversial [57,62]. That DNA-mediated gene transfer of c-onc genes isolated from certain human tumors can transform certain mammalian cells in vitro [34,161] is suggestive of a role for c-onc genes in the etiology of human cancer.

It was proposed [139] that the chromosomal rearrangements which are associated with certain B-cell derived tumors, human Burkitt's lymphoma (BL) and murine plasmocytomas, resulted in the activation of c-onc genes. In the case of human Burkitt's lymphoma, the translocations involve chromosome 8 and the immunoglobulin (lg) gene-bearing chromosomes 14 (lg heavy chain), 2 (lg kappa light chain) or 22 (lg lambda light chain). The observation [270] that the cellular oncogene, c-myc [39], is located on chromosome 8 and is translocated to the lg heavy chain locus in BL cells bearing a t(8;14) chromosome was further suggestive of a role for the juxtaposition of the lg locus and a c-onc gene in Burkitt's lymphoma (BL). The molecular genetics of the BL translocations has been extensively studied and has been the subject of excellent reviews [36,61,138,177,216,229]. The nucleotide sequence analysis of human c-myc genomic and complementary DNA clones [9,13,282,284] indicate that the c-myc gene is composed of three exons and two introns. The c-myc gene is transcribed from two active promoters situated just 5′ of or just within the first exon [243,271]. The c-myc protein coding sequences are contained within the second and third exons while no functional role has clearly been assigned to the first exon which is not translated into protein [243]. In the typical, t(8;14), Burkitt's lymphoma translocation, a breakpoint occurs on chromosome 8 band q24 either within the non-coding first exon or in regions upstream (5′) of the c-myc gene. The c-myc coding (second and third) exons and any flanking upstream DNA are then transposed to chromosome 14 band q32 which contains the immunoglobulin heavy (IgH) chain locus [1,40,68,96,271]. The c-myc and IgH genes are arranged in a head to head (5′ to 5′) configuration on the translocated (derivative 14) chromosome. Chromosomal translocations involving the immunoglobulin light (IgL) chain loci are observed less frequently in Burkitt's lymphoma. In these variant translocations, the breakpoint on chromosome 8 occurs 3′ (downstream) of the c-myc gene. The immunoglobulin light chain loci, kappa on chromosome 2 band p12 and lambda on chromosome 22 band q11, are translocated to the region 3′ of the c-myc gene. The IgL locus and c-myc gene are arranged, in the variant case, head to tail (5′ to 3′) on the translocated (derivative 8) chromosome [37,46,67,110].

Flow karyotypic analysis has been used to study the typical and variant Burkitt lymphoma translocations [269,291]. Taub et al. [269] were able to identify and flow sort the derivative 2 and derivative 8 chromosomes in a BL cell line displaying the variant t(8;2) translocation. By Southern blot analysis of the flow sorted DNA from the two fractions, it was possible to assign the rearranged (putatively activated) c-myc gene to the derivative 8 chromosome, thereby demonstrating that the rearranged c-myc DNA sequence remains on chromosome 8, the translocation breakpoint occurring, in this cell line, some 20 kb downstream of the c-myc gene.

Role of c-myc. In spite of these detailed molecular genetic analyses of the Burkitt lymphoma translocations, the role of c-myc in neoplastic transformation has been difficult to elucidate and has led to what is termed the paradox of c-myc. The great variability and location of BL translocation breakpoints, which are upstream of c-myc in the typical case and downstream of c-myc in the variant translocation, are observations which are difficult to reconcile with a unifying mechanism of c-myc activation. The opposite transcriptional orientation (head to head) of the IgH and c-myg genes in the typical BL translocations is different from the head to tail orientation of the two genes in the variant cases of BL. Mutations within the c-myc gene coding or noncoding sequences [229], deregulation of the translocated c-myc gene allowing its increased expression [68,208], and the apparent involvement of IgH gene enhancer sequences in c-myc transcriptional activation [104] are not common to all Burkitt's lymphomas and, hence, cannot fully explain the mechanism of c-myc activation. Differential usage of the c-myc gene dual promoters in malignant relative to normal B cells has also been observed and assumed to be associated with the activation of c-myc [271].

Investigations of how c-myc gene expression varies with the growth and differentiation of normal cells are providing clues to the possible role of the c-myc gene in neoplastic transformation. Levels of c-myc mRNA were observed to increase by 16- to 40-fold within hours of activation of normal lymphocytes by mitogen or normal fibroblasts by platelet derived growth factor, and then to decrease at later times after the application of the growth stimulus [135]. Treatment of normal resting lymphocytes with the protein synthesis inhibitor, cyclohexamide, in the presence or absence of a growth stimulus also induced high levels of c-myc mRNA [135,136]. These data were taken as an indication that the c-myc gene might normally be repressed by a transacting protein and that upon induction by a growth stimulus, c-myc gene expression becomes deregulated and, hence, elevated to the lability of the repressor protein [135,177].

c-myc mRNA Lifetime. Two important observations have subsequently demonstrated that the control of c-myc mRNA degradation might also be of importance in regulating the level and the timing of c-myc gene expression. Firstly, the c-myc mRNA is extremely unstable with a half-life of less than 30 min in normal cells [44,196]. Secondly, the dramatic increase in the levels of c-myc mRNA observed at short times after growth stimulation are not due to changes in the rate of c-myc mRNA transcription [18]. Although the absolute levels of c-myc mRNA are barely detectable in resting cells, the c-myc mRNA is transcribed at a high rate [18]. Post-transcriptional processes such as the degradation of c-myc messenger RNA are, therefore, contributing in large part to the observed differences in c-myc mRNA levels between stimulated and resting cells. In terms of the role of the c-myc gene in malignant transformation, the aberrant (i.e., translocated or truncated) c-myc genes from neoplastic cells have been shown to produce mRNA species with half lives of one to several hours [65,227] as compared to the 15 to 30 min half-life of the mRNA species produced from the c-myc gene associated with the normal chromosome of the same cell. Changes in the stability of c-myc mRNA resulting from a chromosomal rearrangement of the c-myc gene may therefore be directly involved in the development of tumors such as Burkitt's lymphoma.

Cell Cycle Studies. The demonstration of elevated levels of c-myc mRNA following the stimulation of resting lymphocytes and quiescent fibroblasts [135] does indicate that the c-myc gene, at least in terms of changes in the stability of the c-myc mRNA, may be important in the control of the G_0 to G_1 transition in normal cells.

The stimulation of normal B lymphocytes from the resting to the proliferative state can be divided into at least two steps [263]. Crosslinking of the B cell surface Ig receptors renders the cell competent to respond to B-cell growth factor (step 1). The progression of the competent B cell to DNA synthesis is dependent on the presence of the growth factor (step 2). Normal human B cells that were blocked in step one did not progress into division cycle although they did demonstrate high levels of c-myc mRNA [263]. Elevated levels of c-myc mRNA were also detected in B cells, which were allowed to proceed through step 1 and enter S phase in response to B-cell growth factor [263]. These data not only indicate that the c-myc gene product may play a functional role in the early G_1 phase of the B cell stimulation process but also imply that elevated levels of c-myc mRNA are not sufficient for progression of B cells into cell cycle. Similar results have been obtained in the case of fibroblasts in which microinjected c-myc protein stimulated DNA synthesis in quiescent cells but only in the presence of a growth factor, and the cells did not enter S phase when the growth factor alone was added to the cell culture [124]. It is also of interest to determine how the levels of c-myc mRNA and protein vary with phases of the division cycle other than early G_1. The expression of the c-myc gene as a function of cell cycle phase has been greatly elucidated by flow cytometric [24,66,98, 217,228,273] analysis of DNA content which permits rapid determination of the cells in each phase of the cell cycle under various growth conditions. By combining flow cytometric techniques with the molecular methods for measuring levels of expression of c-myc mRNA or protein, much can be learned of the regulatory role of c-myc in cell growth control. Centrifugal elutriation and flow cytometric analysis of DNA content were used to separate and analyse cells in the different phases of the cell cycle for avian cells [174]. The levels of c-myc mRNA were observed to increase transiently after serum stimulation of serum-deprived chicken embryo fibroblasts but were constant as a function of cell cycle in cells grown in the presence of serum [174]. Similar results have been obtained for other cell types, using antibodies to the c-myc protein in order to study changes in levels of c-myc protein throughout the division cycle [98,217,228].

The production of antibodies specific for the human c-myc proteins [70,97,218,232] has permitted a partial characterization of the major c-myc gene products. Two phosphoproteins with molecular weight ranges of 62,000 to 65,000 and 64,000 to 68,000, respectively, have been identified by precipitation with anti-myc antibodies [70,97,218,232]. Cloning of the c-myc gene in an expression vector system has been used to produce a 64,000 molecular weight protein [283]. The human c-myc gene product is associated with the nucleus [69,97,219] and binds to single- and double-stranded DNA [218,283]. The c-myc gene is strongly conserved in evolution [16] and antibodies to the human c-myc protein can be used to identify the c-myc encoded proteins of other species such as mouse [228].

An elegant use of an anti-c-myc antibody has involved dual fluorescence flow cytometry to study the levels of c-myc encoded protein as a function of cell cycle in mouse fibroblasts [228]. The simultaneous analysis of DNA content and levels of c-myc protein was accomplished by measuring the fluorescence intensity of a fluorescein-conjugated antibody to c-myc versus the intensity of the DNA-specific propidium iodide fluorescence [228]. The results of this experiment clearly demonstrated that the amount of c-myc protein does not vary with cell cycle phase. Other proto-oncogenes that display rapid but transient increases in mRNA expression following cellular growth stimulation include c-fos [30,89, 148,202], which is the normal cellular homolog of the FBJ-osteosarcoma virus transforming gene, and c-myb [272], which shares homology with the avian myeloblastosis virus (v-myb) transforming gene. The c-fos and c-myb proto-oncogenes, similarly to c-myc, code for nuclear proteins. It is within the nucleus, therefore, that these three proto-oncogenes act as cell growth regulatory molecules. Another class of proto-oncogenes code for cytoplasmic proteins and, therefore, affect cell growth by a different mechanism. An excellent discussion of the possible mechanisms of action of these two classes of proto-oncogenes has been presented elsewhere [285].

C-oncs and Growth Factors. Among the cytoplasmic proto-oncogene protein products, several are cell surface antigens which are encoded by proto-oncogenes that show striking homology to cell growth factors and growth factor receptors [107]. The best examples of these are the similarities between the $p28^{sis}$ and human platelet-derived growth factor [52,236,281] and between the oncogene product of c-erb B and the receptor for epidermal growth factor (EGF) [54]. Homology between the c-fms proto-oncogene cell surface product and the receptor for mononuclear phagocyte growth factor, CSF-1, has also been demonstrated [256]. In the latter case, it has been shown that the expression of the fms oncogene cell surface product is required for transformation [238]. The relationship between the initiation and maintenance of the transformed phenotype and an oncogene-coded cell surface product has also been demonstrated for the rat neuroblastoma proto-oncogene, neu [56]. The neu onc gene transforms NIH3T3 cells, which, as a result of the in vitro transformation, express a cell surface antigen, p185 [56]. Using a fluorescence activated cell sorter, hybridoma cell lines secreting anti-p185 monoclonal antibodies have been isolated [56]. The anti-p185 monoclonal antibodies have been shown to down modulate the p185 cell surface antigen, in that treatment of neu-transformed NIH3T3 cells with anti-p185 results in a reversible loss of p185 [55]. Neu-transformed 3T3 fibroblasts which are incubated with anti-p185 revert to a non-transformed phenotype [55]. These results imply that the expression of the p185 cell surface antigen which is induced by transformation with the neu proto-oncogene is required for the maintenance of the transformed cell phenotype. The neu onc gene is highly homologous to a subregion of the erb B gene that encodes the EGF receptor [7,247] suggesting that the p185 cell surface protein is a receptor for an as yet undefined cellular growth factor.

c-abl and Chronic Myelogenous Leukemia. Several other human neoplasia have been associated with specific chromosomal translocations [286], and the study of these at the molecular genetic level is well under way. In chronic myelogenous leukemia (CML), the presence of the Philadelphia chromosome can be detected in the vast majority of cases. The Philadelphia chromosome results from a translocation between chromosomes 9 and 22 [239]. The breakpoint on chromosome 9 includes the c-abl cellular oncogene and a variable amount of flanking DNA from the 5′ adjacent region [48,106]. The c-abl gene and its flanking DNA sequences are translocated to chromosome 22 [48,106], in which the breakpoint is restricted to a region of approximately 5.8 kb termed the breakpoint cluster region (bcr) [91]. The c-abl mRNA transcript of 8 kb observed in the leukemic cells of CML patients is larger than the normal 6- and 7-kb transcripts [82]. These data, which correlate the occurrence of an abnormal c-abl mRNA with the presence of the t(9;22) and which indicate that the chromosome 22 breakpoint of the t(9;22) occurs in the limited bcr region, strongly suggest a role for c-abl in the etiology of CML. In support of this notion, it has recently been shown [259] that the 8kb c-abl transcript in the t(9;22) CML cells results from a fusion of the c-abl and bcr genes.

Chromosomal Rearrangements and DNA Sequence Amplification in Human Neuroblastoma

Chromosomal abnormalities such as homogeneously staining regions (HSRs) and double minutes (DMs) are cytogenetic manifestations of the process of DNA amplification which is observed in certain drug resistant cell lines [248]. Several human tumors, particularly neuroblastomas (Nb) [5], also display cytogenetic evidence of DNA amplification. Molecular studies of DNA-sequence amplification have been initiated for a number of different neoplasias [3,83]. Eleven different cloned DNA fragments isolated from a DNA library constructed from HSR-bearing #1 chromosomes flow sorted from a Nb cell line demonstrated a considerable increase in hybridization intensity in Southern blots of neuroblastoma cell line DNA [129]. These cloned DNA fragments were shown to be specific for the HSR region of chromosome 1 by in situ hybridization and could be mapped to normal chromosome 2 in neuroblastoma cell lines [129]. The amplified chromosomal DNA regions appear, therefore, to result from a transposition of DNA from chromosome 2 to the HSR-bearing chromosome. In a separate series of experiments [251], it was shown that a DNA sequence showing partial homology to the myc proto-oncogene was amplified in the DNA of eight of nine neuroblastoma tumor cell lines and one primary neuroblastoma tumor sample. These eight neuroblastoma tumor cell lines showing amplification of the c-myc homologous DNA sequence [251] all displayed cytogenetic evidence of DNA amplification, that is either HSRs involving different chromosomes or DMs. The one neuroblastoma cell line which did not show DNA amplification [251] possessed neither an HSR nor DMs. Similarly, a c-myc homologous DNA sequence, termed N-myc [145], was shown to be amplified in seven Nb tumor cell lines bearing either HSRs or DMs [145]. These results imply that the proto-oncogene, N-myc, is amplified in most Nb tumor cell lines exhibiting cytologic evidence of DNA amplification and that, in the case of HSR-bearing chromosomes of Nb cell lines, the chromosome 2 specific N-myc sequence is transposed to the site of DNA amplification which can be on a variety of different chromosomes [145,254]. The amplified unit of DNA appears, furthermore, to be composed of noncontiguous chromosome 2 DNA segments, which are spliced together by an unknown mechanism [258] and then relocated to a different region of the genome. The N-myc DNA amplification regions of neuroblastoma tumors contain some 300 to 400 kb of DNA and are highly conserved among the different neuroblastoma tumor DNA samples [137]. In neuroblastoma cell lines, in which no N-myc gene amplification is observed, there appears nonetheless to be an increase in N-myc gene expression as measured by the elevation of N-myc mRNA per N-myc gene copy [144]. The involvement of N myc in the etiology of neuroblastoma is further supported by the observation that the N-myc gene is frequently amplified in primary neuroblastoma tumor DNA samples [22,251,260] and that N-myc DNA sequence amplification is associated with advanced clinical stages of the disease. Amplification of N-myc and other DNA sequences specific for amplified chromosomal segments of Nb has also been observed in retinoblastoma tumor cell lines [244,252], and in primary retinoblastoma tumor samples [178]. The role and relationship of N-myc gene amplification in retinoblastoma and neuroblastoma, which are both neuroectodermal tumors, remains to be elucidated. Amplification of the c-myc proto-oncogene has been observed in two neuroblastoma tumor cell lines [145] as well as in the amplified chromosomal regions of two neuroendocrine tumor cell lines derived from a colon carcinoma [2] and in an acute promyelocytic leukemia cell line [41]. The N-myc oncogene has recently been shown to transform rat embryo cells in tissue culture when co-transfected with the c-HA-ras-1 proto-oncogene [253, 294]. The cells transformed by the N-myc/H-ras co-transfection display high levels of N-myc mRNA transcription [253,294] indicating that increased expression of the N-myc gene may be important not only in the maintenance but in the initiation of the neuroblastoma neoplastic transformation.

Chromosomal DNA Deletions in Retinoblastoma and Wilms' Tumor

Retinoblastoma. Retinoblastoma (Rb), an eye tumor occurring in both hereditary and nonhereditary forms, affects children usually before the age of 4 years. Using statistical techniques, it was postulated that two mutational events were involved in the development of retinoblastoma [141]. In hereditary retinoblastoma, one mutation is inherited via the germinal cells and a second mutation occurs in somatic cells whereas, in the non-hereditary form of the disease, both mutational events occur in the somatic cells. One mutational event which is evident in a small fraction of retinoblastoma patients is a deletion of band q14 on chromosome 13 [142,297]. The involvement of a genetic locus associated with the 13q14 subregion in retinoblastoma is further suggested by the gene mapping of the hereditary retinoblastoma disease locus, which has been assigned to band 13q14 [266] due to its demonstrated linkage to the gene coding for the enzyme esterase D which had previously been localized to 13q14 [267]. To study this disorder at the molecular level, DNA probes have been obtained from recombinant phage libraries constructed either from the DNA of hybrid cell lines containing human chromosome 13 on a rodent chromosome background [27,59] or from the DNA of flow-sorted chromosome 13 [158]. Chromosome 13 specific probes from chromosomal subregions adjacent to 13q14 which detect restriction fragment length polymorphisms were used in the Southern blot analysis of the DNA from both the peripheral blood (constitutional tissue) and the primary tumor tissue of retinoblastoma patients [25]. The loss of one or more hybridization bands in the Southern blots of tumor DNA relative to those of the constitutional DNA implies that homozygosity of alleles surrounding band 13q14 has developed during the retinoblastoma tumorigenesis [25,57]. That the tumor has become homozygous for the chromosome 13 bearing the mutant allele and not the homolog containing the normal (wild-type) allele has been demonstrated by the molecular genetic analysis of tumors from patients who suffer from hereditary retinoblastoma in which the chromosome 13 homolog present in the tumors is derived from the affected parent [26]. Using similar techniques and the same DNA probes as in the retinoblastoma investigation, homozygosity of alleles surrounding the retinoblastoma locus has been demonstrated in osteosarcoma DNA samples [58,99]. The loss or mutation of a gene or several genes located within band 13q14 evidently predisposes individuals to other forms of cancer. In this regard, it has been reported that deletions of band 13q14 have been observed in the peripheral blood or bone marrow of patients suffering from malignant hematologic disorders [117,200]. The chromosomal mechanism by which the normal allele at the retinoblastoma genetic locus is eliminated in the tumor can result from either mitotic nondisjunction or recombination [25]. Two very important conclusions can be drawn from the molecular studies of retinoblastoma. Firstly, hereditary retinoblastoma, which behaves as an autosomal dominant disorder by classical genetics is, in fact, recessive when analyzed at the cellular level [10]. Secondly, the two mutation theory of retinoblastoma [141] has been validated.

Wilms' Tumor. Tumorigenesis resulting from homozygosity at an 11p chromosomal locus has been demonstrated for other childhood cancers, including Wilms' tumor (a tumor of the kidney) [75,147,210,234] and two other rare embryonal tumors, rhabdomyosarcoma and hepatoblastoma [146]. The availability of several cloned genes which have been mapped to 11p and which detect restriction fragment length polymorphisms greatly facilitated molecular studies of these embryonal tumors. Homozygosity for the insulin, beta-globin and c-Harvey-ras 1 genes as well as a random genomic DNA fragment, all of which have been localized to 11p, was demonstrated for the rhabdomyosarcoma and hepatoblastoma tumors [146]. The loss of alleles at these four loci, as well as for the gamma-globin and parathyroid hormone genes, had previously been demonstrated in Wilms' tumor samples [75,147,210,234]. A locus at 11p is also involved in other human cancers. Five of 12 patients suffering from bladder cancer displayed loss of one allele of the insulin and/or Ha-ras genes in the DNA of the tumor relative to that of the normal tissue [74]. A translocation involving the T-cell alpha chain gene and a locus in 11p13 has also been observed in several cases of human T-cell leukemia [182].

A possible clue to the cellular mechanisms which might result in the development of Wilms' tumor has recently been demonstrated [233,255]. Transcription of the gene for insulin-like growth factor-II (IGF-II) is elevated 10 to 100 fold in tumor DNA samples relative to the levels of expression of this gene in normal tissue, including normal kidney tissue adjacent to the tumor [233,255]. Interestingly, the levels of IGF-11 gene expression are comparable to those of several fetal tissues including kidney. The elevated expression of IGF-II in Wilms' tumor samples could either reflect the stage of tumor differentiation or, given that IGF-II is an embryonal mitogen, imply its involvement in the etiology of Wilms' tumor. The gene for IGF-II maps to band 11p14 which is adjacent to the band (11p13) that is deleted in the germline of some Wilms' tumor patients and is presumably the site of the chromosomal locus for the Wilms' tumor susceptibility gene(s). Whether the deletion of one copy of the tumor susceptibility locus and subsequent loss of the second allele during Wilms' tumor development results in the deregulation of genes such as IGF-II is an intensive area of current research.

A detailed molecular analysis of several deletions at 11p13 associated with Wilms' tumor has been carried out by Southern blot dosage analysis [279]. The constitutional DNA from five patients displaying cytologically detectable deletion of 11p and suffering from Wilms' tumor was analyzed using the

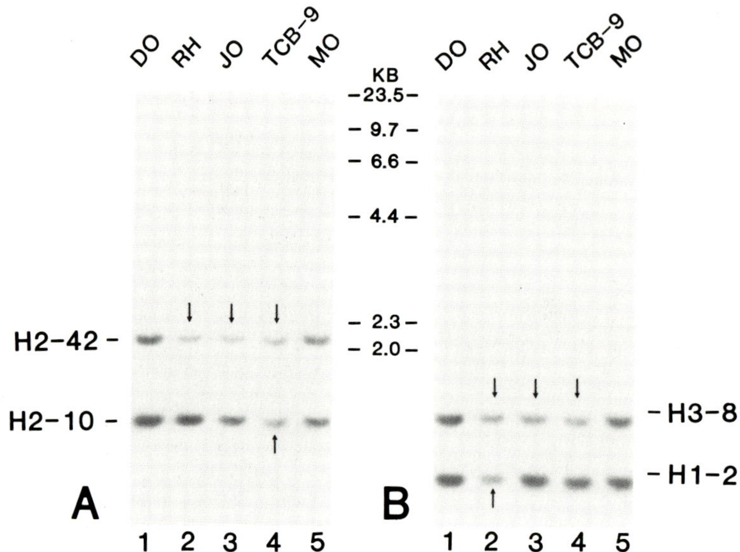

Fig. 15. Southern blot dosage analysis of overlapping retinoblastoma deletions. Pattern of hybridization of chromosome 13-specific DNA probes H2-42 and H2-10 (A) and H3-8 and H1-2 (B) in equal amounts. Hind III-digested DNA isolated from the following individuals: (lane 1, DO) karyotypically and phenotypically normal male (father of JO); (lane 2, RH) male suffering from unilateral retinoblastoma and displaying a deletion encompassing bands 13q13 and 13q14; (lane 3, JO) male suffering from bilateral retinoblastoma and displaying a 13q14.1 .13q21.1 deletion; (lane 4, TCB-9) male suffering from unilateral retinoblastoma and demonstrating 13q deletion mosaicism and from which a T lymphocyte cell line was isolated where all metaphases examined displayed a 13q14.1-21.2 deletion; (lane 5, MO) karyotypically and phenotypically normal female (mother of JO). A two-fold reduction inhybridization intensity, as indicated by the arrows, is molecular evidence that a deletion of the DNA probe has occurred in one of the chromosome 13 homologs. Based on this analysis, one copy per genome of the 2.2 kb H2-42 (A) and the 1.5 kb H3-8 (B) DNA probes are deleted in the retinoblastoma patient DNA samples (lanes 2,3,4). The H2-42 and H3-8 DNA probes can, therefore, be assigned to the common region of deletion overlap shared by the three retinoblastoma patients i.e., 13q14.1 . 13q14.3. One copy per genome of the 1.4 kb H2-10 (A) DNA probe is deleted in the TCB-9 DNA sample (lane 4) indicating that H2-10 is localized to the only subregion (13q21.2) which is deleted in TCB-9 and not in the other individuals of the panel. Similarly, the 1.0 kb H1-2 (B) DNA probe can be assigned to the 13q13-13q14.1 subregion. Based on the data of Lalande et al. [157].

11p-specific probes for the beta-globin, calcitonin, parathyroid hormone and catalase genes [279]. The results of this analysis demonstrated that one copy per genome of the catalase gene probe was deleted in the DNA of four of the five patients, whereas other DNA probes were present in two copies per genome in all patients [279]. These data not only indicate which DNA probe (catalase) maps closest to the Wilms' tumor susceptibility locus but also show which patient displays the smallest deletion, in terms of DNA, associated with the disease. The smallest region of deletion overlap common to affected individuals can thereby be identified and serve as a DNA source for mapping probes which are very tightly linked to the disease locus.

Molecular Analysis of Retinoblastoma Deletions. A similar study has been carried out on DNA samples from retinoblastoma patients displaying overlapping deletions of band 13q14 [157] (Figure 15). In this case, the DNA fragments used as probes were obtained from a chromosome 13 enriched recombinant phage library constructed from flow sorted material [158]. Two of the probes (H3-8 and H2-42) were deleted in three patients with cytogenetically evident deletions involving band 13q14. The common region of deletion overlap in this case encompasses the distal part of 13q14.1 and all of bands q14.2 and q14.3 (Fig. 15). Probes H3-8 and H2-42 are probably within 10^6bp (1% recombination) of the retinoblastoma susceptibility locus. The use of DNA probes such as H3-8 and H2-42 greatly extend chromosome banding studies of deletions such as those involving 13q14 in that a more precise estimate of breakpoints as well as an ordering of the probes along the chromosome can be achieved [157] (Fig. 15).

Molecular analyses of chromosomal probes may also provide a more precise estimate of the frequency of occurrence of deletions in retinoblastoma which, by cytogenetic techniques, has been observed to be 5 to 10%. A better estimate of the proportion of patients displaying a 13q14 DNA deletion is important in studies of the etiology of the disease and will be obtained as more 13q14-specific probes become available.

Enriching Deleted Sequences. An elegant molecular technique has recently been described [153] which permits the construction of recombinant DNA libraries enriched for sequences that are deleted in diseased individuals relative to normal controls. In this procedure, a large excess of DNA isolated from a male patient with a small chromosomal deletion is combined with DNA from an individual bearing no such deletion. The mixture is then subjected to a phenol-enhanced reassociation technique and only those molecules which are appropriately reassociated are then molecularly cloned [153]. In this way, only those DNA sequences which are present in the normal individual's genome and absent from the patient's genome where a deletion has occurred, are isolated and, hence, subjected to molecular cloning. Using DNA from a patient displaying a small deletion involving

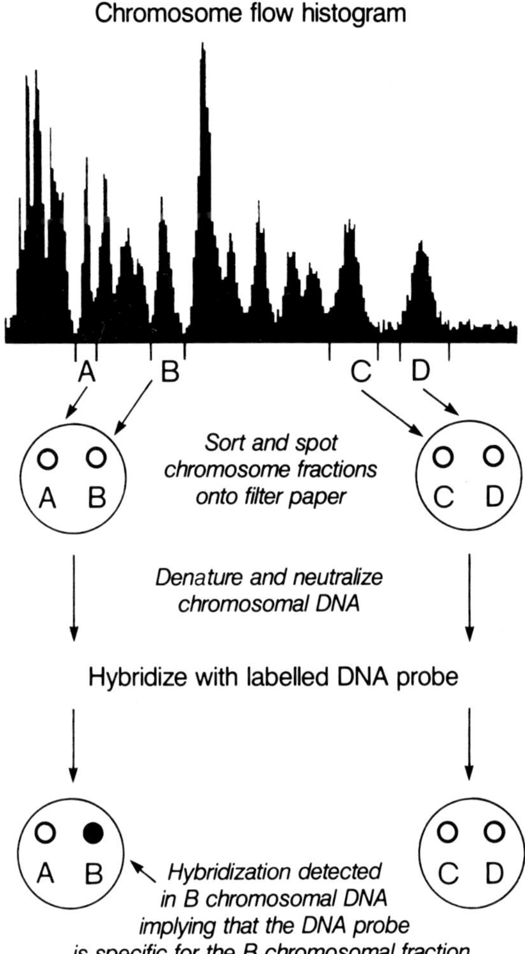

Fig. 16. Principle of DNA spot blot analysis technique. The chromosome flow histogram was obtained by analysing 33258 Hoechst-stained metaphase chromosomes isolated from a 46,XY cell line. The flow cytometer used was an EPICS V instrument. Based on the procedures of Bernheim et al. [14] and Lebo et al. [174].

band Xp21, this technique has yielded Xp21-specific DNA probes which are tightly linked to the Duchenne's muscular dystrophy locus [199]. The DNA reassociation procedure may prove to be more useful for obtaining DNA probes specific for deleted chromosomal subregions which are associated with certain genetic diseases than flow sorting followed by molecular cloning as described above. It should be noted, however, that in addition to the 13q14 specific probes mentioned above, DNA libraries constructed from flow-sorted chromosomes, have yielded DNA probes useful in the molecular analysis of a chromosomal deletion of 15q associated with Prader-Willi syndrome [51] and of Yp associated with gonadal dysgenesis [203].

Note Added in Proof. The overall goal of both approaches for enriching deleted sequences described above is to obtain DNA probes which are close enough to the genetic disease locus so that chromosome walking can be used to isolate the disease gene. The Xp21-specific DNA probes obtained by the DNA reassociation technique have been used in such a manner and, as a result, the Duchenne muscular dystrophy gene has recently been isolated [143] and its genomic organization outlined. It is a huge gene, probably 2 to 2.5 million bp long, with at least 60 exons and a 14-kb message. Similarly, DNA probe H3-8 (Figure 15) obtained from a phage library constructed from flow sorted chromosome 13 has been used to clone the gene that predisposes to retinoblastoma [60,80].

CONCLUSIONS

Flow cytometry and sorting are useful in the following three areas of molecular genetics. First, production of gene libraries from flow-sorted chromosomes for each of the 24 human chromosomal types—these are important sources of cloned DNA sequences for both the study of genetic diseases and also gene structure and function. Second, flow cytometry is useful in monitoring and selecting for the expression of donor genes which have been introduced into host cells by somatic cell genetics methods. These techniques also facilitate studies of the cell surface ranging from the chromosome mapping of a surface antigen gene, to its isolation on a transfected cell for cloning. And it has contributed to the isolation of rare cell variants, and to an understanding of the mechanism and dynamics of gene amplification. Future developments in both recombinant DNA vectors and flow cytometry technology may extend these applications to genes which are not normally expressed at the cell surface. Finally, cloned DNA fragments containing known genes or arbitrary sequences can now be readily assigned to their chromosomal location, and often sub-chromosomal regions, using flow sorting and molecular hybridization. In addition to providing new mapping methods, flow cytometry techniques are proving useful in molecular studies of chromosomal rearrangements such as amplification, deletions, and translocations involving oncogenes. There is very active current research to determine the relationship between these phenomena and neoplasia.

ACKNOWLEDGMENTS

We thank Larry Thompson (Lawrence Livermore National Laboratory), Carol Jones (Eleanor Roosevelt Institute for Cancer Research, Denver), and Steve O'Brien (National Cancer Institute, Frederick, MD) for their crucial aid to this work in supplying us with human–hamster hybrid cell lines. We also thank Lil Mitchell and Sue Lunman for their skillful preparation of this manuscript. Part of the work described here was performed under the auspices of the U.S. Department of Energy by the Lawrence Livermore National Laboratory under contract W-07405-ENG-48 and by the Los Alamos National Laboratory under contract W-7405-ENG-36. Figs. 7, 10, 12, and 13 reproduced by permission from NATURE, Copyright © Macmillan Magazines Ltd.

REFERENCES

1. **Adams JM, Gerondakis S, Webb E, Corcoran LM, Cory S (1983)** Cellular myc oncogene is altered by chromosome translocation to an immunoglobulin locus in murine plasmacytomas and is rearranged similarly in human Burkitt lymphomas. Proc. Natl. Acad. Sci. USA 80:1982–1986.
2. **Alitalo K, Schwab M, Lin CC, Varmus HE, Bishop JM (1983)** Homogeneously staining chromosomal regions contain amplified copies of an abundantly expressed cellular oncogene (c-myc) in malignant neuroendocrine cells from a human colon carcinoma. Proc. Natl. Acad. Sci. USA 80:1707–1711.
3. **Alt FW, Kohl NE, Murphy J, Gee CE (1985)** Amplification of c-myc and N-myc, genes in human and

murine tumors. In Simon M, Herskowitz I (eds) "Genome Rearrangement." New York: Alan R. Liss, pp 233–251.

4. **Athwal RS, Smarsh M, Searle BM, Deo SS (1985)** Integration of a dominant selectable marker into human chromosomes and transfer of marked chromosomes to mouse cells by microcell fusion. Somat. Cell Mol. Genet. 11:(2)177–187.
5. **Balaban-Malenbaum G, Gilbert F (1977)** Double minute chromosomes and the homogeneously staining region of a human neuroblastoma cell line. Science 198:739–741.
6. **Barbosa JA, Kamarck ME, Biro PA, Weissman SM, Ruddle FH (1982)** Identification of human genomic clones coding the major histocompatibility antigens HLA-A2 and HLA-B7 by DNA-mediated gene transfer. Proc. Natl. Acad. Sci. USA 79:6327–6331.
7. **Bargmann CI, Hung M-C, Weinberg RA (1986)** The new oncogene encodes an epidermal growth factor receptor-related protein. Nature (Lond.) 319: 226–230.
8. **Bates PF and Swift RA (1983)** Double cos site vectors: Amplified cosmid cloning. Gene 26:137–146.
9. **Battey J, Moulding C, Taub R, Murphy W, Potter H, Lenoir G, Leder P (1983)** The human c-myc oncogene structural consequences of translocation into the IgH locus in Burkitt lymphoma. Cell 34:779–787.
10. **Benedict WF, Murphree AL, Banerjee A, Spina CA, Sparkes MC, Sparkes RS (1983)** Patient with 13 chromosome deletion: Evidence that the retinoblastoma gene is a recessive cancer gene. Science 219: 973–975.
11. **Benton WD, Davis RW (1977)** Screening λgt recombinant clones by hybridization to single plaques in situ. Science 196:180–182.
12. **Berman JW, Basch RS, Pellicer A (1984)** Gene transfer in lymphoid cells: expression of the Thy-1.2 antigen by Thy-1.1 BW5147 lymphoma cells transfected with unfractionated cellular DNA. Proc. Natl. Acad. Sci. USA 81:7176–7179.
13. **Bernard O, Cory S, Gerondakis S, Webb E, Adams J (1983)** Sequence of murine and human cellular myc oncogenes and two modes of myc transcription resulting from translocation in B lymphoid cells. EMBO J. 2:2375–2383.
14. **Bernheim A, Metezeau P, Guellaen G, Fellous M, Goldberg ME, Berger R (1983)** Direct hybridization of sorted human chromosomes: Localization of the Y chromosome on the flow karyotype. Proc. Natl. Acad. Sci. USA 80:7571–7575.
15. **Bernheim A, Metezeau P, Guellaen G, Weissenbach J, Fellous M, Goldberg M, Berger R (1982)** Une methode d'identification par hybridation moleculaire de chromosomes purifies avec un trieur de cellules. C. R. Acad. Sci. Paris 295:439–442.
16. **Bishop JM (1983)** Cellular oncogenes and retroviruses. Annu. Rev. Biochem. 52:301–354.
17. **Bishop JM (1985)** Viral oncogenes. Cell 42:23–38.
18. **Blanchard J-M, Piechaczyk M, Dani C, Chambard J-C, Franchi A, Pouyssegur J, Jeanteur P (1985)** c-myc gene is transcribed at high rate in G-arrested fibroblasts and is post-transcriptionally regulated in response to growth factors. Nature (Lond.) 317: 443–445.
19. **Blattner FR, Williams BG, Blechi AE, Denniston-Thompson K, Faber HE, Furlong L-A, Grunwald DJ, Keifer DO, Moore DD, Schumm JW, Sheldon EL, Smithies O (1977)** Charon phages: Safer derivatives of bacteriophage lambda for DNA cloning. Science 196:161–169.
20. **Botstein D, White RL, Skolnick M, Davis RW (1980)** Construction of a genetic linkage map in man using restriction fragment length polymorphyisms. Am. J. Hum. Genet. 32:314–331.
21. **Bradshaw HD (1983)** Molecular cloning and cell-cycle specific regulation of a functional human thymidine kinase gene. Proc. Natl. Acad. Sci. USA 80: 5588–5591.
22. **Brodeur GM, Seeger RC, Schwab M, Varmus HE, Bishop JM (1984)** Amplification of N-myc in untreated human neuroblastomas correlates with advanced disease stage. Science 224:1121–1124.
23. **Cami B, Bregegere F, Abastado JP, Kourilsky P (1981)** Multiple sequences related to classical histocompatibility antigens in the mouse genome. Nature (Lond.) 291:673–675.
24. **Campisi J, Gray HE, Pardee AB, Dean M, Sonenshein GE (1984)** Cell-cycle control of c-myc but not c-ras expression is lost following chemical transformation. Cell 36:241–247.
25. **Cavenee WK, Dryja TP, Phillips RA, Benedict WF, Godbout R, Gallie BL, Murphree AL, Strong LC, White RL (1983)** Expression of recessive alleles by chromosomal mechanisms in retinoblastoma. Nature (Lond.) 305:779–784.
26. **Cavenee WK, Hansen MF, Nordenskjold M, Kock E, Maumenee I, Squire JA, Phillips RA, Gallie BL (1985)** Genetic origin of mutations predisposing to retinoblastoma. Science 228:501–503.
27. **Cavenee W, Leach R, Mohandas T, Pearson P, White R (1984)** Isolation and regional localization of DNA segments revealing polymorphic loci from human chromosome 13. Am. J. Hum. Genet. 36:10–24.
28. **Cepko CL, Roberts BE, Mulligan RC (1984)** Construction and applications of a highly transmissible murine retrovirus shuttle vector. Cell 37: 1053–1062.

28a. **Chao MV, Bothwell MA, Ross AH, Koprowski H, Lanahan AA, Buck CR, Sehgal A (1986)** Gene transfer and molecular cloning of the human NGF receptor. Science 232:518–521.

29. **Goring DR, Gupta K, DuBow MS (1987)** Analysis of spontaneous mutations in a chromosomally located HSV-1 thymidine kinase (TK) gene in a human cell line. Somat. Cell. Mol. Genet. 13:47–56.
30. **Cochran BH, Zullo J, Verma IM, Stiles CD (1984)** Expression of the c-fos Gene and of an fos-related gene is stimulated by platelet-derived growth factor. Science 226:1080–1082.
31. **Cohen SN, Chang ACY, Boyer HW, Helling RB (1973)** Construction of biologically functional bacterial plasmids in vitro. Proc. Natl. Acad. Sci. USA 70: 3240–3244.
32. **Collard JG, de Boer JAJ, Janssen JWG, Schijven JF, de Jong B (1985)** Gene mapping by chromosome spot hybridization. Cytometry 6:179–185.
33. **Collard JG, Schijven J, Tulp A, Meulenbroek M (1982)** Localization of genes on fractionated rat chromosomes by molecular hybridization. Exp. Cell. Res. 137:463–469.

34. **Cooper GM (1982)** Cellular transforming genes. Science 218:801–806.
35. **Creagan RP, Ruddle FH (1975)** The clone panel: A systematic approach to gene mapping using interspecific somatic cell hybrids. Cytogenet. Cell. Genet. 14: 282–286.
36. **Croce CM, Klein G (1985)** Chromosome translocations and human cancer. Sci. Am. 252:54–60.
37. **Croce CM, Thierfelder W, Erikson J, Nishikura K, Finan J, Lenoir GM, Nowell PC (1983)** Transcriptional activation of an unrearranged and untranslocated c-myc oncogene by translocation of a CI locus in Burkitt lymphoma cells. Proc. Natl. Acad. Sci. USA 80:6922–6926.
38. **D'Eustachio P, Ruddle FH (1983)** Somatic cell genetics and gene families. Science 220:919–924.
39. **Dalla-Favera R, Gelmann EP, Martinotti S, Franchini G, Papas TS, Gallo RC, Wong-Staal F (1983)** Cloning and characterization of different human sequences related to the onc gene (v-myc) of avian myelocytomatosis virus (MC29). Proc. Natl. Acad. Sci. USA 79:6497–6501.
40. **Dalla-Favera R, Martinotti S, Gallo RC, Erikson J, Croce CM (1983)** Translocation and rearrangements of the c-myc oncogene locus in human undifferentiated B-cell lymphomas. Science 219:963–967.
41. **Dalla-Favera R, Wong-Staal F, Gallo RC (1982b):** onc gene amplification in promyelocytic leukaemia cell line HL-60 and primary leukaemic cells of the same patient. Nature (Lond.) 299:61–63.
42. **Dangl JL, Herzenberg LA (1982)** Selection of hybridomas and hybridoma variants using the fluorescence activated cell sorter. J. Immunol. Methods 52:1–14.
43. **Dangl JL, Parks DR, Oi VT, Herzenberg LA (1982)** Rapid isolation of cloned isotype switch variants using fluorescence activated cell sorting. Cytometry 2: 395–401.
44. **Dani C, Blanchard JM, Piechaczyk M, Sabouty SE, Marty L, Jeanteur P (1984)** Extreme instability of myc mRNA in normal and transformed human cells. Proc. Natl. Acad. Sci. USA 81:7046–7050.
45. **Davies KE, Young BD, Elles RG, Hill ME, Williamson R (1981)** Cloning of a representative genomic library of the human X chromosome after sorting by flow cytometry. Nature (Lond.) 293:374–376.
46. **Davis M, Malcolm S, Rabbits TH (1984)** Chromosome translocation can occur on either side of the c-myc oncogene in Burkitt lymphoma cells. Nature (Lond.) 308:286–288.
47. **Dean PN (1985)** Methods of data analysis in flow cytometry. In Van Dilla MA, Dean PN, Laerum OD, Melamed MR (eds), "Flow Cytometry: Instrumentation and Data Analysis." Orlando, Florida, Academic Press pp 195–221.
48. **de Klein A, van Kessel AG, Grosveld G, Bartram CR, Hagemeijer A, Bootsma D, Spurr NK, Heisterkamp N, Groffen J, Stephenson JR (1982)** A cellular oncogene is translocated to the Philadelphia chromosome in chronic myelocytic leukaemia. Nature (Lond.) 300:765–767.
49. **Disteche CM, Carrano AV, Ashworth LK, Burkhart-Schultz K, Latt SA (1981)** Flow sorting of the mouse cattanach X chromosome, T (X;7) 1 Ct, in an active or inactive state. Cytogenet. Cell. Genet. 29: 189–197.
50. **Disteche CM, Kunkel LM, Lojewski A, Orkin SH, Eisenhard M, Sahar E, Travis B, Latt SA (1982)** Isolation of mouse X-chromosome specific DNA from an X-enriched lambda phage library derived from flow sorted chromosomes. Cytometry 2:(5)282–286.
51. **Donlon TA, Lalande M, Wyman A, Bruns G, Latt SA (1986)** Isolation of molecular probes associated with the chromosome 15 instability in the Prader–Willi syndrome. Proc. Natl. Acad. Sci. USA 83:4408–4412.
52. **Doolittle RF, Hunkapiller MW, Hood LE, Devare SG, Robbins KC, Aaronson SA, Antoniades HN (1983)** Simian sarcoma virus onc gene, v-sis, is derived from the gene (or genes) encoding a platelet-derived growth factor. Science 221:275–276.
53. **Dorman BP, Shimizu N, Ruddle FH (1978)** Genetic analysis of the human cell surface antigenic marker for the human X chromosome in human mouse hybrids. Proc. Natl. Acad. Sci. USA 75:2363–2367.
54. **Downward J, Yarden Y, Mayes E, Scrace G, Totty N, Stockwell P, Ullrich A, Schlessinger J, Waterfield MD (1984)** Close similarity of epidermal growth factor receptor and v-erb-B oncogene protein sequences. Nature (Lond.) 307:521–527.
55. **Drebin JA, Link VC, Stern DF, Weinberg RA, Greene MI (1985)** Down-modulation of an oncogene protein product and reversion of the transformed phenotype by monoclonal antibodies. Cell 41:695–706.
56. **Drebin JA, Stern DF, Link VC, Weinberg RA, Greene MI (1984)** Monoclonal antibodies identify a cell-surface antigen associated with an activated cellular oncogene. Nature (Lond.) 312:545–548.
57. **Dryja TP, Cavenee W, White R, Rapaport JM, Peterswen R, Albert DM, Bruns GAP (1984)** Homozygosity of chromosome 13 in retinoblastoma. N. Engl. J. Med. 310:550–553.
58. **Dryja TP, Rapaport JM, Epstein J, Goorin AM, Weichselbaum R, Koufos A, Cavenee WK (1986)** Chromosome 13 homozygosity in osteosarcoma without retinoblastoma. Am. J. Hum. Genet. 38:59–66.
59. **Dryja TP, Rapaport JM, Weichselbaum R, Bruns GAP (1984)** Chromosome 13 restriction fragment length polymorphisms. Hum. Genet. 65:320–324.
60. **Dryja TP, Rapaport JM, Joyce JM, Pettersen RA (1986)** Molecular detection of deletions involving band q14 of chromosome 13 in retinoblastomas. Proc. Natl. Acad. Sci. USA 83:7391–7394.
61. **Duesberg PH (1985)** Activated proto-onc genes: Sufficient or necessary for cancer? Science 228:669–677.
62. **Duesberg PH (1983)** Retroviral transforming genes in normal cells? Nature (Lond.) 304:219–226.
63. **Dunn IS, Blattner FR (1987)** Charon 36-40: multi-enzyme high capacity recombination deficient replacement vector with polystuffers. Nucl. Acids Res. 15:2677–2698.
64. **Ege T, Ringertz NR (1974)** The preparation of microcells by enucleation of micronucleate cells. Exp. Cell Res. 87:378–382.
65. **Eick D, Piechaczyk M, Henglein B, Blanchard J-M, Traub B, Kofler E, Wiest S, Lenoir GM, Bornkamm GW (1985)** Aberrant c-myc RNAs of Burkitt's lymphoma cells have longer half-lives. EMBO J. 4:3717–3725.

66. **Einat M, Resnitzky D, Kimchi A (1985)** Close link between reduction of c-myc expression by interferon and GO/G1 arrest. Nature (Lond.) 313:597–600.

67. **Erikson J, Nishikura K, ar-Rushdi A, Finan J, Emanuel B, Lenoir G, Nowell PC, Croce CM (1983)** Translocation of an immunoglobulin k locus to a region 3′ of an unrearranged c-myc oncogene enhances c-myc transcription. Proc. Natl. Acad. Sci. USA 80:7581–7585.

68. **Erikson J, ar-Rushdi A, Drwinga HL, Nowell PC, Croce CM (1983)** Transcriptional activation of the translocated c-myc oncogene in Burkitt lymphoma. Proc. Natl. Acad. Sci. USA 80:820–824.

69. **Evan GI, Hancock DC (1985)** Studies on the interaction of the human c-myc protein with cell nuclei: p62c-myc as a member of a discrete subset of nuclear proteins. Cell 43:253–261l.

70. **Evan GI, Lewis GK, Ramsay G, Bishop JM (1985)** Isolation of monoclonal antibodies specific for human c-myc proto-oncogene product. Mol. Cell. Biol. 5:3610–3616.

71. **Evans GA, Ingraham HA, Lewis K, Cunningham K, Seki T, Moriuchi T, Chang HC, Silver J, Hyman R (1984)** Expression of the Thy-1 glycoprotein gene by DNA-mediated gene transfer. Proc. Natl. Acad. Sci. USA 81:5532–5536.

72. **Evans GA, Margulies DH, Camerini-Otero RD, Ozato K, Seidman JG (1982)** Structure and expression of a mouse major histocompatibility antigen gene, H-2Ld. Proc. Natl. Acad. Sci. USA 79:1994–1998.

73. **Evans GA, Margulies DH, Shykind B, Seidman JG, Ozato K (1982)** Exon shuffling: Mapping polymorphic determinants on hybrid mouse transplantation antigens. Nature (Lond.) 300:755–757.

74. **Fearon ER, Feinberg AP, Hamilton SH, Vogelstein B (1985)** Loss of genes on the short arm of chromosome 11 in bladder cancer. Nature (Lond.) 318:377–380.

75. **Fearon ER, Vogelstein B, Feinberg AP (1984)** Somatic deletion and duplication of genes on chromosome 11 in Wilms' tumours. Nature (Lond.) 309: 176–178.

76. **Fellous M, Nir U, wallach D, Merlin G, Rubinstein M, Revel M (1982)**: Interferon-dependent induction of mRNA for the major histocompatibility antigens in human fibroblasts and lymphoblastoid cells. Proc. Natl. Acad. Sci. USA 79:3082–3086.

77. **Fournier REK (1982)** Microcell-mediated transfer. In Shay J (ed), "Techniques in Somatic Cell Genetics." New York: Plenum Press, pp 76–84.

78. **Fournier REK, Ruddle FH (1977)** Microcell-mediated transfer of murine chromosomes into mouse, Chinese hamster and human somatic cells. Proc. Natl. Acad. Sci. USA 74:319–323.

79. **Friend KK, Dorman BP, Kucherlapati RS, Ruddle FH (1976)** Detection of interspecific translocations in mouse-human hybrids in alkaline Giemsa staining. Exp. Cell. Res. 99:31–36.

80. **Friend SH, Bernards R, Rogelj S, Weinberg RA, Rapaport JM, Albert DM, Dryja TP (1986)** A human DNA segment with properties of the gene that predisposes to retinoblastomas and osteosarcoma. Nature (Lond.) 323:643–646.

81. **Fuscoe JC, Clark LM, Van Dilla MA (1986)** Construction of fifteen human chromosome-specific DNA libraries. Cytogenet. Cell. Genet. 43:79–86.

82. **Gall RP, Canaani E (1984)** An 8-kilobase abl RNA transcript in chronic myelogenous leukemia. Proc. Natl. Acad. Sci. USA 81:5648–5652.

83. **George DL, Powers VE (1981)** Cloning of DNA from double minutes of Y1 mouse adrenocortical tumor cells: Evidence for gene amplification. Cell 24:117–123.

84. **Gerhard DS, Kawasaki ES, Bancroft FC, Szabo P (1981)** Localization of a unique gene by direct hybridization in situ. Proc. Natl. Acad. Sci. USA 78: 3755–3759.

85. **Gilliam TC, Healy ST, MacDonald ME, Stewart GD, Wasmuth JJ, Tanzi RE, Roy JC, and Gusella JF** (1987) Nucl. Acids Res. 15:1445–1458.

86. **Goodenow RS, McMillan M, Orn A, Nicolson M, Davidson N, Frelinger JA, Hood L (1982)** Identification of a Balb/c H-2Ld gene by DNA-mediated gene transfer. Science 215:677–679.

87. **Graham FL, Van der Eb AJ (1973)** A new technique for the assay of infectivity of human adenovirus 5 DNA. Virology 52:456–467.

88. **Gray JW, Langlois RG, Carrano AV, Burkhart-Schulte K, Van Dilla MA (1979)** High resolution chromosome analysis: One and two parameter flow cytometry. Chromosoma 73:9–27.

89. **Greenberg ME, Ziff EB (1984)** Stimulation of 3T3 cells induces transcription of the c-fos proto-oncogene. Nature (Lond.) 311:433–438.

89a. **Greve JM, Davis G, Meyer AM, Forte CP, Yost SC, Marlor CW, Kamarck ME, McClelland A** (1989) The major human rhinovirus receptor is ICAM-1. Cell 56:839–847.

90. **Griffith JK, Cram LS, Crawford BD, Jackson PJ, Schilling J, Schimke RT, Walters RA, Wilder ME, Jett JH (1984)** Construction and analysis of DNA sequence libraries from flow-sorted chromosomes: Practical and theoretical considerations. Nucl. Acids Res. 12:4019–4034.

91. **Groffen J, Stephenson JR, Heisterkamp N, de Klein A, Bartram CR, Grosveld G (1984)** Philadelphia chromosomal breakpoints are clustered within a limited region, bcr, on chromosome 22. Cell 36:93–99.

92. **Gross SH, Little PFR (1986)** A cosmid vector for systematic chromosome walking. Gene 49:9–22.

93. **Grunstein M, Hogness DS (1975)** Colony hybridization: A method for the isolation of cloned DNA's that contain a specific gene. Proc. Natl. Acad. Sci. USA 72:3961–3965.

94. **Guan J-L, Rose JK (1984)** Conversion of a secretory protein into a transmembrane protein results in its transport to the golgi complex but not to the cell surface. Cell 37:779–787.

95. **Gusella JF, Keys C, Varsanyi-Breiner A, Kao F-T, Jones C, Puck TT, Housman D (1980)** Isolation and localization of DNA segments from specific human chromosomes. Proc. Natl. Acad. Sci. USA 77:2829–2833.

96. **Hamlyn PH, Rabbits TH (1983)** Translocation joins c-myc and immunoglobulin Y1 genes in a Burkitt lymphoma revealing a third exon in the c-myc oncogene. Nature (Lond.) 304:135–139.

97. **Hann SR, Eisenman RN (1984)** Proteins encoded by

the human c-myc oncogene: Differential expression in neoplastic cells. Mol. Cell. Biol. 4:2486–2497.

98. **Hann SR, Thompson CB, Eisenman RN** (1985) C-myc oncogene protein synthesis is independent of the cell cycle in human and avian cells. Nature (Lond.) 314:366–369.

99. **Hansen MF, Koufos A, Gallie BL, Phillips RA, Fodstad O, Brogger A, Gedde-Dahl T, Cavenee WK** (1985) Osteosarcoma and retinoblastoma: A shared chromosomal mechanism revealing recessive predisposition. Proc. Natl. Acad. Sci. USA 82:6216–6220.

100. **Harper ME, Saunders GF** (1981) Localization of single copy DNA sequences on G-banded human chromosomes by in situ hybridization. Chromosoma 83: 431–439.

101. **Harper ME, Ullrich A, Saunders SF** (1981) Localization of the human insulin gene to the distal end of the short arm of chromosome 11. Proc. Natl. Acad. Sci. USA 78:4458–4460.

102. **Harris H, Hopkinton DA** (1976) "Handbook of Enzyme Electrophoresis in Human Genetics." Amsterdam: North-Holland Publishing Co.

103. **Harris H, Watkins JF** (1965) Hybrid cells derived from mouse and man: Activated heterokaryons of mammalian cells from different species. Nature (Lond.) 205:640–646.

104. **Hayday AC, Gillies SD, Saito H, Wood C, Wiman K, Hayward WS, Tonegawa S** (1984) Activation of a translocated human c-myc gene by an enhancer in the immunoglobulin heavy-chain locus. Nature (Lond.) 307:334–340.

105. **Haynes BF, Hemler ME, Thomas CA, Strominger JL, Fauci AS** (1981) Characterization of a monoclonal antibody (4F2) that binds to human monocytes and to a subset of activated lymphocytes. J Immunol. 126:1409–1414.

106. **Heisterkamp N, Stephenson JR, Groffen J, Hansen PF, de Klein A, Bartram CR, Grosveld G** (1983) Localization of the c-abl oncogene adjacent to a translocation break point in chronic myelocytic leukaemia. Nature (Lond.) 306:239–242.

107. **Heldin CH, Westermark B** (1984) Growth factors: Mechanism of action and relation to oncogenes. Cell 37:9–20.

108. **Hemler ME, Strominger JL** (1982) Characterization of the antigen recognized by the monoclonal antibody (4F2): Different molecular forms on human T and B lymphoblastoid cell lines. J. Immunol. 129: 623–628.

109. **Hohn B** (1979) In vitro packaging of lambda and cosmid DNA. Methods Enzymol. 68:299–309.

110. **Hollis GF, Mitchell KF, Battey J, Potter H, Taub R, Lenoir GM, Leder P** (1984) A variant translocation places the A immunoglobulin genes 3′ to the c-myc oncogene in Burkitt's lymphoma. Nature (Lond.) 307:752–755.

111. **Holtkamp B, Cramer M, Lemke H, Rajewsky K** (1981) Isolation of a cloned cell line expressing variant H-$2K^k$ using fluorescence-activated cell sorting. Nature (Lond.) 289:66–68.

112. **Hsu C, Kavathas P, Herzenberg LA** (1984) Cell-surface antigens expressed on L-cells transfected with whole DNA from non-expressing and expressing cells. Nature (Lond.) 312:68–69.

113. **Hughes SH, Stubblefield E, Payvar F, Engel JD, Dodgson JB, Spector D, Cordell B, Schimke RT, Varmus HE** (1979) Gene localization by chromosome fractionation: Globin genes are on at least two chromosomes and three estrogen-inducible genes are on three chromosomes. Proc. Natl. Acad. Sci. USA 76: 1348–1352.

114. **Jackson DA, Symons RH, Berg P** (1972) Biochemical method for inserting new genetic information into DNA of Simian virus 40: Circulate SV40 DNA molecules containing lambda phase genes and the galactose operon of *Escherichia coli*. Proc. Natl. Acad. Sci. USA 69:2904–2909.

115. **Janssen JWG, Vernole P, de Boer PAJ, Oosterhuis JW, Collard JG** (1986): Sublocalization of c-myb to 6q21+q23 by in situ hybridization and c-myb expression in a human teratocarcinoma with 6q rearrangements. Cytogenet. Cell. Genet. 41:129–135.

116. **Jelinek WR, Toomey TP, Leinwand L, Duncan CH, Biro PA, Choudary PV, Weissman SM, Rubin CM, Houck CM, Deininger PL, Schmid CW** (1980) Ubiquitous, interspersed repeated sequences in mammalian genomes. Proc. Natl. Acad. Sci. USA 77:1398–1402.

117. **Johnson DD, Dewald GW, Pierre RV, Letendre L, Silverstein MN** (1985) Deletions of chromosome 13 in malignant hematologic disorders. Cancer Genet. Cytogenet. 18:235–241.

118. **Johnston RN, Beverley SM, Schimke RT** (1983) Rapid spontaneous dihydrofolate reductase gene amplification shown by fluorescence-activated cell sorting. Proc. Natl. Acad. Sci. USA 80:3711–3715.

119. **Jolly DJ, Okayama H, Berg P, Esty AC, Filpula D, Bohlen P, Johnson GG, Shively J-E, Hunkapillar T, Friedman T** (1983) Isolation and characterization of a full length expressible cDNA for human hypoxanthine phosphoribosyltransferase. Proc. Natl. Acad. Sci. USA 80:477–481.

120. **Jones C, Kao F-T** (1978) Regional mapping of the gene for human liposomal acid phosphatase (ACP2) using a hybrid clone panel containing segments of human chromosome 11. Hum. Genet. 45:1–10.

121. **Jongkind JF, Verkerk A, Schaap GH, Galjaard H** (1980) Flow sorting of in vitro cultured cells in studies on genetic and metabolic interaction. Eur. J. Cell. Biol. 22:594.

122. **Jongkind JF, Verkerk A, Schaap GH, Galjaard H** (1980) Non-selective isolation of fibroblast cybrids by flow sorting. Exp. Cell. Res. 130:481–484.

123. **Joyner AL, Lebo RV, Kan YW, Tjian R, Cox DR, Martin GR** (1985) Comparative chromosome mapping of a conserved homoeo box region in mouse and human. Nature (Lond.) 314:173–175.

124. **Kaczmarek L, Hyland JK, Watt R, Rosenberg M, Baserga R** (1985) Microinjected c-myc as a competence factor. Science 228:1313–1315.

125. **Kam W, Clauser E, Kim YS, Kan YW, Rutter WJ** (1985) Cloning, sequencing, and chromosomal localization of human term placental alkaline phosphatase cDNA. Proc. Natl. Acad. Sci. USA 82:8715–8719.

126. **Kamarck ME, Barbosa JA, Kuhn L, Peters PGM, Shulman L, Ruddle FH** (1983) Somatic cell genetics and flow cytometry. Cytometry 4:99–108.

127. **Kamarck ME, Barker PE, Miller RL, Ruddle FH** (1984) Somatic cell hybrid mapping panels. Exp. Cell. Res. 152:1–14.

128. **Kamarck ME, Macyko CA, Cunningham AC, Ruddle FH** (1983) The gene coding the human S11 surface antigens maps between the loci for HPRT and G6PD on the X-chromosome. Exp. Cell. Res. 149: 325–334.
129. **Kanda N, Schreck R, Alt F, Bruns G, Baltimore D, Latt S** (1983) Isolation of amplified DNA sequences from IMR-32 human neuroblastoma cells: Facilitation by fluorescence-activated flow sorting of metaphase chromosomes. Proc. Natl. Acad. Sci. USA 80:4069–4073.
130. **Kao F-T, Johnson RT, Puck TT** (1969) Complementation analysis on virus-fused Chinese hamster cells with nutritional markers. Science 164:312–314.
131. **Kaufman RJ, Bertino JR, Schimke RT** (1978) Quantitation of dihydrofolate reductase in individual parental and methotrexate-resistant murine cells. J. Biol. Chem. 253:5852–5860.
132. **Kavathas P, Herzenberg LA** (1983) Stable transformation of mouse L cells for human membrane T-cell differentiation antigens, HLA and B2-microglobulin: Selection by fluorescence-activated cell sorting. Proc. Natl. Acad. Sci. USA 80:524–528.
133. **Kavathas P, Herzenberg LA** (1983) Amplification of a gene coding for a human T-cell differentiation antigen. Nature (Lond.) 306:385–387.
134. **Kavathas P, Sukhatme VP, Herzenberg LA, Parnes JR** (1984) Isolation of the gene encoding the human T-lymphocyte differentiation antigen Leu-2 (T8) by gene transfer and cDNA subtraction. Proc. Natl. Acad. Sci. USA 81:7688–7692.
135. **Kelly K, Cochran BH, Stiles CD, Leder P** (1983) Cell-Specific Regulation of the c-myc gene by lymphocyte mitogens and platelet-derived growth factor. Cell 35: 603–610.
136. **Kelly K, Cochran BH, Stiles CD, Leder P** (1984) The regulation of c-myc by growth signals. Curr. Topics Microbiol. Immunol. 113:117–126.
137. **Kinzler KW, Zehnbauer BA, Brodeur GM, Seeger RC** (1986) Amplification units containing human N myc and c-myc genes. Proc. Natl. Acad. Sci. USA 83: 1031–1035.
138. **Klein G** (1983) Specific chromosomal translocations and the genesis of B-cell-derived tumors in mice and men. Cell 32:311–315.
139. **Klein G** (1981) The role of gene dosage and genetic transpositions in carcinogenesis. Nature (Lond.) 294: 313–318.
140. **Klobutcher LA, Ruddle FH** (1979): Phenotype stabilization and integration of transferred material in chromosome-mediated gene transfer. Nature (Lond.) 280:657–660.
141. **Knudson AG** (1971) Mutation and Cancer: Statistical study of retinoblastoma. Proc. Natl. Acad. Sci. USA 68:820–823.
142. **Knudson AG, Meadows AT, Nichols WW, Hill R** (1976) Chromosomal deletion and retinoblastoma. N. Engl. J. Med. 295:1120–1123.
143. **Koenig M, Hoffman EP, Bertelson CS, Monaco AP, Feener C, Kunkel LM** (1987) Complete cloning of the Duchene muscular dystrophy (DMD) cDNA and preliminary genomic organization of the DMD gene in normal and affected individuals, Cell 50:509–51.
144. **Kohl NE, Gee CE, Alt FW** (1984) Activated expression of the N-myc gene in human neuroblastomas and related tumors. Science 226:1335–1337.
145. **Kohl NE, Kanda N, Schreck RR, Bruns G, Latt SA, Gilbert F, Alt FW** (1983) Transposition and amplification of oncogene-related sequences in human neuroblastomas. Cell 35:359–367.
146. **Koufos A, Hansen MF, Copeland NG, Jenkins NA, Lampkin BC, Cavenee WK** (1985) Loss of heterozygosity in three embryonal tumours suggests a common pathogenetic mechanism. Nature (Lond.) 316: 330–334.
147. **Koufos A, Hansen MF, Lampkin BC, Workman ML, Copeland NG, Jenkins NA, Cavenee WK** (1984) Loss of alleles at loci on human chromosome 11 during genesis of Wilms' tumour. Nature (Lond.) 309:170–172.
148. **Kruijer W, Cooper JA, Hunter T, Verma IM** (1985) Platelet-derived growth factor induces rapid but transient expression of the c-fos gene and protein. Nature (Lond.) 312:711–716.
149. **Krumlauf R, Jeanpierre M, Young BD** (1982) Construction and characterization of genomic libraries from specific human chromosomes. Proc. Natl. Acad. Sci. USA 79:2971–2975.
150. **Kuhn LC, Barbosa JA, Kamarck ME, Ruddle FH** (1983) An approach to the cloning of cell surface protein genes. Mol. Biol. Med. 1:335–352.
151. **Kuhn LC, McClelland A, Ruddle FH** (1984) Gene transfer, expression and molecular cloning of the human transferrin receptor gene. Cell 37:95–103.
152. **Kunkel LM, Lalande M, Monaco AP, Flint A, Middlesworth W, Latt SA** (1985) Construction of a human X-chromosome-enriched phage library which facilitates analysis of specific loci. Gene 33:251–258.
153. **Kunkel LM, Monaco AP, Middlesworth W, Ochs HD, Latt SA** (1985) Specific cloning of DNA fragments absent from the DNA of a male patient with an X chromosome deletion. (1985) Proc. Natl. Acad. Sci. USA 82:4778–4782.
154. **Kunkel LM, Smith KD, Boyer SH, Borgaonkar DS, Wachtel SS, Miller OJ, Breg WR, Jones Jr HW, Rary JM** (1977) Analysis of human Y chromosome-specific reiterated DNA in chromosome variants. Proc. Natl. Acad. Sci. USA 74:1245–1249.
155. **Kunkel LM, Tanravahi U, Eisenhard M, Latt A** (1982) Regional localization on the human X of DNA segments closed from flow sorted chromosomes. Nucl. Acid Res. 10:1557–1578.
156. **Kunkel LM, Umadevi T, Eisenhard M, Latt SA** (1982) Regional localization on the human X of DNA segments cloned from flow sorted chromosomes. Nucl. Acids Res. 10:1557–1578.
157. **Lalande M, Donlon T, Petersen RA, Liberfarb R, Manter S, Latt SA** (1986) Molecular detection and differentiation of deletions in band 13q14 in human retinoblastoma. Cancer Genet. Cytogenet. 23:151–157.
158. **Lalande M, Dryja TP, Schreck RR, Shipley J, Flint A, and Latt SA** (1984) Isolation of human chromosome 13-specific DNA sequences cloned from flow sorted chromosomes and potentially linked to the retinoblastoma locus. Am. Genet. Cytogenet. 13:283–295.
159. **Lalande M, Kunkel L, Flint A, Latt SA** (1984) Development and use of metaphase chromosome flow-sorting methodology to obtain recombinant phage li-

braries enriched for parts of the human X chromosome. Cytometry 5:101–107.

160. **Lalande M, Schreck RR, Hoffman R, Latt SA (1985)** Identification of inverted duplicated #15 chromosomes using bivariate flow cytometric analysis. Cytometry 6:1–6.

161. **Land H, Parada LF, Weinberg RA (1983)** Cellular oncogenes and multistep carcinogenesis. Science 222: 771–778.

162. **Landegent JE, De Wal NJ, Baan RA, Hoeijmakers JHJ, Van Der Ploeg M (1984)** 2-Acetylaminofluorene-modified probes for the indirect hybridocytochemical detection of specific nucleic acid sequences. Exp. Cell. Res. 153:61–72.

163. **Langer PR, Waldrop AA, Ward DC (1981)** Enzymatic synthesis of biotin-labeled polynucleotides: Novel nucleic acid affinity probes. Proc. Natl. Acad. Sci. USA 78:6633–6637.

164. **Langlois RG, Yu LC, Gray JW, Carrano AV (1982)** Quantitative karyotyping of human chromosomes by dual beam flow cytometry. Proc. Natl. Acad. Sci. USA 79:7876–7880.

165. **Latt SA, Alt FW, Schreck RR, Kanda N, Baltimore D (1982)** In Schimke RT (ed), "Gene Amplification." Cold Spring Harbor Laboratory, New York: Cold Spring Harbor Laboratory, pp 283–289.

166. **Lau Y-F, Kan YW (1983)** Versatile cosmid vectors for the isolation, expression, and rescue of gene sequences: Studies with the human alpha-globin gene cluster. Proc. Natl. Acad. Sci. USA 80:5225–5229.

167. **Law DJ, Frossard PM, Rucknagel DL (1984)** Highly sensitive and rapid gene mapping using miniaturized blot hybridization: Application to prenatal diagnosis. Gene 28:153–158.

168. **Lawn RM, Fritsch EF, Parker RC, Blake G, Maniatis T (1978)** The isolation and characterization of linked δ- and β-globin genes from a cloned library of human DNA. Cell 15:1157–1174.

169. **Leary JJ, Brigati DJ, Ward DC (1983)** Rapid and sensitive colorimetric method for visualizing biotin-labeled DNA probes hybridized to DNA or RNA immobilized on nitrocellulose: Bio-blots. Proc. Natl. Acad. Sci. USA 80:4045–4049.

170. **Lebo RV (1982)** Chromosome sorting and DNA sequence localization. Cytometry 3:145–154.

171. **Lebo RV, Bastian AM (1982)** Design and operation of a dual laser chromosome sorter. Cytometry 3: 213–219.

172. **Lebo RV, Carrano AV, Burkhart-Schultz K, Dozy AM, Yu L-C, Kan YW (1979)** Assignment of human δ-, Y-, and 6-globin genes to the short arm of chromosome 11 by chromosome sorting and DNA restriction enzyme analysis. Proc. Natl. Acad. Sci. USA 76: 5804–5808.

173. **Lebo RV, Cheung M-C, Bruce BD, Riccardi VM, Kao F-T, Kan YW (1985)** Mapping parathyroid hormone β-globin, insulin, and LDH-A genes within the human chromosome 11 short arm by spot blotting sorted chromosomes. Hum. Genet. 69:316–320.

174. **Lebo RV, Gorin F, Fletterick RJ, Kao F-T, Cheung M-C, Bruce BD, Kan YW (1984)** High-resolution chromosome sorting and DNA spot-blot analysis assign McArdle's Syndrome to chromosome 11. Science 225:57–59.

175. **Lebo RV, Kan YW, Cheung MC, Carrano AV, Yu L-C, Chang JC, Cordell B, Goodman HM (1982)** Assigning the polymorphic human insulin gene to the short arm of chromosome 11 by chromosome sorting. Hum. Genet. 60:10–15.

176. **Lebo RV, Tolan DR, Bruce BD, Cheung M-C, Kan YW (1985)** Spot-blot analysis of sorted chromosomes assigns a fructose intolerance disease locus to chromosome 9. Cytometry 6:478–483.

177. **Leder P, Battey J, Lenoir G, Moulding C, Murphy W, Potter H, Stewart T, Taub R (1983)** Translocations among antibody genes in human cancer. Science 222: 765–771.

178. **Lee W-H, Murphree AL, Benedict WF (1984)** Expression and amplification of the N-myc gene in primary retinoblastoma. Nature (Lond.) 309:458–469.

179. **Lemonnier FA, Dubreuil PC, Layet C, Malissen M, Bourel D, Mercier P, Jakobsen BK, Caillol DH, Svejgaard A, Kourilsky FM, Jordan BR (1982)** Transformation of LMTK- cells with purified class I genes. II. Serologic characterization of HLA-A3 and HLA-CW3 molecules. Immunogenet 16:407–424.

180. **Lemonnier FA, Malissen M, Golstein P, LeBoutellier P, Rebai N, Damotte M, Birnbaum D, Caillol D, Trucy J, Jordan BR (1982)** Expression of human class I histocompatibility antigens at the surface of DNA-transformed mouse L cells. Immunogenetics 16:355–361.

181. **Levine M, Rubin GM, Tjian R (1984)** Human DNA sequences homologous to a protein coding region conserved between homeotic genes of drosophila. Cell 38:667–673.

182. **Lewis WH, Michalopoulos EE, Williams DL, Minden MD, Mak TW (1985)**: Breakpoints in the human T-cell antigen receptor α-chain locus in two T-cell leukaemia patients with chromosomal translocations. Nature (Lond.) 317:544–546.

183. **Liesegang B, Radbruch A, Rajewsky K (1978)** Isolation of myeloma variants with predefined variant surface immunoglobulin by cell sorting. Proc. Natl. Acad. Sci. USA 75:3901–3905.

184. **Lin P-F, Zhao S-Y, Ruddle FH (1983)** Genomic cloning and preliminary characterization of the human thymidine kinase gene. Proc. Natl. Acad. Sci. USA 80:6528–6532.

185. **Littlefield JW (1964)** Selection of hybrids from matings of fibroblasts in vitro and their presumed recombinants. Science 145:709–710.

186. **Littman DR, Thomas Y, Maddon PJ, Chess L, Axel R (1985)** The isolation and sequence of the gene encoding T8: A molecule defining functional classes of T lymphocytes. Cell 40:237–246.

187. **Loenen WAM, Blattner FR (1983)** Lambda charon vectors (Ch 32, 33, 34, and 35) adapted for DNA cloning in recombination-deficient hosts. Gene 26: 171–179.

188. **Look AT, Peiper SC, Rebentisch MB, Ashmun RA, Roussel ME, Lemons RS, le Bean MM, Rubin CM, Sherr CJ (1986)** Transfer and expression of the gene encoding a human myeloid membrane antigen (GP150). J. Clin. Invest. 77:914–921.

189. **Loyter A, Scangos G, Ruddle FH (1982)**: Mechanisms of DNA uptake by mammalian cells: Fate of exogenously added DNA monitored by use of fluorescent dyes. Proc. Natl. Acad. Sci. USA 79:422–426.

190. **Malcolm S, Barton P, Murphy C, Ferguson-Smith**

MA (1981) Chromosomal localization of a single copy gene by in situ hybridization–human β-globin genes on the short arm of chromosome 11. Ann. Hum. Genet. 45:135–141.

191. **Malissen M, Malissen B, Jordan BR** (1982) Exon–intron organization and complete nucleotide sequence of an HLA gene. Proc. Natl. Acad. Sci. USA 79:893–897.

192. **Maniatis T, Fritsch EF, Sambrook J** (1982) "Molecular Cloning." Cold Spring Harbor, New York: Cold Spring Harbor Laboratory.

193. **Maniatis T, Hardison RC, Lacy E, Lauer J, O'Connell C, Quon D, Sim GK, Efstratiadis A** (1978) The isolation of structural genes from libraries of eucaryotic DNA. Cell 15:687–701.

194. **McBride OW, Ozer HL** (1973) Gene transfer with purified mammalian chromosomes. Proc. Natl. Acad. Sci. USA 70:1258–1262.

195. **McCluskey J, Germain RN, Margulies DH** (1985) Cell surface expression of in vitro recombinant class II/Class I major histocompatibility complex gene product. Cell 40:247–257.

196. **McCormack JE, Pepe VH, Kent RB, Dean M, Marshak-Rothstein A, Sonenshein GE** (1984) Specific regulation of c-myc oncogene expression in a murine B-cell lymphoma. Proc. Natl. Acad. Sci. USA 81: 5546–5550.

197. **Meyne J, Bartholdi MF, Travis G, Cram LS** (1984) Counterstaining human chromosomes for flow karyology. Cytometry 5:580–583.

198. **Mitelman F** (1984) Restricted number of chromosomal regions implicated in aetiology of human cancer and leukaemia. Nature (Lond.) 310:325–327.

199. **Monaco AP, Bertelson CJ, Middlesworth W, Colletti C-A, Aldridge J, Fischbeck KH, Bartlett R, Pericak-Vance MA, Roses AD, Kunkel LM** (1985) Detection of deletions spanning the Duchenne muscular dystrophy locus using a tightly linked DNA segment. Nature (Lond.) 316:842–845.

200. **Morgan R, Hecht F** (1985) Deletion of chromosome band 13q14: A primary event in preleukemia and leukemia. Cancer Genet. Cytogenet. 18:243–249.

201. **Muller CR, Davies KE, Cremer C, Rappold G, Gray JW, Ropers HH** (1983) Cloning of genomic sequences from the human Y chromosome after purification by dual beam flow sorting. Hum. Genet. 64: 110–115.

202. **Muller R, Bravo R, Burckhardt J** (1984) Induction of c-fos gene and protein by growth factors precedes activation of c-myc. Nature (Lond.) 312:716–720.

203. **Müller U, Lalande M, Disteche CM, Latt SA** (1986) Construction, analysis, and application to 46, XY gonadal dysgenesis, of a recombinant phage DNA library from flow sorted human Y chromosomes. Cytometry 7:418–424.

204. **Mulligan RC, Berg P** (1980) Expression of a bacterial gene in mammalian cells. Science 209:1422–1427.

205. **Nabholz M, Miggiano V, Bodmer WF** (1969) Genetic analysis with human somatic cell hybrids. Nature (Lond.) 223:358–363.

206. **Neuberger MS, Rajewsky K** (1981) Switch from hapten specific immunoglobulin M to immunoglobulin D secretion in a hybrid mouse cell line. Proc. Natl. Acad. Sci. USA 78:1138–1142.

207. **Newman R, Domingo D, Trotter J, Trowbridge I** (1983) Selection and properties of a mouse L-cell transformant expressing human transferrin receptor. Nature (Lond.) 304:643–645.

208. **Nishikura K, ar-Rushdi A, Erikson J, Watt R, Rovera G, Croce CM** (1983) Differential expression of the normal and of the translocated human c-myc oncogenes in B cells. Proc. Natl. Acad. Sci. USA 80:4822–4826.

209. **Olsen AS, McBride OW, Otey MC** (1980) Isolation of unique sequence human × chromosomal deoxyribonucleic acid. Biochemistry 19:2419–2428.

210. **Orkin SH, Goldman DS, Sallan SE** (1984) Development of homozygosity for chromosome 11p markers in Wilms' tumour. Nature (Lond.) 309:172–174.

211. **Ozato K, Evans GA, Shykind B, Margulies DH, Seidman JG** (1983) Hybrid H-2 histocompatibility gene products assign domains recognized by alloreactive T cells. Proc. Natl. Acad. Sci. USA 80:2040–2043.

212. **Parker RC, Mardon G, Lebo RV, Varmus HE, Bishop JM** (1985) Isolation of duplicated human c-src genes located on chromosomes 1 and 20. Mol. Cell. Biol. 5:831–838.

213. **Parks DR, Bryan VM, Oi VT, Herzenberg LA** (1979) Antigen-specific identification and cloning of hybridomas with a fluorescence-activated cell sorter. Proc. Natl. Acad. Sci. USA 76:1962–1966.

214. **Partridge C, Francke U, Kidd K, Ruddle FH;** Human Gene Library. Supported by NIH grant GENE-GM29111.

215. **Perlman J, Fuscoe JC** (1986) Molecular characterization of human chromosome-specific DNA libraries. Cytogenet. Cell. Genet. 43:87–96.

216. **Perry RP** (1983) Consequences of myc invasion of immunoglobulin loci: Facts and speculation. Cell 33: 647–649.

217. **Persson H, Gray HE, Godeau F** (1985) Growth-dependent synthesis of c-myc-encoded proteins: Early stimulation by serum factors in synchronized mouse 3T3 cells. Mol. Cell. Biol. 5:2903–2912.

218. **Persson H, Hennighausen L, Taub R, Degrado W, Leder P** (1984) Antibodies to human c-myc oncogene product evidence of an evolutionarily conserved protein induced during cell proliferation. Science 225: 687–693.

219. **Persson H, Leder P** (1984) Nuclear localisation and DNA binding properties of a protein expressed by human c-myc oncogene. Science 225:718–720.

220. **Perucho M, Hanahan D, Wigler M** (1980) Genetic and physical linkage of exogenous sequences in transformed cells. Cell 22:309–317.

221. **Peters PGM, Kamarck ME, Hemler ME, Strominger JL, Ruddle FH** (1982) Genetic and biochemical characterization of a human surface determinant on somatic cell hybrids: The 4F2 antigen. Somat. Cell. Genet. 8:825–834.

222. **Peters PM, Kamarck ME, Hemler ME, Strominger JL, Ruddle FH** (1984) Genetic and biochemical characterization of human lymphocyte cell surface antigens. J. Exp. Med. 159:1441–1454.

223. **Pinkel D, Gray JW, Straume T** (1986) Cytogenetic analysis using quantitative, high-sensitivity fluorescence hybridization. Proc. Natl. Acad. Sci. USA 83: 2934–2938.

224. **Ploegh HL, Orr HT, Strominger JL** (1981) Major

histocompatibility antigens: The human (HLA-A,OB,-C) and murine (H-2K, H-2D) Class 1 molecules. Cell 24:287–299.

225. **Potter H, Weir L, Leder P** (1984) Enhancer-dependent expression of human kappa immunoglobulin genes introduced into mouse pre-B lymphocytes by electroporation. Proc. Natl. Acad. Sci. USA 81:7161–7165.

226. **Poustka A, Rackewitz HR, Frischauf AF, Hohn B, and Lehrach H** (1984) Selective isolation of cosmid clones by homologous recombination in *E. coli.* Proc. Natl. Acad. Sci. USA 81:4129–4133.

227. **Rabbits PH, Forster A, Stinson MA, Rabbits TH** (1985) Truncation of exon 1 from the c-myc gene results in prolonged c-myc mRNA stability. EMBO J. 4:3727–3733.

228. **Rabbits PH, Watson JV, Lamond A, Forster A, Stinson MA, Evan G, Fischer W, Atherton E, Sheppard R, Rabbits TH** (1985) Metabolism of c-myc gene products: c-myc mRNA and protein expression in the cell cycle. EMBO J. 4:2009–2015.

229. **Rabbits TH, Forster A, Hamlyn P, Baer R** (1984) Effect of somatic mutation translocated c-myc genes in Burkitt's lymphoma. Nature (Lond.) 309: 592–597.

230. **Rabin M, Hart CP, Ferguson-Smith A, McGinnis W, Levine M, Ruddle FH** (1985) Two homoeo box loci mapped in evolutionarily related mouse and human chromosomes. Nature (Lond.) 314:175–178.

231. **Radbruch A, Liesegang B, Rajewsky K** (1980): Isolation of variants of mouse myeloma X63 that express changed immunoglobulin class. Proc. Natl. Acad. Sci. USA 77:2909–2913.

232. **Ramsay G, Evan GI, Bishop JM** (1984) The protein encoded by the human proto-oncogene c-myc. Proc. Natl. Acad. Sci. USA 81:7742–7746.

233. **Reeve AE, Eccles MR, Wilkins RJ, Bell GI, Millow LJ** (1985) Expression of insulin-like growth factor-II transcripts in Wilms' tumour. Nature (Lond.) 317: 258–260.

234. **Reeve AE, Housiaux PJ, Gardner RJM, Chewings WE, Grindley RM, Millow LJ** (1984) Loss of a Harvey ras allele in sporadic Wilms' tumour. Nature (Lond.) 309:174–176.

235. **Robbins DM, Ripley S, Henderson A, Axel R** (1981) Transforming DNA integrates into the host chromosome. Cell 23:29–39.

236. **Robbins KC, Antoniades HN, Devare SG, Hunkapiller MW, Aaronson SA** (1983) Structural and immunological similarities between simian sarcoma virus gene product(s) and human platelet-derived growth factor. Nature (Lond.) 305:605–608.

237. **Rosa F, LeBouteiller PP, Abadie A, Mishal Z, Lemonnier FA, Bourrel D, Lamotte M, Kalil J, Jordan B, Fellous M** (1983) HLA class I genes integrated into murine cells are inducible by interferon. Eur. J. Immunol. 13:495–499.

238. **Roussel MF, Rettenmier CW, Look AT, Sherr CJ** (1984) Cell surface expression of v-fms-coded glycoproteins is required for transformation. Mol. Cell. Biol. 4:1999–2009.

239. **Rowley JD** (1973) A new consistent chromosomal abnormality in chronic myelogenous leukaemia identified by quinacrine fluorescence and Giemsa staining. Nature (Lond.) 243:290–293.

240. **Ruddle FH** (1981) A new era in mammalian gene mapping: Somatic cell genetics and recombinant DNA methodologies. Nature (Lond.) 294:115–120.

241. **Ruddle FH, Creagan RP** (1975) Parasexual approaches to the genetics of man. Ann. Rev. Genet. 9:407–486.

242. **Ruddle FH, Kamarck ME, McClelland A, Kuhn LC** (1984) DNA-mediated gene transfer in mammalian gene cloning. In Setlow JK, Hollaender A (eds), "Genetic Engineering," Vol. 6. New York: Plenum Press, pp 319–338.

243. **Saito H, Hayday A, Wiman K, Hayward WS, Tonegawa S** (1983) Activation of the c-myc gene by translocation: a model for translational control. Proc. Natl. Acad. Sci. USA 80:7476–7480.

244. **Sakai K, Kanda N, Shiloh Y, Donlon T, Schreck R, Shipley J, Dryja T, Chaum E, Chaganti RSK, Latt S** (1985) Molecular and cytologic analysis of DNA amplification in retinoblastoma. Cancer Genet. Cytogenet. 7:95–112.

245. **Scangos G, Huttner KM, Juricek DK, Ruddle FH** (1981): DNA-mediated gene transfer in mammalian cells: Molecular analysis of unstable transformants and their progression stability. Mol. Cell. Biol. 1: 111–120.

246. **Scangos G, Ruddle FH** (1981) Mechanisms and applications of DNA-mediated gene transfer in mammalian cells—A review. Gene 14:1–10.

247. **Schechter AL, Hung M-C, Vaidyanathan L, Weinberg RA, Yang-Feng TL, Francke U, Ullrich A, Coussens L** (1985) The neu Gene: An erbB-homologous gene distinct from and unlinked to the gene encoding the EGF receptor. Science 229:976–978.

248. **Schimke RT** (1984) Gene amplification in cultured animal cells. Cell 37:705–713.

249. **Schmeckpeper BJ, Smith KD, Dorman BP, Ruddle FH, Talbot CC Jr** (1979) Partial purification and characterization of DNA from the human X chromosome. Proc. Natl. Acad. Sci. 76:6525–6528.

250. **Schroder J, Nikinmaa B, Kavathas P, Herzenberg LA** (1983) Fluorescence-activated cell sorting of mouse-human hybrid cells aid in locating the gene for the Leu 7 (HNK-1) antigen to human chromosome 11. Proc. Natl. Acad. Sci. USA 80:3421–3424.

251. **Schwab M, Alitalo K, Klempnauer KH, Varmus HE, Bishop JM, Gilbert F, Brodeur G, Goldstein M, Trent J** (1983) Amplified DNA with limited homology to myc cellular oncogene is shared by human neuroblastoma tumour. Nature (Lond.) 305:245–248.

252. **Schwab M, Ellison J, Busch M, Rosenau W, Varmus HE, Bishop JM** (1984) Enhanced expression of the human gene N-myc consequent to amplification of DNA may contribute to malignant progression of neuroblastoma. Proc. Natl. Acad. Sci. USA 81:4940–4944.

253. **Schwab M, Varmus HE, Bishop JM** (1985) Human N-myc gene contributes to neoplastic transformation of mammalian cells in culture. Nature (Lond.) 316: 160–162.

254. **Schwab M, Varmus HE, Bishop JM, Grzeschik KH, Naylor SL, Sakaguchi AY, Brodeur G, Trent J** (1984) Chromosome localization in normal human cells and neuroblastomas of a gene related to c-myc. Nature (Lond.) 308:288–291.

255. **Scott J, Cowell J, Robertson ME, Priestley LM,**

Wadey R, Hopkins B, Pritchard J, Bell GI, Rall LB, Graham CF, Knott TJ (1985) Insulin-like growth factor-11 gene expression in Wilms' tumour and embryomic tissues. Nature (Lond.) 317:260–264.

256. **Sherr CJ, Rettenmier CW, Sacca R, Roussel MF, Look AT, Stanley ER** (1985) The c-fms proto-oncogene product is related to the receptor for the mononuclear phagocyte growth factor, CSF-1. Cell 41: 665–676.

257. **Shih C, Weinberg RA** (1982) Isolation of a transforming sequence from a human bladder carcinoma cell line. Cell 29:161–169.

258. **Shiloh Y, Shipley J, Brodeur GM, Bruns G, Korf B, Donlon T, Schreck RR, Seeger R, Sakai Kazuo, Latt SA** (1985) Differential amplification, assembly, and relocation of multiple DNA sequences in human neuroblastomas and neuroblastoma cell lines. Proc. Natl. Acad. Sci. USA 82:3761–3765.

259. **Shtivelman E, Lifshitz B, Gale RP, Canaani E** (1985) Fused transcript of abl and bcr genes in chronic myelogenous leukaemia. Nature (Lond.) 315:550–554.

260. **Shulman LM, Kamarck ME, Barbosa JA, Ruddle FH** (1982) Interferon induced surface expression of HLA in LTK-cells transfected with HLA genomic clones. In "The Third Annual International Congress for Interferon Research." p 86 (abst.)

261. **Sillar R, Young BD** (1981) A new method for the preparation of metaphase chromosomes for flow analysis. J. Histochem. Cytochem. 29:74–78.

262. **Singer DS, Camerini-Otero RD, Satz ML, Osborne B, Sachs D, Rudikoff S** (1982) Characterization of a porcine genomic clone encoding a major histocompatibility antigen: Expression in mouse L cells. Proc. Natl. Acad. Sci. USA 79:1403–1407.

263. **Smeland E, Godal T, Ruud E, Beiske K, Funderud S, Clark EA, Pfeifer-Ohlsson S, Ohlsson R** (1985) The specific induction of myc protooncogene expression in normal human B cells is not a sufficient event for acquisition of competence to proliferate. Proc. Natl. Acad. Sci. USA 82:6255–6259.

264. **Smith HO, Wilcox KW** (1970) A restriction enzyme from *hemophilus influenzae*. I. Purification and general properties. J. Mol. Biol. 51:379–391.

265. **Southern E** (1975) Detection of specific sequences among DNA fragments separated by gel electrophoresis. J. Mol. Biol. 98:503–517.

266. **Sparkes RS, Murphree AL, Lingua RW, Sparkes MC, Field LL, Funderburk SJ, Benedict WF** (1983) Gene for hereditary retinoblastoma assigned to human chromosome 13 by linkage to esterase D. Science 219:971–973.

267. **Sparkes RS, Sparkes MC, Wilson MG, Towner JW, Benedict WF, Murphree AL, Yunis JJ** (1980) Regional assignment of genes for human esterase D and retinoblastoma to chromosome band 13q14. Science 208:1042–1044.

268. **Stark GR, Geoffrey GM** (1984) Gene amplification. Annu. Rev. Biochem. 53:447–491.

269. **Taub R, Kelly K, Battey J, Latt S, Lenoir GM, Tantravahi U, Tu Z, Leder P** (1984) A novel alteration in the structure of an activated c-myc gene in a variant t(2;8) Burkitt lymphoma. Cell 37:511–520.

270. **Taub R, Kirsch I, Morton C, Lenoir G, Swan D, Tronick S, Aaronson S, Leder P** (1982) Translocation of the c-myc gene into the immunoglobulin heavy chain locus in human Burkitt lymphoma and murine plasmacytoma cells. Proc. Natl. Acad. Sci. USA 79: 7837–7841.

271. **Taub R, Moulding C, Battey J, Murphy W, Vasicek T, Lenoir GM, Leder P** (1984) Activation and somatic mutation of the translocated c-myc gene in Burkitt lymphoma cells. Cell 36:339–348.

272. **Thompson CB, Challoner PB, Neiman PE, Groudine M** (1986) Expression of the c-myb proto-oncogene during cellular proliferation. Nature (Lond.) 319: 374–380.

273. **Thompson CB, Challoner PB, Neiman PE, Groudine M** (1985) Levels of c-myc oncogene mRNA are invariant throughout the cell cycle. Nature 314:363–366.

274. **Thompson LH, Mooney CL, Burkhart-Schultz K, Carrano AV, Siciliano MJ** (1985) Correction of a nucleotide–excision–repair mutation by human chromosome 19 in hamster–human hybrid cells. Somat. Cell. Mol. Genet. 11:87–92.

275. **Trask B, Van Den Engh G, Landegent J, De Wal NJ, van der Ploeg M** (1985) Detection of DNA sequences in nuclei in suspension by in situ hybridization and dual beam flow cytometry. Science 230:1401–1403.

276. **van den Engh GJ, Trask B, Cram S, Bartholdi M** (1984) Preparation of chromosome suspensions for flow cytometry. Cytometry 5:108–117.

277. **van den Engh GJ, Trask BJ, Gray JW, Langlois RG, Yu, L-C** (1985) Preparation and bivariate analysis of suspensions of human chromosomes. Cytometry 6: 92–100.

278. **Van Dilla MA, Deaven LL, Albright KL, Allen NA, Aubuchon MR, Bartholdi MF, Browne NC, Campbell EW, Carrano AV, Clark LM, Cram LS, Crawford BD, Fuscoe JC, Gray JW, Hildebrand CE, Jackson PJ, Jett JH, Longmire JL, Lozes CR, Leudemann ML, Martin JC, McNinch JS, Meincke LJ, Mendelsohn ML, Meyne J, Moyzis RK, Munk AC, Perlman J, Peters DC, Silva AJ, Trask BJ** (1986) Human chromosome-specific DNA libraries: Construction and availability. Biotechnology 4:537–552.

279. **Van Heyningen V, Boyd PA, Seawright A, Fletcher JM, Fantes JA, Buckton KE, Spowart G, Porteous DJ, Hill RE, Newton MS, Hastie ND** (1985) Molecular analysis of chromosome 11 deletions in aniridia-Wilms tumor syndrome. Proc. Natl. Acad. Sci. USA 82:8592–8596.

280. **Wahl GM, Lewis KA, Rinz JC, Rothenberg B, Shao J, and Evans GA** (1987) Cosmid vectors for rapid genomic walking, restriction mapping, and gene transfer. Proc. Natl. Acad. Sci. USA 84:2160–2164.

281. **Waterfield MD, Scrace GT, Whittle N, Stroobant P, Johnsson A, Wasteson A, Westermark B, Heldin C-H, Huang JS, Deuel TF** (1983) Platelet-derived growth factor is structurally related to the putative transforming protein $p28^{sis}$ of simian sarcoma virus. Nature (Lond.) 304:35–39.

282. **Watt R, Nishikura K, Sorrentino J, Ar-Rushdi A, Croce CM, Rovera G** (1983) The structure and nucleotide sequence of the 5′ end of the human c-myc oncogene. Proc. Natl. Acad. Sci. USA 80:6307–6311.

283. **Watt RA, Shatzman AR, Rosenberg M** (1985) Expression and characterization of the human c-myc DNA-binding protein. Mol. Cell. Biol. 5:448–456.

284. **Watt R, Stanton LW, Marcu KB, Gallo RC, Croce**

CM, Rovera G (1983) Nucleotide sequence of cloned cDNA of human c-myc oncogene. Nature (Lond.) 303:725–728.

285. **Weinberg RA** (1985) The action of oncogenes in the cytoplasm and nucleus. Science 230:770–774.

286. **Weiss MC, Green H** (1967) Human–mouse hybrid cell lines containing partial complements of human chromosomes and functioning human genes. Proc. Natl. Acad. Sci. USA 58:1104–1111.

287. **Wensink PC, Finnegan DJ, Donelson JE, Hogness D** (1974) A system for mapping DNA sequences in the chromosomes of *Drosophila melanogaster.* Cell 3: 315–325.

288. **White R, Leppert M, Bishop DT, Barker D, Berkowitz J, Brown C, Callahan P, Holm T, Jerominski L** (1985) Construction of linkage maps with DNA markers for human chromosomes. Nature (Lond.) 313:101–105.

289. **Wigler M, Silverstein S, Lee L-S, Pellicer A, Cheng T, Axel R** (1977) Transfer of purified herpes virus thymidine kinase gene to cultured mouse cells. Cell 11: 223–232.

290. **Wigler M, Sweet R, Sim G-K, Wold B, Pellicer A, Lacy E, Maniatis T, Silverstein S, Axel R** (1979) Transformation of mammalian cells with genes from procaryotes and encaryotes. Cell 16:777–785.

291. **Wirschubsky Z, Perlmann C, Lindsten J, Klein G** (1983) Flow karyotype analysis and fluorescence-activated sorting of Burkitt-lymphoma-associated translocation chromosomes. Int. J. Cancer 32:147–153.

292. **Wolfe J, Erickson RP, Rigby PWJ, Goodfellow PN** (1984) Cosmid clones derived from both euchromatic and heterochromatic regions of the human Y chromosome. EMBO J. 3:1997–2003.

293. **Wyman A, White R** (1980) A highly polymorphic locus in human DNA. Proc. Natl. Acad. Sci. USA 77:6754–6758.

294. **Yancopoulos GD, Nisen PD, Tesfaye A, Kohl NE, Goldfarb MP, Alt FW** (1985) N-myc can cooperate with ras to transform normal cells in culture. Proc. Natl. Acad. Sci. USA 82:5455–5459.

295. **Young BD, Ferguson-Smith MA, Sillar R, Boyd E** (1981) High-resolution analysis of human peripheral lymphocyte chromosomes by flow cytometry. Proc. Natl. Acad. Sci. USA 78:7727–7731.

296. **Yunis JJ** (1983) The chromosomal basis of human neoplasia. Science 221:227–236.

297. **Yunis JJ, Ramsay N** (1978) Retinoblastoma and subband deletion of chromosome 13. Am. J. Dis. Child. 132:161–163.

29

Flow Cytometric Studies of Microorganisms

Harald B. Steen

Department of Biophysics, Institute for Cancer Research, The Norwegian Radium Hospital, Montebello, Oslo 3, Norway

During the past decade, flow cytometry has contributed significantly to the biology of eukaryotic, and primarily mammalian, cells. In comparison, it has had remarkably little impact on microbiology, in spite of the fact that microbiology, and bacteriology in particular, presents a number of basic problems that in practice can be solved only by the kind of precise measurement of individual cells in large numbers facilitated by flow cytometry. This situation is primarily attributable to the technical problems arising because of the small size of many microorganisms. But it may also have to do with the fact that the type of data generated by flow cytometry is not always readily comprehended by people who study cells in bulk by biochemical methods. As a result, there has been little pressure from microbiologists to develop the appropriate instrumentation and methodology.

Although various types of microorganisms have been measured by flow cytometry, the number of papers on the subject is limited, and much of the work that has been published gives the impression of pilot studies intended to demonstrate the feasibility of the method rather than to apply it to specific biological problems.

Most of the work has been on yeast and bacteria; yeast are technically simpler to study because of their larger size and DNA content, yet they are roughly 200 times smaller than a diploid human cell [30]. The cell cycle of yeast has been studied by several investigators measuring DNA and RNA as well as protein and light scattering [1,16–24,46]. The feasibility of such measurements on other eukaryotic microorganisms, including protozoa, algae, and molds has also been demonstrated [5,11,16]. Hutter et al. measured effects of various heavy metals on such organisms [19,20,24]. They also showed that different strains of yeast may be distinguished by flow cytometry using FITC-conjugated antiserum [22].

Paau et al. measured the nucleic acid content and size of three types of bacteria, including *Escherichia coli,* using ethidium bromide to stain the cells [36,37]. They found that RNA-bound dye contributed significantly to the fluorescence. Treatment of the cells with RNase had an appreciable effect but did not produce histograms of sufficient resolution to facilitate cell-cycle analysis. A few other papers reported similar results [2,12,15]. None of these provided any new information about the bacteria, primarily because the quality of the histograms did not justify quantitative analysis. The inability to obtain histograms with satisfactory resolution may be the reason why these studies were not continued.

This chapter deals only with flow cytometric work done on prokaryotic organisms and on bacteria in particular. Flow cytometry of bacteria has several obvious applications in both basic and applied bacteriology. Applications in basic research include primarily studies of cell-cycle kinetics under various conditions and the regulatory mechanisms that govern cell growth and replication. The cell cycle of bacteria is not as well understood as that of mammalian cells. This is partly because the cell cycle of bacteria is both more complicated than that of mammalian cells and is much more dependent on the growth conditions. It is also attributable to a lack of appropriate experimental methods. Thus, several basic problems with regard to how bacteria grow and multiply have not been resolved. And the knowledge that we do have is limited almost exclusively to one species: *E. coli.* Our recent flow cytometric studies of *E. coli* have shown that this knowledge has to be modified on several points. These studies also indicate why it has been difficult to reach conclusive results in studies of the bacterial cell cycle by conventional methods. Thus, it appears that reproducible results are more dependent on strictly controlled growth conditions than is the case in corresponding experiments with mammalian cells. Although this complicates experimental work, it does suggest a field of study that has interesting aspects in both basic and applied bacteriology. Furthermore, we have found that the staining of bacterial DNA is somewhat dependent on the state of the cells when they are harvested for investigation. This may create problems in some experiments. By contrast, it points to the interesting possibility of studying the distribution and/or conformation of the chromatin in individual cells.

In clinical bacteriology, flow cytometry may replace some of the rather slow and cumbersome assays currently used for detection and identification of bacteria, for testing of susceptibility to antibiotics and assessment of infectiousness. The

Flow Cytometry and Sorting, Second Edition, pages 605–622

time required for these various assays, and for testing of antibiotic susceptibility in particular, may be reduced significantly. Flow cytometry has an obvious potential for automation of several of the work intensive manual assays now in use, and could yield additional information in the process. It also has interesting possibilities in several areas outside biomedicine. For example, it may be used to monitor bacteria, as well as other microorganisms, employed in industrial processes or occurring as contamination in materials from drinking water to crude oil.

Technically, the main problem with flow cytometric measurement of bacteria is their small size and the correspondingly high demands on instrument sensitivity. The DNA content of the *E. coli* genome is about 4×10^6 base pairs (bp) [13], which is about 10^3 times smaller than that of a diploid human cell, or some 40 times smaller than the DNA content of an average human chromosome. Its volume is also roughly three orders of magnitude smaller than that of human leukocytes [9], which is to say that the surface area to be characterized by antibodies is about two orders of magnitude below that of human cells. Hence, to measure the contents of DNA and other constituents of bacteria by flow cytometry calls for the brightest dyes and highly sensitive instruments. The obvious choice for a DNA specific dye may thus be ethidium bromide or propidium iodide, excited preferentially by the 514-nm line of a large argon laser. This combination was indeed employed in the first flow cytometric studies of bacteria [2,36,37]. However, these dyes also have some affinity for RNA. And since the RNA content of bacteria may be relatively higher than that of mammalian cells [7,8], RNA associated fluorescence will tend to blur the DNA fluorescence histograms. So far, it has proved impossible to eliminate this problem by RNase treatment of the cells prior to staining. Consequently, highly DNA-specific dyes, such as Hoechst 33258, DAPI, or mithramycin, must be used. However, the argon laser does not have sufficient intensity at the wavelengths required to excite these dyes. Another possible reason why it has been difficult to obtain highly resolved DNA histograms of bacteria, may be the nonspherical shape of many species. The rodlike shape of *E. coli* and other species may produce orientational artefacts especially in laser-based instruments.

Nevertheless, the last few years have shown that high-resolution DNA histograms of bacteria can be obtained. The data recorded in these recent experiments make it reasonable to expect that as appropriate instruments become available flow cytometry may become as important in microbiology as it has been in mammalian cell biology. The following discussion concentrates primarily on these recent data.

INSTRUMENT AND METHODS

Instrument

The main reason flow cytometry has not penetrated into bacteriology is the lack of sufficiently sensitive instruments. The major emission lines of the argon laser used in laser-based flow cytometers do not coincide well with the absorption of the highly DNA-specific dyes that must be used in order to avoid fluorescence from dye bound to RNA. The high-pressure mercury lamp does have strong emission lines close to the wavelength of the peak absorption of several of these dyes, such as Hoechst 33258, chromomycin A_3, and mithramycin. However, the various types of commercial flow cytometers using such lamps do not appear to have the sensitivity required to yield histograms of sufficient resolution.

A laboratory-built instrument with arc lamp excitation, which has yielded high-resolution DNA and light-scattering histograms of various types of bacteria, is shown in Figure 1.* The main reason for the high sensitivity of this instrument is the low level of background fluorescence and scattered light produced by the type of flow chamber employed [47]. The result is a high signal-to-noise [S/N] ratio, further enhanced by the use of adjustable slits that eliminate background more efficiently than the fixed "pinhole" employed in other instruments. Excitation intensity is maximized by the use of illumination optics, which give critical illumination. Light-scattering detection is facilitated by a dark-field configuration [48,49]. This device produces a dark-field image of the cells, which exhibits very high contrast. The resulting S/N ratio is sufficient to measure particles with diameters to below 0.2 μm. At the signal level produced by *E. coli,* light-scatter resolution corresponds to a coefficient of variation (CV) below 2%. The light-scattering signal in this instrument has been found to be approximately proportional to the total protein content of the cells [6,51], as shown in Figure 2, and may thus be taken as a measure of protein content. The commercial version of this instrument also facilitates measurement of light scattering separately at low and high scattering angles. This makes it possible to distinguish different types of bacteria on the basis of light scattering only.

Staining

With respect to staining of DNA, the best results have been obtained with a combination of mithramycin and ethidium bromide [3,41] subsequent to fixation in 70% ice cold ethanol. The excitation light is mainly the 436-nm mercury line, which coincides with the peak absorption of mithramycin, whereas ethidium bromide bound to RNA contributes negligibly to the fluorescence. Van Dilla et al. [54] obtained histogram peaks with good resolution using Hoechst 33258 and chromomycin A_3, a close analogue to mithramycin. We have found that Hoechst 33258, excited by 366-nm light, yields approximately the same fluorescence intensity as the combination of mithramycin and ethidium bromide, but the histogram peaks are not so narrow. With Hoechst 33258, the smallest peak widths correspond to CV values around 6%, which may be compared with about 3% in the best histograms obtained with mithramycin–ethidium bromide.

Except for the high demands for instrument sensitivity, bacteria have proved quite suitable for flow cytometry. Typically, they are easy to obtain in large numbers, they are rugged, simple to prepare, do not easily aggregate and keep well in storage.

CELL-CYCLE KINETICS

Cell-Cycle Kinetics of *E. coli*

Whereas the natural growth parameters of mammalian cells are strictly controlled by the particular type of cell, bacteria naturally adapt to a wide variety of growth conditions. This may be the reason why the cell cycle of bacteria, as is known mainly from studies of *E. coli,* is much more variable and flexible than that of mammalian cells. For ex-

*A commercial version of this instrument, Argus, is now available from Skatron, Tranby, Norway.

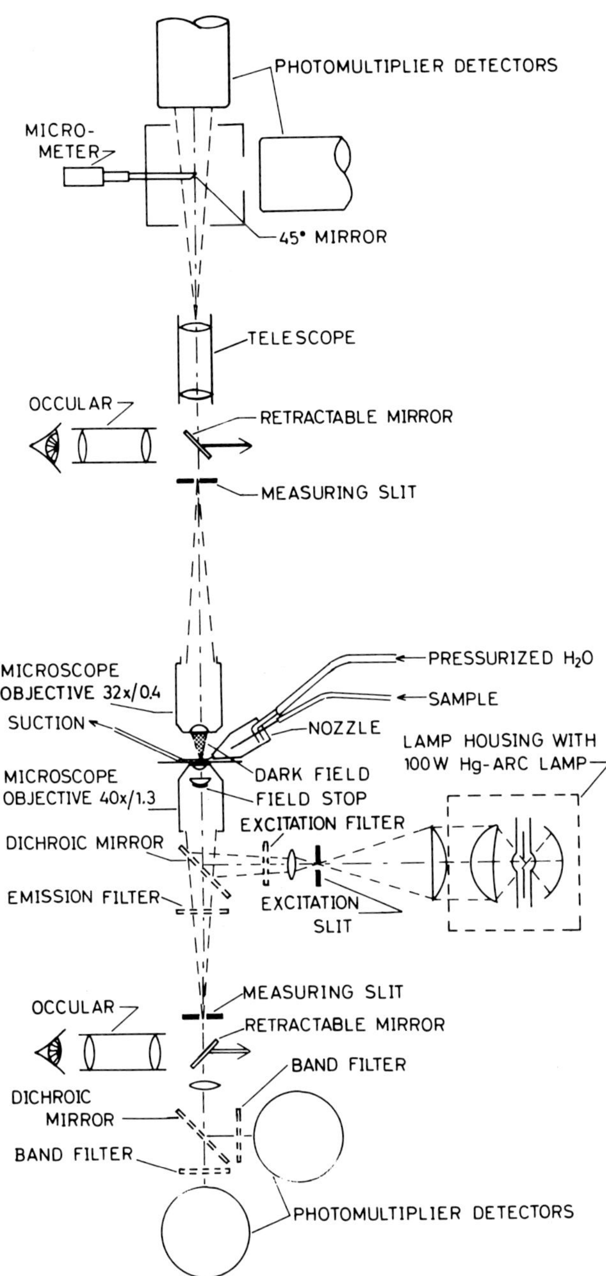

Fig. 1. Arc lamp-based flow cytometer used for bacteria measurements. The arc of the mercury lamp is imaged on the excitation slit, which in turn is imaged in the object plane with the sample flow. The excitation slit may thus be set such that the illumination field is limited to the area immediately around the sample flow, reducing background. The fluorescence and dark-field image of the sample flow are formed on measuring slits set so as to eliminate light from other parts of the object field. The telescope in the dark-field microscope forms an image of the entrance pupil of the dark-field microscope objective. Being behind the measuring slit, this image exhibits the angular distribution of the light scattering of the cells, whereas background is largely eliminated. Thus, the small mirror in the central part of this image reflects light scattered from angles ≥ 15° onto one photomultiplier, whereas the rest of the light, which is dominated by low scattering angles (≥ 2°) falls on the other photomultiplier. For details, see refs. 50 and 53 (see also Chapter 2, this volume). A commercial version of this instrument is now available.

ample, the cell doubling time of *E. coli* may vary from about 20 min under optimal growth conditions to many hours in a poor medium.

The cell cycle of bacteria is divided into phases corresponding to the G_1, S, and G_2 phases used to describe the cell cycle of mammalian cells. The period of chromosome replication is denoted C phase, the prereplication phase B, and the postreplication phase D. In contrast with the case of mammalian cells, the duration of each of these phases may vary greatly with growth conditions. Moreover, under certain conditions (i.e., during rapid growth), the phases may overlap considerably, that is so that the C and D phase of one cell cycle may extend into the next cycle. This is possible because the replication of the bacterial chromosome does not have to be complete before a new round of replication of the two new chromosomes commences. In other words, replication forks may form on each branch of a replication fork. The bacterial genome is contained in one "circular" chromosome. The replication starts at a point called *ori* C and runs at the same rate in both directions until the point of terminus, which is just opposite *ori* C.

Cooper and Helmstetter Model

The currently accepted model of the cell cycle of bacteria, as represented by *E. coli*, was proposed by Cooper and Helmstetter in 1968 [10]. Considering the scarce and partly indirect data available at the time, their work, in my opinion, was a remarkable achievement. Their model was intended to cover *E. coli* growing under near-optimal conditions, i.e., with doubling times below 60 min. The basic assumptions of the model may be summarized as follows: First, the rate of movement of the chromosome replication forks is independent of the cell cycle time, T, and it is the same in all replication forks. For *E. coli*, this rate of DNA synthesis corresponds to replication of the chromosome in about 40 min. In other words, the C phase lasts about 40 min independent of T. Second, after termination of the replication of a chromosome, there is always a period, the D phase, before cell division can occur. The duration of this period is independent of T. For *E. coli* it is about 20 min. Third, initiation of chromosome replication may commence before the previous round of replication is completed. Finally, initiation of replication in the two arms of a replication fork is synchronous.

Examples of chromosome replication patterns according to this model are depicted in Figure 3. As may be seen from these patterns, chromosome replication in cells having T below 40 min goes on throughout the entire cell cycle. For example, in cells in which T = 40 min, replication of the chromosome is initiated in the middle of one cycle and continues until the middle of the following cycle. As a consequence, DNA histograms of such cultures should not exhibit distinct peaks representing the B and/or D phase. It also follows from the model of Cooper and Helmstetter that the average DNA content of the cells increases with decreasing cell cycle time. The maximum number of chromosomes, according to this model, should be 6, that is, when T is the minimum: 20 min.

We have tested the model designed by Cooper and Helmstetter quantitatively by fitting the computer-simulated histograms based on the model to observed ones [42], as shown in Figures 4 and 5. The results suggest that under the growth conditions used in these experiments the model is largely correct. The best-fit values of the cell cycle parameters C and D obtained in this study (Table 1) are in quantitative accordance with the results of Cooper and Helmstetter [10]. In

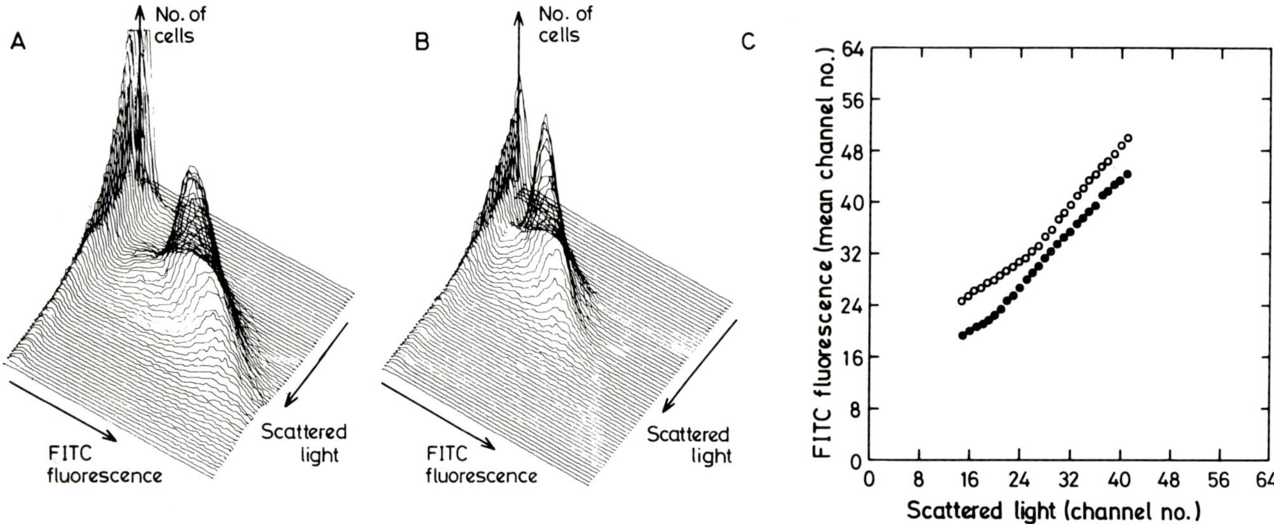

Fig. 2. Correlation between light scattering and protein content of *Escherichia coli* grown in the absence **(A)** and presence **(B)** of chloramphenicol, which inhibits protein synthesis and thereby reduces the protein–DNA ratio of the cells. Cells were fixed in ethanol and labeled with FITC under conditions in which the amount of bound FITC is proportional to total protein content [33]. **C:** Graphs showing that light scattering is roughly proportional to protein content.

some histograms, however, there is a noticeable difference between the experimental and theoretical histograms. This difference, which occurs around DNA values corresponding to the point of initiation of new chromosomes, may be explained if it is assumed that the rate of DNA synthesis is somewhat lower during the first few minutes of the chromosome replication cycle.

Synchrony of Chromosome Replication

One of the assumptions underlying the model of Cooper and Helmstetter is that initiation of replication of parallel chromosomes within the same cell is synchronous. The validity of this assumption may be checked by the use of substances that inhibit initiation while replication is allowed to continue. Therefore, if such a drug is added to an asynchronously growing culture, all cells should eventually end up with an even number of chromosomes, if the assumption is correct, provided cell division is inhibited as well. Figure 6A shows such a histogram as obtained for *E. coli* B/rA treated with rifampicin [43]. The great majority of the cells are indeed in the two and four chromosome peaks, indicating that the assumption of synchronous initiation is largely correct. However, the histogram exhibits a noticeable three-chromosome peak, suggesting that the initiation synchrony in these cells is not perfect. On the basis of the percentage of the cells in the three-chromosome peak we calculate that initiation of parallel chromosomes in these cells occurs within about 3 min [43]. Similar results were found for other strains of *E. coli*.

However, not all strains exhibit such synchrony, as shown for several dna A-deficient mutants (Fig. 6B–D). Note that the degree of asynchrony may vary from a few percent (Fig. 6A) through intermediate cases (Fig. 6C,D) to almost total asynchrony (Fig. 6B). Detailed flow cytometric studies of this phenomenon in a variety of dna A and rec A mutants of *E. coli* have given new information on the molecular mechanisms of control of chromosome replication in bacteria [44,45].

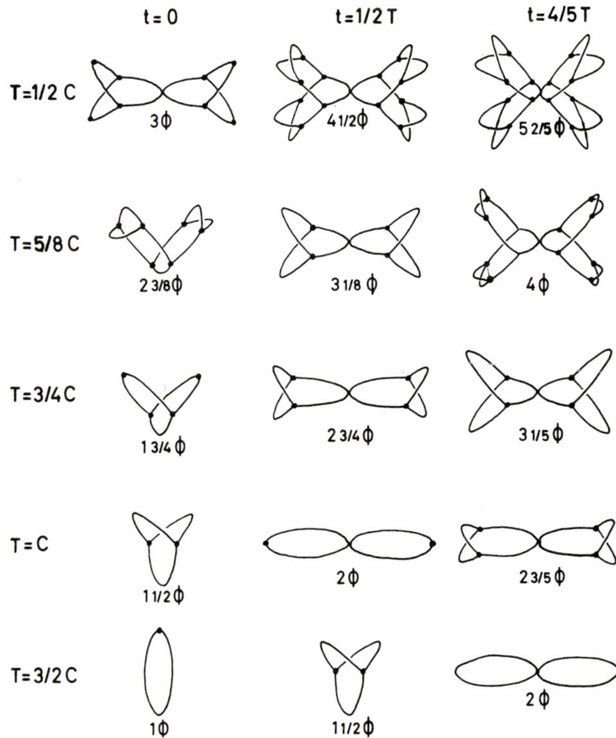

Fig. 3. Chromosome replication patterns according to the model of Cooper and Helmstetter [10]. The patterns are drawn for three different positions, t, in the cell cycle and for various cell-cycle times, T, given in units of the time, C, required to replicate one chromosome. In accordance with refs. [10] and [42], it was assumed that the cells require a postreplication period D = C/2 after completion of a chromosome before cell division can occur. The DNA content in units of chromosome equivalents, ø, is given for each pattern. Replication forks are represented by black dots.

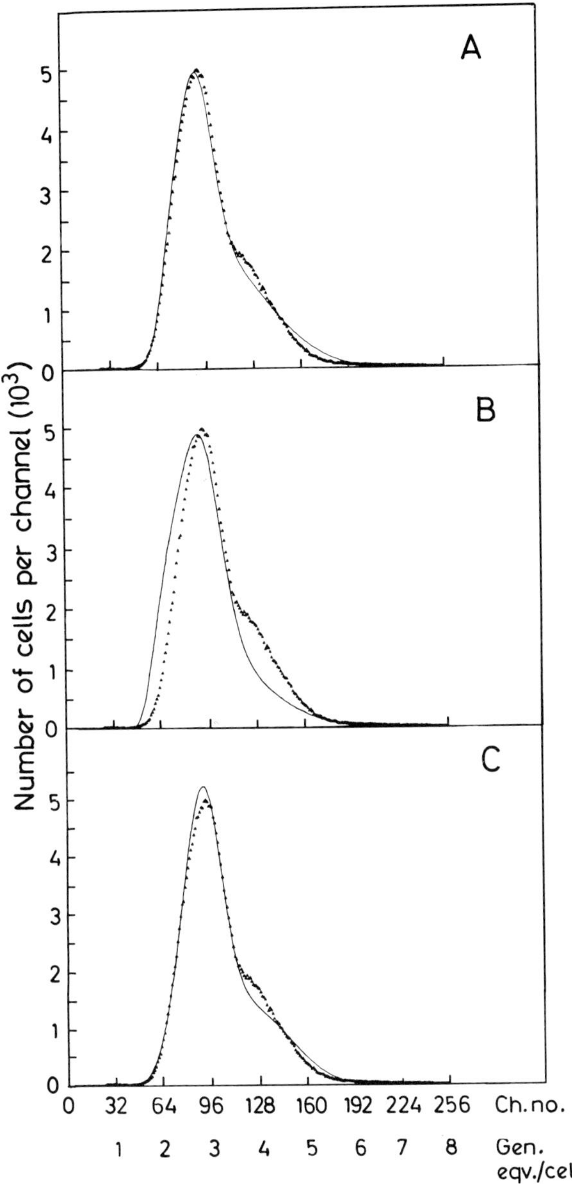

Fig. 4. DNA histogram (▲), as obtained for a batch culture of *Escherichia coli* B/rA growing exponentially in a synthetic (K-glucose) medium with a doubling time of 27 min, compared with three different histograms (solid line) generated by a computer simulation based on the model designed by Cooper and Helmstetter [10]. **A,C:** Best-fit histograms with two different models of biological variation (see Table 1). **B:** Simulated histogram generated with cell-cycle parameters 5–10% off the best-fit values to demonstrate the "sensitivity" of the fitting. (Reproduced from Skarstad et al. [42], with permission from the publisher.)

Deviations from the Model of Cooper and Helmstetter

Although the model of Cooper and Helmstetter appears to be largely correct for some experimental conditions it does not seem valid in general, even for rapidly growing cells. Data recorded for cells growing under somewhat different

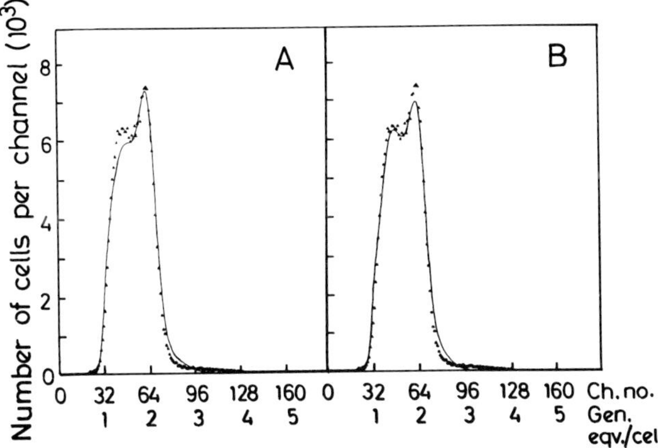

Fig. 5. DNA histogram (▲), obtained for a chemostat culture of *Escherichia coli* B/rA with a cell doubling time of 60 min, compared with computer-simulated histograms (solid line) as generated for two different models of biological variation (see Table 1). (Reproduced from Skarstad et al. [42], with permission from the publisher.)

conditions than those referred to above, seem to be at variance with one of the basic assumptions of the model, namely that the duration of the cell cycle phases B, C, and D is unambiguously determined by the cell cycle time, T. This assumption implies that the cell cycle distribution of an asynchronous exponential culture should remain constant as long as T is constant. As shown in Figure 7, this is not always the case. These data demonstrate a significant variation in the cell cycle distribution during the period of exponential growth. Similar data were obtained under different growth conditions and other cell-doubling times [6,41]. Generally, the average DNA content of the cells decreases during the period of exponential growth. At no point in the experiment of Figure 7, 8, and 9 is the DNA histogram in accordance with that predicted by the model (Fig. 9). Especially, at the higher cell densities (OD > 0.1), which are still well within the exponential range, the experimental data differ dramatically from the predictions of the model. Since the cell doubling time remains constant, the data in Figure 9 implies that the duration of either the C phase or the D phase or both decreases significantly as the cell density increases.

Another aspect of the data in Figure 9 is the gradual reduction in the light scattering of the cells during exponential growth (Fig. 10). Thus, it appears that the average protein content of the cells decreases along with the content of DNA. This result has an important practical implication. The standard method for determination of cell doubling time in bacteria cultures is to measure the optical density at visible wavelengths where absorption is negligible. Thus, optical density is essentially a measure of low-angle light scattering. On the assumption that the light scattering is proportional to cell mass [28] (see also Fig. 2) the OD doubling time is equivalent to mass doubling time, which in turn is usually taken to mean cell doubling time, on the (tacit) assumption that the light scattering per cell remains constant during exponential growth. The data presented in Figures 7 and 10 demonstrate that the latter assumption is not generally correct. According to these data the doubling time based on OD measurements should be larger than the "true" value based on direct measurement of cell numbers. This expectation was borne out by the results in Figure 8, which show that the cell doubling

TABLE 1. Best-Fit Values for Cell-Cycle Parameters of *Escherichia coli* Obtained by Fitting Computer-Simulated DNA Histograms[a] to Experimentally Observed Histograms[b,c]

Variation model[d]	T (min)	C (min)	D (min)	s[e]	h (min)	S (%)	Fig.
1	27	43	23	2.5	3.1	—	5A
1	27	40	21	7.1	3.0	—	5B
2	27	43	22	2.9	—	9	5C
1	60	43	26.5	3.0	7.7	—	6A
2	60	46	22	3.3	—	12	6B

[a]Based on the model designed by Cooper and Helmstetter [10].
[b]Shown in Figures 5 and 6.
[c]Data from Skarstad et al. [42].
[d]Two different models for the biological variation were incorporated into the computer simulation. Variation model 1 assumes that the postreplication period D consists of a constant period subsequent to which the probability (h^{-1}) of division per cell and unit time is constant. Variation model 2 assumes a normal distribution of the cell cycle time, T, having a relative width, S.
[e]s is a measure of the deviation between the experimental and the simulated histogram.

time obtained from cell count is considerably below that found from measurements of optical density, i.e., light scattering. It may also be noted from the results in Figure 8 that the cell doubling time obtained from cell count is well below the 20 min usually considered to be the minimum value for *E. coli.*

The biological mechanisms underlying the variation in the cell-cycle distribution during exponential growth are unknown. General medium depletion may be excluded since even the highest cell density in these experiments was below 1% of that reached at stationary state of growth. Nevertheless, it is conceivable that the medium may be depleted in some specific component which is of importance for the regulation of the cell cycle. Alternatively, it is possible that the cells excrete growth factors which affect the cell-cycle distribution. In any case, further studies of these effects may aid our understanding of the regulation of the cell cycle and open up new possibilities to control bacterial growth for experimental as well as for industrial purposes.

Slow Growth

According to the model of Cooper and Helmstetter, the cell cycle of *E. coli* becomes more similar to that of mammalian cells when the bacteria are growing slowly, i.e., with doubling times above 60 min. In that case, there should be no overlap between the C phase of one cell cycle and the next, the resulting DNA histogram should exhibit a discrete peak representing the D phase and possibly a peak representing the B phase as well. It would appear that this situation should be easier to study by conventional methods than rapid growth. However, especially the duration of the B phase has been a matter of controversy. Thus, it has proved difficult to agree on the quantitative aspect, in particular with regard to the relative duration of the B and D phases. Some workers reached the conclusion that as T increases the B phase remains short or nonexistent while C increases [14,29]. Others have found that the B phase increases more than D phase [31,32]. The flow cytometric data definitely support the latter conclusion. DNA histograms of slowly growing cells (Fig. 11) show that as growth slows down the B phase becomes increasingly dominant (Table 2). The C and D phases appear to remain constant for cell cycle times up to above 90 min. With slower growth, however, also the C and D phase increase. The rate of protein synthesis, as judged from the rate of increase of the light scattering of cells growing very slowly, appears to be the lowest in the B phase and increases through the cell cycle (Table 3). Quantitatively, different strains of *E. coli* may be significantly different in this regard.

Another matter of controversy has been whether *E. coli* in slow-growing cultures require a certain minimum size or protein content before they can proceed from the B phase into the C phase. The histogram in Figure 11 suggests that this is the case. It appears that the great majority of cells do not enter the C phase before they have reached a certain light scattering value, i.e., protein content.

The histogram represented in Figure 11 exhibits another interesting feature, namely the "ridge" extending toward large light-scattering values from the single chromosome peak. Apparently, a certain fraction of the cells grow to much above normal size before they enter C phase. Closer analysis of this histogram shows that these large cells do eventually enter C phase and that they pass through it at approximately the same rate as the smaller cells. The origin of this subpopulation of larger cells is not known. However, it is one of many examples seen in the course of our flow cytometric studies of bacteria that cells within the same culture appear to exhibit distinctly different modes of growth. It may be speculated that such differentiation of the culture may have some survival value. Conceivably, the different modes of growth may provide alternative escape routes in case of crisis. In any case, this aspect of the histogram in Figure 11 exemplifies why it has been so difficult to obtain consistent data by means of the conventional, biochemical assays which yield data only in terms of average values for large numbers of cells.

Stationary Cultures

Many experiments in bacteriology have been based on the use of cells grown to stationary phase, assuming that under such conditions the majority of cells end up with one chromosome. As shown in Figure 12, this assumption does not hold true. In this particular case, there were more cells with two chromosomes than with one. Moreover, a significant fraction of the cells had an intermediate DNA content. Obviously, the assumption that growth into stationary phase leads to synchronization of the cells is not correct under these conditions. In our experience, DNA histograms of stationary phase cultures depend on growth conditions etc. However,

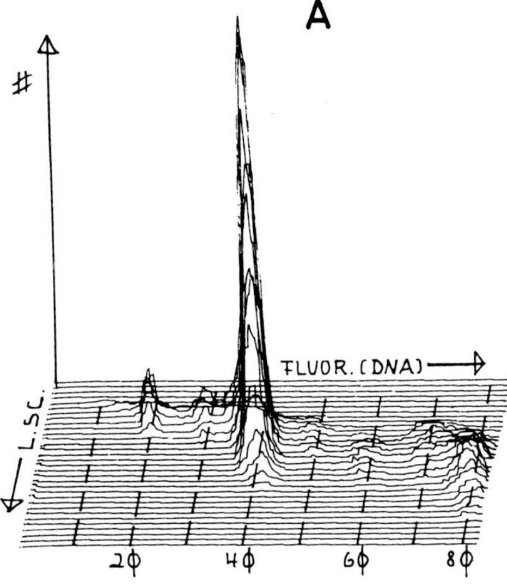

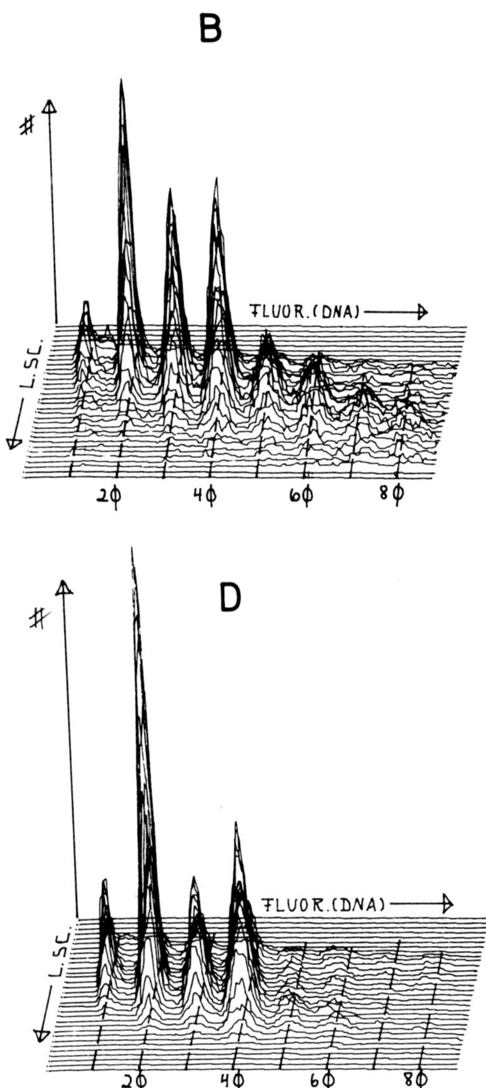

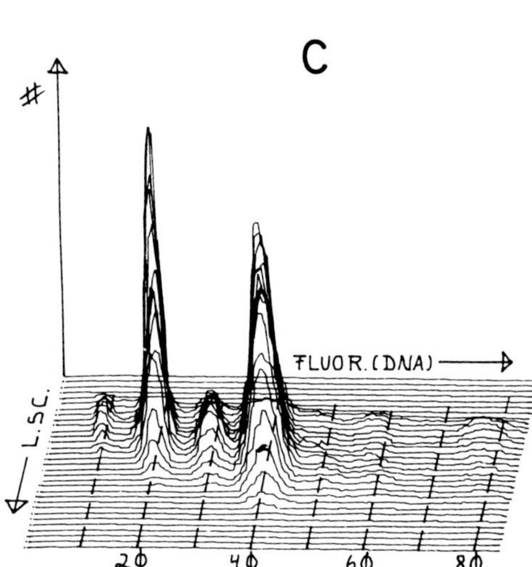

Fig. 6. Dual-parameter fluorescence (DNA)/light-scattering (protein) histogram of various strains of *E. coli* grown in the presence of rifampicin for 1.5 hours in Luria broth medium at 37°C. **A:** *Escherichia coli* B/rA. **B–D:** Different strains of DNA-deficient mutants. Integral chromosome numbers are indicated by broken lines. Perfect synchrony in the initiation of parallel chromosomes within the same cell should yield histograms with peaks only at 2, 4, and 8 chromosomes, whereas with total asynchrony, peaks with odd and even numbers of chromosomes should have approximately the same size.

we have found no case in which the cells were dominantly in the one chromosome peak.

EFFECTS OF ANTIBIOTICS

Flow cytometry has proved an invaluable tool in studies of how various agents, from radiation to cytostatic drugs, affect the cell cycle of mammalian cells. Experiments that we have carried out to study effects of antibiotics on bacteria suggest that in the future flow cytometry may play a similar role in corresponding studies of bacteria and other microorganisms.

Figure 13 shows dual parameter DNA/protein histograms of *E. coli* harvested from an exponential culture exposed to the antibiotic rifampicin for various periods of time. It may be seen that the drug gives very clear effects on the cell cycle distribution of these cells within 10 min of culture, which is about one-third of the cell doubling time (T = 27 min). Including the time required for sample preparation, a flow cytometric test of this kind can easily be completed within a half-hour. This time may be compared to the typical culture period of 24 hours required in the plaque assay currently used in clinical laboratories. In addition, flow cytometry yields much more information than the plaque assay, which only shows whether cells are able to grow or not. For example, the histograms in Figure 13 show that, in the presence of the drug, most cells accumulate in two peaks, representing cells with two and four chromosomes, respectively. By comparison with the histogram of uninhibited cells (Fig. 13A),

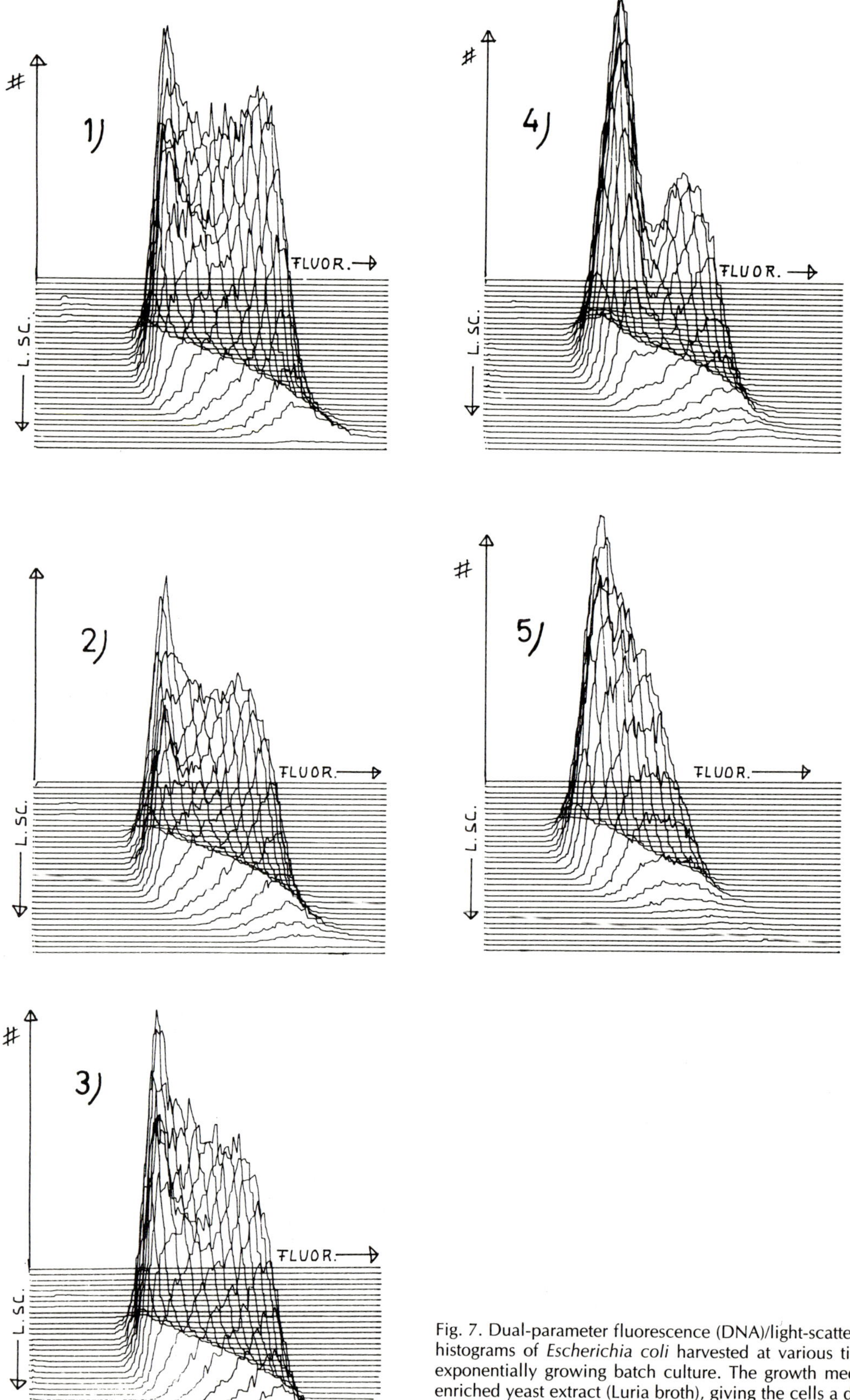

Fig. 7. Dual-parameter fluorescence (DNA)/light-scattering (protein) histograms of *Escherichia coli* harvested at various times from an exponentially growing batch culture. The growth medium was an enriched yeast extract (Luria broth), giving the cells a doubling time close to the theoretical minimum. Panel numbers refer to time of harvest as indicated in Figure 8.

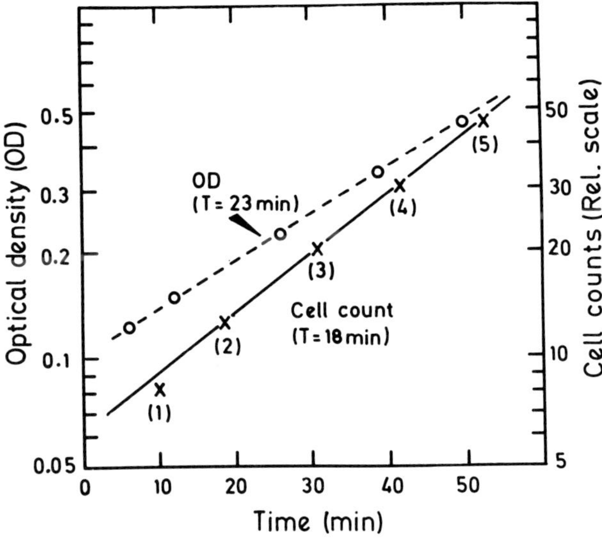

Fig. 8. Growth curves for the *Escherichia coli* culture described in Figure 7, obtained by measurement of the optical density (OD) of the culture at about 500 nm and by cell counting recorded in the flow cytometer.

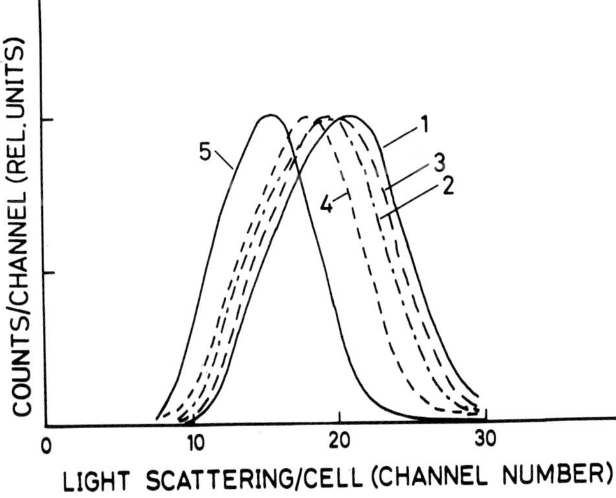

Fig. 10. Light-scattering (protein) distributions of the histograms in Figure 7. The numbers on the curves refer to the time of harvest as indicated in Figure 8.

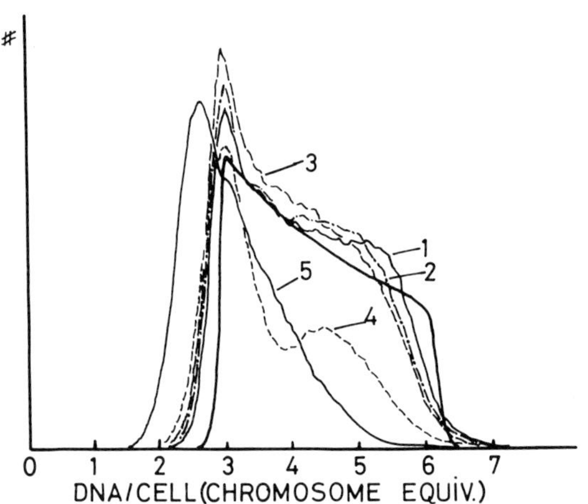

Fig. 9. DNA histograms obtained by integration ("collapsing") the histograms in Figure 7 with regard to light scattering. The numbers of the curves refer to the time of harvest as indicated in Figure 8. The heavy solid line is a theoretical histogram derived from the model designed by Cooper and Helmstetter for T = D = C/2 (the theoretical lower limit of the cell doubling time) assuming an experimental resolution corresponding to CV = 3% and no other biological variation.

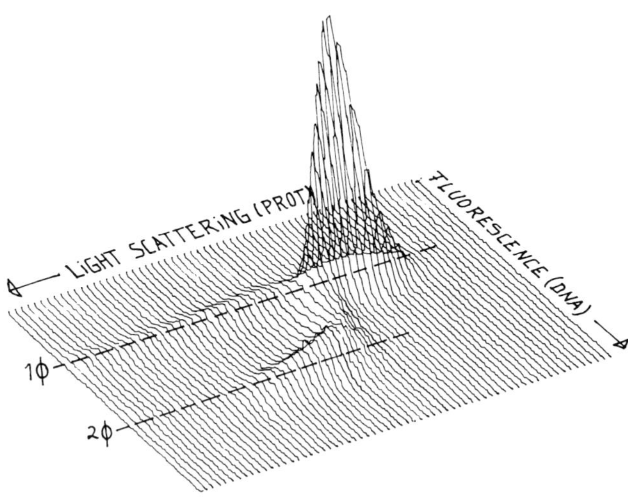

Fig. 11. Dual-parameter fluorescence (DNA)/light-scattering (protein) histogram of *Escherichia coli* B/rA growing exponentially in a chemostat culture with a cell-cycle time of 17 hours. The broken lines represent a DNA content corresponding to 1 and 2 chromosome equivalents, respectively. (Reproduced from Skarstad et al. [41], with permission from the publisher.)

it appears that cells having less than two initiated chromosomes when the drug was given accumulate in the two-chromosome peak, whereas cells with more than two initiated chromosomes accumulate in the four chromosome peak. Cells with more than four initiated chromosomes go on until they reach the eight-chromosome peak. Hence, it appears that the drug inhibits initiation of replication of new chromosomes, while DNA synthesis is allowed to continue. The rate of DNA synthesis, however, is somewhat reduced, as seen from the time required for the histogram peaks to develop. Cell division is also inhibited, as is evident from the absence of cells with only one chromosome, as well as from the observation that the total number of cells in each histogram was the same. Cell death appears to have been negligible, as indicated by the lack of counts close to the origin of the histograms. Such counts are caused by debris from decaying cells (see Fig. 15). Furthermore, it appears that the light scattering, i.e., the protein content of the cells, decreases in the presence of the drug. Most of this information may be superfluous or irrelevant in routine clinical tests of antibiotic susceptibility. However, in screening of new drugs and studies of their mechanisms of action, such data may be quite valuable.

All the data in Figure 13 and the above interpretation are in accordance with current knowledge about the action of

TABLE 2. Duration of Cell-Cycle Periods for *Escherichia coli* B/r A and B/r K Cells[a]

Strain	T	B	C + D	B (min)	C + D (min)	C (min)[b]	D (min)[b]
B/r A	73 min	0.15 T	0.85 T	11	62	—	—
	74 min	0.15 T	0.85 T	11	63	—	—
	95 min	0.25 T	0.75 T	24	71	—	—
	113 min	0.3 T	0.7 T	34	79	—	—
	236 min	0.5 T	0.5 T	118	118	—	—
	293 min	0.5 T	0.5 T	147	147	—	—
	17 hours	0.8 T	0.2 T	816	204	102	102
B/r K	231 min	0.4 T	0.6 T	92	139	—	—
	400 min	0.55 T	0.45 T	220	180	60	120
	16 hours	0.6 T	0.4 T	576	384	144	240

[a]Reproduced from Skarstad et al. [41], with permission of the publisher.
[b]Separate estimates of C and D are not given when T < 5 hours; in these histograms, the two-chromosome peak was not discernable.

TABLE 3. Average Rate of Mass Increase in the Cell-Cycle Periods of *Escherichia coli* B/r K (T = 16 hours) Cells[a]

Strain	Period	Avg. rate of mass increase (channel/hour)
B/r A	B	0.88
	C	2.4
	D	9.4
B/r K	B	3.1
	C	3.8
	D	6.3

[a]Reproduced from Skarstad et al. [41], with permission of the publisher.

rifampicin [27]. Rifampicin inhibits protein synthesis by binding to a subunit of RNA polymerase and thereby blocks initiation of RNA chains. Hence, the formation of the RNA primer at *ori* C is prevented and the initiation of chromosome replication cannot occur. In similar experiments we have studied the effects of several other antibiotics which act as inhibitors of protein synthesis at the ribosomal level, including chloramphenicol, erythromycin, doxycyclin, and streptomycin [6,51,52].

The absence of debris (counts near origin) in the histograms of Figure 13 indicates that the drug in this case acts as a bacteriostatic rather than a bactericidal agent. This contention is confirmed by the data in Figure 14. Some 45 min after the drug is removed, the first signs of renewed growth and DNA synthesis can be seen in the largest cells of the four chromosome peak. Cell growth and DNA synthesis appear to commence at about the same time. Cell division, giving rise to new cells below the four chromosome peak, is not evident until 90 min. For cells having the same number of chromosomes the smaller cells take much longer to resume growth than do the larger ones. Even after 150 min, the smallest four chromosome cells still remain quiescent. Eventually, however, essentially all cells grow again with a cell cycle distribution similar to that recorded for the exponential culture to which the drug was administered.

An example of flow cytometric analysis of another type of antibiotic, ampicillin, is depicted in Figure 15. Apparently, this drug causes the cells to grow to very large sizes and DNA contents. There is no sign of cell division. Presently, however, cell death becomes substantial, as indicated by the increasing amount of debris from decaying cells seen as counts at low channel numbers, and eventually there is no evidence of live cells at all (Fig. 15D). These observations are in accordance with the well-known action of ampicillin and other penicillins, which inhibit synthesis of the cell wall [55]. The result is filamentation, because the sequestering of cells at cell division cannot occur, and eventually there is cell death and decay as the cell wall breaks down. Somewhat similar results have been reported for the effects of several β-lactam antibiotics, which also cause filamentation by inhibition of cell division [35].

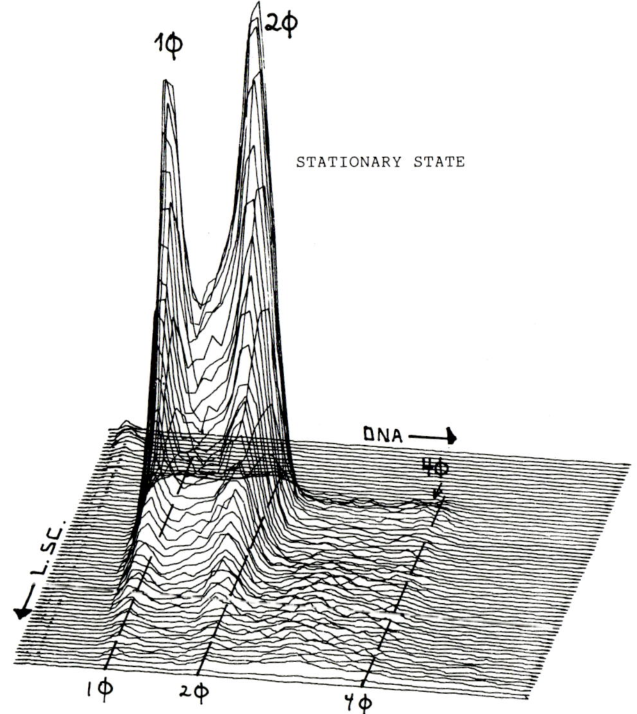

Fig. 12. Dual-parameter fluorescence (DNA)/light-scattering (protein) histogram of a batch culture of *Escherichia coli* about 12 hours after it reached stationary phase. DNA contents corresponding to 1, 2, and 4 chromosome equivalents are indicated by broken lines.

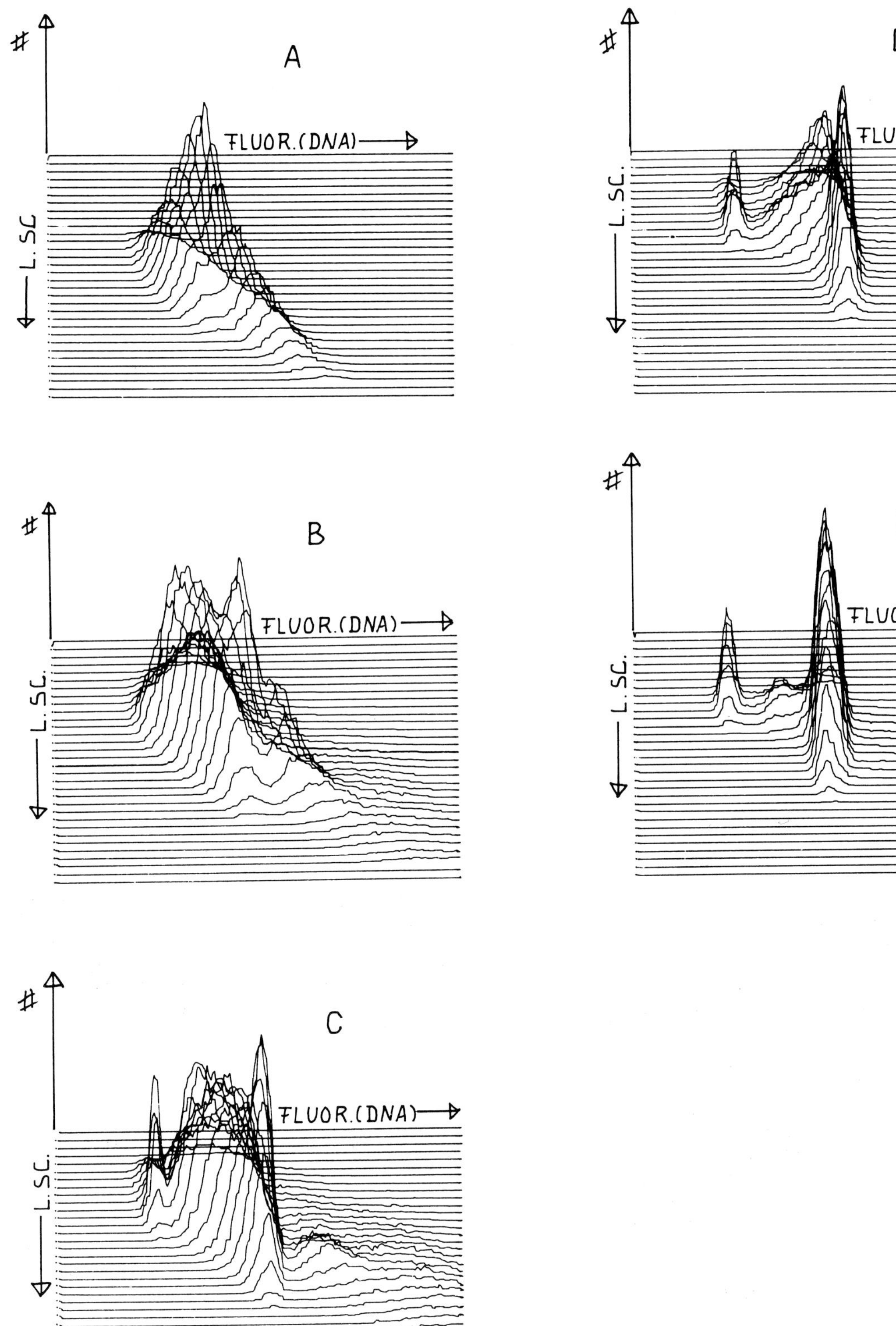

Fig. 13. Dual-parameter fluorescence (DNA)/light-scattering (protein) histograms of *Escherichia coli* grown in the presence of the antibiotic rifampicin for various periods of time. **A:** 0 min. **B:** 10 min. **C:** 20 min. **D:** 30 min. **E:** 90 min. The cells were growing exponentially with 27 min doubling time when the drug was given.

DETECTION AND IDENTIFICATION OF BACTERIA

Clinical Applications

Flow cytometry has several potential applications in other areas of clinical and practical bacteriology. One of these is the detection of bacteria in body fluids such as blood, cere-

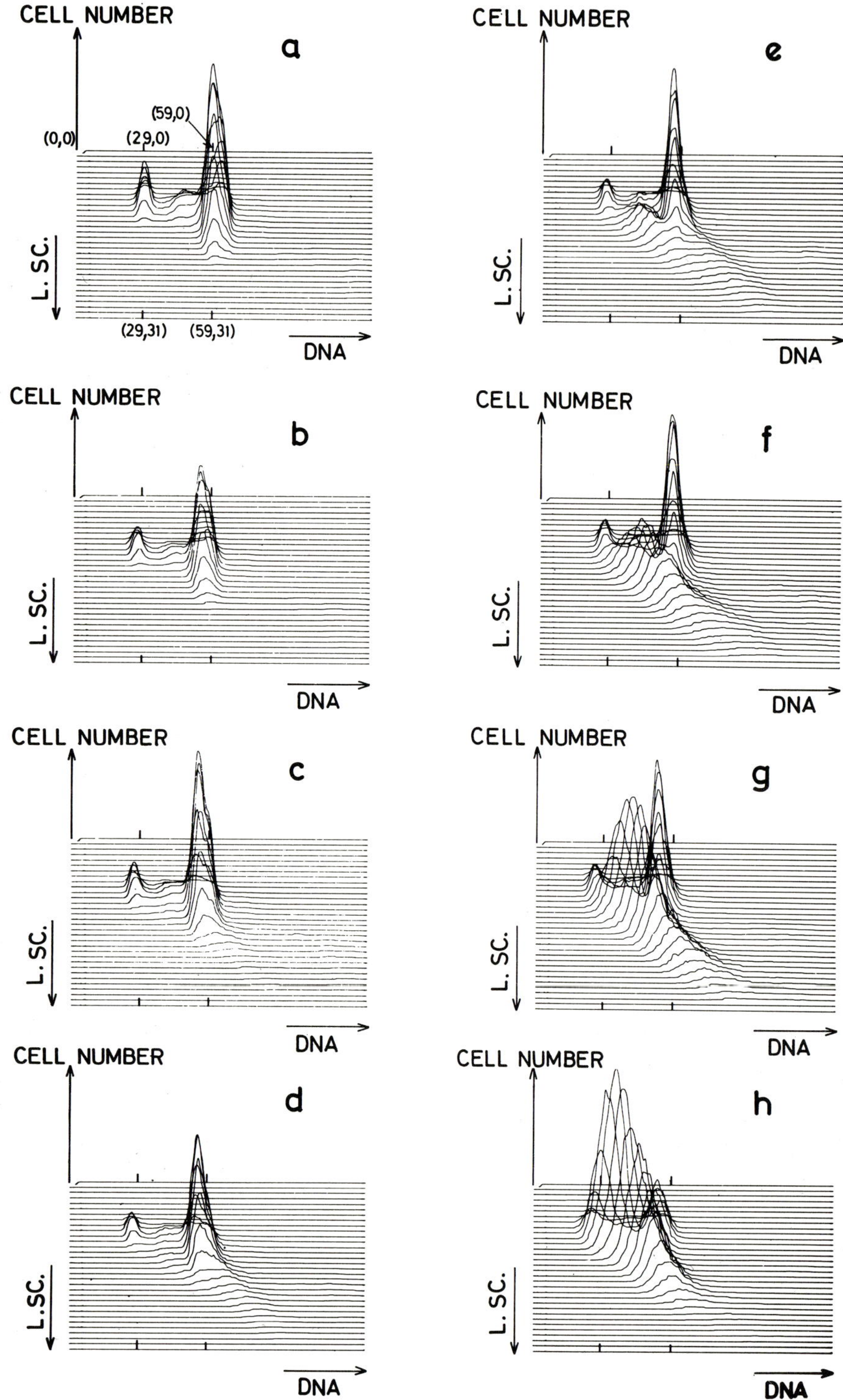

Fig. 14. Dual-parameter fluorescence (DNA)/light-scattering (protein) histograms of *Escherichia coli* grown in the presence of rifampicin for 90 min before being washed and subsequently grown in fresh medium for various periods of time. **a:** 0 min. **b:** 45 min. **c:** 60 min. **d:** 75 min. **e:** 90 min. **f:** 105 min. **g:** 120 min. **h:** 150 min.

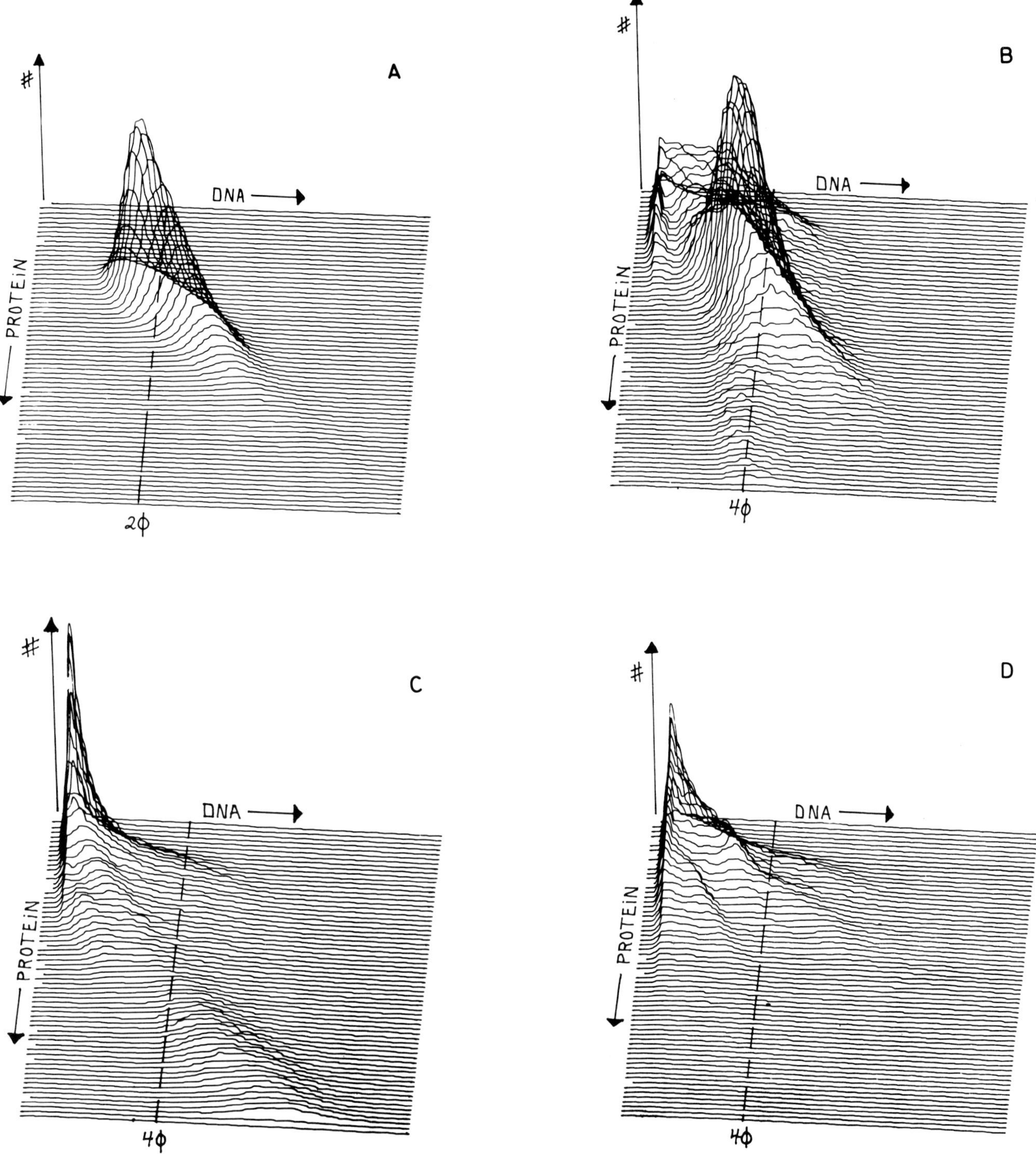

Fig. 15. Dual-parameter fluorescence (DNA)/light-scattering (protein) histograms of *Escherichia coli* grown in the presence of ampicillin for various periods of time. **A:** 0 hours. **B:** 1 hour. **C:** 2 hours. **D:** 4 hours. Counts close to origin are due to degenerating cells and debris. Note the difference in the scale of the fluorescence axis between **A** and **B–D.**

brospinal fluid, and urine. Sepsis, blood infection, is one condition which requires rapid assessment of antibiotic susceptibility. Current methods require that blood samples be inoculated into growth medium followed by an incubation period that may be several days before the bacteria have multiplied sufficiently to be detected and tested. This is longer than it may take for sepsis to reach a critical stage. Flow cytometry may alleviate this problem by its ability to detect a low number of organisms of particular cell type in the presence of large numbers of other cells. The limit of detection of bacteria is determined primarily by the number of flow cytometric parameters used to identify the bacteria and distinguish them from other cells and particles present in the sample. Using a cell-lysing agent to reduce the number of blood cells, Mansour et al. [34] were able to detect less than 100 bacteria per ml blood from animals infected with *E. coli,* and the detection limit for blood infected *in vitro* was of the order of 10 cells/ml. The bacteria were stained with ethidium bromide and identified by light scattering and two fluorescence components. This detection limit is still above the bacteria count in some cases of sepsis. However, it is several orders of magnitude below that achievable by conventional microscopic methods. Thus, although direct bacteria detection in fresh blood may be difficult, it seems possible that flow cytometry may at least facilitate much earlier analysis of the blood cultures used in the current assays. Moreover, it may be possible to lower the detection limit even further by the use of more specific markers, such as antibodies.

Urine is simpler to analyze partly because pathologic concentrations of bacteria are quite high, i.e., above 10^5 cells/ml, and partly because the concentration of other cells is relatively low. Although urinary tract infections are typically not fatal and the high number of bacteria makes antibiotic susceptibility testing relatively simple, flow cytometric analysis can provide rapid determination of the appropriate antibiotic and thereby reduce the number of consultations and shorten the clinical period. The feasibility of detection and identification of bacteria in urine by flow cytometry has been demonstrated [54].

In the case of meningitis, rapid identification and testing of bacteria in the cerebrospinal fluid may be of vital importance. The density of bacteria in such cases is relatively high, suggesting that flow cytometric analysis of antibiotic susceptibility as well as identification of the cells, e.g., by fluorescently labelled antibodies, should be feasible.

Identification of bacteria with regard not only to species but to strain, is a significant task for flow cytometry. At least in principle, bacteria may be characterized and identified by the same parameters as currently used for mammalian cells. The most specific method will probably be the use of antibodies, and monoclonal antibodies in particular. As shown in Figure 16, it is technically feasible to measure the amount of a species specific antiserum on individual bacteria. In this case, however, the surface density of the antigen was relatively high. A few other examples of species-specific antibodies detected by flow cytometry have appeared [4,25,38–40]. However, the surface density of strain specific antigens is likely to be considerably lower. If surface densities of relevant antigens on bacteria are comparable to those for mammalian cells, the sensitivity of current flow cytometers (detection limit of the order of 10^3 antibody molecules per cell (see Chapter 17, this volume) may not be sufficient to detect some bacteria surface antigens. By contrast, cellular autofluorescence, which in practice determines the detection limit for FITC-conjugated antibody bound to the cell, appears to be much weaker for bacteria than for mammalian cells. The introduction of new monoclonal antibodies for bacteria, which is now in progress, should give an indication of the feasibility of flow cytometry for this application within the next few years.

Another approach to the identification of bacteria by flow cytometry is based on differences in CG/AT base pair ratio of the bacterial chromosome in different species (5). Using a method originally developed to determine this ratio in eukaryotic chromosomes, Van Dilla et al. stained bacteria with Hoechst 33258 and chromomycin A_3 which bind preferentially to AT and CG, respectively, and measured the fluorescence of the cells with a dual laser flow cytometer in which they were excited sequentially as they passed through the two separate beams of the two large lasers tuned to the absorption maxima of the respective dyes [54]. With this technique, they were able to determine the fluorescence intensity ratio of the two dyes with a precision corresponding to a coefficient of variation between 15 and 20%, which may be compared with a variation in this ratio between different species of about a factor of 4. Thus, it appears that many species of bacteria may be distinguished quite effectively by such measurements. It seems less likely that strains of the same species, which typically differ much less in CG/AT, can be resolved by this method. Although instruments with dual laser excitation are commercially available, it is doubtful whether they are able to yield corresponding data of similar quality.

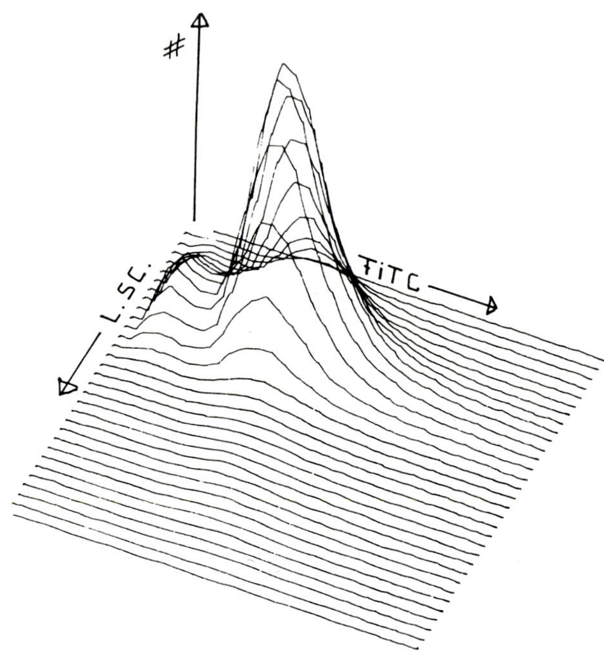

Fig. 16. Fluorescence (FITC)/light-scattering histogram of intact *Escherichia coli* labeled with a rabbit anti-*E. coli* antiserum and subsequently with a FITC-conjugated goat antirabbit antiserum.

In principle, bacteria species may be distinguished also from the size of the genome. However, since the DNA content of growing bacteria varies considerably, i.e., by at least a factor of two (see, e.g., Fig. 9), such measurements will probably be of little value in this regard. By contrast, the DNA–protein ratio remains relatively constant through the cell cycle (Fig. 13A) of exponentially growing cells, and its variation from cell to cell is much less than the variation in the individual parameters. For *E. coli* in exponential growth the variation in this ratio corresponds to a coefficient of variation typically below 10%, which may be compared with a variation between species of the order of 50%. It therefore appears that measurement of the DNA–protein ratio by means of flow cytometry may have some potential for distinction and identification of bacteria species. However, such

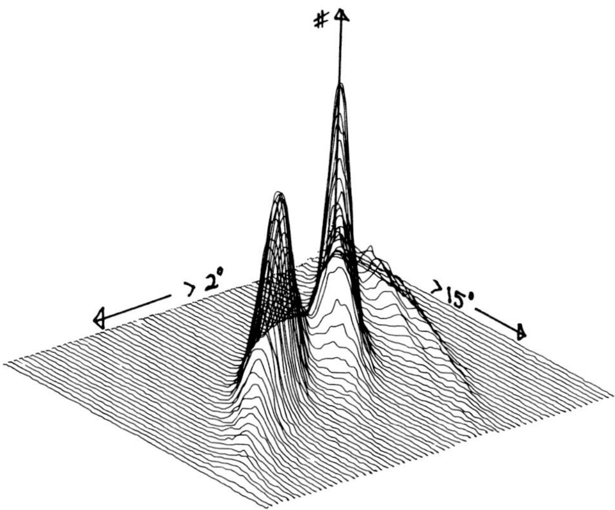

Fig. 17. Example of how different types of bacteria can be distinguished by small-angle and large-angle light scattering in dual-parameter mode. The sample consists of *Staphylococcus aureus* and *Pseudomonas aeruginosa,* both in exponential phase. The cell populations were grown separately and mixed immediately before measurement. The left and right peaks represent *Pseudomonas* and *Staphylococcus,* respectively. (Reproduced from Steen [50], with permission from the publisher.)

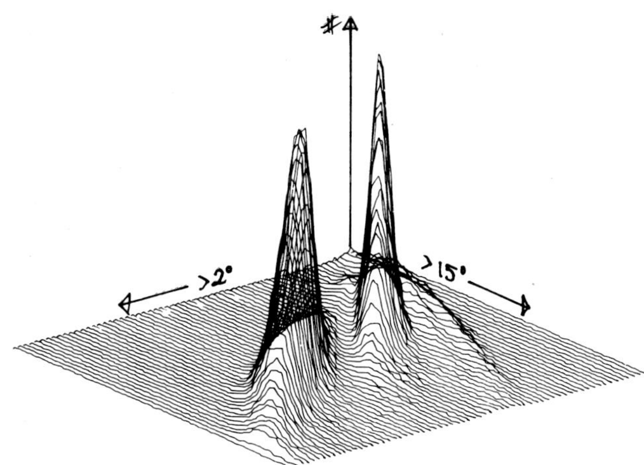

Fig. 18. Example of how bacteria in different states of growth can be distinguished by small-angle and large-angle light scattering in dual-parameter mode. The sample contains *Acinetobacter calcoaceticus.* The left peak represents cells in exponential growth, whereas the right peak represents cells grown to stationary state. The two cell populations were harvested from the same culture at different times and mixed immediately before measurement. The ridge to the right in this histogram and in Figure 17 is due to dead cells and debris.

a method would require well-defined growth conditions, since the DNA–protein ratio may vary considerably with the composition of the medium, cell density, and so forth.

Bacteria species also differ with regard to size, shape, cell wall thickness, and other structural characteristics [26]. Such structural differences are reflected in the angular distribution of the light scattering of the cells [28,57]. Variations in this distribution may be picked up in flow cytometers having facilities for detection of light scattering at two different regions of scattering angle, provided the sensitivity is sufficient. With the arc lamp-based flow cytometer used in our experiments (see Fig. 1), we are able to measure in two different regions of scattering angle, that is, upward from about 2° and upward from 15°, respectively [50,53]. The former region is dominated by diffraction from the main contour of the cell, whereas submicroscopic features, such as the thickness of the cell wall and granularity of the cytoplasm, are more important in the latter region. The relative scattering intensity in the two angle regions should also depend on the shape of the cell, e.g., the ratio of length versus diameter of rod-shaped cells. As demonstrated in Figure 17, different species of bacteria may differ significantly in the ratio of scattering intensity at the different scattering angles. As with the DNA–protein measurements, this light scattering ratio also depends on growth conditions, as shown by the difference between cells in exponential and stationary phase (Fig. 18). Thus, strictly defined growth conditions are a prerequisite for optimal utilization of light scattering for identification of bacteria.

Industrial Applications

Efficient monitoring of microorganisms as contaminants or natural constituents of a wide variety of industrial products has significant practical and economic value. Such products span the range from pharmaceuticals via milk and other beverages to jet fuel and crude oil. Flow cytometry has the potential for detection and identification of microorganisms in fluids faster and with higher precision and more information than is feasible with other methods. The ability of this technique to provide detailed characterization of individual cells in large numbers is likely to make it an indispensable tool for monitoring bacteria used to synthesize various chemicals in the pharmaceutical and other industries. The use of flow cytometry for such purposes is likely to increase rapidly with the growth of these industries expected in the future.

VIRUS

Viruses range in size from about 20 nm (5×10^{-18} g) (parvovirus) to about 180 nm (3×10^{-15} g) (pox virus). Their contents of nucleic acid range from about 3×10^{-18} g to about 3×10^{-16} g, with most types being in the lower part of this range, i.e. below 3×10^{-17} g nucleic acid. These numbers may be compared with the mass and DNA content of small *E. coli,* which are about 1×10^{-12} g and 5×10^{-15} g, respectively. Hence, assuming that virus nucleic acid can be stained with the same efficiency as bacteria DNA (which is not obvious, since the virus nucleic acid is so densely packed) the DNA associated fluorescence of the largest virus (pox) should be some 20 times weaker than that of *E. coli.* This is probably below the detection limit of most of the flow cytometers that are currently commercially available. The light scattering, which in this size range falls off with the square of the particle volume, should be roughly 10^5 times weaker for the pox virus than for *E. coli.* Again, this is probably outside the limit of current commercial instruments. On the other hand, if made sufficiently sensitive flow cytometry could be quite useful for counting and identification of virus. Viruses are usually counted by the number of voids formed in a substrate of living cells, a method which is both time consuming and limited to virus for which a suitable cell substrate exists. Flow cytometry would provide precise counting. Moreover, since most types of virus exhibit

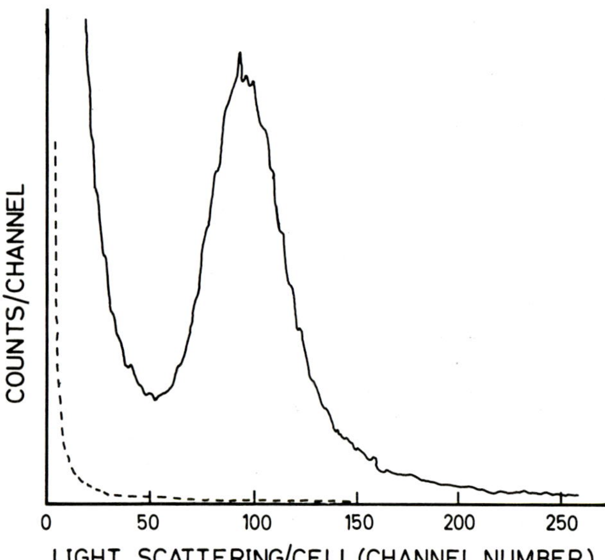

Fig. 19. Light-scattering histogram of "live" pox virus as obtained from a smallpox vaccine. The instrument "background" observed by running pure water is shown by broken line. The "background" in the virus histogram is probably due to debris. The width of the histogram peak corresponds to a coefficient of variance CV_{signal} = 18%, which in turn corresponds to a variation in linear dimension of CV ≈ 3%. The histogram was recorded with an instrument of the type shown in Figure 1, but with a 5-W Ar laser (operated at 488 nm with an output of 1.0 W) replacing the mercury lamp and some modifications in the illumination optics.

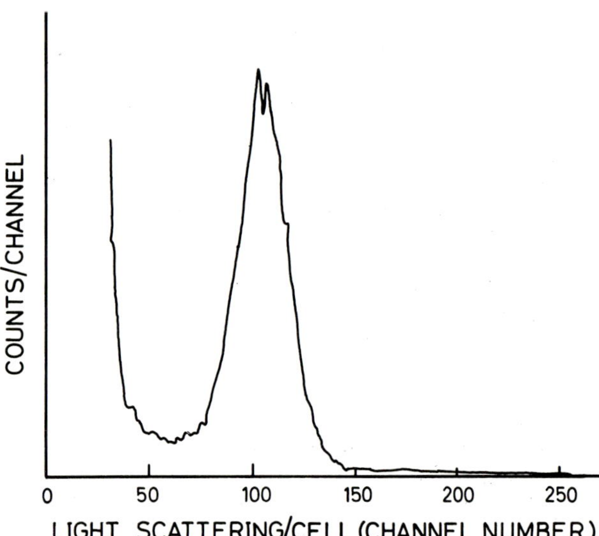

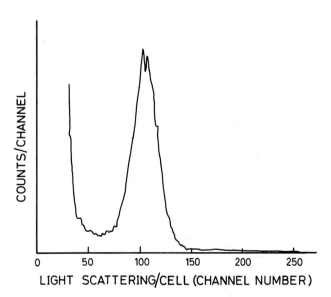

Fig. 20. Light-scattering histogram of "live" T4 phage. The width of the peak corresponds to a precision in the determination of the linear dimensions of the particles corresponding to CV ≈ 2%. The "background" is mainly due to debris. Experimental conditions as for Figure 19.

little variation in size while different types differ greatly in size, measurement of their light scattering should be an efficient method of differentiation. Using a laboratory built flow cytometer Hercher et al. [15] demonstrated the feasibility of this approach for large and intermediate sized virus. Similar data, obtained by the type of instrument shown in Figure 1, are depicted in Figures 19 and 20 (see also Chapter 30, this volume). It has also been shown that by incubating leukocytes with FITC-conjugated virus it is feasible to assay specific virus receptors on the cell surface by means of flow cytometry [56].

REFERENCES

1. **Agar DW, Bailey JE (1982)** Cell cycle operation during batch growth of fission yeast populations. Cytometry 3:123–128.
2. **Bailey JE, Fazel-Makjlessi J, McQuitty DN, Lee LY, Allred JC, Oro JA (1977)** Characterization of bacterial growth by means of flow microfluorometry. Science 198:1175–1176.
3. **Barlogie B, Spitzer G, Hart JS, Johnston DA, Büchner T, Schumann J, Drewinko B (1976)** DNA histogram analysis of human hemopoietic cells. Blood 48:245–257.
4. **Barnett JM, Cuchens MA, Buchanan W (1984)** Automated immunofluorescent speciation of oral bacteria using flow cytometry. J Dent Res 63:1040–1042.
5. **Bonaly J, Mestre JC (1981)** Flow fluorometric study of DNA content in nonproliferative *Euglena gracilis* cells and during proliferation. Cytometry 2:35–38.
6. **Boye E, Steen HB, Skarstad K (1983)** Flow cytometry of bacteria: A promising tool in experimental and clinical microbiology. J Gen Microbiol 129:973–980.
7. **Brandhorst BP, McConkey EH (1974)** Stability of nuclear RNA in mammalian cells. J Mol Biol 85:451–463.
8. **Bremer H, Dennis, PP (1987)** Modulation of chemical composition and other parameters of the cell by growth rate. In Neidhardt FC (ed), "*Escherichia coli* and *Salmonella typhimurium*. Cellular and Molecular Biology." Vol. 2. Washington, DC: ASM, pp 1527–1542.
9. **Buchanan R, Gibbons N (eds) (1974)** "Bergey's Manual of Determinative Bacteriology." Baltimore: Williams & Wilkins.
10. **Cooper S, Helmstetter CE (1968)** Chromosome replication and the division cycle of *Escherichia coli* B/r. J Mol Biol 31:519–540.
11. **Dvorak JA, Hall TE, Crane MSJ, Engel JC, McDaniel JP, Uriegas R (1982)** *Trypanosoma cruzi:* Flow cytometric analysis. I. Analysis of total DNA/organism by means of mithramycin-induced fluorescence. J Protozool 29:430–437.
12. **Fazel-Madjlessi J, Bailey JE, McQuitty DN (1980)** Flow microfluorometry measurements of multicomponent cell composition during batch bacterial growth. Biotechnol Bioeng 22:457–462.
13. **Helmstetter CE (1967)** Rate of DNA synthesis during the division cycle of *Escherichia coli* B/r. J Mol Biol 24:417–427.
14. **Helmstetter CE, Pierucci O (1976)** DNA synthesis during the division cycle of three substrains of *Escherichia coli* B/r. J Mol Biol 102:477–486.
15. **Hercher M, Mueller W, Shapiro HM (1979)** Detection

and discrimination of individual viruses by flow cytometry. J Histochem Cytochem 27:350–352.

16. **Hutter KJ, Eipel HE** (1978) Flow cytometric determinations of cellular substances in algae, bacteria, moulds and yeasts. Antonie van Leewenhoek 44:269–282.
17. **Hutter KJ, Eipel HE** (1978) DNA determination of yeast by flow cytometry. FEMS Microbiol Lett 3:35–38.
18. **Hutter KJ, Eipel HE** (1979) Simultaneous measurements of DNA and protein content of microorganisms by flow cytometry. Eur J Appl Microbiol Biotechnol 6:223–231.
19. **Hutter KJ, Oldiges H** (1978) Untersuchungen zur schadenwirkung von schwermetallverbindungen an microorganismen. In Berichte über Landwirtschaft. 55:724–732.
20. **Hutter KJ, Oldiges H** (1980) Alterations of proliferating microorganisms by flow cytometric measurements after heavy metal intoxication. Ecotoxicol Environ Safety 4:57–76.
21. **Hutter KJ, Stöhr M** (1979) Two-parameter analysis of microbial cell constituents. Microbios Lett 10: 121–128.
22. **Hutter KJ, Eipel HE, Hettwer H** (1978) Rapid determination of the purity of yeast cultures by flow cytometry. Eur J Appl Microbiol 5:109–112.
23. **Hutter KJ, Görtz T, Eipel HE** (1978) Different stages of DNA synthesis during the growth of *Saccharomyces cerevisae.* FEMS Microbiol Lett 3:291–294.
24. **Hutter KJ, Eipel HE, Stöhr M** (1980) Flow cytometric analysis of microbial cell constituents after heavy metal intoxication. In Lærum OD, Lindmo T, Thorud E (eds), "Flow Cytometry." Vol. 4. Oslo: Universitetsforlaget, pp 317–322.
25. **Ingram M, Cleary TJ, Price BJ, Price RL III, Castro A** (1982) Rapid detection of *Legionella pneumophilia* by flow cytometry. Cytometry 3:134–137.
26. **Jawetz E, Melnick JL, Adelberg EA** (1982) "Review of Medical Microbiology." Los Altos, CA: Lange Medical Publications.
27. **Kersten H, Kersten W**: Inhibitors of nucleic acid synthesis. In Kleinzeller A, Springer GF, Wittmann HG (eds), "Molecular Biology Biochemistry and Biophysics." Berlin: Springer pp 108–117.
28. **Koch AL** (1968) Theory of the angular dependence of light scattered by bacteria and similar-sized biological objects. J Theoret Biol 18:133–156.
29. **Koppes LJH, Woldringh CL, Nanninga N** (1978) Size variations and correlation of different cell cycle events in slow growing *Escherichia coli.* J. Bacteriol 134: 423–433.
30. **Kornberg A** (1974) "DNA Synthesis." San Fransisco: WH Freeman.
31. **Kubitschek HE, Freedman ML** (1971) Chromosome replication and the division cycle of *Escherichia coli B/r.* J Bacteriol 107:95–99.
32. **Kubitschek HE, Newman CN** (1978) Chromsome replication during the division cycle in slowly growing, steady-state cultures of three Escherichia coli B/r strains. J Bacteriol 136:179–190.
33. **Lindmo T** (1981) Protein synthesis as a function of protein content in exponentially growing NHIK 3025 cells studied by flow cytometry and cell sorting. Exp Cell Res 133:237–245.
34. **Mansour JD, Robson JA, Arndt CW, Schulte TH (1985): Detection of *Escherichia coli*** in blood using flow cytometry. Cytometry 6:186–190.
35. **Martinez OV, Gratzner HG, Malinin TI, Ingram M** (1982) The effect of some β-lactam antibiotics on *Escherichia coli* studied by flow cytometry. Cytometry 3:129–133.
36. **Paau AS, Lee D, Cowles JR, Oro J** (1977) Flow-microfluorometric analysis of *Escherichia coli, Rhizobium meliloti,* and *Rhizobium japonicum* at different stages of the growth cycle. Can J Microbiol 23:1165–1169.
37. **Paau AS, Lee D, Cowles JR** (1977) Comparison of nucleic acid content in populations of free-living and symbiotic *Rhizobium melitloti* by flow microfluorometry. J Bacteriol 129:1156–1158.
38. **Phillips AP, Martin KL** (1983) Immunofluorescence analysis of bacillus spores and vegetative cells by flow cytometry. Cytometry 4:123–131.
39. **Phillips AP, Martin KL** (1985) Dual-parameter scatter-flow immunofluorescence analysis of bacillus spores. Cytometry 6:124–129.
40. **Sahar E, Lamed R, Ofek I** (1983) Rapid identification of *Streptococcus pyogenes* by flow cytometry. Eur J Clin Microbiol 2:192–195.
41. **Skarstad K, Steen HB, Boye E** (1983) Cell cycle parameters of slowly growing *Escherichia coli* B/r studied by flow cytometry. J Bacteriol 154:656–662.
42. **Skarstad K, Steen HB, Boye E** (1985) DNA distributions of *E. coli* measured by flow cytometry and compared to theoretical computer simulations. J Bacteriol 163:661–668.
43. **Skarstad K, Boye E, Steen HB** (1986) Timing of initiation of chromosome replication in individual *Escherichia coli* cells. EMBO J 5:1711–1717.
44. **Skarstad K, von Meyenburg K, Hansen FG, Boye E** (1988) Coordination of initiation of chromosome replication in *E. coli:* Effects of different *dna A* alleles. J Bacteriol 170:852–858.
45. **Skarstad K, Boye E** (1988) Perturbed coordination of chromosome replication in *rec A* mutants of *Escherichia coli.* J Bacteriol 170:2549–2554.
46. **Slater MI, Sharrow SO, Gart JJ** (1977) Cell cycle of *Saccharomyces* cerevisiae in populations growing at different rates. Proc Natl Acad Sci USA 74:3850–3854.
47. **Steen HB, Lindmo T** (1979) Flow cytometry: A high-resolution instrument of everyone. Science 204:403–404.
48. **Steen HB** (1980) Further developments of a microscope-based flow cytometer: light scatter detection and excitation intensity compensation. Cytometry 1: 26–31.
49. **Steen HB** (1983) A microscope-based flow cytophotometer. Histochem J 15:147–160.
50. **Steen HB** (1986) Simultanous separate detection of low angle and large angle light scattering in an arc lamp-based flow cytometer. Cytometry 7:445–449.
51. **Steen HB, Boye E** (1980) *Escherichia coli* growth studied by dual parameter flow cytophotometry. J Bacteriol 145:1091–1094.
52. **Steen HB, Boye E, Skarstad K, Bloom B, Godal T, Mustafa A** (1982) Applications of flow cytometry on bacteria: Cell cycle kinetics, drug effects, and quantitation of antibody binding. Cytometry 2:249–257.

53. **Steen HB, Lindmo T** (1985) Differential light scattering detection in an arc lamp-based epi-illumination flow cytometer. Cytometry 2:281–285.
54. **Van Dilla MA, Langlois RG, Pinkel D, Yajko D, Hadley WK** (1983) Bacterial characterization by flow cytometry. Science 220:620–622.
55. **Vazquez D** (1979) Inhibitors of protein biosynthesis. In Kleinzeller A, Springer GF, Wittmann HG (eds), "Molecular Biology, Biochemistry, and Biophysics." Vol. 30. Berlin: Springer, pp 108–113.
56. **Wells A, Steen HB, Saeland S, Godal T, Klein G** (1981) A microassay for quantitatively detecting the Epstein–Barr virus receptor on single cells utilizing flow cytometry. J Virol Methods 3:127–136.
57. **Wyatt PJ** (1968) Differential light scattering: A physical method for identifying living bacterial cells. Appl Opt 7:1879–1968.

30

Virus–Cell Interactions Analyzed With Flow Cytometry

John M. Lehman and James W. Jacobberger

Department of Microbiology and Immunology, Albany Medical College, Albany, New York 12208 (J.M.L.); Department of Genetics, School of Medicine, Case Western Reserve University, Cleveland, Ohio 44106 (J.W.J.)

INTRODUCTION

Intracellular parasites and predators (principally viruses, bacteria, and protozoa) usually alter cellular properties in dramatic ways. Viruses use host functions and may alter host metabolism, as in the stimulation of cellular macromolecular synthesis (DNA, RNA, and protein) by certain papovaviruses or the suppression of cellular macromolecular synthesis by the vast majority of viruses. These events produce numerous changes in cellular functions, which can be analyzed by flow cytometry. The complete or partial transcription, translation, and replication of the virus produces virus-specific products (DNA, RNA, and proteins) that can be detected with appropriate antibodies and DNA and/or RNA probes. The intent of this chapter is to describe a multiparametric approach using flow cytometry to study host cell–virus interaction. The literature is reviewed, the ideology delineated, practical and technical aspects discussed, and examples of representative studies presented.

Viruses are small particles (size range 20–400 nm) made up of a nucleic acid genome, associated proteins, glycoproteins, and sometimes a lipid membrane envelope (Table 1). The genome may be DNA or RNA, covered with protein, and membrane coated as a result of the budding of the virus particle through cellular membranes. These viruses incorporate specific virus proteins from the cell membrane as well as cellular proteins, lipids, and glycoproteins. For instance, the papovavirus, Simian virus 40 (SV40), contains only three proteins of viral origin associated with the infectious particle. These are the virus-encoded proteins VP1 (45K), VP2 (42K), and VP3 (30K) covering the viral DNA. The VP1 constitutes 75% of the total virion protein. Associated with the viral DNA are the cellular histones H2A, H2B, H3, and H4, which comprise an internal virion chromatin complex. These proteins can be contrasted to the other virus-encoded proteins not associated with the virion but necessary for transcription, translation, replication, and transformation. They include the tumor antigens, large T antigen (94K), small t antigen (17K), and a highly basic polypeptide, termed the agnoprotein. Much attention has focused on the function(s) of the large T antigen, since this protein is thought to be responsible for the transforming ability of the virus. The T antigen has been assigned the following functions: (1) sequence-specific binding to viral DNA, (2) initiation of viral DNA synthesis, (3) autoregulation of early viral transcription, (4) induction of late viral transcription, (5) ATPase activity, (6) binding to cellular DNA, (7) initiation of cellular DNA synthesis, (8) binding to the cellular p53, and (9) induction of cellular enzyme synthesis.

Table 1 groups the types of viruses by genomic material and organization, coat composition, replication strategy, and major effect on the host cell. Viruses replicate inside the host cell, using various degrees of host machinery and resources. This strategy can be classified as lytic (permissive infection)— the cell dies a short time after exposure to the virus and viral progeny are released, in contrast with a nonlytic virus infection in which virus may be produced continuously over a long period of time without killing the cell. The latter category can be via integration of the virus (retrovirus) into the cell DNA and subsequent replication and continued release of virus. Viral infections of this type occasionally confer additional properties on cells by integrating the viral genome near cell-regulatory genes and increasing the level of transcription of these genes by using the viral promoter encoded in the long terminal repeat (LTR) [36]. Retroviruses excise imperfectly and carry with them these same regulatory genes in an infectious particle defective for replication—a possible mechanism for tumorigenic transformation of normal cells.

Many DNA viruses encode their own regulatory genes, which are capable of activating the cell for growth. This strategy enables a virus like SV40 to infect G_0 cells, stimulate entrance into the cell cycle, and replicate during the S phase using the host DNA synthetic apparatus. While these viruses normally tend to be lytic, integration of the viral genome occasionally takes place through illegitimate recombination [12], and the regulatory genes are constitutively expressed,

Flow Cytometry and Sorting, Second Edition, pages 623–631

TABLE 1. Characteristics of the Major Viruses[a]

Viruses	Size (nm)	Structure	Replication	Host effects
DNA viruses				
Parvoviruses	18–26	Proteins (3) NA-linear, SS, 1.5–1.8 × 10^6 daltons	Nuclear, require host functions	Lytic
Papovaviruses (papillomaviruses)	45–55	Proteins (7–8) NA-supercoiled circular, DS, 3–5 × 10^6 daltons	Nuclear, require host functions	Lytic, transforming
Adenoviruses	65–80	Proteins (11–15) NA-linear, DS, ~20 × 10^6 daltons	Nuclear require host functions	Lytic, transforming
Herpesviruses	120–300	Proteins (15–35) envelope from nuclear membrane NA-linear, DS, 80–150 × 10^6 daltons	Nuclear, require host function	Lytic, latent, transforming
Poxviruses	200–400	Proteins (30–100) enzymes, envelope (Golgi) NA-linear, DS, 85–185 × 10^6 daltons	Cytoplasmic, viral enzymes	Lytic
RNA viruses				
Picornaviruses	24–30	Protein (11–12) NA-linear, SS, 2–3 × 10^6 daltons	Cytoplasmic	Lytic
Reoviruses	70–80	Protein (12) NA (10 segments)-linear, D.S., 15 × 10^6 daltons	Cytoplasmic	Lytic
Togaviruses	40–70	Proteins envelope (host membrane) NA-linear, SS, 4.2–4.5 × 10^6 daltons	Cytoplasmic	Lytic
Rhabdoviruses	60–85	Proteins (5) envelope NA-linear, SS, 4–5 × 10^6 daltons	Cytoplasmic	Lytic
Orthomyxoviruses and paramyxoviruses	80–250	Protein (8–10) envelope NA-linear, SS, 4–6 × 10^6 daltons	Cytoplasmic	Lytic
Retroviruses	80–130	Protein (8–12) envelope, viral enzyme NA-linear, SS, 5–7 × 10^6 daltons	Nuclear	Nonlytic, transforming
Others (RNA)				
Arenaviruses				
Bunyaviruses				
Coronaviruses				
Filoviruses				
Flaviviruses				
Caliciviruses				

[a]Abbreviations: nm = nanometers; proteins = viral; NA = nucleic acid; SS = single stranded; DS = double stranded.

giving rise to a cell population that continues to enter the cell cycle (nonpermissive transforming infection). In all these interactions, new viral genes are expressed, cellular genes may be activated (increased transcription) and/or deactivated, and whole programs of cell behavior initiated (e.g., the cell cycle). These events can be described by correlated quantitative measurement of gene products [messenger RNA (mRNA), proteins)] or cell macromolecules (DNA, RNA, and protein) using flow cytometry. In the following sections, the addition of this information to viral systems under study is evaluated.

With the exception of papers on the papovavirus group, relatively few publications in virology have used flow cytometry. Most reports have been published recently, are preliminary in nature, and do not represent a large contribution to an area of research. However, papers describing SV40 studies have appeared from several laboratories and have elucidated several aspects of host–virus interaction. Perusal of various abstracts suggests that herpesvirus and retrovirus studies will increase in the near future. Published work can be grouped in the following manner:

1. Detection of single virions in solution [14]
2. Analysis of virus binding to cell surface, using fluorescently labeled virions [24,27,39,45]
3. Identification of virus-infected or transformed cells by measurement of viral antigen—specific immunofluorescence [5,6,9,17,19,20,23,47] or light scattering [5,28]
4. Analysis of cell phenotype as a consequence of viral gene expression in transformed cells [4,18,42–44,47] or infected cells [5–7,9,10,15,16,31,38]

Related studies on the expression of immune cell surface antigens during viral infection (e.g., [1]) or as a consequence of infection (e.g., [22,34]) will not be discussed in detail. Surprisingly, there are few reports using flow cytometry in clinical virological assay; however, some are now appearing, including a flow assay for human cytomegalovirus (CMV) [8].

SINGLE VIRIONS

Hercher et al. [14] detected single virions of T2 phage and reovirus capsids by light scattering (LS). Calibration of their instrumentation showed that latex spheres as small as 0.091 μm could be detected and that the relationship between LS and particle size is as the theory predicts. Although Hercher and co-workers did not attempt to size the viral particles by scatter, T2 phage scattered significantly less light than 0.091-μm beads and reovirus capsids scattered significantly less than the T2 phage. These investigators mention the addition of fluorescence as a better or enhanced signal. Fluorescence from lipid, protein, or nucleic acid dyes and/or immunofluorescence are all potential candidates. The ability to detect single virions would seem to be important for (1) comparison of absolute particle number versus plaque-/focus-forming units to characterize viral stocks and/or mutants and correct for efficiency of infection, (2) characterization of the end result of lytic infectious processes (i.e., number of output particles, ratio of empty/full capsids), and (3) protein content distribution measurements of coat proteins. These studies are currently performed with the electron microscope. Flow cytometry should provide better statistics and more information. Electron microscopy establishes structural integrity, whereas flow cytometry could possibly provide a quantitative evaluation of the nucleic acid, protein, and lipid content, as well as specific antigens from which structural integrity could be implied.

VIRUS RECEPTOR STUDIES

Notter et al. [39] measured the binding of Rous sarcoma virus (RSV) to susceptible and resistant chicken embryo fibroblasts. The virus preparations were labeled with R-18, a rhodamine-conjugated octodeconol, which partitions into the lipid bilayer of the virus. As expected [39], Polybrene increased the level of virus binding. By using chicken cells that were genetically susceptible or resistant to infection, Notter and co-workers demonstrate differences in virion binding to the two populations of cells. Saturation of cell receptors was observed and kinetics of virus binding shown for resistant and susceptible cells. Although Notter et al. discuss these binding curves in terms of specific and nonspecific binding, it is not clear from the data presented that the curves do not represent differences in receptor number. The distribution of fluorescence for susceptible cells was broad, but fluorescence was shown in a nonquantitative way to be related to infection potential. Cells of high and low fluorescence were sorted and cultured. Highly fluorescent cells demonstrated a transformed phenotype within 5 days; low fluorescent cells did not. Notter and co-workers state that the ratio of infectious to noninfectious particles is low; therefore, the low fluorescence cells may be largely bound by noninfectious virus. Although this preliminary study demonstrated the usefulness of flow cytometry, it is not clear whether these studies would have been more easily undertaken using more common methods. Isotope-labeling of the virion nucleic acid and/or protein and measuring radioactivity bound when comparing the two genetic populations, measuring the effect of Polybrene, or comparing different virus subgroups would have given similar results. However, provided that the preparation of labeled virus is equivalent labor, kinetic studies almost certainly were easier (and more accurate) than other studies. It would be exciting to see studies such as this incorporate analyses of the virion preparation, quantifying particles per volume and the distribution of fluorescence attributed to single virions, and analysis of viral gene expression. In this manner, one could estimate the number of virions bound per cell across the fluorescence distribution of infected cells and determine the number of successful infections as a function of virions bound.

ANTIGENS

Horan et al. [17] showed that SV40-infected CV-1 cells (African green monkey kidney cell line) could be detected with flow cytometry in an indirect immunofluorescent assay using an antitumor serum (T antigen) raised in Syrian hamsters by injection of SV40-transformed cells. The results demonstrate that at increasing levels of antiserum, there was a saturation of antigen binding accompanied by an equivalent increase in nonspecific binding. The implication of this finding—that immunoglobulin (Ig) is retained intracellularly in fixed cells—is that (1) when staining antigen positive cells in the presence of negative cells, it will be difficult to detect a distinct difference between the two populations; and (2) when developing a particular assay, the optimal protocol will balance background versus specific staining and that rather than being in antibody excess, the cell number, antiserum dilution, and volume of reaction will have to be optimized for greatest difference between the mean fluorescence of positive and negative populations. The degree to which this is a problem is dependent on antigen density, percentage of specific Ig in the antiserum, and finally, the goal of the assay.

Hand et al. [13] used similar methods to describe detection of murine leukemia virus (MLV)-infected cells. In this system, however, a rather broad dilution range (~1/150–1/300) provided a ratio of total mean fluorescence (F_t) to background (F_b) that was quite acceptable. $F_t \sim 100$; $F_b \sim (2–5)$. This corresponds to a percentage specific signal ($F_s = (F_t - F_b)/F_t$, see [20] of 92 to 95%; i.e., 92 to 95% of the fluorescence is attributable to specifically bound antibody. In addition to absorbing the anti-MLV serum with noninfected cells, these investigators further reduced background levels by including bovine IgG in the reaction. They also found that the addition of bovine serum albumin (BSA) increased the F_b. It seems likely that the high F_t/F_b in this system was due to high antigen density/cell. The development of the assay was with Moloney leukemia virus, which was adapted to cell culture and was therefore highly infective for the mouse cell culture system used. When less well-adapted strains were used, the mean fluorescence of the positive population decreased dramatically; however, the ability to detect positive populations by flow cytometry remained comparable to that of visual determinations.

Leary et al. [28] approached the detection of herpes simplex virus type 2 (HSV-2)-infected human cells by an indirect immunoperoxidase technique. In this case, 4-chloro-1-naphthol was used to produce cells stained with a blue polymerized dye, which scattered 632-nm light from a helium–neon laser more effectively than does diaminobenzidine. Scattered light at 1 to 19° (wide forward angle) was measured, and in this case, a decrease in the scattered light intensity correlated with an increase in antigen. Leary et al. noted that staining at high dilutions of primary antibody was necessary because of high background. Nonspecific binding was improved by inclusion of nonconjugated normal serum of the same species with the secondary antibody. The shift in peak channel at 24 hour postinfection was ~42% of the signal for uninfected cells. Using $(LS_i - LS_u)/LS_i$ (i = infected; u = uninfected) as a measure similar to that of F_s, an approximate percentage specific signal of (−) 58% was achieved (our calculations).

Cram and co-workers demonstrated the detection of hog cholera virus within infected PK-15 cells [5,6]. Early results [5] imply ~ 40% F_s for infected cells. These investigators state that "depending on multiplicity of infection and metabolic state of the cells, . . . infected cells . . . measured 4 times brighter." This would represent an F_s of 75%. Since background fluorescence is partially related to cell size [6,10; J.M. Lehman and J.W. Jacobberger, unpublished results], these results could be improved [10] by collecting the fluorescence data as a ratio of fluorescence to Coulter volume in real time. This did not improve F_s (using distribution peak measurements) but did decrease the coefficient of variation (CV) for each distribution, thereby increasing the ability to discriminate each population.

We obtained similar results measuring SV40 T antigen in transformed cells (100% positive) and nontransformed cells (100% negative). However, if one is interested in quantification of antigen in addition to quantifying the number of positive cells, an alternative procedure should be used based on the following argument: Where V is cell volume, F components are as defined previously, and

$$F_t = F_s + F_b \qquad (1)$$

then

$$F_t/V = (F_s + F_b)/V = F_s/V + F_b/V \qquad (2)$$

that is, the specific signal is being reduced by the cell volume as well as the background reduction. Data corrected in this manner would only have meaning if the production of antigen were related to cell size independent of other events. The approach that we took was to use the following argument: if F_b is proportional to V, and in the simplest case

$$F_b = mV + c$$

where m and c are constants representing the slope and intercept, then given equation 1,

$$F_t = F_s + mV + c.$$

Since V is proportional to forward narrow angle LS (FALS), one could determine F_b by measuring nontransformed cells or any population 100% negative and, by measuring FALS simultaneously, m and c. Therefore, for mixed or positive populations, the fluorescence due to nonspecific binding of Ig could be subtracted [21].

Although an analysis routine was developed to determine the relationship of FALS to F_b, in many cases a simpler solution to the F_b problem exists: lower the background until it is insignificant. This is possible largely through the use of monoclonal antibodies. As discussed in detail [20], early results with SV40 transformants that did not produce large amounts of T antigen were plagued with nonspecific binding problems (F_s = 15 to 30%). Although we optimized for primary and secondary serum dilution, reaction volume (ratio of volume to cell number), temperature, washes, and primary serum adsorbed to nontransformed cells, these procedures only minimize F_b. The problem is most likely caused by the Ig present in serum that is not specific for the antigen being measured, and that Ig has an appreciable affinity for the interior of alcohol-fixed cells [20,21]. Three significant steps allowed for low fluorescence backgrounds of 2 to 3% [20]: (1) high affinity monoclonal antibodies, (2) affinity-purified secondary antibody, and (3) inclusion of goat serum (blocking secondary antibody) in all reaction mixtures. The significance of these factors is that one can stain virally infected or transformed cells in a vast excess of antibody; therefore, all available antigenic sites are bound, and the low background (in this case, equivalent to presence of low levels of unbound and bound specific secondary antibody) will not interfere with quantification when test samples have different levels of antigen (e.g., during infection) or variable numbers of cells (e.g., growth curves).

PHENOTYPE ANALYSIS

Cell-Surface Modification

Leary and Todd [29] demonstrated that clustered concanavalin-A (Con A) binding sites on the cell surface of H.Ep.2 cells (human laryngeal carcinoma line) can be detected as an increase in FALS (2–12°) when these cells are treated with Con A, fixed, and then stained by an immunoperoxidase method [26]. On the basis of flow cytometric studies in which Kraemer et al. [26] showed that lectin-binding sites of normal cells are not increased in cell transformation, Leary and Todd hypothesize that Con A binding sites are redistributed as a function of viral or chemical transformation similar to that occurring during lytic herpesvirus infection [28]. Since Con A must be present to observe antigen clustering (cells fixed and then incubated with Con A are equivalent to uninfected cells, i.e., dispersed sites), it seems more correct to hypothesize a change in receptor site mobility as opposed to actual redistribution as a consequence of viral infection. Despite the fact that the "normal" cell type used in these studies was derived from a tumor, Leary et al. cite work demonstrating a high correlation between Con A redistribution and malignancy and state that this may be a useful method for screening potentially carcinogenic agents or as a detection system for cell transformation from the normal to the tumorigenic phenotype. To our knowledge, this method has not been used for either purpose. Nevertheless, they do demonstrate a potentially quantitative method to assay Con A site change as a result of viral infection. This may prove useful when more is known about the consequence of viral infection, and this observation can be attributed to a viral gene. For instance, one may be able to characterize virus strains or mutants by this method.

Cell Fusion

Cram and co-workers [5,6,31] have been able to quantify cell fusion—the principal in vitro cytopathic effect of Newcastle disease virus (NDV). The rationale behind these studies was a practical one. In contrast to earlier reports, Reeve and Poste [40] reported that the degree of syncytia formation could be used to characterize virulence for this virus [2,25]. At the time, Newcastle disease was a costly and emerging disease in the poultry industry. A flow cytometric assay for cell fusion would determine whether the disagreement in results could be attributed to the microscopic counting assay. If virulence were related to cell fusion, the flow cytometry assay could be completed in 3 days, compared with 7–10 days for a plaque assay. Cell fusion was assayed with mithramycin or acriflavin-Feulgen, both staining for DNA content, which detected the multinucleated cells (syncytia). The cell fusion index was the ratio of infected to control cells, i.e., cells with a DNA content greater than G_1 to diploid cells. The kinetics of syncytia formation were studied as well as the effect of cell density and multiplicity of infection. Under conditions in which these parameters did not significantly affect the results, 18 strains of the virus were tested for cell-fusion index. The reproducibility of the assay was as expected quite good. However, results showed a lack of correlation between virulence and ability to induce cell fusion. Although this work did not lead to a clinical assay, it did answer a specific question in a study of many strains that could not have been carried out without flow cytometry.

Stimulation of Cells into DNA Synthesis

In all the following work, a constant feature is that SV40 causes resting cells to cycle. Numerous studies [30,32–34] demonstrated that the cell cycle of cells permissive (African green monkey lines BSC-1 and CV-1), semipermissive (Chinese hamster and human strains), and nonpermissive (mouse) for SV40 viral replication were perturbed within 24 hours of infection.

In all cases, infected cells were induced into one or more rounds of cellular DNA synthesis, with many of the cells (10–30%) exhibiting tetraploid DNA content. Some of the infected monkey cells are blocked in G_2, but some cells cycle to higher DNA content levels (tetraploid S and G_2). All will replicate the viral genome, synthesize late proteins, and die as a result of viral replication. It appears that the stimulation of cellular DNA synthesis provides the necessary environment for the replication of the viral DNA. Gershey [9] demonstrated with cell-cycle data and by collecting mitotic cells as a function of time after infection that infected CV-1 cells did not enter mitosis and were therefore blocked in G_2. Hiscott and Defendi [16] showed similar data from primary African green monkey cells and TC-7 (derived from CV-1) cells infected with a viable deletion mutant of SV40 that produces viral DNA, but slowly. An increase in DNA content greater than 4C seen in DNA histograms was due to an accumulation of viral DNA. This same cell block occurs in semipermissive and in nonpermissive cells [16,18,32]. Hiscott and Defendi [16] found a less pronounced but decided accumulation of cells in the G_2 + M phase of SV40-infected mouse embryo cells (this is not a block so much as an increased residence in G_2). Since similar DNA histograms are observed when dl-884 (an SV40 deletion mutant that does not synthesize small t antigen) was used, both the induction of resting cells to cycle and the G_2 block are a function of large T antigen [4]. In this study on mouse cells, *ts*A58-infected cells at the nonpermissive temperature resulted in a transient accumulation of cells in G_2; however, 48 hours postinfection, the DNA histograms were similar to mock-infected cultures. Autoradiography demonstrated that an increased level of DNA synthesis (at least 2 times) continued in *ts*A58-infected cells at the nonpermissive temperature in this same experiment. Horan et al. [18] showed similar data for Chinese hamster embryo cells. An increase in cycling cells, a G_2 block, and an accumulation of a population of tetraploid cycling cells was clearly demonstrated by 96 hours postinfection. In the isolation of clonal SV40 transformed lines, it is our experience that tetraploid lines are common. Furthermore, DNA analysis of these lines demonstrates the presence of octaploid populations. In one case, we labeled an SV40 transformed Chinese hamster cell line (with a tetraploid subpopulation) with Hoechst 33342, viably sorted the tetraploid G_2 + M population, and reanalyzed the progeny after regrowth. The original distribution of diploid/tetraploid populations was restored (unpublished data). This is consistent with time-lapse cinematography data (unpublished) in which reduction division is observed.

Quiescent cells can be induced to cycle by infection with SV40 virus. This has been demonstrated with flow cytometry [33] and by autoradiography [46]. Since nonpermissive and semipermissive cells are induced to cycle in the absence of viral replication or late gene transcription, this induction must be associated with transcription of SV40 early genes (large T and small t). The results of Hiscott and Defendi [16] demonstrated that small t was not necessary for induction. The most direct evidence that T antigen is the viral gene necessary for induction of cell proliferation is by microinjection [11,50]. In these experiments, purified T antigen was microinjected into cells subsequently incubated with [^{3}H]-thymidine. After fixation and autoradiography, cells injected with T antigen were shown to be cycling, whereas those injected with control solutions were not. Genetic studies had produced mutant virus in the A gene (T antigen), which were temperature sensitive (*ts*A) for viral replication [48]. All the *ts*A mutations were in the coding region for large T antigen. At the temperature permissive for viral replication, nonpermissive (transforming) cells could be infected and transformants selected. These transformants could be shifted to the nonpermissive temperature and the transformation phenotype examined. This body of work represents a lengthy, difficult, and confusing exploration because (1) the mutations are leaky, (2) T antigen is pleiotropic, (3) the transformation frequency of SV40 is low (0.001 to 0.1%), and (4) the cells may express mutations in other genes [51]. The operating wisdom was that *ts*A transformed cells reverted to normal phenotype at the nonpermissive temperature; i.e., they became contact inhibited, stopped dividing at a high rate, and eventually entered quiescence [37]. When Robinson and Lehman [42–44] examined a *ts*A58-transformed line (A58-1) by flow cytometry, surprisingly, the cells were cycling at a rate near that of the same line at the permissive temperature. This finding was followed by time-lapse cinematography studies that showed that cells were dividing, however cell death rates increased and cultures were maintained in a monolayer-like state by a balance of replication and death [44].

Christensen and Brockman [4] expanded this work to include several *ts*A transformants and *ts*A deletion mutants that coded for a temperature-sensitive T protein but did not express small t. The results of this work confirmed that many if not most *ts*A mutants are not temperature sensitive for stimulation of host DNA synthesis and that small t antigen was not necessary for any aspect of the transformed phenotype of these cell lines, i.e., the ability to overgrow a monolayer, the ability to grow in low serum media, and increased growth rate. Currently, much is known about the biochemistry of T antigen [3,41]. Assuming that stimulation of host DNA synthesis takes place through a transactivating DNA binding function, and since T antigen has separate binding domains for viral and host DNA [3,41], it is not surprising that selected *ts* mutations in viral replication are not fully *ts* for stimulation of host DNA synthesis. Since T antigen is autoregulatory, binding to its own promoter through cooperative binding of T antigen molecules inhibits T gene transcription [12,41]; a *ts* mutation in the viral DNA binding domain that inhibited binding would promote excessive T antigen transcription [49] and drive the host DNA binding reaction by mass action even if the affinity for host DNA were decreased by the mutation (secondary folding effects).

MULTIPARAMETRIC ANALYSIS OF CELL AND VIRUS INTERACTION

Much of the recent development in flow cytometry has involved multiparametric correlations of cell phenotype and behavior. Studies coupling antigen and DNA measurements, for example, have demonstrated genes that are cell-cycle regulated or that regulate the cell cycle and can be studied in this manner [20,33]. SV40 is currently the most studied viral system, and quantitative studies of large T antigen expression in both permissive infection and transformation have been initiated. The following summary demonstrates the information obtained by this type of analysis.

Expression of T antigen by SV40-infected CV-1 cells (permissive) was described by Lehman et al. [33] in which T antigen is quantified as a function of the cell cycle, time, and multiplicity of infection during the lytic cycle of the virus. Data were obtained from 6–72 hours postinfection of contact-inhibited CV-1 cells. Primary expression of T antigen was observed in the G_1/G_0 cells of this population; however T-antigen-positive cells were also seen in the G_2 + M and S phase of the cell cycle. Up to 30 hours postinfection, most cells were in the G_0/G_1 phase. By 30 hours postinfection, a tetraploid population was evident and predominated at later time intervals (> 48 hours). When the T-antigen-positive and -negative populations were compared for each cell-cycle phase, a similar percentage was observed up to 30 hours postinfection. This suggests that all phases of the cell cycle were able to induce T-antigen synthesis and are equivalent in their response to SV40 infection. When input virus varied (10 times), the percentage of positive cells increased; however, the levels of T antigen (mean Y) at each phase of the cell cycle were similar. When the G_0/G_1 phase cells were followed at progressive time points, T antigen was synthesized to high levels prior to movement into the S phase of the cell cycle. These data are consistent with the known biochemistry of T antigen and with a hypothesis suggesting that T may be an initiator of $G_1 \rightarrow S$ phase transition [37]. However, other interpretations are possible. For example, T antigen may be transcribed in G_0/G_1 and responsible for the movement from G_0 to G_1, but the level of T could be related to the ratio of DNA to T antigen. In this case, the G_1 exit level would be a function of the length of time in G_0/G_1. This hypothesis could be tested, since cycling cells should exit at a lower level of T antigen than quiescent cells. Furthermore, at later time intervals, the population is induced into high DNA levels (tetraploid) with the beginning of viral DNA and late viral protein (capsid) synthesis. This increase in DNA does not appear to be due to the replication of viral DNA. If infected cells produce 100–1,000 virus particles per cell, this would be approximately 5×10^6 extra base pairs (bp) at the high particle ratio [11]. If a value of 3×10^9 base pairs is assumed for the cellular DNA, then viral genomes (5,243 bp) would increase the signal by 10^{-4} to 10^{-3}, provided the chromatin structure remains the same during infection [35]. This is probably not the case, since it is known that host DNA is incorporated into the virions; therefore, it is probable that single- and double-strand breaks occur in the DNA, and the bound dye to base pair ratio increases (intercalating dyes) during the terminal stages.

EXPRESSION OF SV40 T ANTIGEN IN NONPERMISSIVE CELLS (TRANSFORMING)

T antigen is expressed in G_1, S, and G_2 in most transformed cell lines examined. The following interpretation is the current working model. T antigen is synthesized in G_1 to a high level, and the cell enters S phase; net synthesis in S phase has not been detected. In general, G_2 cells contain twice the average T antigen level (mean Y) of G_1 cells. T antigen is either degraded or the PAB 101 epitope is masked as the cell proceeds to M. Some G_2 cells enter G_1 (tetraploid) without going through mitosis; these cells synthesize higher levels of T antigen to reach the tetraploid threshold value for G_1 exit. In one group of rat cell transformants, different expression patterns have been demonstrated [19]. In these transformants, T antigen was apparently not synthesized or present in the 2C population but was found in the 4C population. Imbert et al. suggested that this is evidence for expression limited to G_2/M. However, another explanation may be that T antigen is expressed normally, but only in the 4C population. Analysis of a large number of SV40-transformed cells demonstrates that the transformants express different levels of T antigen, which may relate to other phenotypic characteristics of transformation (Fig. 1 and 2). Figures 1 and 2 demonstrate the level of staining of the control cells (B-1) and SV40 transformants (WT-HP and WT-4). The average fluorescence (mean Y) of the total populations is 1.25 for the B-1 cells and for the transformants are 6.3 for WT-HP and 21.38 for WT-4. This demonstrates a major difference in populations and single cells for T-antigen quantity. The transformants are 6-fold (WT-HP) and 21-fold (WT-4) higher than are the uninfected cells. Furthermore, the transformants have different levels of T antigen (6.3 vs. 21.38), approximately a threefold difference. These studies are under analysis to determine relationship of T-antigen quantity to transformation.

CLINICAL VIROLOGY

Recently, studies [8] have used flow cytometry to assay an early viral antigen (α) in cells infected with human CMV. Specific monoclonal antibodies demonstrated the presence of

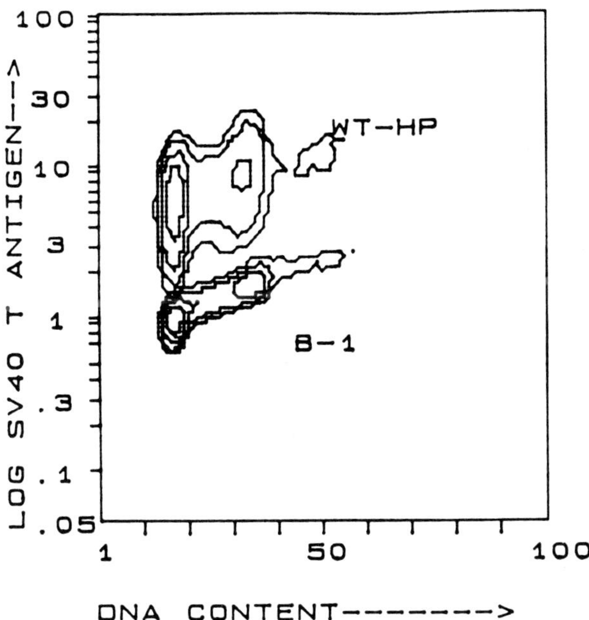

Fig. 1. DNA content and log of SV40 T antigen expression in B-1 (uninfected) and WT-HP SV40 transformed Chinese hamster cells. The cells were fixed, stained, and analyzed by flow cytometry as described [20,33]. The mean of green fluorescence (mean Y) of the B-1 cells is 1.25, and the WT-HP is 6.30, a sixfold difference.

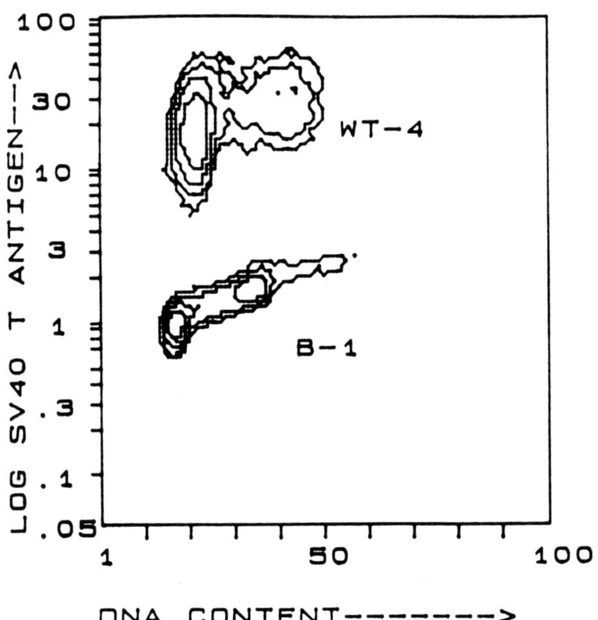

Fig. 2. See Figure 1. The WT-4 and SV40 transformed Chinese hamster cell has a mean Y of 21.38; the B-1 cell has a mean Y of 1.25, a 21-fold difference.

the antigen in a cell-culture model system and in cells obtained from bronchial lavages of suspected CMV-infected patients. The results of this study suggest that flow cytometry may be a useful early diagnostic procedure when used with specific monoclonal antibodies to detect viral proteins. This technology with multiparameter analysis will also be useful for study of the pathogenesis of many viral infections.

SUMMARY

Numerous studies of virus–cell interaction have used flow cytometry. Studies with papovavirus and SV40 have provided some insights into the events in lytic infection, as well as the relationship of the viral protein, T antigen, to transformation. These studies have been made possible by the development of monoclonal and polyclonal antibodies to the specific viral proteins. The use of multiparameter analysis has permitted a comparison, correlation, population analysis, and single-cell analysis of the appearance and quantity of specific viral proteins (T antigen) and a variety of other cell properties in an infected cell population. In many virus–cell interactions, flow cytometry has and will provide a synopsis of the cellular and viral events leading to cell lysis or transformation. The transformation of cells with SV40 is inefficient, with approximately 0.001–0.1 transformed cells observed in nonpermissive cell systems, such as mouse, Chinese hamster and human cells. These studies have provided methodology to detect, determine, and quantify the presence of infected cells and their evolution to the transformed phenotype. Specific cellular proteins (i.e., p53), cell cycle, and population interactions may relate to the phenomena of cellular transformation. Molecular biology studies have amassed considerable information on the virus structure, replication, transcription, and translation, as well as the molecular interactions of viral and cellular proteins. However, these studies are unable to follow and characterize these events on a single-cell or population basis. Flow cytometry with the appropriate reagents (antibodies, DNA, and RNA probes) is providing an approach to this problem. The ability to separate and isolate specific cells via the sorting mode of the flow cytometer will also provide an opportunity to study these virus–cell interactions. As experience and reagents become available, flow cytometry will be useful in clinical assays of virus–cell interaction and will possibly provide a new quantitative and qualitative appraisal of the pathogenesis of many virus–host interactions.

ACKNOWLEDGMENTS

This work was supported in part by grant CA 41608 from the National Cancer Institute. We wish to thank Ms. Judith Laffin for reading the manuscript and for insights into the viral studies, Ms. JoAnn D'Annibale and Maureen Cavanaugh for typing the manuscript, and Mr. David Fogleman for his practical expertise and philosophical insight into many aspects of flow cytometry.

REFERENCES

1. **Anderson J, Byrne JA, Schreiber R, Patterson S, Oldstone MB** (1985) Biology of cloned cytotoxic T lymphocytes specific for lymphocytic choriomeningitis virus: clearance of virus and in vitro properties. J Virol 53:552–560.
2. **Bratt MA, Gallaher WR** (1969) Preliminary analysis of the requirements for fusion from within and fusion from without Newcastle disease virus. Proc Natl Acad Sci USA 64:536–543.
3. **Butel JS, Jarvis DL** (1986) The plasma-membrane associated form of SV40 large tumor antigen: biochemical and biological properties. Biochim Biophys Acta 865:171–195.
4. **Christensen JB, Brockman WW** (1982) Effects of large

and small T antigens on DNA synthesis and cell division in Simian virus 40-transformed BALB/c 3T3 cells. J Virol 44:574–585.

5. **Cram LS, Brunsting A (1973)** Fluorescence and light scattering measurements on hog-cholera-infected PK-15 cells. Exp Cell Res 78:209–213.
6. **Cram LS, Forslund JC, Horan PK, Steinkamp JA (1975)** Application of flow microfluorimetry (FMF) and cell sorting techniques to the control of animal diseases. In Henden CC, Illeni T (eds), "Automation in Microbiology and Immunology." New York: John Wiley & Sons, pp 47–66.
7. **Cram LS, Forslund JC, Jett JC (1978)** Quantification of cell fusion by twenty-one strains of Newcastle disease virus using flow microfluorimetry. J Gen Virol 41:27–36.
8. **Elmendorf S, McSharry J, Laffin J, Fogleman D, Lehman JM (1988)** Detection of an early cytomegalovirus antigen with two color quantitative flow cytometry, Cytometry 9:254–260.
9. **Gershey El (1979)** Simian virus 40–host cell interaction during lytic infection. J Virol 30:76–83.
10. **Gershey EL (1983)** SV40-infected muntjac cells: Cell cycle kinetics, cell ploidy and T antigen concentration. Cytometry 1:49–56.
11. **Graessmann A, Graessmann M, Muller C (1979)** Simian virus 40 and polyoma virus gene expression explored by the microinjection technique. Microbiol Immunol 87:1–22.
12. **Green M (1986)** Transformation and Oncogenesis: DNA Viruses. In Fields BN, Knipe DM (eds), "Fundamental Virology" New York: Raven Press, pp 183–234.
13. **Hand RE Jr, Tennant RW, Yang W, Lavelle GC (1978)** Immunofluorescent analysis of murine leukemia virus-infected cells by microfluorometry. J Immunol Methods 23:175–186.
14. **Hercher M, Mueller W, Shapiro HM (1979)** Detection and discrimination of individual viruses by flow cytometry. J Histochem Cytochem 1:350–352.
15. **Hiscott JB, Defendi V (1979)** Simian virus gene A regulation of cellular DNA synthesis. I. In Permissive cells. J Virol 30:590–599.
16. **Hiscott JB, Defendi V (1981)** Simian virus 40 gene A regulation of cellular DNA synthesis. II. In nonpermissive cells. J Virol 37:802–812.
17. **Horan M, Horan PK, Williams CA (1975)** Quantitative measurement of SV40 T-antigen production. Exp Cell Res 91:247–252.
18. **Horan PK, Jett JH, Romero A, Lehman JM (1974)** Flow microfluorimetry analysis of DNA content in Chinese hamster cells following infection with simian virus 40. Int J Cancer 14:514–521.
19. **Imbert J, Jean-Jaques L, Birg F (1983)** Simian virus 40 antigen is detected only in cells in the G2 phase of the cell cycle in one group of rat transformants. Virology 126:711–716.
20. **Jacobberger JW, Fogleman D, Lehman JM (1986)** Analysis of intracellular antigens by flow cytometry. Cytometry 7:356–364.
21. **Jacobberger JW (1989)** Cell cycle expression of nuclear proteins. In Yen A (ed), "Flow Cytometry and Applications." Boca Raton, FL: CRC Press.
22. **Jacobs RM, Boyce JT, Kociba GJ (1986)** Flow cytometric and radioisotopic determinations of platelet survival time in normal cats and feline leukemia virus-infected cats. Cytometry 7:64–69.
23. **Kimura T, Tanimura E, Yamamoto N, Ito T, Ohyama A (1985)** The quantitative kinetic study of dengue viral antigen by flow cytometry. An in vitro study. Virus Res 2:375–390.
24. **Kimura Y, Yokochi T, Miyadai T, Yoshida K, Yokoo J, Matsumoto K (1985)** Characterization of a porcine kidney cell line resistant to influenza virus infection. J Virol 53:980–983.
25. **Kohn A, Fuchs P (1969)** Cell fusion by various strains of Newcastle disease and their virulence. J Virol 3: 539–540.
26. **Kraemer PM, Tobey RA, Van Dilla MA (1973)** Flow microfluorimetric studies of lectin binding to mammalian cells. J Cell Physiol 81:305–314.
27. **Leary JF, Notter MF (1982)** Kinetics of virus adsorption to single cells using fluorescent membrane probes and multiparameter flow cytometry. Cell Biophys 4: 63–76.
28. **Leary JF, Notter MFD, Todd P (1976)** Laser flow cytophotometric immunoperoxidase detection of herpes simplex virus type 2 antigens in infected cultured human cells. J Histochem Cytochem 24:1249–1257.
29. **Leary JF, Todd P (1977)** Laser cytophotometric detection of variations in spatial distribution of concanavalin-A cell surface receptors following viral infection. J Histochem Cytochem 7:908–912.
30. **Lehman JM, Defendi V (1970)** Changes in deoxyribonucleic acid synthesis regulation in Chinese hamster cells infected with simian virus 40. J Virol 6:738–749.
31. **Lehman J, Horan PK, Cram LS (1979)** Analysis of viral effects on cells utilizing flow cytometry. In Melamed MR, Mullaney PF, Mendelsohn ML (eds), "Flow Cytometry and Sorting." New York: John Wiley & Sons, pp 409–419.
32. **Lehman JM, Klein IB, Cram LS (1979)** Flow cytometry analysis of early DNA content changes in human and monkey cells following infection with Simian virus 40. J Supramol Struct 10:25–31.
33. **Lehman JM, Laffin J, Jacobberger JW, Fogleman D (1988)** Analysis of simian virus 40 infection of CV-1 cells by quantitative two-color fluorescence with flow cytometry. Cytometry 9:52–59.
34. **Lehman JM, Mauel J, Defendi V (1971)** Regulation of DNA synthesis in macrophages infected with simian virus 40. Exp Cell Res 67:230–233.
35. **Lewin B (1980)** Gene Expression 2, 2nd ed. New York: John Wiley & Sons.
36. **Lowry DR (1986)** Transformation and oncogenesis: Retroviruses. In Fields BN, Knipe DM (eds), "Fundamental Virology." New York: Raven Press, pp 235–263.
37. **Martin RG (1981)** The transformation of cell growth and transmogrification of DNA synthesis by Simian virus 40. Adv Cancer Res 34:1–68.
38. **Murray JD, Berger ML, Taylor IW (1981)** Flow cytometric analysis of DNA content of mouse liver cells following in vivo infection by human adenovirus type 5. J Gen Virol 57:221–226.
39. **Notter MF, Leary JF, Balduzzi PC (1982)** Adsorption of Rous sarcoma virus to genetically susceptible and resistant chicken cells studied by laser flow cytometry. J Virol 41:958–964.
40. **Reeve P, Post G (1971)** Studies on the cytopathogenic-

ity of Newcastle disease virus: Relation between virulence, polykaryocytosis and plaque size. J Gen Virol 11:17–24.

41. **Rigby PWJ, Lane DP** (1983) The structure and function of simian virus 40 large T antigen. In Klein G (ed), "Advances in Viral Oncology," Vol. 3. New York: Raven Press, pp 31–57.
42. **Robinson CC, Lehman JM** (1978) Simian virus 40 A gene function. DNA content analysis of Chinese hamster cells transformed by an early temperature-sensitive virus mutant. Proc Natl Acad Sci USA 75:4389–4393.
43. **Robinson CC, Lehman JM** (1982) Simian virus 40 A gene function: Further characterization and growth of tsA transformed Chinese hamster cells. J Cell Physiol 111:225–231.
44. **Robinson CC, Swartzendruber DE, Lehman JM** (1980) Replication of Chinese hamster embryo cells transformed by temperature-sensitive T antigen mutants of simian virus 40. J Virol 35:246–248.
45. **Sinangil F, Harada S, Purtilo DT, Volsky DJ** (1985) Host cell range of adult T-cell leukemia virus I. Viral infectivity and binding to various cells as detected by flow cytometry. Int J Cancer 36:191–198.
46. **Smith HS, Scher CD, Todaro GJ** (1971) Induction of cell division in medium lacking serum growth factor by SV40. Virology 44:359–370.
47. **Stenman S, Zeuthen J, Ringertz NR** (1975) Expression of SV40 T antigen during the cell cycle of SV40-transformed cells. Int J Cancer 15:547–554.
48. **Tegtmeyer P** (1975) Function of Simian Virus 40 gene A in transforming infection. J Virol 15:547–554.
49. **Tenen D, Martin R, Anderson J, Livingston D** (1977) Biological and biochemical studies of cells transformed by simian virus 40 temperature-sensitive gene A mutants and A mutant revertants. J Virol 22:210–218.
50. **Tjian R, Fey G, Graessmann A** (1978) Biological activity of purified simian virus 40 T antigen proteins. Proc Natl Acad Sci USA 75:1279–1313.

31

Application of Flow Cytometry and Sorting to Higher Plant Systems

Michael H. Fox and David W. Galbraith

Department of Radiology and Radiation Biology, Colorado State University, Fort Collins, Colorado 80523 (M.H.F.); Department of Plant Sciences, University of Arizona, Tucson, Arizona 85721 (D.W.G.)

DIFFERENCES BETWEEN PLANT AND ANIMAL CELLS

In considering the application of flow cytometry and cell sorting to plant systems, it is worth remembering that these techniques were developed for animal cells. Animal and plant cells differ in some fundamental ways, and these differences affect the application of flow cytometry and cell-sorting techniques to plant cells. In this chapter, we have chosen to restrict ourselves to the consideration of cells of vascular plants, particularly the seed plants. For a complete review of plant anatomy and morphology, the reader is referred to Esau [35] and to Foster and Gifford [44].

In vascular plants, the cells exist as complex three-dimensional tissues rather than as single cells. Tissues are formed as a function of controlled cell division and elongation during differentation, the individual cells being linked by the presence of a common cellulose cell wall [49]. The cell wall is fundamental to cell division. Cell division in animal cells occurs following the development of an ingrowing cleavage furrow, whereas cytokinesis in vascular plant cells occurs by cell plate formation, except in special cases, such as microspore and generative cell formation.

Different plant cells vary considerably in size. In vascular plant tissues, this situation is complicated further by differences in the state of cellular differentiation, ranging from small, relatively uniform, densely cytoplasmic cells at the meristematic regions (regions of active cell division), to larger cells containing mature chloroplasts typical of leaf tissues, to extremely elongated, fused cells containing little or no cytoplasm, typical of the vascular transport cells of the xylem and phloem. These size ranges in many cases grossly exceed the diameters of the flow cell orifices commonly employed for flow cytometry and sorting of animal cells.

The organized growth of plant cells in tissues is influenced by many factors. The predominant and widespread influences of phytohormones in the control of cell division and expansion are well established, if not well understood. The major phytohormones are divided into five main classes: auxins, cytokinins, gibberellins, abscisic acid, and ethylene. Under the influence of different phytohormone levels, and in the presence of nutritional factors, plant cells can be induced to proliferate aseptically in the form of tissue cultures, either on solid or in liquid (suspension) media [4,5,43,120]. Growth is for the most part heterotrophic, although photoautotrophic cultures can also be successfully established [73].

One spectacular difference between plant and animal cells is the relative ease of regeneration of complete plants from single cells or protoplasts (totipotency). Most attention has focused on the angiosperms, and particular success has been obtained with the dicotyledonous genera. Regeneration of plants from disorganized callus or cell suspension tissues is recognized as occurring through one of two different pathways. The first, organogenesis, involves the induction of shoot meristematic regions. These develop into organized primordia from which bud or shoot-stem structures subsequently emerge. The shoots can then be transferred onto a medium that induces the development of lateral root primordia and the subsequent emergence and elongation of roots [43]. The second developmental pathway, somatic embryogenesis, involves a recapitulation of zygotic ontogeny within the callus tissues [4,5,120]. Through the use of appropriate sectioning techniques, structures corresponding to globular, heart, and torpedo stage embryos can be identified, although cotyledon development is never as complete as within the seed. Somatic embryos germinate through emergence of the radicle and the shoot tip and, in the case of the monocotyledons, the coleoptile [4,5]. Morphogenesis in callus cultures occurs in general in response to removal of auxins from the tissue culture medium. However, the callus must possess morphogenic potential. In certain cases, this potential is linked to genotype or to the type of tissue used as the primary explant, and may require the presence of auxin for its development. Furthermore, callus growth can be accompanied by emergence of morphogenic and nonmorphogenic sectors, so

Flow Cytometry and Sorting, Second Edition, pages 633–650

selective subculturing may be required for the preservation of the morphogenic potential [5,120].

The organogenic and embryogenic pathways both lead to plants that can survive and reproduce under photoautotrophic conditions. However, culture-induced variation (sometimes termed "somaclonal" variation) is frequently observed [99]. The mechanism underlying this variation is not well understood at the molecular level, but may involve, at least in part, the movement of transposable elements. It should be pointed out that plant regeneration through tissue culture is not restricted to the diploid phase. For many plant species, culture of anthers or pollen leads to the emergence of haploid callus which subsequently gives rise to haploid plants through organogenesis. Alternatively, the embryogenic pathway can be followed directly. Thus, the greatly reduced haploid phase typical of seed plants (the gametophytic tissues of the anther and ovary) can be artificially extended through the application of tissue culture techniques [8,82,109,124,125]. Somewhat surprisingly, haploid plants appear, for the most part, quite normal, although typically smaller than the diploid parent. They can be propagated indefinitely by vegetative means.

The successful growth of haploid plants is a specific example of a general observation that plants exhibit an unusual tolerance toward alterations in DNA and ploidy. Thus, plants exhibit a broad interspecies range of nuclear DNA contents, varying from C = 0.2 pg *(Arabidopsis thaliana)* to 127.4 pg *(Fritillaria assyriaca)* [10]. Many individual species also exist in nature as polyploid series. Furthermore, interspecies hybridization is commonly observed and thus blurs the distinction of true species. Aneuploids are viable and can occupy ecological niches, especially when accompanied by vegetative modes of propagation [83]. Finally, the tracing of phylogenetic relationships re-emphasizes the role that polyploidization has played as an evolutionary mechanism. The role of the additional DNA generally present in plant nuclei is not well understood. Much of it is middle-repetitive, for example, in *Zea mays,* in which more than 30% of sequences of the nuclear genome fall into this category [68] and may be the result of transposable element activity [48]. Interestingly, variation in genome size is also observed within the mitochondrion. For the cucurbits, for example, the mitochondrial genome varies in size from 220 to 1,600 Mdal [86]. Once again, the underlying reason for this variation is not clear. By contrast, the chloroplast genome seems to be more highly conserved, being around 99 Mdal in size for most higher plant species [126].

At the cellular level, other unique features of higher plants include the presence of organelles not found in animal cells. Two that are particularly important in our consideration of the use of flow cytometry and cell sorting are the plastids and the vacuole. Plastids are probably the most characteristic organelles of eukaryotic plant cells, categorized on the basis of the type of pigmentation that they contain. They are subdivided into three main categories: the chloroplasts, chromoplasts, and leucoplasts [35]. Chloroplasts are the predominant organelle involved in photosynthesis. They contain chlorophyll and are particularly well developed in leaves. Chromoplasts contain other pigments, such as carotenoids, and are found in colored plant parts such as petals. Leukoplasts are nonpigmented and include undifferentiated plastids such as amyloplasts. Plastids are semiautonomous organelles, in that they contain a genome and ribosomes. Their inheritance is most commonly uniparental (maternal) in nature, though notable exceptions include *Oenothera* and *Pelargonium.* The vacuole frequently comprises the largest single organelle within the plant cell [121]. It possesses a simple ultrastructure. A single bilayer membrane, the tonoplast, bounds an aqueous space that can occupy as much as 90-95% of the cellular volume. The body of the vacuole is frequently traversed by cytoplasmic strands.

The particular importance of plastids and the vacuole to flow cytometry and sorting lies in the observation that many of the natural fluorophores and chromophores of plant cells are localized within these organelles. The chlorophylls are predominant examples of these fluorophores. Others are compounds generally classified as secondary products, such as alkaloids, isoprenoids, and phenolics [9]. Secondary products are defined as classes of molecules that do not play a role in the primary metabolic processes required for the maintenance of life in an individual plant. However, many are essential for the survival of the species in its natural habitat, such as through allelopathy. They are frequently of pharmaceutical, industrial, or cosmetic value. Hydrophilic secondary compounds are often sequestered in the vacuole [9], whereas hydrophobic secondary compounds accumulate in membranes, particularly those of the plastids, or are secreted into the extracellular milieu. Secondary products are frequently concentrated in specific tissues. Therefore, gene expression required for their synthesis is almost always under developmental control and will not normally be observed in relatively undifferentiated plant tissue culture systems. However, somaclonal variation in culture, coupled to the sorting of individual protoplasts on the basis of endogenous fluorescence, has led to the possibility of plant cultures that produce specific secondary products in higher yields. Unfortunately, the wide and continuously expanding spectrum of potentially important secondary compounds has not been paralleled by a comprehensive compilation of the spectroscopic properties of these molecules, although some information can be found [9,25,104]. This has hindered the identification of plant protoplast systems appropriate for cell sorting for enrichment purposes.

THE CELL WALL

The initial problem encountered in the application of flow cytometry and cell sorting to higher plant systems is a consequence of the cell wall. First, it is necessary to dissociate cell clusters to obtain a single-cell suspension. Second, plant cells have a variety of shapes, which may result in positional artifacts in a flow cytometer or lead to unacceptably high coefficients of variation in the signals. Essentially the only way to bypass these problems is to remove the plant cell wall under conditions of osmotic stability. This can be achieved for a large variety of plant cell types through the use of commercial polysaccharidases (cellulases, hemicellulases, and pectinases), in the presence of an inert osmoticum such as mannitol or sorbitol. These enzymes degrade the complex polysaccharides contained in the cell wall, leaving the protoplast within a plasma membrane [37,38,90,119]. The plant plasma membrane differs from that of most animal cells. It lacks an extensive glycocalyx and apparently has less interaction with the cytoskeleton. Thus, the plasma membrane is more fragile than its animal counterpart. Large, dense organelles such as chloroplasts and their cellular segregation in the cytoplasm by an equally fragile tonoplast combine to make protoplasts exceptionally delicate structures. Protoplasts are typically 30–90 μm in diameter, although they can range from 15 μm to 200–300 μm, depending on species and tissue type.

Techniques have been developed for the successful culture of isolated protoplasts, particularly in the dicotyledonous taxa. Under suitable conditions, cell wall regeneration and cell division can be observed after 24–48 hours [37,119]. A subsequent reduction of the osmotic pressure in the culture medium usually promotes the formation of cell clusters and calli. These can be transferred onto solid medium and, for many species, subsequently induced to undergo organogenesis or embryogenesis, giving rise to photoautotrophic plants. Since protoplasts possess a plasma membrane that is exposed to the culture medium, it has been possible to induce fusion between protoplasts of differing species [26,36,79]. This has led to the artificial combination of genomes not possible through sexual means. The role of fluorescence-activated cell sorting in this area is discussed in detail later in this chapter.

INSTRUMENT CONSIDERATIONS AND MODIFICATIONS

The major modification to instrumentation is the use of large flow cell tips, required because of the size and fragility of protoplasts [69,74,106]. Large flow cell tips (range: 100–200 μm) require other adjustments within the flow system. First, the sheath pressure must be lowered, both to maintain conditions of laminar flow through the flow cell tip [78], and to observe droplet formation. Second, the frequency of excitation of the piezoelectric crystal must be reduced to obtain precise droplet breakoff with a larger diameter stream [71,74].

With the current commercially available cell sorters, there is a practical limitation on flow cell tip diameters. The larger droplet size and lower droplet breakoff point require lowering the deflection assembly to view the droplets. The lower flow rate required for laminar flow and the reduced frequency of droplet breakoff results in fewer droplets generated per second and slower sorting. Thus, considerable modifications to the deflection assembly may be required in order to increase the overall sort rate without loss of accuracy or protoplast viability.

An alternative approach to the protoplast size problem involves sample modification. Reducing the size of the protoplasts would enable the use of small flow cell tips, allowing more rapid sorting. This can most easily be achieved through the use of a hypertonic sample solution, which shrinks the protoplast. Protoplasts appear quite tolerant of changes in osmotic environment, particularly under chilled (4°C) conditions [54]. Another possibility is to use miniprotoplasts [40,87,122].

Minor problems have been encountered with plant protoplasts and pollen sample preparation due to sample settling and clumping. Cell settling can be alleviated by gentle agitation, for example by magnetic stirring devices attached to the sample introduction vessel. Cell clumping can be avoided by proper filtering of the sample.

The concept of sample modification in order to avoid size-related effects can also be applied to studies in which protoplast viability is not required. An example is DNA content measurement and analysis of the cell cycle. The plant nucleus is comparable in size to that of animal cells, about 5–10 μm in diameter. However, in a plant protoplast, it occupies an area on the periphery (the perivacuolar cytoplasm) of a much larger sphere. Since one cannot predict, a priori, the orientation of the protoplast as it passes through the laser illumination, there will be substantial variation in the excitation intensity depending on the various paths traversed by the nuclei, inevitably broadening the fluorescence histograms. This problem can be avoided by lysing the unfixed protoplasts under hypertonic conditions in which the nuclear membrane remains intact [53,58]. In this case, the stained nuclei are analyzed separately as small fluorescent spheroids, corresponding to their natural size, with a dramatic improvement in precision of measurement.

An alternative method of flow sorting is available that does not depend on generating droplets (see Chapter 8). Instead, the flow cell consists of two channels, one of which contains a piezoelectric crystal that causes pressure variations briefly diverting the sorting stream to the other channel [18]. This may be a more gentle method for sorting protoplasts, but Bromova and Knopf reported that 23% of the protoplasts were damaged. Protoplast viability >90% has been reported with conventional flow cells with larger orifices [71].

REVIEW OF PAST AND CURRENT WORK

The use of flow cytometry in studying higher plants has a very short history compared with that of animal cells. Nevertheless, several types of measurements have been made in plant systems using flow cytometry. These include measurements of DNA and RNA content, cellulose biosynthesis, chromosome analysis, alkaloid production, fluorescent antibody labeling, binding of lectins to mitochondria, and dual staining of two separate protoplast populations to detect fused heterokaryons. Much of the work is very recent, and the quality is variable. A substantial part is not yet published, although submitted and in press or presented at meetings. Thus, although this chapter reviews the developments to date, it is clear that work in this area is proceeding very rapidly, and inevitably this review will soon be out of date.

DNA Content

Much of the published work involving flow cytometry of plant systems has centered on attempts to measure DNA content and cell-cycle distributions. Galbraith et al. [62] set the stage for DNA measurements by using Hoechst 33258 to study ploidy levels of cultured tobacco protoplasts *(Nicotiana tabacum* cv. *Xanthi).* These investigators used fluorescence microphotometric techniques to quantitate the changes in DNA content of protoplast nuclei as a function of time in culture.

After 6 days in culture, nuclei were measured with DNA levels ranging from diploid to octaploid. However, on further culture, cell clusters were formed in which all nuclei were at the diploid level. This reduction, accompanied by unequal or incorrect chromosome segregation might explain, at least in part, the emergence of tissue culture-induced variation.

Flow cytometry was applied to this problem in a study of the effects of inhibitors of cell-wall synthesis on tobacco protoplast development [63]. The inhibitors used were 2,6-dichlorobenzonitrile (DB) and coumarin. Protoplasts were fixed in 1% sorbitol in ethanol–acetic acid–water (18 : 3 : 2 v/v/v), followed by 70% ethanol, then stained with mithramycin. The DNA content of growing protoplasts was measured with a log amplifier, such that peaks were present at 2C, 4C, 8C, and 16C (Fig. 1). This is because growing protoplasts do not undergo cytokinesis in the presence of inhibitors of cell wall synthesis. Coumarin inhibited the development of these multinucleate protoplasts, but DB did not. These results were correlated with microscopic measure-

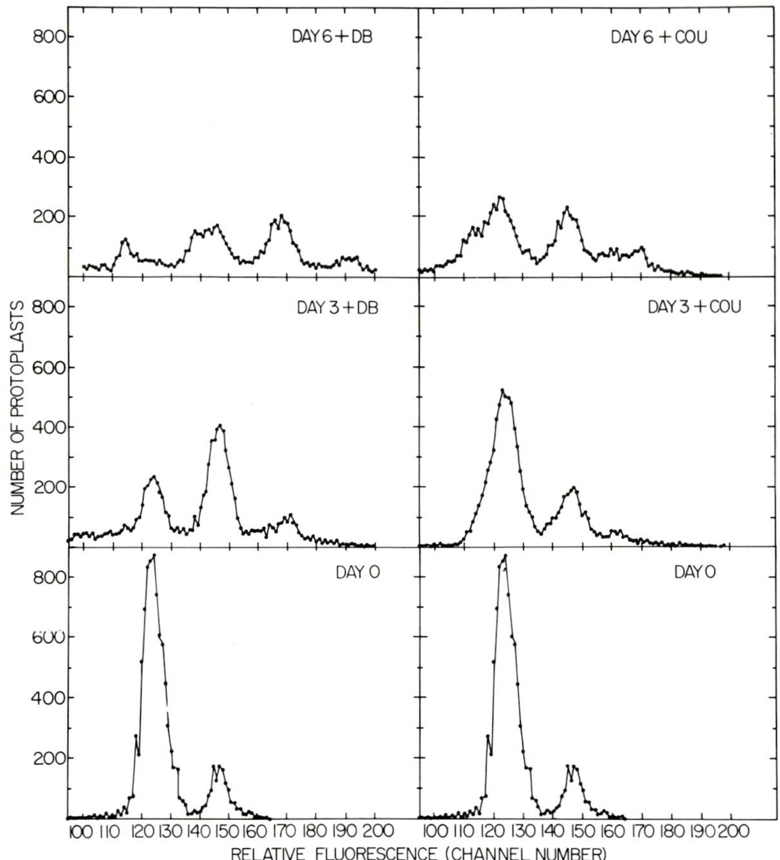

Fig. 1. Changes in nuclear DNA contents of the developing protoplasts as a function of time in culture. DB, + dichlorobenzonitrile; COU, + coumarin. Log fluorescence is plotted on the ordinate. (Reproduced from [63] with permission of the publisher.)

ments of cells with 1, 2, or >2 nuclei. DB inhibited cell wall biosynthesis as measured by fluorescence microphotometry of Calcofluor White, a cell wall specific stain [51,90]. It was concluded that DNA synthesis and cell wall synthesis are not tightly coupled processes.

Another approach to this problem has been reported by Firoozabady [41]. Different protoplast preparations of *Gossypium hirsutum L.* were analyzed for cell cycle parameters, while parallel samples were analyzed by Calcofluor white for cell wall regeneration after 4 days in culture. There was an inverse correlation between %G1 and subsequent cell wall regeneration and a direct correlation between %G2 and cell wall regeneration. Since no histograms were shown, the quality of the results could not be determined.

In a study of DNA synthesis and cell cycle activity in the B73 maize cell line [88], protoplasts were stained with Hoechst 33258 after various times in culture. However, the protoplasts were treated in a highly unusual manner, employing 50 μg/ml DNAase and 0.006% Triton X-100, prior to staining. The histograms were presented on a log display and were impossible to interpret accurately in terms of conventional cell-cycle components.

The first report of an attempt to stain the nuclei of plant protoplasts under viable conditions using Hoechst 33342 was by Puite and Ten Broeke [96]. They employed protoplasts from suspension culture of *Haplopappus gracilis* and *Solanum tuberosum* cv. *Bintje* to study the synchronizing effects of hydroxyurea (HU) and colchicine. Both fixed (ethanol–acetic acid 3 : 1) and unfixed protoplasts were labeled with Hoechst 33342 (1 μg/ml) and analyzed by flow cytometry. However, coefficients of variation (CV) were quite large, and only a small number of protoplasts were analyzed, giving histograms of poor quality, even for the fixed protoplasts. Protoplasts stained viably gave histograms similar in shape to light scatter rather than DNA content. Much better distributions were obtained when nuclei were isolated with Triton X-100 treatment, probably due to the elimination of positional errors and more quantitative staining. Colchicine did block a portion of the population in G_2/M, but HU treatment resulted in cells spread throughout S phase. This was probably because the block was for 20 hours, which has been shown to allow cells to move into S phase (R.G. Alexander, M.E. Wilder, C.M. Naranjo, K.R. Richards, and P.J. Jackson, personal communication, 1985). Similar results have been obtained with HeLa cells when they are blocked with HU for 20 hours or more (M.H. Fox, unpublished results).

Galbraith et al. [58] developed a general technique for obtaining nuclei from plant tissues, involving chopping in a buffer containing 45 mM magnesium chloride, 30 mM sodium citrate, 20 mM 4-morpholine–propane sulfonate (MOPS), and 1 mg/liter Triton X-100. This process released

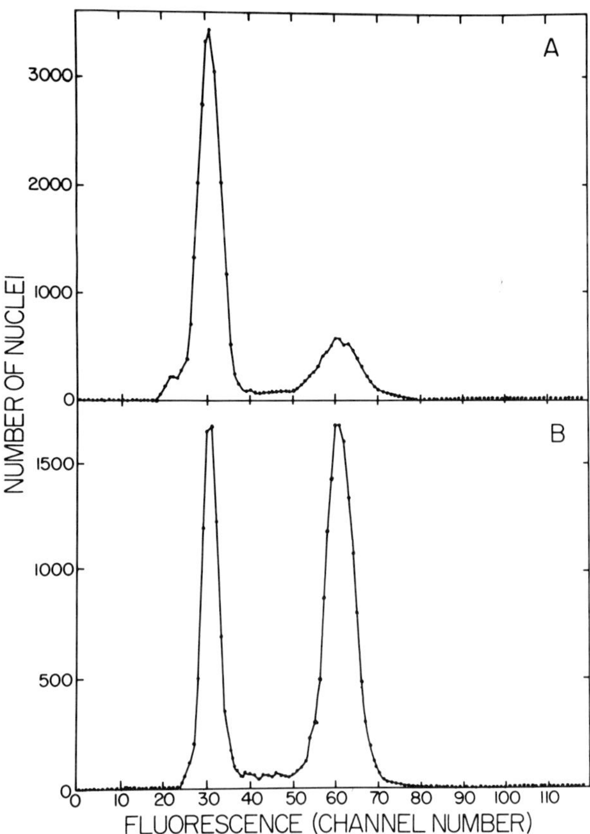

Fig. 2. Flow cytometric analysis of nuclei released by chopping of leaf **(A)** or terminal root **(B)** tissue of *N. tabacum*. The nuclei were stained with mithramycin before analysis. The coefficients of variation for the G_1 peaks were 8.1 and 6.0%, respectively. **A:** G_1, 71.1%; S, 6.6%; and G_2, 22.3%. **B:** G_1, 30.5%; S, 13.6%; and G_2, 55.9%. (Reproduced from [58], Copyright 1983 by the AAAS, with permission of the publisher.)

nuclei in an intact state. After filtration, the nuclei were stained with mithramycin (100 μg/ml) and were analyzed by flow cytometry. Histograms of high quality were obtained for nuclei of leaf or root tissue of *Nicotiana tabacum* (Fig. 2). DNA content was measured for 26 plant species with CV's ranging from 3.8% to 10.8%. Chick red blood cells (CRBC) were used as a standard. The results compared fairly well with those obtained using conventional microdensitometry reported by other workers. An interesting application of DNA analysis was the demonstration that the distribution of cells between G_1 and G_2 phases changed considerably as a function of leaf age. The percentage of cells in G_1 decreased from around 80% for cells from new leaves (apex of the plant) to about 45% for cells from leaves at the bottom of the plant, with a concomitant change in G_2 from 15% to 40%. The percentage of cells in S phase changed only slightly. The one disadvantage of this technique is that mitotic cells would not be analyzed, since mitotic cells do not contain a nuclear membrane and would be lost from the population of nuclei.

This method of isolating nuclei for DNA flow cytometric measurements was also used in a study of plants derived from anther culture of *Nicotiana particulata* and *N. sylvestria* [101]. It was demonstrated by flow cytometry that the leaf nuclear DNA content of anther-derived plants was haploid by comparing the histograms with those obtained on nuclei from diploid plants as well as CRBC. The technique was used in a similar way to discriminate haploid and diploid plants produced in anther culture of *Saintpaulia ionantha* [15] and to examine genotypic variability in nuclear DNA content in maize [7].

Galbraith [53] reviewed their work on the use of flow cytometry for DNA content measurements of higher plants and gave detailed descriptions of appropriate buffers for preparing leaf protoplasts to obtain high-quality DNA histograms.

Another approach to DNA analysis of plant tissue involved fixation with 4% formaldehyde in a Tris buffer with 0.1% Triton X-100, followed by DAPI staining [100]. Nuclei were released from the fixed tissues with higher yield and better CVs (around 3%) than from fresh tissue. They also showed large variations in cell cycle parameters for cells taken from different zones of roots. The ability to use fixed tissue gives greater flexibility by allowing samples to be stored prior to analysis. Presumably other DNA stains could also be used with fixed tissue and give good results.

Protoplasts have also been lysed in water and the nuclei stained with DAPI [116] with relatively good CV, or lysed with detergent/trypsin procedures followed by propidium iodide staining, with CV's of about 5% [107].

Flow cytometric cell-cycle analysis of *Datura innoxia* and *Petunia hybrida* suspension cultures were reported by R.G. Alexander et al. (personal communication, 1985). Isolated protoplasts were filtered and fixed with cold 95% ethanol, then stained with 100 μg/ml mithramycin for DNA analysis. Cell cycle parameters obtained by computer analysis of the FCM histograms were compared with "classic" techniques [23] with excellent results. They also used radioactively labeled precursors of DNA or protein synthesis, [^{3}H]thymidine or [^{3}H]leucine, respectively, combined with cell sorting to analyze the effects of HU or colchicine on protoplast culture.

Another example of DNA measurements to characterize plant heterogeneity is a report of large interspersed populations of isogenic polyploids of *Andropogon gerardii* within natural habitats [81]. Flow cytometry was also used to determine whether embryonic cell-suspension cultures of *Panicum maximum* and *Pennisetum purpureum* remained diploid or became aneuploid [80]. It was concluded that there was a strong selection for euploidy, based on serial DNA distributions over a period of 8–9 days growth in culture. Further work on *Pennisetum purpureum* showed that there were no significant differences in DNA content between leaves of low and high embryogenic competence [111].

Anderson et al. [6] used flow cytometry to compare the DNA content of protoplasts derived from pollen tetrads from 10 species of plants with widely varying DNA content. Protoplasts were fixed in 70% ethanol and stained with chromomycin A3. The DNA content was compared with the length of the synaptonemal complex in pachytene, with a high linear correlation (r = 0.97). The DNA content also compared well with values in the literature obtained by Feulgen staining. An unusual result reported by Fox et al. [45] (also M.H. Fox, L.K. Anderson, and S.M. Stack, personal communication, 1986) was that of 13 plant species analyzed, only 3 showed the single peaks expected of these haploid cells, while the rest showed double peaks, with 10–20% differences in DNA content. When the two peaks were sorted and reanalyzed, they appeared in the same respective chan-

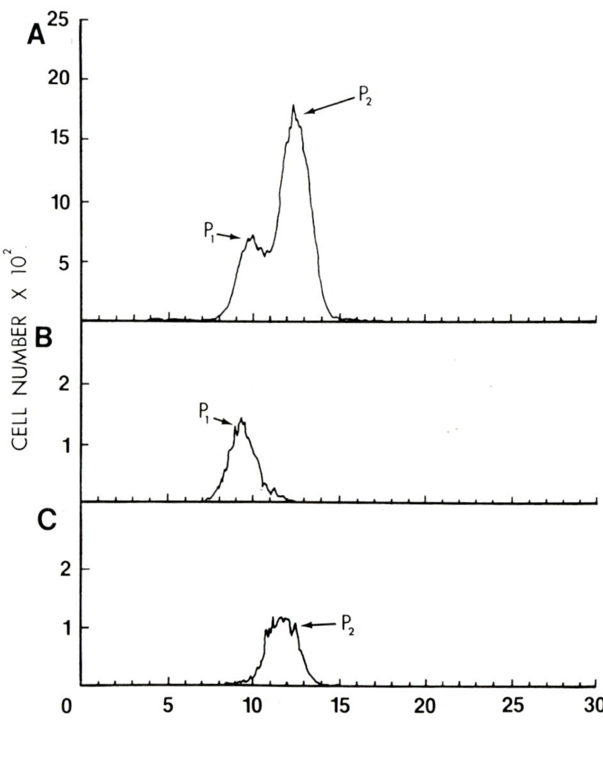

Fig. 3. Distribution of *L. esculentum* pollen tetrad protoplasts based on DNA content determined by flow cytometry. **A:** The cells of this species are distributed in two peaks (P1 and P2). **B:** Cells were sorted from P1, collected, and reanalyzed. **C:** Cells were sorted from P2, collected, and reanalyzed. (Reproduced from [6] with permission of the publisher.)

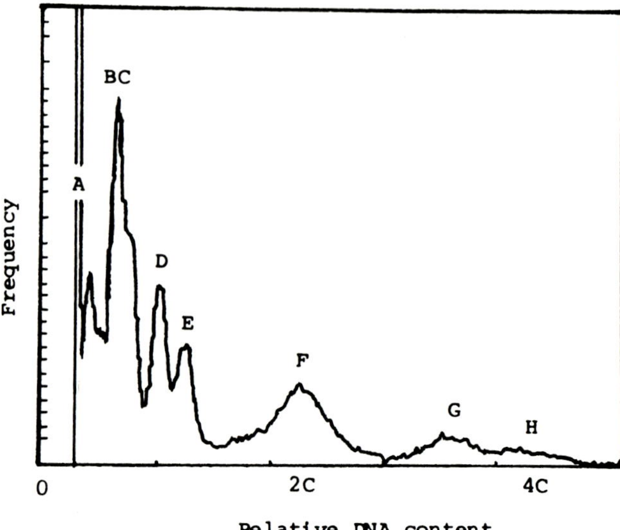

Fig. 4. Frequency distribution of the relative DNA content of ethidium bromide-stained chromosomes and nuclei from mechanically disrupted protoplasts. By flow sorting, peaks were identified as A, debris; B,C, acrocentric; and D,E, metacentric metaphase chromosomes; and F, G_1, nuclei; G, mitotic nuclei; and H, G_2 nuclei. (Reproduced from [30] with permission of the publisher.)

nels, showing that the differences were stable (Fig. 3). Whether these peaks represent an artifact of pollen protoplasts or whether they represent true differences in DNA content is not certain.

In a further example of the value of cell-cycle analysis, *Petunia hybrida* protoplast nuclei stained with ethidium bromide were used to show that cytoplasmic male sterility factor (cms) slows cell growth [12]. Finally, flow cytometry was used to analyze hybrids derived from fused parental protoplasts differing in DNA content [107]. Unexpectedly, the hybrids had twice the DNA content as the sum of the parental protoplasts.

Chromosomes

Relatively little work has been done on analysis of plant chromosomes with flow cytometry. de Laat and co-workers [29–31] partially synchronized suspension cultures of *Haplopappus gracilis* (2n = 4) with HU followed by colchicine to obtain mitotic cells. Protoplasts were swelled in 300 mM mannitol, lysed with a buffer of pH 7.0 containing 15 mM HEPES, 1 mM EDTA, 15 mM DTT, 0.5 mM spermine, 80 mM KCl, 20 mM NaCl, 300 mM sucrose, 1% Triton X-100, and 50 μg/ml ethidium bromide, then forced through a 25-gauge hypodermic needle two times. The resulting histograms (Fig. 4) showed that the individual chromosomes were present, as well as whole nuclei. Interestingly, the homologous chromosomes differed in DNA content, giving two peaks. When the chromosome lengths were measured, similar results were obtained. If this is generally true, it might explain the results of Fox et al. [45] (also M.H. Fox and colleagues, personal communication, 1986). Figure 4 also shows mitotic and G_2 nuclei that are distinguishable on the basis of DNA content (peaks G and H). Mitotic nuclei should not exist, since there is no nuclear membrane in mitotic cells, but in fact peak H consists of aggregates of four chromosomes. A distinction between mitotic and G_2 cells based on DNA staining has not been seen to our knowledge in animal cell culture and was not reported in any of the plant work reviewed by us. While Ethidium bromide was not used in the other plant studies, G_2 and mitotic cells of both HeLa and Chinese hamster ovary cultures give only a single peak when stained with ethidium bromide (M.H. Fox, unpublished results). On the other hand, it has been demonstrated that mitotic and G_2 cells can be resolved with acridine orange staining after heat or acid shock [27]. If the measurable difference between mitotic and G_2 cells, reported by de Laat and Blaas [29–31], can be confirmed it would suggest a very interesting difference between plant and animal cells.

Preliminary results of flow cytometric analysis and sorting of *Petunia hybrida* chromosomes were recently presented [24]. Although this plant species contains seven pairs of chromosomes, only three chromosome peaks were resolved in the histogram. Clearly, further work must be done with plant chromosome preparations to approach the resolution obtained with animal chromosomes.

RNA Content

The use of acridine orange (AO) to measure both RNA and DNA simultaneously was pioneered by Darzynkiewicz and co-workers [27]. This technique was used recently by Bergounioux et al. [11] to analyze nuclei obtained from protoplasts. Nuclei from freshly isolated protoplasts had a low RNA content, which increased significantly by 24 hours in

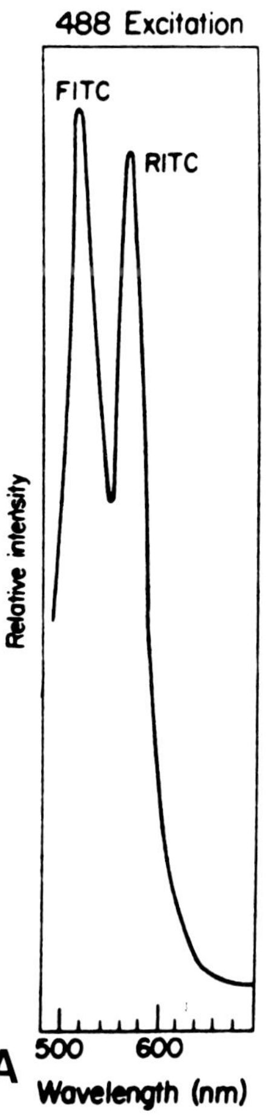

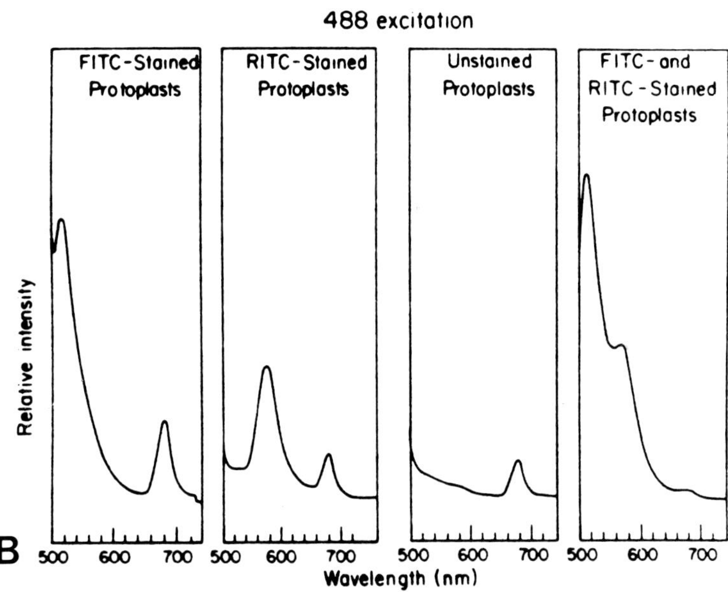

Fig. 5. **A:** Emission spectra for separate 5-mg/ml solutions of fluorescein isothiocyanate (FITC) and rhodamine isothiocyanate (RITC) dissolved in absolute ethanol, at 488-nm excitation. **B:** Emission spectra for differentially stained *E. lathyris* protoplasts at 488-nm excitation. Stain concentrations per ml of enzyme–protoplast solution were 3.6 μl for FITC-stained protoplasts, 7.2 μl for RITC-stained protoplasts, and 3.6 μl RITC for double-stained protoplasts. (Reproduced from [97] with permission of the publisher.)

culture, when the cells were actively cycling. The patterns were very similar to those obtained in lymphoid mammalian populations, showing a quiescent G_0 population initially but a higher RNA content upon stimulation into the cell cycle.

Protoplast Fusion

One of the most promising applications of flow cytometry and cell sorting for plant studies involves the ability to sort viable protoplasts and to regenerate plants from these protoplasts. Of particular interest is the possibility of applying fluorescence-activated sorting for the isolation of heterokaryons formed by protoplast fusion. This can lead to the development of combinations of genomes in plants that would be difficult or impossible to obtain by classical genetic techniques. Much effort has been devoted to developing methods to do this. Instrumentation factors play a significant part in sorting viable protoplasts, as discussed in the previous section.

In order to sort heterokaryons, two cell populations must be differentially labeled and then fused. Plant cells grown in suspension culture typically lack high levels of endogenous fluorochromes, such as chlorophyll. Galbraith and Galbraith [55] approached this problem in suspension cultures of *Glycine max.* using lipophilic fluorescent probes of fluorescein isothiocyanate (FITC) and rhodamine B conjugated to octadecylamine and octadecanol, respectively. When these probes (termed F18 and R18) were added to freshly isolated protoplasts, little labeling occurred, but when cell cultures were preincubated for 12 hours in the presence of the probes, the membranes became labeled. It was then possible to prepare labeled protoplasts. They demonstrated that the presence of F18 or R18 did not inhibit protoplast growth, but that the ethanol in which the probes were dissolved could inhibit growth if the concentration was >0.04 μl/ml. They also demonstrated that the labeled cells exhibited normal morphology, cytoplasmic streaming occurred, labeled protoplasts could be fused, and hybrids could be identified by fluorescence microscopy. No flow cytometry was done in this study.

Galbraith and Mauch [61] followed up this work by labeling cellular proteins using FITC, rhodamine isothiocyanate (RITC), tetramethyl rhodamine isothiocyanate (TRITC) and XRITC. *Nicotiana* mesophyll protoplasts were labeled with these dyes while in the protoplast isolation enzyme solution for 18–21 hours. This method gave uniform cellular staining with all of the dyes. The protoplast yield was not affected by the labeling and plantlets were regenerated from labeled protoplasts, demonstrating that viable protoplasts were obtained. Furthermore, protoplast growth was normal as measured by packed cell volume and protein increases as a function of time in culture. Fusion of labeled protoplasts with polyethylene glycol (PEG) resulted in heterokaryons at a frequency ranging from 0.1 to 2.0%. No flow cytometry or sorting was used and fusion products were not selected individually for growth. These techniques were used later with flow cytometry to demonstrate their applicability and were patented [50,56]. Patnaik et al. [92] used similar procedures to label protoplasts with FITC and manually isolated heterokaryons with a micromanipulator, based on FITC and endogenous chlorophyll fluorescence.

Redenbaugh et al. [97] used FITC and RITC, as well as endogenous chlorophyll fluorescence and light scatter, to characterize protoplasts from five species—*Euphorbia lathyris, Nicotiana glauca, N. langsdorfii, Petunia parodii,* and *P. inflata*—using flow cytometry and cell-sorting techniques. Protoplasts were stained by inclusion of the fluorescent dyes in the enzyme soltuion, as described above. Because of the presence of endogenous chlorophyll, which has a fluorescence emission maximum of 680 nm, a bandpass filter has to be used to exclude chlorophyll fluorescence from the RITC signal. Figure 5 shows fluorescence spectra of these chromophores. It is obvious that spectral overlap between FITC and RITC, as well as chlorophyll, can cause some problems since all three are excited by the 488-nm line used in this study. These spectral overlaps can be partially corrected on most commercial instruments by special circuitry and by selection of appropriate filters. Redenbaugh et al. [97] used a 540-nm dichroic, a 526-nm bandpass (bandwidth of 22.6 nm), and a 577-nm bandpass (bandwidth of 25.5 nm). They were able to distinguish double-stained protoplasts from FITC-stained or RITC-stained protoplasts on a two-parameter histogram using flow cytometry. They also sorted protoplasts based on light scatter and chlorophyll fluorescence and reported the recovery of intact protoplasts. Protoplast viability in culture was not measured. They could not sort on FITC and RITC labeling because the cell sorter they had available had only one photomultiplier tube (PMT).

Galbraith [52] reviewed their work on labeling protoplasts with either the lipophilic probes F18 and R18 or FITC and RITC and gave detailed procedures for isolating and labeling protoplasts. Heterokaryons from PEG-induced fusion of FITC- and RITC-labeled protoplasts were sorted and analyzed microscopically for the presence of both labels. Purities ranging from 78% to 100% were obtained. Protoplasts were also sorted based on chlorophyll fluorescence to determine the effects on viability. Sorting efficiencies were around 50%, with plating efficiencies of 100%. Sorted protoplasts exhibited normal growth in culture.

Harkins and Galbraith [69] used endogenous chlorophyll fluorescence of *Nicotiana tabacum* cv. *Xanthi* leaf protoplasts to evaluate instrumental conditions necessary to obtain the most successful sorting of viable protoplasts. They sorted 1,000–8,000 protoplasts into wells of a microtiter plate with the Coulter EPICS Autoclone system and observed growth in all wells containing more than 3,300 protoplasts (Fig. 6). Growth of the protoplasts into calli was observed and plants were regenerated from those calli. A novel method was developed to determine the proportion of sorted protoplasts that were broken by comparing the number of free chloroplasts and whole protoplasts (Fig. 7). It was determined by light microscopy that these protoplasts contained 72 ± 23 chloroplasts, so the percentage of broken protoplasts could be directly measured by integration of the areas on the flow histograms representing intact protoplasts and free chloroplasts. With this method, they determined that in most cases >97% of the protoplasts remained intact in the flow stream using either a 76-μm or 100-μm flow cell. However, the sort efficiency was significantly better with the 100-μm orifice, yielding a viable recovery of 36% based on vital staining with fluorescein diacetate (FDA). Viability of up to 46% was achieved when hypertonic medium and low temperatures were used to reduce protoplast fragility. Approximately 50% of these grew in protoplast culture, giving a final yield for viable protoplasts of approximately 25%. Similar results were obtained with FITC-labeled protoplasts [54], with viable recovery of 30%. Sorted protoplasts grew up into calli with subsequent shoot induction (Fig. 8).

More recent work with larger flow cells demonstrated that protoplasts could be sorted with an efficiency and viability >90% with a 204-μm diameter flow cell [71]. The droplet frequency must be reduced to accommodate the larger droplets. Nevertheless, adequate sort rates of 200–500/sec can be obtained.

Bergounioux et al. [13] reported sorting protoplasts with 60% recovery of intact protoplasts for *Petunia* or *Medicago* mesophyll and 83% recovery for suspension protoplasts of *Catharanthus*. They also reported that when the suspension buffer was suboptimal (0.65 M instead of 0.75 M), cell recovery declined.

Alexander et al. [3] used either FDA or endogenous chlorophyll as markers for populations of protoplasts from *Barley* cv. *Aramir* mesophyll and *Datura innoxia* suspension culture. Protoplasts were stained with 4 μg/ml FDA in the protoplast isolation enzyme mixture. Protoplasts were fused with PEG, resulting in a fusion frequency of 0.2–1%. Sorting was done with a 200-μm orifice obtained from Becton Dickinson [74] to permit sorting of larger heterokaryons. The piezoelectric crystal frequency was 4.5 kHz. Protoplast diameters were 20 μm for mesophyll and 30 μm for suspension populations, and the fused heterokaryon diameters were approximately 40 μm. The parental protoplast populations were found to differ in light scatter signals as well. Figure 9 shows the results comparing an unfused mixture of protoplasts with a fusion mixture, with the same total number of protoplasts analyzed. Clearly, a larger number of protoplasts appear in the region indicating both red and green fluorescence after fusion. When these protoplasts were sorted and analyzed by fluorescence microscopy, 45% of them were heterokaryons, compared with a background of 0.2% for unfused protoplast mixtures. This indicates that 55% of the protoplasts with both red and green fluorescence were presumably clumped rather than fused. Viability after sorting, and the capability of the fusion products to proliferate into callus, was not tested. It should be pointed out that these heterokaryons were formed between a monocotyledonous and a dicotyledonous plant species, for which development is unlikely to be compatible. Other labeling procedures have been used in the sorting of heterokaryons, including exoge-

Fig. 6. Protoplasts of N. tabacum cultured within microtiter plate wells after flow sorting. **A:** Immediately following sorting. **B:** After 1 day. **C:** After 2 days. **D:** After 6 days. **E:** After 9 days. **F:** After 16 days. Bar = 100 μm. (Reproduced from [69] with permission of the publisher.)

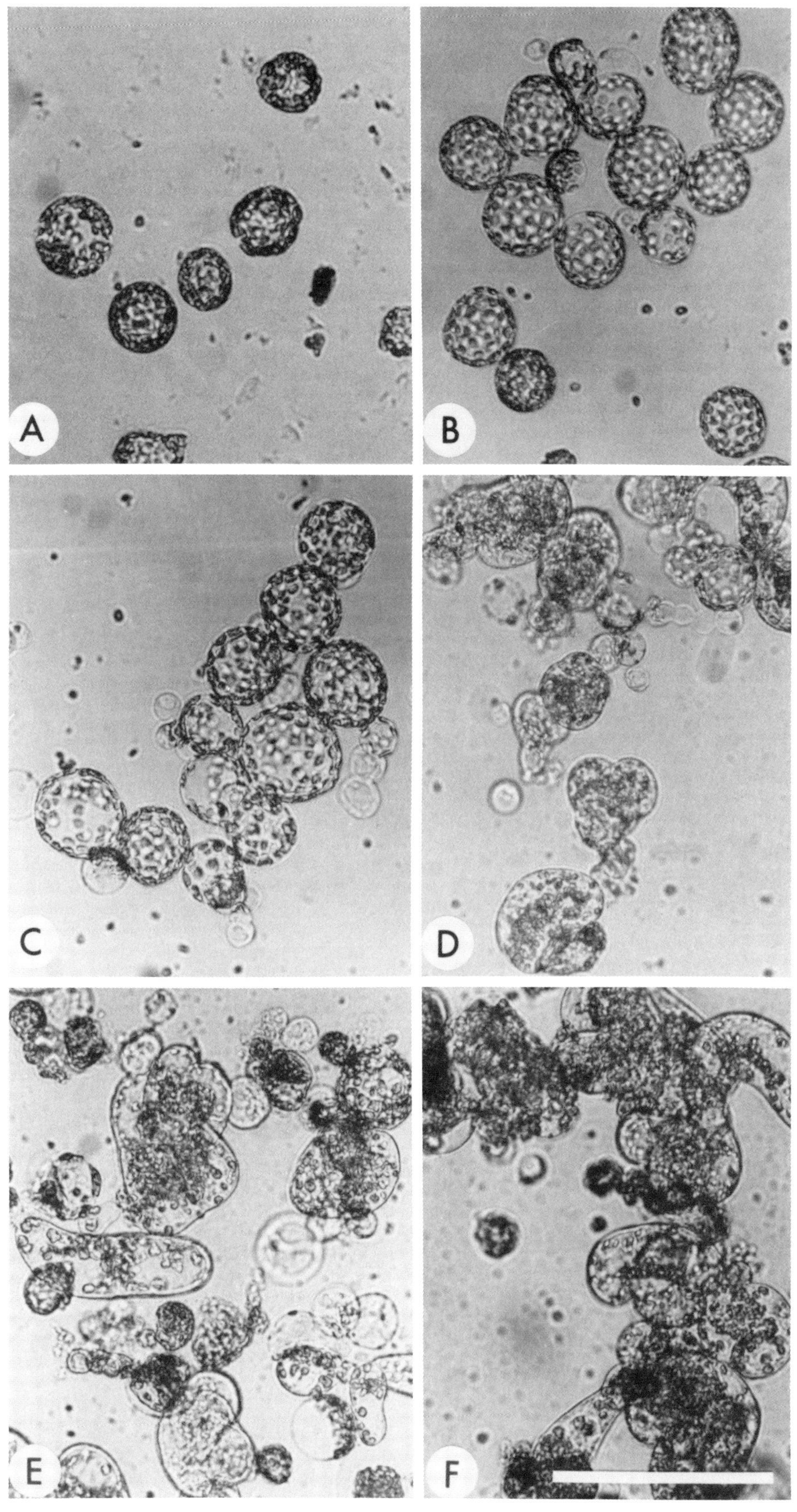
A
B
C
D
E
F

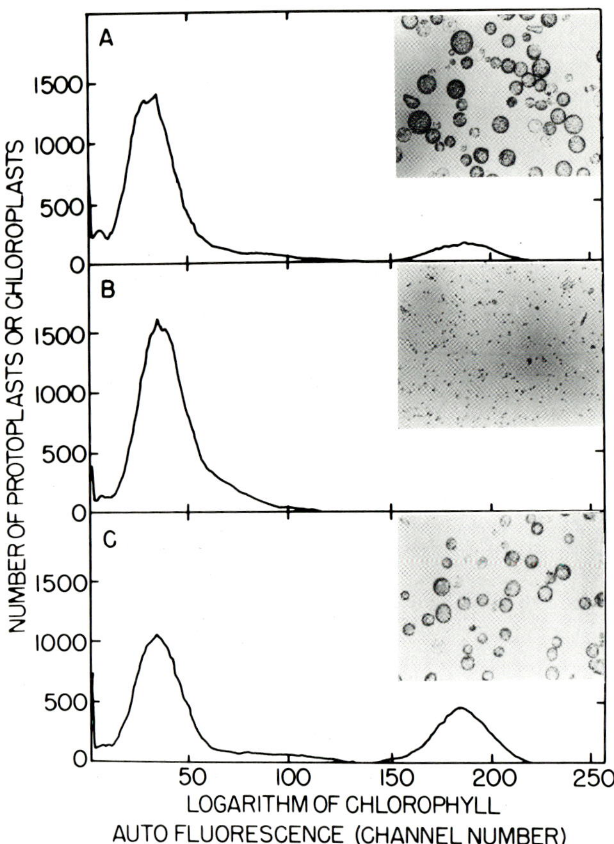

Fig. 7. Histograms obtained by flow analysis of populations of protoplasts and chloroplasts based on chlorophyll autofluorescence. **A:** Freshly prepared and purified protoplasts. **B:** Protoplasts lysed by passage through a 15-μm filter. **C:** Freshly prepared and purified protoplasts subjected to chilling and hyperosmotic conditions. (Reproduced from [69] with permission of the publisher.)

nous carboxyfluorescein and endogenous chlorophyll [64], and FITC with endogenous chlorophyll [93].

The large size of protoplasts is, in many cases, accompanied by the presence of the vacuole. It has been shown [87,122] that viable miniprotoplasts, consisting of the nucleus and some cytoplasm surrounded by a plasma membrane, can be produced by removing the vacuole. Fellner-Feldegg and colleagues [40,72] reported a much higher yield of viable miniprotoplasts than normal protoplasts following sorting with a 76-μm flow cell orifice. The sorted miniprotoplasts developed into the callus stage and regenerated shoots. However, because the yield for making miniprotoplasts is only 25–50%, better results are obtained using normal protoplasts with a larger (120 μm) flow cell tip (H. Fellner-Feldegg, personal communication).

The results summarized here indicate that all the necessary steps have been taken to detect and sort viable heterokaryons and grow them into hybrid plants. This has been done using protoplasts from *N. tabacum, N. sylvestris,* and *N. nesophila* [1]. Protoplasts were prepared from leaf tissue or suspension culture, labeled by incubation in enzyme solution containing FITC or RITC, then fused in a high pH $CaCl_2$ osmoticum. The initial frequency of heterokaryons ranged from 0.2 to 1.5%, whereas the frequency after sorting varied from 43 to 90%. Fused heterokaryons were sorted into wells of 96-well microtiter plates containing growth medium and tissue culture feeder cells to overcome the problem of culturing at low population density. Alternatively, protoplasts were sorted into wells containing growth medium, then sealed with paraffin until cell clusters developed.

A total of 127 plants were regenerated to the greenhouse stage. They were characterized, at the point of flowering, for nuclear DNA content, karyotype and morphology, pollen viability, and isoenzyme pattern. At least 52% of the plants exhibited hybrid characteristics, which was similar to the initial purity of the sorted heterokaryons. The hybrids could be divided into subclasses which exhibited traits characteristic of one or the other parental types, or intermediate in phenotype.

Another study recently compared the use of micromanipulation and cell sorting for selecting heterokaryons derived from the fusion of *Brassica campestris* with *Brassica oleracea* [108]. The parental protoplasts were labeled with either carboxyfluorescein diacetate or endogenous chlorophyll. Heterokaryons were regenerated into plants. They used isozyme analysis for definite characterization of hybrids. All the plants from heterokaryons obtained by micromanipulation were hybrids, while 87% of the plants derived from flow-sorted heterokaryons were hybrids. However, only 30% of the hybrids had symmetrical genetic contributions from each parent.

These results point to a very significant role for flow cytometry and cell sorting in plant hybridization research. One limitation of the technique is the inability to distinguish between adhesion and fusion of protoplasts of opposite type. Probably the best approach is to eliminate clumping to the maximum extent possible by mechanical means prior to sorting and also to employ methods of fusion that minimize adhesion.

Protoplast Characterization

Flow cytometry has been used to measure several properties of plant protoplasts, including chlorophyll content and size. Fluorescence emission from endogenous chlorophyll was used to quantitate chlorophyll content [57]. Nonoverlapping windows of chlorophyll fluorescence were used to sort cell subpopulations, which were then analyzed biochemically for chlorophyll content. A linear correlation was obtained between fluorescence and chlorophyll content, with a correlation coefficient of 0.983.

Because of the large size of protoplasts, forward-angle light scatter may not be a reliable indicator of size. However, time- of-flight (TOF) [85] measurements can be used effectively to measure protoplast diameter [57], since the large size of the protoplasts relative to the laser beam size reduces

Fig. 8. Development of protoplasts following flow sorting. **A:** Protoplasts immediately after sorting. ×210; bar = 100 μm. **B:** Cell colonies derived from sorted protoplasts, after 12 days. ×210; bar = 100 μm. **C:** Green calli produced from cell clusters, 8 weeks after transfer. ×2. **D:** Shoot induction in calli. ×1; bar = 2.0 cm. (Reproduced from [54] with permission of the publisher.)

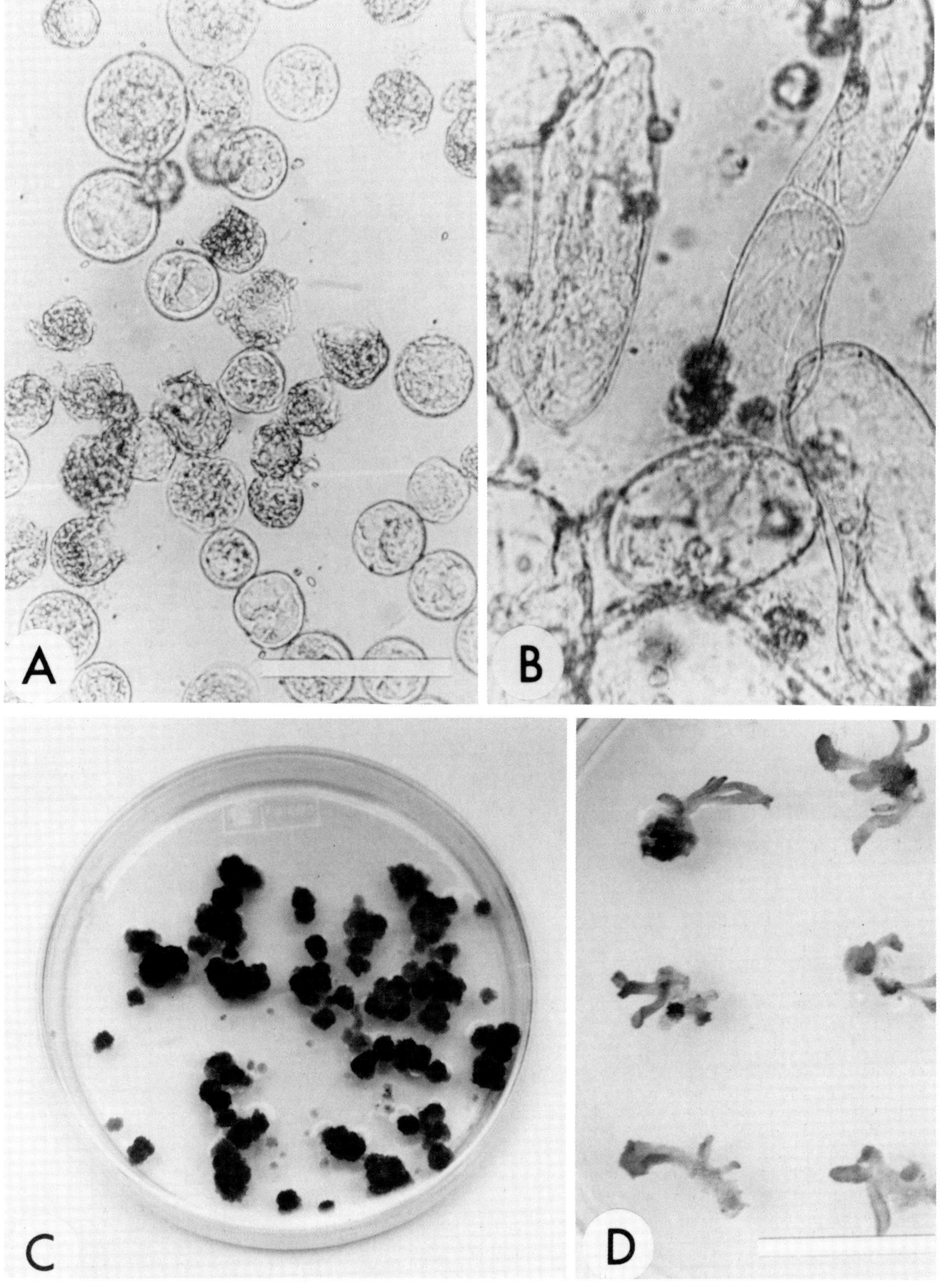

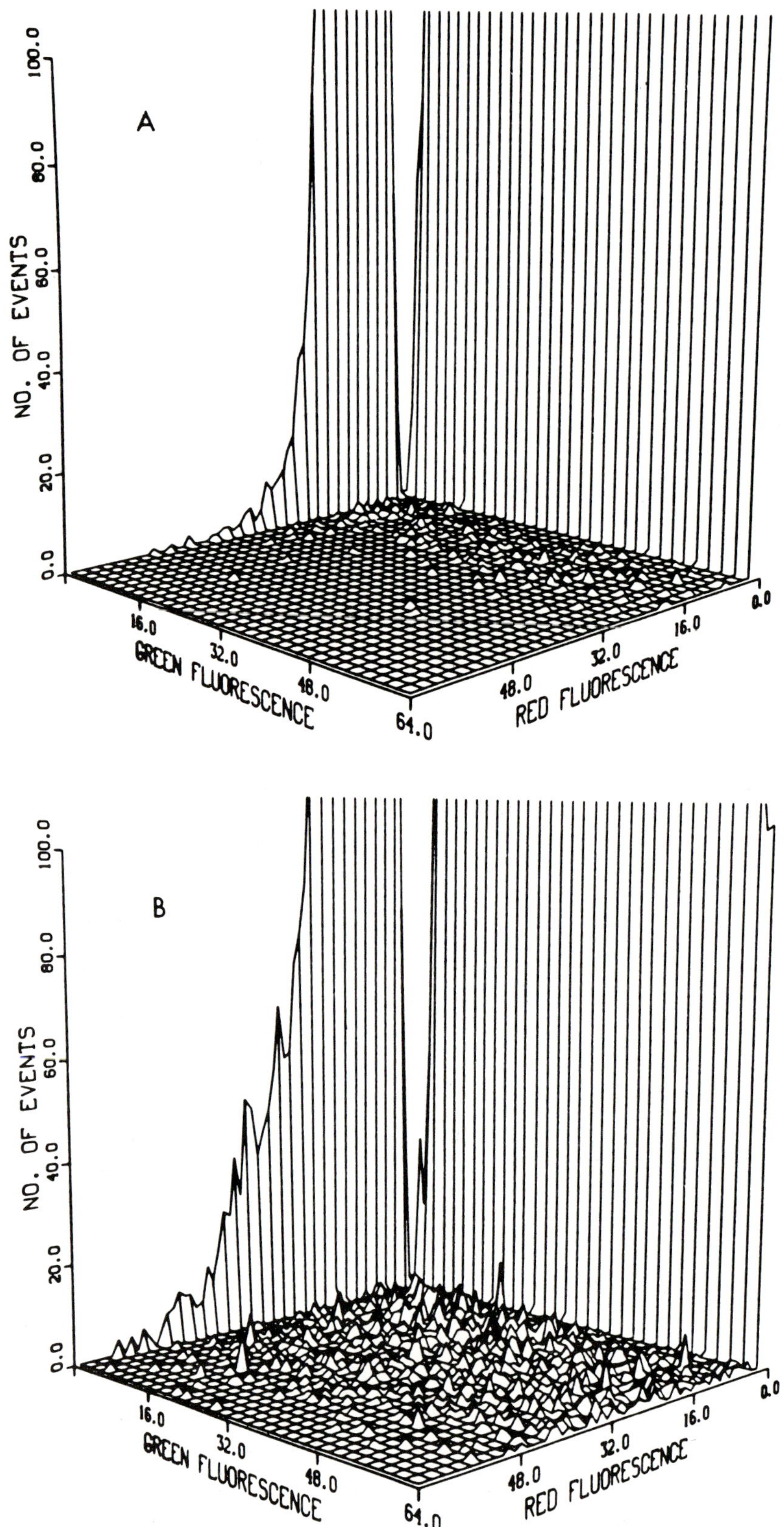

Fig. 9. Correlated red and green fluorescence for unfused and fused mixed populations of protoplasts. The same number of protoplasts were analyzed in each case. **A:** Unfused mixed protoplasts. **B:** Fusion mixture. (Reproduced from [3] with permission of the publisher.)

the nonlinearity. Fluorescence based on endogenous chlorophyll or uptake of aniline dye was used to trigger the TOF signal. A highly linear relationship between TOF and particle diameter was obtained for pollen and protoplasts within the range of 15–55 μm diameter. When fluorescein diacetate (FDA) was used for fluorescence, the relationship was nonlinear in protoplasts containing chlorophyll but was linear for protoplasts lacking chloroplasts.

Protoplast–Microbe Interactions

Interactions of bacterial spheroplasts *(Escherichia coli* and *Agrobacterium tumefaciens),* and plant protoplasts have been studied by flow cytometry [89]. The bacterial spheroplasts were pre-labeled with FDA, then incubated with plant protoplasts. Binding of spheroplasts was accompanied by an increased fluorescence of the protoplasts, determined by volume and fluorescence measurements. Fluorescence increased linearly with the number of bacteria added, and appeared to be a nonspecific phenomenon, since it was independent of the type of bacterial cells and protoplasts. This work is of interest because of the use of *A. tumefaciens* in DNA transfection of plants.

In related work, the uptake of ethidium bromide-labeled plasmid DNA by *Petunia hybrida* protoplasts was analyzed by flow cytometry [110]. Several different methods for DNA uptake were evaluated, including electroporation, PEG treatments, and heat shock.

Cell Wall Biosynthesis

The presence of β-linked glucosides, in particular cellulose or chitin, can be detected by the stain Calcofluor White [51,90]. This provides a probe for cell-wall biosynthesis that can be readily adapted to flow cytometric analysis of protoplast growth in culture. The absorption maximum is 350 nm, with the fluorescence emission maximum at 435 nm. Galbraith [52] demonstrated that the Calcofluor emission intensity was linearly correlated with cellulose content measured by the anthrone reagent. Calcofluor White was also used to study the inhibition of cell wall biosynthesis by coumarin and 2,6-dichlorobenzonitrile (DB) using fluorescence microphotometry [63]. Meadows [88] used Calcofluor White with flow cytometry to analyze cell wall regeneration of *Zea mays* protoplasts. Several distinct peaks were seen on a log plot of fluorescence that changed in intensity with time in culture. Clearly, the need for careful controls of the initial protoplast size and considerations regarding anisotropic growth and the onset of cell division may limit the use of this type of approach.

Alkaloid Fluorescence

The production of the alkaloid serpentine by *Catharanthus roseus* has been studied [21,22] using flow cytometry to detect serpentine fluorescence. This alkaloid can be excited by the argon ultraviolet (UV) laser lines at 351 and 364 nm, and it has a fluorescence emission maximum at 443 nm (Fig. 10). The intracellular pH of the vacuole was measured with the fluorescent probe 9-aminoacridine (9AA) using fluorescence microphotometry. The serpentine concentration per protoplast appeared to vary linearly with diameter, although the authors did not carry out a regression analysis. The distribution of serpentine fluorescence within the population was log-normal, with a broad range. This is consistent with other results [33]. When intracellular pH was compared with cell diameter, there also appeared to be an inverse linear relationship. Again, a precise analysis requires linear regression, since there was substantial scatter in the data. By sorting and comparing populations with either high or low serpentine contents, or high or low pH, they deduced that serpentine content and intracellular pH were positively correlated. The sort rate was 3,000/sec, and the purity of sorting was only around 50%, due most likely to substantial coincidence effects. The observed leakage of serpentine from protoplasts, possibly related to the nonphysiological pH of the sheath fluid, is of some concern. The effects of pH on alkaloid retention have been studied further by Renaudin et al. [98].

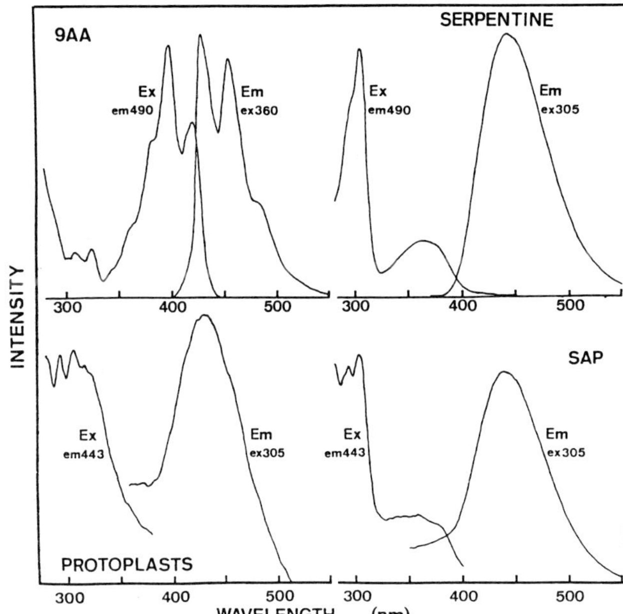

Fig. 10. Excitation and emission spectra. 9AA: 0.5 μM aqueous 9-aminoacridine; SERPENTINE: 1 μM aqueous serpentine tartrate; PROTOPLASTS: 20,000 protoplasts/ml; SAP: 10-fold dilution of cell sap in 0.25% sulfuric acid. (Reproduced from [21] with permission of the publisher.)

Antibodies

Polyclonal antibodies directed against alfalfa mosaic virus were used for immunofluorescence labeling of tobacco and cowpea leaf protoplasts following infection with alfalfa mosaic virus [118]. The patterns of fluorescence labeling, analyzed by flow cytometry, are difficult to interpret due to the unconventional method of data display.

Monoclonal antibodies have been produced against protoplasts from *Petunia hybrida* and *Lycopersicon esculentum* [14] and *Nicotiana tabacum* [59]. These results demonstrate that specific antibodies against various antigens can be fluoresently labeled and used to identify different types of protoplasts. Monoclonal antibodies against plasma membrane antigens have also been used to identify somatic hybrids after protoplast fusion [42,91].

Mitochondria

A novel use of flow cytometry to measure binding of lectins to mitochondria was reported by Petit et al. [94]. Fluorescein-labeled concanavalin A (Con A) or wheat germ agglutinin (WGA) was mixed with mitochondria isolated from potato tubers. Binding to either the outer or inner mitochondrial membrane was measured. There was little binding of WGA, but Con A showed specific binding to about 50% of the mitochondria, with the rest showing little or nonspecific binding.

Pollen

Fox et al. [45] (also M.H. Fox et al., personal communication, 1986) measured the total DNA content of protoplasts derived from pollen tetrads. Pinkel [95] has discussed the limitations and possibilities of using flow cytometry to detect pollen mutants. Tyrer [115] used flow cytometry to compare pollen from maize expressing a wild-type (starchy) phenotype or a mutant (waxy) phenotype. Light-scatter distributions showed slight differences. Both pollen types exhibited similar intensities of autofluorescence when excited at 488 nm, although the starchy phenotype had a broader distribution. When the pollen were incubated in iodine potassium iodide, the autofluorescence of wild-type pollen was quenched significantly compared with the waxy mutant, although there was still considerable overlap.

FUTURE DIRECTIONS

The results that have been obtained with flow cytometry of plant cells demonstrate that it is indeed feasible to label and sort viable protoplasts using various probes. Because of their size and fragility, however, protoplasts require instrument systems with large orifices and low pressures. We believe that improvements in the production of protoplasts will be essential for future flow cytometric studies of higher plants, and that successful culture of protoplasts at low population densities will be necessary. Nevertheless, it is certain that most of these technical details will eventually be resolved and that application of a wide variety of additional probes and techniques can be envisaged. Recent comprehensive reviews of flow cytometry of animal cells by Steinkamp [105] and by Traganos [112,113] illustrate many techniques that could be applied to plants. Brown and colleagues [19,20] have discussed some additional parameters that could advance knowledge of plant cell biology.

Several DNA stains have been used on plant protoplasts, with mithramycin giving the best results. However, this is only because the fixing and staining procedures have been worked out more carefully for mithramycin, and its spectral output does not overlap that of chlorophyll. Since mithramycin is specific for GC regions of DNA [123], the use of dyes specific for AT regions, such as Hoechst 33342 or DAPI, would allow measurements of GC/AT ratios for various plant species, which can vary substantially [16]. Intercalators such as propidium iodide or ethidium bromide should give good results without being influenced by the GC/AT ratio if chlorophyll fluorescence is filtered out. Dual labeling of DNA could also improve resolution of chromosomes [84] for chromosome analysis and sorting. Finally, much better results should be possible with viable staining of protoplasts using Hoechst 33342, if staining concentrations and conditions can be optimized. This would permit selection of populations of viable protoplasts synchronized at various points in the cell cycle.

The study of cell cycle kinetics of mammalian cells has been greatly expanded by the availability of monoclonal antibodies against bromodeoxyuridine [65,66]. Cells undergoing DNA synthesis take up BrdUrd as a thymidine analogue, and dual staining by propidium iodide and anti-BrdUrd antibodies permits accurate measurements of cell-cycle parameters. This methodology has not yet been applied to plant systems (see Chapter 23, this volume).

The possible existence of quiescent (Q) or G_0 cells in plants has been reported. The use of acridine orange to measure RNA and DNA content simultaneously was pioneered by Darzynkiewicz et al. [27] and has provided the capability for distinguishing G_0 and Q cells. This technique has been applied to plant protoplasts in one study [11], and similar characteristics of low RNA content for Q cells were obtained. Chu and Lark [23] report that Q cells in plants are smaller than cycling cells. Further work combining the use of light scattering and acridine orange could help characterize the development of quiescent protoplasts during culture and plant morphogenesis (see Chapter 24, this volume).

The production by higher plants of a variety of medicinal products has great commercial potential. It has been estimated that at least 23% of all prescriptions in industrial countries contain plant products from different sources [33]. The use of flow sorting to select protoplasts containing higher concentrations of these substances is clearly in its infancy. We have indicated that protoplasts can be sorted on the basis of endogenous fluorescence of many of these compounds, if such fluorescence exists. If not, it might be possible to correlate the expression of plasma membrane antigens with the accumulation of nonfluorescent secondary products, in which case monoclonal antibodies might be used for immunofluorescent sorting. There is great potential in this area. However, endogenous substances can also create problems by causing autofluorescence that overlaps with the fluorescence spectrum of the molecules that are of interest. In that case, caution has to be exercised even though optical filters or different excitation lines may help resolve some overlapping spectra.

We anticipate that studies of plant cell biology will be expanded considerably by multiparameter flow cytometry. It is well known that intracellular pH is critical in the regulation of enzyme activity; this has been specifically correlated with serpentine accumulation [21]. The accumulation of other alkaloids is probably strongly dependent on intracellular pH as well. The pH probe ADB has proved useful for animal cells [2,46,117] and can be highly accurate and reproducible in a pH range of 6.5–8.0, giving an accurate discrimination of less than 0.1 pH units [46]. Membrane fluidity can be measured by fluorescence polarization [47,76] of diphenylhexatriene [17] or of the much better charged analogue trimethyl ammonium diphenylhexatriene [83], or R18. Fluorescence polarization can also be used to study the viscosity of the cytoplasm (using FDA) or the sequestration of alkaloids in vacuoles (using endogenous autofluorescence).

Free Ca^{2+} levels play a large role in regulating many cellular events. The fluorescent probe Quin2 AM [114] measures free Ca^{2+} and can be used with flow cytometry when excited in the UV range. An even better Ca^{2+} probe for flow cytometry measurements is Indo-1 [67]. Rhodamine 123 [28,75] has been used in animal cell systems to stain mitochondria and to measure mitochondrial proliferation during the cell cycle. Enzyme kinetics have been studied by measuring the increasing concentrations of a fluorescent enzyme product as a function of time [34]. New substrates could be developed for studying plant enzyme systems. Additional possibilities exist for measuring membrane potential, protein content, and cell volume. The development of specific monoclonal antibodies against antigens is just beginning. Cell wall biosynthesis can be studied by using specific stains such as

Calcofluor or developing antibodies against cell wall components.

In summary, the potential applications of flow cytometry to plant cells encompass virtually all the techniques applied to animal cells, with some additional possibilities related to unique properties of plant cells. As more researchers begin to employ the broad-ranging capabilities of multiparameter flow cytometry on plant cells, we predict a virtual explosion in results leading to an increased understanding of plant cell biology, as well as commercial applications in plant hybridization and selection for specific plant products.

REFERENCES

1. **Afonso CL, Harkins KR, Thomas-Compton MA, Krejci AE, Galbraith DW (1985)** Selection of somatic hybrid plants through fluorescence-activated sorting of protoplasts. Biotechnology 3:811–816.
2. **Alabaster O, Clagett Carr K, Leonaridis L (1984)** Tumor cell heterogeneity: Its determination by flow cytometric analysis of intracellular pH. Methods Achiev Exp Pathol 11:96–110.
3. **Alexander RG, Cocking EC, Jackson PJ, Jett JH (1985)** The characterization and isolation of plant heterokaryons by flow cytometry. Protoplasma 128: 52–58.
4. **Ammirato PV (1983)** Embryogenesis. In Evans DA, Sharp WR, Ammirato PV, Yamada Y (eds): "Handbook of Plant Cell Culture," Vol. 1. New York: Macmillan, pp 82–123.
5. **Ammirato PV (1984)** Induction, maintenance and manipulation of development in embryogenic cell suspension cultures. In Vasil IK (ed): "Cell Culture and Somatic Cell Genetics of Plants." New York: Academic Press, pp 139–151.
6. **Anderson LK, Stack SM, Fox MH, Chuanshan Z (1985)** The relationship between genome size and synaptonemal complex length in higher plants. Exp Cell Res 156:367–378.
7. **Baer GR, Schrader LE (1985)** Seasonal changes and genetic variability for DNA concentration and cellular contents of soluble protein, chlorophyll, ribulose bisphosphate carboxylase and pyruvate Pi dikinase activity in maize leaves. Crop Sci 25:909–916.
8. **Bajaj YPS (1983)** In vitro production of haploids. In Evans DA, Sharp WR, Ammirato PV, Yamada Y (eds): "Handbook of Plant Cell Culture," Vol. 1. New York: Macmillan, pp 82–123.
9. **Bell EA, Charlwood BV (1980)** "Secondary Plant Products." Berlin: Springer-Verlag.
10. **Bennett MD, Smith JB, Heslop-Harrison JS (1982)** Nuclear DNA amounts in angiosperms. Proc R Soc Biol 216:179–199.
11. **Bergounioux C, Perennes C, Brown SC, Gadal P (1988)** Nuclear RNA quantification in protoplast cell-cycle phases. Cytometry 9:87–87.
12. **Bergounioux C, Perennes C, Miege C, Gadal P (1986)** The effect of male sterility on protoplast division in *Petunia hybrida*. Cell cycle comparison by flow cytometry. Protoplasma 130:138–144.
13. **Bergounioux C, Perennes C, Miege C, Brown S (1984)** Osmotic pressure of the cell buffer and conductivity of the sheath fluid influence flow sorting performance with protoplasts of lucerne, *Petunia*, and *Catharanthus*. Biol Cell 52:136a.
14. **Bergounioux C, Perennes C, Tabaeizadeh Z, Prevot C, Gadal P (1983)** Characterization and separation of *Petunia hybrida* and *Lycopersicon esculentum* protoplasts based on natural fluorescence or FITC added to labelled antibodies in immunoreaction with the protoplast membrane. In Potrykus I, Harms CT, Hinnen A, Hutter R, King PJ, Shillito RD (eds): "Protoplasts 1983: Poster Proceedings." Basel: Birkhauser, pp 262–263.
15. **Bhaskaran S, Smith RH, Finer JJ (1983)** Ribulose bisphosphate carboxylase activity in anther-derived plants of *Saintpaulia ionantha* cultivar Shag. Plant Physiol 73:639–642.
16. **Bonner J (1976)** The nucleus. In Bonner J, Varner JE (eds): "Plant Biochemistry." 3rd ed. New York: Academic Press, pp 37–64.
17. **Bouchy M, Donner M, Andre JC (1981)** Evolution of fluorescence polarization of 1,6-diphenyl-1,3,5-hexatriene (DPH) during the labelling of living cells. Exp Cell Res 133:39–46.
18. **Bromova M, Knopf UC (1987)** Separation of intact and damaged plant-protoplasts by using a cell sorter equipped with a two-channel piezo valve chamber. Plant Sci 52:91–97.
19. **Brown S (1984)** Analysis and sorting of plant material by flow cytometry. Physiol Veg 22:341–349.
20. **Brown S, Jullien M, Coutos-Thevenot P, Muller P, Renaudin J-P (1987)** Present developments of flow cytometry in plant biology. Biol Cell 58:173–178.
21. **Brown SC, Renaudin J-P, Prevot C, Guern J (1984)** Flow cytometry and sorting of plant protoplasts: Technical problems and physiological results from a study of pH and alkaloids in *Catharanthus roseus*. Physiol Veg 22:541–554.
22. **Brown S, Prevot C, Renaudin J-P, Guern J (1983)** Flow cytometry and cell sorting on pH and alkaloids in Catharanthus protoplasts. In Potrykus I, Harms CT, Hinnen A, Hutter R, King PJ, Shillito RD (eds): "Protoplasts 1983: Poster Proceedings." Basel: Birkhauser, pp 228–229.
23. **Chu Y, Lark KG (1976)** Cell cycle parameters of soybean (*Glycine max* L.) cells growing in suspension culture: Suitability of the system for genetic studies. Planta 132:259–268.
24. **Conia J, Bergounioux C, Perennes C, Muller P, Brown S, Gadal P (1987)** Flow cytometric analysis and sorting of plant chromosomes from *Petunia hybrida* protoplasts. Cytometry 8:500–508.
25. **Conn EE (1981)** Secondary plant products. In Stumpf PK, Conn EE (eds), "The Biochemistry of Plants, A Comprehensive Treatise." Vol 7. New York: Academic Press, pp 1–798.
26. **Constabel F (1984)** Fusion of protoplasts by polyethylene glycol (PEG). In Vasil IK (ed): "Cell Culture and Somatic Cell Genetics of Plants." New York: Academic Press, pp 414–422.
27. **Darzynkiewicz Z, Traganos F, Melamed MR (1980)** New cell cycle compartments identified by multiparameter flow cytometry. Cytometry 1:98–108.
28. **Darzynkiewicz Z, Traganos F, Staiano-Coico L, Kapuscinski J, and Melamed MR:** Interactions of Rhodamine 123 with living cells studied by flow cytometry. Cancer Res 42, 799–806, 1982.
29. **de Laat AMM, Blaas J (1984)** Isolation, purification and sorting by flow cytometry of metaphase chromosomes of *Haplopappus Gracilis*. In Potrykus I, Harms

CT, Hinnen A, Hutter R, King PJ, Shillito RD (eds): "Protoplasts 1983: Poster Proceedings." Basel: Birkhauser, pp 228–229.

30. **de Laat AMM, Blaas J (1984)** Flow cytometric characterization and sorting of plant chromosomes. Theor Appl Genet 67:463–467.
31. **de Laat AMM, Schel JHN (1986)** The integrity of metaphase chromosomes of *Haplopappus gracilis* (Nutt.) gray isolated by flow cytometry. Plant Sci 47: 145–151.
32. **de Latt AMM, Verhoeven HA, Sree Ramulu K, Dijkhuis P (1987)** Efficient induction by amiprophosmethyl and flow-cytometric sorting of micronuclei in *Nicotiana plumbaginifolia.* Planta 172:473–478.
33. **Deus B, Zenk MHY (1982)** Exploitation of plant cells for the production of natural compounds. Biotechnol Bioeng 24:1965–1974.
34. **Dolbeare FA, Smith RA (1979)** Flow cytoenzymology: Rapid enzyme analysis of single cells. In Melamed MR, Mullaney PF, Mendelsohn ML (eds): "Flow Cytometry and Sorting." New York: John Wiley & Sons, pp 317–333.
35. **Esau K (1977)** "Anatomy of Seed Plants." New York: John Wiley & Sons.
36. **Evans DA (1983)** Protoplast fusion. In Evans DA, Sharp WR, Ammirato PV, Yamada Y (eds): "Handbook of Plant Cell Culture." Vol 1. New York: Macmillan, pp 291–321.
37. **Evans DA, Bravo JE (1983)** Protoplast isolation and culture. In Evans DA, Sharp WR, Ammirato PV, Yamada Y (eds): "Handbook of Plant Cell Culture." Vol. 1. New York: Macmillan, pp 124–176.
38. **Evans PK, Wilson VM (1984)** Plant somatic hybridization. Adv Appl Biol 10:1–69.
39. **Fahleson J, Dixelius J, Sundberg E, Glimelius K (1988)** Correlation between flow cytometric determination of nuclear DNA content and chromosome number in somatic hybrids within *Brassicacaea.* Plant Cell Rep 7:74–77.
40. **Fellner-Feldegg H, Hurlbut P, Glimelius K (1984)** Sorting of miniprotoplasts by flow cytometry. Presented at the Tenth International Conference on Analytical Cytology, June 3–8, 1984, Asilomar, California (abst).
41. **Firoozabady E (1986)** The effects of cell cycle parameters on cell wall regeneration and cell division of cotton protoplasts (*Gossypium hirsutum* L.). J Exp Bot 37:1211–1217.
42. **Fitter MS, Norman PM, Hahn MG, Wingate VPM, Lamb CJ (1987)** Identification of somatic hybrids in plant protoplast fusions with monoclonal antibodies to plasma-membrane antigens. Planta 170:49–54.
43. **Flick CE, Evans DA, Sharp WR (1983)** Organogenesis. In Evans DA, Sharp WR, Ammirato PV, Yamada Y (eds): "Handbook of Plant Cell Culture." Vol. 1. New York: Macmillan, pp 13–81.
44. **Foster AS, Gifford EM (1974)** "Comparative Morphology of Vascular Plants." San Francisco: WH Freeman.
45. **Fox MH, Anderson LK, Stack SM (1984)** Flow cytometric quantitation of DNA content in plant protoplasts. Presented at the Tenth International Conference on Analytical Cytology, June 3–8, 1984, Asilomar, California (abst).
46. **Fox MH, Cook JC (1988)** Intracellular pH measurements using flow cytometry with 1,4-diacetoxy-2,3-dicyanobenzol. Cytometry 9:441–447.
47. **Fox MH, Delohery TM (1987)** Membrane fluidity measured by fluorescence polarization using an EPICS V cell sorter. Cytometry 8:20–25.
48. **Freeling M (1984)** Plant transposable elements. Annu Rev Plant Physiol 35:277–298.
49. **Furuya M (1984)** Cell division patterns in multicellular plants. Annu Rev Plant Physiol 35:349–373.
50. **Galbraith DW (1981)** Identification and sorting of plant heterokaryons. U.S. Patent 4,300,310.
51. **Galbraith DW (1981)** Microfluorometric quantitation of cellulose biosynthesis by plant protoplasts using Calcofluor White. Physiol Plant 53:111–116.
52. **Galbraith DW (1984)** Selection of somatic hybrid cells by fluorescence-activated cell sorting. In Vasil IK (ed): "Cell Culture and Somatic Cell Genetics of Plants." Vol. 1. New York: Academic Press, pp 433–447.
53. **Galbraith DW (1984)** Flow cytometric analysis of the cell cycle. In Vasil IK (ed): "Cell Culture and Somatic Cell Genetics of Plants." Vol. 1. New York: Academic Press, pp 765–777.
54. **Galbraith DW, Afonso CL, Harkins KR (1984)** Flow sorting and culture of protoplasts: Conditions for high-frequency recovery, growth and morphogenesis from sorted protoplasts of suspension cultures of nicotiana. Plant Cell Rep 3:151–155.
55. **Galbraith DW, Galbraith JEC (1979)** A method for identification of fusion of plant protoplasts derived from tissue cultures. Z Pflanzenphysiol 93:149–158.
56. **Galbraith DW, Harkins KR (1982)** Cell sorting as a means of isolating somatic hybrids. In Fujiwara A (ed): "Plant Tissue Culture, 1982." Tokyo: Maruzen Press, pp 617–618.
57. **Galbraith DW, Harkins KR, Jefferson RA (1988)** Flow cytometric characterization of the chlorophyll contents and size distributions of plant protoplasts. Cytometry 9:75–83.
58. **Galbraith DW, Harkins KR, Maddox JM, Ayres NM, Sharma DP, Firoozabady E (1983)** Rapid flow cytometric analysis of the cell cycle in intact plant tissues. Science 220:1049–1051.
59. **Galbraith DW, Maddox JM (1983)** Production of monoclonal antibodies directed against developmentally-regulated protoplast antigens. In Potrykus I, Harms CT, Hinnen A, Hutter R, King PJ, Shillito RD (eds): "Protoplasts 1983: Poster Proceedings." Basel: Birkhauser, pp 234–235.
60. **Galbraith DW, Maddox JR, Harkins KR (1984)** The application of monoclonal antibody and flow sorting techniques to higher plant protoplast systems. In Vitro 20:273 (abst).
61. **Galbraith DW, Mauch TJ (1980)** Identification of fusion of plant protoplasts. II. Conditions for the reproducible fluorescence labelling of protoplasts derived from mesophyll tissue. Z Pflanezenphysiol 98: 129–140.
62. **Galbraith DW, Mauch TJ, Shields BA (1981)** Analysis of the initial stages of plant protoplast development using 33258 Hoechst: reactivation of the cell cycle. Physiol Plant 51:380–386.
63. **Galbraith DW, Shields BA (1982)** The effects of inhibition of cell wall synthesis on tobacco protoplast development. Physiol Plant 55:25–30.

64. **Glimelius K, Djupsjobacka M, Fellner-Feldegg H** (1986) Selection and enrichment of plant protoplast heterokaryons of *Brassicacaea* by flow sorting. Plant Sci 45:133–141.
65. **Gratzner HG, Leif RC** (1981) An immunofluorescence method for monitoring DNA synthesis by flow cytometry. Cytometry 1:385–389.
66. **Gray JW** (ed) (1985) Monoclonal antibodies against bromodeoxyuridine. Special Issue. Cytometry 6: 499–674.
67. **Grynkiewicz G, Poenie M, Tsien RY** (1985) A new generation of Ca^{2+} indicators with greatly improved fluorescence properties. J Biol Chem 260: 3440–3450.
68. **Hake S, Walbot V** (1980) The genome of *Zea mays*, its organization and homology to related grasses. Chromosoma 79:251–270.
69. **Harkins KR, Galbraith DW** (1984) Flow sorting and culture of plant protoplasts. Physiol Plant 60:43–52.
70. **Harkins KR, Galbraith DW** (1984) Flow cytometry and cell sorting: Applications to higher plants. Presented at the Tenth International Conference on Analytical Cytology, June 3–8, 1984, Asilomar, California (abst).
71. **Harkins KR, Galbraith DW** (1987) Factors governing the flow cytometric analysis and sorting of large biological particles. Cytometry 8:60–70.
72. **Hurlbut P, Fellner-Feldegg H, Glimelius K** (1985) Sorting of mini-protoplasts by flow cytometry. Hereditas Suppl 3:142.
73. **Husemann W** (1984) Photoautotrophic cell cultures. In Vasil IK (ed): "Cell Culture and Somatic Cell Genetics of Plants." New York: Academic Press, pp 182–191.
74. **Jett JH, Alexander RG** (1985) Droplet sorting of large particles. Cytometry 6:484–486.
75. **Johnson LV, Walsh ML, Chen LB** (1981) Localization of mitochondria in living cells with rhodamine 123. Proc Natl Acad Sci USA 77:990–994.
76. **Jovin TM** (1979) Fluorescence polarization and energy transfer: Theory and application. In Melamed MR, Mullaney PF, Mendelsohn ML (eds): "Flow Cytometry and Sorting." New York: John Wiley & Sons, pp 137–165.
77. **Junker S, Pedersen S** (1981) A universally applicable method of isolating somatic cell hybrids by two-color flow sorting. Biochem Biophys Res Commun 102: 977–984.
78. **Kachel V, Menke E** (1979) Hydrodynamic properties of flow cytometric instruments. In Melamed MR, Mullaney PF, Mendelsohn ML (eds): "Flow Cytometry and Sorting." New York: John Wiley & Sons, pp 41–59.
79. **Kameya T** (1984) Fusion of protoplasts by dextran and electrical stimulus. In Vasil IK (ed): "Cell Culture and Somatic Cell Genetics of Plants." New York: Academic Press, pp 423–427.
80. **Karlsson SB, Vasil IK** (1986) Growth, cytology and flow cytometry of embryogenic cell suspension cultures of *Panicum maximum* Jacq. and *Pennisetum purpureum* Schum. J Plant Physiol 123:211–227.
81. **Keeler KH, Kwankin B, Barnes PW, Galbraith DW** (1987) Polyploid polymorphism in *Andropogon gerardii*. Genome 29:374–379.
82. **Keller WA** (1984) Anther culture of *Brassica*. In Vasil IK (ed): "Cell Culture and Somatic Cell Genetics of Plants." New York: Academic Press, pp 302–310.
83. **Kuhry J-G, Fonteneau P, Duportail G, Maechling C, Laustreat G** (1983) TMA-DPH: A suitable fluorescence polarization probe for specific plasma membrane fluidity studies in intact living cells. Cell Biophys 5:129–140.
84. **Langlois RG, Yu L-C, Gray JW, Carrano AV** (1982) Quantitative karyotyping of human chormosomes by dual beam flow cytometry. Proc Natl Acad Sci USA 79:7876–7880.
85. **Leary JF, Todd P, Wood JCS, Jett JH** (1979) Laser flow cytometric light scatter and fluorescence pulse width and pulse rise-time sizing of mammalian cells. J Histochem Cytochem 27:314–310.
86. **Leaver CJ, Gray MW** (1982) Mitochondrial genome organization and expression in higher plants. Annu Rev Plant Physiol 33:373–402.
87. **Lorz H** (1984) Enucleation of protoplasts: Preparation of cytoplasts and miniprotoplasts. In "Cell Culture and Somatic Cell Genetics of Plants." Vol. 1. New York: Academic Press, pp 448–452.
88. **Meadows MG** (1982/1983) Characterization of cells and protoplasts of the B73 maize cell line. Plant Sci Lett 28:337–348.
89. **Millman RA, Lurquin PF** (1985) Study of plant protoplast–bacterial spheroplast and cell interactions by flow cytometry. J Plant Physiol 117:431–440.
90. **Nagata T, Takebe I** (1970) Cell wall regeneration and cell division in isolated tobacco mesophyll protoplasts. Planta 92:301–308.
91. **Norman PM, Wingate VPM, Fitter MS, Lamb CJ** (1986) Monoclonal antibodies to plant plasma membrane antigens. Planta 167:452–459.
92. **Patnaik G, Cocking EC, Hamill J, Pental D** (1982) A simple procedures for the manual isolation and identification of plant heterokaryons. Plant Sci Lett 24: 105–110.
93. **Pauls KP, Chuong PV** (1987) Flow cytometric identification of *Brassica napus* protoplast fusion products. Can J Bot 65:834–838.
94. **Petit P, Diolez P, Muller P, Brown SC** (1986) Binding of concanavalin A to the outer membrane of potato tuber mitochondria detected by flow cytometry. FEBS 196:65–70.
95. **Pinkel D** (1981) On the possibility of automated scoring of pollen mutants. Environ Health Persp 37:133–136.
96. **Puite KJ, Ten Broeke WRR** (1983) DNA staining of fixed and nonfixed plant protoplasts for flow cytometry with Hoechst 33342. Plant Sci Lett 32:79–88.
97. **Redenbaugh K, Ruzin S, Bartholomew J, Bassham JA** (1982) Characterization and separation of plant protoplasts via flow cytometry and cell sorting. Z Pflanzenphysiol 107:65–80.
98. **Renaudin JP, Brown SC, Guern J** (1985) Compartmentation of alkaloids in a cell suspension of *Catharanthus roseus*: A reappraisal of the role of pH gradients. In Neuman, et al. (eds): "Primary and Secondary Metabolism of Plant Cell Cultures." Berlin: Springer-Verlag, pp 124–132.
99. **Reisch B** (1983) Genetic variability in regenerated plants. In Evans DA, Sharp WR, Ammirato PV, Yamada Y (eds): "Handbook of Plant Cell Culture." Vol. 1. New York: Macmillan, pp 748–769.

100. **Sgorbati S, Levi M, Sparvoli E, Trezzi F, Lucchini G (1986)** Cytometry and flow cytometry of 4′,6-diamidino-2-phenylindole (DAPI)-stained suspensions of nuclei released from fresh and fixed tissues of plants. Physiol Plant 68:471–476.
101. **Sharma DP, Firoozabady E, Ayres NM, Galbraith DW (1983)** Improvement of anther culture in *Nicotiana*: Media cultural conditions and flow cytometric determination of ploidy levels. Z Pflanzenphysiol 111:441–451.
102. **Smith JA (1982)** The cell cycle and related concepts in cell proliferation. J Pathol 136:149–166.
103. **Sree Ramulu K, Dijkhuis P (1986)** Flow cytometric analysis of polysomaty and *in vitro* genetic instability in potato. Plant Cell Rep 3:234–237.
104. **Stahl E (1969)** "Thin-Layer Chromatography, a Laboratory Handbook." New York: Springer-Verlag.
105. **Steinkamp JA (1984)** Flow cytometry. Rev Sci Instrum 55:1375–1400.
106. **Stoval RT (1977)** The influence of particles on jet break-off. J Histochem Cytochem 25:813–820.
107. **Sundberg E, Glimelius K (1986)** A method for production of interspecific hybrids within *Brassiceae* via somatic hybridization, using resynthesis of *Brassica napus* as a model. Plant Sci 43:155–162.
108. **Sundberg E, Landgren M, Glimelius K (1987)** Fertility and chromosome stability in *Brassica napus* resynthesized by protoplast fusion. Theor Appl Genet 75:96–104.
109. **Sunderland N (1984)** Anther culture of *Nicotiana tabacum*. In Vasil IK (ed): "Cell Culture and Somatic Cell Genetics of Plants." New York: Academic Press, pp 283–292.
110. **Tagu D, Bergounioux C, Perennes C, Brown S, Muller P, Gadal P (1987)** Analysis of flow cytometry of the primary events of *Petunia hybrida* protoplast transformation. Plant Sci 52:215–223.
111. **Taylor MG, Vasil IK (1987)** Analysis of DNA size, content and cell cycle in leaves of Napier grass (*Pennisetum purpureum* Schum.). Theor Appl Genet 74: 681–686.
112. **Traganos F (1984)** Flow Cytometry: Principles and applications. I. Cancer Invest. 2:149–163.
113. **Traganos F (1984)** Flow cytometry: Principles and applications. II. Cancer Invest 2:239–258.
114. **Tsien RY, Pozzan T, Rink TJ (1982)** Calcium homeostasis in intact lymphocytes: Cytoplasmic free calcium monitored with a new intracellularly trapped fluorescent indicator. J Cell Biol 94:325–334.
115. **Tyrer HW (1981)** Technology for automated analysis of maize pollen used as a marker for mutation. 1. Flow-through systems. Environ Health Persp 37: 137–142.
116. **Ulrich I, Ulrich W (1986)** Flow cytometric DNA-analysis of plant protoplasts stained with DAPI. Z Naturforsch 41c:1052–1056.
117. **Valet G, Raffael A, Moroder L, Wunsch E, Ruhenstroth-Bauer G (1981)** Fast intracellular pH determination in single cells by flow cytometry. Naturwissenschaften 68:265–266.
118. **van Klaveren P, Slats J, Roosien J, van Vloten-Doting L (1983)** Flow cytometric analysis of tobacco and cowpea protoplasts infected *in vivo* and *in vitro* with alfalfa mosaic virus. Plant Mol Biol 2:19–25.
119. **Vasil IK, Vasil V (1980)** Isolation and culture of protoplasts. Int Rev Cytol Suppl 11B:1–19.
120. **Vasil V, Vasil IK (1984)** Isolation and maintenance of embryogenic suspension cultures of Gramineae. In Vasil IK (ed): "Cell Culture and Somatic Cell Genetics of Plants." New York: Academic Press, pp 152–158.
121. **Wagner G (1983)** Higher plant vacuoles and tonoplasts. In Hall JL, Moore AL (eds): "Isolation of Membranes and Organelles from Plant Cells." New York: Academic Press, pp 83–118.
122. **Wallin A, Glimelius K, Eriksson T (1978)** Enucleation of plant protoplasts by Cytochalasin B. Z Pflanzenphysiol 87:333–340.
123. **Ward DC, Reich E, Goldberg IH (1965)** Base specificity in the interaction of polynucleotides with antibiotic drugs. Science 149:1259–1263.
124. **Wenzel G, Foroughi-Wehr B (1984)** Anther culture of *Solanum tuberosum*. In Vasil IK (ed): "Cell Culture and Somatic Cell Genetics of Plants." New York: Academic Press, pp 293–301.
125. **Wenzel G, Foroughi-Wehr B (1984)** Anther culture of cereals and grasses. In Vasil IK (ed): "Cell Culture and Somatic Cell Genetics of Plants." New York: Academic Press, pp 311–327.
126. **Whitfield PR, Bottomley W (1983)** Organization and structure of chloroplast genes. Annu Rev Plant Physiol 34:279–310.

32

Measurement of Intracellular Ionized Calcium and Membrane Potential

Peter S. Rabinovitch and Carl H. June

Department of Pathology, University of Washington, Seattle, Washington 98195 (P.S.R.); Naval Medical Research Institute, Bethesda, Maryland 20814 (C.H.J.)

INTRACELLULAR IONIZED CALCIUM

Measurement of intracellular ionized calcium concentration ($[Ca^{2+}]_i$) in living cells is of considerable interest to investigators over a broad range of cell biology. Calcium has an important role as a mediator of transduction of signals from the cell membrane; elevations in $[Ca^{2+}]_i$ are part of the regulation of diverse cellular processes. Eukaryotic cells have an internal calcium ion concentration that is usually maintained at ~100 nM, far below the extracellular environment, by regulation of calcium channels within the plasma membrane and storage and release of Ca^{2+} from sites such as calciosomes, endoplasmic reticulum and mitochondria [90,92,118]. Plasma membrane calcium influx is thought to be initiated by membrane depolarization which opens voltage-gated channels or by the binding of ligands to receptor operated channels. In the latter case, the binding of agonist to its specific membrane receptor causes the activation of a guanine nucleotide-binding (G) protein that, in turn, activates phospholipase C [35] (Fig. 1). Phospholipase C causes the hydrolysis of a membrane phospholipid, phosphatidylinositol 4,5-bisphosphate (PIP2), which yields a water-soluble product, inositol-1,4,5-trisphosphate (IP3) and a lipid, 1,2-diacylglycerol (DAG) [10]. IP3 and perhaps additional products of IP3 metabolism, then causes the release of calcium from intracellular stores, while DAG in conjunction with calcium ions activates protein kinase C. Ca^{2+} itself has a broad range of effects, activating a variety of enzyme systems, both as a cofactor and in conjunction with the calcium-binding protein calmodulin.

Indicators of Intracellular Calcium

An optimal indicator of $[Ca^{2+}]_i$ should span the range of calcium concentrations from 100 nM to above a micromolar, with greatest sensitivity to small changes at the lower end of that range. The response to transient changes should be rapid. The indicator should freely diffuse throughout the cytoplasm but be easily and stably loaded. Finally, the indicator itself should have little or no effect upon $[Ca^{2+}]_i$ or on other cellular functions. Radiolabeled Ca (^{45}Ca) permits sensitive detection but is otherwise completely unsuited, as only the ionized portion of intracellular Ca is of greatest interest, and this can be a small proportion of the total.

Calcium-activated bioluminescent indicators, photoproteins, have been used as indicators of $[Ca^{2+}]_i$, the most widely used being aequorin [12]. Their greatest limitation is the necessity for loading into cells by microinjection or other membrane disruption. Ease of loading was achieved with the development of quin2, using the technique of incubating the cells in the presence of the acetoxymethyl ester of quin2 [20,114]. This uncharged form is cell permeant and diffuses freely into the cytoplasm where it serves as a substrate for esterases. Hydrolysis releases the tetraanionic form of the dye which is trapped inside the cell [112]. Quin2 has several disadvantages, however, especially for use by flow cytometry. A relatively low extinction coefficient and quantum yield have made detection of the dye at low concentrations difficult; at higher concentrations, quin2 itself buffers the $[Ca^{2+}]_i$ and may affect cellular functions [56]. In addition, quin2 at intracellular concentrations commonly employed may block plasma membrane sodium-calcium transport [1], and quin2 may be quenched by heavy metals which are found in the cytoplasm of some cell lines [3].

More recently, Grynkiewicz et al. [40] described a new family of highly fluorescent indicators of $[Ca^{2+}]_i$ that overcome most of these difficulties. One of these dyes, indo-1 ([1-[2 amino-5-[carboxylindol-2-yl]phenoxy]-2-2′-amino-5-methylphenoxyl]ethane *N,N,N′N′*-tetraacetic acid) (Fig. 2), has spectral properties that make it especially useful for analysis by flow cytometry. In particular, when excited by ultraviolet (uv) light, indo-1 exhibits large changes in fluorescence emission wavelengths upon calcium binding (Fig. 3). The use of the ratio of intensities of fluorescence at two wavelengths permits calculation of $[Ca^{2+}]_i$ independent of variability in intracellular dye concentration. Recently, Tsien and colleagues [76] reported two new calcium indicator dyes with excitation and emission spectra similar to fluorescein (fluo-3) and rhodamine (rhod-2), but that do not undergo significant emission or excitation wavelength shifts on Ca^{2+} bind-

Flow Cytometry and Sorting, Second Edition, pages 651–668

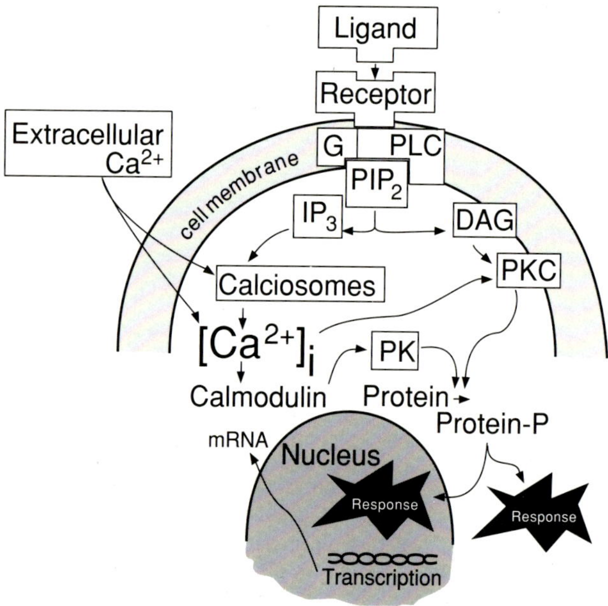

Fig. 1. Intracellular calcium plays a key role as a mediator of transmembrane signaling (see text). G, G protein; PLC, phospholipase C; PIP_2, phosphatidylinositol 4,5-bisphosphate; IP_3, inositol 1,4,5-triphosphate; DAG, diacylglycerol; Calciosomes, organelles of intracellular calcium storage, principally endoplasmic reticulum and mitochondria; PKC, protein kinase C; PK, other calmodulin-activated protein kinases.

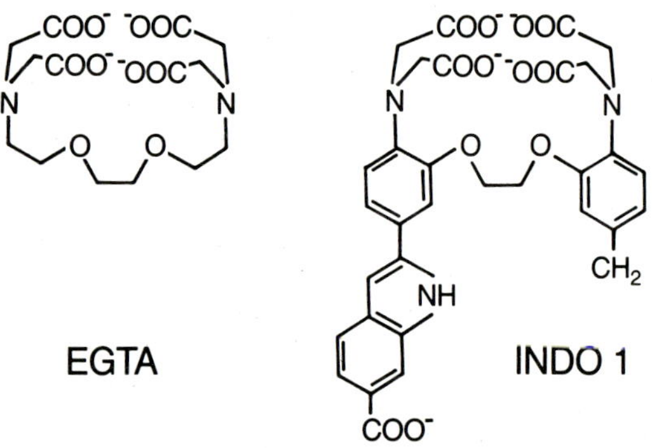

Fig. 2. Molecular structures of EGTA and indo-1. Note the similarities in structure in those portions of the molecules involved in calcium binding.

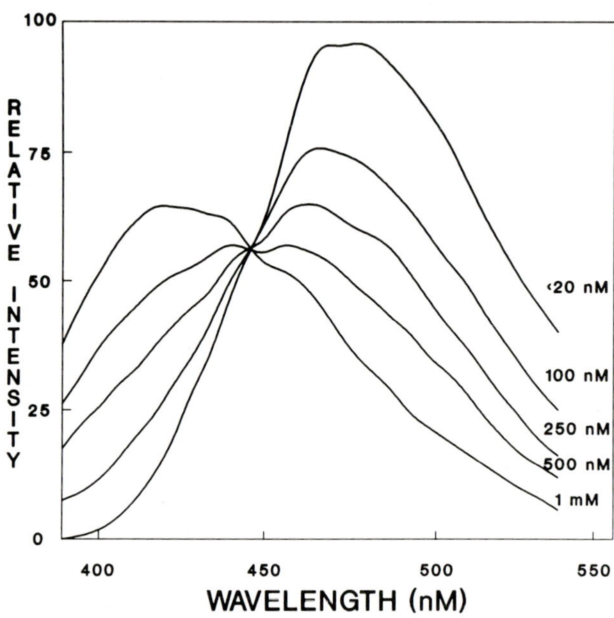

Fig. 3. Emission spectra of indo-1 as a function of calcium concentration. At wavelengths below 400 nm, emission in the absence of Ca drops to a few percent of that seen in the presence of Ca. Above 500 nm, there is a fourfold converse difference. The $[Ca^{2+}]$ of EGTA buffer solutions used is shown at the right. Analyses were performed on a Tracor Northern Fluoroplex III.

ing. The use of these dyes is thus more similar to quin2 and does not share the benefits of ratio analysis; if a uv light source is not available, however, useful results can be obtained with these dyes, as shown subsequently.

Flow Cytometric Assay with Indo-1

Loading of cells with indo-1. Uptake and retention of indo-1 is facilitated by the use of the acetoxymethyl ester of indo-1, using the scheme described above. Approximately 20% of the total dye is trapped in this manner during typical loadings. After loading, the extracellular indo-1 should be diluted 10- to 100-fold before flow cytometric analysis [91]. One incidental benefit of this loading strategy is that this procedure, like the more familiar use of fluorescein diacetate or carboxyfluorescein diacetate, discriminates between live and dead cells. The latter will not retain the hydrophilic impermeant dye and should be excluded during subsequent analysis.

The lower limit of useful intracellular loading concentrations of indo-1 is determined by the sensitivity of fluorescence detection of the flow cytometer and the upper limit is determined by avoidance of buffering of $[Ca^{2+}]_i$ by the presence of the calcium-chelating dye itself. Fortunately, indo-1 has excellent fluorescence characteristics (30-fold greater quantum yield than quin2 at a given dye concentration [40]) and useful ranges of indo-1 loading are relatively low. For human peripheral blood T cells, we have found adequate detection at or above 1 μM loading concentrations under conditions that achieve ~5 μM intracellular indo-1. Buffering of $[Ca^{2+}]_i$ was observed as a slight delay in the rise in $[Ca^{2+}]_i$ and a retarded rate of return of $[Ca^{2+}]_i$ to baseline values seen at loading concentrations above 3 μM (22 μM intracellular concentration), and a reduction in peak $[Ca^{2+}]_i$ seen at even higher indo-1 concentrations [91]. Chused et al. [19] observed slightly greater sensitivity of murine B cells to indo-1 buffering, recommending a loading concentration of no greater than 1 μM. For human platelets, a 2-μM loading concentration has been reported [23]. Rates of loading of the indo-1 ester can be expected to vary between cell types, for example, as a consequence of variations in intracellular esterase activity. In peripheral human blood, more rapid rates of loading are seen in platelets and monocytes than in lymphocytes. Even within one cell type, donor or treatment-

TABLE 1. Effect of Wavelength Choice on Calcium-Sensitive Indo-1 Ratio Shifts[a]

Wavelength Pair (nM)	R_{max}	R_{min}	R	S_{fo}/S_{bo}	R_{max}/R_{min}	R_{max}/R
475/395	2.33	0.040	0.352	3.3	58.2	6.62
475/405	2.38	0.100	0.410	3.3	23.8	5.80
495/395	3.51	0.048	0.429	4.2	73.1	8.18
495/405	3.59	0.119	0.501	4.2	30.2	7.17
515/395	5.75	0.070	0.644	4.63	82.1	8.93
515/405	5.88	0.176	0.752	4.63	33.4	7.82
530/395	9.68	0.117	1.073	4.68	82.7	9.02
530/405	9.89	0.292	1.252	4.68	33.9	7.90

[a]By spectrofluorimetry, 2-nm slit-width; uncorrected fluorescence.

specific factors may affect loading; for example, lower rates of indo-1 loading are seen in splenocytes from aged than from young mice [89].

The analysis of $[Ca^{2+}]_i$ using indo-1 is predicated on distribution of the dye uniformly within the cytoplasm. In several cell types, the related dye fura-2 has been reported to be compartmentalized within organelles [24,69]. In bovine aortic endothelial cells, fura-2 has been reported to be localized to mitochondria, however, indo-1 remains diffusely cytoplasmic [108]. In lymphocytes, there is as yet no evidence of subcellular localization of indo-1, consistent with the concurrence of calibration experiments with predicted results. It is possible that there will be fewer problems with compartmentalization of indo-1 than with fura-2; however, it seems advisable to examine the cellular distribution of indo-1 microscopically, and in each new application to confirm the expected behavior of the dye using the calibration procedures described subsequently.

It has been suggested that both fura-2 and indo-1 may be incompletely deesterified within some cell types [67,98]. Since the ester has little fluorescence spectral dependence on Ca^{2+}, the presence of this dye form could lead to false estimates of $[Ca^{2+}]_i$. Again, the results of calibration experiments are helpful in assessing this possibility.

Indo-1 has been found to be remarkably nontoxic to cells subsequent to loading. Analysis of the proliferative capacity of either human T lymphocytes [91] or murine B lymphocytes [19] has shown unaltered behavior of cells after loading with indo-1. This is especially pertinent to applications of sorting of cells based on $[Ca^{2+}]_i$, as described subsequently.

Instrumental technique. The choice of medium in which the cell sample is suspended for analysis can be dictated primarily by the metabolic requirements of the cells, subject only to the presence of mM concentrations of calcium (to permit calcium agonist-stimulated calcium influx) and reasonable buffering. The use of phenol red as a pH indicator does not impair the detection of indo-1 fluorescence signals. Although the new generation of Ca^{2+} indicator dyes are not highly sensitive to small fluctuations of pH over the physiologic range [40], unbuffered or bicarbonate buffered solutions can impart uncontrolled pH shifts. If analysis of release of Ca^{2+} from intracellular stores, independent of extracellular Ca^{2+} influx is desired, addition of 5 mM EGTA to the cell suspension (final concentration) will reduce Ca^{2+} from several mM to ~20 nM, thus abolishing the usual extracellular to intracellular gradient.

Regulation of the temperature of the cell sample is essential, as transmembrane signaling and calcium mobilization are temperature-dependent and active processes. Most applications will require analyses at 37°C. If cells are allowed to cool before they flow past the laser beam, $[Ca^{2+}]_i$ will often decline, so that either the sample input tubing should be warmed, or narrow-gauge tubing and high flow rates (e.g., ≥50 μl sample/minute) should be used to keep transit times from warmed sample to flow cell minimized.

The absorption maximum of indo-1 is 330–350 nm, depending on the presence of calcium [49]; this is well suited to excitation at either 351–356 nm from an argon ion laser or to 337–356 nm excitation from a krypton ion laser. Laser power requirements depend upon the choice of emission filters and optical efficiency of the instrument; however, it is our observation that although 100 mW is often routinely employed, virtually identical results can be obtained with 10 mW. This should allow, in principal, the use of helium–cadmium lasers. Stability of the intensity of the excitation source is less important in this application than many others, because of the use of the ratio of fluorescence emissions. Analysis with indo-1 has also been performed using excitation by a mercury arc lamp [115].

An increase in $[Ca^{2+}]_i$ is detected with indo-1 as an increase in the ratio of a lower to a higher emission wavelength. The choice of filters used to select these wavelengths is dictated by the spectral characteristics of the shift in indo-1 emission upon binding to calcium, as shown in Figure 3 [40]. Bandpass filters can be chosen to be centered on the "violet" peak emission of the calcium-bound indo-1 dye (400 nm) and free indo-1 dye "blue" emission (485 nm). However, wavelengths nearer the isobestic point do not exhibit as large a dependence upon calcium binding. This effect is summarized in Table 1 by values of R_{max}/R, which indicates the range of change in indo-1 ratio observed from resting intracellular calcium to saturated calcium; a larger dynamic range in the ratio of wavelengths is obtained if "blue" emission below 485 nm is not collected and the center of the "blue" emission bandpass filter is moved upward. Similarly, the "violet" bandpass filter should be chosen to minimize the collection of wavelengths above 405 nm.

If the indo-1 fluorescence emission is displayed on the flow cytometer as a bivariate plot of "violet" versus "blue" signals, the increase in ratio seen with increased $[Ca^{2+}]_i$ will be observed as a rotation around the axis through the origin. This method of ratio analysis is cumbersome, however, and fortunately commercial flow cytometers all have some provision for a direct calculation of the value of the fluorescence ratio itself, either by analog circuitry or by digital computation. Some analog ratio circuits are limited in their range of

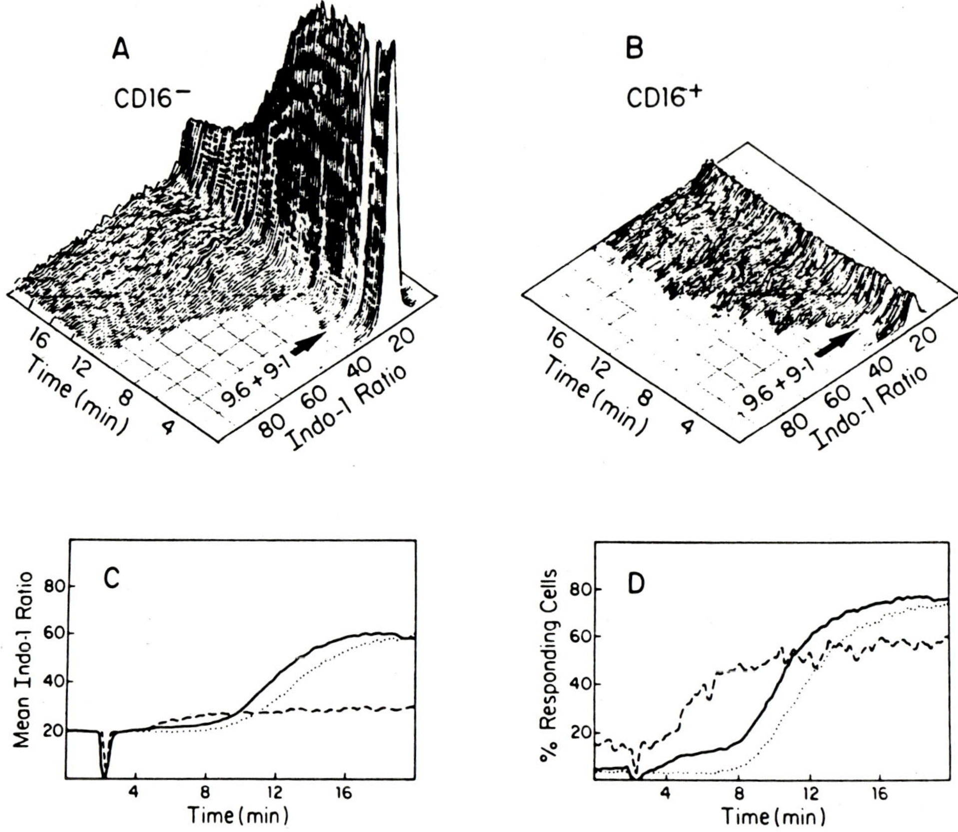

Fig. 4. Indo-1 analysis with simultaneous immunofluorescence staining: the effect of CD2 stimulation on $[Ca^{2+}]_i$ of CD16⁻ or CD16⁺ lymphocyte subsets. Indo-1-loaded nylon-wool-purified T cells were stained with anti-CD16 (PE-conjugated antibody FC-2), which identifies a population of LGL that do not express the CD3 antigen, and stimulated with CD2 antibodies 9.6 and 9.1. **A, B:** Indo-1 analyses of PE negative and positive cells. **C:** Time course for the development of increased mean indo-1 ratio. **D:** Percentage responding cells above a threshold ratio of 26. ——, total cell population analyzed; - - -, $CD16^+$ cells; ····, $CD16^-$ cells. Antibody FC-2 had no effect on the basal $[Ca^{2+}]_i$. The phenotype of the cells used for this experiment was 99% $CD2^+$, 95% $CD3^+$, and 2% $CD16^+$. (Reproduced from June et al. [48] by copyright permission of the American Society for Clinical Investigation.)

acceptable inputs; for example, that the "violet" signal never be greater than the "blue," yielding a ratio of greater than one; in this case the signal gains must be initially set such that subsequent rises in the ratio will not exceed the permitted value. By either analog or digital calculation, it is important that no artifactual offset be introduced in the ratio; this can be quickly tested by altering the excitation power over a broad range of values in an analysis of a nonperturbed indo-1-loaded cell population—a correctly calculated ratio will not show any dependence upon excitation intensity. It can similarly be shown that loading of cells with a broad range of indo-1 concentrations results in a constant value of the "violet"-to-"blue" ratio [91].

The above approach results in a linear display of indo-1 blue and violet fluorescence. If cellular indo-1 loading is extremely heterogeneous, it may be desirable to work with a logarithmic conversion of "violet" and "blue" emission intensities in order to observe a broader range of cellular fluorescence. In this case, the hardware must permit the logarithm of the ratio to be calculated by *subtraction* of the log "blue" from the log "violet" signals [91].

Plotted as a histogram of the ratio values, unperturbed cell populations show narrow distributions of ratio, even when cellular loading with indo-1 is very heterogeneous [91], and coefficients of variation of less than 10% are not uncommon. The effects of perturbation of $[Ca^{2+}]_i$ by agonists can be noted by changes in the ratio histogram profile during sequential analysis and histogram storage. A more informative and elegant display is obtained by a bivariate plot of ratio versus time. The bivariate histogram can be stored, and subsequently subjected to further analysis, or presentation as "isometric plots" in which the x axis represents time, the y axis $[Ca^{2+}]_i$, and the z axis, number of cells (Fig. 4). In the bivariate plot, kinetic changes in $[Ca^{2+}]_i$ are seen with much greater resolution, limited only by the number of channels on the time axis, and the interval of time between each channel. Changes in the fraction of cells responding, in the mean magnitude of the response, and in the heterogeneity of the responding population are easily observed with these displays. For example, it can be seen in Figure 4 that not all cells, even of a specific immunophenotypic subset, respond to a particular signal, and of those that do, the values of $[Ca^{2+}]_i$ are quite heterogeneous. Using simultaneous immunofluorescence with indo-1 analysis (discussed subsequently), some of this heterogeneity was shown to be due to difference between cell subsets; CD4-positive T cells have the highest proportion of cells responding, with less response in the CD8 cells and least in large granular lymphocytes [91]. It is not known to what extent the residual heterogeneity in $[Ca^{2+}]_i$ represents the effect of oscillations of $[Ca^{2+}]_i$ within individual cells as a function of time; since the same cells are not repeatedly examined by flow cytometry, other techniques must be used to address this question. A study using fura2 with digital microscopy has shown that under certain circumstances, anti-immunoglobulin stimulation of B cells produces repetitive, transient oscillations of $[Ca^{2+}]_i$ rather than sustained elevations of $[Ca^{2+}]_i$ [123].

Calculation of the mean y-axis value for each x-axis time interval permits presentation of the data as mean ratio versus time [91]. Calibration of the ratio to $[Ca^{2+}]_i$ permits data presentation in the same manner as traditionally displayed by spectrofluorimetric analysis; i.e., $[Ca^{2+}]_i$ versus time. While this presentation yields much of the information of interest in an easily displayed format, data relating to heterogeneity of the $[Ca^{2+}]_i$ response is lost. Some of this information can be displayed by a calculation of the "proportion of responding cells"—if a threshold value of the resting ratio distribution is chosen, e.g., one at which only 5% of control cells are above, the proportion of cells responding by ratio elevations above this threshold vs. time yields a presentation informative of the heterogeneity of the response [91]. Presentation of both the mean $[Ca^{2+}]_i$ versus time and the percent responding cells versus time facilitates visual comparison of results of different stimuli or treatment.

Calibration of ratio to $[Ca^{2+}]_i$. Before the development of indo-1 and fura-2, $[Ca^{2+}]_i$ determination by measurement of quin2 fluorescence was sensitive to changes in intracellular dye concentration as well as $[Ca^{2+}]_i$. This made calibration necessary at the end of each individual assay by determination of the fluorescence intensity of the dye at zero and saturating $[Ca^{2+}]_i$. By contrast, with indo-1 use of the $[Ca^{2+}]$-dependent shift in dye emission wavelength permits the ratio of fluorescence intensities of the dye at the two wavelengths to be used to calculate $[Ca^{2+}]$:

$$[Ca^{2+}] = Kd \cdot \frac{(R - R_{min})}{(R_{max} - R)} \cdot \frac{S_{f2}}{S_{b2}} \qquad (1)$$

where K_d is the effective dissociation constant (250 nM); R, R_{min}, and R_{max} are the fluorescence intensity ratios of violet to blue fluorescence at resting, zero, and saturating $[Ca^{2+}]$, respectively; and S_{f2}/S_{b2} is the ratio of the blue fluorescence intensity of the calcium-free and -bound dye, respectively [40]. Because this calibration is independent of total intracellular dye concentration and instrumental variation in efficiency of excitation or emission detection, it is not necessary to measure the fluorescence of the dye in the calcium-free and saturated states for each individual assay. In principle, it is sufficient to calibrate the instrument once after set-up and tuning by measurement of the constants R_{max}, R_{min}, S_{f2} and S_{b2}, after which only R is measured for each subsequent analysis on that occasion.

One strategy to obtain the R_{max} and R_{min} values for indo-1 is to lyse cells in order to release the dye to determine fluorescence at zero and saturating $[Ca^{2+}]$, as is carried out with quin2. However, this is not possible with flow cytometry, due to the loss of cellular fluorescence. Another strategy is the use of an ionophore to saturate or deplete $[Ca^{2+}]_i$, in order to allow approximation of the true endpoints without cell lysis. For this approach the ionophore ionomycin is best suited, due to its specificity and low fluorescence. When flow cytometric quantitation of fluorescence from intact cells treated with ionomycin or ionomycin plus EGTA was compared with spectrofluorimetric analysis of lysed cells in medium with or without EGTA, the indo-1 ratio of unstimulated cells (R) and the ratio at saturating $[Ca^{2+}]_i$, R_{max} were similar by both techniques [91]. The latter indicates that ionomycin-treated cells reach near-saturating levels of $[Ca^{2+}]_i$. The value of R_{min}, which can be obtained by treatment of cells with ionomycin in the presence of EGTA, however, is substantially higher than either that predicted from the spectral emission curves (Fig. 3) or that obtained by cell lysis and spectrofluorimetry. Spectrofluorimetric quantitation with either quin2 or indo-1 indicates that $[Ca^{2+}]_i$ remains at approximately 50 nM in intact cells treated with ionomycin and EGTA. Thus, due to the inability to obtain a valid flow cytometric determination with calcium-free dye, we have used for calibration the values of R_{min} and S_{f2}/S_{b2} derived from spectrofluorimetry, either of the indo-1 pentapotassium salt or after lysis of indo-1-loaded cells in the presence of EGTA. It is essential that the same optical filters be used for flow cytometry and spectrofluorimetry, since the standardization is very sensitive to the wavelengths chosen. Typical values of R_{max}, R_{min}, R and S_{f2}/S_{b2} are shown for different emission wavelength combinations in Table 1. Even if careful calibration of the fluorescence ratio to $[Ca^{2+}]_i$ is not being performed for a particular experiment, ordinary quality control can include a determination of the value of R_{max}/R. Unperturbed cells will usually be found to have a reproducible value of $[Ca^{2+}]_i$, and day-to-day optical variations in the flow cytometer are usually minimal (with the same filter set); a limited range of R_{max}/R values should therefore be obtained.

As an alternative to the use of a spectrofluorimeter, a procedure which may be cumbersome or the instrument unavailable, Parks et al. [85] have proposed that the flow cytometer may be used as a spectrofluorimeter with minor modification. In essence, the fluorescence of a steady stream of dye in a buffer of known $[Ca^{2+}]$ is measured by the flow cytometer and the photomultiplier voltage is analyzed directly (as in a standard spectrofluorimeter), or the steady-state fluorescence signal is converted to a pulse for processing by the flow cytometer electronics. The latter can be performed either by interrupting the laser beam periodically, or by converting the photomultiplier tube output from a steady current/voltage level to a pulse. With this technique, it should be possible to determine all constants necessary for direct calibration of indo-1 on the flow cytometer.

Chused et al. [19] suggested that a cocktail of ionomycin, nigericin, high concentrations of potassium, 2-deoxy glucose, azide, and carbonyl cyanide m-chlorophenylhydrazone can be used to collapse the calcium gradient across the cell membrane to zero. To avoid the difficulty of assessing R_{min} in cells, the calibration is based on a regression formula that relates R to ionomycin-treated cells suspended in a series of calcium buffers. Thus, this technique allows one to estimate $[Ca^{2+}]_i$ without the need to determine R_{min}, S_{f2} or S_{b2}, although it is subject to the necessity to prepare Ca^{2+} buffer solutions precisely.

Accuracy of prediction of Ca^{2+} in buffer solutions is dependent upon a variety of interacting factors so that care must be exercised in formulating Ca^{2+} standards. The determination of the ionized calcium concentration in an EGTA buffer system is dependent upon the magnesium concentration; other metals such as aluminum, iron and lanthanum also bind avidly to EGTA [72]. In addition, the dissociation constant of Ca^{2+}–EGTA is a function of pH, temperature, and ionic strength [12]. For example, in a Ca^{2+}/Ca^{2+}–EGTA buffer system, changing the pH from 7.4 to 7.1 can result in the ionized calcium changing from 110 to 375 nM. Finally, it is important to prepare the buffers using the "pH metric technique" [78] because of the varying purity of commercially available EGTA [74].

If, for a particular cell type loaded with indo-1, the values of R and R_{max} obtained by flow cytometry are in good agreement with the values obtained by spectrofluorimetry, then it would be unlikely that the dye is in a compartment inaccessible to cytoplasmic Ca^{2+}, in a form unresponsive to $[Ca^{2+}]_i$ (e.g., still esterified) or in a cytoplasmic environment in which the spectral properties of the dye were altered. With regard to the second condition, it has been proposed that because indo-1 fluorescence, but not that of the indo-1 ester, is quenched in the presence of mM concentrations of Mn^{2+}, Mn^{2+} in the presence of ionomycin can be used as a further test of complete hydrolysis of the indo-1 ester within cells [67].

Simultaneous analysis of $[Ca^{2+}]_i$ and other fluorescence parameters. Although the broad spectrum of indo-1 fluorescence emission will likely preclude the simultaneous use of a second UV-excited dye, the use of two or more excitation sources allows additional information to be derived from visible-light-excited dyes. Perhaps the most usual application will be determination of cellular immunophenotype simultaneously with the indo-1 assay, allowing alterations in $[Ca^{2+}]_i$ to be examined in, and correlated with, specific cell subsets. For example, Figure 4 shows that after CD2 (T11) stimulation of peripheral blood lymphocytes, both $CD3^+$ T cells and $CD16^+$ $CD3^-$ large granular lymphocytes (LGL) mobilized calcium. The pattern of the calcium signal after CD2 stimulation differs, however, in that calcium mobilization in LGL is early in onset and low in magnitude; while in T cells, the calcium signal is delayed and high in magnitude. After CD3 (antigen-specific) stimulation, more than 90% of T cells responded, and as was expected, LGL did not respond [48].

In contrast to T lymphocytes, in human B cells after stimulation by anti-immunoglobulin, only ~50% of cells respond. The CD22 antigen is expressed on a subset of 60–80% of mature B cells, and it was found that all cells responding to anti-immunoglobulin stimulation could be accounted for by cells having the phenotype of IgD^+ IgM^+ $CD22^+$, while cells having the phenotype IgM^+ $CD22^-$ did not respond [87]. Thus, the presence of CD22 rather than of antigen receptor correlates with calcium mobilization after anti-immunoglobulin stimulation, suggesting that the antigen receptor is in some way uncoupled from signal transduction in a subset of B cells. The demonstration of such heterogeneity in $[Ca^{2+}]_i$ signals would have been impossible to discern in conventional assays carried out in a fluorimeter where only the mean calcium response is recorded.

Combining the use of FITC, PE and indo-1 analysis permits determination of $[Ca^{2+}]_i$ in complex immunophenotypic subsets. On instruments without provision for analysis of four separate fluorescence wavelengths, detection of both FITC and the higher indo-1 wavelength using the same filter element may allow successful implementation of these experiments. Gating the analysis of indo-1 fluorescence on windows of FITC versus PE fluorescence enables information relating to each identifiable cellular subset to be derived from a single sample. It has already been mentioned that the $CD4^+$ subset shows a more vigorous response to anti-CD3 and lectin stimulation than does the $CD8^+$ subset [39,91].

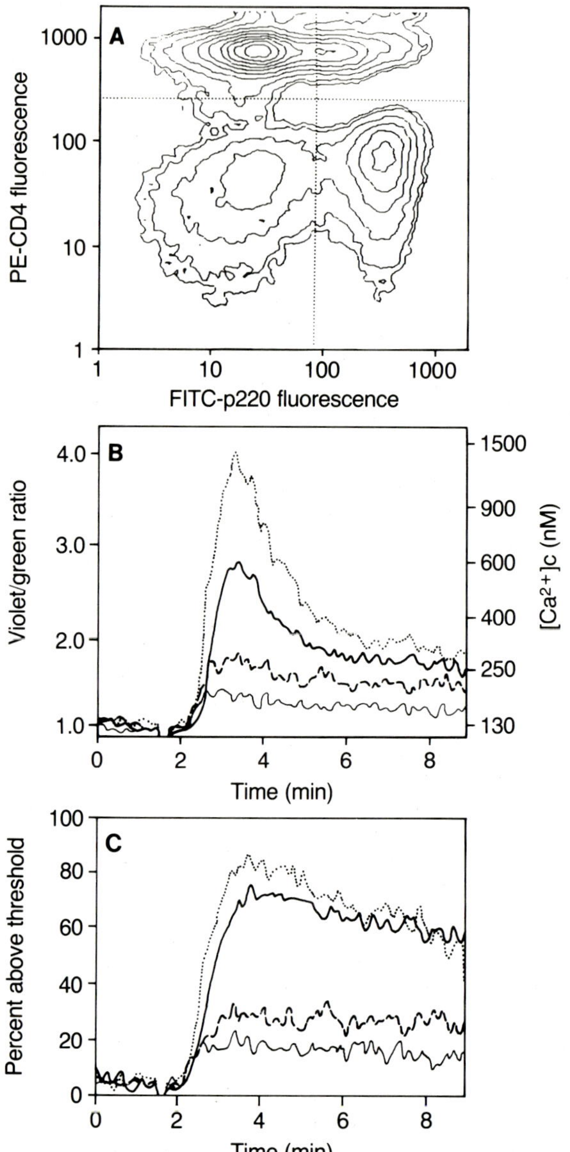

Fig. 5. Analysis of indo-1 fluorescence of subsets of cells identified by FITC/PE staining. A CD2 stimulus was delivered by incubating cells for 10 min with biotinylated anti-CD2 mAb 9.6 before analysis. At time 2 min, the 9.6 mAb was crosslinked by the addition of avidin [63]. **A:** Contour plot of FITC-CD45R (p220, mAb 3AC5) fluorescence (abscissa) vs. PE-CD4 (G19-2) fluorescence (ordinate). The mean indo-1 violet/green fluorescence ratio **(B)** and percentage responding cells above a 2 SD threshold **(C)** were examined by list-mode processing of the data, gating the analysis to display data for only $CD45R^+$ $CD4^+$ cells (heavy solid line), $CD45R^-$ $CD4^+$ cells (dotted line), $CD45R^+$ $CD4^-$ cells (dashed line), or $CD45R^-$ $CD4^-$ (light solid line). These subpopulations were 13%, 24%, 33%, and 30% of the total cells, respectively.

Figure 5 shows that by simultaneous staining of CD4 and CD45R within the CD4 subset, the $[Ca^{2+}]_i$ response to CD2 stimulation is more rapid in the $CD45R^+$ subset ($LFA3^-$ cells with memory function [97]) than in the $CD45R^-$ subset. This difference is similar to, but even greater in magnitude, than the differences previously shown between these subsets following stimulation with anti-CD3 [91].

In the simultaneous analysis of $[Ca^{2+}]_i$ and immunofluorescence, the caution should be considered that use of the antibody probe can itself alter the cellular $[Ca^{2+}]_i$. It is becoming increasingly clear that binding of mAb to cell surface proteins can alter $[Ca^{2+}]_i$, even when these proteins are not previously recognized as part of a signal transducing pathway [2,47,49,59,86,87,91]. For example, antibody binding to CD4 will reduce CD3-mediated $[Ca^{2+}]_i$ signals; if the anti-CD4 mAb is cross-linked, as with a goat-antimouse mAb, the CD3 signals are augmented [59,61]. Antibody binding to CD8 has similar effects (unpublished data).

As a consequence of these concerns, a reciprocal staining strategy can be used whenever possible, so that the cellular subpopulation of interest is unlabeled, while undesired cell subsets are identified by mAb staining. The $CD4^+$ subset in PBL may be identified, for example, by staining with a combination of CD8, CD20, and CD11 mAbs [64], and the $CD5^+$ subset can be identified by staining with CD16, CD20 and anti-HLA-DR [49]. Finally, it is important when staining cells with mAb for functional studies that the antibodies be azide-free, in order that metabolic processes be uninhibited. Commercial antibody preparations may thus require dialysis before use.

Using other probes excited by visible light, it may be possible to analyze additional physiologic responses in cells simultaneously with $[Ca^{2+}]_i$. The simultaneous analysis of membrane potential and $[Ca^{2+}]_i$ is discussed later in this chapter. Using similar studies, it may be possible to analyze intracellular pH simultaneously with $[Ca^{2+}]_i$.

Sorting on the basis of $[Ca^{2+}]_i$ responses. The ability of the flow cytometric analysis with indo-1 to observe small proportions of cells with different $[Ca^{2+}]_i$ responses than the majority of cells suggests that the flow cytometer may be useful to identify and sort variants in the population for their subsequent biochemical analysis or growth. Results of artificial mixing experiments with Jurkat and K562 T-cell and myeloid leukemia cell lines indicated that subpopulations of cells with variant $[Ca^{2+}]_i$ comprising <1% of total cells could be accurately identified [91]. Goldsmith and Weiss [36,37] recently reported the use of sorting on the basis of indo-1 fluorescence to identify mutant Jurkat cells which fail to mobilize $[Ca^{2+}]_i$ in response to CD3 stimulus, in spite of the expression of antigenically normal CD3/Ti complexes. These experiments suggest that sorting on the basis of indo-1 fluorescence can be an important tool for the selection and identification of genetic variants in the biochemical pathways leading to Ca^{2+} mobilization and cell growth and differentiation.

Flow Cytometric Assay with Fluo-3 or Rhod-2

These dyes, very recently developed [76], allow calcium analyses to be performed using visible-light excitation, with spectral characteristics analogous to fluorescein and rhodamine. These dyes do not, unfortunately, exhibit wavelength shifts upon Ca^{2+} binding, so that dual-wavelength ratioing is not possible [76]. The dissociation constants for these dyes (0.4–2.3 μM) are higher than those of indo-1 and fura-2, which may be of use in the measurement of high $[Ca^{2+}]_i$ levels in some stimulated cells. Fluo-3 will probably prove most useful for flow cytometry, both because it is excited by the 488-nm Argon laser line, and because it exhibits a greater fluorescence enhancement upon Ca^{2+} binding than does rhod-2. According to data published in the Molecular Probes, Inc., catalog, fluo-3 shows 8-fold greater fluorescence at 60 μM Ca^{2+} than at 0 μM Ca^{2+} and 3.5-fold greater fluorescence at 60 μM than at typical resting $[Ca^{2+}]_i$ levels, 130 nM.

Cells can be efficiently loaded with these dyes using the acetoxymethyl ester, using the same protocol as described for indo-1. Calibration can be carried out as is performed with quin2; because calibration for nonratiometric measurements is sensitive to loading concentration, each batch of stained cells must be separately calibrated.

Figure 6 shows a direct comparison of results obtained with fluo-3 and indo-1; in this example, both dyes were loaded and analyzed simultaneously in the same cells. The spectral characteristics of fluo-3 also allowed simultaneous staining of cells with PE-conjugated antibody, although compensation for fluorescence crossover (fluo-3 into the PE detector) is required. Both dyes showed a readily measurable response to the agonist, although the change in indo-1 fluorescence ratio was proportionally slightly greater than the increase in fluo-3 fluorescence intensity. Figure 6 also shows that due to heterogeneity in the intracellular concentration of fluo-3 in the loaded cells, the determination of $[Ca^{2+}]_i$ in any particular cell is less accurate, and discrimination of heterogeneity in the cellular response to agonist is less clear using fluo-3. Nevertheless, the ability to utilize fluo-3 with 488-nm excitation, and to perform simultaneous immunofluorescence with the same laser, will make this dye useful in many laboratories.

Applications of Study of $[Ca^{2+}]_i$ in Lymphocytes Using Flow Cytometry

The flow cytometric assay of cellular calcium concentration has already been applied to a wide variety of cells, providing interesting and sometimes unexpected results (Table 2). Using indo-1, a small coefficient of variation (often <10%) was found in lymphocytes for the distribution of the basal calcium level; this value appears to reflect physiologic variation because the instrumental variation was only 4.5% [91]. It has already been mentioned that, in stimulated cells, there is heterogeneity in the responding cells, some of which is correlated with cell surface antigen expression. Surprisingly, there is also heterogeneity in the response of lymphocytes to calcium ionophore [45,91].

Current evidence strongly indicates that calcium mobilization may be necessary for antigen-induced B cell activation. In small resting murine splenic B cells, crosslinking the IgD (delta) receptor causes a brief response onset at ~5 sec that reaches high magnitude, while stimulation of IgM (mu) receptors causes a reaction of later onset, lower magnitude and is more prolonged [19]. The crosslinking of surface immunoglobulin with Fc receptors for Ig on B cells inhibits calcium mobilization when compared to that observed after ligation of the B-cell receptor (by $F(ab)'_2$ anti-immunoglobulin antibodies) that is independent of the Fc receptor [125]. Similarly, the binding of mAb to the CD19 (B4) antigen or treatment with phorbol esters, presumably through the activation of protein kinase C, can also inhibit anti-Ig-induced calcium signaling [86,96]. Stimulation of cells with IL-4 (BSF-1) and lipopolysaccharide (LPS) appears to be independent of the calcium signaling system [50], although the low-molecular-weight form of B-cell growth factor (and not IL-1 IL-2, G-CSF or GM-CSF) is able to cause or potentiate antigen-receptor-induced calcium mobilization in human tonsillar B cells and in some leukemic cells [64].

In both B and T cells, crosslinking of the antigen receptor enhances signal transduction [15,60,71]. The mitogenic effects of monoclonal antibody ligation of antigen receptor is

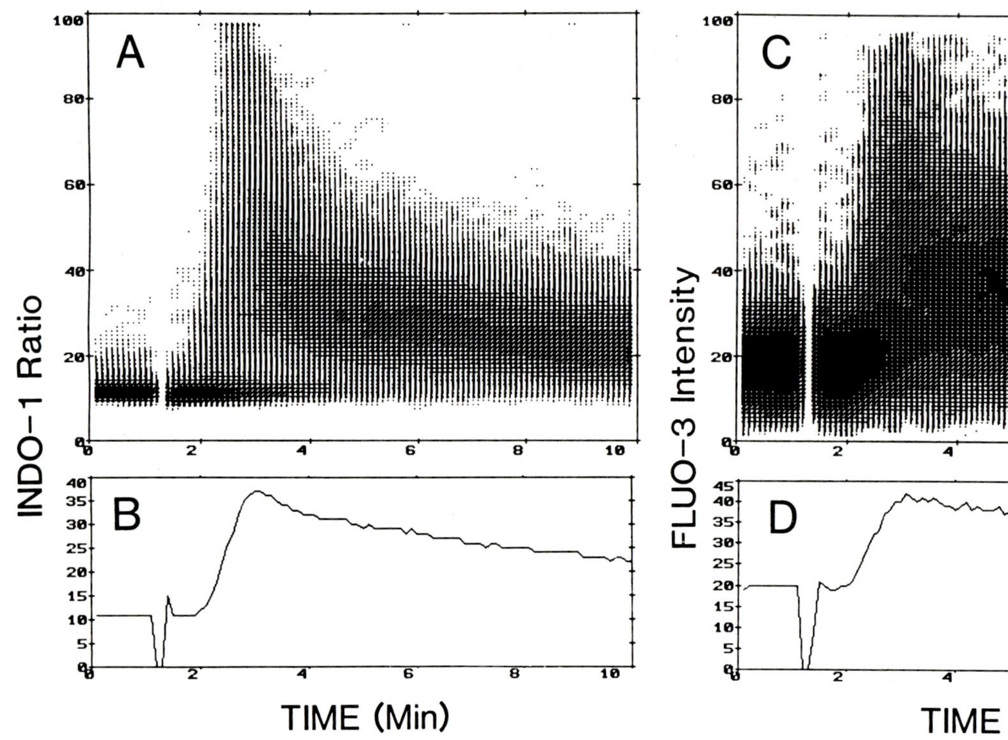

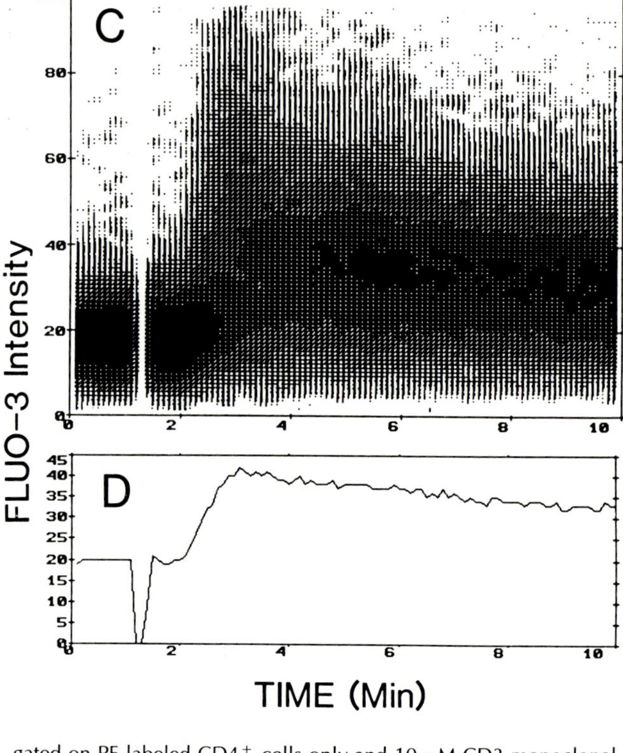

Fig. 6. Simultaneous analysis of indo-1 and fluo-3 in CD4$^+$ T cells. Peripheral blood lymphocytes were loaded with both 3 μM indo-1 and 3 μM fluo-3, stained with PE-CD4 mAb(G19-2) and analyzed using the same optical arrangement as used in Figure 5. Analysis was gated on PE-labeled CD4$^+$ cells only,and 10 μM CD3 monoclonal antibody (G19-4) was added at 1.5 min. The indo-1 green/violet fluorescence ratio **(A)** indo-1 mean ratio **(B),** fluo-3 intensity **(C),** and mean fluo-3 intensity **(D)** are shown as a function of elapsed time.

strongly enhanced by coupling the antibody to a solid support, presumably through prevention of receptor internalization and enhanced signal transduction [59–61]. In studies using indo-1 in T cells, it was found that crosslinking of CD2, CD3, CD4, CD5, CD8, CD28, and MHC class I molecules all can result in calcium mobilization that involves phosphoinositide turnover, while perturbation of other surface molecules such as CD7, CD11, CD25, and CD45 does not cause calcium mobilization [34,49,59]. Most, if not all of the receptor molecules associated with calcium mobilization are members of the immunoglobulin gene superfamily. As mentioned previously, binding of mAb to some of these receptor molecules can also alter signal transduction via the antigen receptor [49,59,61,62,67,86,91]. There is some evidence that the cytoskeleton is involved in the regulation of cytoplasmic calcium [5,111].

The role of calcium signaling in T-cell maturation is being explored. There are qualitative differences between immature and mature thymocytes in the pattern of calcium mobilization [27]. The results suggest that immature T cells have fewer calcium channels or that the channels are less efficiently activated after antigen receptor ligation.

One function of mature T cells is to kill foreign or virus infected cells. Mentzer et al. [73] have used indo-1 and flow cytometry to show that calcium mobilization occurs in the effector T cells during the lytic stage of killing but not during the adhesion stage.

Monocytes. Regulation of cellular calcium concentration in mononuclear phagocytes is complex and only partially understood. In studies using indo-1 and flow cytometry, leukotriene D4 has been shown to increase cytosolic calcium in differentiated HL-60 cells, a leukemic cell line that has characteristics of monocytes [7]. Studies of calcium metabolism in resting nonadherent monocytes have not been reported.

Neutrophils. Lazzari et al. [57] reported the first application of flow cytometric analysis of neutrophils using indo-1. As mentioned below, this analysis was performed simultaneously with measurement of membrane potential. These investigators observed a rapid uniphasic $[Ca^{2+}]_i$ response of neutrophils to *N*-formyl-methionyl-leucyl-phenylalanine (FMLP) over a broad range of FMLP concentration. Interestingly, the membrane potential changes were biphasic at intermediate FMLP concentrations, with hyperpolarization followed by depolarization. Growth factors GM–CSF and G–CSF do not appear to cause calcium mobilization in neutrophils [109].

Platelets

The indo-1-based flow cytometric assay of cells as small as platelets is eminently feasible. Using this technique, Davies et al. [23] found heterogeneity in the response of platelets to thrombin stimulation. At low doses of thrombin, the response consisted of a large Ca^{2+} rise in a subpopulation of cells and a second population of platelets that failed to respond, whereas at high doses of thrombin, all cells responded. It remains to be determined whether a population of platelets exhibits differential thresholds for a response to an agonist based upon different requirements for receptor occupancy or whether all platelets respond and the observed heterogeneity is due to oscillatory changes in Ca^{2+}.

TABLE 2. Summary of Studies of $[Ca^{2+}]_i$ in Hematopoietic Cells Using Indo-1

Cell type	Study aims/conclusions	References
B lymphocytes	Effects of intracellular calcium buffering on antigen-receptor stimulation	[19]
	Effects of cytochalasins on $[Ca^{2+}]_i$	[5,111]
	Phorbol esters and Fc receptors inhibit anti-Ig $[Ca^{2+}]_i$ transients	[31,77,96,124]
	Effects of growth factors on $[Ca^{2+}]_i$	[32,50,64]
	CD19 inhibits anti-Ig stimulation	[86]
	Expression of CD22 correlates with anti-Ig induced $[Ca^{2+}]_i$ transients	[87]
T lymphocytes	Correlation of surface antigen expression with $[Ca^{2+}]_i$ transients	[48,91,39]
	Heterogeneity of $[Ca^{2+}]_i$ response after calcium ionophore	[45,91]
	Aluminum fluoride causes $[Ca^{2+}]_i$ elevation	[84]
	Effects of CD2 on $[Ca^{2+}]_i$	[14,48,63]
	Effects of CD3 on $[Ca^{2+}]_i$	[36,37,58,91,116]
	Effects of CD4 on $[Ca^{2+}]_i$	[13,61]
	Effects of CD5 on $[Ca^{2+}]_i$	[11,49,62]
	Effects of crosslinking on $[Ca^{2+}]_i$	[59,61]
	Diminished $[Ca^{2+}]_i$ response associated with aging	[75,89]
	Cytotoxic T-lymphocyte function and effects on $[Ca^{2+}]_i$	[73]
	HIV-1 infection inhibits CD3-induced $[Ca^{2+}]_i$ transients	[66]
	Cyclosporine does not prevent $[Ca^{2+}]_i$ transients	[93]
	Thymic ontogeny	[27,120]
Large granular lymphocytes	Anti-CD2 causes $[Ca^{2+}]_i$ mobilization	[48]
	Effects of CD16 on $[Ca^{2+}]_i$ after anti-CD2 stimulation	[2]
Monocytes	Effects of leukotriene D4 on $[Ca^{2+}]_i$	[7]
Neutrophils	Effects of FMLP on $[Ca^{2+}]_i$ and membrane potential	[57]
Platelets	Heterogeneity of thrombin effects on $[Ca^{2+}]_i$	[23]

Clinical Applications

Several applications of cytosolic calcium measurement that have direct clinical relevance have been reported. It is possible that the measurement of cytoplasmic free calcium may be useful for the diagnosis of several diseases, such as malignant hyperthermia [54,55], congenital disorders of the immune system [9], cystic fibrosis [65,99,104,117], and the Lambert–Eaton syndrome [53].

The measurement of cellular calcium has proved useful in the investigation of the pathogenesis of certain acquired disorders of the immune system. There is a decline in immune function with aging, and in mice, calcium signaling in splenic lymphocytes after lectin stimulation or after anti-CD3 stimulation is substantially less vigorous in aged mice compared to young mice [75,89]. We recently found that infection with HIV-1, the pathogenic retrovirus that causes acquired immune deficiency syndrome (AIDS), causes a profound signaling defect after CD3 but not after CD2 stimulation [66]. These results provide evidence that cellular activation can occur through distinct biochemical mechanisms and that one mechanism of immunosuppression in AIDS is through an impairment of antigen-specific T-cell signaling.

ANALYSIS OF MEMBRANE POTENTIAL

Resting cells maintain large gradients between intracellular and extracellular concentrations of a variety of ions, including Ca^{2+}, K^+, Na^+, and Cl^-. Potassium ions, for example, are concentrated within cells by action of the Na–K ATPase. The relative permeability of the membrane to K^+ ions is greater than that of other ions; the leakage of K^+ ions establishes an electron countergradient and the cytoplasm becomes electron negative with respect to the external medium. This K^+ electrochemical gradient is the most significant contribution to the negative membrane potential of most mammalian cells. Maintenance of a large negative transmembrane potential has been postulated to be a control mechanism to arrest cells in an inactive stage, and changes in cell membrane potential that occur in various cell types rapidly after binding of ligands to transmembrane receptors have been suggested to be mediators of subsequent physiologic cellular responses. Detailed investigation of membrane potential has been possible in nerve, muscle, and other large cells by direct measurement with implanted microelectrodes. This approach is impractical or impossible in small cells, so that our understanding of the role of membrane potential in cell activation of smaller cells has been dependent on the development and utilization of potential sensitive indicator probes.

Indicators of Membrane Potential

A charged lipophilic molecule can serve as an indicator of membrane potential, as it is expected to partition between the cell and surrounding medium according to the Nernst equation:

$$\frac{C_c}{C_O} = e^{-nEF/RT}$$

where C_c and C_o are the cytosolic and extracellular indicator concentrations, n is the charge of the indicator (positive for a cation), E is the membrane potential and F, R, and T are

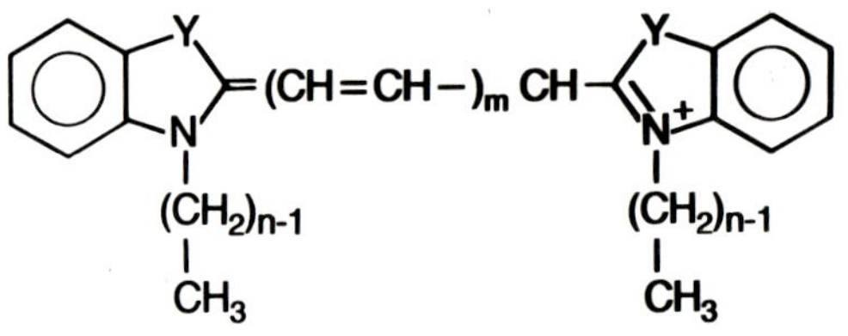

Di Y C_n (2m+1)

Fig. 7. Structure of cyanine dyes. The variable substituents Y, m, and n are described in the text.

TABLE 3. Commonly Used Dyes in Flow Cytometric Analysis of Membrane Potential

	Excitation[a] (nm)	Emission[a] (nm)
Cyanine		
$DiOC_5(3)$	488	510–540
$DiOC_6(3)$	488	520–540
$DiSC_3(5)$	568, 633	>590, >680
$DiIC_5(3)$	488, 514	540–580
Oxonol		
$diBaC_4(3)$	488	510–540
$diBaC_4(5)$	568–595	610–640
$diSBaC_2(3)$	568	590–630

[a]Commonly used laser lines and emission-detection ranges.

the Faraday and gas constants and temperature. For a cationic indicator, the cellular concentration falls as the membrane potential declines towards zero, and rises if the cell hyperpolarizes (i.e., the cytosol becomes more electronegative with respect to the medium). In practice, a useful indicator should not itself perturb the membrane potential, either by its very presence or by cellular toxicity, and changes in the partitioning of the indicator should be readily detectable.

Cyanine dyes. A family of cationic cyanine dyes were described by Hoffmann and Laris [44] to be useful indicators for cells in solution, in the manner described above. These fluorescent dyes have a single negative charge delocalized over an extensive pi-electron system in a highly symmetric molecule (Fig. 7). Sims et al. [106] introduced the shorthand nomenclature $DiYC_n(2m + 1)$ for these dyes, where the Y member of the ring structure may be oxygen (O), sulfur (S), or isopropyl (I). The length of the alkyl side chains, n, affects the lipid solubility, and m, the number of methene groups, affects the fluorescence spectral characteristics.

The response time for detection of membrane potential changes of these indicators is limited by the rate of re-equilibration of dye partitioning. For the cyanine dyes commonly used, the alkyl substituents are of ethyl length or greater, membrane permeability is high, and response times are on the order of 10^{-2} sec [95]. These rates are quite satisfactory for flow cytometry, although they have been termed *slow-response dyes*, in contrast to a group of dyes with time constants of 10^{-5} sec, or less, that are useful for monitoring changes as rapid as action potentials in excitable membranes [119]. Unfortunately, the optical characteristics of the latter group of dyes limits their application to large cells.

Earlier studies of membrane potential with cyanine dyes employed bulk measurements and high dye concentrations. Under these conditions, cellular uptake of the cation during cell hyperpolarization results in a counterintuitive *reduced* fluorescence of the suspension due to fluorescence quenching of the intracellular dye secondary to formation of dye aggregates, and red-shifting of excitation and emission spectra [119]. The use of more fluorescent dyes, such as those shown in Table 3, and flow cytometry allows adequate fluorescence signal detection of single cells with dye concentrations below 10^{-7} M. Under these conditions hyperpolarization is accompanied by *increased* cellular fluorescence and depolarization by decreased fluorescence. The lower dye concentrations also help to minimize the toxicity of these agents, as described subsequently.

Oxonol dyes. The oxonol dyes are chemically unrelated to the cyanine dyes, but are similarly symmetric, membrane-permeant molecules with a highly delocalized charge. This charge is negative, in contrast to the cyanine dyes, so that the changes in partitioning in response to altered membrane potential are in the opposite direction to cyanine dyes: depolarization of the membrane transfers the anion from the external medium onto binding sites within the cell, and hyperpolarization results in dye exclusion and decreased cellular fluorescence.

Cellular dye binding results in red-shifting of absorption and emission and/or an increase in quantum efficiency [6]. For some of the bis-oxonol dyes, the increase in fluorescence quantum efficiency in the nonpolar environment can be as high as 20-fold [94].

Oxonol dyes have not been used as extensively as the cyanine dyes for measurement of membrane potential, in part because only a small proportion of dye is membrane bound. For flow cytometry, however, the properties of the oxonol dyes are especially attractive [18,122]: analysis of cells takes place almost completely apart from dye in the external medium (increasing the S/N ratio). In addition, the lower proportion of bound dye increases the buffering of the external dye concentration and the negative charge of the dye forces exclusion from the highly negatively charged mitochondria, minimizing a complication encountered with cyanine dyes.

Flow Cytometric Assay of Membrane Potential

Flow cytometry was first demonstrated to be applicable to analysis of membrane potential by Shapiro et al. [105], and the techniques used subsequently are fundamentally unaltered. Cells at 37°C are equilibrated with low concentrations of indicator dye, generally ≤50 nM cyanine dyes and ≤150 nM oxonol dyes (toxicity and other artifacts being reduced at lower concentrations of indicator). The cell concentration should be kept low, usually near 2–5 × 10^5/ml, so that in spite of low dye concentrations, the dye/cell ratio is kept high and constant [122]; 5–15 min later, the solution is introduced into the flow cytometer, the chamber (and in some cases the fluidics tubing) being warmed to 37°C. Various indicator dyes permit a range of choices in excitation and emission wavelength combination (Table 3); however, for single-parameter measurements, excitation at 488 nm using $DiOC_5(3)$ or $diBaC_4(3)$ is most common. Adsorption of the dye onto the sample tubing may reach equilibration very slowly, and can result in slowly rising baseline cellular fluorescence (Fig. 8B). This may be minimized by pretreatment of

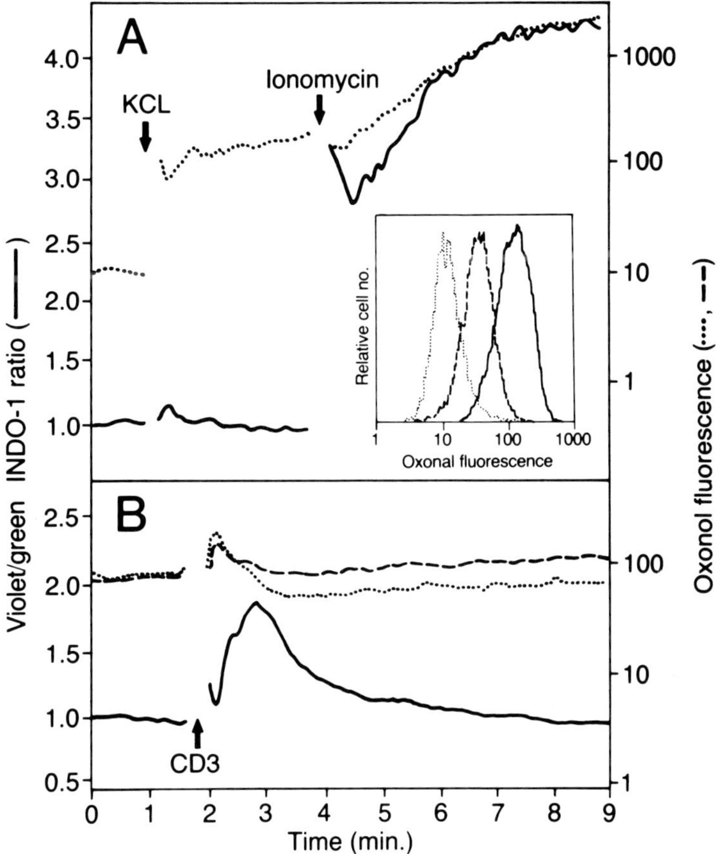

Fig. 8. Simultaneous analysis of bis(1,3-dibutylbarbiturate)-trimethene oxonol [diBaC$_4$(3)] fluorescence (dotted and dashed lines) and indo-1 fluorescence (solid lines). CD4$^+$ cells were isolated using immunomagnetic beads and placed in culture with PHA for 72 hours. After loading with indo-1 (3 μM) cells were incubated in 100 nM oxonol dye at 37°C, 10 min before analysis. Panel A shows control additions of 50 mM KCl (final concentration) at 1 min and 5 μM ionomycin at 4 min. The former results in membrane depolarization without effect on $[Ca^{2+}]_i$, while the latter increases $[Ca^{2+}]_i$ and produces further membrane depolarization. The inset shows profiles of oxonol fluorescence at 1 min (dotted line), 3 min (dashed line) and 8 min (solid line). Panel B shows that after stimulation by simultaneous crosslinking of CD3 and CD4 by addition of CD3/CD4 heteroconjugate [61] at time 2 min, the increase in $[Ca^{2+}]_i$ (solid line) is accompanied by membrane hyperpolarization (dotted line) compared to the oxonol fluorescence of control mock addition (dashed line). The slight gradual increase in baseline of the mock addition is due to the effects of adsorption of oxonol dye onto the sample tubing (see text).

the tubing with the dye solution without cells [122], or measurements can be made at a constant time after addition of the cell suspension to the flow cytometer. Forward-angle scatter is generally used to facilitate discrimination of dead cells and cell aggregates; the former are obviously depolarized and will otherwise confuse the analysis. Analysis can also be gated on light scatter to examine only cells of a uniform size; unlike a ratio measurement, large cells which bind more dye will appear more fluorescent, independent of their membrane potential. As with indo-1 analysis, the resulting fluorescence profiles can be analyzed and displayed as histograms, bivariate plots of fluorescence vs. time, or as mean fluorescence vs. time. Figure 8 shows an example of simultaneous analysis of membrane potential [diBaC$_4$(3)] and $[Ca^{2+}]_i$ (indo-1) in T lymphocytes, illustrating depolarization which occurs in medium with elevated K^+ and hyperpolarization which accompanies receptor–ligand interaction.

Calibration. Because the fluorescence changes of indicator dyes for a given membrane potential change are highly dependent on the dye concentration, dye-to-cell ratio, and cell type used, calibration must ordinarily be established for each experiment. The most commonly used technique was originally described by Hoffmann and Laris [44] and employs the K^+ ionophore valinomycin to increase membrane K^+ permeability in the presence of various external K^+ concentrations. To find the membrane potential by the "null point" method, the concentration of external K^+ is determined, at which no change in the membrane potential-sensitive signal takes place. The membrane potential at this null point can then be calculated from the Nernst equation. A modification of this procedure is to add valinomycin to low K^+ medium, and then elevate the K^+ of the medium using concentrated KCl [94].

The oxonol dyes form complexes with valinomycin and this precludes the use of this ionophore for calibration. An alternative approach has been to take advantage of a calcium ionophore-induced increase in conductance of a Ca^{2+}-sensitive K^+ channel, establishing the membrane potential from the K^+ equilibrium potential (given by the Nernst equation), as above [113]. In this manner, the oxonol fluorescence of A23187-treated T cells was found to vary linearly with the log of the external K^+ over a broad range [122].

Limitations affecting the membrane potential assay. The cyanine dyes are recognized to have a variety of toxic and inhibitory effects, as well as certain limitations inherent in the use of cationic probes (recently reviewed by Chused et al. [18]). These dyes act to uncouple oxidative phosphorylation, deplete cellular ATP, block Ca^{2+}-dependent K^+ conductances, and cause depolarization of the resting membrane potential of lymphocytes and neutrophils [4,83,94,102,107]. These toxicities can be reduced by use of lower dye concentrations but may have selective effects upon different cell types [122].

Because mitochondrial potentials are highly negative, cyanine dye association with mitochondria can be a substantial component of the total fluorescence signal. Elaborate treatments to eliminate the mitochondrial potential have been described [88]; however, these may also affect the plasma membrane potential. A high ratio of dye to cells has been reported to decrease the partitioning of cyanine dye in mitochondria [122]. The complications from mitochondrial dye associations are less in mitochondria-poor polymorphonuclear leukocytes.

Finally, the dye concentration in suspensions of cells in a cyanine dye may be poorly buffered, in that the major fraction of dye may be cell associated. Extrusion of dye from a subpopulation of cells may thus increase the available extracellular dye, which may then be taken up by another cell population independent of changes in membrane potential and merely due to changes in the concentration of dye in the medium [122].

Most of the above limitations do not apply to the anionic oxonol dyes. The smaller fraction of dye associated with the cell results in reduced cellular fluorescence, but this has not been a problem with flow cytometry, and it results in better dye buffering. The negative charge of the oxonol dyes tends to exclude them from mitochondria, reducing both toxicity and this component of the total fluorescence signal.

Application of Study of Membrane Potential by Flow Cytometry

An increasing fraction of studies of membrane potential are being performed by flow cytometry, taking advantage of the sensitivity of this methodology, the ability to recognize heterogeneity in cellular responses, and opportunities for multiparameter analysis. These studies encompass a broad range of cell types (Table 4).

A confusing aspect of the study of lymphocyte membrane potential following antigen receptor activation has been the finding of either hyperpolarization [110,113] (see Fig. 8) or depolarization [16,21,52,79–82,105]. This may be related to variability in activity of a Ca^{2+}-sensitive K^+ channel between cell types. Activity of this channel is detectable in T cells and results in hyperpolarization after treatment with a calcium ionophore, but is reported to be undetectable in murine B cells [122]. Recently, the use of intracellular calcium chelators allowed the detection of both depolarization and hyperpolarization after crosslinking surface Ig on human B cells; these effects were due to an inwardly directed calcium current (causing depolarization) and to activation of calcium-dependent K^+ channels (causing hyperpolarization) [68]. Other reports indicate that there are species differences in the Ca^{2+}-sensitive K^+ channel activity; for example, that Con A-induced activity of the calcium-dependent K^+ channel hyperpolarizes mouse thymocytes, but cannot do so in pig lymphocytes because the channel is already maximally activated [26]. Thus, differential activities of the inward calcium current and outward calcium-dependent K^+ channel in different cell types may be a major explanation of the apparent discrepancies in the reported literature. Simultaneous measurement of $[Ca^{2+}]_i$ may be of use in establishing relationships between these Ca^{2+}-dependent processes and membrane potential.

Flow cytometric analysis of membrane potential has been useful to demonstrate heterogeneity among neutrophils after activation, with subpopulations which hyperpolarize and others which depolarize [101,102]. Heterogeneity and dosage effects may help to explain some observations of hyperpolarization following FMLP activation [57]. In addition to hematopoietic cells, the membrane potential of cells isolated from solid tissues have also been analyzed by flow cytometry. Table 4 provides references to studies of spinal cord cells, pneumocytes, and endothelial cells.

Multiparameter analyses with membrane potential. Given the broad range of excitation and emission characteristics available with membrane potential-sensitive probes (see Table 3), it is perhaps not surprising that a variety of multiparameter assays have been performed. Lazzari et al. [57] analyzed cells loaded with $DiOC_5(3)$ and indo-1 simultaneously, showing that the membrane potential changes in neutrophils stimulated with the oligopeptide chemoattractant FMLP occurred just as rapidly as did changes in $[Ca^{2+}]_i$.

Figure 8 shows an example of the simultaneous analysis of indo-1 and $DiBaC_4(3)$ fluorescence in $CD4^+$ lymphoblasts. Elevations in the $[K^+]$ of the external medium depolarize cells without altering $[Ca^{2+}]_i$, but subsequent exposure to ionomycin causes further depolarization which accompanies increased $[Ca^{2+}]_i$ (Fig. 8A). Activation of the same cells by crosslinking of CD3 and CD4 receptors [61] results in hyperpolarization of the lymphoblasts, which accompanies elevations in $[Ca^{2+}]_i$; the former change is sustained, however, while the latter is transient (Fig. 8B).

Seligmann et al. [101] analyzed binding of fluoresceinated FMLP oligopeptide to neutrophils simultaneously with $DiIC_5(3)$ fluorescence. These studies elegantly demonstrated that heterogeneity in membrane potential after oligopeptide exposure was related to differential binding of the chemoattractant. Simultaneous studies of forward and 90° scatter demonstrated that heterogeneity in $DiOC_5(3)$ fluorescence was not explained by heterogeneity in cells size [30].

Witkowski and Micklem [125] used Texas Red-labeled antibodies to allow estimation of membrane potential with $DiOC_6(3)$ in Lyt-2^+ and Lyt-2^- T-lymphocyte subsets and B cells, demonstrating that in aged mice the two former, but not the latter cell types showed reduced membrane potential. Wilson and Chused [122] use FITC antibodies and $DiSBaC_2(3)$ to distinguish splenic T cells from B cells during analysis. As with analysis of indo-1, simultaneous membrane potential measurement and discrimination of monoclonal antibody fluorescence should be a useful approach in elucidating subset-specific differences in cell responses.

TABLE 4. Summary of Studies of Membrane Potential by Flow Cytometry

Cell type	Study aims/conclusions	Dye	References
B lymphocytes	Receptor crosslinking	$DiOC_5(3)$	[16,21,79–82]
	Depolarization and cell activation	$DiOC_6(3)$	[42,127]
	BCGF—causes depolarization but not cell activation	$DiOC_5(3)$	[127]
T lymphocytes	Early hyperpolarization after Con A is related to elevated $[Ca^{2+}]_i$	$DiSC_3(5)$, $DiOC_6(3)$, $DiBaC_4(5)$, $DiTBaC_2(3)$	[110]
	Hyperpolarization by Con A in mouse but not pig cells	$DiBaC_4(5)$	[26]
	Decreased membrane potential in aged murine T cells	$DiOC_6(3)$	[125]
B and T lymphocytes	First analyses by flow cytometry (PBL, CEM cell line)	$DiOC_6(3)$, $DiOC_2(3)$, $DiOC_6(3)$	[105,103]
	Cyclosporins depolarize B and T cells	$DiOC_6(3)$	[22]
	Oxonol calibration, a Ca^{2+}-sensitive K channel in T but not B cells	$DiBaC_4(3)$	[122]
	Cyanine dye limitations, toxicity in PBL subsets	$DiOC_5(3)$, $DiIC_5(3)$	[122]
Neutrophils	Depolarization by zymosan	$DiOC_5(3)$	[121]
	FMLPLF fluorescence shows differential binding, accounts for membrane potential heterogeneity	$DiIC_5(3)$	[101]
	Depolarization versus light scatter	$DiOC_5(3)$	[30]
	Heterogeneity of potential, comparison of probes	$DiOC_5(3)$	[102]
	Cell recruitment by leukotriene B_4	$DiOC_5(3)$	[28]
	Cell recruitment by GM–CSF	$DiOC_5(3)$	[29]
	Simultaneous measurement of $[Ca^{2+}]_i$ and membrane potential	$DiOC_5(3)$	[57]
	Simultaneous measurement of oxidative metabolism and membrane potential	$DiIC_5(3)$	[100]
	Flow cytometric parameters of neutrophil function	$DiOC_5(3)$	[25]
	Defect in neutrophil activation in malignant infantile osteopetrosis	$DiOC_5(3)$	[8]
Neoplastic cell lines	Daudi and other lines show depolarization by α-interferon	$DiOC_6(3)$	[38]
	L1210 cells, antitumor drug NME uptake vs. membrane potential	$DiOC_6(3)$	[17]
	L1210 cells, heterogeneous potential changes after ionophore	$DiOC_6(3)$	[43]
	Friend cell hemoglobin synthesis vs. mitochondrial membrane potential	rhodamine-123	[126]
Other cell types	Embryonic rat spinal cord cells, batrachotoxin effects	$DiBaC_4(3)$, $DiBaC_4(5)$	[70]
	Effects of cytosine arabinoside on mitochondrial membrane potential in hematopoietic cells	$DiOC_5(3)$	[41]
	Lysophosphatidyl choline in calibration of membrane potential of type II pneumocytes and HL60 cells	$DiOC_5(3)$	[33]
	Membrane potential in endothelial cells after osmotic shock	$DiOC_6(3)$	[51]
	Macrophage membrane potential calibration	$DiOC_6(3)$	[46]

ACKNOWLEDGMENTS

This work was supported in part by National Institutes of Health Grant AG01751, and by the Naval Medical Research and Development Command, Research Task No. MR04120.001-1011. The opinions and assertions expressed herein are those of the authors and are not to be construed as official or reflecting the views of the Navy Department or the naval service at large.

REFERENCES

1. **Allen TJ, Baker PF (1985)** Intracellular Ca indicator quin-2 inhibits Ca^{2+} inflow via Na_i/Ca_o exchange. Nature (Lond) 315:755–756.

2. **Anasetti C, Martin PJ, June CH, Hellstrom KE, Ledbetter JA, Rabinovitch PS, Morishita Y, Hellstrom I, Hansen JA (1987)** Induction of calcium flux and enhancement of cytolytic activity in natural killer cells by cross- linking of the sheep erythrocyte binding protein (CD2) and the Fc-receptor (CD16). J Immunol 139:1772–1779.

3. **Arslan P, Di Virgilio F, Beltrane M, Tsien RY, Dozzan T (1985)** Cytosolic Ca^{2+} homeostasis in Ehrlich and Yoshida carcinomas. J Biol Chem 260: 2719–2725.

4. **Azzone GF, Pietrobon D, Zoratti M (1984)** Determination of the proton electrochemical gradient across

biological membranes. Curr Top Biophys Bioeng 13: 1–77.

5. **Baeker TR, Simons ER, Rothstein TL** (1987) Cytochalasin induces an increase in cytosolic free calcium in murine B lymphocytes. J Immunol 138: 2691–2697.
6. **Bashford CL, Chance B, Smith JC, Yoshida T** (1979) The behavior of oxonol dyes in phospholipid dispersions. Biophys J 25:63–85.
7. **Baud L, Goetzl EJ, Koo CH** (1987) Stimulation by leukotriene D4 of increases in the cytosolic concentration of calcium in dimethylsulfoxide-differentiated HL-60 cells. J Clin Invest 80:983–991.
8. **Beard CJ, Key L, Newburger PE, Ezekowitz RAB, Arceci R, Miller B, Proto P, Ryan T, Anast C, Simons ER** (1986) Neutrophil defect associated with malignant infantile osteopetrosis. J Lab Clin Med 108: 498–505.
9. **Beau P, Marechaud R, Matuchansky C** (1986) Familial defect of CD3 (T3) expression by T cells associated with rare gut epithelial cell autoantibodies. Lancet 1:1274–1275.
10. **Berridge MJ** (1987) Inositol trisphosphate and diacylglycerol; two interacting second messengers. Annu Rev Biochem 65:159–193.
11. **Bierer BE, Nishimura Y, Burakoff SJ, Smith BR** (1988) Phenotypic and functional characterization of human cytolytic T cells lacking expression of CD5. J Clin Invest 81:1390–1397.
12. **Blinks JR, Wier WG, Hess P, Prendergast FG** (1982) Measurement of Ca^{2+} concentrations in living cells. Prog Biophys Mol Biol. 40:1–114.
13. **Blue M-L, Hafler DA, Daley JF, Levine H, Craig KA, Breitmeyer JB, Schlossman SF** (1988) Regulation of T cell clone function via CD4 and CD8 molecules. Anti-CD4 can mediate two distinct inhibitory activities. J Immunol 140:376–383.
14. **Breitmeyer J, Daley JF, Levine HB, Schlossman SF** (1987) The T11 (CD2) molecule is functionally linked to the Tc/Ti T cell receptor in the majority of T cells. J Immunol 139:2899–2905.
15. **Brunswick M, June CH, Finkelman FD, Dintzis HM, Inman JK, Mond JJ** (1989) Surface immunoglobulin mediated B cell activation in the absence of detectable elevations in intracellular ionized calcium: a model for T-cell-independent B-cell activation. Proc Nat Acad Sci (USA) 86:in press.
16. **Cambier JC, Heusser CH, Julius MH** (1986) Abortive activation of B lymphocytes by monoclonal anti-immunoglobulin antibodies. J Immunol 136:3140–3146.
17. **Charcosset J-Y, Jacquemin-Sablon A, Le Pecq J-B** (1984) Effect of membrane potential on the cellular uptake of 2-N-methyl-ellipticinium by L1210 cells. Biochem Pharmacol 33:2271–2275.
18. **Chused TM, Wilson HA, Seligmann BE, Tsien R** (1986) Probes for use in the study of leukocyte physiology by flow cytometry. In Taylor DL, Waggoner AS, Lanni F, Murphy RF, Birge RR (eds) "Applications of Fluorescence in the Biomedical Sciences." New York: Alan R. Liss, pp 531–544.
19. **Chused TM, Wilson HA, Greenblatt D, Ishida Y, Edison LJ, Tsien RY, Finkelman FD** (1987) Flow cytometric analysis of murine splenic B lymphocyte cytosolic free calcium response to anti-IgM and anti-IgD. Cytometry 8:396–404.
20. **Cobbold PH, Rink TJ** (1987) Fluorescence and bioluminescence measurement of cytoplasmic free calcium. Biochem J 248:313–28.
21. **Coggeshall KM, Cambier JC** (1985) B cell activation. VI. Effects of exogenous diglyceride and modulators of phospholipid metabolism suggest a central role for diacylglycerol generation in transmembrane signaling by mIg. J Immunol 134:101–107.
22. **Damjanovich S, Aszalos A, Mulhern S, Balazs M, Matyus L** (1986) Cytoplasmic membrane potential of mouse lymphocytes is decreased by cyclosporins. Mol Immunol 23:175–180.
23. **Davies TA, Drotts D, Weil GJ, Simons ER** (1988) Flow cytometric measurements of cytoplasmic calcium change in human platelets. Cytometry 9:138–142.
24. **De Virgilio F, Steinberg TH, Swanson JA, Silverstein SC** (1988) Fura-2 secretion and sequestration in macrophages. A blocker of organic anion transport reveals that these processes occur via a membrane transport system for organic anions. J Immunol 140: 915–920.
25. **Duque RE, Ward PA** (1987) Quantitative assessment of neutrophil function by flow cytometry. Anal Quant Cytol Histol 9:42–48.
26. **Felber SM, Brand MD** (1983) Early plasma–membrane-potential changes during stimulation of lymphocytes by concanavalin A. Biochem J 210:885–891.
27. **Finkel TH, McDuffie M, Kappler JW, Marrack P, Cambier JC** (1987) Both immature and mature T cells mobilize Ca^{2+} in response to antigen receptor crosslinking. Nature (Lond) 330:179–181.
28. **Fletcher MP** (1986) Modulation of the heterogeneous membrane potential response of neutrophils to N-formyl-methionyl-leucyl-phenylalanine (FMLP) by leukotriene B_4: Evidence for cell recruitment. J Immunol 136:4213–4219.
29. **Fletcher MP, Gasson, JC** (1988) Enhancement of neutrophil function by granulocyte-macrophage colony-stimulating factor involves recruitment of a less responsive subpopulation. Blood 71:652–658.
30. **Fletcher MP, Seligmann BE** (1985) Monitoring human neutrophil granule secretion by flow cytometry: Secretion and membrane potential changes assessed by light scatter and a fluorescent probe of membrane potential. J Leukocyte Biol 37:431–447.
31. **Francois DT, Katona IM, June CH, Wahl LM, Feuerstein N, Huang K-P, Mond JJ** (1988) Anti-Ig mediated proliferation of human B-cells in the absence of protein kinase C. J Immunol 140:3338–3343.
32. **Francois DT, Katona IM, June CH, Wahl LM, Mond JJ** (1988) Examination of the inhibitory and stimulatory effects of interferon-alpha, beta and gamma on human B-cell proliferation induced by various B-cell mitogens. Clin Immunol Immunopathol 48:297–306.
33. **Gallo RL, Wersto RP, Notter RH, Finkelstein JN** (1984) Lysophosphatidylcholine cell depolarization: Increased membrane permeability for use in the determination of cell membrane potentials. Arch Biochem Biophys 235:544–554.
34. **Geppert TD, Wacholtz MC, Davis LS, Lipsky PE**

(1988) Activation of human T4 cells by cross-linking class I MHC molecules. J Immunol 140:2155–2164.

35. **Gilman AG** (1987) G proteins: Transducers of receptor-generated signals. Annu Rev Biochem 56:615–649.

36. **Goldsmith MA, Weiss A** (1987) Isolation and characterization of a T-lymphocyte somatic mutant with altered signal transduction by the antigen receptor. Proc Natl Acad Sci USA 84:6879–6883.

37. **Goldsmith MA, Weiss A** (1988) Early signal transduction by the antigen receptor without commitment to T cell activation. Science 240:1029–1031.

38. **Grimley PM, Aszalos, A** (1987) Early plasma membrane depolarization by alpha interferon: Biologic correlation with antiproliferative signal. Biochem Biophys Res Commun 146:300–306.

39. **Grossmann A, Rabinovitch PS** (1987) Flow cytometry with indo-1 reveals variation in intracellular free calcium within T-cell subsets and between donors after mitogen stimulation. In Burger G, Ploem JS, Goerttler K (eds), "Clinical Cytometry and Histometry." London: Academic Press, pp 192–194.

40. **Grynkiewicz G, Poenie M, Tsien RY** (1985) A new generation of Ca^{2+} indicators with greatly improved fluorescence properties. J Biol Chem 260:3440–3450.

41. **Haanen C, Muus P, Pennings A** (1986) The effect of cytosine arabinoside upon mitochondrial staining kinetics in human hematopoietic cells. Histochemistry 85:609–613.

42. **Heikkilä R, Iversen J-G, Godal T** (1985) No correlation between membrane potential and increased cytosolic free Ca^{2+} concentration, $^{86}Rb^{+}$ influx or subsequent [^{3}H]thymidine incorporation in neoplastic human B cells stimulated with antibodies to surface immunoglobulin. Acta Physiol Scand 124:107–115.

43. **Hickman JA, Blair OC, Stepanowski AL, Sartorelli AC** (1984) Calcium-induced heterogeneous changes in membrane potential detected by flow cytofluorimetry. Biochim Biophys Acta 778:457–462.

44. **Hoffmann JF, Laris PC** (1974) Determination of membrane potentials in human and Amphiuma red blood cells by means of a fluorescent probe. J Physiol (Lond) 239:519–552.

45. **Ishida Y, Chused TM** (1988) Heterogeneity of lymphocyte calcium metabolism is caused by a T cell-specific calcium-sensitive potassium channel and sensitivity of calcium extrusion activity to membrane potential. J Exp Med 168:839–852.

46. **Jenssen H-L, Redmann K, Mix E** (1986) Flow cytometric estimation of transmembrane potential of macrophages—A comparison with microelectrode measurements. Cytometry 7:339–346.

47. **June CH, Ledbetter JA, Rabinovitch PS, Hellstrom KE, Hellstrom I** (1987) Calcium mobilization and enhanced natural killer function in large granular lymphocytes result from crosslinking the CD2 E rosette and CD16 Fc-receptor. In McMichael, AJ (ed), "Leukocyte Typing. Vol. III." Oxford: Oxford University Press, pp 127–131.

48. **June CH, Ledbetter JA, Rabinovitch PS, Martin PJ, Beatty PG, Hansen JA** (1986) Distinct patterns of transmembrane calcium flux and intracellular calcium mobilization after differentiation antigen cluster 2 (E rosette receptor) or 3 (T3) stimulation of human lymphocytes. J Clin Invest 77:1224–1232.

49. **June CH, Rabinovitch PS, Ledbetter JA** (1987) Anti-CD5 antibodies increase cytoplasmic calcium concentration and augment CD3-stimulated calcium mobilization in T cells. J Immunol 138:2782–2792.

50. **Justement L, Chen Z, Harris L, Ransom J, Sandoval V, Smith C, Rennick D, Roehm N, Cambier J** (1986) BSF1 induces membrane protein phosphorylation but not phosphoinositide metabolism, Ca^{2+} mobilization, protein kinase C translocation, or membrane depolarization in resting murine B lymphocytes. J Immunol 137:3664–3670.

51. **Kempski O, Spatz M, Valet G, Baethmann A** (1985) Cell volume regulation of cerebrovascular endothelium in vitro. J Cell Physiol 123:51–54.

52. **Kiefer H, Blume AJ, Kaback HR** (1980) Membrane potential changes during mitogenic stimulation of mouse spleen lymphocytes. Proc Natl Acad Sci USA 77:2200–2204.

53. **Kim YI, Heher E** (1988) IgG from patients with Lambert–Eaton syndrome blocks voltage-dependent calcium channels. Science 239:405–408.

54. **Klip A, Britt BA, Elliott ME, Pegg W, Frodis W, Scott E** (1987) Anaesthetic-induced increase in ionised calcium in blood mononuclear cells from malignant hyperthermia patients. Lancet 1:463–466.

55. **Klip A, Ramlal T, Walker D, Britt BA, Elliott ME** (1987) Selective increase in cytoplasmic calcium by anesthetic in lymphocytes from malignant hyperthermia-susceptible pigs. Anesth Analg 66:381–385.

56. **Lanza F, Beretz A, Kubina M, Cazenave JP** (1987) Increased aggregation and secretion responses of human platelets when loaded with the calcium fluorescent probes quin2 and fura-2. Thromb Haemost 58: 737–743.

57. **Lazzari KG, Proto PJ, Simons ER** (1986) Simultaneous measurement of stimulus-induced changes in cytoplasmic Ca^{2+} and in membrane potential of human neutrophils. J Biol Chem 261:9710–9713.

58. **Ledbetter JA, Gentry LE, June CH, Rabinovitch PS, Purchio AF** (1987) Stimulation of T cells through the CD3/T cell receptor complex: role of cytoplasmic calcium, protein kinase C translocation and phosphorylation of $pp60^{c\text{-}src}$ in the activation pathway. Mol Cell Biol 7:650–656.

59. **Ledbetter JA, June CH, Grosmaire LS, Rabinovitch PS** (1987) Crosslinking of surface antigens causes mobilization of intracellular ionized calcium in T lymphocytes. Proc Natl Acad Sci USA 84:1384–1388.

60. **Ledbetter JA, June CH, Martin PJ, Spooner CE, Hansen JA, Meier KM** (1986) Valency of CD3 binding and internalization of the CD3 cell-surface complex control T cell responses to second signals: distinction between effect on protein kinase C, cytoplasmic free calcium, and proliferation. J Immunol 136:3945–3952.

61. **Ledbetter JA, June CH, Rabinovitch PS, Grossman A, Tsu TT, Imboden JB** (1988) Signal transduction through CD4 receptors: Stimulatory versus inhibitory activity is regulated by CD4 proximity to the CD3/T cell receptor. Eur J Immunol 18:525–532.

62. **Ledbetter JA, Parsons M, Martin PJ, Hansen JA, Rabinovitch PS, June CH** (1986) Antibody binding to

CD5 (Tp67) and Tp44 molecules: Effects in cyclic nucleotides, cytoplasmic free calcium, and cAMP-mediated suppression. J Immunol 137:3299–3305.

63. **Ledbetter JA, Rabinovitch PS, Hellstrom I, Hellstrom KE, Grosmaire LS, June CH (1988)** Role of CD2 crosslinking in cytoplasmic calcium responses and T cell activation. Eur J Immunol 18:1601–1608.
64. **Ledbetter JA, Rabinovitch PS, June CH, Song CW, Clark EA, Uckun FH (1988)** Antigen-independent regulation of cytoplasmic calcium in B cells with a 12 kDa B cell growth factor and anti-CD19. Proc Natl Acad Sci USA 85:1897–1901.
65. **Lin PY, Gruenstein E (1987)** Identification of a defective cAMP-stimulated Cl-channel in cystic fibrosis fibroblasts. J Biol Chem 262:15345–15347.
66. **Linette GP, Hartzman RJ, Ledbetter JA, June CH (1988)** HIV-1 infected T cells exhibit a selective transmembrane signalling defect through the CD3/antigen receptor pathway. Science 241:573–576.
67. **Luckhoff A (1986)** Measuring cytosolic free calcium concentration in endothelial cells with indo-1: The pitfall of using the ratio of two fluorescence intensities recorded at different wavelengths. Cell Calcium 7:233–248.
68. **MacDougall SL, Grinstein S, Gelfand EW (1988)** Detection of ligand-activated conductive Ca^{2+} channels in human B lymphocytes. Cell 54:229–234.
69. **Malgawli A, Milani D, Meldolesi J, Pozzan T (1987)** Fura-2 measurement of cytosolic free Ca^{2+} in monolayers and suspensions of various types of animal cells. J Cell Biol 105:2145–2155.
70. **Mandler RN, Schaffner AE, Novotny EA, Lange GD, Barker JL (1988)** Flow cytometric analysis of membrane potential in embryonic rat spinal cord cells. J Neurosci Methods 22:203–213.
71. **Manger B, Weiss A, Imboden J, Laing T, Stobo JD (1987)** The role of protein kinase C in transmembrane signaling by the T cell antigen receptor complex. Effects of stimulation with soluble or immobilized CD3 antibodies. J Immunol 139:2755–2760.
72. **Martell AE, Smith RM (1968)** "Critical Stability Constants." Vol. 1: "Amino Acids." New York: Plenum Press, pp 269–272.
73. **Mentzer SJ, Smith BR, Barbosa JA, Crimmins MA, Hermann SH, Burakoff SJ (1987)** CTL adhesion and antigen recognition are discrete steps in the human CTL–target cell interaction. J Immunol 138:1325–1330.
74. **Miller DJ, Smith GL (1984)** EGTA purity and the buffering of calcium ions in physiological solutions. Am J Physiol 246:C160–C166.
75. **Miller RA, Jacobson B, Weil G, Simons ER (1987)** Diminished calcium influx in lectin-stimulated T cells from old mice. J Cell Physiol 132:337–342.
76. **Minta A, Harootunian AT, Kao JPY, Tsien RY (1987)** New fluorescent indicators for intracellular sodium and calcium. J Cell Biol 105:89a.
77. **Mizuguchi J, Ji YY, Nakabayaschi H, Huang KP, Beaven MA, Chused T, Paul WE (1987)** Protein kinase C activation blocks anti-IgM-mediated signaling BAL17 B lymphoma cells. J Immunol 139: 1054–1059.
78. **Moisescu DG, Pusch H (1975)** A pH-metric method for the determination of the relative concentration of calcium to EGTA. Pflugers Arch 355:R122.
79. **Monroe JG, Cambier JC (1983)** B cell activation. I. Anti-immunoglobulin-induced receptor cross-linking results in a decrease in the plasma membrane potential of murine B lymphocytes. J Exp Med 157:2073–2086.
80. **Monroe JG, Cambier JC (1983)** B cell activation. II. Receptor cross-linking by thymus-independent and thymus-dependent antigens induces a rapid decrease in the plasma membrane potential of antigen-binding B lymphocytes. J Immunol 131:2641–2644.
81. **Monroe JG, Cambier JC (1983)** B cell activation. III. B cell plasma membrane depolarization and hyper-Ia antigen expression induced by receptor immunoglobulin crosslinking are coupled. J Exp Med 158: 1589–1594.
82. **Monroe JG, Cambier JC (1984)** B cell activation. IV. Induction of membrane depolarization and hyper I-A expression by phorbol diesters suggests a role for protein kinase C in murine B lymphocyte activation. J Exp Med 132:1472–1478.
83. **Montecucco C, Poznan T, Rink TJ (1979)** Dicarbocyanine fluorescent probes of membrane potential block lymphocyte capping, deplete cellular ATP, and inhibit respiration of isolated mitochondria. Biochim Biophys Acta 552:552–557.
84. **O'Shea JJ, Urdahl KB, Luong HT, Chused TM, Samelson LE, Klausner RD (1987)** Aluminum fluoride induces phosphatidylinositol turnover, elevation of cytoplasmic free calcium, and phosphorylation of the T cell antigen receptor in murine T cells. J Immunol 139:3463–3469.
85. **Parks DR, Nozaki T, Dunne JF Peterson LL (1987)** Flow cytometer adaptation for quantitation of immunofluorescent reagents and for calibration of dyes for measuring cellular Ca^{2+} and pH. Cytometry Suppl 1:104.
86. **Pezzutto A, Dürken B, Rabinovitch PS, Ledbetter JA, Moldenhauer G, Clark EA (1987)** CD19 monoclonal antibody HD37 inhibits anti-immunoglobulin-induced B cell activation and proliferation. J Immunol 138:2793–2799.
87. **Pezzutto A, Rabinovitch PS, Dorken B, Moldenhauer G, Clark EA (1988)** Role of the CD22 human B cell antigen in B cell triggering by anti-immunoglobulin. J Immunol 188:1791–1795.
88. **Philo RD, Eddy AA (1978)** The membrane potential of mouse ascites tumour cells studied with the fluorescent probe 3,3′-di propyloxadicarbocyanine. Biochem J 174:801–810.
89. **Proust JJ, Filburn CR, Harrison SA, Buchholz MA, Nordin AA (1987)** Age-related defect in signal transduction during lectin activation of murine T lymphocytes. J Immunol 139:1472–1478.
90. **Putney JW (1986)** A model for receptor-regulated calcium entry. Cell Calcium 7:1–7.
91. **Rabinovitch PS, June CH, Grossmann A, Ledbetter JA (1986)** Heterogeneity of T cell intracellular free calcium responses after mitogen stimulation with PHA or anti-CD3. Use of indo-1 and simultaneous immunofluorescence with flow cytometry. J Immunol 137:952–961.
92. **Rasmussen H (1986)** The calcium messenger system. N Eng J Med 314:1094–1101, 1164–1170.
93. **Redelman D (1988)** Cyclosporin A does not inhibit the PHA-stimulated increase in intracellular Ca^{2+}

concentration but inhibits the increase in E-rosette receptor (CD2) expression and appearance of interleukin-2 receptors (CD25). Cytometry 9:163–165.

94. **Rink TJ, Montecucco C, Hesketh TR, Tsien RY (1980)** Lymphocyte membrane potential assessed with fluorescent probes. Biochem Biophys Acta 595: 15–30.

95. **Ross WN, Salzberg BM, Cohen LB, Grinwald A, Davila HV, Waggoner AS, Wang CH** (1977) Changes in absorption, fluorescence, dichromism and birefringence in stained giant axons: Optical measurement of membrane potential. J Membr Biol 33:141–183.

96. **Rothstein TL, Kolber DL, Simons ER, Baeker TR (1986)** Inhibition by phorbol esters of antiimmunoglobulin-induced calcium signalling and B-cell activation. J Cell Physiol 129:347–355.

97. **Sanders ME, Makoba MW, Sharrow SO, Stephany D, Springer TA, Young HA, Shaw S (1988)** Human memory T lymphocytes express increased levels of three cell adhesion molecules (LFA-3, CD2, and LFA-1) and three other molecules (UCHL1, CDw29, and Pgp-1) and have enhanced IFN-gamma production. J Immunol 140:1401–1407.

98. **Scanlon M, Williams DA, Fay FS (1987)** A Ca^{2+}-insensitive form of fura-2 associated with polymorphonuclear leukocytes. Assessment and accurate Ca^{2+} measurement. J Biol Chem 262:6308–6312.

99. **Schöni MH, Schoni-Affolter F, Jeffry D, Katz S (1987)** Intracellular free calcium levels in mononuclear cells of patients with cystic fibrosis and normal controls. Cell Calcium 8:53–63.

100. **Seeds MC, Parce JW, Szejda P, Bass DA (1985)** Independent stimulation of membrane potential changes and the oxidative metabolic burst in polymorphonuclear leukocytes. Blood 65:233–240.

101. **Seligmann B, Chused TM, Gallin JI (1984)** Differential binding of chemoattractant peptide to subpopulations of human neutrophils. J Immunol 133:2641–2646.

102. **Seligmann BE, Gallin JI (1983)** Comparison of indirect probes of membrane potential utilized in studies of human neutrophils. J Cell Physiol 115:105–115.

103. **Shapiro HM (1981)** Flow cytometric probes of early events in cell activation. Cytometry 1:301–312.

104. **Shapiro BL, Lam LF (1987)** Intracellular calcium in cystic fibrosis heterozygotes. Life Sci 40:2361–2366.

105. **Shapiro HM, Natale PJ, Kamentsky LA (1979)** Estimation of membrane potentials of individual lymphocytes by flow cytometry. Proc Natl Acad Sci USA 76:5728–5730.

106. **Sims PJ, Waggoner AS, Wang CH, Hoffmann JS (1974)** Studies on the mechanism by which cyanine dyes measure membrane potential in red blood cells and phosphatidyl cholinic vesicles. Biochemistry 13: 3315–3330.

107. **Simons TJB (1979)** Actions of a carbocyanine dye on calcium-dependent potassium transport in human red cell ghosts. J Physiol (Lond) 288:481–507.

108. **Steinberg SF, Bilezikian JP, Al-Awqati Q (1987)** Fura-2 fluorescence is localized to mitochondria in endothelial cells. Am J Physiol 253(pt 1):C744–747.

109. **Sullivan R, Griffin JD, Simons ER, Schafer AI, Meshulam T, Fredette JP, Maas AK, Gadenne AS, Leavitt JL, Melnick DA (1987)** Effects of recombinant human granulocyte and macrophage colony-stimulating factors on signal transduction pathways in human granulocytes. J Immunol 139:3422–3430.

110. **Tatham PER, Delves PJ (1984)** Flow cytometric detection of membrane potential changes in murine lymphocytes induced by concanavalin A. Biochem J 221:137–146.

111. **Treves S, Di Virgilio F, Vaselli GM, Pozzan T (1987)** Effect of cytochalasins on cytosolic-free calcium concentration and phosphoinositide metabolism in leukocytes. Exp Cell Res 168:285–298.

112. **Tsien RY (1981)** A non-disruptive technique for loading calcium buffers and indicators into cells. Nature (Lond) 290:527–528.

113. **Tsien RY, Pozzan T, Rink TJ (1982)** T-cell mitogens cause early changes in cytoplasmic free Ca^{2+} and membrane potential in lymphocytes. Nature (Lond) 295:68–71.

114. **Tsien RY, Pozzan T, Rink TJ (1982)** Calcium homeostasis in intact lymphocytes: cytoplasmic free calcium monitored with a new, intracellularly trapped fluorescent indicator. J Cell Biol 94:325–334.

115. **Valet G, Raffael A, Russmann L (1985)** Determination of intracellular calcium in vital cells by flow-cytometry. Naturwissenschaften 72:600–602.

116. **Van Lier RA, Boot JH, Verhoeven AJ, de Groot ER, Brouwer M, Aarden LA (1987)** Functional studies with anti-CD3 heavy chain isotype switch-variant monoclonal antibodies. Accessory cell-independent induction of interleukin 2 responsiveness in T cells by epsilon-anti-CD3. J Immunol 139:2873–2879.

117. **Van Woerkom AE (1987)** The end organ defect in cystic fibrosis; a hypothesis: Disinhibited inositol cycle activation? Med Hypoth 23:383–392.

118. **Volpe P, Krause KH, Hashimoto S, Zorzato F, Pozzan T, Meldolesi J, Lew DP (1988)** "Calciosome," a cytoplasmic organelle: The inositol 1,4,5-trisphosphate-sensitive Ca^{2+} store of nonmuscle cells? Proc Natl Acad Sci USA 85:1091–1095.

119. **Waggoner AS (1979)** Dye indicators of membrane potential. Annu Rev Biophys Bioeng 8:47–68.

120. **Weiss A, Dazin PF, Shields R, Fu SM, Lanier LL (1987)** Functional competency of T cell antigen receptors in human thymus. J Immunol 139:3245–3250.

121. **Whitin JC, Ryan DH, Cohen HJ (1985)** Graded responses of human neutrophils induced by serum-treated zymosan. Blood 66:1182–1188.

122. **Wilson HA, Chused TM (1985)** Lymphocyte membrane potential and Ca^{2+}-sensitive potassium channels described by oxonol dye fluorescence measurements. J Cell Physiol 125:72–81.

123. **Wilson HA, Greenblatt D, Poenie M, Finkelman FD, Tsien RY (1987)** Crosslinkage of B lymphocyte surface immunoglobulin by anti-Ig or antigen induces prolonged oscillation of intracellular ionized calcium. J Exp Med 166:601–606.

124. **Wilson HA, Greenblatt D, Taylor CW, Putney JW, Tsien RY, Finkelman FD, Chused TM (1987)** the B lymphocyte calcium response to anti-Ig is diminished by membrane immunoglobulin cross-linkage

to the Fc gamma receptor. J Immunol 138:1712–1718.

125. **Witkowski J, Micklem HS** (1985) Decreased membrane potential of T lymphocytes in ageing mice: Flow cytometric studies with a carbocyanine dye. Immunology 56:307–313.

126. **Wong W, Robinson SH, Tsiftsoglou AS** (1985) Relationship of mitochondrial membrane potential to hemoglobin synthesis during Friend cell maturation. Blood 66:999–1001.

127. **Yokoyama WM, Chien MM, Engardt SE, Aguiar SW, Ashman RF** (1988) Membrane depolarization of human B cells follows stimulation by either anti-μ or B-cell growth factor, but only anti-μ causes cell volume changes. Hum Immunol 21:155–164.

33

Analysis and Sorting of Blood and Bone Marrow Cells

Jan W.M. Visser
Radiobiological Institute TNO, Rijswijk, The Netherlands

INTRODUCTION

The determination of blood cell composition has been a valuable tool in diagnosing many diseases and in monitoring the effects of therapy. Measurements of red blood cell (RBC) number, size, shape, and color are among the earliest diagnostic tests. The discovery of staining techniques to recognize several types of white blood cells (WBC) about 100 years ago made possible the widespread use of differential WBC in the clinic. The hospital laboratory daily examines between 100 and 1,000 or more blood samples. These are generally still stained using methods first described around the beginning of the twentieth century [16]. However, a large number of laboratories have obtained automated devices to assist their technicians in performing this task. The electric resistance counter, which is widely accepted for counting and sizing of RBC, has been available for many years [4,24,33,70,88] (see also Chapter 4, this volume). Platelet counting and WBC counting (after lysing the RBC) also can be performed by these devices.

Optical measurements for differential counts of stained blood cells by clinical flow cytometers were introduced more recently and reported to be useful and reliable [2, 3,26,32,55,69,72,96,99,100,107]. However, the first flow cytometric devices dedicated to WBC differential counting were not widely distributed in the clinical laboratories, and the new generation of commercial flow cytometers have been used for too short a time to evaluate their accuracy and cost effectiveness. Furthermore, the optimal parameters and stains for automated cell recognition are still not established but are different from the "classic" ones. The development of new stains and their acceptance by the hematologist has just started.

In general terms, flow cytometry has two important advantages over other techniques for measurements of hematopoietic cells. First, these cells are easily brought into suspension, as is necessary for flow measurements. Second, the hematopoietic system contains important rare cells, the stem cells, which can be detected only if large numbers of cells are analyzed. The analysis of large numbers of cells is a standard feature of flow cytometry. Prospects for the application of flow cytometers in hematology may be anticipated from applications of the instrumentation in experimental hematology. Therefore, this chapter describes applications of flow cytometers in the field of hematology which are mainly experimental. A number of applications in clinical hematology are described first.

CLINICAL HEMATOLOGY

An average laboratory that routinely examines about 300 blood samples per day will process less than ten bone marrow samples. Therefore, the application of automated cytometers in clinical hematology mainly concerns the analysis of peripheral blood cells. Morphological examination of blood smears by microscopy led to the classification given in Table 1, which also shows the normal range of counts for the various cell types in peripheral blood. The wide range in normal values is due to biological variation. Some cell types, e.g., basophils and reticulocytes, normally occur at low frequencies, but can still be detected because of their outstanding morphology with the classic stains. Apart from the need to automate the laborious differential count, the advantage of better statistics especially for counting of rare cell types may justify the acquisition of a flow cytometer for the clinical laboratory.

In addition, an increasing number of clinical laboratories routinely determine lymphocyte subpopulations using monoclonal antibodies and flow cytometers. The accuracy here depends mostly on the specificity of the antibody and the reproducibility of the staining procedure. The flow cytometers have a quantitative advantage over other methods for this purpose. A properly calibrated flow cytometer determines not only the fraction of all lymphocytes that belongs to a certain subpopulation, it also measures the amount of label on these cells and thereby the antigen density. Both values may be clinically relevant. The determination of lymphocyte subpopulations is described in Chapter 34 (this volume).

Two kinds of answers can be obtained from automated devices for blood cell classification. First, the distribution of the observed cells is either abnormal or within the normal range with respect to healthy individuals. Second, abnormal

Flow Cytometry and Sorting, Second Edition, pages 669–683

TABLE 1. Normal Range of Blood Cells in Healthy Adults

	Number per mm^3
Platelets	150,000–400,000
Red cells	
Men	4.5–6.5 × 10^6
Women	3.9–5.6 × 10^6
Reticulocytes	39,000–130,000
Total leukocytes	4,000–11,000
Differential leukocytes	
Neutrophils	2,500–7,500
Lymphocytes	1,500–3,500
Monocytes	200–800
Eosinophils	40–440
Basophils	0–100

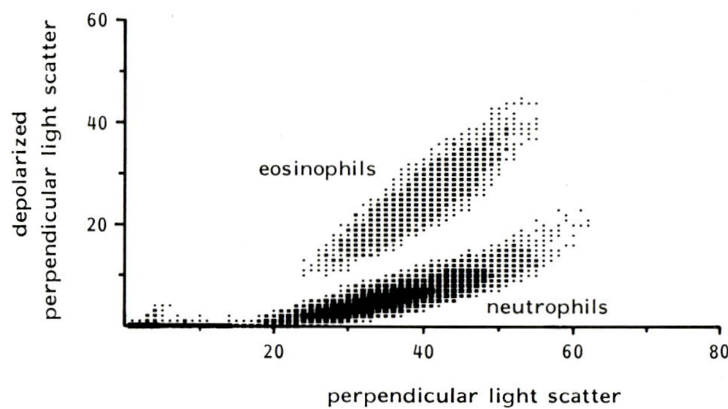

Fig. 1. Discrimination between eosinophils and neutrophils by measurement of the intensity of depolarized perpendicular light scatter. (Reproduced from ref. [50], with permission of the publisher.)

cells may be detected that do not fit the description or measurement of the known normal cell types. Accurate counts of normal cell types depend on the accuracy of the device, on the ability of the staining technique to distinguish between the known cell types, and on the choice of flagging system. The identification of abnormal cells depends heavily on specialized staining techniques and on minimizing artifacts that may yield false-positive or -negative counts. The clinical use of flow cytometers deals mainly with counts of known cell types. Experimental methods for the detection and classification of malignant cell types are described in other chapters.

The cost effectiveness of general purpose flow cytometers in the clinical laboratory depends on the number and type of examinations performed by these machines. Since clinical acceptance is rapidly increasing, and new staining methods and preparatory techniques are being devised for special purposes, it is likely that these instruments will continue to increase in cost effectiveness.

Differential Counting of Blood Cells by Flow Cytometers

Flow cytometers require single cell suspensions. The sampling of peripheral blood, therefore, needs to be performed using an anticoagulant. EDTA, ammonium and potassium oxalate mixtures, trisodium citrate, as well as heparin may all serve this purpose without affecting platelet counts or other enumerations. For experimental hematology, the presence of anticoagulants may influence the measurements because of their effects on pH or ion concentrations.

The total blood cell count is dominated by the presence of RBC and platelets. Analysis of the WBC is normally improved by prior removal of the erythrocytes. Several hemolytic agents are commercially available that have no effect on the WBC enumeration. Some machines (ELT, formerly Ortho Diagnostic Systems Inc., Westwood, MA; H.1, Technicon Instruments Corporation, Tarrytown, NY) split the samples automatically into fractions for RBC and WBC counts, add lysing agent to the latter, and dilute the former. Samples generally have to be diluted according to standard procedures before being measured. Automated diluters can be obtained commercially for this purpose.

Flow cytometers for the hematological laboratory are more specialized than those for experimental research laboratories. The Hemalog D (Technicon) uses a Tungsten Halogen lamp light source and measures only axial light loss and near forward light scatter of WBC stained cytochemically in the device. Technicon's new instrument, the H.1, combines laser light scattering measurements with the type of measurements made by the Hemalog D. The ELT series (Ortho) determine blood cell composition by measuring and processing only the forward and lateral laser light scatter signals of unstained cells. The Coulter counters have been known in the clinic for several decades. Their principles based on electrical resistance are described in Chapter 4 (this volume).

Several machines are in use for routine measurements of immunofluorescence to determine the lymphocyte subpopulations. The Spectrum (Ortho), the EPICS C (Coulter, Hialeah, FL), and the FACSCAN (Becton Dickinson, Sunnyvale, CA) are marketed for this purpose. All combine scatter and immunofluorescence measurements for cell classification. Further development of these specialized devices depends on the practical value of routine lymphocyte subpopulation determination for diagnosis and for the monitoring of therapy.

A number of abnormalities are not detected by most of the available automated differential counters: eosinophilia, low numbers of monocytes or circulating normoblasts and leukoerythroblasts. De Grooth and Terstappen [50] recently devised a method to discriminate between eosinophils and other leukocytes, measuring the depolarization of perpendicular light scatter (Fig. 1). Such measurements will probably be incorporated in the automated counters, permitting detection of eosinophilia. Similarly, new methods will be needed to count low numbers of monocytes and normoblasts differentially. By contrast, since hematopoiesis is a balanced system, in many cases it may be sufficient for the automated devices simply to flag abnormalities based upon such other parameters as anemia, thrombocytopenia, or the presence of unlysed or fragmented erythrocytes or clumps of platelets. The clinician can then order a manual examination of peripheral smears, indicated by these abnormalities or clinical symptoms.

Electric Resistance Pulse Sizing

Coulter counters have been used for more than twenty years for semiautomatic blood cell counting. The principle of electric sizing and counting is now employed in specialized machines for platelets, red blood cells, MCV, etc. The optimum orifice size of "Coulter" counters depends on the particle size counted and generally needs to be changed if platelets and other blood cells are measured on the same device. This is not necessary using optical flow cytometers. The electric size signal is linearly proportional to the volume of the cell, whereas the forward light scatter (FLS) intensity is closely proportional to the cross sectional area in most machines [129]. If one corrects for this difference (assuming that the cells are spherical) the histograms of electric size and FLS of blood and bone marrow cells closely resemble each other. Only the granulocytes show somewhat lower FLS than expected; these cells give relatively more perpendicular light scatter (PLS) perhaps at the cost of FLS or due to a different refractive index.

Cytochemical Staining

The Hemalog D flow cytometer (Technicon Instruments Corporation, Tarrytown, NY) measures scatter and axial light loss, the latter related to absorption. Cytochemical staining of specific blood cell types is used to increase absorption of those cells [12,13]. After the red cells are lysed, the remaining cells are fixed and stained in three aliquots, one for monocytes (esterase), one for basophils (heparin) and the third for neutrophils and eosinophils (peroxidase). The lymphocytes are recognized as unstained cells. All staining procedures are done automatically. Sample throughput is about one per minute; normally, 10^4 cells are counted per sample. The H.1 (Technicon) also determines narrow (2 to 5°) and wider (10 to 15°) angle light scatter to analyse and quantitate red blood cells and hemoglobin.

Light Scatter Patterns

Measurements of light scattered at different angles reveals information about size, refractive index, and structuredness of mammalian cells (see Chapter 5, this volume). Light scatter can be measured with relatively inexpensive devices, and a priori staining is not required. For these reasons research to apply light scatter to the enumeration of blood and bone marrow cells dates back to the early days of flow cytometry [67,94,121,122,128,129].

Figure 2 shows bivariate plots of the forward versus perpendicular light scatter intensities of human peripheral blood cells before and after lysing the erythrocytes. The measurements were performed on a slightly modified FACS II (Becton Dickinson, Sunnyvale, CA) as described in Visser et al. [129]. Similar results are obtained with most other flow cytometers. Without lysing the erythrocytes (Fig. 2A) the plot is dominated by these cells. After partial lysing (Fig. 2B) some of the white blood cells can be distinguished. The white blood cell clusters are best resolved after lysing all erythrocytes (Fig. 2C): lymphocytes are identified by intermediate forward light scatter (FLS) and low perpendicular light scatter (PLS), monocytes by high FLS and low PLS, and granulocytes by high FLS and high PLS. By proper data processing, a limited differential count is obtained, and a sample can be automatically classified as normal or abnormal on the basis of these measurements. The ELT series of flow cytometers (Ortho) employs these measurements for automated blood cell analysis, and most other flow cytometers make use of such light scatter measurements to identify lymphocytes for immunofluorescence assays [51,104].

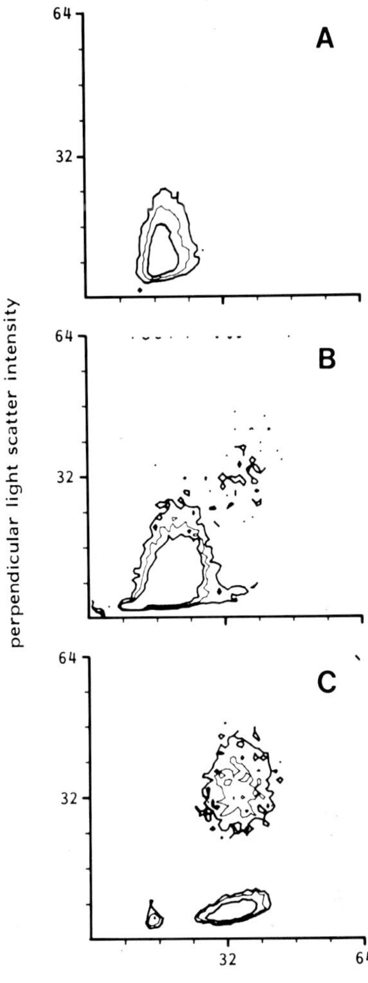

Fig. 2. Bivariate plots of forward and perpendicular light scatter intensities of human peripheral blood cells before (A) and after partial (B) and complete (C) lysis of the erythrocytes using NH_4 Cl.

ENUMERATION OF RARE CELLS

Reticulocytes

Disturbances in the dynamic equilibrium of hematopoiesis can be monitored by flow cytometry at various levels. The kinetics of blood cell formation and removal can be measured by quantitation of cells in various stages of differentiation and maturation or by measurement of cellular function or of cellular products.

For the erythroid differentiation lineage, the simplest measurement is the haematocrit, the packed cell volume (PCV) measurement. If the hematocrit is lower than normal, it may be due to a low rate of erythrocyte formation or to enhanced removal. The rate of formation is checked by counting the immediate precursor cells, the reticulocytes. Several flow cytometric procedures have been described to identify and quantitate reticulocytes. They are based on the presence of RNA in reticulocytes; RNA is absent in mature erythrocytes. Tanke et al. [111] used pyronin Y to stain and measure the RNA content of peripheral blood cells. Four major cell types could be distinguished simultaneously, measuring forward light scatter and pyronin Y fluorescence: mature red blood cells, reticulocytes, leukocytes, and thrombocytes. The fluorescence signals from the leukocytes were much higher than those of the other cell types. Thrombocytes could be eliminated by thresholding on forward light scatter. Figure 3 shows fluorescence histograms of the erythrocyte/reticulo-

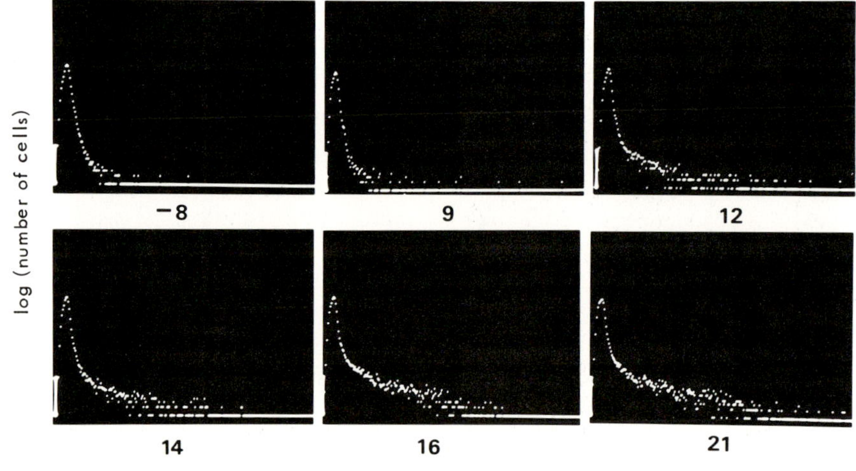

Fig. 3. Frequency distributions of pyronine Y fluorescence in erythrocytes and reticulocytes of a patient at 8 days before (−8) and at 9, 12, 14, 16, and 21 days after bone marrow transplantation. (Reproduced from ref. [111], with permission of the publisher.)

cyte cluster of a patient with aplastic anemia treated with a bone marrow transplantation. At 8 days before transplantation no fluorescent cells are seen. Starting at 9 days after transplantation fluorescent reticulocytes become detectable. Reticulocytes loose their RNA during maturation to erythrocytes; therefore, a continuous transition of just formed reticulocytes with much RNA to mature red cells without RNA can be expected. Quantitation of the number of reticulocytes requires mathematical analysis of the histogram showing transition of reticulocytes to mature red blood cells [111]. Jacobberger et al. [61] described a similar analysis for the quantitation of reticulocytes after staining with $DiOC_1(3)$. Sage et al. [93] similarly introduced thioflavin T. Seligman et al. [98] used a combination of acridine orange and transferrin receptor immunofluorescence measurements to distinguish various stages of erythroid maturation. The precision of the reticulocyte counts by flow cytometric procedures is comparable to or better than that of manual counting, and more reproducable. However, the method using $DiOC_1(3)$ somewhat underestimates the proportion of older, more mature reticulocytes [61]. The pyronin Y and acridine orange methods give results similar to manual counts. Some artifacts may occur in media containing divalent cations which allow platelets to clump [61] or in buffers which strongly affect the shape and size of erythrocytes [111].

Sickle Cells and Somatic Blood Cell Mutants

Although there are several ways by which flow cytometers can detect sickle cells (hemoglobin S containing red blood cells) or hemoglobin AS heterozygotes for genetic counseling, the literature on this application is scarce and focusses on detecting the very rare sickle cells in individuals with normal hemoglobin A. Red blood cells very rarely contain hemoglobin S or hemoglobin C as somatic mutants. In normal individuals the incidence of hemoglobin S containing RBCs is estimated to be one per 10^7 [85].

Enumeration of these mutants in blood samples from persons exposed to mutagens may be of use for biological dosimetry. Mendelsohn et al. [73] and Bigbee et al. [18] described a method to label human red blood cells in suspension with hemoglobin-specific antibodies. Using a membrane-permeable crosslinking reagent, dimethyl suberimidate, intracellular hemoglobin is bound to the membrane. Subsequently, the cells are lysed and washed to yield erythrocyte ghosts with bound hemoglobin, which is then accessible to antibodies. In reconstruction experiments using mixtures of hemoglobin AS and AA cells and antihemoglobin S, quantitative recovery of AS ghosts at an incidence of 3×10^5 was possible with a commercially available flow cytometer. Further development of rare event analysis is expected in the near future and will facilitate this application of flow cytometry. Using other antibodies other mutations can also be detected. The research group in LLNL (Livermore) recently demonstrated up to 10 fold increases in glycophorin A (M and N form) mutants in blood samples from individuals undergoing chemotherapy (see Chapter 27, this volume).

Detection of Parasites in Blood

A number of investigators have described the detection and quantification of parasites in red blood cells [25,58–60,132]. Since mature erythrocytes lack DNA and RNA, they are normally not stained using dyes such as Hoechst 33342 and 33258 or acridine orange. The parasitized red cells, however, contain DNA and RNA and will therefore bind these dyes and fluoresce in proportion to the amount of parasite in the cell. Also early trophozoite forms, so-called rings, which have low quantities of DNA, could be correctly enumerated with a commercially available flow cytometer [25,132]. Figure 4 shows a bivariate distribution of the fluorescence versus forward light scatter intensities of H33258 stained mouse peripheral blood infected with Plasmodium berghei [58]. Red blood cells containing different numbers of parasites as well as nonparasitized cells can be clearly distinguished. Free parasites are also detected at a low angle scatter of about 6° and relative fluorescence 0.8. Jayawardena et al. [62] studied the expression of H-2K, H-2D, and Ia on reticulocytes and erythrocytes infected with nonlethal and lethal variants of the 17XNL strain of *Plasmodium yoelii* using flow cytometry. Their results are of use to understand the immune response against malarial parasites.

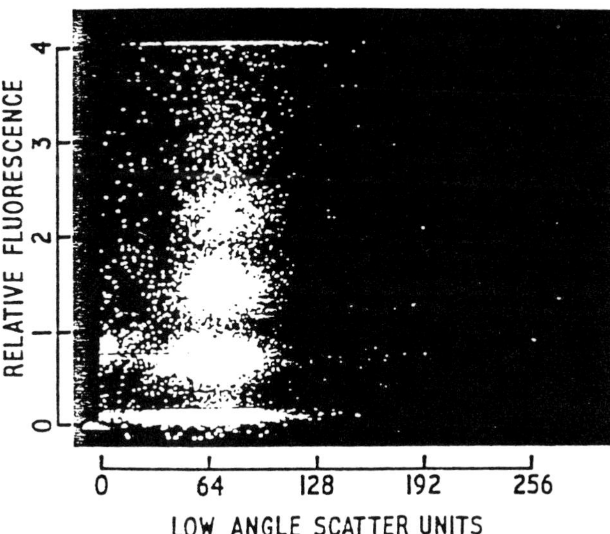

Fig. 4. Bivariate dot plot of Hoechst 33258 fluorescence and forward light scatter intensities of mouse erythrocytes infected with *Plasmodium berghei*. (Reproduced from ref. [58], with permission of the publisher.)

Fetal Cells in the Peripheral Blood

Among the very rare cells in the peripheral blood are fetal cells, which can be found in the blood of pregnant women. Herzenberg and co-workers [54] used fluorescence-activated cell sorting to enrich for leukocytes from maternal blood that bound antiserum specific for paternal antigens. They identified such cells in maternal circulation as early as the 15th week of gestation. Cupp et al. [34] described problems and possible solutions for detecting and sorting fetal erythrocytes in the maternal circulation. Since the incidence of fetal cells is on the order of one per 20,000 to 50,000 maternal erythrocytes, a rare-event analysis technique is obligatory. These workers argue that the prenatal screening of fetal erythrocytes for Rh factor may improve the medical treatment of Rh-negative mothers (with Rh-positive fetuses, which have a Dd Rh-positive father). In addition, inherited single gene disorders that are expressed in erythrocytes or that are genetically linked to marker genes expressed on these erythrocytes, could be diagnosed prenatally by modification of this flow cytometric technique.

FUNCTIONAL ASSAYS

Phagocytosis and Degranulation

Apart from detection and enumeration, flow cytometers can also be used to quantitate responses of cells. Lymphocyte reactions can be studied in a number of ways, as described below. Neutrophils, which are involved in inflammatory and hypersensitivity reactions, respond by a release of enzymes from their intracellular granules. Abrams and co-workers [1] described a quantitative assay of neutrophil degranulation using acridine orange. This fluorochrome accumulates in azurophilic granules, which also contain elastase. It is released from dog neutrophils concommitant with elastase activity in response to the ionophore A23187. Release of the fluorochrome can be measured by flow cytometry. Sklar et al. [103] demonstrated that degranulation of human neutrophils also can be determined by measuring the perpendicular light scatter intensity. The kinetics of the release of elastase are similar to the decrease in perpendicular light scatter. These assays provide means to rapidly evaluate the function of cells involved in inflammatory disease.

Another example is the assay for phagocytosis. A number of investigators [9,10,19,38,108] described methods to quantitate phagocytic capability of leukocytes. Basically, fluorescent particles or bacteria are incubated with the cells and after some time the fluorescence of each cell is measured flow cytometrically. The amount of fluorescence is then a measure of phagocytic activity.

Lymphocyte Activation

Lymphocytes can be stimulated in vitro by specific growth factors, mitogens, antibodies and allogeneic stimuli. Quantitation of the response can be difficult since only some of the lymphocytes respond and often other lymphocyte subsets or other cell types also play a role in the response patterns.

Early measures of lymphocyte activation relied on cell counts or [^{3}H]thymidine incorporation to quantitate proliferation, and ^{51}Cr-release to evaluate cell kill in mixed lymphocyte reactions or NK cell assays. By using flow cytometry the response of lymphocytes can be measured earlier. Increased DNA, RNA, or protein content and changed chromatin structure were the first parameters employed in the analysis of lymphocyte activation by flow cytometry [21–23,27,31,35,36,46,64]. Some of those measurements revealed nonproliferative lymphocyte responses that could be detected without much delay. It was observed, for example, that the stainability of lymphocytes with Hoechst 33342 changed within 12 hours after activation [65]. Although this dye is DNA specific, the change was not due to changes in DNA content but to chromatin structural changes affecting access to DNA binding sites (see Chapt. 16). The same dye could be used to label target cells to quantitate their conjugation with fluorescein labelled killer cells [110]. With the development of fluorescent methods for the determination of intracellular pH [113,130] and the mobilization of Ca^{2+} [86,89,116] still other parameters became available for detection and quantitation of the very early responses of lymphocytes by flow cytometry [29,79,117,130] (see Chapter 32). These have been used also for studies of responses of other hemopoietic cells such as human neutrophils [37,95] and rat erythroid cells [124].

ANALYSIS OF BONE MARROW

Bone marrow is the major site of blood cell production in man. About 2×10^{11} red blood cells, the same number of platelets, 10^{11} granulocytes and monocytes and probably similar numbers of lymphocytes are daily produced in adult human beings. Each of these blood cell types has its own maturation and differentiation pathway (see Chapter 34). The red blood cells arise from reticulocytes, which are formed by maturation from normoblasts, which are themselves formed by a series of cell divisions of early erythroid progenitor cells. The platelets are released from mature megakaryocytes, which are produced through maturation and division with nuclear multiplication (up to 32 nDNA) from megakaryocyte progenitor cells. Similarly, the granulocytes, monocytes and lymphocytes have their own progenitor cells. The immediate precursors of the blood cells viz. normoblasts, megakaryocytes, and so forth, can be recognized and studied by microscopic examination. Their precursors, however, cannot easily be recognized in smears of bone marrow because they are so few and so undifferentiated. The number of divisions to a normoblast from the first committed progenitor cell of the erythroid pathway is about 10 and therefore the progenitor cell occurs about 2^{10} or 1,000 times less frequently in the bone marrow than the normoblast. The existence of such a precursor can be demonstrated using in vitro culture systems, which make use of the capability of the precursor cell to produce a colony of blood cells. The disadvantage of the culture systems is the time required: only about 1 week after the start of the culture can the colonies be analyzed. Furthermore, the plating efficiencies of the colony assays for the various precursors is unknown and, until now, there is no colony assay for the earliest committed lymphoid precursors. In some of the assays several blood cell types can be found within a single colony, demonstrating the existence of a common precursor cell for the committed progenitors of several differentiation lineages. This common precursor is called the pluripotent hematopoietic stem cell (PHSC). In mice and rats, this stem cell can be detected by the spleen colony assay (CFU-S, colony-forming unit spleen [114]). Transplantation of stem cells into lethally irradiated recipients who are devoid of their own blood cell-forming system, leads to the repopulation of hemopoietic tissues and rescue from death. Allogeneic transplantation may lead to graft-versus-host disease if the stem cells are coinjected with lymphocytes. Therefore, purification of stem cells and depletion of lymphocytes from bone marrow grafts is of utmost importance for allogeneic bone marrow transplantation.

The purification of stem cells is also of interest for the preparation of grafts in the autologous situation, e.g., to remove leukemic cells. The antigenic properties of leukemic cells in many patients are variable, whereas the pluripotent stem cells are all antigenically similar. Only the stem cells are required for the repopulation of hemopoiesis. Therefore, for purging of bone marrow the purification of stem cells is the method of choice for most cases with antigenic differences between leukemic and normal stem cells.

Flow cytometers are too slow to sort a sufficient number of stem cells for transplantation in adult human beings. About 2 × 10^8 bone marrow cells per kg body weight are normally required for a successful take of the graft in a matched transplantation. It would take a fast flow cytometer; analyzing at 10^4 cells per second, almost 400 hours to sort the stem cells for a 70-kg recipient. Assuming that the stem cells can be preenriched by a factor of 5 by batch processes such as density centrifugation, the flow cytometer still would need to run for several days. For mismatched transplantations the situation is as yet unthinkable.

Flow cytometers are often used, however, to determine the optical and antigenic properties of the stem cells and other bone marrow cells in order to find differences that may be used to purify the stem cells by batch procedures. Since the stem cells can only be detected by in vitro culture assays or by analysis of the in vivo repopulation after transplantation, the staining and sorting procedures have to be supravital and sterile. The first applications of flow cytometers for this purpose involved unstained mouse bone marrow from which cells with different light scatter intensities and Coulter sizes were sorted and subsequently injected into lethally irradiated recipients to detect CFU-S [44,121,129], or cultured in vitro to find the fractions containing committed granulocyte/monocyte progenitor cells [128,129]. Figure 5 gives an example. By this procedure, the size distribution of these stem cells could be accurately determined for use in a large scale separation method such as elutriation [123,129]. Several supravital stains also were tested for their stem cell specificity.

TABLE 2. Vital Staining and Immunofluorescence of Murine Pluripotent Hematopoietic Stem Cells

	Fluorescence intensity*		
Stain/label	day-8 CFU-S	day-12 CFU-S	Ref.
Hoechst 33342	+	−/+	8, 84
Rhodamine 123	+ +	+/+ +	15, 77
Tetracycline	+ +	n.d.	127
Wheat germ agglutinin	+ +	+ +	126, 127
Anti-H-2K	+	+ +	123, 125
Anti-Qa-2	+	+ +	52
Anti-Thy-1	−/+	+	7, 13, 97

*−, negative; +, dull; + +, bright.

Table 2 lists labels and supravital stains that have been evaluated in stem cell studies. The use of lectins and antibodies turned out to be most successful, and has been applied in large-scale separation methods using lectin or antibody-bound magnetic spheres, panning, columns and so on.

Correction for Blood Cell Admixture

Aspirated bone marrow always contains peripheral blood. Although the percentage of nucleated cells in peripheral blood is relatively low, their presence may disturb quantitative measurements of the bone marrow cells significantly. Holdrinet et al. [57] developed a simple method to determine the fraction of peripheral blood in the suspension and to correct flow cytometer data for this admixture. The method is based on the observation that all erythrocytes and nearly all hemoglobin in the bone marrow aspirate originate from the peripheral blood. The fraction of peripheral nucleated cells in a bone marrow sample (F_{pb}) may then be calculated from the nucleated cell count (NC) and the hemoglobin (Hb) level in both bone marrow aspirate (bm) and venous blood sample (pb):

$$F_{pb} = \frac{Hb_{bm}/Hb_{pb} \times NC_{pb}}{NC_{bm}} \times 100\%$$

The average admixture of peripheral blood in 25 bone marrow samples of healthy volunteers was 14 ± 8%. However, the admixture varied between 6 and 93% of the nucleated cells in the bone marrow aspirate of 20 patients with hematological malignancies.

Labeling Affects In Vivo Assays

The detection of murine and rat pluripotent hemopoietic stem cells is based on their ability to reconstitute hemopoiesis in lethally irradiated recipients after transplantation. The homing of stem cells in recipient mice was found to be significantly affected by labeling with lectins or antibodies. This hampered studies that investigated the presence of membrane structures and differentiation antigens on the stem cells. However, lectins often can be removed from cells by incubation and washing with a competitive sugar [125,126]. Van den Engh and Platenburg [119] were able to study in vivo spleen colony formation by antibody-labeled stem cells after treatment of the labeled cells with papain, which removed the Fc part from bound antibody. Goldschneider et al. [47] used (Fab)2 fragments of Thy-1 to purify rat pluripotent stem cells. Later, it was demonstrated [28] that these cells could be purified by very low concentration of antibody without additional treatment because the stem cells are the most Thy-1 positive cells of rat bone marrow. Although they can be transplanted into syngeneic recipients the labeled stem cells apparently are killed in vivo by a Host-versus-Graft reaction. The amount of label on injected stem cells was found to be of importance and injection of carrageenan before transplantation improved the detection of stem cells in vivo. Thus opsonization of the stem cells by macrophages is probably involved in this HvG reaction. Bauman et al. [11] discovered that the use of biotin–avidin to conjugate the antibodies with fluorochrome also improved the in vivo detection of labeled stem cells. They explain this phenomenon by postulating that the Fc part of the antibody is covered by the large avidin molecule, so that the antibody is not detected by the in vivo defense mechanism. Their finding was of key importance for the development of our method to purify the mouse pluripotent hemopoietic stem cells [125]. It may be speculated, however, that improved techniques for isolation of cells to be administered in vivo will employ antibodies that bind all cells but those that have to be injected or cultured. Recently, such a method for purifying stem cells was published by Muller-Sieburg et al. [78].

Differentiation Antigens

Differentiation and maturation in hemopoiesis is partly reflected in morphological differences, but more generally in membrane antigen modulation. Several approaches have been followed to detect differentiation antigens. Antibodies against mature cell types can be rapidly screened and recognized using immunoperoxidase labeling on slides or by enzyme-linked immunosorbent assay (ELISA) techniques [66]. Their specificity against precursors of the mature cells then can be determined using a combination of flow cytometry, sorting and culturing. In our laboratory antibodies against mature cells were tested in that way using mouse bone marrow cells [120]. An example is given in Figure 6, which shows the forward light scatter and fluorescence intensities of mouse bone marrow cells labeled with anti-T200-biotin and avidin-FITC. This antibody (clone I 3/2 [115]) had been raised against lymphocytes. All bone marrow cells were brightly labeled except for the late erythroid cells: the normoblasts, reticulocytes and erythrocytes [118]. The presence of T200 on the stem cells, however, cannot be detected from such a bivariate plot because of the low incidence of those cells. Sorting of cells based on differences in fluorescence intensity, and subsequent culturing of the sorted fractions revealed that T200 was present on the pluripotent and most committed stem cells and that it was lost late during erythroid differentiation. By similar experiments, the Forssman antigen (clone M1/22.25 [106]) was shown to be present only during erythroid differentiation [118].

Another line of antigen research makes use of antibodies against known antigens for which a function is expected during hematopoiesis. One example is presented in Figure 7, which shows red and green fluorescence intensities of low density, human bone marrow cells labeled with antiglycophorin (fluorescein) and antitransferrin-receptor (phycoerythrin) antibodies [68]. The three-dimensional plot correlating these two markers indicated that transferrin receptor appeared before glycophorin. Sorting experiments showed that the cells in transition were erythroblasts. Glycophorin was then expressed and reached a maximum on the early normoblasts. The amount of glycophorin per cell remained constant on the erythroid cells throughout the rest of their maturation. By contrast, the amount of transferrin receptor progressively declined after reaching a maximum on the

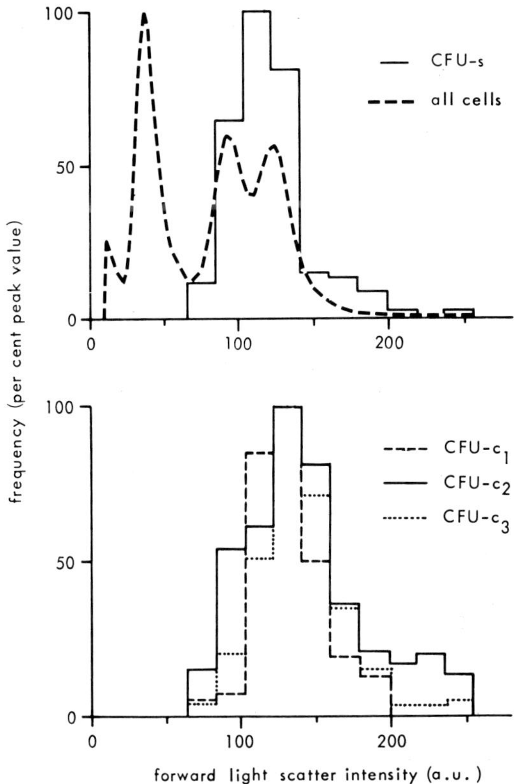

Fig. 5. Frequency distribution of the forward light scatter intensities of mouse bone marrow cells (dashed curve in upper panel), of mouse hemopoietic stem cells (CFU-S) and of three types of progenitor cells which form myeloid colonies in vitro (CFU-C1, −C2, and −C3; for details, see refs. [20] and [129]).

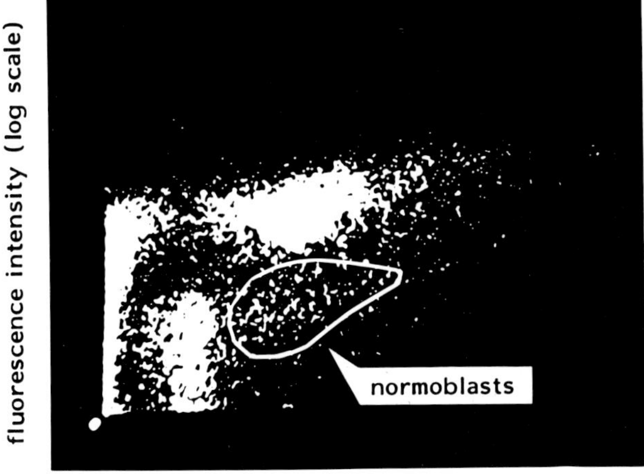

Fig. 6. Bivariate dot plot of anti-T200-biotin/avidin FITC fluorescence and forward light scatter intensities of mouse bone marrow cells [118].

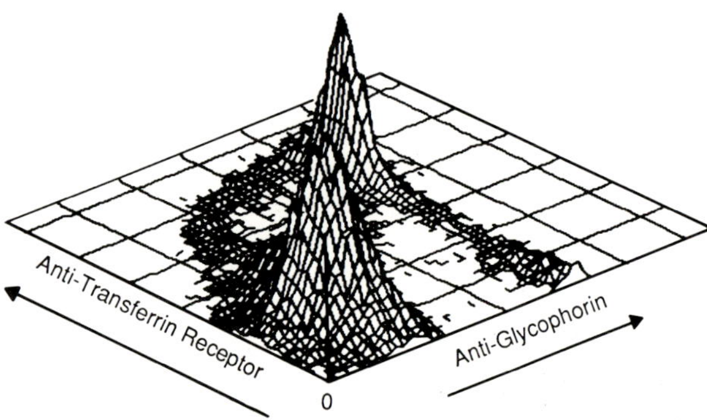

Fig. 7. Bivariate frequency distribution of the antiglycophorin/FITC and antitransferrin receptor/PE flourescence intensities of human bone marrow cells. (Reproduced from ref. [68], with permission of the publisher.)

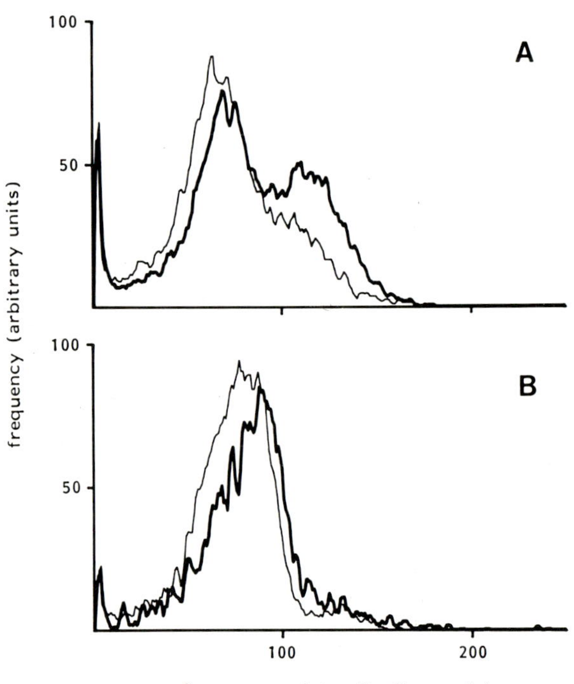

Fig. 8. Frequency distributions of the fluorescence intensities of stimulated (**A**) and unstimulated (**B**) mouse peritoneal macrophages labeled with an anti-*fms* product rat derived antibody and rabbit-antirat Ig/FITC (thick lines) or with rabbit-antirat Ig/FITC only (thin lines).

early normoblasts. Transferrin receptor was lost from the reticulocyte before all the RNA was removed from the cytoplasm.

A recent approach to detect differentiation antigens makes use of antibodies, which are directed against the receptors of growth factors. Molecular biology has been of help here. The growth factor for the formation of monocytes (M-CSF or CSF-1) is the product of the proto-oncogene *c-fms* [92,101]. Antibodies directed against the part of the *c-fms* product that is extracellular have been developed. Figure 8A,B shows fluorescence histograms of stimulated and resident mouse peritoneal macrophages labeled with such an antibody (clone SM3 [5]; Ab1 from Oncogene Science Inc., Mineola, NY) and with a rabbit-antirat Ig/FITC (RaRA/FITC) second layer. Expression of *c-fms* on the stimulated macrophages is clear, no CSF receptor was detected on unstimulated peritoneal macrophages. Expression of *c-fms* in cat macrophages by flow cytometry was demonstrated earlier by Sherr et al. [101]. This indicates that normal expression of growth factors can be detected using flow cytometry. Since specific growth factors are required for the regulation of hemopoietic differentation, this approach of detecting lineage-specific cells is very promising. The antibody which was used to detect the CSF receptors in Figure 8 was raised against the product of the v-onc gene expressed in cat cells. The expression of *c-fms* in human cells had been demonstrated earlier [83]. A second advantage of this approach seems to be that the same antibodies against oncogene products may be useful to study hemopoiesis in experimental animal model systems as well as in human.

The use of lectins and monoclonal antibodies combined with fluorescence-activated cell sorting and culturing revealed subpopulations within the pluripotent stem cell compartment [14,53,56,63,87,131]. It is not clear whether these labels detect different states, such as quiescence or proliferation, of the same cell type, or cells of different developmental or differentiation lineage that are all still pluripotent. One way to analyze these possibilities in more detail is to sort the stem cells and apply biochemical, functional, and molecular biology techniques to the purified suspensions.

Stem Cell Sorting and Analysis

Morphological analysis of bone marrow cells is often restricted to the more mature cell types, the immediate precursors of the peripheral blood cells. These bone marrow cells,

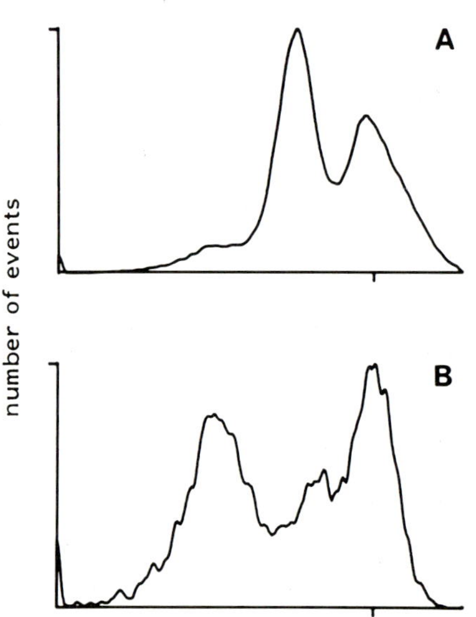

Fig. 9. Frequency distributions of the fluorescence of mouse bone marrow cells. Bone marrow cells were labeled with anti-H-2K biotin/avidin-phycoerythrin (PE), WGA-pyrene, and antibody 45 D8 labeled with goat-antirat-FITC. **A:** All nucleated cells. **B:** Cells (selected from the list mode file) which are brightly H-2K biotin/avidin-PE and WGA-pyrene positive and have low perpendicular and high forward light scatter intensities (stem cell windows). (Courtesy of P. de Vries.)

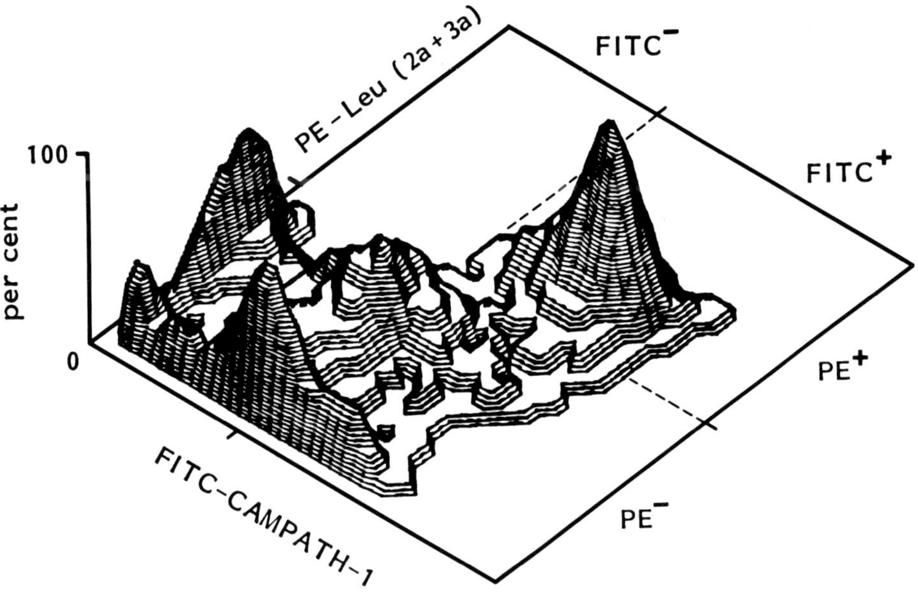

Fig. 10. Bivariate frequency distribution of the red (PE) and green (FITC) fluorescence intensities of Rhesus monkey bone marrow cells labeled with anti-Leu-2a/PE, anti-Leu-3a/PE and CAMPATH-1/FITC which have relatively low forward and low perpendicular light scatter intensities (lymphocyte window). (Courtesy of W. Gerritsen.)

however, only reflect the state of the stem cells or their regulation at an earlier time. It takes about 7 to 14 days before such morphologically recognizable cells are formed out of a pluripotent stem cell. Since during that time a number of cell divisions are needed, the recognizable cells form the majority and the stem cells are rare. Therefore, flow cytometry is the method of choice to examine the early cells if specific labels for these cells is known.

Several stem cells can be purified using light-activated cell sorting. Goldschneider et al. [47] described the purification of pluripotent hematopoietic stem cells from hydrocortisone treated rats using an antibody against the Thy-1 antigen. We have confirmed and combined this method with equilibrium density centrifugation to speed up the sorting [28]. McCarthy et al. [71] recently reported further analysis of rat stem cells which may be of use for purification without cortisone treatment. Murine fetal stem cells can be sorted efficiently using lectins and antibodies [80,81].

The best studied stem cell in experimental hematology is the pluripotent hemopoietic stem cell from adult mouse bone marrow. The purification of that cell type can be combined with a number of defined in vitro and in vivo assays. Mouse bone marrow stem cells can be sorted using a three-step procedure [125]. Firstly, the bone marrow cells are centrifuged on a discontinuous metrizamide gradient and the low density cells are labeled with fluoresceinated wheat-germ agglutinin (WGA-FITC). Subsequently, the WGA-FITC-positive cells with perpendicular and forward light scatter characteristics of blasts are sorted. The WGA-FITC is removed from the cells using N-acetyl-D-glucosamine and the cells are again labeled, now with anti-H-2K biotin antibody and subsequently with avidin–FITC. The brightest cells are again sorted. This fraction is 130- to 200-fold enriched for pluripotent stem cells. It contains about 40% of the pluripotent hematopoietic stem cells, which were present in the starting material. Between 60 and 100% of the sorted cells are pluripotent stem cells. During the second sorting step the cells can be deposited in the wells of Terasaki culture trays in known numbers, and subsequently cultured [125]. With one sorted cell per well 80–100% of the cells gave rise to daughter cells during 4 days of culture with recombinant IL-3 and purified M-CSF. The first divisions of this cell could be directly studied.

Our method can be modified in several ways. Lord [105] gets considerable enrichment for stem cells by sorting twice for WGA-FITC positive cells after density separation. The double sorting is done with the same windows. There are several reasons for better enrichment with two sort runs. One is because the sorting of cells is based in part on perpendicular light scatter; cells with intracellular irregular structures (e.g., an indented nucleus) may pass the laser beam during the first sort run in an orientation such that the irregular structure is not recognized. During a second sorting the orientation may be different so that the cell is recognized and not deflected. Another reason may be that the purity of the sorted cells is improved because of coincident unwanted cells in the deflected droplets during the first run at high speed and the lower frequency of coincidence during the second run at a lower speed. Although the enrichment obtained using Lord's modification is not as high as that with our procedure it is more easily applicable and useful for different strains of mice. We recently modified our procedure by sorting anti-H-2K- biotin/avidin-PE, and WGA-FITC positive cells simultaneous with selection of blasts based on light scatter. The enrichment by this procedure is also somewhat lower than the original one, suggesting an advantage in the purification of stem cells by several rather than a single sort. Other modifications that should be of use are simultaneous labeling with anti-GM1.2 [12], with rhodamine 123 [15,77], and with Hoechst 33342 [8,84], and replacement of anti-

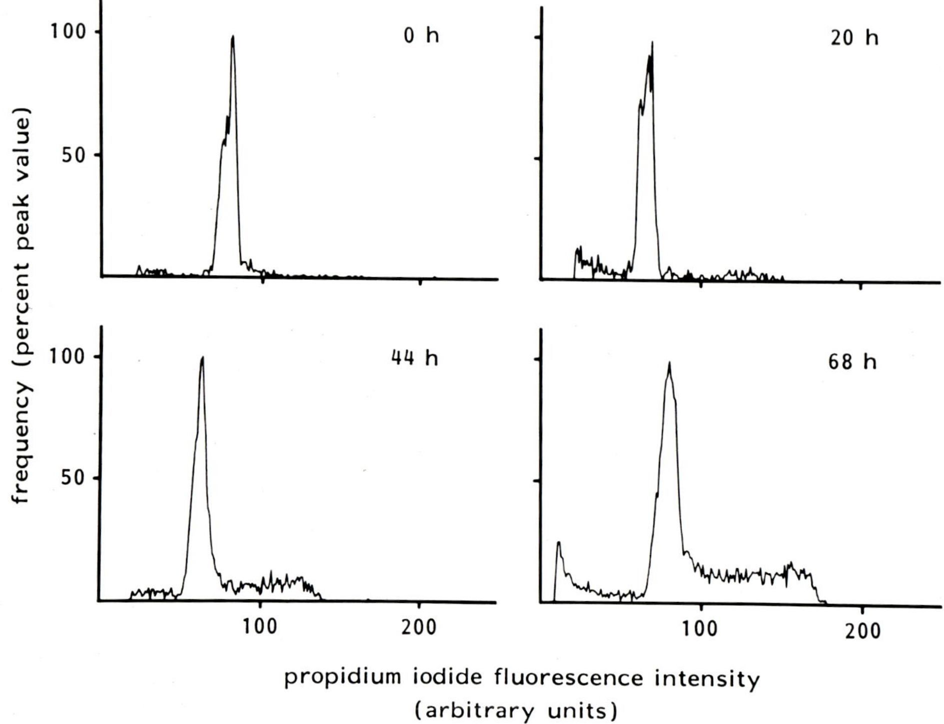

Fig. 11. Frequency distributions of the propidium iodide fluorescence intensity of purified mouse hemopoietic stem cells cultured in the presence of recombinant mouse interleukin-3 under serum-free conditions for 0, 20, 44, and 68 hours, respectively.

H-2K antibody by anti-Qa2 [52]. Figure 9 shows the histogram of FITC fluorescence of all cells from mouse bone marrow (Fig. 9A), and cells labeled with a monoclonal antibody 45 D8 (FITC-labeled), which are also H-2K (phycoerythrin) and WGA (pyrene) positive and are simultaneously within the forward and perpendicular light scatter (488-nm) windows of blast cells (Fig. 9B). The antibody distinguishes three subpopulations of H-2K positive, WGA positive blast cells. Sorting reveals that the day-12 CFU-S are not labeled with the antibody. About 2% of low density nucleated cells from adult mouse bone marrow are in the nonfluorescent subpopulation. Since the density fractionation enriches about fourfold for day-12 CFU-S [125,127], the stem cell content of mouse bone marrow detected this way is about 0.5%. This is in agreement with the outcome of calculations using spleen colony forming cells and the seeding efficiency of these cells to the spleen. It can be envisaged that simpler labeling procedures will be developed to detect and analyze pluripotent stem cells and their characteristics without sorting. It is also likely that this will be developed for human stem cells, in which case it will be a valuable addition for diagnosing hematopoietic malignancies and for monitoring therapy.

A number of lectins and antibodies have been described that discriminate between human colony forming and more mature bone marrow cells [6,17,39–43,48,76,82,90,102]. Some of these recognize stem cell types that occur with an incidence of less than 1%. Unfortunately, it is still not possible to identify and quantitate human pluripotent hematopoietic stem cells either by flow cytometry or other methods, although in principle this is clearly possible and justifies further research. Transplantation of antibody and complement treated cells in a monkey showed that the nonhuman primate rhesus monkey pluripotent hemopoietic stem cells are DR positive [45]. This has now been confirmed by positive cell sorting using flow cytometry (J. Wielenga, personal communication). Unfortunately, the anti My-10 antibody does not react with monkey bone marrow cells, and although this antibody probably binds to human pluripotent stem cells [30] it cannot be proven by transplantation in the monkey model.

T-Cell Depletion

Graft-versus-host (GvH) disease after bone marrow transplantation is caused by the presence of T lymphocytes in the graft. Several procedures for removal of the T cells exist. These are so efficient that the remaining T cells cannot be enumerated by conventional microscope-based techniques. Still GvH reaction occurs using some of these protocols. The question is whether remaining T cells or other cells cause this GvH. Using a flow cytometer, the remaining T cells can be quantitated at levels as low as 1 per 10^5 other cells. This discrimination is possible because of the availability of specific monoclonal antibodies with which the T cells can be brightly labeled. Figure 10 shows a bivariate frequency distribution of the red versus green fluorescence of anti-Leu (2+3)/PE and CAMPATH-1/FITC stained monkey bone marrow cells which have light scatter (forward and perpendicular) characteristics of lymphocytes. The double positive cluster contains 20% of all cells before and 0.02% after T-cell depletion. Control cells which were incubated without label show bivariates with about one per 10^5 cells within the area of this cluster. This method is being used to evaluate T cell depletion for mismatched bone marrow transplantation [45].

Analysis of Cultured Hematopoietic Cells

The progenitor cells of mature blood cells can often be studied indirectly using culture systems. Flow cytometric methods have been applied to analyse the differentiating and maturating offspring of these progenitor cells [49,91]. Even the pluripotent stem cells, at least of the mouse, now can be cultured reproducibly these days using recombinant interleukin-3. Using purified stem cell suspensions the response of these cells to IL-3 upon stimulation in vitro can be studied directly [75]. Figure 11 shows DNA-histograms of sorted stem cells [125] after 0, 24, and 48 hours of culture. The cells were stained with PI in the culture tube (1 ml serum-free modified α-medium [74]), using the method described by Taylor [112]. The cells were simultaneously counted by measuring the ratio of admixed fluorescent beads and cells as described by Stewart and Steinkamp [109]. From the counts in combination with the DNA-histograms, it can be concluded that IL-3 triggered 80% of the sorted cells into cycle, that the lag phase took about 24 hours and that the doubling time was less than 48 hours under these culture conditions. It may be expected that other methods will be developed to detect the response of cells rapidly to their growth factors, based perhaps on pH change measurements, transmembrane potential changes or detection of Ca^{2+} mobilization, and that they will soon be applied to sorted or specifically labeled hematopoietic progenitor cells.

REFERENCES

1. **Abrams WR, Diamond LW, Kane AB (1983).** A flow cytometric assay of neutrophil degranulation. J. Histochem. Cytochem. 31:737–744.
2. **Adams LR, Kamentsky LA (1971).** Machine characterization of human leukocytes by acridine orange fluorescence. Acta Cytol. (Praha) 15:289–291.
3. **Adams LR, Kamentsky LA (1974).** Fluorimetric characterization of six classes of human leukocytes. Acta Cytol. (Praha) 18:389–391.
4. **Akeroyd JH, Gibbs MB, Vivano S, Robinette RW (1959).** On counting leukocytes by electronic means. Am. J. Clin. Pathol. 31:188.
5. **Anderson SJ, Furth M, Wolff L, Ruscetti SK, Sherr CJ (1982).** Monoclonal antibodies to the transformation-specific glycoprotein encoded by the feline retroviral oncogene v-*fms*. J. Virol. 44:696–702.
6. **Andrews RG, Torok-Storb B, Bernstein ID (1983).** Myeloid-associated differentiation antigens on stem cells and their progeny identified by monoclonal antibodies. Blood 62:124–132.
7. **Basch RS, Berman JW (1982).** Thy-1 determinants are present on many murine hematopoietic cells other than T cells. Eur. J. Immunol. 12:359–364.
8. **Baines P, Visser JWM (1983).** Analysis and separation of mouse bone marrow stem cells by H33342 fluorescence-activated cell sorting. Exp. Hematol. 11:701–708.
9. **Bassøe C-F (1984).** Processing of *Staphylococcus aureus* and *Zymosan* particles by human leukocytes measured by flow cytometry. Cytometry 5:86–91.
10. **Bassøe C-F, Laerum OD, Glette J, Hopen G, Haneberg B, Solberg CO (1983).** Simultaneous measurement of phagocytosis and phagosomal pH by flow cytometry: Role of polymorphonuclear neutrophilic leukocyte granules in phagosome acidification. Cytometry 4:254–262.
11. **Bauman JGJ, Mulder AH, Engh GJ van den (1985).** Effect of surface labeling on spleen colony formation: Comparison of the indirect immunofluorescence and the biotin–avidin methods. Exp. Hematol. 13:760–767.
12. **Bauman JGJ, Wagemaker G, Visser JWM (1986).** A fractionation procedure of mouse bone marrow cells yielding exclusively pluripotent stem cells and committed progenitors. J. Cell. Physiol. 128:133–142.
13. **Berman JW, Basch RS (1985).** Thy-1 antigen expression by murine hematopoietic precursor cells. Exp. Hematol. 13:1152–1156.
14. **Berridge MV, Ralph SJ, Tan AS, Jeffery K (1984).** Changes in cell surface antigens during stem cell ontogeny. Exp. Hematol. 12:121–129.
15. **Bertoncello I, Hodgson GS, Bradley TR (1985).** Multiparameter analysis of transplantable hemopoietic stem cells. I. The separation and enrichment of stem cells homing to marrow and spleen on the basis of Rhodamine-123 fluorescence. Exp. Hematol. 13: 999–1006.
16. **Bessis M (1977).** "Blood Smears Reinterpreted." Berlin: Springer International.
17. **Beverley PCL, Linch D, Delia D (1980).** Isolation of human haematopoietic progenitor cells using monoclonal antibodies. Nature (Lond.) 287:332–333.
18. **Bigbee WL, Branscomb EW, Weintraub HB, Papayannopoulou Th, Stamatoyannopoulos G (1981).** Cell sorter immunofluorescence detection of human erythrocytes labeled in suspension with antibodies specific for hemoglobin S and C. J. Immunol. Methods 45:117–127.
19. **Bjerknes R, Laerum OD, Knapp W (1984).** Inhibition of phagocytosis by monoclonal antibodies to human myeloid differentiation antigens. Exp. Hematol. 12: 856–862.
20. **Bol S, Visser J, Van den Engh G (1979).** The physical separation of three subpopulations of granulocyte/macrophage progenitor cells from mouse bone marrow. Exp. Hematol. 7:541–553.
21. **Braunstein JD, Melamed MR, Darzynkiewicz Z, Traganos F, Sharpless TK, Good RA (1975).** Quantitation of transformed lymphocytes by flow cytofluorometry. I. Phytohaemagglutinin response. Clin. Immunol 4:209–215.
22. **Braunstein JD, Melamed MR, Sharpless TK, Hansen JA, Dupont B, Good RA (1976).** Quantitation of lymphocyte proliferative response to allogeneic cells and phytohemagglutinin by flow cytofluorometry. II. Comparison with ^{14}C-thymidine incorporation. Clin Immunol. Immunopathol 5:326–332.
23. **Braunstein JD, Schwartz G, Good RA, Sharpless TK, Melamed MR (1979).** Quantitation of lymphocyte response to PHA by flow cytofluorometry. III. Heterogeneity of induction period. J. Histochem. Cytochem. 27:474–477.
24. **Brecher G, Schneiderman M, Williams GZ (1956).** Evaluation of electronic red blood cell counter. Am. J. Clin. Pathol. 26:1439.
25. **Brown GV, Battye FL, Howard RJ (1980).** Separation of stages of *Plasmodium falciparum*-infected cells by means of a fluorescence-activated cell sorter. Am. J. Trop. Med. Hyg. 29:1147–1149.
26. **Cairns JW, Healy MJF, Stafford DM, Vitek P, Waters**

DAW (1977). Evaluation of the Hemalog D differential leucocyte counter. J. Clin. Pathol. 30:997–1004.
27. **Cassidy M, Yee C, Costa J (1976).** Automated analysis of antigen stimulated lymphocytes. J. Histochem. Cytochem. 24:373–377.
28. **Castagnola C, Visser J, Boersma W, Bekkum DW van (1981).** Purification of rat pluripotent hemopoietic stem cells. Stem Cells 1:250–260.
29. **Chused TM, Wilson AH, Greenblatt D, Ishida Y, Edison LJ, Tsien RY, Finkelman FD (1987).** Flow cytometric analysis of murine splenic B lymphocyte cytosolic free calcium response to anti-IgM and anti-IgD. Cytometry 8:396–404.
30. **Civin CI, Strauss LC, Brovall C, Fackler MJ, Schwartz JF, Shaper JH (1984).** Antigenic analysis of hematopoiesis. III. A hemopoietic progenitor cell surface antigen defined by a monoclonal antibody raised against KG-la cells. J. Immunol. 133:157–165.
31. **Cram LS, Gomez ER, Thoen CO, Forslund JC, Jett JH (1976).** Flow microfluorometric quantitation of the blastogenic response of lymphocytes. J. Histochem. Cytochem. 24:383–387.
32. **Cranendonk E, Abeling NGGM, Bakker A, Jong ME de, Gennip AH van, Behrendt H (1984).** Evaluation of the use of the Hemalog D in acute lymphoblastic leukemia and disseminated non-Hodgkin's lymphoma in children. Acta Haematol. 71:18–24.
33. **Crosland-Taylor P, Stewart JW, Haggis G (1958).** An electronic blood-cell-counting machine. Blood 13: 398–409.
34. **Cupp JE, Leary JF, Cernichiari E, Wood JCS, Doherty RA (1984).** Rare-event analysis methods for detection of fetal red blood cells in maternal blood. Cytometry 5:138–144.
35. **Darzynkiewicz Z, Traganos F. Andreeff M, Sharpless T, Melamed MR (1979).** Different sensitivity of chromatin to acid denaturation in quiescent and cycling cells as revealed by flow cytometry. J. Histochem. Cytochem. 27:478-485.
36. **Darzynkiewicz Z, Traganos F, Sharpless TK, Melamed MR (1976).** Lymphocyte stimulation: A rapid multiparameter analysis. Proc. Natl. Acad. Sci. 73:2881–2884.
37. **Davis BH, McCabe E, Langweiler M (1986).** Characterization of f-Met-Leu-Phe-stimulated fluid pinocytosis in human polymorphonuclear leukocytes by flow cytometry. Cytometry 7:251–262.
38. **Dunn PA, Tyrer HW (1981).** Quantitation of neutrophil phagocytosis, using fluorescent beads. Correlation of microscopy and flow cytometry. J. Lab. Clin. Med. 98:374–381.
39. **Falkenburg JHF, Van der Vaart-Duinkerken N, Veenhof WHJ, Goselink HM, Van Eeden G, Parlevliet J, Jansen J (1984).** Complement-dependent cytotoxicity in the analysis of antigenic determinants on human hematopoietic progenitor cells with HLA-DR as a model. Exp. Hematol. 12:817–821.
40. **Ferrero D, Broxmeyer HE, Pagliardi GL, Venuta S, Lange B, Pessano S, Rovera G (1983).** Antigenically distinct subpopulations of myeloid progenitor cells (CFU-GM) in human periphal blood and marrow. Proc. Natl. Acad. Sci. USA 80:4114–4118.
41. **Fitchen JH, Foon KA, Cline MJ (1981).** The antigenic characteristics of hematopoietic stem cells. N. Engl. J. Med. 305:17–25.
42. **Fitchen JH, Le Fèvre C, Ferrone S, Cline MJ (1982).** Expression of Ia-like and HLA-A,B antigens on human multipotential hematopoietic progenitor cells. Blood 59:188–190.
43. **Foon KA, Schroff RW, Gale RP (1982).** Surface markers on leukemia and lymphoma cells: Recent advances. Blood 60:1–19.
44. **Fulwyler MJ (1969).** Electronic volume analysis and volume fractionation applied to mammalian cells. Thesis, University of Colorado.
45. **Gerritsen WR, Wagemaker G, Jonker M, Kenter MJH, Wielenga JJ, Hale G, Waldmann H, Van Bekkum DW (1988).** The repopulation capacity of bone marrow grafts following pretreatment with monoclonal antibodies against T-lymphocytes in rhesus monkeys. Transplantation 45:301–307.
46. **Gill C, Fischer CL, Wilkins B, Nakamura R (1976).** Quantitation of lymphocyte blastoid transformation by flow cytofluorography. Med. Instr. 9:56.
47. **Goldschneider I, Metcalf D, Battye F, Mandel T (1980).** Analysis of rat hemopoietic cells on the fluorescence-activated cell sorter. I. Isolation of pluripotent hemopoietic stem cells and granulocyte–macrophage progenitor cells. J. Exp. Med. 152:419–437.
48. **Griffin JD, Ritz J, Beveridge RP, Lipton JM, Daley JF, Schlossman SF (1983).** Expression of MY7 antigen on myeloid precursor cells. Int. J. Cell Cloning 1:33–49.
49. **Griffin JD, Sullivan R, Beveridge RP, Larcom P, Schlossman SF (1984).** Induction of proliferation of purified human myeloid progenitor cells: A rapid assay for granulocyte colony-stimulating factors. Blood 63:904–911.
50. **Grooth BG de, Terstappen LWMM, Puppels GJ, Greve J (1987).** Light scattering polarization measurements as a new parameter in flow cytometry. Cytometry 8:539–544.
51. **Hansen WP, Hoffmann RA, Ip SH, Healey KW (1982).** Light scatter as an adjunct to cellular immunofluorescence in flow cytometric systems. J. Clin. Immunol. 2:328–418.
52. **Harris RA, Hogarth PM, Wadeson LJ, Collins P, McKenzie IFC, Penington DG (1984).** An antigenic difference between cells forming early and late haemopoietic spleen colonies (CFU-S). Nature (Lond.) 307:638–641.
53. **Harris RA, Sandrin MS, Sutton VR, Hogarth PM, McKenzie IFC, Penington DG (1985).** Variable expression of Qa-m7, Qa-m8, and Qa-m9 antigenic determinants on primitive hemopoietic precursor cells. J. Cell. Physiol. 123:451–458.
54. **Herzenberg LA, Bianchi DW, Schröder J, Cann HM, Iverson MG (1979).** Fetal cells in the blood of pregnant women: Detection and enrichment by fluorescence-activated cell sorting. Proc. Natl. Acad. Sci. USA 76:1453–1455.
55. **Hinchliffe RF, Lilleyman JS, Burrows NF, Swan HT (1981).** Use of the Hemalog D automated leucocyte differential counter in the diagnosis and therapy of leukemia. Acta Haematol. 65:79–84.
56. **Hoang T, Gihmore D, Metcalf D, Cobbold S, Watt S, Clark M, Furth M, Waldmann H (1983).** Separation of hemopoietic cells from adult mouse marrow by use of monoclonal antibodies. Blood 61:580–588.
57. **Holdrinet RSG, Egmond J van, Wessels JMC, Haa-**

nen C (1980). A method for quantification of peripheral blood admixture in bone aspirates. Exp. Hematol. 8:103–107.

58. **Howard RJ, Battye FL, Mitchell GF (1979).** *Plasmodium*-infected blood cells analyzed and sorted by flow fluorimetry with deoxyribonucleic acid binding dye 33258 Hoechst. J. Histochem. Cytochem. 27: 803–813.

59. **Jackson PR, Winkler DG, Kimzey SL, Fisher FM (1977).** Cytofluorograf detection of *Plasmodium yoelii, Trypanosoma gambiense* and *Trypanosoma equiperdum* by laser excited fluorescence of stained rodent blood. J. Parasitol. 63:593–598.

60. **Jacobberger JW, Horan PK, Hare JD (1983).** Analysis of malaria parasite-infected blood by flow cytometry. Cytometry 4:228–237.

61. **Jacobberger JW, Horan PK, Hare JD (1984).** Flow cytometric analysis of blood cells stained with the cyanine dye $DiOC_1(3)$: Reticulocyte quantification. Cytometry 5:589–600.

62. **Jayawardenen AN, Magis R, Murphy DB, Burger D, Gershon RK (1983).** Enhanced expression of H-2K and H-2D antigens on reticulocytes infected with Plasmodium yoelii. Nature (Lond.) 302:623–626.

63. **Johnson GR, Nicola NA (1984).** Characterization of two populations of CFU-S fractionated from mouse fetal liver by fluorescence-activated cell sorting. J. Cell. Physiol. 118:45–52.

64. **Kmieck PJ, Bagewell CB, Hudson JL, Irvin GL III (1979).** Multiparameter kinetic analysis of killer cell initiation by using immune RNA. J. Histochem. Cytochem. 27:491–495.

65. **LaLande ME, Miller RG (1979).** Fluorescence flow analysis of lymphocyte activation using Hoechst 33342 dye. J. Histochem. Cytochem. 27:394–397.

66. **Lansdorp PM, Oosterhof F, Astaldi GCB, Zeijlemaker WP (1982).** Detection of HLA antigens on blood platelets and lymphocytes by means of monoclonal antibodies in an ELISA technique. Tissue Antigens 19:11–19.

67. **Loken MR, Sweet RG, Herzenberg LA (1976).** Cell discrimination by multiangle light scattering. J. Histochem. Cytochem. 24:284–291.

68. **Loken MR, Shah VO, Dattilio K, Civin CI (1987).** Flow cytometric analysis of human bone marrow. I. Normal erythroid development. Blood 69:255–263.

69. **Mannsberg HP, Saunders AM, Groner W (1974).** The Hemalog D white cell differential system. J. Histochem. Cytochem. 22:711–724.

70. **Mattern CFT, Brackett FS, Olson BJ (1957).** Determination of number and size of particles by electronic gating. I. Blood cells. J. Appl. Physiol. 10:56.

71. **McCarthy KF, Hale ML, Fehnel PL (1985).** Rat colony-forming unit spleen is OX7 positive, W3/13 positive, OX1 positive and OX22 negative. Exp. Hematol. 13:847–854.

72. **Melamed MR, Adams LR, Zimring A, Marnick JG, Mayer K (1972).** Preliminary evaluation of acridine orange as a vital stain for automated differential leukocyte counts. Am. J. Clin. Pathol. 1:95–102.

73. **Mendelsohn ML, Bigbee WL, Branscomb EW, Stamatoyannopoulos G (1980).** The detection and sorting of rare sickle hemoglobin containing cells in normal human blood. In Laerum OD, Lindmo T, Thornd E (eds), "Flow Cytometry." Vol. IV. Bergen: Universitetsforlaget, pp 311–313.

74. **Merchav S, Wagemaker G (1984).** Detection of murine bone marrow granulocyte/ macrophage progenitor cells (GM-CFU) in serum-free cultures stimulated with purified M-CSF and GM-CSF. Int. J. Cell. Cloning 2:356–367.

75. **Migliaccio AR, Visser JWM (1986).** Proliferation of purified murine hemopoietic stem cells in serum-free cultures stimulated with purified stem-cell-activating factor. Exp. Hematol. 14:1043–1048.

76. **Morstyn G, Nicola NA, Metcalf D (1980).** Purification of hemopoietic progenitor cells from human marrow using a fucose-binding lectin and cell sorting. Blood 56:798–805.

77. **Mulder AH, Visser JWM (1987).** Separation and functional analysis of bone marrow cells separated by rhodamine 123 fluorescence. Exp. Hematol. 15:99–104.

78. **Muller-Sieburg CE, Whitlock CA, Weissman IL (1986).** Isolation of two early B lymphocyte progenitors from mouse marrow: A committed pre-pre-B cell and a clonogenic Thy-1^{lo} hematopoietic stem cell. Cell 44:653–662.

79. **Musgrove E, Rugg C, Hedley D (1986).** Flow cytometric measurement of cytoplasmic pH: A critical evaluation of available fluorochromes. Cytometry 7: 347–355.

80. **Nicola NA, Burgess AW, Staber FG, Johnson GR, Metcalf D, Battye FL (1980).** Differential expression of lectin receptors during hemopoietic differentiation. Enrichment for granulocyte–macrophage progenitor cells. J. Cell. Physiol. 103:217–237.

81. **Nicola NA, Metcalf D, Von Melcher H, Burgess AW (1981).** Isolation of murine fetal hemopoietic progenitor cells and selective fractionation of various erythroid precursors. Blood 58:376–386.

82. **Nicola NA, Morstyn G, Metcalf D (1980).** Lectin receptors on human blood and bone marrow cells and their use in cell separation. Blood Cells 6:563–579.

83. **Nienhuis AW, Bunn HF, Turner PH, Gopal TV, Nash WG, O'Brien SJ, Sherr CJ (1985).** Expression of the human c-fms proto-oncogene in hematopoietic cells and its deletion in the $5q^-$ syndrome. Cell 42: 421–428.

84. **Pallavicini MG, Summers LJ, Dean PN, Gray JW (1985).** Enrichment of murine hemopoietic clonogenic cells by multivariate analysis and sorting. Exp. Hematol. 13:1173–1181.

85. **Papayannopoulou T, McGuire TC, Lim G, Garzel E, Nute PE, Stamatoyannopoulos G (1976).** Identification of haemoglobin S in red cells and normoblasts, using fluorescent anti-Hb antibodies. Br. J. Haematol. 34:25–31.

86. **Poenie M, Alderton J, Tsien RY, Steinhardt RA (1985).** Changes of free calcium levels with stages of the cell division cycle. Nature (Lond.) 315:147–149.

87. **Ralph SJ, Tan SA, Berridge MV (1982).** Monoclonal antibodies detect subpopulations of bone marrow stem cells. Stem Cells 2:88–107.

88. **Richar WJ, Breakell ES (1959).** Evaluation of an electric particle counter for the counting of white blood cells. Am. J. Clin. Pathol. 31:384.

89. **Rink TJ, Tsien RY, Pozzan T (1982).** Cytoplasmic

pH and free Mg^{2+} in lymphocytes. J. Cell. Biol. 95: 189–196.

90. **Robinson J, Sieff C, Delia D, Edwards PAW, Greaves M** (1981). Expression of cell-surface HLA-DR, HLA-ABC and glycophorin during erythroid differentiation. Nature (Lond.) 289:68–71.

91. **Ronot X, Adolphe M, Hecquet C, Fontagne J, Jaffray P, Lechat P** (1983). DNA flow cytometric analysis of monogranulocytic colony forming cells from mouse bone marrow in vitro. Cytometry 3:387–389.

92. **Roussel MF, Rettenmier CW, Look AT, Sherr CJ** (1984). Cell surface expression of v-*fms*-coded glycoproteins is required for transformation. Mol. Cell. Biol. 4:1999–2009.

93. **Sage BH Jr, O'Connell JP, Mercolino TJ** (1983). A rapid, vital staining procedure for flow cytometric analysis of human reticulocytes. Cytometry 4:222–227.

94. **Salzman GL, Crowell JM, Martin JC, Trujillo TT, Romero A, Mullaney PF, LaBauve PM** (1975). Cell classification by laser light scattering: Identification and separation of unstained leukocytes. Acta Cytol. (Praha) 19:374–377.

95. **Satoh M, Nauri H, Takeshige K, Minakami S** (1985). Pertussis toxin inhibits intracellular pH changes in human neutrophils stimulated by N-formyl-methionyl-leucyl-phenylalaline. Biochem. Biophys. Res. Commun. 131:64–69.

96. **Saunders AM, Scott F** (1974). Hematologic automation by use of continuous flow systems. J. Histochem. Cytochem. 22:707–710.

97. **Schrader JW, Battye F, Scollay R** (1982). Expression of Thy-1 antigen is not limited to T cells in cultures of mouse hemopoietic cells. Proc. Natl. Acad. Sci. USA 79:4161–4165.

98. **Seligman PA, Allen RH, Kirchanski SJ, Natale PJ** (1983). Automated analysis of reticulocytes using fluorescent staining with both acridine orange and an immunofluorescence technique. Am. J. Hematol. 14: 57–66.

99. **Shapiro HM** (1977). Fluorescent dyes for differential counts by flow cytometry: Does histochemistry tell us much more than cell geometry? J. Histochem. Cytochem. 25:976–989.

100. **Shapiro HM, Schildkraut ER, Curbelo R** (1976). Combined blood cell counting and classification with fluorochrome stains and flow instrumentation. J. Histochem. Cytochem. 24:396–401.

101. **Sherr CJ, Rettenmier CW, Sacca R, Roussel MF, Look AT, Stanley ER** (1985). The c-fms proto-oncogene product is related to the receptor for the mononuclear phagocyte growth factor, CSF-1. Cell 41: 665–676.

102. **Sieff C, Bicknell D, Caine G, Robinson J, Lam G, Greaves MF** (1982). Changes in cell surface antigen expression during hemopoietic differentiation. Blood 60:703–713.

103. **Sklar LA, Oades ZG, Finney DA** (1984). Neutrophil degranulation detected by right angle light scattering: Spectroscopic methods suitable for simultaneous analysis of degranulation or shape change, elastase release and cell aggregation. J. Immunol. 133:1483–1487.

104. **Smart YC, Cox J, Murphy B, Enno A, Burton RC** (1985). Flow cytometric enumeration of absolute lymphocyte number in peripheral blood using two parameters of light scatter. Cytometry 6:172–174.

105. **Spooncer E, Lord BJ, Dexter TM** (1985). Defective ability to self-renew in vitro of highly purified primitive haemopoietic cells. Nature (Lond.) 316:62–64.

106. **Springer T, Galfre G, Secher DS, Milstein C** (1979). Mac-1: A macrophage differentiation antigen identified by monoclonal antibody. Eur. J. Immunol. 9: 301–306.

107. **Steinkamp JA, Romero A, Van Dilla MA** (1973). Multiparameter cell sorting: Identification of human leukocytes by acridine orange fluorescence. Acta Cytol. (Praha) 17:113–117.

108. **Steinkamp JA, Wilson JS, Saunders GC, Stewart CC** (1982). Phagocytosis: Flow cytometric quantitation with fluorescent microspheres. Science 215:64–66.

109. **Stewart CC, Steinkamp JA.** (1982). Quantitation of cell concentration using the flow cytometer. Cytometry 2:238–243.

110. **Storkus WJ, Balber AE, Dawson JR.** (1986). Quantitation and sorting of vitally stained natural killer cell-target cell conjugates by dual beam flow cytometry. Cytometry 7:163–170.

111. **Tanke HJ, Nieuwenhuis IAB, Koper GJM, Slats JCM, Ploem JS** (1981). Flow cytometry of human reticulocytes based on RNA fluorescence. Cytometry 1:313–320.

112. **Taylor IW** (1980). A rapid single step staining technique for DNA analysis by flow microfluorometry. J. Histochem. Cytochem. 28:1021–1024.

113. **Thomas JA, Buchsbaum RN, Zimnick A, Recker E** (1979). Intracellular pH measurements in Ehrlich ascites tumor cells utilizing spectroscopic probes generated in situ. Biochemistry 18:2210–2218.

114. **Till JE, McCulloch EA** (1961). A direct measurement of the radiation sensitivity of normal mouse bone marrow cells. Radiat. Res. 14:213–222.

115. **Trowbridge IS** (1978). Interspecies spleen–myeloma hybrid producing monoclonal antibodies against mouse lymphocyte surface glycoprotein, T200. J. Exp. Med. 148:313–323.

116. **Tsien RY** (1980). New calcium indicators and buffers with selectivity against magnesium and protons. Design, synthesis and properties of prototype structures. Biochemistry 19:2396–2404.

117. **Valet G, Raffael A, Moroder L, Wunsch E, Ruhenstroth-Bauer G** (1981). Fast intracellular pH determination in single cells by flow cytometry. Naturwissenschaften 68:265–266.

118. **Van den Engh G, Bauman J, Mulder D, Visser J** (1983). Measurement of antigen expression of hemopoietic stem cells and progenitor cells by fluorescence-activated cell sorting. In Killmann Sv-Aa, Cronkite EP, Muller-Berat CN, (eds.), "Haemopoietic Stem Cells." Copenhagen: Munksgaard, pp 59–74.

119. **Van den Engh GJ, Platenburg M** (1979). Suppression of CFU-S activity by rabbit anti-mouse brain serum can be overcome by treatment with papain. Exp. Hematol. 6:627–630.

120. **Van den Engh G, Trask B, Visser J** (1981). Surface antigens of pluripotent and committed haemopoietic stem cells. In Neth R, Gallo RC, Graf T, Mannweiler K, Winkler K(eds.), "Haematology and Blood Transfusion." Vol. 26: "Modern Trends in Human Leuke-

mia." Vol. IV. Heidelberg: Springer-Verlag, pp 305–308.

121. **Van den Engh G, Visser J (1979).** Light scattering properties of pluripotent and committed haemopoietic stem cells. Acta Haematol. 62:289–298.

122. **Van den Engh G, Visser J (1984).** Flow cytometry in experimental hematology. Bibl. Haematol. 48: 42–62.

123. **Van den Engh G, Visser J, Bol S, Trask B (1980).** Concentration of hemopoietic stem cells using a light-activated cell sorter. Blood Cells 6:609–623.

124. **Veldman A, Heul C van der, Kroos MJ, Eijk HG van (1986).** Fluorescence probe measurement of the pH of the transferrin microenvironment during iron uptake by rat bone marrow erythroid cells. Br. J. Haematol. 62:155–162.

125. **Visser JWM, Bauman JGJ, Mulder AH, Eliason JF, Leeuw AM de (1984).** Isolation of murine pluripotent hemopoietic stem cells. J. Exp. Med. 159: 1576–1590.

126. **Visser JWM, Bol SJL (1981).** A two-step procedure for obtaining 80-fold enriched suspensions of murine pluripotent hemopoietic stem cells. Stem Cells 1: 240–249.

127. **Visser JWM, Bol SJL, Engh G van den (1981).** Characterization and enrichment of murine hemopoietic stem cells by fluorescence activated cell sorting. Exp. Hematol. 9:644–655.

128. **Visser JWM, Cram LS, Martin JS, Salzman GL, Price BJ (1978).** Sorting of a murine granulocytic progenitor cell by use of laser light scattering measurements. In Lutz D (ed), "Pulse-Cytophotometry." Ghent, Belgium: European Press, pp 187–192.

129. **Visser JWM, Engh GJ van den, Bekkum DW van (1980).** Light scattering properties of murine hemopoietic cells. Blood Cells 6:391–407.

130. **Visser JWM, Jongeling AAM, Tanke HJ (1979).** Intracellular pH-determination by fluorescence measurements. J. Histochem. Cytochem. 27:32–35.

131. **Watt SM, Gilmore DJ, Metcalf D, Cobbold SP,Hoang TK, Waldmann H (1983).** Segregation of mouse hemopoietic progenitor cells using the monoclonal antibody, YBM/42. J. Cell. Physiol. 115:37–45.

132. **Whaunn JM, Rittershaus C, Ip SHC (1983).** Rapid identification and detection of parasitized human red cells by automated flow cytometry. Cytometry 4: 117–122.

34

Applications in Immunology and Lymphocyte Analysis

Kenneth A. Ault
Maine Cytometry Research Institute, Portland, Maine 04102

INTRODUCTION

This chapter discusses the role of flow cytometry and cell sorting in immunology research and medical immunology, emphasizing techniques related to the preparation, surface labeling, analysis, and separation of lymphoid cells. It is the study of the lymphocyte that has brought together the power of flow cytometry and cell sorting with the power of monoclonal antibody technology. This confluence of technologies, which began in the research laboratory, is now in the process of moving into clinical research laboratories and into standard clinical practice. However, there are debates on the relative merits of different techniques, their limitations, the interpretation of results, and so forth; it will be necessary to standardize these techniques before flow cytometric analysis achieves wide clinical application. At the same time, there is pressure from the research laboratory to expand the power of this technology using new monoclonal antibodies, new fluorochromes, and more cellular parameters and to analyze the data in much more sophisticated ways.

We begin with a discussion of the immunological problems that are amenable to solution using the techniques of flow cytometry and focus on those that are currently most useful in clinical and research applications. We then consider several methods of sample preparation that are currently in use, focusing primarily on the preparation of lymphoid cells from blood and tissues and on various techniques of cell labeling.

Finally, this section describes strategies to be used in the sorting of cells, considering frequency of the desired cells, level of purity, and desired number of cells.

IMMUNOLOGICAL PROBLEMS AMENABLE TO FLOW CYTOMETRIC TECHNIQUES

Flow cytometry has had a major impact on immunology research, largely because of the importance of lymphocyte subsets that differ in their surface antigenic expression but are very similar in morphological and physical characteristics. The advent of monoclonal antibody technology, which has been closely coupled with the use of flow cytometers, has made it possible to classify, differentially count and sort out those subsets of lymphocytes for further study. In general, there are two types of classification: lineage specific and functional.

The lineage-specific antigens are expressed on functionally related cells arising from a single progenitor. In the ideal case, they are expressed on all the developmentally related cells of that lineage. However, most such antigens are actually differentiation antigens that appear at a certain stage of differentiation within a lineage and then may disappear at another stage. Nevertheless, such antigens can be used to define a cell lineage and are very useful for that purpose.

An extreme case of a lineage-specific antigen is the clonal marker. This is an antigen that is common to a single cell and all its progeny. It has special significance in the world of immunology because of the principle of clonal selection. The immune response to a single antigenic determinant (epitope) is determined by a clone of lymphocytes expressing receptors for that epitope. In the case of B lymphocytes, the receptor is a surface immunoglobulin (sig). Some of the antibody that a particular B cell is capable of making is not secreted but remains on the surface of the cell and serves as a receptor for antigen. The unique structures of this sig are themselves antigens (idiotypes) that specifically define the B cells of a single clone responsive to one antigenic determinant [29]. Similar structures are present on T cells [42]. Thus, a lineage can be defined at any of several levels (leukocyte > lymphocyte > B lymphocyte > idiotype). A complete description of a cell obviously involves the specification of each of these, although usually specifying the lowest known level may be adequate.

One constraint that must be considered in the use of lineage specific antigens is that they may not be expressed at all stages of differentiation. Thus, antigens may define limited spans of the lineage [47]. This point must be fully understood and lineage antigens investigated before they can be used with confidence to define unknown cell types. A second constraint is that many of these antigens have unexpected distribution on unrelated cells, although they are widely used to define lymphocyte subsets [6,8].

A recurring theme of this review is the advantage of multiple markers in the analysis of lymphocytes [5,26,46]. An-

Flow Cytometry and Sorting, Second Edition, pages 685–696

tigenic structures are likely to be shared, if not with a related cell type, then with entirely unrelated cells. Thus, while it would be naive to expect a given cell type to be uniquely defined by a single antigenic structure, it is frequently possible to define a lineage by combinations of antigens, often a relatively small number of antigens. Even monoclonal antibody technology, which has held out promise of manufacturing antibodies against virtually any desired antigenic structure, may not be enough to dissect the cellular structure of the organism. Nowhere is the use of multiple markers better illustrated than in the case of the human natural killer (NK) cell. These cells are quite well defined, although there is not one unique antigenic structure on their surface. All the antigens used to define NK cells are shared with other cell lineages. Nevertheless, by using a combination of markers, it is still possible to define and study this lymphoid subset [26].

The second type of cell classification and the second major group of markers that are extremely useful in the study of immunology are those associated with the functional state of a cell. These are frequently referred to as activation antigens, although in the broadest sense they may be associated with any number of physiological states of the cell [36,46]. In most cases, these antigens are not lineage specific. That is, they may be found on cells of very different types. Examples include the antigens associated with receptors for hormones or messenger molecules such as the interleukins [50]. Because they are usually not lineage specific, it becomes extremely important to use multiple markers. In a typical application, one or more lineage specific markers may be needed to define a cell type of interest and then evaluate the presence or absence of one or more activation antigens in those cells.

It is useful to think of these antigens in terms of time relative to a functionally significant change in the state of the cell. For example, a lymphocyte encountering an activation signal may first change its membrane potential over a period of a fraction of a second. In response to this signal, it may begin to synthesize a series of new surface structures that appear in a matter of hours and then persist with varying half-lives. For some purposes, it is suitable to think of the change in membrane potential as a signal of activation. However, the membrane potential may change frequently in response to a number of different stimuli, making its significance ambiguous. Thus one may prefer to observe an activation antigen appearing on the cell surface. Some of these antigens reveal an activation event that occurred within the last few hours or days; others signify an event that occurred days or weeks previously. Such antigens are only beginning to be explored on a number of cell types. We will be in a much better position to interpret the results of such experiments as we learn the true functional significance of the various activation antigens.

It is now becoming clear that some antigens previously thought to be lineage specific actually reflect a state of activation of one or more cell types [40]. Since all the surface structures of lymphocytes that we identify as antigens obviously have some functional significance for the cell, the distinction between lineage antigens and activation antigens will continue to blur as we learn more.

A very important role of cell sorting in immunology research is the correlation of cell-surface marker expression with cell function. As monoclonal antibodies are developed that recognize what appear to be new subsets of cells, their significance depends on whether these cells have unique functional attributes. This requires sorting the cells bearing a specific marker (or markers), followed by their analysis in functional assays. Once such a functional correlation is firmly established, the presence of the marker becomes sufficient to infer the function of a cell type. Nowhere is this better illustrated than in the human T-cell subsets. Interestingly, it may be necessary to go through several cycles of re-evaluation to finally establish the correlation of markers with function. For example, when human T cells were resolved into two subsets, first using polyclonal and then monoclonal antibodies, the two subsets were associated with helper and suppressor/cytotoxic functions [12]. After this correlation seemed firmly established and what are now known as the CD4 and CD8 antigens were associated with these two functions, respectively, it was discovered that there were clones of cytotoxic T cells that expressed the CD4 antigen (24). In a re-evaluation, it was discovered that these two antigens were in fact correlated with the ability of the T cell to recognize different classes of MHC antigens on the surface of target cells. The original distinction between helper and suppressor/cytotoxic cells had resulted from the original choice of functional assay systems and had been misleading.

Thus, physically separating cell subsets based on the expression of surface markers, i.e., cell sorting, will remain a critical part of cell biology for a very long time. As we become more sophisticated in our understanding of how cells function, and how they interact in the organism, we will constantly have to re-evaluate the functional significance of both old and new surface antigenic structures.

Another major application of flow cytometry in immunology and biotechnology is in the selection of variant cell types. In essence this is using the cell sorter to select a cell with particular characteristics, usually an infrequent mutation, to apply selective pressure in favor of that mutation. The best example is in the selection of isotype variants of antibody secreting hybridoma cell lines which is increasingly useful in biotechnology [15]. In this application, one has a growing population of hybridoma cells that are secreting a useful monoclonal antibody, but the antibody may be of an isotype (e.g., IgM or IgG2) that is not desirable for a particular application. One option for solving this problem is to search for another hybridoma secreting an antibody with the same specificity but a different isotype (e.g., IgG1). However, using the cell sorter, it is possible to label the original hybridoma with an antibody directed against the desired isotype and sort for the rare hybridoma mutants that happen to be expressing this isotype. When these cells are sorted and grown, perhaps after several cycles of such selective pressure, a new cell line can be established secreting an antibody with the same specificity but with the new isotype.

This use of the cell sorter to drive the selection of mutants may have very wide general application in biotechnology in the preparation of organisms having useful properties that must be maintained against the effects of natural selection. It is applicable any time the desired property of the cell is reflected in the presence or absence of a cell-surface antigen without any selective growth advantage.

In another example, cell sorting is used for transfection experiments. In these experiments new genetic material is transferred into cells, and the goal is to select those cells that have successfully incorporated the new DNA into their genome and begun to express it [23]. The necessity for cell sorting results from the following considerations. First, the transfer of DNA into cells is essentially a random process so that it is necessary to select those cells that have taken up the DNA sequence for the desired gene. Second, only a small fraction of the cells that have taken up a desired gene will stably incorporate it into their genome and express it. Since in most cases, the presence of a foreign gene will confer no selective advantage for the cell, the use of the cell sorter to supply the selective pressure becomes attractive.

This is all the more true because of a curious phenomenon in which many of the newly acquired genes are expressed transiently [11] before the new DNA has been incorporated into the genome of the cell. Thus, the first cells are sorted within a few days of transfection to enrich for those that have taken up the appropriate piece of DNA containing the gene of interest. This preliminary sort is followed later by another sorting to select those cells that have proceeded to stably express the gene.

Obviously, the timing of the sorts must take into consideration the growth characteristics of the cells, the frequency of cells with the desired property, and how well the desired cells can be labeled and distinguished from the rest of the population. This question of sorting strategy must be solved individually for each experiment, based on the best compromise between several mutually antagonistic factors.

A final application of flow cytometry that is seeing increasing application in basic and clinical research in immunology is the detection of very rare cell types (see also Chapter 27). In the situations described above, the cells of interest were capable of growing and replicating and were being selected because they expressed certain properties in a stable fashion. This permits repeated sorting and gradual enrichment of very rare cell types. For example, if the frequency of the desired cell is 10^{-8}, and a single sort is capable of enriching the cells by only a factor of 100, it may require three or four cycles of sorting each followed by culturing of the intermediate enriched population before a pure population is obtained. However, if one is interested in detecting, counting, or purifying a cell type that will not grow or that cannot be trusted to stably express the characteristic that makes it interesting, some other strategy is needed. In this situation, two major problems limit our ability to count very rare cells. First, there must be enough of the rare cells to obtain reasonable statistics in a reasonable length of time. Second, a very high signal to noise ratio is needed for rare event detection. Even cells that are clearly resolvable from the background noise level of unlabeled cells may be lost in the tail of the noise peak when they are present at very low frequency [38]. There is a great deal of interest in developing new generations of cell sorters and new methodology to purify rare cells in a single pass. Examples include the use of high-speed flow rates and marking cells with fluorescent beads to achieve the very high signal-to-noise (S/N) ratios needed to count rare cells [13,22]. Another approach is to sort rare cells onto slides at known locations [1], so that they can be studied in terms of their morphology, genetic content, and other characteristics.

UNIQUE LYMPHOID SUBSETS DEFINED BY FLOW CYTOMETRY

In a field that is developing as rapidly as cellular immunology, any description of the most recent results will surely be quickly outdated. Nevertheless, it may be useful to describe some of the latest achievements of flow cytometry in the identification of new lymphoid cell types, if only to serve as examples of the kind of work that can be done using this technology.

Table 1 lists lymphoid antigens as currently agreed upon by the World Health Organization (WHO)-sponsored Leukocyte Typing Workshops. These cluster designations (CD) are the best way we have at the moment of overcoming the confusion due to the different names given to similar clones by different laboratories and companies.

$CD5^+$ B cells: The existence of B cells that express the CD5 marker has become very clear from a number of lines of evidence. The CD5 marker was originally thought to be pan-T cell. It was then found that it was expressed in most cases of B-cell chronic lymphocytic leukemia [10] (Fig. 1). Similar cells were identified in normal tissues [18]. These cells were found to be the predominant B-cell subset in many patients recovering from marrow transplantation [5], and the same subset was identified among human fetal lymphocytes [3]. They are also present in small numbers in normal adult blood [16]. As this story was unfolding in man, others were finding the analogous subset (the Ly-1 B cell) in the mouse. The murine work shows that these cells are abundant in the fetus and in certain strains of mice that are susceptible to autoimmune diseases [20]. In vitro experiments have shown that these cells may be responsible for most autoantibody production in mice [19] and that they may have an immunoregulatory role [37]. Recently it was suggested that these cells use a restricted family of variable region genes for their immunoglobulin receptors [43].

This entire fascinating story of what appears to be an important lymphoid subset has arisen because these cells can be identified and separated using two-color flow cytometry. Certainly without flow cytometry, and probably without the use of two fluorochromes simultaneously, very little of this work would have been possible.

T cells expressing CD8 and Leu7: This is another example of a lymphoid subset that can only be identified and counted using two simultaneous markers. The presence of CD8 on MHC class 1 restricted T cells that are predominantly suppressor/cytotoxic in function is well known. Leu7 (also known as HNK-1) was originally described as a marker for NK lymphocytes. It is now known to be an interesting and complex marker with a number of useful properties. When it is combined, in two color immunofluorescence, with CD8 there proves to be a population of cells expressing both markers (Fig. 2).

Careful study of these cells, using two- and three-color analysis has shown that cells expressing high levels of CD8 and Leu7 are true T cells in that they also express CD3. These cells can only be identified using this combination of antibodies and they have proved of considerable clinical interest. Although they are present in normal individuals, their numbers are small and quite variable (0–15% of lymphocytes in our hands). In cases of HIV infection, and in some other acute viral illnesses, they are markedly elevated [30,34]. Phillips and Lanier have shown that these are very active cytolytic cells with marked activity against viral infected target cells [41]. Their full clinical significance has yet to be determined, but this is a classic example of the use of multiple markers to identify functional distinct lymphoid subsets.

T cells that are dim or negative for CD5: In addition to being found on some B cells, the CD5 marker is now known to be expressed at low levels, or not at all, on a small population of T cells ($CD3^+$ cells). These cells first came to attention during studies of marrow transplant patients, and patients with an unusual lymphoproliferative disease [44]. They are present in very small numbers in normal blood and appear to have the phenotype ($CD3^+$, $CD8^+$, $Leu7^+$) (Fig.

TABLE 1. Summary of World Health Organization Recommended Nomenclature for Human Leukocyte Differentiation Antigens[a]

Antigen	m.w.	Similar antibodies	Target cells
CD1	p45/12	Leu6, OKT6, T6	Thymocytes
CD2	p50	Leu5, OKT11,T11	E rosette receptor, T and NK cells
CD3	p19/29	Leu4, OKT3, T3	T cells
CD4	p56	Leu3, OKT4, T4	Helper–inducer T cells, monocytes
CD5	p67	Leu5, T1, T101	T cells, B-cell subset
CD6	p120	T12	T cells
CD7	p41	Leu9, 3A1, WT1	T cells, NK cells
CD8	p32–33	Leu2, OKT8, T8	Suppressor-cytotoxic T cells, NK subset
CD9	p24	BA-2	lymphoid leukemias
CD10	p100	CALLA, J5	PMN, pre-B cells
CD11	p170/95	CR3, OKM1, MO1	PMN, monocytes, NK, T-cell subset
CDw13	p150	MY7	PMN, monocytes
CD14	p55	LeuM3, MY4, MO2	Monocytes
CD15	LNFPIII	LeuM1, MY1	PMN, monocytes
CD16	p50–60	Leu11, 3G8	Fc g receptor, NK, PMN
CD19	p95	Leu12, B4	B cells
CD20	p35	Leu16, B1	B cells
CD21	p140	CR2, B2	B cells
CD22	p135	Leu14	B cells
CD23	p45	Blast-2	Fc e receptor, B-cell subset
CD24	p45/55/65	BA-1	B Cells
CD25	p55	IL-2R	IL-2 receptor, activated T cells
CDw29		4B4	T-cell subset
CD33	p67	MY9	Early myeloid
CD34	p115	HPCA1, MY10	Marrow progenitors
CD35	p220	CR1, TO5	PMN, monocytes, B cells
CD38	p45	Leu17, OKT10, T10	Activated T cells, plasma cells
CDw41	gpIIb/IIIa	Many	Platelets
CDw42	gpIb	Many	Platelets
CD45	T200	HLE-1	Pan leukocyte
CD45R	p220	Leu18, 2H4	T subset

[a]Abbreviations: IL-2 = interleukin-2; NK = natural killer (cell); PMN = polymorphonuclear lymphocyte.

3). Morphologically, they fall into the broad category of large granular lymphocyte. In the lymphoproliferative syndrome, they appear to be associated with marrow failure syndromes and neutropenia [9]. Here we have another example of a subset of potential clinical interest whose definition depends on flow cytometry using two markers.

CD4$^-$, CD8$^-$, but CD3$^+$ T Cells: These cells, recently described by Lanier and co-workers [25], appear to be T cells that have not used the usual α/β-genes to form the T-cell receptor but have rearranged the γ-gene. They have been found in normal blood, and Lanier has suggested that they may represent a separate line of differentiation for T cells, of as yet undetermined significance. Recently, there have been reports of lymphoproliferative diseases and immunodeficiency states that may be associated with these cells.

This set of examples clearly illustrates that there is much to be learned about the expression of cell-surface markers on lymphocytes and that we are still in the process of defining functionally significant subsets. Many of these subsets will certainly be of clinical relevance, in addition to their importance in understanding the immune system.

TECHNIQUES OF CELL SEPARATION

Although blood is by far the simplest source of lymphoid cells for study, it has several drawbacks for analysis using flow cytometry. First is the fact that the leukocytes are outnumbered in blood by erythrocytes by a factor of 10^3. Platelets outnumber leukocytes by a factor of about 10, and lymphocytes are only 30–50% of the leukocytes in human blood. Thus, it becomes necessary to distinguish lymphocytes among a large number of irrelevant particles.

There are two basic solutions to this problem: (1) elimination of all the irrelevant particles, and (2) use of physical or immunological parameters that are recognizable by flow cytometry to distinguish the lymphocytes. In practice, both approaches are taken in varying degrees.

Another issue that we must deal with at this time is the storage of blood before processing for the flow cytometer. This major issue for clinical studies has been addressed in the literature. The evidence suggests that the best method of handling blood is to keep it anticoagulated at room temperature and, if it has to be kept for a significant length of time, to dilute it with an equal volume of tissue culture medium. Under these conditions, most workers agree that blood lymphocytes can be successfully analyzed after a delay of about 24 hours [21]. The choice of an anticoagulant has some bearing in that heparin, which is strongly negatively charged, may interfere with some later staining protocols, especially if it is used in excess amounts. By contrast, anticoagulants that chelate calcium may be unintentionally

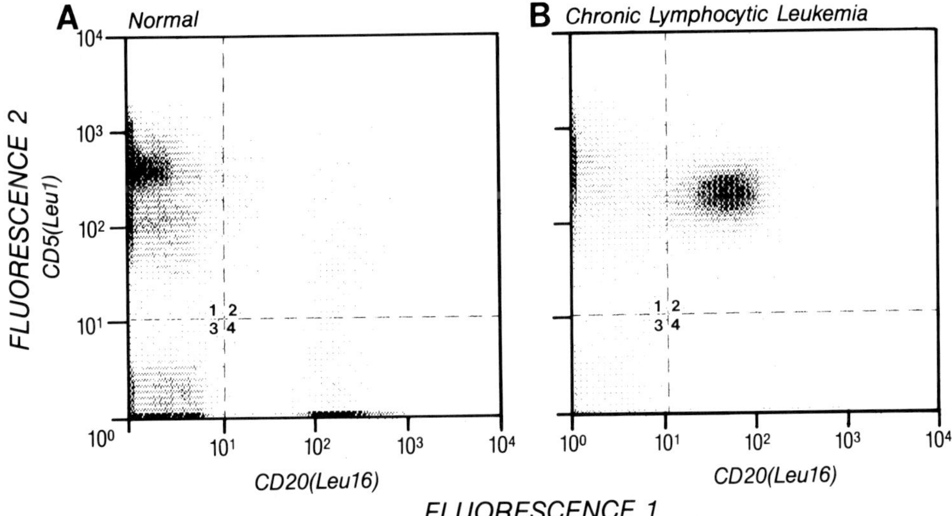

Fig. 1. Expression of CD5 (Leu1) on cells of chronic lymphocytic leukemia. Two-color immunofluorescence analysis of blood cells labeled with FITC-Leu16 (anti-CD20), which recognizes all B cells, and PE-Leu1 (CD5), which is found on most T cells. **A:** Normal blood in which the two markers are mutually exclusive. **B:** From a case of typical chronic lymphocytic leukemia, the leukemic cells express both markers. They have less CD5 antigen than do the residual normal T cells in this patient. These data were obtained on a Becton-Dickinson FACScan instrument.

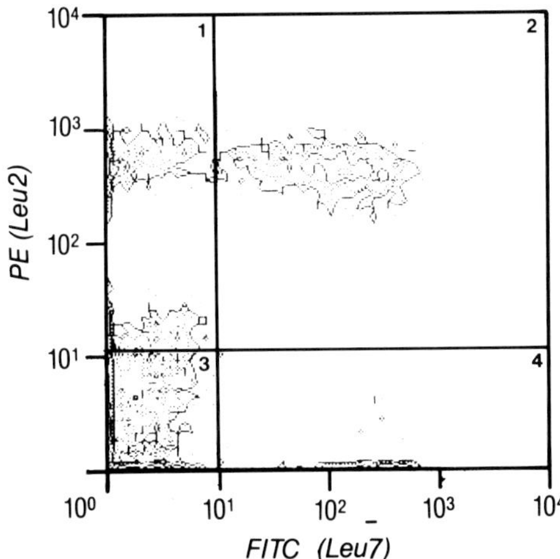

Fig. 2. Expression of Leu7 on T cells that also express CD8 (Leu2). Analysis of blood lymphocytes from a patient with an acute viral infection showing an increase in the proportion of cells labeling with both Leu2 (CD8) and Leu7. Although Leu7 is also expressed on other cells types, these cells are characterized by the presence of large amounts of CD8 antigen. Other studies have shown that these are T cells, as described in the text. Data obtained with a Becton-Dickinson FACScan (TM).

reversed by the addition of too much calcium containing tissue culture medium, and it is well to remember that all cells require divalent cations to maintain their metabolic integrity. Thus, it is advisable to use the minimum amount of anticoagulation that is necessary.

The reason for maintaining blood at room temperature is that nonspecific binding of immunoglobulin to Fc receptors on some lymphocyte subsets is greatly augmented by exposure to cold [31]. In addition, there is evidence that some lymphocyte subsets may be preferentially lost on exposure to cold, although the mechanism of the loss is not clear.

For the reasons described above, it is desirable to reserve all the available fluorescence measuring sensors for immunological information about the lymphocytes. This means that light scatter and electronic volume parameters must be relied on in the discrimination of lymphocytes from other blood elements. Whereas these parameters may be adequate for the resolution of lymphocytes from erythrocytes, platelets, and other leukocytes when they are present in comparable numbers, very large numbers of unwanted cells can make it difficult to resolve the lymphocytes [48]. In an attempt to solve these problems, two basic approaches for the preparation of blood have come into common use (Fig. 4), although both have numerous variations.

The first involves a density-gradient separation, usually centrifugation of the blood over a layer of material having a density sufficient to allow the lymphocytes to float, but the erythrocytes and some of the other leukocytes to sink [17,39]. By far the most common is a mixture of Ficoll (R) (poly-sucrose) and Hypaque (R) (sodium metrizoate) (F–H). The blood is usually diluted with culture medium before being layered on top of the F–H layer and then centrifuged. The cells that float on the F–H are harvested. This technique consistently removes the vast majority of erythrocytes, polymorphonuclear leukocytes, and dead lymphocytes. The majority of monocytes, platelets, and basophils remain with the lymphocytes. Since dead cells are a major source of nonspecific labeling with fluorescent antibodies, their removal is a major advantage. Perhaps the major disadvantage is that monocytes are retained with the lymphocytes, and monocytes are also a source of non-specific labeling. The method is also relatively time consuming.

Variations on this method include pretreatment of the blood to remove platelets by defibrination and pretreatment

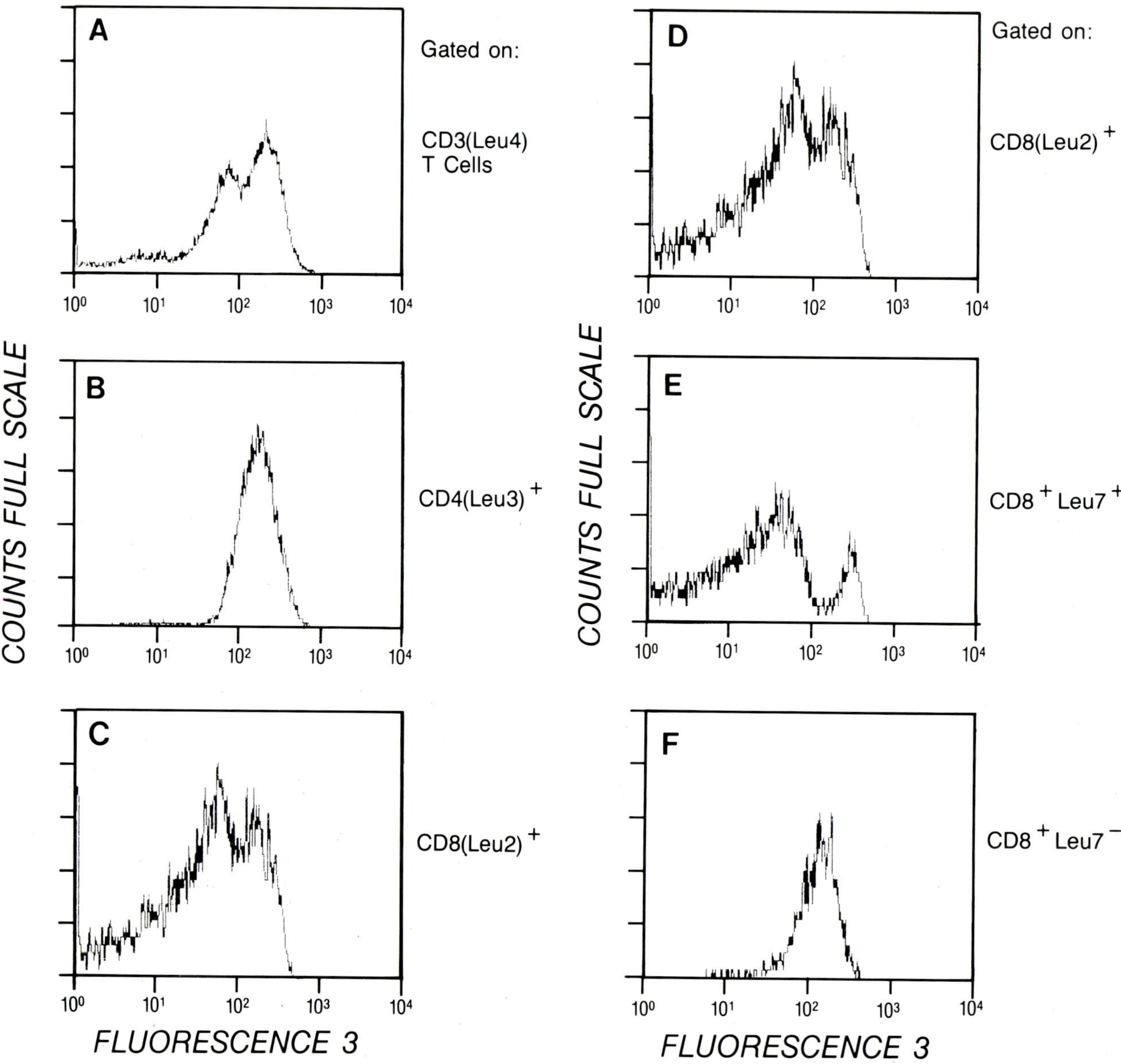

Fig. 3. Examples of a three-color immunofluorescence experiment demonstrating the presence of T cells with low levels of CD5. These data are taken from an experiment in which normal blood cells were labeled with Leu1 (CD5) conjugated with a new phycoerythrin derivative combined with Texas Red. This third color was measured as fluorescence 3 in these experiments and was combined with other reagents to illustrate the levels of CD5 expression on various lymphocyte subsets. It can be seen (**A**) that there is a distinct population of T cells ($C3^+$) that have lower levels of CD5. **B–C:** These Leu1 dim T cells are confined to the $CD8^+$ subset. **D–F:** In experiments in which all three colors are combined it can be shown that the Leu1 dim T cells have the phenotype $CD8^+$, $Leu7^+$. Interestingly, there appears to be an additional subset also expressing CD8 and Leu7 that have high levels of CD5. Data obtained with a Becton-Dickinson FACScan (TM).

by incubating the blood with very small iron particles, which are taken up by the phagocytes (monocytes) and allow them to be removed [7]. These steps add to the complexity and time involved, but they eliminate platelets and monocytes from the floating layer of cells. The result is a nearly pure lymphocyte preparation. Alternatively, the monocytes can be removed after the F–H step by adherence to plastic or other substrates.

The second major method for preparing blood lymphocytes involves lysis of the erythrocytes. This can be done either as a separate step before labeling the lymphocytes, or it can be combined with the labeling procedure to yield a much condensed method [35]. The advantages are the elimination of the erythrocytes and the simplicity of the procedure. Disadvantages include the fact that all other leukocytes remain in the sample, as do any dead or damaged cells. Unless carefully controlled, the lysis procedure can induce unwanted damage of leukocytes, or it can result in incomplete

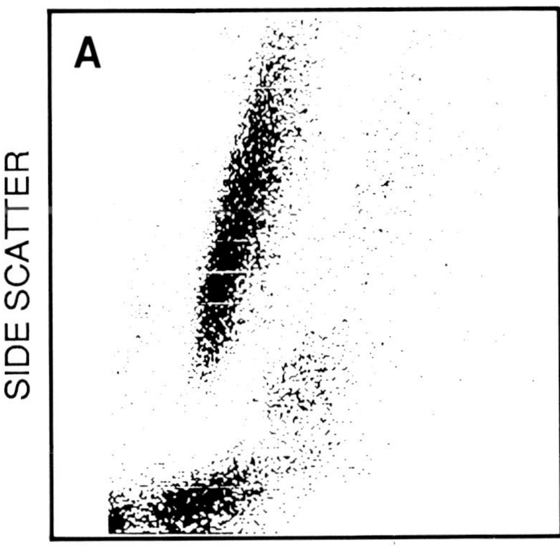

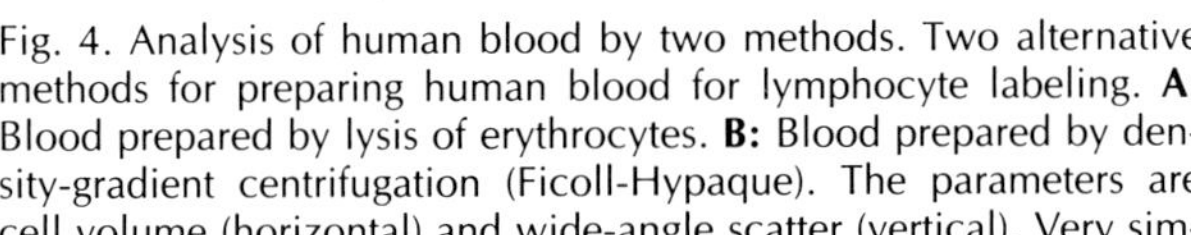

Fig. 4. Analysis of human blood by two methods. Two alternative methods for preparing human blood for lymphocyte labeling. **A:** Blood prepared by lysis of erythrocytes. **B:** Blood prepared by density-gradient centrifugation (Ficoll-Hypaque). The parameters are cell volume (horizontal) and wide-angle scatter (vertical). Very similar patterns are obtained when forward light scatter is used in place of cell volume. Note the presence of large numbers of granulocytes with high wide-angle scatter in the lysed sample and the enrichment of monocytes in the Ficoll-Hypaque sample. These data were obtained on a Becton-Dickinson FACS Analyzer.

lysis of erythrocytes. This latter problem is especially pronounced in abnormal blood samples. There are a large number of conditions, both hematological and nonhematological, in which erythrocytes have increased surface to volume ratio and thus decreased sensitivity to lysis. One particularly difficult problem is the formation of erythrocyte ghosts. These are erythrocytes that have been sufficiently damaged to lose their hemoglobin, hence appear to be lysed, but they are still intact cells and may cause spurious signals by both light scatter and electronic volume.

The preparation of lymphocytes from lymphoid tissues other than blood entails a different set of problems. In general lymphoid tissues are relatively easy to dissociate by teasing or forcing the tissue through a filter. The resulting cell suspension can be freed of debris by filtering or by allowing the debris to settle out. Such cell suspensions, made from thymus or lymph nodes, consist largely of lymphocytes. However, suspensions made from spleen do contain relatively large numbers of erythrocytes and thus benefit from treatments similar to those discussed above for blood, either density-gradient separation or lysis.

Unlike lymphocytes prepared from blood, tissue lymphocytes (with the exception of thymocytes) tend to be very heterogeneous in size. When cells are of uniform size, light scatter can be used to distinguish dead cells [32]. With heterogeneous cell suspensions, however, this is not possible. Dead cells are usually best removed by density gradient methods, but another strategy is to label them in a distinctive manner. A DNA-binding fluorescent dye such as propidium iodide will be excluded from living cells but will penetrate dead cells and brightly label the nucleus. The extremely bright fluorescence of such cells is easily distinguished from most immunofluorescent labeling simply on the basis of intensity, and they can be excluded from analysis. In addition, since the dye fluoresces in the red region, it does not interfere with fluorescein measurements [14].

SELECTION OF ANTIBODIES FOR LABELING

Since the advent of a large number of monoclonal antibodies that are specific for lymphocyte surface markers, the role for conventional polyclonal antibodies has decreased markedly. Nevertheless, there are some situations in which a conventional antibody may still be preferable, even when monoclonals with identical specificities are available. The major advantages of conventional antibodies are that, in general, they have higher avidity than monoclonals and, in situations that require balanced labeling, it may be much easier to find two conventional antibodies that have the same average properties than to find two monoclonals with the same affinity and other physical properties.

The major advantage of monoclonals over conventional antibodies with the same specificity is that labeling with monoclonals is usually much cleaner, with less background. This is because monoclonals are usually used at protein concentrations in the range of 1 to 10 μg/ml, whereas conventional antibodies are used at concentrations on the order of 10 times higher. The higher concentration of immunoglobulin protein leads to higher levels of nonspecific interactions, Fc receptor binding, and so forth.

The quality of labeling with conventional antibodies is usually markedly improved by the use of F(ab′)2 fragments. Removal of the Fc segment decreases the tendency of the antibodies to aggregate on storage and eliminates binding to cellular Fc receptors. Conversely, the preparation of F(ab′)2 fragments of monoclonal antibodies is usually not necessary because they are used at low concentrations in which aggregation and Fc receptor binding are less likely. In addition, the preparation of F(ab′)2 fragments of murine monoclonal antibodies is technically difficult.

Another consideration in the choice of a monoclonal for labeling is the isotype of the monoclonal. In general IgGs are preferable to IgMs from the point of view of stability and nonspecific labeling. Within the IgG class, the various subclasses differ in their affinity for Fc receptors and their tendency to aggregate. It is now customary, especially in the preparation of commercial monoclonal antibodies, to select desirable isotypes for production. If a useful monoclonal has undesirable physical properties it is possible to isotype switch a clone by selection of switch mutants by means of cell sorting [15]. This has already been done in the case of several research and commercial monoclonal antibodies.

As a result of our increasing awareness of the importance of their physical properties, the quality of monoclonals available for lymphocyte identification has been steadily increasing. The possibility of genetic engineering of monoclonals may make further improvement possible for specific purposes.

TECHNIQUES FOR LABELING LYMPHOID CELLS

There are two major techniques for the labeling of cells using single markers: (1) direct, in which the fluorochrome is directly attached to the antibody; and (2) indirect, in which the marker is attached to a second step reagent, which in turn binds to the antibody. Each of these techniques has advantages and disadvantages for flow cytometry. Direct labeling is technically simpler and in general results in less nonspecific labeling. However, it lacks the amplification that can be achieved by the use of the indirect method. Using indirect labeling, it is usually possible to attach several times more fluorochromes to each antigen on the cell surface. Another disadvantage of the direct method is that it is more difficult to define the controls to determine the level of nonspecific labeling. A good control for direct labeling should consists of an irrelevant antibody of the same isotype, having the same ratio of fluorochrome to protein. This is not easy to achieve in practice. Monoclonal antibodies differ in their propensity to bind nonspecifically, even within the same isotype. By contrast, indirect labeling permits better control through the use of an irrelevant first-step antibody followed by the same second-step reagent. Since most of the nonspecific labeling is usually due to inappropriate binding of the second step reagent, this method results in a good control.

A number of other issues must be considered at the time of labeling the cells, including the use of saturating amounts of all reagents, the possible effects of temperature on the expression of surface antigens, the effects of metabolic activity of the cells, and the time required to achieve equilibrium binding. For the most part, it is far preferable to label lymphocytes in the cold or in the presence of inhibitors of metabolism such as sodium azide. It is generally agreed that equilibrium binding occurs rapidly (15 to 30 min) under the conditions used for labeling cells.

Since some subsets of lymphocytes (notably NK cells) and macrophages have receptors capable of binding immunoglobulins via their Fc portion (Fc receptors), it is sometimes advisable to use the F(ab')2 fragments of antibodies to decrease this nonspecific binding. This is especially a problem when using rabbit polyclonal antibodies; it is less a problem with goat antibodies and even less with most monoclonals [2]. Although it is relatively easy to make F(ab')2 fragments of rabbit antibodies, it is more difficult with monoclonal antibodies. For this reason, most monoclonal antibodies are used in their intact form. However, the possibility of excessive nonspecific binding by the Fc receptors must always be kept in mind. It can be minimized by using reagents that are rendered free of aggregates by centrifuging just before use. One particular value of F(ab')2 reagents is that they no longer tend to aggregate, hence are much more stable on long storage.

Washing of labeled cells is especially critical in indirect staining. The unbound first-step antibody must be removed completely before the second step. If it is not, obviously some of the second-step antibody will bind to soluble first-step antibody and not be available to bind to the cells. In addition, the soluble combination of first and second step constitutes an immune complex that is very likely to bind nonspecifically to several other cell types. Washing at the end of the labeling procedure is important, especially if the cells are to be fixed. The usual fixation method employed is 1 to 2% paraformaldehyde [27]. If there is soluble fluorochrome present at the time of fixation, it may be fixed to the cell or it may diffuse into the cell after fixation and lead to markedly increased background. The choice of fixation methods is restricted because some fixatives result in increased autofluorescence of cells. This is especially true for glutaraldehyde and less so for formaldehyde.

It is becoming increasingly common to use multiple labeling protocols that permit the detection of two or more cell-surface markers simultaneously. There are several considerations that apply specifically to this situation. It is much easier to perform multiple labeling if all the labeling is direct. Whenever indirect labeling is used with multiple antibodies, one must be very careful that the second-step reagent(s) are specific, in that they can distinguish the two primary antibodies. The cleanest situation for the use of indirect labeling is to use one antibody as a direct label and the second antibody as a biotin conjugate. The second-step reagent is avidin conjugated to the second fluorochrome. The avidin will bind only to the biotinylated first-step antibody [28]. Alternatively, if the two first-step antibodies are from different species, or of different isotypes, it may be possible to use specific second-step antibodies that can distinguish them. The best examples of this are the use of mouse and rat first-step antibodies followed by antimouse specific and antirat specific immunoglobulins or the use of IgG and IgM first steps followed by isotype specific second steps. In both cases, it is necessary to be very certain that the second-step antibodies are specific; appropriate crossed controls are essential.

For immunofluorescence, the choice of fluorochromes that can provide double or triple labeling has been changing rapidly. Until recently the only practical choices were fluorescein and an analogue of rhodamine known as Texas Red [49]. This combination required the use of two lasers for maximum excitation of both dyes and for complete separation of the emissions. Fluorescein has a long tail of emission extending into the red region, and Texas Red is not as efficient a fluorochrome (lower quantum yield) as fluorescein. In addition, all the rhodamine-like fluorochromes are much more difficult to conjugate than fluorescein and have a grater tendency to cause nonspecific staining of cells. Despite these difficulties, excellent data continue to be generated using this combination.

The introduction of the phycobiliprotein family of fluorochromes has resulted in a dramatic change in our ability to perform two- and three-color immunofluorescence. The first of these, which is now in wide use, is phycoerythrin. This amazing protein, extracted from blue-green algae, can be excited at the same wavelength as fluorescein (488 nm) and emits far enough into the red (580 nm) that it can be used simultaneously with fluorescein. Most importantly, it has a

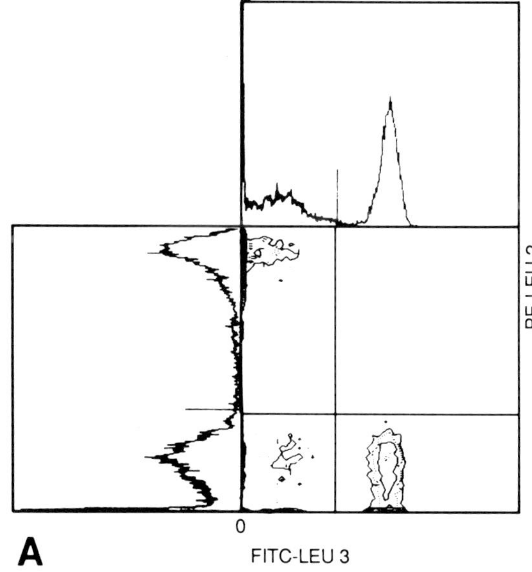

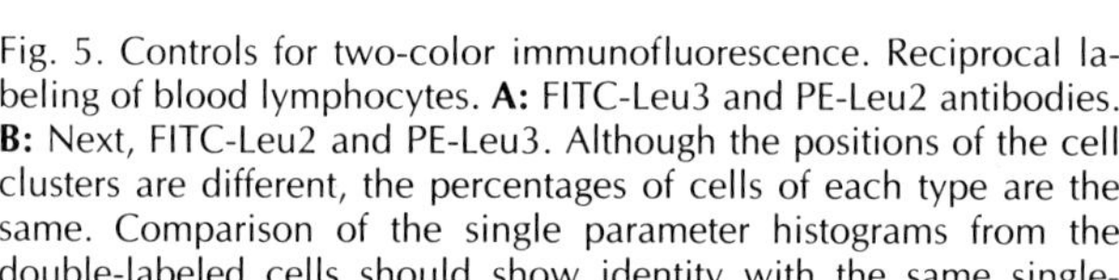

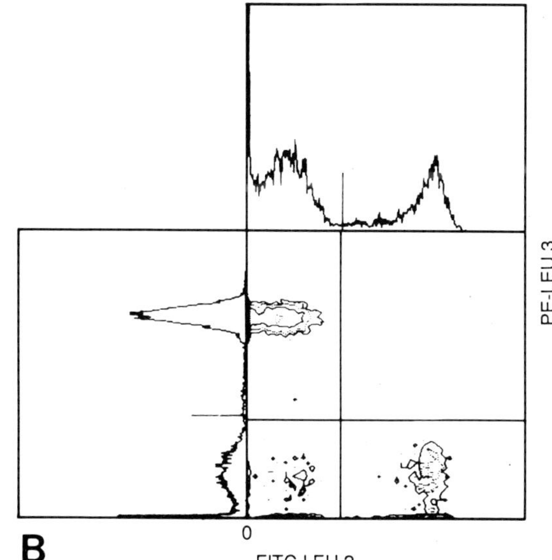

Fig. 5. Controls for two-color immunofluorescence. Reciprocal labeling of blood lymphocytes. **A:** FITC-Leu3 and PE-Leu2 antibodies. **B:** Next, FITC-Leu2 and PE-Leu3. Although the positions of the cell clusters are different, the percentages of cells of each type are the same. Comparison of the single parameter histograms from the double-labeled cells should show identity with the same single-parameter data obtained when the cells are labeled with only one antibody. Such controls rule out nonspecific interactions of the fluorochromes with the cells and interactions between the fluorochromes themselves. These data were obtained on a Becton-Dickinson FACS Analyzer (TM).

quantum yield comparable to that of fluorescein. Although not as easily conjugated to antibodies as fluorescein, it is relatively free of the problems of nonspecific labeling that have plagued the use of rhodamine analogues. When fluorescein and phycoerythrin are used together, it is necessary to correct the red signal for the spill of fluorescein emission into the red region, but this is easily performed electronically [33] and results in excellent dual color immunofluorescence with a single light source (Fig. 5). (See also Chapter 17.)

Other members of this family of fluoresent proteins are being investigated and will likely soon be in regular use, making three-color immunofluorescence practical. Recently, new fluorochromes consisting of pairs of fluorochromes chemically coupled have been introduced. These tandem dyes provide very large Stoke's shifts through the use of internal energy transfer and are rapidly coming into routine use for multicolor immunofluorescence. When performing experiments using multiple immunofluorescence labels, the choice of controls becomes both difficult and crucial. Necessary controls include the use of each fluorochrome separately. Single parameter analysis of double- or triple-labeled cells should show the same fluorescence pattern as the cells labeled with the single fluorochrome. In addition, these single-labeled cells can serve as controls for any spillover of fluoresence emission from one color channel to the other. This can only be tested and corrected by the use of single-labeled cells.

In the case of indirect labeling, it is obviously necessary to control for any nonspecific labeling by the second-step reagent in the presence of all of the primary antibodies. Finally, if possible, it is advisable to rotate the fluorochromes among the various antibodies used to label the cells because of possible nonspecific interactions between the fluorochromes themselves and the antibodies. Thus, the use of fluoroescein with antibody A and phycoerythrin with antibody B should ideally be controlled by the use of fluorescein with antibody B and phycoerythrin with antibody A (Fig. 5).

Several specialized labeling methods are of particular use in the detection of rare cells. The definition of a rare cell is somewhat vague because it depends to large extent on our ability to label the cell. If a cell type can be labeled with a very high S/N ratio, on the order of 10^3, it is certainly possible to detect cells with a frequency of 10^{-6} or less. On the other hand, if the cells are detected by immunofluorescence with a typical S/N ratio of 10, a cell with a frequency of 10^{-2} may be considered rare. It follows that accurate detection of rare cells depends critically on the labeling method.

One method that has been employed successfully for rare cell detection is the use of fluorescent beads [13,38] (see Chapter 19). These can be conjugated to antibodies and provide a very high level of fluorescent signal per molecule of antibody bound. If the antibodies and beads do not bind nonspecifically to unwanted cells, the S/N ratio provided by these beads is adequate for rare cell detection to the level 10^{-6}. The primary drawback to the use of such beads in routine immunofluorescence work is that the signal is no longer necessarily proportional to the amount of antigen on the cell. A single bead may have a great many antibody molecules bound to it and thus may bind to a large number of antigenic sites on the cells. The significance of the number of beads bound to the cell is thus difficult to determine. This method is only useful when one is interested in cells that are either positive or negative for a marker. One of the primary assumptions of flow cytometry—that the signal is proportional to the amount of ligand bound to the cell—is no longer true in this case.

A second method that has been applied to rare cell detection is the use of two-color immunofluorescence to improve the S/N ratio [45]. In this technique, the rare cells are labeled with one fluorochrome, e.g., fluorescein. Then as many as possible of the irrelevant cells are labeled with a second fluorochrome. The use of the second fluorochrome permits gating out a large proportion of irrelevant cells so that the frequency of the rare cells is essentially increased, simplifying their detection and counting. This has been applied to the problem of detecting rare leukemic lymphoid cells in bone marrow. The leukemic cells bear the CALLA antigen, which is fluorescein labeled; most of the nonlymphoid marrow cells are then labeled with a mixture of phycoerythrin-conjugated antibodies. The CALLA-positive cells can be counted as a fraction of marrow lymphoid cells, rather than as a fraction of all marrow cells.

Another technique that has been successfully applied to the problem of labeling rare cells is that of balanced labeling, or clonal excess [4]. Since B lymphocytes express either κ- or λ- light chain on their surface, but never both, it is possible to obtain fluorescence distributions for those B cells expressing κ light chains, and those expressing λ light chains. The fact that these two distributions are nearly identical for normal populations of B cells can be exploited in the detection of relatively small numbers of abnormal (monoclonal) B cells. The presence of the monoclonal B cells disturbs one of the fluorescence distributions, but not the other. This change in the balance of the labeling can be detected mathematically as a significant difference in the shape of the distribution curves. Detection of the abnormal cells in this way is more sensitive to the presence of small numbers of cells than is simply counting the fraction of cells labeled by anti-κ or anti-λ. Unfortunately, other instances of such balanced expression of two antigenic markers are not immediately apparent. Thus, this technique, although interesting because it makes use of the shape of the fluorescence distribution rather than its integral, appears to have very limited application.

INTERPRETATION OF DATA

The analysis and interpretation of flow cytometric data is obviously a very large subject (see Chapter 22). From the limited point of view of interpreting lymphocyte subset data, it can be reduced to three basic operations: (1) determining percentage positive cells, (2) determining the intensity of labeling of a population, and (3) comparing two histograms. Most lymphocyte subset data are reduced to percentage positive by integrating a histogram. This operation requires that the positive cells be clearly resolved from the negative cells. If they are not, there are a variety of strategies for estimating the percent positive. The choice of these strategies depends primarily on assumptions about the nature of the negative population.

Control samples are generally used to define the negative population. In the case of indirect immunofluorescence, the control may be an irrelevant first antibody or no first antibody followed by the second antibody. In the case of direct labeling, the control will usually consist of a directly labeled irrelevant monoclonal, preferably of the same isotype, and with the same fluorochrome-to-protein ratio. High levels of nonspecific labeling may be due to unwanted binding of the antibodies to all, or to a subset, of the cells. This is illustrated by binding to the Fc receptor on those cells (B cells and NK cells) that have a strong FC receptor. Alternatively, nonspecific labeling may be due to binding to dead cells, or debris. In general, it is far preferable to work out a labeling protocol that permits complete or nearly complete resolution of the positive population.

There is one instance in which it is simply not possible to resolve a positive population. This is illustrated by certain antigens whose expression does not define a discrete subpopulation, but rather reflects the physiological state of the cell. A good example is the expression of HLA-DR on T cells. Such data can still be interpreted in terms of percentage positive; however, the distinction between positive and negative cells will of necessity be arbitrary. Perhaps a better way to think about such data is in terms of mean fluorescence intensity for the whole population. This can be thought of as a measure of the average level of activation of all the cells.

Finally, there are some situations in which it is desirable to compare two histograms statistically. In this situation, the question is whether (or to what degree) the two histograms differ. A number of goodness of fit analyses can be applied. Perhaps the most versatile is the Kolmogorov–Smirnov test because it is nonparametric [51]. Other, such as χ^2, can also be used. One major area of application for this type of analysis is in balanced labeling or clonal excess testing [4]. When used as a statistical test, these methods suffer from one major weakness. In order to derive a P value, it is necessary to assume the number of degrees of freedom in the data. For flow cytometric data, the number of cell features measured is sometimes used for this. However, it is clear that such an assumption is incorrect and results in a much too sensitive test. To my knowledge an appropriate solution to this problem has not yet been offered.

SORTING STRATEGIES

There are two critical issues that affect the choice of methods for cell sorting: (1) the required purity of the cells, and (2) the required yield. Both are limited by the number of starting cells, the amount of time available for sorting, and the quality of the label that will permit identification of the cells of interest. The interactions between these variables are frequently complex; often, the best sorting strategy is not immediately obvious. In many situations, it is necessary to try several different strategies before deciding on the best for a given project.

When one needs large numbers of relatively rare cells, the direct approach of labeling the starting population and then simply sorting is frequently not the best. The overall yield of sorting, when one considers the losses in labeling the cells, sorting them, and then recovering them from a dilute solution after sorting, is often in the vicinity of 50%. With careful attention to detail, this can be raised, but as the cells of interest become rarer, the yield will decrease even further. For these reasons, there must be some means of preselecting the cells in order to increase the proportion of desirable cells before sorting. Unfortunately, most other methods of separating lymphoid cells, including density or sedimentation technique, immunoadsorption, and rosetting, also incur losses that approach 50%. As a general rule of thumb, if the desired cells are present at a frequency of less than 1%, and if they can be raised to a frequency of 1% or more by a single-step preselection procedure, it is probably advisable to do so. Sorting will be much more efficient at the level of 1% or higher, and losses from the single preselection step will be acceptable compared with the losses resulting from sorting cells in the less than 1% range. By contrast, if the desired cells are present at a frequency of less than 1%, and it is unlikely that a single preselection step will achieve enrichment to the 1% or higher level, it may be wiser to proceed with sorting directly. One useful strategy here is a preliminary high-speed sort (with the sorter set to ignore coincidences of cells) to achieve purities above 1%, followed by a second sort to achieve final purity. In this case, the losses involved in the first sort will be comparable to, or perhaps much less than, the losses suffered by one or more preselection steps.

Obviously, sorting strategy must be customized to fit the situation. One factor not often considered is that many cell-separation methods do not offer the controls that cell sorting does. An affinity column, for example, may take some time to set up, and may, for unknown reasons, work well sometimes and poorly at other times. If such a technique is applied as a preselection step, one may not know that the column did not work well until it is time to begin sorting, wasting considerable time and effort. The key advantage of flow cytometry sorting is the control that the operator has over the process, with instant feedback if things are not going well. This may be a very important consideration and may lead to a strategy involving sorting even when alternative separation methods are available.

One potential problem that is frequently critical to success in sorting relatively rare cells is the detection of dead cells. Even the best cell preparations usually contain a few percentage dead cells, and dead cells will usually take up fluorescent labels nonspecifically. Thus, if the desired cells are also present at the level of a few percent, one finds that dead cells make up a very large proportion of the sorted cells. In some cases, dead cells can be excluded on the basis of their apparently small size using forward-angle light scatter [43]. In practice, this is only useful if the cell population is relatively uniform in size to begin with. In the case of lymphocyte suspensions from lymphoid tissues, in which there is considerable size heterogeneity, it will not be possible to distinguish dead cells this way.

A better alternative is to use a dye such as propidium iodide, which will not enter living cells but does stain the nuclei of dead cells as already described [34]. The fluorescence of propidium iodide-stained cells, which is in the red region (above 620 nm), is much more intense and easily distinguished from any immunofluorescence signal. Thus, the dead cells can be excluded from the sorted population.

CONCLUSION

The steady increase in the proportion of scientific papers that make use of flow cytometry is testimony to the wide application of this technology in immunology and cell biology. While the study of lymphocytes has progressed most rapidly, increasing attention and new applications to the study of other cell types have been noted. It is clear that, for the forseeable future, the powerful combined technologies of flow cytometry and monoclonal antibodies will be used with increasing effectiveness not only in basic immunological research but in clinical practice as well.

REFERENCES

1. **Alberti S, Stovel R, Herzenberg LA (1984)** Preservation of cells sorted individually onto microscope slides with a fluorescence activated cell sorter. Cytometry 5:644–647.
2. **Alexander EL, Sanders SK (1977)** F(ab′)2 reagents are not required if goat rather than rabbit antibodies are used to detect human surface immunoglobulin. J Immunol 119:1084–1088.
3. **Antin JH, Emerson SG, Martin P, Gadol N, Ault KA (1986)** Leu1+ (CD5+) B cells: A major lymphoid subpopulation in human fetal spleen: Phenotypic and functional studies. J Immunol 136:505–510.
4. **Ault, KA (1979)** Detection of small numbers of monoclonal B lymphocytes in the blood of patients with lymphoma. N Engl J Med 300:1401–1405.
5. **Ault KA, Antin JH, Ginsburg D, Orkin SH, Rappeport JM, Keohan ML, Martin P, Smith BR (1985)** Phenotype of recovering lymphoid cell populations after marrow transplantation. J Exp Med 161:1483–1502.
6. **Ault KA, Springer TA (1981)** Cross reaction of a rat anti-mouse phagocyte specific monoclonal antibody (anti-Mac-1) with human monocytes and natural killer cells. J Immunol 126:359–364.
7. **Ault KA, Unanue ER (1977)** Comparison of two types of immunoglobulin bearing lymphocytes in human blood with regard to capping and motility. Clin Immunol Immunopathol 7:394–404.
8. **Basch RS, Berman JW (1982)** Thy-1 determinants are present on many murine hematopoietic cells other than T cells. Eur J Immunol 12:359–364.
9. **Beverly PCL, Linch DC, Callard RE, Worman CP, Cawley JC (1982)** Cytopenia and T cell proliferation. J Clin Immunol 2:135–140.
10. **Boumsell L, Coppin H, Pham D, Raynal B, Lemerle J, Dausett J, Bernard A (1980)** An antigen shared by a human T cell subset and B cell chronic lymphocytic leukemic cells. J Exp Med 152:229–234.
11. **Chang LJA, Gamble CL, Izaguirre CA, Minden M, Mak TW, McCulloch EA (1982)** Detection of genes coding for human differentiation markers by their transient expression after DNA transfer. Proc Natl Acad Sci USA 79:146–150.
12. **Chess L, Schlossman SF (1977)** Functional analysis of distinct human T cell subsets bearing unique differentiation antigens. Contemp Topic Immunobiol 7:363–379.
13. **Cupp JE, Leary JF, Cernichiari E, Wood JCS, Doherty RA (1984)** Rare event analysis methods for detection of fetal red blood cells in maternal blood. Cytometry 5:138–144.
14. **Dangl JL, Herzenberg LA (1982)** Selection of hybridomas and hybridoma variants by fluorescence activated cell sorter. J Immunol Methods 52:1–14.
15. **Dangl JL, Parks DR, Oi VT, Herzenberg L (1982)** Rapid isolation of cloned isotype switch variants using fluorescence activated cell sorting. Cytometry 2:395–401.
16. **Gadol N, Ault KA (1986)** Phenotypic and functional characterization of human Leu1 (CD5) B cells. Immunol Rev 93:23–34.
17. **Gadol N, Nakamura G, Saunders A (1985)** A new method for separating mononuclear cells from whole blood. Diagn Immunol 3:145–154.
18. **Gobbi M, Caligaris-Cappio F, Janossy G (1983)** Normal equivalent of cells of B cell malignancies: Analysis with monoclonal antibodies. Br J Haematol 54:393–403.
19. **Hayakawa K, Hardy RR, Honda M, Herzenberg LA, Steinberg AD, Herzenberg LA (1984)** Ly-1 B cells: Functionally distinct lymphocytes that secrete IgM autoantibodies. Proc Natl Acad Sci USA 81:2494–2498.
20. **Hayakawa K, Hardy RR, Parks DR, Herzenberg LA (1983)** The "Ly-1 B'cell subpopulation in normal, immunodefective, and autoimmune mice. J Exp Med 157:202–218.
21. **Hensleigh PA, Waters VB, Herzenberg LA (1983)** Human T lymphocyte differentiation antigens: Effects of

blood sample storage on LEU antibody binding. Cytometry 3:453–455.

22. **Iverson GM, Bianchi DW, Cann HM, Herzenberg LA** (1981) Detection and isolation of fetal cells from maternal blood using the fluorescence activated cell sorter. J Prenatal Diagn 1:61–73.
23. **Kavathas P, Herzenberg LA** (1982) Stable transformation of mouse L cells for human membrane T cell differentiation antigens, HLA and β2-microglobulin: Selection by fluorescence activated cell sorting. Proc Natl Acad Sci USA 80:524–528.
24. **Krensky AM, Reiss CS, Mier JW, Strominger JL, Burakoff SJ** (1982) Long term human cytolytic T cell lines allospecific for HLA-DR6 antigen are OKT4$^+$. Proc Natl Acad Sci USA 79:2365–2369.
25. **Lanier LL, Federspiel NA, Ruitenberg JJ, Phillips JH, Allison JP, Littman D, Weiss A** (1987) The T cell antigen receptor complex expressed on normal peripheral blood CD4$^-$, CD8$^-$, T lymphocytes: A CD3 associated disulfide linked gamma chain heterodimer. J Exp Med 165:1076–1094.
26. **Lanier LL, Loken MR** (1984) Human lymphocyte subpopulations identified by using three color immunofluorescence and flow cytometry analysis. J Immunol 132:151–156.
27. **Lanier LL, Warner NL** (1981) Paraformaldehyde fixation of hematopoietic cells for quantitative flow cytometry (FACS) analysis. J Immunol Methods 47:25–30.
28. **Ledbetter JA, Rouse RV, Spedding Micklem H, Herzenberg LA** (1980) T cell subsets defined by expression of Ly-1,2,3 and Thy-1 antigens: Two parameter immunofluorescence and cytotoxicity analysis with monoclonal antibodies modifies current views. J Exp Med 152:280–295.
29. **Levy R, Warnke R, Dorfman RF, Haimovich J** (1977) The monoclonality of human B cell lymphomas. J Exp Med 145:1014–1028.
30. **Lewis DE, Puck JM, Babacock GF, Rich RR** (1985) Disproportionate expansion of a minor T cell subset in patients with lymphadenopathy syndrome and acquired immunodeficiency syndrome. J Infect Dis 151: 555–559.
31. **Lobo PI, Westervelt FB, Horowitz DA** (1975) Identification of two populations of immunoglobulin bearing lymphocytes in man. J Immunol 114:116–119.
32. **Loken MR, Herzenberg LA** (1975) Analysis of cell populations with a fluorescence activated cell sorter. Ann NY Acad Sci 254:163–171.
33. **Loken MR, Parks DR, Herzenberg LA** (1977) Two color immunofluorescence using a fluorescence activated cell sorter (FACS). J Histochem Cytochem 25: 899–907.
34. **Maher P, O'Toole CM, Wreghitt TG, Spiegelhalter DJ, English AH** (1985) Cytomegalovirus infection in cardiac transplant recipients associated with a chronic T cell subset ratio inversion with expansion of a Leu7$^+$ Ts-c$^+$ subset. Clin Exp Immunol 62:515–524.
35. **Mishell BB, Shiigi SM (eds)** (1980) "Selected Methods in Cellular Immunology." San Francisco: WH Freeman, pp. 22–23.
36. **Nairn RC, Rolland JM** (1980) Fluorescent probes to detect lymphocyte activation. (Review.) Clin Exp Immunol 39:1–13.
37. **Okumura K, Hayakawa K, Tada T** (1982) Cell to cell interaction controlled by immunoglobulin genes. Role of Thy-1$^-$, Lyt-1$^+$ (B') cells in allotype restricted antibody production. J Exp Med 156:443–453.
38. **Parks DR, Bryan VM, Oi VT, Herzenberg LA** (1979) Antigen specific identification and cloning of hybridomas with a fluorescence activated cell sorter. Proc Natl Acad Sci USA 76:1962–1966.
39. **Peper RJ, Tina WZ, Mickelson MM** (1968) Purification of lymphocytes and platelets by gradient centrifugation. J Lab Clin Med 72:842–848.
40. **Phillips JH, Lanier LL** (1985) A model for the differentiation of human natural killer cells: Studies on the in vitro activation of Leu-11$^+$ granular lymphocytes with a natural killer sensitive tumor cell, K-562. J Exp Med 161:1464–1482.
41. **Phillips JH, Lanier LL** (1986) Lectin dependent and anti-CD3 induced cytotoxicity are preferentially mediated by peripheral blood cytotoxic T lymphocytes expressing Leu7 antigen. J Immunol 136:1579–1585.
42. **Reinherz EL, Acuto O, Fabbi M, Bensussan A, Milanese C, Royer HD, Meuer SC, Schlossman SF** (1984) Clonotypic surface structure on human T lymphocytes: Functional and biochemical analysis of the antigen receptor complex. Immunol Rev 81:95–129.
43. **Reininger L, Ollier P, Poncet P, Kaushik A, Jaton J-C** (1987) Novel V genes encode virtually identical variable regions of six murine monoclonal anti-bromelein treated red blood cell autoantibodies. J Immunol 138: 316–323.
44. **Reynolds CW, Foon KA** (1984) T_γ lymphoproliferative disease and related disorders in humans and experimental animals: A review of the clinical, cellular, and functional characteristics. Blood 64:1146–1158.
45. **Ryan D, Mitchell S, Hennessy L, Bauer K, Horan P, Cohen H** (1984) Improved detection of rare CALLA positive cells in peripheral blood using multiparameter flow cytometry. J Immunol Methods 75:115–128.
46. **Shapiro HM** (1983) Multistation multiparameter flow cytometry: A critical review and rationale. Cytometry 3:227–243.
47. **Springer TA** (1980) Cell surface differentiation in the mouse. Characterization of "jumping" and "lineage" antigens using xenogeneic rat monoclonal antibodies. In Kennett RH, McKearn TJ, Bechtol KB (eds), "Monoclonal Antibodies." New York: Plenum Press, pp 185–217.
48. **Thompson JM, Gralow JR, Levy R, Hay JB** (1985) The optimal application of forward and ninety degree light scatter in flow cytometry for the gating of mononuclear cells. Cytometry 6:401–406.
49. **Titus JA, Haugland R, Sharrow SV, Segal DM** (1982) Texas red, a hydrophilic, red emitting fluorophore for use with fluorescein in dual parameter flow microfluorometric and fluorescence microscopic studies. J Immunol Methods 50:193–204.
50. **Uchiyama T, Broder S, Waldmann TA** (1981) A monoclonal antibody (anti-Tac) reactive with activated and functionally mature human T cells. J Immunol 126:1393–1397.
51. **Young IT** (1977) Proof without prejudice: Use of the Kolmogorov-Smirnov test for the analysis of histograms from flow systems and other sources. J Histochem Cytochem 25:935–941.

35

Flow Cytometry of Leukemia

Michael Andreeff
Memorial Sloan-Kettering Cancer Center and Cornell University Medical College, New York, New York 10021

INTRODUCTION

The advent of flow cytometric techniques in the last quarter century has created numerous opportunities for investigations of the hematopoietic system, which can be considered as an in vivo suspension culture. Flow cytometry became the preferred tool for the investigation of complex hematopoietic cell populations and has now achieved an undisputed role in the study of leukemia. Investigations of clonality, cell kinetics, and immunologic phenotyping resulted in the detection of a hitherto unknown heterogeneity, some of great theoretical and clinical relevance. Recently, the development of new probes, in particular monoclonal antibodies (mAB) to cell surface antigens, has improved our understanding of cell phenotypes while the molecular biology and molecular genetics techniques of gene sequencing and cloning, Western (protein), Northern (RNA), and Southern (DNA) blotting have made great contributions to the etiology, pathogenesis, and pathophysiology of leukemia. But these latter techniques yield averages for large populations and do not analyze single cells, with the possible exception of the DNA polymerase chain reaction (PCR) [116]. Adaptation of the molecular probes to flow cytometry permits rapid multiparameter molecular analysis of single cells. This chapter surveys recent flow cytometric studies of human leukemia and summarizes the applications of flow cytometric techniques in the diagnosis, treatment, and monitoring of leukemia.

NORMAL HEMATOPOIESIS

The electronic impedance technique for determination of cell number and volume, introduced by Coulter in 1956 [53], is still the most widely used application of flow cytometry in hematology (see Chapter 4, this volume). Improvement of this technique came with the development of hydrodynamic focusing, i.e., aligning cells in the center of the flow channel by differential pressures between sample and sheath fluid (see Chapter 3, this volume). In modern multiparameter flow cytometers, cell number and cell size may be determined by measurements of laser light scattered at small angles (0.5 to 2°), which is porportional to cell volume [150], while scatter at 90° is related to granularity (see Chapter 5, this volume). These two measurements alone, on unstained samples, permit discrimination of debris/erythrocytes/platelets from lymphocytes, monocytes, and granulocytes [180]. In another early approach, supravital staining with the metachromatic dye acridine orange (AO) allowed Melamed et al. in 1972 to automate blood cell differential counts by flow cytometry [144].

The complex binding of AO to nucleic acids was investigated by Darzynkiewicz, who with Traganos introduced the now standard techniques for simultaneous quantitation of cellular DNA and RNA content [200] (see Chapter 15, this volume). This 60-sec-procedure has been invaluable for the determination of cell cycle phases and DNA ploidy in hematopoietic tissues, since it permits the discrimination of cells in G_0, G_{1A}, G_{1B}, S, and G_2M (see Chapter 24, this volume). In non-proliferating peripheral blood cells, systematic differences were found in the RNA content of B, T, and non-B, non-T cells: B and T (suppressor) cells had the lowest RNA content, with T (helper) and null cells having increased RNA content [6]. Enriched peripheral blood granulocyte progenitor cells were characterized by a highly increased RNA content [70]. The AO two-step technique, which employs Triton-X to make cells permeable for AO, also results in damage to the cytoplasmic membrane. In particular, granulocytes are affected. They may be stripped of their cytoplasm and elongated by the shearing force of the fluidic system. The resulting elongation of the nucleus can be detected by an increase in DNA flourescence duration (pulse width) (Fig. 1). Cells characterized by nuclear elongation and loss of cytoplasmic RNA have a characteristic increase in integrated green (DNA) fluorescence, which can easily be misinterpreted as hyperdiploidy. Gating on DNA fluorescence pulse width corrects for this. Pulse width is also very useful for discrimination of cell aggregates. Figure 2 demonstrates that gating on pulse width is crucial in determining the true number of cells in G_2M and in excluding cell aggregates that mimic polyploidy (arrow).

Staining the small amounts of RNA in reticulocytes with Pyronin Y was used by Tanke to automate the counting of these immature erythrocytes [193]; acridine orange and thiazole orange have also been used for this purpose.

A further modification of the AO technique permits the measurement of double-stranded (native) versus single-

Flow Cytometry and Sorting, Second Edition, pages 697–724

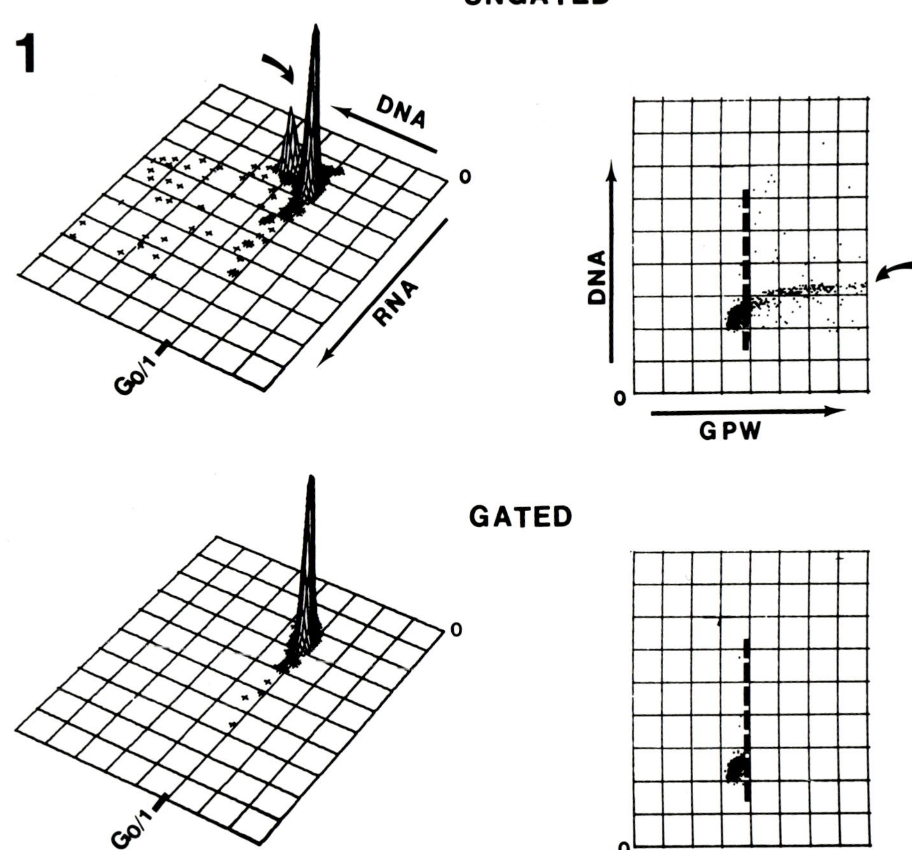

Fig. 1. Peripheral blood cells stained with acridine orange after Ficoll–Hypaque separation. DNA (green fluorescence), RNA (red fluorescence), and green pulse width (GPW) were measured simultaneously. Upper panels show the correlated histograms. Cells that give rise to high GPW values (arrow) have increased (pseudohyperdiploid) DNA content and very low RNA content. After gating on GPW, this cell population disappears. The cells are granulocytes and are regularly seen in peripheral blood after Ficoll separation in chronic myelogenous leukemia. The polymorphic nuclei of granulocytes are sheared of cytoplasm in flow, unfolded and elongated to give the increased GPW.

stranded (denatured) nucleic acids in the intact cell or nucleus (see Chapter 16, this volume). As shown by Darzynkiewicz [58], there are changes in denaturation of DNA (by heat or acid) as cells traverse the different compartments of the cell cycle so that with this technique all proliferating and quiescent cell compartments can be distinguished. In particular, G_0 cell chromatin is more sensitive to denaturation than that of G_1 cells, and mitotic cells are more sensitive than G_2 cells. This technique permits the best assessment of proliferative response in the hematopoietic system and is surpassed in sensitivity only by the measurement of intracellular Ca^{2+} concentration, which changes within minutes after cellular activation (see Chapter 32, this volume).

The development of specific monoclonal antibodies (mAB) to cell surface antigens was probably the single most important factor in the development of FCM as a hematological–immunological technique. Table 1 lists the names of mAB and their cluster designation (CD). The initial antibodies to lymphocyte antigens have now been complemented by those directed against antigens of the erythroid, myeloid, and megakaryocytic lineages [76,143]. These mAB have become indispensible tools in leukemia classification. Examples are given below for the specific leukemia subtypes.

While flow cytometry techniques have been successful in identifying a variety of blood cells, the heterogeneity of bone marrow cells has precluded an automated approach to the assessment of bone marrow. Differentiation antigens are expressed over rather prolonged periods of cell development and are not sufficiently restricted to even match the accuracy of morphological evaluation.

However, first steps toward a better understanding of cell differentiation have been made, notably for the erythroid [131] and myeloid [91] cell lineages. Loken et al. [131] measured the expression of the transferrin receptor, of glycophorin A and of the HLe-1 antigen and were able to correlate proliferation and differentiation for the erythroid cell lineage. In the myeloid series, the earliest antigen identified so far is CD-34 (MY10), followed by CD-33 (MY9), CD-13 (MY7), MY8, CD-11 (Mo1, OKM-1), and CD-14 (MY4, Mo2). However, such antigens as MY9 and MY7 are expressed on CFU-GM, myeloblasts, promyelocytes, myelocytes, metamyelocytes, monoblasts, and monocytes and are therefore not helpful in identifying discrete steps of cellular differentiation.

DNA and DNA/RNA measurements are hampered by the same problem of lineage heterogeneity. It can be demon-

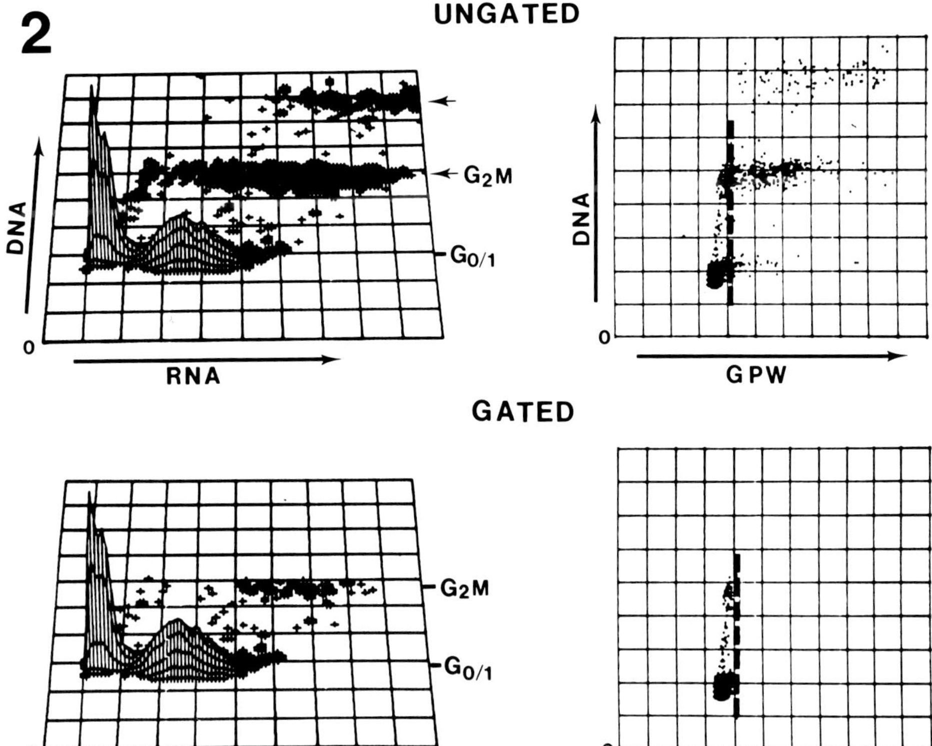

Fig. 2. Recognition of doublets by DNA pulse width gating. The duration (width) of the nuclear fluorescence pulse as a cell (or coincident cells) passes through a narrow beam of excitation light is proportional to the nuclear diameter in the axis of flow (see Chapter 6, this volume). Bone marrow from an acute myeloblastic leukemia was stained with acridine orange. DNA (green fluorescence), RNA (red fluorescence), and green pulse width or pulse duration (GPW) were measured simultaneously. Upper panels show the ungated measurements, suggesting a very high G_2M population, which could also be interpreted as a polyploid (tetraploid) G_1 population. Lower panels show the results of gating: reduction of G_2M and complete disappearance of cells with higher ploidy, thereby revealed as an artifact due to aggregation or coincidence of two cells.

strated, however, that the cellular RNA content decreases when cells undergo differentiation. The RNA index, i.e., the cell cycle corrected RNA content of G_1 cells [10], was 27.2 for normal myeloblasts and promyelocytes, 22.4 for myelocytes, and 21.3 for metamyelocytes using normal bone marrow fractionated by a 1g velocity sedimentation gradient. The number of cells in S phase also decreases with differentiation (Fig. 3). The same was found in experimental cell systems when cells were treated with inducers of differentiation, including DMSO and TPA [47,60].

Although FCM measurements of DNA can provide information regarding the number of cells in different cell cycle phases, the important cell cycle transit times cannot easily be obtained from flow cytometric measurements and are calculated from autoradiographic data using labeled thymidine. Table 2 lists the DNA synthesis times (T_S) for the different hematopoietic cells. A review of hematopoietic cell kinetics was recently published [3]. The introduction of the BUdR technique into FCM has so far not provided complete kinetic analysis of normal hematopoietic cells, but the combination of mAB to more restricted antigens and kinetic double-labeling techniques such as Dolbeare's BUdR/IUdR method [124] may finally result in a more complete picture of the kinetics of developing hematopoietic cells. Pallavicini's elegant investigations of DNA content, DNA synthesis (BUdR), and agglutinin binding in hematopoietic cell populations point in this direction [155] (see Chapter 23, this volume).

Numerous attempts have been made to identify the putative human hematopoietic stem cell [48] and flow cytometry has become an important means to that end. Atzpodien et al. sorted human progenitor cells based only on light scattering properties and obtained three- to fourfold enrichment of CFU_{GM} and BFU_E [26]. Lu et al. used mAB to MY10 and HLA-DR to enrich and phenotype human marrow progenitor cells; these workers achieved a cloning efficiency (CE) of up to 47% in fractions of cells expressing high MY10 and low DR antigens [137]. By comparison, CE was 0.01% in $MY10^-DR^-$ cell fractions. In the mouse system, a group from Stanford succeeded in isolating a pluripotent hematopoietic stem cell by depletion of unwanted cells with magnetic immunobeads and mAB and subsequent cell sorting [187]. The 1,000-fold enriched CFU were mostly in $G_{0/1}$ (>97%), and only 30 of these cells were needed to rescue one-half of a group of lethally irradiated mice. Once the most immature, truly pluripotent stem cells can be isolated, they will permit a much better understanding of hematopoietic development and of the mechanisms underlying leukemogenesis. They will also be used in the reconstitution of the human hematolymphoid system after myeloablative therapy and in gene insertion therapy.

ACUTE MYELOBLASTIC LEUKEMIA

The acute leukemias are usually defined as an accumulation of immature cells "frozen" at an early stage of differ-

TABLE 1. Cluster Designation Nomenclature of Human Leukocyte Differentiation Antigens According to the Third International Workshop*†

Cluster designation group	Molecular weight (kD)	Reactivity	Examples of antibodies	Antigen designation
CD1a	49	Thy, LC	(NA1/34, OKT6)	T6
CD1b	47	Thy, LC	(NU-T2, 4A76)	T6
CD1c	43	Thy, LC, some B	(M241)	T6
CD2	50	T	(OKT11, NU-T1)	T11, E_R, LFA-2
CD3	19–29	T (mature)	(UCHT1, OKT3)	T3
CD4	55	Th_{HI}, M	(anti-Leu-3a, OKT4)	T4
CD5	67	T, B-CLL	(anti-Leu-1, Tu71)	T1
CD6	120	T (mature)	(Tu33, OKT17)	T12
CD7	41	T, T-ALL	(Tu14, 3A1)	
CD8	32–33	$T_{s/c}$	(anti-Leu-2a, OKT8)	T8
CD9	24	nT-nB, B, M, G, Th	(BA-2, J 2, FMC8)	
CD10	100	nT-nB	(BA-3, J 5, NU-N1)	CALLA
CD11a	180, 95	T, B, G, M	(CC5-07, 25.1.2)	LFA-1 α-chain
CD11b	160, 95	G, M, DRC, LGL	(OKM1)	MAC1 α-chain ($C3bi_R$)
CD11c	150, 95	M, HCL	(KB90, anti-Leu-M5)	p150, 95 α-chain
CDw12		M, G, Th		Resting
CD13	150	G, M	(MY7, MCS2)	
CD14	55	(G), M, IDR, DRC	(UCHM1, MO2, MY4)	
CD15	180, 110, 68, 50	G, (M), R-S	(3C4, anti-Leu-M1)	X-hapten
CD16	50–60	G, NK, M	(anti-Leu-11)	Fc_R
CDw17		G, M, Th	(T5A7, G-035)	Lactoceramide
CD18	95	B, T, G, M	(M232, 68-5A5)	β-chain of CD11a,b,c
CD19	95	B	(anti-B4, HD37)	
CD20	37, 32	B, M	(anti-B1, NU-B2, G28-2)	
CD21	140	B, DRC	(HB5, B2)	$C3d_R$
CD22	140, 130	B	(To15, HD39, HD6)	
CD23	45	B, DRC	(MHM6, Tul, Blast-2)	B_{act}, IgE–Fc_R
CD24	42	B, G, M	(BA-1, HB8)	
CD25	55	T, (B), M, NK	(anti-Tac, Tu69)	$IL2_R$
CDw26	200, 120	T_{act}	(TS145, 4EL1C7)	Elusive
CD27	120–55	T (mature), (B), Pla	(S152, VIT14)	
CD28	44	T	(J ohn9-3, Kolt-2)	T-subset (T8 cytotoxic)
CDw29	135	CD4-subpop, B, M	(K20, 4B4)	
CD30	90–110	T_{act}, B_{act}, R-S	(Ki-l, Ber-H2)	
CD31	130–140	G, M, Th, (T)	(SG134, TM3)	
CDw32	40	G, M, Th, B	(2E1, CIKM5, IV3)	40 kD Fc_R
CD33	67	Myeloid progenitors	(L4F3, L1B2, MY9)	
CD34	115	(G), (M)	(MY10, B1-3C5)	Immature G
CD35	220	G, M, B, RBC, DRC, glomeruli	(E11, To5, J 3B11)	$C3b_R$
CD36	85	M, Th	(5F1, ClMegl, 4C7)	gpIV
CD37	40–45	B, (T), G, M	(BL14, HD28, G28-1)	
CD38	45	Thy, B_{act}, Pla, G, M	(OKT10, HB-7)	T10
CD39	80	B, (T_{act}), G, M	(G28-10, G28-8)	B_{act}
CDw40	50	B, M, IDR	(G28-5, S2C6)	B_{act}
CDw41	130, 115	Th	(J15 etc.)	gpIIb/IIIa
CDw42	150	Th	(HPL14, AN51)	gpIb
CD43	95	NL (T, G, brain)	(84-3C1, G10-2)	
CDw44	65–85	NL	(1-173, F10-44-2)	
CD45	220, 205, 190, 180	NL (Leukocytes)	(2D1, 72-5D3)	LCA
CD45R	220	NL ((T), (B), (G))	(RFB5, 4KB5, RFB2)	LCA-restricted

*Reproduced from McMichael et al. [143], with permission of the publisher.

†B lymphocyte: B_{act} = B cell activation antigen; B-CLL = B-type lymphatic leukemia: CALLA = common acute lymphatic leukemia antigen; DRC = dendritic reticulum cell; E_R = sheep erythrocyte receptor; FC_R = Fc-receptor (for IgG); G = granulocyte and precursors; HCL = hairy cell leukemia; IDR = interdigitating reticulum cell; $IL2_R$ = interleukin-2-receptor; LC = Langerhans cell; LCA = leukocyte common antigen; LFA = lymphocyte function-associated antigen; LGL = large granular lymphocyte, M = monocyte/macrophage; NK = natural killer cell; NL = nonlineage specific; nT-nB = non-T, non-B cell; Pla = plasma cell; RBC = red blood cell; R-S = Reed-Sternberg cell; T = T lymphocyte; T_{act} = T cell activation antigen; T-ALL = T-type acute lymphatic leukemia; Th = thrombocyte; $T_{H/I}$ = T helper/inducer cell; Thy = thymocyte; $T_{s/c}$ = T suppressor/cytotoxic cell.

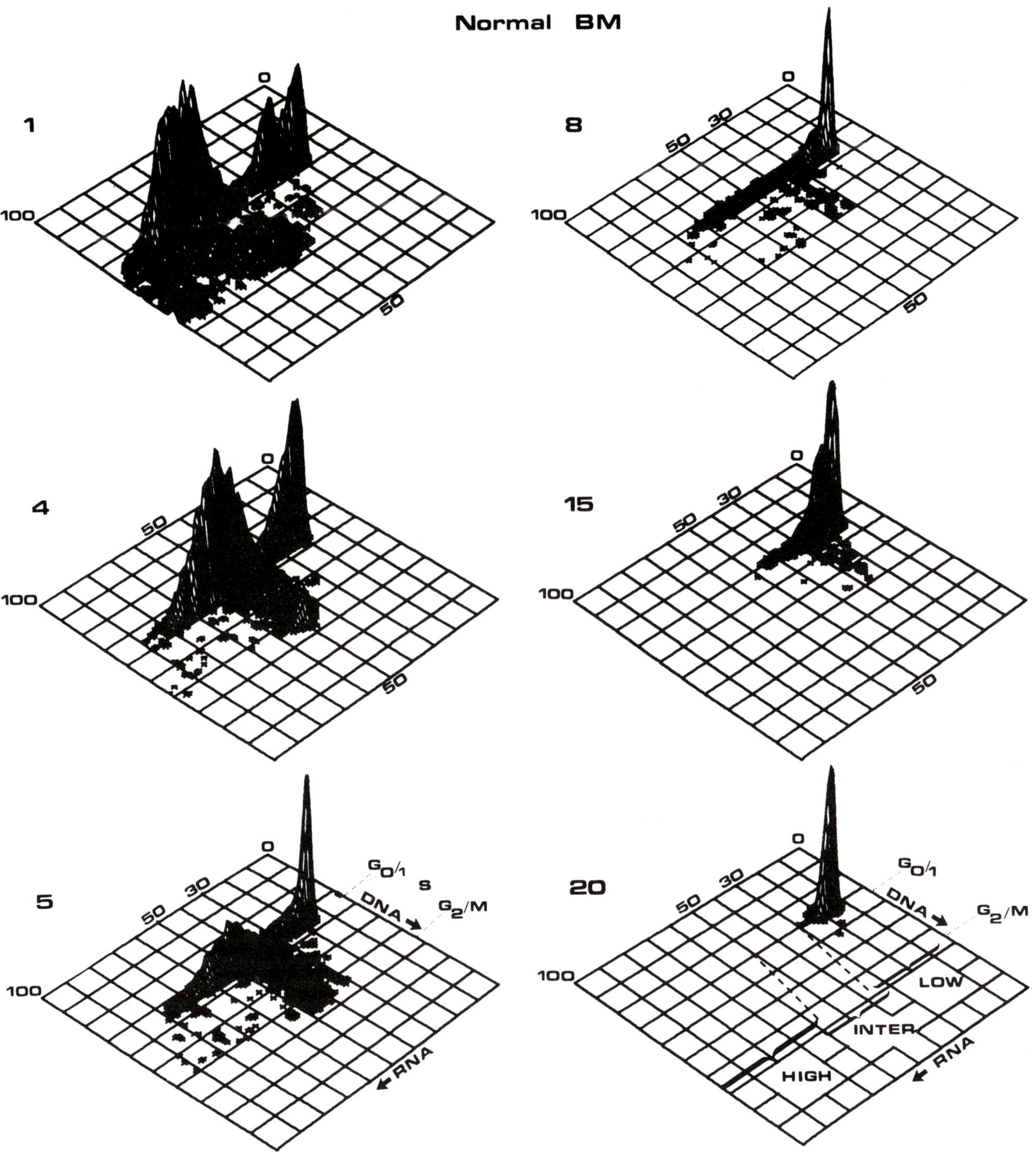

Fig. 3. Velocity sedimentation of normal bone marrow cells. Bone marrow cells were separated by 1XG velocity sedimentation and fractions 1, 4, 5, 8, 15, and 20 were subjected to DNA/RNA flow cytometry using acridine orange. Fraction 1 contains mostly myeloblasts and promyelocytes (75%) with high RNA content (RNA index 27.2) and high proliferation. Fraction 4 contains mostly myelocytes (75%) with high RNA content (RNA index 22.4) and high proliferation. Fraction 5 contains mostly metamyelocytes (70%) with high RNA content (RNA index 21.3) and proliferating cells. Fractions 8 to 20 contain progressively smaller cells with decreasing RNA content and decreasing proliferation.

entiation. The cells arise in the marrow and cause clinical consequences by way of quantitative or functional defects in hematopoiesis. The diagnosis of acute leukemia is made by morphological and histochemical criteria as defined in the French–American–British (FAB) classification [35], which has recently been revised [37]. It encompasses acute myeloid

TABLE 2. DNA Synthesis Times (Ts) of Normal and Leukemic Cell Populations

	T_S (hours)
Normal hematopoiesis	
Myeloblasts	13.2 ± 1.8
Promyelocytes	13.4 ± 0.9
Myelocytes	14.1 ± 1.1
Proerythroblasts	9.4 ± 1.9
Basophilic erythroblasts	11.0 ± 2.3
Polychromatic erythroblasts	17.0 ± 3.4
Lymphoblasts	13.5
Acute myeloblastic leukemia myeloblasts	11–29
Acute lymphoblastic leukemia lymphoblasts	18.3 [85]
	42–47 [87]

leukemia (M1 = myeloblastic leukemia without maturation, M2 = myeloblastic leukemia with maturation, M3 = promyelocytic leukemia, M4 = myelomonocytic leukemia, M5A = monocytic leukemia without maturation, M5B = monocytic leukemia with maturation), erythroblastic leukemia (M6), megakaryoblastic leukemia (M7) [38] and three subtypes of acute lymphocytic leukemia (L1, L2, L3). The reproducibility of this taxonomic system is not very good, but it provides at least morphology-based criteria that are used universally. The clinically important discrimination between AML and ALL can be done with a high degree of accuracy by well-trained hematologists, although FAB M1 and L2 cases can easily be confused. Treatment protocols and results differ greatly for AML and ALL. ALL in children has a rate of complete remissions (CR) in excess of 90%; 70% of children are long-term survivors. In adults, the CR rate is in the 70 to 80% range; 30% to 35% of patients survive 5 years or longer [14]. The response rates are lower in patients with AML, with only 10% of adults surviving 5 years [50]. In AML, the FAB classification is of no prognostic importance, with the possible exception of M3, which may be associated with longer remission duration (RD) [50]. Age, in all studies, is the main prognostic variable. Cytogenetics has recently emerged as another major prognostic factor. FCM has contributed greatly to the unraveling of the heterogeneity of AML. In particular, nucleic acid content measurements have identified DNA stemlines (aneuploidy and multiploidy) and provided cell kinetic information.

Cell Cycle Kinetics

Autoradiographic studies have shown that AML is characterized by a high rate of production of myeloblasts [24]. It is important to understand that an increase in proliferation, a decrease in differentiation or decrease in cell loss can all result in the accumulation of immature blast cells characteristic of leukemia. Since it is very difficult to estimate cell loss in patients, the relative contributions of the different mechanisms is not known. The average T_S (DNA synthesis time) is estimated between 11 and 29 hours [65], but the range reported in the literature varies from less than 10 hours to more than 40 hours [51]. Cell cycle times determined autoradiographically vary from 49 to 95 hours but, by eliminating quiescent cells, a probably more accurate generation time of 24 hours was estimated.

The size of the quiescent compartment is difficult to determine, since cells are constantly entering and leaving it. Continuous labeling with radiolabeled DNA precursors in patients demonstrated the presence of 1 to 12% unlabeled (i.e., quiescent) cells after 8 to 21 days of labeling. Another way of determining the growth fraction is by estimation of percent cells having DNA polymerase and primer-template DNA (TTP index) [127]. A recalculation of the published data shows a mean TTP index of 42.6% for AML and of 28.2% for ALL [3]. The pretreatment bone marrow labeling index for myelocyte leukemia is approximately comparable to that of normal myeloblasts. Flow cytometric determinations of S-phase (DNA synthesizing) cells vary: 5.1 ± 3.8% [10], 6.9 ± 3.6% [20], 11.3% [112], 5.06 ± 3.79 [4], and 9.9% [174]. Variability in results can be attributed to (1) sample error, i.e., quality of the marrow aspirate; (2) DNA dyes used; (3) coefficient of variation of measurement; and (4) DNA histogram analysis program used. A major source of error is the composition of the sample studied: the more peripheral blood cells present in the marrow sample, the lower the measured percentage of cells in S phase. This is due to dilution with nonproliferating normal and low-proliferating leukemic blood cells. Consequently, S phase is consistently higher in marrow biopsies as compared with (diluted) marrow aspirates and peripheral blood samples in the same patient [68,99,101]. Several methods were developed by Hiddemann et al. [102] and Holdrinet et al. [111] to subtract the peripheral blood component as estimated by the number of erythrocytes (or hemoglobin) presumed to be present in the blood but not in the true marrow compartment.

Since the development of antibodies to bromodeoxyuridine (BUdR) by Gratzner [84,85] and others [61,67], the identification of S-phase cells has been greatly improved. It is now not only possible to enumerate the cells synthesizing DNA, but also to calculate T_S [181,210]. Raza et al. [166,168] developed a double-labeling technique for leukemic cells by combining the BUdR technique with [^{3}H]thymidine labeling and autoradiography. Short-term intravenous infusion of BUdR into patients with leukemia was used for in vivo labeling. The number of S cells in biopsies was 20%, in aspirates 9%. Double-labeling experiments gave T_S values from 4 to 49 hours (medium 17 hours), and T_C (cell cycle time) values from 16 to 292 hours (median 76 hours). The surprisingly short T_S values estimated in some samples can be explained by a change in DNA synthesis rate ex vivo; when the synthesis rate decreases between the application of the two labels (sample manipulation) the calculated T_S will be shortened. Nevertheless, the extreme heterogeneity in cell cycle parameters appears to be of clinical significance, as patients with very short and with very long T_C had a poor response to therapy [154,166]. Preisler et al. [162] reported that low pretreatment S phase was correlated with poor response in AML patients treated with high-dose ara-C. The prognostic importance of the number of cells in S phase in AML was uncertain in other studies and may have to be reinvestigated using the BUdR technique. A very elegant modification was recently developed by Crissman et al. [54] and Kubbies et al. [123], who determine BUdR incorporation not by antibody binding, but by the quenching of Hoechst 33258 fluorescence. This approach avoids the denaturation of DNA and may permit simultaneous identification of incorporated BUdR, and measurement of RNA and fluorescein isothiocyanate antibody labeling. With this technique, it would be possible to measure proliferation of antibody-specified cell lineages in the hematopoietic system. BUdR was also shown to suppress the green (DNA) fluorescence of AO-stained stimulated lymphocytes [57].

No significant cell kinetic differences were found among the different FAB subgroups [4], except in one series in which promyelocytic leukemia was found to have the lowest and erythroleukemia the highest S phase [139]. Unclassified leukemia (M0) also had a high S phase.

The introduction of AO flow cytometry, first for RNA measurements [16] and subsequently for simultaneous measurements of DNA and RNA [200], added a new dimension to FCM of leukemia [9,11]. The method permitted discrimination of AML from ALL [10], based on a calculation of RNA index (RNA index (RI) = mean RNA content of $G_{0/1}$ cells of sample × 10, divided by the median RNA content of normal control lymphocytes) (Fig. 4). The RNA index of AML (20.6 ± 4.0) is significantly higher than that of ALL (11.2 ± 2.0). These results were confirmed by Maddox et al. [139] and by Barlogie et al. [31]. No clear correlation was found between RNA index and FAB subgroups [4,208].

The additional measurement of RNA content also proved useful in detecting subtle changes in the $G_{0/1}$ compartment, which could not be appreciated by DNA FCM. Figure 5 demonstrates that G_0, G_1, S, and G_2M as well as DNA aneuploidy can be discriminated. The importance of RNA lies also in the identification of populations of quiescent G_0 cells with low RNA content, which are conceptually of great interest since they probably harbor those leukemic stem cells that survive the initial chemotherapy and lead to relapse. Studies of the crucial transition between G_0 and the different G_1 compartments (G_{1A}, G_{1B}) are also greatly aided by their discrimination with AO staining. In cell kinetic terms, high RNA content is associated with active proliferation. Of interest is also the correlation between the percentage of bone marrow blasts and the number of cells actively in cycle as determined by AO flow cytometry (G_1 + S + G_2, excluding cells in G_0 with low RNA content) (Fig. 6). It is not surprising that acute leukemia with high pretreatment RNA index (i.e., high proliferative fraction) was associated with high response rate and long remission duration [4]. None of 16 patients with low RNA content (RI $G_1<16$) at diagnosis survived 12 months, but 35 of 81 patients with RNA index $G_1>16$ did. This difference was also apparent in the response to induction therapy, and RNA index remained an independent variable in Cox multivariate analysis for survival ($p = 0.026$). The prognostic importance of RNA content was only apparent when specific indices for G_0 (RI G_0) and G_1 (RI G_1) were used. The prognostic importance of the RNA index was confirmed in Maddox's analysis [139] in which AML patients with RNA index $G_{0/1} > 20$ had longer CR durations ($p = 0.04$) than did patients with an RNA index of <20. Another series from M.D. Anderson Hospital, which did not discriminate RNA index subgroups, failed to confirm the prognostic significance of RNA content [208].

Ffrench et al. [73] studied protein content in acute myeloid leukemia by staining with FITC and propidium iodide. Cells with low protein content, assumed to be quiescent, were correlated with FAB groups in that the peak value of protein content histograms was significantly lower in myeloid (M1–M3) than in monocytic (M4, M5) leukemias. The fraction of cells with low protein content (LPC) was higher in the CR group than in nonresponders ($p<0.01$). In a second report, the LPC fraction was lower for patients older than age 50 who did not achieve CR and higher for CR patients over age 50 and for younger nonresponders [74].

A modification of DNA FCM allows for the cellular quantitation of double-stranded RNA (ds-RNA). DNA is treated with DNase and the remaining ds-RNA is stained with propidium iodide (PI) [115]. High numbers of cells with ds-RNA were found in AML and the persistence of ds-RNA excess in the marrow of patients with AML in remission predicted a short remission duration (7 versus 22 months; $p = 0.05$).

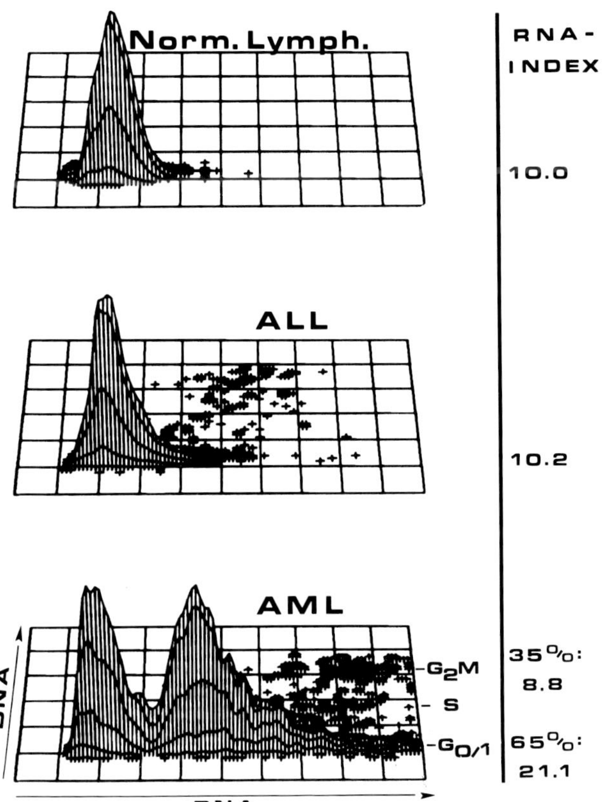

Fig. 4. DNA/RNA histogram of normal control lymphocytes, acute lymphoblastic leukemia and acute myeloblastic leukemia cells. Staining with acridine orange. Normal control lymphocytes have RNA index = 10.0. ALL cells have RNA index = 10.2 with some cells in S and G_2M. AML has two populations of cells, nonleukemic cells (35%) with low RNA index and myeloblasts with high RNA index (65%, RI = 21.1).

When RNA is digested with RNase, and DNA is partially denatured in situ, the metachromatic properties of AO permit selective staining of single-stranded (ss) and double-stranded (ds) DNA [58]. All phases of the cell cycle can be distinguished with this method (see Chapter 24, this volume) and significant differences were found between AML and ALL [7,8,12]: the chromatin of ALL G_1 cells is more sensitive to standardized denaturation than that of AML, leading to higher α_t (ratio of denatured to total DNA) values for ALL than for AML. Besides their diagnostic utility these results may suggest basic kinetic differences between AML and ALL; the high α_t associated with ALL may indicate a more quiescent cell cycle state than that found in AML (low α_t). With BUdR now available for in vivo studies of leukemia kinetics, questions related to growth fraction and recruitment in leukemia will be reassessed during the next few years. As noted, BUdR not only quenches Hoechst 33258 but also suppresses the AO green fluorescence of stimulated lymphocytes: thus, cycling cells in the presence of BUdR will have diminished AO green fluorescence and cells entering G_0

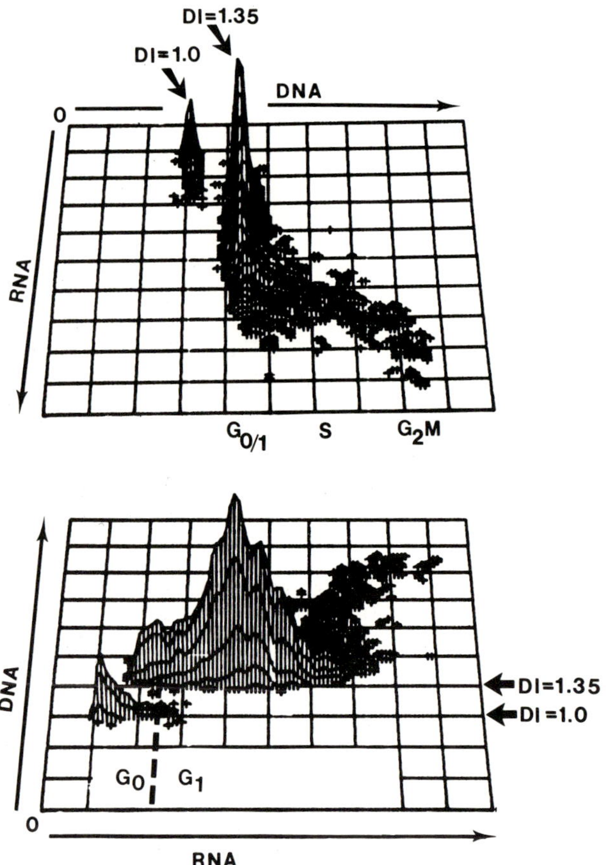

Fig. 5. DNA/RNA flow cytometry of DNA–hyperdiploid leukemic cells. Staining with acridine orange. DNA index of diploid cells = 1.0, of hyperdiploid cells = 1.35. Hyperdiploid cells have high RNA content and high proliferation. Lower panel (same data, rotated) shows discrimination between cells in G_0 with low RNA content and cells in G_1 with high RNA content.

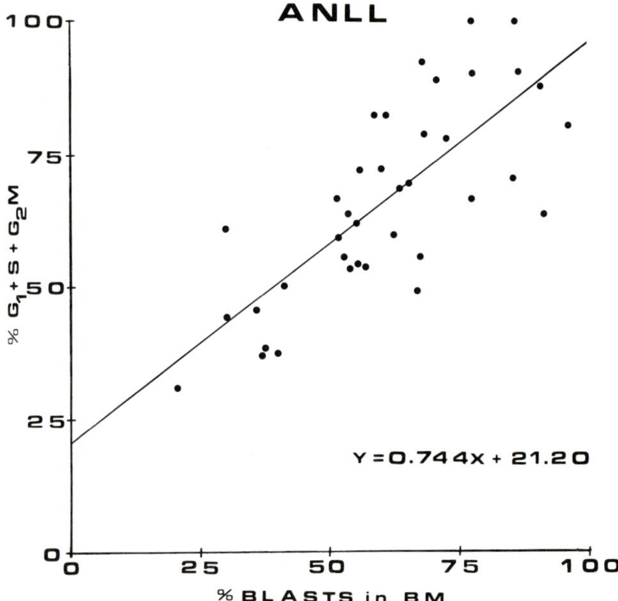

Fig. 6. Correlation between percentage of blast cells in bone marrows of untreated acute myeloblastic leukemia (X axis) as compared with proliferating cells in cell cycle: G_1 + S + G_2 (Y axis).

after having progressed through DNA synthesis are distinguishable from cells that remained quiescent during the period of BUdR exposure [57].

Clonal Abnormalities

Detection of DNA aneuploidy by FCM permits identification of malignant cells without being dependent on dividing cells, as required for cytogenetic analysis. It is important to acknowledge the limitations of FCM of DNA stemlines. First, the sensitivity is limited by the coefficient of variation (CV) of the measurement. Under optimum conditions, some instruments can provide a CV value of <1% and distinguish male and female cells. Even under those ideal conditions, balanced translocations with no gain or loss in DNA cannot be detected, by definition. Many leukemias have only small abnormalities in DNA content; monosomies of chromosomes 5 or 7, for example, will usually remain undiscovered by one-parameter DNA measurements (see Table 3). This abnormality is more easily detected by in situ hybridization using chromosome-specific DNA sequences as markers [160]. Nevertheless, among 44 patients with AML exhibiting balanced translocations and inversion of chromosome 16, 36% had hyperdiploid DNA stemlines and 20% of patients with diploid karyotype displayed an abnormal DNA index [208]. Second, FCM measures dye binding to DNA in chromatin and all DNA may not be stained [59]; accessibility of DNA is affected by chromatin proteins and varies in different cell types and during differentiation. Appropriate controls are therefore required from the same (normal) tissue, individual, or species as outlined in the Convention of Nomenclature for DNA Cytometry [104]. The degree of DNA aneuploidy in AML is usually small (DNA index (DI) ≤1.15) compared with ALL (DI = 1.20 ± 0.04); and DNA aneuploidy is less frequent in AML (27%) [208], 21% [4], and adult ALL (26%) [208], 20% [4], than in pediatric ALL (39%) [106,135,175,192]. Once an aneuploid DI is determined, it can serve as an excellent leukemia cell marker for double staining with mAB to differentiation antigens, to specific proteins (including oncogene proteins and other markers of interest). It is this ability of DNA FCM to positively identify the leukemic cell that makes it invaluable in leukemia research, whereas most other methods are unable to positively identify individual leukemic cells in large cell populations.

Multiple DNA stemlines are not infrequent and permit the investigation of polyclonality and lineage fidelity (multilineage leukemias). Redner et al. [171] observed a patient with four DNA stemlines, all of which were leukemic, as evidenced by their expression of the common leukocyte antigen (CALLA) and pre-B cell markers. If a specific marker is present, only a small number of DNA aneuploid cells are sufficient to identify them positively as leukemic. Of particular value is the detection of small numbers of aneuploid cells after induction therapy for leukemia as a sign of incomplete response, or later, relapse [20]. In central nervous system (CNS) leukemia, small numbers of hyperdiploid cells are sufficient for diagnosis, since there is no background of proliferating normal cells as in bone marrow samples. CNS relapse was detected by FCM up to 6 months before clinical

TABLE 3. Chromosomal DNA as a Percentage of Total DNA*

Chromosome	DNA	
No.	Male (%)	Female (%)
1	4.14	4.07
2	4.05	3.99
3	3.37	3.31
4	3.20	3.15
5	3.07	3.03
6	2.87	2.83
7	2.67	2.63
8	2.44	2.40
9	2.30	2.26
10	2.27	2.24
11	2.26	2.23
12	2.23	2.20
13	1.82	1.79
14	1.72	1.69
15	1.67	1.65
16	1.56	1.53
17	1.42	1.40
18	1.35	1.33
19	1.05	1.04
20	1.11	1.09
21	0.79	0.78
22	0.87	0.86
X	2.55	2.51
Y	0.94	—
	51.72	50.01

*Calculated from Mayall BH, Carrano AV, Moore DH, Ashworth LD, Bennett DE, Mendelsohn ML (1984) The DNA-based human karyotype. Cytometry 5:376–385.

diagnosis [170,173] and the doubling time of leukemic lymphoblasts in the CNS could be calculated as 20 days in one case. No prognostic importance of DNA aneuploidy is evident for AML, in marked contrast to the significant contribution of cytogenetics.

Myeloid Differentiation Markers

Markers of differentiation were initially defined by reactions of histochemical stains on slides. They still are useful in identifying myeloid cells by myeloperoxidase (POX) or Sudan Black (SB), monocytic cells by esterases (naphthyl AS-D acetate esterase inhibited by fluoride), and erythroleukemia cells (M6) by periodic acid-Schiff (PAS) positivity. One commercial FCM instrument measures light scatter and peroxidase content to provide WBC differentials of good quality [71]. In the rare and underdiagnosed acute megakaryoblastic leukemia [38] (FAB: M7), platelet peroxidase (PPO) detected by electron microscopy and mAB to platelet glycoprotein IIb/IIIa are the best diagnostic tests [117]. Morphological criteria alone are insufficient and would probably result in a misdiagnosis of M7 as undifferentiated leukemia (M0).

The development of mABs reactive with lineage- and differentiation-associated cell surface antigens has provided specific reagents for the discrimination and subdivision of leukemic phenotypes and was first successfully applied to ALL. The understanding of myeloid-associated antigens has developed more slowly, but a large number of myeloid mAB are now available. Three International Workshops on Human Leukocyte Differentiation Antigens were conducted to assess the interrelationships between mAB and to determine their reactivities.

As a result, groups of mAB that recognize the same or similar antigens have been identified and designated cluster differentiation groups (CD) (see Table 1). Some groups of antibodies appear to identify the same or similar antigens by immunofluorescence; however, when not confirmed, or when different molecules appear to be recognized, the CD is designated a provisional working group (CDW). The expression of several myeloid antigens corresponds to stages of differentiation within the myeloid lineage [90]. Cell sorting can be used to determine antigen expression by multipotent stem cells. For instance, CFU-GEMM express MY10 and MY9; CFU-GM (d14) are positive for MY10, MY9, MY7, p67, and CD15; and CFU-GM (d7) react with MY9, MY7, p67, CD15, VIM2, and CDW17. BFU-E have antigens for MY10, MY9, and p67; CFU-E are positive for p67; and FCU-meg are positive for the MY10, p67, and VIM 2 antigens.

In the diagnosis of leukemia, mAB are most useful in discriminating ALL and AML, although none detects leukemia-specific determinants. Majdic et al. [141] studied 208 leukemias and found that 91% of AML reacted with mAB VIM2, while all nonmyeloid leukemias were negative. Chan et al. [46] found that 86% of AML and no ALL were positive with MY9. Mirro and Behm [147] summarize the results from several large series as shown in Table 4.

The prognostic importance of mAB in AML is not entirely clear. Griffin et al. [89] reported that MY4 and MY7 predicted for low complete remission (CR) rates ($MY4^+$: 53%, $MY4^-$: 69%, $p = 0.03$; $MY7^+$: 55%, $MY7^-$: 73%, $p = 0.01$). The paired combination of antigen expression enhanced the prognostic significance: CR rate in $MY4^-$ $MY7^-$ was 87% and in all other cases was 54% ($p = 0.001$). Expression of HLA-DR, MY8, and Mo1 was associated with decreased remission duration ($p<0.05$), and expression of MY8 was associated with decreased survival ($p = 0.03$). Interestingly and possibly limiting the prognostic value of mAB phenotyping are observed differences between the phenotype of the total leukemic population and that of the clonogenic cells [126]. It will be a chemotherapy-resistant subpopulation of these cells that repopulates the marrow and leads to relapse. Several ways to investigate drug resistance are discussed later (see "Drug Effects and Drug Resistance in Leukemic Cells"), but so far it is not possible to identify drug-resistant clonogenic cells by FCM. These results were confirmed by Lowenberg and Bauman [136].

The classic definition of acute leukemia as a block in differentiation deserves amendment: recent studies using the molecular restriction fragment length polymorphism (RFLP) method have unequivocally documented clonal abnormalities in the granulocytes of patients with AML [72], suggesting terminal differentiation of a small number of leukemic cells. Induction of differentiation in vitro of DNA aneuploid myeloid leukemic cells resulted in macrophage-like cells with persistently aneuploid DNA content [47].

The same can be concluded from studies using the technique of premature chromosome condensation [109] and in situ hybridization using chromosome-specific DNA probes [86]. In the near future, the clonal origin of cells characterized by specific mAB and sorted in a cell sorter, or identified by their nuclear morphology (granulocytes), will be ascer-

TABLE 4. FAB Category*

Antibody	M1 + M2 (%)	M3 (%)	M4 + M5 (%)	Overall (%)
MY9	77	76	75	76
MY7	71	73	63	67
MY8	33	32	70	52
My-1	32	33	54	46
Mol	38	40	84	46
MY4/Mo2	17	2	54	35

*From Mirro and Behm [147].

tained by subsequent in situ hybridization. This will undoubtedly result in more precise phenotyping of leukemia, since the leukemic origin of specific cells will be proved beyond doubt and morphological ambiguities.

Differentiation of myeloid cells is also accompanied by changes in chromatin structure that result in differences in the accessibility of fluorochromes to DNA [59], in differences in the sensitivity of chromatin to denaturation [60] and in a decrease in cellular RNA content [20], all amenable to flow cytometric measurement.

ACUTE LYMPHOBLASTIC LEUKEMIA

The acute lymphoblastic leukemias (ALL) are also defined by morphological criteria as an accumulation of immature blast cells in the bone marrow. The FAB classification discriminates three different subgroups: L1 is characterized by small cells with homogeneous nuclear chromatin, regular nuclear shape, invisible or small nucleoli, and scanty cytoplasm; L2 is characterized by larger, heterogeneous cells with more variable nuclear chromatin, irregular nuclear shape with common indentations, one or more nucleoli, and often moderately abundant cytoplasm; L3 consists of large and homogeneous cells with finely stippled chromatin, regular nuclear shape, one or more prominent nucleoli, and abundant cytoplasm with deep basophilia and frequent vacuolization. Cytochemical stains are essentially negative, except for PAS, which can also be positive in acute nonlymphocytic leukemia (ANLL). In children, the presence of L2 or L3 ALL, or a significant number of L2 cells in otherwise L1 disease, has been described as conferring a poor prognosis. In adult ALL, morphology has no prognostic significance, except for L3, which is associated with poor prognosis. White blood cell (WBC) count and immunophenotype as well as the time to achieve remission are now universally accepted prognostic factors in both children and adults [14].

Cell Cycle Kinetics

Cell kinetics of ALL has been studied extensively. The reported mean DNA synthesis time is 18.3 hours and no systematic differences were seen between T-ALL and common ALL [66]. The DNA synthesis time was significantly longer in ALL than in normal lymphoblasts from various tissues (thymus 7.9 hours, spleen 10.5 hours, lymph nodes 12.7 hours, and bone marrow 13.5 hours) [66]. ALL, therefore, appears to be the only type of leukemia in which a prolonged DNA synthesis time has been documented in comparison to normal cells. DNA synthesis time and labeling index (LI) of hyperdiploid ALL were increased over diploid ALL, consistent with equal DNA synthesis rates in both groups. S phase in 65 patients with diploid DNA stemline was 4.7%; and 9.5% in 44 cases with aneuploid stemline, respectively [192]. Because the speed of DNA synthesis is constant for both groups, T_S is prolonged in hyperdiploid ALL, which has more DNA to be synthesized before cells can enter the next mitotic division. Peripheral blood lymphoblasts have a lower LI or fraction of cells in S-phase as compared to bone marrow blasts from the same patient [192,201]. Using the percent-labeled mitosis method, Wagner and Hirt [205] found a T_C of 52 to 55 hours, T_S of 42 to 47 hours, T_{G2} of 6 hours, and T_M of 1 hour. This estimate of T_{G2} leaves very little time for G_1. Clarkson et al. [51] found intermitotic times at 2 to 12 days. The grain count halving time of interphase cells was twice that of mitotic cells, indicating that many daughter cells entered a quiescent state after their parents had divided. The TTP index for ALL bone marrow was 28.3% and 5.4% for peripheral blood [127]. This result is in concordance with studies of chromatin structure by AO, indicating a more quiescent state for ALL as compared with AML [3], and the observation that RNA content is significantly lower in ALL [10,208] gives further support to this hypothesis. RNA index of G_0/G_1 cells in ALL was 11.2 ± 2.0, only slightly higher than that of normal control lymphocytes (RI = 10.0), and no difference was seen between L1 or L2 subtypes [192]. A higher S phase was found for L2 lymphoblasts (12.4%) as compared to L1 (5.8%) ($p = 0.01$) [192]. As in AML, better cell kinetic data can be obtained from marrow biopsy material, since it avoids contamination with low-proliferating peripheral blood cells. In pediatric ALL, bone marrow aspirate S phase was 5.5 ± 0.3%, as compared with 8.0 ± 0.5% in biopsies [169]. The RNA index $G_{0/1}$ was 11.6 ± 0.4 in bone marrow aspirates and 10.6 ± 0.2 in biopsies. Based on RNA measurements, most $G_{0/1}$ cells are quiescent (G_0). A definitive assessment of the growth fraction in ALL will probably be possible when studies of continuous BUdR infusions in patients are combined with second lables of cell proliferation, including cell cycle specific proteins. Figure 7 emphasizes the point that new techniques may be useful in reevaluating old concepts. BUdR incorporation in vitro into leukemic blood blasts in lymphoblastic leukemia was much higher than incorporation into bone marrow blasts (mean fluorescence 134 and 47, respectively), despite the much smaller number of S-phase cells in the blood, indicative of a higher DNA synthesis rate (J. Haimi and M. Andreeff, unpublished data).

Most studies have failed to demonstrate a significant effect of pretreatment S-phase fraction on response in ALL. In one study, patients with marrow LI over 6% or blood LI over 4% were found to have significantly shorter remission durations than did patients with lower LI [69]. In pediatric and adult ALL at Memorial Hospital, no significance of pretreatment S phase was apparent [4,169]. Pretreatment RNA content of cells in G_0/G_1, however, appears to be of prognostic importance with regard to achievement of remission, achievement of early remission, and remission duration. Extensive multivariate analysis in a larger series of adult ALL showed that a low G_0 RNA index was the fourth most important unfavorable prognostic characteristic ($p = 0.011$), after L3 or AUL morphology, WBC count $>10{,}000/mm^3$, and weight loss, for achieving complete remission.

In analysis of remission duration, a high RNA index (RI>14) was the third most important factor ($p = 0.017$). In pediatric lymphoblastic leukemia, as in adult ALL, the RNA index $G_{0/1}$ was significantly lower (11.66 ± 0.44) than in myeloblastic leukemia (14.24 ± 0.51) [169]. Early response to treatment, the single most important factor for remission duration, was influenced by four main variables identified in stepwise logistic regression analysis: WBC ($p = 0.0221$), age ($p = 0.0158$), platelet count ($p = 0.0006$), and RNA index G_0 ($p = 0.04$). The higher the RNA content, the more likely was early response to therapy. The lower the RNA content, the more quiescent the cells and the slower they will be killed by chemotherapy. Consequently, recruitment into the cell cycle prior to chemotherapy should make cells more susceptible to some of the drugs used in combination chemotherapy. In adult ALL, once remission is achieved, low pretreatment RNA index (<12.7) was associated with longer remission duration ($p = 0.017$) [4]. Recent cell kinetic analysis has also shed some light on the kinetic and prognostic importance of high WBC in pediatric ALL [95]: WBC was found to be inversely correlated with the $G_{0/1}$ RNA index but not with the S phase of marrow cells, indicating a more quiescent state of leukemic marrow cells in patients with high WBC at diagnosis ($r = 0.51$, $p = 0.02$). For these reasons, RNA index appears to be a more sensitive indicator of cell kinetic activity than standard S-phase measurements and will be useful in evaluating the effects of cytokines on leukemic cell populations.

A new indicator of kinetic activity appears to be the nuclear protein detected by Gerdes' antibody Ki67 [83]. Ki67 is not detectable in nonproliferating peripheral blood G_0 lymphocytes but is expressed and can be quantitated by flow cytometry in G_1, S, and G_2M cells (Fig. 8). A systematic evaluation of Ki67 in leukemia has not yet been published, but expression of Ki67 in non-Hodgkin's lymphoma was of prognostic importance in one study (see Chapter 36, this volume). The proliferating cell nuclear antigen (PCNA)/cyclin has likewise not yet been systematically investigated in leukemia. Recent findings suggest that PCNA is an auxiliary protein of DNA polymerase delta. Kurki et al. [125] developed a flow cytometric method to identify cells expressing PCNA and showed a pattern similar to that of BUdR, i.e., staining of cells in S phase. The potential of PCNA for cell cycle analysis may therefore be limited to the identification of cells undergoing DNA synthesis.

Clonal Abnormalities

Approximately 40% of pediatric ALL [106,135,175,192] and 30% of adult ALL [4,208] have DNA aneuploid stemlines detectable by standard DNA or DNA/RNA flow cytometry with CV values of 2% to 3%. The frequency of DNA aneuploidy is related to FAB morphology; in pediatric ALL, 34% of L1 and 71% of L2 ALL were aneuploid [192]. It is likely that the frequency of detectable DNA aneuploidies will increase with more precise measurements, because many karyotypic abnormalities are undetectable with low or "standard" resolution DNA measurements. As demonstrated by Look and co-workers [133,135], DNA ploidy can be directly correlated with differentiation antigen expression, e.g., with the common ALL antigen (CALLA), which was found to be expressed on diploid, hypotetraploid, and tetraploid cells in the same patient. Redner et al. [171] reported a case with four DNA stemlines and differences in the expression of antigens for each stemline.

In addition to being a marker for malignancy, DNA ploidy may be a prognostic factor in ALL. Look et al. [133,135] reported in children treated at St. Jude's Hospital that a DNA index of ≥1.15 (≥53 chromosomes) was associated with better prognosis than DI <1.16. The markedly hyperdiploid group generally had B-cell-precursor phenotype with expression of CALLA and HLA-DR. Combining two prognostic variables, children with hyperdiploid leukemia, DI ≥ 1.16, and leukocyte count ≤25 × 10^9/L were found to have the lowest probability of relapse. The near-diploid group included many chromosomal translocations, e.g., t(9;22), t(4:22), t(8;14), which are known to be unfavorable, based on the analysis of the Third International Workshop on Chromosomes in Leukemia [198] and results from the Pediatric Oncology Group [55]. These results are not uniformly accepted. Our series of Memorial Hospital patients did not show prognostic importance of DI, emphasizing that prognostic models frequently are valid only in the context of the therapeutic regimen from which they were derived.

Improving therapeutic modalities will successively eliminate prognostic factors until all prognostic factors become insignificant when all patients are cured. Whether DNA aneuploidy may be an important prognostic factor in the natural history of the disease is an hypothesis beyond verification. Supporting evidence can be found in the German BFM trials of pediatric ALL, which have the best therapeutic results reported to date: an initial study (ALL protocol 79/81) showed a tendency for longer remissions for patients with DNA aneuploidy ($p = 0.053$) [106], but the follow-up study (ALL protocol 81/83) did not confirm significant differences between DNA diploid and DNA aneuploid patients [103].

A group of particular importance are the markedly hypodiploid (DI < 0.8) or near-haploid ALL. This rare entity consists of only 19 cases reported so far (17 detected by cytogenetics and 2 by flow cytometry) but was recently enlarged by our study of 23 such cases, of which 19 patients had ALL or lymphoblastic lymphoma. Many cases have a secondary DNA stemline, with twice the DNA content of the primary stemline. The median survival from the diagnosis of hypodiploid ALL was only 6.0 ± 3.5 months; survival from initial diagnosis was 16.1 ± 2.9 months (some patients were identified as hypodiploid at relapse) [19]. In one such case high expression of the multidrug resistance associated P-glycoprotein (gp180) was found [172] and the bone marrow cells were deeply quiescent and responded only minimally to chemotherapy.

Small structural and numerical cytogenetic abnormalities that are not evident by conventional DNA flow cytometry may be detectable by fluorescence in situ hybridization (FISH) using chromosome-specific DNA probes [160]. In cases with known abnormal DNA content, this technique adds information on whether specific types of chromosomal abnormalities are present. The method was adapted for flow cytometry by Gray et al. [86] and has been successfully used in the detection of minimal numbers of blast cells in sex-mismatched bone marrow transplantation and also for identifying the translocation t(4;11) in leukemia cells [159]. Once the CML-specific Philadelphia translocation t(9;22) is detectable, FISH could become a valuable routine method in many clinical laboratories.

Lymphoid Differentiation Markers

ALL cells are "frozen" at early stages of differentiation, but most cases can be identified as pre-B or T-cell ALL, with very few expressing a true null cell phenotype. The most useful marker is the common ALL antigen CALLA, detectable by several antibodies. Both Smets et al. [183] and Look et al. [135] reported a close association between DNA hyperdiploidy and CALLA expression that remains to be fully understood.

Foon and Todd [76] recently reviewed the phenotyping of ALL with panels of monoclonal antibodies directed against

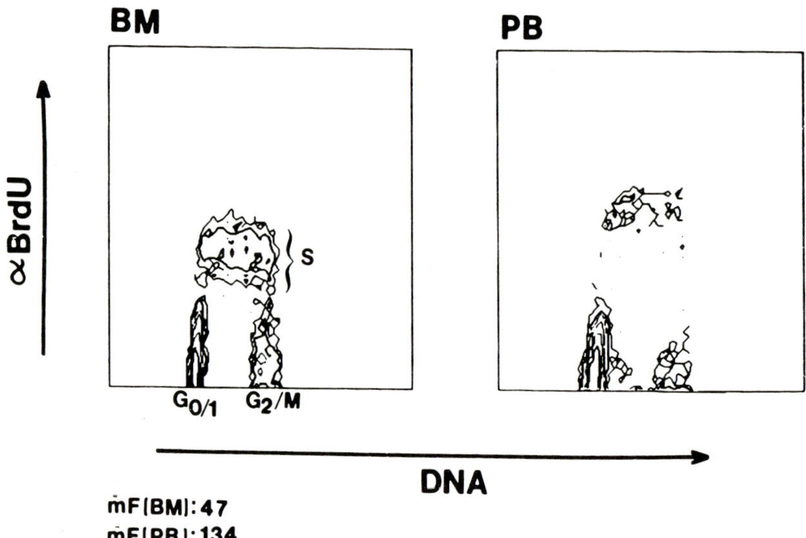

Fig. 7. DNA/BRdU flow cytometry of bone marrow (BM) and blood cells (PB) of a patient with ALL. Cells were incubated with BRdU for 1 hour in vitro and then fixed and stained with IU4 antibody. An isotypic control was used to define background fluorescence. Positivity is indicated by the bracket (S phase). Counterstaining with propidium idodide for total DNA content (X axis). Y axis has a logarithmic scale. Number of cells in S phase is significantly lower in PB as compared with BM, but the BUdR incorporation is three times higher in PB as compared to BM, indicative of a higher DNA-synthesis rate.

cell surface and cytoplasmic antigens expressed during B [2,132] and T-cell differentiation (Fig. 9).

Non-T-ALL were classified into six groups based on their degree of differentiation (Table 5). T-ALL markers were categorized in three groups (Table 6).

The rare subgroup of L-3 ALL is almost invariably classified as B-cell, but exceptions have been found. A special entity is the Adult T-cell leukemias (ATL) described in Japan and the Caribbean [142]. They are virus-induced (HTLV1) and essentially nonresponsive to chemotherapy. They are also characterized by a high number of interleukin-2 (IL-2) receptors. Contrary to initial reports, T-cell phenotype in adult ALL (excluding ATL) is associated with a better prognosis than the pre-B phenotype [82,110]. Of interest is the absence of detectable IL-2 receptors in common and pre-B ALL, which can be induced in vitro by the lectin phytohemagglutinin or by phorbol ester [199], resulting in proliferation and colony formation. Our own studies have confirmed the inducibility of IL-2 receptors on aneuploid leukemic lymphoblasts by IL-2, TPA (phorbol 12-myristate 13-acetate), and OKT3. Recruitment into G_1, but not into S phase, was observed in 80% of patients studied [13]. This approach may be useful in rendering the leukemic cells more sensitive to cell cycle-specific drugs as discussed.

The typical pre-B ALL are not completely frozen in early stages of B-cell differentiation. Cossman et al. and Korsmeyer et al. [52,118,119] were able to differentiate these cells with TPA in vitro and using the concept of clonal excess (CE), Koziner et al. [120] found evidence of κ-light-chain expression in six patients and λ CE in two patients in a series of 35 patients studied. They also found SIg λ expression in the REH cell line, which is derived from a common ALL patient, and rearrangement of the lambda gene by Southern blotting. Although CE, which is indicative of clonal predominance of restricted B cells, cannot be found in normal blood and marrow, additional markers to identify leukemic cells would increase confidence in the existence of fully differentiated leukemic B cells. Double staining for DNA and κ- or λ-light-chain expression revealed CE in 16/18 aneuploid ALL studied [25]. An estimated 5% to 8% of cells expressed restricted light chains. Because of the frequently observed discrepancy between the (higher) number of aneuploid cells and the (lower) number of blast cells in ALL [112], cell sorting was carried out to test the hypothesis that the differentiated leukemic cells were also characterized by morphological features that would exclude them from being identified as blasts. Figure 10 shows the morphology of sorted κ-light-chain expressing cells: the top panel shows κ (solid line) as compared with λ (dotted line) expression. Cells were sorted, and the κ-positive fraction was remeasured (middle). While the κ-negative cells have morphological features typical of lymphoblasts, κ-positive cells have a more mature phenotype. This is evidence for morphological changes of leukemic lymphoblasts, associated with differentiation along the B cell lineage. Flow cytometric evaluation of immunological phenotype has become the standard method for the classification of ALL; results are incorporated in prognostic models that are used to develop more efficacious therapies [82,110].

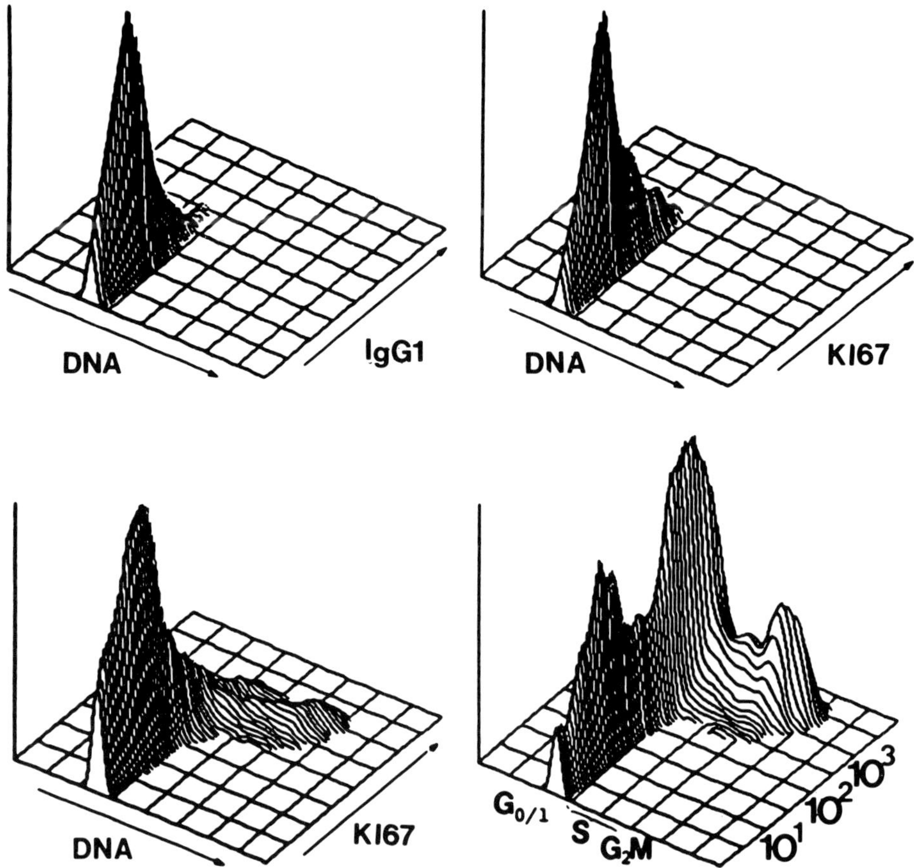

Fig. 8. DNA/Ki67 expression of stimulated peripheral blood lymphocytes. Peripheral blood mononuclear cells were stimulated with OKT3 + IL2. Staining for Ki67 (Y axis, logarithmic scale) and DNA (propidium iodide, X axis) was performed according to Larsen et al. (personal communication). Upper panels show cells after 24 hours of stimulation: upper left shows isotypic control (IgG_1), which defines unspecific background fluorescence; upper right panel shows only weak Ki67 positivity. Left lower panel shows Ki67 expression after 48 hours and lower right panel after 72 hours. Cells in G_0, G_1, S, and G_2M are clearly distinguishable. All cells in G_1, S, and G_2M are positive for Ki67.

MULTILINEAGE LEUKEMIA

With increasingly sophisticated techniques to identify hematopoietic lineages affected by leukemic transformation, a surprisingly complex picture is evolving. A considerable number (22%) of acute nonlymphocytic leukemias were found to exhibit deoxynucleotidyltransferase (TdT) activity, as measured in a biochemical assay, and TdT^+ patients had lower remission incidence (48% versus 68% for TdT^- patients) and experienced shorter remission durations (p = 0.003) [153]. TdT was thought to be a marker specific for the lymphoid lineage in leukemia; two alternative explanations were postulated: (1) myeloid leukemia samples contain a small number of cells with lymphoid phenotype that express TdT; or (2) immature cells in AML are TdT positive but also have myeloid characteristics. The term *biphenotypic* leukemia has been used for these *hybrid* leukemias; the expression of *aberrant* markers was interpreted as *lineage infidelity* (i.e., the cell is committed to a specific lineage and expresses aberrant markers) or "lineage promiscuity" (the cell has not been committed to a specific lineage and expresses markers of two or more differentiation pathways [39,145,152,156,163,184].

In some cases, lymphoid and myeloid markers were expressed on different cells, in others on the same cells indicating malignant transformation of a pluripotent stem cell. In certain human myeloid leukemic cell lines, a rearrangement of the immunoglobulin (Ig) heavy-chain genes that typically occurs during the early stages of B cell differentiation was detected [156]. In some cases, a lineage switch was identified when the patient relapsed [190,207], usually from lymphoid to myeloid. Two alternative explanations are possible: (1) chemotherapy eradicates the dominant clone present at diagnosis, permitting expression of a secondary clone with a different phenotype; or (2) drug-induced changes in the original clone may either amplify or suppress differentiation programs resulting in a phenoyptic shift. Indeed, lineage infidelity could bc induced in vitro by exposing MOLT-3 cells to a single dose of 5-azacytidine [185].

Additional complexity is introduced when antigen expression is examined with regard to different DNA stemlines in cases with aneuploidy or polyploidy. Expression of TdT or CALLA on diploid cells in cases of hyperdiploid ALL indicates that at least some of the DNA diploid cells are leukemic as well [132,171]. Large variations in the amount of antigen

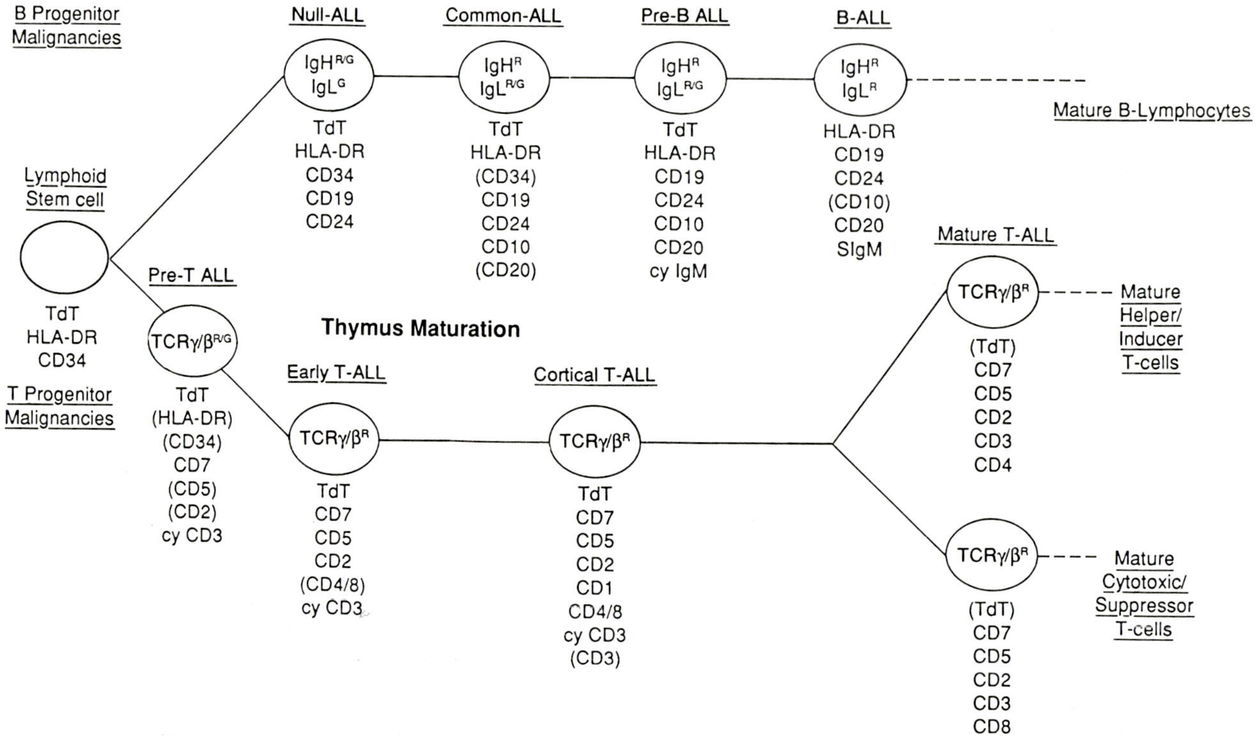

Fig. 9. Schema of lymphoid differentiation: normal B-T-cell development and corresponding immunological types of acute lymphoblastic leukemia (reproduced from Ludwig [138], personal communication).

TABLE 5. Classification of Non-T-ALL*

Group	Antigens				Cytosplasmic (μm)	Surface membrane Ig
	Ia	B4	CALLA	B1		
I	+	−	−	−	−	−
II	+	+	−	−	−	−
III	+	+	+	−	−	−
IV	+	+	+	+	−	−
V	+	+	+	+	+	−
VI	+	−	+/−	+	−	+

*From Foon and Todd [76].

expressed can be found on cells with similar DNA ploidy [171]. Clonal evolution may be the mechanism responsible for many, if not all, cases with multiple DNA stemlines.

Some cases express myeloid and lymphoid markers on the same or on different aneuploid DNA stemlines. A particularly interesting case expressed TdT on diploid, near-triploid, and tetraploid cells, but only the tetraploid cells also expressed the myeloid-specific M2 antigen. In vitro treatment with retinoic acid induced the expression of the M2 antigen on diploid cells as well. Cytogenetic analysis revealed four clones: 46XX, 92XXXX, 92XXXX ($2q^-$), and 65–73 chromosomes (near-triploid) all with $2q^-$. It is possible that the deletion of 2q after cells became tetraploid introduced clonal instability and that all DNA-stemlines and karyotypically defined clones originated from the same cell [153]. This example demonstrates the not yet elucidated complexity of clonal and phenotypic evolution in leukemia. Another case is shown in Figure 11. Diploid cells (2c) with low RNA content and tetraploid cells (4c) with high RNA content were found. Both populations were proliferating and when separated by 1XG velocity sedimentation the diploid cells were lymphoblasts by morphology with low RNA index (9.7), TdT positivity, and Sudan black negativity, while the tetraploid cells had myeloid morphology and high RNA index (29.9), and stained positive for Sudan black [15]. Multilineage leukemias with heterogeneous phenotypes (lineage infidelity or promiscuity [87]) and clonal variability may be appropriate targets of differentiation therapy in the future. These are unsolved challenges to the hematologist. Only a persistent multiparametric approach using flow cytometry with simultaneous staining of DNA, RNA, and at least two to three differentiation antigens coupled with cytogenetic and molecular investigations of particular subpopulations will shed light on the not yet fully appreciated complexity of the acute leukemias.

MYELODYSPLASTIC SYNDROMES

The myelodysplastic syndromes (MDS) comprise a number of hematopoietic malignancies with poor prognosis. Five different entities are defined by the FAB classification [36]: (1) refractory anemia (RA), (2) refractory anemia with ring

TABLE 6. Classification of T-ALL

	Antigens						
Group	Leu-9*	Leu-1	T11/Leu-5	T3/Leu-4	T3/Leu-3	T8/Leu-2	T6
I	+	+(90%)	+(75%)	−	−	−	−
II	+	+	+(25%)	+(90%)	+(90%)	+	
III	+	+	+	+	+/−	+/−	−

Foon and Todd [76].
*Found on virtually all T-ALL cells.

sideroblasts (RARS), (3) refractory anemia with excess of blasts (RAEB), (4) refractory anemia with excess blasts in transformation (RAEBIT), and (5) chronic myelomonocytic leukemia (CMMoL). The FAB subtype is highly predictive of survival with median survivals of 64 months for RA, 71 months for RARS, 7 months for RAEB, 5 months for RAE-BIT, and 8 months for CMMoL [77]. MDS involve two or more hematopoietic cell lineages, indicating involvement of a more undifferentiated stem cell than in many acute leukemias [164]. Chromosomal abnormalities in MDS include monosomies of chromosomes 5 and 7 and deletions of the long arms of chromosomes 5, 7, and 20 [62,128]. A number of genes crucial for hematopoietic growth and differentiation are localized on 5q, including the genes for interleukin-3 (5q23–31), granulocyte–macrophage colony-stimulating factor (5q23–32), macrophage-stimulating factor (5q23–31), and the c-*fms* protooncogene, which encodes the receptor of M-CSF (5q23–34).

Because of the cellular heterogeneity, few flow cytometric studies of MDS have been reported. In a study of 24 patients by Montecucco et al. [149], two patients had aneuploid DNA stemlines, proliferation was higher in RARS and RA than in RAEB and CMMoL and regardless of diagnosis, low S phase predicted short survival. Maiolo et al. found no cell kinetic differences between MDS and AML [140]. Pedersen-Bjergaard et al. [157] analyzed patients with secondary preleukemia by flow cytometry and cytogenetics and demonstrated DNA hypodiploidy in the granulocytes of 4/9 karyotypically hypodiploid (monosomy 7) samples, indicative of the leukemic origin of these granulocytes. Peters et al. [158] described both hyperdiploidy and hypodiploidy in patients with RA, and Clark et al. [49] reported DNA aneuploidy in 34/70 patients with MDS. In their study, hypodiploidy was noted in RA and RAEB and was associated with poor prognosis ($p = 0.001$). It was found to be a better prognostic factor than blast percentage.

CHRONIC LYMPHOCYTIC LEUKEMIA

The chronic lymphocytic leukemias are characterized by a uniform population of lymphoid cells with distinct morphology. They are mostly (95%) of B-cell origin with rearrangement of Ig heavy- and light-chain genes. Clonality of CLL has been demonstrated by expression of a single Ig light chain on the cell surface. Typical is the relatively weak fluorescence of surface Ig, which can be used to discriminate CLL from lymphocytic lymphomas and from prolymphocytic leukemia. Ig isotype analyses indicate that most CLL display a single heavy-chain class, typically μ or μ and δ [75]. Cytoplasmic Ig heavy chains have also been reported. B-CLL cells display receptors for mouse erythrocytes, IgG, and complement. Antigens expressed include Ia, BA1, B1, B2, and B4 and surprisingly a 65-kd glycoprotein (T1) thought to be restricted to T lymphocytes (detected by the T101 and Leu 1 antibodies). The lineage infidelity discussed for the acute leukemias is also evident in CLL: rearrangement of the T receptor has been reported in 10% of B-CLL and in 25% of pre-B ALL. This again emphasizes that Ig and receptor rearrangments alone are not adequate to assign lineage. In one patient with CLL, a second population of CML cells became evident, and flow cytometric marker analysis permitted the determination of different immunological phenotypes [96].

Prolymphocytic leukemia (PLL) differs from B-CLL both morphologically and clinically [78]. Like B-CLL, PLL express Ia, B1, B2 and B4, but surface Ig expression is more intense and these cells usually lack the T1 receptor. There is high binding of the FMC-7 antibody [75].

Hairy cell leukemia is characterized by distinct mononuclear cells with hairy cytoplasmic projections. Histochemically, cells are positive for tartrate-resistant acid phosphatase. The surface markers expressed and Ig gene rearrangements confirm the monoclonal B-cell nature of this disease. Antigens detected include Ia, B1, FMC-1, FMC-7, and (BA1), and surprisingly the plasma cell antigen PCA-1. In addition, the Tac antigen (IL-2 receptor) is expressed in most cases (for review, see ref. [75]).

The cell kinetics of CLL has been reviewed elsewhere [3]. Characteristically, S phase is very low, almost undetectable in peripheral blood and marrow cells (0.2%), but has been found to be as high as 60% in CLL lymph nodes [196]. In studies of continuous infusion of [^{3}H]thymidine over 7 days, about 95% of small peripheral blood lymphocytes were long-lived, with turnover times of more than 1 year. CLL can be considered to have an accumulative type of growth [64]. RNA content determined by DNA/RNA flow cytometry is characteristically low (RNA index = 8), comparable to that of normal B cells, and DNA aneuploidy is usually not detectable by standard resolution DNA measurements (CV 2 to 3%) [7]. Of interest may be an unexpectedly large variability of α_t in AO measurements of chromatin structure and cell cycle position (M. Andreeff, unpublished results): α_t values vary over a large range from deep G_0 to G_1. Supporting the notion of considerable proliferative activity in some CLL or in subpopulations of CLL cells is the observation of a high expression of the cell cycle related proto-oncogene p53 in the blood of patients with CLL [189]. Better understanding of the molecular and cell biology of CLL may ultimately result in improved therapy for this previously incurable disease.

CHRONIC MYELOGENOUS LEUKEMIA

Chronic myelogenous leukemia (CML) is usually easy to diagnose by virtue of increased production of granulocytes. CML often involves multiple cell lineages and is cytogenetically characterized by the Philadelphia translocation (9;22) [186]. This results in rearrangement of the abl proto-onco-

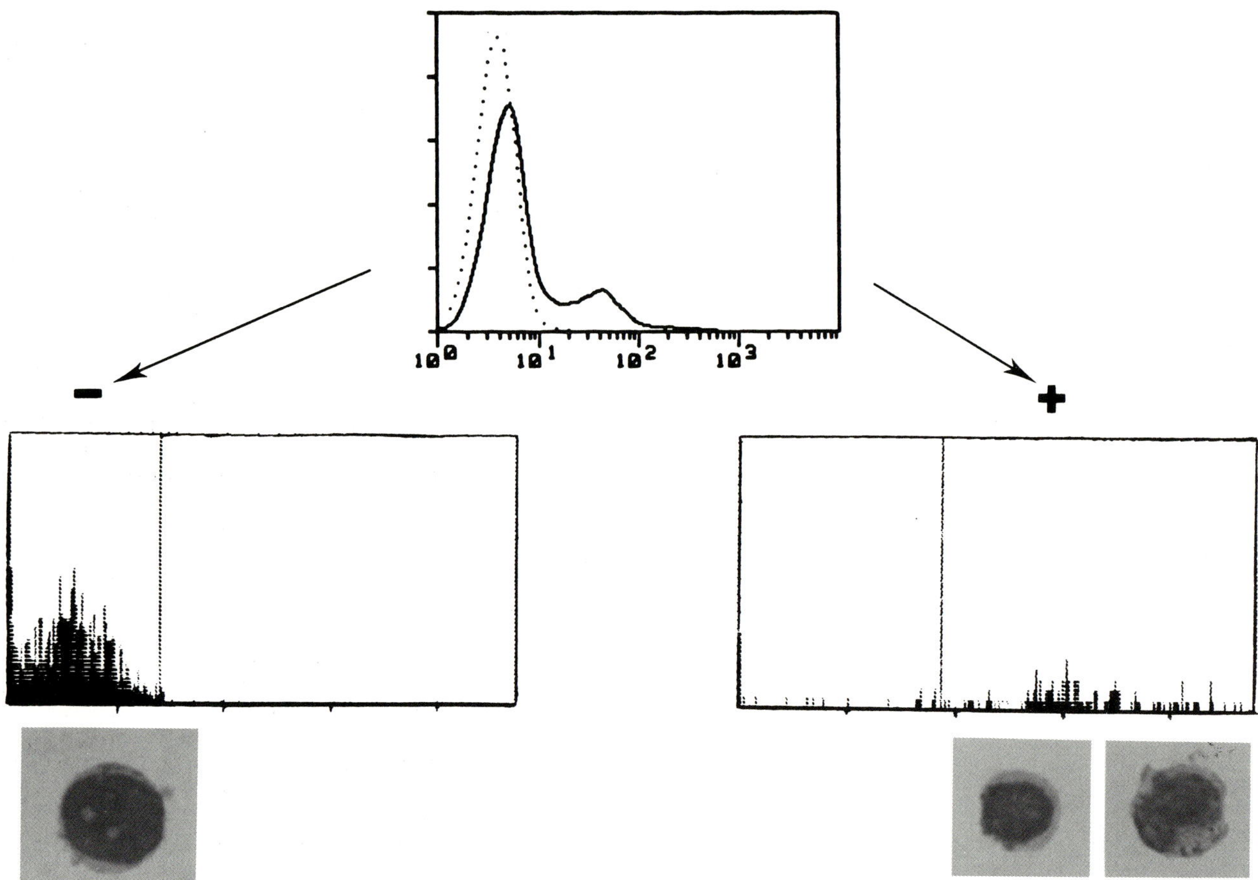

Fig. 10. Differentiation of pre-B acute lymphoblastic leukemia. Cells from a primary ALL (more than 95% lymphoblasts) were stained for kappa (solid line) and lambda (dotted line) light chains. Top panel shows excess of κ-light-chain expressing ALL blast cells. The sample was sorted in an EPICS C Cell Sorter and remeasured for purity (middle panel). Histogram on the left shows κ-negative, on the right κ-positive cells. Sorted cell populations were evaluated on slides. κ-negative cells had the typical morphology of lymphoblasts, while κ-positive leukemic cells showed various morphological degrees of differentiation.

gene [98] and formation of a new hybrid gene (bcr-abl) encoding the CML-specific protein p210 with demonstrated tyrosine kinase activity [56]. The cellular defect leading to CML was described by Strife and Clarkson as "discordant maturation" [191]. Griffin et al. [88] isolated progenitor cells from the peripheral blood of CML patients and obtained cells of which 47% formed myeloid colonies or clusters in culture. The kinetics of bone marrow cell production has been analyzed by Doermer et al. [63].

AO flow cytometry with measurement of green pulse width (GPW) can identify CML by a characteristic pattern similar to the one shown in Figure 1: Ficoll–Hypaque separation of mononuclear cells fails to remove all granulocytes, identified by their unique GPW pattern. This in conjunction with high S phase in the blood is characteristic of CML in AO flow cytometry.

Large discrepancies between flow cytometric S phase and autoradiograph-labeling index determinations were observed by Raza et al. [165]. Holdrinet et al. [113] analyzed CML cells and found no consistent differences in cell kinetics as compared with normal hematopoiesis but observed DNA aneuploidy in 4/11 patients undergoing blastic transformation. Following a 3- to 4-year chronic phase, CML eventually evolves into a blastic phase, with either myeloblastic or lymphoblastic features. As in the case of acute leukemias, myeloblastic transformation of CML is usually characterized by high RNA index [34], and lymphoblastic transformation by low RNA index [5,7]. Ligler et al. [130] tried to predict the transformed phenotype by studying cell surface markers, but early detection of transformation has not yet been achieved with consistency by any method. Some patients have mixed-lineage blastic transformations [92].

With improving therapeutic options, which now include bone marrow transplantation, the need for predicting transformation at an early stage has increased and flow cytometric techniques could be helpful. The detection of minimal disease in CML at present is clearly the domain of the PCR technique, which can identify less than 1 CML cell per 100,000 normal cells [116].

MONITORING OF LEUKEMIA/MINIMAL RESIDUAL DISEASE

Monitoring of leukemia is conventionally done by serial morphological evaluations of bone marrow and blood samples. The clinician's aim of achieving complete remission

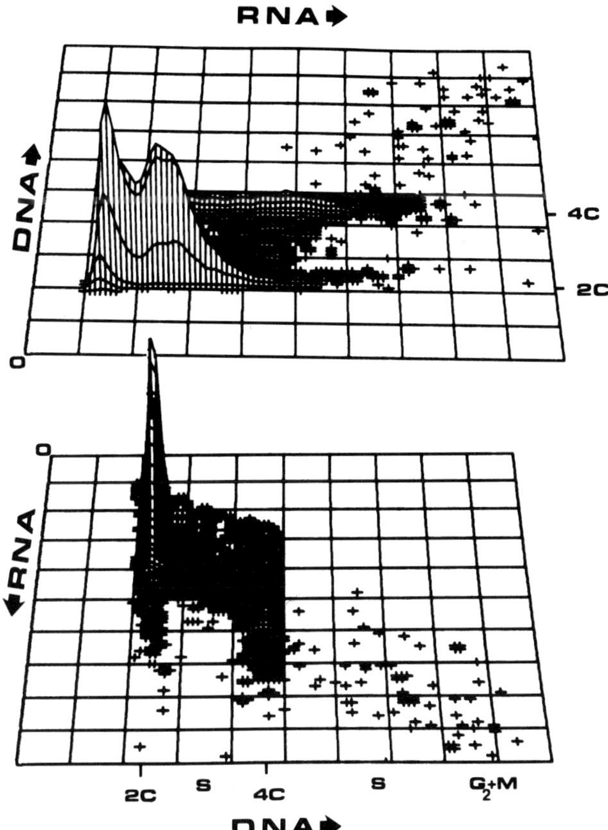

Fig. 11. DNA/RNA histograms from the bone marrow of a patient with biophenotypic leukemia. Sample was stained with acridine orange and shows DNA diploid cells (2C), S-phase cells, and cells with tetraploid DNA content (4C), which are also proliferating. RNA content is higher for tetraploid as compared to diploid cells. Separation on 1XG velocity sedimentation gradient identified diploid cells as lymphoblasts and tetraploid cells as myeloblasts.

(CR) is thought to be reached when the leukemic cell mass is reduced to less than 5% of marrow blasts and when subsequent normalization of peripheral blood counts indicates repopulation of the marrow with normal progenitors, which are able to proliferate and differentiate. These clinical criteria for CR are insufficient, since most patients with ALL or AML will relapse with the same leukemic clone. Flow cytometry can improve the detection of minimal disease.

Monitoring of leukemia encompasses four areas:

1. Cell kinetic monitoring
2. Monitoring of cytoreduction (cell kill kinetics)
3. Monitoring of minimal residual disease (MRD) using DNA aneuploidy as a marker
4. Immunological monitoring of MRD

Because of technological difficulties, flow cytometry has not yet been widely used for monitoring leukemia: it is easier to measure the expression of CALLA at diagnosis than to detect 0.01% abnormal cells by double staining of CALLA and TdT.

Cell Kinetic Monitoring

The introduction of flow cytometry made cell kinetic analysis much faster and easier and a number of reports appeared during the 1970s, beginning with the work of Buechner, Barlogie, Hiddemann, Hillen, and others [28,29,32,33,41,42, 44,107,108]. In particular, effects of the cell cycle-specific drug cytosine arabinoside (ara-c) [41] on leukemic cells in vivo were investigated [1,79]. Low doses of (ara-c) result in an increase of S-phase cells due to a block of DNA synthesis, whereas high ara-c doses kill cells in S, and DNA histograms will show a depletion of S-phase cells. In a study of AML patients treated with a daunomycin/ara-c-containing regimen, responding and nonresponding patients showed different effects of induction therapy on the kinetics of the leukemic cells. In patients achieving CR, S phase decreased to 1.33 ± 0.99%; and to 2.65 ± 2.3% in patients failing therapy ($p = 0.04$) [20]. Another study found a decreasing RNA index (mean 1.4) in responders, but not in nonresponders (mean 2.5) [139]. In the same study, S-phase recovery was different between patient groups: S phase was 16.9% for responders and 12.0% for nonresponders ($p < 0.001$) between days 18 and 22 of treatment: persistent leukemic cells probably prevented recovery of normal hematopoiesis. Future studies of BUdR incorporation will better define the role of kinetic changes in the evaluation of response to chemotherapy.

Cell Kill Kinetics

Quantitation of cytoreduction in terms of cell number per volume of bone marrow was made possible by two techniques, developed by Hiddemann et al. [102] and by Blumenreich et al. [40]. Hiddemann's method requires simultaneous bone marrow aspirate, biopsy, and blood samples and corrects for the admixture of blood cells in the marrow aspirate. Cell kinetic differences identify the *number* of contaminating blood cells, and the ratio of red cell hematocrit in blood and aspirate corrects for the volume of contaminated blood. The total cell number found in acute leukemia marrows was determined to be 10^5 to 10^6 blast cells/mm^3, which amounts to a total of 1 to 5 × 10^{12} bone marrow blasts per patient. Cell kill with daunomycin, ara-c, and 6-thioguanine in 5 days was 1.1 to 3.3 log^{10} and was significantly higher in responding patients than in nonresponders ($p < 0.01$). Patients treated with high-dose thymidine had up to 3.6 log^{10} cell kill. The method also permits determination of cell kill for the different cell lineages present in the marrow [7,182]. The efficacy of different drug schedules can be objectively quantitated [100]. Blumenreich's method is technically much easier and requires the complete disaggregation and precise counting of cells in bone marrow biopsy samples. In a recent study, it was used to assess ALL induction therapy in children. Continuous infusion of daunomycin induced a 1 log^{10} higher cell kill than did the same total dose given IV push [95]. This important information cannot be obtained with other methods, since almost all children achieve CR. It is expected, but remains to be proved, that rapid initial cell kill results in prolonged remission duration. Clinical information certainly supports this hypothesis [82,169].

Monitoring of Minimal Residual Disease by DNA Aneuploidy

The high rate of DNA aneuploidy in ALL (40% in children) makes DNA flow cytometry an excellent technique for monitoring therapy-induced cytoreduction. The sensitivity of detecting small numbers of DNA aneuploid leukemic cells is only 1% to 3%, but in hypocellular marrows after cytotoxic therapy, and in regenerating marrows with high proliferative activity of normal hematopoietic cells, monitoring of DNA aneuploid cells is clearly superior to morphological assessment. Patients with persistent aneuploid cells invariably relapse within three months. Long-term follow-up of 93 children in remission whose ALL cells were characterized by DNA aneuploidy identified only 3/15 relapsed patients prior to clinical relapse, with a lead time of 1 to 6 months [20,169]. Despite the usefulness of DNA aneuploidy monitoring during induction therapy, more sensitive methods are clearly needed for long-term follow-up studies. Fluorescence in situ hybridization (FISH) using chromosome-specific DNA markers has already been shown to detect as few as 2 of 2,000 abnormal cells, when slides were counted manually [86]. Flow cytometry may increase the sensitivity of FISH by one or two orders of magnitude and may help in detecting recurrent disease early enough for successful clinical intervention.

Immunological Monitoring of Minimal Residual Disease

Since no leukemia-specific immunological markers have been identified, the use of flow cytometry in this context appears limited to the early stages of therapy, when reduction and/or persistence of the abnormal phenotype can give clues regarding residual disease. In mixing experiments using $CD1^+$ malignant T lymphoblasts and thymocytes, the detection limit was 1% to 2%. Likewise, measurements of TdT activity and CALLA were not useful in predicting ALL relapses (reviewed in ref. 138), since both markers are present in normal marrow [114] and blood cells. When cell populations were gated by light-scatter properties and irrelevant cells were tagged with phycoerythrin-labeled antibodies, the sensitivity for detecting CALLA-positive lymphoblasts was 1 per 100,000 in the peripheral blood [179]. Enumeration of $CALLA^+/TdT^+$ cells by double staining in children with ALL on maintenance therapy showed that levels >0.03% were indicative of imminent relapse. The technique of κ–λ clonal excess [27,209] could be useful in B-cell lymphomas and B-ALL, but has not been used extensively in leukemia monitoring. The following table [138] summarizes the present possibilities for following MRD in ALL using immunological markers (Table 7).

In myeloid leukemia, the rat AML model BNML was investigated by Martens, Hagenbeek, and Visser; the high expression of a myeloid antigen permitted detection of small numbers of leukemic cells [93,204]. It was estimated that 1 per 10^4 to 1 per 10^5 cells can be detected.

Other techniques to detect minimal residual disease in leukemia include the detection of chromosomal abnormalities by premature chromosome condensation, which identifies cells with high or low proliferative potential (PPI, or cells in late G_1). Fourteen of 16 patients with high PPI relapsed after 23 weeks on average, and 19/44 patients with low PPI relapsed after 68 weeks [109]. Patients with specific translocations, including those with t(9;22), t(8;14), and t(4;11) will be excellent candidates for monitoring with the DNA polymerase chain reaction technique (PCR) [116], which can detect <1 abnormal cell per 100,000. A combination of PCR, FISH, and flow cytometry is conceivable and may provide the quantum jump in sensitivity urgently needed for the detection of truly minimal residual leukemia.

DRUG EFFECTS AND DRUG RESISTANCE IN LEUKEMIC CELLS

Drug effects on leukemic cells have been studied extensively in vitro and in vivo and the cell kinetic effects are described elsewhere (see Chapter 39, this volume). Here we highlight the effects of the three drugs most used in leukemia, i.e., ara-c, methotrexate, and the anthracyclines.

Cytosine Arabinoside

Cytosine arabinoside is the classic blocker of DNA synthesis. Early studies by Buechner et al. [41,43,44] Barlogie et al. [29] and Yataganas et al. [211] confirmed the expected cell cycle specificity of ara-c. Using the AO techniques, recruitment of cells into the cell cycle could be demonstrated in patients receiving therapy with high-dose ara-c [18]. Of potential clinical value is Waldmann's observation that cells resistant to ara-c can be identified based on their persistent ability to incorporate BUdR. The sensitivity of the method is remarkable and approaches 10^{-3} [206]. Figure 12 shows the inhibition of DNA synthesis of HL-60 cells by different concentrations of ara-c: 10^{-9} M ara-c has no detectable effect as compared with controls, 10^{-6} M ara-c blocks DNA synthesis in all but 10/10,000 cells measured and 10^{-3} M ara-c blocks DNA synthesis in all cells. Persistent DNA synthesis of a few leukemic cells in patients would probably lead to relapse. The clinical value of this approach to the identification of ara-c-resistant cells remains to be established.

Methotrexate

Methotrexate (MTX) has been studied extensively with regard to its cell cycle effects (S-phase block), its uptake and metabolism. The uptake of fluorescinated MTX was studied in MTX-sensitive and -resistant leukemic cells [161,176]. A fluorescent MTX analogue was used as a dihydrofolate reductase (DHFR) probe by Rosowsky et al. [177] and can be utilized in a flow cytometry assay to detect methotrexate resistance based on DHFR overproduction. Taylor and Tattersall [195] investigated cell cycle effects using DNA/RNA flow cytometry. Flow cytometry has clear advantages in the study of MTX resistance, since it can identify small numbers of MTX-resistant cells that are undetectable in pharmacological assays.

Anthracyclines

The three main anthracyclines, daunomycin, adriamycin and idarubicin fluoresce upon excitation with a 488-nm argon-ion laser, permitting measurement of their intracellular concentration, as pioneered by Krishan [121]. Drug uptake and modulation of uptake were extensively studied in vitro [81,122]. A comparison of anthracyclin concentrations in human hematopoietic cells measured by flow cytometry and high-pressure liquid chromatography showed good correlation [188]. Comparative studies of daunomycin and adriamycin uptake in sensitive and resistant leukemia cells were also conducted [194] and effects of calmodulin inhibitors were studied [80]. These studies of leukemia cell lines showed effects of calcium channel blockers on anthracycline uptake, but a study by Ross et al. [178] failed to demonstrate an effect of verapamil on the uptake of daunomycin in cells from clinically resistant AML. Additional studies to evaluate the clinical value of this technique are needed.

Multidrug resistance

Multidrug resistance describes the phenomenon of cells selected for resistance to a single cytotoxic agent developing resistance to a group of unrelated agents, which include colchicin, daunomycin, doxorubicin, actinomycin D, vinca alkaloids, and epipodophyllitoxins. The gene conferring multidrug resistance (MDR 1) and its product, a P-glycopro-

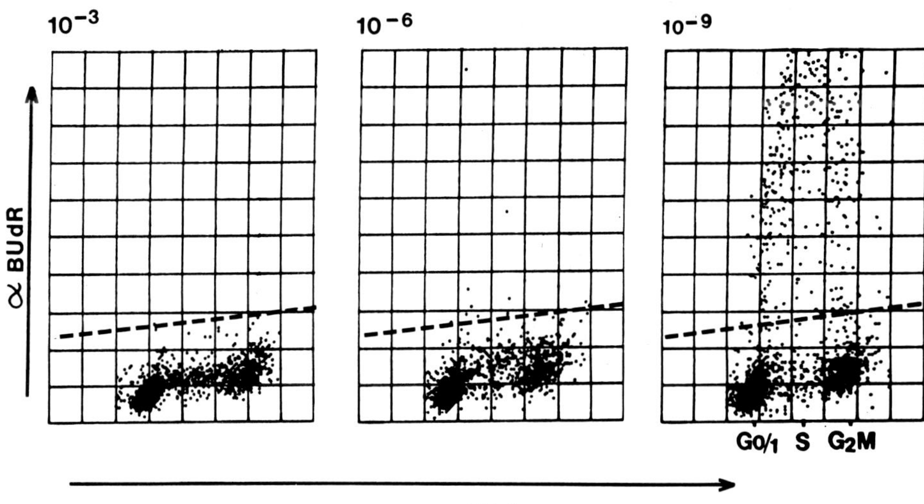

Fig. 12. DNA (X-axis)/BUdR histograms from HL-60 leukemic cells after incubation with cytosine arabinoside (ara-c) for 1 hour. Right panel shows effect of 10^{-9} M ara-c: no effect on S cells is noted as compared with controls (data not shown). Middle panel shows 10^{-6} M ara-c: 10 of 10,000 cells are still incorporating BUdR. Left panel shows 10^{-3} M ara-c inhibits DNA synthesis in all leukemic cells.

TABLE 7. Detection of Minimal Residual Disease in ALL by Flow Cytometry*

B-progenitor cell ALL:
TdT and/or CALLA in peripheral blood (double immunofluorescence (IF) staining):
Detection limit: $TdT^+/CALLA^+$ 0.03%
pre-T/T-ALL:
TdT and TdT/T-cell marker (CD2, CD5, CD7) in bone marrow or peripheral blood (double IF staining)
Detection limit:
TdT^+/T-cell $marker^+$ 0.01–0.04% (bone marrow) 0.01–0.03% (peripheral blood)
B-ALL/B-NHL
κ–λ labeling technique in peripheral blood
Detection limit: 1–10% monoclonal cells

*From Ludwig [138], personal communication.

tein, have been identified [197] and antibodies to detect the MDR phenotype were developed. One such antibody [146] detects a 180-kd cell surface P-glycoprotein and has been used in the study of primary human leukemia cells [97]. Relapsed leukemias expressed twice the amount of P-glycoprotein as newly diagnosed leukemias, and patients failing anthracycline/vinca-containing treatment regimens expressed much higher levels of (gp180) than did responding patients.

Flow cytometric studies of drug uptake, retention, and metabolism and of drug effects on cell cycle progression and survival in vitro combined with investigations of drug resistance may be useful in designing better therapeutic regimen. A particular advantage of flow cytometric techniques is the ability to identify the small number of resistant cells that will result in relapse and ultimately determine the fate of the patient.

ONCOGENE PRODUCTS IN LEUKEMIA

During the past few years advances in molecular biology have provided unprecedented insight into the regulation of cell growth and differentiation. Emerging from this work is the concept of oncogenes and proto-oncogenes: normal cellular genes (proto-oncogenes) controlling growth, development, and differentiation change (to oncogenes) and contribute to the malignant transformation characteristic of cancer cells [45]. In leukemia, four different mechanisms of (proto-) oncogene activation are known and are of help in recognizing the potential of flow cytometric studies in this field (22) (Table 8).

It is not yet possible to identify any of these alterations by flow cytometry directly, although the polymerase chain reaction has the potential of amplifying specific DNA and/or RNA sequences and making them amenable to fluorescence in situ hybridization. Sometimes the oncogene product is expressed in such large quantities that it can be detected, e.g., amplified *myc* mRNA in HL-60 cells. When new proteins are generated, as is the case in the *bcr–abl* translocation of CML, where a novel protein of 210kD (p210) is found, or in *ras* mutations where one amino acid is substituted as the consequence of a single point mutation, the potential for specific and unique monoclonal antibodies exists. However, the antigenic differences of these proteins may be too small to be identified by specific antibodies. To date, no antibody exists that recognizes p210 only without binding to the normal 145-kD *abl* product. Likewise, antibodies detecting mutated *ras* proteins are not available for flow cytometric applications. An additional problem is the cellular localization of many of these proteins, which can reside in the cell nucleus ($p65^{c\text{-}myc}$) or on the inner side of the plasma membrane ($p21^{c\text{-}ras}$).

TABLE 8. Mechanisms of Oncogene Activation in Leukemia

Translocation oncogene	Localization chromosome	Translocation	Leukemias/lymphomas
myc	8q24	t(8;14)	Burkitt's
		t(8;22)	Burkitt's
		t(2;8)	Burkitt's
abl	9q34	t(9;22)	CML, ALL
ets 1	11q23	t(4;11)	AUL
		t(9;11)	AML (M5)
ets 2	21q23	t(8;21)	AML (M2)
bcl 2	18q21	t(14;18)	nodular lymphoma
bcl 1	11q13	t(11;14)	CLL, diffuse lymphoma
mos	8q22	t(8;21)	AML
Ha *ras*	11p15.1 p15	t(11;14)	T-ALL
Deletion			
myb	Adjacent to oncogene		ALL, AML
fms	Monosomy 5		AML, MDS
Amplification			
myb	M1 (AML) cell line		
myc	HL-60 (AML) cell line		
abl	K562 (CML) cell line		
Mutation			
ras	Codon 12,13,61		AML, MDS

A flow cytometric technique for the detection of $p21^{c\text{-}ras}$ in leukemic cells was developed that used a pan-*ras* antibody and provided information regarding changes in *ras* expression in different cell cycle phases identified with propidium iodide [21]. The low level of *ras* protein required several amplification steps, and careful use of control reagents is necessary in these studies. Figure 13 shows $p21^{ras}$ expression in marrow cells from a patient with myelodysplastic syndrome and diploid and hyperdiploid DNA stemlines: both diploid and the hyperdiploid cells express $p21^{c\text{-}ras}$. An improvement in this technique discriminates G_0-G_1-S-G_2M cells by triple staining for DNA (Hoechst), RNA (pyronin Y), and $p21^{c\text{-}ras}$ (FITC) in dual-laser three-color flow cytometry [23]; G_0 cells appear to express lower levels of $p21^{c\text{-}ras}$ than G_1, S or G_2M cells in primary acute lymphoblastic leukemia.

Measurements of oncogene expression are clearly emerging as a novel field of flow cytometry. Unfortunately, some studies have been poorly controlled, yielding implausible results that are inconsistent with those obtained by molecular techniques. Still, the advantage of single-cell measurements of oncogene expression by flow cytometry will most likely be a welcome addition and expansion of molecular techniques.

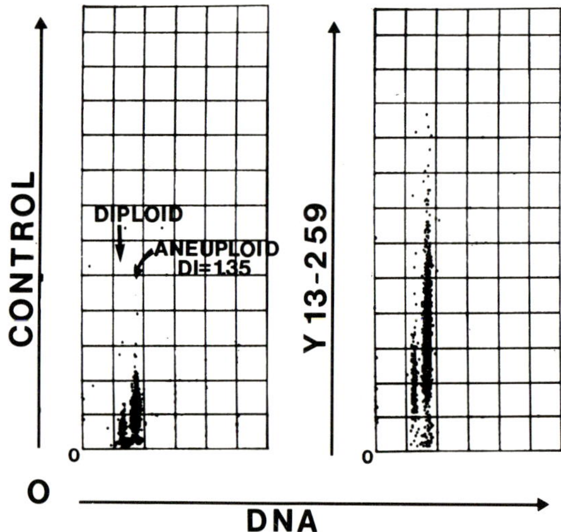

Fig. 13. DNA/$p21^{c\text{-}ras}$ histogram of bone marrow cells from a patient with myelodysplastic syndrome. Control (left) shows immunofluorescence of an isotypic control antibody. X axis shows DNA–diploid and DNA–hyperdiploid cells (DI = 1.35). Right panel shows positivity for p21 c-ras, identified by antibody Y13-259 in both DNA–diploid and hyperdiploid cells. Cytogenetic analysis revealed loss of the long arm of chromosome 5 (5q-), characteristic for MDS, and additional hyperdiploid abnormalities.

OUTLOOK

Flow cytometry has undergone rapid technological development over the past 20 years; complex instruments are now available that are surprisingly easy to use. As a tumor in suspension, leukemias attracted the attention of investigators working with flow cytometry early on, simply because there was no need for cell disaggregation as in solid tissues and tumors. Studies of clonality and cell proliferation initiated in Europe were paralleled by studies of cellular differentiation initiated in the United States. These have greatly aided our understanding of leukemia cell proliferation and differentiation and have become routine hematologic procedures in many centers.

What lies ahead? Multiparameter flow cytometry will employ more dyes, improved fluorochromes, and more user-friendly machines and software. Multicolor fluorescence will probably become routine, and an improved understanding of the developmental steps at which leukemic cells are arrested will evolve. Molecular probes will be applied at the single cell level and it may be possible to look at certain genes, at their RNA message and proteins in the same cells. Hence, flow cytometry of leukemias will merge with immunology and molecular biology and will open the door to single cell analysis for these areas of research. More automated diagnosis and treatment monitoring of leukemia and a better understanding of the biology of the disease should evolve from these studies.

ACKNOWLEDGMENT

I wish to thank Bayard Clarkson, Zbigniew Darzynkiewicz, and Myron Melamed for many years of friendship, support, and stimulating exchanges of the topics discussed in this chapter. Frank Traganos and Thomas Sharpless gave precious help and generous advice to make many of our studies possible. Wolfgang Hiddemann, Carlos Suarez, Joseph Haimi, Susanna Hegewisch-Becker, Agostino Tafuri, and especially Arlene Redner have worked on, and persistently thought about, the problems related to the biology of leukemia and the application of flow cytometry and made significant contributions. Jim Steinmetz, Barry Eagel, Suzanne Swartwout, Edith Espiritu, Gordon Assing, Jan Bressler, James Squires, Loren Godfrey, Walter Verbeek, Carsten Bokemeyer, and Barbar Rehermann have worked extremely hard to fill conceptual frameworks with results. Finally, I am indepted to Linda Tatum for her excellent assistance and patience in the preparation of this manuscript. This work was supported by grants CA 20194, CA 41305, and CA 38980 from the National Cancer Institute.

REFERENCES

1. **Aardal NP, Talstad I, Lerum OD (1979)** Sequential flow cytometric analysis of cellular DNA-content in peripheral blood during treatment for acute leukaemia. Scand. J. Haematol. 22:25–32.
2. **Al-Katib A, Wang CY, Bardales R, Koziner B (1985)** Phenotypic characterization of "non-T, non-B" acute lymphoblastic leukemia by a new panel (BL) of monoclonal antibodies. Hematol. Oncol. 3:271–281.
3. **Andreeff M (1986)** Cell kinetics of leukemia. Semin. Hematol. 23:300–314.
4. **Andreeff M, Assing G, Cirrincione C (1986)** Prognostic value of DNA/RNA flow cytometry in myeloblastic and lymphoblastic leukemia in adults: RNA content and S-phase predict remission duration and survival in multi-variate analysis. Ann. NY Acad. Sci. 468:387–406.
5. **Andreeff M, Beck JD, Darzynkiewicz Z, Arlin Z, Melamed MR, Clarkson B (1979)** DNA, RNA and chromatin structure assayed in situ by flow cytometry: A means for the classification of human leukemias. Verh. Dtsch. KrebsGes. 2:348.
6. **Andreeff M, Beck J, Darzynkiewicz ZD, Traganos F, Gupta S, Melamed MR, Good RA (1978)** RNA content of human lymphocyte subpopulations. Proc. Natl. Acad. Sci. USA 75:1938–1942.
7. **Andreeff M, Darzynkiewicz Z (1981)** Multiparameter flow cytometry. Part II. Application in hematology. MSKCC Clin. Bull. 11:120–121.
8. **Andreeff M, Darzynkiewicz Z (1982)** Chromatin structure in leukemia: Classification of subtypes and cell kinetics during induction therapy. Cell. Tissue Kinet. DD :679.
9. **Andreeff M, Darzynkiewicz Z, Melamed MR, Gee T, Clarkson B (1978)** Nucleic acid content and chromatin structure of leukemic cells *in-situ* as studied by flow cytometry. Proc. Am. Assoc. Cancer Res. 19:786.
10. **Andreeff M, Darzynkiewicz Z, and Sharpless TK, et al. (1980)** Discrimination of human leukemia subtypes by flow cytometric analysis of cellular DNA and RNA. Blood 55:282–293.
11. **Andreeff M, Darzynkiewicz Z, Sharpless TK, Melamed MR, Clarkson B (1977)** Cytofluorometric DNA and RNA measurements in the classification of human leukemias. Blood 50 (suppl. 1):181.
12. **Andreeff M, Darzynkiewicz Z, Steinmetz J, Melamed MR, Clarkson BD (1981)** Chromatin structure and nucleic acid content determined in-situ by flow cytometry (FC) in acute leukemia. Blood 58:127a.
13. **Andreeff M, Espiritu E, Bressler J, Welte K (1990)** Induction of Interleukin-2 receptor and recruitment of aneuploid lymphoblastic leukemia by Interleukin-2 (in preparation).
14. **Andreeff M, Gaynor J, Chapman D, Little C, Gee T, Clarkson BD (1987)** Prognostic factors in acute lymphoblastic leukemia in adults: The Memorial Hospital Experience. In Buechner T, Schellong G, Hiddemann W, Urbanitz D, Ritter W (eds), "Haematology and Blood Transfusion." Vol. 30: "Acute Leukemias." Berlin: Springer-Verlag, 111–124.
15. **Andreeff M, Gee T, Mertelsmann R, McKenzie S, Steinmetz J, Chaganti R, Koziner B, Clarkson B (1980)** Biclonal lymphoblastic and myeloblastic acute leukemia. Proc. Am. Assoc. Cancer Res. 21:313 (abst.).
16. **Andreeff M, Haag D (1975)** Fluorochromierung mit Acridinorange in der Impulscytophotometrie. In Andreeff M (ed), "Impulscytophotometrie." Berlin: Springer-Verlag, pp 31–37.
17. **Andreeff M, Hansen H, Cirrincione C, Filippa D, Thaler H (1986)** Prognostic value of DNA/RNA flow cytometry in B-cell non-Hodgkin's lymphoma; development of laboratory model and correlation with four taxonomic systems. In Andreeff M (ed), "Clinical Cytometry." Ann. NY Acad. Sci. 468: 368–386.
18. **Andreeff M, Kempin S, Arlin Z, Mertelsmann R, Espiritu E, Gee T (1983)** High dose cytosin-arabinoside (HARAC) in acute leukemia (AL). Correlation of clinical response and cell kinetics. Proc. Am. Assoc. Cancer Res. 24:166.
19. **Andreeff M, Redner A, Jhanwar SC, Miller D, Melamed MR, Clarkson BD, Chaganti RSK (1990)** Severe hypodiploidy in leukemia: A new entity with lymphoid phenotype and poor prognosis (in preparation).
20. **Andreeff M, Redner A, Thongprasert S, Eagle B, Steinherz P, Miller D, Melamed MR (1985)** Multiparameter flow cytometry for determination of ploidy, proliferation and differentiation in acute leukemia: Treatment effects and prognostic value. In Buechner T (ed), "Tumor Aneuploidy." Berlin: Springer-Verlag, pp 81–105.
21. **Andreeff M, Slater DE, Bressler J, Furth M (1986)** Cellular ras oncogene expression and cell cycle measured by flow cytometry in hematopoietic cell lines. Blood 67:676–681.
22. **Andreeff M, Squires J, Bokemeyer C, Verbeek W, Bressler J, Redner A (1989)** Oncogenes in human leukemia and lymphoma: chromosomal localization, activation and expression of p21 ras and p53. In Hiddemann W (ed), "Chemotherapy of Infectious

Diseases and Malignancies." Munich: Futura, in press.

23. **Andreeff M, Verbeek W, Bokemeyer C, Bressler J, Redner A, DeAnglis P, Traganos F, Darzynkiewicz Z** (1990) Flow cytometric analysis of cellular ras oncogene expression in human hematopoietic malignancies (in preparation).
24. **Andreeff M, Welte K** (1989) Hematopoietic colony-stimulating factors. Semin. Oncol. 16:211–229.
25. **Andreeff M, Wong G, Koziner B, Chaganti RSK, Al-Katib B, Espirtu E, Espiritu B, Hermann TK, Clarkson B** (1984) Evidence for complete B-cell differentiation of a subpopulation of DNA aneuploid leukemic cells in "Non-B, Non-T" acute lymphoblastic leukemia (ALL). Blood 64 (suppl. 1):185a.
26. **Atzpodien J, Gulati SC, Kwon JH, Wacheter M, Fried J, Clarkson BD** (1987) Human bone marrow CFU-GM and BFU-E localized by light scatter cell sorting. Exp. Cell Biol. 55:265–270.
27. **Ault KA** (1979) Detection of small numbers of monoclonal B lymphocytes in the blood of patients with lymphoma. N. Engl. J. Med. 300:1401–1405.
28. **Bakkeren JAJM, Vaan Gam DE, Hillen HFP** (1979) Interrelationship of immunologic characteristics, proliferation pattern and prednisone sensitivity in acute lymphoblastic leukemia of childhood. Blood 53: 883–891.
29. **Barlogie B, Buechner T, Asseburg U, Kamanabroo D** (1974) Zellkinetischer Effekt von Cytosin-arabinosid im klinischen Test. Med. Welt 25:1532.
30. **Barlogie B, Latreille J, Freireich EJ, Fu CT, Mellard D, Meistrich M, Andreeff M** (1980) Characterization of hematologic malignancies by flow cytometry. Blood Cells 6:719–744.
31. **Barlogie B, Maddox AM, Johnston DA, Raber MN, Drewinko B, Keating MJ, Freireich EJ** (1983) Quantitative cytology in leukemia research. Blood Cells 1:35–55.
32. **Barlogie B, Roessner A, Asseburg U, Kamanabroo D, Hiddemann W, Buechner T** (1974) Untersuchungen zur Teilsynchronisation von Zellen der experimentellen Rattenleukaemia L5222. Verh. Dtsch. Ges. Inn. Med. DD :80.
33. **Barlogie B, Spitzer G, Hart JS, Johnston DA, Buechner T, Schumann J, Drewinko B** (1976) DNA histogram analysis of human hemopoietic cells. Blood 48: 245–257.
34. **Beck JD, Andreeff M, Mertelsmann R, Haghbin M, Tan C, Miller DR, Good RA, Gupta S** (1980) Childhood CML in blastic stage: An analysis of cell markers and cell kinetics. Am. J. Hematol. 9:337–344.
35. **Bennett JM, Catovsky D, Daniel MT, Flandrin G, Galton DAG, Gralnick HR, Sultan C** (1976) Proposals for the classification of the acute leukaemias. Br. J. Haematol. 33:451–458.
36. **Bennett JM, Catovsky D, Daniel MT, Flandrin G, Galton DAG, Gralnick HR, Sultan C** (1982) Proposals for the classification of the myelodysplastic syndromes. Br. J. Haematol. 51:189–199.
37. **Bennett JM, Catovsky D, Daniel MT, Flandrin G, Galton DA, Gralnick HR, Sultan C** (1985) Proposed revised criteria for the classification of acute myeloid leukemia. Ann. Intern. Med. 103:626–629.
38. **Bennett JM, Catovsky D, Daniel MT, Flandrin G, Galton AG, Gralnick HR, Sultan C** (1985) Criteria for the diagnosis of acute leukemia of megakaryocyte lineage (M7). Ann. Intern. Med. 103:460–462.
39. **Bettelheim P, Paietta E, Majdic O, Gadner H, Schwarzmeier J, Knapp W** (1982) Expression of a myeloid marker on TdT-positive acute lymphocytic leukemia cells: Evidence by double-fluorescence staining. Blood 60:1392–1396.
40. **Blumenreich MA, Woodcock TM, Andreeff M, Hiddemann W, Chou T-C, Vale K, O'Hehir M, Clarkson BD, Young CW** (1984) Effect of very high-dose thymidine infusions on leukemia and lymphoma patients. Cancer Res. 44:2203–2207.
41. **Buechner T, Barlogie B, Hiddemann W, Kamanbroo D, Goehde W** (1974) Accumulation of S-phase cells in the bone marrow of patients with acute leukemia by cytosin-arabinoside. Blut 28:399.
42. **Buechner T, Dittrich W, Goehde W** (1971) Die Impulscytophotometrie in der Haematologischen Cytologie. Klin. Wochenschr. 49:1090–1092.
43. **Buechner T, Goehde W, Schumann J, Barlogie B** (1976) In vivo kinetic response of human leukemia cells to antileukemic drugs as seen in the DNA histogram. Exerpta Medica Int. Cong. Ser. No. 415, Topics Hematol., pp 265–269.
44. **Buechner T, Hiddemann W, Schneider R, Kamanabroo D, Goehde W** (1973) Untersuchungen ueber zellkinetische Effekte der Leukaemiebehandlung in der Klinik anhand des DNS-Histogramms mittels Impulscytophotometrie. Med. Welt. 24:1616–1617.
45. **Burck KB, Liu ET, Larrick JW** (1988) "Oncogenes: An Introduction to the Concept of Cancer Genes." New York: Springer-Verlag.
46. **Chan LC, Pegram SM, Greaves MF** (1985) Contribution of immunophenotype to the classification and differential diagnosis of acute leukemia. Lancet 1: 475–479.
47. **Chiao JW, Andreeff M, Freitag WB, Arlin Z** (1982) Induction of in vitro proliferation and maturation of human aneuploid myelogenous leukemic cells. J. Exp. Med. 155:1357–1369.
48. **Civin CI, Strauss LC, Brovall C, Fackler MJ, Schwartz JF, Shaper JH** (1984) Antigenic analysis of hematopoiesis. III. A hematopoietic progenitor cell surface antigen defined by a monoclonal antibody raised against KG-la cells, J. Immunol. 133:157–165.
49. **Clark R, Peters S, Hoy T, Smith S, Whittaker K, Jacobs A** (1986) Prognostic importance of hypodiploid hemopoietic precursors in myelodysplastic syndromes. N. Engl. J. Med. 314:1472–1473.
50. **Clarkson B, Gee T, Arlin Z, Mertelsmann R, Kempin S, Dinsmore R, O'Reilly R, Andreeff M, Berman E, Higgins C, Little C** (1984) Cirrincione C and Ellis S: Current status of treatment of acute leukemia in adults: An overview. In Buechner T, Urbanitz D, Van de Loo J (eds), "Therapie der akuten Leukaemien." Berlin: Springer-Verlag, pp 1–31.
51. **Clarkson B, Strife A, Fried J, Gulati S** (1986) Cytokinetics and cancer treatment. In Freireich EJ, Frei T (eds), "Proceedings of the General Motors Cancer Research Foundation." Philadelphia: J.B. Lippincott, pp 131–190.
52. **Cossman J, Neckers LM, Arnold A, Korsmeyer SJ** (1982) Induction of differentiation in a case of common acute lymphoblastic leukemic. N. Engl. J. Med. 307:1251–1254.

53. **Coulter WH (1956)** High speed automatic blood cell counter and cell size analyzer. Proc. Natl. Electron. Conf. 12:1034–1042.
54. **Crissman HA, Oishi N, Steinkamp JA (1988)** Analysis of DNA replication patterns in BrdU-labeled cell populations and conditions for correlated measurement of RNA and FITC-labeled components. Cytometry Suppl. 2:18.
55. **Crist W, Pullen J, Boyett J, Falletta J, van Eys J, Borowitz M, Jackson J, Dowell B, Russell C, Quddus F, Ragab A, Vietti T (1988)** Acute lymphoid leukemia in adolescents: Clinical and biologic features predict a poor prognosis. A Pediatric Oncology Group Study. J. Clin. Oncol. 6:34–43.
56. **Daley GQ, McLaughlin J, Witte ON, Baltimore D (1987)** The CML-specific p210 bcr/abl protein, unlike v-abl, does not transform NIH/3T3 fibroblasts. Science 237:532–536.
57. **Darzynkiewicz Z, Andreeff M, Traganos F, Sharpless T, Melamed MR (1978)** Discrimination of cycling and noncycling lymphocytes by BUdR-suppressed acridine orange fluorescence in a flow cytometric system. Exp. Cell. Res. 115:31–35.
58. **Darzynkiewicz Z, Traganos F, Andreeff M, Sharpless T, Melamed MR (1979)** Different sensitivity of chromatin to acid denaturation in quiescent and cycling cells as revealed by flow cytometry. J. Histochem. Cytochem. 27:478–485.
59. **Darzynkiewicz Z, Traganos F, Kapuscinski J, Staiano-Coico L, Melamed MR (1984)** Accessibility of DNA in-situ to various fluorochromes: Relationship to chromatin changes during erythroid differentiation of Friend leukemia cells. Cytometry 4:355–363.
60. **Darzynkiewicz Z, Traganos F, Sharpless T, Friend C, Melamed MR (1976)** Nuclear chromatin changes during erythroid differentiation of Friend virus induced leukemia cells. Exp. Cell. Res. 99:301–309.
61. **Dean PN, Dolbeare F, Gratzner H, Rice GC, Gray JW (1984)** Cell-cycle analysis using a monoclonal antibody to BrdUrd. Cell Tissue Kinet. 17:427–436.
62. **Dewald G, Davis MP, Pierre RV, O'Fallon JR, Hoagland HC (1985)** Clinical characteristics and prognosis of 50 patients with a myeloproliferative syndrome and deletion of part of the long arm of chromosome 5. Blood 66:189–197.
63. **Doermer P, Lau B, Wilmanns W (1980)** Kinetics of bone marrow cell production in human acute and chronic myeloid leukemia. Leukemia Res. 4: 231–237.
64. **Doermer P, Theml H, Lau B (1983)** Chronic lymphocytic leukemia: A proliferative or accumulative disorder. Leukemia Res. 7:1–10.
65. **Doermer P, Ucci G, Hershko CH, et al. (1989)** Cell kinetics in leukemia and preleukemia. In Buechner T, Schellong G (eds), "Acute Leukemias." New York: Springer-Verlag.
66. **Doermer P, Ucci D, Lau B, et al. (1984)** In-vivo production of childhood acute lymphoblastic leukemia cells in relation to ploidy and immunological subtype. Leukemia Res. 8:587–595.
67. **Dolbeare F, Gratzner H, Pallavicini MG, Gray JW (1983)** Flow cytometric measurement of total DNA content and incorporated bromodeoxyuridine. Cell Biol. 80:5573–5577.
68. **Dosik GM, Barlogie B, Goehde W, Johnston D, Tekell JL, Drewinko B (1980)** Flow cytometry of DNA content in human bone marrow: A critical reappraisal. Blood 55:734–740.
69. **Dow LW, Chang LJA, Tsiatis AA, et al. (1982)** Relationship of pretreatment lymphoblast proliferative activity and prognosis in 97 children with acute lymphoblastic leukemia. Blood 59:1197–1202.
70. **Drapkin RL, Andreeff M, Koziner B, Strife A, Wisniewski D, Darzynkiewicz Z, Melamed MR, Clarkson B (1979)** Subpopulations of human peripheral blood cells: Analysis of granulocytic progenitor cells by flow cytometry and immunologic markers. Am. J. Hematol. 7:163–172.
71. **Drewinko B (1987)** Flow cytochemistry in the diagnosis of acute leukemia. In Stass SA (ed), "The Acute Leukemias: Biologic, Diagnostic and Therapeutic Determinants." New York: Marcel Dekker, pp 131–152.
72. **Fearon ER, Burke PJ, Schiffer CA, Zehnbauer BA, Vogelstein B (1986)** Differentiation of leukemia cells to polymorphonuclear leukocytes in patients with acute nonlymphocytic leukemia. N. Engl. J. Med. 315:51–58.
73. **Ffrench M, Byron PA, Fiere D, Vu-Van H, Gentilhomme O, Adeleine P, Viala JJ (1985)** Cell-cycle, protein content and nuclear size in acute myeloid leukemia. Cytometry 6:47–53.
74. **Ffrench M, Bryon PA, Fiere D, Vu-Van H, Guyotat D, Extra JM, Viala JJ (1986)** Cell cycle prognostic value in adult acute myeloid leukemia. The choice of the best variables. Leukemia Res. 10:51–57.
75. **Foon KA, Gale RP, Todd RF (1986)** Recent advances in the immunologic classification of leukemia. Semin. Hematol. 23:257–283.
76. **Foon KA, Todd RF (1986)** Immunologic classification of leukemia and lymphoma. Blood 68:1–31.
77. **Foucar K, Landgon II RM, Armitage JO, Olson D, Carroll TJ (1985)** Myelodysplastic syndromes. Cancer 56:553–561.
78. **Freedman AS, Nadler LM (1987)** Cell Surface markers in hematologic malignancies. Semin. Oncol. 14: 193–221.
79. **Fried J, Arlin Z, Clarkson BD (1976)** Flow microfluorometry of bone marrow cells from a patient with acute lymphocytic leukemia during chemotherapy. In Gohde, Schuman, Buchner (eds), "Pulse Cytophotometry." Ghent: European Press, pp 350–361.
80. **Ganapathi R, Grabowski D, Turinic R, Valenzuela R (1984)** Correlation between potency of calmodulin inhibitors and effects on cellular levels and cytotoxic activity of doxorubicin (adriamycin) in resistant p388 mouse leukemia cells. Eur. J. Cancer Clin. Oncol. 20:799–806.
81. **Ganapathi R, Reiter W, Krishan A (1982)** Intracellular adriamycin levels and cytotoxicity in adriamycin-sensitive and adriamycin-resistant P388 mouse leukemia cells. J. Natl. Cancer Inst. 68:1027–1032.
82. **Gaynor J, Chapman D, Little C, Andreeff M, Gee T, Clarkson B (1988)** A cause-specific hazard rate analysis of prognostic factors among adult patients with acute lymphoblastic leukemia: The Memorial Hospital Experience. J. Clin Oncol. 6:1014–1030.
83. **Gerdes J, Lemke H, Balach HM, Backer HH, Schwab U, Stein HJ (1984)** Cell cycle analysis of a cell proliferation-associated human nuclear antigen defined

by the monoclonal antibody Ki67. J. Immunol. 133: 1710–1715.

84. **Gratzner HG (1982)** Monoclonal antibody to 5-bromo- and 5-iododeoxyuridine: A new reagent for detection of DNA replication. Science 218:474–475.

85. **Gratzner HG, Leif RC (1981)** An immunofluorescence method for monitoring DNA synthesis by flow cytometry. Cytometry 1:385–389.

86. **Gray JW, Pinkel D, Trask B, van den Engh G, Segraves R, Collins C, Andreeff M, Haimi J, Perle MA (1988)** Application of flow karyotyping and fluorescence in-situ hybridization to detection of chromosome aberrations. In Karr JP, Coffey DS, Gardner W Jr (eds), "Prognostic Cytometry and Cytopathology of Prostate Cancer." New York: Elsevier, pp 304–313.

87. **Greaves MF, Chan LC, Furley AJW, Watt SM, Molgaard HV (1986)** Lineage promiscuity in hematopoietic differentiation and leukemia. Blood 1:1–11.

88. **Griffin JD, Beveridge RP, Schlossman SF (1982)** Isolation of myeloid progenitor cells from peripheral blood of chronic myelogenous leukemia patients. Blood 60:30–37.

89. **Griffin JD, Davis R, Nelson DA, Davey FR, Mayer RJ, Schiffer C, McIntyre OR, Bloomfield CD (1986)** Use of surface marker analysis to predict outcome of adult acute myeloblastic leukemia. Blood 68:1232–1241.

90. **Griffin JD, Larcom P, Schlossman SF (1983)** Use of surface markers to identify a subset of acute myelomonocytic leukemia cells with progenitor cell properties. Blood 62:1300–1303.

91. **Griffin JD, Ritz J, Nadler LM, Schlossman SF (1981)** Expression of myeloid differentiation antigens in normal and malignant myeloid cells. J. Clin. Invest. 68: 932–941.

92. **Ha K, Freedman MH, Hrincu A, Petsche D, Poon A, Gelfand EW (1985)** Separation of lymphoid and myeloid blasts in the mixed blast crisis of chronic myelogenous leukemia: No evidence for Ig gene rearrangement in CALLA-positive blasts. Blood 66: 1404–1408.

93. **Hagenbeek A, Martens AC (1985)** Detection of minimal residual disease in acute leukemia: Possibilities and limitations. Eur. J. Cancer Clin. Oncol. 21:389–395.

94. **Haimi J, Dolbeare F, Andreeff M (1988)** Cell kinetics of low-dose cytosine arabinoside therapy in acute non-lymphocytic leukemia and myelodysplastic syndromes. Proc. AACR 29:27.

95. **Haimi J, Redner A, Steinherz P, Andreeff M (1988)** Quantitative cytoreduction and cell kinetics during induction therapy in acute lymphoblastic leukemia (ALL). Blood 72 (suppl. 1):202a.

96. **Hashimi L, Al-Katib A, Mertelsmann R, Mohamed AN, Koziner B (1986)** Cytofluorometric detection of chronic myelocytic leukemia supervening in a patient with chronic lymphocytic leukemia. Am. J. Med. 80: 269–275.

97. **Hegewisch S, Andreeff M, Bertino J (1988)** Correlation of P-glycoprotein (Pgp180) expression and response to therapy in human leukemia. Blood 72 (suppl. 1):204a.

98. **Heisterkamp N, Stam K, Groffen J, Klein A, Grosveld G (1985)** Structural organization of the bcr gene and its role in the pH′ translocation. Nature (Lond.) 316: 758–761.

99. **Hiddemann W, Buechner T, Andreeff M, Woermann B, Melamed MR, Clarkson BD (1982)** Cell kinetics in acute leukemia. A critical re-evaluation based on new data. Cancer 50:250–258.

100. **Hiddemann W, Buechner T, Andreeff M, Arlin Z, Woermann B, Clarkson BD (1983)** Vergleich der Therapieeffekivitaet von zwei Induktionsprotokollen bei akuter myeloischer Leukaemie (AML) mittels exakter Quantifizierung der Knochenmarkzellularitaet. Onkologie 6:179–183.

101. **Hiddemann W, Buechner T, Andreeff M, Wormann B, Melamed MR, Clarkson BD (1982)** Bone marrow biopsy instead of "marrow juice" for cell kinetic analysis. Comparison of bone marrow biopsy and aspiration material. Leukemia Res. 6:601–612.

102. **Hiddemann W, Clarkson BD, Buechner T, Melamed MR, Andreeff M (1982)** Bone marrow cell count per mm^3 bone marrow: A new parameter for quantitating therapy induced cytoreduction in acute leukemia. Blood 59:216–255.

103. **Hiddemann W, Ritter J, Woermann B, Budde M, Creutzig U, Schellong G, Buechner T, Riehm H (1985)** DNS-Aneuploidie bei Kindern mit akuten Leukaemien. I. Inzidenz und klinische Bedeutung im Rahmen der BFM-Studien. Klin. Paediatr. 197:215–220.

104. **Hiddemann W, Schumann J, Andreeff M, Barlogie B, Herman CJ, Leif RC, Mayall BH, Murphy RF, Sandberg AA, Schumann J (1984)** Convention of nomenclature for DNA cytometry. Cytometry 5:445–446; Cancer Gene Cytogenet. 13:181–183.

105. **Hiddemann W, Woermann B, Goehde W, Buechner T (1986)** DNA aneuploidy in adult patients with acute myeloid leukemia. Incidence and relation to patient characteristics and morphologic subtypes. Cancer 22:2146–2152.

106. **Hiddemann W, Woermann B, Ritter J, Thiel E, Goehde W, Lahme B, Henze G, Schellong G, Riehm H, Buechner T (1986)** Frequency and clinical significance of DNA aneuploidy in acute leukemia. Ann. NY Acad. Sci. 468:227–240.

107. **Hillen H, Haanen C, Wessels J (1975)** Pulse-cytophotometry as a monitor for treatment in acute leukemia. In Hannen CAM, Hillen HFP, Wessels JMC (eds), "First International Symposium of Pulse Cytophotometry." Ghent: European Press Medikon, pp 315–332.

108. **Hillen H, Wessels H, Haanen C (1975)** Bone marrow proliferation patterns in acute myeloblastic leukemia determined by pulse cytophotometry. Lancet 1:609–611.

109. **Hittelman WN, Menegaz SD, McCredie KB, Keating MJ (1984)** Premature chromosome condensation studies in human leukemia. 5. Prediction of early relapse. Blood 64:1067–1073.

110. **Hoelzer D, Gale RP (1987)** Acute lymphoblastic leukemia in adults: Recent progress, future directions. Semin. Haematol. 24:27–39.

111. **Holdrinet RSG, Egmond JV, Wessels JMC, Haanen C (1980)** A method for quantification of peripheral blood admixture in bone marrow aspirates. Exp. Hematol. 8:103–107.

112. **Holdrinet RS, Pennings A, Crenthe-Schonk AM,**

vanEgmon J, Wessels JM, Haanen C (1983) Flow cytometric determination of the S-phase compartment in adult acute leukemia. Acta Haematol. (Basel) 70:369–378.

113. **Holdrinet RS, Pennings A, van-Egmond J, Wessels JM, Haanen C (1983)** DNA flow cytometry of blood and bone marrow in chronic myelogenous leukemia. Acta Haematol. 69:98–105.

114. **Janossy G, Bollum FJ, Bradstock KF, Ashley J (1980)** Cellular phenotypes of normal and leukemic hemopoietic cells determined by analysis with selected antibody combinations. Blood 56:430–441.

115. **Kantarjian HM, Barlogie B, Pershouse M, Swartzendruber D, Keating MJ, McCredie KB, Freireich EJ (1985)** Preferential expression of double-stranded ribonucleic acid in tumor versus normal cells: Biological and clinical implications. Blood 66:39–46.

116. **Kawasaki ES, Clark SS, Coyne MY, Smith SD, Champlin R, Witte ON, McCormick FP (1988)** Diagnosis of chronic myeloid and acute lymphocytic leukemias by detection of leukemia-specific mRNA sequences amplified in vitro. Proc. Natl. Acad. Sci. 85:5698–5702.

117. **Koike T (1984)** Megakaryoblastic leukemia: Characterization and identification of megakaryocytes. Blood 64:683–692.

118. **Korsmeyer SJ, Arnold A, Bakhshi A, Ravetch JV, Siebenlist U, Hieter PA, Sharrow SO, LeBien TW, Kersey JH, Poplack DG, Leder P, Waldmann TA (1983)** Immunoglobulin gene rearrangement and cell surface antigen expression in acute lymphocytic leukemias of T and B-cell precursor origins. J. Clin. Invest. 71: 301–313.

119. **Korsmeyer SJ, Hieter PA, Ravetch JV, Poplack DG, Waldmann TA, Leder P (1981)** Development hierarchy of immunoglobulin gene rearrangements in human leukemic pre-B cells. Proc. Natl. Acad. Sci. USA 78:7096–7100.

120. **Koziner B, Stavnezer J, Al-Katib A, Gebhard D, Mittelman A, Andreeff M, Clarkson BD (1986)** Surface immunoglobulin light chain expression in pre-B cell leukemias. Ann. NY Acad. Sci. 468:211–226.

121. **Krishan A (1986)** Flow cytometric monitoring of anthracycline transport in tumor cells. Ann. NY Acad. Sci. 468:80–84.

122. **Krishan A, Sauerteig A, Gordon K, Swinkin C (1986)** Flow cytometric monitoring of cellular anthracycline accumulation in murine leukemic cells. Cancer Res. 46:1768–1773.

123. **Kubbies M, Poot M, Schindler D, Seyschab H, Hoehn H, Rabinovitch P (1988)** Flow cytometric analysis of continuous BrdUrd-labeling with Hoechst 33258/ ethidium bromide for cell cycle kinetics and clinical diagnosis. Cytometry 2 (suppl.):18.

124. **Kuo WL, Dolbeare F, Vanderlaan M, Thomas CB, Gray JB (1988)** Simultaneous cytometric analysis of bromo- and iodo-deoxyuridine labeled BN myelocytic leukemia. Cytometry 2 (suppl.):90.

125. **Kurki P, Ogata K, Tan EM (1988)** Monoclonal antibodies to proliferating cell nuclear antigen (PCNA)/ cyclin as probes for proliferating cells by immunofluorescence microscopy and flow cytometry. J. Immunol. Methods 109:49–59.

126. **Lange B, Ferrero D, Pessano S, Palumbo A, Faust J, Meo P, Rovera G (1984)** Surface phenotype of clonogenic cells in acute myeloid leukemia defined by monoclonal antibodies. Blood 64:693–700.

127. **Lange-Wantzin G (1977)** Nuclear labelling of leukaemic blast cells with triated thymidine triphosphate in 35 patients with acute leukaemia. Br. J. Haematol. 37:475–482.

128. **LeBeau MM, Albain KS, Larson RA, Vardiman JW, Davis EM, Blough RR, Golomb HM, Rowley JD (1986)** Clinical and cytogenetic correlations in 63 patients with therapy-related myelodysplastic syndromes and acute nonlymphocytic leukemia: Further evidence for characteristic abnormalities of chromosomes No. 5 and 7. J. Clin. Oncol. 4:325–345.

129. **LeBien TW, Bollum FJ, Yasmineh WG, Kersey JH (1982)** Phorbol ester-induced differentiation of a non T-, non B-leukemic cell line: Model for human lymphoid progenitor cell development. J. Immunol. 128: 1316–1320.

130. **Ligler GS, Brodsky I, Schlam ML, Fuscaldo KE (1985)** Cytogenetics and cell surface marker analysis in CML. 1. Prediction of phenotype of acute phase transformation. Leukemia Res. 9:1093–1098.

131. **Loken MR, Shah VO, Dattilio KL, Civin CI (1987)** Flow cytometric analysis of human bone marrow. I. Normal erythroid development. Blood 69:255–263.

132. **Loken MR, Shah VO, Dattilio KL, Civin CI (1987)** Flow cytometric analysis of human bone marrow. II. Normal B lymphocyte development. Blood 70:1316–1324.

133. **Look T (1985)** The emerging genetics of acute lymphoblastic leukemia: Clinical and biologic implications. Semin. Oncol. 12:92–104.

134. **Look AT, Melvin SL, Brown LK, Dockter ME, Roberson PK, Murphy SB (1984)** Quantitative variation of the common acute lymphoblastic leukemia antigen (gp 100) on leukemic marrow blasts. J. Clin. Invest. 73:1617–1628.

135. **Look AT, Melvin SL, Williams DL, Brodeur GM, Dahl GV, Kalwinsky DK, Murphy SB, Mauer AM (1982)** Aneuploidy and percentage of S-phase cells determined by flow cytometry correlate with cell phenotype in childhood acute leukemia. Blood 60:959–967.

136. **Lowenberg B, Bauman JG (1985)** Further results in understanding the subpopulation structure of AML: clonogenic cells and their progeny identified by differentiation markers. Blood 66:1225–1232.

137. **Lu L, Walker D, Broxmeyer HE, Hoffman R, Hu W, Walker E (1987)** Characterization of adult human marrow hematopoietic progenitors highly enriched by two-color cell sorting with MY10 and major histocompatibility class II monoclonal antibodies. J. Immunol. 139:1823–1829.

138. **Ludwig W-D (1989)** Possibilities and limitations of immunologic marker analysis for the detection of minimal residual disease in childhood ALL. In Riehm H (ed), "Childhood Leukemia: Why Not Cure All?" (personal communication).

139. **Maddox AM, Johnston DA, Barlogie B, Youness E, Keating M, Freireich EJ (1985)** DNA–RNA measurements in patients with acute leukemia undergoing remission induction therapy. J. Clin. Oncol. 3:799–808.

140. **Maiolo AT, Foa P, Mozzana R, Chiorboli O, Ciani A, Maisto A, Starace G (1982)** Flow cytometric anal-

ysis of cellular DNA in human acute nonlymphatic leukemias and dysmyelopoietic syndromes. Cytometry 2:265–267.

141. **Majdic O, Liszka K, Lutz D, Knapp W (1981)** Myeloid differentiation antigen defined by a monoclonal antibody. Blood 58:1127–1133.

142. **Matsuoka M, Hattori T, Chosa T, Tsuda H, Kuwata S, Yoshida M, Uchiyama T, Takatsuki K (1986)** T3 surface molecules on adult T-cell leukemia cells are modulated in vivo. Blood 67:1070–1076.

143. **McMichael AJ (ed)** (1987) Leucocyte typing III. White cell differentiation antigens. Oxford: Oxford University Press, pp 1050.

144. **Melamed MR, Adams LR, Zimring A, Murnick JG, Mayer K (1972)** Preliminary evaluation of acridine orange as a vital stain for automated differential leukocyte counts. Am. J. Clin. Pathol. 57:95–102.

145. **Mertelsmann R, Koziner B, Ralph P, Filippa D, McKenzie S, Arline ZA, Gee TS, Moore MAS, Clarkson BD (1978)** Evidence of distinct lymphocytic and monocytic populations in a patient with terminal transferase-positive acute leukemia. Blood 51:1051–1056.

146. **Meyers MB, Rittmann-Grauer L, O'Brien JP, Safa AR (1989)** Characterization of monoclonal antibodies to a 180 kDa P-glycoprotein: differential expression of the 180 kDa and 170 kDa P-glycoproteins in multidrug-resistant human tumor cells. Cancer Res. 49:3209–3214.

147. **Mirro J Jr, Behm FG (1987)** Immunophenotype in the characterization and Diagnosis of acute leukemia. In Stass S (ed), "Myeloid markers in the Acute Leukemias." New York: Marcel Dekker, pp 271–298.

148. **Mirro JJR, Sitchingman GR, Stass SA (1987)** Lineage heterogeneity in a acute leukemia. Acute mixed lineage leukemia and lineage switch. Acute Leukemias 6:383–402.

149. **Montecucco C, Riccardi A, Traversi E, Giordano P, Mazzini G, Ascari E (1983)** Proliferative activity of bone marrow cells in primary dysmyelopoietic (preleukemic) syndromes. Cancer 52:1190–1195.

150. **Mullaney PF, Van Dilla MA, Coulter JR, et al. (1969)** Cell sizing: A light scattering photometer for rapid volume determination. Rev. Sci. Instr. 40:1029–1034.

151. **Nadler LM, Ritz J, Bates MP, Park EK, Anderson KC, Sallan SE, Schlossman SF (1982)** Induction of human B-cell antigens in Non T-cell acute lymphoblastic leukemia. J. Clin. Invest. 70:433–442.

152. **Neame PB, Soamboonsrup P, Browman G, Barr RD, Saeed N, Chan B, Pai M, Benger A, Wilson EC, Walker IR, McBride JA (1985)** Simultaneous or sequential expression of lymphoid and myeloid phenotypes in acute leukemia. Blood 65:142–148.

153. **Paietta E, Andreeff M, Papenhausen P, Gucalp R, Wiernik P (1989)** Distinct antigen expression related to DNA ploidy in a case of biphenotypic leukemia. Leukemia 3:76–78.

154. **Paietta E, Mittermayer K, Schwarzmeier J (1980)** Proliferation kinetics and cyclic AMP as prognostic factors in adult acute leukemia. Cancer 46:102–108.

155. **Pallavicini MG, Summers LJ, Giroud FJ, Dean PN, Gray JW (1985)** Multivariate analysis and list mode processing of murine hemopoietic subpopulations for cytokinetic studies. Cytometry 6:539–549.

156. **Palumbo A, Minowada J, Erikson J, Croce CM, Rovera G (1984)** Lineage infidelity of a human myelogenous leukemia cell line: Blood 64:1059–1063.

157. **Pedersen-Bjergaard J, Vindelov L, Philip P, Ruutu P, Elmgreen J, Repo H, Christensen IJ, Killmann S-A, Jensen G (1982)** Varying involvement of peripheral granulocytes in the clonal abnormality -7 in bone marrow cells in preleukemia secondary to treatment of other malignant tumors: Cytogenetic results compared with results of flow cytometric DNA analysis and neutrophil chemotaxis. Blood 60:172–179.

158. **Peters SW, Clark RE, Hoy TG, Jacobs A (1986)** DNA content and cell cycle analysis of bone marrow cells in myelodysplastic syndromes (MDS). Br. J. Haematol. 62:239–245.

159. **Pinkel D, Landegent J, Collins C, Fuscoe J, Segraves R, Lucas J, Gray J (1988)** Fluorescence in situ hybridization with human chromosome-specific libraries: Detection of trisomy 21 and translocations of chromosome 4. Proc. Natl. Acad. Sci. USA 85: 9138–9142.

160. **Pinkel D, Straume T, Gray JW (1986)** Cytogenetic analysis using quantitative, high-sensitivity, fluorescence hybridization. Proc. Natl. Acad. Sci. USA 83: 2934–2938.

161. **Poppitt DG, McGown AT, Fox BW (1984)** The use of a fluorescent methotrexate probe to monitor the effects of three vinca alkaloids on a mixed population of parental L1210 and gene-amplified methotrexate-resistant cells by flow cytometry. Cancer Chemother. Pharmacol. 13:54–57.

162. **Preisler HD, Epstein J, Barcos M, et al. (1984)** Prediction of response of acute nonlymphocytic leukaemia to therapy with "high dose" cytosine arabinoside. Br. J. Haematol. 58:19–32.

163. **Prentice AG, Smith AG, Bradstock KF (1980)** Mixed lymphoblastic–myelomonoblastic leukemia in treated hodgkin's disease. Blood 56:129–133.

164. **Raskind WH, Tirumali N, Jacobson R, Singer J, Fialkow PJ (1984)** Evidence for a multistep pathogenesis of a myelodysplastic syndrome. Blood 63:1318–1323.

165. **Raza A, Bhayana R, Ucar K, Kirshner J, Preisler HD (1985)** Differences between labeling index and DNA histograms in assessing S-phase cells from a homogenous group of chronic phase CML patients. Cytometry 6:445–451.

166. **Raza A, Maheshwari Y, Preisler HD (1987)** Differences in cell cycle characteristics among patients with acute nonlymphocytic leukemia. Blood 69:1647–1653.

167. **Raza A, Preisler HD (1985)** Double labeling of human leukemic cells using ^{3}H-cytarabine and monoclonal antibody against bromodeoxyuridine. Cancer Treatm. Rep. 69:195–198.

168. **Raza A, Spiridonidis C, Ucar K, Mayers G, Bankert R, Preisler HD (1985)** Double labeling of S-phase murine cells with bromodeoxyuridine and a second DNA-specific probe. Cancer Res. 45:2283–2287.

169. **Redner A, Andreeff M (1989)** Clinical applications of acridine-orange flow cytometry to the diagnosis and treatment of pediatric acute leukemia (in preparation).

170. **Redner A, Andreeff M, Miller DR, Steinherz P,**

Melamed MR (1984) Recognition of CNS leukemia by flow cytometry. Cytometry 5:614–618.

171. **Redner A, Grabowski E, Al-Katib A, Andreeff A** (1986) Common acute lymphoblastic leukemia characterized by four different DNA stemlines with heterogeneity in DNA content, antigen expression and sensitivity to chemotherapy. Leukemia Res. 10:671–676.

172. **Redner A, Hegewisch S, Haimi J, Steinherz P, Jhanwar S, Andreeff M** (1989) A study of multidrug resistance and cell kinetics in a child with near-haploid acute lymphoblastic leukemia (in preparation).

173. **Redner A, Melamed MR, Andreeff M** (1986) Detection of central nervous system relapse in acute leukemia by multiparameter flow cytometry of DNA, RNA and CALLA and comparison with conventional cytology. In Andreeff M (ed), "Clinical Cytometry." Ann. NY Acad. Sci. 468:241–255.

174. **Riccardi A, Danova M, Montecucco G, Ucci G, Cassano E, Giordano M, Mazzini G, Giordano P: Adult acute non-lymphoblastic leukaemia** (1986) Reliability and prognostic significance of pretreatment bone marrow S-phase size determined by flow cytofluorometry. Scand. J. Haematol. 36:11–17.

175. **Ritter J, Hiddemann W, Woermann B, Buechner T, Schellong G** (1985) DNA aneuploidy in children with acute leukemia. II. Correlation with the phenotype of blasts, clinical picture and course of disease. Klin. Paediatr. 3:221–224.

176. **Rosowsky A, Wright JE, Cucchi CA, Boeheim K, Frei E** (1986) Transport of a fluorescent antifolate by methotrexate-sensitive and methotrexate-resistant human leukemic lymphoblasts. Biochem. Pharmacol. 35:356–360.

177. **Rosowsky A, Wright JE, Shapiro H, Beardsley P, Lazarus H** (1982) A new fluorescent dihydrofolate reductase probe for studies of methotrexate resistance. J. Biol. Chem. 257:14162–14167.

178. **Ross DD, Joneckis CC, Schiffer CA** (1986) Effects of verapamil on in vitro intracellular accumulation and retention of daunorubicin in blast cells from patients with acute nonlymphocytic leukemia. Blood 68:83–88.

179. **Ryan DH, Mitchell SJ, Hennessy LA, Bauer KD, Horan PK, Cohen HJ** (1984) Improved detection of rare CALLA-positive cells in peripheral blood using multiparameter flow cytometry. J. Immunol. Methods 74:115–128.

180. **Salzman GC, Crowell JM, Martin JC, et al.** (1975) Cell identification by laser light scattering: Identification and separation of unstained leukocytes. Acta Cytol. (Praha) 19:374–377.

181. **Sasaki K, Murakami T, Ogino T, Takahashi M, Kawasaki S** (1986) Flow cytometric estimation of cell cycle parameters using a monoclonal antibody to bromodeoxyuridine. Cytometry 7:391–395.

182. **Shank B, Andreeff M, Li D** (1983) Cell survival kinetics in peripheral blood and bone marrow during total body irradiation for marrow transplantation. Int. J. Radiat. Oncol. Biol. Phys. 9:1613–1623.

183. **Smets LA, Slater RM, Behrendt H, Vant-Veer MD, Homan-Block J** (1985) Phenotypic and karyotypic properties of hyperdiploid acute lymphoblastic leukaemia of childhood. Br. J. Haematol. 61:113–123.

184. **Smith LJ, Curtis JE, Messner HA, Senn JS, Furthmayr H, McCulloch EA** (1983) Lineage infidelity in acute leukemia. Blood 61:1138–1145.

185. **Smith LJ, McCulloch EA** (1984) Lineage infidelity following exposure of T-lymphoblasts (MOLT-3 cells) to 5-azacytidine. Blood 63:1324–1330.

186. **Sokal J, Gomez GA** (1986) The Philadelphia chromosome and Philadelphia chromosome mosaicism in chronic granulocytic leukemia. J. Clin. Oncol. 4: 140–111.

187. **Spangrude GJ, Heimfeld S, Weissman IL** (1988) Purification and characterization of mouse hematopoietic stem cells. Science 241:58–62.

188. **Speth PA, Linssen PC, Boezeman JB, Wessels HM, Haanen G** (1985) Quantitation of anthracyclines in human hematopoietic cell subpopulations by flow cytometry correlated with high pressure liquid chromatography. Cytometry 6:143–150.

189. **Squires J, Redner A, Andreeff M** (1988) P53 expression in human leukemia: Restriction to lymphoid cell lineage and potential prognostic importance in acute lymphoblastic leukemia. Proc. AACR 29:454.

190. **Stass S, Mirro J, Melvin S, Pui C-H, Murphy SB, Williams D** (1984) Lineage switch in acute leukemia. Blood 64:701–706.

191. **Strife A, Clarkson B** (1988) Biology of chronic myelogenous leukemia: Is discordant maturation the primary defect? Semin. Hematol. 25:1–19.

192. **Suarez C, Miller D, Steinherz PG, Melamed MR, Andreeff M** (1985) DNA and RNA determinations in 111 cases of childhood acute lymphoblastic leukemia (ALL) by flow cytometry: Correlation of FAB classification with DNA stemline and proliferation. Br. J. Haematol. 60:677–686.

193. **Tanke HJ, Niewenhuis IAB, Koper GJM, et al.** (1983) Flow cytometry of reticulocytes applied to clinical hematology. Blood 61:1091–1097.

194. **Tapiero H, Fourcade A, Vaigot P, Farhi JJ** (1982) Comparative uptake of adriamycin and daunorubicin in sensitive and resistant friend leukemia cells measured by flow cytometry. Cytometry 2:298–302.

195. **Taylor IW, Tattersall MH** (1981) Methotrexate cytotoxicity in cultured human leukemic cells studied by flow cytometry. Cancer Res. 41:1549–1558.

196. **Theml H, Trepel F, Schick P, et al.** (1973) Kinetics of lymphocytes in chronic lymphocytic leukemia: Studies using continuous ^{3}H-thymidine infusion in two patients. Blood 42:623–636.

197. **Thiebaut F, Tsuruo T, Hamada H, Goettesman MM, Pastan I, Willingham MC** (1987) Cellular localization of the multidrug-resistance gene product P-glycoprotein in normal human tissues. Proc. Natl. Acad. Sci. USA 84:7735–7738.

198. **Third International Workshop on Chromosomes in Leukemia** (1983) Chromosomal abnormalities and their clinical significance in acute lymphoblastic leukemia. Cancer Res. 43:868–873.

199. **Touw I, Delwel R, Bolhuis R, van-Zanen G, Lowenberg B** (1985) Common and pre-B acute lymphoblastic leukemia cells express interleukin 2 receptors and interleukin-2 stimulates in vitro colony formation. Blood 66:556–561.

200. **Traganos F, Darzynkiewicz Z, Sharpless T, Melamed MR** (1977) Simultaneous staining of ribonucleic and deoxyribonucleic acids in unfixed cells using acridine

orange in a flow cytofluorometric system. J. Histochem. Cytochem. 25:46–56.

201. **Ucci G, Riccardi A, Doermer P, Danova M** (1986) Rate and time of DNA synthesis of human leukaemic blasts in bone marrow and peripheral blood. Cell Tissue Kinet. 19:429–435.

202. **Uckun FM, Jaszxz W, Ambrus JL, Fauci AS, Gajl-Peczalska K, Song CW, Wick MR, Myers DE, Waddick K, Ledbetter JA** (1988) Detailed studies on expression and function of CD19 surface determinant by using B43 monoclonal antibody and the clinical potential of anti-CD19 immunotoxins. Blood 71:13–29.

203. **Uckun FM, Ledbetter JA** (1990) Immunobiologic differences between normal and leukemic human B-cell precursors. Proc. Natl. Acad. Sci. USA (in preparation).

204. **Visser JW, Martens AC, Hagenbeek A** (1986) Detection of minimal residual discasc in acute leukemia by flow cytometry. Ann. NY Acad. Sci. 468:268–275.

205. **Wagner HP, Hirt A** (1980) Sequential surface marker and cytokinetic studies on individual cells from children with acute lymphoid leukemia. Blood 56:648–652.

206. **Waldman FM, Dolbeare F, Gray JW** (1985) Detection of cytosine arabinoside resistant cells at low frequency using the bromodeoxyuridine/DNA assay. Cytometry 6:657–662.

207. **Walker LMS, Sandler RM** (1978) Acute myeloid leukaemia with monosomy-7 follows acute lymphoblastic leukaemia. Br. J. Haematol. 38:359–366.

208. **Walters RS, Johnston DA, Dixon DO, Stass SA, Keating MJ, Trujillo JM, McCredie KB, Barlogie B** (1987) Nucleic acid flow cytometry: An aid to diagnosis and prognosis in acute leukemia in adults. In Stass SA (ed), "The Acute Leukemias." 6:203–245.

209. **Weinberg DS, Pinkus GS, Ault KA** (1984) Cytofluorometic detection of B cell clonal excess: A new approach to the diagnosis of B cell lymphoma. Blood 63:1080–1087.

210. **White RA, Meistrich ML** (1986) A comment on "a method to measure the duration of DNA synthesis and the potential doubling time from a single sample." Cytometry 7:486–490.

211. **Yataganas X, Strife A, Perez A, Clarkson B** (1974) Microfluorimetric evaluation of cell kill kinetics with 1-β-D-Arabinofuranosylcytosine. Cancer Res. 34: 2795–2806.

36

Flow Cytometry of Lymphoma

Michael Andreeff

Memorial Sloan-Kettering Cancer Center and Cornell University Medical College, New York, New York 10021

INTRODUCTION

Lymphoma is the eighth most common type of cancer in the United States (29,700 cases) and the seventh most common cause of cancer death (14,200 cases). The assessment of lymph nodes suspected to contain malignant lymphoma has developed into a fine art for specialized pathologists. Years of training enable the hematopathologist to interpret subtle histologic patterns, cytologic textures and clinical associations to diagnose and classify Hodgkin's (HD) and non-Hodgkin's (NHL) lymphomas. A number of taxonomic systems were developed, most notable those proposed by Rappaport (Table 1), Lukes and Collins, Dorfman, Lennert (Kiel classification), and the World Health Organization (WHO). These classifications are difficult to compare; in 1982 a Working Formulation was established to develop a common system that could be translated into, and compared with, the existing classifications [116,130,169]. It identified 10 major types using morphologic criteria only, grouped into low-, intermediate-, and high-grade lymphomas. While these groups may reflect the natural history of NHL [96], it is a remarkable paradox that the low-grade lymphomas cannot be cured, whereas a significant number of high-grade lymphomas is amenable to curative therapy. Two additional problems are to be considered: (1) the lack of immunological, cell kinetic, and molecular features in the working formulation; and (2) its questionable reproducibility. It is of interest that no data regarding the inter- and intraobserver reproducibility of the Working Formulation were published in the original paper [169]. Berard et al. [19] reported unacceptable reproducibility. Furthermore, the remarkable advances in our understanding of the molecular events underlying NHL and their immunological and kinetic consequences have not yet been incorporated into a unified, biology-based taxonomic system. Flow cytometry (FCM) has been used predominantly for immunological phenotyping and cell cycle/ploidy (DNA) analysis. As expected for leukemias, it is hoped that the availability of new probes and techniques, particularly those reflecting molecular events, will permit the development of a biology-based taxonomic system for NHL in the future.

Few FCM studies have been conducted on HD, mainly due to the rare presence of the Reed–Sternberg (RS) cell, but the node infiltrating lymphocytes have been characterized and Hodgkin cell lines have been analyzed with the aim of identifying the origin and function of the RS cell [69].

It is hoped that prognostic models [26,52,129,149,169] will permit a much more targeted therapeutic approach than is possible today. This chapter discusses the advances in immunophenotyping of lymphomas, the role of cell cycle analysis and aneuploidy, the detection of minimal disease in staging and treatment monitoring, studies of drug effects and drug resistance, and finally the emerging investigations of oncogene expression in the lymphomas.

IMMUNOPHENOTYPING OF LYMPHOMA

The recognition of lineages and discrete steps in lymphocyte differentiation permits an immunological classification of NHL [117]. A large panel of monoclonal antibodies is now available for the identification of B cells, T cells, and other rare classes of lymphocytes [78] (Table 2). Attempts have been made to identify malignant cells with marker profiles identical to those of normal cells (Fig. 1) [5,158]. Thus, antigens shown to be B-cell-restricted on normal lymphocytes are, by definition, expressed only on B-cell lymphomas. These are divided into three major subgroups: pre-B cell stage (non-T acute lymphoblastic leukemia), mid-B-cell stage (poorly differentiated lymphomas), and secretory B-cell stage (large cell lymphomas and plasma cell tumors).

Panels of antigens defining stages of B- and T-cell maturation were recently tabulated by Foucar et al. [79] (Tables 3,4,5). The CD nomenclature is explained in chapter 35 of this volume. Monoclonal antibodies used for antigen recognition are listed in Table 2. The stepwise acquisition and loss of antigenic determinants during T- and B-cell differentiation is shown in Figure 9 of Chapter 35 (this volume).

The diagnosis of B-cell lymphoma is established when there is a significant increase in the percentage (>50%) of pan-B antigen-bearing cells [115], accompanied by a decrease in T cells. In some cases, staining for cytoplasmic immunoglobulin (Ig) by immunoperoxidase can be helpful. Intracytoplasmic Ig can also be detected by FCM [11]. A

Flow Cytometry and Sorting, Second Edition, pages 725–743

TABLE 1. Classification, Incidence, Clinical Presentation, and Survival of Non-Hodgkin's and Hodgkin's Lymphomas

A. Classification of Malignant Lymphomas[a] (Rappaport)

Nodular (follicular) lymphomas	Diffuse lymphomas	Hodgkin's disease
Lymphocytic, poorly differentiated (NPDL)	Lymphocytic, well-differentiated (DWDL)	Nodular sclerosis
Mixed lymphocytic–histiocytic (NML)	Lymphocytic, intermediate differentiation (DIL)	Lymphocyte predominant
Histiocytic (NHL)	Lymphocytic, poorly differentiated (DPDL)	Mixed cellularity
	Mixed lymphocytic–histiocytic (DML)	Lymphocyte depleted
	Histiocytic (DHL)	
	Undifferentiated—Burkitt's	
	Undifferentiated—non-Burkitt's	
	Lymphoblastic	
	Unclassified	

B. Incidence of Non-Hodgkin's Lymphoma by Histology

		Percentage	Percentage 5-year survival
Nodular	40%		
Lymphocytic, poorly differentiated (NPDL)		30	55–75
Mixed lymphocytic–histiocytic (NML)		10	65
Histiocytic (NHL)		<5	25
Diffuse	60%		
Lymphocytic, well-differentiated (DWDL)		<5	40–80
Lymphocytic, poorly differentiated (DPDL)		20	5–35
Mixed lymphocytic–Histiocytic (DML)		5	25
Histiocytic (DHL)		30	20
Undifferentiated (DUL)		<5	<20

C. Clinical Presentation

Nodular lymphomas	Diffuse large cell ("histiocytic") lymphoma
Stage I and II (<10%)	Stage I and II (~30%)
Widespread nodal involvement at presentation (greater than 90% of cases)	Only 40% of patients have disease on both sides of the diaphragm
Extranodal sites of disease rare	Extranodal disease present in 40% of patients (GI tract, skin, bone)
Bone marrow involvement very common at presentation	Bone marrow involvement uncommon at presentation (when present, risk of lymphomatous meningitis)
Liver and spleen: small, uniform miliary nodules	Liver and spleen: large destructive tumor masses

[a]Modified by Berard [19].

comparison of immunohistochemical and FCM assessment of cell-surface immunoglobulin (SIg) was published by Gebhard et al. [81]. Other groups have also published their experience in lymphoma/phenotyping [58,62,83,86,137,153] (Table 5).

Normal lymph node B cells are polyclonal, encompassing both κ- and λ-light-chain-bearing cells. B-cell lymphoma can be diagnosed when (1) light-chain restriction is evident ("clonal excess" of κ/λ) [177], or (2) a restricted heavy chain is present in a >3:1 ratio over the total percentage of cells expressing other heavy chains. However, κ/λ ratios vary greatly, and variations can occur with other processes [115]. Liendo et al. [112] examined lymph node suspensions from 80 patients and found monoclonal SIg in 43 of 47 (91%) of nodes from NHL, but also in 3 of 11 (27%) of hyperplastic lymph nodes. De Martini et al. [57] analyzed 271 cases of B-cell NHL and found 13% SIg-negative cases. Light-chain monoclonality was most frequent in low-grade and follicular center cell lymphomas, whereas SIg-negative cases were most common in high-grade NHL. Low-grade NHL had high median percentages of neoplastic cells and low T-cell numbers, whereas high-grade NHL exhibited lower and less uniform median percentages of B cells, with higher numbers of T cells and lower CD4/CD8 ratios than did low-grade lymphomas.

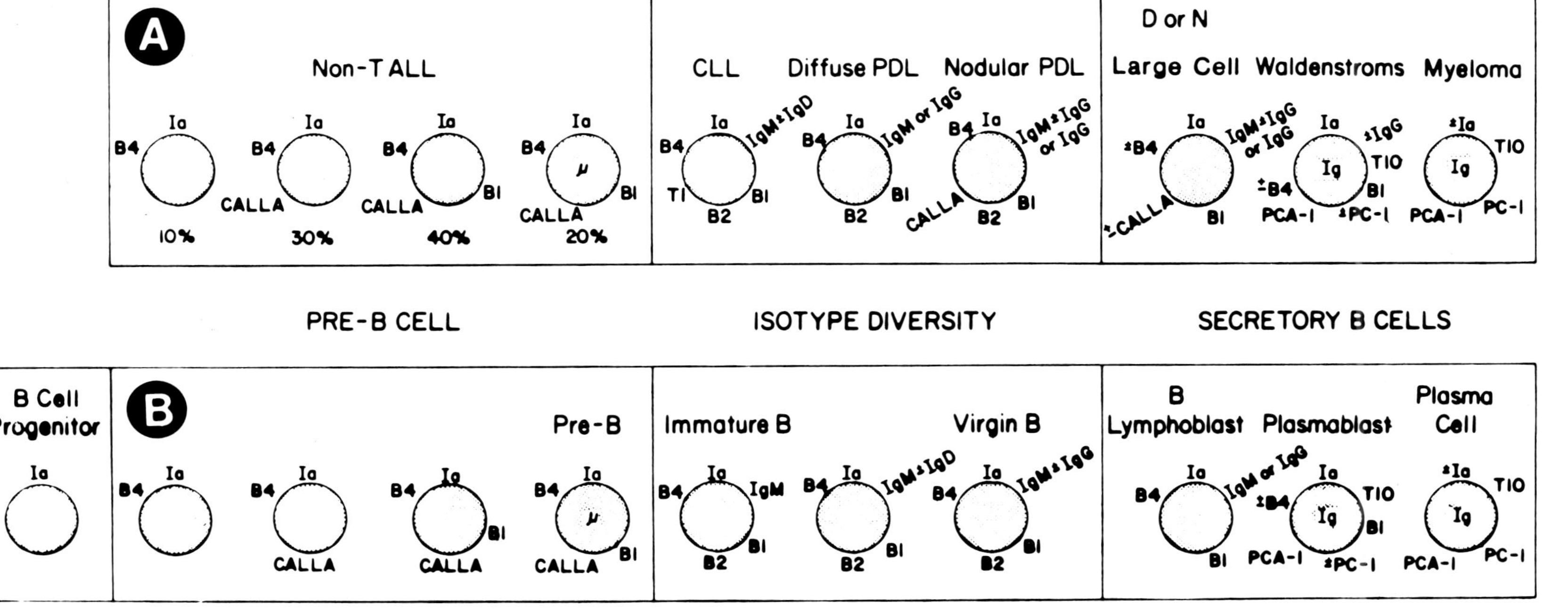

Fig. 1. Comparison of normal and malignant B-cell phenotypes. Hypothetical model relating the malignant B cell to its normal cellular counterpart on the basis of cell-surface phenotype. **A:** Human B-cell tumors. **B:** Normal B-cell counterpart. (Reproduced from Anderson et al. [5], with permission of the publisher.)

TABLE 2. CD Antigens*

CD design.	Selection of assigned monoclonal antibodies	Main cellular reactivity	CD design.	Selection of assigned monoclonal antibodies	Main cellular reactivity
CD1a	NA1/34; T6; VIT6; Leu6	Thy, DC, B subset	CD40	G28-5	B, carcinomas
CD1b	WM-25; 4A76; NUT2	Thy, DC, B subset	CD41	PBM 6.4; CLB-thromb/7; PL273	Plt
CD1c	L161; M241; 7C6; PHM3	Thy, DC, B subset	CD42a	FMC25; BL-H6; GR-P	Plt
CD2	9.6; T11; 35.1	T	CD42b	PHN89; AN51; GN287	Plt
CD2R	T11.3; VIT13; D66	activated T	CD43	OTH 71C5; G19-1; MEM-59	T, G, brain
CD3	T3; UCHT1; 38.1; Leu4	T	CD44	GRHL1; F10-44-2; 33-383; BRIC35	T, G, brain, RBC
CD4	T4; Leu3a; 91.D6	T subset	CD45	T29/33; BMAC 1; AB187	Leucocytes
CD5	T1; UCHT2; T101; HH9; AMG4	T, B subset	CD45RA	G1-15; FB-11-13; 73.5	T subset, B, G, M
CD6	T12; T411	T, B subset	CD45RB	PT17/26/16	T subset, B, G, M
CD7	3A1; 4A; CL1.3; G3-7	T	CD45RO	UCHL1	T subset, B, G, M
CD8	alpha-chain; T8; Leu2a; M236; UCHT4; T811	T subset	CD46	HULYM5; 122-2; J4B	Leucocytes broad
	beta-chain; T8/2T8-5H7		CD47	BRIC 126; CIKM1; BRIC 125	Broad
CD9	CLB-thromb/8; PHN200; FMC56	Pre-B, M, Plt	CD48	WM68; LO-MN25; J4-57	Leukocytes
CD10	J5, VILA1, BA-3	Lymph.Prog., cALL, Germ Ctr. B, G	CDw49b	CLB-thromb/4; Gi14	Plt, cultured T
CD11a	MHM24; 2F12; CRIS-3	Leukocytes, broad	CDw49d	B5G10; HP2/1; HP1/3	M, T, B, (LHC), Thy
CD11b	Mo1; 5A4.C5; LPM19C	M, G, NK	CDw49f	GoH3	Plt, (T)
CD11c	B-LY6; L29; BL-4H4	M, G, NK, B sub	CDw50	101-1D2; 140-11	Leucocytes broad
CDw12	M67	M, G, Plt	CD51	13C2; 23C6; NKI-M7; NKI-M9	(Plt)
CD13	MY7, MCS-2, TUK1, MOU28	M, G	CDw52	097; YTH66.9; Campath-1	Leucocytes
CD14	Mo2, UCHM1, VIM13, MoP15	M, (G), LHC	CD53	HI29; HI36; MEM-53; HD77	Leucocytes
CD15	My1, VIM-D5	G, (M)	CD54	7F7; WEHI-CAMI	Broad, Activ.
CD16	BW209/2; HUNK2; VEP13; Leu11c	NK, G, Mac.	CD55	143-30; BRIC 110; BRIC 128; F2B-7.2	Broad
CDw17	GO35, Huly-m13	G, M, Plt	CD56	Leu19; NKH1; FP2-11.14, L185	NK, activ. lymphocytes

CD18	MHM23; M232; 11H6; CLB54	Leucocytes broad
CD19	B4; HD37	B
CD20	B1; 1F5	B
CD21	B2; HB5	B subset
CD22	HD39; S-HCL1; To15	cytopl. B/surface B subset
CD23	Blast-2, MHM6	B subset, act.M, Eo
CD24	VIBE3; BA-1	B, G
CD25	TAC; 7G7/B6; 2A3	activated T, B, M
CD26	134-2C2; TS145	activated T
CD27	VIT14; S152; OKT18A; CLB-9F4	T subset
CD28	9.3; KOLT2	T subset
CD29	K20; A-1A5	broad
CD30	Ki-1; Ber-H2; HSR4	activated T, B; Sternberg-Reed
CD31	SG134; TM3; HEC-75; ES12F11	Plt, M, G, B, (T)
CDw32	CIKM5; 41H16; IV.3	M, G, B
CD33	My9; H153; L4F3	M, Prog., AML
CD34	My10, BI-3C5, ICH-3	Prog
CD35	TO5, CB04, J3D3	G, M, B
CD36	5F1, CIMeg1; ESIVC7	M, Plt, (B)
CD37	HD28; HH1; G28-1	B, (T, M)
CD38	HB7; T16	Lymph. Prog., PC, Act. T
CD39	AC2; G28-2	B subset, (M)
CD57	Leu7; L183; L187	NK, T, B sub, Brain
CD58	G26; BRIC 5; TS2/9	Leucocytes, Epithel
CD59	Y53.1; MEM-43	Broad
CDw60	M-T32; M-T21; M-T41; UM4D4	T sub
CD61	Y2/51; CLB-thromb/1; VI-PL2; BL-E6	Plt
CD62	CLB-thromb/6; CLB-thromb/5; RUU-SP1.18.1	Plt activ.
CD63	RUU-SP2.28; CLB-gran/12	Plt activ., M, (G, T, B)
CD64	Mab32.2; Mab22	M
CDw65	VIM2; HE10; CF4; VIM8	G, M
CD66	CLB gran/10; YTH71.3	G
CD67	B13.9; G10F5; JML-H16	G
CD68	EBM11; Y2/131; Y-1/82A; Ki-M7; Ki-M6	Macrophages
CD69	MLR3; L78; BL-Ac/p26; FN50	activated B, T
CDw70	Ki-24; HNE 51; HNC 142	activated B, -T, Sternberg-Reed cells
CD71	138-18; 120-2A3; MEM-75; VIP-1; Nu-TfR2	Proliferating cells, Mac.
CD72	S-HCL2; J3-109; BU-40; BU-41	B
CD73	1E9.28.1; 7G2.2.11; AD2	B subset, T subset
CD74	LN2; BU-43; BU-45	B, M
CDw75	LN1; HH2; EBU-141	mature B, (T subset)
CD76	HD66; CRIS-4	mature B, T subset
CD77	38.13(BLA); 424/4A11; 424/3D9	restr. B
CDw78	Anti Ba; LO-panB-a; 1588	B, (M)

*For abbreviations, see Table 1 footnote, Chapter 35, this volume.

TABLE 3. Antigenic Profiles of Stages of Normal T-Cell Maturation*

	Blood and other sites			Peripheral T cells of either helper or suppressor phenotype	
	Immature thymocyte	Common thymocyte	Mature thymocyte		
CD2	+	+	+	+	+
CD38	+	+	+		
T9	+				
CD5	+	+	+	+	+
CD7	+	+	+	+	+
CD1		+			
CD4		+	+	+	
			or	or	
CD8		+	+		+
TdT	+	+	±		
CD3			+	+	+

*Reproduced from Foucar et al. [79], with permission of the publisher.

Johnson et al. [98] studied clonal excess in fine-needle aspirates from NHL patients. In primary diagnostic procedures, light-chain analysis established a diagnosis of lymphoma in 5 of 14 (36%) aspirates from patients with poorly differentiated tumors. In aspirates performed as part of staging procedures, 7 of 19 (37%) showed abnormal light-chain analysis.

Lindh et al. [114] studied blood from 69 patients with NHL and found monoclonal B cells in 29 cases, 8 high-grade and 21 low grade; 17 of 29 had normal lymphocyte counts. De Paoli and Santini [58] found interleukin-2 (IL-2) receptors expressed on well-differentiated B-cell lymphomas.

No single criterion defines T-cell monoclonality, but T lymphoma is suggested when there is (1) an elevated expression of pan-T antigens, (2) a skewed helper/suppressor ratio, or (3) T cell expression of aberrant phenotypes, such as helper and/or suppressor antigens in the absence of pan-T antigens.

T cells predominate in HD and in reactive lymph nodes, which could lead to a mistaken immunophenotype for T-NHL [115]. FCM determinations of terminal deoxynucleotidyl transferase positive cells (TdT) can be helpful [93, 161].

Stonesifer et al. [166] identified the T cells in Lennert's lymphoma as helper cells (T11+, T3+, T4+). A large series (n = 63) of T-NHL was described by Coiffier et al. [49] who found an immature T phenotype in 11 cases, CD4 in 26 cases, CD8 in 13 cases, and undefined immunophenotype (CD4 + CD8) in 10 cases. CD4-positive patients seemed to have a better prognosis than did those with other phenotypes. Haynes et al. [88] found cutaneous T-cell lymphomas to be positive for helper T-cell markers (CD4) and negative for CD8. The circulating Sézary T cells had identical markers.

The immunophenotype, using an indirect immunoalkaline phosphatase (APAAP) method, was compared with rearrangement of the genes for Ig and β-chain of the T cell (β-TCR) in 91 specimens [91]. Ig gene rearrangement was correlated with a monotypic (κ/λ) phenotype in 32 of 33 lymphoma samples, and β-TCR rearrangement was found in 19 of 19 malignant T-cell lymphomas. In some samples, DNA analysis was the only way to demonstrate clonality, and many groups now perform both tests when the patient's lymphoma is initially diagnosed.

Antigens on Reed–Sternberg cells are almost impossible to determine because of the low frequency of these cells in Hodgkin's disease. Some results favor a myelocyte/macrophage origin based on cytochemical staining; however, consistent reactivity with antimonocyte antibodies has not been demonstrated [78]. Paietta et al. [135] found expression of the macrophage growth factor, CSF-1, and its receptor, c-fms, on Hodgkin's derived cell lines. Gerdes et al. [82] found coexpression of the Ki-1 membrane antigen and of the nuclear proliferation-related antigen Ki-67 by an immunoenzymatic double-labeling method. Other investigations regarding the origin of RS cells were conducted on cell lines derived from HD patients without conclusive determination of the origin of the RS cells [69]. The well-known T-cell defect seen in patients with Hodgkin's disease was found by DeAngelis et al. [56] to be treatment induced, since normal T- cell numbers and helper/suppressor cell ratios were seen in untreated patients. Following therapy, decreased helper/suppressor ratios were observed both in patients with HD and NHL, caused by a significant decrease in helper cells.

DNA ANEUPLOIDY IN LYMPHOMA

Abnormal DNA stem lines as determined by DNA FCM can serve as an excellent marker and are highly specific for malignant cells. Unlike solid tumors, DNA stem lines in a lymphoma are frequently close to diploid and their detection depends largely on the quality of the FCM measurement, i.e., on the coefficient of variation (CV). As is the case in leukemias, DNA aneuploidy is a reflection of numerical cytogenetic abnormalities and the frequency of detected DNA aneuploidies is inversely related to the CV. Nowell and Croce [134] recently reviewed and summarized our present knowledge of these abnormalities. These investigators concluded that (1) most tumors have chromosomal abnormalities that are not present in other cells of the body; (2) in a given tumor, all neoplastic cells often have the same cytogenetic change or related changes; (3) chromosomal abnormalities are more extensive in advanced tumors; and (4) although chromosomal alterations often differ between tumors, they are nonrandom patterns. These statements are also valid for NHL. Unfortunately, translocations such as t(8;14), t(8;22), and t(2;8) characteristic for diffuse small noncleaved cell lymphomas, t(14;18) found in follicular lymphomas, deletion 6(q21) of diffuse large cell lymphoma, and trisomy 12 and t(11;14) seen in diffuse small lymphocytic (well-differentiated) lymphomas [37–39] are not detectable by DNA

TABLE 4. Antigeneic Profiles of Stages of Normal B-Cell Maturation*

		B-Cell	Precursor		Pre-B	B-Cell	Plasma Cell
HLA-DR	+	+	+	+	+	+	+
Tdt	+	+	+	+	±		
CD19		+	+	+	+	+	
CD38		+	+	+	+		+
CD10			+	+	+		
CD20				+	+	+	
Cmu					+		
SIg						+	
CD21						±	
PCA 1							+
PCA 2							+
PC 1							+

*Reproduced from Foucar et al. [79], with permssion of the publisher.

TABLE 5. Antigenic Profiles of Subtypes of Non-Hodgkin's Lymphoma*,†

Type of NHL	Subtypes with usual antigenic profile	Comments
T NHL	Lymphoblastic lymphoma: CD7, T9, CD5, CD38, CD1, CD2, CD4, CD8, variable T9	Same phenotypic spectrum as T ALL
	Peripheral T-cell lymphoma CD2, CD3, CD5, CD7, TCR-1, CD4 or CD8	May need molecular studies to confirm T-cell lineage; aberrant expression of pan T-cell antigens common
B NHL	FCC lymphomas: HLA-DR, CD19, CD20, CD21, SIg, variable CALLA	SIg negative B-cell lymphomas and other types of aberrant antigen expression described
	Mantle zone lymphomas: HLA-DR, CD19, CD20, SIg, variable CALLA	Same antigenic spectrum as CLL
	Small lymphocytic lymphomas: HLA-DR, CD19, CD20, CD21, CD5, weak SIg	
	Immunoblastic HLA-DR, CD20, SIg, variable CD19	With more differentiation toward plasma cells may lose HLA-DR and SIg

*Abbreviations: FCC = follicular center cell; SIG = surface immunoglobulin.
†Reproduced from Foucar et al. [79], with permission of the publisher.

FCM. These translocations will hopefully be detected by fluorescence in situ hybridization (FISH) techniques, similar to those already described for t(4;11), characteristic of a specific type of acute leukemia or for t(4;3) induced by radiation [140]. Nevertheless, a series of studies of lymphomas have been performed using DNA FCM. Christensson et al. [46] analyzed 208 cases of NHL. They found that in aneuploid B-cell lymphomas, the frequency of monoclonal light chain restricted cells correlated well with the frequency of aneuploid cells. Of the 208 patients analyzed, 146 had NHL with near-diploid DNA content (Table 7). The frequency of aneuploid tumors was markedly higher in the centroblastic, centrocytic, follicular, and diffuse group as compared with the follicular group. Juneja et al. [101] measured DNA content in 115 cases of NHL using chromomycin A3 and found 33% DNA aneuploidy. The incidence of aneuploidy increased significantly with increasing histological grade ($P = 0.0002$). Low-, intermediate-, and high-grade lymphomas had 14%, 47%, and 62.5% aneuploid cases, respectively. The median S-phase fraction in the three groups was 1.0%, 4%, and 27%, respectively. Wooldridge et al. [180] studied 52 untreated patients with newly diagnosed diffuse large cell (n = 48) or mixed cell (n = 4) NHL. DNA aneuploidy was detected in 56% and predicted a significantly longer duration of complete remission ($P = 0.01$) in these uniformly treated patient groups. Likewise, 2-year survival was 60% for patients with aneuploid versus 36% for those with diploid tumors ($P = 0.01$).

Aneuploid NHL may have higher proliferation than near-diploid NHL, and the most aneuploid component of a polyploid sample generally has the highest S fraction [50,157], favoring the concept of clonal selection and clonal evolution of tumors. In an attempt to increase detection of near-diploid DNA aneuploidy, some investigators have suggested that a broadening of the G_1 peak (increased CV) or shoulders on the distribution of G_1 cells could be indicative of underlying aneuploidy [64]. Double staining for DNA and a second immunological marker [105], such as surface Ig for the monoclonal B-cell population, has also been suggested to bring out DNA aneuploidy. While deserving serious consideration, these approaches do not meet the requirements laid down at the "Convention on Nomenclature for DNA Cytometry, [92] which postulates distinct peaks for the identification of DNA aneuploid cells.

Figure 2 shows a hyperdiploid case of DHL, permitting not only determination of DNA ploidy, but also cell-cycle analysis of the DNA aneuploid population. Within this context, it is interesting to compare today's FCM DNA histograms with those obtained by single-cell microphotometry little more than a decade ago [147]. Some of the conclusions drawn at that time are a reflection of the poor quality of measurements rather than lymphoma biology.

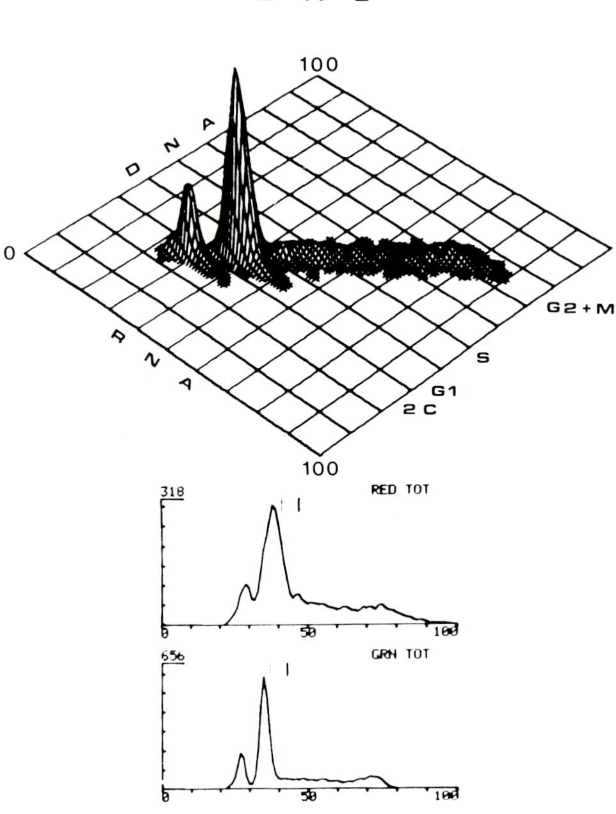

Fig. 2. DNA/RNA flow cytometry of non-Hodgkin's lymphoma. Diploid and hyperdiploid cells can be separated and cell-cycle analysis of the hyperdiploid population is possible. **Lower:** RED TOT = RNA; GRN TOT = DNA.

Algorithms for nuclear DNA analysis from scanning cytophotometry were developed by Boecking et al. [25], who analyzed 95 NHL and identified three groups of patients with significantly different survival, based solely on DNA measurements. These results are awaiting independent confirmation. In a large survey, Barlogie et al. [17] found DNA aneuploidy in 53% of 360 published cases of lymphoma investigated by FCM. Braylan's group pioneered multiparameter FCM for NHL [27,28,31,105] and measured DNA, cell volume, light scatter, and cell-surface Ig light-chain expression. Comparison of cellular DNA of Ig light-chain-bearing neoplastic cells with that of nonneoplastic cells from the same tumor enabled them to detect DNA aneuploidy in almost 80% of NHL and to establish an association between DNA aneuploidy and proliferative state (S phase) [28] (Figs. 3,4,5).

Bauer et al. [18] studied paraffin-embedded surgical biopsies from 50 patients with newly diagnosed diffuse large cell lymphoma, of which 62% were aneuploid, and found no prognostic importance for DNA aneuploidy. In a study of 177 cases of NHL, Srigley et al. [164] from the M.D. Anderson group reported similar rates of aneuploidy for follicular and diffuse large-cell NHL and also concluded that aneuploidy was not an independent prognostic factor. Andreeff et al. [9] found less DNA aneuploidy in low-grade as compared with high-grade NHL (DWDL 0%, DPDL 18%, NPDL 20%, NML 43%, LBL 21%, DHL 45%, undifferentiated NHL 50%), and Cox regression analysis identified DNA ploidy as significant for survival ($P = 0.0264$). Multivariate analysis, however, eliminated DNA ploidy in favor of cellular RNA content. Likewise, Costa et al. [53] found 61% aneuploidy in NHL, more in high-grade (72%) than in low-grade lymphoma (55%). In patients with mycosis fungoides (MF) and Sézary syndrome, 63% had DNA aneuploidy as determined by FCM [36], and aneuploidy was associated with poor prognosis. These results confirmed a smaller study by Wantzin et al. [175,176], who found 7 of 17 skin biopsies of patients with MF to be aneuploid. Another report noted aneuploidy after transformation in 4 of 4 samples [74]. A number of studies have been conducted on paraffin-embedded material [41,90,97,120,181]. In 29 cases of T-NHL, only four cases (13%) were aneuploid, and aneuploidy was not related to histological grade [74]. In a study of body cavity fluids, aneuploidy was identified in some samples reported cytologically negative from patients with lymphoma (and other malignancies) [167]. DNA aneuploidy was also found in 6 of 11 highly malignant NHL with primary bone manifestation [173].

DNA/RNA FCM of 363 lymph nodes using the acridine orange techniques revealed an overall frequency of 69 cases with DNA aneuploidy. In 51 of these cases, results of simultaneous bone marrow studies were available (Table 6, Fig. 6). Marrow involvement was documented by FCM only in 4 of 18 (23%) of aneuploid cases [8]. These results have implications for staging of patients with NHL: marrow involvement assigns patients to stage IV, and this in turn may affect the therapeutic approach. Also, autologous bone marrow transplantation requires that the transfused marrow be free of tumor cells after ablative therapy. Thus, all available methods, including DNA and SIg FCM, should be used to screen for submicroscopic bone marrow involvement with lymphoma cells. Furthermore, methods to "purge" bone marrow of tumor cells depend crucially on sensitive methods for the detection of "minimal disease."

Hodgkin's disease has been neglected by FCM researchers, probably because of the scarcity of Reed–Sternberg cells. One study investigated 115 cases of HD and found 13 aneuploid cases (11%) [127], but aneuploidy and cell kinetics were not of prognostic importance.

CELL KINETICS OF NHL

A very large number of samples from patients with NHL have been investigated with cytokinetic techniques. Initially, the ^{3}H-TdR labeling index (LI) and mitotic index (MI) provided evidence of significant cell kinetic heterogeneity, which was correlated with response to therapy, remission duration, and survival. The advent of FCM permitted cell kinetic analysis in a shorter time, but in one-parameter histograms the results were representative of all cells measured, not just lymphoma cells. Two-color analysis of SIg and DNA circumvents this problem but has not yet become the standard technique in the study of NHL.

Table 7 lists cell kinetic papers on NHL. A total of 2,916 patients have been investigated. The list is probably incomplete, and some patients may have been reported in more than one study. Nevertheless, these reports document the important contribution of cell kinetic parameters, measured as mitotic index (MI), labeling index (LI), or FCM S-phase fraction. In general, an association between lymphoma grade and proliferation has been clearly established in many papers. Low-grade lymphomas are characterized by low numbers of S-phase cells, and high-grade lymphomas by high

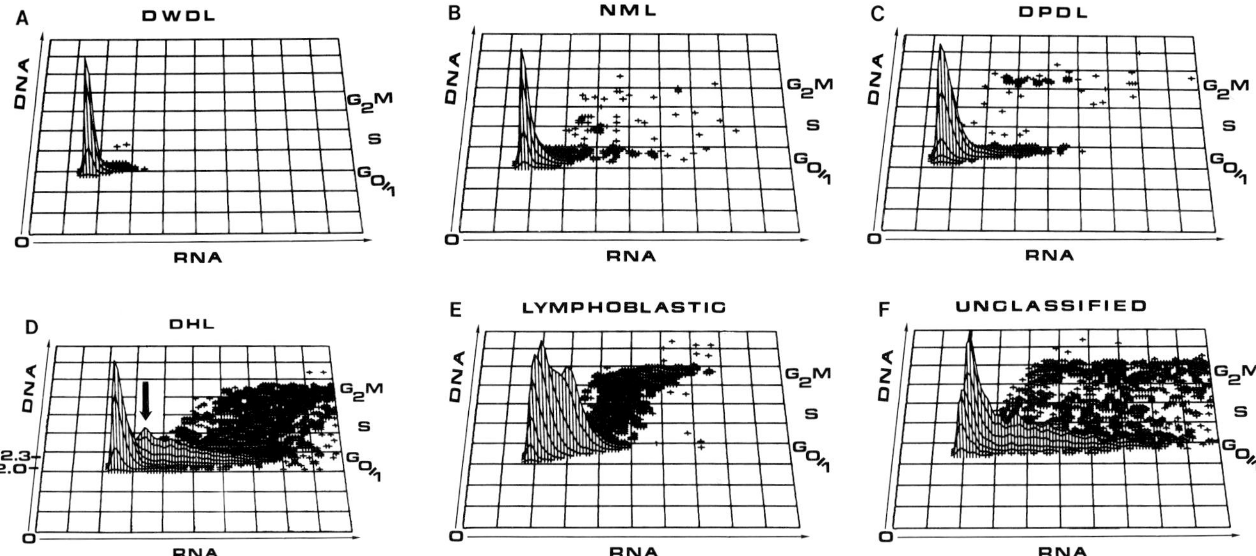

Fig. 3. DNA/RNA flow cytometry histograms of non-Hodgkin's lymphoma lymph node cells. **a:** Diffuse well-differentiated lymphocytic lymphoma, (DWDL). Note the very limited evidence of proliferation. **b:** nodular mixed lymphocytic–histiocytic lymphoma (NML). Note the small increase of G_1 RNA and of S-phase cells. **c:** Diffuse poorly differentiated lymphocytic lymphoma (DPDL). **d:** diffuse histiocytic lymphoma (DHL). Hyperdiploid cells are prominent (arrow). Note the high proportion of proliferating cells, many with increased RNA. **e:** Lymphoblastic lymphoma. There is increased RNA content and active proliferation of many cells. **f:** Unclassified lymphoma. High proliferation and RNA content are apparent.

percentages of cells in S phase (Figs. 3,4,5). There is, however, significant heterogeneity particularly in the high-grade group [1,28,31,61,75,101,106,111,131,148,156,164,174]. Since DNA aneuploidy is more frequent in high-grade lymphomas, it is no surprise that correlations between proliferation and aneuploidy have been reported. Several studies also show a correlation of high proliferative activity with high response rate [30]. High S phase found in low-grade lymphoma can predict subsequent transformation into high-grade lymphoma [118]. Important correlations between labeling index and growth fraction and an inverse correlation between labeling index and cell cycle time or doubling time was established by Hansen et al. [87]. Since aggressive chemotherapy can induce remissions in patients with highly proliferative lymphomas, a correlation between high S phase and high incidence of CR can be expected. These same lymphomas, however, also have a high probability of recurrence and one can expect a correlation between high S phase and short remission duration and survival [148]. A direct comparison of different studies is difficult, since different chemotherapeutic regimens are used. In particular, comparisons of the prognostic value of cell kinetics across the board in high- and low-grade lymphomas are of limited value, since the therapeutic approaches differ considerably. Studies of histological or cell kinetic subgroups with standardized treatment are still needed to elucidate their clinical value in NHL.

The addition of RNA as a novel parameter of cell-cycle analysis proved valuable [9]. Andreeff et al. [9] found that S phase was not prognostic for their group of NHL but that high RNA index was a good indicator of poor prognosis independent of other prognostic parameters ($P = 0.001$). Likewise, McLaughlin [123] found that RNA was the most important prognostic factor in multivariate analysis of large cell lymphoma. Differential measurements of double-stranded RNA added additional information [102,164]. A significant increase in ds-RNA was found in the progression from low to intermediate to high-grade lymphomas, and this parameter alone predicted grade in 59% of cases. A combination of DNA aneuploidy, ($S + G_2M$) percent, RNA index, and ds-RNA improved prediction of grade to 85% [164].

It is now possible to analyze DNA distributions in paraffin-embedded samples of lymphomas [29,41,90,97,120,172,181]. Although the coefficient of variation of DNA measurements is inferior to that obtained from fresh material and usually results in a lower number of detected DNA aneuploidies, the technique is sufficient for cell-cycle analysis. Existing publications strongly suggest that FCM assays of lymphoma kinetics will provide significant and clinically useful contributions to our understanding of lymphoma biology [16,29,45,54,65,70,71,110,121,122,154,155,181]. There are a number of reports on the taxonomy and the prognostic potential of proliferation associated nuclear antigens in NHL [48,85,138,178]. Schrape et al. [150] compared the measurement of growth fraction in NHL using three different methods (Ki67, BrdU, and transferrin receptor). In low-grade lymphomas, 3.5 ± 1.6% of cells were Ki67 positive and 1.2 ± 0.9% were BUdR positive. The values for high-grade lymphomas were 22.5 ± 18.7 and 8.9 ± 7.8, respectively. Surprisingly, high numbers of transferrin receptor-positive cells were found in low-grade lymphomas of all histological types, whereas Ki67 positivity correlated closely with grading. Grogan et al. [85] found Ki67 positivity to be an independent prognostic factor for diffuse large cell lymphomas, and Medeiros et al. [124] found that it was significant for diffuse small cell lymphomas.

Del Bino et al. [60] studied 73 patients with bivariate DNA/protein flow cytometry. Protein heterogeneity was more frequently observed in diffuse than in nodular lymphomas and in high-grade than in low-grade lymphomas but was not related to DNA ploidy or cell proliferation.

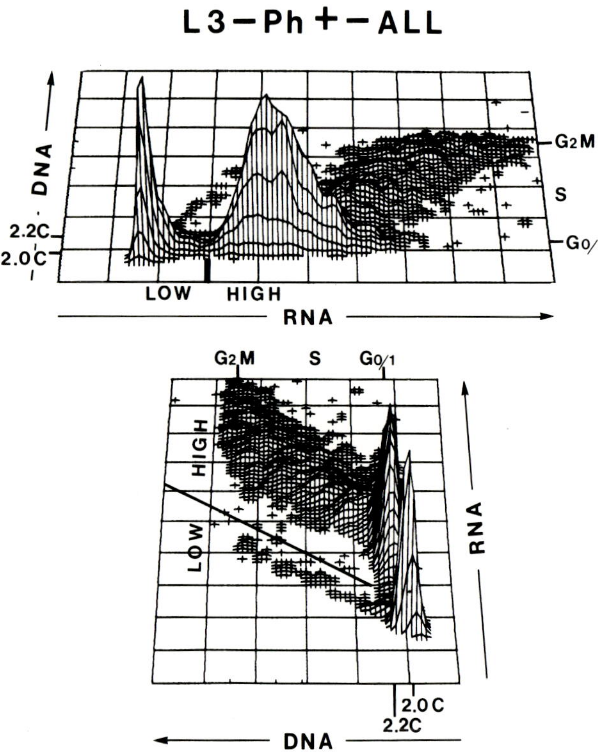

Fig. 4. DNA/RNA flow cytometry of a case of Burkitt's lymphoma/ FAB L3 leukemia. Diploid (2.0C) cells with low RNA and low proliferation can be discriminated from hyperdiploid (2.2C) cells with high RNA and high proliferation [42]. Lower: Rotated 90° from upper panel.

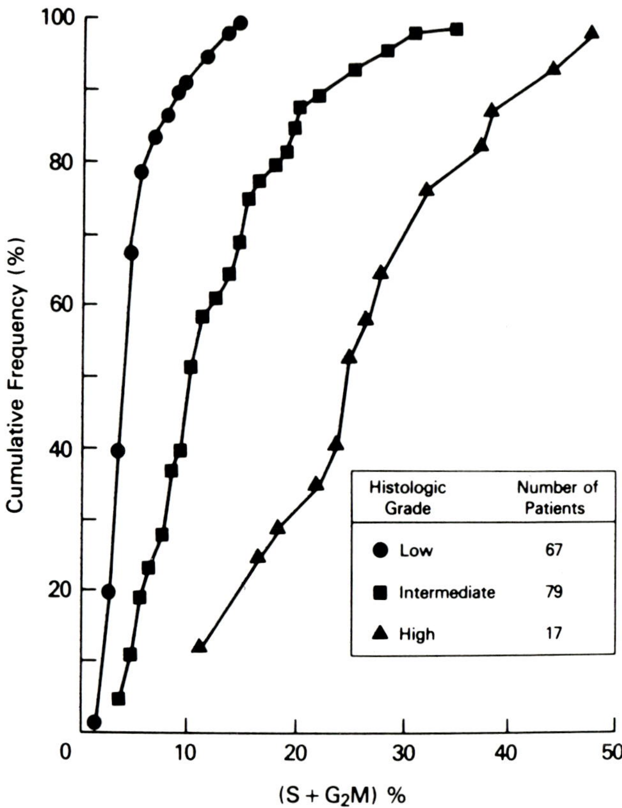

Fig. 5. Cumulative frequency distribution of proliferative activity in non-Hodgkin's lymphoma. Eighty percent of low-grade lymphomas showed ($S + G_2M$) < 5% and the highest proliferative activity was 13%. By contrast, 80% of intermediate-grade tumors showed ($S + G_2M$) >5%. All patients with high-grade lymphoma showed proliferative activity >10%. (Reproduced from Srigley et al. [164], with permission of the publisher.)

The advent of the Ki67 antibody [151,159,179] and other such markers as nucleolar organizer regions (NORs) [55] and nuclear oncoproteins (e.g., c-myc, p53) signal a new era of molecular studies, which will elucidate the growth potential of a variety of normal and malignant cell populations.

MONITORING OF LYMPHOMA/MINIMAL RESIDUAL DISEASE

Successful therapy of NHL results in decreased lymphadenopathy, organomegaly, bone marrow infiltration, and circulating monoclonal malignant lymphocytes. Clinical methods of restaging, including computed tomography (CT) and magnetic resonance imaging (MRI) scans, are insufficient for the detection of minimal residual disease in NHL, as evidenced by a high relapse rate of patients achieving complete remission. In selected cases of T-NHL [4] and Burkitt's lymphoma, clonogenic cells can be grown in vitro in methylcellulose. Results obtained by DNA/RNA flow cytometry are shown in Table 6, indicating minimal disease in some patients with normal marrow morphology. This approach requires DNA aneuploidy as a marker, however.

The detection of B-cell clonal excess has made a significant clinical difference, not only in staging, but also in monitoring of minimal disease in B-NHL [113,177] (see Chapter 34, *this volume*). Weinberg et al. [177] found B-cell clonal excess in 24 of 25 cases of B-NHL using this flow cytometric method. Laemhuis et al. [107] used two-color fluorescence clonal-excess analysis for the successful detection of bone marrow infiltration by B-cell lymphoma. The sensitivity of this approach was sufficient to detect 1–5% lymphoma cells. Johnson et al. [98–100] studied 96 patients with NHL. Blood involvement was found in 24% of patients by routine hematological methods and in an additional 27% by FCM κ/λ light-chain clonal excess. In patients with active disease but without morphological evidence of leukemia, 37% showed abnormal κ/λ distributions, as did 18% of patients in complete remission.

Standard molecular techniques investigating specific gene rearrangements are not better than FCM clonal excess analysis, but the introduction of the DNA–polymerase chain reaction (PCR) [145] has expanded our detection of minimal residual disease significantly, in particular in follicular lymphoma carrying the translocation t14;18 [165]. However, PCR can detect less than one abnormal cell in 100,000 normal cells, and this sensitivity may indicate residual disease without clinical significance. Follow-up investigations are presently underway [32].

DRUG EFFECTS IN LYMPHOMA

Drug effects have only sporadically been studied in vivo, since serial lymph node biopsies can not be obtained routinely [80]. Only in lymphomas in leukemic phase would

TABLE 6. Bone Marrow Involvement in Aneuploid Non-Hodgkin's Lymphoma

Flow Cytometry (DNA/RNA) of 363 Lymph Nodes	N	%	(%)
Lymph nodes with aneuploid DNA stem line	69		
Aneuploid LN with bone marrow data	51	100	
Bone marrow involved			
By morphology and/or flow cytometry	18 =	35.3	(100)
By morphology (no FC available)	8 =	15.7	(44)
By morphology and FC	6 =	11.8	(33)
By FC only (morph negative)	4 =	7.8	(23)

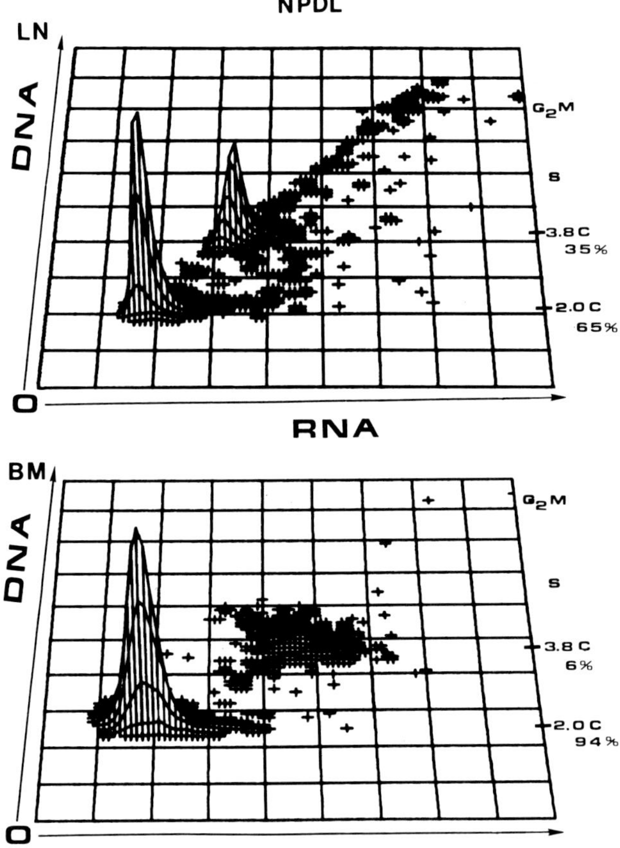

Fig. 6. Detection of minimal bone marrow involvement of lymphoma cells by DNA/RNA flow cytometry. **Top:** DNA/RNA histogram showing 35% hypotetraploid lymphoma cells (3.8C) in a lymph node sample of nodular poorly differentiated lymphocytic lymphoma (NPDL). These cells are proliferating. **Bottom:** DNA/RNA histogram showing bone marrow from same patient: most cells are diploid, but 6% hypotetraploid cells are identified.

such studies be feasible. Hedley [89] studied the effects of 5-fluorouracil in four patients with end-stage NHL by DNA FCM and found changes in the S-phase compartment.

Drewinko and Barlogie [68] studied cell kinetic effects of 34 chemotherapeutic agents in human cell lines and concluded that the position in the cell cycle at the time of exposure was an important determinant of killing rate. Others studied Adriamycin analogues [133,142], hydroxyurea [77,160], thymidine [76], N-methylformamide [20], interferon-α (IFN-α) [144], IFN-γ [126], ricin–antibody conjugate [170], deoxyadenosine [2], α-difluoromethylornithine (DFMO) [3], ricinus agglutinin [34], transforming growth factor (TGF)-β [44,162], and cyclic adenosine monophosphate (cAMP) [119].

Such studies in lymphoma cell lines are of limited clinical value, since cell lines have highly increased proliferation as compared to the tumors from which they originate. They do cast light on the mechanisms of action of certain drugs. Scott et al. [152] demonstrated that anti-μ causes a block in the transition of WEHI-231 B-cells from G_1 to S phase. Exposure to anti-μ for only 2 hours prevented purified G_1 cells from entering S phase. TGF-β also inhibited G_1 to S transition, but not the activation of human B lymphocytes [162]. TGF-β did not block the induction of transferrin receptor (TF) expression, which normally occurs in late G_1; it was concluded that TGF-β arrests cells in middle to late G_1 before TF receptor expression [162]. These studies will certainly continue and may be useful models for investigating mechanism of drug action (see Chapter 39, *this volume*).

The only study of multidrug resistance in NHL so far was reported by Salmon et al. [146], who correlated the sensitivity to doxorubicin with the expression of *p*-glycoprotein. These and similar studies have potential clinical value in choice of therapy.

ONCOGENE EXPRESSION IN LYMPHOMA

The role of oncogenes in hematopoietic malignancies was described in some detail in Chapter 35. In lymphoma, three oncogenes have been of particular interest, i.e, c-myc, bcl-2, and c-myb. The role of c-myc in Burkitt's lymphoma has been studied extensively, and it is postulated that the activation of c-myc by the characteristic translocation (8;14) represents an essential step in the genesis of this tumor [104]. The transposed myc gene becomes constitutively activated and resistant to normal regulatory mechanisms. Leder et al. [109] point out that c-myc activation may be necessary for some malignant processes but is not sufficient. Holte et al. [95] studied c-myc expression in malignant B-cell lymphomas using a fluoresceinated polyclonal anti-myc antibody and propidium iodide and found a strong correlation with other parameters of cell activation. The oncoprotein level was largely unvarying from late G_1 through the rest of the cell cycle. The same was found for p21 ras in other hematopoietic cell systems [7,10]. Yunis et al. [182] investigated the expression of bcl-2 in NHL, and Barletta et al. [12] studied the relationship between 6q− and c-myb. No other FCM data regarding oncogene expression in NHL have been reported in the literature to date.

TABLE 7. Cell Kinetics and DNA Ploidy in Lymphoma*†

No. of patients	Aneuploidy incidence %	Prolif–histol.	Ploidy–histol.	Prolif.–prognosis	Ploidy–prognosis	Reference
208	29	Y				Christensson et al. [46]
115	33	Y ($P<0.001$)	Y ($P=0.0002$)			Juneja et al. [101]
74	61	Y		N		Costa et al. [53]
50	62			Y	N	Bauer et al. [18]
52	80	Y		Y		Braylan et al. [28]
65		Y				Braylan et al. [31]
52	56			Y ($P<.02$)	Y ($P<.01$)	Wooldridge et al. [180]
27				Y (S in 17-p)		Cabanillas et al. [38]
80		Y				Hansen et al. [87]
88				Y		Costa et al. [52]
40		Y (B>T>non-T-non-B)				Murphy et al. [128]
28		S-OKT9				Porwit-Ksiazek et al. [141]
27				Y ($P=0.01$)		Braylan et al. [30]
15		Y				Meyer and Higa [125]
64		Y				Kvaloy et al. [106]
69		Y ($P=0.02$)		Y ($P=0.015$)		Scarffe and Crowther [148]
17	59	Y				Walle et al. [174]
81		Y				Mccartney et al. [118]
105		Y		N		Ellison et al. [75]
65	22			Y ($P=0.0004$)	N	Roos et al. [143]
123		Y		Y		Del Bino et al. [61]
69				Y		Del Bino et al. [59]
140				Y		McLaughlin et al. [123]
154	41 unfavorable			Y		Christensson et al. [47]
	15 favorable					Hirt et al. [94]
101		Y		Y		Akerman et al. [1]
177		Y		Y	Y	Srigley et al. [164]
96		Y	Y	N (RNA:Y)	Y	Andreeff et al. [9]
220		Y				Shackney et al. [156]
26		Y				Katz et al. [103]
279				Y		Donhuijsen [66]
80				Y		Lenner et al. [111]
49		Y				Naus [131]
58				Y		Brooks et al. [33]
22		Y				Palutke et al. [136]
2,916 patients reported						

*Abbreviations: Y = positive, N = negative correlation.
†Review of literature on cell kinetics and DNA ploidy in non-Hodgkin's lymphoma. Correlations between proliferation and histological type, DNA ploidy and histology, proliferation and prognosis, and ploidy and prognosis are indicated for each report referenced.

FLOW CYTOMETRY OF MULTIPLE MYELOMA

Multiple myeloma is the major malignant tumor of plasma cells [72]. The malignant cells are usually present in the bone marrow, and monoclonal Ig is present in serum or in the urine of virtually all patients. Important prognostic factors relate to plasma cell mass, cell kinetics, and tumor cell sensitivity to chemotherapeutic agents (for review, see Durie [72]). Circulating B lymphocytes included a subpopulation of myeloma precursors that were found to express CALLA and HLA-DR but that were negative for surface and cytoplasmic Ig. After isolation, these cells could be transformed into plasma cells by stimulation with the phorbol ester 12-O-tetradecanoyl-phorbol-13 acetate and synthesized the same heavy and light chains as myeloma cells [40,139]. CALLA-positive marrow plasma cells were found to be a sign of poor prognosis [73]. Plasma cell acid phosphatase (AP) was also related to survival and to labeling index (LI) and identified a subgroup of patients with very high tumor mass, low LI, but high AP score who had very poor survival [21,22]. The determination of β_2-microglobulin has become one of the most powerful determinants of prognosis in myeloma [84].

Myeloma kinetics was initially investigated using ^{3}H-thymidine and provided a framework of kinetic data: $S + G_2$ was found to be >60 hours [67], the generation time 8 days, and the growth fractions (GF) 19% and 47% in two patients studied with continuous infusion of ^{3}H-thymidine [67]. Another study found a GF of only 4–6% [24]. Several independent studies showed that a low pretreatment LI ($\leq$1%)

heralded longer survival than LI>1%, independent of tumor burden [23,108].

DNA aneuploidy was detected in as many as 80% of cases [15,35,108], although significantly lower rates of aneuploidy were reported as well [168]. Cytogenetic analysis revealed abnormal karyotypes in about 50% of cases [63], which were associated with shorter survival ($P = 0.0089$).

Dual-parameter measurements of cytoplasmic Ig and DNA [14,43] permit kinetic studies specifically of abnormal cells and increase confidence in low-degree DNA aneuploidy determinations. Likewise, double staining for DNA and plasma cell antigens such as PC-1 enhances diagnostic precision and provides insights into the precise stage of B-cell differentiation [6]. DNA/RNA flow cytometry identified RNA content as an important independent prognostic factor in one study [15], but results were not confirmed in patients treated with the M-2 protocol [168]. These and other studies, such as measurements of protein and nucleolar antigen content in myeloma, are extensively discussed by Barlogie et al. [15], Neckers and Nordan [132], and Smith et al. [163]. The expression of the oncoprotein p21 ras was studied in conjunction with DNA; higher p21 ras levels were found in aneuploid myeloma as compared with normal diploid cells [171].

Several excellent reviews summarize the state of flow cytometric analysis of myeloma [13,15,24]. Phenotyping, cell-cycle analysis, and the detection of DNA aneuploidy are all useful applications of FCM in the investigation of myeloma.

OUTLOOK

Flow cytometry of NHL has reached a mature stage of investigation. DNA aneuploidy is an excellent marker of malignant cells, when present, and is found more often in high-grade NHL as compared with low-grade lymphomas. Immunophenotyping of lymphoma cells is best done by monoclonal antibody studies in conjunction with flow cytometry and has become a routine procedure in many institutions. These studies will usually determine the lineage without too many problems.

Finally, there is growing evidence that measures of cell proliferation, including S-phase fraction, RNA content, double-stranded RNA content, and nuclear antigen expression, including Ki67 and PCNA, will help determine the aggressiveness of NHL in individual cases.

The limitations of FCM result from an absence of direct correlation with tissue architecture, which is lost in the preparation of cell suspensions. The advantages of FCM in lymphoma studies derive from quantitative correlations of cell measurements on a cell-by-cell basis. Panels of antibodies used in conjunction with DNA and RNA analysis provide a biochemical, physiologic, and genetically controlled immunologic basis for the classification, prognosis and treatment monitoring of lymphomas. This technology has become an important adjunct and in some cases is a very attractive alternative to other diagnostic procedures in lymphoid neoplasia [29]. As is the case in leukemia research, new probes, including recently developed in situ hybridization methods for the detection of clonal abnormalities and gene expression, can be expected to further enhance our understanding of lymphoma biology and pathophysiology in the future.

ACKNOWLEDGMENTS

This work was supported by grants CA20194, CA41305, and CA38980 from the National Cancer Institute. I wish to acknowledge the encouragement and help I received from Dr. M.R. Melamed in the conduct of these studies. Dr. D. Filippa, Dr. B. Koziner, and Dr. B. Clarkson were helpful in our early studies of NHL by flow cytometry. Ms. Linda Tatum provided excellent and dedicated assistance in the preparation of this manuscript.

REFERENCES

1. **Akerman M, Brandt L, Johnson A, Olsson H (1987)** Mitotic activity in non-Hodgkin's lymphoam. Relation to the Kiel classification and to prognosis. Br J Cancer 55:219–213.
2. **Albert DA, Nodzenski E (1988)** Deoxyadenosine toxicity and cell cycle arrest in hydroxyurea resistant S49 T-lymphoma cells. Exp Cell Res 179:417–428.
3. **Allen ED, Natale RB (1985)** Effect of alpha-difluromethylornithine (DFMO) as a combination agent on the growth of p3J, a Burkitt's lymphoma cell line. Fed Proc 44:1142.
4. **Allouche M, Georgoulias V, Bourinbaiar A, et al (1987)** Abnormal proliferation of T-colony-forming cells from peripheral blood of patients with T-cell acute lymphoblastic leukemias and lymphomas in complete remission: Potential prognostic value. Br J Haematol 65:411–418.
5. **Anderson KC, Bates MP, Slaughenhoupt BL, et al (1984)** Expression of human B cell-associated antigens on leukemias and lymphomas: A model of human B cell differentiation. Blood 63:1424–1433.
6. **Anderson KC, Bates MP, Slaughenhoupt BL, Schlossman SF, Nadler LM (1984)** A monoclonal antibody with reactivity restricted to normal and neoplastic plasma cells. J Immunol 132:3172–3179.
7. **Andreeff M, Bressler J, Higgins P (1985)** Oncogenes and cancer. Review and a new method for measuring the gene expression in relation to the cell cycle. Dtsch Med Wochenschr 110:30–35.
8. **Andreeff M, Darzynkiewicz Z, Clarkson BD, Miller D, Melamed MR (1982)** Role of flow cytometry (FC) of nucleic acid content and chromatin structure in classification and monitoring of acute leukemia (AL) and non-Hodgkin's lymphoma. Proc 13th Int Cancer Congr p. 589.
9. **Andreeff M, Hansen H, Cirrincione C, Filippa D, Thaler H (1986)** Prognostic value of DNA/RNA flow cytometry of a B-cell non-Hodgkin's lymphoma: Development of laboratory model and correlation with four taxonomic systems. Ann NY Acad Sci 468: 368–386.
10. **Andreeff M, Slater DE, Bressler J, Furth ME (1986)** Cellular ras oncogene expression and cell measured by flow cytometry in hematopoietic cell lines. Blood 67:676–681.
11. **Bardales RH, Al-Katib AM, Carrato A, Koziner B (1989)** Detection of intracytoplasmic immunoglobulin in by flow cytometry in B-cell malignancies. J Histochem Cytochem 37:83–89.
12. **Barletta C, Pelicci P-G, Kenyon LC, Smith SD, Dalla-Favera R (1987)** Relationship between the c-myb locus and the 6q− chromosomal aberration in leukemias and lymphomas. Science 235:1064–1067.
13. **Barlogie B, Alexanian R, Gehan EA, Smallwood L, Smith T, Drewinko B (1983)** Marrow cytometry and prognosis in myeloma. J Clin Invest 72:853–861.

14. **Barlogie B, Alexanian R, Pershouse M, Smallwood L, Smith L** (1985) Cytoplasmic immunoglobulin content in multiple myeloma. J Clin Invest 76:765–769.
15. **Barlogie B, Latreille J, Alexanian R, et al** (1982) Quantitative cytology in myeloma research. Clin Haematol 11:19–46.
16. **Barlogie B, McLaughlin P, Alexanian R** (1987) Characterization of hematologic malignancies by flow cytometry. Anal Quant Cytol Histol 9:147–155.
17. **Barlogie B, Raber MN, Schumann J, et al** (1983) Flow cytometry in clinical cancer research. Cancer Res 43:3982–3997.
18. **Bauer KD, Merkel DE, Winter JN, et al** (1986) Prognostic implications of ploidy and proliferative activity in diffuse large cell lymphomas. Cancer Res 46: 3173–3178.
19. **Berard CW, Bloomfield C, Bonadonna G, et al** (1985) Classification of non-Hodgkin's lymphomas: Reproducibility of major classification systems. Cancer 55:91–95.
20. **Bill CA, Gescher A, Hickman JA** (1987) The effects of N-methylformamide (NMF, NSC 3051) on the growth and glutathione status of murine TLX5 lymphoma cells in vitro. Proc Annu Meet Am Assoc Cancer Res 28:46.
21. **Boccadoro M, Gallamini A, Fruttero A, et al** (1985) Plasma cell acid phosphatase activity as prognostic factor in multiple myeloma: Relationship to the thymidine-labeling index. J Clin Oncol 3:1503–1507.
22. **Boccadoro M, Gavarotti P, Ciaiolo C, et al** (1986) Cytokinetic studies in human myeloma. Basic Appl Histochem 30:233–237.
23. **Boccadoro M, Marmont F, Tribalto M, et al** (1989) Early responder myeloma: Kinetic studies identify a patient subgroup characterized by very poor prognosis. J Clin Oncol 7:119–125.
24. **Boccadoro M, Pileri A** (1987) Cell kinetics of multiple myeloma. Hematol Pathol 1:137–142.
25. **Bocking A, Adler CP, Common HH, Hilgarth M, Granzen B, Auffermann W** (1984) Algorithm for a DNA-cytophotometric diagnosis and grading of malignancy. Anal Quant Cytol 6:1–8.
26. **Bonadonna G, Jotti GS** (1987) Prognostic factors and response to treatment in non-Hodgkin's lymphomas. (Review.) Anticancer Res 7:685–694.
27. **Braylan RC, Benson NA, Nourse V, Kruth HS** (1982) Correlated analysis of cellular DNA, membrane antigens and light scatter of human lymphoid cells. Cytometry 2:337–343.
28. **Braylan RC, Benson NA, Nourse VA** (1984) Cellular DNA of human neoplastic B-cells measured by flow cytometry. Cancer Res 44:5010–5016.
29. **Braylan RC, Benson NA** (1989) Flow cytometric analysis of lymphomas. Arch Pathol Lab Med 113: 627–633.
30. **Braylan RC, Diamond LW, Powell ML, Harty-Golder B** (1980) Percentage of cells in the S phase of the cell cycle in human lymphoma determined by flow cytometry: Correlation with labeling index and patient survival. Cytometry 1:171–174.
31. **Braylan RC, Fowlkes BJ, Jaffe ES, Sanders SK, Berard CW, Herman CJ** (1978) Cell volumes and DNA distributions of normal and neoplastic human lymphoid cells. Cancer 41:201–209.
32. **Bregni M, Siena S, Dalla-Favera R, Gianni AM** (1987) High sensitivity and specificity assay for detection of leukemia/lymphoma cells in human bone marrow. Ann NY Acad Sci 511:473–482.
33. **Brooks JJ, Enterline HT** (1983) Primary gastric lymphomas. A clinicopathologic study of 58 cases with long-term follow-up and literature review. Cancer 51:701–711.
34. **Brossmer R, Bohn B, Sauer A, zur-Hausen HA** (1985) Comparison of established human lymphoma lines by flow cytometry: Quantitation of *Ricinus communis* agglutinin binding and the effect of specific glycosidases. Eur J Cancer Clin Oncol 21:825–831.
35. **Bunn PA, Krasnow S, Makuch RW, Schlam ML, Schechter GP** (1982) Flow cytometric analysis of DNA content of bone marrow cells in patients with plasma cell myeloma: Clinical implications. Blood 59:528.
36. **Bunn PA, Whang-Peng J, Carney DN, Schlam ML, Knutsen T, Gazdar AF** (1980) DNA content analysis by flow cytometry and cytogenetic analysis in *Mycosis fungoides* and Sézary syndrome. J Clin Invest 65: 1440–1448.
37. **Cabanillas F, McLaughlin P** (1988) New methods of investigation for the characterization of lymphomas. In Fuller LM (ed), "Hodgkin's Disease and Non-Hodgkin's Lymphomas in Adults and Children." New York: Raven Press, pp 102–114.
38. **Cabanillas F, Pathak S, Trujillo J, et al** (1988) Frequent nonrandom chromosome abnormalities in 27 patients with untreated large cell lymphoma and immunoblastic lymphoma. Cancer Res 48:5557–5564.
39. **Cabanillas F, Trujillo JM, Barlogie B, et al** (1986) Chromosomal abnormalities in lymphoma and their correlations with nucleic acid flow cytometry. Cancer Genet Cytogenet 21:99–106.
40. **Caligaris-Cappio F, Janossy G, Bergui L, et al** (1985) Identification of malignant plasma cell precursors in the bone marrow of multiple myeloma. J Clin Invest 76:1241–1251.
41. **Camplejohn RS, Macartney JC** (1985) Comparison of DNA flow cytometry from fresh and paraffin embedded samples of non-Hodgkin's lymphoma. J Clin Pathol 38:1096–1099.
42. **Chaganti RSK, Jhanwar JC, Arlin Z, et al** (1983) Do lymphoid neoplasia arise in multipotential stem cells? Evidence from translocations in Burkitt's lymphoma. Cancer Genet Cytogenet 10:95–104.
43. **Chan CSP, Wormsley SB, Peter JB, Schechter GP** (1989) Dual parameter analysis of myeloma cells by flow cytometry: DNA content of cells containing monotypic cytoplasmic immunoglobulin. Am J Clin Pathol 91:12–17.
44. **Chasserot-Golaz S, Schuster C, Dietrich JB, Beck G, Lawrence DA** (1988) Antagonistic action of RU38486 on the activity of transforming growth factor-beta in fibroblasts and lymphoma cells. J Steroid Biochem 30:381–385.
45. **Christensson B, Biberfeld P, Ost A, Tribukait B** (1984) Flow-cytofluorometric DNA analysis in non-Hodgkin's lymphomas. In Cavalli F, Bonadonna G, Rozencweig M (eds), "Second International Conference on Malignant Lymphoma." Boston: Nijhoff Press, pp 101–114.
46. **Christensson B, Tribukait B, Biberfeld P** (1985)

DNA content and proliferation in non-Hodgkin's lymphoma: Flow cytofluorometric DNA analysis in relation to the kiel classification. Dev Oncol 31:101–104.

47. **Christensson B, Tribukait B, Linder IL, Ullman B, Biberfeld P** (1986) Cell proliferation and DNA content in non-Hodgkin's lymphoma. Flow cytometry in relation to lymphoma classification. Cancer 58: 1295–1305.

48. **Clevenger CV, Bauer KD, Epstein AL** (1985) A method for simultaneous nuclear immunofluorescence and DNA content quantitation using monoclonal antibodies and flow cytometry. Cytometry 6: 208–214.

49. **Coiffier B, Berger F, Bryon PA, Magaud JP** (1988) T-cell lymphomas: Immunologic, histologic, clinical and therapeutic analysis of 63 cases. J Clin Oncol 6:1584–1589.

50. **Colly LP, Peters WG, Hermans J, Arentsen-Honders W, Willemze R** (1987) Percentage of S-phase cells in bone marrow aspirates, biopsy specimens and bone marrow aspirates corrected for blood dilution from patients with acute leukemia. Leuk Res 11:209–213.

51. **Coon JS, Landay AL, Weinstein RS** (1987) Advances in flow cytometry for diagnostic pathology. Lab Invest 5:453–479.

52. **Costa A, Bonadonna G, Villa E, Valagussa P, Silvestrini R** (1981) Labeling Index as a prognostic marker in non-hodgkins lymphomas. J Natl Cancer Inst 66: 1–5.

53. **Costa A, Mazzini G, Del Bino G, Silvestrini R** (1981) DNA content and kinetic characteristics of non-Hodgkin's lymphoma: Determined by flow cytometry and autoradiography. Cytometry 2:185–188.

54. **Costa A, Silvestrini R, Bonadonna G, et al** (1988) Prognostic relevance of cell kinetics in non-Hodgkin's lymphomas. Proc Ann Meet Am Soc Clin Oncol A917.

55. **Crocker J, Macartney JC, Smith PJ** (1988) Correlation between DNA flow cytometric and nucleolar organizer region data in non-Hodgkin's lymphomas. J Pathol 154:151–156.

56. **DeAngelis P, Kimmel M, Lacher M, Filippa D, Melamed MR** (1988) T-cell subsets in peripheral blood of patients with newly diagnosed and posttreatment Hodgkin's and non-Hodgkin's lymphomas. Analysis by monoclonal antibody staining and flow cytometry. Anal Quant Cytol Histol 10:235–242.

57. **De Martini RM, Turner RR, Boone DC, Lukes RJ, Parker JW** (1988) Lymphocyte immunophenotyping of B-cell lymphomas: A flow cytometric analysis of neoplastic and nonneoplastic cells in 271 cases. Clin Immunol Immunopathol 49:365–379.

58. **De Paoli P, Santini GF** (1987) Well-differentiated B cell lymphomas/leukaemias express IL-2 receptors identified by immunofluorescence and flow cytometry. Clin Exp Immunol 68:223–224.

59. **Del Bino G, Bruni C, Koch G, Mazzini G, Costa A, Silvestrini R.** (1985) Validation of a mathematical procedure for computer analysis of flow cytometric DNA data in human tumors. Cytometry 6:31–36.

60. **Del Bino G, Silvestrini R, Costa A, Mazzini G, Giordano P** (1987) DNA–protein flow cytometric analysis in non-Hodgkin lymphoma. Basic Appl Histochem 31:183–190.

61. **Del Bino G, Silvestrini R, Costa A, Veneroni S, Giardini R** (1986) Morphological and clinical significance of cell kinetics in non-Hodgkin's lymphomas. Basic Appl Histochem 30:197–202.

62. **Delia D, Villa S, De-Braud F, Giardini R** (1984) Monoclonal antibodies and flow cytometry in the study and diagnosis of non-Hodgkin's lymphomas. Sanofi Rech 219–230. (Abst).

63. **Dewald GW, Kyle RA, Hicks GA, Greipp PR** (1985) The clinical significance of cytogenetic studies in 100 patients with multiple myeloma, plasma cell leukemia, or amyloidosis. Blood 66:380–390.

64. **Diamond LW, Braylan RC, Bearman RM, Winberg CD, Rappaport H** (1980) The determination of cellular DNA content in neoplastic and non-neoplastic lymphoid populations by flow cytofluorometry. In Laerum OD, Lindmo T, Thorud E (eds), "Flow Cytometry." Belgen: Universitetsforlaget, pp 478–482.

65. **Diamond LW, Nathwani BN, Rappaport H** (1982) Flow cytometry in the diagnosis and classification of malignant lymphoma and leukemia. Cancer 50: 1122–1135.

66. **Donhuijsen K** (1987) Mitoses in non-Hodgkin's lymphoams. Frequency and prognostic relevance. Pathol Res Pract 182:352–357.

67. **Drewinko B, Alexanian R** (1977) Growth kinetics of plasma cell myeloma. J Natl Cancer Inst 58:1247.

68. **Drewinko B, Barlogie B** (1984) Cell cycle perturbation effects. Handb Exp Pharmacol 72:101–141.

69. **Drexler HG, Amlot PL, Minowada J** (1987) Hodgkin's disease derived cell lines-conflicting clues for the origin of Hodgkin's disease? Leukemia 1: 629–637.

70. **Duque RE, Braylan RC** (1988) Flow cytometry in the diagnosis of non-Hodgkin's lymphomas. Clin Immunol Newsl 9:13–16.

71. **Duque RE, Stoolman LM, Hudson JL, Ward PA** (1985) Multiparameter analysis of immunohematological disorders by flow cytometry. Surv Synth Pathol Res 4:323–340.

72. **Durie BG** (1986) Staging and kinetics of multiple myeloma. Semin Oncol 13:300–309.

73. **Durie BGM, Grogan TM** (1985) CALLA positive myeloma: An aggressive subtype with poor survival. Blood 66:229–232.

74. **Egerter DA, Said JW, Epling S, Lee S** (1988) DNA content of T-cell lymphomas. A flow cytometric analysis. Am J Pathol 130:326–334.

75. **Ellison DJ, Nathwani BN, Metter GE, et al** (1987) Mitotic counts in follicular lymphomas. Hum Pathol 18:502–505.

76. **Eriksson S, Skog S, Tribukait B, Jaderberg K** (1984) Deoxyribonucleoside triphosphate metabolism and the mammalian cell cycle. Effects of thymidine on wild-type and dCMP. Exp Cell Res 155:129–140.

77. **Eriksson S, Skog S, Tribukait B, Wallstrom B** (1987) Deoxyribonucleoside triphosphate metabolism and the mammalian cell cycle. Effect of hydroxyurea on mutant and wild-type mouse S49 T-lymphoma cells. Exp Cell Res 168:79–88.

78. **Foon KA, Todd RF** (1986) Immunologic classification of leukemia and lymphoma. Blood 68:1–31.

79. **Foucar K, Chen IM, Crago S** (1989) Organization

and operation of a flow cytometric immunophenotyping laboratory. Sem Diagn Pathol 6:13–36.

80. **Fried J, Arlin Z, Alikpala A, Tan CTC, Clarkson B (1976)** Kinetic response of human leukemic and lymphoma cells in vivo to combination chemotherapy using flow microfluorometry. Cancer Treatm Rep 60:166–175.

81. **Gebhard DF Jr, Mittelman A, Cirrincione C, Thaler HT, Koziner B (1986)** Comparative analysis of surface membrane immunoglobulin determination by flow cytometry and fluorescence microscopy. J Histochem Cytochem 34:475–481.

82. **Gerdes J, Schwarting R, Stein H (1986)** High proliferative activity of Reed Sternberg associated antigen Ki-1 positive cells in normal lymphoid tissue. J Clin Pathol 39:993–997.

83. **Glassman AB, Self S, Christopher J (1987)** Lymphomas membrane markers and cell flow cytometric diagnosis. Ann Clin Lab Sci 17:1–7.

84. **Greipp PR, Katzmann JA, O'Fallon M, Kyle RA (1988)** Value of beta-2 microglobulin level and plasma cell labeling indices as prognostic factors in patients with newly diagnosed myeloma. Blood 72: 219–223.

85. **Grogan TM, Lippman SM, Speir CM, et al (1988)** Independent prognostic significant of a nuclear proliferation antigen in diffuse large cell lymphomas as determined by the monoclonal antibody Ki-67. Blood 71:1157–1160.

86. **Grogan TM, Spier CM, Richter LC, Rangel CS (1988)** Immunologic approaches to the classification of non-Hodgkin's lymphomas. Cancer Treatm Res 38:31–148.

87. **Hansen H, Koziner B, Clarkson B (1981)** Marker and kinetic studies in the non-Hodgkin's lymphomas. Am J Med 71:107–123.

88. **Haynes BF, Metzgar RS, Minna JD, Bunn PA (1981)** Phenotypic characterization of cutaneous T-cell lymphoma: Use of monoclonal antibodies to compare with other malignant T cells. N Engl J Med 304: 1319–1323.

89. **Hedley DW (1987)** DNA flow cytometric study of 5-fluorouracil used to treat end stage non-Hodgkin's lymphoma. Br J Cancer 55:107–108.

90. **Hedley DW, Friedlander ML, Taylor IW, Rugg CA, Musgrove EA (1983)** Method for analysis of cellular DNA content of paraffin-embedded pathological material using flow cytometry. J Histochem Cytochem 31:1333–1335.

91. **Henni T, Gaulard P, Divine M, et al (1988)** Comparison of genetic probe with immunophenotypic analysis in lymphoproliferative disorders: A study of 87 cases. Blood 72:1937–1943.

92. **Hiddemann W, Schumann J, Andreeff M, et al (1984)** Convention on nomenclature for DNA cytometry. Cytometry 5:445–446.

93. **Hirata M, Okamoto Y (1987)** Enumeration of terminal deoxynucleotidyl transferase positive cells in leukemia/lymphoma by flow cytometry. Leuk Res 11: 509–518.

94. **Hirt A, Baumgartner C, Imbach P, Luethy A, Wagner HP (1984)** Differentiation and cytokinetic analyses of normal and neoplastic lymphoid cells in B and T cell malignancies of childhood. Br J Haematol 58:241–248.

95. **Holte H, Strokke T, Smeland E, et al (1989)** Levels of myc protein, as analyzed by flow cytometry, correlate with cell growth potential in malignant B-cell lymphoma. Int J Cancer 43:164–170.

96. **Horning SJ, Rosenberg SA (1984)** The natural history of initially untreated low-grade non-Hodgkin's lymphomas. N Engl J Med 311:1471–1475.

97. **Joensuu H, Klemi PJ, Eerola E (1988)** Diagnostic value of DNA flow cytometry combined with fine needle aspiration biopsy in myeloma. J Pathol 154: 237–245.

98. **Johnson A, Akerman A, Cavallin-Stahl E (1987)** Flow cytometric detection of B-clonal excess in fine needle aspirates for enhanced diagnostic accuracy in non-Hodgkin's lymphoma in adults. Histopathology 11:581–590.

99. **Johnson A, Cavalin-Stahl E (1985)** Detection of small amounts of monoclonal lymphoma cells in peripheral blood by flow cytometry. In Cavalli F, Bonadonna G, Rozencweig M (eds), "Second International Conference on Malignant Lymphoma." Boston: Nijhoff Press.

100. **Johnson A, Cavallin-Stahl E, Akerman M (1985)** Flow cytometric light chain analysis of peripheral blood lymphocytes in patients with non-Hodgkin's lymphoma. Br J Cancer 52:159–165.

101. **Juneja SK, Cooper IA, Hodgson GS, et al (1986)** DNA ploidy patterns and cytokinetics of non-Hodgkin's lymphoma. J Clin Pathol 39:987–992.

102. **Kantarjian HM, Barlogie B, Pershouse M, et al (1985)** Preferential expression of double-stranded ribonucleic acid in tumor versus normal cells: Biological and clinical implications. Blood 66:39–46.

103. **Katz RL, Raval P, Manning JT, McLaughlin P, Barlogie B (1987)** A morphologic immunologic, and cytometric approach to the classification of non-Hodgkin's lymphoma in effusions. Diagn Cytopathol 3:92–101.

104. **Klein G (1986)** Constitutive activation of oncogenes by chromosomal translocations in B-cell derived tumors. AIDS Res S167–176.

105. **Kruth HS, Braylan RC, Benson NA, Nourse VA (1981)** Simultaneous analysis of DNA and cell surface immunoglobulin in human B-cell lymphomas by flow cytometry. Cancer Res 41:4895–4899.

106. **Kvaloy S, Godal T, Marton PF, Steen H, Brennhovd IO, Abrahamsen AF (1981)** Spontaneous [^{3}H]-thymidine uptake in histological subgroups of human B-cell lymphomas. Scand J Haematol 26:221–234.

107. **Leemhuis T, Srour E, Hanks S, Smith BR, Jansen J (1987)** Detection of infiltration of bone marrow by B-cell-lymphoma using two-color fluorescence clonal-excess analysis. Blood 70:217a. (Suppl 1)

108. **Latreille J, Barlogie B, Johnston D, Drewinko B, Alexanian R (1982)** Ploidy and proliferative characteristics in monoclonal gammopathies. Blood 59:43–51.

109. **Leder P, Stewart T, Kuo A, Pattengale P (1985)** Growth and differentiation of cells in defined environment. In Murakami H, Yamane I, Barnes DW, Mather JP, Hayashi I, Sato GH (eds) , "Cell Growth and the New Genetics." New York: Springer-Verlag, pp 473–480.

110. **Lehtinen T, Aine R, Kallioniemi OP, et al (1989)**

Flow cytometric DNA analysis of 199 histologically favourable or unfavourable non-Hodgkin's lymphomas. J Pathol 157:27–36.

111. **Lenner P, Roos G, Johansson H, Lindh J, Dige U** (1987) Non-Hodgkin's lymphoma. Multivariate analysis of prognostic factors including fraction of S-phase cells. Acta Oncol 26:179–183.

112. **Liendo C, Danieu L, Al-Katib A, Koziner A** (1985) Phenotypic analysis by flow cytometry of surface immunoglobulin light chains and B and T cell antigens in lymph nodes involved with non-Hodgkin's lymphoma. Am J Med 79:445–454.

113. **Ligler FS, Smith RG, Kettman JR, et al** (1980) Detection of tumor cells in the peripheral blood of nonleukemic patients with B-cell lymphoma: Analysis of "clonal excess." Blood 55:792–801.

114. **Lindh J, Lenner P, Roos G** (1984) Monoclonal B cells in peripheral blood in non-Hodgkin's lymphoma. Correlation with clinical features and DNA content. Scand J Haematol 32:5–11.

115. **Little JV, Foucar K, Horvath A, Crago S** (1989) Flow cytometric analysis of lymphoma and lymphoma-like disorders. Semin Diagn Pathol 6:37–54.

116. **Lukes RJ, Collins RD** (1975) A functional classification of malignant lymphomas. In Rebuck JW, Berard CW, Abell MR (eds), "The Reticuloendothelial System." Huntington, NY: Williams and Wilkins, p213.

117. **Lukes RJ, Taylor CR, Chir B, et al** (1978) A morphologic and immunologic surface marker study of 299 cases of non-Hodgkin lymphomas and related leukemias. Am J Pathol 90:461–486.

118. **McCartney JC, Camplejohn RS, Alder J, Stone MG, Powell G** (1986) Prognostic importance of DNA flow cytometry in non-Hodgkin's lymphomas. J Clin Pathol 39:542–546.

119. **McConlogue JC, Marton LJ, Coffino P** (1983) Growth regulatory effects of cyclic AMP and polyamine depletion are dissociable in cultured mouse lymphoma cells. J Cell Biol 96:762–767.

120. **McIntire TL, Goldey SH, Benson NA, Braylan RC** (1987) Flow cytometric analysis of DNA in cells obtained from deparaffinized formalin-fixed lymphoid tissues. Cytometry 8:474–478.

121. **McLaughlin P, Osborne B, Johnston D, et al** (1987) Prognostic value of nucleic acid flow cytometry (FCM) in diffuse large cell lymphoma. Proc Annu Meet Am Assoc Cancer Res 28:37.

122. **McLaughlin P, Osborne B, Johnston D, et al** (1987) Prognostic value of nucleic acid flow cytometry (FCM) in diffuse large cell lymphoma. Third International Conference on Malignant Lymphoma, p 38.

123. **McLaughlin P, Osborne BM, Johnston D, et al** (1988) Nucleic acid flow cytometry in large cell lymphoma. Cancer Res 48:6614–6619.

124. **Medeiros LJ, Picker LJ, Gelb AB, et al** (1989) Numbers of host "helper" T-cells and proliferating cells predict survival in diffuse small-cell lymphomas. J Clin Oncol 7:1009–1017.

125. **Meyer JS, Higa E** (1979) S-phase fractions of cells in lymph nodes and malignant lymphomas. Arch Pathol Lab Med 103:93–97.

126. **Mohammed RM, Clark C, Mohamed AN, Al-Katib A** (1988) Analysis of gamma-interferon and phorbol ester-treated human leukemia and lymphoma cell lines by 2-dimensional gel electrophoresis: Correlation with phenotypic changes. Proc Annu Meet Am Assoc Cancer Res 29:A298.

127. **Morgan KG, Quirke P, O'Brien CJ, Bird CC** (1988) Hodgkin's disease: A flow cytometric study. J Clin Pathol 41:365–369.

128. **Murphy SB, Melvin SL, Mauer AM** (1979) Correlation of tumor cell kinetic studies with surface marker results in childhood non-Hodgkin's lymphoma. Cancer Res 39:1534–1538.

129. **Nabholtz JM, Friedman S, Collin F, Guerrin J** (1987) Modification of kiel and working formulation classifications for improved survival prediction in non-Hodgkin's lymphoma. J Clin Oncol 5:1634–1639.

130. **Nathwani BN** (1979) A critical analysis of the classifications of non-Hodgkin's lymphomas. Cancer 44: 347–384.

131. **Naus GJ** (1985) Proliferative activity in malignant non-Hodgkin's lymphoma. Fed Proc 44:909.

132. **Neckers LM, Nordan RP** (1988) Regulation of murine plasmacytoma transferrin receptor expression and G1 traversal by plasmacytoma cell growth factor. J Cell Physiol 135:495–501.

133. **Nooteboom GN, Chuang LF, Israel M, Chuang RY** (1985) N-trifluoroacetyladriamycin-14-0-hemiadipa te (AD143) effects on nucleoside incorporation and cell cycle in human AW ramos lymphoma cells. Fed Proc 44:1141.

134. **Nowell PC, Croce CM** (1987) Cytogenetics of neoplasia. In Greene MI, Hamaoka T (eds), "Development and Recognition of the Transformed Cell." New York: Plenum, pp 1–19.

135. **Paietta E, Racevskis J, Stanley ER, Andreeff M, Papenhausen P, Wiernik PH** (1990) Expression of the macrophage growth factor, CSF-1, and its receptor, c-fms by a Hodgkin's disease derived cell line and its variants. Cancer Res (in press).

136. **Palutke M, Schnitzer B, Dresner D, Tabaczka P** (1986) Comparison of morphologic features and mitotic rate to cytometrically determined DNA content of poorly differentiated lymphocytic lymphomas. Ann NY Acad Sci 468:178–194.

137. **Parker JW** (1985) Flow cytometry in lymphoma/leukemia diagnosis. Dev Oncol 31:126–142.

138. **Pastolero GC, Mori S** (1987) Expression of proliferation-associated nuclear antigen in human malignancies. Jpn J Exp Med 57:193–198.

139. **Pilarski LM, Mant MJ, Ruether BA** (1985) Pre-B cells in peripheral blood of multiple myeloma patients. Blood 66:416–422.

140. **Pinkel D, Landegent J, Collins C, et al** (1988) Fluorescence in situ hybridization with human chromosome-specific libraries: Detection of trisomy 21 and translocation of chromosome 4. Proc Natl Acad Sci USA 85:9138–9142.

141. **Porwit-Ksiazek A, Christensson B, Lindemalm C, et al** (1983) Characterization of malignant and non-neoplastic cell phenotypes in highly malignant non-Hodgkin lymphomas. Br J Cancer 32:667–674.

142. **Potmesil M, Levin M, Traganos F, et al** (1983) In-vivo effects of adriamycin or N-trifluoroacetyladriamycin-14-valerate on a mouse lymphoma. Eur J Cancer Clin Oncol 99:109–122.

143. **Roos G, Dige U, Lenner P, Lindh J, Johansson H** (1985) Prognostic significance of DNA analysis by

flow cytometry in non-Hodgkin's lymphoma. Hematol Oncol 3:233–242.

144. **Roos G, Leanderson T, Lundgren E (1984)** Interferon-induced cell cycle changes in human hematopoietic cell lines and fresh leukemic cells. Cancer Res 44:2358–2362.

145. **Saiki RK, Gelfand DH, Stoffel S, et al (1988)** Primer-directed enzymatic amplication of DNA with a thermostable DNA polymerase. Science 239:487–491.

146. **Salmon SE, Grogan TM, Miller T, Scheper R, Dalton WS (1989)** Prediction of doxorubicin resistance in vitro in myeloma, lymphoma, and breast cancer by p-glycoprotein staining. J Natl Cancer Inst 81:696–701.

147. **Sandritter W, Grimm H (1977)** DNA in non-Hodgkin-lymphoma—A cytophotometric study. Beitr Pathol 160:213–230.

148. **Scarffe JH, Crowther D (1981)** The pre-treatment proliferative activity of non-hodgkins lymphoma cells. Eur J Cancer 170:99–108.

149. **Schneider RJ, Seibert K, Passe S, et al (1980)** Prognostic significance of serum lactate dehydrogenase in malignant lymphoma. Cancer 46:139–143.

150. **Schrape S, Jones DB, Wright DH (1987)** A comparison of three methods for the determination of the growth fraction in non-Hodgkin's lymphoma. Br J Cancer 55:283–286.

151. **Scott CS, Ramsden W, Limbert HJ, Master PS, Roberts BE (1988)** Membrane transferrin receptor (TfR) and nuclear proliferation-associated Ki-67 expression in hemopoietic malignancies. Leukemia 2:438–442.

152. **Scott DW, Livnat D, Pennell CA, Keng P (1986)** Lymphoma models for B cell activation and tolerance. III. Cell cycle dependence for negative signalling of WEHI-231 B lymphoma cells by anti-mu. J Exp Med 164:156–164.

153. **Self SE, Burdash NM, Ponzio AD, Lavia MF (1986)** Lymphocyte subsets in lymph node hyperplasias and B cell neoplasms as determined by fluoresceinated antibodies and flow cytometry. Ann NY Acad Sci 468:195–210.

154. **Shackney S, Levine A, Simon R, et al (1983)** Dual parameter flow cytometry (FCM) studies in 220 cases of non-Hodgkin's lymphoma (NHL). Proc Annu Meet Am Soc Clin Oncol 2:C-32.

155. **Shackney SE (1986)** The use of flow cytometry in the diagnosis and biological characterization of the non-Hodgkin's lymphomas. Ann NY Acad Sci 468:171–177.

156. **Shackney SE, Levine AM, Fisher RI, et al (1984)** The biology of tumor growth in the non-Hodgkin's lymphomas: A dual parameter flow cytometry study of 220 cases. J Clin Invest 730:1201–1204.

157. **Shackney SE, Skramstad KS, Cunningham RE, et al (1980)** Dual parameter flow cytometry studies in human lymphomas. J Clin Invest 66:1281–1294.

158. **Shawler DL, Wormsley SB, Dillman RO, et al (1985)** The use of monoclonal antibodies and flow cytometry to detect peripheral blood and bone marrow involvement of a diffuse, poorly differentiated lymphoma. Int J Immunopharmacol 7:423–432.

159. **Silvestrini R, Costa A, Veneroni S, Del-Bino G, Persici P (1988)** Comparative analysis of different approaches to investigate cell kinetics. Cell Tissue Kinet 21:123–131.

160. **Skog S, Tribukait B, Wallstrom B, Eriksson S (1987)** Hydroxyurea-induced cell death as related to cell cycle in mouse and human T-lymphoma cells. Cancer Res 47:6490–6493.

161. **Slater DE, Mertelsmann R, Koziner B, et al (1986)** Lymphoblastic lymphoma in adults. J Clin Oncol 4:57–67.

162. **Smeland EB, Blomhoff HK, Holte H, et al (1987)** Transforming growth factor type beta (TGF beta) inhibits G1 to S transition, but not activation of human B lymphocytes. Exp Cell Res 171:213–222.

163. **Smith L, Hall R, Pershouse M, Alexanian R, Barlogie B (1984)** Quiescent cells in S-phase: A cytokinetic sanctuary in myeloma. Proc Annu Meet Am Soc Clin Oncol 3:11.

164. **Srigley J, Barlogie B, Butler JJ, et al (1985)** Heterogeneity of non-Hodgkin's lymphoma probed by nucleic acid cytometry. Blood 65:1090–1096.

165. **Stetlet-Stevenson M, Raffeld M, Cohen P, Cossman J (1988)** Detection of occult follicular lymphoma by specific DNA amplification. Blood 72:1822–1825.

166. **Stonesifer KJ, Benson NA, Ryden SE, Pawliger DF, Braylan RC (1986)** The malignant cells in a Lennert's lymphoma are T lymphocytes with a mature helper surface phenotype. A multiparameter flow cytometric analysis. Blood 68:426–429.

167. **Stonesifer KJ, Xiang JH, Wilkinson EJ, Benson NA, Braylan RC (1987)** Flow cytometric analysis and cytopathology of body cavity fluids. Acta Cytol (Praha) 31:125–130.

168. **Tafuri A, Myers J, Lee B, Andreeff M (1988)** DNA and RNA flow cytometry in multiple myeloma (MM): Clinical correlations. Proc Annu Meet Am Soc Clin Oncol 7:A172.

169. The Non-Hodgkin's Lymphoma Pathologic Classification Project National Cancer Institute sponsored study of classifications of non-hodgkin's lymphomas. Cancer 49:2112–2135.

170. **Tonevitsky AG, Trakht IN, Rudchenko SA (1984)** Action of a chain ricin–antibody conjugate on Burkitt lymphoma cells correlation with the change of cell population distribution on the cell cycle. Presented at the Sixteenth Meeting of the Federation of European Biochemical Societies p. 197

171. **Tsuchiya H, Epstein J, Selvanayagam P, et al (1988)** Correlated flow cytometric analysis of H-ras p21 and nuclear DNA in multiple myeloma. Blood 72:796–800.

172. **Vassallo J, Mellin W, Pill C, Roessner A, Grunchmann E (1987)** Flow cytometric DNA analysis of malignant lymphomas with primary bone manifestation. J Cancer Res Clin Oncol 113:249–252.

173. **Vassallo J, Mellin W, Pill C, Roessner A, Grundmann E (1987)** Flow cytometric DNA analysis of malignant lymphomas with primary bone manifestation. J Cancer Res Clin Oncol 113:249–252.

174. **Walle AJ (1986)** Identification of L3 leukemia and Burkitt's lymphoma cells by flow cytometric quantitation of nuclear and cellular RNA and DNA content. Leuk Res 10:303–312.

175. **Wantzin GL, Larsen JK, Christensen J, Ralfkiaer E, Thomsen K (1984)** Early diagnosis of cutaneous T-cell lymphoma by DNA flow cytometry on skin biopsies. Cancer 54:1348–1352.

176. **Watzin GL, Thomsen K, Larsen JK, Christensen IJ,**

Rasmussen BB (1983) DNA analysis by flow cytometry in lymphomatoid papulosis. Clin Exp Dermatol 8:505–512.

177. **Weinberg DS, Pinkus GS, Ault KA** (1984) Cytofluorometric detection of B cell clonal excess: A new approach to the diagnosis of B cell lymphoma. Blood 63:1080–1087.

178. **Weiss LM, Strickler JG, Medeiros LJ, Gerdes J, Stein H, Warnke PA** (1987) Proliferative rates of non-Hodgkin's lymphomas as assessed by Ki67 antibody. Hum Pathol 18:1155–1159.

179. **Weiss LM, Strickler JG, Medeiros LJ, Gerdes J, Stein H, Warnke RA** (1987) Proliferative rates of non-Hodgkin's lymphomas as assessed by Ki67 antibody. Hum Pathol 18:1155–1159.

180. **Wooldridge TN, Grierson HL, Weisenburger DD, et al** (1988) Association of DNA content and proliferative activity with clinical outcome in patients with diffuse mixed cell and large cell non-Hodgkin's lymphoma. Cancer Res 48:6608–6613.

181. **Young GA, Hedley DW, Rugg CA, Iland HJ** (1987) The prognostic significance of proliferative activity in poor histology non-Hodgkin's lymphoma: A flow cytometry study using archival material. Eur J Cancer Clin Oncol 23:1497–1504.

182. **Yunis JJ, Mayer MG, Arnesen MA, Aeppli DP, Oken MM, Frizzera G** (1989) bcl-2 and other genomic alterations in the prognosis of large-cell lymphoma. N Engl J Med 320:1047–1054.

37

DNA Flow Cytometry of Human Solid Tumors

Martin N. Raber and Bart Barlogie

Department of Medical Oncology, University of Texas M.D. Anderson Cancer Center, Houston, Texas 77030 (M.N.R.); Department of Hematology, University of Texas M.D. Anderson Cancer Center, Houston, Texas 77030 (B.B.)

Over the past decade, DNA content analysis has developed from a technique limited to a relatively small number of research laboratories into one available at many hospitals around the world and performed by a growing number of commercial laboratories. Nevertheless, the interpretation and importance of these measurements remains controversial. This chapter reviews our current understanding of DNA analysis in solid tumors with respect to the diagnosis of malignancy, differential diagnosis, and prognosis. The advantages and limitations of the technique are discussed, as well as how these measurements may be applied to the management of patients with cancer. In most cases, data have been drawn from clinical studies in which firm conclusions can be made based on a clear definition of the patient population studied according to stage of disease and treatment employed, with an adequate description of the technical aspects of measurement and analysis.

The focus of research in applied tumor cytometry has been on understanding the morphologic heterogeneity of human malignancies, related to patterns of metastatic spread and prognosis [12]. In addition to an inherent pleomorphism of tumor cells, most neoplasms contain a varying proportion of normal cells which at times are not readily distinguishable from their malignant counterparts. This cellular diversity underlines the need for an unequivocal tumor marker before trying to assess the phenotypic heterogeneity of tumor cells. The presence of an abnormal DNA stemline remains the most reliable tumor marker available to flow cytometric analysis [7].

TECHNICAL ASPECTS

Having developed in a variety of laboratories worldwide, it is not surprising that there is considerable variation in techniques of cell preparation, staining, instrumentation, and data interpretation. However, there is broad agreement on how the results of analysis should be displayed and reported. To facilitate communication among laboratories a convention on nomenclature has provided guidelines for expressing abnormalities of DNA content [37]. The DNA index (DI) is calculated by dividing the modal DNA content of the population being studied by the DNA content of the corresponding normal cells. Thus, samples with a DNA content equal to that of normal cells have a DI equal to 1, and are referred to as diploid. Populations with a DNA content not equal to that of normal cells are referred to as aneuploid and are further subdivided into hypodiploid (DI <1), hyperdiploid (DI >1) and occasionally tetraploid (DI $= 2$). Most authorities follow this convention, and it is adhered to in the figures and tables presented in this chapter.

While the techniques for processing, staining, and measuring DNA content are standardized and reproducible, certain limitations are apparent when studying solid tumors. When evaluating aneuploidy, care should be taken that the reference diploid standard be from the same species, preferably from the same tissue. This is particularly important in dealing with minimum deviation aneuploidy (DI 0.95–1.05). In situations in which one peak in the diploid range with a high coefficient of variation (CV) is found, consideration should be given to repeating the analysis using different fluorochromes. The interpretation of these broad peaks is controversial because of the possibility of hidden near-diploid, aneuploid populations, but in themselves these peaks should not be considered aneuploid. Some authors have chosen to avoid the controversy by lumping together diploid and "near diploid" tumors as a single group [10,39]. This approach appears to be valid.

Similarly, interpretation of the S-phase fraction (SPF) of the histogram is controversial. The techniques for estimating SPF are numerous, and are beyond the scope of this chapter (see Chapter 22). Our own approach is to combine S + G2M, which reduces the problems of modeling to one (GI–S) interface. One also must be concerned that differences in fluorescent intensity may result from differences in chromatin structures and/or nuclear protein binding, rather than differences in cellular DNA content (see Chapter 16). We recommend that laboratories engaged in clinically applied cytometry test the stoichiometry of staining in their system, either by using populations of cells with defined differences in DNA content or relating DNA modal values to defined chromosomal abnormalities.

Flow Cytometry and Sorting, Second Edition, pages 745–754

There have been two important advances in processing solid tumor samples that merit particular attention because they have considerable impact on solid tumor cytometry. The first is the technique for analysis of paraffin embedded samples. Described by Hedley et al. in 1983, this technique employs thick sections of paraffin embedded tumor which are dewaxed in xylene, rehydrated through a series of alcohol solutions of increasing strength, then pepsinized and prepared for analysis in a conventional manner [34]. Its impact on our understanding of the value of DNA analysis has been considerable. It has taken us from an era in which only fresh samples could be analyzed, which meant prospective studies with small patient numbers and limited follow-up, to one in which whole pathology archives are available for retrospective analysis. This has allowed investigators to focus on homogeneous patient populations selected after enough time has passed to evaluate the possible relationship of DNA measurements to the natural history of the tumor. While this approach has certain limitations, its impact has already been considerable.

A second interesting technique meriting comment is that of Dressler and her colleagues, who analyzed breast cancer samples obtained previously and stored frozen for steroid receptor assay [22]. The cell sample is prepared by homogenization of pulverized frozen tumor tissue, progressive filtration and removal of debris. It permits flow cytometry of small tumor samples without compromising the tissue needed for receptor analysis, and could make DNA analysis readily available for all breast tumors biopsied or removed. Like Hedley's technique, it has been applied retrospectively to a large population with apparent success. Based on these techniques, investigators are beginning to make significant progress in assessing the prognostic value of DNA content analysis.

DNA CONTENT AS A MARKER OF MALIGNANT TUMORS

To date, well over 7,000 solid tumor samples have been analyzed for DNA content and reported in the literature. Table 1 is an incomplete listing of studies, focusing on large series. As can be seen, an aneuploid population is consistently reported in approximately 70% of the tumors analyzed. Aneuploidy appears to be an excellent marker for the malignant cell population [7,8]. The DNA index correlates well with chromosome number as determined by karyotypic analysis both in the leukemias and solid tumors, as has been elegantly demonstrated by investigators at the Karolinska Institutet [79]. One might ask, if this is the case, why a higher percentage of tumors are not aneuploid? The explanation lies in the present limits of resolution of DNA content analysis (a variation of 0.05 in the DI would represent 1 chromosome), and in the tumors that have undergone translocation or other genomic changes not reflected in DNA content or chromosome number.

A major question pertains to the nature of diploid subpopulations in aneuploid tissue samples. It is a general assumption that most human tumors display a single DNA stemline, as the incidence of bi- or multiclonality is generally <10%. In extensive studies of human myeloma using bivariate DNA-cytoplasmic immunoglobulin (CIG) analysis, we have noted monoclonal light chain expression in diploid cells of approximately 10% of otherwise aneuploid samples [6]. To date, no meticulous study has been conducted to determine by flow sorting or other means whether diploid subpopulations in aneuploid human solid tumors are also tumor cells or a benign component. This would be of considerable importance, as there is experimental evidence that neoplastic clonogenic cells may display a diploid DNA content, whereas their differentiated progeny are aneuploid [73].

Is DNA content a good marker of malignancy? It is certainly easily and reproducibly measured. Also, DNA content can be determined on biopsy or aspirated material and results correlate well with analysis of the entire tumor [67]. It appears to be stable over time and in samples from different sites within a tumor [10,13]. In extensive studies by many investigators an abnormal DNA index has been reported to be highly specific for neoplasia, particularly in the case of solid tumors [12]. Since some normal tissues such as liver exhibit tetraploid cells, with mononuclear or binuclear morphology, tetraploidy is considered different from aneuploidy, and most investigators require that tetraploid cells comprise at least 15% of the population for a diagnosis of neoplasia. While a number of tumors are hypodiploid, haploid cells are found normally only in testis tissue.

There are also a number of early malignant or premalignant lesions associated with distinctly abnormal DNA stemlines, including preleukemia [9], benign monoclonal gammopathy [11], angioimmunoblastic lymphadenopathy [8], ulcerative colitis [32], and low-stage bladder tumors [44]. This is an area of great interest biologically, as it permits study of the transition from incipient to uncontrolled tumor growth with invasion and metastasis. Extensive studies of ulcerative colitis demonstrated progressively higher proportions of aneuploid lesions with increasing morphologic abnormalities progressing toward overt adenocarcinoma of the colon [32]. The absence of clear-cut morphologic stigmata of carcinoma in DNA-aneuploid lesions does not compromise the tumor marker value of abnormal DNA content. We interpret this as indicating a malignant genome in tissue still capable of apparently normal differentiation. It is the clonal proliferation of cells with an abnormal DNA stemline that confers the attribute of neoplasia.

There are a number of diagnostic problems for the histopathologist making the distinction between reactive proliferation, benign tumors and malignant neoplasms. These include some thyroid [4,53] and brain tumors [28,38], as well as some body cavity fluids [81] and spinal fluid (see chapter 38). For these specimens, aneuploidy may be of benefit in arriving at the correct diagnosis.

RELATIONSHIP OF DNA CONTENT TO THE TYPE OF MALIGNANCY AND STAGE OF DISEASE

Among the reported studies as a whole, and in our experience in particular, the distribution of DNA index values does not differ between primary and metastatic sites of a given tumor type. While this observation is statistically relevant, few studies have actually examined DNA ploidy in primary and metastatic tumors as they present either concurrently or sequentially [13]. In an autopsy study of 12 patients, we noted conformity of DNA stemline at metastatic sites and primary tumor, although there were differences in the relative proportions of tumor subpopulations in cases with multiple clones [62]. The uniqueness of DNA stemline for a given patient's tumor is further emphasized by the exceptional ploidy change accompanying the development of drug resistance in patients with leukemia and myeloma whom we have followed [5]. With the introduction of Hed-

TABLE 1. DNA Aneuploidy in Solid Tumors

Tumor type	No. of patients	% Aneuploidy	Investigators Author
Bladder	53	70	Barlogie (personal communication)
	98	95	Klein et al. [45]
	123	94	Klein et al. [44]
	200	85	Tribukait et al. [78,79]
	474	88	
Brain	18	66	Barlogie (personal communication)
	11	90	Frederikson and Bichel [27]
	18	78	Hoshino et al. [38]
	47	77	
Breast	1184	57	Dressler et al. [22]
	473	60	Hedley et al. [36]
	166	70	Cornelisse et al. [16]
	275	71	G. Fraschini (personal communication)
	70	44	Kute et al. [48]
	65	80	Moran et al. [58]
	92	92	Olzeweski et al. [60]
	79	78	Thornthwaite and Coulson [75]
	565	71	Cornelisse et al. [17]
	2969	65	
Cervix	90	68	Adelson et al. [1]
	105	80	Jakobsen [39]
	40	98	Linden et al. [49]
	235	78	
Colon	77	65	Kokal et al. [46]
	134	55	Armitage et al. [2]
	264	52	Scott et al. [68]
	279	62	Schutte et al. [66]
	45	65	Barlogie (personal communication)
	78	72	Frankfurt et al. [26]
	33	39	Wolley et al. [85]
	910	59	
Head and neck	73	86	Johnson et al. [42]
Lung	100	55	Zimmerman et al. [87]
	200	83 small cell	
		85 non-small cell	Bunn et al. [14]
	64	90	Moran and Melamed [57]
	35	86	Teodori et al. [74]
	30	79	Vindelov et al. [82]
	187	84	Volm et al. [83]
	616	78	
Melanoma	72	78	Frankfurt et al. [26]
	38	79	Hansson et al. [33]
	605	76	J. Schumann (personal communication)
	715	76	
Ovarian	84	61	Blumenfeld et al. [13]
	128	73	Friedlander et al. [29]
	55	66	Frankfurt et al. [26]
	31	45	J. Schumann (personal communication)
	298	65	
Prostate	91	58	Winkler et al. [84]
	30	67	Barlogie (personal communication)
	31	36	Rotors et al. [64]
	300	71	Tribukait et al. [80]
	40	55	Zetterberg and Esposti [86]
	492	64	
Sarcomas	73	68	Barlogie (personal communication)
	88	61	Frankfurt et al. [26]
	34	67	Kreicbergs et al. [47]
	41	98	J. Schumann (personal communication)
	236	67	
Squamous cell	220	82	J. Schumann (personal communication)
Testicular	74	93	J. Schumann (personal communication)
Miscellaneous tumors	125	40	Dixon and Carter [19]
	190	66	Frankfurt et al. [26]
Total	7674	68	

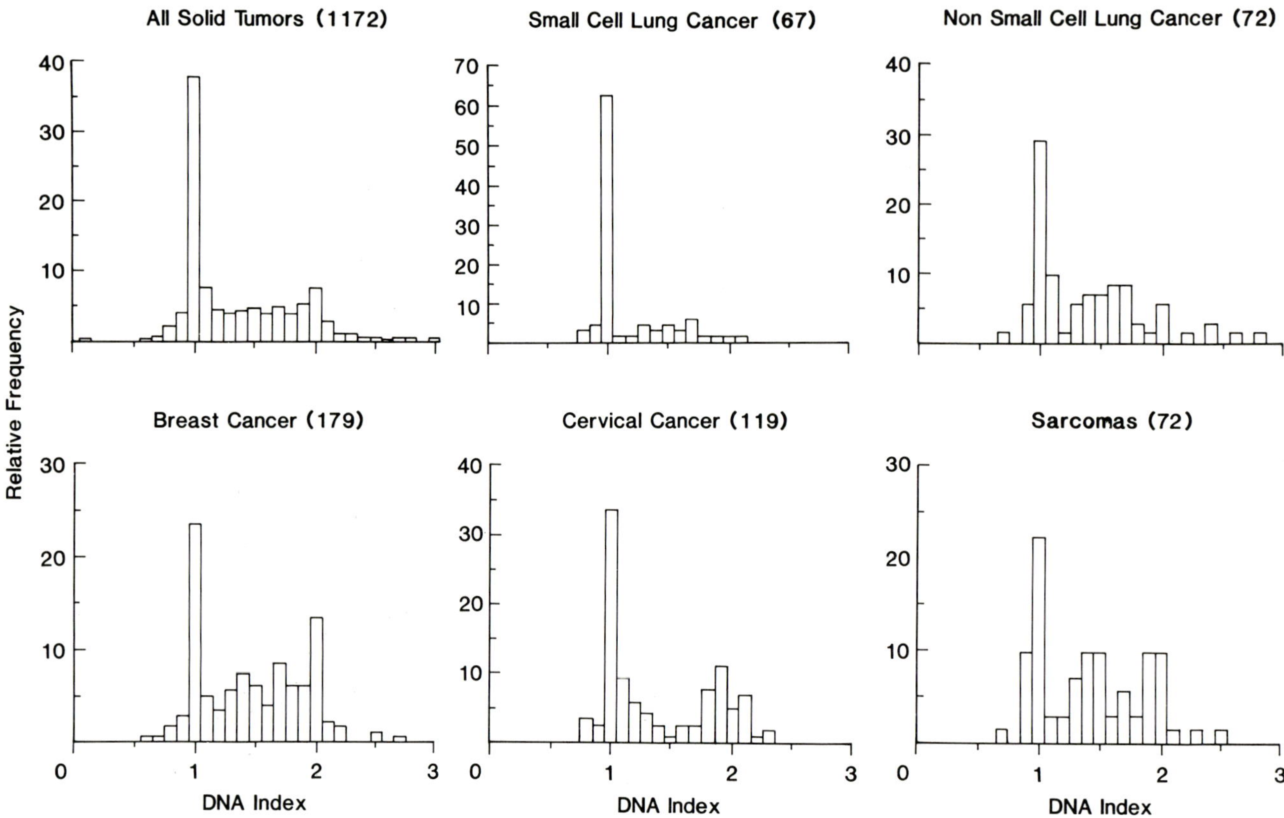

Fig. 1. Relative frequency of distribution of DNA index values in a series of 1,172 solid tumors, according to dominant diagnostic subgroups.

ley's technique for paraffin embedded samples, this question is being addressed, and the stability of DNA content over time confirmed at least in preliminary studies [13].

Figure 1 illustrates the relative frequency distributions of DNA index values among the dominant diagnostic categories of more than 1,000 tumors analyzed in our laboratories. A trimodal distribution pattern with frequency peaks at near-diploid, near-triploid, and near-tetraploid DNA index values was observed among most histopathologic categories. A notable exception was small cell carcinoma of the lung with a preponderance of near-diploid tumors.

In summary, abnormal DNA content appears to be a common, stable, reproducible, easily measured marker of the malignant cell. Given these characteristics, it is not surprising that its measurement has come under intensive study and evaluation.

CELL-CYCLE DISTRIBUTION IN HUMAN SOLID TUMORS

Figure 2 presents cell-cycle distribution data as a function of diagnosis. As is true of DNA ploidy, there was remarkable heterogeneity in the cytokinetic parameters. No disease-specific distribution patterns of proliferative activity could be discerned. Nor was there significant difference between primary and metastatic sites of tumor involvement [25]. Unlike the DNA stemline measurements, which do not differ in different sections or different metastatic sites of a given tumor in the same patient, cell cycle distribution patterns do vary as a function of tumor size and site of sampling within the tumor [76]. In experimental tumors, there is an increased proportion of cells in $[S+G_2M]$ with progression from the center to the periphery, and this probably is true of human tumors as well. Given the pitfalls of cell cycle distribution analysis of both diploid and aneuploid human tumors, comparisons with other assays of proliferative activity are important. Unfortunately, there are few case-by-case comparisons of the tritiated thymidine labeling index (in vitro and in vivo) and DNA-derived S-phase fraction. Excellent correlation between these two parameters has been demonstrated in one report of bone marrow specimens from normal individuals and from patients with various types of leukemia [9], but not in another study of patients with chronic myelogenous leukemia [65]. Meyer et al [56] carefully studied labeling index in breast cancer and noted lower values for thymidine labeling than were obtained for percent S-phase cells by flow cytometry.

Table 2 compares the reported labeling indices and flow cytometry S phase percentage. While there are several possible explanations for the discrepancies, including diffusion of the radioisotope throughout the tumor tissue, there is reason to believe that some tumor cells with S-phase DNA content may have low or even absent DNA synthetic activity, comparable to S_0-phase cells during plateau growth in vitro. Evidence supporting this concept has been reported in DNA

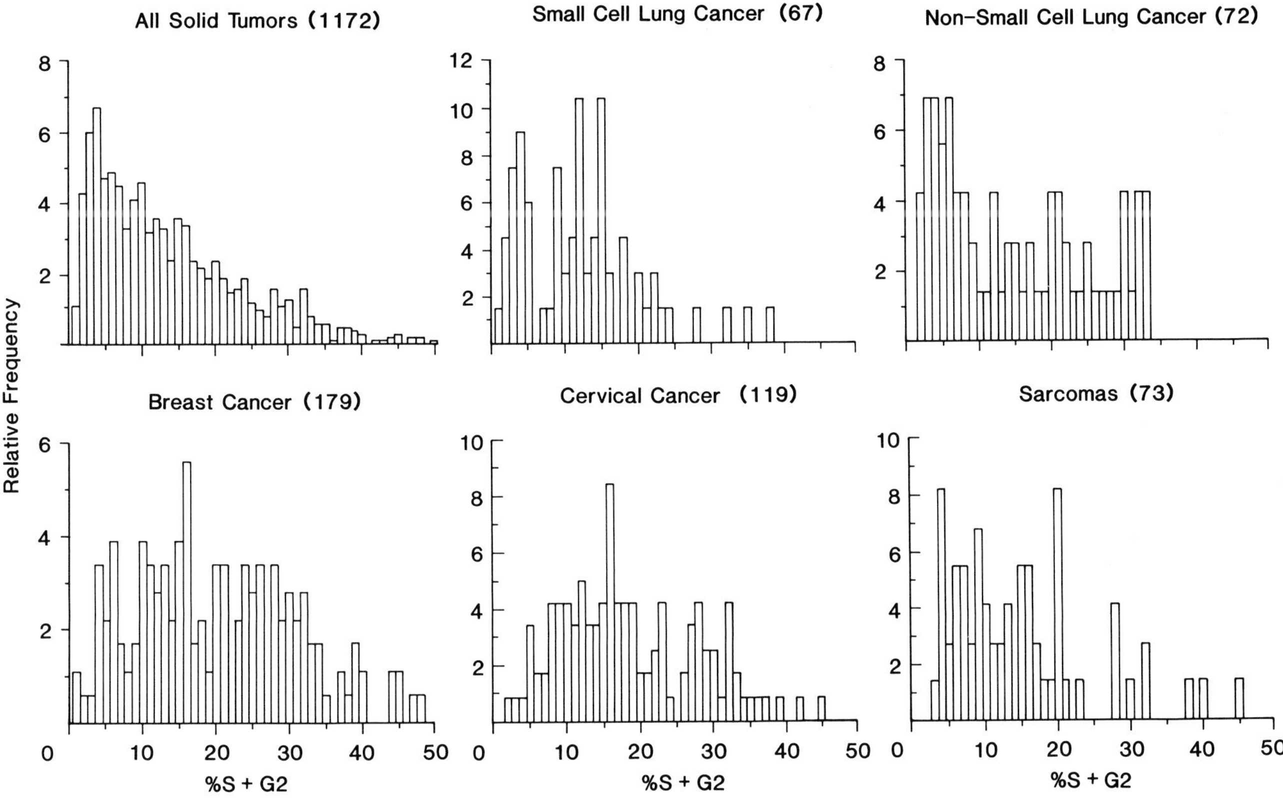

Fig. 2. Relative frequency distribution of the S plus G2 phase compartments in a series of 1,172 solid tumors according to dominant diagnostic subgroups.

TABLE 2. S-Phase Compartment in Primary Breast Cancer Comparison of Autoradiographic and Flow Cytometric Analysis

Investigators	No. of patients	$[^3H]$-TdR labeling index	Mean S %—FCM
Kute et al. [48]	70	—	13
Olzeweski et al. [60]	92	—	9
Meyer et al. [56]	27	6.5	15
Meyer et al. [54]	170	6.0	—
Raber et al. [61]	80	—	13
Silvestrini et al. [69]	199	4.3	—
Sklarew et al. [70]	56	2.4	—
Dressler et al. [22]	1,084	—	5.8

denaturation studies employing acridine orange [18] (see Chapter 24). There is additional evidence from studies of multiple myeloma, in which the average plasma cell labeling index of 1% at diagnosis was markedly lower than the S phase fraction of 4% derived from bivariate DNA–CIg studies [6]. This issue is of considerable biological importance, as the sensitivity to most chemotherapeutic agents and ionizing radiation is dependent not only on cell-cycle distribution but on cycle traverse rate as well. Thus, quiescent cells in S would constitute a cytokinetic sanctuary [24]. This important question deserves further study, which may come from application of the bromodeoxyuridine monoclonal antibody technique, in conjunction with DNA flow cytometry [31] (see also Chapter 23).

RELATIONSHIP BETWEEN PLOIDY AND CELL-CYCLE DISTRIBUTION

Several studies have noted a relationship between the S-phase compartment size and the degree of DNA content abnormality. In our own laboratory, we demonstrated a progressive increase in the proportion of tumor cells in S with increasing DNA index among more than 1,000 tumors investigated [10], which was in excellent agreement with studies by Johnson et al. [41] in spontaneous canine tumors. An update of the literature and of our own experience confirms such a relationship for tumors with DNA indices between 1 and 2 (Table 3). Tetraploid tumors with some exceptions appear to have a smaller S-phase compartment than that of

TABLE 3. Median Percentage Cells in S Phase According to DNA Index

DNA index	All tumors (n = 1,102)	Cervix (n = 119)	Lung (n = 140)	Sarcomas (n = 73)	Breast (n = 248)
0–1.1	7	13	7	11	12
>1.1–1.5	12	18	16	15	19
>1.5–2.0	18	26	20	12	19
>2	16	22	15	14	20

highly aneuploid tumors. Look et al. [51] suggested that in childhood acute leukemia a longer time is required for completion of DNA synthesis in cells with a higher DNA complement. However, it must be pointed out that the increased SPF in aneuploid compared with diploid tumors is in part an artifact produced by the inclusion of nonproliferating normal cells in the diploid peak. The different relationships between ploidy and proliferative activity may point to fundamental biological differences between various solid tumors. In the case of breast cancer, this notion is supported by differences in estrogen receptor (ER) expression and proliferative activity among pre- and post-menopausal patients (Fig. 3). While there is an inverse relationship between these two parameters in premenopausal patients, well known from tritiated thymidine autoradiographic studies [54], such relationship is not apparent for postmenopausal women who have consistently higher ER values and more spread in proliferative activity.

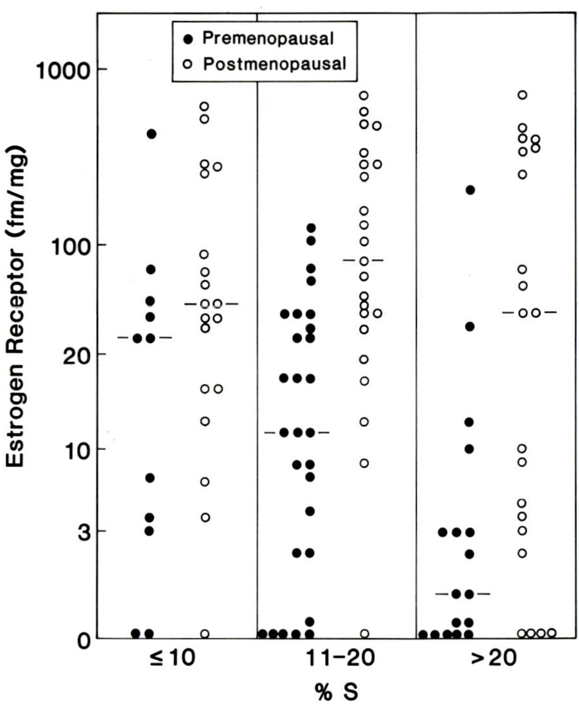

Fig. 3. Relationship between estrogen receptor content and the percentage of cells in the S-phase compartment of the cell cycle in pre- and postmenopausal patients. The lines represent median values.

CLINICAL RELEVANCE

The concept that aneuploidy would predict for a poor clinical outcome was founded on the earlier work of Atkin [3] and Caspersson [15] using measurements based on the Feulgen technique. Given their observations, it is at first surprising that it has taken so long to assess the prognostic impact of DNA ploidy by flow cytometry. However, until Hedley's description of a technique for extracting nuclei from paraffin embedded tissue and measuring DNA content by flow cytometry, investigators were limited to prospective studies. Therefore, most early studies had small numbers of patients, limited follow-up, and were weighted in favor of patients with advanced disease.

With the introduction of techniques for the analysis of archival material, it has been possible to analyze discrete patient populations, and quantitate the impact of aneuploidy on relapse free and overall survival. While breast cancer and colon cancer have been studied most intensively and will be considered separately, there are some observations that apply generally to all tumors.

In patients with advanced metastatic disease, ploidy has no impact on prognosis. This has been demonstrated in many studies, including breast [72], colon [2], and ovarian cancer [29,30]. It explains in part the failure of early investigators to demonstrate the prognostic importance of DNA content analysis and most likely reflects the fact that tumor burden is the most important prognostic variable for solid tumors. Large tumor burden overshadows the impact of DNA ploidy. By contrast, in those cases in which investigators focused on earlier stage disease, and had the opportunity to observe the natural history of the tumor, aneuploidy was consistently associated with a poorer prognosis. This has been demonstrated in many tumor types, including breast [21], colorectal [2,46,66,67,85], prostate [71,77,84], ovary [13,29,63], non-small cell lung [83,87], endometrium [59], cervix [39,40], and melanoma [33]. In these patients with a small tumor burden and long natural history, we can more easily see the impact of DNA ploidy.

In patients with intermediate tumor burden, or when treatment alters the natural history of the tumor, the relationship between DNA ploidy and clinical course is controversial. An example of this situation would be Stage II breast cancer [17,20,35,36].

It is extremely important that in any analysis of groups of patients they be homogeneous with respect to tumor type, stage and treatment. Failure to match for stage and treatment may account for seemingly contradictory results in otherwise similar studies conducted by different investigators. Paradoxically, Look has reported that in two childhood cancers, acute lymphocytic leukemia and neuroblastoma, aneuploidy predicts for prolonged survival [50,52], possibly because newer chemotherapies are more effective against (aneuploid) tumors that would have a more aggressive natural history if untreated. Thus, aneuploidy becomes a predictor of responsive disease and overall improved survival.

While it appears that tumors can be satisfactorily grouped as diploid, or near-diploid versus aneuploid, the importance of tetraploidy remains controversial. Tetraploid tumors often have an SPF lower than that of other aneuploid tumors. Some investigators consider that tetraploidy and diploidy have similar prognostic significance [43], while others classify tetraploid with aneuploid tumors [10,84].

An important consideration is whether aneuploidy is of independent predictive value with respect to other known prognostic variables. This issue remains controversial. In many series, highly aneuploid tumors are also the "most aggressive" when histologic and cytologic criteria are applied [20,58,84]. These tumors are often referred to as "high grade" or "poorly differentiated." Since we would expect highly aneuploid tumors to have high SPF, numerous mitoses, and cytologic atypia, it is possible that DNA content is in effect providing a reproducible, objective measurement of these subjective pathologic distinctions.

CELL-CYCLE ANALYSIS

While we have made great strides in understanding the impact of DNA ploidy on the natural history of many solid tumors, the case for proliferative activity as measured by DNA distribution is much less clear. The reasons for this are numerous. As with aneuploidy, there is good reason to believe that high proliferative activity would be a marker for tumors with poor prognosis. Studies of breast cancer proliferation by the Thymidine labeling index (TLI) have demonstrated that a high TLI predicts for poor survival [55,69]. Nevertheless, the importance of DNA-based measurements of the SPF remains unclear. In part, this is due to the technical difficulties in calculating SPF for specimens with overlapping DNA distributions of aneuploid and diploid populations, mixtures of benign and near-diploid malignant cells, and less than optimally preserved archival tissue in paraffin blocks. In addition there is the problem of distinguishing So from S-phase cells as has already been commented on. Thus, while there are reports in the literature that suggest SPF is of value in determining prognosis of breast cancer [35,22], SPF data are reported much less often than DI, and an accurate assessment of proliferation will probably require measurement with other probes.

ANALYSIS OF SPECIFIC TUMORS

Breast Cancer

There has been much interest in DNA content analysis of breast cancer for some time. Although the results of some studies are contradictory, certain conclusions are apparent. As in other solid tumors, 60–70% of breast tumors are aneuploid [17,22,60]. Abnormal DNA content is not significantly related to stage of disease or other prognostic factors, although some investigators have suggested a trend toward higher frequency of aneuploidy in ER-negative tumors and in tumors associated with a greater number of positive nodes [23,60].

SPF appears to be higher in aneuploid tumors than in diploid and, like TLI, inversely related to ER status, particularly in premenopausal women (see Fig. 3). It is not significantly related to stage of disease [22,55].

In patients with advanced disease, as demonstrated by Stuart-Harris et al. [72], aneuploidy does not impact on survival; tumor burden is probably the overwhelming prognostic variable. However, in node negative breast cancer, preliminary results from Dressler et al. [21] indicate that ploidy is an important prognostic variable and may well serve to identify patients with stage I disease who have a poor prognosis. In stage II disease (positive nodes), the data are less clear. Hedley and others suggest that in homogeneously matched subsets of patients aneuploidy predicts for poor survival, with its impact decreasing as clinical stage increases [23,35], but not all investigators agree [17,20]. Differences in treatment may explain some of these discordant results.

The impact of SPF is unclear, although it has been suggested that high SPF also predicts for poor prognosis in node positive patients [35,54,55].

Colon Cancer

Early prospective studies of DNA ploidy in colon cancer were reported by Wolley et al. in 1982 [85]. These workers demonstrated in a small series of patients that those with diploid tumors had a significant survival advantage when compared to patients with aneuploid tumors. In fact, colon cancer has proved an excellent model for clinical correlative studies of aneuploidy, since the patients are almost all treated by resection of the tumor and surgically staged, usually with no other therapy.

Using paraffin-embedded material, a number of investigators have extended these early observations in larger retrospective studies of patients. Armitage and colleagues [2] reported that 5-year survival was significantly improved in diploid patients with Dukes' B & C colon cancer, but not in patients with Dukes' D disease. More recently, Kokal et al. [46], Schutte et al. [66] and Scott et al. [68] report similar results in more than 500 patients. There is little informaton available on the importance of SPF in these patients.

USING DNA CONTENT AS A PROGNOSTIC TOOL

From the numerous studies reported, there is growing evidence that in solid tumors aneuploidy is a sign of poor prognosis. Where the evidence is strong as in breast and colon cancer, it has been suggested that adjuvant therapy trials be stratified for aneuploidy [68]. Nevertheless, DNA content measurements should not be used in the management of individual patients without a clear understanding of their limitations. In comparison with other known prognostic variables, particularly tumor burden and stage, DNA ploidy is a weak predictor of clinical course. But within well-defined subsets of patients (e.g., stage I ER negative breast cancer; Dukes' B colon cancer; level III melanoma, aneuploidy may help define patients with poor prognosis. Finally, it must be noted that a new effective therapy may completely change the prognostic significance of these measurements, depending on their correlations with cell features that predict response to treatment. As new treatment strategies are introduced, the value of DNA flow cytometry measurements will have to be re-examined.

ACKNOWLEDGMENTS

Original work described in this chapter was supported in part by grants CA 28771 and CA 16672 from the National Cancer Institute, National Institutes of Health, Bethesda.

REFERENCES

1. **Adelson M, Johnson TS, Sneige N, Williamson K, Freedman RS, and Peters LJ (1987)** Cervical carcinoma: DNA content S-fraction and malignancy grad-

ing. II Comparison in the clinical staging. Gynecol Oncol (26)57–70.
2. **Armitage NC, Robins RA, Evans DF, Turner DR, Baldwin RW, Hardcastle JD (1985)** The influence of tumor cell DNA abnormalities on survival of colorectal cancer. Br J Surg 72:828–830.
3. **Atkin NB, Kay R (1979)** Prognostic significance of modal DNA value and other factors in malignant tumors based on 1465 cases. Br J Cancer 40:210–221.
4. **Auer GU, Backdahl M, Forsslund G, Askensten U (1985)** Ploidy levels in non-neoplastic and neoplastic thyroid cells. Ann Quant Cytol 7:97–105.
5. **Barlogie B, Alexanian R, Gehan E, Smallwood L, Smith L, Drewinko B (1983)** Marrow cytometry and prognosis in myeloma. J Clin Invest 72:853–861.
6. **Barlogie B, Alexanian R, Pershouse M, Smallwood L, Smith L (1985)** Cytoplasmic immunoglobulin content in multiple myeloma. J Clin Invest 76:765–769.
7. **Barlogie B, Drewinko B, Schumann J, Gohde W, Dosik G, Johnston DA, Freireich EJ (1980)** Cellular DNA content as a marker of neoplasia in man. Am J Med 69:195–203.
8. **Barlogie B, Gohde W, Johnston DA, Smallwood L, Schumann J, Drewinko B (1978)** Determination of ploidy and proliferative characteristics of human solid tumors by pulse cytophotometry. Cancer Res 38: 3333–3339.
9. **Barlogie B, Hittelman W, Spitzer G, Hart JS, Trujillo JM, Smallwood L, Drewinko B (1977)** Correlation of DNA distribution abnormalities with cytogenetic findings in human adult leukemia and lymphoma. Cancer Res 37:4400–4407.
10. **Barlogie B, Johnston DA, Smallwood L, Raber MN, Maddox AM, Latreille J, Swartzendruber DE, Drewinko B (1982)** Prognostic implications of ploidy and proliferative activity in human solid tumors. Cancer Genet Cytogenet 6:17–28.
11. **Barlogie B, Latreille J, Swartzendruber DE, Smallwood L, Maddox A, Raber MN, Drewinko B, Alexanian R (1982)** Quantitative cytometry in myeloma research. In Schmidt WR (ed), "Clinics in Hematology," Philadelphia: WB Saunders, pp. 19–46.
12. **Barlogie B, Raber MN, Schumann J, Johnson TS, Drewinko B, Swartzendruber DE, Gohde W, Andreeff M, Freireich EJ (1983)** Flow cytometry in clinical research. Cancer Res 43:3982–3997.
13. **Blumenfeld D, Braly PS, Ben-Ezra J, Klevecz RR (1987)** Tumor DNA content as a prognostic feature in advanced epithelial ovarian carcinoma. Gynecol Oncol 27(3):389–402.
14. **Bunn PA, Carney DN, Gazdan AF, Whang Peng J, Matthews MJ (1984)** Diagnostic and biological implication of flow cytometric DNA content analysis in lung cancer. Cancer Res 43:3982–3997.
15. **Caspersson T (1979)** Quantitative tumor cytochemistry. G.H.A. Clowes Memorial Lecture. Cancer Res 39: 2341–2355.
16. **Cornelisse CJ, deKong HR, Moolenaar AJ, van de Valde CJ, Ploem JE (1984)** Image and flow cytometric analysis of DNA content in breast cancer, relation to estrogen receptor content and lymph node involvement. Anal Quant Cytol 6:9–18.
17. **Cornelisse CJ, van de Velde CJ, Caspers RJ, Moolenaar AJ, Herman J (1987)** DNA ploidy and survival in breast cancer patients. Cytometry 8(2): 225–234.
18. **Darzynkiewicz Z, Traganos F, Andreeff M, Sharpless T, Melamed M (1979)** Different sensitivity of chromatin to acid denaturation in quiescent and cycling cells revealed by flow cytometry. J Histochem Cytochem 27:478–485.
19. **Dixon B, Carter CJ (1980)** DNA microdensitometry and flow cytofluorometry of human tumor biopsies. In Laerum OD, Lindmo T, Thorud E (eds), "Flow Cytometry." Vol. IV. Oslo: Universitetsflorlaget, pp. 427–430.
20. **Dowle CS, Owainati A, Robins A, Burns K, Ellis IO, Elston CW, Blamey RW (1987)** Prognostic significance of the DNA content of human breast cancer. Br J Surg 74(2):133–136.
21. **Dressler L, Clark G, Owens M, Pounds G, Oldaker T, McGuire W (1987)** DNA flow cytometry predicts for relapse in node negative breast cancer patients. Proc Am Soc Clin Oncol 6:57.
22. **Dressler LG, Seamer LC, Owens MA, Clark GM, McGuire WL (1988)** DNA flow cytometry and prognostic factors in 1331 frozen breast cancer specimens. Cancer 61:420–427.
23. **Ewers SB, Langstrom E, Baldetorp B, Killander D (1984)** Flow cytometric DNA analysis in primary breast carcinoma and clinicopathologic correlations. Cytometry 5:408–419.
24. **Ford SS, Shackney SE (1977)** Lethal and sublethal effects of hydroxyurea in relation to drug concentration and duration of drug exposure to sarcoma 180 *in vitro*. Cancer Res 37:2628–2637.
25. **Frankfort OS, Greco WR, Slocum HK, Arbuck SG, Gamarra M, Pavelic ZP, Rustum YM (1984)** Proliferative characteristics of primary and metastatic human solid tumors by DNA flow cytometry. Cytometry 5: 629–635.
26. **Frankfurt OS, Slocum HK, Rustum YM, Arbuck SG, Pavelic ZP, Petrelli N, Huben RP, Pontes EJ, Greco WR (1984)** Flow cytometric analysis of DNA aneuploidy in primary and metastatic human solid tumors. Cytometry 5:71–80.
27. **Frederikson P, Bichel PL (1980)** Sequential flow cytometric analysis of the single cell DNA content in recurrent human brain tumors. In Laerum OD, Lindmo T, Thorud E (eds), "Flow Cytometry." Vol. IV. Oslo: Universitetsflorlaget, pp 398–402.
28. **Frederikson P, Knudsen V, Reske-Nielsen E, Bichel P (1979)** Flow cytometric DNA analysis in human medulloblastomas. In Paoletti P, Walker G, Burtti G, Knerich (Eds), "Multidisciplinary Aspects of Brain Tumor Therapy." Amsterdam: Elsevier/North-Holland, pp 363–367.
29. **Friedlander ML, Hedley DW, Swanson C, Russell P (1988)** Prediction of long-term survival by flow cytometric analysis of cellular DNA content in patients with advanced ovarian cancer. J Clin Oncol 6(2):282–290.
30. **Friedlander MD, Hedley DW, Taylor IW, Russell P, Tattersall MH (1984)** Influence of cellular DNA content on survival in advanced ovarian cancer. Cancer Res 44:397–400.
31. **Gratzner HG (1982)** Monoclonal antibody to 5 bromo- and 5 iododeoxyuridine: A new reagent for detection of DNA replication. Science 218:474–475.

32. **Hammarberg C, Slezak P, Tribukait B (1984)** Early detection of malignancy in ulcerative colitis: A flow-cytometric DNA study. Cancer 53:291–295.
33. **Hansson J, Tribukait B, Lewensohn R, Ringborg O (1982)** Flow cytofluorometric DNA analysis of metastasis of human malignant melanoma. Anal Quant Cytol 4:99–104.
34. **Hedley DW, Friedlander ML, Taylor IW, Rogg CA, Musgrove EA (1983)** Methods for analysis of cellular DNA content of paraffin embedded pathologic material using flow cytometry. J Histochem Cytochem 31: 1333–1335.
35. **Hedley DW, Rugg CA, Gelber RD (1987)** Association of DNA index and S phase fraction with prognosis of nodes positive early breast cancer. Cancer Res 47: 4729–4735.
36. **Hedley DW, Rugg CA, Ng ABP, Taylor IW (1984)** Influence of Cellular DNA content on disease-free survival of Stage II breast cancer patients. Cancer Res 44:5395–5398.
37. **Hiddemann W, Schumann J, Andreeff FM, Barlogie B, Herman C, Leif R, Mayall B, Murphy R, Sandberg A (1984)** Convention on nomenclature for DNA cytometry. Cytometry 5:445–446.
38. **Hoshino T, Nomura K, Wilson CB, Knebel KD, Gray JW (1978)** The distribution of nuclear DNA from human brain tumor cells. J Neurosurg 49:13–21.
39. **Jakobsen A (1984)** Prognostic impact of ploidy level in carcinoma of the cervix. Am J Clin Oncol 7:475–480.
40. **Jakobsen A (1984)** Ploidy level and short time prognosis of early cervix cancer. Radiother Oncol 1:271–275.
41. **Johnson TS, Raju MR, Giltman PK, Gillete EL (1981)** Ploidy and DNA distribution analysis of spontaneous dog tumors by flow cytometry. Cancer Res 41:3005–3009.
42. **Johnson TS, Williamson KD, Cramer MM, Peters LJ (1985)** Flow cytometry analysis of head and neck carcinoma DNA index and S fraction from paraffin-embedded sections: Comparison with malignancy grading. Cytometry 6:461–470.
43. **Kallioniemi OP, Punnonen R, Mattila J, Lehtinen M, Koivula T (1988)** Prognostic significance of DNA index, multiploidy, and S-phase fraction in ovarian cancer. Cancer 61:334–339.
44. **Klein FA, Whitmore WF, Herr H, Melamed M (1982)** Flow cytometry follow up of patients with low stage bladder tumors. J Urol 126:88–92.
45. **Klein FA, Herr HW, Whitmore WF Jr, Sogani PC, Melamed MR (1982)** An evaluation of automated flow cytometry (FCM) in detection of carcinoma *in situ* of the urinary bladder. Cancer 50:1003–1008.
46. **Kokal W, Sheibani K, Terz J, Harada R (1986)** Tumor DNA content in the prognosis of colorectal carcinoma. JAMA 255:3123–3127.
47. **Kreicbergs A, Silvferswaro C, Tribukait B (1984)** Flow DNA analysis of primary bone tumors. Cancer 53: 129–136.
48. **Kute TE, Moss HB, Anderson D, Crumb K, Miller B, Burns D, Dube L (1981)** Relationship of steroid receptor cell kinetics and clinical status in patients with breast cancer. Cancer Res 41:3524–3529.
49. **Linden WA, Baisch H, Ochlich K, Scholz KU, Mauss HJ, Stegner HE, Joshi D, Wu CT, Koprowska I, Nicolini C (1979)** Flow cytometric prescreening of cervical smears. J Histochem Cytochem 27:529–535.
50. **Look AT, Hayes FA, Nitschke R (1984)** Cellular DNA content as a predictor of response to chemotherapy in infants with unresectable neuroblastoma. N Engl J Med 311:231–235.
51. **Look A, Melvin SL, Williams DL, Brodeun GM, Dahl GV, Kalwensky DH, Murphy SB, Mauer AM (1982)** Aneuploidy and percent of S phase cells determined by FCM correlated with cell phenotype in childhood acute leukemia. Blood 60:959–967.
52. **Look AT, Roberson RK, Williams DL (1985)** Prognostic importance of blast cell DNA content in childhood ALL. Blood 63:1079–1086.
53. **Lukas GL, Miko TL, Fabian E, Nagy IZ, Csaky G, Balazs G (1983)** The validity of morphologic methods in the diagnosis of thyroid malignancy. Acta Chir Scand 149:759–766.
54. **Meyer JS, Bauer WC, Rao BR (1978)** Subpopulations of breast carcinoma defined by S phase fraction, morphology and estrogen receptor content. Lab Invest 39: 225–235.
55. **Meyer JS, Lee JW (1980)** Relationship of S-phase fraction of breast carcinoma in relapse to duration of remission, estrogen receptors content, therapeutic responsiveness and duration of survival. Cancer Res 40: 1890–1896.
56. **Meyer JS, Micko S, Craven JL, McDivitt RW (1984)** DNA flow cytometry of breast carcinoma after acetic acid fixation. Cell Tissue Kinet 17:1–13.
57. **Moran R, Melamed M (1984)** Flow cytometric analysis of human lung cancer: Correlation with histologic type and stage. Anal Quant Cytol 6:99–104.
58. **Moran RE, Stauss MJ, Black M, Alvarez EM, Evans MD, Evans R (1982)** Flow cytometry DNA analysis of human breast tumors and comparison with clinical and pathologic parameters. Proc Am Assoc Cancer Res 23:33.
59. **Mosberger B, Auer G, Forsslund G, Mosberger G (1984)** The prognostic significance of DNA measurements in endometrial carcinoma. Cytometry 5:430–436.
60. **Olzeweski W, Darzynkiewicz Z, Rosen PP, Schwartz MK, Melamed MR (1981)** Flow cytometry of breast carcinoma. I. Relation of DNA ploidy level to histology and estrogen receptor. Cancer 48:980–984.
61. **Raber MN, Barlogie B, Latreille J, Bedrossion C, Fritsche H, Blumenschein G (1982)** Ploidy, proliferative activity and estrogen receptor content in human breast cancer. Cytometry 3:36–41.
62. **Raber MN, Barlogie B, Luna M (1984)** Flow cytometric analysis of DNA content on postmorten tissue. Cancer 53:1705–1707.
63. **Rodenburg CJ, Cornelisse CJ, Heintz PA, Hermans J, Fleuren GH (1987)** Tumor ploidy as a major prognostic factor in advanced ovarian cancer. Cancer 59(2): 317–323.
64. **Rotors M, Laemmel A, Kastendieck H, Becker H (1980)** DNA distribution pattern of prostatic lesions measured by flow cytometry. In Laerum OD, Lindmo T, Thorud E (eds), "Flow Cytometry." Vol IV. Oslo: Universitetsflorlaget, pp 431–435.
65. **Roza A, Bhayana R, Ucar K, Kirshnin J, Preisler H (1985)** Differencc between labeling index and DNA histograms in assessing S-phase cells from a homoge-

neous group of chronic phase CML patients. Cytometry 6:445–451.

66. **Schutte B, Reynders MM, Wiggers T, Arends JW, Volovics L, Bosman FT, Blijham GH** (1987) Retrospective analysis of the prognostic significance of DNA content and proliferative activity in large bowel carcinoma. Cancer Res 47(20):5494–5496.

67. **Scott NA, Grande UP, Weiland LH, Pemberton JH, Beart RW, Lieber M** (1987) Flow cytometric DNA patterns from colorectal cancers—How reproducible are they? Mayo Clin Proc 62:331–337.

68. **Scott NA, Wieland HS, Moertel CG, Cha SS, Beart RW, Lieber MM** (1987) Colorectal Cancer. Dukes' stage, tumor site, preoperative plasma CEA level, and patient prognosis related to tumor DNA ploidy pattern. Arch Surg 122:1375–1379.

69. **Silvestrini R, Daidone MG, Gaspanni G** (1985) Cell kinetics as a prognostic marker in node negative breast cancer. Cancer 56:82–87.

70. **Sklarew RJ, Hoffman J, Post J** (1977) A rapid in vitro method for measuring cell proliferation in human breast cancer. Cancer 40:2299–2302.

71. **Stephenson RA, James BC, Gay H, Fair WR, Whitmore WF, Melamed MR** (1987) Flow cytometry of prostate cancer: Relationship of DNA content to survival. Cancer Res 47:2504–2507.

72. **Stuart-Harris R, Hedley DW, Taylor IW** (1985) Tumor ploidy, response and survival in patients receiving endocrine therapy for advanced breast cancer. Br J Cancer 51:525–559.

73. **Swartzendruber DE, Cox KZ, Wilder ME** (1980) Flow cytoenzymology of the early differentiation of mouse embryonal carcinoma cells. Differentiation 16:23–30.

74. **Teodori L, Tirindelli-Danesi D, Mauro F, DeVita R, Uceelli R, Botti C, Modine C, Nervi C, Stipa S** (1983) Non-small cell lung carcinoma: Tumor characterization on the basis of flow cytometrically determined cellular heterogeneity. Cytometry 4:174–183.

75. **Thornthwaite JT, Coulson PB** (1982) Comparison of steroid receptor content, DNA levels, surgical staging and survival in human breast cancer. Cytometry in the Clinical Laboratory, Engineering Foundation, Santa Barbara, CA.

76. **Thornthwaite JT, Sugarbaker EV, Temple WJ** (1980) Preparation of tissues for DNA flow cytometric analysis. Cytometry 1:229–237.

77. **Tribukait B, Esposti P** (1976) Comparative cytofluorometric and cytomorphologic studies in non-neoplastic and neoplastic human urothelium. In Gohde W, Schumann J, Buchner T (eds), "Second International Symposium on Pulse-Cytophotometry." Ghent, Belgium: European Press Medikon, pp 176–187.

78. **Tribukait B, Esposti PL** (1978) Quantitative flow microfluorometric analysis of the DNA in cells from neoplasms of the urinary bladder: Correlation of aneuploidy with histological grading and cytological findings. Urol Res 6:201–205.

79. **Tribukait B, Gramberg-Ohman I, Wijkstrom H** (1986) Flow cytometric DNA and cytogenetic studies in human tumor: A comparison and discussion of the differences in modal values obtained by the two methods. Cytometry 7:194–199.

80. **Tribukait B, Ronstlom L, Esposti PL** (1983) Quantitative and qualitative aspects of flow DNA measurements related to the cytologic grade of prostate carcinoma. Anal Quant Cytol 5:107–111.

81. **Unger K, Raber M, Bedrossian C, Stein D, Barlogie B** (1983) Analysis of pleural effusion using automated flow cytometry. Cancer 52:873–877.

82. **Vindelov LL, Hansen HH, Christensen IJ, Spang-Thomsen M, Hirsch FR, Hansen M, Nissen MI** (1980) Clonal heterogeneity of small cell anaplastic carcinoma of the lung demonstrated by flow cytometric DNA analysis. Cancer Res 40:4295–4300.

83. **Volm M, Mattern J, Sanka J, Vogt-Schaden M, Wayss K** (1985) DNA distribution in non small cell lung carcinoma and its relationship to clinical behavior. Cytometry 6:348–356.

84. **Winkler HZ, Rainwater LM, Myers RP, Farrow GM, Therneau TM, Zincke H, Lieber MM** (1988) Stage D1 prostatic adenocarcinoma: Significance of nuclear DNA ploidy patterns studied by flow cytometry. Mayo Clin Proc 63:103–122.

85. **Wolley RC, Schreiber K, Koss LG, Karas M, Sherman A** (1982) DNA distribution in human colon carcinoma and its relationship to clinical behavior. J Natl Cancer Inst 69:15–22.

86. **Zetterberg A, Esposti P** (1980) Prognostic significance of nuclear DNA levels in prostatic carcinoma. Scan J Urol Nephrol 55:53–58.

87. **Zimmerman PV, Hawson GA, Bint MH, Parsons PG** (1987) Ploidy as a prognostic determinant in surgically treated lung cancer. Lancet 2:530–533.

38

Flow Cytometry in Clinical Cytology

Myron R. Melamed and Lisa Staiano-Coico

Departments of Pathology, Memorial Sloan-Kettering Cancer Center and Cornell University Medical College, New York, New York 10021 (M.R.M.); and Department of Surgery, Cornell University Medical College, New York, New York 10021 (L.S.)

Much of the initial impetus to develop an instrument for measuring and analyzing cells came out of the desire to automate the cytologic detection of uterine cervical cancer. During the 1950s and 1960s, trained cytotechnologists were few. Then, as now, the diagnosis of cancer by cytology was highly labor intensive and greatly dependent on the skill and experience of the cytotechnologist and cytopathologist. Early work by Caspersson [11,12], Mellors et al. [89], Leuchtenberger et al. [80], Atkin and Richards [3], Sandritter et al. [102], and others had demonstrated that a wide range of carcinomas were characterized by an abnormal increased content of nuclear DNA. Nuclear DNA content was and still is the single most important objective descriptor of cancer cells.

The major applications of clinical cytology, excluding needle aspirates of solid tumors (see Chapter 37, this volume), are in examination of the uterine cervix, the urinary bladder, sputum and bronchial secretions, effusions, and spinal fluid. The status of flow cytometry differs for each of these different clinical applications.

UTERINE CERVICAL CYTOLOGY

The first efforts to automate cytologic examinations by flow cytometry were directed to the Papanicolaou (Pap) test. They were designed to identify cancer cells in samples of uterine cervical epithelium by the presence of abnormal nuclei with increased DNA content. (see Figs. 1,2). This was an adaptation of the approach that had been used for at least two decades to scan Pap smears on glass slides. The first field trial, reported by Koenig et al. [76], used specimens of unstained cells scraped from the uterine cervix and put into suspension to be measured simultaneously for total nucleic acid absorption in the ultraviolet (2,537 Å) and cell size by visible light scatter. A minimum of 100,000 cells/specimen were examinated; 45% of 1,155 examinations were machine readable, and 85% of the cancers were identified in 35% of the cases defined as abnormal.

With thc introduction of fluorescent stains, including stains specific for DNA, measurements of much greater precision were possible. In a series of publications, Sprenger, Bohm, Sandritter, and associates described a methodology for preparing and examining cervical scrape cytology specimens by flow cytometry using a fluorescent Acriflavine Feulgen staining technique [105]. The DNA histograms they obtained were quite good. Measurements of normal specimens yielded DNA values distributed symmetrically around the diploid value, whereas specimens from cervical carcinoma yielded histograms of DNA skewed toward higher values, including some tetra- and hexaploid measurements (Fig. 3A). In one series of 60 cases comparing flow cytometry with conventional cytologic examinations, 26 of 44 satisfactory examinations were positive by flow cytometry, including 20 with positive and 6 with negative conventional cytology; 18 were negative and all had negative cytology [105].

Further technical improvements were introduced by Gohde and associates [49,99,133], who simplified specimen preparation, performed in a single tube without centrifugation and without any solution change. Thus there was minimal possibility of cell loss [133]. Gohde also introduced a coincidence circuit to avoid falsely high DNA measurements due to coincidences of cells, increasing the sensitivity of the method to detect smaller numbers of aneuploid cells [50] (Fig 3B). One of the major goals of specimen preparation in the case of uterine cervical cytology, as with solid tissues in general, is to obtain a monodispersed cellular specimen. In the method used by this group and developed by Zante et al. [133], the cellular specimen is digested in 0.5% pepsin, preserving and dissociating epithelial cell nuclei, which are then stained by a combination of the DNA-specific fluorochromes ethidium bromide and mithramycin in the presence of RNase. Linden et al. [83], Jakobson et al. [62], and Dudzinski et al. [32] have confirmed the presence of aneuploid cells in cervical carcinoma and precancerous lesions.

While the DNA parameter is crucial to identify cancer cells, a second feature may be of value in selecting for analysis the subset of cells of interest within a mixed population of cells. Thus, cervicovaginal cytology specimens commonly contain leukocytes that can obscure small numbers of aneuploid cancer cells (see Fig. 2C) and endocervical or endometrial glandular cells that are problems in differential diagno-

Flow Cytometry and Sorting, Second Edition, pages 755–772

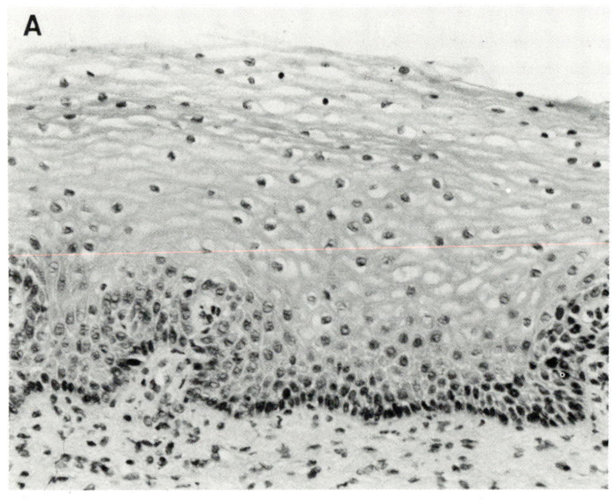
A

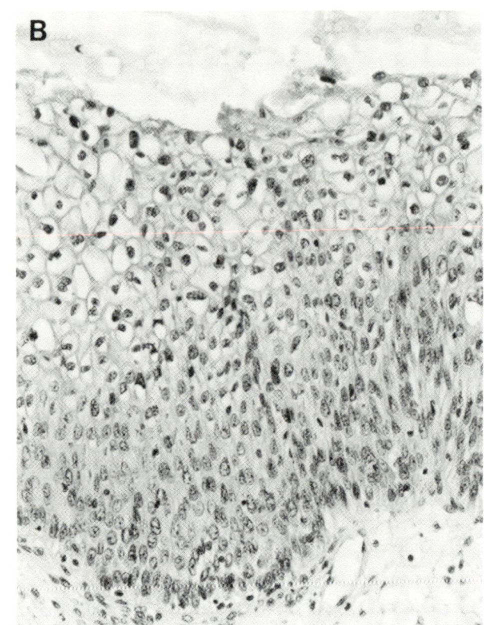
B

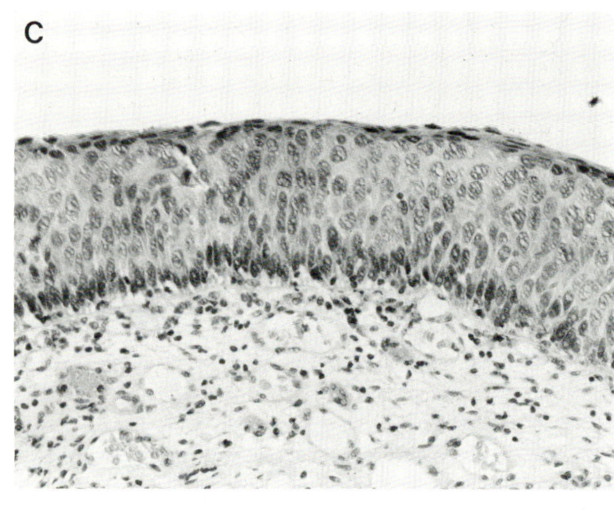
C

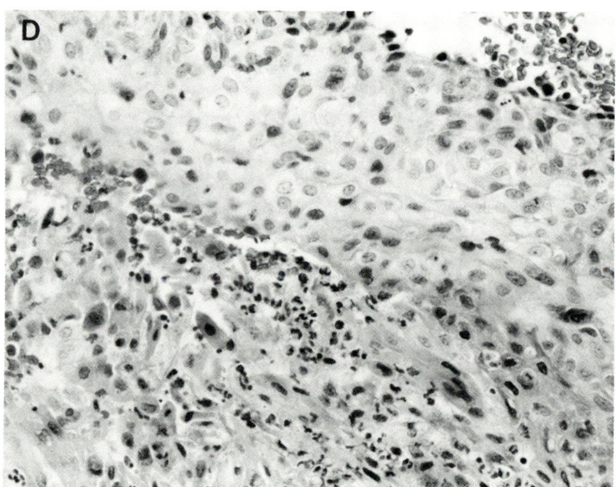
D

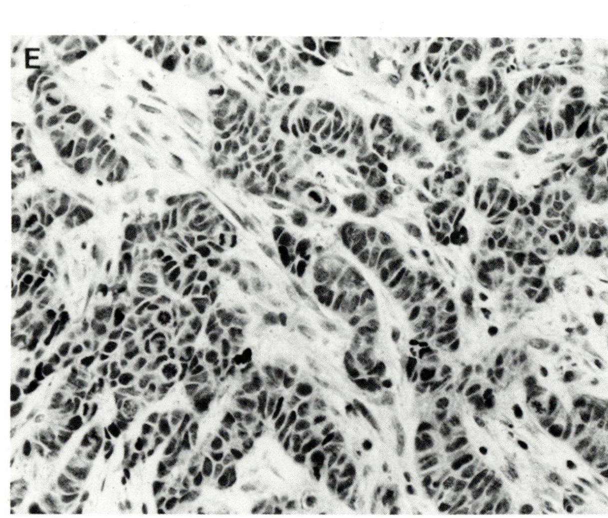
E

sis (see Fig. 2D). To identify and exclude the leukocytes, which have much less cytoplasm than epithelial cells, two-parameter DNA/protein measurements have been proposed. Steinkamp and Crissman [111] collected normal human cervicovaginal epithelium with a Dacron swab, which was then immersed in saline solution and subsequently treated with collagenase to yield a suspension of single cells. The cells were fixed briefly in 70% ethanol, treated with ribonuclease, and then stained sequentially for DNA with propidium iodide (PI) and total protein with fluorescein isothiocyanate (FITC). The emission spectra differ for these two florescent dyes, making it possible to measure DNA and total protein simultaneously for each cell. In addition, the time duration of each fluorescence signal (time of flight) was measured, (Fig. 3B), giving the relative nuclear (PI) and cellular (FITC) diameter. Foulkes et al. [42] used the combined PI/FITC staining technique to study a series of specimens from patients with cervical neoplasia and were able to distinguish a subpopulation of presumed cancer cells with increased nucleic acid content and increased ratio of nucleic acid to protein. They examined 70 cases and reported 11.8% false-positive and 5.8% false-negative rates in a nonblinded study. Barrett et al. [9] quantified DNA by chromomycin A_3 fluorescence and cell size by light scatter.

A technique for measuring and analyzing multiangle light scatter was described by Salzman et al. [100] and applied to the study of suspensions of mechanically dissociated, unfixed cells scraped or swabbed from the uterine cervix or vagina. A 5-mW helium neon laser was focused on the cell stream in the flow channel, and the scattered light from each cell measured by a detector consisting of 32 separate concentric rings of photodiode material covering 0° to 22°. A mathematical clustering algorithm was developed for analysis of these 32-parameter light-scatter data. Normal specimens yielded three different clusters, believed to represent squamous cells and leukocytes. In the samples from patients with dysplasia or invasive carcinoma a fourth cluster was identified.

Following these early, very rapid strides toward a flow cytometry-based system for detecting cervical carcinoma, there has been little recent progress. Johnson et al. [64], and Adelson et al. [2] have speculated on the prognostic importance of S-phase fraction as well as DNA index but have found these parameters to be independent of known clinical predictive variables. The most innovative contributions came from Wheeless et al. [129], who introduced a set of morphologic parameters derived from two- and three-dimensional slit scanning (see Chapter 6, this volume). Although potentially useful, the highly sophisticated instrumentation required for these measurements will have to be simplified for clinical application. Ideally, slit-scan morphometry could be coupled with DNA and either protein, RNA, or other measurements using currently available specific and stoichiometric fluorescent stains (see Chapters 13–16, this volume). Mention should be made also of the possible value of monoclonal or polyclonal antibodies to tumor associated or other antigens [41,55,77,116], and to cDNA or cRNA probes as for example in identifying human papillomavirus associated with precancerous lesions of the cervix [134] (Fig. 4). As yet it is too early to anticipate the practical value of these new reagents (see also Coleman and Ormerod (16), for a review of the role of antibodies as markers in clinical cytology).

The limited progress in developing an instrument for automated screening of cervical cytology may reflect a decrease in the perceived need for such a system. The shortage of skilled cytotechnologists has been met by an expansion of training programs, and Papanicolaou-stained smears provide information not yet available by flow cytometry, including hormonal status, the presence of inflammation or blood, and the type of micro-organisms present.

Finally, the identification of cases with very small numbers of malignant cells can be a problem for flow cytometry. Nevertheless, there is reason to believe that with an optimized method for collecting cell samples multiparameter flow cytometry would be capable of identifying neoplastic lesions of the cervix with acceptable sensitivity and specificity. The immediate need is for a simple, efficient means of sampling and preparing monodispersed cell suspensions from the cervical epithelium, after which the new markers can be re-evaluated and clinical diagnostic trials considered again.

UROLOGIC CYTOLOGY

In contrast to the experience with uterine cervical cytology, flow cytometry of the urinary bladder has been extremely effective not only in the detection of carcinoma but in monitoring the effects of conservative treatment. In many respects the urinary bladder is an ideal model for flow cytometry. It is surfaced by a relatively simple epithelium (urothelium) which is easily accessible for sampling (Fig. 5). There is no difficulty in the preparation of a monodispersed cell suspension, and the desquamated urothelial cells are a uniform population (Fig. 6), more easily analyzed than the complex mixture of glandular and squamous cells of the uterine cervix. Normal specimens are "clean", there are no inflammatory cells, no micro-organisms, no mucus. Even in cases of carcinoma at an early or in situ phase of development there is little inflammation and a relatively pure population of urothelial cells (Figs. 7,8). Only when the carcinoma invades and ulcerates is there appreciable associated inflammation and bleeding. Interestingly, there are other differences between the superficial or noninvasive and deeply invasive tumors that offer the possibility of distinguishing early from advanced disease, e.g., chromosomal number and modal DNA values near diploid (normal) in the low-grade noninvasive tumors and quite abnormal (aneuploid or tetraploid) in high-grade invasive carcinomas [34,78,101]. Also,

Fig. 1. Series of histologic sections showing progressive stages in the development of uterine cervical carcinoma. **A:** Normal epithelium. Note that the epithelial cells have uniform nuclei with delicate chromatin texture. **B:** A precancerous or "borderline" neoplastic lesion (dysplasia). The cytologic abnormalities are most evident in superficial and intermediate cell layers. Nuclei are enlarged, hyperchromatic (aneuploid), and often irregular with coarse chromatin texture. Many of the cells also show perinuclear halolike clearing of cytoplasm (koilocytosis), resembling the viral infected cells of the common wart. These cells bind monoclonal antibodies and DNA probes to human papillomavirus (HPV), strongly suggesting that the "warty" or "koilocytotic" precancerous lesions are in fact caused by HPV(see Fig. 4). **C:** Carcinoma in situ. Nuclei are larger and darker staining than in normal epithelium, consistent with an increased (aneuploid) DNA content, with coarse nuclear texture attributable to chromatin structural changes. Occasional mitoses (not shown here) indicate increased proliferation. Cellular abnormalities are evident throughout all layers of the epithelium. **D,E:** Superficially invasive **(D)** and deeply invasive **(E)** carcinomas of the cervix showing the same nuclear abnormalities as above, but typically with greater variability of nuclear size, staining, and configuration, disarranged growth pattern, and surface ulceration with bleeding and inflammation.

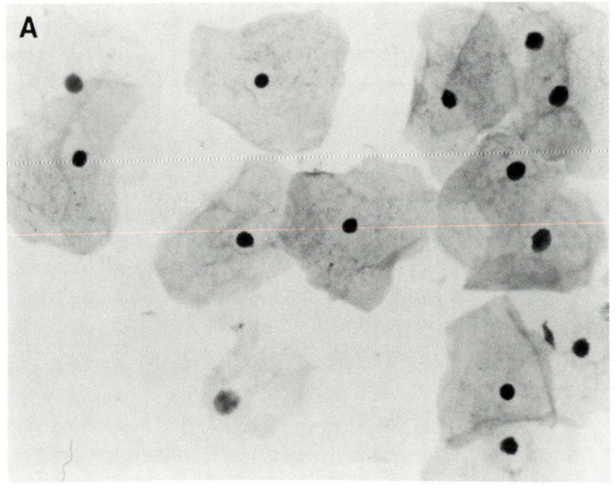

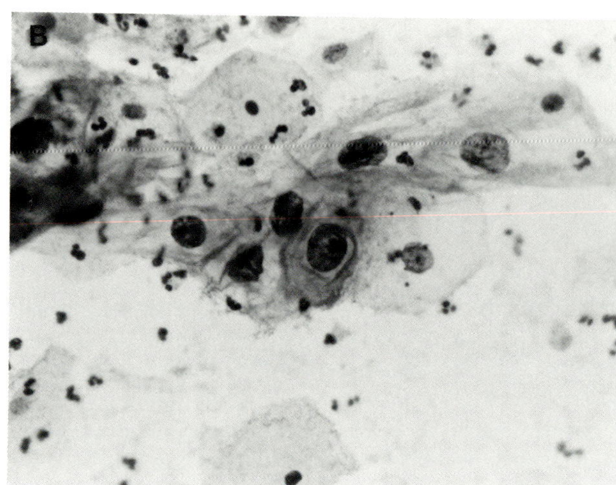

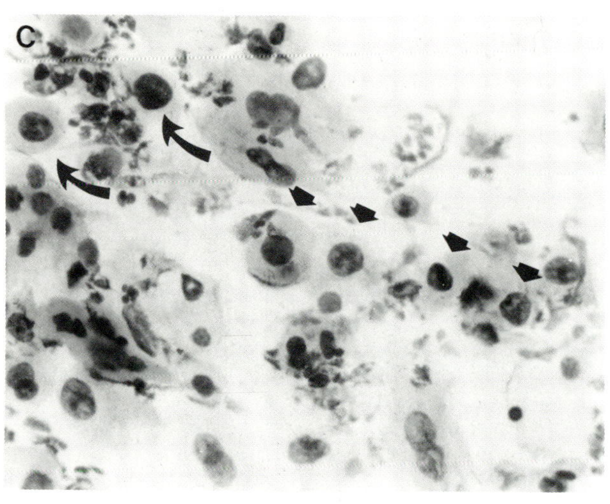

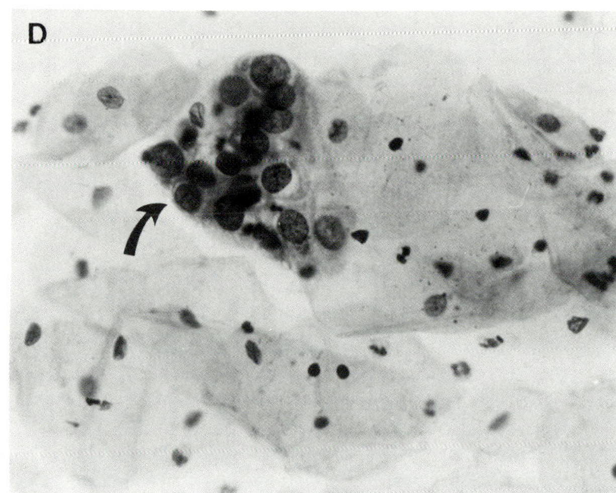

Fig. 2. Papanicolaou-stained uterine cervical smears. **A:** From a normal, healthy young woman at midcycle showing benign superficial squamous cells with small, uniform nuclei and abundant thin cytoplasm. (cf. Fig. 1A). **B:** Precancerous "borderline" lesion of cervix (dysplasia) showing a group of six or seven superficial squamous cells with enlarged hyperchromatic coarsely textured nuclei and variable perinuclear halo. Also present are normal superficial squamous cells and many degenerating polymorphonuclear leukocytes. **C:** Carcinoma in situ. Malignant cells (arrows) are present in a smear showing rather marked inflammation with atypia of accompanying benign epithelial cells due to the inflammation. **D:** Normal smear. There is a cluster of benign, metaplastic endocervical cells present (arrow). Normal smears also may contain endometrial cells, histiocytes and even pollen or other extrinsic cells that can cause difficulty in diagnosis.

the malignant cells comprising carcinoma in situ do not as a rule show the great variability that characterizes invasive carcinoma (126). (Figs. 7–10). Finally, there are very recent reports suggesting that the cells of bladder cancer undergo a change in cell surface antigen expression as they become more deeply invasive [44].

The first reports of flow cytometry of bladder epithelial cells quickly established the feasibility of this technique [46,87,118,122]. Tribukait and Esposti showed that whereas non-neoplastic urothelium had diploid DNA, at least some bladder tumors were aneuploid [118]. Melamed et al studied voided urine, bladder irrigation specimens and tumor cell suspensions and found differences between tumor and normal urothelium in RNA as well as DNA content [87]. The next few years saw a series of studies from these and other laboratories that refined the technique and established flow cytometry of bladder irrigation specimens as a reliable means of identifying bladder cancer [18–20,92]. Tribukait and Esposti demonstrated that the presence of aneuploid cell populations correlated with histologic and cytologic grade, increasing from 7% of grade 0,1 tumors to 79% of grade 3 tumors [119]. In a later study they reported that all grade 3 tumors were aneuploid and that S-phase fractions, which were about 6% in diploid tumors, were about 17% in aneuploid tumors, indicating more rapid proliferation of the more anaplastic tumors [120]. Chin et al. [13] found all grade I tumors were diploid, whereas 30% of grade II and 77% of grade III tumors were aneuploid, and the frequency of aneuploidy in early-stage tumors was similar to the expected incidence of progression. Helander et al. [57] found

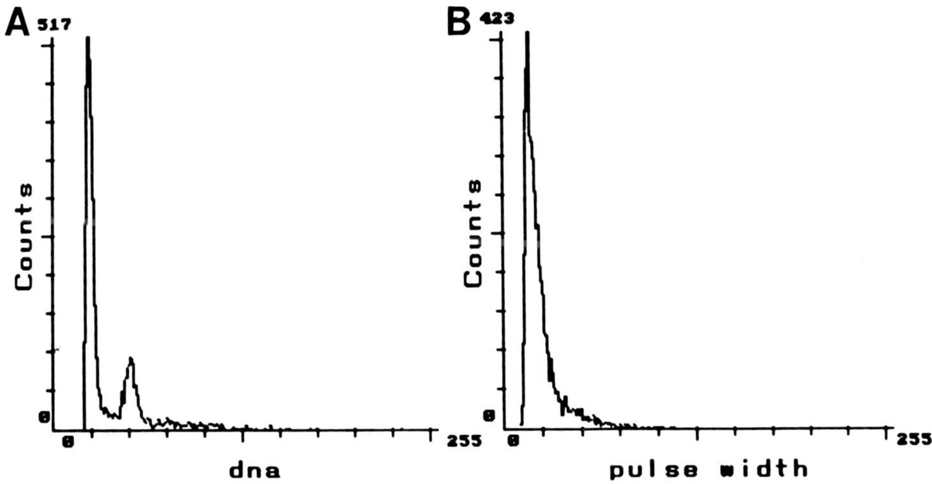

Fig. 3. **A:** DNA histogram of cells from an invasive carcinoma of uterine cervix. Acridine orange stain. The larger peak at lower DNA value represents diploid cells (channel 24), probably mostly lymphocytes and admixed benign epithelium. The smaller peak at higher DNA value is hypertetraploid (channel 52), and composed of aneuploid tumor cells. Cell doublets and larger cell groups have been excluded by "gating" DNA measurements on nuclear pulse width. **B:** Histogram of nuclear pulse width measurements showing a unimodal distribution skewed to the right, indicating a population of single cells that includes a small number with enlarged nuclei.

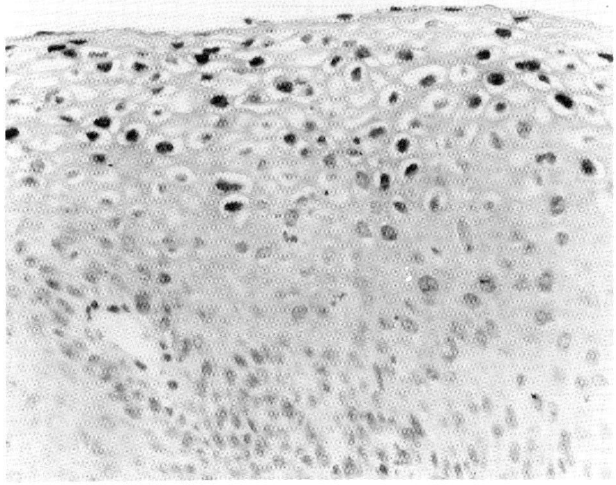

Fig. 4. A precancerous lesion of uterine cervix. The darkly stained koilocytotic cell nuclei are infected by human papilloma virus (HPV), as demonstrated by binding of a peroxidase-labeled HPV cDNA probe. It is possible that markers of this kind could be used to identify HPV infected cells by flow cytometry, increasing the specificity and perhaps the sensitivity of uterine cervical cancer detection.

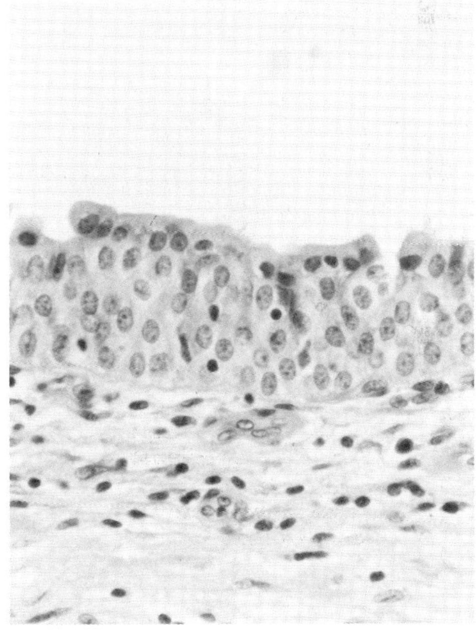

Fig. 5. Normal bladder mucosa. The epithelium is relatively simple, composed of several layers of uniform cells; the surface layer of "umbrella" cells has somewhat more cytoplasm and a unique asymmetric cell membrane that protects against hyperosmotic contents of urine.

all grade III carcinomas to be aneuploid. Granberg-Ohman and associates compared chromosomal karyotypes with DNA distribution in 15 bladder tumors and found higher chromosome counts in high-grade tumors, with good correlation between actual chromosomal number and predicted number from flow cytometry [51]. A much larger, more recent study by Wijkstrom et al. confirmed that tumor grade, chromosomal count, and modal DNA values were highly correlated [130]. While diploid and near-diploid tumors could not be detected by flow cytometry, they were as a group biologically more benign than tumors with higher chromosome counts.

Most of this early work was carried out by single-parameter flow cytometry of cells stained for DNA, usually with propidium iodide or ethidium bromide, and more recently with DAPI [45]. In 1980, Collste et al. [18] introduced the two-step acridine orange (AO)-staining technique of Darzynkiewicz and Traganos (see Chapter 15, this volume) to stain the DNA and RNA of bladder epithelial cells differ-

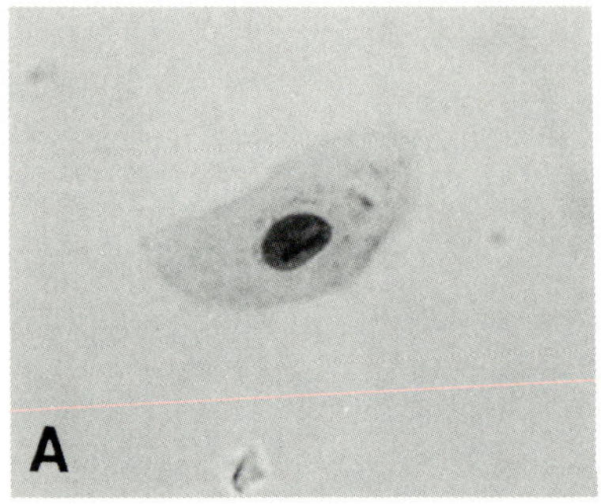

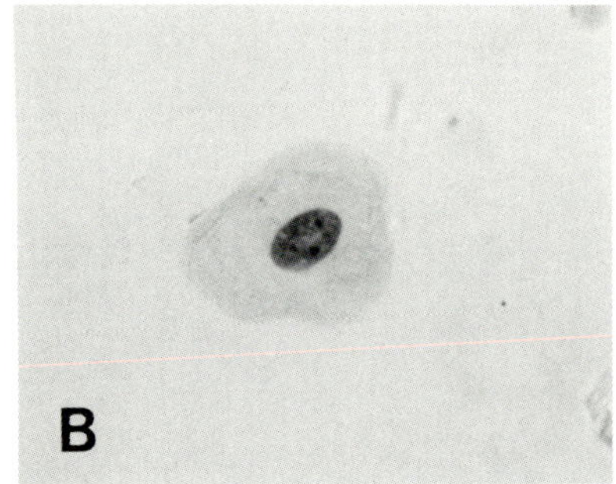

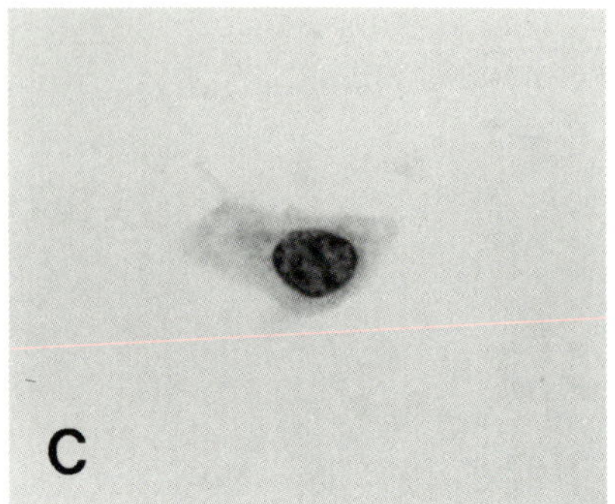

Fig. 6. Normal desquamated urothelial cells; two superficial "umbrella" cells with abundant cytoplasm **(a,b)** and a deeper, intermediate cell **(c)**. These are found in voided urine but are more abundant and usually better preserved in irrigation specimens. The background is "clean."

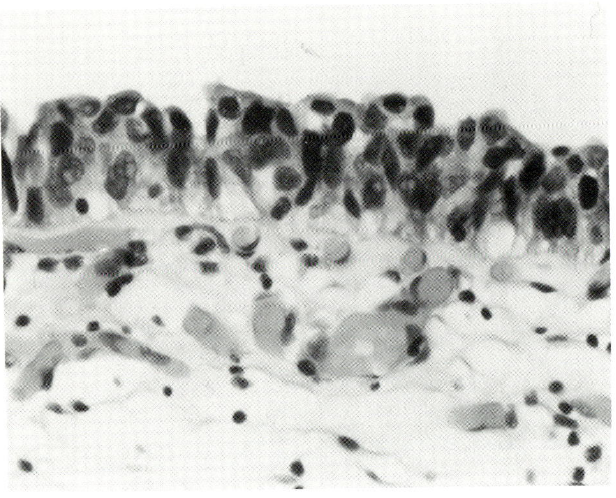

Fig. 7. Carcinoma in situ of urinary bladder (cf. Fig. 5). The epithelium is composed of cancer cells with enlarged hyperchromatic nuclei, but there is no invasion into the underlying stroma.

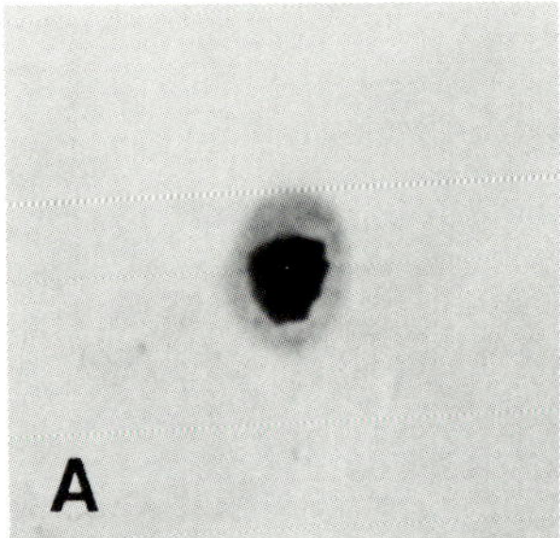

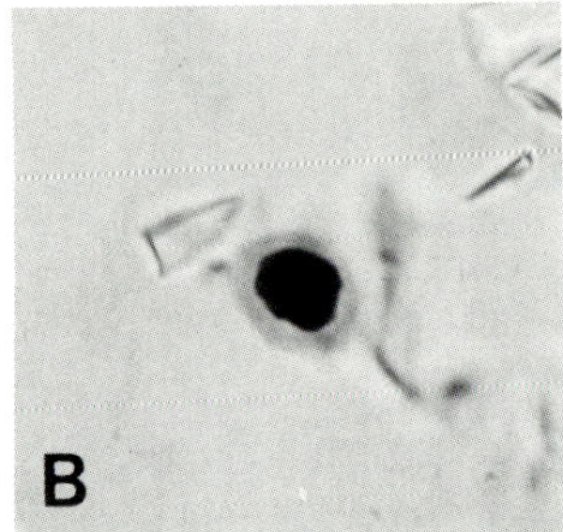

Fig. 8. **A,B:** Single desquamated cells from carcinoma in situ of bladder. The cells are comparable to normal in size, or even somewhat smaller, with relatively large, hyperchromatic, irregular nuclei and coarse chromatin structure.

entially in irrigation specimens (Fig. 11). With this two-parameter stain and nuclear size estimates by pulse width measurements (time of flight), these investigators were able to identify and exclude from analysis polymorphonuclear leukocytes, squamous cells, and cell doublets or larger groups [19]. They then demonstrated that at least some bladder tumors without a well-defined aneuploid peak could be identified by the presence of an increased number of hyperdiploid cells (>15%). The hyperdiploid compartment contains aneuploid tumor cells that are too few or too variable to form an identifiable peak, as well as cycling cells that are more numerous in tumors than in normal mucosa [17,37]. Klein et al. [69,71] and Badalament et al. [8], respectively, summarized the early and more recent Memorial Sloan-Kettering experience with flow cytometry of the bladder, which was diagnostic in 80 to 90% of carcinomas and 50% of papillomas (Table 1). It was also effective in a small series of adenocarcinomas of the bladder [4]. In a new series of 70 patients with biopsy-proven bladder tumors, Badalament reported 83% sensitivity, which was greater than for conventional cytology of voided urine or bladder irrigation specimens on the same patients at the same time [6,7]. Another recent study, by Klein and White, reported 93% agreement between flow cytometry and cystoscopy in 286 patients [73]. Klein et al. [68] also carried out a study of 100 urologic patients with normal or non-neoplastic disease of the bladder to estimate the false-positive rate (specificity) and found only two with positive flow cytometry. Both patients had bladder calculi with severe cystitis and >15% hyperdiploid cells with a tetraploid but not aneuploid population. In these two cases the false-positive flow cytometry was very likely the result of a marked reactive proliferative response to inflammation. Klein and White suggested that the percentage of hyperdiploid cells required for diagnosis by dual-parameter acridine orange-stained specimens might not apply to evaluate the single-parameter flow cytometry histograms of propidium iodide-stained specimens [73]. In many patients with inflammatory disease, they noted that DNA distributions were asymmetric, with a "tail" of higher values. They regarded as suspicious, but not positive, those specimens with 15% hyperdiploid or tetraploid cells but without a defined aneuploid population. Thus, while there is agreement that a distinct aneuploid population is indicative of carcinoma (Fig. 11b) [25,28,30,45,79,84,86,90,110,117,124] criteria for diagnosis based on percentage hyperdiploid cells can be affected by inflammatory disease and by the method of specimen preparation, staining technique, and threshold selection for excluding leukocytes, squamous cells, and degenerated or other unwanted cells.

In our experience with papilloma, only about one-third have an aneuploid population and another 15% have an increased proportion of hyperdiploid cells; thus, about half

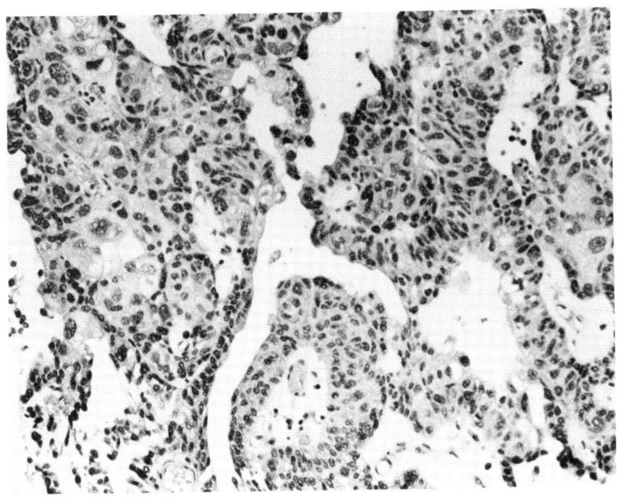

Fig. 9. Fragments of invasive carcinoma of bladder, showing the marked variation in cellular and nuclear size and staining compared with carcinoma in situ.

the cases may be diagnosed by flow cytometry. Since papillomas, by definition, are cytologically benign, low-grade papillary tumors (Fig. 12) to be distinguished from papillary carcinoma (Fig. 13), it is surprising that even one-third of the papillomas have aneuploid populations. Possibly these are cases with foci of carcinoma or carcinoma in situ undetected in the papillary tumor, or present elsewhere in the bladder. It is known that carcinoma in situ may be present in otherwise "normal" bladder mucosa of patients who have bladder tumors. That would explain the findings of Collste et al. [20] and others who have shown that sampling by bladder irrigation may reveal aneuploid populations not identified in cell suspensions of selected site biopsies, and Farsund et al. [36], who found aneuploid populations in biopsies of cystoscopically normal appearing mucosa from patients with bladder tumors. deVere White et al. recently reported that all patients with low grade noninvasive papillary tumors and aneuploid DNA histograms experienced recurrent tumors [28].

The increased proportion of hyperdiploid cells is consistent with an increased proliferative rate in papillomas compared with normal bladder epithelium. Klein et al. [72] also reported an increase in RNA content of cells from papillomas, suggesting increased metabolic activity; and Staiano-Coico et al. found that nuclear RNA also is increased [110].

Flow cytometry is very effective in the detection of carcinoma in situ [71]. Because it requires catheterization and irrigation of the bladder for adequately cellular specimens, however, flow cytometry as a cancer detection technique is limited to high risk populations. These include industrial workers exposed to urinary bladder carcinogens, patients with a past history of conservatively treated bladder tumors and adult patients with unexplained or persisting urologic symptoms. It should be emphasized that the symptoms of carcinoma in situ are those of a nonspecific cystitis—frequency, dysuria, and nocturia—usually with microscopic hematuria [88]. In a study of approximately 30,000 unselected urologic patients at the Mayo Clinic, Farrow et al. found unsuspected carcinoma in situ by cytologic examination of urine in 0.2 to 0.3% [35]. Whether it will eventually be

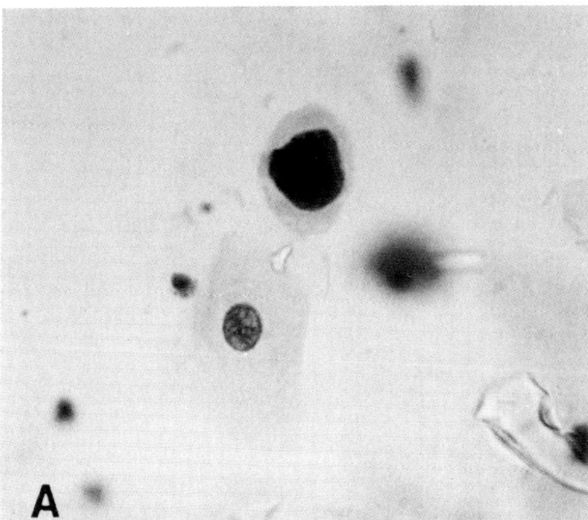

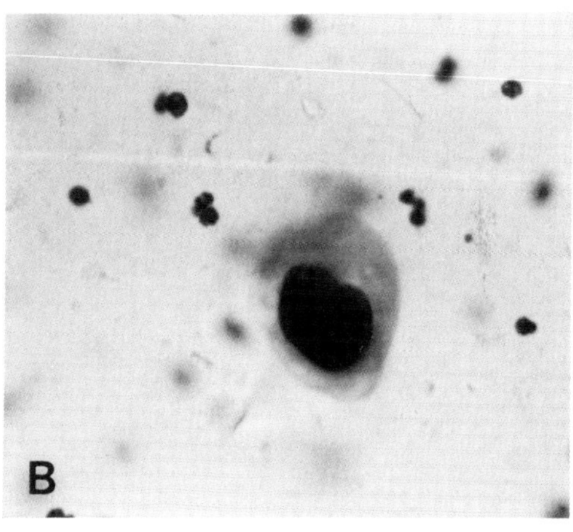

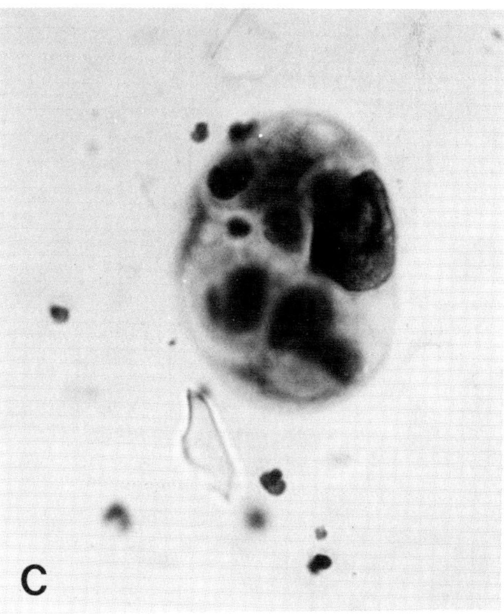

Fig. 10. **A,B:** Desquamated cells from invasive bladder carcinoma. **C:** There is great variation in size and shape of cancer cells, clusters of cancer cells, and inflammation and bleeding.

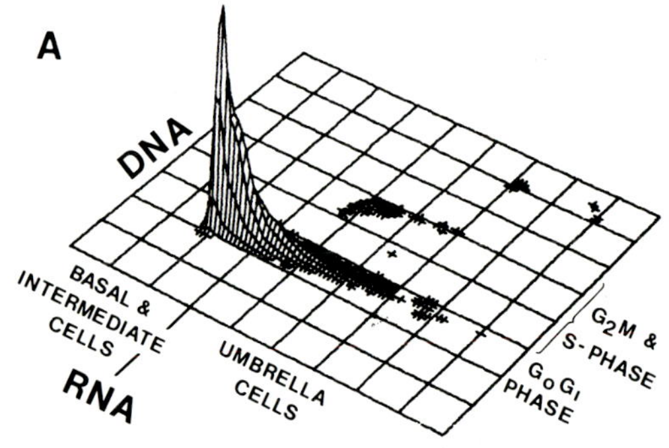

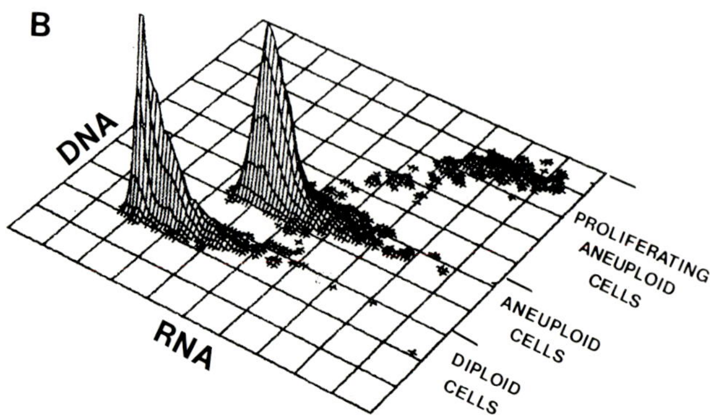

Fig. 11. Two-parameter DNA–RNA acridine orange-stained flow cytometry histograms of irrigation specimens from normal bladder **(a)** and carcinoma **(b)**. The height of the peaks indicates number of cells at each joint distribution of DNA and RNA. Note that in the normal specimen **(a)** most cells (90 to 95%) are diploid and resting (G_0, G_1); only a small number are proliferating (S, G_2M). The superficial or umbrella cells with abundant cytoplasm have increased RNA content, accounting for the skewed RNA distribution of diploid cells. In the bladder irrigation specimen from a patient with bladder carcinoma **(b)**, there is a second prominent population of (aneuploid) cells with increased DNA content, and also an increased number of proliferating aneuploid cells.

TABLE 1. Sensitivity of Flow Cytometry in Histologically Confirmed Bladder Tumors at Memorial Sloan-Kettering*

	Papilloma	Papillary carcinoma	In situ carcinoma	Invasive carcinoma	All cases	All cases excluding papilloma
POS FCM/total cases (% positive)	63/126 (50%)	75/92 (82%)	157/177 (89%)	119/133 (89%)	463/588 (79%)	400/462 (87%)

*From Badalament et al. [8].

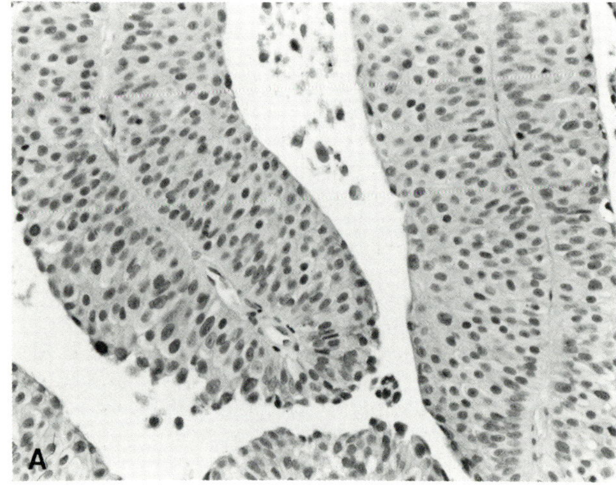

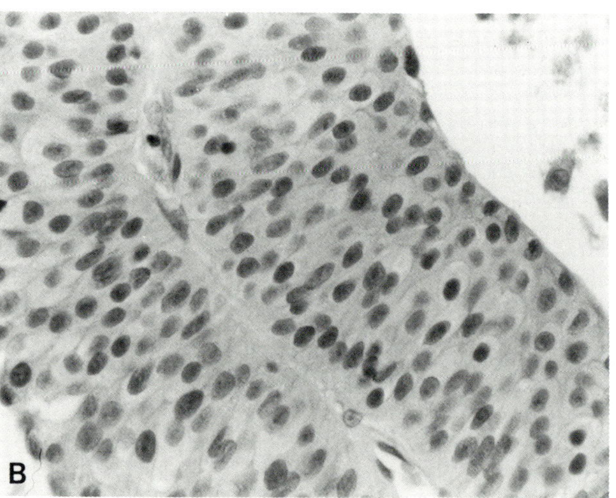

Fig. 12. **A,B:** Bladder papilloma. These tumors are composed of fronds of uniform, cytologically benign or slightly atypical urothelial cells on a central fibrovascular core. By definition, the cells should be diploid. **A:** Medium magnification. **B:** High magnification.

possible to perform flow cytometry successfully on voided urine specimens is still in doubt. deVere White et al. have reported encouraging preliminary results [27]. If it can be achieved, flow cytometry of voided urine certainly will be a major technical advance with important clinical applications.

The most important urologic application of flow cytometry is in following patients with low-stage or superficial tumors who are treated conservatively. Flow cytometry is at least as sensitive as conventional cytology for this purpose and provides quantitative information that can be presented graphically and is easily monitored in sequential examina-

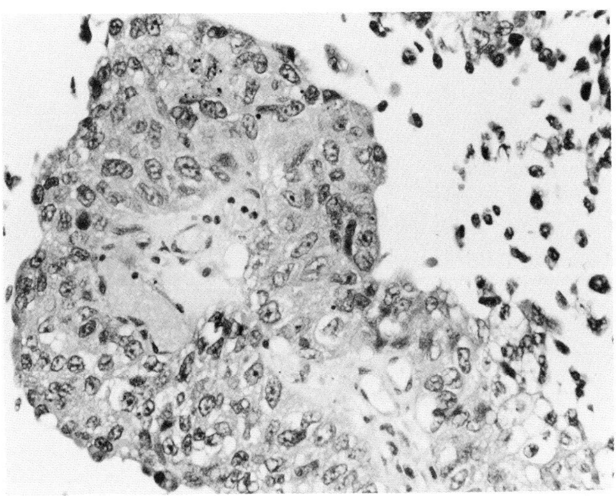

Fig. 13. Papillary bladder carcinoma. These tumors differ from benign papilloma in that the cells making up the tumor fronds are cytologically malignant and less orderly in arrangement. Compare with Fig. 12.

tions (Fig. 14). Thus, flow cytometry has been used to evaluate local resection [29,31,74,84] and intravesical BCG treatment of low-stage tumors [5,70,109] and may help in monitoring radiotherapy [75,119,131] and chemotherapy [39,59]. It also has been effective in examining urethral irrigation specimens for carcinoma [132], including specimens taken postcystectomy [58]. Finally, there is some evidence to suggest that the type of aneuploidy and the proliferative fraction may be useful parameters for subclassifying bladder cancer [13,28,38,51–54, 57,121,131] and may be helpful in selecting therapy or predicting clinical course.

In order to make flow cytometry more widely available to the urologists in small community hospitals or in their office practice it will be necessary to devise some method of specimen preservation for transport to the laboratory. Ratliff et al. reported good results with specimens preserved in 95% ethanol [96]. We found that equal mixtures of 50% ethanol with the irrigation cell suspension gave good preservation for most specimens, but was not consistently reliable. These results are encouraging and suggest that a suitable transport medium will be available within the near future.

A number of new parameters have been proposed to increase diagnostic accuracy or provide additional information on tumor behaviour; most are still under evaluation. Tachibano et al. [114] and Jitsukawa et al. [63] have emphasized the value of DNA variability or heterogeneity. Wheeless et al. [129] have begun to assess multidimensional slit-scan flow cytometry (see Chapter 6, this volume). But for the most part, the proposed new parameters are based on immunocytochemical determinations of cell surface antigens or cytoplasmic constituents. Feitz et al. [40] and Huffman et al. [61] reported using fluoresceinated anticytokeratin antibodies in conjunction with the DNA fluorochrome propidium iodide. They were able to select cytokeratin-containing epithelial cells from among cytokeratin-negative stromal and inflammatory cells and to determine the DNA content of bladder epithelial cells better in specimens with large numbers of inflammatory cells.

The blood group-related antigens also are attractive as potential markers for identifying or characterizing neoplastic bladder epithelium. While it has long been evident that there was a loss of ABH antigen expression in carcinoma and carcinoma in situ [82,128], now thought to be due to suppression or loss of glycosyl transferases, Cordon-Cardo et al. recently reported that normal urothelium in the approximately 20% of individuals who are nonsecretors also lacks expression of ABH, Lewis b, and y antigens [21]. Lack of expression in these individuals is attributable to the regulatory secretor gene and is not a deletion indicating neoplastic transformation. For this and other reasons, it would be desirable to have a positive marker for identification of cancer cells rather than loss of a normally expressed marker. Cordon-Cardo and associates have reported that Lewis X is expressed in neoplastic urothelium but not in the benign urothelial cells, except for slight reactivity on the lumenal surface of umbrella cells [22]. They speculate that this is due to an increase in fucosyl transferases coupled with suppression of glycosyl transferases in the neoplastic cells, altering the carbohydrate composition of membrane glycoproteins and glycolipids. Thus, enhanced expression of X antigen, perhaps accompanied by increased expression of precursor substance or changes in other blood group antigens, is a possible marker of malignant transformation of the urothelium.

Finally, an increasing number of tumor-associated antigens are beginning to yield useful markers [14,43,44]. Some of these differentiate tumors of different histologic grade. Invasive carcinomas, for example, are more likely to express an antigen binding the monoclonal antibody T-138, while low-grade, noninvasive papillary tumors preferentially bind a different monoclonal antibody, Om-5 [44]. A change in antigen expression may indicate transition in stage or grade [95] or may be an indication of response to therapy [60].

EFFUSIONS: PLEURAL, ASCITIC, AND PERICARDIAL FLUIDS

Effusions would seem to be readily adaptable to examination by flow cytometry (Fig. 15a), but there are technical problems. The fluids vary greatly not only in total cell count and relative proportion of inflammatory and reactive mesothelial cells, but also in the number of degenerated versus preserved cells, in the amount of protein present, number of red blood cells, and presence of small fibrin clots. Some mesothelial and tumor cells may be single, but many are in doublets or larger groups and must be dissociated. Finally, since there can be no normal specimens of effusion, there are no normal values for the cellular contents.

All flow cytometry studies of effusions that have been carried out so far rely primarily on DNA distribution of the total population to identify the component of aneuploid cells which, if present, indicates malignant tumor. Unger et al. [124] examined heparinized pleural effusions from 33 patients stained with ethidium bromide and mithramycin and correctly identified 10 of 12 malignant and 20 of 21 benign effusions. Similar results were obtained by Evans et al. [33]. Unfortunately, small numbers of aneuploid (malignant) cells may not be identified in an effusion rich in reactive mesothelial and inflammatory cells. This was well demonstrated by Schneller et al. [104], who identified aneuploid populations in only 16 of 26 cytologically cancerous effusions. Feulgen cytophotometry was carried out on smears of the same fluid specimens in 8 of the 10 cases without a demonstrated aneuploid population. When 500 cells were randomly selected by an automatic device, mimicking flow cytometry, the same

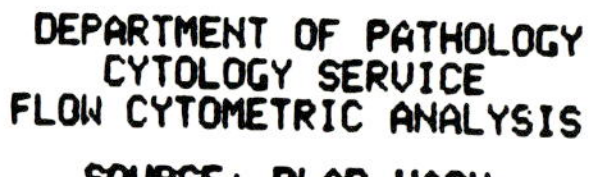

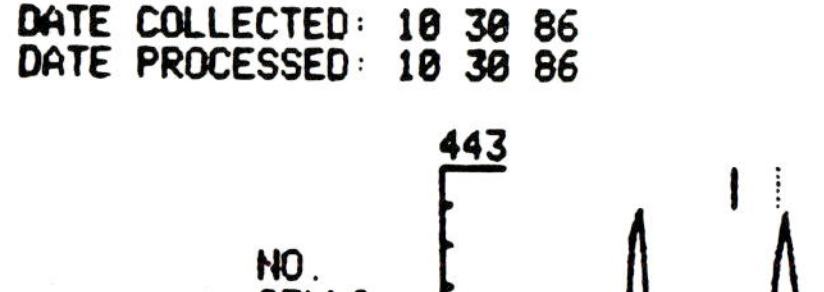

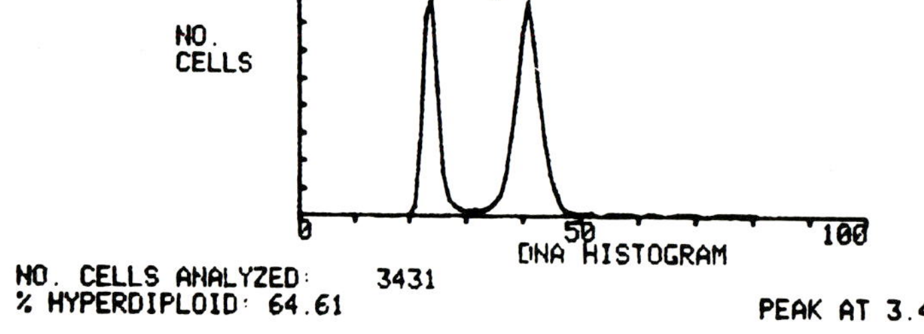

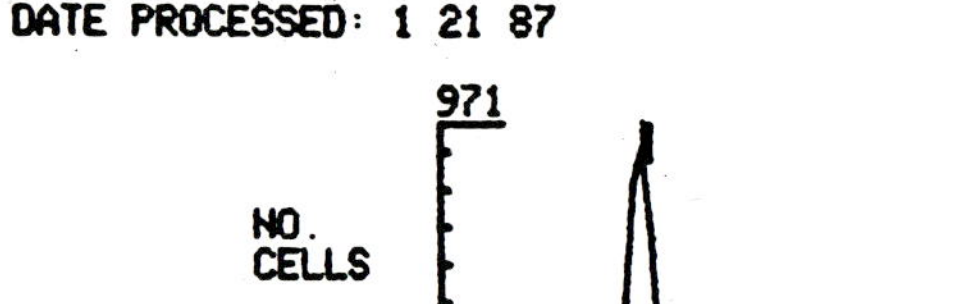

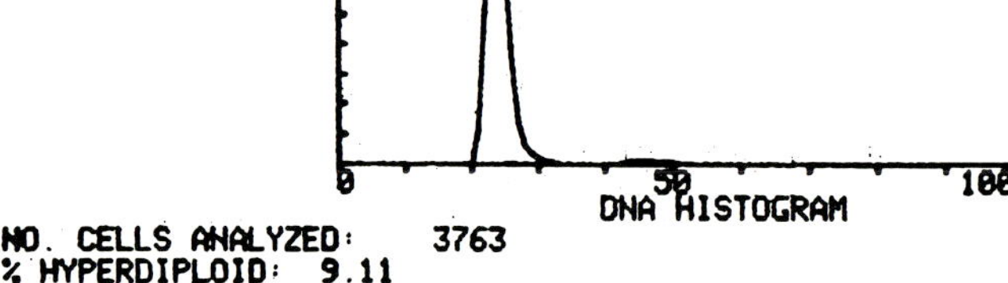

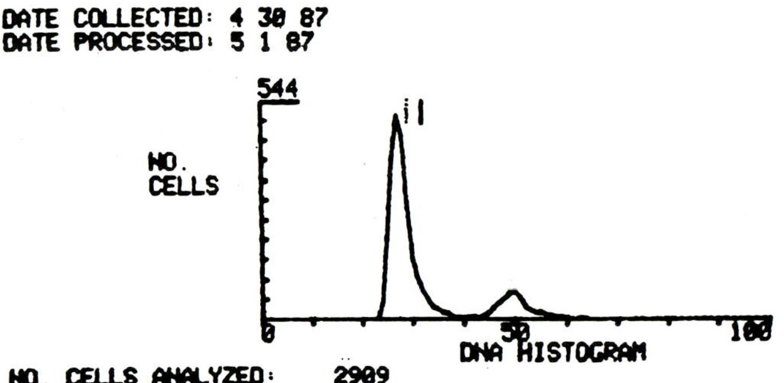

Fig. 14. Sequence of three flow cytometry DNA histograms of bladder irrigation specimens taken at 3-month intervals to show the value of this technique in monitoring therapy. In the initial specimens (top), there was an obvious aneuploid population caused by tumor. Following transurethral resection and a course of intravesical BCG, the DNA histogram showed no residual tumor (middle), but 6 months after the initial examination, there was evidence of recurrence. (lower panel).

diploid distribution was obtained. However, when cytologically malignant cells were selected for measurement by an experienced cytopathologist, the DNA distribution by Feulgen cytophotometry (of this subpopulation) was aneuploid in 6 of the 8 cases. Similarly, Hedley et al. [56] detected aneuploid cells in 23 of 36 effusions with malignant cells present by conventional cytology; 13 were diploid. Thus, they concluded that with presently available techniques flow cytometry is likely to have only limited value in this clinical application.

Stonesifer et al. [113] extended conventional criteria for aneuploidy to include broadened and asymmetrical G_1 distributions, increasing sensitivity but also obtaining a small number of false-positive cases. Katz et al. [65] used the acridine orange DNA/RNA staining technique (see Chapter 15, this volume) and examined effusions of 78 patients. They found that RNA and proliferative indices added little to flow cytometry determinations of DNA ploidy except in the case of hematologic neoplasms (see Chapter 35, this volume).

Thus, the presence of an aneuploid population has been reliably diagnostic of malignant tumor, although there are a substantial number of malignant effusions in which the DNA

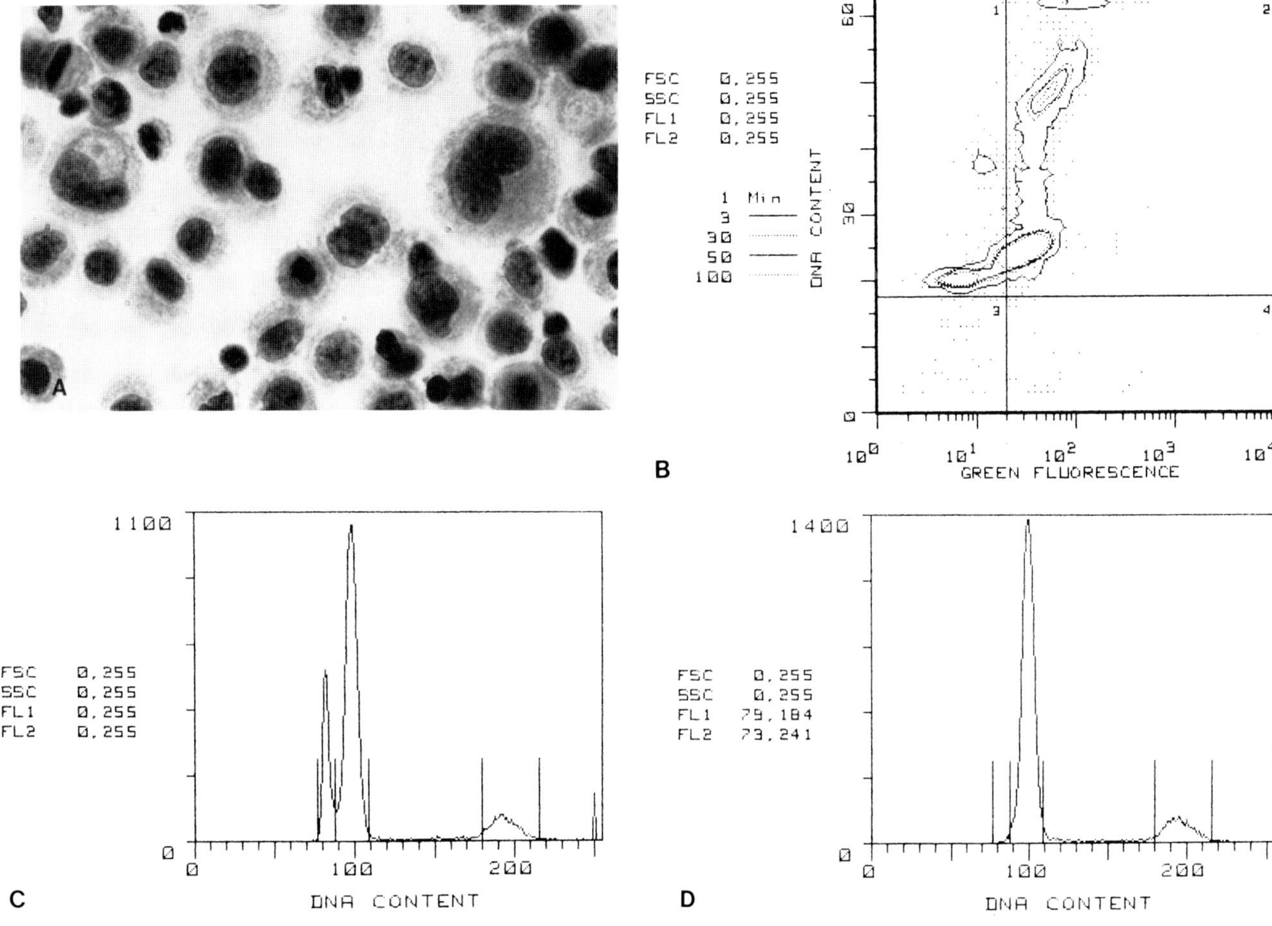

Fig. 15. **A:** Cytocentrifuged preparation of pleural fluid from a patient with a history of adenocarcinoma of the breast. Many of these cells are suspicious for adenocarcinoma, but it is difficult to be certain that they are not aytpical, reactive, but benign mesothelial cells. **B:** Dual-parameter flow cytometry histogram of FITC-monoclonal antikeratin antibody-stained specimen (abscissa), counterstained for DNA with propidium iodide (ordinate). Note that the cytokeratin parameter divides the cell sample into two populations: cytokeratin-negative cells are diploid and nonproliferating (G_0G_1), cytokeratin positive cells (higher FITC green fluorescence) are aneuploid and proliferating. **C:** Ungated DNA histogram showing a mixed diploid and aneuploid population. The latter comprises 79.3% of cells. **D:** Gated DNA histogram of keratin-positive cells showing a pure population of aneuploid (carcinomatous) cells. The diploid mesothelial cells, histiocytes and other reactive cells did not stain for keratin and were excluded. DNA index, 1.20; CV, 3.8%; G_0G_1, 80.8%; S, 5.6%; G_2/M, 13.6%. Case of Dr. Martin L. Kelsten [66].

distribution is diploid. Many of the latter are due to the relatively small number of tumor cells present in an effusion that has abundant inflammatory and reactive mesothelial cells. It should be possible to improve detection of the aneuploid (tumor cell) population in these cases either by electronically gating selected subpopulations of cells based on other feature measurements, or by enriching for the tumor cells through such physical means as differential sedimentation [127] (see also Chapter II, this volume). Since the tumor cells in effusions may be of almost any type from any source, it is difficult to envision a single feature measurement that would distinguish tumor cells as a group from all other cell types. The epithelial cancers, primarily adenocarcinomas, are the most frequent cause of malignant effusions and one possible marker would be the anticytokeratin antibodies, which are specific for epithelial cells. Kelsten et al were encouraged by their success in using anticytokeratin antibodies to identify aneuploid epithelial cells in effusions [66] (see Fig. 15). Yet, even the monoclonal anticytokeratin antibodies may cross-react with mesothelial cells, and they do not react with cells of malignant lymphomas or sarcomas. At this time, preprocessing by differential sedimentation probably is most realistic for enrichment of malignant cells in effusions, even in systems using electronic gating by unique antigen expression or other feature measurements. Finally, it should be remembered that a substantial proportion of carcinomas and lymphomas are diploid by flow cytometry (see Chapter 37, this volume), and these tumors cannot be identified by measurements of DNA alone. Whether antibodies to tumor-associated antigens will be useful markers in those instances is still an open question. Anti-CEA antibody is of possible value, at least for carcinomas known to express CEA. Another antibody (Ca 1), a mouse monoclonal IgM antibody to glycoproteins of human laryngeal carcinoma, was recently reported to correctly identify 14 of 17 malignant effusions [24]. Certainly other antibodies will be developed and evaluated over the coming years, and there is reason to be optimistic that some of these will prove clinically useful.

Having described in detail efforts to diagnose cancer in effusions, we must note that the greatest potential value of flow cytometry lies not in diagnosis but in characterizing the tumor cells present. Valet et al., for example, treated short-term cultures of cancerous effusions with cytostatic drugs and then examined them for drug effect by multiparameter flow cytometry [125]. They measured cell volume to distinguish inflammatory from tumor cells, then used an esterase activated pH-indicator dye to identify living cells and propidium iodide to identify dead cells and measure their DNA content. From those data, they calculated a therapeutic index based on the number of dead and surviving tumor and inflammatory cells. The key to these types of studies is the availability of multiparameter flow cytometry, a selection of dyes that measure a range of cell functions, and our own ingenuity.

CEREBROSPINAL FLUID

We have carried out flow cytometry studies of cerebrospinal fluid (CSF) from patients with a variety of neoplastic and other diseases of the central nervous system (CNS). Normally there are up to 10 cells/mm^3 in the CSF, primarily lymphocytes. Thus, 1 ml of normal CSF would contain up to 1,000 cells, and 3 to 5 ml is sufficient for flow cytometry—not an unreasonable amount of fluid to request for this examination. Patients with inflammatory or neoplastic disease will have increased spinal fluid cellularity.

Flow cytometry has been most accurate in identifying meningeal carcinomatosis and CNS involvement with leukemia or lymphoma [15]. Primary brain tumors and metastatic tumors in the brain do not shed cells into the CSF unless or until tumor extends to involve the meninges. While in cases of meningeal carcinomatosis metastatic carcinoma is readily diagnosed by conventional cytologic technique, this is not true of lymphoma or leukemia. It is in these latter cases that flow cytometry of CSF is most valuable, both for diagnosis [97] and for monitoring results of treatment [98]. The presence of a uniform population of lymphocytes may be highly suggestive of lymphoma in a patient known to have the disease, but complicating viral encephalitis or reaction to treatment can give similar cytologic findings, even including the presence of immature cells [1]. Conversely, the presence of immature lymphomatous cells in a pleocellular specimen may be mistakenly interpreted as reactive. In these instances, demonstration of an aneuploid population of cells by flow cytometry is conclusive evidence of lymphoma (or leukemia).

An estimated 60 to 95% of leukemias, depending on cell type, and 80% of non-Hodgkin's lymphomas are diploid by flow cytometry; these will not be detected by measurements of DNA content. Other parameters are available and have been applied to studies of marrow and peripheral blood (see Chapters 34 to 36, this volume) and to CSF fluid [97], making use of RNA measurements and cell surface antigen expression, but it is not clear that they will be as effective as DNA ploidy in CSF. An important limitation is the volume of fluid available for sequential measurements with different antibody labels. Modifications of the instrumentation to accept and measure all (or nearly all) the cells in smaller volumes of samples may make sequential examinations feasible. Still more useful would be an increase in the number of parameters than can be measured simultaneously. But even with present techniques, CSF flow cytometry is a practical and accurate means of identifying (or excluding) central nervous system

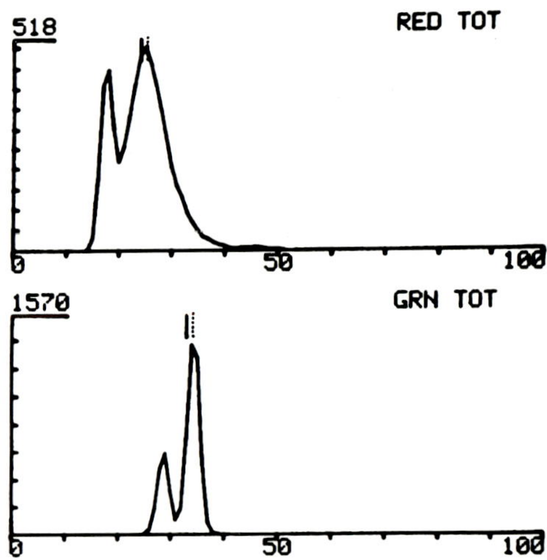

Fig. 16. DNA histogram of spinal fluid from a child with central nervous system relapse of acute leukemia following a short drug-induced remission. The leukemia was aneuploid in prior studies of marrow, as it is here in the spinal fluid. Because the leukemic cellswere uniform, with few blasts, it was not possible by conventional cytology to distinguish this leukemic relapse from an inflammatory reaction. Acridine orange stain: GRN TOT, DNA; RED TOT, RNA.

involvement at least in the case of lymphomas or leukemias known to be aneuploid on examination of peripheral blood or marrow (Fig. 16). Flow cytometry can be of diagnostic value in cases with suspected meningeal carcinomatosis.

Flow cytometry also has been reported to detect an increase in stimulated lymphocytes within the CSF of patients with multiple sclerosis [91] and in the synovial fluid of patients with rheumatoid arthritis [10]. While it can hardly be used as a diagnostic technique in these cases, flow cytometry may provide clues to the activity of these and other inflammatory diseases of immunologic etiology.

SPUTUM

Cough specimens of sputum are not well suited for flow cytometry, and there are few reports of this application. The cells that are dislodged from lung cancer are typically sparse, and may be present only intermittently in the sputum. Thus, it is unlikely that a cell suspension from sputum will contain sufficient malignant cells to be identified as an aneuploid population. Bronchial brush specimens can be expected to contain more tumor cells and have been successfully analyzed by flow cytometry [115]. The tumor cell fraction can be enriched by centrifugation on a density gradient [94,127], but even after enrichment there is not likely to be a detectable aneuploid population in most specimens. Furthermore, the time and cost of processing offer little or no advantage over conventional cytologic examinations of Papanicolaou-stained smears.

An interesting and innovative approach was taken by Frost et al. [47,48,123] who reported enriching the cancer cells in sputum by means of flow cytometry/sorting for subsequent visual examination. Sorting was based on nucleic acid content and light scatter. However, the procedure was time consuming and of little practical value.

Steinkamp et al. [112] and Lehnert et al. [81] applied flow cytometry to the study of alveolar macrophages recovered by bronchoalveolar irrigation. Bronchoalveolar irrigants contain numerous macrophages in addition to bronchial epithelium and inflammatory cells and are well suited to flow cytometry studies. Rather than seeking rare tumor cells in these lavage specimens, they used flow cytometry to assay the effects of toxic agents on the biochemical or functional properties of the component cells. In their hamster model, lavage specimens contained macrophages, leukocytes, and ciliated columnar and basal epithelial cells. Cellular parameters that could be studied and that distinguished the various cell types included nuclear and cellular diameters, DNA content, protein, and esterase activity.

MISCELLANEOUS APPLICATIONS

Some interesting applications have been proposed for flow cytometry of nonconventional cytology specimens. We describe two such applications that are of interest to us. Others will certainly be able to expand this list.

Disorders of Skin Proliferation and Maturation

Dual-parameter RNA and DNA measurements can resolve three subpopulations of normal cutaneous keratinocytes, which differ in cell cycle kinetics and represent successive stages of maturation (Fig. 17, bottom). One subpopulation, containing the lowest cellular RNA content and designated A, comprises slowly cycling keratinocytes that have been identified as basal layer cells by immunochemical staining with specific antibodies. The second subpopulation, designated B, is composed of larger keratinocytes with higher RNA content and significantly shorter cell doubling times. The third subpopulation C is composed of the largest most superficial squamous cells, which are nondividing and exhibit yellowish cytoplasmic fluorescence due to the presence of keratohyalin granules [67,108].

We have used flow cytometry to study a series of patients with psoriasis vulgaris, a chronic hyperproliferative disorder of the skin [106]. Samples were taken before therapy and after 2 to 3 weeks treatment with ultraviolet irradiation and coal tar. Before therapy, both the uninvolved and lesional areas of skin show an increased proportion of cells from the more rapidly dividing B population (Fig. 17, top). Moreover, epidermal cells derived from psoriatic lesions showed a significant elevation of cellular RNA content compared with nonlesional or control skin. Post-treatment cellular RNA content returned to quasinormal levels as patients responded to therapy. By contrast, nonresponsive patients did not show reductions in cellular RNA post-treatment, suggesting that cellular RNA content may have clinical value as a prognostic indicator of treatment response in psoriasis.

We also examined 35 basal cell carcinomas of the face, to compare their keratinocyte subpopulation distributions with that of normal and psoriatic skin. Instead of three cell subpopulations, we found that the basal cell carcinomas were composed primarily of low RNA A cells [107] (Fig. 18A). Moreover, in the larger, more aggressive carcinomas, the keratinocytes exhibited not only differences in RNA content, but also were tetraploid, exhibiting DNA indices of 2.0 [107] (Fig. 18b).

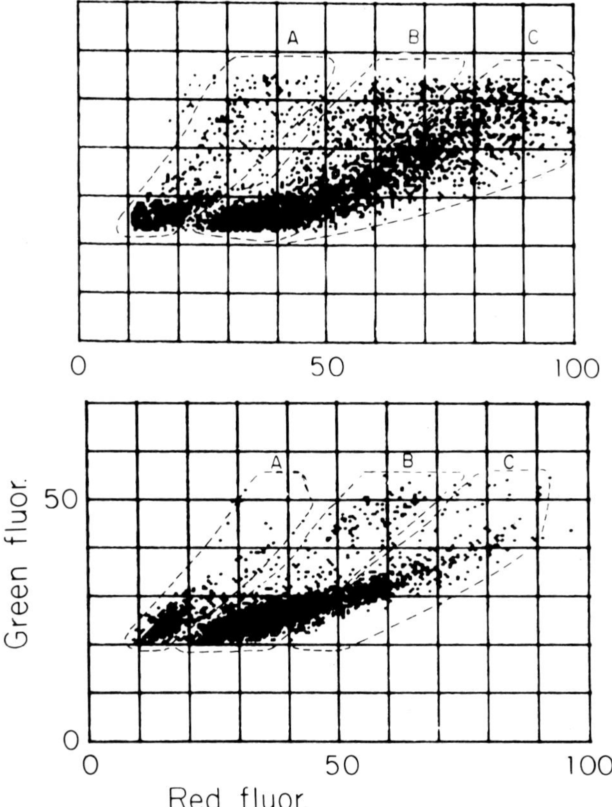

Fig. 17. Dual-parameter RNA (red fluor) / DNA (green fluor) scattergrams of hyper-proliferative keratinocytes from psoriatic skin (top) compared with normal skin (bottom). Three subpopulations are evident in both cases, based on RNA content. **a:** Low-RNA basal cells known to have prolonged cell doubling time. **b:** Intermediate cells with higher RNA content and much shorter cell doubling time. **c:** Superficial cells, which are the largest and most keratinized cells and do not proliferate. The psoriatic skin shows a great increase in DNA synthesizing (proliferating) cells, predominantly in the B (intermediate cell) compartment. It is difficult to assess the **(c)** compartment because there is considerable nonspecific acridine orange staining of these cells, presumably due to large amounts of cytokeratin.

Cancer Cells in Bone Marrow

During the 1960s, considerable effort was invested into developing a technique for the identification of circulating cancer cells in peripheral blood (for review, see Melamed [85]). While it was clear that cancer cells could be recovered in some instances from random samples of blood draining an invasive carcinoma, these were rare occurrences and of little or no prognostic value. However, interest in identifying "circulating" cancer cells was recently restimulated by Dearnaley et al. [26], who reported that fluoresceinated antibody to epithelial membrane antigen could be used to identify otherwise undetectable breast cancer cells within samples of bone marrow from women undergoing breast cancer surgery. Their results have been confirmed in our laboratory by Cote et al. [23] and by others [93,103]. The probability of finding these occult cancer cells in the marrow is increased in women who have axillary lymph node metastases at the time their breast cancer is discovered and resected. Thus there is reason to expect this test to be of clinical value in predicting which patients will eventually develp metastases, and to modify their treatment plan accordingly. No one has yet reported adapting this manual examination to flow cytometry, although it is clearly feasible to do so.

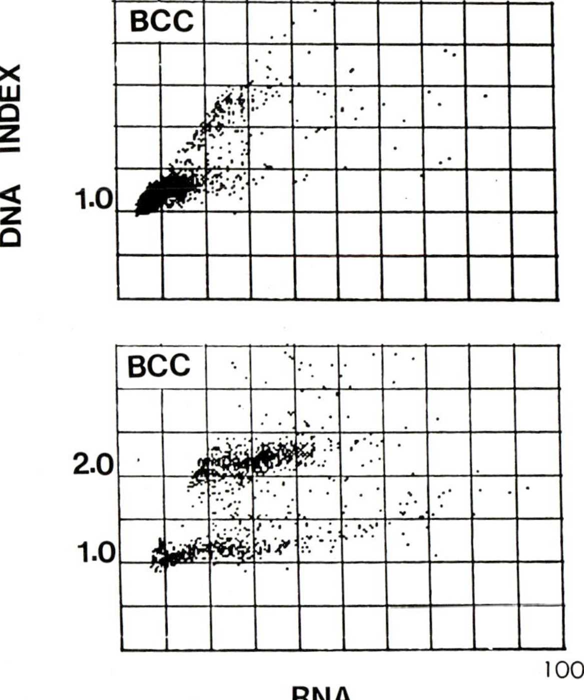

Fig. 18. Dual-parameter RNA/DNA scattergrams of two basal cell carcinomas. Note the increased proportion of low-RNA containing cells (i.e., lack of differentiation) compared with normal (see Fig. 17, bottom). The more marked tetraploidy of the tumor in the lower panel has been observed in association with more aggressive behavior.

REFERENCES

1. **Aaronson A, Hajdu SI, Melamed MR** (1975) Spinal fluid cytology during chemotherapy of leukemia of the central nervous system in children. Am J Clin Pathol 63:528–537.
2. **Adelson MD, Johnson TS, Sneige N, Williamson KD, Freedman RS, Peters LJ** (1987) Cervical carcinoma DNA content, S-fraction and malignancy grading. II. Comparison with clinical staging. Gynecol Oncol 26: 57–70.
3. **Atkin NB, Richards BM** (1956) Deoxyribonucleic acid in human tumors as measured by microspectrophotometry of Feulgen stain. A comparison of tumours arising at different sites. Br J Cancer 10:769–786.
4. **Badalament RA, Cibas ES, Reuter VE, Fair WR, Melamed MR** (1987) Flow cytometric analysis of primary adenocarcinoma of the bladder. J Urol 137: 1159–1162.
5. **Badalament RA, Gay H, Whitmore WF Jr, Herr HW, Fair WR, Oettgen HF, Melamed MR** (1986) Monitoring intravesical bacillus calmette-guerin treatment of superficial bladder carcinoma by serial flow cytometry. Cancer 58:2751–2757.
6. **Badalament RA, Hermansen DK, Kimmel M, Gay H, Herr HW, Fair WR, Whitmore WF Jr, Melamed MR** (1987) The sensitivity of bladder wash flow cytometry, bladder wash cytology, and voided cytology in the detection of bladder carcinoma. Cancer 60: 1423–1427.
7. **Badalament RA, Kimmel M, Gay H, Cibas ES, Whitmore WF Jr, Herr HW, Fair WR, Melamed MR** (1987) The sensitivity of flow cytometry compared with conventional cytology in the detection of superficial bladder cancer. Cancer 59:2078–2085.
8. **Badalament RA, Whitmore WF Jr, Fair WR, Melamed MR** (1988) The relative value of cytometry and cytology in the management of bladder cancer: The Memorial Sloan-Kettering Cancer Center experience. Semin Urol 6:22–30.
9. **Barrett DL, Jensen RH, King EB, Dean PN, Mayall BH** (1979) Flow cytometry of human gynecologic specimens using Log chromomycin A_3 fluorescence and Log 90° light scatter. J Histochem Cytochem 27: 573–578.
10. **Bonvoisin B, Cordier G, Revillard JP, Lejeune E, Bouvier M** (1984) Increased DNA and/or RNA content of synovial fluid cells in rheumatoid arthritis: A flow cytometry study. Ann Rheum Dis 43:222–227.
11. **Caspersson T** (1936) Uber den chemischen Aufbau der strukturen des zellkernes. Scand Arch Physiol 73 (Suppl 8):1–151.
12. **Caspersson T** (1950) "Cell Growth and Cell Function." New York: WW Norton and Co., Inc.
13. **Chin JL, Huben RP, Nava E, Rustum YM, Greco JM, Pontes JE, Frankfurt OS** (1985) Flow cytometric analysis of DNA content in human bladder tumors and irrigation fluids. Cancer 56:1677–1681.
14. **Chopin DK, deKernion JB, Rosenthal DL, Fahey JL** (1985) Monoclonal antibodies against transitional cell carcinoma for detection of malignant urothelial cells in bladder washing. J Urol 134:260–265.
15. **Cibas ES, Malkin MG, Posner JB, Melamed MR** (1987) Detection of DNA abnormalities by flow cytometry in cells from cerebrospinal fluid. Am J Clin Pathol 88:570–577.
16. **Coleman DV, Ormerod MB** (1984) Tumor markers in cytology. In Koss LG, Coleman DV (eds), "Advances in Clinical Cytology," Vol 2. New York: Masson Publishing, USA pp 33–47.
17. **Collste LG, Darzynkiewicz Z, Traganos F, Sharpless TK, Devonec M, Claps ML, Whitmore WF Jr, Melamed MR** (1979) Cell-cycle distribution of urothelial tumor cells as measured by flow cytometry. Br J Cancer 40:872–877.
18. **Collste LG, Darzynkiewicz Z, Traganos F, Sharpless TK, Sogani P, Grabstald R, Whitmore WF Jr, Melamed MR** (1980) Flow cytometry in bladder cancer detection and evaluation using acridine orange metachromatic nucleic acid staining of irrigation cytology specimens. J Urol 123:478–485.
19. **Collste L, Darzynkiewicz Z, Traganos F, Sharpless TK, Whitmore WF Jr, Melamed MR** (1979) Identification of polymorphonuclear leukocytes in cytological samples for flow cytometry. J Histochem Cytochem 27:390–393.
20. **Collste LG, Devonec M, Darzynkiewicz Z, Traganos F, Sharpless TK, Whitmore WF Jr, Melamed MR** (1980) Bladder cancer diagnosis by flow cytometry. Correlation between cell samples from biopsy and bladder irrigation fluid. Cancer 45:2389–2394.
21. **Cordon-Cardo C, Lloyd KO, Finstad CL, McGroarty ME, Reuter VE, Bander NH, Old LJ, Melamed MR**

(1986) Immunoanatomic distribution of blood group antigens in the human urinary tract. Influence of secretor status. Lab Invest 56:444–454.

22. **Cordon-Cardo C, Reuter VE, Lloyd KO, Sheinfeld J, Fair WR, Old LJ, Melamed MR (1988)** Blood group antigens in human urothelium: Enhanced expression of precursor, Le[x] and Le[y] determinants in urothelial carcinoma. Cancer Res 48:4113–4120.
23. **Cote R, Rosen PP, Hakes TB, Sedira M, Bazinet M, Kinne DW, Old LJ, Osborne MP (1988)** Monoclonal antibodies detect occult breast carcinoma metastases in the bone marrow of patients with early stage disease. Am J Surg Pathol 12:333–340.
24. **Czerniak B, Papenhausen PR, Herz F, Koss LG (1985)** Flow cytometric identification of cancer cells in effusions with Ca 1 monoclonal antibody. Cancer 55:2783–2788.
25. **Dean PJ, Murphy WM (1985)** Importance of urinary cytology and future role of flow cytometry. Urology (Suppl 26):11–15.
26. **Dearnaley DP Ormerod MG, Sloane JP, Lumley H, Imrie S, Jones M, Coombes RC, Neville AM (1983)** Detection of isolated mammary carcinoma cells in marrow of patients with primary breast cancer. J R Soc Med 76:359–364.
27. **deVere White RW, Deitch AD, Baker WC, Strand MA (1988)** Urine: A suitable sample for deoxyribonucleic flow cytometry studies in patients with bladder cancer. J Urol 139:926–928.
28. **deVere White RW, Deitch AD, West B, Fitzpatrick JM (1988)** The predictive value of flow cytometric information in the clinical management of stage 0 (Ta) bladder cancer. J Urol 139:279–282.
29. **deVere White RW, Olsson CA, Deitch AD (1986)** Flow cytometry: Role in monitoring transitional cell carcinoma of bladder. Urology 28:15–20.
30. **Devonec M, Darzynkiewicz Z, Kostyrka-Claps ML, Collste L, Whitmore WF Jr, Melamed MR (1982)** Flow cytometry of low stage bladder tumors: correlation with cytologic and cystoscopic diagnosis. Cancer 49:109–118.
31. **Devonec M, Darzynkiewicz Z, Whitmore WF Jr, Melamed MR (1981)** Flow cytometry for followup examinations of conservatively treated low stage bladder tumors. J Urol 126:166–170.
32. **Dudzinski MR, Haskill SJ, Fowler WC, Currie JL, Walton LA (1987)** DNA content in cervical neoplasia and its relationship to prognosis. Obstet Gynecol 69: 373–377.
33. **Evans DA, Thornthwaite JT, Ng ABP, Sugarbaker EV (1983)** DNA flow cytometry of pleural effusions. Anal Quant Cytol 5:19–27.
34. **Falor WH, Ward RM (1977)** Prognosis in well differentiated noninvasive carcinoma of the bladder based on chromosomal analysis. Surg Gynecol Obstet 144:515–518.
35. **Farrow GM, Utz DC, Rife CC, Greene LF (1977)** Clinical observations of sixty-nine cases of in situ carcinoma of the urinary bladder. Cancer Res 37:2794–2798.
36. **Farsund T, Didrick O, Hoestmark L (1983)** Ploidy disturbance of normal-appearing bladder mucosa in patients with urothelial cancer: Relationship to morphology. J Urol 130:1076–1082.
37. **Farsund T, Hoestmark J (1982)** Mapping of cell cycle distribution in normal human urinary bladder epithelium. Scand J Urol Nephrol 17:51–56.
38. **Farsund T, Hoestmark JG, Laerum OD (1984)** Relation between flow cytometric DNA distribution and pathology in human bladder cancer. A report on 69 cases. Cancer 54:1771–1777.
39. **Farsund T, Laerum O, Hoestmark J, et al. (1984)** Local chemotherapeutic effects in bladder cancer demonstrated by selective sampling and flow cytometry. J Urol 131:22–32.
40. **Feitz WFJ, Beck HLM, Smeets AWGB, Debruyne FMJ, Vooijs GP, Herman CJ, Ramaekers FCS (1985)** Tissue-specific markers in flow cytometry of urological cancers: cytokeratins in bladder carcinoma. Int J Cancer 36:349–356.
41. **Flint A, McCoy JP, Asch TR, Beckwith AL, Morley GW (1987)** Simultaneous measurement of DNA content and detection of surface antigens of cervico-vaginal cells by flow cytometry. Anal Quant Cytol Histol 9:419–424.
42. **Foulkes BJ, Herman CJ, Cassidy M (1976)** Flow microfluorometric system for screening gynecologic cytology specimens using propidium iodide–fluorescein isothiocyanate. J Histochem Cytochem 24:322–331.
43. **Fradet Y, Cordon-Cardo C, Thomson T, Daly ME, Whitmore WF Jr, Lloyd KO, Melamed MR, Old LJ (1984)** Cell surface antigens of human bladder cancer defined by mouse monoclonal antibodies. Proc Natl Acad Sci USA 81:224–228.
44. **Fradet Y, Cordon-Cardo C, Whitmore WF JR, Melamed MR, Old LJ (1986)** Cell surface antigens of human bladder tumors: Definition of tumor subsets by monoclonal antibodies and correlation with growth characteristics. Cancer Res 46:5183–5188.
45. **Frankfurt OS, Huben RP (1984)** Clinical applications of DNA flow cytometry for bladder tumors. Urology(suppl 23):29–34.
46. **Freni SC, Reijnders-Warner O, DeVoogt HJ, Beyer-Boon ME, Brussce JAM (1975)** Flow fluorophotometry on urinary cells compared with conventional cytology. In Haanen CAM, Hillen HFB, Wessels JMC (eds), "First International Symposium on Pulse Cytophotometry, Nijmegen," Part III. Ghent: European Press, pp 194–203.
47. **Frost JK, Tyrer HW, Pressman NJ, Albright CD, Vansickel MH, Gill GW (1979)** Automatic cell identification and enrichment in lung cancer. I. Light scatter and fluorescence paarameters. J Histochem Cytochem 27:545–551.
48. **Frost JK, Tyrer HW, Pressman NJ, Adams LA, Vansickel MH, Albright CD, Gill GW, Tiffany SM (1979)** Automatic cell identification and enrichment in lung cancer. III. Light scatter and two fluorescence parameters. J Histochem Cytochem 27:557–559.
49. **Gohde W, Dittrich W (1971)** Impulsfluorometrie—ein neuartiges durchflussverfaheren zur ultraschnellen mengenbestimmung von zillinhaltsstoffen. Acta Histochem Supp X:42–51.
50. **Gohde W, Schumann J, Fruh J (1976)** Coincidence eliminating device in pulse cytophotomoetry, In Gohde W, Schumann J, Buchner TH (eds), "Second International Symposium on Pulse Cytophotometry." Ghent: European Press, pp 79–85.
51. **Granberg-Ohman, Tribukait B, Wijkstrom H, Berlin T, Collste LG (1980)** Papillary carcinoma of the uri-

nary bladder. A study of chromosome and cytofluorometric DNA analysis. Urol Res 8:87–93.

52. **Gustafson H, Tribukait B, Esposti PL** (1982) DNA patterns, histological grade and multiplicity related to reurrence rate in superficial bladder tumours. Scand J Urol Nephrol 16:135–139.

53. **Gustafson H, Tribukait B, Esposti PL** (1982) The prognostic value of DNA analysis in primary carcinoma in situ of the urinary bladder. Scand J Urol Nephrol 16:141–146.

54. **Gustafson H, Tribukait B, Esposti PL** (1982) DNA profile and tumour progression in patients with superficial bladder tumors. Urol Res 10:13–18.

55. **Haines HG, McCoy JP, Hofheinz DE, Ng ABP, Nordqvist SRB, Leif RC** (1981) Cervical carcinoma antigens in the diagnosis of human squamous cell carcinoma of the cervix. J Natl Cancer Inst 66:465–474.

56. **Hedley DW, Philips J, Rugg CA, Taylor IW** (1984) Measurement of cellular DNA content as an adjunct to diagnostic cytology in malignant effusions. Eur J Cancer Clin Oncol 20:749–752.

57. **Helander K, Krikhus B, Iversen OH, Johansson SL, Nilsson S, Vaage S, Fjorduane H** (1985) Studies on urinary bladder carcinoma by morphometry, flow cytometry, and light microscopic malignancy grading with special reference to grade II tumors. Virchows Arch 408:117–126.

58. **Hermansen DK, Badalament RA, Whitmore WF Jr, Fair WR, Melamed MR** (1988) Detection of carcinoma in the post-cystectomy urethral remnant by flow cytometric analysis. J Urol 139:304–307.

59. **Hermansen DK, Reuter VE, Whitmore WF Jr, Fair WR, Melamed MR** (1988) Flow cytometry and cytology as response indicators to M-VAC (methotrexate, vinblastine, doxorubicin and cisplatin). J Urol 140:1394–1396.

60. **Huffman JL, Fradet Y, Cordon-Cardo C, Herr HW, Pinsky CM, Oettgen HF, Old LJ, Whitmore WF Jr, Melamed MR** (1985) Effect of intravesical bacillus calmette-guerin on detection of a urothelial differentiation antigen in exfoliated cells of carcinoma in situ of the human urinary bladder. Cancer Res 45:5201–5204.

61. **Huffman JL, Garin Chesa P, Gay H, Whitmore WF Jr, Melamed MR** (1986) Flow cytometric identification of human bladder cells using a cytokeratin monoclonal antibody. NY Acad Sci 468:302–315.

62. **Jakobson A, Kristensen PA, Poulsen HK** (1983) Flow cytometric classification of biopsy specimens from cervical intraepithelial neoplasia. Cytometry 4:166–169.

63. **Jitsukawa S, Tachibana, M, Nakazono M, Tazaki H, Addonizio JC** (1987) Flow cytometry based on heterogeneity index score compared with urine cytology to evaluate their diagnostic efficacy in bladder tumor. Urology 29:218–222.

64. **Johnson TS, Adelson MD, Sneige N, Williamson KD, Lee AM, Katz R** (1987) Cervical carcinoma DNA content, S-fraction and malignancy grading. I. Interrelationships. Gynecol Oncol 26:41–56.

65. **Katz RL, Johnson TS, Williamson KD** (1985) Comparison of cytologic and acridine-orange flow-cytometric detection of malignant cells in human body cavity fluids. Anal Quant Cytol Histol 7:227–235.

66. **Kelsten ML, Chianese D** (1988) Multiparameter flow cytometric analysis in the detection of malignant effusions. Acta Cytol (Praha) 32:773 (abst).

67. **Kimmel M, Darzynkiewicz Z, Staiano-Coico L** (1986) Stathmokinetic analysis of human epidermal cells *in vitro*. Cell Tissue Kinet 19:289–304.

68. **Klein FA, Herr HW, Sogani PC, Whitmore WF Jr, Melamed MR** (1982) Flow cytometry of normal and nonneoplastic diseases of the bladder: An estimate of the false positive rate. J Urol 127:946–948.

69. **Klein FA, Herr HW, Sogani PC, Whitmore WF Jr, Melamed MR** (1982) Detection and follow-up of carcinoma of the urinary bladder by flow cytometry. Cancer 50:389–395.

70. **Klein FA, Herr HW, Whitmore WF Jr, Pinsky CM, Oettgen HF, Melamed MR** (1981) Automated flow cytometry to monitor intravesical BCG therapy of superficial bladder cancer. Urology 17:310–314.

71. **Klein FA, Herr HW, Whitmore WF Jr, Sogani PC, Melamed MR** (1982) An evaluation of automated flow cytometry (FCM) in detection of carcinoma in situ of the urinary bladder. Cancer 50:1003–1008.

72. **Klein FA, Melamed MR, Whitmore WF Jr, Herr HW, Sogani PC, Darzynkiewicz Z** (1982) Characterization of bladder papilloma by two-parameter DNA–RNA flow cytometry. Cancer Res 42:1094–1097.

73. **Klein FA, White FKH** (1988) Flow cytometry deoxyribonucleic acid determinations and cytology of bladder washings: Practical experience. J Urol 139:275–278.

74. **Klein FA, Whitmore WF Jr, Herr HW, Melamed MR** (1982) Flow cytometry follow-up of patients with low stage bladder tumors. J Urol 128:88–92.

75. **Klein FA, Whitmore WF Jr, Wolf RM, Herr HW, Sogani PC, Staiano-Coico L, Melamed MR** (1982) Presumptive downstaging from preoperative irradiation for bladder cancer as determined by flow cytometry: A preliminary report. Int J Radiat Oncol Biol Phys 9:487–491.

76. **Koenig SH, Brown RD, Kamentsky LA, Sedlis A, Melamed MR** (1968) Efficacy of a rapid cell spectrophotometer in screening for cervical cancer. Cancer 21:1019–1026.

77. **Koprowska I, Zipfel S, Ross A, Herlyn M** (1986) Develpment of monoclonal antibodies that recognize carcinoma. Acta Cytol (Praha) 30:207–213.

78. **Lamb D** (1967) Correlation of chromosome counts with histological appearances and prognosis in transitional cell carcinoma of the bladder. Br Med J 1: 273–277.

79. **Larsen JK, Pedersen T** (1978) Pulse-cytophotometry on bladder washings. In "Proceedings of the Third International Symposium on Pulse Cytophotometry, Vienna." Ghent: European Press pp 519–526.

80. **Leuchtenberger C, Leuchtenberger R, Davis AM** (1954) A microspectrophotometric study of the desoxyribose nucleic acid (DNA) content in cells of normal and malignant human tissues. Am J Pathol 30: 65–85.

81. **Lehnert BE, Valdez YE, Fillak DA, Steinkamp JA, Stewart CC** (1986) Flow cytometric characterization of alveolar macrophages. J Leukocyte Biol 39:285–298.

82. **Limas C, Lange PH, Fraley EE, Vessella RL** (1979) A, B, H antigens in transitional cell tumors of the urinary bladder. Cancer 44:2099–2107.

83. **Linden WA, Ochlich K, Baisch H (1979)** Flow cytometric prescreening of cervical smears. J Histochem Cytochem 27:529–535.
84. **Melamed MR (1984)** Flow cytometry of the urinary bladder. Urol Clin North Am 11:599–608.
85. **Melamed MR (1979)** Circulating cancer cells. In Koss LG (ed), "Diagnostic Cytology and Its Histopathologic Bases." Philadelphia: JB. Lippincott, pp 1105–1119.
86. **Melamed MR, Klein FA (1984)** Flow cytometry of urinary bladder irrigation specimens. Hum Pathol 15:302–305.
87. **Melamed MR, Traganos F, Sharpless T, Darzynkiewicz Z (1976)** Urinary cytology automation. Preliminary studies with acridine orange stain and flow through cytofluorometry. Invest Urol 13: 331–338.
88. **Melamed MR, Voutsa N, Grabstald H (1964)** Natural history of carcinoma in situ of the urinary bladder. Cancer 17:1535–1545.
89. **Mellors RC, Keane JF Jr, Papanicolaou GN (1952)** Nucleic acid content of the squamous cancer cell. Science 116:265–269.
90. **Murphy WM, Emerson LD, Chandler RW, Moinuddin SM, Soloway MS (1986)** Flow cytometry versus urinary cytology in the evaluation of patients with bladder cancer. J Urol 136:815–819.
91. **Noronka ABC, Richman DB, Arnason BGW (1980)** Detection of in vivo stimulated cerebrospinal fluid lymphocytes by flow cytometry in patients with multiple sclerosis. N Engl J Med 303:713–717.
92. **Pedersen T, Larsen JK, Krarup T (1978)** Characterization of bladder tumours by flow cytometry on bladder washings. Eur Urol 4:351–355.
93. **Porro G (1988)** Monoclonal antibody detection of carcinoma cells in bone marrow biopsy from breast cancer patients. Cancer 61:2407–2411.
94. **Pressman NJ, Albright CD, Frost JK (1981)** Centrifugal separation of cells in sputum specimens from patients with adenocarcinoma. Cytometry 1:260–264.
95. **Ramaekers FCS, Huysmans A, Moesker O, Schaart G, Vooijs GP, Herman CJ (1985)** Cytokeratin expression during neoplastic progression of human transitional cell carcinomas as detected by a monoclonal and a polyclonal antibody. Lab Invest 52:31–38.
96. **Ratliff JE, Klein FA, White FKH (1985)** Flow cytometry of ethanol-fixed versus fresh bladder barbotage specimens. J Urol 133:958–960.
97. **Redner A, Andreeff M, Miller DR, Steinherz P, Melamed MR (1984)** Recognition of central nervous system leukemia by flow cytometry. Cytometry 5: 614–618.
98. **Redner A, Melamed MR, Andreeff M (1986)** Detection of central nervous system relapse in acute leukemia by multiparameter flow cytometry of DNA, RNA and Calla. NY Acad Sci 468:241–255.
99. **Reiffenstuhl G, Severin E, Dittrich W, Gohde W (1971)** Die impulscytophotometric des vaginal und cervical smears. Arch Gynakol 211:595–616.
100. **Salzman GC, Crowell JM, Hansen KM, Ingram M, Mullaney PF (1976)** Gynecologic specimen analysis by multiangle light scattering in a flow sytem. J Histochem Cytochem 24:308–314.
101. **Sandberg AD (1977)** Chromosome markers and progression in bladder cancer. Cancer Res 37: 2950–2956.
102. **Sandritter W, Cramer H, Mondorf W (1960)** Zur krebsdiagnostik an vaginalen zellausstrichen mittels cytophotometrischer mesungen. Arch Gynakol 192: 293–303.
103. **Schlimok G, Funke I, Holzmann B, Gottlinger G, Schmidt G, Hauser H, Swierkot S, Warnecke HH, Schneider B, Koprowski H (1987)** Micrometastatic cancer cells in bone marrow: In vitro detection with anti-cytokeratin and in vivo labeling with anti-17-IA monoclonal antibodies. Proc Natl Acad. Sci USA 84: 8672–8676.
104. **Schneller J, Eppich E, Greenebaum E, Elequin F, Sherman A, Wersto R, Koss LG (1987)** Flow cytometry and feulgen cytophotometry in evaluation of effusions. Cancer 59:1307–1313.
105. **Sprenger E, Sandritter W, Bohm N, Wagner D, Hilgarth M, Schoden M (1972)**: Flowthrough-cytophotometry—A step towards automated cytology? Acta Cytol (Praha) 16:297–303.
106. **Staiano-Coicò L, Gottlieb AB, Barazani L, Carter DM (1987)** RNA, DNA and cell surface characteristics of lesional and nonlesional psoriatic skin. J Invest Dermatol 88:646–651.
107. **Staiano-Coico L, Gottlieb AB, Prioleau P (1988)** RNA, DNA and cell surface characteristics of basal cell carcinomas. J Invest Dermatol 90:610.
108. **Staiano-Coico L, Higgins, PJ, Kimmel M, Gottlieb AB, Pagan-Charry I, Madden MR, Finkelstein JL, Hefton JM (1986)** Human keratinocyte culture: Identification and staging of epidermal cell subpopulations. J Clin Invest 77:396–404.
109. **Staiano-Coico L, Huffman J, Wolf R, Pinsky CM, Herr HW, Whitmore WF Jr, Oettgen HF, Darzynkiewicz Z, Melamed MR (1985)** Monitoring intravesical bacillus Calmete–Guerin treatment of bladder carcinoma with flow cytometry. J Urol 133: 786–788.
110. **Staiano-Coico L, Wolf R, Darzynkiewicz Z, Whitmore WF Jr, Melamed MR (1984)** RNA and DNA content of isolated nuclei from bladder irrigation specimens as measured by flow cytometry. Anal Quant Cytol 6:24–29.
111. **Steinkamp JA, Crissman HA (1974)** Automated analysis of deoxyribonucleic acid, protein and nuclear to cytoplasmic relationships in tumor cells and gynecologic specimens. J Histochem Cytochem 22:616–621.
112. **Steinkamp JA, Hansen, KM, Wilson JS, Salzman GC (1977)** Automated analysis and separation of cells from the respiratory tract: Pulmonary characterization studies in hamsters. J Histochem Cytochem 25: 892–898.
113. **Stonesifer KJ, Xiang J, Wilkinson EJ, Benson NA, Braylan RC (1987)** Flow cytometric analysis and cytopathology of body cavity fluids. Acta Cytol (Praha) 31:125–130.
114. **Tachibano M, Chandhury M, Addonizio J, Burson ML, Chiao JW, Nagamatou GR (1985)** Heterogeneity index score (HIS). New computerized method for classification of human bladder carcinomas using flow cytometry. Urology 26:356–361.
115. **Tirindelli-Danesi D, Teodori L, Mauro F, Modini C, Botti C, Cicconette F, Stipa S (1987)** Prognostic sig-

nificance of flow cytometry in lung cancer. A five-year study. Cancer 60:844–851.

116. **To A, Dearnaley DP, Ormerod MG, Canti G, Coleman DV** (1982) Epithelial membrane antigen—Its use in the cytodiagnosis of malignancy in serous effusions. Am J Clin Pathol 78:214–219.

117. **Tribukait B** (1984) Flow cytometry in surgical pathology and cytology of tumors of the genito-urinary tract. In Koss LG, Coleman DV (eds), "Advances in Clinical Cytology", Vol 2. New York: Masson Publishing, USA, pp 163–189.

118. **Tribukait B, Esposti P** (1976) Comparative cytofluorometric and cytomorphologic studies in non-neoplastic and neoplastic human urothelium. In Gohde W, Schumann J, Buchner TH (eds), "Second International Symposium on Cytophotometry, Munster." Ghent: European Press, pp 176–187.

119. **Tribukait B, Esposti PL** (1978) Quantitative flow-microfluorometric analysis of the DNA in cells from neoplasms of the urinary bladder: Correlation of aneuploidy with histological grading and the cytological findings. Urol Res 6:201–205.

120. **Tribukait B, Gustafson H, Esposti P** (1979) Ploidy and proliferation in human bladder tumors as measured by flow cytometric DNA analysis and its relations to histopathology and cytology. Cancer 43: 1742–1751.

121. **Tribukait B, Gustafson H, Esposti PL** (1982) The significance of ploidy and proliferation in the clinical and biological evaluation of bladder tumors: A study of 100 untreated cases. Br J Urol 54:130–135.

122. **Tribukait B, Moberger G, Zetterberg A** (1975) Methodological aspects of rapid-flow cytofluorometry for DNA analysis of human urinary bladder cells. In Haanen CAM, Hillen HFP, Wessels, JMC (eds), "First International Symposium on Pulse Cytophotometry, Nijmegen." Ghent: European Press, pp 50–60.

123. **Tyrer HW, Golden JF, Vansickel MH, Echols CK, Frost JK, West SS, Pressman NJ, Albright CD, Adams LA, Gill GW** (1979) Automatic cell identification and enrichment in lung cancer. II. Acridine orange for cell sorting of sputum. J Histochem Cytochem 27:552–556.

124. **Unger KM, Raber M, Bedrossian CWM, Stein DA, Barlogie B** (1983) Analysis of pleural effusions using automated flow cytometry. Cancer 52:873–877.

125. **Valet G, Warnacke HH, Kahle H** (1984) New possibilities of cytostatic drug testing on patient tumor cells by flow cytometry. Blut 49:37–43.

126. **Voutsa NG, Melamed MR** (1963) Cytology of in situ carcinoma of the human urinary bladder. Cancer 16: 1307–1316.

127. **Walle A, Kodama T, Melamed MR** (1983) A simple density gradient for enriching subfractions of solid tumor cells. Cytometry 3:402–407.

128. **Weinstein RS, Miller AW, Coon JS** (1984) Tissue blood group ABH and Thomsen-Friedenreich antigens in human urinary bladder carcinoma. In Kurth KH, Debruyne FMJ, Schroeder FH, Splinter TAW, Wagener TDJ (eds) "Progress and Controversies in Oncological Urology." New York: Alan R. Liss, Inc., pp 249–262.

129. **Wheeless LL, Berkan TK, Patten SF, Reeder JE, Robinson RD, Eldidi MM, Hulbert WC, Frank IN** (1986) Multidimensional slit-scan detection of bladder cancer. Preliminary clinical results. Cytometry 7:212–216.

130. **Wijkstrom H, Granberg-Ohman I, Tribukait B** (1984) Chromosomal and DNA patterns in transitional cell bladder carcinoma. A comparative cytogenetic and flow-cytofluorometric DNA study. Cancer 53:1718–1723.

131. **Wijkstrom H, Gustafson H, Tribukait B** (1984) Deoxyribonucleic acid analysis in the evaluation of transitional cell carcinoma before cystectomy. J Urol 132:894–898.

132. **Winkler HZ, Lieber MM** (1988) Primary squamous cell carcinoma of the male urethra: Nuclear deoxyribonucleic acid ploidy sutdied by flow cytometry. J Urol 139:298–303.

133. **Zante J, Schumann J, Barlogie B, Gohde W, Buchner Th** (1976) New preparing and staining procedures for specific and rapid analysis of DNA distributions. In Gohde W, et al (eds), "Second International Symposium on Pulse Cytophotometry." Ghent: European Press, pp 97–106.

134. **Zur Hausen H** (1986) Intracellular surveillance of persisting viral infections. Human genital cancer results from deficient cellular control of papilloma virus gene expression. Lancet 2:489–491.

39

Single- and Multiparameter Analysis of the Effects of Chemotherapeutic Agents on Cell Proliferation In Vitro

Frank Traganos

Investigative Cytology Laboratory, Memorial Sloan-Kettering Cancer Center, New York, New York 10021

It is generally assumed that mammalian tumors form as a result of inappropriate, uncontrolled cell growth arising out a defect in the ability of a cell to respond to normal growth control stimuli. The cause(s) of this phenotypic derangement may include some of the following: (1) genetic damage or mutation induced by chemicals or irradiation; (2) expression of abnormal genetic information induced by viral oncogenes; (3) gain or loss of chromosomal material by neoplastic cells; (4) derepression of oncofetal genes that are present but normally silent in adult cells; and/or (5) alteration in post-transcriptional processing of critical cellular macromolecules [125]. Whatever the cause, the tumor cells attain some selective advantage for growth over normal cells, argueably as a result of their inability to differentiate.

Ideally, the ultimate technique for ridding an organism of tumor cells is to use an agent that interferes in some way with the ability of the cell to proliferate. In principle, this is a relatively simple task, since a variety of agents are available that are extremely toxic to proliferating cells. In practice, however, the problem is much more complex. Thus, while tumor cells grow without the restraints of the normal feedback control mechanisms, their growth rate is rarely faster than that of normal cells, a significant proportion of which are cycling in critical tissues such as the bone marrow and gastrointestinal tract. In addition, while some tumors have a high proportion of actively proliferating cells (e.g., leukemias and lymphomas), most malignant neoplasms are "solid" tumors that have a relatively low growth fraction and are therefore less susceptible to treatment with agents that affect, for example, the ability of the cells to synthesize DNA. Thus, it has become all important to both understand the factors controlling cell growth and differentiation, and to analyze the mechanisms by which various prospective agents perturb or inhibit such growth.

THE CELL CYCLE

Based on the early work of Howard and Pelc [90], it has been possible to describe cell growth in terms of specific phases through which a cell cycles. Thus, a cell must duplicate its complement of DNA during the DNA synthesis (S) phase prior to division of its DNA and cytoplasmic material into two daughter cells during mitosis (M phase). In most mammalian cell systems, "gaps" have been observed between the termination of DNA synthesis and mitosis (G_2), and between division and the subsequent reinitiation of DNA synthesis in the daughter cells (G_1). In some instances, either as a result of a genetically programmed event or in reaction to the presence or absence of a stimulant in the environment, the cell will exit the cycle, generally following mitosis, and reside in a quiescent, nondividing compartment variously referred to as G_0, G_{1Q}, G_{1D}, and so forth [14,126] (see also Chapters 23,24, this volume). As a practical matter it is possible to discriminate morphologically between mitotic cells and the remaining interphase population. Using microspectrophotometric techniques to measure cellular DNA content [25], or by visualizing the incorporation of a radioactive component (tritiated thymidine) into cells during DNA synthesis [128], it is possible to discriminate S phase cells within the interphase population. However, these techniques are tedious and time consuming. With the development of flow cytometry [66,94,180] it became possible to measure the DNA content, chromatin structure and other constituents of individual cells rapidly and accurately. This provided for the first time, the flexibility, speed and power to extend kinetic analyses to everyday use in the laboratory and clinic [84].

DNA DISTRIBUTIONS AND THE CELL CYCLE

A single DNA distribution (frequency histogram) contains detailed information about the growth kinetics of an asynchronously growing cell population [84] (see Chapter 23, this volume). With the development and application of techniques for staining cells in suspension with DNA-specific fluorescent probes [33,37,111] it became possible to obtain a sequence of DNA histograms of cells taken from cultures to

Flow Cytometry and Sorting, Second Edition, pages 773–801

TABLE 1. Effect of Continuous Exposure of Exponentially Growing FL cells to CI-921

Drug concn. (nM)	Exposure* (hr)	DNA distribution			
		G_1	S	G_2M	$>G_2M^{\dagger}$
0	0	36.0	47.5	16.5	–
0	4	34.7	47.5	17.8	–
1	4	33.7	50.2	16.2	–
10	4	16.1	59.0	24.9	–
50	4	2.8	70.8	26.4	–
0	8	32.1	50.4	17.4	–
1	8	37.1	44.7	18.2	–
10	8	9.9	40.3	49.8	–
50	8	0.7	50.6	48.8	–
0	24	34.7	51.4	13.9	–
1	24	36.3	49.5	14.2	–
10	24	8.4	28.6	43.8	19.2
50	24	0	5.3	72.9	21.8

*Cells were continuously exposed to the drug, an aliquot removed from culture and analyzed by flow cytometry at the indicated times.
†Polyploid cells.

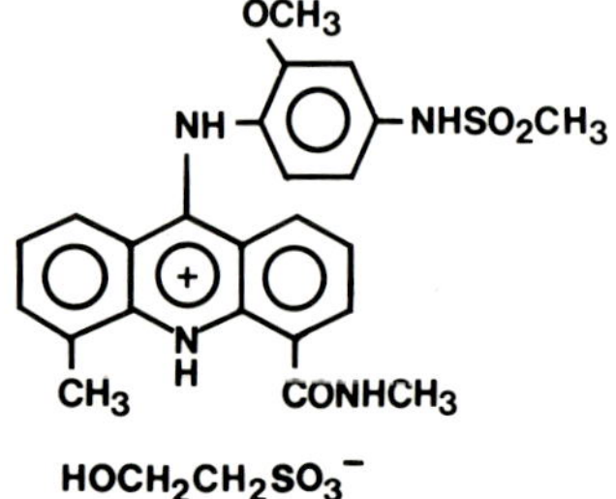

Fig. 1. Chemical structure of CI-921.

which various chemotherapeutic agents had been added. Thus, DNA-distribution sequences could be used to study the dynamics of cell cycle traverse, or to study the kinetic response of a cell population to a perturbing agent or condition.

ANALYSIS OF DRUG ACTION USING SINGLE-PARAMETER DNA DISTRIBUTIONS

While it is not possible to detail all studies that have used flow cytometry to assess the effects of chemotherapeutic agents on cell proliferation, a brief overview with representative examples can be provided. Some of the earliest studies utilized an adaptation of the fluorescent Feulgen–Schiff procedure for staining of cells in culture [158,159,187]. However, the procedure was time consuming and the repeated centrifugations required led to significant cell loss [37]. As a result, simpler staining techniques which could be performed on alcohol-fixed or detergent-treated whole cells [3,11,33, 38,66,80,136,147,150,188] or nuclei [101,182] were developed using DNA intercalating or base specific dyes [33,37,111] (see Chapters 13,14). In most studies employing these techniques, chemotherapeutic agents were administered for varying lengths of time to cell cultures and the DNA distributions analyzed periodically [1,2,6–9,12,13,19,23,26, 40,65, 67–71, 75,77, 79, 86, 93, 96,102–107,116-118,121, 122,124,129,131,134,142,148,151–154,156,157,161,181, 185]. In combination with cell counts and/or clonability studies, the alterations in cell cycle distribution of drug-treated cells can yield information on the cell cycle phase specificity (if any) of the drug.

More detailed cell kinetic information can be obtained by addition of a stathmokinetic agent (i.e., inhibitor of cell division). By blocking cell exit from a particular cell cycle compartment (e.g., mitosis), the cell doubling time, mean residence time in G_1 and the duration of the S and G_2M phases can be determined with efficiency and accuracy by flow cytometry [100,109]. Changes in these values induced by addition of a prospective chemotherapeutic agent may then be used to pinpoint drug action.

Synchronized cell populations may also be useful in the study of drug effects in vitro [130,160]. Asynchronously growing cultures provide a multiplicity of targets for drug action, since cells in various phases may have varying degrees of sensitivity to a drug. The redistribution of an asynchronous population following drug treatment is likely to depend upon the fraction of cells existing in each cell cycle phase upon addition of the drug, the duration of each phase relative to the duration of drug exposure and the point of action of the drug. Exposing relatively uniform populations of cells to a drug (i.e., in terms of cell cycle phase; synchronized cells) simplifies the interpretation of cell cycle distributions following drug exposure.

Another attempt at increasing the information available from a single-parameter DNA histogram analysis of cell proliferation is by use of the thymidine analog bromodeoxyuridine (BUdR) in combination with the fluorescent DNA stain Hoechst 33258 (HO33258) (see Chapters 13,14,23,24, this volume). Whereas HO33258 can be used as a stain for total DNA content, its fluorescence is quenched by the presence of BUdR [20,112] and thus a cell synthesizing DNA will not have the expected increase in fluorescence. Upon division daughter cells will have lowered fluorescence intensity enabling one to follow drug effects from one cycle to the next. Early flow cytometric studies utilized this technique to follow the effects of irradiation on cell cycle progression [18,88,119].

THE MULTIPARAMETER APPROACH

While DNA distributions have proven useful in monitoring the cell cycle effects of drugs, DNA content alone is not sufficient to detect subtle cell cycle effects associated with unbalanced growth or to provide information on cell subpopulations in complex heterogeneous systems. Thus, cells in G_2 phase cannot be discriminated from mitotic cells nor can quiescent or slowly growing populations (e.g., G_0, G_{1Q}) be detected by analysis of DNA content alone. Occasionally, drugs have differential effects on DNA versus RNA or protein synthesis (see below) that lead to unbalanced growth which would be undetectable from single-parameter DNA distributions. Thus, multiparameter measurements of drug effects can provide all the data present in single-parameter DNA distributions and a good deal more. Below is a brief description of some of the multiparameter approaches devel-

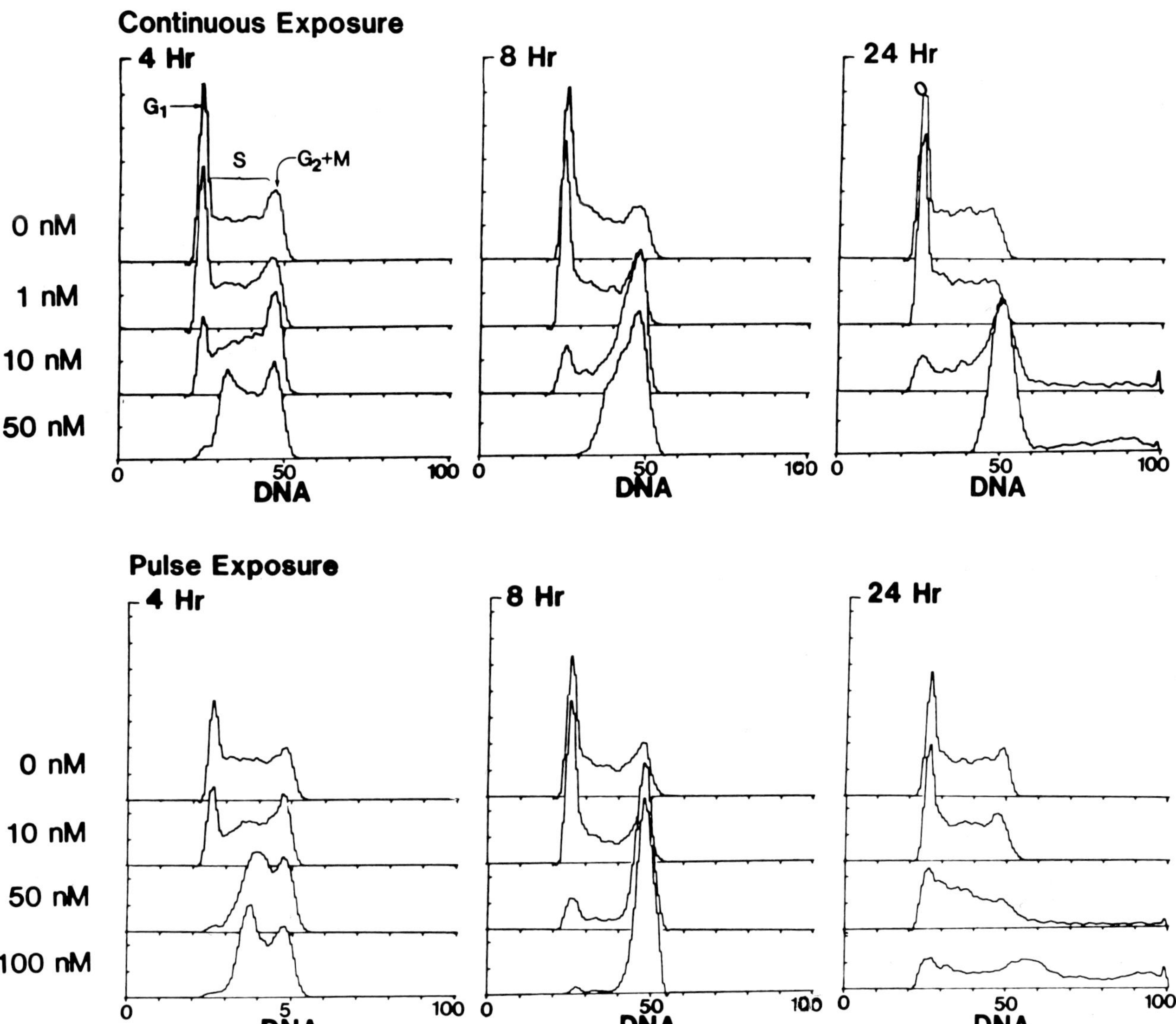

Fig. 2. Single-parameter (DNA) histograms of FL cells exposed for a 1 hr pulse (bottom) or continuously (top) to CI-921. All frequency histograms represent the green fluorescence distribution of AO stained cells. In the case of pulse-exposed cultures, the cells, after a 1 hr incubation with CI-921, were rinsed twice with HBSS and then recultured in fresh medium without drug. The relative proportions of cells in each cell cycle phase at each time point are indicated in Tables 1 and 2.

oped for analysis of cell cycle kinetics by flow cytometry, and their application.

DNA versus Protein Content

A variety of staining techniques have been developed to measure DNA and total protein content in intact cells and isolated nuclei. Most use fluorescein isothiocyanate (FITC), a green fluorescing dye, to stain protein and a red fluorescing DNA stain such as EB or PI [34,36,39,76,81,123,132,144, 146]. With multiple laser systems, one can use such protein stains as rhodamine 640 and 4-acetamido-4′-isothio-cyanate-stilbene-2,2′-disulfonic acid (SITS) in combination with UV- or visible light-excited DNA stains [146]. However, with single as well as multiple laser systems analysis of DNA and various combinations of other parameters is possible, including protein content, nuclear and cytoplasmic diameters, nuclear to cytoplasmic ratio, and phagocytic capacity [36,144–146].

DNA versus RNA Content

Simultaneous measurement of both DNA and RNA content in individual cells has been accomplished by differentially staining the two nucleic acids with the metachromatic dye acridine orange (AO) [41,56,169,174] (see also Chapter 15, this volume) or by dye combinations such as HO33342 for DNA and pyronin Y (PY) for RNA [137,149]. The latter

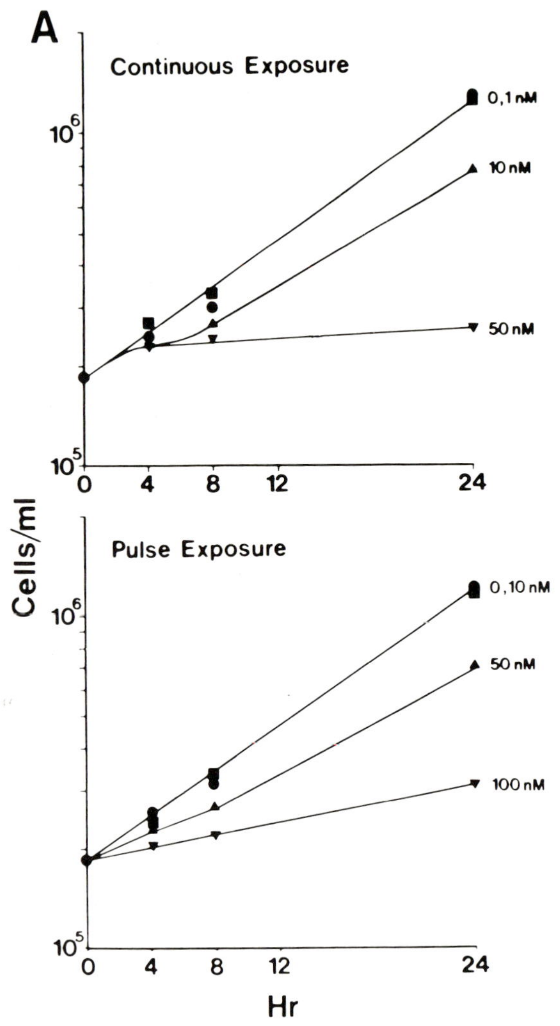

Fig. 3. **A:** Cell counts of FL cells exposed to CI-921 as in Figure 2. Cells were seeded at a concentration of 1.8 × 10^5/ml and growth monitored at the indicated times by counting Trypan Blue negative cells. The dose which inhibited cell growth by 50% at 24 hours (ID_{50}) was 9.5 and 67.5 nM for continuous and pulse exposure, respectively. **B:** Colony formation of exponentially growing and stationary Chinese hamster ovary (CHO) cells exposed to CI-921 for a 1-hour pulse. Cells were treated, washed free of drug, trypsinized and plated at a concentration of 200 cells in 35-mm tissue culture dishes. Following incubation for 1 week, the number of colonies consisting of 50 or more cells were counted and compared to control colony formation (cf. ref [162] for details). The curves represent the drug concentration at which colony formation was inhibited by 50% (ICF_{50}). The ICF_{50} values were 80 and 190 nM for exponential and stationary-phase CHO cells, respectively.

technique requires two-laser excitation for the UV and green excited dyes, respectively [137].

Using the protein or RNA parameter it became possible, for the first time, to characterize different functional states within a single cell cycle phase, i.e., quiescent or differentiated cells with the same DNA content as actively cycling G_1 cells but different RNA content [4,16,17,48,53,56,63, 64,108,137,166,170,174,183,184]. Also, since the rate of cell progression through the cell cycle is related to RNA content [42,46,47,78], differential inhibition of DNA or RNA synthesis, leading to unbalanced growth, could be analyzed flow cytometrically [155,168].

DNA versus Chromatin Structure

The metachromatic properties of AO can also be used to stain DNA partially denatured by heat or acid [50,54,55, 57,58] (see Chapter 16). The green fluorescence associated with binding to double-stranded nucleic acids and the red luminescence due to dye association with single-stranded nucleic acids [95] permits comparison of total DNA (green plus red luminescence) with the degree of DNA denaturation (red divided by green plus red luminescence; α_t) in RNase treated cells. The sensitivity to DNA denaturation (α_t value) is greatest for cells with condensed chromatin and provides an automated technique for scoring such cells (i.e., mitotic cells, G_0 cells) by flow cytometry [41,42,48,49,51,53].

DNA versus Other Cellular Attributes

Other cellular attributes which have been combined with measurement of cellular DNA content to provide multiparametric data include light scatter (complex signals dependent upon combinations of cell size, shape, refractive index, and internal structure) [163], Coulter volume [35,92,145,174,

179], membrane antigens [109,110,113,114,135], total cell surface protein (85), nuclear and nucleolar antigens [10,29] and hormone receptors [10]. Quenching of HO33258 fluorescence by BUdR has also been combined with staining of total DNA by EB to provide two-parameter data, permitting one to follow cell growth over several cell cycles [21]. BUdR incorporation can also be detected using FITC-labeled antibodies directed against the base and, when combined with PI staining of total DNA, provides a powerful tool for cell cycle analysis (see Chapter 23, this volume). Additional multiparameter approaches include correlated analysis of three or more parameters such as DNA, cell surface receptors, and light scatter [22,133] or DNA, RNA, and protein [32,146].

Several of these approaches have been applied to the in vitro analysis of drug effects on cell growth [44,59–62,72, 73,75,98,154,162,164,165,167,171–173,175–178]. These techniques often provide a detailed picture of the cell cycle phase specificity of the agent under study as well as its effects on cellular metabolism and may provide a hint as to whether the agent interacts with DNA in intact cells.

To give the reader a feeling for the kinds of information available from flow cytometric analysis of drug effects on cell cycle kinetics, a rather comprehensive analysis of a single, new prospective antitumor agent follows. A variety of experimental approaches and flow cytometric techniques are presented both to illustrate the quality and quantity of data available and to compare the results of various approaches, consistent with a specific mode of action of the drug. However, before presenting the flow cytometry data, the reader is provided with a brief review of the general classes of antitumor agents presently in use and under study, as well as the cell systems and staining procedures used.

ANTICANCER DRUGS

Generally, antitumor drugs can be divided into five categories: (1) alkylating agents, (2) antimetabolites, (3) antibiotics, (4) antimitotics, and (5) miscellaneous agents, including hormones and enzymes. Alkylating agents act by binding covalently through the alkyl (or substituted alkyl) group to a cellular consistuent, predominantly DNA. There are five classes of alkylating agents: nitrogen mustards, nitrosoureas, triazenes, methane sulfonic acid esters, and ethylenimines [125]. The nitrogen mustard, methclorethamine, was the first chemical agent used to systematically treat cancer patients [82]. In addition to methclorethamine, some of the more clinically useful members of this group of anticancer agents include melphalan, cyclophosphamide, chlorambucil, carmustine (BCNU), comustine (CCNU), busulfan, dacarbazine, and thiotepa. Most alkylating agents are considered cell cycle phase nonspecific. Although these agents interact with the DNA of resting (quiescent) cells as well as with DNA of proliferating cells during all cell cycle phases, it is clear that cells undergoing DNA replication (S phase cells) are the most sensitive. It is also assumed that the longer the exposed cell has before initiation of DNA synthesis, the more time it has to repair the damage to DNA caused by the alkylating agent.

The second group of anticancer agents are the so-called antimetabolites. Antimetabolites structurally resemble a natural metabolite necessary for cell function and interfere with its normal utilization. There are three types of antimetabolites: antifolates, antipurines, and antipyrimidines. The first antimetabolite used clinically was the antifolate aminopterin used by Farber et al. [74] in 1947. Other useful antifolates include methotrexate and dichloromethotrexate. Among the first antipurines in clinical use was β-mercaptopurine. Another active analog is 6-thioguanine. Antipyrimidines were developed in an attempt to synthesize analogues that could inhibit uracil utilization by tumors; tumors use more uracil than orotic acid for nucleic acid synthesis, while the opposite is true of normal tissue. Heidelberger [87] introduced the first substituted pyrimidine into the clinic in the form of 5-fluorouracil (5-Fu). Among the more active of the recently developed antipyrimidines are cytosine arabinoside (ara C) and 5-azacytidine, both mainstays of antileukemic therapy [125]. Antifolates normally act by inhibiting the enzyme dihydrofolate reductase which leads to inhibition of DNA, RNA, and protein synthesis, although their major action is probably inhibition of thymidylate synthesis. Antifolates such as methotrexate, which inhibits RNA and protein synthesis thereby slowing the rate of entry of cells in S phase, is considered a "self-limiting" S phase specific drug. Most antipurines such as 6-mercaptopurine must be converted to the nucleotide to be active. The agents appear to have several mechanisms of action including inhibition of purine biosynthesis. The antipyrimidines such as 5-fluorouracil, like the antipurines, must be converted to the nucleotide for activity. The mechanism of action of antipyrimidines is generally to inhibit DNA synthesis through thymidylate synthesis inhibition [125]. Both antipurines and antipyrimidines tend to be more toxic to proliferating cells and are therefore considered cell cycle specific. They do not appear, however, to be phase specific, although cells in G_1 or S may be somewhat more sensitive than cells in other phases.

Antitumor antibiotics were the next class of drugs to be used in the clinic. Farber in 1954 used actinomycin D (AMD) to treat Wilms' tumor in children [89]. Almost all cytotoxic antibiotics act by either directly or indirectly altering DNA function. Thus, effective antitumor antibiotics such as AMD, Adriamycin, and daunorubicin bind noncovalently to DNA, acting as intercalating agents; mithramycin also binds noncovalently to DNA but does not intercalate; mitomycin C binds covalently to DNA, and bleomycin causes DNA strand scission [125]. Antibiotics that intercalate into DNA tend to inhibit cell growth by interfering with either DNA synthesis or transcription. In addition to altering chromatin structure, these agents can cause chromosome damage (strand breaks). Thus, AMD, an RNA synthesis inhibitor, is neither cell cycle nor phase specific while the anthracycline antibiotics such as adriamycin and daunorubicin appear to be more effective on cycling cells and to be somewhat phase specific (S, G_2M). Both mithramycin and mitomycin C inhibit nucleic acid synthesis. However, mithramycin and its analogs are base specific and block RNA synthesis to a greater extent than DNA synthesis while mitomycin C crosslinks DNA like an alkylating agent and selectively inhibits DNA synthesis. Bleomycin, like mitomycin C, profoundly inhibits DNA synthesis while RNA and protein synthesis are much less affected, resulting in a slowdown in cell progression through S and into G_2 phase.

The vinca alkaloids, vincristine and vinblastine, represent the first clinically useful antimitotic agents to be introduced. These "spindle poisons" interfere with cell division by preventing polymerization of tubulin, the major protein of microtubules that form the mitotic spindle. Thus, both agents inhibit division, although under appropriate conditions the blockade is reversible. Other antimitotics include the anti-inflammatory agent colchicine (colcemid) and the podophyllotoxins VP 16 and VM 26, all of which act by binding to tubulin. However, VP 16 and VM 26 act in late S and/or G_2 to prevent the entry of cells into mitosis [31].

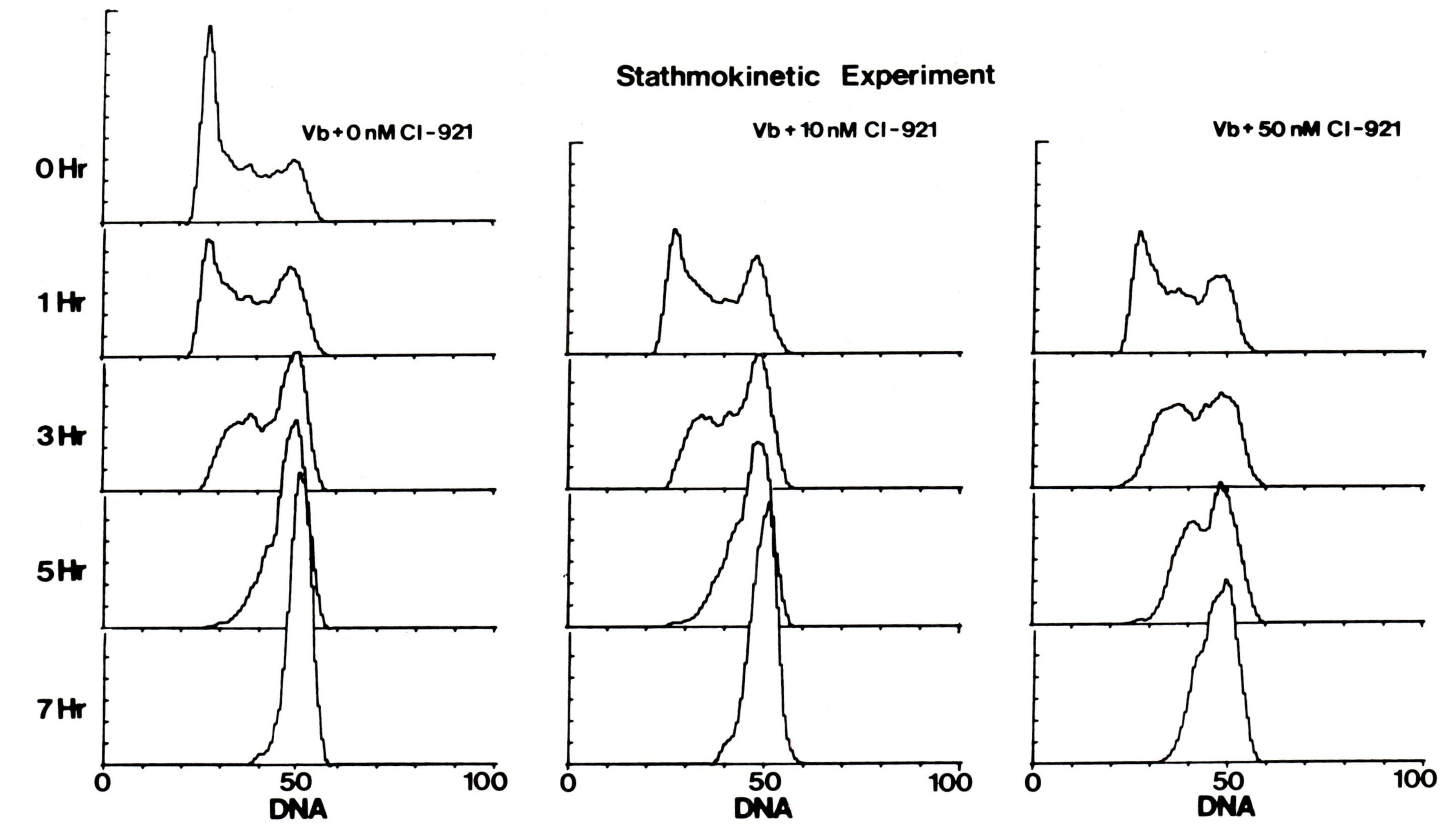
Stathmokinetic Experiment
Vb + 0 nM CI-921
Vb + 10 nM CI-921
Vb + 50 nM CI-921
0 Hr
1 Hr
3 Hr
5 Hr
7 Hr
0
50
100
DNA

Some of the miscellaneous antitumor agents include the enzyme L-asparaginase which is useful against tumors that lack asparagine synthesis activity; hydroxyurea, an inhibitor of ribonucleoside reductase, which inhibits DNA synthesis resulting in cell accumulation at the G_1-S border and in S phase; the platinum compound, cis-diamminodichloroplatinum II, which inhibits DNA synthesis but blocks cells in all phases of the cell cycle; the EDTA-like analogue Razoxane (ICRF-159), which inhibits cell division but allows DNA synthesis to continue, resulting in formation of multinucleated cells; the anthracenedione derivative, mitoxantrone, a powerful intercalator which slows progression of cells through S and blocks them in G_2 phase; m-AMSA, an acridine derivative which also appears to act by intercalation into DNA; and hormones such as prednisolone, limited in use to specific tissues that normally contain cells with receptors to mediate their action [125].

Antitumor Agent CI-921

In order to provide a degree of continuity to the discussion of drug effects on cell proliferation, the majority of data presented in this Chapter will be from the study of a prospective antitumor agent recently synthesized by a group in New Zealand [5]. This new agent is a derivative of the DNA intercalating antitumor drug, amsacrine (NSC 249992) [23]. The agent, N-5-dimethyl-9-[(2-methoxy-4-methylsulfonylamino)phenylamino]-4-acridinecarboxamide (CI-921; NSC 343499) also binds DNA by intercalation, is slightly more lipophilic and has a higher association constant for DNA than amsacrine. CI-921 shows superior activity against several model tumors as compared with amsacrine, daunorubicin, and Adriamycin. Preliminary studies [5] have suggested that, as with most intercalators which are considered S phase active agents, cells become blocked in G_2 phase at subcytotoxic concentrations. The CI-921 used in this study was a generous gift of Dr. Bruce C. Baguley (Cancer Research Laboratory and Department of Pathology, University of Aukland Medical School, Aukland, New Zealand) and its structure is presented in Figure 1. A report on this work was been published [164a].

CELL SYSTEMS

All the data presented in this chapter were derived from experiments on two cell systems, Chinese hamster ovary (CHO) cells and Friend erythroleukemia (FL) cells. Though both are immortal cell lines, they differ in several respects; the CHO cells grow attached to a substrate, are easily synchronized by mitotic detachment and have a high cloning efficiency, while the FL cells grow in suspension and in the presence of an appropriate stimulus will differentiate [170,172].

Fig. 4. Single-parameter (DNA) distribution of exponentially growing FL cells exposed to both a stathmokinetic agent (Vinblastine) and 0, 10 or 50 nM CI-921. The AO (green fluorescence) DNA distributions were obtained as in Figure 2. Vinblastine (0.05 μg/ml) was added to all cultures at t = 0. Two separate cultures received either 10 or 50 nM CI-921 also at t = 0. Samples were removed at 2-hour intervals beginning at 1 hour after drug addition. The accumulation of cells in G_2M in the presence of vinblastine and the absence or presence of CI-921 can be observed.

CHO Cells

A detailed description of the growth conditions employed for CHO cells is presented elsewhere [172]. Because CHO cells grow attached to a substrate it is a simple matter to grow the cells to high density without fear of exhausting the nutrients in culture; they can be readily refed by pouring off the spent medium and adding fresh medium. The result of growth to high cell density is a partial synchronization of most (>80%) of the cells in a subcompartment of G_1 characterized by a lower mean RNA and protein content compared to G_1 cells from exponentially growing cultures. Cell synchronization in mitosis (see below) can be easily accomplished, since mitotic and immediately postmitotic G_1 cells are only loosely attached to the substrate and may be easily dislodged by mechanically shaking the culture vessel. Large numbers of these synchronized cells can be collected and used to provide relatively homogeneous populations of M, G_1, S, or G_2 cells for drug experiments.

FL Cells

This cell line is not easily synchronized except by density gradient centrifugation or elutriation. However, the fact that FL cells grow in suspension facilitates kinetic experiments of drug action, since aliquots can be easily withdrawn from a common culture vessel at numerous times after addition of drug and culture to culture variations are avoided. Additionally, the culture system is uniquely suited to test for agents that cause cellular differentiation in vitro. As an example, FL cells respond by loss of proliferative activity, reduction in RNA and protein content and synthesis of hemoglobin, all of which can be followed flow cytometrically [164,175].

Staining Techniques

DNA and RNA Content. Simultaneous staining of DNA and RNA with AO has been detailed elsewhere [56,169] (see Chapt. 15). Cells removed directly from cultures at the indicated time points are made permeable by addition of 0.2 ml of cell suspension to 0.4 ml of a detergent solution containing 0.08 N HCl, 0.15 M NaCl, and 0.1% Triton X-100. After 30 seconds to allow for the cells to become permeable, chromatographically purified AO is added at a concentration of 6 μg/ml in 0.2M Na_2HPO_4–0.1 M citric acid buffer (pH 6.0) containing 1 mM EDTA. Under these conditions, the dye to DNA molar ratio (in permeable cells) is greater than 2, and the interaction of the dye with DNA results in green luminescence with a maximum emission at 530 nm, whereas interaction with cellular RNA gives red luminescence with maximum emission at 640 nm [95]. Studies have shown that these staining conditions provide green fluorescence proportional to DNA content [30], while the red luminescence, after appropriate subtraction of nonspecific luminescence, is stoichiometric for RNA [15].

DNA and Protein Content. After overnight fixation in 70% ethanol, the cells may be stained for DNA and protein, respectively, in a solution containing 40 μg PI/ml, 0.05 μg FITC/ml, and 40 μg RNase/ml in Tris buffer (pH 8.5), as adapted from a previously published report [36]. PI-FITC stained cells fluoresce red in proportion to their DNA content (after enzymatic digestion of double-stranded RNA) and green in proportion to their protein content.

Acid Denaturation of DNA. Both control and drug-treated FL cells are washed in Hank's balanced salt solution (HBSS), resuspended in 1 ml HBSS, and fixed in 9 ml ice-cold acetone–alcohol (a 1:1 mixture by volume of 70% ethanol and acetone). Following overnight fixation at 4°C, cell aliquots

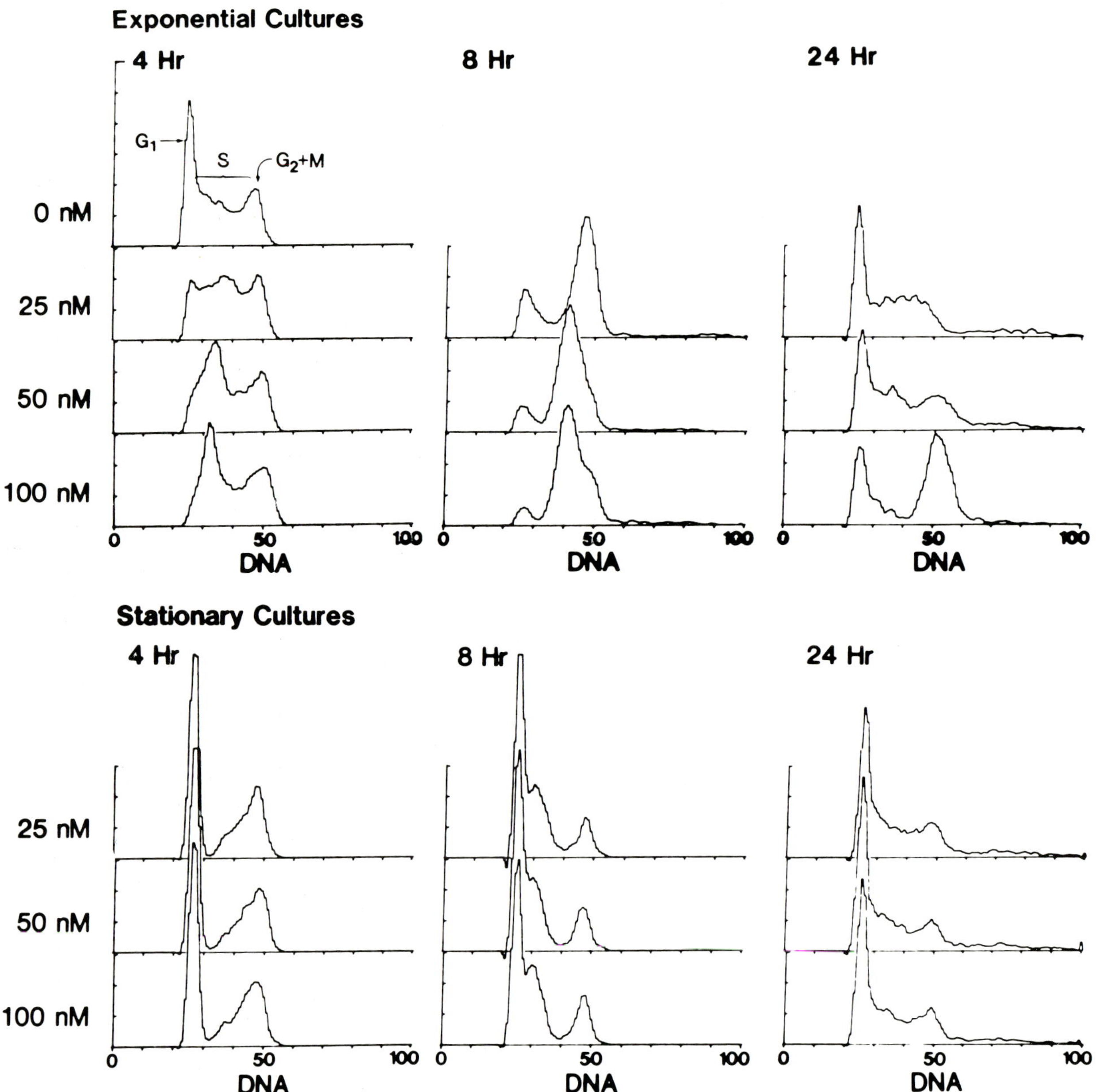

Fig. 5. Single-parameter (DNA) distributions of CHO cells exposed to a 1-hour pulse of CI-921 while in exponential (top) or stationary (bottom) growth. The frequency histograms are the same as in Figure 2. Stationary cultures were obtained by growing CHO cells to confluence and refeeding with fresh medium for 2 days. Next, cells were treated with medium alone or drug for 1 hour, washed free of drug, trypsinized, and replated at approximately 2 × 10^5 cells/ml in multiple dishes. At each time point, cells were harvested by trypsinization, stained with AO and the DNA (green fluorescence) distributions obtained.

are centrifuged twice and resuspended in HBSS to which 1,000 U of RNase is added. After a 1-hour incubation in the enzyme at 37°C, 0.1 ml of cell suspension (0.5–1 × 10^6 cells) is mixed with 0.4 ml of a 0.1 M KCl–HCl buffer (1:1) (pH 1.5) for 30 sec. The cells are then stained by addition of 2 ml of AO solution (4 μg/ml) in 0.2M Na_2HPO_4–0.1 M citric acid buffer at pH 2.6. Under these staining conditions, green fluorescence is proportional to double stranded DNA while red luminescence is proportional to single-stranded DNA, the RNA having been enzymatically digested [41]. Thus, total fluorescence (red plus green) is proportional to DNA content per cell while the α_t value (red divided by red plus green fluorescence) is related to the degree of sensitivity of the cells to acid-induced denaturation (see Chapter 16,

TABLE 2. Effect of a 1-hour Pulse Exposure of CI-921 on Exponentially Growing FL Cells

Drug Concn. (nM)	Measurement* (hr)	DNA Distribution			
		G_1	S	G_2M	$>G_2M^{\dagger}$
0	4	29.0	51.6	19.4	–
10	4	20.1	56.2	23.7	–
50	4	2.0	67.7	30.3	–
100	4	1.7	67.9	30.4	–
0	8	38.0	44.6	17.4	–
10	8	45.5	33.6	20.9	–
50	8	11.1	27.5	61.4	–
100	8	1.5	17.7	80.8	–
0	24	31.8	48.6	19.6	–
10	24	32.5	48.8	18.7	–
50	24	22.1	47.8	15.5	14.6
100	24	13.9	24.6	27.6	33.9

*Cells were exposed to the drug for 1 hour, washed, and replated in fresh drug free medium. An aliquot was removed from culture and analyzed by flow cytometry at the indicated times.
†Polyploid cells.

this volume). All samples were run and analyzed on an Ortho FC 200 flow cytometer interfaced to a Data General Nova minicomputer for which data collection and analysis facilities have been described [139].

SINGLE-PARAMETER ANALYSIS

Continuous Exposure

DNA Distributions. One of the simplest methods for determining the effect of an agent on cell cycle progression is to expose exponentially growing cultures to various concentrations of the drug and then examine the redistribution of cells within the cell cycle after different periods of incubation. It is important in these studies to choose drug concentrations that are not likely to produce large numbers of dead or damaged cells. Partially disrupted cells may manifest lowered DNA fluorescence, skewing the distribution toward G_1 phase. Complete cell disintegration as a result of drug treatment may occur randomly or be cell cycle phase specific such that the redistribution of cells throughout the cycle will be miscalculated (e.g., underestimated for one phase and overestimated for others). In the present instance, prior studies [5] and preliminary experiments established a range of drug concentrations which, under each set of incubation conditions, would not give rise to significant cytotoxicity over the course of the experiment.

The effect of continuous exposure of exponentially growing FL cells to various concentration of CI-921 is shown in the top of Figure 2 and in Table 1. At a concentration of 1.0 nM, CI-921 had virtually no effect on the DNA distribution at any of the measurement times. However, 10 nM CI-921 caused an obvious redistribution of FL cells within 4 hours; cells accumulated in S and G_2M phase. The accumulation of cells in G_2M was even more pronounced by 8 hr. Following 24-hour incubation, the data became more difficult to interpret. As shown in Figure 2 and Table 1, the percentages of cells in S and G_2M phase diminished between 8 and 24 hours, while some cells either remained in G_1 or entered G_1 upon dividing. In addition, a considerable number of higher ploidy level cells appeared.

There are several possible explanations for the observed change in cell cycle distribution: (1) some cells escaped the G_2M block to divide and re-enter the cell cycle, (2) some cells, unable to divide, proceeded to grow at a higher ploidy level, or (3) some combination of both of these possibilities occured. These possibilities can be tested in several ways.

The effects of continuous exposure of exponentially growing FL cells to 50 nM CI-921 are relatively easily interpreted. By 4 hours, cells had accumulated in G_2M as well as in early S phase. Seemingly few, if any, cells divided and by 8 hours, most cells accumulated in late S and G_2M phase. Finally, by 24 hours, most cells were observed in what operationally must be defined as G_2M, with 21.8% of the cells having entered a higher ploidy (Table 1).

Cell Counts. Interpretation of complex changes in cell cycle distributions occurring as a result of drug action can often be simplified by utilizing information derived from cell counts. Figure 3A (top) illustrates the growth of FL cells in the absence and presence of CI-921 at the concentrations and time points displayed in Figure 2 (top). The observation that 1 nM CI-921 had little or no effect on FL cell growth can be confirmed by the cell counts. The growth curve for cells treated with 10 nM CI-921 suggested nearly normal growth, followed by a slowdown between 4 and 8 hours and then more rapid growth between 8 and 24 hours. The observation that the cell number increased over the time course of the experiment suggests that the G_1 cells observed in the histograms are likely to have arisen as a result of cell division (i.e., a leaky G_2M block), albeit at a much slower rate than normal, and were not the result of a block in early G_1 phase, which had frozen some 8 to 9% of the cells in that cell cycle compartment. The fact that the slopes of the growth curves between 8 and 24 hours are lower than for untreated cells was very likely the result of both a continued slowdown in cell progression through S and especially G_2M, and the fact that some cells did not divide but continued to cycle at a higher ploidy level. At 50 nM CI-921, cell growth was somewhat slowed over the first 4 hours, after which growth was almost completely inhibited. This observation would explain the absence of substantial numbers of G_1 and early S phase cells after 4-hour treatment; cells cycling at a higher ploidy level (24 hours), not having divided, would not add to the cell number.

Stathmokinetic Experiment. While it is absolutely essential to obtain cell growth curves when testing for cell cycle effects, the use of this information to explain changes in cell cycle distributions does not always provide unequivocal results. This is especially true when the question concerns partial cell blocks, for instance in G_1 phase. However, under appropriate circumstances, by including a stathmokinetic agent (e.g., a drug which inhibits completion of mitosis) it can be determined whether cells have been blocked in G_1 or have escaped the effects of the drug early on, divided, and re-entered G_1 phase. Thus, the immediate addition of vinblastine, a mitotic inhibitor, to exponentially growing FL cultures grown in the absence or presence of various concentrations of CI-921 provides evidence that some cells escape the G_2 block, divide and enter G_1 at 10 nM but not at 50 nM CI-921 (Fig. 4).

Pulse Exposure

DNA Distributions. Except when administered as a slow infusion, most drugs have a relatively short serum half-life such that a pulse exposure of cells to a drug more closely approximates the in vivo circumstance than does continuous

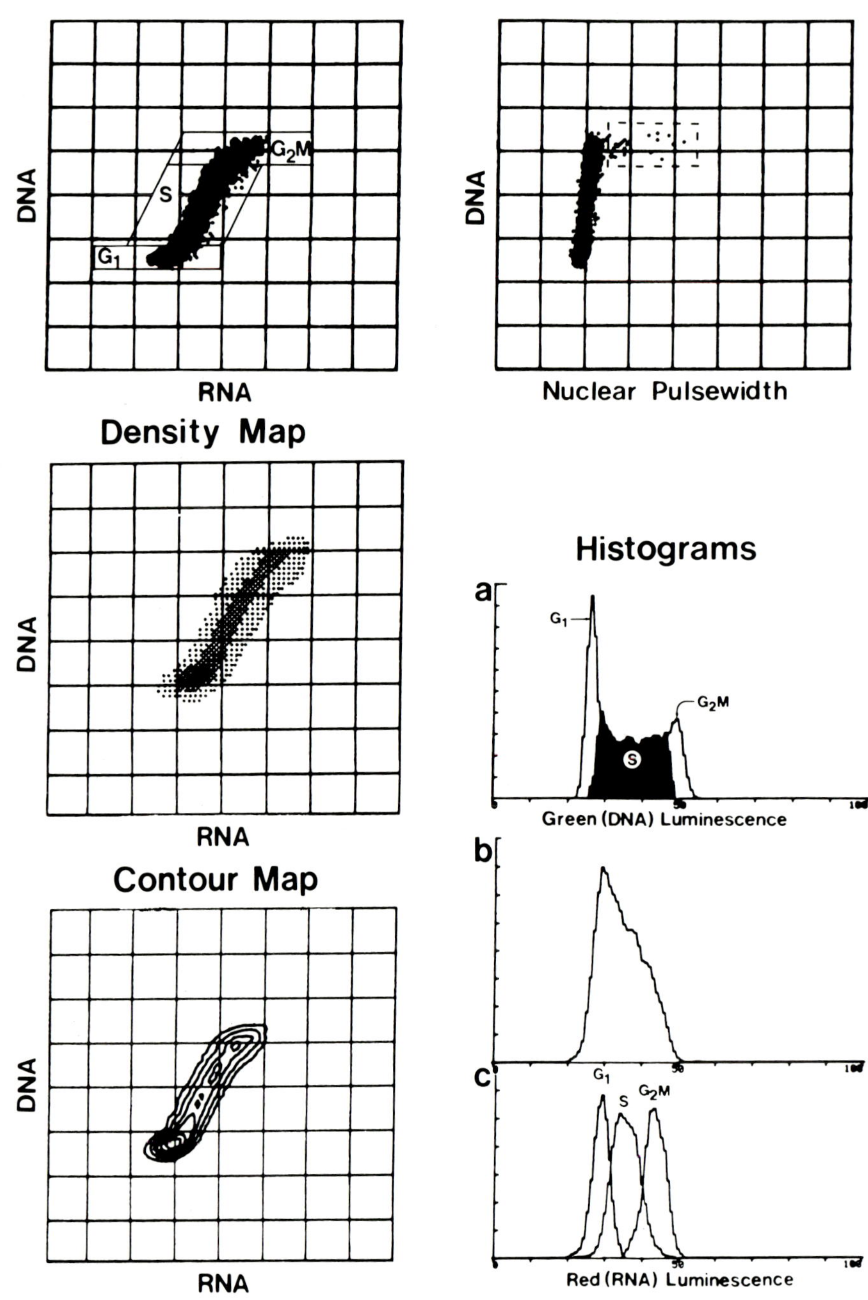
Scattergrams
DNA
RNA
S
G_1
G_2M
DNA
Nuclear Pulsewidth
Density Map
DNA
RNA
Contour Map
DNA
RNA
Histograms
a
G_1
G_2M
S
Green (DNA) Luminescence
b
c
G_1
S
G_2M
Red (RNA) Luminescence

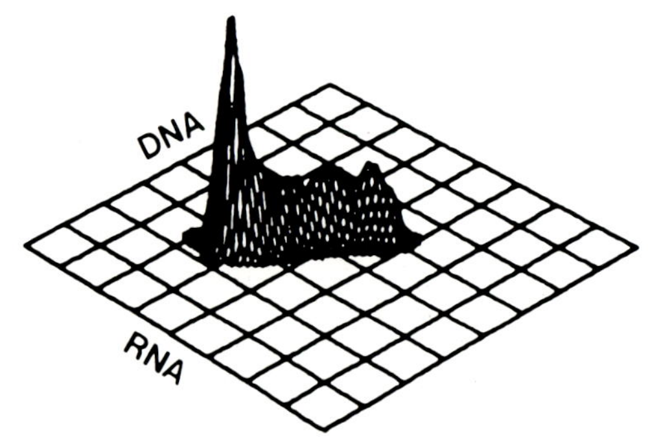
Two Dimensional Histogram
DNA
RNA

exposure. Suspension cultures may be exposed to a drug for a short period of time (generally 1 hour) washed several times and resuspended in drug-free medium. The drug's subsequent effect on cell cycle kinetics can then be analyzed as in the bottom of Figure 2. It should be pointed out, however, that washing cells in drug-free medium does not necessarily remove all traces of drug especially in those instances where the drug has a high binding affinity for some cellular macromolecule. This is generally the case with intercalators and DNA [104]. Nevertheless, under most circumstances such a procedure effectively lowers the drug concentration to below the level at which it would normally be active.

It is clear by comparing the results in the top and bottom of Figure 2, that a higher concentration is necessary to affect cell kinetics when the drug is present for only a short length of time. In addition, perturbations observed soon after removal of some drug concentrations are likely to be transient. Thus, following a 1-hour pulse with 10 nM CI-921, cells accumulated once again in S and G_2M phase (Fig. 2 and Table 2) much as they did at the same concentration following 4 hr in the continuous presence of the drug. By 8 hr, the cell cycle distribution had nearly returned to normal with only a small G_1 accumulation apparent. The control and 10 nM CI-921 treated cultures were indistinguishable by 24 hours. No obvious fluctuations were observed in the growth curve of cells treated with a 10 nM pulse (Fig. 3A, bottom).

A pulse of 50 nM CI-921 had primarily the same effect as continuous exposure to that drug concentration over the first 4 hours of culture. However, by 8 hours the distributions differed, with the appearance of G_1 cells in the pulse exposed cultures. The 24-hour samples differed dramatically, with a majority of the cells having divided in the pulse exposed culture compared with few or no cells cycling at the original ploidy level in the culture exposed continuously to drug. Examination of the growth curve confirms the fact that a 1-hour pulse of 50 nM CI-921 was insufficient to block cells from proliferating; the growth curve very nearly resembles that of the culture exposed continuously to 10 nM CI-921.

Finally, exposure to a 1-hour pulse of 100 nM CI-921 resulted in a redistribution of the cell cycle over the first 8 hours similar to that observed following continuous exposure to 50 nM CI-921. However, the 24-hour cell cycle distribution indicated that at least one cell division took place, as was confirmed by the cell growth curve (Fig. 3A, bottom).

In inspecting the 24-hour samples of cultures exposed for 1 hour to 50 or 100 nM CI-921 one observes a broadening of the peaks of G_1 phase and especially G_2M phase cells. CI-921, like its parent compound Amsacrine, induces single- and double-stranded breaks in DNA [5]. The broadening of the peaks is likely to be due to unequal division of DNA; when cells are blocked before division, the peaks remain relatively sharp (cf. 24-hour sample of cells exposed continuously to 50 nM CI-921). Although not shown, the cell number dropped precipitously between 24 and 48 hours in the culture exposed to a 1-hour pulse of 100 nM CI-921, while continuous exposure to 50 nM CI-921 had much less dramatic an effect on viability. This might suggest that the agent's toxicity is more readily apparent in a shorter time frame when cell division is permitted to occur.

Pulse Exposure of Exponentially Growing versus Stationary Cell Cultures

As stated above, CHO cells grow attached to the substrate and can be maintained in exponential growth by subculturing, or can be grown to a high cell density. Confluent or stationary cultures of CHO cells can be maintained at high viability by providing fresh medium to the cultures periodically without reducing the cell number. The result of such treatment is to accumulate a majority of cells in G_1 phase. Since CHO cells are not normal, untransformed cells, they tend to also accumulate in late S phase and G_2M.

DNA Distributions. Exponentially growing CHO cells exposed to a 1-hour pulse of various concentrations of CI-921 differed slightly in their response as compared with FL cells (Fig. 5). Examining the 4-hour time points, it appeared that the higher the drug concentration: (1) the fewer the cells that re-entered G_1 phase; and (2) the earlier and more distinct was the accumulation of cells in S phase. These same effects were observed in the 8-hour samples. By 24 hours, the cell cycle distribution was near-normal at the lowest concentrations whereas at 100 nM, cells had accumulated predominantly in G_1 and G_2 phase. There were relatively few cells at higher ploidy levels in any of the cultures.

By virtue of their ability to form colonies on plastic, CHO cells can be used in a colony formation assay to determine drug effectiveness. A 1-hour pulse of CI-921 resulted in inhibition of colony formation of exponentially growing CHO cells as illustrated in Figure 3B. The point at which 50% inhibition of control colony formation occurred (ICF_{50}) was 80 nM for exponentially growing CHO cells. In comparison, the dose that inhibited FL cell growth by 50% (ID_{50}) when cells were exposed to a 1-hour pulse of CI-921 was 67.5 nM.

Stationary CHO cells exposed to a 1-hour pulse of CI-921, followed by reculture at low cell densities, were relatively resistant to the effect of the drugs. There is generally a delay following replating of stationary cultures at low cell densities before cells begin cycling. There was no obvious effect of the drug at any concentration within the first 4 hours (Fig. 5, bottom). By 8 hours, a cohort of cells could be observed entering S phase in each culture. A typical cell cycle distribution consistent with exponential growth was observed at 24 hours for each culture. Clearly CI-921 is an S phase active agent. Few cells reside in S phase in stationary cultures.

Fig. 6. Presentation of multiparameter data. Exponentially growing FL cells were stained with the metachromatic dye AO to provide green fluorescence proportional to DNA content and red luminescence proportional to RNA content: The green fluorescence pulse-width (nuclear diameter) as a third measurement provides a means of discriminating against cell clumps [132]. The two scattergrams (top) represent the three measurements. Each dot indicates the position of an individual cell. The position of G_1, S, and G_2M phase cells is indicated based on the DNA content; note the single-parameter (DNA) histogram a. The red luminescence associated with RNA content is presented for the whole population in histogram b, while histogram c contains the RNA histograms for the individual cell cycle phases G_1, S and G_2M selected along the DNA fluorescence histogram shown in the scattergram at the top left. The density map was prepared from the same data file as the scattergram with the relative density of cells in the distribution split into three levels: 1 to 10, 11 to 50, and more than 50 cells. The contour map was also prepared from the same data set and six slices were selected at various intervals. With both the density and contour maps, the location of high concentrations of cells (peaks) can be appreciated. Finally, at the bottom of the figure, a two-dimensional histogram (pseudo-three-dimensional projection) is presented in which the heights of the peaks represent an accumulation of cells with similar attributes; here the major peak represents G_1 cells.

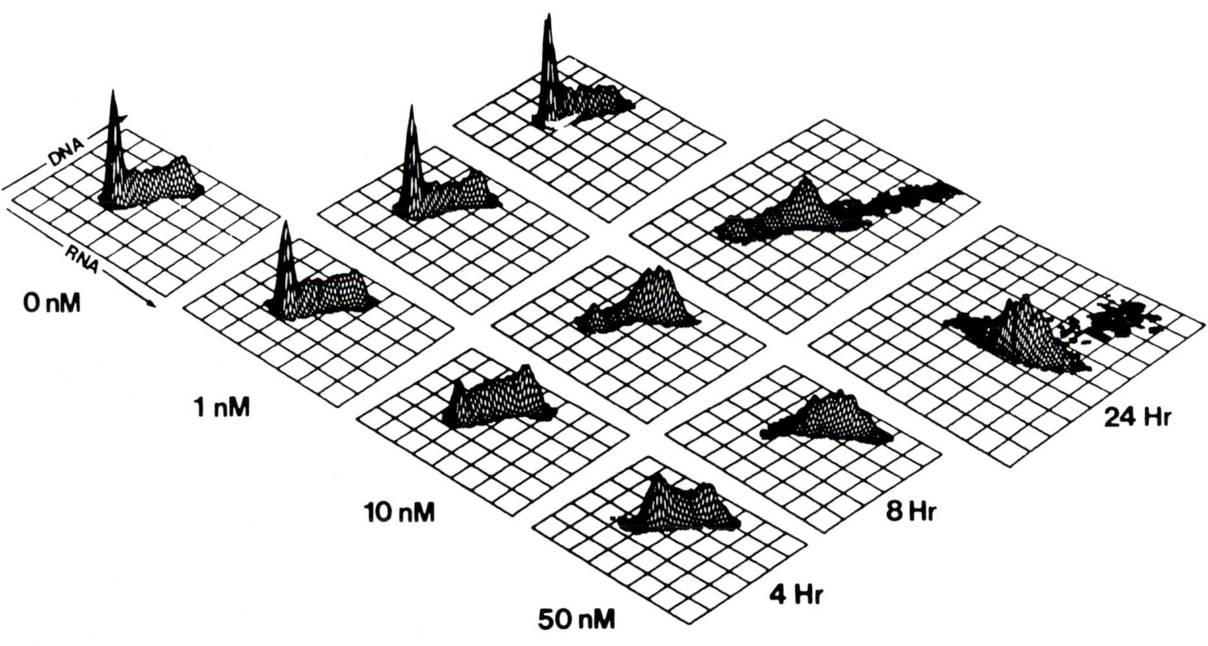

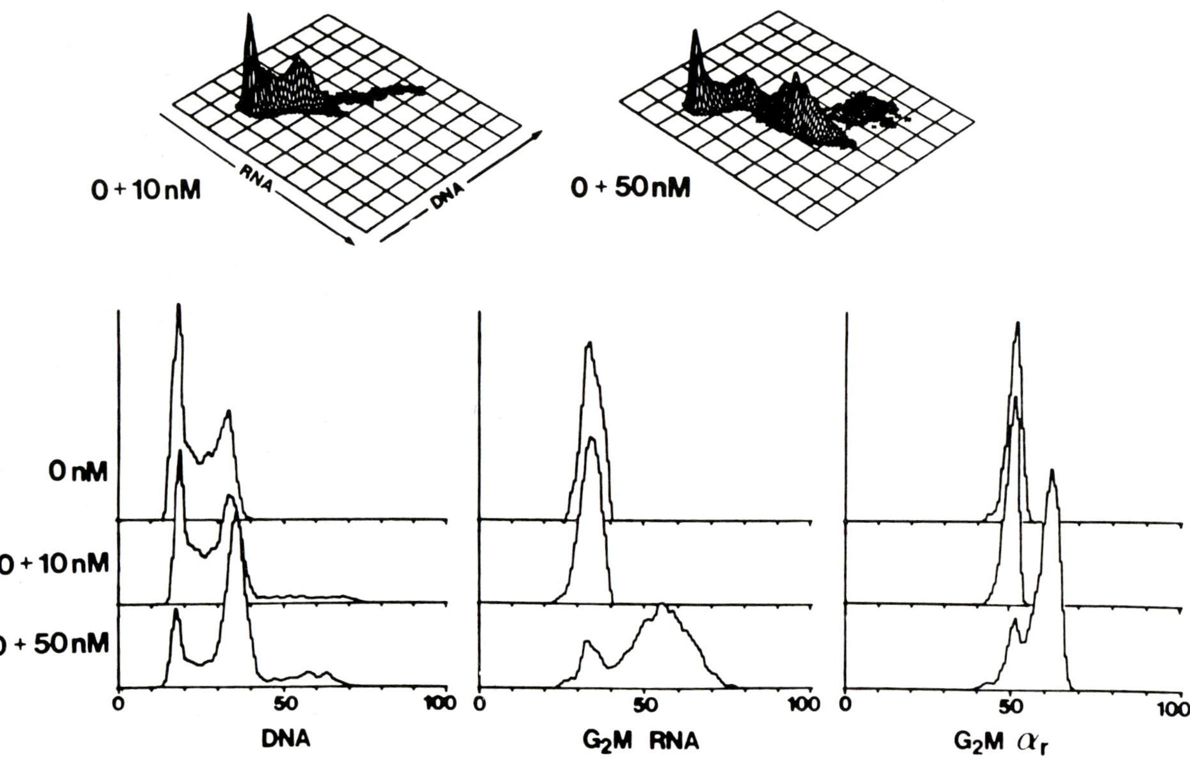

Fig. 7. Two-dimensional (DNA versus RNA) distributions of FL cells treated with CI-921 as in Figure 2. Aliquots were removed from cultures and stained with AO to provide green and red luminescence of individual cells, which is proportional to DNA and RNA content, respectively. In addition, equal aliquots were removed from the 24-hour control and drug-treated cultures, mixed and stained with AO. The single-parameter histograms of green (DNA) fluorescence for the control and the two mixtures are presented at the bottom. Utilizing these distributions to select only G_2M cells, the RNA and α_r (RNA/(RNA + DNA)) distributions were obtained.

Therefore, little effect is expected among these cultures. This observation is supported by the colony formation data (Fig. 3B) in which nearly twice as much drug was required to reduce colony formation by 50% (i.e., 190 nM) as was used in the experiment presented in Figure 5 (bottom). It would appear that at high enough concentrations damage due to drug binding is not reparable in the time between removal of the drug and the signal for the cells to start synthesizing DNA, which occurs when cells are plated at low density (e.g., a duration of approximately 1 hr). At a drug concentration of 100 nM, colony formation in stationary CHO cell cultures was inhibited by about 10%, which was approximately equal to the number of cells in S phase when the drug was added. Since most cells were in G_1 during exposure to 100 nM CI-921 any damage to these cells was probably repaired prior to their entrance into S phase. Thus, there was no observable difference in the cell cycle distributions with increasing drug concentration within the range tested.

MULTIPARAMETER ANALYSIS

Most flow cytometers are capable of simultaneous measurements of more than one cellular attribute (see Chapters 2,9, this volume). Thus, it is possible to measure intrinsic cellular attributes, such as autofluorescence, light absorption, or light scatter, at one or several wavelengths or angles of collection [138] (Chapt. 5). In addition, extrinsic probes such as fluorochromes may be added, which upon binding to specific subcellular constituents will provide fluorescence signals at one or several wavelengths. As a result, combinations of fluorescence and/or absorption and/or light scatter signals can routinely be obtained from flow cytometers with a single light source and appropriate lens, filter, mirror and detector combinations [138]. Advanced or specially designed systems with multiple excitation sources can provide even greater numbers of signals, since the light sources are separated along the flow stream with each interrogation point yielding a unique signal(s) [32,45]. Assuming one has the capability of making multiple measurements on a single cell, it is then only necessary to choose the appropriate signal (dye) combination and to record the data in a format such that the measurements remain correlated (see Chapter 7, this volume). The alternative, a series of uncorrelated single-parameter histograms, often do not permit detection and analysis of important cell subpopulations.

True multiparameter data consist of measurements of two or more independent signals. Often these measurements consist of peak (e.g., the difference between the measurement maxima and background signal) or area (e.g., the integrated value minus background) values of fluorescence, light scatter, absorption, and other factors. Occasionally a single measurement may be processed in several ways to give rise to pseudo two parameter data, examples of which would include peak versus area or area versus pulse width measurements of a single cellular attribute such as DNA fluorescence. Such pairs of measurements can be useful in gating out, for example, cell clumps, resulting in single parameter histograms free of cell doublets that might otherwise distort DNA histograms [141].

An example of true multiparameter data is provided in Figure 6. In this instance, exponentially growing FL cells were stained with the metachromatic dye AO to provide both green and red luminescence signals, which are proportional to cellular DNA and RNA content, respectively. In addition, one of the signals, green (DNA) luminescence, was processed to obtain nuclear pulsewidth data useful in eliminating cell clumps from analysis. Such data can be presented for display and analysis in the form of a scattergram of dots, the position of each dot indicating the relative DNA and RNA content of a particular cell. Two scattergrams each presenting a different pair of the three values recorded for each cell, are displayed at the top of Figure 6. A three-dimensional representation of the data can be obtained by plotting it in the form of a density or contour map, or as a two dimensional histogram in which the number of cells within the distribution with a specific set of attributes are visualized either by increased density, a grouping of contour lines or the height of a peak (Fig. 6).

A pseudo-two-dimensional analysis of the data in Figure 6 would provide histograms such as a and b, which represent the DNA and RNA content distributions, respectively, of the entire FL cell population recorded independently. This information is also available from true multiparameter data. However, using the two-parameter correlated data, it is possible to obtain the distribution and statistics for specific cell subpopulations, e.g., the RNA distributions of individual cell cycle phases identified by their DNA content (Fig. 6, histogram c). This type of data is provided whether a single metachromatic dye is used, as presently, or whether multiple dyes with similar absorption but separable emission spectra are used. With multiple light sources many dyes can be excited and if the excitation is spatially separated and accompanied by appropriate detectors, the presence or quantity of three or more cellular constituents can be obtained for large numbers of individual cells.

DNA versus RNA Content

The two parameter (DNA vs RNA) distributions of FL cells grown in the continuous presence of CI-921 are presented in Figure 7. The single parameter DNA distributions of the same samples have been presented in Figure 2. The heights of the peaks in the two-dimensional histograms indicate the relative frequency of cells with specific RNA and DNA content. In the control (0 nM) distribution, the major peak represents G_1 cells, while cells at twice the DNA content define the locus of G_2M cells. Cells with intermediate luminescence value represent S phase cells. Note that as cells proceed through the cycle from G_1 through S to G_2M, they increase in RNA content (Figure 6). Although not shown, when adjustments are made for nonspecific red luminescence (i.e., if the red luminescence remaining in stained, RNase-treated cells is subtracted) it is clear that the RNA content doubles, on average, as cells proceed from G_1 to G_2M phase.

RNA content is also a sensitive measure of cell growth [42,45–48,78]. Cells grown to high cell densities have reduced metabolic activity generally accompanied by a shift of the population to lower mean RNA content [168]. This phenomenon was observed when cells were exposed to 1 nM CI-921 (Fig. 7). Since this concentration of drug had virtually no effect on cell growth (Fig. 3A), by 24 hours, cells had reached a high cell density (subplateau phase) and the mean RNA content had diminished by about 20% (Table 3). However, when prevented from reaching a high cell density by treatment with 10 nM CI-921, the cells remained in balanced growth (i.e., the RNA content remained equivalent to that of exponentially growing untreated cells) (Table 3). The maintenance of balanced growth is presented graphically at the bottom of Figure 7. When equal aliquots of untreated FL cells and cells exposed for 24 hours to 10 nM CI-921 were mixed and stained simultaneously, no difference could be observed between the RNA distribution of G_2M cells in the

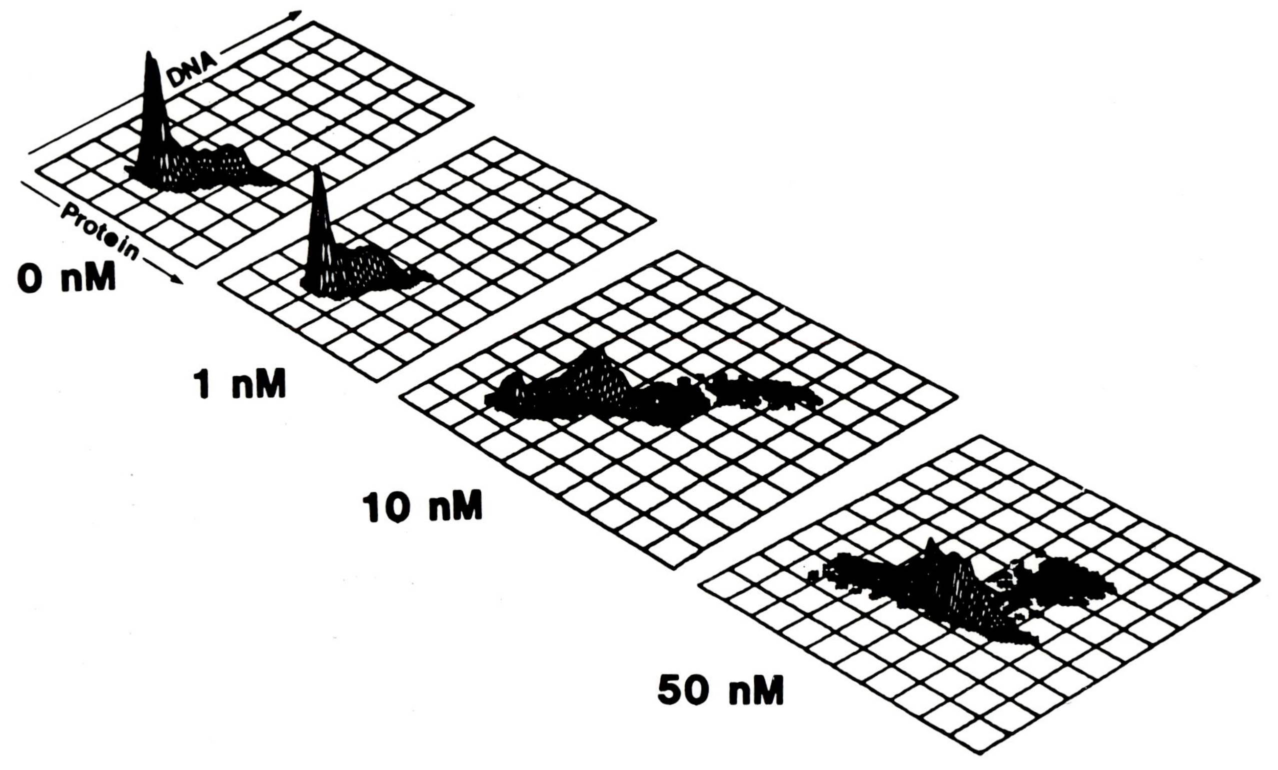

DNA
Protein
0 nM
1 nM
10 nM
50 nM

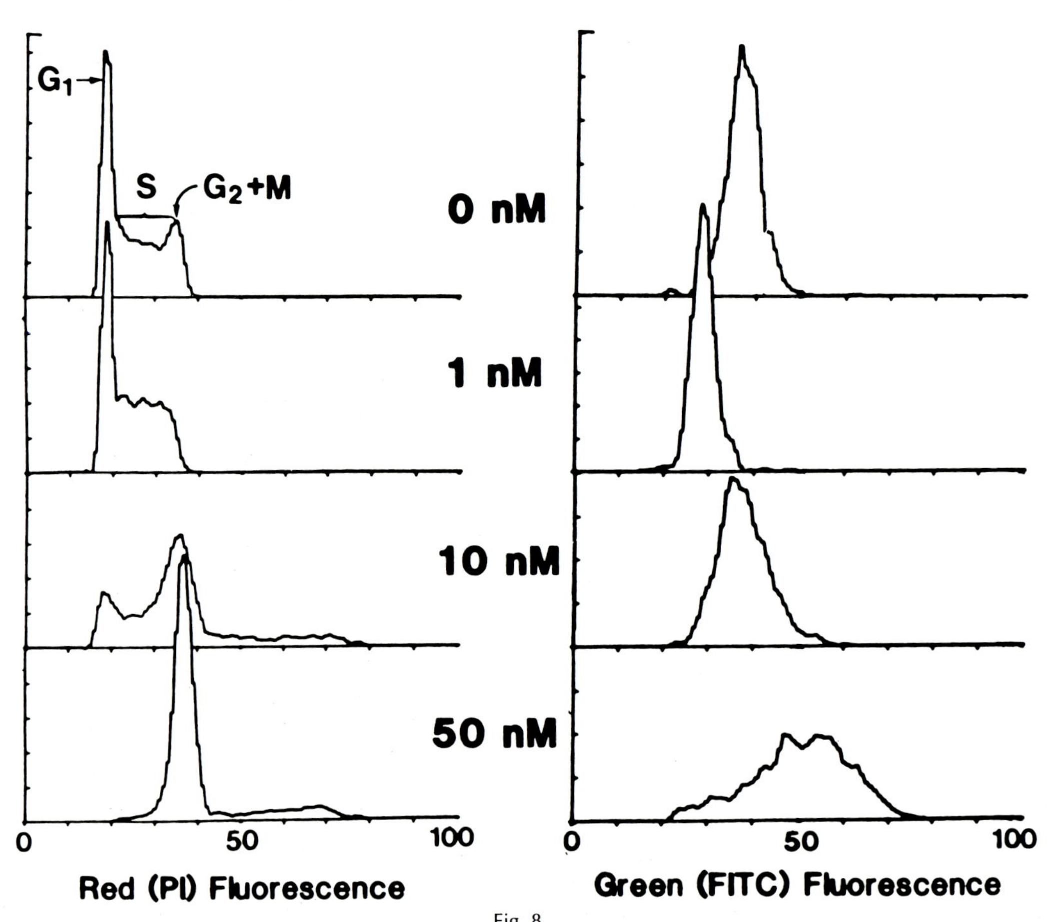

DNA Content
Protein Content
G1
S
G2+M
0 nM
1 nM
10 nM
50 nM
0
50
100
Red (PI) Fluorescence
Green (FITC) Fluorescence

Fig. 8.

TABLE 3. Effect of Continuous Exposure of FL Cells to Varying Concentrations of CI-921

	CI-921 Concn.					
	1.0 nM		10.0 nM		50.0 nM	
hr*	RNA†	Prot.‡	RNA	Prot.	RNA	Prot.
4	1.0	1.0	1.02	0.98	1.07	1.08
8	0.90	1.02	1.00	1.04	1.52	1.19
24	0.81	0.79	0.99	1.03	1.75	1.34

*Time of exposure to drug.
†RNA content determined after subtraction of RNase-insensitive red fluorescence from total red fluorescence of G_2M cells determined from the AO green (DNA) luminescence distribution.
‡Total protein as determined by FITC staining of G_2M cells selected from the PI (DNA) red fluorescence distribution.

control and mixed specimens. The ratio of RNA content to total nucleic acid (DNA plus RNA) content, α_r, which is a useful measure of balanced growth [155,168], was also unchanged when drug treated cells were mixed with controls (Fig. 7, bottom). Again, although α_r takes into consideration the cellular DNA content and thus is not limited to comparisons of cells from the same cell cycle phase, it is possible with multiparameter analysis to provide the ratio distribution of cells selected from a second parameter (using, e.g., the DNA versus α_r distribution). In this instance, the α_r distribution of the G_2M populations were compared and showed no change. This was not true of cells treated with higher drug concentrations. When continuously exposed to 50 nM CI-921, FL cells, which accumulated in G_2M phase, continued to increase in RNA content (Fig. 7; Table 3). A mixture of control cells with a sample treated for 24 hours with 50 nM CI-921 showed two distinct populations in terms of RNA content which would not have been observed from examining the DNA distribution alone (Fig. 7). The extent of the increase in RNA can be appreciated both in the two-dimensional histogram of DNA vs RNA and in the RNA and α_r distribution of the G_2M populations (Fig. 7).

Interestingly, an increase in RNA content of cells blocked in G_2M seems to be common when cells are treated with moderate concentrations of intercalating agents which, on the one hand, are sufficient to prevent cell division but are not so high as to interfere completely with RNA synthesis [172,176]. Clearly, since RNA content is a function of the combination of RNA synthesis and breakdown, accumulation of RNA can only occur when synthesis is allowed to continue, breakdown is not accelerated, and cells are unable to cycle and distribute RNA to daughter cells. It is important to realize that this type of information can only be obtained from measurement of individual cells; bulk measurements provide averages and thus obscure the presence of subpopulations. Whereas microspectrophotometric and microspectrofluorometric measurements of cellular constituents in individual cells on slides are possible, even with the most sophisticated, automated apparatus, at most, several hundred cell measurements are attempted. Flow cytometry provides more accurate measurements on literally thousands of cells in seconds or, at most, minutes.

Fig. 8. Two-dimensional (DNA versus protein content) distributions of FL cells treated with CI-921 as shown in Figure 2. Cells were stained simultaneously for DNA with PI (red fluorescence) and total protein with FITC (green fluorescence). Note the similarity between DNA distributions stained with PI and the AO green (DNA) fluorescence distributions in Figure 2. The 24-hour control sample was replaced with cells growing at a lower density to demonstrate the loss of protein content associated with growth to high cell density (1-nM sample). Again, using the DNA distributions, the protein content of cells with G_2M DNA content were selected and displayed.

DNA versus Total Protein Content

Multiparameter analysis of drug effects in vitro is by no means limited to measurements of DNA versus RNA content. Of several possible combinations of measurements, the analysis of DNA vs protein content was selected to provide a second example of this approach. Thus, the same 24-hour samples analyzed in Figure 7 were fixed and stained with PI to obtain DNA content and FITC for total protein content and are displayed in Figure 8. The two dimensional histograms of DNA vs protein content are clearly very similar to the DNA vs RNA histograms in Figure 7. Comparison of the DNA content histograms obtained with PI to those obtained from the AO green luminescence distributions (Fig. 2, top, last column) illustrates that the two stains yield essentially identical results (note, the untreated control in Figure 8 is from an exponential culture while the control in Figure 2 has reached a high cell density).

The FITC-stained protein distributions presented in Figure 8 are from the G_2M population only, selected according to DNA content from the multiparameter data. As stated above, an exponentially growing control replaced the high density control, which had essentially the same protein content as the cells selected from cultures treated with 1 nM CI-921 (not shown). Thus, protein content decreased upon reaching high cell density, as did RNA content. However, when mean protein content was compared for the same G_2M populations as was RNA content, the patterns were similar (Table 3). Protein content, as a function of balanced growth, increased only in cultures exposed to 50 nM CI-921 (Table 3).

Multiparameter Analysis of Synchronized Cell Populations

Often identification of cell cycle phase specific effects is simplified by utilizing synchronized cell cultures. Care must be exercised, however, in the choice of the method of cell synchronization. The use of drug, nutrient deprivation, or serum starvation may affect cell response to subsequent drug exposure in a variety of complicated ways. Generally, when applicable, mechanical synchronization techniques (e.g., differential centrifugation, elutriation, mitotic shakeoff) have proven the most useful since they avoid some of the interactive and possibly synergistic effects of synchronizing agents.

CHO cells are rather easily synchronized by mitotic detachment [172]. By choosing appropriate procedures for collection of mitotic cells, relatively uniform populations can be obtained for any cell cycle phase. One important consideration is in choosing the time between the "cleanup" shake and the shakeoff to be collected for drug treatment. Since CHO cells roundup during mitosis and require 30 min to 1-hour growth in conditioned medium to reattach following mitosis, an interval of 30 min or longer between shakeoffs will provide carly G_1 cells in addition to cells in various stages of mitosis. However, the shorter the duration between

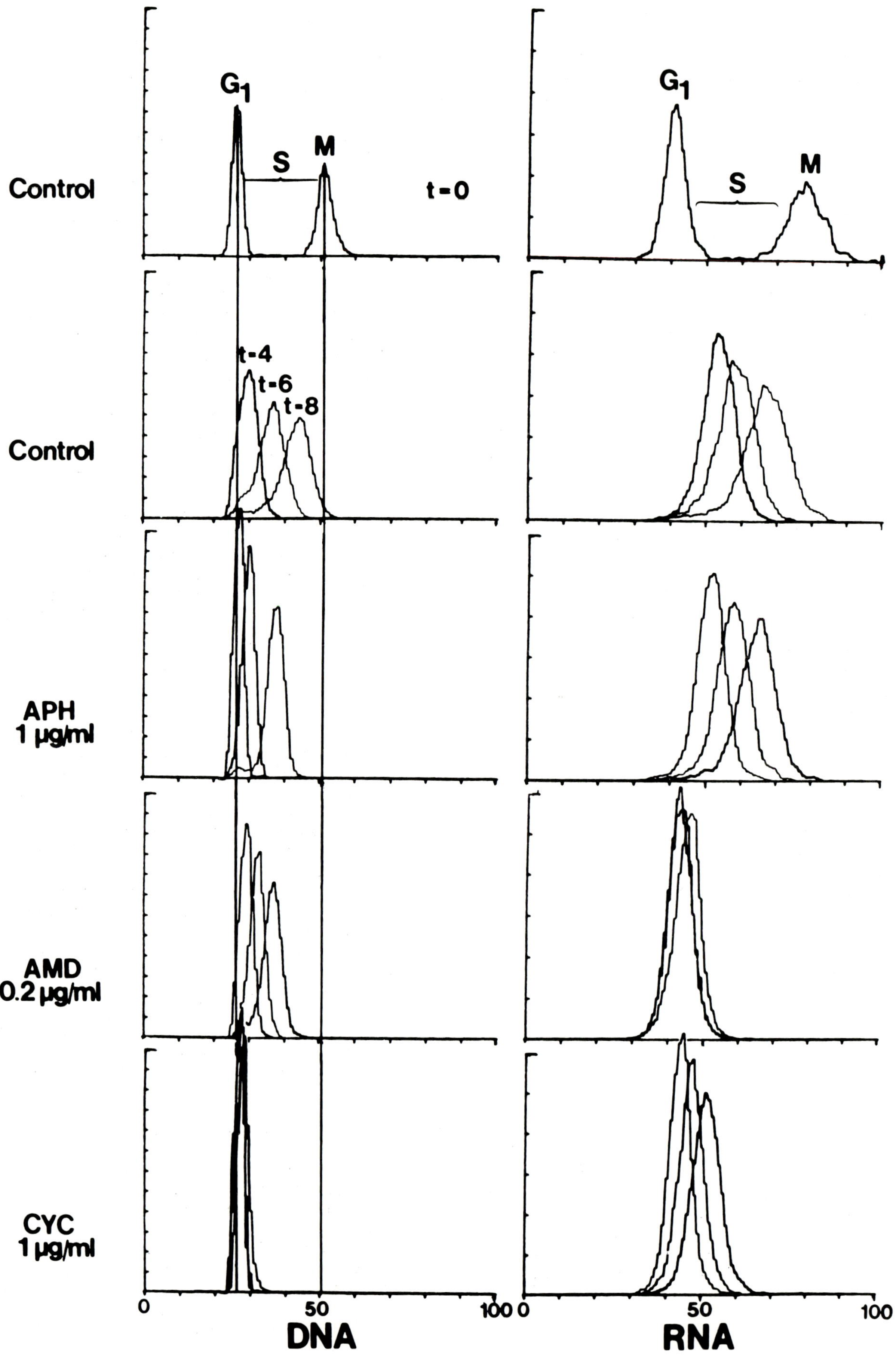

Control
Control
APH
1 µg/ml
AMD
0.2 µg/ml
CYC
1 µg/ml
G1
S
M
t=0
t=4
t=6
t=8
G1
S
M
0
50
100
DNA
0
50
100
RNA

shakeoffs, the lower the yield of cells per shakeoff. Since large numbers of cells are often required for drug studies, several shakeoffs performed sequentially on the same set of cultures can be combined, collected and kept at 0 to 4°C.

To illustrate the application of synchronized cultures in examining drug effects, several agents that are known to affect the progression of cells from G_1 to S phase were tested on mitotically synchronized CHO cells as illustrated in Figure 9. Thus, large numbers of mitotic and early G_1 cells could be collected (samples shaken off at 30-min intervals), which in the absence of drug treatment proceed synchronously through the cell cycle with a doubling time of approximately 11.5 hours and cell cycle phase duration of 3.5, 5.5, 2.0, and 0.5 hours for G_1, S, G_2, and M, respectively. Using AO staining, the DNA and RNA content of the population was obtained at various times of culture (Fig. 9). Three drugs, aphidicolin (APH), a DNA polymerase inhibitor, actinomycin D (AMD), an RNA polymerase inhibitor, and cycloheximide (CYC), an inhibitor of protein synthesis, were added to the synchronized populations for 4 hours at t = 0. After 4 hours, the drugs were washed out and the DNA and RNA content determined for the populations at several time intervals.

APH, at a concentration of 1.0 μg/ml should cause the accumulation of cells at the G_1–S boundary (i.e., allow transit through G_1) but has only a limited effect on accumulation of RNA content. Thus, APH-treated cells should enter a state of unbalanced growth in the sense that their RNA content increases in the absence of a proportional increase in DNA content. Upon release from the drug, the cells should proceed through S phase after a slight delay. Figure 9 illustrates that the DNA content of APH-treated cells was more uniform during their delayed transit through S than control cells as a result of being synchronized again at the G_1–S boundary. The RNA content of APH-treated cells increased at nearly the same rate as control, untreated CHO cells (Fig. 9).

Relatively low concentrations of AMD (0.1 to 0.2 μg/ml) will inhibit RNA synthesis but fail to entirely block cell progression through S phase (Fig. 9). This is an interesting phenomenon inasmuch as the continuous presence of higher AMD concentrations blocked cells in G_1 and led to continuous reduction in RNA content (Fig. 10B). Thus, long term exposure to low concentrations of AMD led to unbalanced and abortive growth; cells require 18 hours to reach G_2 phase in the continuous presence of 0.1 μg/ml AMD. By subtracting the nonspecific red luminescence (subtraction of the red luminescence remaining following RNase treatment) and scaling the value to 1.0 for immediately post-mitotic G_1 cells, the trajectory of a synchronized population through the cell cycle can be plotted (Fig. 10B). A comparison of AMD-treated to control values in Figure 10B illustrates that cells can enter and traverse S phase, albeit extremely slowly, with very small increases in RNA content at low drug concentrations. Higher concentrations of AMD completely block cell progression (Fig. 10B).

CYC at 1.0 μg/ml completely inhibited cell entrance into S phase but allowed some increase in RNA content (Fig. 9). Apparently, 4-hour treatment with 1.0 μg/ml CYC is not reversible. This, as others have suggested [24], may result from the breakdown of a labile protein required for cell progression, which must be synthesized to initiate DNA synthesis. Once this protein is broken down, the cell enters a state in which several hours are required before the signal for reinitiation of cell cycle traverse is triggered [120].

The drug effects observed in Figures 9 and 10 would also occur in exponentially growing cell culture but be significantly more difficult to visualize and interpret. Thus, S phase cells are likely to be variously affected by each of the agents, while G_2 phase cells are generally more resistant, especially to the action of APH and AMD. With exponential cultures, one would be faced with a complicated redistribution of cells among the various cell cycle stages that would depend on such factors as (1) the relative susceptibility of cells in each phase to the effect of the drug; (2) the length of exposure to the drug times the transit time through each phase; (3) the proportion of cells residing in each phase at the time of drug addition; and (4) drug concentration.

Fig. 9. Multiparameter analysis of the effect of various agents on the progression of synchronized CHO cells through the cycle. CHO cells were synchronized in mitosis and immediately postmitotic G_1 phase by mechanical shakeoff of the loosely attached cells (interphase cells require trypsinization for removal). An aliquot was stained with AO to provide the DNA and RNA distributions of mitotic and early G_1 phase cells (top). The second set of the histograms demonstrate the change in DNA and RNA content as the synchronous cohort of cells traversed S phase (4, 6, and 8 hours after shakeoff). The remaining aliquots received APH (an inhibitor of DNA synthesis), AMD (an inhibitor of RNA synthesis) or CYC (a protein synthesis inhibitor) for 4 hours, were then washed free of the drug, replated with fresh medium and harvested at 4, 6, and 8 hours. See text for a description of drug effects.

A MULTIPARAMETER ANALYSIS OF A STATHMOKINETIC EXPERIMENT IN VITRO

Considerable evidence exists that both subtle and dramatic changes occur in the gross and molecular structure of chromatin in situ (DNA and its attendant nuclear proteins) during the cell cycle and during transitions of cells from proliferation to quiescence or upon differentiation [42]. Many of these changes in structure, however, are abolished upon extraction and solubilization of chromatin in vitro. Recently, it has become possible by means of flow cytometry to observe and quantitate changes in chromatin structure occurring during the cell cycle [42] (see Chapter 24, this volume). By exposing cells to partially denaturing condition (i.e., temperature, pH) followed by quantitation of the remaining amounts of double and single-stranded DNA using the metachromatic properties of AO, it appears that the sensitivity of DNA in situ to denaturation correlates with chromatin condensation. Thus, the DNA most sensitive to denaturation is contained in condensed chromosomes of mitotic cells and in immediately postmitotic G_1 cells (Fig. 11), quiescent cells and cells which have undergone differentiation [49–51,53,55,57,58,175]. Within a population of exponentially growing cells, late G_1 (G_{1B}) and early S phase cells are the most resistant to denaturation (Fig. 11).

In fixed, RNase-treated cells, AO binding to native and denatured regions of DNA result in green and red luminescence, respectively. With appropriate staining and instrumentation, the combined intensities of green and red fluorescence indicate total DNA content of a cell and thus the cell cycle (DNA) distribution, whereas the ratio of red luminescence (denatured DNA) to total luminescence (expressed as α_t) indicates the relative sensitivity of cells to denaturation (Fig. 11). Thus, mitotic cells, which have undergone rather

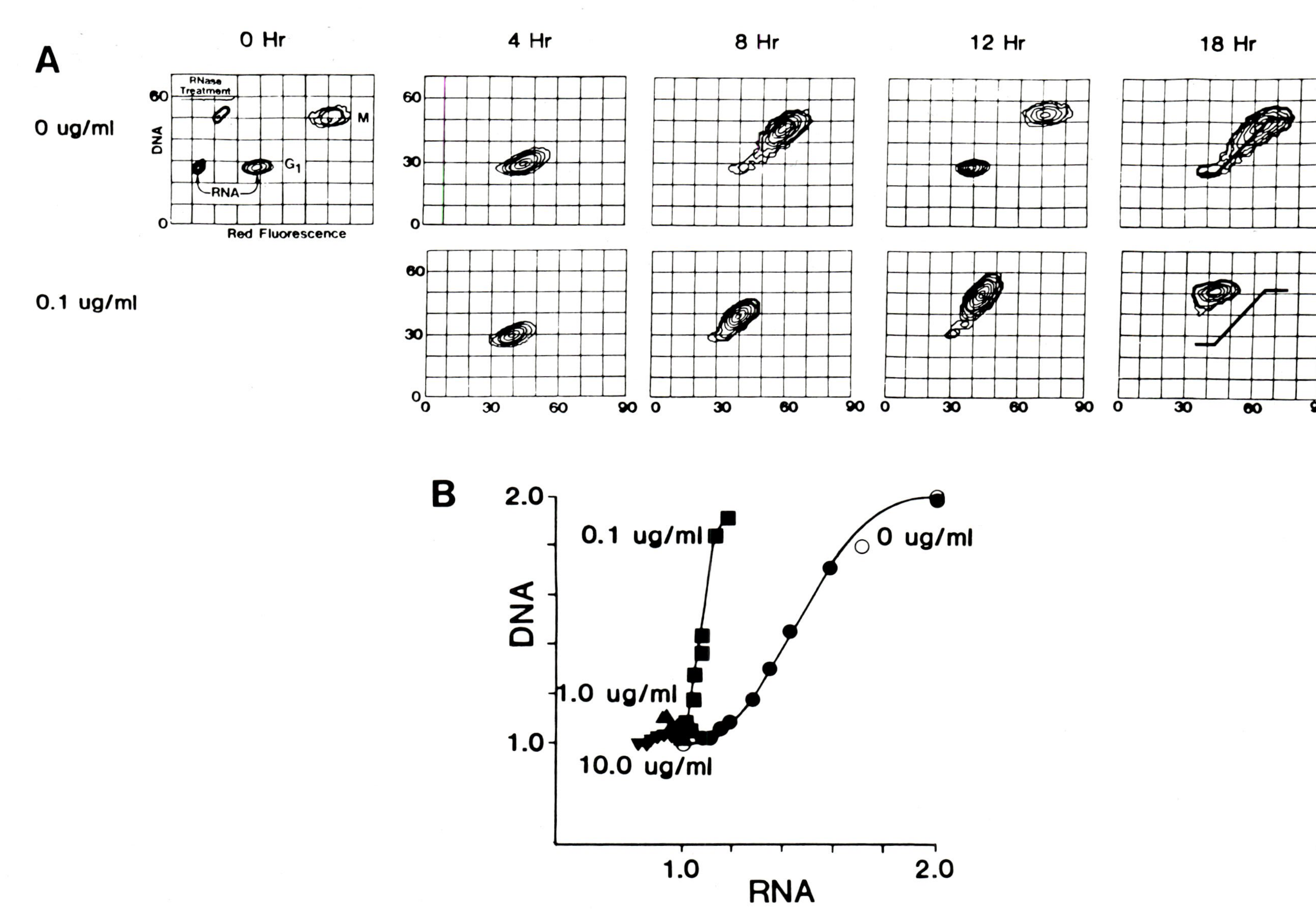
Actinomycin D
A
0 Hr
4 Hr
8 Hr
12 Hr
18 Hr
0 ug/ml
0.1 ug/ml
RNase Treatment
DNA
RNA
G1
M
Red Fluorescence
0
30
60
90
B
DNA
RNA
0.1 ug/ml
1.0 ug/ml
10.0 ug/ml
0 ug/ml
1.0
2.0

obvious and dramatic DNA condensation in the process of forming chromosomes, actually appear as a clearly identifiable, distinct cluster separate from G_2 cells. The latter contain an equivalent amount of DNA but in a much less condensed state (Fig. 11). Immediately post mitotic G_1 cells (G_{1A}) form a shoulder on the G_1 distribution with an intermediate degree of denaturability (α_t value) compared with M and early S phase cells. Using this staining technique, it is possible to describe 5 cell subpopulations in exponentially growing cultures, i.e., G_{1A}, G_{1B}, S, G_2, and M as opposed to the three apparent from typical single parameter DNA distributions (e.g., G_1, S, and G_2M).

In the presence of a stathmokinetic agent such as vinblastine, cells will accumulate in mitosis. As observed in Figure 5, a single-parameter analysis of the effect of a drug in the presence of a stathmokinetic agent can indicate if cell transit through a particular cell cycle phase is perturbed. With the addition of a second parameter, in this case α_t, a great deal more information becomes available. Two-parameter contour maps of selected time points following addition of Vinblastine and either 0, 10, or 50 nM CI-921 are presented in Figure 11.

In the absence of CI-921, the exit of cells first from G_1 and then early S phase is concomitant with cell accumulation in mitosis. The addition of 10 nM CI-921 appears to diminish cell accumulation in M and slows traverse through S phase. Both effects are even more dramatic in the culture exposed to 50 nM CI-921. These results can be quantitated with high precision by multiparameter analysis of the distributions obtained from such experiments.

The addition of the stathmokinetic agent at $t = 0$ almost immediately prevents the entrance of cells into G_1 phase. By setting appropriate thresholds along the total fluorescence (DNA) distribution, it is possible to estimate the percentage of cells remaining in G_1 phases as a function of time (Fig. 12A). The curve of emptying of the G_1 phase has both a linear and exponential component [52]. It is believed, for a variety of reasons, that the exit of cells from the G_{1A} compartment is exponential or stochastic in nature as is evident from its straight-line character on the semilogarithmic plot, while transit through G_{1B} is relatively deterministic (i.e., linear) and responsible for the curved portion of the G_1 exit curve [43]. A more detailed discussion of this point is presented in Chapter 24. Nevertheless, CI-921 has relatively little effect on cell exit from G_1 or G_{1A}; the slight "slowdown" between 5 and 7 hr in the G_1 exit curve is based on 1% or less cells which is not sufficient to be considered significant.

Fig. 10. Multiparameter analysis of the progression of synchronized CHO cells through the cell cycle in the absence and presence of AMD. **A:** Contour maps (DNA versus RNA) of synchronized CHO cell growth in the presence of 0 or 0.1 μg/ml AMD. Note, the normal doubling time for these cells is 11.5 to 12 hours. **B:** The mean green (DNA) and red (RNA) luminescence was calculated for each population presented in the top as well as for cultures treated with 1.0 and 10.0 μg/ml AMD (not shown). The nonspecific red fluorescence was then calculated based on enzymatic digestion of RNA and subtracted from the red luminescence value to give RNA-specific red luminescence. At this point, all fluorescence values were normalized to that of immediately postmitotic G_1 phase cells, which were assigned a value of 1.0 for both DNA and RNA. The trajectories for each set of cultures were then plotted. Note that the control trajectory contains points (open circles) from the second cell cycle (12 and 18 hours).

Cell traverse through S phase can be studied in several ways. In the present instance, equal slices were obtained along the total (DNA) fluorescence distribution corresponding to equivalent areas of early-, mid-, and late-S phase cells. In the absence of drug treatment, the percentage of cells should increase in each S phase window (due to the exponential nature of the culture), reach a maximum at a point where the effect of the stathmokinetic agent on cell entrance into G_1 becomes apparent, and then decrease according to a specific function [52]. The time at which the curves reach a maximum should be later for the cells in mid- and late-S phase. This effect can be observed for the family of curves of non-drug-treated cultures presented in Figure 12B. Thus, the curves peak at 3, 4, and 5 hours for early-, mid-, and late-S phase cells, respectively. (Note, however, that these points are affected by the frequency of sampling such that more frequent sampling (e.g., at 10 min intervals) may reveal somewhat different intervals of time between the peak values).

There was clearly an effect of the drug on S phase traverse. At the lower concentration, CI-921 caused a slight increase in accumulation of early-S phase cells though their rate of traverse (slope of the decrease) seemed unaffected. Cells were delayed about 1 hour in their traverse of mid-S phase by 10 nM CI-921. This delay became approximately 2 hours, when late-S phase cells were analyzed. At the higher concentration, CI-921 caused a 1-hour delay in transit through early S phase, which became approximately 3 hours for mid-S phase cells, while late S phase cells appeared to still be accumulating at the last time point (7 hours). A separate peak in late S phase can be observed in the contour map of the 7-hour sample exposed to 50 nM CI-921 (Fig. 11), which would agree with the data in Figure 12B as well as with the observations made earlier in Figures 2 and 4.

The results of cell accumulation in G_2M and M phase are generally the most interesting and dramatic when dealing with subcytotoxic concentrations of intercalating agents. This is clearly the case with CI-921. The mitotic cell accumulation curve should be exponential under optimal growth and experimental conditions. This is also true for the G_2M accumulation curve over the portion that is as yet unaffected by the stathmokinetic agent (i.e., the duration of G_2, S and most if not all of G_1 phase). A deviation from the straightline exponential slope of the M phase accumulation curve that is not accompanied by a change of equivalent slope of the G_2M accumulation curve indicates cells are blocked in G_2 phase. Note the slopes of both the M and G_2M accumulation curves provide the doubling time (T_d) of the culture while the distance between them along the time axis approximates the duration of G_2 phase. If both the M and G_2M slopes are affected, it would suggest that entrance of cells from S to G_2 phase is slowed, which would affect entrance into M. The timing of any changes in slope and the actual values of those changes provide evidence of whether the drug effectively blocks cells in a specific phase, lengthens the time required to traverse a phase, or some combination of both.

Within the first hour of treatment, CI-921 dramatically affects cell entrance into mitosis (Fig. 12C). At 10 nM CI-921, cells continued to divide albeit much more slowly than normal, while no cells divide except perhaps for the first 30 min in the culture exposed to 50 nM CI-921 (cf. Figs. 2 and 4). Very late S phase cells appeared not to be affected by either drug concentration as is evident from the G_2M accumulation curve. However, after 1 hour there was a dramatic slowdown in entrance of cells into M phase at both drug

STATHMOKINETIC EXPERIMENT

0 Hr

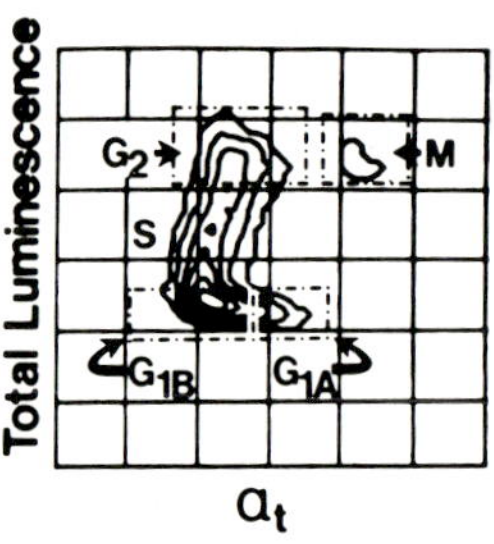

0 nM CI-921 | **10 nM CI-921** | **50 nM CI-921**

1 Hr

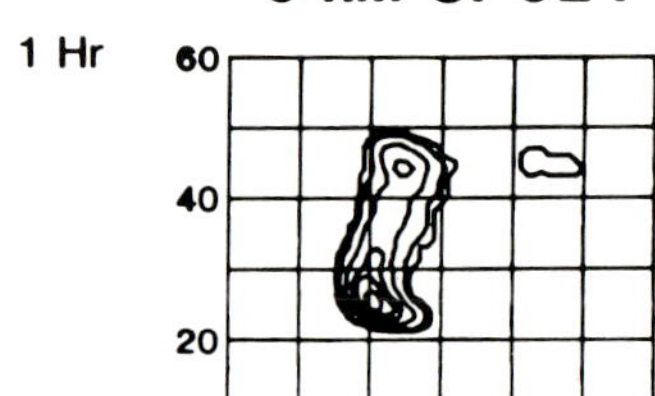

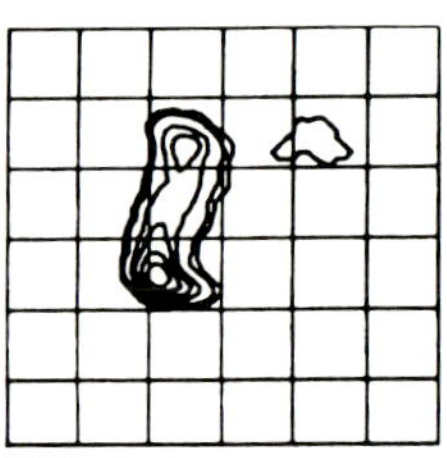

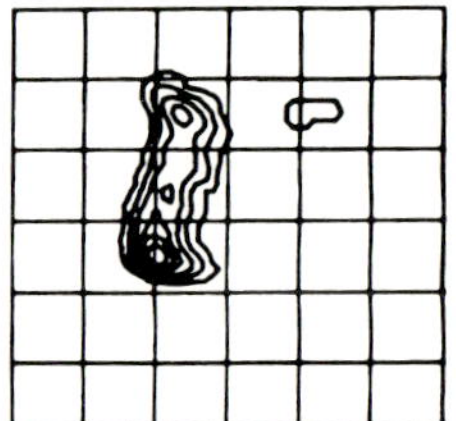

3 Hr

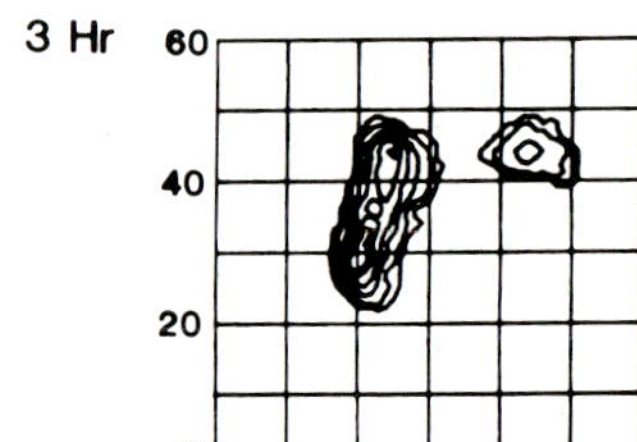

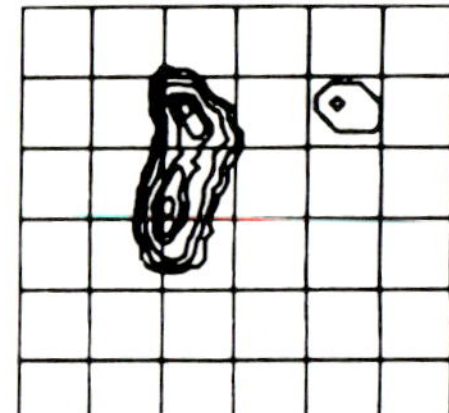

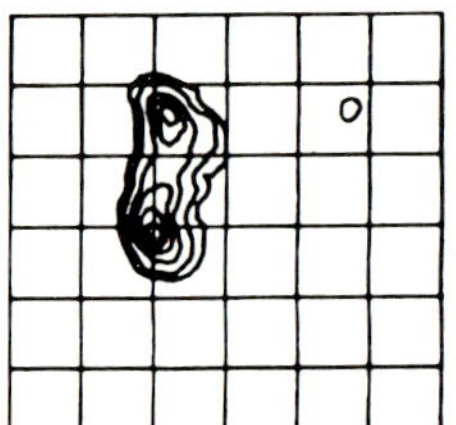

5 Hr

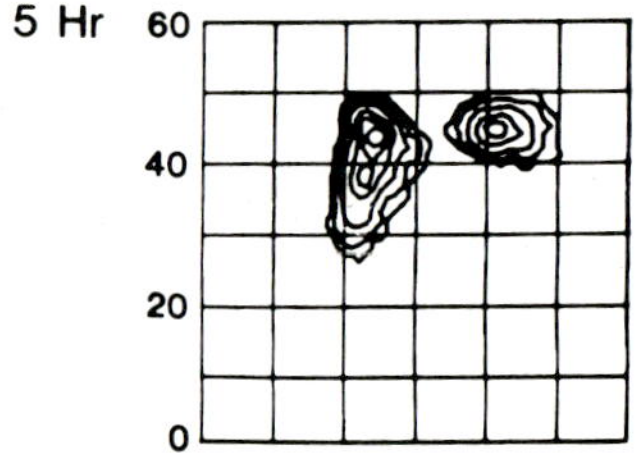

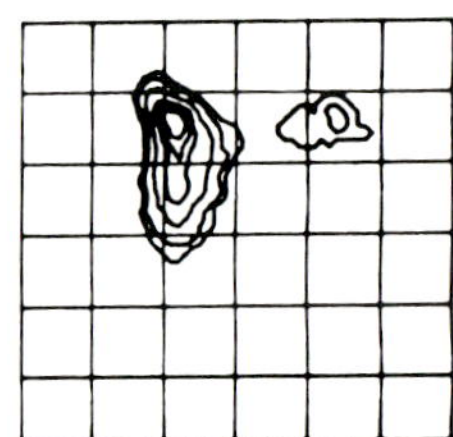

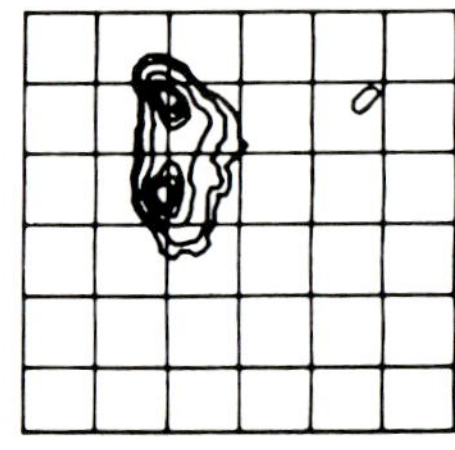

7 Hr

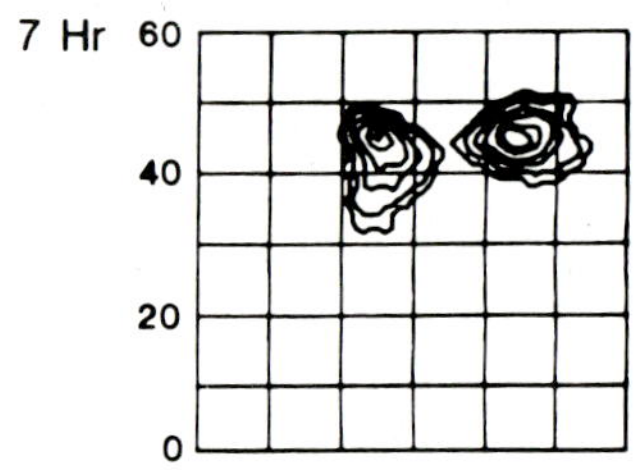

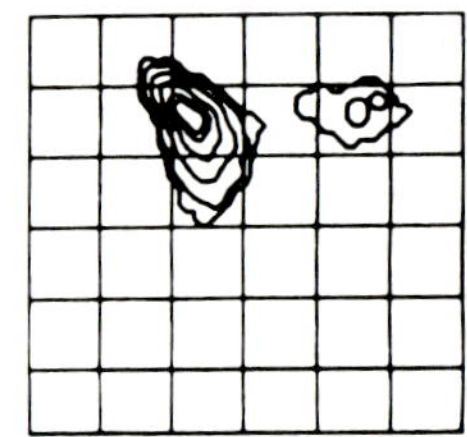

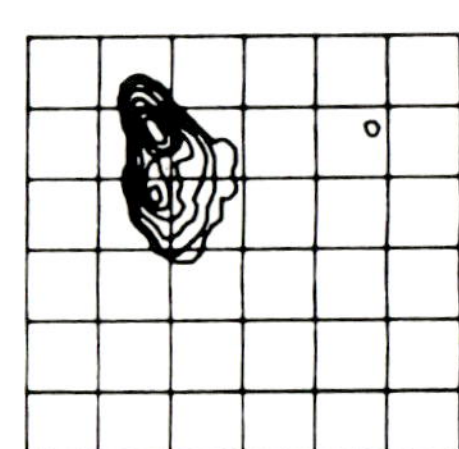

Mixture of Untreated Control and 7 Hr Samples

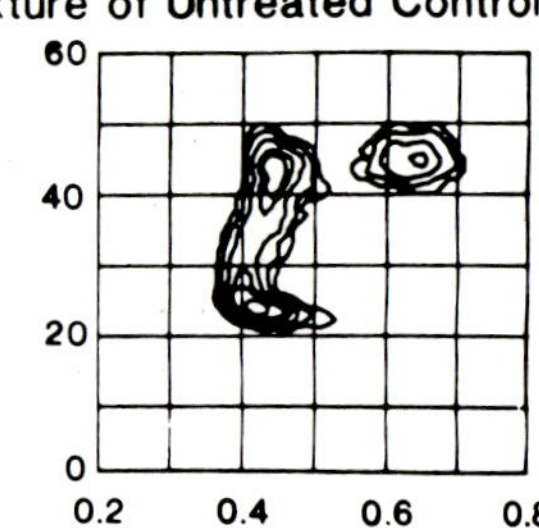

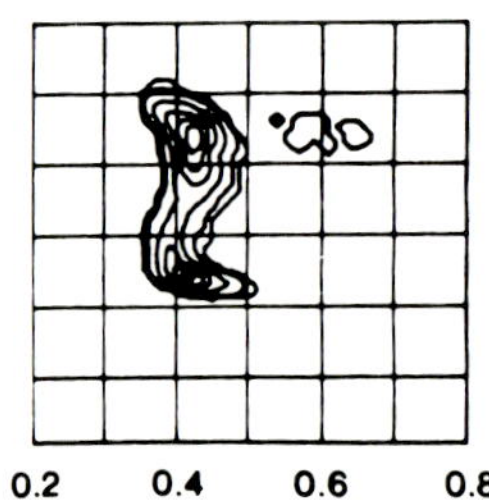

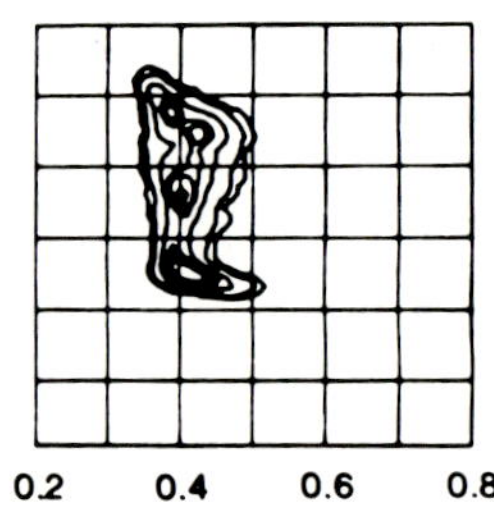

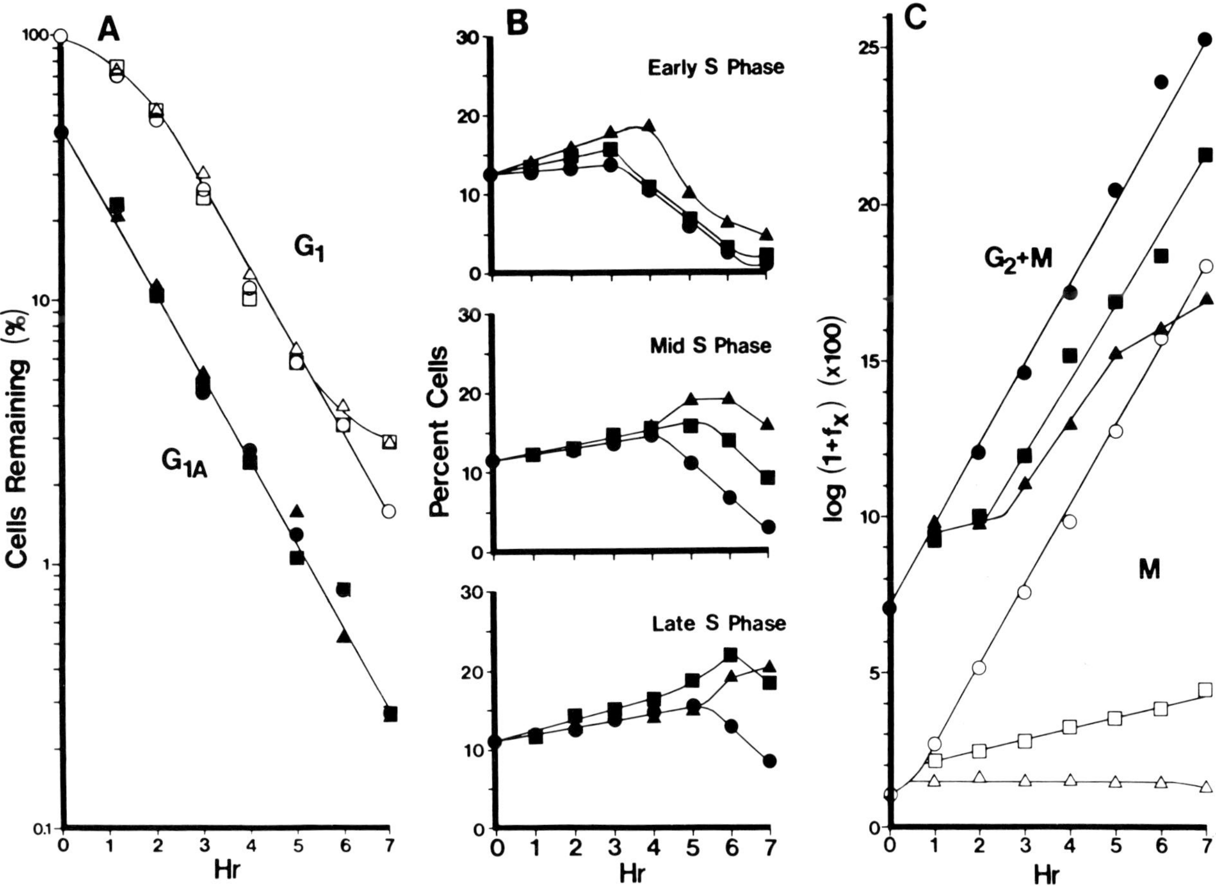

Fig. 12. Analysis of the cell cycle specific action of CI-921 obtained from the stathmokinetic experiment presented in Figure 11. **A:** Exit of cells from G_1 (stathmokinetic agent prevents cell division) calculated for control (○), 10 (□) and 50 (△) nM CI-921 cultures versus time. Exit of cells from G_{1A} (see text) also plotted (solid symbols). **B:** Passage of cells through S phase. Three equivalent "windows" in early-, mid-, and late-S phase were determined (e.g. windows of equal channel width). The percentage of cells in each window at each time point was determined (symbols as in A). **C:** Accumulation of cells in mitosis (open symbols) and G_2M (closed symbols). The action of the stathmokinetic agent should result in a logarithmic accumulation of cells in mitosis (straightline on log plot) if the culture is growing exponentially. In the absence of a drug effect, the G_2M accumulation curve should be parallel to the mitotic curve and separated along the time axis by an amount equal to the G_2 transit time (see text for detailed discussion of the data).

Fig. 11. Contour maps of selected measurements from a stathmokinetic experiment to determine the point of action of CI-921. Exponentially growing FL cells were treated with 0.05 μg/ml vinblastine and split into three separate cultures, two of which received CI-921 (10 and 50 nM). At each time point, an aliquot was collected, washed with HBSS and fixed in acetone–alcohol. When all specimens were collected, they were rehydrated in HBSS, treated with RNase, exposed to a pH 1.5 buffer, and stained with AO. The acid treatment acts to partially denature DNA in situ. Using AO to stain the remaining double-stranded DNA green and the single-stranded (denatured) DNA red, the total DNA (green plus red luminescence) and sensitivity of the DNA to denaturation, α_t (red/green plus red luminescence), could be plotted. Five distinct populations can be identified in such distributions: G_1 cells plus a subcompartment G_{1A} characterized by high α_t values, S phase cells, G_2 phase cells and a separate mitotic population (top). The bottom row of contour maps illustrate the distributions obtained by combining equal numbers of cells from the 0-hour control and the 7-hour samples treated with 0, 10, or 50 nM CI-921, respectively. Both the redistribution among cell cycle compartments and any shift in stability of DNA to denaturation sometimes caused by interaction of the drug with DNA (e.g., intercalation) can be discerned in such a display.

concentrations. After 3-hour treatment at the lower drug concentration, cells quickly began accumulating in G_2 phase at nearly the control rate. At the higher drug concentration, the situation is more complex with a slowdown from 1 to 3 hours followed by slow but steady entrance of cells from S to G_2 for the next 2 hours and then another slowdown in cell entrance into G_2. Thus, it appears that the major action of the drug is to cause an accumulation of cells in G_2 phase, though the higher the drug concentration the slower the traverse of cells through S and into G_2 phase.

In addition to an effect on cell cycle traverse, many agents, especially intercalators, also affect chromatin structure which is manifested as a shift in the α_t distribution [62]. Such is the case with CI-921 (Fig. 11). If one notes the α_t value for the 7-hour time points, the remaining interphase (G_2) and the mitotic cells have approximate α_t values of 0.46 and 0.65, respectively. The low concentration of the drug had no apparent effect on the few mitotic cells present but caused a shift in the interphase population toward lower α_t values (0.42). The shift in acid denaturability was more dramatic for interphase cells following treatment with 50 nM CI-921.

In this instance, the shift in α_t was about 19% from 0.48 to 0.39 (Fig. 11). That these shifts were not due to random machine or staining variations could be demonstrated by mixing cells from each 7-hour specimen with an equal number of 0-hour control cells (Fig. 11, bottom). The α_t shift among the G_2 cells in the sample treated with 10 nM CI-921 could still be demonstrated. At the higher concentration, four peaks can be observed: The G_1 cells; cell accumulated in mid to late S phase; G_2 cells from the control culture (slightly shifted due to the overlap of drug-treated cells) and the new "G_2" population, accumulated as a result of drug treatment, which had low α_t and slightly increased total fluorescence values (Fig. 11, bottom).

The documented shift in α_t values in the drug-treated samples is frequently observed with intercalating agents that stabilize chromatin structure [62]. Note that the intercalating agent for the most part has little effect on the binding of the intercalating dye; the total fluorescence values were not diminished. However, the slight increase in total fluorescence observed among a portion of G_2 cells in the drug treated cultures was of interest. Some cells within this population may represent tetraploid "G_1" cells which were preparing to cycle at a higher ploidy level as was observed at prolonged culture times (see Fig. 2).

Mathematical Analysis of Stathmokinetic Experiments

With appropriate sampling, flow cytometric analysis can provide information on the long-term (i.e., greater than one generation time) effects of subcytotoxic concentrations of chemotherapeutic agents on in vitro cell proliferation. Alternatively, short-term (i.e., less than one generation time) analysis of drug effects, investigated with the aid of a stathmokinetic experiment, can be accomplished more conveniently and provides greater detail. Thus, stathmokinetic experiments performed on asynchronous, exponentially growing cell lines, in addition to providing information about the basic cell kinetic characteristics of the culture (e.g., the duration of the S, G_2, and M phases and the kinetics of emptying of the G_1 and G_{1A} compartments), can, with the addition of the prospective chemotherapeutic agent, provide a detailed analysis of the point(s) of drug action. Whereas the data obtained from stathmokinetic experiments on unperturbed cell populations have proved useful in determining the relative utility and appropriateness of various theoretical models of cell proliferation [27,28,83,91,97,115, 127,140,143,186], new mathematical approaches have been developed to utilize the perturbations in the stathmokinetic data caused by different drugs to predict their long-term effects [98,99].

An attempt has been made [98] to test whether data obtained from a short-term stathmokinetic experiment (6-hour) involving two different G_2-blocking agents each tested at two concentrations could predict the DNA distributions obtained by flow cytometry measured 24 to 48 hours after drug addition. A set of feasible parameter values were obtained that characterized the action of the drugs and when substituted into a model of cell proliferation could, to some extent, reproduce the observed continuous exposure effects of the drugs. In this regard, a more refined mathematical analysis has been developed using a nonparametric approach that, although general enough to be applicable in principle to any chemotherapeutic agent whose action is distributed throughout the cell cycle, successfully predicted the long-term (24-hour) effects of continuous exposure of FL cells to CI-921 [99]. Thus, a complete "drug action curve" was constructed that predicts the percentage of cells affected as a function of their cell cycle phase.

One assumption of the model described above is that the blocking action of the drug is permanent. Modeling transient blocks and/or cell death is more difficult and requires estimating additional coefficients [99]. This represents a significant complication in predicting long-term drug effects from data obtained from only a portion of a single cell cycle. Clearly, the long-term effects of many agents will depend upon a number of factors including: drug concentration; length of drug exposure (e.g., continuous versus pulse exposure); the percentage of cells in the "sensitive" phase of the cell cycle when the drug is added; and the length of exposure of a cell to the agent within the "sensitive" cell cycle phase. Nevertheless, the information obtained by these techniques can be useful in predicting long-term drug effects and, as the mathematical models improve, it is likely that information useful for chemotherapeutic treatment schedules will be forthcoming.

CONCLUSIONS

In this review, an attempt has been made to provide the reader with an overview of some of the approaches and flow cytometric techniques currently available to study drug effects on cell cycle kinetics. It should be emphasized that, while flow cytometry is a powerful tool and can provide useful information concerning a drug's point of action, care must be exercised in extrapolating drug activity data from in vitro systems to the situation existing in human tumors in vivo. By their very nature, tissue culture cell lines are likely to provide drug activity data that are more applicable to fast growing disseminated tumors (e.g., lymphoproliferative disorders, ascites tumor) than for slow-growing, solid tumors in which drug penetration and tumor heterogeneity are difficult to model. Nevertheless, the advantages of flow cytometry in the study of drug effects on cell proliferation are numerous and include the ability: (1) to monitor both synchronized and exponentially growing cultures swiftly and accurately; (2) to study unusual populations not amenable to classical cell kinetic studies (e.g., samples devoid of S phase cells); (3) to distinguish cells within discrete subpopulations (e.g. early-, mid-, late-S phase cells); (4) to detect slowly growing or quiescent subpopulations that may exist originally or arise as a response to drug action; (5) to detect abnormalities in cell division (e.g., polyploidy or nondisjunction); (6) to determine whether and to what extent unbalanced growth is induced; (7) to monitor the ability of a drug to induce differentiation; (8) to establish (under certain circumstances) whether a drug binds to a specific subcellular target (specifically DNA); and (9) to analyze, with great precision, the point or points in the cell cycle where a drug exerts its primary action. Perhaps one of the unique advantages of flow cytometry is that, since culture conditions can be monitored almost instantaneously, one has the ability to modify ongoing experiments. Clearly, with the continued development of new and more sophisticated multiparameter flow cytometric approaches to cell cycle analysis, it will be possible to provide a great deal more information about a prospective antitumor agent prior to its introduction into clinical trials.

REFERENCES

1. **Adams EG, Crampton SL, Bhuyan BK (1981)** Effect of 7-Con-O-Methylnogarol on DNA synthesis, sur-

vival, and cell cycle progression of Chinese hamster ovary cells. Cancer Res. 41:4981–4987.

2. **Alabaster O (1978)** Cytokinetic effect of cyclophosphamide and cytosine arabinoside given concurrently and sequentially to L1210 ascites in exponential growth. In Lutz D (ed), "Pulse Cytophotometry." Ghent, Belgium: European Press, pp 709–718.
3. **Arndt-Jovin D, Jovin T (1977)** Analysis and sorting of living cells according to deoxyribonucleic acid content. J. Histochem. Cytochem. 25:585–589.
4. **Ashihara T, Traganos F, Baserga R, Darzynkiewicz Z (1978)** A comparison of cell cycle related changes in post-mitotic and quiescent AF8 cells as measured by flow cytometry after acridine orange staining. Cancer Res. 38:2514–2518.
5. **Baguley BC, Denny WA, Atwell GJ, Finlay GJ, Rewcastle GW, Twigden SJ, Wilson WR (1984)** Synthesis, antitumor activity, and DNA binding properties of a new derivative of amsacrine, N-5-dimethyl-9-((2-methoxy-4-methyl-sulfonylamino) phenylamino)-4-acridinecarboxamide. Cancer Res. 44: 3245–3251.
6. **Bakke O, Eik-Nes KB (1981)** Cell cycle-specific glucocorticoid growth regulation of a human cell line (NHIK 3025). J. Cell. Physiol. 109:489–496.
7. **Bakke O, Jakobsen K, Eik-Nes KB (1984)** Concentration-dependent effects of potassium dichromate on the cell cycle. Cytometry 5:482–486.
8. **Barlogie B, Drewinko B (1978)** Cell cycle stage dependent induction of G_2 phase arrest by different antitumor agents. In Lutz D (ed), "Pulse Cytophotometry." Ghent, Belgium: European Press, pp 545–552.
9. **Barlogie B, Drewinko B, Buchner T, Gohde W, Schumann T, Freireich E (1976)** VP-16 and Yoshi 864: Effects on the cell cycle progression in a human lymphoma cell line. In Gohde W, Schumann J, Buchner T (eds), "Pulse Cytophotometry." Ghent, Belgium: European Press, pp 226–235.
10. **Barlogie B, Raber MN, Schumann J, Johnson TS, Drewinko B, Swartzendruber DE, Gohde W, Andreeff M, Freireich E (1983)** Flow cytometry in clinical cancer research. Cancer Res. 43:3982–3997.
11. **Barlogie B, Spitzer G, Hart JS, Johnston DA, Buchner T, Schumann J, Drewinko B (1976)** DNA-histogram analysis of human hemopoietic cells. Blood 48:245–258.
12. **Barranco SC, May JT, Boerwinkle W, Nichols S, Hokanson KM, Schumann J, Gohde W, Bryant J, Guseman LF (1982)** Enhanced cell killing through the use of cell kinetics-directed treatment schedules for two-drug combinations *in vitro*. Cancer Res. 42:2894–2898.
13. **Bartholomew JC, Pearlman AL, Landolph JR, Straub K (1979)** Modulation of the cell cycle of cultured mouse liver cells by benzo(a)pyrene and its derivatives. Cancer Res. 39:2538–2543.
14. **Baserga R (1968)** Biochemistry of the cell cycle: A review. Cell Tissue Kinet. 1:167–191.
15. **Bauer KD, Dethlefsen LA (1980)** Total cellular RNA content. Correlation between flow cytometry and ultraviolet spectroscopy. J. Histochem. Cytochem. 28: 493–498.
16. **Bauer KD, Dethlefsen LA (1981)** Control of proliferation in HeLa-S3 suspension cultures. I. Simultaneous determination of cellular RNA and DNA content by acridine orange staining and flow cytometry. J. Cell. Physiol. 108:99–112.
17. **Bauer KD, Keng PC, Sutherland RM (1982)** Isolation of quiescent cells from multicellular tumor spheroids using centrifugal elutriation. Cancer Res. 42:72–78.
18. **Beck HP (1981)** Proliferation kinetics of perturbed cell populations determined by the bromodeoxyuridine-33258 technique. Radiotoxic effect of incorporated (^{3}H)thymidine. Cytometry 2:170–174.
19. **Bhuyan BK, Adams EG, Johnson M, Crampton SL (1985)** Synergistic combination of menogarol and melphalan and other two drug combinations. Invest. New Drugs 3:233–244.
20. **Bohmer RM (1979)** Flow cytometrical cell cycle analysis using the quenching of 33258 Hoechst fluorescence by BUdR incorporation. Cell Tissue Kinet. 12: 101–110.
21. **Bohmer RM, Ellwart J (1981)** Cell cycle analysis by combining the 5-bromodeoxyuridine/33258 Hoechst technique with DNA-specific ethidium bromide staining. Cytometry 2:31–34.
22. **Braylan RC, Benson NA, Nourse V, Kruth HS (1982)** Correlated analysis of cellular DNA, membrane antigens and light scatter of human lymphoid cells. Cytometry 2:337–343.
23. **Buchner T, Barlogie B, Hofschroer J, Metz J, Ortheil U, Hortebusch H, Gohde W, Schumann J (1976)** Cell kinetic effects of cytosine arabinoside. I. Experimental studies using *in vitro* and *in vivo* systems. In Gohde W, Schumann J, Buchner T (eds), "Pulse Cytophotometry." Ghent, Belgium: European Press, pp 221–225.
24. **Campisi C, Medrano EE, Morreo G, Pardee AB (1982)** Restriction point control of cell growth by a labile protein: Evidence for increased stability in transformed cells. Proc. Natl. Acad. Sci. USA 79: 436–440.
25. **Casperson T (1936)** Uber den Chemischen Aufbau der struktusen des Zellkernes. Scand. Arch. Physiol. 73(suppl. 8):1–151.
26. **Charcosset J-Y, Bendridjian J-P, Lantieri M-F, Jacquemin-Sablon A (1985)** Effects of 9-OH-ellipticine on cell survival, macromolecular synthesis and cell cycle progression in sensitive and resistant Chinese hamster lung cells. Cancer Res. 45:4229–4236.
27. **Chuang S-N, Lloyd HH (1975)** Mathematical analysis of cancer chemotherapy. Cell Math. Biol. 37:147–160.
28. **Chuang S-N, Soong TT (1978)** Mathematical analysis of cancer chemotherapy: The effects of chemotherapeutic agents on the cell cycle traverse. Bull. Math. Biol. 40:499–512.
29. **Clevenger CV, Bauer KD, Epstein AL (1985)** A method for simultaneous nuclear immunofluorescence and DNA content quantitation using monoclonal antibodies and flow cytometry. Cytometry 6: 208–214.
30. **Coulsen PB, Bishop AO, Lenarduzzi R (1977)** Quantitation of cellular deoxyribonucleic acid by flow microfluorometry. J. Histochem. Cytochem. 25:1147–1153.
31. **Creasey WA (1981)** The vinka alkaloids and similar compounds. In Crooke ST, Prestayko AW (eds), "Cancer and Chemotherapy." Vol. III. Orlando, Florida, Academic Press, pp 79–96.

32. **Crissman HA, Darzynkiewicz Z, Tobey RA, Steinkamp JA** (1985) Correlated measurements of DNA, RNA and protein in individual cells by flow cytometry. Science 228:1321–1324.
33. **Crissman HA, Mullaney PF, Steinkamp JA** (1975) Methods and applications of flow systems for analysis and sorting of mammalian cells. In Prescott DA (ed), "Methods in Cell Biology." New York: Academic Press, pp 179–246.
34. **Crissman HA, Oka MS, Steinkamp JA** (1976) Rapid staining methods for analysis of deoxyribonucleic acid and protein in mammalian cells. J. Histochem. Cytochem. 24:64–71.
35. **Crissman HA, Ovlicky DJ, Kissane RJ** (1979) Use of the two parameter (DNA and cell size) flow cytometric analysis for determining the effects of potential chemotherapeutic agents. In Laerum OD, Lindmo T, Thorud E (eds), "Flow Cytometry. Vol. IV. Oslo: Universitetsforlaget, pp 526–530.
36. **Crissman HA, Steinkamp JA** (1982) Rapid, one step staining procedure for analysis of cellular DNA and protein by single and dual laser flow cytometry. Cytometry 3:84–90.
37. **Crissman HA, Stevenson AP, Kissane RJ, Tobey RA** (1979) Techniques for quantitative staining of cellular DNA for flow cytometric analysis. In Melamed MR, Mullaney PF, Mendelsohn ML (eds), "Flow Cytometry and Sorting." New York: John Wiley & Sons, pp 243–262.
38. **Crissman HA, Tobey RA** (1974) Cell cycle analysis in 20 minutes. Science 184:1297–1298.
39. **Crissman HA, Van Egmond J, Holdrinet RS, Pennings A, Haanen C** (1981) Simplified method for DNA and protein staining of human hematopoietic cell samples. Cytometry 2:59–62.
40. **Crowther PJ, Cooper IA, Woodcock DM** (1985) Biology of cell killing by 1-β-D-arabinofuranosylcytosine and its relevance to molecular mechanisms of cytotoxicity. Cancer Res. 45:4291–4300.
41. **Darzynkiewicz Z** (1979) Acridine orange as a molecular probe in studies of nucleic acids *in situ*. In Melamed MR, Mullaney PF, Mendelsohn ML (eds), "Flow Cytometry and Sorting." New York: John Wiley & Sons, pp 285–316.
42. **Darzynkiewicz Z** (1983) Molecular interactions and cellular changes during the cell cycle. Pharmacol Ther 21:143–188.
43. **Darzynkiewicz Z** (1984) Metabolic and kinetic compartments of the cell cycle distinguished by multiparameter flow cytometry. In Skehan P, Friedman SJ (eds), "Growth, Cancer and the Cell Cycle." Clifton, NJ: Humana Press, pp 249–278.
44. **Darzynkiewicz Z, Carter S, Kimmel M** (1984) Effect of (^{3}H)-UdR on the cell cycle progression of L1210 cells. Cell Tissue Kinet. 17:641–655.
45. **Darzynkiewicz Z, Crissman H, Traganos F, Steinkamp J** (1982) Cell heterogeneity during the cell cycle. J. Cell. Physiol. 113:465–474.
46. **Darzynkiewicz Z, Evenson DP, Staiano-Coico L, Sharpless T, Melamed MR** (1979) Correlation between cell cycle duration and RNA content. J. Cell. Physiol. 100:425–438.
47. **Darzynkiewicz Z, Evenson D, Staiano-Coico L, Sharpless T, Melamed MR** (1979) Relationship between RNA content and progression of lymphocytes through the S phase of the cell cycle. Proc. Natl. Acad. Sci. USA 76:358–362.
48. **Darzynkiewicz Z, Traganos F** (1982) RNA content and chromatin structure in cycling and noncycling cell populations studied by flow cytometry. In Padilla GM, McCarty KS (eds), "Genetic Expression in the Cell Cycle." Orlando, Florida: Academic Press, pp 103–128.
49. **Darzynkiewicz Z, Traganos F, Andreeff M, Sharpless T, Melamed MR** (1979) Different sensitivity of chromatin to acid denaturation in quiescent and cycling cells as revealed by flow cytometry. J. Histochem. Cytochem. 27:478–485.
50. **Darzynkiewicz Z, Traganos F, Arlin Z, Sharpless T, Melamed MR** (1976) Cytofluorometric studies on conformation of nucleic acids *in situ*. II. Denaturation of deoxyribonucleic acid. J. Histochem. Cytochem. 24:49–58.
51. **Darzynkiewicz Z, Traganos F, Kapuscinski J, Melamed MR** (1985) Denaturation and condensation of DNA *in situ* induced by acridine orange. Relationship to chromatin changes during differentiation of Friend erythroleukemic cells. Cytometry 6: 195–207.
52. **Darzynkiewicz Z, Traganos F, Kimmel M** (1986) Assay of cell cycle kinetics by multiparameter flow cytometry using the principle of stathmokinesis. In Gray JW, Darzynkiewicz Z (eds), "Techniques for Analysis of Cell Proliferation." Clifton, NJ: Humana Press, pp 291–336.
53. **Darzynkiewicz Z, Traganos F, Melamed MR** (1980) New cell cycle compartments identified by multiparameter flow cytometry. Cytometry 1:98–108.
54. **Darzynkiewicz Z, Traganos F, Sharpless T, Melamed MR** (1974) Thermally-induced changes in chromatin of isolated nucleic acids and of intact cells as revealed by acridine orange staining. Biochem. Biophys. Res. Commun. 59:392–399.
55. **Darzynkiewicz Z, Traganos F, Sharpless T, Melamed MR** (1975) Thermal denaturation of DNA *in situ* as studied by acridine orange staining and automated cytofluorometry. Exp. Cell. Res. 95:143–153.
56. **Darzynkiewicz Z, Traganos F, Sharpless T, Melamed MR** (1976) Lymphocyte stimulation: A rapid multiparameter analysis. Proc. Natl. Acad. Sci. USA 73: 2881–2884.
57. **Darzynkiewicz Z, Traganos F, Sharpless T, Melamed MR** (1977) Interphase and metaphase chromatin: Different stainability of DNA with acridine orange after treatment at low pH. Exp. Cell. Res. 110:201–214.
58. **Darzynkiewicz Z, Traganos F, Sharpless T, Melamed MR** (1977) Different sensitivity of DNA *in situ* in interphase and metaphase chromatin to heat denaturation. J. Cell. Biol. 73:128–138.
59. **Darzynkiewicz Z, Traganos F, Staiano-Coico L, Kapuscinski J, Melamed MR** (1982) Interactions of rhodamine 123 with living cells studied by flow cytometry. Cancer Res. 42:799–806.
60. **Darzynkiewicz Z, Williamson B, Carswell EA, Old LJ** (1984) Cell cycle-specific effects of tumor necrosis factor. Cancer Res. 44:83–90.
61. **Darzynkiewicz Z, Traganos F, Xue S, Melamed MR** (1981) Effect of n-butyrate on cell cycle progression and *in situ* chromatin structure of L1210 cells. Exp. Cell. Res. 136:279–293.

62. **Darzynkiewicz Z, Traganos F, Xue S, Staiano-Coico L, Melamed MR (1981)** Rapid analysis of drug effects on the cell cycle. Cytometry 1:279–286.
63. **Dethlefsen LA (1979)** In quest of the quaint quiescent cells. In Meyn RF, Withers HR (eds), "Radiation Biology in Cancer Research." New York: Raven Press, pp 415–435.
64. **Dethlefsen LA, Bauer KD, Riley M (1980)** Analytic cytometric approaches to heterogeneous cell populations in solid tumors. Cytometry 1:89–97.
65. **Dethlefsen LA, Gray JW, George YS, Johnson S (1976)** Flow cytometric analysis of the perturbed cellular kinetics of solid tumors: problems and promises. In Gohde W, Schumann J, Buchner T (eds), "Pulse Cytophotometry." Ghent, Belgium: European Press, pp 188–200.
66. **Dittrich W, Gohde W (1969)** Inpulsefluorometrie bei Einzelzellen in Suspensionen. Z. Naturforsch. 24B: 360–361.
67. **Drewinko B, Barlogie B (1976)** Survival of cultured human lymphoma cells after treatment with 3′,3′-iminodi-1-propanol, dimethanesuflonate (ester), hydrochloride (Yoshi 864) and 4′-demethylepipodophyl-lotoxin 9-(4,6-0(ethylidene-β-D-glucopyranoside) (VP-16). In Gohde W, Schumann J, Buchner T (eds), "Pulse Cytophotometry." Ghent, Belgium: European Press, pp 236–243.
68. **Drewinko B, Patchen M, Yang L-Y, Barlogie B (1981)** Differential killing efficacy of twenty antitumor drugs on proliferating and nonproliferating human tumor cells. Cancer Res. 41:2328–2333.
69. **Drewinko B, Roper P, Barlogie B (1979)** Patterns of cell survival following treatment with anti-tumor agents *in vitro*. Eur. J. Cancer. 15:92–98.
70. **Drewinko B, Yang L-Y, Barlogie B (1982)**: Lethal activity and kinetic response of cultured human cells to 4′(9-acridinylamine) methanesufon-m-anisidine. Cancer Res. 42:107–111.
71. **Esnault K, Roques BP, Jacquemin-Sablon A, Le Pecq JB (1984)** Effects of new antitumor bifunctional intercalators derived from 7H-pyridocarbazole on sensitive and resistant L1210 cells. Cancer Res. 44: 4355–4360.
72. **Evenson DP, Darzynkiewicz Z, Staiano-Coico L, Traganos F, Melamed MR (1979)** Effects of 9,10-anthracenedione, 1-4 bis ({2[(hydroxyethyl)amino]-ethyl}amino)-diacetate on cell survival and cell cycle progression in cultured mammalian cells. Cancer Res. 39:2574–2581.
73. **Evenson DP, Traganos F, Darzynkiewicz Z, Staiano-Coico L, Melamed MR (1980)** Effects of 9,10-anthracenedione, 1-4 bis ({2-[(hydroxyethyl)amino] ethyl}amino)-diacetate on cell morphology and nucleic acids of Friend leukemia cells. J. Natl. Cancer Inst. 64:857–866.
74. **Farber S, Diamond LK, Mercer RD, Sylvester RF, Wolff JA (1948)** Temporary remissions in acute leukemia in children produced by folic acid antagonist, 4-amino-pteroyl-glutamic acid (aminopterin). N. Engl. J. Med. 238:787–793.
75. **Fox RM, Kefford RF, Tripp EH, Taylor IW (1981)** G_1 phase arrest of cultured human leukemic T-cells induced by deoxyadenosine. Cancer Res. 41:5141–5150.
76. **Freeman DA, Crissman HA (1975)** Evaluation of six fluorescent protein stains for use in flow microfluorometry. Stain Technol. 50:279–284.
77. **Fried J, Perez AG, Doblin JM, Clarkson BD (1981)** Cytotoxic and cytokinetic effects of thymidine, 5-fluorouracil, and deoxycytidine on HeLa cells in culture. Cancer Res. 41:2627–2632.
78. **Fujikawa-Yamamoto K (1982)** RNA dependence in the cell cycle of V79 cells. J. Cell. Physiol. 112:60–66.
79. **Gehring U, Gray JV, Tomkins GM (1976)** Corticosteroid effect on the cell cycle. In Gohde W, Schumann J, Buchner T (eds), "Pulse Cytophotometry". Ghent, Belgium: European Press, pp 284–290.
80. **Gohde W, Schumann J, Zante J (1978)** The use of DAPI in pulse-cytophotometry. In Lutz D (ed), "Pulse Cytophotometry." Ghent, Belgium: European Press, pp 229–232.
81. **Gohde W, Spies I, Schumann J, Buchner T, Kliene-Dopke G (1976)** Two parameter analysis of DNA and protein content of tumor cells. In Gohde W, Schumann J, Buchner T (eds), "Pulse Cytophotometry." Ghent, Belgium: European Press, pp 27–32.
82. **Gilman A, Philips FS (1946)** The biological actions and therapeutic applications of β-chloroethyl amines and sulfides. Science 103:409–415.
83. **Gray JW (1983)** Quantitative cytokinetics: Cellular response to cell cycle specific agents. Pharmacol. Ther. 22:163–197.
84. **Gray JW, Dean PN, Mendelsohn ML (1979)** Quantitative cell-cycle analysis. In Melamed MR, Mullaney PF, Mendelsohn ML (eds), "Flow Cytometry and Cell Sorting." New York: John Wiley & Sons, pp 383–407.
85. **Hawkes SP, Bartholomew JC (1977)** Quantitative determination of transformed cells in a mixed population by simultaneous fluorescence analysis of cell surface and DNA in individual cells. Proc. Natl. Acad. Sci. USA 74:1626–1630.
86. **Hecquet C, Nafziger J, Ronot X, Marie J-P, Adolphe M (1985)** Flow cytometric analysis of pentakis(aziridino)-thiatriazadiphosphorine oxide (SOAz)-induced changes in cell cycle progression of HeLa and HL-60 cells. Cancer Res. 45:552–554.
87. **Heidelberger C (1965)** Fluorinated pyrimidines. In Davidson JN, Cohn WE (eds), "Progress in Nucleic Acid Research and Molecular Biology." New York: Academic Press, pp 1–50.
88. **Hieber L, Beck H-P, Lucke-Huhle C (1981)** G_2-delay after irradiation with particles as studied in synchronized cultures and by the bromodeoxy-uridine-33258 H technique. Cytometry 2:175–178.
89. **Hollstein U (1974)** Actinomycin. Chemistry and mechanism of action. Chem. Rev. 74:625–652.
90. **Howard A, Pelc SR (1953)** Synthesis of deoxyribonucleic acid in normal and irradiated cells and its relation to chromosomal breakage. Heredity 6(suppl.):261–273.
91. **Jagers P (1983)** Stochastic models for cell kinetics. Bull. Math. Biol. 45:507–519.
92. **Kachel V, Glossner E, Kordwig E, Ruhenstroth-Bauer G (1977)** FLUVO-Metrical, a combined cell volume and cell fluorescence analyser. J. Histochem. Cytochem. 25:804–812.
93. **Kaiser TN, Lojewski A, Dougherty C, Juergens L, Sahar E, Latt SA (1982)** Flow cytometric character-

ization of the response of Fanconi's anemia cells to mitromycin C treatment. Cytometry 2:291–297.

94. **Kamensky LA, Melamed MR (1969)** Instrumentation for automated examination of cellular specimens. Proc. IEEE 57:2007–2016.

95. **Kapuscinski J, Darzynkiewicz Z, Melamed MR (1982)** Luminescence of the solid complexes of acridine orange with RNA. Cytometry 2:201–211.

96. **Kefford RF, Taylor IW, Fox RM (1983)** Cytometric analysis of adenosine analogue lymphocytotoxicity. Cancer Res. 43:5112–5119.

97. **Kimmel M (1985)** Nonparametric analysis of stathmokinesis. Math. Biosci. 74:111–123.

98. **Kimmel M, Traganos F (1985)** Kinetic analysis of a drug-induced G_2 block *in vitro*. Cell Tissue Kinet. 18:91–110.

99. **Kimmel M, Traganos F (1986)** Estimation and prediction of cell cycle specific effects of anticancer drugs. Math. Biosci. 80:187–208.

100. **Kipp JBA, Jongsma APM, Barendsen GW (1979)** Cell cycle phase durations derived by a flow cytofluorometric method using the mitotic inhibitor vinblastine. In Lareum OD, Lindmo T, Thorud E (eds), "Flow Cytometry." Vol. IV. Oslo: Universitetsforlaget, pp 341–344.

101. **Krishan A (1975)** Rapid flow cytofluorometric analysis of mammalian cell cycle by propidium iodide staining. J. Cell. Biol. 66:188–193.

102. **Krishan A, Dutt K, Israel M, Ganopathi R (1981)** Comparative effects of adriamycin and N-trifluoroacetyladriamycin-14-valerate on cell cycle kinetics, chromosomal damage, and macromolecular synthesis *in vitro*. Cancer Res. 41:2745–2750.

103. **Krishan A, Frei E III (1976)** Effect of adriamycin on the cell cycle traverse and kinetics of cultured human lymphoblasts. Cancer Res. 36:143–150.

104. **Krishan A, Ganapathi RN, Israel M (1978)** Effect of Adriamycin and analogs on the nuclear fluorescence of propidium iodide-stained cells. Cancer Res. 38: 3656–3662.

105. **Krishan A, Paika K, Frei E III (1975)** Cytofluorometric studies on the action of podophyllotoxin and epipodophyllotoxins (VM-26, VP-16-213) on the cell cycle traverse of human lymphoblasts. J. Cell. Biol. 66:521–530.

106. **Krishan A, Paika KD, Frei E III (1976)** Cell cycle synchronization of human lymphoid cells *in vitro* by 2,3-dihydro-1H-imidazo(1,2b)pyrazole. Cancer Res. 36:138–142.

107. **Krishan A, Pitman SW, Tattersall MHN, Paika KD, Smith DC, Frei E III (1976)** Flow microfluorometric patterns of human bone marrow and tumors cells in response to cancer chemotherapy. Cancer Res. 36: 3813–3820.

108. **Kurland J, Traganos F, Darzynkiewicz Z, Moore MAS (1978)** Macrophage mediated cytostasis of neoplastic hemopoietic cells: Cytofluorometric analysis of the reversible cell cycle block. Cell. Immunol. 36: 318–330.

109. **Langen P, Graetz H, Lehmann E (1979)** The use of cytosine arabinoside and colchicin in FCM analysis of cell cycle effects of inhibitors in an asynchronous cell culture. In Laerum OD, Lindmo T, Thorud E (eds), "Flow Cytometry." Vol. IV. Oslo: Universitetsforlaget, pp 362–366.

110. **Lanier LL, Warner NL (1981)** Cell cycle related heterogeneity of Ia antigen expression on a murine B lymphoma cell line: Analysis by flow cytometry. J. Immunol. 126:626–631.

111. **Latt SA (1979)** Fluorescent probes of DNA microstructure and synthesis. In Melamed MR, Mullaney PF, Mendelsohn ML (eds), "Flow Cytometry and Sorting." New York: John Wiley & Sons, pp 263–284.

112. **Latt SA, George YS, Gray JW (1977)** Flow cytometric analysis of bromo-deoxyuridine substituted cells stained with 33258 Hoechst. J. Histochem. Cytochem. 25:927–934.

113. **Leong SPL, Bolen JL, Chee DO, Smith VR, Taylor JC, Benfield JR, Klevecz RR (1984)** Cell-cycle-dependent expression of human melanoma membrane antigen analysed by flow cytometry. Cancer 55:1276–1283.

114. **Lindmo T, Davies C, Rofstad EK, Fodstad O, Sundan A (1984)** Antigen expression in human melanoma cells in relation to growth conditions and cell cycle distribution. Int. J. Cancer 33:167–171.

115. **Macdonald PDM (1981)** Towards an exact analysis of stathmokinetic and continuous-labelling experiments. In Rotenberg M (ed), "Biomathematics and Cell Kinetics." Amsterdam: Elsevier/North-Holland Biomedical Press, pp 125–142.

116. **Meck RA, Clubb KJ, Allen LM, Yunis AA (1981)** Inhibition of cell cycle progression of human pancreatic carcinoma cells *in vitro* by L-(αS,5S)-α-amino-3-chloro-4,5-dihydro-5-isoxazoleacetic acid, Acivicin (NSC 163501). Cancer Res. 41:4547–4553.

117. **Miller-Faures A, Michel N, Aguilera A, Blave A, Miller AOA (1981)** Laser flow cytofluorometric analysis of HTC cells synchronized with hydroxyurea, nocodazole and aphidicolin. Cell Tissue Kinet. 14:501–514.

118. **Murgo AJ, Fried J, Burchenal D, Vale KL, Strife A, Woodcock T, Young CW, Clarkson BD (1980)** Effects of thymidine and thymidine plus 5-fluorouracil on the growth kinetics of a human lymphoid cell line. Cancer Res. 40:1543–1549.

119. **Nusse M (1981)** Cell cycle kinetics of irradiated synchronous and asynchronous tumor cells with DNA distribution analysis and BrdUrd-Hoechst 33258-technique. Cytometry 2:70–79.

120. **Pardee AB, Campisi J, Gray HE, Dean M, Sonenshein G (1985)** Cellular oncogenes, growth factors and cellular growth control. In Ford RJ, Maizel AL (eds), "Mediators in Cell Growth and Differentiation." New York: Raven Press, pp 21–29.

121. **Pedrali-Noy G, Miller-Faures A, Miller AOA, Kruppa J, Koch G (1980)** Synchronization of HeLa cell cultures by inhibition of DNA polymerase with aphidicolin. Nucleic Acid Res. 8:377–387.

122. **Pettersen EO, Nome O, Dornish JM, Oftebro R (1985)** Mitotic arrest and interphase inhibition induced by the pyrimidine sulfoxide NY4138. Invest. New Drugs 3:245–253.

123. **Pollack A, Moulis H, Block NL, Irvin GL III (1984)** Quantitation of cell kinetic responses using simultaneous flow cytometric measurements of DNA and nuclear protein. Cytometry 5:473–481.

124. **Potmesil M, Levin M, Traganos F, Israel M, Darzynkiewicz Z, Khetarpal VK, Silber R (1983)** *In*

vivo effects of adriamycin or N-trifluoroacetyl-adriamycin-14-valerate on a mouse lymphoma. Eur. J. Clin. Oncol. 18:109–122.

125. **Pratt WB, Ruddon RW (1979)** "The Anticancer Drugs." New York: Oxford University Press.

126. **Prescott DM (1976)** "Reproduction of Eukaryotic Cells." New York: Academic Press.

127. **Puck TT, Steffen J (1963)** Life cycle analysis of mammalian cells. Part I. Biophys. J. 3:379–397.

128. **Quastler H, Sherman FG (1959)** Cell population kinetics in the intestinal epithelium of the mouse. Exp. Cell. Res. 17:420–438.

129. **Reddel RR, Murphy LC, Hall RE, Sutherland RL (1985)** Differential sensitivity of human breast cancer cell lines to the growth inhibitory effects of tamoxifen. Cancer Res. 45:1525–1531.

130. **Ronning OW, Lindmo T, Pettersen EO, Seglen PO (1979)** Kinetics of entry into S phase and into G_1 phase of the subsequent cell cycle for synchronized NHIK 3025 cells. In Laerum OD, Lindmo T, Thorud E (eds), "Flow Cytometry." Vol. IV. Oslo: Universitetsforlaget, pp 350–353.

131. **Ronot X, Adolphe M, Kuch D, Jaffray P, Lechat P, Deysson G (1982)** Effect of sodium cis-β-4-methoxybenzoyl-β-bromacrylate (Cytembena) on HeLa cell kinetics. Cancer Res. 42:3193–3195.

132. **Roti Roti JL, Higashikubo R, Blair OC, Uygur N (1982)** Cell-cycle position and nuclear protein content. Cytometry 3:91–96.

133. **Rudolph NS, Ohlsson-Wilhelm BM, Leary JF, Rowley PT (1985)** Single-cell analysis of the relationship among transferrin receptors, proliferation and cell cycle phase in K562 cells. Cytometry 6:151–158.

134. **Samy TSA, Siegel PJ, Hopper WE Jr, Krishan A (1984)** Experimental pharmacology of auromycin in L1210 tumor cells *in vitro* and *in vivo*. Cancer Res. 44:3202–3207.

135. **Sarkar S, Glassy MC, Ferrone S, Jones OW (1980)** Cell cycle and the differential expression of HLA-A,B and HLA-DR antigens on human B lymphoid cells. Proc. Natl. Acad. Sci. USA 77:7297–7301.

136. **Schnedl W, Lutz D (1978)** DNA measurements by using the A-T specific fluorochrome DIPI and the G-C specific Mithramycin. In Lutz D (ed), "Pulse Cytophotometry." Ghent, Belgium: European Press, pp 233–237.

137. **Shapiro H (1981)** Flow cytometric estimation of DNA and RNA content in intact cells stained with Hoechst 33342 and pyronin Y. Cytometry 2:143–150.

138. **Shapiro HM (1983)** Multistation multiparameter flow cytometry: A critical review and rationale. Cytometry 3:227–243.

139. **Sharpless TK (1979)** Cytometric data processing. In Melamed MR, Mullaney PF, Mendelsohn ML (eds), "Flow Cytometry and Sorting." New York: John Wiley & Sons, pp 359–379.

140. **Sharpless TK, Schlesinger FH (1982)** Flow cytometric analysis of G_1 exit kinetics in asynchronous L1210 cell cultures with the constant transition probability model. Cytometry 3:196–201.

141. **Sharpless T, Traganos F, Darzynkiewicz Z, Melamed MR (1975)** Flow cytofluorimetry: Discrimination between single cells and cell aggregates by direct size measurement. Acta Cytol. (Praha) 19:577–581.

142. **Smets LA, Homan-Blok J (1985)** S1-phase cells of the leukemic cell cycle sensitive to 1-β-D-arabinofuranosyl-cytosine at a high dose level. Cancer Res. 45: 3113–3117.

143. **Staudte RG (1981)** On the accuracy of some estimates of cell cycle time. In Rotenberg M (ed), "Biomathematics and Cell Kinetics." Amsterdam: Elsevier/North-Holland Biomedical Press, pp 233–239.

144. **Steinkamp JA, Crissman HA (1974)** Automated analysis of deoxyribonucleic acid, protein and nuclear to cytoplasmic relationships in tumor cells and gynecologic specimens. J. Histochem. Cytochem. 22:616–621.

145. **Steinkamp JA, Hansen KM, Crissman HA (1976)** Flow microfluorometric and light-scatter measurement of nuclear and cytoplasmic size in mammalian cells. J. Histochem. Cytochem. 24:292–297.

146. **Steinkamp JA, Stewart CC, Crissman HA (1982)** Three-color fluorescence measurements on single cells excited at three laser wavelengths. Cytometry 2:226–231.

147. **Stohr M, Petrova L (1975)** The alkaline hydrolysis of nucleic acid for removal of RNA associated fluorescence in phenanthridium related flow through cytofluorometry. Histochemistry 45:95–99.

148. **Sutherland RL, Hall RE, Taylor IW (1983)** Cell proliferation kinetics of MCF-7 human mammary carcinoma cells in culture and effects of tamoxifen on exponentially growing and plateau-phase cells. Cancer Res. 43:3998–4006.

149. **Tanke HJ, Niewehuis IAB, Koper GJM, Slats JCM, Ploem JS (1980)** Flow cytometry of human reticulocytes based on RNA fluorescence. Cytometry 1:313–320.

150. **Taylor IW (1980)** A rapid single step staining technique for DNA analysis by flow microfluorimetry. J. Histochem. Cytochem. 28:1021–1024.

151. **Taylor IW, Bleehen NM (1977)** Changes in sensitivity to radiation and ICRF 159 during the life of monolayer cultures of EMT6 tumor line. Br. J. Cancer 35:587–594.

152. **Taylor IW, Hodson PJ, Green MD, Sutherland RL (1983)** Effects of tamoxifen on cell cycle progression of synchronous MCF-7 human mammary carcinoma cells. Cancer Res. 43:4007–4010.

153. **Taylor IW, Slowiaczek P, Francis PR, Tattersall MHN (1982)** Biochemical and cell cycle perturbations in methotrexate treated cells. Mol. Pharmacol. 21:204–210.

154. **Taylor IW, Tattersall MHN (1981)** Methotrexate cytotoxicity in cultured human leukemic cells studied by flow cytometry. Cancer Res. 41:1549–1558.

155. **Taylor IW, Tattersall MHN, Fox RM (1979)** Flow cytometry analysis of unbalanced cell growth. In Laerum OD, Lindmo T, Thorud E (eds), "Flow Cytometry." Vol. IV. Oslo: Universitetsforlaget, pp 516–521.

156. **Thornthwaite JT, Allen LM (1980)** The effect of the glutamine analog, AT-125, on the cell cycle of MCF-7 and BT-20 human breast carcinoma cells using DNA flow cytometry. Res. Commun. Chem. Pathol. Pharmacol. 29:393–396.

157. **Thorud E, Clausen OPF (1978)** The effects of bleomycin on murine epidermal cell kinetics. In Lutz D

(ed), "Pulse Cytophotometry." Ghent, Belgium: European Press, pp 553–560.

158. **Tobey RA, Crissman HA** (1972) Use of flow microfluorometry in detailed analysis of effects of chemical agents on cell cycle progression. Cancer Res. 32: 2726–2732.

159. **Tobey RA, Oka MS, Crissman HA** (1975) Differential effects of two chemotherapeutic agents, streptozotocin and chlorozotocin, on the mammalian cell cycle. Eur. J. Cancer 11:433–441.

160. **Tobey RA, Oka MS, Crissman HA** (1979) Analysis of effects of chemotherapeutic agents on cell-growth kinetics in cultured cells. In Melamed MR, Mullaney PF, Mendelsohn ML (eds), "Flow Cytometry and Sorting." New York: John Wiley & Sons, pp 573–582.

161. **Tomasovic SP, Higashikubo R, Cohen AM, Brown AD** (1982) Comparative kinetics of poly (6-thioinosinic acid) and 6-mercaptopurine treated L1210 cells. Cytometry 3:48–54.

162. **Traganos F** (1982) Dihydroxyanthraquinone and related bis(substituted) amino anthraquinones: A novel class of antitumor agents. Pharmacol. Ther. 22:199–214.

163. **Traganos F** (1984) Flow cytometry: Principles and applications. I. Cancer Invest. 2:149–163.

164. **Traganos F, Bueti C, Darzynkiewicz Z, Melamed MR** (1984) Effects of retinoic acid *versus* dimethyl sulfoxide on Friend erythroleukemia cell growth. I. Cell proliferation, RNA and protein content. J. Natl. Cancer Inst. 73:193–204.

164a. **Traganos F, Bueti C, Darzynkiewicz Z, Melamed MR** (1987) Effects of a new amsacrine derivative, N-5-dimethyl-9-(2-methoxy-4-methylsulfonylamino)-phenylamino-4-acridnecarboxamide, on cultured mammalian cells. Cancer Res. 47:424–432.

165. **Traganos F, Darzynkiewicz Z, Bueti C, Melamed MR** (1984) Effects of a prospective antitumor agent, 1,4-bis(2′-chloroethyl)-1,4-diazabicyclo-(2.2.1) heptane diperchlorate, on cultured mammalian cells. Cancer Invest. 2:1–13.

166. **Traganos F, Darzynkiewicz Z, Melamed MR** (1979) RNA content and the cell cycle. In Laerum OD, Lindmo T, Thorud E (eds): "Flow Cytometry." Vol. IV. Bergen: Universitetsforlaget, pp 331–335.

167. **Traganos F, Darzynkiewicz Z, Melamed MR** (1981) Effects of the L isomer (+)-1,2 bis (3,5-dioxopiperazine-1-yl) propane on cell survival and cell cycle progression of cultured mammalian cells. Cancer Res. 41:4566–4576.

168. **Traganos F, Darzynkiewicz Z, Melamed MR** (1982) The ratio of RNA to total nucleic acid content as a quantitative measure of unbalanced cell growth. Cytometry 2:212–218.

169. **Traganos F, Darzynkiewicz Z, Sharpless T, Melamed MR** (1977) Simultaneous staining of ribonucleic and deoxyribonucleic acids in unfixed cells using acridine orange in a flow cytofluorimetric system. J. Histochem. Cytochem. 25:46–56.

170. **Traganos F, Darzynkiewicz Z, Sharpless T, Melamed MR** (1979) Erythroid differentiation of Friend leukemia cells as studied by acridine orange staining and flow cytometry. J. Histochem. Cytochem. 27:382–389.

171. **Traganos F, Darzynkiewicz Z, Staiano-Coico L, Evenson D, Melamed MR** (1979) Rapid, automatic analysis of the cell cycle point of action of cytostatic drugs. In Laerum OD, Lindmo T, Thorud E (eds), "Flow Cytometry." Vol. IV. Oslo: Universitetsforlaget, pp 511–515.

172. **Traganos F, Evenson DP, Staiano-Coico L, Darzynkiewicz Z, Melamed MR** (1980) The action of dihydroxyanthraquinone on cell cycle progression and survival of a variety of cultured mammalian cells. Cancer Res. 40:671–681.

173. **Traganos F, Evenson DP, Staiano-Coico L, Darzynkiewicz Z, Melamed MR** (1980) Ellipticine-induced changes in cell growth and nuclear morphology. J. Natl. Cancer Inst. 65:1329–1336.

174. **Traganos F, Gorski AJ, Darzynkiewicz Z, Sharpless T, Melamed MR** (1977) Rapid multiparameter analysis of cell stimulation in mixed lymphocyte culture reactions. J. Histochem. Cytochem. 25:881–887.

175. **Traganos F, Higgins PJ, Bueti C, Darzynkiewicz Z, Melamed MR** (1984) Effects of retinoic acid *versus* dimethyl sulfoxide on Friend erythroleukemia cells. II. Induction of quiescent, nonproliferating cells. J. Natl. Cancer Inst. 73:205–218.

176. **Traganos F, Staiano-Coico L, Darzynkiewicz Z, Melamed MR** (1980) Effects of ellipticine on cell survival and cell cycle progression in cultured mammalian cells. Cancer Res. 40:2390–2399.

177. **Traganos F, Staiano-Coico L, Darzynkiewicz Z, Melamed MR** (1980) Effects of prospidine on cell survival and growth of mammalian cells in culture. J. Natl. Cancer Inst. 65:993–999.

178. **Traganos F, Staiano-Coico L, Darzynkiewicz Z, Melamed MR** (1981) Effects of aclacinomycin on cell survival and cell cycle progression of cultured mammalian cells. Cancer Res. 41:2728–2737.

179. **Valet G, Fischer B, Sundergeld A, Hanser G, Kachel V, Rubenstroth-Bauer G** (1979) Simultaneous flow cytometric DNA and volume measurements of bone marrow cells as sensitive indicator of abnormal proliferation patterns in rat leukemias. J. Histochem. Cytochem. 27:398–403.

180. **VanDilla MA, Trujillo TT, Mullaney PF, Coulter JR** (1969) Cell microfluorometry: A method for rapid fluorescence measurement. Science 169:1213–1214.

181. **Vilarem M-J, Charcosset J-Y, Primaux F, Gras M-P, Calvo F, Larson C-J** (1985) Differential effects of ellipticine and aza-analogue derivatives on cell cycle progression and survival of BALB/c 3T3 cells released from serum starvation of thymidine double block. Cancer Res. 45:3906–3911.

182. **Vindelov L** (1978) A new method for rapid isolation and staining of nuclei for FMF-analysis of nuclear DNA in cells from solid tissues and cell suspensions. In Lutz D (ed), "Pulse Cytophotometry." Ghent, Belgium: European Press, pp 483–488.

183. **Watson JV, Chambers SH** (1977) Fluorescence discrimination between diploid cells on their RNA content: A possible distinction between clonogenic and monoclonogenic cells. Br. J. Cancer. 38:592–600.

184. **Watson JV, Chambers SH** (1978) Nucleic acid profile of EMT6 cell cycle *in vitro*. Cell Tissue Kinet. 11: 415–422.

185. **Wheeler RH, Clauw DJ, Natale RB, Ruddon RW** (1983) The cytokinetic and cytotoxic-effects of ICRF-

159 and ICRF-187 *in vitro* and ICRF-187 in human bone marrow *in vivo*. Invest. New Drugs 1:283–295.

186. **Wright NA, Appleton DR (1980)** The metaphase arrest technique. A critical review. Cell Tissue Kinet. 13:643–663.

187. **Yataganas Z, Clarkson BD (1974)** Flow microfluorometric analysis of cell killing with cytotoxic drugs. J. Histochem. Cytochem. 22:651–659.

188. **Zante J, Schumann J, Barlogie B, Gohde W, Buchner T (1976)**: New preparation and staining procedures for specific and rapid analysis of DNA distributions. In Gohde W, Schumann J, Buchner T (eds), "Pulse Cytophotometry." Ghent, Belgium: European Press, pp 97–111.

Index